SAMPLE REPORT LAYOUT

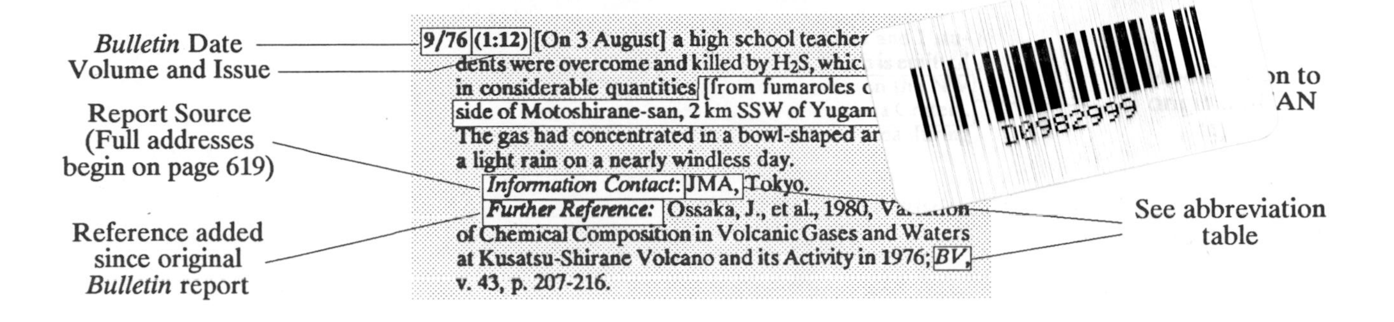

ERUPTION CHRONOLOGY INFORMATION

Eruption Dates

Code Before Date

?	=	Eruption Uncertain	S	=	SOFAR (hydrophonic)

Uncertainty Code After Date

a	=	± 1 day	*q*	=	± 45 days	
b	=	2 days	*r*	=	60	(2 months)
c	=	3	*s*	=	75	
d	=	4	*t*	=	90	(3 months)
e	=	5	*u*	=	120	(4 months)
f	=	6	*v*	=	150	(5 months)
g	=	7	*w*	=	180	(1/2 year)
h	=	8	*x*	=	270	
i	=	9	*y*	=	365	(1 year)
j	=	10	*z*	=	545	
k	=	12	*	=	>730	(2 years)
l	=	15	<	=	before date listed	
m	=	20	>	=	after date listed	
n	=	25	?	=	Uncertain date	
p	=	30 (1 month)				

Eruption Date Examples

1975 1017<	=	started before 17 Oct
1976 0115e	=	started between 10 and 20 Jan
?1975 11..	=	eruption uncertain
1975 0122?	=	date uncertain (range unknown)

Codes following durations (in days) indicate range, and > symbol indicates eruption continuing at last report date used in duration calculation.

Eruption Characteristics

Central Vent eruption
Flank Vent, or parasitic crater
Radial Fissure eruption
Regional Fissure eruption

Submarine eruption
New Island formation
Subglacial eruption
Crater Lake eruption

Explosive eruption
Pyroclastic Flow, or nuée ardente
Phreatic, or water-driven, eruption
Sulfur (molten) flow

Lava Flow(s)
Lava Lake eruption
Dome extrusion
Spine extrusion

Fatalities, casualties
Damage to land, property
Mudflow, or lahar
Tsunami long-period sea wave

(See pages 12-14 for discussion.)

Volume of Products

VV	=	Volume of lava erupted (left column)
LT	=	Volume of tephra (right column)
7 -	=	10^7 m^3 lava, no recorded tephra volume
- 9	=	10^9 m^3 tephra, no recorded lava volume
7 8	=	10^8 m^3 lava, 10^8 m^3 tephra

Volcanic Explosivity Index (VEI)

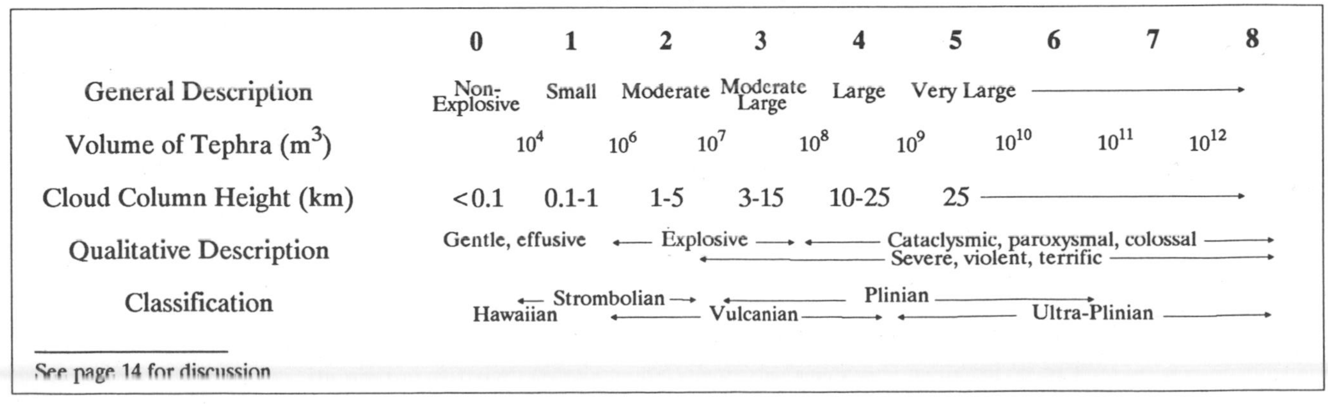

	0	1	2	3	4	5	6	7	8
General Description	Non-Explosive	Small	Moderate	Moderate Large	Large	Very Large →			
Volume of Tephra (m^3)		10^4	10^6	10^7	10^8	10^9	10^{10}	10^{11}	10^{12}
Cloud Column Height (km)	<0.1	0.1-1	1-5	3-15	10-25	25 →			
Qualitative Description	Gentle, effusive	← Explosive →			← Cataclysmic, paroxysmal, colossal → / ← Severe, violent, terrific →				
Classification	Hawaiian	← Strombolian →		← Vulcanian →	← Plinian →		← Ultra-Plinian →		

See page 14 for discussion

GLOBAL VOLCANISM 1975-1985

GLOBAL VOLCANISM 1975-1985

The First Decade of Reports from the
Smithsonian Institution's
Scientific Event Alert Network (SEAN)

Editors:

Lindsay McClelland
Tom Simkin
Marjorie Summers
Elizabeth Nielsen
Thomas C. Stein

National Museum of Natural History
Smithsonian Institution
Washington, DC 20560

Prentice Hall, Englewood Cliffs, New Jersey
American Geophysical Union, Washington, DC

Library of Congress Cataloging-in-Publication Data

Global volcanism 1975–1985.

Includes index.

1. Volcanoes—History—20th century. I. McClelland, Lindsay. II. Scientific Event Alert Network (National Musuem of Natural History (U.S.))
QE522.G57 1989 551.2′1′0904 88-32287
ISBN 0-13-357203-X

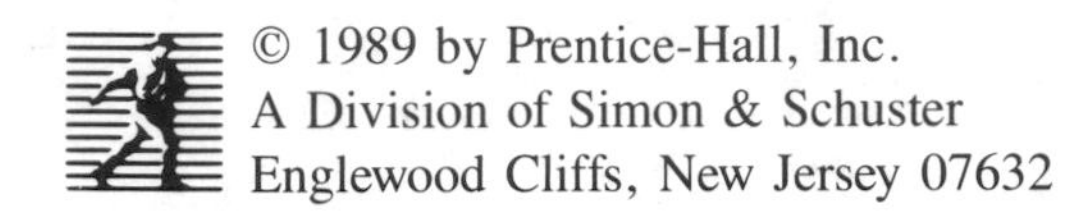

A Division of Simon & Schuster
Englewood Cliffs, New Jersey 07632

Printed in the United States of America

10 9 8 7 6 5 4 3 2 1

0-13-357203-X

Prentice-Hall International (UK) Limited, *London*
Prentice-Hall of Australia Pty. Limited, *Sydney*
Prentice-Hall Canada Inc., *Toronto*
Prentice-Hall Hispanoamericana, S.A., *Mexico*
Prentice-Hall of India Private Limited, *New Delhi*
Prentice-Hall of Japan, Inc., *Tokyo*
Simon & Schuster Asia Pte. Ltd., *Singapore*
Editora Prentice-Hall do Brasil, Ltda., *Rio de Janeiro*

CONTENTS

PART III

INSIDE COVERS

PART I

Telegram

western union

1985 JAN 22 AM 10 12

1-003606A022 01/22/85
ICS IPMWGWD WSH
00511 01-22 0817A EST
ICS IPMWGWK
1-002012A022 01/22/85
ICS IPMMLAJ MGY
01791 01-22 0401A CST MLAC
ICS IPMWGWB
1-055906G022 01/22/85
ICS IPMIIHA IISS
IISS F M RCA 22 0501
PMS WASHINGTON DC
WUC5201 MWR583 M8/17 011
URAX CO SUMX 145
PETROPAVLOVSKKAMCHATSKII URSS 145/141 22 0910
LINDSAY MC... AND SEAN NATIONAL MUSEUM OF
NATURAL HI... AIL STOP 129
WASHINGTO... USA
IN NOVEM... HE TERMINAL ERUPTION AT
KLYUCHE... INUED STOP ISSUED LAVA FLOWS
REACHE... NG STOP VOLCANIC BOMBS EJECTED
TO 30... MN ROSE TO 2-4 KM STOP AT

UP153
R I
PM-VOLCANO SKED 5-16
ALL ISLANDERS RESCUED FROM VOLCANIC EXPLOSION
AGANA, GUAM (UPI) -- A JAPANESE FREIGHTER RESCUED ALL 53 RES...
ON THE TINY PACIFIC ISLAND OF PAGAN TODAY, PICKING THEM OFF A BE...
WHERE THEY HAD FLED TO ESCAPE THE SUDDEN ERUPTION OF A VOLCANO T...
SPEWED OUT MOLTEN LAVA, POISONOUS GASES AND ASH.
NO INJURIES WERE REPORTED.
THE VOLCANO ERUPTED WITHOUT WARNING FRIDAY, SHAKING THE ENTIRE
ISLAND IN THE U.S. PACIFIC TRUST TERRITORY AND CUTTING OFF RADIO
COMMUNICATION TO THE OUTSIDE WORLD.
RESCUERS INITIALLY THOUGHT THE EXPLOSION MIGHT HAVE KILLED
...ERYONE ON PAGAN, AN ISLAND 3.5 MILES WIDE AND 8.5 MILES LONG IN T...
...ANA CHAIN, 450 MILES NORTH OF GUAM.
...COMMERCIAL PILOT FLYING NEAR THE ISLAND AFTER THE ERUPTION
...ED THE ASH CLOUD WAS 60,000 FEET HIGH AND RESIDENTS ON AGRIHAN
...65 MILES AWAY SAID ASH WAS RAINING DOWN ON THEM.
...S. COAST GUARD ON GUAM SAID JAPANESE CREWMEN FROM THE
...HOYO MARU EVACUATED ALL 53 PEOPLE FROM THE BEACH, INCLUDING
...AND SIX WOMEN.
...OTHER RESIDENT OF THE ISLAND, ITS MAYOR, HAD JOURNEYED TO
...D ON BUSINESS BEFORE THE EXPLOSION.
...REPORTS INDICATE THE ISLANDERS FLED TO THE SOUTHWEST
...AND IN THE OPPOSITE DIRECTION OF THE LAVA FLOW, WHICH
...THE MOUNTAIN AND POURED INTO THE SEA.
...UARD SPOKESMAN IN GUAM SAID HE UNDERSTOOD FROM THE
...S WERE IN GOOD HEALTH.
...THE FREIGHTER THAT THERE WERE NO INJURIES AND ALL THE
...HOYO MARU WAS TO RENDEZVOUS WITH THE FENTRESS, A TRUST
...ITORY SHIP, WHICH WILL TRANSPORT THE SURVIVORS TO SAIPAN, A TRIP
...ING ABOUT 12 TO 14 HOURS.
THE DRAMA BEGAN FRIDAY MORNING WHEN PEDRO CASTRO, THE ISLAND'S
RADIO OPERATOR, BROADCAST A MESSAGE TO COMMONWEALTH OFFICIALS SAYING
"THE WHOLE ISLAND IS SHAKING."
...SHORTLY THEREAFTER CASTRO SHOUTED, "THE
...ND THEN COMMUNICATIONS WITH THE ISLAND CEASED.
...ROUTE TO AUSTRALIA AND WAS SOME 137 MILES
...THE VOLCANO ERUPTED. THE SHIP'S CAPTAIN
...FROM THE JOINT RESCUE CENTER IN GUAM LATE
...EDIATELY SPED TO THE ISLAND IN THE DARKNESS
...THROUGHOUT THE NIGHT. AT DAWN IT MOVED IN TO
...NDERS.
...IN GUAM SAID REPORTS FROM THE ... MARU
...STOPPED SMOKING AND IT APPEA...
...THE TIME BEING.
...FICIAL OF THE COMMONWEA...
...ON PAGAN HAD BEEN...
...EMBER.
...CAUSED SPECULA...
...OTHING RECEN...
...AS ABOUT...

THE WASHINGTON POST Saturday, June 26, 1982 A11

Volcanic Ash Chokes Jet's Engines; Pilot Revives Them After 5-Mile Fall

JAKARTA, Indonesia, June 25 (AP)—A British jetliner, its engines choked off by ash from an erupting volcano, plunged five miles in a terrifying dive before the pilots restarted the engines and saved more than 200 lives, aviation and airline officials reported today. No injuries were reported.

"It seemed to go on for an eternity," Australian passenger Gerry Middleton said later. "Everybody was petrified. There was no noise. "By the time we pulled out . . . everybody was just about on their knees praying."

The British Airways Boeing 747 was carrying 224 passengers and a crew of 16 on a flight last night from Kuala Lumpur, Malaysia, to Pe... Australia. It was the final ... flight from London.

About 100 mil... flying at 37... a thi...

...tered "mild turbulence" and there was nothing to worry about. But "we knew it was more serious," said Middleton.

Some children cried at first, but then there was silence in the cabin, said another passenger, Douglas Cross, 40.

"One elderly woman looked out of the window and commented on the terrible storm," he said. "She did not realize she was seeing the flashbacks from the engines."

"The crew acted in th... emplary manner, m... cabin to calm ... oxygen ...

...roughly, and he turned back to Jakarta, where he made a safe emergency landing despite the fact that the ash had "sandblasted" the cockpit window, hampering visibility.

"Indications are that the crew did an absolutely splend... job in very difficult circu... pilot of ... the chief Ca... Airways, ...ondon. ... alert

Caribbean Volcano Erupts

KINGSTOWN, St. Vincent—Explosions sounding like gunfire reverberated from Mt. Soufriere volcano, sending a huge column of ash into the air and forcing evacuation of residents within a 10-mile radius.

The 4,000-foot volcano, which killed 2,000 persons in an eruption in 1902, erupted at 5 a.m. Prime Minister Milton Cato declared a state of emergency and ordered evacuation of all villages surrounding the volcano 30 miles north of Kingstown.

western union

Telegram

TIMES LAVA
COL 129 DC20560 300M 2...
PAGE2
SPOUTING WAS OBSERVED ABOVE THE CRATER STOP ON
NOVEMBER 23-24 ASH COLUMN ROSE TO 4KM HEIGHT STOP IN JANUARY
THE ERUPTION CHARACTER REMAINS PREVIOUS STOP AFTER THE PAROXYSM
OF 10-14 OCTOBER BEZYMIANNY VOLCANO REMAINS TO BE IN A STAGE
OF MODERATE ASH-GASEOUS ACTIVITY STOP ON DECEMBER 27 GORELY
VOLCANO BEGAN
COL 23-24 4KM 10-14 27
PAGE3
TO ERUPT STOP DURING INITIAL STAGE PHREATIC
EXPLOSIONS WERE THREE POINT FIVE KM HEIGHT STOP ON DECEMBER
30 ASH EXPLOSIONS BEGAN THEIR HEIGHT REACHED 3 KM PLUME
EXTENDED 100 KM TO EAST STOP ON JANUARY 8 THE ERUPTION CEASED
DR IVANOV
COL 30 3 100 8
NNN
0502 EST

0806 EST

W.U. 1201-SF (R5-69)

western union

1016 EST

Telegram

Western Union International, Inc.

International

N

RCA AUG 22 2307*

SMITHSONIA WSH

ZCZC YWB1561 RMM7727 RMB4399 ECN812 SC1926

URWN HL EDQO 109

QUITOECUADOR 109/105 22 1830 PAG/1 CK/53/50

LT

SEAN

SMITHSONIAN INSTITUTION NHB ROOM 9

WASHINGTONDC

ERUPTION FERNANDINA GALAPAGOS BEGAN 8...

TOURIST BOATS MOORE/SNELL PARTY REA...

AUGUST LAVA ISSUED FROM FISSURES ...

TERRACE FLOWED TO LAKE OVER DESCENT ROO...

APPARENTLY GREATER THAN 1973 ERUPTION ASH

INTRODUCTION

Volcanism is one of nature's primal spectacles. It commands our attention, whether for its awesome power, its danger, or its beauty. The years 1975–85 brought dramatic examples of volcanism—from the 1980 blasts of Mt. St. Helens, through the continuing lava rivers of Hawaii, to the tragic eruption of Colombia's Nevado del Ruiz in 1985—and these have been abundantly documented by the news media as well as scientists. However, the decade also brought less spectacular volcano stories—from eruptions that never took place (despite months of increasingly ominous warnings), through pumice (source unknown) floating on vast expanses of open ocean, to small eruptions of special concern only to those who happened to be too close to them. Most of these nonsensational events were ignored by the news media, and a surprisingly high proportion have not yet been adequately documented in the scientific literature. This book reports the decade's full spectrum of volcanism: small eruptions as well as large, and subtle warning signs as well as notorious disasters. By treating this wide range of volcanic activity, we attempt to provide a baseline of global volcanism: a context into which events of the future can be placed for comparison or measurement.

The heart of this book, Part II, is a reorganization of past volcano reports from the *SEAN Bulletin*, a monthly publication of the Smithsonian Institution's Scientific Event Alert Network. This global network consists of more than 1000 correspondents, including professional volcanologists, scientists in other specialties, travelers, and other careful observers of the world around them. They contribute news of volcanoes seen from field investigations, volcanological observatories, private homes, airplanes, and remote sensing laboratories. After some combining and editing, the reports then go out to interested readers around the world in the hope that timely information will both increase volcanological understanding and facilitate study while events are still new.

The SEAN reports, though, are necessarily preliminary. They lack the ruminative evaluation and cross-checking of facts that characterize careful retrospective studies of recent volcanism. Such studies have been published for less than a third of the events described in this book, but we have asked our correspondents to correct errors in their original reports and to provide references to published work on the events that they have described. These changes have been incorporated in the text—with square brackets ([ ]) identifying new wording and ellipses (. . .) marking deletions—and we have added further references to the original SEAN reports where appropriate. Nonetheless, many original preliminary reports remain unchanged, and readers must be cautioned against treating the accounts in this book as definitive.

The original *Bulletin* information has been reorganized geographically, making it easier to find reports from a particular volcano, or group of volcanoes, and learn what has happened there in the last decade. The geographic sequence is that adopted 40 years ago for the *Catalogue of Active Volcanoes of the World*[1] and used in our 1981 book, *Volcanoes of the World*[2]. Chapter numbers in Part II correspond to the 20 major volcanic regions in these references[3] and are shown in the index map that makes up the inside rear cover of this book.

In addition to the editing and further references described above, we have introduced each volcano with a paragraph of geographical and historical background. A tabular chronology of the decade serves as a temporal index to the reports and facilitates searches for eruptions of a certain size or type. And finally, a detailed index opens up the full decade of reports to inquiries by subject, person, or place.

Making room for these additions, and reducing 1460 original pages to 571, has been difficult. Adopting a two-column format with proportionally-spaced type has helped, as has deletion of introductory text and illustrations that were necessarily repeated from month to month in the original *Bulletin*. We have also trimmed, redrafted, or reduced many illustrations to save space. By listing the full address of each correspondent only in Part III, we have saved unnecessary repetition while providing current rather than outdated addresses for our information sources.

In the introductory pages that follow, we add background information on SEAN, summarize highlights of the decade, provide some quantitative information on its activity, introduce the event chronology, and describe both editing conventions and cautions needed before embarking on the reports of Part II.

The record of global volcanism is a growing document, essential to understanding the eruptions that are certain to continue in the future. You may have the opportunity to contribute to that growing document if you happen to be in the right place at the right time, or if you know of events inadequately described here. We strongly encourage interested readers to contact us. A better record of global volcanism depends on a continuation of international cooperation between many different people from many different backgrounds. We solicit your help in building this better record.

---

1 This group of catalogs, hereinafter referred to as *CAVW*, was published from 1950 to 1975 by the International Association of Volcanology and Chemistry of the Earth's Interior (IAVCEI).

2 Smithsonian Institution, 1981, *Volcanoes of the World.* Hutchinson Ross, Stroudsburg, 240 pp.

3 There were no reported eruptions during the decade in Regions 18 (Atlantic Ocean S of Iceland) or 20 (Arctic Ocean). Therefore, Region 19 of *CAVW* (Antarctica) becomes Chapter 18 here. Our Chapter 19 covers global atmospheric effects of volcanism, and Chapter 20 examines the extraterrestrial volcanism discovered in 1979 on one of Jupiter's moons.

ACKNOWLEDGMENTS

Our first and strongest thanks go to our correspondents. Without them, there would be no SEAN and certainly no *SEAN Bulletin*. The volume and quality of information in the *Bulletin* is a testimonial to the dedication and cooperation of people all over the world who share an interest in active volcanoes. Our need for news inevitably comes at their busiest times, under the pressure of ongoing eruptions, yet they have repeatedly provided detailed and timely data that have stimulated research on active volcanism. The *Information Contacts* section that concludes every report in Part II includes 696 names. Some appear only once, but many are repeated, report after report, and often volcano after volcano. All can be found in Part III, both in the index and in the address list. However, many people have provided information relayed to us by someone else, and the number of actual "information contacts" extends well beyond 696. To all of them, we extend our warmest thanks, and our hope that strong communication will continue into the years ahead.

The *Bulletin*, however, does not spring effortlessly from neatly typed reports spontaneously generated by meticulous and punctual correspondents. There are illustrations to redraft, handwritings to decipher, telexes to ungarble, telephone calls to make, foreign languages to translate, specialist information to solicit, divergent reports to assimilate, inquiries to answer, obscure references to find and digest, camera-ready *Bulletin* copy to prepare, print, and mail each month, followed by summaries for *EOS*, *Geotimes*, and *Bulletin of Volcanology*. It makes for busy days, particularly when a newsworthy eruption is in progress. All these tasks and more have been accomplished with dedication and hard work by a small SEAN staff through the years. The roles of those who are no longer with us are described in the following section (SEAN History), and we thank them for their valuable contributions to SEAN.

The *Bulletin* has drawn on more Smithsonian resources than the SEAN staff alone. The SEAN Advisory Committee must be thanked, along with the overlying administrative structure (Department of Mineral Sciences, Museum of Natural History, and Smithsonian Institution), for their support through the years. Volcanologist curators Tom Simkin, Dick Fiske, and Bill Melson have provided much helpful advice, Gene Jarosewich has helped with translations, and Lee Siebert has been a constant source of information through the Volcano Reference File (VRF, our database of global volcanism through the last 10,000 years). Beyond our department, field biologists, anthropologists, and librarians have aided information searches in parts of the world unfamiliar to us, and the Smithsonian Institution Press has arranged printing of the *Bulletin* since 1980. The Institution's computer specialists (including Dave Bridge, Bruce Daniels, Gary Gautier, Pete Kauslick, Kathy Lawson, and Anne Quade) have assisted with databases such as the VRF, SEAN addresses, and bibliographic files. Luigi Mancini and other friends have repeatedly helped translate communications to and from English.

Preparing this book from the monthly *Bulletins*, however, requires a special set of acknowledgments that go with the developmental sequence of the book. The start was the Congressional funding, in late 1984, of our Global Volcanism Program. This provided the support essential to reformat the *Bulletins*, index them, and put them between hard covers as a useful reference work on global volcanism.

Marjorie Summers deserves full credit for the next step, the huge job of assembling 10 years of *Bulletins* into a coherent electronic data set. This began with capture of original typescript by optical character reader, and continued with manipulation by desktop computer, for which all of us thank Jon Dehn. Marge was ably assisted by summer interns Susan Harrington (1987) and Dana Bahar (1988). Mailing of text for review by original correspondents was another large job, which was handled by Toni Duggan and Dana Bahar. Responses from these mailings were then incorporated into the text by Lindsay McClelland. References to the scientific literature—detailing and updating original *Bulletin* reports—were gathered by Tom Simkin, Toni Duggan, and Lisa Wainger before final assembly (with those provided by SEAN correspondents) by McClelland.

Elizabeth Nielsen produced not only the comprehensive index, but also the decadal data summarized later in Part I. She was assisted by the powerful indexing capabilities of our word-processing software, Nota Bene, and support from its developer, Steve Siebert. The Chronology, which serves as a temporal and process index to the reports, was produced from the VRF by Lee Siebert and computer specialist Tom Stein using the Pick Operating System.

Illustrations were another large responsibility of Marge Summers. She drafted or redrafted some, resized or cropped others, and produced final camera-ready figures with the help of volunteer Terry Shumaker and interns Dana Bahar and Katherine Duncker. The world map (inside rear cover) was drafted by volunteers Felix Nienstadt and Ken Cavagnaro, with help from Tim O'Hearn and Ken McCormick. Final page layout of camera-ready copy was prepared by Tom Stein, using Ventura Desktop Publisher, in consultation with Marge Summers. Patricia Rayner and Patrice LaLiberté of the American Geophysical Union (AGU), and Michael Hays of Prentice Hall, provided helpful advice on layout.

The new material in Part I and volcano introductions in Part II were written largely by Simkin and McClelland, but editing of the full text has been a group effort of all 5 editors (with 5 very different views of grammar, punctuation, and style!). Additional editorial help has come from Dick Fiske (Smithsonian) and

Sophie Papanikolaou (Prentice Hall). Through all of this, Ellen Thurnau has been an administrative mainstay, keeping track of orders, paychecks, and budgets with cheerful efficiency.

The overall production has been very much a team effort, and we could not have done it without the help of all those named above and the correspondents named in the individual reports. We thank them for their essential assistance in a job that seemed small when we started. We now know that it was not small, but we think it was worth it. We hope you do, too.

SEAN HISTORY

The Smithsonian Institution has long had a special interest in volcanology. Even before there was an institution bearing his name, James Smithson described a new volcanic sublimate from the slopes of Vesuvius. In 1912, meteorologist C. G. Abbot, fifth Secretary of the Institution, recorded the atmospheric effects of a major eruption near Katmai, Alaska, and became one of the early investigators of volcanic influences on climate. And William Foshag, who documented the growth and development of the Mexican volcano Paricutin in the 1940s, is an example of the many Smithsonian scientists with active research programs in volcanology.

During the years 1963–67, the Surtsey eruption off the south coast of Iceland built a new island amid considerable multidisciplinary interest. Volcanologists were joined by atmospheric scientists studying eruptive cloud dynamics, oceanographers studying marine interactions, and biologists studying the arrival of life on this new island. Many of Surtsey's unique research opportunities required swift scientific observation, but these opportunities were often lost in other, less-publicized events. Smithsonian administrators (notably Assistant Secretary for Science Sidney Galler) recognized the need for a global communications hub to report natural events swiftly and facilitate prompt scientific attention. At the end of 1967, the Institution established the Center for Short-Lived Phenomena (CSLP) in Cambridge, Massachusetts. The Center was located at the Smithsonian Astrophysical Observatory (SAO) to take advantage of the international communication network then in use there for satellite tracking. Director Bob Citron and Lee Cavanaugh developed a network of correspondents around the world and mailed individual postcards describing natural events shortly after information was received. The Center grew rapidly and was soon reporting on a wide variety of biologic and anthropologic, as well as volcanic, events. By 1975, however, the distance between the Center and the Washington-based Smithsonian scientists most interested in its work had become a problem. Furthermore, the SAO communications network had ceased with the end of their satellite tracking contract. Therefore, the Institution transferred key CSLP employees to Washington, and established the Scientific Event Alert Network (SEAN) in the National Museum of Natural History.

SEAN started operations on 1 October 1975. Its initial staff consisted of geographer David Squires (Operations Manager), biologist Shirley Maina, John Whitman, and administrative assistant Betty Grier. SEAN was administratively part of the Museum Director's office, and oversight was provided by a Scientific Advisory Committee chaired successively by Henry Setzer, William Melson (from October 1976), and Tom Simkin (from February 1978). In keeping with the interests of museum scientists, reporting of man-made events (such as oil spills) was discontinued, biological reporting emphasized marine mammals, and reporting of both volcanic and meteoritic phenomena was greatly expanded. Rather than mailing individual reports on postcards, information from various sources was synthesized for each event and published in a monthly *SEAN Bulletin* mailed to correspondents. Our studies showed that report receipt was not substantially delayed by the new system, and our savings in publication time allowed us to devote more attention to individual notification by telephone and telex of urgent information to correspondents needing to act on a particular event.

CSLP's list of correspondents was systematically expanded, both for better geographic coverage and for more information sources in the new areas of scientific emphasis. The list grew from a few hundred in 1975 to more than a thousand by the end of 1985. This growing network of correspondents, primarily scientists actively engaged in field-oriented research, has been the lifeblood of SEAN's operations.

Initial reports, particularly of events with a severe human impact, have often come from the press, and monitoring of the various news services has been a daily routine since SEAN's beginning. Armed with such preliminary (and not always reliable) news, confirmation and detailed information about an eruption's progress has been sought. Smithsonian scientists with varied field experience, as well as colleagues from the USGS and the university community, have helped direct us to additional information sources. Establishing contact with scientists working at the eruption site is the most important (and often most challenging) initial task. They are generally best placed to observe a developing event, but are often under the intense pressures that attend life-threatening crises, far from communication facilities, and *always* enormously busy. The high quality of reports from correspondents under such trying circumstances is most impressive.

Reports from the eruption site are then passed by

SEAN to other specialists who are in turn consulted for additional pertinent information. Satellite specialists are among the first notified, for they can often provide data on the timing of activity, as well as height, dispersal, and SO_2 content of eruption clouds. Such information is of course passed back to the scientists at the erupting volcano, and this timely exchange helps catalyze research on rapidly developing events.

This generalized response developed rapidly during SEAN's early years. John Whitman resigned after 5 months, and Lindsay McClelland, who had previously worked with our computerized Volcano Reference File before leaving for graduate work in volcanology, joined SEAN in August 1976. He arrived in the midst of the volcanic crisis at Soufrière de Guadeloupe and has been the mainstay of SEAN's volcano reporting ever since. Having a volcanologist at SEAN's end of the communication line has been essential to our success.

In addition to volcanism, SEAN has regularly reported earthquakes, fireballs, and meteorite falls. Earthquake data are received from the U.S. Geological Survey's National Earthquake Information Center[4], and only the larger (Richter magnitude greater than 6.5) or more damaging events are tabulated. Fireball reports cover 3 pages or less per average issue and cannot pretend to be complete, but they provide a valuable link to meteorite falls and recovery of these scientifically precious samples of extraterrestrial material.

To bring SEAN's reports to a broader geologic audience, since February 1977 we have summarized the *Bulletin* reports in a monthly column for *Geotimes*, published by the American Geological Institute. Starting with the June 1979 *Bulletin*, we forwarded volcano and earthquake reports to the American Geophysical Union (AGU), at the same time that they were sent to our printer, for use in the next available issue of *EOS*, AGU's weekly newsletter. Additional reports have been sent directly to *EOS* when major events significantly preceded our monthly *Bulletin* schedule. Since mid-1981 only a selection of reports have been published in *EOS* each month along with one-line summaries of all. Starting in early 1986, volcanological summaries have also been prepared 6 times a year for the *Bulletin of Volcanology*, the journal of the International Association of Volcanology and Chemistry of the Earth's Interior (IAVCEI).

The *SEAN Bulletin* has always been mailed gratis to the worldwide network of correspondents who supply us with information, but growing interest in the *Bulletin* soon led to increasing requests for subscriptions. Not being equipped to handle these requests, we approached the National Technical Information Service (NTIS), a clearinghouse for government publications. Starting in January 1979, they made the *SEAN Bulletin* available to subscribers. Since mid-1981, a less expensive subscription to the *Bulletin* has been offered by the AGU[5]. Readers interested in swifter receipt of the *Bulletin* are now able to access it by computer. Since late 1987, the text portion has been posted on two electronic bulletin boards—KOSMOS and OMNET[6].

The middle years of SEAN's first decade brought substantial personnel and administrative changes (Fig. 1). Betty Grier left Washington in mid-1980, ending the swift and accurate typing of the monthly *Bulletin*, and David Squires was asked to take over the Natural History Museum's shipping office at the end of that year. In January 1981, SEAN was moved—physically and administratively—into the Museum's Department of Mineral Sciences. At the same time, Lindsay McClelland took over as Operations Manager and two half-time assistants were added for geology (Janet Crampton) and biology (Paula Rothman). Government-wide budget cuts seriously affected the Smithsonian in 1981, however, and the decision was reluctantly made to end SEAN's biological reporting in July 1982. Shirley Maina and Paula Rothman were transferred to fill open biological positions in the Museum, and tabulation of marine mammal and turtle data was taken over by other groups[7]. Although SEAN made the switch from typewriter to word processor in 1981, the office was still seriously understaffed, and in December 1982, Elizabeth Nielsen joined the department, with half her time going to SEAN and half to research assistance.

With the announced end of SEAN's biological reporting, its Scientific Advisory Committee was disbanded in April 1982. Oversight of SEAN's meteoritical reporting and final editing of the *Bulletin*'s fireball section remained with Roy Clarke, a member of the Advisory Committee through its full lifetime, until assumed by Glenn MacPherson in 1986. Tom Simkin continued this role for the remaining parts of SEAN, and has been final editor of the volcano and earthquake reports since the *Bulletin*'s first issue.

In late 1984, the Congress approved the Museum's Global Volcanism Program, and this new support has allowed expansion in several areas (not least of which is the preparation of this book). Upon Janet Crampton's resignation in 1985, we hired Emily Wegert—again on a half-time basis—and soon added Marge Summers and Toni Duggan. All had substantial SEAN involvement, continuing the growth in volcano reporting and partially freeing Elizabeth Nielsen for development of our new

4 National Earthquake Information Center, U.S. Geological Survey, Stop 967, Denver Federal Center, Box 25046, Denver, CO 80225, (303/236-1500) (TWX 510 6014123).

5 Subscriptions are on a calendar year basis at $18/year to U.S. addresses and $28/year to all other countries. Single back-issues are also available from AGU. Orders must be prepaid, and checks made payable to AGU. American Geophysical Union, 2000 Florida Ave. NW, Washington, DC 20009, (202/462-6903) (800/424-2488).

6 Both networks are accessed through GTE Telenet. KOSMOS is operated by the AGU (address and telephone numbers in footnote above): our bulletin board is named SEAN. OMNET's address is 70 Tonawanda Street, Boston, MA 02124 (617/265-9230) and our bulletin board is named *VOLCANOES.ETC*. Volcanological news will reach us via *SMITHSONIAN.SEAN* on KOSMOS or *SEAN.SMITHSONIAN* on OMNET.

7 Marine Mammal Events Program (cetaceans only), c/o NHB Stop 108, Smithsonian Institution, Washington, DC 20560. Sea Turtle Stranding and Salvage Network, National Marine Fisheries Service, Southeast Fisheries Center, 75 Virginia Beach Drive, Miami, FL 33149.

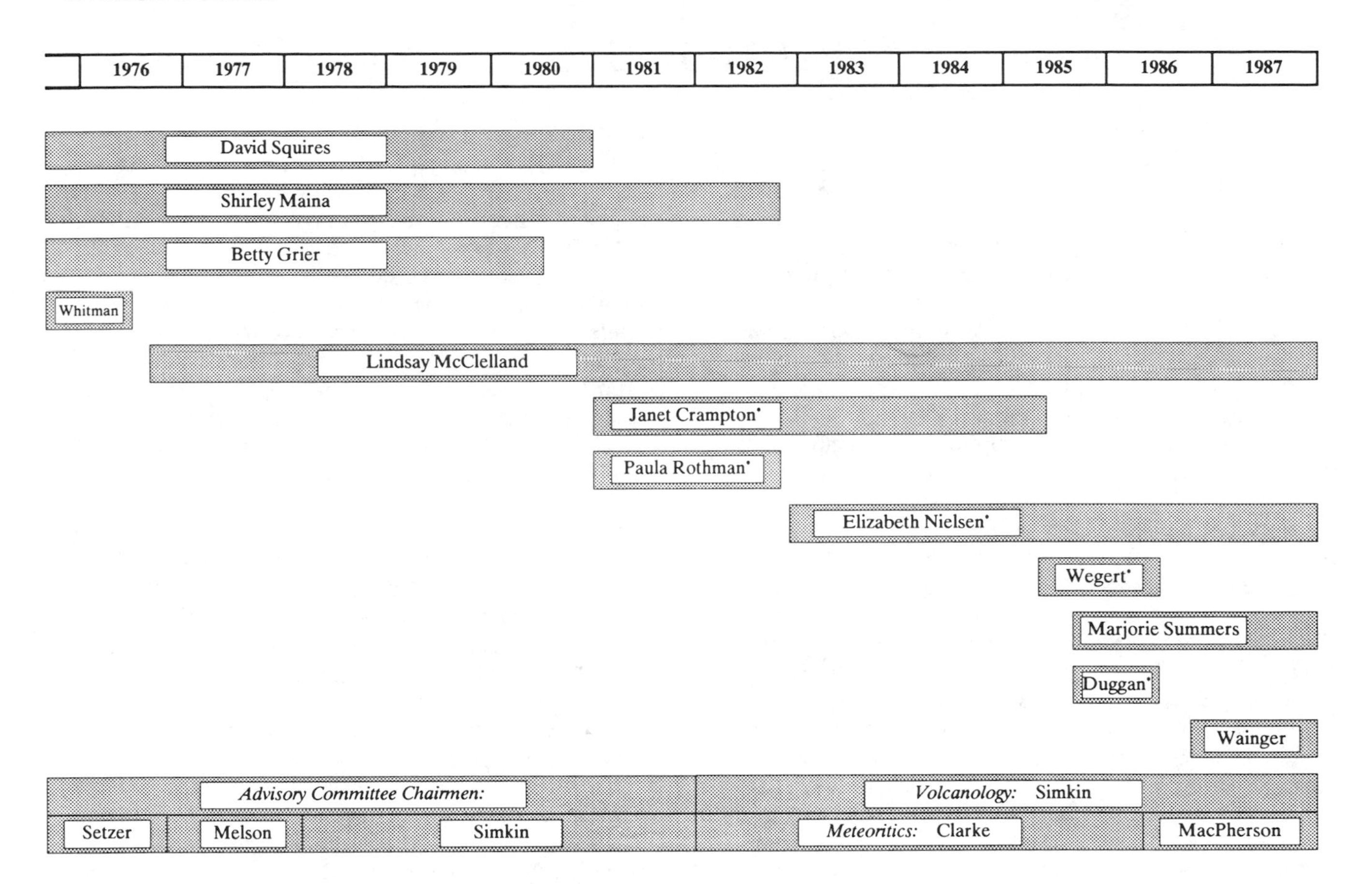

Fig. 1. SEAN personnel from the program's Oct 1975 start through 1987. Asterisks denote part time SEAN employees.

Volcanological Archives. Lisa Wainger, who replaced Toni Duggan in 1986, worked full time with SEAN, and freed more time for Summers and McClelland to work on the book.

The primary emphasis of the Global Volcanism Program has been the improvement of our Volcano Reference File—a database of the world's volcanoes and their eruptions of the last 10,000 years—and its conversion from mainframe to desktop computers. We believe that the elements of this new program are of a piece—complementing and cross-fertilizing each other toward a better understanding of how volcanoes work. SEAN, in addition to facilitating study of contemporary volcanism, provides an essential baseline record of volcanic activity—the small as well as the large events. By extending this record historically, the VRF database gives context to today's events and aids identification of past events that may help illuminate tomorrow's. The Volcanological Archives and our Museum collections provide safe places for the photographs, maps, specimens, and other documentation needed for the future studies essential to a better understanding of volcanism.

In looking back at SEAN's history, we can see many positive trends. In 1975, during our first 3 months, an average of 5 volcanoes were covered in each issue and the reports totaled little over a page in each (see collage on p. 39 for the entire volcano section of our first issue). These averages grew rapidly to 10 volcanoes and nearly 10 pages per issue in 1977. Growth has since continued, but at a more gradual rate, with averages in recent years of at least 20 volcanoes and nearly 15 pages per issue. We have worked hard to follow up on initial reports: less than one-fourth of the volcanoes covered in this book were reported in only one issue, and most of these were simply status reports on infrequently visited volcanoes. Often these follow-up reports carry information on important post-eruptive processes missed in descriptions of the more spectacular eruptive activity. Another area of concentration has been increased communication with the satellite and atmospheric science communities. By alerting weather satellite specialists to an eruption, we have at times obtained data that might otherwise have been lost, and the tracking of volcanic aerosols has both aided understanding of their distribution and increased interest in the far-reaching effects of large eruptions. Occasional reports of atmospheric effects appeared in the *SEAN Bulletin* after the 1980 St. Helens eruption and—following the widespread effects of El Chichón—this section has been a regular feature since mid-1983.

The most important trend, though, and the one that pleases us the most, is the continuing increase in good correspondents. Better reporting of global volcanism depends ultimately on the number and quality of our correspondents. We value enormously the many friends who take time to pass on to us, and to *SEAN Bulletin* readers, their observations of volcanoes near to them but very distant from us.

HIGHLIGHTS

Two events in SEAN's earliest months—an earthquake in Hawaii and an eruption in Iceland—developed into two of the decade's most interesting long-running stories; and both highlighted the increased monitoring capability that has been the decade's outstanding advance.

Iceland sits astride the Mid-Atlantic Ridge, offering the opportunity for land-based observations of processes that generally take place beneath the sea: processes accompanying the spreading apart of two of the earth's major crustal plates. A brief eruption in December 1975 at Iceland's Krafla volcano marked the onset of a fascinating, seven-year period of rifting showing that plates do not separate at a steady few centimeters per year but in intense surges of several meters over only a few years followed by perhaps a century of quiet. In the first major activity at Krafla since 1729, repeated cycles of inflation—swelling of the earth's surface over central magma reservoirs—were punctuated by brief episodes of volcanic rifting. Some of the early rifting events were accompanied by eruptions, but most magma moved below the surface, and extruded lava volumes were small compared to the volume of magma intruding the rift zone. These episodes included history's smallest documented eruption, when 26 m^3 of basaltic scoria emerged from a 1200-m geothermal drillhole in September 1977. Since March 1980, there have been 5 fissure eruptions, most recently in September 1984, and more than 0.2 km^3 of lava has emerged from the rift zone north of the magma reservoirs. Rifting measured since 1975 along the 80-km zone has reached 5-7 meters.

In a quite different tectonic setting on the island of Hawaii, remarkably similar relationships between rifting and volcanism followed the large (magnitude 7.2) earthquake of 29 November 1975. Much of the huge structural block south of Kilauea's East and SW rifts moved seaward by several meters, making room for magma to move laterally as subsurface intrusions. From that date through the end of 1981, the USGS Hawaiian Volcano Observatory detected 15 intrusive episodes but only 2 eruptions at the surface. Few intrusive events had previously been detected anywhere in the world, and it has only been the development of sensitive monitoring techniques in recent years that has allowed this subsurface activity to be documented as it happens. The character of Kilauea's activity changed dramatically with the 3 January 1983 onset of the first of 48 eruptive episodes from East Rift vents. Summit deflation and harmonic tremor with each episode indicated drainage of magma from the summit chamber, with gradual reinflation between episodes. By the end of 1985, the cumulative lava volume was approaching 0.5 km^3, making it the largest of Kilauea's historic rift eruptions.

Considerable property damage resulted from both examples of rift volcanism; lava overran more than 50 houses on Kilauea's south flank, and seismicity associated with rifting at Krafla halted a costly geothermal power project. At both volcanoes, however, relatively quiet lava production plus close cooperation between volcanologists and public officials have allowed timely evacuations and minimized the risk to human life.

Also in SEAN's early months, volcanologists on the Caribbean island of Guadeloupe were coming to grips with what was to become the other major theme of the decade— the social consequences of volcanism and attendant problems of prediction. Local seismicity had begun to increase in November 1975 but caused little concern until a small phreatic eruption from Soufrière Volcano on 8 July 1976. Other ash eruptions in August prompted authorities to order the evacuation of 72,000 residents. However, activity declined in September and deformation studies detected no significant inflation of the volcano. An international committee convened by the French government concluded that the probability of a dangerous eruption was low. Officials then ended the evacuation, and the last of Guadeloupe's displaced residents were allowed to return by 1 December.

The decisions at Guadeloupe were controversial and acrimonious but the 1983 eruption on Indonesia's Una Una Island proved the importance of timely evacuation of island populations. At least 10 days of seismicity preceded the start of explosions on 18 July. On the recommendation of the Volcanological Survey of Indonesia, the island's 7,000 residents were evacuated by 22 July. The next day an explosion swept nearly all of the island with pyroclastic flows, destroying most homes and 700,000 coconut trees. No human lives were lost.

Extraterrestrial volcanism was detected for the first time in March 1979, as the Voyager 1 spacecraft passed Jupiter's moon, Io. Several of Io's 8 vents erupted repeatedly during the 4 days of observations, feeding plumes that rose hundreds of kilometers. No active lava flows were evident, although fresh-looking flows appeared to radiate from several volcano-like features. Four months later, Voyager 2 data showed that 6 of the 8 vents remained active.

Volcanologists have only rarely been killed in eruptions, but on the same day that volcanism was detected on Io a terrestrial volcano took the lives of Robin Cooke, senior volcanologist at Papua New Guinea's Rabaul Volcanological Observatory, and his assistant Elias Ravian. The eruption of Karkar volcano began in January 1979, after 6 months of seismicity, and on the night of March 8 a directed blast devastated the Cooke-Ravian campsite on the caldera rim. A year later, volcanologist David Johnston was killed by a much larger directed blast at a volcano on the opposite side of the Pacific.

The 1980 eruption of Mount St. Helens was the event of the decade in bringing volcanism to the attention of the Western world. Of a magnitude that occurs only about once every 10 years, it killed 57 people, devastated dense forests 30 km away, sent a cloud to 24 km altitude, and dropped measureable ash over many states. Although the United States ranks third (behind Indonesia and Japan) in number of historically active volcanoes, three generations had passed since the last eruption in the conterminous states. The St. Helens eruption swiftly focused the attention (and funding) of a large, industrial nation on a single volcanic event. It has unquestionably become history's most studied eruption, yielding important new data about debris avalanches, directed blasts, dome growth, and more. Interdisciplinary efforts, particularly in biology and atmospheric science, were major elements of the investigation. Tracking of the eruption cloud by aircraft, balloon, satellite, and lidar (laser radar) documented its initial dispersal at various altitudes and its continued presence in the stratosphere.

This growing cooperation between volcanologists and atmospheric scientists promises to improve substantially our understanding of the atmospheric and climatic effects of volcanism. Increased scrutiny with a variety of instruments can also help to define the character and size of eruptions, especially those poorly observed from the ground. In early 1982, a new layer of stratospheric aerosols began to be observed by lidar stations around the globe. Soon named the Mystery Cloud, its source long remained unknown. However, later processing of data from NASA's Total Ozone Mapping Spectrometer (TOMS) on the Nimbus-7 polar orbiting satellite showed a large sulfur dioxide cloud originating from eastern Zaire in late December 1981. The initial explosive phase of Nyamuragira's eruption that month was almost certainly the source of the Mystery Cloud, and the eruption later fed a 26-km lava flow, the longest from that volcano in historic time.

The atmospheric effects of that eruption were soon dwarfed by a series of major explosions from México's El Chichón in 1982. The 28 March explosion, its first in recorded history, produced a column 15 km high and heavy tephra falls. Stronger explosions during 3-4 April sent gas-rich material well into the stratosphere. Pyroclastic surges destroyed 8 villages within 8.5 km of the summit. Although heavy tephra falls from the initial explosion had driven many people from flank villages, about 2,000 people who had remained or returned to their homes were killed by the surges. Only about 0.5 km^3 of tephra was deposited, comparable to that at St. Helens in 1980, but the eruption columns were far richer in sulfur gases. The amount of aerosol injected into the stratosphere was at least ten times that from St. Helens, and its long residence in the stratosphere has drawn attention to the variation in sulfur content of different eruptions. The aerosol cloud—sulfuric acid droplets formed by the combination of volcanic SO_2 gas with water vapor and so small that they fall very slowly—soon circled the globe and dispersed over the northern hemisphere, causing striking sunsets and other atmospheric effects. Strongly enhanced stratospheric aerosols were documented for at least 3 years by lidar, balloon, and satellite. Such aerosol layers are known to filter solar radiation and therefore cool climates below them after major eruptions, but the climatic effects of El Chichón, despite intense study by atmospheric scientists, remain uncertain. A powerful oceanic warming event ("El Niño") that began in the eastern Pacific later that year caused substantial perturbations in weather patterns, making it difficult to separate El Niño effects from those caused by volcanic aerosols.

A day after the strongest activity at El Chichón, explosions began at Java's Galunggung. Although not as large as those from El Chichón, the explosions continued through 1982, forcing tens of thousands of residents from their homes on the densely populated flanks and causing tephra falls that disrupted a wide area. Galunggung sits astride major international air routes, and on 2 occasions jetliners carrying hundreds of passengers flew through tephra clouds that temporarily stalled their engines. Although both landed safely, these terrifying incidents catalyzed growing concern in the aviation community about the hazards posed by eruption clouds. Several of the most-affected nations, led by Indonesia, Australia, and the United States, have taken steps to improve communications between volcanologists and aviation interests, while international regulatory agencies are developing ways to gather and rapidly distribute warnings of dangerous eruption clouds.

The events at St. Helens, El Chichón, and Galunggung dramatically set the stage for strong international interest in geophysical unrest at 3 volcanic centers in the early 1980s: California's Long Valley, Italy's Campi Flegrei, and Papua New Guinea's Rabaul. All 3 are calderas—large, broadly circular depressions formed when a volcanic edifice collapses into a large magma chamber from which the magma has just been erupted. Recorded history has fortunately not seen eruptions of the scale that produced these calderas, and all 3 are now centers of substantial human populations. Their potential for large-scale devastation lent new urgency to understanding the earthquakes, elevation changes, and thermal anomalies that were recognized in the early 1980s.

At Long Valley, on the eastern front of the Sierra Nevada east of San Francisco, 4 moderate (magnitude 5.5-6.1) earthquakes around the time of the 1980 St. Helens eruption heralded several years of unrest. A series of earthquake swarms accompanied by spasmodic tremor, uplift of the resurgent dome, and increased fumarolic activity prompted the USGS to issue a notice of potential volcanic hazard in May 1982. The most intense swarm, in January 1983, included more than 1000 shocks in its first 12 hours, and analysis suggested that it was associated with magma intrusion along the caldera's south margin. Seismicity declined in succeeding months and the USGS reduced its alert in 1984. Uplift continued, however, reaching a maximum of nearly $^1/_2$ meter (since 1975) by summer 1985, and a detailed geophysical monitoring program continues. Only 3000 people live within the caldera, but the ski season brings tens of thousands of daily visitors. A new highway has been constructed that would facilitate their evacuation if necessary.

Slow vertical changes of more than 10 m have been measured since Roman times at Campi Flegrei, on the Bay of Naples about 25 km west of Vesuvius. Periods of more rapid change included several meters of uplift preceding the caldera's most recent eruption, in 1538, and 1.7 m of nearly aseismic inflation in 1969 – 72. Vigorous uplift resumed in mid-1982, and seismicity began to build in November with earthquakes centered only 2-3 km below the surface, many large enough to be felt by the local population. Uplift rates reached a maximum of 5 mm per day in October 1983, accompanied by increasing seismicity. But the crisis then passed. In late 1984, inflation and seismicity declined to pre-crisis, background levels. Stress from the earthquakes and as much as 1.8 m of uplift weakened buildings, forcing out 60% of Pozzuoli's 70,000 residents. New housing was constructed within the caldera but outside the present zone of uplift and seismicity. However, many residents chose to return to Pozzuoli after activity ceased, continuing several millenia of uneasy coexistence between humans and volcanoes in this densely populated region.

Rabaul caldera forms a natural harbor at the north end of New Britain Island, and 70,000 people live within 15 km of the caldera center. After about 500 died in the 1937 eruption, the Rabaul Volcanological Observatory was founded, and long-term monitoring by this group has documented increased seismicity there since 1971. Uplift has continued since the first measurements in 1973. In late August 1983, however, a major increase in seismicity followed by new uplift prompted a warning from volcanologists that an eruption was possible within a few months. Frequent vigorous seismic swarms associated with strong local uplift were thought to indicate intrusions of magma to within 1 km of the surface. Activity peaked in April 1984 with nearly 14,000 recorded events, then declined gradually. By November, volcanologists were able to reduce the hazard warning, and seismicity had declined to pre-crisis levels by August 1985. Thus, all 3 caldera crises ended without eruption, but careful monitoring continues and few volcanologists would be surprised by a renewal of unrest.

Near the end of the decade, the dangers of volcanic gases—unspectacular, yet lethal gases—were again brought to the attention of a volcanological community preoccupied by the threat of large eruptions. In February 1979, on Java's Dieng Plateau, carbon dioxide had flowed down a hillside from two small craters, killing 149 persons fleeing a small predawn explosive eruption. But it was in the African nation of Cameroon that two lakes, not previously recognized as volcanic, killed residents with no warning. At Lake Monoun in August 1984, a gas cloud was emitted from the lake after an audible explosion, and 37 people in low-lying areas were suffocated. Although details of the episode remained uncertain and there was no evidence of a true volcanic eruption, victims had skin damage and survivors on the fringes of the whitish, smoke-like cloud reported that it smelled bitter and acidic. The region is volcanically young, and vents at the lake bottom are believed to have gradually fed CO_2 into the denser bottom waters. An overturn of the density-stratified lake, perhaps triggered by an earthquake or internal seiche, could then have suddenly exsolved lethal quantities of CO_2. Support for this hypothesis comes from a similar but much more deadly episode two years later and only 100 km to the WNW. In August 1986 (beyond the period covered in this volume), 1700 persons around Lake Nyos were suffocated by a dense cloud of CO_2 that flowed down several valleys and through streamside villages. Although some scientists advocate a sudden influx of CO_2 from vents in the bottom of the lake, rather than rapid lake overturn releasing exsolved gas from CO_2-rich bottom water, geologists and limnologists are working together to understand this newly recognized hazard before the tragedy is repeated.

The report decade began with a volcanic threat leading to a controversial evacuation of people in Guadeloupe. It came full circle and ended with the tragic consequences of a warning that was never heeded in Colombia. The 13 November 1985 eruption of Nevado del Ruiz killed about 22,000 people, history's fourth largest death toll from a single eruption (after 92,000 at Tambora in 1815, 36,000 at Krakatau in 1883, and 28,000 at Pelée in 1902). Felt seismicity near the summit had begun a year earlier, increased fumarolic activity was noted by January 1985, and seismic and thermal activity continued through the spring and summer. A small explosion on 11 September caused light ashfalls and spawned a mudflow that moved 27 km down one valley. Geologists prepared a hazards map that was released in preliminary form on 7 October and received front-page news coverage. The map warned of severe mudflow risk to river valleys as much as 5000 vertical meters below the summit should an eruption suddenly melt some of the summit's heavy icecap. A large mudflow in 1845 had killed about 1000 people and probably overran the site of Armero, the town shown by the map to be in the greatest danger. But experience in Guadeloupe and elsewhere has shown that long-term evacuations of large numbers of people are difficult, unpopular, and economically disruptive. Most residents of Armero and other threatened areas did not move.

The 13 November eruption was preceded by high-frequency earthquake swarms that began 6 days earlier and by 3 days of continuous volcanic tremor, but the seismicity was weaker than that associated with the September explosion. Small explosions began in midafternoon and by evening, ash began falling during a rainstorm on the doomed town of Armero. The paroxysmal explosion began shortly after 9 P.M., with hot pyroclastic flows turning summit ice to water that swiftly picked up ash and mud, roared down steep valleys and spilled out onto the plains below. Armero was inundated by 11:30. About 21,000 of its 25,000 citizens died, and several thousand perished in other valleys around the volcano. The eruption itself was not particularly large—it was considerably smaller than St. Helens in 1980—yet that serene and scenic snow cap, high above the valley floors, was a deadly time bomb. More than 2 years later, detailed monitoring continued to document vigorous seismicity and gas emission, with some deformation. However, installation and maintenance of a short-term warning system for future mudflows have proven to be very difficult, and the evacuation of a populated area on short notice remains a formidable challenge.

Such challenges must be met. More attention must be paid to understanding phenomena as different as CO_2-laden lakes, snow-capped volcanoes, and restless calderas. And attention must also be paid to the human problems of translating a hazard map into effective evacuation procedures. Enormous progress has been made in monitoring volcanoes and forecasting their activity, but only a small fraction of the world's dangerous volcanoes are adequately monitored. The human population of our planet continues to rise—inexorably and exponentially—yet the world's volcanoes show no sign of slowing down their activity. The successes of the last decade have been encouraging, but the death toll that ended it is a somber reminder of how much remains to be done.

CHRONOLOGY

This section provides a listing that is both a temporal index and tabular summary of the events reported in the book. To compress a decade of information into a 14-page table requires some abbreviations and conventions, which must first be introduced.

The chronology is adapted from a similar table in *Volcanoes of the World*, to which the reader may turn for more detailed description (and pre-1975 data). Along with *Date*, *Volcano*, and *Region* for each eruption, its location in the text is shown by Chapter (*CH*) and Page (*PP*), together with coded eruptive *Characteristics* and indicators of magnitude (*VEI* and *Vol*ume of both *L*ava and *T*ephra). In addition to finding the volcanic events of a given time period, it should be possible to scan the decade for events of a particular type or size. However, the reader interested only in the eruptions of a specific time period may safely ignore the coded complexities of the right-hand side of the table.

The chronology incorporates some information not found in the SEAN reports that make up this book. Some dates, volumes, and characteristics have been modified from our VRF database, which utilizes information received after publication of the intial SEAN reports. VEI assignments (see discussion below) are made only after receipt of all initial reports. The chronology also shows (in parentheses) the volcano names used in both the VRF and *Volcanoes of the World* when they differ significantly from those used in the SEAN reports.

ERUPTIONS

Individual eruption durations vary from the minutes-long Krafla borehole eruption to the centuries-long activity of Stromboli. Eruptions lasting many years are common in the volcanic record, and are often overlooked in listings of shorter eruptions. Therefore, this chronology lists all eruptions *continuing* from the preceding year at the start of each new year (unless an individual event at that volcano appears later in the same year). Thus every volcano known to be active in a given year should be listed in the entries for that year.

Significant explosive events are also listed individually, even if part of a longer eruption. To assist the reader in locating the principal events, each eruption is listed once in **boldface** type whereas its additional phases (including its start, in some cases) or continuation into subsequent years are shown in lightface *italics*. Normally the strongest event is the one shown in boldface. For example, the Mt. St. Helens eruption of May 18, 1980, is boldfaced, whereas the eruption's start and smaller explosive phases are shown in italics. Its nonexplosive dome-building activity of later years is shown only by "continuing" entries at the start of those years. Several volcanoes (e.g., Erta Ale, Yasour) have been erupting continuously for decades before the start of the 1975–85 report period. These are included in the chronology with italicized "continuing" entries but no boldfaced "principal event" line. Brief entries have been added to Part II for these volcanoes—for completeness of the decade's record—even when they were not covered by original SEAN reports. Of the 160 volcanoes represented in the chronology, events at 6 of them (Karua 1977, Lokon-Empung 1976-80, Didicas 1978, Ekarma 1980, Callaqui 1980, and Kick-'em-Jenny 1977) were not reported to SEAN, but have been added here.

Many of SEAN's reports have covered geophysical unrest that did not result in an eruption. This non-eruptive unrest, when described in the text, is also included (in italics) in the chronology.

DATES AND UNCERTAINTIES

Dates are listed in year/month/day format. Many uncertainties surround historic eruptions—even in the last decade—and codes both before and after the eruption date are used to show these uncertainties.

A **?** *before* the date indicates uncertainty about the eruption itself (as in questionable submarine eruption reports or subaerial activity that may have been only increased fumarolic emission), while the same symbol *after* the date indicates that only the date is uncertain. When the location of the eruption is the main uncertainty (as when the activity may have been from any one of several neighboring volcanoes), a commercial "at" symbol (@) is used before the date. The only other code that precedes a date is the letter **S**, which indicates a probable submarine eruption detected by geophysical means (generally seismically or acoustically) but not directly witnessed. Seven of the 8 episodes designated with this code were sequences of T-Phase signals from Macdonald Seamount, recorded by the Polynesian Seismic Net. Macdonald was discovered in 1967 by direct submarine acoustic monitoring, but data from such defense-related programs are no longer available to volcanologists.

Codes *after* dates are normally lowercase letters indicating the range of uncertainty surrounding that date, but a **?** is used when no range is known (yet the date is questionable). The letters **a** through **j** indicate 1 through

10 days. Additional uncertainty codes used are at longer intervals, corresponding to longer time units, and are shown in Table 1 (and on the inside front cover). The only other code used after eruption dates is the "less than" symbol (<), indicating that the eruption was already in progress at the time of the first report.

Duration information (shown here in days) normally carries larger uncertainties, since the end of an eruption is inherently less dramatic than its beginning and consequently is not as well reported. When the length of an eruption is known, or can be approximated, it is shown in the duration column. The end of an eruption may be known to the month but not to the day, or only a range of possible dates for the eruption's end is known ("still going" at one observation, but "ended" by the next): This is shown by one of the uncertainty codes described above. If even the range is unknown, a **?** follows the uncertainty letter code. When several phases of a single eruption are listed, the duration appears only with the first entry (and is boldfaced even if the rest of the initial entry is not). Some 53 durations were less than a full day (**<1**), and 11 were greater than 999 days (**>3yr**). Finally, there are the eruptions that were continuing at the time of the last report: These are indicated by the > symbol at the end of the duration entry.

This decade's data shows that eruption record-keeping has been improving. In *Volcanoes of the World*, 62% of the 5,564 eruptions listed had no termination date, and hence no duration, but that figure has dropped to only 22% for the eruptions of this book's report period. Durations for 59% of the eruptions were recorded to ±1 day.

CHARACTERISTICS: INTRODUCTION

Forty years ago, when organizing the *Catalogue of Active Volcanoes of the World*, the International Association of Volcanology established 20 specific eruptive characteristics to be checked for each eruption. These were tabulated in *Volcanoes of the World* and are shown in the same format here as a handy way of finding eruptions of a particular type. The presence, or possible presence, of a characteristic is shown by an **X** or **?** in the appropriate column. A dash (-) in that column, however, does not necessarily mean the absence of that characteristic; only that it is not clear from the reported information. The characteristics for the *full* eruption are repeated for each phase or continuing year, but appear in boldface uppercase type only with the principal entry (thus any reader counting occurrences of a particular characteristic should ignore those appearing in non-bold, lowercase, italicized type). Readers interested in any of these 20 characteristics should use these columns as an index to them. Because so much additional information is available in the chronology—to guide readers to the most appropriate part of the text—we have worked to make these columns complete and have not included these subject words in the index (Part III).

Code		Value	Code		Value
a	=	± 1 day	*n*	=	± 25 days
b	=	2 days	*p*	=	30 (1 month)
c	=	3	*q*	=	45
d	=	4	*r*	=	60 (2 months)
e	=	5	*s*	=	75
f	=	6	*t*	=	90 (3 months)
g	=	7	*u*	=	120 (4 months)
h	=	8	*v*	=	150 (5 months)
i	=	9	*w*	=	180 (½ year)
j	=	10	*x*	=	270
k	=	12	*y*	=	365 (1 year)
l	=	15	*z*	=	545
m	=	20	*	=	>730 (2 years)

Table 1. Uncertainty codes used with Chronology dates and durations, to cover accounts such as "the eruption began sometime between the 10th and 12th of May and ended in July" (shown in the chronology as a 0511*a* start and 65*l* duration).

We have divided the 20 standard characteristics into 5 sets, each with 4 related codes. They are described below, with references to illustrative figures in Part II where appropriate.

CHARACTERISTICS I: LOCATION

The first set indicates where the eruption took place:

- *Central* crater activity was reported for more than ⅔ of the decade's 388 eruptions, and nearly 85% of those for which an eruption site was given.
- *Flank* activity not associated with well-developed radial or regional fissures account for only about 10% of the decade's eruptions. An example is the SW foot of Santa María volcano (Guatemala), where extrusion of viscous lava has continued since 1922, building Santiaguito Dome within the flank crater left by the enormous 1902 eruption (Fig. 14-5).
- *Radial fissure* eruptions characterize the behavior of several frequently active sites, with fully half from just 3 volcanoes: Etna (Sicily), Piton de la Fournaise (Réunion I.), and Kilauea (Hawaii). Such activity may originate from a series of fissures that form a distinctly radial pattern around a central crater, as elegantly shown by the French Geological Survey's 1985 *Carte des Coulées Historiques du Volcan de la Fournaise* and the Italian National Research Council's 1979 *Carta Geologica del Monte Etna*, or occur repeatedly from different segments of well-defined rift zones as at Kilauea.
- *Regional fissure* eruptions are concentrated along linear rifts, many tens of kilometers long, that show no orientation radial to local volcanic centers. Along the Ardoukoba Rift (Djibouti), for example, lava issued from a new 500-m fissure in 1978 (Fig. 2-2). Although activity soon concentrated at a single vent, about 25 fissures, parallel to the rift trend, opened during the activity. Of the entries in this column, all but Plosky Tolbachik are from Icelandic or East African rifts.

CHARACTERISTICS II: WATER-RELATED

The next set of 4 characteristcs addresses eruption interactions with water:

- *Submarine* eruptions account for only 15% of the total documented during the period, but civilian monitoring of the deep oceans is almost nonexistent and detects only a small proportion of the actual activity. French seismic monitoring on Tahiti recorded acoustic waves (T-phase) from 17 eruptions at Macdonald Seamount since its discovery in 1967, as well as seismicity that may have been associated with eruptions or intrusive episodes in the sea off Mehetia and at Teahitia Seamount. Extensive rafts of floating pumice drifted thousands of kilometers from eruptions in the Tonga Islands at Metis Shoal (1979) and Home Reef (1984).

• *New island* formation was reported with only 4 of the submarine eruptions: at Kavachi (Solomon Islands) in 1976 and 1978 (Fig. 5-12), at Metis Shoal, and at Home Reef. Islands built primarily of tephra are not long able to resist the ocean's pounding, and most disappear below sea level within months.

• *Subglacial* eruptions seem uncommon, with only 5 reported between 1975 and 1986, but most occur in remote regions and many may remain undocumented. One of the world's most active sites of subglacial volcanism, Grímsvötn caldera in Iceland, has been the source of more than 20 jökulhlaups (glacier bursts, or sudden catastrophic flooding from beneath the ice) since 1332.

• *Crater lakes* bring water into close, potentially explosive proximity with underlying magma, and, like glaciers, can supply the large volumes of water that feed lethal mudflows. Of the decade's 24 eruptions that occurred through a crater lake, more than $^2/_3$ were from New Zealand's Ruapehu (Fig. 4-9) and Costa Rica's Poás. Both volcanoes combine persistent minor activity with good local monitoring, and activity at more remote crater lakes has no doubt gone unreported.

CHARACTERISTICS III: EXPLOSIVE

The next set of 4 characteristics emphasize the explosive fragmentation of gas-charged magma, in contrast to the more gentle effusion of lava flows.

• *Explosive* activity occurred during more than $^3/_4$ of the decade's eruptions. An X in this column indicates that tephra fall occurred, even when the explosive activity itself was not witnessed, but does not necessarily indicate that fresh magma was present in the ejecta. Measurable tephra from the 1980 St. Helens eruption fell more than 500 km away (in Montana), but several of its prehistoric eruptions were significantly more voluminous.

• *Pyroclastic flows* are hot avalanches that move downslope at hurricane speeds, often leaving nearly total destruction in their wake (Fig. 6-14). Pyroclastic flows are often channeled down stream valleys, or surges may spread outward along a broad front from the source of a violent explosive eruption, devastating everything within several kilometers of the crater (as at El Chichón in 1982 and Una Una in 1983). Pyroclastic flows are associated with the more violent eruptions: they were reported in over half of the 49 most explosive eruptions of the decade (VEI ≥ 3—see VEI explanation below) but in only 4 of the 194 eruptions with VEI <2.

• *Phreatic* eruptions are those in which cold water interacts with hot (but not molten) rock, driving explosive activity that ejects fragmented older rock rather than fresh magma. Noteworthy examples include the 1976 activity at Guadeloupe, the pre-May 18 explosions at St Helens, and the 1977 formation of Ukinrek Maars in Alaska. When such eruptions include a significant proportion of fresh magma in the ejecta they are termed phreato-magmatic, and designated here by an X in the "explosive" column. An X in both columns, as found in more than $^3/_4$ of the phreatic eruptions of the decade, indicates only the presence of significant tephra fall, and does not necessarily mean that fresh magma was involved in the eruption. Less than 24% of the decade's eruptions had reported phreatic activity, and $^2/_3$ of these were in the VEI = 1 group.

• *Sulfur* does not designate venting of sulfur-rich gases, a virtually universal characteristic of volcanism. Instead, an X in this column marks the comparatively rare emission of molten sulfur from volcanic vents. During the report period, such flows were reported at only Kirishima (Japan) and Poás (Costa Rica). On Jupiter's moon Io, sulfur is apparently volcanism's dominant volatile component, forming substantial surface deposits of airfall tephra and flows (Fig. 20-4).

CHARACTERISTICS IV: LAVA

Nonexplosive lava production is marked by the next group of characteristics:

• *Lava flows* were reported from only $^1/_3$ of the decade's eruptions, but are the dominant eruption product at the world's rift systems, most of which are submerged in the deep ocean basins. The most spectacular lava production of the decade occurred in Zaire, when the lava lake that had persisted for 49 years at Nyiragongo suddenly drained through flank fissures on 10 January 1977. Within 1 hour, 20,000,000 m^3 of extremely fluid melilite nephelinite lava had poured downslope at speeds up to 60 km/hour, leaving coatings only millimeters thick in places (Fig. 2-9). Some flows moved through populated areas, where they destroyed several villages, killing ~180 people and leaving 800 homeless.

• *Lava lakes* have also remained active for many years at Erebus (Antarctica), Erta Ale (Ethiopia), and Masaya (Nicaragua). Short-lived lava-lake activity was reported at 8 other volcanoes during the decade.

• *Domes* of lava are built from much more viscous material that forms a steep-sided body instead of flowing downslope. Gas pressure and structural instability make lava domes prone to partial collapse and sudden explosions that often generate dangerous pyroclastic flows. The May 1980 eruption of Mt. St. Helens was followed by repeated cycles of lava dome extrusion and destruction before magma was sufficiently degassed to allow growth of the current composite dome (Fig. 12-25). Some dome growth takes place completely below the surface. These are called cryptodomes, or endogenous domes, and are recognized by large uplift, such as that following a week of strong explosions at Usu (Hokkaido, Japan) in 1977. During almost 5 years of endogenous dome growth, some points were uplifted more than 180 m. Dome activity was reported at 18 of the 203 volcanoes described in this book.

• *Spines* are tabular extrusions of particularly viscous lava usually squeezed up from the surface of growing domes. After the disastrous 1902 eruption of Mont Pelée, a massive spine grew to 230 m in height before crumbling, but such features are structurally unstable and generally topple before attaining such size. Several spines were described during the growth of St. Helens' dome, two reaching heights of 60 m in early 1983 (Fig. 12-23).

CHARACTERISTICS V: HUMAN EFFECTS

The last set of characteristics emphasizes the human effects of eruptions:

- *Fatalities* are listed for 4% of the 1975–85 eruptions, and total close to 30,000. Most are from the 1985 Ruiz disaster (22,000) and El Chichón (1-2000), but the remaining 20 total nearly 1000 (Table 6). Large as those numbers are, we must count ourselves fortunate that they were not much larger; the modern world has been spared the enormous eruptions seen in the very recent geologic record, but such eruptions are inevitable in the future. Even eruptions like several in the last century would have far greater tolls today because of the exponential increases in the number of people threatened by volcanoes. In the last decade, however, evacuations of residents from densely populated agricultural areas have saved many thousands of lives during several volcanic crises, particularly in the Philippines and Indonesia where cooperation between government volcanological surveys and public safety officials is long established.
- *Damage* to property was reported for only 16% of the decade's eruptions, a minimum figure reflecting particularly destructive activity. Modest ashfalls are disruptive but do not often cause permanent damage, so are not usually assigned this characteristic unless crop damage was reported. Damage is understandably associated more often with the more explosive eruptions. Half of those larger than VEI = 2 included damage, compared with only 3 of the VEI = 0 eruptions. One of these 3, however, was the continuing East Rift eruption of Kilauea that has caused such well-publicized damage since 1983. Successful lava diversion on Etna in 1983 prevented some damage to property.
- *Mudflows* (or lahars) form when fragmental deposits on steep slopes are mobilized by water. These dense flows are particularly dangerous because their momentum often takes them far beyond the base of the volcano. Eruptions not only provide new ejecta for mudflows but can also release the water needed to mobilize it, by sudden drainage of a crater lake or melting of a snow/ice cap (as at Ruiz). For years after an eruption, unconsolidated tephra remains a hazard; heavy rains can trigger damaging mudflows such as those at Mayon (Philippines) during typhoon Daling in 1981. Mudflows were reported in 35 events—only 9% of the decade's eruptions, but 8 caused fatalities.
- *Tsunamis*, or devastating giant sea waves, can be generated when large masses of water are suddenly displaced by earthquake, landslide, or eruption. The most notorious volcanic tsunamis, at Krakatau in 1883, killed more than 34,000 people. In 1979 another Indonesian volcano, Iliwerung, produced a tsunami that killed hundreds when a landslide entered the sea. The only other fatal tsunami during the report period, generated by a major earthquake at Kilauea in 1975, killed two people on the south coast of Hawaii.

VOLCANIC EXPLOSIVITY INDEX (VEI)

One of the most commonly asked questions about eruptions is "how big was it?" Unfortunately, volcanologists have no global network of instruments like that used in earthquake studies to determine Richter Magnitude. There are several measures of eruption "bigness"—volume of tephra, eruptive cloud height, explosive energy, distance traveled by ballistic ejecta, etc.—but data are missing for many of today's eruptions, and those of the historic record may not even carry subjective descriptors like "huge," "small," or "major." To counter this problem, Newhall and Self[8] devised the Volcanic Explosivity Index (VEI) to provide a relative measure of the explosive vigor of eruptions. The VEI combines volume of products, eruption cloud height, and more qualitative information to yield an explosivity value on an open-ended scale ranging from 0, for nonexplosive events, to 7, for history's largest explosive eruption (Table 2). A VEI can be assigned with more confidence if several types of information yield the same value, but volume of tephra, if known, supersedes atmospheric data. Only one event of the decade, the 18 May 1980 St. Helens eruption, rated a VEI of 5. The two closely spaced El Chichón eruptions, both of which rated 4s, had a combined volume that just exceeded 10^9 m^3, qualifying the full eruption for a VEI of 5. Five other individual eruptions rated 4s during the decade: Augustine 1976, Alaid 1980, Pagan 1980, Galunggung 1982, and Una Una 1983. Plosky Tolbachik's major explosive activity had declined by the time SEAN started reporting, but its total 1975–76 activity also rated a 4. Table 2 includes the count of eruptions in other VEI groups both for the 1975–85 decade and for all historic eruptions in our files.

In the Chronology, VEI values are listed in 2 columns if the eruption has more than one entry (i.e., multiphase or continuing year). The lefthand column lists the VEI for the eruptive phase on that line, or date. The righthand column contains an asterisk (*) on the line describing the most vigorously explosive phase, while in lines tabulating weaker phases it shows the strongest VEI of the eruption. The start of the 1980 St. Helens eruption, for example, has "3 5" in the VEI columns, indicating a VEI of 3 for the initial explosions and 5 for the pyroxysmal event. That event, on May 18, is shown by "5 *." Thus the righthand column swiftly shows whether there are other dates in the chronology for that particular eruption and the value of its maximum VEI.

VOLUME OF LAVA AND TEPHRA

No system comparable to the VEI has been developed for the relative vigor of nonexplosive eruptions. Lava production rates have been directly measured in a few instances, and can be derived in other cases when

8 Newhall, C. G., and Self, S., 1982, The Volcanic Explosivity Index (VEI): An Estimate of Explosive Magnitude for Historical Volcanism; *JGR (Oceans & Atmospheres)*, v. 87, p. 1231-38.

	Volcanic Explosivity Index								
	0	1	2	3	4	5	6	7	8
General Description	Non-Explosive	Small	Moderate	Moderate Large	Large	Very Large →			
Volume of Tephra (m^3)	10^4	10^6	10^7	10^8	10^9	10^{10}	10^{11}	10^{12}	
Cloud Column Height[1] (km)	<0.1	0.1-1	1-5	3-15	10-25	25 →			
Qualitative Description	Gentle, effusive	← Explosive →		← Cataclysmic, paroxysmal, colossal → / ← Severe, violent, terrific →					
Classification	Hawaiian	← Strombolian →		← Vulcanian → / ← Plinian →		← Ultra-Plinian →			
Total Historic Eruptions	487	623	3176	733	119	19	5	2	0
1975-1985 Eruptions	70	124	125	49	7	1	0	0	0

[1] For VEI's 0-2, data are km above crater; for VEI's 3-8, data are in km above sea level.

Table 2. Relationship between Volcanic Explosivity Index (VEI) and other measures of eruption magnitude.

the timing and volume of an eruption are well constrained. However, eruptive volumes can often be measured long after the end of an eruption. Although rapid erosion of tephra and uncertain lava flow thicknesses cause problems in generating accurate values, order-of-magnitude determinations provide a useful comparison between eruptions. When volume estimates have been made (only 25% of the 388 individual eruptions listed here), they are shown in the Chronology simply as order-of-magnitude values in the final columns. The exponent (10^x) of the volume in m^3 is listed: for example, 5.6×10^8 m^3 of lava would be shown as an 8 in the L column under VOL. Volumes of the decade ranged from history's smallest, 26 m^3 from an Icelandic borehole in 1977, to 2.2×10^9 m^3 (2.2 km^3) in the 1975–76 Plosky Tolbachik eruption on the Kamchatka peninsula. The decade fortunately lacked any of the gigantic eruptions of the recent geologic record. A Yellowstone eruption only 2 million years ago, for example, produced 2500 km^3 in a single, unbroken event[9]—5000 times the magmatic production of St. Helens in 1980.

Total volumes are listed and tephra values are bulk, rather than magmatic (or dense-rock equivalent) volumes. As with "Characteristics," volumes for the entire eruption are repeated for individual phases or continuing years, where they appear as lightfaced italics. Readers scanning the volume data should pay attention to the boldfaced single values for each eruption, rather than the italicized repeat entries. But more importantly, readers of the volume data must keep in mind the uncertainties surrounding these estimates and the many events for which no estimates at all have been made.

9 Christiansen, R. L., 1984, Yellowstone Magmatic Evolution: Its Bearing on Understanding Large-Volume Explosive Volcanism, *in* Boyd, F. R. (ed.), *Explosive Volcanism: Inception, Evolution, and Hazards*; National Academy of Sciences Press, Washington, p. 84-95.

DATE Year	MoDy	DURA-TION	VOLCANO NAME (Subregion)	CH	PP	CHARACTERISTICS (CENTRAL VENT, FLANK VENT, RADIAL FISSURE, REGIONAL / SUBMARINE, NEW ISLAND, SUBGLACIAL, CRATER LAKE / EXPLOSIVE, PYRO FLOW, PHREATIC, SULFUR / LAVA FLOW, LAVA LAKE, DOME, SPINE / FATALITIES, DAMAGE, MUDFLOW, TSUNAMI)	VEI	VOL L/T
1975	....	.?..	**KRAKATAU** (Indonesia)	6		X--- ---- X--- X--- ----		
1975	....	.?..	**NISHINO-SHIMA** (Volcano Is-Japan)	8		---- ?--- ---- ---- ----	0	
1975	....	....	**EREBUS, MOUNT** (Antarctica)	19	573	X--- ---- X--- XX-- ----	2 *	
1975	01..	>3yr	**MARAPI** (Sumatra)	6		X--- ---- X-X- ---- XXX-	2 *	
1975	0212	11*a*	**NGAURUHOE** (New Zealand)	4		X--- ---- XX-- ---- --X-	3	-/6
1975	0224	186*a*	**ETNA** (Italy)	1		--X- ---- X--- X--- ----	1	6/-
1975	03..	91*p*	**LOPEVI** (Vanuatu-SW Pacific)	5		XX-- ---- X--- ---- ----	2	
1975	03..	60*p*	**RAUNG** (Java)	6		X--- ---- X-X- ---- ----	1	
1975	04..	....	**BAGANA** (Bougainville-SW Pac)	5		X--- ---- XX-- X-X- ----	2 *	
1975	04..	.?..	**ESMERALDA BANK** (Mariana Is-C Pac	8		---- X--- ---- ---- ----	0	
1975	*04..*	*....*	*SANTIAGUITO (SANTA MARIA)*	*14*	*475*	*-x-- ---- xxx- x-xx xxx-*	*2 3*	*8/8*
1975	0424	3	**RUAPEHU** (New Zealand)	4	123	---- ---X XXX- ---- -XX-	2	
1975	0528	166*a*	**FUEGO** (Guatemala)	14	480	X--- ---- XX-- ---- ----	2	
1975	*0617*	*....*	*ARENAL (Costa Rica)*	*14*		*xxx- ---- xxx- x--- xxx-*	*2 3*	*7/7*
1975	*0628*	*535a*	*PLOSKY TOLBACHIK (Kamchatka)*	*10*	*323*	*xx-x ---- x--- x--- ----*	*1 4*	*8/9*
1975	0705	<1	**MAUNA LOA** (Hawaiian Is)	13	454	X-X- ---- ---- X--- ----	0	7/-
1975	0706	....	**PLOSKY TOLBACHIK** (Kamchatka)	10	323	XX-X ---- X--- X--- ----	4 *	8/9
1975	*0723*	*142a*	*COTOPAXI (Ecuador)*	*15*	*526*	*(Increased fumarolic activity)*		
1975	08..	14*c*	**KAVACHI** (Solomon Is-SW Pac)	5		---- X--- X--- ---- ----	0	
1975	0801	119*a*	**ETNA** (Italy)	1		X--- ---- X--- X--- ----	2	
?1975	0824?	83*l*	**LOIHI SEAMOUNT** (Hawaiian Is)	13		---- ?--- ---- ---- ----	0	
1975	0825	.?..	**MINAMI-HIYOSHI SEAMOUNT** (Volcano	8	290	---- X--- ---- ---- ----	0	
?1975	09..	.?..	**CLEVELAND** (Aleutian Is)	11		---- ---- ---- ---- ----		
?1975	0910	.?..	**UNNAMED** (Izu Is-Japan)	8		---- ?--- ---- ---- ----	0	
?1975	0910	.?..	**UNNAMED** (Mariana Is-C Pac)	8		---- ?--- ---- ---- ----	0	
?1975	0910	.?..	**UNNAMED** (Mariana Is-C Pac)	8		---- ?--- ---- ---- ----	0	
1975	0913	549*l*>	**PAVLOF** (Alaska Peninsula)	11	334	X--- ---- X--- ?--- --?-	2 *	
1975	0913	43*e*	**SHISHALDIN** (Aleutian Is)	11	332	X--- ---- X--- ?--- --?-	2	
1975	10..	89*l*	**ASO** (Kyushu-Japan)	8		X--- ---- X--- ---- ----	1	
1975	10..	16*l*	**PACAYA** (Guatemala)	14	489	X--- ---- ---- X--- ----	0	
1975	1002	.?..	**TORI-SHIMA** (Izu Is-Japan)	8	288	---- X--- ---- ---- ----	2	
1975	1017<	.?..	**RUAPEHU** (New Zealand)	4		---- ---X --X- ---- ----	1	
?1975	11..	.?..	**KASUGA SEAMOUNT** (Volcano Is)	8	290	---- ?--- ---- ---- ----	0	
1975	11..	218*l*	**COLIMA, VOLCAN DE** (Mexico)	14	465	X--- ---- -X-- X-X- ----	2	8/-
1975	1104	20*a*	**STROMBOLI** (Italy)	1	50	X--- ---- X--- X--- ----	1	6/5
1975	1104	154*a*	**FOURNAISE, PITON DE LA** (Indian O	3	91	-XX- ---- X--- XX-- ----	1	7/5
1975	1129	406*a*	**ETNA** (Italy)	1	52	--X- ---- X--- X--- ----	2	
1975	1129	<1	**KILAUEA** (Hawaiian Is)	13	406	X-X- ---- ---- X--- ----	0	5/-
1975	1220	<1	**KRAFLA** (Iceland-N)	17	557	---X ---- ---- X--- ----	0	5/-
1976	*Continuing*		*STROMBOLI (Italy)*	*1*		*x--- ---- x--- ---- -x--*	*1?*	
1976	*Continuing*		*ETNA (Italy)*	*1*		*x--- ---- x--- ---- ----*	*2 **	
1976	*Continuing*		*ERTA ALE (Ethiopia)*	*2*		*x--- ---- ---- xx-- ----*	*0*	
1976	*Continuing*		*NYIRAGONGO (Africa-C)*	*2*		*x--- ---- ---- xx-- ----*	*0*	
1976	*Continuing*		*MANAM (New Guinea-NE of)*	*5*		*x--- ---- xx-- x--- -x--*	*2 **	
1976	*Continuing*		*LANGILA (New Britain-SW Pac)*	*5*		*xx-- ---- xxx- x-x- ----*	*2 3*	*6/-*
1976	*Continuing*		*RABAUL CALDERA (New Britain)*	*5*		*(Seismicity, deformation)*		
1976	*Continuing*		*BAGANA (Bougainville-SW Pac)*	*5*		*x--- ---- xx-- x-x- ----*	*1 2*	
1976	*Continuing*		*YASOUR (Vanuatu-SW Pacific)*	*5*		*x--- ---- x-x- ---- ----*	*2 3*	
1976	*Continuing*		*MARAPI (Sumatra)*	*6*		*x--- ---- x-x- ---- xxx-*	*1 **	
1976	*Continuing*		*DUKONO (Halmahera-Indonesia)*	*6*		*x--- ---- x--- x--- -x--*	*2 3*	
1976	*Continuing*		*SAKURA-JIMA (Kyushu-Japan)*	*8*		*x--- ---- xxx- -x-- xxx-*	*2 3*	
1976	*Continuing*		*ASO (Kyushu-Japan)*	*8*		*x--- ---- x--- ---- ----*	*1*	
1976	*Continuing*		*KARYMSKY (Kamchatka)*	*10*		*x--- ---- xx-- x-x- --x-*	*2 **	*7/6*
1976	*Continuing*		*PLOSKY TOLBACHIK (Kamchatka)*	*10*		*xx-x ---- x--- x--- ----*	*1 4*	*8/9*
1976	*Continuing*		*COLIMA, VOLCAN DE (Mexico)*	*14*		*x--- ---- -x-- x-x- ----*	*2*	*8/-*
1976	*Continuing*		*MASAYA (Nicaragua)*	*14*		*x--- ---- x--- -x-- -x--*	*0 1*	
1976	*Continuing*		*GALERAS (Colombia)*	*15*		*x--- ---- x--- ---- ----*		
1976	*Continuing*		*CHILLAN, NEVADOS DE (Chile-C)*	*15*		*-x-- ---- x--- x-x- ----*	*2 **	*5/-*
1976	*Continuing*		*EREBUS, MOUNT (Antarctica)*	*19*		*x--- ---- x--- xx-- ----*	*1 **	

DATE Year	MoDy	DURA-TION	VOLCANO NAME (Subregion)	CH	PP	CHARACTERISTICS (CENTRAL VENT, FLANK VENT, RADIAL FISSURE, REGIONAL " / SUBMARINE, NEW ISLAND, SUBGLACIAL, CRATER LAKE / EXPLOSIVE, PYRO FLOW, PHREATIC, SULFUR / LAVA FLOW, LAVA LAKE, DOME, SPINE / FATALITIES, DAMAGE, MUDFLOW, TSUNAMI)	VEI	VOL L/T
1976	....	.?..	**IWO-JIMA** (Volcano Is-Japan)	8	290	X--- ---- --X- ---- ----	1	
1976	....	....	**SANGAY** (Ecuador)	15	526	X--- ---- X--- X--- X---	2	4/-
1976	0102u	.?..	**LONG ISLAND** (New Guinea-NE of) .	5	163	X--- ---- X--- ---- ----	1	
1976	0104	1331	**REVENTADOR** (Ecuador)	15	524	X--- ---- XX-- X--- ----	2	7/-
1976	0115e	.?..	**GAUA** (Vanuatu-SW Pacific)	5		X--- ---- X--- ---- ----	2	
1976	0122?	477a?	**AUGUSTINE** (Alaska-SW)	11	345	X--- ---- XX-- --X- -XX-	4 *	7/8
?1976	02..	.?..	**MINAMI-HIYOSHI SEAMOUNT** (Volcano	8		---- ?--- ---- ---- ----	0	
1976	0203?	101e?	**PACAYA** (Guatemala)	14	489	X--- ---- X--- ---- ----	1	
1976	*0204*	*....*	*MERAPI (Java)*	*6*	210	*x--- ---- xx-- x-x- --x-*	2 3	
1976	0206	>3yr	**LOKON-EMPUNG** (Sulawesi-Indonesia	6	228	X--- ---- X?-- --X- ----	2 *	4/-
1976	*0210*	*105e*	*MICOTRIN (W Indies)*	*16*	550	*(Seismic swarm)*		
1976	0302	<1	**KUSATSU-SHIRANE** (Honshu-Japan) .	8	279	X--- ---- X-X- ---- ----	2	
1976	0309	7a	**SAN CRISTOBAL** (Nicaragua)	14	493	X--- ---- X--- ---- ----	1	
1976	0325	.?..	**BEZYMIANNY** (Kamchatka)	10		X--- ---- XX-- X-X- --X-	2	
1976	0402	<1	**KUCHINOERABU-JIMA** (Ryukyu Is) ..	8	247	X--- ---- X-X- ---- ----	2	
1976	0406	<1	**RUAPEHU** (New Zealand)	4	123	X--- ---X --X- ---- ----	1	
1976	0406	175a?	**SHISHALDIN** (Aleutian Is)	11	333	X--- ---- X--- ---- --?-	2 *	
1976	*05..*	*857 >*	*KADOVAR (New Guinea-NE of)*	*5*	143	*(Increased thermal activity)*		
1976	0501	126a	**LOPEVI** (Vanuatu-SW Pacific)	5		X--- ---- X--- ---- ----	1	
1976	*0507*	*....*	*SANTIAGUITO (SANTA MARIA)*	*14*	476	*-x-- ---- xxx- x-xx xxx-*	3 *	*8/8*
1976	*0524*	*....*	*AMBRYM (Vanuatu-SW Pacific)*	*5*		*xx-- ---- x--- -x-- ----*	2 3	
1976	*06*	*....*	*MAKIAN (Halmahera-Indonesia)* ...	*6*	232	*(Evacuation, no eruption)*		
1976	0607	167a	**RAUNG** (Java)	6		X--- ---- X--- ---- ----	2	
1976	*0621*	*....*	*KILAUEA (Hawaiian Is)*	*13*	407	*(Magma intrusion)*		
1976	0621	1471	**POAS** (Costa Rica)	14	512	X--- ---X X--- ---- ----	2	
1976	0703	....	SUWANOSE-JIMA (Ryukyu Is)	8	243	x--- ---- x--- ---- ----	2 3	
1976	0708	236a	**SOUFRIERE GUADELOUPE** (W Indies)	16	541	X-X- ---- X-X- ---- -XX-	2	-/5
1976	*0714*	*....*	*KILAUEA (Hawaiian Is)*	*13*	407	*(Magma intrusion)*		
?1976	08..	>3yr	**FUKUTOKU-OKANOBA (SHIN-IWO-JIMA)**	8	290	---- ?--- ---- ---- ----	0	
?1976	0802	262a	**FUKUJIN SEAMOUNT** (Volcano Is) ..	8	292	---- ?--- ---- ---- ----	0	
1976	0823	5a	**NGAURUHOE** (New Zealand)	4		X--- ---- X--- ---- ----	1	
1976	0824	50a?	**KAVACHI** (Solomon Is-SW Pac)	5	192	---- XX-- X--- X--- ----	2	
1976	0829	<1	**SAN CRISTOBAL** (Nicaragua)	14	494	X--- ---- X--- ---- ----	1	
1976	*0831*	*....*	*SEMERU (Java)*	*6*	213	*x--- ---- xx-- x-x- xxx-*	2 3	
1976	0903	44a	**TAAL** (Luzon-Philippines)	7	239	-X-- ---- X-X- ---- -X--	2	-/6
1976	0910	....	**PAVLOF** (Alaska Peninsula)	11	335	X--- ---- X--- ?--- --?-	2 *	
1976	0915	3651	**API SIAU** (Sangihe Is-Indonesia)	6	229	-X-- ---- X--- X--- XX--	2	7/-
1976	0923	9a	**SARYCHEV PEAK** (Kuril Is)	9	318	X--- ---- X--- X--- ----	2	6/-
1976	*0925e*	*5a*	*KRAFLA (Iceland-N)*	*17*	557	*(Magma intrusion)*		
1976	0927	....	**SHISHALDIN** (Aleutian Is)	11	333	X--- ---- X--- ---- --?-	2 *	
1976	*1012*	*....*	*ARENAL (Costa Rica)*	*14*	505	*xxx- ---- xxx- x--- xxx-*	2 3	*7/7*
1976	1015q	205q>	**AKUTAN** (Aleutian Is)	11	329	---- ---- X--- ---- ----	2	
1976	*1021*	*....*	*LA LORENZA (Colombia)*	*15*	517	*(Mud volcano eruption)*		
1976	*1031*	*1a*	*KRAFLA (Iceland-N)*	*17*	557	*(Magma intrusion)*		
1976	11..	17d	**RUAPEHU** (New Zealand)	4	123	X--- ---X X-X- ---- ----	1	
1976	1102	<1	**FOURNAISE, PITON DE LA** (Indian O	3	92	--X- ---- X--- X--- ----	1	5/4
1976	1103	438l>	**TELICA** (Nicaragua)	14	495	X--- ---- X-X- ---- ----	1	
?1976	1127	.?..	**MATTHEW ISLAND** (SW Pacific)	5	197	X--- ---- --?- ---- ----	1?	
?1976	12..	102a	**MINAMI-HIYOSHI SEAMOUNT** (Volcano	8	291	---- ?--- ---- ---- ----	0	
1976	1202	77a	**SAN MIGUEL** (El Salvador)	14	493	X--- ---- X--- ---- -X--	1	6/-
1976	*1218*	*>3yr*	*WHITE ISLAND (New Zealand)*	*4*	105	*x--- ---- xxx- ---- ----*	2 3	*-/6*
1976	1223	143e	**NYAMURAGIRA** (Africa-C)	2	80	-XX- ---- X--- XX-- ----	1	6/-
1977	*Continuing*		*STROMBOLI (Italy)*	*1*		*x--- ---- x--- ---- -x--*	1?	
1977	*Continuing*		*ERTA ALE (Ethiopia)*	*2*		*x--- ---- ---- xx-- ----*	0	
1977	*Continuing*		*NYAMURAGIRA (Africa-C)*	*2*		*-xx- ---- x--- xx-- ----*	1	*6/-*
1977	*Continuing*		*KADOVAR (New Guinea-NE of)*	*5*		*(Increased thermal activity)*		
1977	*Continuing*		*MANAM (New Guinea-NE of)*	*5*		*x--- ---- xx-- x--- -x--*	2 *	
1977	*Continuing*		*LANGILA (New Britain-SW Pac)* ...	*5*		*xx-- ---- xxx- x-x- ----*	2 3	*6/-*
1977	*Continuing*		*RABAUL CALDERA (New Britain)* ...	*5*		*(Deformation)*		

DATE Year	MoDy	DURA-TION	VOLCANO NAME (Subregion)	CH	PP	CHARACTERISTICS (Central vent, Flank vent, Radial fissure, Regional; Submarine, New island, Subglacial, Crater lake; Explosive, Pyro flow, Phreatic, Sulfur; Lava flow, Lava lake, Dome, Spine; Fatalities, Damage, Mudflow, Tsunami)	VEI		VOL L/T
1977	*Continuing*		*BAGANA (Bougainville-SW Pac) ...*	*5*		x--- ---- xx-- x-x- ----	1	2	
1977	*Continuing*		*YASOUR (Vanuatu-SW Pacific) ....*	*5*		x--- ---- x-x- ---- ----	2	3	
1977	*Continuing*		*MARAPI (Sumatra) ....*	*6*		x--- ---- x-x- ---- xxx-	1	*	
1977	*Continuing*		*MERAPI (Java) ....*	*6*		x--- ---- xx-- x-x- --x-	1	*	
1977	*Continuing*		*SEMERU (Java) ....*	*6*		x--- ---- xx-- x-x- xxx-	2	3	
1977	*Continuing*		*LOKON-EMPUNG (Sulawesi-Indonesia*	*6*		x--- ---- x?-- --x- ----	2	*	*4/-*
1977	*Continuing*		*API SIAU (Sangihe Is-Indonesia)*	*6*		-x-- ---- x--- x--- xx--	1		*7/-*
1977	*Continuing*		*DUKONO (Halmahera-Indonesia) ...*	*6*		x--- ---- x--- x--- -x--	2	3	
1977	*Continuing*		*SUWANOSE-JIMA (Ryukyu Is) ......*	*8*		x--- ---- x--- ---- ----	2	3	
1977	*Continuing*		*SAKURA-JIMA (Kyushu-Japan) .....*	*8*		x--- ---- xxx- -x-- xxx-	2	3	
?1977	*Continuing*		*FUKUTOKU-OKANOBA (SHIN-IWO-JIMA)*	*8*		---- ?--- ---- ---- ----	0		
?1977	*Continuing*		*MINAMI-HIYOSHI SEAMOUNT (Volcano*	*8*		---- ?--- ---- ---- ----	0		
1977	*Continuing*		*KARYMSKY (Kamchatka) ....*	*10*		x--- ---- xx-- x-x- --x-	2	*	*7/6*
1977	*Continuing*		*AKUTAN (Aleutian Is) ....*	*11*		---- ---- x--- ---- ----	2		
1977	*Continuing*		*PAVLOF (Alaska Peninsula) ......*	*11*		x--- ---- x--- ?--- --?-	2	*	
1977	*Continuing*		*SAN MIGUEL (El Salvador) ....*	*14*		x--- ---- x--- ---- -x--	1		*6/-*
1977	*Continuing*		*TELICA (Nicaragua) ....*	*14*		x--- ---- x-x- ---- ----	1		
1977	*Continuing*		*MASAYA (Nicaragua) ....*	*14*		x--- ---- x--- -x-- -x--	0	1	
1977	*Continuing*		*ARENAL (Costa Rica) ....*	*14*		xxx- ---- xxx- x--- xxx-	1	*	*7/7*
1977	*Continuing*		*GALERAS (Colombia) ....*	*15*		x--- ---- x--- ---- ----			
1977	*Continuing*		*SANGAY (Ecuador) ....*	*15*		x--- ---- x--- ---- ----	2		
1977	*Continuing*		*CHILLAN, NEVADOS DE (Chile-C) ..*	*15*		-x-- ---- x--- x-x- ----	2	*	*5/-*
1977	*Continuing*		*SOUFRIERE GUADELOUPE (W Indies)*	*16*		x-x- ---- x-x- ---- -xx-	1		*-/5*
1977	*Continuing*		*EREBUS, MOUNT (Antarctica) .....*	*19*		x--- ---- x--- xx-- ----	1	*	
1977	0110	<1	**NYIRAGONGO** (Africa-C)	2	83	X-X- ---- X--- X--- XX--	1		7/4
S1977	0114	<1	**KICK-'EM-JENNY** (W Indies)	16	554	---- X--- ---- ---- ----	0		
1977	*0120*	*<1*	*KRAFLA (Iceland-N) ....*	*17*	557	*(Magma intrusion)*			
1977	0120e	8*e*?	**AMBRYM** (Vanuatu-SW Pacific)	5		X--- ---- X--- ---- -X--	2		
1977	*0130*	*1a*	*NASU (Honshu-Japan) ....*	*8*	281	*(Seismic swarm)*			
?1977	0201	.?..	**KARUA** (Vanuatu-SW Pacific)	5	197	---- ?--- ---- ---- ----	0		
1977	*0208*	*1a*	*KILAUEA (Hawaiian Is) ....*	*13*	407	*(Magma intrusion)*			
1977	*0209*	*....*	*SANTIAGUITO (SANTA MARIA) ......*	*14*	476	-x-- ---- xxx- x-xx xxx-	3	*	*8/8*
1977	0222	.?..	**KAVACHI** (Solomon Is-SW Pac)	5	194	---- X--- X--- ---- ----	1		
1977	0303	47*a*	**FUEGO** (Guatemala)	14	480	X--- ---- X--- ---- ----	1		
1977	0306	2*a*?	**SEGUAM** (Aleutian Is)	11	328	--X- ---- X--- X--- ----	1		
1977	*0311*	*....*	*WHITE ISLAND (New Zealand) .....*	*4*	106	x--- ---- xxx- ---- ----	2	3	*-/6*
1977	0319	9*a*?	**PURACE** (Colombia)	15	523	X--- ---- X--- ---- ----	2		
1977	0323	4*a*	**FERNANDINA** (Galapagos)	15	527	X--- ---- X--- X--- ----	1		
1977	0324	23*a*	**FOURNAISE, PITON DE LA** (Indian O	3	92	--X- ---- X--- X--- -X--	0		8/-
1977	0325	.?..	**BEZYMIANNY** (Kamchatka)	10	323	X--- ---- XX-- X-X- --X-	3		5/7
1977	0330	10*a*	**UKINREK MAARS** (Alaska Peninsula)	11	344	-X-- ---- X-X- -XX- ----	3		5/7
1977	0404	29*a*	**CONCEPCION** (Nicaragua)	14	503	X--- ---- X--- ---- ----	2		
1977	0405	5*a*	**KARTHALA** (Indian O.-W)	3	91	--X- ---- ---- X--- -X--	0		
1977	0411	102*a*	**ASO** (Kyushu-Japan)	8	266	X--- ---- X--- ---- -X--	2		
1977	*0411*	*....*	*AUGUSTINE (Alaska-SW) ....*	*11*		x--- ---- xx-- --x- -xx-	1	4	*7/8*
1977	0413	.?..	**GAUA** (Vanuatu-SW Pacific)	5		X--- ---- X--- ---- ----	2		
1977	0427	134	**KRAFLA** (Iceland-N)	17	558	---X ---- X-X- X--- ----	1		6/1
1977	05..	60*p*?	**POAS** (Costa Rica)	14		X--- ---X X-X- ---- ----	1		
1977	0609	21*a*	**RAUNG** (Java)	6		X--- ---- XX-- ---- ----	2		
1977	0704	<1	**NGAURUHOE** (New Zealand)	4	122	X--- ---- X--- ---- ----	1		
1977	0716	256*a*	**ETNA** (Italy)	1	52	X--- ---- X--- X--- -X--	3		
1977	0719	.?..	**KAVACHI** (Solomon Is-SW Pac)	5	194	---- X--- X--- ---- ----	1		
1977	*0802*	952*a*	*KLIUCHEVSKOI (Kamchatka) ....*	*10*		x-x- ---- x--- x--- ----	2	3	*6/-*
1977	0807	>3yr	**USU** (Hokkaido-Japan)	8	298	X--- ---- X-X- --X- XXX-	3	*	-/7
1977	0808?	86*a*?	**RUAPEHU** (New Zealand)	4	123	---- ---X XX-- ---- --X-	1		-/5
1977	0819	56*1*>	**PACAYA** (Guatemala)	14	490	X--- ---- X--- ---- ----	2		
1977	0825	....	**WHITE ISLAND** (New Zealand)	4	108	X--- ---- XXX- ---- ----	3	*	-/6
1977	0911	696*a*	**FUEGO** (Guatemala)	14	480	X--- ---- XX-- X-X- ----	2		-/6
1977	0913	17	**KILAUEA** (Hawaiian Is)	13	407	--X- ---- ---- X--- ----	0		7/-
1977	0930	<1	**AMBRYM** (Vanuatu-SW Pacific)	5		---- ---- ---- ---- ----			
1977	1003	39*a*?	**TAAL** (Luzon-Philippines)	7	241	-X-- ---- X-X- ---- ----	2		

DATE Year	MoDy	DURA-TION	VOLCANO NAME (Subregion)	CH	PP	CHARACTERISTICS (CENTRAL VENT, FLANK VENT, RADIAL FISSURE, REGIONAL ?; SUBMARINE, NEW ISLAND, SUBGLACIAL, CRATER LAKE; EXPLOSIVE, PYRO FLOW, PHREATIC, SULFUR; LAVA FLOW, LAVA LAKE, DOME, SPINE; FATALITIES, DAMAGE, MUDFLOW, TSUNAMI)	VEI	VOL L/T
?1977	1014	161*a*	**FUKUJIN SEAMOUNT** (Volcano Is) ..	8	292	---- X--- ---- ---- ----	0	
1977	1016	<1	**SAN CRISTOBAL** (Nicaragua)	14	494	X--- ---- X--- ---- ----	2	
1977	1017	549 >	**MONOWAI SEAMOUNT** (Kermadec Is) .	4	136	---- X--- ---- ---- ----	0	
1977	1024	24	**FOURNAISE, PITON DE LA** (Indian O	3	92	--X- ---- X--- X--- ----	1	7/-
1977	*1030*	*18a*	*OSHIMA (Izu Is-Japan) ....*	*8*	282	*(Seismic swarm)*		
1977	*1102*	*<1*	*KRAFLA (Iceland-N) ....*	*17*	560	*(Magma intrusion)*		
1977	1108	206*a*	**ASO** (Kyushu-Japan)	8	267	X--- ---- X-X- ---- ----	1	
1977	12..	>3yr	**COLIMA, VOLCAN DE** (Mexico)	14	465	X--- ---- XX-- X-X- ----	1	5/-
1977	1207	<1	**AZUMA** (Honshu-Japan)	8	282	X--- ---- X-X- ---- ----	1	
S1977	1210	5	**MACDONALD** (Pacific-C)	13	460	---- X--- ---- ---- ----	0	
1977	1218	159*e*	**POAS** (Costa Rica)	14	512	X--- ---X X--- ---- ----	2	
1978	*Continuing*		*STROMBOLI (Italy) ....*	*1*		*x--- ---- x--- ---- -x--*	*1?*	
1978	*Continuing*		*ERTA ALE (Ethiopia) ....*	*2*		*x--- ---- ---- xx-- ----*	*0*	
1978	*Continuing*		*WHITE ISLAND (New Zealand) .....*	*4*		*x--- ---- xxx- ---- ----*	*2 3*	*-/6*
1978	*Continuing*		*MONOWAI SEAMOUNT (Kermadec Is) .*	*4*		*---- x--- ---- ---- ----*	*0*	
1978	*Continuing*		*KADOVAR (New Guinea-NE of) .....*	*5*		*(Increased thermal activity)*		
1978	*Continuing*		*MANAM (New Guinea-NE of) ....*	*5*		*x--- ---- xx-- x--- -x--*	*2 **	
1978	*Continuing*		*LANGILA (New Britain-SW Pac) ...*	*5*		*xx-- ---- xxx- x-x- ----*	*2 3*	*6/-*
1978	*Continuing*		*RABAUL CALDERA (New Britain) ...*	*5*		*(Seismicity, deformation)*		
1978	*Continuing*		*BAGANA (Bougainville-SW Pac) ...*	*5*		*x--- ---- xx-- x-x- ----*	*1 2*	
1978	*Continuing*		*YASOUR (Vanuatu-SW Pacific) ....*	*5*		*x--- ---- x-x- ---- ----*	*2 3*	
1978	*Continuing*		*MERAPI (Java) ....*	*6*		*x--- ---- xx-- x-x- --x-*	*1 **	
1978	*Continuing*		*SAKURA-JIMA (Kyushu-Japan) .....*	*8*		*x--- ---- xxx- -x-- xxx-*	*2 3*	
1978	*Continuing*		*ASO (Kyushu-Japan) ....*	*8*		*x--- ---- x-x- ---- ----*	*1*	
?1978	*Continuing*		*FUKUTOKU-OKANOBA (SHIN-IWO-JIMA)*	*8*		*---- ?--- ---- ---- ----*	*0*	
1978	*Continuing*		*KARYMSKY (Kamchatka) ....*	*10*		*x--- ---- xx-- x-x- --x-*	*2 **	*7/6*
1978	*Continuing*		*KLIUCHEVSKOI (Kamchatka) ....*	*10*		*x-x- ---- x--- x--- ----*	*2 3*	*6/-*
1978	*Continuing*		*COLIMA, VOLCAN DE (Mexico) .....*	*14*		*x--- ---- xx-- x-x- ----*	*1*	*5/-*
1978	*Continuing*		*SANTIAGUITO (SANTA MARIA) ......*	*14*		*-x-- ---- xxx- x-xx xxx-*	*1 **	*8/8*
1978	*Continuing*		*FUEGO (Guatemala) ....*	*14*		*x--- ---- xx-- x-x- ----*	*2*	*-/6*
1978	*Continuing*		*TELICA (Nicaragua) ....*	*14*		*x--- ---- x-x- ---- ----*	*1*	
1978	*Continuing*		*MASAYA (Nicaragua) ....*	*14*		*x--- ---- x--- -x-- -x--*	*0 1*	
1978	*Continuing*		*ARENAL (Costa Rica) ....*	*14*		*xxx- ---- xxx- x--- xxx-*	*1 **	*7/7*
1978	*Continuing*		*SANGAY (Ecuador) ....*	*15*		*x--- ---- x--- ---- ----*	*2*	
1978	*Continuing*		*CHILLAN, NEVADOS DE (Chile-C) ..*	*15*		*-x-- ---- x--- x-x- ----*	*2 **	*5/-*
1978	*Continuing*		*EREBUS, MOUNT (Antarctica) .....*	*19*		*x--- ---- x--- xx-- ----*	*1 **	
1978	*....*	*....*	*FUSS PEAK (Kuril Is) ....*	*9*	319	*(Increased thermal activity)*		
1978	*01..*	*....*	*LOKON-EMPUNG (Sulawesi-Indonesia*	*6*	228	*x--- ---- x?-- --x- ----*	*1 2*	*4/-*
1978	01..	700*p*>	**RAUNG** (Java)	6		X--- ---- X-?- ---- ----	1	
1978	*01..*	*....*	*SEMERU (Java) ....*	*6*		*x--- ---- xx-- x-x- xxx-*	*1 3*	
1978	0106	3*a*	**DIDICAS** (Luzon Is-N of)	7	242	-X-- ---- X--- ---- ----	2	
1978	*0107*	*19a*	*KRAFLA (Iceland-N) ....*	*17*	560	*(Magma intrusion)*		
1978	*0113*	*....*	*USU (Hokkaido-Japan) ....*	*8*	302	*x--- ---- x-x- --x- xxx-*	*2 3*	*-/7*
?1978	0126?	57*a*?	**MINAMI-HIYOSHI SEAMOUNT** (Volcano	8	291	---- ?--- ---- ---- ----	0	
1978	0204	5*a*	**WESTDAHL** (Aleutian Is)	11	330	X--- ---- X--- ---- -XX-	3	
1978	0208	<1	**SHISHALDIN** (Aleutian Is)	11	333	X--- ---- X--- ---- ----	2	
1978	0222	299*a*	**API SIAU** (Sangihe Is-Indonesia)	6	229	X--- ---- X--- ---- ----	1	
1978	0307	316*a*	**RUAPEHU** (New Zealand)	4	124	X--- ---X X-X- ---- ----	1	
1978	*0307*	*1931?*	*MAYON (Luzon-Philippines) ......*	*7*	236	*x--- ---- x--- x--- ----*	*1 2*	*7/-*
1978	0330<	35*d*>	**CONCEPCION** (Nicaragua)	14	503	X--- ---- X--- ---- ----	2	
?1978	0331	.?..	**UNNAMED** (Kuril Is)	9	318	---- ?--- ---- ---- ----	0	
1978	*0419*	*....*	*SUWANOSE-JIMA (Ryukyu Is) ......*	*8*	244	*x--- ---- x--- ---- ----*	*2 3*	
1978	0429	39*a*	**ETNA** (Italy)	1	54	--X- ---- X--- X--- ----	2	7/-
1978	0507	7*a*	**ULAWUN** (New Britain-SW Pac)	5	174	X-X- ---- XX-- X--- -X--	3	6/7
1978	0507	.?..	**MAYON** (Luzon-Philippines)	7	236	X--- ---- X--- X--- ----	2 *	7/-
1978	0514	3*a*	**TARUMAI** (Hokkaido-Japan)	8	311	X--- ---- XX-- ---- ----	1	-/4
1978	0621	31*f*	**KAVACHI** (Solomon Is-SW Pac)	5	194	---- XX-- X--- X--- ----	2	
1978	0627	67*a*	**CANLAON** (Philippines-C)	7	233	X--- ---- X-X- ---- ----	2	
1978	07..	....	**DUKONO** (Halmahera-Indonesia) ...	6	231	X--- ---- X--- X--- -X--	3 *	
1978	*0707*	*....*	*KIRISHIMA (Kyushu-Japan) ....*	*8*	264	*(Seismic swarm)*		

DATE Year	MoDy	DURA-TION	VOLCANO NAME (Subregion)	CH	PP	CHARACTERISTICS (CENTRAL VENT, FLANK VENT, RADIAL FISSURE, REGIONAL / SUBMARINE, NEW ISLAND, SUBGLACIAL, CRATER LAKE / EXPLOSIVE, PYRO FLOW, PHREATIC, SULFUR / LAVA FLOW, LAVA LAKE, DOME, SPINE / FATALITIES, DAMAGE, MUDFLOW, TSUNAMI)	VEI	VOL L/T
1978	0710	97l>	**KRAKATAU** (Indonesia)	6	200	X--- ---- X--- ---- ----	1	-/4
1978	*0710*	*3a*	*KRAFLA (Iceland-N)*	*17*	561	*(Magma intrusion)*		
?1978	0720?	.?..	**TIATIA** (Kuril Is)	9	317	---- ---- --?- ---- ----		
1978	0729	16a	**BULUSAN** (Luzon-Philippines)	7	234	X--- ---- X-X- ---- -X--	2	-/5
1978	*0801*	*47e*	*SHASTA, MOUNT (US-Calif)*	*12*	398	*(Seismic swarm)*		
1978	0808	18a	**FERNANDINA** (Galapagos)	15	527	X--- ---- X--- X--- ----	2	
1978	0823	6a	**ETNA** (Italy)	1	55	--X- ---- X--- X--- ----	2	6/-
?1978	0824	1a	**FUKUJIN SEAMOUNT** (Volcano Is) ..	8	291	---- ?--- ---- ---- ----	0	
1978	*0908*	*....*	*MARAPI (Sumatra)*	*6*	199	x--- ---- x-x- ---- xxx-	2 *	
1978	0908p	.?..	**BEZYMIANNY** (Kamchatka)	10		X--- ---- X--- ---- --X-	2	
1978	0922	84l	**POAS** (Costa Rica)	14	512	X--- ---X X-XX ---- ----	1	
1978	0925e	.?..	**AKUTAN** (Aleutian Is)	11	329	X--- ---- X--- X--- ----	2	
1978	1107	7a	**ARDOUKOBA** (Djibouti)	2	77	---X ---- X--- XX-- ----	1	7/-
1978	*1107*	*<1*	*ILIAMNA (Alaska Peninsula)*	*11*	346	*(Steam emission)*		
1978	*1110*	*5a*	*KRAFLA (Iceland-N)*	*17*	561	*(Magma intrusion)*		
?1978	1116	.?..	**NISHINO-SHIMA** (Volcano Is-Japan)	8	288	---- ?--- ---- ---- ----	0	
1978	1118	11a	**ETNA** (Italy)	1	55	--X- ---- X--- X--- ----	2	7/-
1978	*1122*	**1131**	*LOPEVI (Vanuatu-SW Pacific)*	*5*		xx-- ---- x--- x--- ----	0 2	
1978	*1210*	*....*	*LAWU (Java)*	*6*	213	*(Seismic swarm)*		
1978	1211	<1	**IWO-JIMA** (Volcano Is-Japan)	8	290	X--- ---- --X- ---- ----	1	
1978	1212	150a	**TARUMAI** (Hokkaido-Japan)	8	312	X--- ---- X--- ---- ----	1	
1979	*Continuing*		*STROMBOLI (Italy)*	*1*		x--- ---- x--- ---- -x--	1?	
1979	*Continuing*		*ERTA ALE (Ethiopia)*	*2*		x--- ---- ---- xx-- ----	0	
1979	*Continuing*		*WHITE ISLAND (New Zealand)*	*4*		x--- ---- xxx- ---- ----	2 3	-/6
1979	*Continuing*		*MONOWAI SEAMOUNT (Kermadec Is)* .	*4*		---- x--- ---- ---- ----	0	
1979	*Continuing*		*MANAM (New Guinea-NE of)*	*5*		x--- ---- xx-- x--- -x--	2 *	
1979	*Continuing*		*LANGILA (New Britain-SW Pac)* ...	*5*		xx-- ---- xxx- x-x- ----	2 3	6/-
1979	*Continuing*		*RABAUL CALDERA (New Britain)* ...	*5*		*(Deformation)*		
1979	*Continuing*		*BAGANA (Bougainville-SW Pac)* ...	*5*		x--- ---- xx-- x-x- ----	1 2	
1979	*Continuing*		*YASOUR (Vanuatu-SW Pacific)*	*5*		x--- ---- x-x- ---- ----	2 3	
1979	*Continuing*		*MARAPI (Sumatra)*	*6*		x--- ---- x-x- ---- xxx-	1 *	
1979	*Continuing*		*MERAPI (Java)*	*6*		x--- ---- xx-- x-x- --x-	1 *	
1979	*Continuing*		*LAWU (Java)*	*6*		*(Seismic swarm)*		
1979	*Continuing*		*SEMERU (Java)*	*6*		x--- ---- xx-- x-x- xxx-	2 3	
1979	*Continuing*		*RAUNG (Java)*	*6*		x--- ---- x-?- ---- ----	1	
1979	*Continuing*		*LOKON-EMPUNG (Sulawesi-Indonesia*	*6*		x--- ---- x?-- --x- ----	2 *	4/-
1979	*Continuing*		*DUKONO (Halmahera-Indonesia)* ...	*6*		x--- ---- x--- x--- -x--	2 3	
1979	*Continuing*		*SUWANOSE-JIMA (Ryukyu Is)*	*8*		x--- ---- x--- ---- ----	2 3	
1979	*Continuing*		*SAKURA-JIMA (Kyushu-Japan)*	*8*		x--- ---- xxx- -x-- xxx-	2 3	
?1979	*Continuing*		*FUKUTOKU-OKANOBA (SHIN-IWO-JIMA)*	*8*		---- ?--- ---- ---- ----	0	
1979	*Continuing*		*USU (Hokkaido-Japan)*	*8*		x--- ---- x-x- --x- xxx-	0 *	-/7
1979	*Continuing*		*TARUMAI (Hokkaido-Japan)*	*8*		x--- ---- x--- ---- ----	1	
1979	*Continuing*		*KARYMSKY (Kamchatka)*	*10*		x--- ---- xx-- x-x- --x-	2 *	7/6
1979	*Continuing*		*KLIUCHEVSKOI (Kamchatka)*	*10*		x-x- ---- x--- x--- ----	2 3	6/-
1979	*Continuing*		*COLIMA, VOLCAN DE (Mexico)*	*14*		x--- ---- xx-- x-x- ----	1	5/-
1979	*Continuing*		*FUEGO (Guatemala)*	*14*		x--- ---- xx-- x-x- ----	2	-/6
1979	*Continuing*		*MASAYA (Nicaragua)*	*14*		x--- ---- x--- -x-- -x--	0 1	
1979	*Continuing*		*ARENAL (Costa Rica)*	*14*		xxx- ---- xxx- x--- xxx-	1 *	7/7
1979	*Continuing*		*SANGAY (Ecuador)*	*15*		x--- ---- x--- ---- ----	2	
1979	*Continuing*		*CHILLAN, NEVADOS DE (Chile-C)* ..	*15*		-x-- ---- x--- x-x- ----	2 *	5/-
1979	*Continuing*		*EREBUS, MOUNT (Antarctica)*	*19*		x--- ---- x--- xx-- ----	1 *	
1979	*0112?*	*209a?*	*KARKAR (New Guinea-NE of)*	*5*	157	x--- ---x x-x- ---- x-x-	2 *	-/6
1979	*0126*	*....*	*LONG ISLAND (New Guinea-NE of)* .	*5*	163	*(Increased fumarolic activity)*		
1979	0129	36a>	**AZUL, CERRO** (Galapagos)	15	532	XXX- ---- X--- X--- ----	2?	
1979	02..	.?..	**LOPEVI** (Vanuatu-SW Pacific)	5		XX-- ---- X--- X--- ----	2 *	
1979	02..	.?..	**SHISHALDIN** (Aleutian Is)	11	333	X--- ---- X--- ---- ----	2	
1979	0207	4a?	**AMBRYM** (Vanuatu-SW Pacific)	5	196	X--- ---- X--- ---- -X--	2	
?1979	0208	.?..	**WESTDAHL** (Aleutian Is)	11	331	---- ---- ?--- ---- ----	3?	
1979	0211	.?..	**BEZYMIANNY** (Kamchatka)	10	323	X--- ---- XX-- X--- ----	3	5/7

DATE Year	MoDy	DURA-TION	VOLCANO NAME (Subregion)	CH	PP	CHARACTERISTICS: CENTRAL VENT, FLANK VENT, RADIAL FISSURE, REGIONAL	SUBMARINE, NEW ISLAND, SUBGLACIAL, CRATER LAKE	EXPLOSIVE, PYRO FLOW, PHREATIC, SULFUR	LAVA FLOW, LAVA LAKE, DOME, SPINE	FATALITIES, DAMAGE, MUDFLOW, TSUNAMI	VEI	VOL L/T
1979	0216r	.?..	KIRISHIMA (Kyushu-Japan)	8	264	X---	----	--XX	----	----	1	
1979	0220	<1	DIENG VOLCANIC COMPLEX (Java) ..	6	209	-X--	----	--X-	----	XXX-	1	-/5
1979	0308	....	KARKAR (New Guinea-NE of)	5	158	X---	---X	X-X-	----	X-X-	2?*	-/6
1979	04..	543l	CHIRINKOTAN (Kuril Is)	9	318	X---	----	X---	X---	----	2	
1979	0413	196e	SOUFRIERE ST. VINCENT (W Indies)	16	551	X---	---X	XXX-	--X-	-XX-	3	7/7
1979	0414	36a	CARRAN-LOS VENADOS (Chile-C) ...	15	540	X---	----	X---	X---	----	2	6/6
?1979	0426	382a	FUKUJIN SEAMOUNT (Volcano Is) ..	8	291	----	?---	----	----	----	0	
1979	*0430*	*<1*	*MARAPI (Sumatra) ....*	*6*	199	*(Landslides, lahar)*						
1979	0510<	72 >	METIS SHOAL (Tonga-SW Pacific) .	4	136	----	XX--	X---	----	----	2	
1979	*0513*	*5a*	*KRAFLA (Iceland-N) ....*	*17*	562	*(Magma intrusion)*						
1979	0514	.?..	CURACOA (Tonga-SW Pacific)	4	138	----	X---	----	----	----	1	
1979	0528	47a	FOURNAISE, PITON DE LA (Indian O	3	94	--X-	----	X---	X---	----	1	5/-
1979	*0529*	*....*	*KILAUEA (Hawaiian Is) ....*	*13*		*(Magma intrusion)*						
1979	0531	<1	API SIAU (Sangihe Is-Indonesia)	6	230	-X--	----	X---	----	----	1	
1979	0606	101l?	AMBRYM (Vanuatu-SW Pacific)	5		X---	----	X---	----	----	2	
1979	0612	269a	ASO (Kyushu-Japan)	8	267	X---	----	X-X-	----	XX--	2	
?1979	0630	15a	RUAPEHU (New Zealand)	4	124	----	---X	--X-	----	----	1	
1979	*07..*	*<1*	*ILIWERUNG (Lesser Sunda Is) ....*	*6*	222	*(Landslide, tsunami)*						
1979	0702	72a	LOPEVI (Vanuatu-SW Pacific)	5	197	-X--	----	X---	----	----	2	
?1979	0712	.?..	NIKKO SEAMOUNT (Volcano Is-Japan	8	292	----	?---	----	----	----	0	
?1979	0713	.?..	MYOJIN-SHO (BAYONNAISE ROCKS) ..	8	287	----	?---	----	----	----	0	
1979	0716	25a	ETNA (Italy)	1	56	X-X-	----	X---	X---	-X--	2	
1979	0727	81l	KRAKATAU (Indonesia)	6	201	X---	----	X---	X---	----	2	4/-
1979	*0812*	*....*	*KILAUEA (Hawaiian Is) ....*	*13*		*(Magma intrusion)*						
1979	*0823*	*....*	*SANTIAGUITO (SANTA MARIA) ......*	*14*	476	*-x--*	*----*	*xxx-*	*x-xx*	*xxx-*	*2 3*	*8/8*
1979	0908	129l	POAS (Costa Rica)	14	512	X---	---X	X-X-	----	----	1	
1979	0912	33l	ETNA (Italy)	1	58	X---	----	X-X-	----	XX--	2	
1979	0918	.?..	BEZYMIANNY (Kamchatka)	10		X---	----	XX--	X---	----	2	5/6
S1979	0930	<1	MACDONALD (Pacific-C)	13	460	----	X---	----	----	----	0	
1979	1015	40 >	LLAIMA (Chile-C)	15	535	X---	----	X---	X---	--X-	2	
1979	1028	179a	ON-TAKE (Honshu-Japan)	8	274	X---	----	XXX-	----	-X--	2	
1979	1113	62 >	NEGRA, SIERRA (Galapagos)	15	530	----	----	X---	X---	-X--	3	
?1979	1115	.?..	FARALLON DE PAJAROS (Mariana Is-	8	292	----	X---	----	----	----	0	
1979	1116	<1	KILAUEA (Hawaiian Is)	13	408	--X-	----	----	X---	----	0	5/-
1979	*1203*	*5a*	*KRAFLA (Iceland-N) ....*	*17*	562	*(Magma intrusion)*						
1979	1205	132a	RUAPEHU (New Zealand)	4	124	----	---X	X-X-	----	----	1	
1979	*1227*	275a	*BULUSAN (Luzon-Philippines) ....*	*7*	234	*x-x-*	*----*	*x-x-*	*----*	*----*	*2 3*	
1980	*Continuing*		*STROMBOLI (Italy) ....*	*1*		*x---*	*----*	*x---*	*----*	*-x--*	*1?*	
1980	*Continuing*		*ERTA ALE (Ethiopia) ....*	*2*		*x---*	*----*	*----*	*xx--*	*----*	*0*	
1980	*Continuing*		*WHITE ISLAND (New Zealand) .....*	*4*		*x---*	*----*	*xxx-*	*----*	*----*	*2 3*	*-/6*
1980	*Continuing*		*MANAM (New Guinea-NE of) ....*	*5*		*x---*	*----*	*xx--*	*x---*	*-x--*	*2 **	
1980	*Continuing*		*RABAUL CALDERA (New Britain) ...*	*5*		*(Seismicity, deformation)*						
1980	*Continuing*		*BAGANA (Bougainville-SW Pac) ...*	*5*		*x---*	*----*	*xx--*	*x-x-*	*----*	*1 2*	
1980	*Continuing*		*YASOUR (Vanuatu-SW Pacific) ....*	*5*		*x---*	*----*	*x-x-*	*----*	*----*	*2 3*	
1980	*Continuing*		*MERAPI (Java) ....*	*6*		*x---*	*----*	*xx--*	*x-x-*	*--x-*	*1 **	
1980	*Continuing*		*SEMERU (Java) ....*	*6*		*x---*	*----*	*xx--*	*x-x-*	*xxx-*	*2 3*	
1980	*Continuing*		*LOKON-EMPUNG (Sulawesi-Indonesia*	*6*		*x---*	*----*	*x?--*	*--x-*	*----*	*2 **	*4/-*
1980	*Continuing*		*DUKONO (Halmahera-Indonesia) ...*	*6*		*x---*	*----*	*x---*	*x---*	*-x--*	*2 3*	
1980	*Continuing*		*SUWANOSE-JIMA (Ryukyu Is) ......*	*8*		*x---*	*----*	*x---*	*----*	*----*	*2 3*	
1980	*Continuing*		*SAKURA-JIMA (Kyushu-Japan) .....*	*8*		*x---*	*----*	*xxx-*	*-x--*	*xxx-*	*2 3*	
1980	*Continuing*		*ON-TAKE (Honshu-Japan) ....*	*8*		*x---*	*----*	*xxx-*	*----*	*-x--*	*2*	
?1980	*Continuing*		*FUKUTOKU-OKANOBA (SHIN-IWO-JIMA)*	*8*		*----*	*?---*	*----*	*----*	*----*	*0*	
?1980	*Continuing*		*FUKUJIN SEAMOUNT (Volcano Is) ..*	*8*		*----*	*?---*	*----*	*----*	*----*	*0*	
1980	*Continuing*		*USU (Hokkaido-Japan) ....*	*8*		*x---*	*----*	*x-x-*	*--x-*	*xxx-*	*0 **	*-/7*
1980	*Continuing*		*CHIRINKOTAN (Kuril Is) ....*	*9*		*x---*	*----*	*x---*	*x---*	*----*	*2*	
1980	*Continuing*		*KARYMSKY (Kamchatka) ....*	*10*		*x---*	*----*	*xx--*	*x-x-*	*--x-*	*2 **	*7/6*
1980	*Continuing*		*COLIMA, VOLCAN DE (Mexico) .....*	*14*		*x---*	*----*	*xx--*	*x-x-*	*----*	*1*	*5/-*
1980	*Continuing*		*MASAYA (Nicaragua) ....*	*14*		*x---*	*----*	*x---*	*-x--*	*-x--*	*0 1*	
1980	*Continuing*		*ARENAL (Costa Rica) ....*	*14*		*xxx-*	*----*	*xxx-*	*x---*	*xxx-*	*1 **	*7/7*

DATE Year	MoDy	DURA-TION	VOLCANO NAME (Subregion)	CH	PP	CHARACTERISTICS (CENTRAL VENT, FLANK VENT, RADIAL FISSURE, REGIONAL; SUBMARINE, NEW ISLAND, SUBGLACIAL, CRATER LAKE; EXPLOSIVE, PYRO FLOW, PHREATIC, SULFUR; LAVA FLOW, LAVA LAKE, DOME, SPINE; FATALITIES, DAMAGE, MUDFLOW, TSUNAMI)	VEI		VOL L/T
1980	*Continuing*		*SANGAY (Ecuador)*	*15*		x--- ---- x--- ---- ----	2		
1980	*Continuing*		*NEGRA, SIERRA (Galapagos)*	*15*		---- ---- x--- x--- -x--	0		
1980	*Continuing*		*CHILLAN, NEVADOS DE (Chile-C)*	*15*		-x-- ---- x--- x-x- ----	2	*	*5/-*
1980	*Continuing*		*EREBUS, MOUNT (Antarctica)*	*19*		x--- ---- x--- xx-- ----	1	*	
1980	....	....	**LANGILA** (New Britain-SW Pac)	5	165	XX-- ---- XXX- X-X- ----	3	*	6/-
1980	01..	.?..	**KLIUCHEVSKOI** (Kamchatka)	10	324	X-X- ---- X--- X--- ----	3	*	6/-
?1980	0107	10a?	**KARKAR** (New Guinea-NE of)	5	161	X--- ---- --?- ---- ----	1		
1980	0110	1a	**TUPUNGATITO** (Chile-C)	15	534	X--- ---- X--- ---- ----	2		
1980	0111	2481	**ETNA** (Italy)	1	58	X--- ---- X--- ---- ----	2		
1980	*0122*	....	*SANTIAGUITO (SANTA MARIA)*	*14*	477	-x-- ---- xxx- x-xx xxx-	2	3	*8/8*
1980	0130	25	**NYAMURAGIRA** (Africa-C)	2	80	-XX- ---- X--- X--- -X--	3		7/7
1980	*02..*	....	*BABUYAN (Luzon-Philippines)*	*7*	242	*(Landslide forces evacuation)*			
1980	0207	....	**BULUSAN** (Luzon-Philippines)	7	234	X-X- ---- X-X- ---- ----	3	*	
1980	*0210*	*5a*	*KRAFLA (Iceland-N)*	*17*	563	*(Magma intrusion)*			
S1980	0212	<1	**MACDONALD** (Pacific-C)	13	460	---- X--- ---- ---- ----	0		
1980	*0302*	*10a*	*KILAUEA (Hawaiian Is)*	*13*	410	*(Magma intrusion)*			
1980	0311	<1	**KILAUEA** (Hawaiian Is)	13		--X- ---- ---- X--- ----	0		0/-
1980	0313	<1	**IWO-JIMA** (Volcano Is-Japan)	8		X--- ---- --X- ---- ----	1		
1980	0315e	276e	**KRAKATAU** (Indonesia)	6	202	X--- ---- X--- X--- ----	2		
1980	0316	221	**KRAFLA** (Iceland-N)	17	563	---X ---- ---- X--- ----	0	*	7/-
1980	0324	173a	**API SIAU** (Sangihe Is-Indonesia)	6	230	X--- ---- X--- ---- ----	1		
1980	*0327*	>3yr	*ST. HELENS, MOUNT (US-Wash)*	*12*	347	xx-- ---- xxx- --x- xxx-	2	5	*7/9*
1980	0329	.?..	**MARAPI** (Sumatra)	6	199	X--- ---- X--- ---- ----	1		
1980	0415	127a	**LOPEVI** (Vanuatu-SW Pacific)	5		XX-- ---- X--- X--- -X--	3		
1980	0418	1a	**BEZYMIANNY** (Kamchatka)	10		X--- ---- XX-- X-X- --X-	3		5/7
1980	0501s	.?..	**MAKUSHIN** (Aleutian Is)	11	329	X--- ---- X-X- ---- ----	1		
1980	*0516*	*94a*	*AMBRYM (Vanuatu-SW Pacific)*	*5*		x--- ---- x--- ---- ----	2	3	
1980	0518	....	**ST. HELENS, MOUNT** (US-Wash)	12	349	XX-- ---- XXX- --XX XXX-	5	*	7/9
1980	0524	.?..	**EKARMA** (Kuril Is)	9	319	---- ---- X--- ---- ----	1		
1980	*0525*	>3yr	*LONG VALLEY CALDERA (US-Calif)*	*12*	399	*(Seismicity, deformation)*			
1980	*0525*	....	*ST. HELENS, MOUNT (US-Wash)*	*12*	355	xx-- ---- xxx- --x- xxx-	3	5	*7/9*
1980	0605a	107b	**BROMO (TENGGER CALDERA)** (Java)	6	215	X--- ---- X--- ---? ----	2		
1980	*0612*	....	*ST. HELENS, MOUNT (US-Wash)*	*12*	355	xx-- ---- xxx- --x- xxx-	3	5	*7/9*
1980	0615z	.?..	**SEAL NUNATAKS GROUP** (Antarctica)	19	580	XX-- ---- X--- X--- ----			
1980	*0620*	378a?	*GORELY (Kamchatka)*	*10*	322	x--- ---- xxx- ---- ----	2	3	*-/7*
1980	0620<	96 >	**VILLARRICA** (Chile-C)	15	537	X--- ---- XX-- ---- ----	2?		
1980	*07..*	*60p*	*MIYAKE-JIMA (Izu Is-Japan)*	*8*	283	*(Seismic swarm)*			
1980	0703<	5a?	**AKUTAN** (Aleutian Is)	11	329	X--- ---- X--- ?--- ----	2		
1980	*0706*	....	*HOOD, MOUNT (US-Oregon)*	*12*	397	*(Seismic swarm)*			
1980	0706a	.?..	**PAVLOF** (Alaska Peninsula)	11	335	X--- ---- X--- ---- ----	2		
?1980	0707	.?..	**NISHINO-SHIMA** (Volcano Is-Japan)	8		---- ?--- ---- ---- ----	0		
1980	0708	80a	**ETNA** (Italy)	1	60	X--- ---- X--- X--- ----	3		
1980	*0710*	....	*KRAFLA (Iceland-N)*	*17*	563	---x ---- ---- x--- ----	0	*	*7/-*
1980	*0722*	....	*ST. HELENS, MOUNT (US-Wash)*	*12*	357	xx-- ---- xxx- --x- xxx-	3	5	*7/9*
1980	0723	....	**AMBRYM** (Vanuatu-SW Pacific)	5		X--- ---- X--- ---- ----	3	*	
1980	0729	<1	**MALINAO** (Luzon-Philippines)	7	239	-X-- ---- --X- ---- -X--	1		
1980	0731	.?..	**GORELY** (Kamchatka)	10	322	X--- ---- XXX- ---- ----	3	*	-/7
1980	0807	41a	**GARELOI** (Aleutian Is)	11	327	X--- ---- X--- ---- ----	3?		
1980	*0807*	....	*ST. HELENS, MOUNT (US-Wash)*	*12*	358	xx-- ---- xxx- --x- xxx-	3	5	*7/9*
1980	0817	3a	**HEKLA** (Iceland-S)	17	555	---X ---- X--- X--- -X--	3		8/7
1980	0821	5d	**BEZYMIANNY** (Kamchatka)	10		X--- ---- XX-- X--- ----	2		5/-
1980	0823	403a	**SHEVELUCH** (Kamchatka)	10	326	X--- ---- XX-- --X- ----	1		7/-
1980	*0827*	*<1*	*KILAUEA (Hawaiian Is)*	*13*	410	*(Magma intrusion)*			
1980	09..?	.?..	**MARION ISLAND** (Indian O.-S)	3	104	-XX- --X- X--- X--- ----	1		6/-
1980	0904	19a	**GAMALAMA** (Halmahera-Indonesia)	6	232	X--- ---- X--- ---- -X--	2		-/6
1980	0911	<1	**POAS** (Costa Rica)	14	513	X--- ---X X-X- ---- ----	1		
1980	0924	<1	**ASO** (Kyushu-Japan)	8	271	X--- ---- X-X- ---- ----	1		
1980	0928	<1	**KUCHINOERABU-JIMA** (Ryukyu Is)	8	247	X--- ---- X-X- ---- ----	2		-/6
1980	10..	<1	**CALLAQUI** (Chile-C)	15	535	X--- ---- X-X- ---- ----	1		
1980	1006	<1	**ULAWUN** (New Britain-SW Pac)	5	175	X--- ---- XX-- ---- -X--	3		-/7

	DATE Year	MoDy	DURA-TION	VOLCANO NAME (Subregion)	CH	PP	CHARACTERISTICS (CENTRAL VENT, FLANK VENT, RADIAL FISSURE, REGIONAL / SUBMARINE, NEW ISLAND, SUBGLACIAL, CRATER LAKE / EXPLOSIVE, PYRO FLOW, PHREATIC, SULFUR / LAVA FLOW, LAVA LAKE, DOME, SPINE / FATALITIES, DAMAGE, MUDFLOW, TSUNAMI)	VEI		VOL L/T
	1980	1007	141a	KAVACHI (Solomon Is-SW Pac)	5	195	---- X--- X--- ---- ----	1		
	1980	1016	....	ST. HELENS, MOUNT (US-Wash)	12	361	xx-- ---- xxx- --x- xxx-	3	5	7/9
	1980	1018	15a	RUAPEHU (New Zealand)	4	125	---- ---X X-X- ---- ----	1		
	1980	1018	....	KRAFLA (Iceland-N)	17	564	---x ---- ---- x--- ----	0	*	7/-
	1980	1022	<1	KILAUEA (Hawaiian Is)	13	411	(Magma intrusion)			
	1980	1029<	.?..	GAUA (Vanuatu-SW Pacific)	5		X--- ---- --?- ---- ----	1?		
	1980	11..	228w	KIRISHIMA (Kyushu-Japan)	8	265	(Increased fumarolic activity)			
	1980	11..	1211	TARUMAI (Hokkaido-Japan)	8	313	(Increase in seismicity)			
	1980	1101q	.?..	PACAYA (Guatemala)	14	490	X--- ---- X--- ---- ----	1		
	1980	1102	<1	KILAUEA (Hawaiian Is)	13	411	(Magma intrusion)			
	1980	1105	3141	PALUWEH (Lesser Sunda Is)	6	220	X--- ---- XX-- --X- -X--	2		6/-
	1980	1108	4	PAVLOF (Alaska Peninsula)	11	335	x--- ---- x--- x--- ----	1	3	
S	1980	1110	97	MACDONALD (Pacific-C)	13	460	---- X--- ---- ---- ----	0		
	1980	1111	....	PAVLOF (Alaska Peninsula)	11	335	X--- ---- X--- X--- ----	3	*	
?	1980	1115	41 >	MYOJIN-SHO (BAYONNAISE ROCKS)	8	288	---- ?--- ---- ---- ----	0		
	1980	1225	3a	KRAFLA (Iceland-N)	17	564	(Magma intrusion)			
	1980	1226	<1	POAS (Costa Rica)	14	513	X--- ---X X-X- ---- ----	1		
	1981	Continuing		STROMBOLI (Italy)	1		x--- ---- x--- ---- -x--	1?		
	1981	Continuing		ERTA ALE (Ethiopia)	2		x--- ---- ---- xx-- ----	0		
	1981	Continuing		WHITE ISLAND (New Zealand)	4		x--- ---- xxx- ---- ----	2	3	-/6
	1981	Continuing		MANAM (New Guinea-NE of)	5		x--- ---- xx-- x--- -x--	2	*	
	1981	Continuing		LANGILA (New Britain-SW Pac)	5		xx-- ---- xxx- x-x- ----	2	3	6/-
	1981	Continuing		RABAUL CALDERA (New Britain)	5		(Deformation)			
	1981	Continuing		BAGANA (Bougainville-SW Pac)	5		x--- ---- xx-- x-x- ----	1	2	
	1981	Continuing		YASOUR (Vanuatu-SW Pacific)	5		x--- ---- x-x- ---- ----	2	3	
	1981	Continuing		MERAPI (Java)	6		x--- ---- xx-- x-x- --x-	1	*	
	1981	Continuing		SEMERU (Java)	6		x--- ---- xx-- x-x- xxx-	2	3	
	1981	Continuing		PALUWEH (Lesser Sunda Is)	6		x--- ---- xx-- --x- -x--	2		6/-
	1981	Continuing		DUKONO (Halmahera-Indonesia)	6		x--- ---- x--- x--- -x--	2	3	
	1981	Continuing		SUWANOSE-JIMA (Ryukyu Is)	8		x--- ---- x--- ---- ----	2	3	
	1981	Continuing		SAKURA-JIMA (Kyushu-Japan)	8		x--- ---- xxx- -x-- xxx-	2	3	
?	1981	Continuing		FUKUTOKU-OKANOBA (SHIN-IWO-JIMA)	8		---- ?--- ---- ---- ----	0		
	1981	Continuing		USU (Hokkaido-Japan)	8		x--- ---- x-x- --x- xxx-	0	*	-/7
	1981	Continuing		GORELY (Kamchatka)	10		x--- ---- xxx- ---- ----	2	3	-/7
	1981	Continuing		KARYMSKY (Kamchatka)	10		x--- ---- xx-- x-x- --x-	2	*	7/6
	1981	Continuing		SHEVELUCH (Kamchatka)	10		x--- ---- xx-- --x- ----	1		7/-
	1981	Continuing		ST. HELENS, MOUNT (US-Wash)	12		xx-- ---- xxx- --x- xxx-	1	5	7/9
	1981	Continuing		LONG VALLEY CALDERA (US-Calif)	12		(Seismicity, deformation)			
S	1981	Continuing		MACDONALD (Pacific-C)	13		---- x--- ---- ---- ----	0		
	1981	Continuing		COLIMA, VOLCAN DE (Mexico)	14		x--- ---- xx-- x-x- ----	1		5/-
	1981	Continuing		ARENAL (Costa Rica)	14		xxx- ---- xxx- x--- xxx-	1	*	7/7
	1981	Continuing		SANGAY (Ecuador)	15		x--- ---- x--- ---- ----	2		
	1981	Continuing		CHILLAN, NEVADOS DE (Chile-C)	15		-x-- ---- x--- x-x- ----	2	*	5/-
	1981	Continuing		EREBUS, MOUNT (Antarctica)	19		x--- ---- x--- xx-- ----	1	*	
	1981	01..	....	POAS (Costa Rica)	14	513	(Increased thermal activity)			
?	1981	0107	1a	FUKUJIN SEAMOUNT (Volcano Is)	8	291	---- ?--- ---- ---- ----	0		
	1981	0120	19a	KILAUEA (Hawaiian Is)	13	412	(Magma intrusion)			
	1981	0125	187a?	KLIUCHEVSKOI (Kamchatka)	10		X--- ---- X--- ---- ----	1		
	1981	0126e	41j	ETNA (Italy)	1	60	X--- ---- X--- X--- ----	2		6/-
	1981	0130	4	KRAFLA (Iceland-N)	17	564	---X ---- ---- X--- ----	0		7/-
	1981	02..	....	SANTIAGUITO (SANTA MARIA)	14	478	-x-- ---- xxx- x-xx xxx-	2	3	8/8
	1981	02..?	.?..	TELICA (Nicaragua)	14	496	X--- ---- X--- ---- ----	1		
	1981	0203	91a	FOURNAISE, PITON DE LA (Indian O	3	94	--X- ---- X-X- X--- ----	1		7/-
	1981	0209	>3yr	PACAYA (Guatemala)	14	490	xx-- ---- x--- xxx- -x--	1	2	6/-
	1981	0220	222a	AMBRYM (Vanuatu-SW Pacific)	5		X--- ---- X-X- X--- -X--	2		
	1981	0227<	.?..	TARUMAI (Hokkaido-Japan)	8	313	X--- ---- --X- ---- ----	0		-/2
	1981	03	....	MIYAKE-JIMA (Izu Is-Japan)	8	283	(Seismic swarm)			
	1981	03..	>3yr	ETNA (Italy)	1	61	X--- ---- X--- ---- ----	2		
	1981	0304	143a	GAMKONORA (Halmahera-Indonesia)	6	231	X--- ---- X-X- ---- ----	1		-/5

DATE Year	MoDy	DURATION	VOLCANO NAME (Subregion)	CH	PP	CHARACTERISTICS (CENTRAL VENT, FLANK VENT, RADIAL FISSURE, REGIONAL ∴ / SUBMARINE, NEW ISLAND, SUBGLACIAL, CRATER LAKE / EXPLOSIVE, PYRO FLOW, PHREATIC, SULFUR / LAVA FLOW, LAVA LAKE, DOME, SPINE / FATALITIES, DAMAGE, MUDFLOW, TSUNAMI)	VEI	VOL L/T
?1981	0305	286l	**MEHETIA** (Pacific-C)	13	461	---- ?--- ---- ---- ----	0	
?1981	0306	13a	**BAM** (New Guinea-NE of)	5	145	---- ?--- ---- ---- ----	0	
1981	*0306*	*13a*	*KADOVAR (New Guinea-NE of)*	*5*	145	*(Offshore water discoloration)*		
1981	0317	6	**ETNA** (Italy)	1	61	--X- ---- X--- X--- -X--	1	7/5
1981	0409	18a	**BULUSAN** (Luzon-Philippines)	7	235	X--- ---- X--- ---- ----	3	
1981	0409	7	**HEKLA** (Iceland-S)	17	556	--X- ---- X--- X--- ----	2	7/-
1981	0424	178	**KRAKATAU** (Indonesia)	6	203	X--- ---- X--- ---- ----	1	
1981	*0427*	*39a*	*ALAID (Kuril Is)*	*9*	319	x--- ---- x--- ---- -xx-	3 4	-/8
1981	0430	.?..	**ALAID** (Kuril Is)	9	319	X--- ---- X--- ---- -XX-	4 *	-/8
1981	0515	>3yr	**PAGAN, MOUNT** (Mariana Is-C Pac)	8	293	XXX- ---- XX-- X--- -X--	4 *	7/8
1981	0610	15a	**TIATIA** (Kuril Is)	9	317	??-- ---- X--- ---- ----	2?	
1981	0612	724a	**BEZYMIANNY** (Kamchatka)	10	323	X--- ---- XXX- X-XX --X-	3 *	-/7
1981	0615	<1	**ASO** (Kyushu-Japan)	8	271	X--- ---- X-X- ---- ----	1	
1981	*0625*	*<1*	*KILAUEA (Hawaiian Is)*	*13*	413	*(Magma intrusion)*		
1981	0709	<1	**GAUA** (Vanuatu-SW Pacific)	5		X--- ---- ?-?- ---- ----	1?	
1981	0801x	.?..	**FERNANDINA** (Galapagos)	15	529	X--- ---- ---- X--- ----	0	
1981	*0810*	*1a*	*KILAUEA (Hawaiian Is)*	*13*	413	*(Magma intrusion)*		
1981	0831e	442m?	**GUAGUA PICHINCHA** (Ecuador)	15	525	X--- ---- X-X- ---- ----	1 *	-/4
?1981	0915e	.?..	**KAVACHI** (Solomon Is-SW Pac)	5	195	---- ?--- ---- ---- ----	0	
1981	*0920*	*....*	*PALLAS PEAK (KETOI) (Kuril Is)* .	*9*	318	*(Intense fumarolic activity)*		
1981	0925	2a	**PAVLOF** (Alaska Peninsula)	11	336	X--- ---- X--- X--- ----	3	6/-
?1981	0925	.?..	**SHISHALDIN** (Aleutian Is)	11	333	---- ---- ?--- ---- ----		
1981	1025?	171l	**RUAPEHU** (New Zealand)	4	127	---- ---X X-X- ---- ----	1	
1981	*1117*	*281*	*DIENG VOLCANIC COMPLEX (Java)* ..	*6*	210	*(Seismicity)*		
1981	1118	5	**KRAFLA** (Iceland-N)	17	565	---X ---- ---- X--- ----	0	7/-
1981	1125	.?..	**ALAID** (Kuril Is)	9		X--- ---- X--- ---- ----	2	
1981	*1125e*	*97e*	*TELICA (Nicaragua)*	*14*	496	x--- ---- x--- ---- -x--	1 2	
1981	*12..*	*231*	*KIRISHIMA (Kysushu-Japan)*	*8*	265	*(Increased fumarolic activity)*		
1981	1216	....	**MASAYA** (Nicaragua)	14	501	X--- ---- X--- -X-- -X--	1 *	
?1981	1221	.?..	**KLIUCHEVSKOI** (Kamchatka)	10	325	---- ---- ---- ---- ----		
1981	1225	20a	**NYAMURAGIRA** (Africa-C)	2	81	-XX- ---- X--- X--- -X--	3	8/8
1982	*Continuing*		*STROMBOLI (Italy)*	*1*		x--- ---- x--- ---- -x--	*1?*	
1982	*Continuing*		*ETNA (Italy)*	*1*		x--- ---- x--- ---- ----	2	
1982	*Continuing*		*ERTA ALE (Ethiopia)*	*2*		x--- ---- ---- xx-- ----	0	
1982	*Continuing*		*NYAMURAGIRA (Africa-C)*	*2*		-xx- ---- x--- x--- -x--	3	*8/8*
1982	*Continuing*		*RUAPEHU (New Zealand)*	*4*		---- ---x x-x- ---- ----	1	
1982	*Continuing*		*RABAUL CALDERA (New Britain)* ...	*5*		*(Seismicity, deformation)*		
1982	*Continuing*		*BAGANA (Bougainville-SW Pac)* ...	*5*		x--- ---- xx-- x-x- ----	1 2	
1982	*Continuing*		*YASOUR (Vanuatu-SW Pacific)*	*5*		x--- ---- x-x- ---- ----	2 3	
1982	*Continuing*		*MERAPI (Java)*	*6*		x--- ---- xx-- x-x- --x-	1 *	
1982	*Continuing*		*SEMERU (Java)*	*6*		x--- ---- xx-- x-x- xxx-	2 3	
1982	*Continuing*		*DUKONO (Halmahera-Indonesia)* ...	*6*		x--- ---- x--- x--- -x--	2 3	
1982	*Continuing*		*SUWANOSE-JIMA (Ryukyu Is)*	*8*		x--- ---- x--- ---- ----	2 3	
1982	*Continuing*		*SAKURA-JIMA (Kyushu-Japan)*	*8*		x--- ---- xxx- -x-- xxx-	2 3	
1982	*Continuing*		*KIRISHIMA (Kysushu-Japan)*	*8*		*(Increased fumarolic activity)*		
?1982	*Continuing*		*FUKUTOKU-OKANOBA (SHIN-IWO-JIMA)*	*8*		---- ?--- ---- ---- ----	0	
1982	*Continuing*		*USU (Hokkaido-Japan)*	*8*		x--- ---- x-x- --x- xxx-	0 *	*-/7*
1982	*Continuing*		*KARYMSKY (Kamchatka)*	*10*		x--- ---- xx-- x-x- --x-	2 *	*7/6*
1982	*Continuing*		*LONG VALLEY CALDERA (US-Calif)* .	*12*		*(Seismicity, deformation)*		
1982	*Continuing*		*COLIMA, VOLCAN DE (Mexico)*	*14*		x--- ---- xx-- x-x- ----	1	*5/-*
1982	*Continuing*		*SANTIAGUITO (SANTA MARIA)*	*14*		-x-- ---- xxx- x-xx xxx-	1 *	*8/8*
1982	*Continuing*		*PACAYA (Guatemala)*	*14*		xx-- ---- x--- xxx- -x--	1 2	*6/-*
1982	*Continuing*		*ARENAL (Costa Rica)*	*14*		xxx- ---- xxx- x--- xxx-	1 *	*7/7*
1982	*Continuing*		*POAS (Costa Rica)*	*14*		*(Increased thermal activity)*		
1982	*Continuing*		*SANGAY (Ecuador)*	*15*		x--- ---- x--- ---- ----	2	
1982	*Continuing*		*CHILLAN, NEVADOS DE (Chile-C)* ..	*15*		-x-- ---- x--- x-x- ----	2 *	*5/-*
1982	*Continuing*		*EREBUS, MOUNT (Antarctica)*	*19*		x--- ---- x--- xx-- ----	*1 **	
1982	....	.?..	**API SIAU** (Sangihe Is-Indonesia)	6		X--- ---- ?--- ---- ----	1	
1982	*....*	*....*	*PLOSKY (Kamchatka)*	*10*	326	*(Advance of glacier in caldera)*		

DATE Year	MoDy	DURATION	VOLCANO NAME (Subregion)	CH	PP	CHARACTERISTICS (Central vent, Flank vent, Radial fissure, Regional; Submarine, New island, Subglacial, Crater lake; Explosive, Pyro flow, Phreatic, Sulfur; Lava flow, Lava lake, Dome, Spine; Fatalities, Damage, Mudflow, Tsunami)	VEI	VOL L/T
1982	01..	67a?	KILAUEA (Hawaiian Is)	13	415	(Magma intrusion)		
?1982	0112	64a	FUKUJIN SEAMOUNT (Volcano Is) ..	8	291	---- ?--- ---- ---- ----	0	
1982	0115	<1	GARELOI (Aleutian Is)	11	327	X--- ---- X--- ---- ----	3	
1982	0115e	.?..	CONCEPCION (Nicaragua)	14	503	X--- ---- X-X- ---- ----	2	
1982	0128	24a	GRIMSVOTN (Iceland-S)	17	556	(Jokulhlaup)		
?1982	0210	4a	TIATIA (Kuril Is)	9		X--- ---- --?- ---- ----	1	
1982	0212	.?..	TELICA (Nicaragua)	14	496	X--- ---- X--- ---- -X--	2 *	
1982	0213	....	LANGILA (New Britain-SW Pac) ...	5	168	xx-- ---- xxx- x-x- ----	3 *	6/
S1982	0301	84	MACDONALD (Pacific-C)	13	460	---- X--- ---- ---- ----	0	
1982	0309	1	IWO-JIMA (Volcano Is-Japan)	8	290	X--- ---- --X- ---- ----	1	
1982	0310<	66l>	MARAPI (Sumatra)	6	200	X--- ---- X--- ---- ----	1	
?1982	0316	64a	TEAHITIA (Pacific-C)	13	462	---- ?--- ---- ---- ----	0	
1982	0319	....	ST. HELENS, MOUNT (US-Wash)	12	374	xx-- ---- xxx- --x- xxx-	3 5	7/9
1982	0319	....	DESCABEZADO GRANDE (CHILE-C) ...	15	534	(Increased fumarolic activity)		
1982	0321	26e	AKAN CALDERA (Hokkaido-Japan) ..	8	315	(Increased seismicity)		
1982	0324	39a	KLIUCHEVSKOI (Kamchatka)	10	325	X--- ---- X-?- ---- ----	1	
1982	0327	....	MANAM (New Guinea-NE of)	5	149	x--- ---- xx-- x--- -x--	3 *	
1982	0328	167a	CHICHON, EL (Mexico)	14	467	X--- ---- XXX- ---- XXX-	4 *	-/8
1982	0329	<1	ALAID (Kuril Is)	9	321	X--- ---- X--- ---- ----	2	
?1982	04..	15l	NISHINO-SHIMA (Volcano Is-Japan)	8	288	---- ?--- ---- ---- ----	0	
1982	0404	....	CHICHON, EL (Mexico)	14	468	x--- ---- xxx- ---- xxx-	4 *	-/8
1982	0405	278a	GALUNGGUNG (Java)	6	205	x--- ---- xx-- x--- xxx-	3 4	-/8
?1982	0406	.?..	ESMERALDA BANK (Mariana Is-C Pac	8	298	---- ?--- ---- ---- ----	0	
1982	0407	58a?	KAVACHI (Solomon Is-SW Pac)	5		---- X--- X--- ---- ----	2	
1982	0418	<1	GAUA (Vanuatu-SW Pacific)	5		X--- ---- ?-?- ---- ----	2	
1982	0426	<1	ASAMA (Honshu-Japan)	8	277	X--- ---- XXX- ---- -X--	2	
1982	0430	<1	KILAUEA (Hawaiian Is)	13	415	X-X- ---- ---- X--- ----	0	5/-
1982	0517	....	GALUNGGUNG (Java)	6	206	X--- ---- XX-- X--- XXX-	4 *	-/8
1982	0528k	....	GUAGUA PICHINCHA (Ecuador)	15	525	x--- ---- x-x- ---- ----	1 *	-/4
1982	0610	....	BEZYMIANNY (Kamchatka)	10		x--- ---- xxx- x-xx --x-	2 3	-/7
1982	0621	116l?	NYIRAGONGO (Africa-C)	2	85	X--- ---- --X- X--- ----	1	7/-
1982	0622	2a	KILAUEA (Hawaiian Is)	13	416	(Magma intrusion)		
1982	0701	<1	WHITE ISLAND (New Zealand)	4	119	X--- ---- X--- ---- ----	2	
1982	0715q	898w	CAMPI FLEGREI (Italy)	1	41	(Seismicity, deformation)		
?1982	0715q	.?..	PAVLOF (Alaska Peninsula)	11		X--- ---- --?? ---- ----		
1982	0718	2a	RAUNG (Java)	6	216	X--- ---- X--- ---- -X--	3	
1982	0826	76a	SOPUTAN (Sulawesi-Indonesia) ...	6	226	X--- ---- X--- ---- -X--	3	-/6
1982	0828	9a?	WOLF, VOLCAN (Galapagos)	15	529	X-X- ---- X--- X--- ----	1	
1982	09..	....	FUSS PEAK (Kuril Is)	9	319	(Increased fumarolic activity)		
1982	0923	6a	HAROHARO VOLC COMPLEX (New Zeal)	4	122	(Seismic swarm)		
1982	0925	<1	KILAUEA (Hawaiian Is)	13	416	X--- ---- ---- X--- ----	0	6/-
1982	0930j	....	PAGAN, MOUNT (Mariana Is-C Pac)	8	296	xxx- ---- xx-- x--- -x--	3?4	7/8
1982	1002	<1	ASAMA (Honshu-Japan)	8	278	X--- ---- X--- ---- ----	1	
1982	1005d	221m	AKUTAN (Aleutian Is)	11		X--- ---- X--- ---- ----	2	
1982	1007	69l	KLIUCHEVSKOI (Kamchatka)	10	325	X--- ---- X--- ---- ----	1	
1982	1007	....	MASAYA (Nicaragua)	14	502	x--- ---- x--- -x-- -x--	1 *	
1982	1016	27a	CAMEROON, MOUNT (Africa-W)	2	90	--X- ---- X--- X--- -XX-	2	7/5
1982	1024	1a	LOPEVI (Vanuatu-SW Pacific)	5	197	X--- ---- X--- ---- ----	2	
1982	1026	64	KUSATSU-SHIRANE (Honshu-Japan) .	8	279	X--- ---X X-X- ---- ----	1	
1982	1117	<1	ILIBOLENG (Lesser Sunda Is)	6	221	X--- ---- X--- ---- ----	2	
1982	1122	....	CHIRPOI GROUP (Kuril Is)	9	318	X--- ---- ?--- ---- ----	2	
1982	1128	.?..	IWO-JIMA (Volcano Is-Japan)	8	290	X--- ---- --X- ---- ----	1	
1982	12..	15l	MARAPI (Sumatra)	6		X--- ---- X--- ---- ----	1	
1982	1209	<1	KILAUEA (Hawaiian Is)	13	419	(Magma intrusion)		
?1982	1215	.?..	FUKUJIN SEAMOUNT (Volcano Is) ..	8	291	---- ?--- ---- ---- ----	0	
1982	1227	....	MIYAKE-JIMA (Izu Is-Japan)	8	283	(Seismic swarm)		
1983	Continuing		CAMPI FLEGREI (Italy)	1		(Seismicity, deformation)		
1983	Continuing		STROMBOLI (Italy)	1		x--- ---- x--- ---- -x--	1?	
1983	Continuing		ERTA ALE (Ethiopia)	2		x--- ---- ---- xx-- ----	0	

DATE Year	DATE MoDy	DURATION	VOLCANO NAME (Subregion)	CH	PP	CHARACTERISTICS (CENTRAL VENT, FLANK VENT, RADIAL FISSURE, REGIONAL; SUBMARINE, NEW ISLAND, SUBGLACIAL, CRATER LAKE; EXPLOSIVE, PYRO FLOW, PHREATIC, SULFUR; LAVA FLOW, LAVA LAKE, DOME, SPINE; FATALITIES, DAMAGE, MUDFLOW, TSUNAMI)	VEI		VOL L/T
1983	Continuing		MANAM (New Guinea-NE of)	5		x--- ---- xx-- x--- -x--	2	*	
1983	Continuing		RABAUL CALDERA (New Britain) ...	5		(Seismicity, deformation)			
1983	Continuing		BAGANA (Bougainville-SW Pac) ...	5		x--- ---- xx-- x-x- ----	1	2	
1983	Continuing		YASOUR (Vanuatu-SW Pacific)	5		x--- ---- x-x- ---- ----	2	3	
1983	Continuing		GALUNGGUNG (Java)	6		x--- ---- xx-- x--- xxx-	1	4	-/8
1983	Continuing		MERAPI (Java)	6		x--- ---- xx-- x-x- --x-	1	*	
1983	Continuing		SEMERU (Java)	6		x--- ---- xx-- x-x- xxx-	2	3	
1983	Continuing		DUKONO (Halmahera-Indonesia) ...	6		x--- ---- x--- x--- -x--	2	3	
1983	Continuing		SUWANOSE-JIMA (Ryukyu Is)	8		x--- ---- x--- ---- ----	2	3	
?1983	Continuing		FUKUTOKU-OKANOBA (SHIN-IWO-JIMA)	8		---- ?--- ---- ---- ----	0		
1983	Continuing		PAGAN, MOUNT (Mariana Is-C Pac)	8		xxx- ---- xx-- x--- -x--	2	*	7/8
1983	Continuing		AKUTAN (Aleutian Is)	11		x--- ---- x--- ---- ----	2		
1983	Continuing		LONG VALLEY CALDERA (US-Calif) .	12		(Seismicity, deformation)			
1983	Continuing		PACAYA (Guatemala)	14		xx-- ---- x--- xxx- -x--	1	2	6/-
1983	Continuing		MASAYA (Nicaragua)	14		x--- ---- x--- -x-- -x--	0	1	
1983	Continuing		ARENAL (Costa Rica)	14		xxx- ---- xxx- x--- xxx-	1	*	7/7
1983	Continuing		SANGAY (Ecuador)	15		x--- ---- x--- ---- ----	2		
1983	Continuing		CHILLAN, NEVADOS DE (Chile-C) ..	15		-x-- ---- x--- x-x- ----	2	*	5/-
1983	Continuing		EREBUS, MOUNT (Antarctica)	19		x--- ---- x--- xx-- ----	1	*	
1983	....	.?..	MARAPI (Sumatra)	6		X--- ---- X--- ---- ----	1		
1983	0101	>3yr	LENGAI, OL DOINYO (Africa-E) ...	2	79	X--- ---- X--- XX-- ----	2		
1983	0103	>3yr	KILAUEA (Hawaiian Is)	13	419	--X- ---- ---- XX-- -X--	0		8/-
1983	0126	....	SAKURA-JIMA (Kyushu-Japan)	8	258	x--- ---- xxx- -x-- xxx-	3	*	
1983	0129	....	SANTIAGUITO (SANTA MARIA)	14	478	-x-- ---- xxx- x-xx xxx-	2	3	8/8
1983	0202	....	ST. HELENS, MOUNT (US-Wash)	12	379	xx-- ---- xxx- --x- xxx-	2	5	7/9
1983	0206	15a	RINCON DE LA VIEJA (Costa Rica)	14	504	X--- ---X X-X- ---- --X-	1		
1983	0211	4a	COLIMA, VOLCAN DE (Mexico)	14		X--- ---- X--- X-X- ----	1		
1983	0308	111a	KLIUCHEVSKOI (Kamchatka)	10	325	--X- --X- X-X- X--- --X-	1		8/-
S1983	0314	68a	MACDONALD (Pacific-C)	13	460	---- X--- ---- ---- ----	0		
1983	0316	9a	CONCEPCION (Nicaragua)	14	504	X--- ---- X--- ---- ----	2		
1983	0317	25a	MOMOTOMBO (Nicaragua)	14	499	(Volcanic tremor)			
1983	0328	131a	ETNA (Italy)	1	62	--X- ---- X-X- X--- -X--	1		8/5
1983	0408	<1	ASAMA (Honshu-Japan)	8	278	X--- ---- X--- ---- ----	2		
1983	0415	<1	NIIGATA-YAKE-YAMA (Honshu-Japan)	8	276	X--- ---- X-X- ---- ----	1		
?1983	0415e	74f	BROMO (TENGGER CALDERA) (Java) .	6		X--- ---- --?- ---- ----	1		
1983	0418	....	LANGILA (New Britain-SW Pac) ...	5	170	xx-- ---- xxx- x-x- ----	3	*	6/-
1983	05..	9491>	API SIAU (Sangihe Is-Indonesia)	6		xx-- ---- xxx- x--- ----	1	3	
1983	0511	196a	ILIBOLENG (Lesser Sunda Is)	6	222	X--- ---- X--- ---- ----	1		
1983	0516	....	NGAURUHOE (New Zealand)	4	122	(Seismic swarm)			
1983	0517	....	LANGILA (New Britain-SW Pac) ...	5	170	xx-- ---- xxx- x-x- ----	3	*	6/-
1983	0522	....	BEZYMIANNY (Kamchatka)	10	324	X--- ---- XXX- X-XX --X-	3		
1983	0528	4	GRIMSVOTN (Iceland-S)	17	556	X--- --XX X--- ---- ----	2		
1983	06..	....	TANGKUBAN PARAHU (SUNDA CALDERA)	6	203	(Seismicity, deformation)			
1983	0604	317a	VENIAMINOF (Alaska Peninsula) ..	11	337	X--- ---- X--- X--- ----	3		7/-
1983	0615	1a	UNNAMED (New Britain-SW Pac) ...	5	163	---- X--- ---- ---- ----	0		
1983	0625	4	BULUSAN (Luzon-Philippines)	7	235	X--- ---- X-X- ---- ----	2		
1983	0708	<1	OKMOK (Aleutian Is)	11	328	X--- ---- X--- ---- ----	2		
1983	0711	7a	PAVLOF (Alaska Peninsula)	11	337	X--- ---- X--- ---- ----			
?1983	0712	14a	TEAHITIA (Pacific-C)	13	463	---- ?--- ---- ---- ----	0		
1983	0715	<1	FOURNAISE, PITON DE LA (Indian O	3	97	(seismic swarm)			
1983	0718	1501	UNA UNA (COLO) (Sulawesi)	6	222	x--- ---- xxx- ---- -x--	2	4	
1983	0723	....	UNA UNA (COLO) (Sulawesi)	6	223	X--- ---- XXX- ---- -X--	4	*	
1983	0726	147	KUSATSU-SHIRANE (Honshu-Japan) .	8	281	X--- ---- X-X- ---- ----	1		
1983	0809	2	GAMALAMA (Halmahera-Indonesia) .	6	232	X--- ---- X--- ---- -X--	3		
1983	0817	<1	ILIWERUNG (Lesser Sunda Is)	6	222	-X-- X--- ---- ---- X--X	1		
1983	1003	<1	MIYAKE-JIMA (Izu Is-Japan)	8	284	-XX- ---- XXX- X--- -X--	3		6/6
1983	1014	.?..	VILLARRICA (Chile-C)	15	537	X--- ---- XX-- ---- ----	2?		
S1983	1027	68a	MACDONALD (Pacific-C)	13	461	---- X--- ---- ---- ----	0		
1983	1106	128a	ULAWUN (New Britain-SW Pac)	5	178	X--- ---- --X- ---- ----	1		
1983	1114	34a?	PAVLOF (Alaska Peninsula)	11	336	X--- ---- X--- ---- ----	3		

DATE Year	MoDy	DURA-TION	VOLCANO NAME (Subregion)	CH	PI	CHARACTERISTICS (CENTRAL VENT, FLANK VENT, RADIAL FISSURE, REGIONAL; SUBMARINE, NEW ISLAND, SUBGLACIAL, CRATER LAKE; EXPLOSIVE, PYRO FLOW, PHREATIC, SULFUR; LAVA FLOW, LAVA LAKE, DOME, SPINE; FATALITIES, DAMAGE, MUDFLOW, TSUNAMI)	VEI		VOL L/T
1983	1204	76a	FOURNAISE, PITON DE LA (Indian O	3	97	--X- ---- X--- X--- ----	1		7/-
?1983	1218	118a	TEAHITIA (Pacific-C)	13	463	---- ?--- ---- ---- ----	0		
1983	1221	<1	BROMO (TENGGER CALDERA) (Java) .	6		X--- ---- X--- ---- ----	1		
1983	1226e	48e	WHITE ISLAND (New Zealand)	4	120	X--- ---- X--- ---- ----	2		
1984	Continuing		CAMPI FLEGREI (Italy)	1		(Seismicity, deformation)			
1984	Continuing		STROMBOLI (Italy)	1		x--- ---- x--- ---- -x--	1?		
1984	Continuing		ERTA ALE (Ethiopia)	2		x--- ---- ---- xx-- ----	0		
1984	Continuing		LENGAI, OL DOINYO (Africa-E) ...	2		x--- ---- x--- xx-- ----	1		
1984	Continuing		FOURNAISE, PITON DE LA (Indian O	3		--x- ---- x--- x--- ----	1		7/-
1984	Continuing		WHITE ISLAND (New Zealand)	4		x--- ---- x--- ---- ----	2		
1984	Continuing		LANGILA (New Britain-SW Pac) ...	5		xx-- ---- xxx- x-x- ----	2	3	6/-
1984	Continuing		RABAUL CALDERA (New Britain) ...	5		(Seismicity, deformation)			
1984	Continuing		BAGANA (Bougainville-SW Pac) ...	5		x--- ---- xx-- x-x- ----	1	2	
1984	Continuing		YASOUR (Vanuatu-SW Pacific)	5		x--- ---- x-x- ---- ----	2	3	
1984	Continuing		SEMERU (Java)	6		x--- ---- xx-- x-x- xxx-	2	3	
1984	Continuing		DUKONO (Halmahera-Indonesia) ...	6		x--- ---- x--- x--- -x--	2	3	
1984	Continuing		SUWANOSE-JIMA (Ryukyu Is)	8		x--- ---- x--- ---- ----	2	3	
?1984	Continuing		FUKUTOKU-OKANOBA (SHIN-IWO-JIMA)	8		---- ?--- ---- ---- ----	0		
1984	Continuing		PAGAN, MOUNT (Mariana Is-C Pac)	8		xxx- ---- xx-- x--- -x--	2	*	7/8
1984	Continuing		ST. HELENS, MOUNT (US-Wash)	12		xx-- ---- xxx- --x- xxx-	1	5	7/9
1984	Continuing		LONG VALLEY CALDERA (US-Calif) .	12		(Seismicity, deformation)			
1984	Continuing		KILAUEA (Hawaiian Is)	13		--x- ---- ---- xx-- -x--	0		8/-
S1984	Continuing		MACDONALD (Pacific-C)	13		---- x--- ---- ---- ----	0		
?1984	Continuing		TEAHITIA (Pacific-C)	13		---- ?--- ---- ---- ----	0		
1984	Continuing		SANTIAGUITO (SANTA MARIA)	14		-x-- ---- xxx- x-xx xxx-	1	*	8/8
1984	Continuing		SANGAY (Ecuador)	15		x--- ---- x--- ---- ----	2		
1984	Continuing		CHILLAN, NEVADOS DE (Chile-C) ..	15		-x-- ---- x--- x-x- ----	2	*	5/-
1984	0109	22a?	GALUNGGUNG (Java)	6	209	X--- ---- X-X- ---- ----	1		
1984	0205	3441>	BEZYMIANNY (Kamchatka)	10		x--- ---- xx-- --x- ----	2	3	-/7
1984	0217	....	MANAM (New Guinea-NE of)	5	153	x--- ---- xx-- x--- -x--	3	*	
1984	0223	20a	NYAMURAGIRA (Africa-C)	2	83	-X-X ---- X--- X--- ----			
1984	03..	2751>	KLIUCHEVSKOI (Kamchatka)	10	325	X--- ---- X--- X--- ----	3		
1984	0301	4a	HOME REEF (Tonga-SW Pacific) ...	4	139	---- XX-- X--- ---- ----	3?		
1984	0307	222a?	KAITOKU SEAMOUNT (Volcano Is) ..	8	288	---- X--- X--- ---- ----	0		
1984	0325	21a	MAUNA LOA (Hawaiian Is)	13	454	X-X- ---- ---- X--- ----	0		8/-
1984	0330	12a?	FERNANDINA (Galapagos)	15	528	X--- ---- X--- X--- ----			
1984	0331	161	RINCON DE LA VIEJA (Costa Rica)	14		X--- ---X X-X- ---- --X-	1		
1984	04..	....	MASAYA (Nicaragua)	14	502	x--- ---- x--- -x-- -x--	1	*	
1984	0405d	....	EL MISTI (Peru)	15	532	(Increased vapor emission)			
1984	0420	<1	LLAIMA (Chile-C)	15	536	X--- ---- X--- ---- ----	2		
1984	0427?	174	ETNA (Italy)	1	67	X--- ---- X--- X--- ----	2		6/-
1984	0509	12a	PALUWEH (Lesser Sunda Is)	6		X--- ---- X--- ---- ----	2		
1984	0515	....	PACAYA (Guatemala)	14	491	XX-- ---- X--- XXX- -X--	2	*	6/-
1984	0521	10a	BROMO (TENGGER CALDERA) (Java) .	6		X--- ---- X--- ---- ----	1		
1984	0522	<1	SHEVELUCH (Kamchatka)	10	326	X--- ---- X--- ---- ----	1		
1984	0524	99	SOPUTAN (Sulawesi-Indonesia) ...	6	227	X--- ---- X--- ---- ----	3		-/7
1984	0603	....	ARENAL (Costa Rica)	14	510	xxx- ---- xxx- x--- xxx-	2	3	7/7
1984	0603<	375 >	TINAKULA (Santa Cruz Is-SW Pac)	5	196	X--- ---- X--- ---- ----	2		
?1984	0605d	163m	LOKON-EMPUNG (Sulawesi-Indonesia	6	228	X--- ---- --?- ---- ----	1		
1984	0615	....	MERAPI (Java)	6	212	x--- ---- xx-- x-x- --x-	3	*	
1984	0701*	.?..	ALAID (Kuril Is)	9		X--- ---- X--- ---- ----	2?		
1984	0720	6041	ETNA (Italy)	1	69	X--- ---- X--- ---- ----	2		
1984	0721	....	SAKURA-JIMA (Kyushu-Japan)	8	261	x--- ---- xxx- -x-- xxx-	3	*	
1984	0806	....	UNZEN (Kyushu-Japan)	8	265	(Seismic swarm)			
1984	0811	4611>	VILLARRICA (Chile-C)	15	537	X--- ---- X--- X--- -XX-	1		6/-
1984	0816	<1	LAKE MONOUN (BAMBOUTO VOLC FIELD)	2	88	(Toxic gas cloud)			
1984	0823	19a	ULAWUN (New Britain-SW Pac)	5	178	X--- ---- X-X- ---- ----	1		
1984	0904	14	KRAFLA (Iceland-N)	17	567	---X ---- ---- X--- ----	0		
1984	0905	....	API SIAU (Sangihe Is-Indonesia)	6	230	XX-- ---- XXX- X--- ----	3	*	

DATE Year	MoDy	DURATION	VOLCANO NAME (Subregion)	CH	PP	CENTRAL VENT / FLANK VENT / RADIAL FISSURE / REGIONAL	SUBMARINE / NEW ISLAND / SUBGLACIAL / CRATER LAKE	EXPLOSIVE / PYRO FLOW / PHREATIC / SULFUR	LAVA FLOW / LAVA LAKE / DOME / SPINE	FATALITIES / DAMAGE / MUDFLOW / TSUNAMI	VEI		VOL L/T
1984	0909	27	MAYON (Luzon-Philippines)	7	237	X---	----	XX--	X---	-XX-	3	*	7/7
1984	0913	....	MAYON (Luzon-Philippines)	7	237	x---	----	xx--	x---	-xx-	3?	*	7/7
1984	0913	....	EREBUS, MOUNT (Antarctica)	19	579	x---	----	x---	xx--	----	2	*	
1984	1013	....	BEZYMIANNY (Kamchatka)	10	324	X---	----	XX--	--X-	----	3	*	-/7
1984	1024	243a	ASO (Kyushu-Japan)	8	271	X---	----	X-X-	----	----	1		
?1984	1111	71a	LOIHI SEAMOUNT (Hawaiian Is)	13	405	----	?---	----	----	----	0		
1984	1115	<1	MARAPI (Sumatra)	6	200	X---	----	--X-	----	----	1		
1984	1129	.?..	VENIAMINOF (Alaska Peninsula)	11	343	X---	----	X---	----	----	2		
1984	12	171	CONCEPCION (Nicaragua)	14	504	x---	----	x-x-	----	-X--	1	*	
1984	1222	87a?	RUIZ (Colombia)	15	517	X---	----	X-X-	----	----	1		
1984	1227	366 >	GORELY (Kamchatka)	10	322	X---	----	--X-	----	----	2		
1984	1230	28a	ULAWUN (New Britain-SW Pac)	5	179	X---	----	X---	X-X-	----	1		
1985	Continuing		ERTA ALE (Ethiopia)	2		x---	----	----	xx--	----	0		
1985	Continuing		LENGAI, OL DOINYO (Africa-E)	2		x---	----	x---	xx--	----	1		
1985	Continuing		MANAM (New Guinea-NE of)	5		x---	----	xx--	x---	-x--	2	*	
1985	Continuing		LANGILA (New Britain-SW Pac)	5		xx--	----	xxx-	x-x-	----	2	3	6/-
1985	Continuing		RABAUL CALDERA (New Britain)	5		(Seismicity, deformation)							
1985	Continuing		BAGANA (Bougainville-SW Pac)	5		x---	----	xx--	x-x-	----	1	2	
1985	Continuing		TINAKULA (Santa Cruz Is-SW Pac)	5		x---	----	x---	----	----	2		
1985	Continuing		YASOUR (Vanuatu-SW Pacific)	5		x---	----	x-x-	----	----	2	3	
1985	Continuing		MERAPI (Java)	6		x---	----	xx--	x-x-	--x-	1	*	
1985	Continuing		SEMERU (Java)	6		x---	----	xx--	x-x-	xxx-	2	3	
1985	Continuing		API SIAU (Sangihe Is-Indonesia)	6		xx--	----	xxx-	x---	----	2	3	
1985	Continuing		DUKONO (Halmahera-Indonesia)	6		x---	----	x---	x---	-x--	2	3	
1985	Continuing		SUWANOSE-JIMA (Ryukyu Is)	8		x---	----	x---	----	----	2	3	
1985	Continuing		ASO (Kyushu-Japan)	8		x---	----	x-x-	----	----	1		
?1985	Continuing		FUKUTOKU-OKANOBA (SHIN-IWO-JIMA)	8		----	?---	----	----	----	0		
1985	Continuing		PAGAN, MOUNT (Mariana Is-C Pac)	8		xxx-	----	xx--	x---	-x--	2	*	7/8
1985	Continuing		GORELY (Kamchatka)	10		x---	----	--x-	----	----	2		
1985	Continuing		ST. HELENS, MOUNT (US-Wash)	12		xx--	----	xxx-	--x-	xxx-	1	5	7/9
1985	Continuing		LONG VALLEY CALDERA (US-Calif)	12		(Seismicity, deformation)							
?1985	Continuing		LOIHI SEAMOUNT (Hawaiian Is)	13		----	?---	----	----	----	0		
1985	Continuing		KILAUEA (Hawaiian Is)	13		--x-	----	----	xx--	-x--	0		8/-
1985	Continuing		PACAYA (Guatemala)	14		xx--	----	x---	xxx-	-x--	1	2	6/-
1985	Continuing		CONCEPCION (Nicaragua)	14		x---	----	x-x-	----	-x--	2		
1985	Continuing		SANGAY (Ecuador)	15		x---	----	x---	----	----	2		
1985	Continuing		CHILLAN, NEVADOS DE (Chile-C)	15		-x--	----	x---	x-x-	----	2	*	5/-
1985	Continuing		VILLARRICA (Chile-C)	15		x---	----	x---	x---	-xx-	1		6/-
1985	Continuing		EREBUS, MOUNT (Antarctica)	19		x---	----	x---	xx--	----	1	*	
1985	....	....	ARENAL (Costa Rica)	14	511	xxx-	----	xxx-	x---	xxx-	2	*	7/7
1985	0102	....	CONCEPCION (Nicaragua)	14	504	X---	----	X-X-	----	-X--	2	*	
1985	0106	3a	BEERENBERG (Atl-N-Jan Mayen)	17	570	-X--	----	X---	X---	----	2		
?1985	0110	15a	TEAHITIA (Pacific-C)	13	463	----	?---	----	----	----	0		
1985	0114	.?..	HEARD VOLCANO (Indian O.-S)	3	103	X---	--X-	X---	X---	----	2?		
1985	0124	....	SANTIAGUITO (SANTA MARIA)	14	479	-x--	----	xxx-	x-xx	xxx-	2	3	8/8
1985	0203	<1	PALUWEH (Lesser Sunda Is)	6		X---	----	X---	----	----	1		
1985	0224	....	SAKURA-JIMA (Kyushu-Japan)	8	261	x---	----	xxx-	-x--	xxx-	3	*	
1985	0308	127a	ETNA (Italy)	1	71	X-X-	----	X-X-	X---	-XX-	2		7/-
1985	0313	<1	CANLAON (Philippines-C)	7	233	X---	----	X-X-	----	----	1		
1985	0321	<1	NIUAFO'OU (Tonga-SW Pacific)	4	141	X---	---X	X---	----	----	0		-/2
1985	04..	....	MASAYA (Nicaragua)	14	502	x---	----	x---	-x--	-x--	1	*	
1985	0421	....	TECAPA (El Salvador)	14	493	(Seismicity, graben formation)							
1985	0424	....	VULCANO (Italy)	1	51	(Increase in seismicity)							
1985	0503?	.?..	KARYMSKY (Kamchatka)	10	322	X---	----	----	X---	----	0		
1985	0519	<1	SOPUTAN (Sulawesi-Indonesia)	6	228	X---	----	X---	----	-X--	2		-/6
1985	0521<	19 >	RUAPEHU (New Zealand)	4	134	X---	---X	--X-	----	----	1		
1985	0526	152a	SHEVELUCH (Kamchatka)	10	326	X---	----	X-X-	----	----	2		
1985	0612	185a	BEZYMIANNY (Kamchatka)	10		x---	----	xx--	x-x-	-x--	1	*	
1985	0614	394	FOURNAISE, PITON DE LA (Indian O	3	99	X-X-	----	----	X---	-X--	1		7/-

DATE Year	MoDy	DURA-TION	VOLCANO NAME (Subregion)	CH	PP	CHARACTERISTICS: CENTRAL VENT / FLANK VENT / RADIAL FISSURE / REGIONAL	SUBMARINE / NEW ISLAND / SUBGLACIAL / CRATER LAKE	EXPLOSIVE / PYRO FLOW / PHREATIC / SULFUR	LAVA FLOW / LAVA LAKE / DOME / SPINE	FATALITIES / DAMAGE / MUDFLOW / TSUNAMI	VE[I]		VOL L/T
1985	0619	1	TOKACHI-DAKE (Hokkaido-Japan) ..	8	314	X---	----	X-X-	----	----	1		
1985	0629	47l?	BEZYMIANNY (Kamchatka)	10		X---	----	XX--	X-X-	-X--	3	*	
1985	0730	625l>	SANGEANG API (Lesser Sunda Is) .	6	217	X---	----	XX--	X---	--X-	3	*	
1985	*08..*	*....*	*UMBOI (New Guinea-NE of) ....*	*5*	*163*	*(Seismicity, subsidence)*							
1985	*0807<*	*....*	*EL MISTI (Peru) ....*	*15*	*533*	*(Increased fumarolic activity)*							
1985	*0816*	*333l*	*KLIUCHEVSKOI (Kamchatka) ....*	*10*		*xx--*	*----*	*x---*	*x---*	*--x-*	*2*	*3*	
1985	0823	127a	RAUNG (Java)	6	216	X---	----	X---	----	----	2		
?1985	0902	.?..	FARALLON DE PAJAROS (Mariana Is-	8	293	----	?---	----	----	----	0		
?1985	0902	.?..	SAN CRISTOBAL (Nicaragua)	14	495	X---	----	--?-	----	----			
1985	*0909*	*3a*	*NASU (Honshu-Japan) ....*	*8*	*282*	*(Seismic swarm)*							
1985	*0911*	*52a*	*RUIZ (Colombia) ....*	*15*	*517*	*x---*	*----*	*xx--*	*----*	*xxx-*	*2*	*3*	*-/7*
1985	0921	1a	NII-JIMA (Izu Is-Japan)	8	283	(Seismic swarm)							
1985	0929?	16l?	HEARD VOLCANO (Indian O.-S)	3	104	XX--	--X-	----	X---	----	0		
1985	*1002*	*13a*	*LAMONGAN (Java) ....*	*6*	*215*	*(Seismic swarm)*							
1985	1005	2a?	CANLAON (Philippines-C)	7	233	X---	----	X---	----	----	1		
1985	1015q	.?..	RINCON DE LA VIEJA (Costa Rica)	14		X---	----	X-X-	----	----	1		
1985	11..	91p>	SAN MIGUEL (El Salvador)	14		X---	----	X-X-	----	----	1		
1985	1113	....	RUIZ (Colombia)	15	518	X---	----	XX--	----	XXX-	3	*	-/7
?1985	1115	<1	TANGKUBAN PARAHU (SUNDA CALDERA)	6	204	X---	----	-X--	----	----	1		
1985	*1117*	*4*	*ULAWUN (New Britain-SW Pac) ....*	*5*	*179*	*x---*	*----*	*x---*	*x---*	*----*	*2*	*3*	
1985	1120	....	ULAWUN (New Britain-SW Pac)	5	180	X---	----	XX--	X---	----	3	*	6/6
1985	12..?	82l?	AMBRYM (Vanuatu-SW Pacific)	5		X---	----	X---	----	----	2?		
?1985	1201	.?..	GUALLATIRI (Chile-N)	15	533	----	----	--?-	----	----			
?1985	1202	<1	NISHINO-SHIMA (Volcano Is-Japan)	8	288	----	?---	----	----	----	0		
1985	1202	....	KLIUCHEVSKOI (Kamchatka)	10		XX--	----	X---	X---	--X-	3	*	
1985	1202a	139b?	CONCEPCION (Nicaragua)	14	504	X---	----	X---	----	-X--	1		
1985	1206	112a	STROMBOLI (Italy)	1		X---	----	XXX-	X---	----	2		6/4
1985	1209	81c	KAVACHI (Solomon Is-SW Pac)	5	195	----	X---	X---	----	----	1		
?1985	1210	.?..	CLEVELAND (Aleutian Is)	11	328	X---	----	--?-	----	----	1		
1985	1219	13a	ETNA (Italy)	1	75	X-X-	----	X---	X---	----	2		

DECADE SUMMARY AND STATISTICS

The reports in this volume represent, we believe, history's best-documented decade of global volcanism. Here we attempt to summarize that decade's data and compare it with the previous volcanic record.

VOLCANOES ACTIVE

Part II presents reports for 205 volcanoes, only 158 of which actually erupted during the decade. In part this difference indicates increased attention to the signs of volcanic unrest, the subsurface activity that is so important to understanding surface volcanism, and in part our interest in reporting baseline information from non-erupting but potentially violent volcanoes. Table 3 presents data for each year during the decade. Multiple eruptions at single volcanoes account for the difference between the annual figures and the 158 volcanoes responsible for the totals.

The number of active volcanoes in the world very much depends upon definition of the word "active." Does this number mean those erupting at the time of the statement? Those with historically documented eruptions? Those that could erupt in the near future? In recent years, at least 15 have been in virtually constant eruption (on-going eruptions such as Stromboli, Sakurajima, Erta Ale) and an average of 56 erupted each year during the decade. At least 380 have erupted during this century, 534 have had historically documented eruptions, and more than 1500 have erupted in the Holocene (post-glacial time, or the last 10,000 years). Because the dormant intervals between major eruptions at a single volcano are commonly thousands of years long and the historic record in many parts of the world is very short, this latter figure of 1500 must be taken as a minimum number of volcanoes with eruptive potential. This number does not include the many thousands of young volcanoes on the sea floor[10].

10 Several recent studies have placed the number of identifiable sea floor volcanoes in the millions, but these include many Cenozoic volcanoes that are no longer active. Assessing eruptive potential is even harder with seamounts than with subaerial volcanoes, and the number of "active" submarine volcanoes is not known. One recent study is: Fornari, D. J. et al., 1987, Seamount Abundances and Distribution near the East Pacific Rise 0°-24°N Based on Seabeam Data, *in* Keating, B. H. et al. (eds.) *Seamounts, Islands, and Atolls*, Geophysical Monograph 43, American Geophysical Union, p. 13-21.

CHANGES IN VOLCANISM THROUGH TIME

Any discussion of the numbers of active volcanoes raises the question of whether global volcanism has changed through time. Historical records document only 1 or 2 volcanoes active each year through most of the fifteenth century, but that number jumped to 3-5 early in the sixteenth century (with the great Spanish/Portugese explorations and the invention of the printing press) and continued to increase through the middle of this century. These increases, we believe, are closely related to increases in the world's human population and communication. We are seeing an increased *reporting* of eruptions, rather than increased frequency of global volcanism. The last 130 years (Fig. 2) show this generally increasing trend along with some major "peaks and valleys" that suggest global pulsations. A closer look at the two largest "valleys," however, shows that they coincide with the two World Wars, when people (including editors) were preoccupied with other things. Many eruptions were probably witnessed during those times, but reports do not survive in the scientific literature.

Year	Continuing	New	Questionable	Total
1975	22	28	4	54
1976	19	36	4	59
1977	29	29	4	62
1978	23	25	5	53
1979	28	23	5	56
1980	25	35	5	65
1981	25	24	4	53
1982	22	30	6	58
1983	19	32	2	53
1984	20	30	4	54
1985	25	22	5	52

Table 3. Number of volcanoes active, 1975 – 85. Eruptions continuing from the previous year, volcanoes with new eruptions during the year, and volcanoes with only questionable activity are all listed. Subsurface activity, or unrest, is not included in these data.

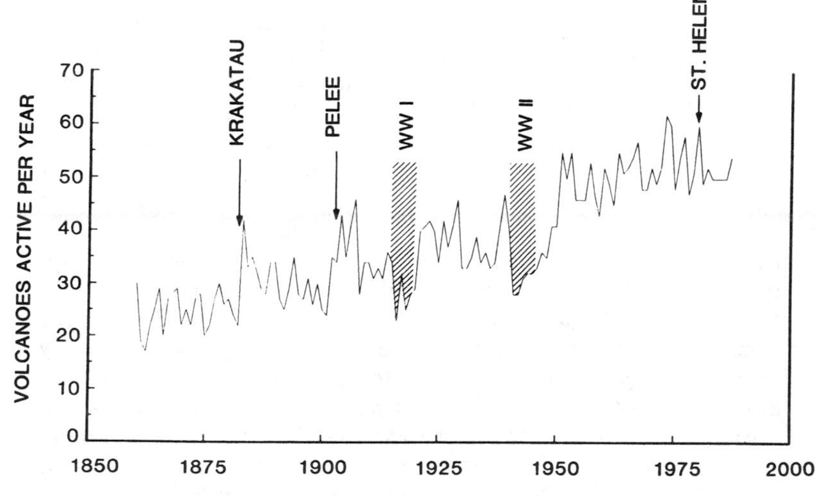

Fig. 2. The total number of volcanoes active each year, 1850-1987. Volcanoes with only uncertain activity (normally about 5% of the total) are included in these counts.

If these apparent drops in global volcanism are caused by *de*creased human attention to volcanoes, then it is reasonable to expect that *in*creased attention after major, newsworthy eruptions should result in higher-than-average numbers of volcanoes being reported in the historic literature. The 1902 disasters at Mont Pelée, St. Vincent, and Santa María (Fig. 2) were highly newsworthy events. They represent a genuine pulse in Caribbean volcanism, but we believe that the higher numbers in following years (and following Krakatau in 1883) result from increased human interest in volcanism. People reported events that they might not otherwise have reported, and editors were more likely to print those reports.

We felt so confident in this interpretation that we predicted[2] a similar increased reporting of volcanism after the 1980 eruption of Mount St. Helens. This obviously did not happen. We would like to think, however, that a partial explanation is that we are finally approaching reasonably comprehensive reporting of global volcanism. Note (in Fig. 2) that the number of volcanoes active per year shows encouraging consistency during the last 30 years. The most active year in recent decades was 1974, when 5 of 7 western Papua New Guinean volcanoes erupted: apparently another genuine pulsation of activity along a plate margin[11].

VOLCANO DISTRIBUTION AND REGIONAL VARIATIONS

Mentioning regional pulses of volcanism in 1902 and 1974 prompts a short discussion of volcano distribution and a look for regional variations during the report decade.

The map on the inside rear cover shows that the world's known subaerial volcanoes are strongly concentrated into a relatively small number of linear belts. These account for >94% of known historic eruptions and total 32,000 km in length, but they cover less than 0.6% of the earth's surface. Most are adjacent to deep oceanic trenches, and it is clear that subduction of one plate beneath another—old crust being consumed at converging plate margins—causes most of the volcanism that we see. The complementary tectonic process of rifting—the creation of new crust at *di*verging plate margins—takes place largely on the deep ocean floor and is not witnessed by humans, but accounts for an estimated ³/₄ of the new lava reaching the earth's surface each year[12]. Rift volcanism dominates the planet and, although most of it poses no threat to humans, this dominance must be remembered for global perspective.

11 Cooke, R. J. S. et al., 1976, Striking Sequence of Volcanic Eruptions in the Bismarck Volcanic Arc, Papua New Guinea, in 1972 – 75, *in* Johnson, R. W. (ed.), *Volcanism in Australasia*; Elsevier, Amsterdam, p. 149-172.

12 Crisp, J. A., 1984, Rates of Magma Emplacement and Volcanic Output; *JVGR*, v. 20, p. 177-211

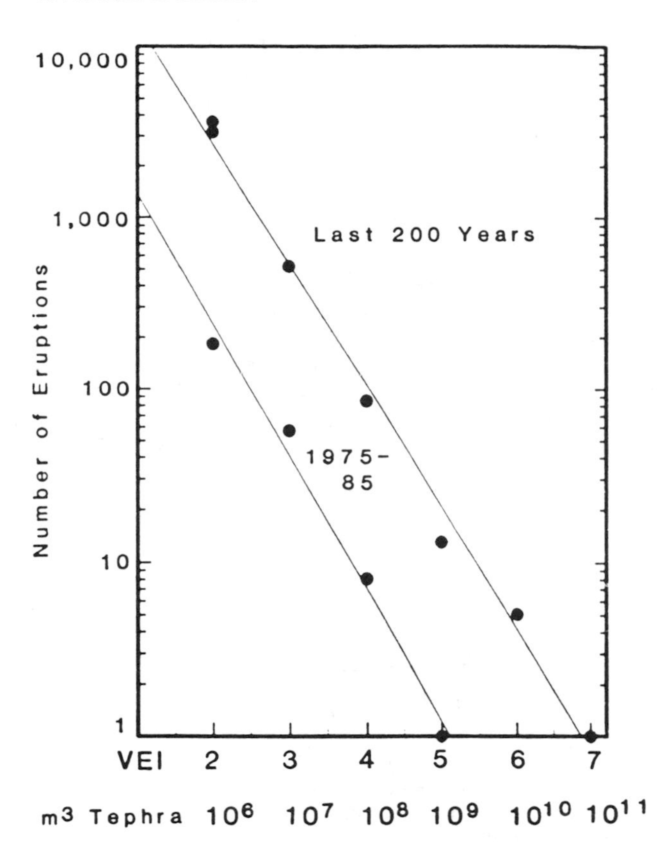

Fig. 3. Frequency of volcanic eruptions in different VEI groups over the last 200 years and the 10-year report decade 1975–85. VEI is described above and is proportional to tephra volume (also shown on horizontal axis). Eruptions of low explosivity (VEI 0 and 1) are not shown, and a logarithmic scale is used to allow resolution of the lower frequencies found with the larger eruptions. Best-fit lines determined by an exponential regression model.

Several years ago, we published a study[13] comparing different volcanic belts and the basis of their recent eruptive records. Although the recent record for many belts is too short for meaningful measurements, and comparisons are complicated by differing definitions of words such as *volcano* and *eruption*, some striking variations between belts were obvious. Here (Table 4), we reprint some of that information for comparison with data from the decade 1975–85.

Eight of these volcano belts experienced no reported volcanism at all during the 1975–85 decade: Greece, the New Guinea highlands, the Banda arc, Taiwan, S Chile, and all Atlantic Ocean archipelagoes south of Iceland. As shown by the VRF eruption data, none of these have been particularly vigorous in the previous 100 years, but their combined total for that century averaged over 5 eruptions per decade, so their quiet during this last decade is surprising. Other broad areas of scattered, nonlinear volcanism, like mainland Asia, were likewise quiet during the decade. Most belts showed a somewhat higher number of eruption years in the last decade than in previous years, probably reflecting better reporting than the average over the previous 100 years. Two regions, however, show dramatically higher numbers during the last decade. The Cascades increase is understandable as 6 years of Mount St. Helens eruptions are superimposed on a volcano belt that was exceedingly quiet during the previous 100 years, but the Bismarck arc increase in Papua New Guinea may be a genuine in-

		Volcanoes			Eruptions		
#	Belt Name	1975 -85	pre-1981	%	1975 -85		pre-1981
1	Italy-S	4	7	57	24		24
	Greece	0	6	0	0	<	1
2	Africa-E	3	82	4	14+	>	6
3	Comores	1	2	50	1		1
	Reunion	1	1	100	8		6
4	New Zealand	4	31	13	18+		14+
	Kermadecs	1	5	20	2+	>	1-
	Tonga	3	10	30	3+		3
5	Bismarck-W	8	11	73	30	>	12+
	Bismarck-E	2	12	17	14	>>	2
	New Guinea	0	10	0	0	<	1
	Solomons-NI	2	13	15	17		>
	Vanuatu	4	13	31	23		18
6	Sumatra	2	30	7	13		15
	Java	9	36	25	38		28
	Sunda	4	30	13	9		11
	Banda	0	6	0	0	<	1
	Sangihi	2	17	12	13		10
	Halmahera	4	9	44	14		8+
7	Philippines	5	30	17	12	>	6-
	Philippines-N	1	9	11	1		1
	Taiwan-N	0	7	0	0		0+
8	Ryukyu-Kyushu	6	14	43	36	>	18-
	Honshu	6	36	17	11		14+
	Izu-Bonin	12	19	63	38	>	12-
	Marianas	3	11	27	6+		4-
	Hokkaido	4	15	27	8+		7
9	Kuriles	7	50	14	14	>	7
10	Kamchatka	7	47	15	39		20+
11	Aleutians	8	40	20	19		13
	Alaska Peninsula	4	33	12	12		10
12	Cascades	3	38	8	6	>>	1
13	Hawaii	3	6	50	13		9-
14	Mexico	2	22	9	8		8
	Central America	13	79	33	55		34-
15	Colombia-Ecuador	7	23	30	27		19-
	Galapagos	4	15	27	8		4+
	Peru-Chile-N	2	63	3	3		4+
	Chile-Central	7	26	27	20		15-
	Chile-S	0	11	0	0	<	1-
16	West Indies	4	17	24	5		3-
17	Iceland	3	29	10	15	>	4-
18	Azores	0	10	0	0	<	1-
	Canaries	0	7	0	0		0+
19	S Sandwich	0	9	0	0	<	1

Table 4. Activity of the world's volcanic belts, comparing the 1975-85 decade with a 100-year (pre-1981) record from the VRF[13]. Number of volcanoes active during the decade is compared with number of known Holocene volcanoes (and given as a percentage of that total). Eruption years are the combined total of calendar years that each volcano in the belt was active in the decade. For the longer period, the total for 1880–1981 is divided by 10, so a comparison of these two columns shows the last decade's activity against the average decade's activity during the last 100 years. "Greater than" (>) and "less than" (<) symbols are used when the difference between the two values is more than 2:1 and are doubled when more than 5:1. Chapter numbers precede belt names

13 Simkin, T. and Siebert, L., 1984, Explosive Eruptions in Space and Time: Durations, Intervals, and a Comparison of the World's Active Belts, *in* Boyd, F. R. (ed.), *Explosive Volcanism: Inception, Evolution, and Hazards*; National Academy of Sciences Press, Washington, p. 110-21.

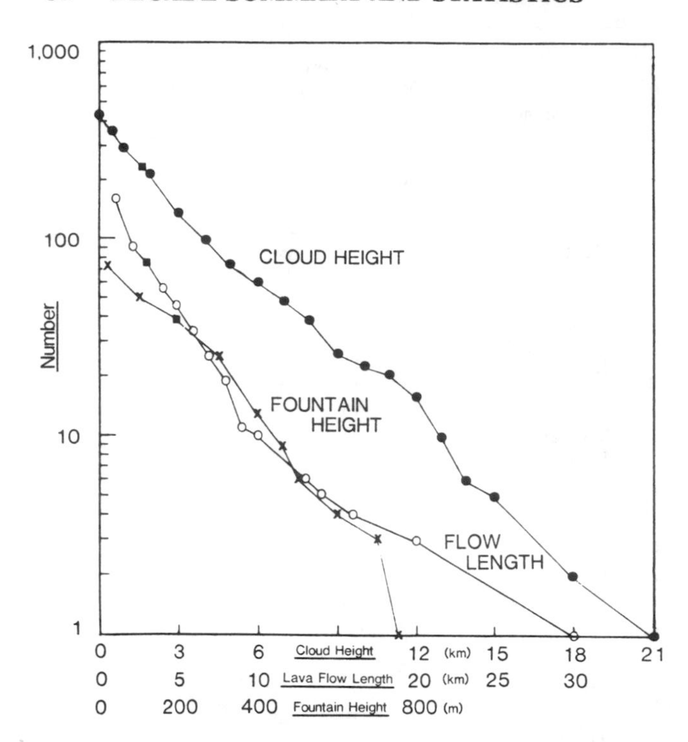

Fig. 4. Cloud height, lava flow length, and fountain height measurements, 1975–85. Cumulative frequency of maxima are plotted on a logarithmic scale as in Fig. 3 above. One cloud of 21 km is listed, 5 of ≥ 15 km, 22 of ≥ 10 km, 100 of ≥ 4 km, and so forth. The median for each data set is marked by a square. Cloud heights include satellite and radar measurements, pilot's sightings, and estimates from ground observers. Some purely vapor clouds may be included in the data set although our attempt was to count only ash clouds. Flow length data are dominated by Kilauea and Etna (half of all measurements) but 27 volcanoes are represented. Projectile heights are not included in lava fountain data, but can be found under "trajectory" in the index. Kilauea accounts for 34 of the 74 fountain heights.

crease following the 1974 pulsation discussed above.

ERUPTION FREQUENCY AND MAGNITUDE

Like earthquakes, the frequency of eruptions decreases with increasing size—there are many small ones, fewer medium-sized ones, and very few large ones. We

	Unit	Min.	Median	Max.	N
Eruptive Cloud Height*	km	0.1	2.0	20	474
Lava Fountain Height*	m	10	200	750	74
Lava Flow Length*	km	1	3	30	112
Lava Flow Velocity	m/hr	1	200	60,000	26
Lava extrusion rate	m^3/sec	1	16	600	28
Lava Flow Volume	$m^3 \times 10^6$	0.2	11.7	460	62
Tephra Volume	$m^3 \times 10^6$	0.1	4.0	1000	21
SO_2 Emission Rate	tons/day	20	170	24,000	67
Fumarole Temperature	°C	60	400	1000	101
Fatalities	total	1	27	>22,000	20

Table 5. Volcanological measurements 1975–85. Range, median, and number of measurements (N) for various rates, distances, volumes, and temperatures listed in this book. Only the larger value was tabulated when multiple values were listed for a single event, but events spanning several issues of the *Bulletin* may have different maxima for successive issues. Full data sets for parameters marked by an asterisk are plotted in Figure 4. See text for discussion (and limitations). Fatality data appear in Table 6.

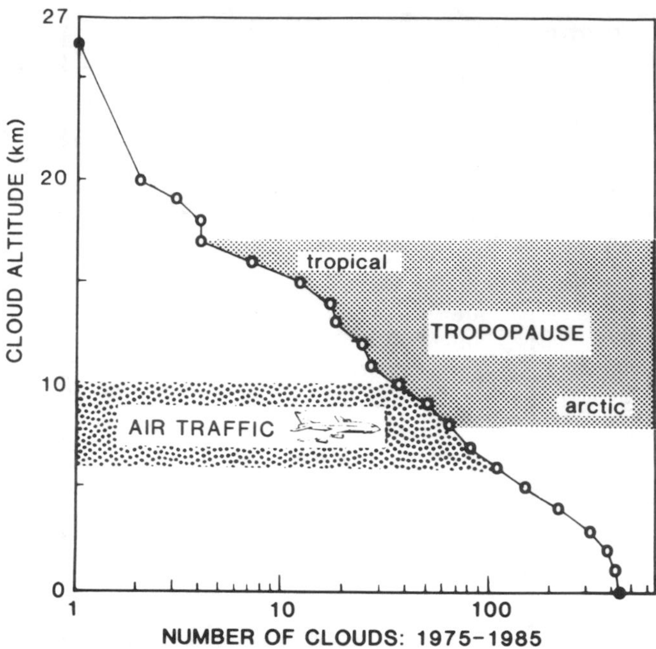

Fig. 5. Volcanic cloud altitudes, 1975–85. Data as in Figure 4, but recalculated as altitude above sea level and with axes switched for ease of visualizing elevation data. Ranges of tropopause[14] and most commercial air traffic are shown. During the decade, over 100 eruptive clouds are known to have penetrated the air traffic range, some with near-disastrous results (see Chapter 7, Galunggung 7:6-7), and at least 9 have passed through the tropopause into the stratosphere, where volcanic products are easily dispersed around the world.

have borrowed from the seismologists' use of frequency/magnitude plots, taking the VEI (described above under "Chronology"), which closely parallels the volume of erupted tephra, as our best measure of explosive magnitude. In Fig. 3, the number of eruptions recorded over the last 200 years is plotted for each VEI class. We have had one VEI 7 event (Tambora, 1815) in the last 200 years, 5 ≥ VEI 6, 13 ≥ VEI 5 (Mt. St. Helens size), and so forth. The upper data point at VEI 2 represents the more accurate figure (based on the last 10, rather than 200, years) since it is the smaller VEI 2 events that are least likely to have been fully reported in the last century. The number of eruptions in each VEI class during the 1975–85 decade appear in the VEI table (Table 2) and are plotted on Figure 3 to show that the frequency/magnitude relationship during the decade was comparable to that of the longer historic record.

Another way of reading this plot is that eruptions producing at least 10^6 m^3 of tephra take place at an average of once or twice a month somewhere on Earth. Those producing 10^7 m^3, like the Ruiz event that generated devastating mudflows in 1985, take place 2 or 3 times a year. Eruptions the size of St. Helens 1980 (10^9 m^3, or 1 km^3) occur perhaps once a decade, and those like Krakatau 1883 (≥ 10 km^3) 2 or 3 times a century. The historic record, however, fails to prepare us for the much larger eruptions of the

14 Reiter, E. R., 1975, Stratospheric-Tropospheric Exchange Processes; *Reviews of Geophysics and Space Physics*, v. 13, p. 459-74.

past. The Toba eruptions, only 75,000 years ago, were 1000 times larger than the St. Helens eruption, and even larger eruptions are known from the geologic record.

ERUPTION MEASUREMENTS

As a part of the indexing process for this book, we noted the value, as well as type, of various volcanological measurements. This allowed us to break these indexed measurements into numerical groups, so that a reader interested in unusually high lava fountains, or short lava flows, can quickly find them through the index. But it also allows us to make some generalizations on a decade's gathering of such values—the small as well as the large—toward a better understanding of typical as well as extreme volcanism on the planet. Ten of these measures are listed in Table 5 and 3 are plotted in Fig. 4.

The great variety of volcanic behavior is illustrated by the large range in values, and the generally low medians remind us of the prevalence of smaller events (despite the volcanological literature's emphasis on the larger events). Some measurements, such as lava fountain heights, must obviously be made by the time that preliminary reports are submitted to SEAN, but others, like flow volume, are normally calculated after such reports. Thus tephra volumes are listed for fewer than 1 out of 15 explosive eruptions, and 71% of the lava volumes tabulated are from 2 volcanoes (Kilauea and Etna). Clearly many of these measurements are not representative of global volcanism during the decade, but others may be helpful. Figure 4 plots data frequency for 3 such parameters.

SEAN has sought volcanic cloud height data wherever possible and worked to help communication between the volcanological and meteorological communities. Cloud height is a useful measure of eruptive vigor[15] and is presented in Fig. 4 as height above the volcano. Some measurements, however, are absolute altitude measurements, without regard for the volcano's own elevation, and we have converted between altitudes and heights using the appropriate volcano summit elevation. It is the altitude values, however, that are of interest to the meteorological community, and this interest is now shared by

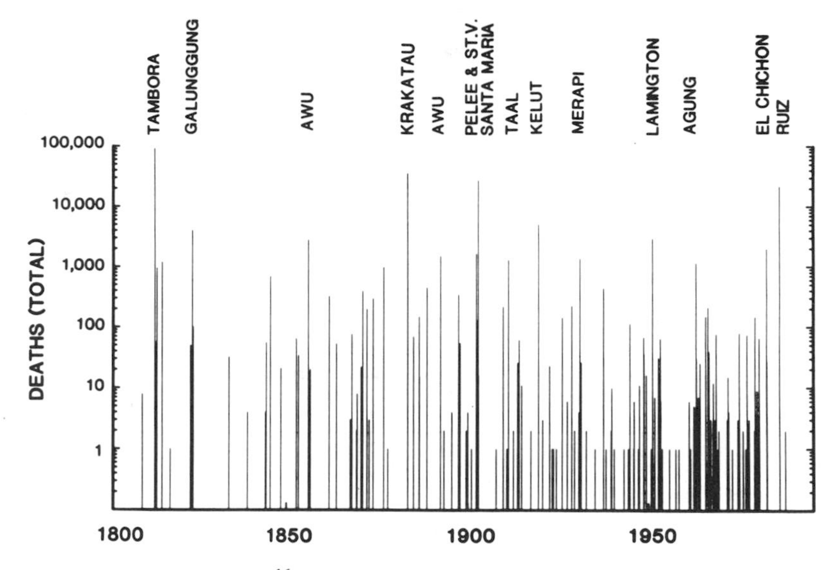

Fig. 6. Volcanic death toll[16] for the years since 1800. The total number of fatalities—indirect as well as direct—are plotted on a logarithmic axis and all eruptions with >1,000 fatalities are labeled.

Date			Volcano	VEI	Evacuees	Deaths	Cause(s)
1975	Nov	29	Kilauea (Hawaii)			2	tsunami (earthquake induced)
1976	Jul	8	Soufriere (Guadeloupe)	2	72,000	0	
	Aug	12	Sangay (Ecuador)	2		2	ballistics
	Sep	3	Taal (Philippines)	2	thousands	0	
	Oct	15	Api Siau (Sangihe Is)	2	1800	1	pyroclastic flow
1977	Jan	10	Nyiragongo (Zaire)	1		>75	lava flows
	Aug		Usu (Japan)	3	27,000	3	mudflows
	Oct	24	Piton de la Fournaise	1	2500	0	
	Nov	05	Karthala (Comore Is)	0	4000	0	
1978	May	7	Mayon (Philippines)	2	8000	0	
	Sep	2	Santiaguito (Guatemala)	1		1	mudflows
1979	Feb	20	Dieng (Java)	1		149	gas
	Mar	8	Karkar (PNG)	2		2	blast
	Apr	30	Marapi (Sumatra)			80	landslides (rain induced)
	Jul	7	Iliwerung (Indonesia)			500	tsunami (landslide induced)
	Sep	6	Aso (Japan)	2		3	ballistics
		12	Etna (Italy)	2		9	ballistics
1980	May	18	Mt. St. Helens (USA)	5	>3-400	57	blast, mudflow
	Jul	22	Ulawun (PNG)	4	2000	0	
	Sep	4	Gamalama (Indonesia)	2	40,000	0	
1981	Mar	29	Semeru (Java)	2		1	pyroclastic flow
	May	14	Semeru (Java)			252	mudflows (rain induced)
		15	Pagan (Marianas)	4	53	0	
	Jun	30	Mayon (Philippines)			40	mudflows (typhoon induced)
1982	Mar	28+	El Chichón (Mexico)	4		2000	pyroclastic flows
	Apr	05+	Galunggung (Java)	4	75,000	27	mudflows (lahars)
	Aug	25	Santiaguito (Guatemala)		hundreds	0	
1983	Jul	23	Una Una (Indonesia)	4	7000	0	
	Oct	4	Miyake-jima (Japan)	3	1400	0	
1984	Aug	16	Lake Monoun (Cameroon)			37	gas
1985	Nov	13	Ruiz (Colombia)	3		>22,000	mudflows

Table 6. Volcano-related fatalities, 1975–85, with dominant cause of death. Explosive magnitude (VEI) is listed for eruptive events, and number of evacuees is included—for some nonfatal eruptions as well as fatal. Many of these numbers are only approximate.

15 Wilson, L. et al., 1978, The Control of Volcanic Column Heights by Eruption Energetics and Dynamics; *JGR*, v. 83, p. 1829-36.
16 Many of these data are from Blong, R. J., 1984, *Volcanic Hazards*, Academic Press, Sydney, 427 pp.

1988	Anak Ranakah (Indonesia)
1987	Kupreanof (Alaska Peninsula)
1983	Unnamed (New Britain)
1982	El Chichón (Mexico)
	Teahitia (French Polynesia)
1981	Mehetia (French Polynesia)
1980	Marion Island (Indian Ocean)
	Malinao (Philippines)
1979	On-Take (Honshu, Japan)
	Nikko Seamount (Kazan Is, Japan)
1978	Ardoukoba (Djibouti)
	Unnamed (Kuril Is)
1977	Monowai Seamount (Kermadec Is)
	Ukinrek Maars (Alaska Peninsula)
1975	3 Unnamed seamounts (Izu/Marianas)
	Minami-Hiyoshi Seamount (Kazan Is)

Table 7. Eruptions from volcanoes with no prior historic activity.

Year	Volcano	First Historic Eruption?	Deaths
1982*	El Chichón (Mexico)	yes	2000
1980	St Helens (USA)	no	57
1956	Bezymianny (Kamchatka)	yes	0
1932	Cerro Azul (Chile)	no	0
1912	Novarupta (Katmai, AK)	yes	2
1907	Ksudach (Kamchatka)	yes	0
1902	Santa María (Guatemala)	yes	>5000
1886	Tarawera (New Zealand)	yes	>150
1883	Krakatau (Indonesia)	no	36,417
1875	Askja (Iceland)	yes	0
1854	Sheveluch (Kamchatka)	yes	0
1835	Cosiguina (Nicaragua)	yes	5-10
1822	Galunggung (Indonesia)	yes	4011
1815	Tambora (Indonesia)	yes	92,000

Table 8. Largest explosive eruptions of the nineteenth and twentieth centuries (VEI ≥ 5, or tephra volume ≥ 1 km^3). All but 3 were the first historic eruption known from the volcano and the high death tolls (in heavily populated regions) reflect that fact. El Chichón is included because tephra from its two VEI 4 eruptions, separated by only a week, total 1 km^3.

those concerned with air traffic safety near active volcanoes. We therefore replot the data as altitude above sea level in Fig. 5.

VOLCANIC HAZARD

Table 6 lists fatalities suffered during the decade reported here. Also listed are the VEI for each eruption, and the number of evacuees where appropriate. Many more evacuations could be listed, but we wanted to temper the fatality table with a reminder of the many eruptions in which timely evacuations have prevented loss of life.

Among the eruption fatalities, a wide variety of causes is demonstrated. Although the more explosive eruptions (VEI ≥ 3) have been the major killers, note that more than half of the fatal eruptions have been relatively small (VEI ≥ 2) events.

Fig. 6 shows the numbers of fatalities during the 19th and 20th centuries. Large death tolls continue from Tambora's 92,000 to Ruiz in 1985. There is no clear pattern to the larger numbers, but the frequency of fatal eruptions clearly increases with time.

One of the certain contributors to large volcanic death tolls is the fact that the repose interval between eruptions is very much longer than the historic records in many parts of the world. Most eruptions of low explosivity follow repose intervals of only 1 to 10 years, while the more explosive events, of VEI ≥ 5, follow repose intervals measured in hundreds or thousands of years[17]. In regions with short historic records, the human population is commonly unprepared for such events and the results are often tragic.

For at least 2 centuries we have had an average of 1 or 2 eruptions per year from volcanoes with *no* previous historic activity (Table 7 lists eruptions over the last 13 years from such volcanoes). Other volcanoes have erupted after hundreds of years of quiet. Because their danger was not widely recognized, and because unusually violent eruptions are known to follow unusually long periods of quiet, these events included some of history's worst natural disasters. Table 8 lists the most violent eruptions of the last 2 centuries. Of these 14 most explosive eruptions, 11 were the first historic eruption at that volcano (and for others, like Krakatau, earlier activity was not widely known). Attention is commonly focused on volcanoes that have produced history's longest-running and best-documented eruptions, but we must not neglect the nonfamous, thickly vegetated volcanoes that have had *no* historic activity. They may be the most dangerous of all.

EDITING AND FORMAT CONVENTIONS

The original *SEAN Bulletin* coverage of volcanism during this ten-year period totaled more than 1,400 pages. To reduce this to manageable (and affordable) book length, without sacrificing information, has required careful but substantial trimming. The conventions adopted for these—and other—changes are described in this section, along with an introduction to the conventions of the original *Bulletin* itself.

SEAN BULLETIN CONVENTIONS

Because of SEAN's emphasis on swift reporting, the *Bulletin* has normally been restricted to field observations and measurements—the timing and sequence information so important to understanding the dynamics

17 See Figure 6, *Volcanoes of the World*[2].

of volcanism but so easily lost if not captured on paper at the time. Most of the analytical, theoretical, and interpretive aspects of these eruptions—requiring laboratory equipment, library access, or quiet reflection—are normally absent from the *Bulletin*. In short, the *Bulletin* does not compete with scientific journals, to which the reader is referred for more complete coverage of recent volcanism. We have added many such references in this book (see "General References" section below), but it is important for the reader to understand the fundamental limitation of the *Bulletin*'s preliminary nature.

The conventions followed by SEAN through the years, and retained here, are as follows:

- **Volcano names** are those used in *Volcanoes of the World*, with only a few exceptions (e.g., "Santiaguito Dome" for the continuously active feature of Guatemala's Santa María). These exceptions are mentioned with the appropriate reports in this book and both names, along with other common synonyms, are treated alphabetically in the index. Some names (e.g., Ukinrek Maars) were not known at the time of the initial SEAN report, but all reports are here combined under the current name.
- **Latitude and longitude** are listed in degrees and decimal parts of a degree (rather than minutes and seconds), and again are from *Volcanoes of the World*. These normally locate the volcano's (or volcanic field's) center point and thus may not accurately locate the site of a noncentral eruption.
- **Local time** is used unless otherwise specified. The conversion factor for Greenwich Mean Time (GMT, or universal time) is noted for each volcano (if times are listed in the report). GMT time is needed for close comparison with seismic or other global data sets, but local time helps the reader to place an observation in relation to the local solar (and human) cycle. Local hours of darkness are often difficult to recognize in GMT yet knowing them aids interpretation of a report's visual details (e.g., incandescence, cloud heights).
- **24-hour time** is used, rather than A.M. and P.M. (e.g., 1:30 P.M. is 1330).
- **Metric units** are used throughout. Measurements in other units are converted when they appear in incoming reports. After conversion, metric measures are rounded to reflect the number of significant figures in the original value (i.e., 1000 ft becomes 300 m, not 304.8 m). Abbreviations (and some conversion factors) are shown on the inside front cover of this book.
- **16 point compass direction** is used for most bearings and these standard abbreviations are also shown on the inside front cover. Directions are therefore accurate, theoretically, to ±11.25 degrees, although many reports come from observers making more general estimates (e.g., using SE when SSE would be more accurate).
- **Issue date** has been, since the fourteenth issue, the last day of the month summarized (the first 13 were dated the first day of the following month). The deadline for receipt of reports has generally been about the tenth day of the following month: aiming to provide enough time for correspondents to compile and mail a month's summary, but also allowing last-minute updates by telephone and telex. Thus it is common to see, for example, early June data in the May *Bulletin*.

TEXT EDITING

Reorganizing the reports on a volcano-by-volcano basis has allowed **removal of repetitive material** that was necessary in a monthly bulletin. Many of the original reports included an introductory section briefly describing activity from previous months, plus a closing comment summarizing the volcano's historic eruptions. Only a single statement summarizing a volcano's eruption history has generally been retained. Similarly, a single map has typically been selected for a given volcano, although sequences of maps have been kept where they show progressive changes to the volcanic edifice. Where possible, cumulative graphs of deformation and seismic data have been used, as at New Zealand's White Island, where a single figure illustrating changes over several years includes many smaller data sets from earlier *Bulletins*. Repeated monthly tables were replaced either by a cumulative table (as with explosion counts at Sakurajima) or by graphics. Volcanic aerosols, the persistent stratospheric remnants of major eruption clouds, were measured at several sites around the globe and regularly tabulated in the *Bulletin*. However, several reporting groups have combined these data into multi-year graphs, which provide a clearer picture of long-term changes in stratospheric aerosol loading. These replace the tables here, at considerable savings in book length.

Abbreviations and acronyms are used extensively for words, organizations, and publications that appear repeatedly. The first time that a given acronym is used, it appears in parentheses after the full name. Readers not moving sequentially through the book will often miss those initial definitions and should seek them on the inside front cover, or in the List of Acronyms. Months with short names (March – July) are spelled out; others were shortened to 3 or 4 characters. Units of length and mass are expressed as mm, km, etc., if they are preceded by a number. Compass directions are generally abbreviated. (See inside front cover and above section on "SEAN Bulletin Conventions.")

Corrections of original text have been done cautiously, attempting to walk the line between intrusive editorial devices and unmarked changes that leave a careful reader, comparing the book with the original *Bulletin*, wondering whether a change is a correction or a new error. Obvious typographical errors in the original text (where found!) have been corrected with no additional marks, but any other changes have been marked by **square brackets** ([ ]). Deletions are marked by **ellipses** (. . .). We have generally resisted the urge to improve upon the prose of the original text, and have not even considered the **addition** of more recent information (e.g., chemical analyses of products, refinement of original estimates, petrochemical interpretation) that could triple the size of the book (and more rightfully belongs in scientific journals). However, we have sent copy for **review by original information contacts**, and have done our best to incorporate their comments. Some have submitted corrections, some have asked for deletions, and many have responded to our main request for references to more detailed work in the published literature.

ILLUSTRATIONS AND LAYOUT

In addition to the removal of repeated maps as described above, many illustrations have been reduced in size. The pressure of SEAN's monthly deadline often precludes both experimentation with smaller size and the minor redrafting of newly received illustrations that makes such reduction possible. The use of smaller illustrations is also facilitated by the two-column format adopted for this book. Illustrations have been renumbered by chapter, so, for example, the tenth figure of Chapter 12 is numbered 12-10.

Volcanoes are listed in the sequence adopted by IAVCEI in the late 1930s for its *Catalog of Active Volcanoes of the World* series, and revised for the Smithsonian's database and *Volcanoes of the World*. Each region is assigned a separate chapter, and these regional divisions are shown on the map that makes up the inside rear cover of this book. Following each volcano entry, we list latitude, longitude, summit elevation, and (when reports carry times) time conversion to GMT. Current elevations (and depths) are used, but these measurements are inherently changeable (e.g., Mt. St. Helens which started the decade at 2950 m and ended it at 2549 m). Depths to shallow seamounts are particularly variable.

A few sentences then introduce most volcanoes, attempting to provide geographic and volcano-historic context if not present in the initial report. Original SEAN reports follow chronologically, with *Bulletin* date shown (month/year) in the left margin, followed by volume and issue in parentheses (e.g., "**1/77 (2:1)**") indicates Volume 2, issue 1, of January 1977)[18]. Reports in a particular issue, however, often deal with observations made at other times, so the issue date is not necessarily that of the information contained in the report. To aid recognition of their sequence, reports in consecutive *Bulletin*s are separated by a single blank line, brief gaps between *Bulletin*s are shown by an extra blank line, and different eruptive (or intrusive) events are separated by a horizontal line.

INFORMATION CONTACTS

In the *SEAN Bulletin* the sources for each report are rigorously listed, along with full address. This information is essential, both to acknowledge the correspondents who are so essential to SEAN and to provide a source for readers interested in follow-up data. Addresses change, however, and the geographic reorganization of the book commonly brings together multiple reports from the same source. Therefore, we have saved space by reducing each information contact reference to name (with the briefest organizational affiliation needed for identification) and compiled a separate list—with the most recent known full address—of every contributor to the *Bulletin*. This list can be found in Part III and should be a useful compilation of volcano information sources. As with other parts of the book, we welcome updates and corrections to the address list from readers.

REFERENCES

SEAN reports, as we repeatedly emphasise, are by their very nature preliminary. They remain the only available descriptions of an unfortunately large number of eruptions, but for others they have been complemented and in some cases superseded by more detailed reports in the scientific literature. We can neither duplicate nor incorporate these papers—a recent bibliography[19] of the St. Helens eruption alone lists over 1500 references without touching the thousands of articles in the popular press—but we consider it essential to guide readers toward some selected reports in the published literature.

These additions appear under the heading "Further References" with the original SEAN reports. At the end of the report decade, we asked all correspondents for references to published work detailing the events that they originally reported to SEAN. In addition to these references, many written by SEAN correspondents themselves, we searched our own bibliographies and the GeoRef[20] database, selecting references where available and appending them to the appropriate SEAN reports. We then sent review copy to each correspondent asking them to pay special attention to the references in reviewing their portion of the text. In the end, we were unable to find references for many events, but found an abundance for others. Space limitations have forced us to choose only a few papers for each of these events. We have tended to select compilations over individual papers, longer over shorter, and more recent over older papers (in the expectation that readers can find earlier references among those cited by more recent papers).

References describing a single eruption are listed after that eruption's report, and papers describing several eruptions follow the report of the most recent. General overviews of a volcano group or region can provide a valuable counterpoint to SEAN's descriptions of individual eruptions. When available, we have included these (under the heading "General References") at the beginning of a volcano section or regional chapter. Special compilations of papers are generally listed as a single reference (under the editors' names) but their nature is noted. Those few papers that were cited in the original *Bulletin*s are listed (as they were then) under the heading of "References." This scheme of interspersing references throughout the text precludes a standard,

---

18 Volume 1 includes 15 issues—from SEAN's first, in October 1975, through 1976—but in all subsequent volumes the issue number corresponds to the report month.

19 Manson, C. J., Messick, C. H., and Sinnott, G. M., 1987; Mount St. Helens—A Bibliography of Geoscience Literature, 1882–1986; *USGS Open File Report* 87-292, 205 pp.

20 GeoRef is produced by the American Geological Institute, 4220 King Street, Alexandria, VA 22302.

alphabetically arranged bibliography, but all are so sequenced, under the senior author's name[21], in the detailed index of Part III. We believe that the references themselves are where they are most needed.

While far from comprehensive, we hope that this small selection of references will be a useful starting point for all those who seek additional information about contemporary volcanism.

GENERAL REFERENCES

Above we have described our conventions for referenced citations in Part II. Here we list a variety of general references that might be helpful for the interested reader. These cover the volcanoes of the world, and should be considered as information sources complementing the "General References" listed at the start of some sections of Part II.

Three scientific journals are so frequently cited in volcanology that we have used their acronymns in text citations. They carry specific papers on a wide variety of volcanological subjects. The *Bulletin of Volcanology* (or *BV;* formerly *Bulletin Volcanologique*) has been the journal of the International Association of Volcanology and Chemistry of the Earth's Interior (IAVCEI) since 1932. Since 1985, it has been published by Springer-Verlag. The *Journal of Volcanology and Geothermal Research* (*JVGR*) has been published since 1975 by Elsevier. The *Journal of Geophysical Research* (*JGR*) is published in 3 sections by the American Geophysical Union (AGU): references cited here are to the Solid Earth section unless otherwise specified. Although the majority of its papers are non-volcanological, its volcanological contributions are substantial.

Three works provide comprehensive reference information on the world's volcanoes. From 1950 to 1975, the IAVCEI published 22 regional volumes in a series entitled *Catalogue of the Active Volcanoes of the World* (*CAVW*). Catalogs for Alaska and Iceland are still in progress, and some catalogs are seriously out of date, but the series is a most valuable guide and handy reference to information not easily found elsewhere. Another IAVCEI project, the *Data Sheets on Post-Miocene Volcanoes*, was partially published in the mid 1970s. These sheets contain data on a much larger population of volcanoes, those believed to have been active in the last 15 million years, but the data for each are limited. The complete set is being reprinted on microfiche by the American Geophysical Union. Our 1981 book, *Volcanoes of the World*[2] contains the essence of our VRF coverage on volcanoes active in the last 10,000 years. Its formats are regional, chronologic, gazetteer, and bibliographic. We are at work on a second edition that will incorporate the many additions and corrections of the last 7 years.

Temporal reviews can also be found in 3 places. First and most important is the *Bulletin of Volcanic Eruptions*, compiled annually by the Volcanological Society of Japan, and published in the *Bulletin of Volcanology*. This normally appears 2-3 years after the close of the report year and this provides a valuable retrospective complement to the *SEAN Bulletin*. The monthly journal *Geotimes* has included an annual review issue to which we have contributed articles[22] covering each of the years 1975–85. On a longer review basis, national committees to the International Union of Geodesy and Geophysics (IUGG) provide reports every 4 years, several of which deal specifically with volcanology[23].

A map of the world's volcanoes is now in press at the U.S. Geological Survey. It is a joint USGS/Smithsonian publication at a scale of 1:30 million and includes global physiography, earthquake epicenters, and plate tectonics features. A set of tectonic maps published by the American Association of Petroleum Geologists, as part of their Circum-Pacific map project, consist of 5 overlapping maps at a scale of 1:10 million. It covers most of the world's volcanoes, and a variety of other data are also shown.

Finally, we list some general references to aid the reader interested in learning more about volcanology. Textbooks by Decker and Decker[24], Francis[25], Macdonald[26], and Williams and McBirney[27] are aimed at introductory to more technical readers (in the sequence listed). All provide helpful, authoritative information. Many other texts are available in other languages as well as in English. A monograph by Sawada[28] describes satellite imaging of many eruption clouds, and an impor-

21 In the case of two-author papers, the junior author's name is also indexed.

22 *Geotimes*, v. 21(1), p. 38; v. 22(1), p. 42-43; v. 23(1), p. 49-50; v. 24(1), p. 50-51; v. 25(2), p. 47-49; v. 26(2), p. 57-58; v. 27(2), p. 57-59; v. 28(2), p. 39-41; v. 29(7), p. 18-20; v. 29(11), p. 15-16; v. 31(4), p. 14-17.

23 National reports are published by different groups in different nations. U.S. reports are published by the AGU and cover the periods 1971–78 (Simkin, T.), 1979–82 (Swanson, D., and Casadevall, T.), and 1983–86 (Self, S., and Francis, P. W.).

24 Decker, R, and Decker, B., 1981, *Volcanoes;* W. H. Freeman Co., San Francisco, 244 pp. Second edition in preparation.

25 Francis, P. W., 1976, *Volcanoes;* Penguin Books, Harmondsworth, England, 368 pp.

26 Macdonald, G. A., 1972, *Volcanoes*; Prentice Hall, Englewood Cliffs, 510 pp. Second edition in preparation by J. P. Lockwood.

27 Williams, H., and McBirney, A. R., 1979, *Volcanology;* Freeman, Cooper & Co., San Francisco, 397 pp.

28 Sawada, Y., 1987, Study on Analysis of Volcanic Eruption Cloud Image Data Obtained by the Geostationary Meteorological Satellite (GMS); *Technical Reports of the Meteorological Research Institute (Japan)*, no. 22, 335 pp.

tant compilation has just been published detailing geophysical unrest (with and without accompanying volcanism) at active calderas around the world[29]. Several other volumes referenced elsewhere in this book (e.g., Blong[16], Boyd[9]) deal with volcanoes on a global basis. And a series of books by Maurice and Katia Krafft[30] provide beautiful photographs to illustrate many aspects of volcanism.

CAUTIONS

The reports in this volume constitute, we like to think, the best record available for a full decade of global volcanism. It is essential to emphasise, though, that this record is far from complete—even for the visible volcanoes above sea level—and it barely touches the sea floor eruptions that are the overwhelmingly dominant type of global volcanism.

As emphasised in the "Decade Summary" section above, estimates of the magma volume reaching the earth's surface each year indicate that at least $^3/_4$ is added to the sea floor, either at oceanic spreading centers or at intraplate "hot spots." A very few of these seamounts have built themselves into islands—like Hawaii and Réunion—and some spreading center volcanism is visible on Iceland, but the vast majority of the world's eruptions take place unnoticed under 1-3 km of sea water. The last decade has included major advances in understanding this style of volcanism[31], but the many eruptions that must take place on the deep ocean floor are totally undocumented. Events are reported in this book from only 21 submarine volcanoes, and many reports are of uncertain activity (e.g., discolored water, floating pumice from an unknown source). Furthermore, all but 3 of the 21 are shallow vents along active volcanic chains at converging plate boundaries. The 3 exceptions—Loihi, Macdonald, and Teahitia—are seamounts close to seismographic networks, and their eruptive record comes largely from interested seismologists who monitor their activity instrumentally. Recent estimates suggest that the number of active volcanoes on the deep ocean floor mayh be in the tens of thousands[10]. However adequate the SEAN record may be for subaerial volcanism, it is woefully *in*adequate for the submarine eruptions that dominate the world's volcanism.

The "Decade Summary" section emphasized that many of volcanic record's apparent trends reflect changes in human reporting rather than real changes in volcanic activity. Human-induced variations in the volcanic record are paralleled in SEAN's history, and are probably enhanced by the preliminary nature of the data in the *Bulletin*. As SEAN gradually built a network of information sources, the volume and quality of reports have grown, and the number of volcanoes that could erupt without our knowledge has diminished. Superimposed on these trends are changes in reporting caused by the arrivals and departures of dedicated observers at individual volcanoes or volcanic regions[32].

We have made little attempt to smooth these variations either in the original *SEAN Bulletin*s or in this book. Minor activity at closely monitored sites has therefore sometimes been reported in more detail than larger eruptions at more remote volcanoes. While we regret our limited coverage of some eruptions, volcanoes, and regions, we see no merit in deleting information in order to de-emphasize lesser, but better reported activity. Such activity is often far from unimportant, as demonstrated by the year of seismicity and increased gas release prior to the November 1985 catastrophe at Ruiz, and similar activity that apparently began well before El Chichón's powerful 1982 eruption. Details of the behavior that preceded gas releases at Lakes Monoun and Nyos would go far toward dispelling uncertainties about their triggering mechanisms and help in evaluating risks at other possibly dangerous lakes. The absence of baseline information that seems to characterize most volcanic crises testifies to the value of detailed data from the world's comparatively few well-monitored volcanoes. We salute (and are pleased to report) the continuing hard work by many volcanologists at sites of unspectacular, low-level activity. Any one of these sites could become the focus of worldwide attention in the near future, adding enormous importance to today's baseline information. And if the focus is on some other, distant, volcano, the baseline data at these sites will provide valuable comparative information.

Finally, the preliminary nature of these SEAN reports must again be emphasized. They remain the only available descriptions for many eruptions, but all were necessarily compiled quickly and without the benefit of scholarly checking and cross-checking of information. The reader should consult the more recent references cited for many eruptions and exercise caution in drawing conclusions from the preliminary reports presented here.

---

29 Newhall, C. G., and Dzurisin, D., 1988, Historical Unrest at Large Calderas of the World; *U.S. Geological Survey Bulletin* 1855, 1108 pp.
30 Krafft, M., and Krafft, K., 1975, *Volcano;* Abrams, New York, 174 pp., and many more in their native French language.
31 For example, the 1977 discovery of a completely unsuspected group of organisms thriving on sulfur emanations 1500 m below the limit of photosynthesis.
32 In Alaska, for example, note the drop in reporting of Pavlof and Shishaldin when Sergeants Sventek and Dean were assigned elsewhere, or the increased reporting of Alaskan volcanoes when John Reeder began to gather data from local aircraft pilots.

PART II

NATURAL SCIENCE EVENT
BULLETIN

November 1, 1975

Vol. 1, No. 1

Scientific Event Alert Network
Smithsonian Institution - Museum of Natural History
Room 9 - Mail Stop 103
Washington, D.C. 20560 U.S.A.

Telex: 89599

Telephone: (202)381-4174

Natural Science Events - October, 1975

GEOPHYSICAL EVENTS

Volcanic Activity

Cotopaxi Volcano, Ecuador, South America. (00°50'S. latitude, 78°26'W. longitude). Cotopaxi last erupted in 1944. On 23 July 1975 small grayish puffs of smoke were observed emanating from the crater. In mid-September a 1,000-foot vapor plume rose above the crater, and a small earthquake shook the volcano on 24 September. In mid-October vapor plumes were reported to be increasing in volume and frequency.

Information contact: Dr. Minard Hall, Escuela Politecnica Nacional, Casilla 2759, Quito, Ecuador.

Fuego Volcano , Guatemala, Central America (14°29'N. latitude, 90°53'W. longitude). Unusual activity on Fuego, aside from minor vapor emissions, started the night of September 19. Since that time there have been occasional emissions of dark gray to black ash clouds. The ash was either dissipated as dust in the atmosphere or fell on the upper slopes of the cone. A trace fell on some populated areas but was insufficient for collection. A large ash cloud was observed on 2 October at 0830 (L.T.), and on 11-12 October ash, brownish-gray in color, fell on Antigua.

Information contacts: Don Willever, 7 Calle O. No. 18, Antigua, Guatemala.

Samuel Bonis, Instituto Geografico Nacional, Guatemala City, Guatemala.

...tude, ...Pavlof has ...,400 meters high. ...st up the Alaska Peninsula. ...prevent observation of the volcano, ...was active for 30 minutes out of the one ...s visible. On 31 October, for 1.5 hours, lava ...ted streaming down the north side of the cone. (This was probably ash and mud flow.)

Information contacts: Sgt. Paul Sventek, Box 96, 714 ACW SQ, APO Seattle, Washington, U.S.A.

Tom Miller, Branch of Alaskan Geology, U.S.G.S., 1209 Orca Street, Anchorage, Alaska, U.S.A.

Shishaldin Volcano, Aleutian Is., Alaska, U.S.A. (54°45'N. latitude, 163°58'W. longitude). Shishaldin, during the few times it could be observed from Cold Bay, Alaska, was seen to be continually active in September and into October. At 0815 GMT, 17 September, the NOAA research vessel Millard Freeman, located at 55°33'N. latitude, 163°49'W. longitude, experienced a rainfall which contained ash. This continued for 15 minutes while the ship was headed on a course 250° true. The ash eruptions had apparently ceased by the end of October.

Information contacts: Same as for Pavlof Volcano.

ITALY

CAMPI FLEGREI

Bay of Naples, S Italy
40.83°N, 14.14°E, summit elev. 458 m

1/84 (9:1) The following reports are largely from Giuseppe Luongo, Roberto Scandone and Franco Barberi.

"Campi Flegrei (Phlegraean Fields) is a large caldera some 12-14 km across, roughly 25 km W of Vesuvius and 5 km WSW of Naples. The caldera formed after a huge eruption 35,000 years ago that produced 80 km^3 of dense rock [as finely fragmented ash distributed over much of the eastern Mediterranean]. Several other eruptions of decreasing intensity have occurred since then. In the past 10,800 years at least 22 different centers are recognizable. The last eruption occurred in 1538.

"Campi Flegrei has been the site of slow vertical movements since at least Roman times. A slow subsidence had occurred since the last eruption in 1538. An uplift observed in 1970 continued until 1972 without significant seismic activity. The inferred maximum uplift with respect to previous levellings was 170 cm. Slow ground oscillations observed between 1972 and 1982 had an annual period with a range of about 10-15 cm per year in the zone of maximum uplift. Since the summer of 1982 oscillation has not reversed as in previous years. The overall uplift amounted to 110 cm between Jan 1982 and Dec 1983 in the zone of maximum movement, within the town of Pozzuoli in the center of the caldera. Repeated levelling surveys in the area have given evidence of an area of uplift of about 6 km radius with a fairly circular symmetry.

"In Nov 1982 moderate seismic activity was observed by the permanent seismic network that has been operating since 1972 (fig. 1-1). The level of activity was slightly above the microseismic background in the area. In Jan 1983, public officials were notified of the anomalous trend of the phenomenon and the possibility of increasing seismic and volcanic hazard. In March, a distinct increase in seismic activity was observed with the first magnitude 3 earthquake. Since then, ground uplift has continued with a velocity that reached 5 mm/day during Oct. After Oct, oscillations in the rate of uplift were observed, with a range between 1 and 4 mm per day. The seismic activity increased, following a trend similar to that of the uplift velocity. A magnitude 4 earthquake occurred on 4 Oct when the rate of uplift reached 5 mm per day. This earthquake caused some building collapses (without injuries) in the town of Pozzuoli. Downtown Pozzuoli was evacuated after this event because of concern about the increasing seismic hazard. The main part of the town is built of old brick houses that were increasingly affected by the continuous seismic activity. On 13 Oct a seismic swarm of >250 shocks occurred in 5 hours. Maximum magnitude was 3.0.

"The people evacuated from Pozzuoli were temporarily resettled in the resort areas surrounding Campi Flegrei. A new settlement has already been planned on the border of the more vulnerable area. The choice of its location was made by public authorities to minimize the social consequences of evacuating people from their residences. The new settlement is relatively safe from a seismic point of view but is not safe from a maximum probable volcanic event.

"The permanent surveillance network operating in the area comprises measurements of ground deforma-

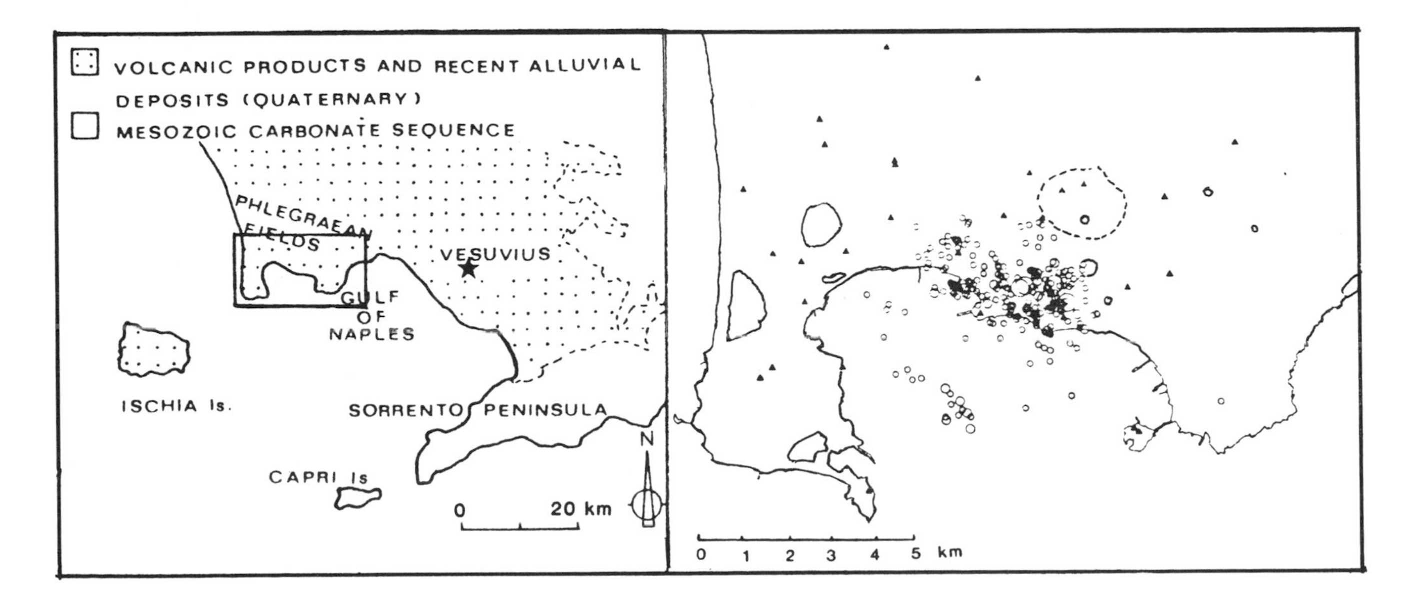

Fig. 1-1. *Left,* index map (after Nunziata and Rapolla, 1981). *Right,* distribution of the 212 best-located earthquakes as of early 1984 (circles) and positions of seismic stations (triangles).

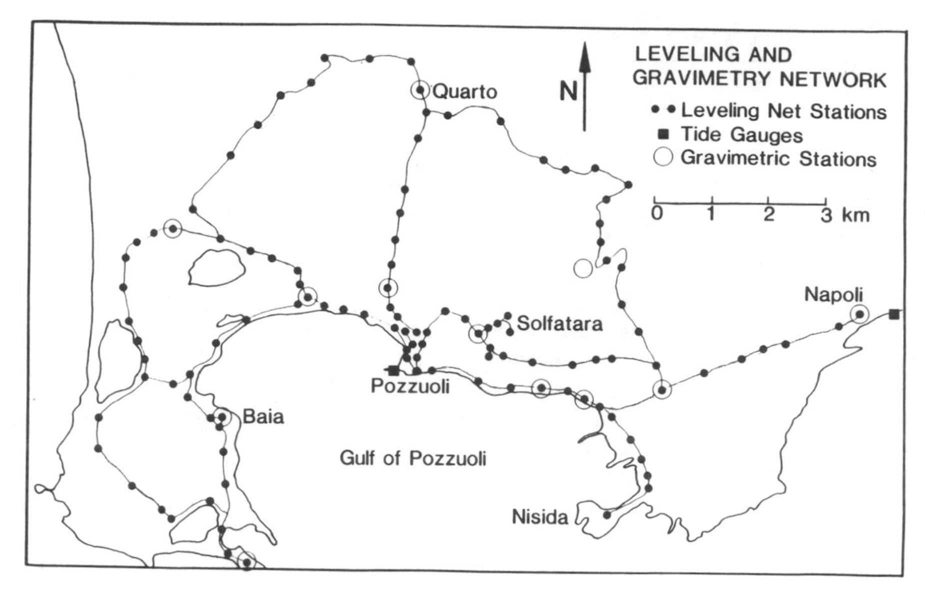

Fig. 1-2. Levelling network in the Campi Flegrei area, early 1984.

tion and seismic activity, and monitoring of gas content and temperatures of fumaroles. Temporary measurements of self-potential have been performed by a French team from the Institut de Physique du Globe (IPG).

"Vertical ground deformation is measured by a repeated levelling of the permanent network (fig. 1-2) and is also checked daily by a tide gauge in Pozzuoli harbor. Measurements are referred to a tide gauge located in the nearest stable place in Naples. Horizontal deformation is also measured on a network covering Campi Flegrei. The data give evidence of a maximum extension of about 40 cm over 4 km, nearly coincident with the area of maximum vertical uplift.

"The permanent seismic network operating in the area is composed of 22 vertical seismometers, 8 of which are operated by the Osservatorio Vesuviano (OV) and 14 by the AGIP company for the initial purpose of monitoring seismic activity connected with the geothermal field. Seven AGIP stations and the 8 OV stations are cable-connected to a central point in Naples. Routine locations are made on these 15 stations using the HYP071 program. Subsequent analysis of data from all 22 stations is made utilizing the program LQUAKE by R. Crosson. The preliminary velocity model is based on data collected from the geothermal wells in the area. The shallow character of the seismic activity does not give any evidence of a zone of anomalous propagation of S waves. A seismic explosion campaign has been planned in the Gulf of Pozzuoli to provide information on the deeper structure of the area.

"The earthquakes of higher magnitude are mainly confined within a restricted area under the Solfatara Crater. They are offset with respect to the area of maximum uplift and their mean depth is about 3 km (E-W profile is shown in fig. 1-3). Preliminary focal mechanisms indicate a predominantly tensile field in this area. The data on temporal distribution of earthquakes indicate a swarm-type character. The event of maximum magnitude (4) occurred 4 Oct 1983, and its epicenter was in the Solfatara area. A close correlation seems to exist between the velocity of uplift and the seismic activity. The more energetic earthquakes seem to coincide with the higher rates of uplift (4-5 mm/day).

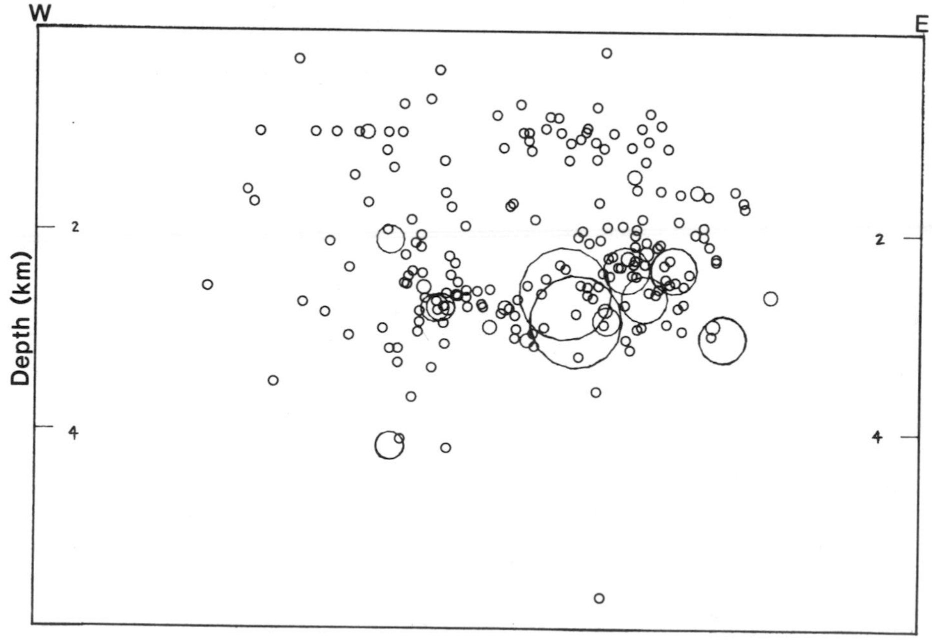

"A cross-cooperation with seimologists from the Univ. of Wisconsin is underway. A temporary network of ten 3-component stations with high dynamic range has already been deployed in the area and will be operating for some months. A temporary network of 3-component stations was also operated during Nov 1983 by IPG seismologists.

"Since April 1983, radon measurements have been made in water wells located in the area. The data are still too preliminary to infer any model. We await a prolonged period of measurements to infer what may be the seasonal trend. Temperatures of the Solfatara fumaroles are also continuously monitored. No significant change has been detected.

"Gas monitoring of the Sol-

Fig. 1-3. Profile (W-E) showing depth distribution of the earthquakes shown in fig. 1-1. Maximum magnitude = 4.0.

fatara fumaroles is carried out by several teams from the Universities of Pisa, Palermo, and Florence, both by continuous measurement and periodic sampling. Preliminary data seem to indicate an increase in the energy flux supplied to the deep water table located at 1.2 km depth by the geothermal wells.

"Two detailed surveys of the helium content of the ground have been performed by a team from the Univ. of Rome. Order of magnitude variations have been detected in a large area NW of Pozzuoli."

Reference: Nunziata, C. and Rapolla, A., 1981, Interpretation of Gravity and Magnetic Data in the Phlegraean Fields Geothermal Area, Naples, Italy; *JVGR*, v. 9, p. 209-225.

Information Contacts: G. Luongo, R. Scandone, OV, Napoli; F. Barberi, Univ. di Pisa.

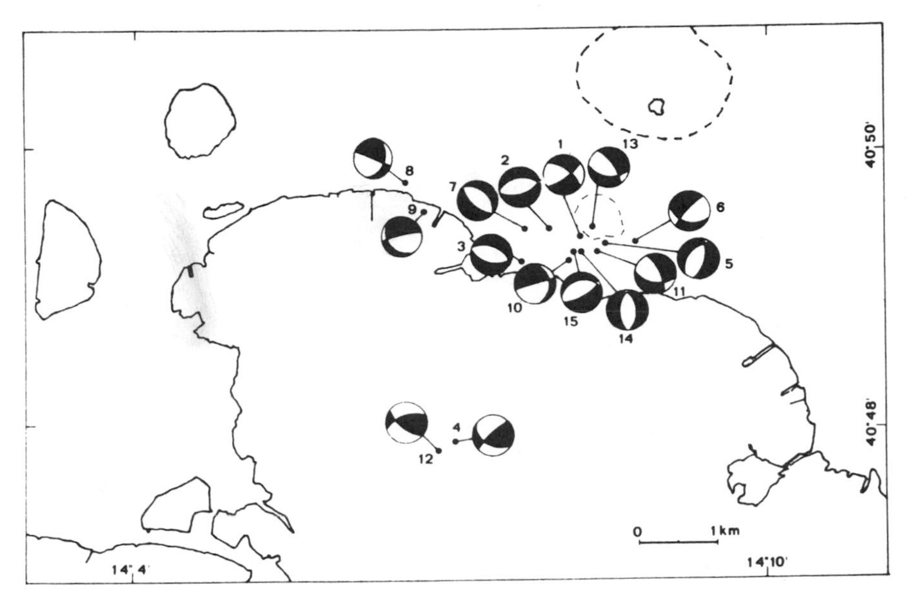

Fig 1-4. Focal mechanisms of 15 selected earthquakes. Courtesy of OV.

3/84 (9:3) "The rate of seismic strain energy release at Campi Flegrei was higher during the first 3 months of 1984 than during 1983. The seismic strain energy released in the first 3 months of 1984 was almost as high as that released during all of 1983, although it must be noted that seismic activity did not begin to be detected until March 1983. Seasonal trends in periods of quiet activity have been observed at Campi Flegrei. The complex interplay between external and internal causes of the present crisis makes prediction of the development of the phenomenon still more difficult.

"A peak in activity occurred during the second week of March and at the beginning of April 1984. A magnitude 3.9 earthquake occurred 9 March and five days later a magnitude 4.0 earthquake caused the roof collapse of a fifteenth century church in Pozzuoli. No injuries occurred. The close association of these two earthquakes caused some concern about the possible development of the continuing crisis. On 1 April, a seismic swarm of 499 events occurred between 0300 and 0800. The maximum magnitude was 3.0. The preliminary location is in an area about 1 km W of Pozzuoli, the same area as another swarm on 13 Oct 1983.

"An analysis of the reliability of hypocentral determinations has been performed using different velocity models. The uncertainties in the velocity model prevent any reliable estimate of the true depth of the earthquakes, even if they are definitely confined within the upper 4 km of crust. The effect of different velocity models does not appreciably change the epicentral determinations. The most energetic events are still confined in a small area around Solfatara Crater. The magnitude 4.0 earthquake of 14 March was located about 1 km SE of Solfatara.

"The focal mechanisms of 37 selected events have been studied. Fifteen reliable solutions were obtained (fig. 1-4). Ten events, located around Solfatara Crater, are of tensional type. Two events, located in the Gulf of Pozzuoli, seem of compressive type. No predominant orientation of the P and T axes is found.

"The previously inferred trend of high deformation rates preceding the largest shocks was not observed for the last 2 large events. The velocity of uplift, as measured by the Pozzuoli tide gauge, was in the range of 2-3 mm per day during the first 2 weeks of March 1984; it increased to about 4 mm per day during the last 2 weeks in March. A survey of the levelling network was completed during March. The pattern of deformation is similar to that observed in Dec 1983. A maximum uplift of 32 cm since Dec 1983 was measured near Pozzuoli. The maximum uplift since Jan 1982 was 142 cm. Geochemical surveillance was implemented in 1983 as a consequence of the uplift. Some variations in the chemical composition of water wells have been detected between Jan and March 1984. In the same period, small variations in the composition of fumarolic gases from Solfatara Crater have been recorded (slight increase of the S/C ratio). The radon content was approximately constant. The reducing capacity of fumarolic gases has been continuously monitored at two fumaroles within Solfatara Crater, showing a broad peak since mid-Feb, and reaching a maximum in mid-March.

Information Contacts: G. Luongo, R. Scandone, OV, Napoli; F. Barberi, Univ. di Pisa; M. Carapezza, Univ. di Palermo.

4/84 (9:4) "Seismic activity at Campi Flegrei decreased in April. Following the earthquake swarm of 1 April and a magnitude 3.5 event that occurred on 3 April, the cumulative seismic strain energy release had a smaller slope than in the previous 3 months (fig. 1-5).

"The distribution of epicenters for the best-located events (number of stations >8, RMS <0.16) until 15 Jan shows an interesting feature (fig. 1-6). Three different clusters are evident. The first, which includes the events with highest energy, is located around the S border of

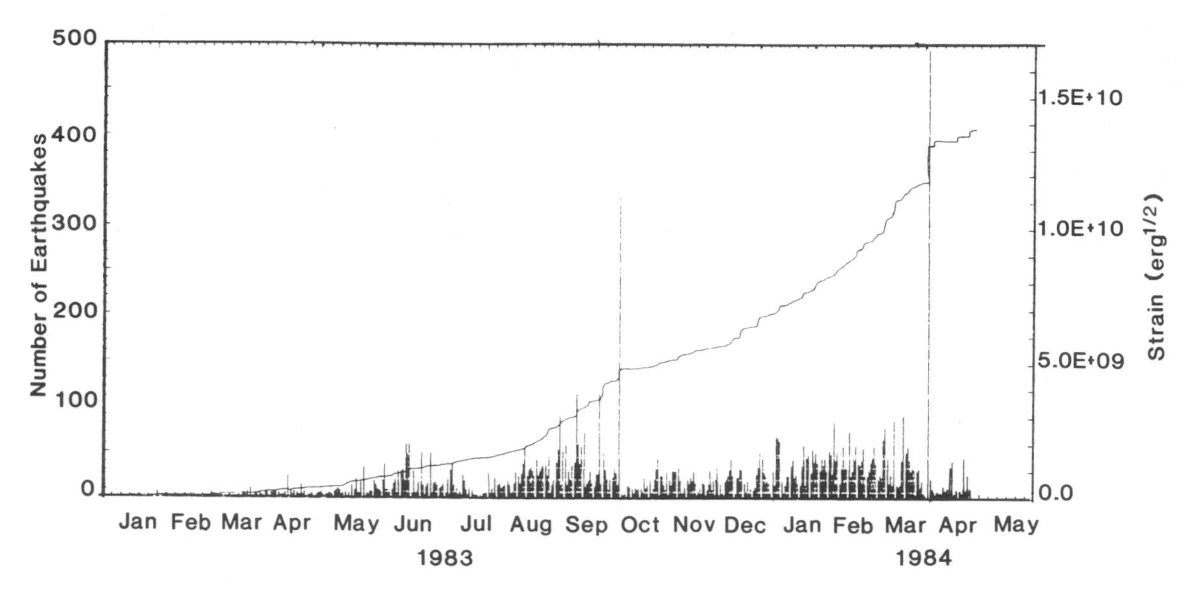

Fig. 1-5. Daily number of earthquakes at Campi Flegrei (vertical lines) and cumulative strain release (curve), Jan 1983–April 1984. Courtesy of OV.

Solfatara Crater. The events that occur during swarms are located on the W edge of the active seismic region. The third cluster includes a few events located in the Gulf of Pozzuoli.

"The mean velocity of uplift during April, as measured by the tide gauge located in Pozzuoli harbor, was 1.6 mm per day. This value is definitely smaller than the previous month. Temperatures of the 2 fumaroles within Solfatara Crater remained at constant values of 110° and 157-159°C.

"The decrease in activity observed after the swarm of 1 April is similar to the decrease observed during Nov 1983. In that case, a diminuition in seismic activity and a decrease in the uplift velocity were observed after a magnitude 4.0 earthquake, and a swarm that occurred 13 Oct 1983."

Information Contacts: G. Luongo, R. Scandone, OV, Napoli; F. Barberi, Univ. di Pisa.

6/84 (9:6) "Seismic activity remained at a low level for most of May and June. The mean velocity of uplift, measured by the Pozzuoli tide gauge, was of the order of 3 mm per day in May.

"We were puzzled by the low level of seismicity and the relatively high rate of ground uplift. Our previous experience indicated that uplift velocities greater than 2-3 mm per day in Puzzuoli are often associated with increased seismic activity. This had not happened during April and May. We were suspicious that a change might have occured to the system. Finally, there was a small swarm of 80 events in 2 hours on 28 May. The maximum magnitude was 2.0. The location of the swarm was in the same area as the swarms of 13 Oct 1983 and 1 April 1984. The significant difference is that this time we were able to identify 3 low-frequency events that occurred at the beginning of the swarms. They had unclear onsets and a typical frequency of about 1 Hz. Two were sufficiently well-recorded on many stations of the network to have a good location, in the same area as the swarm. These events have remained isolated so far, but

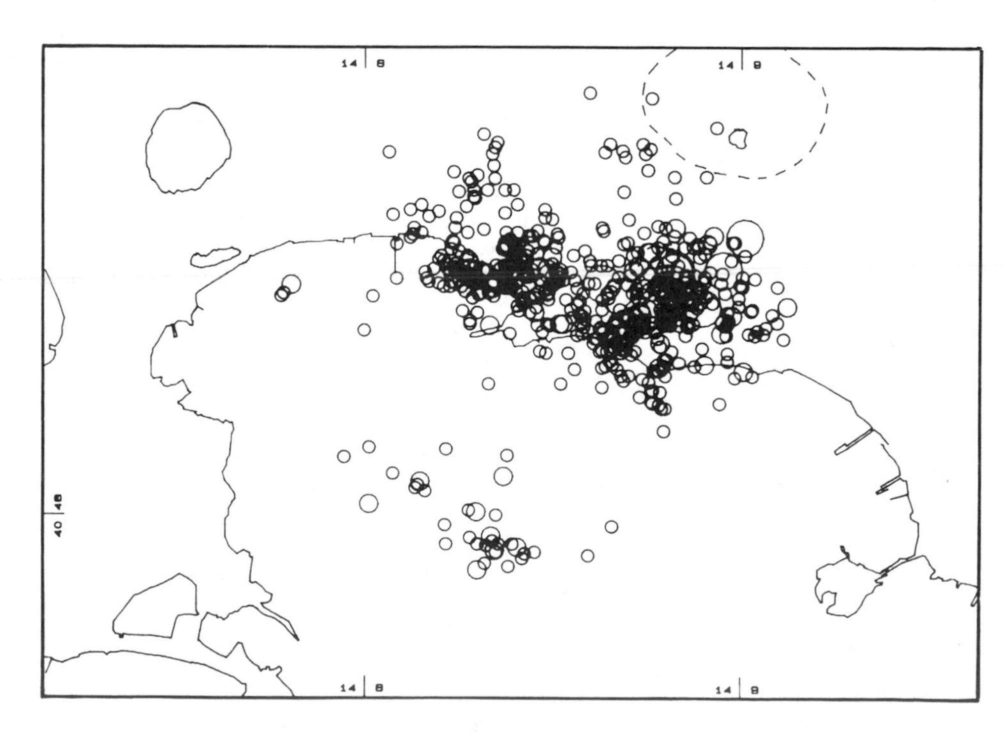

Fig. 1-6. Epicentral distribution of best-located events through 15 Jan 1984. Courtesy of OV.

we have alerted personnel to look carefully for other occurrences. We cannot exclude the possibility that other occurrences may have passed unnoticed. Unfortunately, the bad season, with frequent sea storms and high human background noise, may have completely hidden the low-level seismic activity.

"Four magnitude 3+ earthquakes occurred 1-2 June near Solfatara Crater on the E edge of the active seismic area, where we again observed that the most energetic events occur. Two events of magnitude 3.3 and 2.6 occurred 17 June on the E side of the caldera near the Nisida Peninsula, and a magnitude 3.6 event on 27 June was located in the center of the Gulf of Pozzuoli. They are the most energetic events that have occurred in these areas. On 1 July, a magnitude 3.6 event started a swarm of more than 100 events near Solfatara Crater. Another magnitude 3.5 event occurred during this swarm.

"Uplift measured by the Pozzuoli tide gauge was about 1 mm per day 1-24 June. A new tide gauge was installed on the E border of the caldera (in Nisida) to monitor the differential movement of the caldera. It gave a value of 0.2 mm per day 1-24 June. A levelling survey was completed in June. Preliminary data indicate a maximum uplift of 17 cm with respect to March 1984. The center of maximum uplift remained in the same area as in previous levellings, totaling 160 cm since Jan 1984.

"At the end of May, divers reported a visible increase in submarine fumarolic activity in the Gulf of Pozzuoli, in front of Monte Nuovo Crater (site of the last eruption at Campi Flegrei, in 1538). We still do not know how to monitor this activity. A few pictures have been taken at different times, but we do not know if the increased activity occurs episodically.

"Radon measurements have been routinely performed since April 1983 in 2 water wells. They gave evidence of a strong peak (of probable seasonal origin) during the winter (fig. 1-7). Fluctuations of smaller period are superimposed on the trend. Fumarole temperatures in Solfatara Crater remained at 158° and 110°C, with variations of only a few degrees.

"Discussions have been carried out on the character of recognizable short-term precursors, if any, of a possible volcanic event in the area. We were strongly relying on a significant increase in seismic activity days before an eruption. We based this assumption on historic chronicles of the eruption of Monte Nuovo. The occurrence of 3 swarms without any evident volcanic activity has reduced our confidence in recognizing a swarm as a possible precursor of volcanic activity. The maximum duration of a swarm has been 6 hours. We are now relying on the total energy release during a swarm and on the character of the seismic activity. We have never observed harmonic tremor or a significant number of low-frequency events. The deformation rate is another possible precursor, but we could recognize it as a precursor if a significant uplift occurred in a matter of days before a volcanic event. The magnitude of the uplift should be at least the same as the daily sea tide (tens of cm). We are planning the installation of more tide gauges, to be able to detect more localized uplift if it occurs. Geochemical monitoring is presently carried out mainly in Solfatara Crater; plans have been made to extend this monitoring to other areas.

"The beginning of the summer season is increasing the demographic pressure on the area. The evacuation order in the central part of the caldera is not strictly enforced. The high density of population of the area (250,000 people live within the caldera) is posing a tremendous problem with respect to the measures that should be taken to reduce the hazard.

"The length of the present crisis (2 years since the beginning of uplift) has produced a social impact and economic loss comparable to that of an actual eruption.

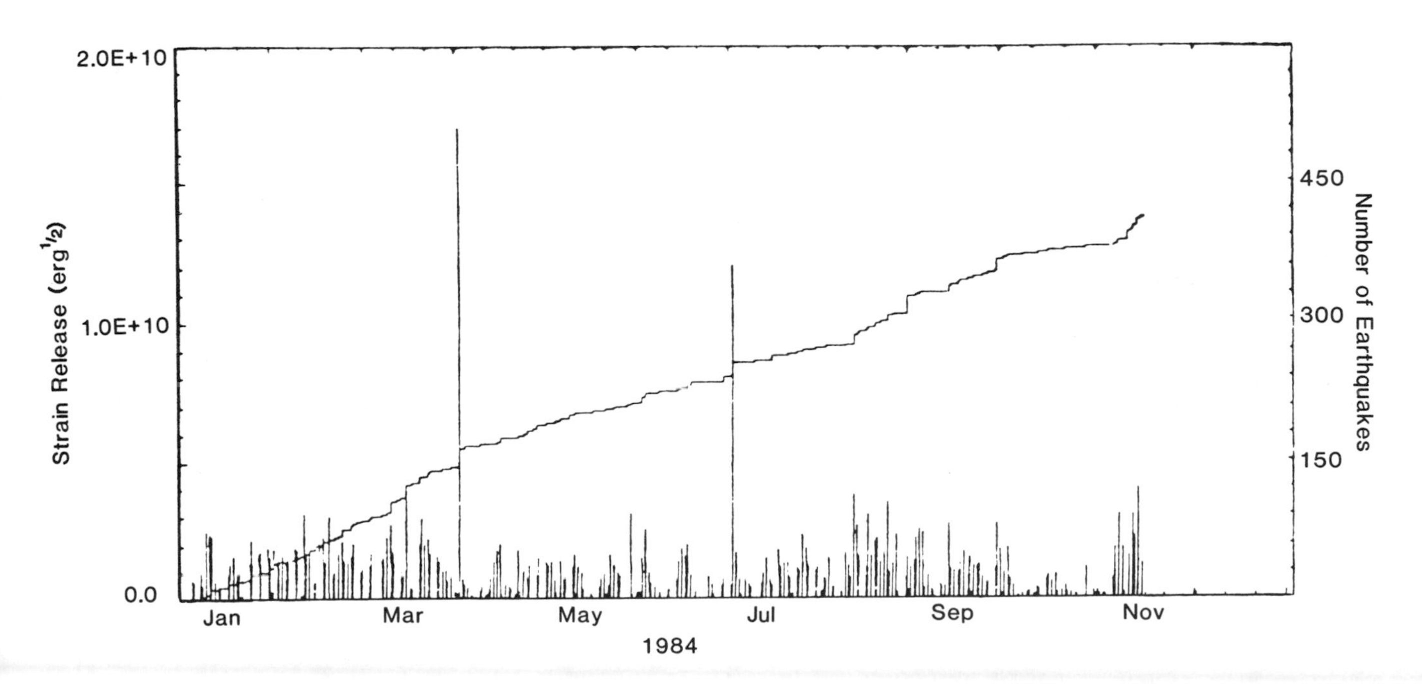

Fig. 1-7. Daily number of earthquakes at Campi Flegrei (vertical lines) and cumulative strain release (curve), Jan-Nov 1984. Courtesy of OV.

The researchers devoted to the difficult task of monitoring and understanding development of the phenomenon suffer from the accumulated stress of being responsible for the lives of thousands of people."

Information Contacts: Same as 9:4.

9/84 (9:9) "Activity at Campi Flegrei displayed some changes July-Sept. The monthly mean uplift velocity was in the range of 1.2 mm per day at Pozzuoli and less than 1 mm per day at Nisida. A new tide gauge is now operating on the W side of the caldera near Cape Miseno. The mean velocity of uplift in that area is comparable to that at Nisida. The velocity of uplift during the last 3 months is lower than that during the winter and spring.

"Seismic energy release remained at the same level as in the past year. We observed a higher number of the most energetic events (M 3) than in past months (table 1-1). They are also no longer confined within the restricted area around Solfatara Crater, but instead occur all over the seismically active area.

"The temperatures of fumaroles in Solfatara Crater remained stable. The radon content of the water wells reached a minimum in the last year, with a slight increase again at the end of Sept.

"The decrease in the uplift velocity during constant seismic energy release does not support the idea that uplift velocity is closely related to the occurrence of larger earthquakes. On the contrary, in this last period

Date		Magnitude
July	1	3.7, 3.5
	11	3.8
Aug	22	3.6
	29	3.9, 3.8
Sep	12	3.3
	13	3.2
	28	3.5, 3.7

Table 1-1 Magnitudes of the most energetic events in the Campi Flegrei area, July–Sept 1984.

we have observed higher energy events. They often occur as double events (1 July, 29 Aug, and 28 Sept; table 1-1). The area that was previously affected by swarms of low-energy events (1 km W of Pozzuoli) is now also affected by high-energy events. We do not know if this pattern of activity indicates a higher risk of eruption."

Information Contacts: Same as 9:4.

11/84 (9:11) "Activity at Campi Flegrei declined in Oct. The velocity of uplift, as measured by the tide gauge at Pozzuoli, was null for the entire month and seismic activity was very low. This was the first substantial decrease in activity observed since the beginning of the crisis in the summer of 1982 (figs. 1-7, 1-10). A new increase in

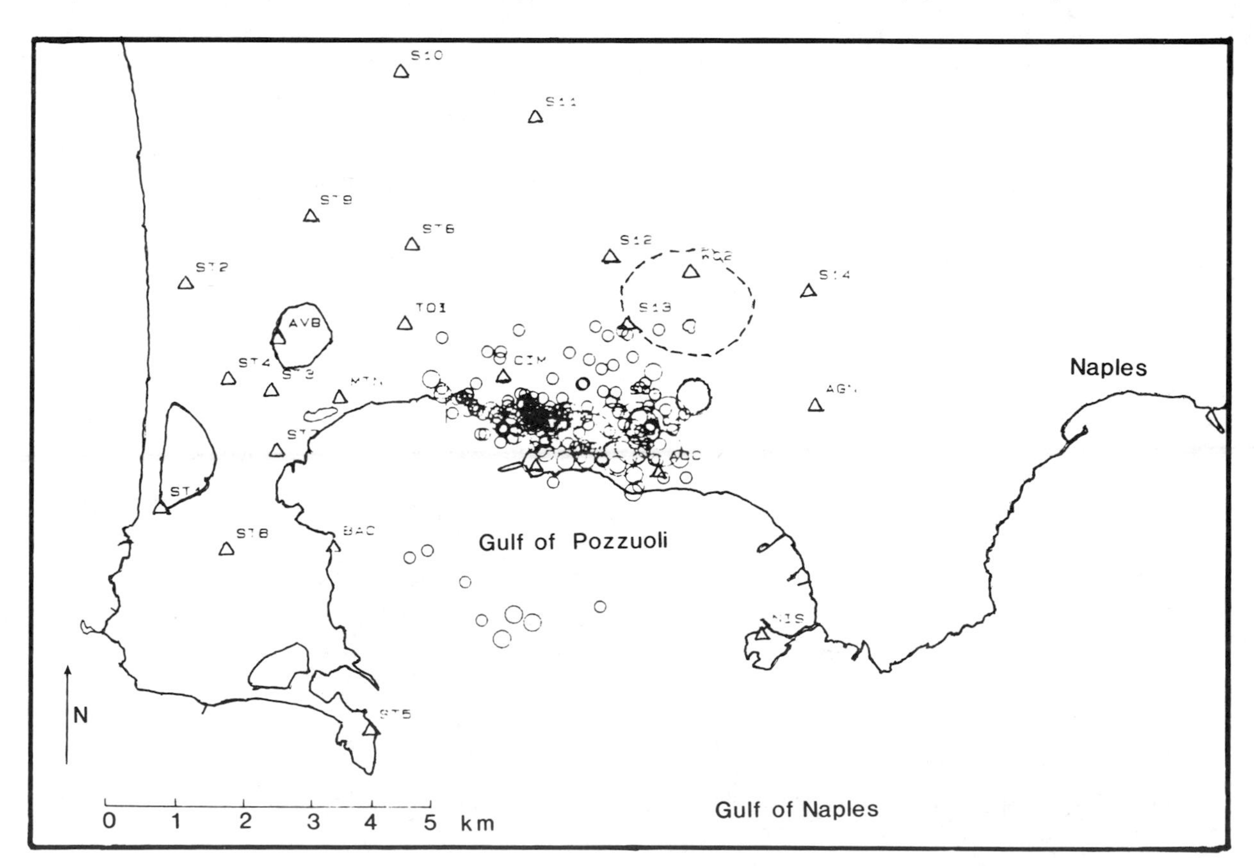

Fig. 1-8. Distribution of the epicenters of the 181 best-located seismic events, Jan 1983–May 1984. Triangles mark seismic stations. Courtesy of OV.

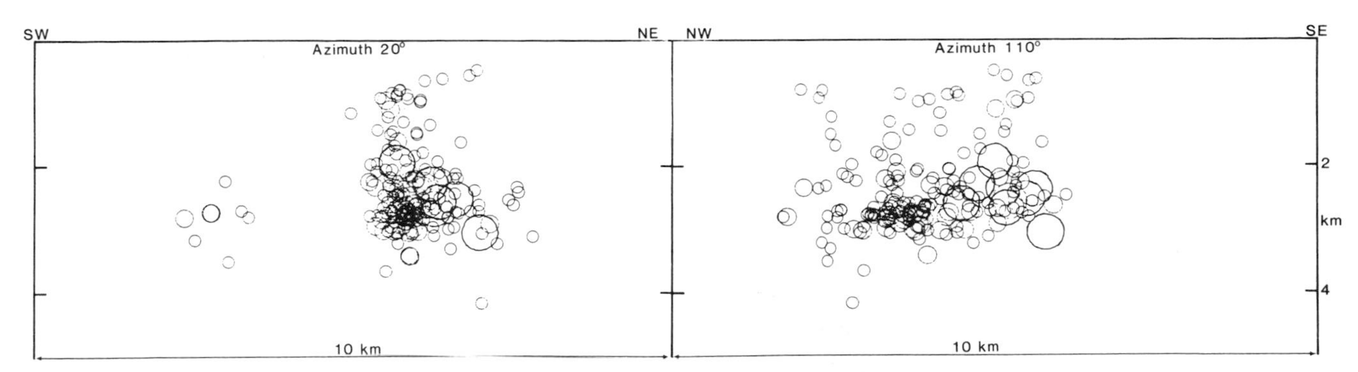

Fig. 1-9. Epicenters in fig. 1-8 projected into cross sections oriented N20°E, *left*, and N110°E, *right*. Courtesy of OV.

seismic activity was observed during the first week of Nov with a magnitude 2.7 earthquake on the 5th. During the second week, uplift started again with a mean velocity of 1 mm per day. On 8 Nov, events of magnitude 3.2, 3.2, 2.8, and 3.3 occurred in the Gulf of Pozzuoli. Other events occurred on the 10th (M 2.9) and 11th (M 3). Minor events occurred during the third week, but uplift velocity remained at 1 mm per day. During the last week in Nov, uplift velocity rose to 2.5 mm per day with two earthquakes (M 3.1, 2.9) on 26 Nov and several others of lower magnitude.

"The temperatures of fumaroles in Solfatara Crater remained constant at 156°C. A new fumarolic vent opened in the crater during the night of 16-17 Nov with a violent emission of steam.

"Fig. 1-8 shows the distribution of the 181 best-located events between Jan 1983 and May 1984. These events were recorded at more than 12 stations (and have an rms less than 0.1). The figure shows an alignment of the Jan 1983-May 1984 epicenters on land along a N110°E direction with a clustering of the stronger earthquakes around Solfatara Crater. The earthquake area is limited on the E by the remains of the old Agnano Crater and on the west by Monte Nuovo Crater. A striking feature is that the earthquakes occur mainly along the coastline and the remains of an old marine terrace, which uplifted some 40 m about 5000 years BP.

"Fig. 1-9 shows two perpendicular sections of the same data set. The N20°E section shows two clusters of events, on land and in the sea, with an aseismic area in the middle. The same section shows the rather abrupt boundary between the seismically active areas on land and in the sea. Most of the earthquakes occur between 2 and 3 km depth, but these data are still preliminary due to the uncertainties of the velocity model."

Information Contacts: Same as 9:4.

1/85 (10:1) "A magnitude 3.8 earthquake occurred on 8 Dec and was located on the E side of Solfatara Crater. After this event, the seismic activity was very low during Dec and Jan. During Dec the tide gauges in the Gulf of Pozzuoli recorded no uplift of the ground in the area. In the second week of Jan, a slight deflation was measured

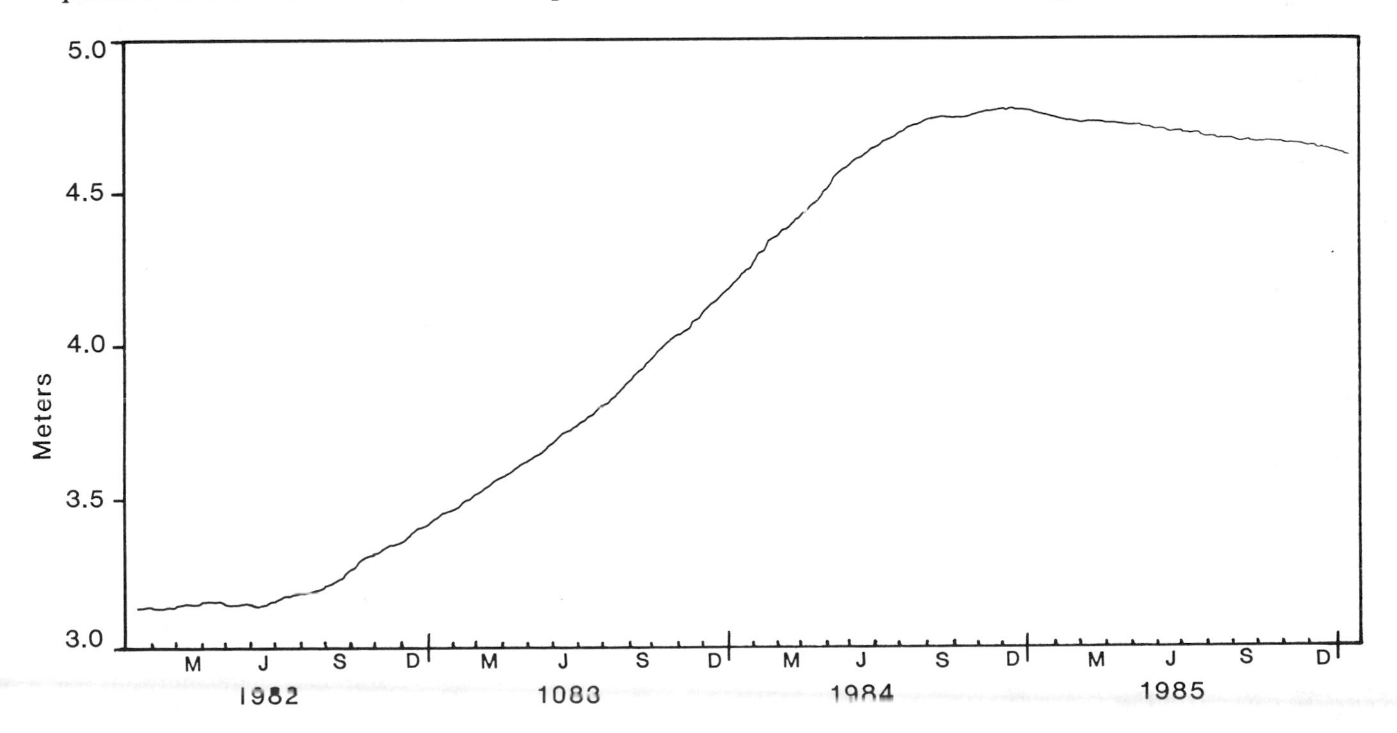

Fig. 1-10. Ground movements measured by the tide gauge in Pozzuoli Harbor, Jan 1982 – Dec 1985. Courtesy of Roberto Scandone.

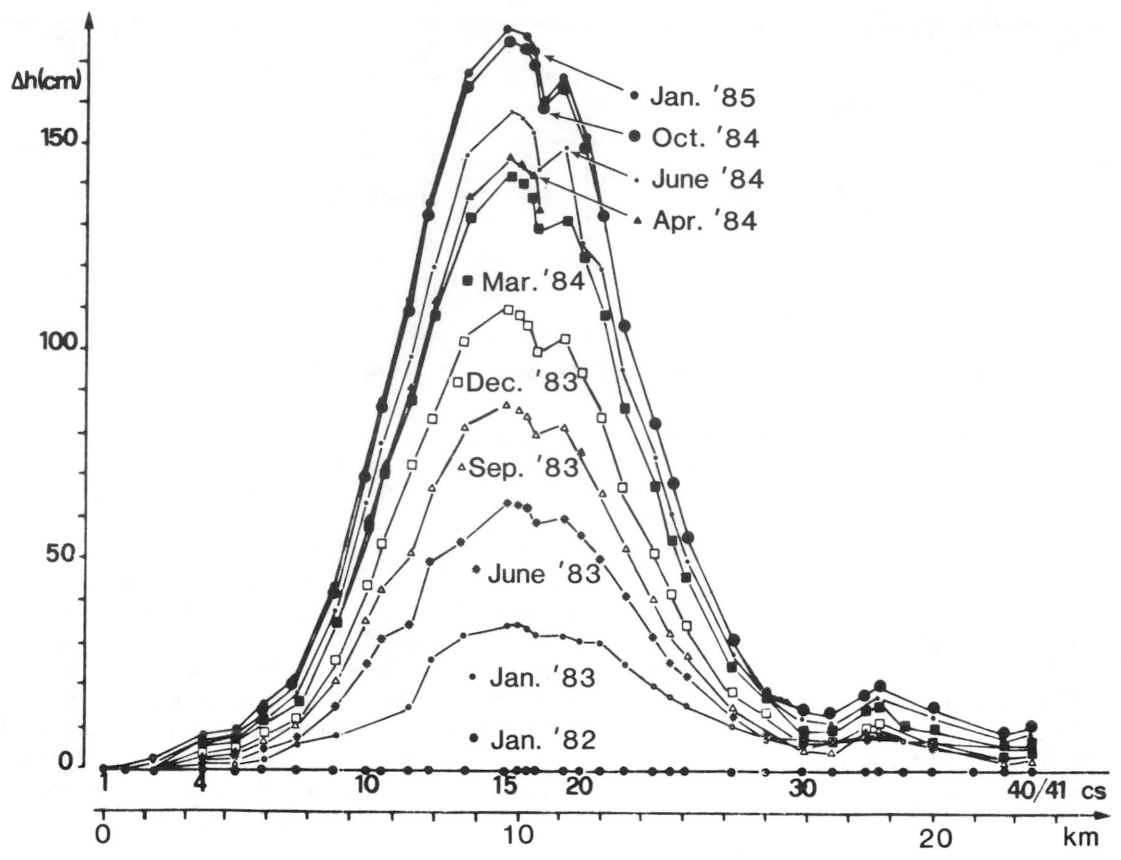

Fig. 1-11. Results of 10 surveys of the levelling network, Jan 1982 – Jan 1985. Benchmarks are oriented along an approximately E–W line (see fig. 1-1 for their locations).

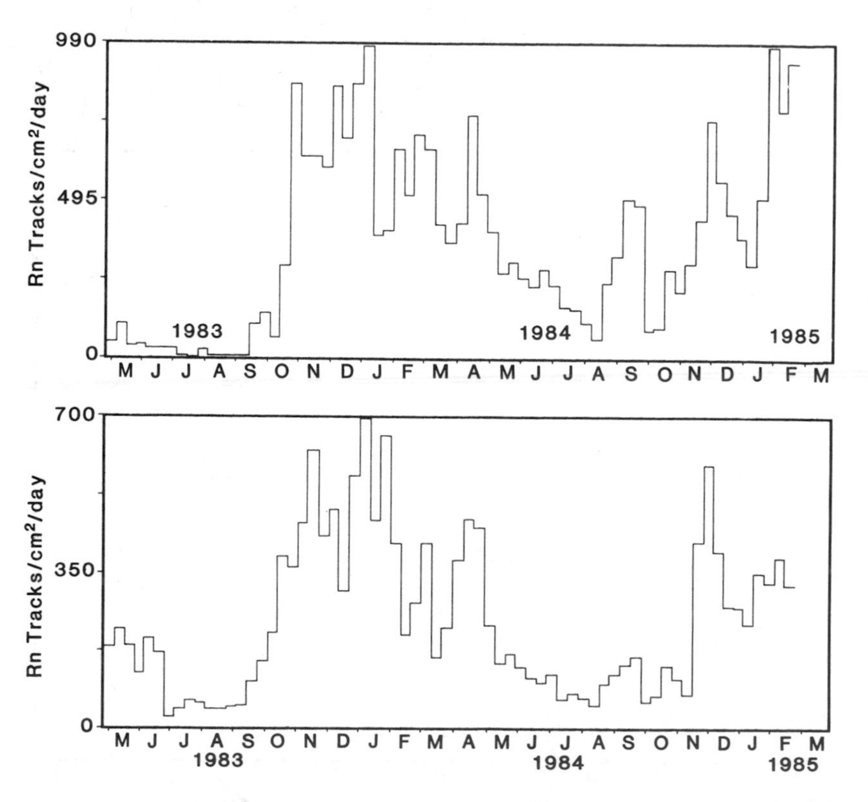

Fig. 1-12. Results of radon measurements at 2 water wells: *top*, near Solfatara, 26 April 1983 – 2 March 1985 (Tortorelli); *bottom*, near Monte Nuovo, 26 April 1983 – 25 Feb 1985 (Damiani). Courtesy of OV.

by the Pozzuoli tide gauge. Only the radon contents of water in monitored wells showed an increase. In mid-Nov the values were similar to those in the corresponding period of Dec 1983-Jan 1984. We are waiting for a longer quiescent period before releasing a ceased alert warning."

Information Contacts: Same as 9:4.

2/85 (10:2) "Activity at Campi Flegrei has decreased substantially after 2 1/2 years of continuous ground uplift and seismic activity. The tide gauge in Pozzuoli harbor measured 2 cm of ground subsidence during Feb (fig. 1-10). Seismic activity remained at a very low level with only a few low-energy events in the Gulf of Pozzuoli. The periodic survey of the entire vertical deformation network was completed in Jan. An uplift of 3 cm since Oct 1984 was measured on the benchmark in the area of maximum deformation (fig. 1-11).

"Radon and temperature measurements are currently made in 4 water wells. The pattern of radon content of 2 of them, near Solfatara (Tortorelli well) and Monte Nuovo (Damiani well), shows good correlation (fig. 1-12). At the present time, we observe high radon emission, similar to observations in the same period last year. The 2 years of continuous monitoring give evidence of a seasonal trend in radon emission.

A possible increase, due to influx of hot fluids from depth, may have occurred Dec 1983-April 1984. During this period, a slight but significant increase in water temperature was observed (fig. 1-13), which matches well with the period of highest seismic energy release. A more detailed study is under way.

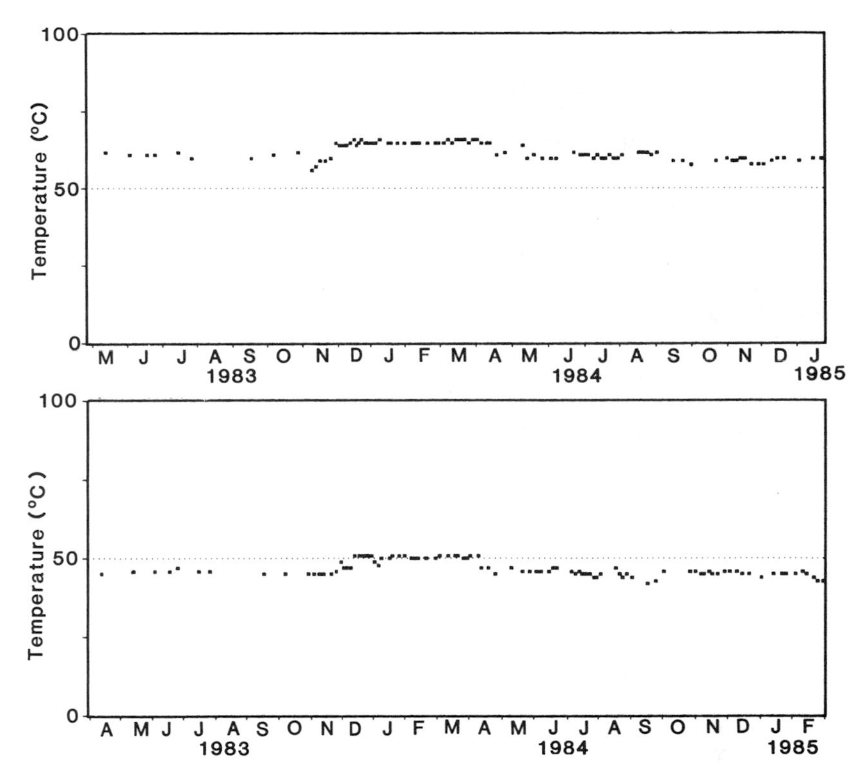

Fig. 1-13. Temperature measurements at the same wells as in fig. 1-12: *top,* 26 April 1983 – 23 Jan 1985 (Tortorelli); *bottom,* 30 March 1983 – 2 March 1985 (Damiani). Courtesy of OV.

"A maximum uplift of 180 cm has occurred since June 1982, and ~320 cm since 1969 (fig. 1-14). The moderate seismic activity during the present crisis has caused extensive damage to old buildings. We consider this period of unrest as the major geological event at Campi Flegrei in the last 180 years (since the first measurements of sea level in the area). The order of magnitude of the ground uplift during the last 16 years is comparable to that which occurred before the eruption of Monte Nuovo in 1538, and we consider this phase as a possible precursor (on a time scale of years) to new volcanic activity.

"If the present crisis follows the same pattern as that of 1969-72, we must expect a period of slight subsidence, followed by a period of slight oscillations of the ground level, until a new crisis eventually occurs. The reversal of the secular pattern of subsidence of the area (the most significant in 450 years) that occurred in 1969, along with the relatively long period of quiescence of Vesuvius (since 1944; 6 times longer than any repose period since 1694) may all be related to a change in the stress field controlling volcanic activity in the Naples area."

Information Contacts: G. Luongo, R. Scandone, R. Pece, OV, Napoli; F. Barberi, Univ. di Pisa.

Further References: Barberi, F., Hill, D., Innocenti, F., Luongo, G., and Treiul, M., (eds.), 1984, The 1982-1984 Bradyseismic Crisis at Phlegraean Fields (Italy); *BV,* v. 47, no. 2, p. 173-370 (21 papers).

Rosi, M. and Sbrana, A., eds., 1987, Phlegrean Fields; *Quaderni de "La Ricerca Scientifica" 114,* v. 9, 176 pp.

Lirer, L., Luongo, G., and Scandone, R., 1987, On the Volcanological Evolution of Campi Flegrei, Italy; *EOS,* v. 68, no. 16, p. 226-234.

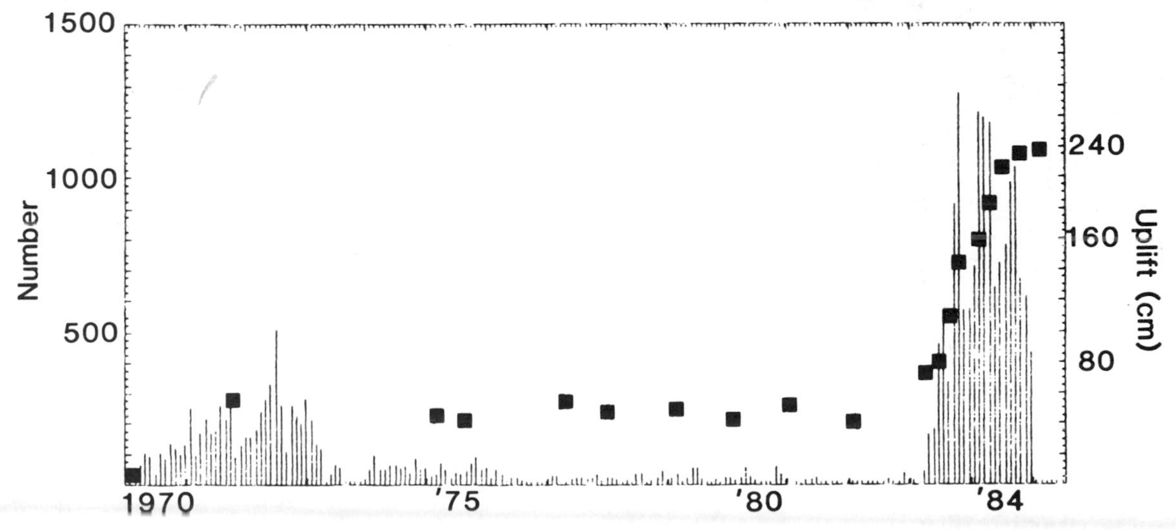

Fig. 1-14. Uplift and seismicity at Campi Flegrei, 1970 – 1984. Squares show relative height of benchmark no. 15 of the levelling network (see fig. 1-2 and 9:1), vertical lines show monthly number of seismic events. Courtesy of OV.

STROMBOLI

Æolian Islands, Italy
38.78°N, 15.22°E, summit elev. 926 m
Local time = GMT +1 hr

This small island N of Sicily has been in almost continuous eruption for over 2,000 years. Its small "Strombolian" explosions, hurling incandescent scoria above the crater rim, occur several times an hour, but larger eruptions are less frequent.

11/75 (1:2) For the first time since the spring of 1971, a new effusive phase began at [1200] on 5 Nov. Three summit craters were active, and lava flowing down the Sciara del Fuoco reached the N coast. By mid Nov effusive activity was decreasing below the 700 m level on the Sciara. The eruption had no adverse effects on the island's population or agriculture (fig. 1-15).

Information Contact: G. Nappi, IIV, Catania.

Further References: Villari, L., ed., 1980, The Aeolian Islands, an Active Volcanic Arc in the Mediterranean Sea; *Rendiconti Soc. Ital. Mineral. Petrol.,* v. 36, 185 pp. + refs.

Capaldi, G., et al., 1978, Stromboli and its 1975 Eruption; *BV,* v. 41, p. 259-285.

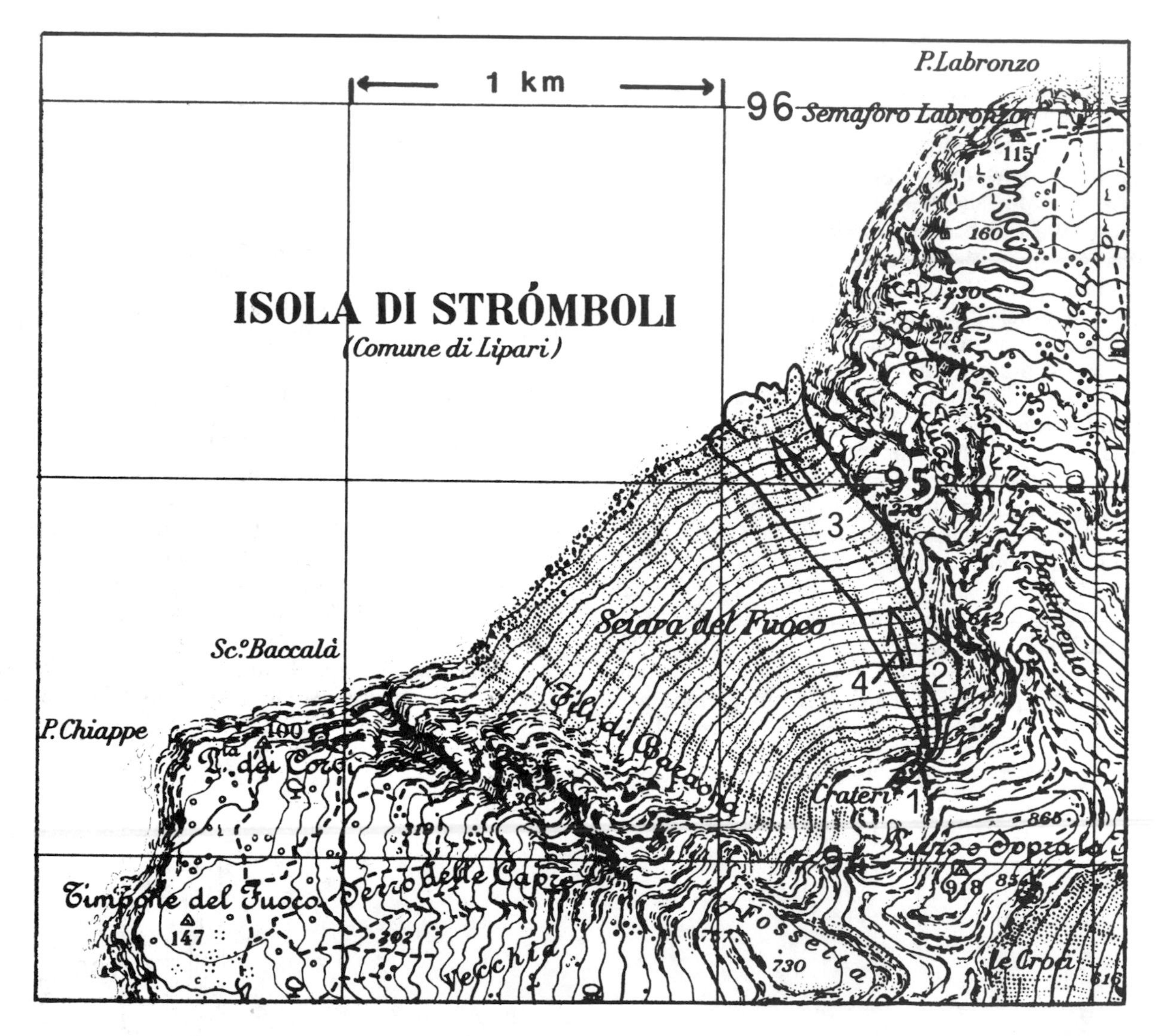

Fig. 1-15. Map of NW Stromboli. Locations: *1,* newly formed eruptive gorge; *2,* first lavas (morning of 6 Dec); *3,* lava flows of 6 Dec 1985 – 1 Jan 1986; *4,* path of nuée ardente (6 Dec). Courtesy of Mauro Rosi and Alessandro Sbrana.

VULCANO

Æolian Islands, Italy
38.40°N, 14.96°E, summit elev. 500 m

The southern island in the Æolian group has been quite active in the past, particularly in the last few centuries BC. As the legendary home of Vulcan's forge, it has given its name to all volcanoes.

6/85 (10:6) "Daily earthquake frequency in the Vulcano area showed, from 24 April, a siginficant increase (fig. 1-16). Seismic energy may also be considered unusual even though magnitudes have not exceeded 2.5. Epicenters lay predominantly in the Gran Cratere area or very close to it (fig. 1-17); focal depths were generally less than 1 km. The timing of event and energy distribution reveals the swarm character of the seismicity. Waveforms lead us to hypothesize that both degassing and fracturing phenomena occurred. A levelling survey carried out in May showed a slight uplift (1 cm) of the epicentral area with respect to the S part of the island."

Vulcano's most recent eruption began in [1888] and lasted until 1890. A submarine cable about 5 km east of Vulcano was broken several times during the eruption, usually accompanied by violent boiling of the sea and the appearance of pumice or scoria on the surface. The cable broke again in December 1892.

Information Contacts: S. Falsaperla, G. Neri, IIV, Catania.

Further References: Frazzetta, G., Gillot, P.Y., La Volpe, L., and Sheridan, M.F., 1984, Volcanic Hazards at Fossa of Vulcano: Data from the Last 6000 Years; *BV,* v. 47, p. 105-125.

Falsaperla, S. and Neri, G., 1986, Seismic Monitoring of Volcanoes: Vulcano (Southern Italy); *Periodico di Mineralogia,* v. 55, p. 143-152.

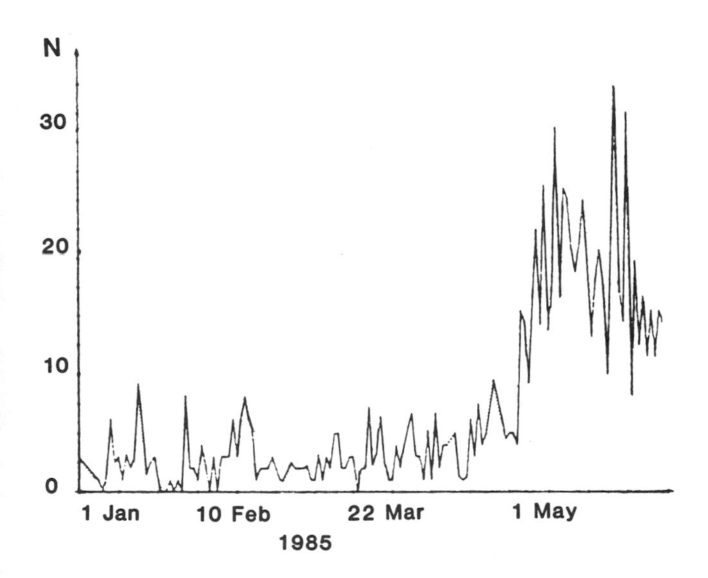

Fig. 1-16. Daily frequency of local earthquakes recorded 1 Jan-30 May by the Vulcano Cratere seismic station on the N flank.

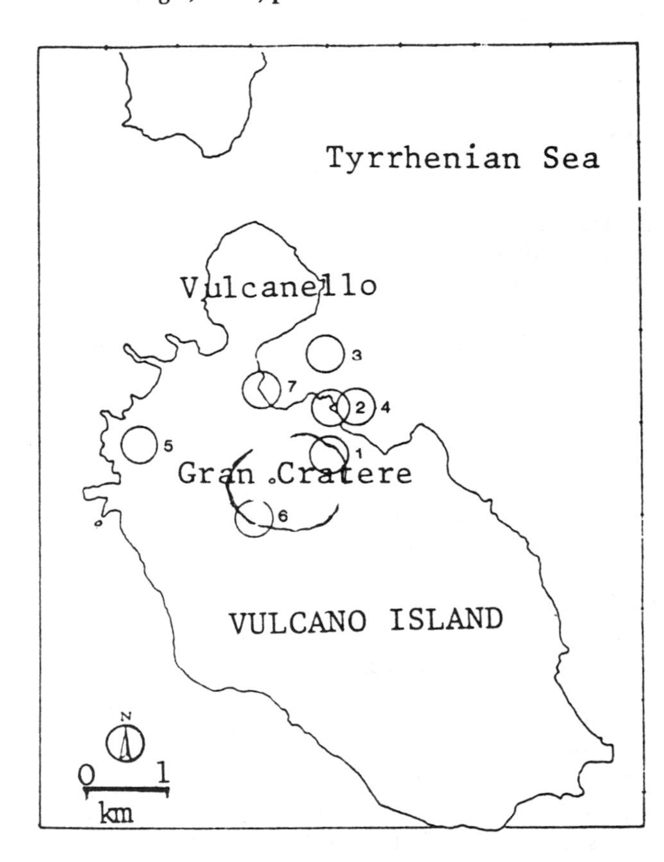

Fig. 1-17. Epicenters of the 7 highest energy earthquakes that occurred in the Vulcano area 24 April – 30 May.

ETNA

Sicily, Italy
37.73°N, 15.00°E, summit elev. 3290 m
Local time = GMT +1 hr winter (+2 summer)

Etna's documented record of volcanism is the longest of the world's volcanoes, with the first historic eruption in 1500 BC. Etna's frequent activity is evidenced by the following reports from 60 SEAN Bulletins. Its physiographic dominance of Catania, Sicily's largest city (with over one million inhabitants), is shown in fig. 1-18.

General References: Romano, R., ed., 1982, Mount Etna Volcano: A Review of Recent Earth Sciences Studies; *Memorie della Società Geologica Italiana,* v. XXIII, 205 pp. + maps.

Chester, D.K., Duncan, A.M., Guest, J.E., and Kilburn, C.R.J., 1985, *Mount Etna: The Anatomy of a Volcano;* Stanford University Press, 404 pp.

11/75 (1:2) Continually active throughout most of 1975, lava began flowing from a new opening on the NW side of the volcano on [29] Nov.

Information Contact: G. Nappi, IIV, Catania.

Further Reference: Pinkerton, H. and Sparks, R.S.J., 1976, Formation of the 1975 Subterminal Compound Lava Flow, Mount Etna; *JVGR,* v. 1, p. 167-182.

5/76 (1:8) In mid-May, Etna [temporarily] ceased erupting lava. For several days prior to 15 May, the volcano's two lateral craters were inactive, and only a thin column of vapor rose from the central crater. A team of scientists, led by Haroun Tazieff, was on the scene conducting investigations.

Information Contact: IIV, Catania.

6/76 (1:9) Observations were made by the University College London group and R. Romano of the IIV, 16 May-4 June 1976. Until this time, weather conditions had been very bad on the volcano, and reports of activity were sporadic. In the summit region, Bocca Nuova (the W vent of the central crater) was fuming strongly and had a depth estimated to be in excess of 200 m. This pit was exploding throughout the period of observation, at rates varying from one explosion per 3 minutes to 2 or 3 explosions per minute. Incandescent material was occasionally thrown above the level of the crater rim and vesicular scoria and Pele's hair fell close to the crater downwind. The Chasm (the larger E vent of the central crater) was also fuming strongly, and although its bottom was not seen, it had a depth of several hundred meters. During the afternoon of 25 May the Chasm started deep violent explosions at a rate of 20-25 per minute. No bombs were seen from these explosions, and the activity died down over the next few days. The new pit on the W side of Northeast Crater was fuming quietly but not exploding. The area around this new pit contained large amounts of sublimates and its total depth was around 50 m. The Northeast Crater itself is now inactive, the vent being plugged by scree; but there is still heavy fumarolic activity high on the W inside wall.

The effusive activity during the period was occurring from new vents on the N side of the mountain (near Punta Lucia). A 40 m-high cone formed at about the 2,900 m level. Inside this cone was a conelet from which mild Strombolian explosions of fresh gassy lavas were occurring on 26 May, though it was quiet on 31 May. Lava emissions were taking place farther downslope in the new lava field, from a number of boccas with positions that changed from day to day in an area above the 2,500 m contour. The rate of emission from one bocca was measured to be 0.4 m^3/sec. and the total emission for the whole field was about 2 m^3/sec.

Information Contacts: J. Guest, J. Murray, S. Scott, W. O'Donnell, Univ. of London; R. Romano, IIV, Catania.

7/77 (2:7) Lava effusion and explosive activity, from a new vent at 3200 m elevation just N of the Northeast Crater, began during the early morning of 16 July. A small lava lake, first observed at 0400, occupied the new vent, which was the source of 2 lava flows. The larger moved E into the Valle del Leone, and the smaller to the N. By 18 July, the flows had reached about 2 km and 800 m length respectively. Strombolian activity began at 0515 on 16 July and rapidly increased in frequency and power. Ejecta reached 600 m in height and fell over an area about 1 km in diameter. The explosions had ended by 20 July and lava extrusion ended on 23 or 24 July.

Precision levelling 2 weeks before the eruption by John Guest and co-workers showed a 1 cm inflation of the S flank since Sept 1976, but strong deflation under aa fields on the N flank.

Information Contacts: J. Guest, Univ. of London; R. Romano, G. Frazzetta, IIV, Catania.

8/77 (2:8) "After a period of quiescence since the beginning of this year, Etna erupted again on 16 July. New explosive and effusive boccas opened on the lower flank of the Northeast Crater Cone at an altitude of 3200-3270 m. The first signs of eruption occurred in early July when a small pit opened on the N edge of the Northeast Crater Cone. Initially this pit emitted high-temperature gas which during the night was seen to be incandescent.

"The first sign of lava emission was seen by a guide at 0400 on 16 July when he observed a small 'lake' of lava in the bottom of the pit. An hour later weak explosive activity started and increased to become stable at about 15 explosions per minute by 1000. Incandescent material was thrown about 300 m above the vent, with a maximum height of about 500 m. The maximum range of the bombs was 400-500 m. During this early period of explosive activity, a fracture opened at ~3240 m on the lower slopes of the cone and flows were emitted from the end of the fracture (at 3220 m), first towards the NE and later to the E. At the same time ephemeral boccas (3200 m) opened below the main fracture and fed a small flow in a northerly direction. Explosive activity reached its peak during the night of 17 July. During the early hours of 18 July collapse occurred in the Northeast Crater itself, causing great clouds of dust and ash. The explosive activity at the new vent then diminished both in frequency and intensity, becoming extremely variable.

"On 18 July the flows had reached a length of 800 m to the N and 2 km to the E. From then onwards the explosive activity continued to diminish, as did the effusion of lava to the N flow. The E flow reached the edge of Valle del Leone and continued as the principal flow, reaching a total length of 4 km by the end of the eruption. The N flow stopped with a total length of 1 km. Explosive activity ended at 1000 on 22 July and the lava flows stopped in the late afternoon.

"The area covered by lava was 0.16 km^2, the volume of lava was 4.8 x 10^5 m^3 and volume of pyroclastic material 500 m^3. Microseismic activity was noted before and during the eruption.

"The Northeast Crater started erupting again on 5 Aug at 1430. The activity began with strong explosions from the July vent, with lava fountains being thrown to 400 m in the early stages. One lava flow formed, advancing N from a fracture on the N slope of the Northeast Crater cone. The fracture was oriented approximately NNE. The lava was fluid and travelled about 3 km. Seismic activity was observed just before and during the eruption, which ended at 0630 on 6 Aug."

UPI reported that a third eruption began early 14 Aug and lasted only 14 hours. Fountains or Strombolian ejecta rose 200 m above the vent, and two lava flows, each about 200 m wide, moved about 4 km down the volcano.

Information Contacts: R. Romano, G. Frazzetta, D. Condarelli, IIV, Catania; J. Guest, Univ. of London; UPI.

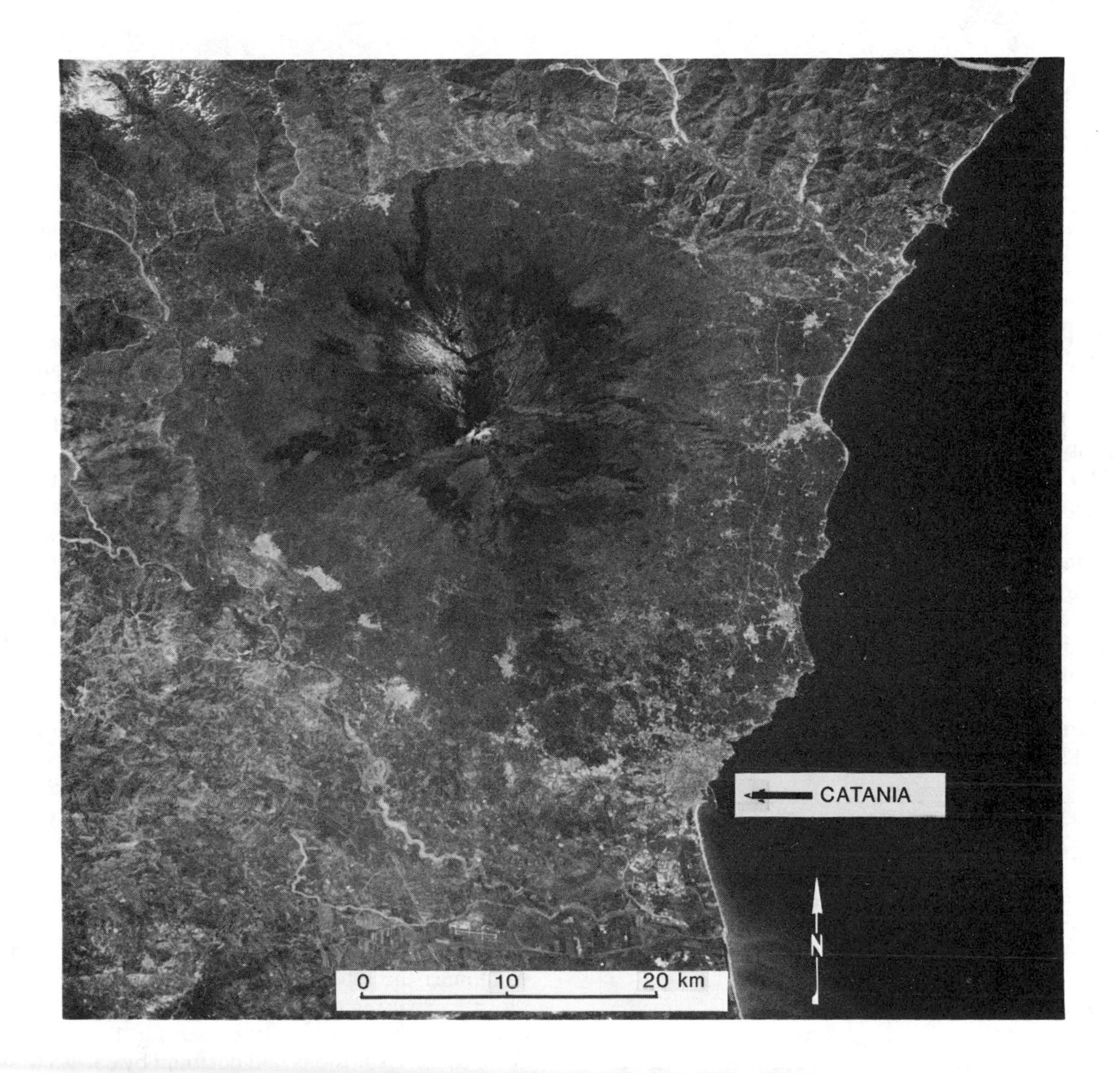

Fig. 1-18. Photograph of Mt. Etna and vicinity taken from the Space Shuttle in April 1981. Photo no. STS 1-13-444, courtesy of Charles A. Wood.

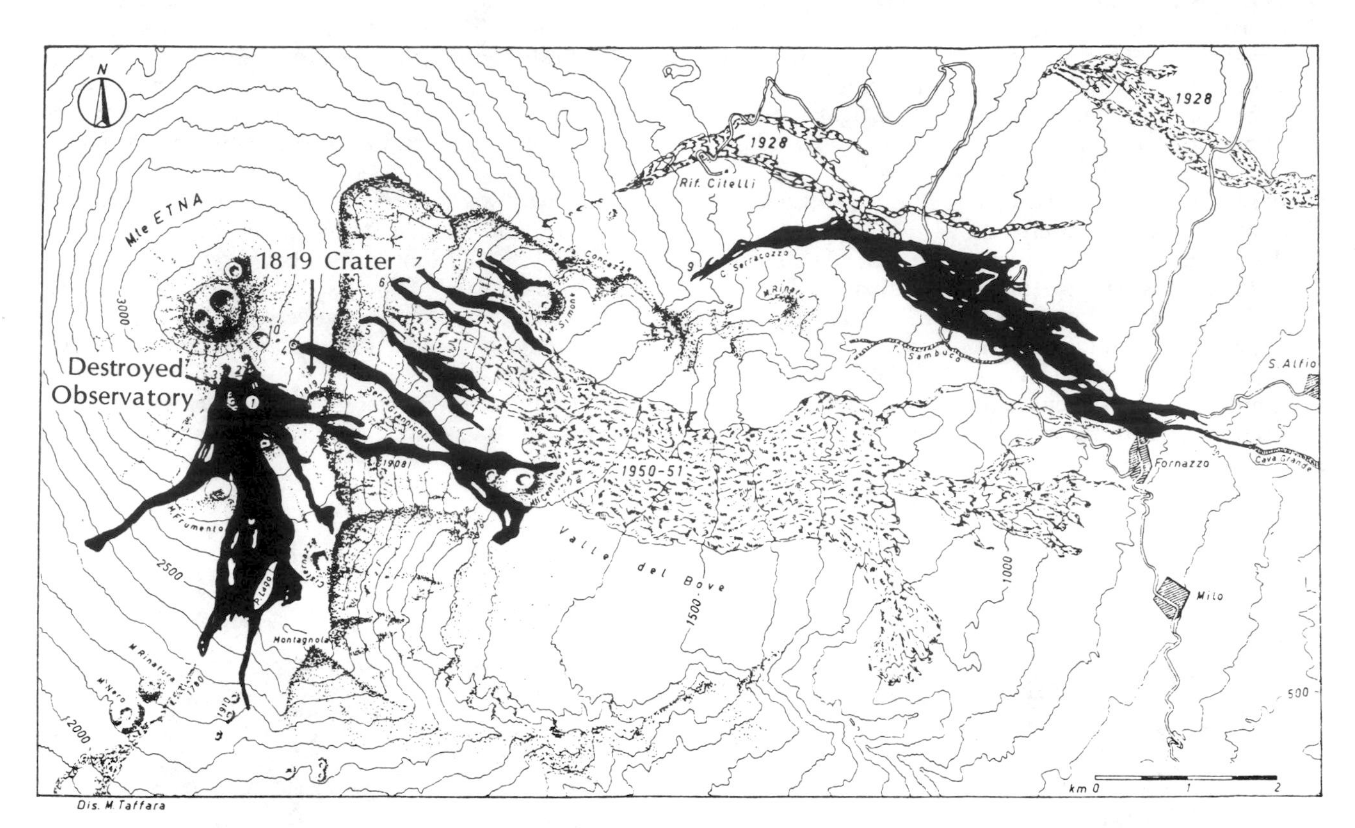

Fig. 1-19. Map of Mt. Etna, showing 1971 lava flows in black and earlier flows in other patterns. Contour interval, 100 m. From Rittmann, A., Romano, R., and Sturiale, C., L'Eruzione Etnea dell' Aprile-Giugno 1971; *Atti della Accademia Gioenia di Scienze Naturali in Catania,* Serie Settima, v. 3, 1971.

11/77 (2:11) Etna's fourth eruption since mid-July began during the night of 2 Nov from the Northeast Crater cone, site of the 3 previous events. Two lava flows were extruded, but (in contrast to the earlier events) explosive activity was very weak. The larger flow traveled about 2.5 km NW from the vent and the smaller about 1 km to the N, before the eruption ended about noon on 4 Nov.

Information Contacts: R. Romano, IIV, Catania; J. Guest, Univ. of London.

12/77 (2:12) Frequent eruptions from Etna's Northeast Crater continued through early Jan (table 1-2). Since the 2-4 Nov activity, progressively briefer 1-day eruptions occurred on 7, 22, and 25 Nov. On 6 Dec, extrusion of a single lava flow began at about 1100 from a NNE-trending fissure on the N side of the Northeast Crater. The flow traveled about 4.5 km down the E flank before the eruption ended at 2200. On 24 Dec, tephra was thrown 1000 m above the vent, and renewed activity 5 days later, accompanied by small earthquakes felt in nearby villages, projected tephra several hundred m above the vent. Lava extrusion resumed 2 Jan and the flow had advanced about 1 km on a 150 m-wide front by evening. The eruption was continuing as of the morning of 4 Jan.

Information Contacts: R. Romano, L. Villari, IIV, Catania; J. Guest, Univ. of London; NY Times; UPI.

1977-78	Dates of Eruption
Nov	2-4, 7-8, 22 (10 hrs), 25 (7-8 hrs), 27 (10 hrs)
Dec	6 (8 hrs), 10-(?)13, 18, 24-25, 29
Jan	2-3, 4, 5, 7
Mar	25-26, 27-28

Table 1-2. Summary of Northeast Crater activity since Nov 1977. Courtesy of R. Romano.

3/78 (3:3) Etna erupted for the first time since Jan on [25] March, extruding a lava flow, destroying trees on the flank, and emitting ash, that was blown S towards Catania (25 km from the summit) by a strong wind.

Information Contact: UPI.

4/78 (3:4) An eruption began 29 April and continued as of 3 May. Lava was extruded from 4 new vents on the SE flank, and had reached the 1700 m level near Monte Centenari, the 1852 cone, by the morning of the 3rd. Explosive activity ejected bombs to 300 m above the largest of the vents (near the 1819 crater; fig. 1-19) and built a 50-m cone. Explosions had declined by early 3 May, but vigorous lava effusion continued.

Frequent brief eruptions from Etna have occurred

since July 1977, but all have originated from vents on the Northeast Crater cone. The last SE flank eruption occurred 5 April-12 June, 1971, destroying the Etna Observatory.

Information Contacts: J. Guest, Univ. of London; R. Romano, IIV, Catania.

5/78 (3:5) R. Romano reported the eruption started on 29 April at about 2000-2030 from bocca 1 (fig. 1-20). Explosive activity built a cone that was about 50 m high by 2 May. Lava flowed from the E side of the cone into the Valle del Bove. Fissures opened, extending into the 1819 crater, where they intersected another fissure set which runs along the wall of the Valle del Bove.

"During the afternoon of 1 May new boccas 2 and 3 opened on the existing fissures, then early the next afternoon bocca 4 opened, emitting a small lava flow that stopped the same day. Activity at bocca 1 also ended on 2 May, but the main flow front advanced at 100 m/hr. Bocca 3's activity diminished 6 May and ended by 7 May."

John Guest and J.B. Murray arrived at the volcano on 10 May and reported: "Lava effusion was limited to bocca 2, marked by a hornito about 10 m high. The rate of eruption was about 10 m^3/sec through midday 13 May when bomb ejection began at a rate that increased to 40-50 m^3/sec, but decreased to 20-30 m^3/sec by the next day and had returned to about 10 m^3/sec by 15 May.

"Although bocca 1 was not emitting lava during this period, there were several collapses in the vent, producing billowing brown smoke. Occasional big explosions began 14 May, throwing bombs as much as 100 m above the vent. Explosions intensified on 27 May, but activity quickly returned to the 15 May level, and the eruption was continuing on 31 May.

"These eruptions have completely changed the Northeast Crater and surrounding area; the highest point on the volcano is now the Northeast Crater. Strong explosive activity during several eruptions has covered much of the summit area with ash, and during the Easter eruptions (25-28 March) there was a light ashfall as far away as Catania (25 km SE). Lava flows were extensive, one reaching as far down as 1700 m elevation of the NW flank."

Information Contacts: R. Romano, IIV, Catania; J. Guest, Univ. of London.

7/78 (3:7) Etna's SE flank eruption [which began on 29 April] stopped for about 12 hours on 26 May, then resumed and continued until the evening of 5 June. When visited by John Guest and others in late July, the main SE flank vents contained glowing red fissures and emitted jets of gas at high pressure. Occasional deep explosions could be heard inside Bocca Nuova, accompanied by rumbling and frequent collapse activity. The Chasm, normally continuously active, was largely filled with ash and snow, and showed no signs of activity. Steam emission continued from the Northeast Crater, site of a series of brief eruptions between July 1977 and March 1978 (2:8, 3:5).

Information Contact: J. Guest, Univ. of London.

8/78 (3:8) Activity resumed on Etna's SE flank during the night of 24-25 Aug. The initial activity consisted of ejection of spatter bombs and ash from one of the 1971 eruption craters, at 3000 m altitude on the SE flank of the summit cone. Lava extrusion from this crater began the night of 25-26 Aug and had ended by the next morning. Lava flowed eastward into the Valle del Bove, traveling 2.5 km to 2000-2100 m elevation. During the afternoon of the 26th, a second vent opened at 2725 m altitude, on the wall of the Valle del Bove. Its flow moved about 3 km in 12 hours, reaching 1650 m altitude.

The explosive activity that started 24-25 Aug began to decrease on 27 Aug, but 7 more vents opened that afternoon on the walls of the Valle del Bove, between 2800 m and 2500 m. By 29 Aug, the number of active vents had decreased to 4 with a notable diminution of lava effusion, and explosions had ended.

Information Contact: R. Romano, IIV, Catania.

9/78 (3:9) The SE flank eruption ended on the morning of 30 Aug. Activity since then has been confined to occasional explosions from Bocca Nuova. Two distinct sets of fissures formed during the eruption. The active vents trended NE-SW, parallel to the 1971 vents. No lava was extruded from the second set, which trended N-S. Fault throws of up to 3 m were observed.

Information Contact: J. Guest, Univ. of London.

Further Reference: Mackey, M. and Scott, S.C., 1980, The Eruption of Mt. Etna in Aug 1978; *in* Huntingdon, E.T., et al., (eds.), 1980, *U.K. Research on Mt. Etna;* Royal Society of London, p. 43-44.

11/78 (3:11) A new SE flank eruption began on 18 Nov. Initial activity consisted of ejection of ash and wallrock from one of the spatter cones formed in the Aug eruption. Ejection of incandescent ash and larger tephra from this cone started during the morning of 23 Nov, and was accompanied by minor lava effusion on the crater floor. Lava fountains rose 500 m above the Aug cone during the afternoon of 25 Nov and lava began to flow eastward into the Valle del Bove. That night, 2 new vents opened on the wall of the Valle del Bove at about 2600 m altitude, extruding flows that traveled 4 km E, to an altitude of about 1500 m. Strong ash emission from the Aug cone was visible from $\sim$40 km away on 26 Nov.

Two more vents opened in the Valle del Bove on 27 Nov at about 1700 m above sea level, and another vent opened nearby the next day, at about 1800 m altitude. Lava flows from these vents had traveled 4 km into the Valle Calanna, a steep valley extending SE from the E end of the Valle del Bove, by the evening of the 27th. A sixth vent opened that night at 1650 m altitude, extruding lava that advanced slowly toward the town of Zafferana Etnea (population 7000), a few km S of Milo. The eruption ended during the night of 29-30 Nov.

Information Contacts: R. Romano, IIV, Catania; UPI; AP.

7/79 (4:7) By early July, activity had resumed at the Chasm in Etna's central crater. The Chasm had been continuously active for many years but had become quiescent and largely filled with ash during the 1978 eruptions (3:5). John Guest and coworkers arrived at the volcano 11 July and observed small, sharp, Strombolian explosions from a small pit that had formed in the floor of the Chasm. When visited again on 27 July, the pit was filled with lava, covered by a thin crust that swelled prior to frequent Strombolian explosions. The lava lake had grown further by the next day. Large blisters formed in the lake, then burst, throwing bombs 200 m or more high. Some fell 50-60 m outside the rim of the Chasm.

Bocca Nuova, adjacent to the Chasm, was quiet on 11 July. However, collapse activity deep in this crater could be heard 27 July, and billowing clouds of dust were emitted.

Guest and coworkers observed the beginning of activity at a third site during the morning of 16 July, when strong gas emission started from a vent at the bottom of one of the 1978 craters on the upper SE flank (fig. 1-20). Ejection of lithic blocks and a little fresh magma soon commenced, with the proportion of juvenile material increasing steadily. By afternoon, strong Strombolian activity was occurring from the vent. The next day, bombs from many of the spasmodic explosions rose 200-300 m above the rim of the approximately 100 m-deep 1978 crater. Most bombs fell back inside the crater, but a few landed as much as 50 m outside the rim. Another vent, on the side of the 1978 crater, emitted ash, building a small cone. Similar activity continued until the night of 22-23 July, when the explosions became stronger and more frequent. The stronger activity continued through the morning of the 23rd, then declined to the more moderate levels of 17-22 July. Six vents were active at various times, 2 of which were dominant. This activity persisted, with some fluctuations in intensity, through 28 July, when Guest and coworkers left the volcano.

After a series of felt earthquakes 29-30 July, strong explosions from the Chasm began during the night of 3-4 Aug. Heavy ashfall took place in Catania, closing the airport, and ash fell as far away as Syracusa, about 80 km to the SSE. Unusual lightning accompanied the explosions, which were visible from the mainland, 75 km from Etna. Two fissures opened early 4 Aug near crater l, at 2950 m and 2875 m altitude. By afternoon, fluid lava from these fissures had traveled about 13 km down the E flank, threatening the village of Fornazzo and forcing its evacuation. However, about 300 m from Fornazzo the lava changed direction, and damage was limited to about 1000 acres of fruit and nut orchards. By late 4 Aug, summit explosions had apparently ceased.

Several new fissures opened the next day. The first was about 1 km long, located at about 1800 m altitude in the Valle del Bove, on the SE flank. Others opened later on the NE flank, producing lava that flowed down

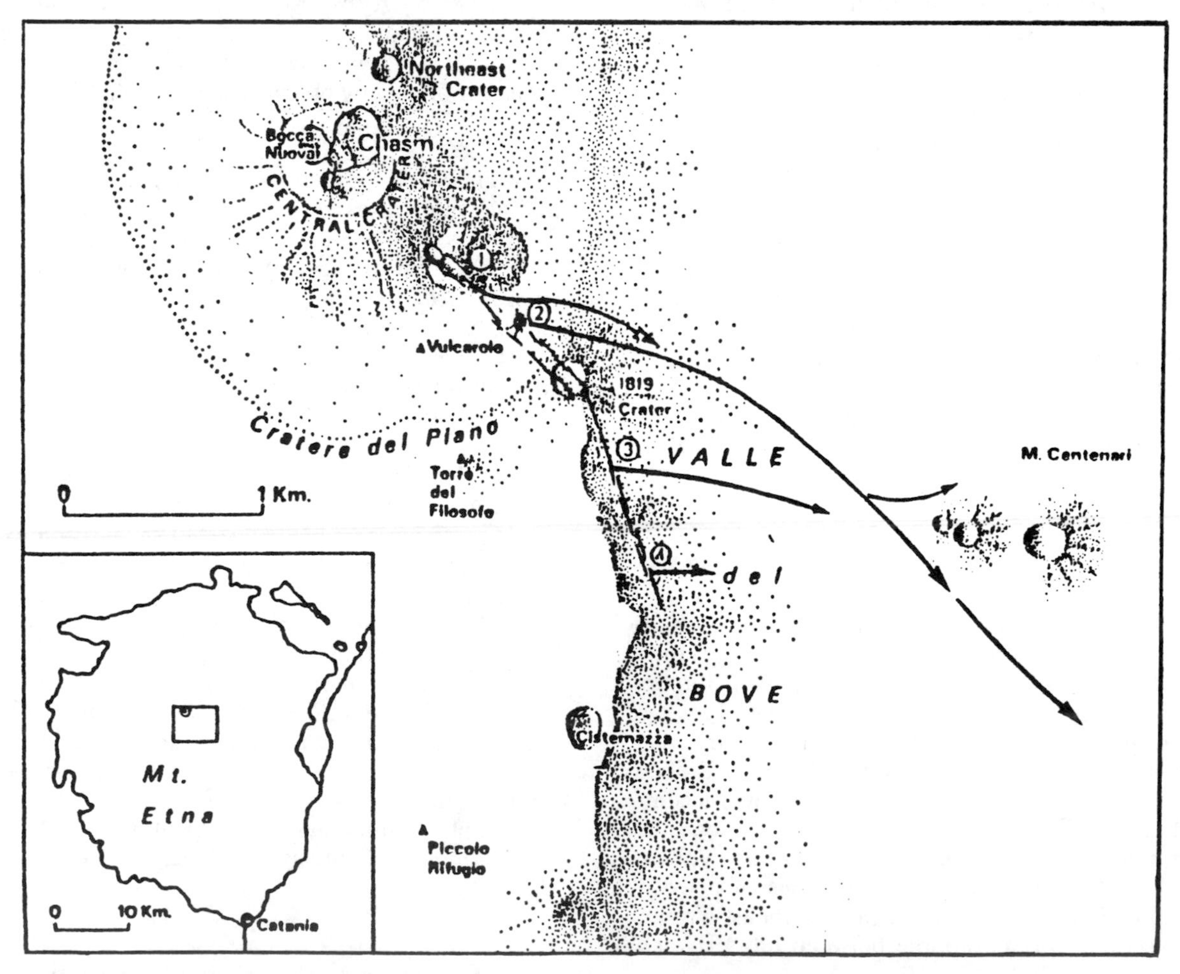

Fig. 1-20. Map showing locations of vents, fissure systems and lava flows of the April – May 1978 eruption. Bocca Nuova is W of the Chasm in the Central Crater; other boccas are identified by number. Prepared by J. Guest and J.B. Murray.

2 valleys. Lava effusion from some of these vents was continuing as of the morning of 7 Aug.

Information Contacts: J. Guest, J. Murray, Univ. of London; R. Romano, IIV, Catania; UPI; Reuters.

8/79 (4:8) The following report was prepared by John Guest and Romolo Romano from observations by E. Lo-Giudice, D. Condarelli, and A. Pellegrino.

"In the afternoon of 3 Aug, lava fountaining up to a height of 300 m started in the 1978 crater, which we are now calling the Southeast Crater. This crater has been active since the middle of July. Ash from this explosive activity fell over the E flank of the volcano, then later that day fell over the SW flank as far as Syracusa. During the evening of 3 Aug, an eruptive fissure opened near the 1819 Crater. Lava erupted that night reached Monte Centenari, some 2 km away. Fifty-four earthquakes were recorded with magnitudes up to 3.5-4.

"At 0545 on 4 Aug, another fissure opened in the Valle del Bove, SE of Monte Simone from 1800 m to about 1700 m elevation, approximately 1 km long. Two flows were erupted quietly from the fissure. The flow from the top of the fissure moved SE (towards Rocca Musarra). The second, from the lower part of the fissure, traveled along the N wall of the Valle del Bove past Rocca Caora and reached the Torrente Fontanelle by midday. At 1430 the flow front was advancing at 100 m/hr and cut the Rifugio Citelli-Fornazzo road. The flow continued to advance, stopping in the evening 50 m from the N-S road through Fornazzo, just N of the town at 870 m above sea level.

"In the central crater area, large explosions had occurred from the Chasm during the end of July. At 1000 on 4 Aug, the magma level in the Chasm dropped, and in the Southeast Crater explosive activity was greatly diminished. At 1130 on 5 Aug there were again large explosions from the Southeast Crater and fountaining resumed at 1345 with heights of up to 400 m.

"At 1615 on 5 Aug, a new fissure opened just NW of the 1819 Crater, with fountaining. A lava flow from this fissure reached the foot of the wall of the Valle del Bove. At 1715, another eruptive fissure opened SE of the 1819 Crater, again with fountaining, and a flow moved into the Valle del Bove. Ash from this eruption also fell in the region of Catania and Syracusa. At 1730, a fissure with a NE trend opened at approximately 2500 m above sea level in the Valle del Leone. During the night of 5-6 Aug, lava from this fissure traveled about 3 km.

"Early in the morning of 6 Aug, many earthquakes of up to magnitude 3.5-4 were recorded until about 1218. In the afternoon, activity increased from the fissure near Monte Simone, which had been active on 4 Aug, and lava flows overlapped the earlier ones from this vent. The 6 Aug flow traveled some 1.5 km. At 2030, yet another fissure opened, with an ENE trend, at an altitude of 2150 m, coinciding almost exactly with the 1928 fissure on the outer flank of the Valle del Bove. A sluggish flow followed the path of the 1928 lava, stopping on 8 Aug, 50 m from the Rifugio Citelli road, having traveled 1 km. Flows from vents in the Valle del Bove stopped on 9 Aug."

Information Contacts: J. Guest, Univ. of London; R. Romano, IIV, Catania.

9/79 (4:9) Activity resumed during the night of 1-2 Sept, with a collapse of part of the wall of Bocca Nuova and small explosions from the neighboring summit crater, the Chasm, the following day. The volcano then re-

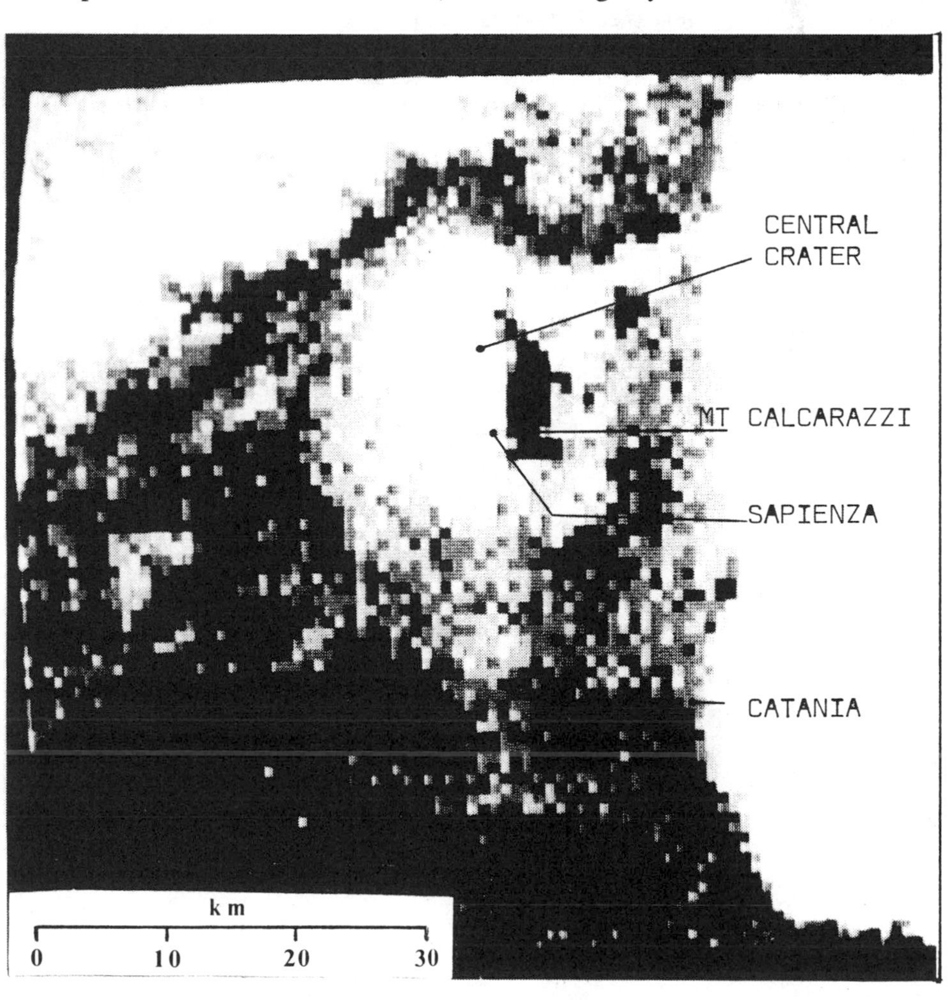

Fig. 1-21. Infrared image taken 16 Sept 1978 by the NOAA 5 weather satellite. The thermal anomaly is shown by the dark area around Monte Calcarazzi.

mained quiet until a 30-sec explosion from Bocca Nuova killed 9 persons and injured 23 at 1747 on 12 Sept. Some of the 150 tourists in the area at the time of the explosion were on the crater rim, and others were at a parking lot 400 m to the NW, where a large number of blocks about 25 cm in diameter fell. The explosion was apparently somewhat directed, because the distribution of blocks was dominantly to the NW of Bocca Nuova and no blocks traveled as far as 200 m in directions other than NW. No fresh magma was ejected by the explosion. The next day, considerable quantities of fine ash were emitted from Bocca Nuova and 1 or 2 small deep explosions were heard, but activity since then has been limited to weak emission of vapor containing SO_2.

Information Contact: J. Guest, Univ. of London; UPI.

Further Reference: Kieffer, G., 1981, Les Explosions Phreatiques et Phreatomagmatiques Terminales a l'-Etna; *BV,* v. 44, p. 655-660.

10/79 (4:10) The following thermal anomaly report is from C. Archambault, J. Stoschek, and J. C. Tanguy.

"The existence of a N-S elongated thermal anomaly about 10 km in length has been determined on the upper half of Mount Etna, both by field measurements and infrared satellite imagery (see *References* below). It is believed that this high-temperature anomaly is symptomatic of storage of magma at shallow depth within the rift zones of Etna. However, the magnitude of the thermal anomaly is expected to change with respect to volcanic activity (4 flank eruptions have occurred from this zone during the past 2 years). Since 26 June, ground temperatures have been continuously recorded at the end of the thermal anomaly (fig. 1-21). The results in this part of the volcano are considered the most significant because no volcanic activity has occurred in this zone for 70 years, therefore, the abnormal temperature cannot be due to cooling of residual magma. In order to eliminate the climatic effect, the data are presented (fig. 1-22) as the difference between recordings of the central part of the thermal anomaly (Monte Calcarazzi Station) and those of a "reference station" (Sapienza) located at the same altitude about 1 km outside the anomalous zone.

"There had been considerable temperature increase (3°C) at the Monte Calcarazzi station during the month that preceded the flank eruption of 3 Aug. On 10 Aug the temperature at 120 cm depth, checked by direct measurements, was 9°C higher than the normal (Sapienza). Such a variation may be either a transient response to the Aug eruptive phenomena on the SE, E, and NE flanks, or the indication of magma motion southward—a problem that should be solved in the months to come."

References: Mise en Évidence de Zones Thermiquement Anormales sur le Massif de l'Etna. *7éme R. ann. Sci. Terre,* Lyon, Avril 1979, p. 15.

Mise en Évidence d'Anomalies Thermiques dans la Basse Zone du Secteur Sud de l'Etna, *Note Technique CNET,* no. 110/79.

Etablissement d'une Carte Thermique du Massif de l'Etna à Partir des Données Transmises par le Satellite Météorologique NOAA V...; *Note Technique CNET,* no. 111/79.

Information Contacts: C. Archambault, J. Stoschek, CNET, France; J. Tanguy, Univ. de Paris VI; PIRPSEV, CNRS-INAG.

Further Reference: Guest, J.E. and Murray, J.B., 1980, Summary of Volcanic Activity on Etna, 1977-1979; *in* Huntingdon, E.T. et al., (eds.), 1980, *U.K. Research on Mt. Etna;* Royal Society of London, p. 50-52.

1/80 (5:1) [Some elements of this report were excised at the authors' request.] "Since the Aug flank eruption, no major volcanic event has occurred on Mount Etna, although on 12 Sept a moderate phreatic explosion resulted in 9 casualties near the W vent Bocca Nuova (4:9). On 11 Jan 1980, ashes were emitted from both Bocca Nuova and the Southeast Crater, where activity has preceded all the flank eruptions for the past two years.

"However, ground temperature measurements made on 21 Jan in the S part of Etna (Sapienza, Monte Silvestri, Monte Calcarazzi; 4:10) show that the thermal anomaly has been reduced to a very low level (1-2 °C, in contrast to 9 °C in Aug 1979. This is the lowest temperature recorded in this zone since temperature measurements were initiated in Sept 1978."

Information contacts: Same as above.

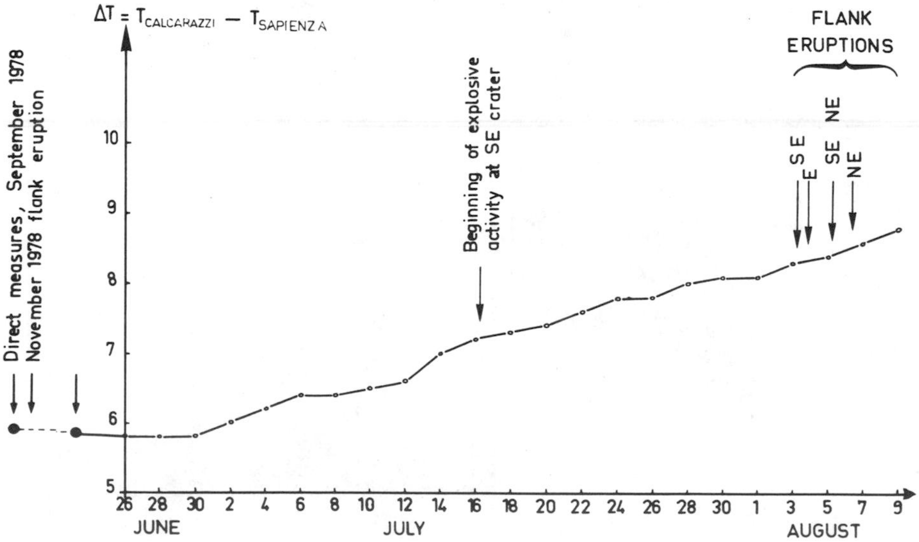

Fig. 1-22. Graph showing the variation between temperatures recorded at Monte Calcarazzi and Sapienza, Sept 1978 – 9 Aug 1979.

2/80 (5:2) During the night of 29 Feb-1 March, red glow was visible in the Southeast Crater. No glow was reported the next

2 nights, but cloud cover may have obscured the crater.

Information Contacts: J. Guest, Univ. of London; R. Romano, IIV, Catania.

4/80 (5:4) During the evening of 14 April, explosions began from Etna's summit area and red-hot gases were emitted from the Southeast Crater. Explosions continued for the next several days. Residents of Zafferana, 11 km SE of the summit, saw large explosions on 16 and 17 April that were especially spectacular at night because of incandescence or perhaps lightning. Poor weather prevented observations from the IIV in Catania.

Etna guides who climbed the volcano, probably on 17 April, saw large fresh bombs near the Chasm and Bocca Nuova craters. Bombs were particularly prominent on the N and W sides of the central crater area. On 27 April at 1705, a large summit explosion produced a 3 km-high ash cloud. Renewed explosions began during the afternoon of 29 April and continued through the night from the Southeast Crater.

Information Contacts: R. Romano, L. Villari, IIV, Catania; J. Guest, Univ. of London; C. Archambault, J. Stoschek, CNET, France; J. Tanguy, Univ. de Paris VI; G. Scarpinati, Acireale.

5/80 (5:5) "After a 3-month period of stabilization, ground temperatures in Etna's S flank fissure zones (6 km S of the summit) began to increase again in March (fig. 1-23). However, station 3, only 25 m from station 2, continued to show a nearly constant temperature. Variations in temperature are calculated by comparison with Sapienza reference station 1, located outside the fissure zones, about 0.5 km WNW of stations 2 and 3, and 1.25 km from station 5 (4:10).

"The increase in temperature preceded renewed activity from the summit crater system. On 14 April, red-hot gases were emitted from the Southeast Crater and on the following days fresh lava lumps were ejected from the Chasm.

"On 20 April, S flank temperatures were stationary or even slightly lower. Between 20 April and 1 May, temperatures strongly increased again. A large explosion on 27 April (probably at Bocca Nuova) and strong lava fountaining on 29 April at the Southeast Crater were followed by moderate magmatic activity until at least 3 May. After a short period of stabilization (1-4 May), temperatures were still increasing, although more slowly."

[Archambault, Stoschek, and Tanguy added the following paragraph to replace explanatory material excised from 5:1. " It was initially believed that the thermal anomaly fluctuations were related to the reopening of cracks caused by increases in volcanic pressure. Further investigations showed this hypothesis to be incorrect, with the systematic seasonal increase of surficial temperatures being mainly the consequence of a microclimate (Bourlet and Bourlet, 1982). A volcanic effect does occur, but its influence cannot be simple and is probably linked to the circulation of hot waters through the S flank. From this standpoint, a striking example is the rapid temperature increase measured in Nov 1982 at 6 m depth (fig. 1-27, 8:4), where a climatic effect cannot be invoked. Such a variation was not observed the following years (Nov 1983 and 1984). It may have resulted from the heating of meteoric waters by an intrusion of magma into the S flank a few months before the March 1983 eruption."]

"The Southeast Crater was intermittently active throughout May. On 31 May, two vents were observed inside the crater with moderate ejections of incandescent magmatic material up to 50 or 100 m, with some bursts occasionally reaching 200 m. As of 3 June, activity had increased noticeably, and explosions were stronger and more frequent. The number of explosions per hour reached 165 and the ejecta reached heights of 200-300 m."

References: Bourlet, Y. and Bourlet, F., 1982, Etude Microclimatique de 5 Stations sur le Versant Sud de l'-Etna; *Bull. PIRPSEV,* no. 63.

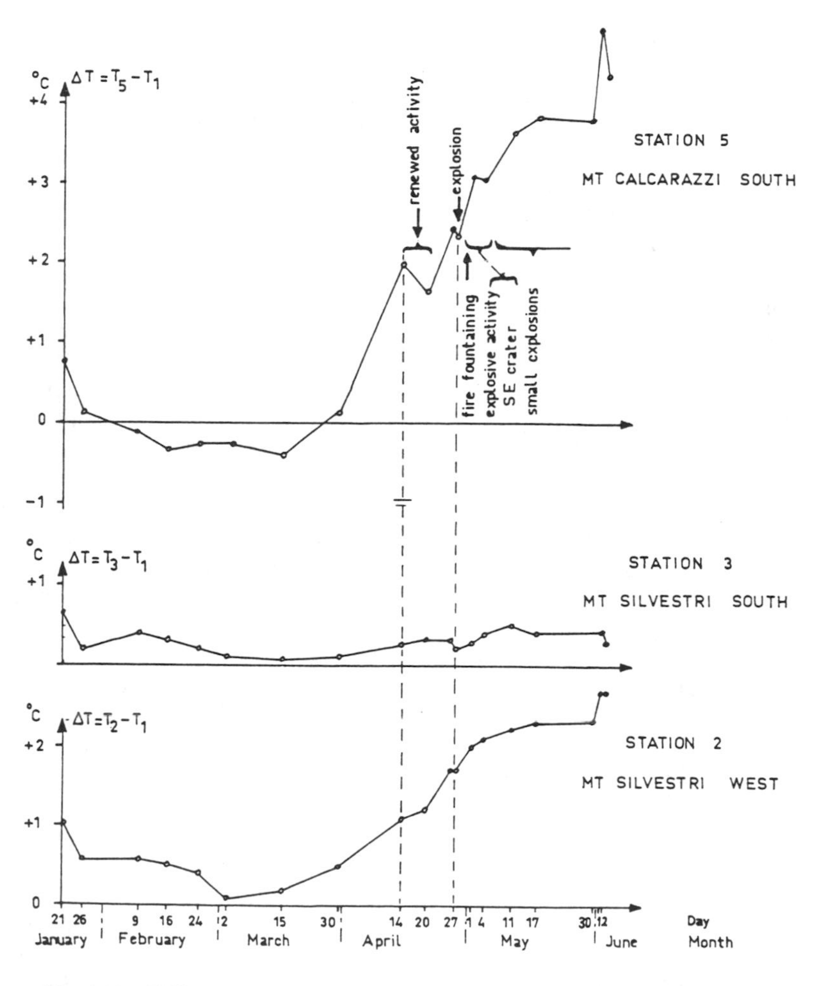

Fig. 1-23. Differences between ground temperatures measured at 120 cm depth at stations 2, 3, and 5, and reference station 1 (T_2-T_1, T_3-T_1, and T_5-T_1), 21 Jan – 2 June, 1980.

Bourlet, Y. and Bourlet, F., 1983, Etude des Anomalies Thermiques et Hydriques sur le Versant NE à Citelli et sur le Versant Sud à la Montagnola; *Bull. PIRPSEV,* no. 73.

Information Contacts: J. Tanguy, Univ. de Paris VI; C. Archambault, J. Stoschek, CNET, France; G. Scarpinati, Acireale (direct measurements and observations); PIRPSEV, CNRS/INAG, France.

7/80 (5:7) Activity increased gradually between 13 and 20 July in the Southeast Crater, site of lava fountaining in late April, intermittent eruptions in May, and stronger explosions in early June. No information was available on activity between 4 June and 12 July [but explosive activity was persistent from the SE Crater and Bocca Nuova]. In mid-July, many small explosions were observed on some days, while other days were characterized by fewer but larger explosions. After a period of poor weather, renewed observations of the Southeast Crater on 31 July revealed 4 active vents, located about 25 m below the lowest portion of the crater rim. The 2 larger vents steamed continuously, and exploded about every 2 seconds, ejecting incandescent tephra to heights of about 25 m. At approximately 2-minute intervals, stronger explosions sent tephra upward about 100 m. A small spatter cone surrounded the vents. The 2 smaller vents ejected juvenile material only occasionally.

Explosions from deep within Bocca Nuova were heard about every 2 seconds on 31 July. No tephra reached the rim. Since last Sept, Bocca Nuova's [diameter] has increased by as much as 80 m, bisecting a small crater formed in 1964. Most of the increase occurred during the winter, but further crater growth took place in June.

The Chasm, E of Bocca Nuova, was filled with solidified lava to within 25 m of its rim on 31 July. Large amounts of spatter and many bombs surrounded the crater as a result of activity in April (5:4).

Gravimeter readings made in late July by Tim Sanderson on the S flank yielded values that were significantly lower than in Sept 1979, indicating a loss of mass on that flank. July 1980 N flank gravity values were very similar to those of the previous Sept.

Information Contacts: J. Guest, J. Murray, C. Kilburn, R. Lopes, Univ. of London; T. Sanderson, Imperial College.

8/80 (5:8) Relatively weak activity similar to that of late July continued through Aug at Etna. A 1-day eruption on 1 Sept deposited ash on the E flank and extruded 2 lava flows.

John Guest climbed the volcano on 18 Aug. As on 31 July, explosions occurred deep within Bocca Nuova. The Chasm remained quiet. Mild Strombolian activity continued from the Southeast Crater, but reportedly weakened the following week. R. Romano reported that a swarm of local seismic events began on 21 Aug.

At 0957 on 1 Sept, a pale brown plume was seen rising from the Northeast Crater, which last erupted in March 1978 (3:5). By 1130, explosions were ejecting large bombs or blocks every 10-15 minutes. An increase in seismicity at about 1700 was followed at 1730 by stronger explosions that were audible in Fornazzo, 10 km E of the crater. A large black eruption column rose to 6 km above the crater. By 1800, ash was falling on Fornazzo and the entire E flank. Geologists reached the eruption area by about 2000 and saw nearly continuous Strombolian explosions from 2 vents in the Northeast Crater ejecting tephra to 500-600 m above the rim. Lava from the Northeast Crater flowed to the N and NW. By the next morning, the eruption had stopped. Heavy fog made mapping of the 2 lava flows difficult, but the NW flow had moved past Punta Lucia, about $^3/_4$ km from the Northeast Crater.

A second brief eruption from the Northeast Crater began early 6 Sept, ending at about 1500 the same day, after an estimated 10 hours of activity. A small lava flow was extruded. No further information was available at press time.

Information Contacts: J. Guest, C. Kilburn, Univ. of London; T. Sanderson, Imperial College; R. Romano, IIV, Catania; G. Kieffer, Univ. de Clermont-Ferrand.

9/80 (5:9) The 6 Sept eruption began at about 0500. Vigorous Strombolian activity continued for 10 hours, ejecting bombs to about 500 m above the same Northeast Crater vent that had erupted 1 Sept. Bombs 20 cm in diameter or larger fell as far as 750 m away. A lava flow, extruded from the same vent at a rate of 10-20 m^3/sec, traveled about 2 km to the N, directly over the main lobe of the bifurcated 1 Sept flow, which extended about $^1/_2$ km farther to the N. Ash fell on the coastal towns of Acireale and Taormina, 20 km SE and 30 km NE of Etna. Observations after the eruption showed the vent completely filled by lava and slumped debris. As of 3 Oct, no further eruptions had been reported.

Rumbling and deep explosions continued in Sept from Bocca Nuova. The Chasm remained inactive. Mild Strombolian activity at the Southeast Crater stopped in Sept, with only infrequent gas emission and small collapse events reported.

Tim Sanderson collected gravity data before the 1 Sept eruption and after the 6 Sept one. Frequent ground temperature measurements by J. C. Tanguy and associates continue (5:5).

Information Contacts: C. Kilburn, Univ. of London; T. Sanderson, Imperial College.

2/81 (6:2) The IIV reported explosions and extrusion of lava from the Northeast Crater. After a period of ash emission at the end of Jan and the beginning of Feb, stronger activity began with intense explosions the evening of 5 Feb. Lava flowed through a breach in the W to NW side of the Northeast Crater cone, forming 3 lobes that moved W, NW, and N, covering the upper NW slope of the volcano. The N lobe, the largest, traveled about 2 km to about 2600 m elevation where it had a 1.2 km front. The eruptive activity stopped the evening of 7 Feb.

Information Contacts: R. Romano, IIV, Catania; J. Guest, Univ. of London.

3/81 (6:3) An eruption 17-23 March extruded lava from several fissures on the NNW flank. Initial estimates indicate that the main flow reached about 7.5 km in length, lava flows covered an area of about 6 km^2, and about 30-35 x 10^6 m^3 of lava were extruded [but later calculations of C. A. Wood from topographic measurements by Murray (1982) yield 18 x 10^6 m^3], at a rate of 58-70 m^3/sec. Damage was estimated at about $10 million. Of the 90 historic eruptions of Etna for which location data are available, only 3 (1614, 1764, and 1918) occurred on the NW or NNW flanks.

Etna began to erupt on 17 March after a 2-day swarm of about 500 earthquakes, including a magnitude 4-5 event during the morning of 16 March. On 17 March at 1337 an eruption fissure opened at about [2550 m] above sea level on the NW flank, trending approximately NW-SE. Lava fountains rose 100-200 m from this fissure and lava flowed rapidly westward. In the next 4 hours, 3 more fissures opened, the first and third also trending NW, the second WNW. All showed strong lava fountaining and were the source of lava flows. As fissures formed at lower altitudes, those higher on the volcano ceased to be active.

At 1855 on 17 March, another fissure opened at 1800 m elevation on the NNW flank, trending NW at its upper end, but after a short distance changed direction to more directly downslope. A large lava flow that originated from this fissure traveled 5 km within 4 hours, cut a railroad and highway (at about 730 m altitude) during the night, and crossed another railway line and road (at about 680 m altitude) early on 18 March. The lava destroyed orchards and farm buildings, and passed very close to the village of Montelaguardia, forcing the evacuation of its 250 residents. The fissure propagated downslope to about 1300 m altitude at 1130 on 18 March. The lower section extruded a small lava flow that briefly threatened Randazzo (population 15,000) but did not force its evacuation. By 1630, the center of the main flow was more than 1 km wide and its front had reached 650 m altitude, about 100 m from the bed of the Alcantara River.

At 2200, another fissure opened between 1235 and 1140 m altitude, extruding flows that moved toward Randazzo. By this time, the system of eruptive fissures had a total length of about 7.5 km. The main flow reached the Alcantara River bed (600 m above sea level) on 19 March at 1100, while the flows extruded from the fissure between 1235 and 1140 m altitude continued to advance slowly. By noon on 20 March, this fissure was characterized by mild spatter ejection that continued to feed slow-moving lava flows. However, the main flow had nearly halted. Sporadic activity between 1235 and 1140 m continued 21-22 March, finally ending during the evening of the 23rd. The longest flow from this fissure stopped at 900 m elevation, about 2 km from Randazzo. More than 25 small earthquakes were recorded on 23 March, centered around the eruption fissures.

Throughout the period of lava extrusion, more or less intense emission of sand-size tephra occurred from Bocca Nuova, enlarging it to the W. Strong winds caused flank ashfalls on 22 March [as the Northeast Crater briefly ejected juvenile cinders; Tanguy and Patané, 1984; citation after 8:4].

Information Contacts: R. Romano, IIV, Catania; UPI; AP.

Further References: Murray, J.B., 1982, Les Déformations de l'Etna à la Suite de l'Eruption de Mars 1981; *Bull. PIRPSEV,* no. 57.

Sanderson, T.J.O., Berrino, G., Corrado, G., and Grimaldi, M., 1983, Ground Deformation and Gravity Accompanying the March 1981 Eruption of Mount Etna; *JVGR,* v. 16, p. 299-316.

Guest, J.E., Kilburn, C.R.J., Pinkerton, H., and Duncan, J.M., 1987, The Evolution of Lava Flow-fields: Observations of the 1981 and 1983 Eruptions of Mt. Etna, Sicily; *BV,* v. 49, p. 527-540.

7/81 (6:7) On 29 July, a dense ash cloud was ejected for more than 30 minutes from Bocca Nuova. The ash emission may have been produced by collapse within the crater; no significant explosions were associated with the activity. Similar events have occurred on several occasions since the March eruption.

Information Contacts: J. Guest, Univ. of London; R. Romano, IIV, Catania.

9/81 (6:9) Collapse activity deep within Bocca Nuova has been frequent since the 17-23 March fissure eruption (6:3). No fissuring or other evidence of surface collapse has been observed around Bocca Nuova. Explosions associated with the collapse activity ejected fine ash, caused strong ground vibrations 300 m from the crater, and could be heard as much as 10 km away. Plumes produced by this activity could sometimes be seen on the satellite images returned once daily by the NOAA 7 polar orbiter. Images returned shortly after noon on 3 and 4 Oct showed narrow, well-defined plumes extending about 75 km downwind from Etna. A smaller, less dense plume extending outward only about 20 km was present on the 6 Oct image.

Information Contacts: J. Guest, Univ. of London; M. Matson, NOAA.

10/81 (6:10) Images returned by the NOAA 7 polar orbiting satellite continued to show occasional plume emission: on 9 Oct at 1442 a plume roughly 75 km long drifted SE, and on 1 Nov at 1329 a much smaller plume, roughly 20 km long, was moving SW. No other activity was seen on the almost daily imagery between those dates.

Information Contact: M. Matson, NOAA.

12/81 (6:12) As of mid-Jan, collapse on Bocca Nuova's inner walls was almost continuous, producing plumes that contained fine ash but no fresh magma. At times of little or no wind, the plumes rose 3-4 km above the crater. There was no evidence of collapse beyond the crater rim. No changes have occurred in seismicity or tilt.

Images from the NOAA 7 satellite showed plumes emerging from the summit area on 4 and 6 Jan. A plume observed on 4 Jan at 1431 extended about 80 km to the SE, beyond the coast. Infrared data showed that the plume's apparent temperature was comparable to that of the sea water beneath it, and thus it probably remained at a relatively low altitude. The next image of the area, at 1408 on 6 Jan, showed a fairly diffuse linear plume about 55 km long, drifting SE. A plume from Etna had last been observed on satellite imagery (available most days for the Etna area) on 1 Nov (6:10).

Information Contacts: L. Villari, IIV, Catania; M. Matson, NOAA.

5/82 (7:5) [Some elements of this report were excised at the authors' request.] Since the fissure eruption of 17-23 March 1981, explosions associated with collapse activity deep within Bocca Nuova have produced small to moderate ash plumes. Magma was observed in the bottom of Bocca Nuova in mid-May, mid-July, and early Sept 1981, and incandescence was seen there in Feb 1982. All activity from Bocca Nuova stopped 8 May, but resumed about a week later around a time of unusual seismic and thermal activity on the NE flank. Incandescent scoria rose above the crater rim 20-22 May.

Seismicity increased during the third week of May and culminated with an explosion, possibly at 1515 on 27 May, when the summit seismometer operated by the Univ. of Catania detected a magnitude 3.5 earthquake that was felt by local residents. The next day, geologists found blocks of old lava that had been ejected more than 300 m from the Chasm. Blocks up to 1.5 m across were found in small impact craters at the rim. Many blocks larger than 10 cm occupied elongated depressions, implying relatively oblique impact, oriented radially to the crater. The greatest concentration of blocks was immediately N of the crater, although the pattern of smaller blocks suggested a NW orientation. The depth of the Chasm had increased from about 50 m in March to about 100 m after the explosion, by removal of material that had filled it since 26 May 1980 [explosive activity had begun in mid April], and vents in its walls were steaming. However, activity 28 May was concentrated in Bocca Nuova, where occasional detonations could be heard and the crater floor was obscured by sulfurous steam. On 29 May, Bocca Nuova ejected black ash, containing fresh magma, every 2-3 minutes. Ash clouds rose 200-300 m above the crater rim and were blown a few hundred m to the south. Only fumarolic activity was observed in the Northeast and Southeast craters.

Information Contacts: R. Romano, IIV, Catania; M. Malin, M. Sheridan, Arizona State Univ.; J. Sheridan, Tempe, AZ; C. Archambault, J. Stoschek, CNET, France; J. Tanguy, Univ. de Paris VI; R. Basile, S. Scalia, G. Scarpinati, Gruppo Ricerca Speleologica; M. Cosentino, S. Gresta, G. Lombardo, G. Patane, Ist. di Scienze della Terra, Catania.

7/82 (7:7) Etna erupted on 8 Aug, when a black tephra column was emitted from Bocca Nuova and lava rose to 150-200 m below the crater rim. The activity was accompanied by a marked increase in minor earth tremors. Sicilian authorities have restricted tourist access to the volcano.

Information Contacts: J. Guest, Univ. of London; Reuters.

9/82 (7:9) Strombolian activity on the floor of Bocca Nuova was observed in Sept. Fine ash fell on the crater rim. Since summer 1981, continuing collapse activity at Bocca Nuova has widened the crater by nearly 100 m, to 250 m in SW-NE dimension. Poor visibility prevented determination of the crater's depth. Just to the E, vapor emission was continuous from the Chasm, site of an explosion in late May that removed about 50 m of debris that had choked this vent for 2 years (7:5).

Information contact: J. Guest, Univ. of London.

11/82 (7:11) A C-141 cargo plane pilot observed tephra from Etna at about 12 km altitude on 6 Dec at 1800. Romolo Romano reported there had been many small explosions from Bocca Nuova throughout the day. No eruption plume was visible on the only satellite image of Etna that was available on 6 Dec, from a NOAA polar orbiter at 1530. Poor weather and heavy snow make access to Etna's summit area difficult during the winter.

Information Contacts: R. Romano, IIV, Catania; M. Matson, NOAA.

12/82 (7:12) Explosive activity of varying intensity continued through early Jan from the floor of Bocca Nuova. Large ash emissions were sometimes observed. During the night of 24-25 Dec, intense explosions ejected incandescent tephra. Most of the tephra fell back within the crater, but some was deposited outside the crater rim.

Press sources reported emission of large quantities of gray and white "smoke" from the Northeast Crater but Romolo Romano noted that the Northeast Crater activity was fumarolic and no ash was ejected. The Northeast Crater last erupted in Feb 1981, producing a lava flow and ash (6:2).

Information Contacts: R. Romano, IIV, Catania; UPI.

Further Reference: Scarpa, R., Patane, G., and Lombardo, G., 1983, Space-time Evolution of Seismic Activity at Mt. Etna during 1974-1982; *Ann. Geophysicae*, v. 1, no. 6, p. 451-462.

3/83 (8:3) A destructive south flank fissure eruption began on 28 March, preceded by a series of strong earthquakes first felt during the night of 26-27 March. At about noon on the 27th, a strong smell of H_2S was noted from an old cone (Monte Silvestri) roughly 2 km S of the initial eruption, although H_2S is not normally present in that area. Seismicity continued through the following night. At about 0845 on 28 March a NNE-SSW-trending eruptive fissure opened from about 2450 to 2250 m altitude, roughly 4 km S (bearing about 170°) of the central crater (between the eruption fissure of AD 1910 and La Montagnola). The base and E side of this fissure fed several lava flows that initially moved to the SSE and SSW then turned S. Weak explosive activity along the entire fissure ejected modest quantities of lava fragments. By evening, the main flow had cut a road and overrun several buildings.

During the morning of 1 April, vigorous emission of gas, ash, and old lava, accompanied by occasional phreatic explosions, began from 2 explosion craters upslope at 2700 m altitude. At the end of the day, explosions from the southern vent ejected lava fragments. On 2 April, nearly constant lava production fed numerous superposed flows that formed a 500 m-wide lava field extending to 1900 m altitude. As of 3 April, the lava had not advanced below 1450 m altitude, 3.5 km from the fissure. At least 4 principal effusive vents were active along the 750 m fissure, and from its upper part strong gas emission with sporadic explosions occurred at about 30 hornitos.

Bands of open fractures, oriented about N-S, extended from the central crater area to the eruptive fissure. A substantial widening was noted at the S rim of Bocca Nuova, site of frequent collapse activity since Etna's last eruption (from N flank fissures in March 1981; 6:3). Strong vapor emissions from Bocca Nuova sometimes included abundant ash. There was no activity from the Northeast and Southeast craters.

The temperature of the lava was less than 1100°C and its chemistry (alkali basalt) [corrected from phonolitic tephrite] was similar to that from some of the more recent eruptions. An area of more than 1 km^2 was covered by lava and the volume emitted was estimated at about 8-10 x 10^6 m^3. The IIV considered the eruption to be a typical slow subterminal type. The last activity of this type on the S flank was in 1780. As of 8 April, effusive activity had diminished, but the eruption had not yet ended.

The lava destroyed ski lifts [the cable car system originally reported destroyed survived until March 1985] and destroyed or seriously damaged 9 privately owned huts and 11 small buildings owned by local authorities, including restaurants, chalets, mountain refuges, and a first aid station. Lava remained 8 km from the village of Nicolosi, its closest approach to a village or town.

Information Contacts: R. Romano, IIV, Catania; M. Krafft, Cernay, France; UPI.

4/83 (8:4) The eruption was continuing as of 6 May. J. C. Tanguy noted that temperature station cables in the S part of the summit zone broke on 27 March between 0530 and 0946, probably because of the opening of the eruption fracture. Fissures extended S from 2700 m (where 2 explosion vents formed on 31 March) to 2450 m, then turned SSW along the trend of the 1910 eruption. Small hornitos and spatter cones formed between 2450 m and 2350 m. The main effusive vent was at 2280 m altitude. On 31 March, the temperature of lava at this vent was 1067°C at 60 cm depth, and the same value was measured in the main flow 30-40 m downslope on 4 April. F. Mousnier-Lompré and G. Scarpinati recorded a lava temperature of 1078°C on 1 May. J.B. Murray and A. Pullen reported that lava flow surface velocities measured 17-30 April were in the 1.17-3.41 m/sec range. Using surface velocities, Murray and Pullen calculated an average effusion rate of 22 m^3/sec if flow thickness was 3 m and 44 m^3/sec if the thickness was 5 m.

Romolo Romano reported that lava flowed S and SW, forming a wide, complex field as much as 1 km across and 40 m thick. On the E side of the field, the fronts of the longest flows were about 6 km from the vent and reached 1150 m altitude before stopping. On 23 April, the principal flow shifted to the W side of the field, and advanced to about 6.5 km from the vent, stopping at 1100 m altitude only 30 m from a road. On 4 May the primary flow was again moving down the E side of the lava field and had reached 1450 m altitude as of early 6 May. Lava flowed from the vent at about 2 m/sec, a rate of about 10 m^3/sec, and had a temperature of about 1030°C [but note 1067° and 1078° above]. The rate of advance of the flow fronts was quite variable and dependent on the gradient; estimated velocities ranged from a very few m per hour to about 60 m per hour. The area covered by the lava was about 4 km^2 on 6 May and the volume of lava emitted was about 40 x 10^6 m^3.

Earlier in the eruption, small hornitos and spatter cones had formed above the main effusive vent and small quantities of lava fragments were ejected. Vapor emission continued from this portion of the eruption fissure but no lava fragment ejection has been noted since 10 April. [Two] explosion vents at 2700 m altitude ejected ash during the morning of [1 April; Tanguy and Patané (1984), Frazzetta and Romano (1984)], but only vapor emission has been observed at this vent since then.

Throughout the eruption, ash was ejected at varying intensity from Bocca Nuova. More vigorous ash ejection 2-5 May sent plumes to 2 km above the crater and caused ashfalls on the mainland (Calabria coast) at least 60 km from the volcano. On 4 May, a polar orbiting satellite image showed the beginning of an ash emission episode at 0453. By the next image at 0606, a narrow plume extended nearly 1000 km SE from Etna at an altitude of about 7.5 km. Venting was continuing at 0945 but the plume was smaller; at 1603, it was 100-150 km long. Another image, at 0705 on 6 May, showed a plume about 100 km long. From the Chasm, weak emission of gas that sometimes contained reddish ash has been observed during the eruption. In the Northeast and Southeast craters, eruptive activity was limited to weak fumarolic emissions, but new concentrations of large fractures were seen in these craters as well as small internal collapses.

Murray and Pullen reported that vertical ground deformation during 1981-2 was characterized by summit deflation and S flank inflation. Reoccupation of a precise leveling network 20-22 April 1983 showed large changes on the upper S flank (near Piccolo Rifugio) since Sept 1982: 42 cm of uplift W of the eruption fissure, 12 cm of uplift E of the fissure, and a drop of 126 cm on the fissure itself. Large and complex movements had occurred elsewhere, including deflation of 76 cm above the eruption fissure at about 2900 m altitude (near Torre del Filosofo). Horizontal measurements across the fissure showed an E-W extension of 1.3 to 2 m since Aug 1981, with contraction of 17 cm and 6 cm at 500 m and 1 km W of the fissure. A network of 25 dry tilt stations occupied 24-29 April 1983 showed no general in-

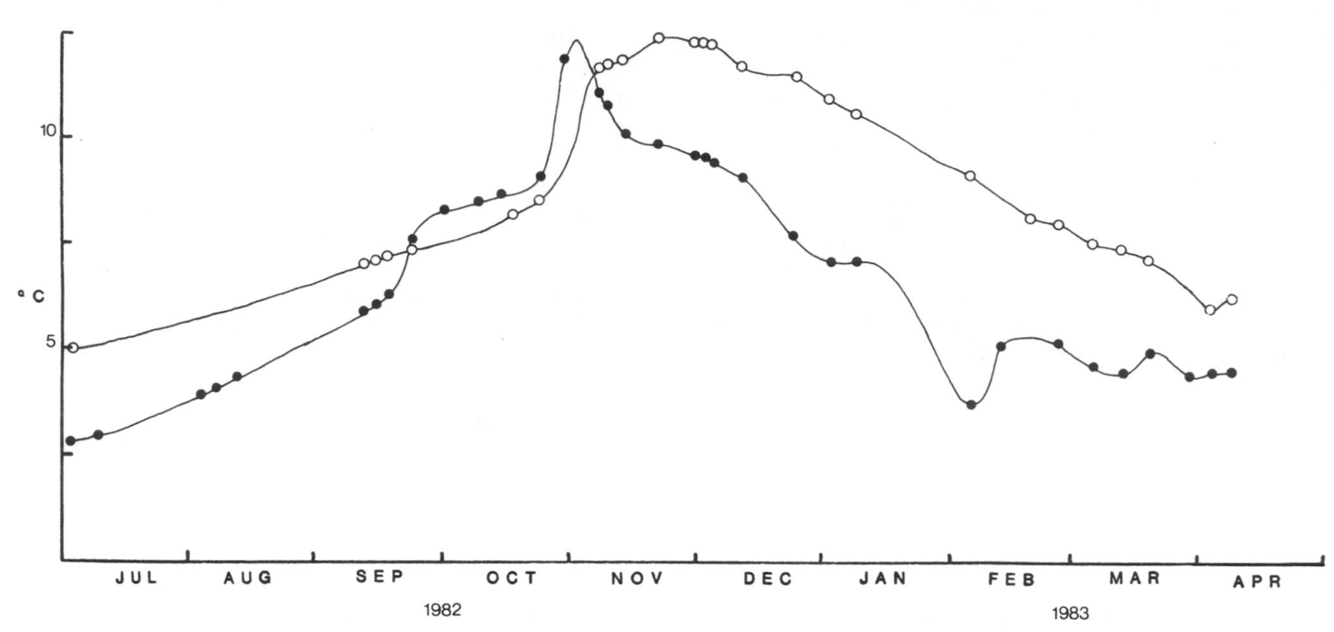

Fig. 1-24. Ground temperatures measured July 1982–March 1983 at 6 m depth, Silvestri station (closed circles) and Calcarazzi station (open circles), south rift zone, 1900 m altitude.

Fig. 1-25. Ground self-potential, *top,* and temperature, *bottom,* at 120 cm depth, recorded at Vulcarolo Station, south summit zone, 2965 m altitude. Note that the very steep rise in self-potential occurred in early April, after the start of the eruption. Telemeasurements by ARGOS system.

flation or deflation of the edifice since 1982.

Ground temperature measurements at 6 m depth on the S flank at 1900 m altitude showed a strong increase in Nov 1982 (fig. 1-24). In the summit zone, ground self-potential increased at roughly the same time and again in mid-Jan. A very steep rise occurred in early April, after the start of the eruption (fig. 1-25). Immediately before the eruption a geophone in the S summit zone registered strong seismic activity, up to 134 events per hour during the morning of 27 March.

Further References: Frazzetta, G. and Romano, R., 1984, The 1983 Etna Eruption: Event Chronology and Morphological Evolution of the Lava Flow; *BV,* v. 47, p. 1079-1096.

Tanguy, J. and Patané, G., 1984, Activity of Mount Etna, 1977-1983: Volcanic Phenomena and Accompanying Seismic Tremor; *BV,* v. 47, p. 965-976. (both in Barberi and Villari, (eds.), 1984, cited after 8:7)

Information Contacts: R. Romano, L. Villari, S. Gresta, O. Consoli, IIV, Catania; J. Tanguy, J. Murray, PIRPSEV; A. Pullen, Imperial College; M. Matson, NOAA; C. Archambault, J. Stoschek, CNET, France; S. Scalia, G. Scarpinati, Acireale; J. and N. Bartaire, St. Maur des Fossés; R. Cristofolini, M. Cosentino, G. Lombardo, G. Patanè, A. Viglianisi and P. Villari, Ist. di Scienze della Terra, Catania; P. Mousnier-Lompré, Servizio Sismico Regionale.

5/83 (8:5) The continuing eruption showed signs of a progressive decline on some days. The velocity of the lava flow from the vent at about 2280 m altitude decreased from about 1.7 m/sec on 28 May to less than 0.5 m/sec in early June, with a reduction in the rate of outflow to about 3 m^3/sec. In early May, the velocity at the vent had been about 2 m/sec and the effusion rate was about 10 m^3/sec. New flows, most of which moved S and SW, continued to add to the S flank lava field that has accumulated during the eruption. On 25 May a flow advanced beyond the edge of the lava field, reaching 1240 m altitude by the 27th, about 5.5 km from the vent. From 28 May, all of the lava flowed toward the interior of the lava field or approached its W side. In early June, frequent and impressive overflows occurred from both sides of the main lava channel at about 2200 m altitude. Within the lava field, effusive pseudo-vents at around 1800-1980 m altitude have remained numerous. As of 8 June the area covered by lava was estimated at 6 km^2 and the volume of lava produced by the eruption was about 75-80 x 10^6 m^3.

Explosions of varying intensity continued, particularly from Bocca Nuova, and primarily ejected reddish ash. On 15 May and 1-4 June the explosions were quite large. A NOAA 7 satellite image at 1529 on 15 May showed a strong plume (as large or larger than the 4 May plume; 8:4) that extended 100-150 km to the SE. Pilots reported that the plume had emerged from the summit area at about 1100. On 1 June at 0800, pilots estimated the height of a plume at about 5 km and reported that it was drifting SSE. In May and early June, only vapor emission was observed from an explosion crater at 2700 m altitude and a fissure between 2450 and 2300 m altitude that were active early in the eruption (8:3). On 3 June, deep felt shocks (M 3.3) occurred near the active vent; these events continued the next day.

Because the advancing lava flows threatened additional property damage, efforts were made to divert the lava. Explosive charges were detonated at 0409 on 14 May to blast a passage from the W side of the main lava flow at 2100 m altitude into a previously prepared artificial channel. Initially 20% of the total lava flux was diverted into the artificial channel, but by 16 May lava had ceased to flow in this channel after reaching 700 m length. After the explosions, however, there were numerous substantial overflows from the main natural channel, particularly from the W side, and these slowed the advance of the most distant flow. Artificial embankments have also been constructed E and W of the main natural channel, allowing control of the overflows from this channel that were moving over earlier flows.

Pham Van Ngoc and D. Boyer obtained self potential data during a PIRPSEV mission to Etna 27 April to 3 May. The following is a report from Pham Van Ngoc.

"Five self potential (SP) profiles were carried out from 2350 to 2700 m elevation. These profiles straddled the open fissure that trends NNE-SSW below Piccolo Rifugio (at 2500 m altitude on the S flank) and the N-S fractures above it. The profiles were located upstream from the lava flow emerging at 2320 m altitude. Fig. 1-26 shows the results of the SP profiles.

"The shape of the SP anomalies was very different S and N of Piccolo Rifugio. To the S, profiles I, II, and III show huge (more than 500 mV) and sharp anomalies suggestive of a superficial origin. It is notable that the maxima of these anomalies were not located just above the open fissure, but 40-50 m westward. The detail of profile III, just below Piccolo Rifugio, indicated clearly that the axis of the underground flow lies 40 m W of the open fissure (under the ski lift building W of Piccolo Rifugio). These results confirm that the pressure exerted by underground flow induced an E-W extension,

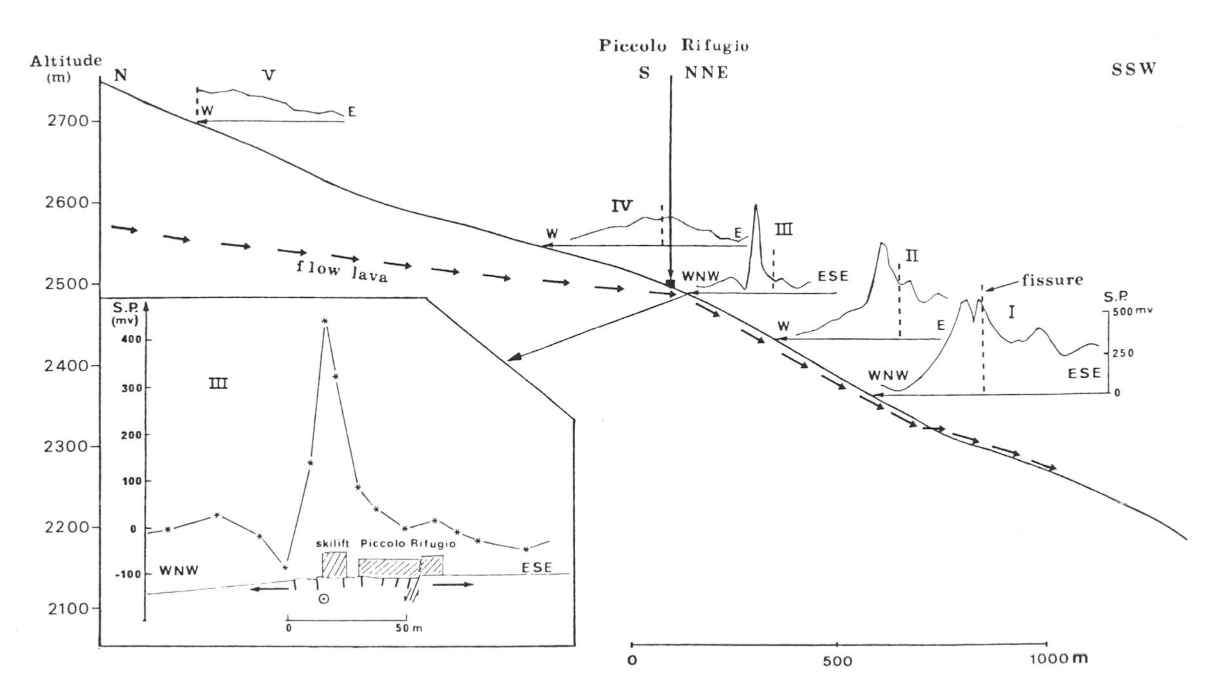

Fig. 1-26. Results of self potential profiles obtained 27 April – 3 May 1983 across open fractures that trended N-S above 2500 m altitude (Piccolo Rifugio) and NNE – SSW (about 200° azimuth) below 2500 m. The topographic profile (vertical exaggeration 2:1) parallels the fissure trends, changing strike at Piccolo Rifugio. Arrows pointing left show positions of self-potential profiles. Arrows pointing right diagrammatically show the inferred subsurface flow of magma, its emergence from the vent at 2320 m altitude, and continuation downslope as flowing lava. The inset at lower left details profile III, indicating that the axis of the subsurface flow was under the ski lift building, 40 m W of the open fissure. Here arrows show local stress field and fault movement. Courtesy of Pham Van Ngoc.

creating a small graben clearly visible in the area of Piccolo Rifugio. The open fissure corresponds to a normal fault on the E edge of the graben and caused the collapse of the W part of Piccolo Rifugio.

"Above Piccolo Rifugio, profiles IV and V show much smaller SP anomalies (about 250 mV). Furthermore, these anomalies widen, thus indicating a deeper origin. It is notable that the shape of the SP anomalies changed completely from profile III to profile IV in a distance of about 300 m.

"SP results suggest that: 1) above Piccolo Rifugio, the magma flowed deeply in a N-S direction; and 2) level with Piccolo Rifugio (2500 m altitude), the magma suddenly approached the surface and followed a shallow underground NNE-SSW channel that ran some 50 m W of the open fissure, then came to the surface at about 2320 m altitude. Arrows in fig. 1-26 indicate the path inferred for the lava.

Information Contacts: R. Romano, IIV, Catania; Pham Van Ngoc, Ecole Nationale Supérieure de Géologie; M. Matson, NOAA.

6/83 (8:6) Romolo Romano reported that the velocity of the lava flow from the main vent, at about 2280 m altitude on the S flank, continued to decrease in early July, from somewhat less than 0.5 m/sec in early June, to an estimated 0.1 m/sec on 7 July. The rate of outflow, about 3 m^3/sec in early June, had dropped to about 0.5 m^3/sec by 7 July. Lava flowed SE, S, and SW onto the S flank lava field that has accumulated since the eruption began, but the flows were smaller than in previous months and reached a maximum length of 1 km. Between 1800 and 1950 m altitude, some flows approached the E edge of the lava field and effusive pseudo-vents remained numerous. Moving lava was visible through many small windows. Between the main vent and the lava field, numerous overflows occurred from the E and W sides of the main lava channel from 2270 to 2100 m altitude. As of 8 July, the volume of lava erupted since 28 March was estimated at 100-110 x 10^6 m^3.

J. C. Tanguy reported that on 14 June, the temperature of a slow (about 0.15 m/sec) lava outflow at 2280 m altitude was determined to be 1071°-1073°C using both Cr-Al and Pt-Rh thermocouples inserted at 40 cm depth. This temperature was identical to that measured in the main lava flow 12 May by P. Mousnier-Lompré. Lava velocity at the vent began to increase 14 June and reached almost 4 m/sec 18-21 June. Gas pressure increased during the same period and new hornitos were built just below the main vent between 2280 and 2260 m altitude. Lava overflows occurred in this zone 18 and 20 June. On 21 June, gas pressure decreased and the lava velocity at the vent decreased to less than 1 m/sec. The level of lava in the main channel dropped again, leaving impressive grottoes where gas combustion produced temperatures as high as 1137°C (thermocouple) to 1165°C (infrared measurements) on 21 June.

Moderate to violent explosions have continued to occur from Bocca Nuova. Explosions on 20, 24, and 26 June, and 8 July were especially strong. On 24 June, emission of white vapor had been increasing since at least 0700, culminating in an explosion at 1015 that ejected old material, including reddish cinders and large blocks, that fell as much as 250 m from the vent, primarily to the W. A polar orbiting satellite image 26 June at 0606 showed a low-altitude plume that extended about 100 km ESE, and a similar plume, about 150 km long, trended SSE on 4 July at 0640. Larger plumes were seen on satellite images 8 July at 1427 and 1609 (more than 500 km long, to the ESE), 10 July at 0615 (500 km, to the SSE) and 2036 (150 km, to the SE), and 11 July at 0554 (300-500 km, to the SSE). On 12 July at 1415, the U. S. Navy reported a cloud extending 250 km to the SE with a base at 1 km altitude and a top at about 5 km altitude. Only weak vapor emission has been observed from the Chasm, and from vents at 2700 m on the S flank active earlier in the eruption.

Information Contacts: R. Romano, IIV, Catania; J. Tanguy, PIRPSEV; R. Clocchiatti, CEN, Saclay, France; F. De Larouziere, CNRS; R. Cristofolini, M. Cosentino, G. Patane, A. Viglianisi, P. Villari, Ist. di Scienze della Terra, Catania; M. Matson, NOAA.

7/83 (8:7) After 131 days of activity, the eruption stopped during the morning of 6 Aug. The July activity was similar to that of the second half of June. The main lava channel was almost completely roofed over, but moving lava was visible through 4 windows in the channel roof. Numerous overflows from the upper windows produced modest lava flows of short duration during the first 10 days of July. Through the end of the month, lava emerged from scattered short-lived pseudo-vents at about 1860-1800 m above sea level and flowed onto the S flank lava field that has accumulated during the eruption (fig. 1-27). These small superposed flows approached the E and W edges of the lava field; one advanced beyond the field's W margin on 13 July but stopped quickly. Efforts to contain the lava flows continued with the construction of new small embankments. None of the July flows moved below 1600 m altitude.

Ash emissions occurred at irregular intervals from Bocca Nuova, but were not as strong as in the previous month. High-altitude winds carried ash to Catania (about 30 km to the SSE) on 9, 10, and 11 July. No significant activity occurred from other vents.

Preliminary estimates suggest that the 131-day eruption extruded about 100 x 10^6 m^3 of lava, at a rate of 10 m^3/sec. Lava flowed a maximum of 7 km from the vent, reaching 1100 m altitude (E of Mt. Mazzo), and covered an area of about 6 km^2.

Information Contact: R. Romano, IIV, Catania.

Further References: Kieffer, G., 1983, L'Eruption de l'Etna Commencée le 28 Mars, 1983: sa Place dans l'Exceptionnel Cycle Eruptif en Cours (1971-1983); *C. R. Acad. Sci. Paris,* Ser. II, v. 296, p. 1689-1692.

Barberi, F. and Villari, L., (eds.), 1984, Special Issue on Mt. Etna and its 1983 Eruption; *BV,* v. 47, no. 2, p. 877-1177 (22 papers).

Lockwood, J.P. and Romano, R., 1985, Diversion of Lava during the 1983 Eruption of Mount Etna; *Earthquake Information Bull.,* v. 17, no. 4, p. 124-133.

1/84 (9:1) The following 3 reports are from R. Romano.

"Beginning in Dec, numerous seismic crises were recorded, mainly connected to the degassing of the magma column through the central vents. In the same period, ejections of reddish ash (old material) or dark ash (fresh material) occurred from the central crater. At times (14, 16, and 28 Jan) these have been rather significant, depositing thin layers of ash on the E flank. Some nights, pulsating flashes, due to the ejection of incandescent material from Bocca Nuova were observed. Tiltmeter variations were also recorded."

Information Contact: Same as 8:7.

4/84 (9:4) "During the night of 27-28 April, an eruptive fissure oriented approximately NW-SE opened inside the Southeast Crater, near its NE margin. This crater, which is at about 3000 m elevation behind the SE side of the central crater, formed in May 1971 and had numerous eruptive episodes in 1978 (April-July, Aug, Nov), 1979 (July-Aug), and 1980 (Jan-Sept, explosive activity only).

"Initially, moderate activity was observed from 3 explosive vents along the fissure, which ejected lava fragments. At the same time, small lava flows emerged from the fissure, remaining inside the Southeast Crater. During the next few days, Strombolian activity increased with the ejection of lava fragments to 250 m or slightly more in height. During the morning of 1 May, lava flowed over the NE rim of the crater then turned SE, covering the E side of the Valle del Bove (fig. 1-19) and quickly reaching a length of 2 km. Feeding of this flow was continuous but of variable volume, resulting in numerous superposed and parallel subflows but little advance of the flow front after its initial rapid movement. On 6 May, lava overflowed the SE rim of the crater, advancing NE and later E. This flow, which was fed until the morning of 8 May, reached a length of about 1 km. As of 10 May, both the Strombolian activity (which formed a scoria cone inside the Southeast Crater) and the effusive activity appeared to be decreasing.

"Starting 5 May, strong ash ejections from the Chasm were observed, while from Bocca Nuova only emission of gas under pressure was detected. In the past months, a lava lake 150 m from the rim of Bocca Nuova has been noted."

Information Contact: Same as 8:7.

5/84 (9:5) "The Southeast Crater eruption was continuing in early June. The explosive Strombolian activity from the small new cone within the Southeast Crater had been diminishing, and stopped almost completely 13 May. Starting that day, ash ejections have been observed at more or less regular intervals, while slow emission of gas and vapor usually occurred. The Strombolian activity started again in late May; at times (25 May and 4 June) it was particularly violent.

"The effusive activity has been continuous, with alternating phases of greater or lesser vigor. The lava field has grown noticeably toward the S (reaching a maximum dimension of more than 500 m) and by early June had in its interior many ephemeral effusive vents, which generated small lava flows that advanced over earlier ones. The main flows (generally one to the S and another to the N), which originated from convergence of the small flows, barely got below 2700 m elevation.

"At irregular intervals, more or less violent ejections of reddish ash from the Chasm have been noted, while from Bocca Nuova there have only been gas emissions."

Information Contact: Same as 8:7.

6/84 (9:6) Tanguy and Clocchiatti reported that in late May and early June, explosions, usually 10-30 per minute, ejected lava fragments and scoria to 50-200 m above the inner cone that had formed inside the Southeast Crater. Explosive activity sometimes declined to weak puffs of gas without much tephra, but at other times the ejecta were rich in large fragments of magmatic material. During periods of more vigorous activity, occasional

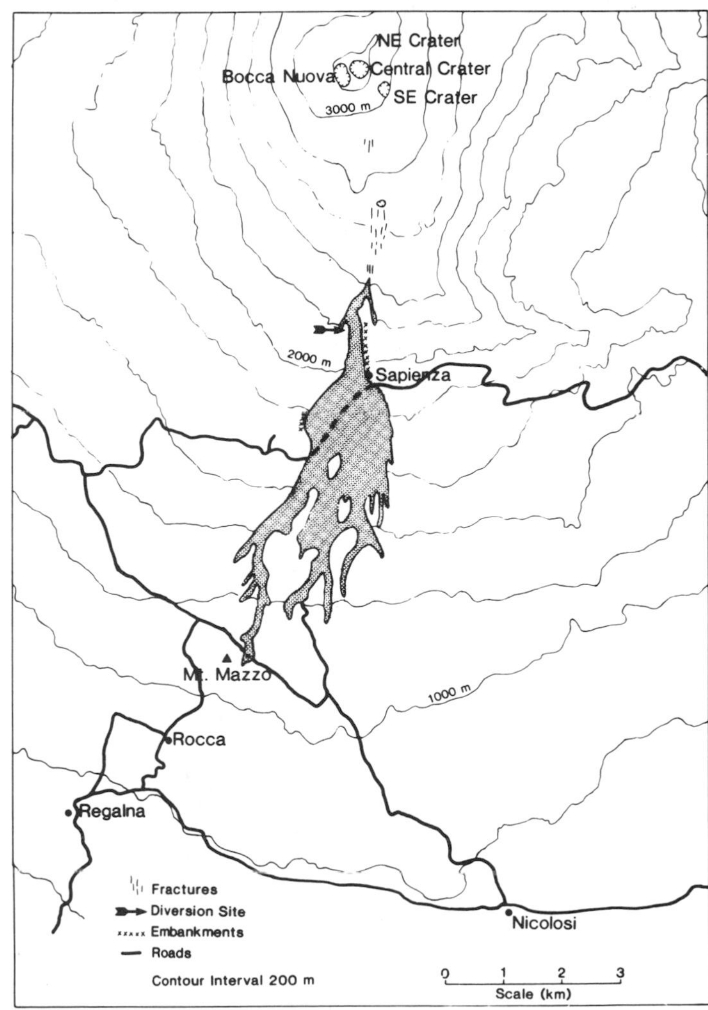

Fig. 1-27. Summit and S flank of Etna, showing the active vents and lava field of the 1983 eruption. Fractures are shown diagrammatically by short N-S lines. Contour interval 200 m. Large arrow on the upper W side of the lava field indicates the site of the partially successful attempt to divert lava into an artificial channel 14 May (8:5). Embankments constructed to limit the lava's spread are shown by x's. Several roads and villages in the area are shown (Sapienza is an inn, and Mt. Mazzo is an old vent). Nicolosi, Regalna, and Rocca cover larger areas than indicated. Courtesy of Romolo Romano.

bursts hurled the smallest tephra to 500-600 m height. In contrast, lava effusion occurred at a steady rate of a few (perhaps 2-5) m^3 /sec, significantly lower than in paroxysmal eruptions (> 10 m^3/sec) but probably higher than in typical subterminal persistent activity (1 m^3/sec). The maximum temperature measured at small effusive vents on 31 May and 5 June was 1075-1076°C (at 60 cm depth).

Romolo Romano reported that explosive activity at the Southeast Crater in June and early July was limited to ejection of incandescent tephra at varying intervals, occasionally accompanied by ejection of dark ash. Effusive activity continued, resulting in further enlargement of the lava field, especially to the N. Lava flows did not extend much below the 2800 m level. During the last 10 days of June, lava overflows occurred from the SE side of the Southeast Crater, but did not flow over the edge of the Valle del Bove. On 4 July, an overflow of very thin (0.5 m thick) fast-moving lava occurred from the still-active SE side of the Southeast Crater.

Tanguy and Clocchiatti reported Bocca Nuova was filled with lava to within about 100 m of its rim in April, but by early June the lava column was again very deep (more than 300 m) and activity was limited to quiet emission of large amounts of SO_2. Romano reported that violent ejection of reddish ash from Bocca Nuova was observed beginning in the second half of June. Recently, the ash has been gray (indicating presence of new material). On 3 and 9 July, violent explosions ejected lava fragments that fell outside the crater rim, especially on the SW flank. During this period, strong emission of gas under pressure was noted at the Chasm.

Tanguy reported that in late April seismicity recorded by Christian Archambault from a geophone about 1 km SSE of the Southeast Crater increased from less than 500 to more than 2200 events per day (fig 1-28). A seismic crisis was also recorded Jan-March, accompanied by lava filling and Strombolian activity at Bocca Nuova. The temperature at 3 cm depth at the station about 400 m SSE of the Southeast Crater decreased before the 1983 flank eruption but increased before the current eruption began in April 1984.

Information Contacts: R. Romano, IIV, Catania; J. Tanguy, Univ. de Paris VI; R. Clocchiatti, CEN, Saclay, France; C. Archambault, CNET, France.

7/84 (9:7) Quoted material in the following 2 reports is from Romolo Romano.

"The Southeast Crater eruption was continuing in early Aug with more or less intense Strombolian activity, accompanied at irregular intervals by violent expulsions of dark ash. This activity produced a scoria cone (about 50 m high) higher than the rim of the Southeast Crater. The effusive activity took place from vents around 3000 m above sea level that changed their positions continuously. On 6 Aug, two effusive vents were active along the old rim of the Southeast Crater, one on the NE edge, the other on the S edge. Some rather well-fed flows originated from these vents. The final flow direction was always E, toward the Valle del Bove. During this period, the lava flows never advanced below 2600 m. The lava field that formed from this continuous and variable (in terms of intensity and position) effusive activity was larger than 1 km in extent. The volume of lava emitted can be estimated at around 8-10 x 10^6 m^3.

"An increase in central crater eruptive activity was recorded in July. From Bocca Nuova, violent expulsions of gray ash continued at irregular intervals, while on the vent floor, violent and continuous Strombolian activity continued. At times, incandescent lava rose higher than the crater rim. The Chasm, after showing activity similar to that at Bocca Nuova in mid-July, was the source of violent activity on 19 July between 1300 and 1700. Very violent Strombolian activity ejected incandescent lava fragments about 1 m in diameter to 500 m from the crater rim. The S and N flanks of the central crater were most often impacted by the lava fragments (their average diameter was about 30 m and they fell within an average radius of 300 m)."

The pilot of an aircraft flying near Etna at 1542 on 19 July observed an eruption cloud that reached about 6.5 km altitude. At 1613, the NOAA 7 polar orbiting satellite showed a plume extending 100 km E from Etna.

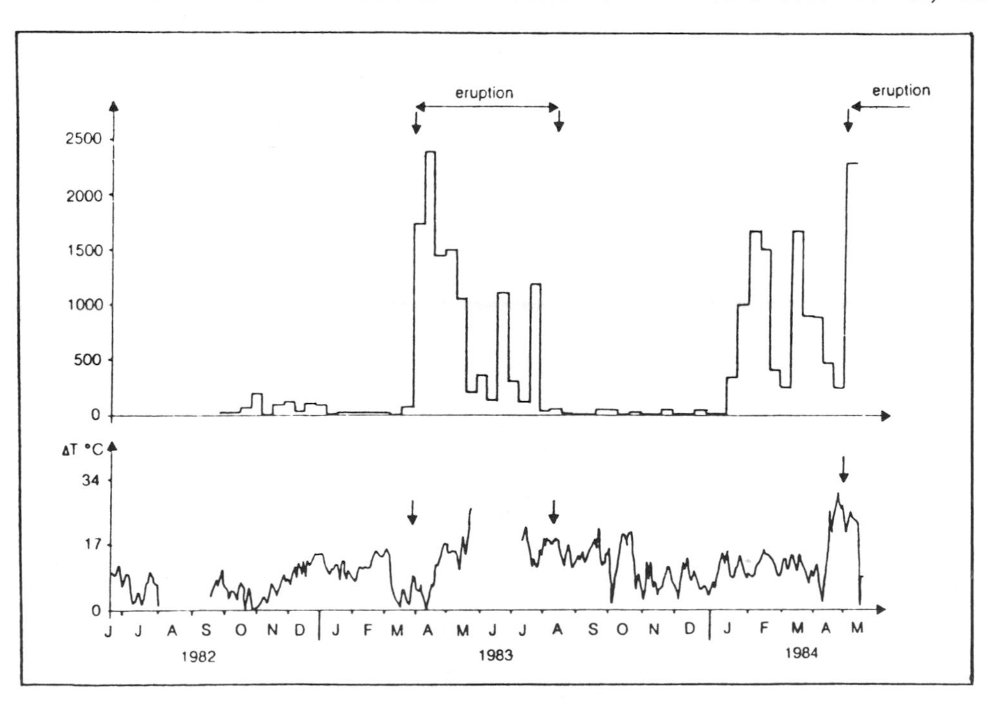

Fig. 1-28. *Top,* number of seismic events recorded per day (averaged over a 10-day period) at the TDF geophone about 1 km SSE of the Southeast Crater. *Bottom,* plot of the difference in subsurface temperature (measured at 3 cm depth) between the CC1 (about 400 m SSE of the Southeast Crater) and TDF stations. Arrows indicate periods of eruptive activity. Courtesy of J.C. Tanguy.

"After this, the Chasm remained obstructed until 1 Aug, when it reopened (at 1900) with the expulsion of old material that fell outside the crater rim. On 6 Aug, this vent was once again obstructed (around 1300) as the result of internal landslides.

"The Northeast Crater, inactive since Feb 1981, had a violent explosion that ejected old material on 20 July at 1715. Since then, strong emissions of gases occurred from the small vent that formed near the summit."

Information Contacts: R. Romano, IIV, Catania; M. Matson, NOAA.

8/84 (9:8) "The Southeast Crater Strombolian activity was intense at times (23 Aug, 1 and 6 Sept) and almost absent at other times (afternoon of 7 Sept). Violent expulsions of dark ash still occurred, at irregular intervals. The effusive activity took place through various vents along the edge of the Southeast Crater (around 3000 m elevation). In Aug the effusive activity occurred mainly on the SE side, producing lava flows variable in number, position, and rate of feeding. At first they were directed toward the S; later they turned E, rarely (on 23 Aug) advancing below the edge of the Valle del Bove (about 2700 m elevation). On 31 Aug, one of these flows advanced to about 200 m from the rifugio Torre del Filosofo at 2910 m elevation. The flow then turned E and stopped 1 Sept at 2780 m elevation.

"Aug activity from Bocca Nuova was similar to that of the previous month, mainly showing emission of gas and steam. The Chasm remained obstructed by landslides within the conduit. During the first few days of Sept, isolated expulsions of reddish ash from the Northeast Crater were noted, always in the afternoon. The last one was observed on 5 Sept. Usually, strong emissions of gas occur from this crater."

On 16 Aug at 0606, a weather satellite image showed a plume extending about 200 km SE from Etna at about 5.5 km altitude. The next morning at 0726, a similar plume was present on the imagery. Low sun angles in the early morning improve the visibility of eruption plumes.

Information Contacts: R. Romano, IIV, Catania; M. Matson, J. Paquette, NOAA.

9/84 (9:9) The following report from Romolo Romano was received 11 Oct.

"Continuing Strombolian activity, particularly intense at times (on the 14th and 20th) and very irregular in intensity and duration, marked the activity in Sept. Beginning 27 Sept, this activity was nearly replaced by expulsions of reddish ash at irregular intervals, rarely accompanied by the ejection of incandescent material (bombs and incandescent lava fragments). Lava flows, rarely thick (20 Sept), usually emerged from effusive vents on the S rim of the Southeast Crater at 3050 m elevation, and were directed mainly SE, E, S, and ENE. A lava flow advanced S to about 100 m from the rifugio Torre del Filosofo. The flows moved across the lava field, rarely advancing below 2700 m elevation, within the Valle del Bove.

"Strombolian activity occurred at the bottom of Bocca Nuova. The activity was particularly violent at times with ejection of incandescent lava fragments above the level of the crater rim (24 and 27 Sept). Generally, only gas emissions were observed. The Chasm usually remained obstructed. Rarely (11 and 24 Sept), incandescent gas under pressure was observed emerging from a small vent. From the Northeast Crater, isolated expulsions of reddish ash continued, but activity was usually limited to quiet gas and vapor emissions."

Michael Matson reported that a weather satellite image on 14 Sept at 0740 showed a plume at 4.5 km elevation or higher extending SE from Etna. Ejection of the plume was estimated to have begun 1-2 hours earlier. No plume was evident on an image at 0946.

The following report by Christopher Kilburn, is based on observations 25 Aug - 1 Oct.

"Since the eruption began in late April, a cone has

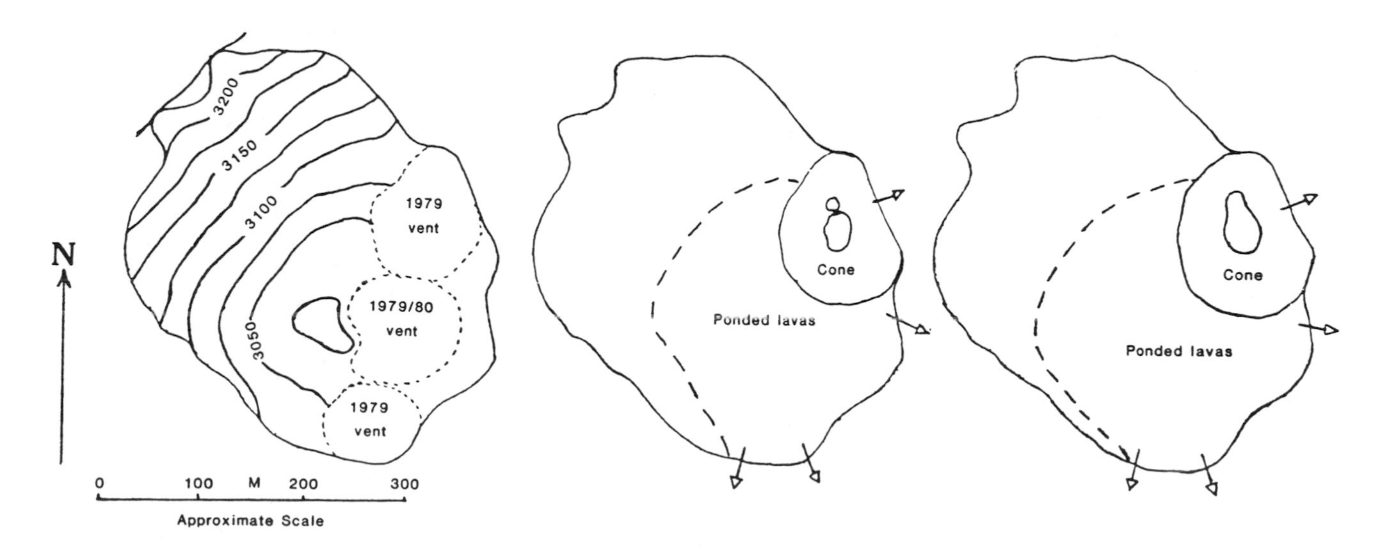

Fig. 1-29. *Left*, morphology of Southeast Crater before the 1984 eruption; contours are approximate, in meters above sea level. *Center and right*, morphology of Southeast Crater 11 Sept and 1 Oct 1984, dashed line indicates the approximate extent of ponded lava, arrows the direction of lava discharge. After Murray, J. B., 1982, Sommit del Mt. Etna, Settembre 1981, 1:5000 scale map, Ordnance Survey, Southampton, U.K. Courtesy of Christopher Kilburn.

developed inside the upper E rim of the Southeast Crater, roughly coincident with the position of the most northerly vent active in 1979 (fig. 1-29). On 1 Oct the cone was about 120 x 100 m at its base (the longer axis oriented roughly NNW) and some 30-40 m high. Until 20 Sept, the Strombolian activity consisted of intermittent violent explosions, with peak frequencies of 14 per minute and bomb trajectories ranging from subvertical to directed. Lower-angle ejections were mainly to the E and S, but occasionally to the N and W. Bombs of the order of 0.04 m^3 had maximum ranges of about 300-400 m. After 20 Sept, the violence of the explosions markedly decreased. Weak ejections every few minutes or hours were interspersed with periods of inferred internal collapse that generated small convection columns of pale brown ash and dust. This change in explosive behaviour followed a brief period of rapid lava discharge to the E on 20 Sept as well as a period of heavy precipitation. Enlargement of the vent at the top of the new cone, which occurred sometime between 13 and 24 Sept, may have been associated with the 20 Sept lava discharge.

"Lava discharge was virtually continuous from ephemeral vents S and E of the base of the new cone. Lava flowed S, SE, and E over low points in the Southeast Crater rim. Activity typically alternated between the S and E vents at intervals of one to several days. The successive aa flows, few of which were longer than 1.5 km with frontal thicknesses rarely exceeding 5 m, created a compound flow field with an estimated area that approached 1.5 km^2 by 1 Oct (fig.1-30). By 23 Sept, ponding inside the Southeast Crater had reached a maximum accumulated thickness of more than 45 m, raising the level of its bottom to about 3065 m above sea level. Assuming a conservative average thickness of 5 m, and including the material within the Southeast Crater, a provisional estimate of the minimum volume of the flow field on 1 Oct is 8.5 x 10^6 m^3, yielding a minimum average effusion rate of about 0.64 m^3/sec. Estimates of the eruption rate, made near the vent 17-28 Sept, were 0.5-1.0 m^3 /sec for lava discharged toward the S. The chief uncertainties in these discharge rates are the velocity profiles across the flow surface and with depth, and the thickness of the flow; for the minimum value, parabolic velocity profiles and a flow depth of 3-4 m have been assumed, the maximum (central) surface velocity and lava channel width being estimated in the field.

"The Strombolian activity at Bocca Nuova was accompanied by the emission of large quantities of SO_2-rich fumes. Until at least 24 Sept, magma was probably roughly 100-200 m below the W rim; observations were obscured by fumes. An apparently fresh bomb was found about 20-30 m from the NW rim 24 Sept, while others, up to 70 cm across, were seen being expelled toward the N and NW 28-29 Sept to a maximum estimated range of about 100-120 m. It is not known whether ejection of material outside the crater indicated a rise in level of magma within Bocca Nuova or an increase in explosive vigour due to changes in the physical state of the magma, notably in its vesicle content."

Information Contacts: R. Romano, IIV, Catania; C. Kilburn, Univ. of London; J. Murray, D. Decobecq, C. Delmotte, Univ. Paris Sud; Pierre Briole, PIRPSEV; M. Matson, NOAA.

10/84 (9:10) Quoted material is from Romolo Romano.

"After 172 days of activity, the eruption in the Southeast Crater stopped during the evening of 16 Oct, although activity continued at the Northeast and central craters. On 17-18 Oct, only violent ejections of ash at rather long intervals occurred from the small cone that had formed inside the Southeast Crater. The ash ejections stopped completely the morning of 18 Oct. During the last week of activity from the Southeast Crater, there was a gradual decrease in lava effusion, which was limited to vents along the S rim, and ash ejection from the small cone inside the crater. The lava flows were directed mainly toward the SE, generally stopping after only a few hundred meters.

"During the late morning of 16 Oct, Strombolian activity began from a vent at the bottom of the Chasm. Activity was extremely violent during the evening. At times, lava fragments were ejected above the crater rim, falling back within a 100-m radius. The Strombolian activity diminished during the night but continued, with alternating phases, through 17 Oct. During the morning of 18 Oct, the crater was obstructed by consolidated lava. It reopened 25 Oct with the ejection of old lava and ash, and more or less

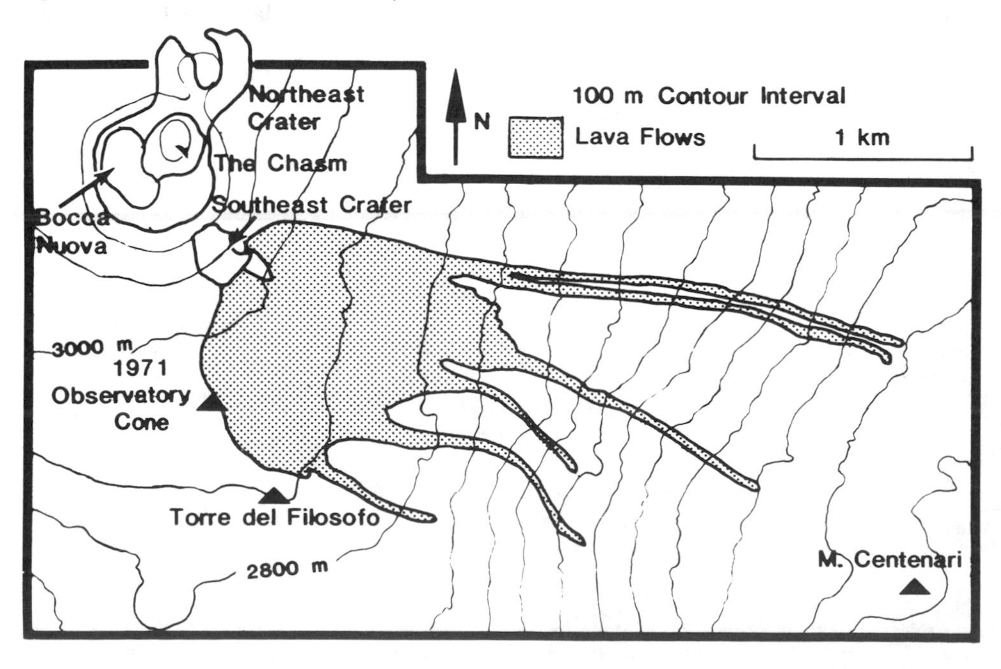

Fig. 1-30. Provisional sketch map of 1984 lava flows as of 1 Oct. Courtesy of Christopher Kilburn.

intense emission of gas has continued since then. Bocca Nuova alternated phases of slow emission of gas with periods of intense emission of vapor mixed, at times, with mainly reddish ash.

"Ejections of vapor mixed with reddish ash that had started at the beginning of Sept from the Northeast Crater intensified during the final phase of the Southeast Crater eruption. During the night of 19 Oct, weak Strombolian activity from the Northeast Crater was recorded at irregular intervals. This activity was succeeded by almost continuous ejection of dark ash, followed during the night of 21 Oct by the collapse of the crater's summit area. This enlarged the central part of the crater and ejected old lava to a distance of about 500 m, mainly toward the W. Discontinuous and more or less intense emission of vapor mixed with reddish ash started 21 Oct from this crater, while continuous emission of dense white vapor occurred from a small vent to the N.

"During the afternoon of 16 Oct a seismic crisis began, with earthquakes mainly occurring in the middle and upper parts of the E flank. The strongest shocks (magnitude about 3.5-4.5), which were felt, were all shallow (around a few km)."

A few of the largest earthquakes in the swarm are described below. Date, time, epicenter, and some casualty and damage data are from Romolo Romano; intensity values and the remaining casualty/damage information are from press sources.

18 Oct, 1258: centered near Piano Pernicana (15.5 km NE of the summit): Intensity reached MM V-VII, causing ground fracturing and some cracking of roofs and walls.

19 Oct, 1843: centered near Zafferana Etnea (11.5 km SE of the summit): 1 person was killed and others injured. Intensity reached MM VII-VIII. More than 400 buildings were damaged, including 50% of the historic district, and about 500 people were left homeless. Damage also occurred in Milo (11 km ESE of the summit), Fornazzo (10.5 km E), Santa Venerina (14.5 km SE), and Giarre (16.5 km ESE).

25 Oct, 0211: centered near Fleri (14 km SE of the summit), where it reached MM VIII, injured 12-15 people, destroyed 80% of the houses, and left 900 homeless. Ground cracking was observed in the area. Mt. Ilice, a 350 m-high prehistoric cone roughly 1.5 km upslope from Fleri, lost about 20 m of height during the earthquake. The shock was also felt in Catania.

7 Nov, 0956: centered near Pedara (15.5 km SE of the summit), where a few buildings were damaged.

Information Contacts: R. Romano, IIV, Catania; La Stampa, Torino; Corriere della Sera, Milano.

11/84 (9:11) No major eruptive activity has occurred since the Southeast Crater eruption ended in mid-Oct. From the Northeast Crater, emission of white vapor was more or less continuous and consistent. Sporadic expulsions of reddish ash were observed 27 Nov and 3 Dec. Ejection of mainly reddish ash observed at Bocca Nuova was particularly violent 22-24 Nov. Ash ejected 23 Nov was mainly dark in color, but on succeeding days was mostly reddish older material that had fallen into the conduit. Ash fell on the lower SE flank. Only weak emission of gas and vapor occurred from the Chasm.

Flank seismicity began as the Southeast Crater eruption ended in mid-Oct (9:10). Isolated tremors continued in Nov. Both felt and located events were mainly on the N and NE flanks. No additional damage was reported.

Information Contact: R. Romano, IIV, Catania.

2/85 (10:2) Weak Strombolian activity started 8-9 March in the Southeast Crater. Lava began to flow from the Southeast Crater the morning of 10 March and advanced E (toward the Valle del Bove), stopping that night as feeding ended. Mudflows were also observed in the Valle del Bove; heavy rains had caused flooding in Sicily during the previous week. On 11 March, fissures opened on the upper S flank (in Piano del Lago Alto). The press reported that temperatures in some of the fissures were high enough to melt plastic at nearby cable car stations. Ash fell on the towns of Acireale and Fiumefreddo (about 20 km SE and 20 km ENE of the central crater). During the morning of 12 March, lava emerged from vents that opened at progressively lower elevations (from 2620 m to 2500 m above sea level) and flowed S and SSW, reaching 2250 m elevation by the next morning. Numerous mudflows preceded the lava flows.

An 11 March newspaper article citing the National Institute of Geophysics reported that microtremors with epicenters in the W part of the Valle del Bove had been recorded for the past few days. A magnitude 3.4 event near the central crater occurred 9 March at 1523, and a shock with a focus at 5 km depth was felt 10 March at 1101 in the towns of Linguaglossa, Milo, and Sant Alfio (16 km NE, 11 km ESE, and 13 km E of the central crater). No magnitude was reported for the 10 March earthquake but both events were said to reach MM IV-V intensity.

Information Contacts: R. Romano, IIV, Catania; Il Progresso, New York; Corriere della Sera, Milano; UPI.

3/85 (10:3) The following is a report from Romolo Romano. Additional information about the eruptive activity from French volcanologists has been inserted in parentheses in this section. Their report of seismicity and tilt associated with the eruption is presented below.

"Explosive and effusive activity occurred 8-10 March from the Southeast Crater (French volcanologists noted that a strong but short-lived phase of the Southeast Crater activity started 10 March at 1100, with vigorous lava fountaining and overflow of lava toward the Valle del Bove). After numerous fractures formed 11 March on the upper S flank between 3000 and 2600 m above sea level, an exclusively effusive vent opened 12 March at 2620 m elevation, with no recorded seismic activity. The same day, three more effusive vents opened at lower elevations (2600, 2510, and 2490 m); only the last two (near Piccolo Rifugio) remained active. The lava flows that originated from these vents moved mainly toward the S and SW, giving rise to numerous individual lobes. Near the vent at 2510 m elevation, weak explosive activity occurred, soon creating several small spatter cones and hornitos.

"On 14 March, the lava flows moving S destroyed two Etna cableway pylons; the cable broke the next day. The longest flow stopped 15 March at 2080 m elevation. Along this trend, superposing lava flows were noted during the following days, until the effusive activity from the vent at 2490 m ceased on 23 March.

"The lava flows that moved SW created an extensive lava field (maximum width 500 m). At 2100 m elevation the lava field split into at least 5 lobes. The longest (and easternmost) flow had descended to 1850 m elevation by 4 April, covering a distance of about 3 km.

"As of early April, the main lava channel had become a lava tube between 2450 and 2300 m above sea level. Short-lived effusive vents opened near the lower end of the tube, with lava flows approaching the E side of the lava field or flowing over it. The velocity of the lava flows varied from a few meters per hour to about 30 m per hour. The surface covered by the lava can be estimated at around 2.5 km^2 and the volume of lava at about 12 x 10^6 m^3. The temperature of lava at the vents was around 1050°C.

"Particularly during the first few days of the eruption, because of the snow cover, several mudflows formed ahead of the lava flows. Phreatic explosions, violent at times (25 March) were also observed. During the eruption, more or less vigorous emission of vapor, gas, and rarely ash occurred from the central crater vents. (French volcanologists reported a strong explosion from the Chasm on 1 April at about 1625). Isolated expulsions of dark ash from Northeast Crater were observed.

"Collection of information on the eruptive activity was possible thanks to the cooperation of the following Italian Alpine Club rescue team guides and volunteers: G. Baglio, A. Cariola, A. Cristaudo, C. Ferlito, A. Nicotra, G. Puglisi, and F. Zipper."

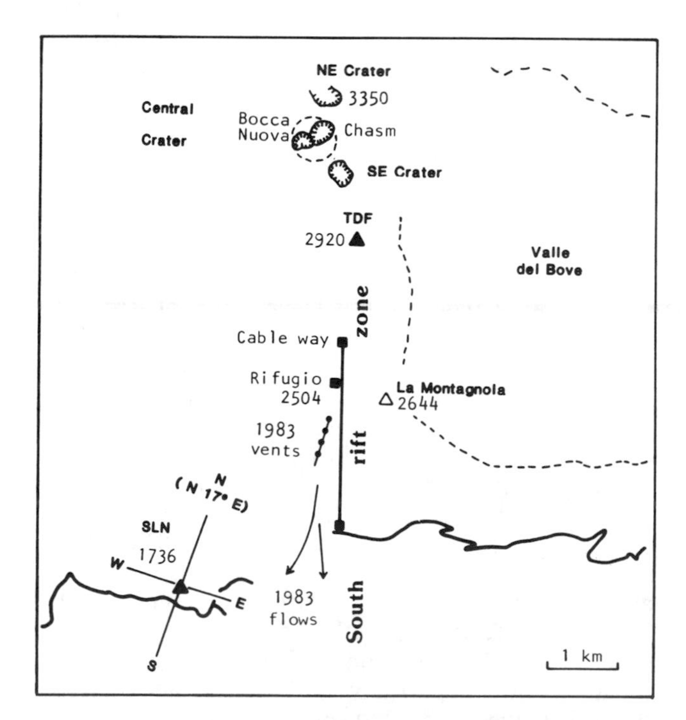

Fig. 1-31. Sketch map of the summit and south flank of Etna, showing locations of geophones at Torre del Filosofo (TDF) and Serra la Nave (SLN), and orientation of the tiltmeter at SLN.

Later newspaper reports described a swarm of about 30 earthquakes, some reaching intensities of MM IV-V, that started 7 April and continued until early on the 9th. The events were felt most strongly in the towns of Santa Venerina, Acireale, and Giarre (14.5 km SE, 20 km SE, and 16.5 km ESE of the central craters). After the swarm, the rate of lava production increased by a factor of about $^1/_3$.

The following report, on activity monitored through the ARGOS system, is from PIRPSEV, CNRS-INAG, in the context of French-Italian cooperation.

"Two geophones and a Blum pendulum inclinometer, connected with ARGOS for telerecording of the data, operate near the summit (TDF) and on the south flank (SLN) (fig 1-31). The geophones record both earthquakes and pulses of volcanic tremor as seismic events, and appear to be good indicators of seismic activity linked to volcanism. The eruption was preceded by an increase in seismic activity (mainly tremor energy) as indicated by the 2 geophones (TDF and SLN, fig. 1-32). However, after the Southeast Crater eruption,

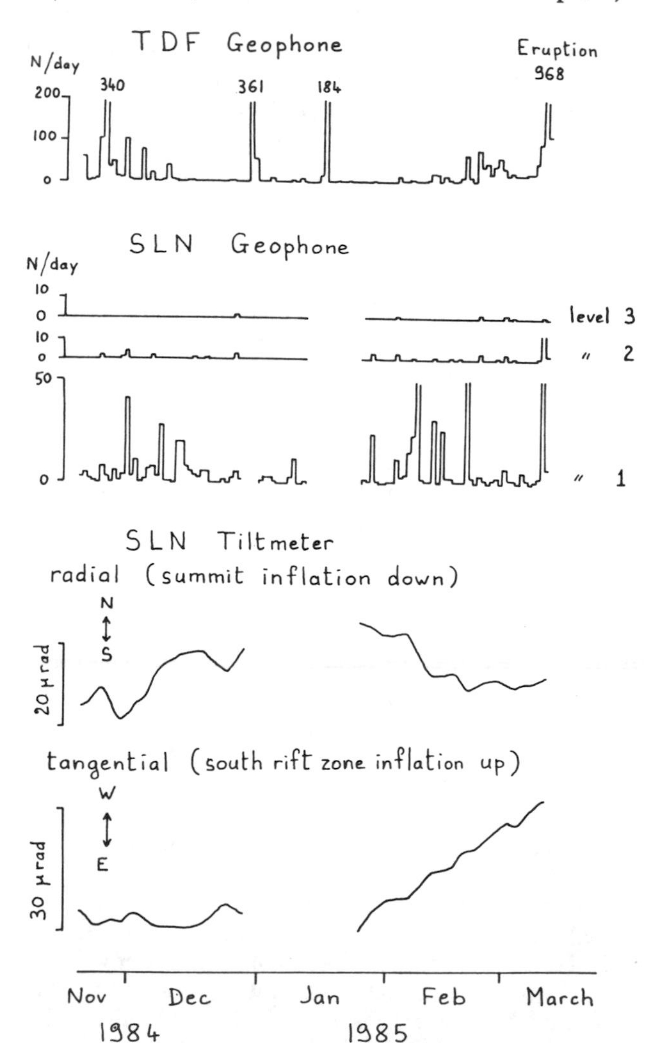

Fig. 1-32. Seismicity (number of events per day), *top*, and tilt, *bottom*, preceding the 10 March eruption, recorded by the geophones at TDF and SLN, and the SLN tiltmeter. The TDF geophone detection level is 6 μm/sec, the three SLN levels are 1.8, 9, and 45 μm/sec.

the tremor energy returned to very low levels (fig. 1-33), although effusive vents subsequently opened on the S flank.

"The SLN inclinometer showed continuous tilt toward the WSW during early Feb, consistent with inflation on the site of the 1983 (and March 1985) eruptions (fig. 1-32). Since mid-Feb, however, tilt has been mainly westward (figs. 1-34, 1-35), thus indicating inflation lower on the S rift zone. This change appears almost synchronous with the appearance of new frequencies of volcanic tremor (S. Gresta, personal communication)."

Information Contacts: R. Romano, IIV, Catania; C. Archambault, C. Pambrun, CNET/PIRPSEV; P. Blum, IPG/PIRPSEV; P. Briole, IIV/IPG/PIRPSEV; G. Kieffer, Centre de Recherches Volcanologiques, Clermont-Ferrand/PIRPSEV; J. Tanguy, PIRPSEV; La Stampa, Torino.

4/85 (10:4) The following reports are from R. Romano.

"The south flank activity that began 12 March was continuing in early May. During April, the main lava channel was transformed to a continuous lava tube (with at least 3 windows) from 2510-2320 m elevation. Around 2300 m elevation numerous ephemeral effusive vents formed, variable in number and location, from which several lava flows originated and advanced over the lava field or along its E edge.

"The lava flows, directed mainly SE, S, and SSW, have not advanced much, generally stopping at 2150-2050 m above sea level. The lowest elevation reached during this period was 1950 m (20 April). At times (19 and 21 April and 6 May), because of an increase in the production rate, lava overflows occurred following roof collapses in the upper parts (2510 and 2485 m elevation) of the lava tube, giving rise to small lava flows of short duration.

"No gas emission was noted from the mid-April hornitos. More or less intense emission of gas and vapor continued from both of the central crater vents; ash emission was very rare. The Northeast Crater generally emitted vapor and rarely (10 April and 9 May) ejected reddish ash.

"The surface covered by lava can be estimated at around 3 km^2 and the volume at about 20 x 10^6 m^3. In this period there has been almost a total absence of earthquakes. However, during the week of 8-14 April, 20 shocks with magnitudes less than 3 were recorded. Afterwards, a variation in the main spectral peaks of the tremor was observed. A gradual increase in the tremor energy was observed during the entire month of April (personal communication by S. Gresta).

"The Etna Guides and volunteers from the Italian Alpine Club rescue team cooperated in the collection of information about eruptive activity."

Information Contact: R. Romano, IIV, Catania.

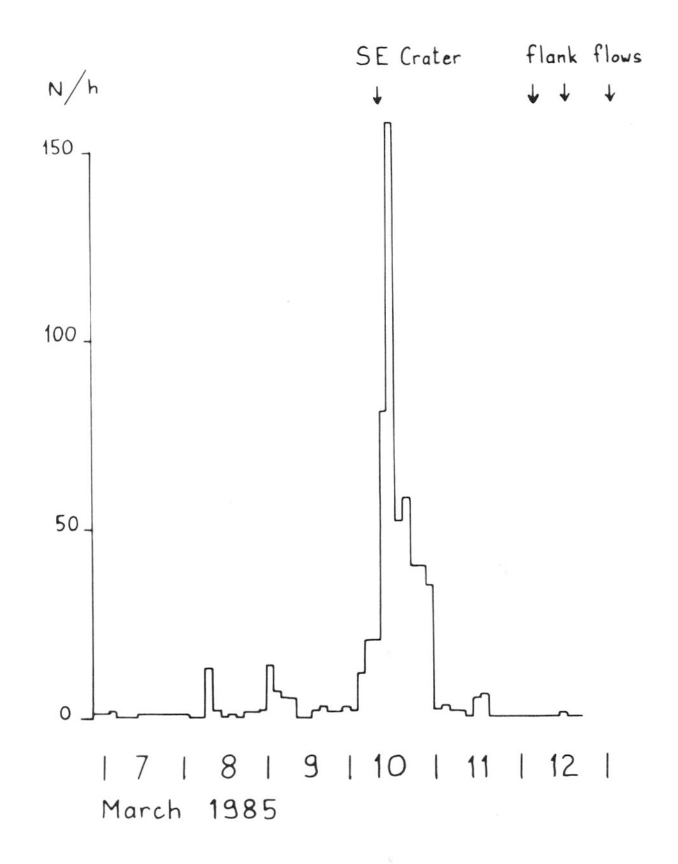

Fig. 1-33. Number of events per hour recorded by TDF geophone, 7-12 March 1985, showing a strong increase during the Southeast Crater eruption. Arrows at top indicate that eruption and the production of flank lava flows.

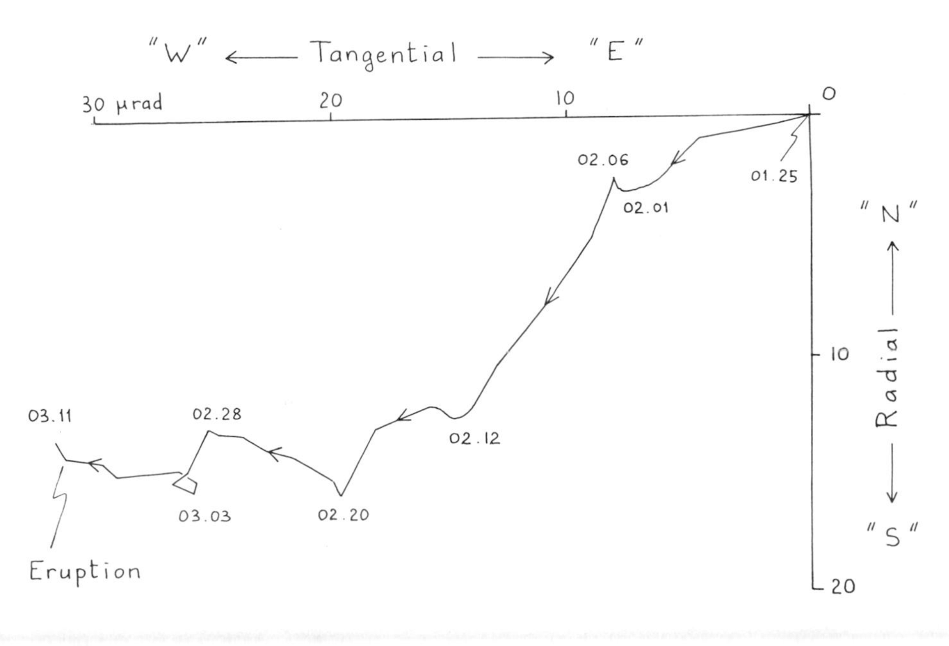

Fig. 1-34. Tangential vs. radial component of the SLN tiltmeter, 25 Jan – 11 March 1985 (S rift zone inflation westward, summit inflation southward).

5/85 (10:5) "The eruption continued through May without significant changes from last month. The main lava channel has been transformed into a lava tube, between 2510 m and 2320 m elevation, that has at least four windows through which it was possible to see the lava flow. The lava has maintained a constant velocity during the last few weeks. On 23 May another lava overflow occurred as a result of roof collapse along the upper portions of the lava tube (2485 m elevation), generating small lava flows of brief duration.

"The numerous short-lived vents inside the lava field (from 2320 m to 2150 m elevation) were variable, as usual, in number and position. The lava flows that originated from these short-lived vents have increased the size of the lava field on both the W and E sides to a maximum width of about 1.5 km.

"Beginning 10 May the lava moved mainly toward the SW (Monte Rinatura and Monte Nero). Around the beginning of June there were numerous lava flows toward the SE (1910 craters area). Lava continued to flow toward the south (Monte Castellazzi area), but these flows were not strongly fed. None of the lava flows descended below 2000 m elevation.

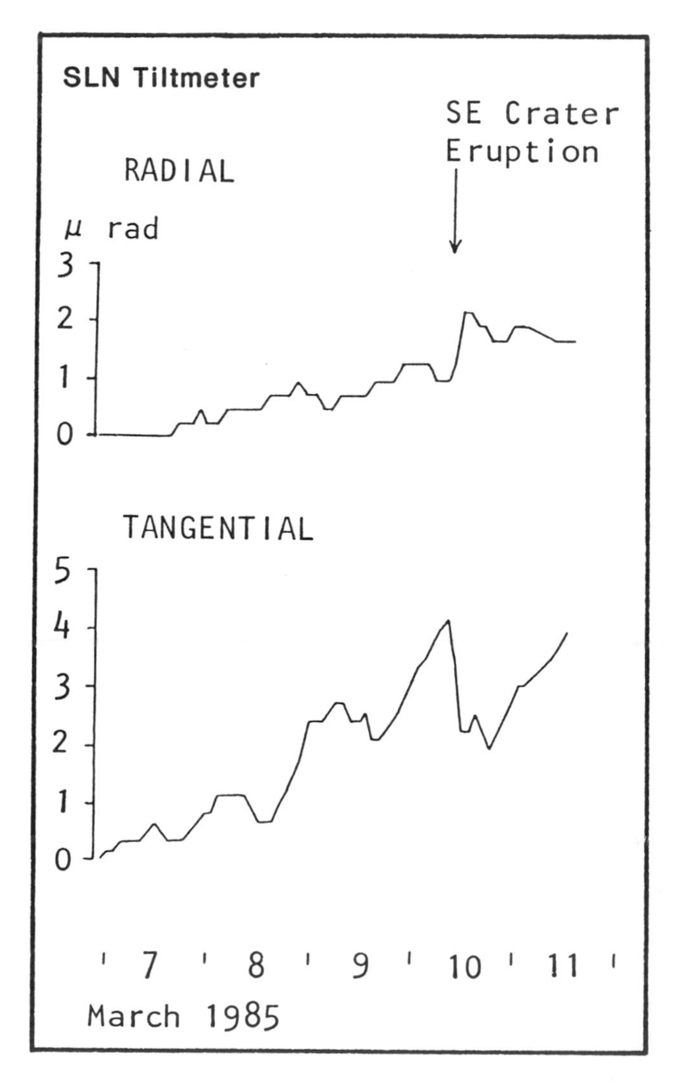

Fig. 1-35. Small changes in tilt accompanying the Southeast Crater eruption, as recorded by the SLN tiltmeter, 7-11 March.

"The more or less intense emission of gas and vapor from both vents of the central crater and from the Northeast Crater continue. Emissions of ash are rare and inconsistent. Gas under pressure emerged from a small opening at the southern base of the 1984 cone inside the Southeast Crater (P. Briole, personal communication). R. Clocchiatti conducted temperature measurements with a thermocouple; the temperature ranged between 1053°C and 1088°C (CEN-SACLAY).

"M. Cosentino and G. Lombardo reported that no particular seismic activity was recorded. Instead, an increase in the average amplitude of harmonic tremor was noticed, probably connected to the degassing of the central and Northeast craters.

"The Etna guides and rescue volunteers from the Italian Alpine Club (A. Cristaudo and A. Nicotra) helped with information on the activity."

Information Contact: R. Romano, IIV, Catania.

6/85 (10:6) "After a brief period of quiescence from 11-13 June, the eruption continued. Only one effusive vent remained active (2485 m elevation) with major lava flows originating from it. From 9 to 11 July, following a brief increase in activity, four effusive vents were present at high elevation (2490-2480 m). The velocity of the lava produced in this period ranged from 20 to 7 cm/sec, volume 2.5 to 0.5 m^3/sec.

"The directions of the lava flows were mainly toward the S, SSE, and SE. In the last few weeks, lava flows of significant size were also present, and were directed toward the SSW, S, and SW. The short-lived vents were still present and numerous, scattered on the lava field from 2350 to 2250 m elevation. As usual, they varied in number and location. Recently, the lava field enlarged on the E side, but the lava flows have not descended below 2100 m elevation. The emission of gas continued and was sustained from the central craters, and the Southeast Crater. At times, expulsion of generally reddish ash from Bocca Nuova was noted. From the Northeast Crater, violent expulsions of old material (28 June), and emission of reddish ash (10 July) were observed.

"The temperature of the lava flow at the main effusive vent was around 1080°C (P. Briole). During this period seismicity remained at low levels. On 12 June at 1848, a seismic event (magnitude 2.5) with an epicenter in the lower E flank (14 km depth) was recorded. On 7 July, three events of low magnitude (<2.8) with epicenters on both the E and W flanks were also recorded.

"On 9 July at 0430, the amplitude of harmonic tremor increased abruptly, coinciding with the increase in effusive activity. During the following 24 hours, the amplitude of the harmonic tremor returned to normal levels (communications with M. Cosentino and G. Lombardo). The Etna guides and rescue volunteers from the Italian Alpine Club have cooperated in the collection of information."

Information Contact: R. Romano, IIV, Catania.

7/85 (10:7) "The eruption ended 13 July. The first phase (12 March-11 June) lasted 92 days, the second phase (13 June-13 July) lasted 31 days.

"The effusive activity that had shown new strength

during the last days of the eruption greatly diminished on 12 July. Only one main effusive vent at 2490 m elevation and 2 short-lived effusive vents in the lava field (at 2300 m and 2600 m) remained active. The flows from the upper vent never descended below 2400 m elevation and those from the lava field vents never below 2100 m. From preliminary estimates, the area covered by lava was about 2.2 km^2 and volume of lava produced was about 30×10^6 m^3. The maximum width of the lava field is 1.5 km, the maximum length of flows, 1830 m (fig. 1-36).

"Bocca Nuova had a relatively sustained gas emission. Inconsistent expulsions of reddish ash were also observed. This activity was related to explosive activity on the floor of the vent (about 300 m below the rim), and was limited to the collapse of the vent's internal walls. The Chasm produced only weak gas emission. Collapses of the walls of the Northeast Crater were observed, and starting 22 July, this crater was partially obstructed, with emission of gas under pressure.

"Except for the isolated seismic events felt in the upper part of the volcano on 19 and 20 July (magnitude 2.8) and 22 July (magnitude 3), the seismicity remained at low levels."

Information Contact: R. Romano, IIV, Catania.

12/85 (10:12) *Eruptive Activity.* (Romolo Romano) "Strombolian activity started from the Southeast Crater on 19 Dec. The activity occurred at irregular intervals, becoming increasingly intense and continuous in successive days. Since the end of the 12 March – 13 July 1985 eruption, more or less intense Strombolian activity had been observed from several explosive vents (variable in number and position) on the floor of Bocca Nuova at variable depths (from 100 to 300 m or more from the rim of the vent).

"On 25 Dec at 0340 an eruptive fissure opened on the W side of the Valle del Bove (on Etna's SE flank), beginning at 2750 m above sea level. The opening of this E-W-oriented fissure was preceded by a seismic crisis (see below). Strombolian activity soon started along this fissure from at least 3 explosive vents, while a lava flow began from the lower end of the fissure, covering, in a period of 18 hrs, a distance of 1.5 km. The lava flow stopped at a point NE of Monte Centenari, at about 1700 m elevation, within the Valle del Bove. Eruptive activity from the fissure stopped early the next morning. Activity resumed from the same place early 28 Dec. This second eruptive phase was characterized by weak Strombolian activity from the 3 explosive vents, and created small spatter cones. The effusive activity decreased; a very viscous lava flow moved about 300 m from the origin, branching at about 2600 m elevation. The eruptive phase ceased during the early morning of 31 Dec. A violent expulsion of ash and lapilli from the summit craters (E and W vents of the central crater, Southeast Crater, and Northeast Crater) during the first hours of the eruption was succeeded by more or less continuous and consistent emission of reddish ash.

"The Etna Guides (S. Carbonaro, O. Consoli, A. Mazzaglia, A. Nicotizza, and volunteers of the Alpine Rescue Team of the Italian Alpine Club (A. Cristaudo) have collaborated in collecting information about the eruptive activity."

Seismic Activity. (M. Cosentino, M. DiFrancesco, and E. Lombardo) "A seismic crisis began during the early morning of 25 Dec with shocks located mainly between Piano Provenzana (NE flank) and the Valle del Bove. The shocks were very shallow (2 km or less). At the same time, the amplitude of harmonic tremor increased sharply. The shock that destroyed the hotel Le Betulle at Piano Provenzana, killing one person and injuring seven others, occurred on 25 Dec at 0338 and had a magnitude of 3.5. During the following 48 hrs, about 200 more tremors with magnitudes of 1-4 were recorded. The strongest, M 4, occurred on 26 Dec at 0334, with its epicenter at Piano Provenzana. Focal depths were 2-3 km. The area of the epicenters remained between Piano Provenzana and Valle del Bove. Beginning 27 Dec, seismicity decreased in frequency, energy, and number of events, and stabilized to values of about 5-6 shocks per day. Their location was mainly in the eruptive area and on the E and W flanks (especially in early Jan). Similar activity was continuing on 10 Jan."

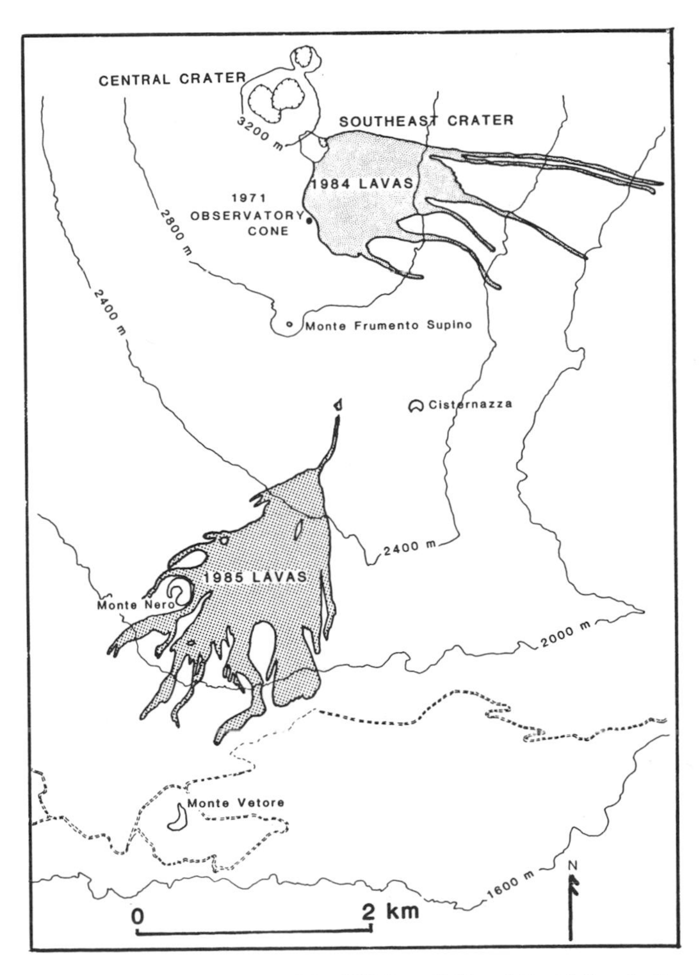

Fig. 1-36. Preliminary map of the 1985 lava field and its relation to the 1984 lavas. Data, provided by Romolo Romano, is provisional and subject to revision. [This map originally appeared in 10:8, with a brief summary of the report from 10:7.]

Newspapers reported that continuing seismicity included an event on 12 Jan at 0037, centered in the Zafferana Etnea area (11 km SE of the summit), that reached MM 6 and damaged some buildings. At least 3-4 minor shocks (one of MM 3) were felt the previous day.

Ground Deformation. (G. Nunnari and R. Velardita) "During the second half of Dec, the tilt stations (Pizzi Deneri, NE flank, elevation 2850 m; and Serra Pizzuta Calvarina, S flank, elevation 1650 m) recorded a progressive inflation of the upper E flank. At the Pizzi Deneri station the variation was 13 microradians (μrad) radially and 22 μrad tangentially between 20 Dec at 0900 and 23 Dec at 0900. The same stations recorded a variation of 45 μrad radially and 42 μrad tangentially between 24 Dec at 0900 and 25 Dec at 0900. The deformation produced during the first hours of 26 Dec reached a level that remained substantially unchanged as of 10 Jan."

Information Contacts: R. Romano, G. Nunnari, R. Velardita, IIV, Catania; M. Cosentino, M. DiFrancesco, G. Lombardo, Ist. di Scienze della Terra, Catania; La Stampa, Torino.

AFRICA

ERTA ALE

N Ethiopia
13.603°N, 40.673°E, summit elev. 613 m

Erta Ale is a shield volcano in the Afar depression, the southerly tectonic extension of the Red Sea rift and one of the few places where oceanic rifting processes can be observed on land. This remote and difficult area is rarely visited, but an active lava lake has been known in Erta Ale's summit crater since at least 1967. SEAN reported no direct observations during the 1975-85 period, but *Bulletin* 12:8 included satellite confirmation of continuing activity and measurement of surface temperatures to 1150°C from the lava lake in January, 1986.

Reference: Rothery, D.A., Francis, P.W., and Wood C.A., 1988, Volcano Monitoring Using Short Wavelength Infrared Data from Satellites; *JGR,* v. 93, p. 7993-8008.

ARDOUKOBA

Ardoukoba Rift, SE of Lake Assal, Djibouti
11.61°N, 42.47°E, elev. -70 m
Local time = GMT + 3 hrs

Djibouti is near the intersection of rift systems forming the Red Sea and the Gulf of Aden (fig. 2-1).

11/78 (3:11) The crew of a French observation aircraft saw an eruption early 8 Nov in a virtually uninhabited area SE of Lake Assal. On the preceding day a series of weak earthquakes was felt in the city of Djibouti and 2 larger shallow events were located by the U. S. Geological Survey's National Earthquake Information Center (USGS/NEIC). The first, m_b 5.2, occurred on 7 Nov at 2006 (felt as a sharp shock in Djibouti) and the second, m_b 5.0, on 8 Nov at 0808. Hypocenters calculated by the USGS for these events are 8 and 17 km WSW of the eruption, or well within the location error for events in this region.

Two basaltic lava flows were reportedly extruded: one flow traveled about 1 km to the SE, the other about 0.5 km to the NW. Lava effusion rates reached an estimated maximum of 1000 m^3/minute [but see 0.5 x 10^6 m^3/hour in 2:12]. Ash clouds rose about 300 m and larger pyroclastics about 70 m. The eruption formed a crater about 30 m in diameter and built a cone about 100 m high.

Activity began to decline 14 Nov and within 2 days only steaming was visible. As of 22 Nov, some vapor emission continued from subsidiary vents. No casualties or damage were reported.

Although no previous eruptions have been reported in the area in historic time, the most recent lava flows are substantially younger than sediments ^{14}C dated at 5300 BP, and are thought to be younger than 3000 BP. Eruptions are also mentioned in local legends (Delibrias et al., 1975).

Reference: Delibrias, G., Marinelli, G., and Stieltjes, L., 1975, Spreading Rate of the Asal Rift *in* Pilger, A. and Rosler, A. (eds.), *Afar Depression of Ethiopia; Inter-Union Commission on Geodynamics Scientific Report No. 14;* E. Schweitzerbart'sche Verlagsbuchhandlung, Stuttgart, p. 214-221

Information Contacts: W. Clarke, U.S. Embassy, Djibouti; M. Krafft, Cernay, France; USGS/NEIC.

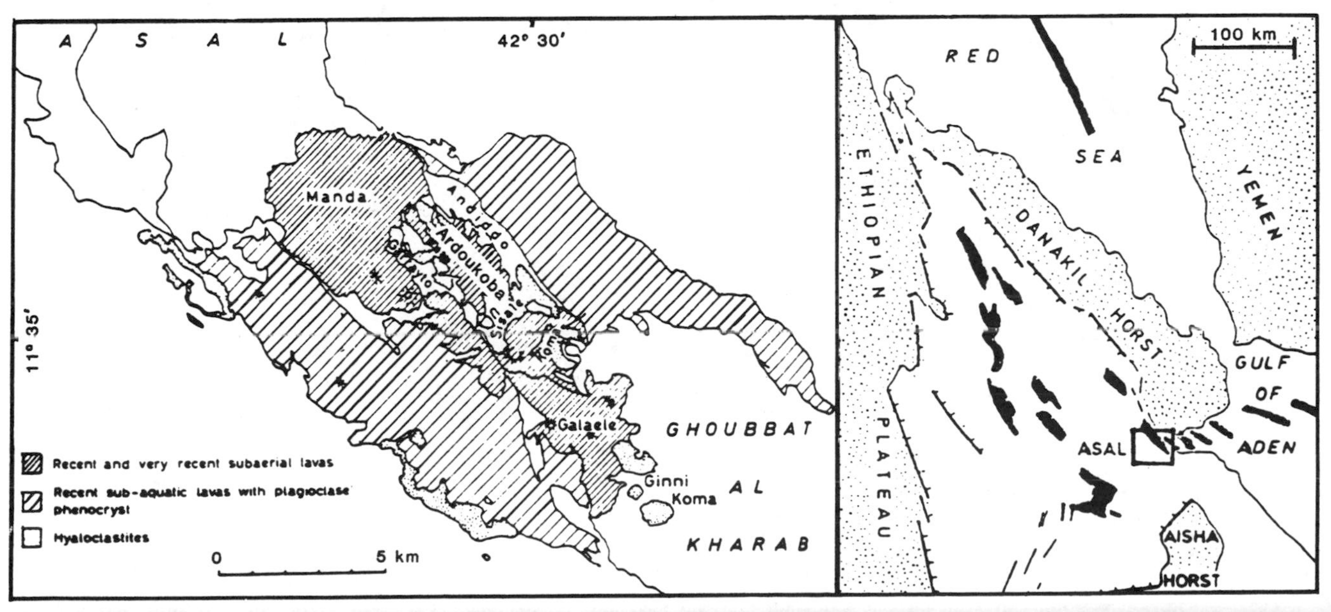

Fig. 2-1. Map of the Ardoukoba Rift area, *left,* indicated by index map, *right,* showing rock types and volcanic centers. Latitudes and longitudes were added by SEAN and are approximate. Both maps after Delibrias et al., 1975.

12/78 (3:12) The eruption of Ardoukoba began on 7 Nov. During the preceding 24 hrs, about 800 earthquakes with M < 3.3 had been recorded. These events occurred at the end of the Ghoubbat al Kharab (fig. 2-2), just S of the rift and about 6 km SE of the eruption site.

After a brief period of gas emission, lava fountaining began from a 500 m-long, newly opened fissure. Three spatter cones formed on this fissure, but the primary activity soon shifted about 1/2 km to the SE. For the first 2-4 hrs as much as 0.5 x 10^6 m^3 of lava was extruded/hr, but the extrusion rate declined rapidly and by the 5th or 6th day had decreased by an order of magnitude.

By 9 Nov, only the main vent was still active. A cone (named Gira-le-Koma) about 30 m high, 200 m long, and 25 m wide, with a 60° external slope, had formed around this vent, which contained a small lava lake. On 13 Nov the level of the lava lake dropped rapidly, then the lake disappeared. Scoria and bombs were ejected briefly before the eruption ended on 14 Nov.

The total volume of lava emitted was estimated to be at least 12 x 10^6 m^3. Lava covered an area of slightly more than 3 km^2. Flow thicknesses reached 25 m but usually ranged from 1 to 2 m. About 25 fissures, parallel to the NW-SE trend of the rift, opened during the activity. Most fissures were from 1-2 mm to about 1 m wide, but there was one short segment of 3 m width. Many of the fissures also showed a component of normal movement, with scarps up to 1/2 m high.

Reference: Needham, H.D. et al., 1976, The Accreting Plate Boundary: Ardoukoba Rift (Northeast Africa) and the Oceanic Rift System; *Earth and Planetary Science Letters,* v. 28, p. 439-453.

Information Contact: M. Krafft, Cernay.

Further References: Abdallah, A. et al., 1979, Relevance of Afar Seismicity and Volcanism to the Mechanics of Accreting Plate Boundaries; *Nature,* v. 282, p. 17-23.

Demange, J., and Tazieff, H., 1978, The "Tectonic" Eruption of the Ardoukoba (Djibouti); *C.R. Acad. Sci. Paris, Ser. D,* v. 287, p. 1269-1272.

Hernandez, J. and Ruegg, J.C., (eds.), 1980, Colloque Rift d'Asal: Réunion Extraordinaire de la Société Géologique de France; *Bulletin de la Société Géologique*

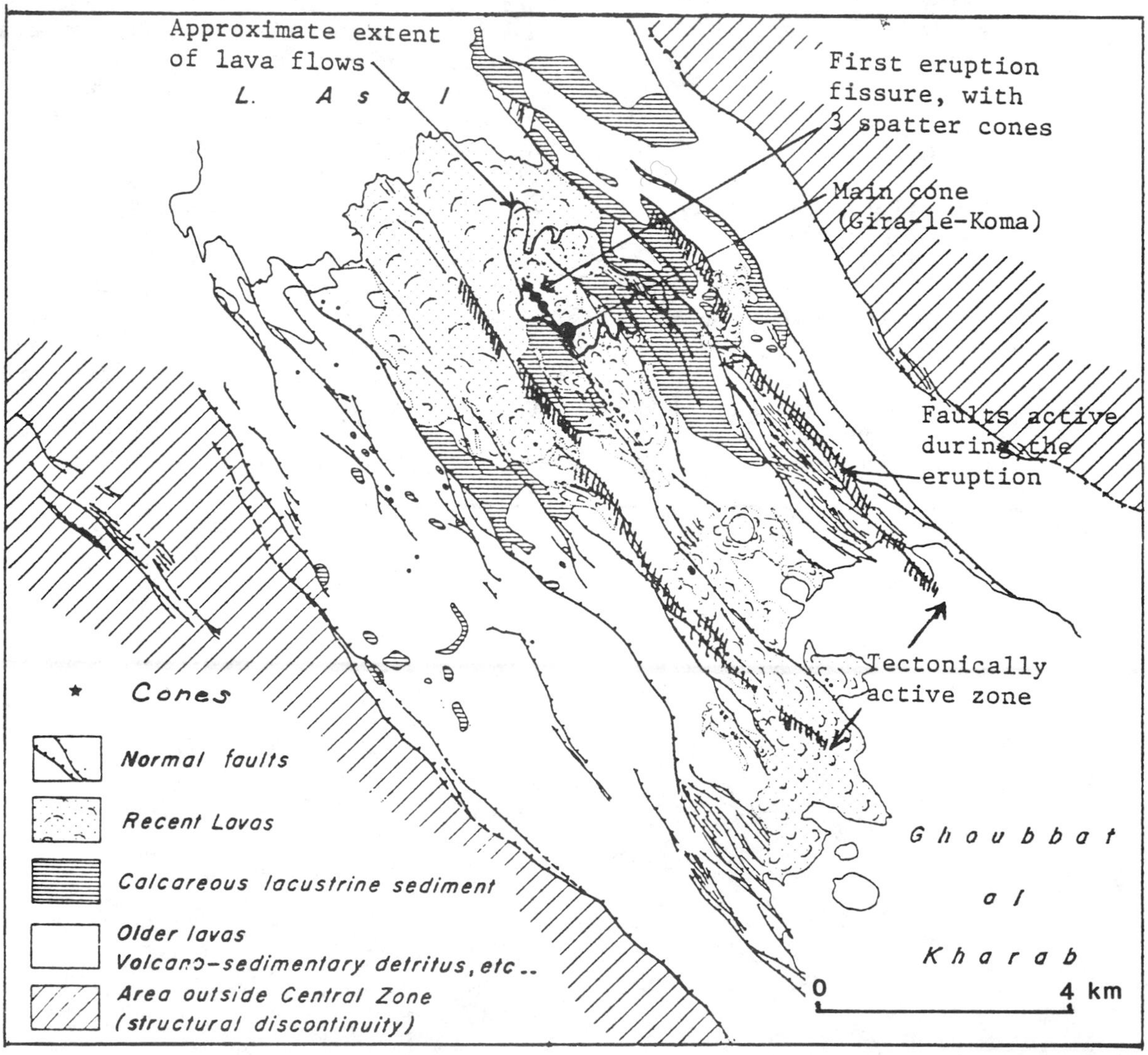

Fig. 2-2. Rock types, faults, and old volcanic cones in the Ardoukoba Rift, after Needham et al., 1976. An overprint, from M. Krafft, shows the active vents and lava flows from the 7-14 Nov eruption and the zone of active faulting accompanying the eruption.

de France, serie 7, t. XXII, no. 6, p. 797-1013.

Tarantola, A., Ruegg, J.C., and Lepine, J.C., 1979, Geodetic Evidence for Rifting in Afar: A Brittle-Elastic Model for the Behavior of the Lithosphere; *Earth and Planetary Science Letters,* v. 45, p. 435-444.

OL DOINYO LENGAI

N Tanzania
2.751°S, 35.902°E, summit elev. 2886 m
Local time = GMT + 3 hrs

190 km SSW of Nairobi and 40 km NE of Olduvai Gorge, this strato-volcano is famous for the unusually high proportion of sodium carbonate (washing soda) in its lavas.

General References: Dawson, J. B., 1962, The Geology of Oldoinyo Lengai; *BV,* v. 24, p. 349-387.

Nyamweru, C., 1980, Rifts and Volcanoes; Nelson Africa, Nairobi, 128 pp.

1/83 (8:1) On 6 Jan at first light (about 0600), Philip Sanders observed a cloud of vapor and fine tephra, not present the previous day, emerging from the volcano. During 12 hours of observation, the size of the plume gradually increased, but there were no audible explosions. Clouds obscured the summit the next morning and no additional observations are available.

Information Contact: J. Sanders, Columbia Univ.

2/83 (8:2) Satellite images and reports from the ground indicated that the activity continued through early March with varying intensity. Imagery from the NOAA 7 polar orbiter showed a small plume over the crater on most days. Although larger than most observed in Feb, a plume on the 9th appeared to be at low altitude and extended less than 10 km NNE. On 2 March, a plume was visible to 80-100 km E of the volcano. Low-level winds were blowing in the opposite direction, but those at 6-7.5 km were moving toward the E at about 10 km per hour. Dennis Haller noted, however, that weather stations in the area are widely separated and that low-velocity winds are often quite variable in direction. On 8 March, the plume extended less than 50 km NE, again in the wrong direction for the low-altitude winds observed elsewhere in the region.

Tanzania National Park Service officials in Arusha (roughly 120 km SE of the volcano) reported that during activity on 2 Feb, an eruption column and incandescence could be seen 150 km away. A second eruption occurred 19 Feb, and as of early March activity was said to be occasional. The active vent was described as "inside the cone." Most eruptions in this century have been from the N crater, which contained a tiny hornito when Maurice Krafft climbed the volcano in 1980. Evidence of flowing lava along the road nearest the volcano was also reported. A passenger aboard a commercial airliner (date of flight unknown) saw lava flowing from the back of the crater and bubbling inside. Peter Swan and Peter Jones reported fine ash around Olduvai (about 70 km from the volcano) 14 Feb. Jones noted that visibility was limited and there was a constant odor of volcanic fume. From Magadi, Kenya (130 km N of Ol Doinyo Lengai), P. R. Ellis saw small plumes rising from the volcano's W side at the beginning of March and noted that there had been several other observations of activity from the Magadi area.

Ol Doinyo Lengai's most recent reported eruption was Aug-Sept 1974, when it ejected tephra from the N crater.

Information Contacts: D. Haller, National Oceanographic and Atmospheric Administration/ National Environmental Satellite Data and Information Service (NOAA/NESDIS); Natl. Park Service, Tanzania; D. Miller, U.S. Embassy, Dar es Salaam; P. Swan, Arusha; P. Jones, Ngorongoro; P. Ellis, Magadi, Kenya; M. Krafft, Cernay; R. Hay, Univ. of Calif., Berkeley.

3/83 (8:3) Marie Benson reported that Ol Doinyo Lengai erupted 1 and 9 Jan, and 17 and 20 Feb but not since then. During the 20 Feb eruption, lava was flowing toward Lake Natron (N of the volcano). Andrew Stirrat, and the Outward Bound expedition of which he was co-leader, were N and W of the volcano 13-16 March. They saw small gray plumes emerging from the summit area on 13-14 March, and the morning of the 15th but none later that day or on the 16th. Clear views of the summit were obtained during the evening of 14 March and throughout the night of 15-16 March but no glow was evident. An apparent thin layer of ash on the upper N flank was visible through binoculars but no fresh ash was recognized in the area visited by the expedition, which approached to within 2.5 km of the foot of the volcano, nor did they observe any new lava. A local guide who was near the volcano 7-10 March reported that the air was hazy and smelly and caused chemical burns on his arms after 6 hours of exposure, the only time he had encountered such problems in the previous 3 months. Madelon Kelly and others visited the area a few km E and N of the volcano 15 March but observed neither lava nor any apparent eruptive activity. Their photographs showed no plumes.

Information Contacts: M. Benson, USAID, Arusha; D. Miller, U.S. Embassy, Dar es Salaam; A. Stirrat, Outward Bound; M. Kelly, Tyrone, PA.

10/84 (9:10) On 22 Jan, climbers observed that the S half of the main crater was about half-filled with talus. South of the center of the main crater, but beyond the margin of the talus pile were 2 small vents that were "burping" at intervals of about 30 seconds to 1 minute. Other fumaroles were observed just below the N rim and within the talus. Fumaroles were still visible on the north rim of the crater in Sept.

Information Contact: G. Lewis, Arusha.

Further Reference: Nyamweru, C., Activity of Ol Doinyo Lengai, Tanzania, 1983-1986; *Journal of African Earth Sciences,* in press.

NYAMURAGIRA

north of Lake Kivu, Zaire
1.42°S, 29.20°E, summit elev. 3053 m
Local time = GMT + 2 hrs

Nyamuragira is a shield volcano in the Virunga Mountains of eastern Zaire (fig. 2-7). Flank radial fissure activity has been common in the 30 eruptions known since 1882.

General Reference: Hamaguchi, H., (ed.), 1983, Volcanoes Nyiragongo and Nyamuragira: Geophysical Aspects; Tohoku University Faculty of Sciences, Sendai, Japan, 130 pp.

12/76 (1:15) An eruption from a new crater on the SW flank of Nyamuragira began on 23 Dec. The eruption was not believed to present any danger to life or property.

Information Contacts: Agence France Presse (AFP); Brussels Domestic News Service (DNS).

2/77 (2:2) The eruption of Nyamuragira began at about 1540 ["about 1200" in 2:3] on 23 Dec, from a new flank crater 8 km SW of the summit. Lava fountains 150-200 m high formed a 70-m spatter cone around the vent. Two lava flows were extruded, one that extended about 3 km to the SW, the other about 1 km to the W. The eruption was continuing as of 31 Dec.

More than 10 times the normal number of volcanic earthquakes were recorded on 11 Dec from Lwiro (about 100 km SSW of the volcano), and significantly more than the normal number on 12, 21, and 22 Dec. The March-May, 1971 eruption of Nyamuragira took place at Rugarama, on the NW flank (Pouclet and Villeneuve, 1972) rather than the SW flank or the summit crater as reported elsewhere.

Information Contact: S. Ueki, Institut de Recherche Scientifiques Afrique Centrale (IRS).

Reference: Pouclet, A. and Villeneuve, M., 1972, L'Eruption du Rugarama (Mars-Mai, 1971) au Volcan Nyamuragira; *Bulletin Volcanologique (BV),* v. 36, p. 200-221.

3/77 (2:3) At about 1200 on 23 Dec, an eruption began from the NE end of a 1 km-long, N45°E-trending fissure on the SW flank of Nyamuragira. Explosive activity was concentrated at 3 locations along the fissure, building 3 small cinder cones. The cones were breached to the W and a little lava was extruded in that direction (flow 1, fig. 2-7). The next night, activity shifted to the SW end of the fissure and a new pyroclastic cone (Murara) began to grow. Cinders and lapilli fell within a 1 km radius of Murara, and a lava flow (flow 2, fig. 2-7) moved southward from a breach in the SW wall of the crater. Extrusion of flows 3 and 4 had begun by 6 and 10 Jan 1977 respectively.

On 18 Jan, Murara was 150 m high, and a flow rate of 10 m/minute was measured 150 m from the vent. The flow rate had slowed to 2-4 m/minute by the next day, and the lava had nearly solidified by evening. Explosions increased in intensity, sending ejecta 500-600 m high every 2 seconds. Lava effusion had resumed from the N foot of Murara by 27 Jan. On 28 Jan, 6 subsidiary craters opened within 200 m of the N foot vent. Lava extruded from these craters rapidly coalesced into a single flow; its front was 6 km from Murara on the 28th. A 60 m-diameter lava lake was observed on 10 Feb in Murara's crater, and small pahoehoe flows were issuing from the SW foot.

Information Contacts: Y. Pottier, IRS; M. Krafft, Cernay; H. Tazieff, CNRS.

Further Reference: Brousse, R., Cochemé, J.J., Pottier, Y., and Vellutini, P.J., 1979, Eruption and Lavas of Murara: New Volcano (December 1976-April 1977) in Kivu (Zaire); *C.R. Acad. Sci. Paris,* serie D, v. 289, p. 809-812.

1/80 (5:1) The Agence Zairoise de Presse (AZAP) news agency reports that an eruption from the N side of Nyamuragira began at 2100 on 30 Jan. As of 1 Feb lava was moving northward toward the agricultural community of Mweso at what was described as a "quite rapid" rate. A "considerable" distance and area had been covered by the lava.

Information Contact: AZAP.

2/80 (5:2) Citing an information bulletin sent to regional authorities by scientists at IRS, AZAP news agency reported that extrusion of a new lava flow began at 1423 on 11 Feb from the N flank vent active since 30 Jan. By 20 Feb the new flow had advanced about 7 km and the extrusion rate had increased. Incandescent tephra and poisonous gas spread by violent winds killed livestock and vegetation, but no loss of human life has been reported.

Information Contacts: AZAP; IRS.

3/80 (5:3) The Lwiro seismological station, 95 km S of Nyamuragira, recorded the onset of volcanic explosions at 1330 on 30 Jan. A new fissure, trending N20°E, opened on the N flank of the volcano. During the first week of activity, lava was extruded from 150 m of the fissure, with fountain heights exceeding 100 m. Beginning 6 Feb, the zone of fountaining along the fissure was reduced in length and a new cone (named Gasenyi) began to form 3 km N of the summit at 2400-2500 m above sea level and adjacent to an older cone (fig. 2-3).

Lava initially flowed NE, covering an area 2 km long and 0.5 km wide, with a maximum thickness of about 1 m. The second (main) flow moved more than 7 km N, then turned in a northeasterly direction, over the 1958 Kitsimbanyi Flow, destroying a 13 km-long section of forest in Rwindi National Park. This flow reached a maximum width of 2.5 km, and was less than 0.5 m thick at its front. Three temporary seismic stations installed by the Geophysical Department of IRS began to record long-period microseisms at 1200 on 10 Feb. The next day at 1623, a new vent opened at the S end of the new cone, extruding a third lava flow that traveled over 7.5

km in a N70°E direction. This flow was about 1.5 m thick near the vent.

Explosions from the 2 vents were vigorous and often very noisy. Occasionally, both vents were active simultaneously, but explosions usually alternated between them, building a double cone with estimated dimensions of 100 m height and 200-300 m width by 17 Feb.

Katia Krafft flew over the eruption on 18 Feb. Fountains rose about 100 m from the 2 vents, which were within an active lava lake contained by the double cone. Extrusion of the small lava flows continued. Similar activity was continuing when she reached the volcano on the ground late 20 Feb. Red glow from the lava moving to the N was visible that night. Fountaining continued 21 Feb and aa lava was extruded slowly from the base of the N vent. More vigorous fountaining began that evening and a third vent was occasionally active. Lava was ejected to 120 m above the S vent. Discontinuous explosions occurred from the N vent, which overflowed during the night.

Microseismic activity recorded by IRS seismic stations declined sharply on 22 Feb. By 0300 on 23 Feb, lava extrusion had stopped and only a few feeble explosions were occurring from the vents. Only weak gas emission was observed the next day. Katia Krafft reports that both aa and pahoehoe flows were observed, and the lavas contained xenoliths of melted granite, pumice, and porcellanite, as in most flank eruptions of Nyamuragira.

Information Contacts: N. Zana, IRS; K. Krafft, Cernay; Y. Pottier, Univ. Paris-Sud.

Further Reference: Brousse, R., Caron, J.P., Kampunzu, A.B., et al., 1981, Eruption et Nature de la Lave du Gasenyi: un Nouveau Volcan (Janvier-Fevrier 1980) au Flanc Nord du Nyamulagira (Kivu, Zaire); *C.R. Acad. Sci. Paris,* serie II, v. 292, p. 1413-1416.

12/81 (6:12) A 400 m-long fissure opened before dawn 26 Dec in the saddle between Nyamuragira and Nyiragongo. The exact location of the fissure was not known at press time, but there are numerous old vents within the 14 km, NNW-trending rift zone between the 2 volcanoes. Activity began, probably at about 0130, with a strong explosion from the upper end of the fissure (at 2300 m altitude) that ejected a 3-4 km-high tephra cloud. After the initial explosion, lava was extruded from the lower end of the fissure and flowed N. Residents fled the area. Images from the NOAA 7 satellite showed a large hot area between the volcanoes at 0230 on 27 Dec that had not been present 24 hours earlier. By the afternoon of 30 Dec, the lava flow was 15-20 km long, but only 300-400 m wide. Lava fountaining was continuing and a cone was growing at the vent. Tephra ejection was also continuing, preventing aircraft from flying near the active fissure, although it was not as vigorous as during the initial explosion.

Shallow earthquakes occurred 18 Nov at 1118 and 1234 at 2.11°S, 22.83°E and 2.22°S, 22.52°E, about 725 km ESE of Nyamuragira in an area that is not normally seismically active. Kampala Domestic Service reported a felt earthquake in the Bushenyi district of Uganda, about 150 km NE of Nyamuragira, at 0300 on 30 Dec.

Information Contacts: M. Krafft, Cernay; M. Matson, NOAA; USGS/NEIS; Kampala Domestic Service, Uganda; WCBS Radio, USA.

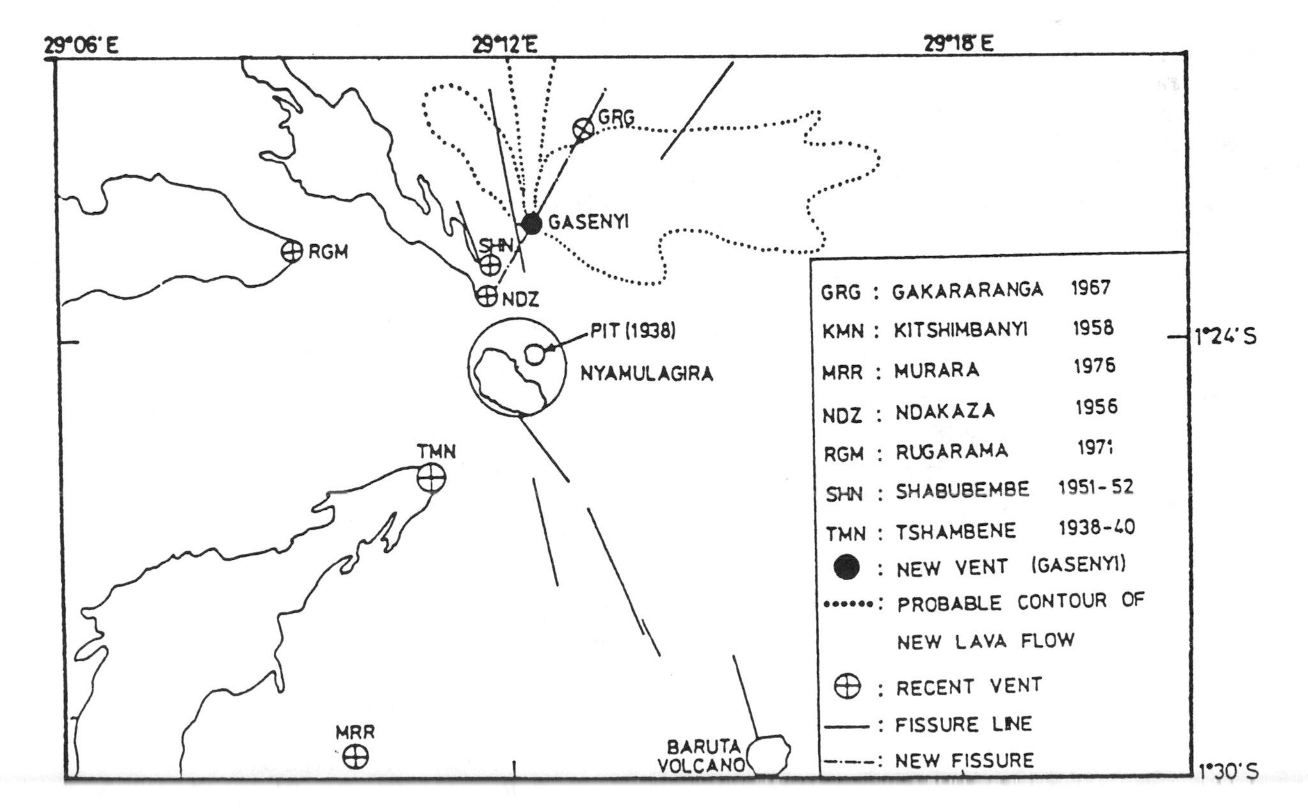

Fig. 2-3. Map of Nyamuragira and vicinity showing approximate outlines of major lava flows from the new 1977 cone. Courtesy of N. Zana.

1/82 (7:1) A fissure eruption from the SE foot of Nyamuragira started during the night of 25-26 Dec. Ground observers first noted activity about 0130 on 26 Dec, but satellite data indicated that the eruption may have begun as much as 2 hours earlier. According to the satellite data, lava flowed slightly N of E for 6-8 km, then turned more to the N. By 0100 on 26 Dec, the flow front appeared to be 12-14 km from the vents, but its advance had slowed substantially from the initial rate of 20-24 km/hr to less than 6 km/hr.

The source of the eruption was a 1.5-km-long zone of en echelon fissures, trending roughly N70°W at 2170 m altitude (fig. 2-4). Lava initially fountained from the entire length of the fissure zone, but activity was soon confined to its SE (and lowest) end. After a few days, 400 m-wide lava flows had advanced as much as 30 km, destroying forest in Virunga National Park. A plume of scoria and ash rose 6 km or slightly more. A large quantity of tephra, including Pele's hairs and tears, fell within 25 km of the vents, damaging vegetation and killing many animals (see below).

When Katia Krafft began field work at the eruption site 3 Jan, a pair of lava fountains, 100-200 m high, was emerging from a small lava lake in the asymmetrical cone that had formed over the SE end of the fissure zone. The E wall of the cone, near the dominant vents, was about 100 m high, but the W wall was only about 30 m high. An aa lava flow a few meters wide cascaded from a breach in one wall of the cone and moved downslope at about 20 km/hr, but within 50 m had widened to 10 m and slowed to 7 km/hr as it passed through a channel between 5 m-high blocks. A scoria column rose about 1 km. Fountaining became more vigorous about 2200 on 3 Jan, but had declined to its previous level by the next morning. A second increase in lava fountaining, to as much as 200 m height, began suddenly about 1300 on 4 Jan. Scoria up to 5 cm in diameter fell 3 km from the vent, as during the first week of the eruption. Explosions occurred every few seconds during the early afternoon. Shock waves from explosions occasionally shook tents 3 km away, but these could not be correlated with increases in lava fountain heights. Strombolian activity from a third vent, in the NE part of the cone, began abruptly during the early evening. Dark plumes of ash and scoria were occasionally ejected from this vent.

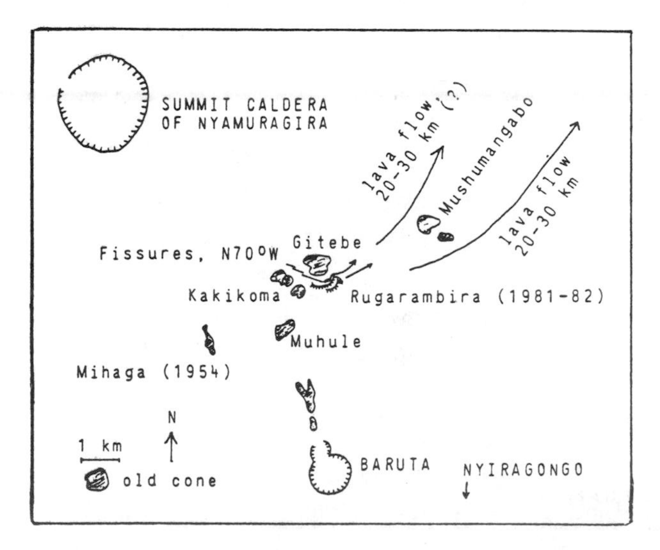

Fig. 2-4. Sketch map showing 1981-82 eruption fissures, directions of lava flows, and surrounding older cones. Courtesy of K. Krafft.

Lava fountaining was much weaker 5 Jan, but the rate of effusion of the lava flow remained constant that day, then increased during the evening of 6 Jan. The lava cascade through the breach in the crater wall widened to at least 10 m and lava flowed away at 20 km/hr, but a crust formed over the flow 50 m from the crater. Incandescent fissures appeared in the NE wall of the cone. During the next 24 hours a large quantity of lava was discharged, forming a mass of blocks in a steep 10 m-high crest at the edge of the flow. A small flow, 1-2 m thick, moved toward the E then turned NE for about 1 km; the lava formed slabs near the vent then changed to an aa flow downslope.

A team of Japanese seismologists measured strong seismic activity 6-9 Jan, then a decrease on the 10th. Residents of Goma, Zaire (30 km from the volcano) and Kigali (the capital of Rwanda, about 120 km away) felt a strong earthquake 9 Jan at 1927, but the event was not noticed by geologists near the volcano and was not recorded by the Worldwide Seismic Stations Network (WWSSN). The effusion rate and speed of the lava flow remained unchanged 9-11 Jan, but the height of the lava fountains decreased. The eruption ended during the night of 14-15 Jan.

Heavy tephra falls occurred N, E, and S of the cone. By the time the eruption ended, 1.5 m of ash and scoria had fallen within 1.5 km of the vents and deposits were > 0.5 m thick 3-4 km away; only a few cm of tephra fell W of the vents. All shrubs and trees in the area were stripped of leaves and bark. Many birds and rodents similar to small rabbits were killed in the tephra fall area. Within ~12 km of the cone, the eruption created a microclimate characterized by numerous violent thunderstorms with torrential rains and hail during Krafft's 8 days of observations.

The scoria was very vesicular and was not found in large fragments because it exploded upon striking the ground. A few bombs weighing 1-2 kg were thrown as far as 100 m from the cone during Krafft's field work. The lava was a basanite with numerous augite and olivine phenocrysts. Numerous xenoliths, coated with a thin glaze of lava, were ejected during the eruption, both by the explosive activity (in fragments 5-20 cm in diameter) and in the lava flows. The xenoliths were whitish to yellowish, and spongy, consisting primarily of büchite (altered quartzite) and porcellanite (altered argillaceous schist).

[Later analysis of Nimbus-7 satellite data by Krueger and associates revealed the emission of a large SO_2 plume at the start of the eruption in late 1981. This plume is thought to be the source of the early 1982 "Mystery Cloud" of stratospheric aerosols detected by atmospheric scientists before the much larger March-April 1982 El Chichon eruption (see Chapter 19).]

Information Contacts: K. Krafft, Cernay; N. Zana, IRS.

Further Reference: Kampunzu, A. B., Lubala, R. T., Brousse, R., et al., 1984, Sur l'Eruption du Nyamulagira de Decembre 1981 a Janvier 1982: Cone et Coulée du Rugarambiro (Kivu, Zaire); *BV,* v. 47, p. 79-105.

2/84 (9:2) A NW flank eruption began 23 Feb at 1013 and was continuing in early March. By the end of Feb, 2 lava flows had extended about 10 km.

Information Contacts: IRS; M. Krafft, Cernay.

3/84 (9:3) The NW flank fissure eruption ended during the evening of 14 March. Lava flows covered a large area and a substantial quantity of tephra was deposited near the vents.

An A-type volcanic earthquake at 0323 on 23 Feb was followed by volcanic tremor. A fissure trending N100°E began opening gradually from E to W at 1013. Lava issued from the entire fissure during the first day of the eruption, but activity soon concentrated at two vents about 400 m from the E end of the fissure and two others about 1.5 km to the W (A and C on fig. 2-5). About a week later, a new vent (B), about 500 m W of the E vents, began to emit lava, but at a lower rate than the other vents.

Lava extrusion was accompanied by explosions that could be heard 25-30 km away, along the W margin of the rift valley. About 2 m of scoria and many spindle bombs were deposited within 600-800 m S of the E vents. A bomb weighing about 12 kg was found 600 m away. N. Zana judged the volume of ejecta to be much more than in the 1976, 1980, or 1982 eruptions.

When Zana visited the eruption site 8-11 March, activity had ended at the W vents and was declining at the new vent. Both aa and pahoehoe were observed between the new vent and the W vents. Cones at the W vents stood about 80 m above the surface of the lava. At about 2300 on 10 March an aa lava flow from the E vents flooded the area of the new vent and carried away its small cone. This flow was still moving S about 0900 on 11 March. By 11 March the composite cone at the E vents had a basal diameter of 300 m and was about 250 m high, but activity was becoming intermittent.

Eruptive activity, including the explosions, ceased in the evening of 14 March. Night glow disappeared 16-17 March. Lava had flowed 20 km to the W and the lava field had an average width of 2.5 km (fig. 2-6). The new cones have been named Kivandimwe, meaning "things running together."

Information Contact: N. Zana, IRS.

NYIRAGONGO

Kivu Province, eastern Zaire
1.48°S, 29.23°E, summit elev. 3465 m
Local time = GMT + 2 hrs

The summit crater of this volcano, 14 km SE of Nyamuragira, was occupied by a famous lava lake for at least 42 years before the event described below. The tragic 1977 eruption was the shortest major eruption known to us (<1 hr), and its contrast to Stromboli's virtually continuous 2400-year record illustrates the huge variation in eruption durations.

General Reference: Tazieff, H., 1979, Nyiragongo, the Forbidden Volcano; J. F. Bernard (trans.); *Barron's*, New York, 287 pp. Originally published as Nyiragongo; *Flaminarion*, Paris, 1975.

1/77 (2:1) Nyiragongo volcano began erupting at about 1000 on 10 Jan, apparently from several flank craters. The eruption was primarily effusive, but a mushroom cloud was reported by the Brussels Domestic Service. Lava flows, mainly on the volcano's SE flank, are reported to have moved as much as 16 km from the volcano, engulfed 2 villages, and cut several roads. The eruption had probably ended by 11 Jan.

Earthquakes were felt in Bukavu, about 125 km SE of Nyiragongo, on 1 and 6 Jan. Most of the 65,000 residents of Goma, 17 km SW of Nyiragongo, fled prior to the eruption because of "incessant" earth tremors. By the night of 10-11 Jan, many were returning to their homes. The 23 Dec eruption of Nyamuragira, 14 km NW of Nyiragongo, had apparently ended. Estimates of casualties range from none to 2000, the latest (26 Jan) being 50-100.

Information Contacts: Kigali (Rwanda) DNS; Reuters; AZAP (Zaire); Brussels DNS; UPI; U.S. Dept. of State.

2/77 (2:2) Activity at Nyiragongo began in early Dec, 1976, when the lava lake rose and covered half of the first platform. Premonitory seismic activity was noted from Dec 11 (see Nyamuragira, 2:2). The eruption began from 5 fissure vents at about 2200 m elevation on the SE flank. Four lava flows were extruded, to the S, SE, SW, and NE. The latter flow blocked a major road, preventing supplies from reaching the area. About 100-150 hectares of land and 400 houses were destroyed by the flows. Estimates of deaths range from 38-60.

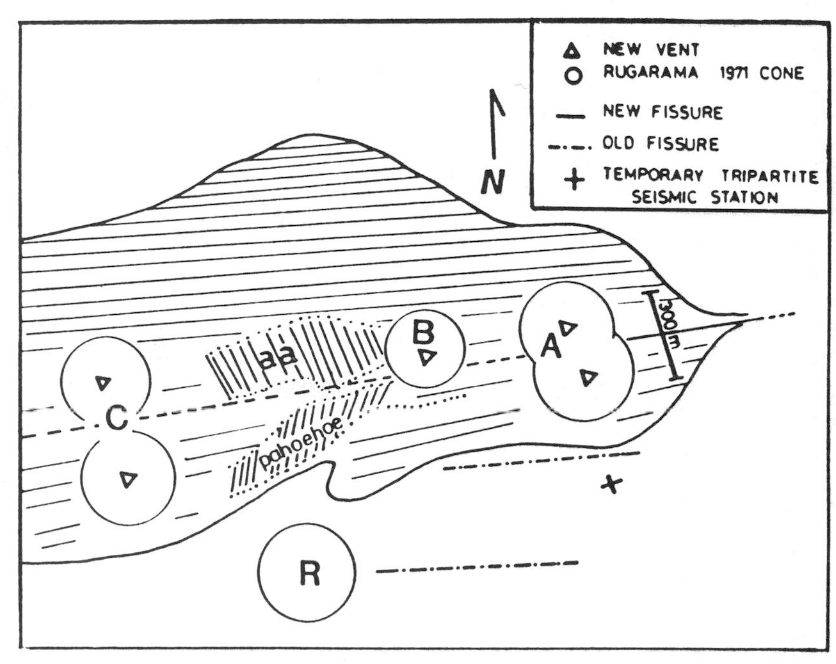

Fig. 2-5. Sketch map of part of Nyamuragira's NW flank, showing the relative positions of vents active in the 1984 eruption, and the location of the 1971 cone, S of the 1984 vents. Courtesy of N. Zana.

Information Contacts: U.S. Dept. of State; S. Ueki, IRS.

3/77 (2:3) At 1015 on 10 Jan, Nyiragongo's lava lake, present in the main crater since 1928, began to drain rapidly through a system of parallel fissures that opened simultaneously on the N (Baruta) and S (Shaheru and Djoga) flanks of the volcano (fig. 2-7). Within 1 hour, 20-22 x 10^6 m^3 of melilite nephelinite lava poured from the fissures. The flows, which moved downslope at up to

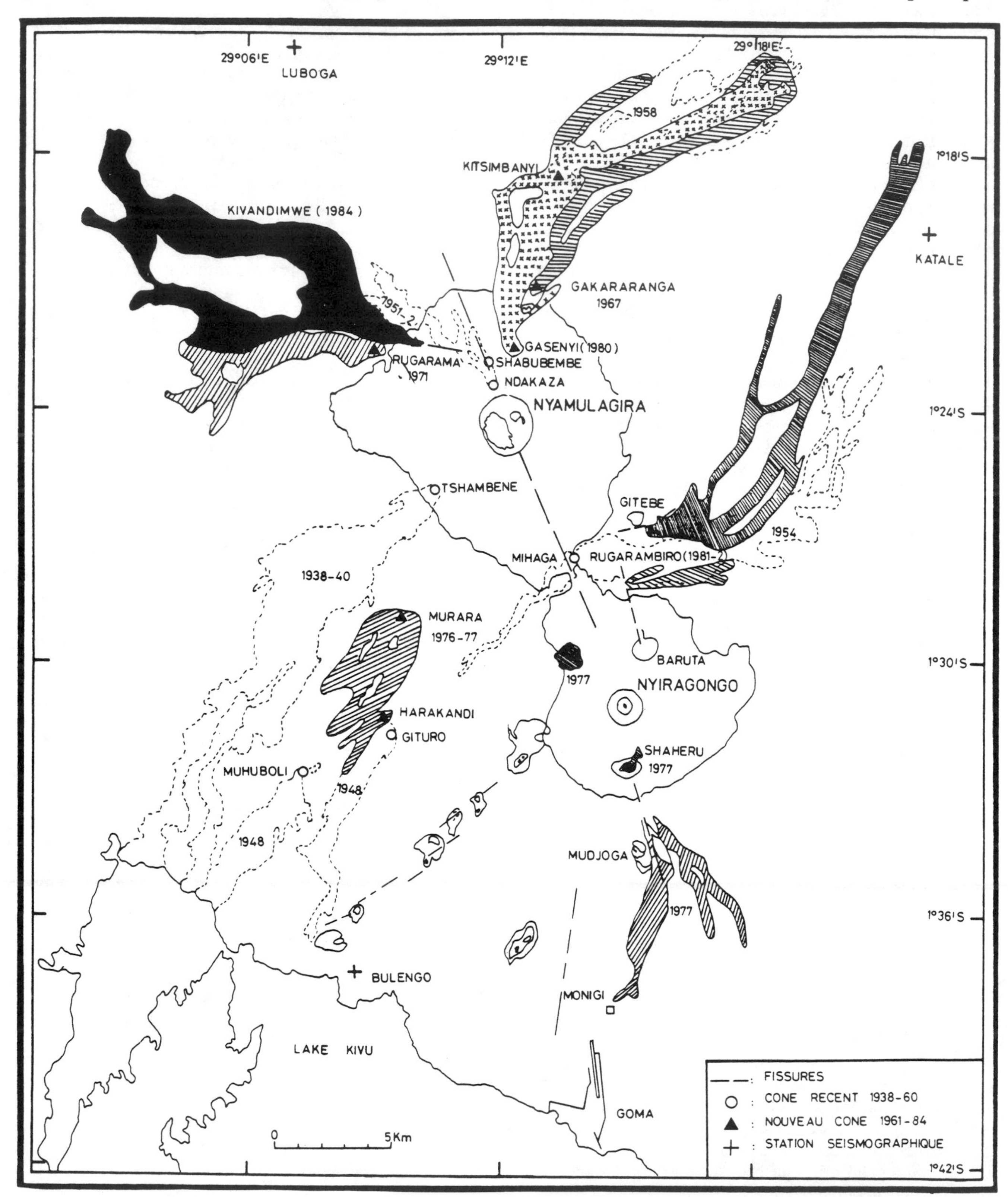

Fig. 2-6. Map of the Nyamuragira-Nyiragongo area showing recent vents and lava flows. The 1984 lava flow is shaded solid black. Courtesy of N. Zana.

60 km/hr, were pahoehoe type for most of their length, but changed to aa at their distal ends. Mean flow thickness was about 1 m, ranging from a few mm near the fissures to about 2 m at flow fronts (figs. 2-8 and 2-9). No spatter ramparts or hornitos were observed. About 70 persons were killed and 800 left homeless by the lava, which destroyed ~1200 hectares of agricultural land.

Five days before the eruption, the lava lake had risen to its highest recorded level, partially covering the terraces which had surrounded it. The rapid draining of the lava lake caused the partial collapse of the terraces, where groundwater flashed to steam, producing a dark gray cloud that rose more than 1 km before being blown towards Goma. Goma was plunged into semi-darkness and most of its residents fled to neighboring Rwanda.

Seven hours after the eruption began, lava effusion had completely ended, but fumarolic activity continued. At 1500 on 16 Jan, a strong gas eruption from the main crater projected a cloud containing little or no solid material to about 1 km above the volcano. At 1550, earthquakes felt on the crater rim preceded the collapse of what remained of the terraces, terminating major gas emission. Weak fumarolic activity persisted. The crater, with a bottom filled with debris, was about 1200 m in diameter and 1000 m deep after the 16 Jan events.

Information Contacts: H. Tazieff, CNRS; M. Krafft, Ensisheim; Y. Pottier, IRS.

Further References: Nakamura, Y. and Aoki, K., 1980, The 1977 Eruption of Nyiragongo Volcano, Eastern Africa, and Chemical Composition of the Ejecta; *Bulletin of the Volcanological Society of Japan*, v. 25, no. 2, p. 17-32.

Pottier, Y., 1978, Première Eruption Historique du Nyiragongo et Manifestations Adventives Simultanées du Volcan Nyamulagira (Chaine des Virunga–Kivu–Zaire: Dec. 76–Juin 77); *Musée Royal de l'Afrique Centrale (Tervuren, Belgium): Rapport Annual du Département de Géologie et de Minéralogie (1977)*, p. 157-175.

Tazieff, H., 1976/7, An Exceptional Eruption: Mt. Nyiragongo, January 10th, 1977; *BV*, v. 40, p. 189-200.

6/82 (7:6) Lava fountaining began 26 June in Nyiragongo's central crater, and, by 7, July a lava lake covered the crater floor. No activity had been reported at Nyiragongo since the lava lake drained on 10 Jan 1977. On 26 and 27 June, two 5-10 m-high lava fountains were observed at the bottom of the crater. By 30 June, only one fountain was active, feeding a very small lava pool. However, when a geologist climbed to the crater rim 7 July, a lava lake ~1/2 km across covered the crater floor. In the center of the lake was a domical lava fountain 30-50 m high and 150 m in diameter. The fountain and the rim of the lake were bright orange, a color similar to that seen during periods of vigorous fountaining before 1977. About 1/3 of the lake was covered by a fissured black skin. Based on comparisons with known pre-eruption features in the crater, the lake level was estimated to have risen 100-150 m between 26 June and 7 July. Most of the lava seemed to be entering the lake from below, but a very small amount of lava was emerging from a 10 m-diameter vent in the N wall of the crater, about 50 m above the lake surface. This vent had apparently been more active a few days earlier. From Goma, 17 km to the SSW, a glow was still visible over the crater 13 July. Earthquakes during the night of 4-5 July shook furniture and formed fissures in old houses in Goma.

Information Contact: M. Krafft, Ensisheim.

7/82 (7:7) The eruption began between 0400 and 0430 on 21 June when an explosion was heard. A vent on the NW wall of the crater, about 10 m above the floor in a mass of fallen rock (A in fig. 2-10), was fountaining lava to 50 m. Lava was flowing from the vent and forming a small lake in the crater bottom, which had been about 800 m below the rim before the eruption. By 1400 a high, wide, pine-tree-shaped column of white vapor was visible over the crater. The initial period of eruption was apparently phreatic, and was accompanied by continuous explosions.

When observers visited the crater on 26 June (fig. 2-10, top right), the lava lake was about 250 m wide; its surface was 730 m below the crater rim. The initial vent had been submerged, and the lava lake surface was domed 20 m high over it with a 10 m-high fountain in

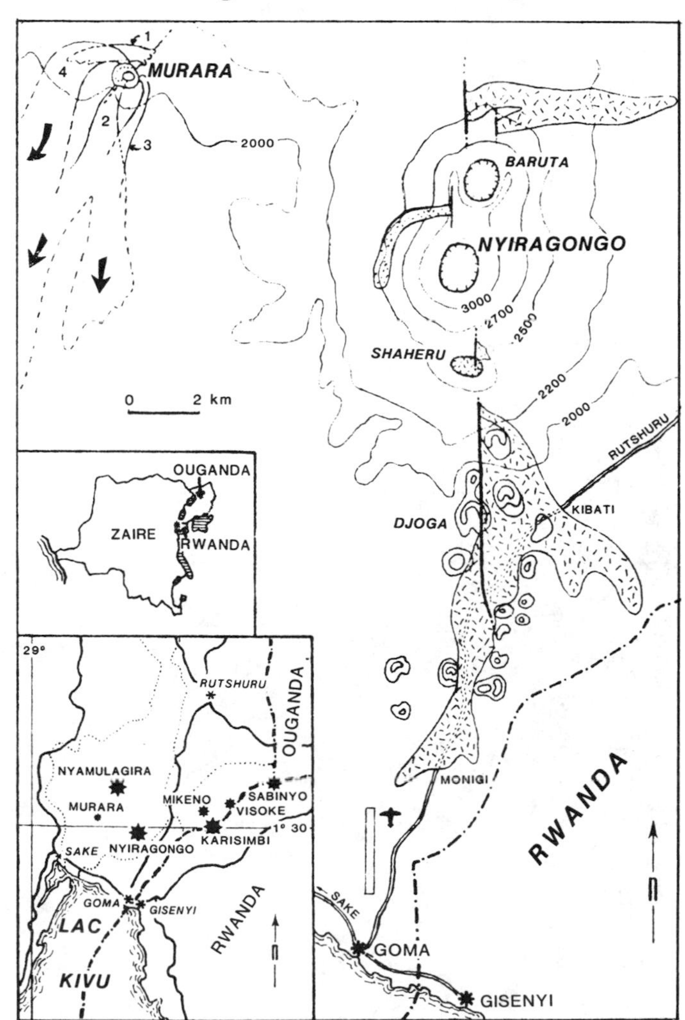

Fig. 2-7. Map of eruption sites and lava flows at Nyiragongo and Nyamuragira's Murara vent. The location of the stippled flow on Nyiragongo was provided by M. Krafft. Courtesy of Y. Pottier.

the dome's center. Two new vents were active (B and C, top right), one about 60 m above the lake level and N of the now-submerged vent, the other, bright red inside, 220 m above the lake level and NE of its companion. Both had formed hornitos, and were steaming vigorously and ejecting lumps of fluid lava. A 15 m-wide lava flow descended from the first new vent, a 60 m-wide flow from the second. The flow surfaces were chilled but lava was moving through the tubes into the lava lake. Narrow streams of fluid red lava were running over the surface of the second flow into the lake. The S half of the lake, already chilled, was covered by a fissured black crust. The N half had a moving, striped gray skin, the movements starting where the flow from the first new vent entered the lake.

The lake level continued to rise by several meters per day, and by 30 June it was 680 m below the crater rim and about 300 m wide. The bubbling dome of the submerged vent was not visible. All of the central and S parts of the lake were covered by a fissured black crust. The first new vent, now only a few meters above the lake level, was fountaining lava to 50 m and emitting a lava flow that entered the lake from the NE. The second new vent was 170 m above the lake level, still red inside and steaming strongly, but had no lava flow. The edge of the lava lake was molten, and bright orange at night.

The activity of the original vent had ceased by 3 July (fig. 2-10, bottom left). The first new vent had built a cone about 50 m high since it became active. Observers on 4 July found the first new vent submerged and forming a large domical lava upwelling about 100 m in diameter and 10-20 m high in the N-central part of the lake. The rest of the lake surface was covered with a fissured black crust. The second new vent was steaming and sometimes emitting yellow flames; it had no lava flow.

The lava lake surface lay 550 m below the crater rim and was 500 m wide on 7 July. The domical lava upwelling over the submerged first new vent was 160 m in diameter and 30-50 m high. Concentric fluid lava waves traveled from the dome's center to its edge; lava tongues overflowed to the E and S onto the $^{2}/_{3}$ of the lake that was covered by crust. The upwelling, the tongues, and a thin line around the edge of the lake were bright orange at night. The second new vent stood 15 m high and was 40 m above the lake level. It was still steaming strongly and was bright orange at night.

During the next 10 days the lake continued to rise. The lava upwelling over the submerged, first new vent flattened and narrowed to about 120 m in diameter. By 15 July the second new vent was only 10 m above the lake level and had breached to the S. A small lava flow from it entered the lake; lumps of fluid lava projected around its cone. By 17 July the lake level was 510 m below the crater rim. The second new vent was submerged and making a second lava upwelling in the lake,

Fig. 2-8. Nyiragongo (background) and the 10 Jan lava flow with destroyed vehicle (VW Combi) in foreground. Photograph by M. Krafft. [Originally from 2:4, without accompanying text.]

about 100 m in diameter with a central fountain 40 m in diameter and about 20 m high. This upwelling had connected with the one over the first new vent.

By 23 July the lake level appeared to have slowed its rise. It was 500 m below the crater rim and 600 m wide. The NW part was occupied by the 2 active lava upwellings (fig. 2-10, bottom right), the SW one 140 m in diameter with a 40 m-wide, 10-20 m-high bubbling in the center, the NE one 100 m in diameter with a 40 m-wide, 10-30 m-high bubbling in the center. The remaining 2/3 of the lake surface was covered by a fissured black crust. A line of moving red lava was visible along the W edge of the lake. Before the eruption, the crater bottom was about 300 m wide and covered with eroded lava blocks 30-40 m high. Lava had submerged all the blocks by 3 July, by 15 July the crater bottom was completely filled, and by 23 July the lake was 300 m deep. Maurice Krafft estimated the volume of lava emitted 21 June-23 July at 36 x 10^6 m^3. When the volcano's lava lake drained in 1977, 22 x 10^6 m^3 were emitted.

The eruption apparently was not preceded or accompanied by noticeable seismic activity. A seismic observation on 3 July showed continuous harmonic tremor, interpreted as lava rising in the conduit. Precursor events that were observed included fumarole activity in the crater and along the southern fissure—which had increased significantly since Jan—and an apparent 100-150 m uplift of the crater bottom.

Information Contacts: M. Krafft, Cernay; N. Zana, IRS.

10/82 (7:10) By 23 July, lava from several vents had formed a lake about 300 m deep with an estimated volume of 36 x 10^6 m^3. By mid-Sept, the volume of the lava lake had approximately doubled, but the eruption rate was declining slowly (table 2-1).

Geologists climbed to the crater rim on 31 July, 4 and 17 Aug, and 3 and 15 Sept. Between 24 and 31 July, the level of the lake surface rose about 20 m. On 4 Aug, however, the domical upwellings of lava that had marked the location of the 2 vents beneath the NW part of the lake on 23 July were not active. The entire surface of the lake was chilled and no glow was visible. On 17 Aug, an area of upwelling about 100 m in diameter and 10 m high was again active over one of the vents. A geologist descended to the surface of the lava lake 3 Sept and took samples. In hand specimen, these appeared to have a composition very similar to the melilite nephelinite of the 1928-1977 lake. A glowing red fissure that averaged about 30 cm wide and 20 cm deep ringed the lake about 2 m from its edge, emitting gases.

Dates		Eruption Rates: (x 10^6 m^3/day)
June	21-26	0.2
	27-30	0.6
July	1-7	2.5
	8-21	1.0
	21-31	0.5
Aug	1 - Sept 15	0.4

Table 2-1. Daily eruption rates at Nyiragongo calculated by M. Krafft over various time periods between 21 June and 15 Sept 1982.

Fig. 2-9. Jan 1977 Nyiragongo pahoehoe lava, with flow texture showing its extreme fluidity. Photograph by M. Krafft. [Originally from 2:4, without accompanying text.]

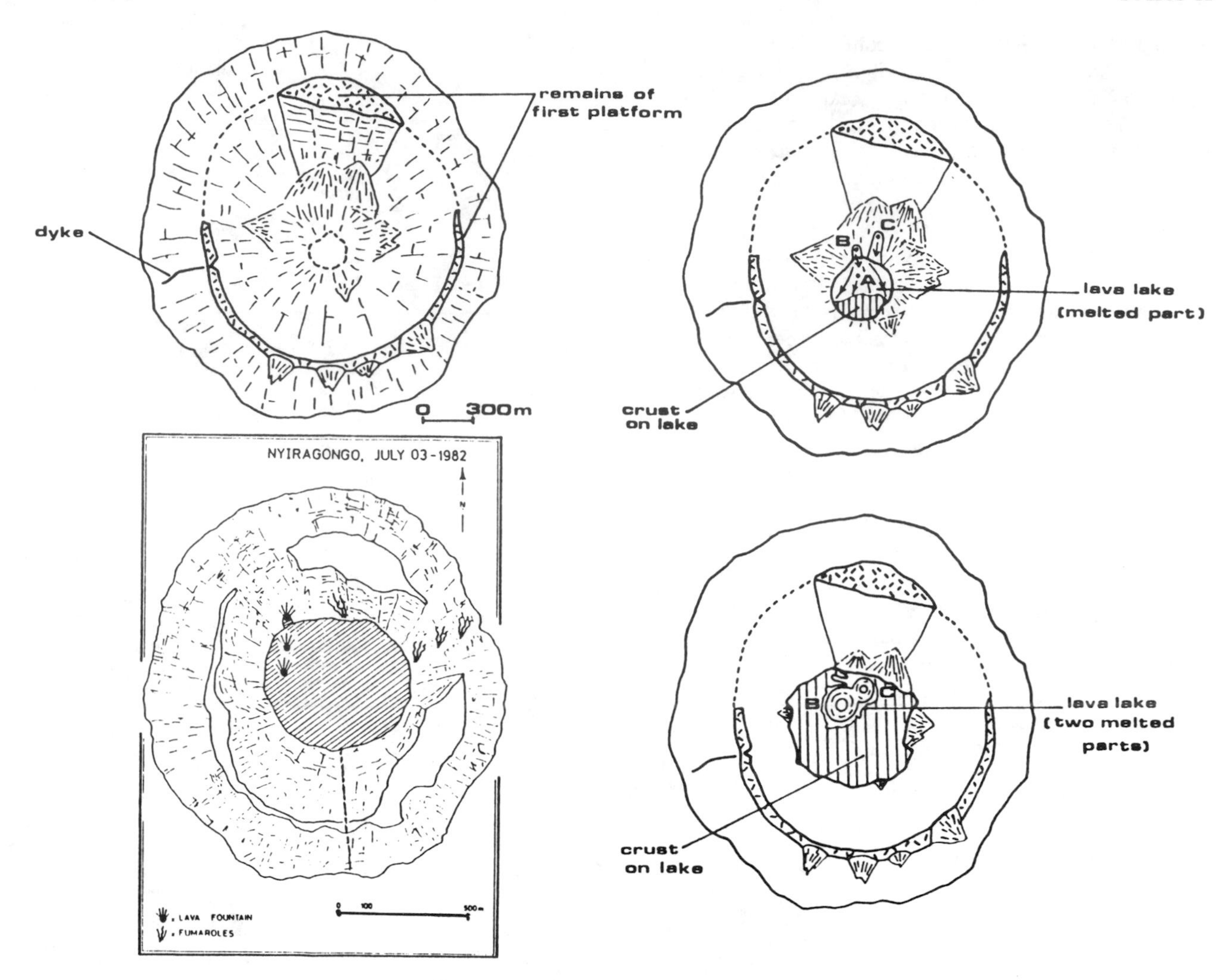

Fig. 2-10. Sketch maps of Nyiragongo's crater: *top left*, May 1982, before the eruption; *top right*, 26 June, showing the initial vent (A) now submerged, 2 new vents (B and C) N of it, and the crust on the S part of the lava lake; *bottom left*, 3 July, with the first of the new vents also submerged by the lava lake (diagonal pattern) – sites of lava fountaining and fumarolic activity are indicated; *bottom right*, 23 July with all 3 vents submerged – most of the lava lake is crusted over, but there are 2 upwellings. Map at bottom left is from N. Zana; others from M. Krafft.

Between late July and 15 Sept, the lava lake continued to grow, raising its surface an additional 100 m to about 400 m below the highest point on the crater rim. The lake's average diameter on 15 Sept was 700 m, its depth was 400 m, and it contained about 70 x 10^6 m^3 of lava. The zone of upwelling over the active vent was 250 m across and lava fountains reached a maximum height of 70 m. At the time of the 1977 eruption, the lava lake volume was about 20 x 10^6 m^3 and the surface stood about 200 m from the crater rim. The surface of the more voluminous 1982 lava lake was roughly 200 m lower than that of the 1977 lake because morphological changes associated with the 1977 eruption considerably increased the volume of the crater.

Information Contact: M. Krafft, Cernay.

Further References: Krafft, M. and Krafft, K., 1983, Le Reapparition du Lac de Lave dans le Cratère du Volcan Nyiragongo de Juin a Septembre 1982 (Kivu-Zaire). Histoire, Dynamisme, Debits et Risques Volcaniques; *C.R. Acad. Sci. Paris,* serie II, v. 296, p. 797-802.

Tazieff, H., 1984, Mt. Nyiragongo: Renewed Activity of the Lava Lake; *JVGR,* v. 20, p. 267-280.

LAKE MONOUN (BAMBOUTO FIELD)

Cameroon, W Africa
5.57°N, 10.58°E, summit elev. 1080 m.
Local time = GMT + 1 hr

Located in the Cameroon Line of young volcanoes (fig. 2-11), Lake Monoun is part of the Bambouto Volcanic Field.

General Reference: Fitton, J.G. and Dunlop, H.M., 1985, The Cameroon Line, West Africa, and its Bearing on the Origin of Oceanic and Continental Alkali Basalt; *Earth and Planetary Science Letters,* v. 72, p. 23-38.

8/84 (9:8) Before dawn on 16 Aug an explosion in Lake Monoun (about [7 km] from the village of Fuombot and 218 km NE of Mt. Cameroon) produced a 5-m water wave [shock waves and burned vegetation were originally reported by the embassy], flattening vegetation around the lake. A white cloud in the vicinity of the lake at daybreak looked like typical fog, but contained poisonous gas that killed 37 people. Victims suffered vomiting, paralysis, and very rapid death; some lost the outer layer of their skin. The Cameroon government sent a team of investigators to the scene, who collected samples of rock and damaged vegetation. Residents fled the area but most had returned to their homes within a few days. As of late Aug, the lake water was still reddish.

No previous eruptions are known from Lake Monoun. A 1:1,000,000 geologic map shows basalts and some rhyolites and trachytes in the area but the rocks are not dated (Le Marechal, 1975a). Lake Monoun is located at the intersection of an E-W trending zone of mylonites and the Cameroon Line (Le Marechal, 1975b), a chain of Tertiary to Recent generally alkaline volcanoes extending from Annobon Island in the Atlantic Ocean northeastward through Cameroon, dividing into two branches at its northeastern end (Fitton, 1980, and fig. 2-11). Mt. Cameroon, the only known site of Recent volcanism along the Cameroon Line, last erupted Oct-Nov 1982. Rocks as old as 31 million years have been dated along the Cameroon Line (on Principe Island; Fitton, 1980).

Information Contacts: C. Twining, U.S. Embassy, Douala; Yaoundé Domestic Radio Service.

References: Fitton, J. G., 1980, The Benue Trough and Cameroon Line—A Migrating Rift System in West Africa; *Earth and Planetary Science Letters*, v. 51, p. 132-138.

Le Marechal, A., 1975a, *Carte Geologique de l'Ouest du Cameroun et de l' Adamaoua*, 1:1,000,000; *ORSTOM*.

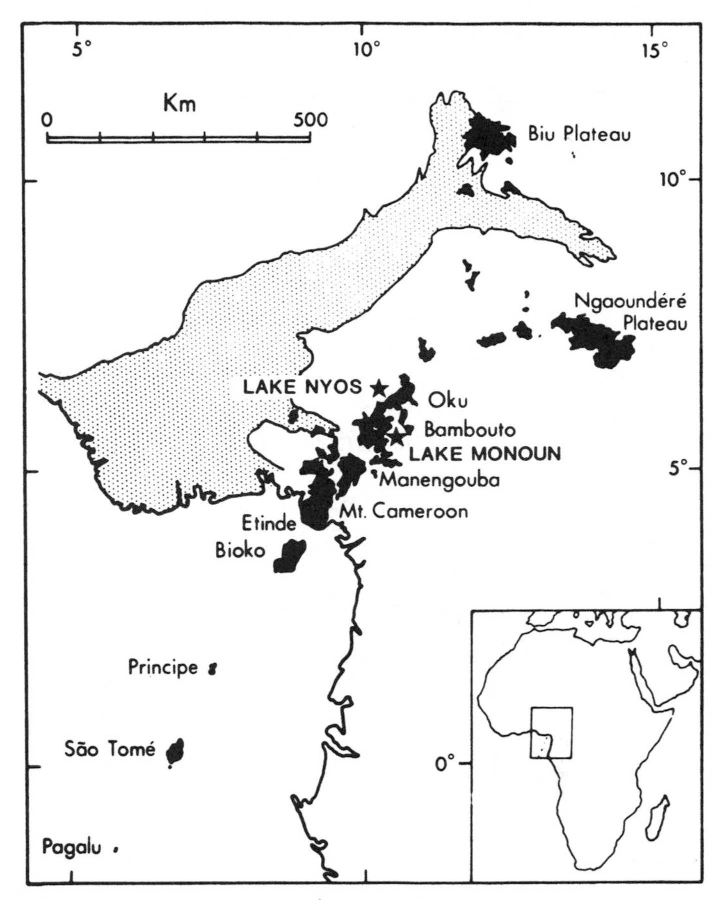

Fig. 2-11. Map of the Cameroon Line and Benue Trough, after Fitton and Dunlop, 1985. Volcanic rocks are shown in black. The approximate position of Lakes Monoun and Nyos are indicated by stars.

Le Marechal, A., 1975b, *Directions Structurales de l'Ouest du Cameroun et de l'Adamaoua*, 1:1,000,000; *ORSTOM*.

8/85 (10:8) In March 1985, Haraldur Sigurdsson, J. D. Devine, and F. Tchoua investigated the August 1984 event. The following briefly summarizes their findings; Sigurdsson and others have submitted a paper for publication and they will also present their data in Dec at the American Geophysical Union meeting in San Francisco.

On 15 Aug at about 2330, several people heard a loud noise or explosion from the Lake Monoun area and there were unconfirmed reports that an earthquake was felt that day at a town 6 km N of the lake. The gas cloud was emitted from the E part of the lake, where a crater about 350 m in diameter and at least 96 m deep is located. Victims of the cloud were in a low-lying area and had apparently died between 0300 and dawn. No autopsies were performed and the exact causes of death are unknown; all bodies had suffered skin damage [corrected from first-degree burns]. Persons on the fringes of the cloud reported that it smelled bitter and acidic. From 0630 until it dissipated by 1030, the whitish, smoke-like cloud remained 0-3 m above the ground. Vegetation was flattened within 100 m of the lake's east end, indicating that a water wave as much as 5 m above lake level was associated with the event.

Lake Monoun is near the center of a volcanic field that includes at least 34 recent craters, and there is evidence that eruptive activity has occurred there as recently as a few hundred years ago. However, the chemistry of the lake water and sediments, the uniformly low lake temperature (23-24°C), and the absence of new tephra in or around the lake suggested to the research team that the Aug 1984 event was not the result of an eruption or a sudden ejection of volcanic gas from the lake. Gradual emission of CO_2 from volcanic vents is thought to have led to a buildup of HCO_3 in the lake. An earthquake or internal seiche is thought to have upset the density stratification of the lake, triggering its overturn and catastrophic exsolution of CO_2, which suffocated the victims. Explanations of the cloud's acid odor and the agent of victims' skin damage are uncertain.

[Two years later, 1700 people perished 100 km to the NNW (fig. 2-11) when CO_2 was suddenly exsolved from Lake Nyos and flowed down several inhabited valleys. Results of initial research on this newly recognized hazard are summarized by Sigurdsson (1987), drawing substantially on discussions at the March 1987 international conference in Yaoundé, Cameroon.]

Information Contact: H. Sigurdsson, Univ. of Rhode Island.

Further References: Tchoua, F.M., 1983, The Phreatomagmatic Explosions of Monoun; *Rev. Sci. Tech.*, v. 3, p. 87-97.

Sigurdsson, H., Devine, J.D., Tchoua, F.M., et al., 1987, Origin of the Lethal Gas Burst from Lake Monoun, Cameroon; *JVGR,* v. 31, p. 1-16.

Sigurdsson, H., 1987, Lethal Gas Bursts From Cameroon Crater Lakes; *EOS,* v. 68, no. 23, p. 570-573.

MT. CAMEROON

Cameroon, W Africa
4.20°N, 9.17°E, summit elev. 4070 m.
Local time = GMT + 1 hr

Rising 4 km above the Gulf of Guinea, this massive strato-volcano has a long historic record. It is the only volcano outside the Mediterranean to have a documented eruption before the time of Christ.

General References: Déruelle, B., N'Ni, J., and Kambou, R., 1987, Mount Cameroon: An Active Volcano of the Cameroon Line; *Journal of African Earth Sciences,* v. 6, p. 197-214.

Fitton, J.G., 1987, The Cameroon Line, West Africa: A Comparison Between Oceanic and Continental Alkaline Volcanism; *in* Fitton, J.G. and Upton, B.G.J., (eds.), *Alkaline Igneous Rocks;* Geological Society Special Publication no. 30, p. 273-291.

10/82 (7:10) On 16 Oct, residents of Buea, about 10 km SE of the summit, were frightened by rumbling sounds and saw a red glow on the W flank. The next day, a helicopter crew making a survey flight over the area observed an eruption along a fissure about 6.5 km SW of the summit (on the SW border of the terminal plateau, at 4.15°N, 9.13°E). Steam was emitted from the upper part of the fissure, which began at 2600 m altitude and trended NNE-SSW, parallel to the general tectonic trend. Basaltic lava emerged from the fissure at 2500 m altitude and moved downslope at 20 km/hr in a 3 m-wide flow that reached forested areas. At night, glow from the lava was visible 50 km away. By 19 Oct, the active vent had built a cone 20 m high and 50 m in long dimension, and was emitting steam, ash, and lava fountains. Activity fluctuated 20-22 Oct, with periods of vigorous degassing lasting about 50 minutes. The crater included 2 active vents that ejected gas, ash, and bombs to 300-400 m height. Short, noisy explosions every 5-10 seconds made talking difficult 200 m from the crater. At 2450 m altitude, a lava flow poured out of the fissure at 20 km/hr. Newspapers reported some slowing of the lava advance during the night of 21-22 Oct, and by then the lava had moved through heavy rain forest to about 6 km from the seacoast village of Bakingili (population 300) on the SSW flank. Geologists reported that similar vent activity continued 22-30 Oct, and was audible 15-20 km away. On 30 Oct, the flow was about 200 m wide and 20 m thick, and its front had apparently stopped at 1200 m above sea level. However, the eruption was reported to be continuing. Total lava output was estimated at 10^7 m^3. Gas and aerosols were collected by a CNRS team. The eruption plume reached 4000 m altitude. Total SO_2 output was about 1000 metric tons/day (t/d).

At press time, Fitton, Hughes, and Kilburn reported that the moderate lava outflow and Strombolian activity observed 7-8 Nov had declined to a very sluggish trickle on the 9th, and the eruption had apparently ended 10 Nov.

Most of the 380 inhabitants of Bakingili were evacuated, as were the residents of Tsonko (Isongo), a nearby coastal town. Larger towns and an oil refinery were not endangered by the current activity, but palm oil plantations have reportedly been damaged.

Mt. Cameroon last erupted Feb-March 1959, producing a large E flank lava flow. Increased seismicity was recorded in Nov 1975, but no eruption occurred.

Information Contacts: F. Le Guern, D. Polian, CNRS; B. Déruelle, F. Tchoua, Univ. of Yaoundé; Drs. Betah and Ngungi, Ministère des Mines et Energie, Yaoundé; Drs. Kergomard, Lorfsher, and Rousset, Elf Serepca, Douala; P. Vincent, Univ. de Clermont; Dr. Lonne, Ardic Cameroun Aerienne, Douala; G. Fitton, Univ. of Edinburgh; C. Kilburn, Univ. of London; D. Hughes, Portsmouth Polytechnic; Ambassador H. Horan, U.S. Embassy, Yaoundé; L. Matteson, U.S. Consulate, Douala; British Consulate, Douala; J. Baade, T. Humphrey, Gulf Oil, Douala; R. Shaw, International Air Transport Assoc. (IATA), Montreal; D. Haller, NOAA/NESS; Cameroon Tribune.

11/82 (7:11) The upper SW flank fissure eruption that began 16 Oct ended by 10 Nov. When French volcanologists departed at the end of Oct, lava was continuing to emerge from the lower part of the fissure while Strombolian activity built a scoria cone less than 1 km upslope.

Between 31 Oct and the 7 Nov arrival of British volcanologists, there was extensive landsliding around the lower part of the fissure. An elongate depression had formed, trending radially downslope, about 500 m long, 200 m wide, and as much as 150 m deep. An estimated 5 x 10^6 m^3 of old lava and pyroclastics moved a maximum of 1.5 km. Lava from later stages of the eruption flowed over and around the debris pile, entering the rain forest 2 km from the vent.

When Fitton, Kilburn, and Hughes arrived at the volcano 7 Nov, activity from the lower part of the fissure had stopped. However, moderate Strombolian activity was continuing from the upper part of the fissure, ejecting bombs as much as 50 cm in diameter to 100 m above the rim of the scoria cone, which was 50 m high and 150 m across. Lava flowed from a vent below the SW side of the cone at 2700 m altitude. The next day, a velocity of 2 m/sec was measured in a 2 m-wide flow from the vent. At 2600 m altitude, this lava formed an aa flow 3 m wide, advancing 1-4 m/sec. At the flow front, 1 km from the vent, the temperature of the lava was 1070°C. On 9 Nov, lava emission had slowed to a very sluggish trickle and activity at the cone was reduced to gas emission with a period of 5-15 minutes. The eruption had apparently ceased by 10 Nov.

Information Contacts: G. Fitton, Univ. of Edinburgh; C. Kilburn, Univ. of London; D. Hughes, Portsmouth Polytechnic.

Further Reference: Fitton, J.G., Kilburn, C.R.J., Thirlwall, M.F., and Hughes, D.J., 1983, 1982 Eruption of Mount Cameroon, West Africa; *Nature,* v. 306, p. 327-332.

INDIAN OCEAN

KARTHALA

Grand Comoro Island, Indian Ocean
11.75°S, 41.05°E, summit elev. 2361 m
Local time = GMT + 3 hrs

Karthala is the most active volcano in the Comores Archipelago, between Africa and the island of Madagascar. It has erupted at least 22 times since 1828.

3/77 (2:3) An effusive eruption was reported on 6 April. About 4000 people were evacuated from the path of the lava, which was flowing towards the sea. No deaths have occurred.

Information Contact: G. Beauchamp, Jr., Office of Foreign Disaster Assistance (OFDA), Washington, DC.

4/77 (2:4) The eruption began about noon on 5 April from a SW flank vent, after a series of local tremors during the morning. Basaltic lava was extruded, which divided into 2 flows about 300 m wide and 3-15 m thick, separated by several hundred meters. The flows reconverged downslope and reached the sea on 6 April. Strong earthquakes were felt on the SE flank on 8 April, but were not accompanied by surface activity. Lava extrusion had ended on 10 April, although heavy fuming from nearby fissures continued as late as 17 April, preventing close approach to the vent, which was surrounded by up to 6 m of lapilli. No casualties were reported, but 4000 people were evacuated and 3 villages damaged or destroyed. Karthala last erupted Sept-Oct, 1972.

Information Contacts: P. de Saint Ours, St. Maurice, France; G. Beauchamp, Jr., OFDA, Washington, DC.

Further Reference: Krafft, M., 1982, L'Eruption Volcanique du Kartala en Avril 1977 (Grande Comore, Ocean Indien); *C.R. Acad. Sci. Paris,* serie II, v. 294, p. 753-758.

PITON DE LA FOURNAISE

Réunion Island, Indian Ocean
21.23°S, 55.71°E, summit elev. 2631 m
Local time = GMT + 4 hrs

This basaltic shield volcano forms the SE half of Réunion Island, 700 km E of Madagascar. It has been one of the most active oceanic volcanoes, with over 100 eruptions in the last 300 years.

11/75 (1:2) After a 31-month repose, the volcano began erupting on 4 Nov at the S end of Brulant Crater, 2300 m above sea level. A 400 m-long fissure opened, trending NW-SE, and 5 cinder cones formed along its length. On 5 Nov, one cone, 50 m high, was still active at the SE end of the fissure. A 10 m-wide lava lake formed. During the following 10 days 180,000 m^3 of aphyric basalt aa was emitted. By 14 Nov the eruption was decreasing, the crater was degassing, and the lava flows were reported as small.

Information Contact: L. Montaggioni, Univ. de la Réunion.

Further Reference: Krafft, M. and Gerente, A., 1977,

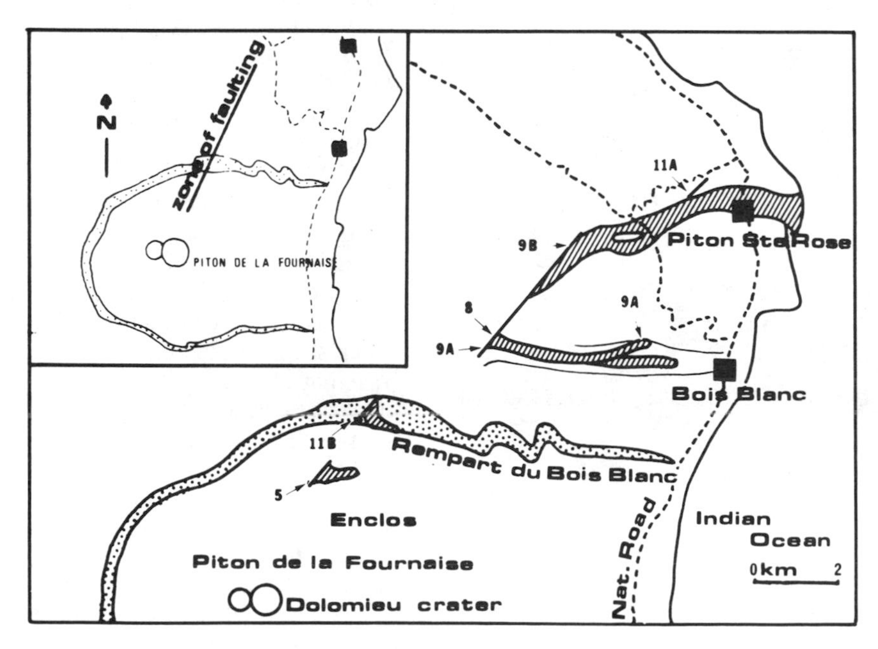

Fig. 3-1. Map, by Maurice Krafft, of the fissures and lava flows at Piton de la Fournaise 5-16 April 1977. The number beside each flow represents the date (in April) on which extrusion began. Vents 8, 9A, and 9B form a continuous fissure, and flows 8 and 9B overlap for most of their lengths.

LAVA EXTRUSION

Flow	Start	Stop
5	5; 1700	7; 1200
8	8; 1900	9; 1000
	12; 1400	12; 2400
9A	9; 0700	9; 1200
9B	9; 0930	10; 1000
	13; 0100	16; 1200
11A	11; 1200	11; 1800
11B	11; 1630	11; 2100

Table 3-1. April 1977 lava flows and their durations of extrusion. Flow numbers are from fig. 3-1. Dates (in April) are separated from start and stop times by semicolons.

L'Activité du Piton de la Fournaise entre Novembre 1975 et Avril 1976 (Ile de la Réunion, Ocean Indien); *C.R. Acad. Sci. Paris,* serie D, v. 284, p. 2091-2094.

11/76 (1:14) An effusive eruption began at 1300 on 2 Nov from a 300 m-long fissure N of Dolomieu Crater. The eruption ended at 0400 the next day after producing a lava flow 1 km long. No casualties or damage were reported. Piton de la Fournaise had been dormant since the end of 5 months of activity on 6 April, 1976 [*Bulletin of Volcanic Eruptions (BVE),* no. 16].

Information Contact: M. Krafft, Ensisheim.

4/77 (2:4) A new eruption of Piton de la Fournaise included its first flank activity since 1800. The following is a summary of events.

[24] March: 4 fissures opened at 2000 m altitude on the SE flank of the main crater (Dolomieu) and emitted lava for half a day.

4 April: Felt tremors began.

5 April: At 1700 a 500-m fissure opened at 1900 m altitude in the NE quarter of the caldera (fig. 3-1 and table 3-1) and extruded lava until the morning of 7 April.

8 April: At 1900 an explosion was heard and a fissure opened at 1300 m altitude on the N flank, producing lava fountains, gas, and a lava flow. The flow ceased 500 m from the village of Boisblanc during the night of 9 April.

9 April: A new fissure formed at 0700 near the 8 April fissure, extruding a lava flow that reached 700 m altitude. At 1100 another fissure opened (3 km N of the 2 previous ones) at 600 m altitude, from which a 50 m-wide flow moved during the night through the village of Sainte Rose, destroying 12 houses (fig. 3-2). It widened to 250 m and reached the sea between 0230 and 0300 on 10 April.

11 April: A new fissure opened 500 m N of Sainte Rose but emitted only gas. During the afternoon, lava flowed towards the sea from the caldera (L'Enclos) [see 2:5].

12 April: Earlier flows stopped, but new activity, lasting from afternoon until about midnight, began at 1500 m altitude above Boisblanc, near the 8-9 April eruption sites.

13 April: Lava again flowed from the NE quarter of the caldera during the morning [but see 2:5]. Lava extrusion resumed at 0100 from the fissure that had opened 9 April above Sainte Rose. The new flow reached the village at 1830, destroyed 21 houses and a church (figs. 3-3, 4), and entered the sea at 2200. About 1000 people were evacuated, but no casualties were reported.

Information Contacts: M. Krafft, Ensisheim; P. de Saint Ours, St. Maurice.

5/77 (2:5) The eruption ended about noon on 16 April with the cessation of lava extrusion from the fissure above Piton Sainte Rose (fig. 3-1). Maurice Krafft reports that the lava described last month as flowing towards the sea on 11 April was a small flow (11 B on fig. 3-1) originating from a fissure in the caldera wall (Rempart de Bois Blanc) and that the extrusion of a flow from the NE quarter of the caldera on 13 April is "very contested now."

Information Contact: M. Krafft, Ensisheim.

Further References: Bout, P., 1979, Observations sur les Coulées de Basalte (Oceanites) des Eruptions des 9-17 Avril 1977 de Piton Sainte-Rose (Réunion); 4ème Colloque de Geomorphologie Volcanique; Problemes du Volcanisme Explosif; *Clermont-Ferrand, Univ., Fac. Lett. Inst. Geogr.* v. 57, p. 47-52.

Kieffer, G., Tricot, B. and Vincent, P.M., 1977, Une Eruption Inhabituelle (Avril 1977) du Piton de la Fournaise (Ile de la Réunion): Ses Enseignements Volcanologiques et Structuraux; *C.R. Acad. Sci. Paris,* serie D, v. 285, p. 957-960.

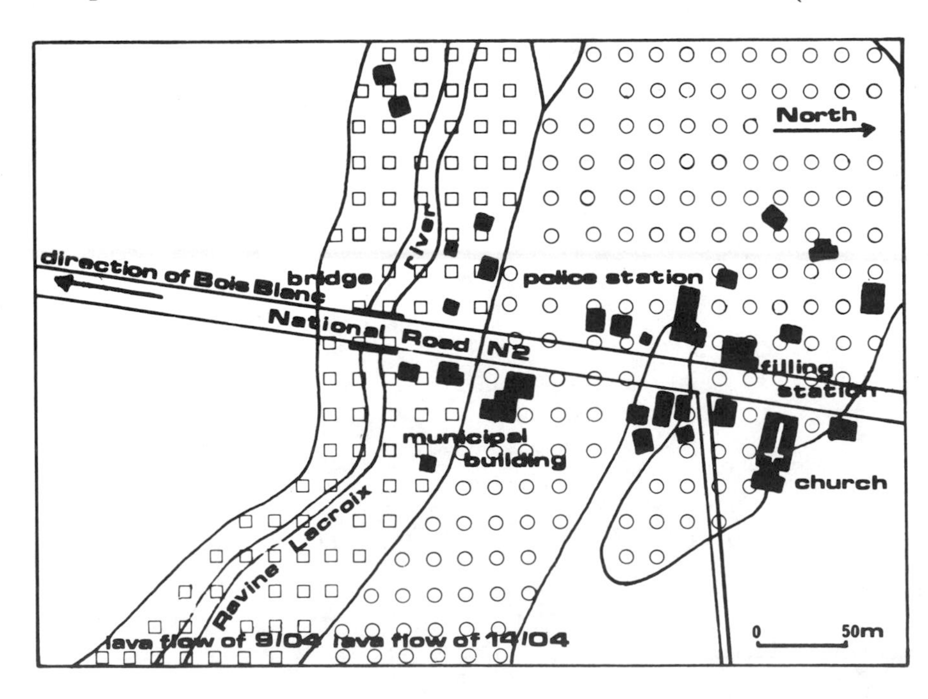

Fig. 3-2. Map, by Maurice Krafft, of lava flows through the village of Piton Ste. Rose, April 1977.

11/77 (2:11) Two fissures opened during the morning of 24 Oct on the E slope of Dolomieu Crater, within the caldera. The first fissure opened at 2180 m altitude at 0830, and emitted an aa lava flow, which stopped the next day after traveling to 500 m altitude. The second fissure opened 1 hour after the first at 1920 m altitude. Moderate explosive activity from this fissure built a single tephra cone until 26 Oct, when a second, immediately adjacent cone began to grow. Activity from the 2 cones remained fairly constant until lava fountaining began on 1 Nov. Fountains initially rose 200 m above the cones, but gradually declined until 10 Nov, when new

Fig. 3-3. Oblique airphoto by Maurice Krafft looking SW. Flow 9B, of 9 April 1977 (fig. 3-1) is outlined in white, and the caldera is at left.

Fig. 3-4. Destruction, by the lava flow at left, of the church of Piton Ste. Rose. April 1977 photo by M. Krafft.

lava covered about 4 km^2. Fountaining re-intensified 10-13 Nov, and was succeeded on 14 Nov by effusion of basaltic lava flows, with velocities reaching 60 km/hr. No casualties or damage occurred. The eruption ended at 1250 on 17 Nov.

Information Contact: M. Krafft, Cernay.

Further Reference: Kieffer, G. and Vincent, P.M., 1978, The October-November 1977 Eruption of Piton de la Fournaise (Réunion Island): A Terminal Eruption Without Terminal Crater; *C.R. Acad. Sci. Paris*, serie D, v. 286, p. 1767-1770.

5/79 (4:5) A small eruption began during the evening of 28 May and ended about noon the next day. A 100 m-long fissure opened about 1.5 km SE of Cratère Dolomieu, extruding a lava flow about 150 m in length (fig. 3-5). Three spatter cones formed on the fissure.

Information Contact: Same as above.

7/79 (4:7) A small eruption began when 2 radial fissures opened almost simultaneously on the N and S flanks of Cratère Dolomieu (the central crater) at about 1845 on 13 July. A line of three small fountains, each about 50 m high, formed along the N flank fissure and aa lava flowed about 400 m downslope. N flank activity ended at about 2200. Ten spatter cones were generated by the 0.5 km-long S flank fissure and a second aa flow traveled 1.5 km before the eruption stopped at 1130 on 14 July.

Information Contact: Same as above.

1/81 (6:1) A summit-area eruption began on 3 Feb after 12 days of local earthquakes and 17 cm of summit inflation. After a fairly sudden onset of seismicity 23 Jan, about forty magnitude 2 events were recorded daily by the newly established Observatoire Volcanologique du Piton de la Fournaise (OVPDLF). The day before the start of the eruption, 73 earthquakes were recorded, with foci about 1 km beneath Cratère Bory, the smaller of the 2 summit craters. Seismicity intensified in the hour prior to the first eruptive activity on 3 Feb. About 250 small discrete events were followed by 5 minutes of harmonic tremor. At 2030 a small fissure opened in Cratère Bory. A minor lava flow was extruded during 2 hours of activity along this fissure and a 6 m-high hornito formed at the vent. During the second hour of the eruption, a small amount of aa lava flowed from a vent about 200 m below the rim separating the larger Cratère Dolomieu from Bory. This lava covered about 3/4 of a small crater ruin (Enclos Velain) between Bory and Dolomieu. After about 2 hours, 2 or 3 small fissures opened on the NE side of Cratère Dolomieu, each extruding a lava flow about 100 m long.

The next morning at about 0400, a 300 m-long N-S trending fissure formed lower on the NE side of Dolomieu. Three spatter vents were active initially, but within an hour fountaining (15-30 m high) was limited to the lower portion of the fissure. Lava flowed downslope through channels and tubes onto the caldera floor.

As of early 6 Feb, lava fountaining as much as 70 m high was continuing from a 30 m-long segment of the lower end of the fissure. The activity had built a small, elongated cone with 3 vents. The lava flow, composed of aphyric basalt, was 1.5-2 km in length and covered several thousand m^2 of the caldera floor (fig. 3-6). Seismicity beneath Cratère Bory had stopped a few hours after the eruption began, but small events were occurring 6 Feb beneath Nez Coupé de Ste. Rose, on the caldera's N rim.

This eruption produced more lava than the 2 most recent eruptions, 28-29 May and 13-14 July, 1979. However, the 1981 volume is of the same order of magnitude as has been extruded in most of its numerous lava flow eruptions from the summit area in the past 50 years.

Information Contacts: L. Stieltjes, Bureau de Recherches Géologiques et Miniéres (BRGM), Réunion; OLPDLF, Réunion; M. Krafft, Cernay.

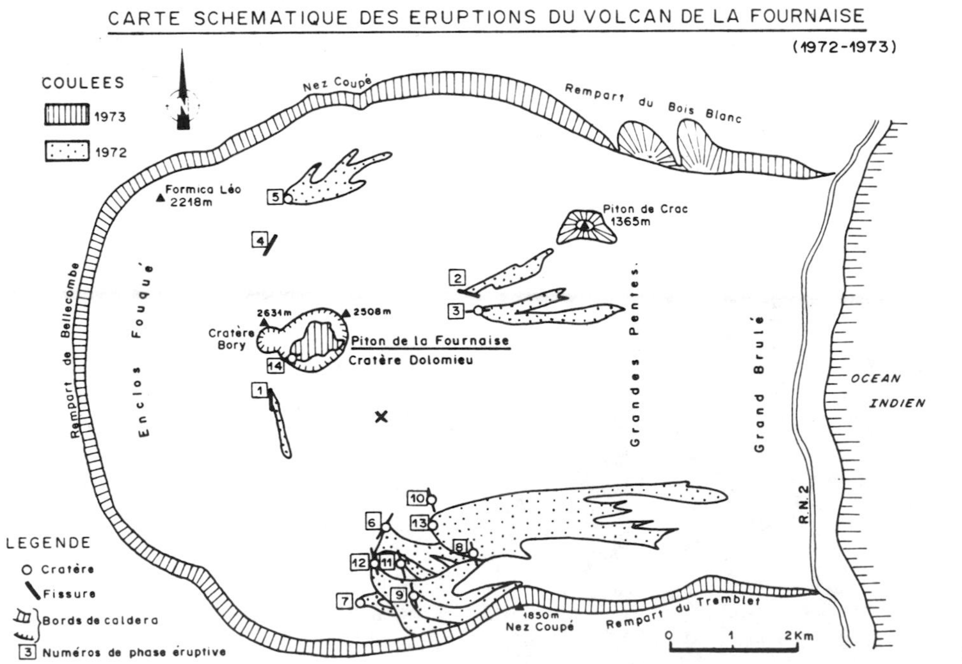

2/81 (6:2) Lava extrusion that began 3 Feb from the N side of the updomed summit region that surrounds Bory and Dolomieu craters con-

CARTE SCHEMATIQUE DES ERUPTIONS DU VOLCAN DE LA FOURNAISE (1972-1973)

Fig. 3-5. Map of the caldera of Piton de la Fournaise, from Krafft, M. and Gerente, A., 1977, L'activité du Piton de la Fournaise entre Octobre 1972 et Mai 1973; *C.R. Acad. Sci. Paris,* Serie D, v. 284, p. 607-610. The approximate site of the May 1979 activity is indicated by an X.

tinued until 25 Feb. After about 13 hours of seismicity, fissures opened on the SW side of the summit area and began to eject lava. The eruption was continuing as of 3 March.

Activity N of the Summit, 3-25 Feb. During the first few days of the eruption, lava was extruded from a series of radial fissures in the N summit region. By 6 Feb, lava fountaining was confined to a spatter cone at 2350 m altitude at the lower end of a fissure that opened 4 Feb. Lava flows emerged from 1 or 2 vents about 300 m down slope from the active spatter cone and moved about 1.3 km to the E. Fountaining was most intense 10 Feb (30 m high) and 18 Feb (100 m high). About 19 Feb, a small lava lake formed inside the active cone. Lava fountains rose a few m above the lake surface. A 2 m-diameter vent high on the cone emitted blue and yellow flames 3-4 m high. The spatter cone partially collapsed 20 Feb. Lava overflowing the collapsed area formed a front 100 m wide.

Fountaining and extrusion of lava flows began a rapid decline on 23 Feb and stopped on the 25th. Several million m^3 of lava were extruded 3-25 Feb.

Activity SW of the Summit Beginning 26 Feb. Seismographs at OVPLDF began to record a series of small (about magnitude 1) local earthquakes around midnight of 25-26 Feb. Earthquakes became increasingly frequent that morning and by 1230 were occurring once every 15 seconds under the summit's Bory Crater. Harmonic tremor started at 1300 and the beginning of eruptive activity was observed at 1306. Two minutes later, a large black cloud rose to 2 km height. Two en echelon radial fissures, trending N74°E, opened on the SW side of the updomed summit region. The upper fissure, 200-300 m long, extended from 2400 m to 2250 m altitude. The lower fissure, offset about 100 m from the base of the upper fissure, extended about 100 m farther downslope. Lava fountains rose to 15 m height from the entire length of the upper fissure, while fountains from the lower fissure were 50-60 m high. After half an hour, lava from the 2 fissures had merged into a single aa flow 2 km long that spread onto the caldera floor and moved toward the S caldera wall. Mid-afternoon outflow rates from the 2 fissures were about 300 m^3/sec (about 1 x $10^6 m^3$/ hr), much higher than at any time during the N summit region activity earlier in the month. The lava was an aphyric basalt, as was the 3-25 Feb material. By about 1800, lava fountaining along the upper fissure was concentrated at its lower end, where a cone was growing. Seismicity ended within a few hours of the start of eruptive activity on 26 Feb, a pattern similar to that observed at the beginning of the eruption 3 Feb.

Lava fountaining along the entire lower fissure continued until 0200 on 27 Feb, then was limited to the middle of this fissure, where a cone formed. The rate of lava outflow declined to 60 m^3/sec by the morning of the 27th and 10 m^3/sec the following day. Fountaining from the upper fissure stopped 28 Feb but continued from the lower fissure, building a 15 m-high spatter cone. Two other spatter cones formed along the lower fissure 1 March, with activity concentrating at one of these, also about 15 m high, on 2 March. The rate of lava production remained at about 10 m^3/sec as of 2 March [but see 6:4], feeding a slow-moving lava flow that was incandescent for the upper 1.5 km of its length.

Information Contacts: M. Krafft, Cernay; L. Stieltjes, BRGM, Réunion; OVPDLF, Réunion.

Further References: Bachelery, P., Blum, P.A., Cheminée, J.L., et al., 1982, Eruption at le Piton de la Fournaise Volcano on 3 February 1981; *Nature,* v. 297, p. 395-397.

Blum, P., Gaulon, R., Lalanne, F., and Ruegg, J., 1981, Sur l'Evidence de Précurseurs de l'Eruption du Volcan Piton de la Fournaise a la Réunion (Fevrier 1981); *C. R. Acad. Sci. Paris,* v. 292, serie II, p. 1449-1455.

3/81 (6:3) The activity SW of the summit that began 26 Feb continued until 25-26 March. A new eruption on 1 April was preceded by a swarm of local earthquakes, starting at 1923. The seismographs at OVPDLF registered 72 discrete events in the next few hours, before the onset of harmonic tremor and the start of an eruption at 2141. Observatory personnel reported that lava extruded from a vent in the north-central area of the caldera, about 3 km ENE of the summit, flowed toward the N caldera wall, reaching it during the night. By the early afternoon of 2 April, the flow front was about 1 km W of the coast highway [but see 6:4], but the lava's rate of advance had

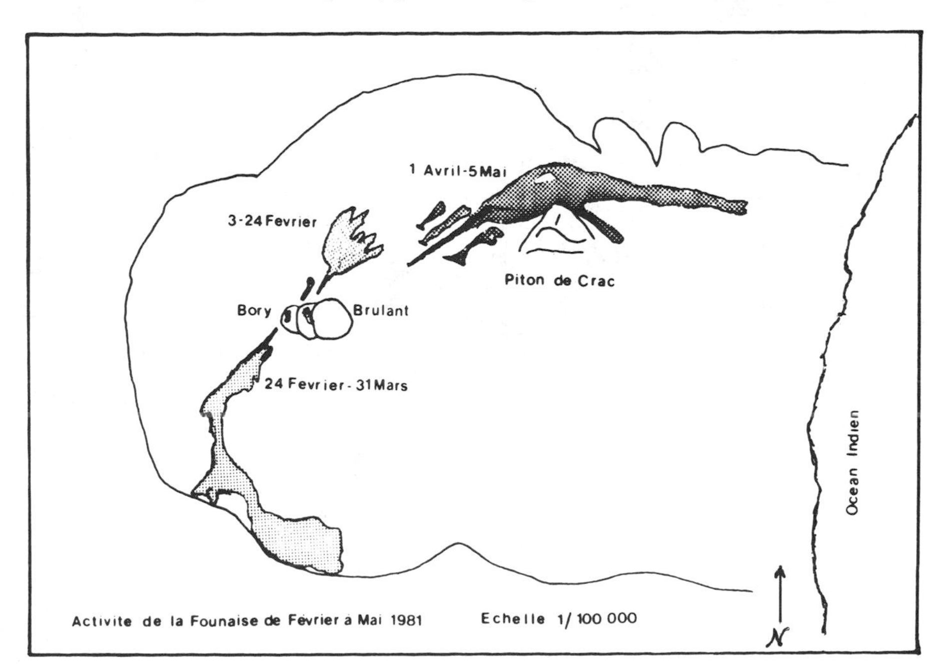

Fig. 3 6. Map prepared by F. X. Lalanne, Observatoire Volcanologique de la Réunion, showing the eruption fissures and lava flows from the 3 phases of the 3 Feb-5 May 1981 eruption of Piton de la Fournaise.

slowed considerably.

Information Contacts: J. Le Mouel, J.L. Cheminée, Institut De Physique Du Globe (IPG), Paris.

4/81 (6:4) As lava extrusion N of the summit was ending, seismic activity resumed on 25 Feb. By 0400 on the 26th, 50 events had been recorded, 20 with magnitudes greater than 1. Seismographs registered an additional 521 local earthquakes, 111 of magnitude 2 or greater, between 0400 and 1304. Harmonic tremor began as 2 en echelon fissures, trending 215°, opened about 800 m SW of Cratère Bory at 1304. Aphyric basalt flows had reached 2 km in length by the next morning when lava was pouring from the fissures at a rate of 600 m^3/sec (about 2 x 10^6 m^3/hr). By 1 March, the outflow rate had declined to about 20 x 10^6 m^3/sec and a 400 m-long segment of the flow was moving through a lava tube. Lava fountains reached an average height of 20-40 m. Blue and green flames were nearly always visible over the fissures. Three vents remained active on 2 March, but by 4 March fountaining continued from only 1 vent, reaching an average height of about 20 m. By the 4th, lava had reached the break in slope above the caldera wall, and most of the flow travelled through tubes. The flow reached the S caldera wall, at 1900-2000 m altitude, on 7 March.

Seismicity, which had remained quiet since 26 Feb, resumed on the morning of 7 March when 500 events were recorded, all with magnitudes less than 1.0. On 9 March, 700 more events, again with magnitudes less than 1.0, were recorded between 0819 and 1651 from a seismic station on the N side of the caldera. Between the 13th at 1000 and the 15th at 2344, the same station recorded 50 more events, 22 of which had magnitudes of 1 or more.

The lava flows spread laterally as the rate of outflow slackened 16-21 March. Lava reached 1800 m altitude and the flow front was about 5 km from the vents. The activity declined progressively, ending completely 25-26 March. A levelling network W of the summit was reoccupied 11 March, indicating that a net inflation of 11 μrad had occurred since the last survey on 13 Feb.

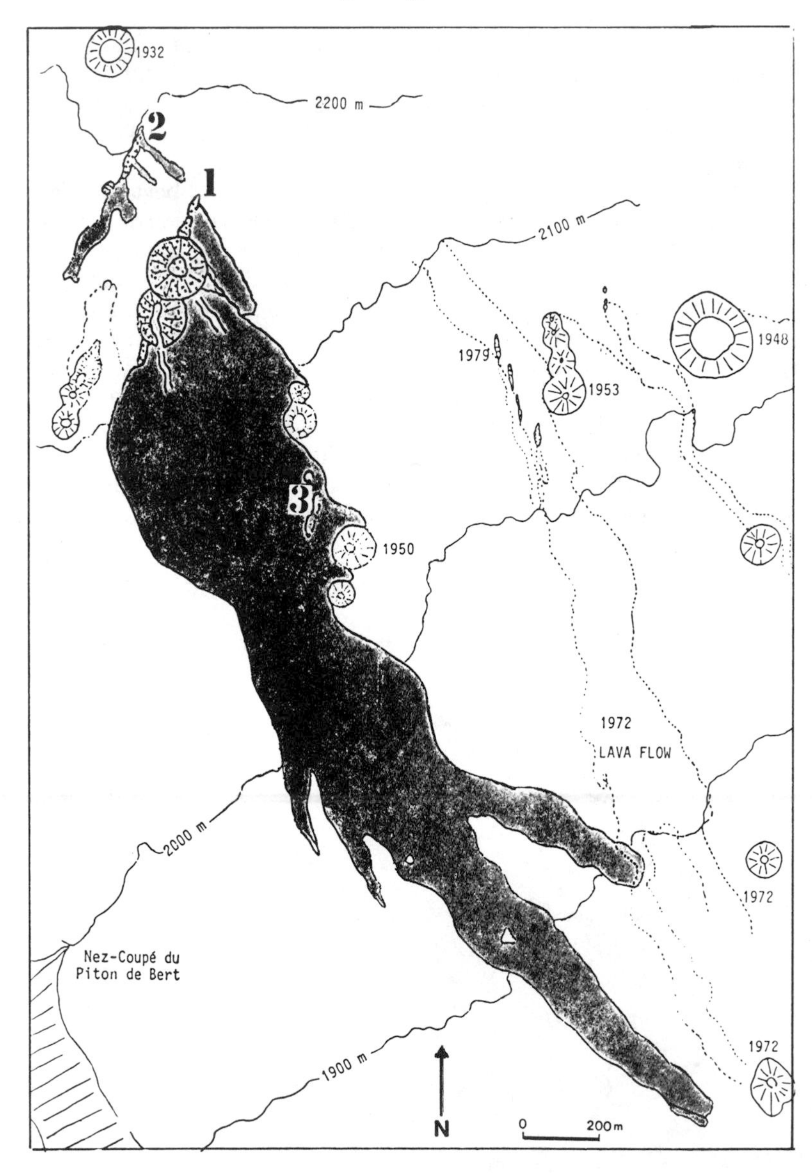

Fig. 3-7. Map showing the 1983 lava flows (shaded) and eruption fissures as of 14 Dec. Older craters and fissures are dated and the 1972 lava flow is outlined with dots. Activity stopped on 4 Dec at about 1400 along fissure zone 2 and about 0700 the next day along fissure zone 3. Courtesy of OVPDLF.

After about 2 hours of local seismicity, a 1 km-long fissure trending 060° opened on 1 April at 2141, between 1900 and 1600 m altitude about 2 km NE of the updomed summit crater area (fig. 3-6). Olivine basalt lava flowed rapidly downslope 1 and 2 April, stopping at 480 m altitude about 2.3 km W of the coast highway. On 3 April, effusive activity was limited to 2 vents along the fissure at 1650 and 1600 m altitude. Intense explosive activity ejected lava fragments to an average height of 50 m, building 2 cones, while numerous flows (as many as a dozen at once, 200-300 m in length) moved as much as 1 km toward the N caldera wall. Activity at the upper vent stopped 6 April. Many phreatomagmatic explosions occurred 7 April, of which 2 were observed between 1200 and 1700. A large quantity of cauliflower bombs and ash was ejected.

Poor weather prevented observations for the next 13 days. By 22-24 April, when weather had cleared, tephra ejections were reaching an average altitude of only 30 m. Lava flowed from the ends of tubes, forming numerous small tongues of aa and pahoehoe tens of meters in length. A 2 km x 300 m field of lava had formed by the 24th.

On 3 and 4 May the explosive activity ceased as effusive activity weakened. Only 3 aa flows persisted in the lava field, one 300 m long, the other two only 10 m in length. By the night of 4-5 May, all eruptive activity appeared to have ended.

Heavy rains stopped some of the seismometers from working, but the 1-2 instruments that remained functional recorded no seismicity after the premonitory swarm on 1 April. No significant changes in tilt were measured.

Information Contacts: J. Le Mouel, J.L. Cheminée, IPG, Paris; M. Krafft, Cernay.

5/81 (6:5) An earthquake swarm beneath the N caldera rim began suddenly on 18 May at 1000. By 2200, 800 small events had been recorded, but no eruption followed.

Information Contacts: J.L. Cheminée, IPG, Paris; M. Krafft, Cernay.

7/83 (8:7) A 12-hour earthquake swarm occurred 15 July at Piton de la Fournaise, the first seismic crisis there since shortly after the 3 Feb-5 May 1981 eruption, which produced 10 x 10^6 m^3 of lava during 3 active phases. Since then, background seismicity had been less than 0.5 events per day. The 21 events between 0830 and 2015 on 15 July occurred in the central area at shallow depth but were poorly located because they were recorded on only 1-3 stations, had emergent onsets, and had poorly-defined phases. Event durations ranged from 15 to 150 seconds. The seismic crisis prompted the resurvey of deformation networks, but no significant changes were measured.

The OVPLDF noted that for the past 50 years the mean eruption frequency has been once every 12-14 months. The 27 months since the last eruption is one of the longer repose intervals during that period.

Information Contact: OVPDLF, Réunion.

11/83 (8:11) The following quoted reports are from the OVPDLF.

"Eruptive activity started 4 Dec at 0900, preceded by an increase in seismicity that began [20] Nov. From 20 Nov, the seismic events clustered at depths of about 1.5-5 km beneath the summit area. The maximum number of earthquakes was recorded 1 and 2 Dec, with 10 events ($M < 1$) per day. The strain release was not very large. A seismic swarm began 4 Dec at 0642. About 300 events were recorded by 0859, when harmonic tremor began, and a few minutes later the eruption was seen. The initial eruption fissure was situated on the SSW flank of the central cone (Cratère Dolomieu). Activity at 2 segments of this fissure stopped during the evening. A second fissure zone opened at 1027 and stopped erupting about 1400; a third fissure opened at 1319 and ceased erupting about 0700 the next morning. The amount of lava extruded 4 Dec was estimated at 1 x 10^6 m^3. The initial fissure was the most productive. Only weak deformation was measured before and during the eruption."

Information Contacts: J. Lenat, F. Lalanne, OVPDLF, Réunion; Univ. de la Réunion; J. LeMouel and J.L. Cheminée, IPG, Paris.

12/83 (8:12) "The eruption was continuing as of 5 Jan, but the effusive and explosive activity and tremor amplitude had decreased. A rough estimate of the total magma output through 5 Jan was 10 x 10^6 m^3, about double the volume that had reached the surface by 14 Dec (fig. 3-7).

"A resurvey of the deformation networks during the days following the onset of the eruption has shown that ground deformation has taken place in the summit area. Tilt and distance measurements indicate an inflation centered S of the summit craters with a shallow focus. Records from a continuously recording strainmeter in the summit area indicate that the major part of the ground deformation happened within ½-¾ hour (fig. 3-8). [Rapid deformation started about two hours before the eruption and just after the beginning of the seismic swarm associated with the intrusion toward the surface (fig. 3-9).] No other significant ground deformation has been observed during the course of the eruption through 21 Dec.

"Since the beginning of the eruption, the only recorded seismic activity has been volcanic tremor with amplitude that varied with time."

Information Contacts: J. Lenat, A. Bonneville, P. Tarits, H. Delorme, OVPDLF, Réunion; P. Bachelery, J. Bougeres, Univ. de la Réunion.

1/84 (9:1) "A second eruptive phase began on 18 Jan at 0454, preceded by inflation of the summit area that began in

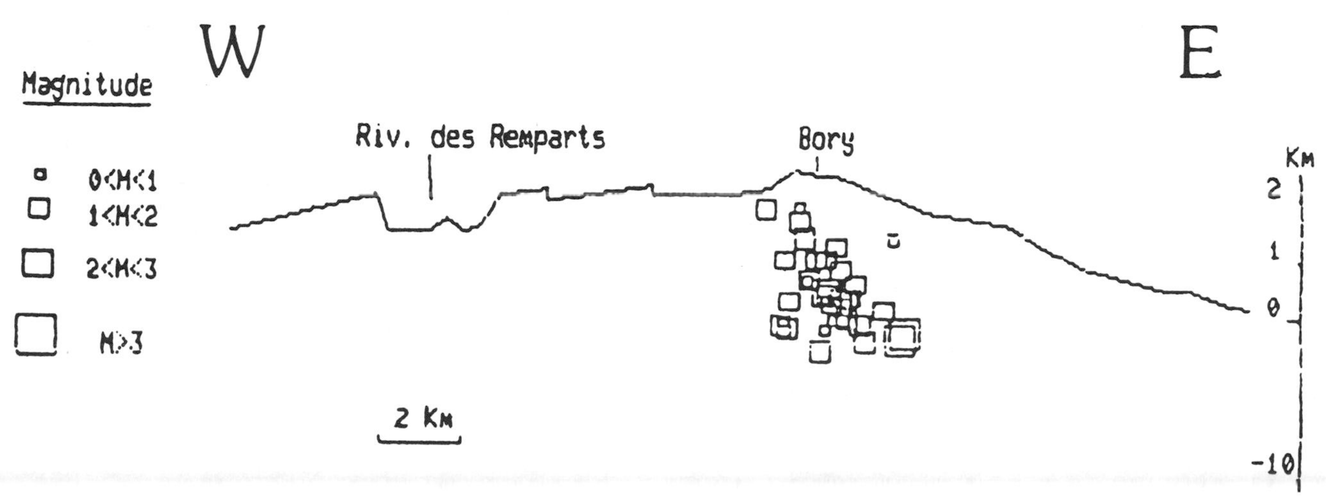

Fig. 3-8. Extension in mm, detected 4 Dec 1983 between midnight and 1400 [local time] by a continuously recording strainmeter in the summit area. Courtesy of OVPDLF.

early Jan. A seismic swarm of about 50 low-energy events occurred between 0313 and 0454 on 18 Jan, when harmonic tremor began. The eruptive fissure of the first phase, which began on 4 Dec, was still active (but with virtually no explosive activity and a low level of effusive activity) when the second phase began, but apparently ceased during the day on 18 Jan.

"Two new eruptive fissures formed about 400 m NNW of the main first phase fissure. Activity at the upper one rapidly decreased, and stopped at 0200 on 19 Jan. The other, about 200 m long, sustained lava fountaining more than 80 m high 18-19 Jan. The fountains produced a large amount of Pele's hair that was transported SW by wind and deposited on inhabited areas, causing a potential hazard for livestock grazing in the area. Emergency measures were taken by local authorities; fortunately, heavy rains and wind 19-23 Jan washed away most of the tephra that remained on the grass.

"During 18 Jan, the lava discharge was vigorous (up to 100 m^3/sec). At 1200 the flow extended 4 km from the vents. Inflation, possibly related to emplacement of an intrusion, was measured on 18 Jan, showing the same pattern as after the start of the first phase on 4 Dec.

"On 24 Jan, the eruption was localized at 2 vents that were building 2 cinder cones. On 27 Jan, only one of the vents was still active. Eruptive activity was continuing as of 8 Feb."

Information Contacts: J. Lenat, A. Bonneville, C. Hemond, F. Lalanne, P. Tarits, OVPDLF, Réunion; P. Bachelery, J. Bougeres, Univ. de la Réunion.

2/84 (9:2) The following is from J. F. Lenat.

"The eruption ended 18 Feb. During the previous days tremor amplitude had become irregular with periodic bursts. In the last 12 hours of the eruption, the tremor was intermittent, with bursts occurring less and less frequently. The sporadic tremor progressively died away during the afternoon of 18 Feb and an observation from the rim of the previously active vent brought confirmation that lava was no longer present at the bottom of the crater although it was still red hot."

Information Contact: J. Lenat, OVPDLF, Réunion.

Further References: Bachelery, P., 1984, L'Eruption du Piton de la Fournaise (Réunion) 12-83/02-84; *Bulletin du Laboratoire de Géographie Physique, Univ. de la Réunion,* no. 1, sommaire, p. 2-14.

Lenat, J.F., Bachelery, P., Bonneville, A., Tarrilo, P., Cheminée, J.L., and Delorme, H., The December 4, 1983 to February 18, 1984 Eruption of Piton de la Fournaise (La Réunion, Indian Ocean): Description and Interpretation; *JVGR,* in press.

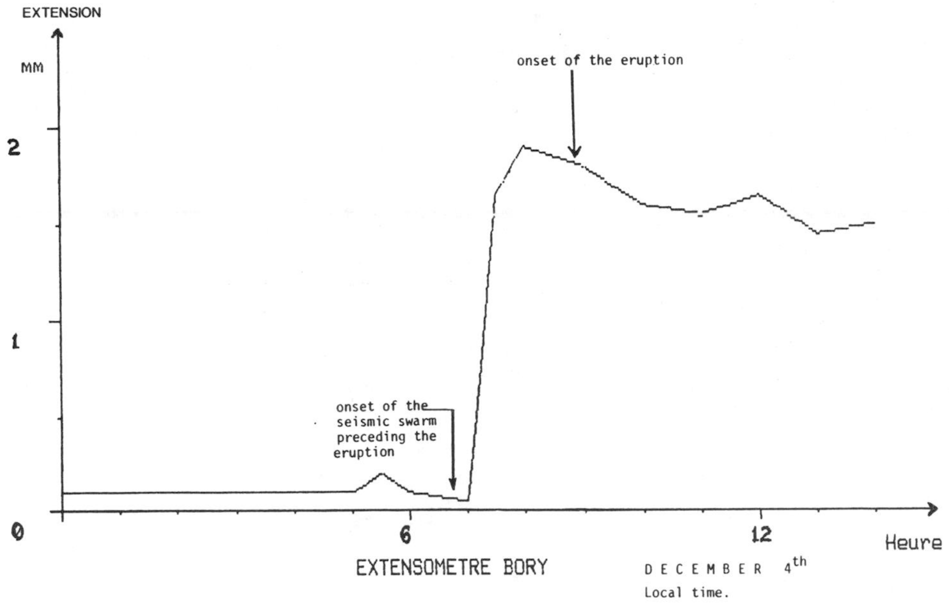

5/85 (10:5) The following quoted reports are from OVPDLF.

"Since the last eruption, seismicity and deformation measured at the 4 summit stations had remained at very low levels (0-5 earthquakes and 1-7 μrad of tilt per month). The pattern of tilt vectors implied a progressive deflation of the area that had undergone large deformation during the last eruption.

"Beginning on 15 May, several small earthquakes (M <1) were recorded at depths of 1.5-2.5 km beneath the summit. Simultaneously, the summit dry tilt stations began to show an inflationary pattern. The number of seismic events progressively increased. Seismic activity peaked on 8 June with 13 events, and decreased to a low level on the 13th. No significant migration of the earthquakes was observed. Inflation appeared to have been almost continuous except during 2 episodes when deformation slowed (3 June) or reversed (8 June) corresponding to periods of largest seismic energy release.

"This activity differed from the 1983 pre-eruptive crisis. The late 1983-early 1984 eruption was preceded by 2 weeks of seismic activity (48 events) with no associated deformation. The most recent activity lasted for at least 4 weeks. The earthquakes are located in the same area as before, but summit tilt stations show inflation of 15-40 μrads."

Information Contacts: J. Lenat, OVPDLF, Réunion; M. Kasser, Inst. Geographique National (IGN), Paris; A. Nercessian, IPG, Paris; R. Vie le Sage, Délégation aux Risques Majeurs (DRM), Paris; P. Bachelery, Univ. de la Réunion; A. Bonneville, Univ. du Languedoc; G. Boudon, Obs. Volc. de la Mt. Pelée; M. Halbwaks, Univ. de Chambéry.

EXTENSION MM 2 1 0 — onset of the eruption — onset of the seismic swarm preceding the eruption — 6 12 Heure — EXTENSOMETRE BORY — DECEMBER 4th Local time.

Fig. 3-9. Pre-eruption seismic cross-section through the summit, with hypocenters and magnitudes of seismicity between 20 Nov and the onset of the pre-eruption swarm of 4 Dec 1983. The 3 initial swarm events are also shown. Courtesy of OVPDLF.

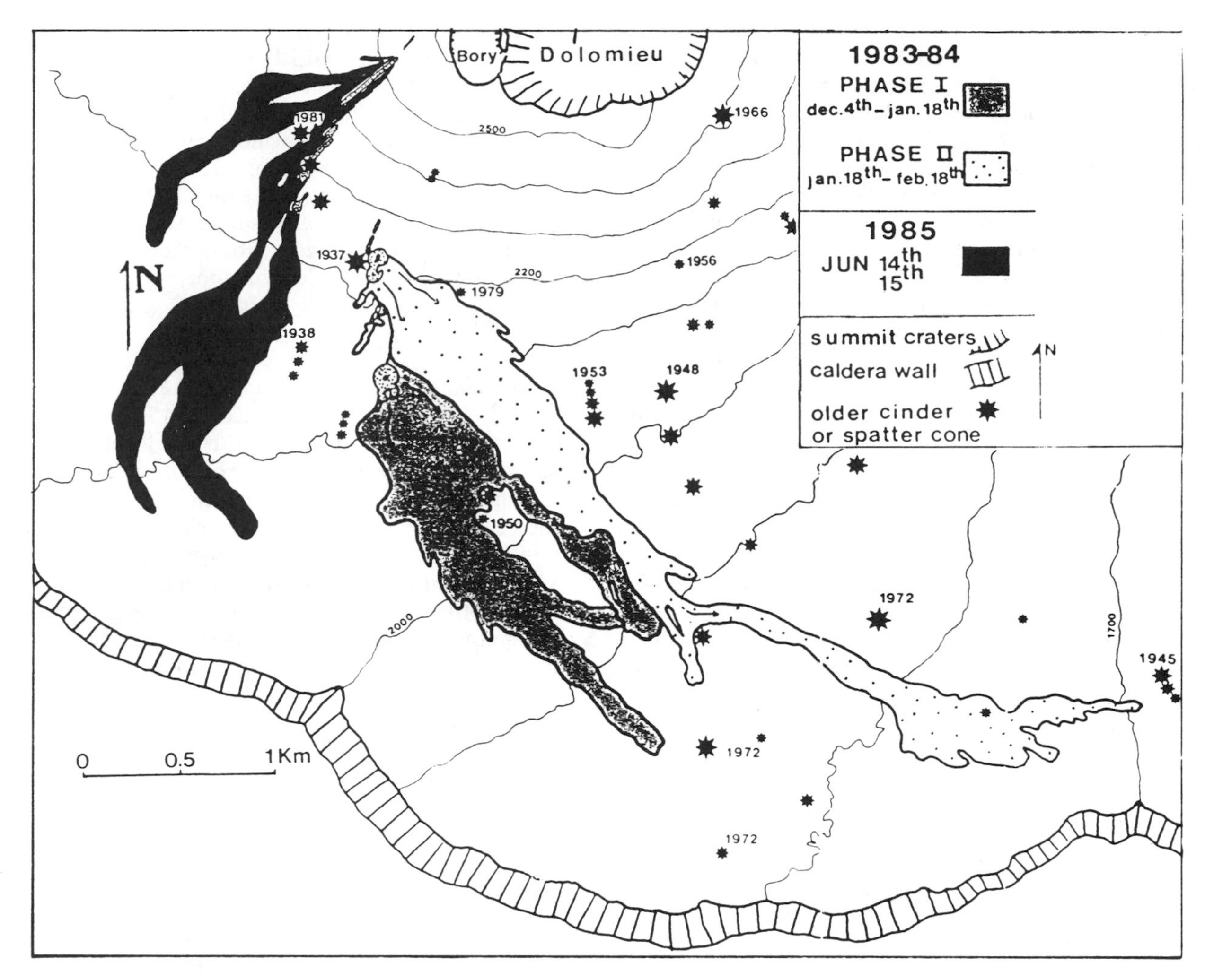

Fig. 3-10. Sketch map of the central and south parts of the caldera, showing lava flows from the 1983-4 and 1985 eruptions, and vents from earlier eruptions. Courtesy of OVPDLF.

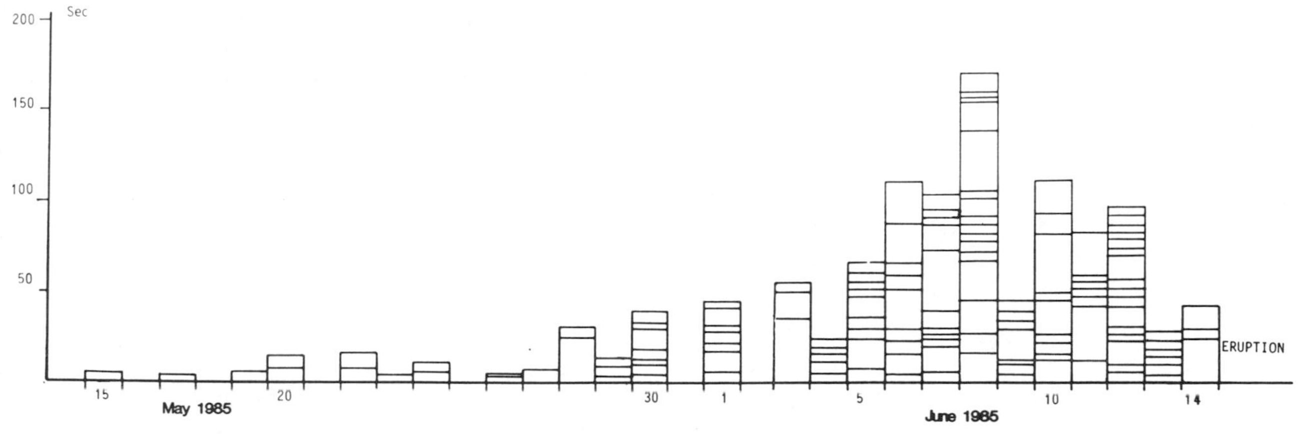

Fig. 3-11. Chronology of the 1985 pre-eruption seismic crisis. Each box represents one earthquake. Box heights represent earthquake durations. Courtesy of OVPDLF.

6/85 (10:6)"An eruption began on 14 June at 1604 on the SW flank of the central cone (fig. 3-10), just as in the July 1979, Feb 1981, Dec 1983, and Jan 1984 eruptions. The main eruptive fissure, oriented 230°, was about 1 km long and composed of 4 main en echelon fractures. A small fissure oriented 330° was located NW of the summit. The NW fissure and the upper part of the SW fissure were active for 2-6 hours after the beginning of the eruption. Activity stopped on the lower part of the SW fissure on 15 June at about 1600. Flows were emitted all along the fissure, the longest extending about 2 km from it. No calculation of emitted volume has yet been made, but a crude approximation is 1×10^6 m^3.

"This eruption was preceded by a 30-day pre-eruptive seismic crisis and by a 40-minute intrusive seismic swarm. Seismicity had been at background levels since the end of the 1983-84 eruption. The pre-eruptive seismic crisis began on 15 May and the number of recorded earthquakes has increased since then (fig. 3-11). All were located beneath the summit at a mean depth of 1.5-2.5 km. The duration of the pre-eruptive seismic crisis and the number of events have been greater than for the Dec 1983 eruption, but the location of events has been the same.

"Since the end of the 1983-84 eruption, the deformation has been small, which could be interpreted as a relaxation of the area that had undergone large displacements during the two intrusive crises related to the 4 Dec 1983 and 18 Jan 1984 outbreaks. A pattern of inflation of the central area has been measured since May 1985. This inflation developed during the pre-eruptive crisis and was in the range of 20-50 μrads at the dry tilt stations located near the summit. Inflation was not seen at the stations more than 2-3 km from the summit.

"Figure 3-12 shows the chronology of seismicity 14-15 June and detailed plots of the 40-minute intrusive period. As the intrusion approached the surface, local surface movements occurred, as shown by strainmeters across an open fissure and by the tiltmeter about 200 m

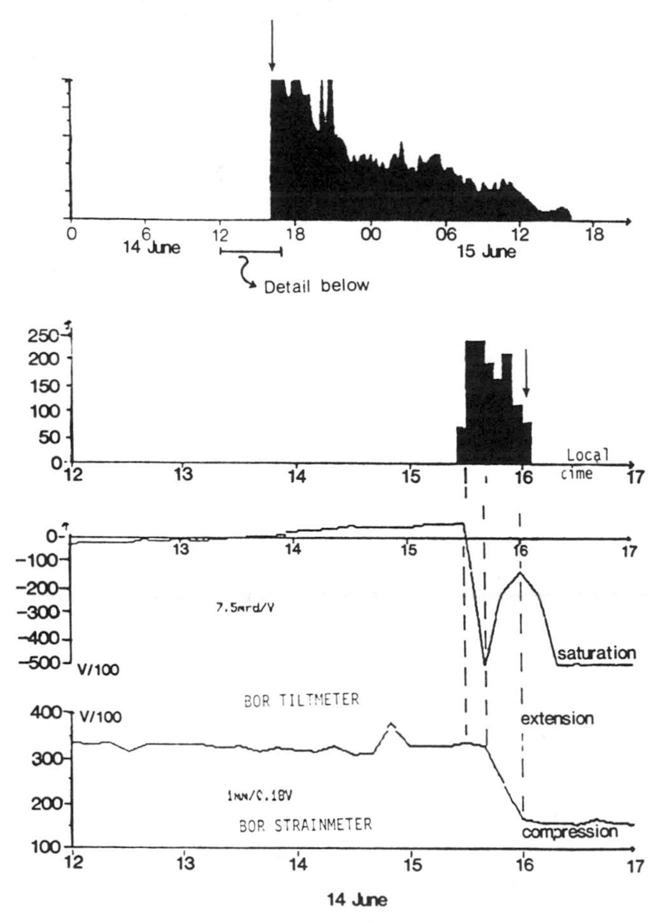

Fig. 3-12. *Top,* relative amplitude of harmonic tremor at BOR station, about 200 m from the fissure, 14-15 June; *center,* cumulative duration of seismic activity (in seconds) during 5-minute intervals, 14 June between 1200 and 1700 hours; *bottom,* tiltmeter and strainmeter data at BOR station, 14 June, 1200-1700. Arrows at top and center mark the start of the eruption, as does the right-hand vertical dashed line in the tilt/strain section of the figure. Courtesy of OVPDLF.

from the eruptive fissure.

"The lava of the 14-15 June eruption is a transitional aphyric basalt with a composition (Table 3-2) similar to that of the 1983-84 basalt. This supports the hypothesis that the shallow magma chamber could not have been replenished before this eruption.

Information Contacts: J. Lenat, OVPDLF, Réunion; P. Bachelery, Univ. de la Réunion; A. Bonneville, Univ. du Languedoc; G. Boudon, Obs. Volc. de la Mt. Pelée; M. Halbwachs, Univ. de Chambéry; M. Kasser, IGN, Paris; A. Nercessian, IPG, Paris; R. Vie le Sage, DRM, Paris.

7/85 (10:7) The following is from J. F. Lenat.

"After the 24-hour eruption of 14-15 June, seismicity and deformation had returned to low levels. Sustained seismic activity began suddenly on 9 July. The intrusive seismic crisis, similar to the ones that have preceded all the outbreaks in the central area since 1981, continued until about 2250. The earthquakes were located between 1 and 2.5 km beneath the summit. The intrusion failed to reach the surface. The deformation pattern shown by the summit dry-tilt stations seems to indicate that the intrusion was emplaced E of the summit area.

"After declining for about 1 hour, the seismic activity resumed with a new swarm of earthquakes that were mostly centered 3-6 km WNW of the summit area at depths of 1-4 km. This crisis peaked on 10 July with more than 1200 events, and progressively decreased during the following days. From 13 July to the end of the month the rate of seismicity decreased from 5-10 events per day to about 1 event per day.

"This second swarm is the first to be observed outside of the central area since the Observatory was established in 1980. It is also by far the largest recorded seismic crisis at Piton de la Fournaise. A large number of epicenters are also found along the 'Grandes Pentes' line which is the inland boundary of the 'Grand Brule' slump structure that is thought to result from the instability of the free flank of the volcano.

	A	B
SiO_2	49.50	49.20
Al_2O_3	14.60	14.60
Fe_2O_3	4.14	4.89
FeO	7.53	6.18
MgO	7.10	7.10
CaO	11.60	11.60
Na_2O	2.50	2.50
K_2O	0.80	0.80
TiO_2	2.90	2.90
MnO	0.18	0.18
TOTAL	100.85	100.58

Table 3-2. Composition of two samples of the 14-15 June 1985 basalt. Courtesy of OVPDLF.

"On 5 Aug at 2250 a new intrusive seismic swarm began at shallow depth beneath the summit area and more than 200 events were recorded in less than 3 hours. An eruption started at 2340 N of the summit at the base of the central cone (E of Cratère Magne). Deformation related to this eruption affected the summit area and the descent to the Plaine des Osmondes, corresponding to a major inflation on both sides of a N15°E fissure. From the beginning of the eruption until 9 Aug, 4 x 10^6 m^3 of aphyric basalt have been emitted at a rate of 10-15 m^3/sec. Temperatures of 1100-1120°C have been recorded. The eruption was continuing as of 11 Aug."

Information Contacts: J. Lenat, H. Delorme, J. Delarue, OVPDLF, Réunion; A. Hirn, J. Delattre, IPG, Paris; P. Bachelery, Univ. de la Réunion.

8/85 (10:8) The following quoted reports are from OVPDLF.

"The eruptive episode that began on 5 Aug ended 1 Sept. The major part of the aphyric basalt lava flow was emitted during the first 10 days. The lava front stayed at the bottom of the Grandes Pentes inside the Plaine des Osmondes (near the N caldera wall, fig. 3-13). During the last week, small amounts of pahoehoe were

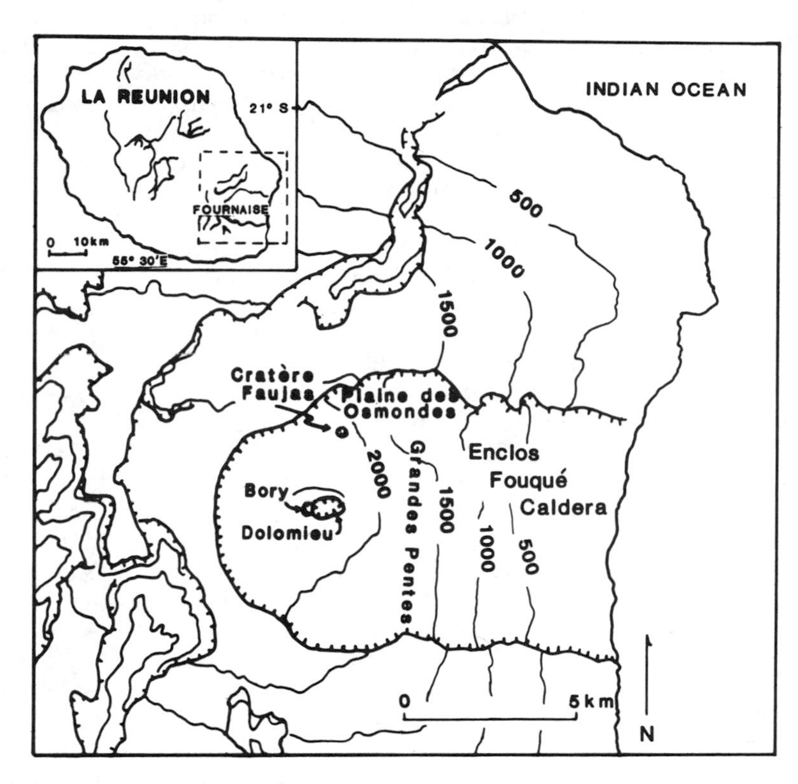

Fig. 3-13. Sketch map of the summit caldera region of Piton de la Fournaise after Blum, et al. (1981).

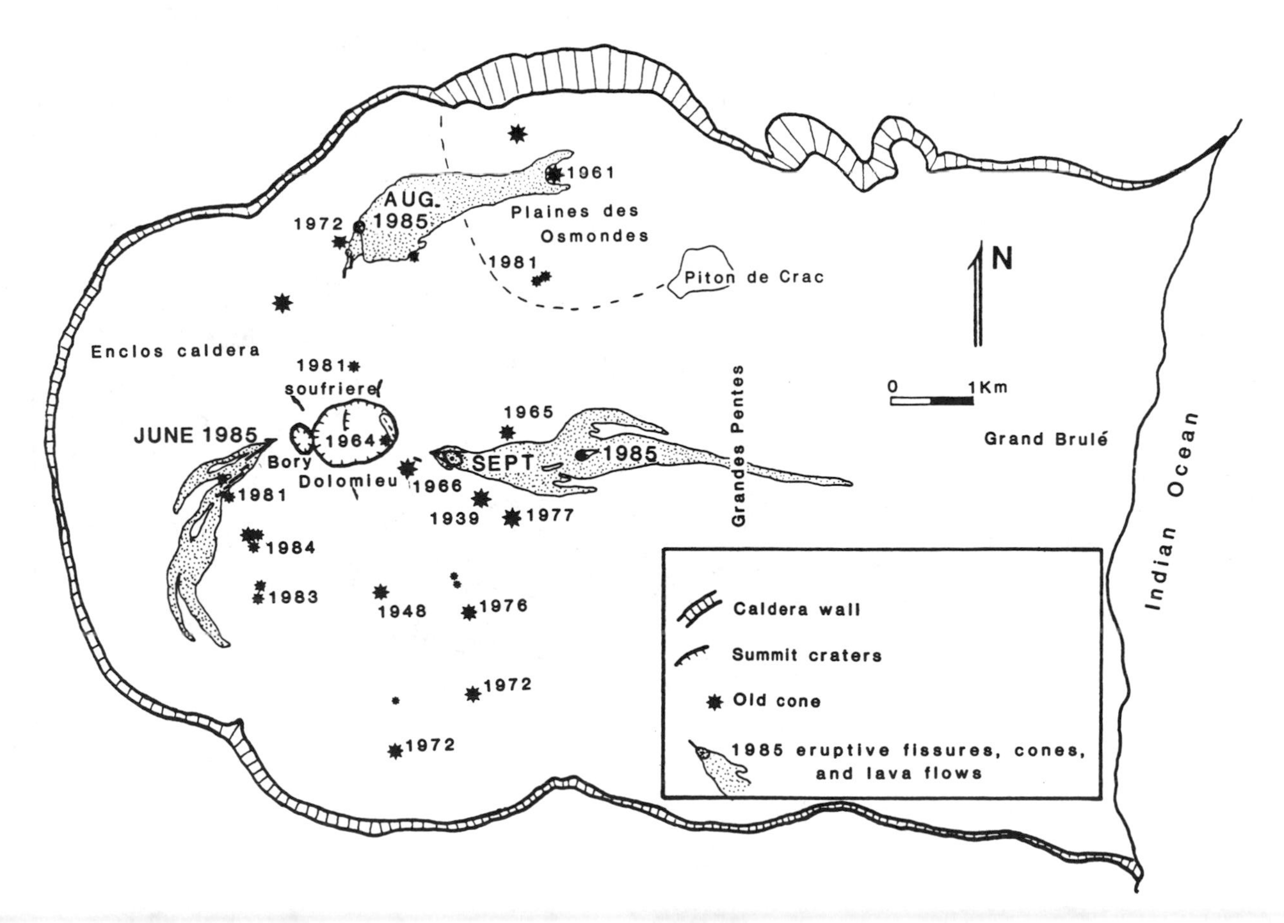

Fig. 3-14. Provisional map of the 1985 eruptive fissures and lava flows, as of 30 Sept. Courtesy of OVPDLF and the Université de la Réunion.

emitted from tunnels at the foot of Cratère Faujas, about halfway between the central cone and the N caldera wall. Seismic tremor lasted for 10 days at a very low level.

"Deformation related to the opening of the 5 Aug fissure affected the summit and the flank of the Plaine des Osmondes. No significant deformation was recorded during Aug, but minor summit deflation was noted around the fissure zone."

Information Contact: H. Delorme, J. Delarue, OVPDLF, Réunion; P. Bachelery, Univ. de la Réunion.

9/85 (10:9) A very short seismic crisis began on 6 Sept at 1408. As for the June and Aug eruptive episodes, seismic events were centered under Dolomieu and Soufrière, at 1 km depth (10:5-8). After 30 shallow summit seismic events, 3 eruptive fissures opened at 1520. The first fissure (near Soufrière) was 50 m long and continued erupting for 2 hours. The second, 250 m long, opened inside the Dolomieu summit crater and erupted until 0500 the next morning. The third, E-oriented and 500 m long, opened on the E flank of Dolomieu (fig. 3-14). [As of 7 Sept at 1400, the eruption was continuing from this fissure and the lava flow along the Grandes Pentes had reached 1000 m altitude (10:8)].

Eruptive activity was then limited to Dolomieu's E flank fissure at 2200 m altitude. A major cone (Thierry crater) has formed, emitting lava that advanced to an altitude of 550 m, following 1977 flows. After a few days, active flow fronts reached only 1700 m altitude, about 2 km from Thierry crater. The upper part of the fissure system was then reactivated 15-17 Sept, emitting a small amount of scoria and a small lava flow. Since 20 Sept, lava flows have been emitted through a tube at 1900 m altitude. The volume of the lava was estimated to be 14-17 x 10^6 m^3 as of 30 Sept. The lavas are basalts with variable amounts of olivine phenocrysts. Major degassing took place, rich in sulfur compounds.

Before the 6 Sept eruption, a small summit inflation was recorded. Geodetic and leveling measurements show that horizontal and vertical deformation related to the new fissures reached 40 cm, coinciding with inflation of the E zone of Dolomieu. Deformation has been recorded in the whole W side of the Enclos caldera. During the eruption, no significant movement was recorded, except on the E coast.

Information Contacts: H. Delorme, J. Delarue, OVPDLF, Réunion; P. Bachelery, Univ. de la Réunion.

10/85 (10:10) "The eruption ended during the second week of Oct. After major lava extrusion in Sept, lava flows were emitted at low rates through a tube at 1800 m altitude [but 1900 in 10:9]. The total volume of lava for the Sept episode is estimated to be 17-20 x 10^6 m^3. Degassing was very important, and eruptive cones were covered with sulfur deposits.

"Since the beginning of the month, a major deflation was detected on the E coast. This may correspond to the end of the inflation that began on 10 July (10:7). In mid-Oct, at the end of the eruption, this deflation was recorded both at the summit station and in the caldera. After some sporadic tremors, seismic activity associated with this eruption ended on 16 Oct."

Information Contacts: H. Delorme, J. Delarue, OVPDLF, Réunion; P. Bachelery, Univ. de la Réunion.

11/85 (10:11) "Seismic activity was very low during all of Nov and was at shallow depth (1-2 km) under the sum-

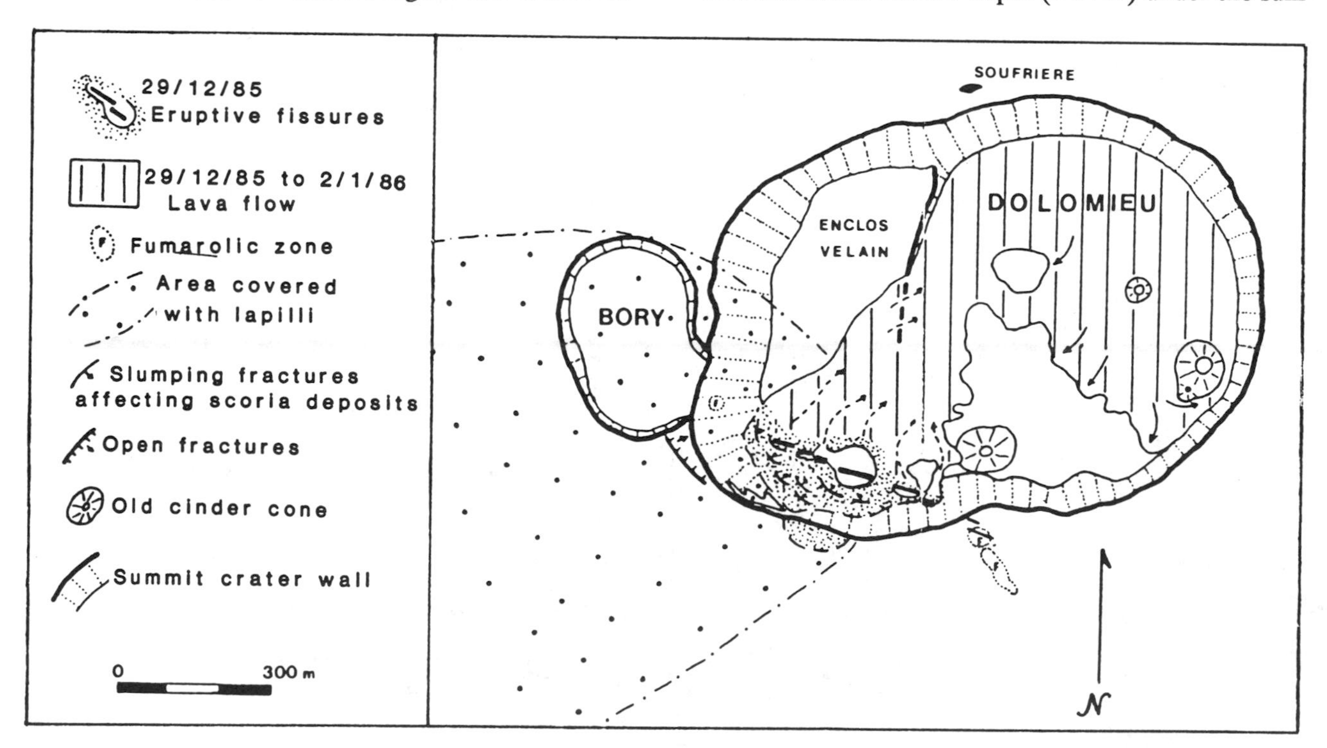

Fig. 3-15. Map of the summit craters of Piton de la Fournaise as of 2 Jan 1986, courtesy of OVPDLF. Active fissures and lava flows produced since the beginning of the eruptive episode on 28 Dec are shown.

mit. Small deformation was measured only by summit stations. Continuously-recording tiltmeters indicated progressive deformation on Bory Crater since 28 Nov.

"During the night of 1-2 Dec, a very short seismic crisis occurred. For 17 minutes, very shallow low-magnitude events occurred under Dolomieu crater at depths of 0.5-1.5 km. A 1.5-km fissure opened from the top of Bory crater down to the S flank of the central cone. This eruption lasted only 28 hours, from 2 Dec at 0215 to 3 Dec at 0600. The amount of basaltic lava emitted was very small.

"A major inflation pattern was recorded only on the summit stations. No deformation was found in the rest of the geodimeter net."

Information Contacts: H. Delorme, J. Delarue, OVPDLF, Réuion; P. Bachelery, Univ. de la Réunion.

12/85 (10:12) "After the very short (28-hour) eruption on 2 Dec, seismic activity was limited to very small shallow events in the summit zone for a few days.

"Since 25 Dec some deeper events have occurred under the summit zone (1-3 km depth). On the 28th two strong events (20 s) were recorded on the whole seismic network (11 stations) followed by a few events on the 29th. During the evening of the 28th very small events were noticed at the summit station, followed by a very short crisis (1836-1850) and opening of fractures inside Dolomieu crater (1854-1857). Aphyric basalt began to cover a large part of the crater floor (fig. 3-15).

"The opening of the fractures was sudden and rocks, cinder, and lava were ejected to heights of as much as 250 m. Lava fountains were 100-150 m high for one day, than activity was limited to one main cone (1 Jan). On 9 Jan, the main cone was still active, emitting important volumes of gases, mainly SO_2. Lava temperatures were between 1140° and 1150°C. Some pahoehoe was observed in tubes.

"After the 2 Dec eruption, deformation indicated no relaxation of the summit area. From 6-27 Dec, a progressive deformation was recorded (30 μrad) on the Bory permanent tiltmeter. Tilt stations around the summit indicated a small summit inflation. The summit area and the S and W flanks of Bory were affected by deformation. Nothing has been detected anywhere else in the Enclos caldera."

Information Contacts: Same as above.

HEARD ISLAND

S Indian Ocean
53.105°S, 73.513°E, summit elev. 2745 m
Local time = GMT + 5 hrs

Heard is an uninhabited and rarely visited island volcano 1500 km N of Antarctica on the Kerguelen Ridge.

2/85 (10:2) During the night of 14-15 Jan, personnel of the French oceanographic mission *Sibex* observed an eruption on Heard Island. A lava flow emerged from a vent at about 2750 m altitude, on the upper S flank (between Big Ben and Mawson, fig. 3-16). For 2 days, a plume was

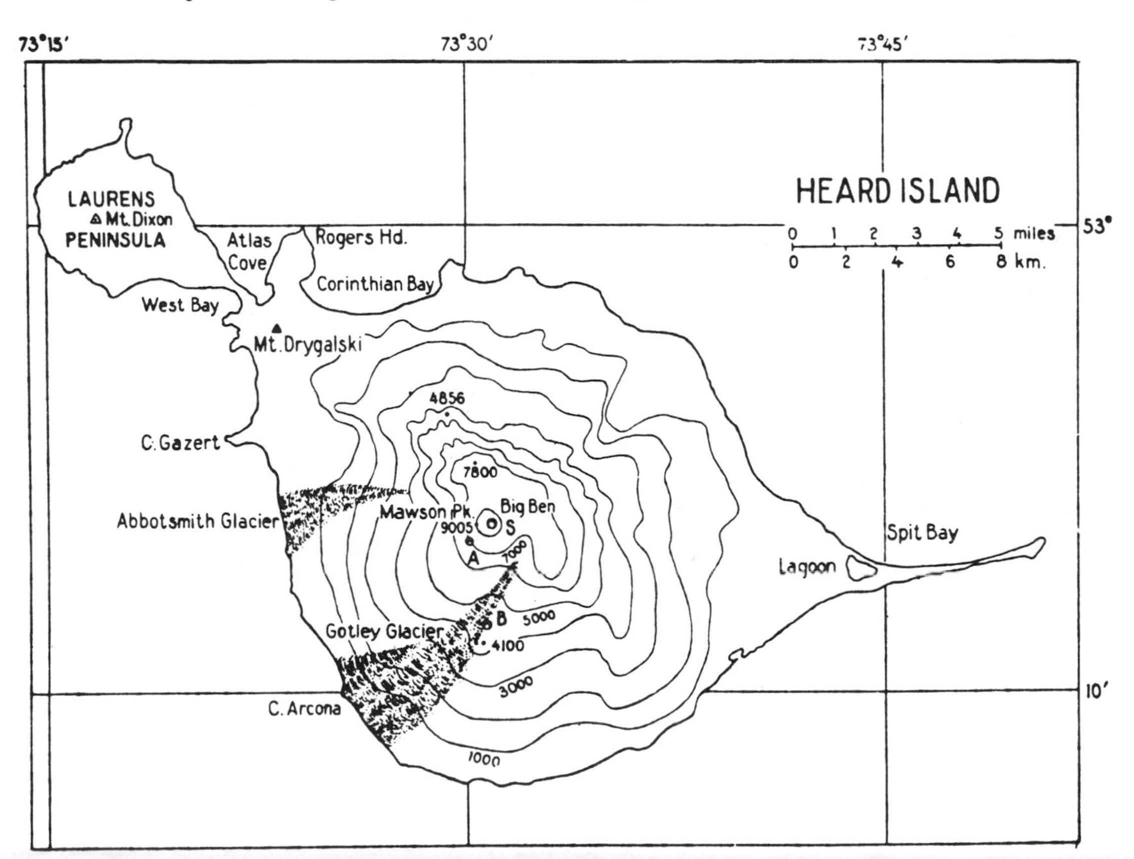

Fig. 3-16. Sketch map of Heard Island, from Neumann van Padang, M., 1963, *CAVW;* Part XVI.

visible from 32 km away. An image from the NOAA 7 polar orbiting satellite showed a diffuse plume extending SE from the island on 14 Jan at 1712. The plume was narrow over the island, widening over the ocean until obscured by weather clouds 20-25 km to the SE. The next day at 1658, NOAA 7 imagery showed that the island was free of weather clouds, but no volcanic plume was evident.

Heard Island is uninhabited and 80% of its land area is covered by glaciers. Eruptive activity was observed there in 1881, 1910, and intermittently from 1950 through 1954. Weak explosions were reported by most observers, but lava flows were seen only in 1950, 1951, and 1952. Much of the activity has been subglacial.

Information Contacts: J. Bull, Terres Australes et Antarctiques Françaises; A. Giret, Expeditions Polaires Françaises; J.L. Cheminée, IPG, Paris; W. Gould, NOAA.

10/85 (10:10) No further reports of volcanic activity on this remote, uninhabited island were received until 29 Sept, when scientists from the MV *Nella Dan* observed steam rising from a vertical crack in the Gotley Glacier (on the SW flank). On 4 Oct at about 2100, from 11.5 km offshore, 2 distinct areas of glow were observed: at the summit, and at a point between 1500 and 2000 m altitude in Gotley Glacier, 6.5-7 km from the summit. The observers believed the flank glow to be a second source of lava, although they noted the possibility that lava originating at the summit might have flowed under the glacier and emerged at that point. Space Shuttle astronauts observed the volcano emitting vapor every day during their 30 Oct-5 Nov mission and took a photograph (no. 61A-49-047), but there was no sign of fresh lava or ash on the summit..

Information Contacts: R. Varne, Univ. of Tasmania; M. Helfert, NASA, Houston.

Further Reference: Quilty, P., 1985, New Volcanic Vent on Heard Island?; *ANARE News,* Dec. 1985, p. 11

MARION ISLAND

Prince Edward Islands, Indian Ocean
46.90°S, 37.75°E, summit elev. 1185 m

This shield volcano forms a 20 km-diameter island midway between Africa and Antarctica. A meteorological station is maintained there by the South African government.

1/81 (6:1) During the first week in Nov, research station personnel visiting the W side of Marion Island observed 2 new cinder cones, 3 small lava flows, and fresh tephra deposits, none of which were present when the scientists were last in the area in Feb.

Russell and Berruti traveled to the eruption site in late Nov. Regrowth of burnt vegetation indicated that the activity had probably occurred at least 2 months earlier. The smaller of the 2 cinder cones, about 6 m high with a crater 15 m in diameter, had formed at the summit of Kaalkoppie, an eroded, 100 m-high tuff cone. A lava flow that apparently originated from the W (seaward) flank of the summit cone had poured over nearby cliffs 50-70 m high and ponded in a small amphitheater-like area at their base. About 10 m of lava remained in the amphitheater in November, but caves above this level were partially filled with lava. Some of the lava had drained from the amphitheater and continued about 100 m seaward, flowing into the ocean and forming a front about 120 m wide and 10 m high. A lava tube seen at the S edge of this flow in early Nov had collapsed by the time Russell and Berruti saw it on the 26th, forming a 4 m-wide trench. This flow covered about 2 hectares, including the portion between the summit cone and the cliffs.

A second lava flow occupied a few hundred m^2 of the promontory above the amphitheater. A small amount of this lava had spilled through a fissure onto the first flow, but most remained on the promontory or poured over its concave N cliff face into the sea.

On the flank of Kaalkoppie, E of the new summit cone and near its base, a larger tephra cone had formed around a 35 m-diameter crater. The E side of the cone was breached by a lava flow, 35 m wide as it emerged from the crater, that eventually reached 50 m width before diverging into 2 lobes. One lobe flowed about 350 m to the NW, the second about 200 m to the S along a shallow valley. The total area covered by this flow was about 7 hectares.

Irregular blocks and spheroidal bombs nearly 1 m in diameter were found on the flank cone. Fusiform and ribbon bombs fell as much as 350 m from the cone with heaviest tephra fall extending from its E, breached, side. A continuous layer of ash and lapilli covered an area extending several hundred m to the E and 40 m S of the 2 cones, with scattered fragments found 250 m to the S and much farther to the SE.

No other eruptions have been reported in historic time from Marion Island. Some unvegetated lava flows appear no more than a few hundred years old (Verwoerd, 1967).

Reference: Verwoerd, W. J., 1967, Marion and Prince Edward Islands; *Nature,* v. 213, no. 5073, p. 230-232.

Information Contacts: A. Berruti, Univ. of Cape Town; S. Russell, Univ. of Orange Free State; M. du Plessis, Geological Survey of South Africa; C. Hide, South African Embassy, Washington, DC.

Further Reference: Verwoerd, W.J., Russell, S., and Berruti, A., 1981, 1980 Volcanic Eruption Reported on Marion Island; *Earth and Planetary Science Letters,* v. 54, p. 153-6.

NEW ZEALAND and TONGA

Most of SEAN's original information about New Zealand volcanoes was from the *Immediate Reports*, a limited-circulation publication of the New Zealand Geological Survey (NZGS), the Geophysics Division of the Department of Scientific and Industrial Research (DSIR), and Victoria University of Wellington. Compiled within days of frequent field visits to several moderately active sites, the reports that follow include preliminary data more detailed than, but otherwise typical of, what is available to SEAN from many volcanic regions.

Activity is summarized annually in the *New Zealand Volcanological Record*, and we recommend nos. 5-15 (1975-1985) for the perspective that immediate reports cannot provide. At the request of New Zealand geologists, we have made some changes to correct errors of fact and emphasis, but as in other regions, we refer readers to the published literature for additional detail and interpretation.

General References: Gregory, J.G. and Watters, W.A. (eds.), 1986, Volcanic Hazards Assessment in New Zealand; *New Zealand Geological Survey Record 10,* 104 pp. (9 papers).

Houghton, B.F. and Weaver, S.D. (eds.), 1986, Taupo Volcanic Zone: Tour Guides C1, C4, C5, and A2; and North Island Volcanism: Tour Guides A1, A4, and C3; *New Zealand Geological Survey Records 11-12*, 212 and 138 pp.

Smith, I.E.M. (ed.), 1986, Late Cenozoic Volcanism in New Zealand; *Royal Society of New Zealand Bulletin 23*, 371 pp. (18 papers).

WHITE ISLAND

Bay of Plenty, New Zealand
37.50°S, 177.23°E, summit elev. 321 m
Local time = GMT +12 hrs winter (+13 hrs summer)

This 2 x 2.4-km uninhabited island, 50 km off the coast of North Island, has been New Zealand's most active volcano during the report decade.

1/77 (2:1) A visit on 13 and 15 Dec by a team of 10 scientists resulted in the following observations: "Continued bulging in the Noisy Nellie/1971 Crater area, the vigorous high-temperature fumaroles in the back wall of 1971 Crater, and generally higher ground temperature in the active part of the main crater all suggest further activity may occur. The situation is similar to that in 1967 prior to the eruption of Rudolf, and a phreatic eruption could occur in the next few months."

18 Dec: A fishing vessel reported an explosion from White Island at 1545. The White Island seismograph recorded a tremor.

20-21 Dec: Over 50 local earthquakes per day were recorded. At 1100 on 21 Dec a large steam plume rose 700 m.

25 Dec: Moderate-heavy tremor began, continuing for at least 3 days.

26 Dec: Discolored steam and some "rocks" were emitted for several hours prior to the eruption of a large, very dark ash cloud at 1545.

27 and 29 Dec: Steam and ash eruptions.

30 Dec: C. Hewson (DSIR Geophysics Division), S. Nathan, and E. Lloyd (NZGS) flew over White Island between 1033 and 1058. The E part of the island was sprinkled with mustard-green ash, which appeared thickest within the crater. Ash thickness may have exceeded 1 m near the active vent; several cm of ash coated the S rim. No impact craters were observed in the recent ash. The vent was located near the 1967 (Rudolf) crater (fig. 4-1) and was about 20 m in diameter. Two nested craters appeared to have formed: the larger was saucer shaped and occupied the approximate position of the 1971 crater; the smaller inner crater was erupting a medium to dark gray steam and ash cloud, from which a very light ashfall was visible for 2-4 km downwind. The sea along the N coast was discolored by extensive orange-brown and milky precipitates. Several streams of water were visible on the crater floor and a particularly vigorous spring emerged at the base of the E crater wall.

6-7 Jan: Commercial pilots reported increased activity.

8 Jan: C. Hewson, S. Nathan, and 2 television reporters flew over White Island between 1450 and 1510. Light gray-brown ash was vigorously erupted. The vent appeared larger than on 30 Dec. Noisy Nellie was steaming strongly, but Donald Mound appeared less active, with only distinct fumaroles emitting steam. All visible parts of the island were mantled by light gray ash, estimated to be at least 1 cm thick on coastal vegetation; ash was much more widespread than on 30 Dec. The area within about 50 m of the vent was plastered with pale gray mud. No bomb craters were observed.

11 Jan: Victoria University and NZGS scientists flew over White Island between 1110 and 1135. Steam, colored pale brown-gray by ash, billowed slowly to 800-900 m above sea level. The ash content of the eruption cloud was less than on 30 Dec. Since then, no more than 1 cm of ash appeared to have been deposited on the island. The active vents were larger and deeper, and covered the area of the 1971, Rudolf, and part of the 1933 craters [but see 2:2]. Minor fumaroles were noted on the outer slope of the volcano.

15 Jan: Two large eruptions were reported at 0830 and 1200, the eruption cloud from the latter reaching 1300 m.

17 Jan: J. Healy, B. Scott, and E. Lloyd (NZGS) overflew White Island between 1133 and 1146 and observed 1000 m steam clouds containing some ash. The active vent seemed about the same size (150 m in diameter) as on 11 Jan. Although the observation period was brief, it appeared that the nature of the eruption had changed from continuous emission to one characterized by brief periods of little or no emission. The last erup-

tion occurred 19-20 July, 1971.

Information Contacts: J. Cole, Victoria Univ., Wellington; D. Shackelford, Villa Park, CA.

2/77 (2:2) White Island was still erupting vigorously when visited on 25 Jan and inspected from the air on 12 Feb. The new crater (Christmas Crater) had increased slightly in size, but had not, as previously thought, engulfed Rudolf Crater.

25 Jan: Weak to moderate emission of ash in a moderately convoluting eruption cloud was occurring from a small vent (about 10 m in diameter) within the new 160 m-diameter collapse crater (Christmas Crater). No incandescence was observed. Two scientists enveloped in the eruption cloud for about 30 seconds were covered by fine, dry ash and reported the cloud to be only "pleasantly warm." More than 1 m of fine- to moderate-grained tephra had been deposited on the E rim of Christmas Crater, but no large ejecta were observed. Noisy Nellie was much more active than on 13 Dec (2:1), emitting steam and SO_2 under high pressure.

12 Feb: A dense buff-brown plume of convoluting steam and ash rose 600-750 m from a small vent near one wall of Christmas Crater (fig. 4-2). The eruption cloud was considerably darker than on previous inspections because of a higher ash content, and ash cover within the crater appeared much thicker than on 25 Jan. Christmas Crater had increased 10 m in diameter and was 100 m deep, with steep, buttressed walls and a generally flat floor. Slump blocks had collapsed into the crater and pools of discolored water were on its floor. Mudflows covered the steep slopes around the crater.

Rudolf Crater was only a few meters deep, having been nearly filled with ash and slump debris. Noisy Nellie was steaming strongly, emission from Donald Mound was normal, and a little steam was issuing from the S wall of Gilliver. The crater floor around Noisy Nellie and Donald Mound was coated with sulfur.

Information Contacts: I. Nairn, NZGS, Rotorua; D. Shackelford, Villa Park, CA.

4/77 (2:4) White Island was inspected from the air on 25 March and 14 April, and visited on 4 April. On 25 March, a tan gas and ash cloud with an orange base was emitted from the vent, which had migrated from the wall to the N base of Christmas Crater, allowing access to the vent by runoff water for the first time. A glow had been observed during an 11 March overflight, but could not be confirmed on 25 March because of the large quantity of gas filling the crater. Many impact craters and large blocks, not present on 11 March, were seen S and E of Christmas Crater and on the floor of 1933 Crater, indicating that a major explosion had taken place between 11 and 25 March.

Observers reported a deep red glow above White Island during the night of 26 March, and clouds, frequently blackish, rising to 2000 m on 26-27 March.

On 4 April, a voluminous, moderately convoluting cloud of incandescent ash was rising to 600 m in brief

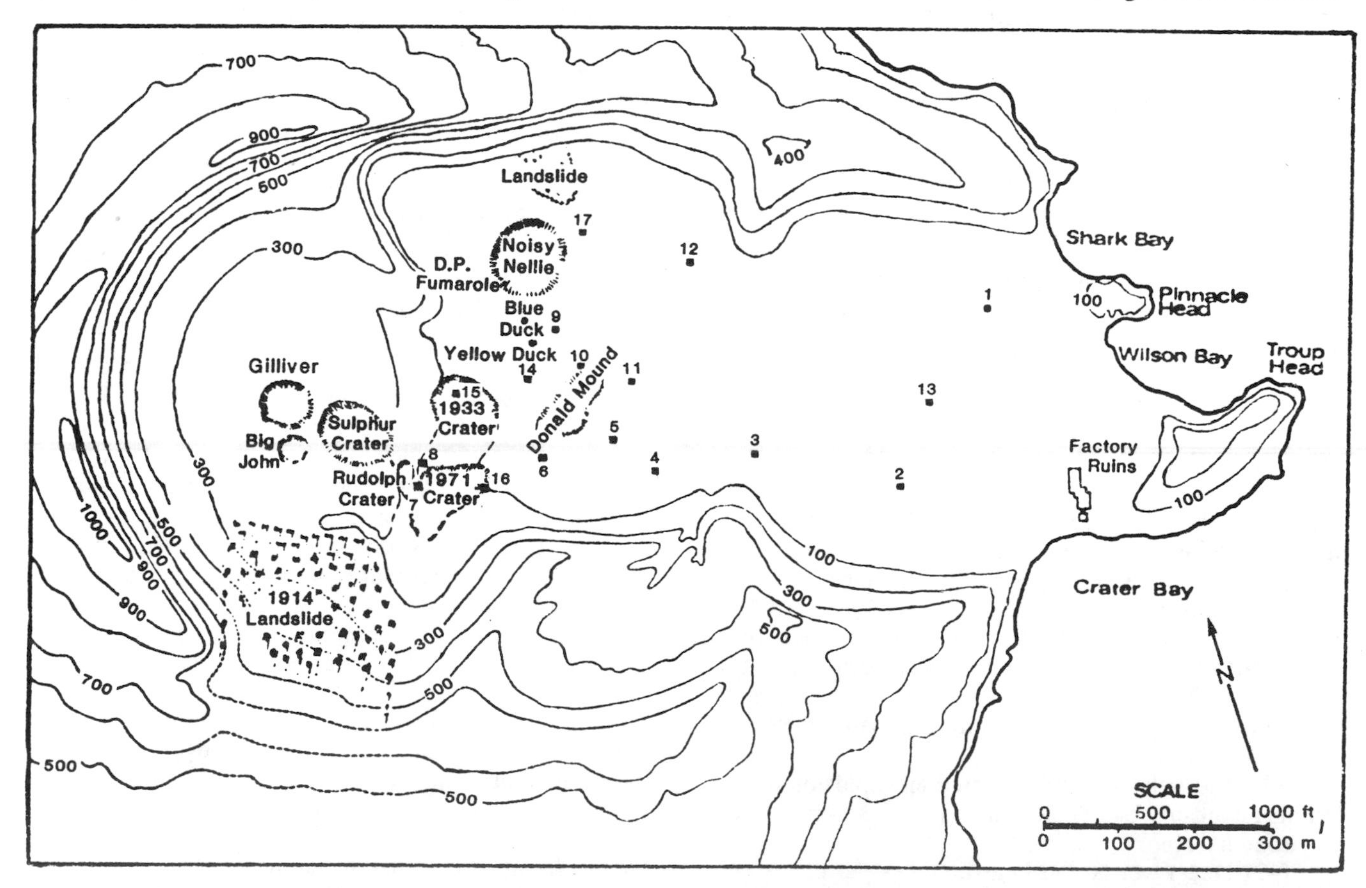

Fig. 4-1. Main crater of White Island Volcano, showing the locations of various vents and fumaroles. After Browne, P.R.L. and Cole, J.W., 1973, Surveillance of White Island Volcano, 1968-1972; *New Zealand Journal of Geology and Geophysics*, v. 16, no. 4, p. 959-963.

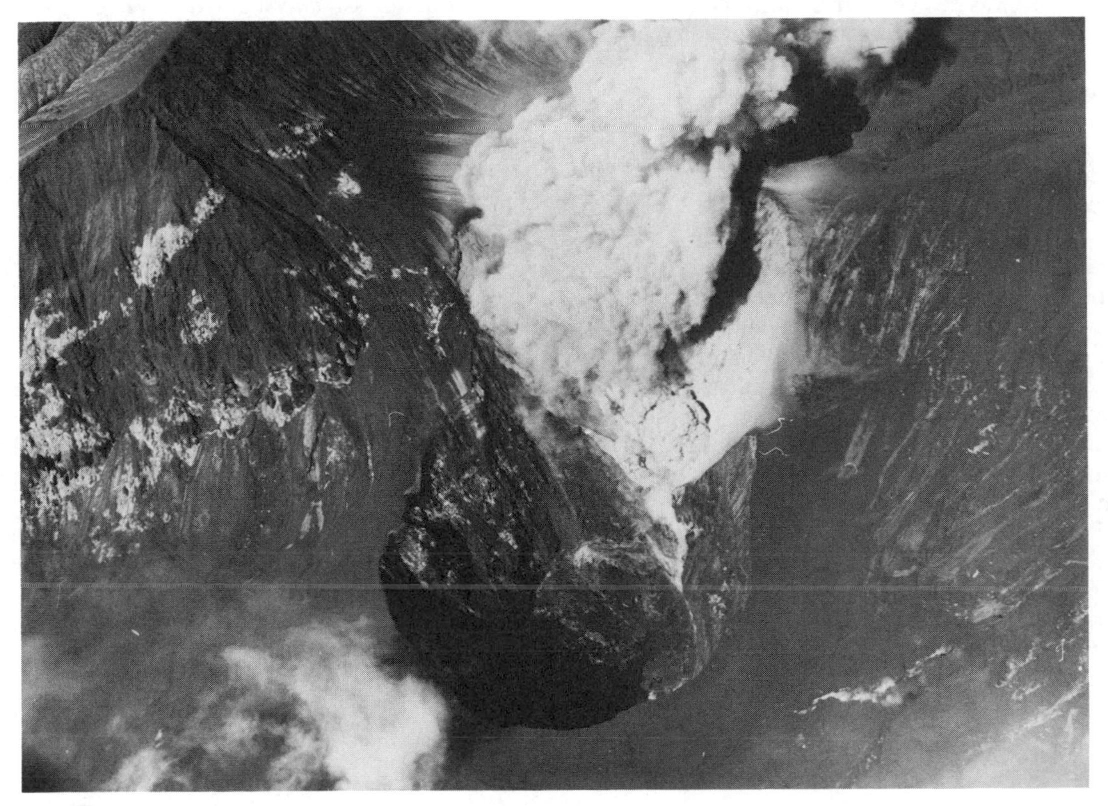

Fig. 4-2. Airphotos of Christmas Crater, taken 12 Feb 1977, showing steam and ash emission from the active vent. Courtesy of Ian Nairn.

puffs, and drifting to the SE. A comparison of 25 March airphotos with 4 April ground observations indicated that there had been no eruption of large ejecta since 25 March. The largest blocks from the March eruption were composed of accidental material, but most of the tephra consisted of scoriaceous essential lava ranging in size from ash, to blocks and bombs up to several m across. The March eruption was the largest 20th-century explosion at White Island and the first to produce essential ejecta, but no eyewitness reports of a large eruption have been received. [A careful search by J.H. Latter of the records of regional seismic stations failed to detect any earthquakes at White Island during this period.]

By the 14 April overflight, activity had declined to low-volume emission of a dark fawn-colored, slowly convoluting steam cloud, containing a little ash. There was no evidence of major explosive activity postdating the March eruption. A linear fumarole zone had developed, extending from the N end of Wilson Bay across the W end of Shark Bay to the crater wall.

Information Contacts: I. Nairn, B. Scott, NZGS, Rotorua; J. Latter, DSIR, Wellington; R. Clark, Victoria Univ., Wellington.

Further Reference: Clark, R.H., Cole, J.W., Nairn, I.A., and Wood, C.P., 1979, Magmatic Eruption of White Island Volcano, New Zealand, December 1976-April 1977; *New Zealand Journal of Geology and Geophysics*, v. 22, no. 2, p. 175-190.

5/77 (2:5) The volcano was inspected from the air on 5 May. A steam and ash column, light tan and slowly convoluting, rose 600-900 m above Christmas Crater. No incandescence was observed, nor were there any new large ejecta, new impact craters, or major new ash deposits on the main crater floor. Christmas Crater appeared to be unchanged from 25 April.

Information Contact: NZGS, Rotorua.

6/77 (2:6) Another aerial inspection on 24 May revealed 2 new ash deposits: the E half of the island was covered by gray-green ash erupted about 10 May, and the main crater contained red-pink ash that had fallen during the night of 23-24 May. Steam was being emitted at low velocity from Christmas Crater. On another overflight 1 June, Christmas Crater was emitting a steam column containing some ash which rose slowly to 1100 m, interspersed with occasional higher velocity ash pulses. Some ash adhered to the aircraft, for the first time since 28 March. No new large ejecta were observed during either inspection.

Observers in Whakatane (about 55 km from White Island) reported eruption clouds during the early morning of 28 May and at 0627 on 1 June. An earthquake was felt in the Bay of Plenty at 0653 on 1 June.

Information Contact: Same as above.

8/77 (2:8) There were inspections from the air on 5 and 26 July, and from the ground on 3 Aug. At 1117 on 5 July, Mr. Harvey, Civil Defence Whakatane, observed an ash column that rose rapidly to an estimated 1200 m. The crater and the base of the column appeared to be glowing. Mr. D. Thursby viewed White Island at 1150 from Te Kaha, about 50 km away, and reported a dense, red ash cloud that dispersed after rising about 300 m. The aircraft reached White Island at 1534. A pink, moderately convoluting cloud, interspersed with higher velocity ash pulses, was rising 300-400 m. A thick layer of deeply channeled red ash covered the main crater floor. No large ejecta or bomb craters were observed, nor was there any incandescence.

During the 26 July aerial inspection, ash content of the plume was reduced to an orange tint near the base. Brick-red ash, thought to be several days old, covered the island. No new impact craters were noted.

The 3 Aug ground inspection revealed that the gullied red ash seen from the air on 5 and 26 July had been covered by coarse gray tephra, field identified as glassy basaltic andesite, ranging from ash to blocks (up to 1 m in diameter) and bombs. Some of the gray tephra initially deposited on the main crater walls had been remobilized to form debris flows, about 3 m wide at their fronts and usually less than 100 m long, that had moved down pre-existing debris fans towards the crater floor. Pits dug at several locations in the main crater revealed that more than 82 cm of ash had fallen near Christmas Crater since 5 May. The gray tephra layer comprised 37 of the 82 cm, and was overlain by 2-5 mm of fine red ash.

The ash column on 3 Aug was light tan, and was emitted at moderate velocity to about 600 m. No incandescence was observed.

Information Contacts: J. Cole, Victoria Univ., Wellington; B. Scott, B. Houghton, NZGS, Rotorua.

9/77 (2:9) White Island was inspected from the air on 25 Aug and from the ground on 28 Aug. A large but brief ash eruption from Christmas Crater began at 1445 on 25 Aug. Steam emission had been moderate to voluminous during the 2 previous weeks and a deep red glow was reported by pilots beginning about 1 week before the eruption. Pilot Bruce Black reported that the ash column, which included a few jets of incandescent material, rose rapidly to about 4500 m before being blown NNW by a 40 km/hr wind. A few minutes later, a second explosion produced a cloud that reached 6000 m altitude, but contained less incandescent material. Base surges from this cloud moved across the main crater, some of them against the wind. By 1515, activity had declined to steam emission.

Ground inspection revealed light gray ash, about half the thickness of the late July deposits, containing lithic blocks up to 1 m in diameter. Impact craters with subdued outlines were common, the largest about 3 m in diameter and 1 m deep. Base surges had deposited moderately well-developed cross bedded transverse dunes. Slumping and a mudflow deposit were observed NW of Christmas Crater, and strong steam emission was occurring from the associated concentric cracks.

Information Contacts: I. Nairn, E. Lloyd, B. Houghton, P. van der Werff, NZGS, Rotorua.

10/77 (2:10) No major explosions have occurred since the 25 Aug event. An aerial inspection on 9 Oct revealed a pink, moderately convoluting ash column rising about 900 m from Christmas Crater. Some rainwater channeling of the 25 Aug ash deposit had taken place.

Information Contact: B. Scott, NZGS, Rotorua.

11/77 (2:11) A 2000-m brown ash cloud was observed over White Island on 7 Nov and eruption of additional ash was reported the next morning. An aerial inspection on 10 Nov revealed a white 600-m non-convoluting steam column emerging from a new fumarole in the N wall of Christmas Crater. Strong steaming, depositing sublimates, continued from several fumaroles.

White Island was visited on 20 Nov. Poorly sorted accessory ash and lapilli SE of Christmas Crater reached a maximum thickness of 330 mm above the 25 Aug deposits. A tongue of debris flow tephra, including blocks up to 1.6 m in diameter, extended 400 m SE from Christmas Crater. Base surge dunes were observed N of Christmas Crater.

Information Contact: E. Lloyd, NZGS, Rotorua.

12/77 (2:12) White Island seismograms for the period 30 Aug-15 Nov were analyzed (table 4-1). [About 2700] small volcanic earthquakes have been counted and separated into 4 groups [for which the magnitude-frequency coefficient b was determined.] All the events interpreted as explosion earthquakes occurred during a [strong] swarm of A-type events between 1722 on 21 Oct and 0724 on 22 Oct. The largest, M_L 2.6, occurred at 0104. [An eruption is known to have taken place between 9 Oct and 10 Nov (date unknown). From the explosion earthquakes, J.H. Latter suggests that the peak of this activity probably occurred at 0104 on 22 Oct.]

Information Contact: J. Latter, DSIR, Wellington.

Number of Events	Earthquake Type (30 Aug–5 Nov 1977)
1584	A-type (high-frequency).
873	Minakami B-type (low frequency); produced in or very close to magma or in areas of hot gas.
218	Intermediate frequency ["C-type"] Generally closely associated with periods of volcanic tremor [which has a dominant frequency of $2^1/_2$ to $3^1/_2$ Hz, but sometimes as high as 4-5 Hz.]
20	Explosion earthquakes.

Table 4-1. Number [revised] and type of events recorded by White Island seismic stations, 30 Aug–5 Nov 1977.

1/78 (3:1) White Island was inspected from the air on 19 Jan, the first observation since 20 Nov. No significant activity had been reported during the intervening period.

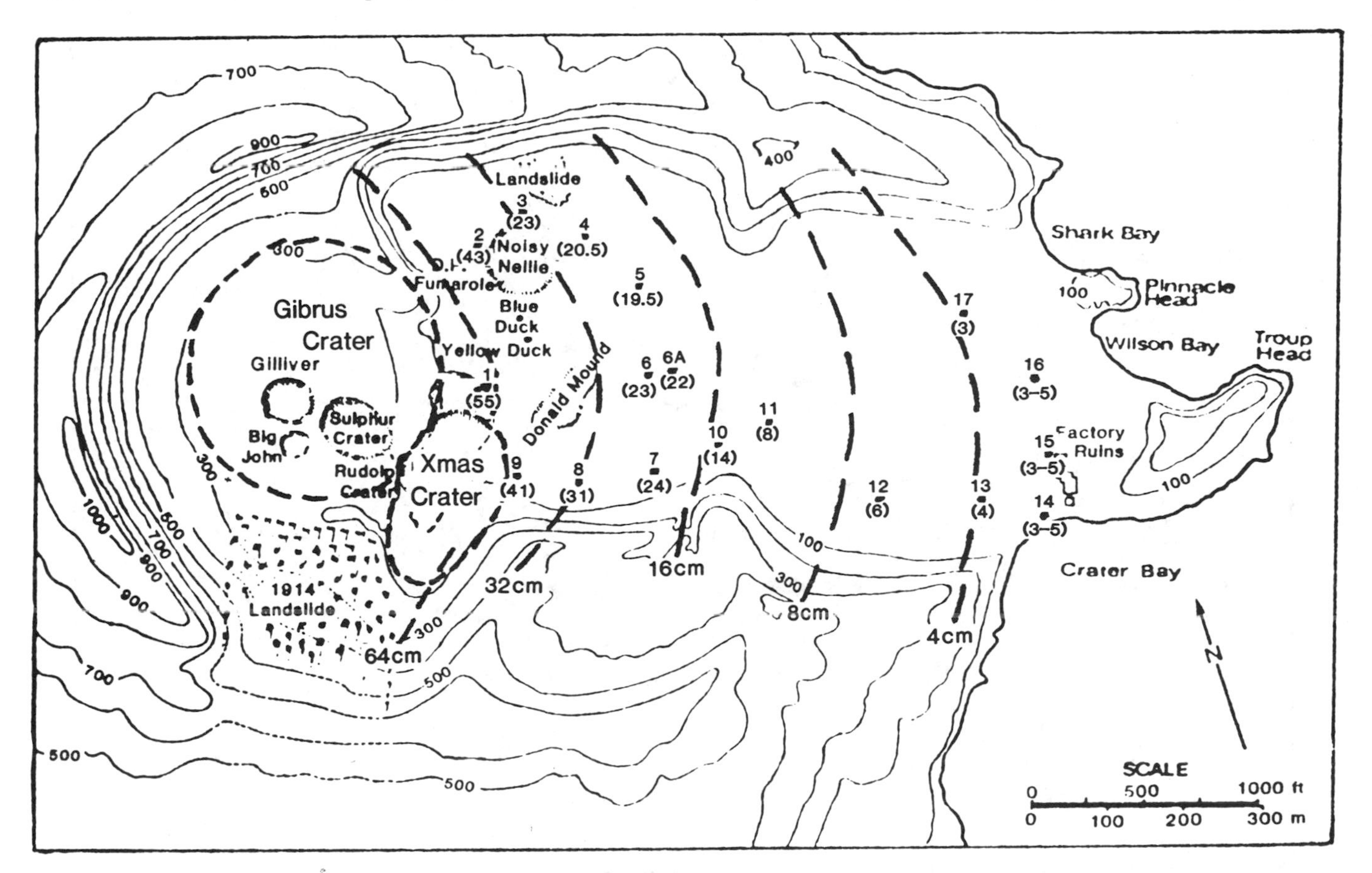

Fig. 4-3. March 1978 map showing approximate boundaries of the new crater (Gibrus) and Christmas crater, tephra isopachs, and sites of measured sections (thicknesses in cm, measured 21 March, in parentheses). Courtesy of the NZGS.

A 600-m column of weakly convoluting white steam was continuously emitted from Christmas Crater during the 20-minute overflight. A yellow-green pond, first noted in Nov, had grown to occupy much of the central floor of Christmas Crater, but was not steaming or otherwise active. Strong steaming continued from the new vent in the NW wall of Christmas Crater and from older fumaroles within the large main crater.

Information Contact: I. Nairn, NZGS, Rotorua.

3/78 (3:3) White Island was visited on 16 Feb, and was viewed from the air on 23 & 27 Feb and 2 & 16 March. Gas and a little ash were emitted at high pressure from a new vent in Christmas Crater on 16 Feb. . . . On 20 Feb between 1500 and 1800, observers about 55 km from the volcano noted a black ash column rising to about 1500 m, the largest seen since Nov 1977. During this activity, vigorous tremor and trains of apparent B-type earthquakes were recorded. Tremor intensified 21-22 Feb, accompanied by large-amplitude long-duration earthquakes.

During the 23 Feb overflight, a moderate amount of gray ash rose to about 1200 m altitude from the new Christmas Crater vent. Strong fumarolic activity continued from various vents. Ash emission from Christmas Crater was continuing on 27 Feb, and had begun from Gilliver (the Nov 1966-March 1967 eruption crater), about 200 m to the NW, which previously had only fumarolic activity during the current eruption.

Between 27 Feb and 2 March, the seismograph recorded periods (45 minutes to 6 hours) of semi-continuous 3-4 Hz vibration and several low-frequency, probable B-type shocks. Gray ash covered White Island on 2 March, but no impact craters were seen near the Christmas Crater vents, which emitted a moderate amount of dark brown ash that rose to about 900 m altitude at low velocity. Some ash was in the steam column from Gilliver Crater, but less than on 27 Feb. Strong fumarolic activity continued at other vents.

Semi-continuous 3-5 Hz vibration was recorded 2-10 March, accompanied by numerous A-type and some B-type shocks. After the 10th, seismic activity was limited to a few 45-80-minute trains of A-type events recorded 13-16 March.

A thick blanket of light gray ash covered all of White Island on 16 March. A 350 m-diameter collapse crater had formed since the 2 March overflight, bounded by concentric cracks that developed in the spring of 1977. The collapse crater occupied the area of the new Christ-

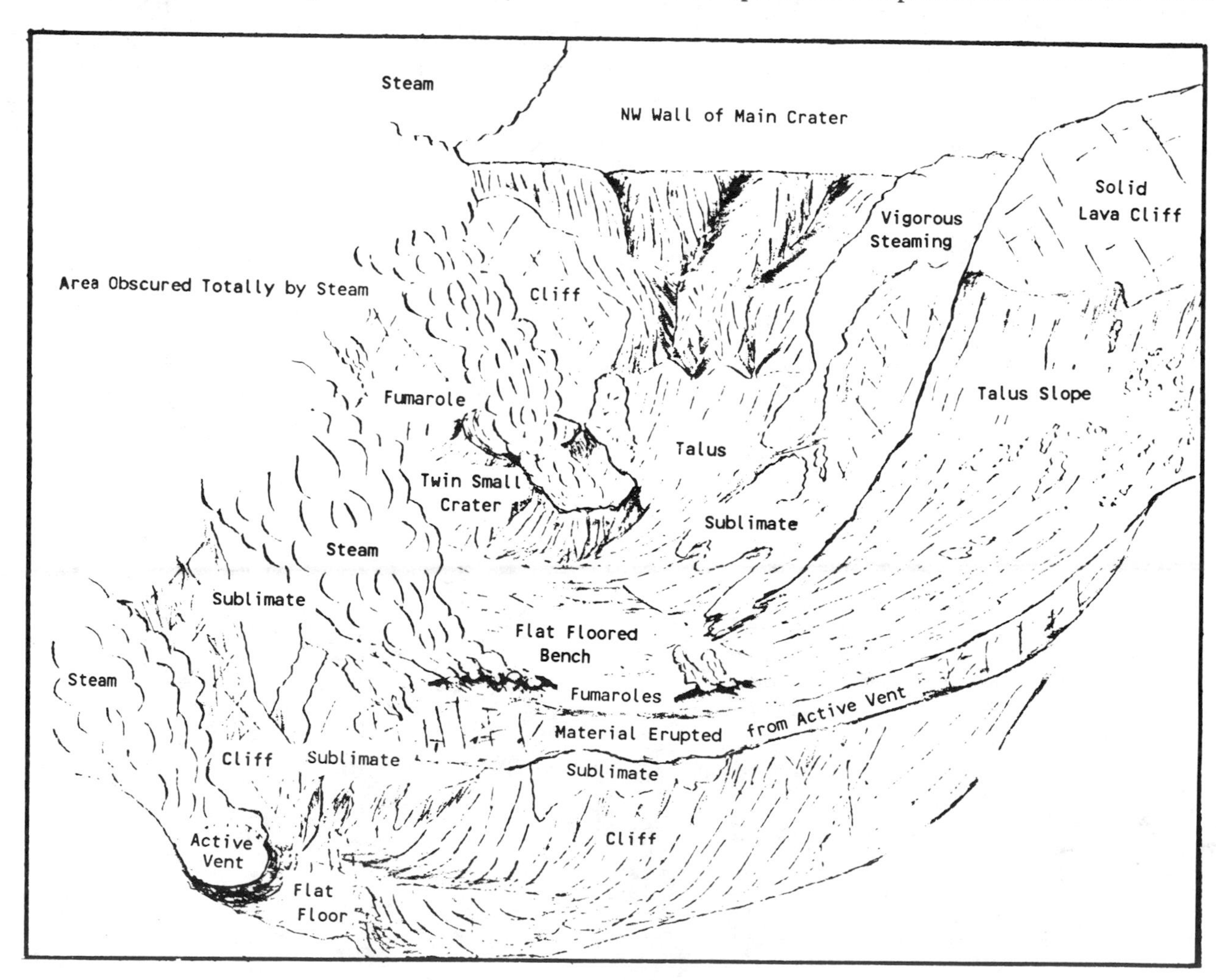

Fig. 4-4. Sketch by B. J. Scott showing the interior of Gibrus Crater on 21 March 1978, looking W from the E crater rim.

mas Crater vent and older craters Gilliver, Sulphur, and Big John. Strong steaming obscured the crater floor, but it appeared to have a maximum depth of about 100 m. Episodic emission of ash and blocks (to a maximum height of about 200 m) was occurring from the collapse crater, at the approximate former location of Sulphur Crater.

Information Contacts: I. Nairn, B. Scott, NZGS, Rotorua.

4/78 (3:4) White Island was visited on 21 March and inspected from the air on 5 April. The new collapse crater, provisionally named Gibrus Crater, is the largest new topographic feature to form on White Island since 1914 (average diameter is 325 m, depth about 100 m; fig. 4-3). Its approximate volume was 6.7 x 10^6 m^3, more than 3 times that of adjacent Christmas Crater.

Steam rising to 400 m obscured the new crater during most of the 21 March visit, but cleared briefly revealing steam vents, including the "active vent," which ejected ash and blocks during the 16 March overflight. A pair of 20 m-diameter vents near the NW wall had built small ash cones (fig 4-4). Hot gas (>550°C) was emitted at high velocity from a new fumarole 50 m NNE of Noisy Nellie and temperatures of 370°C and 250°C were recorded in H_2S-rich gas from strongly active fumaroles on Donald Mound. Strong fumarolic activity deposited sulfur on the S wall of Christmas Crater.

Numerous impact craters containing accessory blocks up to 0.6 m in diameter were found on 21 March up to 500 m from the new crater. Tephra thickness exceeded 85 cm near the crater rim. All tephra had been deposited as airfall.

Large black eruption clouds were observed at 0950 on 30 March (rising 3000-4500 m) and during the early morning of 4 April (visible from Tauranga, about 100 km away). However, no major new ashfall was seen on 5 April, nor were there any apparent changes in crater morphology (figs. 4-5 and 4-6). Steam from Gibrus and Christmas craters rose 750 m, but other fumarolic activity had declined.

The formation of Gibrus Crater between 2 and 16 March was not observed, but numerous local earthquakes were recorded during this period. Larger events are tabulated below (table 4-2).

[J. H. Latter has rewritten the following paragraph based on more up-to-date information.

"Volcanic tremor was conspicuous until 17 March. It was strong on 8 and 17 March, and ended immediately after the explosion earthquake at 2146 on 17 March."]

Information Contacts: B. Scott, E. Lloyd, I. Nairn, NZGS, Rotorua; J. Latter, DSIR, Wellington.

5/78 (3:5) Eruptions accompanied by seismicity resumed on 13 May after more than a month of weak to moderate fumarolic activity. Voluminous dark eruption clouds rising to about 3000 m were reported by persons living near the Bay of Plenty (50-60 km from White Island) on 13 May, but poor weather hampered observations for the next several days. Visibility improved on the morning of 17 May, and a large cloud was seen from Whakatane, 55 km S of White Island. An aerial inspection early

Fig. 4-5. Oblique airphoto taken by J. Perrin on 11 March 1977, looking W, showing much of the W portion of White Island's main crater before the formation of the new (Gibrus) crater in March 1978. Compare with April photograph (fig. 4-6). Courtesy of B. J. Scott, NZGS.

1978	Time	Magnitude (M_L)	Type	Remarks
Mar 2	0125	~3.1	A	
4	0316	2.4	E	
5	0112	2.6	A	2 events
	0355	2.4	E	
	1743	2.4	C	
6	2006	2.7?	A	
8	0302	2.4	E	
	1805	2.5?	C	
15	1700	2.4	C	
16	0320	2.3	C	
17	2111	2.6	C	largest C-type shock during the period
	2136	2.4	A	
	2146	2.5	E	probably the largest explosion earthquake during the period

Table 4-2. Larger seismic events 2-17 March 1978. Earthquake type A is Minakami A-type (>3 Hz); type B, low frequency (<1.75 Hz); C type, intermediate in frequency between A and B types. Type E indicates an explosion event. J. H. Latter has corrected elements of this table.

that afternoon revealed a 2000-m steam cloud, containing little or no ash, fed by vigorous fumarolic activity from several sites, including a new, nearly horizontal fumarole on the SW wall of Christmas Crater. Fine dark gray ash mantling the main crater floor had been disrupted by rainfall on 15 May and probably 16 May. No coarse debris or impact craters were visible.

Subsidence, not seen during the previous overflight (2 May), was defined by arcuate scarps extending from the W wall of the new (Gibrus) crater to the SE wall of Christmas Crater. Vigorous steaming occurred from the collapse scarps, which were displaced about 1 m down to the NE.

Low-amplitude, low-frequency tremor began at about 2100 on 12 May and amplitude increased for ~22 hours. A similar tremor episode started on 15 May at about 0300, increased in amplitude until around 1900, then gradually subsided 16-18 May. Low-frequency B-type shocks accompanied the first period of tremor, but decreased noticeably during the second tremor episode.

Information Contacts: B. Scott, E. Lloyd, B. Houghton, NZGS, Rotorua.

7/78 (3:7) Two brief periods of volcanic tremor, each fol-

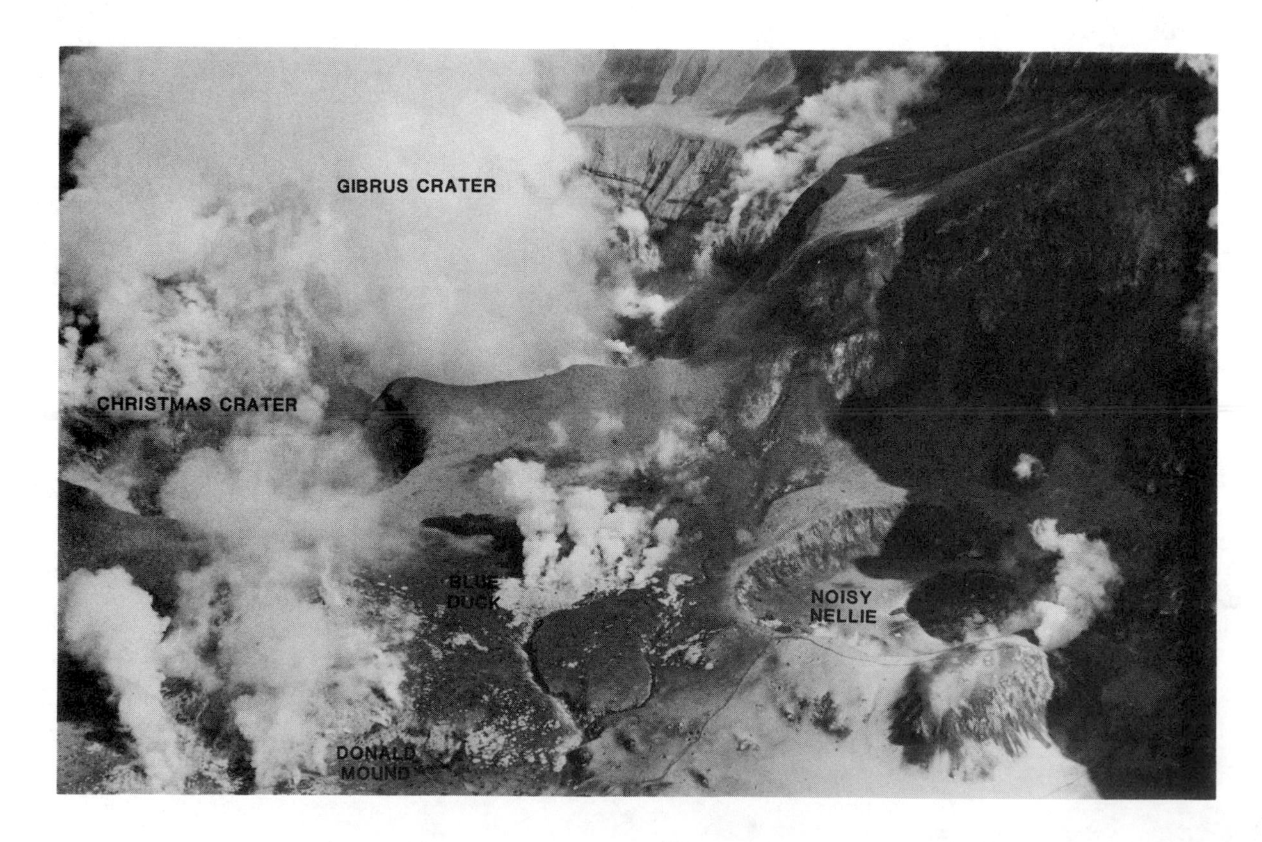

Fig. 4-6. Oblique airphoto taken by E. F. Lloyd on 5 April 1978, looking WNW. The large, vigorously steaming Gibrus Crater is behind the shallower, gently steaming Christmas Crater.

lowed by multiple earthquakes, were recorded during June. After low-frequency tremor between 0100 and 2000 on 8 June, multiple shocks occurred at 002645 on the 9th. Medium-frequency tremor lasting from 1200 on 22 June until 1100 on 24 June was succeeded by a high-frequency tremor episode beginning at 1518 on the 24th and culminating in a series of earthquakes at about 1646. Local seismicity was otherwise limited to a few small earthquakes per day.

White Island was inspected from the air on 28 June. No evidence of major eruptive activity since 17 May was observed. Voluminous steam emission from the new collapse crater (Gibrus) produced a convoluting steam column that rose about 600 m. Some new fumaroles were visible inside Gibrus, but other fumarolic activity had declined. A formerly hot (> 550°C in late March) and vigorous fumarole NW of Noisy Nellie had declined to weak steaming. Two new pits were observed within Christmas Crater. The larger, 20-25 m in diameter, contained a pond of green water; the smaller, 15 m in diameter, was surrounded by a small amount of grayish ejecta.

Information Contacts: P. van der Werff, I. Nairn, B. Scott, NZGS, Rotorua.

8/78 (3:8) White Island was inspected from the air on 26 July and visited on 6 and 9 Aug. Impact craters not present on 28 June were visible in 26 July airphotos. The time of the eruption that produced the craters is uncertain. However, a 4500-m black eruption column was seen from the mainland by one observer, about 50 km away, at 1045 on 17 July.

Numerous impact craters from 0.5 to 2.0 m in diameter were found during the Aug visits, extending E from the active vent in Gibrus crater and reaching a concentration of about 1 crater/4 m^2 area. Geologists excavated a few of the craters, each of which contained an angular andesite block. Poorly sorted ash- to lapilli-sized gray tephra, 10-20 mm thick near the vent, mantled the surface. The tephra deposit contained no evidence of fresh magma. According to NZGS geologists, the tephra was probably deposited by a pyroclastic flow or base surge. The impact craters contained tephra deposits only on their distal rims, indicating that the tephra's movement included a substantial horizontal component.

The narrow ridge separating Christmas and Gibrus craters was virtually destroyed between 28 June and 26 July. The NW portion of the old Christmas Crater had deepened, and was occupied by vigorous fumaroles and a series of yellowish-green ponds. Continuous fumarolic activity was observed at other sites within this new composite crater, and weaker fumaroles were active NW of it.

Information Contacts: B. Houghton, E. Lloyd, B. Scott, NZGS, Rotorua.

10/78 (3:10) At least 50 discrete high-frequency earthquakes were recorded between 0034 and 0608 on 5 Oct. Large events occurred at 0037, 0250, and 0424. With the possible exception of the second event, the shocks had very sharp onsets and were interpreted as A-type events. The earthquake swarm was followed by bursts of low-amplitude volcanic tremor, which continued sporadically through the rest of the day.

Time	Description
1849:45	explosion-type event, lasting a few seconds.
1849:50-1901	continuous high-frequency tremor.
1901	brief explosion event, slightly larger than the first.
1901-1906	continuous high-frequency tremor, with more than twice the amplitude of the tremor before 1901.
1903:30-1906	at least 8 high-amplitude multiple events.
1906	Tremor rapidly died away.

Table 4-3. Sequence of seismicity accompanying the 24 Dec 1978 eruption.

An aerial inspection on 5 Oct revealed no new tephra or impact craters. A white, gently convoluting steam cloud rose from fumaroles in 1978 crater and at the base of the main crater wall, north of Noisy Nellie. Yellow and white fumarole sublimates covered the walls of 1978 Crater.

Information Contact: E. Lloyd, NZGS, Rotorua.

1/79 (4:1) A period of renewed ash emission that began on 24 Dec 1978 has been one of the longest and most voluminous since the eruption began in Dec 1976. The ash emission was preceded by considerable local seismicity recorded since early Nov. The seismicity consisted of periods of continuous high-frequency tremor lasting up to 30 hours, separated by quiet periods of similar length, and other periods of similar total duration characterized by up to 9 bursts of noise per minute.

Shortly after 1900 on 24 Dec, persons on the mainland (about 50 km from White Island) observed the ejection of a dark, billowing cloud to an altitude of about 3 km. This cloud drifted E and was followed by several smaller dark ash clouds. The eruption was accompanied by a brief seismic sequence (table 4-3).

The next morning, NZGS personnel flew over the volcano and observed new gray ash near the rim of 1978 Crater. Vapor rose from a cluster of small vents on the 1978 Crater floor, near the former site of Rudolf vent. Some block-sized ejecta surrounded the active vents, but no large ejecta could be seen outside 1978 Crater. Later that morning, pilot Graeme Bell saw blocks and ash ejected from these vents, with blocks rising 30-50 m.

Numerous small, high-frequency earthquakes occurred on 25 and 28 Dec, followed by periods of semi-continuous high-frequency tremor on the 28th and 29th. A few low-frequency B-type shocks were also recorded. Tremor became nearly continuous after about 0800 on 29 Dec. On the 31st, pilots reported block and ash eruptions.

During overflights on 1 and 9 Jan, columns of gas and ash rose 600-700 m from 1978 Crater. A few blocks and impact craters were visible near the crater rim on 1 Jan, but block ejection was not observed on either occasion. Between the 1st and 9th, long periods of semi-continuous 3-7 Hz tremor were recorded on 6 days. Explosion-type earthquakes were recorded on 2 and 6 Jan and two large A-type events occurred on 8 Jan. Eruptive activity was visible from the mainland (about 50 km away) on 4 Jan and five ash clouds, each rising about 1 km, were ejected in a 40-minute period the next morning. A 2 km-high cloud was seen during the late afternoon of 7 Jan, followed by 7 more large clouds in the next 35 minutes.

Ground inspection on 12 Jan showed four separate layers of fresh ash totaling 32 cm thick mantling the crater rim, underlain by a blocky layer that may have predated the 24 Dec activity. Ash thicknesses decreased to 6 cm about 200 m to the E. An ash column rose to about 450 m altitude from the active vent, about 70 m in diameter on the 1978 Crater floor.

Information Contacts: E. Lloyd, I. Nairn, B. Scott, NZGS, Rotorua; H. Palmer, Post Office, Rotorua.

2/79 (4:2) Increased ash emission and local seismicity were continuing in late Jan. Long periods of semi-continuous, moderate- to high-frequency tremor continued from the 12th, but were punctuated on 16 Jan by 19 possible explosion-type earthquakes between 0052 and 1738 (the largest at 0451). On 22 Jan there were 6 explosion events (maximum amplitude less than half that of the largest 16 Jan event) from 1158 to 1222, and tremor gradually declined to low levels after 0800 on the 23rd. Eleven hours later, however, tremor amplitude increased for 2 1/4 hours, then remained continuous at moderate to high frequency for over 5 days. A single explosion-type event occurred at 1015 on 26 Jan and tremor declined to a low level at 0500 on the 29th. [The largest B-type shock of the period, magnitude 2.45, occurred during a 3-hour swarm of about 50 similar events on 24 Feb.]

On 23 and 25 Jan an ash column rose to about 600 m above sea level from the active vent, which was at least 50 m deep. About 40 cm of ash was measured at the rim of 1978 Crater on the 25th, 8 cm more than on 12 Jan. Blocks up to 1 m across, not observed on the 12th, were found about 200 m to the E. Fumarole temperatures of up to 535°C were recorded, similar to those measured on 7 Dec.

Information Contacts: B. Scott, I. Nairn, NZGS, Rotorua.

4/79 (4:4) The morning of 12 April, a light tan, mildly convoluting eruption column rose from a deep vent in the floor of 1978 Crater. The column reached an altitude of about 500 m before drifting E and producing a fine ash fallout. No fresh impact craters or blocks were observed near the 1978 Crater rim, but brown ash covered the main crater floor. Strong vapor emission also occurred from a fumarole E of 1978 Crater. Almost continuous medium- to high-frequency tremor, had persisted for the past several weeks.

Information Contact: B. Scott, NZGS, Rotorua.

5/79 (4:5) At least 2 moderate explosive eruptions seemed to have occurred since 12 April. R. R. Dibble and E. Hardy observed recently ejected tephra deposits on 21-22 April. [J.H. Latter notes that the strongest period of volcanic tremor recorded between Aug 1977 and Sept 1979 started at 1800 on 29 April and continued until 1 May at 0856.] A newspaper article (*Whakatane Beacon*, 9 May) reported that an eruption at 2030 on 1 May produced a shock wave recorded by a vessel anchored off White Island and a thin layer of ash fell on the same vessel. At 1737 the next day, S. Harvey (Civil Defence, Whakatane, about 50 km from White Island) observed an eruption column that rose to 4-5 km altitude.

NZGS personnel overflew White Island on 18 May. A low, white gas column from the active vent in 1978 Crater obscured the W portion of the main crater floor. Strong steaming also occurred from fumaroles N and E of 1978 Crater. Reddish brown tephra covered much of the visible area of the main crater, and there were post-12 April impact craters up to 2 m in diameter 1/2 km E of the active vent.

Information Contacts: B. Houghton, I. Nairn, NZGS, Rotorua; R. Dibble, E. Hardy, Victoria Univ., Wellington; S. Harvey, Civil Defence, Whakatane; The *Whakatane Beacon*.

6/79 (4:6) A NZGS team visited White Island on 28 May. The active vent, in the smaller SE section of the dumb-bell-shaped 1978 Crater, emitted a weakly convoluting, ash-poor eruption column, punctuated by occasional pulses of more vigorous ash ejection. Bombs up to 1 m in diameter were infrequently thrown above the vent, breaking open to reveal incandescent interiors. Sharp detonations were heard as often as every 3 to 5 seconds, but appeared to be related to gas release, not bomb ejection. The bottom of the steep-sided active vent was not visible, but it appeared to be more than 100 m deep. Fumarolic activity continued at several locations, but appeared to be at lower pressures and temperatures than during other recent visits. The temperature at one fumarole a few hundred meters E of 1978 Crater had decreased to 410°C from 535°C on 25 Jan.

The entire main crater floor was covered by red tephra from the April-May eruptions. A pit dug 5 m from the edge of 1978 Crater (but not along the E-W axis of heaviest deposition) revealed 230 mm of April-May tephra overlying 110 mm of older compacted gray ash. Large ejecta were abundant within about 300 m of the vent, forming a discontinuous layer along the axis of deposition. Most of the large ejecta were low-density, dark brown, flattened or fusiform bombs with a slaggy breadcrust surface and cowdung or ribbon shapes. Bomb sizes were commonly about 0.3 m, but reached 2 m in longest dimension. Some apparently fresh, dense, and vesiculated glassy bombs were also observed. Large dense blocks of altered lithic material were often associated with impact craters.

A levelling survey over the main crater floor showed deflation of both 1978 Crater (at the W end of the main crater) and the SE portion of the main crater, relative to the NE section. This represents a continuation of the trend seen between the surveys of 7 Dec 1978 and 21-22

April 1979.

Information Contacts: B. Houghton, I. Nairn, NZGS, Rotorua.

7/79 (4:7) Mr. B. Adams reported a dark, pulsating tephra column that rose intermittently from White Island to about 1 km altitude. He first observed the eruption at about 0800 on 13 July from 50 km to the SSE. Ash appeared to be falling downwind from the vent. A white plume had been present over White Island for the previous several weeks.

NZGS personnel overflew the volcano on 16 July. An eruption cloud containing red ash emerged from the active vent in the SE section of 1978 crater, reaching about 600 m altitude before being blown N by a moderate wind. Much of White Island was covered by red ash, which was thick enough near the vent to cover the bombs ejected in April and May (4:5-6). No fresh impact craters were observed. Vigorous fumarolic activity continued at several sites within the main crater.

Information Contact: E. Lloyd, NZGS, Rotorua.

8/79 (4:8) NZGS personnel visited White Island on 6 Aug. Rapidly convoluting clouds of gas and fresh ash were ejected every 5-30 seconds, apparently from a magma column deep in the active vent. Ashfall rates at the 1978 Crater rim reached 1 mm/10-20 seconds. During the 3 1/2 hours of observations, a single more violent explosion threw 1-10 m-diameter blocks about 100 m above the crater.

Samples of the 6 Aug tephra were crystal-rich, containing plagioclase, pyroxene, olivine (?), both a pumiceous and nonvesicular (low-silica) andesitic glass, and limited amounts of altered lithic material.

Temperatures measured at the hottest accessible fumaroles (using a thermocouple) reached a maximum of 400°C, similar to those recorded during the last ground inspection, on 28 May. Since then, 213 mm of new tephra had fallen at a sampling site 20 m from the 1978 Crater rim, along the axis of maximum airfall.

A new level survey indicated that deflation in the direction of the vent continued (fig. 4-7) although the deflation rate had slowed from a maximum of 40 μrad/month between Dec 1978 and April 1979 to 17 μrad/month since 28 May. A magnetic resurvey showed a substantial decrease in near-vent values, consistent with the change to magmatic activity since the last survey, in Dec 1978.

Seismic activity since 7 June (when the seismometer resumed operation) included long periods of medium-frequency volcanic tremor, interspersed with quiet periods. [An estimated 1150 events occurred in a 6-hr period on 22 July.] Five low-frequency earthquakes (apparently B-type) were recorded 30-31 July. [Microearthquakes continued, with only minor breaks, until at least 30 Sept.]

Information Contacts: B. Houghton, I. Nairn, NZGS, Rotorua.

10/79 (4:10) NZGS personnel flew over White Island on 28 Sept. Since the previous inspection on 6 Aug, moderate ash emission had apparently been almost continuous. On 3-4 Sept, Sam Harvey of Whakatane observed a stronger ash eruption that produced a 2 km-high cloud. Seismic activity between the 2 inspections was characterized by prolonged bursts of high-frequency tremor [the most vigorous starting on 25 Sept at 1239 and lasting about 4 1/2 hours], interspersed with infrequent quieter periods of up to a few hours duration.

Eruptive activity during the 10-minute overflight consisted of spasmodic ejection of a pale brown, weakly convoluting gas and ash column that rose about 400 m before being blown NE by a light wind. Reddish-brown tephra appeared to uniformly mantle the main crater floor. No impact craters or other evidence of coarse ejecta were observed. Steam rose from a gully that enters 1978 Crater from the W, but pre-existing fumaroles on the main crater floor seemed unchanged.

Information Contact: E. Lloyd, NZGS, Rotorua.

11/79 (4:11) White Island was overflown 25 Oct, about a month after the previous aerial inspection. A weakly convoluting pink fume cloud rose to about 500 m from the eruption vent in 1978 Crater, and moderate steam emission occurred from fumaroles elsewhere in the main crater. Brown ash mantled the main crater floor, and a few blocks and apparent impact craters were visible near 1978 Crater.

Seismic activity between 28 Sept and 25 Oct typically consisted of intermittent to semicontinuous bursts of high-frequency tremor, with few quiet periods. Strong local earthquakes were rare.

Information Contact: I. Nairn, NZGS, Rotorua.

Further Reference: Houghton, B.F., Scott, B.J., Nairn, I.A., and Wood, C.P., 1983, Cyclic Variation in Eruption Products, White Island Volcano, New Zealand, 1976-1979; *New Zealand Journal of Geology and Geophysics*, v. 26, p. 213-216.

2/80 (5:2) The press reported that an eruption occurred during the late afternoon of 6 Feb. Records telemetered by the White Island seismograph showed a substantial increase in seismicity during the preceding week. Shortly after 1648 on the 6th, the seismograph recorded a high-amplitude, high-frequency event lasting about 3 minutes.

Personnel from the NZGS and Victoria Univ. flew over White Island on 7 Feb. A 500-m steam column emerged from deep within the active vent of 1978 Crater. Other fumarolic activity was at its weakest level since the eruption began. Extensive erosion, caused by heavy rainfall in mid-Jan, could be seen on the main crater floor. No fresh tephra was visible.

A second overflight on 25 Feb revealed fresh, pale brown ash mantling the main crater. Impact craters to about 1 m in diameter reached an estimated density of 1 per 10 m^2 at the E and SE rim of 1978 Crater, and could be seen, at reduced density, to about 200 m from the rim. Activity from 1978 Crater was similar to that observed on 7 Feb. High pressure fumarolic emission had begun at 2 sites on the main crater floor, but remained weak elsewhere.

Information Contacts: B. Houghton, E. Lloyd,

NZGS, Rotorua.

4/80 (5:4) S. Harvey reported that continuous ash eruptions were visible from Whakatane beginning 9 April. The ash clouds were voluminous, but did not rise to great heights. During the first 2 days of the activity, the clouds were pinkish brown and were accompanied by emission of white vapor from a vent to the E. On 11 April, the ash clouds were black and vapor emission had apparently ceased.

At 1114 on 12 April, the White Island seismograph recorded a large-amplitude, high-frequency event with virtually instantaneous onset, against a background of the almost continuous high-frequency tremor that has characterized long periods of the seismic record in recent months. At about the same time, persons on the mainland saw a large ash eruption.

NZGS personnel flew over the volcano about 2 hours later. Red-brown ash rose from deep within 1978 Crater to about 700 m height. Impact craters were seen inside 1978 Crater and blocks littered the ground within 20 m of the SE rim. Much of the main crater floor was mantled by ash, including the area where the blocks were deposited. However, there was no ash on the blocks, which were therefore assumed to have been ejected quite recently, probably by the late-morning explosion. The interior of 1978 Crater has changed little since the present active vent was established in late 1978.

Information Contact: E. Lloyd, NZGS, Rotorua.

5/80 (5:5) Geologists visited White Island on 26 May. During the first half of the 5-hour visit, eruptive activity was limited to emission of a white vapor column. After ash darkened the vapor column briefly, a small explosion ejected accessory blocks to a few tens of meters above the rim of 1978 Crater. A vigorously convecting gas and ash column deposited ash and lapilli-sized accessory material around the crater. Acidic, ash-charged water droplets fell within 400 m of the crater. A smaller explosion occurred about 40 minutes later. Both explosions were recorded by the White Island seismograph, which had recorded 2 similar but much larger events during the 2 previous days. One of the earlier explosions was observed from near Motiti Island, about 70 km away.

Since the previous ground inspection on 18 April, a line of 3 new fumaroles had formed NE of 1978 Crater, along a trend of pre-existing intense fumarolic activity. Temperatures in the new fumaroles ranged from 475°-615°C. The largest vent, 10 m in diameter and 15 m deep, was surrounded by 2 small lobes of ejecta. Fumarolic activity elsewhere on the main crater floor had become substantially stronger, and several other new gas vents were observed. Divers found warm springs just off the NE and NW coasts of White Island, measuring a temperature of 45°C at a depth of 0.3 m at one of the sites.

A levelling survey revealed much more extensive and rapid uplift in the main crater since 18 April than had occurred in the 5 previous months. The maximum inflation values of 22-27 mm were measured in the portion of the survey area nearest 1978 Crater and the zone of intense fumarolic activity mentioned above. The inflation recorded since Nov 1979 was a reversal of steady deflation between Feb 1978 and Aug 1979. A magnetic survey of the main crater yielded no significant changes since 18 April. A pit dug about 200 m E of 1978 Crater went through 270 mm of tephra before reaching Aug 1979 deposits. Very little tephra had accumulated more than 400 m from 1978 Crater since Aug 1979.

Information Contacts: B. Houghton, E. Lloyd, I. Nairn, NZGS, Rotorua; R. Dibble, Victoria Univ., Wellington.

6/80 (5:6) NZGS personnel flew over White Island on 13 June. A vapor cloud produced by several fumarolic areas rose to a maximum height of about 1 km. A thin layer of gray ash mantled the main crater floor, obliterating footprints left by the 26 May survey party. A band of grey ash trending NE from 1978 Crater was particularly conspicuous on the outer slopes of White Island. Impact craters, numerous on the E side of 1978 Crater, extended several hundred meters to the ESE, farther

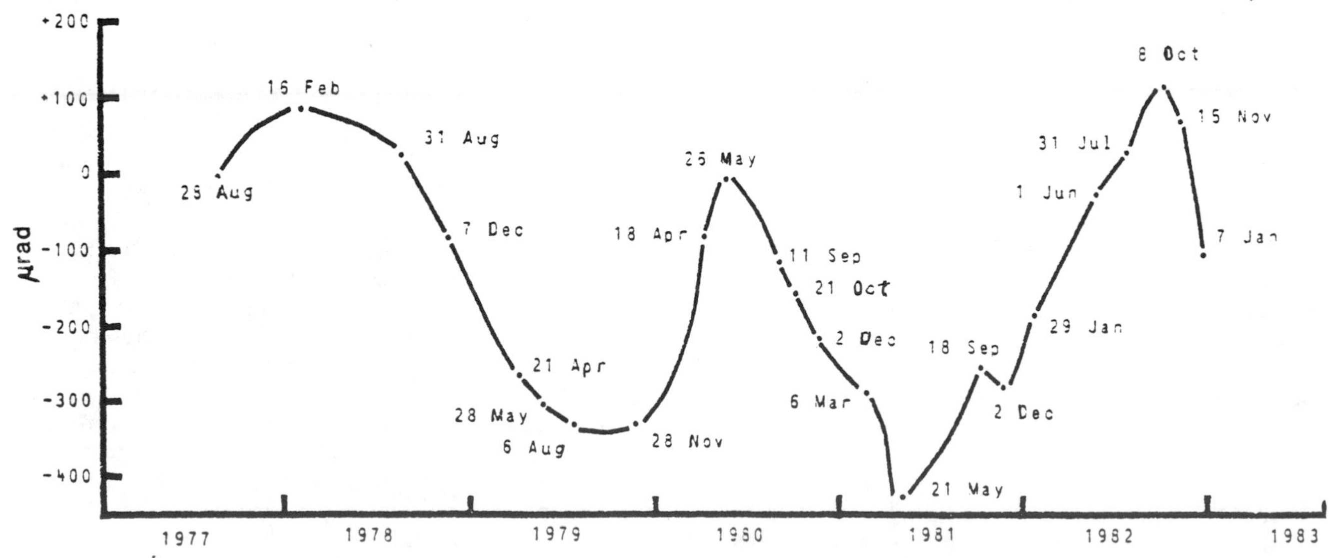

Fig. 4-7. Tilt data 50 m E of Donald Mound, 28 Aug 1977-7 Jan 1983 [originally from 8:1]. Courtesy of NZGS.

than at any time during 1980. The distant impact craters appeared large (no size estimate given) but were widely spaced. Seismicity during the few days prior to the overflight was characterized by periods of moderate-amplitude, high-frequency tremor interspersed with longer periods when tremor was low in both amplitude and frequency or was entirely absent. Brief episodes of high-amplitude, high-frequency tremor were recorded on 11 June (4 minutes) and 13 June (1 1/2 minutes). Two A-type shocks occurred about 2 1/2 minutes apart on 13 June, about 3 hours before the 1 1/2 minute episode of intense tremor. None of the strong seismicity occurred during the overflight.

Information Contacts: B. Houghton, E. Lloyd, I. Nairn, NZGS, Rotorua.

8/80 (5:8) On 19 July, residents of Whakatane observed a black ash column that rose about 3 km above the volcano. However, a 28 July NZGS overflight found it little changed from 13 June. The island was covered by red-gray ash, deeply gullied in places on the crater floor. Partly eroded impact craters extended up to 200 m from the SE rim of 1978 Crater. The latter was hidden by voluminous clouds of white steam and gas while the recently formed subsidiary gas vents to the E (5:2) were obscured by vapor from adjacent fumaroles. Other fumarolic activity farther to the N appeared to have declined since the previous inspection. The sea immediately off the NW coast was colored white, probably due to suspension of sulphur (and other precipitates from underwater hot springs; 5:5).

Seismicity during July was characterized by relatively low tremor levels and occasional earthquakes, some of low frequency (B-type). Periods of high-amplitude, high-frequency tremor were recorded 11-16 July. Some larger, high-frequency, probably A-type or regional earthquakes were recorded on 27 July. The seismograph was not operational during the 19 July eruption.

Information Contact: I. Nairn, NZGS, Rotorua.

10/80 (5:10) During 5 hours of observation by geologists and geophysicists on 21 Oct, the active vent in 1978 Crater emitted an ash-poor gas column of varying intensity which deposited no tephra. Gray ash postdating the 11 Sept inspection was 1 cm thick on the 1978 Crater rim, and less than 2 mm thick beyond 200 m from the rim. No fresh impact craters were observed. Fumarolic activity continued from the gas vents NE of 1978 Crater that were first observed in May and other gas vents remained vigorous. Fumarole temperatures were generally somewhat lower than in May and Sept with a maximum of 600°C.

A levelling survey showed that the relatively rapid rate of subsidence (as much as 2.3 mm/week) between May and Sept, which had followed 6 weeks of localized inflation, decreased to less than 1 mm/week in the 6 weeks since 11 Sept. Magnetic field intensity appeared to have fallen near the 1978 Crater rim since 11 Sept.

Information Contacts: B. Houghton, I. Nairn, NZGS, Rotorua.

1/81 (6:1) NZGS personnel flew over White Island on the morning of 6 Jan. In the 10 minutes they were over the island, the voluminous convoluting emissions of white steam and gas clouds obscured their view around and into 1978 Crater. The lower portion of the 600-750 m-high eruption column was slightly ash-charged. The main crater was thickly covered with eroded brown-green ash. Impact craters extended a few hundred meters NE from 1978 Crater. Conspicuous blue fumes were associated with the steam-gas column rising in the 1914 landslide area just SE of 1978 Crater.

Seismicity since ground inspections in early Dec was characterized by 4 distinct periods of marked increase. Intervals of high-frequency, high-amplitude tremor were recorded for 32 hours on 15-16 Dec, for 35 hours on 22-23 Dec, and for 26 hours on 27-28 Dec. Strong ash emissions were likely during these periods. Large discrete earthquakes were recorded on 14 Dec and 2 Jan.

Information Contact: B. Scott, NZGS, Rotorua.

3/81 (6:3) NZGS personnel visited White Island on 17 Jan and 6 March, and overflew the volcano on 24 Feb after reports of explosive eruptions in Jan. On 17 Jan, a weakly convoluting gas column charged with light brown ash was issuing from the active vent in the S part of 1978 Crater. Gray tephra covered the main crater floor. Since the magmatic activity in early Nov, about 500 mm of ash had been deposited near the 1978 Crater rim, about 230 mm since 2 Dec. Two or possibly three generations of impact craters, the youngest perhaps less than a day or two old, were found 250-600 m NE of the active vent, with concentrations of up to 3 craters per m^2 300-400 m away. Crater diameters ranged from 30 mm to 1.2 m. Blocks up to 0.7 m in diameter were found in some of the impact craters, but no fresh magma appeared to have been ejected. The apparent near-vertical final trajectory of the blocks in the impact craters was striking.

Seismic records showed that periods of high-frequency tremor occurred 7-13, 17-19, 22-23 and 25-28 Jan, and 30 Jan-11 Feb. Short bursts of harmonic tremor were recorded after the high-frequency tremor declined. Large discrete events (eruption sequences) were recorded on 24 and 29 Jan and 6, 12, 17, 23 and 24 Feb. The eruption accompanying the 24 Jan event produced ashfall at Cape Runaway (75 km E of White Island, on the mainland) and was witnessed by P. M. Otway (NZGS).

So little gas and tephra were being emitted during the 24 Feb overflight that viewing was excellent. The main crater floor appeared to be covered by a thick layer of red-brown, principally fine-grained tephra. Impact craters formed since 17 Jan pocked the floor to 600 m E of the active vent, the greatest range since the March 1977 eruptions.

On 6 March, emissions from the active vent, entrenched in a subcrater at least 200 m deep, were very low. The rim of 1978 Crater showed no major changes, but a large portion of the crater floor N of the active vent had been raised several meters by rapid accumulation of tephra between 17 Jan and 24 Feb. At the E edge of 1978 Crater rim, 410 mm of new tephra overlay earlier deposits. Within a kilometer to the E, the new tephra thinned to about 8 mm. The surface layer was a fine pink ash (mean diameter about 63 μm) containing abundant lithic clasts, subordinate crystals of pyroxene and olivine, and minor amounts of glassy, weakly to moderately vesicular, essential, low-silica andesite. This layer was underlain by a finer green ash containing a greater percentage of essential clasts.

A few new impact craters had formed just outside 1978 Crater's E rim, some occupied by lithic blocks that were not coated with ash. Some of the older impact craters scattered across the main crater floor contained buried scoriaceous bombs, apparently of recent magmatic origin.

Fumaroles checked from the air in Feb were monitored in March. The inspection team measured minimum temperatures of 490°-650°C at 3 fumaroles formed within the past year E of 1978 Crater rim. One is the most energetic feature on the island other than the active vent.

The March levelling survey showed that subsidence had continued after the previous survey on 2 Dec. The volcano had deflated about 300 μrad since May, the peak of a 6-month period of inflation (fig. 4-7). The greatest deformation, 12-13 mm since Dec and a total of 60 mm since May, was about in the center of the main crater, near the zone of fumarolic activity just E of 1978 Crater.

Information Contacts: B. Scott, B. Houghton, I. Nairn, NZGS, Rotorua.

7/81 (6:7) A routine surveillance flight by NZGS personnel was made on 13 July from 1034 to 1057. Weak gas and steam emissions from the fumaroles and vents rose to 600 m and showed no sign of ash. The main crater floor appeared dark gray-brown near the active vent, but reddened away from it. Distinct yellow-green areas were visible both on the S side of the crater and on the N outer slopes. A tan area was also on the S side. Numerous impact craters of more than one generation extended about 600 m E of the active vent and were concentrated about 300 m from it. All the impact craters had subdued margins. Tephra deposits were extensively gullied. Although discrete explosive events had occurred, there had been little tephra emission since the previous visit on 21 May.

Seismic records revealed a continued decline in activity, apparent since early this year. Since 21 May a marked decline was evident in the number of low-frequency (B-type) events from more than 30/day to about 5/day. Volcanic tremor was recorded on 26 and 30 May; 3-4, 6, 12, and 28 June; and 2, 5-6, and 8 July. High-frequency (volcano-tectonic) events numbered fewer than 10/day except during 14-15 June, when several hundred per day were recorded. The increased high-frequency activity was accompanied by distinctive, medium-frequency seismic signatures which were symmetric and had emergent onsets. NZGS personnel interpreted these as volcanic earthquakes and, possibly, intrusive events.

Information Contact: B. Scott, NZGS, Rotorua.

9/81 (6:9) On 18 Sept, NZGS personnel found that little eruptive activity had occurred in the last 6 months. The fine tephra evenly mantling the main crater floor showed rain wash, pitting, and considerable erosion along the main water courses. A small landslide near the midpoint of the main crater's S rim had sent several tongues of muddy, sulfur-rich, hydrothermally altered material onto the crater floor. A distinctive pink ash that had formed the surface on 6 March (6:3) was at about 100 mm depth on the rim of 1978 Crater, but was only 30 mm below the surface in the center of the main crater 300 m to the E.

The active subcrater was 200 m wide and 150 m deep. A shallow green pond occupied most of its floor. A 20-30 m-wide vent on the NW side of its floor was emitting pink-tinged gas at high velocity. Throughout the visit this gas and other vapor from 1978 Crater formed a white column.

Fumaroles in a 300 m-wide zone across the main crater floor E of 1978 Crater appeared to be slightly less active than in March. Temperatures were 560°C in a large, reactivated vent; 550°C (100° lower than in March) in a smaller one nearby; and 340°C at a vent now so enlarged that the thermocouple could reach no more than halfway into the pit.

Except for inflation near the center of the main crater, the levelling survey showed no tilt change since May, in contrast to general subsidence earlier in the year. Significant inflation (115 μrad), had occurred just E of active fumaroles about 300 m E of 1978 Crater. The levelling survey team noted that this tilt reversal bears some resemblance to that in April 1980, which preceded the formation of 3 fumaroles in the same area.

Seismic records 1 Aug-20 Sept showed a marked increase in the number of low-frequency (B-type) events from fewer than 5 to more than 30 per day, reversing a decline from late May through mid-July. High-frequency (volcano-tectonic) events numbered fewer than 5 per day except on 11 and 12 Sept, when 10 were recorded each day. On 26 Aug, and 8 and 10 Sept, single distinctive seismic signatures (symmetric with emergent onsets) were recorded. The NZGS interpreted these as medium-frequency volcanic earthquakes, probably intrusive events. Volcanic tremor was recorded on 10 Sept.

Information Contacts: B. Houghton, I. Nairn, NZGS, Rotorua.

2/82 (7:2) Observers from Victoria Univ. and NZGS spent $4^1/_2$ hours the afternoon of 29 Jan inspecting the island and reactivating the seismograph. No tephra deposition had occurred since the last visit 2 Dec 1981. Tephra deposited between 18 Sept and Dec measured 46 and 23 mm at two sites on the crater rim, and was absent beyond 250 m from the active vent. Its fraction coarser than 250 μm was composed of roughly equal amounts of fresh glassy low-silica andesite (52 wt.%) and accessory, hydrothermally altered lithic material (48 wt.%).

Activity in 1978 Crater appeared less intense than in Sept. The depth of the active subcrater was 105 m. Vents on its floor and W wall emitted white steam. Numerous geysers jetted black muddy water to several meters in the N part of the shallow bright-green pond that covered about half of the subcrater's flat floor. Many large lava blocks had fallen from the W walls; several rockfalls occurred during the observation period. The high-pressure gas vent that had been on the NW side of the subcrater floor in Sept was not present.

Fumaroles around Donald Mound, a 100 × 150-m area in the middle of the main crater, appeared more active than in Sept. Several had strong discharges of high-pressure, clear gas and some blue fume. Temperatures just E of the mound were 590°, 640°, and 655°C. Low-pressure gas in a fumarole W of it, where blue fume was condensing, had a temperature of 560°C. A large fumarole that had developed earlier on the N side of the main crater was at least 5 m deep and was emitting large volumes of white, superheated steam. Ephemeral minerals were collected on the last three visits. X-ray diffraction analysis indicated that these were predominantly unoxidized sulfur and anhydrous forms in the throats of fumaroles, and sulfate-hydrate and sulfate-hydroxide-hydrates surrounding fumaroles and their outwash areas.

The levelling survey found little tilt change between Sept and Dec, and none since Dec except in the vicinity of Donald Mound, where an 11-mm rise had been recorded. The inflated area was about 400 m across, centered at the E side of the Mound, about 200 m E of 1978 Crater. After a similar localized inflation, 3 small gas vents formed N of the Mound in May 1980.

Seismic records 21 Sept-19 Jan (when the seismograph stopped working) showed a gradual decline in the number of low-frequency (B-type) events from more than 40 to fewer than 5 per day by 10 Dec. The level has remained low. High-frequency (volcano-tectonic) events numbered fewer than 5 per day except on: 25 Nov, 7 events; 16 Dec, 16; 21 Dec, 6; 8 Jan, 35; and 14 Jan, 7. Distinctive seismic signatures (symmetric, with emergent onsets) were recorded on 29 Nov, 1, 13, and 19 Dec, and 16 Jan. The NZGS interpreted these as medium-frequency volcanic earthquakes, probably intrusive events. Volcanic tremor was recorded weakly on 25-26 Sept, 8, 24, and 25 Oct, and 15 Nov.

Information Contacts: A. Cody, B. Houghton, I. Nairn, P. Otway, B. Scott, C. Wood, NZGS, Rotorua.

6/82 (7:6) NZGS personnel visited White Island on 1 June. No significant eruptive activity had occurred since Jan, but fumarolic activity in 1978 Crater appeared more intense. Vents on the walls and floor of the active subcrater emitted large white steam columns, and numerous small geysers jetted black muddy water. Rockfalls from the W wall had formed a fan onto the subcrater floor. A green lake occupied much of the SE sector of the subcrater floor.

Inflation localized around Donald Mound (a 100-m × 150-m area in the middle of the main crater and about 150 m E of the edge of 1978 Crater) had continued, with a rise of 13 mm measured about 50 m to the E, and 18 mm about 100 m to the SE. The inflation started a year ago, but paused between Sept and Dec 1981. Fumaroles near the Mound and about 200 m N of it were emitting large white steam columns. Fumarole temperatures around Donald Mound showed a general increase from 630°C on 2 Dec 1980, to 650°C in May 1981 and Jan 1982, and 690°C in June 1982. More sulfur appeared to have been deposited at the ground surface. For the area between 1978 Crater and Donald Mound and extending about 200 m to the N, magnetometer data showed increases in total magnetic field of between 64 and 326 nanoteslas, compared to changes of less than 25 nanoteslas over the rest of the main crater since Jan.

Between 27 Feb and 2 June, more than 150 (and occasionally as many as 400) usually small, low-frequency (B-type) events were recorded per day during 3 periods of increased seismicity: 15-20 March, 26 March-8 April, and 22 May. No more than 39 low-frequency events were recorded on other days. High-frequency (volcano-tectonic) events usually numbered fewer than 5 per day except for 6 on 4 and 13 March, 44 on 3 April, and 6 on 2 June. Volcanic tremor was recorded on 25-26 March, and 5 and 12 May.

Information Contacts: I. Nairn, C. Wood, B. Scott, NZGS, Rotorua; P. Otway, NZGS, Wairakei; D. Sheppard, DSIR, Gracefield.

7/82 (7:7) On 1 July at 1413 a wide-band earthquake sequence indicative of an eruption was recorded, with peak-to-peak amplitude exceeding Full-Scale Deflection (FSD). From Whakatane, S. Harvey reported that a dark eruption column and ash fallout accompanied the earthquakes. Harvey observed an eruption column until 1517, by which time the seismicity had declined to 2-3 mm peak-to-peak, low-frequency, tremor-like activity. Two smaller wide-band earthquake sequences were recorded on 3 July at 0234 and 5 July at 0424.

About noon on 8 July NZGS personnel flew over the volcano. There was no sign of any new ash deposition or impact craters. A steam plume was rising to about 900 m from the SE part of 1978 Crater, and much smaller steam plumes were being emitted from the fumarole area in the center of the main crater. Emission was more intense than during the 1 June visit and during a photographic flight on 13 June. The green lake observed in the SE part of 1978 Crater during several previous visits was still present.

A sharp-walled pit had formed in the N part of 1978 Crater. It was filled with steam and opened into the gully that drains into the SE part of 1978 Crater. It had not been seen on March airphotos, but a small depression was noted in this area on 13 June photographs.

Between 2 June and 5 July the daily number of low-frequency (B-type) seismic events declined. Only on 24, 28, and 30 June were there more than 20 (maximum of 37 on 28 June), as compared to 27 Feb-2 June when the number of events often exceeded 150 per day. All were very small.

High-frequency (volcano-tectonic) seismic events numbered fewer than 3/day after 2 June. All events were small except on 2 July at 2027 when they exceeded FSD. On 21-22 June, 4 medium-frequency volcanic earthquakes were recorded. NZGS personnel interpret these as possible intrusive events. Tremor-like seismic activity was recorded 18-19 and 30 June, and 2, 3, and 4 July.

Information Contacts: I. Nairn, B. Scott, NZGS,

Rotorua; P. Otway, NZGS, Wairakei.

10/82 (7:10) When NZGS personnel inspected the volcano on 8 Oct, 2 months after the last visit on 31 July, they found no new tephra deposits. There has been no significant tephra accumulation on the main crater floor since Jan 1982, the longest interval without deposition since the eruption began in Dec 1976. Much of the tephra cover in the W half of the main crater, where the fumaroles are, has been subjected to intense hydrothermal alteration.

Substantial steam columns were being emitted from vents on the floor and walls of 1978 Crater, and from Donald Mound and Noisy Nellie fumaroles. The strongest fumarolic activity was on the N side of 1978 Crater, near the new collapse pit. The small green pond in the SE part of the crater was still present; in a muddy black pond N of it, there was vigorous geysering. Fumarole temperatures measured at 4 places on the main crater floor ranged from 470°C in vents with low gas pressure to 695°C in high-pressure vents near Donald Mound.

Inflation of the Donald Mound area has continued at a steady rate. The E side had risen 9 mm since 31 July, and 58 mm since 21 May 1981 (fig. 4-7), equivalent to 500 μrad of tilt. A deflationary trend was continuing about 200 m N of the Mound. The NZGS interpreted the inflation as a possible precursor to eruptive activity in this area.

Between 5 July and 1 Oct low-frequency (B-type) seismic events did not exceed 10/day except in early July, when they increased to 31/day on the 9th and 10th, then declined. All the low-frequency events were small. High-frequency (volcano-tectonic) events usually numbered fewer than 5/day in this period, except as shown in table 4-4. These events were also generally small.

Moderately high frequency volcanic tremor began about 2200 on 28 July and gradually increased in amplitude until 0800 on 30 July, when a distinct decline in amplitude was apparent; tremor ceased by 1600 that day. Wide-band earthquake sequences indicative of eruptions were recorded on 9, 15, and 26 Sept, but no eruptions were reported. Medium-frequency volcanic earthquakes were recorded on 16, 17, and 26 July, 14 Aug, and 4 and 18 Sept. The NZGS interpreted these as possible intrusive events.

Information Contacts: I. Nairn, B. Scott, NZGS, Rotorua.

1/83 (8:1) Field work by NZGS personnel 7 Jan revealed no evidence of eruptive activity since their previous visit on 15 Nov. Only minor changes were observed in 1978 Crater. Temperatures were measured at 3 fumaroles. At vents E and NE of Donald Mound temperatures were 620° and 630°C, 60-65° lower than on 8 Oct. West of Donald Mound, a vent formed in 1980 had a temperature of 556°C, 86° warmer than in Oct.

	July	Aug				Sept	
Date	18	7	15	18	28	6	14
Events	67	9	15	8	10	11	18

Table 4-4. Days between 5 July and 1 Oct 1982 when more than 5 high-freqency events were recorded.

The center of the deflating area, near the E edge of 1978 Crater in Nov, had deepened and moved E several hundred meters to the Donald Mound area, which had been inflating from mid-1981 until Nov 1982. A site on the E side of the Mound area had subsided 16 mm since Nov. A nearby tiltmeter measured deflation of 50 μrad 8 Oct-15 Nov, and 170 additional μrad by 7 Jan (fig. 4-7).

The inflation in the Donald Mound area had been interpreted by the NZGS as a possible precursor of minor eruptive activity, as in May 1980 when 3 new vents formed between there and the 1978 Crater. But the inflation rate was only 1/3 that of 1980, and the recent abrupt deflation is now thought to be the end of the 15-month inflationary episode.

Information Contacts: B. Scott, NZGS, Rotorua; P. Otway, NZGS, Wairakei.

3/83 (8:3) Aerial inspection by NZGS personnel on 10 March revealed no evidence of eruptive activity since 7 Jan. A white steam plume was rising to about 600 m altitude from the SE part of 1978 Crater. For 200 m to the N, there were moderate emissions from vents in deep gullies and from 2 fumaroles. Very little emission was originating from Donald Mound. Most of the Mound was covered with yellow sublimates, but a central zone was gray.

Since 7 Jan the number of low-frequency (B-type) events has increased, especially 9-15 Feb (more that 25/day; maximum, 42) and 22 Feb-4 March (more than 21/day). High-frequency (volcano-tectonic) events usually numbered fewer than 5/day, except for 6 on 29 Jan, 7 on 6 Feb, and 10 on 21 Feb. Wide-band seismic events were recorded on 19 and 24 Feb, and 2 and 6 March. They lasted 4-40 minutes with peak-to-peak amplitudes up to 70 mm.

Information Contact: B. Scott, NZGS, Rotorua.

4/83 (8:4) When NZGS personnel visited the island on 14 April, they observed few differences in 1978 Crater. The small green ponds present during the previous visit on 22 March had enlarged and merged to cover 60-70% of the crater floor. Color ranged from orange adjacent to the fumaroles to lime green. Fumarolic activity continued on the W wall and up the NW-trending gully system. Activity at Donald Mound appeared to have declined, but no fumarole temperatures were measured. Very small changes in tilt were recorded: +8 mm about 150 m W of Donald Mound, and +9 mm about 100 m N of it. The NZGS noted that these changes indicated an end to the deflation of the Donald Mound area recorded between Nov 1982 and Jan 1983 (fig. 4-7).

Information Contacts: P. Otway, NZGS, Wairakei; B. Scott, NZGS, Rotorua; G. Sorrell, DSIR, Wellington.

2/84 (9:2) A tephra eruption began at White Island in late Dec 1983. No significant activity had been reported since late 1981.

NZGS personnel reported that a ground inspection on 23 Nov and an aerial inspection on 27 Nov 1983 had yielded no evidence of eruptive activity. Airborne COSPEC measurements on the 27th showed an SO_2 emission rate of 1200 t/d. Voluminous steam columns above the island were observed from Pukehina Beach (about 65 km SE) in late Dec-early Jan by 2 NZGS geologists. A pilot reported that the eruption column had changed from white to gray about 20 Jan. Photographs taken by a yachtsman who landed on White Island on 27 Jan showed a dense ash column.

During an aerial inspection on 2 Feb, crater conditions appeared similar to those shown in the 27 Jan photographs. Geologists who flew over the island (from 0944 to 1000) observed a gray-green ash layer on the main crater floor and the island's outer slopes. Ash appeared thickest on the N slopes. A white steam plume with a little light gray, fine ash was rising to 1.2-1.5 km above sea level. All eruptive activity appeared to come from a new vent (about 20 x 30 m and at least 50 m deep) that had formed at the site of a vigorous fumarole on the N margin of the older (SE) portion of the dumbell-shaped 1978 Crater complex. A small tuff ridge had been built up from the floor of 1978 Crater to the rim of the new vent. Impact craters and blocks up to 1 m in diameter were visible on the tuff ridge and throughout the complex, extending E to Donald Mound. Impact crater density decreased rapidly away from the vent. Fumarolic activity in other areas appeared to have declined from the level observed in Nov. At about 1200, after geologists left the area, a vigorous, dark eruption column was observed from many points along the coast of North Island, more than 50 km away.

During field work on 6 Feb, NZGS personnel observed little apparent change since their previous visit. No significant new ashfall was evident outside the rim of 1978 Crater. Small blocks (less than 10 cm in diameter) on the floor of the 1978 Crater complex were not ash-coated, suggesting recent ejection. All the blocks consisted of lithic andesite; some were altered. Moderate gas emission from the active vent was punctuated by occasional pulses of fine gray ash. More voluminous emissions occurred at about 1200 and 1530. Acid rain and a little ash fell on the geologists.

Post-Nov 1983 tephra thicknesses ranged from 60 mm about 150 m N of the new vent, to 15 mm about 350 m E, and 2 mm about 800 m E. Sand-sized material in new ash NE of the vent was dominated by abundant fresh plagioclase, pyroxene, some possible olivine crystals, and magnetite, with subordinate glass and minor amounts of lithic fragments. Considerable altered silt- and clay-sized material was also present in the ash. At a site closer to the vent, tephra from an earlier phase of the 1983-4 activity had the same crystal and glass components but a much larger proportion of altered lithic fragments.

At 762°C, the temperature of a fumarole E of Donald Mound was similar to that measured on 23 Nov 1983; temperatures below 700°C were measured on all other visits in 1983, 1982, and 1981. Inflation had continued in the Donald Mound area. The localized uplift area was about 400 mm across and centered on Donald Mound.

NZGS geologists returned to White Island on 17 Feb. Except for a small quantity of blocky ejecta around the E rim of the new vent, there was little evidence of recent eruptive activity. Fumarole temperatures remained high, and the Donald Mound area was still inflated.

Telemetry of seismic data, which stopped in late Nov, resumed 6 Feb. Between the 6th and 23rd, 6-30 low-frequency B-type volcanic earthquakes were recorded daily. There were more than 20 events per day, 9-15 Feb, and fewer than 15 per day, 16-23 Feb. High-frequency volcano-tectonic earthquakes numbered fewer than 3 per day except on 6, 7, 8, and 13 Feb when 13, 20, 12, and 11 events were recorded. Long-duration, wide-band, multiple-frequency earthquake sequences were recorded on 8, 9, 10, 11, 14, and 17 Feb. Similar sequences had previously been correlated with eruptive activity, but there was no visual confirmation of any eruption clouds associated with these events. Weak low- or medium-frequency tremor was recorded for 8-9 hours on 11 Feb.

Information Contacts: I. Nairn, NZGS, Rotorua; W. Rose, Michigan Tech. Univ.

Further Reference: Rose, W.I., Chuan, R.L., Giggenbach, W.F., Kyle, P.R., and Symonds, R.B., 1986, Rates of Sulfur Dioxide and Particle Emissions from White Island Volcano, New Zealand, and an Estimate of the Total Flux of Major Gaseous Species; *BV*, v. 48, p. 181-188.

11/85 (10:11) When geologists visited White Island on 13 Nov, there was no evidence that any eruptive activity had occurred since their visit on 21 May. Deflation of the Donald Mound area, roughly 100 m E of the 1978 Crater, continued. The area of subsidence was a NW-SE ellipsoid about 400 m long by 250 m wide, centered on Donald Mound. One station had dropped 112 mm since a small ash eruption in Feb 1984; stations immediately W of the area, which had dropped 15-25 mm May 1984-May 1985, had fully recovered by the Nov visit.

Magnetic data showed a small but high-amplitude anomaly centered N of Donald Mound, suggesting to geologists that substantial near-surface cooling had occurred in the area since the May magnetic survey. At one vent, fumarole temperatures had declined to 390°C from 523° in May.

No low-frequency (B-type) events were recorded from Feb until late Sept, when they resumed following a magnitude 7 event, 350-400 km NE of White Island. Since 25 Oct, 6-35 low-frequency events have occurred per day, with unusually large amplitudes (up to 50 mm peak to peak). The number of high-frequency (volcano-tectonic) earthquakes remained relatively constant in 1985, with 3-5 recorded on most days.

Information Contacts: I. Nairn, A. Cody, B. Scott, C. Wood, W. Davis, NZGS, Rotorua; P. Otway, NZGS, Wairakei; D. Christoffel, E. Hardy, Victoria Univ., Wellington; W. Giggenbach, DSIR, Wellington.

HAROHARO VOLCANIC COMPLEX

North Island, New Zealand
38.092°S, 176.508°E, summit elev. 914 m
Local time = GMT +12 hrs winter (+13 hrs summer)

This complex of overlapping lava domes, midway between White Island and Lake Taupo, has had no historic activity, but lies just north of Mt. Tarawera, the scene of a particularly violent and fatal eruption in 1886 (2 km^3 of basalt were ejected in only 4 hours).

11/82 (7:11) A series of shallow earthquakes occurred 23-29 Sept a few km SE of Haroharo Dome, in the Okataina Volcanic Center (fig. 4-8). The main earthquake was at 1423 on 23 Sept. A foreshock preceded it by about 3 minutes, and 3 of the 4 large aftershocks followed at 1429, 1440, and 1452 (table 4-5). Many other aftershocks were recorded, the last at 0530 on 29 Sept.

I. A. Nairn, working on the N side of Tarawera Volcano (about 11 km S of the epicenters) on 23 Sept, felt shocks and heard rockfalls nearby. He estimated the Modified Mercalli intensities of the foreshock and main shock as IV, and of the aftershocks at 1440 and 1452 at III-IV. He described the ground vibrations as low-frequency but relatively large-amplitude. Other nearby NZGS personnel noted the relatively low frequency of the felt shocks compared to typical local felt earthquakes. Observers noted that although they did not feel the shocks strongly outdoors, houses and vehicles resonated to large-amplitude vibrations. A small seiche was recorded on the N side of Lake Tarawera (7-8 km SW of the events). Three tilt networks around Tarawera Volcano showed no significant changes.

J. H. Latter placed the hypocenter for the 27 Sept event [at 38.129°S, 176.531°E, fig. 4-8] about 5 km SE of Haroharo Dome at a depth of about 2 km. Nairn reported that this location coincides with a small area of surface faulting and geothermal activity. The 23 Sept earthquakes could not be located because of the lack of any nearby seismic records, but epicenters were estimated to be within 6 km of the 27 Sept event. Latter noted that the slow propagation of energy from the earthquakes and the low frequency of the felt shocks might suggest that they were "roof rock" events generated by activity in an underlying magma body. However, no volcanic tremor was detected during or after the earthquake sequence.

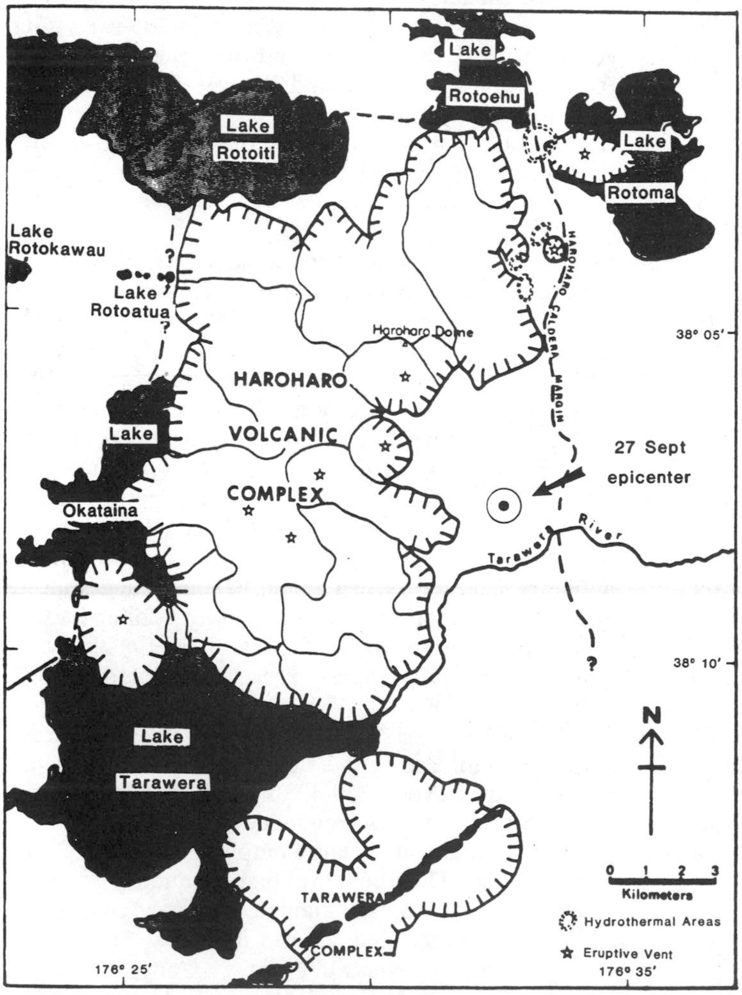

Although Haroharo has not been historically active, 5 eruptions in the last 10,000 years have been dated by ^{14}C or tephrochronological methods. Very large explosive eruptions occurred roughly 4000, 4800, 7000, and 9000 years before the present (BP). Dome extrusion occurred at 4400 years BP (± 400 years).

Information Contacts: J. Latter, DSIR, Wellington; I. Nairn, B. Scott, NZGS, Rotorua; P. Otway, NZGS, Wairakei.

Fig. 4-8. Map showing the Haroharo Volcanic Complex and part of Tarawera Volcano in the Okataina Volcanic Center. From Cole, J.W. and Nairn, I.A., 1975, *CAVW*, Part XXII.

1982	23 Sept					27 Sept
Time	1420	1423	1429	1440	1452	1806
Mag.	[3.3]	4.1	2.6	3.2	[2.9]	[2.7]

Table 4-5. Earthquakes of M > 2.5, 23-29 Sept 1982.

NGAURUHOE

North Island, New Zealand
39.158°S, 175.63°E, summit elev. 2291 m

Ngauruhoe is a perfectly shaped cone on the SW flank of the Tongariro massif. It has been New Zealand's most active volcano in historic time, with 61 eruptive episodes since its first recorded eruption in 1839.

General Reference: Latter, J.H., 1981, Volcanic Earthquakes and Their Relationship to Eruptions at Ruapehu and Ngauruhoe Volcanoes; *JVGR*, v. 9, p. 293-310.

The NZGS reports that cigar-shaped eruption columns rose to 900 m above the crater on 4 July 1977 at 0800 and 1030.

5/83 (8:5) [The following seismic data, from Balsillie and Latter (1985) replaces the first two sentences of the original report, which dealt only with 16 May seismicity.

Three swarms of A-type earthquakes were recorded during the first half of 1983. The first and smallest included 14 events to magnitude 1.8 on 16-19 Feb. About 20 A-type and 4 B-type (including the largest, at magnitude 2.4) shocks occurred 1-7 April, plus 2 episodes of possible volcanic tremor (for 7 hours on 2 April and 10 hours on 4 April). The largest swarm began gradually on 7 May, peaked 14-16 May (maximum magnitude 2.1), and ended about 29 May, accompanied by 4 more episodes of possible volcanic tremor. During the strongest activity, 15-20 events of magnitude ≥1.0 were recorded daily.]

On 10 May, NZGS personnel measured temperatures of the hot gases issuing from the crater bottom. Rockfalls from the overhanging E crater wall and growing talus fans had reduced the degassing area. Temperatures of 320°-350°C were measured 1-2 m from the accessible edge of the hot area. These reflected a nearly steady decrease: 620° (1 July 1978), 520° (24 Feb 1979), 478° (25 June 1981), 418° (21 Jan 1982), 458° (10 June 1982), [349° (4 May 1983) and 105° (7 Feb 1985)]. Ngauruhoe last erupted 12-23 Feb 1975, when strong explosive activity sent eruption plumes to 10 km and pyroclastic flows moved down the flanks (Nairn and Self, 1978).

Reference: Nairn, I.A. and Self, S., 1978, Explosive Eruptions and Pyroclastic Avalanches from Ngauruhoe in February 1975; *JVGR*, v. 3, p. 39-60.

Information Contact: W. Giggenbach, DSIR, Wellington.

Further Reference: Balsillie, F.H. and Latter, J.H., 1985, Volcano-seismic Activity at Ngauruhoe during 1983; *New Zealand Volcanological Record*, no. 13, p. 66-71.

RUAPEHU

North Island, New Zealand
39.27°S, 175.58°E
summit elev. 2796 m
Local time = GMT +12 hrs
winter (+13 hrs summer)

North Island's highest summit, 13 km SSW of Ngauruhoe, is marked by an acidic crater lake of highly variable temperature. This has been the site of more historic eruptions (43) than any other crater lake in the world. The DSIR maintains a seismic observatory at the foot of the volcano, 9 km NNW of the crater lake.

A major phreatic eruption in 1975, the second largest in history, preceded SEAN's start by only 5 months. The NZGS notes that, in contrast to 1975, 1976 was a year of minor volcanic activity at Ruapehu, with relatively small geyser-like phreatic eruptions observed in April and Nov.

Reference: Nairn, I.A., Wood, C.P., Hewson, C.A.Y., and Otway, P.M., 1979, Phreatic Eruptions of Ruapehu: April 1975; *New Zealand Journal of Geology and Geophysics*, v. 22, p. 155-173.

General Reference: Houghton, B.F., Latter, J.H., and Hackett, W.R., 1987, Volcanic Hazard Assessment for Ruapehu Composite Volcano, Taupo Volcanic Zone, New Zealand; *BV*, v. 49, p. 737-751.

11/77 (2:11) A moderate phreatomagmatic eruption from Crater Lake occurred at 1350 on 2 Nov, accompanied by a magnitude [3.4] earthquake. Two eruptive pulses were observed, the second producing an ash cloud that rose 1800 m above the summit before being blown NE.

NZGS personnel inspected the volcano from the air less than 2 hours after the eruption and conducted ground investigations on 5 and 9 Nov. Airfall tephra was deposited in a narrow zone, extending several km NE of the vent (fig. 4-9). The margins of the deposits consisted of 40-150 mm of normally graded ejecta, ranging in size from coarse lapilli at the base to ash at the surface. Accessory material was estimated to comprise > 99% of the deposit, which included a substantial quantity of sulfur spherules and gypsum. Within about 600 m of the vent, numerous impact craters were observed, most

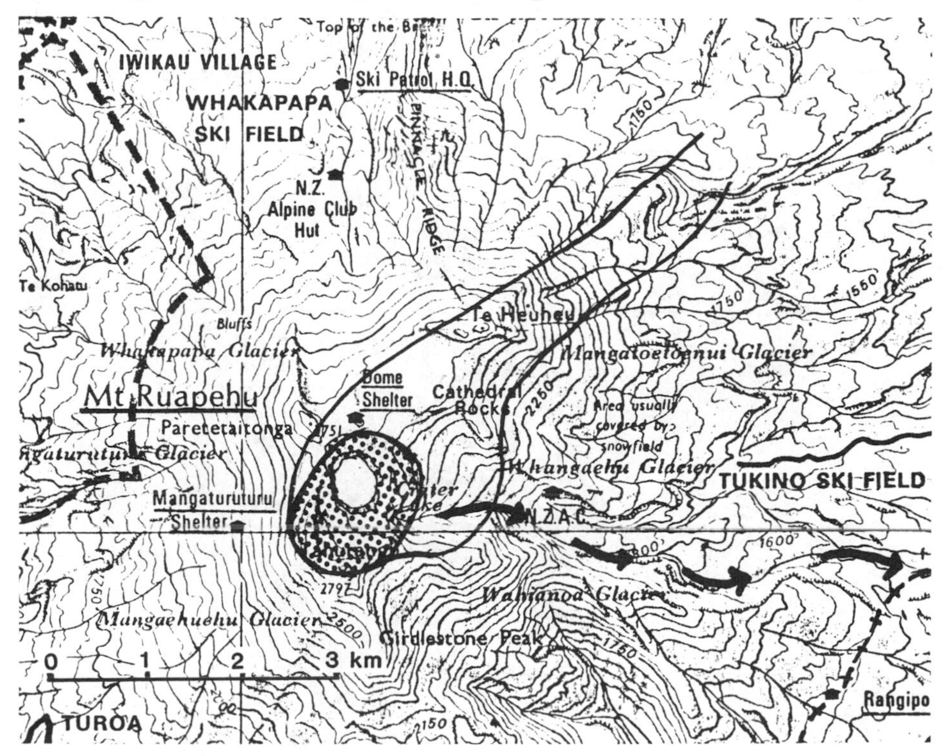

Fig. 4-9. Summit region of Mt. Ruapehu, showing area of 2 Nov 1977 impact craters (stippled) and ash deposits (enclosed by solid line). Crater Lake is 500 m in diameter. Arrows indicate the upper portion of the path of the lahar down the Whangaehu River. Contour interval is 50 m.

0.15-0.2 m in diameter, cylindrical, and steeply inclined (fig. 4-10). The largest crater in this area was 1 m in diameter, but a single isolated crater, 5 m in diameter, was discovered 1.2 km from the vent. Craters were formed by dense accessory andesite blocks that were also scattered across the surface of the deposits, accompanied by rare (< 5%) fresh pumiceous andesite breadcrust bombs (fig. 4-11) reaching 0.8 x 0.4 x 0.4 m in size. A lahar traveled 21 km down the Whangaehu River valley, depositing a narrow band of debris.

Several months of small-scale phreatic activity in Crater Lake preceded the 2 Nov event. During the aerial survey, vigorous steam emission was occurring from the lake, which contained a black sulfur slick near its center, indicating convective upwelling. Several small steam eruptions were observed between 4 and 9 Nov, none of which produced new tephra deposits. The volume of 2 Nov ejecta is estimated to be at least as great as that of the 1971 phreatomagmatic eruption.

Information Contacts: B. Houghton, E. Lloyd, NZGS, Rotorua; P. Otway, NZGS, Wairakei.

Further Reference: Wood, C.P., 1978, Bombs from the Ruapehu Eruption, 2 November, 1977; *New Zealand Volcanological Record*, no. 7, p. 39.

[Summary of 1978-79 activity from NZGS: "A dark column of water was erupted to about 610 m above Crater Lake on 7 March 1978 at about 1240. Only fine grey airfall ash, similar to the material suspended in the lake water, was ejected. This tephra fell within a 500 m-wide band that extended 2.3 km NNE of Crater Lake at a thickness that nowhere exceeded 5 mm. Small geyser-like eruptions were observed in Crater Lake in June, when the lake temperature rose to 48.3°C. Tephra, presumably deposited by a phreatic eruption, was observed around Crater Lake on 2 Sept, and further eruptions occurred on 7 and 9 Sept. Small geyser-like phreatic eruptions were also observed in early Oct.

"Following a period of quiescence since mid Oct 1978, small phreatic eruptions occurred on 7 and 17 Jan 1979. The temperature of the lake rose to reach a maximum recorded value of 33.5°C on 26 March, while the discharge also increased to a maximum of 185 liters/sec on the same date but had decreased to 24 liters/sec by 2 April. Further small phreatic episodes were reported in late June and July. Lake temperatures rose to 26.5°C during this time, then gradually declined for the rest of the year."]

3/80 (5:3) Small phreatic explosions, accompanied by summit inflation and an increase in crater lake temperature, have occurred at Ruapehu, probably beginning in late Jan (fig. 4-12). During each several-hour summit visit by NZGS personnel (from late Jan through mid-March), 2-4 explosions took place from the crater lake.

Although explosion sizes varied, their characteristics were similar. A dark eruption slug of water and particulate matter broke through the lake surface (sometimes preceded by updoming of the lake water), rose a maximum of a few tens of meters, then collapsed to form a toroidal ring. Surtseyan jets formed, distinct from the main mass of ejecta. A steam column was sometimes produced, from which base surges moved radially outward. Waves generated by the explosions usually overtopped the lake outlet, sending pulses of water eastward down the Whangaehu River. Trains of black or yellow sulfur were seen floating on the lake surface after some explosions. Dark floating objects about 1 m across were also observed, but it was not possible to determine whether these were pumiceous blocks or frothy sulfur.

Lake temperature rose from 22°C on 31 Jan to 43°C on 22 Feb, then declined to 37.5° by 14 March. Six μrad of summit inflation took place between surveys on 15 Jan and 12 Feb, but no inflation has been measured in the 3 surveys since then.

The NZGS notes that similar periods of phreatic activity have occurred several times in the past 12 years, including 1968, 1971, 1974, and 1978. The 1971 activity culminated in a major phreatomagmatic eruption, but the eruptions of 1969 and 1975 took place up to a year after such active periods, and the 1977 eruption was not preceded by renewed phreatic events.

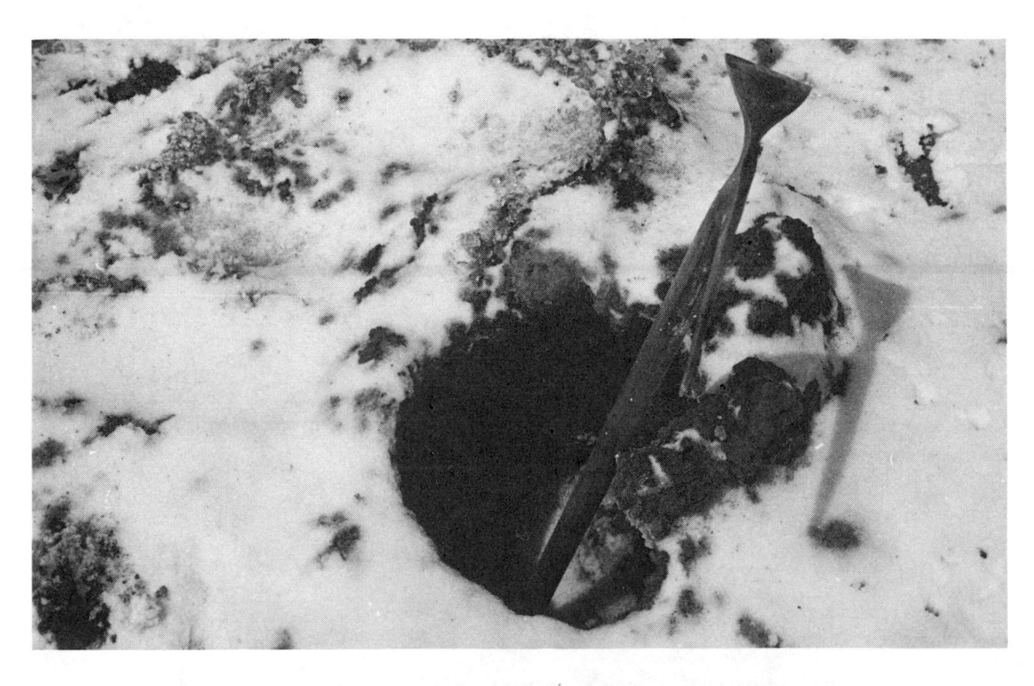

Fig. 4-10. Impact Crater, northern Crater Lake basin, 9 Nov 1977. Courtesy of NZGS.

Information Contacts: B. Houghton, E. Lloyd, B. McG. Simpson, NZGS, Rotorua; P. Otway, NZGS, Wairakei.

4/80 (5:4) Geologists visited Ruapehu on 27 March, and 12-13 and 16 April. During the 27 March visit, 3 small phreatic explosions took place through Crater Lake in 4 hours and 40 minutes, continuing the activity that probably began in late Jan. The largest explosion produced 20 m-high water jets and a few waves big enough to overtop the lake outlet, sending very small pulses of water down the Whangaehu River. A strong H_2S odor was noted after this explosion. The lake water temperature near the outlet was 40°C, 2.5° higher than on 14 March, but close to the average value recorded since 20 Feb. Strong upwelling occurred near the center of the lake, but was less distinct in the N area. There was no evidence of recent ashfall around the crater.

Activity was similar during the 12-13 April inspection, with 6 explosions in about 23 hours. One, with a muffled booming sound, ejected jets of water 30 m high, formed 1-2 m waves, and sprinkled dark ash on the observers. There was a strong H_2S odor downwind from all of the explosions. The lake color remained battleship gray, but lake level had risen more than $^1/_2$ m since 27 March and was overflowing at about 20 liters/sec down the Whangaehu River. The lake temperature at the outlet had dropped 3° to 37°C since 27 March. A leveling survey indicated that less than 1 μrad of tilt had occurred since mid-March. There has been minimal deformation during the last 2 $^1/_2$ months.

No explosions took place during a 2 $^1/_2$ hour visit on 16 April, but a thin layer of ash had fallen on snow NNE of the crater, perhaps from an eruption cloud seen from a distance early 14 April.

Information Contact: P. Otway, NZGS, Wairakei.

5/80 (5:5) NZGS personnel visited Ruapehu on 7 May. Although occasional small phreatic explosions had occurred through Crater Lake during previous visits beginning in late Jan, no explosions were observed on 7 May. There was no ash on snow that had fallen around the summit area on 29 April.

Large, yellow-green sulfur slicks floated at the N end of the battleship-gray lake. Upwelling at the center of the lake was only intermittent, in contrast to the continuous upwelling seen during earlier visits. Lake water temperature was at 39°C at the outlet, 2° higher than on 13 April, but within the range of temperatures recorded since mid-Feb.

The seismometer recorded continuous low-level tremor [on 7 May, as on many other days].

Information Contact: B. McG. Simpson, NZGS, Rotorua.

6/80 (5:6) NZGS personnel observed reduced activity during a 1 $^1/_2$ hour visit on 30 May. Upwelling in Crater Lake had been continuous and fairly vigorous when visited from mid-Feb through mid-April, and intermittent on 7 May. However, upwelling was not evident on 30 May, although faint yellow sulfur slicks were visible near the center of the lake. The water temperature near the lake outlet was 31°C, 7.5° lower than 22 days earlier. The small explosions through the lake that had occurred intermittently in previous months were not observed on 30 May, and there was no ash on snow surrounding the lake. Tiltmeters recorded 4 μrad of deflation since 13 April, in contrast to less than 1 μrad of change 12 Feb-13 April.

Information Contact: P. Otway, NZGS, Wairakei.

10/80 (5:10) Visits by F. Greenhall on 19 Oct and P. Otway on 20 Oct revealed a thin layer of pale gray ash in a 400 m-wide area S of Crater Lake. The ash appeared to be composed entirely of lake sediments and contained no coarse particles. It had been deposited in the upper layers of snow known to have fallen between 17 Oct and the late afternoon of 18 Oct. Only a minor wave surge of less than 1 m appeared to have been associated with the ash ejection.

J. H. Latter reported that a period of low-frequency volcanic seismicity occurred on 18 Oct at 1435, reaching a maximum magnitude of 2.5. [Similar activity occurred

Fig. 4-11. Broken face on large breadcrust bomb, revealing scoriaceous interior, Crater Lake basin, 9 Nov 1977. Courtesy of NZGS.

13 and 15 Sept and 3 Nov without associated ash emission.] The temperature of the lake on 20 Oct was 31°C, a 6° increase in 13 days. The lake was a turbid gray, with large slicks of dark sulfur floating near its center and much steam rising from the surface. Upwelling near the center appeared strong, although steam partially obscured this area.

Occasional explosions from the crater lake began in late Jan and continued through mid-April. No evidence of additional activity had been observed since April.

Information Contacts: P. Otway, NZGS, Wairakei; J. Latter, DSIR, Wellington; F. Greenhall, Ohakune.

10/81 (6:10) Ruapehu's Crater Lake has cooled from 32.5°C in late April to 11°C in early Oct. The temperatures recorded since late Aug were the lowest since detailed measurements began. Periods of low Crater Lake temperature preceded strong eruptions in June 1969 and April 1975.

The following report, from J. H. Latter, describes recent seismicity.

"The period of declining temperatures up to 7 Sept was seismically quiet, apart from minor volcanic tremor and associated B-type earthquakes that ended on 6 July, a short swarm of roof rock shocks from 11 to 13 July, and less than an hour of strong volcanic tremor on 24 Aug. In particular, there were no large B-type earthquakes, as after the April 1975 eruption, which might have suggested activity at the volcanic focus that was unable to break out at the surface. This might be interpreted as evidence that the low lake temperatures were due either to a blockage deeper than the volcanic focus (thought to be about 1 km below Crater Lake), or that heat transfer had declined or stopped, for some different reason.

"Beginning gradually on 7 Sept, a swarm of roof rock earthquakes marked the end of seismic quiescence. The swarm was unusually prolonged, ending on 2 Oct with an earthquake of M_L 2.4 (i.e. quite small), which was not a true roof rock earthquake. This was well-recorded on the Tongariro National Park network and could be accurately located at 2 km below Crater Lake, which, in the model used, marks the top of the Tertiary sedimentary rocks below the volcano. Rather than being a roof rock earthquake, it was therefore a 'wall rock', or perhaps even 'floor rock' earthquake, for the known volcanic focus; although it is possible that it may represent a roof rock earthquake for a deeper volcanic focus (magma chamber?) below the Tertiary sedimentary rocks.

"A train of weak B-type earthquakes preceded the 2 Oct earthquake by about 54 hours. Volcanic tremor began about 8 hours later but lasted only a few hours. Tremor did not become common on the records until after the 2 Oct earthquake. Since then it has built up steadily on the records, becoming quite strong and almost continuous since 14 Oct. Clearly the roof rock swarm and terminating earthquake took place in response to stress caused by blockage of some kind and had the effect of removing the blockage. Since 2 Oct, more typical volcano-seismic activity, as shown by the tremor, has resumed.

"An unusual feature has been that some of the volcanic tremor has been of a much higher frequency than measured previously at Ruapehu. In the past, 3 Hz tremor has been thought to have been correlated with eruptions. On 7 Oct, tremor with dominant frequencies as high as 4.5 Hz was recorded, probably because of intrusion at shallow levels high in the volcano in very restricted or narrow spaces [see also 28 Oct in 6:12].

"In spite of the interesting type of recent seismic activity, it is stressed that the amount of energy released is small. Roof rock activity has peaked at a level well below that reached in Nov 1976 and March 1977. Furthermore, B-type earthquakes have all been small (the largest M_L 2.3), and the volcanic tremor, which peaked on 18 Oct at about 1.9×10^6 J [and again on 25 Oct (about 3.5×10^6 J)], was exceeded in Nov 1976, Oct 1977, Oct 1978, and Aug and Oct 1979. It is far below the level that prevailed before the 1971 eruptions.

"It seems likely that the present moderately high level of tremor will be associated with reheating of Crater Lake, and that eruptions will take place. But the volcano can perhaps be considered in a less dangerous state than

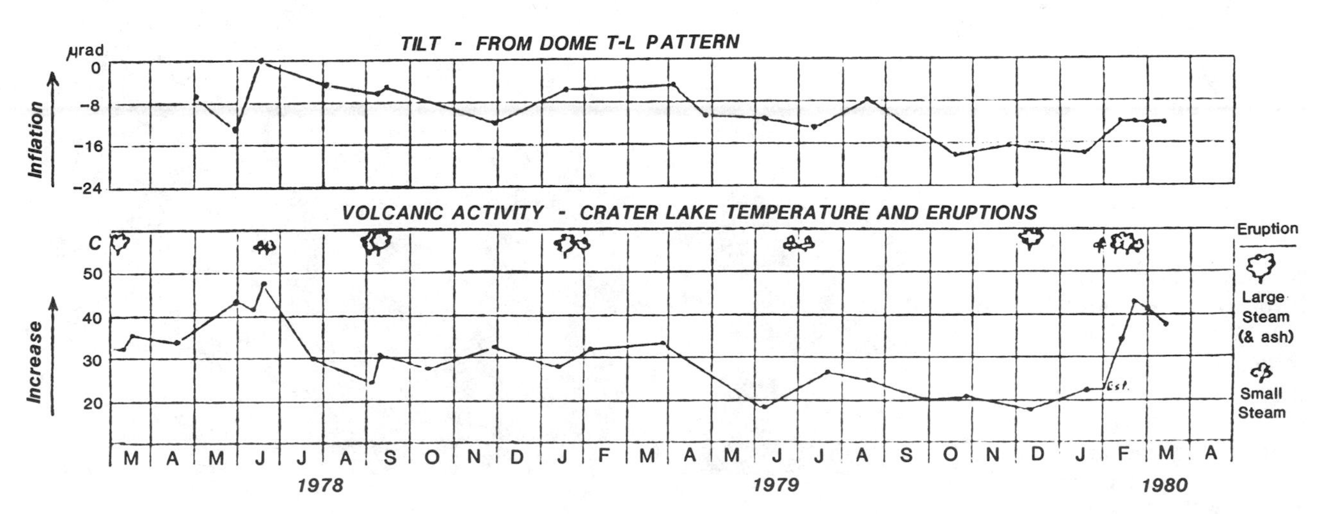

Fig. 4-12. Summit tilt in µrad, *top*, and Crater Lake temperature with eruptions shown schematically, *bottom*, March 1978 – March 1980.

during its recent period of low temperatures and marked roof rock activity."

Information Contact: J. Latter, DSIR, Wellington.

11/81 (6:11) The increased seismicity reported last month was followed by resumption of upwelling in Crater Lake, a sharp increase in its temperature, and a small ash eruption.

Geologists recorded a Crater Lake temperature of only 10.5°C on 8 Oct, a decline of 3° since 28 Aug and the lowest since detailed measurements began (although the lake was reportedly frozen in 1886 and 1926). The lake showed no sign of sulfur slicks or steam and only minor upwelling was evident, at the N end. When geologists returned late 12 Oct, sulfur was upwelling from the center of the lake at 10-20 minute intervals and the temperature had risen slightly, to 11.8°C. Tilt measurements indicated that about 8 μrad of inflation had occurred since 17 Aug. The distance across the N part of the crater rim had increased 21 mm between surveys 1 May and 28 Aug, also indicating inflation, but this increase had been entirely reversed by 13 Oct.

Float plane pilot K. Newton reported that he saw a patchy layer of ash extending southward from the crater during the week prior to 30 Oct, probably about 27 Oct. Geologists inspecting the crater area 4 Nov observed a thin layer of ash extending 500 m S of the lake. Snow that fell 1 Nov had covered the ash, but subsequent melting exposed some of the distal end of the deposit. Tilt measurements 4 Nov indicated that the 8 μrad of inflation noted 13 Oct had been completely reversed. The lake temperature had increased to 24°C, its color had changed from blue-green to its normal battleship gray since mid-Oct, and upwelling in the center produced black sulfur slicks.

Information Contact: P. Otway, NZGS, Wairakei.

12/81 (6:12) Increased seismicity and higher crater lake temperatures preceded a small ash eruption in late Oct. About 3 weeks later, a second small explosion from the crater lake ejected tephra that may have included fresh magma. Minor explosive activity was continuing in late Dec. Volcanic tremor may indicate shallow intrusions of magma beneath the summit crater lake, or lava extrusion onto the lake bottom.

Volcanic tremor started to increase in early Oct, and had become quite strong and almost continuous by the 14th. For the next 2 weeks, seismographs recorded moderately strong tremor with frequencies between 1.1 and 2.7 Hz, a normal range for Ruapehu during periods of activity [but see 6:10]. Tremor declined on 28 Oct, stopping completely for 5 1/2 hours. Tremor resumed at 2230 on 28 Oct, at the highest frequency (5 Hz) recorded at Ruapehu since a seismograph was installed near the summit in May 1976. The high-frequency tremor continued for about 10 hours and was interpreted by J. H. Latter as indicating shallow intrusive activity. Between 29 Oct and 14 Nov, seismographs recorded occasional normal-frequency (2.0-2.8 Hz) tremor and a few weak low-frequency (B-type) volcanic earthquakes, the strongest a magnitude 2.4 event on 6 Nov. A 3-hour episode of high-frequency (3-3.5 Hz) tremor was recorded late 17 Nov, and 3-5 Hz tremor that started late the next day lasted about 10 hours.

Park Ranger Pat Sheridan observed ash on snow in the Crater Lake area just before noon on 19 Nov. However, an Air New Zealand pilot flew over the volcano 4 hours later and saw no evidence of a recent eruption. Geologists have not been able to resolve the conflicting observations, but their overflight the next morning revealed dark gray mud extending about 150 m SW of the lake. NZGS personnel visited the summit area 24 Nov and saw dark gray ash to 700-800 m down the valley of the Whangaehu River, which flows down the E flank from Crater Lake. The maximum thickness of the deposit was less than 1 cm. C. P. Wood analyzed the tephra, primarily precipitated lake sediment (including yellow sulfur spheres to 1 mm in diameter) and altered andesite, but containing many angular chips of fresh-looking dense black glassy andesite, particularly in the coarser size fraction. Wood suggested that the glassy andesite may have been ejected directly as fresh magma or may have been fragments of lava extruded very recently onto the crater lake floor. A [very small (M 1.7)] B-type earthquake at 0123 on 20 Nov, at the end of an episode of high-frequency tremor, seemed the seismic event most likely to have accompanied the tephra eruption. The crater lake temperature was 42°C on 24 Nov, up from 36.5° 3 days earlier and 24° during the last visit by geologists on 4 Nov (6:11). Tilt stations were reoccupied on 24 Nov, but no significant changes had occurred since measurements 20 days before.

Between 18 and 27 Nov (the last day for which detailed seismic records were available at press time), 10-20 hours of volcanic tremor were recorded on most days, at frequencies of 4.5-5 Hz until 22 Nov, 3.5-5 Hz until the 24th, and 3-3.5 Hz thereafter. There were only about 2 hours of tremor on the 24th and 25th. None of the tremor was strong, but the highest amplitudes were recorded 27 Nov. Small low-frequency volcanic earthquakes began early 23 Nov, apparently centered less than 1 km below the crater lake, at about the level where roof rock events normally occur (6:10). Latter noted that this suggests magma has intruded the roof rock beneath the lake. Low-amplitude tremor was continuing as of late Dec.

Geologists returned to the volcano 28 Dec. They observed 3 large vapor plumes during their climb to the summit, and while at the crater lake saw a vigorous explosion that produced a 500-m steam column and waves more than 2 m high. No ashfall was noted although the initial jets of water were darkened by tephra. A nearby seismograph recorded a 3 Hz signal during the explosion. The only indications of recent tephra emission were small 1-2 mm-thick lobes of dark gray ash and sulfur that extended about 100 m from the lake. The temperature of the lake had risen further since the previous measurement 24 Nov, to nearly 47°C. Reoccupation of tilt stations showed about 7 μrad of inflation since 24 Nov, but there had still been a net deflation of 4 μrad since 13 Oct.

Information Contacts: J. Latter, DSIR, Wellington; B. Scott, I. Nairn, C. Wood, NZGS, Rotorua.

1/82 (7:1) Frequent explosions from Crater Lake continued through late Jan. The lake temperature continued to rise, and volcanic tremor and shallow volcanic earthquakes were frequent. NZGS personnel carried out field work in the summit area 12, 18, and 21 Jan.

Eruptive Activity, Temperature, and Ion Concentration of Lake Water. During the roughly 12-hour visit on 12 Jan, 23 small hydrothermal explosions from the crater lake were noted, separated by quiet periods of 4-127 minutes. Water jets from these events rose as much as 70 m, and 2 generated large steam columns. A dark-colored deposit on snow extended from Crater Lake onto the lower flanks. The phreatic explosion that produced this deposit was not observed, but possibly accompanied a moderate B-type volcanic earthquake recorded the previous day. Crater Lake temperature was about 57°C, 10° hotter than when it was last measured, on 28 Dec. Geologists were hampered by poor weather 18 Jan, but observed 4 hydrothermal explosions, similar to those of 6 days earlier, during brief periods of good visibility. During better weather on 21 Jan, 20 hydrothermal explosions were seen in 7 hours, again similar to those of 12 Jan. The temperature of the lake remained at 57°C. Depth soundings at 2 points in the center of Crater Lake indicated that no major lava dome growth has occurred, but did not rule out the extrusion of small quantities of lava onto the lake floor. Concentrations of Mg and Cl, and the Mg/Cl ratio of lake water have both shown large increases after falling during early 1981. The NZGS noted that these increases demonstrate that interaction between fresh rock and lake water has been occurring, but it is uncertain whether the fresh rock is new magma or older rock exposed to lake water for the first time because of explosive activity.

Deformation. Tilt surveys in the summit dome area measured only 3 μrad of apparent deflation between 28 Dec and 12 Jan, and no additional deflation was detected on 18 Jan. Since inflation peaked 25 June, very little change in tilt has been detected in this area. However, horizontal deformation measurements showed that 16 mm of expansion had occurred 13 Oct and 21 Jan along a 600-m line across the N side of the crater. Horizontal movements across the dome during this period did not exceed 10 mm.

Seismic Activity. The increased seismic activity that started 18 Nov peaked 30 Nov-2 Dec, then declined. [Although reported in considerable detail here, Nov-Dec seismicity was substantially weaker than that associated with October's minor eruptive activity.] High-frequency (3.5 Hz) tremor, strongest 28 Nov, was succeeded by low-frequency (1-2 Hz) tremor of deeper origin that peaked 30 Nov and declined 1 Dec. A magnitude 2.3 B-type (low-frequency volcanic) earthquake was recorded on 1 Dec and a magnitude 2.4 low-frequency volcanic event occurred the next day at 1221, centered in the roof rock overlying the usual focus of B-type shocks. Until the first roof rock event of clearly intrusive or magmatic character occurred on 23 Nov, all of the roof rock seismicity had been tectonic and of high frequency. Volcanic tremor remained at a low level until 26 Dec, except for a minor peak on the 3rd. However, tremor was detected every day through this period, with dominant frequencies of 3.5-4.5 Hz through 15 Dec, and more normal frequencies (2-2.5 Hz) after that date.

A sequence of small earthquakes on 22 Dec was probably accompanied by an explosive eruption. A magnitude 2.0 B-type shock at 1208 was followed $2^1/_2$ min later by a magnitude 2.2 low-frequency volcanic event located in the roof rock zone. A weak air wave, detected by a nearby microbarograph, had a probable origin time of 8 sec before the second earthquake. Similarly, a small explosion witnessed by geologists on 28 Dec was accompanied by the onset of very weak low-frequency tremor, and was followed 9 and 12 sec later by magnitude 1.25 and 1.5 volcanic events in the roof rock zone.

Tremor strengthened in late Dec and early Jan, accompanied by discrete low-frequency volcanic earthquakes, the strongest at 0722 on 24 Dec (in the roof rock zone, magnitude 2.2), 1 Jan at 0613, and 2 Jan at 0104 (both slightly deeper, magnitude 2.3, B-type events). This period of increased seismicity was interpreted by J. H. Latter to suggest that high-level intrusion of magma was accelerating, probably accompanied by extrusion onto the crater lake floor.

High-frequency tremor continued until 5 Jan, then diminished by about a factor of 4. Tremor increased again on the 14th and reached a peak 17-18 Jan. Dominant frequencies ranged from 3 to 5 Hz, indicating that the activity continued to be very shallow. Few significant discrete volcanic earthquakes were recorded during the first half of Jan, but explosions observed on 12 Jan were accompanied by very small B-type events. Portable seismographs were operated by NZGS geologists visiting the crater 18 and 21 Jan. On the 18th, they recorded weak high-frequency tremor but detected no earthquakes accompanying the 4 observed hydrothermal explosions. Increases in tremor amplitude preceded some of the explosions observed 21 Jan by as much as 5 minutes, but again, none were accompanied by discrete earthquakes.

Information Contacts: J. Latter, DSIR, Wellington; D. Sheppard, Chemistry Division, Wellington; B. Scott, P. Otway, I. Nairn, NZGS, Rotorua.

2/82 (7:2) NZGS personnel returned to Ruapehu 5 Feb and observed an apparent decline from the fairly vigorous Jan activity. During $3^1/_2$ hours of field work the geologists saw only 1 explosion; a geysering of muddy black water from near Crater Lake's center that lasted at least 25 seconds. Sounds that may have been produced by 2 additional explosions were heard during cloudy periods that obscured Crater Lake for most of the last 2 hours of the NZGS visit. In contrast, 2-3 explosions per hour were noted 12 and 21 Jan. The amplitude of volcanic tremor, measured for 2 hours by a portable seismograph on 5 Feb, had decreased considerably since 21 Jan [but J. Latter notes that such fluctuations are common]. Tremor frequency was about 3 Hz. Crater Lake temperature dropped from 57°C on 12 and 21 Jan to

49°C on 5 Feb. However, preliminary analyses indicated that both Mg and Cl concentrations and the Mg/Cl ratio continued to increase, consistent with increasing interaction between lake water and magma or rock not previously exposed to lake water. Depth soundings in the central area of Crater Lake indicated that no major lava dome growth had occurred on that portion of the lake bottom.

Information Contact: I. Nairn, NZGS, Rotorua.

3/82 (7:3) Seismic activity, Crater Lake temperature, and strength and frequency of the lake's hydrothermal eruptions declined in Feb and early March, but increased again in mid-March.

Summit-area monitoring by NZGS personnel 11 Feb showed little change since the visit 6 days earlier. Only 4 small explosions from Crater Lake were noted in 8 1/2 hours. The largest, lasting about a minute, ejected three 30 m-high columns of muddy black water, which collapsed onto the lake surface to form small base surges. The temperature of the lake water had risen slightly, from 49° to 50.5°C. Distance-measuring and tilt surveys showed no significant changes. The next visit by geologists, on 5 March, lasted 4 hours, but no explosions were observed nor was there any evidence of new ash around the lake. However, climbers saw 2 very small explosions the next day. The lake temperature had dropped almost 10°, to 41°C, in about 3 weeks. Only minor tilt changes were observed.

Park rangers received a report of an eruption at about 1215 on 16 March that generated a steam cloud filling the entire crater area to an estimated height of 1 km. NZGS personnel saw one steam explosion during a 2 1/2 hour visit 18 March. Continuous steaming of Crater Lake was reported during the early morning of 20 March. Geologists returned 23 March and observed 5 explosions from Crater Lake in 10 hours. Four were relatively small, producing columns of water 5-30 m high. However, a larger explosion at about 1430 produced large waves, and jets of black water that rose more than 100 m above the lake surface. Lake temperature had increased 6° since 5 March, to 47°C. No significant tilt changes were detected during surveys 23 and 26 March. A single Crater Lake explosion was observed during 5 hours of NZGS fieldwork 26 March.

The following is from reports by J. H. Latter. [For his later analysis of this activity, see *New Zealand Volcanological Record,* no. 12, p. 31-37].

A period of higher amplitude volcanic tremor began about 1600 on 14 Jan (7:1), climaxed 26 Jan, and ended 30 Jan. Since then, strong tremor has been recorded only during an 8-hour period 10-11 Feb. Through 25 Jan, the tremor was dominantly high-frequency (3-4 Hz), suggesting that its origin was very shallow, but since then the strongest tremor has been mainly low-frequency (1-2 Hz). The focus of activity has evidently moved down to a lower level within the volcano. Latter notes that this could either be due to a process of withdrawal of magma, which up to now has been standing at a high level, or to the arrival of fresh magma from greater depths at the normal volcanic focus about 1 km below Crater Lake.

Only small volcanic earthquakes occurred between mid Jan and the end of Feb. A marked swarm of low-frequency volcanic earthquakes (B-type) took place, at about the normal focus, 20-22 Feb; activity peaked about 1200 on 20 Feb with several magnitude 2.1 earthquakes. This magnitude was relatively low, and it was not known whether the events were accompanied by eruptions. Latter notes that it was likely that the B-type swarm represented a minor stoppage in the volcano's conduits, but that the stoppage must have been rather weak since it was evidently overcome by quite small-magnitude earthquakes. Similar but smaller events took place 21-22 Jan (when no eruptions took place), and 3 and 14 Feb.

Shallower seismic activity peaked 23-25 Jan, when high-frequency tremor was fairly strong, preceded by the largest magnitude volcanic earthquakes at this level since 24 Dec (the so-called C-types, two M_L 2.0 events). A smaller C-type earthquake (M_L 1.8) occurred 28 Jan; since then there have been few, the largest only M_L 1.6 (on 26 Feb). During the declining stages of activity 24-25 Jan, 31 Jan, and 24-26 Feb (after the B-type swarm mentioned above), high-frequency roof rock earthquakes with magnitudes between 1.6 and 1.9 have been detected.

Latter notes that "the best fit for B-type earthquake data suggests a mean depth of origin of 0.77 km beneath the floor of Crater Lake. Adopting an explosion model for the earthquakes, and equating the travel time (origin time of earthquake minus observed eruption time) of 8.5 seconds with upward movement of gas from this depth, gives an average velocity of the gas column of about 90 m/sec. Applying the same velocity to the onsets of C-type earthquakes yields a depth of origin of about 250 m below the floor of the lake. This estimate, though crude, is probably of the right order, and suggests that magma had risen during the increased activity (since Sept 1981) by about 500 m.

"The decline in seismic activity at the end of Jan, and the change to tremor of deeper origin, appears more likely to have been due to withdrawal of magma than to a major blockage of conduits within the volcano. Although lake temperature has declined, partly no doubt because of the accelerated melt around Crater Lake during the long spell of fine weather, the volcano still gives the impression of being 'open vent'. The small magnitude (M_L 2.1) of the largest earthquakes occurring since activity declined suggests that only minor blockages have formed, and have been fairly quickly overcome."

High-level (high-frequency) tremor continued 1-23 March, although none was recorded 4 or 7-10 March. Tremor was strong 11-16 March, peaking on the 13th, but remained much weaker than in late Jan. Occasional episodes of low-frequency tremor were recorded during the first 3 weeks in March, some lasting for several hours. These were interpreted by Latter as indicating movement at the base of the magma column, at least 500 m tall, that may extend from 200-300 to 700-800 m below Crater Lake. A swarm of B- and C-type earthquakes began on 15 March, culminating in a 6-minute B-type sequence 21 March that reached a magnitude of 2.7, the largest volcanic earthquake at Ruapehu since 2 Jan. Clouds obscured the volcano 21 March, so it was impossible to determine if an eruption accompanied this event. The swarm was continuing as of 23 March.

Information Contacts: J. Latter, DSIR, Wellington; I. Nairn, B. Scott, NZGS, Rotorua; P. Otway, NZGS, Wairakei; R. Beetham, NZGS, Turangi.

4/82 (7:4) NZGS personnel observed increased explosive activity from Crater Lake in mid to late March, but no explosions occurred during visits to the summit area 15 and 21 April. The temperature of the lake water declined from 47°C on 23 March to 39° on 15 April, then increased slightly to 42° six days later. On 15 April, lake-surge deposits could be seen on 6-day-old snow as much as 2 m above the lake surface, but on 23 April there was no evidence of additional surges or recent ash emission. Deformation surveys indicated that about 12 mm of inflationary expansion had occurred across Crater Lake 23 March-15 April, but 10 mm of contraction of the same line was measured on 23 April. However, this line remained 15 mm longer than it had been a year earlier.

Information Contacts: A. Cody, I. Nairn, NZGS, Rotorua; P. Otway, NZGS, Wairakei.

6/82 (7:6) NZGS personnel surveying Ruapehu on 29 May observed no hydrothermal eruptions from Crater Lake during their 8-hour stay, nor did they see evidence of recent large surges of lake water in the heavy snow cover within 1-2 m of the water's edge. The lake level was lower than on the 15 and 21 April visits; measurements indicated it was 2 m below overflow, the lowest since the April 1975 eruption. Bathymetric surveys had shown a decrease in the depth to the vent area of about 250 m between 1965 and 1970 but data collected from remote depth sounding buoys in 1982 indicated that the vent area was about 150 m deeper than in 1970. Water temperature in the outlet area was 27°C, 15° lower than on 21 April (fig. 4-13). A large yellow-black sulfur slick usually occupied the center of the battleship gray lake. Chemical analysis of lake water showed small increases in Mg and Cl concentrations, but no change in the Mg/Cl ratio.

Deformation surveys of the volcano showed no measurable tilt, but an increase of about 10 mm, since 21 April, in a precisely measured 600-m line across the crater. About 10 μrad of inflation have occurred since the latter half of 1981, and the crater was about 20 mm wider than a year ago (fig. 4-14). NZGS interpreted the slight inflation as either non-elastic expansion possibly related to the eruptive period that began late in 1981 (6:10-12), or still-elevated gas pressure within the vent.

Information Contacts: P. Otway, NZGS, Wairakei; I. Nairn, NZGS, Rotorua.

7/82 (7:7) No hydrothermal eruptions from Crater Lake were observed on 5 or 26 July, when NZGS personnel worked at the volcano. The lake, colored its usual battleship gray, was steaming moderately on the 5th; on the 26th there was upwelling from the center of the lake but only a little steam was rising.

On 5 July the water temperature in the Outlet area was 33°C, 6° higher than on 29 May. Snow within 1-2 m of the water's edge showed no signs of recent large surges. Considerable sulfur coated the lake margin in the outlet area. A preliminary water analysis showed no change in Mg concentration but a small increase in Cl: Mg/Cl = 0.119. A triangulation survey showed that the length of a precisely measured 600-m line across the crater had decreased 7 mm since 29 May. NZGS interpreted the 1981 crater width measurements as showing virtually no deformation following the 20 mm expansion recorded during the onset of the eruptive period in Jan (fig. 4-14).

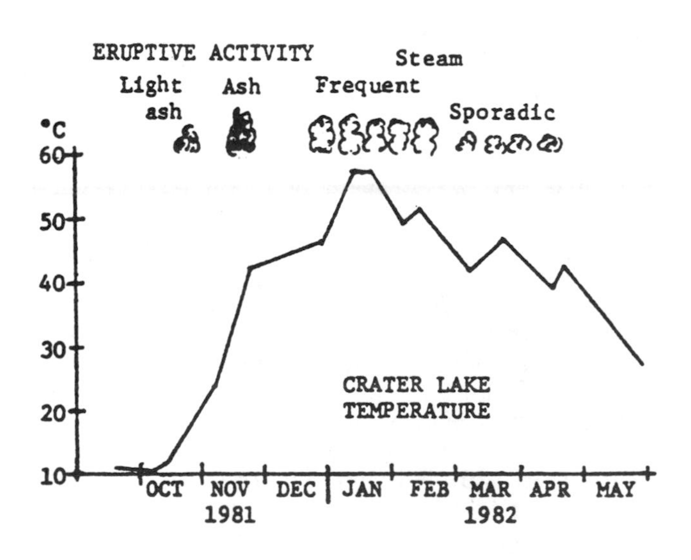

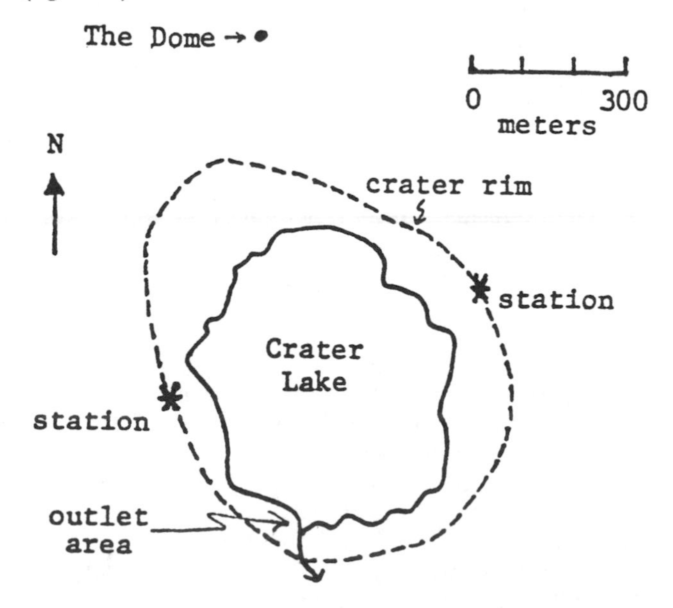

Fig. 4-13. *Left,* temperature of Crater Lake as measured in the Outlet area. *Right,* sketch of the crater area showing measurement sites used to collect data. Courtesy of P. Otway.

On 26 July the water temperature had dropped to 24.5°C. Snow lay within 0.5 m of the edge and showed no signs of recent ash deposits or surging. Mg concentration in the lake water was down slightly (by 4 ppm), but was only 1 ppm less than the mean Jan-April value.

Information Contacts: I. Nairn, B. Scott, NZGS, Rotorua; P. Otway, NZGS, Wairakei.

10/82 (7:10) Few changes have been observed since the end of July. NZGS personnel visited Ruapehu on 19-20 and 24 Aug, 17-18 Sept, and 21 Oct.

On all the visits the gray Crater Lake had yellow sulfur patches on its surface and minor to moderate upwelling in the center and near the N shore. Snow lay within 0.5 m of the water's edge and showed no sign of water surge. Water temperatures were 24°C on 24 Aug, 29° on 17 Sept, and 25° on 21 Oct. Mg concentration in the water has remained unchanged since Feb, but Cl concentration has increased. The Mg/Cl ratio has gradually declined from 0.130 in Feb to 0.115 on 24 Aug and 0.113 on 17 Sept. Deformation surveys indicated no apparent summit inflation.

Volcanic tremor has been recorded for some time and peaked in early Sept. The NZGS interpreted the low [normal] tremor frequency and the unchanged Mg concentration to indicate that magma is still deep beneath Crater Lake. The increased Cl concentration suggested that gas can freely vent into the bottom of the lake, and the slow rate of lake refilling suggested that no shallow magmatic intrusion had occurred.

Information Contacts: P. Otway, NZGS, Wairakei; I. Nairn, B. Scott, NZGS, Rotorua; A. Cody, F.R.I., Rotorua.

12/82 (7:12) A NZGS deformation survey on 15 Dec showed a moderate degree of inflation. The distance between two stations on opposite sides of the 600-m wide crater was 15-20 mm longer than on 17 Sept. About half the increase had occurred since 9 Nov.

The temperature of Crater Lake as measured at the outlet was 13°C, the lowest of the year. The lake temperature had been falling since the eruptive activity in Jan (7:1, 6). Some intermittent upwelling was observed in the N part of the lake, but little or none in the center. Due to meltwater inflow, the lake was greener and less turbid than in Oct, and the Mg and Cl concentrations were lower. The Mg/Cl ratio was 0.115 on 21 Oct, 0.113 on 9 Nov and 0.109 on 15 Dec. The decline was attributed to lack of interaction between lake water and lava (lower Mg values), but some continued fumarolic input of Cl.

The NZGS team noted that similar periods of cooling lake water in 1979, 1980, and 1981 ended with rapid lake temperature increase accompanied by hydrothermal eruptions. They also noted that the inflation was moderate but probably significant, indicating that Ruapehu may enter a phase of renewed activity within the next few months.

Information Contact: I. Nairn, NZGS, Rotorua.

1/83 (8:1) NZGS personnel visited Ruapehu on 3 and 24 Jan. Crater Lake was clearer than in Dec, and was not steaming on either day. On 3 Jan, no upwelling was apparent over the central vent, but moderately strong upwelling was occurring from one of the vents at the N end of the lake, radiating discolored water and yellow-gray sulfur slicks. Lake temperature measured at the outlet was 22°C, 7° higher than on 15 Dec. The NZGS ascribed this to the significantly increased upwelling from the N vent, but also to reduced meltwater inflow. Reduced heat flow from the main vent was credited with the steady color change from gray to blue-green, as had occurred in the past.

On 24 Jan, upwelling was slight over the main vent and moderate over the N vent. A considerable number of sulfur globules were floating in the outlet area, where lake temperature was 20°C.

Although concentrations of magnesium and chlorine were higher on 3 Jan than in Dec, and lower on 24 Jan, the Mg:Cl ratio remained stable at 0.110 and 0.109 respectively, reflecting the generally low level of activity in the lake.

The 3 Jan horizontal deformation survey showed that the Nov and Dec inflationary trend had reversed. The 20-mm extension of the 600 m-wide crater that had developed between the 17 Sept and 15 Dec surveys had disappeared. The tilt-levelling survey detected no significant changes since the previous measurements on 17 Sept. However, a second deformation survey on 24 Jan revealed renewed extension. The distance between 2 stations on opposite sides of the crater was only 10 mm less than on 19 Oct. Analysis of deformation measurements since Sept showed greater changes than had been estimated, the most rapid changes recorded in recent years.

According to an NZGS tentative interpretation, the changing conditions at the volcano (table 4-6) indicate "strain release as a deep blockage (at about 1 km) of the vent was overcome in the latter part of Dec. The main vent may now be in an open state, and will possibly allow magma or gas to rise relatively freely. . . . The degree of

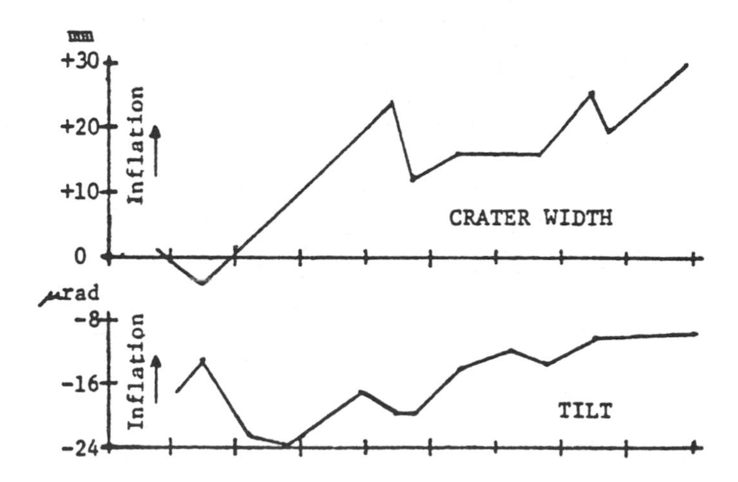

Fig. 4-14. *Top,* crater width as measured Oct 1981-May 1982 between two stations (fig. 4-13) about 600 m apart on the crater rim. *Bottom,* tilt as measured from the Dome, about 200 m N of the crater rim. Courtesy of P. Otway.

Interval	Deformation	Temperature Change °C	Main Vent Upwelling	North Vent Upwelling	Mg (ppm)	Cl⁻ (ppm)	Mg/Cl⁻ Ratio
17 Sept-19 Oct	None	-4° (29°-25°)	None	Slight	1022	8850	0.115
19 Oct-9 Nov	+8 mm	-4° (25°-21°)	None	None	988	8725	0.113
9 Nov-15 Dec	+22 mm	-8° (21°-13°)	None	Slight	900	8240	0.109
15 Dec-3 Jan	-50 mm	+9° (13°-22°)	None	Strong	918	8350	0.110
3 Jan-24 Jan	+10 mm	-2° (22°-20°)	Slight	Moderate	850	7785	0.109

Table 4-6. Changing conditions at Ruapehu between Sept 1982 and Jan 1983. Deformation is measured between 2 stations on opposite sides of the crater (see fig. 4-13). Lake temperature is measured at the outlet. Upwelling and Mg and Cl⁻ measurements are on the last day of each interval.

activity at the N vent is complex and is related to the degree to which the main vent is blocked as well as to the supply of heat from depth. . . .

"The pattern of low lake temperatures with no visible upwelling above the main vent combined with temporary inflationary expansion of about 20 mm is similar to that of July 1980 and Aug 1981. In both cases heating from the main vent recommenced 6-7 weeks later, followed by hydrothermal eruptions from the same site about a month after that. If the same sequence occurs on this occasion we can expect to see strong convection start from the main vent by the end of Jan or early Feb, and (provided excessive heat is not released by the N vent) eruptions may commence by early March. (However) . . . due to the now apparently relaxed state of main vent, the onset of renewed activity . . . may start with relatively quiet lake reheating."

Information Contacts: P. Otway, NZGS, Wairakei; I. Nairn, NZGS, Rotorua.

2/83 (8:2) When NZGS personnel returned to Ruapehu 10-11 Feb, they found the lake turbid. It had been clear on their previous visit, 24 Jan. Upwelling was slight over the central vent with a trace of dark sulfur, and minor from 2 or 3 cells at the N end of the lake.

Thick strands of gray sulfur spheroids and some yellow teardrop shapes floated near the outlet. Fine-grained glass-foam fragments also were present in the floating material. Glass-foam appeared in May 1973; was produced in abundance during the April 1975 eruption; was found during the Oct-Nov 1977 eruptive period; and appeared in sulfur slicks on 21 Feb 1978. No glass-foam was found during the Feb 1980 or Nov 1981-Jan 1982 eruptive periods.

Lake water temperature at the outlet was 19°C, 1° cooler than on 24 Jan. The water's Mg concentration had remained stable, but Cl concentration had risen by 250 ppm, indicating to the NZGS a resumption of fumarolic activity. The Mg/Cl ratio was 0.106.

The horizontal deformation survey showed a 12-mm extension of the 600 m-wide crater as measured between 2 stations on opposite sides of the rim. After a period of rapid inflation, then deflation, the distance across the crater had returned to that of 19 Oct.

When NZGS personnel flew over Ruapehu 6 days later, the lake was relatively clear and a pale blue-green. Upwelling was absent over the central vent, but moderate at the N end of the lake, where 3 brownish cells were visible. The NZGS attributed the lake's rapid clearing (by sediment settlement) to cessation of heat flow from the main vent.

On 22 Feb the NZGS found the slightly steaming, calm lake a bright blue-green, with no upwelling over the main vent. Yellow and gray sulfur strands were drifting S from moderate upwelling over at least 3 locations at the lake's N end. Water temperature measured at the outlet was 23.5°C, up 4.5° from 10 Feb. The horizontal deformation survey showed shortening of 5 mm across the crater. Only 2 μrad of tilt had occurred since the last measurements on 3 Jan.

The level of volcanic tremor and B-type earthquakes was moderately high throughout Jan and [tremor peaked on 2 Feb]. Activity rapidly declined to a very low level 10-15 Feb. It remained low until 0845 on 23 Feb, when a B-type earthquake sequence with events of M_L 3.0-3.1 was triggered by a magnitude 2.1 roof rock earthquake. On 24 Feb, the NZGS noted that "The increased seismicity and the recent changes in the appearance of the lake indicate that Ruapehu has entered a phase where the probability of eruption is now at a relatively high level. Visitors to the crater are being advised not to approach the lake too closely." A similar sequence of B-type earthquakes occurred 26 Feb at 2356. The series of magnitude 3.0-3.1 events was again triggered by a high-frequency roof rock earthquake, of magnitude 2.0. On 1 March at 0757, a third sequence of B-type events reached magnitude 2.9. Depths for the 1 March events were estimated at 300-600 m beneath Crater Lake, somewhat shallower than usual. Weak volcanic tremor began 2 March at about 0500, at perhaps 300 m below Crater Lake. J. H. Latter noted that this probably represented gas moving toward the surface.

The Chief Ranger, Tongariro National Park, and pilot K. Newton both reported that the lake was gray early 24 Feb, but there were no signs of ash deposits or upwelling from the main vent. The NZGS interpreted the color change "as being due to a sudden, strong upwelling, possibly in the form of a hydrothermal eruption, following the shallow seismic events at 0845 on 23 Feb." By the 28th, the lake temperature had risen a further 3.5° in 6 days, to 27°C, and an additional 9 mm of shortening (deflation) was measured across the 600 m-wide crater. Moderate upwelling was noted from the N end of the lake but the water became noticeably clearer during 4 hours of observations. The lake was still milky gray when observed from the air 2 March, but only slight upwelling was occurring and there were no signs of recent eruptions.

Information Contacts: P. Otway, NZGS, Wairakei; J. Latter, DSIR, Wellington.

3/83 (8:3) Viewing Crater Lake from the air, pilot K. Newton had reported the color was gray on 5 March, but was reverting to blue-green 3 days later. When NZGS personnel visited Ruapehu on 17 March, Crater Lake was gray. They found no evidence of recent eruptions;

neither ash deposits nor surge marks. Upwelling over the N vent area was slight. Over the central vent no upwelling was visible, but thin black sulfur strands appeared in midafternoon.

Lake water temperature measured at the outlet was 23°C, 4° lower than on their last visit, 28 Feb. Concentrations of both Cl and Mg had risen slightly; the Mg/Cl ratio remained 0.104.

The horizontal deformation survey showed that the distance between 2 stations on opposite sides of the crater had decreased an additional 8 mm since 28 Feb, for a total contraction of 22 mm since 10 Feb.

Since seismicity increased 23 Feb, there have been 9 B-type earthquake sequences, the 3 reported last month plus others on 4, 7, 8, 9, 12, and 14 March. These sequences typically began with a high-frequency roof rock (tectonic) earthquake of about magnitude 2 at a relatively shallow depth. Within a minute or so, this was followed by a deeper B-type (volcanic) earthquake of magnitude 2.9-3.4, at a depth between the focus of normal magmatic events (about 1 km depth) and those in the roof rock. No significant volcanic tremor has occurred since the earthquake series began. The lake's color changes appeared to correlate with the earthquake series.

The largest B-type earthquake in the series, [M_L 3.25], occurred at 1406 on 12 March. According to J.H. Latter, earthquakes at Ruapehu have not in the past exceeded this magnitude in a closed-vent situation, as this appears to be, without an accompanying eruption.

The seismicity and deflation were tentatively interpreted by the NZGS as indicating a decreasing magmatic or gas pressure at a deep level below the N vents (or, less likely, intrusion occurring beyond the crater, resulting in compression of the crater rim). As long as the present seismicity persists, they consider the probability of eruption to remain higher than usual.

Information Contacts: P. Otway, NZGS, Wairakei; J. Latter, DSIR, Wellington.

4/83 (8:4) When W. W. Chadwick visited the volcano on 18 April, the lake was mostly green, with extensive floating yellow and gray sulfur slicks. Minor upwelling, marked by a light gray patch, was occurring over the N vents; none was observed over the central vent. There was no evidence of recent eruptions or surging. Lake temperature measured at the outlet was 20°C, 3° lower than on the previous visit, 17 March. . . .

On 25 March the New Zealand Railways Communications Section, Taumaranui, reported an abnormally low pH for the Whangaehu River, which drains Crater Lake. Records showed a pH of 3.5 on 3 March, 4.0 on 11 March, 3.5 on 18 March, but 1.5 on 25 March. Chadwick interpreted the data as showing that Crater Lake was overflowing continuously during March and that a substantial increase in flow occurred toward the end of the month.

Information Contact: W. Chadwick, NZGS, Wairakei.

5/83 (8:5) On 15 May, when NZGS personnel visited the volcano, Crater Lake was uniformly battleship gray. Faint upwelling over the central (main) vent was marked by black sulfur slicks. Weak upwelling was observed over the N vents. At the outlet there was no evidence in the new snow surrounding the lake of recent ash deposits or surging. Lake temperature measured at the outlet was 16.5°C, 3.5° lower than on the last visit, 18 April.

The horizontal deformation survey showed that since 24 April the maximum change in distance between pairs of stations around the crater was an increase of 17 mm. This distance, between stations on opposite sides of the 600 m-wide crater, had decreased 22 mm 10 Feb-17 March, then stayed virtually unchanged until 24 April. There has been little cumulative movement of the several survey stations during the past year, but 3 stations S of the lake have moved about 18 mm SE. Only 4.5 μrad of apparent inflationary tilt were recorded on the Dome since 22 Feb.

Sequences of B-type earthquakes occurred every 2-4 days in late Feb and March (8:3), but have been less frequent since then; only 2 B-type earthquake sequences were recorded 18 April-15 May. Small shocks, interpreted as tectonic, have occurred since 29 April. The largest was magnitude 1.8 on 5 May. There have also been occasional episodes of weak tremor.

Information Contact: P. Otway, NZGS, Wairakei.

6/83 (8:6) When NZGS personnel visited Ruapehu on 15-16 and 22 June, conditions at Crater Lake were very similar to those observed on 15 May. The lake color was battleship gray, and water temperature measured at the outlet remained 16.5°C. Concentrations of both Mg and Cl (750 and 7455 ppm) were also virtually unchanged. Discontinuous sulfur slicks indicated intermittent upwelling over the main vent. Upwelling was also observed over 3 sites at the N end of the lake. According to the NZGS, the volcano has entered a phase of steady, but low activity.

Gray mud and yellow sulfur covering the outlet suggested higher recent rates of outflow, but snow lay within 30 cm of the water, indicating no recent large surges. Varying pH values measured by the New Zealand Railways Communications Section, Taumaranui, on the Whangaehu River, which drains Crater Lake, marked periods of strong overflow (low pH) alternating with little or no overflow (high pH).

The horizontal deformation survey on 22 June revealed a 20 mm decrease since mid May in the distance between stations on opposite sides of the 600 m-wide crater. Since the stations were installed in 1976, this distance had shortened to its 22 June length on only a few occasions (in early 1976 and several times in 1980-81). The NZGS interpreted the shortening as deflationary contraction indicating low magmatic or gas pressure.

No volcanic (B-type) earthquakes were recorded [8 June through 21 Aug]. Bursts of possible moderate-frequency, moderate-amplitude tremor were recorded during the first half of June.

Information Contacts: I. Nairn, NZGS, Rotorua; P. Otway, NZGS, Wairakei.

8/83 (8:8) Inspections by NZGS personnel revealed no significant changes. On 15 and 21 July, and 16 Aug, Crater

Lake appeared its usual battleship gray color. Moderate upwelling was observed over the vents at the N end of the lake, none over the central vent. Water temperature measured at the outlet was 18.7°C in July (2.2° higher than on 15 June), and 18.3° in Aug. Concentrations of Mg (778 and 744 ppm) and Cl (7310 and 7470 ppm) were virtually unchanged for the 2 months. Horizontal deformation surveys showed that by Aug the distance between 2 stations on opposite sides of the 600 m-wide crater had increased 16 mm from a 2-year minimum value measured in June (7:6).

Information Contacts: P. Otway, NZGS, Wairakei; B. Scott, A. Cody, NZGS, Rotorua.

[Summary of 1984 activity from NZGS: "1984, as with 1983, was characterized by a low level of eruptive activity from Crater Lake. The relatively high lake temperatures measured at the end of 1983 continued until 8 Feb when a maximum of 31°C was recorded. Lake temperature dropped gradually to 11.6°C in July, accompanied by the cessation of convection above the central vent area. Convection continued above the northern vent areas. When convection resumed from the central area in Sept, the lake temperature began to rise, reaching 21°C in Oct. The cessation of convection from the central vent area in Dec was accompanied by a temperature decline.

"The lake level was recorded at 0.5 m below overflow at the beginning of 1984 but was overflowing by 8 Feb. The discharge remained very constant (about 50 liters/sec) through the year, until the summer thaw induced a significant increase in the flow. Only one hydrothermal eruption (2 April) was positively reported during 1984."]

5/85 (10:5) After 3 years of quiescience, small hydrothermal eruptions began on or shortly before 21 May. On 16 May, major overflow of Crater Lake into the Whangaehu River began, as shown by pH and conductivity measurements downstream (at Tangiwai). Seismic activity, characterized by increased high-frequency tremor and local earthquakes, began [in early May, with some minor tremor on the 5th and small volcanic earthquakes on the 11th]. Eruptions were first observed in Crater Lake on 21 May and were also seen on 25 May.

On 28 May NZGS personnel noted 15 small hydrothermal eruptions during a 10-hour period. All the eruptions occurred in the lake center, where resumption of weak upwelling had first been observed on 26 April. The eruptions were characterized by updoming of the central lake surface, noisy ejection of water and mud jetting to 10 m above the lake surface, and waves radiating from the lake center. Clouds limited deformation measurements across the crater but a pair of horizontal angle observations indicated no major change in crater diameter.

The lake was battleship-gray in color, with clean snow around the water margin about 1 m above lake level. The lake temperature had increased to 45°C from the 20-25°C range of recent months. Interim analyses of the lake water show a progressive dilution throughout the summer months. Cl and Mg contents have increased since activity began, but the Mg/Cl ratio has not changed significantly.

Information Contacts: I. Nairn, B. Scott, NZGS, Rotorua; P. Otway, NZGS, Wairakei.

6/85 (10:6) Small hydrothermal eruptions from Crater Lake, accompanied by increased seismicity, continued through early June. Geologists returned to the volcano on 4 June and found activity similar to that observed during the previous inspection on 28 May. However, the frequent minor hydrothermal eruptions that began on or before 21 May appear to have concluded, perhaps as early as 9 June.

Eight small hydrothermal eruptions from the central vent were noted during 8.5 hrs of observations on 4 June. Water spurted to as much as 15 m above the vent, and waves surged onto the lake shore and through the outlet. One eruption generated a 100-m steam plume. Convective upwelling between eruptions was observed over an area approximately 50 m in diameter in the center of the main vent area, but was often obscured by steam. No activity was observed from the N vent. Lake temperature remained at 44-45°C but the eruptive activity was accompanied by a major drop in outflow rate to about 5 liters/sec, from 200 liters/sec on 28 May. The lake surface remained battleship-gray in color, and new snow was trimmed back to about 0.5-1 m above lake level.

Geologists returned on 28 June and observed neither upwelling nor evidence of recent eruptions. Dark green sulfur slicks appeared from time to time in the central area but were constantly visible in the N vent area, accompanying faint upwelling there. Steam columns were formed by the large temperature differential (40°C) between the lake surface and the air. Lake temperature had declined 7°C since 4 June. The water had dropped to 0.2 m below overflow level, and the channel was full of clean snow. Water samples were taken during both the June inspections.

The start date of this eruptive episode remains uncertain. A major overflow from Crater Lake began 16 May. From Whakapapa (about 9 km NNW of Crater Lake) at noon the next day (± 50 minutes), Quentin Forman of Auckland University heard a noise and saw steam rising from the direction of Ruapehu, but a ridge blocked his direct view of Crater Lake. The first observed eruptive activity was on 21 May. Uncertainty also surrounds the end of the eruption. After the 4 June eruptions reported above, P. M. Otway had a clear view of the volcano from Taupo (roughly 75 km from Ruapehu) on 9 and 11 June, but saw no plume. Steam clouds were seen over Ruapehu by geologists on 25 June (the day before they arrived at the crater) but were thought to have most likely been caused by atmospheric effects.

The eruptive activity was preceded and accompanied by weak seismicity in a wide vertical range beneath Crater Lake. Volcanic tremor associated with heating of Crater Lake began fairly strongly on 5 May at 0452 in the roof rock zone below the lake. The tremor had dominant frequencies of 1.5-2.5 Hz, usual at Ruapehu. A number of small volcano-tectonic earthquakes occurred around that time. The largest, magnitude 2.15 on

6 May, was at 2 km depth. Their amplitudes at the summit station were anomalously high [relative to those measured at the volcano's foot, indicating that they were unusually shallow or near the station]. Other similar events were associated with high-frequency volcanic tremor, suggesting to seismologists that magma was intruding the wall-rock north of Crater Lake conduit. Volcanic earthquakes (multiple events showing characteristic features attributed to an origin in magma or pockets of hot gas), which had begun to occur at shallower foci in Sept 1983, reached maximum magnitudes (2.25) on 11 May.

High-frequency volcanic tremor (superimposed on the 2 Hz tremor that started 5 May) first became conspicuous on 19 May at 2209, immediately after a magnitude 2.9 earthquake about 10.5 km ESE of Crater Lake. This was the smallest of four earthquakes at the same focus: two others occurred on 19 May (both magnitude 3.0), and one magnitude 3.25 event on 20 May. The earthquakes were shallow, probably less than 2 km deep, on or very close to the prominent fault bounding Ruapehu on the SE side. A composite focal mechanism determination suggested right-lateral strike-slip on this fault, with a compressional component in the direction of the summit. Low-frequency volcanic earthquakes began on 25 May at 1519, culminating in a magnitude 2.4 roof rock event at 1725, and were continuing 9 June. J. H. Latter noted that the episode is typical of an open-vent period of energy release at Ruapehu. He also noted that the data suggest that the present state of increased activity is similar to, but less intense than, the 1981-1982 period when similar events continued for 3-4 months without any major eruptions (8:1).

Preliminary results of deformation surveys indicate that no significant crater deformation has occurred in the last few months. NZGS geologists noted that this appears to be in accordance with the seismic interpretation of open-vent conditions during this episode.

Information Contacts: A. Cody, I. Nairn, B. Scott, NZGS, Rotorua; J. Latter, S. Sherburn, DSIR, Wairakei; P. Otway, NZGS, Wairakei.

7/85 (10:7) Geologists returned to Crater Lake on 11 July and 5 Aug to monitor trends in lake temperature and crater deformation following June's increased activity.

In July, the lake was the normal battleship-gray color with some sulfur slicks at the central vent area, unchanged from late June. The lake temperature was 31.5°C (a 5.5° decline since late June) and the air carried a strong gas odor toward one lake outlet. The lake was about 0.5 m below overflow level and there was no evidence of recent large surges. No hydrothermal eruptions occurred during the 4 $^1/_2$ hour observation period.

By 5 Aug, the lake color had changed to light gray and its temperature was nearly unchanged. There was no sign of convection at either of the two vents, and snow and ice were within 0.5 m of the lake surface. The lake had begun overflowing at 3 liters/sec.

Only a 15 ± 10 mm expansion across the N rim of the crater since June was measured during the 11 July inspection. Rapid, but minor deformations were occurring in July, but their significance was uncertain.

Information Contacts: I. Nairn, NZGS, Rotorua; P. Otway, NZGS, Wairakei.

9/85 (10:9) Geologists visited Ruapehu on 13, 14, and 20 Sept. Crater Lake's temperature had declined by 10°C since 5 Aug. On 20 Sept the lake was 0.4 m below overflow level (it was overflowing on 5 Aug). There were no clear signs of any eruptions having occurred.

Information Contact: P. Otway, NZGS, Wairakei.

11/85 (10:11) Park Ranger Paul Dale witnessed a large upwelling and surging of the lake on 31 Oct at 1342, lasting 30-40 seconds and producing $^1/_4$ m waves on the shore. The lake temperature at 1405 was 26.5°C with an outflow rate of 150 liters/sec, a significant increase from the 20.0°C temperature and water level 0.1-0.2 m below overflow recorded on 17 Oct. Fresh ash was not present on the snow. Pilot Bruce Williams, flying just E of the mountain on 11 Nov at 1100, reported steam accompanying strong upwelling in the lake center. He considered the lake to be in the most agitated state he had observed since witnessing a relatively large eruption 4 years ago; the activity was also observed by a Geyserland Airways pilot.

During a 15 Nov inspection by NZGS personnel, the lake was gray with scattered yellow sulfur slicks, in contrast to the 17 Oct observation of dark concentric rings of sulfur. An upwelling of muddy water occurred at 1027, doming to 2 m above the lake surface, and generating steam and $^1/_2$-m waves near the center of the lake. At 1336 water domed to 4-5 m, producing 1 m waves. The waves did not further erode the low-lying snow or generate significant surges in the outlet channel, illustrating the difficulty of detecting evidence for small events. The upwellings observed during the Nov inspection were smaller than some seen during a period of apparently similar activity in May (10:5). A minor inflationary extension measured on 17-18 Oct had reversed by 15 Nov. Tremor and volcanic earthquakes increased in magnitude in Oct. Seismic records 4-10 Nov included both low- (about 2 Hz) and high-frequency (about 5 Hz) tremor of generally low amplitude. Some small B-type events were recorded on 4 Nov and very shallow earthquakes (L-shocks) were common. During the geologists' 15 Nov observations, seismometers recorded low-frequency (2 Hz) tremor.

Information Contact: P. Otway, NZGS, Wairakei.

MONOWAI SEAMOUNT

Kermadec Islands, S Pacific Ocean
25.92°S, 177.15°W, submarine depth -120 m
Local time = GMT + 12 hrs

Unnamed at the time of SEAN's original report, this volcano is about 350 km NE of Raoul Island, the largest of the Kermadec group and the nearest known island volcano.

10/77 (2:10) Royal New Zealand Air Force (RNZAF) personnel observed submarine volcanic activity at a site above the Tonga Ridge on 13 Oct at 1430. Discoloration stretched about 5 km SW from a patch of brown, gaseous, turbulent water 200 m in diameter. A sonar buoy dropped into the turbulent water detected pulsating rumbles and an explosion, believed to originate from a source about 4000 m deep. No records exist of previous activity in the area.

Information Contact: J. Latter, DSIR, Wellington.

Further Reference: Davey, F.J., 1980, The Monowai Seamount: An Active Submarine Volcanic Centre on the Tonga-Kermadec Ridge; *New Zealand Journal of Geology and Geophysics,* v. 23, p. 533-536

[Note the unknown submarine eruption reference following the next set of reports. Although encountered west of the Kermadecs, the source of this floating pumice is unknown, and the report is grouped with other problematic pumice sightings after the extensive pumice reports from Metis Shoal.]

METIS SHOAL

Tonga Islands, S Pacific Ocean
17.18°S, 174.87°W, submarine depth -4 m
Local time = GMT + 12 hrs

This shallow submarine volcano in the Tonga group produced the first eruption reported by the Smithsonian's Center for Short-Lived Phenomena in early 1968. Eight eruptions are known from Metis; the first in 1851. The early pumice reports that follow were originally listed by SEAN as "Tonga Islands" before activity at Metis and Curacoa was known.

5/79 (4:5) A large body of floating pumice was first reported on 10 May by a Soviet vessel, in the area from 17.80°-18°S and 176.45°-176.67°W. On 14 May, a U.S. Naval Research Laboratory aircraft flew over continuous 30-45 m-wide patches of dirty rust-brown pumice from 17°-17.45°S and 174.13°-177.60°W. Four days later, the same aircraft flew over Late Island (18.810°S, 174.65°W), last reported to erupt in 1854. White material, apparently ash, covered the cone, and streaks of pumice extended back to the island from the main body of the raft [but see 4:12].

While enroute from Nandi, Fiji NE to Apia, Western Samoa on 22 May, Captain Gallagher of Polynesian Airlines observed pumice extending about 30 km SE from 16.2°S, 178.38°W. On the 24th, Captain Gallagher again observed pumice, in large areas at about 15.8°S, 177.5°W. An Air Pacific plane flew over pumice that extended about 25 km N-S and about 45 km E-W from 16.37°S, 177.63°W on 28 May.

Captain F.R. Sutherland of Polynesian Airlines reported pumice between about 15.58°S, 176.50°W and 16.10°S, 178.18°W on 30 May, as he flew SW from Apia to Nandi. He first saw several long, thin, mustard-brown streaks, then the pumice became more concentrated. The largest area of pumice, about 5 x 1 km, extended NW-SE and tapered to the SE.

During the afternoon of 4 June, Fijian naval vessels collected samples of the pumice from just inside the Tonga entrance to Vanua Balavu (17.3°S, 178.9°W), an island of the Lau Group about 135 km ESE of Vanua Levu, one of the two main Fiji islands. Pumice fragments in this area were greenish-gray, ranged from 2.5 to 500 cm across, and carried coarse barnacles 0.5-0.7 cm long. There have been no reports of pumice washing ashore at any location but the samples were collected within the reef surrounding Vanua Balavu.

The rate of drift of the pumice is estimated by Ronald Richmond at about 3 km/day, to the NW. Its source was apparently Late Island (see above), but residents of nearby islands in the Tonga chain did not report an eruption. A thick, odorless haze covering the Tonga and Fiji Islands early in the week beginning 6 May reduced visibility to 3 km or less in Nandi, Fiji. Such haze is very unusual in this area and may have been caused by volcanic activity. On 10 May, what appears to be a volcanic plume could be seen on NOAA weather satellite imagery, drifting SE from about 18°S, 174°W between 1200 and 2100-2200, when the source of the plume was cut off. However, an earlier eruption must be postulated as the source for at least some of the pumice, given the sighting on 10 May.

Information Contacts: R. Richmond, Mineral Resources Dept., Fiji; Capt. Gallagher, T. Sutherland, Polynesian Airlines; N. Cherkis, US Naval Research Lab; A. Krueger, NOAA.

6/79 (4:6) The pumice described in 4:5 continued to drift in a generally westward direction in June. In mid-June, patches of pumice were reported near Vanua Balavu, Katafaga, and Yacata, small islands at the N end of the Lau Ridge (fig. 4-15). The pumice at Yacata formed a strip 5 km long but only about 5 m wide. Large quantities of pumice were seen during the same period around Late Island (18.81°S, 174.65°W), about 400 km ESE of the Lau Group.

The source of the pumice remains uncertain. In mid-May, streaks of pumice extended back to Late Island from the main raft and the island's cone was covered with white material, which was thought to be fresh ash (4:5). However, activity has been reported at 3 other sites along the Tonga Arc in May and June.

A Tongan vessel reported a strong eruption between 0700 and 1200 on 19 June at 19.18°S, 174.83°W, in the area of Metis Shoal, a frequently active submarine volcano that last erupted Dec 1967-Jan 1968. The vessel's crew observed a mushroom cloud climbing to about 1.5 km and saw the sea steaming. At about the same time, the *M.V. Niuvakai* (enroute from Auckland, New Zealand to Samoa) passed Metis Shoal, reporting an eruption column a little less than 200 m high at 0754. Captain Robert Jones of Air Tonga flew over Metis Shoal late the next morning and saw rocks ejected to about 150 m altitude as water boiled vigorously around the center of the eruption. During subsequent flights, Captain Jones observed a growing area of tephra, either an island or a mass of floating pumice, around the eruption site. He estimated its diameter at 3 km on 21 June, 8 km on 22

June, and 16 km on 24 June.

A fisherman saw an eruption near Tafahi [380 km NNE of Metis Shoal] on 14 May, the only day he was in the area. A message received from the ship *M.V. Marama* at 1445 on 31 May stated that the W side of frequently active Tofua (19.80°S, 175.07°W) was emitting smoke, but that no activity was visible on Late.

A research vessel, carrying geologists from Fiji and Tonga, left Fiji on 3 July. Each of the reported eruption sites along the Tonga Arc will be investigated and pumice floating in the area will be sampled.

1979	Time (GMT)	Magnitude	Latitude	Longitude	Depth of Focus	# of Stations
April 28	0054	5.7	18.136°S	174.855°W	Shallow	93
	0303	5.2	19.252°S	175.669°W	244 km	101
29	0851	4.7	15.906°S	173.046°W	Shallow	33
May 6	0907	–	15.932°S	173.882°W	Shallow	25
12	2332	4.2	16.347°S	174.325°W	282 km	8
28	0003	5.3	17.512°S	175.187°W	274 km	81
30	1645	5.1	15.390°S	173.174°W	Shallow	39
June 1	0210	5.0	15.398°S	173.180°W	Shallow	69
	2057	4.7	20.340°S	173.799°W	Shallow	38

Table 4-7. Events recorded by the WWSSN between late April and early June with epicenters near one of the reported eruption sites. Data are from the USGS-NEIS, published in their *Preliminary Determination of Epicenters.*

Many moderate seismic events, most shallow, occur each month in the Tonga Arc. Nine events between 28 April and 1 June had epicenters near one of the reported eruption sites (table 4-7).

Information Contacts: R. Richmond, Mineral Resources Dept., Fiji; J. Latter, DSIR, Wellington; S. Tongilava, Lands Survey and Natural Resources, Tonga; R. Jones, Air Tonga; USGS/NEIS, Denver; AP.

Further Reference: Woodhall, D., 1979, Cruise of the *R.V. Bulikula* to Investigate Recent Volcanic Activity in Tonga, July 11-18, 1979; Fiji Ministry of Lands & Mineral Resources, *Mineral Resources Division Report* 14.

7/79 (4:7) Geologists aboard the *R. V. Bulikula* inspected the volcanoes along the Tonga Arc between 3 and 21 July. A new island, named Late Iki, had formed above Metis Shoal at 19.18°S, 174.85°W, site of strong activity in June. The new island, comprised of tephra ranging in size from ash to large bombs, was about 300 m long, 120 m wide, and 15 m high in mid-July. Emission of hot ash continued from the E end of the island.

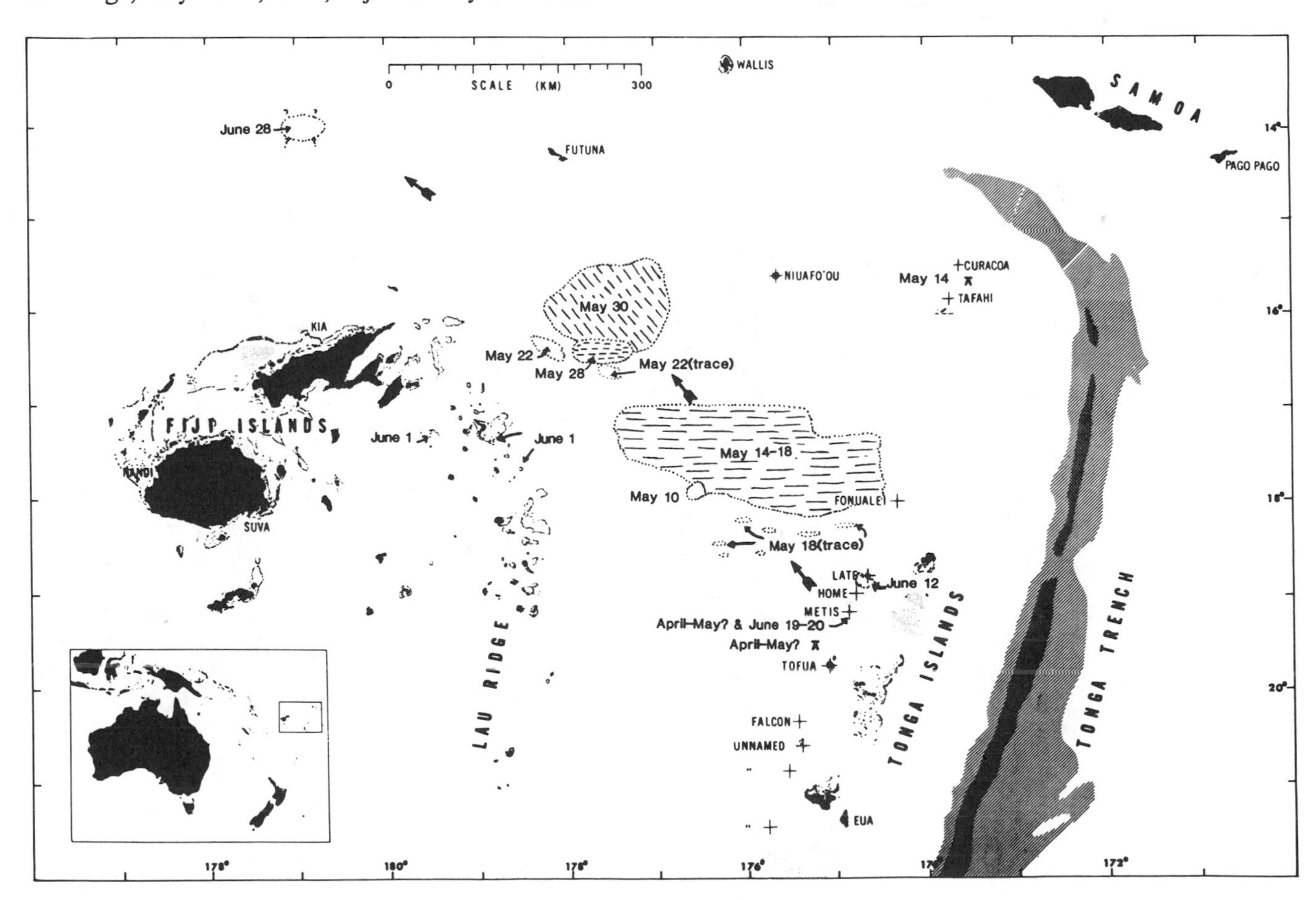

Fig 4-15. Map by T. Simkin adapted by Woodhall (1979) to show the distribution of floating pumice between Tonga and Fiji, and the locations of volcanic activity in Tonga, May-June, 1979. Historically active volcanic centers are indicated by crosses. Trench depths > 6 and 8 km are indicated by diagonal and cross-ruled patterns.

Short-lived islands were also formed over Metis Shoal during the eruptions of 1858 and 1967-8. The position determined (by satellite navigation) for the new island is approximately 1 km E of the one formed in 1967-68 (Melson and others, 1970). That island, estimated to have been 700 x 100 m and 15 m high, was eroded away within a month of the eruption's end.

Fonualei, about 140 km NNE of the new island, was emitting steam when viewed from the *R.V. Bulikula* in mid-July. No new pumice was being ejected by any of the volcanoes along the Tonga Arc.

Reference: Melson, W., Jarosewich, E., and Lundquist, C., 1970, Volcanic Eruption at Metis Shoal, Tonga, 1967-1968, Description and Petrology; *Smithsonian Contributions to the Earth Sciences,* No. 4, 18 pp.

Information Contact: R. Richmond, Mineral Resources Dept., Fiji.

8/79 (4:8) Some fumarolic activity continued to take place during Aug on Late Iki. Geologists do not expect the island to persist for more than a few months.

Reports from Rotuma Island (12.5°S, 177.08°E, about 1150 km NW of Late Iki) state that a thick blanket of pumice washed up on the S side of the island in mid-Aug. Fish that jumped out of the sea and landed on the pumice were unable to return to the water. No other reports of pumice have been received by the Fiji Mineral Resources Department in Aug.

Information Contact: Same as above.

10/79 (4:10) Floating pumice, presumably from the recent island-forming eruption of Metis Shoal was reported at 2 widely separated locations during Oct. On 6 Oct, Samuel Iko, aboard the *M. V. Independence*, observed a blanket of pumice floating between the islands of Guadalcanal, the Florida Group, and Malaita (Solomon Islands), with the greatest concentration at about 9.5°S, 160.5°E, approximately 3000 km WNW of Metis Shoal. Other reports indicated that pumice washed ashore at Marau, an islet at 8.5°S, 159.5°E, as the raft moved southward through the Solomons.

On 25 Oct, John Cårney sighted dark friable pumice on the beach on the S coast of Efate Island, Vanuatu (17.8°S, 168.5°E), about 1850 km W of Metis Shoal. This is the first known observation of pumice from Metis Shoal in Vanuatu.

The last observations of Late Iki, the island that formed over Metis Shoal, were made in early Oct. At that time, the island had nearly disappeared beneath the ocean's surface.

Information Contacts: D. Tuni, Ministry of Natural Resources, Solomon Islands; J. Carney, A. MacFarlane, Geological Survey, Vanuatu; R. Richmond, Mineral Resources Dept., Fiji.

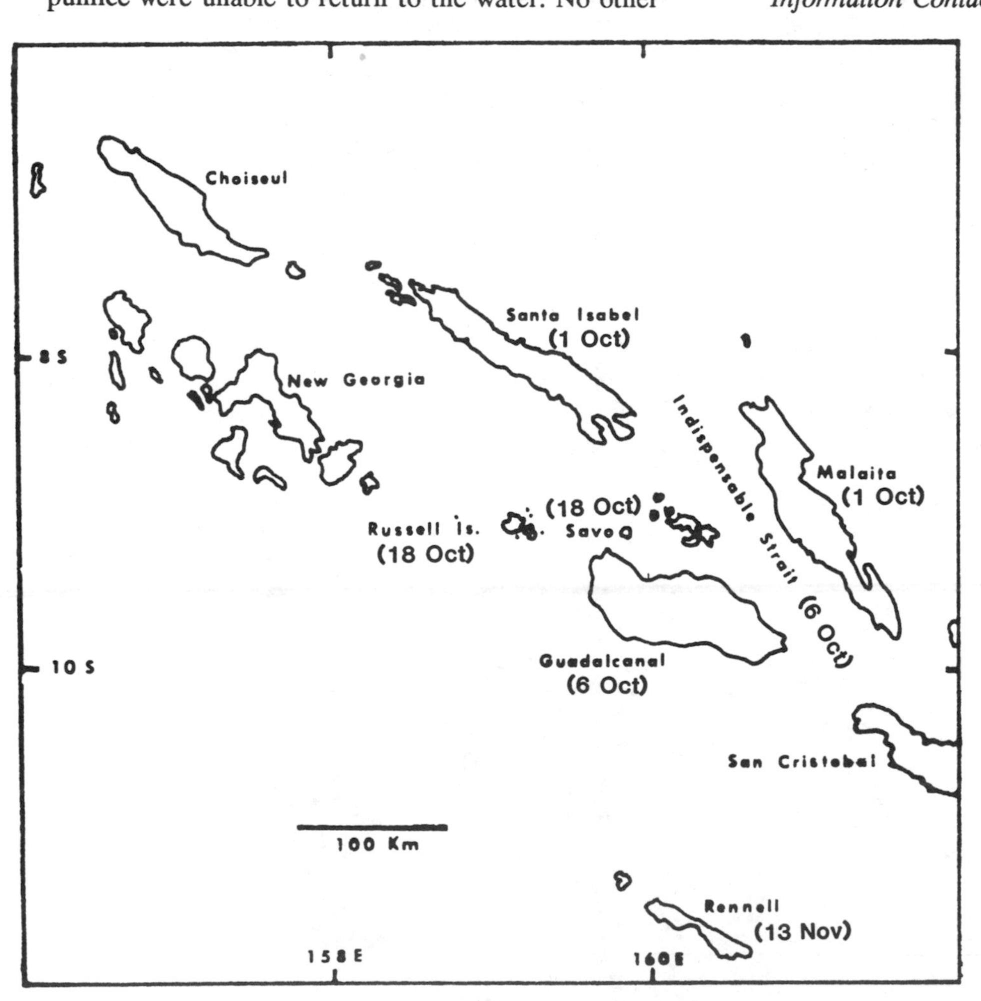

Fig. 4-16. Sketch map of the Solomon Islands, showing locations and dates of pumice arrivals, Oct-Nov, 1979.

12/79 (4:12) Most of the reported eruption sites were investigated during the 11-18 July cruise of the Fijian research vessel *R. V. Bulikula*. An inspection of Late Island, where whitish material thought to be ash reportedly covered the summit cone on 18 and 21 May, revealed no trace of a whitish covering other than scattered patches of white lichen, and no evidence of a recent eruption. Large fluctuations in the intensity of fumarolic activity were observed at both Fonualei and Tofua.

Further details from Tongan authorities indicated that submarine activity reported by a fisherman occurred 14 May about 13 km N of Tafahi at the N end of the Tonga Islands. [This location is 24 km S of Curacoa, a shallow submarine volcano that produced abundant pumice in 1973, and the 14 May 1979 activity has been entered in the Chronology Table (and our records) as Curacoa]. Thick black "smoke" reached a height of about 100 m, and the eruption may have been rhythmic or spasmodic. At about the same time, a small earthquake was felt on Nuiatoputapu, about 25 km from the eruption site.

Deni Tuni reported that pumice arrived at the following Solomon Islands locations (fig. 4-16) on the following dates: Malaita and the N and E shores of Santa Isabel, 1 Oct; Indispensable Strait and Guadalcanal, 6 Oct; Savo and the Russells, 18 Oct; and the SW shore of Rennell, 13 Nov.

Information Contacts: D. Woodhall, R. Richmond, Mineral Resources Dept., Fiji; D. Tuni, Ministry of Natural Resources, Solomon Is.

UNKNOWN SUBMARINE VOLCANO

SW Pacific Ocean
Local time = GMT + 12 hrs

The source of the floating pumice in the following reports is not known. The reports are placed here with the known pumice rafts from Metis Shoal (earlier) and Home Reef (later).

4/83 (8:4) While sailing on his yacht *Cookoos' Nest*, Captain J. McInnis passed through the edge of a roughly 1-hectare area of small pieces of pumice on 6 April at 1206. He reported some "bubbling" but noted no smells. His location at the time of the sighting was fixed by satellite at 27.58°S, 177.40°E, several hundred kilometers W of the Kermadec volcanic trend. Previous pumice rafts in the area have drifted generally toward the W. Water depths in the area of the sighting are in excess of 4000 m. Seismic records in New Zealand (roughly 1400 km from the site) showed no earthquakes associated with the activity, and nothing was reported by the observer on Raoul Island (in the Kermadecs and about 460 km from the site).

Information Contacts: I. Everingham, Mineral Resources Dept., Fiji; W. Smith, DSIR, Wellington.

8/83 (8:8) The source of the pumice remains unknown. Analysis of March and April records from the Réseau Sismique Polynésien (RSP) revealed no acoustic waves (T-phase) from eruptions other than that of Macdonald Seamount. However, the numerous small islands in the area of the Kermadecs, Tonga, Samoa, and Fiji interfere with acoustic waves, preventing effective T-phase monitoring of volcanic activity in some parts of the S Pacific. J. Talandier notes that measurements of surface currents in French Polynesia and similar latitudes suggest that pumice from Macdonald should drift eastward, away from the 6 April site.

Pumice came ashore at both the SE and NW ends of the Tuamoto Archipelago, on the Gambier Islands (23.15°S, 134.97°W) and at Rangiroa (15.00°S, 147.67°W), 4800 km E and 3900 km ESE of the 6 April observation. No information on the amount of pumice or the date of its arrival at these locations was available. Talandier noted that Rangiroa is very remote from known active volcanoes other than those in the Mehetia region, where eruptions occur at depths that are too great for production of pumice.

Information Contact: J. Talandier, Lab. de Géophysique, Tahiti.

HOME REEF

Tonga Islands, S Pacific
18.99°S, 174.78°W
Local time = GMT + 13 hrs

Home Reef is a shallow submarine volcano 23 km NNE of Metis Shoal.

2/84 (9:2) An eruption in the vicinity of Home Reef was reported on 2 March at 1107. Intense submarine activity ejected a plume to an altitude estimated by an airline pilot at more than 7.5 km. A surface layer, probably pumice, extended 60 km to the NE and was 20-30 km wide, enveloping Late Island (25-30 km to the NE). Surface discoloration of the sea covered a larger area. Another report at about the same time described a pumice raft of the same dimensions drifting SW. South Pacific Islands Airways (SPIA) reported that the activity was at 19.0°S, 174.80°W.

Gerald Dion piloted Pan American World Airways flight 811 (Honolulu to Auckland) over the area on 3 March at about 0730. From about 18 km upwind, the eruption was visible through broken weather clouds for about 1 minute. A medium-dark reddish-brown eruption column rose from a submarine vent within a horseshoe-shaped island open to the E. The eruption column reached slightly more than 12 km altitude (several hundred meters above the aircraft) where winds carried its top at least 15 km NE.

During the morning of 4 March, an SPIA pilot reported that an eruption cloud was still visible, rising high above the sea surface. He saw floating pumice drifting away from the eruption site but no island appeared to have formed. However, before the eruption had ended, by 5 March at 1030, two small islands had formed with a maximum elevation of about 20 m, enclosing a crater about 1500 by 500 m. Island-forming eruptions of Home Reef occurred in 1852 and perhaps in 1857.

Information Contacts: R. Krishna, Fiji Meteorological Service; J. Latter, DSIR, Wellington; G. Dion, Pan American World Airways; Meteorological Office, Tonga; W. Smith, T. Kossarias, FAA, Washington.

4/84 (9:4) Tonga government geologist David Tappin reported that brown discolored water preceded the eruption, which started 1-2 March. The new island was visible by 2 March. When Captain Jeff Heard of SPIA flight 607 flew over the eruption site on 5 March at 1030, explosive activity had declined. Weak steaming occurred from a submarine crater surrounded by the new island.

In mid-March, a cargo vessel traveling from Tonga

to Fiji at 12 km/hr took 9 hours to pass through a zone of pumice. Samples were collected from this vessel about 150 km W of Tonga. Pumice rafts were reportedly sighted at Oneata Island, Lau Group (18.45°S, 178.50°W, roughly 500 km WNW of Home Reef) on 5 April. On 1 May, ships between Tonga, Fiji, and Samoa reported that floating pumice was so thick that it was clogging their seawater intake systems.

RNZAF personnel flew over the new island 23 March. They gave its location as 19.02°S, 174.73°W, about 10 km S of Late Island. Dimensions of the new island were estimated at 1500 m by 500 m, with cliffs about 30-50 m high (fig. 4-17). Discolored water just NW of the island suggested submarine activity. Photographs taken from upwind showed the island to be yellowish brown in color, but atmospheric haze caused it to appear dark brown from downwind. David Tappin reported that activity was continuing in early April.

The RSP did not record any seismicity from the eruption. Islands and deep water between Tahiti and Tonga prevented RSP stations from recording any acoustic waves (T-phase).

Information Contacts: D. Tappin, Ministry of Lands, Survey, and Natural Resources, Tonga; P. Shepherd, RNZAF; J. Latter, DSIR, Wellington; J. Lum, Ministry of Energy and Mineral Resources, Fiji; R. Krishna, Fiji Meteorological Service; J. Talandier, Lab. de Geóphysique, Tahiti; N. Banks, HVO, Hawaii.

7/84 (9:7) Large rafts of floating pumice probably from the early March eruption of Home Reef (9:2, 4), were reported in the Fiji area through early July. The press reported that pumice was first sighted in waters near the Lau island group (roughly 400 km WNW of Home Reef) in April. By early May, large pumice rafts in several regions of Fiji were forcing ships to return to port, and pumice covered the shoreline of many islands. The Fiji press also reported that pumice had reached Vanuatu and the Solomon Islands by early May. From a ship on 2 May, David Tappin observed pumice rafts about 100 m long and 20 m wide just outside the harbor at Suva (Fiji's capital, about 750 km WNW of Home Reef). Flying from Suva to Tonga on 12 May, Tappin saw 5-10 similar en echelon rafts between Suva and about 19°S that appeared to be moving WNW or W. A large kill of deep-sea fish was reported at Oni-i-Lau (at the south end of the Lau islands) at the end of May. Thick layers of pumice had accumulated at the shoreline, strong enough to support the weight of adults. Tides and winds carried a thick blanket of pumice into Suva harbor 24 June, but officials said that by then there was less pumice in the Fiji area than in April.

Information Contacts: P. Rodda, Mineral Resources Dept., Fiji; D. Tappin, Ministry of Lands, Survey, and Natural Resources, Tonga; R. Krishna, Fiji Meteorological Service; *Fiji Times*.

8/84 (9:8) Large quantities of pumice, probably from the early March eruption of Home Reef, began to arrive about 10 April at Futuna Island (14.42°S, 178.33°W) and Alofi Island (14.45°S, 178.08°W), roughly 700 km NW of Home Reef. ORSTOM geologists collected samples of pumice fragments that were typically 3-4 cm in diameter but occasionally reached 15-20 cm in largest dimension. During fieldwork on Futuna and Alofi islands in July, ORSTOM personnel saw pumice accumulations as much as 30 cm thick on the upper parts of some beaches. People aboard a ship that left Fiji in early May saw pumice as far west as 100 km from Vanuatu about 8 May. Arrival of pumice in Vanuatu was reported in late June. It was apparently found mainly in the central part of the island group in the vicinity of Efate (17.75°S, 168.3°E)

Fig. 4-17. RNZAF photo of the new island formed by the Home Reef eruption, taken 23 March 1984 from about 300 m altitude. The island trends approximately N-S, with N at right. Courtesy of W/O P. J. R. Shepherd and J. H.

and the Shepherds, about 1800 km WNW of Home Reef. The pumice seemed to move as discontinuous "streamers", but was as much as 0.25 m thick on parts of some beaches.

Information Contacts: P. Maillet, J. Eissen, M. Monzier, ORSTOM, New Caledonia; A. Dahl, Noumea, New Caledonia; A. McCutchan, Dept. of Geology, Mines, and Rural Water Supplies, Vanuatu; G. Greene, USGS, Menlo Park CA.

9/84 (9:9) Pumice that was probably from the March eruption of Home Reef was collected in Aug at Beautemps-Beaupré (northern Loyalty Islands, 20.4°S, 166.2°E) and on 3 Sept at Yate (southern New Caledonia, 22.15°S, 167.00°E), both roughly 1300 km WSW of Home Reef. Fishermen reported that pumice arrived at Yate by 19 Aug or 1-2 days earlier during a few days of rough seas. The largest pumice fragment found was 9 cm long but the pumice averaged 1-2 cm in diameter. Small shells up to 1 cm long were frequently attached to the pumice. As of mid-Sept, no pumice had arrived in the Solomon Islands, roughly 2600 km NW of Home Reef.

Information Contacts: P. Maillet, J. Eissen, M. Monzier, ORSTOM, New Caledonia; D. Tuni, Ministry of Lands, Energy, and Natural Resources, Solomon Is..

10/84 (9:10) While traveling southeast of Fiji in May, June, and July, Dutch ships encountered rafts of floating pumice, probably from the March eruption of Home Reef.

On 15 May between 0100 and 0530, the *M.V. Amanda Smits* traveled through floating pumice from 18.10°S, 178.90°E to 18.78°S, 178.17°E, about 675 and 740 km WNW of Home Reef. The pumice ranged in size "from grit to large dice".

A volunteer meteorological observer aboard the *M.V. Nedlloyd Alkmaar* reported that the vessel first encountered pumice on 8 June at 0800 at 21.58°S, 177.85°W (about 430 km SW of Home Reef). Steaming on a course of 282° (slightly N of W), the ship continued to pass through pumice rafts of varying density for nearly 280 km, with the last observation on 8 June at 1800 at 21.03°S, 179.53°E (about 635 km WSW of Home Reef). Pumice fragments ranged in size from "fine grit" to about 10 cm in diameter. Live shellfish up to 1 cm across that looked like mussels were attached to some pumice fragments.

The *M.V. Nedlloyd Barcelona* sailed through small amounts of pumice 21 July; her positions at 0700 and at 1300 were 22.4°S, 178.7°E and 24.0°S, 178.7°E, about 775 and 875 km SW of Home Reef. Pumice fragments reached a maximum size of 2-3 cm and were aligned in E-W strands up to several hundred meters long.

Information Contact: L. Mahieu, Royal Netherlands Meteorological Inst.

3/85 (10:3) Fragments of pumice were found in early 1985 at 3 sites in the Solomon Islands. Pumice fragments found near the high water mark were without attached shellfish, but those found lower on the beach were encrusted with barnacles of the genus *Lepas* that were up to about a cm in length, suggesting that they had been in the water for a few months or more. Since the March 1984 eruption of Home Reef, floating pumice has been found in numerous locations to the SW and NW, as far away as 1800 km (Efate Island, Vanuatu in late June; 9:8).

Large quantities of pumice were found by Henry Isa on the SE coast of Alu Island at Aleana village (7.1°S, 155.75°E, about 3450 km WNW of Home Reef). Mr. Isa saw the pumice by 31 Jan, but its date of arrival was not known. A large amount of pumice was also present at Koela village on the SW coast of Savo Island (9.1°S, 159.8°E, about 2950 km WNW of Home Reef) when Alison Papabatu visited the island 27 Feb. Lesser quantities of pumice were found 19 March and collected by Deni Tuni on Savo's SE coast at Kolika village. Residents of the island did not know when the pumice had arrived. No fresh pumice was found along the E side of Savo Island, nor had any been seen at Honiara, Guadalcanal Island, about 30 km SE of Savo.

Information Contact: D. Tuni, Ministry of Natural Resources, Solomon Is.

Further Reference: Rodda, P., 1986, Home Reef Pumice in Fiji (second edition); *Mineral Resources Dept. Note* BP1/58, 5 pp.

NIUAFO'OU

Tonga Islands, S Pacific
15.60°S, 175.63°W, summit elev. 260 m
Local time = GMT + 13 hrs

This basaltic shield volcano lies west of the Tonga-Kermadec-New Zealand line of volcanoes as shown in fig. 4-15. Niuafo'ou has 11 known (and 3 uncertain) eruptions since 1814.

9/85 (10:9) During the night of 21-22 March, an earthquake swarm with Modified Mercalli intensities as high as VII was felt on the island of Niuafo'ou. A large crack formed on the NE flank, and floating pumice appeared in the caldera lake.

The first felt earthquake (MM IV) was at 2050 on 21 March. Strong tremors, lasting about 2 seconds each, continued for 30 minutes. A thunderous rumbling noise was heard, but it was difficult for island residents to locate its origin. Seismicity continued from 2130 to 2200 with intensities of II-IV; the tremors appeared to have a W to E motion. Between 2200 and 2300, the motion of the tremors appeared to change to E to W and intensities increased to III-V. Rumblings increased until [about] midnight when a particularly loud noise was heard and earthquake intensities increased to VI-VII. By 0100, the number of felt earthquakes had decreased and intensities had declined to III-IV. A small event of intensity II was felt at 0156. None of the earthquakes were large enough to be recorded at the Afiamalu seismic station, Upolu, Western Samoa (the nearest seismograph, about 500 km ENE of the volcano). [Villagers at Betani (Petani) and Tongamama'o (SE side of the island) reported very loud rumblings like thunder "which they estimated to have come from the reef SE (of the

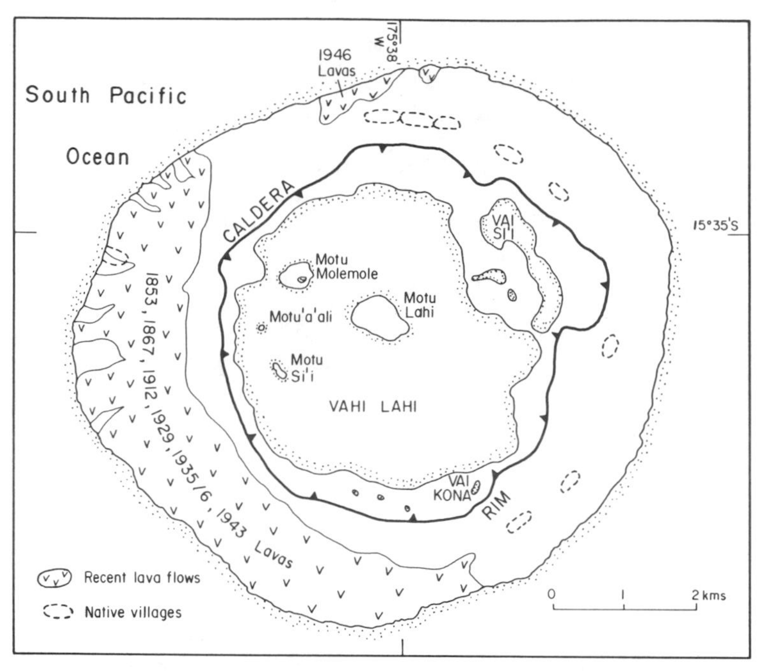

Figure 4-18. Map of Niuafo'ou, showing the distribution of historic lava flows. The outlines of caldera lakes Vai Lahi and Vai Sii, and the solfataric area Vai Kona, are stippled. Courtesy of Paul Taylor.

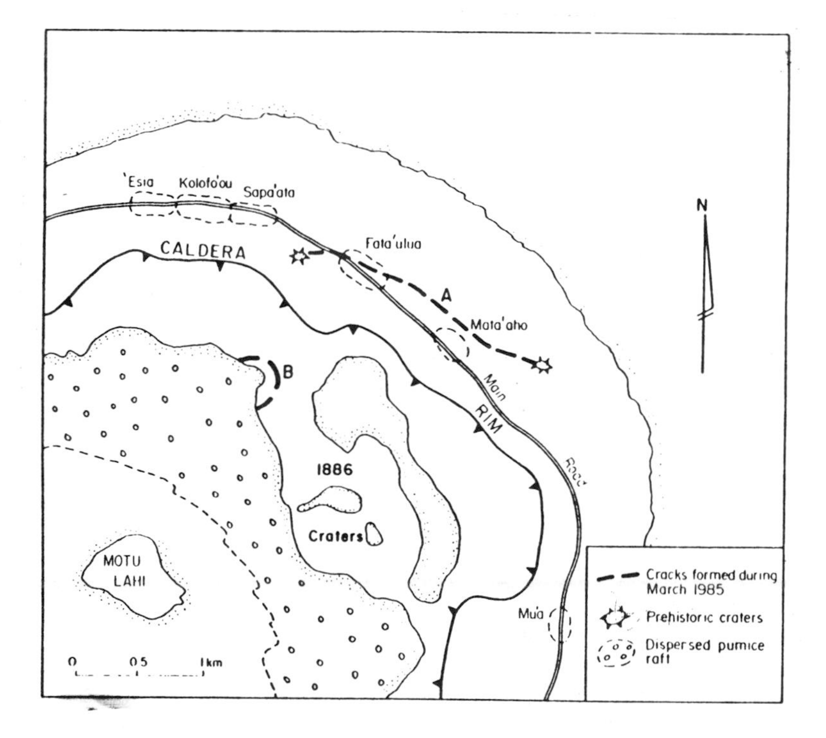

Fig. 4-19. Map of the NE part of Niuafo'ou, showing the location of the March 1985 cracks (A and B); the villages affected by the earthquakes; and the approximate distribution of pumice floating in the crater lake in late May. Courtesy of Paul Taylor.

villages) and not from underground" (New Zealand Foreign Affairs Report Telegram, 23 March 1985, p. 2).—J. Latter]

During the night, a 250 m-long crack was discovered near the village of Fata'ulua (fig. 4-19), extending inland from the shore. It appears that nothing was erupted from it. The RNZAF flew over the island on 23 March, and reported that it appeared normal and no destruction was observed.

On 24 March, Mr. Fifita, the Chief Meteorological Officer of Tonga, gave the following information to John Latter: No earthquakes were felt on Niuafo'ou after 1132 on the 22nd. Residents saw fresh black pumice on the shore of Motu Lahi island in the caldera lake (fig. 4-19). [A sample examined at Victoria University of Wellington was a very vesicular basaltic andesite.] From reports that he had received, Fifita estimated the extent of the pumice as 100 m long, 7 m wide, and 7-10 cm thick. There was no sign of a new crater, bubbles were not seen in the lake, nor was the lake water hot.

The 8 km-diameter island of Niuafo'ou is a shield volcano that rises 260 m above sea level; its summit caldera, about 4 km wide, is filled with a lake (Vai Lahi) about 80 m deep that contains several island cones (Jaggar, 1931). In 1946, an eruption of the volcano forced the evacuation of almost all the island's 1200 inhabitants (Macdonald, 1948). Additional eruptions may have occurred in 1947 and 1959.

References: Jaggar, T. A., 1931, Geology and Geography of Niuafo'ou Volcano; *Volcano Letter,* no. 318, p. 1-3.

Macdonald, G.A., 1948, Notes on Niuafo'ou; *American Journal of Science,* v. 246, p. 65-77.

Information Contacts: J. Latter, DSIR, Wellington; Mr. Fifita, Chief Meteorological Officer, Tonga; R. Blong, P. Taylor, Macquarie Univ.

Further References: Rogers, G., 1981, The Evacuation of Niaufo'ou, an Outlier in the Kingdom of Tonga; *Journal of Pacific History,* v. 16, p. 149-163.

Taylor, P.W., 1986, Geology and Petrology of Niuafo'ou Island, Tonga; Subaerial Volcanism in an Active Back-arc Basin; *New Zealand International Volcanological Congress Abstracts,* p. 123.

MELANESIA

General References: Johnson, R.W. (ed.), 1976, *Volcanism in Australasia*; Elsevier, Amsterdam, 405 pp. (28 papers).

Johnson, R.W. (ed.), 1981, *Cooke-Ravian Volume of Volcanological Papers*; Geological Survey of Papua New Guinea Memoir 10, 265 pp. (25 papers).

Lowenstein, P.L., 1982, Problems of Volcanic Hazards in Papua New Guinea; *Geological Survey of Papua New Guinea Report* 82/7, 62 pp.

Lowenstein, P.L. and Talai, B., 1984, Volcanoes and Volcanic Hazards in Papua New Guinea; *in* Shimazaki, Y. (ed.), *Geologic Evolution, Resources, and Geologic Hazards*; Proceedings of the International Centennial Symposium of the Geological Survey of Japan; Geological Survey of Japan Report 263, p. 315-331.

KADOVAR

off N coast of New Guinea
3.63°S, 144.63°E, summit elev. 365 m

Kadovar and Blupblup are the westernmost active volcanoes in the Bismarck Arc, a group of islands paralleling the N coast of New Guinea and roughly 10-50 km offshore.

11/76 (1:14) The following is excerpted from a report by R.J.S. Cooke.

Kadovar is a 365 m-high, 1.4 km-diameter volcanic island. A 1700 report of smoke seen briefly there was believed at the time to be a volcanic eruption. Since then no eruptions are known, and no traces of thermal activity remained in early 1976, although there was supposedly some such activity several generations ago.

The island has the form of a steep-sided cone 250 m high, with a 600 m-wide crater breached to the SSE. The breach probably extended at least to sea level, but the crater and breach are now occupied by a high-standing, conical lava dome. The island is completely vegetated, and supported a village of more than 300 people built around the crater rim. Most of the village gardens are on the side of the dome.

In early Aug 1976, the first reports were received of activity there, the precise nature of which has not been clearly established. A summary of events to mid-Nov, and of investigations so far carried out, follows.

May-June 1976: The precise nature of the early events is not clear. Some sort of disturbance seems to have taken place in the sea just off (50 m?) the S coast, which may have been a small hydrothermal eruption or, more likely, a vigorous ebullition of gas bubbles. Weak earth tremors were felt, apparently during the preceding few days, and an explosive sound was heard. Scum was noticed on the water, persisting for a few days, and reddish coloration appeared on the rocks in the tidal zone at one point on the S coast.

3 Aug: An investigation was made by Rabaul Volcano Observatory (RVO) volcanologists D. Wallace and R. Almond. The only effect of the earlier activity was a 100 m-long reddish zone of iron hydroxide discoloration at sea level, and associated sea discoloration. No thermal activity or other unusual phenomena were observed, and no definite volcano-seismic activity was recorded during 5 hours operation of a portable seismograph.

Mid-Sept: Hot ground was first noticed midway up the side of the lava dome about 16 Sept.

26 Sept: An investigation was made by volcanologist R. Cooke. The coastal discoloration was more extensive than in Aug, and ground temperatures up to 49°C were present in this zone. The reported hot ground on the mid-slope of the dome was about 30 × 15 m in area, and was producing vigorous emissions of SO_2 and HCl gases (indicated by Dräger tubes) from one main vent and numerous minor vents, at temperatures up to 99.7°C. The white gas column was visible above the treetops from the boat, but only minor vegetation damage had occurred. A newly formed small, weak patch of hot ground was found 100-200 m upslope of the main fumarole area. No volcano-seismic activity was recorded during 5 1/2 hrs with a portable seismograph.

14 Oct: Government officers R. Allen and D. Mahar from Wewak inspected the fumarole area, which had intensified in activity and was about 150 × 50 m in area. The expansion had proceeded principally upslope.

23 Oct: D. Mahar again inspected the fumarole area, which had further enlarged. All vegetation in the area had been killed.

10-12 Nov: An investigation was made by volcanologist V. Dent, physicist D. Norris (University of Papua New Guinea, Port Moresby), and D. Mahar. The main fumarole area had enlarged again and was estimated at about 150 × 70 m, and the area of dead vegetation was more prominent. A second main fumarole vent had developed 20 m downslope of the original main vent of Sept, but the principal expansion since then had been upslope, and to the E. The maximum temperature measured was again about 99°C. Collections of fumarole gas were made in evacuated glass tubes containing silica gel, and gas condensates were also collected. The maximum ground temperature in the S coast discolored zone was 49.5°C, similar to Sept. About 50 hours recording by a portable seismograph produced 5-10 very small and close A-type events. A seismic event counter was installed on the island. D. Norris carried out a total magnetic field survey of the island using a proton magnotometer.

It is intended to continue regular inspection visits to the volcano, but a full-time observation post is not planned at this time. Such a post will be established, probably on Blupblup Island, 13 km north of Kadovar, when an eruption is believed to be imminent, or once an eruption has commenced. All residents of Kadovar have been evacuated to Blupblup Island.

Information Contact: R. Cooke, RVO, Rabaul.

Further Reference: Wallace, D.A., Cooke, R.J.S., Dent, V., Norris, D., and Johnson, R., 1981, Kadovar Volcano and Investigations of an Outbreak of Thermal Activity in 1976; *in* Johnson, R.W. (ed), *Cooke-Ravian Volume of Volcanological Papers;* Geological Survey of Papua New Guinea Memoir, p. 1-12.

4/77 (2:4) The following is from R.J.S. Cooke.

"Since the previous report, 5 more ground inspections have been made, and a sixth is planned for the last week in April.

"Complete investigations, including temperature measurement, collection of gases and gas condensates, measurement of magnetic field, and seismic recording, were made during visits on 15-16 Dec (Cooke, Norris), and 16-18 Feb (Dent, Norris), and are planned for the forthcoming visit (Cooke, Norris). Partial investigations were made on 3 April (Wallace), when temperatures and gases were investigated, and on 26 Jan (Mahar) and 14 Feb (volcanological assistant J. Kuduon), when temperatures were measured. Vertical aerial photographs were taken by a survey firm on 15 Nov, and aerial obliques were taken during the Dec inspection. Another aerial inspection was carried out on 7 Jan, the day after a shallow magnitude 6.5 earthquake about 30 km WSW of Kadovar on 6 Jan, 0611 GMT (preliminary location by USGS). This earthquake had no apparent affect on the volcano at the time.

"During the period covered by this report, the level of activity seemed to have stabilized. Maximum temperatures have been steady at 99-100°C, marked expansion of the main thermal area has ceased (although weak isolated gas vents are still occasionally found in new areas), and the quantity of gas emitted may even have declined slightly. The thermal area was not as unpleasant to the investigators as it was last Nov-Dec, although as some gas samples have not yet been analyzed quantitative information on the changing gas content is not available.

"No significant magnetic field changes have been detected. A few volcano-seismic events were recorded in both Nov and Dec, but such events appeared to be absent in Feb. Felt earthquakes were noted by inhabitants of nearby islands on several occasions, but there is no strong reason to associate these with Kadovar volcano. Unusually high seismic event counts on Kadovar are suspect because of a malfunctioning event counter.

"Although the initiation and early rapid development of this thermal activity led to the belief in a forthcoming Kadovar eruption, the stabilization (or even slight decline) in activity suggests the possibility that the event may be confined to thermal activity. Such purely thermal events have been reported elsewhere. However, the event will continue to be treated as a possible precursor to an eruption, and the former inhabitants will be advised to maintain the evacuation for the present. Only a small number of men are presently living on Kadovar in order to maintain the original gardens, to supply the evacuees on Blupblup Island. It is interesting to speculate that the 6 Jan earthquake may have been connected with the levelling-off of activity."

Information Contact: Same as above.

7/77 (2:7) The following is from D.A. Wallace.

"Three detailed investigations have been made since the March report: 25-28 April (Cooke, Norris)—72 hours of seismic recording, gas condensates collected, temperatures measured, and magnetic field re-surveyed; 7-8 June (Dent)—30 hours seismic recording, temperatures measured, and gas and gas condensates collected; and 7-8 July (Wallace, Norris, Emeleus)—30 hours seismic recording, temperatures measured, gas and gas condensates collected, and magnetic field re-surveyed.

"The main thermal area has been little changed during this period. The original fumarole of last Sept shows the same degree of activity as it did then. Volcano-seismic activity was virtually nil. Maximum temperatures were still 99-100°C but there may have been slight changes in the magnetic field pattern. The only features presently indicating continued development are weak thermal areas that are still being established in new areas on other parts of the island. Some of the population have returned, contrary to a previous report, but they have established a new settlement in a comparatively safe area, having abandoned their old village site."

Information Contact: D. Wallace, RVO, Rabaul.

11/77 (2:11) The following is from R.J.S. Cooke.

"Volcanologists have made 3 detailed investigations at Kadovar since the report in July. On each occasion, temperatures were measured, gases and gas condensates were collected, and seismic recording was carried out; a magnetic resurvey was made during the most recent inspection. Dates and personnel were: 18-19 Aug, V.F. Dent; 30 Sept, C.O. McKee; 13-14 Nov, C.O. McKee and D. Norris. On 14 Sept, photographs were taken during an aerial inspection.

"The main thermal area was much the same size as before, apart from some expansion at the SW corner, and spanned about 250 m upslope, and about 130 m maximum around the slope. It is now much more sharply defined and clear because of progressive fall of dead vegetation. Overall vapour emission seemed much less than in earlier periods, and was more concentrated in a few main areas, with fewer of the weak but widespread vents evident in early 1977. The original main fumarole may be a little reduced in intensity, although its temperature has not declined, remaining near 100°C. No measured temperatures exceeded 100°C, although the upper part of the main zone has become hotter, approaching 100°C. Vapour emission was still most profuse in the upper part.

"No volcano-seismic activity has been recorded during the report period, and although significant relative changes in total magnetic field have taken place, no simple pattern is apparent."

Information Contact: R. Cooke, RVO, Rabaul.

5/78 (3:5) Thermal activity was continuing in late March. An aerial inspection 21 March and a ground inspection the next day revealed that although principal fumaroles remained at the same level of activity as before, a slight decrease of activity in the main (S flank) thermal zone is indicated by minor revegetation. The odor of acid gases was no longer present at several groups of fumaroles and temperatures have dropped slightly (from 100°C to 97-97.5°C). However, another area of thermal activity (high on the E side of the lava dome) had produced a significant vegetation kill in the past few months.

Information Contact: Same as 2:11.

9/78 (3:9) Kadovar's thermal activity was continuing as of 20 Sept, but had shown a further slight decline since March.

Information Contact: Same as 2:11.

3/81 (6:3) "An area of orange discolouration of the sea at the NE shore was observed during aerial inspections on 6 and 19 March. Previously (1976-77) sea discolouration was present at the S shore and was related to the development of a new thermal area on the S flank of the island. This thermal area was now observed to have been reduced in size by regrowth of vegetation."

Information Contact: Acting Senior Volcanologist, RVO, Rabaul.

BAM

off N coast of New Guinea
3.60°S, 144.85°E
summit elev. 600 m

Bam is 25 km E of Kadovar and has had 23 reported eruptions in the last 120 years.

3/81 (6:3) During aerial inspections 6 and 19 March, a 1 km-long zone of orange sea discoloration was noted at the S shore of Bam Island. Bam's last eruption, in 1960, consisted of explosions from the central crater.

Information Contact: Acting Senior Volcanologist, RVO, Rabaul.

MANAM

off N coast of New Guinea
4.10°S, 145.06°E, summit elev. 1807 m
Local time = GMT + 10 hrs

Manam is 60 km SE of Bam and has been the most active volcano in the Bismarck Arc during historic time. The first of its 39 recorded eruptions was in 1616.

General References: de Saint Ours, P., 1982, Potential Volcanic Hazards at Manam Island; *Geological Survey of Papua New Guinea Report* 82/22, 19 p.

Scott, B.J. and McKee, C.O., 1986, Deformation, Eruptive Activity, and Earth Tidal Influences at Manam Volcano, Papua New Guinea, 1957-1982; *Royal Society of New Zealand Bulletin* 24, p. 155-171.

4/77 (2:4) A minor eruption has been in progress from Main and Southern craters (fig. 5-1) since mid-Feb. Weak intermittent lava fountaining has been observed 7 times at Southern crater, while brief phases of ash ejection, and on one occasion lava fountaining, were seen at Main crater during Feb and March. Low-level volcanic tremor has been recorded, but no significant tilt effects preceded the eruption.

Volcano-seismic events normally occur at the rate of about 1 per minute beneath Manam. Minor eruptive phenomena occur intermittently between major eruptions (such as 1974-75). [2:4 erred in adding that these phenomena were usually confined to Main Crater.]

Reference: Palfreyman, W.D. and Cooke, R.J.S., 1976, Eruptive History of Manam Volcano, Papua New Guinea *in* Johnson, R.W. (ed.) *Volcanism in Australasia*; Elsevier, Amsterdam, p. 117-131.

Information Contact: R. Cooke, RVO, Rabaul.

5/78 (3:5) A long phase of relatively weak ash ejection from Southern crater in 1977 ended in mid-Jan 1978. Similar activity briefly resumed at the end of Feb and for a few days in mid-March. Emission of moderate amounts of white vapor from Main and Southern craters has occurred since then.

Information Contact: Same as above.

9/78 (3:9) Minor ash ejection from Southern crater has been reported at intervals since mid-March: on 8 and 19-20 June; 11-13, 16-17, 19, and 27 July; 24, 26-28, and 31 Aug; and 2, 8-11, 14-16, and 18-22 Sept. Ashfall has been noted a number of times, particularly since mid-Sept.

Information Contact: Same as above.

10/78 (3:10) Small amounts of ash were emitted from Southern crater on most days in Oct. Incandescent fragments were seen briefly on 3 nights. Ash ejection from Main crater began in mid-Oct, and glow was seen there late in the month.

Information Contact: Same as above.

12/78 (3:12) Intermittent ash emission from Southern crater was less frequent in Nov and early Dec than in Oct. Ejection of incandescent lava fragments was seen on 7 nights in Nov and glow was observed on 7 nights in mid-Dec. For most of the second half of Dec, white and occasionally blue vapor issued from Southern crater. Ash ejection from Main crater was reported on 9, 13, 22, and 29 Nov, and on 6 Dec. Glow was seen over Main crater on 11 Nov, and 17, 21, and 23 Dec.

Information Contact: C. McKee, RVO, Rabaul.

8/79 (4:8) The following is from C. O. McKee.

"Eruptive activity strengthened in Aug and it appears that the level of lava has risen within the volcano. Glow was visible above both Main and Southern craters on most nights, and ejections of incandescent lava were observed on several nights. The lava ejections were stronger from Main crater, rising occasionally 200-300 m above the summit. Brown ash was emitted from Main crater on many days and grey ash emissions were seen rarely. Southern crater discharged brown or grey ash on only a few days. Otherwise both vents released white and blue vapours. Eruptive sounds included rumbling, roaring, booming, and sharp detonations. Seismic activity showed a slight intensification, but tilts were steady.

"An aerial inspection was made on 19 Aug. Main crater was a circular pit several hundred meters wide and contained a broad mound of black lava (fragments?). The surface of the mound was about 50-80 m below the crater rim. A central, circular vent was discharging blasts of red incandescent lava fragments at 1-10 second intervals. Southern crater was a deep, funnel-shaped structure with a pool of red incandescent lava at its base. Copious quantities of blue vapour were being released from both craters at the time of the inspection."

Information Contact: Same as 3:12.

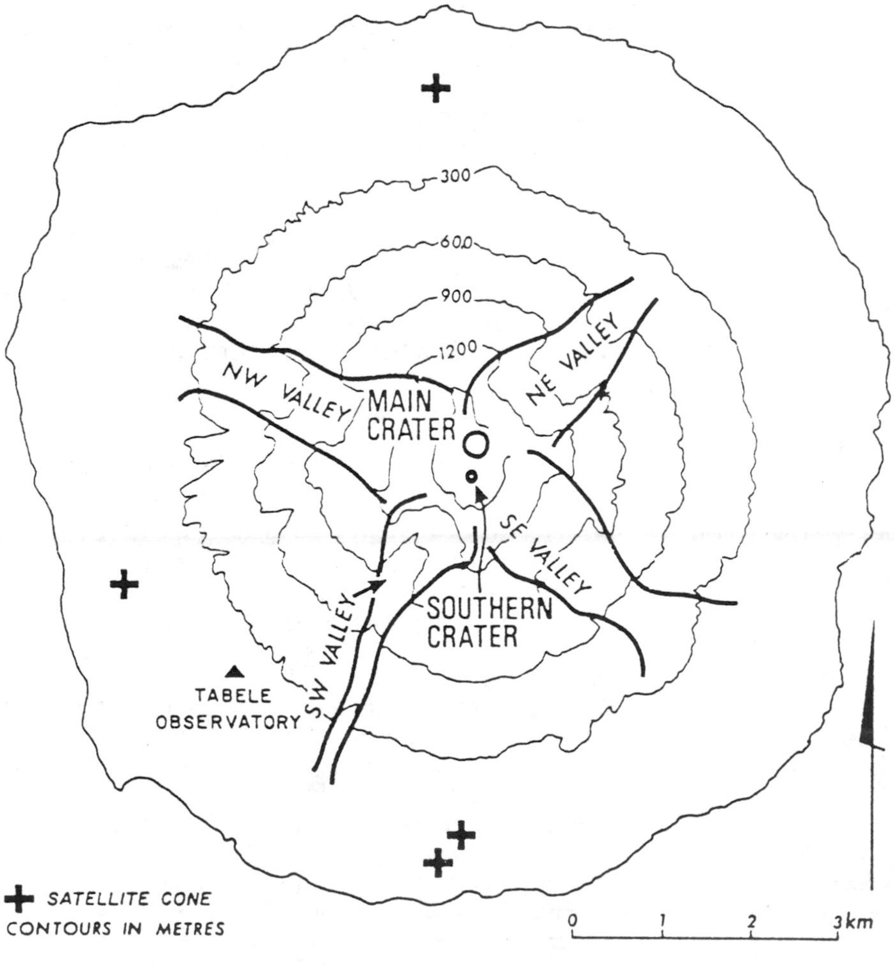

9/79 (4:9) The following is from Richard Almond.

"The period of strong eruptive activity that developed at the beginning of Aug terminated on 24 Aug, and the level of B-type seismic activity dropped back to that of July. Observations were infrequent during Sept because of bad weather but it is likely that emissions from both Main and Southern craters consisted solely of vapour. Weak glow from the summit was seen on 2 nights. The only audible effects were weak roaring noises."

Information Contact: R. Almond, RVO, Rabaul.

10/79 (4:10) The following 8 reports are from C. McKee.

"During the reporting period 12-22 Oct, white vapour was emitted from both Main and Southern craters. Additionally, Main crater released grey ash on 13 Oct and brown ash 16-18 Oct. Glow was seen from both vents on most nights during the reporting period, and sparse lava ejections from Southern crater were seen on the 16th. Roaring and rumbling noises were heard 13-15 Oct. There appeared to be no trend in tilt, and seismic activity remained steady at the same level as in Sept. Reports from Manam have been interrupted by the breakdown of Tabele Observatory's electrical generators.

"A brief aerial inspection on 26 Oct revealed that only thick white vapour emission was occurring at both vents. No clear views into Southern crater were obtained, but a dark surface of lava (fragments?) was seen in Main crater."

Information Contact: C. McKee, RVO, Rabaul.

2/80 (5:2) "Moderate activity occurred in Jan and Feb. Grey ash emissions from Southern crater were seen commonly in mid-Jan together with crater glow and occasional lava ejections. Activity weakened somewhat in Feb when mostly white and blue emissions were seen, and glow, with or without lava ejections, was seen on 7 nights evenly distributed through the month. Main crater was usually obscured, but grey or brown ash emissions were seen sporadically in mid-Jan. No sounds of eruptive activity from either vent were heard in the 2-month period."

Information Contact: Same as above.

4/80 (5:4) "During March, Manam's summit was often

Fig. 5-1. Sketch Map of Manam Island, after Palfreyman and Cooke, 1976.

obscured by clouds, but when visible, both vents were usually emitting white vapours. Weak rumbling noises were heard on the E side of the volcano on 7 March. Activity strengthened in April, when glow and lava ejections from Southern crater were commonly seen. The highest ejections rose to about 50-60 m above the summit. No eruptions of ash were seen, but white and blue vapours were emitted. A blue haze was occasionally seen in the upper parts of some of the major valleys that descend from the summit. Main crater was usually obscured, but white emissions were observed on a few days. Seismic activity remained at its usual level and tilt readings indicated no trends."

Information Contact: Same as above. [Information Contacts for this and following RVO reports were corrected by the RVO staff in Jan 1988.]

5/80 (5:5) "Activity strengthened near the end of the first week in May. Brown or grey ash emissions from Southern crater were observed on most days in May, and grey ash emissions from Main crater were also observed on several days when the vent was not obscured. Light ashfalls were experienced on the W side of the island. This ash emission often occurred with little accompanying sound. However, deep booming, rumbling, and roaring noises were occasionally heard. Orange-red glows above both vents were seen on a few days, and lava fragment ejections from Southern crater were observed on 1 and 18 May. No trends were observed in tilts, but seismic amplitudes increased by a factor of about two over normal levels at the onset of the phase of ash emission."

Information Contact: Same as above.

6/80 (5:6) "Manam's activity returned to a lower level in June. Occasional brown or grey ash emissions from Southern crater were observed, and light grey emissions from Main crater were seen on 5 June. However, no explosive sounds and no glows were observed from either crater. No trends were registered by the tiltmeters, and seismicity remained at its normal level."

Information Contact: Same as above.

7/80 (5:7) "Manam's July activity was at a level similar to that of June, although brief periods of stronger seismicity were recorded at the beginning and end of the month. Pale grey/brown or thick white emissions from Southern crater were commonly observed, and blue vapours were occasionally seen. Main crater was often obscured but several observations of pale grey/brown emissions were made. Occasional rumbling sounds were heard, but no crater glows were seen. No trends were registered by the tiltmeters."

Information Contact: Same as above.

8/80 (5:8) "Activity during Aug continued at the same level as that of June and July. Emissions from Southern crater were mostly white vapour and brown ash, but grey ash emissions were observed on 20 Aug and blue vapours were occasionally seen. Roaring and rumbling sounds were associated with stronger pulses of ash emission. Main crater was obscured most of the time, but on several occasions white vapour emissions were observed. Brown ash emissions from Main crater were observed on 20 Aug. No crater glows or lava fragment ejections were observed from either crater. Seismicity and tilt remained steady."

Information Contact: Same as above.

9/80 (5:9) "Activity continued at the same level as that of the previous 3 months. Southern crater usually emitted white vapour, but brown, grey, or blue emissions were occasionally seen. Weak rumbling noises were heard on a few days at the beginning of Sept. Emissions from Main crater were usually white, but grey emissions were seen on 2 occasions, on 2 and 10 Sept. No glows or lava fragment ejections from either crater were observed. Seismic activity and tilts were steady."

Information Contact: Same as above.

1/81 (6:1) The following is from Ben Talai and Peter Lowenstein.

"Moderate to strong light brown to grey ash-laden vapour and, rarely, dark brown dust were sporadically ejected from Southern crater. Main crater occasionally emitted weak white vapour. Light ashfall from Southern crater was recorded at nearby Tabele on 2 Dec. Low rumbling noises were heard on 20 and 25 Dec. A weak glow from Southern crater was observed at night 26-29 Dec. Seismic activity was at its normal level. [Tilt measurements indicated radial inflation of about 3 μrad in Dec. This followed a net inflationary radial tilt of about 2 μrad in Oct.]"

Information Contacts: B. Talai, P. Lowenstein, RVO, Rabaul.

3/81 (6:3) Reports through (7:11) are from C.O. McKee.

"During the first 2 months of 1981 a low level of activity prevailed. White and occasionally brown emissions were observed from both craters. In March, moderate to strong brown and light grey ash-laden emissions were common from Southern crater. Main crater emissions were also grey on several occasions. Explosive sounds from the summit were rarely heard in Jan and Feb but became noticeable in the second half of March. Night observations of the volcano in Jan and Feb indicated no lava fragment ejections above the craters, although weak glow above Southern crater was reported for 2 Jan. Sparse ejections of lava fragments from the crater were observed overnight on 14-15 March, and glow above Southern crater was observed on 30 March.

"Background volcano-seismic levels remained fairly steady Jan-March, but a significant change in seismic activity was the occurrence of strong local earthquakes, possibly of volcanic origin. Preliminary analysis of seismic records showed that 5 such events were recorded in Feb and 14 in March.

"The tiltmeters at Manam continued to show a trend of northerly uplift. After the last major eruptive period in 1974 a pattern of summit deflation prevailed until early 1978. Total deflation was about 14 μrad. A definite trend of inflation began in the second half of 1979. The accumulated tilt during the last 2 years was about 6 μrad.

"Aerial inspections were made on 6 and 19 March. Cloud cover prevented detailed observations of summit activity on 6 March but a distinct blue vapour haze was drifting down the N and NW flanks. On 19 March, brown ash-laden ejections from Southern crater were occurring at about half-minute intervals. Main crater continuously released white vapours. Again a blue vapour haze was present, extending about 1 km N of the summit."

Information Contact: C. McKee, RVO, Rabaul.

4/81 (6:4) "Visible activity intensified somewhat in April. On most days when the summit was not obscured, moderate volumes of variously white, and light to dark brown or grey emissions were seen rising above Southern crater. Occasional rumbling, roaring, and booming noises from the summit were heard at Tabele Observatory (about 4 km away). A weak glow above Southern crater was seen on 13 and 16 April, and ejections of incandescent lava fragments from there were seen on 28 and 29 April. Main crater was usually obscured, but thick white emissions were noted on 7 and 28 April. The intensity of background volcano-seismicity remained steady. A preliminary analysis indicates that only one A-type volcanic earthquake occurred in April. Tilt measurements were steady."

Information Contact: Same as above.

5/81 (6:5) "Activity was fairly stable during May. Southern crater released white vapours and emitted moderate volumes of brown and grey ash. Main crater was usually obscured, but in clear conditions it was seen to release thick white vapours. Main crater glow was seen on several nights 20-23 May. Rumbling and booming noises from the summit were heard occasionally. Background seismic activity and tilts remained steady."

Information Contact: Same as above.

6/81 (6:6) "Activity in June was similar to that observed in May. Southern crater continued to emit white vapours and brown to grey ash in moderate volumes. Main crater usually released thick white vapours. Low rumbling sounds from the summit, heard frequently during the second half of the month, probably originated from Southern crater. No summit glows or incandescence were observed. A light ashfall was recorded at Manam's [Tabele] Observatory, about [4] km from the summit, on 4 June. Seismic activity and tilts remained steady."

Information Contact: Same as above.

7/81 (6:7) "Visible activity in July remained similar to that observed in June. Emissions from Southern crater varied between white, brown, light grey, and dark grey. Main crater released grey emissions on several days. Occasional weak rumbling sounds, and deep booming sounds on 30 July, probably originated from Southern crater. No observations were made of summit glow or ejections of incandescent lava fragments. Occasional ashfalls were registered on the SW to NW flanks. Seismic activity and tilts showed no significant changes from previous trends [but see 6:8]."

Information Contact: Same as above.

8/81 (6:8) "Brown or grey ash emissions from Southern crater were seen less frequently in Aug than in July. Main crater usually released white vapours, but grey or brown ash emissions were more common in Aug than at other times in 1981. Sound effects from the volcano ranged from weak rumbling to loud deep booming. Static dull orange-red glow at both craters overnight on 31 July-1 Aug was occasionally disturbed by flashes of bright orange-yellow glow, and sprays of incandescent lava fragments from Southern crater were also seen occasionally. Glow and ejections of incandescent lava fragments from Southern crater were observed 14 July, and glow from Main crater was seen on the 25th.

"Detailed analysis of seismic records indicates that, contrary to reports for June and July (6:6-7), seismicity began a slight but apparently steady intensification in early June. For most of Aug, average seismicity was probably twice as strong as in the first half of 1981. The character of seismicity, however, is unchanged, and consists of volcanic B-type or explosion earthquakes. Tilt observations were steady in Aug."

Information Contact: Same as above.

9/81 (6:9) "There were fewer instances of ash emission from both craters in Sept. Usually, Southern crater released moderate volumes of thick white vapour. However, rumbling sounds from the volcano were common, and weak glow and ejections of incandescent lava fragments from Southern crater were observed on 15 Sept. No trends were evident from tiltmeter measurements. Seismicity remained at a level higher than normal for most of the month, but declined near the end of the month to levels prevailing in the first half of 1981."

Information Contact: Same as above.

10/81 (6:10) "Both craters showed stronger activity in Oct. Incandescent tephra ejections from Southern crater were seen on 4 nights at the beginning of the month, and on most nights from 12 Oct onwards. Brown or grey ash emissions from Southern crater were seen 1-9 and 21-30 Oct. Main crater emitted brown or grey ash on most days 1-22 Oct. Main crater glow was seen on 1 Oct, and incandescent tephra ejections on the 14th. A substantial ash plume from Main crater (reportedly several tens of kilometers long) was observed on 14 Oct. White and blue vapours were emitted from both craters during non-explosive intervals throughout Oct, and rumbling sounds from the volcano were heard on most days. No trends were shown by tiltmeter measurements, and seismic activity remained steady."

Information Contact: Same as above.

11/81 (6:11) "Visible activity appeared to intensify somewhat in Nov. Brown or grey emissions from Southern crater were observed on most days, and crater glow or ejections of incandescent tephra were seen on most nights. Sound effects associated with this activity varied from low rumbling to thudding, booming, and occasional detonations. Ashfalls in coastal locations 4-5 km from the summit were reported early in Nov. Main crater was less active, usually emitting white vapours, although grey and brown emissions were observed on 2 days late in the month. Instrumental readings remained steady."

Information Contact: Same as 6:3.

12/81 (6:12) "Activity further intensified in Dec. Southern crater emissions became darker after 9 Dec when spearheaded projections of tephra were first reported. Rumbling, thudding, and roaring sounds also intensified after the 9th. Larger tephra was visible beginning 13 Dec. Incandescent tephra ejections were seen throughout the month. Daytime incandescence, visible from Tabele Observatory, was reported occasionally, but became more common at the end of the month. These ejections rose to a maximum height of about 0.7 km. Spearheaded tephra projections became common at the end of the month. Light ashfalls in coastal areas 4-5 km from the summit were reported on several days in the second half of the month.

"Main crater was less active than Southern crater. Weak to moderate volumes of brown to grey emissions were reported occasionally in the first 10 days of Dec. However, activity intensified later in the month, and tephra emissions were reported every day 21-30 Dec. Weak glow from Main crater was seen on 29 Dec. Seismicity strengthened throughout Dec, reaching Aug and Sept levels by the end of the month. Distinct inflation was evident from tiltmeter measurements."

Information Contact: Same as 6:3.

1/82 (7:1) "Dark tephra emissions from Southern crater continued throughout Jan, although periods of white vapour emission became more common from the 16th. Roaring, rumbling, booming, and thudding sounds were more pronounced in the first half of the month. Incandescent tephra ejections were seen on most nights, occurring at intervals of 1-3 minutes. Ashfalls were reported on the E coast of Manam 7-15, 21-22, and 29-30 Jan.

"Activity at Main crater was subdued, consisting of weak white and brown emissions. Seismicity declined from a peak between mid-Dec and early Jan, and by mid-Jan seismicity was back to normal levels. Tilts were stable during Jan following the inflationary phase of Dec 1981."

Information Contact: Same as 6:3.

2/82 (7:2) "Weak to moderate ejections of grey or brown tephra from Southern crater were observed on most days. Frequency of eruptions varied, and occasionally they resembled Strombolian activity. Incandescent tephra ejections were visible on the nights of 6, 10, 11, 14, 15, and 17 Feb. Ashfalls in coastal areas were reported on most days 1-20 Feb. The only emissions observed from Main crater were white vapours. Seismicity increased during Feb, and by the end of the month the level was similar to peaks reached in Aug-Sept and Dec 1981. Tilts remained stable."

Information Contact: Same as 6:3.

3/82 (7:3) "Strong eruptive activity occurred in 2 intervals in March, the first during several days at the beginning of the month. Spearheaded projections of tephra from Southern crater were observed on 2 and 3 March. Tephra ejections were less intense 4-7 March, but instability of the rapidly accumulated tephra caused avalanches of this material to descend from the summit into the SW valley. Inspections by volcanologists on 10 and 11 March suggested these avalanches were small. No significant changes in tiltmeter readings accompanied this eruptive phase, but seismicity showed a marked intensification on 5 March.

"Much stronger activity occurred near the end of the month. A paroxysmal eruption was observed at 1207 on 27 March. The dark grey-brown Vulcanian eruption cloud ascended to 6-7 km. Lightning flashes were seen in parts of the cloud. Strong Strombolian explosive activity followed the paroxysmal eruption at about 1215. The E side of the island experienced a brief period of darkness and tephra falls were locally severe, but the maximum thickness of the tephra deposit was probably only a few millimeters. Fragments up to 7 cm in size were collected at one village. Vegetation was strongly affected by the tephra fall and water supplies were polluted, but no structural damage was done to houses. A pyroclastic flow descended the SE valley during the eruption, but stopped about halfway to the coast.

"Seismicity was very strong at the time of the eruption and was still high at month's end. Before and after the eruption discrete B-type earthquakes occurred at the rate of about 1 per minute. For about 15 hours from the commencement of visible activity, discontinuous seismic tremor was recorded. No significant changes were evident in tiltmeter readings."

Information Contact: Same as 6:3.

4/82 (7:4) "Activity was weaker in April. Southern crater was relatively quiet for most of the month, emitting brown-grey tephra clouds of small to moderate volume. However, 2 discrete explosions were observed on 24 and 25 April, with clouds reaching heights of about 600 m and 1500 m respectively. Main crater was reactivated on about 10 April when small grey tephra clouds were observed. The period of strongest activity was 15-20 April and included louder explosive sound effects, increased emission of tephra, and incandescent tephra ejections to about 130 m. Seismic activity was generally low, but higher 15-20 April. A change in the character of the seismicity, from discrete events occurring at rates of several per minute to more or less continuous tremor, was noticed 7-15 April. A rapid intensification in seismic amplitude occurred 15-18 April, and an abrupt decrease took place after the 20th. No significant variations were registered by the tiltmeters."

Information Contact: Same as 6:3.

5/82 (7:5) "After reactivation of Main crater on 10 April, moderate volumes of dark grey-brown ash-laden emissions from this crater continued to be observed on most days in May. This activity was stronger in the first half of the month when roaring, rumbling, and explosion sounds accompanied the ash emissions. Fine ashfalls, probably from Main crater, took place on the E coast on 11, 13, 14, and 16-17 May. No crater glows or incandescent tephra ejections from Main crater were seen in May.

"Southern crater showed a low level of activity throughout May, releasing small to moderate volumes of grey-brown ash-laden emissions on most days. Blue vapour emissions were common. Weak crater glow was observed occasionally from the 14th.

"Seismicity was at a low (normal) level throughout May. Tilt measurements were steady in the first half of the month, but a slight inflation (about 1 μrad) accumulated during the second half."

Information Contact: Same as 6:3.

7/82 (7:7) "Fairly steady levels of activity prevailed at both craters in June and July. Southern crater has released moderate volumes of grey-brown ash-laden emissions and occasionally blue vapours. Rumbling and roaring sounds accompanying these emissions indicated a relatively mild eruptive condition. No glows from this crater were seen in June or July.

"Main crater emitted grey to dark grey and rarely brown ash clouds in usually moderate volumes, also accompanied by roaring and rumbling sounds. Crater glows were seen on 18 June and 17 July. Light ashfalls in coastal areas (about 5 km from the craters) were reported on 28 June and 14, 15, 17, 18, and 24 July. A persistent blue vapour haze on the upper flanks of the volcano was observed on 19 and 24 June and 6, 11, and 30 July.

"Seismic activity was remarkably steady in June and July at a low level. The tiltmeters at Tabele Observatory on the SW side of Manam have registered 3-4 μrad of northerly down-tilt since early March."

Information Contact: Same as 6:3.

8/82 (7:8) "Main crater erupted impressively from 8 Aug when thick dark grey ash clouds were emitted and weak to strong crater glow was observed at night. Sharp ejections were seen occasionally. Rumbling and roaring sounds accompanied similar activity until a decline became evident on 23 Aug. Voluminous blue vapour emissions were seen on several days between 19 and 31 Aug. Ashfalls in coastal areas were reported on about 30% of days, evenly distributed throughout the month.

"Southern crater activity was very mild during the first half of Aug. An increase occurred in the second half of the month, and weak to moderate pale grey-brown ash emissions were reported on most days after the 14th. On a few days at mid-month, weak to strong roaring and rumbling sounds were heard from this crater. No Southern crater incandescent activity was observed."

Information Contact: Same as 6:3.

11/82 (7:11) "Both craters displayed steady, moderate levels of activity in Nov. Southern crater was somewhat more active with rumbling sounds commonly accompanying the usually pale grey to brown ash-laden emissions. Main crater emissions were usually white and pale grey, but occasional brown ash-laden emissions were seen. Light ashfalls were reported in coastal parts of Manam on most days in Nov. No crater incandescence was observed. Seismic activity was steady at normal levels throughout Nov, and tiltmeter measurements showed no trends."

Information Contact: Same as 6:3.

12/82 (7:12) The remaining Manam reports are from the RVO staff.

"Both craters displayed a moderate level of activity through Dec. A moderately thick white to grey plume was released from Main crater while Southern crater produced grey to brown ash-laden emissions. These emissions, accompanied by weak and low rumbling noises, resulted in occasional light ashfalls on coastal parts of the island.

"The level of seismicity increased during the month (number of recorded shocks from 1500 to 2000/day; background tremor becoming continuous from 19 Dec). The average amplitude of the seismic events increased on 13-17 Dec and again on 27-28 Dec. Emissions from Main crater increased in volume and pressure, but instead of explosions or crater glow the seismic activity dropped again suddenly (from 2200 on 28 Dec) to a low level and tremor stopped."

Information Contacts: P. de Saint Ours, P. Lowenstein, RVO, Rabaul.

Further Reference: Scott, B.J. and McKee, C.O., 1984, Deformation, Earth Tidal Influences, and Eruptive Activity at Manam Volcano, Papua New Guinea, 1957-82; *Geological Survey of Papua New Guinea Report* 84/3.

1/83 (8:1) "Manam Volcano was essentially quiet during Jan. Moderately thick white vapour was released from Main crater, and a white-grey plume from the Southern crater. Hardly any noise was heard. Only very light ashfalls were reported from downwind coastal areas of the island. Seismic activity remained steady, with about 2000 moderate-amplitude events recorded per day. Tilt curves at both stations remained flat."

Information Contacts: P. de Saint Ours, B. Talai, RVO, Rabaul.

2/83 (8:2) "Main crater [experienced increased activity] for a few days in mid Feb. Ash-laden emissions from Southern crater also increased.

"White vapour was first observed over Main crater 8 Feb and increased on the 9th. This was accompanied by a change in the seismic pattern, with a progressive decrease in the daily number of B-type events (from 2100) but an increase in amplitude of the shocks.

"Activity stayed at low level until 15-16 Feb, when low to loud rumbling noises from Main crater were heard at 5-minute intervals. Harmonic tremor, formerly in bands, became continuous. Large amounts of blue vapour were observed with the white plume. On the night of 16 Feb red glow was seen over Main crater. The activity lasted 4 days and was accompanied by an increase in the daily number of recorded seismic events from 1400 to 1800.

"Beginning 20 Feb, night glow and blue vapour emissions disappeared, and rumbling noises and plume volume decreased. In the last days of the month seismic activity fluctuated between [1100] and 1300 daily events, but some explosions from Southern crater were again heard and recorded."

Information Contacts: P. de Saint Ours, P. Lowenstein, RVO, Rabaul.

3/83 (8:3) "Southern crater became more active in March after showing increasing activity in late Feb. Weak explosion sounds were heard on most days until 14 March, accompanying weak to moderate white-grey vapour and ash emissions. Weak ejections were reported on 12 and 13 March, when weak crater glow was seen. A period of somewhat stronger emissions was reported 19-24 March, including low to moderate explosion sounds on 23 and 24 March, and continuous vapour and ash emissions were noted on 26 and 29 March. Weak rumbling was heard from the 29th to the month's end, and deep booming sounds were reported on the 31st.

"Generally steady activity of weak to moderate white-grey emissions occurred at Main crater until about 23 March. These emissions were usually not accompanied by sound effects, but on 2 and 3 March weak explosion sounds were heard. From the 24th to the end of the month emissions were reported to be moderate, and included brown ash from the 27th. Light ashfalls were reported on about 30% of days from locations on the E and SE flanks.

"A steady increase in the daily number of volcanic earthquakes took place in March, from about 1200 at the beginning of the month to about 2100 at month's end. Event amplitudes showed a slight stepwise increase at mid-month. A marked brief increase in seismic amplitudes was also noted on 4 and 5 March.

"Tilt measurements showed steady changes of about 1 μrad down to the NW from the observatory, on the SW flank (fig. 5-1)."

Information Contact: C. McKee, RVO, Rabaul.

4/83 (8:4) "A slight intensification of the ongoing mild eruptive activity was evident in April. Increased seismicity and continuous, voluminous blue vapour emissions occurred at both summit craters from mid-month. Both craters continued to emit pale grey or brown ash clouds that reached maximum heights of about 1000 m above the summit. Emissions from Southern crater were accompanied by sharp detonations, booming, and rumbling of varying intensities, at intervals of 5-25 minutes. Similar sound effects from Main crater were occasionally heard.

"Southern crater glow was seen on most nights in the month, although a hiatus occurred 14-16 April. Ejections of incandescent lava that reached maximum heights of about 250 m above the crater were seen on about 30% of nights. Weak or fluctuating Main crater glow was observed 5-8 and 14-19 April. Ashfalls were recorded in coastal areas, about 5 km from the summit, on about 35% of days.

"No significant tilt changes were recorded in April, but seismicity showed a mild intensification. Seismic amplitudes increased near the beginning of the month and remained double normal size. Daily earthquake totals varied from 2500 at the beginning and end of the month to a peak of 3500 on 23 April."

Information Contact: Same as above.

5/83 (8:5) "Further intensification of activity in May was shown by increased seismicity, stronger ash emissions, and more frequent sightings of nighttime incandescence.

"Both craters continued to emit pale grey ash clouds, but the emissions from Southern crater were frequently reported to be strong and thick. The emission column from Southern crater usually rose several hundred meters above the summit, and a plume several tens of kilometers in length was reported on a few days. Intervals between discrete explosion and rumbling sounds varied from about 3 to 35 minutes. Glow and incandescent lava ejections were observed on most nights. Ballistic trajectories of incandescent tephra reached heights of up to 600 m above the crater rim.

"A brief period of intensified Main crater activity took place 12-19 May. Stronger ash emissions were accompanied by explosion and rumbling sounds, and crater glow and incandescent lava ejections were reported 12-16 May. Incandescent tephra fragments reached a maximum height of about 500 m.

"Volcano-seismic amplitudes continued to increase from the elevated April levels and reached a peak of about 5 times normal 22-24 May. A slight reduction was evident during the last week of May. At the month's end, amplitudes were still about 3 times normal. The number of volcano-seismic events declined from the peak of about 3500 per day reached in late April, and stabilized at about 2500 per day in May.

"A small but distinct tilt change was evident in May on the E-W component of the water tube tiltmeters at Tabele Observatory, about 4 km SW of the summit. The amplitude of the change was about 1 μrad, and its orientation was consistent with inflation of the volcano. No corresponding changes were shown by the N-S component."

Information Contact: Same as above.

6/83 (8:6) "Despite continuing high levels of seismicity, audible and visible activity tended to be less intense in June. Both craters produced pale grey ash-laden emissions in moderate volumes, accompanied by rumbling and booming sounds at intervals of 3-25 minutes. Most of the sounds seemed to originate from Southern crater.

"Blue vapours were also emitted from both craters, but particularly from Southern crater, resulting in a haze of blue vapour on the downwind flank of the volcano. Probably as a result of strong winds, the emission column rarely rose higher than about 100 m above the summit. On several days the bluish emission plume stretched out for several tens of kilometers. Fallout of ash was common in coastal areas, but usually light.

"Daily totals of volcanic earthquakes rose from about 2400 at the beginning of the month to about 2900 at the month's end. The amplitudes of discrete volcanic earthquakes were about 3 times normal at the beginning of the month, but rose to about 5 times normal 11-24 June. By month's end amplitudes had returned to early

June levels. Tiltmeter measurements showed no trends in June."

Information Contact: Same as 8:3.

7/83 (8:7) "Most observed parameters showed little or no change in July. Visible activity at Southern crater was unchanged from that of June: pale grey ash-laden emissions in moderate volumes ejected to no more than about 100 m above the crater rim. Occasional weak rumbling and booming sounds accompanied these emissions. Blue vapour emissions also continued at the same rate. On several days a bluish emission plume stretched several tens of kilometers downwind. The crater was often obscured at night, and incandescence was seen only on the nights of 2, 4, 12, and 13 July, as weak fluctuating glow. On 2 July ejections of incandescent tephra to heights of about 250 m were also seen.

"Main crater activity was weaker in July than in June. Usually, pale grey ash-laden clouds were emitted in small to moderate volumes. Blue vapour emissions were seen on 12 and 14 July, and 26-31 July. No eruption sounds could be detected from Main crater, and no instances of nighttime incandescence were reported.

"A helicopter inspection on 26 July revealed that Main crater was a deep funnel-shaped structure having a central vent from which weakly ash-laden, blue-tinged clouds were being released. Abundant fumaroles were noted on the crater walls. Views of Southern crater were obscured by steady emission of ash-laden clouds that filled the entire crater.

"Amplitudes of B-type volcanic earthquakes were remarkably steady throughout July at about double non-eruptive levels, but representing a distinct decline from the high levels of mid-June. Daily totals of seismic events were about 2000 at the beginning and end of the month, but varied up to about 3000 in mid-month. Tilt measurements showed a continuation of the flat trend evident in June."

Information Contacts: C. McKee, P. de Saint Ours, RVO, Rabaul.

8/83 (8:8) "A slight intensification of the moderate eruptive activity at Southern crater took place in the second half of Aug. Activity at Main crater remained essentially unchanged from July. In the first half of Aug a pale grey or brownish plume of moderate volume was produced at Southern crater, accompanied by blue emissions. Similar emissions were released from Main crater, although in lesser amounts. Seismic recordings in this period indicate a daily average of about 2000 B-type earthquakes of small amplitude.

"From about 16 Aug, brown or dark grey ash clouds were ejected from Southern crater. These convoluting ash-laden clouds usually rose 100-200 m over the crater rim. Daily totals of B-type earthquakes increased to as many as 2800 events, of slightly larger amplitude. Crater glow and sprays of incandescent lava fragments were observed on 27 and 28 Aug, rising to 150-200 m above Southern crater at 2-3-minute intervals. Weak glow at Main crater was observed on 28 Aug. Near the end of the month the daily totals of B-type events declined to about 2300. No significant tilts were registered by the water-tube tiltmeters at Tabele Observatory."

Information Contact: Same as above.

Volcano	Date	t/d SO_2
Bagana	Sept 8	3100
Langila	Sept 11	74
	12	1300
Manam	Sept 12	920
Ulawun	Sept 11	71

Table 5-1. Rates of SO_2 emission at 4 volcanoes, Sept 1983. Airborne COSPEC data from R. Stoiber, S. Williams and C. McKee.

9/83 (8:9) "Apart from a short period of stronger activity at mid-Sept, Manam continued its mild eruptive activity which has been characterized by moderate emissions of pale grey-brown ash clouds from both craters. The stronger activity occurred 10-13 Sept, when impulsive ejections of dark tephra-laden clouds took place from Southern crater at intervals of several minutes. The ejections rose to 150 m above the crater and fed a plume that extended several tens of kilometers downwind. However, this stronger activity had no distinctive seismic expression. Throughout the month amplitudes of volcanic earthquakes remained steady at 2-3 times normal, and the average daily total of events was about 2500. Tilt measurements from Tabele Observatory indicated a gradually accumulating [upward tilt] of about 2 μrad to the NNW."

Richard Stoiber, Stanley Williams, and Chris McKee used a COSPEC to measure the rate of SO_2 emission from several volcanoes in Papua New Guinea during Sept (table 5-1). Plumes at Bagana and Manam were strong, and Ulawun's plume was small. Activity at Langila was weak 11 Sept, but had intensified during measurements the next day. The quiet-phase Langila data were collected from the ground; all other data were acquired while flying under the plumes.

Information Contacts: C. McKee, RVO, Rabaul; R. Stoiber, S. Williams, Dartmouth College.

10/83 (8:10) "Activity intensified somewhat at Manam's Main crater in Oct, while Southern crater activity was generally mild with forceful ejections of tephra on 2 days. The intensified Main crater activity took place in several brief periods. Explosive eruption sounds (rumbling, roaring, and booming) were heard about 4-5 km away on 14-15, 23-25, and 29-31 Oct, and crater glow was seen on 14, 23-24, and 29-30 Oct. Moderate, pale grey-brown emissions were observed during these periods, but at other times only small to moderate amounts of tephra-poor vapour were emitted.

"For most of the month, Southern crater emissions consisted of pale grey-brown ash clouds, but on 16 and 31 Oct impulsive, dense ejections of tephra and vapour to 200-250 m above the crater occurred. On 20, 22, and

31 Oct the emissions from both craters combined to form a plume several tens of kilometers long.

"Amplitudes of volcanic earthquakes were mainly steady at about 2-3 times normal levels, but increased 15-25 Oct. Daily earthquake totals remained around 2700 for the first half of the month, decreased to about 1900 at mid-month, and approached 2600 on 31 Oct. A small down-tilt of about 1 μrad to the N (deflation) accumulated steadily during the month."

Information Contact: C. McKee, RVO, Rabaul.

11/83 (8:11) "Activity at Manam's Main crater showed a further intensification in Nov but Southern crater activity remained mild. The stronger Main crater activity commenced about mid-month, and was marked by moderate to strong, pale grey-brown, tephra-laden emissions, accompanied by louder rumbling and roaring sounds, and a doubling of the amplitudes of explosion and B-type volcanic earthquakes. Daily totals of earthquakes showed a corresponding decrease from about 2500 to 1500. No significant tilting accompanied the intensified eruptive activity. Fluctuating or dull glow from Main crater was seen occasionally throughout the month. Southern crater continued to release emissions that usually contained little or no tephra. Weak rumbling or roaring sounds were heard occasionally. Blue vapours were emitted from both craters on most days of Nov."

Information Contact: Same as above.

12/83 (8:12) "The moderate level of activity from Main crater reported in Nov decreased early in Dec. After a comparative lull 3-13 Dec, vapour emission from Southern crater increased, and from the 17th onwards night glow and ejection of incandescent fragments were observed. Glow from Main crater was also observed from 18 Dec onwards. The combined activity was accompanied by deep rumblings, explosion noises, and the discharge of a moderately large, strongly coloured, ash-laden vapour plume. Light ashfalls were experienced daily on the SW side of the island for the remainder of the month."

Information Contact: P. Lowenstein, RVO, Rabaul.

1/84 (9:1) "The eruptive activity continued through Jan. It consisted of night glow at both Main and Southern craters, and frequent Strombolian ejections of glowing lava fragments up to several hundred meters above Southern crater. Explosion noises and sub-continuous rumbling sounds were heard. Scoria and bombs ejected from Southern crater avalanched down the SW and SE valleys.

"Beginning 13 Jan, the amplitude of recorded B-type earthquakes started to increase considerably, although their number remained about 2000/day. From 18-26 Jan, the amplitudes of events increased to about 4 times normal. At 1155 on 26 Jan, a large explosion from Southern crater produced a voluminous, dark ash-laden plume rising to 3.5 km. The amplitude of the sub-continuous tremor and B-type events then returned rapidly to normal. Up to 3 mm of ash were deposited on the coastal areas. For the remainder of Jan, the Strombolian activity continued at the same level as at the beginning of the month."

Information Contact: Same as above.

2/84 (9:2) "A phase of major eruptive activity commenced at Manam's Southern crater in mid-Feb when a series of pyroclastic avalanches was discharged into the SE valley. Moderate Strombo-vulcanian explosive activity took place at Southern crater during the first half of the month, but an intensification was noted from 12 Feb, and on the 17th the first pyroclastic avalanche was discharged. This and the succeeding avalanche on the 21st descended about 4 km from the summit. Smaller avalanches were produced on most days after the 21st, usually terminating about 2 km from the summit.

"Ground and aerial inspections near the end of the month revealed that the numerous avalanches had obliterated most of the pre-existing surface in the upper half of the valley. Trees were flattened and had lost limbs and foliage. Scorching of vegetation had taken place on the 200 m-high valley walls and beyond to distances of 100 m. In addition to these hot pyroclastic avalanches, numerous flows of loose scoria from rapidly accumulated airfall deposits around the vent were also noted. These scoria flows descended into both the SE and SW valleys, terminating within 2 km from the summit.

"Vertical explosion activity at Southern crater produced an impressive eruption column that rose 5-8 km above the vent on several days. Incandescent pyroclasts were ejected to heights of about 700 m, 17-29 Feb. Ashfalls in coastal areas were generally light, although the accumulated thickness may have been up to several centimeters in places, resulting in the loss of branches from some trees.

"Main crater was moderately active throughout the month. The rate of ash and vapour emission was generally weak to moderate. Weak, fluctuating glow at night indicated small ejections of incandescent lava within the crater.

"Seismicity showed a strong increase at mid-month corresponding with the intensified visible explosive activity. Between 14 and 19 Feb, the amplitude of B-type events was about 8 times normal. During the remainder of the month a slight reduction to about 5 times normal was noted. Daily totals of volcanic earthquakes were steady at about 1700 (1-12 Feb), rose to 2100 (13-25 Feb), then returned to 1700. Tiltmeter measurements indicated a steady deflationary change of about 2 μrad.

"The stage-1 volcano alert, declared on 24 Jan in anticipation of increased activity, was maintained in force throughout the month. Warnings were issued to the local population to stay out of the SE and SW valleys."

Information Contact: C. McKee, RVO, Rabaul.

Further Reference: Johnson, R.W., 1984, Volcanological Inspections in Papua New Guinea, February 1984; *Geological Survey of Papua New Guinea Report* 84/8.

3/84 (9:3) "The phase of major Southern crater eruptive activity continued until mid-March, with the same pattern of high Strombolian projections resulting in scoria flows and glowing avalanches in the SE valley. Sub-continuous vertical jets of incandescent fragments (up to 10 per minute) commonly reached 300-500 m above the crater rim. The Strombolian jets appeared to originate without recognizable synchronization from 2 and possibly 3 vents within Southern crater.

"Under the influence of the seasonal NW wind, the fragments, mainly scoria, accumulated on the SE side of the crater on 35° slopes. Approximately 20-30% of the scoria rolled down to the base of the talus fan or, gathering in channels, formed scoria-fed lava flows which progressed at about 100 m per day. The high rate of scoria accumulation prevented cooling and consolidation of the deposits. Their instability resulted in debris flows and glowing avalanches on 6-9 and 11-12 March. The avalanches, occurring in series of 10-30 within periods of 20-90 minutes, generally came to rest at the base of the talus fan, a descent of about 900 m from the summit at 1800 m. The most voluminous avalanches, however, on 8 and 11 March, had enough momentum to travel another kilometer down the SE valley to about 300 m elevation. Pyroclastic avalanches ended on 12 March with the decrease in intensity of the summit Strombolian explosions. During the last 2 weeks of March, they averaged 1 per minute and reached heights of 100-300 m above the crater. Main crater activity remained unchanged, consisting of thick white vapour emission, illuminated at night by weak fluctuating glow.

"Seismicity was high throughout the month with noticeable peaks. The amplitude of B-type events was up to 10 times normal 15-16 March, and 7 times normal 23-26 March. Background harmonic tremor was strong 6-17 March. The daily number of volcanic earthquakes reached 2800 on the 9th before decreasing to 1400 on 16-19 March, and rising to over 2000 after the 20th. Tiltmeter measurements at Tabele Observatory continued to register a steady deflationary change of about 2 μrad per month."

Information Contact: P. de Saint Ours, RVO, Rabaul.

4/84 (9:4) "The intensity of the eruption remained at a moderate-high level throughout April. Moderate-strong Strombolian explosions from 2 vents in Southern crater produced an eruption column about 0.5 km high until 18 April. Incandescent lava ejections took place at relatively steady rates of up to about 4 per minute throughout the month, but an increase in average height of ejections from about 190 m to about 260 m was noted beginning 15 April. This tended to correlate with a change in eruption sound effects from roaring and rumbling in the first half of the month to sharp detonations beginning 15 April. Ashfalls in coastal areas were reportedly light throughout the month.

"Several cycles of waxing and waning seismicity were noted, with an overall peak (about 15 times normal amplitude) at the beginning of the month. Succeeding seismic peaks (18-19 April and early May) were less intense. Daily earthquake totals correlated with trends in seismic amplitudes. Up to 2500 events per day were recorded when amplitudes were high, and about 1300 per day in the intervening lulls. Cyclicity of the seismicity has been noted since Jan, with a period ranging from 2.5 to 3.5 weeks. The highest seismic peak in 1984 was at the start of April.

"The sustained high rate of explosions in April resulted in constant feeding of the cinder apron around Southern crater. These deposits were highly unstable, and frequent avalanches of debris flowed down the upper parts of the SE and SW valleys, usually coming to rest at elevations greater than 1000 m. A breach developed in the S side of Southern crater about 27 March. This appeared to channel flowing lava to the precipitous headwall of the SW valley until early April when the crater rim was reconstructed. The lava flow fragmented completely in its descent into the SW valley.

"Main crater produced pale grey-brown, lightly ash-laden emissions in low-moderate volumes through April. These emissions were usually silent. No crater incandescence was noted.

"Fairly steady deformation of the volcano since late March has resulted in an accumulation of about 2 μrad of radial inflation up to the end of April."

Information Contact: C. McKee, RVO, Rabaul.

Further Reference: Scott, B.J., 1985, Manam Volcano, Papua New Guinea: Eruptive Activity 26 March-17 April 1984; *New Zealand Geological Survey Report* G88.

5/84 (9:5) "Eruptive activity remained at a moderate-high level until 12 May; for the remainder of the month, activity was reduced. Moderate-strong tephra ejections at Southern crater, at rates of up to about 7 per minute, were observed until 12 May. The activity produced an eruption column 1-2 km high. At night, incandescent tephra ejections rose about 500 m. Main crater was also more active in this period; emissions had a higher ash content, and crater glow was visible on 1 May. Eruption sound effects included loud roaring and sharp detonations.

"Beginning 13 May, visible activity was significantly weaker, with Southern crater ejections occurring at rates of 1-2 per minute. Sound effects changed to muffled detonations and weak rumbling. Main crater emissions were usually thick white vapour, or vapour with light ash content. Throughout the month, ejecta from Southern crater was channelled into the SW valley, where the headwall was often obscured by dust clouds.

"Seismicity 1-12 May was the highest for the entire eruption, peaking at about 16 times normal amplitude on the 10th. A peak in daily earthquake totals of 2800 on 7 May also occurred in this period. The amplitude of seismic events during May continued to show a marked correlation with solid-earth tides, with maximum amplitudes being recorded when the daily tidal variations were greatest. The number of earthquakes per day showed a similar but less distinct relationship. No significant tilt changes were recorded in May."

Information Contact: Same as above.

6/84 (9:6) "Moderate Strombolian eruptive activity continued throughout June, with slightly more intense activity at the beginning and end of the month. The plume formed by emissions from both craters rose to a height of about 1-1.5 km above the summit.

"Southern crater ejections took place at a fairly steady rate of 1-2 per minute. Pale grey-brown clouds usually rose about 250 m, although from 3 to 7 June the average height of ejection clouds was 400 m. The ejections were rich in incandescent material. Much of the ejecta was channelled into the SW valley where it cascaded down the precipitous headwall of the valley in debris slides. Main crater activity was weaker, consisting of lightly ash-laden emissions that were released relatively gently. Fluctuating glow was seen on 4 June, indicating lava ejections within the deep funnel-shaped crater.

"Peaks in seismic amplitude (up to 10 times normal) occurred on 2 and 28 June. Daily earthquake totals averaged about 2400, 3-13 June; for the remainder of the month the daily average was about 1750. No significant tilt changes took place in June."

Information Contact: Same as 9:4.

7/84 (9:7) "Moderate Strombolian activity continued in July. Explosive activity at Southern crater showed little change throughout the month, with ejections at a rate of 1-2 per minute. The ejections continued to be rich in incandescent material and formed pale grey-brown ash clouds rising 200-300 m above the crater. Observers at 4-5 km distance noted rumbling and roaring sounds from the Southern crater on most days, but sharp detonations were heard on 24-25 June. Debris avalanches in the SW valley were semi-continuous. Main crater usually released dense white vapours or very pale, lightly ash-laden emissions. Blue vapour emissions were noted on several days at the end of the month, and no Main crater incandescence or eruption sounds were observed.

"Seismic activity was relatively steady compared with previous months, although mild fluctuations occurred. Seismic amplitudes were up to about 10 times normal. Daily totals of earthquakes averaged about 1500. No significant tilt changes were recorded."

Information Contact: Same as 9:4.

8/84 (9:8) "Mild Strombolian activity, which had persisted at Southern crater since mid-May, was interrupted on the evening of 25 Aug by a series of strong Vulcanian explosions. At 1830 a dense column of ash was ejected about 2 km above the summit by the first and strongest explosion. Fallout of incandescent ejecta from the eruption column produced glowing avalanches that descended the SW and SE valleys. Activity remained high for about 1 hour as the explosions continued, then declined rapidly. Seismicity, which had increased dramatically during the strong explosive phase, had returned to 'normal' within a few hours.

"Light ashfalls were experienced in inhabited coastal areas on the W side of the volcano but no damage was caused to property or gardens. However, some people were alarmed by the sudden onset and strong intensity of the eruption. No distinct precursors to the eruption were observed."

Information Contact: Same as 9:4.

9/84 (9:9) "Following the brief intense eruption of 25 Aug, activity was steady at a weak-moderate level in Sept. Both summit craters continued to release ash-laden emissions. Incandescent activity was observed at Southern crater at the beginning and middle of the month, and Main crater incandescence was seen for a few days at mid-month. Seismic amplitudes declined slightly to levels about 3 times normal. Slight radial inflation of the volcano took place early in the month."

Information Contact: Same as 9:4.

11/84 (9:11) "No strong eruptive activity took place in Oct and Nov, but both craters continued to release ash-laden emissions, usually in small-to-moderate volumes. Summit incandescence was seen only once, from Main crater on 7 Oct.

"Daily earthquake totals decreased from about 2300 at the beginning of Oct to 1500 around 19 Oct and have remained steady since then. Seismic amplitudes declined slightly in early Oct, then stabilized at about 2-3 times normal. No significant tilts were recorded."

Information Contact: Same as 9:4.

3/85 (10:3) "Activity has remained stable at a low level since the 25 Aug 1984 eruption. However, seismicity began increasing on 5 March and tremor amplitude reached a peak about 10 times normal on the 20th before decaying sharply to about 2-3 times normal on the 23rd. No definite changes were evident in the low-density ash and vapour cloud emissions from the summit craters, but explosion sounds were heard from 19 March after a period of silence starting in late Feb. The tiltmeters at Tabele Observatory registered a 3 μrad radial inflation in March."

Information Contact: Same as 9:4.

4/85 (10:4) "A period of intensified activity took place about 12-24 April. Stronger explosions occurred at both summit craters, particularly from the 19th to the 22nd. Weak glows from Main crater were observed on the 19th and 22nd, and from Southern crater on the 21st. Seismicity during mid-April reached a peak 5-10 times normal. No tilting was recorded in April following the 3 μrad of inflation in March."

Information Contact: Same as 9:4.

6/85 (10:6) "Following the brief period of intensified activity 12-24 April, activity was mild at both summit craters until late June. Activity during this 2-month interval was characterized by gentle release of white or pale grey emissions with low ash content. No summit incandescence was observed, no tilt changes were recorded, and seismicity was weak.

"Seismicity began to increase in mid-June and seismic amplitudes reached a peak about 4 times non-eruptive levels on the 26th, then appeared to decline at the end of the month. The increase was apparently associated with Southern crater activity; its emissions became more voluminous from the 26th and consisted of dark brown ash. Fine ashfall was reported 4-5 km downwind 2 days later. Activity at Main crater did not appear to be affected and continued to consist of weak to moderate, pale grey, low-density ash and vapour clouds."

Information Contact: Same as 9:4.

7/85 (10:7) "Seismicity remained at somewhat elevated levels (seismic amplitudes about twice non-eruptive levels) during the first half of July as Southern crater continued to emit dark brown ash clouds. At mid-month activity re-intensified with stronger Southern crater ash emissions, resulting in ashfalls in coastal areas. Seismic amplitudes rose to about 4 times non-eruptive levels and daily earthquake totals peaked at about 2500, up from about 1700. A slight decline in activity was evident from about 20 July and seismicity returned to the level of early July. Throughout the month, Main crater released weak to moderate, pale grey, low-density ash and vapour clouds. No significant tilt changes were recorded in July."

Information Contact: Same as 9:4.

8/85 (10:8) "A decline in activity was evident in Aug. Although brown ash clouds were normally emitted from Southern crater, on several days only white vapours were released. Emissions from Main crater were usually white. Seismic amplitudes declined somewhat, but were still slightly above non-eruptive levels at the end of Aug. Daily earthquake totals were usually about 1600, but rose briefly to about 2400 between 24 and 26 Aug. Tilt measurements fluctuated over a 2 μrad range, with no clear trends."

Information Contact: Same as 9:4.

9/85 (10:9) "The decline in activity noted in Aug continued into Sept, with only a few reports of brown ash clouds from Southern crater toward the end of the month. Seismic amplitudes returned to the non-eruptive levels recorded during late 1984 and early 1985. Daily numbers of earthquakes also remained at low levels throughout the month."

Information Contact: J. Mori, RVO, Rabaul.

11/85 (10:11) "A slight increase in activity occurred in Nov with reports of brown ash clouds from Southern and Main craters. Ashfalls were reported along the NW and SW parts of the island. On the 26th, small incandescent ejections were reported from Southern crater and rumbling was heard at the observatory 7-9, 25-26, and 29 Nov.

"Seismic amplitudes remained at non-eruptive levels with a slight increase toward the end of the month. Daily numbers of earthquakes remained low throughout the month."

Information Contact: B. Talai, RVO, Rabaul.

12/85 (10:12) "Activity was at reduced levels. Emissions from Southern and Main craters were occasionally lightly laden with ash. Light ashfalls were noted 17-20 December at the Observatory, 4 km SW of the summit. The number of earthquakes and seismic amplitudes remained at non-eruptive levels."

Information Contact: J. Mori, RVO, Rabaul.

KARKAR

off N coast of New Guinea
4.649°S, 145.964°E, summit elev. 1840 m
Local time = GMT + 10 hrs

This 19 × 25 km island volcano is 100 km ESE of Manam. It has erupted about 9 times since 1643, most often from Bagiai Cone in its 3 km-diameter summit caldera (fig. 5-2).

9/78 (3:9) The following is from R.J.S. Cooke.

"Unusual periodic bands of seismic noise were first recorded at Karkar Observatory on 10 July, and continued virtually unchanged until mid-Aug, when the amplitude began to increase slowly (fig. 5-3). A strong increase in amplitude took place on 28 Aug, but after a time of relatively steady level a slight decline has occurred since mid-Sept. The volcanic tremor at its peak was almost as strong as during the peaks of the 1974 and 1975 eruptions.

"Investigations with a portable seismograph during Aug determined that the seismicity originated near the center of the island. Minor explosive sounds may have been heard on 25 Aug and possible minor ash was observed in the normal vapor column on 30 Aug. An extensive new area of thermal activity was noted on 31 Aug, but it is not clear when this commenced. Karkar residents were first advised on 4 Sept that an eruption was forecast. On 27 Sept, Karkar's activity increased. Red glow was observed and explosion sounds were heard, but no major eruption occurred."

Information Contact: R. Cooke, RVO, Rabaul.

10/78 (3:10) Activity continued to increase, but no major eruption had occurred by 30 Oct. Incandescence, first observed 27 Sept, has been visible each night since then in an area about 20 m in diameter at the SE foot on Bagiai Cone (in the W part of the caldera). Temperatures of 750-850°C in the incandescent area were measured by optical pyrometer. The nature of the incandescent area was uncertain, but it did not appear to be newly extruded lava.

The explosion sounds reported on 27 Sept are of uncertain origin, but volcanic explosions were frequently heard in mid-Oct. At the end of Oct, a voluminous column of white vapor was rising from the flank of Bagiai Cone and from the SE part of the caldera floor, and some blue vapor was issuing from the incandescent area. After 27 Sept, periods of nearly continuous tremor were felt on many occasions at an observation post 2 km from the volcano. Recorded seismic activity reached its highest level at the end of Oct.

Information Contact: Same as above.

11/78 (3:11) Strong seismic activity persisted through Nov.

Strongly marked periodic bands on the seismograph records were again evident between the 1st and 17th. Tremor was felt at the observation post, 2 km from the volcano, throughout the month. Seismicity began to increase on 10 July (3:9-10), peaked 23 and 24 Oct, then remained steady or declined slightly 29 Oct-29 Nov. Incandescence and strong vapor emission also continued through Nov.

Information Contact: Same as 3:9.

12/78 (3:12) Strong volcano-seismic activity persisted in Dec, but with lower intensity than in Nov. After 19 Nov, periodic banding of stronger and weaker seismicity became indistinct and at times almost nonexistent. Incandescence and strong vapor emission continued, and brief sharp explosion sounds were heard on several days.

Information Contact: C. McKee, RVO, Rabaul.

1/79 (4:1) The following is from R.J.S. Cooke.

"An eruption commenced from a new crater late 12 or early 13 Jan. The crater was on the floor of the inner caldera, SE of Bagiai cone, very close to but not precisely at the previously reported incandescent zone. Thick but fairly light-colored ash clouds were erupted continuously for about 10 days from the new crater, which was 100-150 m in diameter. Eruptive activity ceased about 22-23 Jan and a voluminous column of white vapour has been emitted from the new crater since then, although possible very light ash content was reported on the 30th. A temporary observation post (established in early Sept and occupied continuously since then in expectation of the eruption) was in operation throughout the month, but observations have been severely hampered by heavy monsoon weather and the precise times of commencement and cessation are not known.

"The commencement of the eruption was not marked by any unusual events, audible or instrumental. Volcano-seismic activity continued at a moderately high

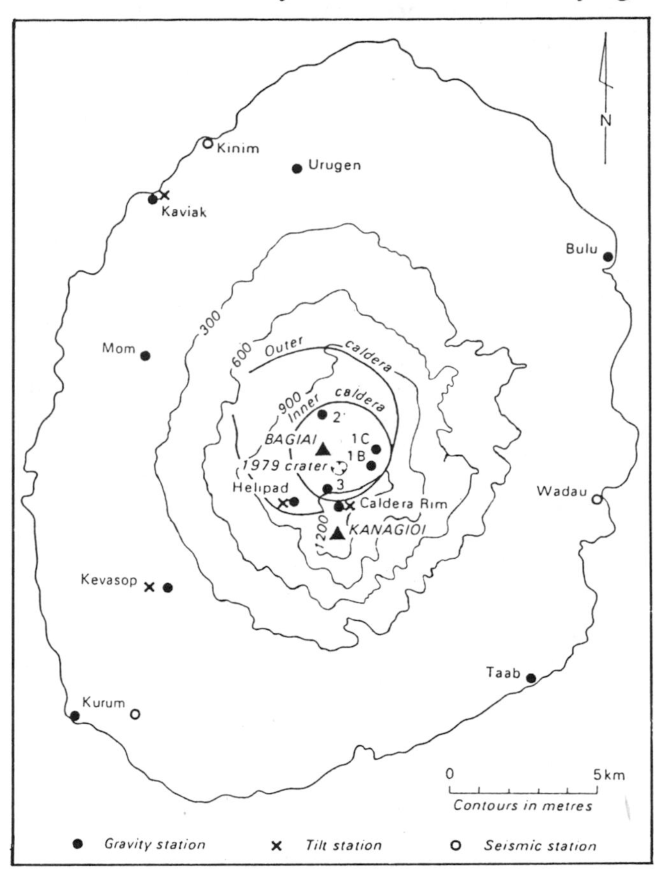

Fig. 5-2. Map of Karkar Island after McKee and others, 1976, showing recording stations. [Originally from 4:3.]

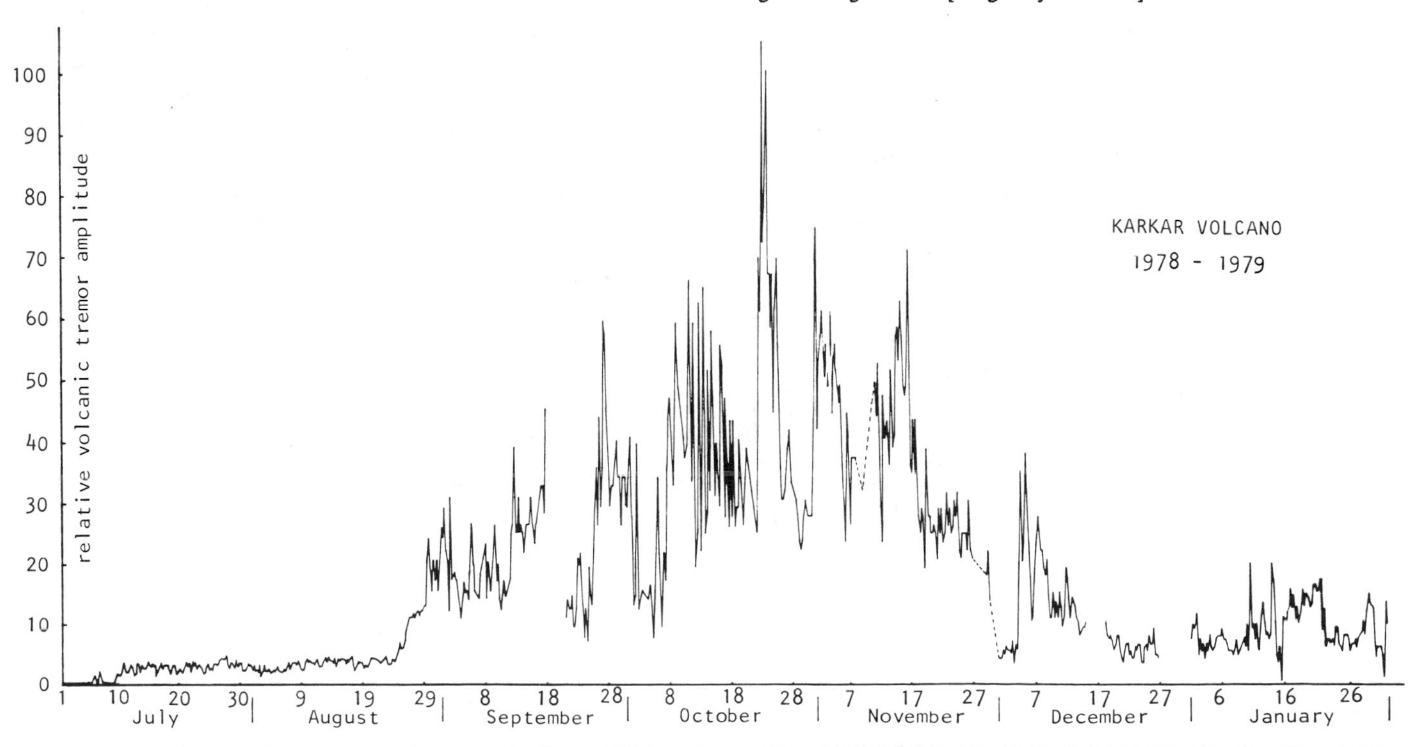

Fig. 5-3. Graph, courtesy of C. O. McKee, showing the relative amplitude of volcanic tremor at Karkar, 1 July 1978-31 Jan 1979. For the most part, the amplitude is from the periodic stronger bands of tremor.

level throughout Jan, although it was considerably weaker than the peak activity of Oct-Nov. The eruption was not marked by any distinctive volcano-seismic features. Periodic stronger and weaker phases of ash emission corresponded with banding on seismic records, similar to that previously reported. At times of stronger ash ejection, flashing arcs were often seen at a rate of several per second, but noises were not prominent.

"Heavy ashfall blanketed the upper levels of forest on Karkar Island, downwind from the volcano, but ashfall was light in inhabited and cultivated areas and there was no significant damage reported. Fouling of streams by ash has affected water supply in some villages."

Information Contact: R. Cooke, RVO, Rabaul.

2/79 (4:2) During Feb, 4 additional phases of heightened eruptive activity took place. Audible explosions ejected ash and rock fragments on 1, 8, 15-16, and 24 Feb, accompanied by felt earthquakes or periods of felt tremor. Frequent flashing arcs, associated with strong ejections of ash and blocks, were observed 7-9 Feb. Between the phases of strong activity, Karkar usually emitted moderately dense white vapor with intermittent light ash content, usually accompanied by weak rumbling and roaring.

Since the eruption began on 12 or 13 Jan, seismic activity has been generally steady, at levels much lower than during the peak of pre-eruptive seismicity in Oct 1978. Banding of periodic stronger and weaker tremor had been sporadically evident. Discrete (B-type) seismic events began to be recorded in greater numbers during the second half of Feb.

Information Contact: C. McKee, RVO, Rabaul.

3/79 (4:3)

We are saddened to report that R.J.S. Cooke, 40, and Elias Ravian, 34, were killed on 8 March by a directed blast of debris from Karkar volcano. Robin Cooke came to Rabaul Volcanological Observatory in 1971 and was named Senior Volcanologist 2 years later. His contributions to volcanology were many, particularly in seismic monitoring and in generously sharing his reports of local volcanism with scientists of the world. Elias Ravian had been a highly respected worker at the Observatory for 9 years. Both men devoted much of themselves to better understanding of the volcanism that took their lives.

The following is from a report by Richard Almond. Seismic activity at the caldera rim camp, about 1.2 km from the vent, decreased slightly 1-2 March, then remained at a relatively low level through the 4th. Dense billows of white vapor from the vent were accompanied by occasional mild ash ejections on 1 March, but there was little or no ash emission 2-4 March. Seismicity increased gradually 5-7 March, but only one weak ash ejection, on the 5th, accompanied the continuing dense vapor emission.

At 0125 on 8 March, most of Karkar Island's 23,000 residents were awakened by the sound of a strong eruption. Continuous ejections of dense ash blanketed the W side of the island, where there were heavy ash and scoria falls for 45-60 minutes. Ash and blocks were scattered over the caldera floor. Accompanying the vertical eruption was a blast of hot gas and ejecta directed towards the SE. The blast devastated the entire SE quadrant of the caldera rim, killing Cooke and Ravian, who were camped near the S rim. Ash ejection, as observed from the coast, continued at a reduced level throughout 8 and 9 March. Seismic activity increased slightly during the major eruption, then decreased markedly.

Poor weather prevented observations on 10 and 11 March. From 12 to 16 March, aerial inspections revealed continuous emission of light brown ash from the NW portion of the vent, which had been enlarged by the 8 March eruption from 75 m to 300 m in diameter, and was about 150 m deep. Density of ash emission increased towards the end of the period,

On the 17th, emissions had changed to white vapor containing very little ash. Background seismicity was relatively low, but there were numerous discrete events, probably B-type. These had ceased by 0200 on the 18th and there was little or no vapor or ash. The source of previous emissions was then clearly visible; a cavern at the bottom of the crater extending in a NW direction beneath Bagiai Cone. There was a strong H_2S smell when flying downwind of the volcano, even several kilometers from the vent. On 19 and 20 March, helicopter inspection revealed strong jets of vapor being emitted from the cavern with an extremely loud, continuous roaring noise. The angle of ejection was about 30° from the horizontal and toward the SE. These conditions continued until the morning of the 23rd, when a grey ash column was observed rising vertically from the vent. The ash cloud streamed to about 15 km SW of the island. Moderate ash emission continued through the end of March, accompanied by a moderately high level of harmonic tremor and periods of occasional B-type events.

McKee and others (1976) note that Bagiai "seems to have been the main source of recent activity, and was the source of the 1974 and 1975 eruptions".

Reference: McKee, C.O., Cooke, R.J.S., and Wallace, D.A., 1976, The 1974-75 Eruptions of Karkar Volcano, Papua New Guinea *in* Johnson, R.W. (ed.) *Volcanism In Australasia*, Elsevier, 1976, p. 173-190.

Information Contacts: R. Almond, C. McKee, RVO, Rabaul.

4/79 (4:4) The following is a report from C.O. McKee and W.G. Melson.

"Activity fluctuated during April, but was generally weaker than in March. Strong grey ash emission occurred more or less continuously 7-18 April. Tightly convoluted ash clouds rose to several hundred meters above the vent. The ash plume reached as much as 30 km downwind (mainly eastward) on some days. Ashfalls early in the month were heavy enough at times to prevent work on plantations and village gardens on the E side of the island. The ash changed in dominant color from grey and brownish-grey to black around 18 April. [Patches of incandescent rocks] were observed inside the crater on 10 April and intermittently since that time.

"A churning, steaming, brown-colored lake was first observed 23 April in the bottom of the new explosion crater, which opened on 12 or 13 Jan to about 75 m in diameter and greatly enlarged, to about 300-400 m in diameter, in the fatal explosions of 8 March. Eruptions occurred through this lake, producing black ash ejections reminiscent of shallow marine volcanic explosions. Most had insufficient energy to leave the estimated 150-200 m-deep crater except on rare occasions, when a few blocks fell near but outside the crater. Lake level appeared to fluctuate greatly. Although clearly visible on 23 April, the lake was not visible the next day, following a period of heavy black ash emission, probably during the night, that deposited about a centimeter of ash on the SW side of the crater. Sounds at the crater edge during subaqueous explosions were like those of some geysers, and were accompanied by a tinkling noise probably produced by fallback. Aerial inspection on 28 April revealed very large black ash emissions to 200 m above the crater rim, with some blocks falling outside the crater. Frequency of explosions was nearly continuous, every second or less.

"Seismic activity consisted of irregular distinct periods of strong tremor 1-14 April. From 15 April, seismic activity consisted of steady low-amplitude tremor, with a few distinct B-type shocks on most days. Gravity stations were established on 13 April [in an effort to detect any possible subsurface mass changes or elevation changes]. Twelve remeasurements were made during the month. Three stations are on the caldera floor, 2 on the rim, 3 on the mid-flanks, and 4 along the coast. The measurements indicated that no major [subsurface mass changes], rapid uplift, or subsidence of the caldera floor was occurring. The sensitivity of these measurements is estimated at ± 10 cm [for elevation changes]. Between 18 and 22 April, 4 dry tilt stations were installed, 1 on the caldera rim, 1 on the coast, and 2 at intermediate points. As of the end of April, relevelling of these had not revealed significant changes in tilt."

Information Contacts: C. McKee, RVO, Rabaul; W. Melson, SI, Washington DC.

5/79 (4:5) The following reports are from C.O. McKee.

"The eruption continued through May, strengthening considerably at the end of the month. At 1900 on the 30th, following reports of heavy ash eruption during the day, Civil Defence and government authorities were informed that activity had intensified by the declaration of an alert. No other action is to be taken at this stage.

"A trend of increasing activity began on 6 May, and from the 16th, significantly stronger activity persisted. White vapours crowned the summit until 22 May, forming a column about 1 km high. Grey emissions began to color the base of this column on the 7th. Much stronger activity occurred on the 16th and 17th when heavy ashfalls took place in the summit region and light ash deposits (less than 1 mm) were registered at the NW coast. At this time, convoluted grey ash clouds were rising from the crater.

"During the evening of 19 May a period of about 30 minutes of strong volcanic tremor was recorded. It was discovered during an aerial inspection of the crater on the 20th that 5 new vents had opened. Mixed activity was observed, ranging from steady, convoluted, pale grey ash and vapor emission, to continuous black ash jets, to frequent Surtseyan ash and vapor projections from vents in a shallow lake. Activity continued in a similar fashion over the next few days with variable numbers of vents active.

"On the 24th, a new higher level of activity commenced. Ash emissions were markedly more voluminous, resulting at times in the formation of an ash haze in the caldera. A plume of ash driven by the wind stretched out about 10-15 km from the summit. Ashfalls were heavy in the summit region but only light at the W and NW coasts.

"Similar activity persisted through the remainder of the month, but on 30 May much denser ash emission took place, resulting in moderate to heavy ashfalls on the lower W flanks. An ash plume 50-60 km long formed W of Karkar. At one village 2 km from the W coast an ash deposit 1 cm thick was measured on the morning of the 31st. Severe damage to vegetation was reported from W coast areas.

"During the last week of May the stronger ash eruptions were accompanied by a higher level of seismic activity, and explosive sounds were heard at villages on the lower W flanks. However, levels of volcanic tremor remained relatively low compared with pre-eruptive seismic activity in 1978. Discrete volcanic earthquakes were registered on most days in May.

"The crater lake, which had been observed intermittently since 23 April, has not been seen since 22 May. Rapid infilling of the crater with ejecta began on 24 May. On 31 May part of Bagiai Cone collapsed into the crater, reducing its depth by about 60 m to an estimated 50-80 m.

"A program of gravity measurements which commenced on 13 April has shown a trend of steadily increasing gravity values at the 3 stations on the caldera floor. The gravity change averaged 0.03 mgal. None of the other stations, situated at coastal, mid-flank, and summit locations, showed any trends. This suggests that the causative agent is spatially restricted to the environs of the inner caldera. Levelling measurements at 4 dry tilt stations spread from the coast to the inner caldera rim showed no evidence of flank deformation. It is uncertain at present whether gravity effects are caused by elevation changes of the caldera floor relative to the rest of the volcano, or to a change in the density structure of the volcano.

"Copious volumes of muddy water have been observed flowing off the slopes of Bagiai on several days since 13 May, and mudflows on parts of the caldera floor have blanketed underlying lava flows. When smaller, similar streams of water were running off Bagiai in April, it was considered that this was condensation from the emission cloud. However, the large volume of liquid observed on several days in late May suggests that the source in this case may be springs."

Information Contact: C. McKee, RVO, Rabaul.

6/79 (4:6) "Through most of June, a moderate-sized white vapor plume stood over the summit. This plume was usually 1-2 km long and rose 0.5 km above the crater. Surtseyan-like explosions of dark ejecta and vapors were seen from several vents in the crater floor on most days. These explosions were mainly contained within the 1979 crater, which now has horizontal dimensions of 450 × 300 m and is 300 m deep. The crater was significantly enlarged by the strong explosions at the end of May and beginning of June. Other periods of moderate to strong activity occurred on 8, 17, and 23 June. At these times continuous grey ash emissions formed an ash plume stretching for several kilometers, and heavy ashfall occurred on the W part of the summit region. Explosive sounds from the summit were heard on the W flank and coast during the stronger activity. Flashing arcs were seen on 16, 18, and 22 June. Horizontally moving ground-hugging clouds were seen advancing slowly from the crater rim in late May and early June, and on 23 June. Thin lobe-shaped deposits of ash about 100-200 m long were formed by the ash clouds on the W rim of the crater on 23 June. A pond of grey water was briefly visible in the bottom of the crater after the strong explosions. Sulfur deposits were present for several days after 20 June on the N and S walls of the 1979 crater, on the summit of Bagiai Cone, and on a small part of the W floor of the caldera, but these were obliterated by the strong ash eruptions of 23 June.

"A significant intensification of seismic activity commenced in mid- or late May and reached the highest sustained level for the year at mid-June."

Information Contact: Same as 4:5.

7/79 (4:7) "For most of July the eruption remained at low intensity. Characteristically, a white vapour plume several kilometers long trailed away from the summit. A sulphurous odour was detected downwind from the new crater during early and late July. A haze of pale blue vapour was formed in the W part of the summit caldera on several days late in July.

"Ponds of grey water occupied the base of the new crater, and dark, Surtseyan-like explosions of solid ejecta, vapour, and water occurred at high frequency. Numerous different explosion sites were observed, although at some only mild intermittent bubbling was taking place.

"Stronger explosive activity occurred on 6, 9, and 15-16 July, when dark clouds of ash and rocks were ejected above the crater rim. Fields of impact craters were formed on the caldera floor to distances of several hundred meters from the rim of the crater. Most of the ash in these ejections fell in the W summit region, but weak ashfalls at the coast on 15 and 16 July were reported. An association between these explosive phases and prior heavy rainfalls is evident.

"The intensity of volcano-seismic activity declined slightly from the higher June level. Discrete volcanic earthquakes were recorded on many days, and bands of strong tremor were evident on a few days.

"Gravity and tilt measurements are continuing. Several caldera floor gravity stations have been irretrievably lost because of ashfalls and mudflows. A trend towards increasing gravity values on the caldera floor seems to have ceased, and 2 stations on the caldera rim are now showing values lower by a few hundredths of a mgal. Of the 4 tilt stations, only the 2 caldera rim ones are showing definite trends. These stations, on the S part of the summit, are showing downwards tilting in an approximately NW direction."

Information Contact: Same as 4:5.

8/79 (4:8) "During the first 9 days of Aug, moderate explosive activity occurred. The heavier solid fragments in dark ejections were contained mainly within the crater, but on the 1st blocks rose about 200 m above the crater rim, and some landed on the crater rim. Several vents were active but the most productive were those at the centre and the W extremity of the crater floor. Types of explosive activity, which varied from one vent to another, included frequent dense black convoluting ejections, grey ash-laden vapour emission, streaky black and white jets, and weaker grey-brown ash and vapour emission. Eruption sounds were not loud and ranged from low roaring and rumbling to banging. The emission cloud was usually grey, and sometimes brown, between the 1st and 9th. Occasionally it rose to 500-800 m above the crater rim. Ashfalls in this period were largely restricted to the summit and were heaviest on the 1st.

"Grey water was seen in a pond in the W extremity of the crater floor on the 4th, and from the 10th a lake covered most of the crater floor. Profuse steaming at the surface of the lake partly obscured the explosive activity. Ejections through the lake occasionally resembled Surtseyan blasts, but commonly they only rose several tens of meters above the lake surface and had the appearance of water fountains.

"After the 9th the emission cloud consisted of white vapour in small to moderate volume. Sulphurous odours were noticed around and downwind from the crater for most of the month.

"Direct measurements of the crater on the 24th showed it to be 700 m long and 440 m wide, elongated in the E-W direction. No depth measurement was made but depth is estimated to be 300-400 m."

Information Contact: Same as above.

Further Reference: McKee, C.O., Wallace, D.A., Almond, R.A., and Talai, B., 1981, Fatal Hydro-eruption of Karkar Volcano in 1979: Development of a Maar-like Crater; *in* Johnson, R.W. (ed.), *Cooke-Ravian Volume of Volcanological Papers*; Geological Survey of Papua New Guinea Memoir 10, p. 63-84.

9/79 (4:9) The following is from Richard Almond.

"The relatively low level of activity at the end of Aug continued until the end of Sept. Daily observations by helicopter took place until 12 Sept. The activity continued to consist of small, black, Surtseyan-like bursts breaking through the surface of the small crater lake. Seismic activity continued at a low level, although slight increases in activity appeared as faint bands in the records, a characteristic of this eruption. Several discrete A-type events occurred. Tilt and gravity observations until 12 Sept showed no significant changes.

"Continuous surveillance using a helicopter will no longer take place unless activity increases again. Instead, a helicopter will be chartered once every 2 weeks to enable detailed visual observations, and tilt and gravity measurements to be made. Seismic monitoring via the permanent surveillance station and 2 temporary stations will continue as before."

Information Contact: R. Almond, RVO, Rabaul.

10/79 (4:10) The following reports are from C.O. McKee.

"Observations from the Kinim Observatory at the NW coast of Karkar, 11 km from the 1979 crater (fig. 5-2), indicated that no strong explosive activity occurred during Oct. Usually a small to moderate white vapour cloud crowned the summit.

"Inspection trips were made 12-15 and 26-29 Oct. Activity in the 1979 crater continued to consist of small, fountain-like ejections of dark material through the crater lake. The sources of ejections define an approximately E-W zone across the centre of the crater floor. Most ejections rose to a maximum of 30 m above the lake surface. Vent diameters are estimated to be 10-20 m. The entire surface of the lake 'steamed' profusely during both inspection visits, and a small to moderate white vapour cloud was produced. Occasionally the vapour cloud rose to 600-800 m above the crater rim. H_2S and SO_2 gases were detected.

"White and pale yellow sublimates have been deposited strongly on the N crater wall, and areas of mainly white deposits were scattered on the S and W walls. A pink tint was present in some deposits on the N wall. Temperatures in a fumarolic area on the W part of the caldera floor have dropped to 75-83°C, after remaining at 85-96°C between April and Aug.

"Seismic activity consisted of discrete volcanic events and 1 or 2 periods of tremor per hour. The tremor bursts usually lasted 3-4 minutes. Tilt and gravity observations showed no significant changes."

Information Contact: C. McKee, RVO, Rabaul.

11/79 (4:11) "As observed from the Kinim Observatory, no strong explosive activity occurred during Nov. During inspections of the summit and the crater lake floor on 17 and 18 Nov, it was found that fountaining of dark muddy water has virtually ceased and that several islands have been formed. The main island was [at the end of a peninsula] joined to the base of the S wall of the crater, and the other island was near the E edge of the lake. Several large boulders were present on the main island and it was bordered by narrow, smooth shores. Orange-brown discolouration was present over the central part of the island. At its N edge a vigorous geyser continuously jetted yellow-tinged water to heights of several tens of meters. The E half of the lake was green and the W half retained the same grey-brown colour seen previously."

Information Contact: Same as above.

12/79 (4:12) Minor activity continued through Dec. No ash explosions have been observed since early Aug.

Information Contact: R. Almond, RVO, Rabaul.

1/80 (5:1) The following is from B.J. Scott and C.O. McKee.

"Observations from the Kinim Observatory at the NW coast of Karkar indicated that no strong explosive activity occurred during Jan. Unconfirmed reports were received of weak ash emissions on 7 and 17 Jan.

"During a summit inspection 28 Jan, the lake was of a uniform light grey/brown colour, and there was no sign of the peninsula reported in Nov and Dec. Convection was occurring in the lake. Fumarolic activity on Bagiai Cone and on the SE part of the caldera floor was extensive but fumaroles were not strong and vapour was emitted under low pressure.

"Gravity observations on 27 and 28 Jan indicated increased values at stations on the flanks and summit, a reversal of previous trends. Tilt observations for the same period are incomplete. The new gravity trend may be interpreted as deflation, and an indication that the eruption is over. However, a further round of measurements will be required to confirm this."

Information Contacts: B. Scott, C. McKee, RVO, Rabaul.

4/80 (5:4) The following reports are from C.O. McKee.

"Observations from Kinim Observatory indicated only weak white vapour emission during March. Inspections of crater activity on 11, 12, and 24 March indicated that phreatic explosions were continuing from several sources in the lake occupying the 1979 crater. These events, occurring at intervals of a few minutes on 11 March, involved updoming of the lake surface and projection of spear-headed columns of dark crater-floor mud to heights of about 20 m. The lake, occupying the W half of the crater floor, was a uniform pale brown colour, and steam emission from the lake surface was profuse. Fumarolic activity from the walls of the crater continued unchanged, but fumaroles on Bagiai Cone and the caldera floor were weaker. The E crater margin had migrated 3-5 m by collapse of the crater wall, leaving a near-vertical escarpment.

"No changes were observed in summit emissions in April. An aerial inspection on 26 April revealed that only a small pond remained in the base of the 1979 crater, on its W side. About 6 closely spaced vents on the crater floor were releasing white emissions under low to moderate pressure. The floor of the crater was flat and smooth. From the air, white sublimates on the E wall of the crater appeared quite conspicuous. Fumaroles on the caldera floor continued to show weak activity, but those on Bagiai may have strengthened since late March.

"Gravity measurements in March suggested that a deflationary trend indicated in Jan had eased. Seismic activity continued to show discrete events and traces of tremor recorded as irregular bands through April."

Information Contact: C. McKee, RVO, Rabaul.

5/80 (5:5) "Summit observations 10 May revealed that a small lake of brown water occupied the NW part of the 1979 crater. Silting had formed a flat crater floor some 300 m square. Near the centre of the crater floor was a circular pool of brown water about 25 m in diameter,

Continuous upheavals of vapour, water, and dark crater-floor material were observed rising only slightly above the level of the water surface. A line of fumaroles stretched W from this pool. Extensive weak fumarolic activity had deposited white and yellow sublimates over the lower ²/₃ of the E crater wall, and diffuse vapour emission on the W side of the crater was slightly stronger. Vapour emission continued from sources on Bagiai Cone. Weak fumaroles extended in a line from the crater to the S caldera wall, and several others on the SW caldera floor were also observed.

"Gravity observations between 9 and 11 May were similar to previous sets in 1980, and suggested a cessation of inflation of the upper part of the volcano. Dry tilt observations seemed to agree with this trend. Seismic activity remained unchanged."

Information Contact: Same as 5:4.

6/80 (5:6) "An aerial inspection on 28 June indicated that conditions in the 1979 crater were similar to those during the previous inspection on 10 May. Fumarolic activity on Bagiai Cone may have strengthened since the May inspection, while fumarolic activity on the caldera floor appeared to have weakened over the same time interval. Seismic activity remained unchanged."

Information Contact: Same as 5:4.

7/80 (5:7) "Possible ash emissions were reported on 13, 14, and 15 July. However, an aerial inspection on 29 July did not detect any morphological changes to the crater that might have accompanied the reported ash emissions. Fumarolic activity inside the 1979 crater may have increased. Fumaroles on the E part of the caldera floor may have weakened while those on the W side may have intensified. White sublimates were conspicuous on Bagiai Cone. The seismic recording level was unchanged, and the seismic activity continued to consist of small, probably B-type, shocks, and periods of tremor."

Information Contact: Same as 5:4.

8/80 (5:8) "White vapour continued to be emitted from summit sources. An aerial inspection on 12 Aug, and aerial and ground inspections on the 14th and 15th, revealed copious white vapour emissions from the E side of Bagiai Cone and from sources on the E part of the caldera floor. These sources are believed to be stronger than they were at the time of the previous inspections (May 1980). Conditions inside the 1979 crater were similar to those in May. A descent was made into the crater and its depth was measured at 150 m. Silting of the crater floor has resulted in a flat floor consisting of coarse sandy sediment and some small boulders. Ejections of dark, sediment-rich water to heights of about 0.4 m were taking place over the entire surface of a pool of water at the centre of the crater floor. Vigorous gas ebullition was occurring from the muddy area immediately surrounding the pool. Numerous fumaroles on the E wall of the crater released vapour under low pressure. No unusual seismic activity was recorded in Aug and the intensity of the seismicity was unchanged."

Information Contact: Same as 5:4.

9/80 (5:9) "Local observers made 4 inspections of the summit area 8-13 Sept. They observed small ejections of crater floor material and water from scattered sources in the previously inactive W pool of water at the base of 1979 crater. Small dark ejections continued from the central pool, but a new gas vent was present on the crater floor nearby. The water in the pools was reported to be yellow on 13 Sept. Vapours rising from 1979 crater were white, and small to moderate in volume. Seismic activity intensified during Sept. Single volcanic earthquakes became more numerous early in the month and the tremor level showed a general strengthening at mid-month."

Information Contact: Same as 5:4.

1/81 (6:1) The following is from B. Talai and P. Lowenstein.

"A transient increase in hydrothermal and fumarolic activity for 2-3 days at the beginning of Dec coincided with the onset of seasonal heavy rains. Minor geysers were observed on the floor of 1979 crater. There were voluminous emissions of white vapour from a landslide on the Bagiai side of 1979 crater floor. Fumarolic activity was strong on the W side of Bagiai Cone and on the E side of the caldera floor right up to the caldera wall. Weak to moderate vapour emissions at these localities continued for the rest of the month."

Information Contacts: B. Talai, P. Lowenstein, RVO, Rabaul.

3/81 (6:3) The following 2 reports are from C.O. McKee.

"Aerial and ground inspections were made 6-8 March and other aerial inspections were carried out on 19 and 26 March. Conditions in the caldera appeared similar to those during inspections in Nov and Dec 1980. Hydrothermal activity was continuing at the base of the 1979 crater and maximum measured temperatures were 97.5°C. The other main source was on the W part of Bagiai Cone. During the 26 March aerial inspection the volume of emission was reportedly greater than previously observed.

"Gravity measurements and levelling were carried out 6-8 March. The gravity measurements were consistent with previous sets in 1980, and might indicate summit deflation. Levelling up to the end of 1980 showed possible small deflationary trends of several μrad at the 3 mid- and upper-flank tilt arrays. However the changes were very small and similar in size to the limits of error in making the measurements."

Information Contact: C. McKee, RVO, Rabaul.

7/81 (6:7) "Aerial and ground inspections by volcanologists 18-20 July confirmed local reports that significant development of fumaroles on the W flank of Bagiai Cone has taken place. H_2S and SO_2 gases were identified in the fumarolic emissions. New hot water springs have been created near Bagiai's summit and on its W flanks. Temperatures in the fumaroles and springs were about 97°C.

"Hydrothermal activity in 1979 Crater has declined. No active mud or water pools were seen in the crater, and only tenuous vapour emission was taking place from sources on the crater floor and walls. Temperatures in fumaroles were about 97°C. Gravity and levelling measurements showed no significant changes. Seismic activity remained steady."

Information Contact: Same as 6:3.

LONG ISLAND

off N coast of New Guinea
5.36°S, 147.12°E, summit elev. 1304 m

Long Island is between Karkar and New Britain. A major eruption around 1660, as large as Krakatau's 1883 event, is documented largely by oral legends and its extensive ash deposits in the New Guinea Highlands.

Motmot Island is a small cone within Long Island's 11 km-diameter caldera lake. A small new crater was formed on Motmot between aerial observations on 5 Sept 1975 and 29 April 1976 (*BVE*, no. 15, p. 19).

General Reference: Blong, R.J. 1982, *The Time of Darkness*, Australian National University Press, Canberra, 257 pp.

1/79 (4:1) The following is a report from R.J.S. Cooke.

"A report was received on 26 Jan of a supposed eruption of Motmot cone about mid-Jan. A flypast by a volcanologist on 15 Jan had revealed nothing unusual. However, aerial inspections on 29 Jan indicated considerable new vapor emission, including 1 powerful fumarole that was jetting white vapor and possibly water at the W shoreline of Motmot Island (350 m in diameter). Vapor from this jet obscured details nearby, but no clear evidence of recent eruptive activity was noted. No new craters or cones were present, and no new ejecta could be recognised on the island. Pending further information, it is assumed that only vapor activity has occurred so far."

Information Contact: R. Cooke, RVO, Rabaul.

2/79 (4:2) The following is from C.O. McKee.

"An aerial inspection of Long Island on 23 Feb confirmed that no eruption had taken place in Jan. Only small volumes of vapour were seen emanating from the summit craters and from a fumarole at the W shoreline. This fumarole is in a different position from the powerful one seen on 29 Jan."

Information Contact: C. McKee, RVO, Rabaul.

UMBOI

W of New Britain Island, Papua New Guinea
5.59°S, 147.88°E, summit elev. 1655 m

No historic eruptions are known from this volcano.

9/85 (10:9) The following is a report from J. Mori.

"Following damaging earthquakes and changes at a thermal area on Umboi Island during Aug, local people were concerned about possible volcanic activity. A brief visit was made to Umboi by R. W. Johnson (Bureau of Mineral Resources, Canberra) on 5 Sept. He reported no signs of imminent volcanic threat but recommended that an officer from RVO visit Umboi for further investigations.

"On 12 Sept, inspection of the thermal area on the W flank of Talo volcano (on the W side of Umboi Island) indicated that no increase in temperature had occurred. However, local subsidence had taken place. This was probably due to strong shaking of the ground during the strongest earthquake on 19 Aug.

"Nine local earthquakes were recorded 11-14 Sept, 2 reportedly felt. These earthquakes are probably local to Umboi Island, but it is uncertain whether they are directly related to the volcano. Seismic records indicate that earthquakes were continuing at a low level in mid-Sept."

Information Contact: J. Mori, RVO, Rabaul.

SUBMARINE VOLCANO (unnamed)

off the W coast of New Britain Island
5.2°S, 148.57°E
Local time = GMT + 10 hrs

Although submarine activity has been reported between Karkar and Long Island, no previous historic activity is known at this site 40 km NNE of Langila.

7/83 (8:7) The following report is from P. de Saint Ours and C.O. McKee.

"An unnamed seamount, 30 km NNE of Cape Gloucester, W New Britain, may have been the site of a short-lived eruption on 15-16 June. A subcontinuous swarm of long-period earthquakes was registered by several seismic stations in Papua New Guinea at 1913-2001 on 15 June and 0427-0450 on 16 June. The swarm was recognized when the records were analyzed at RVO in early July. Preliminary determinations indicated shallow origins over a broad area at the W extremity of New Britain.

"Inquiries with the local people resulted in accounts of sounds like a jet plane coming from the sea, and glow in the sea a long distance from the coast. Northeastward migration of the incandescence was also reported, possibly suggesting a fissure eruption. Airborne observations on 28 July failed to find water discolouration or any other evidence of the 6-week-old event.

"Until further information is obtained, the most likely source for these phenomena is a large seamount mapped in the general area of earthquake locations and visible reports."

Information Contacts: P. de Saint Ours, C. McKee, RVO, Rabaul.

LANGILA

New Britain Island, Papua New Guinea
5.53°S, 148.42°E, summit elev. 1189 m
Local time = GMT + 10 hrs

This volcano, on the W tip of New Britain, has been

described in 68 SEAN reports over the decade, a number exceeded only by St. Helens (69) and Sakura-jima (84). Langila marks the E end of the Western Bismarck Arc, 470 km ESE of Kadovar.

5/78 (3:5) The following two reports are from R.J.S. Cooke.

"Last year's mild Strombolian eruption at Crater 2 [NE crater] (figs. 5-4 and 5-5) ended in mid-Dec, after a peak in early Sept during which a small amount of lava rose into the crater without overflowing. A new inner crater was blown through this lava a few weeks later and seems to have become considerably enlarged between aerial inspections on 21 March and 18 May. An unusually thick column of white vapour has been issuing from this crater during the last few weeks and a number of explosions were reported during May. A new mild Strombolian eruption [but see 3:9] from Crater 2 began 27-28 May and is continuing."

Information Contact: R. Cooke, RVO, Rabaul.

Further References: Johnson, R.W., Davies, R.A., and Palfreyman, W.D., 1971, Cape Gloucester Area, New Britain: Volcanic Geology, Petrology, and Eruptive History of Langila Craters through 1970; *Australia Bureau of Mineral Resources, Geology, and Geophysics Record* no. 1971/14, 34 pp.

Palfreyman, W.D., Wallace, D.A., and Cooke, R.J.S., 1981, Langila Volcano: Summary of Reported Eruptive History, and Eruption Periodicity from 1961 to 1972 *in* Johnson, R.W. (ed.), *Cooke-Ravian Volume of Volcanological Papers;* Geological Survey of Papua New Guinea Memoir 10, p. 125-133.

9/78 (3:9) "An eruptive phase which began in late April and intensified markedly in late May has continued to the present (25 Sept). However, a volcanological party observed the activity from 30 May-9 June and found that it was not Strombolian as suggested by early reports, but Vulcanian in character. Typically, 1 or 2 larger explosions occurred per day, with a number of lesser ones. A lull was noted during mid-July and from late Aug. Since the first week of Sept, there has been little activity."

Information Contact: Same as above.

10/78 (3:10) Only 6 explosions were reported during Oct, (7, 9, 11, 19, 25, and 29 Oct) but some were much larger than usual.

Information Contact: Same as above.

12/78 (3:12) The character of activity changed in late Oct, after 6 strong explosions had occurred earlier in the month. Explosions became more frequent, but only a little ash was present in the eruption clouds. Ash contents increased on 27 Nov and remained high through Dec while frequent explosions persisted. Since 27 Nov, a gray ash plume has been continuously visible over the volcano.

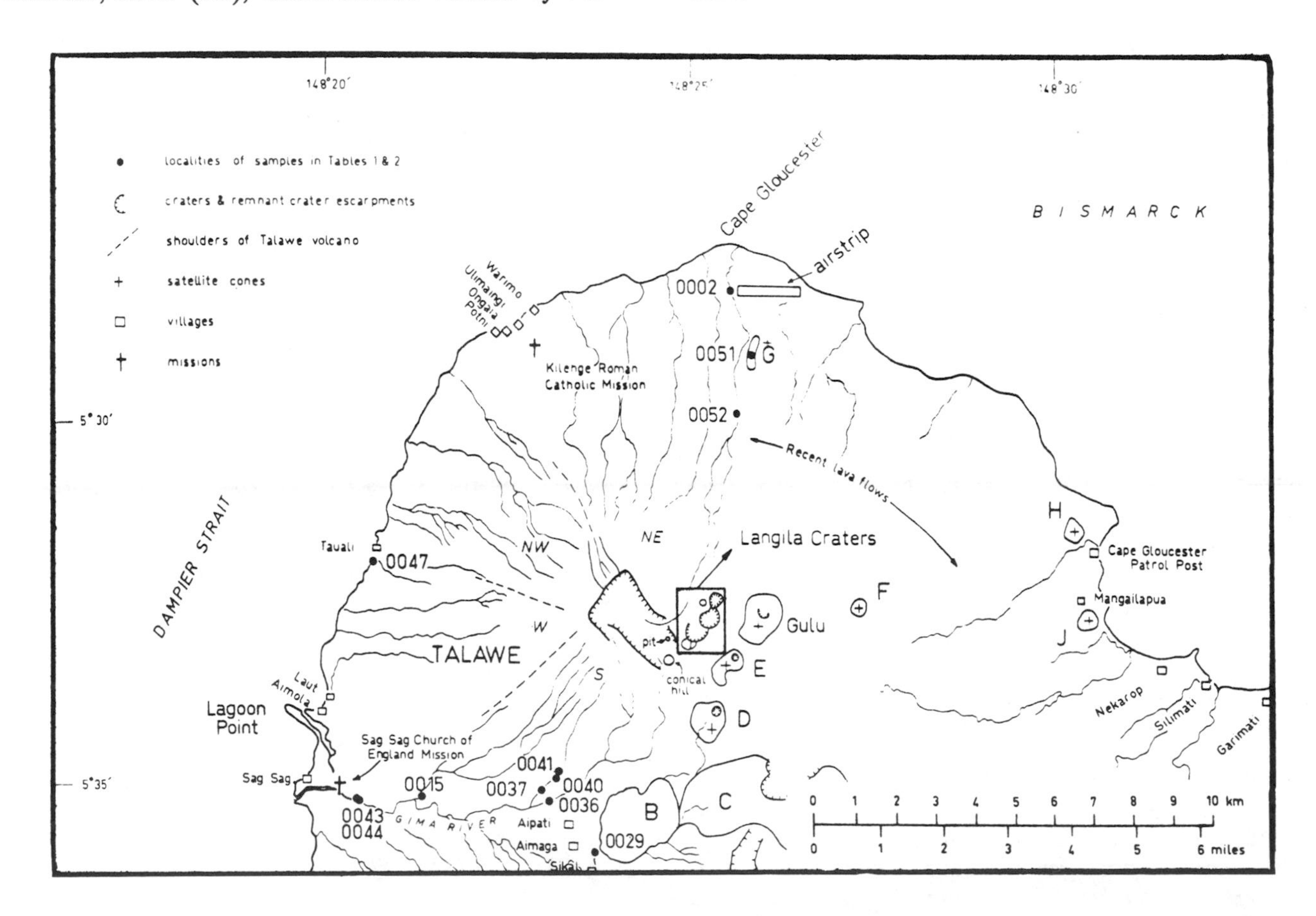

Fig. 5-4. Sketch map of the NW tip of New Britain Island, showing Langila volcano and surrounding features. The portion of Langila's summit outlined by a box is shown in detail in fig. 5-5. After Johnson and others, 1971.

Aerial inspections on 7 and 15 Dec revealed an active lava dome in Crater 2 (Langila's NE crater). Incandescence was visible in the dome's surface fissures during these (daytime) overflights. Glow could be seen at night from 10 km away on 13 and 21 Dec.

Information Contact: C. McKee, RVO, Rabaul.

1/79 (4:1) Frequent ash ejection continued through Jan. Aerial inspections on 4, 15, and 29 Jan showed that the lava dome discovered in Dec has risen to the rim of Crater 2 without overflowing. By the end of Jan, a new craterlet had formed in the fresh lava surface.

Information Contact: R. Cooke, RVO, Rabaul.

9/79 (4:9) Quoted reports below are from RVO staff.

"For most of Sept the activity consisted of weak vapour emission accompanied by occasional small explosions. Only a very little ash was emitted. On 27 Sept a series of 10 loud explosions was accompanied by the emission of a dense ash cloud. A heavy ashfall occurred at Kilenge mission, 10 km NW of the volcano. No further details are yet available."

Information Contact: R. Almond, RVO, Rabaul.

10/79 (4:10) "No further strong explosions have been observed since those of 27 Sept. The ash deposit from the 27 Sept explosions was several millimeters thick 10 km from the source. Grey ash clouds were seen on a few days at the beginning and end of Oct, and white vapour emissions were observed occasionally. Seismic activity remained at a low level."

Information Contact: C. McKee, RVO, Rabaul.

11/79 (4:11) No ash emission was observed during the first half of Nov. Gray clouds were occasionally ejected from the 16th through the end of the month.

Information Contact: Same as above.

1/80 (5:1) "A new crater was formed on the flank of Crater 3 on 19 Jan. Moderately thick brown/grey ash was explosively ejected until 20 Jan. Weak-moderate ash emissions resumed 24 Jan and continued intermittently until 29 Jan. Incandescence and rumbling were observed 28 Jan.

"During an overflight 28 Jan a brown-coloured deposit was observed extending up to 600 m from the new vent. Seismic activity consisted of numerous discrete volcanic events and possibly periods of tremor."

Information Contacts: B. Scott, C. McKee, RVO, Rabaul.

2/80 (5:2) "Sometime between overflights on 28 Jan and 15 Feb, effusion of lava commenced on the upper W flank of Crater 3, from the new vent formed on 19 Jan. When first observed on 15 Feb, the flow was about 700 m long. Explosions and rumbling were heard throughout the month, and glow and lava fragment ejections from the new vent were seen occasionally. Little ash emission was observed, but blue and white emissions were common. Crater 2 emissions consisted solely of white vapour.

"The seismograph was shifted from Kilenge Mission (10 km W of Langila) to Cape Gloucester airstrip (8 km N of the volcano) to enable joint direct visual and seismic observations. Seismic activity associated with the eruption was weak, consisting of tremor-like signals and probable explosion events."

Information Contact: C. McKee, RVO, Rabaul.

4/80 (5:4) "Langila's eruption continued but observations have been prevented at times by poor weather. Incomplete reports for March indicated that incandescent explosive activity was continuing in the new crater, formed on 19 Jan. Blue and white vapour emissions from the new crater were commonly observed in April. Brown ash emissions were seen 9 and 10 April, and at night on those dates incandescent lava fragment ejections occurred. Crater glow was seen occasionally during the month, and rumbling and explosion sounds were heard daily. The 1980 lava flow appeared to have grown little since it was first seen on 15 Feb. Crater 2 usually emitted white vapour, and brown emissions were reported on 2 days in April. Explosive activity was expressed seismically as small discrete events that had the appearance of discontinuous tremor."

Information Contact: Same as above.

5/80 (5:5) "The eruption continued at moderate to low intensity. From the observation post about [10] km away, roaring, rumbling, and detonations were heard almost incessantly. Small Vulcanian explosions were occasionally observed from the active vent in Crater 3 (formed on 19 Jan).

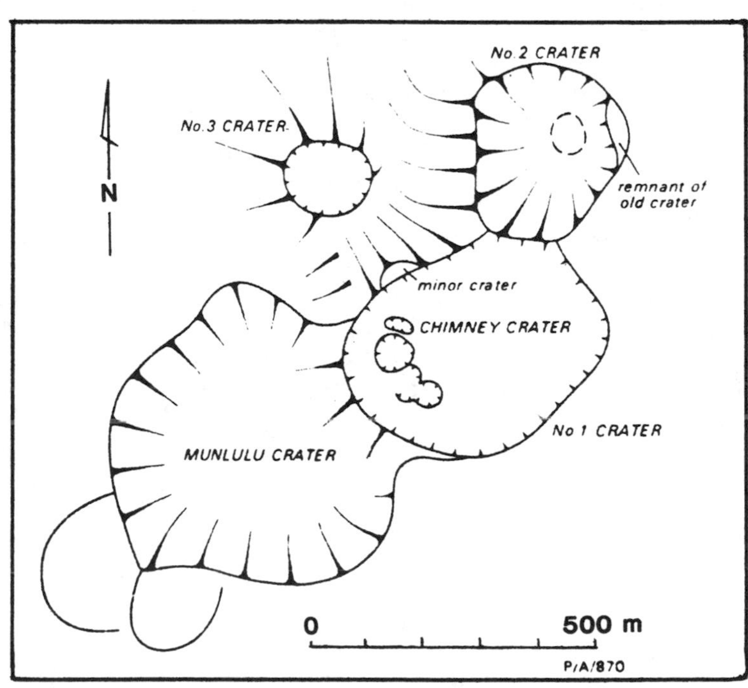

Fig. 5-5. Summit area of Langila in 1970, from Palfreyman and others, 1981. The area shown is outlined in fig. 5-4.

"During a ground inspection on 8 May, detailed observation of several explosions revealed that blocks were commonly ejected to a height of about 600 m, accompanied by loud roaring. The concluding stages of explosions were characterized by streaming of translucent vapour followed by conspicuous emission of blue vapour. A larger explosion on 9 May produced a small base surge which travelled about 300 m W, leaving a pale brown deposit.

"The 1980 lava flow is blocky, and similar in hand specimen to lavas produced in the 1970's. The lava flow had ceased moving, and its length was about 800-1000 m. Since it was first observed in Feb, three distinct lobes have entered adjacent valleys, the E lobe being the longest. A preliminary estimate of flow volume is 3×10^5 m^3."

Information Contacts: B. Scott, C. McKee, RVO, Rabaul.

6/80 (5:6) "Eruptive activity continued apparently without change during June. Light grey explosion clouds were occasionally observed rising from the active vent in Crater 3. Steady glow from this vent was observed on 5 June. Seismicity continued at the same intensity as previously."

Information Contact: C. McKee, RVO, Rabaul.

7/80 (5:7) "Vulcanian explosions continued in July from the active vent at Crater 3, and crater glow was seen on several nights. Eruption clouds were usually brown/grey, but otherwise emissions were white and rarely blue. Emissions from Crater 2 were usually white, but possible ash contents in the emissions were reported on several days. Weak glow from Crater 2 was reportedly observed on one occasion. The level of seismic activity remained steady, and consisted of moderate-amplitude volcanic earthquakes, probably of explosion origin."

Information Contact: Same as above.

8/80 (5:8) "The eruption continued at the same intensity in Aug. Emissions from the active vent in Crater 3 were mostly white and blue, but during a more active phase at mid-month grey emissions were observed. Lava fragment ejections and glows were observed on four consecutive nights 14-17 Aug. Rumbling and explosion sounds were heard on most days during the month. Crater 2 usually emitted white vapour, but on 4 occasions in the first half of Aug, brown emissions were observed. Glow at Crater 2 was observed on 4 Aug. The level and character of seismic activity was unchanged."

Information Contact: Same as above.

9/80 (5:9) "Observations of glow and lava fragment ejections from the active vent at Crater 3 were more numerous than in previous months. White and blue vapours were the common emission products, but grey emissions were seen on several days in the second half of Sept. Rumblings were heard throughout, occasionally augmented by explosive sounds. Occasional brown or grey emissions from Crater 2 were observed, but usually this vent released only white vapours. Seismic activity was reportedly unchanged."

Information Contact: Same as above.

10/80 (5:10) "Langila was inspected between 1630 on 28 Oct and [0915] on 29 Oct, following reports of increased activity. Large Vulcanian explosions took place from Crater 2 at intervals of several minutes or more and produced dark grey convoluting ash clouds reaching heights of 3-5 km. Minor ejections of incandescent lava fragments also took place, producing impact craters to a distance of at least 200 m. Crater 3 contained a body of plastic incandescent lava that was upheaved frequently by Strombolian explosions. Incandescent lava fragments were ejected to 300 m in horizontal distance. The crater was topped by a pale grey emission cloud. A blocky lava flow [first noticed on 13 Oct] had advanced about 2 km on a broad front from Crater 3. The lava looked similar to previous flows, which have consisted of low-silica andesite. Seismic activity (volcanic tremor and B-type events) was greatly intensified in Oct, necessitating an increase in seismograph attenuation of 18 decibels."

Information Contact: Same as above.

11/80 (5:11) "Intensified eruptive activity that began in mid-Oct (5:10) continued until 8 Nov. Dark emission clouds continued to be released from Crater 2, and emission clouds from Crater 3 were pale grey. Ejections of incandescent lava fragments from both craters were accompanied by rumblings and explosion sounds. The lava flow from Crater 3 was reported to be still active on 11 Nov.

"A decline in the intensity of the eruption was evident on 8 Nov, when seismograph attenuation was reduced by 18 decibels. However, glows and ejections of incandescent lava fragments continued from both craters, and grey ash and vapour clouds continued to be emitted."

Information Contact: Same as above.

1/81 (6:1) "Vapour emissions continued from Craters 2 and 3. Some small ejections of brown-grey ash rose from Crater 2. The lava flow from Crater 3 was still active and had almost reached the terminus of the 1975 flow."

Information Contacts: B. Talai, P. Lowenstein, RVO, Rabaul.

3/81 (6:3) Reports through 10/81 are from C.O. McKee.

"The first 3 months of 1981 have shown a steady decline of eruptive activity. Occasional brown ash-laden emissions from Crater 2 were observed in Jan, but in Feb and March emissions from Crater 2 were white and apparently of declining volume. Crater 3 released blue and white vapours in Jan and Feb; in March only small volumes of white vapour were emitted. The last time glow was observed was on 21 Jan from Crater 2. Seismic activity from Langila was at a low level Jan-March. Small tremor-like signals continued to be recorded."

Information Contact: C. McKee, RVO, Rabaul.

4/81 (6:4) "Eruptive activity, which had been declining since a peak in Oct 1980, increased in April. Explosions from Crater 2 produced dark ash clouds on 13 and 29 April.

Weak glow above this crater was seen on 25 and 26 April. Pale grey emissions from Crater 3 were observed on 2, 14, 16, and 17 April. Blue emissions from the same vent were noticed on 8 days. No explosive sounds were detected at the observation post about 10 km away. The intensity of seismic activity remained low."

Information Contact: Same as 6:3.

5/81 (6:5) "On most days of observation, Crater 2 emitted white vapour, and brown or rarely grey ash in moderate volumes. Crater 3 released small volumes of white, blue, and occasionally grey emissions. No glows or lava fragment ejections from either crater were observed, and no volcanic sounds were heard."

Information Contact: Same as 6:3.

6/81 (6:6) "A further intensification of activity took place in June. Moderate to strong white and brown emissions from Crater 2 were commonly seen. Ashfalls were reported on several days from locations about 10 km from the volcano. Rumbling and/or explosive sounds were heard on most days. Crater glow or ejections of incandescent lava fragments from Crater 2 were seen on 5 days in the second half of the month. Crater 3 was less active, commonly releasing white or blue vapours, but weak grey emissions were occasionally seen. Seismic activity strengthened considerably. Large-amplitude, multiple explosion-type earthquakes and prolonged periods of tremor clearly represented tephra explosions and bouts of gas venting at Crater 2."

Information Contact: Same as 6:3.

7/81 (6:7) "Vulcanian explosions occurred at Crater 2 throughout July. Explosion and rumbling sounds were heard at an observation post about 10 km N of Langila. Brief aerial inspections on 16 and 20 July revealed that eruptions at Crater 2 were taking place at intervals of about 10 minutes. The maximum height reached by the eruption clouds was probably about 300-400 m. Glow from Crater 2 was seen on 4 consecutive nights, 25-28 July. Crater 3 was less active, usually releasing white and blue vapours in small volumes, but grey emissions were occasionally seen.

"The overall level of seismic activity intensified near the end of July, but its character remained similar to that observed in June. Periods of strong tremor were generated by explosions and prolonged gas jetting at Crater 2."

Information Contact: Same as 6:3.

8/81 (6:8) "Eruptive activity at Crater 2 declined near the middle of Aug; frequent Vulcanian explosions and periods of ash emission gave way to white and blue vapour emission occasionally punctuated by intervals of brown or grey ash emission. Ashfalls were noted at coastal locations about 10 km N and NW of the volcano until 14 Aug. Ejections of incandescent material were seen on 6, 7, 8, and 14 Aug. Sounds of detonations, rumbling, and roaring were heard during the first half of the month. Crater 3 released white vapours throughout Aug, and blue emissions were seen rarely in the second half of the month. Seismic activity declined from a peak reached near the end of July, but the character of the activity was unchanged."

Information Contact: Same as 6:3.

9/81 (6:9) "Ash emission from Crater 2 resumed near the beginning of Sept after about 2 weeks of mainly white and blue vapour emission in late Aug. Significant ashfalls occurred W and N of the volcano, particularly in the first 3 weeks of Sept. Glow and ejections of incandescent lava fragments from Crater 2 were occasionally observed, and rumbling and explosion sounds were heard throughout the month. Crater 3 activity continued to be weak, consisting of weak emissions of white vapour.

"Seismicity fluctuated in strength. Several seismic events had the appearance of Vulcanian explosion earthquakes (from Crater 2), but the main feature of the seismicity in Sept was prolonged periods of discontinuous tremor, probably representing periods of ash emission. Frequent small brief seismic events began to be recorded at mid-month."

Information Contact: Same as 6:3.

10/81 (6:10) "Eruptive activity intensified in Oct when Crater 3 became strongly active, emitting incandescent tephra and a new lava flow. After a brief interval of ash emission on 3 Oct, Crater 3 began a sustained period of explosive activity on 9 Oct. Ejections of incandescent tephra were seen on 15 Oct and continued from 17 Oct throughout the month. The production of a new lava flow from Crater 3 was suspected on 27 Oct and confirmed on the 28th. The flow rate was apparently low.

"Activity at Crater 2 may have declined since late Sept. Vulcanian explosions occurred at intervals of several days, and periods of less-explosive ash emission were observed at the beginning and end of Oct. At other times, Crater 2 released white and blue vapours. Crater 2 glow was observed on 1, 3, 17, and 18 Oct.

"Ashfalls were reported in inhabited areas about 10 km N and W of the volcano on about 50% of days during the first half of Oct. Seismicity remained at a fairly low level, consisting mainly of numerous small events probably originating from Crater 3."

Information Contact: Same as 6:3.

11/81 (6:11) The remaining Langila reports are from RVO staff.

"Strong eruptive activity continued in Nov. The present eruptive episode is the most intense since Oct-Nov 1980 (5:10-11). Detailed observations were made by a volcanologist 17-26 Nov.

"Crater 2 produced a wide variety of activity. Vulcanian explosions rising to about 2 km were commonly observed. Frequency of explosions varied from minutes to hours, and periods of near-continuous (Strombolian) activity were also noted. Crater 2 glow was seen during the first half of the month, followed by apparently more intense activity including ejections of incandescent tephra occasionally rising to 0.5 km above the source. Commonly the larger Crater 2 explosions registered a seismic airwave. Rumbling and detonations from Crater 2 were heard throughout the month.

"Small nuées ardentes were produced by backfall of ejecta from some of the Crater 2 explosions. The largest of the nuées travelled about 2 km down the E flank. Periods of continuous harmonic tremor appeared to be related to resonance effects in the Crater 2 lava conduit. Continuous fluctuating glow, incandescent tephra ejections, and associated rumbling and booming sounds were noted from Crater 2 during some of the periods of harmonic tremor.

"Crater 3 activity may have declined during the month. Incandescent tephra ejections or crater glow were common during the first half of the month, but only occasionally observed later in Nov. Several large Vulcanian explosions were observed. One on 13 Nov was reported by a passing aircraft pilot as rising to about 6-7 km altitude, and having a columnar appearance. Crater 3 eruptions later in the month were often single events. Some were accompanied by loud detonations and others were soundless. Some of the soundless eruptions were registered seismically as large-amplitude, low-frequency events.

"The new lava flow from Crater 3 appeared to become inactive in mid-Nov. The flow rate was evidently low, as only a small volume of lava was extruded during the 2-week period of lava effusion."

Information Contacts: P. Lowenstein, C. McKee, RVO, Rabaul.

12/81 (6:12) "Strong eruptive activity continued during Dec although a decline was apparent in the last week of the month. Activity was centred at Crater 2, which produced dark brown or dark grey tephra emissions throughout Dec. On several days in the first half of the month the emissions were reported as being thick. Significant ashfalls at inhabited areas about 10 km to the N and W occurred 6, 8-12, 14, 17, 22, and 29 Dec. Sounds of explosions and rumbling were heard 1-14 Dec, and detonations were frequent 9-10 and 15-17 Dec. Incandescent tephra ejections or crater glow were seen 1-4, 8, 13-18, and 20 Dec. Dark emissions from Crater 3 were observed once (on 1 Dec); otherwise, the only products from this crater were white and blue vapours. Seismic activity showed a good correlation with visible activity. Periods of stronger visible activity were usually accompanied by large-amplitude volcanic earthquakes and periods of harmonic tremor."

Information Contact: C. McKee, RVO, Rabaul.

1/82 (7:1) "Occasional Vulcanian explosions from Crater 2 were reported. One explosion on 11 Jan was accompanied by a loud detonation, and an associated earthquake was felt 10 km away. Crater 2 glow was seen on 14 and 16 Jan.

"Crater 3 produced white and blue emissions, 1-17 Jan. From the 18th, pale grey Strombolian eruption clouds were seen. Rumbling and explosion sounds were associated with these eruptions. Crater 3 glow or ejections of incandescent tephra were seen beginning then. An active lava flow from Crater 3 was first observed 21 Jan.

"Seismicity recorded 11-23 Jan contained large-amplitude Vulcanian explosion shocks originating from Crater 2, and other smaller scale activity which probably originated at Crater 3. The apparent Crater 3 seismicity consisted of numerous, brief, small-amplitude events 14-17 Jan, and larger amplitude tremor envelopes 18-23 Jan which probably were an expression of the visible Strombolian eruptive activity that commenced 18 Jan."

Information Contact: Same as above.

2/82 (7:2) "Both craters produced occasional Vulcanian explosions during Feb. Activity appeared to be stronger in the first half of the month. Explosions from Crater 3 rose to 3-4 km on 9 and 11 Feb, and from Crater 2 to 6-7 km on the 13th and 14th. Glow and incandescent tephra ejections were observed at Crater 2 most nights the first half of the month, occasionally the second half. At Crater 3 incandescent tephra emission was reported only on 7 Feb.

"The main feature of Feb seismicity was the registration of several Vulcanian explosion earthquakes per day. Larger events were recorded 8-17 Feb. Two periods of frequent, brief, small-amplitude seismic events occurred 6-9 and 13-16 Feb. This activity became more intense on the 14th and 15th, giving rise to signals resembling discontinuous seismic tremor."

Information Contact: Same as above.

3/82 (7:3) "A fairly low level of activity prevailed in early March, but in the second half of the month activity at both craters intensified. Crater 3 erupted incandescent tephra 18-22 March, accompanied by frequent explosive detonations and loud rumbling. From 22 March until the end of the month, glow and incandescent tephra ejections from Crater 2 were seen on most nights. Dark eruption clouds were occasionally seen, and loud explosions and rumblings were heard. Seismicity was stronger from 18 March, and correlated with the intensified visible activity."

Information Contact: Same as above.

4/82 (7:4) "A steady level of mild eruptive activity prevailed at Crater 2 during the first part of April. Distinctly stronger activity was evident on the 23rd. Eruption clouds became more heavily charged with tephra, audible explosions more intense, and ejections of incandescent lava more frequent. Activity at Crater 3 was subdued, consisting usually of weak emissions of white and rarely grey vapour, and occasional emissions of blue vapour. Relatively small explosions producing dark brown or grey tephra clouds were observed on 15 and 26 April. Ashfalls at the observation post, about 10 km N of the craters, occurred on 15, 20, 25, and 30 April. Seismicity was steady at moderate to low levels during the first half of April, but intensified on the 20th. The increased seismicity was directly associated with the stronger visible explosive activity at Crater 2."

Information Contact: Same as above.

5/82 (7:5) "Following a period of moderately intensified explosive activity from Crater 2 in the second half of April, strong eruptive activity took place from this vent 2-16 May. During peak activity, 5-9 May, strong Strombolian activity was observed. Frequent ejections of bright red incandescent lava to maximum heights of 250-300 m above the crater were seen. An eruption column rose to about 4 km on 6-7 May. Ashfalls were experienced in inhabited areas about 10 km N and NW of the volcano during most of the first half of the month. The total thickness of these deposits was probably several mm.

"A significant event in this phase of Crater 2 activity was the production of a substantial lava flow. Several lobes developed NE of the crater, the largest extending to about 1 km in length. The volume of the flow is estimated to be 0.5-1.0 × 10^6 m^3. Occasional Vulcanian explosions took place 17-31 May at Crater 2, and crater glow was seen once, on the 19th. Throughout the month Crater 3 emitted white and blue vapours at a low rate.

"Seismicity corresponded closely with the intensified visible activity. During the peak eruptive period seismic tremor was produced by the Strombolian explosive activity. As eruption intensity waned, discrete earthquakes associated with Vulcanian explosions at Crater 2 became prominent."

Information Contacts: B. Talai, C. McKee, RVO, Rabaul.

7/82 (7:7) "Following a strong eruption from Crater 2 in early May, activity declined sharply. From about 17 May occasional Vulcanian explosions took place at this crater, but usually the only emissions were white vapours and occasionally blue vapours in small volumes. Ashfalls were reported 10 km N of the crater at the beginning of June and on 8, 10, and 31 July.

"Crater 3 has shown declining activity since April when occasional explosions were seen. Weak white and blue emissions were seen until about mid-June, but since then the crater has been completely inactive except for weak white vapour emissions on 18 July. Seismic activity has remained low since mid-May."

Information Contact: C. McKee, RVO, Rabaul.

8/82 (7:8) "A resurgence of activity was evident in Aug as Vulcanian explosions from Crater 2 became more common. Explosive eruptions occurred on 5, 12, 14, 15, 17, 24, and 31 Aug, and light ashfalls were experienced at the observatory post 10 km to the N after most of these eruptions. The cloud from the largest explosion, on 15 Aug, reached a height of about 6 km. Crater 3 continued to show little or no activity. Seismic activity generally remained low, although earthquakes associated with the Vulcanian explosions were recorded."

Information Contact: Same as above.

9/82 (7:9) "Activity in Sept was similar to that seen in Aug. However, Crater 2 showed a more intense phase of activity in the second half of the month which included more frequent Vulcanian explosions and periods of continuous ash emission. Light to moderate ashfalls at the observation post 10 km N of the volcano were recorded on 5 days but mostly in the second half of the month. No crater incandescence was seen. Blue vapour emissions from Crater 2 were observed on several days at the beginning and end of the month.

"Crater 3 showed little or no activity apart from weak emissions of white-grey vapour. The greyish color is not considered to be caused by entrained ash but to be an effect of peculiar light conditions [during early-morning periods of observation].

"Seismicity was generally at a low level, although explosion earthquakes were associated with the Vulcanian activity at Crater 2."

Information Contact: Same as above.

10/82 (7:10) "The level of activity in Oct was similar to that observed in Sept. Grey and brown ash emissions from Crater 2 were seen on about 50% of days; otherwise, white vapours were emitted. Ashfalls in areas about 10 km from the volcano were reported on 4, 5, and 6 Oct. Vulcanian explosions were observed on 7 and 30 Oct. On the 7th a dark brown eruption column rose to about 3.5 km, and incandescent ejecta started fires in vegetation at the foot of the volcano. The explosion on the 30th was much smaller, the ash column rising only a few hundred meters. Crater 2 was more active from the 22nd. Crater glow was seen on 22, 24, and 25 Oct, and weak to strong rumbling sounds from the crater were heard on 22, 23, 24, 26, and 29 Oct.

"Crater 3 continued its low level of activity, releasing white and blue vapours in small volumes. Seismic activity remained at a low level. A few Vulcanian explosion earthquakes were recorded."

Information Contact: Same as above.

11/82 (7:11) "A generally low level of activity prevailed during Nov, when both craters usually emitted tenuous white vapours, and usually no crater incandescence or sound effects were observed. However, 2 large Vulcanian explosions from Crater 2 were reported, on 12 and 22 Nov. A dense ash cloud was erupted at about 1930 on the 12th. Brilliant lightning displays were reported, and the eruption cloud contained incandescent tephra. On the 22nd a 3 km-high eruption column was produced by an explosion at about 0200. Light ashfalls 10 km N and W of the volcano were reported. Seismic activity was generally low, although several explosion earthquakes were recorded.

"A detailed aerial and ground inspection was carried out on 29 Nov. Both craters were releasing thin white and blue vapours. At the S rim of Crater 2 a strong odour of SO_2 was detected. The strong eruptive activity at Crater 2 in May resulted in significant accumulation (about 100 m) of coarse red-brown tephra mainly on the W rim of the crater. Vulcanian explosions since then have reamed out a funnel-shaped crater 250-300 m wide and about 200 m deep. No vents were visible in the floor of the crater. Crater 3 was about 150 m across and shallow, and most emissions seemed to originate from sources scattered around the crater walls. The nose of the westernmost lobe of the May Crater 2 lava flow was visited. The flow thickness at that point was ~20 m."

Information Contacts: C. McKee, P. Lowenstein, RVO, Rabaul.

12/82 (7:12) "Activity increased in Dec. Crater 2 produced thick white emissions while Crater 3 released white or blue vapours. Seismicity was generally low. Vulcanian explosions of increasing violence occurred from Crater 2 on 1, 14, 16, 25, and 26 Dec. On these occasions, the development of a dark grey ash-laden column up to 9 km was accompanied by loud rumbling noises and explosion earthquakes, and followed by light ashfalls. On 16 and 26 Dec harmonic tremor accompanied the subsequent escape of gases and tephra. The large explosion on the 26th at 1400 was accompanied by vivid lightning, ejected incandescent boulders up to 2 km from the crater, and started bush fires on the NW flank of the volcano. Twenty explosion shocks were recorded over the background tremor during a 25-minute period after the eruption, and Crater 2 remained incandescent for 45 minutes."

Information Contacts: P. de Saint Ours, P. Lowenstein, RVO, Rabaul.

1/83 (8:1) "Volcanic activity increased markedly in Jan and was concentrated in Crater 2. Several Vulcanian explosions were recorded every day. The most violent, 11-14 and 22-24 Jan, ejected incandescent boulders and tephra, illuminating the dark ash-laden column and setting fire to vegetation on the volcano's flanks. Consecutive ashfalls were blown SE over unpopulated areas. Crater 3, meanwhile, was quietly exhaling weak white vapours. In the sub-continuous seismic background, each discrete explosion was recorded as a sharp, large-amplitude event with a period of 1.5 Hz."

Information Contacts: P. de Saint Ours, B. Talai, RVO, Rabaul.

2/83 (8:2) "The increased Vulcanian activity of Crater 2 in Jan culminated in a rise of the magma column, with an eruptive phase maximum 11-16 Feb. The 3-11 Feb build-up of the eruption consisted of approximately hour-long periods of loud rumbling noises, with deep explosion sounds at 5-30-second intervals. Several times per day at irregular intervals, individual explosions produced black ash-laden columns that rose as much as 3-4 km before being dissipated by the NW winds. Night glow, observed 3 Feb, became more intense during this period. Low Strombolian fountaining was visible 3-5 and 9 Feb.

"During the 6 days of maximum activity, Crater 2 simultaneously displayed continuous Strombolian fountaining to 100 m and intermittent powerful Vulcanian explosions. Most of the Vulcanian explosions were laterally directed, while the continuous moderate vapour emissions and Strombolian fountaining were central and vertical, leading to the conclusion that Crater 2 may contain two more or less independent vents.

"Seismic activity consisted of a sub-continuous background of harmonic tremor and Strombolian B-type earthquakes. Each individual Vulcanian eruption produced large-amplitude low-frequency explosion events. The most powerful explosions occurred 12-13 and 15 Feb at the rate of 2-5/hr.

"During the eruption, Crater 3 (a separate composite cone 300 m W of Crater 2) released only weak white vapours. However, the volume of emission increased to moderate or large during the first 10 days of Feb, the time of the activity buildup at Crater 2."

Information Contacts: P. de Saint Ours, P. Lowenstein, RVO, Rabaul.

1983		Height (km)
April	9	0.3
	10	0.3
	15	3.0
	18	8.0
	26	5.0

Table 5-2. Heights of Vulcanian explosion columns above Crater 2, Langila, April 1983.

3/83 (8:3) "After the increased explosive activity from Crater 2 in Feb, the level of activity returned to normal. White-grey vapour and ash emissions were seen on a few days, and occasional relatively small Vulcanian explosions were observed. Explosion and rumbling sounds were heard occasionally. The only volcano-seismic activity recorded was from Vulcanian explosions at Crater 2 which occurred at intervals of a few hours to about 4 days."

Information Contact: C. McKee, RVO, Rabaul.

4/83 (8:4) "A generally low level of activity prevailed during April. However, Crater 2 produced some discrete Vulcanian explosions (table 5-2). The explosion on the 18th was witnessed from an aircraft about 70 km away. The eruption cloud rose steadily for about 15 minutes, then began spreading at its top to become mushroom-shaped. Weak emissions of grey ash clouds were also reported on 8, 14, 20, 21, and 29 April. Crater 3 was quiet for most of the month, releasing weak white vapour, although grey emissions were reported on 21 and 26 April. On the 30th a Vulcanian explosion sent ash and vapour to a height of about 3 km. Seismic activity was generally at a low level, with occasional explosion earthquakes."

Information Contact: Same as above.

5/83 (8:5) "A generally steady moderate level of Vulcanian explosive activity prevailed in May. On most days when the summit was clear, pale grey or brown emissions were reported from Crater 2. Vulcanian eruption columns rose to about 300 m on 16 May, and 7 km on 17 May. Explosion or rumbling sounds were heard on about 60% of days. Weak glow from Crater 2 was seen on 4 nights: 8, 11, 13, and 25 May. Crater 3 continued to show very mild activity, usually releasing thin white vapours and rarely emitting pale grey clouds.

"Volcano seismicity consisted of occasional Vulcanian explosion earthquakes at an average rate of several per day. On 3, 5, and 16 May explosion earthquakes were followed by periods of about 30 minutes of tremor probably produced by prolonged powerful degassing at Crater 2."

Information Contact: Same as above.

6/83 (8:6) "Activity increased somewhat in June as Vulcanian explosions became more frequent. Ash emissions from Crater 2 were observed on most days and Vulcanian explosions usually occurred at rates of 0-5 per day. However, 2 periods of more intense activity took place; 4-17 June, and starting 26 June and continuing at month's end.

"In the first period of stronger activity, rates of recorded Vulcanian explosions reached about 30 per day on 7 June, and about 15 per day on 8 and 10-14 June. Weak red glow from Crater 2 was seen on the nights of 10 and 11 June. Light ashfalls were reported from about 10 km W of the crater.

"In the second period of stronger activity, from 26 June onwards, ash emissions from Crater 2 became more frequent, and sounds of detonations and rumbling more noticeable. Ejection of incandescent tephra was seen the night of 26 June, and continuous glow was observed early on the morning of the 27th. In the last few days of the month the volcano was obscured by the ash emissions from Crater 2. Ashfalls of several millimeters were reported at locations 10 km N and W of the volcano.

"Crater 3 continued to show a low level of activity, usually releasing tenuous white vapours. However, thick dark brown ash clouds were emitted on 6 June, and pale grey emissions were reported on 12 June. Volcano seismicity at Langila was dominated by the earthquakes produced by Vulcanian explosions."

Information Contact: Same as 8:3.

7/83 (8:7) "Stronger activity that commenced at Crater 2 on 26 June continued into July. During the first week of July, ash emissions blown down the volcano's flanks by strong winds occasionally obscured the active vent. A few large Vulcanian explosions were observed, and associated detonations and rumblings were heard frequently at the beginning of the month. Ashfalls continued during this period in inhabited coastal areas about 10 km to the NW and N. Weak crater glow was noted on 2 July.

"Seismic records indicate an average of about 5 Vulcanian explosions per day in the first week of July, accompanied by large-amplitude harmonic tremor on most days. A decline in activity was evident after 7 July as emissions became less voluminous and less ash-rich, and explosive sounds less frequent.

"Activity re-intensified somewhat from 15 July. Greater quantities of ash were ejected, resulting in renewed ashfalls in coastal areas. On a few days the volcano was again obscured by its own ash emissions. Vulcanian explosions continued to register on seismograms at an average of about 3 per day.

"Crater 3 remained relatively inactive, mainly releasing white vapours. However, pale grey and blue emissions were reported on 9 July.

Information Contacts: C. McKee, P. de Saint Ours, RVO, Rabaul.

8/83 (8:8) "A definite strengthening of Vulcanian activity at Crater 2 has been observed since late June. This phase of activity is the strongest since Jan-Feb.

"On most days in Aug ash-laden vapour emissions from Crater 2 were blown down the volcano's flanks by strong winds, obscuring both the active vent and Crater 3. Strong activity in the first 8 days of the month consisted of up to 15 Vulcanian explosions per day, and periods of harmonic tremor recorded 4-8 Aug. Light to moderate ashfalls were common in coastal inhabited areas about 10 km NW and N. A decline of activity was observed 9-20 Aug as emissions became less voluminous and less ash-rich, and sound effects were heard less frequently. Occasional explosion earthquakes were recorded. More frequent Vulcanian explosions, loud sound effects, and ashfalls resumed on 21 Aug.

"In contrast to activity in Jan and Feb 1983, when incandescent lava ejections took place, the more explosive activity at Crater 2 in Aug was rarely accompanied by incandescence. Weak crater glow was observed on only 2 nights (17 and 29 Aug). Crater 3 remained relatively inactive, usually releasing thin white vapours."

Information Contacts: Same as above.

9/83 (8:9) "Explosive eruptive activity continued at Crater 2, and was stronger 2-6 Sept. Up to 12 explosions per day were recorded, and periods of strong volcanic tremor were produced by prolonged gas venting. Activity declined 8-11 Sept, but re-intensified on the 12th. Moderate activity persisted for the remainder of the month, including patchy and weak tremor and 1-9 explosions per day. Weak red glow was observed over Crater 2 on 25 and 29 Sept. Light to heavy ashfalls were common in coastal inhabited areas about 10 km to the NW and N."

[See 8:9 Manam report (above) for SO_2 measurements at Langila.]

Information Contact: C. McKee, RVO, Rabaul.

10/83 (8:10) "Moderate explosive activity persisted at Crater 2 during the first week of Oct, but from the 8th until the end of the month, eruptive activity was at a low level.

"During 1-7 Oct, activity at Crater 2 consisted of weak-strong ash emission accompanied by explosion and rumbling sounds heard 10 km away. Weak red night glow from the crater was observed 3-7 Oct. This activity was represented seismically by occasional large Vulcanian explosion earthquakes (2-5 per day), numerous smaller explosion shocks, and rare periods of continuous and discontinuous harmonic tremor. Ashfall was experienced in coastal inhabited areas about 10 km N and NW of the volcano on 2 Oct.

"From the 8th until the end of the month, Crater 2 released white emissions in small to moderate quantities. However, Vulcanian explosions accompanied by loud detonation sounds and ejections of incandescent material took place on 24 and 27 Oct. Seismicity was at a low level 8-31 Oct, but rare Vulcanian explosion events were also recorded. Crater 3 released small volumes of white and occasionally blue vapours throughout the month."

Information Contact: Same as above.

12/83 (8:12) "Vulcanian activity continued from Crater 2 with a pronounced increase toward the end of the month that culminated in the production of a small lava flow on the NE flank.

"From 1-20 Dec, activity consisted of occasional emissions of white-grey tephra rising to 0.6-1.2 km above the summit, accompanied by weak to low rumbling noises and explosions. Glow was visible 12-14 Dec, with ejection of glowing fragments to 270 m on the 13th. Fine ashfalls occurred 10 km to the N on 6, 7, 11, and 15 Dec.

"From 21-27 Dec, activity became much stronger, with continuous ejection of thick tephra-laden vapour to 1.8-2.7 km above the summit accompanied by continuous loud rumbling and explosion sounds. Continuous glow and ejections of lava fragments were observed at night with overflow of lava to the NE 23-28 Dec. This was accompanied by long periods of continuous harmonic tremor which increased in amplitude during the periods of maximum activity. Crater 3 remained inactive throughout, with only weak emissions of white vapour."

Information Contact: P. Lowenstein, RVO, Rabaul.

1/84 (9:1) "Activity remained high at Crater 2, but Vulcanian explosions replaced the more continuous activity that produced the lava flow on the NE flank in Dec. For most of the time, Crater 2 produced moderate amounts of white to brown ash-laden vapour, accompanied by discontinuous rumbling and explosion sounds, while the seismic station at Cape Gloucester airstrip, 9 km away, recorded discontinuous tremors and large explosion earthquakes. Peaks of activity occurred on 7, 12, and 25 Jan with emission of columns of thick dark tephra-laden vapour to heights of 1.5-2.5 km above the crater. Large blocks were ejected as far as 2 km from the vent by the more powerful explosions, and ashfalls were experienced on the coast, 10 km downwind. Activity at Crater 3 was confined to the emission of white and blue vapours."

Information Contact: Same as above.

2/84 (9:2) "Activity was substantially reduced during Feb. The main activity at Crater 2 was the release of white vapours and, rarely, pale grey ash clouds. One Vulcanian explosion was observed on 7 Feb, producing a thick dark ash cloud that rose about 3 km above the vent. Crater glow was seen on 3 and 5 Feb. Crater 3 continued its usual activity of weak to moderate white and blue vapour emission. Seismicity was at a low level, with only a few recorded explosion earthquakes."

Information Contact: C. McKee, RVO, Rabaul.

Further Reference: Johnson, R.W., 1984, Volcanological Inspections in Papua New Guinea, February 1984; *Geological Survey of Papua New Guinea Report* 8/84.

3/84 (9:3) "A relatively low level of activity persisted during March, although a slight intensification was noted after the 19th. Pale grey ash emissions from Crater 2 were occasionally observed 1-18 March. A strong Vulcanian explosion took place on 19 March, and on the 20th further explosions were observed. The seismic record for 20 March indicates a total of 10 explosions. A lull in Crater 2 activity was noted 21-25 March but weak ash emission began on the 26th, and explosive activity resumed on the 27th. Two or three explosion earthquakes were recorded on 27, 28, and 30 March. Crater 3 released white and blue vapours at low rates throughout the month."

Information Contact: Same as above.

4/84 (9:4) "For most of April, activity was at a low level with weak to moderate white and occasionally grey emissions from Crater 2, and weak white and rarely blue emissions from Crater 3.

"However, between 7 and 17 April stronger activity took place at Crater 2. Occasional strong Vulcanian explosions producing eruption columns up to about 5 km high were accompanied by detonations and rumblings, and ashfalls were recorded in coastal areas 10 km to the N and NW. Characteristic impulsive earthquakes also accompanied these explosions."

Information Contact: Same as above.

5/84 (9:5) "Activity in May was similar to that in April. Intermittent weak-moderate ash emissions from Crater 2 were reported during the first 8 days of May. During the remainder of the month Crater 2 usually released white vapours at low rates. However, strong Vulcanian eruptions accompanied by loud detonations took place on 13, 25, and 28 May. These eruptions resulted in reportedly heavy ashfalls 10 km downwind. Weak emissions of white and blue vapours continued at Crater 3."

Information Contact: Same as above.

6/84 (9:6) "Activity was generally very weak at both craters in June. The emissions were usually white or occasionally blue vapours. However, lightly ash-laden clouds were emitted from Crater 2 on 16-18 and 21 June. Seismicity was at a very low level throughout, with occasional low-amplitude volcanic earthquakes."

Information Contact: Same as above.

8/84 (9:8) "Activity remained at the generally low level that has persisted for several months. Occasional grey or brown ash emissions from Crater 2 were reported. Seismicity was at a very low level throughout, with only occasional small volcanic earthquakes."

Information Contact: Same as above.

11/84 (9:11) "For the period Sept to Nov, activity remained at a low level. Occasional grey or brown ash emissions from Crater 2 were reported. Seismicity was at a very low level with only a few volcanic earthquakes recorded."

Information Contact: Same as above.

2/85 (10:2) "Activity intensified in Feb after being at a very low level for about 1 year. Several large Vulcanian explosions at Crater 2 in late Jan marked the commencement of this new eruptive phase. Explosions occurred at rates of up to about 20 per day in Feb. The largest explosions produced eruption columns reportedly about 3-4 km high. No significant ashfalls took place in inhabited areas. Weak glow was frequently observed from Crater 2 up to about mid-Feb, and ballistic incandescent lava fragments were seen at the bases of some eruption columns. Seismicity accompanying this activity not only included the usual discrete explosion earthquakes, but also a number of periods of harmonic tremor, thought to be caused by sequences of prolonged gas discharge at Crater 2. Crater 3 remains inactive, except for mild fumarolic activity from the crater walls."

Information Contact: Same as 9:2.

3/85 (10:3) "Langila returned to a low level of activity in early March after a period of moderate-to-strong explosive activity at Crater 2 in Feb. From 3 March, activity at Crater 2 usually consisted of weak emission of white vapour, although occasional Vulcanian explosions took place. Crater 3 remained quiet."

Information Contact: Same as 9:2.

6/85 (10:6) "A new phase of increased activity at Crater 2 began about 8 June. This followed several months of very weak activity punctuated by occasional mild Vulcanian explosions. From mid to late May a more prolonged period of slightly intensified activity consisted of grey ash emissions, occasional explosion sounds, and several periods of crater incandescence.

"Rumbling and explosion sounds, large quantities of blue vapor, and occasional grey ash clouds were reported from Crater 2 between 8 and 12 June, marking the onset of a new phase of activity. A further intensification took place mid-month and an eruption column about 1 km high was reported on the 17th. Weak glow from Crater 2 was observed on the night of the 20th, and explosion sounds were more frequently heard on the 22nd. Light ashfall was reported 10 km downwind on the 27th. Seismicity was comprised of characteristic Vulcanian explosion earthquakes. This higher level of activity was continuing at the end of June."

Information Contact: Same as 9:2.

7/85 (10:7) "Further intensification of the eruption at Crater 2 took place in July. Increased activity was noticeable from 19 July when the frequency of Vulcanian explosions increased from about 2 to 10 per day. These explosions were stronger, as indicated by ashfalls as far as 10 km downwind, louder detonations, and rumbling. Incandescent lava fragments were ejected on the night of 19 July. A further increase in activity was apparent from 22 July when daily totals of Vulcanian explosions climbed to about 30 and periods of continuous harmonic tremor were recorded."

Information Contact: Same as 9:2.

8/85 (10:8) "The stronger eruptive activity reported in late July declined rapidly at the beginning of Aug, but this was followed by two more spasms of intensified activity. More frequent Vulcanian explosions from Crater 2, accompanied by characteristic volcanic earthquakes and sequences of continuous harmonic tremor, took place 10-14 and 19-23 Aug. Weak red incandescence at the crater was observed from locations about 10 km downwind on 10 and 11 Aug."

Information Contact: Same as 9:2.

9/85 (10:9) "Activity intensified in Sept. A moderate phase of eruption began on 6 Sept when an ashfall was noted at Kilenge Mission (10 km NW of Langila). From 9 to 13 Sept, incandescence was noted at Crater 2 and there was an increase in audible explosions and rumblings (heard at the observation post, 10 km NW). Small incandescent ejections were reported on the night of 9 Sept. Seismic tremor lasting from a few minutes to a few hours was recorded during this period, with the strongest bursts on the 9th and 10th.

"Incandescence from Crater 2 was again noted 17-26 Sept with an increase of audible explosions and rumblings. During this period there were several ashfalls at the observation post and Kilenge Mission. Ash was also reported at Siassi, Umboi Island (30 km W) on 24 and 25 Sept. The brightest glow and highest level of tremor during this period were on 23 and 24 Sept."

Information Contact: J. Mori, RVO, Rabaul.

10/85 (10:10) "The eruption of Crater 2 persisted through Oct with periods of increased activity from 4-7 and 14-25 Oct. When inspected on 2 and 3 Oct, the 200 m-wide crater was only 50 m deep, plugged with a magma column covered by debris. High-pressure degassing was occurring from 6 fissures in the crater floor. Occasional explosions that may have been phreatomagmatic produced ash-laden plumes.

"Intensification of magmatic activity on the 4th resulted in sub-continuous Strombolian ejections to 250 m above the crater and intermittent Vulcanian explosions. A light to dark grey ash-laden plume rose 1500 m above the crater, and produced ashfall 10 km downwind. Loud rumbling noises and sharp explosions were heard at an observation post 9 km N of the volcano. Periods of seismic tremor lasting from a few minutes to a few hours were recorded during this time.

"Activity returned to the former degassing mode 7-18 Oct, but eruptions resumed after the 19th. The strongest seismic activity was recorded on the 19th and 20th but harmonic tremor lasted until the 24th. Strombolian eruptions were discontinuous 19-24 Oct and a few Vulcanian explosions expelled dark ash-laden columns to 2000 m above the crater. By the end of the month, the eruptive activity had declined to persistent degassing, with a few explosion shocks recorded daily."

Information Contact: P. de Saint Ours, RVO, Rabaul.

11/85 (10:11) "Activity was at a low level during Nov with occasional reports of grey ash clouds from Crater 2. A column of dark cloud was reported rising to about 200-300 m above the summit on the 30th. Two audible explosions (heard at the observation post, 10 km NW of the crater) were reported on the 17th and 30th. Crater 3 remained inactive throughout the month. Seismic activity remained at a low level throughout the month with only 2 explosion shocks recorded, on the 17th and the 30th."

Information Contact: B. Talai, RVO, Rabaul.

12/85 (10:12) "Activity was at a low level for most of Dec. Starting the 29th there were daily reports of rumbling sounds and light ashfalls at the observation post 10 km NW of the summit. Glow was noted from Crater 2 on the night of the 29th. Seismic activity also increased from the 29th, with several Vulcanian explosions recorded per day and some harmonic tremor.

Information Contact: J. Mori, RVO, Rabaul.

ULAWUN

New Britain Island, Papua New Guinea
5.04°S, 151.34°E, summit elev. 2300 m
Local time = GMT + 10 hrs

This strato-volcano, 130 km SW of New Britain's NE tip (Rabaul), is the highest in the Bismarck Arc. Of its 23 known eruptions since 1700, the last 6 are described here.

General References: Johnson, R.W., Davies, R.A., and White, A.J.R., 1972, Ulawun Volcano, New Britain; *Australia Bureau of Mineral Resources, Geology and Geophysics Bulletin* 142, 42 pp.

McKee, C.O., 1983, Volcanic Hazards at Ulawun Volcano; *Geological Survey of Papua New Guinea Report* 83/13, 21 pp.

5/78 (3:5) The following is a report from R.J.S. Cooke.

"A new eruption was first observed just after 1900 on 7 May, when intermittent glow was seen at the summit. This rapidly developed into explosions of incandescent lava fragments from the summit crater, which increased in intensity until by 2100-2200 a full-scale eruption was in progress. At this stage, new ejections were visible every few seconds and explosive sounds were prominent; the intensity was probably as strong as that during the peaks of the 1970 and 1973 eruptions.

"Aerial inspections the following morning showed powerful ejections of red incandescent bombs every few seconds, feeding an eruption column 1-1.5 km high. The ash content in the ejections was fairly small, and the eruption column was relatively thin in texture. Only one source was noted and this appeared to be towards the N part of the 1973 'chasm', which had been deepened considerably by collapse sometime between May and July 1976.

"Similar activity continued through the night of 8-9 May, probably with a slight decline in intensity. However, beginning about 0700 on 9 May the vigor of ejections increased and the ash content built up very rapidly, until by mid-afternoon large quantities of black ash were being erupted in a column about 2.5 km high, and ashfall was accumulating rapidly at the downwind coast. A layer about 1 cm thick was deposited in a few hours 12 km WNW of the summit, and the ash plume extended tens of kilometers out to sea. An aerial inspection early 9 May revealed a weak small second vent on the SE side of the summit. This strong phase slackened off noticeably after about 1630 on 9 May.

"One or more nuées ardentes traveled down the SE flank at the peak of this phase, starting at about 1545, but were not directly observed, being on the opposite side of the volcano to the observation post. Fallen trees were visible only at the edges of the extensive area of forest destroyed by the nuées ardentes. The devastation extended 6-7 km from the summit and exhibited several adjoining lobes (fig. 5-6).

"A spectacular but so far unidentified phenomenon was observed by pilots of two aeroplanes at about 1045 on 9 May, although nothing unusual was noted at that time from the observation post. This had some of the characteristics of both nuées ardentes and lava flows, to judge by the descriptions, and the mountain was [reported] to have split from top to bottom. It is hoped that photographs taken from both planes will assist in interpreting this event.

"Summit activity during the night of 9-10 May was reduced in comparison to that of the previous nights, with much less incandescence, and intermittent spells of inactivity up to 30 seconds long. Little activity was evident by about 0800 on 10 May; rumblings were often heard, but only a thin wispy eruption column was present.

"Later that morning, ash emission again increased strongly, a thick black column of ash was present through 11 May, and additional ashfalls were recorded at the coast. During the night of 10-11 May, incandescent lava fragments were still occasionally ejected from the summit crater, but the next night, only occasional weak glow was noted.

"A morning aerial inspection on the 11th showed that a thick ash column was billowing up rapidly and continuously from virtually the whole area of the summit, although individual projections of dense black ash with little solid content could be seen periodically. No incandescence was visible. A lava flow was active on the lower E flank of the volcano; its source was apparently a radial fissure about 500 m long, some 6 km from the summit and 2000 m vertically below it, along which a number of individual lava vents could be seen. Some issued lava as brightly incandescent but quiet flows, others as vigorous continuous fountains of bright orange lava estimated (under poor conditions) to be gushing lava to heights of at least 30-40 m. It was thought probable that this fissure had opened sometime the previous day (10 May).

"The lava flow sources remained vigorously active 12-13 May. More than a dozen individual vents were present, 5 of them fountains. The cumulative effusion rate was probably 100-200 m^3/sec (provisional estimate), and the lava velocity was about 2-2.5 m/sec near the sources. During this time however, summit explosive activity weakened and eventually ceased in the late afternoon of 13 May. Although ash emission was still fairly strong on the 12th and the morning of the 13th, the ash column was no longer black but a mid-grey/brown color. During the afternoon of the 13th, intermittent periods of only white vapor emission were noted during the final stages of explosive activity.

"By 14 May, the lava flow sources were fewer in number and much weaker; sources higher up the fissure had stopped and only one fountain was still active, at the bottom end of the fissure. Lava flow ceased altogether, probably during the night of 14-15 May, just as the flow front had finished blocking the main channel of the Pandi River (about 11 km from the volcano), diverting it into an existing older channel. Lava had passed less than 50 m from Naisapuna, a hamlet of 4 houses and the only inhabited spot on the whole E flank, and had cut the only road on that flank (a rough track). Provisionally, the flow may contain 20×10^6 m^3 of lava. The lava appears to be basalt, broadly similar to that of the 1970 and 1973 eruptions, although no analyses have yet been carried out.

"A short-period, vertical-component seismograph has been in use at Ulamona Catholic Mission (11 km WNW of Ulawun's summit) since Dec 1976. Between then and Aug 1977, a number of brief swarms of B-type volcanic events were recorded. During the last few days of April 1978, a number of barely discernible events, probably of the same kind, were recorded (mostly on 27-29 April) and a few were seen every day until the start of the eruption. Somewhat larger events resembling short bursts of volcanic tremor occurred on 2-3 May. During the few hours before the start of the visible eruption on 7 May, patchy, weak tremor was recorded. This tremor became strong and continuous with the commencement of visible activity, and maximum amplitudes were recorded during the first night. Tremor declined slightly after that and from mid-week, amplitudes fluctuated. Tremor ceased altogether with the end of explosive activity on 13 May. Some large and unusual events were recorded on the 9th and 10th that may be volcanic, but their interpretation is unclear at present."

Information Contact: R. Cooke, RVO, Rabaul.

Further Reference: McKee, C.O., Almond, R.A., Cooke, R.J.S., and Talai, B., 1981, Basaltic Pyroclastic Avalanches and Flank Effusion from Ulawun Volcano in 1978; *in* Johnson, R.W. (ed.), *Cooke-Ravian Volume of Volcanological Papers*; Geological Survey of Papua New Guinea Memoir 10, p. 153-66.

9/80 (5:9) Australian radio reports that Ulawun began to erupt during the night of 6-7 Oct. Pilots saw ash "billowing" to about 6 km above the crater during the morning of the 7th. A police spokesman said ash was thought to be falling more than 30 km from the volcano. Police were considering the evacuation of Ulamona Catholic mission, the settlement closest to the volcano.

Information Contact: Melbourne Overseas Service.

10/80 (5:10) The following reports are from C.O. McKee.

"A brief but powerful eruption took place at Ulawun between 6 and 7 Oct. The visible commencement of the eruption was late at night on 6 Oct. At about 2300, faint glow was seen at the summit, intensifying at intervals of 1-2 minutes, presumably signifying weak ejections of incandescent lava fragments. At 0400 on the 7th the lava ejections were more frequent, occurring at intervals of 1 minute or less. No sounds of the eruption had been heard up until this time by observers at Ulamona Catholic Mission, and no significant ash emission had been observed.

"After 0600, weak emissions of dark ash were seen from Ulamona. Observers in aircraft approaching from the NE noted that at about 0640 the emission cloud above Ulawun was slightly more voluminous than normal, and was reported as pale to dark. At about 0700, a series of strong explosions commenced, heard as deep rumbling at Ulamona. Within 10 minutes the top of the eruption column had reached about 4 km above sea level. Red incandescence was seen at the base of the column. By about 0715 the eruption column had grown to about 7-10 km in height. It was vertical and straight-sided right up to its top, where slight lateral expansion had begun.

"Small pyroclastic avalanches were seen shortly after the beginning of the strong explosive activity. At about 0720 they were reportedly much larger and descended all flanks of the volcano, particularly the N and SW flanks. All observers reported that these avalanches originated directly from the crater, and were not formed by collapse of the eruption column. They were described as not moving quickly down the volcano's flanks. Between 0720 and 0730 several particularly strong explosions occurred, accompanied by visible shaking of the volcano, likened to the initial shaking seen in quarry blasts. The eruption column was reinforced by these explosions, and the reports of directly associated pyroclastic avalanches may be interpreted as base surges. However, pyroclastic avalanching was also reported to

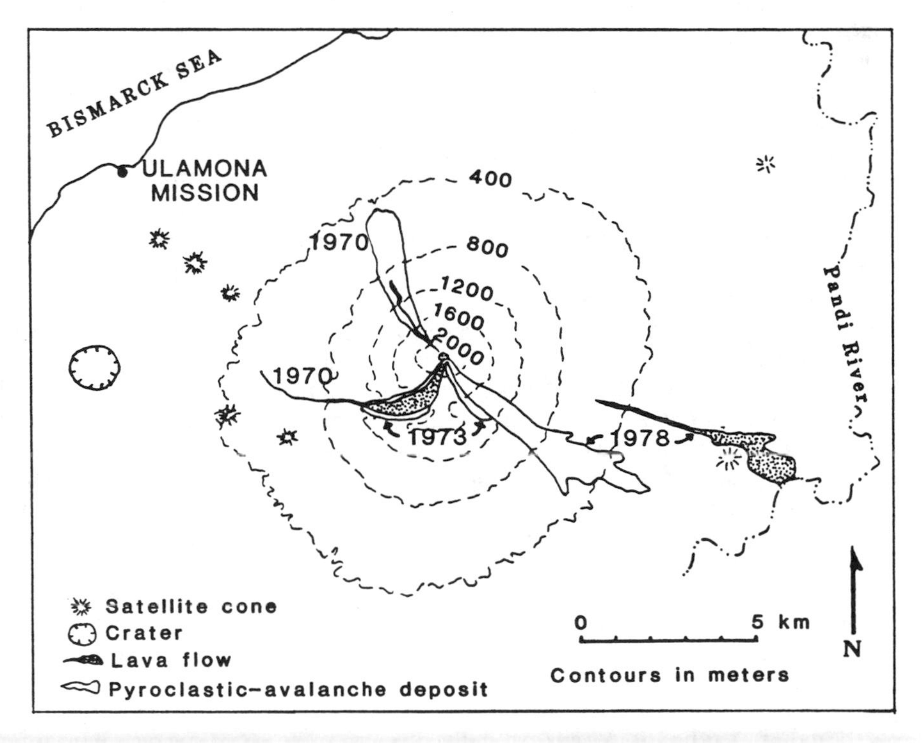

Fig. 5-6. Sketch map of Ulawun, after McKee and others, 1981. 1970-78 lava flows and pyroclastic avalanche deposits are shown. Contour interval is 400 m. [Originally from 10:11.]

have been more or less continuous during this period.

"By about 0735 the upper part of the eruption column was spreading out more noticeably and the clouds on the volcano's flanks had become more voluminous. The diameter of the 'mushroom' top of the cloud was estimated at 50-60 km. The eruption column continued to be fed by apparently frequent explosions in the crater but only slight upward growth was evident. The volcano gradually became obscured as the eruption cloud began to dissipate. By 1000 the volcano was totally obscured down to its base, and at Ulamona the darkness was total for several hours. The ash cloud drifted slowly towards the SW and was thick enough to cause 1 hour of total darkness in mid-afternoon at Bialla, 45 km from the crater.

"Observations from the flank of a neighbouring volcano indicated that strong explosive activity continued until about 1215. Until this time a violent electrical storm had prevailed in the eruption cloud. The cessation of electrical discharges coincided with the cessation of strong explosions.

"At about 1800 the ash had cleared sufficiently to allow observations of the summit from Ulamona. No ash emission was seen, but through the night a faint glow was present above the crater. Bursts of glow on the upper NE flank and associated clouds of ash on the same flank were observed from Ulamona.

"No further ash emissions from the summit crater were seen. However, ash clouds were seen occasionally for several days on the upper N flanks, and at night spots of incandescence were seen in the same place on the volcano. The ash clouds may have originated as slides of unstable parts of the cone, and in one case a large ash cloud may have been produced by explosive interaction of meteoric water and hot parts of the cone.

"No lava flows were produced during the eruption. Estimation of the total volume of fragmental flows awaits receipt and analysis of aerial photographs, but the volume is probably greater than that of the 1978 pyroclastic avalanche deposits (about 17×10^6 m^3 of juvenile material). A preliminary estimate of the volume of airfall ash is $10\text{-}20 \times 10^6$ m^3. The lava appears similar to the basaltic material produced in previous eruptions.

"Apart from the addition of a veneer of deposits from the pyroclastic avalanches and airfall ash, the topographic changes to the volcano brought about by the eruption include the formation of a series of gouges on the upper NE flank and the reaming out of an enlarged summit crater. The crater is now slightly elliptical with its larger diameter, estimated at 100-150 m, oriented E-W. The inner walls of the crater are steep, and the deepest part of the crater floor is about 60-100 m below the rim.

"Since the installation of a short-period vertical component seismograph at Ulamona in Dec 1976, B-type volcanic earthquakes have been common, sometimes occurring in swarms. During 1980 these events often occurred at intervals of less than 2 minutes, but during the night of 6 Oct they became more frequent, resembling patchy volcanic tremor. Several strong local earthquakes, probably A-type volcanic events, were recorded on 3, 5, and 6 Oct.

"At the commencement of the strong, visible activity, the seismic activity intensified dramatically, becoming continuous tremor, which persisted until about 1215 on 7 Oct. After that time, tremor ceased altogether, signifying cessation of the eruption." [See Ch. 19, 6:10 for data on eruption cloud dispersal.]

Information Contact: C. McKee, RVO, Rabaul.

1/81 (6:1) "The volcano was very quiet throughout Dec with only continuous moderate emission of white vapour from the summit crater."

Information Contact: Same as above.

12/82 (7:12) The remaining quoted reports are from the RVO staff.

"A very large white vapour cloud was emitted from the summit for several hours during the morning of 24 Dec, and moderate to strong vapour emission continued for 3 days. This coincided with a decline in seismicity which had been at a higher than usual level since mid-Nov. During the last week of the month seismic activity remained below the usual level of 1000 to 1500 B-type events per day. The vapour emission on the 24th is the most visible sign of activity since the volcano last erupted in Oct 1980 (5:10)."

Information Contacts: P. de Saint Ours, P. Lowenstein, RVO, Rabaul.

1/83 (8:1) "After the 24 Dec increase in vapour emission, Ulawun was back to a rather low level of activity. The summit crater released a low-pressure, sulfur-laden vapour plume, and daily seismicity included fewer than 1000 B-type recorded events. In Jan the seismicity steadily increased again in both amplitude and frequency, to its former moderate level of 1000-1300 events per day."

Information Contacts: P. de Saint Ours, B. Talai, RVO, Rabaul.

3/83 (8:3) "A notable seismic crisis occurred 21-23 March. From mid-Jan until 21 March, seismicity had been fluctuating between 400 and 1200 B-type volcanic shocks of moderate amplitude per day, in an apparently cyclic manner with a period of 16 days. On 21 March, shocks occurred at the rate of about 800/day until 1700, when their amplitude decreased sharply. At 1930 their frequency started to increase, and from 2000 they became subcontinuous (1800/day) and moderate in amplitude. This high level of seismicity continued for about 38 hours before gradually decreasing again.

"The volcano was obscured by heavy cloud for most of this period. The first visual observations were made at 0730 on 22 March when 4-6 almost perfect smoke rings, reportedly pale-grey in colour, were ejected rapidly to about 300 m above the summit crater. Consistently strong white vapour emissions from the summit crater commenced on 25 March, contrasting with the usually weak to moderate white emission. A report on 26 March suggests a brief interval of dark grey emission, but seismic records show no changes that could indicate eruptive activity. The decline in seismicity on 23 March was followed by almost 2 weeks of exceptionally low levels interrupted only by a brief increase, 30-31 March.

"It is interesting to note that the high level of seismicity closely followed a heavy rainfall of 65 mm. It also occurred only 3 days after a major tectonic earthquake (M_L 7.7) in the Solomon Sea, felt at MM V in the vicinity of Ulawun. It is possible that the earthquake opened microfissures in the volcanic edifice allowing ground water to penetrate and interact phreatically with shallow magma. The cyclic variations in seismicity since mid-Jan and increasing instability in March are believed to indicate that an eruption may occur in the near future."

Information Contact: C. McKee, RVO, Rabaul.

4/83 (8:4) "Intriguing seismicity, possibly indicating an eruption in the near future, continued in April and included periods of volcanic tremor. Amplitudes of discrete events were generally low, although a degree of cyclicity in amplitudes was apparent, with a period of about 8-11 days. Daily earthquake totals increased from about 600 to about 1500.

"After the 21-23 March seismic crisis, Ulawun's seismicity showed a fairly steady decay, reaching a very low level in early April. One clear A-type event was recorded on 7 April. A new seismic crisis, preceded by a lull about 2 hours long, began on 10 April at about 0310. The initial, strong continuous tremor changed to discontinuous tremor within a few hours. The entire period of tremor lasted about 6 hours. After this crisis, a steady decline was evident until 17 April. Small A-type events were recorded 11-16 April.

"On 17 April, 5 periods of tremor occurred. After about 3 hours of very low seismicity, the first began at 1645. A distinct lull also preceded the third period. Tremor was mostly continuous, with total duration of about 280 minutes. Individual periods lasted about 29-106 minutes, and were followed by about 5 hours of frequent discrete shocks and discontinuous tremor. Beginning 18 April a gradual decay in amplitude and frequency of occurrence of the shocks was recorded. Possible small A-type events were recorded 20-29 April.

"No unusual visible activity directly accompanied the seismic crisis. However, ejection of one or more smoke rings, seen to rise rapidly to about 500 m above the summit, was reported 11-18 April. Blue vapour emission was seen 14 April. Ulawun's usual white vapour emissions were moderate to strong throughout the month, but increased toward the end."

Information Contact: Same as above.

5/83 (8:5) "Seismicity continued to show interesting variations in May. Very low levels of seismicity were recorded at the beginning and end of the month, but somewhat stronger seismicity took place 5-23 May. On 19 May a new seismic crisis was recorded. Commencing at about 1000, the crisis consisted first of continuous tremor which lasted for about 7 hours. The strongest part of the crisis was in the first few hours. The tremor became discontinuous from about 1700, gradually giving way over the following few hours to frequent discrete shocks.

"Throughout the month, emissions from the summit crater were reported to be strong white vapours. During an aerial inspection 24 May a white emission column rose several hundred meters above the summit, but the vapours were not being emitted with any force. A plume 5-10 km long was formed by the emissions."

Information Contact: Same as above.

6/83 (8:6) "Ulawun's seismicity continued to show interesting variations, although there appears to be no correlation between the seismicity and the steady, moderate to strong white vapour emission from the summit crater.

"A number of seismic crises took place in June in which periods of volcanic tremor were recorded. The first, on 10-11 June, lasted for about 15 hours and was followed by 4 more: 15 June (7 hrs), 16 June (7 hrs), 17-18 June (17 hrs), and 30 June-2 July (48 hrs). In the last, several periods of tremor were recorded. Other periods of tremor-like signals were recorded on 14, 27, and 28 June, but these were probably the effects of rainstorms on the volcano.

"Amplitudes of discrete earthquakes were generally low in June, although slightly higher amplitudes were recorded 10-18 June in relation to the first 4 periods of tremor. A brief interval of low amplitudes followed the crisis of 17-18 June, but a steady rise in amplitudes was recorded beginning 20 June.

"At times other than seismic crises, daily totals of volcanic earthquakes were about 1500, although fluctuations, from 1000 to 2000 per day, took place at the end of June."

Information Contact: Same as above.

7/83 (8:7) "The moderate white vapour plume released at Ulawun's summit crater was undisturbed by the volcano's continuing unstable seismicity. A seismic crisis that started on 26 June was the longest since March when this pattern of activity started. It consisted of several periods of tremor up to 13 hours long, and sub-continuous volcanic earthquakes. This activity declined progressively 2-3 July to return to a rate of 1000-1500 B-type events per day. However, the average amplitude of discrete events remained fairly high (about 3 times normal levels) until 20 July. Further seismic crises on 16, 17, and 19 July marked the end of this particular period of stronger seismicity. No significant tilt changes were evident in July."

Information Contacts: C. McKee, P. de Saint Ours, RVO, Rabaul.

9/83 (8:9) [See 8:9 Manam report (above) for SO_2 measurements at Ulawun.]

11/83 (8:11) "About 2 months of steady mild activity consisting of weak-moderate white vapour emission and low seismicity ended abruptly with a very intense seismic crisis 3-8 Nov.

"Weak discontinuous tremor began at 1900 on 3 Nov but showed a dramatic change at 2300 when frequent volcanic earthquakes of large amplitude started. No change was noted in the summit crater emissions.

Numerous volcanic earthquakes continued over the next 2 days and a stage-1 alert warning of an increased risk of an eruption was issued on 5 Nov. At approximately 0200 on 6 Nov a new period of continuous large-amplitude earthquakes occurred and a stage-2 alert was issued, warning that an eruption was likely in the near future. A series of large events was recorded between 0800 and 1000 while observers near the base of the volcano reported dark summit emissions; one report suggested that they contained incandescent material. A stage-3 volcano alert was issued, warning of the possible development of a full-scale eruption later in the day, and the authorities stood by in case people in the most dangerous areas had to be moved.

"Reports from passing aircraft late in the day indicated that tephra emissions may have recurred in early or mid-afternoon. An aerial inspection at 1700 by an RVO volcanologist failed to confirm the presence of new ejecta on the volcano's flanks, but faint haziness possibly due to earlier ejection of fine dust was noted downwind. The emission plume at the summit was found to be white and normal in size.

"At 0530 on 7 Nov, shock waves were seen radiating from the summit crater through the emissions, indicating further mild explosive activity. The plume was larger than normal (about 300 m high) and extended 5-10 km from the summit. Seismic activity and emissions remained at high levels until 8 Nov when they abruptly declined to normal again, allowing a return to a stage-1 alert.

"At the peak of the crisis, over 3000 volcanic earthquakes per day were recorded, declining to about 1300 per day between 8-15 Nov, and 700 at the end of the month. Four A-type events were recorded on 22 Nov."

Information Contact: C. McKee, RVO, Rabaul.

12/83 (8:12) "Volcano-seismic activity continued at a low level throughout most of Dec but increased suddenly again with the appearance of continuous harmonic tremor at 0200 on 24 Dec, less than 2 days after 2 strong regional earthquakes (M_L 6.4 and 6.5, MM IV-V) about 150 km ESE of Ulawun at 0932 and 1102 on 22 December.

"The first period of harmonic tremor, which was low in amplitude, lasted less than 2 hours. It was followed by 3 days (23-26 Dec) of high seismic activity consisting of further periods of low-amplitude, continuous and discontinuous harmonic tremor and numerous larger-than-normal B-type volcanic events. No changes to the normal white vapour emission were observed during the seismic crisis. Disturbances consisting of loud bangs and the production of vapour rings were observed from Ulamona Catholic Mission on 15-16 and 20-21 Dec, but were not accompanied by any events on the seismic records."

Information Contact: P. Lowenstein, RVO, Rabaul.

3/84 (9:3) "Following increased seismicity and reports of audible explosions and vapour rings in Dec 1983, a phase of mild explosive activity took place in Jan. The eruptions were probably phreatic or phreatomagmatic as some correlation with rainfall is evident. Explosions producing tephra clouds were observed 12-14 and 19-20 Jan. These emissions rose as high as 2000 m above the summit. Seismicity throughout this period was of low intensity and characterized by occasional bursts of tremor. A major seismic crisis began at about 1400 on 20 Jan. Tremor rapidly became strong and continuous. Large-amplitude volcanic earthquakes were discernible in the tremor by 2300. Soon afterwards tremor began to subside. It is not known whether eruptive activity accompanied this seismicity as the volcano was obscured by clouds. For the remainder of Jan and throughout Feb, no eruptive activity took place, and seismicity was low.

"An explosion sound from Ulawun was reported on 4 March, and tephra emissions were observed on the 5th and 13th. The emissions on the 13th rose to about 2000 m above the summit. Seismicity was generally at a low level in March. On most days, 100-300 volcanic earthquakes were recorded, but a peak of about 750 events was reached on 9 March. Earthquake amplitudes were slightly higher 9-11 March."

Information Contact: C. McKee, RVO, Rabaul.

8/84 (9:8) "Apart from a few minor steam blast explosions earlier in the year, the level of activity in 1984 has been low. However, in mid-Aug the amplitudes of volcanic earthquakes increased markedly, and by late Aug the frequency of occurrence of volcanic earthquakes was increasing.

"Since 23 Aug occasional explosion sounds from the summit have been heard and occasional small ash clouds have been observed rising 400-500 m above the summit. Weak summit crater glow was reported on 25 Aug, and on the 28th provincial authorities were alerted to the possibility of further developments."

Information Contact: Same as above.

9/84 (9:9) "After several weeks of intensified seismicity and occasional small phreatic explosions, a mild Strombolian eruption began on 4 Sept and ended on 11 Sept.

"Previous Ulawun eruptions developed rapidly (within hours) to full-scale eruptions, including lava effusion and paroxysmal explosive activity culminating in the formation of pyroclastic flows. However, the 1984 eruption developed slowly from the first appearance of summit glow on 4 Sept, and no effusion or paroxysmal activity took place. The eruption reached a peak 6-9 Sept when mild, rhythmic, almost ash-free, Strombolian explosions took place every few seconds.

"During an aerial inspection on 7 Sept, a pool of fluid lava was observed in the bottom of the summit crater, about 50-100 m below the rim; explosions took place from a small circular central vent. Most ejecta were contained by the crater walls, but occasional larger explosions showered the upper flanks of the volcano with incandescent tephra. Visible shock waves were generated by the stronger explosions. A white eruption plume several tens of kilometers long was formed at the peak of the eruption. Seismic amplitudes were 30-40 times normal on 7 and 8 Sept, and the seismicity was characterized by irregular tremor produced by the frequent explosions.

"Regular reports and advice were transmitted by RVO to provincial government authorities throughout, and reactions to the eruption were controlled and rational, resulting in minimal disturbance to daily life."

Information Contact: Same as 9:3.

11/84 (9:11) "No further eruptive activity has occurred since the short-lived Strombolian eruption, 4-11 Sept. The only visible activity has been moderate emissions of white vapours. However, volcano seismicity started to increase again on 9 Nov. Beginning 14 Nov, daily totals of B-type events averaged about 600, a marked increase from daily totals that usually numbered less than 50, 14 Sept-8 Nov. The amplitudes of these events also increased substantially."

Information Contact: Same as 9:3.

12/84 (9:12) "Seismicity, which had intensified in Nov, decreased to a normal level by early Dec. A re-intensification began on 17 Dec and reached a peak near the end of the month, when seismic amplitudes were about 10 times normal.

"Weak summit crater glow was observed on 30 Dec, indicating that an eruption was probably in progress. Fluctuating glow was observed on 3 Jan at about 0400, and occasional small dark tephra emissions from the summit crater were noted between 0800 and 0900. Seismic amplitudes began to increase at about 2300 on 3 Jan and steadily climbed to a peak of about 20 times normal on 8 Jan.

"Aerial inspections by volcanologists on 4 and 5 Jan revealed that the eruptive activity consisted of weak ejections of incandescent tephra from several vents in a mound of fresh lava within the summit crater. The ejections occurred at a rate of 1-2 per minute, and the largest rose about 100 m. The ash content of the emissions was low, and the plume was only a few kilometers long.

"A change in visible activity was evident from 9 Jan when the emissions became rich in white vapour, and ash ejections were less frequent. Crater glow, which had been observed consistently from 3 Jan, was absent from 9 Jan. Despite the weaker visible activity, seismicity persisted at about the same level as that of 8 Jan."

Information Contact: Same as 9:3.

1/85 (10:1) "The eruption, characterized by generally low-intensity Strombolian activity, terminated on 27 Jan. At the peak of seismicity (about 20 times normal levels, on 8 Jan), weak ejections of incandescent tephra occurred at a rate of 1-2 per minute from 1 or 2 vents in a mound of fresh lava in the summit crater. From 9 Jan, seismicity declined steadily, and nighttime incandescence from the crater was absent.

"Seismicity stabilized on about 17 Jan at about 10 times normal levels. Despite the reduced seismicity, summit crater incandescence returned on 16 Jan and persisted until the 25th. Ejections of incandescent tephra were more frequent than earlier in the month, occurring at rates of up to about 10 per minute, and rising to about 130 m above the crater rim. The ash content of ejections was insignificant, and the eruption plume was only 1-2 km long. Seismicity started to decline on 25 Jan and dropped sharply to normal levels on the 27th, marking the end of the eruption.

"An aerial inspection on 30 Jan revealed complex topographic changes in the summit crater. The lava dome formed early in the eruption was surmounted by a small cinder cone, and the flanks of the dome were draped with a mantle of scoria and ash. Lava flows formed an almost continuous moat around the base of the dome. There was no overflow of lava onto Ulawun's flanks, but the lava had almost reached the lowest point in the crater rim, at the head of the NW valley. A small pit crater was present near the SW edge of the summit crater. The volume of new lava and tephra in the summit crater is provisionally estimated to be 10^5-10^6 m^3."

Information Contact: Same as 9:3.

10/85 (10:10) On 17 Nov, Ulawun began to erupt ash and lava, producing 2 slow-moving lava flows. On 21 Nov, one flow was moving N of the summit; the other to the NW. Officials issued evacuation warnings to the [area's] 700 inhabitants, although there was no immediate danger.

Information Contacts: UPI; AP.

11/85 (10:11) "A brief, spectacular Strombolian eruption took place 17-22 Nov, developing rapidly after about 5 days of precursory seismicity.

"Seismic activity began to increase on 12 Nov with the occasional appearance of small discrete B-type volcanic earthquakes. These increased in size and number over the following 3 days, resulting in an official notification on the 15th of an increased risk of an eruption. Seismic activity continued to increase over the following 2 days with the appearance of low-amplitude continuous harmonic tremor on 17 Nov at 1600.

"At 2000, notification was received by radio from Ulamona Catholic Mission (fig. 5-6) of a summit Strombolian eruption in progress, with ejections of incandescent lava fragments to heights of 300-500 m above the crater every few seconds. The eruption was reported to have begun at about 1830, following, or in association with, a rapid increase in the amplitude of the harmonic tremor.

"The intensity of the eruption increased with the emission of 2 large, dark clouds of ash at about 2045 before settling down to a more steady pattern of Strombolian explosions every few seconds, accompanied by a fairly constant level of strong continuous harmonic tremor.

"Volcanologists carried out an aerial inspection at about 0700 the next day and observed regular Strombolian ejections of incandescent lava every few seconds to heights of about 200 m above the summit crater. A small cone was being constructed in the summit crater, but some ejecta were falling outside the crater rim. Small debris slides were occurring intermittently around the N side of the crater. The eruption column at that time was about 2 km high and was lightly laden with ash.

The eruption plume was about 10 km long and trended approximately S. During the day the rate of ash production increased, resulting in a dense pall of ash on the E side of the volcano.

"A lava flow started to descend the N slope in the early evening of the 18th. This flow originated from a fissure about 70 m below the summit crater, and although it moved rapidly at first on the steep upper slopes of the volcano (it may have advanced about 3 km downslope in the first few hours), its progress became very slow when it reached the volcano's gentler middle slopes.

"Spectacular 'fire-fountaining' at the summit crater was observed beginning the night of the 18th. The subcontinuous showering of explosion debris around the crater built up an apron of highly unstable material. Intermittent slides of this material, mainly into the NW valley, produced impressive ash clouds rising from the volcano's slopes. The first of these moderate-sized avalanches was observed moving down the N flank at about 0715 on the 19th. Several more slides occurred later the same day. Throughout the 20th, debris slides were common on the N flank and NW valley. Most advanced less than 4 km from the crater, but a few travelled about 5 km downslope.

"By early morning of the 19th the lava flow had bifurcated, with the E lobe slightly longer. Progress of the flow was slow on the 19th and 20th. At 1400 on the 20th, the nose of the E lobe was about 4 km from the summit, and was about 100 m longer than the W lobe. Their widths were 20-40 m.

"The eruption column was very impressive on the morning of the 20th. Dark grey and convoluted at its base, it paled upward, rising to an altitude of 7-8 km, and was crowned by an elliptical pale grey vapour and ash plume extending W. Most of the ash fallout was controlled by the low-level wind system (below 4 km altitude) blowing from the NW.

The intensity and mode of activity remained unchanged on the 21st, but seismicity began increasing during the early hours of the 22nd. After reaching a peak at about 0800 on the 22nd, seismicity suddenly declined and within 2 hours, the eruption ended.

"When the eruption stopped, the most distal part of the lava flow was about 5.5 km from the crater. Samples collected from the flow are coarsely porphyritic with conspicuous plagioclase phenocrysts. The lava has a similar appearance to previous Ulawun lavas, which are quartz tholeiites.

"During the eruption, on the 18th and 20th, measurements were made at a number of dry tilt arrays around Ulawun. The readings on these days showed that little or no tilting was occurring. Unfortunately, no base line measurements are available to check whether deformation had occurred prior to the eruption."

Information Contacts: C. McKee, P. Lowenstein, RVO, Rabaul.

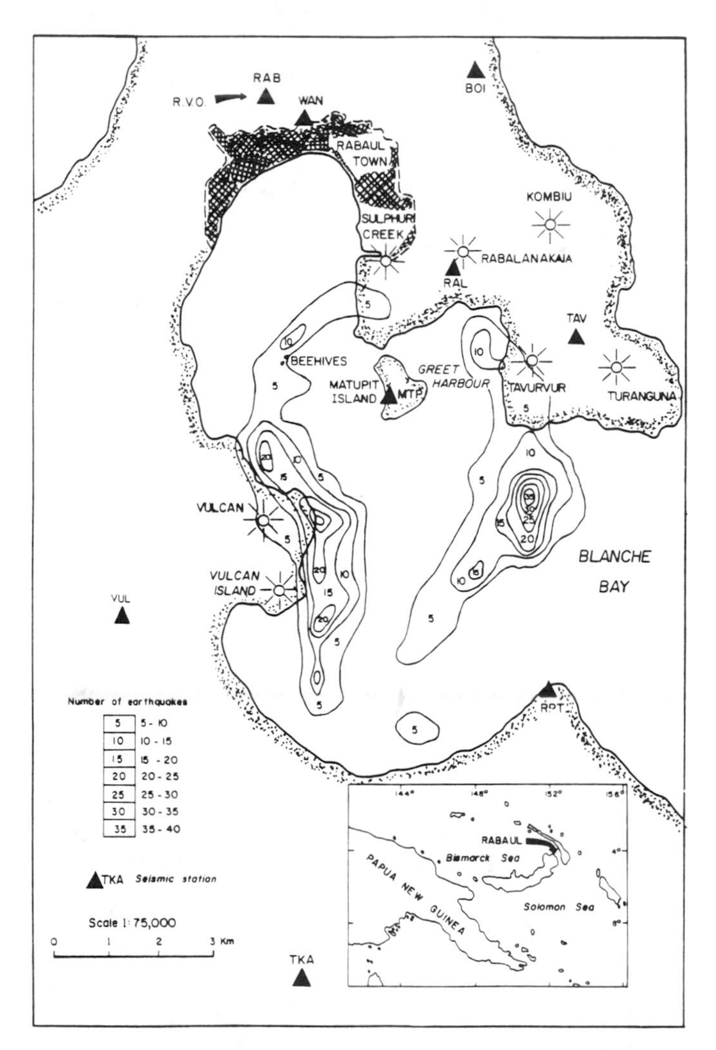

Fig. 5-7. Pattern of seismicity in Rabaul Caldera, Feb 1977-June 1982. Solid triangles represent seismic stations (including the RVO), sunbursts represent volcanic centers. Isoseismal lines show numbers of earthquakes as listed in the table. Arrow on inset index map shows location of Rabaul. Courtesy of C. McKee.

RABAUL

New Britain Island, Papua New Guinea
4.271°S, 152.203°E, summit elev. 229 m
Local time = GMT + 10 hrs

This caldera forms a beautiful natural harbor on the NE tip of New Britain Island. Only 9 historic eruptions are known. The 1937 eruption killed over 500 people, but led directly to the founding of Rabaul Volcano Observatory. This small but able team watches over the many active volcanoes of Papua New Guinea.

General References: Greene, H.G., Tiffin, G.L., and McKee, C.O., 1986, Structural Deformation and Sedimentation in an Active Caldera, Rabaul, Papua New Guinea; *JVGR*, v. 30, p. 327-356.

Johnson, R.W. and Threlfall, N.A., 1985, *Volcano Town: the 1937-43 Eruptions at Rabaul*, Robert Brown and Associates, Bathurst, 151 pp.

McKee, C.O., Johnson, R. W., Lowenstein, P., Riley, R., Blong, R., de Saint Ours, P., and Talai, B., 1985, Rabaul Caldera, Papua New Guinea: Volcanic Hazards and Eruption Contingency Planning; *JVGR*, v. 23, p. 195-237.

8/82 (7:8) Reports through 8:11 are from C.O. McKee.

"The results of geophysical surveillance in Rabaul caldera since the early 1970's have been forming an intriguing scenario which can be interpreted as the prelude to an eruption.

"Seismic surveillance began in Rabaul in 1940 but it was not until 1967, when a network of seismic stations was installed around the N part of Blanche Bay (fig. 5-7), that a reasonable appreciation of local seismicity became possible. Since that time, the apparent seismicity of Rabaul caldera has consisted essentially of shallow, short-period, volcano-tectonic earthquakes originating from about 6 km to near-surface depths. Other, longer-period, harmonic events of possibly local origin are much less common.

"In the early 1970's, the seismic network was extended S and now consists of 9 stations. As the most sensitive stations of the present network are some of those established in 1967, a consistent appreciation of the frequency of occurence of caldera earthquakes is therefore possible. Between 1967 and 1971 seismicity of the caldera was quite stable and the rate of occurrence of local earthquakes varied from about 20 to slightly more than 100 events per month (fig. 5-8). Since late 1971, distinctly higher numbers of earthquakes have been registered in what have come to be recognized as swarms of caldera earthquakes. Typically, these swarms last for periods of about 10 minutes to several hours in which individual earthquakes may occur at such short intervals that resolution on seismograms is commonly prevented.

"Fig. 5-8 clearly shows the effects of a number of highly active periods (seismic crises) since 1971, in which sequences of seismic swarms took place. A trend of progressively higher numbers of events in successive seismic crises is discernable. In the 2 most recent crises, in Sept-Oct 1980 and Jan-March 1982, the strongest earthquakes registered at M_L 5.2 and 5.1 respectively.

"A clear picture of the seismically active parts of Rabaul Caldera has emerged for the period 1977-1982, for which computer locations of caldera earthquakes have been obtained. A majority of the events located in this period have been shallow ones within about 2 km of the surface. In practice, the computer-locatable proportion of the total number of detected caldera earthquakes is about 10%, as reliable locations are only achievable when events are registered on 5 or more stations of the network. A further 10-15% of the total of detected events are registered by the group of 3 stations MTP, RAL, and TAV, (fig. 5-7) suggesting that these (smaller) events originate from the Greet Harbour area.

"The broad pattern of Rabaul caldera seismicity is seen to be 2 arcuate zones, near the mouth and in the W part of Blanche Bay. A markedly elongated concentration of seismicity skirts the headland formed by the historically active Vulcan-Vulcan Island volcanoes (active 1878 and 1937), and an intense zone of seismicity lies about 3 km S of Tavurvur volcano, also active historically (1767(?), 1791, 1878, 1937, and 1941-43). Other distinct concentrations within the main seismic zones lie near the Beehives and in the Greet Harbour area. The shape of the main zones combined could be linked to a major, caldera-modifying eruption about 1300 BP; that is, the seismicity may define the fault(s) along which caldera deepening or widening took place. In addition, there appears to be an intimate relationship between the most recently active volcanoes in the caldera and some of the concentrations of seismicity. The most notable of these associations is in the Vulcan-Vulcan Island area, and weaker seismic zones appear to be related to Tavurvur (Greet Harbour) and the Beehives. The zone of strongest seismicity, near the entrance to Blanche Bay, is not closely associated with any well-known centres, although there is some evidence suggesting the possible existence of several submarine volcanoes nearby. Some clarification of these relationships and interpretations is provided by the results of deformation studies.

"The methods used to monitor deformation in Rabaul caldera include routine tiltmetry, optical levelling, gravity measurements, and sea level monitoring. Remarkable deformation in the Sulphur Creek-Matupit Island area has been monitored closely since 1973, principally by optical levelling and gravity techniques. Since 1973, the S end of Matupit Island has been uplifted almost 1 m. Fig. 5-9 shows the results of optical levelling surveys carried out since 1973 along a line from a stable bench mark (BM 21) near Rabaul Observatory to Matupit Island. It appears that along this level line, elevation changes are only significant S of Sulphur Creek. Other incomplete levelling data on stations immediately N and 1-2 km E and SE of Greet Harbour indicate that over the same time interval, minor uplift, probably less than 10 cm, has occurred in these areas. However, significant uplift is believed to have taken place at the SW coast of Tavurvur volcano. Partial contours defining these changes indicate that the focus of uplift probably lies S of the entrance to Greet Harbour. Results from dry tiltmetry around Greet Harbour since early 1981 and measurements from a spirit level tiltmetry at TAV seismometer site since early 1972 confirm the existence of an uplift source in this area. Dry tilt data from a station on the S end of Matupit Island indicate a current rate of uplift (to the SE) of about 4-8 μrad per month. Level and gravity results from the W

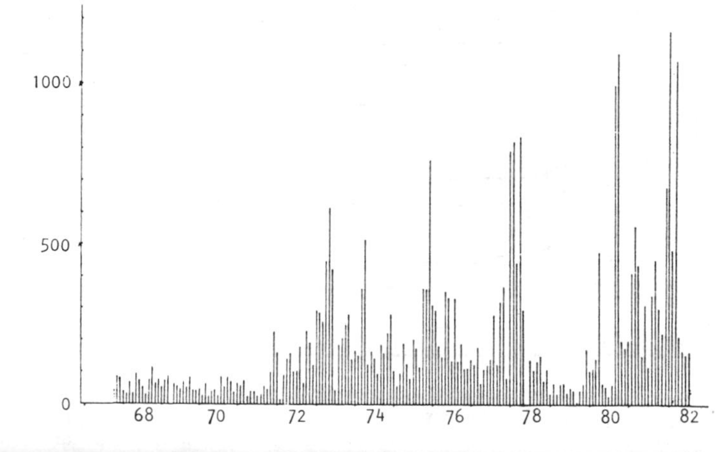

Fig. 5-8. Rabaul Caldera seismicity, showing monthly numbers of earthquakes, 1967-1982.

and S shores of Blanche Bay indicate relatively minor uplift there.

"The localized nature of the strongly deformed area is indicative of a shallow source of inflation. Preliminary determinations, comparing the uplift over different time intervals, suggest a focal depth of about 1-3 km. A noteworthy feature of the level changes shown in fig. 5-9 is the steady rate of deformation. These aspects of the uplift suggest that the focus of the deformating source is relatively static in a vertical sense and that the deformation may be due to enlargement of the source.

"The developments in seismicity and deformation in Rabaul caldera appear ominous, although the rate of change in each of these parameters is steady, indicating a slowly evolving situation. It appears that uplift of Matupit Island was taking place before the regular program of gravity and levelling surveys commenced in 1973, but the time of onset of this effect is not known. A tiltmeter at Rabaul Observatory underwent a net change of about 40 μrad of southwards uplift between 1964 and 1981, but a tiltmeter at Sulphur Creek, much closer to the present focus of deformation, has shown a complex record of SW uplift since its installation in 1972. The seismicity has a much clearer history, with unequivocal changes commencing in late 1971.

"Significantly, the centroid of the 2 arcuate seismic zones has a similar location to the epicentre of the deformation. It could be concluded that the seismic zones outline a rising area in agreement with the known deformation. A possible cause for this apparently related activity could be accumulation of magma at shallow depth beneath the centre of Blanche Bay. This may eventually lead to an eruption."

Information Contact: C. McKee, RVO, Rabaul.

Further Reference: Cooke, R.J.S, 1977, Rabaul Volcanological Observatory and Geophysical Surveillance of the Rabaul Volcano; *The Australian Physicist*, Feb. 1977, p. 27-30.

9/83 (8:9) An exponential increase in seismic activity in Rabaul caldera began in late Aug and culminated in an intense crisis with 621 earthquakes on 19 Sept. The strongest event had an M_L of 4.2. Since then seismicity has remained high at 40-120 events per day and has included several minor crises. The total number of caldera earthquakes in Sept was 2135, a significant increase over the previous highest monthly totals of 1170 and 1079 in Jan and March 1982.

"The earthquakes have been concentrated at depths of 0-3 km near Tavurvur Volcano, a small post-caldera

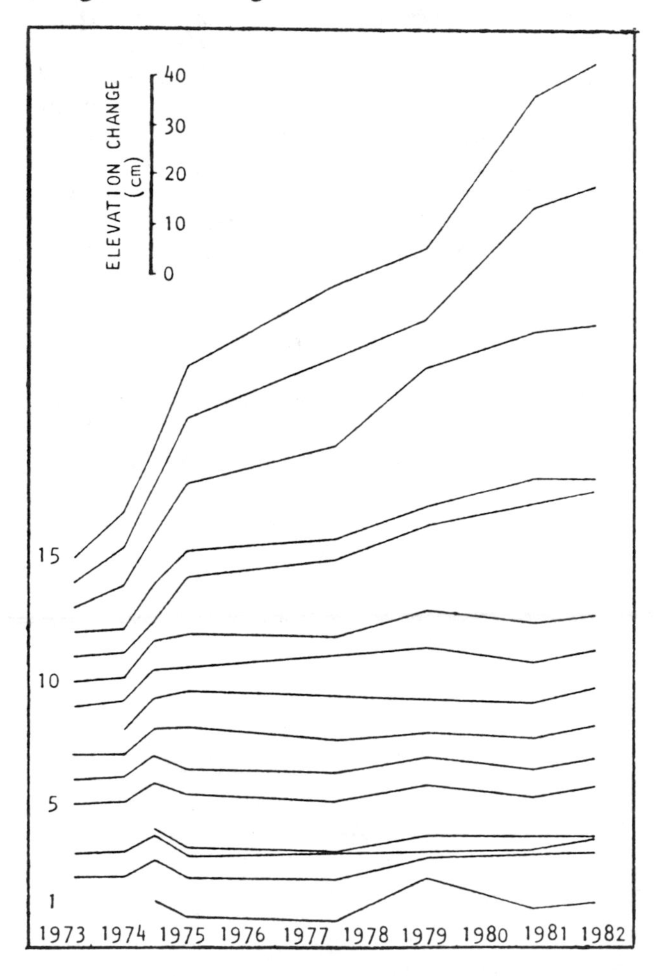

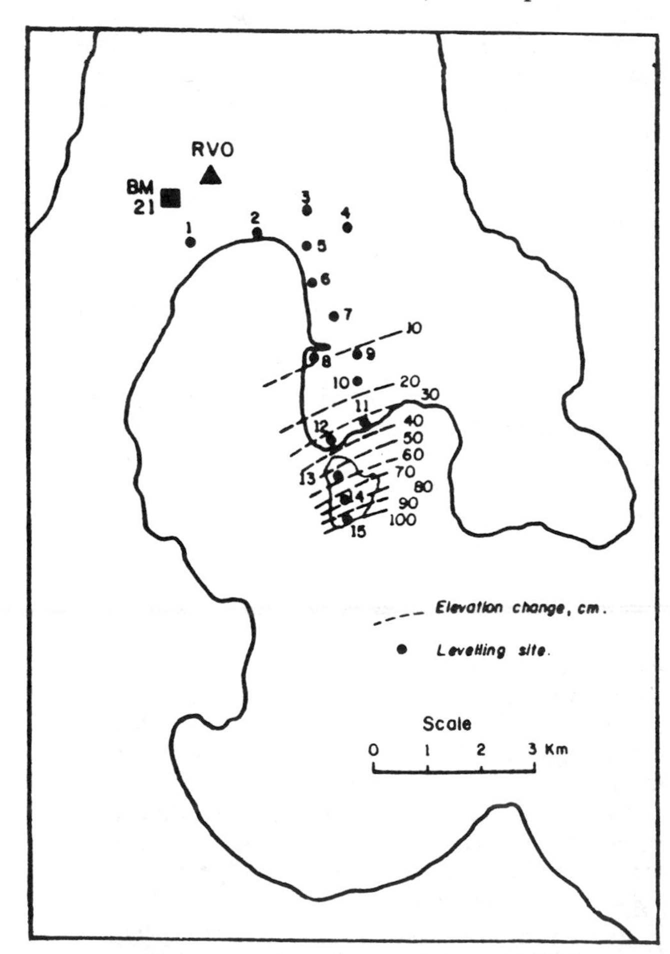

Fig. 5-9. Elevation changes at Rabaul, 1973-1982, S of RVO. *Left,* map showing gravity network stations used as levelling sites and partial contour lines of elevation changes relative to stable benchmark BM 21. *Right,* results of optical levelling on gravity network stations. Elevation changes for each station are relative to the first measurement for each station (left end of each line) and can be read from the vertical scale in the upper left corner. Courtesy of C. McKee.

cone on the E section of the elliptical caldera bounding fault, but other sections of this fault have also been seismically active (7:8-9).

"Tilt measurements showed distinct uplift centered 1.5 km S of Tavurvur. Uplift commenced in early Sept in relation to increasing seismicity. A sharp tilt change of up to 49 μrad accompanied the seismic crisis of 19 Sept, but tilt rates have since returned to normal. The depth and increase in volume of the source of ground deformation are estimated to be about 1.7 km and 1.9 × 10^2 m^3."

Information Contact: Same as 7:8.

10/83 (8:10) "Further dramatic increases in seismicity and ground deformation rates took place in Oct. The total number of caldera earthquakes for the month was 5198, about 2.5 times the total for Sept and about 35 times the average monthly total since the last peak of activity in March 1982. A large proportion of events occurred in seismic crises on 15-16 Oct (daily earthquake totals of 868 and 305) and 28-29 Oct (daily totals of 338 and 513 events). The strongest earthquakes had magnitudes (M_L) of about 4, and subterranean rumblings accompanied many of the felt events. The earthquakes continued to be concentrated in the E and NE parts of the caldera at depths of about 0-3 km.

"Tilt measurements showed maximum values around Tavurvur Cone in the NE part of the caldera. Tilts of up to about 20 μrad accumulated gradually until the seismic crisis at the end of the month, when step-wise changes of up to about 40 μrad were measured. Ground cracks were found on the W flank of Tavurvur after the seismic crisis of 28-29 Oct.

"Interpretation of the tilting using a point-source model indicates that the centre(s) of deformation could be immediately offshore about 1 km below the SW flank of Tavurvur. The increase in volume of the deforming source(s) is estimated to be of the order of 1 × 10^6 m^3."

Information Contact: Same as 7:8.

11/83 (8:11) "A stage-2 volcano alert, implying a possible eruption within a few months, was declared by the RVO on 29 Oct, in response to the increased seismic activity and ground deformation in Sept and Oct.

"In Nov the seismic activity continued to intensify but the rate of ground deformation remained the same. The total number of caldera earthquakes for the month was 5748, an increase of 550 over October's total. Seismic crises became more common but less intense, occurring on 5, 9, 15, 17, 18, 26, 29, and 30 Nov. Daily totals of earthquakes on these days ranged from 220 to 538. During the month, there was a linear increase in background seismic activity from about 100-170 events per day. Early in Nov, earthquakes appeared to be concentrated on the NE part of the caldera, near Tavurvur, but the area around Vulcan, on the W side of the caldera, became active later in the month. The strongest earthquakes had magnitudes of about 3.5.

"The pattern of ground deformation was similar to that seen in the previous 2 months. Dry tilt stations on the NE side of the caldera continued to indicate that the centre of uplift is near the mouth of Greet Harbour. The maximum measured tilt was 43 μrad at a station on the coast of Tavurvur's SW flank. Tilts accumulated gradually during the month at most stations, although an offset of about 10 μrad was recorded at a station on Matupit Island after the seismic crisis of 5 Nov.

"EDM data from networks newly established in Nov showed that horizontal movements were near or within noise levels but that slight expansion may have been occurring within about 3 km of the inflationary centre on the NE side of the caldera."

Information Contact: Same as 7:8.

12/83 (8:12) The following 2 reports are from P. Lowenstein.

"Although the monthly total number of earthquakes was higher than in any of the previous months, the percentage of stronger earthquakes (those with sufficient energy to be recorded by 5 or more of the 9 harbour seismic stations) was lower than in any of the previous months (table 5-3). The distribution of earthquakes was similar to that in Nov with the highest concentration in the NE part of the caldera near Tavurvur, but most of these were small, recorded by only the 3 closest seismic stations.

"The maximum measured tilt was 17 μrad, by the station on the coast, on Tavurvur's SW flank. Tilts again accumulated gradually throughout the month, and no offsets were associated with any of the minor swarms of harbour earthquakes. Ground deformation rates were less than half those in Nov, and indications are that there was considerably less energy release in Dec than in any previous month during the current crisis.

"It is still too early to determine whether the decrease of unrest in the Rabaul Caldera in Dec is just temporary, or marks the onset of a longer-term improvement in the situation."

Information Contact: P. Lowenstein, RVO, Rabaul.

Further Reference: McKee, C.O., Lowenstein, P.L., de St. Ours, P., Talai, B., Itikarai, I., and Mori, J., 1984, Seismic and Ground Deformation Crises at Rabaul Caldera: Prelude to an Eruption?; *BV*, v. 47, p. 397-411.

1/84 (9:1) "There was a marked increase in the amount of unrest in Rabaul Caldera during Jan, with a total of 8372 volcanic earthquakes recorded, an increase of 1255 over the Dec total.

"A major seismic crisis took place on 15 Jan when

1983	Number	% Stronger
Sept	2135	13.3
Oct	5199	6.9
Nov	5748	4.2
Dec	7117	3.8

Table 5-3. Earthquakes at Rabaul, Sept-Dec 1983. 'Stronger' events were recorded by 5 or more of the 9 harbor stations.

942 earthquakes occurred, including several strongly felt events. The maximum magnitude earthquake (M_L 4.9) was accompanied by underground rumbling sounds. This crisis was accompanied by a maximum tilt change of 32.5 μrad at [a station immediately N of Greet Harbour]. . . .

"The overall distribution of earthquakes in Jan was similar to that in Dec, with high concentrations on the NE (Greet Harbour) and W (Karavia Bay) sides of the harbour. . . . Steady inflation of the Karavia Bay and Greet Harbour magma reservoirs continued throughout the month. . . .

"As a result of the increased activity in Jan, a warning was issued to the authorities to the effect that the eruption, which was previously thought to be only a possibility when the Stage-2 volcanic alert was declared on 29 Oct, was now much more likely to occur within the next few months."

Information Contact: Same as 8:12.

2/84 (9:2) The remaining reports are from the RVO staff.

"Seismicity in Rabaul caldera showed a further intensification in Feb. Although the total number of volcanic earthquakes recorded (8339) was slightly lower than in Jan (8372), there were more stronger events in Feb. The entire caldera seismic zone was active during the month. However, seismicity was concentrated in the NE, N, and W parts of the caldera. Seismic crises took place on 1, 5, 7, 8, 11, 12, 13, 18, and 27 Feb, the most energetic on the 7th, 13th, and 18th, affecting the NE, N, and W parts of the caldera, respectively. The strongest caldera earthquake in these crises was a magnitude (M_L) 4.3 event on the 18th.

"An interesting development in Feb was the increase in rates of tilting around Vulcan cone on the W side of the caldera. The tilt change at 1 station immediately SW of Vulcan was about 40 μrad, about 4 times the tilting seen there in Jan. By contrast, tilting around Greet Harbour in the NE part of the caldera was reduced compared with that of Jan, and the pattern of tilts there was more complex. Relatively minor tilting occurred during seismic crises. The largest crisis-related tilts were about 15 μrad near Vulcan.

"EDM data over the last 3 months show that significant horizontal deformations are taking place at Rabaul and are most pronounced in the Greet Harbour area. Extension of about 60 microstrain (change in length divided by original length $\times 10^6$) has occurred across Greet Harbour in this period. There is no evidence of widespread horizontal movements."

Information Contact: C. McKee, RVO, Rabaul.

3/84 (9:3) "Seismicity continued to intensify in March. The total number of caldera earthquakes was 8729, as compared to 8339 in Feb. More significant than actual numbers of earthquakes, however, was the continued increase in the proportion of stronger earthquakes. This change appears to be related to the increased incidence of somewhat deeper (2-4 km), more energetic earthquakes in the Vulcan area. The remainder of the caldera seismic zone continued to be active, highlighted by the usual strong concentration of very shallow, relatively low-energy events under the W flank of Tavurvur.

"Major seismic crises took place on 3 and 25 March. The totals of caldera earthquakes on those days were 932 and 726, respectively. The crisis of 3 March involved the E part of the caldera seismic zone, at the mouth of Blanche Bay, and included an event of M_L 5.1. Strong ground deformation was associated with this crisis. Tilts of up to about 50 μrad indicated inflation centred near Sulphur Point at the mouth of Greet Harbour. [Horizontal distance measurements near the time of the crisis were affected by movement of one of the caldera rim base stations.] After this crisis, expansion of Greet Harbour resumed at the same rate as before, about 25 microstrain per month. The crisis of 25 March was centred immediately NE of Vulcan, and the strongest earthquake was an M_L 3.7. Waterspouts up to about 3 m high were observed briefly near the NE shore of Vulcan during this crisis. No significant ground deformation accompanied the seismicity.

"Five levelling surveys carried out between late Nov 1983 and mid-March 1984 showed that the area of maximum measured uplift in the caldera is at the S end of Matupit Island. The rate of uplift in this period was steady at about [50 mm per month]. This compares with an uplift rate of about [8 mm per month] for the period 1973-1983."

Information Contact: Same as above.

4/84 (9:4) "A further intensification of seismic activity took place in April. The total number of caldera earthquakes was 13,749, 60% more than in March. Seismicity was concentrated on the E side of the caldera, in Greet Harbour and at the entrance to Blanche Bay.

"Major seismic crises occurred on 21 and 22 April, when 1011 and 1717 events were recorded. The crisis on the 21st was centred at the mouth of Blanche Bay, and the strongest earthquake was a magnitude (M_L) 3.6 event. Only minor ground deformation was associated with this crisis.

"On the 22nd at 1100 an M_L 4.8 earthquake heralded the most energetic crisis to date, which was centred at the head of Greet Harbour. Structural damage in this and the Sulphur Creek area included cracking, and in one case collapse, of masonry walls, cracks in concrete floors, a burst water main, and burst household water tanks. Tilts around Greet Harbour ranged from 30 to 50 μrad, generally showing a pattern of radial inflation centred in the harbour. Measurements of horizontal deformation indicated expansion of the Greet Harbour area by 20-30 microstrain.

"The overall pattern of ground deformation in April indicated that the strongest tilting, of up to 80 μrad, was in the Greet Harbour area. Rates of horizontal deformation indicated expansion was about double that in any previous month (40-50 microstrain).

"Levelling surveys from Rabaul Township to Matupit Island and around Greet Harbour showed that between mid-March and mid-April the S end of Matupit Island rose 76 mm. Further uplifts of about 50 mm on Matupit Island and at the head of Greet Harbour accompanied the 22 April seismic crisis, making the total uplift in April about double that in any previous month."

Information Contacts: C. McKee, P. Lowenstein, RVO, Rabaul.

5/84 (9:5) "Overall, a substantial decline in seismicity took place in May. The total number of earthquakes was 8938, compared to 13,794 in April.

"Following the intense seismic and deformation crisis of 21-22 April, seismicity decayed exponentially and by mid-May the daily earthquake totals were averaging about 215. Small swarms of volcanic earthquakes that occurred on 4 and 27 May were followed by a major seismic and deformation crisis on 29 May. The earthquake swarms affected the NW part of the caldera seismic zone on the 4th and the Greet Harbour area on the 27th, but no significant ground deformation accompanied them.

"The seismic crisis of the 29th was centered near Vulcan; the main event had a magnitude (M_L) of 3.8. An unusual feature of this crisis was that deflation of up to about 30 μrad accompanied the seismicity in the Vulcan area and that apparently aseismic inflation of the same magnitude took place at the head of Greet Harbour. Levelling measurements about 1 week after the crisis indicated maximum uplift of 43 mm at the S end of Matupit Island.

"The main features of ground deformation before the crisis of the 29th were mild steady inflation (10-20 μrad) immediately SE of Vulcan, deflation (up to 40 μrad) and slight subsidence (a few mm) near Rapindik at the head of Greet Harbour, and uplift of Matupit Island (about 41 mm at the SE coast). Horizontal deformation continued unchanged throughout May.

"The effects of the crisis of the 29th suggested direct interplay between different parts of the caldera. It is noteworthy that eruptions occurred simultaneously at Vulcan and Tavurvur in 1878 and 1937, and that disturbances along a line connecting these 2 centres were also observed at the same time."

Information Contact: C. McKee, RVO, Rabaul.

6/84 (9:6) "Overall, June was a relatively quiet month. A further decline in seismicity took place. The total number of earthquakes for the month was 5304, and average daily earthquake totals were steady at about 160. The only perturbation to the steady rate of seismic energy release was a swarm of earthquakes in the Greet Harbour area on 25 June. The total number of earthquakes that day was 450, and the largest event had a magnitude (M_L) of 2.8. No significant ground deformation accompanied this seismicity.

"Uplift continued throughout June in the Matupit Island-Greet Harbour area, but at variable rates. Between 23 May and 6 June the S end of Matupit Island rose 43 mm. From 6 to 26 June uplift in the same area was only 10 mm. Inflationary tilt around Greet Harbour persisted but at a reduced rate, with maximum changes of 20 μrad at Sulphur Point and Rapindik. Slight deflationary tilt occurred in the Vulcan-Karavia Bay area, with a maximum change of 20 μrad at Karavia. Horizontal distance measurements across Greet Harbour indicated E-W expansion at the same rate as previously, but a decline in the rate of N-S expansion."

Information Contact: Same as above.

7/84 (9:7) "Seismicity continued to decline in July (fig. 5-10), but some parameters of ground deformation showed an increase after a lull in June.

"The total number of caldera earthquakes for July was 4404, compared to 13,794 in April, 8938 in May, and 5304 in June (fig. 5-11). Daily earthquake totals were fairly steady, ranging from 56 to 280. Two small swarms of earthquakes, including 1 or 2 felt events, took place 10 and 13 July. The swarm on the 10th occurred in the Vulcan-Karavia Bay area, while the one on the 13th was restricted to the Vulcan area. However, the total numbers of earthquakes on these days were 193 and 280, only slightly above background. Neither of these swarms was accompanied by ground deformation.

"Although some parameters of ground deformation

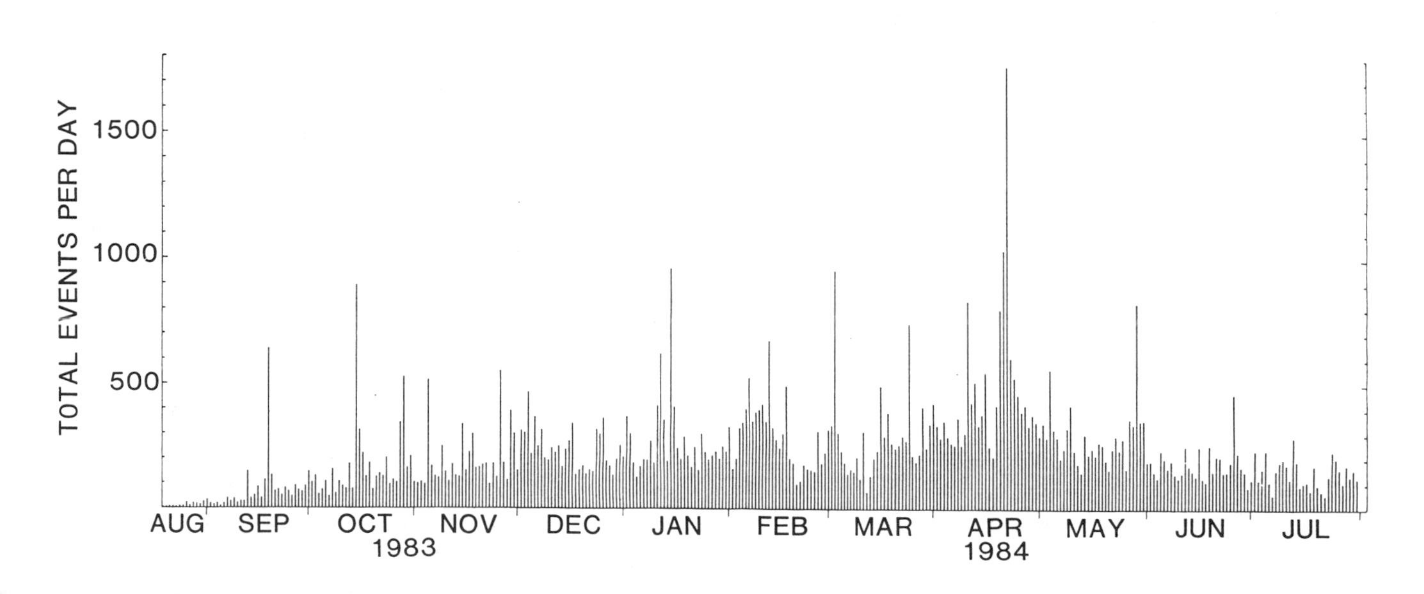

Fig. 5-10. Number of earthquakes per day at Rabaul Caldera, Aug 1983-July 1984. Courtesy of RVO.

tended to be stronger in July than in June, none of them showed above average rates of change. Levelling measurements carried out in the last week of July indicated 38 mm of uplift at the S end of Matupit Island since late June. Only 10 mm of uplift occurred in this area 6-26 June. Horizontal distance measurements indicated a return to fairly uniform expansion across Greet Harbour, at rates of about 20-25 ppm per month. In general, tilt changes were smaller in July than in June, with most tilts less than 8 μrad. However, the station at the NW shore of Greet Harbour continued to show inflation to the SE (in Greet Harbour) at a rate of about 20 μrad per month. The pattern of tilts indicated continued weak inflation of the Vulcan and Greet Harbour areas."

Information Contact: Same as 9:6.

8/84 (9:8) "The state of unrest continued in Aug at about the same level as in July. The total number of earthquakes for the month was 5285, compared to 8938 in May, 5304 in June, and 4404 in July. But the total [seismic energy] release was about [2.4×10^{16}] ergs, compared to [1.4×10^{17}] in May, [1.9×10^{15}] in June, and [1.3×10^{17}] in July. [These and subsequent energy release values for Rabaul have been corrected by RVO.]

"Seismicity in Aug was concentrated in the N half of the caldera seismic zone with a crisis consisting of 628 earthquakes, including a magnitude 3.4 event, on 3 Aug. Six smaller earthquake swarms took place in the Greet Harbour area, one on 11 Aug, and five in the period 25-27 Aug.

"None of the seismic activity was accompanied by any sudden ground deformation changes. Ground deformation measurements showed continuing slow inflation, mainly in the Matupit-Greet Harbour area, at a similar rate to that in July. The maximum measured vertical uplift at the SE end of Matupit Island was about 35 mm, and expansion across Greet Harbour amounted to about 10-20 ppm.

"Most of the evidence obtained in the past 2 months suggests that the volcano has settled into a fairly steady and linear rate of progress toward the anticipated eruption. In the absence of any unexpected changes, the situation could continue at the present rate for several months to a few years before an eruption occurs."

Information Contact: Same as 9:6.

9/84 (9:9) "Declining seismicity and rates of ground deformation were evident in Sept. The total number of caldera earthquakes for the month was 4048 and the seismic energy released amounted to about [8×10^{14}] ergs (compared to 5285 earthquakes and about [2.4×10^{16}] ergs in Aug). Seismicity was concentrated in the N

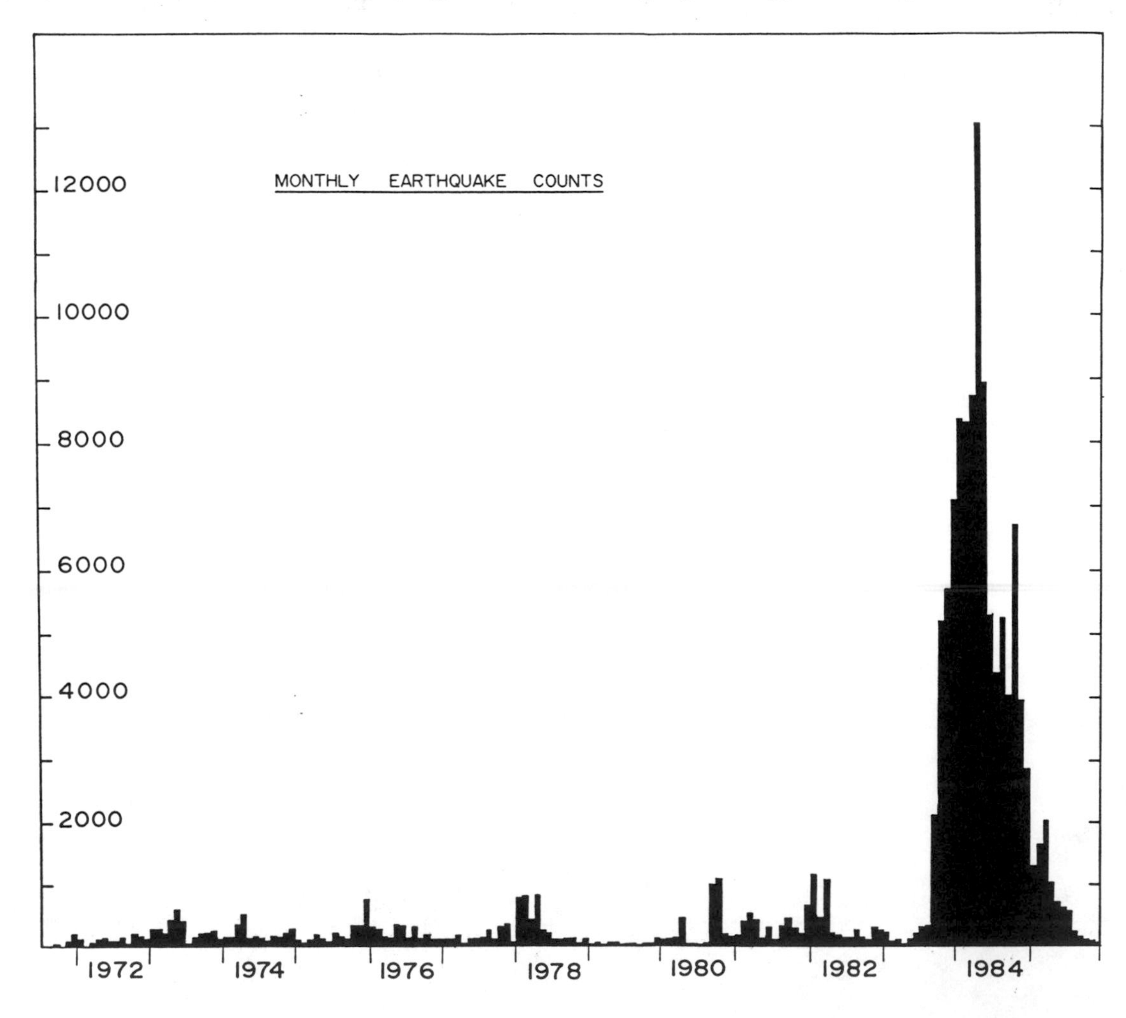

Fig. 5-11. Monthly earthquake totals, Rabaul caldera, 1971–1985. Courtesy of RVO.

part of the caldera seismic zone. Measured vertical uplift continued to be greatest along the SE coast of Matupit Island. However, the maximum uplift in Sept was only 14 mm (compared to 35 mm in Aug). Tilt and horizontal distance measurements continued to indicate slow steady inflation and expansion in the Greet Harbour area."

Information Contact: Same as 9:6.

10/84 (9:10) "Seismicity and rates of ground deformation reintensified in Oct. The total number of caldera earthquakes for the month was 6749, and seismic energy released was [5.5×10^{18}] ergs.

"The increased level of activity was due mainly to a seismic and ground deformation crisis on 18 Oct. The seismicity was concentrated in the Blanche Bay area and included 4 moderate-to-strong earthquakes (M_L 4.9, 3.6, 3.35, and 3.3). The seismic energy released during the crisis amounted to . . . about 90% of the month's total seismic energy. Tilt changes measured soon after the crisis indicated a deformation source immediately offshore (W) from Sulphur Point, at the N edge of Blanche Bay. The maximum measured tilt change was about 90 µrad. Using a point-source model, the deformation source was calculated to be about 1.2 km deep, and the volume change at the source about 1×10^6 m^3. The uplift at Sulphur Point was about 100 mm. The ground deformation associated with the crisis was very localized. At the SE coast of Matupit Island, about 1.5 km from the deformation source, the uplift was only 33 mm; at the N shore of Greet Harbour, about 2.5 km away, it was only 5 mm. No marked horizontal deformation took place in association with the crisis.

"In addition to the crisis on the 18th, there were a number of seismic swarms and a few moderate-to-strong discrete earthquakes. The most notable was a swarm at Greet Harbour on 8 Oct (maximum M_L 3.8), a moderate-to-strong earthquake (M_L 3.8) at the entrance to Blanche Bay about 10 hours after the crisis on the 18th, and seismic swarms from around the Vulcan headland on 24 (maximum M_L 2.8) and 26 Oct (maximum M_L 3.2).

"Most of the Oct ground deformation took place on the 18th, but tilting and uplift continued at a reduced rate around Greet Harbour for the remainder of the month. An offset of about 25 µrad was registered at 1 station on Vulcan after the seismic swarm on the 26th. The maximum ground deformation recorded for the month was 130 µrad of tilt and 100 mm of uplift at Sulphur Point. Horizontal deformation was mostly insignificant, although a distinct N-S dilation was evident at the mouth of Blanche Bay. This was due largely to a northward shift (about 50 mm) of Sulphur Point."

Information Contact: Same as 9:6.

11/84 (9:11) "The level of activity subsided again following the seismic and ground deformation crisis on 18 Oct. Caldera seismicity in Nov consisted of 3985 earthquakes with total energy of about [6×10^{15}] ergs (maximum M_L, [2.9]). Most of the energy was released during small swarms on 6, 8, 10, 18, 19, and 27 Nov. Seismicity was concentrated in the N half of the caldera seismic zone. Meanwhile, ground deformation measurements reflected slow steady inflation at both shallow magma reservoirs, under the mouth of Greet Harbour and immediately E of Vulcan. In the Greet Harbour area, maximum uplift was 19 mm, maximum tilt was 30 µrad, and maximum horizontal strain was 20 ppm.

"In view of the general decrease in activity at Rabaul since the beginning of June 1984, the RVO advised government authorities on 22 Nov that the situation was considered to have returned to a stage-1 volcanological level of alert in which the anticipated eruption is not now expected to occur before several months to a few years."

Information Contacts: C. McKee, P. Lowenstein, RVO, Rabaul.

12/84 (9:12) "A further decrease in activity was evident in Dec. Seismicity reached its lowest level since the beginning of the current crisis period in Aug 1983. Seismic energy release for the month was [1.3×10^{15}] ergs; 2887 earthquakes were recorded. The largest earthquake for the month was an M_L 2.5 event that was part of a small seismic swarm in the Vulcan area on 22 Dec. The month's seismicity was concentrated in the Vulcan-Matupit Island area. A mild rate of ground deformation persisted. The greatest changes were in the Greet Harbour area, where the maximum tilts were about 20 µrad on each side of the mouth of the harbour."

Information Contact: C. McKee, RVO, Rabaul.

2/85 (10:2) "Activity at Rabaul remained at a low level in Jan and Feb. Monthly caldera earthquake totals were 1297 and 1672, and seismic energy outputs were about [4.4×10^{14}] and [3.4×10^{15}] ergs. In Jan the strongest earthquake was an M_L [2.3] event on the 25th, and only 6 events had magnitudes exceeding 2.0. The strongest earthquake in Feb was an M_L 3.0 event on the 24th, and 13 events had magnitudes exceeding 2.0. Jan's seismicity was concentrated in the Tavurvur and Sulphur Creek-Matupit Island areas, while in Feb, seismicity was strongest in the Vulcan-Matupit Island area and at the head of Greet Harbour.

"Mild ground deformation continued, with changes being above uncertainty levels only in the Greet Harbour area. The largest tilt and horizontal distance changes were of the order of 15 µrad and 10 ppm per month. Levelling completed at the end of Jan showed maximum uplift since early Dec of only 15 mm at the S end of Matupit Island. This is only slightly higher than the average pre-crisis uplift rate from 1971-1983."

Information Contact: Same as above.

3/85 (10:3) "After about 4 ½ months of relatively low activity, a moderate seismic and deformation crisis took place in Rabaul Caldera on 3 March. The part of the caldera affected was the region between Davapia Rocks and Rabalanakaia Volcano (fig. 5-12). The strongest earthquake was an M_L 3.6 event; altogether there were 6 earthquakes of magnitude 3 or greater. More than 600 small earthquakes were also detected, most less than 2 km deep. Associated measured ground deformation was mild; the largest tilt change was 16 µrad at the NW shore of Greet Harbour, and no significant changes were noted in EDM lines across the affected area.

"Apart from the crisis, activity was at low levels in March. The total number of earthquakes for the month (including the crisis of the 3rd) was 2052, and the average background number of events was 45 per day. The seismic energy output was about [6.9×10^{16}] ergs, most released during the crisis. The Matupit Island-Greet Harbour area continued to be the focus of ground deformation, with the largest tilts about 30 μrad at Sulphur Point. Levelling carried out 11-18 March indicated that since the previous survey (21-29 Jan) greatest uplift (45 mm) had taken place at the S end of Matupit Island. Maximum rates of horizontal deformation continued to be about 10 ppm per month."

Information Contact: Same as 9:12.

4/85 (10:4) "A very low level of activity prevailed in April. The total number of caldera earthquakes in April (1041) is the lowest monthly total for the whole crisis period (starting Sept 1983). The seismic energy output was only about 6×10^{13} ergs and the strongest caldera earthquake was only M_L 2.1. Seismicity was concentrated in three areas within the caldera seismic zone: at the entrance to Blanche Bay, around the Vulcan headland, and in the NE part of Greet Harbour.

"Rates of ground tilt were barely above noise levels. The biggest tilt changes, about 6 μrad, were recorded at stations around Greet Harbour. Maximum rates of horizontal deformation continued to be about 10 ppm per month, near the entrance to Greet Harbour."

Information Contact: Same as 9:12.

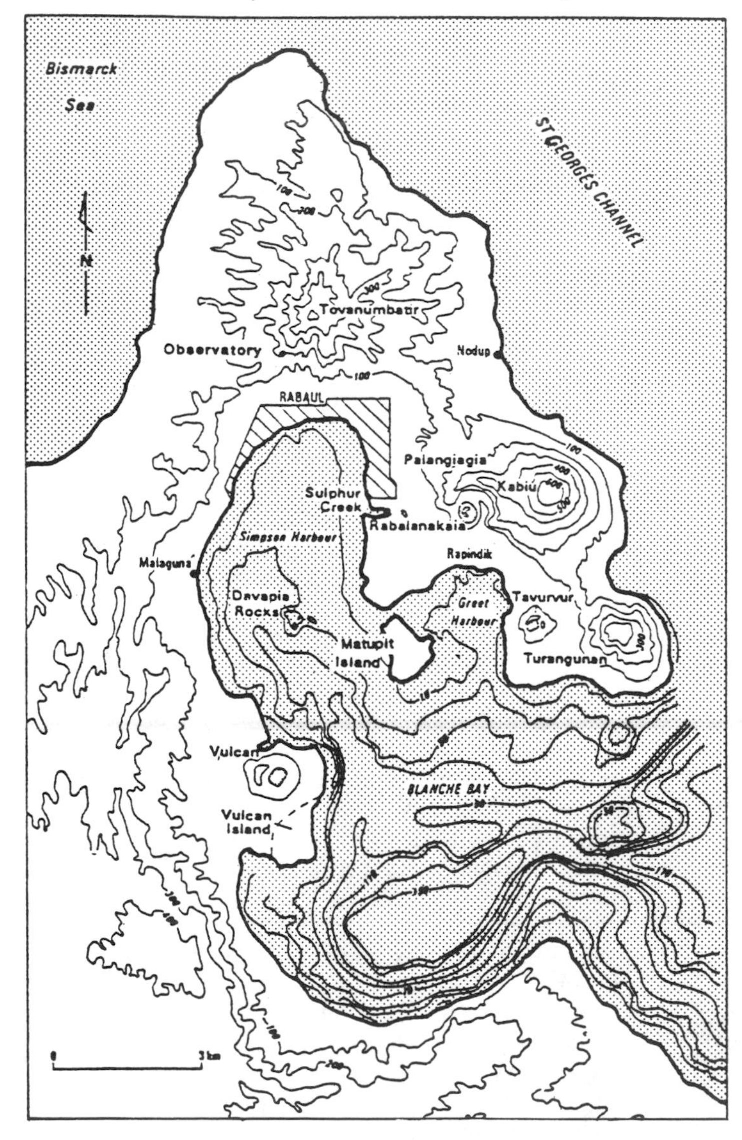

Fig. 5-12. Topography and bathymetry of Rabaul Caldera, including intra-caldera and satellite cones. Contours on land are in meters and isobaths are in fathoms, at about 37 m intervals. After McKee and others, 1983.

6/85 (10:6) "Generally low levels of volcano seismicity and ground deformation have prevailed since the 3 March crisis. However, two recent major regional earthquakes, both magnitude 7.1 (M_L), caused strong ground movements at Rabaul: intensities were MM V on 10 May and MM VI-VII on 3 July. The 10 May event triggered a swarm of caldera earthquakes. During the following 5 hours, over 100 caldera earthquakes were recorded, mainly from the Vulcan area. The largest was a magnitude 3.7 (M_L) earthquake about 1 hour after the regional event. There was little measurable ground deformation associated with this seismic swarm. No comparable caldera seismic response was detected following the 3 July regional earthquake, suggesting that rates of stress accumulation in the caldera are low. However, a close watch will be maintained for any long-term effects of the present intensified regional seismicity on the Rabaul volcano.

"Monthly caldera earthquake totals were 710 in May and 644 in June. In May, the pattern of seismicity was dominated by a broad zone of activity in the Vulcan area with a clustering of events at the E extremity of the Vulcan Headland. In June, events were scattered around the caldera seismic zone with no strong concentrations. The caldera seismic energy output for the two-month period was about [6×10^{16}] ergs, most of which was related to the seismic swarm induced by the central New Britain earthquake of 10 May.

"Horizontal distance measurements have continued to show gradual rates of change, the maximum about 10 ppm per month in the Greet Harbour area. Levelling measurements carried out between 8 and 10 May showed that uplift rates were low since the previous survey in mid-March. The largest change in this two-month interval was 13 mm at the S end of Matupit Island. Since the March crisis, maximum monthly tilt rates have been generally low, ranging from about 6 to 17 μrad per month. The area of biggest tilt changes has continued to be around Greet Harbour. In June a higher tilt rate (30 μrad per month) was recorded at the NW shore of Greet Harbour, but the significance of this is uncertain at present."

Information Contact: Same as 9:12.

7/85 (10:7) "On 3 July, a major regional earthquake (M_L 7.1) in S New Ireland (about 80 km from Rabaul) caused large local ground deformations in Rabaul caldera. The main effects were tilt offsets of up to several hundred μrad. Tilt changes were greatest at the mouth of Greet Harbour. At Sulphur Point and the SE end of Matupit Island, tilt on the day of the earthquake may have reached 330 and 120 μrad, respectively. The tilt vectors at these locations and those of much smaller deformations at the head of Greet Harbour appear to radiate outward from a point near the mouth of the harbour. Tilting also occured on the Vulcan headland in response to the 3 July earthquake. Tilts of about 45-50 μrad were registered at 2 stations in the W-central and SE parts of the headland, but other nearby tilt stations showed only minor changes, and no clear pattern of tilt was evident. Other ground deformation measurements (levelling and EDM) did not show any notable changes associated with the earthquake.

"It is uncertain what significance should be attached to the fact that the earthquake-induced tilting in the Greet Harbour area conforms to some extent with effects produced by processes purely internal to the caldera. Such large tilt changes might be expected to have been accompanied by caldera seismicity, but as far as we know the caldera did not respond seismically to the 3 July earthquake. This contrasts with the effects of the [10] May earthquake, which included a seismic swarm and minor tilt changes in the Vulcan area. The apparently aseismic nature of the deformation on 3 July may be an indication of merely local surface movements.

"Apart from the effects of the 3 July earthquake, a low level of activity prevailed in Rabaul caldera in July. The total number of caldera earthquakes for the month was 595, down from 639 in June. A decrease in the daily rate of earthquakes was noted, from about 25 per day during the first half of the month to about 8 per day in the last week of the month. Caldera earthquakes were concentrated in the Rapindik area, immediately N of Greet Harbour. There were no strong caldera earthquakes in July; the strongest events had magnitudes (M_L) of less than 2.

"Levelling measurements along the line from Rabaul town to Matupit Island in late July showed that uplift in the caldera was continuing at a low rate. The greatest change since the previous survey in May was 26 mm at the SE coast of Matupit Island. This change indicates a rate of uplift probably similar to the pre-crisis rate if some of the uplift was associated with the 3 July regional earthquake.

"Horizontal distance measurements indicated continuing slow dilation in the caldera, with the greatest changes (5-10 ppm per month) occurring in the Greet Harbour area.

"Tilt changes after the 3 July earthquake were evident in the Greet Harbour area and at one station on the Vulcan headland. The largest changes were about 43 μrad about 1 km SW of the Vulcan cone and about 23-33 μrad at the mouth of Greet Harbour. These tilt changes accumulated at an exponentially decreasing rate and are believed to be related to the earthquake-induced tilting on 3 July.

Information Contact: Same as 9:12.

8/85 (10:8) "Activity showed a further reduction in Aug, and appeared to be at pre-crisis levels. The total number of caldera earthquakes for the month was 236, as compared to 595 in July. The located earthquakes were mostly from the N part of the caldera seismic zone, and an unusual concentration of events was detected 1-2 km N of that zone. Tilt measurements generally showed a continuation of the exponentially decreasing tilts observed following the New Ireland earthquake of 3 July. Steady, slow dilation (up to 10 ppm per month), centred near the mouth of Greet Harbour, continued to be observed by horizontal distance measurements."

Information Contact: Same as 9:12.

9/85 (10:9) "Seismic activity, which had returned to pre-crisis levels in Aug, continued to decline. A total of 176 events were recorded in the caldera in Sept compared to 236 in Aug. Tilt changes continued to show an exponential decrease following the large change at the time of a strong regional earthquake on 3 July. Levelling and horizontal distance measurements showed insignificant changes."

Information Contact: J. Mori, RVO, Rabaul.

10/85 (10:10) Seismicity and ground deformation continued to decline at Rabaul.

Information Contact: P. Lowenstein, RVO, Rabaul.

11/85 (10:11) "Seismicity and ground deformation rates were in further decline from previous months. The total number of caldera earthquakes in Nov was 115, of which only 1 was locatable. On 26 Nov, no caldera earthquakes were recorded for the first time since 24 May 1983. Ground deformation was minimal."

Information Contact: B. Talai, RVO, Rabaul.

12/85 (10:12) "Seismicity continued to decline in Dec with only 48 events of small magnitude (M_L 0.5), compared to 115 events in Nov. Levelling 11 and 12 Dec showed consistent subsidence on Matupit Island for the first time since the levelling program was begun in 1973. The southern tip of Matupit Island had dropped 1.4 cm since the last levelling in early Oct. This may indicate the beginning of a deflationary period within the caldera. There were no significant changes in the tilt, EDM data, or sea level."

Information Contact: J. Mori, RVO, Rabaul.

BALBI

Bougainville Island, Solomon Islands
5.295°S, 154.98°E, summit elev. 2591 m

Bougainville, the largest and westernmost of the Solomon Islands, is part of Papua New Guinea. Balbi is 330 km ESE of Rabaul and (with Bagana and Loloru below) under the jurisdiction of RVO.

10/84 (9:10) Balbi, Bougainville's highest point, is in the N-central part of the island on the axial volcanic chain. In [what is possibly an eroded caldera at the summit of the composite cone, 7 craters] lie on a N-S ridge about 3 km long. [There are also two cones about 2 km W of the line of craters.] Only Crater B, about 600 m in diameter and second from the S, shows activity. Anthropological evidence suggests an explosive eruption accompanied by nuées ardentes and fatalities, sometime between 1800 and 1850 [but recent geological work does not support this].

"A boiling mud pool and up to a dozen large, very active fumaroles flanked the lineament of craters. Large collapses have occured into Crater B, and further extensive tension cracks were visible around the crater's rim. Many small solfataras were still active in the W wall of Crater B. The lake in Crater C had diminished in size." [Recent RVO investigations show that the fumaroles are aligned orthogonally to the 7-crater lineament and that the mud pool is no longer active.]

Information Contact: K. McCue, Bougainville Copper Ltd., Panguna [with Jan 1988 additions from RVO based on their recent work].

BAGANA

Bougainville Island, Solomon Islands
6.14°S, 155.19°E, summit elev. 1702 m
Local time = GMT + 10 hrs

The most active volcano on Bougainville (and 97 km SE of Balbi), Bagana has 25 known eruptions since the mid-1860s. The current eruption has continued since 1972.

5/78 (3:5) A sluggish lava flow remained active on the N flank of Bagana, but no explosions have been reported there in the past 2 years.

Information Contact: R. Cooke, RVO, Rabaul.

9/78 (3:9) An aerial inspection on 14 Sept showed that lava continued to flow down the N flank, as it has since late 1976. No explosive activity has been reported during this phase of lava flow extrusion.

Information Contact: Same as above.

3/81 (6:3) "Moderate to strong emission of white vapour continued throughout March. An active lava flow descending the [established lava channel on] the N slope had reached ²/₃ of the way down the mountain. Small nuée ardente-type avalanches caused by collapse of the flow front were observed during an aerial inspection on 9 Jan."

Information Contact: C. McKee, RVO, Rabaul.

5/83 (8:5) From 5-7 March, the weak white vapor emissions from the summit increased to a thick high-pressure plume rising to 2000 m above the summit. Vapor release remained high until 21 March, but no glow was reported. Vapor emission was again strong at the end of the month.

Aerial inspections on the mornings of 15, 16, and 17 March revealed a thick but apparently normal plume being released from the lava dome occupying the summit crater. The viscous blocky lava flow on the N flank appeared to be moving extremely slowly, perhaps a few meters per week. At the source of this flow, the maximum lava temperature (measured by a portable infrared optical pyrometer from a helicopter) was only 175°C on the slow-moving, blocky surface.

During 24 hours of seismic monitoring from the W flank at 1100 m altitude, 540 B-type events and 1 or 2 sharp, impulsive, shallow events were recorded.

After an explosive phase that culminated in a nuée ardente on 30 May 1966 and destroyed the existing lava dome, a new dome began to form in the SE part of the summit crater. Lava flowed intermittently down the S flank of the volcano from then until sometime between April and early June 1975, when explosive activity destroyed that dome. A new dome began growing in the NW part of the crater, and lava has flowed sluggishly down Bagana's N and NW slopes since then (Bultitude and Cooke, 1981).

Reference: Bultitude, R.J., and Cooke, R.J.S., 1981, Note on Activity from Bagana Volcano from 1975 to 1980, *in* Johnson, R.W. (ed.),*Cooke-Ravian Volume of Volcanological Papers,* Geological Survey of Papua New Guinea Memoir 10, p. 243-248.

Information Contact: P. de Saint Ours, RVO, Rabaul.

[See 8:4 Manam report (above) for SO_2 measurements at Bagana.]

12/83 (8:12) Reports through 8/84 are from the RVO staff.

"A marked increase in activity was observed from Bagana in Dec, with an increase in vapour emission and darkening of the plume early in the month. Bright glow was observed at night on 7 Dec followed by explosion and rumbling noises on the 8th and the emission of abundant blue vapour. On the 14th, incandescent boulders were observed tumbling down the upper NW flank. By the end of the month activity had decreased again, with no glow at night and the emission of moderate amounts of slightly brownish vapour."

Information Contact: P. Lowenstein, RVO, Rabaul.

1/84 (9:1) "The increase of summit activity noted in Dec resulted in [pulses of lava down the channel on the N flank] of the volcano from 5 Jan onwards. This lava remained active throughout the month and produced plumes of white and grey vapour."

Information Contact: Same as above.

3/84 (9:3) "The increased activity continued in Feb and March. Occasional brown and grey tephra emissions were observed, and rumbling and explosion sounds were heard 17 km away. Nighttime summit glows were occasionally seen.

"New lava flows were reported in Jan, but aerial inspections have failed to confirm these reports. They indicated a relatively static body of lava extending about 200 m from the summit, but this is an old lava flow. The main development of the known active lava flow at Bagana in recent times has been a sharp change in direction of flow on the lower slopes. The nose of the flow is now abutting the dome on the W foot of Bagana after completing a 60° turn toward the W from the established flow channel on the N flank."

Information Contact: C. McKee, RVO, Rabaul.

8/84 (9:8) "The predominately effusive eruption continued. When last observed (21 July), the viscous blocky lava flow on the NW flank had reached an altitude of 1000 m and had an estimated volume of 1.3×10^6 m^3. A moderate plume of dense, white, SO_2-rich gases continued to be emitted from the summit crater. Seismicity from the volcano was at a low level, with only a few B-type and explosion earthquakes per day."

Information Contact: P. de Saint Ours, RVO, Rabaul.

10/84 (9:10) The following is from Kevin McCue.

"An uncharacteristic swarm of shallow tectonic-like earthquakes together with banded low-amplitude harmonic tremor commenced on 19 Oct about 2000 and continued through Oct. During the previous week the NW lava flow collapsed to form a well-defined lava channel below the point where the flow turns sharply W. The toe of the lava flow continued to encroach on a satellite dome at the W foot of Bagana. Numerous solfataras have given a distinctive facia to the ESE summit."

Information Contact: K. McCue, Bougainville Copper Ltd.

11/84 (9:11) This report is from the RVO and Kevin McCue.

"A new eruptive phase started in Oct 1984. The extrusion of a fresh batch of andesitic magma into the summit crater was accompanied by a marked increase in the volume of vapour rising above the crater, an increase in the area of fumarolic activity in and around the crater, night glows, and incandescent material tumbling down the flanks. Simultaneously, volcanic seismicity increased from fewer than 10 B-type events per day to over 100 per day by 19 Oct; harmonic tremor appeared on the 12th and became sub-continuous after the 15th. A relative drop in volcano seismicity (24-27 Oct) was followed by re-intensification. The daily frequency of events was about 1000 by 11 Nov, and consistently above this after 15 Nov. Strong tremor was recorded for periods of several hours on 4, 5, 9, 13, 18, 20, and 22 Nov. Explosion earthquakes were occasionally recorded.

"An aerial inspection by Bougainville Copper Ltd. geologists revealed that the dome of viscous andesite had bulged to about 15 m above the crater rim and lava was spilling over the N, E, and W parts of the rim. Debris on these 3 flanks corroborated the observations of incandescent material avalanching down the sides of the volcano, presumably from collapse of parts of the dome. Paradoxically, the long-established lava flow channel on the N flank of the volcano seems to have been drained, leaving an empty lava channel from the crater rim down to about 1100 m altitude."

Information Contact: P. de Saint Ours, RVO, Rabaul; K. McCue, Bougainville Copper Ltd.

12/84 (9:12) "Eruptive activity was at about the same intensity as in Nov. Slow effusion of viscous andesitic lava continued in the summit crater, and unstable parts of the crater dome collapsed, causing avalanches of incandescent lava blocks. [Strong fumarolic activity was continuing] on the upper E flank to about 200 m below the crater rim."

Information Contacts: C. McKee, RVO, Rabaul; Bougainville Copper Ltd.

1/85 (10:1) The remaining reports are from RVO. "Bagana's summit crater dome continued growing through Jan, to fill 95% of the crater and bulge up to about 30 m above the crater rim. A thick, ash-laden plume was fed by numerous sources in the dome. A large solfataric area on the upper E flank of the volcano was also contributing vapours to the plume. An ash haze was observed stretching horizontally more than 100 km to the NE at about 2000 m altitude. The steep flanks of the volcano were covered with thin ash deposits and a large number of blocks from avalanches of unstable parts of the dome.

"The long-established N lava flow was still active, and was broadening and thickening on the NW basal slopes at about 900 m altitude. Frequent avalanching was taking place from the [edges of the flow near its terminus]."

Information Contact: P. de Saint Ours, RVO, Rabaul.

2/85 (10:2) "Eruptive activity continued at Bagana in Feb, although summit explosive activity was absent, or sporadic and very weak. Strong vapour emission continued, and the very light ash content in the plume was considered to be an effect of entrainment of dust from occasional avalanches on the sides of the summit lava dome. Frequent rockfalls from the edges of the active lava flow on the [N and] NW flanks of Bagana continued."

Information Contact: C. McKee, RVO, Rabaul.

3/85 (10:3) "Activity at Bagana in March was back to normal, after a period of vigorous activity that began in Oct 1984. The daily totals of volcanic earthquakes ranged from 0 to 3, and summit glow was observed on only 3 nights: 16, 25, and 31 March. Moderate-to-strong white vapour emission continued."

Information Contact: Same as above.

10/85 (10:10) "Inspection by Bougainville Copper, Ltd. geologists, confirmed the ongoing mild eruptive activity, with extremely slow but sustained progress of the blocky andesite flow active (since [1975]) on the [N and] NW flanks. Weak glow was occasionally observed at night above the summit crater. The viscous lava dome in the crater continued to gently release a faint vapour plume. Seismicity remained weak with a few B-type earthquakes occurring daily."

Information Contact: Same as 8:10.

LOLORU

Bougainville Island, Solomon Islands
6.52°S, 155.62°E, summit elev. 1894 m

Loloru is 63 km SE of Bagana.

10/84 (9:10) Quoted material is from Kevin McCue.

"The strato-volcano Loloru in S-central Bougainville has not been historically active. A prehistoric lava dome fills the W part of the 2.5 km-diameter summit [caldera and there is a lake] between the dome and the E wall.

"Aerial inspection showed that dome and flank solfataras were active as normal. Water temperature most recently recorded in a pool by the stream draining the caldera lake was 80°C."

Information Contact: K. McCue, Bougainville Copper Ltd.

KAVACHI

Solomon Islands, SW Pacific
9.02°S, 157.95°E, submarine depth variable
Local time = GMT + 11 hrs

This remote submarine volcano, two-thirds of the way from Bougainville to Guadalcanal, has been the site of at least 8 island-forming eruptions since 1939. The islands were made of finely fragmented volcanic materials and none has long withstood erosion by the open sea.

No other Solomon Island volcanoes S of Bougainville were reported active during the decade, but pumice was found on several beaches in early 1985 (see 10:3 report under Home Reef, Tonga).

8/76 (1:11) A new eruption of Kavachi volcano was sighted on 24 Aug at 0800 by Solair pilot Bruce Kirkwood. A fountain of water and volcanic ejecta reached 30 m height. The eruption was continuing as of 2 Sept. Seismicity was being monitored by Deni Tuni at Vakambo, 45 km north of the volcano. The last eruption of Kavachi ended in Feb 1970.

Information Contact: R. Thompson, Geological Survey, Honiara.

9/76 (1:12) Much of the following information was received by R.B.M. Thompson from Solair pilots Bruce Kirkwood, Robert Snape, and Eric Cooper.

Fountains of water and rock continued at least through 4 Sept, reaching about 60 m height. No information is available for 5 Sept.

6 Sept: A resident of Gatukai Island (NE of Kavachi) noted a change of the eruption column from white steam and spray to "smoke" during the afternoon (fig. 5-13).

7 Sept: A morning overflight revealed that a rubble pile had reached the surface. Heavy swells were breaking over the rubble, which continuously emitted steam and "smoke", accompanied by bursts of rocks. The maximum height reached by the ejecta was about 90 m. By

Fig. 5-13. Airphoto of Kavachi, taken 6 Sept, 1976 (airplane wing at left of photo). Courtesy of R.B.M. Thompson.

1600, a cone, extruding lava, extended 1.5-3 m above the surface.

8 Sept: At 1300 the cone was about 9 m above sea level and 40 m across (at sea level). A red glow from the cone's central vent was clearly visible in bright sunlight. Lava was pouring over the NW rim, which was about 3 m lower than the rest of the cone. Eruptions of blocks, reaching more than 150 m height, occurred at approximately 1-minute intervals. Some of the blocks were fairly large and some were glowing. Brown and white "smoke" was emitted continuously.

9 Sept: At 1030 the volcano was less active than the day before. Lava surged up and down, overspilling the NW rim into the sea on the larger upsurges. Eruptions of blocks, ash, and vapor, occurring every 1-2 minutes, were less intense than on the 8th. The island appeared to be breaking up, most rapidly on the NW, the prevailing wave direction. The maximum height of the cone above sea level was about 6 m. The surrounding water was discolored by a brownish scum. Photographs were taken by seismological observer Deni Tuni (fig. 5-14).

Information Contact: R. Thompson, Geological Survey, Honiara.

10/76 (1:13) The lava flow, which began on 7 Sept when the cone first built itself above sea level, ceased between 9 and 11 Sept. Pyroclastic eruptions continued, with small bursts every 30 seconds and larger ones every 2 minutes. "Smoke" emission was continuous. Tephra deposition had filled the breach in the SE wall of the crater by 15 Sept. Photographs taken on that date show an elongate island (NW-SE) about 100 m in length. Pyroclastic activity lasted until at least 1 Oct, maintaining its 30-second and 2-minute periodicity. No pumice was reported. Cone height stabilized at 15-20 m, with accretion of tephra compensating for erosion by the sea. An observer

Fig. 5-14. Airphoto of Kavachi, taken 9 Sept, 1976 by Deni Tuni. View is from the E. Note the bombs being emitted from the vent. Courtesy of R.B.M. Thompson.

24 km away noted no activity on 6 Oct. A survey to determine the size and position of the new cone was planned for 12 Oct.

A temporary seismological station installed on Vakambo Island (8.37°S, 157.85°E, near New Georgia Island 73 km from Kavachi) began recording on 24 Sept. Twenty earthquakes were recorded 24 Sept-5 Oct, all with S-P separations corresponding to Kavachi epicenters, but they cannot be conclusively associated with the eruption since they were not recorded by another station. The station was operated by Deni Tuni.

Information Contact: R. Thompson, Geological Survey, Honiara.

3/77 (2:3) The following is from R.B.M. Thompson.

Captain Stewart Evans approached Kavachi at 1500 on 13 Oct observing subsurface activity, but no island. He reports:

"There was a definite eruption cycle during the first 2 hours of observation. Eruptions were fairly regular at approximately 15-minute intervals, and every 4th or 5th eruption was of noticeably greater magnitude. The cycle broke down about 1730. Frequent tremors were then heard and felt from the government survey vessel *Wakio*, stationed about 1 km from the vent. The tremors were felt at 5-10-minute intervals, but eruptions became less frequent and less violent. No eruptions were sighted for a 40-minute period while steaming away from the area at about 1800.

"The crest of the volcano was estimated to be less than 15 m below the surface and the sea was stained brown over an area approximately 100 m in diameter. The average height attained by ejecta (water and volcanic debris) was 10-15 m above the sea surface. The maximum height attained following the highest ejection was estimated at 30 m. Small waves were generated by each eruption. The average wave was 1 m high and radiated to about 500 m from the volcano. No bottom was found in a 365-m range on the echo sounder around a 1-km perimeter of the volcano."

"On 23 Oct, I observed a circular area of shallow (yellow-blue) sea surrounding a deep (dark blue) area, from an airplane above Kavachi. A yellow-blue area looking like a reef extended to the SE."

A study of 8 mm movies taken 9, 14, and 30 Sept, 1976 shows that red-hot blocks reached a maximum dimension of more than 1 m.

No other activity was reported until 0945 on 22 Feb, 1977 when Solair Chief Pilot Bruce Kirkwood reported plumes of spray and rock debris thrown upwards to about 60 m at intervals of 10 minutes.

Information Contact: Same as above.

7/77 (2:7) Renewed activity was first observed on 19 July at 1000 by Solair Pilot Bruce Kirkwood, 3 days after the previous overflight. Eruptions were seen on 20 July, and at about 30-second intervals on 22 July. Ash was ejected by the vent, which remained below the ocean surface.

Information Contact: Same as above.

7/78 (3:7) On 21 June, Bruce Kirkwood observed an eruption at Kavachi. Geologists from the Solomon Islands Ministry of Natural Resources flew over the site the next

Fig. 5-15. Airphoto taken 22 June 1978 showing the island formed by Kavachi's eruption. Courtesy of Frank Coulson.

day and observed a circular island 30-50 m in diameter and 1-2 m in height (fig. 5-15). Occasional weak explosions ejected ash and scoria, and the perimeter of the island was steaming.

Kirkwood flew over the volcano again on 14 July and reported a N-S elongate island with a central cone estimated to be 30 m high. Bombs and blocks were thrown from the central vent to as much as 400 m altitude, and lava poured down the S flank. A small subsidiary vent appeared to be forming near sea level on the E coast.

By 16 July (Kirkwood's next overflight) the S half of the island, including the central vent, had dropped along an E-W fault with a scarp that formed the new S edge of the island. The subsided area appeared to be just below sea level and eruptive activity there continued. During the 10-minute overflight, 1 explosion was observed.

Information Contact: F. Coulson, Ministry of Natural Resources, Honiara.

8/78 (3:8) When observed on 28 July, the eruption had ended. A small rocky island, about 15 m long, 5 m wide and no more than 3 m high remained at the eruption site. The island was expected to be destroyed by wave action within a few days to a few weeks.

Information Contacts: F. Coulson, D. Tuni, Ministry of Natural Resources, Honiara.

10/80 (5:10) Bruce Kirkwood observed a small submarine eruption on 7 Oct. A mixture of water, steam, and blocks rose intermittently to more than 100 m above the ocean surface. Sea water surrounding the eruption site was slightly discolored.

Information Contacts: D. Tuni, Ministry of Natural Resources, Honiara; B. Kirkwood, Solair.

11/80 (5:11) Solair pilots flying over Kavachi Volcano on 14 Oct observed a submarine eruption similar to that reported by Kirkwood on 7 Oct, although there appeared to be more mud in the surrounding seas. By 23 Oct, activity had decreased to occasional bursts of hot water at the surface.

Information Contact: D. Tuni, Ministry of Natural Resources, Honiara.

2/81 (6:2) On 11 Nov at 1215 a Solair flight diverged from its normal route to observe the volcano. Drs. Hughes and Dunkley of the Geological Division, Ministry of Natural Resources, reported that a dense, nearly vertical steam jet was billowing to approximately 300 m, but dissipated as the plane approached. The eruption site was marked by white water, and a stream of muddy, turbid, pale brown water extending several km NE from the volcano. On 3 Dec Dunkley observed an area of discolored water several hundred meters wide extending NW (down current) about 4 km. No eruption was in progress.

Information Contact: Same as above.

3/81 (6:3) Solair Captain Peter Cox overflew Kavachi on 9 Jan at 1540. At 30- to 60-minute intervals steam fountains rose to a height he estimated at 150 m. The sea was stained light brown for as much as 15-20 km from the volcano, but Cox saw no floating pumice.

Ministry of Natural Resources geologist A. Smith observed minor eruptive activity on 25 Feb. Submarine explosions apparently originating at 5-10 m depth transmitted shock waves to the surface. Some gas bubbles could be seen but no ejecta were evident. The prevailing wind drove seas to the NE, carrying an expanding plume of yellow-brown to yellow-green water visible on the surface for at least 2 km from the volcano.

Information Contact: Same as above.

9/81 (6:9) Solair pilots who overflew Kavachi in mid-Sept reported gas bubbling and discolored sea water. No eruption columns were observed.

Information Contact: F. Coulson, Ministry of Natural Resources, Honiara.

Further Reference: Johnson, R.W., and Tuni, D., 1987, Kavachi, an Active Forearc Volcano in the Western Solomon Islands: Reported Eruptions Between 1950 and 1982; *in* Taylor, B. and Exon N.F., (eds.), *Marine Geology, Geophysics and Geochemistry of the Woodlark Basin, Solomon Islands;* CircumPacific Council for Energy and Mineral Resources, Earth Science Series, v. 7, Houston, p. 89-112.

12/85 (10:12) On the morning of 30 Dec, Solair Captain Brian Smith observed dirty water over Kavachi while flying from Honiara to Munda. When he returned to Honiara that afternoon, Kavachi was ejecting rock debris, mud, steam, and water to a height of 30 m. The eruption was observed by the captain and crew of the MV *Iu Mi Nao* the same afternoon. On 31 Dec, the eruption was intensifying, with one explosion per minute ejecting material to 200-300 m height. By 1 Jan, both eruptive intensity and height of the ejecta column had decreased. On 7 Jan, Deni Tuni observed explosions while en route to Honiara on the MR *Compass Rose II.* Two explosions ejected black plumes that probably contained ash to about 20 m height. A third explosion, which appeared to consist mainly of steam and perhaps other gases, spread along the horizon, forming a pyramid-shaped cloud before disappearing. As of mid-Jan, no island had formed.

Information Contact: D. Tuni, Ministry of Natural Resources, Honiara.

TINAKULA

Santa Cruz Islands, SW Pacific Ocean
10.47°S, 165.75°E, summit elev. 650 m
Local time = GMT + 11 hrs

The Santa Cruz Islands lie about 600 km E of Guadacanal and N of the NNW-trending Vanuatu group. Tinakula, the only historically active volcano in the group, has known 13 eruptions since 1595.

6/84 (9:6) The following is from the cruise report of the USGS research vessel *S. P. Lee*, engaged in multichannel seismic profiling in the Vanuatu and Solomon Islands areas. The report was written by H. G. Greene, A. Macfarlane, and other members of the scientific party.

"On 3 June, Tinakula could be seen 'smoking' in the distance, some 25 km away. As the ship approached the island, large billowing clouds of steam were observed emanating from the summit of the volcano. Occasionally a large, billowing, dark gray, ash-laden plume was observed rising to several kilometers above the island without being disturbed by the prevailing SE trade winds, unlike the steam which was rapidly dissipated by those winds. The emanation of the ash appeared to take place at fairly regular intervals, about every 2 hours.

"The *Lee* passed within 400 m of Tinakula, on a W-E path along the N side of the island, which presented a good vantage point for observing the active vent at the head of the landslide scarp on the NW side of the volcano. As the ship drew abeam of Tinakula, rumbling sounds could be heard from within the active vent, immediately N of the central crater. Boulder-size (football-size) rocks were ejected from the vent, and were still steaming as they rolled and skipped down the steep, N-dipping scree slope and splashed into the sea. At least a dozen of the boulders and much more material of cobble size (65-250 mm in diameter) were seen being thrown from the vent every minute. Much of this debris was accumulating on the scree slope, which is actively infilling the void left by the 1971 landslide.

"Geophysical data collected by the *Lee* showed that another volcanic cone is present about 90 m beneath the surface of the water some 5 km W of Tinakula. This submarine volcano is a little smaller than Tinakula and in consequence not projecting above the sea surface. It has a very youthful geomorphic profile with sharp steep flanks, and appears from its morphology to be active. This volcano lies on the W flank of Tinakula and could be a vent associated with the same magmatic processes that are building the island today. This volcano or volcanic vent has not been identified before and is not on any bathymetric map."

Tinakula's last reported eruption started 6 Sept 1971, preceded by a small tsunami at the island. Intermittent explosive activity built a small summit cone, and incandescent blocks rolled down the volcano's flanks into the sea. A slow-moving lava flow extended about 300 m down the NW flank. About 160 people were evacuated from the island. The eruption ended in Dec 1971.

Information Contacts: A. Macfarlane, Dept of Geology, Mines, and Rural Water Supplies, Vanuatu; H. G. Greene, USGS, Menlo Park, CA.

6/85 (10:6) On 13 June at 0946 geologists (about 4 km from Tinakula in a boat) observed a vapor and ash cloud rising slowly from the crater. The ash emission was followed about 5 minutes later by whitish vapor, and the eruption cloud drifted slowly toward the W from the crater.

Stopping 50 m from the breached NW side of the volcano, the geologists observed rising steam, and large boulders that rolled down the flank before splashing into the sea. There was no lava flow.

During the following two hours, no further eruption was observed as the summit was nearly always covered with clouds. A possible site for a telemetering seismometer was selected on the E side of the volcano at Mendana Cone. Local people reported that the volcano has exhibited similar low-level eruptive behavior since 1984.

Information Contact: D. Tuni, Ministry of Natural Resources, Honiara.

AMBRYM

Ambrym Island, Vanuatu
16.25°S, 168.08°E, summit elev. 1334 m

This large strato-volcano lies 690 km SSE of Tinakula. Explosive eruptions have been frequent since its discovery in 1774, mostly from young cones within the large 9 × 12 km central caldera.

3/79 (4:3) According to press reports, an eruption from Benbow Crater occurred on 10 Feb. Gases from the eruption caused acid rainfall on the SW portion of Ambrym Island, destroying most vegetation within 24 hours, contaminating water supplies, and burning some inhabitants. Jean-Luc Saos, Director of Mineral Resources for the New Hebrides government, reported a high concentration of HCl and sulfur compounds in the volcanic gases. Although heavy ashfalls have occurred in the area in the past, this is the first report of acid rains.

Information Contacts: A. L. Dahl, S Pacific Commission, Noumea; *Les Nouvelles Caledoniennes*, Noumea.

4/79 (4:4) The following report, written by A. Macfarlane using information from J.-L. Saos, supplements the above.

"An increase in activity from Benbow Crater (in the SW portion of Ambrym's caldera) involving the copious emission of cinders, lapilli, and gases was first noted on 7 Feb. On the night of 10-11 Feb, ash and gas columns apparently combined with local rainstorms to produce a fallout of dilute sulfuric acid, which affected an area of some 90 km^2 across SW Ambrym Island between and inland of the coastal villages of Baiap and Calinda (about 16 and 10 km SSW of Benbow Crater). The most extreme effects were observed in the Lalinda-Port Vato area (a 5-km zone along the coast) where there was extensive burning of pastures and yellowing of cash crops such as cocoa and coconut palms. In addition, the local population suffered from gastric upsets and from burning of the skin by contact with the acid rainwater. The local water supply was contaminated to the extent that pH values in cisterns had decreased to 5.2-5.5 when subsequently measured and were probably even lower at the time of the acid rains.

"Owing to heavy cloud cover, it has not been possible to determine whether these eruptions mark a significant change in the configuration of vents within Benbow Crater."

Information Contacts: A. Macfarlane, Geological Survey, Vanuatu; J.-L. Saos, Dept. of Mineral Resources and Rural Water Supply, Vanuatu.

LOPEVI

Vanuatu, SW Pacific
16.51°S, 168.35°E, summit elev. 1447 m
Local time = GMT + 11 hrs

Forming a small island 40 km SE of Ambrym, Lopevi has a history of 20 eruptions since 1863.

10/82 (7:10) A. Macfarlane reported that Lopevi ejected a black ash cloud on 24 Oct that rose to an altitude of about 6 km and drifted NW. Weather clouds prevented satellite observations until late that day, when no plume was visible. On 25 Oct, the area's first daylight image (at 0645) from the GOES West geostationary weather satellite showed a relatively diffuse plume extending about 250 km S from the volcano. By 0900, the far end of the cloud was roughly 500 km S of Lopevi and feeding appeared to be ending. The cloud had clearly begun to dissipate on the next image, an hour later. As of early Nov, no additional activity had been reported either in Vanuatu or by Solair pilots (who fly near Lopevi 3 days a week). Macfarlane notes that such activity at Lopevi is not unusual. Lopevi was last reported active on 12 Sept 1979, when small amounts of ash were emitted from cones on the SW flank.

Information Contacts: A. Macfarlane, Geological Survey, Vanuatu; F. Coulson, Ministry of Natural Resources, Solomon Islands; D. Haller, NOAA/NESS; R. Shaw, IATA.

KARUA

Vanuatu, SW Pacific
16.83°S, 168.54°E, summit depth variable

Submarine eruptions have produced ephemeral islands several times since 1897 at this shallow submarine volcano between Tongoa and Epi, the islands immediately S of Lopevi. Although not reported to SEAN at the time, bubbling of sea water and yellow water discoloration were observed on 1 Feb 1977 (*Bulletin of Volcanic Eruptions* no. 17, 1979).

YASOUR

Tanna Island, Vanuatu
19.52°S, 169.43°E, summit 350 m

Continuous mild eruptions have been reported from this volcano since it was visited by Captain Cook in 1774. Strombolian bomb ejection and occasional Vulcanian explosions (sending ash clouds to 2000 m altitude) were reported in the *Bulletin of Volcanic Eruptions* (no. 15-20, 1977-82) and activity is listed as continuing in this book's Chronology table. Tanna is near the S end of the Vanuatu group, 390 km SSE of Ambrym.

MATTHEW ISLAND

New Caledonia, SW Pacific
22.33°S, 171.32°E, summit elev. 200 m

Matthew lies near the S end of the New Hebrides Trench and E of New Caledonia Island. Four eruptions are known since 1949. A line of active volcanoes runs nearly 1500 km to the NNW through Vanuatu to Tinakula.

12/76 (1:15) On 27 November, the crew of a RNZAF P3 aircraft noted extensive dust clouds above Matthew Island, and discoloration of the surrounding water. The New Zealand Defense Scientific Establishment is investigating this activity.

Information Contacts: J. H. Latter, DSIR, Wellington; Defense Scientific Establishment, Auckland; J. Barnes, N.Z. Defense Staff, Washington DC.

1/77 (2:1) The following is additional information obtained by the RNZAF from the crew of the aircraft that flew over Matthew Island on 27 Nov.

During the overflight, the volcano emitted a gray dust cloud that merged with a layer of stratus clouds (held to about 350 m altitude by a temperature inversion). A strong sulfur odor was present in the area. Muddy gray discolored water was carried SE by the current. Because of poor weather and light conditions, no photographs were taken. More recent information on the activity is not available.

Information Contacts: Defense Scientific Establishment, Auckland; J. Barnes, N.Z. Defense Staff, Washington DC.

7/77 (2:7) [An RNZAF crew flew over Matthew Island on 11 Feb and photographed the volcano. The vent was weakly emitting vapor (fig. 5-16).]

Information Contact: I. Nairn, NZGS, Rotorua.

3/83 (8:3) A Vanuatu government team arrived at Matthew Island on 10 March at 0700. The only activity noted was emission of wispy, white vapor from the central crater in the island's W (main) edifice.

[A report in this original issue described an apparent eruption of Hunter Island, 85 km E of Matthew, observed by a team of geologists on 9 March. Later information revealed that human activity had started fires and no eruption had taken place.]

Information Contact: A. Macfarlane, Dept. of Geology, Mines, and Rural Water Supplies, Vanuatu.

Further Reference: Maillet, P., Monzier, M., and Lefevre, C., 1987, Petrology of Matthew and Hunter Volcanoes, South New Hebrides Island Arc (Southwest Pacific); *JVGR*, v. 30, p. 1-29.

Fig. 5-16. RNZAF photograph, taken from the N on 11 Feb 1977, showing the vent and about $^2/_3$ of Matthew Island.

INDONESIA

General Reference: Kusumadinata, K. (ed.), 1979, *Data Dasar Gunungapi Indonesia*; Volcanological Survey of Indonesia, Bandung, 820 pp.

MARAPI

Sumatra
00.38°S, 100.47°E, summit elev. 2891 m
Local time = GMT + 7 hrs

Marapi has been the most active volcano in Sumatra during historic times, with 58 eruptions known since 1770, and the only one with recorded eruptions during the report decade. It lies just S of the equator in the middle of Indonesia's largest, 1700 km-long, island.

10/78 (3:10) At 1830 on 8 Sept Marapi ejected a thick blackish-gray cauliflower-shaped cloud to 1500 m above the crater, accompanied by glow and a roaring noise. Andesitic ash and lapilli fell on a 30 km^2 area. This explosion was preceded by a number of smaller ones that produced 300-500 m-high clouds. Fumarolic emissions, rising as much as 700 m and containing some ash, were continuing as of 18 Sept. No seismicity was felt in villages around the volcano. The activity originated from the lateral extension of a small, pre-existing, summit area crater (fig. 6-1), about 300 m E of the central crater. When visited on 13 Sept the active crater was an elongate feature 95 m long, 50 m wide, and about 50 m deep.

Information Contact: F. Suparban Mitrohartono, Volcanological Survey of Indonesia (VSI), Bandung.

4/79 (4:4) According to press reports, 60 persons were killed by an eruption of Marapi during the morning of 30 April [but see 4:5], and rescue workers searched for 19 others believed trapped by "landslides." The volcano was said to have ejected "stones" and "mud" or "lava," causing damage in at least 5 villages. The deaths were apparently caused by large airfall tephra.

Information Contacts: Agence France Presse (AFP); *Kompas*, Jakarta.

5/79 (4:5) Press reports describing a tephra eruption of Marapi on 30 April were incorrect. About 300 mm of rainfall remobilized an old lahar and other volcanic material on Marapi's N and E flanks, producing several landslides. The largest began at 2400 m altitude on 30 April, and traveled as much as 20 km downslope to about 700 m altitude, leaving a deposit 20-150 m wide and 1-3 m thick. Eighty people were killed, 5 villages were damaged, and several acres of farmland were destroyed.

A VSI team inspected Marapi's crater on 8 May. Fumaroles emitted thin white vapor columns that had a slight sulfur odor and a temperature of 90°-104° C.

Information Contacts: F. Suparban Mitrohartono, A. Sudradjat, VSI, Bandung.

9/79 (4:9) A small eruption of Marapi occurred on 11 Sept. The eruption column rose 700 m and deposited ash to about 3 km W of the volcano. An inspection revealed that 3 summit-area craters (Verbeek, C, and Tuo) had been active. No activity has been reported since then.

Information Contact: A. Sudradjat, VSI, Bandung.

3/80 (5:3) Plumes rose 700 m from Marapi at 0627 and 0755 on 29 March, then were blown S by the wind. No further activity had occurred as of 2 April.

Information Contacts: A. Sudradjat, L. Pardyanto, VSI, Bandung.

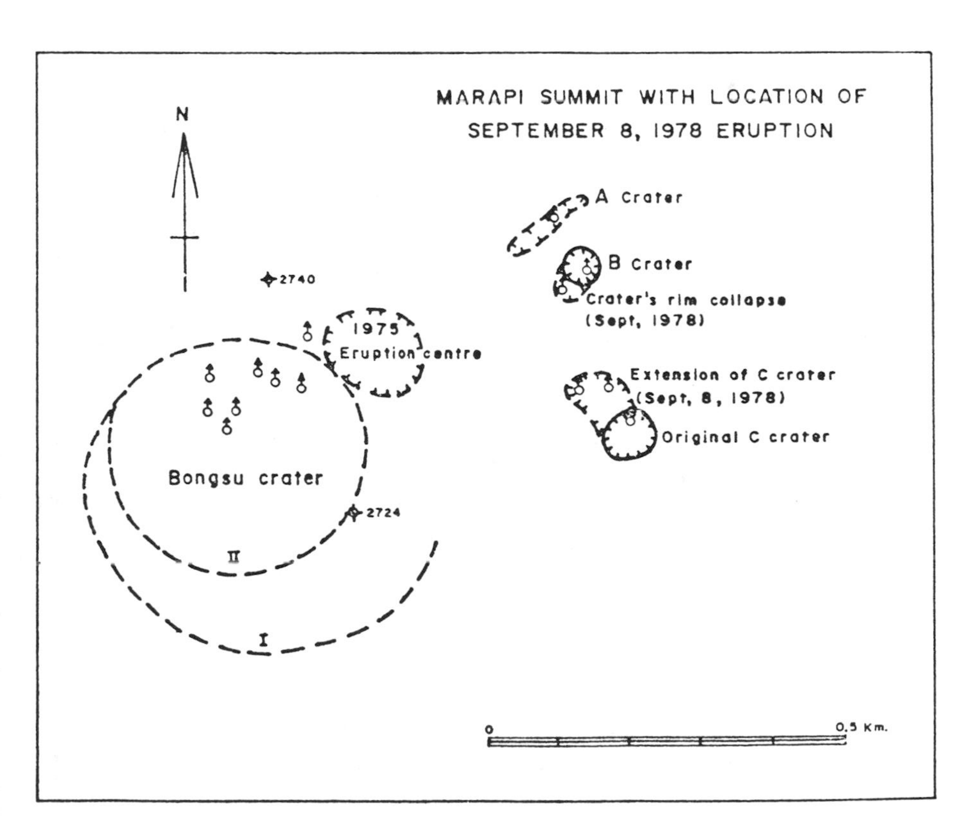

Fig. 6-1. Sketch map, courtesy of VSI, showing Marapi's summit area and the 1978 eruption crater.

	1971	1976	1978
Solfataras	110°	156°	123°
Crater Lake	45°		34°

Table 6-1. Temperatures (in °C) of solfataras and a crater lake at Kaba.

7/82 (7:3) Indonesian newspapers reported that Marapi ejected a black eruption column for about 30 minutes beginning at 0700 on 10 March. The governor of West Sumatra noted that there have been 2 previous small eruptions in 1982. Camps have been prepared to receive evacuees if a large eruption occurs.

Information Contact: M. Krafft, Cernay.

11/84 (9:11) On 15 Nov at 0500 Marapi emitted a white to brownish plume. A small, possibly phreatic, eruption occurred at 0830, ejecting a blackish plume to about 400 m height. No additional activity was reported. No residents were evacuated.

Information Contact: A. Sudradjat, VSI, Bandung.

KABA

Sumatra
3.52°S, 102.62°E, summit elev. 1952 m

12/79 (4:12) Some press reports indicated that increased activity from Kaba and neighboring Bukitdaun volcanoes preceded a destructive 17 Dec earthquake. Adjat Sudradjat and Suparto Siswowidjoyo report that no unusual volcanic activity took place. Thick white solfataric vapor was emitted from 2 craters (Kaba Lama and Kaba Besar) on Kaba, which last erupted in 1956. Temperatures of solfataras and a crater lake at Kaba were measured on several occasions in the past 9 years (table 6-1).

Active fumaroles are present on Bukitdaun, but the volcano has not erupted in historic time.

Information Contacts: A. Sudradjat, VSI, Bandung; Suparto S., Java and Sumatra Observatory.

KRAKATAU

Sunda Strait
6.10°S, 105.42°E, summit elev. 813 m
Local time = GMT + 7 hrs

In the Sunda Strait, between Sumatra and Java, Krakatau is famous for its devastating 1883 eruption. At least 36,417 people were killed, largely by tsunamis that reached 40 m heights; explosions were heard 4650 km away; and atmospheric effects were recognized around the world. Anak Krakatau, or "Child of Krakatau," is a young cone that has been growing by repeated eruptions in the center of the collapsed caldera since 1927.

Reference: Simkin, T. and Fiske, R.S., 1983, *Krakatau 1883: The Volcanic Eruption and its Effects*; Smithsonian Institution Press, Washington, 464 pp.

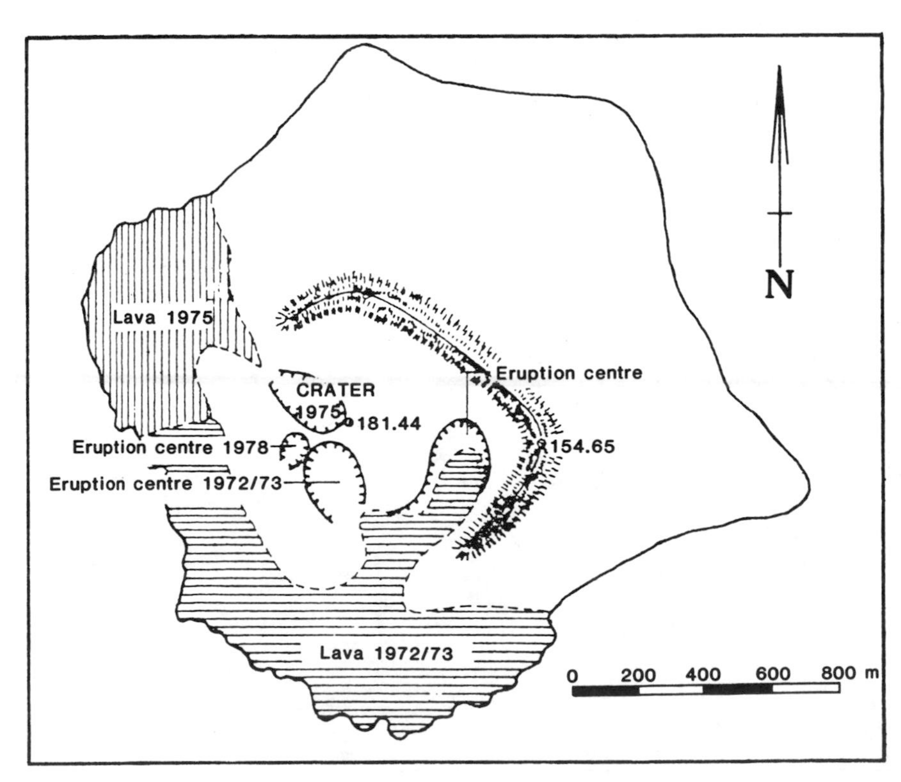

Fig. 6-2. 1978 map of Anak Krakatau, courtesy of VSI.

7/78 (3:7) Activity began in mid-July and continued through early Aug. On 2 Aug, the volcano ejected a "huge" column of incandescent material, visible from the W Java coast, about 50 km away.

Information Contacts: Reuters; D. Shackelford, Villa Park, CA.

10/78 (3:10) An eruption from a summit crater of Anak Krakatau (fig. 6-2) began on 10 July and was continuing in Oct. Lightning over the summit was seen from a nearby village on 10 July and small amounts of basaltic ash were ejected. Other explosions occurred on 14, 18, 21-23, and 30 July, and on 3 Aug. During the largest explosions (on 14 and 22 July) tephra clouds, including some bombs, rose 400-500 m above the crater. Activity was confined to vapor emission 11-13, 15-17, and 26-29 July.

VSI scientists visited the volcano 21-23 July. Tephra emission occurred at intervals of 15 minutes to 6 hours. Ash was always ejected by the explosions but larger tephra was only occasionally present in the eruption clouds.

The eruption had declined by the third week of Aug but had returned to July levels when the volcano was revisited 2-5 Oct. Activity was Strombolian, consisting of discrete groups of explosions. Each explosion group lasted an average of 9 minutes, with an observed time interval between first explosions of successive groups ranging from 5 to 27 minutes. The first explosion of each group was always the largest, typically ejecting bombs 75-100 m above the crater. Some bombs fell back into the crater and others described parabolic arcs, falling 300-400 m away and forming impact craters averaging 40 cm in diameter and 10-15 cm deep. Dark gray, ash-laden, cauliflower-shaped clouds ejected with the bombs rose 200-400 m at an average velocity of 5 m/sec. Lightning was visible in the ash clouds. Coarse ash fell 500 m from the crater and finer material was blown into Sunda Strait. Water vapor was emitted from cracks and fissures formed along the inner wall of the active crater. Vapor emission appeared to increase 3-5 seconds before the first explosion of each group.

A single component (vertical) Hosaka electromagnet seismograph recorded 554 explosion earthquakes during 77 hours of observation, using a 0.3 second transducer 1 km from the crater. Using the minimum amplifier magnification (about 2000 ×) a maximum double amplitude of 15 mm was recorded.

Information Contacts: R. Hadisantono and Suratman (July-Aug activity), L. Pardyanto (Oct activity), VSI, Bandung.

7/79 (4:7) Activity from Anak Krakatau's 100 m-diameter 1978 crater resumed in mid-July. Bombs (average diameter 1 m), lapilli, and ash were ejected every 5-15 minutes, rising 200 m and covering the area within about 700 m of the crater (fig. 6-3). Lava flowed W, reaching the coast about 450 m away. A danger zone has been delineated within 3 km of the crater.

The 1979 eruption was stronger than that of 1978, when ash and lapilli were ejected, but no bombs or lava flows.

Information Contact: A. Sudradjat, VSI, Bandung.

9/79 (4:9) Lava extrusion had ended by early Sept, but tephra emission continued. Activity fluctuated during 10 days of observations in early and mid Sept, but usually consisted of discrete explosions at intervals ranging from 20 seconds to 40 minutes. Ash clouds rose to as much as 2 km above sea level and incandescent tephra formed

Fig. 6-3. Oblique airphoto of Anak Krakatau, in the center of Krakatau caldera. Photo was taken 12 Sept 1979 by Maurice Krafft, looking approximately NW. Verlaten Island (caldera rim) is in the background. An eruption column rises from the 1978 crater. [Originally in 4:10 with a brief summary of previous reports.]

fountains that reached several hundred meters height. Some of the explosions were audible up to 50 km away. Activity continued at the end of Sept, with ejection of 1-2 m bombs and finer pyroclastics taking place about every 2 1/2 minutes.

Maurice Krafft visited Anak Krakatau 5-8 and 13-15 Sept and flew over the volcano for 3 hours on 12 Sept. A 3-man team (Rudy Hadisantono, Stephen Self, and Michael Rampino) investigating the products of the 1883 eruption observed the volcano 10-12 Sept.

The following is from Maurice Krafft.

5 Sept: 3 vents, aligned NE-SW, were active in the 1978 crater. The SW vent emitted clouds that rose 200 m; an ash cloud rose 300 m from the middle vent; and incandescent bombs (average diameter about 1/2 m) from the NE vent reached 400-500 m in height, covering the area within about 1000 m of the vent. Ash and gases from the NE vent rose 1000 m. Explosion frequency averaged one every 3-4 minutes.

6 Sept: The SW vent had become quiescent, but explosions from the other 2 vents occurred every 5-10 minutes. Bombs reached 400-500 m above the crater and fell as much as 1200 m away, on the E end of the island. At about 2100, activity began to weaken, and continued to decline during the night. Explosions on 5 and 6 Sept were heard within 50 km of Krakatau, on Java and Sumatra.

7 Sept: Only the NE vent remained active. An eruption cloud containing considerable ash but very few bombs was ejected every 20-30 minutes, rising about 500 m. Most of the explosions were not audible, but noisy explosions ejected bombs at about 2-hour intervals.

8 Sept: Weaker activity; explosion frequency declined to every 30-40 minutes and audible events that ejected bombs were less common than the day before.

The following is from Stephen Self.

"During the night of 9-10 Sept, the explosions could be heard, accompanied by volcanic tremor, on mainland W Java, 45 km from Krakatau.

"The activity observed on 10 and 11 Sept consisted of periodic explosive ejection of juvenile bombs, non-juvenile lithic blocks and large amounts of fine ash. The interval between explosions varied from 20 seconds to 20 minutes with no obvious pattern of periodicity. The explosions were often frequent enough to maintain an eruption column of fine ash and gases to a maximum of 2000 m above sea level; winds blew the column WNW.

"The estimated initial velocity of the most powerful explosions was 150-170 m/sec based on timed plume-rise during the gas thrust phase. Convective rise velocities varied from 10-20 m/sec; large blocks (1-2 m diameter) were ejected into the sea up to about 1 km from the active crater.

"At night, the ejecta were incandescent, forming spectacular lava fountains up to 200 m above the vent. The activity, therefore, has the characteristics of Strombolian explosions, but produces much more fine ash than in more basic Strombolian activity.

"A new cone has been built around the 1978 crater and reaches about 150 m above sea level. At times, deposition on the cone was so heavy that it was 50% coated by glowing bombs at night. Fine gray deposits were accumulating on the older islands of the Krakatau group, with a total of 3 cm on the N and central parts of Sertung Island (about 5-6 km NNW of the crater).

"On the afternoon of 11 Sept, the activity had dwindled to occasional weak convective plumes and the team landed on Anak Krakatau. They ascended the 1930-40 crater rim on the E side of the island and collected fresh bombs ejected from the active vent (on the W side of Anak Krakatau). The bombs were andesitic and varied from massive and glassy to poorly vesicular lava with a plagioclase phenocryst content of 15-20%. The phenocrysts were up to 3-5 mm in length.

"The coarse ejecta were purely magmatic and it appeared that there was no contact between sea water and the rising magma. However, the large quantity of fine ash may suggest some phreatomagmatic mechanism.

The team last observed the volcano late on 12 Sept, when the activity was slightly less than on 10 and 11 Sept."

The following is from Maurice Krafft:

12 Sept: A 3-hour morning overflight revealed explosions every 4-5 minutes, producing 800 m-high ash clouds. After some explosions, bombs fell into the sea at the W coast of the island.

13 Sept: Explosions occurred at about 20-minute intervals. Ash clouds were voluminous and rose about 1500 m, but few bombs were ejected.

14 Sept: Activity was similar to the previous day. In addition, a considerable amount of lightning was observed in the ash clouds.

15 Sept: Explosion frequency dropped to one each 30-40 minutes, but ash clouds continued to rise 1000-1500 m.

The following is from Adjat Sudradjat.

At the end of Sept bombs 1-2 m in diameter fell as much as 400 m from the crater, and finer pyroclastics fell as much as 700 m away. Two eruption columns were visible, indicating that there were 2 active vents. Quiet intervals between explosions were about 2 1/2 minutes long.

Information Contacts: M. Krafft, Cernay; R. Hadisantono, A. Sudradjat, VSI, Bandung; S. Self, M. Rampino, NASA, New York.

12/79 (4:12) The average number of eruption tremors/day recorded at Pasauran, [about 45] km from the volcano, was 70 in July, 50 in Aug, and 100 in Sept, then dropped sharply to 2 in Oct and Nov. No tremors were recorded in Dec, and detonations and ejection of incandescent materials were no longer observed.

Information Contacts: A. Sudradjat, VSI, Bandung; Suparto S., Java and Sumatra Observatory.

4/80 (5:4) Activity began to increase at the end of March. Detonations from explosions were heard 50 km away and window glass trembled on the W coast of Java. Incandescent material rose 200 m above the vent, which approximately coincided with the 1975 eruption center. About 65 explosion events were recorded on 13 April, and 290 on 16 April. The strongest activity occurred during a 5-hour period on 19 April, when 200 explosions were recorded. There were 335 explosions on 20 April, but activity was declining the next day. Rough seas prevented a landing on the island.

Information Contacts: A. Sudradjat, L. Pardyanto, Suparto S., VSI, Bandung.

9/80 (5:9) Increased activity from Anak Krakatau began in March, when detonations were heard from Pasauran on the W coast of Java. Incandescent material was thrown to 200 m height during a period between 19 April at 2300 and 20 April at 0415, when 200 explosions were recorded. Activity declined after 20 April but continued intermittently through Sept.

A stronger explosion occurred on 9 Sept at 0039. It rattled windows and shook houses in Pasauran, and a 3-4 cm-amplitude explosion event was recorded on the seismograph there. Ash clouds reached about 1.5 km altitude in Sept and explosions continued at the end of the month. The source of the 1980 activity is a new vent that approximately coincides with the 1975 crater and is about 250 m NW of the 1978-79 eruption center.

A research group consisting of a biologist, oceanographer, environmentalist, and volcanologist are studying Krakatau under the auspices of the centennial commemoration of the 1883 eruption. The Centennial Committee invites scientists worldwide to participate in a 3-year period of research at Krakatau.

Information Contacts: A. Sudradjat, L. Pardyanto, VSI, Bandung; M. Krafft, Cernay; *Kompas*, Jakarta.

10/81 (6:10) Explosions resumed 20 Oct after several months of fumarolic activity. Guy Camus and Pierre Vincent visited the volcano for 4 hours during the afternoon of 19 Oct, but noticed no premonitory activity. Explosions began between 0300 and 0400 the next morning. From Rakata Island (about 3 km SE of Anak Krakatau), Camus and Vincent noted 19 explosions in the 2 hours just after sunrise, before leaving the island. They had seen several others by midafternoon during discontinuous observations from a boat. Most were initiated by a "cannon-like" explosion from the main cone, followed by convective growth of an eruption column (typically to 400-600 m, but occasionally to [2] km in height). No noise could be heard on Rakata Island. The explosions usually lasted one to several minutes, but the last one observed by Camus and Vincent as they left the area began at 1511 and continued until 1525. Most of the eruption columns were dark, containing abundant ash but few blocks and no incandescent material. Water vapor could be seen condensing at the top of several eruption columns and lightning was occasionally observed. Ash fell on Sertung Island, about 2 km W of Anak Krakatau.

Information Contacts: G. Camus, P. Vincent, Univ. de Clermont-Ferrand.

Further References: Camus, G., Gourgaud, A., and Vincent, P.M., 1987, Petrologic Evolution of Krakatau (Indonesia): Implications for a Future Activity; *JVGR*, v. 33, p. 299-316.

Siswowidjoyo, S., 1985, The Renewed Activity of Krakatau Volcano After its Catastrophic Eruption in 1883 *in* Indonesian Institute of Sciences, Jakarta, *Proceedings of the Symposium on 100 Years Development of Krakatau and its Surroundings;* p. 192-198.

Sudradjat, A., 1985, The Morphological Development of Krakatau Volcano, Sunda Strait, Indonesia; *Ibid*., p. 141-146.

TANGKUBANPARAHU (SUNDA CALDERA)

W Java

6.77°S, 107.60°E, summit elev. 2084 m

This strato-volcano has been formed within the Holocene Sunda Caldera by repeated eruptions during the last several thousand years. Fifteen eruptions are known since 1826. The volcano overlooks the city of Bandung, home of 1.6 million people (and the VSI), 30 km to the S.

8/83 (8:8) Tangkubanparahu has shown increased seismicity, thermal activity, and inflation during the past several months. A VSI seismograph recorded as many as 7 shallow volcanic tremors/day in early June, 19/day by early Aug and 25/day the last week in Aug. At the beginning of Sept, seismicity increased sharply from an average of 14 to 60-74 events/day and by the middle of the month reached a daily average of 106-110. Epicenters were in the vicinity of Baru fumarolic field, on the W wall of Crater 3. No deep volcanic earthquakes have been detected. A telemetering seismograph (obtained through the VSI-USGS cooperative program) permitted monitoring from Bandung, 30 km S of the volcano.

Little change in surface activity was observed until Sept. Three fumarole fields (Baru, Ratu, and Upas) were monitored, with highest temperatures recorded at Baru, increasing to 98°C from an average of 94°C. On 13 Sept, a pair of white steam columns rose from Baru field for about 20 minutes each, reaching heights of 100-150 m. SO_2 emission measured by COSPEC increased from 70 t/d in Aug to 80 t/d in Sept. Since July, dry tilt measurements have shown a consistent inflation centered on Baru fumarole field totaling about 20 μrad.

Civil authorities warned of the volcanic hazard and the National Park has issued an alert to tourists. Within a 3-km danger zone, camping and auto parking were forbidden.

Information Contact: A. Sudradjat, VSI, Bandung.

10/83 (8:10) Seismicity continued to increase through mid-Oct but no surface changes have been noted. Tectonic earthquakes and both A- and B-type microtremors were recorded. A-type events occurred irregularly, usually at 1-3 per day, but as many as 5 were detected on several days. B-type earthquakes increased substantially, as shown in table 6-2.

Ground deformation did not show regular changes. From 10 to 25 mm of irregular inflation and deflation were detected but their significance is doubtful. Fuma-

Month		Average	Maximum
May-July		3	7
Aug,	first weeks	10	19
	last week	14	25
Sept,	first half	70	120
	second half	57	103
Oct,	first half	72	127

Table 6-2. Number of B-type events per day at Tangkubanparahu, May to mid-October 1983.

role temperatures remained stable at 96°C in the 3 fumarole fields (Baru, Ratu, and Upas). The previously declared forbidden zone of 3 km radius remained in effect.

Tangkubanparahu's most recent reported eruption was in 1969, when phreatic activity produced a thin ash layer on all sides of the volcano. Increased thermal activity in 1971 ejected small columns of mud.

Information Contact: Same as 8:8

[Minor steam explosions were reported (*Bulletin of Volcanic Eruptions* no 25, 1988) from Kawah Baru crater on 15 November 1985. The explosions were described as essentially phreatic, producing a 200-m high eruption cloud. There was no mention of tephra ejection. After this event, temperatures in the Kawah Baru solfatara field increased from 95°C to 168°.]

11/85 (10:11) "Temperatures in Kawah Baru, a small vent on the W side of the summit crater active during the 1896 eruption, rose steadily from 90°C to 140° in the 2 weeks prior to 9 Dec. No changes have been detected in the volcano's continuing low-level background seismicity."

Information Contacts: T. Casadevall, L. Pardyanto, VSI, Bandung.

12/85 (10:12) "Fumarole temperatures at Kawah Baru continued to climb through Dec, reaching a maximum of 171°C on 31 Dec. In Jan, temperatures have gradually declined to 148°C as of the 11th. No seismicity was recorded."

Information Contacts: Suparto S., T. Casadevall, VSI, Bandung.

GALUNGGUNG

W Java
7.25°S, 108.05°E, summit elev. 2168 m
Local time = GMT + 7 hrs

Galunggung has only 5 historic eruptions, but 2 have been devastating, killing more than 4000 people in 1822,

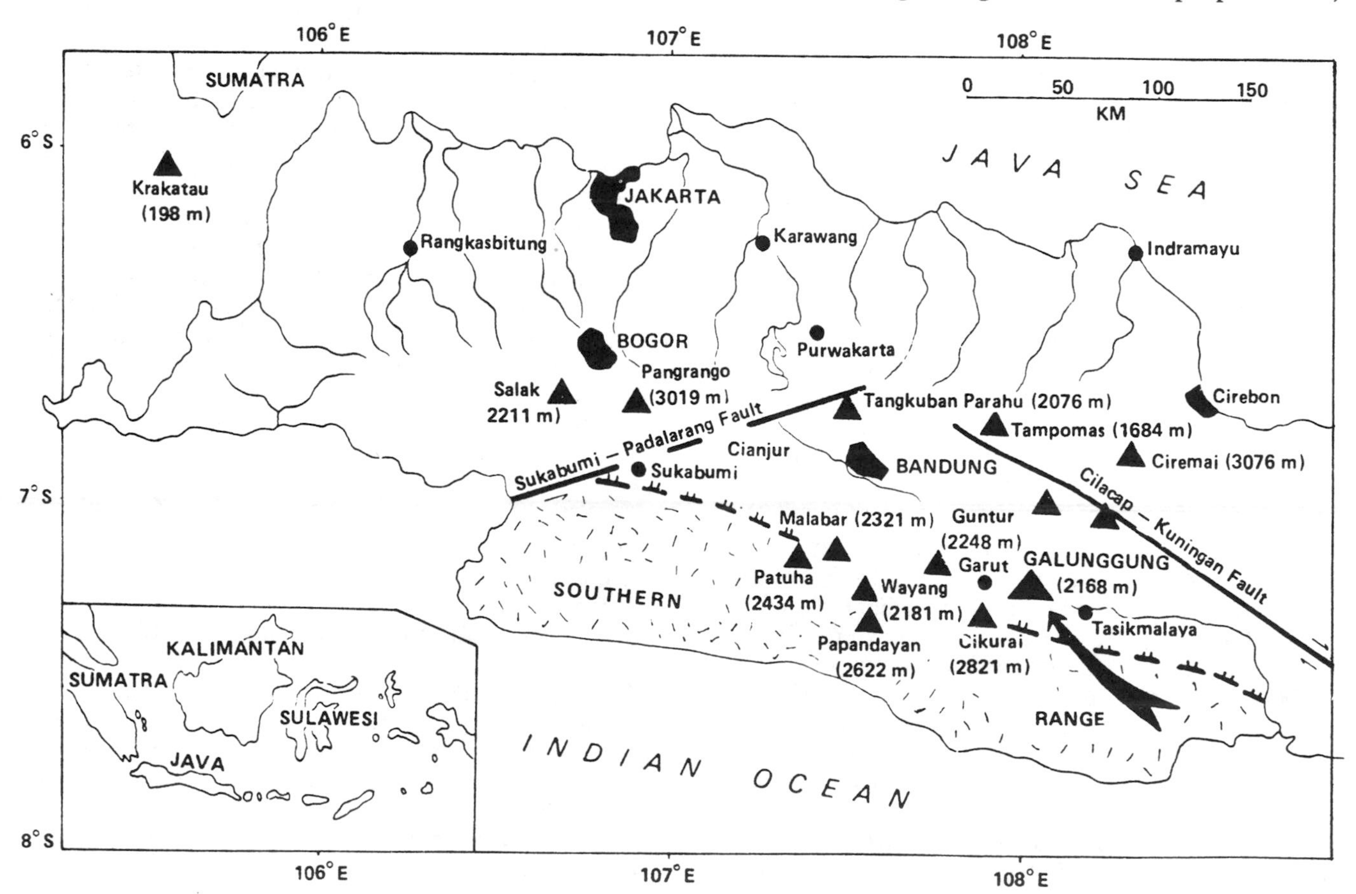

Fig. 6-4. Map of western Java and vicinity, showing some of the numerous volcanoes (triangles; summit elevations in parentheses) and major fault systems of the region. From Katili and Sudradjat, 1984.

and causing extensive property loss in 1894. The volcano is 75 km SSE of Tangkubanparahu (fig. 6-4).

General Reference: Katili, J.A. and Sudradjat, A., 1984, *Galunggung: The 1982-1983 Eruption,* Volcanological Survey of Indonesia, Bandung, 102 pp.

3/82 (7:3) A brief explosive eruption began before dawn 5 April, ejecting incandescent tephra and "stones as big as a human head" according to press reports. An image returned at 0700 by the Japanese GMS showed an eruption column about 50 km in diameter. The next available image, at 1410, showed that feeding of the eruption column had stopped and the plume had drifted about 250 km to the N. As much as 25 cm of ash fell on the flanks and ashfalls were reported from as far away as Garut, 35 km to the NW. The activity was accompanied by strong felt seismicity, and felt events continued in midafternoon. Two persons were killed and as many as 31,000 were evacuated, but most of the evacuees returned home within a few hours.

A second explosive eruption occurred during the night of 8-9 April, associated with at least 1 felt earthquake. Hot mud flowed at 60 km/hr as far as 11 km down the SE flank, buried houses in at least 6 villages, and destroyed a bridge over the Cikunir River, which emerges from a large breach in the SE side of the crater (fig. 6-5). Officials said that only about half of the 8.6×10^6 m^3 of material in the crater had been ejected and feared that the steady rain falling on the area could trigger more mudflows. AFP reported 8 persons dead, 3 missing, and 22 injured. UPI reported that many were burned or suffering from the effects of toxic volcanic gases. Authorities have forbidden entry into several areas where gases were seeping from cracks in the ground. The rice crop, within a month of its harvest, was destroyed.

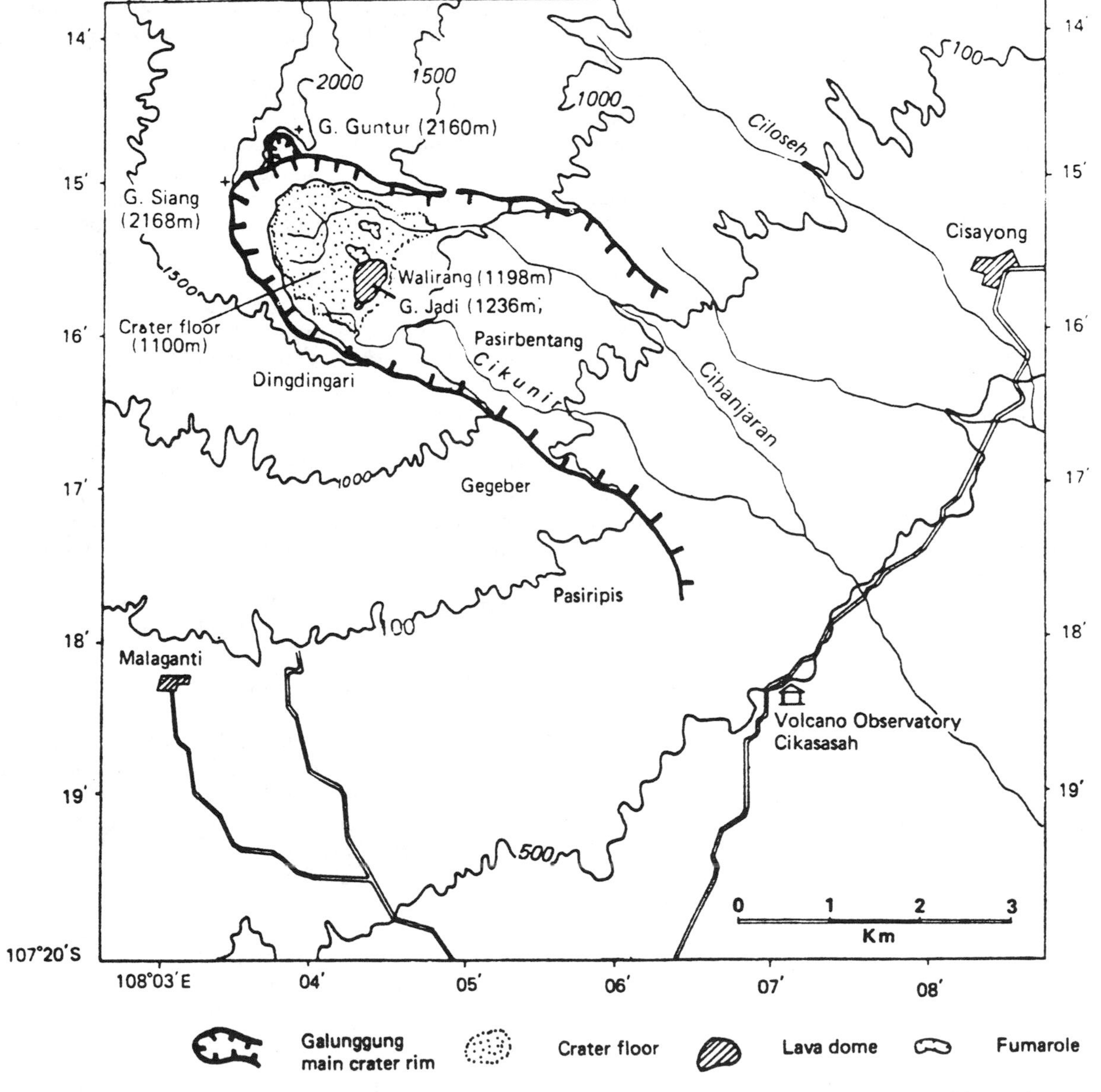

Fig. 6-5. Sketch map of Galunggung crater and vicinity, showing Gunung Jadi lava dome, Walirang ridge, major drainages, and flank towns. A temporary volcano observatory was established in Cikasasah, 7 km SE of the crater. From Katili and Sudradjat, 1984.

During Galunggung's last large eruption in 1918, a lava dome was extruded. In 1822, more than 4000 people were killed by lahars that emerged from the large breach in the SE side of the crater. Tens of thousands of people live on voluminous mudflow deposits SE of the volcano.

Information Contacts: D. Haller, NOAA/NESS; C. D. Miller, USGS, Denver, CO; Jakarta DRS; AFP; UPI.

4/82 (7:4) The following is primarily from VSI.

After 64 years of quiescence, a series of explosions from Galunggung began 5 April. The most recent of these, which are reported to have claimed 30 lives, occurred the night of 17 May and at 1600 on 18 May. There is no permanent observatory at the volcano, but seismic monitoring has been conducted there for 3 months of every year since 1976. No significant seismic activity was recorded during the last monitoring period, Sept-Nov 1981.

Residents of villages 3 km from the summit felt an earthquake about 2200 on 4 April. At midnight, subsurface rumbling was heard, followed by emission of thick white fume. Detonations began at about 0200 on 5 April, and at 0400 a larger explosion ejected a black cloud. Explosive activity continued until about 1600, depositing ash on Bandung, 65 km to the NW. Persons living near the volcano fled the activity. The next day, VSI recommended to local officials that the SE flank remain evacuated within 6 km of the summit, but other residents could return to their homes.

An explosion at 2108 on 8 April, 23 minutes after a magnitude 3.8 felt earthquake, was described by VSI on 1 May as the main eruption of the series. No local earthquakes had been reported since 6 April. A pyroclastic flow traveled 5 km from the crater down a SE flank river valley (the Cibanjaran), destroying small villages in the evacuated zone. Large volcanic ejecta killed 3 people 7 km from the volcano and ash fell as far as 23 km away. After this explosion, both eruptive and seismic activity ceased for several days. Earthquakes resumed 11 April. The number of events varied, but magnitudes and total energy release increased. Based on the seismicity, residents were warned 20 April of the danger of more eruptions.

Local newspapers reported moderate explosions 20 April. A thick black and white cloud ejected at 0808 rose about 2 km. Louder explosions at 0815 were followed by the ejection of a dense black cloud. Ash fell on nearby villages. At 0850 explosions were occurring every 15 seconds. By late afternoon 6 explosive episodes had been detected from nearby villages and rain mixed with ash was falling. Small eruption columns were observed 21-22 April. In accordance with VSI recommendations construction of dams to protect inhabited areas from lahars had begun by mid-April.

Renewed explosive activity began 25 April at 0455. A pyroclastic flow again moved down the Cibanjaran River valley, reaching 4.7 km from the crater. Tephra fell to the E and SE, with lapilli falling about 7 km away. As of 30 April, no additional eruptions had occurred, but VSI expected activity to resume within a few days.

AFP reported that an explosion occurred 6 May at 0105. Ash fell 160 km away and 2 persons were injured by hot tephra in Garut. On 17 May hot tephra showered the town of Tasikmalaya (17 km ESE) and injured 4 persons; on the 18th, 2 cm of tephra accumulated there.

Information Contacts: VSI, Bandung; M. Krafft, Cernay; Jakarta DRS; *Kompas* and *Sinar Harapan*, Jakarta; AFP.

6/82 (7:6) Intermittent explosions began 5 April, destroying about 90% of Gunung Jadi, the lava dome extruded into Galunggung's large breached summit crater during the last eruption in 1918. Lahars and nuées ardentes flowed SE through the breach onto the upper portion of the major prehistoric landslide deposit. The highest reported ash column reached 16.5 km, pitting the windshield of a passing airplane. Ash fell as far away as the Yogyakarta-Solo area, about 300 km to the E.

The 10 eruptive episodes that had occurred as of late June were separated by quiescent periods that ranged from 3 days (early in the eruption) to 3 weeks (before the 24-27 June explosions). Nuées ardentes produced by the first 5 explosions (5, 8, 20-21, and 24 April, and 6 May) traveled a maximum of 5 km. After the 6th eruptive episode, on 13 May, about 90-95% of the 1918 lava dome was still intact, with destruction limited to weak zones on its NW and SE margins, but the next explosions, on 17-19 May, left only about 10% of the old dome in the crater.

The 24-27 June explosions dropped 8-18 cm of ash and lapilli on villages 7-10 km W of the volcano, destroying hundreds of houses. Indonesian newspapers reported that people a few kilometers from the crater heard thundering sounds and saw glow over the volcano before the explosions began. Residents of Tasikmalaya (17 km ESE of the volcano) saw incandescent tephra ejection at 1900, 1910, and 1930. At 2050, a British Airways jumbo jet with 240 persons on board, flying roughly 150 km WSW of Galunggung at about 11 km altitude, encountered an ash cloud that stalled all 4 of its engines and abraded its windshield and wing surfaces. St. Elmo's fire was noted in the cockpit as the jet flew through the ash cloud. The aircraft lost 7.5 km of altitude before the engines could be restarted, but it landed safely in Jakarta. Light ashfall began about midnight in Bandung, stopping by morning, but the city remained in semi-darkness into the afternoon. Images from the Japanese GMS and

1982	Time	Temp (°C)	Altitude (km)
25 June	1400	-61	12.5
14 July	0100	-72	16 (at tropopause)
	1400	-72	16 (at tropopause)
29 July	0200	-53	11.5
13 Aug	1500	-72	15 km
16 Aug	1400	-58	no radiosonde data
29 Aug	1500	-72	15.5 km

Table 6-3. Cloud top temperatures determined by Michael Matson from NOAA 7 polar orbiting satellite images, with altitudes calculated from nearby radiosonde temperature/altitude profiles.

the NOAA 7 polar orbiter showed that the eruption cloud moved W then curved toward the S with its distal end reaching a point roughly 850 km S of Galunggung.

A similar explosion occurred at about 1000 on 25 June. Five hours later, data from infrared satellite imagery showed that the top of the cloud produced by this explosion had a temperature of -65°C, corresponding to an altitude of about 13.5 km, well below the tropopause at 15.3 km (-76°C). Winds at 13.5 km were from the NNE, and satellite images showed that the cloud moved toward the SSW to roughly 900 km from the volcano, before dissipating by 1500 the next day.

Chemical analyses show a consistent decline in SiO_2 content, from 55.7 weight % in samples from the first explosion to 49.2% in the June tephra. A significant increase in MgO, FeO, and CaO content was also noted between early April and late June. The seismic record has shown continuous tremor (amplitude 2 mm at magnification 2000 ×) and specific precursory signatures before explosions. According to a preliminary estimate of the VSI, magma was located about 3 km below the surface in late June. The VSI indicated that the interval from 27 June until the next explosive episode might last 1-3 weeks but that quiescent intervals may tend to increase in length.

Three additional explosions were seen on satellite imagery 13-15 July. On a GMS image returned 13 July at 2300, a plume that appeared to be several hours old extended S from Galunggung, then curved SE. A Singapore Airlines jumbo jet carrying 230 passengers flew into this cloud at about 9 km altitude. Three of its 4 engines stalled, but after losing about 2.4 km of altitude, pilots restarted one engine and landed safely in Jakarta. Air traffic has now been rerouted away from Galunggung. A much smaller cloud emerged from the volcano shortly before 0700 the next day and moved SE. By 1900, the distal end of the first plume was about 850 km ESE of Galunggung and the second, narrower, plume extended nearly 600 km SE. A third explosion at about 1800 on 15 July appeared comparable in size to the first. By 2300, the cloud had moved about 500 km to the S.

A volcanic hazard map prepared in 1974 has been used successfully by volcanologists and civil authorities to delineate danger zones at Galunggung, including areas at risk from lahars produced by rainfall on recent ejecta. After the 24-27 June activity, the number of evacuees had climbed to 40,000, living in temporary government barracks. Of the 27 deaths associated with the eruption, only 3 have been caused directly by pyroclastic material.

Historic activity at Galunggung has shown cyclical formation and destruction of lava domes. The large 1822 eruption destroyed a pre-existing dome, then extruded a new one. The 1822 dome was destroyed in the 1894 eruption, then replaced in 1918 by Gunung Jadi, of which the 1982 explosions have left only an estimated 10% in the summit crater.

Information Contacts: A. Sudradjat, VSI, Bandung; D. Haller, M. Matson, J. Paquette, NOAA/NESS; W. Smith, FAA; M. Krafft, Cernay; *Sinar Harapan*, Jakarta.

7/82 (7:7) Intermittent eruptive activity continued to shower tephra on the area around the volcano and to send avalanches of pyroclastic material down river valleys draining SE.

Beginning at 1940 on 13 July and culminating with a fatal eruption on 16 July, explosions sent ash and lapilli as far as 40 km, destroying homes and refugee facilities. The first explosion was heard up to 50 km away, and rumbling sounds continued during the episode.

By 14 July at least 2 cm of ash had fallen in Tasikmalaya. The daytime sky was dark, the smell of sulfur noticeable. Explosions were heard as often as every 2 minutes, and the eruption column rose to 3-4 km. Six hot avalanches had traveled SE down the Cibanjaran and Cikunir Rivers, some as far as 2 km, by 15 July. At least 18 volcanic earthquakes had been recorded by the seismograph at Cikasasah, about 7 km SE of the volcano. When Galunggung erupted at 1500 on 16 July, at least 10 persons were killed, dozens injured, and many settlements destroyed. No information was available for 16-28 July.

Eruption plumes extending S and W were noted on GMS images 28, 29, 30, and 31 July, and 1 Aug. Images between 2000, 28 July and 0400, 29 July, showed a plume extending 800-900 km SSW from the volcano. Cloud-top temperatures of -40°C to -45°C indicated an altitude of 10-10.5 km. This high plume was accompanied by a smaller, lower plume extending W. A new plume that had traveled WSW was seen on the image at 2300, 29 July. Indonesian radio reported that the 29 July eruption began at 1930 with thunderous explosions that continued for at least 3 hours and resulted in scorched forest areas. Tephra fell in Garut.

Satellite images recorded eruption plumes extending SW on 7, 9, 11, and 13 Aug. On 14 Aug a plume extended about 110 km S, then 110 km W. The Indonesian news agency Antara said the 9 Aug eruption began at 0935 with roars heard 50 km away and lasted over 3 hours. It produced "explosions of glowing magma," a 3-4 km-high eruption column, and tephra fall over a 40-km radius.

According to Antara, seismographs registered slight tremors of 2 mm amplitude on 13 Aug, followed by larger shocks of 8-12 mm amplitude at 1124. The eruption began at 1215 with rumbling and explosions that sent tephra and vapor to 2 km. At 1230 another tremor was recorded, accompanied by rumbling and tephra fall through a slight drizzle as far as Garut, where residents likened the sky to "gloomy twilight." The eruption lasted until 1455.

Ash from the larger explosions has been reported as reaching Australia. Brilliant sunsets and twilight enhancement have been observed in Sydney.

Galunggung's explosive eruptions have forced the evacuation of 62,000 persons from the densely populated area close to the volcano. The horseshoe-shaped crater opens SE. Glowing avalanches have traveled 5 km, and mudflows generated by heavy rains 10 km down main river valleys draining in that direction. Houses, rice fields, fish ponds, and roads have been destroyed. The government of Indonesia asked UNDRO to coordinate international assistance and to help evaluate the hazards and relief needs.

Information Contacts: M. Matson, S. Arnette, R. Borneman, D. Haller, J. Hawkins, NOAA/NESS; D. Anderson, Singapore Airlines; *Kompas* and *Sinar Harapan*, Jakarta; Jakarta DRS; Antara News Agency, Jakarta; AFP; *UNDRO News* (July 1982).

8/82 (7:8) Frequent periods of explosive activity continued through early Sept at Galunggung. A GMS image on 16 Aug at 1000 showed a plume that extended S then curved W to about 8°S, 100°E. By the time of the next available image, 5 hours later, feeding of the plume had ended and only remnants remained, S of Java.

After about 10 days of quiescence, explosions resumed early 26 Aug. Antara radio reported that "fire flashes" were ejected from the crater for about 6 minutes starting at 0311, followed by roaring sounds that caused panic in Tasikmalaya. Around dawn, grayish ash rose 7-8 km above the summit and glowing material appeared to slide 200-500 m from the crater. "Flames" rose 50-100 m around the crater. Between 5 and 10 cm of ash fell in Garut. Ash darkened Bandung until 1030, reducing visibility to 3 m even in lighted areas. Residents of Jakarta (Indonesia's capital, about 170 km NW of Galunggung) remained indoors as ash thick enough to cover car windshields fell within a few minutes. GMS images indicated that ash ejection began about 0500 and lasted about 2 hours. Feeding had clearly stopped by 1000. A smaller explosion started just before midnight and appeared to be continuing on an image returned at 0100. On 30 Aug at 1015, Galunggung ejected a 5-6 km-high dense black and white eruption cloud followed by a series of thundering sounds and flashes. Heavy ashfall was reported near the volcano, and lighter ashfall occurred in Garut and in Bandung, where it began about 1300. Loud explosions accompanied by flashes of incandescence started again at 0353 on 31 Aug and continued until about 0930. Heavy tephra fall darkened villages a few km from the volcano and lapilli as large as marbles caused the collapse of 2 buildings. Satellite images showed a cloud, somewhat larger than many previously ejected by Galunggung, moving SSW. An explosion 2 Sept at 0400 was followed by a second, more powerful one at 0425. Heavy ashfalls disrupted traffic between Bandung and Cirebon (about 100 km from Bandung and 75 km NNE of the volcano) between 0900 and 1600. No additional explosions had been reported as of early Sept.

Cloud top temperatures determined from NOAA 7 polar orbiter infrared images have been compared with radiosonde temperature/altitude profiles above 5.95°N, 116.05°E to yield approximate cloud top altitudes, as shown in table 6-3.

Information Contacts: D. Haller, O. Karst, M. Matson, NOAA/NESS; Antara News Agency, Jakarta; AFP; BBC.

9/82 (7:9) Occasional eruption clouds could be seen on satellite imagery in Sept. Explosive activity that probably began about 0000-0030 on 19 Sept produced a moderate-sized plume that appeared on an image at 0100. Six hours later, the plume was considerably more diffuse and feeding had stopped. At 1900 on the 22nd, an image showed an eruption cloud beginning to emerge from the volcano. No other explosions have been reported through the end of Sept.

Information Contacts: M. Matson, D. Haller, J. Hawkins, NOAA/NESS.

10/82 (7:10) Occasional explosions continued through early Nov. On 14 Oct, emission of thick grayish clouds was accompanied by intermittent thunderous rumbling sounds. An explosion at 1145 ejected ash that fell on the area around the volcano. At 1300, a GMS image showed a small plume moving SE. By the next image, 3 hours later, the plume was dispersing. After roughly 2 weeks of quiet, strong earthquakes were felt and several flashes of light preceded an explosion during the night of 3-4 Nov that ejected a thick cloud. Ash again fell near the volcano. GMS images showed a moderate to small plume from activity that probably began about 1700 on 4 Nov. The plume, which drifted SSE, was visible on the 1900 image but had dissipated 4 hours later. Inspection of satellite data revealed no other explosions, but the rainy season has begun in Indonesia and weather clouds are making eruption plumes considerably more difficult to detect by satellite.

Local authorities and villagers were building small dams near the volcano to prevent lahars from causing damage in river valleys during the rainy season.

Information Contacts: D. Haller, E. Hooper, A. Smith, NOAA/NESS; AFP; Antara News Agency, Jakarta.

11/82 (7:11) Press sources reported that an explosion on 3 Dec ejected ash and incandescent material. Strong winds blew fine ash to Garut. Within an hour, lahars were flowing down 2 river valleys. Heavy clouds that cover Java for much of the day during the rainy season prevented satellite observation of the eruption cloud. No explosions had been reported since 4 Nov and the 3 Dec explosion was said to follow "a period of relative calm."

Since the eruption began 5 April, about 40×10^6 m^3 of tephra and mud have accumulated on the flanks of the volcano. Monsoon rains threaten to remobilize this material and form destructive lahars. In the primary danger zone, residents have been given 40,000 plastic bags, which are to be filled with sand and used as a protection against flooding. Authorities have warned residents of Tasikmalaya of the danger of lahars.

Information Contacts: D. Haller, NOAA/NESDIS; AFP; *Sinar Harapan*, Jakarta.

12/82 (7:12) VSI reported that tephra ejection continued through mid-Jan, but with consistently decreasing intensity. Three explosions were reported in Nov, 2 in Dec, and 1 in the first half of Jan. Ash columns rose to about 3-5 km in Nov and Dec and to about 2 km in Jan. Glows, frequently observed during the Dec and Jan explosions, lasted for several hours to several days. Indonesian air traffic control authorities reported explosive activity on 4 Jan at 1900 and issued an alert to aviation.

Within the 1982-83 crater, a tephra cone about 200 m in diameter and 70-80 m high has developed in the area once occupied by the 1918 lava dome (Gunung Jadi). A 20-30 m-diameter vent in the new cone ejected pyroclastics to several tens of meters height. Strombolian activity occurred from this vent in Jan.

The Department of Public Works has controlled lahars along the Cikunir, Cibanjaran, and Ciloseh rivers by constructing a series of dikes and pocket dams 5-7 km long, concentric and radial to the active crater. As of mid-Jan, the dikes had stopped lahars from reaching the city of Tasikmalaya, population 700,000.

Press sources reported that on 9 Oct at 1111 Galunggung ejected a thick grayish cloud to 800 m height. At a VSI observation post on the flank, tremor amplitude reached a maximum of 45 mm. About 15 minutes later, a second explosion produced a 1 km-high cloud. The much larger explosion that followed generated a dense eruption column that reached a height of 8 km and was accompanied by lightning and thunderous sounds. At about 1610 a fourth explosion ejected sand-sized tephra that caused the collapse of 60 houses in a nearby village. Rainfall later that afternoon washed ash from house roofs in Tasikmalaya.

The 3 Dec explosion began at 1730. The activity began with a thunderous sound and ejection of incandescent material from the crater. Tremor amplitude at the VSI observation post increased from 25 mm to 41 mm. By 1820 lahars were flowing down 2 rivers. Large glows and frequent lightning were visible from Tasikmalaya. Ash blew to the W and 2 cm was deposited in the vicinity of Garut.

Information Contacts: A. Sudradjat, O. Mandraguna, VSI, Bandung; W. Smith, FAA, Washington; M. Krafft, Cernay; *Kompas* and *Sinar Harapan*, Jakarta.

2/84 (9:2) A phreatic explosion 9 Jan produced a cloud that was mostly steam with a little ash. Steam emission remained more vigorous than usual for about 2 days. Local volcanic tremor accompanied the activity but no significant increase in deeper seismicity was noted before or during the increased steam emission. Explosive activity was also reported 31 Jan but no details were available from the ground, and weather clouds prevented satellite observations.

Explosive activity stopped after extrusion of a small lava flow onto the crater floor in early Jan 1983.

Information Contacts: A. Sudradjat, VSI, Bandung; W. Smith, FAA, Washington.

Further References: Hanstrum, B.N. and Watson, A.S., 1983, Case Study of Two Eruptions of Mount Galunggung and an Investigation of Volcanic Eruption Cloud Characteristics Using Remote Sensing Techniques; *Australian Meteorological Magazine*, v. 31, p. 171-177.

Sawada, Y., 1987 (see Part I—References).

DIENG VOLCANIC COMPLEX

Central Java
7.20°S, 109.90°E, summit elev. 2565 m
Local time = GMT + 7 hrs

This volcanic plateau, 200 km E of Galunggung, consists of many cones and craters. Fifteen historic eruptions are known in the last 200 years, and another from the 14th or 15th century. Five have caused fatalities.

2/79 (4:2) Gas containing CO_2 and H_2S was ejected from Sinila Crater at 0330 on 20 Feb, killing (at latest report) 182 persons and injuring about 1000, over 100 of whom had to be hospitalized [but see 4:3]. A police colonel said that 17,000 people had been evacuated from 6 villages near Sinila. A number of livestock were also killed by the gas, and ponds near the crater were covered with dead fish.

An AP wirephoto showed bodies of some of the victims, who did not appear to be burned, nor were they covered with ash. Nearby vegetation also appeared unburned.

Many of the fatalities occurred in the village of Bandjarnegara, located in a valley on Dieng's SW flank. The Dieng complex consists of a caldera containing 2 large craters, plus a NW-SE trending, 14 × 6 km zone of small (100-300 m-high) cones.

The gas emission from Sinila was preceded by 7 felt earthquakes and followed by ejection of "smoke." UPI reported that "lava continued to pour over the slopes of the Dieng Plateau" on 20 Feb. Sikidang Crater is located in the summit caldera, 27 km NE of Bandjarnegara, and is said to be 11 km from Sinila Crater.

Information Contacts: AFP; Jakarta DRS; Antara News Agency, Jakarta; AP; UPI; Reuters; S. Russell, USGS, Reston, VA.

3/79 (4:3) The following information is compiled largely from interviews with Subroto Modjo, Chief of the Geothermal Section, VSI.

Nomenclature for this area presents some difficult problems, because of its complexity and the uncertain relationships between its numerous and closely spaced volcanic features. The best general name for the region is "Dieng Volcanic Complex," which includes the volcanoes listed in the *CAVW* as "Dieng" and "Butak Petarangan." [SEAN's initial report listed this volcano as "Prahu (Mt. Dieng)."]

Poisonous gas emission, accompanying a phreatic eruption from the SW portion of the Dieng volcanic complex, killed 149 persons early on 20 Feb. Nearby residents felt earthquakes at 0155, 0240, and 0400, after which an explosion at 0504 ejected a dark gray cloud from Sinila Crater (fig. 6-6), a small, water-filled pre-existing vent. Less than 1 hour later, after the ejection of a second dark gray cloud, a hot lahar from Sinila flowed down a gully, cutting a road. At 0650, a new crater, named Sigluduk, opened 250 m W of Sinila and began emitting vapor. Sinila produced a second, larger lahar a few minutes later. The total volume of Sinila's 2 lahars was about 15,000 m^3.

Residents of Koputjukan village fled the activity in several directions. Many of those who headed W towards Batur were killed by gas containing CO_2 and H_2S, emitted from many small vents and fissures S and W of the 2 craters, ponded in a low-lying area beside the road. Others were killed near an elementary school, to which they had retreated after observing the deaths along the Koputjukan-Batur road. Both areas had been known to emit poisonous gases in the past. Four rescue workers were also killed. About 15,000 were evacuated from 6 nearby villages, and 2 danger zones were delineated around the gas vents (fig. 6-6); entry into the inner zone was completely prohibited, and use of the outer zone was limited to daylight hours.

Activity from the 2 explosion craters declined after ejection of a dark gray cloud from Sigluduk at 1100 on 20 Feb. By dawn the next day, Sinila was emitting dense white vapor, but only a little white vapor was escaping from Sigluduk. VSI personnel visited the craters early on the 22nd. Sinila's vapor emission had ended, although the bottom of the crater, 60-70 m across and 40 m deep, was filled with bubbling mud. Sigluduk, 25 m in diameter and 10 m deep, contained an active solfatara with a temperature of 71°C.

The Dieng complex has a long history of fatal activity, including eruptions in 1918 (1 killed), 1939 (10 killed), and 1944 (114 killed), as well as a geothermal well blowout that killed 3 persons in late 1978.

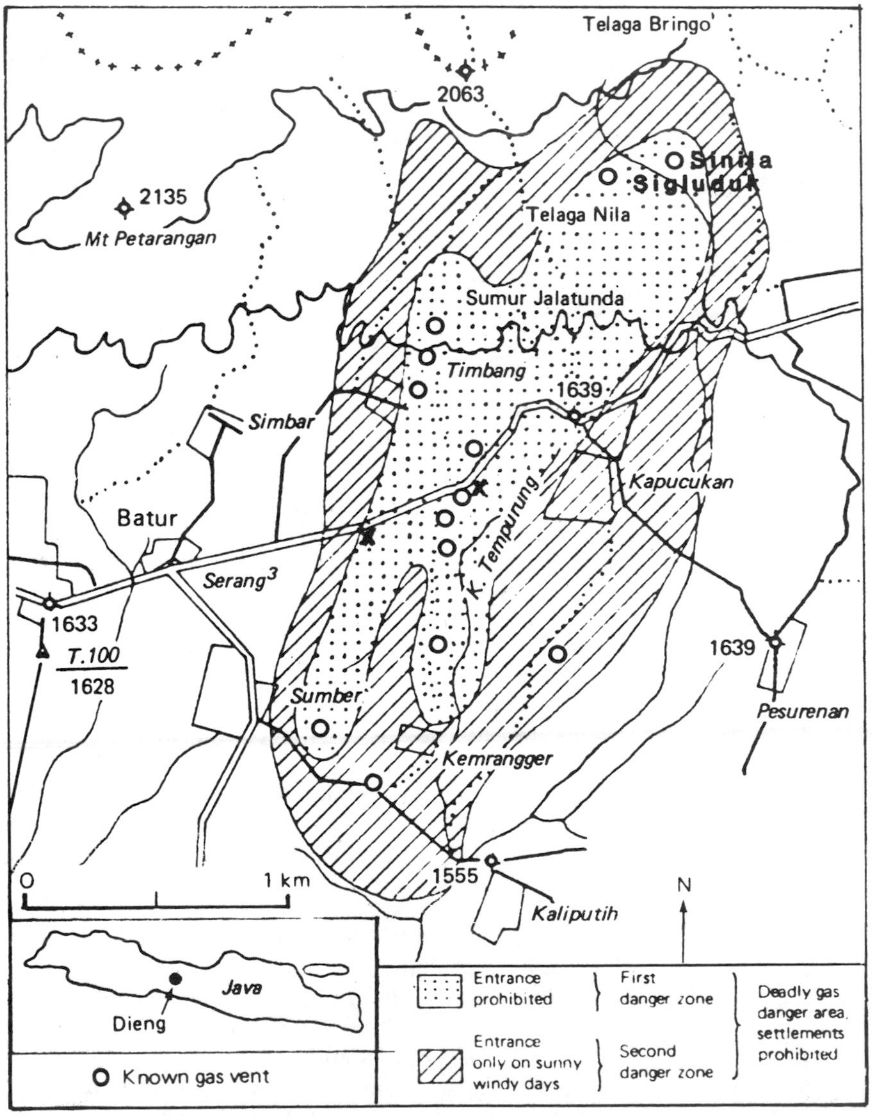

Fig. 6-6. Hazard zonation map, courtesy of VSI, showing the SW portion of the Dieng Volcanic Complex. Entry is prohibited to the first danger zone, indicated by a dotted pattern; entry is permitted to the second danger zone (diagonal lines) only on sunny, windy days. Known gas vents are represented by circles; fatalities are marked by x's. Courtesy of Adjat Sudradjat.

Information Contacts: S. Modjo, VSI, Bandung; M. Krafft, Cernay; *Kompas* and *Sinar Harapan*, Jakarta.

Further Reference: LeGuern, F., Tazieff, H., and Faivre Pierret, R., 1982, An Example of Health Hazard: People Killed by Gas During a Phreatic Eruption: Dieng Plateau, Java, Indonesia; *BV*, v. 45, p. 153-156.

12/81 (6:12) On 17 Nov an area of the Dieng volcanic complex with a history of poisonous gas emission experienced an MM 2-3 felt earthquake. A tripartite electromagnetic seismometer located the shocks in a zone of fractures and gas emission 1-2 km E of Batur. Newspapers and radio broadcasts reported that 3-12 seismic events were recorded during each of the 10 days ending 6 Dec. However, the number of events began to decrease in early Dec, and seismicity was continuing to decline as of the 17th.

There were no indications of volcanic activity. The poisonous gas content of vapor sampled at several known emission sites had not changed more than 15% from previous measurements. The temperature of the pool of boiling mud at Sileri cone and the thermal feature Sikidang, a few km E of the zone, remained unchanged at 90°C. Government officials issued an alert, but removed it after receiving a VSI report.

Information Contacts: A. Sudradjat, VSI, Bandung; Jakarta DRS.

MERAPI

Central Java
07.54°S, 110.44°E
summit elev. 2911 m
Local time = GMT + 7 hrs

This strato-volcano with an active summit lava dome has a grim history. It is located 70 km SE of Dieng and immediately N of Yogyakarta, a city of half a million persons. 32 of its 67 historic eruptions are known to have been accompanied by nuées ardentes—more than any other volcano in the world—and 11 of them have caused fatalities. The volcano is carefully watched by several VSI observatories and monitored by at least 6 telemetering seismometers.

3/76 (1:6) Activity increased in Feb

compared to the previous month. The number of volcanic earthquakes increased, the lava dome grew higher, and red glares were observed at 2 points on the dome. The number of glowing avalanches decreased from Jan, however, and there was only one nuée ardente, on 4 Feb, that traveled about 750 m SW into the Batang River.

On 5 March at 1503, reddish white smoke emitted with strong gas pressure was observed to about 350 m above the summit. The volcano was hidden behind clouds. On 6 March, small avalanches took place continuously from 0135 to 0348, then the volcano was covered by clouds. At 1702, three avalanches were observed. From 1733 on, bigger nuées ardentes took place, interspersed with minor ones, until 13 March (table 6-4).

From the evening of 8 March until the evening of 9 March, when the volcano became visible from the Ngepos Volcano Observatory 11 km SW of the summit, it could be observed that the lower part of the lava dome, estimated to be about 400,000 m^3, or almost $^1/_3$ of its volume (1.4×10^6 m^3), had slid away. During the events on 6 and 7 March nuées ardentes moved SW, entering a tributary of the Batang River and continuing much farther downstream into the rivers Sat, Blongkeng, and Bebeng. The prevailing wind blew E, depositing finer volcanic products on the E and SE flanks to a maximum distance of 37.5 km. Thickest ash falls were on the ESE flank. At 6 km from the summit the deposit apparently measured 5 mm and temporarily panicked the villagers.

Nuées ardentes caused forest fires on the upper SW flank near the tributaries of the rivers Sat, Blongkeng, and Bebeng. Roughly 580,000 m^3 of ash were deposited on the E flank, and around 300,000 m^3 of avalanche material were deposited on the upper SW flank.

Information Contact: D. Hadikusumo, Volcanology Division, Geological Survey of Indonesia (GSI), Bandung.

11/76 (1:14) Merapi was unusually active on 6 and 7 Nov, emitting 3000-m ash clouds, and nuées ardentes that moved 2.5 km down the SW flank. No casualties or serious property damage were reported. The new dome, which began growing after the March activity, had an estimated volume of 20,000 m^3 in June.

Information Contacts: D. Hadikusumo, Volcanology Division, GSI, Bandung; D. Shackelford, Villa Park, CA; UPI.

Date	Large	Small	Maximum Distance
6	10	19	4 km
7	17	16	5.5 km
8	5	2	3.5 km
9	5	6	6 km
10	1	4	2.75 km
11	1	9	5.5 km
12	-	6	1.5 km
13	-	3	1.5 km

Table 6-4. Daily number of nueés ardentes and maximum distances traveled from the vent, 6-13 March 1976.

3/77 (2:3) The new lava dome had increased in volume from about 2×10^3 m^3 in June, 1976 to about 2.5×10^6 m^3 in late Feb, 1977. Avalanches continued to occur at intervals of 15 minutes or more; sometimes, but not every day, nuées ardentes accompanied the avalanches, especially after a heavy rainfall.

Information Contacts: J. Matahelumual, Volcanology Division, GSI, Bandung; D. Shackelford, Villa Park, CA.

8/78 (3:8) Merapi was visited by Haroun Tazieff in July. No eruption was occurring, but seismicity was increasing, and fumarole temperatures on the N flank reached 980°C, as compared to 400-500°C during previous periods of quiescence.

Information Contacts: H. Tazieff, CNRS, Gif-sur-Yvette; P. Jezek, SI, Washington.

9/78 (3:9) François LeGuern and Haroun Tazieff provided the following fumarole temperatures and gas compositions (table 6-5), obtained during their July visit to Merapi. Gases were analyzed with a field gas chromatograph.

Information Contacts: F. Le Guern and H. Tazieff, CNRS, Gif-sur-Yvette.

7/79 (4:7) Newspaper reports state that hot avalanches from the lava dome decreased in frequency during May, after as many as 40/day had occurred in April. The May

Location	Max. Temp.	Analysis Temp.	Gas composition, in volume %: CH_4	CO_2	H_2O	H_2S	SO_2	H_2	O_2+Ar	N_2	CO
Gendol	820°C	720°C	0.013	3.28	94.37	0.24	0.25	1.12	0.15	0.55	0.03
		720°C	0.017	4.02	93.08	0.30	0.31	1.36	0.19	0.68	0.03
		819°C	<0.001	4.57	93.71	0.47	0.51	0.56	0.005	0.13	0.01
Crater near active dome	901°C	744°C	0.06	5.26	90.75	1.15	1.14	1.46	0.02	0.12	0.02
Woro	760°C	580°C	3.21	94.27	0.12	0.06	Tr	Tr	2.34	0.02	

Table 6-5. Fumarole temperatures and gas compositions at Merapi, July 1978, courtesy of François LeGuern and Haroun Tazieff.

avalanches traveled as much as 1.75 km. Growth continued at the new dome, which first appeared in Jan 1979 and now covers remnants of the 1973 and 1978 domes.

Information Contacts: *Kompas*, Jakarta; M. Krafft, Cernay.

1/80 (5:1) Lava dome extrusion continued through 1979 from a SW flank vent about 200 m below the summit. A lobe of lava extending a short distance down the upper SW flank gave the dome an asymmetrical form and occasionally spawned nuées ardentes d'avalanche. The last significant nuée ardente traveled about 6 km in Aug 1979, but remained within the forbidden zone, where human access is prohibited.

The maximum thickness of the active dome was about 100 m in Oct. In Nov, VSI estimated its volume at 1.14×10^6 m^3 and its extrusion rate at about 10^5 m^3/month.

Information contacts: A. Sudradjat, L. Pardyanto, VSI, Bandung.

2/81 (6:2) The lava dome that began to emerge at the summit in 1979 was still growing in Feb and had reached an altitude of 2947 m. Lava fragments from the E and central part of the cone had moved 2.0 km toward the Batang River, and 250-500 m farther in Dec. Personnel at the Ngepos Observatory have counted 34 larger and 468 smaller lava avalanches in recent months; the time interval was not reported. Nuées ardentes d'avalanche occurred in Nov and Dec but were confined to the summit area. Two Minakami A-type earthquakes, the first in several months, were recently recorded by the seismograph at the Babadan Observatory (4 km from the summit). No important lahars have occurred along the Putih, Bebeng, and Krasak Rivers since the beginning of this year's rainy season.

Information Contact: Same as above.

Further Reference: Tazieff, H., 1983, Monitoring and Interpretation of Activity on Mt. Merapi, Indonesia, 1977-80: A Practical Example, "Civil Defense" *in* Tazieff, H. and Sabroux, J.C. (eds.), *Forecasting Volcanic Events*, Elsevier, Amsterdam, p. 485-494.

12/81 (6:12) Landsliding on the summit lava dome 29 Nov was followed by a nuée ardente d'avalanche that flowed 3-4 km down the Batang River valley. Ash fell 15-20 km to the NW. An avalanche of incandescent lava from the dome set fire to 15 hectares of tropical mixed forest. On 1 Dec approximately $^3/_5$ of the lava dome slid down the upper flank, flooding the Krasak and Batang Rivers and adding 1.2×10^6 m^3 to the deposits from previous debris flows. As of early Dec, there were about 6.9×10^6 m^3 of unstable material in the summit area. Authorities feared that monsoon rains could cause large cold lahars, and officials of Magelang District, W of Merapi, remained on constant alert in early Dec. AFP reported that the volcano continued to emit ash in mid-Dec.

A tripartite electro-seismometer at Babadan Observatory and 2 single-detector seismographs near the volcano recorded continuous trains of tremors and very shallow earthquakes that may have been landslide events. Continuous records of magnetic values showed no significant changes. No sharp variations in CO_2 and H_2S emissions were detected. Tilt measurements showed deformation of 10-20 μrad, within the normal variation.

Information Contacts: A. Sudradjat, VSI, Bandung; AFP; Jakarta DRS.

1/82 (7:1) Heavy ashfall that began 6 Feb damaged crops and halted traffic in the nearby Boyolai region, according to a press report dated 8 Feb.

Information Contact: AFP.

6/84 (9:6) The quoted material is a report from Adjat Sudradjat.

"Merapi erupted 15 June between 0215 and 0600, accompanied by nuées ardentes that extended 7 km down rivers (the Batang, Bebeng, and Krasak) on the SW side of the volcano. An eruption plume rose to 6 km height and caused ashfall in Muntilan, Ambarawa, and Semarang, approximately 60 km N of the volcano. The eruption was accompanied by detonations. The first explosion was followed by a milder eruption producing a plume to 2 km height and a nuée ardente to 6 km distance at noon. The frequency of nuées ardentes progressively decreased until the morning of 16 June. No eruptions were observed the following day. Lahar material estimated to exceed 4×10^6 m^3 along the Bedeng, Krasak, and Putih Rivers may threaten Magelang city (population about 125,000)."

An image from the NOAA 7 polar orbiting satellite on 15 June at 1441 showed a hazy area that did not seem to be weather-related extending from 7°S to 5°S and from 110°E to the edge of the image at 106°E. The hazy area did not extend back to Merapi, nor was any tephra emission apparent at the volcano. No evidence of activity could be seen on the image returned the next day at about the same time.

Newspapers reported ashfalls at Magelang (30 km NW of Merapi) and Salatiga (35 km NE of the volcano). Visibility near Salatiga was limited to 10 m and more than a cm of ash covered roads, slowing traffic. More than 2 cm of ash fell at Solo (45 km E of the volcano) and ashfall was reported at Cilacap, on the coast 160 km SW of Merapi.

"Seismographs detected a progressive increase in seismicity from 4 counts per day on 8 June to 59 on 12 June. A warning was issued 13 June, and the evacuation of 1000 persons from forbidden zone section VI (Kemiren village) was immediately implemented. The tong-tong warning system was tested again to be sure that it was operational. The eruption was preceded by an intense lava avalanche on 13 June that caused a nuée ardente d'avalanche (nuée ardente of Merapi type).

"As of 22 June, no volcanic A-type earthquakes had been recorded. The dominant seismicity has been continuous tremors, very shallow volcanic earthquakes, and avalanche events with multiphase signatures. Judging from this evidence, 2 possibilities are: (1) the 15 June eruption is the final phase of the 1969-1984 lava dome growth (effusive stage – see below); or (2) the 15 June eruption is evidence of the gas phase of a new cycle. Both possibilities would show quiescence of A-type earthquakes. The rate of dome growth, which was reported to have sharply increased 20-21 June, may tend to support the latter hypothesis if it continues.

"After the 1969 gas explosion, Merapi continued to build a lava dome in its summit crater. Since the orifice is not symmetrical, the lava dome becomes unstable as it grows, and portions slide away, producing avalanches and nuées ardentes d'avalanche. The maximum growth of the dome was 3.6×10^6 m^3 and the rate of growth was seemingly constant at about 0.1×10^6 m^3 per month. Intensive sliding usually occurs in rainy seasons, in Nov annually, and removes 20-30% of the dome's total volume. The lava blocks that slide away from the dome may originate lahars, affecting a large area of the SW sector of the volcano."

Information Contacts: A. Sudradjat, VSI, Bandung; M. Matson, NOAA/NESDIS; *Jakarta Times*.

Further References: Bardintzeff, J.M., 1984, Merapi Volcano (Java) and Merapi-type Nuée Ardente; *BV* v. 47, p. 433-46.

Zen, M.T., Siswowidjoyo, S., Djoharman, L., and Harto, S., 1980, Type and Characteristics of the Merapi Eruption; *Buletin Dept. Teknik Geologi*, Inst. Teknologi Bandung, v. I, p. 34-46.

LAWU

Central Java
7.63°S, 111.19°E, summit elev. 3265 m
Local time = GMT + 7 hrs

Lawu is 85 km E of Merapi and has had no historic eruptions.

5/79 (4:5) The Indonesian newspaper *Kompas* reported that the first earthquakes of a swarm in the vicinity of the Lawu volcanic complex were felt on 10 Dec 1978. Area residents reported 14 felt shocks in Dec, 5 in Jan, 2 in Feb, 6 in March, and 8 in April. The earthquakes were usually preceded by thunder-like rumbling from the direction of Lawu.

Seismicity became more frequent in late April and early May. At least 4 felt events occurred on 26 April, including a 10-second earthquake at 1900 that damaged a temple and a transmitting station. On 4 May a landslide in Lawu's Candradimuka Crater (in the S part of the complex) was followed by emission of a thick vapor cloud that was accompanied by a sulfur odor. Between the evening of 4 May and 0700 the next morning, 9 events were felt. A total of 27 felt shocks occurred on 5 May, 37 on the 6th, and 35 on the 8th. A series of 5 earthquakes lasting 4-6 seconds each took place at about 1230 on 9 May. During a 12-hour period 14-15 May, there were more than 1000 recorded events, more than 50 of which were felt. A VSI team is investigating the seismicity. Only solfataric activity has been reported from Lawu in historic time.

Information Contact: *Kompas*, Jakarta.

Further Reference: Tjia, H.D. and Hamidi, S., 1981, An Earthquake Swarm Around Lawu Volcano in Java; *Berita Geologi*, v. 13, p. 108-111.

SEMERU

E Java
8.11°S, 112.92°E, summit elev. 3676 m

Semeru, the highest and one of the most active volcanoes in Java, has had 56 eruptions since 1818, most of them relatively mild. It is in Eastern Java, 280 km ESE of Merapi.

Further Reference: Suryo, I., 1986, Semeru; *Bulletin of the Volcanological Survey of Indonesia*, no. 111, 52 pp.

11/76 (1:14) The eruption that began on 31 Aug 1967 continued with ejection of ash and nuées ardentes, and growth of the lava dome.

Information Contacts: D. Hadikusumo, Volcanology Division, GSI, Bandung; D. Shackelford, Villa Park, CA.

3/77 (2:3) Semeru continued its activity essentially unchanged during Feb, with lava avalanches, occasional nuées ardentes, and emission of ash and gases.

Information Contacts: J. Matahelumual, Volcanology Division, GSI, Bandung.

1/80 (5:1) Extrusion of a summit dome began in 1967 and was continuing in late 1979. Until 1974, dome growth was concentrated in the S portion of the summit, shifting since then to the SE but remaining in the summit area. Lava avalanches, nuées ardentes (traveling a maximum of about 7 km), and ash explosions have accompanied the dome extrusion.

During 1979, the rate of dome growth slowed, and ash explosions became weaker and less frequent, occurring at intervals of 45 minutes to 1 hour in contrast to 30-35 minutes in 1977-78. Although almost all of the 1977-78 explosions were recorded by a seismograph 7 km S of the summit, few of the most recent explosions can be detected by this instrument.

Information Contacts: A. Sudradjat, L. Pardyanto, VSI, Bandung.

9/80 (5:9) Ash eruptions were continuing in late Sept. However, no additional dome growth has been observed, and nuées ardentes have not been reported since Jan.

Information Contacts: A. Sudradjat, VSI, Bandung; D. Shackelford, Fullerton, CA.

2/81 (6:2) Ash emission continued at an average rate of once every 56 minutes in Nov and Dec. Ash columns typically rose 500-700 m above the crater rim. Some clouds were less ash-rich, as indicated by a grayish color. Incandescent lava fragments were sometimes visible at

1985	Apr	May	June	July	Aug	Sept	Oct	Nov	Dec
Explosions	2529	3832	3748	3321	3192	2357	2315	1464	1797
Rockfalls	179	323	437	364	341	303	475	224	320
Nuées Ardentes	0	0	5	18	15	12	10	6	5

Table 6-6. Monthly numbers of explosions, rockfalls and nueés ardentes d'avalanche at Semeru, April-Dec 1985. Explosion and rockfall counts are from seismic records; nuées ardentes are counted from extended rockfall signals on seismic records and from observations from Gunungsawur Observatory at 650 m altitude, 12 km SE of the summit.

night. Strombolian-type eruptions have accompanied the formation of the lava dome since extrusion began in 1967.

Lava avalanches from the dome have usually been contained at about 3 km altitude on the S flank of the volcano, in the upper reaches of the Kembar River, but one traveled farther down the river valley in early Dec. Before this year's monsoon rains VSI has alerted local authorities to the S and SE of the danger of lahars along the Kembar, Kobokan, Rejali, Sat, and Glidil Rivers.

Information Contacts: A. Sudradjat, L. Pardyanto, VSI, Bandung.

3/81 (6:3) Activity increased 28 March. The first nuée ardente moved about 4 km from the summit down the Kembar and Kobokan Rivers (on the S flank) at 1755. During the following days, increasingly intense nuées ardentes reached a distance of more than 7 km from the summit. Four nuées ardentes and 19 lava avalanches (presumably accompanied by nuées ardentes of eruptive origin) were reported on 29 March and 4 more nuées ardentes and 36 lava avalanches were observed the next day. As of 31 March, tremors were being continuously recorded by the VSI seismograph about 10 km from the summit.

One person was killed by a nuée ardente and 272 others were evacuated. The ongoing rainy season may cause lahars and associated flooding.

Information Contacts: Same as 6:2.

SiO_2	56.8
Al_2O_3	19.9
Fe_2O_3*	7.61
MgO	2.28
CaO	8.09
Na_2O	3.45
K_2O	1.22
TiO_2	0.70
P_2O_5	0.18
MnO	0.18
loi**	0.01
Total	100.45

* Total iron expressed as Fe_2O_3
** Loss on ignition at 900°C

Table 6-7. Whole rock analyses of breadcrust bomb sample collected from Semeru, 17 Aug 1985. X-ray fluorescence analyses by the USGS Laboratory, Denver, Colorado.

5/81 (6:5) Thirty cm of rain in 2 hours on 14 May dislodged pyroclastic deposits from the upper flanks. Approximately 5-6 × 10^6 m^3 of breccia, volcanic sands, ash, surficial cover, and vegetation flowed down the 40-60° E flank into the valleys of the Tunggeng and Sat Rivers. The mudflow killed 252 persons, left 152 injured and 120 missing, and flooded 626 hectares of rice fields and 16 villages along the river's banks. It eroded old mudflow deposits and washed away a dike built in 1912 after a similar event had destroyed the city of Lumajang (40 km E of the volcano) in 1901. The devastated area appears to be within the alert zone on VSI's 1973 hazard map.

Fresh nuée ardente deposits on the upper S flank, combined with the onset of the monsoon season, had prompted the VSI to warn local authorities in Jan of the danger of mudflows S and SE of Semeru. Although a mudflow also moved down the S flank on 14 May, no casualties were reported there.

Activity at Semeru was normal during May, with about 80 gas eruptions each day. The lava dome continued to grow at about 100 m^3/day.

Information Contacts: Same as 6:2.

8/85 (10:8) The quoted material in the next 2 reports is from VSI.

"Semeru has been continuously active during at least the past 2 years. Activity consists of a blocky lava flow from the Semeru vent accompanied by Vulcanian explosions that occurred 6-10 times per hour during Aug, sending tephra-rich plumes to 1 km above the vent."

On 4 Aug, Space Shuttle astronauts observed a plume extending 100 km W of the summit.

Information Contacts: J. Matahelumual, T. Casadevall, VSI, Bandung; C. Wood, NASA, Houston.

12/85 (10:12) "Semeru has been active throughout 1985. The principal activity consisted of frequent small Vulcanian explosions and slow lava extrusion from the summit vent, small rockfalls from the steep-sided walls of the summit lava plug, and occasional nuées ardentes d'avalanche from the lava flow, where it rests on the steep SE flank of the summit cone. These small pyroclastic flows generally remain within 5 km of the summit vent. Typical maximum cloud height for the small Vulcanian explosions is 1000 m or less. The frequency of explosions decreased steadily from 3832 in May (125 per day) to 1797 in Dec (60 per day). The monthly average number of rockfalls was about 320. The maximum number of nuées ardentes d'avalanche was 18 in July, declining from 15 in Aug to 5 in Dec (table 6-6). The composition of a breadcrust bomb sample collected on 17 Aug is shown in table 6-7.

Information Contacts: Suparto S., T. Casadevall, VSI, Bandung.

TENGGER CALDERA

E Java
7.94° S, 112.95°E, summit elev. 2329 m
Local time = GMT + 7 hrs

Bromo, the historically active pyroclastic cone in Tengger Caldera, has erupted frequently in the 19th and 20th centuries. It is just 20 km N of Semeru.

6/80 (5:6) Vapor emission from Bromo cone, on the floor of the 5 km-diameter Tengger Caldera, began 2 June. Several days later, Bromo started to eject incandescent tephra. Glowing bombs 1 m in diameter fell as much as 400 m from the crater rim, while ash-laden vapor rose 600-800 m, accompanied by rumbling. Crowds of tourists approached to within 500 m of the vent during the first 2 weeks of the eruption.

Activity intensified during the 3rd week, with explosions every few seconds to 2 minutes. By 21 June, some cauliflower-shaped ash clouds rose 900 m. VSI moved tourists away from the immediate vicinity of the eruption to the caldera rim, where a resort hotel is located about 4 km from Bromo.

By 24 June, explosive episodes, accompanied by loud detonations, had become more vigorous but less frequent, occurring at 4-5 minute intervals. Although about half of the ejecta fell back into the crater, 1 m-diameter bombs set bushes afire at the foot of Batok cone, 1.5 km away.

Information Contacts: A. Sudradjat, L. Pardyanto, Suparban, VSI, Bandung.

9/80 (5:9) Activity at Bromo cone peaked on 20-21 June when 480 explosions were recorded during a 24-hour period. Ash-laden steam columns rose 1.2 km and bombs fell 1.5 km away, setting vegetation afire. Activity declined in late June and remained weak until mid-July. Frequent explosions resumed 15 July, when about 1000 were recorded in 24 hours. Explosions remained frequent for more than a week, then declined steadily after 24 July to about 1 every 48 hours in Aug. *Kompas* newspaper reported a minor eruption on 9 Sept at 0825 and added that tourists were not being allowed to descend to the crater. However, by the end of Sept, activity had weakened sufficiently to allow tourists into the crater area.

Information Contacts: A. Sudradjat, L. Pardyanto, VSI, Bandung; M. Krafft, Cernay; *Kompas*, Jakarta.

Further Reference: Siswoyo, S., 1978, *Report on Seismic Activity at Bromo and Lamongan (1977-78);* Geological Survey of Indonesia, Bandung.

LAMONGAN

E Java
8.00°S, 113.342°E, summit elev. 1651 m

Lamongan is a strato-volcano 45 km E of Tengger with 42 eruptions in the last 2 centuries.

10/85 (10:10) These reports are from VSI.

"Lamongan experienced a seismic swarm with ground breakage 2-15 Oct. More than 3000 earthquakes were recorded at the VSI observation post at Gunungmeja, near Klakah, W of Lamongan (fig. 6-7). Two additional seismometers were installed in the epicentral area on 5-6 Oct. Ground breakage occurred in a region about 5 km W of the summit and consisting of numerous ground cracks, some with vertical offsets of 15 cm, striking ENE. No changes were noted in the temperatures or character of the many lakes in the Lamongan area, which fill maars created by prehistoric eruptions. The 1985 activity is in the same general area as two seismic swarms that occurred in 1924-25 and in 1978 but did not culminate in eruptions.

"Lamongan was quite active during the 19th century with 40 eruptions, including more than a dozen that produced lava flows. Most were from vents on the W flank above about 400 m altitude. It is notable that the maars on the W slope of Lamongan all occur at elevations below the historic eruption vents, suggesting that the position of the groundwater table controls its eruptive style. Eruptions from above 400 m altitude form fissure vents and cinder cones that pose little threat to the local inhabitants. Activity from vents below 400 m may give rise to explosive, maar-forming eruptions. The 1985 epicentral area has a surface elevation of 200-380 m. Plans are underway to increase the level of surveillance.

Reference: Van Bemmelen, R., 1949, *The Geology of Indonesia*, v. 1B; M. Nijhoff, The Hague.

Information Contacts: J. Matahelumual, Suparto S., T. Casadevall, VSI, Bandung.

11/85 (10:11) "Nov activity consisted of an average of 1 earthquake per day in the Oct swarm area W of the volcano. Detailed geological and monitoring efforts are now in progress by VSI to evaluate the possibility of a future eruption from the epicentral area. This area contains numerous prehistoric maars and has a population of about 100,000.

Information Contacts: T. Casadevall, L. Pardyanto, VSI, Bandung.

RAUNG

E Java
8.125°S, 144.042°E, summit elev. 3332 m
Local time = GMT + 7 hrs

Raung is near the eastern end of Java, just 50 km from Bali and 80 km E of Lamongan. Its first historic

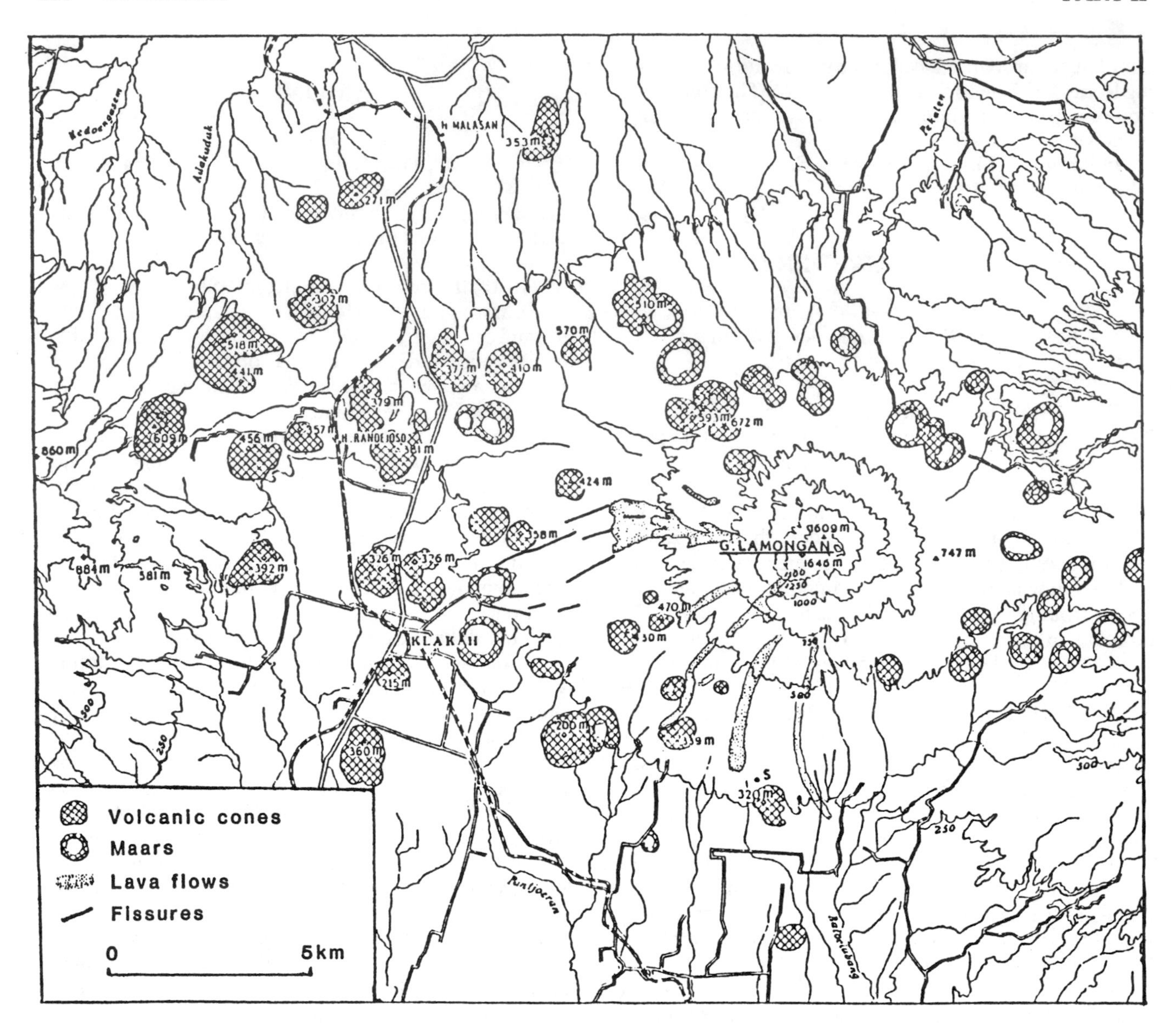

Fig. 6-7. Map of Lamongan, after Van Bemmelen, 1949. Contour interval is 250 m. Contours on the W side of the map mark the lower E flank of Tengger Caldera. Fissures shown are from the 1924-25 earthquake swarm.

eruption, in 1586, caused fatalities as did 5 more through 1817. Recorded eruptions since then have been more frequent (52) but only moderately explosive.

7/82 (7:7) Raung erupted 18 July, emitting dark clouds of tephra to 6000 m. Activity apparently began with a weak earth tremor at about 0300, felt by inhabitants of the tea and coffee plantations on the slopes and in the areas around Kalibaru and Glenmore, at the S foot of the volcano. On 19 July at about 0500 an earthquake was felt, and inhabitants at the foot of Raung reported hearing two consecutive explosions. A seismograph post for Ijen Volcano (8.058°S, 114.242°W) at Pal Tuding, which also monitors Raung, registered vibrations with a maximum amplitude of 3.9 mm.

Two eruptions on 19 July were reported by a Garuda Airlines pilot flying from Jakarta (about 840 km NW of Raung) to Denpasar (on Bali, about 140 km SE) and back. One eruption occurred around 0511 and the other around 1130. The pilot estimated that columns of "smoke" reached 0.6-0.9 km height. Flights have been diverted from near the volcano.

Ash fell lightly all day, leaving a whitish coat on the leaves of tea and coffee plants. An employee of one tea plantation said part of the top leaves were damaged and withered. Light ashfall was reported until 20 July.

Raung's last activity was an explosive central crater eruption with pyroclastic flows, 9-30 May 1977.

Information Contacts: *Kompas,* Jakarta; AFP.

11/85 (10:11) "Raung resumed activity during Nov with a series of small explosions. The summit crater has not been visited since the start of the latest activity. However, observations begun 1 Nov from the new Raung observation post about 15 km SE of the summit at 700 m altitude (at Mangaran) indicated that the explosions have been centered along the E side of the large summit crater, near the recently active eruptive vent on the crater floor. At least 44 explosion clouds were observed during Nov, mostly whitish in color but dark gray ash-laden clouds were also seen. On 15 Nov, light ashfall was reported from SE flank villages (Bejong, Mangaran, and Seragi) and from Banjuyangi City, 35 km ESE of the volcano. The Mangaran seismometer recorded 52 explosion earthquakes during Nov. Activity was continuing as of 12 Dec.

"The last eruption of Raung that produced a lava flow was in 1973. That flow was confined to the summit crater. Explosions similar to the Nov activity have frequently been reported over the last decade."

Information Contacts: T. Casadevall, L. Pardyanto, VSI; Antara News Service, Jakarta.

12/85 (10:12) No explosions were recorded in Dec.

Information Contacts: Suparto S., T. Casadevall, VSI, Bandung.

SANGEANG API

Nusa Tenggara [formerly Lesser Sunda] Islands
8.18°S, 119.06°E, summit elev. 1949 m
Local time = GMT + 8 hrs

This small island (13 km diameter) lies 400 km E of Bali and 60 km NW of Komodo. The first of its 17 historic eruptions was in 1512.

7/85 (10:7) "Sangeang Api Volcano, on the island of the same name NE of Sumbawa Island, began to erupt on 30 July, with a series of explosions from the summit crater (Doro Api). The first explosion occured at about 0900 and produced a plume of tephra and gas to about 3500 m altitude. Additional explosions occurred at 1130 (6500 m), 1320 (6500 m), and 1800 (1500 m). Ashfall in Bima (50 km SW of the volcano), the capital of Sumbawa, totaled about 2 mm, and ashfall along the NE coast of Sumbawa, in the vicinity of the village of Wera, totaled about 2 cm. Between 30 July and 1 Aug, the 1242 inhabitants of Sangeang Api island were evacuated to Sumbawa. Numerous additional Vulcanian explosions took place during the following week, reaching a maximum altitude of 2500 m. Poor weather prevented systematic observations 1-5 Aug. On 6 Aug at 1939, a 0.7 km-long aa flow was observed advancing W from the region of Doro Api crater, toward the village of Doro Sangeang. By 9 Aug, the flow was 1.7 km long.

"The most recent eruption of Sangeang Api was in 1966; earlier eruptions occurred in 1512, 1715, 1821, 1860, 1911, 1912, 1927, 1953, [1954-1958 (6 eruptions)], and 1964. Typical activity begins with strong Vulcanian explosions followed within a few days by Strombolian explosions. Eruptions in 1953 and 1964 ended with a lava flow and weak Strombolian explosions. Lavas from Sangeang Api are of basaltic composition (49-51% SiO_2)."

Information Contact: VSI, Bandung.

Time	Length (km)	Width (km)	Min. Temp. (°C)	Max Altitude (km)	Remarks
1400	135	105	-71°	14.1	emerging
1700	315	100	-50°	11.8	detached
1800	385	150	37°	10.2	detached
1830	405	160	30°	9.4	detached
1900	425	130	24°	8.5	detached

Table 6-8. Sangeang Api eruption cloud data from GMS satellite images on 30 July, 1985. Maximum altitude of the eruption cloud is estimated by comparing the GMS digital temperature data with temperature/altitude data profiles from nearby radiosondes. No plume was evident on the image returned at 1100. A plume could be detected from 30 July at 1400 until 31 July at 0200, but none was visible 3 hours later. Maximum plume extent was reached at 1900.

8/85 (10:8) "Sangeang Api continued to erupt in early Sept. Activity consisted of a single lava flow that has moved approximately 4.5 km W from Doro Api. The frequency of explosions has declined steadily from about 150 per day in early Aug to about 50 per day in early Sept. The maximum explosion cloud height was about 1 km. From 9 to 14 Aug, the number of small ($M < 1$) shallow volcanic earthquakes ranged from 190 to 400 per day. By 6-11 Sept the number of volcanic earthquakes had decreased to 12-40 per day.

"A gas plume has been continuously emitted from the volcano since activity began on 30 July. In late Aug, the gas plume rose less than 0.5 km and was carried W by steady winds. The plume was translucent with a bluish gray to bluish white color. The larger explosions contributed dark gray clouds of fine tephra to the plume. Its SO_2 content was in the range of 50-100 metric tons per hour when measured on 23 Aug.

"The evacuation of the approximately 1250 inhabitants from Sangeang Island was completed in early Aug. Inhabitants are in the process of moving their homes to Sumbawa Island."

Information Contacts: J. Matahelumual, T. Casadevall, VSI, Bandung.

9/85 (10:9) GMS images showed plumes from the initial 30 July activity (fig. 6-8). Digital data analyses of temperatures at the tops of the plumes compared with radiosonde altitude profiles suggested that they reached a maximum altitude of 14.1 km (table 6-8), below the tropopause. However, Yosihiro Sawada notes that for both this and other eruptions, altitudes calculated with this technique seem to be consistently higher than estimates from ground observers.

Information Contact: Y. Sawada, Meteorological Research Institute (MRI), Tsukuba.

10/85 (10:10) "Sangeang Api entered into eruption on 30 July. The main features of the activity included an initial phase of mild Vulcanian explosions, followed by Strombolian explosions, a lava flow, and a persistent plume of water vapor and gas. The initial explosions on 30 July were accompanied by a small nuée ardente that followed the drainage of the Sori Kawangge 4 km down

	No. 1	No. 2
SiO_2	53.8	49.3
Al_2O_3	18.9	17.9
Fe_2O_3*	7.23	11.1
MgO	2.51	4.60
CaO	7.30	10.2
Na_2O	4.33	3.07
K_2O	3.60	2.40
TiO_2	[0.66]	1.03
P_2O_5	0.38	0.32
MnO	0.22	0.23
loi**	1.09	0.05
Total	[100.02]	100.2

* Total iron expressed as Fe_2O_3
** Loss on ignition at 900°C

Table 6-9. Whole rock analyses of samples collected from Sangeang Api. X-ray fluorescence analyses by the USGS Laboratory, Denver, Colorado. *Sample 1,* pumice from eruption on 30 July 1985. *Sample 2,* ejected block, dense lava explosion mid-Aug 1985.

the SSW flank to about 200 m altitude, and produced a dark-colored, hornblende-bearing pumice. The eruption has continued through the first week of Nov but at a greatly reduced level from Aug. The frequency of explosions has declined steadily from about 150 per day in early Aug to about 50 per day in early Sept and fewer than 6 per day in early Nov. The maximum height of the explosion cloud was about 1 km above the vent in Aug and less than 300 m above the vent by early Nov.

". . . The daily earthquake count shows a steadily decreasing trend, and in early Nov the number of earthquakes was less than 30 per day. The lava flow, first observed on 6 Aug flowing W from the region of Doro Api crater (toward the village of Doro Sangeang), had advanced to about 4.7 km from the summit by 1 Oct. The flow showed no additional growth during Oct.

Photographs taken by astronauts from Space Shuttle mission 61A (30 Oct-5 Nov) showed a fairly dense plume extending roughly 50-100 km from Sangeang Api (fig. 6-9).

Information Contacts: J. Matahelumual, S. Supartuo, T. Casadevall, VSI, Bandung; C. Wood, NASA, Houston.

11/85 (10:11) "Sangeang Api continued to erupt with 10-30 Strombolian explosions per day in Nov and the first week in Dec. The 1985 lava channel had not lengthened from the 4.7 km observed in late Sept. However, the volume of the main channel has grown because of a considerable increase in height during the past 2 months. A slow-

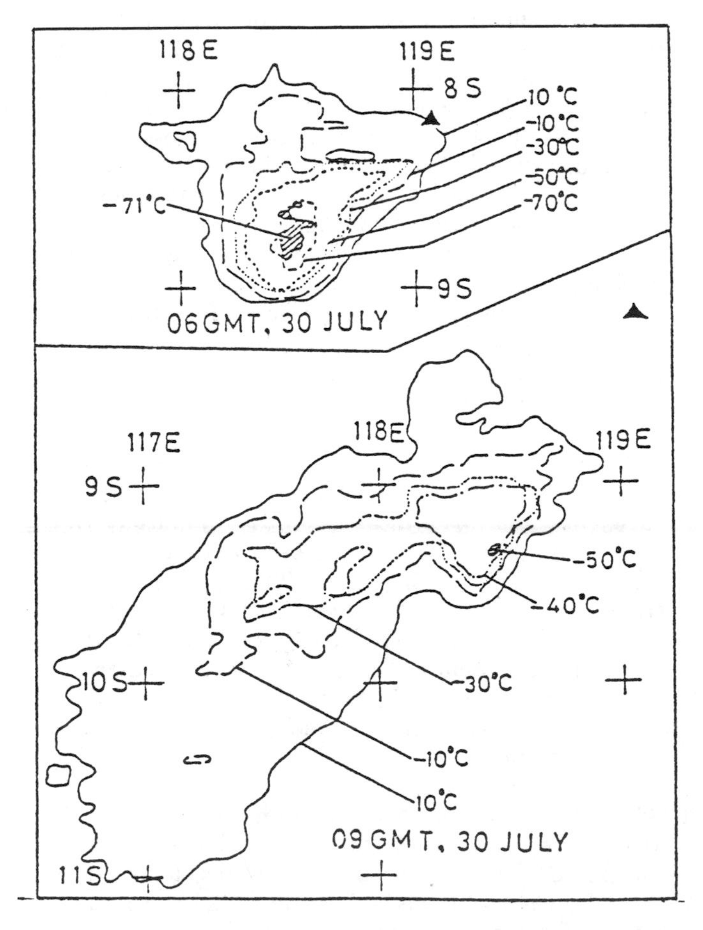

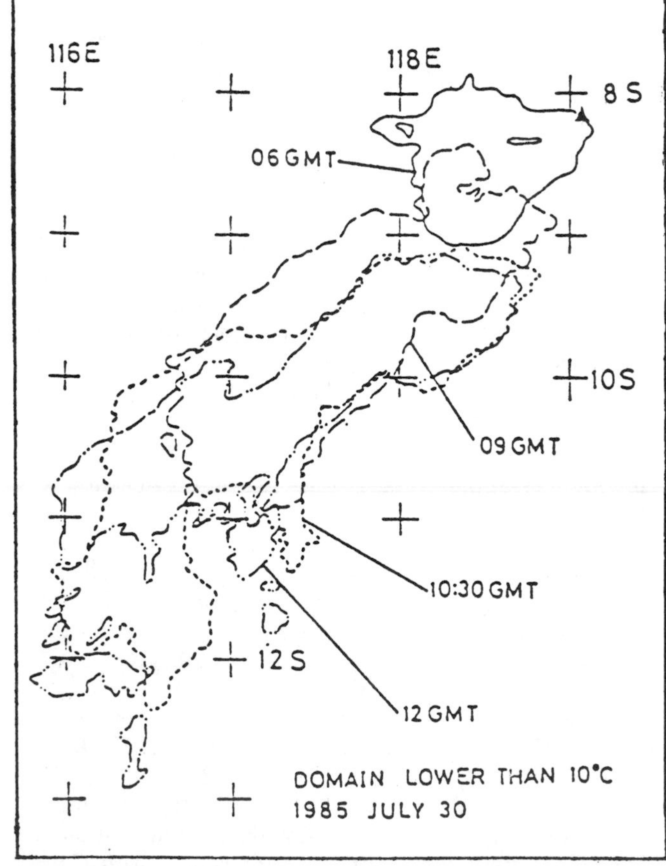

Fig. 6-8. *Left,* surface temperatures (in °C) of eruption clouds from Sangeang Api; 30 July 1985, 1400 and 1700. The volcano is indicated by a solid triangle. *Right,* zones of plume surface from Sangeang Api with temperatures less than 10°C on 30 July at 1400, 1700, 1830, and 2000. Courtesy of Yosihiro Sawada.

moving lava flow in the central channel is located atop the main feeder tube. Periodic overflows of this channel add both fluid lava and rubble to the outer flanks of this tube. On 5 Dec, four separate overflows of fluid lava and rock rubble were observed rolling down the sides of the central channel.

"A brief visit was made to the central crater on 4 Dec during a period of quiet. The active vent was a large cinder cone in Doro Api Crater. This central cone, which previously rose about 40 m from the floor of Doro Api, was estimated to be approximately 180-200 m above the floor during the 4 Dec visit. Thundering detonations were heard almost continuously during the 15 minutes spent in the crater and 1 mild Strombolian explosion hurled incandescent blocks. At night, a continuous reddish glow at the bottom of the gas plume over the crater suggested that a small lava lake may exist within Doro Api or the central cone. Fountaining of fluid lava, sheets, or ribbons was observed to accompany some of the larger Strombolian explosions at night.

"We also confirmed the existence of a small-volume pyroclastic flow that was probably produced during the initial activity of 30 July. Local residents reported that the activity caused a number of small fires in the vicinity of Doro Montoy crater (just N of Doro Api) and the Sori Mbere drainage on the S flank of the volcano. A small block and ash flow, still hot to the touch, was found in the Sori Mbere drainage in Dec. Lahars generated by heavy rainfall and unconsolidated material on the upper slopes of the volcano have been common during and immediately after previous episodes. Outcrops along the shoreline indicate that a number of lahars and possibly also pyroclastic flows have entered the sea. Small mudflows were produced by heavy rains on 2 Dec. One travelled down the Sori Mbere drainage while a second mudflow entered the sea at Oi Nono Jara on the S side of Sangeang Island.

"This pattern of activity conforms to that of previous 20th century eruptions of Sangeang Api. The 1911 activity included numerous explosions and a lava flow from the summit crater that moved more than 6 km down the W flank. The activity that began in March 1953 produced a lava flow and intermittent explosions through 1954. The 1964 eruption began on 29 Jan with strong explosions from the 1954 crater, and a lava flow first observed on 3-4 Feb moved N and E to about 750 m elevation. Explosions and outflow of lava continued at least several months and possibly until the end of 1965."

Information Contacts: T. Casadevall, L. Pardyanto, VSI, Bandung.

12/85 (**10:12**) "Sangeang Api continued to erupt in Dec. Compositions of 30 July pumice and mid-Aug dense lava bombs are given in table 6-9."

Information Contacts: Suparto S., T. Casadevall, VSI, Bandung.

PALUWEH

Nusa Tenggara Islands
8.32°S, 121.71°E, summit elev. 875 m
Local time = GMT + 8 hrs

Paluweh is an even smaller island (8 km diameter) than Sangeang, and 180 km E of it, lying just off the N

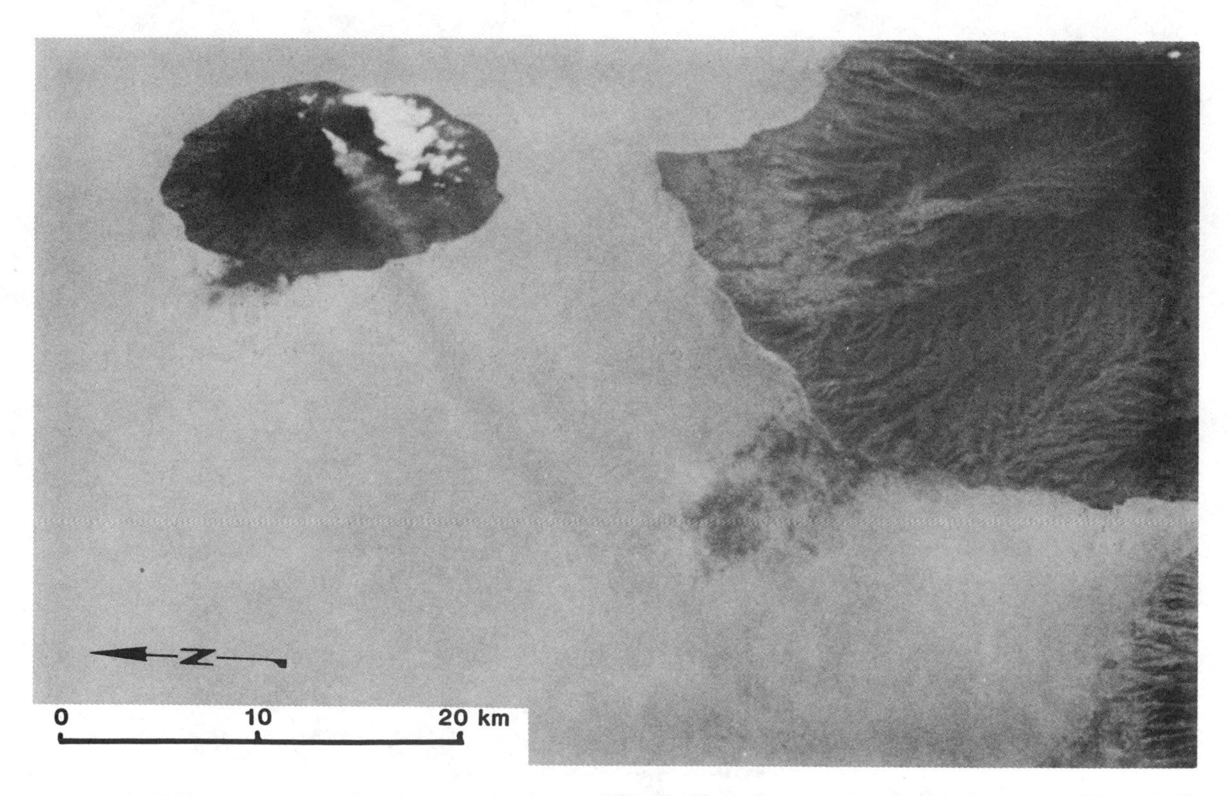

Fig. 6-9. Space Shuttle photograph (no. 61A-40-062), taken 5 Nov 1985, showing a plume emerging from Sangeang Api, plus a portion of Sumbawa Island. Note that north is to the left. Courtesy of Charles A. Wood.

coast of Flores Island. Seven eruptions are known in this century and one in the 17th century.

1/81 (6:1) Activity began to increase on 5 Nov and continued intermittently through the end of Jan. On 9 Nov, an eruption column rose 1 km from the summit crater. Bombs fell nearby and 2 mm of ash were deposited 1 km to the W. Bombs and ash were ejected for about 15 minutes starting at 1115 on 13 Nov, from a summit crater vent [but see 6:2] 40 m in diameter. The tephra column reached 700 m in height. On 27 Jan ejecta set bushes afire near a flank village. Detonations from explosions on 31 Jan were heard at Kota Baru, Flores Island (50-60 km from the volcano) at 0740, 0803, 0807, 0913, 1030, and 1215. No additional activity had occurred as of 5 Feb.

Information Contacts: A. Sudradjat, L. Pardyanto, VSI, Bandung.

2/81 (6:2) VSI provided further details about the intermittent explosive activity. The 40 m-diameter vent was formed during one of the early Nov eruptions and is situated on the upper NNE flank. Bombs from the 1 km-high eruption column on 9 Nov measured up to 60 cm in diameter. Beginning 18 Jan renewed activity was reported. A hot air wave was felt by the inhabitants of 2 E flank villages. About 1850 persons were evacuated from the danger zone. After the explosions on 31 Jan a new lava dome was observed in the crater. Activity declined gradually, and the volcano appeared to be normal again on 1 Feb at 1200. No casualties from Paluweh's Nov-Jan activity were reported.

Information Contacts: Same as above.

8/81 (6:8) Indonesian Meteorological Institute personnel reported that Paluweh has erupted for the second time this year. The meteorology bureau on Flores Island (about 55-60 km away) said it recorded 809 "strong tremors" 17-22 Aug. "Lava and burning rocks" were repeatedly erupted on 22 Aug. On 24 Aug the bureau recorded ocean water temperatures as high as [98°C] around the island. No casualties were reported, but inhabitants evacuated the area. Crops and fishing boats were damaged.

Information Contact: UPI.

9/81 (6:9) No pyroclastic flows were observed during the growth of the lava dome first seen 31 Jan (although some sliding occurred), but it generated blasts of hot air felt by residents of a flank village. They were evacuated by the end of Feb, after VSI had issued a volcanic hazard warning. By July, the lava dome was 200 m high, its volume exceeded 8.5×10^6 m^3, and its summit had be-

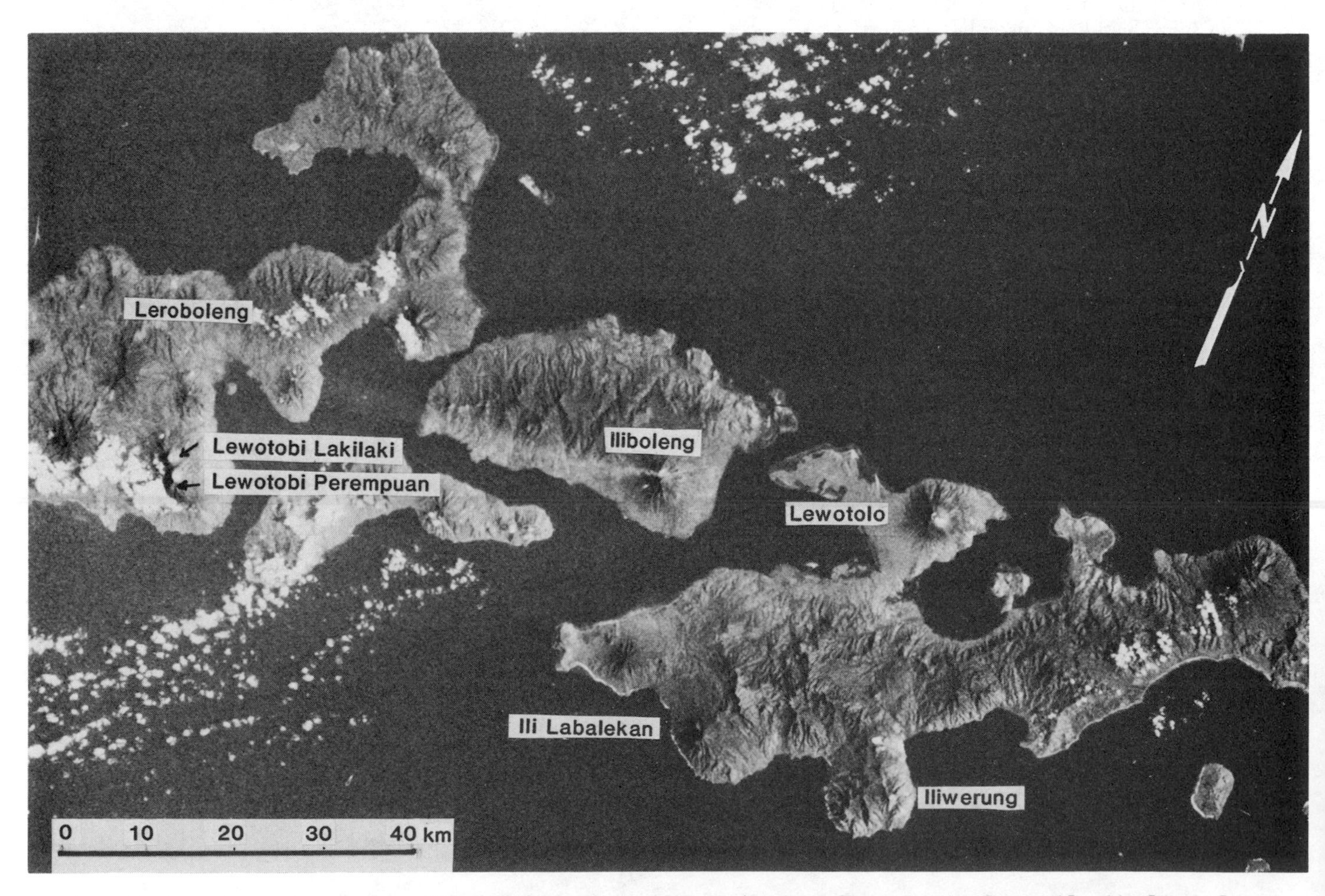

Fig. 6-10. Space Shuttle photograph (no. STS8-46-960) taken 31 Aug 1983, showing eastern Flores, Adonara, Solor and Lomblen Islands. Some of the islands' numerous volcanoes are labeled. A plume emerges from Iliboleng volcano. Courtesy of Charles A. Wood.

come the highest point on the volcano at 875 m above sea level. Explosive activity resumed on 5 Sept between 2010 and 2105, producing a 1 km-high plume. This activity was followed by the destruction of the lava dome. Pyroclastic flows and nuées ardentes d'avalanche moved downslope, depositing 5-20 cm of tephra at one village, and starting fires at 36 structures, including a church and 5 shelters, at another. Because residents had previously been evacuated, there were no casualties. Since the destruction of the dome, the 3-component seismograph monitoring the volcano has recorded shallow earthquakes which VSI believes may be generated by sliding from remnants of the dome.

Information Contacts: A. Sudradjat, L. Pardyanto, VSI, Bandung.

ILIBOLENG

Nusa Tenggara Islands
8.342°S, 123.258°E, summit elev. 1659 m
Local time = GMT + 8 hrs

170 km E of Paluweh, on the island of Adonara in the Solar group (fig. 6-10), this strato-volcano has had 18 historic eruptions since 1885.

12/82 (7:12) VSI reported that an ash eruption began on 17 Nov at 0615. The ash column rose about 1 km and was blown N, causing thin ashfalls on nearby villages. The

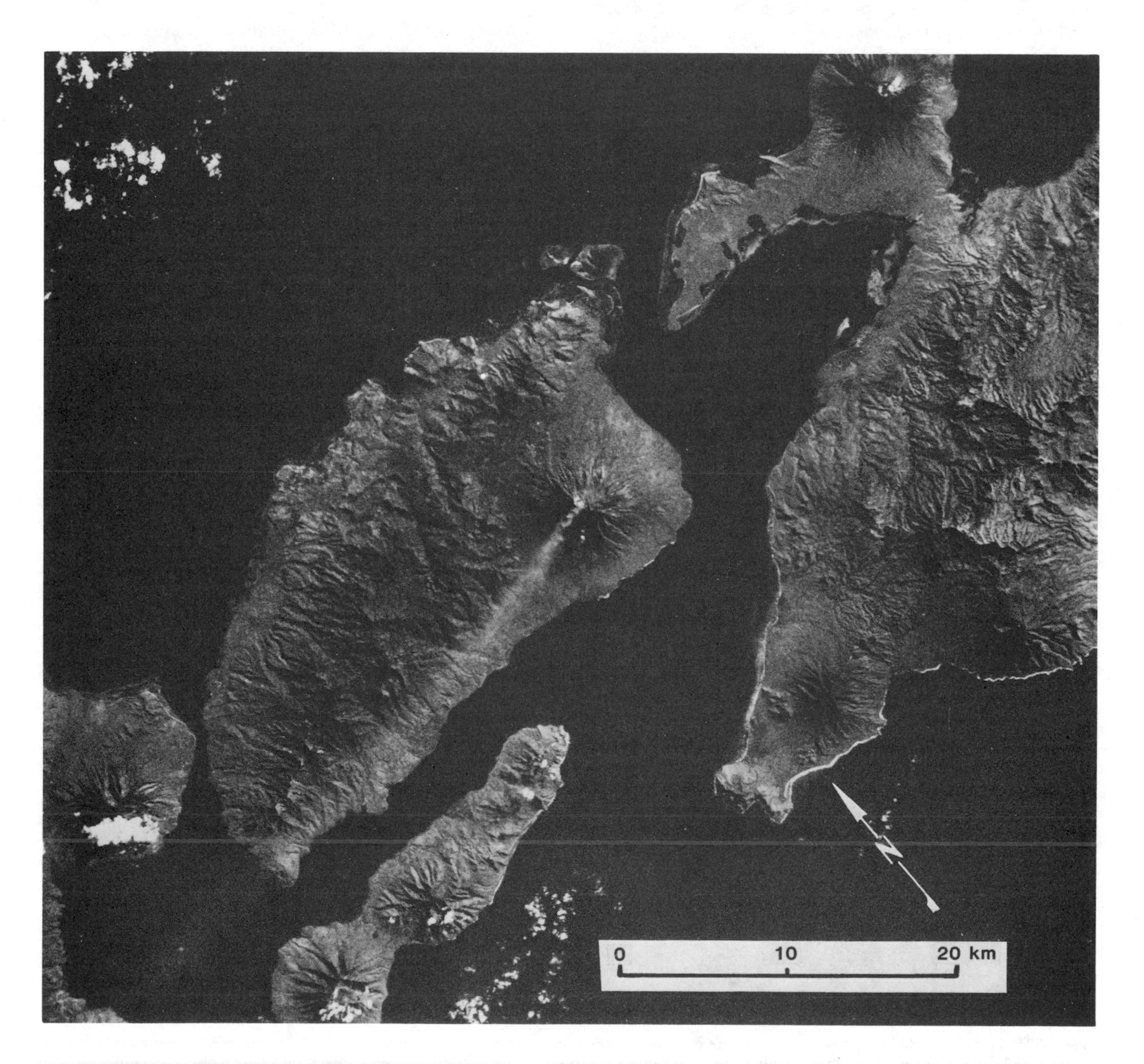

Fig. 6-11. Space Shuttle photograph (no. STS8-60-1840) taken 31 Aug 1983 at virtually the same time as fig. 6-10, but showing a more detailed view of Iliboleng emitting a diffuse plume. Note that north is to the upper left. Courtesy of Charles A. Wood.

eruption ended after several hours. As of mid-Jan, no additional activity had been reported. Press sources noted that the eruption was preceded by a moderately strong, felt earthquake on 15 Nov at about 2300. Iliboleng last erupted April 1973-April 1974, ejecting tephra from the central crater that destroyed flank vegetation.

Information Contacts: A. Sudradjat, O. Mandraguna, VSI, Bandung; M. Krafft, Cernay; *Sinar Harapan*, Jakarta.

8/83 (8:8) [Photographs taken 31 Aug (figs. 6-10 and 6-11) and] 4 Sept by astronauts on Space Shuttle mission STS-8 showed small, relatively diffuse plumes emerging from Iliboleng. [The 31 Aug plume extended at least 40 km and the 4 Sept plume was roughly 50-70 km long.]

Information Contacts: C. Wood, R. Underwood, NASA, Houston.

ILIWERUNG

Nusa Tenggara Islands
8.54°S, 123.59°E, summit elev. 1018 m

Iliwerung, 43 km SE of Iliboleng on Lomblen Island, has a similar history, with 11 eruptions known since 1870.

8/83 (8:8) Explosions occurred 17 and 18 Aug from a submarine vent (Gobal) S of Iliwerung. No casualties were reported. Its previous eruption began with submarine activity in Dec 1973; by July 1974, there were 2 small islands at the eruption site and a third emerged in Aug. All have subsequently been eroded away. In July 1979, about 50×10^6 m^3 of Iliwerung were removed by a landslide. About $1/3$ of this material entered the sea, and (probably aided by additional submarine slumping) generated tsunami that killed hundreds of people.

Information Contacts: L. Pardyanto, VSI, Bandung; AFP.

UNA UNA

Sulawesi
0.17°S, 121.61°E, summit elev. 508 m
Local time = GMT + 8 hrs

Nearly 1000 km N of Iliwerung, Una Una Island (13 km diameter) lies in the middle of the huge bay formed by the Minahassa Peninsula on N Sulawesi. It had another large eruption in 1898 and a minor one sometime between 1928 and 1948. The 40 m-high summit cone (site of the 1983 eruption) was called "Colo" (originally "Tjolo") and the name is often applied to the volcano itself.

7/83 (8:7) An explosive eruption produced pyroclastic flows that destroyed most homes, vegetation, and animal life on 40 km^2 Una Una Island and probably injected tephra into the stratosphere. Initial activity prompted evacuation of everyone on the island before the devastating explosions.

The eruption was preceded by seismicity that increased from 9-11 felt events per day on 8 July to 30-40/day on 15 July. The number of recorded events was 33 on 14 July, increasing the following days to 49, 53, and 73 then to an average of more than 90/day 18-21 July. The strongest earthquake was felt 400 km away on 18 July. That morning, a 1-km column of ash and incandescent material was ejected. AFP reported that a strong explosion occurred 19 July, and thick gray clouds containing incandescent tephra were visible from Ampana,

Fig. 6-12. Portions of 3 GMS images showing the expansion of the cloud produced by the explosions of 23 July 1983, when hot avalanches devastated Una Una island shortly after residents had been evacuated. An arrow points to the eruption plume on each image. Land areas are outlined, from Sumatra and the Malay Peninsula at left to Timor and Halmahera at right. Image scans began at 1631, *upper left;* 1831, *lower left;* and 1931, *lower right,* with x marking position of aircraft 84 min later (see text). Courtesy of Yosihiro Sawada. [Originally from 8:9.]

more than 100 km to the S, the next day.

By the 20th, almost all houses and buildings in the 8 villages near the volcano had been destroyed and nearly half of the residents of the island had been evacuated. All had left by the time of a major explosion on 21 July at 1623 that subjected 80% of the island to temperatures of up to 200°C. Tephra as large as 5-10 cm in diameter fell near a VSI observation vessel and the monitoring team reported flames on parts of the island. A government geologist estimated that all 700,000 coconut trees and all livestock on the island must have been burned, probably by pyroclastic flows. Ash darkened much of the region. People in Falu, 250 km away, were forced to protect themselves from ashfall until late 23 July. A VSI field party arriving on the island 22 July at 0100 felt 10 earthquakes during their 15-hour stay and observed a 1.5-km eruption column at 1649.

On 23 July at 2055, a British Airways jet (en route from Singapore to Perth) flying at 10.6 km altitude encountered an eruption cloud at 1.4°S, 120.71°E, about 150 km S of Una Una (fig 6-12). Pilots noted a volcanic smell, lack of visibility, and St. Elmo's Fire around the windshield. The aircraft returned immediately to Singapore and suffered no damage. On 24 July at 1930, a satellite image showed a cloud about 120 km wide, extending about 600 km S from Una Una. Earlier in the eruption, weather clouds had obscured the Una Una area. Press reports quoted a local government official who said that 80% of the island was covered by volcanic clouds on 24 July, burning vegetation and destroying trees. On 26 July at 0000, the Japanese GMS satellite showed what appeared to be a dense eruption column rising from the island. On the next image, 2 hours later, a fan-shaped plume was visible, probably in or near the stratosphere. High-altitude material was blowing SW and W, while low and mid-level debris was drifting slowly S to SSE.

On 28 July at 0200 the GMS satellite showed a small plume over the island. By 0500 a plume about 60 km wide extended about 200 km WSW from the volcano. The plume appeared denser at 0800 and by 1100 vigorous activity fed a cloud that reached 118°E and at least 13.5 km altitude. At 1400 the plume stretched about 500 km to the WSW and was very dense within 250 km of the volcano. Temperatures and wind directions at the tropopause (15 km altitude) were consistent with the plume's direction of movement and coldest temperature (-76°C) from a NOAA 7 image at 1430 (fig. 6-13). By the next image, at 2000, the plume had dissipated. The GMS satellite showed the beginning of another eruptive episode on 30 July at 1630. At 2000, a NOAA 7 image contained a WSW-drifting plume, similar to the one on 28 July but not as spectacular. Feeding of this plume was continuing at 2300; it drifted SW, then W toward Sulawesi. It extended from the volcano about 200 km to 1.5°S, 119.5°E on 31 July at 0200, but was dissipating 3 hours later. At 2000 an image showed what appeared to be an eruption column, but little activity was visible 3 hours later.

Another explosive episode first appeared on the imagery 2 Aug at 0500. Before activity ended at 1700, a plume had moved about 200 km to the SW and reached roughly 9-12 km altitude. A dense eruption column appeared over the island 3 Aug at 0000 and extended roughly 120 km to the W and SW 2 hours later. The plume was relatively diffuse and appeared to have

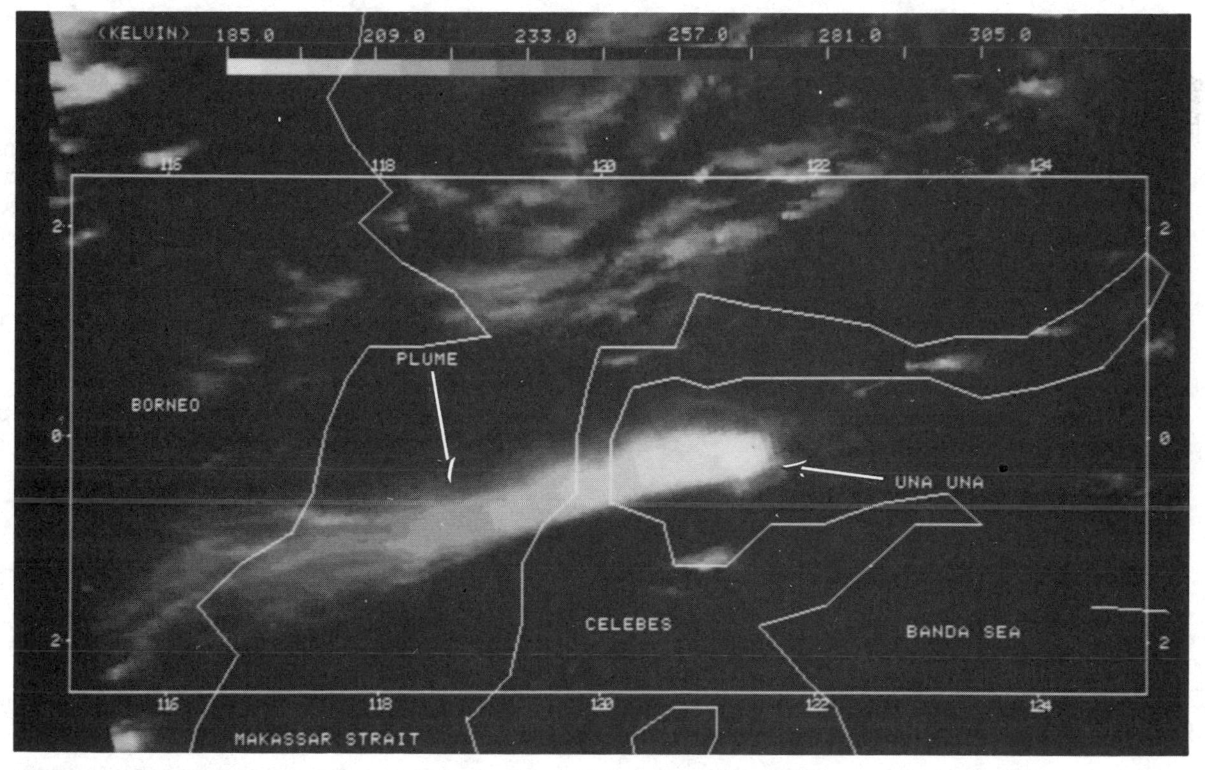

Fig 6-13 NOAA 7 thermal infrared satellite image, courtesy of Michael Matson, showing an 800 km-long eruption plume from Una Una 28 July 1983 at 1430. White areas are coldest (see gray scale at top of figure). The coldest part of the plume had a temperature of -76°C, indicating that it had penetrated the stratosphere.

1983	Time	Length (km)	Width (km)	Temp (°C)
23 Jul	1631	100	90	-74
	1801			-76
	1831	320	200	
	1931	560	180	
25 Jul	↑2331	30	30	-70
26 Jul	0131	210	160	
	0431	(520)	(160)	
27 Jul	↑1931	60	20	
28 Jul	0131	160	20	
	0431	170	50	
	0631	290	70	
	0731	260	40	
	1031	500	70	
	1331	550	60	
	1631	710	80	
	1831	650	70	
	1931	500	30	
	2331	(340)	(20)	
30 Jul	↑1631	80	40	-73
	1801			-80
	1831	260	110	
	1931	370	120	
	↑2331	190	50	
		(340)	(320)	
31 Jul	0131	250	130	
	0431	(260)	(100)	
	↑1801			-70
	1831	60	30	-71
	1901			-71
	1931	160	50	-71
	2331	(150)	(80)	
1 Aug	0131	(160)	(110)	
	0431	(130)	(50)	
2 Aug	0431	70	50	-72
	0601			-65
	0631	160	70	-63
	0701			-72
	0731	240	80	-75
	1031	420	100	-79
	↑1331	50	15	-60
		(340)	(240)	
	1631	60	20	-56
		(420)	(320)	
	1831	(130)	(20)	
	1931	(450)	(320)	
	2331	80	30	-71
3 Aug	0131	(120)	(60)	
	0431	(110)	(20)	
	0631	(130)	(15)	
4 Aug	1031	80	40	-73
	1331	150	110	-73
	1631	190	120	-63
	1831	(190)	(160)	
	1931	(210)	(130)	
	↑2331	90	100	-79
5 Aug	0131	170	100	-81
	0431	280	200	-73
	0601			-61
	0631	170	170	-60
	0701			-64
	↑0731	30	30	-70
	1031	60	60	-61
	1331	100	90	-63
6 Aug	↑0131	30	20	-64
	0431	110	60	
	0631	190	0	
	0731	200	130	
	↑1631	60	50	-73
	1801			-69
	1831	110	70	
	1931	(130)	(80)	
7 Aug	↑1331	190	80	-75
	1631	320	110	-79
	1801			-77
	1831	190	150	-75
	1931	240	160	
	2331	150	70	
8 Aug	0131	(190)	(60)	
	↑0431	30	15	
	0631	60	15	
	0731	(70)	(20)	
10 Aug	0131	30	20	(obscure)
	0431	30	10	(obscure)
	0631	50	15	(obscure)
11 Aug	1331	290	80	-69
	1631	(510)	(110)	
	1831	(680)	(160)	
12 Aug	↑0131	140	60	-73
	0431	460	100	
22 Aug	1331	20	15	(obscure)
	1631	150	110	(obscure)
	1831	(260)	(110)	
	1931	(300)	(140)	
26 Aug	↑1331	(almost circular)		
	1631	(cloud detected)		

Table 6-10. Una Una eruption cloud data extracted from GMS satellite images by Yosihiro Sawada. Data are tentative; some apparent plumes may have been weather clouds. Times are the beginnings of image scans, which are completed in about 25 minutes. Images are returned 14 times per day at intervals ranging from 30 minutes to 3 hours. New explosions are indicated by an arrow to the left of the time. Data shown in parentheses are for plumes that are detached from the volcano because explosive activity had (apparently) stopped. A new plume was sometimes ejected before remnants of the previous explosive pulse had dissipated; dimensions of the old plume are then listed in parentheses below data on the new activity. Coldest temperatures at the tops of plumes are shown.

reached only the mid-troposphere. Satellite images indicated that another explosion started 4 Aug at about 1000, feeding a plume that moved about 350 km to the NNW. The different direction of drift was the result of a weather change; this plume probably remained in the troposphere. AFP reported an eruption on 9 Aug at 0835 that ejected a gray plume to 3 km. No activity was evident on satellite images until 12 Aug at 0130, when a plume was observed that was not visible 2 hours earlier. At 0300, NOAA 7 data showed a dense plume, similar to that of 28 July, extending about 300 km SW to central Sulawesi.

Una Una's only known large historic eruption occurred from its crater lake in 1898, producing mudflows and more than 10^7 m^3 of tephra.

Information Contacts: A. Sudradjat, VSI, Bandung; N. Banks, HVO, Hawaii; M. Matson, J. Hawkins, O. Karst, S. Kusselson, NOAA/NESDIS; AFP; Antara News Agency, Jakarta; UPI.

8/83 (8:8) Satellite images continued to show occasional eruption plumes through late Aug. After the 12 Aug plume described last month, activity was next observed on an image from the GMS satellite on 22 Aug at 1400, when a very small fairly bright area was present directly over the island, a feature typical of the initial stages of an explosive episode. This event developed rapidly, with a NOAA 7 image 35 min later showing a bright plume extending about 60 km to the W. By 1700 (GMS data), it had moved about 120 km W from the volcano, and its leading edge had just reached the coast of Sulawesi. On the next image, at 2000, feeding from the volcano appeared to have ended and the plume was dissipating to the W. On 26 Aug at 1100, a GMS image showed a bright, newly-ejected plume about 40 km wide that appeared to have reached the tropopause. At 1400, a very dense high-level cloud about 80 km wide had spread W then SW about 250 km, but on a NOAA 7 image 1 hour later the cloud appeared to be dispersing and the eruption had clearly ceased by the next GMS image at 1700. On 29 Aug, GMS imagery showed the beginning of an explosive episode at about 1930. By midnight, a moderately dense plume extended WSW along the equator, then turned abruptly to the SSW, reaching 120°E at 1-2°S. On the next image, at 0500, feeding from the volcano had stopped and the plume was nearly dissipated.

Government officials noted that several villages have been completely destroyed by the eruption and that all of the island's coconut trees had been killed. All of the people living on Una Una were evacuated before the devastating explosions 23 July. Officials anticipated that it would be several years before the island would again be habitable, so residents have been resettled on other islands until they can return.

The WWSSN noted 66 events in the vicinity of Una Una beginning late 16 July. No earthquakes smaller than magnitude 4.5 were tabulated, and most magnitudes were between 4.8 and 5.4. Of these, the 21 recorded by 20 or more stations had a mean epicenter of 0.09°S, 121.70°E (standard deviations for both latitude and longitude, 0.05°), about 15 km NE of the volcano. Depths of the same 21 events averaged 55 km (standard deviation 7.5 km). Earthquakes recorded by the WWSSN had become less frequent by the time of the largest explosion 23 July; few were recorded after 26 July and none after 1 Aug. Other events in the region included single M 5.0-5.3 shocks on 27, 28, and 31 July, about 200 km ENE of Una Una at roughly 40 km depth; and M 5.5 and 5.1 earthquakes 20 and 31 July at depths of 299 and 272 km, 300 km WNW of the volcano.

Information Contacts: M. Matson, J. Hawkins, S. Kusselson, NOAA/NESDIS; USGS/NEIC; Antara News Agency, Jakarta.

1983		Time	Height (km)
Jul	23	1623	10
	25-6	2325-0021	7.5
	27	0400-0605	7.5
		1500-2010	7
	28	0002-0045	8
		1630-1730	8
	30	1615-?	6
Aug	1	1834-2000	7
	1-2	2130-0230	6
	2	0314-0600	8
		0800-0900	8
	2-3	1905-0200	5
	4	0915-1100	6
	6	1520-?	6
	7	1100-1900	10
	11	1115-1135	8
	12	0047-0147	9
	18	1013-1240	12
	22	1203-?	8
	24	2148-2220	4
	25	1847-2000	5.5
	26	1023-1139	10

Table 6-11. Times of Una Una eruption clouds with heights estimated by VSI geologists.

9/83 (8:9) Since late Aug, no explosions have been reported by ground observers or seen on satellite imagery. Yosihiro Sawada searched all July and Aug images from the GMS satellite and provided table 6-10.

Information Contact: Y. Sawada, MRI, Tsukuba.

10/83 (8:10) A VSI team monitored the eruption from near the island [beginning 23 July, and observed 22 distinct explosions (table 6-11). Many, but not all of these explosions were detected by satellite (table 6-10).]

Maurice Krafft visited Una Una in mid-Aug. He observed and photographed the 22 Aug explosion (table 6-11 and fig. 6-14) and pyroclastic flow deposits from previous explosions (fig. 6-15). The entire island had been devastated except for a narrow strip of undamaged vegetation and villages along the E coast.

Information Contacts: A. Sudradjat, VSI, Bandung; M. Krafft, Cernay.

Further Reference: Katili, J.A. and Sudradjat, A., 1984, The Devastating 1983 Eruption of Colo Volcano, Una Una Island, Central Sulawesi, Indonesia; *Geologisches Jahrbuch*, v. A75, p. 27-47.

Fig. 6-14. Explosion photographed from the S on 22 Aug 1983 by Maurice Krafft. Pyroclastic flows from this explosion continued $^1/_2$ km beyond the SSW coast of Una Una and 1 km beyond the NNW coast.

Fig. 6-15. Coconut trees uprooted by pyroclastic flows on the SE side of the island, photographed 19 Aug 1983 by Maurice Krafft. Pyroclastic flow deposits from the major 23 July explosion were 5 m thick on the island's SW side. A plume from the summit is in the background.

SOPUTAN

Sulawesi
1.108°N, 124.725°E, summit elev. 1784 m
Local time = GMT + 8 hrs

Soputan is on N Sulawesi Island, 375 km ENE of Una Una, where the Minahassa Peninsula curves northward toward the Philippines. It is a strato-volcano with 25 historic eruptions known since 1785.

8/82 (7:8) An explosive eruption began 26 Aug at 1142. Ash rose to 3 km above the summit crater at 1300. Ash fell on a village 7 km SE of the volcano and lightning was observed at night. Detonations and roaring sounds followed the explosions. The 850 inhabitants of a village on the flanks of the volcano moved to nearby towns. Indonesian authorities issued a warning notice to aircraft.

An image returned by the GMS satellite at 1400 showed a fairly dense plume, about 120 km in diameter, emerging from Soputan. Four hours later, a Cathay Pacific Airlines pilot estimated that the top of the cloud was at about 15 km, well above his flight altitude. At 2000, a satellite image showed continued vigorous feeding of the plume, which extended about 700 km W from the volcano to 119°E, where it was about 350 km wide. On this image, the plume appeared to be rising to roughly the cirrus cloud level, in the upper troposphere. By midnight, feeding of the plume was weakening, and on the next image, at 0200 on 27 Aug, the plume was detached from the volcano.

Satellite imagery showed renewed activity shortly before 0800. A plume about the same size as the one ejected the previous day moved W. Light ashfall started at 0915 on a town about 40 km N of the volcano. Feeding of the plume continued until about 1400. An additional seismograph was installed 27 Aug at a site 5 km S of Soputan.

Antara radio reported a third, smaller, explosive episode 28 Aug. Hot ash, but no large tephra, was ejected from dawn to about midday, accompanied by thunderous sounds. Ash fell on 12 villages around the volcano, but there were no casualties. No additional explosions were reported until 16 Sept, when an eruption plume appeared on satellite imagery. Antara reported 5 explosions that day.

Geologists had visited the volcano 2 days before the eruption began but saw no increased surface activity. They measured fumarole temperatures of 76°, 79°, and 84°C. Soputan last erupted in 1973.

Information Contacts: A. Sudradjat, VSI, Bandung; T. Baldwin, D. Haller, M. Matson, NOAA/NESS; Antara News Agency, Jakarta.

9/82 (7:9) At 1700 on 16 Sept the GMS satellite showed a plume extending WSW from the volcano, at about the cirrus cloud level. By midnight, the plume was less dense, but feeding from the volcano appeared to be continuing. Antara radio reported that an explosion at 0120 on 17 Sept ejected hot ash, pebbles, and rocks 10-40 cm in diameter. Other explosions occurred at 1014, 1129, 1132, and 1715 the same day. Ejecta sometimes rose to 2 km above the summit. An ash cloud remained over the area for 18 hours. At 1100 on 18 Sept, a GMS image showed the cloud from a moderate to intense explosion that began about 1000 and ended around 1300-1400. No additional activity has been reported (table 6-12).

Information Contacts: M. Matson, D. Haller, T. Baldwin, NOAA/NESS; Antara News Agency, Jakarta.

10/82 (7:10) VSI reported that explosive activity resumed at 2045 on 9 Nov. Ash rose to about 5 km altitude and was blown NW, falling in the city of Amurang, 20 km away. A light ashfall was also reported at the VSI's Kakaskasen Observatory, 30 km N of Soputan. Newspapers reported that ash fell 40 km from the volcano and that within hours streets near the volcano were covered by as much as 10 cm of ash, halting traffic in many areas. VSI reported that rumbling and detonations accompanied the activity, and lightning flashes were observed at 4 km altitude. The eruption ended at about 1800 on 10 Nov, and no rumbling was heard that night.

1982		Time	Temp (°C)	Altitude
Aug	26	1500	-72°	15 km
	27	1500	-37°	10 km
Sept	18	1330	-65°	14 km

Table 6-12. Cloud top temperatures for 3 plumes from Soputan determined by Michael Matson from NOAA 7 polar orbiting satellite images, with altitudes calculated from nearby radiosonde temperature/altitude profiles.

A VSI seismograph at Silian, 6 km S of the volcano, recorded tremors for 4 hours before the eruption. The Teledyne seismograph, set at a magnification of 2000, indicated a maximum amplitude of 6 mm during the early morning of 10 Nov and 4 mm in the afternoon. Deep (tectonic) earthquakes and 2 shallow volcanic events also preceded the explosive activity.

The GMS satellite showed no activity from Soputan at 1700, but on the next image, at 2000, there was a plume extending about 150 km W from the volcano. From the plume's development and rate of drift, satellite specialists estimated that gas emission had begun shortly after 1700, or well before the first explosive activity noticed on the ground at 2045. By 0200 on 10 Nov, relatively diffuse, apparently low-level material was drifting N and NW, while higher altitude ejecta formed a dense plume that moved almost directly W. Satellite data continued to show feeding of the cloud through 1400 on 10 Nov. On the next image, at 1700, the cloud had separated from the volcano. By 2000, a plume with dense and diffuse patches extended from about 3°N, 123°E to 4°S, 117°E, a length of more than 1000 km, and had a maximum width of about 500 km, considerably larger than the Aug and Sept plumes.

Information Contacts: A. Sudradjat, Suratman, VSI, Bandung; D. Haller, M. Matson, NOAA/NESS; *Sinar Harapan*, Jakarta; AFP.

Further Reference: Sawada, Y., 1983, Attempt on Surveillance of Volcanic Activity by Eruption Cloud Image from Artificial Satellite; *Bulletin of the Volcanological Society of Japan*, v. 28, p. 357-373.

5/84 (9:5) Soputan erupted from 24 May at 2243 until 26 May at 0300. An ash column rose to 4 km and moved W. Ash and sand-sized tephra fell on the area W of the volcano, forming a deposit more than 10 cm thick over about 75 km and 1-10 cm thick over an additional 125 km. Although there were no people within the danger zone, about 3000 were in the alert zone. About 350 spontaneously evacuated from the area, where the primary cultivation is of coconut palms. Manado and Gorontalo airports (about 50 km NNE and about 200 km SW of Soputan) were closed 26 and 27 May.

As of 30 May, no volcanic earthquakes had been recorded, although 2 tectonic events were detected. No premonitory activity was observed.

Information Contact: A. Sudradjat, VSI, Bandung.

1984	Hour	Density	Width (km)	Length (km)	Movement direction	Minimum plume temperature
May 25	23	dense	60	60	circular	-79°C
26	05	dense	190	240	NW	-82°C
	11	dense	160	330	NW	-70°C
	17	diffuse	460	440	W	
Aug 31	08	dense	30	40	circular	-74°C
	14	dense	350	420	W	-75°C
	20	diffuse	430	660	W	
Sept 1	02	diffuse	430	360	W	

Table 6-13. Dimensions of plumes from Soputan determined from GMS images by Yosihiro Sawada.

8/84 (9:8) A 5-hour explosive eruption occurred at Soputan on 31 Aug, the first activity since the 24-26 May tephra ejection. VSI seismic instruments recorded a progressive increase in local seismicity beginning 6 Aug. On 14 Aug, a sequence of tremors appeared between 0400 and 0800, with amplitude increasing to 25 mm (at 2000 magnification). VSI issued a warning to civil authorities and an alert was put into effect on the 14th. Seismicity continued 15-25 Aug with an irregular number of A- and B-type events, averaging 1-2 per day. From 25 Aug until the time of the eruption, seismicity totally stopped, increasing suspicion among VSI scientists that an eruption was possible.

The eruption started at 0709 on 31 Aug and lasted until about noon. An ash column rose to about 6 km and moved NE. Authorities and area residents were well-prepared, and neither casualties nor an evacuation were reported. Press sources reported that the ash cloud could be seen from Manado, the provincial capital 5 km to the NNE. The ash cloud covered a large area and disrupted traffic on the trans-Sulawesi highway.

On 31 Aug at 1457, a visible-band image from the NOAA 7 polar orbiting satellite showed a plume extending about 450 km W from the volcano. The plume was quite dense and about 120 km wide.

Information Contacts: A. Sudradjat, VSI, Bandung; M. Matson, NOAA/NESDIS; UPI.

2/85 (10:2) Yosihiro Sawada observed a series of plumes from the May and Aug 1984 eruptions of Soputan (table 6-13) on images from the GMS satellite. In 9:5, Adjat Sudradjat reported that Soputan erupted from 24 May at 2243 until 26 May at 0300, depositing tephra west of the volcano. A GMS infrared image 25 May at 2300 shows a nearly circular plume rising from the volcano. By the time of the next image, at 0500 the next day, a large eruption cloud was evident, but feeding from the volcano had ended 2 hours before. Six hours later, the plume was clearly detached from the volcano. A 5-hour explosive eruption of Soputan started 31 Aug at 0709 and lasted until about noon. A circular eruption column was evident less than an hour after the start of the activity (fig. 6-16, left) and a large plume was visible 6 hours later, about 2 hours after the eruption ended (fig. 6-16, right).

Information Contact: Y. Sawada, MRI, Tsukuba.

5/85 (10:5) An ash eruption from Soputan's main crater occurred 19-20 May from 1815 to 0130. The eruption column rose to 4 km altitude, and about 2 cm of ash (fine to coarse) accumulated at villages (Kawangkoan, Langoan, Noongan, and Ratahan) 9-12 km from the crater. There were no casualties and no evacuations were necessary. The volcano has remained quiet since 22 May.

On 20 May at 0058, a jumbo jet en route from Hong Kong to Sydney, Australia with 267 passengers and 16 crew members encountered the ash cloud about 80 km SSE of the volcano (approximately 0.5°N, 124.54°E). An orange glow discharged from the nose of the aircraft and orange sparks passed over the windshield. Engine inlets were illuminated by a white light. A light haze that smelled like burnt dust filled the cabin, and ash accumulated on flat surfaces. These effects continued for 7-8 minutes, while the aircraft remained on course at 0.85 times the speed of sound, for a distance of roughly 120-135 km. The aircraft continued to Sydney, arriving 4 hours after exiting the ash cloud, and landed uneventfully. Because of damage caused by the ash cloud, it was necessary to replace all 4 of the aircraft's engines, other navigational components, and more than a dozen windows.

The TOMS instrument on the NIMBUS 7 polar orbiting satellite detected an area of SO_2 enhancement SE of Soputan during its pass at local noon on 20 May. The area of enhancement extended from about 124.5°E to 126°E near the equator and from about 125°E to 127°E at 1.5°S with the maximum at about 1°S, 126°E.

Information Contacts: VSI, Bandung; Boeing, Seattle, WA; A. Krueger, NASA/GSFC.

LOKON-EMPUNG

North Sulawesi
1.36°N, 124.79°E, summit elev. 1579 m

The twin volcanoes Lokon and Empung lie 30 km NE of Soputan. Intermittent gas and ash eruptions from Tompaluan crater, in the saddle between the two volcanoes, took place from 1976 into 1980 (*Bulletin of Volcanic Eruptions* no. 16-19, 1978-81). In 1984 gas eruptions were reported from early June to November, with plumes reaching 400-500 above the crater (*Bulletin*

Fig. 6-16. GMS infrared satellite images with arrows pointing to eruption clouds from Soputan 31 Aug 1984 at 0800, *left*, and 1400, *right*. Courtesy of Yosihiro Sawada.

of the Volcanological Survey of Indonesia, no. 113, 1986). This crater has erupted at least 17 times since 1829, and eruptions are known from Empung in the late 14th and 18th centuries.

API SIAU

Sangihe Islands
2.78°N, 125.48°E, summit elev. 1784 m
Local time = GMT + 8 hrs

On the island of Ulu, 200 km NNE of Soputan, Api Siau is the highest and most active volcano in the Sangihe island chain. Its first historic eruption in 1675 has been followed by 40 known eruptions.

General Reference: Manalu, L., 1986, Karangetang (Api Siau); *Bulletin of the Volcanological Survey of Indonesia*, no. 109, 48 pp.

10/76 (1:13) An eruption of lava and pyroclastics from a new vent on the SW flank began on 15 Sept at about 0700, accompanied by loud rumbling. For 11 days before the eruption, approximately 120 earthquakes/day had been felt by inhabitants of the area. Continuous volcanic tremor was recorded at Tarorane Volcano Station from 15 Sept onwards. Through 5 Oct, an andesitic block lava flow, 20-40 m thick and 100-200 m wide, moved about 70 m/day, threatening the villages of Bubali and Salili, from which 1800 people were evacuated. By 12 Oct, the flow rate had slowed to 20 m/day, but one edge of the flow was within 650 m of Ulu City, the capital of Siau Island. One spectator was killed and another badly injured on 19 Sept by a small avalanche, caused by the collapse of one flank of the flow as it reached the edge of a steep valley. The lava flow destroyed a bridge, 24 homes (44 others in the path of the flow were dismantled to avoid destruction), and killed an estimated 37,500 coconut, clove, and nutmeg trees. The flow is presently being mapped by a team of surveyors. Api Siau's last eruption was in early 1974.

Information Contact: D. Hadikusumo, Volcanology Division, GSI, Bandung.

11/76 (1:14) The following additional information is from a report by scientists from the Volcano Observation Section, Geological Survey of Indonesia.

The eruption was preceded by an earthquake swarm that began on 4 Sept. The number of felt earthquakes reached a maximum of 120/day early in the swarm, then gradually declined. At 0700 on 15 Sept, a minor pyroclastic eruption began from a new vent (A in fig. 6-17) at about 1100 m elevation on the S flank, producing a thick ash cloud. This activity was succeeded by lava effusion from vent A. Another pyroclastic eruption occurred on 17 Sept from a second new vent (B in fig. 6-17), about 300 m S of vent A. Lava effusion began shortly afterwards from B. The 2 flows coalesced near vent B and moved downslope 200-300 m/day. The rate of movement gradually decreased to 10 m/day as the flow front reached more level terrain. By 21 Oct, the flow was about 50 m thick near its source, about 6 km long, and was only 400 m from Ulu City, but lava extrusion had apparently ended. The flow front continued to advance 5-10 m/day. Thick "smoke" was still being emitted from the summit crater Karangetang. Lava volume was estimated at 2×10^7 m^3.

Information Contact: Same as above.

1/79 (4:1) Three explosions from the summit crater occurred during Oct and 2 others were observed at night during Nov and early Dec. Ashfalls reached Ulu (6 km SE of the summit, fig. 6-17), where about 0.5 mm of ash was deposited.

A 10-minute eruption beginning at 1730 on 15 Dec

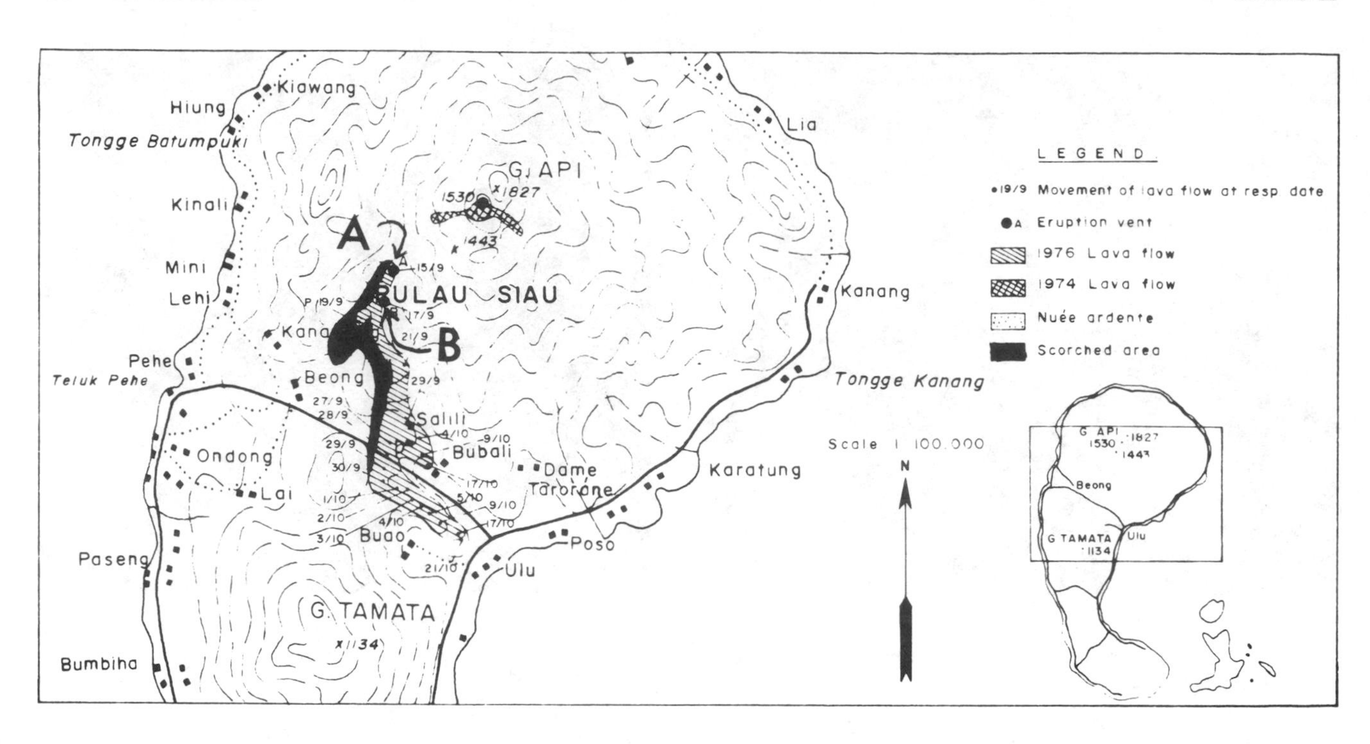

Fig. 6-17. Map of Api Siau, showing the 1974 and 1976 lava flows and the 1976 nuée ardente deposits. Courtesy of VSI.

produced a thick tephra column that rose 600-700 m above the summit and dropped more than 0.5 mm of ash on Ulu. Sharp explosion sounds were clearly audible from the temporary observatory at Tarorane, about 5 km from the summit. Volcanic tremors were frequently felt, but none were recorded at Tarorane.

Vapor emission from the 1976 vents has stopped, but a solfatara about 50 m above the upper vent emitted a thin stream of bluish gas.

Information Contacts: A. Sudradjat, M. Kamid, VSI, Bandung.

8/79 (4:8) On 31 May at about 0200 an explosion produced a new crater at about 1300 m elevation on the NNW flank. A 600-m eruption column and a loud noise were reported. Gas emission from 3 vents in the new crater was continuing in mid-Aug and the 784 inhabitants of nearby villages remained on alert.

Information Contact: A. Sudradjat, VSI, Bandung.

3/80 (5:3) Incandescent tephra was ejected to 200 m above the main crater on 24 March at 1905. As of 27 March, no additional activity had been observed.

Information Contact: Same as above.

9/80 (5:9) A seismograph recorded an explosion from Api Siau on 12 Sept at 0410. The next day at 1140, a cauliflower-shaped eruption column rose 1200 m above the crater. Similar explosions had occurred on 3 July and 3 times in Aug, depositing incandescent tephra over an area 3 km in diameter.

Information Contacts: A. Sudradjat, L. Pardyanto, VSI, Bandung.

8/84 (9:8) Press sources reported that explosions 28 Aug and 5 Sept ejected dense ash clouds that rose to 4 km altitude. Ash and larger tephra fell over a wide area. Lava flows and lahars destroyed terraced rice fields and nutmeg orchards on the volcano's upper flanks. Magnitude 3-5 earthquakes had been felt since the beginning of Sept. About 25 shocks were felt near the volcano on 5 Sept, and some were also felt on nearby islands. By late 11 Sept, about half of Siau Island's 40,000 residents had been evacuated and officials warned people living on the lower flanks to be ready to leave quickly.

Information Contacts: UPI; DPA; AFP.

9/84 (9:9) An explosive eruption on 5 Sept was preceded by seismicity and 9 months of minor tephra ejection. Rumblings were heard on 4 Jan, followed by an explosion that ejected ash. From Feb through April, rumbling preceded episodes of ash emission. On 31 May at 0724, an ash column rose to 1.5 km above the summit. During the night of 7-8 June, glowing lava fragments were ejected from the main crater. On 20 July, ash emission was accompanied by rumbling. The number of local seismic events increased through the first half of 1984 (table 6-14). Volcanic tremors were recorded 24 Aug, although no surface activity was seen. Ash emission occurred 3 Sept at 0447, producing an eruption column that rose 600 m. Glowing lava fragments were occasionally ejected. Rumbling accompanied the activity.

On 5 Sept at 0905, an ash column rose 4 km from the main crater. Nuées ardentes flowed 2 km S and 1 km W,

with estimated volumes of 1.5 and 0.5 × 10^6 m^3. One week later, ash emission was continuing and weak rumbling was heard. Ten volcanic and 5 tectonic earthquakes were recorded daily through 16 Sept. About 4500 people were temporarily evacuated from the S and W sides of the danger zone but were allowed to work in their fields during the day. No casualties were reported.
Information Contact: A. Sudradjat, VSI, Bandung.

2/85 (10:2) A 4-day eruption began 24 Feb. Effusive material was erupted from a S flank new vent, about 300-400 m from the main crater. A lava flow extended about 350 m S (along the Batuawang River), stopping after 4 days. Inhabitants downslope of the lava flow were alerted, including the villages of Kola, Bola, and Kopi. Harmonic tremor indicated lava movement a few hours before effusive activity began, but no other significant changes in seismicity were detected during the activity.

Api Siau has been the site of nearly continuous activity for the last 20 years.
Information Contact: Same as 9:9.

11/85 (10:11) "Api Siau erupted from the main crater (Kawah Utama) on 6 Nov. A 1.5 km-high ash column covered villages S of the volcano (Salili, Beong, Kanawong, and Lehi) with 1-3 mm of ash. Detonations were heard from the observation post at Muaralawa, 4.5 km SW of the volcano. Additional detonations were reported throughout the month. Possible precursory signs of this activity included a darkening of the normally whitish quiescent plume beginning on 4 Nov.
Information Contacts: T. Casadevall, L. Pardyanto, VSI, Bandung.

12/85 (10:12) Api Siau was quiet during Dec with only a small gas plume continuing to be emitted.
Information Contacts: Suparto S., T. Casadevall, VSI, Bandung.

DUKONO

Halmahera Island
1.70°N, 127.87°E, summit elev. 1087 m

Halmahera lies E of the Sangihe Islands across the Molucca Sea. Dukono is the northernmost volcano in the group, 200 km ESE of Api Siau. The first of its 5 historic eruptions was in 1550, and it has been erupting virtually continuously since 1933.

8/78 (3:8) Geologists Muslim Monoarta and Peter Jezek visited Dukono from 20-25 July. Activity had begun to intensify 4-7 days before their arrival and explosions of varying strength occurred about once every 10 seconds during their visit, ejecting bombs as large as 4 m in longest dimension. The bombs fell 200-250 m from the crater rim. Several 6-8-hour periods of quiescence interrupted the explosions, always at night. Ash emission was

1984	Tectonic Earthquakes	Volcanic Earthquakes
Jan	62	18
May	82	57
June	200	139
July	456	85

Table 6-14. Number of local earthquakes per month recorded at Api Siau.

nearly continuous, with clouds rising as high as 10 km above the crater. The ash ranged from black sand-size particles to very fine light gray clay-size material.

On 21 July, fine light gray ash was blown N beyond the town of Galela, 15 km from the volcano. Moderate ashfall continued through most of the day on the 21st and during the morning of the 22nd. During the night of 23-24 July, the loudest explosions in many years rattled windows in Galela and incandescent ejecta were visible.
Information Contacts: M. Monoarta, Volcanology Division, GSI, Bandung; P. Jezek, SI, Washington.

1/80 (5:1) VSI personnel visited Dukono in Aug 1979. Explosive activity had increased since the previous visit 13 months earlier. Ash was ejected every 9-15 minutes but there were no audible detonations.
Information Contacts: A. Sudradjat, L. Pardyanto, VSI, Bandung; D. Shackelford, Brea, CA.
Further Reference: Kusumadinata, K., 1977, Data on the Dukono Volcano; *Berita Direktorat Geologi*, v. 9, no. 16, p. 183.

GAMKONORA

Halmahera Island
1.38°N, 127.52°E, summit elev. 1635 m
Local time = GMT + 9 hrs

Gamkonora, 50 km S of Dukono, has erupted 12 times since 1564.

7/81 (6:7) Explosive activity from the summit crater began about 0900 on 19 July. The eruption apparently began with the ejection of incandescent tephra, followed by about 1½ hours of ash emission. An eruption cloud rose about 700 m and ash fell 5 km S of the summit, where 1-1.5 mm accumulated. Occasional felt earthquakes continued after the 19 July ash ejection ended. More than 3500 people fled the area.

Smaller explosions occurred on 22 July at about 0400 and 1800, accompanied by booming noises heard in a village at the NNW foot of the volcano, 5 km from the summit. Glow was visible over the crater at night. A VSI team arrived on the island immediately after the second explosion. After the team issued an evaluation, the evacuees returned to their homes

Local officials reported that the summit crater had occasionally emitted thick "smoke" since March. Gamkonora's last eruption, from mid-July through early Oct 1952, also consisted of explosive activity from the summit crater.

Information Contacts: A. Sudradjat, S. Suparto, Suratman, VSI, Bandung; Antara News Agency, Jakarta; AFP.

GAMALAMA

Ternate Island
0.80°N, 127.325°E, summit elev. 1715 m
Local time = GMT + 9 hrs

Gamalama, or Peak of Ternate, is a strato-volcano forming an 11 km-diameter island just W of Halmahera Island. It is the highest volcano in the northern Moluccas and a historic center of the spice trade. Over 67 eruptions are known since 1538.

8/80 (5:8) An incandescent tephra eruption from the central crater began with an explosion on 4 Sept at 1430, followed by a second about 2 hours later, and others at 0030, 0330, and 1120 the next day. Incandescent material fell 500-750 m from the crater, starting brush and forest fires. Ash fell on the entire island, accumulating to a depth of 10 cm by the second day of the eruption at Ternate City, 7-8 km E of the crater. Ash thicknesses reached 15 cm on some parts of the island, according to AFP.

Two earthquakes were felt by persons remaining on the island on 6 Sept as the eruption continued. By 7 Sept, activity had declined. Ash clouds rose about 1 km and were blown N by the prevailing wind, keeping ash away from the S half of the island, including Ternate City.

About 40,000 of the approximately 60,000 residents fled Ternate Island for Tidore Island, 5 km to the S, during the first 2 days of the eruption. No casualties have occurred according to the VSI, AFP, and Reuters, although some of the broadcast press apparently incorrectly reported casualties. A hazard map previously prepared by VSI delineates a danger zone of 33 km^2 (population 2500) in the summit area and an alert zone of 30 km^2 on the N, NW, and NE flanks (population 2500). The S and E parts of the island are considered to be safe by VSI. The volcano's last eruption, 31 Dec 1962-2 Jan 1963, was also from the central crater. Numerous other eruptions have been recorded since 1538.

Information Contacts: A. Sudradjat, VSI, Bandung; AFP; Reuters.

9/80 (5:9) The eruption continued vigorously into late Sept. Seven explosions were recorded during the first week of activity, 17 the second week, and 15 the third week. Eruption clouds ranged from about 500 m to 1800 m in height. Bombs fell as much as 1 km away, most heavily to the N and NE. By early Oct, eruptions had ended, but seismicity continued, with episodes of volcanic tremor lasting up to 4½ hours. Most evacuees had returned home, but some small villages in the NE sector red zone remained evacuated.

Kompas newspaper reported that a new crater had formed ENE of the summit. Both the summit crater (2 vents) and the new crater have ejected incandescent tephra.

Information Contacts: A. Sudradjat, L. Pardyanto, VSI, Bandung; M. Krafft, Cernay; *Kompas*, Jakarta.

7/83 (8:7) AFP reported that an eruption began 9 Aug. Residents of villages closest to the volcano were awakened at 0445 by the activity. A thick black eruption column containing incandescent material rose 1.5 km and "red-hot lava" moved down the N flank, destroying scores of homes and plantations. Ash fell W of the volcano, closing an airport. Explosions on 10 Aug at about 1000 and 1200 produced 1.5-km ash columns. A wind shift threatened to cause ashfalls E of Gamalama. Earthquakes centered on the volcano accompanied the eruption.

More than 5000 persons living near the volcano evacuated to the town of Ternate, capital of North Moluccas regency. No casualties were reported. Despite bad weather, vessels were standing by in case Ternate required evacuation.

Gamalama has had numerous small explosive eruptions in historic time, some accompanied by lava flows.

Information Contact: AFP.

MAKIAN

North Maluku Islands, Halmahera
00.32°N, 127.40°E, summit elev. 1357 m

Makian is a small island 55 km S of Gamalama, with 8 historic eruptions (4 of them fatal) since its first around 1550.

General Reference: Matahelumual, J., 1986, Makian; *Bulletin of the Volcanological Survey of Indonesia*, no. 110, 37 pp.

7/76 (1:10) The Jakarta Domestic Service reported on 13 June that "the Social Affairs Dept. has designated Malifut County in Kao Subdistrict, North Maluku District, as a resettlement area for 3250 families being evacuated from Makian Island, because there is a possibility that the volcano on the island will erupt at any time."

Information Contact: Jakarta DRS.

9/76 (1:12) Fears of renewed activity at Makian, which last erupted in 1890, stemmed from earthquakes felt in the area in 1972. Investigations carried out between Jan 1973 and July 1976 by GSI volcanologists indicated that the earthquakes recorded were of tectonic origin and that the volcano shows no signs of renewed activity.

Information Contact: G. de Nève, GSI, Bandung.

PHILIPPINE ISLANDS

General References: Catalog of Philippine Volcanoes and Solfataric Areas (3rd rev.); Commission on Volcanology, Quezon City, 1981, 132 pp.

Geologic Hazards and Disaster Preparedness Systems; Philippine Institute of Volcanology and Seismology (PHIVOLCS), Quezon City, 1987, 128 pp. (6 papers).

CANLAON

Negros Island
10.41°N, 123.13°E, summit elev. 2465 m
Local time = GMT + 8 hrs

Canlaon, the highest point on the central island of Negros, is 510 km ESE of Manila and 890 km N of Api Siau, the nearest Indonesian volcano described in the last chapter. Fourteen eruptions are known since 1866.

6/78 (3:6) A brief ash eruption began at [0610] on 27 June. A grayish eruption column rose about [1500] m above the summit before being blown S by strong prevailing winds. Ash fell on S flank villages within 8 km of the summit. The eruption ended after about 20 minutes, and was succeeded by voluminous but low-pressure emission of white vapor.

Information Contact: G. Andal, Commission on Volcanology (COMVOL), Quezon City.

7/78 (3:7) Intermittent ash emission continued through July. The most recent ash emission occurred at 1253 on 31 July when a brownish column rose 700 m above the crater, accompanied by volcanic tremor.

Information Contact: Same as above.

8/78 (3:8) Canlaon ejected steam and a little ash for 4 1/2 hours on 2 Aug. The cloud rose 800-1400 m above the crater rim. Seismographs recorded local activity. A smaller eruption occurred at 0610 on 9 Aug, producing a 250-m cloud. [The eruption ended with small explosions on 14 Aug at 0615 and 2 Sept at 1600.]

Information Contact: Same as above.

Further Reference: Oanes, A., 1978, 1978 Canlaon Volcano Eruption; COMVOL report (unpub.).

5/80 (5:5) COMVOL reported that an average of 60 earthquakes per day were recorded near Canlaon in mid-May, reaching a maximum of 160 on 17 May [but see 5:6]. COMVOL installed 3 additional seismic stations around the volcano and extended the danger zone from 4 km to 10 km. A similar earthquake swarm preceded the 1978 eruption.

Information Contact: AFP.

6/80 (5:6) An earthquake swarm began 6 May in the vicinity of Canlaon. The number of earthquakes recorded per day on nearby seismographs reached a peak of 108 on 19 May, then seismicity began a gradual decline. In late June, volcanic shocks continued to occur at a rate of about 40 per day. As of 27 June, a total of 2652 events had been recorded. No surface volcanism has been reported with this earthquake swarm.

Information Contact: O. Peña, COMVOL, Quezon City.

11/80 (5:11) Seismic activity has lessened considerably, but remained above normal as of late Nov.

Information Contact: Same as above.

2/85 (10:2) A small steam and ash eruption 13-14 March was preceded by local seismic activity. Three high-frequency volcanic tremors were recorded on 9 March and one each on 10, 11, and 12 March by a PHIVOLCS seismograph 9 km from the vent. Sustained seismic activity started suddenly on 13 March at 1312, 4 1/2 hours before the onset of the eruption, and totaled 139 recorded events, 60 of which were high-frequency and 79 low-frequency types. At the beginning of the seismic crisis, only high-frequency tremors were recorded, then low-frequency types predominated between 1400 and 1600. The number of events per minute declined before the eruption began.

Mild ejection of steam with a minimal amount of ash started 13 March at 1745, producing a plume that reached 300-500 m height. At 1515 the next day, the summit was steaming voluminously, and the vapor column rose 500-700 m. On 15 March at 1700, "moderate" steaming was reported. The volcano was cloud-covered and only occasionally visible. During the eruption's first 24 hours (until 14 March at 1759) 51 high-frequency and 9 low-frequency tremors were recorded.

Information Contact: PHIVOLCS, Quezon City.

9/85 (10:9) On 5 Oct a small eruption produced an 800-m eruptive cloud, depositing ash on 4 nearby villages. A strong booming sound was reported to have preceded the eruption and earthquakes were felt during the eruption. Activity had declined to voluminous emission of white vapor on 8 Oct and weakened further the next day, but seismographs continued to record low-frequency events. PHIVOLCS declared a 4-km diameter danger zone 2 weeks before the eruption, and residents of the zone were ordered to evacuate when the eruption began.

Information Contact: AFP.

BULUSAN

Luzon Island
12.77°N, 124.05°E, summit elev. 1559 m
Local time = GMT + 8 hrs

Bulusan is the southernmost volcano on Luzon and has 13 known eruptions since 1886. It is 280 km NNE of Canlaon and 70 km SSE of Mayon.

7/78 (3:7) Bulusan ejected a grayish steam and ash cloud for about 30 minutes on 29 July, beginning at 1155. The cloud rose several hundred meters above the crater and deposited 1-7 mm of ash NE of the volcano. The eruption was accompanied by rumbling.

Steam emission persisted after the eruption, and as of 3 Aug, white to dirty white vapor was being continuously emitted to about 100 m above the crater. COMVOL delineated a danger zone within 5 km of the summit, to be evacuated in the event of a major eruption. Bulusan last erupted in Dec 1933.

Information Contacts: G. Andal, COMVOL, Quezon City; AFP.

8/78 (3:8) A voluminous ash-laden cloud rose 2500-3000 m above the crater rim on 14 Aug, then was blown NE by the prevailing wind. The eruption, accompanied by hissing and rumbling sounds, began at 0545 and lasted about 30 minutes. A seismograph 5 km SSE of the crater was saturated by an explosion-type earthquake.

Information Contact: G. Andal, COMVOL, Quezon City.

Further Reference: Aguila, L.G., 1978, The 1978 Eruptions of Bulusan Volcano; *The COMVOL Letter*, v. 10, nos. 3-4, p. 4-8.

1/80 (5:1) Bulusan began to erupt at 1301 on 27 Dec, ejecting significant amounts of ash to about 1 km above the summit. Ash was blown WSW by the prevailing wind. The eruption lasted for 1 hour and 13 minutes, and was accompanied by hissing and rumbling. As of early Jan, some activity was continuing.

Information Contact: O. Peña, COMVOL, Quezon City.

2/80 (5:2) The eruption continued with a second explosion on 12 Jan at 1640. Ash rose about 500 m above the crater rim, then drifted NNW. COMVOL had predicted further activity in the *Bulusan Volcano Bulletin* dated 4 Jan, based on patterns exhibited during previous eruptions, especially the 1878 eruption.

Citing a COMVOL official, Kyodo radio reported that a third explosion occurred on 7 Feb, apparently ejecting ash to 6 km. Ash emission was continuing early the next day. Residents left the area, reportedly because of 28 volcanic earthquakes and 2 tectonic events.

Information Contacts: O. Peña, COMVOL, Quezon City; Kyodo radio, Tokyo.

3/80 (5:3) Occasional explosions continued through March. The 7 Feb explosion began at 0215, after an increase in the number of volcanic earthquakes. Cauliflower-shaped, ash-laden clouds rose about 6 km above the crater rim before drifting SSW. Rumbling and hissing sounds accompanied the activity. Milder ejection of ash-laden steam clouds started at 1102 the next day, with light ashfall continuing until late afternoon on the volcano's SW sector. Another weak ash ejection occurred 9 Feb.

Clouds and rain made observation of the summit and collection of ashfall data difficult in March. A strong odor of H_2S was detected at San Benon station, 6 km SSW of the crater, at 0945-0950 and 1030-1200 on 10 March, but no ashfall was reported. On 22 March, light ashfall occurred from 1120 to 1135.

Three weak ejections of ash-laden steam clouds took place on 26 March. At 0449, a cloud rose to approximately 400 m above the crater rim before winds blew it NW. A second steam and ash column reached 500-700 m above the rim about 1200, and a third was ejected to 300 m height 2 hours later. About 2 mm of ash fell on a settlement 8 km NNW of the crater.

At 1525 the next day, a volcanic earthquake of intensity I on the Modified Rossi-Forel (MRF) scale was felt at Lake Bulusan station, 4.5 km SE of the crater. Rumbling and hissing sounds accompanied the earthquake. Ejection of ash-laden clouds was visible through weather clouds at the summit.

Information Contact: O. Peña, COMVOL, Quezon City.

4/80 (5:4) Almost daily mild ejection of ash-laden steam clouds characterized activity from 2 April until late in the month. The clouds rose 100-1500 m above the summit. On 4, 5, 15, 17, and 22 April, light ashfalls took place on villages as much as 8 km from the crater in the volcano's NW and SW quadrants. Activity was sometimes accompanied by volcanic earthquakes, including B-type events at 1243 on 2 April and 0722, 0950, 1007, and 1627 on 4 April, and by hissing sounds.

At 1950 on 29 April, Bulusan erupted a series of dark, cauliflower-shaped ash clouds, forming a column that reached 3-4 km above the summit. About 2.5 mm of ash fell on SW flank villages. Hissing and intermittent rumbling sounds were heard during the ash ejection. The 29 April activity was preceded by about 3 days of increased seismicity, when seismographs recorded 170 volcanic earthquakes. Two of these were felt at intensity I on the MRF scale at a COMVOL station.

Information Contact: Same as above.

6/80 (5:6) Occasional mild emission of ash-laden steam clouds continued through June. During June, some light ashfalls occurred on the S and SW flanks.

Information Contact: Same as above.

7/80 (5:7) An earthquake swarm began on 6 July, when 108 volcanic events were recorded in a 24-hour period. Six events were felt at intensities of I-IV on the MRF scale and were accompanied by rumbling. Seismicity peaked on 10 July with 189 recorded shocks, 10 felt at intensities I-IV. The swarm declined gradually after the 10th.

On 19 July, ash-rich steam clouds rose about 6 km above the summit. About 3 mm of ash fell on villages to the SE. Seismicity increased again on 22 July, with 235 volcanic earthquakes recorded in 24 hours. Ten were felt at intensities up to IV. AFP reported a 24 July COMVOL announcement that persons living within 4 km of Bulusan had been instructed to evacuate because of the earthquakes. On 27 July, 323 events were recorded and 11 felt, again at intensities of as much as IV. The swarm was continuing as of 29 July.

Information Contacts: O. Peña, COMVOL, Quezon City; AFP.

8/80 (5:8) Occasional explosions continued through late Aug. Four explosions have occurred since last month, on 30 July (the strongest) and 1, 5, and 24 Aug. Ash-laden steam clouds rose 4-7 km above the summit, depositing ash as much as 14 km from the crater. The maximum measured ashfall thickness was 3.5 mm. The explosions were preceded by earthquakes of intensities I-IV on the MRF scale, continuing the seismicity that began on 6 July. As of 28 Aug, occasional felt events were continuing.

Information Contact: O. Peña, COMVOL, Quezon City.

9/80 (5:9) Only one explosion was reported during Sept, a mild ash ejection on the 28th at 1155. The summit was covered by weather clouds, but seismographs in the area recorded the activity and villages in the SE sector of the volcano received traces of ashfall.

Information Contact: Same as above.

11/80 (5:11) The most recent ash eruption, on 28 Sept, was followed by a series of volcanic earthquakes that became less frequent with time. Felt events of intensity I-II on the MRF scale have also occasionally been recorded.

Information Contact: Same as above.

4/81 (6:4) Bulusan began to eject ash on 9 April at 1008, its first such activity since 28 Sept 1980. Eruption clouds reached 4 km height and deposited 4 mm of ash on villages on the W side of the volcano. AFP reports that the 9 April eruption lasted about 8 hours.

A second episode of ash ejection started 15 April at 1858. Lightning flashes were observed within the eruption clouds, which reached an approximate height of 8 km. Most of the ashfall was again on villages W of Bulusan, with accumulations of as much as 6 mm.

An earthquake swarm began 20 April. During the next 7 days, 1812 events were recorded, 64 of which were felt at MRF I-V and were accompanied by rumbling. On 27 April at 1745, renewed explosive activity produced ashfalls on villages SW of the crater. Clouds obscured the summit, preventing determination of the eruption column height. Seismicity has declined since then. No additional eruptive activity had occurred as of 11 May.

Information Contacts: O. Peña, COMVOL, Quezon City; AFP.

6/81 (6:6) Seismographs registered a local earthquake swarm 30 June-4 July. Of the 700 recorded events, 11 were felt at intensities I-II on the MRF scale and 7 others were accompanied by rumbling. No eruption took place.

Information Contact: O. Peña, COMVOL, Quezon City.

7/81 (6:7) No eruption followed the 30 June-4 July earthquake swarm, and seismic activity at the volcano has remained relatively quiet since then.

Information Contact: Same as above.

6/83 (8:6) Mild phreatic explosions from the summit crater occurred at 0515 on 25 June and 0515 on 29 June. Eruption clouds reached heights of 1000 and 1200 m. Between 1 and 2 mm of ash fell on the NW quadrant of the volcano but no damage was reported.

An intermittent increase in steaming from the summit crater and vents on the upper W flank was noted about a month before the eruption. A week before the first explosion, the temperature of hot springs 4 km S of the summit crater increased by 2°C and their discharge rate increased by 4 liters/sec. No increase in seismicity was noted.

Information Contact: Same as above.

MAYON

SE Luzon Island
13.26°N, 123.68°E, summit elev. 2462 m
Local time = GMT + 8 hrs

This classically conical strato-volcano has the most active historic record in the Philippines (47 eruptions since 1616). Nuées ardentes have been recorded from 18 eruptions, and at least 12 eruptions have caused fatalities. It is 300 km SE of Taal.

11/77 (2:11) Summit crater glow was observed between 2002 and 2125 on 6 Nov from COMVOL observation stations. Glow was continuous and deep red for the first 3 minutes, then became intermittent and yellowish in color. The next day, steam emission, under increased pressure, occasionally varied in volume and intermittently changed from the normal white color to brown. Several new steam vents had formed outside the crater, on the upper flank. Glow was again observed during the night of 8-9 Nov and volcanic tremor was recorded.

Mayon's last major eruption, in 1968, killed at least 6 persons, covered roughly 100 km^2 with more than 5 cm of airfall ash, and produced a 3½ km-long aa lava flow as well as numerous nuées ardentes.

Reference: Moore, J. G. and Melson, W. G., 1969, Nuées Ardentes of the 1968 Eruption of Mayon Volcano, Philippines; *BV*, v. 33, no. 2, p. 600-620.

Information Contact: G. Andal, COMVOL, Quezon City.

2/78 (3:2) The following reports are from Gregorio Andal.

"Crater glow at the summit, observed on 6 Nov, was followed by increased volcanic tremor 2 days later. Aerial investigations disclosed bluish fumes in the SW portion of the crater. At 0116 on 12 Nov an imperceptible volcanic earthquake with a rather large double amplitude was recorded, lasting more than one minute. Drying of vegetation on the upper flank (about 1700 m elevation) was noted on 18 Nov. On 22 Dec, 13 large-amplitude volcanic tremors were recorded, accompanied by increased steam emission forming a cauliflower-shaped cloud. Glow, accompanied by volcanic tremor, was continuing as of late Jan."

Information Contact: Same as 2:11.

5/78 (3:5) "An eruption started at about 2030 on 7 May and gradually increased in intensity, reaching a maximum on 22 May. The event was characterized by weak lava flow extrusion at the start, and at the height of the eruption there was a fascinating night display as incandescent basaltic [andesite] lava flowed down the SW flank. Strong earthquakes were felt at the Mayon Rest House Observatory (on the N flank, elevation 760 m, fig. 7-1)

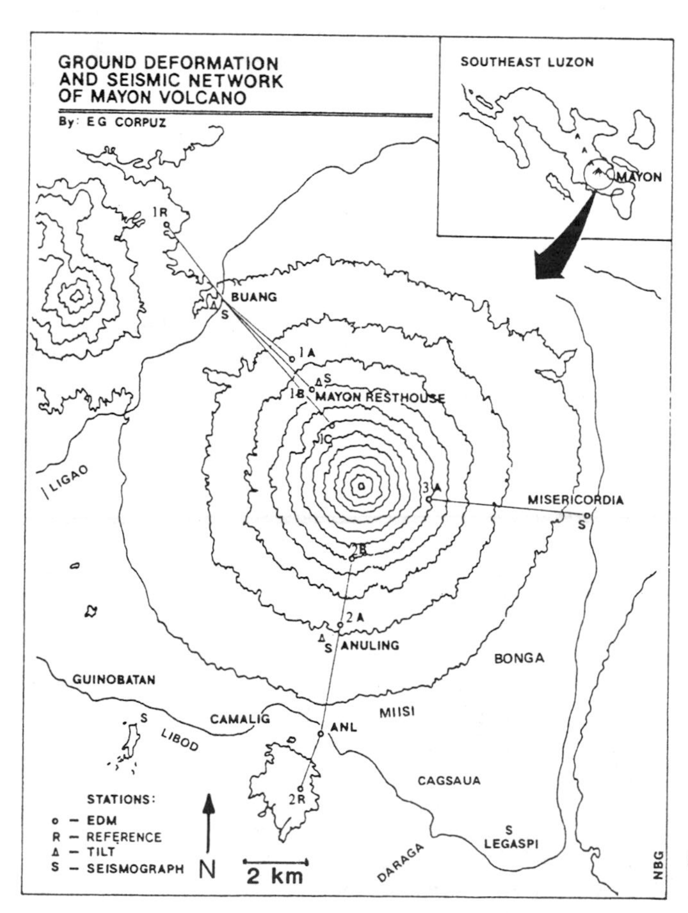

Fig. 7-1. Location map of Mayon Volcano showing ground deformation and seismic network, after Corpuz, 1985. [Not in original *Bulletin*.]

on 26 May, accompanied by voluminous ejection of ash-laden clouds and spattering of incandescent pyroclastic materials. From then on, activity began to decline, punctuated at first by short lulls, which became longer each day until ash ejections, rumblings, and volcanic tremor ceased on 29 May. Small amounts of lava continued to be extruded as of 2 June. This decline in activity suggests that the critical period of the eruption has apparently passed.

"No casualties were reported during the eruption, but 8000 people on the SW sector of the volcano and within an 8 km radius had to be evacuated. An additional 15,000 people evacuated voluntarily due to their fear of the eruption, even though they were not within the declared danger zone."

Activity briefly intensified early on 7 March, when ash-laden clouds and a little incandescent material were ejected.

The following information was provided by Chris Newhall.

The new aa lava flow emerged from a breach in the summit crater wall and traveled down the SW flank directly over the 1968 flow. By 26 May, the nose of the new lava was about 200 m beyond the end of the 1968 flow. The breach grew to about $1/3$ of the crater's circumference and about 100 m depth before it began to fill in during the later stages of the eruption.

Maximum eruption cloud height was about 3 km above the summit. Ash contents were low and maximum ashfalls were only a few millimeters. Harmonic tremor was nearly continuous during the period of maximum activity and shallow explosion earthquakes were also recorded.

Information Contacts: G. Andal, COMVOL, Quezon City; C. Newhall, Dartmouth College.

8/78 (3:8) Lava extrusion ended on 6 July, more than a month after activity began to decline. Harmonic tremor had accompanied the extrusion, increasing in magnitude when the lava flow rate increased. Hollow rumbling sounds were occasionally heard. Some ash puffs continued to be ejected, the most recent at 1727 on 21 Aug, saturating a nearby seismograph. Chemical analysis of the ash and petrographic analysis of the lava both indicated a basaltic andesite composition. COMVOL recommended the return of the last of the evacuees on 7 July, after recommending a partial end to the evacuation several weeks earlier.

Information Contact: G. Andal, COMVOL, Quezon City.

Further Reference: Peña, O., 1978, Notes on the Mayon Eruption from May 3-July 4, 1978 and COMVOL's Role; *COMVOL Letter*, v. 10, no. 3-4, p. 1-3.

8/79 (4:8) Seismic activity and crater glow were observed at Mayon 14-28 July. The same type of activity preceded the 1978 eruption. COMVOL has established a close watch on the volcano and will monitor any changes.

Information Contacts: G. Andal, COMVOL, Quezon City; J. Wolfe, Pan Asean Technical Services, Manila.

8/80 (5:8) Short-duration harmonic tremor began to be recorded at Mayon on 16 Aug. Bursts of tremor continued, but have become less frequent since the 16th. Similar seismicity preceded the 1978 eruption and accompanied crater glow in July 1979.

Information Contact: O. Peña, COMVOL, Quezon City.

11/80 (5:11) Occasional tremor continued through Nov, and as of the 30th, 214 episodes had been recorded.

Information Contact: Same as above.

12/80 (5:12) A moderate quantity of dirty white steam rose weakly to 200 m above the crater rim on 4 Dec at 1247, accompanied by short-duration harmonic tremor on the Mayon Resthouse Observatory seismograph. Faint crater glow was first noted at 2315 the same day. Additional steam emission was observed 12 and 14 Dec. Episodes of tremor and discrete earthquakes continued through Dec.

Information Contact: Same as above.

6/81 (6:6) At about 2000 on 30 June, mudflows triggered by continuous rains accompanying Typhoon Daling swept villages in the S and E sectors of Mayon. Preliminary estimates set casualties at about 100 persons with many more missing [but see 6:7]. Mayon has not erupted since 1978.

Information Contact: Same as above.

7/81 (6:7) Updated casualty figures indicate that the S and E flank mudflows triggered 30 June by typhoon Daling killed 40 persons, injured 9, and left 7 missing. Other reported casualties were caused by the typhoon itself and associated flooding.

Information Contact: Same as above.

Further Reference: Gianan, O., 1982, A Volcano Disaster Preparedness Plan: Mechanics of Implementation of "Operation Mayon"; *Proceedings of the First Seminar Workshop on Philippine Volcanoes and Volcanic Terranes*, Quezon City, Dec. 1982, p. 88-97.

8/84 (9:8) Slow summit lava production started 10 Sept at 0821, after at least 12 hours of harmonic tremor [but eruptive activity started the previous evening (9:9)]. Activity was dominantly Strombolian 10-11 Sept. On the 10th, aerial observers reported that lava from the summit crater was slowly spilling over the rim. Incandescent blocks rolled 600 m down the NW flank, destroying or incinerating trees in their path. Ash-laden clouds rose several hundred meters above the summit. PHIVOLCS recommended the evacuation of people living in a danger zone within 8 km of the crater. The next day, ash-laden steam clouds were ejected to heights of as much as 3 km at intervals of 1-5 minutes. The strongest explosion, accompanied by loud detonations, began at about noon, producing a steam and ash column that rose 3 km and deposited as much as 2 cm of ash. Two lava flows about 200 m wide advanced at about 3 m per minute to about 2 km from the summit crater by evening.

Activity intensified and became more Vulcanian in character beginning on the 12th. Pyroclastic flows reached several kilometers from the summit. Nine explosions occurred within 3 hours starting around 1100, sending ash to 14.5 km altitude. The day's strongest explosion, at 1553, sent ash to nearly 15 km altitude. Incandescent tephra was evident in the eruption columns. Smaller explosions took place at 1-5-minute intervals. Sixteen additional "fairly strong" explosions occurred in a 17-hour period ending midmorning 13 Sept. Activity on the 13th was characterized by persistent ash emission to 3 km above the crater but no large eruption columns were observed. A total of 26 strong explosions ejecting incandescent tephra were recorded in a 12-hour period ending the morning of 14 Sept. Rain falling on nearby towns deposited 7.5 cm of wet ash. Three lava flows continued to advance down the NW, N, and SW flanks.

More than 16,000 people were evacuated from 36 villages, most of which were within the 8-km danger zone. Nine people were reported killed, 8 by burial in volcanic debris and one by hot steam [but note that PHIVOLCS reported in 9:9 that no casualties were directly attributed to the eruption and mudflows].

Information Contacts: R. Punongbayan, PHIVOLCS, Quezon City; UPI; AP; DPA.

9/84 (9:9) Quoted material is from PHIVOLCS.

"Eruptive activity started 9 Sept at 1923. Initial activity was dominantly Strombolian, with incandescent spattering at the summit and production of small lava flows. A mound of solidified lava inside the crater blocked the 1968 notch at the SW rim, so the small lava flows

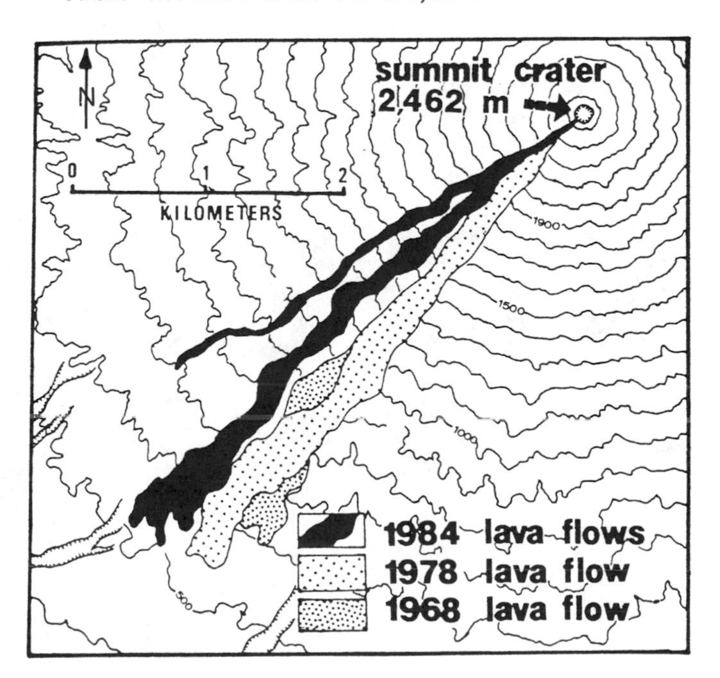

Fig. 7-2. Map, after Corpuz, 1985, showing the extent of the 1984 lava flows. [Not in original *Bulletin*.]

1984	Hour	Density	Width (km)	Length (km)	Movement Direction
Sept 9	?14	diffuse	30	60	W
11	?14	diffuse	50	70	WSW
12	?14	diffuse	20	60	W
13	?00	diffuse	70	110	SW
	?08	diffuse	20	70	SW
	14	diffuse	20	90	SW
15	08	diffuse	20	70	NW
	14	diffuse	30	40	NW
16	08	diffuse	30	70	W
	14	diffuse	30	40	W
	20	diffuse	30	80	W
17	02	diffuse	30	40	W
18	02	diffuse	30	40	SW
	08	diffuse	40	50	W
	14	diffuse	30	60	W
	20	diffuse	30	60	W
19	02	diffuse	30	60	W
20	02	diffuse	40	70	NW
22	02	diffuse	40	60	W
23	08	dense	40	120	W
	14	dense	80	260	W
	20	dense	40	140	W
24	02	dense	40	220	W
	08	dense	80	190	W
	14	dense	70	100	SW
	20	diffuse	70	80	SW
25	02	diffuse	40	40	SW
	08	diffuse	40	80	W
	14	diffuse	40	60	NW

Table 7-1. Series of plumes from the Sept 1984 eruption of Mayon observed on images from the GMS satellite. Plume lengths and widths were measured from images rather than digital data. Courtesy of Yosihiro Sawada.

and initial pyroclastic flows (see below) moved predominantly NW.

"A fairly strong eruption 10 Sept at 2300 marked the start of Vulcanian activity. Ash-laden steam clouds rose 5 km above the summit and a pyroclastic flow moved down to the NW, reaching 700 m elevation. Stronger explosions 11 Sept reopened the notch at the SW rim, so more of the later lava and pyroclastic flows moved SW than NW. The eruption continued to intensify, peaking 13 Sept. Cauliflower-shaped, ash-laden steam clouds accompanied by rumbling sounds reached a maximum height of 15 km before drifting SW, W, and NW. Continuous volcanic tremors with increasing amplitude were recorded, punctuated by explosion earthquakes. Two lava flows emerged through the SW breach. One reached 500 m elevation, adjacent to and W of the 1978 flow. The other, a little farther W, advanced to 1400 m elevation (fig. 7-2). The new lava is porphyritic augite-hypersthene andesite.

"Activity gradually declined 14-21 Sept. A mild eruption 22 Sept at 0502 was accompanied by a volcanic earthquake felt at intensity II on the MRF scale at the Mayon Resthouse Observatory. A relatively quiet period followed. A very strong explosion 23 Sept at 0433 ejected voluminous ash-laden steam clouds that reached 10 km in height. Incandescent tephra rose 2 km above the summit and spread in all directions, covering the summit area with red-hot tephra to about 1500 m elevation. A large notch was formed in the SE rim of the crater and a smaller one in the E rim. Subsequent pyroclastic flows were directed predominantly SE and E, although some moved in other directions along gullies. Ash fell within about 50 km SW, W, and NW of the summit. Areas E and NE of the volcano received most of the fine airfall tephra generated by pyroclastic flows. The eruption continued to intensify until the 24th. Voluminous ash emission, sometimes sustained for 5 minutes, occurred at intervals of 2-15 minutes, accompanied by strong detonations and at times by electrical discharges. Maximum height of the eruption clouds was 15 km. On 24 Sept at 1614, a nuée ardente [almost] reached the nearest village. A large volume of pyroclastic flow material was deposited on the SE flank. The eruption started to decline 25 Sept. By 5 Oct activity was limited to weak steaming and faint to moderate crater glow, accompanied by volcanic tremors and discrete earthquakes." Press sources reported reintensification of the eruption 6 Oct. Ash-laden steam clouds rose as much as 1.7 km above the summit and lava flowed 1 km from the crater.

"Mudflows generated by rain destroyed 3 sections of the Legaspi-Santo Domingo highway ~8 km SE of the volcano. Larger mudflows on the 27th overran the same portion of highway. Two bridges were destroyed along the Malilipot-Santo Domingo highway, roughly 8 km E of Mayon." As of 30 Sept, press sources reported that 6500 hectares of farmland had been covered by mudflows.

"Implementation of the Mayon preparedness plan was fairly smooth. On 10 Sept, the area within 6 km of the summit was declared off-limits and all residents were recommended for evacuation. On 12 Sept, the danger zone was extended to 8 km from the summit on the S and SW flanks. About 26,000 people were evacuated during

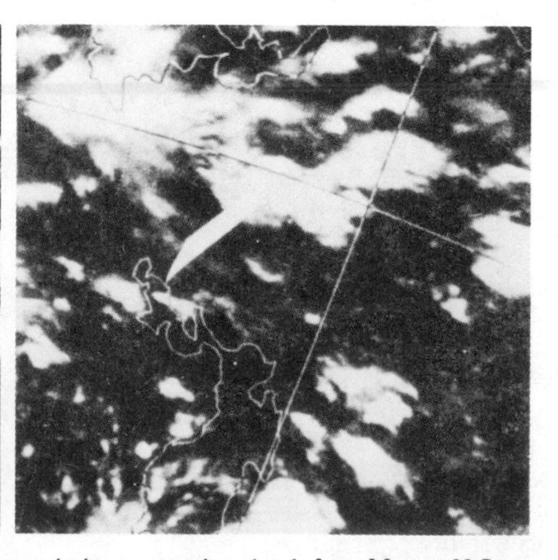

Fig. 7-3. GMS infrared satellite images with arrows pointing to eruption clouds from Mayon, 23 Sept 1984 at 0800, *left*, and 1400, *right*. Land areas are outlined with fine white lines. Courtesy of Yosihiro Sawada.

the first phase of the eruption. On 23 Sept, the danger zone was expanded again, to 10 km from the summit on the SE side and 8 km from the summit around the rest of the volcano. All residents of that area were recommended for evacuation and the number of evacuees swelled to more than 73,000 at 50 centers. No casualties were attributed directly to the eruption or mudflows."

Reference: Corpuz, E., 1985, Chronology of the September-October 1984 Eruption of Mayon Volcano, Philippines *in* Punongbayan, R.S. (ed.), 1985, Special issue on the 1984 eruption of Mayon; *Philippine Journal of Volcanology*, v. 2, p. 36-51 (9 papers, 205 pp.).

Information Contacts: PHIVOLCS, Quezon City; DPA; AP.

Further Reference: Umbal, J., 1987, Recent Lahars of Mayon Volcano *in* National Science and Technology Authority, *Geologic Hazards and Disaster Preparedness Systems;* p. 56-76.

2/85 (10:2) Yosihiro Sawada observed a series of plumes from the Sept 1984 eruption of Mayon (table 7-1) on images from the GMS satellite. Problems with the scanning system of the GMS limited images to every 6 hours during the spring and summer, and at times prevented data returns from its southern zone of coverage.

Eruption clouds from Mayon's intense activity 23-24 Sept appeared much larger and denser on satellite imagery than those from the early- to mid-Sept activity. A moderate plume on 23 Sept at 0800 had grown much larger 6 hours later (fig. 7-3) and plumes remained large and dense through 1400 the next day. Declining activity remained visible until 26 Sept at 1400.

Information Contact: Y. Sawada, Meteorological Research Institute, Tsukuba, Japan.

Further Reference: Sawada, Y., 1987 [see Part I - References].

MALINAO

Luzon Island
13.42°N, 123.60°E, summit elev. 1548 m

Malinao, only 20 km N of Mayon, had no historic activity prior to 1980.

8/80 (5:8) A small phreatic explosion took place on 29 July from one of the pools of hot water in the Naglagbong thermal area, Tiwi geothermal field, at the E foot of Malinao. The explosion ejected hot mud and blocks up to $^2/_3$ m in diameter. The ejecta reached heights of 150 m and fell as much as 350 m from the vent. One person received second-degree burns and 2 buildings were damaged, one a COMVOL seismic station.

Before the explosion the pool was 15 m in diameter and 4.3 m deep, with clear emerald-green water at a temperature of 85°C. As early as 6 July, the seismograph recorded unusual microseisms. Two hours before the explosion, geysering of muddy water was observed.

Information Contact: O. Peña, COMVOL, Quezon City.

TAAL

Luzon Island
14.02°N, 121.00°E, summit elev. 400 m
Local time = GMT + 8 hrs

Taal is a strato-volcano in a 15 × 22-km caldera lake (fig. 7-4) 60 km S of Manila. Thirty four historic eruptions are known since 1572; 6 have been fatal, with many deaths caused by tsunamis in the lake.

9/76 (1:12) COMVOL scientists monitoring Taal during late Aug noted the following signs of increased activity: profuse steaming from the junction of the SW portion of the 1966-70 eruption cone and the western rim of the pre-1965 lagoon (fig. 7-5); visible ground heating at some points above the pre-1965 lagoon; and a rise in the temperature of some ground probes in the area to 95°C. As a result of these observations, the evacuation of several thousand persons from Taal Island ("Volcano Island" on figs. 7-4 and 7-5) and some lakeshore towns was recommended on 31 Aug.

2 Sept: Steaming increased at the 1966-70 eruption site and at the main crater. The area of steaming had expanded considerably, particularly to the SW. A fissure widened from 5 to 20 cm and lengthened from 50 to 100 m. Seismic activity was low.

3 Sept: An eruption of steam and ash began at about

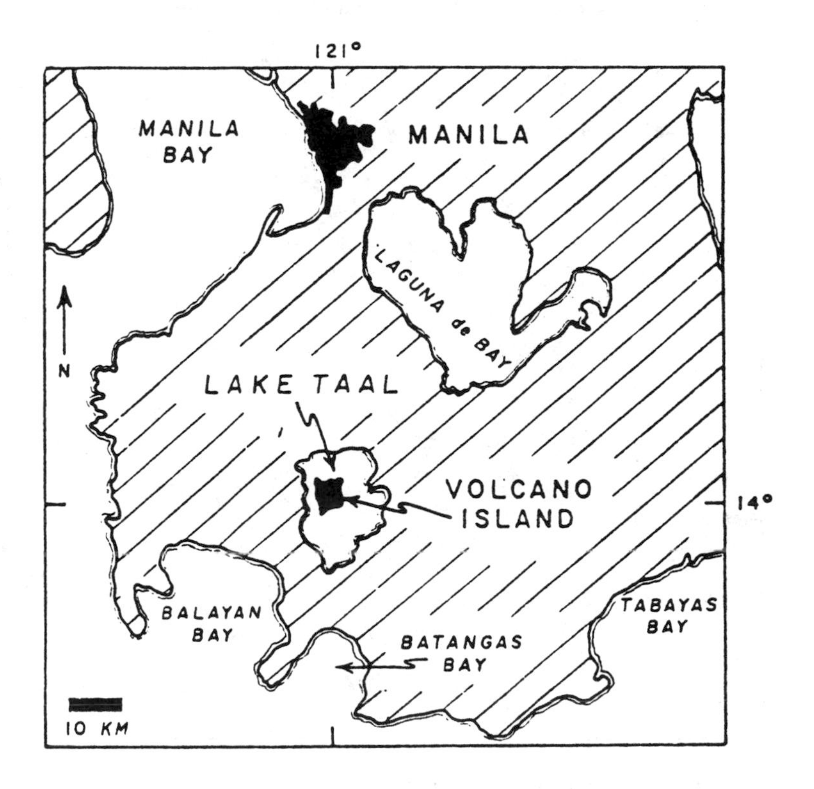

Fig. 7-4. Index map showing the area around Taal, after Moxham, 1967. Land area is cross-ruled.

0430 from Mt. Tabaro vent, a few meters SW of the site of 1966-70 activity. Eruptions occurred at about 4-minute intervals, producing clouds 1500-3000 m high. Harmonic tremor accompanied the activity. Evacuation had been completed at 0400.

4 Sept: Eruptive activity ceased at 0500. Renewed steam and ash emission occurred between 0900 and 0920 and again at 1300; the latter included some large fragments and produced a cloud 300-400 m high. Harmonic tremor was recorded during eruptions. Brownish ash blanketed the S half of the island to an unknown depth.

5 Sept: Individual eruptive bursts contained a greater proportion of ash and large fragments than those of previous days. No incandescent material or rumbling was observed. During the morning, eruption periods were longer than the repose periods separating them; by late afternoon, eruptions lasted about 3 minutes, followed by about 5 minutes of repose. Eruptions were nearly vertical with a slight southward component; the resulting clouds reached more than 2000 m in the morning, but were limited to 300 m by strong SW winds in the afternoon. Late-afternoon rain caused steaming to intensify. Continuous harmonic tremor of 4 mm amplitude was recorded. The new vent was enlarged by coalescence with the adjacent cone.

6 Sept: Harmonic tremor amplitudes began to increase at 0330 to 3-10 mm. Emission of black, grayish-black, or reddish-brown ash-laden eruption clouds was continuous, accompanied by faint rumbling and occasional brief lightning flashes just above the vent. The eruption clouds, which contained some large rock fragments, averaged 200-300 m, reaching a maximum of 2200 m in the early morning. Activity had slackened by afternoon, when maximum eruption cloud height was slightly more than 500 m, and the diameter was about 50 m. Lower ejection velocity and light wind caused much of the erupted material to collapse and form base surges at the foot of the 1966 cone. Rock fragments up to 30 cm in diameter produced dust clouds on the 1966 cone by impact and downslope rolling. By 1700, activity was still continuous, but the eruption cloud had diminished to 30 m across and less than 500 m high. Maximum harmonic tremor amplitude had declined.

7 Sept: Steam and ash eruptions continued, generating a 3000-m cloud and base surges over the SW flank. Harmonic tremor was continuous and had a maximum amplitude of 5 mm. A new fissure opened, measuring 100 × 0.2 m.

8 Sept: After a midmorning period of quiescence, a new vent opened a few meters SW of the active vent of the past several days. Ash and larger fragments were ejected to 250 m height. Maximum harmonic tremor amplitude increased to 8 mm.

9 Sept: The strongest explosion of the eruption produced base surges that moved over the 1965 and 1966 craters. Harmonic tremor amplitudes rose from 5 to 9 mm shortly before the explosion, and reached a maximum of 12 mm during the explosion. By late afternoon, eruptive force had declined, but ash emission was still continuous and voluminous. Strong west winds carried the ash across Lake Taal to its E shore.

10 Sept: At 2331, incandescent gases and some incandescent fragments were noted for the first time, accompanied by a continuous hollow sound. Steam and ash emission continued.

11 Sept: Minor fountaining and incandescent ash clouds were observed. Steam and ash emission continued.

Reference: Moxham, R.M., 1967, Changes in Surface Temperature at Taal Volcano, Philippines 1965-1966; *BV*, v. 31, p. 215-234.

Information Contacts: G. Andal, COMVOL, Quezon City; W. Ward, HQ 13th Air Force (PACAF); D. Shackelford, Villa Park, CA.

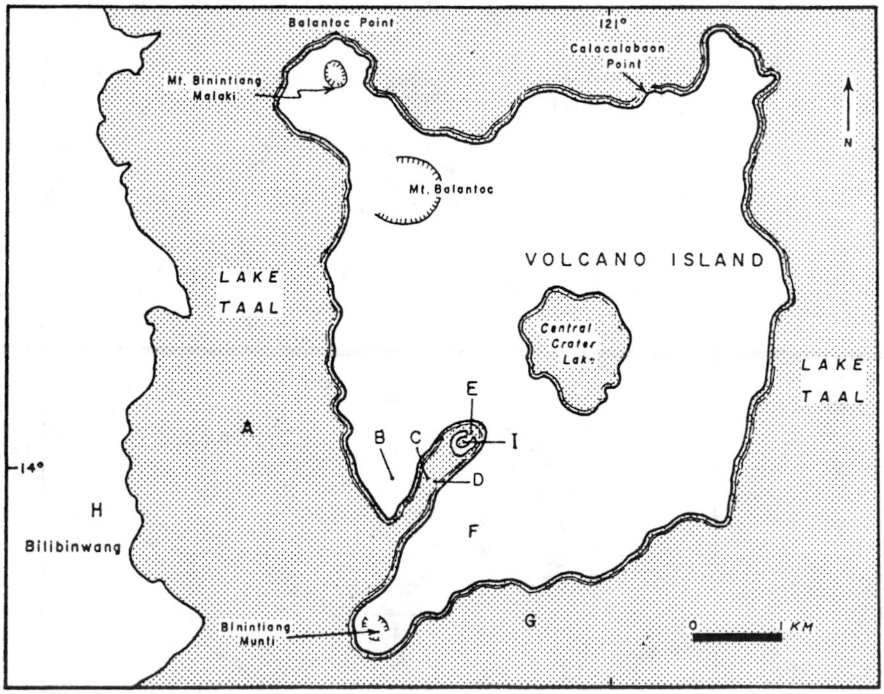

Fig. 7-5. Map of Taal Island and vicinity, Oct 1965, after Moxham, 1967. The 1965 explosion crater extends SW from I; the 1966 explosion crater formed immediately to the NW. Water is shaded.

11/76 (1:14) Activity continued through Sept and early Oct.

13 Sept: Voluminous quantities of steam and ash were emitted, causing ashfalls at towns up to 10 km E of the vent. Harmonic tremor was continuous. Intermittent incandescence was reported during the previous week.

14 Sept: Strong steam and ash eruptions continued until evening, when activity slackened.

15 Sept: Activity increased around 0700. Just before noon, a

strong explosion ejected rock fragments 150 m above the vent and produced an 1800-m ash cloud. Ash fell on towns surrounding Lake Taal, up to 16 km E and N of the vent. One mm of ash fell in a 20-minute period and a strong stench of sulfur was noted at the volcano station, 4 km from the vent.

17 Sept: Ash fell continuously on towns East and North of Lake Taal, frequently mixing with rain to fall as mud. Heavy ashfall and sulfur fumes forced the evacuation of Taal Island, except for a skeleton staff remaining at the volcano station. Many work animals had died and many others were ill, probably because of sulfur-contaminated food and water. Harmonic tremor gradually increased in amplitude.

18 Sept: Ash and sulfur emission decreased slightly in the evening.

19 Sept: Diminished activity allowed an inspection on the vent area. The new crater formed on the first day of the eruption was steaming weakly. Most activity was from a 100 m-diameter vent within the 1965-1970 cone, on which scattered impact craters a few centimeters to 0.7 m across were present up to 500 m from the vent. Harmonic tremor amplitude had declined. Ash emission weakened further in the evening, with occasional brief pauses. Ash clouds took 2-3 minutes to rise 700 m.

20 Sept: Continuous eruption resumed in the morning. No incandescence was observed. Harmonic tremor amplitude was unchanged. About 100 people had eruption-related diseases: asthmatic bronchitis, gastroenteritis, respiratory tract infection, sore eyes, and allergy.

21-22 Sept: The eruption weakened considerably.

23 Sept: Steam and ash emission increased in intensity, but no incandescence was observed. Harmonic tremor amplitude increased.

4 Oct: The eruption resumed after a "short lull". Ash clouds rose more than 2100 m, accompanied by lightning, thunderstorms, and harmonic tremor.

5 Oct: Ash emission, in black to brownish clouds, increased. By afternoon, heavy ashfall, which forced cars to use their headlights, had deposited 2.5 cm of ash in one town near Lake Taal.

7 Oct: Continuous heavy ashfall forced the evacuation of 5 towns up to 16 km E and N of the vent. The eruption was accompanied by hollow sounds from within the crater and by harmonic tremor. Profuse emission of white steam was occurring from the source of the 1968 lava flow. A new step fault was noted near the summit of the vent.

8 Oct: Ash-laden steam clouds were ejected continuously to a height of 500-600 m, depositing ash around Lake Taal. Activity had migrated back to the elongate main vent. Harmonic tremor amplitude began to increase at 0330.

10 Oct: A reconaissance at 1200 noted ejection of coarser fragments, accompanied by hollow reverberations, lightning, and thunder. Ash fell SW of the vent (most ash had previously fallen to the N and E). Ash eruption ended at 1650.

11 Oct: Activity resumed at 0100 with the ejection of brownish ash clouds, accompanied by harmonic tremor.

Information Contact: *Manila Times-Journal.*

12/76 (1:15) By 30 Oct, Taal's eruption had declined to weak steaming, and harmonic tremor had ended. Harmonic tremor resumed at 1925 on 30 Oct and steam eruptions began at about 2130. During the early morning of 31 Oct, cauliflower-shaped steam clouds reaching 100 m height were observed, containing only minimal amounts of ash. This activity continued through the day, the clouds gradually decreasing in height and volume. No further activity had been reported by 16 Nov.

Information Contact: Same as above.

Further References: Alcaraz, A. and Datuin, R., 1977, Notes on Taal Volcano Prognostics; *Journal of the Geological Society of the Philippines*, v. 31, no. 2, p. 18-20.

Andal, G.A. and Aguila, L.G., 1977, Prognostic Criteria of the 1976 Eruption of Taal Volcano; *COMVOL Letter*, v. 9, no. 1-2, p. 1-6.

10/77 (2:10) A weak phreatic eruption from the NE portion of the 1976 crater began at 1515 on 3 Oct, and had ended by 1400 the next day. Brownish to grayish basaltic ash clouds initially rose 300 m, and reached a maximum height of 500 m after a few hours. Activity then gradually declined to strong steaming. The eruption was accompanied by harmonic tremor. Residents of Taal Island were evacuated, and had not returned by late Oct because of the possibility of renewed activity.

Information Contact: G. Andal, COMVOL, Quezon City.

11/77 (2:11) Taal ejected voluminous ash-laden steam clouds at 1200 on 9 Nov. The clouds, similar to those of 3 Oct, rose about 250 m before being blown toward the mainland by a moderately strong NE wind. Ground probe temperatures have increased, and harmonic tremor continued at about the same amplitude. Initial evacuation procedures remained in effect.

Information Contact: Same as above.

2/78 (3:2) "The renewed activity of Taal on 9 Nov resulted in the formation of a circular conelet protruding a few meters from the floor of the elongated 1976 crater. Voluminous ash-laden clouds rose to a height of 500 m with a roaring sound audible on the lakeshore. Bluish fumes were emitted from the NE inner wall of the 1976 crater. Eruptive activity ended by the early morning of 12 Nov, but volcanic tremor of varying amplitude continued. Tremor with a maximum double amplitude was recorded at 2210 on 24 Nov and at 0504 on 25 Nov. The initial evacuation procedure remained in effect in late Jan."

Information Contact: Same as above.

Further References: Ruelo, H.B., 1983, Morphology and Crater Development of the Mt. Tabaro Eruption Site, Taal Volcano, Philippines; *Philippine Journal of Volcanology*, v. 1, no. 2, p. 19-68.

Wolfe, J.A., 1980, Eruptions of Taal Volcano 1976-1977; *EOS*, v. 61, p. 57-58.

DIDICAS

North of Luzon Island
19.08°N, 122.202°E, summit elev. 228 m

Several island-forming eruptions have occurred at this remote site, 50 km N of Luzon and 280 km N of Taal, since its first historic eruption in 1773. Lava dome growth in 1952-53 produced an island more than a kilometer in diameter that has remained. An eruption from 6 to 9 Jan 1978 covered the island with fresh volcanic ash (*Bulletin of Volcanic Eruptions*, no. 18, 1980).

BABUYAN CLARO

Babuyan Islands
19.50°N, 121.95°E, summit elev. 1160 m

Fifty-two km NNW of Didicas, Babuyan Claro is one of two adjacent volcanoes on Babuyan Island. Nine eruptions are known from the island since 1652, but few reports specify which volcano was active.

2/80 (5:2) Hot mudflows from Babuyan Claro [but see below] damaged rice fields and roads, and forced the evacuation of nearby residents according to a Kyodo radio report dated 8 Feb. PNA radio reported that a 3 man COMVOL was sent to study the volcano. . . .

Information Contacts: Kyodo radio, Tokyo; PNA radio, Manila.

3/80 (5:3) A COMVOL investigation has determined that press reports of hot mudflows at Babuyan Claro were incorrect. COMVOL found that a landslide, probably triggered by continuous rainfall, had occurred at 600 m above sea level on the NW flank.

Information Contact: O. Peña, COMVOL, Quezon City.

JAPAN and MARIANA ISLANDS

SUWANOSE-JIMA

Ryukyu Islands, Japan
29.53°N, 129.72°E, summit elev. 799 m
Local time = GMT + 9 hrs

This small island is 200 km SSW of Kyushu (Japan's third-largest island) and 1160 km SW of Tokyo. It's first historic eruption was in 1813, and it has displayed virtually continuous Strombolian activity since 1956.

Generic names such as "jima" or "shima" ("island") are treated in different ways by different sources (and in different parts of the original *SEAN Bulletins*). Here they are generally hyphenated but not capitalized.

General Reference: Hirasawa, K. and Matsumoto, H., 1983, Volcanic Geology of Suwanose-jima, the Tokara Islands, Kagoshima Prefecture; *Bulletin of the Volcanological Society of Japan*, v. 28, p. 101-115.

6/76 (1:9) Mild Strombolian eruptions continued, with smoke columns reaching heights of 100-1000 m, rumblings, and occasional ash falls.

Information Contact: Japan Meteorological Agency (JMA), Tokyo.

9/76 (1:12) TOA (domestic) airline pilots noted unusually high "smoke" columns from On-take summit crater in July (table 8-1). [This crater name, used in original SEAN reports (and *Volcanoes of the World*), should not be confused with the volcano of the same name on Honshu (see below).]

Information Contact: Same as above.

5/77 (2:5) Frequent explosions from the summit crater occurred 26-29 April. Considerable ash fell nearby from eruptive clouds that rose 3000 m above the vent. Explosion sounds and air shocks were often detected and glow was seen on 1 or 2 occasions.

Information Contact: Same as 1:12.

6/77 (2:6) Strombolian activity continued into May, but explosions were less frequent.

Information Contacts: JMA, Tokyo; D. Shackelford, CA.

8/77 (2:8) Eruptions from On-take summit crater continued through July, with ash clouds rising 2000-3000 m and considerable ash deposition.

Information Contact: Same as above.

10/77 (2:10) Strombolian activity continued during Aug and Sept. An eruption at 1000 on 14 Sept produced clouds that rose 2000 m above On-take.

Information Contact: JMA, Tokyo.

8/79 (4:8) Activity has been observed only once since mid-March (table 8-2), in contrast to the pattern of the preceding 26 months. From Jan 1977-March 1979 there were 1-4 periods of tephra ejection per month, each lasting up to a few days.

Information Contact: Same as above.

9/79 (4:9) "On-take vent exploded on 5 Sept, after a few small explosions on 20 July. Explosive activity continued from about 2000 on 5 Sept to 0100 on 6 Sept. Incandescent columns rose 500 m above the crater and explosions occurred every 10 seconds during the most active stage on the 5th. Explosive sounds (rumblings) were heard at Yaku-shima Island, 90 km NE of Suwanose-jima, and windows and doors on Yaku-shima were rattled by air vibrations. Ash fell in the sea E of Suwanose-jima.

"A village of 65 people lies on Suwanose-jima Island. People there said that the activity on 5 Sept was one of the strongest of the many explosive periods since 1956. No damage was caused by the explosions. Explosions at the volcano had become less frequent this year than before."

Information Contact: Same as above.

1/80 (5:1) A loud explosion from On-take summit crater occurred at about 0700 on 10 Dec, after 3 months of quiescence. The initial explosion produced a 1.5 km-high cloud, and smaller explosions continued for about 3 hours. Ash clouds then decreased, but ejections of incandescent material and reflected glow were seen that night. Activity decreased further to weak ash emission the next day, although glow in the summit crater were seen from the air on 12 Dec. The explosions caused no damage on the island.

Information Contact: JMA, Tokyo.

7/80 (5:7) Aerial observers reported that tephra clouds rose to about 1.8 km above On-take vent on 18-19 May.

Information Contacts: JMA, Tokyo; D. Shackelford, CA.

12/80 (5:12) Strombolian explosions have occurred almost every month since Nov 1956 from On-take, the highest point on Suwanose-jima Island. Eruptive activity has typically lasted from one to a few days. The only damage from the 1980 explosions was caused by minor ashfalls on crops. Between explosive periods, white vapor rose a few hundred meters above the vent.

Information Contact: JMA, Tokyo.

10/81 (6:10) During 1981, explosions from Suwanose-jima have been recorded every month through Aug. Observations were made from 3 km S of the active B crater (fig. 8-1) and from Nakano-shima Island, 26 km NE. An explosion is registered when visual observation of an eruption cloud is correlated with sound of an explosion. Aircraft crews reported 3 eruption clouds: 28 June, cloud height 1.2 km; 17 July, 2.4 km; and 10 Aug, 2.7 km.
Information Contact: Same as 5:12.

11/81 (6:11) After 2 months of quiesence, Crater B was active 25-28 Nov. Ash was ejected on 25 Nov. Explosive sounds were recorded from about 0200 on the 26th. Activity intensified to register 5-6 explosions/minute from

1975		Plume Height	Activity
Oct	4	700	sometimes moderate explosions
	5	1000	sometimes large explosions, ashfall
	6	1500	large explosions, ashfall
	7		explosion
	8	1000	large explosion
	9	500	explosion
Nov	9	1500	large explosion
	10	600	explosion
	19	500	large explosion
Dec	7	200	sometimes small explosions
	21	700	frequent large explosions
	22	400	remarkable incandescent column, frequent large explosions
	23	700	explosion
	24	400	sometimes large explosions
	25	700	explosion
	26	1000	frequent large explosions
	27	700	large explosion
	28	700	moderate explosion
	31	400	

1976		Plume Height	Activity
Jan	8	300	large explosion
	24	1000	ashfall
	25	1000	large explosion, ashfall
	26		ashfall
	27	500	large explosion, ashfall
	28-29	1500	
Feb	14		moderate explosion
	15	800	moderate explosion, ashfall
	16		sometimes explosion sounds, ashfall
	17		sometimes explosion sounds
	18	300	sometimes explosion sounds
	19	700	explosion
	20	500	moderate explosion
	26		explosion
Mar	12	500	frequent large explosions
	13	700	frequent large detonations
	14	500	sometimes moderate explosions
Apr	13		moderate detonation
	14	300	ashfall
	15	300	ashfall, moderate explosion
	16	1000	ashfall
May	10	500	explosion
Jul	3	4200	moderate detonation
	4	3600-4200	TOA airline pilot observed ash cloud
	7	3600	TOA airline pilot observed ash cloud
	23	3000	explosion, large amount of ash cloud, ashfall
	24	3000	explosion, ashfall
Oct	4	500	ashfall
	12	500	ashfall
Nov	1		ashfall

1977		Plume Height	Activity
Jan	11	1200	1 explosion
Mar	19-21	700	ash ejection and ashfall
	24-25		2 explosions
Apr	1-3	1000	1 explosion, ash ejections
	10-11	3000	ash ejections and ashfall
	14-20	1000	2 explosions, ash ejections and ashfall
	27-29	3000	explosions, ash ejections
May	8		1 explosion
	13-18	2000	several explosions, ashfall
	26	1000	1 explosion
Jun	3-5	3000	frequent explosions, incandescent column, ashfall
	21-22	3000	frequent explosions
Jul	9-10	4000	ash ejections, 1 explosion
	23		explosions
Aug	8-10	3000	frequent explosions, incandescent ejecta
Sep	10-15	3000	frequent explosions, ash ejections, ashfall
	18	3000	ash ejections and ashfall
	23	1000	1 ash ejection
Oct	26-30	2000	ash ejections
	14		1 explosion
	19	2000	ash ejections
	6		1 explosion

1978		Plume Height	Activity
Jan	23		3 explosions
Mar	19	500	2 ash ejections
	22-25	1000	ash ejections
Apr	9	1000	ash ejections
	19-22	3000	ash ejections and ashfall
May	1-3	500	ash ejections and ashfall
	31	500	frequent explosions, incandescent column
Jun	1	500	frequent explosions, incandescent column, ashfall
	16-19		frequent explosions
Jul	2-5	700	frequent explosions, incandescent column, ashfall
	19-21	3000	ash ejections and ashfall
Aug	7-10	3000	frequent explosions, incandescent column, ashfall
	24	2000	ash ejections and ashfall
	7	3000	frequent explosions, ashfall
	17		1 explosion
Sep	23-24	700	frequent explosions, incandescence
Oct	10-12	5000	frequent explosions, ashfall
	24-28	3000	frequent explosions, incandescence, ashfall
Nov	6-10	700	explosions, incandescence, ashfall
	19-22	500	explosions, incandescent column, ashfall
Dec	4-5	300	explosions

Table 8-1. Explosions and tephra emissions from Suwanose-jima, Oct 1975-1978. White vapor was emitted during days not listed in the table for 1977-78. Plume heights are given in meters. In this and following tables, data have been combined from monthly *Bulletin* tables and JMA has added data missing from original reports. Courtesy of JMA.

1979		Plume Height	Activity
Feb	6	500	explosions, incandescence
	14-15	500	explosions, ashfall
	19	500	1 explosion
	22		1 explosion
Mar	11-13	500	4 explosions
Jun	5		incandescent column, explosions
Jul	20	500	7 explosions, ash emissions
	21	200	ash emissions
Sep	5-6	2000	frequent, strong explosions, incandescent columns, ashfall to sea
	7	1000	ash emissions
	8	800	ash emissions, incandescence
	9	2000	frequent explosions
	12	1000	10 explosions
Dec	10	1500	explosions, incandescent blocks, reflected glow
	11		ash emissions
	12		"flames" in crater
	18	300	3 explosions

1980		Plume Height	Activity
Feb	5-6	1500	about 10 explosions; incandescent column
Mar	21-22	1000	many explosions
Apr	25-26	1500	explosions: ashfall on inhabited areas
May	13		3 explosions
	18	500	6 explosions; persistent ash ejection
Jun	4-5	500	more than 25 explosions
Jul	16-19		many explosions
Aug	3-8	1500	several tens of explosions; incandescent columns
	21-23	1000	more than 20 explosions; incandescent column
Sep	8-9	1000	more than 1000 explosions
	20		3 explosions
	24-27	2000	more than 1000 explosions
Oct	25-27	500	persistent ash ejection
Nov	8-10	1500	more than 1000 explosions
	29	500	persistent ash ejection
Dec	13		explosions

1981		Plume Height	No. of Explosions
Jan	4-8	100-1000	continuous ash cloud
	29	500	10
	30	300	2
	31	–	2
Feb	12	1000	1
	20	1000	3
Mar	26	500	16
	17	1000	20
Apr	15	–	many
	16	–	16
	26	500	many
May	14	500	10
Jun	12	1500	3
	13	–	2
	15	–	3
	16	–	3
	29-30		many
Jul	1-4	–	many
	5	500	10
	6	–	10
	13	–	3
	14	2000	many
	15	–	6
Aug	9-12		no observation
Nov	26	–	many
	27	1500	many
Dec	13	500	40-50; R
	14	500	many; R
	15	500	30-40; R
	16	500	2
	23	–	2

1982		Plume Height	No. of Explosions
Jan	2	1500	4
	23	1000	2-3/hour
	25	200	10
	26	500	many
	27	300	12
	28	300	3
Feb	3	–	5
	6	200	5
	7	200	1
	8	200	1
	13	1000	1; R
	14-15	500	many; R
	16-17	–	many; R
	18	500	many; R
	19	–	6
	22	1000	3
	23	–	4
	24	–	3
	25	1000	7; R
	26	500	5
Mar	9	1000	many; R
	10	1000	many; R
	11	500	10
	28	–	10; R
	29	–	20-25; R
	30	1000	4
	31	–	6
Apr	1	1000	2
	2	1000	10
	3	–	3
	15	500	18-20; R
	17	500	5-6/hr
	18	–	2-3/hr
	19	–	10; R
May	5	200	1
	6	1000	many; R
	27	1000	1;
June-July		–	quiet
Aug	16	1000	4
	17	500	4
Sep	2	500	5
	16	–	many
	25	200	9
	26	200	3
Oct	5	500	3; R
	6	–	6
	7	–	7
	8	–	10; R
	9	300	7
	17	–	1

Table 8-2 Explosions and tephra emissions from Suwanose-jima, 1979-Oct 1982. A dash indicates no visual observation due to bad weather. An R indicates rumbling. Plume height is given in meters.

1230 to 1700 that day, then declined to about 10/hr. From 1700 on the 27th to 0200 on the 28th about 4 explosions/hr were recorded. On 28 Nov activity was limited to continuous emission of white vapor. The ash and blocks ejected during the activity caused no damage.

Information Contact: Same as 5:12.

11/82 (7:11) No damage was reported from 1982 activity, but there was heavy ashfall on the inhabited, S end of the island on 7 Oct. The eruptive activity typically lasted from 1 to a few days. Between active periods, white vapor rose a few hundred meters above the vent.

Information Contact: Same as 5:12.

1982		Plume Height	Explosions	Observation Time	Note
Nov	3	500	3-4/hr	0500-2200	
	4	500	5-6/hr	0500-2200	
	5	500	5-6/hr	0700-2100	
	6	500	3/hr	0500-2200	
	7	500	1-2/hr	0600-2200	
	23	500	5	0900-1300	
Dec	15	500			A
	17	1000	15-16	1300-2000	
	18	500			A
	19	1000	2-3/min	0900-2100	
	20	1000	1-2/5 min	0600-2100	
	29	500	2	1000-1300	

1983		Plume Height	Explosions	Observation Time	Note
Jan	17	500	4-6/min	0630-0800	
			3-4/hr	0800-2000	
	18	unknown	3-5/hr	0600-2200	
	19	unknown	15-20	0600-1700	
	28	500	25-30	1700-2200	
	29	700	2-3/min	0500-1800	
			4-5/hr	1800-2200	
	30	1000	4-5/hr	0500-1600	
Feb	10	200	a few		
	11	500	1-2/min	1900-2100	
	26	1000	5-6/min	1530-2300	
	27	1000	10	0500-1200	
Mar	4	unknown	1		
	11	1000	2-3/min	1400-2300	
	12	unknown	12	0600-1200	
	13	-	3	0800-1200	
Apr	28	unknown	3-5/min	1400-1600	
	29	500	3	1500-1900	
May	8	200	6	0700-1500	
	11	200	3	1400-1500	
	12	200	3	1300-1500	
Jun	8	1500	7	1400-1900	
	9	1500	4	1500-1800	
	10	500	7	0900-2000	
Jul	5-6	-		2100-0300	I
	27	500	2	1200-1730	
Sep	30	1500	7-9/hr	0500-0900	
Oct	1	1000	17	0500-0900	
	17	unknown	15	0600-1800	
	18	1000	5	0530-1130	
	27	500		1400-2300	R
	28	1000		0500-1330	R
	29	1000		0500- ?	R
Nov	22	500	10-15/min	1000-2200	
	23	500	15-20/min	0920-1100	
			15-20/min	1400-1700	
	25	300	16	0600-1800	
Dec	3	500	2-10/min	1500-2300	
	4	300	6	1000-2200	
	18	500	2-3/min	0400-1500	
			10	1500-2200	
	19	500	15-20/min	0500-0900	
			20	1500-2300	
	20	500	2-3/min	0400-0700	
			20	0700-2100	

Table 8-3. Summary of explosions at Suwanose-jima, Nov 1982–Dec 1983. *Notes:* A = Ashfall; I = Incandescent column; R = Rumbling. Plume heights are given in meters.

1984		Plume Height	Explosions	Observation Time
Jan	2	300	13	0500-2000
	26	300	13	0600-2200
	27	500	5-7/hr	1600-2200
	28	500	30-40/hr	0800-2300
	29	500	7	0500-1200
	30	500	10	0400-1500
Mar	13			
	14-16		4-5/min	
	17			
	18		10	0500-2300
Apr	12	1000	7	2100-2300
	15		7	0730-1300
		-	5-6	1700-2200
	16	-	6	0700-1700
	17	-	12	0500-1000
	18	-	3	0600-1000
May	15		12-13	

1985		Plume Height	Comments
June	28	2000	airplane pilot saw the ash cloud, light ashfall
Aug	3		airplane pilot saw the ash cloud
Sep	14-15		frequent explosion sounds
Oct	1		airplane pilot saw the ash cloud
Nov	17-21		frequent explosion sounds
Dec	5-6		frequent explosion sounds

Table 8-4. Explosions at Suwanose-jima, 1984–85. Plume height is given in meters. May 1984 and 1985 data were added by JMA.

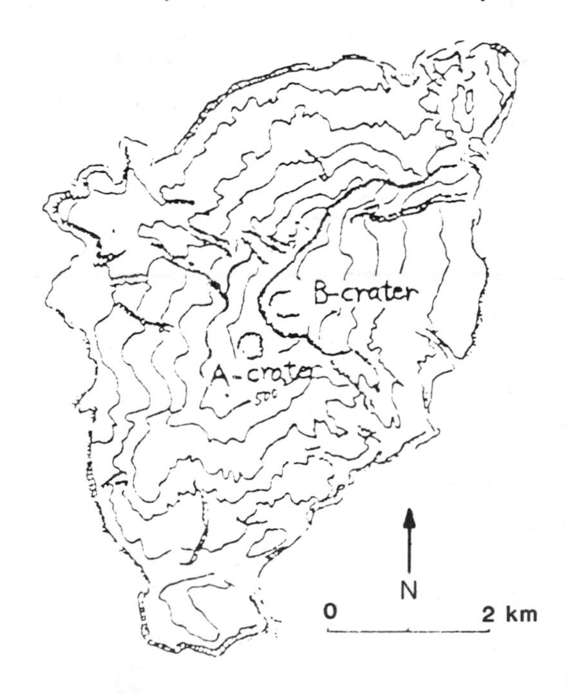

Fig. 8-1. Topographic map of Suwanose-jima Island; the active vent is in Crater B.

3/84 (9:3) Strombolian activity has been recorded almost every month, Nov 1982-Dec 1983 (table 8-3). No damage was reported, although there were often heavy ashfalls on the inhabited area of the island, along the shore 3.5 km SSW of the active vent.

Information Contact: Same as 5:12.

7/85 (10:7) Occasional Strombolian activity continued until May 1984, but no explosion sounds had been reported since June (table 8-4).

During the afternoon of 28 June 1985, an aircraft pilot flying near Suwanose-jima saw a plume rising to an altitude of 2-2.5 km (table 8-4). At Nakano-shima, about 25 km NE of the volcano, slight ashfall was observed, but no explosions were heard.

Information Contact: Same as 5:12.

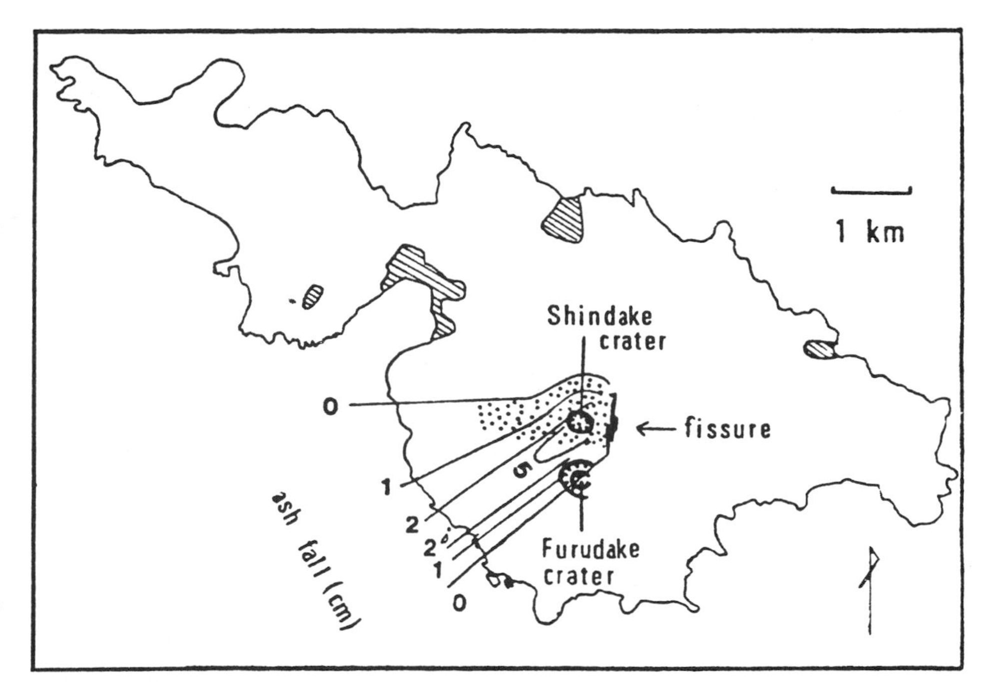

Fig. 8-2. Sketch map of Kuchinoerabu-jima Island, 28 Sept 1980. Ashfall isopachs are in centimeters. Blocks fell in the stippled area. Shaded zones are inhabited. Courtesy of JMA.

KUCHINOERABU-JIMA

Ryukyu Islands, Japan
30.43°N, 130.22°E, summit elev. 649 m
Local time = GMT + 9 hrs

Shin-dake is the summit cone of this small island strato-volcano, midway between Suwanose-jima and Kyushu. It has been the site of all 13 eruptions known in the last 150 years.

Early SEAN reports spelled the volcano differently, reflecting confusion between two different styles of transliteration from Japanese to Roman characters. Here (and in *Volcanoes of the World*) we have have used the Hepburn style (e.g. "Fuji" rather than "Huzi", and "Kuchinoerabu-jima" rather than "Kutinoerabuzima").

4/76 (1:7) An explosion at Shin-dake cone yielded columns of smoke 3000 m high at about [1540] on 2 April. After 30 minutes the height of the smoke column decreased to about 200 m (twice that of usual emissions). Egg-sized volcanic ejecta fell, and ash 2 cm deep was measured in a village 3 km from the crater. The last explosion took place on 3 June 1974. [JMA reported that there was no damage.]

Information Contact: T. Tiba, National Science Museum, Tokyo.

9/80 (5:9) After 4 years of quiet, a brief, weak explosion produced a [2-3] km-high ash cloud on 28 Sept at 0510. Ash fell on the sea [SW] of the volcano, missing the homes of the 12 × 5 km island's 300 residents. Activity after the explosion was limited to emission of white vapor through the end of Sept. Minor ash explosions have occurred in 7 different years since 1966. Kuchinoerabu-jima's last major eruption, in 1933-4, destroyed a village, killing 8 persons and injuring 26.

Information Contact: JMA, Tokyo.

12/80 (5:12) The following, from JMA, supplements the reports of the 28 Sept eruption.

"After the 28 Sept eruption, which lasted for 1/2 hour, no additional eruptions had occurred as of the end of Dec. Eight scientists from Kyoto University, Kagoshima University, and the JMA observatory arrived at the island on 1 Oct, installing portable seismometers at 5 sites. The next day, they climbed to the new fissure, which was 0.6-6.0 m wide and 750 m long, trending N-S near Shindake crater (fig. 8-2), active in historic time. A considerable amount of white vapor was emitted from the fissure.

"The SW sector of the volcano was covered with gray ash, 1 m thick near the fissure and 2 cm thick at the base of the volcano, on the coast. Blocks were scattered N and W of the fissure, the largest block measuring about 2 m in diameter. No essential ejecta were observed. The volume of ejecta was estimated at about 10^5 m^3. Steaming decreased gradually during Oct, and was restricted to 10 small craters on the fissure by mid-Oct.

"Seismicity was relatively weak in Oct and Nov except on 4 and 9 Oct when swarms of small B-type earthquakes were recorded (fig. 8-3). The JMA's seismometer was removed on 15 Nov because the volcano was quiet. People on the island reported no felt earthquakes, and decreasing steam activity through Dec. Life returned to normal for the island's 300 inhabitants soon after the 28 Sept eruption."

Information Contact: JMA, Tokyo.

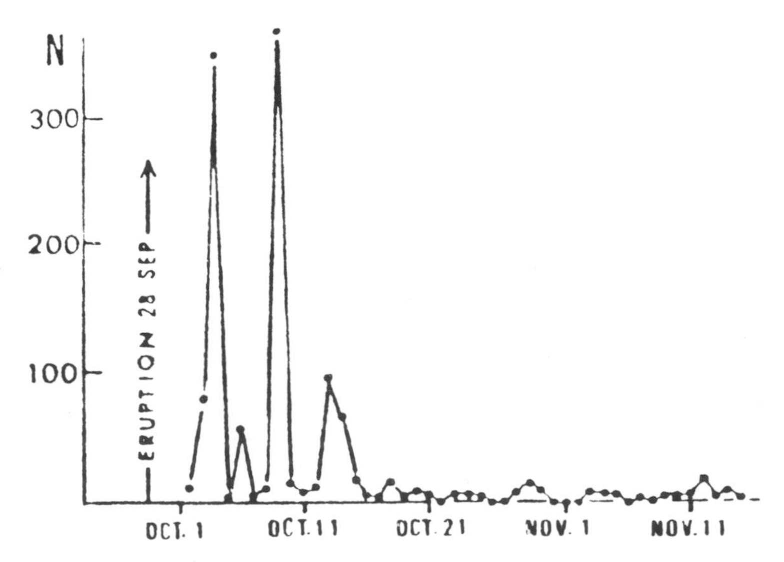

Fig. 8-3. Number of seismic events/day at Kuchinoerabu-jima, 1 Oct – 15 Nov 1980. Courtesy of JMA.

SAKURA-JIMA

Kyushu, Japan
31.58°N, 130.65°E, summit elev. 1118 m
Local time = GMT + 9 hrs

Located in the large, caldera-formed bay at the S tip of Kyushu, Sakura-jima is one of the world's most active volcanoes. Its first historic eruption was in 708 AD, and its continuing eruption (since 1955) is unusually well-monitored. Reports of Sakura-jima's activity have appeared in more *SEAN Bulletins* (84) than any other volcano's during the decade. Monitoring is both by JMA's Kagoshima Meteorological Observatory, where a seismograph was installed at the early date of 1888, and by the Sakura-jima Volcano Observatory of Kyoto University, which maintains one of the world's most accurate deformation monitoring programs (fig. 8-4).

JMA added the following note about their documentation of Sakura-jima's frequent tephra emissions.

"It is very difficult to count explosions at the volcano. Some of the eruptions cause strong air shocks, sounds, and/or seismic shocks, while others emit ash, sometimes a copious amount, without any shocks. JMA counts the explosions using the following rules: (1) an explosive sound is heard, air shock is felt, or ejection of blocks is recognized by observers at the JMA Kagoshima Meteorological Observatory, 10 km W of the volcano, or (2) a typical explosion earthquake is recorded by seismographs on the island, or (3) a typical explosion air shock is recorded by a microbarograph at the Observatory. We have reported to SEAN the number of explosions documented under the above rules."

General References: Aramaki, S., Kamo, K., and Kamada, M., 1988, *A Guide Book for Sakura-jima Volcano;* Kagoshima International Conference on Volcanoes, Kagoshima Prefectural Government, 88 pp.

Kobayashi, T., 1982, Geology of Sakura-jima Volcano: A Review; *Bulletin of the Volcanological Society of Japan*, v. 27, p. 277-292.

10/75 (1:1) During the summer of 1975 Sakura-jima was normally active. There were [27] explosions in Sept (table 8-5). The highest eruption cloud reached a few thousand meters altitude.

Information Contact: JMA, Tokyo.

5/76 (1:8) Two notable explosions occurred during May. Volcanic activity had been increasing during 1976 and dark smoke had been frequently observed since an explosion on 7 April. On 13 May at 0738, an explosion sent a cloud to 2000 m height. This moderate explosion was the 29th of 1976 at the Minami-dake summit crater.

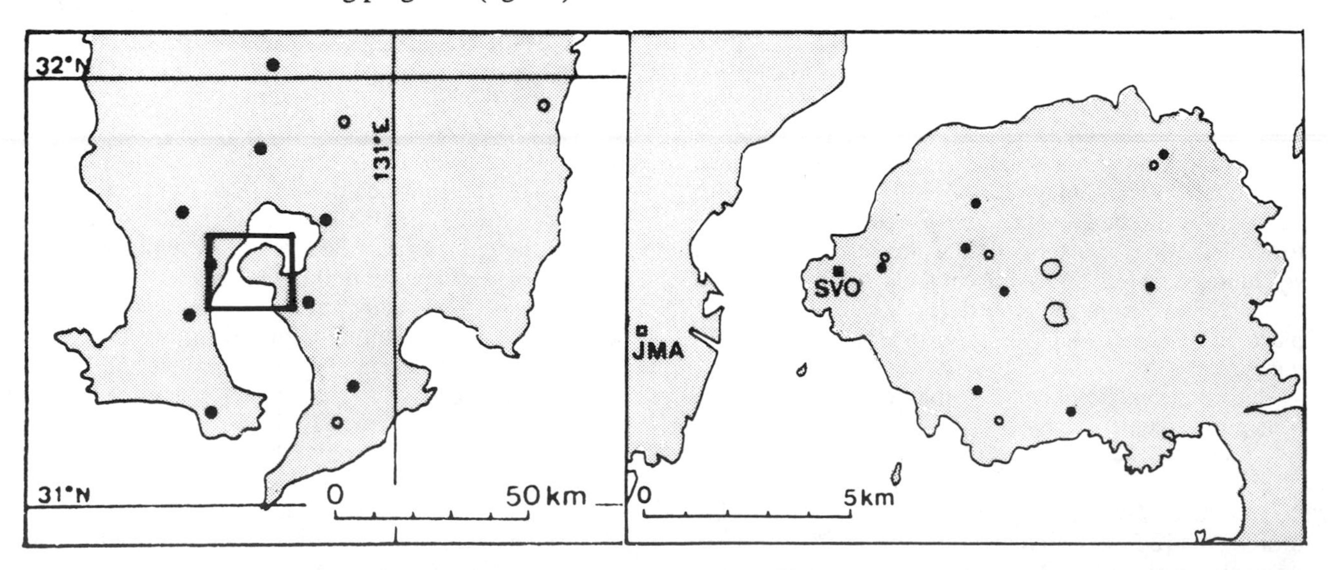

Fig. 8-4. Sakura-jima, *left,* and Southern Kyushu, *right,* showing locations of JMA Observatory, Sakura-jima Volcano Observatory (SVO), and the seismic stations maintained by them.

Large amounts of pumice and lapilli fell onto the E slope of the crater. At Sakura-jima-guchi, 5.5 km ESE of the crater, up to 3 cm of pumice was deposited (as measured on a road), and maximum ejecta size was 25 × 25 × 4 cm. Automobiles and crops in the area were damaged by the ashfall.

On 17 May at 1342, an eruption cloud reached 2700 m height (not 16 miles as reported by UPI in the press) following the 31st and largest explosion recorded in 1976. Tiba reported blocks up to 1 m across falling 2.5 km from the vent. Shimozuru reported cinders 2-6 cm in size at Arimura, 3 km SSE of the crater. At Furusato, 3 km S of the crater, and at a school 7 km SE of the crater, window glass was broken by the strong airshock which measured 0.34 millibars on a microbarograph 10 km W of the crater. Volcanic tremor occurred 32-36 hours before this major eruption, then volcanic earthquakes continued up to 22 hours before its onset.

Both explosions took place at the newly-opened (since 10 Dec 1975) 40 m-diameter crater SE of crater A at Minami-dake. An estimated 170,000 tons of molten lava filled the bottom of crater A from 90 to 50 m depth.

Information Contacts: JMA, Tokyo; D. Shimozuru, ERI, Tokyo; T. Tiba, National Science Museum, Tokyo.

6/76 (1:9) Sixteen explosions were recorded between 1-5 June, but none 6-[19] June.

Information Contact: JMA, Tokyo.

9/76 (1:12) An explosion at 0727 on 5 June deposited a small amount of lapilli (to 1 cm in diameter) at Kurokami, 5 km E of the crater. Maximum June ash cloud height exceeded 3000 m. Ashfalls were frequent during late June at Kagoshima City, 10 km W of the crater. Volcanic earthquakes occurred frequently. July explosions were small, but ash emission was heavy. The maximum height of ash clouds was 3000 m. Kagoshima City continued to experience frequent ash falls. The July 26th explosion occurred at 0357, producing a felt air shock. Vibrations continued to be felt after the explosion at Higashi-Sakura-jima, 3 km SSW of the crater. Detonations, air shocks, cinder falls, [an incandescent column], and frequent ashfalls were noted during Aug. The maximum eruption cloud height 1-10 Sept was 2000 m.

Information Contact: Same as above.

11/76 (1:14) Most of the explosions 10 Sept-28 Oct were small. [Ash cloud] emission was also observed, reaching a maximum of 3000 m [above the crater]. Volcanic earthquakes were frequent. The explosion at 1142 on 10 Sept deposited a large quantity of ash and lapilli at Kagoshima City. The 23 Sept explosion at 2015 was accompanied by a large detonation, an air shock, [an incandescent column], thunder, and rumbling. A large amount of ash and lapilli were deposited on the S flank, 3 km from the crater. Activity on 2 Oct began at 1238 and lasted about 3 hours, causing a heavy ashfall W of the vent. A large detonation and air shock were observed at 0028 on 6 Oct. This explosion caused a cinder fall on the middle flank.

Information Contact: Same as above.

12/76 (1:15) Almost all 1 Nov-20 Dec explosions were small. Strong detonations, air shocks and [incandescent columns] were sometimes observed at the Kagoshima meteorological observatory. Emission of [ash] clouds, some of which rose 3000 m above the crater, was almost continuous, and ashfalls around the crater were frequent. At the time of one of these (21 Oct at 1228), a large quantity of ash moved down the W slope, but it was not possible to confirm it as a nuée ardente. On 4 Nov at 1700, an overflight revealed a red-hot lava mass [or mound] (40 m across) at the bottom of the summit crater. A weak [reflection of] glow, probably caused by the lava, was sometimes seen during Oct.

Information Contact: Same as above.

1975	1	2	3	4	5	6	7	8	9	10	11	12	13	14	15	16	17	18	19	20	21	22	23	24	25	26	27	28	29	30	31	Total
Oct	-	-	-	-	-	-	-	-	-	-	-	2	-	-	-	-	-	-	-	-	1	-	-	1	-	1	1	2	3	2	2	15
Nov	-	-	1	-	2	-	-	-	1	-	1	-	1	1	1	1	1	-	5	1	-	5	-	-	-	1	1	-	1	-		24
Dec	1	-	-	-	-	1	-	1	1	-	-	-	-	-	-	-	-	-	-	-	-	1	2	-	-	-	-	1	-	-	1	9

1976	1	2	3	4	5	6	7	8	9	10	11	12	13	14	15	16	17	18	19	20	21	22	23	24	25	26	27	28	29	30	31	Total
Jan	1	-	-	-	-	-	-	-	1	-	-	-	-	-	-	-	-	-	1	1	-	-	-	-	1	-	-	-	-	-	-	5
Feb	-	-	-	-	-	-	-	-	-	-	-	-	-	-	-	-	-	-	-	-	-	-	-	-	-	1	-	1	2			4
Mar	-	-	1	1	-	1	2	1	-	-	-	-	-	-	-	-	-	-	-	-	-	-	-	-	-	-	-	-	-	-	-	6
Apr	-	-	-	-	-	-	1	-	-	-	-	-	1	-	1	-	-	-	-	-	-	-	-	1	-	-	1	1	2	1		9
May	1	1	-	-	1	-	-	-	-	-	1	-	1	-	-	1	2	-	2	-	-	1	2	-	2	1	2	4	-	5	4	31
Jun	5	1	4	3	3	-	-	-	-	-	-	-	-	-	-	-	-	-	-	2	4	-	-	-	-	-	-	1	-	-		23
Jul	-	-	-	-	-	-	-	-	-	-	-	1	-	-	-	-	-	-	-	-	1	-	2	1	-	1	-	-	-	-	-	6
Aug	-	-	1	-	-	-	-	-	-	1	-	-	-	1	1	-	-	-	-	-	-	-	1	4	1	2	3	2	2	-	-	19
Sep	1	4	-	1	3	-	8	-	1	1	-	-	-	2	-	2	-	1	-	-	-	-	1	-	-	-	1	-	-	1		27
Oct	-	1	-	-	-	1	-	-	-	1	1	3	-	1	2	-	-	-	-	-	2	-	-	-	1	-	-	1	-	-	-	14
Nov	1	1	-	-	-	-	2	2	-	-	-	-	-	1	3	1	-	-	-	-	-	-	1	1	-	-	-	-	1	1		15
Dec	-	1	-	-	1	1	3	-	-	-	2	1	1	1	-	1	-	-	-	-	-	-	-	1	2	1	-	1	-	-	-	17

Table 8-5. Summary of explosions from Minami-dake crater Oct 1975 – Dec 1976. [Oct 1975 – May 1976 data are newly added to this table and errors in the June – Dec 1976 portion have been corrected.]

2/77 (2:2) 6 Dec, 2149: A moderately strong detonation, air shock, [incandescent column] and volcanic [lightning] were reported. Incandescent tephra fell on the middle flank.

11 Dec, 0345: Considerable fall of tephra, including [scoria] (4 cm maximum diameter) and non-scoriaceous lapilli (2 cm maximum diameter) E of the summit. Two cm of ash fell 3 km SE of the summit.

13 Dec, 0106: Lapilli (4.5 cm maximum diameter) broke car windows 3.5 km E of the summit.

[On 6 Dec] a swarm of about 500 volcanic earthquakes occurred during a 3-hour period. . . .

Information Contact: Same as 1:9.

3/77 (2:3) The explosion at 1632 on 31 Jan (table 8-6) deposited a considerable quantity of lapilli [S of the crater]. At 1005 the next day, a strong air shock broke [65 windowpanes at a school in Tarumizu City, about 10 km SE of the crater]. After the 1 Feb explosions, [ash] emission from the crater ceased until 6 March [but white vapor emission continued from 2 Feb until the 6 March explosions]. The number of volcanic earthquakes decreased in Feb and March.

Information Contact: Same as 1:9.

4/77 (2:4) A slight air shock from the 15 March explosion was felt at the Kagoshima Meteorological Observatory.

Information Contact: Same as 1:9.

5/77 (2:5) The 30 April eruption cloud rose 2700 m above the crater and caused a minor ashfall on the SE flank.

Information Contact: Same as 1:9.

6/77 (2:6) Sixteen explosions were recorded 1-16 May. Local earthquakes continued.

Information Contacts: JMA, Tokyo; D. Shackelford, CA.

7/77 (2:7) June explosions were accompanied by air shocks, rumbling of about 10 seconds duration, and explosion sounds. Eruption columns rose to 2700 m from the crater and contained some cinders.

Information Contact: JMA, Tokyo.

8/77 (2:8) Ashfalls, scoria ejection, incandescence, air shocks, and rumbling frequently accompanied the July explosions. Eruption clouds rose more than 2800 m on 3 July, more than 3000 m on 6 July, and about 2700 m on 20 July. . . .

Information Contacts: JMA, Tokyo; D. Shackelford, CA.

9/77 (2:9) Ash clouds rose to a maximum of 2800 m above the crater in Aug, but ashfalls were minor. Falling scoria from an explosion at 1317 on 29 Aug caused [grass fires at 8 points] on the SW flank. [The fires were extinguished in a few minutes.]

Information Contact: JMA, Tokyo.

10/77 (2:10) Ash clouds rose to a maximum of 2100 m above the crater during Sept. Explosions were accompanied by air shocks and scoria ejection.

Information Contact: Same as above.

11/77 (2:11) Maximum ash cloud height during Oct was 2500 m above the crater. Some incandescent material was ejected and reflected glow was occasionally observed. The air shock from an explosion at 0347 on 30 Nov broke [102 windowpanes] in villages about 3 km S of the summit. Tephra started grass fires, but these were quickly extinguished. No [injuries] resulted from the explosion, the 21st to occur in Nov.

Information Contacts: JMA, Tokyo; Japanese Press.

12/77 (2:12) . . . Maximum Nov ash cloud height was 2200 m above the crater, on the 2nd. Volcanic thunder was heard on more than 20 occasions during the month.

Information Contact: JMA, Tokyo.

1/78 (3:1) Tephra from explosions at 1220 on 8 Dec, 1834 on 20 Dec, 0621 and 1507 on 22 Dec, and on 27 Jan, caused [grass] fires. A 300-m column of incandescent ejecta was observed at 0534 on 1 Dec, and reflected glow was seen between 0322 and 0325 on 10 Dec. The air shock from an explosion at 2140 on 8 Dec broke 3 [windowpanes] in a nearby village. Ash clouds rose 2000 m above the crater on 6, 8, and 22 Dec.

Information Contact: Same as above.

2/78 (3:2) Tephra from an explosion on 25 Dec cracked the

1977	1	2	3	4	5	6	7	8	9	10	11	12	13	14	15	16	17	18	19	20	21	22	23	24	25	26	27	28	29	30	31	Total
Jan	1	2	1	1	3	-	-	-	-	-	-	-	1	-	-	-	-	-	-	-	-	-	-	-	-	-	-	-	-	-	1	10
Feb	2	-	-	-	-	-	-	-	-	-	-	-	-	-	-	-	-	-	-	-	-	-	-	-	-	-	-	-				2
Mar	-	-	-	-	-	1	-	-	-	-	-	-	1	-	1	-	-	-	-	-	-	-	-	-	-	-	-	-	1	-	-	4
Apr	-	-	1	-	-	-	-	-	-	-	-	-	-	-	-	-	-	-	-	-	-	-	-	-	-	-	-	-	1	1		3
May	2	1	5	1	1	-	1	-	1	-	-	1	2	-	-	1	-	-	-	-	1	-	-	-	3	-	1	1	1	2	4	29
Jun	-	5	2	3	3	1	1	1	-	1	1	-	1	-	-	-	-	-	-	-	-	-	1	1	-	-	-	1	-	-		22
Jul	1	1	5	2	-	2	-	-	2	-	1	-	4	2	2	-	3	-	-	2	-	-	-	-	-	-	-	1	-	-	-	28
Aug	-	-	1	-	-	4	-	-	-	-	1	5	2	-	-	2	1	6	-	-	-	2	-	-	1	1	1	1	3	2	2	35
Sep	2	-	-	-	1	-	2	1	-	-	-	-	-	-	3	-	1	1	1	-	3	2	-	2	1	-	2	-	-	1		23
Oct	-	-	-	-	1	-	-	-	1	2	-	-	-	2	2	4	-	-	-	3	1	-	-	-	1	1	1	-	-	-	-	19
Nov	3	2	1	-	1	-	1	-	-	-	-	-	-	2	-	1	2	-	-	-	1	-	2	-	1	-	1	-	2	1		21
Dec	1	-	-	-	-	1	-	2	1	1	-	1	-	-	-	2	-	-	3	2	1	5	4	1	2	-	-	-	-	-	-	27

Table 8-6. Summary of explosions from Minami-dake crater during 1977. [Monthly data from this and following tables are summarized in table 8-20, covering the full eruption from 1955 through 1985.]

windshield of an All Nippon Airways aircraft passing over the Hayato area, about 20 km NE of the crater. Maximum Jan ash cloud height was 1900 m above the crater on the 23rd. Incandescence, scoria ejection, air shocks, and rumbling accompanied the activity. Aerial observations in mid-Jan confirmed the continued growth of the lava mass present in the crater since Sept. Several earthquake swarms . . . were recorded during Dec and Jan.

Information Contacts: JMA, Tokyo; D. Shackelford, CA.

3/78 (3:3) February explosions were accompanied by air shocks and rumbling. Scoria from explosions at 0024 on 8 Feb and 0218 on 26 Feb caused [grass] fires. The maximum Feb cloud height was 1700 m above the crater, at 0850 on the 25th.

Information Contact: JMA, Tokyo.

5/78 (3:5) Powerful explosions from the summit crater of Minami-dake have become more frequent since summer 1977 (tables 8-6 and 8-7). Beginning in Aug, explosions have been preceded by earthquake swarms lasting several days. . . . This pattern has often occurred 4-5 times/month and has enabled scientists at the JMA's Sakura-jima Observatory to [forecast] the explosions. The frequent property damage that has occurred near the volcano since last summer continued in March and April. Many windowpanes and a car windshield were broken by airshocks and tephra during March. Incandescence was also observed during March and April.

Information Contact: Same as above.

6/78 (3:6) The air shock from the 22 May explosion broke 5 windowpanes. Earthquake swarms . . . occurred in mid and late May, correlated with an increase in the number of explosions during the same period. A swarm of exceptionally long duration (16 hours) on 30 May was followed by 4 explosions on the 31st.

Information Contact: Same as above.

7/78 (3:7) About 20 earthquakes . . . were recorded in June. . . . Consequently, JMA scientists believe that a large volume of magma remained on or just under the crater bottom in early June. Reuters reported a loud explosion on 21 July, producing a 1600-m eruption column, and 2 explosions on 23 July, throwing incandescent ejecta 2500 m above the vent. UPI reported that strong winds from a typhoon spread about 2.5 cm of ash within 5 km of the summit after an explosion on 31 July.

Information Contacts: JMA, Tokyo; Reuters; UPI; D. Shackelford, CA.

8/78 (3:8) During July, 25 explosions from the summit crater of Minami-dake were recorded. Large quantities of tephra (blocks, lapilli, and ash) were ejected by 3 explosions on 30-31 July. Strong [SE] winds from a typhoon carried the ejecta to [an inhabited area] 10 km [NW of] the crater. Three persons were slightly injured, and [windowpanes] were broken in 62 houses and 45 cars. More than 50 earthquake swarms, each lasting several hours, occurred during July. . . .

Information Contact: JMA, Tokyo.

9/78 (3:9) . . . [There were] 32 recorded explosions from the summit crater of Minami-dake during Aug. The explosions produced ash clouds that rose about 2000 m above the crater. Emission of . . . ash clouds was continuous between explosions. Ash fell every day around the volcano, primarily to the NW, causing slight damage to crops, electric wires, and homes over a broad area. Total Aug ashfall was estimated at a few cm (20 kg/m^2) in a village 5 km from the volcano.

Information Contact: Same as above.

11/78 (3:11) The summit crater of Minami-dake exploded 25 times in Sept and 15 times in Oct. Since June, dense ash clouds have frequently been emitted from Minami-dake between recorded explosions. Ashfalls near the volcano were almost continuous June-Oct, damaging . . . crops, and impeding traffic. However, there was no damage from coarse tephra as in late July nor from explosion air shocks, which broke [windowpanes] in late May.

Information Contact: Same as above.

12/78 (3:12) Only one explosion from the summit crater of Minami-dake was recorded during Nov, on the 15th. The ash ejection that had frequently occurred between explosions since the spring ended about 25 Nov. Activity has been limited to steam emission since then.

1978	1	2	3	4	5	6	7	8	9	10	11	12	13	14	15	16	17	18	19	20	21	22	23	24	25	26	27	28	29	30	31	Total
Jan	-	1	-	-	-	-	1	-	2	-	-	-	-	-	-	-	-	1	1	1	1	-	2	2	-	-	1	1	2	1	-	17
Feb	-	-	-	-	1	1	-	1	-	-	-	-	-	-	-	-	-	-	1	-	-	-	-	1	1	1	-	-				7
Mar	-	-	-	2	1	1	1	1	1	5	1	1	2	1	1	-	-	1	-	-	-	-	-	-	1	1	1	1	-	1	1	25
Apr	1	-	1	-	-	-	-	-	-	-	-	-	-	2	1	-	1	-	-	-	-	-	3	-	-	1	1	2	1	-		14
May	-	-	-	-	-	-	-	-	-	-	1	-	1	-	-	-	-	2	-	-	-	1	2	1	-	1	4	-	3	-	4	20
Jun	2	4	-	4	1	3	2	-	2	-	-	1	-	1	-	-	-	2	1	-	-	1	2	1	2	1	4	1	2	2		39
Jul	-	1	2	1	-	-	-	-	-	1	-	-	-	-	1	-	1	-	-	2	3	1	5	-	-	-	1	-	1	2	3	25
Aug	-	-	-	-	3	1	-	-	-	1	1	2	2	-	2	4	1	1	2	-	-	5	1	-	-	-	-	1	-	1	4	32
Sep	-	-	3	-	1	2	1	1	-	3	-	1	-	1	1	-	1	-	-	-	1	3	5	-	-	-	1	-	-	-		25
Oct	-	2	2	3	-	1	1	1	1	-	-	1	-	-	-	-	-	1	1	-	-	-	-	-	1	-	-	-	-	-	-	15
Nov	-	-	-	-	-	-	-	-	-	-	-	-	-	-	1	-	-	-	-	-	-	-	-	-	-	-	-	-	-	-		1
Dec	1	-	-	1	-	-	-	-	-	-	-	-	-	-	1	-	-	-	-	-	1	1	-	-	2	-	1	-	1	-	2	11

Table 8-7. Summary of explosions from Minami-dake crater during 1978.

Information Contact: Same as 3:9.

Further Reference: Kamo, K., 1979, The Recent Activity of Sakura-jima Volcano; *in* Report on Volcanic Activities and Volcanological Studies in Japan for the Period from 1975 to 1978; *Bulletin of the Volcanological Society of Japan*, v. 24, no. 4, p. 26-34.

1/79 (4:1) Eleven explosions from the summit crater of Minami-dake were recorded in Dec. The number of explosions in 1978 was 231, little changed from 223 in 1977. Lapilli ejected on 4 Dec cracked 2 windshields of All Nippon Airways airplanes. Similar damage to aircraft above Sakura-jima occurred 8 April 1975 and 25 Dec 1977.

Information Contact: Same as 3:8.

2/79 (4:2) Fifteen explosions from the summit crater of Minami-dake were recorded during Jan (table 8-8). On 5 Jan, lapilli broke [windshields] in 7 cars. Ash emission was observed between explosions, but not as frequently as in the autumn of 1978.

Information Contact: Same as 3:8.

3/79 (4:3) Sixteen explosions occurred from the summit crater of Minami-dake during Feb. Ash emission between explosions remained infrequent, as it has since Dec. Yosihiro Sawada reports that high-amplitude volcanic tremor ended in Nov 1978, and shallow B-type earthquakes began to be recorded. An earthquake swarm in mid-Dec preceded the emplacement of a new incandescent lava [mound] on the [crater] floor of Minami-dake, after which explosions became more frequent.

Information Contacts: JMA, Tokyo; Y. Sawada, MRI, Tokyo; D. Shackelford, CA.

4/79 (4:4) Seven explosions from the summit crater of Minami-dake were recorded in March. Tephra from the explosions caused no damage.

Information Contact: JMA, Tokyo.

5/79 (4:5) Seven explosions from the summit crater of Minami-dake were recorded in April. Reflected glow was observed above the crater on 22 April. Earthquake swarms . . . were frequently recorded in April, indicating that the andesitic lava [mound] in the bottom of the crater was increasing in volume. The strongest explosions of the current eruption have taken place during periods when the lava dome was most voluminous.

Information Contact: Same as above.

6/79 (4:6) No explosions from the summit crater of Minami-dake were recorded during May, the first time in 6 years that a month without explosions has occurred there [but see 4:9]. Sakura-jima's current eruption began in Oct 1955.

Information Contact: Same as above.

7/79 (4:7) For the second consecutive month, no explosions from the summit crater of Minami-dake were recorded in June. The last explosions occurred on 30 April. Sakura-jima's current eruption has included several quiet phases, the most recent from Dec 1972 to April 1973.

Information Contact: Same as above.

8/79 (4:8) No explosions have occurred from Sakura-jima since [2 were recorded on] 30 April, a long quiet phase for this eruption.

Information Contact: Same as above.

9/79 (4:9) The following is a report from Manabu Komiya.

"Sakura-jima exploded on 19 Aug after 110 days of quiescence, producing an ash cloud 1 km high. It was the 46th explosion of this year. Five more explosions occurred in early Sept, but none caused any damage.

"The fact that no explosion had occurred in 110 days does not immediately suggest decreasing volcanism, because intermittent ash emission without explosive shocks had occurred through the explosion-free period, as frequently as before. Daily ash emission often caused ashfalls at cities and towns near the volcano. Volcanic gas that flowed down the flanks damaged vegetation, adding to the damage caused by falling ash.

"Reflected glow above the crater was often observed at night during July and early Aug. Swarms of B-type earthquakes . . . (1 burst consisted of hundreds of earthquakes) were recorded a few times/month in June and Aug. These facts indicate that the lava mound on the crater floor persisted or grew during this period. An aerial inspection on 30 July revealed a large mound that had reached a volume of 7×10^5 m^3, twice that of May."

Information Contact: Same as above.

1979	1	2	3	4	5	6	7	8	9	10	11	12	13	14	15	16	17	18	19	20	21	22	23	24	25	26	27	28	29	30	31	Total
Jan	4	2	-	-	2	-	1	-	1	-	-	1	-	-	-	-	-	-	1	-	-	-	-	-	-	-	1	-	-	-	2	15
Feb	-	-	-	-	-	-	-	-	1	-	-	-	-	-	-	-	1	1	1	1	4	1	1	1	1	1	1	1				16
Mar	1	1	2	-	-	-	-	1	-	-	-	-	-	-	-	-	1	-	-	-	-	-	-	-	-	-	-	-	-	-	1	7
Apr	-	-	-	-	-	1	-	-	-	-	1	-	-	1	-	-	-	-	-	-	-	-	-	1	-	1	-	-	-	2		7
May	-	-	-	-	-	-	-	-	-	-	-	-	-	-	-	-	-	-	-	-	-	-	-	-	-	-	-	-	-	-	-	--
Jun	-	-	-	-	-	-	-	-	-	-	-	-	-	-	-	-	-	-	-	-	-	-	-	-	-	-	-	-	-	-		--
Jul	-	-	-	-	-	-	-	-	-	-	-	-	-	-	-	-	-	-	-	-	-	-	-	-	-	-	-	-	-	-	-	--
Aug	-	-	-	-	-	-	-	-	-	-	-	-	-	-	-	-	-	-	1	-	-	-	-	-	-	-	-	-	-	-	-	1
Sep	-	-	1	-	-	-	1	1	1	1	-	-	-	1	-	-	-	-	-	1	2	2	-	-	-	-	-	-	-	2		13
Oct	-	-	-	1	-	-	-	1	2	1	2	2	1	4	1	4	1	2	-	-	-	-	1	-	-	1	1	-	-	1	-	26
Nov	-	1	1	1	-	-	-	1	1	3	-	-	-	2	1	1	-	1	1	3	4	1	-	2	-	-	1	2	1	-		28
Dec	2	-	-	2	-	-	-	-	-	1	1	1	1	2	2	1	3	1	1	-	1	1	4	5	1	3	2	1	-	-	-	36

Table 8-8. Summary of explosions from Minami-dake crater during 1979.

10/79 (4:10) Thirteen explosions were recorded in Sept. The highest eruption cloud rose 2.8 km above the crater, on 8 Sept (table 8-9). No damage from any of the explosions was reported.

Information Contact: Same as 4:4.

11/79 (4:11) Explosions recorded at the JMA's Kagoshima Observatory, 11 km from Sakura-jima, were about twice as frequent in Oct and Nov as they were in Sept. Typically, an "explosion" consisted of a weak shock, registered both seismically and sonically, followed by ash ejection. Ash emission without an accompanying explosion occurred more often than the explosion-triggered events.

[The explosion at 1400 on 10 Nov was followed by about 20 minutes of tephra emission and continuous tremor.] Lightning was frequent in the tephra cloud, which deposited 2 cm of ash during a rainstorm at Furosato, 3 km from the crater. Lapilli cracked a car windshield and 2 cars collided after skidding on the wet ash. The windshields of 2 domestic airliners were cracked as they flew into an eruption cloud near Sakura-jima at 0801 and 0805 on 18 Nov, about 20 minutes after a recorded explosion. In both cases, damage was restricted to the outermost of 3 sheets of glass, and the planes landed safely. Another 2 cm of ash fell at Furosato after this explosion.

Information Contact: Same as 4:4.

12/79 (4:12) Frequent explosions and ejections of dense ash clouds continued through Dec. Between 28 Nov and 28 Dec, 36 explosions were recorded at the JMA's Kagoshima Observatory. The 110-day explosion-free period of 1 May-18 Aug dropped the number of recorded explosions in 1979 to [149], from 231 in 1978 and 223 in 1977. However, the monthly totals for Oct-Dec 1979 are higher than the 1977-8 averages. The windshield of a domestic YS-11 airliner was cracked by tephra at 1.5 km altitude at 1749 on 24 Dec, 9 minutes after an explosion from Sakura-jima (table 8-10). The plane landed safely at Kagoshima airport 7 minutes later.

Information Contact: JMA, Tokyo.

1/80 (5:1) Thirty-six explosions were recorded in Dec by the JMA's Kagoshima Observatory, bringing the total number in 1979 to 149. Tephra cracked airplane windshields on 18 and 24 Dec; both planes landed safely. In Jan, 10 explosions had been recorded by the 27th, none of which caused any damage. Sakura-jima's current eruption began 13 Oct 1955 after 5 years of quiescence. A large eruption in 1946 extruded $8.3 \times 10^7 m^3$ of lava, and subsequent weak explosions continued for several years.

Information Contact: Same as above.

4/80 (5:4) Twenty explosions from Sakura-jima were recorded in Feb and 10 in March (table 8-11), none of which caused any damage. Ash emission without explosions occurred less frequently in Feb and March than

1979	Time	Cloud Ht (km)
Aug 19		1.0
Sep 3		-
7		1.2
8		2.8
9		0.8
10		-
Oct 4	0555	1.2
8	1025	1.2
9	0907	
10	0900	2.0
11	0240	
	1408	2.0
12	0012	
	2116	
13	0310	
14	0623	1.6
	1321	
	1957	
	2247	
15	1403	1.2
16	0550	1.3
	0751	1.5
	1658	
	1846	
17	2245	
18	0509	
	1559	
23	1346	2.5
26	1503	2.7
27	0204	
30	1655	2.0

1979	Time	Cloud Ht (km)	Activity
Nov 2	0246		
3	1038		
4	0804		
8	0807	1.5	
9	1054		
10	0148		
	1400		
	1815		
14	0410		
	1513	2.3	
15	1623		
16	0156		
18	0742		2 airplane windshields cracked
19	1733	1.8	
20	1514	2.3	
	1522	2.7	
	1536	2.0	
21	1458	1.6	
	1646	1.7	
	1806		
	1920		incandescent cinders
22	2106		
24	0834	1.0	
	1955		incandescent cinders, lightning
27	2237		
28	0209		
28	1305	2.0	lightning
29	0142		

Table 8-9. Explosions recorded at the Kagoshima Observatory, 19 Aug–29 Nov 1979.

has been usual over the past 5 years. Two bursts of B-type earthquakes were recorded in the 2-month period. About 1000 events occurred in 11 hours on 13 Feb and about 500 in 5 hours on 15 March. Both of these bursts, thought by JMA geologists to be caused by magma rising in the vent, were followed by increased explosion activity. In addition, columns of incandescent tephra were ejected after the Feb earthquakes. On 14 Feb, 2 columns rose 200 m at 0438 and a 400-m column was ejected at 2137.

Activity beginning late 15 Feb was typical of the more prolonged eruptive periods that have characterized Sakura-jima since explosions resumed in Aug 1979.

1979	Time	Plume Height	Activity
Dec 1	0210		
	0435		
4	0824	800	
	1630	1000	lightning
10	0352		
11	0147	1500	
12	0516		
13	0942	1500	
14	1512	1400	
	1547	2400	
15	1247	1300	
	1944		
16	0002		
17	0038		
	0153		lightning, 3 incandescent columns, rumbling
	2219		incandescent column
18	2118		reflected glow after explosion
19	2307		
21	0135		
22	0119		
23	1138		
	1540		rumbling
	1812		
	2007		rumbling
24	1710	1700	
	1740	2300	lightning
	2218		two 400 m incandescent columns, lasting 20 sec; much lightning
	2227		much lightning
	2231		much lightning
25	1059	800	lightning
26	0827	2000	
	1033	2100	
	1049	2100	lightning
27	0733	1000	
	0915	1700	
28	1151	2000	

1980	Time	Plume Height
Jan 8	1400	2000
10	0929	2500
	1049	2000
18	1322	1200
19	0550	
22	2335	
23	2216	
24	1656	1000
	1707	1300
27	0828	1600

Table 8-10. Explosions recorded at Kagoshima Observatory, 1 Dec 1979 – 27 Jan 1980. Plume heights are given in meters.

Soon after a loud explosion at 2241, 2 columns of incandescent tephra rose 200 m above the summit. Ejection of incandescent blocks was almost continuous until about midnight, then occurred every few minutes until 0110 on the 16th. Lapilli up to 3 cm in diameter fell on inhabited areas, but caused no damage. Much lightning was seen in the eruption clouds. Strong, continuous air vibrations rattled windows in towns at the base of the volcano. A seismograph recorded strong tremor until about 0100 on 16 Feb. By about 0300, eruptive activity had declined, and vapor emission or weak ash ejection continued until the 9 March explosion.

Information Contact: Same as 4:12.

5/80 (5:5) A total of 48 explosions from Sakura-jima were recorded in April, the largest monthly figure in 5 years. Frequent explosions continued into early May. Four of the April explosions produced ash clouds higher than 2 km, and 2 ejected incandescent columns. No damage was reported.

Information Contact: Same as 4:12.

6/80 (5:6) The period of unusually frequent explosions at Sakura-jima that began in late March ended on 7 June. In May, 69 explosions were recorded, the third highest monthly total since the eruption began in October 1955. Most of the May and June explosions produced ash clouds that rose 1-2 km above the crater; the highest clouds were 2.5 km on 17 May and 2.7 km on 7 June. A 100 m-high incandescent column was observed at night on 1 May.

Ash emission without an explosion [shock] has also more frequent than usual from late March through early June. Ash often fell on Kagoshima City, causing various minor disruptions. Fallen ash derailed a streetcar on 2 May, and electric power for 3700 homes was cut off on 8 May after wet, sticky ash caused numerous short circuits. During a heavy rainfall on 12 May, a debris flow damaged a concrete bridge at the foot of the volcano. Ash emission declined in mid and late June.

Information Contact: Same as 4:12.

7/80 (5:7) After a decline from 69 explosions in May to 12 in June, activity remained at a similar level until late July, when explosions became stronger and more frequent. The last July explosion was also the month's largest. A tephra cloud rose about 3 km, and [blocks fell onto] the lower half of the volcano. Ash fell to the NE, reaching [the city of] Miyakonojo, 40 km away, 1 hour after the explosion. No damage was reported.

Information Contact: Same as 4:12.

8/80 (5:8) The number of recorded explosions increased from 16 in July to 34 in Aug. The highest Aug eruption cloud rose 2 km above the summit, on the 2nd. Ash frequently fell NE of the volcano but no damage was reported. . . .

Information Contact: Same as 4:12.

9/80 (5:9) During Sept, 21 explosions were recorded. The highest Sept ash cloud rose 1800 m on the 7th. Blocks fell on the flanks on several occasions. A 200 m-high incandescent column lasted for 10 seconds on 30 Sept. No damage was reported from the Sept activity.

Information Contact: Same as 4:12.

11/80 (5:11) The number of recorded explosions declined from 21 in Sept to 4 in Oct, then increased to 21 in Nov. The highest Oct ash cloud reached 2.0 km, on the 1st. None of the Oct activity caused any damage. Lapilli from the largest Nov tephra cloud, which rose 2.5 km on the 8th, broke 5 car windshields. The air shock from the 28 Nov explosion broke 2 [windowpanes] in a hotel at the base of the volcano. No injuries were reported.

Information Contact: Same as 4:12.

12/80 (5:12) Explosions from the summit crater of Minami-dake were continuing at the end of 1980. The 9 explosions in Dec brought the year's total to 276, the largest number since 362 were recorded in 1974. The highest Dec ash cloud rose 1.8 km on the 3rd. Airshocks and tephra fall from the explosions broke [windowpanes] in buildings, automobiles, and aircraft; disrupted traffic; and interrupted electric power on occasion in 1980; but caused no injuries.

Information Contact: JMA, Tokyo

1/81 (6:1) A burst of B-type earthquakes that began at 0200 on 18 Jan prompted the JMA observatory at Sakura-jima to issue an explosion warning at 0930. Reflected glow was seen over the summit that night. Four strong explosions occurred during the next 2 days (table 8-12). Each of the first 3 produced a 200 m-high incandescent column. The fourth, strongest explosion at 1632 on 20 Jan ejected an incandescent block that formed a 1.3 m-diameter crater when it fell near an inhabited area. Similar occurrences of B-type earthquake bursts, reflected glow of the lava mound in the crater, and explosions were observed in July and Aug 1979. None of the Jan explosions caused any damage.

Information Contact: Same as above.

1980	1	2	3	4	5	6	7	8	9	10	11	12	13	14	15	16	17	18	19	20	21	22	23	24	25	26	27	28	29	30	31	Total
Jan	-	-	-	-	-	-	-	1	-	2	-	-	-	-	-	-	-	1	1	-	-	1	1	2	-	-	1	1	-	1	-	12
Feb	-	-	-	1	1	-	2	2	3	3	-	1	-	4	3	-	-	-	-	-	-	-	-	-	-	-	-	-	-			20
Mar	-	-	-	-	-	-	-	-	1	-	-	-	-	-	-	1	1	1	-	1	3	-	-	-	-	-	-	-	-	1	1	10
Apr	5	1	-	-	-	2	1	-	1	1	2	3	1	5	3	2	1	3	1	-	-	1	1	1	4	6	-	1	1	1		48
May	4	1	2	7	1	-	-	6	-	-	3	6	5	-	1	3	6	3	2	2	-	3	5	1	-	2	2	1	2	1	-	69
Jun	3	1	-	2	-	3	1	-	-	-	-	-	-	-	-	-	-	1	-	-	-	-	-	1	-	-	-	-	-	-		12
Jul	-	1	-	-	-	-	-	-	-	1	-	-	-	-	-	-	1	-	-	1	-	-	-	-	2	2	2	-	-	3	3	16
Aug	-	1	1	1	-	-	1	-	7	2	1	-	1	-	-	2	1	2	-	-	1	2	4	-	-	3	3	-	1	-	-	34
Sep	-	-	-	-	-	-	8	2	1	-	-	1	-	-	-	-	-	-	-	-	-	-	-	-	1	-	1	3	-	4		21
Oct	1	-	-	-	-	-	-	-	-	-	-	-	-	-	-	1	-	-	1	-	-	-	-	1	-	-	-	-	-	-	-	4
Nov	2	1	3	-	-	-	2	2	1	1	-	-	2	-	-	-	-	-	-	-	-	1	2	3	-	-	-	1	-	-		21
Dec	-	-	1	-	2	-	-	-	-	1	-	-	2	-	-	-	-	1	-	1	-	-	1	-	-	-	-	-	-	-	-	10

Table 8-11. Summary of explosions from Minami-dake crater during 1980.

1981		Time	Ground Shock Amplitude (mm)	Plume Height (km)	Air Shock Amplitude (mBars)	Activity and Damage
Jan	1	0621	18	1.5	0.24	Sound heard at Miyakonojo City, 40 km NE
	5	2110	12		0.05	Sound heard at Miyakonojo City, 40 km NE
	6	0747	11		0.11	Sound heard at Miyakonojo City, 40 km NE
	14	1648	20	1.3	0.30	Sound heard at Miyakonojo City, 40 km NE
	19	0530	25		0.30	Incandescent column 200 m high; glow seen at 2200, 18 Jan
		1714	17	0.5	0.15	Incandescent column 200 m high; sound heard at Miyakonojo
	20	0325	12		0.17	Incandescent column 200 m high
		1632	15	2.1	0.25	Sound heard at Miyakonojo
	25	2259	13		0.29	Sound heard at Miyakonojo
Mar	20	0521	6		0.19	Two simultaneous columns from Minami-dake craters A and B
Jun	14	0419	14		0.45	Rumbling for 30 seconds
Sep	13	1401	5	2.6	0.14	Mud-like material ejected
Oct	2	0444	10		0.26	Two successive incandescent columns 100 m high
Nov	3	0440	15		0.27	Weak rumbling for 10 seconds
	8	1558	13	1.9	0.19	Sound heard at Miyakonojo City
	14	0450	4		0.17	Incandescent column 100 m high
	16	1528	10	3.0	0.30	Large ash cloud; large air shock
	21	1322	10	2.5	0.19	Lapilli broke windshields on a few cars, caused forest fire
	22	2113	6		0.25	Incandescent column 100 m high
	29	0558	3	2.3	0.17	Incandescent column 100 m high, rumbling for 20 seconds
	30	0608	3		0.25	Incandescent column 200 m high
Dec	3	1336	13	3.1	0.29	Largest ash cloud in 1981
	9	1836	9	2.0	0.24	Incandescent column 200 m high
	22	0620	10	1.0	0.24	Incandescent column 200 m high

Table 8-12. Notable explosions at Sakura-jima, 1981.

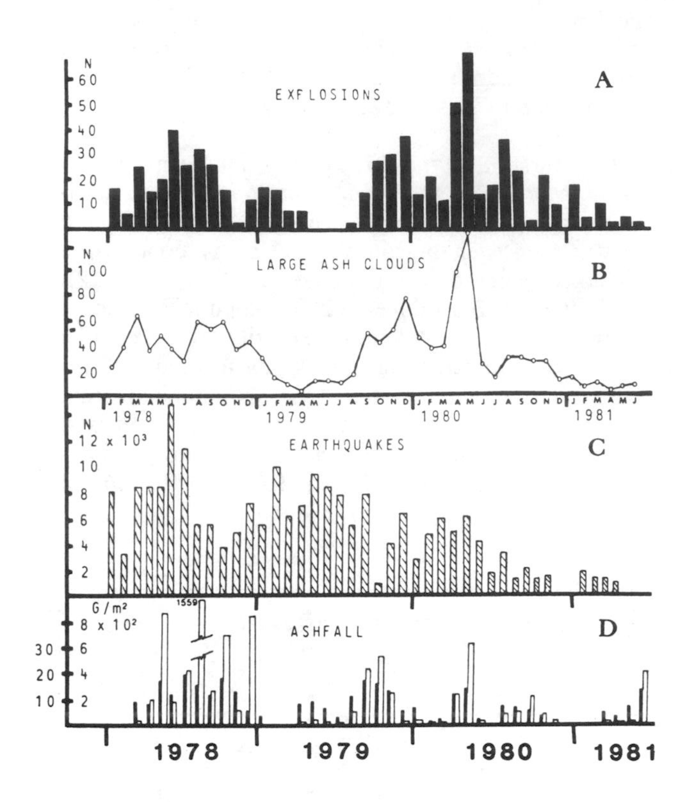

Fig. 8-5. Monthly number of recorded explosions (A), large ash clouds (B), recorded earthquakes (C), days when ashfall was observed at JMA's Kagoshima Observatory (D, black bars), and monthly volume of ash at Kagoshima Observatory (D, white bars) Jan 1978-June 1981. Ashfall is measured in a tray at Kagoshima Observatory. Explosions are counted by visual observation, microbarographs, and seismometers.

2/81 (6:2) After an active month in Jan, only 5 explosions [occurred] in Feb (table 8-13). The highest Feb ash cloud rose 1.2 km on the 21st. The Feb explosions caused no damage.

Information Contact: Same as 5:12.

3/81 (6:3) Eleven explosions occurred in March. The highest ash cloud grew to 2 km on the 18th. Two incandescent columns rose simultaneously from vents in the summit crater of Minami-dake on the 20th. No damage was reported.

Information Contact: Same as 5:12.

4/81 (6:4) Activity declined in April. Only two explosions, on 1 and 21 April, had been recorded by the 27th.

Information Contact: Same as 5:12.

5/81 (6:5) After only 2 explosions from the summit crater of Minami-dake had occurred in April, none were recorded in May until the 18th. As of 28 May, 4 explosions had occurred. The largest produced a 1.2 km-high ash cloud on the 24th. Although explosive activity was limited in April and May, there was continued ash ejection during which explosions were not registered at the JMA's Kagoshima Observatory. Bursts of B-type earthquakes were recorded on 3, 5, 18, and 19 May.

Information Contact: Same as 5:12.

6/81 (6:6) Activity has declined since April. Only 3 explosions, on 1, 8, and 14 June, had been recorded by 28 June. None caused damage. The highest ash cloud rose 2.5 km on 8 June.

Information Contact: Same as 5:12.

7/81 (6:7) During the first 27 days of July, only 1 explosion from the summit crater of Minami-dake was recorded by the JMA's Kagoshima Observatory. Ash ejection without recorded explosions and local seismicity also remained at low levels through July. Activity since Jan 1978 is summarized in fig. 8-5.

Information Contact: Same as 5:12.

8/81 (6:8) The decline in activity first noted in April reversed in Aug, when 34 explosions were recorded. Only one was recorded in July. Most of the explosions were small. The largest eruption cloud reached 2.5 km on 26 Aug. Ash frequently fell NE of the volcano, but no damage was reported. Weak glow was seen above the active crater on 13, 24, and 31 July.

Information Contact: Same as 5:12.

9/81 (6:9) Frequent explosions from the summit crater of Minami-dake continued through Sept, after increasing from 1 in July to 34 in Aug. None of the 38 recorded in Sept caused damage. Ash clouds higher than 2 km above the summit were observed on 13, 14, 18, 20, and 23 Sept. On the 13th an ash cloud that rose to 2.6 km was ejected simultaneously with [dense powder] that covered a 100 × 100 m area on the W slope of the active crater. The highest cloud rose to 2.7 km, also on the 13th.

1981	1	2	3	4	5	6	7	8	9	10	11	12	13	14	15	16	17	18	19	20	21	22	23	24	25	26	27	28	29	30	31	Total
Jan	1	1	-	-	1	1	1	-	1	1	-	-	-	1	-	-	1	-	2	2	1	-	-	-	1	-	-	1	1	-	1	18
Feb	-	-	-	-	-	1	-	-	1	-	-	-	-	-	-	-	1	-	-	1	-	-	-	-	-	-	-	1	-	-		5
Mar	-	-	-	-	1	-	-	-	2	1	-	-	-	-	-	1	-	2	-	1	-	1	-	-	-	-	-	-	1	1	-	11
Apr	1	-	-	-	-	-	-	-	-	-	-	-	-	-	-	-	-	-	-	-	1	-	-	-	-	-	-	-	-	-		2
May	-	-	-	-	-	-	-	-	-	-	-	-	-	-	-	-	-	1	-	-	-	-	-	2	-	1	-	-	-	-	-	4
Jun	1	-	-	-	-	-	-	1	-	-	-	-	-	1	-	-	-	-	-	-	-	-	-	-	-	-	-	-	-	-		3
Jul	-	-	-	-	-	-	-	-	-	-	-	-	-	-	-	-	-	-	-	-	-	-	-	-	-	-	1	-	-	-	-	1
Aug	1	1	-	-	1	-	1	1	2	2	-	1	2	2	2	1	1	2	-	2	2	3	-	1	2	1	-	-	1	-	2	34
Sep	3	-	1	-	2	-	1	3	-	1	-	1	2	1	3	1	2	2	2	1	-	-	3	-	1	-	1	2	2	3		38
Oct	5	3	3	2	1	2	1	-	1	-	-	-	1	-	3	-	5	1	-	-	-	-	-	-	1	1	2	-	1	-	2	35
Nov	-	1	1	2	2	3	1	5	1	3	6	-	3	2	-	2	-	1	2	4	2	2	-	1	1	1	1	1	1	1		50
Dec	1	1	2	1	-	-	-	-	2	-	1	-	-	2	-	2	2	3	3	2	-	2	-	1	2	1	-	-	1	3	-	32

Table 8-13. Summary of explosions from Minami-dake crater during 1981.

Information Contact: Same as 5:12.

10/81 (6:10) Frequent explosions continued through Oct, when 35 were recorded. Two 100 m-high incandescent columns were recorded for 3 seconds at 0444 on 2 Oct. Ash clouds higher than 2 km above the summit were observed on 3, 17 and 31 Oct.

Information Contact: Same as 5:12.

11/81 (6:11) Eruptive activity intensified in Nov. Explosions became more frequent and stronger; 50 were recorded. The explosion rate had been about 35/month Aug-Sept, but fewer than 4/month April-July. An explosion at 1528 on 16 Nov ejected incandescent tephra that caused [grass fires] on the SW flank. The [eruption] column rose 3 km, the greatest height this year. Lapilli from an explosion at 1322 on 21 Nov broke windshields on a few cars passing 3 km S of the summit crater of Minami-dake. Incandescent columns were observed on several occasions: 100 m high for 5 seconds on 7 and 22 Nov, and for 3 seconds on 14 and 29 Nov; 200 m high for 3 seconds on 30 Nov.

Information Contact: Same as 5:12.

12/81 (6:12) The rate of explosive activity at the summit crater of Minami-dake dropped slightly in Dec; 27 explosions had been recorded by the 25th. Incandescent columns 200 m high were observed for 7 seconds at 1836 on 9 Dec and for 5 seconds at 0622 on 22 Dec. Ash clouds higher than 2 km were observed on 3, 4, and 9 Dec.

Information Contact: JMA, Tokyo.

1/82 (7:1) Since last Aug, explosions from the summit crater of Minami-dake have been frequent. In Jan, 27 explosions were recorded (table 8-14). A 200 m-high incandescent column lasting 3 seconds was observed on 20 Jan, and another 300 m high for 5 seconds on the 21st.

Information Contact: Same as above.

3/82 (7:3) The rate of explosions from the summit crater of Minami-dake declined in early and mid-Feb, then increased late in the month. Frequent explosions continued through March. Recorded explosions numbered 15 in Feb, 47 in March.

On 26 Feb, an explosion at 1044 produced a 1600 m-high eruption column, then a continuous ash cloud was observed until 1150, and from 1430 until sunset ended visual observation from the JMA's Kagoshima Observatory. A 1500 m-high eruption column was ejected at 1731. On 28 Feb a continuous ash cloud was observed 0620-1230, and 3 explosions were recorded the next day. On 24 March a 100 m-high incandescent column was observed for 15 seconds, and on the 28th a 200 m-high incandescent column lasting 30 seconds was accompanied by rumbling. Local seismicity was active in the first half of Feb, when explosive activity had declined. JMA scientists have observed that a swarm of B-type earthquakes, which they interpret as possibly caused by magma rising to a shallower depth, is often followed by increased explosive activity. In March local seismic events and continuous ash clouds were frequently observed, but only rarely did an explosion with a large amount of ejecta occur. There was some damage to nearby farm products.

Information Contact: Same as 6:12.

4/82 (7:4) Recorded explosions declined from 47 in March to 15 in April. Ash clouds higher than 2000 m were observed on 12 and 19 April. These 2 explosions ejected incandescent blocks, but no damage was reported. In Feb and March, volcanic gas . . . damaged farm crops at the SW foot of the volcano.

Information Contact: Same as 6:12.

5/82 (7:5) The number of recorded explosions from the summit crater of Minami-dake was 24 in May. After mid-May, most explosions ejected incandescent tephra, but no damage was reported. In late May activity gradually changed to continuous ash ejection without recorded explosions. Wind carried ash toward the W, causing heavy ashfall in and around the city of Kagoshima, on 25-26 May. The explosion on 19 May, at 0727, produced an eruption column higher that 2500 m and was associated with 30 seconds of rumbling. Incandescent columns were observed on several occasions: 400 m high for 8 seconds on 4 May; 200 m high for 3 seconds on 5 May and for 5 seconds on 24 May.

Information Contact: Same as 6:12.

6/82 (7:6) The number of explosions from the summit crater of Minami-dake increased from [24] in May to 34 in June. A 300 m-high incandescent column was ob-

1982	1	2	3	4	5	6	7	8	9	10	11	12	13	14	15	16	17	18	19	20	21	22	23	24	25	26	27	28	29	30	31	Total
Jan	1	1	-	1	-	-	-	1	4	3	3	-	-	-	1	1	1	-	1	1	3	1	1	-	-	1	-	1	-	1	-	27
Feb	1	-	1	1	-	1	-	-	-	-	-	-	-	1	1	-	-	-	-	-	-	1	1	1	1	2	-	3				15
Mar	2	1	1	3	1	-	5	3	2	1	2	-	1	-	1	-	-	2	4	2	2	3	1	1	2	1	2	1	-	2	1	47
Apr	-	2	1	2	-	-	-	1	-	-	2	1	-	-	-	-	-	-	1	-	1	-	-	-	1	-	-	1	1	-		15
May	-	-	-	2	2	-	4	-	-	-	-	-	-	1	-	1	-	3	1	1	1	-	2	3	2	-	-	-	-	-	1	24
Jun	1	1	-	2	3	1	3	2	1	1	-	1	-	1	-	2	2	1	1	3	3	-	-	1	1	-	1	1	-	1		34
Jul	1	-	-	-	-	-	-	-	-	-	-	-	1	-	-	1	1	-	3	-	1	1	2	2	-	-	-	-	-	-	-	13
Aug	-	1	-	-	-	1	2	-	1	-	-	-	-	-	-	-	-	-	-	2	2	1	-	1	1	1	-	1	-	-	-	14
Sep	1	-	-	-	-	-	-	-	-	-	-	-	-	-	-	-	1	-	-	-	-	-	-	-	-	-	-	-	-	-		2
Oct	-	-	-	-	-	1	1	-	1	-	-	-	-	-	-	-	-	-	-	-	-	1	-	-	-	-	-	-	1	-	1	6
Nov	2	-	-	-	-	-	-	-	-	-	-	1	-	-	-	-	-	-	-	-	-	-	1	-	-	-	-	-	-	-		4
Dec	-	-	-	-	-	-	-	1	2	-	2	5	-	-	6	3	1	-	-	-	2	3	1	1	-	-	1	1	-	3	-	32

Table 8-14. Summary of explosions from Minami-dake crater during 1982.

served at 2223 on 8 June, and the highest plume rose 3000 m above the summit on 10 June. Carried by a strong wind, ejecta from the explosion at 0325 on 14 June fell in the area between Kurokami (4.7 km E of the crater) and Sakura-jima-guchi (5.5 km SE). At Kurokami, lapilli broke a car windshield and a building's windowpane.

Information Contact: Same as 6:12.

8/82 (7:8) Thirteen explosions were recorded in July. Activity was low during the first half of the month, then increased 16-24 July. A debris flow [caused by rainfall] carried away a concrete bridge at the S base of the volcano on the 24th. After the 24 July explosion, activity changed to continuous ash ejection without explosion [shocks].

In Aug, 14 explosions were recorded. Activity continued at the July level. On 24 Aug, continuous ash ejection without explosion [shocks] began in the morning, causing heavy ashfall to the NW. Ashfall from 1415-1500 was measured at 285 g/m^2 at the JMA's Kagoshima Observatory (about 10 km W of the crater). At the Kagoshima Prefecture Office, 2 km N of the JMA Observatory, 5522 g/m^2 of ash fell from 0900-1600. The ejection declined in the evening. The ash caused slight damage.

Information Contact: Same as 6:12.

11/82 (7:11) Explosive activity at the summit crater of Minami-dake and local seismicity have declined since Sept. Two explosions were recorded in Sept, 6 in Oct, and 4 in Nov. An explosion at 1130 on 17 Sept was followed by continuous ash ejection, without explosion [shocks], lasting for 98 minutes and resulting in heavy ashfall on the S half of the city of Kagoshima, about 15 km NW. In Oct, ash ejection occurred every day, although few explosions were recorded. The frequent ash ejection was accompanied by an increase in the number of recorded tremor events, which totaled 334 hours in Oct.

Each of the Nov explosions produced lapilli ejection, air shocks, and sounds. At 1523 on 23 Nov, the largest explosion of the 3-month period sent an eruption cloud to 3 km. The cloud was observed from the JMA's Miyakonojo Weather Station, 45 km ENE. An Air Nauru jet carrying 39 passengers and flying at about 3 km altitude entered the cloud 23 km ESE of Sakura-jima at 1545, 6 minutes after leaving Kagoshima Airport. The impact of the lapilli produced hairline cracks in 3 cockpit windows, prompting the pilot to return to Kagoshima, where he landed safely. At the SE foot of the volcano, a car windshield was destroyed.

Information Contacts: JMA, Tokyo; UPI.

12/82 (7:12) Explosive activity at the summit crater of Minami-dake intensified in Dec when 32 explosions were recorded. Only 12 had been recorded from Sept through Nov. Ash ejections lasting less than 25 minutes and not accompanied by instrument-recorded explosions were frequently observed. Although eruptive activity was frequent, large amounts of ejecta were rarely observed. No damage was reported. In early Dec, 124 hours of volcanic tremor were recorded but only 20 hours were recorded during the middle of the month, when explosive activity peaked.

Information Contact: JMA, Tokyo.

1/83 (8:1) Activity at the summit crater of Minami-dake intensified in Jan; 54 explosions (table 8-15) and 35 ash ejections not accompanied by instrument-recorded explosions were observed. In the second half of Jan, explosions were accompanied by larger amounts of lapilli. An explosion at 1059 on 26 Jan (table 8-16) ejected an eruption column that rose 3.8 km above the summit, the highest observed in the past 9 years. This cloud was large enough to be observed from JMA's Aso-san Weather Station, 150 km to the NNE; aircraft crews reported that it rose 7.8 km above sea level. Lapilli from this explosion broke the windshield of a car passing near Arimura, 3 km S of the summit. Incandescent columns were observed during two of the explosions on 31 Jan. A 300-m high column accompanied the explosion at 0228 and lasted 40 seconds; a column 200 m high lasted 30 seconds during the 0545 explosion.

Information Contact: Same as above.

2/83 (8:2) The rate of explosions at the summit crater of Minami-dake has gradually increased since Dec. In Jan, 53 were recorded and in Feb, 73 explosions were recorded, one of the larger monthly totals since the eruption began in 1955.

1983	1	2	3	4	5	6	7	8	9	10	11	12	13	14	15	16	17	18	19	20	21	22	23	24	25	26	27	28	29	30	31	Total
Jan	1	1	-	2	-	5	6	3	1	3	6	-	1	1	3	2	2	-	1	1	-	1	-	1	2	3	3	-	-	1	3	53
Feb	2	4	2	5	2	-	4	4	-	5	6	4	-	3	2	3	3	3	3	3	2	2	2	2	-	3	3	1				73
Mar	-	1	1	2	2	-	2	4	1	-	1	2	3	1	2	1	-	1	-	-	2	2	1	-	2	1	1	-	2	-	1	36
Apr	1	-	1	-	1	1	-	1	1	1	-	1	1	-	-	1	3	1	-	1	1	-	1	2	-	1	1	1	-	-		22
May	-	1	-	3	1	2	-	1	1	-	-	-	-	-	-	-	1	4	1	1	2	1	-	-	1	2	-	-	-	-	-	22
Jun	1	1	1	-	2	-	-	1	-	1	-	1	1	1	1	2	-	2	6	2	3	2	1	-	-	1	-	3	-	-		33
Jul	-	-	-	-	-	3	1	-	-	-	-	1	2	2	2	1	-	1	2	2	3	3	2	3	-	-	-	-	-	-	3	31
Aug	-	2	1	-	1	1	3	1	2	1	-	-	-	1	-	1	2	2	1	1	4	1	3	-	-	-	7	3	-	-	1	39
Sep	1	1	2	1	1	-	-	1	4	1	1	3	1	1	1	-	-	2	-	3	1	3	1	-	-	-	2	-	2	3		36
Oct	1	2	2	-	-	-	-	-	1	3	1	-	2	1	-	-	3	-	-	2	2	-	-	-	-	-	-	-	-	-	1	21
Nov	-	-	1	-	1	-	2	-	1	2	-	-	1	-	-	-	-	-	2	-	1	-	-	-	1	1	-	-	1	2		16
Dec	1	-	-	3	-	-	3	1	3	1	1	2	3	2	-	1	2	-	2	1	-	3	2	1	3	-	-	-	-	2	-	37

Table 8-15. Summary of explosions from Minami-dake crater during 1983.

Ashfall and eruption clouds were observed on 11, 14, and 27 Feb at Miyakonojo Observatory. The air shock from the explosion at 2241 on 5 Feb was large enough to be felt at Miyazaki Observatory, 80 km NE. On 18 Feb a hut at Arimura, 3 km SSE of the summit, was set on fire by an incandescent block, 50-100 cm in diameter. The explosion at 1043 on the 21st was not large enough to be accompanied by any observed explosive sound or felt air shock at Kagoshima Observatory but strong NW winds carried lapilli toward the SE foot of the volcano where 4 car windshields were cracked or broken. Two incandescent columns, rising about 200 m above the crater, were observed in Feb for 10 seconds on the 1st, and for 5 seconds on the 27th.

Rain on 2 Feb triggered debris flows in S flank valleys. One flowed into 9 houses and a hotel after pushing away a 10 m-long sand-trap wall, and covered the adjacent road for about 50 m. The monthly number of recorded seismic events was 4456 in Jan, but decreased to 2410 in Feb.

Information Contact: Same as 7:12.

4/83 (8:4) The number of recorded explosions from the summit crater of Minami-dake declined to [36] in March and 22 in April. Observers saw lapilli ejected by about $^1/_4$ of the explosions in March and $^1/_3$ in April, but no damage was reported. The monthly number of recorded seismic events, including tremors, was 1358 in March and 768 in April. On 2 March, rain-triggered debris flows in S and E flank valleys temporarily blocked a road.

Information Contact: Same as 7:12.

5/83 (8:5) The number of explosions from the summit crater of Minami-dake recorded in May was 22, the same as in April. In the second half of May, most of the explosions ejected lapilli and sent ash and vapor to more than 2 km above the summit. The 4.4 km-high [eruption] column from an explosion at 1320 on 18 May was one of the largest since the eruption began in 1955.

On 22 May an explosion at 1237 that sent a column to more than 4 km was followed by 2 hours of continuous ash and vapor ejection. The activity was accompanied by thunder and temporary interruption of electric and telephone service. Ash fell as far as Aburatsu, on the E coast of Kyushu, 70 km E of the volcano.

An explosion on 26 May was also accompanied by continuous ash ejection. A large amount of ash fell SSE of the volcano; more than 20 car windshields were broken, and the roof of a primary school cracked.

Information Contact: Same as 7:12.

7/83 (8:7) In June, 33 explosions were recorded from the summit crater of Minami-dake and 31 in July. Although the explosion rate from June to mid-July was above its usual level, rarely was a large amount of ejecta observed in any explosion. Only about $^1/_8$ of the explosions ejected much lapilli, or produced eruption columns that rose to more than 2 km above the summit.

Activity intensified slightly 19-24 July. Most explosions produced large amounts of ejecta and ash frequently fell on the cities of Miyakonojo (40 km ENE) and Miyazaki (80 km NE). The end of explosive activity on 25 July was followed by continuous ash ejection. Bad weather limited visual observation, but volcanic tremor that was assumed to be accompanied by ash ejection was recorded until 29 July. The number of large B-type earthquakes suddenly increased about 1800 on 29 July and remained high until 0300 the next day. Earthquake size then returned to its usual level, but the recorded events were still more numerous than usual.

Explosive activity resumed on 31 July, accompanied by a decrease in seismic activity. An explosion at 1445 on 2 Aug ejected large amounts of lapilli, which fell near Kyoto University's Sakura-jima Volcano Observatory (about 1.7 km SW of the summit) and the site of sand trap wall construction, where 1 worker [originally reported as 4] was slightly burned. [Blocks] made many craters [near the University Observatory and the con-

1983	Event	Comment
Jan 26	Explosion at 1059	Windshields broken on 4-5 cars at SE foot
Feb 2	Debris flows	One flow damaged 9 houses and a hotel at S foot after pushing away part of a sand trap wall
18	Explosion at 1326	Ejected blocks as large as 0.5-1 m; one hut at SE foot burned
21	Explosion at 1043	Windshields broken on 4 cars at SE foot
Mar 2	Debris flows	Road at S foot temporarily covered
May 22	Explosion at 1237	Eruption column to 4 km above the summit, 2 hours of continuous ash ejection; electric supply interrupted, equipment damaged at Kurokami, 5 km E
26	Explosion	5 hours of continuous ash ejection; windshields broken on 23 cars, gymnasium roof cracked
Jul 24	3 Explosions	Strong SW wind carried ash to Miyazaki City, 80 km NE, reducing visibility to 4 km
Aug 2	Explosion at 1401	Large amount of lapilli fell near Karutayama Volcano Observatory, about 3 km NW, and at site of sand trap construction, 2 km SE, where 4 workers were slightly burned
14	Explosion at 1614	Windshields on 16 cars, windows in 2 houses and a hut roof broken at Nojiri (SW foot); windshields on 3 cars broken in Kamoike (S part of Kagoshima City, 10 km WSW)
27	Explosion at 1401	Car windshields broken at Arimura, 3 km S
Sep 1	Debris flows	Roads temporarily closed. Total of 5 flows (on Sept 1, 10, 20, and 21)
17-18	Continuous ash ejection	Ash on track derailed streetcar in N Kagoshima City, 10 km WNW
20	Explosion at 1518	A few windows in a temple damaged at Kamoike. Debris flows (see Sept 1)
Oct 10	Explosion at 1351	Windshields broken on 2 cars at SW foot
Dec 7	Explosion at 1702	Car windshield broken in Tarumizu City, 10 km SSE
13	Explosion at 1028	Large air shock broke windows in hotel and house

Table 8-16. Notable eruptive activity of Sakura-jima, 1983.

struction site]; the largest was 1.5 m in diameter and 1 m deep [produced by a block 50 cm in diameter].

Information Contact: Same as 7:12.

12/83 (8:12) Recorded explosive eruptions from the summit crater of Minami-dake were frequent in Aug [33] and Sept (36), fewer in Oct (21) and Nov (16). In mid-Aug strong wind carried a large amount of ejecta to the inhabited area around the volcano, damaging cars and houses. On 14 Aug lapilli as large as 6-7 cm in diameter fell on Nojiri, at the foot of Sakura-jima 4 km SW of Minami-dake, breaking or cracking windshields on 16 cars, a house windowpane, and the roof of a hut. Radio news from Kagoshima reported that windshields on 3 cars were cracked by lapilli as large as 1-3 cm in diameter at Tarumizu Wharf, on the S side of the city. An explosion at 0153 on 16 Aug scattered lapilli up to 2.5 cm in diameter between Mochiki and Yumoto (3.5 km SSW of Minami-dake), where windshields on 26 cars ware cracked.

Explosive activity remained at a high level in Sept. Incandescent blocks from an explosion at 1148 on 12 Sept started a flank [brush] fire. A heavy ashfall on 17-18 Sept deposited 270 g/m^2 of ash at JMA's Kagoshima Observatory (table 8-17); a Kagoshima streetcar derailed on 19 Sept due to ash on the tracks. A large air shock from an explosion at 1518 on 20 Sept broke windows in Kagoshima, and another explosion at 1638 sent lapilli as large as 1 cm in diameter toward the N. Five debris flows moved down S flank valleys on 1, 10, 20, and 21 Sept. Bursts of discrete seismic events were frequent in Sept. Bursts on 19 and 28 Sept lasted for about 10 hours and included large events.

Information Contact: Same as 7:12.

1983	Ashfall (g/m^2)	1983	Ashfall (g/m^2)
Jan	16	Jul	209
Feb	66	Aug	55
Mar	[174]	Sep	297
Apr	224	Oct	43
May	121	Nov	[117]
Jun	530	Dec	60

Table 8-17. Monthly ashfall measured at Kagoshima Observatory, 1983.

1/84 (9:1) Recorded explosive eruptions were fewer in Oct (21) and Nov (16), but about as frequent in Dec (37) as in Aug [33] and Sept (36). About 1/4 of Oct's explosions were accompanied by large quantities of ejecta. On 10 Oct, the last and strongest of explosions at 1001, 1131 and 1351 sent an eruption column to 2.5 km above the summit. A large amount of lapilli broke windshields on 2 cars at Nojiri and Mochiki, at the SW foot of the volcano about 4 km from the summit. Activity remained at a relatively low level from late Oct to mid Nov. In late Nov stronger explosions were again frequently observed. During the first 12 explosions in Dec, observers at the Kagoshima Observatory witnessed lapilli ejection. Ejecta from an explosion at 1702 on 7 Dec broke a windshield at Usine in Tarumizu City (10 km SSE). On 13 Dec an explosion at 1028 generated an air shock that broke 5 windows in a hotel and 1 in a house. There were 413 recorded explosions in 1983, the second largest annual total since the current eruption began in 1955.

Information Contact: JMA, Tokyo.

3/84 (9:3) An explosion at Sakura-jima on 12 April at about 0940, the 95th in 1984, produced an eruption cloud that rose to 2.3 km (table 8-18). According to press reports, ejecta fell over half the volcano and broke windows at the foot.

Information Contacts: Kyodo News Service, Tokyo; UPI.

1984		Time	Result
Jan	4	1928	Lapilli broke car windshield at S foot of volcano.
	10	1457	Air shock cracked a window at E foot.
	11	2140	Air shock cracked 3 windows.
Mar	8	0724	Lapilli cracked car windshield.
	19		Debris flows reached E foot; no damage reported.
	30	1853	Large quantity of lapilli; forest fires started.
Apr	12	0941	Strong air shock cracked 3 windows at E and S foot, 2 windows in Kagoshima.
	19		Debris flows occurred.
	29	1800	Large quantities of incandescent ejecta started forest fires in more than 10 places; strong air shock broke a window in Kamoike. (Damage in Kagoshima is rare.)
May	4		A windowpane was broken in Yasui.
	8		Air shock broke a hospital windowpane at the SW foot.
Jun	23		Heavy ashfall caused a traffic jam and interrupted electric service in Kagoshima.
	3		Strong air shock broke windowpanes at a junior high school and a house in Koike at the W foot. A Sakura-jima Volcano Observatory employee was injured by broken glass in Yokoyama at the W foot.
	7		A large amount of lapilli broke 11 car windshields and a house windowpane at the NE foot.
Jul	21		A large amount of ejecta including pieces up to a few meters in diameter fell at the S foot. Incandescent fragments broke the roofs of 10 houses and a warehouse, burning some slightly. Telephone and electric service was interrupted for a few hours.
Oct	10-11		Volcanic gas damaged farm crops at the SW foot.
Dec	20		Lapilli broke a few car windshields at the S foot.

Table 8-18. Damage caused by Sakura-jima eruptive activity in 1984.

5/84 (9:5) Explosive activity at the summit crater of Minami-dake continued at a higher level, Jan-April (table 8-19). The average monthly number of recorded explosions was 27, including larger ones that caused damage.

Information Contact: JMA, Tokyo.

12/84 (9:12) The number of recorded explosions at Sakura-jima in 1984 declined during July-Nov, but increased in Dec. Eruptive activity at the summit crater of Minami-dake caused damage around the volcano throughout the year.

Information Contact: Same as above.

1/85 (10:1) Explosive activity at the summit crater of Minami-dake declined Aug-Nov, then increased in Dec when 59 explosions were recorded. An explosion at 1820 on 20 Dec was accompanied by an air shock powerful enough to have been recorded at Uwajima (260 km NE); ash and lapilli up to 2 cm in diameter fell on the road at the S foot and broke a car windshield. An explosion at 2132 on the 31st was accompanied by another strong air shock that broke 11 windowpanes in 3 of the hotels at the S foot. The rate of explosions decreased in Jan, when 20 were recorded. A strong air shock recorded at Uwajima, Nobeoka (140 km NE), Kumamoto (135 km N, and other places broke a windowpane in a S-foot hotel on 23 Jan.

Information Contact: Same as above.

2/85 (10:2) An explosion at the summit crater of Minami-dake on 24 Feb at about 1030 ejected a plume to 4 km height. Lapilli 4-5 cm in diameter fell as far as 5 km SE of the crater, damaging 43 cars at the S foot and in Tarumizu City (5 km SE). Brush fires started by hot tephra quickly died. An air shock was recorded in Miyazaki, about 75 km to the NE. In the 4 hours after 1200, 140 swarm earthquakes were recorded.

Information Contact: T. Tiba, National Science Museum, Tokyo; Kyodo News Service, Tokyo.

3/85 (10:3) Explosive activity at the summit crater of Minami-dake intensified in Feb, when 35 explosions were recorded (table 8-19). The largest, at 1030 on the 24th, ejected a plume to an altitude of more than 4 km and was accompanied by a strong air shock that saturated the microbarograph at the Kagoshima Observatory. A strong NW wind carried a large amount of lapilli toward the SE foot. Between Arimura (3 km SSE) and Ushine-fumoto (6 km ESE, in the middle of the city of Tarumizu), 28 car windshields were broken. Tile or slate roofs on 53 houses were slightly damaged at Ushine-fumoto, and a windowpane was broken at Arimura. The number of recorded explosions increased to 54 in March, although the number expelling a large amount of ejecta decreased. Swarms of volcanic earthquakes were observed on 4, 17, and 19 Feb and 2, 8-9, and 24 March. On 31 March, an explosion sent lapilli 4-5 cm in diameter toward the SE foot, where 4 car windshields were cracked.

Information Contact: JMA, Tokyo.

4/85 (10:4) Activity remained at a high level in April, when 37 explosions were recorded, as compared to 35 in Feb and 54 in March. An explosion on 9 April at 1827 ejected a plume to 4 km above the crater, and was followed by 40 minutes of continuous ash emission and volcanic

1984	1	2	3	4	5	6	7	8	9	10	11	12	13	14	15	16	17	18	19	20	21	22	23	24	25	26	27	28	29	30	31	Total
Jan	1	-	-	1	-	-	1	-	1	3	1	-	-	-	-	-	-	1	2	3	-	-	-	-	2	1	1	1	1	1	1	22
Feb	1	5	-	2	-	-	1	1	2	3	-	-	-	-	-	1	1	-	-	2	1	-	2	-	-	1	1	2	-			26
Mar	-	2	-	-	-	1	-	2	1	1	2	-	-	1	-	1	2	1	1	1	1	1	2	3	3	3	2	1	1	3	-	36
Apr	3	1	-	-	-	2	-	1	1	-	2	2	-	-	-	-	-	3	1	1	-	2	1	-	1	-	-	2	2	-		25
May	1	-	1	2	2	3	1	2	1	1	-	2	-	2	4	1	1	2	-	2	1	2	2	2	1	2	2	1	-	2	-	43
Jun	1	5	3	2	1	1	1	-	-	-	-	2	1	-	2	1	1	-	-	3	2	1	1	1	-	2	4	3	2	2		42
Jul	2	-	-	-	-	-	4	-	-	1	-	-	3	2	2	-	-	1	-	-	1	-	-	3	1	-	-	-	1	-	-	21
Aug	-	1	-	-	1	-	2	-	1	2	-	1	1	-	-	-	-	1	1	-	-	-	-	-	-	-	1	-	-	-	-	12
Sep	-	-	-	-	-	1	-	-	-	-	-	1	-	-	2	-	-	1	-	1	-	1	2	1	1	-	-	1	-	1		13
Oct	-	-	-	2	-	2	-	-	-	-	1	1	-	-	-	-	1	1	-	-	1	2	1	-	2	-	-	-	-	-	-	14
Nov	2	1	-	-	-	-	-	-	-	1	1	-	-	1	1	1	-	-	-	-	-	1	5	2	1	2	-	-	-	-		19
Dec	-	-	1	-	2	2	1	1	4	-	3	3	2	2	2	7	1	3	2	4	-	1	-	-	-	3	6	3	2	-	4	59

1985	1	2	3	4	5	6	7	8	9	10	11	12	13	14	15	16	17	18	19	20	21	22	23	24	25	26	27	28	29	30	31	Total
Jan	1	-	3	-	-	-	1	2	-	-	-	1	1	1	-	-	1	1	-	1	-	2	1	3	-	-	-	-	1	-	-	20
Feb	-	-	2	2	6	3	2	1	2	3	-	1	2	-	1	-	1	1	1	1	1	-	-	1	2	-	1	1				35
Mar	1	2	5	3	3	3	-	-	-	3	3	2	1	2	5	1	1	1	1	5	1	-	1	1	3	3	1	1	-	-	1	54
Apr	1	2	3	-	1	-	4	-	1	2	3	2	4	2	-	-	1	-	2	1	-	-	3	-	1	2	1	1	-	-		37
May	-	-	-	-	-	-	-	1	-	1	1	-	2	2	-	-	-	-	-	-	-	-	1	-	1	1	-	-	-	-	-	10
Jun	-	1	-	-	-	-	3	3	-	2	1	1	1	1	2	1	-	-	3	-	-	1	1	2	3	2	1	2	1	1		33
Jul	1	6	2	-	-	1	3	-	1	3	2	2	1	1	2	-	1	3	4	2	4	1	1	4	3	2	-	2	1	4	3	60
Aug	2	-	1	2	-	-	-	1	-	3	2	-	2	1	-	1	-	-	-	-	-	-	-	1	1	-	1	-	2	-	-	20
Sep	-	7	-	4	2	1	1	2	1	2	-	7	3	-	1	-	2	3	2	-	-	4	4	1	1	1	-	-	-	-		49
Oct	1	4	3	2	2	-	3	2	2	1	1	-	1	1	-	3	4	1	5	2	2	1	2	-	-	1	1	-	1	-	1	47
Nov	-	1	1	1	1	-	-	1	-	-	-	1	-	2	2	1	1	1	-	1	2	1	-	4	3	2	2	1	4	1		34
Dec	2	3	6	3	5	5	3	1	2	4	4	3	-	2	2	6	-	3	3	1	1	-	-	5	1	2	1	-	-	3	4	75

Table 8-19. Summary of explosions from Minami-dake crater during 1984–85.

lightning. A small pyroclastic flow moved down the SW flank to about 600 m elevation. In the city of Kagoshima, ashfall began at 1840 and continued until around 0200 the next day. In the 24 hours beginning at 0900 on the 9th, ashfall at the Kagoshima Observatory was 1608 g/m^2, the largest daily total since measurements of ash deposits started in April 1969.

Another event of the same type occurred 13 April at 0722. The explosion earthquake was followed by continuous tremors, which gradually changed into a swarm of B-type earthquakes. Lapilli up to 1 cm in diameter fell on the foot, where 4 car windshields were cracked. On 3 May, Space Shuttle astronauts photographed a whitish, relatively diffuse plume that extended at least [50] km from Sakura-jima (fig. 8-6).

Information Contacts: JMA, Tokyo; C. Wood, NASA, Houston.

6/85 (10:6) Explosive activity declined in May, when only 10 explosions were recorded. The number of recorded volcanic earthquakes and tremors also decreased, from 6580 in April to 2755 in May.

Information Contact: JMA, Tokyo.

7/85 (10:7) The number of recorded explosions increased from 10 in May to 33 in June and 60 in July. Explosions caused damage on 8, 13, 16, 22, and 30 June and on 6, 10, and 21 July. Bombs fell on roads and farms, producing large craters. A bomb from an explosion on 8 June at 1316 made a 1 m-diameter crater in a road near a residential area. On 13 June lapilli up to 3 cm in diameter fell on the S foot, cracking windshields and solar water heaters on rooftops. Ash fell heavily that day at Kagoshima and closed railway crossings. After an explosion on 16 June at 1147, a bomb fell on a farm, producing a crater 4 m in diameter and 0.8 m deep. On 22 June at 1029 an ash cloud rose 3.5 km. The air shock from the explosion broke windows at a school near the foot of the volcano, and falling lapilli cracked a car windshield on the E foot. Windshields of cars at the S foot were cracked by falling lapilli from a 30 June explosion. Total June ashfall was 1510 g/m^2 at JMA's Kagoshima Observatory.

Observations of the eruption column heights and quantity of ejecta were prevented by bad weather in early July. Debris flows from the volcano on 2 July blocked roads at 3 places after a heavy rainfall. On 6 July at 1720 a bomb fell on a house 3 km from the crater, brief-

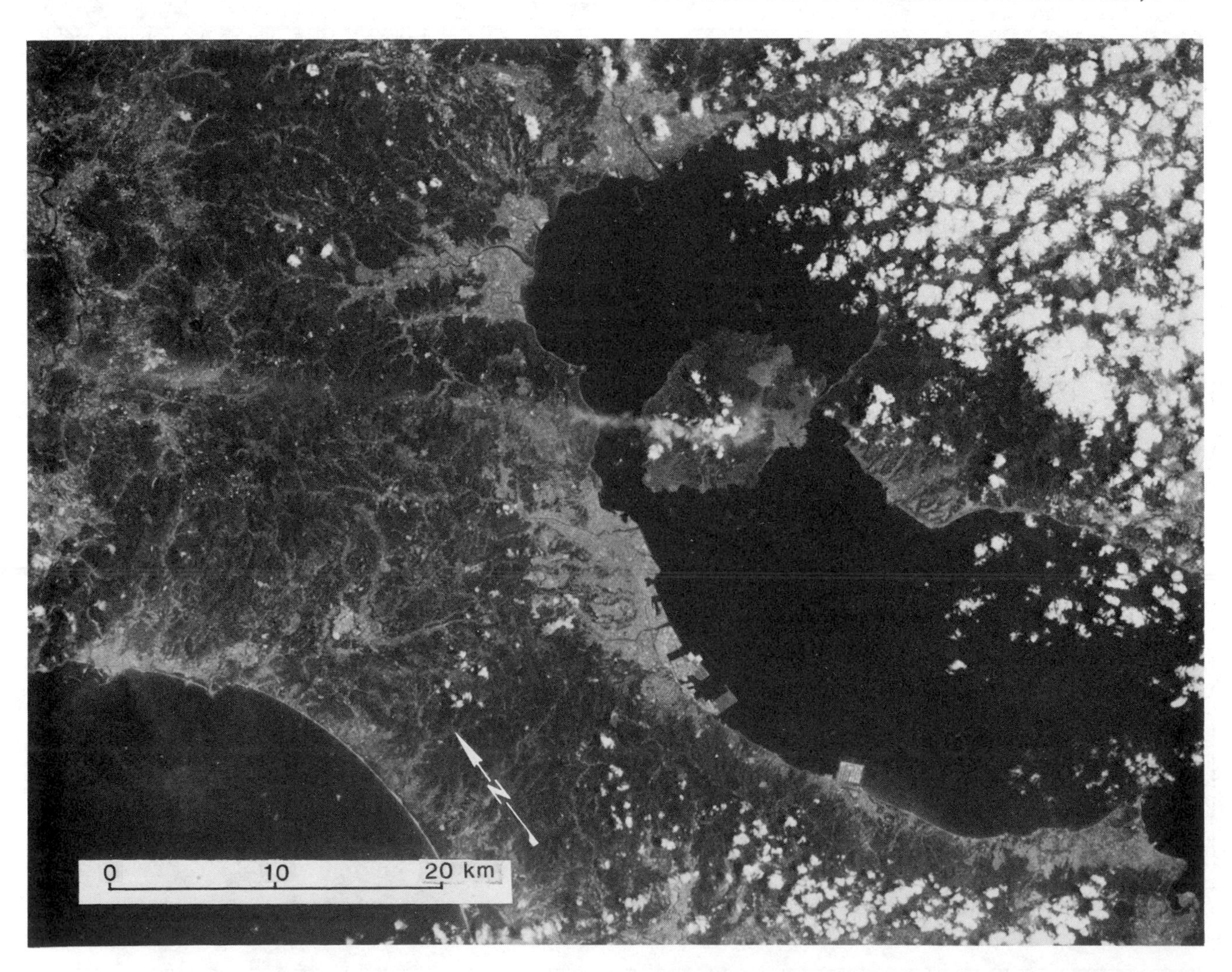

Fig. 8-6. Photograph of Sakura-jima taken by Space Shuttle astronauts on 3 May 1984 (no. 51B-52-74). A plume extends more than 50 km NW. Courtesy of Charles A. Wood.

ly setting it afire and making a 2-m hole in the roof; no casualties or injuries occurred. On 10 July, incandescent rocks fell on an inhabited area in Arimura, at the S foot. One broke into pieces and cracked roof tiles on a house.

An airshock from a 21 July explosion broke windowpanes of a high school and a restaurant in the central part of Kagoshima. The explosion was followed by continuous emission of ash, which was carried toward Kagoshima. Early the next day, a Japan National Railway crossing gate in the N part of the city malfunctioned because of ash deposits on the rails; a car was struck by a train, but the car's driver was only slightly injured. Debris flows on 27 July blocked a road and broke buried water pipes at the S foot.

Ashfall was observed daily at the Kagoshima Observatory in late July. During the 24 hours beginning 28 July at 0900, ashfall was 2476 g/m^2, the largest daily total, raising monthly and yearly totals to the largest since measurements of ash deposits started in April 1969.

On 31 July, 3 explosions with large eruption columns at 0700, 0848, and 0951 were followed by vigorous ash emission that continued until about 1200. Driven by a SE wind, the ash fell over the W coast of Kyushu (fig. 8-7).

Information Contact: Same as 10:6.

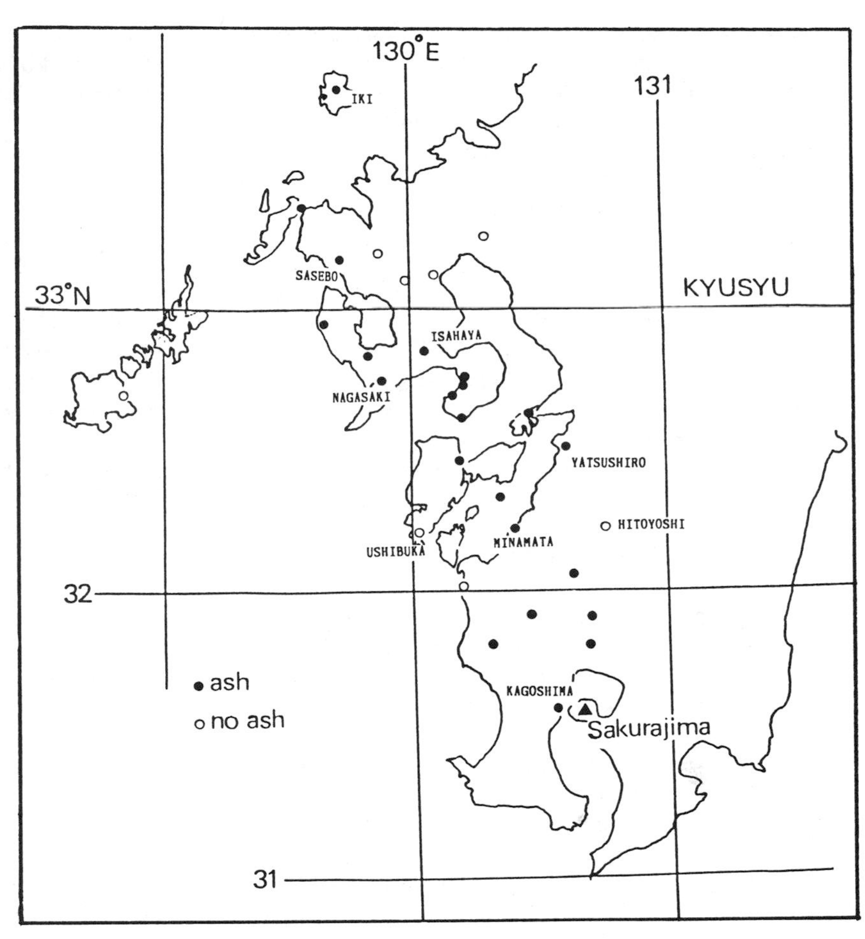

Fig. 8-7. Ashfall from 31 July 1985 explosions at Sakura-jima.

10/85 (10:10) Explosive activity at the summit of Minami-dake declined in Aug, although vigorous ash emission was frequently observed. A large amount of ash fell on the city of Kagoshima, temporarily cutting off electric power and delaying trains. Lapilli broke car windshields and solar water heaters at 1919 on 11 Aug, at 0816 on the 14th, and at 1142 on the 16th.

Minami-dake's activity intensified in Sept, when 49 explosions were recorded. Successive vigorous ash emissions began with an explosion on 4 Sept at 1127, followed by 3 more the same day. An ash cloud moved N and slight ashfall was observed at many points in Kyushu and along the E coast of Honshu (fig. 8-8, left). An eruption column on 9 Sept rose to 4 km above the summit. An explosion on 12 Sept at 0156 ejected lapilli up to 1 cm in diameter; a car windshield was broken at the SE foot. On 22 Sept, activity similar to that on the 4th sent ash over the E coast of Kyushu (fig. 8-8, right).

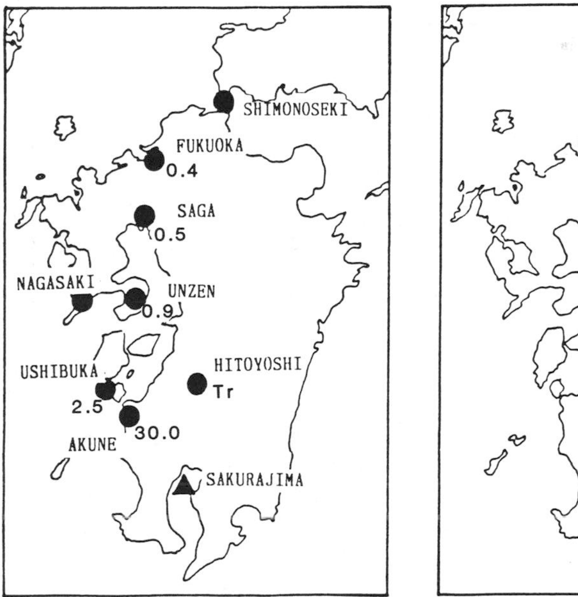

Fig. 8-8. Locations of reported ashfall from explosions of Sakura-jima on 4 Sept, *left*, and 22 Sept 1985, *right*. Numbers beside data points on the left map are ashfall in g/m^2.

Forty-seven explosions were recorded in Oct, but there were no damages. Photographs by Space Shuttle astronauts on 3 and 4 Oct show a V-shaped plume to the SE of Sakura-jima.

Information Contacts: JMA, Tokyo; W. Daley, NASA, Houston.

11/85 (10:11) Eruptive activity remained vigorous in Nov, when 34 explosions were recorded. Frequent powerful explosions produced air shocks and scattered incandescent blocks to about 2 km from the summit. No damages were reported. An explosion on 25 Nov at 1427 was accompanied by a small pyroclastic flow, which only covered part of the summit area.

The frequency of explosions increased further in early Dec. By the 5th, 19 explosions were recorded, bringing 1985's total to 416, the largest since discrete explosive activity began in 1955 (table 8-20). A series of small explosions on 3 Dec shattered windows of several buildings in northern Kagoshima and disrupted telephone service in some areas. The summit crater of Minami-dake erupted on 5 Dec at 0220 and 0648, dropping ash on Kagoshima, but there was no damage. The last major explosive eruption of Sakura-jima was in 1914.

Information Contacts: JMA, Tokyo; UPI.

12/85 (10:12) Weather satellite images on 24 and 31 Dec show eruption plumes from Sakura-jima. On the 24th an approximately 200-km plume extended SW, then turned E for about 60 km. The ash cloud was fairly dense and milky gray in color. On the 31st there was a 120-km V-shaped plume ESE of the volcano. The end of the plume was about 40 km wide and very diffuse.

Information Contacts: M. Matson, NOAA/NESDIS; JMA, Tokyo.

Further Reference: Eto, T., Kamada, M., and Kobayashi, T., 1987, The 1983-1986 Activities of Sakura-jima Volcano *in* XIX IUGG General Assembly, 1987, *Report on Volcanic Activities and Volcanological Studies in Japan for the Period from 1983 to 1986*, p. 18-27.

KIRISHIMA

Kyushu, Japan
31.88°N, 130.92°E, summit elev. 1700 m

Kirishima is 43 km NNE of Sakura-jima. It is formed of more than 20 eruptive centers over a 20 × 30 km area that also includes Japan's first National Park. 66 eruptions are known since 742 AD.

9/78 (3:9) An earthquake swarm in the vicinity of Kirishima, including some felt events, was recorded on 7 July. No surface activity was observed. Kirishima's last eruption, on 5 Aug 1971, ejected accessory tephra and mud. No seismicity was recorded preceding the 1971 eruption.

Information Contacts: JMA, Tokyo; D. Shackelford, CA.

Further Reference: Ida, Y., Yamaguchi, M, and Masutani, F., 1986, Recent Seismicity and Stress Field in Kirishima Volcano; *Journal of the Seismological Society of Japan*, v. 39, p. 111-121.

6/79 (4:6) JMA personnel visited the summit crater of Shinmoe-dake, one of the SE group of volcanic edifices in the Kirishima complex, on 23 April. They found a 10 cm-

Year	Jan	Feb	Mar	Apr	May	Jun	Jul	Aug	Sep	Oct	Nov	Dec	Total
1955	0	0	0	0	0	0	0	0	0	6	0	0	6
1956	3	2	28	15	4	0	5	4	20	13	11	10	115
1957	23	19	5	0	0	1	0	3	0	0	6	0	57
1958	4	4	7	9	0	8	2	4	4	9	9	23	83
1959	33	11	0	0	0	1	3	0	48	4	8	1	109
1960	36	11	19	52	48	39	22	0	69	80	34	4	414
1961	6	10	8	8	15	15	4	10	13	12	57	38	196
1962	15	0	10	10	6	9	7	9	9	7	1	6	89
1963	0	22	5	7	18	8	10	5	24	14	14	9	136
1964	37	13	2	1	9	12	2	2	2	1	4	3	88
1965	4	0	2	3	2	1	3	3	4	3	2	2	29
1966	4	1	0	3	4	7	5	3	2	2	1	12	44
1967	4	7	2	6	5	29	15	19	3	24	13	0	127
1968	0	3	10	2	16	3	1	0	0	2	0	0	37
1969	0	1	2	0	2	1	0	6	9	1	0	0	22
1970	0	2	0	0	0	1	1	4	3	1	3	4	19
1971	2	3	1	4	0	0	0	0	0	0	0	0	10
1972	0	0	5	1	0	0	2	1	8	16	47	28	108
1973	0	0	0	1	2	4	2	17	14	38	35	31	144
1974	30	32	12	1	30	93	49	38	28	15	21	13	362
1975	11	29	41	27	8	7	3	9	16	15	24	9	199
1976	5	4	6	9	31	23	6	19	27	14	15	17	176
1977	10	2	4	3	29	22	28	35	23	19	21	27	223
1978	17	7	25	14	20	39	25	32	25	15	1	11	231
1979	15	16	7	7	0	0	0	1	13	26	28	36	149
1980	12	20	10	48	69	12	16	34	21	4	21	10	277
1981	18	5	11	2	4	3	1	34	38	35	50	32	233
1982	27	15	47	15	24	34	13	14	2	6	4	32	233
1983	53	73	36	22	22	33	31	39	36	21	16	37	419
1984	22	26	36	25	43	42	21	12	13	14	19	59	332
1985	20	35	54	37	10	33	60	20	49	47	34	75	474

Table 8-20. Monthly and yearly number of explosions at Sakura-jima since the summit eruption began in 1955.

wide sulfur flow that had traveled more than 50 m downslope from one of the summit crater fumaroles, which also was surrounded by scattered, explosively ejected sulfur. The sulfur, molten when it was ejected during the winter, was a dark brown solid in April.

Information Contact: JMA, Tokyo.

3/81 (6:3) Fumarolic activity had increased since Nov 1980 in the Iwodani ("Sulfur Valley") area at the W base of the volcano. Personnel from Kagoshima and Tokyo Universities, Tokyo Industrial College, and the JMA observatory monitored the temperatures and the chemical composition of the vapor. The highest temperature measured was 98°C in March, the same as during the past 2 years. The gas content varied from 90% CO_2-10% H_2S to 70% CO_2-30% H_2S.

New fumaroles appeared in a residential area. Because of the dense gas, civil police closed a parking area and part of a road. . . . Landslides in 1959 and 1971 accompanied weak phreatic explosions in the geothermal area but there were no casualties.

Information Contact: Same as 4:6.

1/82 (7:1) On 8 Jan JMA personnel discovered a 6 m-long, 30 cm-wide, 2 cm-thick sulfur flow that had emerged from Fumarole No. 6 in Shinmoe-dake (fig. 8-9), a crater in the summit area of the Kirishima volcanic complex. Pieces of sulfur and tar-like material were scattered around the fumarole.

Increased geothermal activity was observed on the inner wall of Shinmoe-dake in Dec. The measured temperature of Fumarole No. 6 was 184°C, the highest since JMA's Kagoshima Observatory began summit area temperature measurements in 1979. Vegetation on the N inner wall had been damaged by the increased activity. Fumarolic activity in the Iodani area, about 5 km WSW of Shinmoe-dake, had increased from Nov 1980 through 1981. On 8 Jan JMA observers found that this activity had declined to its usual level. The last major activity at Shinmoe-dake was a steam explosion in 1959, when tephra was ejected, mainly from the fissure trending W from the crater.

Information Contact: Same as 4:6.

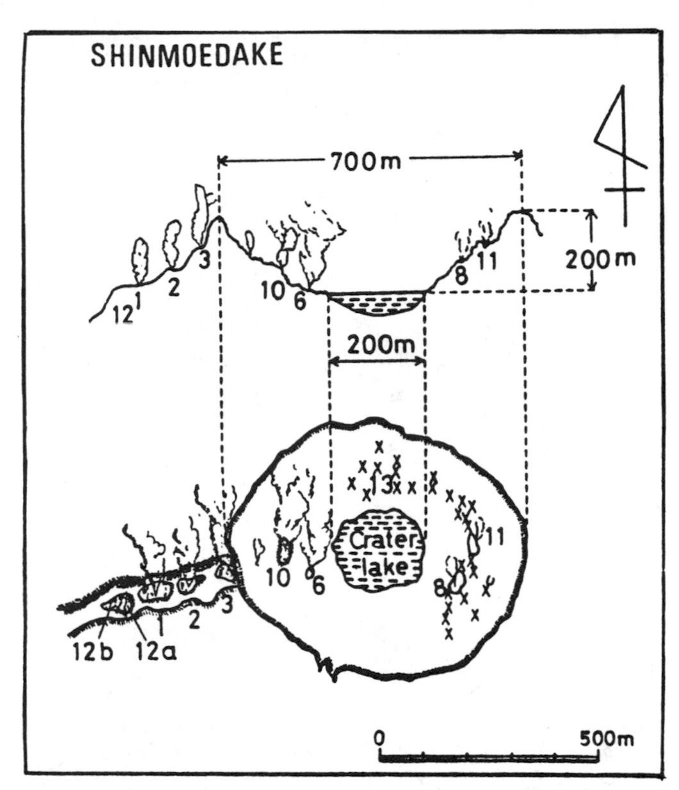

Fig. 8-9. Cross-section and map of Shinmoe-dake crater. Numbers identify individual fumaroles.

UNZEN

Kyushu, Japan
32.75°N, 130.30°E, summit elev. 1360 m
Local time = GMT + 9 hrs

In W Kyushu, the large volcanic complex of Unzen is linked to the city of Nagasaki, 40 km to the W, by a narrow isthmus. Only 5 historic eruptions are known, but that of 1792 was tragic. The volcano is nearly surrounded by well-populated bays, and volcanogenic tsunamis are a major hazard.

9/84 (9:9) Local seismic activity, which had increased since May, peaked in Aug when 6370 seismic events, including 409 felt events, were recorded (fig. 8-10). During the evening of 6 Aug the number of recorded earthquakes increased rapidly (table 8-21). Two large felt shocks and several minor felt shocks occurred within 10 minutes, causing 10 landslides and injuring 2 persons. Seismic activity has since declined, but remained above normal as of 26 Sept.

Epicenters of the swarm, determined by the JMA seismic network, were concentrated on the W coast of Shimabara Peninsula about 8 km W of Fuken-dake, the most recently active peak of the Unzen Volcano Complex (fig. 8-11). The depth of seismic sources deepened

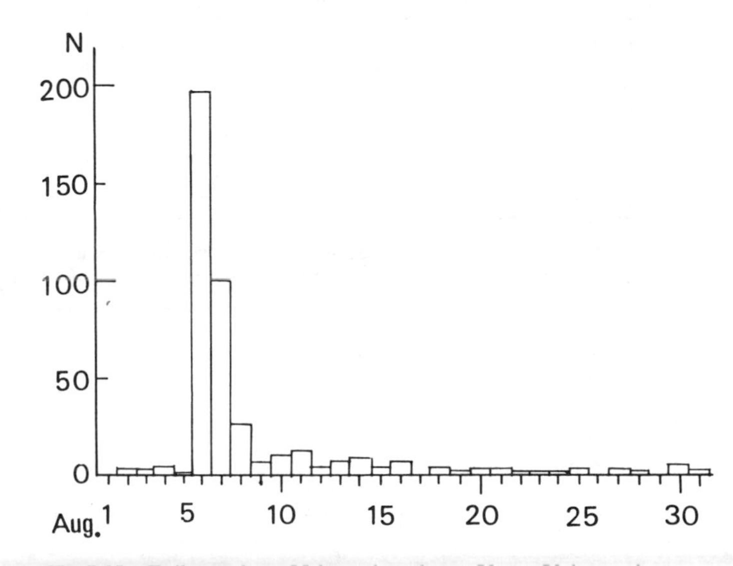

Fig. 8-10. Daily number of felt earthquakes at Unzen Volcano, Aug 1984.

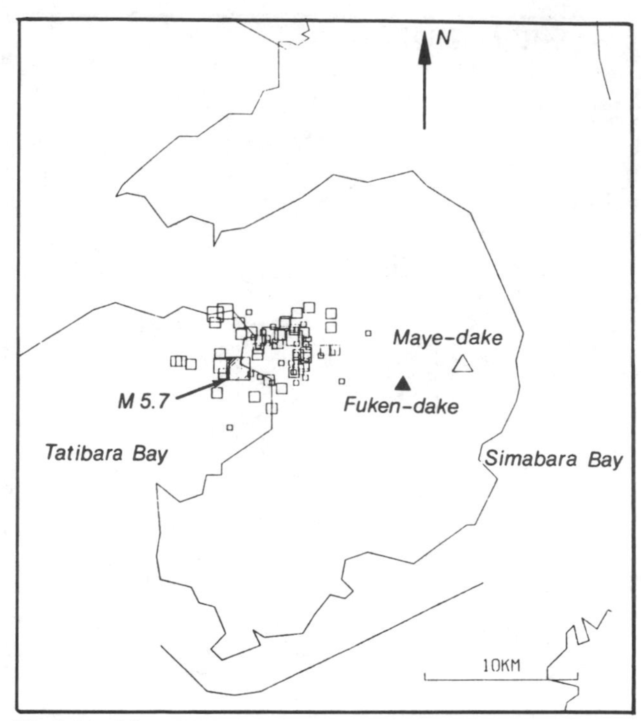

Fig. 8-11. Map of Shimabara Peninsula. The Unzen Volcanic Complex occupies most of the peninsula. Fuken-dake (solid triangle), Maye-dake (open triangle) and epicenters (open squares) of events, 6 Aug, 1300-7 Aug, 0325 are shown.

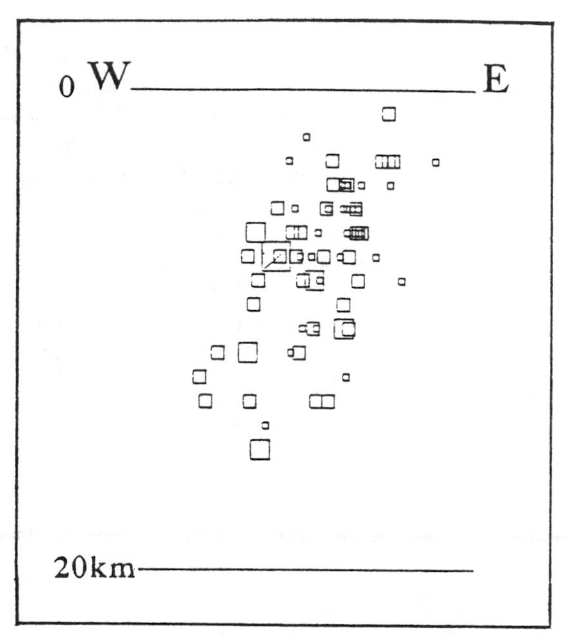

Fig. 8-12. E-W cross section of the W side of Shimabara Peninsula, showing depth distribution of earthquakes in fig. 8-11.

Time	Magnitude	Depth of focus	Intensity
1728	4.8	6 km	IV
1730	5.7	7 km	IV
1735	4.0	15 km	IV
1738	5.0	11 km	V
1841	4.1	10 km	IV
1949	4.0	8 km	IV

Table 8-21. Earthquakes of M $\geq$ 4.0 (JMA), 6 Aug 1984. Intensities were measured at the JMA Unzen-dake Weather Station, using the JMA scale (0-VII).

from E to W (fig. 8-12). The characteristic hypocenter distribution coincides with previous results from the University of Kyoto and JMA. Earthquake swarms without eruption have often occurred at Unzen.

No eruptive activity has been observed since 1792, when explosions at Fuken-dake in Feb were followed by lava extrusion in March, and earthquakes and landslides in April and May. On 21 May an avalanche from Maye-dake (near the E coast), caused by an earthquake, created a major tsunami in Shimabara Bay that caused [15,000] fatalities and much destruction.

Information Contacts: JMA, Tokyo; Kyodo News Service, Tokyo.

Further Reference: Sawada, Y., 1978, Seismic activity at Unzen-dake Volcano and the Unusual Number of its Occurrence and Frequency; *Papers in Meteorology & Geophysics*, v. 29, p. 83-96.

ASO

Kyushu, Japan
32.88°N, 131.10°E, summit elev. 1592 m
Local time = GMT + 9 hrs

75 km E of Unzen and 150 km N of Sakura-jima, Aso has produced more historic explosive eruptions than any other volcano in the world. Many small cones lie within a 20 km-diameter caldera, and one of them (Naka-dake) has erupted 163 times since Japan's first documented eruption in 553 AD.

General Reference: Ono, K., Kubotera, A, and Ota, K., 1981, Aso Volcano *in* Kubotera, A. (ed.), *Symposium on Arc Volcanism Field Excursion Guide to Sakura-jima, Kirishima, and Aso Volcanoes;* Volcanological Society of America, p. 33-52.

5/77 (2:5) Normal fumarolic activity in [Crater 1 of] Naka-dake, one of the central cones within Aso caldera, began to increase on 31 March. Slight rumbling on 4 April was succeeded on 11 April by heavy rumbling, further increase in gas emission, and ejection of some grayish ash. On 12 April, the grayish [plume] contained many fist-sized cinders, which fell on the floor of Naka-dake's crater. Glow was seen 21 April but the [plume] had turned white and contained only a little ash. There was slight [ash] emission from the crater bottom, which had contained a small hot water pool.

Information Contact: JMA, Tokyo.

6/77 (2:6) Minor ash emission, similar to the April activity occurred in mid-May. By late May, the ash emission had abated.

Information Contacts: JMA, Tokyo; D. Shackelford, CA.

7/77 (2:7) Emission of gas and tephra from the [Crater 1] of Naka-dake continued through June. Much ash and sand, and occasional fist- to hand-size cinders were ejected. Activity declined in early July, but an explosion on 20 July at 1321 sent a column of ash . . . about [1500] m above the crater and [ejecta slightly] injured [3] persons nearby.

Information Contacts: JMA, Tokyo; T. Tiba, National Science Museum, Tokyo; UPI.

8/77 (2:8) The increased activity on 31 March was accompanied by continuous large amplitude tremor. A small vent formed on 8 May and emitted ash. During the night, a 2-10 m "flame" was observed. In late May, more fist-sized ejecta fell on the crater floor. An earthquake swarm occurred 3 June, and large-amplitude tremor, some of which could be felt near the crater, was recorded 18 June. Small scale eruptions deposited ash in early June. These eruptions increased in strength 20 – 23 June, depositing fist-sized ejecta inside the crater.

Activity then declined until 20 July, when explosions at 1321, accompanied by airshocks, projected black and gray ash clouds 1500 m above the crater. Ash fell up to 500 m from the crater, reaching a maximum depth of 30 cm, and blocks up to 80 cm across fell on the rim. Similar explosions occurred at 1341 on 22 July, producing a 1300 m cloud and depositing considerable ash around the crater. Activity then declined. No fresh magma was ejected at any time during the eruption.

Information Contacts: JMA, Tokyo; D. Shackelford, CA.

12/77 (2:12) A steam and ash cloud rose more than 300 m from Naka-dake at 0704 on 8 Nov. Ash ejections continued through the end of Nov, producing substantial ashfalls in and near the crater on 14 and 18 Nov.

Information Contact: JMA, Tokyo.

2/78 (3:2) Steam and ash emission from Naka-dake continued in Dec and Jan, producing occasional slight ashfalls nearby. Frequent earthquakes and volcanic tremor were recorded.

Information Contacts: JMA, Tokyo; D. Shackelford, CA.

6/78 (3:6) Increased activity from [Crater 1] of Naka-dake continued through April. The vapor cloud contained little or no ash during March, but on 4 April the emissions increased in volume and a small ashfall was observed. Continuous emission of a grayish-white cloud began 7 April and lasted through the end of the month. Small quantities of mud and fine particles were ejected from vents in Naka-dake crater, but none of this material rose more than 20 m above the crater bottom. None of the ejecta contained evidence of fresh magma. Short-period volcanic tremor recorded by seismographs near the crater continued through April.

Information Contact: Same as above.

9/78 (3:9) Steam and ash emission from [Crater 1] of Naka-dake occurred 29 May – 1 June, producing an ashfall on the N side of the crater. Rumbling accompanying the activity could be heard from [JMA's Aso-san Weather Station] 1 km from [the crater of] Naka-dake. Mud and [block] spattering was observed in mid June, but the ejecta rose only 40-50 m and remained within the crater. No further activity has been reported.

Information Contact: Same as above.

6/79 (4:6) An explosive eruption from Naka-dake crater began at 1510 on 13 June. Activity lasted more than 1 hour, producing a 1500-2000 m-high steam and ash column, and thundering sounds. [Blocks] larger than a man's head were thrown 400 m above the crater rim. Kyodo radio reported that the eruption reintensified during the night of 15 – 16 June. Hot tephra was ejected to 200 m above the crater rim, accompanied by a roaring noise. Doors and windows rattled in nearby houses and some residents fled the area, according to police reports. . . .

Information Contacts: T. Tiba, National Science Museum, Tokyo; Kyodo Radio, Tokyo.

7/79 (4:7) A small explosion occurred at the beginning of June from Crater 1 of Naka-dake (fig. 8-13). Activity then declined for 10 days, although an earthquake was felt on [9] June. Loud rumbling and frequent small explosions began on 12 June and incandescent block ejection was seen that night. At 1510 on the 13th, a larger explosion produced a 2 km-high ash column and the ejection of numerous blocks. Lightning could be seen in the ash column. Ash emission was observed until nightfall and incandescent blocks continued to be ejected every few minutes through the night of 13-14 June. A field investigation on 14 June revealed more than 10 cm of ash and many scoria bombs and blocks (up to 70 cm in diameter) in the summit area.

Frequent periods of activity, consisting of weak and continuous ash emission or explosive block ejection, occurred daily until 25 June. Incandescent scoria was seen at night on the 14th, 15th, and 16th. A new vent (791 pit)

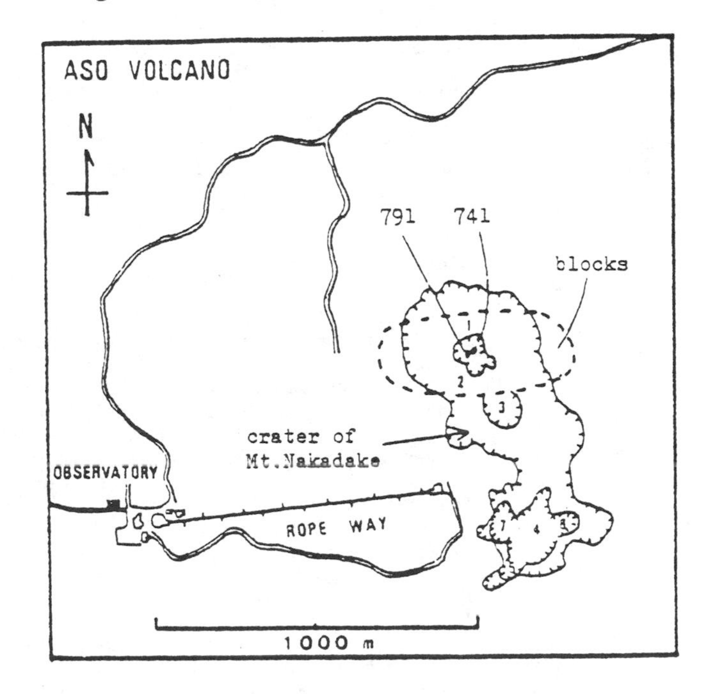

Fig. 8-13. Map of Naka-dake summit area, Aso volcano. Naka-dake's crater includes smaller craters 1-7 (hachured lines). Vents 741 (active since 1974) and 791 (formed during the June activity) are shown in crater 1. The dashed line delineates the area where tephra larger than 3 cm in diameter fell during June, 1979.

was formed during the June eruption at the bottom of Crater 1 near 741 pit, which had been active since 1974 (fig. 8-13). Since the end of June, activity has declined to occasional weak ash emission. Seismicity remained low in June, in contrast to the strongly increased seismicity that accompanied the eruptions of 1977 (fig. 8-14).

Information Contact: JMA, Tokyo.

8/79 (4:8) The following is from JMA.

"The eruption of Aso continued through July (table 8-22). Seismicity remained low. It is empirically known for Aso volcano that the amplitude of continuous tremor becomes large before an eruption and remains large throughout the eruptive period, as shown in fig. 8-15. The local disaster control group for Aso volcano [closed the area within 1 km of the crater] at 1310 on 11 June because high-amplitude continuous tremor had begun to be recorded at JMA's Aso-san Weather Station [originally referred to as Aso Observatory] during the early morning. Civil Defense personnel kept people 1 km from the crater, visited by many persons when the volcano is inactive. The eruption began during the evening of 12 June. No casualties have occurred.

"The second characteristic event of this eruption was the decrease in the amplitude of the continuous tremor just before the largest explosion, on 13 June. The extraordinary decrease in amplitude was observed for 11 and 1/3 hours, from 0336 to 1457 on the 13th. The explosion occurred at 1510, after a steep increase in tremor amplitude for 13 minutes. Many cases of a decrease in tremor amplitude before a larger explosion are known for past eruptions at Izu-Oshima and Aso. For example, a decrease lasting 4 days was recorded before Aso's large explosion of 31 Oct 1965."

Kyodo radio reported that [3] persons were killed and [11] injured by blocks ejected at about 1300 on 6 Sept. The area within 1 km of the active vent remained off limits.

Information Contacts: JMA, Tokyo; Kyodo radio, Tokyo.

9/79 (4:9) The following is from JMA.

"Eruptive activity continued through early Sept. Ash eruptions occurred almost every day from mid-July until 10 Aug (table 8-23) and ash fell on towns near the vol-

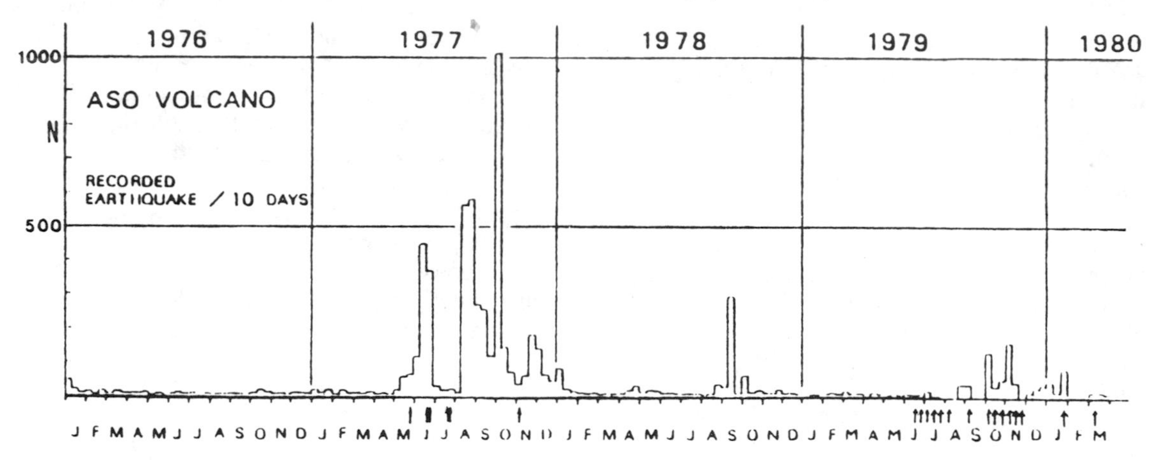

Fig. 8-14. Number of recorded seismic events/10 day period at Aso, Jan 1976-March 1980. Arrows represent eruptions. [Originally from 5:4.]

1979		Activity
late May-early June		Intermittent phreatic ejections; wet ash and blocks fell on crater floor, common activity at Aso.
June	[9]	Felt earthquake [at 0904].
	11	Continuous tremor amplitude increased during the morning; [zone 1 km from crater closed] at 1310.
	12	Further tremor amplitude increase in the morning but sudden decrease in amplitude at 1821; ash eruption began in the evening; a few incandescent blocks seen at night.
	13	Further tremor amplitude decrease at 0336, then steep increase at 1457; large explosion at 1510 produced a 2 km-high ash cloud; blocks fell 350 m from the vent.
	14-15	Intermittent ash eruptions; incandescent blocks; very loud rumbling ([people said] the strongest in 50 years) began at 2340 on 15 June and lasted until the next morning, rattling doors and windows.
	16	Activity decreased briefly in the morning, but loud rumbling resumed at 1110 and explosions started again, ejecting incandescent scoria.
	17-26	Ash ejection every day; 25 cm deposited in the summit area 12-23 June; rumbling declined during the morning of the 18th, then resumed at 0950 on the 20th, continuing through 27 June; block ejection was observed on 3 days, and reflected glow could be seen on 22 and 23 June; lightning was seen on 19 June.
	27-30	Poor weather prevented observations; estimated volume of ejecta in June, 1.4 x 10^6 tons.
July	1-5	Ejection of ash and incandescent blocks, accompanied by rumbling; largest scoria bombs measured at 86 cm in diameter.
	6-19	Ash ejection and weak rumbling; no blocks observed.
	20-31	Ash and incandescent blocks ejected; lightning and reflected glow seen; continuous weak rumbling, punctuated by occasional louder periods; July ejecta volume estimated at 1.64 x 10^6 tons.

Table 8-22. Activity at Aso, late May-early Aug, 1979.

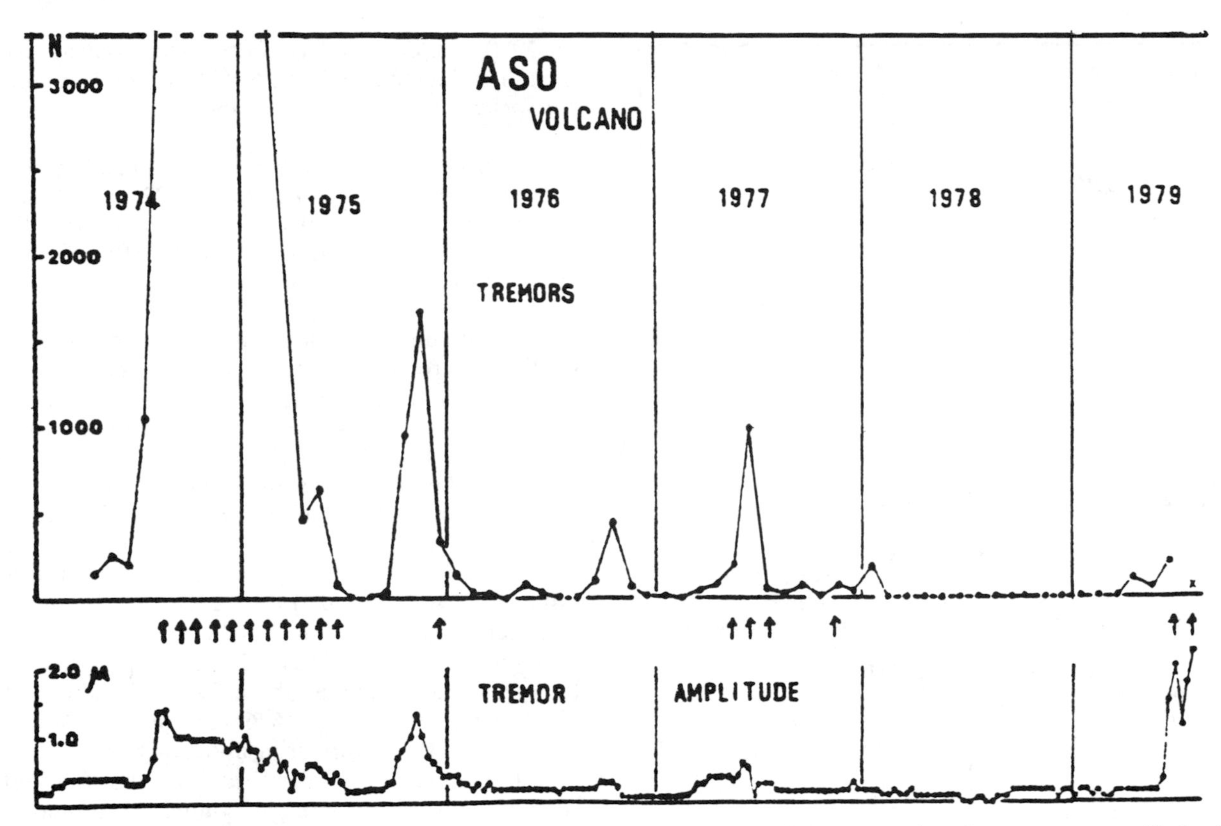

Fig. 8-15. Aso seismicity since 1973. Monthly number of isolated tremor events, *top,* plus 10-day means of continuous tremor amplitudes during the same period, *bottom.* Arrows show months in which Aso was erupting.

cano; ash reached Takeda City, 30 km NE of Aso, in early Aug. Strong rumbling resumed on 10 Aug, but the volcano suddenly stopped emitting ash at about 1300 that day. The rumbling lasted until 26 Aug and was occasionally heard at the towns of Aso-machi and Ichinomiya-machi, 10 km from the crater. A steep decrease in the amplitude of the recorded continuous tremor took place at about 0900 on 27 Aug and the volcano was very quiet (no ash or block ejection, nor any rumbling) until 6 Sept.

"A loud explosion occurred at 1306 on 6 Sept. A dark ash cloud, in which lightning was seen, rose 700 m. The air shock reached 0.8 millibars and the ground shock had an amplitude of 17 μm at the JMA's Aso-san Weather Station, 1.2 km from the crater. Three tourists were killed, 2 injured seriously, and 9 slightly, by falling blocks 10-20 cm in diameter at a site 0.9 km from the vent (fig. 8-16). Numerous blocks pierced the roof of a ropeway station (also 0.9 km from the vent), made of concrete as thick as 25 cm. A few of the lesser injuries occurred inside the station house. People said that the blocks that fell around the station house were hot and the cores of some of them were dimly glowing. Ash reached Oita city, 65 km NNE of Aso. The activity declined to white vapor emission 7 minutes after the explosion and no further eruption had occurred as of 12 Sept.

"The amplitude of recorded continuous tremor remained small (about 2 μm) through the explosion, became large (to 17 μm) 40 minutes after the explosion, then declined gradually to around 5 μm 14 hours after

1979		Activity
Aug	1	Ejection of ash and incandescent blocks; reflected glow; lightning.
	2-3	Ejection of ash and incandescent blocks; lightning.
	4	Ash ejection; lightning.
	5-6	Ejection of ash and incandescent blocks.
	7	Bad weather - no observations.
	8-9	Ejection of ash and incandescent blocks; lightning.
	10	Ejection of ash and incandescent blocks stopped about 1300, then loud rumbling began.
	11-26	Loud rumbling; white steam emission, a few ash ejections.
	27	161 mm of rainfall; rumbling stopped; steep decline in continuous tremor amplitude at 0900.
Aug 28-Sept 5		No eruptive activity [but slight ashfall 5 Sept; 4:10]; tremor remained weak.
Sept	6	Large explosion at 1306, killing 3 persons; tremor amplitude increased sharply at 1350 then decreased gradually over the next 14 hours
	7-12	White steam emission; weak tremor

Table 8-23. Activity at Aso, 1 Aug-12 Sept, 1979.

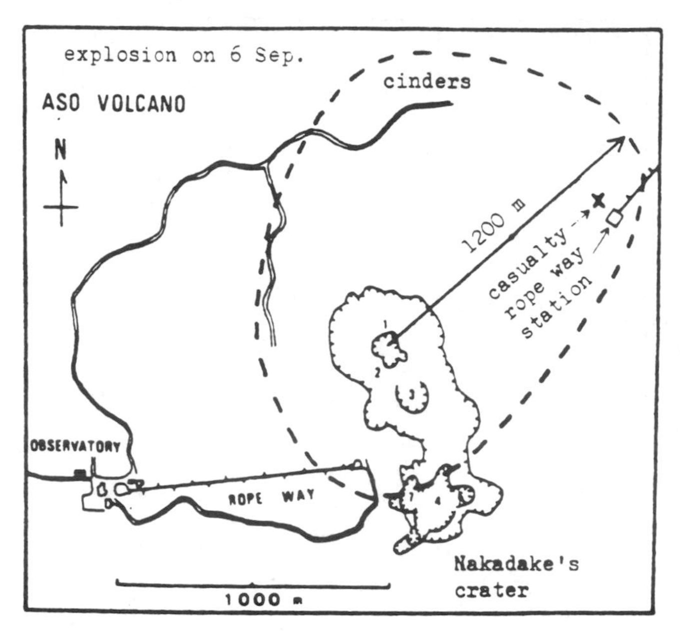

Fig. 8-16. Summit area of Aso. Casualties from the 6 Sept 1979 explosion took place at the site marked X. Cinders larger than 3 cm fell in the area enclosed by a dashed line, a maximum of 1200 m from the vent, Crater 1 of Naka-dake. Craters 1-7 are shown within the larger Naka-dake Crater, by hachured lines.

the explosion. It is not known whether the low-amplitude stage, which lasted from 27 Aug to just after the explosion, was an example of 'amplitude decrease prior to explosion'(4:8). Seismicity remained relatively low both before and after the explosion.

"The restricted area [designated] on 11 June by the local disaster control committee for Aso volcano was still [closed] on 6 Sept. The committee will re-examine the size of the restricted area (within 1 km of the crater), although the casualties occurred in this area.

"The summit area of Naka-dake was surveyed by JMA personnel and by Koji Ono of the Japan Geological Survey 8-11 Sept. Cinders larger than 3 cm were scattered in the area enclosed by a dashed line in fig. 8-16, reaching 1.2 km from Crater 1. A large block found 0.3 km from Crater 1 was 4.6 × 2.6 × 2.6 m and weighed about 50 metric tons. No scoria or other essential fragments were found. The explosion is considered to be a steam explosion, and may have been caused by heavy rainfalls on 27 Aug (161 mm) and 3-4 Sept (127 mm)."

Information Contact: JMA, Tokyo.

Further Reference: Wada, T., Kikuchi, S., and Ono, H., 1980, The Explosion of Naka-dake, Volcano Aso on the 6th of September, 1979; *Bulletin of the Volcanological Society of Japan*, v. 25, p. 245-253.

10/79 (4:10) After the large 6 Sept eruption that killed 3 persons, Aso remained quiet through 22 Sept. Occasional weak ash emission took place 23-29 Sept (table 8-24). All of the eruptive activity occurred from the 6 Sept vent.

Information Contact: Same as 4:9.

11/79 (4:11) Frequent ash ejections resumed on 24 Sept and continued through late Nov. During Oct, no blocks were seen to reach the rim of Naka-dake Crater nor were incandescent blocks observed at night. By the end of Oct, the concentration of ash at the JMA's [Aso-san Weather Station] (1 km from the active vent) had reached more than 10 kg/m^2, equal to about 1 cm of ash thickness. Although continuous tremor amplitude had correlated well with June–Sept eruptive activity, amplitudes remained low (about 0.5) during Oct. The number of local earthquakes also remained low in Oct.

A characteristic decrease in the amplitude of continuous tremor began at about 0900 on 2 Nov, lasting until a large explosion at 1626. An eruption cloud rose 1.5 km above the crater during about an hour of ash ejection. Four mm of ash fell at the [Weather Station]. A survey by [Weather Station] personnel 2 days later found scoria up to 200 m from the vent, overlying 0.6 m of ash that had fallen in the summit area since the eruption began 12 June. The tremor amplitude decrease was the third since June that had preceded a sizeable explosion. An alert was issued from the [Weather Station] one and a half hours before the explosion. No casualties occurred.

Ash emission in Nov was stronger than in Oct, causing heavy ashfalls near the volcano. Slight ashfalls occurred occasionally at Mt. Takachiho (110 km S), Kumamoto city (40 km W), and in Oita Prefecture (50 km E). Ejection of incandescent blocks was observed at night on 11 and 19 Nov, for the first time since 6 Aug. Tremor amplitude increased through most of Nov, but declined late in the month. The Strombolian activity of June, July, and early Aug occurred while tremor amplitude was high.

Information Contact: Same as 4:9.

Further Reference: Tanaka, Y., Tsuchiya, Y., and Yamaura, Y., 1981, Detection of Volcanic Smoke and Ashfall Area at Aso from Landsat Data; *Papers in Meteorology & Geophysics*, v. 32, no. 4, p. 275-291.

1979		Activity
Sept	7-22	Quiet; no ashfalls observed.
	23	Ash eruption, lasting about 10 minutes.
	24	10-minute ash eruption in the morning; continuous ashfall 1450 until night; accumulation less than 0.1 mm.
	25-26	No ashfalls observed.
	27	Continuous ashfall from 0540 to 1730, accumulating about 0.2 mm.
	28	No ashfalls observed.
	29	About 0.1 mm of continuous ashfall between 1510 and 1645.

Table 8-24. Activity at Aso, 5-30 Sept 1979. Ash thicknesses, at about 1 km from the source, were estimated by weighing the small amount of material accumulated on a measured area.

1/80 (5:1) Activity stopped on 28 Nov after strong ash emission through most of the month (table 8-25). Between June and Nov, ash caused about 1 billion yen ($4 million) in damage to crops and forests. No further ashfalls were observed until an explosion at 2107 on 26 Jan deposited 3 cm of ash and fist-sized scoria on the rim of Naka-dake, the source crater. A small amount of ash fell on Aso-machi town, at the base of the volcano. The explosion caused no damage, and the volcano returned to quiescence the next day.

Information Contact: JMA, Tokyo.

1979	Ashfall ($x 10^3$ tons)
June	1420
July	1620
Aug	1590
Sept	300
Oct	970
Nov	3270
Dec	0
Total	9170

Table 8-25. Monthly ashfall in thousands of tons, June-Dec 1979, as estimated by personnel from JMA's [Aso-san Weather Station].

4/80 (5:4) A weak and brief emission of ash from Aso occurred on 8 March, producing ashfall on the S flank. Since strong ash emission stopped on 28 Nov 1979, eruptive activity had been confined to ejection of a small amount of ash and scoria, accompanied by a strong air and ground shock, on 26 Jan, and emission of white vapor at other times.

The amplitude of continuous tremor recorded at [Aso-san Weather Station] declined in Dec 1979 and has remained low through March (fig. 8-17). The number of local earthquakes increased somewhat around the Jan tephra ejection, but declined in Feb and did not increase substantially during the March activity.

Information Contact: Same as above.

5/80 (5:5) Activity at Aso has been confined to weak but steady emission of white vapor since the brief 8 March ash ejection. The number of seismic events per day and the amplitude of the continuous tremor recorded at [Aso-san Weather Station] were both small in April and early May.

Information Contact: Same as above.

9/80 (5:9) A brief, weak explosion on 24 Sept ejected ash to about 800 m above Crater 1 of Naka-dake [after quiescence since the 8 March ash ejection]. The area within 1.5 km of the summit was closed immediately after the explosion but reopened 2 days later.

Information Contact: Same as above.

6/81 (6:6) Ash and block ejection from Crater 1 of Naka-dake, the northernmost of 7 in Naka-dake, was observed at 1230-1300 on 15 June, after 9 months of quiescence. Blocks rose to 30 m, but fell within the 100 m-diameter crater. 1 μm ground shocks were recorded at 1239 and 1244, and a 3.7 μm shock at 1251 [at the Weather Station]. Activity then subsided. The explosions caused no damage. The area within 1 km of the summit, closed immediately after activity began, was reopened 17 June. Aso-san Weather Station personnel [visited the crater on 15 June and] observed that the greenish water pooled in Crater 1 since Oct had become gray tinted [but returned to its usual green the next day]. The [level of some points on the surface of the water] rose intermittently. Naka-dake is the historically active part of the Aso volcanic complex.

Information Contact: Same as above.

10/84 (9:10) A warning of increased volcanic activity was issued by the [Aso-san Weather Station] on 11 Oct. On the morning of 24 Oct, ash was ejected from a fumarole that had formed in mid-Sept on the lowest part of the E inner wall of [Crater 1 of] Naka-dake. A small gray plume rose to 300 m above the fumarole at 0920, but the ejection was too weak to send ash beyond the crater rim.

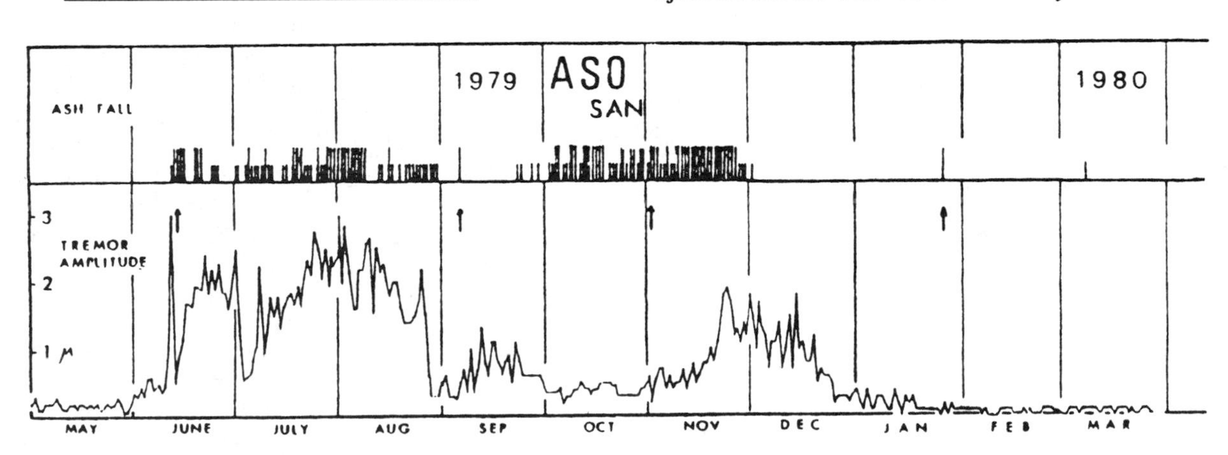

Fig. 8-17. Daily mean amplitude of continuous tremor, *bottom*, and occurrence of ash emission, *top*, observed from the JMA's [Aso-san Weather Station], May 1979-March 1980. Larger ashfalls are represented by longer lines in the top section of the figure. The arrows in the bottom section indicate strong explosions on 13 June, 6 Sept (3 persons killed), 2 Nov 1979, and 26 Jan 1980.

Entry to the area within 1 km of the crater was prohibited shortly before 1000. Another plume rose to 200 m at 1030 the next day, accompanied by intermittent ejection of small blocks, as large as fist size. Activity had subsided by 1550.

The level of water in Crater 1 has gradually decreased since early April (fig. 8-18). A part of the crater bottom could be seen in mid-Sept. As of 26 Oct, only 30% of the bottom was covered by the hot water.

Information Contacts: JMA, Tokyo; A. Izumo, Yokohama Science Center; *Japan Times*; Kyodo News Service, Tokyo.

1985		Ash (g/m^2)
April	2	141
	6	130
	7	14
	12	338
	13	37
May	12	2.5
	14	1550
	[16	757.9]
	22	119.7

Table 8-26. April-May 1985 daily ash accumulation at the JMA's [Aso-san Weather Station], 1.2 km SW of Crater no. 1 of Naka-dake. [Ash was measured each day at 0900 and represents amount deposited the previous 24 hours. There was no accumulation on other days in April and May. Data for May originally included in table 8-27.]

12/84 (9:12) Activity has gradually increased at Crater 1 of Naka-dake. After weak ash emission 2–3 Jan, an ash-laden plume has been occasionally observed since 11 Jan. A gray plume rose to 700 m above the crater rim on the 18th. . . . The ash included juvenile material.

Information Contact: JMA, Tokyo.

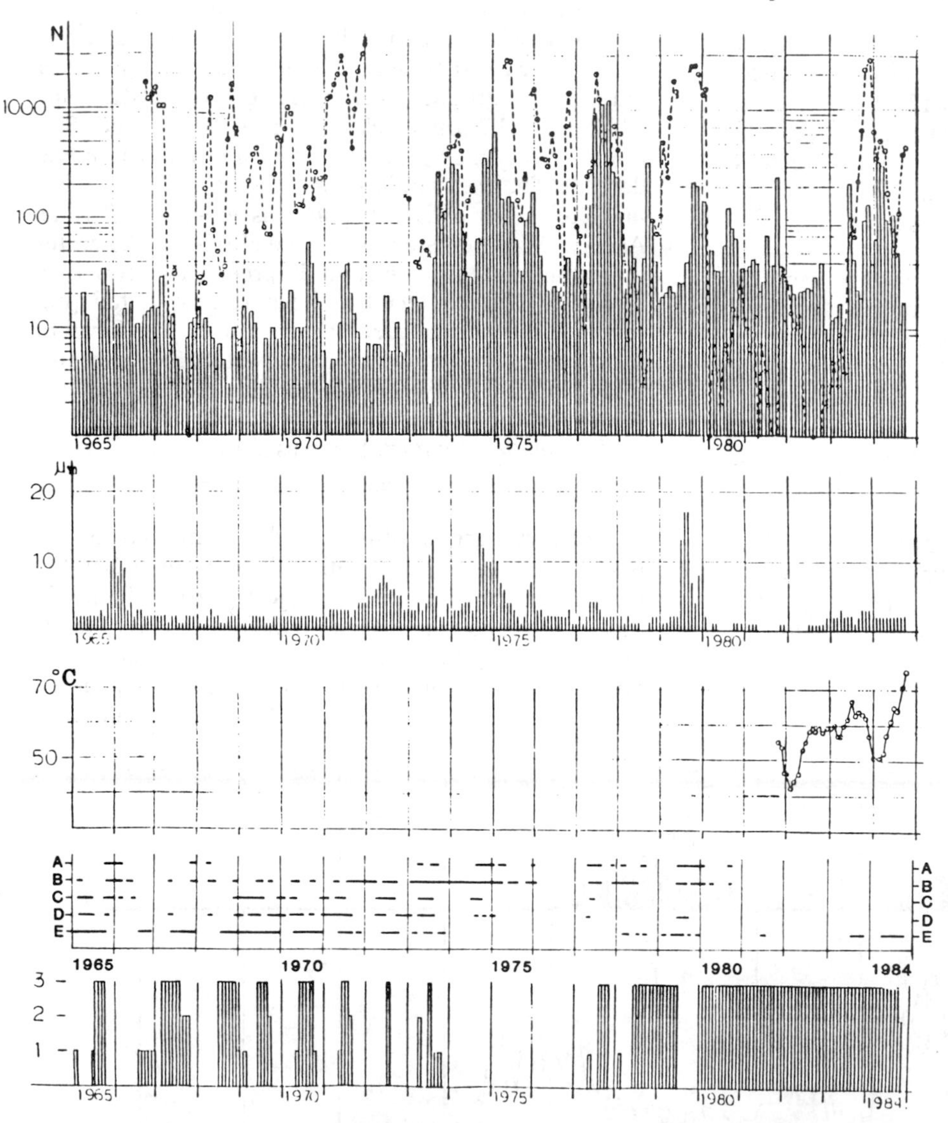

Fig. 8-18. Activity at Aso Volcano, 1965-84, courtesy of JMA. *Top* to *bottom:* monthly number of isolated tremor episodes (open circles connected by dashed lines) and volcanic earthquakes (bars); monthly averaged amplitude of continuous tremor; monthly maximum surface temperature of water in Naka-dake's Crater 1; eruptive phonomena--block ejection (A), ash ejection (B), volcanic flame (C), red-hot lava at the bottom of Crater 1 (D), and ash and mud ejection (E); quantity of water pooled in Crater 1--small amount (1), intermediate amount (2), full (3).

1/85 (10:1) Ash was ejected a little more vigorously on 31 Oct, rising to the crater rim (about 100 m above the vent). On 2 Nov, ashfall was again confined to the crater. On 5–6 Nov weak rumbling was heard at JMA's Aso-san Weather Station, 1.2 km SW of the crater.

Occasional weak ash emission that continued after 2 Nov was interrupted by water flowing into the active vent on 12 Nov. From then until 30 Dec muddy water and small rocks were ejected almost continuously, at varying intensity. During the strongest activity, on 9 Dec, this mixture fountained to an average height of 10 m above the crater floor, sometimes to more than 20 m.

Moderate ejections of ash to 150 m (as high as the crater rim) resumed 30 Dec and ended 2 Jan. On 30 Dec at 0815 an ash-laden plume rose 200 m above the rim. Ash fell on the S slope of the crater. . . . A prohibited zone within 1 km of Crater 1 was established 28 Nov.

On 11 Jan a grayish plume was observed rising 400-500 m above the crater rim, where ashfall was 1 cm thick. A little ash also fell on the S flank. Since the 11th an ash-laden plume was observed almost

May 1985	1	2	3	4	5	6	7	8	9	10	11	12	13	14	15	16	17	18	19	20	21	22	23	24	25	26	27	28	29	30	31
Incandescent Blocks	N	N	N	N	N	Y	-	Y	N	N	N	N	N	N	Y	Y	N	Y	-	-	Y	Y	-	-	N	N	Y	-	-	Y	-
Ash-laden Plume	Y	Y	Y	Y	Y	Y	-	Y	N	N	Y	Y	Y	Y	Y	Y	Y	Y	-	-	Y	Y	N	Y	N	N	Y	-	-	Y	Y
Height Blocks Ejected above Crater floor (m)						50									50	50		70			near floor	50					70			120-130	

Table 8-27. Activity at Aso in May 1985. Y = observed; N = not observed; - = not observed because of bad weather.

every day in Jan. Activity increased slightly on 18 Jan, when a gray plume rose 600 m above the rim. Volcanic flame [from a pit on the crater floor] was observed [during visits] the nights of 21 and 25 Jan.

Information Contacts: JMA, Tokyo; A. Izumo, Yokohama Science Center; *Japan Times.*

3/85 (10:3) Moderate ash-laden emissions from Naka-dake were occasionally observed in Feb and March. Gray plumes were seen on 8, 17-18, and 25-28 Feb. Ash accumulation on 26 Feb was 295 g/m^2 at JMA's [Aso-san Weather Station] (at the SW foot of Naka-dake [1.2 km from the crater]). On 1 March, the emission from the pit that had been active from Jan through Feb declined rapidly, although a new pit formed ~40 m N of the older one. Ash emission resumed the next day. Volcanic flame [from the pit on the crater floor was observed during visits] the nights of 6 and 25 March.

In late Dec, the press reported that volcanic tremors were continuously felt. However, JMA noted that average tremor amplitude was within the range of 0.1-0.3 μm through Dec and Jan. As of 1 April, tremor remained at about the same level.

Information Contact: JMA, Tokyo.

4/85 (10:4) Moderate ash-laden emissions from Crater 1 of Naka-dake were observed on almost every day in April (some daily ash accumulations are listed in table 8-26). Volcanic flame had been observed rising 20-40 m above [the pit in] the floor of Crater 1 on 25 March, and the remainder of the water pool in the crater disappeared on 16 April. Average tremor amplitude remained at around 0.3 μm in April.

On 6 May, a small eruption was observed at a new pit about 10 m E of the one that had formed on 1 March. Rocks several tens of centimeters in diameter rose to a height of 50 m above the floor of Crater 1 and ash-laden emission was almost continuous from the 1 March pit. It was not certain if juvenile material was included in the 6 May tephra.

Information Contact: Same as above.

6/85 (10:6) Since Oct 1984, activity has gradually increased at Naka-dake, site of all of the more than 140 eruptions known from the Aso complex in historic time. Activity at Crater 1 increased in May. Ash accumulated at JMA's Aso-san Weather Station on 12, 14, [16], 22, 23, 29, and 30 May (table 8-26). An ash-laden plume was observed almost daily (table 8-27). On the morning of 6 May, incandescent blocks were ejected at a new vent that had not been present the previous day. Block ejection stopped on 8 May but resumed on 15 May, when the vent increased to 40 m diameter by collapse of a wall between it and another vent that had formed in March. Incandescent blocks, including scoria a few tens of centimeters in diameter, were ejected to 50 m above the crater floor. Ash and mud jetted to the level of the top of the wall that had separated the 2 vents.

Incandescent blocks were occasionally ejected at the vent after 16 May. The most powerful May eruption occurred on the morning of the 30th; larger amounts of incandescent blocks (mainly scoria) as much as 1-2 m in diameter were ejected 120-130 m above the crater floor at 1020. The intensity of ejection then gradually declined.

Information Contact: Same as above.

7/85 (10:7) Activity at Aso has gradually increased since Oct 1984. Moderate ash emissions from Crater 1 of Naka-dake were observed almost daily from the beginning of May (10:6) until 20 June. White steam vapor dominated for the last 10 days of the month. Total June ashfall at JMA's [Aso-san Weather Station] was 1429 g/m^2.

Activity declined in July, when neither ash plumes nor deposits were observed at the [Weather Station]. On the morning of 1 July, [Weather Station] personnel found that the vent on the floor of the crater had been covered by water from heavy late-June rains. The level of water within the crater gradually rose in early July. About 70% of the crater floor has been covered by hot (60°C) water since mid-July. There were many fumaroles along the margin of the pool, and sand and water were ejected at many points on the surface. Seismic activity remained low throughout June and July.

Information Contact: Same as 10:3.

ON-TAKE

Central Honshu, Japan
35.90°N, 137.48°E, summit elev. 3063 m
Local time = GMT + 9 hrs

Near the center of Honshu, On-take is a large complex volcano 675 km NE of Aso and 200 km W of Tokyo.

Among Japan's volcanoes, its height is second only to Fuji, 130 km to its SE. No historic activity was recognized until 1979.

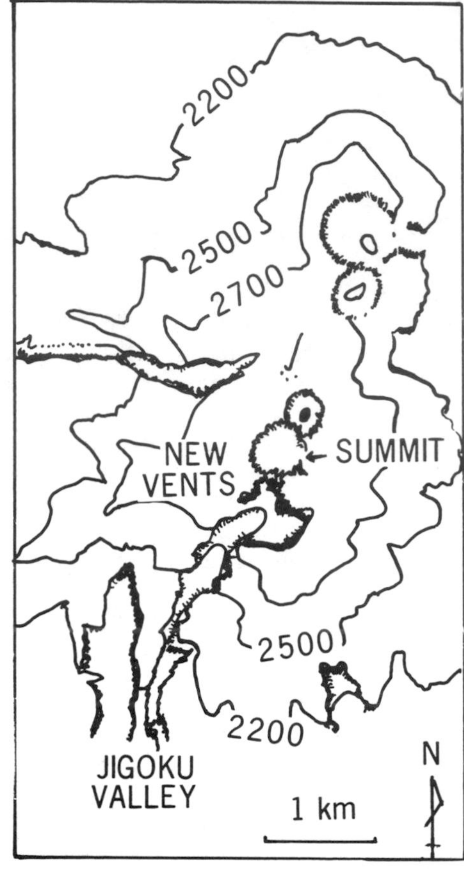

Fig. 8-19. Map of On-take, courtesy of JMA. The 500 m-long zone of 28 Oct 1979 vents is shown by a heavy NW-SE line immediately S of the summit. The 4 large craters trending NNE from the summit were formed during an eruption ^{14}C dated at 23,000 BP. The short north-northeasterly line 1 km N of the summit represents a succession of small prehistoric explosion craters.

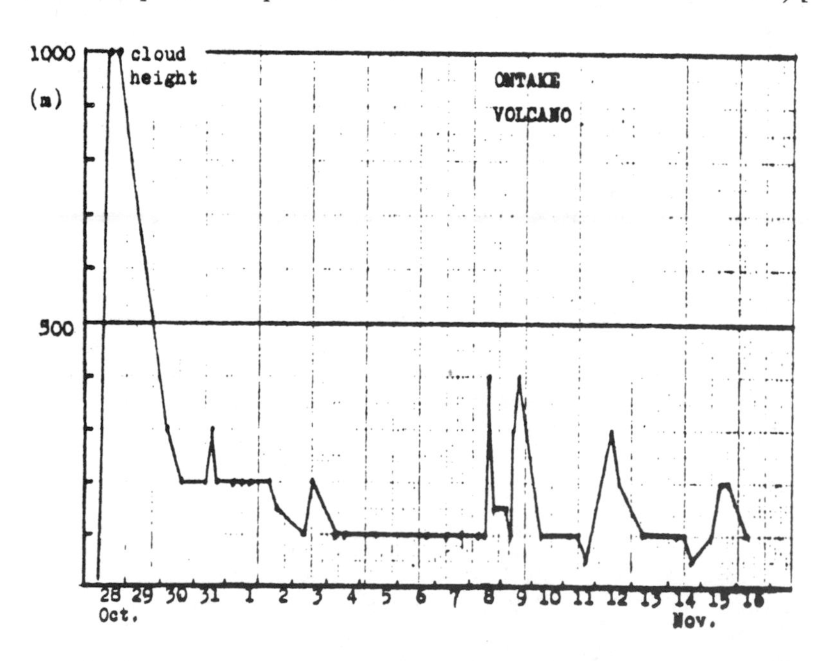

Fig. 8-20. Eruption or vapor cloud height above On-take's summit 28 Oct-16 Nov 1979. Courtesy of JMA.

10/79 (4:10) [JMA and Tokiko Tiba have made substantial revisions to this report.] On-take's first eruption in historic time began early on [28] Oct. Vapor emission apparently started at about 0500, then activity strengthened at about 0800 to emission of an [ash] column . . . [Activity increased further from around 1100.] A tephra column that included ash and lapilli . . . was rising 1500 m, and ash was falling on the E flank. Tephra clouds were ejected [continuously] during the afternoon. [Only white vapor was emitted the next day.]

Tephra was ejected from several craters that [formed in a fissure about 300 m S of the summit (fig. 8-19). There had been no fumaroles in the area]. . . .

The Kyodo radio network reported that officials were assessing damage caused by the eruption; damage was particularly heavy at Kaida, about [12 km NE of the summit (Kenga-mine)]. Some livestock had to be evacuated because of ashfall on pastureland.

An earthquake swarm, including a magnitude [5.3 shallow] event, was recorded in the On-take area in Oct 1978. . . .

Information Contacts: T. Tiba, National Science Museum, Tokyo; Kyodo Radio, Tokyo; USGS/NEIS, Denver CO.

11/79 (4:11) On-take's first eruption in historic time began suddenly before dawn on 28 Oct. No initial explosion was heard and no shock wave was recorded by JMA seismic stations, [but long-duration tremor was recorded from 0520 that may have been eruption tremor]. Ash emission continued through the day, intermittently during the morning but continuously during the afternoon, with strongest activity at about 1400. A dark ash cloud rose about 1 km (fig. 8-20). About 1.5 m of ash fell near the summit and a few tens of centimeters fell on the flank. [A few millimeters] were deposited on [the villages of] Kaida-mura (12 km NE) and Mitake-mura (14 km SE) [the nearest inhabited areas], and slight ashfall covered a wide area as far as 150 km NE of the volcano (fig. 8-21). No ejections of incandescent blocks were seen, but aerial observers reported that blocks were scattered around the vents (see below). Activity declined to white vapor emission before dawn on 29 Oct, although a slight ashfall occurred at a nearby village that morning. Steaming continued at varying intensity through mid-Nov.

During the eruption, numerous vents were active in a 500 m-long NW-SE-trending zone near the summit (figs. 8-19 and 8-22). Along some portions of the zone, curtains of steam were visible from the air on 28 Oct. Kazuaki Nakamura reported that 10 small craters were seen during an aerial inspection on 3 Nov; 4 craters were emitting vapor. Takeshi Kobayashi, who has studied the volcano for 20 years, reported that no fumaroles existed prior to the eruption in the new vent zone.

No historic eruptions are known from On-take, one of Japan's larger strato-volcanoes. Kobayashi reports that the youngest ^{14}C dates from On-take, 23,000 BP, are from scoria and lava flows composed of 2 pyroxene andesite, ejected from at least 5 craters forming a NNE-SSW line on the N flank.

Information Contacts: JMA, Tokyo; JMA also forwarded information from T. Kobayashi (Toyama Univ.); K. Nakamura, ERI, Univ. of Tokyo.

Further References: Aoki, H., 1980, *A Compilation of Reports on the Volcanic Activity and Hazards of the 1979 Eruption of On-take Volcano*; special publication by the research group for the 1979 On-take eruption, no. B-54-3, 168 pp.

Aramaki, S. and Ossaka, J., 1983, Eruption of On-takesan, October 28, 1979 *in* XVIII IUGG General Assembly, Hamburg, *Report on Volcano Activities and Studies in Japan for the Period from 1979 to 1982*, p. 1-7.

12/79 (4:12) In the 2 months since the 1-day eruption of 28 Oct, no ash ejection has been observed and vapor emission has declined. At the end of Dec, the vapor column was only 100 m high.

About 10 local seismic events per day have been recorded since instruments were installed [on 29 Oct]. Epicenters have been distributed along a sharply defined narrow linear zone extending about 2 km N-S on the SE flank. Earthquakes had been felt in the area for about 2 years, but the number of felt events declined to about 1 per week after the eruption.

Information Contact: JMA, Tokyo.

1/80 (5:1) Steam from the 28 Oct vents rose steadily to 100-200 m through Jan. No ashfalls have been observed in inhabited areas on the flanks since Nov, although snow in the summit area has been [slightly] darkened by ash and colored yellow by sulfur during this period. Muddy acidic water has been flowing from the 28 Oct vents since the eruption, killing fish in the Otaki River. The number of dead fish decreased in Jan. Local earthquake activity remained at Nov-Dec levels, about 10 recorded events per day.

Information Contact: Same as above.

4/80 (5:4) Steady vapor emission, to 100-300 m heights, continued through March. Vapor emerged from 4 of the numerous (at least 10) vents formed during the eruption in a linear, 500 m-long, NW-SE-trending zone near the summit. Airphoto data show that 3 of the 4 active vents have increased in size since the eruption. Ash in the vapor columns has repeatedly caused slight darkening of the snow in the summit area.

Information Contact: Same as above.

5/80 (5:5) Steady emission of a 100-300 m-high vapor column continued through early May. Ash had darkened snow in the summit area through March, but the snow melted in April. However, a pale dust cloud, presumably the source of the ash, continued to drift over the summit.

Information Contact: Same as above.

6/80 (5:6) Quiet emission of white vapor continued through June. The vapor rose several hundred meters above 4 summit-area vents. JMA personnel removed temporary seismic equipment installed 29 Oct and discontinued the visual monitoring begun on that date.

Information Contact: Same as 4:12.

Fig. 8-21. Sketch map of S Honshu and Shikoku Islands, Japan. Area of ashfall from On-take is shaded. Courtesy of JMA.

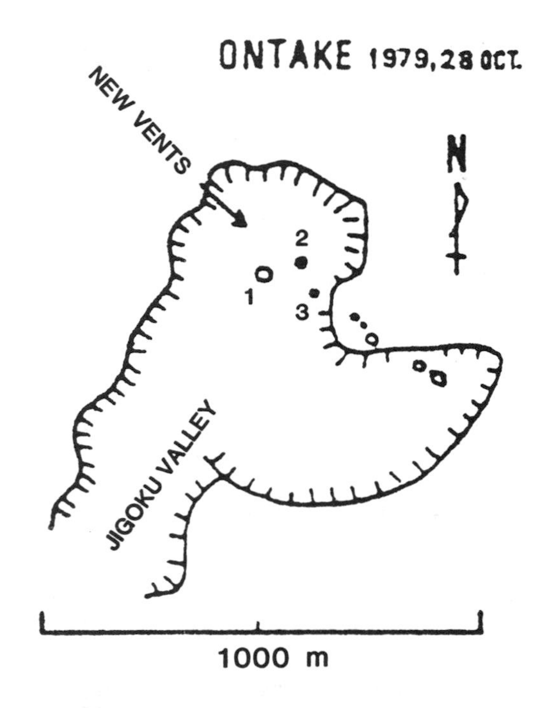

Fig. 8-22. Zone of new vents on On-take, sketched from the air on 3 Nov by Kazuaki Nakamura. Numerous vents that had existed along the fissure on 28 Oct were no longer visible at the time of this flight. The 28 Oct ash cloud was ejected [mainly] by Vent 1, the largest of the vents at 30 m diameter. The other vents emitted mostly white steam during the eruption.

NIIGATA YAKE-YAMA

Central Honshu, Japan
36.92°N, 138.07°E, summit elev. 2400 m

This lava dome, near Honshu's west coast and 125 km N of On-take, erupted in 1361 and 6 times since 1949.

4/83 (8:4) On the morning of 15 April, a local resident noticed gray-tinted snow on the N flank and a plume near the summit area. On 18 April, a joint observation team from the firehouse, the district forestry office, and the city office of [Itoigawa] (20 km NE of the volcano), visited the summit area and found that ash had fallen on the snow over the N flank. Analysis of an ash sample at Niigata University showed that the ash was not freshly magmatic, but was composed of fragmented old rock that contained mainly plagioclase, amphibole, and pyroxene phenocrysts. No seismic activity was recorded, but no JMA seismograph is installed within 30 km of the volcano.

The volcano is a 400 m-high lava dome that rests on 2000 m-high mountains underlain by sedimentary rock. During its last eruption, a phreatic explosion from fissures in the N and W flanks [near the summit] on 28 July 1974, three climbers were killed by ejecta, ash fell as far as 100 km to the NE damaging about 220 km^2 of farmland, and 2 mudflows descended the N flank.

Information Contact: JMA, Tokyo.

ASAMA

Central Honshu, Japan
36.40°N, 138.53°E, summit elev. 2550 m
Local time = GMT + 9 hrs

Asama is 110 km NE of On-take along the high "spine" of central Honshu. Its first historic eruption in 685 AD is the second oldest in Japan, and at least 108 eruptions are known since then. Fatalities have resulted from 13 of these events.

General Reference: Aramaki, S., Shimozuru, D., and Ossaka, J., 1981, Asama Volcano; *in* Aramaki, S. (ed.), *Symposium on Arc Volcanism Field Excursion Guide to Fuji, Asama, Kusatsu-Shirane, and Nantai Volcanoes*; Volcanological Society of Japan, p. 23-48.

9/80 (5:9) Seismicity has increased substantially during the past 6 months, reaching the highest monthly total since Aug 1977 (fig. 8-23). The number of recorded earthquakes peaked in Sept, declining slightly late in the month. No eruptive activity was observed, but monitoring was enhanced at Asama's JMA observatory.

Information Contact: JMA, Tokyo.

11/80 (5:11) Monthly seismicity at Asama increased from 1114 recorded events in Sept to [1365] in Oct, the highest monthly total since Aug 1977. Seismic activity decreased to 897 recorded events in Nov. No eruption or increase in steam emission were observed. Asama last erupted in 1973, when the earthquakes reached 5612 per month.

Information Contact: Same as above.

5/81 (6:5) After a substantial increase in seismicity during the second half of 1980, the number of recorded earthquakes declined Dec-Feb. A sudden increase from fewer than 400 in Feb to nearly 1000 in mid-March was not accompanied by any observed eruption or increase in steam ejection. By the end of March, seismicity had declined to the usual level of fewer than 500 recorded earthquakes per month.

Information Contact: Same as above.

8/81 (6:8) On 10 Aug at about 1000, seismic instruments 1.8 km S of the crater recorded a sudden increase in the number of local earthquakes. The number of recorded events reached 276 on 10 Aug, and 230 were registered the next day. Seismicity began a gradual decline on 12 Aug and had dropped to the normal level of about 10 events/day a few days later (fig. 8-24). No eruptive activity was observed, but vapor emission continued and JMA increased monitoring at its [Karuizawa Weather Station].

There have been several periods of increased seismicity at Asama since its last eruption in 1973.

Information Contact: Same as above.

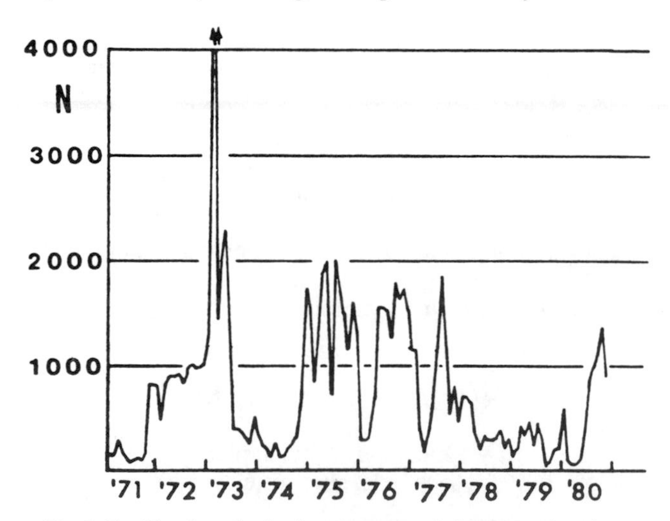

Fig. 8-23. Number of seismic events recorded per month at Asama, Jan 1971 – Nov 1980. The eruptions of Feb and March 1973 are indicated by arrows at the top of the figure.

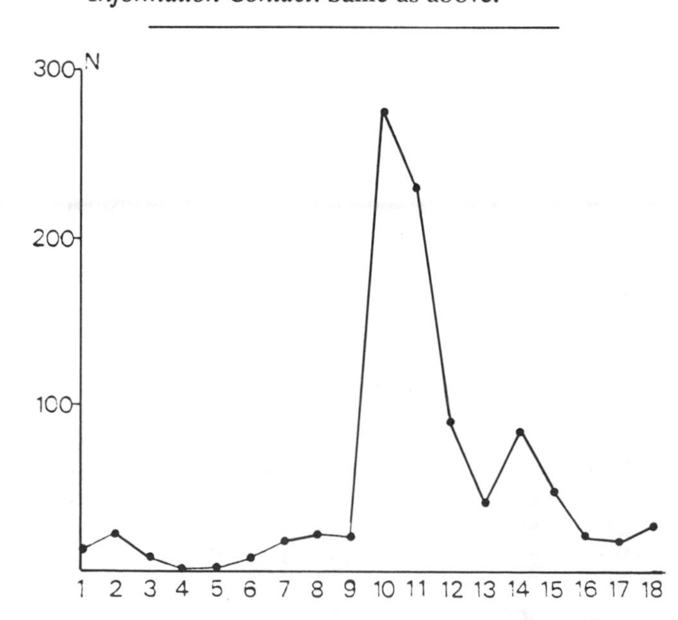

Fig. 8-24. Daily number of earthquakes recorded at Asama, 1-18 Aug 1981.

4/82 (7:4) Asama erupted explosively from the summit crater at 0225 and 0548 on 26 April. No increase in the number of discrete earthquakes was recorded before the eruption, but 55 periods of volcanic tremor were recorded in March, the largest monthly total since Asama last erupted, 1 Feb – 24 May 1973. The white vapor plume normally visible above Asama's summit grew substantially in 1981. Its maximum monthly height declined in late 1981 but had begun to increase again in early 1982 (fig. 8-25).

Local residents heard the first detonation and volcanic rumbling at 0225. A sulfuric smell persisted after the explosion. From the N rim of the summit crater, 2 staff members of the volcano museum at the N foot of Asama witnessed an incandescent eruption column and a pyroclastic flow. Snow on the N slope of Asama melted, and a telemetry cable serving seismic instruments was cut by a debris flow 150 seconds after the eruption began. GMS infrared images showed that an eruption cloud less than 50 km in diameter had risen to about 4.5 km at 0300. The earthquake accompanying the first explosion had a magnitude of 2.0 and a ground-shock amplitude of 35 μm, as recorded at JMA's Karuizawa [Weather Station] 2 km S of the volcano. B-type earthquakes, volcanic tremor, and ash ejection followed the explosion for about 2 hours. Eruptive activity continued intermittently until the second, smaller explosion, which began with an earthquake of 4 μm amplitude and was accompanied by volcanic tremor that lasted 10 minutes. A satellite image did not show this cloud at 0600, nor were remnants of the cloud from the first explosion evident.

Ashfall was noted at the Karuizawa [Weather Station] from 0200 until around 0600. Wind carried the ejecta primarily SE (fig. 8-26). Ash 2-3 cm thick accumulated 12 km SE and SW of Asama at Karuizawa and Komoro. For the first time in 21 years fine ash fell in the metropolitan Tokyo area, 130 km to the SE. The total amount of ash and lapilli was estimated to be about 10,000 tons; no juvenile tephra was found. A gray plume, 500-1000 m high, was observed throughout the day 26 April. All of the day's 89 recorded seismic events occurred after the first explosion. The number of earthquakes per day had been at normal levels since Feb (fig. 8-27) and returned to normal about 1900. By 27 April, activity was limited to a 300 m-high vapor plume. No more explosions had occurred as of 30 April. The kinetic energy of the eruption was estimated at 10^{18} ergs.

Information Contacts: JMA, Tokyo; D. Shimozuru, ERI, Univ. of Tokyo; T. Tiba, National Science Museum, Tokyo; D. Haller, NOAA/NESS; UPI.

Further References: Aramaki, S. and Hayakawa, Y., 1982, Ash-fall during the April 26, 1982 Eruption of Asama Volcano; *Bulletin of the Volcanological Society of Japan*, v. 27, no. 3, p. 203-215.

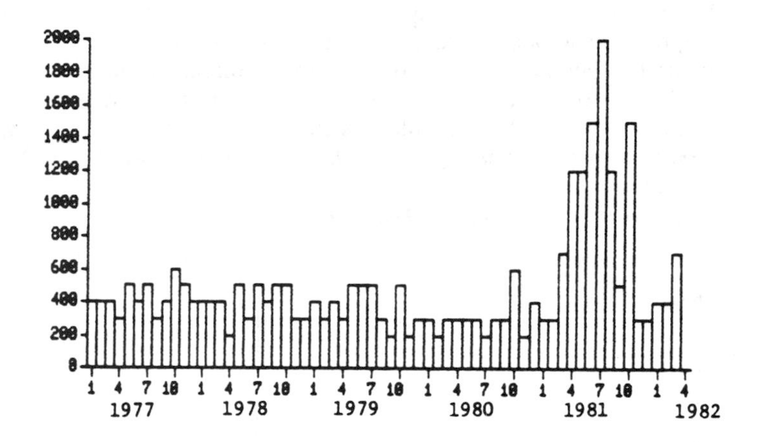

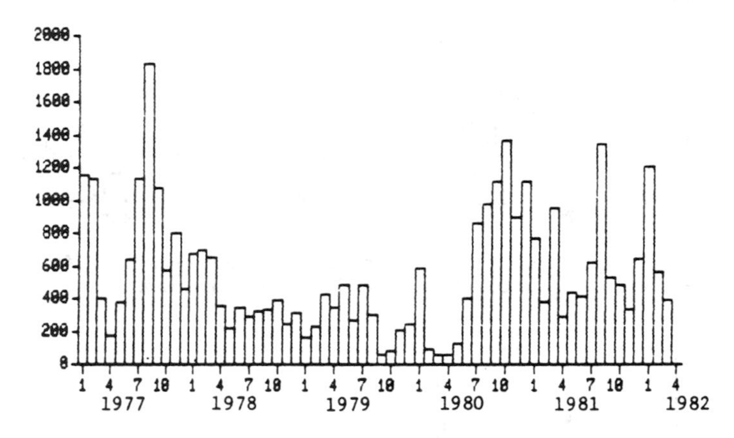

Fig. 8-25. *Top,* maximum height in meters of volcanic plumes in each month, Jan 1977 – March 1982. *Bottom,* monthly number of recorded seismic events at Asama, Jan 1977 – March 1982. Courtesy of JMA.

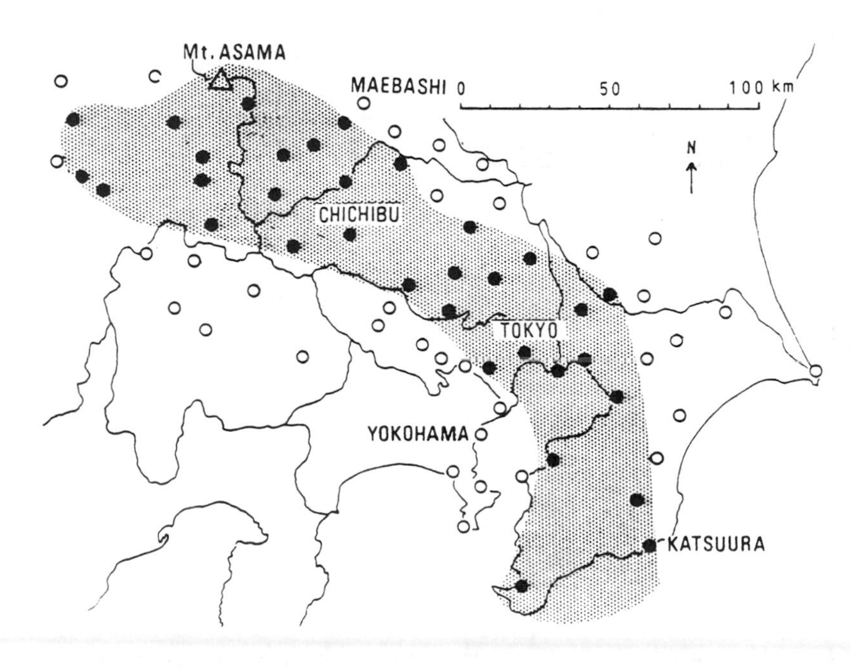

Fig. 8-26. Area of ashfall (shaded) from the eruption at 0225 on 26 April 1982. Courtesy of JMA.

Shimozuru, D., Gyoda, N., Kagiyama, T., Koyama, E., Hagiwara, M., and Tsuji, H., 1982, The 1982 Eruption of Asama Volcano; *Bulletin of the Earthquake Research Institute, Tokyo*, v. 57, no. 3, p. 537-559.

5/82 (7:5) After the 1-day eruption of 26 April, no further eruptions have been observed as of 31 May. Local seismicity has been at its normal level since March, except on the day of the eruption (fig. 8-27). Vapor emission remained at a high level; plumes that rose more than 1000 m above the summit were observed on 9, 15, and 23 May.

Information Contact: JMA, Tokyo.

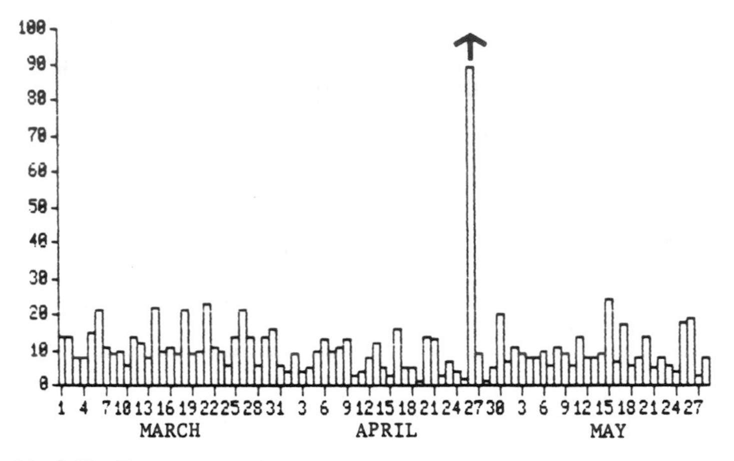

Fig. 8-27. Daily number of recorded seismic events at Asama, March-May, 1982, courtesy of JMA. The 26 April eruption is marked by an arrow.

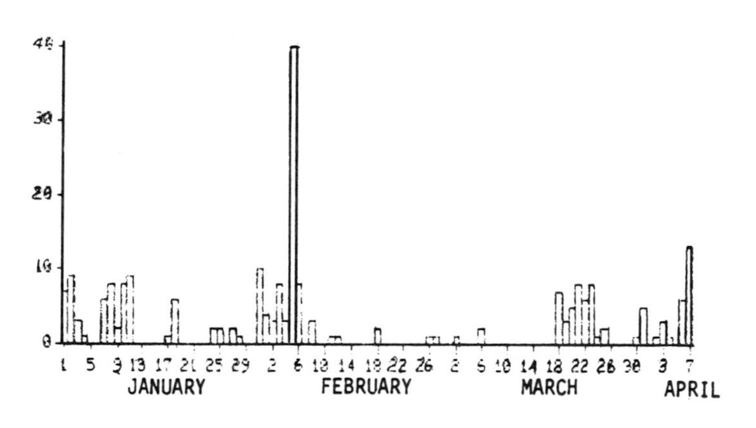

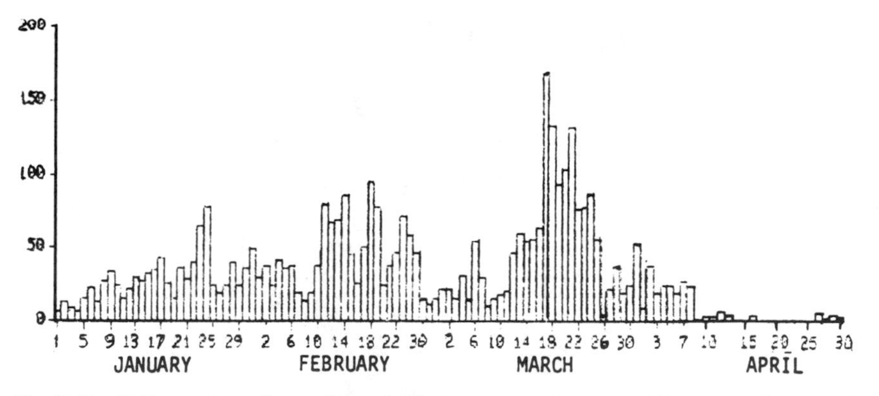

Fig. 8-28. Daily number of recorded volcanic tremor events, *top*, and B-type earthquakes, *bottom*, at Asama, Jan-April, 1983. Courtesy of JMA.

10/82 (7:10) A small explosion occurred at 0958 on 2 Oct, after a sudden increase in seismicity to 108 events on 30 Sept then a decrease to 5 events the next day. The explosion caused slight ashfall on the N side of the volcano, and was accompanied by a ground shock with 2.0 μm amplitude recorded 2.0 km S of the summit.

After the 26 April explosion local seismicity had been at its usual level (7:4-5) until July, when volcanic tremor was recorded for 15 minutes on the 6th. The level was again low until mid-Aug, then increased gradually in Sept, peaking on the 30th. On 7 Oct, 17 volcanic tremor events were recorded, but no explosion occurred; 4 were recorded the next day. As of 4 Nov no further explosions have been recorded, although seismicity has remained at a high level.

Information Contact: Same as above.

3/83 (8:3) Asama ejected incandescent tephra before dawn 8 April. Fine ash carried by W winds fell as much as 250 km away and turned snow-capped mountains gray 80 km from the volcano. Scattered brush fires were started by hot tephra in nearby foothills. During the day, columns of whitish vapor rose from the crater. Visible imagery from the NOAA 7 polar orbiting satellite at 1500 on 8 April showed remnants of a plume extending about 80 km to the ENE, probably at roughly 5 km altitude. No casualties or major damage were reported.

Information Contacts: D. Haller, NOAA/NESDIS; UPI.

4/83 (8:4) JMA scientists have sent details of the explosive, summit crater eruption on 8 April. Local seismic activity had increased in mid-March, but returned to background level in late March. In early April, high-frequency B-type earthquakes and volcanic tremor were observed more frequently than usual (fig. 8-28).

The eruption began at 0159. The air shock (amplitude 0.2 millibars) and eruption earthquake (amplitude, 125 μm) were recorded at the JMA Karuizawa [Weather Station]. JMA personnel heard the thunder-like sound that accompanied the explosion, and observed the ejection of the incandescent tephra column. During the next 11 minutes, 4 more eruption earthquakes were recorded; seismic activity then declined rapidly. Only 2 volcanic earthquakes were recorded between the initial explosion and 0600, when most activity had ceased.

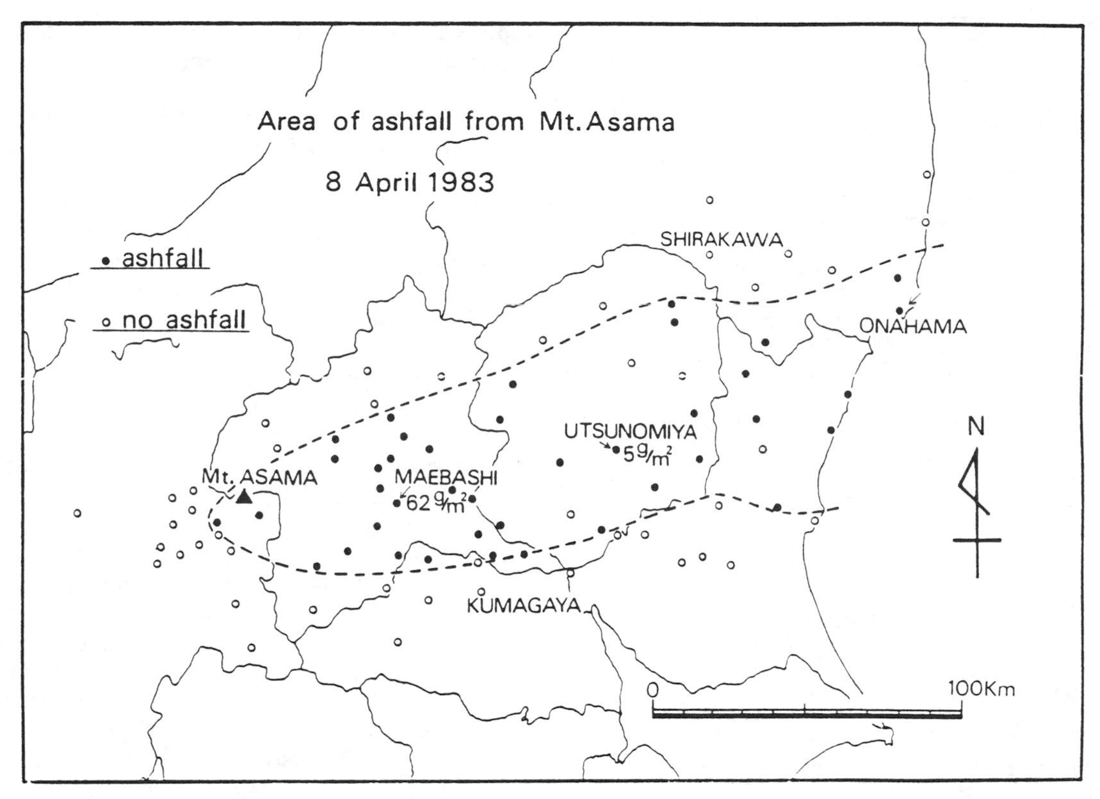

Fig. 8-29. Area of ashfall from Asama, 8 April 1983. Courtesy of JMA.

By 0450, when the summit was first visible from the [Weather Station], a 600 m-high, gray plume was being blown WSW from the summit. As the wind reversed during the eruption, ash was carried ENE (fig. 8-29). Near Ko-Asama, a lava dome about 3 km E of the summit, 2.7 kg of tephra per m^2 accumulated, including lapilli as large as 1 cm in diameter. By 0600 activity was limited to a 300 m-high vapor plume. No further explosions had been recorded by sunset. A forest fire started by the incandescent tephra on the S flank of Asama was extinguished by 0430.

After the eruption, seismic activity declined to below background levels. Only a few seismic events per day were recorded.

Information Contact: JMA, Tokyo.

KUSATSU-SHIRANE

Central Honshu, Japan
36.62°N, 138.55°E, summit elev. 2176 m
Local time = GMT + 9 hrs

Just 25 km N of Asama, this strato-volcano has erupted 17 times since 1805. Most eruptions have been phreatic.

General Reference: Hayakawa, Y., Aramaki, S., Shimozuru, D., and Ossaka, J., 1981, Kusatsu-Shirane Volcano *in* Aramaki, S. (ed.), *Symposium on Arc Volcanism Field Excursion Guide to Fuji, Asama, Kusatsu-Shirane, and Nantai Volcanoes*; Volcanological Society of Japan, p. 49-63.

JMA notes that a small eruption from Mizugama Crater occurred at about 1800 on 2 March 1976. A new crater 50 m in diameter and 10 m deep was formed in the NE part of Mizugama Crater.

9/76 (1:12) [On 3 August] a high school teacher and 2 students were overcome and killed by H_2S, which is emitted in considerable quantities [from fumaroles on the NW side of Motoshirane-san, 2 km SSW of Yugama Crater]. The gas had concentrated in a bowl-shaped area during a light rain on a nearly windless day.

Information Contact: JMA, Tokyo.

Further Reference: Ossaka, J., et al., 1980, Variation of Chemical Composition in Volcanic Gases and Waters at Kusatsu-Shirane Volcano and its Activity in 1976; *BV*, v. 43, p. 207-216.

10/82 (7:10) A brief phreatic explosion from 3 vents on 26 Oct ejected a dark column that rose 100 m above the lakes in 2 summit craters, Yugama and Karagama (fig. 8-30). Onset time of the explosion was uncertain, but the seismograph at the Maebashi District Meteorological Observatory, 1.1 km NE of Yugama, began to record volcanic tremor with an amplitude of 0.2 μm at 0855. Gray plumes were recognized at 0905. The tremor gradually increased in amplitude and peaked about an hour after onset.

Activity at Karagama ended at 0920, only 15 minutes after it was first recognized, and at 0930 activity at Yugama was reduced to white vapor emission (which ended the next day). Volcanic tremor, however, continued at peak levels for a few hours then gradually weakened, ceasing at 0124 on 30 Oct. A swarm of vol-

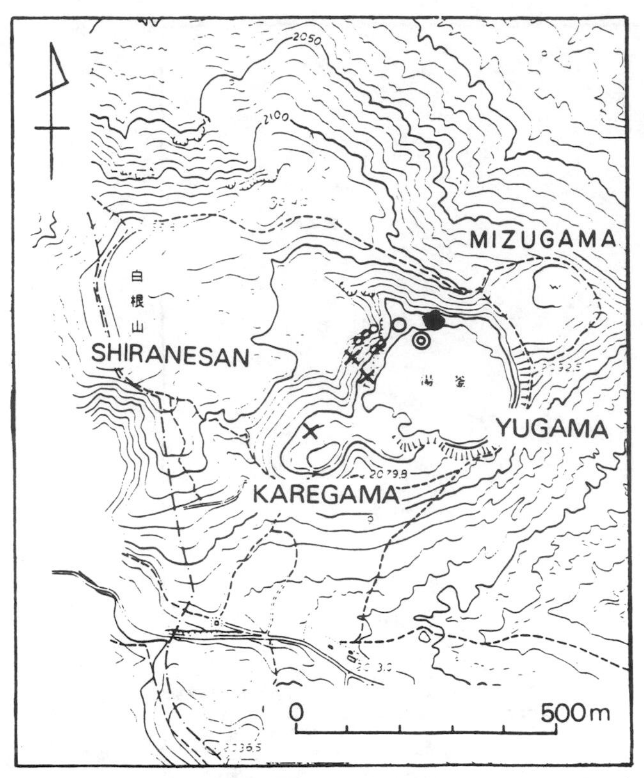

Fig. 8-30. Map of the summit region of Kusatsu-Shirane, courtesy of JMA. Small circles indicate fumaroles; the other marks represent vents active since the 27 Oct 1982 eruption. An 'x' indicates an emission point of the 26 Oct 1982 eruption plume. The large open circle is Pit No. 6; concentric circles are Pit No. 7; the solid circle is a new pit that formed on 13 Nov. Pits No. 2 and 3 (not shown) are on the W inner wall of Yugama. [Originally in 9:1.]

canic earthquakes lasting 5 hours was recorded on 26-27 Oct, and there was another minor swarm on [27] Oct.

JMA reported that a strong wind carried ash from Yugama as far as 3 km to the SE; ash was 1 mm thick on the crater rim. Tokiko Tiba reported a maximum of 3 cm of ash in the summit area. The total volume of ejecta was estimated to be 2800-3400 tons. No damage was reported. As of 2 Nov no further eruptions had been reported, and fumarolic activity in Yugama had declined. . . .

Information Contacts: T. Tiba, National Science Museum, Tokyo; JMA, Tokyo.

OCT	26	27	28	29	30	31
TEMP	14°C	46°C	55°C	56°C	48°C	39°C

Table 8-28. Temperature of water, Yugama Crater lake, 26-31 Oct 1982.

11/82 (7:11) Temperatures measured in the lake in Yugama Crater for 6 days following the 26 Oct explosion are shown in table 8-28.

Information Contacts: Y. Sawada, Meteorological Research Institute, Ibaragi; D. Shackelford, CA.

12/82 (7:12) At 0538 on 29 Dec, after 64 days of quiescence, a phreatic explosion from the pit formed in the 26 Oct explosion ejected a gray plume that rose about 300 m above the rim of Yugama Crater. The plume was 700 m above the rim by 1300, then suddenly declined about 1500. The seismograph at the Maebashi District Meteorological Observatory, 1.1 km NE of Yugama, began to record volcanic tremor 46 minutes before the explosion. Maximum amplitude was 2.1 μm. Tremor ended at 1538, but resumed at 1900 and continued until 1500 the next day. Ash traveled as far as 4 km, mainly NE. Scattered blocks and ejecta carried by spraying water were observed around the pit, including clay material on 30 Dec. No damage was reported. As of 5 Jan, no further explosions had been reported, but tremor had been continuous since 2 Jan.

Information Contact: JMA, Tokyo.

4/83 (8:4) . . . Local seismicity had declined after the Dec activity, but increased in early Jan (table 8-29), when 91 volcanic earthquakes and 44 tremor events were recorded. After mid-Jan no tremor events were recorded, but the number of volcanic earthquakes was greater than background level in early March.

At 0459 on 9 March the seismograph at Maebashi District Meteorological Observatory recorded a swarm of volcanic earthquakes that lasted only a few minutes. JMA scientists noted that these earthquakes were similar in wave form to those recorded during previous eruptions. . . . In April, activity was limited to occasional white vapor ejections. On 10 April, one vapor plume rose to 60 m above Yugama Crater. Volcanic tremor, possibly accompanied by vapor ejections, was observed 9-12 and 19-20 April.

Information Contact: Same as above.

1983	Discrete Seismic Events	Continuous Tremor	Remarks
Jan	Observed swarms	Observed	Rumbling at the summit
Mar	Observed swarms	--	Increased fumarolic activity at Pit no. 2
Apr	--	Observed	White plumes from Pits no. 2, no. 7, and a fumarole on the N inner wall of Yugama; N edge of frozen Yugama lake melted
Jun	Observed swarms	--	--
Jul	Observed swarms	Almost at noise level	Small eruption at Pit no. 6, NW wall of Yugama, on 26 July; small amount of ash
Nov	Observed swarms	Observed during eruptive activity	Explosive eruption at NW inner wall of Yugama and at Karagama on 13 Nov; a larger amount of ejecta
Dec	Observed swarms	Observed during eruptive activity	Eruption on 21 Dec at Pits no. 2 and 3, and at Karagama

Table 8-29. Seismic activity at Kusatsu-Shirane, 1983.

5/83 (8:5) Tokyo Institute of Technology personnel visited the volcano on 12 March to investigate a report of a small phreatic explosion from Pit No. 7 on 9 March [originally reported in 8:4]. A brief swarm of volcanic earthquakes had been recorded at 0459 that day at the Maebashi District Meteorological Observatory. The Institute team found no fresh ash. Although ash was reported to have fallen on the frozen crater lake, another observer said that the lake was not frozen on 9 March. The team concluded that the 9 March activity had been only seismic.

Information Contact: Same as 7:12.

7/83 (8:7) On 26 July a small phreatic explosion occurred at the NW rim of Yugama Crater, at the volcano's summit. JMA personnel observed the eruption, at Pit No. 6, formed during the 26 Oct phreatic explosion and the site of a similar explosion on 29 Dec. Volcanic tremor (amplitude 0.2-0.3 μm) started at 1031. About 1110, the dominant frequency of the tremor decreased and white vapor was ejected. Volcanic rumbling intensified at 1140; about $^1/_3$ of the crater lake was covered with ash by 1150. Accompanied by strong rumbling, a dark plume was ejected at [1204]; it had risen above the crater rim by [1212]. Volcanic tremor returned to a higher frequency at 1213, but shifted back to lower frequency about 20 minutes later. Emission of a white vapor plume that rose to about 100 m above the pit was continuous in the afternoon. Volcanic tremor ended at 1720. The plume weakened suddenly at 1730.

Local seismicity had been at a high level since last autumn. Seismographs recorded 209 volcanic earthquakes in June and 227 in July. On 22 July, swarms of A-type earthquakes occurred. After the 26 July eruptive episode, seismicity declined slightly but remained above background level.

Information Contacts: JMA, Tokyo; T. Tiba, National Science Museum, Tokyo.

10/83 (8:10) UPI reported that thunderous explosions ejected tephra on 13 Nov at 1144. [Blocks fell a few hundred meters from the vent; original press reports of a more distant fall of large tephra were incorrect.] Tephra [reached a town] (Nakanojo) 40 km SE of the volcano. A "secondary" eruption occurred 25 minutes later. No injuries were reported.

Information Contact: UPI.

12/83 (8:12) Summit explosions from Kusatsu-Shirane occurred on 13 Nov and 21 Dec. Precursory seismic activity began with a large-amplitude discrete event on 2 Nov. Tremor was continuous between 0700 and 2000 on thc 10th. Stronger tremor was recorded at 1722 on the 12th. Its mean amplitude on the seismograph, initially 1.9 μm, gradually increased to more than 5.0 μm at 2000, when the wave form of the tremor became similar to one that had accompanied a mud ejection at Aso Volcano. Tremor gradually declined during the early morning of 13 Nov.

Explosions occurred 13 Nov at 1144 and 1208 from small vents (Pits No. 6 and 7) on Yugama Crater's [N] wall and a new pit that appeared about 50 m E of Pit No. 6. [During a 22 Dec inspection] Tokyo Institute of Technology personnel found that a [45] cm-wide, 45 m-long fissure on the N wall of the adjoining Karagama Crater had also ejected ash. JMA noted that the line extending SW from Pit No. 6 across Karagama Crater's N wall has been active since the eruption of Oct 1982.

Lapilli were scattered as much as 600-700 m S of Yugama. Ash was carried 30 km downwind although most fell into the lake in Yugama Crater. The second explosion was accompanied by a ground shock of intensity III (JMA Scale) at the rest area 700 m S of Yugama. Signals from the seismograph at the Maebashi District Meteorological Observatory stopped immediately after this explosion. A field survey revealed that [blocks] had made many craters 1-3 m in diameter and 50 cm deep on the upper SE outer slope of Yugama. The buried seismograph cable seemed to have been cut by one of these [blocks].

Seismic activity increased again 18 – 19 Dec, when 32 and 31 discrete events were recorded. After a brief period of activity from 0954-1007 on the 21st, continuous tremor resumed at 1022. The tremor, accompanied by some large discrete events, saturated the seismograph from 1035-1105, then gradually declined, ending at 1220.

A Kusatsu town employee working S of Yugama first noticed the sound of vapor emission at 1010; 20 minutes later he observed a dark plume containing mud and lapilli rising 300 m above the crater rim. Ash traveled 400-500 m SE, darkening frozen Yugama Lake. Eruptive activity began to decline about 1100. Tokyo Institute of Technology personnel found that the eruption sites were Pits No. 2 and 3 on the W inner wall of Yugama. A small amount of ash was also ejected from the fissure in Karagama Crater that was active in Nov.

Information Contact: JMA, Tokyo.

1/84 (9:1) 1983 activity is summarized in table 8-29; locations are on fig. 8-30.

Information Contacts: JMA, Tokyo.

Further Reference: The 1982 – 1983 Eruptions of Kusatsu-Shirane Volcano *in* XIX IUGG General Assembly, 1987, *Report on Volcanic Activities and Volcanological Studies in Japan for the Period from 1983 to 1986*, p. 5-8.

NASU

N-central Honshu, Japan
37.12°N,139.97°E, summit elev. 1917 m

Nasu is 150 km NE of Asama and 160 km N of Tokyo. Nine historic eruptions are known, the first in 1397.

3/77 (2:3) An earthquake swarm including felt shocks occurred on 30 – 31 Jan, but no surface phenomena were observed.

Information Contact: JMA, Tokyo.

10/85 (10:10) Local seismic activity increased 9–12 Sept at the N foot, about 10 km from the summit. Magnitude [3.5] events recorded on 9 and 11 Sept were felt at intensity I and II at a town 7 km NW of the summit, but were not felt at the JMA Nasu-dake Volcano Observatory, about 19 km to the SSE.

Another swarm of small earthquakes occurred [27–29 Sept] at the NE foot, about 10 km from the summit. Three larger events were felt at Kashi hot spring, 7 km NE of the summit, but again none were felt at the JMA Observatory. The JMA seismic network in this area could not accurately determine hypocenters and magnitudes of small events. The largest events of this swarm were estimated to be roughly magnitude 2.7. Earthquake swarms are common around Nasu.

Information Contact: Same as 2:3.

AZUMA

NE Honshu, Japan
37.73°N, 140.25°E, summit elev. 2024 m

This strato-volcano complex is 70 km NNE of Nasu, and only 20 km NE of the famous volcano Bandai-san. Six eruptions are known since 1844.

1/78 (3:1) The number of volcanic earthquakes at Azuma, including some felt events, began to increase in Sept and continued at an increased rate in Oct. On 26 Oct, the fume cloud rose about 400 m from Oana Crater on the SE flank of Issaikyo, one of the numerous strato-volcanoes that comprise the Azuma complex. Mud and sand spattering began, and fist-sized blocks were thrown 20 m above the crater. Active fuming continued through Nov.

A brief eruption from Oana was observed during the early morning of 7 Dec from Fukushima Meteorological Observatory, about 20 km to the E. The ash cloud rose 500-1000 m above the crater and produced a slight ashfall nearby. Similar ash ejections occurred through the end of Dec.

Information Contacts: JMA, Tokyo; Y. Sawada, Meteorological Research Institute, Tokyo; D. Shackelford, CA.

2/78 (3:2) Steam and ash emissions continued in Jan, with plumes rising 300-500 m from Oana crater.

Information Contacts: JMA, Tokyo; D. Shackelford, CA.

The natural progression of active volcanoes along the Pacific margin must be broken somewhere to accomodate the Izu-Marianas chain, and the organizers of the CAVW decided to make this break between the islands of Honshu and Hokkaido. The next 14 volcanoes are parts of a long arc extending nearly 2000 km S of Tokyo.

OSHIMA

Izu Islands, Japan
34.73°N, 139.38°E, summit elev. 758 m
Local time = GMT + 9 hrs

Oshima, or "big island," lies 100 km S of Tokyo and is often called "Izu-Oshima" to distinguish it from another island volcano of the same name SW of Hokkaido. One or two substantial historic eruptions per century are known from 1112 until 1684, when the summit cone of Mihara-yama was formed in the central caldera. 50 eruptions have followed, all from Mihara-yama.

11/77 (2:11) An earthquake swarm [W] of Oshima Island, accompanied by subterranean rumbling, began in late Oct and was continuing in mid-Nov (table 8-30).

Similar earthquake swarms occurred in Jan 1972 and Nov 1973. Ten minor explosive eruptions have occurred at Oshima since 1962, the latest [in 1974].

Information Contact: T. Tiba, National Science Museum, Tokyo.

Further Reference: Yamashina, K. and Nakamura, K., 1978, Correlations Between Tectonic Earthquakes and Volcanic Activity of Izu-Oshima Volcano, Japan; *JVGR*, v. 4, p. 233-250.

1977		Earthquakes Felt	Earthquakes Recorded
Oct	30	1	79
	31	1	217
Nov	4	3	43
	15	3	86
	16	4	243
	17	3	242

Table 8-30. Number of felt and recorded earthquakes at Oshima, 30 Oct-17 Nov 1977. [JMA replaced the data in the original table.]

[JMA submitted the following description of renewed seismic activity.

A large earthquake shook central Japan at 1224 on 14 Jan 1978. The epicenter was about 10 km W of Oshima island, focal depth was 0 km, and the magnitude was 7.0. A severe shock of intensity 5 (on the JMA scale of 0-7) (fig. 8-31) was felt at JMA's Oshima Weather Station. The earthquake killed 25 persons, injured 139, and totally destroyed 94 houses and damaged 539 others in the Izu Peninsula and Oshima island. In the Izu Peninsula, numerous landslides and rockfalls blocked roads and railways, and destroyed many houses. On Oshima island, houses were destroyed or damaged mainly by strong earthquake motion. Weak tsunami waves were generated. The highest was 70 cm (peak to peak) at Okada port on Oshima island. Aftershocks occurred in a narrow belt elongated from Oshima island to the W (fig. 8-32). Volcanism at Oshima volcano was not affected by the earthquake.]

[After the report decade had ended, in Nov 1986, a major eruption of Oshima attracted widespread attention, both for the evacuation of more than 12,000 people and for the dramatic lava fountaining to record heights of 1.5 km (see 11:11).]

References: Ida, Y. and Kaneoka, I. (eds.), in press, 1986 Eruption of Oshima; *Bulletin of the Volcanological Society of Japan*, special issue.

Special Issue -- The 1986 Eruption of Izu-Oshima Volcano; *Bulletin of the Geological Survey of Japan*, v. 38, p. 601-753, Nov 1987, (12 papers).

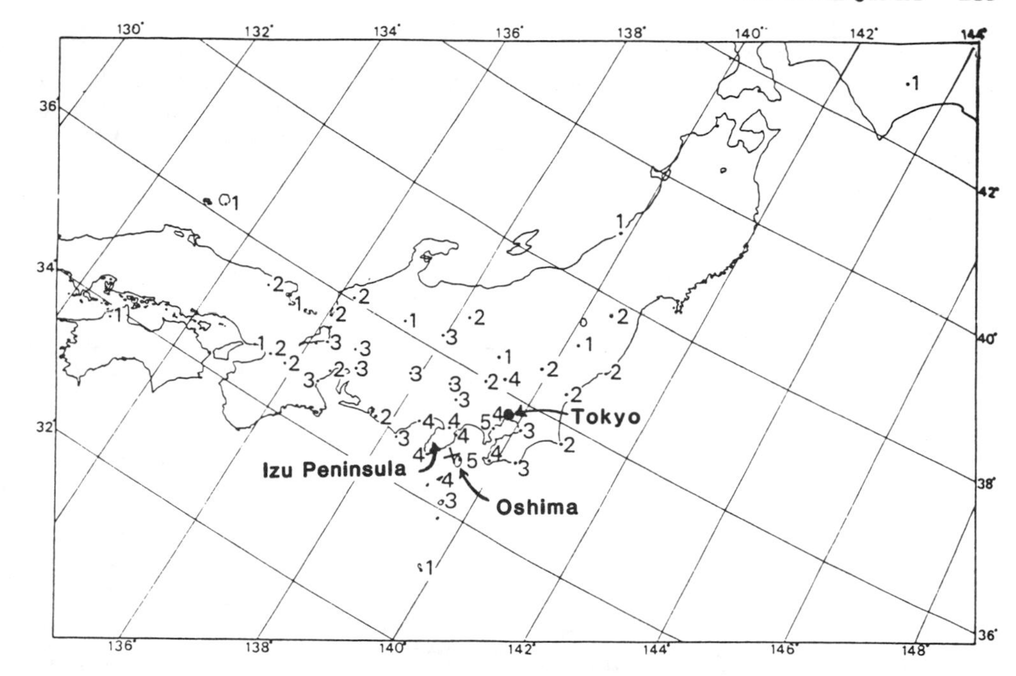

Fig. 8-31. Distribution of seismic intensities (JMA scale of 0-7) of the Oshima earthquake at 1224 on 14 Jan 1978. Courtesy of JMA.

NII-JIMA

Izu Islands, Japan
34.48°N, 139.27°E, summit elev. 429 m

This lava dome forms a small island 40 km S of Oshima. Its only historic eruption was in 1886.

10/85 (10:10) Local seismicity increased in the sea NW of Nii-jima on 21–22 Sept. Residents measured no significant changes in temperatures of the island's hot springs during this period. Because epicenters were close to the island, more than 20 events were felt, the largest of magnitude [3.5 (JMA)]. Earthquake swarms are common around Nii-jima; the most recent was in Aug-Sept 1983 [when the largest event was M 4.2].

Information Contact: JMA, Tokyo.

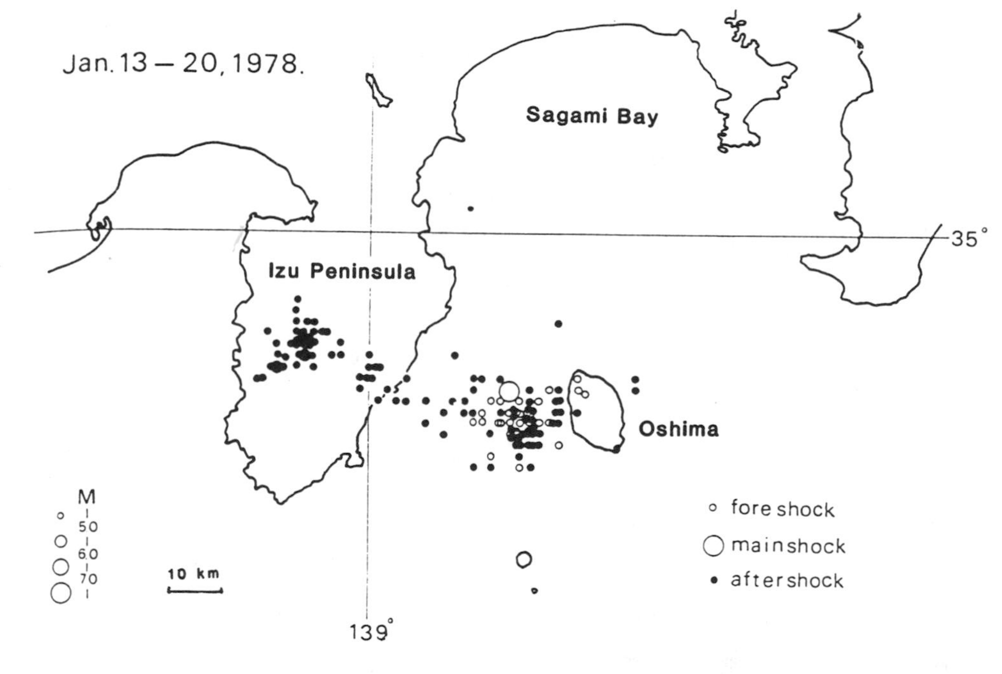

Fig. 8-32. Distribution of foreshocks and aftershocks of the Oshima earthquake at 1224 on 14 Jan 1978. Courtesy of JMA.

MIYAKE-JIMA

Izu Islands, Japan
34.08°N, 139.53°E, summit elev. 815 m
Local time = GMT + 9 hrs

Miyake-jima island, 75 km S of Oshima, has erupted at least 14 times since 1085.

12/82 (7:12) Beginning 27 Dec there have been many recorded earthquakes with epicenters in the ocean ~40 km S of Miyake-jima Island (about 200 km S of Tokyo; figs. 8-33 and 8-34). The first felt shock occurred at 1533. At 1537 on 28 Dec the largest, magnitude [6.4] on the JMA scale, originated at the N edge of the epicentral area. It was followed by gradually decreasing aftershock activity. Earthquake swarms or main shock-aftershock events in the sea around Miyake-jima occurred in July and Sept 1980, and . . . March 1981. The USGS/NEIS recorded 4 shocks of M 5 or greater (table 8-31).

Personnel from the Miyake-jima Weather Station visited the summit area (Oyama) on 29 Dec, but observed no unusual phenomena. The events were interpreted as having been too far from Miyake-jima to be precursors of volcanic activity. A group of shallow earthquakes occurred close to Miyake-jima a few months before the last eruption in 1962.

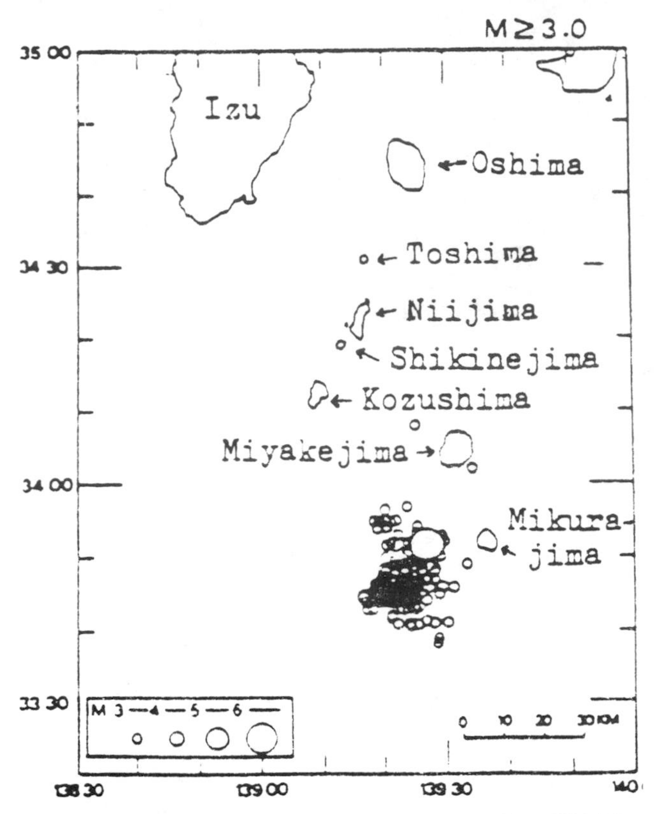

Fig. 8-33. Epicenters of earthquakes M > 3, 27-31 Dec 1982. The epicentral area is 40 km S of Miyake-jima Island. The largest open circle is the main shock that occurred 28 Dec at 1537, 30 km S of Miyake-jima. Focal depths ranged from 0 to 20 km. The Jan 1983 earthquakes were located in the same area. Courtesy of JMA.

Information Contacts: JMA, Tokyo; USGS/NEIS, Denver, CO.

9/83 (8:9) Miyake-jima erupted on 3 Oct after 21 years of quiescence. Two hours of increasing seismicity preceded the eruption onset. A column of tephra and vapor rose to 10 km, and lava flowed down the SW flank.

Small earthquakes began to be recorded at the JMA Miyake-jima Weather Station at [1359]. Weak shocks were felt at the same time in Ako, the largest village on the SW coast. Seismicity increased gradually, and from around 1430 to 1523 as many as 2-3 earthquakes/minute were recorded. The first felt shock (JMA intensity 1) at the weather station occurred at [1447], followed by others at 1500 (JMA 2), 1514 (JMA 1), and 2 at 1522 (both JMA 2). Many more shocks were felt in Ako. JMA personnel judged that the eruption began at 1523, when the amplitude of recorded continuous tremor began to increase. Tremor saturated the seismograph by [1529] and high amplitudes persisted for hours.

The eruption began [near Jinan-yama, 2 km SW of the summit, forming a 4 km-long fissure that extended to the SW coast]. Lava fountains rose to a few hundred meters from more than 9 vents. Lava advanced [mainly] in [4] flows, 300-400 m wide, starting forest fires in many places. The largest flow reached Ako and a smaller one reached Usuki village about 1800; 90% of Ako was destroyed but there were no casualties. Lava reached the

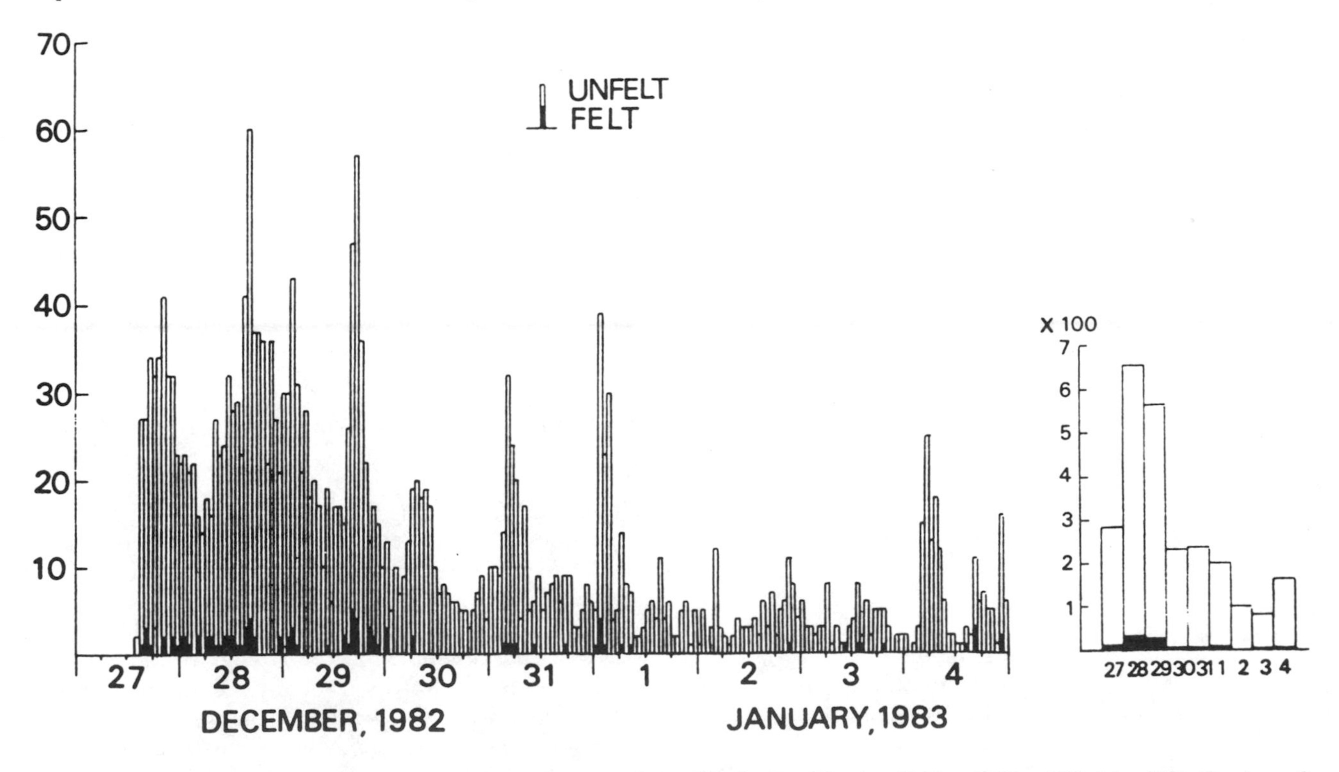

Fig. 8-34. Hourly, *left*, and daily, *right*, number of earthquakes recorded at Miyake-jima Weather Station, 27 Dec 1982-4 Jan 1983. Courtesy of JMA.

sea about 1900.

The pilot of a Japanese airliner reported that an eruption column had reached 10 km altitude around 1600. Tephra covered the entire 55-km^2 island. Tephra was thickest on the E half of the island, where 20-30 cm of ash and lapilli accumulated; many car windshields were broken. In the SW sector, 7-8 cm were reported. The airport was closed by the clouds of tephra and about 7.5 cm of ash and fist-sized tephra on the runway. Rescue planes en route to the island had to return to Tokyo Airport.

Spectacular fountaining and frequent loud explosions continued until midnight. An underwater explosion at the SW end of the fissure was observed from a fishing boat about 2330. Activity subsided during the night and only voluminous white smoke was observed on 4 Oct. [Despite press reports of a new island off the S coast, no new island formed.]

Between the onset of the eruption and 0100 on 4 Oct, 59 earthquakes were felt. Earthquake activity, which had declined at the onset of the eruption, resumed at 1812 and increased gradually. At 2233 a magnitude 5.7 (M_S) [JMA magnitude 6.2] shock struck the island. Preliminary USGS data placed the event at 34.06°N, 139.45°E, at shallow depth. The shock caused landslides at about 10 places along island roads, and was felt weakly in Tokyo and throughout the Kanto district on Honshu, 180 km to the N. After this earthquake, seismic activity decreased through 4 Oct. Three felt earthquakes and a series of weak events of different character from those that preceded the eruption occurred between 1700 and 2100 on 5 Oct. About 3000 earthquakes, including 109 felt shocks, had been recorded on Miyake-jima [27 Dec 1982 – 18 Jan 1983 (7:12)].

When the eruption began, island residents fled to schools and other buildings designated as shelters, but [there was an unconfirmed report that] 30 were forced from the Tsubota town hall (3.5 km SE of the summit) when the roof began to collapse under the weight of tephra. About 2000 residents were moved from the endangered area near the eruption zone to the N coast. There were no casualties. Eleven government ships arrived to stand by in case the entire island population of 4400 needed evacuation. About 10% of the population left the island on 4 Oct.

Although heavy rain on 5 Oct cooled the lava, a stream that threatened the 60 remaining [wooden houses] in Ako continued to advance about $^1/_3$ m/hr. On 6 Oct firemen tried to halt it by spraying water on its front.

Time	Magnitude	Latitude	Longitude	Depth of Focus
1024	5.0 M_s	33.74°N	139.46°E	shallow
1053	5.5 M_s	33.70°N	139.44°E	shallow
1112	5.1 M_s	33.70°N	139.52°E	shallow
1537	6.1 M_s	33.77°N	139.51°E	20 km

Table 8-31. Earthquakes of $M_s \geq 5$, 28 Dec 1982, Izu Islands.

Miyake-jima's most recent eruption was 24-27 Aug 1962, when explosions and lava flows originated from NE flank fissures. Of the 13 recorded eruptions since 1085, Ako was destroyed or badly damaged in 1643, 1712, 1763, and 1835.

Information Contacts: JMA, Tokyo; T. Tiba, National Science Museum, Tokyo; *Japan Times*, Tokyo; Kyodo News Service, Tokyo; AFP; DPA; AP; UPI.

10/83 (8:10) A SW-flank fissure eruption that began 3 Oct in mid-afternoon reached maximum intensity soon after it started and had subsided by the next morning. Frequent explosions and lava fountaining occurred from vents along a 4 km-long fissure extending SW from the island volcano's summit to the coast (fig. 8-35), fountaining along the upper 2 km and phreatomagmatic explosions along the lower 2 km of the fissure. Lava buried 80% of the coast town of Ako, and heavy tephra falls occurred on the SE part of the island.

Pre-eruption Studies. The JMA Mobile Observa-

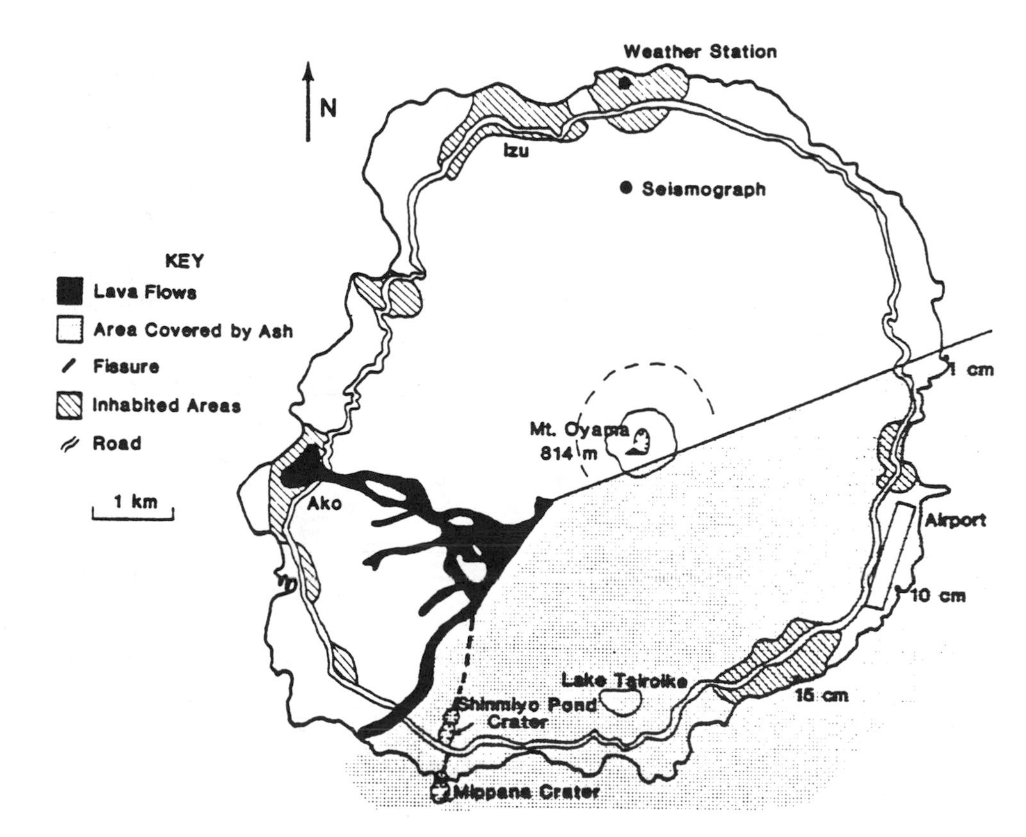

Fig. 8-35. Sketch map of Miyake-jima after the eruption of 3-4 Oct 1983. Craters are indicated by hachured circles. Values in centimeters within the stippled area are thicknesses of new tephra at those locations. Oyama is the summit cone of Miyake-jima volcano. Courtesy of JMA.

Year	Lava	Tephra
1940	11 x 10^6 m^3	greater than 1962
1962	9 x 10^6 m^3	0.3 x 10^6 m^3
1983	[4.7] x 10^6 m^3	[7.4] x 10^6 m^3

Table 8-32. Estimated ejecta volumes of the 3 most recent Miyake-jima eruptions.

tion Team had investigated Miyake-jima 20 Aug – 8 Sept 1983 and found no unusual activity. During this investigation seismometers recorded a swarm of microearthquakes a few tens of kilometers S of the island, and another a similar distance N, but none with epicenters on the island. Temperatures, ground water chemistry and height of the steam column showed no anomalous values. From 27 Dec 1982 to 18 Jan 1983 many earthquakes, the largest of magnitude 6.1 (M_S) [JMA magnitude 6.4] at 1537 on 28 Dec, occurred 40 km S of the volcano, (fig. 8-34 and 7:12), but no relation between them and the eruption is known.

Eruption Chronology and Products. JMA personnel judged that the eruption began at 1523, when the amplitude of recorded continuous tremor began to increase (fig. 8-36). Tremor was saturating instruments by 1527. A dark eruption cloud that had reached 3 km altitude was observed from an [All Nippon Airways] airliner at 1529. People living at the foot of the volcano heard explosions at around 1530, and saw incandescent blocks and voluminous clouds of tephra ejected from points on the SW flank. Ash began to fall at the airport at 1540. The eruption column had reached an altitude of 10 km by around 1600. Lava flowing W was ½ km from Ako by 1630, and began surrounding and igniting houses around 1800. The southernmost lava flow entered the sea about 3.5 km SE of Ako at about 1900.

Phreatomagmatic explosions began at a newly formed explosion crater (named [Shinmyo] Pond Crater) near the S end of the fissure at around 1640. At 1722, a strong explosion near the seashore, perhaps at the same crater, produced an incandescent column. Lapilli started to fall in Tsubota (about 4 km E) at 1646. The lapilli fall was heavy enough to break car windshields at around 1700, but began to decline at 1834 and ceased at 1910 as wet muddy ashfall started. At 2140 a strong explosion occurred near the shore, and about an hour later many incandescent columns were observed there but this activity paused at about 2300. A submarine phreatomagmatic explosion at the SW end of the fissure was observed from a fishing boat at around 2310. At about 0145 occasional submarine explosions were seen . . . in the same area.

A member of the JMA Mobile Observation Team arrived at Miyake-jima on 4 Oct at 0500. Sea and air observations at that time revealed that the fountaining had ended, perhaps about an hour earlier, and observers in aircraft saw neither explosions nor lava extrusion. Four main lava flows had moved down the SW flank. Steam rose weakly from some places on the flows, and voluminously from the SW end of the fissure where phreatic activity had occurred. Small flames seen on flow surfaces were probably from burning trees [or houses]. Lava had virtually stopped flowing, although surveyors who landed at Ako by boat observed a flow front advancing slightly. A series of pits and cinder cones had formed along the fissure in [a slightly curved line] (fig. 8-35). Along the lower 2 km of the fissure there were 4 explosion craters. The largest, 300 m in diameter, was the site of submarine explosions seen during the night. It had added new land at the shore and was given the name "Mippana Crater."

The largest of the lava flows had buried 80% of Ako. It was 50 m wide in valleys, 500 m wide in the town, and 5 m thick at its front. Several homes and 2 schools afire at the flow front were observed from the air at 0500 on 4 Oct. Part of the flow advanced a few meters 4 – 6 Oct; on the 6th sea water was sprayed at 40 places along the front to halt it. At 38 of these, the flow had stopped by the next day, but spraying continued until 8 Oct. The E half of the island was covered with dark tephra, 15 cm deep at Tsubota, 10 cm at the airport, and 1 cm on the NE coast. The coast road around the island was closed at many places by lava flows and tephra.

Volcanologists from the Universities of Tokyo, Tohoku, Chiba, and other institutions arrived on the island 4 – 5 Oct to study

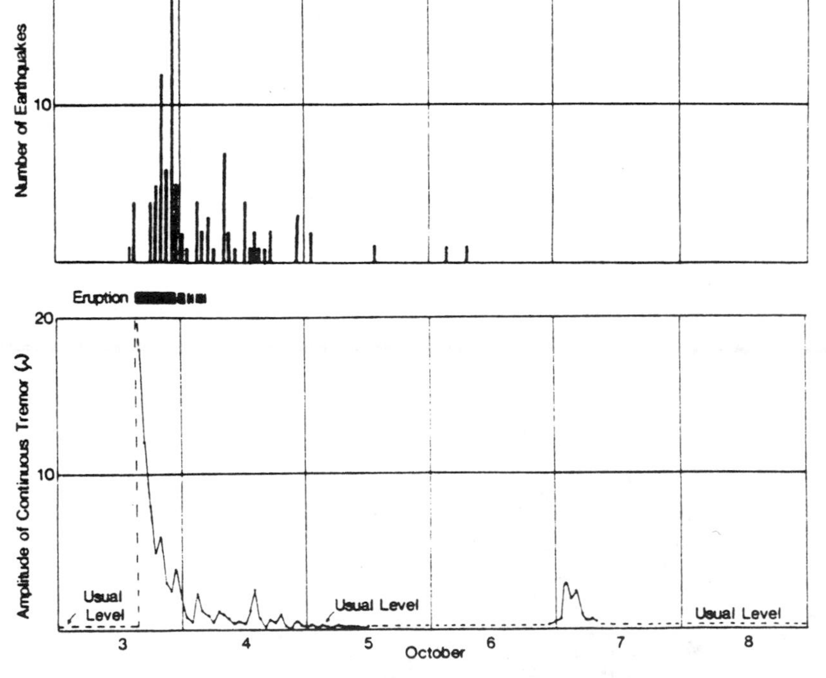

Fig. 8-36. *Top,* number of earthquakes felt per hour at the JMA weather station on the N side of the island. *Bottom,* amplitude of continuous tremor recorded at the JMA weather station 3-8 Oct 1983. The broken line indicates approximately measured data. The heavy line above the graph diagramatically shows the period of eruptive activity. Data courtesy of JMA.

the eruption and its products. Shigeo Aramaki's preliminary estimate of the volume of material ejected was [4.7] $\times 10^6$ m^3 of lava and [7.4] $\times 10^6$ m^3 of tephra (table 8-32).

Seismicity. Five earthquakes were felt between 1448 and 1522 on 3 Oct , but with the steep increase in [continuous volcanic] tremor amplitude at the start of the eruption, felt shocks stopped (fig.8-35). Earthquake activity remained low from 1522-1800, the most vigorous phase of the fissure eruption, then increased gradually from 1812 until 2233 when a 5.7 (M_S) [JMA magnitude 6.2] event shook the island. The few shocks felt 5-6 Oct represent stronger than usual seismicity. The last felt event occurred 15 Oct and the total since the beginning of the activity was 101. Although landslides occurred at about 10 sites along roads, there were no casualties. The number of small earthquakes then gradually decreased, but remained above background level on 27 Oct.... The amplitude of recorded continuous tremor decreased rapidly late 3 Oct, reaching its usual low level on 5 Oct, but increased again briefly early 7 Oct without ... eruptive activity.

Human Effects. The JMA weather station issued an alert on 3 Oct at 1545. Evacuation of Ako's 1400 inhabitants by fishing boat, private car and bus began immediately, as the instruction to go to Izu, on the N shore, was repeated through loudspeakers. Eleven buses moved about 1000 persons in 2 round trips before the road was cut by lava at around 1730. Fishing boats transported the remaining 70. Eleven ships (from JMSA and JSDF) arrived the evening of 3 Oct, in case the entire island population of 4400 needed evacuation. Food, blankets, and tanks of water were shipped from Tokyo.

Electricity was restored to all areas except Tsubota within 24 hours. Water supply to half the island was cut off because of damage to a major conduit (from Tairo-ike Pond). Telephone service was interrupted on 4 Oct when a switching station in Ako failed, but emergency radio service was established the same day. Tephra deposits at the airport were cleared after 4 days, allowing it to reopen. The Tokyo government planned to build temporary housing for evacuees.

Previous Activity. Miyake-jima is a basaltic volcano with a central cone (Oyama) surrounded by a low somma rim about 1 km in diameter. Activity has typically consisted of Hawaiian-type fissure eruptions. In historic time, fissures have been mainly on the NE and SW flanks but older fissures are distributed in a more uniform radial pattern. The last eruption, from a fissure [on the NE slope to] the NE coast, began 24 Aug 1962 and lasted about 30 hours.

Information Contact: JMA, Tokyo.

Further References: Aramaki, S., Hayakawa, Y., Fujii, T., Nakamura, K., and Fukuoka, T., 1986, The October 1983 Eruption of Miyake-jima Volcano; *JVGR*, v. 29, p. 203-230.

Special Issue -- The 1983 Eruption of Miyake-jima; *Bulletin of the Volcanological Society of Japan*, v. 29, Dec 1984, 349 pp. (32 papers).

MYOJIN-SHO (BAYONNAISE ROCKS)

Izu Islands, Japan
31.92°N, 139.92°E, summit elev. 10 m
Local time = GMT + 9 hrs

Myojin-sho, 240 km S of Miyake-jima (fig. 8-37), is a post-caldera lava dome in a mostly submerged 10 km-wide caldera. The Bayonnaise Rocks are the only exposed part of the caldera, but short-lived islands have been built by 3 of the 17 eruptions known since 1896.

7/79 (4:7) After 10 years of inactivity, discolored water was observed on 13 July 1979 at Myojin-sho, where 31 persons aboard the research vessel Kaiyo Maru V were killed by an explosion in 1952.

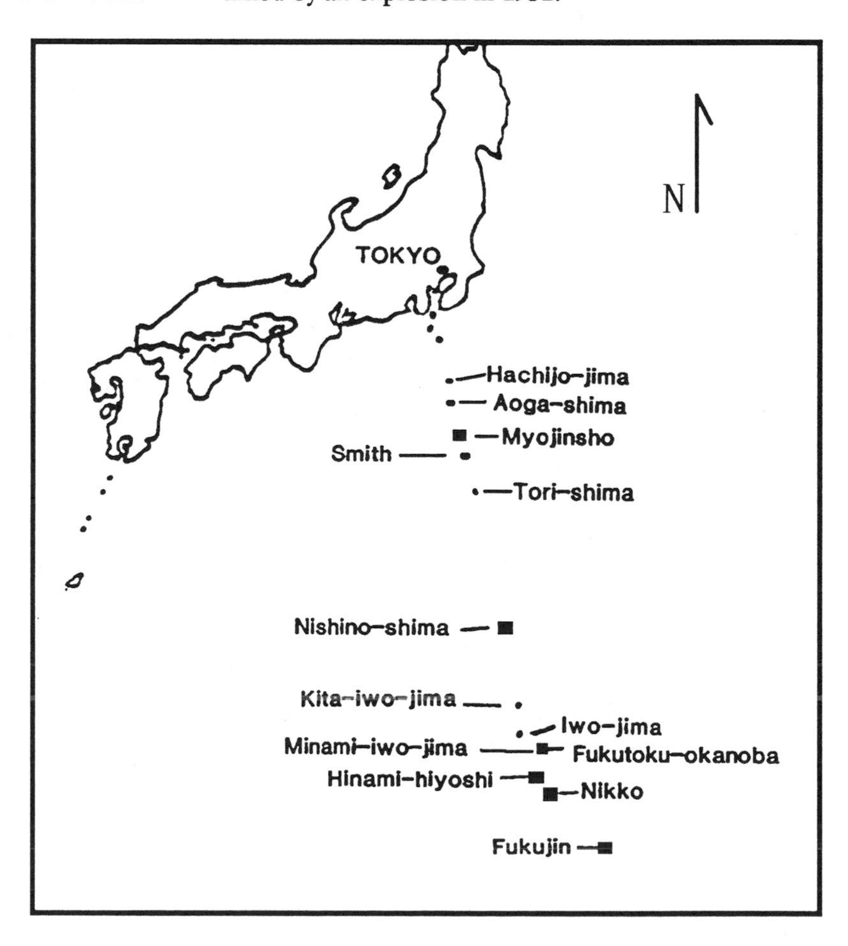

Fig. 8-37. Map showing Japan and the Kazan (Volcano) Island chain. Volcanoes active in 1978 or 1979 are indicated by black squares. Courtesy of JMA. [Originally in 4:7.]

Information Contact: JMSA, Tokyo; JMA, Tokyo.

11/80 (5:11) The crew of the fishing boat *Suitenmaru 11* saw discolored water over Myojin-sho on 15 Nov at around 1530. They reported that no discoloration had been seen there that morning. JMSA personnel flew over the site on 18 Nov and observed three circular areas of pale green water, each 50-80 m across, aligned within a 300-m zone. [Discolored water was] also seen the following day by the crew of the fishing boat *Shinkomaru 26* and again by JMSA personnel on 26 Nov. No floating ejecta or upwelling of water were noted, however.

Information Contact: JMSA, Tokyo; JMA, Tokyo.

12/80 (5:12) [Overflights of Kazan (Volcano) Islands centers on 14 Nov and 18 Dec (see table 8-39 under Fukutoku-okanoba) did not cover Myojin-sho, but that on 23 Dec observed discolored water at the site.]

Information Contact: JMA, Tokyo.

Further Reference: Smoot, N.C., The Growth Rate of Submarine Volcanoes on the South Honshu and East Mariana Ridges, *JVGR* (in press). [Bathymetric data on this and following seamounts throughout the Mariana arc.]

TORI-SHIMA

Izu Islands, Japan
30.48°N, 140.32°E, summit elev. 403 m

This small island, 165 km SSE of Bayonnaise Rocks, is at the S end of the Izu chain. Submarine activity 9 km S of the island was observed in 1975 on SEAN's second day of operation and is reported below (2:1 under Fukutoku-okanoba).

1984		Observation
Mar	7	9 x [28] km of discolored water
	8	Floating lapilli; 0.5-3 x 50 km of discolored water
	9	Sea water temperature at vent 0.5°C higher than floating lapilli; 9-13 x 30 km of discolored water
	12	Floating lapilli; discolored water
	13	Sea water near vent 0.5-1°C higher; floating lapilli; vapor plume
	15	Floating lapilli
	16-19	Floating lapilli; vapor plume
	22	Dark plume; luminescence at night
	23	Vapor plume; luminescence at night
	24, 25	Vapor plume
	26	Floating lapilli
Apr	9	0.3 x 5 km of very weak water discoloration
May	9-10	Area of discolored water 0.3 km in diameter
	15	White upwelling and discoloration, 3.5 km in diameter
	18	Milky white discoloration, 0.18 km wide
June	9	Weak green discoloration, 0.15 km x 0.05 km

Table 8-33. JMSA observations at Kaitoku Seamount, March-June 1984.

NISHINO-SHIMA

Ogasawara (Bonin) Islands, Japan
27.25°N, 140.88°E, summit elev. 52 m

Another small island, 365 km S of Tori-shima, this is at the N end of a linear ridge extending 280 km SSE to Iwo-jima and others in the Kazan (Volcano) Islands. Several new islands were formed during an eruption in 1973–74 and were later joined to Nishino-shima.

7/79 (4:7) On 15 Nov 1978, discolored water was visible at 27.25°N, 140.88°E, 6.5 km NW of Nishino-shima, the first activity there since the explosions of May 1973–summer 1974.

Information Contacts: JMSA, Tokyo; JMA, Tokyo.

Further Reference: Ehara, S., Yuhara, K., and Ossaka, J., 1977, Rapid Cooling of Volcano Nishinoshima-shinto, the Ogasawara Islands; *Bulletin of the Volcanological Society of Japan*, v. 22, p. 75-84 (Part I-Observational Results) and p. 123-131 (Part II-Interpretations).

[Aerial observations (on 2 Dec 1985) of a pale green water discoloration zone, extending 0.6 km SW from the island, were reported by JMSA in the *Bulletin of Volcanic Eruptions* (no. 25, 1988). Water discoloration had last been observed in April, 1982.]

KAITOKU SEAMOUNT

Ogasawara Islands, Japan
26.12°N, 141.10°E
Local time = GMT + 10 hrs

Midway between Nishino-shima and Iwo-jima, this site was not recognized as active before the following reports.

2/84 (9:2) On 7 March at 1230, the crew of a Japan Maritime Self-Defense Force (JMSDF) transport plane flying about 130 km N of Iwo-Jima observed a fan-shaped zone of discolored sea water that extended ~[28] km WSW from a submarine vent. The maximum width of the discolored zone was about 9 km (table 8-33). A helicopter flew over the area shortly thereafter and its crew estimated that the extent of the reddish-brown water was roughly as large as Iwo Jima Island (~5 × 8 km).

The next morning, JMSA personnel observed continuous submarine eruptive activity. Gray or yellowish brown water was ejected every 10 minutes and waves spread outward from the vents. The sea colors included gray, white, yellowish brown, and reddish brown. JMSA observers saw neither plumes nor floating ejecta, although small white plumes and rocks or reefs were seen during a flight by the JMSDF at about noon the same day. On 12 March, personnel aboard a JMSDF patrol plane again saw floating material, and a plume about 100 m above sea level. Only discoloration was found during a JMSA flight 13 March. As of the 13th, no new island had been observed at the eruption site.

The activity was located near the site of an eruption reported in 1543 at 26.00°N, 140.77°E.

Information Contact: JMA, Tokyo.

4/84 (9:4) RSP stations on Rangiroa, Tubuai, and Rikitéa recorded acoustic waves (T-phase) from a strong submarine volcanic eruption in the vicinity of the Izu Islands. Between 25 March and 30 April, more than 500 signals were received, 300 between 2 and 9 April. Most of the events were impulsive and of short duration, indicating explosive volcanic activity at a shallow depth. Some of the events were emergent and of longer duration indicating quiet emission of lava. J.M. Talandier noted that these events appear to be correlated with the submarine activity reported in 9:2.

Information Contact: J. Talandier, Lab. de Géophysique, Tahiti.

5/84 (9:5) JMSA observations beginning 7 March indicated that eruptive activity was strongest in mid to late March, when floating ejecta and vapor plumes were nearly always seen. Beginning in April, activity subsided gradually. The area of discolored water decreased from 13 × 30 km in early March to 300 m in diameter in early May.

On either side of the submarine vent are 2 shallow areas, at 26.13°N, 141.10°E, and 26.05°N, 140.93°E (fig. 8-38). JMSA has formally named the feature Kaitoku Kaizan (Kaitoku Seamount).

Information Contact: JMA, Tokyo.

10/84 (9:10) JMSA reconnaissance revealed few signs of activity since the March eruption. Pumice was observed only around the eruption site; none washed up on shore.

Activity gradually declined. To the crew that overflew the area at noon on 9 April, the water discoloration was barely distinguishable, and no ejecta or plume was

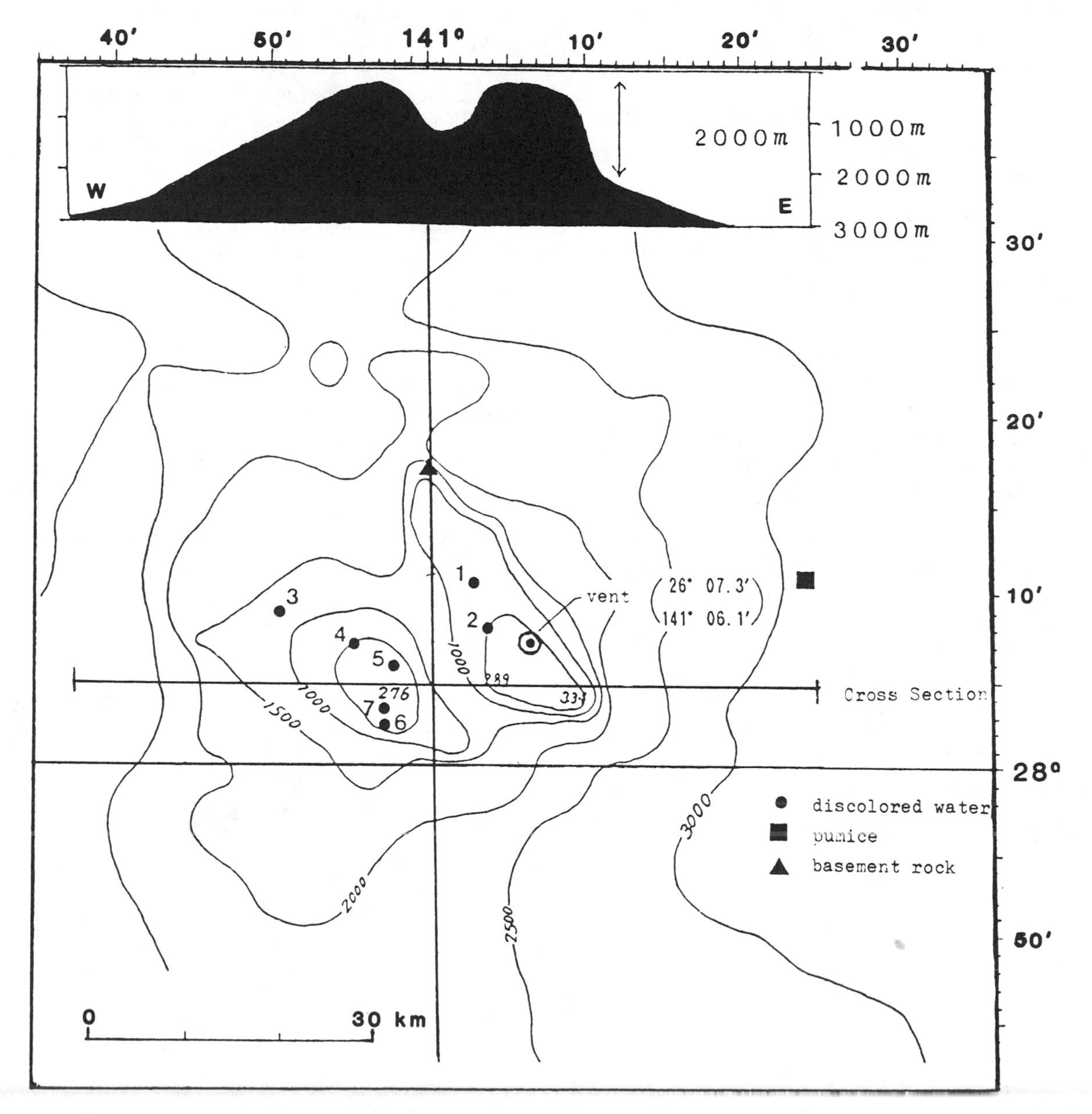

Fig. 8-38. Cross-section, *top*, and bathymetry, *bottom*, at Kaitoku Seamount. Courtesy of JMSA.

seen. JMSA and JMSDF saw no discolored water or other activity on 10 and 31 July, 1 Aug, and 6 and 26 Sept, but the crew of the MV *Emblem* reported a 3.5-5-km area of upwelling and water discoloration on 15 Oct.

Information Contact: Same as 9:5.

Further Reference: Tsuchide, M., Kato, S., Uchida, A., Sato, H., Konishi, N., Osaka, J., and Hirabayashi, J., 1985, Submarine Volcanic Activity at the Kaitoku Seamount in 1984; *Rep. Hydrograph. Res.*, v. 20, p. 47-82.

IWO-JIMA

Kazan (Volcano) Islands, Japan
24.75°N, 141.33°E, summit elev. 165 m

The name means "Sulfur Island", and eruptive activity, mainly small and phreatic, has been reported 12 times since 1930. The island lies within a submarine caldera, and was the scene of one of World War II's worst battles. Continuing volcanic unrest is evident in the 10+ m of uplift that has been recorded here since 1952.

Reference: Newhall, C.G. and Dzurisin, D., 1988, Historical Unrest at Large Calderas of the World; *U.S. Geological Survey Bulletin* 1855, 1108 pp.

9/82 (7:9) Five small phreatic explosions without detonations or recorded seismic events were reported by the Japanese National Research Center for Disaster Prevention. They occurred 9-10 March on the NW shore of the island, producing a new crater N of Asodai Crater (which erupted in 1967, 1969, 1976 and 1978) and "Million Dollar Hill." The 3 craters lie on the Asodai fault. The ejecta were no larger than 25 cm in diameter and were scattered within 300 m of the new crater. No juvenile material was found, only fragmented older rock.

Information Contact: JMA, Tokyo.

4/83 (8:4) The National Research Center for Disaster Prevention reported that 2 weak steam explosions from Asodai Crater 28 – 29 Nov 1982 were associated with an earthquake swarm 25-30 Nov. Among the 1492 recorded seismic events were 13 felt shocks. Before the swarm, the maximum number of recorded seismic events for a 5-day period had been 104. The earthquakes cracked roads in some places.

Information Contact: Same as above.

Volcano	Latitude	Longitude	Date Reported
Kasuga Seamount	21.78°N	143.71°E	Nov 1975
Fukujin Seamount	21.93°N	143.46°E	20 March 1974
Fukutoku-okanoba	24.28°N	141.50°E	Dec 1974
NW of Fukutoku-okanoba	24.42°N	141.32°E	March 1975
NW of Iwo-Jima	24.88°N	141.17°E	Jan 1974
Unnamed	26.13°N	144.48°E	March 1974
S of Tori-Shima	30.38°N	140.32°E	Oct 1975

Table 8-34. Location of some possible submarine eruptions in the Ogasawara-Kazan-Mariana Is. chain since Jan 1974.

FUKUTOKU-OKANOBA (SHIN-IWO-JIMA)

Kazan Islands, Japan
24.28°N, 141.52°E, summit elev. variable

Fukutoku-okanoba is a submerged shoal 55 km SSE of Iwo-jima and about 5 km NE of Minami-Iwo-Jima island. The ephemeral islands of Shin-Iwo-Jima were formed at this location in 1904 and 1914. It has been the most active site of submarine volcanism in the Kazan Island chain during the report decade, and formed another ephemeral island in Jan 1986 (10:12).

Many of the following reports first appeared in the *Bulletin* under the general heading of "Submarine Volcano, Volcano Islands, N Pacific." They cover aerial observations of several sites (see next 3 volcano headings) up to 330 km SSE of Fukutoku-okanoba.

1/77 (2:1) The crew of a JMSA aircraft recently observed that bubbling water near Minami Iwo-Jima had changed to an ashen brown color. Orange and yellow flotsam was seen in the area. [The location given for this observation was 23.48°N, 141.67°E, or 90 km SSE of Fukutoku-okanoba, in the vicinity of Minami-Hiyoshi seamount.] Activity was reported nearby (23.50°N, 141.92°E) in Aug 1975. Possible submarine eruptions have been reported at 7 additional locations in the [Ogasawara-Kazan]-Mariana Is. chain since Jan 1974 (table 8-34).

Information Contacts: AFP; U.S. Defense Mapping Agency.

4/77 (2:4) [A report of surface discoloration in March was removed at JMA's request.] The last reported activity in this area was a [green] discoloration in Aug 1976.

Information Contact: JMA, Tokyo.

8/77 (2:8) Sea surface discoloration was observed from the air on 1, [12, 13, 16, 17, 21, 22, and 25 May].

Information Contacts: JMA, Tokyo; D. Shackelford, CA.

3/79 (4:3) Aerial reconnaissance by the JMSA in late Jan showed slight discoloration of the ocean around this volcano, at 24.30°N, 141.48°E.

Information Contacts: Y. Sawada, Meteorological Research Institute, Tokyo; D. Shackelford, CA.

7/79 (4:7) Activity has been observed at 6 submarine volcanoes in the Volcano Island chain in 1978 and 1979 (fig. 8-37). Discolored water has often been visible over 3 of the volcanoes during frequent overflights by JMSDF and JMSA aircraft (table 8-35). Fukutoku-okanoba, the most active, formed an island in 1904, 1914, [and 1986]. Sea water discoloration was seen there in 1950, 1952–53, 1955–56, [1958–60, 1962–63], 1967–68, and [1973–87]. Minami-Hiyoshi (23.50°N, 141.90°E), formerly named Hiyoshi-okinoba, was active [9 Jan-18 March and 26-28 March] 1977 and [26 Jan-24 March] 1978. Discolored water was occasionally visible over Fukujin (21.93°N, 143.47°E), previously called Hukuzin-okanoba, in 1977, 1978, and 1979.

Information Contacts: JMSA, Tokyo; JMA, Tokyo.

9/82 (7:9) JMSA [and JMSDF have] continued frequent monitoring flights over several known submarine volcanoes. On almost all overflights discolored sea water was seen around Fukutoku-okanoba, the most active submarine volcano in this area. No discolored sea water has been observed near Minami-Hiyoshi since March 1978. Discoloration has been seen occasionally over Fukujin each year since 1977. Analysis of Landsat satellite multi-spectral scanner data revealed discolored water there in Jan 1982, after a year of quiescence.

Information Contact: JMA, Tokyo.

3/84 (9:3) Frequent monitoring of several known submarine volcanoes has continued.... No discolored water has been observed at Fukujin since Dec 1982. Minami-Hiyoshi produced discolored sea water Jan-March 1977 and Jan-March 1978 but no further activity has been observed. [Discolored water was observed at Fukutoku-okanoba in 1983 on 14 Jan, 2 and 15 Feb, 3 and 15-16 March, 24 May, 16 June, 20 July, 25 Oct, 11 Nov, and 12 Dec; and was not visible 12 July, 25 Aug, and 13 Sept.]

Information Contact: Same as above.

10/85 (10:10) A milky white discoloration, about 100 m in diameter, was observed on 17 Sept and discoloration was noted again on 17 Oct.

Information Contact: Same as above.

11/85 (10:11) JMSA has continued frequent aerial monitoring of several known submarine volcanoes. Volcanic activity has often been observed at Fukutoku-okanoba during overflights since late 1983, but no water discoloration has been seen at Kasuga, Nikko, or Fukujin. Discolored water was observed at Fukutoku-okanoba on: 21 Dec 1983; 30 Jan, 23 Feb, 15 March, 6 and 23 April, 18 May, 9 June, 10 July, 1 Aug, 6 Sept, and 16 Nov 1984; and 14 Feb, 15 March, 17 April, 14 May, 17 Sept, and 17 Oct 1985; and was not visible 26 Sept, 24 Oct, and 12 Dec 1984.

Information Contact: Same as above.

[Additional water discoloration in 1985 was reported (*Bulletin of Volcanic Eruptions* no. 25, 1988) on 14 June, 17 July, 13 Aug, 14 Nov, and 23 Dec.]

MINAMI-HIYOSHI SEAMOUNT

Kazan Islands, Japan
23.51°N, 141.91°E, summit depth -130 m

This center, 95 km SSE of Fukutoku-okanoba, was formerly named Hiyosi-okanoba. It has been called Minami-Hiyoshi Seamount (or Kaizan) from *Bulletin* 3:5 onwards.

2/77 (2:2) On 10 Jan, a JMSA aircraft noted 2 adjacent zones of discolored water. One zone was 5.5 km × 1 km, colored gray and milky white at the center. Each color was thought to have been generated by a separate vent.

The second zone, 6.5 km long, was pale green to yellowish-brown. No volcanic ejecta were observed. By 24 Jan, activity had decreased markedly. Depth at the site is reported to be 130 m.

Information Contacts: JMA, Tokyo; Japanese Press.

1978	F-o	M-H	F
Jan 10	D	N	D
24	D	N	D
25	D	-	D
26	-	D	D
Feb 23	D	D	N
Mar 24	D	D	D
Apr 13	D	N	N
May 25	D	N	N
Jun 28	D	-	-
29	D	N	N
Aug 25	D	N	D
Oct 24	N	N	
Nov 15	D	-	-
16	D	N	N
Dec 14	D	N	N

1979	F-o	M-H	F
Jan 11	D	-	-
24	D	N	N
Feb 8	D	N	N
Mar 27	D	[N]	-
Apr 26	D	N	D
Jun 15	N	N	N
Jul 11	N	-	-
12	D	N	D
Aug 24	N	N	-
Sep 9	D	-	-
13	D	N	N
Oct 25	D	N	D
Nov 8	D	N	D
Dec 12	D	N	D

1980	F-o	M-H	F
Jan 29	-	-	D
Feb 15	D	N	D
Mar 19	D	N	D
Apr 24	D	N	N
May 12	D	-	D
Jun 16	D	N	N
Jul 7	D	-	-
8	D	N	N
14	D	N	N
Aug 18	N	N	N
Sep 4	N	N	N
Oct 21	N	N	N
Nov 14	D	N	N
Dec 18	D	N	N

1981	F-o	M-H	F
Jan 7	-	-	D
8	-	-	D
9	D	N	N
29	D	-	-
Feb 12	D	[-]	-
20	D	[N]	N
Mar 12	D	[-]	-
13	D	N	N
16	D	N	N
Apr 20	D	N	N
Jun 17	N	N	N
Jul 16	D	-	-
17	D	[N]	[N]
22	D	N	N
Aug 17	D	N	N
Sep 17	D	N	N
Oct 19	N	N	N
Nov 11	N	N	[-]
Dec 21	D	N	N

1982	F-o	M-H	F
Jan 12	-	-	D
19	D	N	D
Feb 9	D	N	N
28	D	-	-
Mar 16	D	N	D
Apr 1	D	-	N
2	D	N	N
16	D	N	N
May 14	D	N	N
Jun 16	D	N	N
22	D	-	-
Jul 13	D		N
Aug 20	D		N
Sep 27	D		N
Oct 13	D		N
Nov 17	D		N
Dec 15	D		D

Legend

F-o Fukutoku-okanoba
M-H Minami-Hiyoshi
F Fukujin

D Discolored water
N No discolored water
- No overflight

Table 8-35. Volcanic activity at 3 sites in the Kazan Islands, Jan 1978-Dec 1982, detected by JMSA and JMSDF monitoring flights. [This table combines those from 4:7, 5:4, 5:11, 5:12, 6:10 and 7:9.]

4/77 (2:4) A discolored belt was observed on the sea surface at this site on 27 and 28 March. Similar activity had been reported at Minami-Hiyoshi beginning 10 Jan, but had declined by 24 Jan.

Information Contact: JMA, Tokyo.

[The JMA has provided additional chronological information on the 1977 discolorations. They were observed by the crew of JMSA aircraft on Jan 10-14, 17-18, 21, 26, 28-29, and Feb 3–4, 11, 18, 24-25. JMSDF crews continued to see them on Mar 10, 11, and 17, but saw no discolorations on Mar 18, 19, 21, 23, and 25. Discolorations were again visible on the next 3 days (Mar 26–28) during JMSDF flights, but were not seen on 6 Apr by a JMSA flight nor by any subsequent flights that year. (Post-1977 data are shown in table 8-35.)]

5/78 (3:5) Four areas of pale green discolored water were observed on 10 Jan.

Information Contact: Same as above.

[JMA has added the information that JMSA flight crews on 10 and 24 Jan saw no discoloration, but it was seen on 26 Jan and again on both 23 Feb and 24 March by JMSDF aircraft personnel. As noted in table 8-35, discoloration was not seen on 13 April or on any subsequent flights.]

NIKKO SEAMOUNT

Kazan Islands, Japan
23.08°N, 142.32°E, summit depth -612 m

Nikko is 65 km SSE of Minami-Hiyoshi along the Volcano Islands trend.

7/79 (4:7) At Nikko, not previously known to be active, discolored water was seen on 12 July. A JMSA survey vessel recorded a depth of 612 m to Nikko's summit in 1977.

Information Contact: JMA, Tokyo.

11/80 (5:11) Nikko has shown no signs of activity since July 1979.

Information Contacts: JMSA, Tokyo; JMA, Tokyo.

FUKUJIN SEAMOUNT

Kazan Islands, Japan
21.93°N, 143.47°E, summit depth -3 m

Fukujin, earlier called Fukujin- (and Hukuzin-) okanoba, lies 175 km SSE of Nikko at the S end of the Kazan Islands trend. Its first reported activity was in 1951.

2/77 (2:2) A yellowish-green zone of discolored water was observed from the air at this site on 3 Dec 1976. By 11 Feb 1977, activity had declined. The strongest recent activity occurred on 15 Oct 1973, when explosions intermittently ejected a "smoke" cloud containing blocks to 50-80 m above the ocean surface. Explosions were reported by fishing boats on 18 Jan and 20 Feb 1974 and discoloration of water on several other occasions.

The JMSA ship *Syoko* visited the area in Aug 1976 and reported a conical submarine volcano rising 3000 m from the ocean floor. A depth of 3 m was recorded by a fishing boat on 20 March 1974.

Information Contacts: JMA, Tokyo; Japanese Press.

5/78 (3:5) Discolored water was seen 14 Oct and 10, 24, and 26 Jan. Five greenish circular spots were observed on 10 Jan and eruptions occurred about once every 2 minutes on 24 Jan.

Information Contact: JMA, Tokyo.

[From this Jan 1978 activity onwards, aerial observations are included in table 8-39. Discoloration was seen at Fukujin every year from 1977 through 1982. Analysis of Landsat satellite multi-spectral scanner data revealed discolored water there in Jan 1982, after a year of quiescence.]

FARALLON DE PAJAROS

Mariana Islands, N Pacific Ocean
20.53°N, 144.90°E, summit elev. 360 m
Local time = GMT + 10 hrs

In the U.S. Trust Territory of the Pacific, this uninhabited strato-volcano forms a 2 km-diameter island 215 km SSE of Fukujin Seamount. It has erupted at least 19 times since 1864. Like the first report below (originally listed as "Submarine volcano near Uracas Island"), several of these eruptions have been from the volcano's submarine flanks or from nearby vents on the main volcanic ridge.

11/79 (4:11) The crew of the fishing boat *Koyo-maru 5* felt a series of shocks beginning at about 1530 on 15 Nov, followed by the upwelling of water containing sulfur about 15 km SE of Farallon de Pajaros Island at 20.445°N, 145.03°E. Depths in the area of the activity range from 70 to 200 m. The JMSA reports that sea surface discoloration has been observed there in the past.

Since the day before, a larger than normal cloud plume had been observed over Farallon de Pajaros. . . . Submarine eruptions occurred near Farallon de Pajaros (but not at the 15 Nov site) in 1934, 1967, and 1969.

Information Contacts: JMSA, Tokyo; JMA, Tokyo; T. Tiba, National Science Museum, Tokyo.

9/81 (6:9) "A 16 July USN flight also covered the Quaternary volcanoes of the Marianas. Fuming and discolored water were observed at Farallon de Pajaros but were not anomalous conditions."

Information Contacts: N. Banks, HVO, Hawaii.

8/85 (10:8) On 2 Sept, a Continental Air Micronesia pilot observed a zone of brown and light green discolored water about 3 km in diameter, centered about 30 km S of Farallon de Pajaros. No eruption plume was seen by the pilot and inspection of satellite images for several days after 2 Sept revealed no visible plumes. Cloudy weather obscured the area in the days immediately preceding 2 Sept. No other reports of activity have been received.

Farallon de Pajaros is a small island with a single active cone and two small remnants of an older cone. . . . There are no obvious volcanic structures on the ocean bottom near the reported locations of the 1934 and 1967 submarine eruptions, but there is a conical seamount near the 1969 eruption site (20.24°N, 145.02°E) (Corwin, 1971, unpublished manuscript). The location of the 1985 activity is approximate, but is not inconsistent with the position of the 1969 eruption and the seamount.

Information Contacts: N. Banks, CVO, Vancouver, WA; W. Gould, NOAA/NESDIS.

11/85 (10:11) Between 2 Aug and 5 Sept, 109 T-phase events originating in the NW Pacific were received by a high-gain station at Rangiroa, Tahiti. J. M. Talandier notes that their the characteristics are typical of submarine volcanic eruptions, being of shallow (ocean) depth; the timing of the events coincides with the presence of a zone of discolored water near Farallon de Pajaros (10:9). However, a precise origin cannot be determined because the events were of weak amplitude and recorded by only one station.

Information Contact: J. Talandier, Lab. de Geophysique, Tahiti.

PAGAN

Mariana Islands, N Pacific Ocean
18.13°N, 145.80°E, summit elev. 570 m
Local time = GMT + 10 hrs

Pagan is nearly 2,000 km SSE of Tokyo, at the latitude of southern Mexico or the northernmost Philippines, and 280 km S of Farallon de Pajaros. Its largest of 13 historic eruptions, reported below, came only 18 days after another unusually large eruption, at Alaid in the Kuril Islands.

4/81 (6:4) A strong explosive eruption from North Pagan, the larger of the 2 strato-volcanoes that form the Pagan volcano complex, began on 15 May. While reporting strong felt seismicity on the island, radio operator Pedro Castro suddenly announced at 0915 that the volcano was erupting. Communication was then cut off. An infrared image from the GMS at 1000 showed a very bright circular cloud about 80 km in diameter over the volcano. The cloud spread SE at about 70 km/hr, and by 1600 its maximum height was estimated at 13.5 km from satellite imagery. Weakening of activity was evident on the image returned at 1900, and on the next image, at 2200, feeding of the eruption cloud had stopped, with the proximal end of the cloud located about 120 km SE of the volcano. No additional activity has been detected on satellite images, but by 0400 the next morning, remnants of the plume had reached 10°N and 155°E.

Aircraft attempting to land on Pagan Island were prevented from doing so by the eruption. At 1235, pilots reported a mushroom cloud over the island and ashfall over its N and E ends. Ashfalls were also reported from Agrigan Island, 105 km NW. Additional pilot reports at 1410 indicated that the eruption was intensifying and that the cloud had reached more than 7 km altitude. UPI reported that aircraft crews flying past the island at 2 different (but unspecified) times saw ash rising to 10.5 and 18 km. The USN reported that aircraft crews saw lava flowing down the NE and NW flanks toward the sea, and about 2.5 km down the SW flank to within 1 km of the island's village and 0.5 km of the airstrip.

The Japanese merchant ship *Hoyo Maru* rescued all 53 persons on Pagan Island early 16 May. Only 1 minor injury was reported.

The USGS sent a 3-man team from the HVO to Pagan Island. When they overflew the island on 17 May at about 1200, the eruption had ended. Lava had advanced about 1 km down the NE and NW flanks from the summit. The SW flank flow had partially covered the airstrip, but had stopped before reaching the village.

Information Contacts: F. Smigielski, NOAA/NESS; G. Telegadas, NOAA/Air Resources Lab (ARL); R. Tilling, USGS, Reston, VA; U.S. DOD; UPI.

5/81 (6:5) A USGS team of Norman Banks, Robert Koyanagi, and Kenneth Honma carried out ground observations, seismic monitoring, and deformation studies on Pagan Island 20-28 May. The material in quotes is excerpted from their report. Material not between quote marks was abstracted from their tabular data.

"A major eruption of North Pagan started 15 May, preceded by earthquakes first felt in late March or early April. The earthquakes (~3/day) were strong enough to shake houses, and some were felt by persons out of doors; magnitudes were probably between 3 and 4. One resident visited the summit area 13 May and noted new ground cracks, sublimates, and increased gas emissions.

"On 15 May, the first of a series of closely-spaced earthquakes (at least 13 felt) began at 0745 and from descriptions of cracked concrete houses, one of the earthquakes (at 0825) probably exceeded magnitude 4. At 0915, residents heard a loud boom, followed immediately by the beginning of the eruption, first on the N flank then proceeding toward the S part of the summit area. The eruption apparently reached full intensity almost immediately, and lava flows were noted by residents very soon after the appearance of the ash/scoria column. Geologic observations show that ash eruption and lava emission took place simultaneously during most of the eruption.

"Airline and rescue pilots reported that the height of the eruption cloud exceeded 13 km, and Japan-based weather radar reported ash to heights of 18-20 km. Weather satellite images showed that the high-altitude ash cloud traveled SSE (6:4), but ash and scoria deposits on Pagan Island were thickest in the NW sector because of the prevailing SE winds at low altitude."

At 1930 there was a notable decrease in plume height and density. Activity remained weak the next day except for a brief period of vigorous ash ejection reported by the USN around noon. During an overflight between 1000 and 1030 on 18 May, only weak to moderate emission of blue fume was noted. Residents of Alamagan Island, 35 km SSE, reported "fire and smoke on the mountain" 19 May.

"Three vents, oriented about N-S, were active. The northernmost vent, about 1 km N of the summit, was probably the first to open; it built a scoria/ash cinder cone about 80 m high, 0.90 km^2 in area, and about 36 × 10^6 m^3 (12 × 10^6 m^3 of magma, recalculated to a density of 3 g/cm^3) in volume. The central vent was in a notch about 100 m deep in the N rim of the old summit crater. This vent probably ejected most of the material in the large eruption cloud, and it fed flows that went N, NNE, and W. The third and southernmost vent was in a notch about 80 m deep in the S rim of the old summit. This vent fed flows that moved S."

	1981	1925 (max)	1925 (min)
SiO_2	52.0	50.3	50.1
Al_2O_3	15.7	18.1	17.9
FeO	11.3	9.6	8.9
MgO	5.0	4.8	4.8
CaO	9.9	10.8	10.6
Na_2O	2.4	2.7	2.5
K_2O	0.65	0.72	0.70
TiO_2	0.95	1.16	0.98
P_2O_5	0.21	0.16	0.07
MnO	0.22	0.21	0.03

Table 8-36. Chemical analyses of Pagan samples. 1981 values are from microprobe analyses of a fused sample by John Sinton, University of Hawaii; the analyses of 1925 flows are from Larson, et al., 1975.

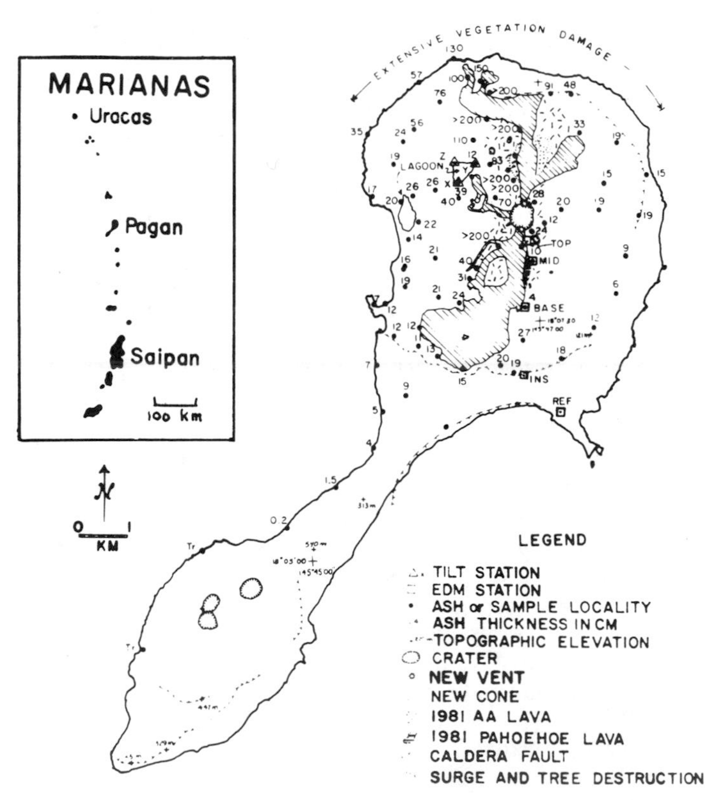

Fig. 8-39. Sketch map showing distribution of Pagan's 15 May 1981 eruption products. Prepared by the USGS.

The USGS team observed intermittent activity during 8 days of field work on Pagan Island. An increase in harmonic tremor level and the number of discrete higher frequency events began 1-2 hours before the extrusion of minor aa lava flows from the S summit vent late 21–early 22 May, and late 23 May. A similar increase in seismicity during the midmorning of 24 May was followed by ash emission from both the N and S summit vents that lasted from early afternoon through the evening. As ash emission was decreasing during the evening of 24 May, extrusion of a small amount of aa lava occurred; lava extrusion also occurred during the evening and predawn hours of 25 and 26 May. Eruptive activity was then limited to minor fuming until USGS personnel left Pagan Island 28 May.

"The volume of eruptive products ejected through 28 May exceeded 50 × 10^6 m^3, and a large part of the island's arable land was covered by lava flows, and airfall ash and scoria (fig. 8-39). Lava flows were predominantly aa and ranged from 3 to 30 m in thickness. Fortunately, extremely devastating phenomena such as widespread pyroclastic flows did not take place. Lithic blocks and juvenile bombs as large as 1 m in diameter were thrown more than 2 km from the summit onto the N flank, and base surge phenomena, evidenced by low-amplitude (4-20 cm) dune and antidune features and preferential upslope tree damage, took place in restricted corridors down to elevations of 200 m on the N and S slopes.

"The events of 15 May caused no injuries to residents, but some livestock were killed outright, and others were starving because of extensive destruction of vegetation. However, the present breaches on the N and S rims of the crater could channel potentially destructive pyroclastic flows into narrow corridors N and S of the summit. Such channeling may have occurred on 15 May and could occur again, even during the current eruptive cycle.

"Ash and scoria deposits on the island, constituted about 4 × 10^6 m^3 of magma (recalculated to a density of 3 g/cm^3) and exceeded 2 m in thickness NW of the summit crater. An unknown volume of ash and scoria was deposited at sea and composed the ash plume that extended S of Guam. The longest lava flow (2.8 km), that issued from the S vent covered 2.5 km^2 and had a volume of about 15 × 10^6 m^3. The N summit vent fed flows (1 km long, 0.2 km^2 area, 1.0 × 10^6 m^3 volume) that spilled over the summit crater on the W and stopped 100 m short of the inner lagoon; it also fed very voluminous flows that advanced N, dissecting and carrying away much of the material in the cinder cone around the vent on the N flank. The NE lobe of these northern flows traveled 1.5 km, covered an area of about 0.2

km^2, and contained 2×10^6 m^3 of lava. The N lobe traveled 3.5 km, locally exceeded 30 m in thickness, covered an area of about 1.3 km^2, and had a volume of about 25×10^6 m^3. The several lava flow and ash-ejecting events after 15 May were all of minor volume.

"Six gas samples were collected from 4 cracks that opened during the 15 May eruption, and 1 sample of airfall scoria was fused and analyzed by electron microprobe by John Sinton (University of Hawaii). Preliminary analyses of the gas samples by Paul Greenland (HVO) suggest that they contain a large portion of air (mostly introduced into the vent system through the porous volcanic edifice). When compared with gases of Kilauea (Hawaii) and Mt. St. Helens (Washington), the Pagan gas had low amounts of SO_2 and high amounts of H_2, CO, CH_4, and COS. The scoria sample was more or less typical of basalt of northern Marianas volcanoes (table 8-36). Hand-specimen examination indicates that the 1981 basalt contains few (1-7%) phenocrysts (1-2 mm long) that are dominantly plagioclase and clinopyroxene.

"A tilt monitor was established 21 May on the inner lagoon to measure possible changes of lake level around the shoreline and tilt of the lakebed. During the 8-day monitoring period, the lake level dropped regularly at a rate of about 24 mm/day, either because of tidal changes, slow drainage of the lake, water evaporation, or uplift of that part of the island. Continued drainage of the lake or increases in temperature of the lake water would cause concern, because such changes could indicate that ground water might be heating in response to the proximity of hot rock in the volcanic edifice. The overall trend of the tilt changes observed can be interpreted in terms of a possible inflation center NE of the lake near the N flank vent. In addition, fluctuations took place that could either be measurement noise or, if real, be associated with the largest eruptive events during our residence on the island (24 May ash event and 25 May lava event).

"A segmented EDM array was installed on the S flank as an additional deformation monitor. In general, changes above "noise" level were not seen in the lines measured from the instrument (INS, fig. 8-38) to the 3 nearest stations, but station TOP (at elevation 540 m) moved steadily southward, 66 mm in 6 days. This movement could have resulted from inflation of the summit due to intrusion of a shallow magma body, gravitational instability of the summit area, or right-lateral movement on the N-S fissure system that formed during the eruption.

"Seismic monitoring 20 – 28 May showed continuous harmonic tremor indicating movement of magma a few kilometers beneath the surface, and short bursts of high-frequency signals, indicating intermittent extrusive events such as degassing and low-level lava fountaining. The ongoing harmonic tremor suggests that more secondary eruptive activity may take place.

"At the same time, no significant earthquake activity was detected, indicating no prominent buildup of stresses typically associated with activities prior to a major explosive eruption. However, the deformation survey indicated possible swelling of the volcano, which may lead to an increase in future microearthquake activity and eventually to an eruptive event more significant than those of 20 – 28 May."

Reference: Larson, E.E., Reynolds, R.L., Merrill, R., Levi, S., and Ozima, M., 1975, Major-Element Petrochemistry of some Extrusive Rocks from the Volcanically Active Mariana Islands; *BV*, v. 38, p. 361-377.

Information Contacts: N. Banks, R. Koyanagi, K. Honma, HVO, Hawaii.

Further References: Banks, N.G., Koyanagi, R.Y., Sinton, J.M., and Honma, K.T., 1984, The Eruption of Mt. Pagan Volcano, Mariana Islands, 15 May 1981; *JVGR*, v. 22, p. 225-270.

Sawada, Y., 1983, Analysis of Eruption Clouds by the 1981 Eruptions of Alaid and Pagan Volcanoes with GMS Images; *Papers in Meteorology and Geophysics*, v. 34, p. 307-324.

6/81 (6:6) Residents of Alamagan Island, [roughly 60 km] SSE of Pagan, observed mild ash ejection and "fire and smoke" emerging from the volcano around 11 June. No additional activity has been reported.

Information Contact: N. Banks, HVO, Hawaii.

9/81 (6:9) The following update is from Norman Banks.

"Overflights of Pagan by the USN on 16 July and the USGS on 12 Sept revealed the formation of a new 60-80 m-diameter crater in the center of the old summit crater. Fume emission rates from the new vent appeared to be significantly greater than the combined rate observed in late May from the 3 vents formed 15 May. However, no new lava flows were identified and ash accumulation downwind of the new vent was not appreciable. The new vent may have formed during the explosive activity observed from a nearby island 11 June.

Information Contact: Same as above.

11/81 (6:11) Residents of Pagan Island, evacuated during the major eruption that began 15 May (6:5), returned for a 1-day visit on 19 Nov. Explosive activity was occurring when they arrived about 0600 and continued through the day, accompanied by booming sounds. Scoria fell on the visitors, who also noted a strong odor of sulfur. At about 1700, a series of booming sounds was followed by ejection of a tephra column that rose about 1.5 km. Since a previous visit in Sept, 1 cm of ash had accumulated on the island's only village, 4 km from the summit.

Information Contact: Same as above.

2/82 (7:2) Residents of Agrigan Island, about 50 km NNW of Pagan, observed voluminous black columns rising from the volcano on 4, 5, and 6 Jan, and white plumes on other days in the first half of Jan. Telefax copies of thermal infrared imagery (8 km resolution), available at 3-hour intervals from the GMS were inspected at the NOAA/NESS, but no eruption cloud was apparent despite clear weather. The activity was preceded by 3 deep earthquakes near Pagan, on 3 Jan at 2005, 2009, and 2015. NEIS located preliminary hypocenters at 18.034°N, 145.633°E, 590 km deep (M 6.1); 17.903°N, 145.530°E, 513 km deep (M 5.8); and 18.166°N, 145.352°E, 596 km deep (M 5.4). The first event was felt on Saipan (about 180 km S of Pagan) and Guam (nearly 300 km SSW of Pagan).

No additional activity was reported until 8 Feb, when the mayor of the northern islands visited Pagan and found it erupting. A minimum of 5 minor eruptions per day were observed through 23 Feb. At one point during this period (date and time not yet determined), a "large smoke cover" from Pagan was observed over Saipan. Images of the area, returned by the NOAA 7 polar orbiting satellite 1-2 times per day, were inspected 25 Jan – 25 Feb, but no eruption plumes were evident. In hand specimen, a sample of ejecta collected 17 Feb appeared to contain at least 30% non-juvenile material. The sample will be analyzed at HVO.

The 53 residents of Pagan, evacuated during the second day of the 1981 eruption, have not yet been able to return for more than brief visits to the island.

Information Contacts: N. Banks, HVO, Hawaii; Cmdr. J. Walker, COMNAVMAR; L. Whitney, Office of the Rep. to the U.S., Washington DC; USGS/NEIS, Denver CO; M. Matson, NOAA/NESS.

11/82 (7:11) Personnel on a USN training flight observed activity at Pagan when they flew near the volcano about 1415 on 10 Dec. A hole roughly 60 m below the crater rim on the NW side was spewing debris and brown smoke. Light-colored vapor was emerging from the center of the crater. Burning was seen along the S and SW slopes, but a cause could not be determined [see 8:3].

Information Contacts: Cmdr. J. Walker and Lt. J. Meyer, COMNAVMAR.

3/83 (8:3) The following is a report from Norman Banks.

"A team of 4 HVO scientists, 5 scientists from the USGS Water Resources Division, plus Civil Defense and other government officials from the Commonwealth of the North Mariana Islands visited Pagan 5 – 15 March.

"There were 2 very minor ash eruptions on 7 and 15 March; ashfall was confined to the summit cone. During the remainder of the visit, activity was limited to degassing. The gases were essentially atmospheric in composition, much different than the May 1981 gases, which had a high magmatic component. Seismic monitors showed varying amounts of B-type events and harmonic tremor. More numerous and stronger seismic events preceded the ash eruptions of 7 and 15 March.

"HVO scientists established a second EDM array (one had been installed in May 1981) and a tilt network, and installed a seismic event counter and two-component tiltmeter. Both EDM arrays showed minor deflation 5 – 15 March. Reoccupation of the original EDM line showed that 25 cm of net inflation had occurred on the higher slopes of the volcano between May 1981 and March 1983, but the lack of other measurements between those dates prevented determination of shorter-term deformation trends.

"Scientists from the Water Resources Division installed equipment to transmit data from the two-component tiltmeter (including periodic temperature measurements), the seismic event counter, and a rain gauge to Hawaii via the GOES West satellite. They also performed a water resource evaluation and sampled volcanic gases.

"Stratigraphy of the tephra deposits indicated that Pagan had erupted at least 4 and perhaps as many as 7 times since May 1981. The volume of the post-May 1981 tephra deposits is minor in comparison to that of the May 1981 deposit. It is difficult to assign eruption dates to each tephra layer because of the sporadic nature of observations on the island. However, it is probable that a single lava flow and one of the tephra layers was produced on 11 June, 1981 (6:6). Other eruptions were observed in Nov 1981 (6:11) and Jan through Feb 1982 (7:2). The date of emplacement of the uppermost and thickest tephra layers is uncertain. However, comparison of Dec 1982 aerial photographs with those taken in Aug 1982 suggested that these layers were emplaced during that interval. In addition, during late Sept – Oct 1982, residents of Saipan (roughly 300 km to the S) reported a dark cloud, similar to the one ejected in May 1981, drifting to the S.

"The most recent eruptive products were slightly richer in phenocrysts than products of the May 1981 eruption. A preliminary microprobe analysis indicated that Feb 1982 eruption material was less differentiated than that of May 1981 but similar in composition to the 1925 magma (6:6).

"The activity seen by USN personnel on 10 Dec 1982 was much less intense than that of March 1983. The burning seen along the S and SW slopes was on another edifice on the opposite (S) end of the island, and was due to a brush fire, unrelated to eruptive activity.

"Early in 1982, there was speculation that Pagan may have been the source for the 'Mystery Cloud' of volcanic aerosols in the stratosphere (Ch. 19, 7:1-3). The volume of individual tephra layers does not by itself suggest that Pagan was the source of the aerosols. However, they contain a large fraction of lithic material, suggestive of powerful gas jetting and erosion of the vent; lidar studies indicated that the source was at the approximate latitude of Pagan; and Pagan is the only volcano thought to have been in eruption at that latitude and at that time. Thus, at present, Pagan remains the best possible source for the 'Mystery Cloud' of early 1982." [Later work on Nimbus 7 satallite data by Arlin Krueger revealed a large SO_2 cloud originating in the vicinity of Nyamuragira Volcano, Zaire, in late Dec 1981. This eruption is now thought to be a more likely source for the 'Mystery Cloud.']

Information Contact: N. Banks, HVO, Hawaii.

8/83 (8:8) On 1 Sept, personnel aboard a Continental Air Micronesia aircraft flying from Saipan to Japan reported "ash and smoke" at an altitude of 6 km within 15 km of the volcano. Scheduled flights pass directly over Pagan Island 1-2 times per day, but no subsequent flights reported increased activity. No eruption clouds were observed on satellite imagery.

Technicians visited Pagan in late July, and saw only minor activity. Inspection of a checkpoint near the volcano on 29 July revealed no indication of major new ashfall since geologists left the island in mid-March. Pagan Island remains uninhabited, as it has since the major eruption of May 1981 (6:4-5), although residents and officials occasionally visit the island. Pagan's mayor, on the island through late Aug, observed no change from the low-level late July activity.

Information Contacts: R. Shaw, IATA, Montreal; R. Koyanagi, HVO, Hawaii; F. Chong, Disaster Control Officer, Saipan; O. Karst, NOAA/NESDIS.

9/83 (8:9) On 8 and 9 Sept USN personnel observed the volcano during pilot-navigation training flights. They described (and photographed) a plume containing a little ash that issued gently from the crater. Although heavy rain, lightning and convective weather cloud activity to more than 9 km obscured the view on the 8th, the plume was seen rising to 3.4 km. A stratus layer extended 30-35 km ENE at 2-2.5 km altitude. In clear conditions on the 9th, the rising plume and the stratus layer were again observed.

Information Contact: Lt. Cmdr. R. Adkerson, COMNAVMARIANAS.

1984	March			April										
Date	29	30	31	1	2	3	4	5	6	7	8	9	10	11
Plume	x	-	x	x	x	x	x	x	x	-	x	-	-	x
	April													
Date	12	13	14	15	16	17	18	19	20	21	22	23	24	25
Plume	x	-	x	-	-	*	*	*	x	x	-	x	x	x
	April					May								
Date	26	27	28	29	30	1	2	3	4	5	6	7	8	
Plume	-	-	x	-	-	x	-	-	-	-	-	x	-	

Table 8-37. Days in 1984 when a plume was visible at Pagan (x = plume visible, - = no plume visible, * = covered by weather clouds).

10/83 (8:10) Several former residents of Pagan, including Mayor Dan Castro, spent Sept on the island. The only activity they observed was an explosive eruption accompanied by glow that began at 0640 on the 26th. That evening, glow remained visible. Ash that fell on the village was collected from concrete slabs. The ash has been identified as magmatic, of medium-sand size, and depleted in fines.

Since Pagan's major eruption in May 1981, 6 or 7 explosions have ejected enough tephra to cause ashfall in

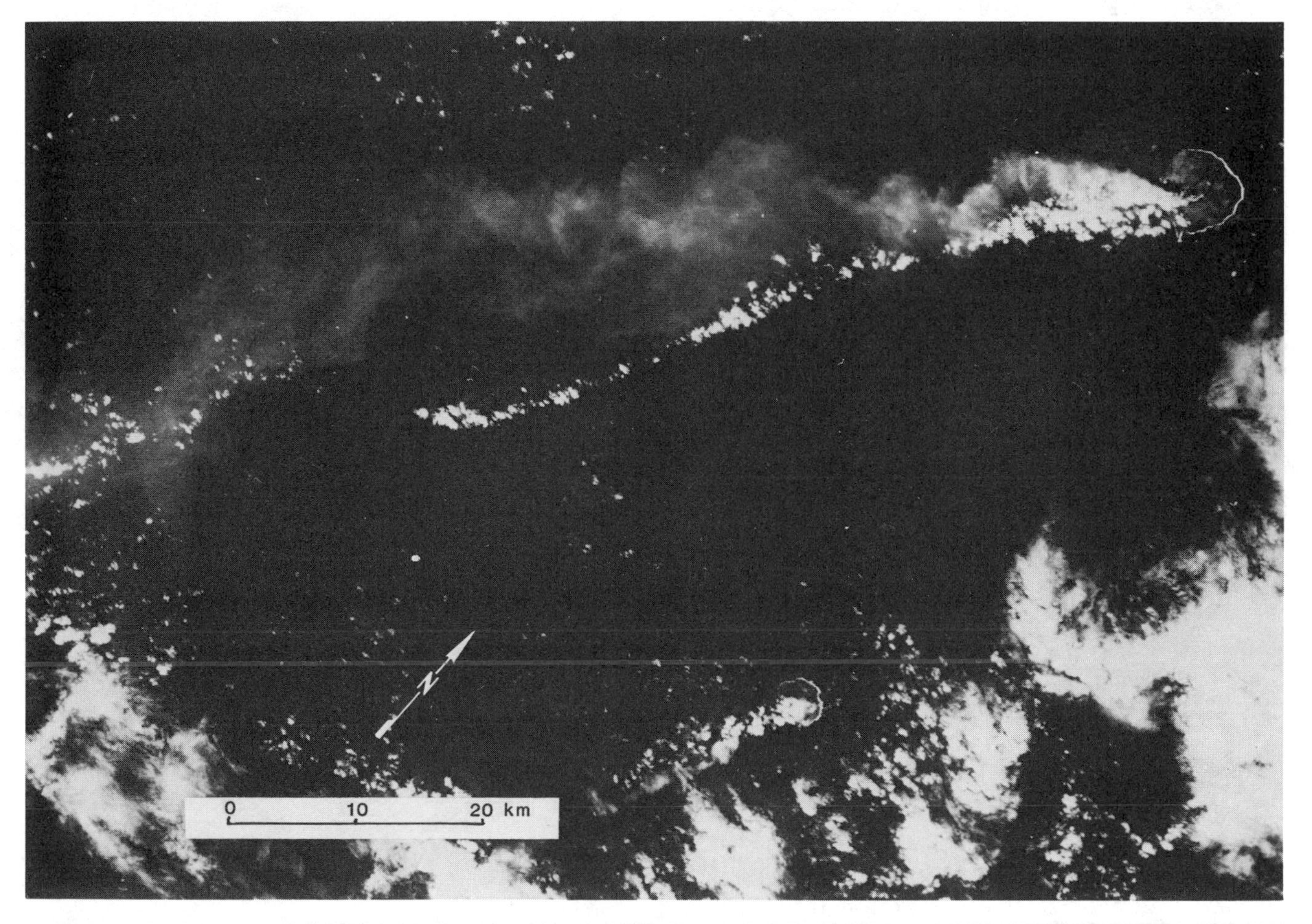

Fig 8-40. Space Shuttle photograph (no 51B 146 030), taken 1 May 1985, showing a plume extending at least 100 km SSW from Pagan. Alamagan, Pagan's neighbor to the SSE, is also shown. Courtesy of C. Wood.

the village. The 26 Sept ash was the first to fall on the concrete slabs since they were swept clean in July.

Information Contact: N. Banks, HVO, Hawaii.

4/84 (9:4) Visible images from NOAA 7 and NOAA 8 polar-orbiting satellites showed Pagan emitting a small, moisture-laden plume on about half of the days between 29 March and 8 May (table 8-37). When present, the plume extended 100-160 km downwind, with its base at about 0.75 km and its top at about 1.8 km. Occasionally the plume appeared as a haze, when its moisture had been lost downwind. Whether the moisture in the plume was vented by the volcano or acquired from the atmosphere has not been determined.

On 15 May, Pagan erupted twice. At 0825 residents of Agrigan Island, 65 km N, felt an earthquake, rushed from their homes, and saw a dark column rising 2-4 km above Pagan. The eruption cloud, about 50 km in diameter, drifted NW. An Air Force weather reconnaissance plane, diverted to inspect the activity, observed no lava emission. A second moderate eruption occured during the afternoon.

Information Contacts: M. Matson, NOAA/NESDIS; N. Banks, HVO, Hawaii.

4/85 (10:4) The crew of a Continental Air Micronesia flight that passed about 25 km W of Pagan at about 6 km altitude on 11 April at 0840 observed a large brown ash cloud obscuring the volcano. The eruption column rose to about 2.5 km altitude and a plume extended about 150 km to the W. On 15 April at 1400, another Continental Air Micronesia flight passed over Pagan and the crew reported a new lava flow. Information on its size and position were not available.

On 1 May at about 1000, Space Shuttle astronauts took 9 photographs that showed a plume originating from near the center of the volcano and extending at least [100] km to the SSW (fig. 8-40). The plume was dense near the volcano and appeared to contain some ash, but diffused rapidly as distance from Pagan increased. Weather clouds obscured the volcano during subsequent orbits.

Information Contacts: N. Banks, HVO, Hawaii; R. D. Morris, Continental Air Micronesia, Guam; W. Dailey, C. Wood, NASA, Houston, TX.

ESMERALDA BANK

Mariana Islands, W Pacific Ocean
15.00°N, 145.25°E, summit depth -60 m

This shallow submarine volcano, 350 km S of Pagan and 175 km N of Guam, is the southernmost historically active site in the Marianas.

6/82 (7:6) In April, "sulfur boil" activity was observed at the submarine volcano Esmeralda Bank from the U.S. National Marine Fisheries Service research vessel *Townsend Cromwell*. The sulfur emission can be seen as an area of strong mixing above the Bank on a 6 April bottom profile. The research vessel's log notes 3 sources of sulfur emission on 21 April, the strongest near a shoal at 50 m depth, a second S of a ridge just S of the main shoal, and the third in the saddle between these 2 shallow areas. The next day, the main source remained strongly active and emission continued from the saddle vent, but by 24 April sulfur emission was only barely visible.

Several submarine eruptions have been reported from Esmeralda Bank, the most recent in April 1975.

Information Contact: L. Eldredge, Univ. of Guam.

Further Reference: Gorshkov, A.P., Gavrilenko, G.M., Seliverstov, N.I., and Scripko, K.A., 1982, Geologic Structure and Fumarolic Activity of the Esmeralda Submarine Volcano; *in* Schmincke, H.-U., Baker, P.E., and Forjaz, V.H. (eds.), *Proceedings of the International Symposium on the Activity of Oceanic Volcanoes*; Arquipélago, Serie Ciéncias da Natureza (Univ. Azores) no. 3, p. 271-298.

Resuming a clockwise progression around the volcanoes of the NW Pacific rim, the northerly trend of Honshu volcanoes turns sharply to the east in SW Hokkaido, starting an unbroken arc through the Kuriles and the Kamchatka peninsula.

USU

Hokkaido, Japan
42.53°N, 140.83°E, summit elev. 725 m
Local time = GMT + 9 hrs

Usu lies 770 km N of Tokyo on the S margin of the Toya caldera. Its largest historic eruption was in 1663 and, like 4 of its subsequent 6 eruptions, was followed by the growth of a substantial lava dome. Four of its 10 historic eruptions have caused fatalities.

8/77 (2:8) A major eruption began at 0912 on 7 Aug from a new vent [on the SE slope of] Ko-Usu dome (fig. 8-41), after [32] hours of premonitory seismicity. About [0.08] km^3 of hypersthene dacite tephra has been ejected, in eruption columns that rose to [9] km or more on 4 occasions (fig. 8-42). Ashfalls forced the evacuation of about 20,000 tourists and 7,000 residents from nearby towns, and caused serious crop damage, particularly N of the volcano (figs. 8-43 and 8-44). Table 8-38 is a summary of the eruption.

Earthquake frequency exceeded 100 events/hr during the premonitory swarm, then ranged between 60 and [120]/hr until the afternoon of 16 Aug, when a decline to 40/hr occurred. [Seismicity declined suddenly when explosions occurred (fig. 8-45).] By 24 Aug, event frequency had declined further but magnitudes had increased to a maximum of 4, resulting in an increase in total energy release. Hypocenters were 1-2 km beneath the [summit].

Uplift and fissuring were observed on 12 Aug between Ko-Usu and O-Usu domes. Ground temperatures in this area were elevated to 33°C from the normal 24°C. Usu last erupted 23 June 1944–30 Sept 1945, producing Showa-Shinzan dome [at the E foot of the volcano].

Information Contacts: T. Tiba, National Science Museum, Tokyo; L. Siebert, SI, Washington DC; J. Bloom, U.S. Embassy, Tokyo.

9/77 (2:9) No explosions have been reported since 14 Aug, but a new cryptodome at the E foot of Ko-Usu, surrounded by a 1.5 km semicircular fissure zone, had risen about 70 m by late Sept. Local earthquake swarms continued.

Information Contacts: Y. Katsui, Hokkaido Univ.; JMA, Tokyo; L. Siebert, SI, Washington DC; *Asahi Evening News*.

10/77 (2:10) Surface activity during Sept was confined to continued cryptodome uplift and weak emission of steam clouds, which rose a maximum of 600 m from summit vents [produced by the Aug eruption]. Daily recorded seismic events, including many felt shocks, declined irregularly through Sept, but remained high at the end of the month.

Information Contact: JMA, Tokyo.

11/77 (2:11) Weak emission of white to grayish-white steam from small craters and fissures near Ko-Usu and O-Usu continued through Oct, reaching a maximum height of 400 m above the vents. Uplift of Ogari-yama [a large block making up most of the 2 km-diameter somma's NE quadrant, that has been uplifted] at the E foot of Ko-Usu since Aug (2:9), continued during Oct at the rate of about 45 cm/day, while Ko-Usu itself subsided about 40 cm/day. Another [unnamed peak of the uplifting block], about 240 m ENE of Ko-Usu's summit, rose 60 cm/day in Oct.

A small steam explosion from 3 new vents began between 0300 and 0400 on 16 Nov (fig. 8-47) and lasted until about 0930. The explosion, the first from Usu since 14 Aug, produced a 300-m cloud and was accompanied by rumbling. The new vents (one is 6 m in diameter, the other two about 2 m in diameter) are clustered atop a 25 m-diameter rise at the edge of the unnamed new cryptodome. Several hundred local earthquakes per day continued to be recorded (fig. 8-48).

Information Contacts: T. Tiba, National Science Museum, Tokyo; JMA, Tokyo.

12/77 (2:12) Weak emission of white to grayish-white steam continued through Nov from small craters and fissures near the summits of Ko-Usu and O-Usu. During Nov, daily rates of uplift declined slightly at both [peaks of the uplifting block] (Ogari-yama and another centered about 240 m ENE of Ko-Usu's summit) to about 40 cm/day, from Oct rates of 45 and 60 cm/day respective-

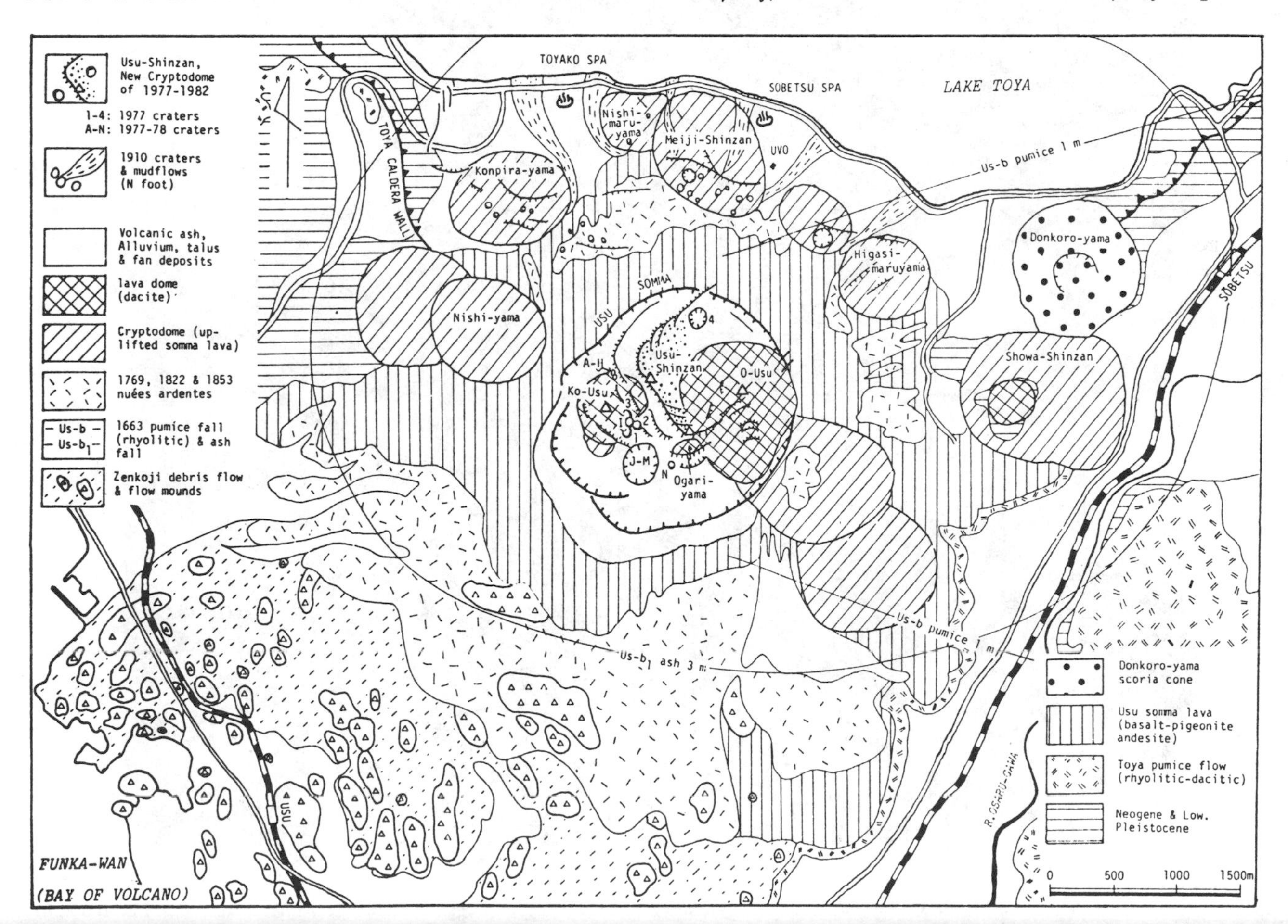

Fig. 8-41. Geologic map of Usu showing vent locations for the 1977–82 eruption. From Katsui et al., 1985.

Fig. 8-42. Ash-laden cloud from Usu at 0950 on 7 Aug 1977, 38 minutes after the eruption began. View approximately to the S. Photo by *The Hokkaido Shinbun,* courtesy of Yoshio Katsui.

Fig. 8-43. Usu volcano on 12 Aug 1977 looking SE across Lake Toya with the town of Toyako-Spa in the foreground. O-Usu, left, and Ko-Usu, right, lava domes lie within the Usu somma. The first three major explosive eruptions 7-9 Aug originated from vents behind Ko-Usu in this view; the fourth major eruption on 9 Aug occurred from a vent in front of O-Usu. Trees on the upper slopes were stripped of leaves by volcanic ejecta. Photo by Lee Siebert.

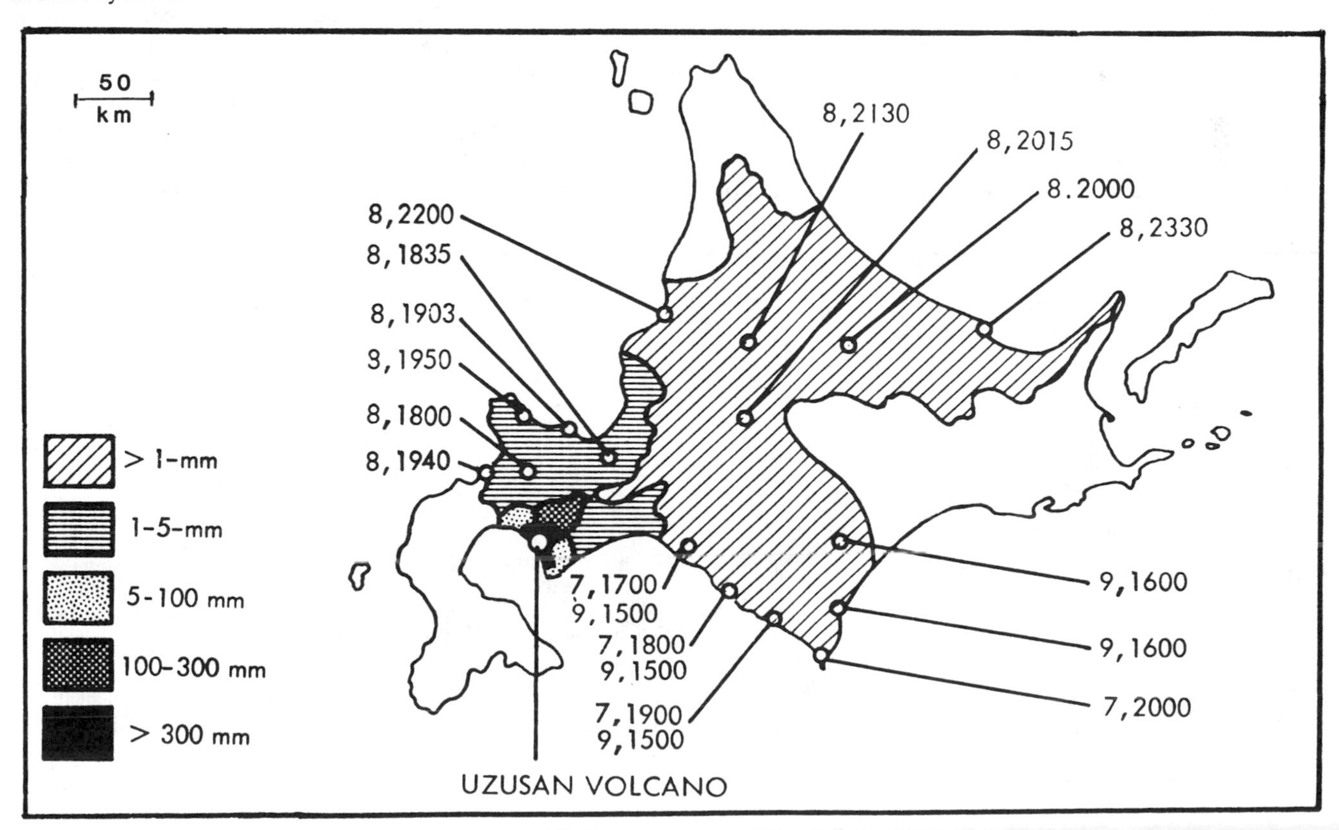

Fig. 8-44. Map showing distribution of Usu's ash on the island of Hôkkaido. Open circles are sampling localities. The associated numbers refer to the date and time of first appearance(s) at each locality of [pumice from] explosions on 7, 8, and 9 Aug 1977. Courtesy of JMA.

1977		Time	Vent	Cloud Ht	Ashfall Azim.	Remarks
Aug	7	0912-1140	1	12	S 60 E	Up to 30 cm of ash and breccia fell 4 km away. Airplane windows were cracked by tephra at 3600 m altitude. Lightning accompanied the eruption cloud.
		1331	1	2		"White smoke".
		1620	1	1.5		
		1822	1	1.5		
	8	1335-1410	2			
		1537-1800	2	10	N 40 W	Ash reached Asahigawa, 180 km NE. Pumice up to 20 cm diameter fell 3 km away.
		1900	2			
	8-9	2340-0215	3	10?	N 45 W	Incandescent material rose several hundred meters
	9	0530-0730	3	6?	N 10 E	Ash reached the Sea of Okhotsk, about 75 km to the N.
		0815-0825	3	1.5		
		0855-0905	3	1.5		
		0908-0925	3	1		
		1020-1105	4	4	N 75 E	
		1120-1420	4	9	S 65 E	
	12	0812-0900	3	2.5	N 35 W	"Black Smoke"
	13-14	2237-0155	3	4?	N 70 E	2 cm of pumice nearby; eruption accompanied by lightning.

Table 8-38. Summary of explosions from Usu, 7-14 Aug, adapted from Katsui, et al. (1978). Vent numbers correspond to those shown in fig. 8-41. Cloud heights (in km) are approximate.

ly. Subsidence of Ko-Usu had also slowed, from 40 cm/day in Oct to about 30 cm/day in Nov. According to the *Japan Times,* a team of Hokkaido University geophysicists has located a shallow ellipsoidal aseismic zone beneath the 2 cryptodomes. The zone has a maximum dimension of 1 km and a minimum dimension of 0.5 km, but it was not possible to determine its depth from the newspaper article.

Reference: Katsui, Y. et al., 1978, Preliminary Report of the 1977 Eruption of Usu Volcano; *Journal of the Faculty of Science, Hokkaido Univ.*Ser. 4, v. 18, p. 385-408.

Information Contacts: Y. Katsui, Hokkaido Univ.; JMA, Tokyo; *Japan Times*.

Further Reference: Yokoyama, I., 1978, The 1977 Eruption of Usu Volcano with a Special Reference to Prediction of Volcanic Activities; *Bulletin of the Volcanological Society of Japan*, 2nd series, v. 23, no. 1, p. 65-82.

Katsui, Y., Komuro, H., and Uda, T., 1985, Development of Faults and Growth of Usu-Shinzan Cryptodome in 1977-1982 at Usu Volcano, North Japan; *Journal of the Faculty of Science, Hokkaido Univ.*, Ser. 4, v. 21, p. 339-362.

1/78 (3:1) Weak white and grayish-white steaming from vents and fissures near the summits of O-Usu and Ko-Usu continued during Dec, with a maximum cloud height of about 300 m. On 13 Jan, white steam clouds rose about 500 m from [a new vent (named B)] at 0740 and at 1125. A thin layer of ash was seen on top of snow at Date City (about 10 km SE of Usu) and Sobetsu (about 5 km NW of Usu).

Cryptodome uplift . . . declined slightly (to about 35

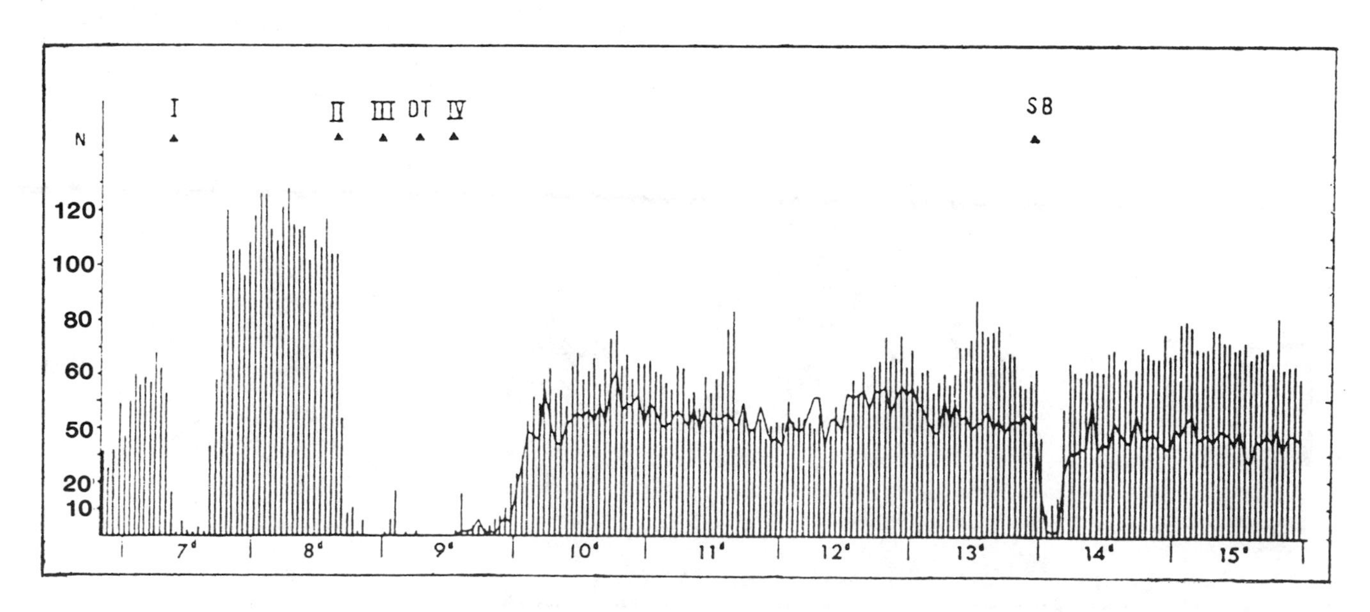

Fig. 8-45. Hourly seismicity at Usu, from the night of 6-7 Aug to 15 Aug 1977. Decreases in seismicity accompanied major explosive episodes, indicated by small triangles. Courtesy of JMA.

cm/day in Dec, from about 40 cm/day the preceding month). Subsidence at Ko-Usu, about 30 cm/day in Nov, decreased to about 15 cm/day in Dec. Daily local seismicity also decreased slightly in Dec.

Information Contacts: JMA, Tokyo; T. Tiba, Nat. Sci. Museum, Tokyo; Y. Katsui, Hokkaido Univ.

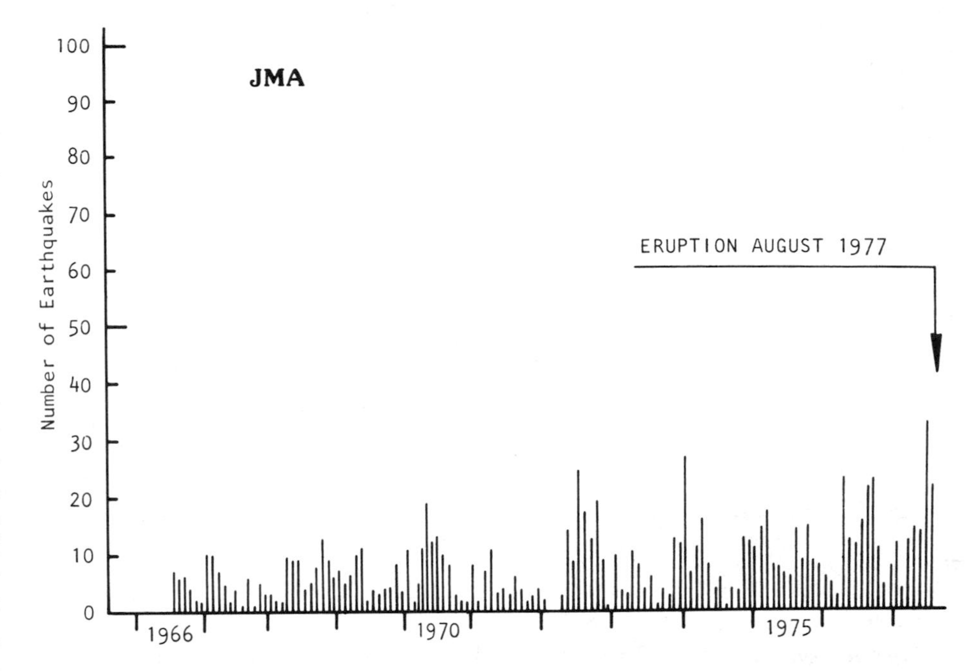

Fig. 8-46. JMA graph of monthly volcanic earthquakes at Usu, July 1966 – July 1977 [showing a gradual increase in seismicity]. The monthly mean is 9.

2/78 (3:2) Weak steaming from vents and fissures on the slopes of . . . Ko-Usu and the new cryptodome NE of Ko-Usu continued during Jan. Steam clouds rose a maximum of 400 m above the vents. The rate of uplift of the new cryptodome briefly increased in late Dec to 38 cm/day (from 35 cm/day in mid-Dec), then declined to 31 cm/day in early Jan and to 26 cm/day by the end of the month. A team from Hokkaido University measured its summit altitude at 568 m in early Jan [compared to about 490 m altitude before the Aug 1977 eruption].

Daily earthquake frequency gradually increased at the end of Jan, and by late Feb had returned to the Sept rate. Epicentral distribution, according to Hokkaido University geophysicists, was unchanged, concentrated beneath Ko-Usu and the N portion of the somma, to slightly N of the somma (2:12).

Information Contacts: JMA, Tokyo; D. Shackelford, CA.

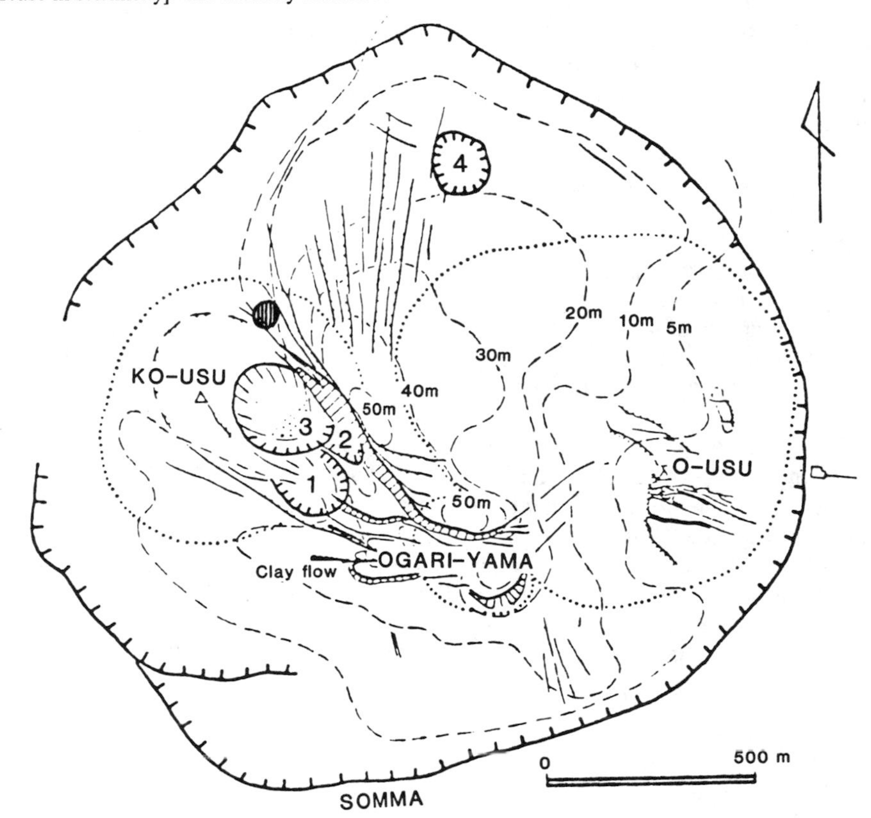

Fig. 8-47. Region within Usu's somma. Solid lines are faults and fractures, dashed lines show amount of new uplift as of 23 Aug, and dotted lines outline pre-existing domes (Ko-Usu, O-Usu, and Ogari-yama). New craters are numbered. The location of the 16 Nov vent (2:11-12) is shown by a small shaded circle N of vent 3; location was provided by JMA. From Katsui, et al., 1978.

3/78 (3:3) Two small eruptions were observed during Feb, originating from vents 250-300 m ENE of the summit of Ko-Usu dome. The first occurred at 1559 on the 25th from a vent 2-3 m in diameter, producing a 500-m gray ash cloud and a trace of ashfall near the volcano. The dark gray ash cloud from the second eruption, at 0740 on 27 Feb, rose about 1200 m from a 15-20 m-diameter crater, causing a 2-3 mm ashfall a few kilometers SE of Usu.

The rate of uplift at the . . . cryptodome increased substantially, from about 23 cm/day in Jan to about 39 cm/day in Feb at Ogari-yama, and from 26 cm/day in late Jan to ~41 cm/day in Feb at the unnamed [peak] about 400 m E of the summit of Ko-Usu. Feb subsidence at Ko-Usu continued at the Jan rate of about 8 cm/day. Daily earthquake counts at Usu began to increase in late Jan and remained relatively high through most of Feb

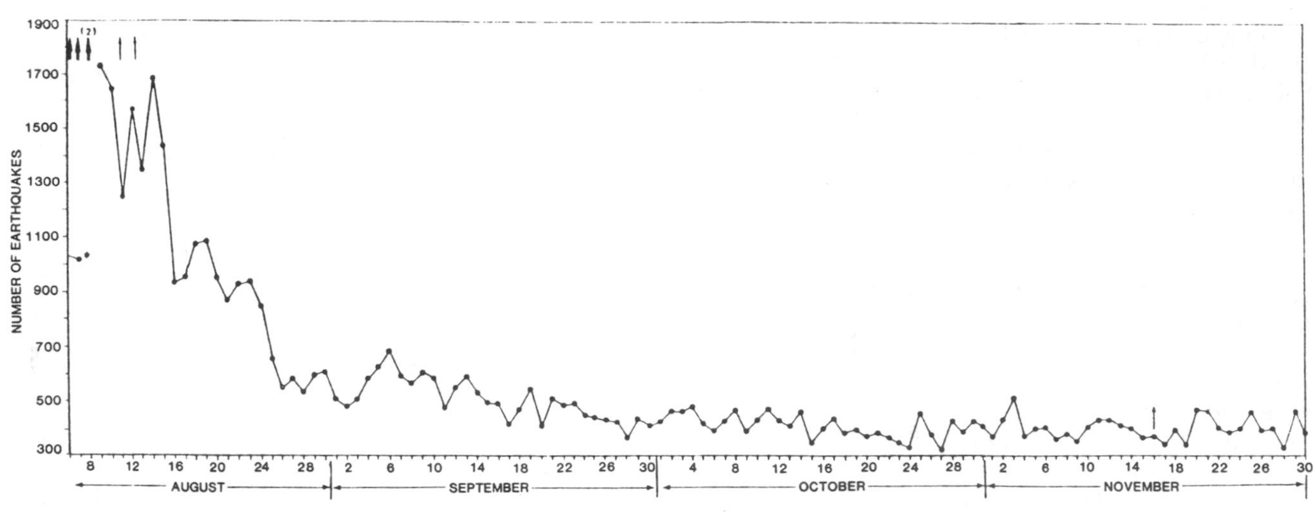

Fig. 8-48. Number of seismic events/day, 7 Aug–30 Nov 1977, recorded by a JMA seismograph 2.3 km N of the summit of O-Usu. Heavy arrows indicate major explosions, lighter arrows smaller explosions [considered to be phreatic]. Continuous [eruption] tremor, represented by *, prevented a [precise] earthquake count on 9 Aug.

(fig. 8-49).

Information Contacts: JMA, Tokyo; T. Tiba, National Science Museum, Tokyo.

5/78 (3:5) The steam explosions of 25 and 27 Feb were followed by similar events, each lasting a few hours, on 2, 3, 4, 5, 6, 9, 11, and 14 March, and on 24 April, causing slight ashfalls around the volcano. All of the explosions originated from a small area NE of Ko-Usu, where geologists found 8 small vents on 5 March.

After a sharp increase to 39 cm/day in Feb, the rate of uplift at Ogari-yama had declined to 26 cm/day by late March and averaged 24 cm/day in April. . . . Daily earthquake counts also increased substantially in Feb to a monthly average of 411/day, but decreased to 345/day in March and 260/day in April. . . .

Information Contact: JMA, Tokyo.

6/78 (3:6) After 30 days of quiescence, steam explosions occurred on 26, 29, and 31 May and 1, 2, 4, and 7 June, causing ashfalls on surrounding villages. The largest explosion produced a 3-mm ash deposit in a village at the foot of the volcano, and blocks were scattered in the summit crater, 500 m from the vent. Scientists from the University of Hokkaido observed the ejection of incandescent blocks during the night of 2 June, the first incandescent ejecta recognized since the initial large [pumice eruptions] of Aug 1977. This suggests that the rising magma column is now near the surface.

. . . The general downtrend in daily seismicity that began in Feb continued through early June.

The rate of uplift at the new cryptodome (centered about 400 km E of the summit of Ko-Usu) increased sharply in Feb (3:3), but began to decrease in March and continued to decrease through May (data from I. Yokoyama).

Information Contacts: Same as above.

7/78 (3:7) Steam explosions continued through late July, with ashfalls around the volcano from those observed on 2, 3, 4, 7, 11, 22, 26, 28, and 30 June, and 2 July. The rate of uplift at the new cryptodome decreased to 17.6 cm/day in May and 13.6 cm/day in June. Subsidence of Ko-Usu dome has virtually stopped.

The *Japan Times* reported a 5-minute eruption on 16 July, with ejecta rising about 1000 m above the vent. Reuters reported that eruption columns reached 2000 m on [29] July. [Ash eruptions occurred frequently 15-31 July.]

Information Contacts: JMA, Tokyo; *Japan Times*; Reuters.

8/78 (3:8) Steam explosions continued, occurring on 2, 9,

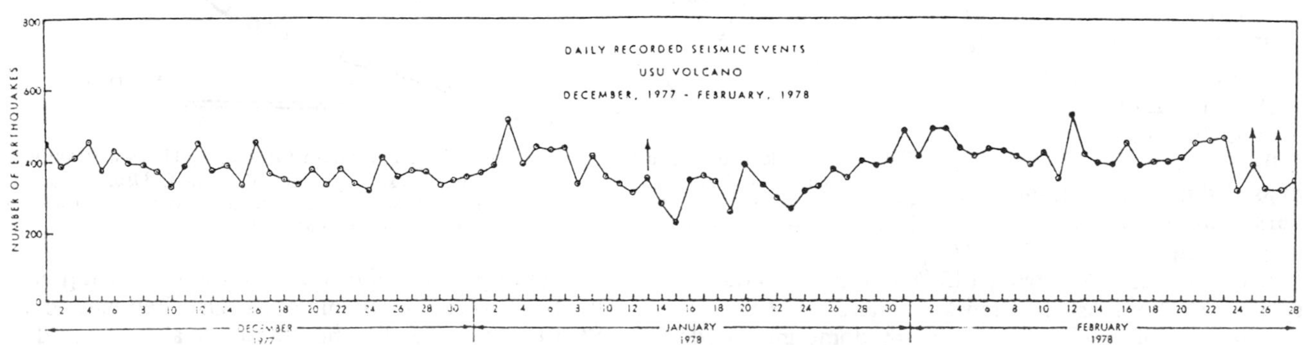

Fig. 8-49. Number of seismic events/day, 1 Dec 1977–28 Feb 1978, recorded by a JMA seismograph 2.3 km N of the summit of O-Usu. Arrows mark steam explosions.

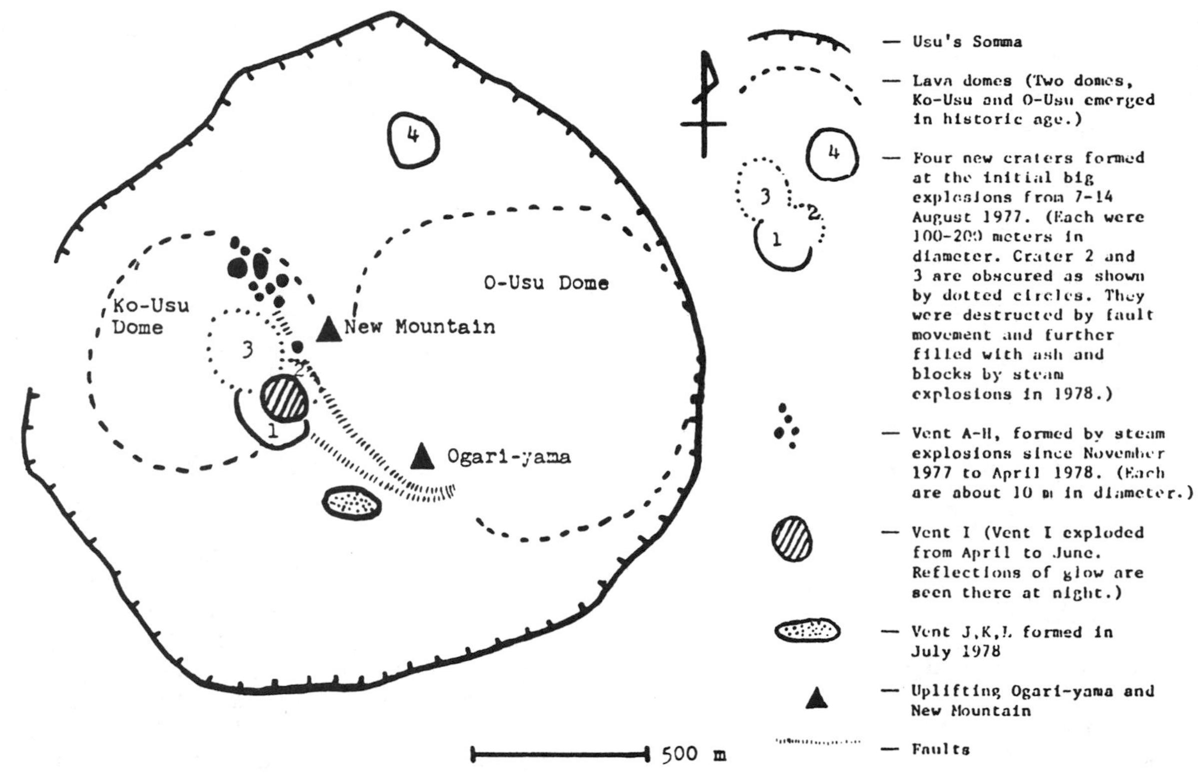

Fig. 8-50. Sketch map of the summit area as of July 1978. Courtesy of JMA.

10, 15-17, 25, and 28-31 July. The 15 July explosions deposited 10 mm of ash on nearby villages, the largest ashfall since Nov 1977. Vent I (fig. 8-50), site of explosions in April, late May, and June became quiescent, but reflected glow could be seen above this area at night in June and July. Vents J, K, and L (about 300 m SE of vent I) were formed by explosions on 9, 15, and 16 July respectively. Vigorous steam emission, accompanied by continuous tremor, enlarged the vents.

Mean daily seismicity fell to 175 events, of which an average of 23/day were felt. Uplift rates at . . . the new cryptodome ENE of Ko-Usu continued to decrease. . . . Subsidence of Ko-Usu dome has virtually ceased.

The Kyodo broadcasting company reported an ex-

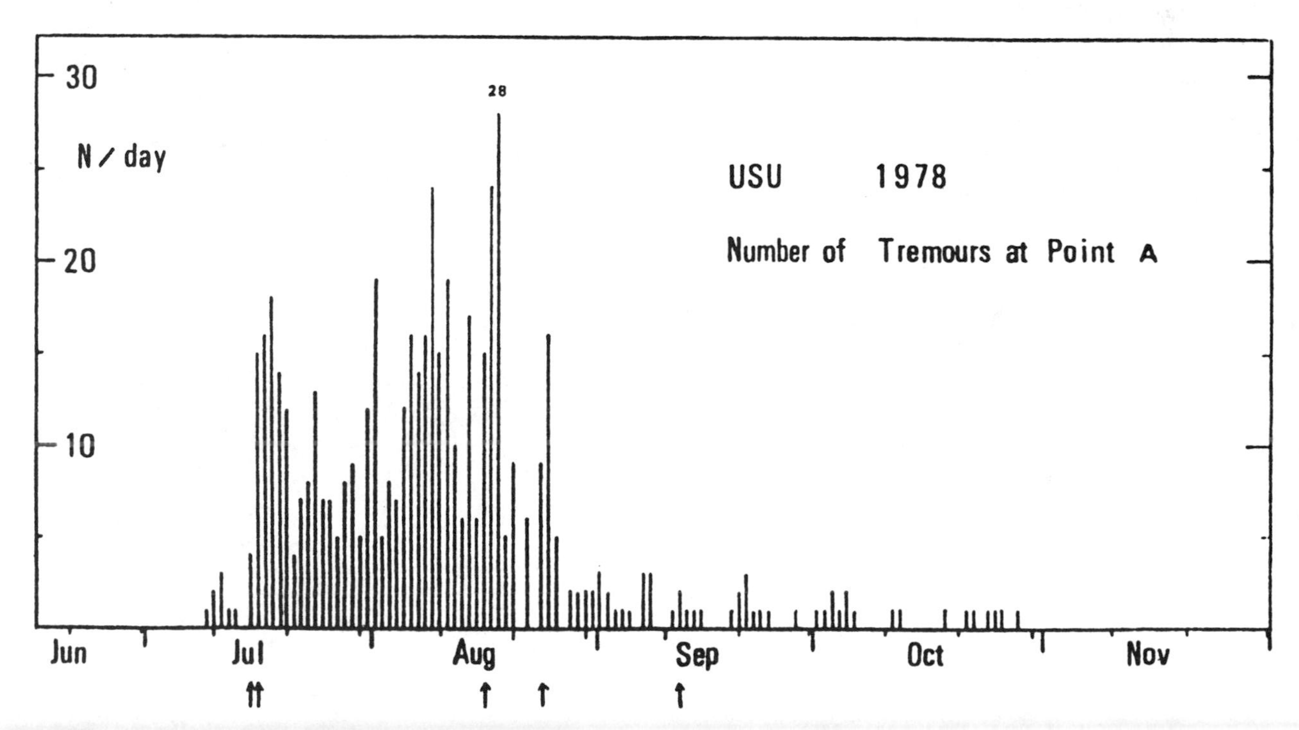

Fig. 8-51. Daily number of continuous tremor episodes, lasting from 1 minute to 1 hour, recorded at Usu July – Oct 1978. Tremor episodes correspond to periods of ash emission. Arrows indicate larger explosions. [Originally from 4:2.]

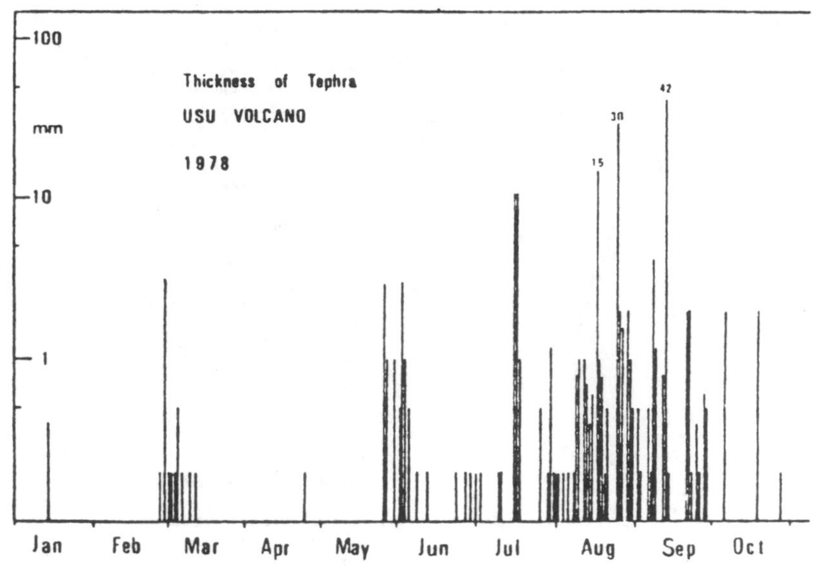

Fig. 8-52. Daily thickness (in mm) of newly deposited tephra at the foot of Usu, Jan–Oct 1978. Note logarithmic scale. [Originally from 4:2.]

plosion of Usu at 0325 on 24 Aug. The ash cloud rose about 800 m and ash fell on Noboribetsu, about 30 km from Usu. Shortly after the explosion, glow was seen above the summit for several minutes. An increase in local seismicity occurred 2 days before the explosion; 345 events were recorded 22 Aug, but only 102 on the 23rd.

Information Contacts: JMA, Tokyo; I. Yokoyama, Y. Katsui, Hokkaido Univ.; Kyodo broadcasting company.

9/78 (3:9) Steam explosions were more frequent in Aug than in July. Explosions, most lasting no more than 15 minutes, occurred on 1, 3, 7-14, 16-20, 24-26, and 28-30 Aug, causing ashfalls in nearby villages. More than 1 cm of ash was deposited on villages a few km E of Usu after explosions on 16 and 24 Aug. During the 24 Aug activity, [an incandescenct column 300 m high] was visible from the base of the volcano for the first time since Usu's major eruption of Aug 1977. The Aug explosions produced a 350 × 200 m crater in the S portion of the summit area, engulfing the smaller vent active in July (3:8). Crater growth persisted through early Sept.

The rates of uplift at the new cryptodome . . . continued to decline, to 7 cm/day during Aug. Local seismicity also continued to decline, to an average of 91 events/day, about 1/6 of which were felt. Tokiko Tiba reported that explosions from 2215-2235 on 12 Sept and 0030-0217 on 13 Sept produced a 3-km eruption cloud,

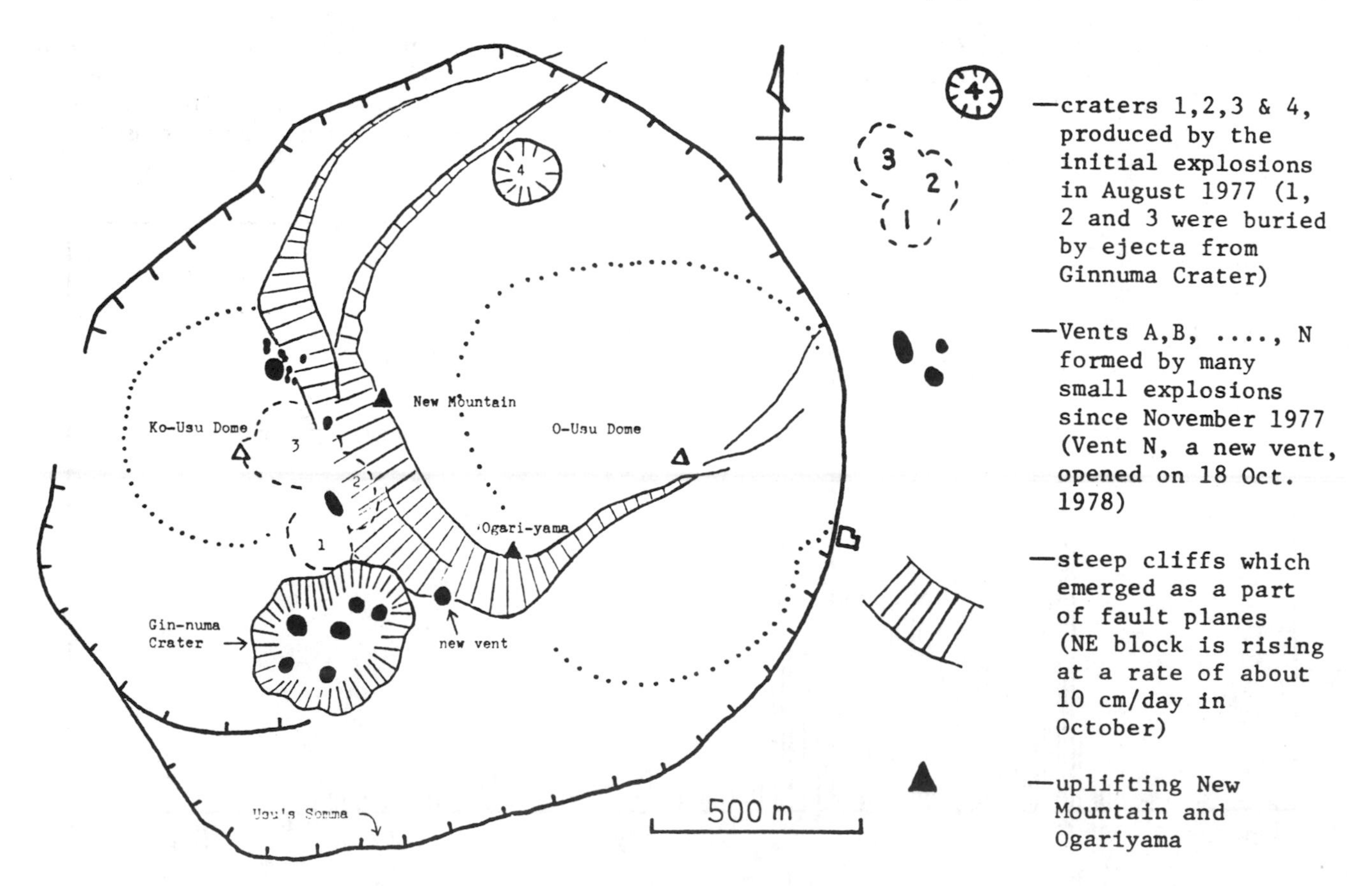

Fig. 8-53. Map of the summit area, late Oct 1978. Locations were surveyed by the Katsui group of the University of Hokkaido and JMA. The figure was adapted by Manabu Komiya from the Katsui group report.

accompanied by lightning, and 1-5 cm of ashfall on Toyako-spa, about 3 km NNW of the active crater.

Information Contacts: JMA, Tokyo; I. Yokoyama, Hokkaido Univ.; T. Tiba, National Science Museum, Tokyo.

10/78 (3:10) Explosions were recorded on 1-2, 5-8, 11-13, 20-22, 24-25, and 27-28 Sept. Most lasted only a few minutes and produced ashfalls of no more than a half millimeter in thickness (figs. 8-51 and 8-52). However, the 2 explosions that occurred during the night of 12-13 Sept were the largest since steam explosions began in Nov 1977. Ash clouds, accompanied by much lightning, rose 2.5 km above the crater and deposited up to 3.5 cm of ash over a broad area N of the volcano. Columns of incandescent tephra, about 600 m high, were present over the crater for three 10-minute periods during the 12-13 Sept activity, dropping blocks in the summit area. All of the Sept explosions occurred from the newly-named Gin-numa Crater, which had grown to a roughly circular feature, 350 m in diameter and 70 m deep, by the end of the month.

Cryptodome uplift rates and the number of local earthquakes per day both increased slightly in Sept, after several months of decline.

Information Contacts: JMA, Tokyo; I. Yokoyama, Hokkaido Univ.

Fig. 8-54. Apartment building near Usu damaged by faulting (and now abandoned). Photo taken in Oct 1978. Courtesy of Manabu Komiya.

Fig. 8-55. Two-meter fault dislocation in a road [2 km NW of] Usu. Note the slight distortion of the flat-roofed building in left-center foreground. Fault movement began in Oct 1977. Photo taken in Oct 1978. Courtesy of Manabu Komiya.

11/78 (3:11) A lahar flowed through the town of Toyako-spa (at Usu's NW foot) on 24 Oct, killing 3 persons, destroying 16 houses, and damaging 20 others. However, the number of summit area explosions declined to 3 in Oct (one each on the 5th, 18th, and 27th) after 1 or more had been recorded on 16 days in Sept. The 5 Oct explosion originated from Gin-numa crater, active since Aug, but a new 30 m-diameter vent (labeled N in fig. 8-53) opened less than 100 m to the NE on 18 Oct and was the source of the explosions on the 18th and 27th.

I. Yokoyama reported that both the average number of earthquakes per day and the rate of cryptodome uplift increased slightly in Oct, continuing a trend that began the preceding month.

Information Contacts: JMA, Tokyo; I. Yokoyama, Hokkaido Univ.

12/78 (3:12) The number of days per month in which one or more explosions were recorded has dropped sharply, from 21 in Aug to 3 in Oct.

As of early Dec, there had been no explosions since 27 Oct. Local seismicity declined to an average of 79 earthquakes per day in Nov, and I. Yokoyama reported that the rate of cryptodome uplift decreased to 6.5 cm/day. Both seismicity and cryptodome uplift rates had increased slightly in Sept and Oct but have been on a general downward trend since Feb.

Since the initial explosions of Aug 1977, the N portion of Usu's somma has been slowly moving outward, causing severe local faulting at the N foot of the volcano. The distance between the N rim of the somma and the shore of Lake Toya decreased by about 65 m Sept 1977–March 1978 (Yokoyama, 1978). Numerous small faults have formed, damaging or destroying buildings in the area (fig. 8-54). Some roads have been cut by faults (fig. 8-55), and underground water pipes, telephone cables, and tubes for hot spring water have been cut at several hundred points.

Information Contacts: JMA, Tokyo; I. Yokoyama, Hokkaido Univ.

Further References: Imagawa, T., 1986, Mud and

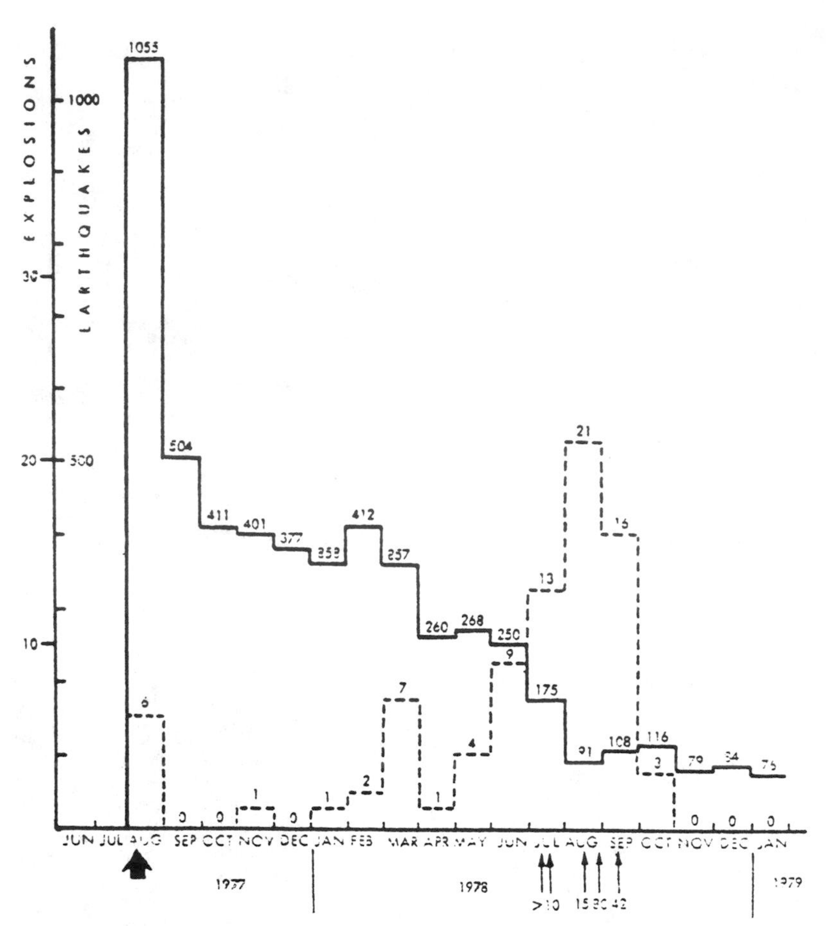

Fig. 8-56. Monthly means of Usu's daily seismicity recorded by a seismograph 2.3 km N of Usu (solid line), and number of days per month in which explosions occurred (dashed line), June 1977–Jan 1979. The double arrow represents the major explosions of 7-14 Aug 1977 and the single arrows indicate the most important of the much smaller 1978 explosions. The number beneath each single arrow is the thickness of tephra, in mm, deposited at the foot of the volcano by that explosion (data from figs. 8-51 and 8-52).

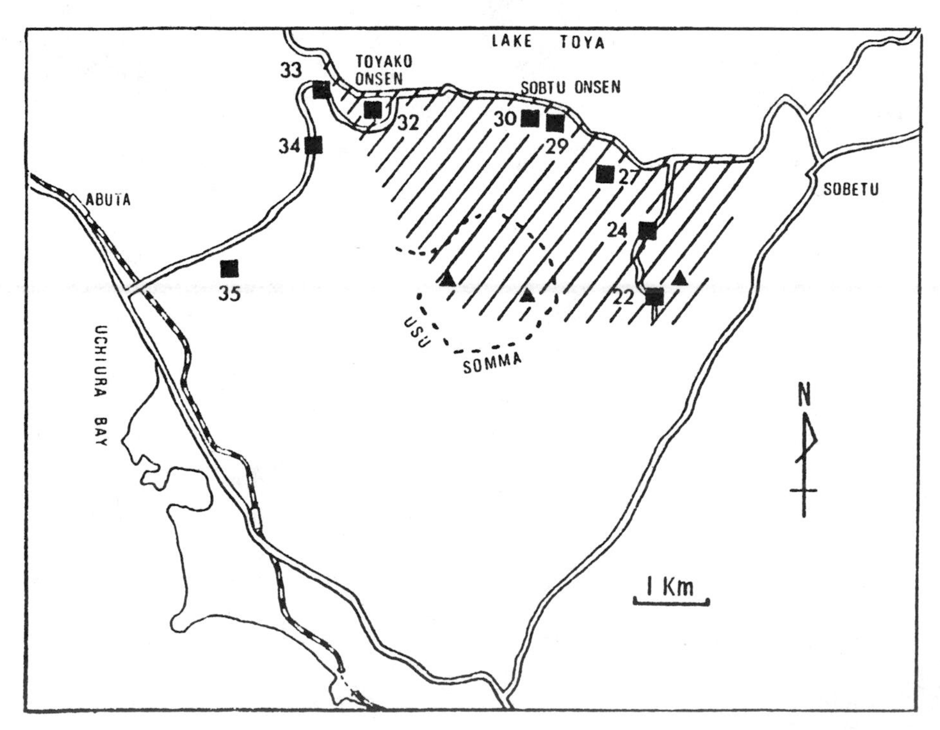

Fig. 8-57. Map of Usu and vicinity, with the area affected by severe ground deformation (3:12) shaded.

Debris Flows on Mt. Usu after the 1977–1978 Eruptions; *Environmental Science, Hokkaido*, v. 9, no. 1, p. 113-135.

Katsui, Y., Yokoyama, I., Watanabe, H., and Murozumi, M., 1981, Usu Volcano *in* Katsui, Y. (ed.) *Symposium on Arc Volcanism Field Excursion guide to Usu and Tarumai Volcanoes and Noboribetsu Spa, Part I*; Volcanological Society of Japan, Tokyo, p. 1-37.

Niida, K., Katsui, Y., Suzuki, T., and Kondo, Y., 1980, The 1977–1978 Eruption of Usu Volcano; *Journal of the Faculty of Science, Hokkaido University*, ser. IV, v. 19, p. 357-394.

Seki, K. (ed.), 1978, *Usu Eruption and its Impact on Environment*; Hokkaido University, Sapporo, 496 pp. (41 papers).

Yokoyama, I., Yamashita, H., Watanabe, H., and Okada, H., 1981, Geophysical Characteristics of Dacite Volcanism—1977–1978 Eruption of Usu Volcano; *JVGR*, v. 9, p. 335-358.

The Eruption of Usu Volcano (August 1977–December 1978); Technical Reports of the JMA, v. 99, 1980, 203 pp.

1/79 (4:1) No explosions have occurred since 27 Oct, although voluminous steam emission has been continuous. The average number of seismic events per day was 84 in Dec, up slightly from 79 per day in Nov.

Information Contact: JMA Tokyo.

2/79 (4:2) No explosions occurred in Jan. The average number of local earthquakes/day declined slightly, from 84 in Dec to 76 in Jan (fig. 8-56), and the rate of cryptodome uplift also showed a small decrease. The ground deformation described in 3:12 continued NW of the volcano, but stopped to the N, NE, and E (fig. 8-57).

Information Contact: Same as above.

3/79 (4:3) The average number of earthquakes per day at Usu declined from 76 in Jan to 56 in Feb, and cryptodome uplift rates also decreased, to about

5 cm/day. No explosions have occurred since Oct.

Information Contact: Same as 4:1.

4/79 (4:4) No explosions have occurred since Oct 1978. In March, seismicity and the rate of cryptodome uplift remained at the same levels as in Feb.

Information Contact: Same as 4:1.

5/79 (4:5) Seismicity declined from an average of 64 events/day in March to 61/day in April. No explosions have occurred since Oct 1978.

Information Contact: Same as 4:1.

6/79 (4:6) Monthly recorded seismicity continued to decline slightly in May, while cryptodome uplift remained constant at about 5 cm/day (fig. 8-58). Ogari-yama and the Usu-Shinzan (Usu "New Mountain"), once thought to be separate cryptodomes, are now considered to be 2 peaks of a single [rising block]. Ogari-yama has risen 160 m since uplift began in Aug 1977.

No explosions have occurred at Usu since 27 Oct 1978.

Information Contact: Same as 4:1.

9/79 (4:9) Yosihiro Sawada provided the following data (table 8-39) on total summit area elevation changes since activity began in Aug 1977 [based on results from Hokkaido University]. The uplift rate at Ogari-yama was 5 cm/day in May and 3 cm/day in July.

Seismicity continued to decline slowly, from an average of 54 events/day in June to 47/day in July. The actual number of events per day sometimes varied considerably from the mean; in June, daily earthquakes ranged from 5 to 293.

When inspected on 31 July, Gin-numa Crater (formed by the activity of Aug – Oct 1978) contained a pool of water. Strong fumaroles on its rim had maximum measured temperatures of 635°C.

Information Contacts: Y. Sawada, MRI, Tokyo; D. Shackelford, CA.

7/80 (5:7) Seismic activity (table 8-40) and cryptodome uplift have continued through May. The rate of cryptodome uplift and outward movement of the N somma wall both averaged about 5 cm/day in Nov 1979 and 4 cm/day in March 1980. In May 1980, white vapor rose from 3 vents. Three new parallel faults with a combined throw of about 60 cm passed through the S portions of the active vents.

Information Contacts: JMA, Tokyo; D. Shackelford, Fullerton CA.

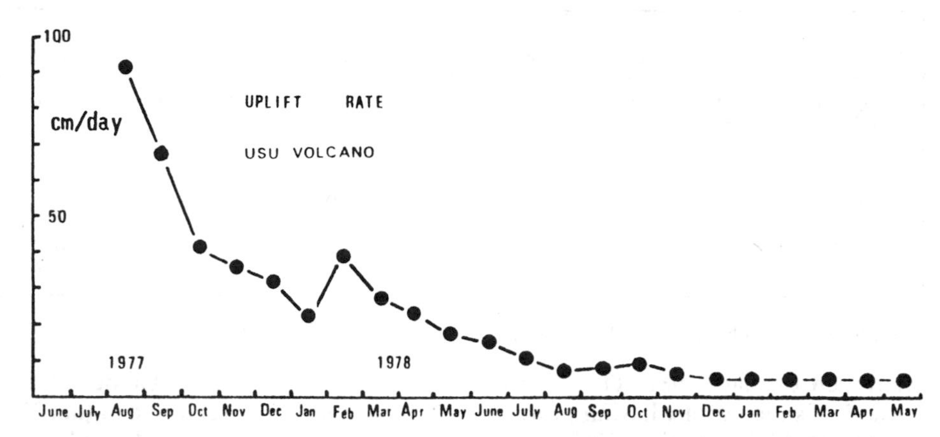

Fig. 8-58. Rate of cryptodome uplift in cm/day, July 1977 – May 1979. Courtesy of JMA.

1/81 (6:1) Cryptodome uplift and local seismicity continued through 1980. Local seismicity continued an irregular decline through 1980 (fig. 8-59 and table 8-41). Felt shocks averaged 3/day in 1980, but swarms of 30-40 felt events in a single day occurred about once a month. . . . Careful correlation of seismic records with observed surface deformation and faulting revealed that larger earthquakes occurred simultaneously with measureable fault movements.

The rate of cryptodome uplift decreased through 1980, from 5 cm/day in Jan to 3-4 cm/day in Dec [at the peak of] Usu-Shinzan . Northward lateral movement of the N flank continued at a similar rate. As a result, compression of the ground N of the volcano also continued, affecting several towns and villages.

Information Contacts: JMA, Tokyo; I. Yokoyama, Hokkaido Univ.

3/82 (7:3) The following is from I. Yokoyama.

"The crustal deformation and local seismicity at Usu

Location	Elevation change
New cryptodome	+ 159.51 m
Ogari-yama	+ 163.26 m
O-Usu	- 23 m
Ko-Usu	- 54.8 m

Table 8-39. Total elevation change at several features on Usu, as of 31 July 1979.

	1979						1980				
Month	Jul	Aug	Sept	Oct	Nov	Dec	Jan	Feb	Mar	Apr	May
Earthquakes/Day	47	36	35	36	36	44	38	35	29	19	22

Table 8-40. Monthly means of the number of recorded earthquakes per day, July 1979 – May 1980.

1980	Jan	Feb	Mar	Apr	May	Jun	Jul	Aug	Sept	Oct	Nov	Dec
Recorded	[1177]	1004	890	582	[674]	[221]	601	486	620	413	604	[571]
Felt	234	216	162	92	121	32	112	82	108	69	106	94

1981	Jan	Feb	Mar	Apr	May	Jun	Jul	Aug	Sept	Oct	Nov	Dec
Recorded	357	289	235	485	153	151	423	317	244	315	[290]	440
Felt	63	49	41	92	35	33	89	64	41	52	54	105

Table 8-41. Monthly number of local seismic events, 1980–81.

continued through 1981. The monthly number of recorded seismic events, having gradually declined since the major 1977 eruption (2:8-9), dropped further to about 308/month in 1981 but remained at about this level through the year (fig. 8-60 and table 8-41). Gradually weakening steam activity from the craters formed in 1978 has been observed. Around these craters, there have been many fumaroles that vigorously emitted white vapor; highest temperature was 643°C in Aug 1981. According to the data from the Usu Volcano Observatory (Hokkaido University) the rate of uplift of the Usu-Shinzan cryptodome decreased from about 2/cm per day in 1980 to about 0.8 cm/day in 1981. The northward lateral movement of the N flank continued at a similar rate."

Information Contact: I. Yokoyama, Hokkaido Univ.

1/83 (8:1) Seismicity and ground deformation at Usu ended in spring 1982, after 58 months of activity. The monthly number of recorded seismic events had gradually declined since the major eruption in 1977 (2:8-9), but remained above background through 1981, when approximately 308 seismic events were recorded per month (7:3). Ground deformation had also continued since the eruption. The rate of uplift of the cryptodome decreased from about 2 cm/day in 1980 to about 0.8 cm/day in 1981. Northward lateral movement of the N flank also continued through 1981.

In 1982, seismic activity decreased to the background level of about 10 events/month by April; 496 events were recorded in Jan, 231 in Feb, 79 in March, 10 in April and 11 in May. Ground deformation has been negligible since April 1982.

Information Contact: JMA, Tokyo.

Further References: Katsui, Y., Komuro, H., and Uda, T., 1985, Development of Faults and Growth of Usu-shinzan Cryptodome in 1977–1982 at Usu Volcano, North Japan; *Journal of the Faculty of Science, Hokkaido University*, ser. IV: Geology and Mineralogy v. 21, no. 3, p. 339-362.

Yokoyama, I. (ed.), 1984, *Report of Joint Geophysical and Geochemical Observations of Usu Volcano in 1982 and Tarumai Volcano in 1983*, 214 pp.

12/83 (8:12) Usu's major explosions of Aug 1977 were followed by rapid (but decreasing) cryptodome growth and seismicity. Smaller explosions started in Nov 1977, became more vigorous and frequent during the following summer, and ended in Oct. The rate of cryptodome growth increased briefly in early 1978 but slowed gradually, as did accompanying seismicity, for the next several years, ceasing rather abruptly in early 1982. Some points on the cryptodome rose as much as 180 m and a baseline between the N foot of the volcano and its N crater rim had shortened by a similar amount by early 1982 (Yokoyama, Katsui, and Abiko, 1983).

The following is a report from S. N. Williams.

"A team of U.S. and Japanese scientists visited Usu's

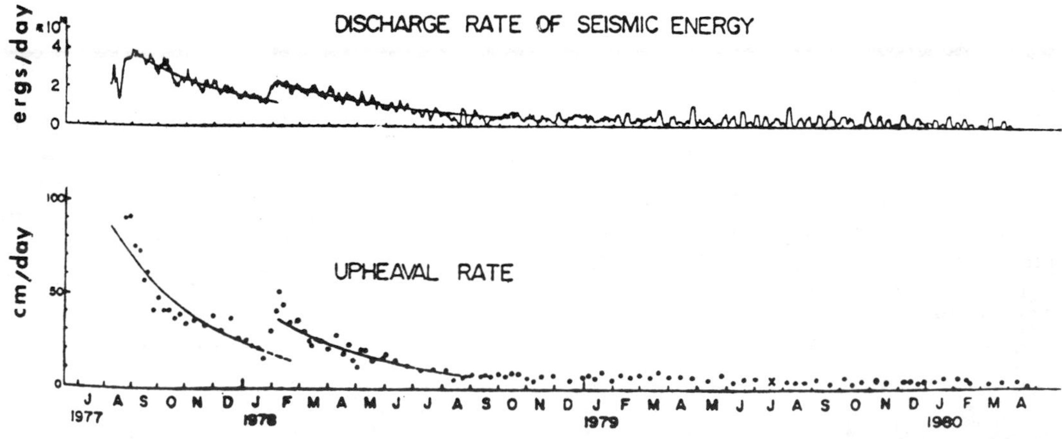

Fig. 8-59. *Top,* discharge rate of seismic energy (ergs/day) from Usu, Aug 1977–April 1980. *Bottom,* uplift rate (cm/day) of the Usu-Shinzan cryptodome for the same period. Note the increase in Feb 1978. [Uplift] data are from I. Yokoyama.

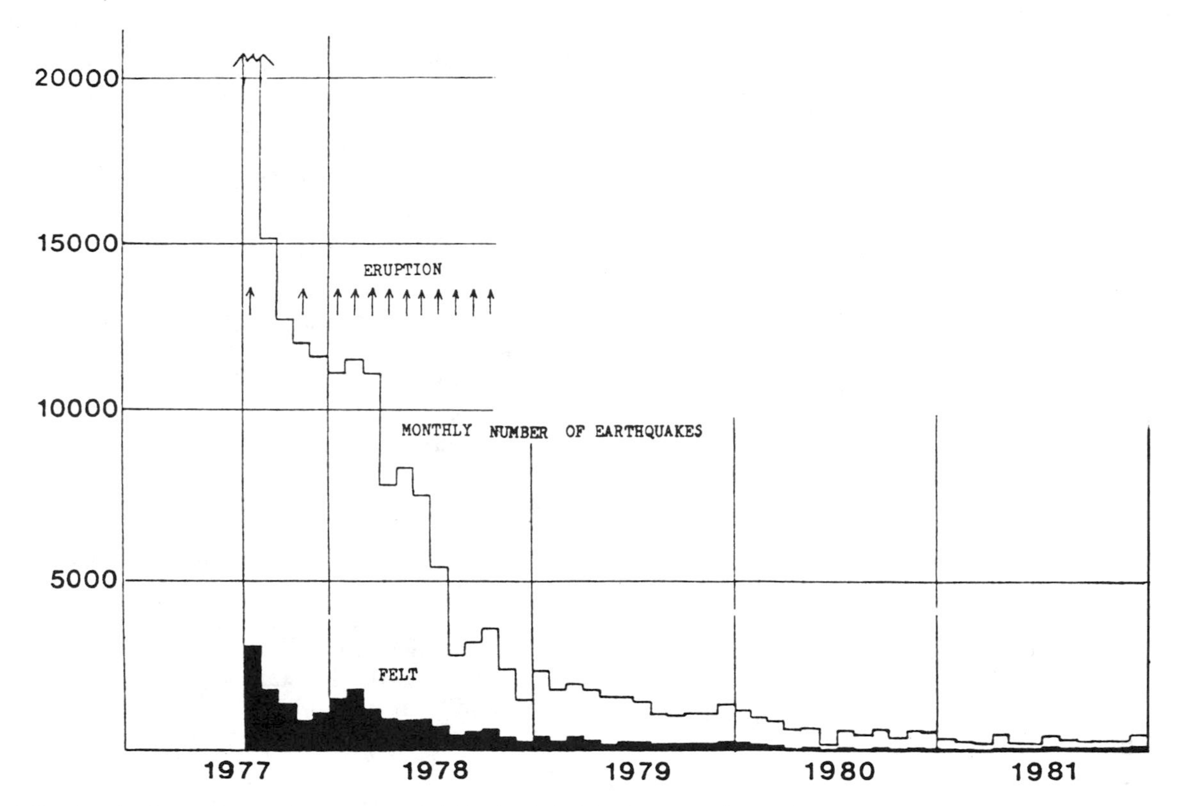

Fig. 8-60. Graph of monthly number of recorded (white bars) and felt (black bars) seismic events at Usu, Aug 1977 – Dec 1981, supplied by I. Yokoyama. [Eruptive] activity during a particular month is indicated by arrows. Earthquakes in Aug 1977 numbered at least 25,000 (2:10).

crater in mid-Aug to continue their study of fumarole mineral geochemistry and mineralogy begun in Sept 1981. Fumarole locations, distribution of fumarole incrustations, and temperatures were mapped. Samples of gas and fumarole sublimates were collected.

"Peak fumarole temperatures have declined somewhat. Fumarole 2, 770°C in 1981, was 730°C in 1983; no. 3 decreased from 690°C to 650°C; no. 1 from 550°C to 545°C. Gas discharge continued at a tremendous rate with the associated roaring jet sound. Extensive deposition of fumarole incrustations continued. All of these fumaroles were located in a narrow valley between the 1977 – 82 cryptodome (Usu-Shinzan) and Ko-Usu, a lava dome extruded in either the 1663 or 1769 eruption.

"Overall, a broad concentric zonation was noted in the temperature distribution of the fumaroles. Large areas of steaming ground on Ko-Usu dome and Usu-Shinzan formed an aureole around the high-temperature fumaroles. These approximately 100°C fumaroles continued to emit large quantities of H_2S. Gin-numa Lake appeared to maintain about the same level as in 1981."

Reference: Yokoyama, I., Katsui, Y., and Abiko, T., 1983, The 1979 – 1982 Activities of Usu Volcano *in* Report on Volcanic Activities and Volcanological Studies in Japan for the Period from 1979 to 1982; *Bulletin of the Volcanological Society of Japan*, v. 28, no. 1, appendix, p. 15-19.

Information Contacts: S. Williams, R. Stoiber, J. Patterson, Dartmouth College; T. Abiko, Muroran Inst. of Tech.

TARUMAI

Hokkaido, Japan
42.68°N, 141.38°E, summit elev. 1024 m
Local time = GMT + 9 hrs

Tarumai, with 35 eruptions since 1667, has been the most active of Hokkaido's volcanoes in historic time. It is 50 km NE of Usu. Large eruptions in 1667 and 1735 formed a 1.5 km-diameter summit caldera. A prominent lava dome grew after the 1909 eruption, and has been the site of most subsequent eruptions.

6/78 (3:6) Tarumai exploded on 14 May at 2253, accompanied by volcanic tremor. A slight ashfall was observed 10 km NNE of the summit crater and a small pyroclastic flow consisting of pumiceous ash and blocks traveled 150 m SE from the vent. Sixteen hours after the eruption, geologists measured a temperature of 200°C at the base of the pyroclastic flow (30 cm below the surface). Smaller explosions occurred on 17 May, mantling the vent area with 1 cm of new ash. No further explosions had been observed as of 7 June.

Tarumai, which last erupted in 1955, is characterized by conspicuous earthquake swarms, followed by explosive eruptions of andesitic magma, then lava dome extrusion. The number of earthquakes at Tarumai has been increasing irregularly for the past 10 years and increased sharply in the first 5 months of 1978 (fig 8-61)

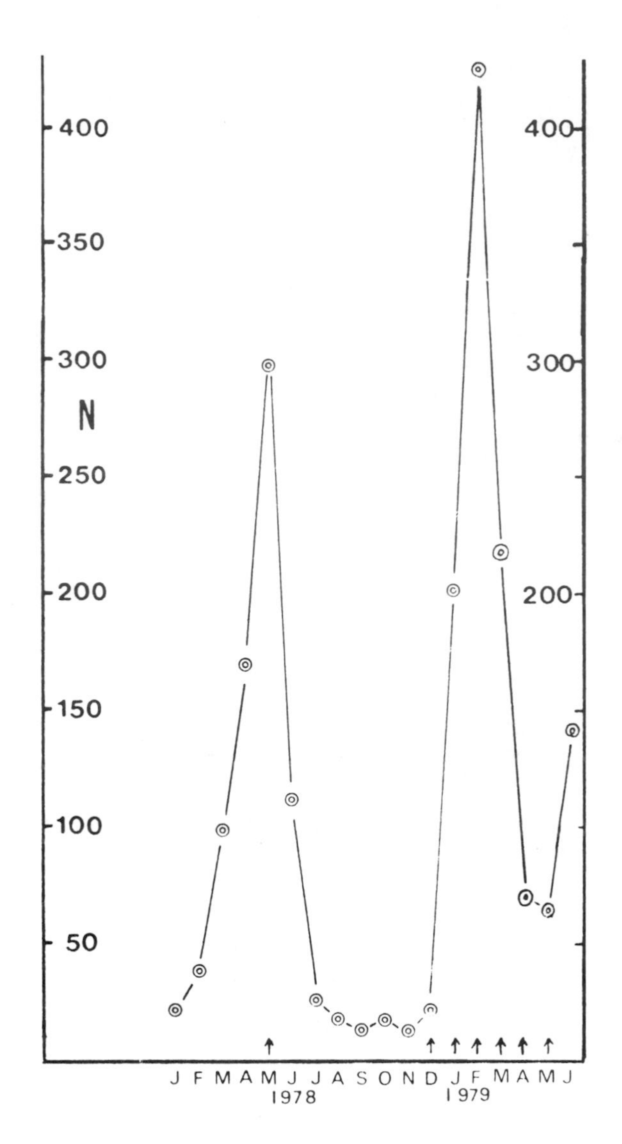

Fig. 8-61. Monthly seismicity Jan 1978–June 1979, recorded at JMA's Tarumai observatory. Arrows represent ash eruptions.

No eruption was associated with the somewhat smaller peak in seismicity that occurred in 1975.

Information Contacts: JMA, Tokyo; D. Shackelford, CA.

Further Reference: Katsui, Y., Onuma, K., Niida, K., Suzuki, T., and Kondo, Y., 1979, Eruption of Tarumai volcano in May 1978; *Bulletin of the Volcanological Society of Japan*, v. 24, no. 2, p. 31-40.

7/78 (3:7) As of early July, no eruptive activity had been observed since the explosions of 14 and 17 May. Seismicity decreased to its normal level in mid-June after a 10-year increase preceding the May explosions.

Information Contact: JMA, Tokyo.

12/78 (3:12) Tarumai erupted ash from 0910 to 0920 on 12 Dec. The ash cloud rose 200 m above the vent, which was also the source of small explosions on 14 and 17 May. Ash slightly darkened the snow in the summit area, but did not reach the foot of the volcano. A second ash eruption occurred on 16 Dec, with ashfall again restricted to the summit area. Seismicity remained at a low level (table 8-42).

Information Contact: Same as above.

1/79 (4:1) Ash eruptions continued through early Jan. Ash was ejected on 16, 26, and 29 Dec and on 5 Jan, falling on the summit area. Seismicity remained at normal levels through Dec.

Information Contact: Same as above.

2/79 (4:2) Ash eruptions continued through Jan. Ash was ejected on 5, 22, 23, and 27 Jan, darkening snow in the summit area, but no ash reached the foot of the volcano.

Local seismicity increased sharply about 10 Jan, after several months of relative quiet. An irregular increase in seismic activity has persisted since 1967. A similar pattern occurred in the 11 years prior to the major 1977 eruption of Usu.

Information Contact: Same as above.

3/79 (4:3) Ash was ejected on 5, 8, 19, and 25–28 Feb, continuing the series of small explosions that resumed 12 Dec. The ash darkened snow in the summit area, but did not reach the foot of the volcano. The number of recorded local earthquakes increased from 201 in Jan to 427 in Feb.

Information Contact: Same as above.

4/79 (4:4) Ash eruptions continued in March. Ash fell on the summit area on 1, 2, 4, 6, and 8 March but did not reach the foot of the volcano. Seismicity declined considerably in March, but remained slightly above the monthly total for Jan.

Information Contact: Same as above.

5/79 (4:5) Eruptive and seismic activity declined substantially in April. Only 1 [ash eruption] took place in April, on the 13th, causing a small ashfall in the summit area. The monthly number of recorded earthquakes dropped from 223 in March to 70 in April.

Information Contact: Same as above.

6/79 (4:6) The decline in activity continued through May. A single small ash eruption occurred in May, on the 11th. Ash fell in the summit area but not reach the foot of the volcano. Monthly recorded earthquakes decreased slightly from 70 in April to 66 in May, but the number of tremor events dropped more sharply, from more than 40 to 9. The tremor events, each lasting a few minutes, were thought to be generated by vigorous activity at a

1978	Jan	Feb	Mar	Apr	May	June	July	Aug	Sept	Oct	Nov	Dec
Events	22	39	95	170	298	112	26	18	13	17	13	[21]

Table 8-42. Monthly seismicity in 1978, Tarumai Volcano.

vent. The strongest of this activity was visible from the base of the volcano.

Information Contact: Same as 3:7.

7/79 (4:7) No explosions or tremor events were recorded in June. However, the number of local earthquakes, which had been declining since March, increased in June.

Information Contact: Same as 3:7.

8/79 (4:8) The number of recorded earthquakes declined to 58 in July after rising sharply to 142 in June. No explosions were observed in July, nor were any tremor events, which are presumably generated by [ash ejection].

Information Contact: Same as 3:7.

11/79 (4:11) An increase in the number of local earthquakes and a resumption of tremor events occurred in mid-Sept, but no eruption was observed. In Oct, seismicity returned to normal levels and no tremor events were recorded.

Information Contact: Same as 3:7.

12/80 (5:12) Seismic activity increased in Nov after about one and a half years of quiet.

Information Contact: Same as 3:7.

1/81 (6:1) Seismic activity increased again to more than 400 recorded events during Jan. No eruption has yet been observed. About 200 events per month were recorded in Nov and Dec, after over a year of fewer than 50 events per month.

Information Contact: Same as 3:7.

2/81 (6:2) In Feb, 1121 seismic events were recorded, the most in any month since 1967, when JMA began routine measurements at the volcano. Seismicity has irregularly but gradually increased in the past 14 years (fig. 8-62). No eruption has occurred during the current increase in seismicity.

Information Contact: Same as 3:7.

3/81 (6:3) Local seismicity began to increase in Nov 1980 and the number of events per month reached 1211 in Feb 1981 (fig. 8-63). Seismicity began a gradual decline in early March and by mid-March had reached the usual average of fewer than 3 recorded earthquakes per day.

Only 87 events were recorded in March. Although the Dec 1978 – May 1979 eruption accompanied the last major increase in seismicity, no eruption has occurred during the current, much larger increase [but see 6:4].

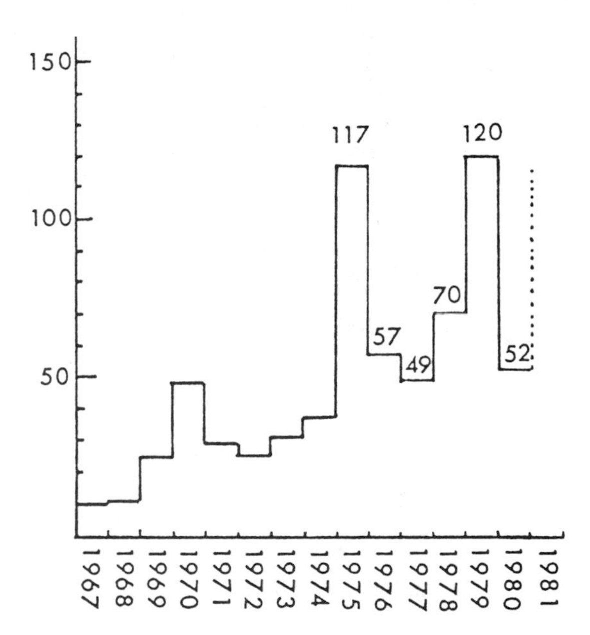

Fig. 8-62. Yearly means of Tarumai's monthly seismicity, 1967-1980.

Information Contact: Same as 3:7.

4/81 (6:4) Seismicity declined to its usual level of fewer than 3 recorded earthquakes per day in March and remained at this level through April. Although no volcanic activity apparently had accompanied the peak of seismicity in Feb (1211 events), subsequent investigation revealed that a weak steam explosion had occurred on 27 Feb or a few days earlier.

On the morning of 27 Feb an All Nippon Airways crew reported that they saw radially darkened snow on the SE part of the summit area. Yoshio Katsui visited the summit on 9 April and found a single layer of gray ash in the snow. The ash layer was only 0.3-0.6 mm thick at the crater rim. The total volume of ejecta was estimated to be no more than 400 m^3.

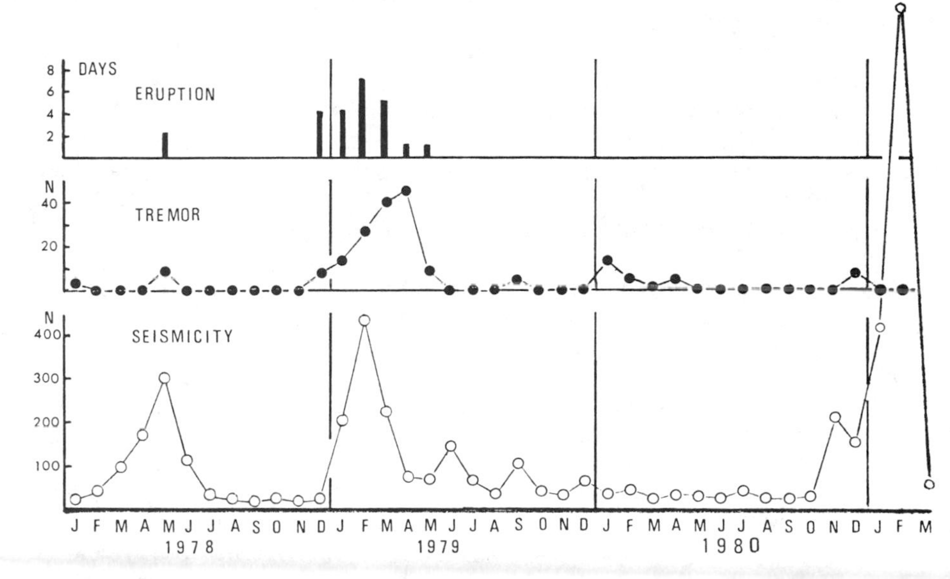

Fig. 8-63. Monthly numbers of days with eruptions, *top*, tremor events, *center*, and recorded earthquakes, *bottom*, at Tarumai, Jan 1978 – March 1981.

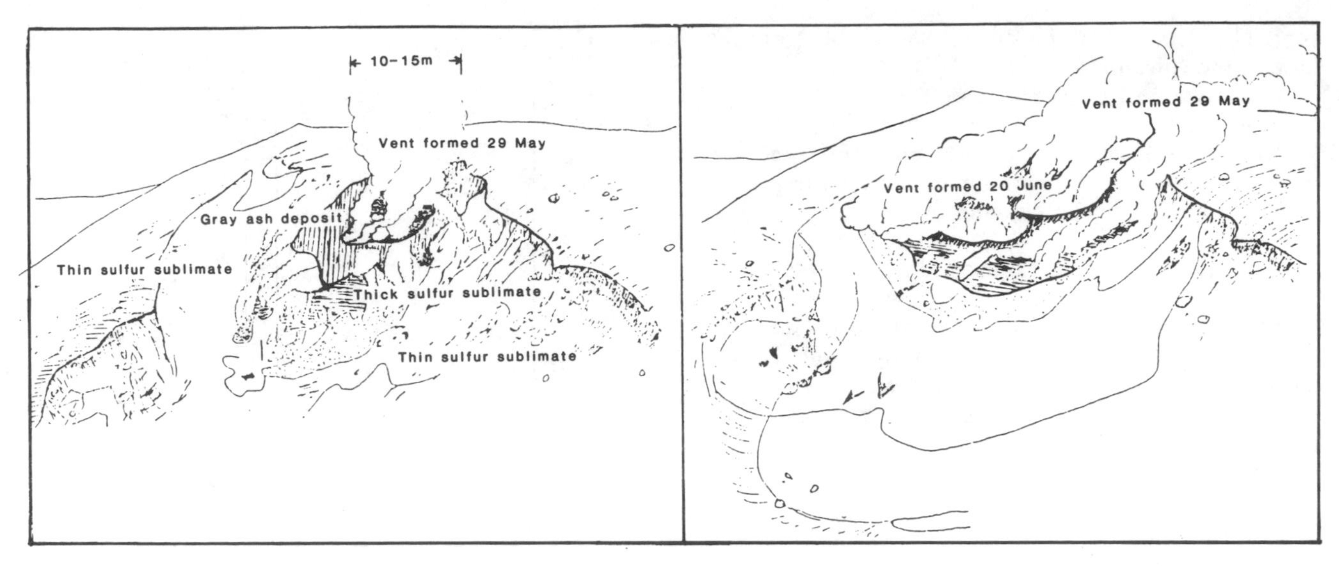

Fig. 8-64. Sketch map of the E wall of the 1962 crater of Tokachi on 7 June, *left*, and 25 June 1985, *right*. Courtesy of JMA.

Information Contacts: JMA, Tokyo; Y. Katsui, Hokkaido Univ.

TOKACHI

Hokkaido, Japan
43.42°N, 142.68°E, summit elev. 2077 m

The highest of Hokkaido's historically active stratovolcanoes, Tokachi is 135 km NE of Tarumai. It has erupted 13 times since 1670, most violently in 1926 and 1962. The suffix "dake", or "mountain" is often added to this volcano's name, as it was in the original SEAN reports that follow.

6/85 (10:6) On the morning of 29 May, personnel of JMA's Tokachi Volcano Observatory saw hot muddy water being ejected from a fumarole on the 1962 crater wall. The next day, they found that a 10-15-m fissure [or an elongated crater] had formed [on the wall of the 1962 crater]. Muddy water was continuously ejected from the fissure during their observations.

This activity stopped on 5 June. Thermal activity at the crater has been gradually increasing since 1983, although local seismic activity has remained at background levels.

Information Contact: JMA, Tokyo.

7/85 (10:7) Thermal activity that formed new vents [on the wall of] the 1962 crater began in late May and continued through July. Beginning on 29 May and ending 5 June, hot muddy water was ejected from a new 10 m-diameter oval vent in the wall of a summit crater formed in 1962. Weak ash emission occurred nearby on 19 June, forming a second new vent, also oval and about 15 m in diameter, on 20 June. Molten sulfur burned in and around the new vents and weak red glow was observed over the 1962 crater from the town of Biei, 7 km away, on the nights of 20-22 June.

The second hot muddy water ejection began on 24 June. JMA personnel from the Tokachi Observatory visited the crater on 9 July and found that this activity had stopped. On 25 July, a third ejection began at the June vent and continued through 31 July. The maximum height of white plumes from the 2 new craters was 80 m [in July]. Seismicity remained low in June and early July, as in previous months. Thermal activity at the 1962 crater has been gradually increasing since 1983.

Information Contact: Same as above.

10/85 (10:10) The third phase of hot muddy ejection began on 25 July. Muddy water was carried 10 m above the [new crater that had formed on the E wall of the 1962 crater] when ejection began, but gradually declined. As of 1 Nov, water ejection had not been observed from the Volcano Observatory since 6 Aug. Fig. 8-64 shows recently formed features at the E wall of the 1962 crater.

Information Contact: Same as above.

Further Reference: Miyakawa, H., Maekawa, T., and Yokoyama, I., 1986, Monitoring of the Temperature by Infrared Thermometry at the Crater Wall of Tokachi Volcano, Hokkaido; *Geophysical Bulletin, Hokkaido University*, v. 47, p. 17-31.

March	1-18	19	20	21	22	23	24	25	26	27	28	29	30	31
Events	None	1	3	9	55	63	26	84	24	30	13	31	50	22

Table 8-43. Seismicity at Me-Akan, March 1982.

AKAN

Hokkaido, Japan
43.38°N, 144.02°E, summit elev. 1503 m
Local time = GMT + 9 hrs

This large caldera, 110 km NE of Tokachi, is one of Japan's many National Parks and a center for preservation of the island's native Ainu culture. Me-Akan is the only historically active cone, with 15 eruptions (13 since 1951).

3/82 (7:3) On 21 March a strong (M 6.9) earthquake occurred near Urakawa, off the S coast of Hokkaido and 182 km SW of Akan. Local seismicity at Me-Akan increased after the earthquake, but the JMA has reported that there is no evidence of a causative relationship. The total number of seismic events recorded in March was 411 (table 8-43). The numbers of recorded seismic events at Me-Akan for 1977–81 are 97, 45, 491, 254, and 194.

Information Contact: JMA, Tokyo

8/82 (7:8) Local seismicity gradually declined, returned to its usual level by mid-April (fig. 8-65), and has remained there through July. Monthly numbers of recorded seismic events are: March-411, April-92, May-16, June-54, July-16.

Information Contact: Same as above.

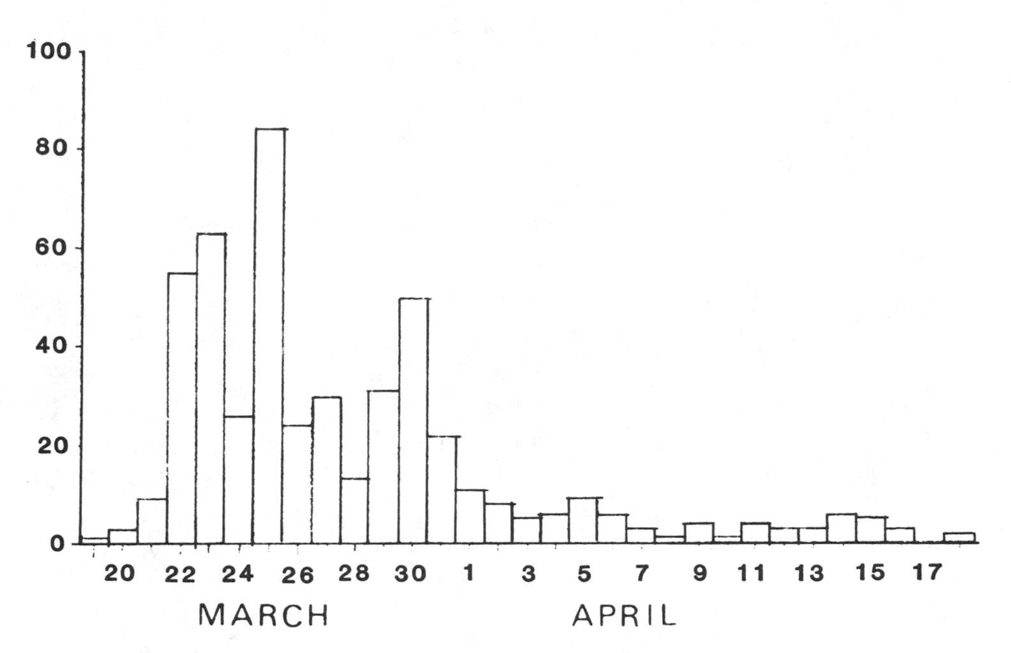

Fig. 8-65. Daily number of recorded seismic events at Me-Akan, March–April 1982.

KURIL ISLANDS

General References: Fedotov, S.A. et al., 1986, The Volcanic Activity in the Kurile-Kamchatka Zone During the Period 1980–1984; *Volcanology and Seismology,* no. 2, p. 3-20.

Ivanov, B.V., Kirsanov, I.T., Khrenov, A.P., and Chirkov, A.M., 1979, Active Volcanoes of Kamchatka and Kurile Islands in 1978–1979; *Volcanology and Seismology*, no. 6, p. 94-100.

TIATIA

Kunashir Island
44.35°N, 146.25°E, summit elev. 1822 m
Local time = GMT + 11 hrs

Tiatia is 210 km NE of Akan, at the NE end of Kunashir Island, in the sparsely populated Kuril Islands. Its first historic eruption was in 1812.

7/78 (3:7) The crew of a JMSA patrol boat observed a white vapor column rising about 600 m above the summit of Tiatia on the morning of [21] July. Tiatia last erupted in 1973, after 161 years of quiet. [Increased thermal activity between 1974 and 1977 melted snow and emitted vapor plumes but produced no tephra (Markhinin, 1984).]

Information Contact: Reuters.

Further Reference: Markhinin, E.K., 1984, On the State of Kunashir Island Volcanoes (March, 1974–May, 1982); *Volcanology and Seismology,* v. 5, no. 1, p. 45-52 (English translation); 1983, no. 1, p. 43-51 (in Russian).

8/78 (3:8) Residents of the E end of Hokkaido heard an explosion on 20 July at 1325. The explosion was not recorded by seismographs or microbarographs in E Hokkaido. Tiatia, approximately 50 km to the E, was obscured by fog.

The next morning, the crew of the JMSA ship *Kunasiri* observed a white cloud rising 600 m from Tiatia, but heard no explosions. No ashfall was found (in Japan) on 20 or 21 July.

Information Contact: JMA, Tokyo.

7/81 (6:7) The crew of a Japanese fishing boat cruising near Kunashir Island observed "smoke" rising from Tiatia on 10 June. During the night of 24 June, an orange glare was observed in the direction of the volcano from [JMA's Nemuro Weather Station], 120 km away. No additional activity has been reported.

Information Contact: Kyodo Radio, Tokyo.

12/81 (6:12) "Aerial inspection on 20 Sept of the volcanoes in the S and central Kuril Islands revealed that Tiatia's summit crater was in a state of moderate fumarolic activity. No individual distinct fumaroles were observed; vapor was being released from the whole crater surface. Numerous vapor sites were also noticed on the outer slopes near the summit crater. There were no remarkable changes near the volcano summit as compared to 1977–78. A certain increase in heat activity was observed near the subordinate vent (formed in 1973) on the S slope. Heat flow measurements made in the vent in 1981 by A. Zemtsov and A. Tron yielded values of q = 7.4 and W/m^2 = 1.77 cal/cm^2s, 1.2 times as large as in 1978. Orange glare was observed by people in Yuzhno-Kurilsk, 50 km SW of the volcano, but we are not sure that it was related to volcanic activity."

Information Contact: G. Steinberg, Sakhalin Complex Institute.

MEDVEZHIA CALDERA

Iturup Island
45.38°N, 148.80°E, summit elev. 1125 m

Kudriavy is the central of three young cones in the 10 km-wide Medvezhia caldera and the only one with historic eruptions. It is 230 km NE of Tiatia, at the NE end of the largest Kuril island, and its last known eruption was in 1958.

12/81 (6:12) The 20 Sept 1981 aerial inspection (see Tiatia, 6:12) also revealed intense fumarolic activity at 5 sites in Kudriavy crater.

Information Contact: G. Steinberg, Sakhalin Complex Institute.

KOLOKOL GROUP

Urup Island
46.05°N, 150.06°E, summit elev. 1330 m

Berg and Trezubetz are strato-volcanoes in the closely spaced Kolokol Group of volcanoes on Urup Island. They are 125 km NE of Medvezhia.

12/81 (6:12) Very weak gas emission was occurring from the summits of extrusive domes at both Berg and Trezubetz during the 20 Sept 1981 flight.

Information Contact: G. Steinberg, Sakhalin Complex Institute.

UNNAMED SUBMARINE VOLCANO

E of Urup Island
46.10°N, 151.50°E
Local time = GMT + 11 hrs

No previous activity has been reported from this site,

60 km NE of the Kolokol Group, but it is directly on the trend of Kuril volcanoes.

5/78 (3:5) The crew of the fishing boat *Shinano Maru* observed a 300 m × 70 m zone of bubbling, light-colored water 7.7 km E of Urup Island between 1730 and 1745 on 31 March. It is not certain whether the activity was caused by submarine volcanism. JMA seismographs recorded no accompanying earthquakes.
Information Contact: JMA, Tokyo.

CHIRPOI GROUP

Urup Island
46.525N, 150.875E, summit elev. 624 m

Snow and Cherny volcanoes are on Chirpoi Island, just 55 km NNE of the unnamed submarine volcano reported above. Seven eruptions are known from the two volcanoes since 1712, the most recent in 1960.

On 22 November 1982 a ship captain saw heavy smoke clouds above the crater of a 395 m peak on the island. Investigation showed that a brief eruption had occurred from Snow volcano.

Reference: Ivanov, B.V., et al., 1988. Active volcanoes of Kamchatka and Kuril Islands: Status in 1982. *Volcanology and Seismology,* v. 6, p. 623-634 (English translation); v. 1984, no. 4, p. 104-110 (in Russian).

KETOI CALDERA

Ketoi Island
47.35°N, 152.475°E, summit elev. 1002 m

Pallas Peak is a young cone on the eastern margin of the Ketoi caldera on the 9 km-wide island of the same name. The last known eruption of this volcano, 250 km NE of the Kolokol Group, was in 1960.

12/81 (6:12) Intense fumarolic activity was occurring on the outer N slope [of Pallas Peak during the 20 Sept 1981 overflight]. There were sulfur deposits near the fumaroles. However, no apparent fumarolic or solfataric activity was observed at **Zavaritski Caldera** (46.925°N, 151.95°E) or **Prevo Peak** (47.02°N, 152.12°E). Both are on Simushir, the island immediately SW of Ketoi.
Information Contact: G. Steinberg, Sakhalin Complex Institute.

USHISHIR CALDERA

Ushishir Island
47.52°N, 152.95°E, summit elev. 401 m

The first historic eruption of this volcano, 30 km NE of Ketoi, was in the early 18th century and one of the first known in the Kuriles. Its most recent eruption was in 1884.

12/81 (6:12) Weak gas release was occurring inside the S part of the caldera [during the 20 Sept overflight].
Information Contact: G. Steinberg, Sakhalin Complex Institute.

SARYCHEV

Matua Island
48.09°N, 153.20°E, summit elev. 1497 m

Sarychev Peak, 70 km NE of Ushishir, has erupted at least 13 times since the 1760s, more than any other volcano in the Kuriles.

10/76 (1:13) Ash emission, accompanied by earth tremors and a booming noise, began in mid-Sept. On 2 Oct, volcanologists observed 2 lava flows moving rapidly down the W flank toward the Sea of Okhotsk. By 16 Oct, eruptive activity was limited to infrequent "smoke" effusion, allowing the staff of the hydro-meteorological station (the only inhabitants of the island) to return. Volcanologists from the Institute of Volcanology, Petropavlovsk (IVP) are presently studying the volcano. Sarychev had a mild ash eruption lasting only 6 minutes on 9 December 1965. Its last major eruption began on 30 Aug, 1960, producing a 5 km-high cloud of andesitic basalt ash.
Information Contact: Tass.
Further Reference: Andreyev, V.N., Shantser, A.Ye., Khrenov, A.P., Okrugin, V.M., and Nechayev, V.N., 1978, Eruption of the Volcanic Peak Sarycheva in 1976; *Byull. Vulkanol. Stn.,* no. 55, p. 35-40.

12/81 (6:12) The W walls of the crater were ruined after the 1976 eruption, and this part of the summit was flat during the 20 Sept overflight. In the central part of the summit crater, there was a shallow but morphologically distinct funnel in which there were many intensely active fumaroles.
Information Contact: G. Steinberg, Sakhalin Complex Institute.

CHIRINKOTAN

Chirinkotan Island
48.98°N, 153.48°E, summit elev. 724 m

This strato-volcano lies NW of the main Kuril trend, 100 km N of Sarychev and 35 km W of Ekarma.

6/80 (5:6) In April 1979, an eruption began from Chirinkotan volcano, which forms an isolated, uninhabited island about 3 km in diameter. Block lava flowed down the SSW flank in April and May. Ash explosions occurred regularly during the summer of 1979 and explosive activity increased in Oct. In Jan and April 1980, moderate explosions occurred 1-2 times per hour.

Chirinkotan last erupted between 1954 and 1961, when a tephra cone was built and a lava flow was extruded, perhaps at the same time as the strong gas emission that occurred in 1955.

Information Contact: G. Steinberg, Sakhalin Complex Institute.

EKARMA

Ekarma Island
48.95°N, 153.94°E, summit elev. 1171 m

Explosions were observed from a passing fishing boat on 24 May 1980. A black eruption column rose to about 1 km height, and ash fell on the ship. Ekarma's slopes were dark gray while those of nearby Chirinkotan were white, as observed from 40-50 km to the SW (Ivanov et al., 1981). Ekarma's only other known eruption was 1767-69.

Reference: Ivanov, B.V., Chirkov, A., Dubik, Yu., Garilov, V.A., Stepanov. V.V., Rulenko, O.P., and Firstov, P.P., 1981. The State of Volcanoes in Kamchatka and Kuril Islands in 1980. *Volcanology and Seismology,* no. 3, p. 99-104.

FUSS PEAK

Paramushir Island
50.22°N, 155.20°E, summit elev. 1772 m

Fuss is on SW Paramushir Island, 170 km NE of Ekarma. Its only historic eruption was in 1854.

10/82 (7:10) Recent aerial infrared surveys and ground investigations have shown increasing thermal activity at Fuss Peak. Gorshkov (1967) reported that there was no fumarolic activity at the volcano, E. K. Markhinin found only very minor signs of it in 1969, and fumaroles were not observed during 1971-1976 overflights. In 1973, an aerial infrared survey detected weak thermal anomalies over an area of about 10^4 m^2 in the N and E parts of the crater and on its E rim (Gusev and Zelenov, 1979). When the volcano was resurveyed in 1978, intense thermal anomalies were measured over most of the 700 m-diameter crater, extending down to its bottom about 200 m below the rim. Weak fumaroles were observed in the E and central part of the crater. The 1978 resurvey also found a 50-80 m-wide zone of anomalously high temperatures extending 250-300 m down the E flank from the crater rim (Ibid.). Fumaroles were seen on the E flank during an overflight in the fall of 1981.

G. S. Steinberg visited the volcano in Sept 1982 and found fumaroles at the base of both the W and E sides of the crater's small median ridge. Activity was stronger at the E base of the ridge, where there were 2 groups of fumaroles, each with 3 powerful vapor jets with temperatures of 95-96°C. Bright yellow sulfur crystals were present in some of the vents. Despite the fumaroles, the majority of the crater floor was snow-covered. Many weak fumaroles were observed in a zone of small fissures on the upper flank in the immediate vicinity of the crater rim, but none had deposited sulfur. About halfway down the cone, in a narrow, shallow canyon that was apparently an extension of the upper flank fissure zone, there were 3 groups of fumaroles, separated by 30-60 m, vigorously emitting a mixture of steam and other gases. Temperatures at these vents were 94-96°C and they had deposited bright yellow sulfur. Within the fissure zone, temperatures at 30-40 cm depth were 9-13°C. Temperatures at similar depths outside the zone were 3-4°C.

References: Gorshkov, G.S., 1967, *Vulkanizm Kurilskoi Ostrovnoi Dugi;* Nauka, Moscow, 288 pp.

Gusev, N.A. and Zelenov, E.N., 1979, The Activization of Heat Regime of the Fuss Peak Volcano According to the Heat Aerial Surveying; *Volcanology and Seismology,* no. 4, p. 102.

Information Contact: G. Steinberg, Sakhalin Complex Institute.

ALAID

Atlasova Island
50.80°N, 155.50°E, summit elev. 2339 m
Local time = GMT + 11 hrs

The northernmost Kuril volcano, Alaid forms an island NW of the main Kuril trend. It is 70 km NNE of Fuss and the same distance W of Kamchatka's S tip. A violent eruption in 1790 was followed by several other large eruptions before the one described here.

4/81 (6:4) Soviet volcanologists reported that an explosive summit eruption from Alaid, on uninhabited Atlasova Island, began after midday on 27 April and intensified the next day. Much of the information on the eruption from both U.S. and Soviet sources is from analysis of satellite imagery. Clouds prevented satellite observations until about 0715 on the 28th, when infrared imagery from the NOAA 6 polar orbiter revealed a distinct V-shaped eruption plume that extended a short distance NE from the volcano before disappearing in heavy weather clouds. An infrared image returned from the GMS at 1100 showed a similar pattern. Microbarographs at Kushiro Weather Observatory (about 1250 km SW of Alaid) recorded 3 distinct pressure waves on 28 April: at 1143 (0.5 millibars), 1153 (0.2 millibars), and 1340 (0.8 millibars).

Vigorous feeding of this cloud could be seen on the satellite imagery for the next 2 days (fig. 9-1). 29 April imagery indicated that the plume consisted of 2 primary layers, at about 9-11 km and 13.5-15 km altitude. The last clear-weather image, on 30 April at 1700, showed a plume at least 120 km wide and 1900 km long. Eighteen hours later (1100 on 1 May, 4 days after the eruption began) partial clearing showed that feeding of the plume had apparently ended.

Significant ashfalls were reported over a wide area. Soviet volcanologists reported that the ash, a pyroxene olivine basalt, fell as much as 1000 km from the volcano, over an area of 150,000 km^2. They noted an accumulation of 30 cm of ash 7 km from Alaid, and Tass reported that 20-25 cm fell on the town of Severokurilsk (45 km

ESE of the volcano), where residents heard roaring noises and saw a glow from the volcano during the night. Schools were closed in Severokurilsk and radio communication was disrupted. Ash mixed with wet snow fell on Petropavlovsk (300 km NE of the volcano) and other inhabited areas on the Kamchatka Peninsula. In the Aleutians, ashfall began 28 April on Shemya (about 1200 km ENE of Alaid), and lasted all day 30 April and 1 May, when roughly 2 mm of ash were measured in very windy weather. Lt. Becker observed intermittent ashfalls and periods of acid rain between 2 and 5 May, always within one and a half hours after low ocean tide. Ash collected at Shemya was sent to the NASA Ames Research Center. Daily precipitation sampling from Adak Island (650 km E of Shemya and 1900 km from Alaid) 1-7 May yielded only a trace of ash, on the 4th.

Tass reported that volcanologists overflew the volcano 29 April and observed an ash column that rose to about 10 km altitude from the summit crater. Soviet volcanologists later reported a maximum eruption cloud height during the activity of 12 km, based on overflights and analysis of satellite imagery.

Soviet volcanologists reported that activity declined 2-4 May. No additional activity was observed on satellite imagery until 8 May at 2300, when the GMS satellite recorded a new eruption column starting to emerge from Alaid. Careful examination of earlier imagery from other satellites indicates that the renewed activity may have started as early as 1930. By 9 May at 0300, a dense plume extended more than 120 km to the ESE. This plume remained shorter and much narrower than the late April clouds, reaching a maximum length of about 400 km ESE from the volcano. GMS imagery continued to show strong feeding of the cloud at 1100, but the eruption seemed to be weakening by 1400 and had apparently ended by the time of the next available image at 2000.

Attempts to observe and sample the ejecta farther downwind continue. During the night of 6-7 May, lidar operated by SRI International near San Francisco, California detected 2 distinct layers of material at 11.9 and 12.8 km altitude, just below the tropopause. However, it was not possible to confirm that this material was of volcanic origin.

A preliminary search for strong seismicity associated with the eruption yielded only a single shallow magnitude 6.0 event at 44.04°N, 149.93°E (860 km SSW of the volcano), on 1 May at 0142.

Alaid's last eruption, in 1972, produced large tephra clouds and lava flows that reached the sea from NW flank vents. Its last summit eruption was in 1894.

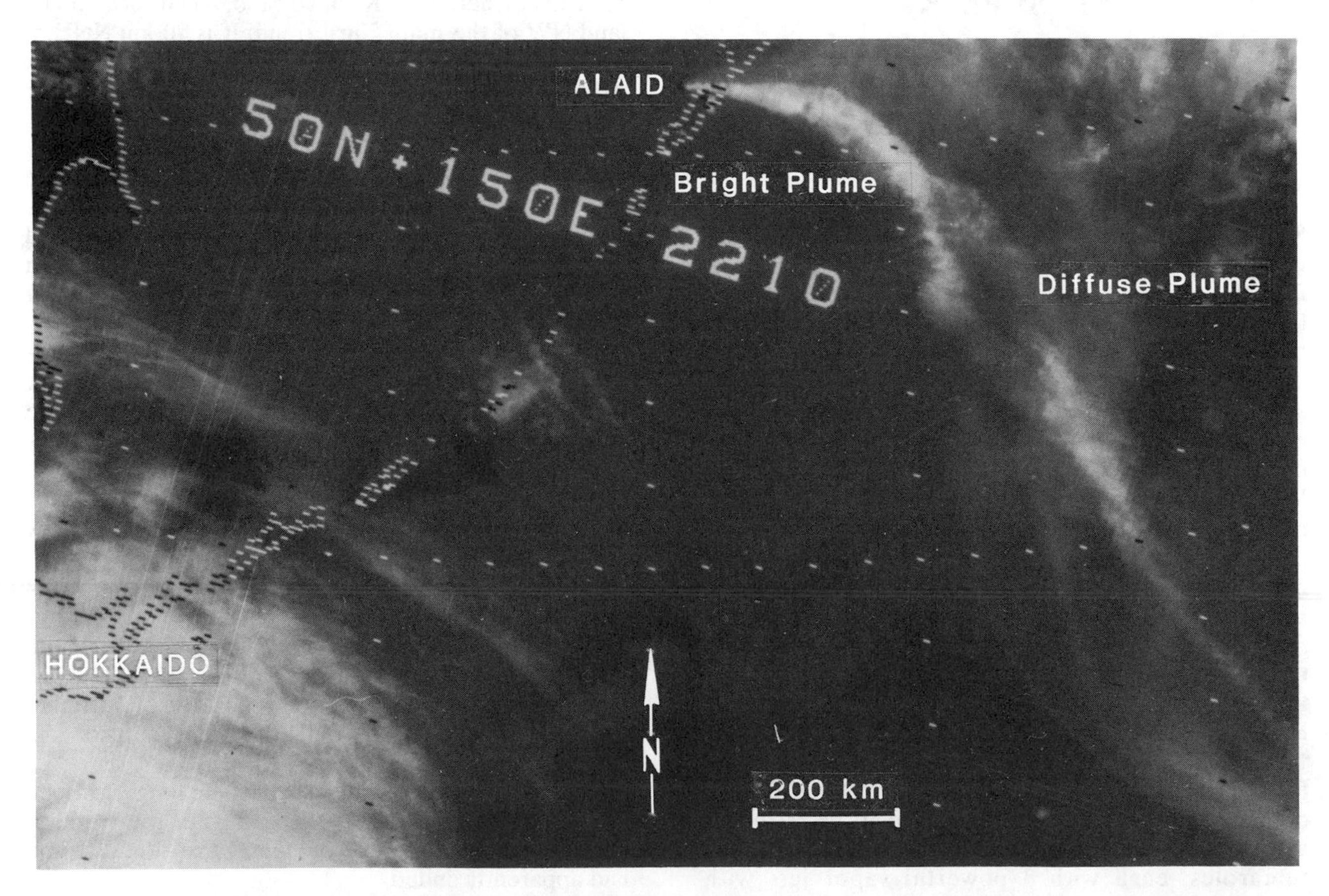

Fig. 9-1. NOAA 6 satellite image, returned 30 April at 0910, showing eruption plumes from Alaid. The image was obtained using the 11 μm thermal infrared sensor. Image resolution is 4 km. Two plumes are evident, one bright and distinct, the other diffuse but well-defined. The plumes are over 1700 km in length, trailing off the right side of the image. Analysis of the imagery shows that the brighter plume has a temperature of about -55°C, and the diffuse plume an apparent temperature of -10°C. Comparison of the temperature of the colder (and therefore higher) of the 2 plumes with radiosonde temperature profiles collected from Petropavlovsk, Kamchatka (300 km NE of the volcano) and Podgornoye, Paramushir Island (50 km S of the volcano) yielded 2 possible altitudes for this plume; 9.8-10.4 km if it had not passed through the tropopause (at 11.6 km), or 13.7-14.3 km if the plume was stratospheric. The image and caption information were provided by Michael Matson.

Information Contacts: S. Fedotov, B. Ivanov, IVP, Kamchatska; F. Smigielski, S. Arnett, M. Matson, NOAA/NESS; G. Telegadas, NOAA/ARL; D. Shimozuru, Univ. of Tokyo; R. Muñoz, NASA, Ames Research Center; M. P. McCormick, NASA, Langley Research Center; P. Russell, SRI International; Lt. Becker, USAF, Shemya AK; Tass; USGS/NEIS.

5/81 (6:5) G. S. Steinberg provided the following additional information on the early stages of the eruption.

Activity began 27 April with the emission of a small white plume. The magnitude of the explosions increased rapidly, soon building a black ash column more than 10 km high. Ashfall was intense at Severokurilsk (45 km ESE of the volcano), and by the end of 27 April as much as 20 cm had been deposited there. Schools and kindergartens were closed the next day and water collecting facilities were disrupted by the ash. The wind changed direction 29 April, blowing the eruption cloud NE. At Petropavlovsk, 300 km to the NE, an estimated 1.5 kg/m^2 of ash had accumulated by the end of 29 April. The eruption reached its maximum intensity 30 April-1 May. Activity began to decrease 2 May, was limited to ejection of ash and larger tephra to less than 100 m height by 4-5 May, and stopped 7 May.

Michael Matson provided the following satellite data.

Plumes of varying langths were intermittently present through late May on imagery (at about 0730 each day) from the NOAA 6 polar orbiting satellite. The minor activity reported by Steinberg after 2 May could not be seen on the images, but a small plume from Alaid appeared 7 May. On 8 May, a plume extended about 350 km from Alaid, but 24 hours later it was only about 60 km long. There was a plume on the 10 May imagery, but no activity was evident 11-14 May. A very small plume reappeared 15 May, had grown to 250 km long by the 17th, and persisted, at lengths that varied from 50 to nearly 600 km, through 27 May (weather clouds obscured the volcano 19 and 25 May). Alaid appeared to emit a plume no more than 25 km long on 1 June. No additional activity was evident on the satellite imagery through 9 June.

[The 1981 eruption of Alaid had been suggested as a possible source of the "Mystery Cloud" of stratospheric aerosols detected in early 1982 before the eruption of El Chichón. Later work on Nimbus-7 satellite data by Arlin Krueger revealed a large SO_2 cloud originating in the vicinity of Nyamuragira Volcano, Zaire, in late Dec 1981. This eruption is now thought to be a more likely source for the "Mystery Cloud."]

Information Contacts: G. Steinberg, Sakhalin Complex Institute; M. Matson, NOAA/ NESS.

Further References: Fedotov, S.A., Ivanov, B.V., Avdeyko, G.P., Flerov, G, et al., 1981, The 1981 Eruption of Alaid Volcano; *Volcanology and Seismology,* 1981, no. 5, p. 82-87.

Sawada, Y., 1983, Analysis of Eruption Clouds by the 1981 Eruptions of Alaid and Pagan Volcanoes with GMS Images; *Papers in Meteorology and Geophysics,* v. 34, p. 307-324.

3/82 (7:3) Imagery from the GMS satellite revealed a narrow, linear eruption plume emerging from Alaid at 1100 on 29 March. The plume extended roughly 100 km to the ESE and was estimated to be roughly 2 hours old. Images returned 3 hours earlier and later showed no evidence of activity.

Information Contact: M. Matson, NOAA/NESS.

KAMCHATKA

General References: Fedotov, S.A. et al., 1986, The Volcanic Activity in the Kurile-Kamchatka Zone During the Period 1980 – 1984; *Volcanology and Seismology,* no. 2, p. 3-20.

Ivanov, B.V. and Khrenov, A.P., 1979, State of Craters of Active Kamchatka Volcanoes in 1977 – 1978; *Volcanology and Seismology*, no. 1, p. 97-101.

Ivanov, B.V., Kirsanov, I.T., Khrenov, A.P., and Chirkov, A.M., 1979, Active Volcanoes of Kamchatka and Kurile Islands in 1978 – 1979; *Volcanology and Seismology*, no. 6, p. 94-100.

GORELY

54.45°N, 158.12°E, summit elev. 1829 m

This caldera, 260 km NE of Alaid, has erupted 12 times since 1828. It has been known as Gorely Khrebet (in the CAVW and other publications), but the Institute of Volcanology in Petropavlovsk now uses the shortened name.

7/80 (5:7) During 1978 – 79, large fumaroles appeared in the central crater and a lake was formed on the crater floor. In Jan 1980, a 1 km-high gas column was first observed over the summit. This remained throughout the spring, and gas vents appeared on the walls and floor of the crater.

In June, intensive gas emission was accompanied by explosions that ejected lithic material. During this presumed phreatic eruption [but see 10:1] the crater lake disappeared. As of late July, activity was increasing and a glow was visible over the crater at night.

The last strong explosive eruption of Gorely took place in 1929 – 31. Minor activity occurred in 1961.

Information Contacts: E. Vakin and I. Kirsanov, IVP, Kamchatka.

Further References: Gavrilov, V.A., Gordeev, E.I., et al., 1984, Volcanic Tremor and the Gorely Volcano Earthquakes During the 1980 – 1981 Eruption; *Volcanology and Seismology*, no. 6, p. 3-17.

Kirsanov, I.T., 1981, Eruption of the Gorely Volcano in Summer 1980; *Volcanology and Seismology*, no. 1, p. 70-73.

1/85 (10:1) Gorely began to erupt 27 Dec. During the initial activity, phreatic explosions ejected clouds that rose to 3.5 km height. Ash explosions began 30 Dec, producing plumes that reached 3 km height and extended 100 km E. The eruption ended 8 Jan. In the previous eruption, June 1980 – July 1981, initial phreatic activity gave way in later phases to phreatomagmatic explosions that ejected alkali-rich basaltic andesite (Kirsanov and Ozerov, 1984).

Reference: Kirsanov, I.T. and Ozerov, A.Yu., 1984, Composition of Products and Energy Yield of the 1980 – 1981 Gorelyi Volcano Eruption; *Volcanology and Seismology*, no. 5, p. 23-43.

Information Contact: Tass.

Further Reference: Gavrilov, et al., 1986, Intensification of Goreli Volcanic Activity in August – September 1984; *Volcanology and Seismology*, no. 5, p. 90-92.

KARYMSKY

54.07°N, 159.60°E, summit elev. 1486 m

Karymsky is a strato-volcano within a large caldera, 200 km NNE of Gorely. It has erupted 31 times since its first historic event in 1771.

General Reference: Khrenov, A.P., et al., 1982, Eruptive Activity of Karymsky Volcano over the Period of 10 Years (1970 – 1980); *Volcanology and Seismology*, no. 4, p. 29-48.

4/76 (1:7) Karymsky erupted during April. Scientists registered 60-80 explosions a day and believed that the plug in the crater's vent was being broken. Ash, slag, and gases were discharged from the crater.

Information Contact: Y. Doubik, IVP, Kamchatka.

Further Reference: Zharinov, I.A. and Firstov, P.P., 1985, Activity, Seismic Regime, and Crust Inclination at Karymsky Volcano During the Summer of 1976; *Volcanology and Seismology*, no. 2, p. 93-95.

8/79 (4:8) Tass reported on 31 July that explosions every 2-3 minutes produced 1 km-high ash clouds, and 2 lava flows were moving down the flanks.

Information Contact: Tass.

[Lava dome growth continued in 1982, filling the crater by mid-year. Small to moderate explosions were frequent Jan – July. Less frequent but more violent explosions occurred in Aug and Sept, then decreased gradually and ended by 10 – 11 Oct. The lava dome was destroyed, leaving a bowl-shaped crater similar to that after an earlier dome was destroyed in Nov 1978.]

Reference: Ivanov, B.V., Chirkov, A.M., Dubik, Yu. M., Khrenov, A.P., Dvigalo, V.N., Razina, A.A., Stepanov, V.V., and Chubarova, O.S., 1988, Active Volcanoes of Kamchatka and Kurile Islands: Status in 1982; *Volcanology and Seismology*, v. 6, p. 623-634 (English translation); 1984, no. 4, p. 104-110 (in Russian).

4/85 (10:4) On 3 May, Space Shuttle astronauts photographed a feature that appeared to be a short, stubby lava flow extending from the summit crater toward the Pacific coast. The feature was a black area with a length (roughly several hundred meters) 2-3 times its width, and boundaries that were sharply defined against the snow cover. No eruption plume was observed.

Karymsky's most recent reported eruption began in 1970, producing moderate explosions with pyroclastic flows and lahars, lava flows, and a period of lava dome growth, all from the summit crater. After the eruption ended in October 1982, only weak fumarolic activity, without accompanying seismicity, was observed during the following 11 months. As of Sept 1983, the summit crater was 160 × 120 m, elongate NE-SW, and post-eruption collapse had deepened it to 60 m (Ivanov, et al., 1984).

Reference: Ivanov, B.V., Gavrilenko, G.M., Dvigalo, V.N., Ovsyannikov, A.A., Ozerov, A.Yu., Razina, A.A., Tokarev, P.I., Khrenov, A.P., and Chirkov, A.M., 1988, Activity of Volcanoes in Kamchatka and the Kurile Islands in 1983; *Volcanology and Seismology*, v. 6, p. 959-972 (English translation); 1984, no. 6, p. 114-121 (in Russian).

Information Contact: C. Wood, NASA, Houston.

PLOSKY TOLBACHIK

55.93°N, 160.47°E, summit elev. 3085 m

This basaltic shield volcano, unusual amidst the andesitic strato-volcanoes and calderas of Kamchatka, is 215 km NNE of Karymsky and 35 km S of Kliuchevskoi. Tolbachik has erupted 31 times since 1740; its most recent being one of the major fissure eruptions of the century. It began on 6 July 1975 and continued into Dec 1976, but the most vigorous activity had ended at the time of SEAN's start. The eruption is well summarized, however, in the following two references:

General References: Fedotov, S.A. and Markhinin, Ye.K. (eds.) 1983, *The Great Tolbachik Fissure Eruption: Geological and Geophysical Data 1975 – 1976*; Cambridge University Press, 341 pp. (English translation of 1978 publication).

Fedotov, S.A. (ed.) 1984, *Large Tolbachik Fissure Eruption: Kamchatka, 1975 – 1976*; Science Press, Moscow, 637 pp. (in Russian).

4/76 (1:7) The eruption of Plosky Tolbachik volcano, which started in July 1975, is continuing.

Information Contact: Y. Doubik, IVP, Kamchatka.

5/76 (1:8) Plosky Tolbachik has been in a state of eruption for almost a year. It was reported on 17 May that a new volcano was developing near there, and a group of IVP scientists is investigating the eruption. They stated that "this is one of the biggest and most interesting volcanic eruptions this century." Reportedly, a chain of new volcanoes was formed during the eruption, four of which are fairly large. "Streams of fire" have been active for nearly a year.

Information Contact: Same as above.

BEZYMIANNY

55.97°N, 160.59°E, summit elev. 2800 m
Local time = GMT + 12 hrs

Between Tolbachik and Kliuchevskoi, the volcano whose name means "no name" erupted violently in 1955 – 56 for the first time in history. It received relatively little international attention until the remarkably similar eruption of Mt. St. Helens in 1980. It is second only to Merapi in historic dome-forming eruptions, with 23 since 1955.

General Reference: Bogoyavlenskaya, G.E. and Kirsanov, I.T., 1981, 25 Years of Activity of Bezymianny Volcano; *Volcanology and Seismology*, no. 2, p. 3-13. An English translation of this paper appears (in 2 parts) in *Volcano News* no. 13, April 1983, p. 6-7, and no. 14, July 1983, p. 4-6.

3/77 (2:3) A column of "stones", ash, and gas was erupted to 6-7 km height on 25 March, following a series of earthquakes the preceding day. Subsequent explosions on 25 and 31 March sent tephra clouds to 15 km altitude and caused heavy ashfall. No damage has occurred to populated areas. The present activity at Bezymianny, which last erupted March 1965 – March 1970, is reportedly the strongest since the great eruption of 1955 – 1957.

Information Contact: Tass.

Further Reference: Bogoyavlenskaya, G.E., Ivanov, B.V., Budnikov, V.A., and Andreev, V.N., 1979, The Eruption of Bezymianny Volcano in 1977; *Byull. Vulkanol. Stn.*, no. 57, p. 16-25.

2/79 (4:2) After a brief period of premonitory seismicity, a series of explosions from Bezymianny began at 1023 on 11 Feb. The explosions produced what was reported as an agglomerate flow about 10 km long and 10 m thick. They also deposited 2 cm of ash on a town several dozen km from the volcano, and destroyed part of the new cone. Lava flowed several hundred meters down the NW flank. The volume of material erupted was about 0.2 km^3.

Bezymianny was thought to be extinct until its major eruption of Oct 1955 – March 1957. The present activity is the strongest since the third and most powerful phase of that eruption, on 30 March 1956, when a catastrophic explosion destroyed the summit and produced large nuées ardentes and lahars.

Information Contacts: N. Kozhemyaka, IVP, Kamchatka; Tass; Sovetskaya Rossiya Radio, Moscow.

6/81 (6:6) In a report dated 16 June, Tass said that Bezymianny had erupted, ejecting an 8 km-high ash column and extruding a lava flow 400 m wide. NESS personnel inspected early and mid-June imagery returned every 3 hours from the GMS satellite, but did not find a large eruption column. Weather is often cloudy over Kamchatka, however, and could have masked evidence of an eruption.

Information Contacts: E. Hooper, NOAA/NESS; Tass.

[Extrusive activity was continuous in 1982, accompanied by ash ejections that sometimes fed small pyroclastic flows.]

Reference: Ivanov, et al., 1988, Active Volcanoes of Kamchatka and Kurile Islands: Status in 1982 (see Karymsky, 4:8 for full reference).

5/83 (8:5) Bezymianny began to erupt 22 May, without premonitory seismicity. Ash was ejected to 5-6 km height and covered the E foot of the volcano. The total area of the ash deposit was about 1500 km^2. Strong explosions destroyed part of the Novy (new) lava dome (see below) and a 4-5 km-long pyroclastic flow was noted at its E base. Andesitic lava was extruded from the dome's summit. Since the beginning of the eruption, the volcano has remained cloud-covered, making observations difficult.

Imagery returned 1 June by the NOAA 7 polar orbiting satellite showed a dark band extending about 250 km ESE from the vicinity of Bezymianny, above a layer of heavy weather clouds. Because of the clouds, it was not possible to locate the dark band's origin more closely than about 56°N, 160°E, or determine if the volcano was feeding the dark band at that time. Continued poor weather has prevented additional satellite observations of eruption plumes.

No eruptions of Bezymianny were known for more than 250 years after the Russian discovery of Kamchatka in 1697. Ash eruptions that began in late 1955, followed by lava dome extrusion and intrusive activity, culminated in a paroxysmal directed explosion on 30 March 1956 that destroyed the summit and formed a large crater, elongate to the E. Lava extrusion then resumed, accompanied by numerous explosive episodes, and has continued through the present, building the Novy dome (Bogoyavlenskaya and Kirsanov, 1981).

Information Contacts: G. Bogoyavlenskaya, IVP, Kamchatka; M. Matson, NOAA/NESS.

10/84 (9:10) The following is a report from G. Ye. Bogoyavlenskaya and P. I. Tokarev.

"Activity increased from late Sept through mid-Oct. On 4 Sept, small surface earthquakes began to be recorded at a seismic station 13 km from the volcano. By 8 Oct, the number of recorded events was 300 per day. On 9 Oct, ash ejections became frequent and rockslides occurred from the dome. On 13–14 Oct the eruption entered its main phase. Volcanic tremor began and an eruption column rose to 5 km height. Several explosions destroyed the E portion of the summit dome. Pyroclastic flows descended along 2 routes, the larger more than 8 km long. Ashfall occurred to the ENE. The ash layer 16 km NE of the volcano was 2 kg/m^2. Weaker activity followed and by 19 Oct the eruption was over."

Information Contacts: G. Bogoyavlenskaya, P. Tokarev, IVP, Kamchatka.

1/85 (10:1) Since the strong explosions in early Oct, moderate gas and ash emission has continued.

Information Contact: B. Ivanov, IVP, Kamchatka.

Further Reference: Malyshev, A.I., 1987, Bezymianny Volcano: Its Eruption in 1981–1984; *Volcanology and Seismology*, no. 2, p. 89-93.

[Explosive activity that began on 29 June 1985 destroyed the E part of the lava dome. A block and ash flow formed a thick deposit that extended 8-10 km to the E, a directed blast covered an area of 10 km^2, and explosive activity fed a series of pyroclastic flows through 1 July. Lava extrusion then began from the new crater and continued for several months. This eruption is described in an extensive report by Bogoyavlenskaya and others in *SEAN Bulletin* v. 11, no. 4, including a summary of Bezymianny's activity since 1956.]

Further Reference: Alidibirov, M.I., Belousov, A.B., and Kravchenko, N.M., 1987, The Phase of Directed Blast During the Bezymianny Eruption in 1985; *Volcanology and Seismology*, no. 2, p. 81-89.

KLIUCHEVSKOI

56.06°N, 160.64°E, summit elev. 4850 m
Local time = GMT + 12 hrs

Massive Kliuchevskoi is the highest volcano on Kamchatka and historically the most vigorous. Its first eruption, in 1697, is the second oldest on the peninsula, and 78 are known since.

General Reference: Fedotov, S.A., Khrenov, A.P., and Zharinov, N.A., 1987, Klyuchevskoi Volcano: Its Activity in 1932–1986 and Prospects of Its Development; *Volcanology and Seismology*, no. 4, p. 3-16.

7/78 (3:7) Activity in the summit crater increased in March, preceded by an earthquake swarm. After a 3-month recess, new lava appeared in the crater and poured through a breach in the NW rim onto the flanks. An 80-m cone was built in the center of the crater and "intense" collapse occurred on the crater walls and the NW Sciarra. Moderate seismicity accompanied the activity.

Information Contact: B. Ivanov, IVP, Kamchatka.

1/80 (5:1) Tass reported, in an article dated 24 Jan, that Kliuchevskoi had begun to erupt. Gas, ash, and "hot rocks" were ejected, and a lava flow was extruded. Heavy snowfall impeded observation of the crater, but the staff of the local IVP station were monitoring the volcano with instruments.

Information Contact: Tass.

3/80 (5:3) Soviet press sources described a renewed eruption at Kliuchevskoi. After a series of volcanic earthquakes, ash was ejected from the summit in bursts that rose as much as 5 km. Ashfall covered an area of more than 6000 km^2. On 6 March, a fissure more than 1 km long opened on the NE flank and began to emit gases. The next day, lava extrusion from the fissure started at 1.5 km altitude. Four small cones, up to 20 m high, formed along the fissure. As of 14 March, lava had flowed 1 km downslope and ash emission from the summit crater was continuing. IVP personnel were investigating the activity at the eruption site.

Information Contacts: Tass; V. Khudyakov in the 14 March edition of *Sovetskaya Rossiya,* Moscow.

Further Reference: Stepanov, V.V. and Chirkov, A.M., 1981, Activity of the Upper Crater of Kliuchevskoi Volcano in January–March 1980; *Volcanology and Seismology*, no. 1, p. 103-106.

12/81 (6:12) A small hot area, apparently centered on the summit, was first noted on the thermal infrared band of a NOAA 6 satellite image on 21 Dec at 1057, and a hot spot was present 22 hours later. An average temperature of 320°C within a given image element (or "pixel", corresponding to a 1.1 × 1.1 km ground area) will saturate the sensors for this band, so it is not possible to determine the actual temperature of the heat source nor its true dimensions. On the next image, at 1233 on 22 Dec, the hot spot was accompanied by a diffuse plume that extended about 350 km SE. The hot spot had decreased in size 24 hours later, and the plume was smaller (detectable only to 60 km to the E) and more diffuse. Clouds partially obscured the volcano for the next several daily images, but a small plume seemed to be present. On the next clear-weather image, at 1400 on 28 Dec, there appeared to be a small warm area at the summit, but no plume was evident. No additional activity has been observed on satellite imagery. There have been no reports from ground observers.

Information Contact: M. Matson, NOAA/NESS.

[Occasional small explosions ejected ash and incandescent bombs 24 March–2 May 1982. A mid-Sept overflight revealed a collapse depression 300 m in diameter in place of the 1977–80 cinder cones. Ash emission resumed 7-8 Oct. Activity became progressively more intense through Nov. and bomb ejection resumed in Dec.]

Reference: Ivanov, B.V., et al., 1988, Active Volcanoes of Kamchatka and Kurile Islands: Status in 1982 (see Karymsky 4:8 for full reference).

3/83 (8:3) An earthquake swarm on the NE flank began 28 Feb. The majority of the events had foci above sea level (Kliuchevskoi's summit elevation is 4850 m) and their maximum magnitude was 3. Based on the swarm's character, the IVP predicted that a flank eruption would start between 4 and 9 March. On 8 March a crater opened at 3000 m altitude on the NE flank. Activity from the crater was purely effusive, producing an andesitic basalt flow that was 3 km long by 18 March.

Information Contact: B. Ivanov, IVP, Kamchatka.

Further References: Special issue on the 1983 eruption of Kliuchevskoi; *Volcanology and Seismology*, 1988, 148 pp. (English translation of *Volcanology and Seismology*, 1985, no. 1) (8 papers).

Panov, V.K. and Slezin, Y.B., 1985, The Mechanism of Formation of a Lava Field During the Predskazanny Flank Eruption; 1983, Klyuchevskoy Volcano, Kamchatka; *Volcanology and Seismology*, no. 3, p. 3-13.

Tokarev, P.I., 1985, Prediction of Lateral Eruption of Kliuchevskoy Volcano in March 1983; *JVGR*, v. 25, p. 173-180.

5/83 (8:5) The NE flank eruption was continuing in early June. As many as 15 lava flows were simultaneously active, some reaching 5 km in length. At the end of May, the maximum discharge rate was 10 m^3/sec. Weak explosions occurred from the summit crater. The activity was not visible on NOAA weather satellite imagery returned in May.

Information Contact: G. Bogoyavlenskaya, IVP, Kamchatka; M. Matson, NOAA/NESS.

5/84 (9:5) Continuous volcanic tremor and night glow over the crater began in March. Tremor and the number of explosive earthquakes increased from late March through May. During this period, the amplitude of volcanic tremor at 14 km distance and the maximum amplitude of explosive earthquakes reached 2 and 5 μm respectively. Since mid-May, a cinder cone has been visible in the central part of the crater. On 22 May, as moderate Strombolian activity continued, lava began to pour into the NW valley.

Information Contact: B. Ivanov, IVP, Kamchatka.

Further Reference: Tokarev, P.I., 1985, Eruption of Kliuchevskoi Volcano in March–April 1984 and Estimation of the Observation Data; *Volcanology and Seismology*, no. 1, p. 106-108.

8/84 (9:8) Eruptive activity continued through Aug. During periods of maximum activity ash was ejected to 5 km and bombs to 1 km above the crater rim. Lava flowed to the NW, NE, and SW from the central crater; the largest flow advanced along the NW valley to about 3 km above sea level and crossed a glacier, forming mud flows. A cinder cone has formed inside the central crater.

On 17 Aug between 0733 and 1027, high-resolution thermal infrared and visual images from polar orbiting weather satellites showed a plume extending about 200 km SE from the volcano below about 6 km altitude. Soviet volcanologists confirmed these observations, reporting that on 16–17 Aug a 15 km-wide ash plume extended 200 km from the volcano.

Information Contacts: B. Ivanov, IVP, Kamchatka; M. Matson, NOAA/NESDIS.

1/85 (10:1) Kliuchevskoi's summit eruption continued in

Nov and Dec. Lava flows reached 1.5 km in length. Gas and ash columns rose 2-4 km above the summit and bombs reached 300 m height. At times lava fountains were observed above the crater rim. Eruption character remained constant through mid-Jan.

Thermal infrared images from the NOAA 6 polar orbiting satellite on 23 Nov at 1852 and 24 Nov at 0839 showed narrow plumes emerging from Kliuchevskoi and extending roughly 60 km E. Soviet volcanologists reported that the ash column rose to about 4 km above the crater 23–24 Nov.

Information Contacts: B. Ivanov, IVP, Kamchatka; M. Matson, W. Gould, NOAA/NESDIS.

Further Reference: Gordeev, E.I., Mel'nikov, Yu.Yu., Sinitsin, V.I., and Chebrov, V.N., 1986, Volcanic Tremor of Kliuchevskoi Volcano (Eruption of Summit Crater in 1984); *Volcanology and Seismology*, no. 5, p. 39-53.

PLOSKY

Kliuchevskoi Volcano Group
56.20°N, 160.63°E, summit elev. 4108 m

A large complex volcano, only 10 km NW of Kliuchevskoi, Plosky has had no historic eruptions.

3/82 (7:3) The volcano's 17 km-long Bilchenok Glacier has begun to advance. The glacier, located in Plosky's caldera, has 3 large ice cascades on its NW flank. Previous surges of this glacier occurred in 1959, 1976, and 1977. Photo reconaissance flights over Kamchatkan glaciers 10–11 March revealed that Bilchenok's front was 1 km from its 1980 position and 500 m from the 1959 maximum surge. Its surface was broken into blocks, and rupture disturbances of the snow cover were observed. Although the Kliuchevskoi group has been quite active, no historic eruptions of Plosky have been reported.

Information Contacts: V. Vinogradov, IVP, Kamchatka.

Further Reference: Ovsyannikov, A.A., Khrenov, A.P., and Murav'yeva, Y.D., 1985, Recent Activity of the Dal'nya Ploskaya Volcano; *Volcanology and Seismology*, no. 5, p. 97-98.

SHEVELUCH

56.78°N, 161.58°E, summit elev. 3395 m

The next major strato-volcano to the N, and 85 km from Kliuchevskoi's summit, Sheveluch's 1854 and 1964 eruptions were particularly violent. The Aleutian arc swings ESE from this general location, with Buldir, the nearest Holocene volcano, approximately 1000 km distant.

Aerophotogrammetric observations documented extrusion of a lava dome in the summit crater beginning in Aug 1980. Growth was most rapid during the first 2 months, but continued through 1981, extruding a dome 180 m high with a volume of about 0.02 km^3. A vent on the SE part of the dome ejected ash and gas, and avalanches advanced 700 m down the dome's S flank.

Reference: Dvigalo, V.N., 1988, Growth of a Dome in the Crater of Shiveluch Volcano in 1980–1981 from Photogrammetry Data; *Volcanology and Seismology*, v. 6, no. 2, p. 307-315 (English translation); 1984, no. 2, p. 104-109 (in Russian).

5/84 (9:5) A single ash ejection to 1 km height occurred 22 May at 2356. No changes to the lava dome were observed. Lava extrusion and explosive activity began in late Aug 1980.

Information Contact: B. Ivanov, IVP, Kamchatka.

[Explosions ejecting andesitic ash were reported in the *Bulletin of Volcanic Eruptions* (no. 25, 1988) on 26 May, 8 August, 19 September, and 25 October, 1985. Rockslides preceded the 19 September explosion, which produced an ash cloud that reached a height of 4 km.]

ALASKA

Local time is used here, as elsewhere throughout the book, but conversion to and from GMT is difficult in this chapter. Alaska covers 2 time zones, Daylight Saving Time is used in the summer, and time zones were shifted ahead 1 hour in 1983. Residents of the eastern Aleutians often use mainland Alaska time, despite officially being within the Bering Time Zone, 1 hr earlier. We have converted to correct local time when aware of a discrepancy. Conversion times given for each volcano are for the time period used in that volcano's reports.

GARELOI

Gareloi Island, Aleutian Islands
51.80°N, 178.80°W, summit elev. 1573 m
Local time = GMT - 11 hrs winter (-10 hrs summer)

The southernmost volcano in the Aleutians, Gareloi forms a small island 1300 km from Kamchatka and 150 km W of Adak. Thirteen eruptions are known since 1760.

8/80 (5:8) During an overflight on 8 Aug, USN pilot Edwin Beech saw vapor rising to about 1 km above the summit crater. The next day, a Northwest Orient Airlines pilot observed a steam and ash column that reached 10.5 km altitude and was blown NNW, away from inhabited areas. Poor weather obscured the summit for the next several days, although large eruption columns would have been visible above the cloud layer from passing aircraft.

On 13 Aug, USN pilots were able to see the top 300 m of the volcano. A light gray eruption cloud that appeared to originate from the NE quadrant of the summit crater rose to about 2.5 km altitude (1 km above the summit), depositing ash to the NW. Both Lt. Beech and David Evans, who operates the USGS seismic station on nearby Adak Island, felt that the summit area had changed significantly during the eruption.

By 23–24 Aug, the activity had declined to weak vapor emission. No lava flows have been observed. Gareloi's last reported eruption was in Jan 1952.

Information Contacts: T. Miller, USGS, Anchorage; Lt. E. Beech, USN, Adak Island; D. Evans, USGS, Adak Island.

9/80 (5:9) USAF pilot Jerry Nelson observed renewed activity from Gareloi during an overflight on 17 Sept. As he approached the volcano at about 1600, he saw a slight wisp of vapor that grew rapidly into a dark brown ash-rich column reaching about 6 km altitude. The eruption column, which appeared to originate from the E side of the summit crater, drifted slightly N or NW. The activity was visible until Nelson left the area about 10 minutes later.

Information Contacts: T. Miller, USGS, Anchorage AK; J. Nelson, USAF, Travis AFB, CA.

11/80 (5:11) On 10 and 11 Aug, SO_2 from a fresh volcanic plume was detected from a research aircraft (flown by NASA under contract from the U.S. Department of Energy) at 19.2 km altitude just S of Anchorage Alaska.

Imagery returned 8 Aug at 1010 by the NOAA 6 satellite shows a high-altitude plume appearing to originate from the vicinity of Gareloi. Using a drift rate of 30 km/hr, Los Alamos National Laboratory (LANL) personnel calculated that the eruption that produced this plume had probably ended about 10 hours earlier. Later visual and infrared images show the plume moving toward the Anchorage area, about 2000 km from Gareloi, at a rate that could have brought it to the sampling area by 10 Aug. The eruption column seen emerging from Gareloi 9 Aug by a commercial pilot (5:8) was also present on satellite images, but clearly was not large enough and did not reach a high enough altitude to have been the source of the material sampled 10–11 Aug. Wind conditions also preclude the 7 Aug eruption clouds from Mt. St. Helens (5:7) as a source for SO_2 in the Anchorage area at this time.

Information Contacts: W. Sedlacek, G. Heiken, E. Mroz, LANL.

Further Reference: Sedlacek, W.A., Mroz, E.J., and Heiken, G., 1981, Stratospheric Sulfate from the Gareloi Eruption, 1980: Contribution to the "Ambient" Aerosol by a Poorly Documented Volcanic Eruption; *Geophysical Research Letters*, v. 8, p. 761-764.

2/82 (7:2) Infrared imagery from the NOAA 7 polar orbiting satellite 15 Jan at 1402 showed an apparent eruption cloud blowing E from the vicinity of Gareloi (fig. 11-1). Analysis of this image yielded a cloud top temperature of -36°C, indicating an altitude of 7-9 km. No eruption clouds were present on images the 2 previous days and the following day, nor was any activity evident on imagery 24 Jan, or 3 and 11 Feb. No ashfall was reported on Adak Island (130 km to the E) or Shemya Island (500 km to the WNW). Navy pilots plan to photograph Gareloi in the near future, for comparison with photos taken during and after its Aug–Sept 1980 eruption (5:8-9,11).

The seismic network on Adak Island recorded an m_b 3.2-3.3 earthquake on 15 Jan at 0521. No unique hypocenter can be determined for this event, but 1 of the 2 possible solutions places it directly under Gareloi. This shock was substantially larger than the m_b 2.6 event that accompanied the Feb 1974 eruption of Great Sitkin, which produced a 3-km plume. Additional events were recorded the following week that may also be centered in the vicinity of Gareloi. An infrasound array at College, Alaska (about 2200 km NE of Gareloi) recorded no acoustic wave associated with the eruption.

Information Contacts: J. Kienle, Univ. of Alaska; S. Billington, NOAA/CIRES, Univ. of Colorado; P. Mutschlecner, LANL.

Fig. 11-1. Infrared image returned 15 Jan 1982 at 1402 by the NOAA 7 polar orbiting satellite. A bright white plume, at left-center, drifts ESE from Gareloi. Dark ripples near the plume and at right-center indicate the position of islands in the Aleutian chain.

SEGUAM

Seguam Island, Aleutian Islands
52.32°N, 172.38°W, summit elev. 1054 m
Local time = GMT - 11 hrs

Seven eruptions have been reported from this strato-volcano 440 km ENE of Gareloi and near the center of the Aleutian chain.

3/77 (2:3) Jürgen Kienle reported that on 6 March at about 0400, the crew of the USCG Cutter *Mellon* observed and photographed eruptive activity at the 750 m level of Pyre Peak, Seguam Island. Eight lava fountains rising to an estimated 90 m were observed along a 1-km radial rift 2.5 km SW of Pyre Peak summit. Activity appeared to be progressing NE (towards the summit).

A lava flow formed and divided into 2 tongues. The larger was approximately 2.5 km long and 1 km wide. The smaller moved toward the S, and was approximately 1 km long and 0.5 km wide. Three dense black clouds were erupted from the vent during 2 hours of observation, after which the *Mellon* departed.

The nearest short-period seismometers, 225 km to the E (Nikolski), and 300 km to the W (Adak), were too far away to register any unusual earthquake activity.

Thomas Miller reported that on 8 March, a Reeve Aleutian Airlines flight passed over Seguam at low altitude. The pilot reported that lava effusion and fountaining had ended, but that a considerable amount of steam, possibly containing some ash, was being emitted from the fissure. Lava flows had not reached the sea. Poor weather during much of the year hampers aerial observation. Several other ship reports of eruptions in the area have been received in the past several years. Seguam is uninhabited.

Information Contacts: J. Kienle, Univ. of Alaska; T. Miller, USGS, Anchorage.

CLEVELAND

Chuginadak Island,
Aleutian Islands
52.82°N, 169.95°W,
summit elev. 1730 m
Local time = GMT - 10 hrs

Mt. Cleveland is 180 km ENE of Seguam, and its first historic eruption was in 1893.

12/85 (10:12) "About midday on 10 Dec, pilot Tom Madsen (president, Aleutian Air) noted an anomalous 400+ m-high eruption column over Mt. Cleveland from the ground at Nikolski, Umnak Island (about 65 km ENE of the volcano). The top of the vertical column had drifted at least 0.5 km to the N. The white eruption cloud probably consisted principally of steam with only minor amounts of ash, if any. Based on observations by Madsen, Mt. Cleveland has been steaming fairly continuously since at least 1982, when he began flying regularly from Dutch Harbor (Unalaska Island) to Atka (Atka Island). Reeder has received several reports about steam-blast and phreatomagmatic eruptions at Mt. Cleveland over the last several years."

Mt. Cleveland is a 1730 m-high strato-volcano that erupted vigorously on 10-12 June 1944. A small detachment of soldiers was stationed on Chuginadak Island at the time. One soldier died in the eruption, and the outpost was subsequently abandoned. Eruptions were reported in 1951 and 1975, but remain unconfirmed.

Information Contact: J. Reeder, Alaska Div. of Geol. and Geophys. Surveys.

OKMOK

Umnak Island, Aleutian Islands
53.42°N, 168.13°W, summit elev. 1073 m
Local time = GMT - 9 hrs

A 12 km-diameter caldera marks this large volcano, 140 km ENE of Cleveland on the NE end of Umnak Island. Eighteen eruptions have been reported since 1805.

8/83 (8:8) Images from the NOAA 7 polar orbiting satellite at 1816 on 8 July showed an elongate plume extending approximately 100 km S from Okmok. Analysis of infrared data yielded a plume temperature of -25°C, corresponding to an altitude of 6.3 km. The plume was not visible on imagery returned 12 hours earlier and later. There were no other reports of activity at Okmok during this period. Umnak Island is not densely populated, but aircraft frequently fly near the volcano. Okmok's last reported eruption June-Dec 1945 produced small eruption clouds and lava flows from the SW part of the caldera.

Information Contacts: M. Matson, NOAA/NESDIS; T. Miller, M. E. Yount, USGS, Anchorage.

MAKUSHIN

Unalaska Island, Aleutian Islands
53.87°N, 166.93°W, summit elev. 2036 m

Makushin, immediately W of Dutch Harbor, forms the prominent N portion of Unalaska Island. The volcano is 90 km from Okmok, and its first historic eruption, in 1768, is among the oldest in the Aleutians.

7/80 (5:7) On 8 July, J. Hauptmann, G. Gunther, and R. [Steuer] visited a seismic station on the E flank and overflew the summit. More than 10 roughly circular vents emitted vapor from the summit area, a flat region about 100 m across. The largest vent was about 30 m in diameter, and others were around 10 m across. An H_2S odor was detected, but no ash or incandescent material was observed.

About 60 m below the summit on the S flank, an explosion vent had recently ejected tephra ranging in size from ash to blocks, deposited in streaks aligned roughly toward the SE. Some impact craters were present in the deposit area, which extended 30-60 m from the vent.

Information Contacts: J. Hauptmann, G. Gunther, R. Steuer, S. McNutt, Lamont-Doherty Geological Observatory (LDGO).

AKUTAN

Akutan Island, Aleutian Islands
54.13°N, 166.00°W, summit elev. 1303 m
Local time = GMT - 10 hrs

This island volcano is 65 km ENE of Makushin. Following several uncertain reports early in the 19th century, the volcano has erupted at least 28 times.

5/77 (2:5) Eruptive activity, reported from this volcano in 1973 and 1974, resumed on 5 May and was continuing 4 days later. Personnel aboard the USCG Cutter *Ironwood* observed eruptions of light brown ash clouds about every 15 minutes between 1700 and 2000 on 5 May. The ash blew to the N but only a thin layer was noted on the snow-covered N side of the island. The ash eruptions were separated by periods of white steam emission. Richard Maloney, an airline pilot, saw incandescent ash during an overflight at 1730 on 6 May and more ash eruptions on 7 May. The *Ironwood* returned to the island at 1900 on 9 May and activity similar to that of 5 May was observed. Villagers on Akutan report sporadic activity since last autumn.

Information Contact: T. Miller, USGS, Anchorage.

10/78 (3:10) Akutan began to erupt in late Sept. Airline pilots reported incandescent fragments, some "as big as a car," rising about 100 m above the crater. The USCG Cutter *Morgenthau* passed N of Akutan during the evening of 6 Oct. Crew members observed incandescent tephra ejection from the summit and glow reflecting upward onto an eruption column. A deep red glow about 1 km long, which appeared to be a lava flow, moved down the flanks.

Information Contacts: M. Compton, USCGC *Morgenthau*; D. Hoadley, Reeve Aleutian Airways; J. Kienle, Univ. of Alaska.

7/80 (5:7) Akutan was observed by J. Davies on 3 July and by J. Hauptmann and G. Gunther on 8 July. The volcano was not active on the 3rd, but Davies saw a fresh-looking lava flow that had moved through a breach in the NNW caldera wall. On 8 July, the volcano was emitting steam and dark brownish-grey ash.

Information Contacts: J. Davies, J. Hauptmann, G. Gunther, S. McNutt, LDGO.

WESTDAHL

Unimak Island, Aleutian Islands
54.52°N, 164.65°W, summit elev. 1560 m
Local time = GMT - 11 hrs winter (-10 hrs summer)

With its larger neighbor Pogromni (only 6 km NW), Westdahl makes up the SW end of Unimak, the largest

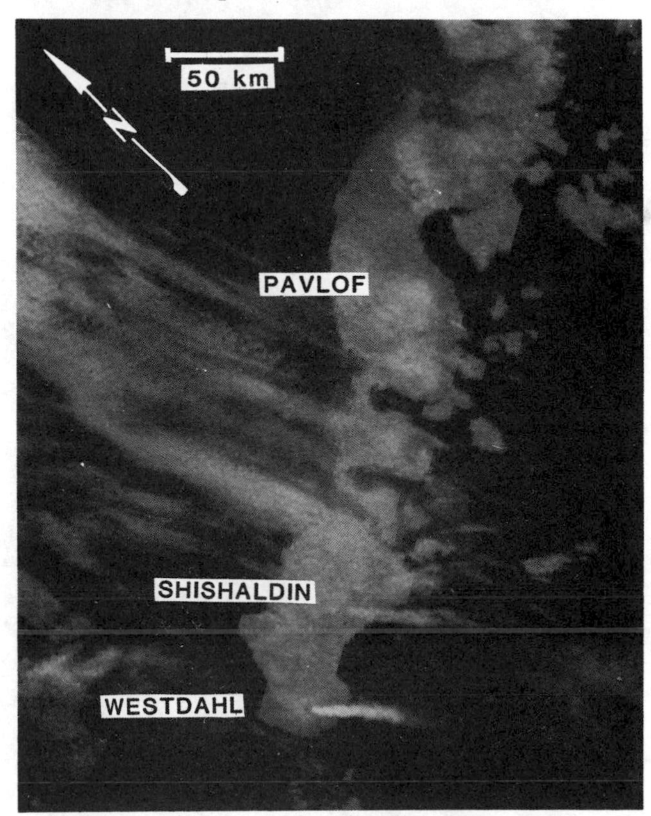

Fig. 11-2. Enlargement of a portion of a NOAA 5 infrared image at 1835 on 8 Feb 1978 (0.9-km ground resolution), showing plumes, each about 60 km long, extending SE from both Westdahl and Shishaldin. Most of the Alaska Peninsula is visible, as is the coast of Alaska NW to the entrance to Kuskokwim Bay. The gray area off the coast is pack ice. Courtesy of Jürgen Kienle and NOAA.

and easternmost island in the Aleutians. Westdahl is 85 km ENE of Akutan, and its only other known eruption was in 1964.

1/78 (3:1) The USCG reported on 6 Feb that ash, accompanied by a sulfur odor, was falling on a station located at the foot of Westdahl. Lightning was observed above the summit, accompanied by thunder and rumbling. The summit is not visible from the Coast Guard station and cloud cover has prevented direct aerial observation of the group of volcanoes which includes strato-volcanoes Westdahl and Pogromni, active in historic time. However, Reeve Aleutian Airways personnel report an ash cloud rising to 8-10 km altitude, including some large blocks visible above the 3-km cloud layer. Satellite images returned 5 Feb show a well-developed ash cloud, but it has not yet been possible to pinpoint the eruption start time [see 3:9].

Information Contact: T. Miller, USGS, Anchorage.

2/78 (3:2) Data from satellite imagery and aircraft observations established that the eruption originated from the summit of Westdahl, one of 5 volcanoes on Unimak Island active in historic times. The eruption column was first visible in 5 Feb satellite imagery (fig. 11-2), but Coast Guard personnel from Scotch Cap (15 km SW of Westdahl's summit) noted a sulfur odor during the evening of the 3rd and reported that ash had begun to fall by the next morning. Eruption column height was estimated at 8-10 km from aircraft observations on 6 Feb (fig. 11-3) and up to 8 km from interpretation of satellite imagery (fig. 11-4). Nearly 1 m of ash fell at Scotch Cap, forcing evacuation of its personnel and damaging Scotch Cap Light. Meltwater caused stream flooding and washed out the coast road. A Reeve Aleutian Airways pilot observed a new cinder cone, about 100 m in diameter, located near the site of Westdahl's last eruption (March–April, 1964). No lava flow has been observed. Activity declined to steaming after 9 Feb.

Information Contacts: J. Kienle, Univ. of Alaska; T. Miller, USGS, Anchorage.

5/78 (3:5) Snow contaminated by dark ash fell on the freight vessel *United Spirit* between 1200 and about midnight on 7 Feb, as it steamed from 48.8°N, 152.5°W to 49.2°N, 156.3°W, about 1000 km SE of Westdahl. Satellite imagery (fig. 11-4) shows that winds were driving the eruption cloud towards the vessel, about 24 hours before the ashfall began. Winds observed from the *United Spirit* during the snowfall were steadily from the NW.

Information Contact: JMA, Tokyo.

Fig. 11-3. Oblique airphoto taken 6 Feb 1978 by George Wooliver (Reeve Aleutian Airways), showing Westdahl's eruption column emerging from the cloud layer at about 3000 m altitude. Wind is from the N at about 130 km/hr. Wooliver estimated that the top of the eruption column was at 8-10 km altitude. Courtesy of Thomas Miller.

9/78 (3:9) On 6-7 Aug a team of 6, including volcanologists Maurice and Katia Krafft and Alain Gerente, climbed Westdahl. The new crater formed by the Feb eruption is about 1.5 km in diameter and 0.5 km deep, at about 1450 m elevation (fig. 11-5). Its upper portion cut through glacial ice, which reached 200 m thickness on the N rim. The bottom of the vertical-walled crater is filled with blocks, ash, ice, and talus.

A lahar deposit, originating on the WSW flank of the new crater, extended down the glacier on Westdahl's flank to the sea, cutting the road from Cape Sarichef to Scotch Cap. The thickness of the upper portion of the deposit averaged about 50 cm, increasing to 1-3 m near the lower end.

A diary from Richard Clark (who was at Scotch Cap during the eruption) has established that the eruption began about 1330 on 4 Feb with the ejection of a steam cloud. Ash emission began soon afterward and ashfall started at Scotch Cap (15 km SW of the crater) at about 1600. A thunderstorm associated with the activity dropped hailstones formed around particles of ash and small lapilli. Tephra fall and the associated thunderstorm continued at Scotch Cap until mid-afternoon on 5 Feb. Clark reported that ash emission was continuing when he left the area during the morning of 8 Feb. A plume was visible in a satellite image at 1129 on 9 Feb (3:2). After the 9th, activity declined to steaming.

Information Contacts: M. Krafft, Cernay; R. Clark, Blaine WA.

2/79 (4:2) A cloud apparently erupted from Westdahl was present on NOAA weather satellite imagery for more than 30 hours on 8 and 9 Feb. The cloud was first observed on an infrared image at 0352 on 8 Feb, about 17 hours after the previous image, on which no eruption cloud could be seen. A cloud was present on successive images on the 8th at 0926 (visible and infrared) and on the 9th at 0842 and 1037 (infrared) but not at 1958 on the 9th or 0952 on the 10th. The cloud was no more than 50 km in longest dimension on any of the images, nor was it elongated into a typical volcanic plume.

The height of the cloud was calculated separately

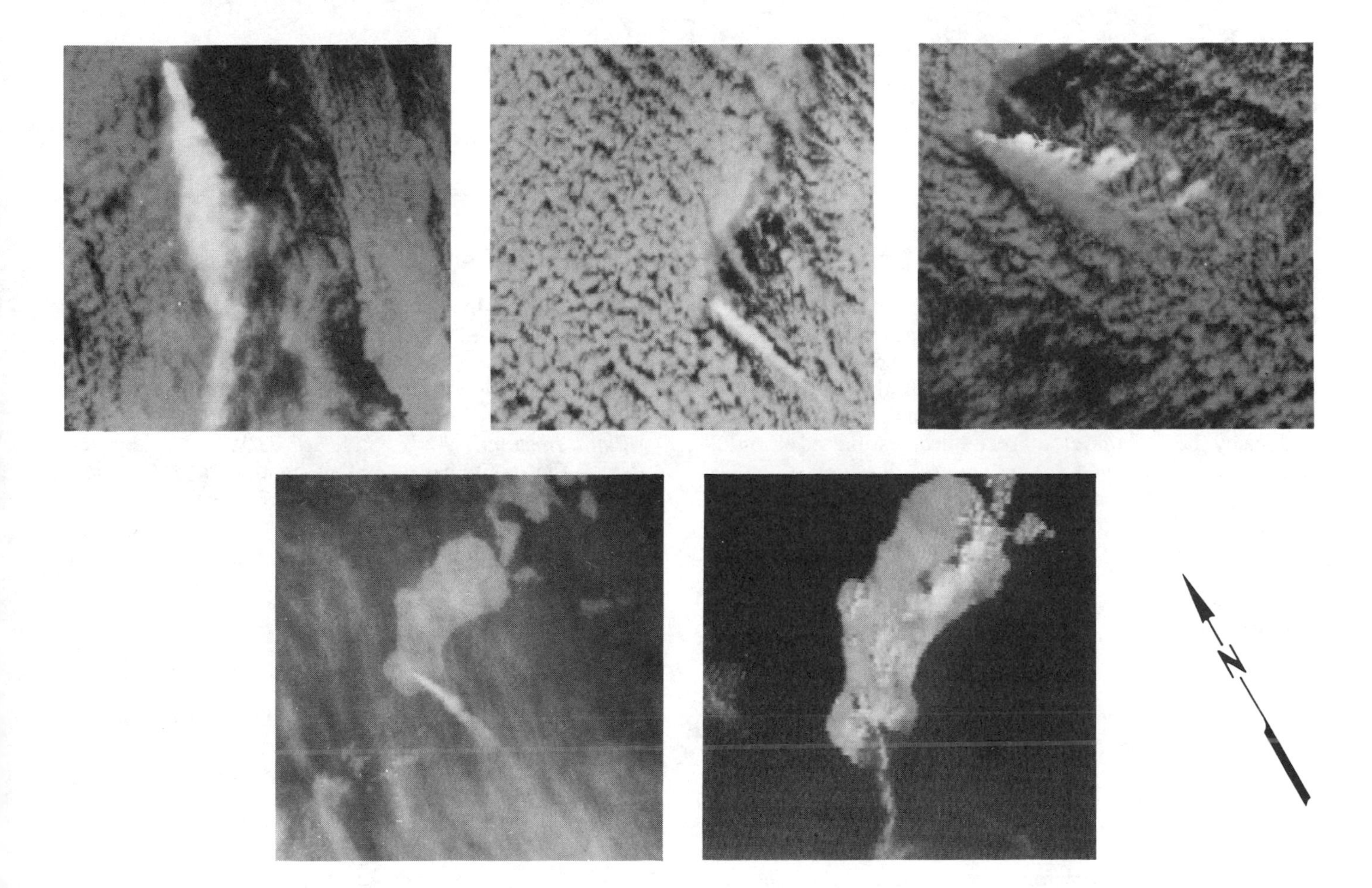

Fig. 11-4. NOAA 5 satellite imagery, 5-9 Feb 1978, showing Unimak Island (110 km long) and vicinity. North orientations (arrow) are the same for each. *A,* (5 Feb 1978, 1036, infrared) Plume extends about 230 km S, rising to an estimated 8 km; black summit hot spot. *B,* (6 Feb 1978, 1141, infrared) Plume extends about 110 km SE; black summit hot spot. *C,* (7 Feb 1978, 1101, infrared) Plume at least 6 km high and about 160 km long, blown primarily SE but some eastward shearing evident, probably by lower altitude winds. *D,* (8 Feb 1978, 1018, infrared) Plume extends 80 km SSE, at lower altitude than in C. *E,* (8 Feb 1978, 1018, visible) Weak plume curves S to SW about 100 km. Shishaldin is no longer active, 16 hours after its plume was visible in fig. 11-2. The dark triangular area S of Westdahl is fresh ashfall, subtending an angle of about 95° and covering an area of about 300 km^2. No ashfall is visible on Shishaldin. Courtesy of Jürgen Kienle and NOAA.

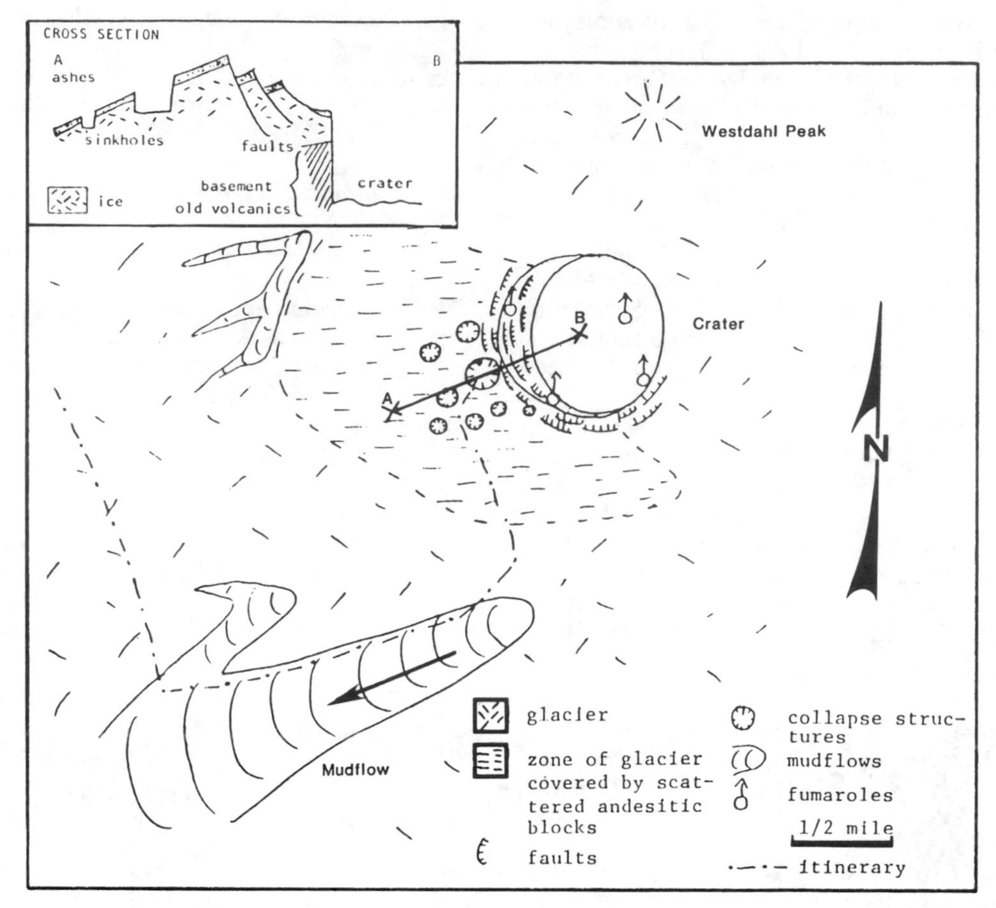

Fig. 11-5. Sketch map with cross section of Westdahl's Feb eruption crater and vicinity, prepared by Maurice Krafft from observations made during a visit 6-7 Aug 1978.

from infrared and visual images taken at 0926 on 8 Feb. Analysis of the infrared image returns a temperature of -53°C at the top of the cloud, corresponding to an altitude of slightly more than 8 km. On the visual image, measurements of the shadow cast on the weather cloud deck by the volcanic cloud result in an estimated altitude of 7.7 ± 1 km.

No activity was reported from the USCG LORAN Station at Cape Sarichef, less than 25 km from Westdahl, nor was there any aircraft confirmation of the activity.

Information Contacts: F. Parmenter, NOAA/NESS, Anchorage, AK; J. Kienle, Univ. of Alaska; USCG LORAN Station, Unimak Island, AK.

SHISHALDIN

Unimak Island, Aleutian Islands
54.75°N, 163.97°W), summit elev. 2857 m
Local time = GMT - 11 hrs winter (-10 hrs summer)

This beautifully symmetrical cone is the highest in the Aleutians. It is 60 km ENE of Westdahl and 90 km WSW of Cold Bay (Fort Randall). Interested observers stationed at Cold Bay during SEAN's first 18 months provided regular reports of Shishaldin and Pavlof. However, weather conditions in this region are frequently bad, and the reports below cover only those times when weather and flights permitted observations.

10/75 (1:1) Shishaldin, during the few times it could be observed from Cold Bay, was seen to be continually active in Sept and into Oct. At 2215 on 16 Sept, NOAA's RV *Millard Freeman*, at 55.55°N, 163.82°W, experienced rainfall that contained ash. This continued for 15 min while the ship was headed on course 250° true. The ash eruptions had apparently ceased by the end of Oct.

Information Contacts: P. Sventek, USAF, Cold Bay; T. Miller, USGS, Anchorage.

12/75 (1:3) White steam was noted on 18 Dec (1200), 23 Dec (0900), and 27 Dec (0900).

Information Contact: P. Sventek, USAF, Cold Bay.

1/76 (1:4) White steam was noted at Shishaldin on 4-5, 11-12, 29-30, and 31 Jan.

Information Contact: Same as above.

3/76 (1:6) Activity during March remained low. 3 March, 1100-1300: steaming intermittently. 8 March, 1700: steaming quietly. 10 March, 1700: steaming quietly. 23 March, 1800: steaming weakly. 28 March, 0600-0615: steaming intermittently.

Information Contact: Same as above.

4/76 (1:7) 2 April, 0600: steaming weakly, snow on cone uniformly white. 6 April, 1500: appeared to be radially covered with ash. 14 April, 0800-1800: steaming intermittently. 25 April, 2200: steaming weakly. 27 April, 1110: A faint, veil-like ash cloud was observed over the volcano. Within ten minutes the cloud had disappeared, the ash having settled on the cone, which appeared to be irregularly covered with radial ash sprays. A dark black streak was observed on the NW flank, extending 5-10 km from the summit to about the 300 m level where the slope of the cone diminishes to about 20°. The volcano was observed until 1900 and no further activity was noted. 30 April, 1100-1800: The cone appeared to be ash-covered. Strong S winds had blown black ash sprays, more noticeable than the day before, onto the N flank.

Information Contact: Same as 1:3.

5/76 (1:8) 14 May, 1400: dark ash visible on cone. 16 May, 2000: dark ash and steaming observed. 20 May, 1900: NE flank from 1600 m to 2500 m appeared to be extensively ash covered. 31 May, 1100-1400: quiet.

Information Contact: Same as 1:3.

6/76 (1:9) 3 June, 1100-1700: volcano quiet; cone darkened by ash or exposed rock due to melted snow. 4 June, 0800-2000: steamed at irregular intervals and occasionally threw out ash that settled onto the slopes. The visible flanks appeared to be about 70% ash covered. 5 June, 1400-2100: no activity noted. 6 June, 1710-1730: no activity noted. 10 June, 1430: emitting black smoke intermittently. At 2000: the cone's upper 600 m, visible above the clouds, was extensively ash-covered; no activity noted. 11 June, 1100-2000: constantly emitting a weak steam plume. 12 June, 0800-2000: emitting a slight steam plume. 15 June, 0500-0600, 2100: steaming constantly. 19 June, 1530: steaming. 22 June, 1300: peak covered by recent snows. NE flank showed ashfall. 27 June, 1100-1400: entire cone from the 1500 m level to the summit was lightly to heavily ash-covered. 28 June, 1700-1730: steaming weakly.

Information Contact: Same as 1:3.

7/76 (1:10) 2 July, 1300-1400: steaming steadily. 3 July, 1200: steaming weakly. 4 July, 0500-2100: steaming weakly. 6 July, 1300-2100: dark ash observed accumulating on N slope. 8 July, 2100-2200: emitting smoke and ash. 18 July, 1500-2100: steaming lightly. 23 July, 1830-2100: steaming heavily.

Information Contact: R. Dean, USAF, Cold Bay.

8/76 (1:11) 31 July, 2045-2130: no activity. 2 Aug, 1133-1145: no activity. 3 Aug, 0755-2130: light steam emission that intensified after 1600, turning to "smoke" after 2015. No new accumulations of ash were noted.4 Aug, 0730-0755: steady plume of light steam. 12 Aug, 1745-1800: heavy steam blown down the W slope by winds. 22 Aug, 0910-1205: no activity. 30 Aug, 0800-0930: no activity; 90% of the ash accumulation had been blown away.

Information Contact: Same as above.

10/76 (1:13) 27 Sept, at about 2025: a 1-minute burst of incandescent gas jetted from the crater and curved S. No further activity was observed during the next several hours. 28 Sept: At 0650 Shishaldin was steaming and emitting occasional ash clouds that were carried away by strong winds.

Information Contact: Same as above.

11/76 (1:14) No activity was observed during Oct.

9 Nov, 0842-0915: steaming, no ash accumulation on the cone. 17 Nov, 1040-1135: steaming, no noticeable ash accumulation. 18 Nov, 0816-1510: steaming. 19 Nov, 0716-1330: steaming. 24 Nov, 1505-1630: bursts of dense "smoke" were emitted at 5-minute intervals from the summit, then moved about 100 m down the E slope.

Information Contact: Same as above.

3/77 (2:3) 12 Feb, 0905-1120: steaming very lightly, no ash deposits. 22-24 Feb: no activity. Snow not discolored.

Information Contact: Same as above.

2/78 (3:2) A NOAA satellite image (fig. 11-2) on 8 Feb at 1835 shows simultaneous 60-km plumes emerging from the summits of Shishaldin and Westdahl, about 50 km apart. No plume from Shishaldin is visible in images taken 8 hours earlier and 16 hours later (fig. 11-4). No ashfall at Shishaldin can be seen in the latter image, but resolution is only 0.9 km and a light or moderate ashfall would probably not be visible.

Information Contact: J. Kienle, Univ. of Alaska.

2/79 (4:2) Peninsula Airways pilots reported unusually strong ash emission during overflights on 14 or 15, and 27 Feb. Ashfall, usually confined to the summit area, was occurring on the upper half of the volcano.

Information Contact: G. Morgan, Peninsula Airways.

9/81 (6:9) Activity accompanied some of the eruption at Pavlof (about 150 km ENE) [see 6:9 Pavlof; Pavlof and Shishaldin are in different time zones]. At 1315 on 25 Sept, NOAA weather satellite imagery revealed plumes emerging from both Shishaldin and Pavlof (Pavlof was also emitting a plume 4 hours earlier when weather clouds had last allowed a clear view of the area). By 1445, Shishaldin's plume had reached an altitude estimated at 6-7.5 km based on cloud top temperatures calculated from infrared imagery. The plume remained evident on the imagery until 1845, drifting E.

However, no activity from Shishaldin accompanied the ejection of a new cloud from Pavlof at 1845. Satellite images next showed a plume from Shishaldin at 0830 the next morning, when plumes from both volcanoes could be seen drifting ESE. On the next image with clear visibility, at 1315, no activity could be seen from Shishaldin. Reports from pilots through this period were very sketchy, but Shishaldin was said to be "steaming hard." No unusual activity was observed after 26 Sept by pilots or on satellite imagery.

Information Contacts: T. Miller, J. Riehle, USGS,

Anchorage; S. McNutt, E. Hauksson, LDGO; W. Younker, NOAA/NESS, Anchorage.

PAVLOF

Alaska Peninsula
55.42°N, 161.90°W, summit elev. 2518 m
Local time = GMT - 10 hrs winter (- 9 hrs summer); after 1982, - 9 hrs winter (-8 hrs summer)

Pavlof is 150 km ENE of Shishaldin and 60 km from Cold Bay. It is the most frequently active Alaskan volcano in historic time, with 41 eruptions reported since 1790.

General References: McNutt, S.R., 1987, Eruption Characteristics and Cycles at Pavlof Volcano, Alaska and Their Relation to Regional Earthquake Activity; *JVGR*, v. 31, p. 239-267.

McNutt, S.R. and Beavan, R.J., 1987, Eruptions of Pavlof Volcano and Their Possible Modulation by Ocean Load and Tectonic Stresses; *JGR*, v. 92, p. 11509-11523.

10/75 (1:1) For the past several weeks Pavlof has been producing columns 2100-2400 m high, carried NE up the Alaska Peninsula. Weather conditions commonly prevent observation of the volcano, but on 26 Oct it was active for 30 minutes of the hour that it was visible. On 31 Oct, for 1.5 hours, "lava" was reported streaming down the N side of the cone; this may have been a mud flow.

Information Contact: P. Sventek, USAF, Cold Bay.

12/75 (1:3) 10 Dec, 0100-0800: black ash and numerous sporadic orange mud/lava bursts. 18 Dec, 1300: white steam. 19 Dec, 1000: grey smoke. 23 Dec, 1000: white steam; fresh snow on N slope was darkened from the ashfall. 23 Dec, 1900: 10-second orange glow. 27 Dec, 1000: white steam; snow was white except for ash on N slope. 28 Dec, 0815: Tephra ejection like "a blowtorch" visible (in daylight) for 30 seconds; 1300: grey smoke. 30 Dec, 1755-1810: energetically emitted surges of glowing red-hot ejecta. In contrast to the sporadic 10 Dec activity, this emission was of a continuous pulsing nature and surges rose at least 150 m above the summit. This was the most energetic activity observed during the last several months. For the next 5 hours Pavlof was visible but quiet.

Information Contact: Same as above.

1/76 (1:4) 3-4 Jan: intermittent steam, grey smoke, and black ash. 5 Jan: black ash. 11-12 Jan: steam. At 2130 on 11 Jan a pilot reported the volcano erupting red-hot mud/lava up to 300 meters above the cone. 17 Jan: grey smoke. 22 Jan: steam. 24-25 Jan: grey smoke and steam. 29-30 Jan: constant steam. 31 Jan: very intermittent steam.

Information Contact: Same as above.

2/76 (1:5) At approximately 1400 on 20 Feb, Pavlof was seen to eject ash clouds at about 3-minute intervals. The ash clouds moved S and dissipated into a layer near the 3000 m level. After sunset, lava emissions became visible, but were not as energetic as those observed on 31 Dec. At 2230 22 Feb, a USAF pilot flying at 12.5 km reported that lava was visible trailing as a rivulet down the entire NW flank. The volcano continuously emitted lava during the 45-minute pilot observation. On 23 Feb at 1530 a pilot reported that the activity was basically the same as on the 22nd. On 24 Feb at 1115 another pilot reported that the lava activity appeared to have ceased, but that strong winds were driving ash, smoke and steam down the S side of the volcano.

Information Contact: Same as above.

3/76 (1:6) Activity during March remained low. 3 March, 1200-1400: steaming continuously. 10 March, 1800: inactive; snow on the W flank was white. 12 March, 1400-1600: steaming quietly; 1730 snow umblemished on all sides of cone visible on fly-by. 19 March, 1600: steaming; snow umblemished on all sides of cone visible on fly-by. 22 March, 1200: steaming, W flank umblemished. 23 March, 1900: steaming weakly. 25 March, 1730: several radial ash sprays visible on N flank. 28 March, 1500-1700: inactive, with ash still visible on N flank.

Information Contact: Same as above.

4/76 (1:7) 6 April, 1118: pilot reported ash extensively covering entire expanse of Pavlof's SE flank. Black smoke and intermittent steam were observed. 14 April, 0900-1900: steaming intermittently; 2 ash sprays visible on NW flank. 15 April, 0600-0700; 25 April, 2300; and 27 April 1200-1900: steam emissions. 29 April, 1500: a pilot reported the volcano steaming. The entire E flank was reported to be ash-covered and very black. 30 April, 1200-1900: Although difficult to differentiate between ash and exposed rock (when viewed from Cold Bay, 60 km to the W) there appeared to be ash over the upper 100 m of the cone and along several radial areas along the near (SW) flank. Pavlof apparently produced extensive ash during the last few days of the month.

Information Contact: Same as above.

5/76 (1:8) 7 May, 1200: all flanks of volcano snow-covered. 14 May, 1500: dark ash visible on cone. 16 May, 2100: dark ash and steam emission visible. 20 May: upper 100 m of cone irregularly darkened, apparently from ash. Steam issuing from cone. At 2000, a thin layer of ash at the 3000-m level was carried S for at least 25 km. At 2045, light grey material was ejected a few hundred meters. 21 May, 0900-0930: steaming. 31 May, 1200-1500: quiet. No longer possible to distinguish (at distance) between new ash darkening snow and dark bedrock exposed by melting snow.

Information Contact: Same as above.

6/76 (1:9) 2 June, 1450: pilot reported volcano steaming, extensive ash deposits visible on top and E flank of cone. 3 June, 1200-1800: volcano quiet, cone darkened by ash, or exposed rock due to melted snow. 4 June, 0900-2100: steamed continuously; occasionally ejected a light tan ash that settled into a thin layer at the 2500-m level and moved NE up the Alaska Peninsula. 5 June, 1500-2200: no activity noted. 6 June 1810-1830: no activity noted. 10 June, 2100: black smoke and ash emitted. Flanks extensively darkened by ash. 11 June, 1500: pilot reported volcano steaming. 12 June, 0900-2100: no activity apparent. 15 June, 0600-0700, 2200: steaming constantly. 19 June, 1630: steaming and emitting grey smoke. 26 June, 1400-2100: vertical steam column reached 300-500 m above cone. 27 June, 1200-1500: intermittent steam and ash emissions formed a thin layer stretching NE at the 2500-m level. 28 June, 1300-2200: ash layer extended NE at the 2500-m level.

Information Contact: Same as 1:1.

7/76 (1:10) 4 July, 0600-2200: black ash noted on the N slope. 15 July, 2000-2300: heavy steam blown down the W slope. 18 July, 1600-2300: heavy steam emitted. 23 July, 2100-2200: no activity.

Information Contact: R. Dean, USAF, Cold Bay.

8/76 (1:11) 31 July, 1730-2130: no activity. 3 Aug, 2200-2232: plume of light steam. 12 Aug, 1845-1900: no activity. 22 Aug, 1010-23 Aug, 1807: no activity.

Information Contact: Same as above.

10/76 (1:13) 10 Sept, 1520-1523: "smoke" emission with some intermixed ash. Ash deposits were noted on the SE slope. 22 Sept, 0745-1200: no activity. Lower slopes were snow-covered, the upper slopes were "dark," and the cone was black. 28 Sept, 0750-0843: a steady 150-m steam plume was observed. The cone was ash-covered, but snow blanketed the rest of the mountain.

Information Contact: Same as above.

11/76 (1:14) 7 Oct, 1852: no activity. Ash covered the top 100 m of the cone. 9 Oct, 0930-1110: heavy steam emission was visible on a rare windless day. The cone was covered with ash. 28 Oct, 1430: a steady "smoke" plume and occasional small ash clouds issued from the vent. 6 Nov, 0330-0500: an eruption was reported by fishermen working in the Bering Sea off Cold Bay. 17 Nov, 1240: "smoking." 18 Nov, 1510, 1715: heavy "smoke" rose about 200 m above the vent. 19 Nov, 0915: steam and ash emission. Ashfall had darkened the cone. 1530: steam emission.

Information Contact: Same as above.

1/77 (2:1) Seismic activity at Pavlof, characterized by [50-1000] events/day Aug-Nov 1976, had declined to 10-15 events/day during a 2-day period in mid-Dec. During periods of high seismic activity, individual events frequently blended together, producing an effect similar, but not identical, to harmonic tremor (1 frequency dominant). Approximately 1 earthquake/week (probably tectonic in origin) is locatable.

Due to poor weather conditions, visual observations of Pavlof were impossible during Jan. When visited on 21 Dec, Pavlof was emitting very small amounts of steam.

Information Contacts: J. Davies, LDGO; R. Dean, USAF, Cold Bay.

3/77 (2:3) 10 Feb, 1010-1640: no activity. Entire volcano covered with snow. 12 Feb, 1105-1355: steaming lightly. Part of cone covered with ash. 22 Feb, 0917-1625: no activity. Entire volcano covered with snow. 23 Feb, 0935-1710: no activity. 13 March, 1525: no activity. 16 March, 1530: overflight in private aircraft piloted by Capt. Young (USAF); steam was issuing from about 20 small vents in an ash-covered area approximately 50 m in diameter, 10 m below the base of the cone on the W slope. An older 30 m-diameter crater, partially filled with snow, was about 20 m below the base of the cone on the N slope.

Information Contact: R. Dean, USAF, Cold Bay.

4/77 (2:4) 22 March, 1915: light "smoke" plume. Ash darkened the top 100 m of the cone; 2305: steaming.

Information Contact: Same as above.

7/80 (5:7) Pavlof was emitting steam when viewed by J. Davies on 3 July. A few days later, according to second-hand reports, ash was present in the steam column.

Information Contacts: J. Davies, S. McNutt, LDGO.

11/80 (5:11) An eruption from Pavlof 11-12 Nov ejected large lava fountains and ash clouds that reached 11 km altitude, and may have produced lava flows.

A seismic station [8.5 km SE] of Pavlof registered a 2.5-minute burst of low-amplitude harmonic tremor beginning 5 Nov at 1351. Emission of steam, ash, and some blocks from a vent high on the NE flank started 8 Nov at 1047 and lasted about 5 minutes, without accompanying seismicity. A second burst of low-amplitude tremor occurred between 0536 and 0541 on 9 Nov.

In contrast to the pattern observed before eruptions in 1973, 1974, 1975, and 1976, virtually no additional seismic activity was recorded until a group of 7 low-frequency volcanic earthquakes occurred at about 2300 on 10 Nov. After an explosion event appeared on seismic records at 0243 on 11 Nov, 10 more low-frequency volcanic earthquakes were recorded between 0300 and 0400. Continuous harmonic tremor, of fairly low amplitude, began at 0608, but amplitude intensified around 0900.

Reeve Aleutian Airways pilot Everett Skinner saw rocks up to 1 m in diameter rising 10-30 m at 1315 on the 11th. An observer in Cold Bay, 60 km to the W, noted an increase in activity about 1600. Skinner returned to the vicinity between 1630 and 1700, reporting lava fountaining from the summit, a black cloud hugging the upper N flank, and an eruption column reaching an estimated 6 km altitude. Between 1800 and 2000, various witnesses reported lava fountaining to a maximum height of 300 m, and incandescent material moving down the N flank. A satellite image returned at 1958 shows a nearly circular plume, 15 km in diameter, N of the volcano. Activity was visible through the night from Cold Bay and the Sand Point area (50-65 km to the ENE).

The next morning, at 0946, a satellite image revealed a plume 160 km long and almost as wide spreading N of Pavlof. Spectral analysis and weather balloon data indicated that the plume reached 8-9 km above sea level. Pilot reports on 12 Nov placed the top of the eruption cloud at 9 km at 1000, 6 km at 1100, and 11 km at 1400. The eruption clouds were described as varying from ash-rich to ash-poor. A helicopter crew from KENI television, Anchorage, videotaped pulses and bursts of lava fountaining, rising 150-300 m between 1600 and 1700. The fountains emerged from a pre-existing vent high on the NE flank, the only vent confirmed active during the eruption.

Very high-amplitude harmonic tremor accompanied the eruption, reaching its strongest levels between 2000 on 11 Nov and 0700 on 12 Nov. Tremor ceased at 1835 on the 12th, when many B-type earthquakes began to be recorded.

By the morning of 13 Nov, the eruption had ended. Several hundred B-type events per day were recorded 14-15 Nov. Renewed high-amplitude tremor began 15 Nov at 1306, lasting until 1711. B-type earthquakes continued 16-19 Nov, but fewer than 100/day were recorded.

Information Contacts: S. McNutt, J. Davies, LDGO; A. Till, USGS, Anchorage; J. Kienle, Univ. of Alaska; G. Roberts, Cold Bay Weather Station; Cmdr. J. Hair, Marine Environmental Branch, Juneau.

9/81 (6:9) NOAA weather satellite images revealed an eruption plume emerging from Pavlof at 1030 on 25 Sept. On the image at 1415, when weather clouds next permitted a clear view of the area, both Pavlof and Shishaldin (~150 km to the SW) were emitting plumes. At 1545, data from infrared imagery showed that the temperature at the top of Pavlof's cloud was -55°C, corresponding to an altitude of about 9 km. This cloud drifted nearly due E and was still visible at 1945 when imagery showed a new plume originating from Pavlof. By 2215, the new plume had reached 9-10.5 km altitude and feeding from Pavlof appeared to be continuing. By 0415 the next morning, the bulk of this plume had drifted SE and appeared to be largely disconnected from its source, although faint traces of plume may have extended back to Pavlof.

Fishermen in Pavlof Bay reported that activity continued through the night, dropping nearly 4 cm of ash on one boat. An ash sample from one of the boats was sent to the USGS in Anchorage. No certain activity could be distinguished on the satellite image returned at 0615, but there were unconfirmed reports of a renewed eruption by about 0700 and by 0930 the imagery again showed plumes from both Pavlof and Shishaldin. From infrared imagery, a temperature of -28°C was determined for the top of Pavlof's plume, indicating that its altitude was approximately 7.5 km. A Reeve Aleutian Airways pilot flying near Pavlof at 1000 observed a black eruption column and estimated the altitude of its top at roughly 6-7 km. He also reported incandescent material on the W flank. A faint plume extended ESE and was still connected to Pavlof on the satellite image at 1415. No eruption clouds have been observed on the imagery since then, and there have been no reports from pilots of renewed activity.

A visit to the volcano 2-3 Oct by Egill Hauksson and Lazlo Skinta revealed that lava had been extruded from a vent about 100 m below the summit and had flowed down the NNW flank to about the 600 m level. The lava covered an area of roughly 3 km^2, and was 6-7 m thick at the thickest portion of the flow front, which was not advancing. A sample of the lava was sent to the LDGO. No ashfall thicknesses could be determined because of redistribution by very strong winds.

A Lamont-Doherty seismic monitoring station [8.5] km SE of the summit recorded occasional periods of harmonic tremor and an increase in the size of B-type events beginning about 2 weeks before the eruption. However, a few days before the eruption began both the number and size of events decreased; only 5 discrete shocks were recorded between 1500 on 22 Sept and 1500 on the 23rd, and only 2 during the next 24 hours, as compared to an average background level of 15-25/day. On 25 Sept, the day Pavlof's eruption was first observed on satellite imagery, the seismographs recorded a few more discrete events and intermittent, very low-amplitude harmonic tremor. Between 2000 on 25 Sept and 0300 on 26 Sept tremor amplitude increased gradually, and by about 0330 tremor was saturating the instruments. The strongest tremor was recorded between 0500 and 0900, then amplitudes began to decrease. However, tremor remained strong and continuous until 1220 on 27 Sept, when it declined to several-minute bursts, between which discrete events could be observed. About 100 discrete events and lower amplitude bursts of tremor were recorded during the 24-hour period ending at 1500 on 28 Sept. As of 5 Oct, B-type events and bursts of harmonic tremor were continuing.

Both the 1980 and the 1981 eruptions occurred from vents high on the N flank, but it was not certain whether these were the same vents.

Information Contacts: T. Miller, J. Riehle, USGS, Anchorage; S. McNutt, E. Hauksson, LDGO; W. Younker, NOAA/NESS, Anchorage.

10/83 (8:10) Strong tremor started to appear on local seismic records on 14 Nov at about 1500 and by 1800 was saturating the instruments. The mayor of Sand Point, about 90 km E of Pavlof, saw glow over the volcano at 2330 that night. At 1220 the next day, an airline pilot reported an eruption column rising to about 5.5 km altitude through weather clouds that covered the summit and obscured the vent area. Twenty minutes later, the column had reached 7.5 km altitude. Tephra emission was continuing at 1300. The plume blew S and SE, and spread to about 50 km width, 50 km S of the volcano. Aircraft were diverted from the area. Tremor continued to saturate the seismic instruments through the afternoon.

Information Contacts: T. Miller, USGS, Anchorage; S. McNutt, LDGO.

11/83 (8:11) On 19 Nov a small vapor cloud rose approximately 100 m above the vent. Bad weather prevented observations until 26 Nov when Pavlof was visible until mid-afternoon from Cold Bay. During the morning, a vapor plume containing a little ash rose to 4.5 km altitude. At intervals of approximately 30 minutes, puffs of dark ash were emitted. The intervals became shorter, and by 1500 ash emission was nearly continuous.

Through Oct and early Nov, an LDGO seismic monitoring station near the volcano recorded background levels of 0-40 (usually 0-30) small low-frequency events per day. A 30-minute burst of volcanic tremor began at 2000 on 4 Nov, and a 6-minute burst at 1757 on 9 Nov. Between 1430 on 11 Nov and 1100 on 13 Nov, 15 explosions were recorded. Several 1-2 minute bursts of tremor occurred between 1700 and 1900, when continuous tremor started. Its amplitude gradually increased, and tremor began to saturate the seismograph at 1100 on 14 Nov. Tremor was strongest between midnight and 1200 on 15 Nov, and continued to saturate the seismograph until 2100 on 15 Nov when its amplitude began to decrease. Tremor remained continuous but at low amplitude between 1300 on 16 Nov and 1200 on 18 Nov. Intermittent low-amplitude tremor and numerous low-frequency (B-type) events recorded after 1200 on 18 Nov were continuing on 21 Nov.

An increase in seismic activity had also been recorded in mid-July. Seismicity remained at background levels until 11 July. During the 24 hours beginning at 1500 on the 11th, 6 explosions were recorded at an LDGO seismic monitoring station near the volcano. The number of recorded events increased to 55 for the same period on 12-13 July, and to 150 on 13-14 and 14-15 July, then decreased to 120 on the 15-16th, 38 on the 16-17th, and 19 on the 17-18th, returning to background after 1500 on 18 July. During the period of increased seismicity, approximately half of the recorded events were explosions and half low-frequency events.

At 1549 on 15 July, a thermal infrared image from the NOAA 7 polar orbiting satellite showed a bright spot over Pavlof and an elongate plume extending approximately 150 km to the E. This plume was distinctly colder (at higher altitude) than the layer of low clouds that covered the area. No activity was visible on other NOAA 7 images returned approximately every 12 hours 11-18 July.

Information Contacts: S. McNutt, LDGO; T. Miller, M. E. Yount, USGS, Anchorage; M. Matson, NOAA/NESDIS.

12/83 (8:12) Activity continued through Dec. At 1400 on 15 Dec, an airline pilot observed a burst of ash from the volcano, producing a plume that drifted NW. Brief periods of ash emission separated by longer quiescent periods were continuing as of 28 Dec. These short eruptions produced plumes that dissipated after a few hours.

Information Contacts: M. E. Yount, T. Miller, USGS, Anchorage.

1/84 (9:1) Six explosions were recorded between 1600 and 2000 on 15 Dec by LDGO's 5-station seismic net 4.5-10 km from the volcano. One station, about [8.5] km from Pavlof, detected bursts of harmonic tremor 17 Dec, 1100-18 Dec, 0330; 18 Dec, 0530-0615 and 1040-1110; 20 Dec, 2200-2245; and 21 Dec, 2035-2048. Seismicity then decreased to the background level of several tens of events per day and remained at that level as of 26 Jan.

Eruption plumes were observed on 3 images returned 15-17 Dec from the NOAA 8 polar orbiting satellite. The images at 2101 on the 15th and 1031 on the 17th showed well-defined, relatively dense plumes extending 225 km E and 400 km NE from Pavlof above the weather cloud layer. A diffuse plume was observed on the image at 2018 on 18 Dec. No volcanic plumes were observed on other images 15-21 Dec, but heavy weather clouds obscured the area. There have been no eyewitness reports of eruptive activity since airline pilots last reported eruption clouds from Pavlof at 1400 on 15 Dec (8:12).

Information Contacts: S. McNutt, LDGO; M. E. Yount, USGS, Anchorage; M. Matson, W. Gould, NOAA/NESDIS.

3/84 (9:3) At 1225 on 16 March, the pilot of Air Pacific flight S27 observed a white vapor plume rising to 6 km altitude from the volcano and drifting NW. There had been no eyewitness reports of activity at Pavlof since 15 Dec 1983 (8:12). After an increase on 17-21 Dec, seismicity decreased to the background level of several tens of events per day and remained at that level as of 2 April.

Information Contact: M. E. Yount, USGS, Anchorage; S. McNutt, LDGO.

VENIAMINOF

Alaska Peninsula
56.17°N, 159.38°W, summit elev. 2507 m
Local time = GMT - 9 hrs winter (- 8 hrs summer)

A large strato-volcano with summit caldera, Veniaminof is 180 km NE of Pavlof. Its first historic eruption, in 1830, lasted for 8 years; 7 events have been reported since.

5/83 (8:5) Pilots began to report eruption clouds from Veniaminof late 4 June, noting that plumes containing some ash rose to about 4.5 km altitude. Residents of Perryville (population 100, about 25 km S of the volcano) saw incandescence and dark skies on 7 June at about 0130. Later that day, USGS personnel flew over Veniaminof. In the S part of the 10 km-diameter, ice-filled caldera, Strombolian activity was occurring from 2 vents on a cone that rises about 300 m above the ice and had previously been the site of fumarolic activity. Cherry-red molten material was ejected several times per minute to roughly 30-60 m height. The lowest 150 m of the eruption column was tephra-rich, but above that height the plume was wispy and light gray. About 50 km^2 of ice, chiefly to the S and SW, was coated with a very thin layer of ash. In the ice just SE of the cone, a prominent set of ring fractures defined a circular depression, produced by melting, that was estimated to be about 0.5 km in diameter and no more than 20 m deep.

By the next USGS overflight, at 1500 on 9 June, Strombolian activity was more vigorous and lava was flowing down the S side of the cone (fig. 17-6). Bombs and scoria were ejected every 3-5 seconds to about 2400

Fig. 11-6. Photograph of Veniaminof from the SW, on 9 June 1983 between 1515 and 1600. The circular collapse depression is shown in right foreground, the meltwater pond to its left. Steam emerges from the hole melted by the lava flow in the ice at the base of the cone. Ash is being ejected from the active cone. The caldera ice is covered by a thin layer of ash. The snow-covered caldera rim is visible in the background, with the notch at right leading to Crab Glacier (fig. 11-7). The steaming ice pit, shown on fig. 11-7 and described above, formed after this photo was taken. Courtesy of USGS.

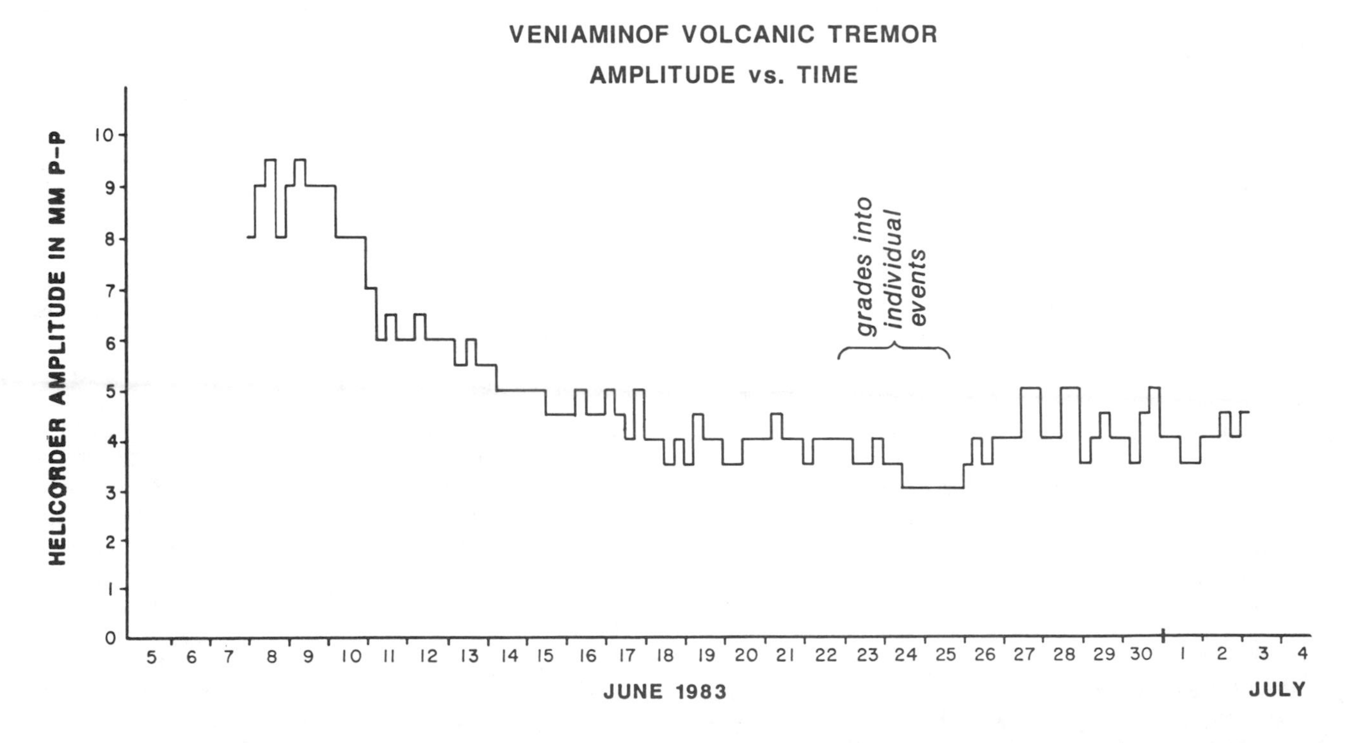

Fig. 11-7. Volcanic tremor amplitude vs. time, 8 June-2 July 1983. Station IVF (Ivanof Bay) has a 1-Hz vertical geophone and is located 31 km SSW of Veniaminof's summit. The signal is radio-telemetered to a central recording site and is recorded on both an event-triggered digital system and on a helicorder. The plot shows the helicorder amplitude of the approximately 1.2 Hz volcanic tremor at 6-hour intervals; measurements were taken for the peak-to-peak waveform with an amplitude that was attained or exceeded at least 3 times during the 6-hour period. On 23-24 June the signal changed gradually from continuous tremor of variable amplitude to discrete individual events (dates are GMT). Courtesy of Stephen McNutt. [Originally appeared in 8:8.]

m altitude, about 250 m above the summit of the cone. Tephra had begun to fill the cone's crater and an increasing number of bombs were falling on its flanks. Lava flowed from the summit down the steeply sloping S side of the cone onto the ice field, where a large E-W-trending dumbbell-shaped depression had formed. Lava flowed into the W part of the dumbbell and sank into the ice, from which billowing clouds of steam were rising. The base of the lava flow was roughly 100 m in diameter. Open water was present in the narrow area between the 2 parts of the dumbbell, and the E part was characterized by an unusual fracture pattern. The E part of this feature closely approached, but did not touch, the depression in the ice observed 2 days earlier. The timing of the beginning of lava flow activity was uncertain, but a pilot who flew past the volcano at 1330 on 8 June saw no steam clouds, so lava probably had not begun to flow onto the ice by then.

A short-period seismic instrument located at Ivanof Bay, about 30 km from the volcano, was operating in its standard event-triggered mode until it was reset to record continuously 8 June. Between 8 and 10 June, it recorded nearly continuous low-amplitude tremor, with occasional larger discrete bursts that reached (preliminary) magnitudes of 1-2 (fig. 11-7). No unusual discharge from streams draining the caldera has been observed. An eruption from Veniaminof was last reported in 1944 [but see 8:6].

Information Contacts: T. Miller, USGS, Anchorage; S. McNutt, LDGO.

6/83 (8:6) The eruption was continuing in early July. Periods of poor weather frequently limited observations, but eruption plumes were sometimes visible above low clouds mantling the volcano's summit.

By the time of a USGS overflight 15 June, the eruption had built a new cinder cone, roughly 150 m high and 500 m in diameter, within the central crater of the pre-existing 300 m-high cone in the S part of the 10 km-diameter ice-filled caldera (fig. 11-8). Every few seconds, rhythmic explosions ejected molten material from the central vent of the new cone. However, the lava flow that had been cascading down the S side of the cone onto the ice on 9 June (8:5 and fig. 11-6) was no longer active. This flow had formed a large pit in the ice, approximately 250-300 m in diameter and roughly 60 m deep. Steam rose from numerous point sources within the pit. In the ice just SE of the cone, a new pit roughly 60 m across had developed, along the trend of what appeared to be fractures on the cone. Vapor clouds from the pit seemed to be emerging from a cavern in the portion nearest the cone, suggesting that a flank eruption may have been occurring beneath the ice. From the central vent, voluminous steam emission fed billowing white clouds that rose to about 3.5 km altitude. Vigorous steaming had been observed after dark on 11 June but the activity was distinctly stronger on the 15th. Only minor ash emission was observed, although thin tephra deposits surrounded the volcano, perhaps reaching a thickness of a few centimeters SW of the cone.

A pilot who flew over the summit on 17 June at 1700 saw no lava fountaining and reported that lava was flowing down the SW side of the cone at a diminished rate. Beginning 21 June, pilots frequently observed ash plumes penetrating the layer of low clouds that usually obscured the crater. Around mid-morning on the 21st, a plume reached an estimated 2.5 km altitude (only about 300 m above the active cone). Early 25 June, ash was moving NE at 3.5-4 km altitude and several ash puffs were observed that afternoon. During a few hours of clear weather on the afternoon of 27 June, lava was fountaining to about 100 m above the cone and an active flow was observed.

A lake about 1.5 km long was seen on the ice filling the caldera. Early 28 June, pilots reported dense ash clouds moving NW at altitudes as high as 6 km. At 1500, an ash burst reached 8 km altitude; ash was rising to 6 km at 1715 and ash and steam to 4.5 km at 2139. Late 29 June, a pilot reported ash and larger tephra being ejected at a high rate, feeding a plume that moved NW and reached 5.5-6 km altitude. During the first week in July, plumes typically rose to a maximum of 3.5-4.5 km, but higher weather clouds often prevented pilots from seeing eruption plumes. Images from the NOAA 7 polar orbiting satellite showed the development of a plume

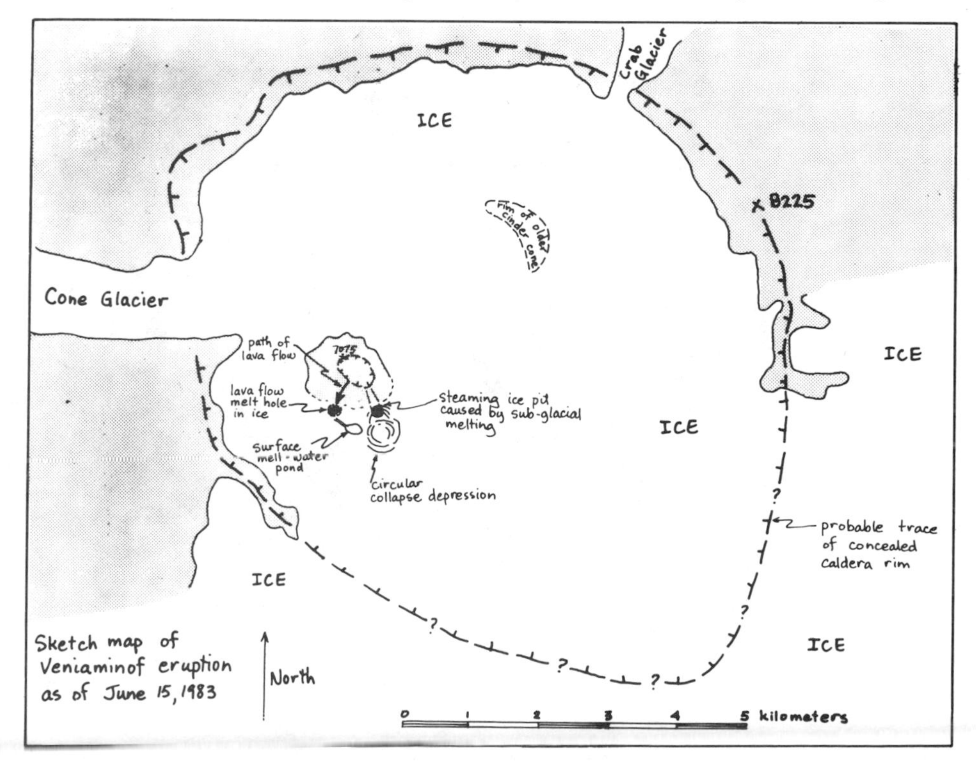

Fig. 11-8. Sketch map by M. Elizabeth Yount showing Veniaminof's crater on 15 June 1983.

that had started forming shortly before an image on 4 July at about 1130. At 1623, the plume extended about 150 km to the SSE and analysis of infrared data yielded a minimum plume temperature of -8°C. Radiosonde temperature/altitude profiles indicated that this temperature corresponded to an altitude of 3.8 km. By 2000, no plume could be seen.

Renewed lava flow activity was observed in early July, from a vent near the summit of the cone, and perhaps from a new vent on its SE flank. On 7 and 8 July, lava fountaining was continuing and bombs rose to about 100 m above the cone.

Last month, we reported that Veniaminof last erupted in 1944. However, a small cone shown in a 1954 airphoto had disappeared by 1973, when USGS personnel carried out field work at Veniaminof, and geologists presume that an eruption must have destroyed it at some time during this 19-year period. The cone was within the 300 m-high cone, active in the current eruption, in the S part of the caldera. Efforts to locate more photographs and narrow the 19-year time window are continuing.

Information Contacts: T. Miller, M. E. Yount, USGS, Anchorage; M. Matson, D. Haller, NOAA/NESDIS.

7/83 (8:7) The eruption remained vigorous through mid-July, but appeared to be declining in late July and early Aug. During an overflight on 13 July, the active cinder cone filled most of the summit crater of the pre-existing intra-caldera cone. From a breach in the S side of the cinder cone, molten material was ejected every 1-2 minutes to 150-300 m height. A blocky lava flow 15-20 m wide moved from the breach down the slope of the intra-caldera cone and ponded at the bottom of a vertically walled ice pit about 1600 m long, 400-800 m wide and 60-100 m deep. The pit, elongate roughly E-W with a slight curvature to the N at its E end, appeared to result from coalescence of smaller ice pits observed in mid-June. It contained a water lake of unknown depth, and white vapor columns rose from the vicinity of the lava flow. The active cone also emitted a thin discontinuous brown-gray eruption column that rose to about 4.2 km altitude, feeding a long narrow plume that extended 30 km or more ENE. Additional tephra had been deposited inside and outside the caldera since the previous month's observations.

Although lava continued to flow down the S side of the cone on 26 July, activity appeared weaker. Yellow sublimates were observed around the vent. By the next overflight, on 3 Aug, no incandescent tephra was being ejected and the lava flow did not appear to be moving. A few bright reddish-brown patches were noted along the lower part of the flow, but it was not possible to determine whether these were incandescent areas or heavily oxidized zones. Yellow sublimates were visible on the N portion of the active cone and the upper part of the lava flow. The nose of the flow was steaming, especially where it was in contact with the ice pit's meltwater lake (fig. 11-9). The flow had advanced farther into the ice pit and was within 50 m of dividing the meltwater lake into 2 parts. No ice was seen falling into the lake, but its S portion was ice-choked. Concentric fractures extended SW from the lake almost to the caldera rim. Ash covered the entire caldera, the slopes outside its rim, and mountains to the S. Glacier ice outside the caldera was colored light chocolate brown by ash. Three fissures, not visible on a 7 June airphoto, extended from an older cinder cone in the N-central part of the caldera about a quarter of the way to the active cone 2.5 km to the SW.

Information Contacts: T. Miller, M. E. Yount, S. Nel-

Fig. 11-9. Photograph taken 3 Aug 1983 by Steven Nelson, showing the nose of the lava flow in the ice pit. Note the concentric fractures in the ice near the edge of the pit. Dark ash covers the ice surface.

son, R. Emanuel, USGS, Anchorage.

8/83 (8:8) During an overflight on 26 July, the active cone emitted small bursts of pink-gray ash, and a white vapor cloud rose from the summit area. Bursts of steam rose from the lava flow which was forming a "delta" in the meltwater lake. Gray ash covered the entire caldera floor. Concentric fractures around the older cinder cone NE of the active cone were also observed.

The eruption appeared to be declining in late July-Aug. No ash emission was observed on 17 Aug at 1930, although a white vapor cloud rose above the active vent. The lava flow did not appear to be moving; no incandescence or steaming was observed on the flow. The water level was lower in the meltwater lake than during the 3 Aug overflight.

Information Contacts: M. E. Yount, T. Miller, USGS, Anchorage; S. McNutt, LDGO.

9/83 (8:9) After declining in Aug and Sept, eruptive activity resumed on 3 Oct, when Perryville residents reported ashfall on the town and saw incandescence at the volcano.

During an overflight on 5 Oct, USGS personnel observed Strombolian bursts of lava from the previously active vent within the summit crater of the intra-caldera cone. An ash cloud rose 300-1200 m above the vent and extended E. On 7 Oct, ash and bombs were ejected 60-90 m above the vent. Lava flowed SW from the vent on top of flows erupted June-July, adding a new lobe to the lava delta at the base of the intra-caldera cone. Steam rose from the active flow fronts. The meltwater lake formed during previous activity remained frozen and had apparently not increased in size.

During the next overflight, from 1345 to 1425 on 13 Oct, a thin wispy ash cloud rose about 60 m above the cone and drifted N, depositing ash on the caldera floor ice N and NE of the intra-caldera cone. Bursts of ash and incandescent bombs were ejected from the summit of the cinder cone at rhythmic intervals of a few seconds early in the overflight, but became more continuous later. Lava flowed from SW flank vents slightly below the summit of the cinder cone down the steep flank of the intra-caldera cone. They extended the lava delta to the SE, remelting the SE portion of the meltwater lake.

Information Contacts: T. Miller, M. E. Yount, USGS, Anchorage.

10/83 (8:10) Eruptive activity continued through early Nov. On the night of 23-24 Oct, Perryville residents observed lava fountains at the summit, and on 30 Oct they observed lava flowing down the SW flank of the intra-caldera cone. On 31 Oct and 1 Nov, an ash cloud rose 1 km above the vent.

Bad weather prevented overflights by USGS personnel during late Oct. During a 4 Nov overflight, a very light-colored vapor plume containing a little ash rose ~100 m and was blown S. Lava flowed down the SW side of the intra-caldera cone, extending the lava delta to the S. They did not observe any water in the large ice pits previously melted into the caldera ice by the lava flows, but their view was obscured by the eruption cloud.

Seismic records available through 8 Oct showed low-amplitude continuous volcanic tremor beginning 1 Oct at 1200. On 2 Oct the amplitude increased to slightly less than half that during the June eruption (8:5, 8). The tremor remained continuous and of about this amplitude through 8 Oct. Some slightly larger bursts of tremor were recorded 4-8 Oct. The eruptive activity reported on 3 Oct by Perryville residents was not distinguishable on the seismic record.

Information Contacts: M. E. Yount, USGS, Anchorage; S. McNutt, LDGO.

11/83 (8:11) Eruptive activity continued through Nov. Perryville residents observed glow over the volcano at night through the week of 13-19 Nov. On 16 Nov and the few nights preceding, the glow was the brightest observed since Strombolian activity resumed in early Oct. They saw steam on 16 Nov and heard rumbling from the volcano. On 18 Nov they observed a large billowing vapor cloud with no ash and again heard rumbling. On 23 Nov, a small amount of ash and steam rose from the intra-caldera cone, but no incandescence was observed. On the evening of 30 Nov, they observed a very small steam cloud with no ash, but they saw no glow over the volcano. Bad weather has prevented any overflights by USGS personnel since 4 Nov.

Information Contacts: M. E. Yount, T. Miller, USGS, Anchorage.

12/83 (8:12) Eruptive activity was continuing in early Jan. Perryville residents saw glow over the volcano on the night of 9 Dec. On the 10th, a small vapor cloud rose from the summit area, and on the 11th a larger cloud rose to more than 4 km altitude. During the night of 26 Dec, they observed incandescent material being ejected from the summit area. Ash and vapor erupted from a vent on the intra-caldera cone and another to the SW. Earthquakes were felt in Perryville at 1045 on 26 Dec and 1415 on 27 Dec. On 8 Jan, Perryville residents saw an ashy vapor cloud rise from the intra-caldera cone, but they saw no glow over the volcano. Bad weather has prevented any overflights by USGS personnel since 4 Nov.

Information Contact: M.E. Yount, USGS, Anchorage AK.

Further Reference: Yount, M.E., Miller, T.P., et al., 1985, Eruption in an Ice-Filled Caldera, Mount Veniaminof, Alaska Peninsula; *USGS Circular* 945, p. 59-60.

1/84 (9:1) Perryville residents observed glow over the summit of the volcano during the evenings of 31 Dec, and 3, 4, 11, and 13 Jan. Activity intensified on 17 Jan, when they observed lava fountains as high as 200 m. The fountains were active for 30-45-minute periods, waned for indefinite intervals, then resumed.

During an overflight by USGS personnel between 1345 and 1410 on 23 Jan, billowing white vapor clouds rose from the summit area of the intra-caldera cone to approximately 3-4 km altitude and were blown E (figs. 11-10 and 11-11). There was no ash in the eruption plume or on the area surrounding the cone. Fountains rose 10-20 m and lava flowed from a vent about 100 m below the summit of the cinder cone that has grown during the eruption to nearly fill the crater of the intra-caldera cone. The lava flow, estimated at 10-20 m wide, was confined by steep levees and extended more than 200 m onto the lava delta formed by earlier flows that covered the entire floor of the ice pit. Although the surface of the lava delta was irregular, its average thickness was estimated at about 30 m. USGS personnel estimated that approximately 45×10^6 m^3 of lava has filled the ice pit since June.

The ice pit had increased in size since the last overflight on 4 Nov to more than 2 km by about 1 km (compare fig. 11-10 and fig. 11-7). Steam rose from numerous sites on the perimeter of the pit where lava contacted ice. Perryville residents observed a very large vapor cloud, but not much incandescent material, over the summit of the volcano 4-6 Feb. On the evening of the 6th, lava fountains rose to heights of several tens of meters for approximately half-hour periods, separated by 45-minute intervals. On 7 Feb a small vapor cloud rose from the summit area, and that evening Perryville residents saw a faint glow but no lava fountains.

Information Contacts: M. E. Yount, T. Miller, USGS, Anchorage.

2/84 (9:2) Eruptive activity continued through early March. Perryville residents observed incandescent lava flowing down the intra-caldera cone on clear evenings between 13 Feb and early March. During the day, they saw a small vapor cloud rise from the cone. On the morning of 27 Feb, a small dark ash cloud that dissipated after about 2 hours was visible from Perryville. On the evenings of 2 and 3 March, low lava fountains were active for about 1-hour periods, declined, then resumed. Bad weather has prevented any overflights by USGS personnel since 23 Jan.

Information Contact: M. E. Yount, USGS, Anchorage.

3/84 (9:3) Eruptive activity continued through mid-March, but declined late March-early April. Weather clouds obscured the volcano for most of March; however, Perryville residents were able to make the following observations. On 7, 11, 14, and 16 March, a vapor plume rose above the intra-caldera cone. Glow was seen over the summit of the volcano on the evenings of the 7th and 16th. Between 1600 and 1700 on 22 March, an eruption cloud with a small amount of ash rose to approximately 4 km altitude, and an earthquake was felt in Perryville at 2345 that evening. A large vapor cloud was observed on the 23rd and a dark ash cloud was "glimpsed" on the 28th. Another vapor plume rose ~60 m above the intra-caldera cone on 30 March. Weather clouds continued to obscure the volcano through 8 April. Perryville residents observed a vapor cloud above the summit during the day 9-10 April but no incandescence during the evening. During an overflight by USGS personnel on 11 April, a vapor cloud containing little or no tephra rose about 90 m from the summit area of the intra-caldera cone. No incandescent ejecta was observed. The lava flow did not appear to be advancing and it was covered by a light dusting of snow. Vapor was emitted from the edges and top of the lava flow.

Information Contact: Same as above.

4/84 (9:4) After declining late March-early April, eruptive activity continued through late April. On 12-15 April, Perryville residents observed a small continuous vapor cloud of variable volume above the intra-caldera cone. On the 16th, a similar cloud, but of constant volume, was observed. Between 1230 and 1400 on 17 April, a very dark continuous ash plume rose to more than 2 km altitude from the intra-caldera cone. On the 18th, a small vapor cloud of variable volume rose from the cone. No observations were made on the 19th-21st, and weather clouds obscured the volcano on the 22nd. On 23-24 April, Perryville residents again observed a small steam cloud that varied in volume. No incandescence was observed over the summit between 11 and 26 April; glow was last observed on 16 March.

Information Contact: Same as above.

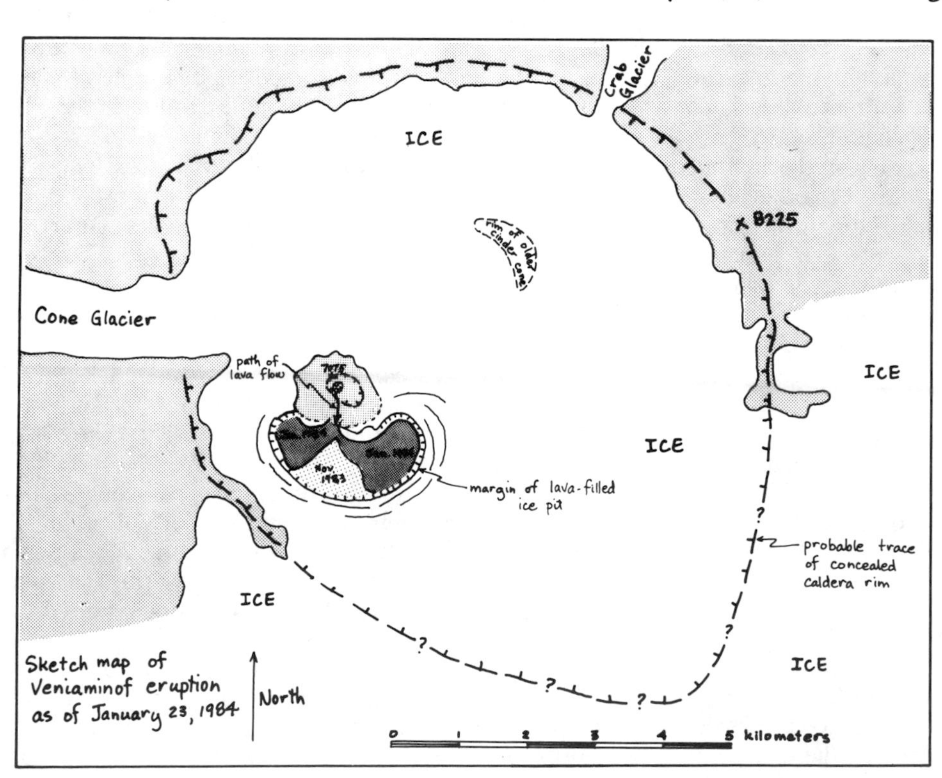

Fig. 11-10. Sketch map by M. Elizabeth Yount of Veniaminof on 23 Jan 1984. The intra-caldera cone, lava-filled ice pit, and lava flows of Nov and Jan are shown. Compare with fig. 11-8 (7 months earlier).

5/84 (9:5) Only weak activity at Veniaminof was observed by Perryville residents between late April and late May. No ash was seen in the plume after 17 April (9:4). Small white vapor plumes were visible 9 and 16 May and a larger white plume was noted on the 19th. During some periods of clear weather, no plume was seen. No glow was evident through the period.

Information Contact: Same as 9:2.

8/84 (9:8) No ash or lava emission from Veniaminof was observed during the summer. USGS personnel saw a vapor cloud, which contained no visible ash, emerging from the intra-caldera cone during an overflight on 15 June. They saw vapor plumes rising above the volcano several times during June from observation points 20-60 km SE to SSW of the volcano.

During USGS fieldwork in the caldera 13-14 Aug, vapor clouds rose from the top and the base of the new cone in the summit crater of the intra-caldera cone. The lava delta in the ice pit was still quite warm and steam rose from it. A repetitive cycle of "vent clearing" was heard; about 15 minutes of a roaring noise like a jet engine was followed by about 20 minutes of quiet, then the roaring resumed. Vapor emission did not appear to increase during the roaring period, but the observers did not have a good view of the cloud. Felt earthquakes occurred during both the roaring and quiet periods. USGS personnel placed control points for aerial photography and sampled material extruded during the 1983-4 eruption.

Information Contacts: T. Miller, M. E. Yount, USGS, Anchorage.

11/84 (9:11) Eruptive activity resumed on 29 Nov. At about 0400, Perryville residents were awakened by rumbling noises from the volcano. By 0800, a black ash cloud was rising to about 3.5-4 km altitude. At 1000, a second plume rose to about 4 km, followed by smaller bursts that were occurring at approximately 5-minute intervals as of about 1020. Pilots reported an ash plume to about 4.5 km altitude at 1045, very little activity at 1100, and another ash plume to about 5.4 km at 1115. No incandescent material was observed from Perryville or by the pilots.

A pilot who flew over the volcano on the morning of 5 Dec reported a white vapor plume, containing only a small amount of ash, rising from 2 small pits on the E side of the previously active cone. One of the pits was

Fig. 11-11. Photograph of Veniaminof from the SE taken 23 Jan 1984 by M. Elizabeth Yount. A vapor plume rises from the new cinder cone built within the summit crater of the intra-caldera cone. Steam rises from the perimeter of the ice pit where lava contacts ice.

steaming more vigorously than the other, and a brownish haze drifted downwind from the volcano. He observed no incandescent material or recent lava extrusions. On 6 Dec, Perryville residents observed large vapor plumes of varying intensity that contained very minor amounts of ash. They saw no incandescent material, and had heard no rumbling noises during the previous several days. On 7-8 Dec the volcano was obscured by weather clouds; however, small intermittent vapor plumes with no ash were observed from Perryville on the 9th. No incandescent material was seen. On the 10th and 11th, the volcano was not visible from Perryville.

Before the eruption, on 25 Nov, an LDGO seismic monitoring station about 30 km from the volcano recorded 3 events (either low-frequency volcanic events or tremor). However, other stations of the LDGO network are triggered by earthquakes greater than about magnitude 2.5-3, and no such events had been recorded as of 0200 on 29 Nov.

Information Contacts: M. E. Yount, T. Miller, USGS, Anchorage; J. Taber, LDGO.

12/84 (9:12) Weather clouds obscured the volcano from mid-Dec through early Jan. A pilot who flew over the volcano on 2 Jan observed no activity.

The following is from John Taber and Ken Hudnut. "The most recent eruptive activity at Veniaminof had little associated seismicity as compared to the previous eruption. During the recent activity an LDGO seismic monitoring station 30 km from the caldera recorded only small amounts of tremor, which lasted a total of about 20 minutes. Only 4 events of 1-2 minutes duration were recorded in the 5 days before the observed eruption. Seismic activity increased at 0420 on 29 Nov with 8 minutes of tremor roughly corresponding to first reports of rumbling noises by Perryville residents. Fourteen low-frequency events, possibly tremor, were recorded during the first 5 hours of the eruption, then recorded seismicity decreased. There is no notable correlation of timing of these seismic events with observations of eruptive activity."

Information Contacts: M. E. Yount, USGS, Anchorage; J. Taber, K. Hudnut, LDGO.

UKINREK MAARS

Alaska Peninsula
57.75°N, 156.35°W, summit elev. 95 m
Local time = GMT - 10 hrs

No previous activity was known from this site, 250 km NE of Veniaminof, and it was titled "Unnamed Vent" in SEAN's first report. The new maars (low-rimmed explosion craters) are directly N of the Hawaiian Islands, and 100 km SW of Katmai, scene of this century's largest explosive eruption (in 1912).

3/77 (2:3) On 30 March, explosions were reported at 1500 and 1720 from a vent 13 km [but see 2:4] NW of Peulik Volcano. The latter explosion was viewed from close range from a Wien Air Alaska plane. Its crew reported that the dense, ash-laden eruption cloud rose higher than 6000 m.

Peulik, an 1844-m strato-volcano with a dacite lava dome, last erupted in 1852. It is surrounded by several small olivine andesite scoria cones.

Information Contact: T. Miller, USGS, Anchorage.

4/77 (2:4) Two new maars formed in tundra terrain, 15 km NW of Peulik Volcano, between 30 March and 9 April. Explosions were first observed on 30 March from 70 km SW of the eruption site. Pilots who overflew the eruption at 1725 and 1800 reported a single vent, 20-30 m in diameter, that emitted white steam, then a dark, ash-laden cloud that rose 6000-7500 m. Fine ash fell 135 km ESE of the vent, and a sulfurous haze layer lay over Kodiak (250 km E of the vent) all day. More ash clouds were seen on 1 and 2 April.

On 2 April, the original crater had filled with water and become quiescent, and a new 60 m-diameter vent had formed 500 m to the E. By the early afternoon of 3 April, the E crater had grown to about 100 m in diameter and contained a yellowish-orange lava lake. Fragments up to 1 m across were being ejected to 300 m height. Later in the afternoon, 15-20-m lava fountains were observed.

An ash cloud rising more than 4000 m deposited traces of ash 95 km to the N on 5 April, but by 6 April activity had declined to steam emission and some ash explosions, which sent tephra to more than 1000 m above the lava lake. Similar activity, including 30-m orange-red lava fountains, was reported on 7 April. No further eruptions were reported until the early morning of 9 April, when violent explosions of incandescent material were seen 30 km away.

A team of volcanologists from the University of Alaska and Dartmouth College visited the eruption site 14-21 April. The W crater was oblong (150 × 65 m) and filled with lukewarm, slightly acidic water. The E crater was about 250 m in diameter and 100 m deep. About $^2/_3$ of its floor was occupied by a lava dome up to 40 m high that was degassing and was coated with sulfur and hematite. Ground water emerged from the crater walls at 50-70 m depth and cascaded onto the dome, where it flashed to steam. Occasional ash puffs were created by the caving of the steep crater walls. Blocks and boulders of highly variable composition and various degrees of rounding, and olivine basalt bombs with lithic cores, decreased in size from 1.5 m in diameter near the crater rims to about 50 cm diameter a few hundred meters away. Fist-sized cinders fell as far as 2 km away. Stripped bark, and mud with imbedded scoria plastered against tree trunks 500 m from the vents, indicate at least minor base surge activity during 1 or more explosions of the E crater.

Two portable short-period seismograph systems, which operated from 15-20 April, recorded a high level of microearthquake activity and 3 distinct earthquake swarms of several hours duration. More than 1 event per minute was recorded during the swarms. Many of the smaller events were shallow, but some of the larger ones showed S-P times indicating hypocenter depths between a few km and 20 km. Some of the larger events were also recorded by a permanent University of Alaska seismic station 25 km N of the eruption site.

Information Contact: J. Kienle, Univ. of Alaska.

5/77 (2:5) When visited in late May, the E vent was filling with water. A glow could be seen through fractures in the E vent dome, which had ceased growing and was degassing less vigorously than in mid-April. The vents are located on the Bruin Bay Fault. Its projected trend cuts an area of about 12 new hot springs that have formed about 3 km to the N in Becharof Lake. Periodic earthquake swarms continued to occur in the vicinity of the new vents. A third seismograph station was added in late May to the two installed in April.

Information Contact: Same as above.

Further References: Kienle, J., Kyle, P.R., Self, S., Motyka, R.J., and Lorenz, V., 1980, Ukinrek Maars, Alaska, I. April 1977 Eruption Sequence, Petrology and Tectonic Setting; *JVGR*, v. 7, p. 11-38.

Self, S., Kienle, J., and Huot, J.P., 1980, Ukinrek Maars, Alaska, II. Deposits and Formation of the 1977 Craters; ibid., p. 39-66.

Kienle, J. and Swanson, S.E., 1983, Volcanism in the Eastern Aleutian Arc: Late Quaternary and Holocene Centers, Tectonic Setting, and Petrology; *JVGR*, v. 17, p. 393-432. (This reference also contains data on Augustine and Iliamna.)

AUGUSTINE

Augustine Island
59.37°N, 153.42°W, summit elev. 1227 m
Local time = GMT - 10 hrs

Augustine Island is in southern Cook Inlet, 150 km NE of Katmai and 275 km SW of Anchorage. Only 7 eruptions are known since 1812, but 3—1883, 1976 (below), and 1986 (11:2-8)—are among the 7 largest in Alaska's history (VEI ≥ 4).

1/76 (1:4) Explosive activity, as measured by University of Alaska Geophysical Institute infrasonic stations, began on the afternoon of 22 Jan, and at 0745 on 23 Jan the first major ash eruption occurred. A second major explosion and ashfall followed at 1645 that afternoon. At least 5 major eruptions took place during the following 3 days. Ash clouds penetrated the tropopause, reaching heights of 14 km as measured by ANR height-finding radars. A light dusting of ash (approximately 1.5 mm) fell at Anchorage. Ash also fell at Iliamna, Homer, and Seldovia, Alaska.

Microearthquake swarms and occasional explosions have been noted on the volcano since mid Oct. Island seismographs, presumably damaged by premonitory activity, stopped telemetering earthquake data about one week prior to the main eruptions. A strong increase of earthquake activity was recorded, however, on 22 Jan, on the University of Alaska and USGS seismic stations located on the mainland, W of the island volcano. Intense swarm activity accompanied the main eruptive phase. Lahars, mudflows and pyroclastic flows descended the flanks and some reached the sea. Vent-clearing and subsequent explosions removed much of the 1963-64 dome, resulting in a crater breached to the N.

The Burr Point Research Station (N tip of the island) was severely damaged by blast and thermal effects from one or more nuées ardentes, and scoria and ashfall. Temperatures greater than 400°C were measured 9 feet below the surface of a pyroclastic flow E of the research station.

Chemical and petrographic analyses of the first 23 Jan ashfall, sampled at Seldovia and Iliamna, indicate that initial melt accompanying vent breaching explosions was dacitic andesite as documented in the following partial chemical analysis: SiO_2 = 63.8%, Fe_2O_3 = 2.1%, FeO = 2.0%, MgO = 2.1%, CaO = 5.1%, Na_2O = 3.9%, K_2O = 1.3%. A relatively quiet period extended from 27 Jan to the end of the month.

Information Contacts: R. Forbes, Univ. of Alaska; P. Sventek, USAF, Cold Bay.

2/76 (1:5) Explosive activity resumed at 0442 on 6 Feb, as documented by seismic and infrasonic signals, and by another ashfall along the NW margin of the Kenai Peninsula. In the morning a muddy rain fell in Kenai and Ninilchik along with considerable ash. Another eruption occurred at 1230, and a Wien Airlines pilot reported the top of the cloud at 7.5-9 km. At 1800 on 6 Feb a blizzard-like ash storm occurred at Homer, becoming so dark that vehicular traffic stopped. This fall of very fine tan-colored ash was the heaviest that Homer had received during the eruption sequence.

The volcano erupted almost continuously until 16 Feb, when the activity primarily returned to steam explosions. During that period there were an abundance of seismic and infrasonic signals. Cloud tops commonly reached heights of 3-4.5 km. The Geophysical Institute made aircraft observations almost daily 8-18 Feb except the 10th and 14th. The pilot observed that ash was erupted until 16 Feb.

Geophysical Institute scientists visited Augustine on 18 Feb and observed occasional ash puffs but not the continuous outpouring of ash that had occurred earlier. During the period of continuous strong ash activity there had also been new nuées ardentes and the NE part of the island was again enlarged. Temperatures of 604°C were measured in the new nuée ardente, which overlay one from the first phase. This temperature was reached at a depth of 2 m, away from fumaroles. Ambient temperature was 0°C. A new lava dome was extruded, possibly on 11-12 Feb, and the top was about 250 m above the base. From 18 Feb to the end of the month the volcano was fairly quiet.

Information Contact: J. Kienle, Univ. of Alaska.

3/76 (1:6) A seismic array continued in operation on Augustine Island during March. The only known activity during this period was normal degassing and several earthquake swarms, with most magnitudes in the 1-2 range.

Information Contact: Same as 1:5.

4/76 (1:7) Passing aircraft noted extensive steaming on 23 April (1015) and 29 April (1500), but no eruptive activity was reported during the month [but see 1:8].

Information Contact: Same as 1:5.

5/76 (1:8) Augustine was very active for a considerable part of April. The eruptions produced mostly nuées ardentes, and during the peak of this period explosions were recorded on the University of Alaska's seismic system 10-15 minutes each hour—about 100 events per day. Deposits on the N and NE sides increased considerably, including those from numerous nuées ardentes. Since the activity was nuées ardentes, the troposphere was not affected.

The eruptions occurred as follows in April: 6th-9th, 6 to 12 per day; 10th-11th, about 1 every hour; 12th-14th, almost continuous; 15th-17th, very intense; 18th-22nd, 12 per day; 23rd, nuée ardente activity stopped.

Through the rest of April and all of May the volcano quieted, pouring out large quantities of gas and steam in a steady-state situation. Some of the plumes were reported to be rather spectacular.

Information Contact: Same as 1:5.

6/76 (1:9) The volcano continued its activity throughout June, continually pouring out large quantities of gas and steam. Some of the plumes were spectacular.

Information Contact: Same as 1:5.

8/76 (1:11) The Augustine investigation team reports that: "Degassing of ash-flow deposits and the new summit lava dome continued during July and Aug. There were only a few local seismic events and no new eruptive activity.

"Field work during mid-Aug disclosed that the initial Jan 1976 eruptions had been the most explosive and extensive of three cycles that occurred in late Jan, mid-Feb, and mid-April. Pyroclastic flows of the Jan cycle extended in all directions away from the summit, reaching the sea on the S and NE sides of the volcano. One of the hot pyroclastic clouds overran Burr Point Research Station, at the northernmost tip of the island 5 km from the summit, damaging it severely. The velocity of the cloud was high enough to dent the aluminum structures. Thermal and blast effects on the structures indicate that they were hit from the side facing away from the volcano. We infer that the structures must have been damaged by a back eddy of the turbulent pyroclastic cloud that continued out to sea.

"The decreasing explosivity of eruptions and extent of deposits during the Feb and April activity reflects a change in the mechanism of eruption, from pyroclastic explosions of Jan and 6 Feb, to block and ash flows avalanching off the newly emerging viscous andesitic lava dome in the crater during the latter part of Feb and April. Emplacement of the lava dome has now nearly obscured the Jan crater, but has not yet restored the summit to its pre-1976 elevation.

"Gas samples have been collected from active fumaroles at the margin and on the flanks of the new lava dome. Temperatures up to 457°C were measured in these fumaroles. A maximum of 430°C was measured in fumaroles emitted from the Feb ash flow deposits, as compared to maximum fumarole temperatures of 603°C measured soon after their emplacement. Fumarolic emission from the Jan deposits has ceased entirely.

"On the basis of the recent work and comparison to the past eruptive history of Augustine, we now believe that the explosive phase of the 1976 eruptive cycle is over and that activity in the near future is likely to be limited to local debris fall and mudflows."

Information Contacts: Augustine Field Party: Geophys. Institute, Univ. of Alaska, and Univ. of Washington; D. Johnston, H.-U. Schmincke, J. Kienle, M. Utting, Univ. of Alaska.

Further Reference: Kienle, J. and Shaw, G.E., 1979, Plume Dynamics, Thermal Energy and Long-Distance Transport of Vulcanian Eruption Clouds from Augustine Volcano, Alaska; *JVGR*, v. 5, p. 139-164.

ILIAMNA

SW Alaska
60.03°N, 153.10°W, summit elev. 3053 m
Local time = GMT - 10 hrs

Iliamna, and its northern neighbor Redoubt, tower over the W side of Cook Inlet. Eleven eruptions have been reported since 1768, but Johnston (1979) notes that Iliamna's high rate of gas emission often results in steam plumes—particularly on cold, clear days—that are easily mistaken for eruption clouds or "steam eruptions" (as in the report below).

Reference: Johnston, D.A., 1979, Volcanic Gas Studies at Alaskan Volcanoes *in* The U.S. Geological Survey in Alaska: Accomplishments during 1978; *USGS Circular* 804-B, p. 83-84.

12/78 (3:12) A brief steam emission began at about 1050 on 7 Nov. Puffs of steam, ejected every 1-5 minutes, rose an estimated 3 km above the summit. No ash was visible in the steam puffs. The activity ended at about 1330. Iliamna's last reported activity occurred 1952-3. A USGS seismic station 20 km NNE of Iliamna recorded no unusual seismicity.

Information Contacts: J. Proffett, Anaconda Co., Anchorage; J. Kienle, Univ. of Alaska; J. Lahr, USGS, Menlo Park, CA.

CASCADES and SIERRA NEVADA

MT. ST. HELENS

Cascade Range, S Washington
46.20°N, 122.18°W, summit elev. 2549 m
Local time = GMT - 8 hrs winter (- 7 hrs summer)

This formerly conical strato-volcano is 165 km S of Seattle and 80 km NNE of Portland, Oregon. Mid-19th century eruptions had been documented, and the volcano was recognized as dangerous (e.g. Crandell and Mullineaux, 1978), but it did not capture widespread attention until 1980.

Reference: Crandell, D.R. and Mullineaux, D.R., 1978, Potential Hazards from Future Eruptions of Mount St. Helens Volcano, Washington; *U.S. Geological Survey Bulletin* 1383 C, 26 pp.

Selected references from the voluminous literature on Mt. St. Helens are listed in the following reports. Note especially those following the 5:12 report summarizing 1980 activity.

3/80 (5:3) After [more than] a week of local seismicity, Mt. St. Helens began to erupt steam and ash on 27 March, the first eruption in the contiguous USA since the 1914–17 activity of Lassen Peak, California. Steam and ash explosions and earthquakes were continuing on 6 April, but no fresh magma had reached the surface.

Seismic activity [began to increase by 16 March and stronger activity was initiated] on 20 March at 1548 by a M [4.2] earthquake centered [about 4 km beneath] Mt. St. Helens. Events of M 3.5 on 22 March, 3.4 on 23 March, and 4.2, 3.4, and 3.4 on 24 March punctuated smaller events that were occurring every few minutes by the 24th at depths of less than 5 km. USGS and Univ. of Washington seismologists installed an array of portable seismographs. USGS personnel who flew over the volcano on 24 March saw no new snow-free patches or fumaroles, but observed a number of snow avalanches triggered by earthquakes. Seismicity became more vigorous on 25 March, when 11 earthquakes of M 3.4-3.8 were recorded, and continued to increase the next day, when 14 shocks of M 3.4-4.0 accompanied numerous smaller events that continued to occur.

At 1236 on 27 March, a loud noise heard more than 15 km from the summit heralded the initial steam and ash explosion. Ash emission continued until a M 4.5 earthquake was recorded at 1401. The explosion prompted the evacuation of about 360 persons from [near] Mt. St. Helens. A team of USGS volcanologists was sent to monitor the volcano. Helicopter reconaissance revealed a new crater 60-75 m in diameter and about 50 m deep, located within the summit crater [that predated] the previous eruption, in 1857. A second explosion occurred at about 0300 the next morning, producing an ash-laden cloud that reached 2 km above the summit and a non-incandescent ash avalanche that flowed down the NW flank. Shortly thereafter, a steam

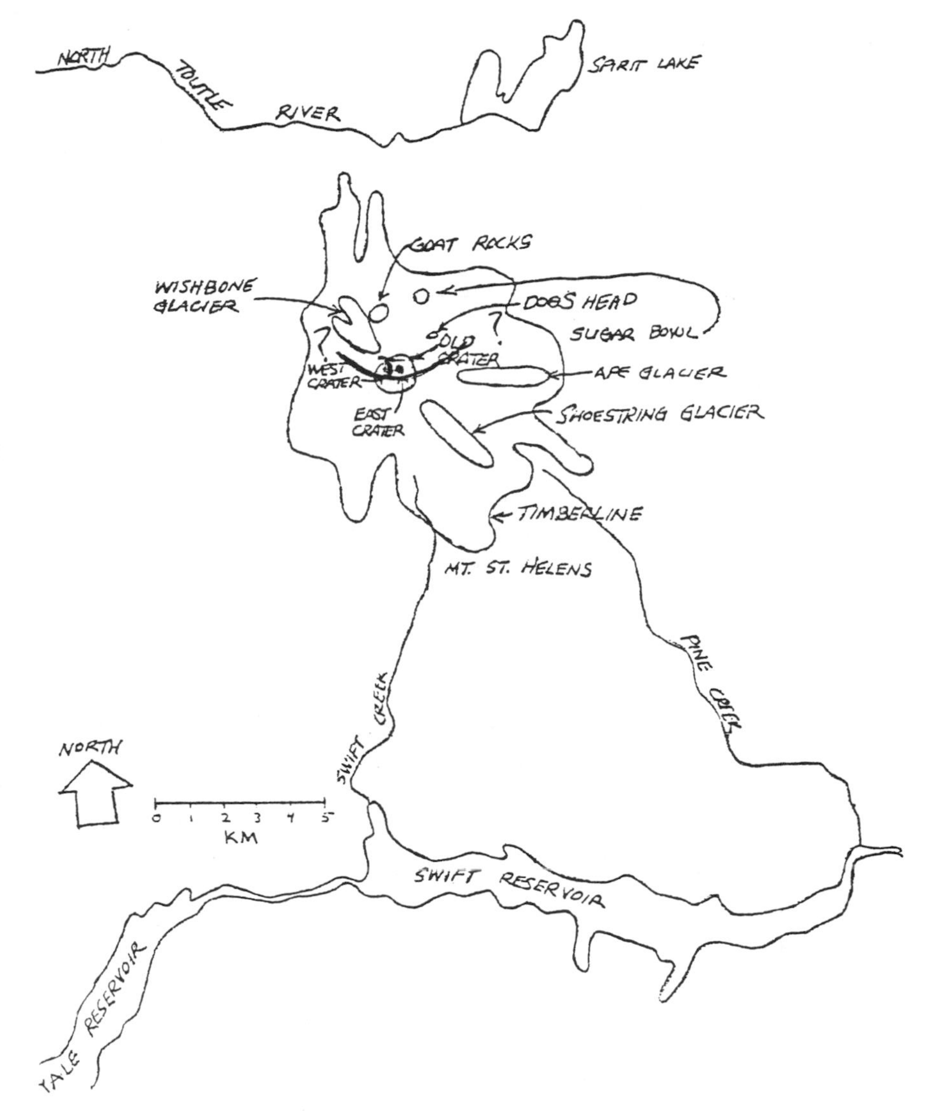

Fig. 12-1. Sketch map of Mt. St. Helens. Positions and lengths of the arcuate summit fissures are approximate, as indicated by question marks at each end. The old summit crater is shown as larger than its actual diameter of about 0.5 km. The 2 active vents have coalesced at the surface, but maintain their separate identities. Courtesy of Robert Christiansen and Robert Tilling.

plume rose to more than 3 km height. Aerial observers reported 2 nested arcuate fissures [first seen the afternoon of 27 March], one about 5 km long, the other about 1.5 km long, trending approximately E-W across the summit, S of and above the new crater (fig. 12-1).

By dawn on 28 March, pulses of dark, dense ash were rising to 3 km above the summit and some blocks were being ejected. Small mudflows moved in pulses and surges down the NE flank, reaching timberline by midafternoon. Occasional periods of rhythmic ash emission, lasting 45 minutes to 1 hour, occurred through the day, and more low-temperature ash avalanches traveled down the N and NE slopes. Many seismic events occurred, including one of magnitude 4.2, concentrated in a zone about 2 km deep in the volcano's NW quadrant. That evening, the water level in Swift Lake Reservoir was lowered by at least 8 m as a precautionary measure, to accommodate any eruption-induced snowmelt runoff or mudflows.

A new 30-50 m-diameter crater, about 10 m from the one formed [on 27 March], was discovered during overflights 29 March. Blue flame was observed in the vents, at times flickering and jumping from one crater to the other. No strong ash pulses were reported between explosions on 28 March at 2300 and 30 March at 0410. During this period, seismic events appeared to migrate to the SSE along a 25 km-long linear trend, extending from 2 km depth in the volcano's NW quadrant to 15-20 km depth below Swift Reservoir, at the S foot of Mt. St. Helens. However, continuing analysis of the seismic data indicated that the apparent migration may be an artifact of data reduction and the crustal model used. Refinement of epicenter determinations is in progress.

Strong activity resumed on 30 March. At 0740 an anvil-shaped steam and ash cloud grew, producing ashfall as far away as Bend, Oregon, about 250 km to the S. The cloud could be seen on high-resolution NOAA weather satellite imagery, but the altitude of its top could not be estimated from the satellite data. An AP photo, probably of this explosion, clearly shows [an ash veil] moving most of the way down the SE flank. [D. Swanson notes that this phenomenon, described as a "sizeable ash avalanche" in the original report, was in fact only a veil of ash that moved slowly by gravity and deposited very little material.] Six more explosions projected ash to more than 1.5 km above the summit on the 30th.

A wind shift on 31 March sent ash from the continuing explosions W of the volcano onto more populated areas. Ashfall began about noon in the Kelso-Longview area (population 75,000), about 65 km W, leaving a thin layer of light-colored, abrasive material. Only minor venting occurred during the night of 31 March-1 April; 3 earthquakes of magnitude 4.5-4.7 were recorded, with foci only 1 km beneath the summit, but the number of events was declining. Ash from a large explosion on 1 April was collected in Spokane, 500 km to the E.

Harmonic tremor was recorded for the first time from 1925 to 1930 on 1 April. This brief tremor episode was weak and poorly-defined, but vigorous, high-amplitude tremor, registered on nearly all seismic stations in W Washington, lasted from 1940 to 1955 on 2 April. Two bursts of harmonic tremor were recorded the next day, from 1840 to 1900 and 2100 to 2111; 11 individual earthquakes took place between the 2 tremor episodes on 3 April and a larger event occurred at 2115. Additional bursts of harmonic tremor occurred on 4 April from 0759 to 0826 and 1110 to 1142, and on 5 April between 1112 and 1130. Steam and ash emission continued intermittently through 6 April. Despite the bursts of tremor — evidence of magma movement within [or beneath] the volcano — there has been no evidence of juvenile material in any of the ejecta studied. To date, analysis of tilt and gravity data from the N flank has yielded no statistically significant deformation trends.

Information Contacts: D. Mullineaux, USGS, Denver, CO; R. Christiansen, USGS, Menlo Park, CA; S. Malone, R. Crosson, E. Endo, Univ. of Washington; R. Tilling, USGS, Reston, VA.

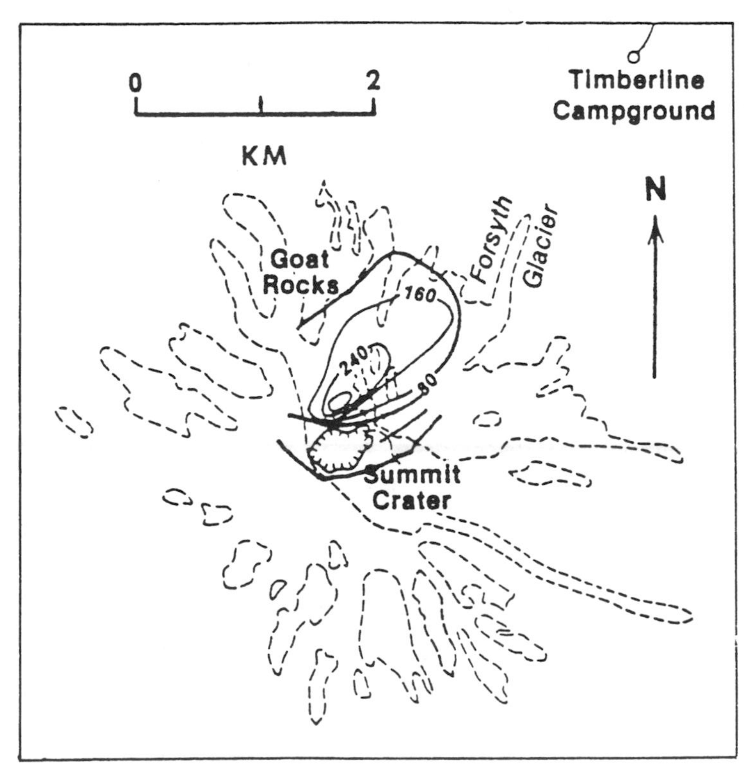

Fig. 12-2. USGS sketch map showing the summit and upper flanks. The N flank uplift is delineated by contours with an interval of 80 feet (about 25 m), generated by measuring differences in elevation shown on airphotos taken in Aug 1979 and April 1980. The boundary faults that define the summit graben wrap around the N flank uplift. Within the graben, the active crater is shown by a hachured line. Glaciers are represented by dashed lines.

4/80 (5:4) Numerous debris flows descended the N and E flanks early in the eruption, reaching nearly to timberline in some areas. These were composed of ice blocks and sparse blocks of rock in a matrix of dark-colored mixed ash and snow. Non-incandescent ash avalanches occasionally flowed down the flanks during the early part of the eruption.

By 8 April, the 2 initially separate active

vents had coalesced into a single crater (fig. 12-2), at least 500 m long x 350 m wide. The crater had deepened to 300 m by 12 April and sometimes contained small quantities of water or ice. Most mid-April eruption clouds were small and consisted primarily of vapor, but ash-rich clouds were occasionally ejected to 1 km or more above the summit. Large ice blocks were sometimes included in the ejecta. The tiltmeter at Timberline, about 4 km NE of the summit, recorded episodes of deformation of as much as a few tens of μrad in a few minutes to a few hours but these episodes oscillated between uplift and subsidence, resulting in little or no net tilt. Periods of uplift appeared to precede explosions, which were followed by subsidence. Explosions decreased in size and ash content through mid-April, with the last confirmed ejection of ash on 22 April. No fresh magma has been identified in any of the ejecta. Episodic explosions were replaced by continuous steaming, which was continuing in early May.

A comparison by James Moore of the 1952 topographic map of Mt. St. Helens with airphotos taken 7 and 12 April 1980 indicated that a substantial net uplift had taken place on the upper N flank. Cartographers from the USGS National Mapping Division comparing Aug 1979 airphotos with the April 1980 sets confirmed Moore's findings, delineating a feature with an area of nearly 4 km^2 that had risen at least 25 m and had a maximum uplift of about 100 m. [D. Swanson notes that the "uplift" was more apparent than real. It resulted from horizontal displacement of high terrain over low terrain, not vertical displacement of the low terrain.] Subsequent geodimeter readings showed that one area of the uplift had [moved outward] as much as 6 m 24-29 April.

The upper portion of Forsyth Glacier has been extensively distended and cracked by the uplift, and USGS personnel warned of the danger of a major avalanche down the N or NE flank. As a result, Washington's Governor Ray ordered the entire Spirit Lake area N and NE of the volcano, plus the zone immediately W and S of the volcano, closed to all but government workers and scientists. Access to a second zone, extending 8 km beyond the inner zone, was allowed to landowners during daylight hours only. About 300 loggers and 60 permanent residents of the area around Spirit Lake had been evacuated earlier, just after the initial explosions in late March.

The arcuate fissures described last month defined a graben extending more than 3 km through the summit, including the active crater. Both ends of the graben died out as gentle sags after wrapping around the uplift just to the N. The graben contained a median fault, E of and trending toward the active crater. This fault was downdropped on the N side by an amount estimated by Hopson and Melson at less than 30 m. Total subsidence of the graben was difficult to estimate because of its proximity to the uplift.

Although the number of local earthquakes has fluctuated (fig. 12-3), total seismic energy release has remained relatively constant through the eruption. Refinement of the crustal model used to locate the earthquakes has clustered the hypocenters beneath the volcano, 1-10 km below the summit. No tremor has been recorded since 12 April.

The USGS continues to monitor the activity in cooperation with the Univ. of Washington, the United States Forest Service (USFS), and others. Numerous instruments, including seismographs, tiltmeters, and gravimeters have been placed around the volcano (fig. 12-4), supplementing continuing visual observations from ground stations and aircraft.

Information Contacts: J. Moore, R. Christiansen, USGS, Menlo Park, CA; D. Mullineaux, D. Crandell, USGS, Denver, CO; R. Tilling, USGS, Reston, VA; C. Hopson, Univ. of California, Santa Barbara; W. Melson, SI; S. Malone, R. Crosson, E. Endo, Univ. of Washington; USGS Newport Geophysical Observatory.

5/80 (5:5) A major eruption destroyed the summit of Mt. St. Helens, projected ash into the stratosphere, devastated the N and NW flanks, and killed dozens of people on 18 May. The initial explosion was heard more than 350 km away. Substantial ashfalls occurred hundreds of kilometers downwind, closing roads, schools, and businesses, and threatening crops in the NW USA.

Explosions similar to those of early to mid-April

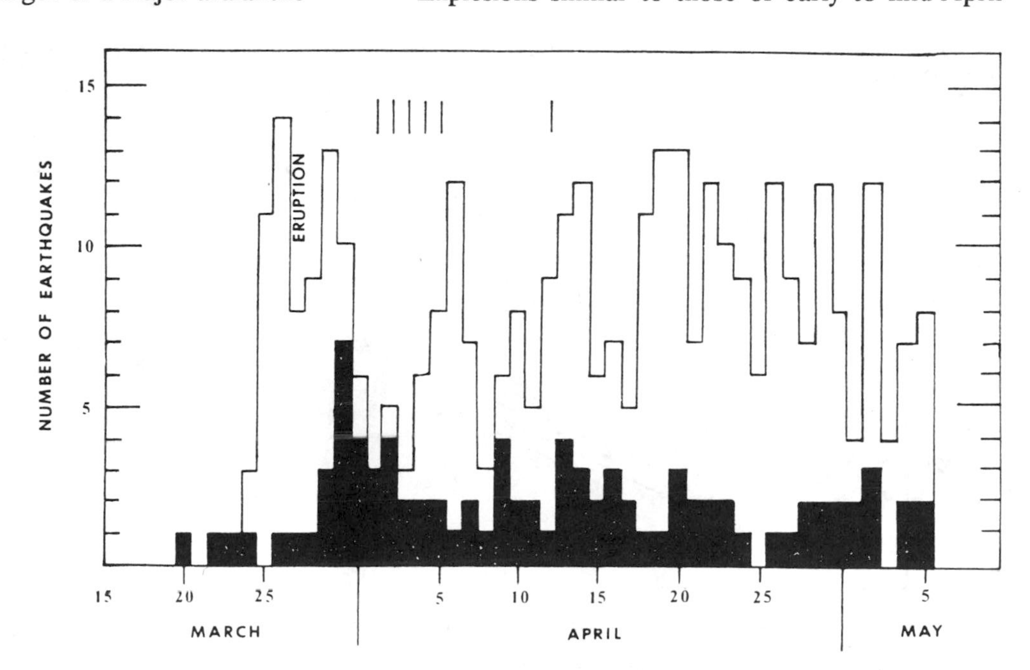

Fig. 12-3. Number of earthquakes per day in the Mt. St. Helens area with magnitudes > 3.4 (upper line) and > 4.0 (lower line, with area below shaded for emphasis) recorded at the Newport Observatory (360 km from the volcano) 20 March-3 May 1980. Short vertical lines above the earthquake graph indicate 1 or more periods of harmonic tremor on that day. Data provided by Robert Tilling and the Newport Observatory.

resumed [7] May and continued until 14 May. Several tens of earthquakes per day of magnitude 3 or greater continued to be recorded through 17 May. Total seismic energy release remained relatively constant through late April, then declined slightly.

The USGS began daily [geodetic] measurements on the periphery of the N flank bulge on 25 April, recording consistent outward [displacement] of 1.5-2 m/day through 17 May (fig. 12-5). The direction of movement was nearly horizontal, to the NNW.

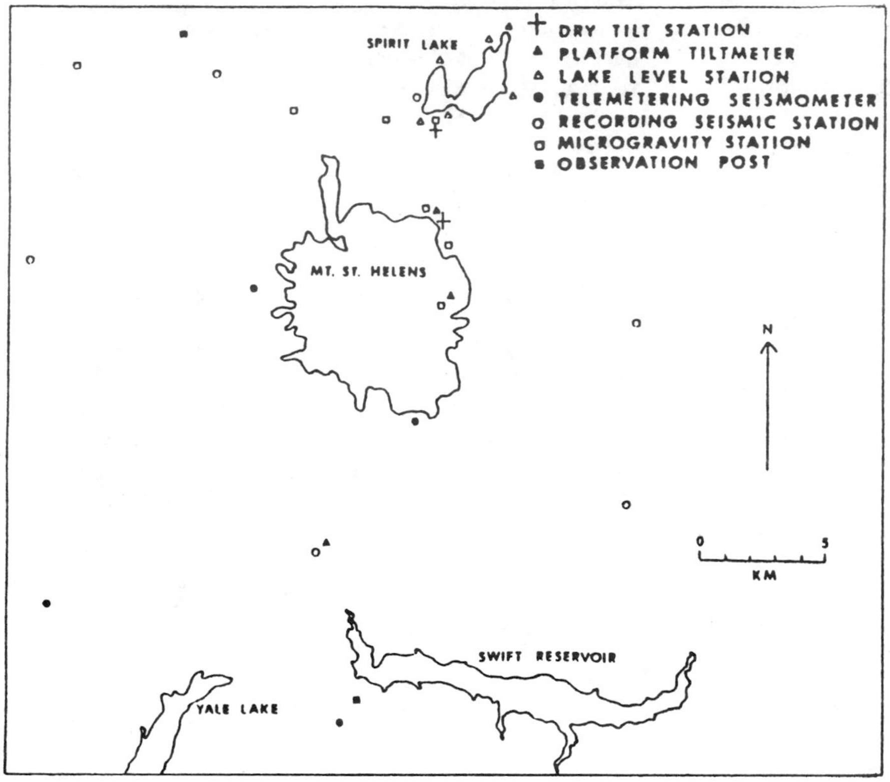

18 May Eruption. Much of the information on the 18 May eruption is from Robert Christiansen. His detailed narrative of the eruption will appear in the news section of *Nature*, [v. 285, p. 531-533.]

At 0832 on 18 May, seismographs recorded an earthquake of about magnitude 5 (its unusual wave characteristics prevented a straightforward magnitude calculation). A remarkable series of photographs by Vern Hodgson shows that the entire N flank bulge immediately began to separate from the volcano along a fissure that opened across its upper section. The bulge quickly formed a massive avalanche that raced downslope, displaced the water of Spirit Lake, and struck a ridge about 8 km to the N. Most of the avalanche material then turned W and flowed down the N fork of the Toutle River (the outlet of Spirit Lake).

. . . The mudflows destroyed 123 homes and most of the bridges crossing the Toutle River for tens of kilometers downstream, then continued down the Cowlitz River into the Columbia, where suspended sediment, logs, and other debris filled the ship channel, stranding many vessels in Portland harbor.

A powerful laterally-directed blast emerged from the area formerly occupied by the bulge and overtook the avalanche within seconds. The blast, carrying lithic ash and lapilli, devastated a zone extending 30 km E-W and more than 20 km outward from the volcano in an arc encompassing almost 180° of the N flank (fig. 12-6). Destruction was virtually total in an inner zone nearly 10 km wide, where no trees remained in the previously thickly forested area. Beyond the inner zone, all trees were blown to the ground, pointing outward from the source of the blast in a nearly uniform radial pattern. In the outer few hundred meters of the blast area, trees were seared but remained standing.

Almost simultaneously with the ejection of the lateral blast, a large vertical cloud rose rapidly from the pre-existing summit crater to more than 19 km above sea level (as measured by Portland airport's weather radar), passing through an unusually high tropopause at 13.5-14 km (fig. 12-7). Vigorous feeding of the vertical column continued for

Fig. 12-4. Monitoring equipment around Mt. St. Helens as of mid-April 1980. Data provided by Robert Tilling.

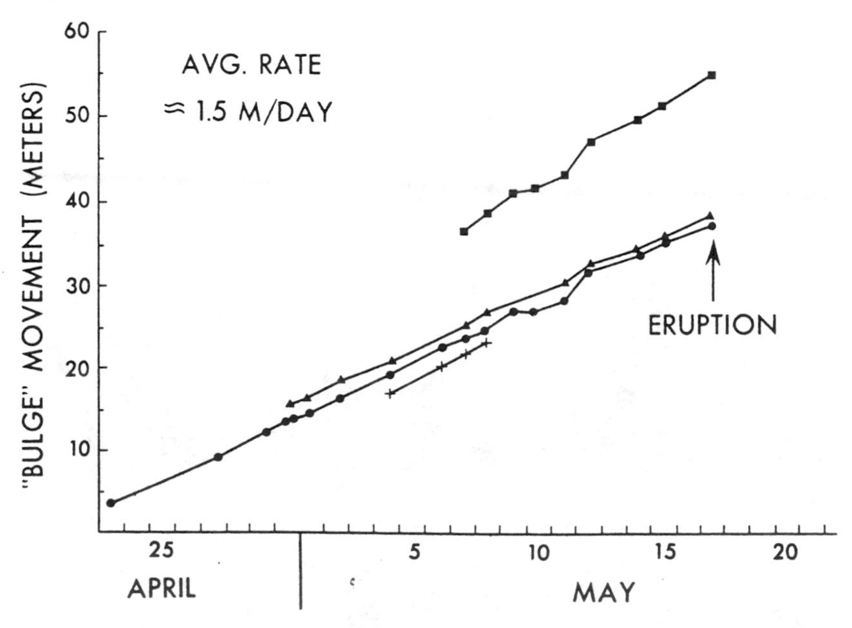

Fig. 12-5. Cumulative outward movement of 4 different points on the N flank bulge, 23 April-18 May 1980. Courtesy of Robert Tilling. [Originally from 5:6.]

more than 9 hours, before declining gradually during the late afternoon. Ash clouds moved rapidly NE and E. Large quantities of ash fell on a wide area of Washington, N Idaho, and W and central Montana. Ashfall at Ritzville, Washington, more than 300 km from Mt. St. Helens, totaled at least 7 cm (fig. 12-8). In Spokane, 500 km NE of the volcano, visibility was briefly reduced to only 3 m at about 1500. A trace of ash fell in Denver about noon the next day, and USGS hydrologists detected slight ashfall in parts of Oklahoma.

Pyroclastic flows, generated both by collapse of the vertical column and direct emission through the large northward breach produced by the directed blast, left a fan-shaped pumiceous deposit extending [into] Spirit Lake [and] the Toutle River, overlying debris flow deposits in that area.

Ash Cloud. NOAA weather satellite imagery clearly recorded the rise and dissemination of ash clouds, and at least 2 distinct major pulses of ash ejection (fig. 12-9). Measurements from the images showed that the cloud from the main explosion initially expanded in all directions, with the bulk of the ejecta moving E. [Measurements from satellite images indicated that the rate of horizontal advance of

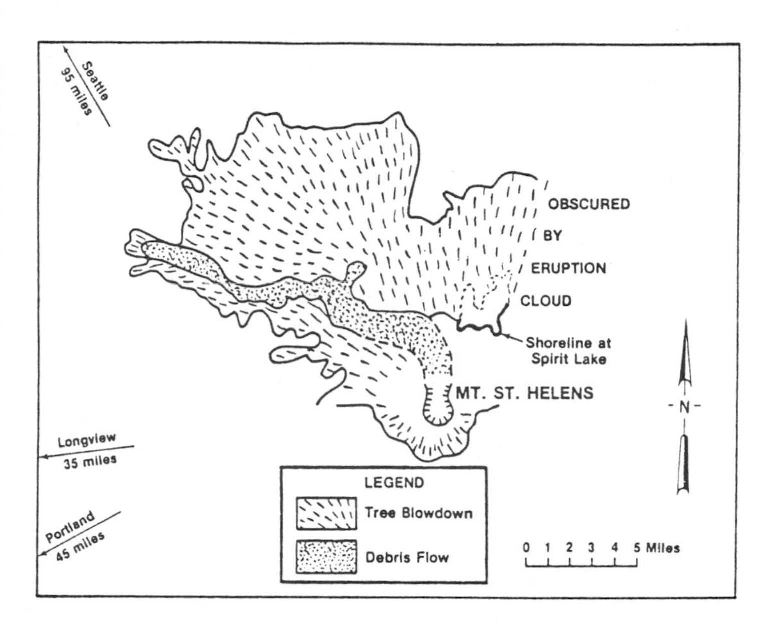

Fig. 12-6. USGS sketch map of Mt. St. Helens. The area devastated by the 18 May 1980 directed blast is shown schematically by sketching radial tree blowdown. The deposit left by the avalanching N flank is stippled.

Fig. 12-7. Oblique airphoto showing Mt. St. Helens erupting at about 1130 on 18 May 1980. View is approximately to the N. Courtesy of Austin Post, USGS.

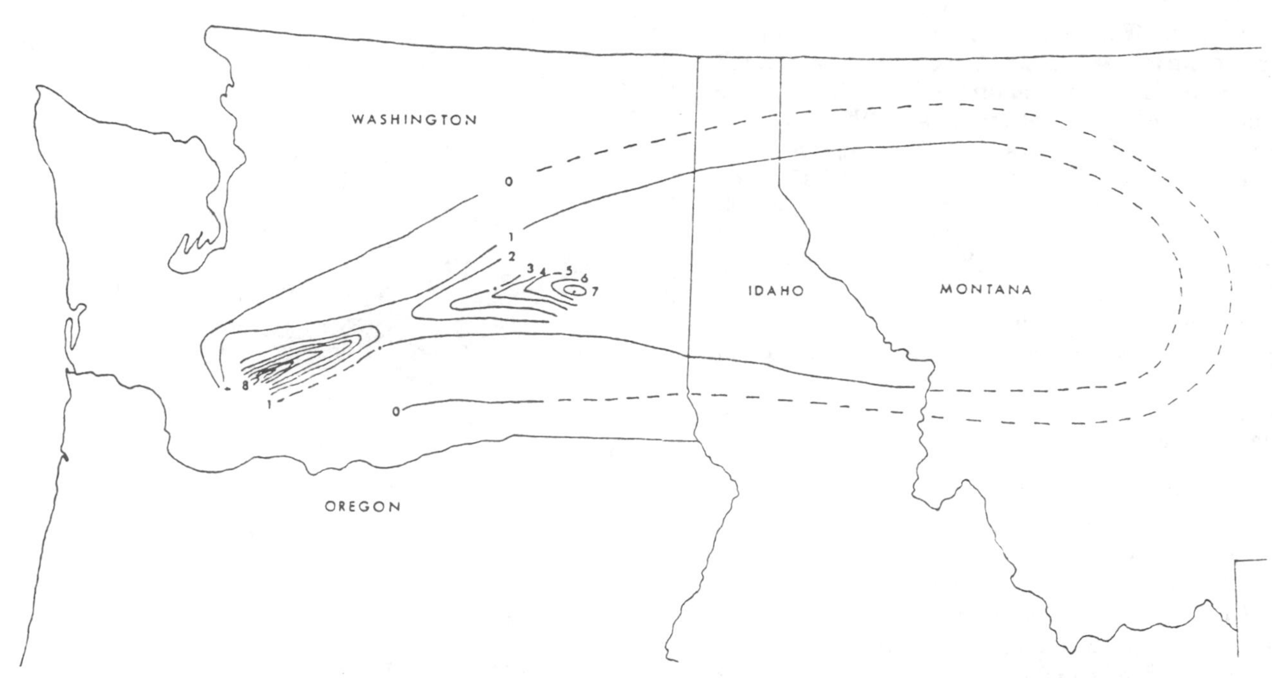

Fig. 12-8. Isopach map of ashfall from the 18 May 1980 eruption, prepared from data provided by Albert Eggers and the USGS. Thicknesses are in centimeters.

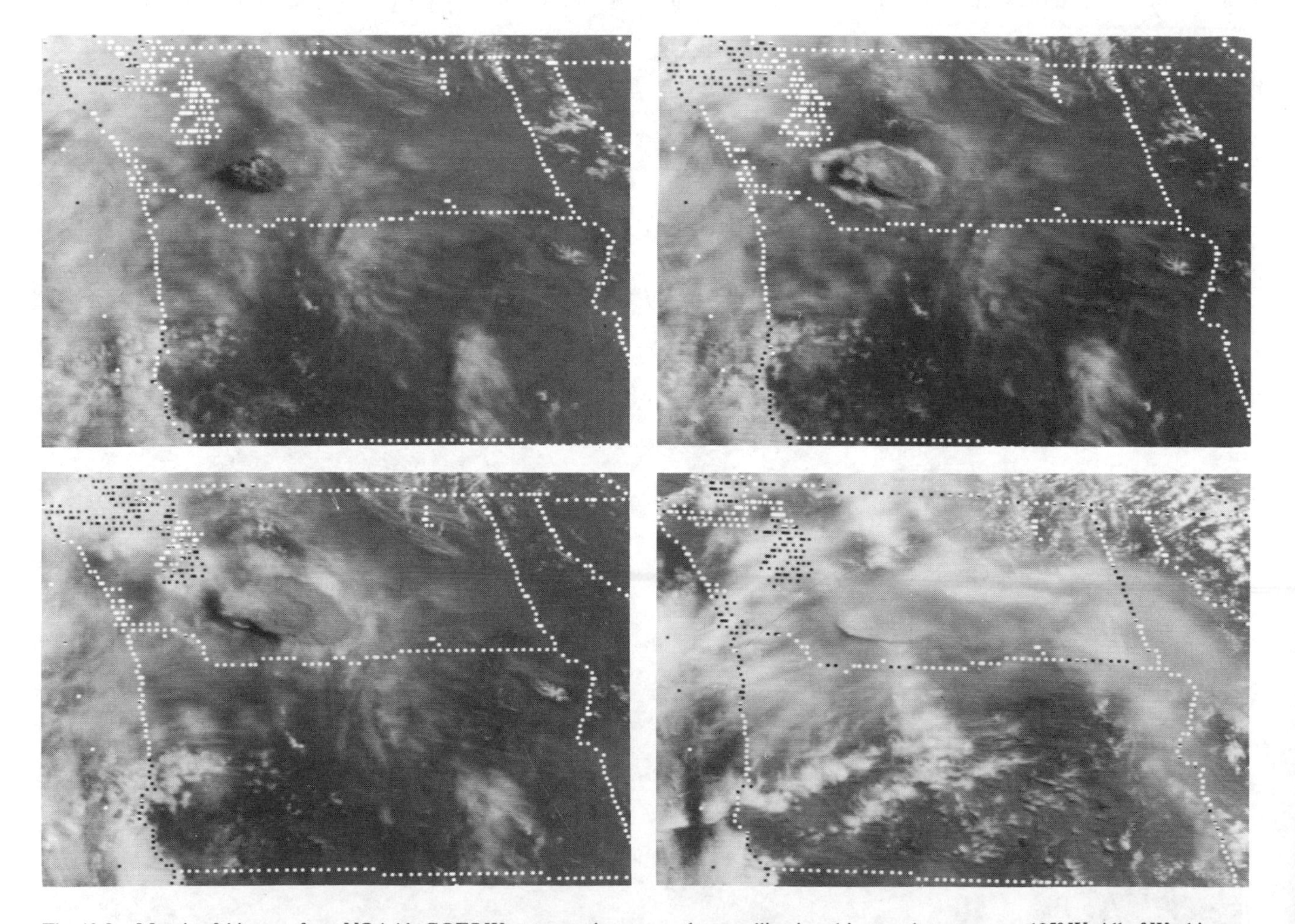

Fig. 12-9. Mosaic of 4 images from NOAA's GOES West geostationary weather satellite, in orbit over the equator at 135° W. All of Washington and Oregon, most of Idaho, and parts of Montana, California, Nevada, and S Canada are shown. Image times (all 18 May 1980) are: *upper left,* 0845 (13 minutes after the eruption began), *upper right,* 0915, *lower left,* 0945, *lower right,* 1315. The first 3 images show the rapid dissemination of the ash cloud after the initial explosion. The last image shows the eruption column from the second major burst, about 1 hour after it was ejected. Ash from earlier activity had spread to the Idaho-Montana border. Courtesy of Arthur Krueger and Andy Horvitz.

the cloud front averaged 250 km/hr for the first 13 minutes after the eruption's onset. Horizontal velocity soon decreased, remaining at about 100 km/hr for the first 1000 km of its dispersal to the ENE.] Portland airport reported wind speeds of only 120 km/hr toward the E at 12 km altitude. The second pulse could be seen on the image returned at 1215. [From an aircraft, D. Swanson observed that] the color of the column [gradually] changed from dark gray to pale gray [between about 1200 and 1220].

Ash was widely dispersed in the atmosphere because of varying wind directions at different elevations (fig. 12-10). Murray Mitchell reported that ash had made a complete circuit of the globe by 29 May. Most of the tropospheric material had fallen out by mid-June, but a diffuse dust veil remained in the stratosphere from the latitude of Mt. St. Helens N to the polar region. Bernard Mendonça reported that as of 9 June, NOAA's solar radiation and lidar equipment in Hawaii had detected no Mt. St. Helens ejecta. Seasonal arctic haze precluded observations from the Barrow, Alaska station. Stratospheric circulation patterns make aerosol movement to the S very unlikely before autumn.

High-altitude studies of the eruption cloud were carried out using aircraft from several NASA installations, including the Ames and Langley research centers; LANL; and NCAR. The pilot of the NASA Ames aircraft saw ash at nearly 23 km altitude while flying E of the volcano 18 May. NASA's SAGE satellite measured particle densities over S Canada on 22 May and the United States 23-27 May. "Ground truth" for the SAGE measurements was gathered by balloon from the Univ. of Wyoming and other locations. Most data from these studies have not yet been analyzed. However, Grant Heiken reported that preliminary results from the LANL aircraft, sampling at about 15 km, show recovery of 1-11 μm particles (the bulk of which were 3-4 μm) that were virtually all glass shards. William Smith will gather a series of reports for the Upper Atmospheric Programs Bulletin, published jointly by the FAA and NASA.

Pollution-monitoring equipment operated by the Alexandria, Virginia Health Department collected an unusual quantity of particulate matter during a rainstorm late 20 May. The 20 May rain was unusually acidic for that area, with a pH of 4-4.5, and occurred as Mt. St. Helens ash passed overhead.

Because of poor weather conditions, few brilliant sunrises or sunsets were reported in the United States after the 18 May eruption, although Grant Heiken was awakened by a gaudy, blood-red sunrise in Los Alamos, New Mexico. Charles Van Zant observed a ring around the sun from Cancún, México (on the Yucatán Peninsula) at about noon on 23 May. The ring filled about $^1/_4$ of the sky and was rainbow-colored at its edge.

Energetics. The acoustic pressure wave produced by the initial 18 May explosion was recorded on microbarographs operated by NOAA in Boulder, Colorado and Washington, DC, and on infrasound equipment operated by William Donn north of New York City at

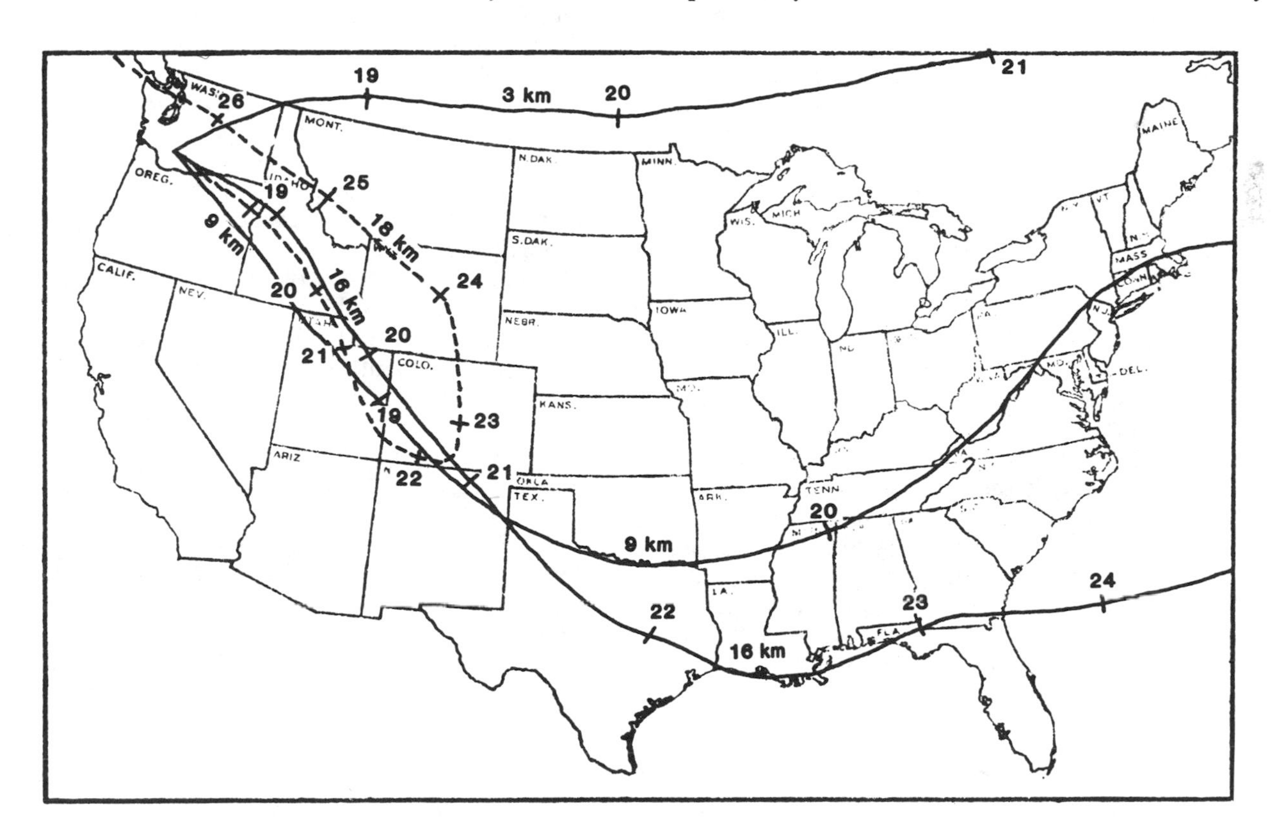

Fig. 12-10. Paths traveled over North America by 18 May 1980 ash at 3, 9, 16, and 18 km altitudes. Tick marks along each line show position of ash cloud front at that altitude every 24 hours, at noon GMT (0500 local time at Mt. St. Helens). The date at each tick mark is indicated. For clarity, the 18 km path is shown as a dashed line. The 21 May cloud front at 9 km altitude is just off the map to the E. Data provided by NOAA's Air Resources Laboratory.

LDGO. The waves recorded on these instruments were comparable to those generated by previous 10 megaton nuclear tests. Wave frequencies were very low, about 1 cycle per 5 minutes (0.003 Hz).

Morphologic Changes. The [debris avalanche], lateral blast, and vertical explosion created a crater, breached to the N, with a N-S dimension of about 3 km and an E-W dimension of about 1.5 km (somewhat wider at the base of the breach). The summit was destroyed. The maximum elevation of the volcano, on the crater rim, was about 350 m less than the previous summit altitude of 2975 m. The lower end of the breach extended downward nearly to the 1500 m level.

Volume. Volume calculations for the eruption were very preliminary, based on the size of the new crater and the amount of ash deposited. Most estimates are in the 1-2 km^3 range, but some are as high as 4 km^3. Comparison with previous eruptions in the VRF indicated that explosions of this size occur only about once a decade.

Socioeconomic Effects. At least 24 people are known to have been killed by the 18 May explosion, most by the avalanche and directed blast. Thirteen others known to have been in the area of maximum devastation at the time of the explosion are missing and presumed dead, including USGS geologist David Johnston (obituary at the end of this report). In addition, 37 persons believed to have been near the volcano on 18 May remain unaccounted for, bringing the probable death toll to 74. [Later estimates yield lower death tolls as the number of missing declined; the most recent data cited by Blong (1984) total 57 deaths.] State of Washington officials estimated the financial losses to private enterprise and state and local government to be at least $2.7 billion. [Blong (1984) tabulates about $1.5 billion in losses and cleanup costs.]

The following is from a report by R.J. Blong, who visited Yakima, Washington (140 km ENE of the volcano) 4 weeks after the 18 May explosion to study the socioeconomic effects of the tephra fall.

"Ashfall at Yakima from the 18 May eruption amounted to 12-18 mm. Automatic street lighting systems came on about 1115 and did not switch off until the following morning, although there was some lightening of the sky just before dusk. No deaths can be directly attributed to the ashfall. Two people died from cardio-pulmonary disease during the clean-up operation. Some people experienced headaches and gastroenteritis the day after exposure to the ash, the symptoms recurring in some of those involved in the cleanup. This may have been a stress reaction, but a virus with similar symptoms was prevalent in the area before the tephra fall. The cleanup operation created a great deal of camaraderie but some hostility developed where cars sped through streets stirring up the ash and where some roofs remained uncleaned. Anxiety but not depression developed.

"The Yakima airport and the airspace above was closed for 7 days. Twenty thousand tons of ash were removed from the 40 hectares of hard surface area by a team of up to 150 people working with 40 pieces of earthmoving equipment. The airport cleanup cost about $75,000 for equipment hire, labour (including overtime), and fuel.

"Pacific Power and Light experienced no problems meeting the peak demand when the darkness fell, probably because it was a warm day and a Sunday. Four or five older-style transformers caught fire. Subsequently, during a rainfall, several poles caught fire either as a result of lightning strikes or from shorting out by ash across contacts. Generally the wind removed the tephra from poles, insulators and transformers and there was less trouble than anticipated by the company.

"Pacific Northwest Bell telephone service experienced an unprecedented demand. The toll network was designed to handle with little delay the busiest days (Mothers Day and Christmas) but there were 70% more attempts to make calls than occurred on Mothers Day, the previous Sunday. The company also had to take steps to keep the tephra out of the electro-mechanical system. Maintenance on public phones has doubled since the tephra fall whereas no increase in maintenance has been necessary for semi-public phones.

"The cost of cleaning the ash from Yakima's approximately 350 km of streets has been estimated at $2-4 million. Downtown businesses suffered a serious loss of revenue through being closed for up to a week. The Greater Yakima Chamber of Commerce estimated on 30 May the cost of the ashfall at $95 million, of which equipment damage and automotive maintenance and repair amounted to $42 million. Motel occupancy rates were down from 80-90% to less than 50%. The tourist and convention industry will continue to suffer severely unless steps are taken to assure townsfolk and potential visitors alike that 20 years of Mt. St. Helens activity does not mean 20 years of 18 May and its aftermath."

Petrology. The following preliminary petrologic data are from William Melson.

"An early dark-colored and later light-colored tephra layer have been noted in airfall from the 18 May explosions (R. Kienle, personal communication). Near the volcano, the upper layer contains large essential ejecta of pumice. These products have been analyzed on the electron microprobe. Pumice from 3 separate localities provided by R. Kienle and Bruce Nolf have essentially identical compositions with the following average: SiO_2 = 63.35, Al_2O_3 = 18.38, FeOT = 3.87, MgO = 1.83, CaO = 5.22, K_2O = 2.01, Na_2O = 4.31, TiO_2 = 0.37, P_2O_5 = 0.09, Cl = 0.1-0.2, F and S $< 0.05\%$. Analyses of glass inclusions in plagioclase typically have consistently low sums, averaging about 93%, suggesting about 7% dissolved H_2O. Cl in these is typically around .15% and F and S less than .05%. The dominant phenocryst in the pumice is plagioclase (average An 47-70), which makes up $\sim$35% by weight. Accessory phases, in order of abundance, are hornblende, hypersthene (Fs 35), and subequal amounts of titan-magnetite and ilmenite.

"Tephra from the lower dark layer of the 18 May explosion layer is similar in composition to the pre-18 May tephra, reflecting the presence of material derived largely from the explosive destruction of the central part of the cone. The upper, lighter colored tephra contains much more pumiceous material, visible megascopically and revealed in bulk analyses, reflecting, finally, the breaching of hornblende hypersthene dacite magma. Bulk analyses of the light-colored tephra show an increasing concentration of glass with distance from the volcano.

"Pumiceous tephra from the 25 May explosion (see below) contains accessory augite, according to C. A. Hopson, a phase absent or extremely rare in the 18 May pumice. It appears that the explosions are successively tapping deeper, less water-rich portions of a zoned magma chamber."

Seismicity. Preliminary analysis of seismic and deformation data indicates that there was no immediate warning of the imminence of a large explosion. After the magnitude 5.0 earthquake that apparently triggered the eruption, a brief period of harmonic tremor was recorded, followed by the absence of any earthquakes with magnitudes greater than 2 until about 1145. Seismic activity increased rapidly after 1145, and almost continuous magnitude 3.5-4 seismicity was recorded from 1400 to 1630 at the USGS Newport Geophysical Observatory. After 1630, seismicity declined.

Tilt. Records of the only surviving tiltmeter, on the S flank, show that rapid inflation began at the same time as the explosion at 0832. Rapid inflation lasted only 10 minutes, succeeded by deflation that continued until about 1630. Moderate inflation then began and has continued.

Post-18 May Activity. Eruptive activity declined after 18 May, and by the 21st was limited to episodic ejections from the crater, mostly of vapor. Large fumaroles and secondary explosions were generated from the debris flow deposit, occasionally producing columns of material as high as 2 km. Between 19 and 24 May, only a few earthquakes with M > 3 were recorded, in contrast to the several tens of events that had occurred each day since late March (fig. 12-11). However, harmonic tremor began during that period (exact date not known to SEAN).

At 0232 on 25 May the amplitude of harmonic tremor began to increase. Within minutes, an ash-rich eruption column had been seen from a surveillance aircraft. By 0245, NWS radar at Portland recorded the top of the column at nearly 14 km. A swarm of small earthquakes, centered about 8 km beneath the volcano, began at 0249 and continued at a rate of 1-2/hr.

The density of ash in the eruption column started to decrease within 5 minutes, and the height of the column was declining within the first hour. Winds were quite variable, but much of the ash blew toward the W half of the compass. By 0600, ash was falling in the Portland-Vancouver area (80 km SW). Ashfall darkened the early morning in the Kelso-Longview area (55 km W) and the ash cloud extended as far as the Olympic Peninsula of NW Washington. Heavy rain during the eruption mixed with the ash to drop mud on much of the affected region. Many airports were closed and ground travel was difficult.

By 0800, harmonic tremor amplitude had declined and the earthquake swarm had begun to subside. However, the eruption continued through most of the day, with the altitude of the top of the column ranging from 4 to 6 km. The eruption declined during the evening, and activity was limited to emission of steam clouds containing varying amounts of ash by 0100 the next morning.

Most of the tephra ejected 25 May was juvenile material. Some pyroclastic flow deposits were emplaced on the N flank. H.H. Lamb's preliminary estimate for the total Dust Veil Index from the 18 and 25 May eruptions is 600-1500.

The ash content of the plume declined during the next several days, and by late 28 May, the plume was entirely composed of vapor. Small incandescent areas were seen on the crater floor during the night of 28-29 May and on several occasions thereafter. Careful inspection showed that the incandescence was caused by the heating of parts of the crater floor by venting gases, not the presence of magma at the surface. Vigorous steaming continued through early June, with vapor rising to about 3.5 km altitude. SO_2 content of the plume continued to be an order of magnitude greater than before the 18 May eruption. An L-shaped lake about 1 km across was observed in the crater 10 June [but see 5:6], away from the area of active steam venting. Harmonic tremor continued, at varying amplitudes, through early June, but earthquake activity remained at very low levels.

At press time, a third large explosion occurred. An Eastern Airlines pilot observed the ejection of a dense ash column at 2045 on 12 June. The cloud blew S and SW, dropping marble-sized tephra on Cougar (18 km

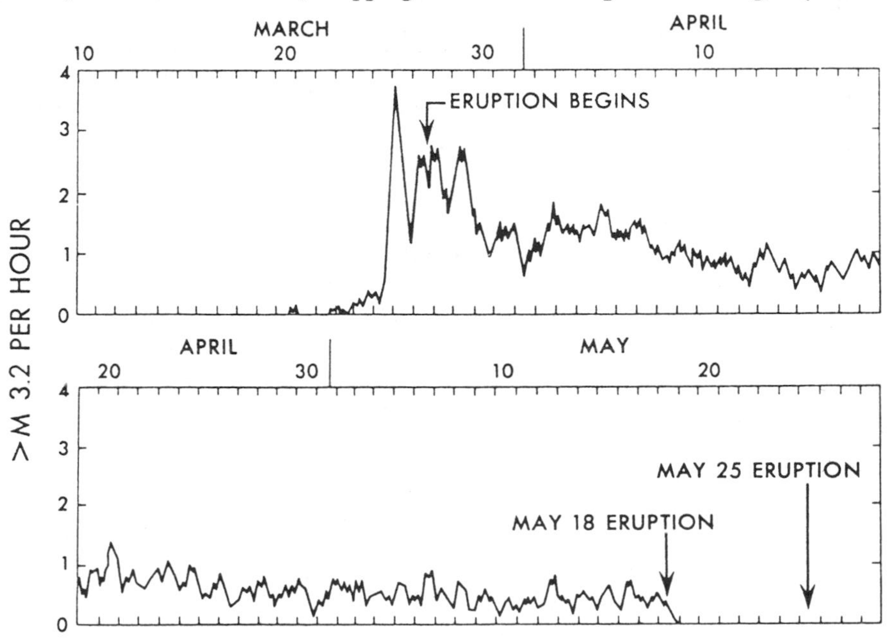

Fig 12-11 Number of seismic events per hour with magnitudes > 3.2, 20 March 28 May 1980. Courtesy of Robert Tilling.

SW of the pre-18 May summit). Ashfall began in Portland (about 80 km SW) by 2300, and more than 1/2 cm accumulated. Portland airport radar recorded pulsating echoes to altitudes of 10.6 to nearly 17 km. New bursts of ash were observed at about 2-minute intervals from a USFS monitoring aircraft. An earthquake of about magnitude 4.0 was recorded by Univ. of Washington seismographs at 2110. About 1500 people were evacuated without injury from a designated danger zone within about 30 km of the volcano. Ash and accompanying rain made roads in NW Oregon and SW Washington muddy and treacherous. Portland airport was closed. After the explosion, on 15 June, the presence of a growing lava dome in the center of the crater was confirmed by the USGS.

Historic Activity. Mt. St. Helens was last active between 1831 and 1857, when a series of eruptions were separated by intervals of up to 7 years. Most were small explosive events, and none approached the size of the 18 May activity. Crandell and Mullineaux (1978) describe a 4000-year-old eruption, [considerably greater than] that of 18 May, in their definitive paper. Pyroclastic flows and mudflows extended more than 30 km down the Toutle River, and more than 20 cm of tephra fell tens of kilometers NE of the volcano.

> Very few volcanologists throughout history have lost their lives by eruption, but last year Robin Cooke and Elias Ravian were killed at Karkar and now we must report the death of David Johnston at Mt. St. Helens. At the time of the 18 May eruption, Dave was monitoring the volcano from a position just 8 km NNW of the summit. No one knew better than Dave the risk involved in his St. Helens work, and no one contributed more to the understanding of this volcano's eruptive mechanisms. Although only 30 years old, his PhD work on Augustine, and subsequent work with the USGS had already established his position among the leading young volcanologists in the world. His enthusiasm and warmth will be missed at least as much as will his scientific strength.

Information Contacts: R. Christiansen, USGS, Menlo Park, CA; R. Tilling, USGS, Reston, VA; D. Mullineaux, D. Crandell, USGS, Denver, CO; A. Krueger, NOAA/NESS; M.P. McCormick, NASA Langley Research Center; W. Smith, FAA; G. Heiken, LANL; M. Mitchell, NOAA, Silver Spring, MD; R. Dalton, Alexandria, VA Health Dept.; B. Mendonça, L. Machta, NOAA/ARL; W. Donn, Lamont-Doherty Geological Observatory; C. Hopson, Univ. of California, Santa Barbara; B. Nolf, Central Oregon Community College; R. Blong, MacQuarie Univ.; A. Eggers, Univ. of Puget Sound; D. Dzurisin, USGS HVO, HI; H.H. Lamb, Univ. of East Anglia; W.G. Melson, SI; S. Malone, R. Crosson, E. Endo, Univ. of Washington; Newport Geophysical Observatory; C. Van Zant, Richardson, TX; UPI; *New York Times*.

Further Reference: Blong, R.J., 1984, *Volcanic Hazards: A Sourcebook on the Effects of Eruptions*; Academic Press, Sydney, 424 pp. [See also references following 5:12.]

6/80 (5:6) After the 12 June explosion, a lava dome began to grow in the crater, accompanied by occasional weak bursts of harmonic tremor. No significant explosions have taken place since the start of dome emplacement, nor has the dome growth generated any nuées ardentes.

Richard Janda calculated that about 2 km^3 of debris was emplaced in the N fork of the Toutle River and Spirit Lake by the 18 May eruption. Richard Waitt estimated the total 18 May airfall component as 1.1 km^3 of unconsolidated material, with a mean bulk density of 0.5 g/cm^3. Poor weather has prevented the aerial photography necessary for the production of a topographic map of the post-18 May volcano. After a map is produced, the amount of material removed from the edifice on 18 May can be accurately determined.

After the 25 May eruption most of the few individual earthquakes recorded on local seismographs were centered 10-15 km N of Mt. St. Helens and were thought likely to be related to regional crustal stresses rather than volcanic activity. Harmonic tremor continued, increasing in amplitude early 3 June, but dropping to a very low level by 5 June before ending shortly thereafter. A persistent steam column rose from the crater to altitudes of 3-4.5 km in early June. A small amount of ash was occasionally included in the lower portion of the plume. Improved visibility on 10 June revealed that a roughly crescent-shaped lake, about 300 m long and 100 m wide, had formed on a portion of the [S] part of the crater floor within the past 5 days. Evidence of recent rock avalanches from the crater walls was also visible.

At about 1300 on 12 June, Steven Malone noted the onset of weak harmonic tremor and an increase in small discrete events on seismographs operated by the Univ. of Washington and USGS. A marked increase in harmonic tremor amplitude occurred at 1905 and an eruption plume rising to 4 km altitude was observed 5 min later. Portland airport radar tracked the eruption column to 11 km altitude shortly thereafter. Eruptive activity fluctuated in intensity for the next 2 hours. Harmonic tremor continued at roughly the same amplitude until about 2000, when it declined to a low level. A dramatic increase in tremor amplitude at 2111 was accompanied by an eruption column that quickly reached almost 16 km altitude, producing pulsating echoes on Portland airport radar. An explosion was heard more than 200 km away in Toledo, Oregon. Activity remained vigorous until about midnight, when cloud height and tremor amplitude began to decline. By 0200, the altitude of the top of the plume had dropped to about 4.5 km.

Infrared images returned by NOAA's GOES West weather satellite show the eruption cloud drifting SW. Marble-sized pumice and ash fell on Cougar accompanied by a strong sulfur odor. Ashfall in Portland began about 2300, reaching accumulations of around 1/2 cm. Infrared imagery shows that the cloud reached the Pacific coast of N Oregon by 2330 and continued to drift SW for several more hours, before becoming impossible to distinguish from heavy weather clouds in the area. Ash fell on more than 11,000 km^2 of NW Oregon and SW Washington, reaching Lake Oswego, Oregon, 110 km SW of Mt. St. Helens.

About 1500 people were evacuated from a zone within about 30 km of Mt. St. Helens. No deaths or injuries were reported near the volcano. Portland airport was closed for several hours during the night, but few flights were affected. Rain before and after the ashfall in the Portland area turned the ash into a slippery mud, as it did to much of the 25 May ash. However, blowing ash continued to cause occasional disruptions to travel more than a week after the eruption.

Several pumice and ash flows were emplaced on the N flank by the 12 June eruption, one stopping roughly 20 m from the shore of Spirit Lake, which was still partially covered by floating debris. The deposits ranged from 2 to 10 m in thickness, and internal temperatures as high as 600°C were measured.

The explosions of 25 May and 12 June were of similar size. Both sent most of their ash toward the heavily populated areas W of the Cascade Range, although analysis of wind directions at various altitudes indicates that prevailing winds blow toward the E half of the compass almost 90% of the time (Crandell and Mullineaux, 1978).

On 15 June, USGS personnel confirmed the presence of an active lava dome, roughly 200 m in diameter and 40 m high, at the bottom of the crater. Water vapor and other gases rose from a shallow lake surrounding the dome. Red glow was visible at night through cracks in the dome's surface. Poor weather often made observations of the dome and its rate of growth difficult. Clear weather on 18 and 19 June revealed that the dome was growing upward about 6 m/day, reaching a height of 65 m by the 19th. Harmonic tremor had stopped by 15 June and did not resume until 2 episodes of very weak tremor, lasting 30 and 65 minutes, were recorded during the night of 24-25 June. Sporadic tremor continued through late June, but very few discrete earthquakes were recorded. Clear visibility 28-29 June indicated that no vertical dome growth had occurred since 19 June. The dome was about 200 m in diameter and was crossed by a 12 m-long fissure in which cherry red incandescence could be seen.

Information Contacts: R. Christiansen, D. Peterson, R. Waitt, USGS, Menlo Park, CA; R. Tilling, USGS, Reston, VA; D. Mullineaux, D. Crandell, USGS, Denver, CO; R. Janda, USGS, Tacoma, WA; A. Krueger, NOAA/NESS; S. Malone, R. Crosson, E. Endo, Univ. of Washington; UPI; AP; *New York Times*.

7/80 (5:7) An eruption similar to those of 25 May and 12 June destroyed the June dome on 22 July. After a slightly smaller eruption on 7 Aug, growth of a new dome began.

Activity was weak from late June through mid-July. A vapor plume rose to a maximum of 3.5-4 km altitude. SO_2 emission, measured by remote sensing equipment, ranged from 1800 to 2600 t/d. No lava dome growth was detected.

After several episodes of harmonic tremor on 27 June, weaker than the tremor preceding any of the eruptions, seismicity recorded by the USGS-Univ. of Washington seismic net was limited to a few small shallow earthquakes per day. In an 11-minute period on 6 July, 4 earthquakes of magnitude 2.0-3.2 occurred about 8 km below Marble Mountain, 14 km SE of Mt. St. Helens. The first of these shocks was recorded only 17 minutes after the start of an earthquake swarm beneath Mt. Hood, 100 km SSE of Mt. St. Helens (see Mt. Hood section). Six events of magnitude 1.4-2.9, centered about 17 km SE of Mt. St. Helens, were recorded in 13 1/2 hours 19-20 July. A few hours later, one seismograph recorded 50 minutes of very weak harmonic tremor, the first since 27 June. On the day of the eruption, 22 July, shallow events centered under the N flank began at about 1000. Four earthquakes were recorded between 1400 and 1500, nine from 1500 to 1600, and 20 from 1600 to 1700. An increase in SO_2 emission just prior to the eruption caused a sharp drop in the CO_2/SO_2 ratio after a steady increase over the preceding week.

The initial tephra ejection phase began at 1714 and lasted 6 minutes. NWS radar at Portland airport measured the top of the eruption cloud at slightly less than 14 km. Clear weather allowed the cloud to be seen from as far away as Seattle, Washington and Salem, Oregon, both about 150 km distant. An overflight by Donald Swanson revealed that the activity had blown a hole in the lava dome that had begun to grow in the crater shortly after the 12 June eruption, but had not destroyed it.

A second, larger, tephra cloud emerged from the volcano at 1825 and was fed vigorously for 22 minutes, destroying the lava dome. Pyroclastic flows moved down the N flank, nearly reaching Spirit Lake. A NOAA weather satellite infrared image at 1845 (2 minutes before the end of this active phase) yielded a temperature of -52°C at the top of the column, corresponding to an altitude of about 13 km (fig. 12-12). Portland airport radar recorded a maximum cloud altitude of more than 18 km during this phase. Arthur Krueger estimates the base of the stratosphere to have been at about 16 km altitude.

The final series of eruptive pulses began at 1901, lasting about 2 hours and 40 minutes. The highest cloud ejected during this period reached almost 14 km at 1907 and pyroclastic flows were most numerous around that time. NOAA weather satellite images show that feeding of the eruption column had declined substantially by 2115, and that the ash plume was definitely detached from Mt. St. Helens on the next image, at 2145. The plume moved NE across Washington into British Columbia, Canada, remaining visible on the images until about 0100, when it merged with a cold front over W Alberta. Ash fell as far away as Montana and Alberta. Ashfalls were light, the thickest reported at Colville, Washington (nearly 450 km NE of Mt. St. Helens) where 1.5 mm were measured.

The USGS informed federal, state, and local government agencies of the earthquake swarm during the early afternoon of 22 July. People working on the flanks of the volcano, including about 120 USFS personnel, were evacuated before eruptive activity began. No casualties were reported.

Gary Heckman reports that a class M-8 solar flare that occurred on 22 July began 37 minutes after the volcano's first explosion (at 1751), peaked at 1803, and ended at 1909. Class M-8 flares, near the top of the intermediate intensity scale, normally occur about once a month. However, solar activity is presently near the peak of its 22-year cycle and intense flares have become more frequent.

Aerial observers reported glowing red areas in the crater during the night of 22-23 July. Overflights the next morning revealed a new crater, about 250-300 m in longest dimension and 100 m deep, in the area formerly occupied by the lava dome. A thick blanket of ash surrounded the new crater, which emitted a small vapor column similar to those of mid-July. SO_2 emission dropped to about 800 t/d on 23 July, but returned to mid-July levels the following day. Temperatures of up to 705°C were measured at 1.5 m depth in the new pyroclastic flows on 24 July, while surface temperatures were 70-80°C.

On 28 July at 0608, a little ash rose slowly from the new crater to 3.5 km altitude. Harmonic tremor accompanied the activity, which lasted about 20 minutes. No unusual pre-eruption seismicity was recorded, but glow was observed a few hours earlier from a USFS monitoring aircraft.

SO_2 emission dropped to 700 t/d on 28 July and did not exceed 1100 t/d through the end of the month. Red glow continued to be visible in the crater as of early Aug. The USGS attributed the glow to superheated rock, not a new lava dome.

At press time, another eruption was followed by renewed lava dome growth. At about noon on 7 Aug, volcanic tremor began, increasing in intensity as a magnitude 2.3 earthquake occurred at 1238 in the vicinity of Marble Mountain, site of earthquakes in early and mid-July. Remote sensing of CO_2 and SO_2 gas emissions showed that the CO_2/SO_2 ratio had dropped from 12-14 on 6 Aug to about 3 on 7 Aug. However, the lower ratio was caused by a decrease in CO_2 rather than the increased SO_2 emission that caused the similar ratio decline before the 22 July eruption. Because of the changes in seismicity and gas emission, the USGS withdrew its personnel from the area closest to the volcano and notified the USFS Emergency Coordination Center that an eruption might be imminent. Other persons working on the volcano's flanks were evacuated, including fire fighters battling persistent smoldering debris in the 18 May eruption deposits.

Tremor intensity continued to increase, and another magnitude 2.3 earthquake was recorded in the Marble Mountain area at 1458. An eruption began at 1623, producing an ash-laden column that rose quickly to about 13.5 km. A small pyroclastic flow moved down the N flank, leaving a thin deposit, but did not reach Spirit Lake. At 1910, an increase in tremor amplitude coincided with a series of pulsating ash ejections that sent clouds to a maximum altitude of less than 7 km. Seismic and eruptive activity waned briefly about 2200, then a stronger explosion at 2232 sent tephra to 11 km altitude, accompanied by more tremor. Variable winds sent ash over a wide area, but ashfalls were very light. Ash was reported in Grays Harbor County, about 150 km NW; Mt. Vernon, 250 km N; Wenatchee, 200 km NE; and Portland, 80 km SW. Small pulses were seen on the USGS monitoring camera the next day at 0935, 1007, and 1734. Ash rose to about 3.5 km altitude as surrounding steam columns reached nearly 5 km. A new lava dome was visible in the vent on 8 Aug, with its top about 40 m below the vent rim. About 8 m of vertical growth appeared to have taken place by 9 Aug, but no further growth was evident on 11 Aug.

Tracking of the 18 May eruption cloud continues at several locations worldwide. Ronald Fegley reported that NOAA's lidar equipment at Mauna Loa, Hawaii (19.5°N, 155.6°W) detected a significant volcanic layer at 18.4 km altitude on 15 July. Weaker layers had been detected on 10 and 13 June and 2 and 8 July at similar altitudes.

M.P. McCormick provided the following informa-

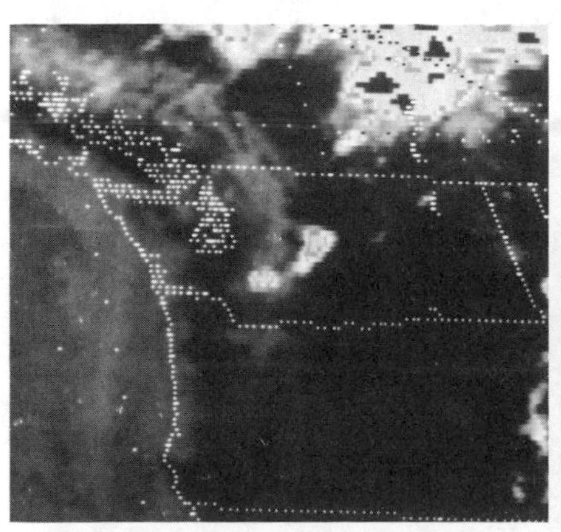

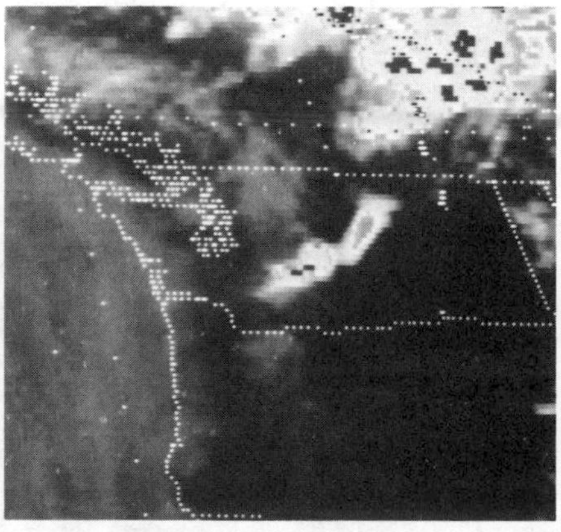

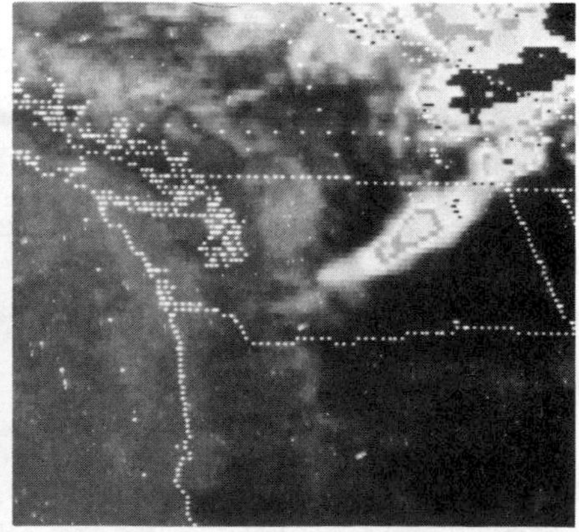

Fig. 12-12. Series of 3 enhanced infrared images returned 22 July 1980 by NOAA's GOES West geostationary weather satellite. *Left,* image at 1845 shows the cloud from the explosion at 1714 drifting NE as the column from the explosion that began at 1825 rises from the volcano. A temperature of -52°C at the top of the latter column was calculated from this image, corresponding to an altitude of about 13 km. *Center,* image 1 hour later, at 1945, shows material from the eruptive pulses that began at 1901 rising from the volcano, as 2 distinct earlier clouds move NE. *Right,* image at 2145 shows that feeding of the cloud has nearly ended, and the distal end of the plume has crossed into Canada at the NE corner of Washington. Courtesy of Arthur Krueger and Andy Horvitz.

tion. On 4 June, the ground-based lidar at NASA's Langley Research Center in SE Virginia (37.1°N, 76.3°W) first detected a persistent layer at 18 km altitude (no data were obtained 30 May-3 June). H. Jäger and R. Reiter observed layers near 12 km altitude (slightly above the local tropopause) over Garmisch-Partenkirchen, Germany (47.5°N, 11.0°E) on several days beginning on 25 May. Material extending from 13 to 15.5 km altitude was detected there on 12 June. Beginning 11 June, G. Visconti and G. Fiocco reported material between 15 and 20 km, and possibly at 14 km, over L'Aquila, Italy (42.4°N, 13.4°E). M. Fujiwara of Kyushu University, Japan (33.6°N, 130.4°E) reported a layer on 3 June just below 15 km, clearly higher than the local tropopause. NASA's SAGE satellite will move through the mid-latitudes of the Northern Hemisphere in Aug and will collect data on ejecta remaining in the atmosphere. H.H. Lamb reports that he observed a colored ring, which he believes to be an example of Bishop's Ring, around the sun at sunset on 12 June at Norwich, England (52.6°N, 1.3°E).

Issue No. 80-3 of the *Upper Atmospheric Programs Bulletin*, published jointly by NASA and the FAA, contains 10 pages of reports of high-altitude atmospheric research and satellite study of the 18 May eruption cloud.

JMA reports that many of their stations, from the NE to the SW ends of Japan, recorded the pressure wave from the 18 May explosion on microbarometers. Arrivals ranged from 1447 to about 1628 (local time at Mt. St. Helens) on 18 May. Calculated apparent wave velocities ranged from 305 to 309 m/sec.

Strong electrical effects were associated with the heavy ashfall of 18 May. Rocke Koreis reported observations of ball lightning at elevations from near ground level into the ash cloud over Yakima, Washington (140 km ENE of the volcano). Strong thunder continued throughout Yakima's ashfall, which was entirely dry and lasted from 1115 to 2300 on 18 May. A distinct second pulse of ash was noted in late afternoon. Koreis also reported that a light plane fleeing the initial 18 May explosion at an air-speed of more than 270 km/hr was briefly being overtaken by the eruption cloud moving at an estimated 320 km/hr.

Information Contacts: R. Decker, USGS HVO, Hawaii; R. Tilling, C. Zablocki, USGS, Reston, VA; D. Swanson, R. Christiansen, USGS, Menlo Park, CA; A. Krueger, NOAA/NESS; S. Malone, R. Crosson, E. Endo, Univ. of Washington; M.P. McCormick, NASA Langley Research Center; R. Fegley, NOAA/ERL; H. Jäger, Fraunhofer-Institut für Atmosphärische Umweltforschung; G. Visconti, Univ. dell'Aquila; G. Fiocco, Istituto di Fisica dell'Atmosfera; M. Fujiwara, Kyushu Univ.; H.H. Lamb, Univ. of East Anglia; Seismological Division, JMA; W.S. Smith, FAA; R. Koreis, Yakima, WA; AP; UPI.

8/80 (5:8) USGS personnel measured temperatures in the 7 Aug pyroclastic flow deposits of 647°C near the crater and 639° near their distal margin the day after emplacement. By 8 Aug a lava dome had filled the inner crater formed during the 22 July explosions to about half its former depth of almost 100 m. More than 20 m of additional dome growth had taken place by the morning of 9 Aug. Occasional bursts of vapor and ash rose to 3-6 km altitude on 8 Aug, accompanied by small seismic events. SO_2 emission increased from about 900 t/d on 8 Aug to at least 2000 t/d on the 9th, but explosive bursts had ended and seismic activity was very low.

During the next several days, the rates of CO_2 and SO_2 emission fluctuated substantially (table 12-1). The dome appeared to have risen slightly between 9 and 11 Aug, but no growth has been observed since. The deformation monitoring line between the crater and a ridge W of Spirit Lake shortened about 3 cm/day 8-13 Aug, a rate typical of previous inter-eruption periods since measurements resumed in mid-June. No significant changes in ground tilt around the base of the volcano have been recorded since June.

On 15 Aug at 1437, an ash-rich cloud rose to about 1 km above the volcano. The cloud became gradually less ash-laden, and dissipated after less than 15 minutes. Volcanic tremor was recorded during the eruption, declining as the eruption waned, but there was no premonitory seismicity. The activity blasted a small crater in the W side of the dome but did not destroy it. A similar eruption had occurred 28 July.

The surface of the lava dome was about 7 m lower on 17 Aug than it had been before the 15 Aug explosion. No shortening of the deformation monitoring line occurred between 13 and 17 Aug, but shortening resumed 19 Aug, indicating renewed inflation. Seismicity was limited to infrequent very small shallow events, many of which probably represented rockfalls in the crater.

On 22 Aug, a small quantity of water from Maratta Creek, a tributary of the North Fork of the Toutle River, breached a portion of the nearly 30 km-long debris dam left in the Toutle by the 18 May eruption. Water and debris flowed about 3 km downstream where it formed a small lake with an estimated volume of 3.8×10^5 m^3. Five days later this lake overflowed, moving nearly 10

	CO_2 (t/d)	SO_2 (t/d)	CO_2/SO_2
Aug 10	3100	600	5.2
11	5100	900	5.4
12	2100	650	3.2
13	19,000	3400	5.6
14	8700	1600	5.4
15*	2400	800	3.2
16	<3000	≈1000	<3.0
17	4200	1500	2.8
19	3700	1300	2.8
20	3300	1900	1.7
21	6900	2600	2.7
22	11,000	2000	5.5
23	5500	1800	3.1
24	6800	1250	5.4
25	2100	520	4.0
26	3900	1400	2.8
29	6000	1000	6.0

*Not including gas released by the 15 Aug explosion.

Table 12-1. Daily rates of CO_2 and SO_2 emission from Mt. St. Helens, measured in t/d by remote sensing equipment. No data are available for 18, 27, and 28 Aug.

km down the Toutle valley to Camp Baker, site of a partially completed dam project designed to control the much larger floods that could occur in the debris-clogged valley when heavy rains resume in the autumn.

Some equipment was damaged at the construction site. Much of the water was contained by the unfinished check dam, but some continued about 15 km farther to the town of Toutle, damaging or destroying some temporary bridges and access roads. No casualties were reported.

Mt. St. Helens remained quiet through early Sept. Gas emission continued to fluctuate but no explosions occurred. Deformation monitoring lines began to lengthen slightly on 25 Aug, indicating slight deflation of the volcano, but an average of 2 cm/day of contraction by early Sept showed a return to more typical gradual inflation. On 30 Aug, the lava dome was the same size as on 9 Aug. Incandescence could be seen through deep cracks in its surface and through the small crater formed by the 15 Aug explosion. Cracks in the walls and floor of the inner crater containing the dome also revealed incandescent material. Seismicity remained generally weak in late Aug and early Sept. A brief earthquake swarm on 4 Sept began with a magnitude 2.5 event at 2046, followed by 4 shocks within the next 9 minutes. All were centered 2-4 km beneath an area about 8 km NNW of Mt. St. Helens and were believed by the USGS to be of tectonic origin.

Monitoring of material ejected into the stratosphere by the 18 May eruption continues. Results from NASA's lidar at Wallops Island, Virginia show that the layer at 18 km had become more diffuse by Aug, occupying a zone between 16 and 20 km. In Tucson, Arizona, Aden and Marjorie Meinel continued to observe a weak layer at this altitude during Aug sunrises and sunsets. A NASA P-3 aircraft will collect data on the 18 May material during a mid-Sept cross-country flight, timed to coincide with information-gathering by NASA's SAGE satellite as it passes over the Northern Hemisphere.

Two reports in the 5 Sept issue of *Science* present data on the petrology, chemistry, and size distribution of the 18 May tephra. Hooper and others emphasize the bimodal character of ash deposited about 400 km ENE of Mt. St. Helens. The ash changed abruptly from a relatively dark, glass-poor silicic andesite to a lighter-colored glass-rich rhyodacite $3^1/_4$ - $3^1/_2$ hours after ashfall began. This interval corresponds quite closely to the timing of changes observed in the character of the eruption column at the volcano (5:5). Fruchter and others present bulk analyses of ash collected at numerous locations in Washington. In addition to petrology and major and trace element chemical analyses, the report focuses on the tephra's toxic and radioactive components, which do not appear to have been abundant enough to have a significant effect on animal or plant life.

The July 1980 issue of the *Washington Geologic Newsletter* lists Mt. St. Helens research projects being carried out by 25 groups at 20 institutions and government agencies, in addition to giving sources for pre- and post-eruption maps and airphotos of the volcano. In addition, names and addresses of USGS personnel with the particular Mt. St. Helens study on which each was working as of 20 June are listed. An eruption chronology, and reports on ashfall distribution and petrography are also presented in this publication.

Pre- and post- 18 May photographs and satellite images of Mt. St. Helens and vicinity are available from the EROS Data Center, Sioux Falls SD 57198. Among these are about 1500 photographs taken by USGS personnel in low-altitude aircraft; color infrared photographs taken at 18 km altitude from NASA's U-2 aircraft on 1 May and 19 June; and cloud-free LANDSAT images.

Information Contacts: D. Peterson, USGS, Vancouver, WA; R. Tilling, USGS, Reston, VA; S. Malone, R. Crosson, E. Endo, Univ. of Washington; M.P. McCormick, NASA Langley Research Center; A. and M. Meinel, Univ. of Arizona; UPI.

9/80 (5:9) Sept activity was limited to slow outward movement of the N flank, minor seismicity, and weak vapor emission. No eruption was associated with the brief earthquake swarm of 4 Sept, and seismicity was limited to a few very minor events through mid-month. Outward movement of the rampart N of the inner crater continued, averaging 1.2 cm/day 26 Aug-9 Sept (in contrast to the 1.5 m/day movement of the N flank "bulge" prior to 18 May). Several small jets of ash were ejected from a vent near the base of the lava dome on 9 Sept. Each burst lasted about 30 minutes, but consisted of only a few cubic meters of ash. Weather and instrument problems limited data on CO_2 and SO_2 emission rates. Early Sept CO_2/SO_2 ratios ranged from 4.7 on the 5th to 2.4 on the 9th, with a maximum CO_2 emission rate of 5000 t/d, on the 6th.

Many large avalanches occurred on the crater walls during the afternoon and early evening of 12 Sept, but no accompanying seismic activity was recorded. Avalanching declined during the night of 12-13 Sept. Ground deformation measurements indicated that outward movement of the crater's N rampart had slowed to less than 1 cm/day by mid-Sept. Several radial fissures in the lava dome within the inner crater widened slightly in the 2 weeks since they first developed about 9 Sept. The USGS believes that this widening was probably related to the N rampart movement. No dome growth has been detected since 11 Aug, but some incandescent areas were visible on the dome on 23 Sept. No significant seismic or eruptive activity was recorded in mid-Sept.

On 24 Sept at 0917 a gray gas plume rose from the volcano to about 3 km altitude, just clearing the crater rim, then drifted to the S. Vigorous gas emission lasted about an hour, but there was little or no ash in the plume. No seismicity accompanied this event, but low-amplitude harmonic tremor began a few hours later at about 1400. The tremor was intermittent, with episodes lasting from less than a minute to 15 minutes, separated by 2- to 15-minute quiet intervals. Tremor episodes declined in number and duration after about 2 hours, and had nearly ended by 1800. On 26 Sept, very low-level harmonic tremor began at 0740. Minor steam emission started at 0747, lasting 9 minutes. The tremor ended by 0800. No additional tremor had been recorded as of 30 Sept. Outward movement of the N crater rim had virtually stopped 24 Sept, but measurements on the 26th indicated that very slow deformation had resumed. Although rates of outward movement have varied, the average rate for the most active portion of the N flank was slightly less than 1 cm/day during Sept.

In mid-Sept, scientists aboard a NASA P-3 aircraft studied the plume at Mt. St. Helens and material remaining in the upper atmosphere from the 18 May eruption. The aircraft left Hampton, Virginia on 17 Sept, flew S to Georgia, traveled W along the 32nd parallel to about Tucson, Arizona, then continued NW to Portland, Oregon. During this transcontinental flight and the return trip across the northern plains the following week, 2 wavelengths of lidar backscatter data showed a generally consistent broad layer of volcanic material at 14 to 22 or 23 km altitude. Peak backscatter typically occurred from 2 levels, 18.5-19 km and about 21.5 km. The lower was usually significantly stronger. Although the 21.5 km peak sometimes approached that at 18.5-19 km, the upper peak disappeared suddenly in certain areas.

Simultaneous data were gathered on 2 occasions from the P-3 aircraft and NASA's SAGE satellite, once over Sacramento, California and once over Portland. A dust sonde released from Laramie, Wyoming by James Rosen collected information on upper atmosphere particulates as the NASA aircraft made lidar measurements in the same area on its return flight.

Scientists with the NASA-funded Research on Atmospheric Volcanic Emissions (RAVE) project collected a variety of data on the plume from Mt. St. Helens during a 4-hour flight in the NASA P-3 on 22 Sept. Filter samples were collected and the plume was analyzed directly for SO_2, H_2S, OCS, CS_2, O_3, NO, and total S.

A Univ. of Washington-USGS report provided additional information on the seismicity associated with the 7 Aug explosions and minor ash emission 8-10 Aug (5:7-8). Harmonic tremor began at 1207 on the 7th, occurring as 10- to 20-second bursts at randomly spaced intervals. Low-frequency events were recorded at 1529 and 1554. At about 1622, records show a gradual increase in tremor amplitude, followed by a large seismic event at 1626:45 that marked the onset of eruptive activity. An ash-laden cloud rose to 13.5 km altitude and a small pyroclastic flow moved down the N flank. By 1730, tremor amplitude had returned to pre-1207 background levels. A series of pulsating ash ejections reaching a maximum altitude of less than 7 km coincided with a gradual increase in tremor amplitude starting at 1910. Tremor amplitude declined substantially by 2045. A second amplitude increase beginning at 2130 was followed by a large seismic event at 2232:22 as a stronger explosion sent tephra to 11 km altitude. Peak tremor levels were similar to those associated with the first 7 Aug explosion. At 2328, the first of eight 9-11 km-deep earthquakes occurred, all with magnitudes less than 1.7. Tremor amplitudes declined slowly. Between 8 Aug at 0325 and 9 Aug at 0129, approximately 18 events similar to large bursts of tremor and lasting 20-130 seconds were recorded. These can be correlated with discrete periods of ash emission. The last significant ash emission event was recorded on 10 Aug at 0813. No changes in tremor amplitude preceded these events.

Information Contacts: D. Peterson, USGS, Vancouver, WA; R. Tilling, USGS, Reston, VA; S. Malone, R. Crosson, E. Endo, Univ. of Washington; M.P. McCormick, NASA Langley Research Center; J. Friend, Drexel Univ.

10/80 (5:10) *Minor Activity — Early Oct.* A gradual increase in activity began 7 Oct with barely detectable harmonic tremor that started a few minutes after midnight and lasted less than an hour. During the day on the 7th, a few plumes, containing no appreciable ash, rose to about 3 km altitude before drifting NE. No harmonic tremor was recorded 8 Oct, but shallow earthquakes of magnitudes 1.6 and 1.8 occurred at 1535 and 1537. Some 8 Oct plumes contained smaller amounts of ash. Several minor gas clouds containing a little fine ash were emitted 10 Oct between 0915 and 1100, accompanied by intermittent low-level harmonic tremor. The initial column on the 10th reached 4.5-6 km altitude and drifted NNE. Two more gas plumes were emitted during the afternoon, again accompanied by minor seismicity. Gas emissions decreased in frequency and intensity 11-13 Oct. Seismic records during poor weather on the 14th indicated that occasional minor gas emission continued. On 15 Oct, some cracks that had not been present the preceding week were visible on the crater floor. Three small volcanic earthquakes were recorded, but no harmonic tremor was detected.

Explosive Eruptions — 16-18 Oct. Shallow earthquakes, most with magnitudes less than 1, became increasingly frequent on 16 Oct, reaching 1 event every few minutes by early evening. A magnitude 3 earthquake, centered about 1 km beneath Mt. St. Helens, occurred at 1902. An hour later, the USGS and the USFS notified state and local officials of the possibility of an eruption, prompting the successful evacuation of 92 persons from around the volcano.

Observers in a USFS aircraft saw strong incandescence in the inner crater area at 2157, then an explosion 1 minute later. Tephra was ejected for only 5-10 minutes, accompanied by strong, regular seismicity probably related to explosive gas release. The top of the eruption column reached 13.5 km, as measured by Portland airport's radar. Ash and pea-sized pumice fell on Cougar. Pumice was reported at Amboy (40 km SW) and a trace of ash fell on the Portland-Vancouver area. A few flights to Portland were rerouted or cancelled and an air pollution alert was issued, but there were no serious disruptions of auto traffic the next morning.

Seismic activity stopped immediately after the 16 Oct explosion, but low-level seismicity resumed about 0900 the next morning. At 0928, an ash-laden eruption cloud began to rise rapidly from the crater, reaching 9 km altitude at 0932 and its maximum altitude of 14.3 km by 0938. This explosion produced a single pyroclastic flow, at about 0935, that traveled 3-4 km down the N flank to the break in slope, coming to rest after about 5 minutes. Ash emission began a gradual decline about 0940 then decreased abruptly at 0954, although sporadic eruptive activity continued until about 1015, when seismicity ceased. Wind directions differed with altitude, but the main plume drifted SE to SSE, reaching north-central

Oregon. Ash from the explosion of late 16 Oct and early 17 Oct fell as far away as Eugene Oregon, 280 km SSW. These explosions destroyed the lava dome that was extruded after the 7 Aug eruption.

Weak seismicity resumed at about 2045 on the 17th, gradually increasing in intensity. Observers in a USFS aircraft noted strong incandescence in the inner crater at 2108, and the emergence of an eruption cloud at 2112 as tremor amplitude increased sharply. An incandescent pyroclastic flow descended the N flank at 2116. The eruption column reached its maximum altitude of 13.7 km at 2119. Eruption intensity declined gradually after 2119, with activity becoming intermittent by 2200. Emission of steam and ash pulses continued until about 2350, accompanied by intermittent seismicity. Ash blew SE.

Occasional seismic activity, consisting of very low-level signals lasting a few seconds to a few minutes, continued until the early afternoon of 18 Oct. A seismic episode that began on the 18th at 1232 abruptly intensified 3 minutes later as an eruption cloud emerged from the crater, reaching a maximum altitude of 6 km at 1239. A second burst, ejected at 1246, reached 7.6 km altitude by 1249. Vigorous tephra ejection stopped 1 minute later, but intermittent weak emissions continued for another 15 minutes. Eruptive activity gradually weakened, but sporadic low-level seismicity continued. A new tephra plume was ejected at 1428, reaching 6 km altitude at 1432, but activity declined quickly and the episode had essentially ended by 1500. Light ashfalls from the 18 Oct explosions were reported from as far as N-central Oregon.

Lava Dome Growth. By the time visibility returned at 1520, a new lava dome about 30 m across and 6 m high was growing in a saucer-shaped depression in an area formerly occupied by the inner crater and the recently destroyed, post-7 Aug lava dome. An hour later, the dome was 40 m in diameter and 9 m high. By the next morning it had grown to an elliptical feature 270 m in largest dimension and 50 m high, but it had essentially ceased increasing in volume, with sagging of the summit compensating for continuing increases in width. Fragments continued to spall from the dome's irregular breadcrust-like surface. Noisy, episodic gas emission occurred around the base of the dome. Occasional low-level seismicity continued, but these episodes became briefer and less frequent through 19 and 20 Oct.

Minor Activity — Late Oct. No significant seismicity was recorded during the next several days. When visible through heavy weather clouds, the summit of the lava dome appeared to have sagged further and the margins widened slightly, but there was no apparent increase in volume. On 25 Oct, a series of small, shallow seismic events between 1100 and 1130 accompanied the ejection of individual vapor plumes that reached 3.6 km altitude. The next day at 1720, a shallow volcanic earthquake of about magnitude 2 was followed by several smaller shocks in the next few hours but no plumes were ejected. When weather conditions permitted in late Oct, measurements showed continued outward movement of the N crater rampart. Gas vents near the dome fumed vigorously through the end of Oct and cracks in the surrounding crater floor continued to widen very slowly.

Information Contacts: D. Peterson, J. Dvorak, D. Mullineaux, C. Newhall, USGS, Vancouver, WA; R. Tilling, S. Russell-Robinson, USGS, Reston, VA; S. Malone; R. Crosson; E. Endo, Univ. of Washington.

11/80 (5:11) Activity was limited to vapor emission and occasional seismic activity through early Dec. Most early Nov seismic events were caused by rockslides from the crater walls. No significant local earthquakes or harmonic tremor were recorded until mid-Nov, when brief episodes of harmonic tremor began, barely within the detection limits of sensitive seismographs on and near the volcano. Intermittent low-level tremor continued through early Dec. Stronger tremor started on 25 Nov at 2054, gradually fading into background noise about 35 minutes later. Observers in a USFS aircraft reported a slightly brighter glow in the dome area after this event. A second burst of stronger tremor began 27 Nov at 2034, continuing for about an hour, and several more such episodes, lasting only a few minutes each, were detected through 30 Nov.

USGS monitoring of the N crater rampart revealed a maximum net outward movement of about 23 cm between the Oct explosions and 26 Nov. A major reversal to inward movement had occurred in late Oct before an outward trend resumed in Nov. Outward growth accelerated in mid-Nov to slightly more than 1.5 cm/day at times, a rate similar to that recorded during the summer. About 20 cm of expansion was measured between 12 and 26 Nov.

No major changes have taken place in the volume or ratio of gases emitted by the mountain. Two large fumaroles opened in the crater floor, very close to the margin of the lava dome, one on 18 or 19 Nov, the other on the 25th. As they opened, both ejected mud (containing no fresh magma) that coated snow on the flank. As of early Dec, the new fumaroles were 2-3 m across, glowed cherry red, and puffed noisily at half-second intervals.

No.	SiO_2	Al_2O_3	FeO*	MgO	CaO	K_2O	Na_2O	TiO_2	P_2O_5	Total	Date	No. analyses
1	64.13	17.61	4.04	1.88	4.90	1.26	4.63	.58	.15	99.18	18 May	9
2	64.19	17.92	3.99	1.91	5.06	1.29	4.83	.60	.15	99.94	25 May	11
3	63.72	18.04	4.24	1.99	5.16	1.25	4.70	.64	.15	99.89	12 June	9
4	63.49	17.87	4.44	2.20	5.22	1.26	4.97	.57	.15	100.17	22 July	7
5	63.28	17.51	4.39	2.17	5.30	1.23	4.89	.64	.16	99.57	7 Aug	10

* Total Iron calculated as FeO.

Table 12-2. Averages for each eruptive episode of Mt. St. Helens, 18 May-7 Aug. Electron microprobe analyses of fused powders by W. Melson, T. O'Hearn, and J. Nelen, SI. Samples collected by D. Swanson, C. Hopson, W. Melson, R. Fiske, and C. Kienle.

The following is a report from W.G. Melson:

"A small but definite trend toward andesite compositions is revealed by major element analyses of the 18 May-7 Aug eruptives. A total of 46 samples of probable essential ejecta have been analyzed (table 12-2 and fig. 12-13), a minimum of 5 such samples from each eruptive episode. The trend is an irregular one and is most pronounced with regard to MgO and CaO when plotted against time of eruption."

Information Contacts: T. Casadevall, C. Newhall, D. Swanson, USGS, Vancouver, WA; R. Tilling, USGS, Reston, VA; S. Malone, R. Crosson, E. Endo, Univ. of Washington; W. Melson, SI.

12/80 (5:12) Renewed dome growth took place in late Dec, without the large explosions that immediately preceded previous dome-building episodes in June, Aug, and Oct. Activity was limited to minor seismicity and weak vapor emission for about a month after the 16-18 Oct explosions and dome extrusion. Frequent periods of very low-level harmonic tremor, lasting a few minutes to several hours, began to appear on seismic records 19 Nov. Bursts of higher level tremor, similar to explosion events seen earlier at Mt. St. Helens, could often be correlated with ejections of vapor columns that sometimes contained ash. A few discrete shallow earthquakes were recorded, but remained infrequent until late Dec.

A series of vapor plumes marked the volcano's behavior throughout much of Dec. A few minutes of stronger tremor accompanied emission of a vapor plume that rose to 3 km altitude 7 Dec, and one of several bursts of higher amplitude tremor on 9 Dec occurred as a plume was ejected to 2.7 km altitude at 1325. A new thin deposit of ash was noted on the upper S flank early on 12 Dec. Emission of this ash was not observed, but a burst of increased tremor had begun at 0417, lasting about 30 minutes. On 13 Dec at 2017, a plume reached 5.5 km altitude as higher level tremor was recorded. Inspection of the dome 15 Dec revealed a new small [explosion] crater in its [SE] edge. Adjacent to the new crater, a roughly triangular section of the dome had been [exploded away], extending about 15 m along its outer edge and 30 m toward the center of the dome. Plumes associated with increased tremor rose to 3.3 km altitude 16 Dec at 0800 and 17 Dec at 1520. A plume reaching 6 km altitude was briefly visible through clouds on 21 Dec at 1409, accompanied by a short burst of tremor. Two days later, at 1258, seismic activity and vapor emission increased simultaneously. Gas and a little tephra rose to almost 3 km, but activity continued for only a few minutes. New crater floor cracks were apparent after this event.

Deformation measurements 6 and 7 Dec showed a halt or possibly a reversal of the outward movement of the N crater rampart that had resumed in Nov (5:11). Measurements 18 Dec revealed little or no change. However, observations on the 23rd showed renewed northward displacement. Fissures in the crater floor appeared to be widening as well as extending radially from the inner crater.

On 25 Dec, the number of discrete shallow earthquakes began to increase. Seismicity peaked before noon 27 Dec, averaging 5 events per hour and occasionally reaching 8/hour during the next 30 hours. Univ. of Washington geophysicists located about 2 dozen of these events. All were centered at 2 km depth or less and were within 1 km NW of the Oct dome. No migration of events was evident.

Aerial observations 26 Dec were hampered by poor visibility, but there were no apparent changes in the crater. Bad weather prevented additional observations of the crater until 28 Dec at 0900, when USGS and USFS

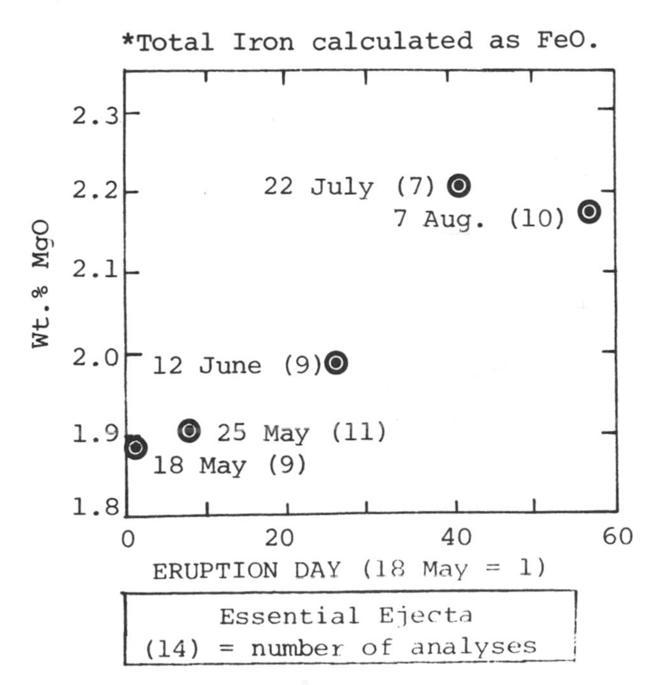

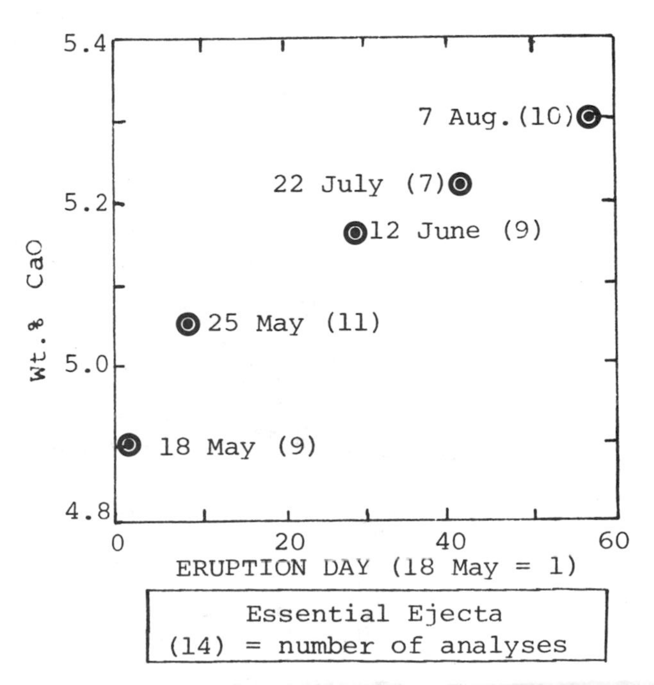

Fig. 12-13. Average of MgO, *left*, and CaO, *right*, concentration in rocks from each eruptive episode at Mt. St. Helens, 18 May-7 Aug 1980, plotted against time of each episode. Analyses are of fused powders by the electron microprobe. Numbers in parentheses are the number of samples analyzed and included for average value.

personnel saw a new extrusion, about $^1/_4$ the size of the Oct dome and emerging from its SE edge. A spine-like structure protruding 30-60 m from the center of the Oct dome was noted an hour later. All but 8 m of this structure toppled the next day at 1540. Growth of the new SE lobe and another much smaller new lobe on the NW edge of the Oct dome had apparently stopped by 3 Jan but crater floor deformation has continued. The SE lobe measured at least 225 m in an E-W direction and reached a maximum height of about 100 m above the crater floor, although by 6 Jan the crest was subsiding somewhat. The NW lobe was about 100 m across. The elliptical Oct dome had been about 230 m in largest horizontal dimension and 50 m high on 19 Oct. A collapse pit formed on the Oct dome during the growth of the new lobes, but the pit's dimensions were unavailable.

Deformation measurements showed that the N crater rampart had moved outward about 85 cm 23-28 Dec and another 1.5 m by 2 Jan. Since then, the crest of the rampart has been uplifted and thrust northward dramatically, as much as 5 m by 6 Jan. Other thrusts have been observed in relatively level terrain on the crater floor.

By the afternoon of 29 Dec, seismicity had declined to a rate of 1 or fewer events per hour. As of 7 Jan, no harmonic tremor and very few discrete earthquakes were being recorded.

Information Contacts: D. Swanson, C. Newhall, J. Dvorak, USGS, Vancouver, WA; S. Malone, E. Endo, C. Weaver, Univ. of Washington; R. Tilling, USGS, Reston, VA.

Further References: Foxworthy, B.L. and Hill, M., 1982, Volcanic Eruptions of 1980 at Mt. St. Helens: The First 100 Days; *USGS Professional Paper* 1249, 125 pp.

Lipman, P.W. and Mullineaux, D.R. (eds.), 1981, The 1980 Eruptions of Mt. St. Helens, Washington; *USGS Professional Paper* 1250, 844 pp. (62 papers).

1/81 (6:1) Lava extrusion resumed 5 Feb, adding a substantial quantity of new material to the dome that grew in the crater after the 16-18 Oct explosions and the 2 new lobes produced in late Dec and early Jan (fig. 12-14).

Minor Activity — Jan. After growth of the Dec-Jan lobes ceased between 2 and 4 Jan, outward movement of the N crater rampart gradually declined to an average of about $^1/_2$ cm/day, although rates were vari-

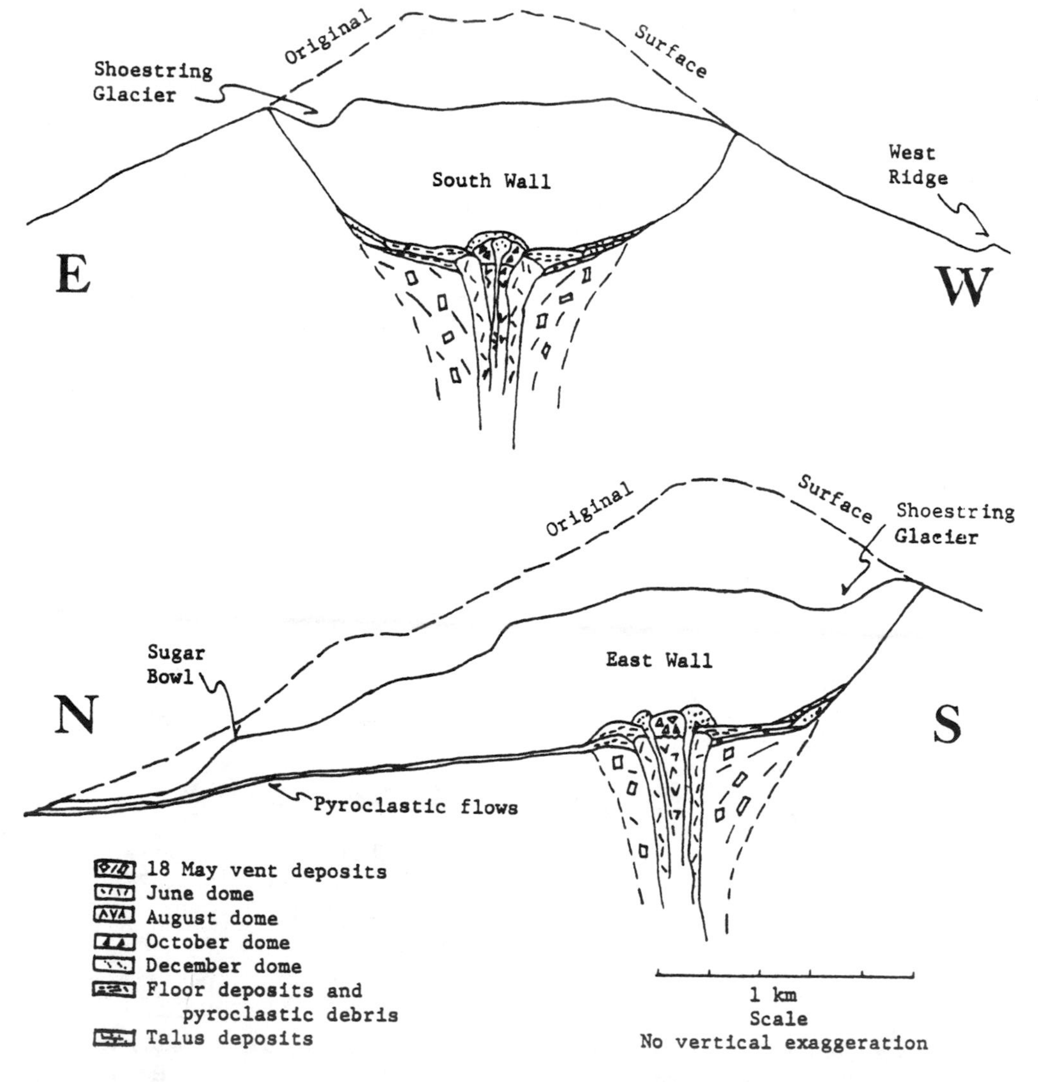

Fig. 12-14. E-W and N-S cross sections through Mt. St. Helens, Jan 1981, by Michael Doukas, USGS.

able and data were limited. Jan seismicity was the quietest of any period since earthquakes began 20 March. Only 40 discrete events were large enough to be recorded on 3 or more stations of the USGS-Univ. of Washington seismic net, in contrast to 136 in Dec and 74 in Nov. Of the Jan earthquakes, about 10 were low-frequency events associated with dome growth early in the month, many others were rock avalanche events, and a few accompanied ejection of steam plumes. A new fumarole opened 9 Jan on the E margin of the lava dome. This fumarole was the probable source of small steam and ash plumes on 16 Jan at 1152 (to 3 km altitude) and 20 Jan at 1204 (to at least 3 km), both accompanied by bursts of seismicity. Similar seismic activity was recorded 24-25 Jan and field parties saw light ash deposits on fresh snow. Several similar bursts occurred 31 Jan-1 Feb, two of which could be correlated with steam and ash emission. However, another steam plume was ejected without accompanying seismicity.

Increased Deformation and Seismicity. Deformation and seismic activity began to increase at the beginning of Feb. Radial fissures in the crater floor began to widen at a noticeably faster rate and movement of thrust faults accelerated. A larger number of glowing cracks in the surface of the lava dome indicated that its temperature was increasing. On 2 Feb at 0336, a 4-minute burst of seismicity was followed by a magnitude 2 earthquake at 0340, then low-level harmonic tremor were recorded until 0630. Occasional bursts of seismic activity continued through the day, and 35 minutes of low-level tremor was recorded that night. A gradual increase in discrete earthquakes began 3 Feb. Occasional low-level tremor was recorded, as were several bursts of seismicity, one of which was associated with a small plume at 1220. By midnight 4-5 Feb, the number of discrete events had reached 4-5/hour and continued at that rate for about 6 hours.

Lava Extrusion. Just before 0500, the USGS and Univ. of Washington issued an advisory predicting an eruption within the next 12 hours. Seismicity began to decline about 0600, probably signaling the beginning of lava extrusion. By 0800, earthquakes were occurring at a rate of only about 1/hr. Very heavy steaming obscured the crater, but new lava could be seen on the Oct dome during about 30 seconds of visibility. The number of discrete seismic events decreased further by mid-afternoon, remaining at many fewer than 1/hr through 8 Feb. However, bursts of unusual seismic signals were recorded, possibly caused by lava extrusion.

Improved visibility revealed that the new lava was extruded through the collapse pit in the center of the Oct dome. The new material appeared to have both uplifted and overridden the Oct dome, leaving this area about 35 m higher by the time growth apparently stopped during the night of 6-7 Feb. The small NW lobe emplaced during the Dec-Jan activity was pushed about 12 m N and partially overridden by new lava. New thrust faulting also occurred in the SW part of the crater, but was much less extensive than the thrusting associated with the Dec-Jan activity. The increase in dome volume produced by the Feb extrusion was roughly equal to the volume of lava produced by each of the 2 previous events, but at press time it was not possible to determine how much volume was of new lava on the surface and how much was caused by uplift of pre-existing lobes.

Information Contacts: D. Swanson, C. Newhall, J. Dvorak, USGS, Vancouver, WA; S. Malone, C. Boyko, E. Endo, C. Weaver, Univ. of Washington; R. Tilling, USGS, Reston, VA.

2/81 (6:2) Mt. St. Helens remained quiet as of 10 March, as it has since the end of the lava extrusion episode of 5-7 Feb. The Feb lava approximately doubled the volume of the composite dome, adding about 5×10^6 m^3 of new material to the 1.5×10^6 m^3 extruded 18-19 Oct and the 3.5×10^6 m^3 extruded 27 Dec-4 Jan. All of the pre-existing dome, except for a portion of the Dec-Jan SE lobe, was covered by the Feb lava. Between 8 and 21 Feb, the Feb lobe spread 12 m while sagging 3 m, resulting in dimensions for the new lava of 281 m in E-W direction and 119 m in maximum height above the crater floor.

Low-frequency volcanic earthquakes associated with the Feb lava extrusion ended 9 Feb. Occasional bursts of seismicity continued to be recorded. One, on 10 Feb at 0915, coincided with the emission of a steam cloud, containing a minor amount of ash, that rose to 4 km altitude. Field crews reported hearing a boom prior to this event. Some rock avalanche events were also recorded after dome emplacement ended. A magnitude 5.5 tectonic earthquake occurred late 13 Feb about 12 km N of Mt. St. Helens. As of 28 Feb, about 175 aftershocks stronger than magnitude 1 had been recorded. Through the end of Feb, seismographs continued to record a few rock avalanche events and bursts of seismicity of the type that has sometimes been associated with steam explosions. Clouds prevented observations of the crater for much of Feb, but clear weather on the 26th revealed evidence of numerous minor steam explosions on the N side of the lava dome.

Geodetic measurements showed a few centimeters of horizontal contraction of the Mt. St. Helens edifice between 4 Feb and 5 March. No significant movement of the N crater rampart occurred after the early Feb dome emplacement, nor has there been any measurable deformation of the crater floor during this period.

The following, from W. G. Melson, is based on microprobe analyses performed on the 1980-81 eruptives.

"The SiO_2 content of the essential ejecta underwent a slight increase in the 7 Aug eruption, peaked in the 17-19 Oct eruption, but remained lower than for the 18 May tephra. This temporarily reversed a prior trend toward more basic compositions, which resumed with the Dec-Jan and Feb dome enlargements. CaO, FeO, and MgO show an inverse relationship to SiO_2 (fig. 12-15), an expected relationship in a 'normal' fractionation sequence."

Information Contacts: D. Swanson, C. Newhall, USGS, Vancouver, WA; C. Boyko, S. Malone, E. Endo, C. Weaver, Univ. of Washington; R. Tilling, USGS, Reston, VA; W. Melson, SI.

3/81 (6:3) March eruptive activity was limited to occasional emission of small steam clouds, at least one of which contained ash. However, significant deformation was measured within the crater, and there was a slight increase in volcanic seismicity during the second half of March. Geologists announced that another eruptive episode was likely if the deformation and seismic trends continued, but none had occurred by press time.

The USGS-Univ. of Washington seismic net recorded 15 bursts of seismicity in March and 5 more bursts during the first 6 days of April. In the past, similar signals have often been correlated with episodes of steam emission, but because of poor weather, correlations with only 2 such episodes could be confirmed in March: a minor puff on 9 March at 1549, and a steam cloud containing some ash on 27 March at 1441. Newly fallen ash (made up of reworked dome material) observed NE of the volcano 25 March may have been ejected during a burst of seismicity the previous day.

The seismic net began to detect small, low-frequency, shallow events on Mt. St. Helens on 21 March. Fifteen of these discrete volcanic events were recorded by the end of March. Numerous aftershocks of the magnitude 5.5 tectonic event that occurred 13 Feb about 12 km N of Mt. St. Helens, continued to appear on seismic records through March.

Deformation measurements showed that outward movement of the N crater rampart resumed in March. Between 9 and 17 March, the rampart moved 7 cm to the N; by 22 March it had advanced 6 cm farther northward; and an additional 3.5 cm of movement was measured by 24 March.

A newly-established levelling net on the crater floor showed pronounced uplift near the lava dome, indicating that the dome was rising. Increasing crater-floor deformation was also demonstrated by accelerations in the rate of widening of a fissure from 3 mm/day to 1 cm/day and the rate of movement of a thrust fault from $^1/_2$ cm/day to 1 cm/day by late March.

Addendum. On 9 April at about 1800, local seismicity began to increase, to about 1 event per hour at first and to about 2/hr after midnight. The USGS-Univ. of Washington team issued an advisory about midnight stating that an eruption was likely within the next day if seismicity continued to increase.

Periods of constant low-frequency seismicity became more common and by 0230 on 10 April low-frequency activity was continuous. Individual events superimposed on this activity had increased to an average of 6-8/hr by 0600 and remained at that level through the day. At 0821, a small explosion produced an ash-bearing plume that rose to 4.5 km altitude. A light ashfall was reported at a ranger station 40 km NE. Although clouds prevented observation of the crater, a USGS helicopter crew could see that this explosion had generated no pyroclastic flows.

About 1900, the pattern of seismic activity started to change. The number of discrete events dropped to 4-6/hr, but these events were slightly stronger and total seismic energy release briefly stayed about the same. However, by midnight there had been a notable decline in both the number of events and seismic energy release, and by 0200 only 1-2 events were being recorded per hour. Seismicity had essentially ended by 2100-2200 on 11 April.

The weather cleared somewhat late 12 April, and geologists were able to view the crater between 1800 and 1900. New lava extended roughly 75 m NNW from the pre-existing dome. Television station videotape taken between 1900 and 1930 showed significant additional lava extrusion.

Information Contacts: D.

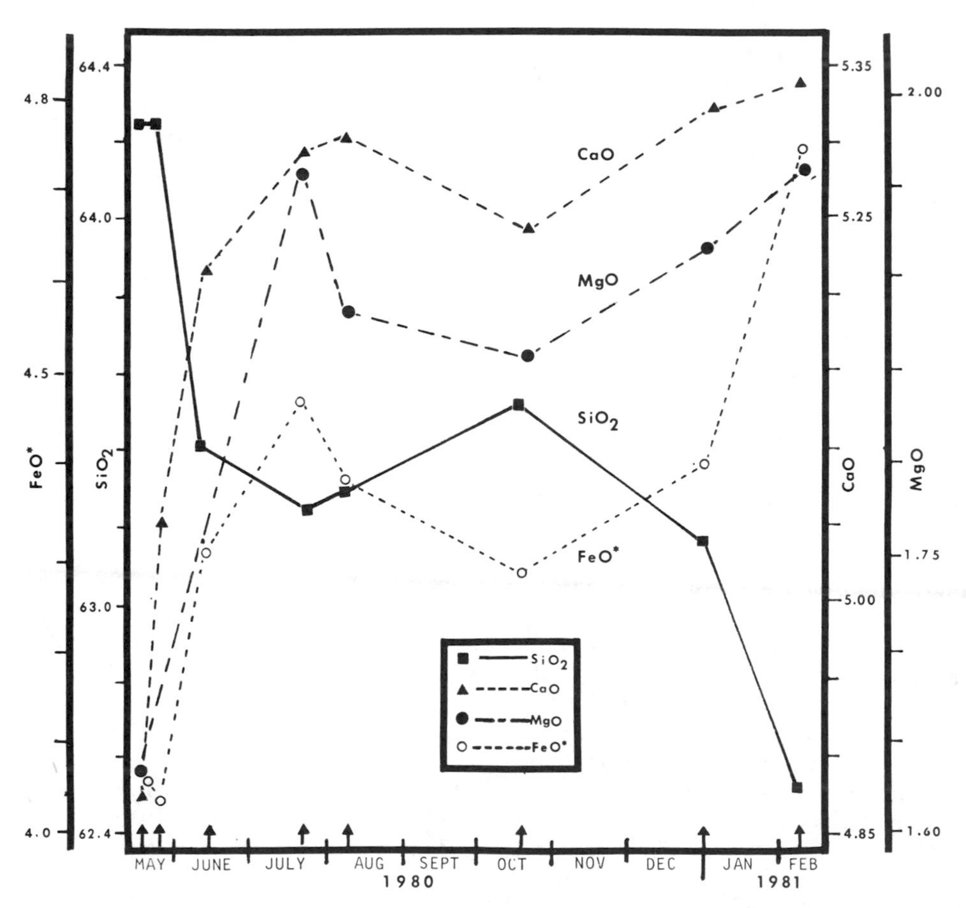

Fig. 12-15. Change of SiO_2, CaO, FeO, and MgO as a function of time of eruption. Electron microprobe analyses of fused powders were performed at SI by W. Melson, J. Nelen and T. O'Hearn. Each data point is the average of the following number of individual analyses of essential ejecta: 18 May, 9; 25 May, 11; 12 June, 9; 22 July, 7; 7 Aug, 10; 17-18 Oct, 11; Dec-Jan dome enlargement, 6; Feb dome enlargement, 1 (sample from D. Swanson, USGS). Analytical precision for each analysis is about a 2σ of: SiO_2=0.62, FeO* (all Fe as FeO)=0.43; MgO=0.33, CaO=0.17.

Swanson, C. Newhall, S. Russell-Robinson, USGS, Vancouver, WA; C. Boyko, A. Adams, S. Malone, E. Endo, C. Weaver, Univ. of Washington; R. Tilling, USGS, Reston, VA.

4/81 (6:4) About 5 × 10^6 m^3 of lava were added to the composite dome in the crater of Mt. St. Helens in early April. Because of poor visibility, no precise time for the start of extrusion could be determined, but a small explosion that ejected an ash-bearing plume to nearly 5 km altitude on 10 April at 0821 may have marked the beginning of the episode. Much of the lava had already been emplaced by the time geologists had their first view of the crater early 12 April, and extrusion was essentially complete by evening. Most of the associated seismicity had ended by late 11 April, but occasional discrete low-frequency events continued to be registered by the USGS-Univ. of Washington net through 17 April.

Although deformation measurements showed that the magma rose through a conduit beneath the central collapse pit of the pre-existing dome, the April lava emerged from a vent somewhat N of the central pit, covered roughly the N quarter of the older material, and extended about 160 m NNW from it previous margin. After the April event, the dome had a volume of about 15 × 10^6 m^3, maximum and minimum lateral dimensions of 630 m (NNW-SSE) and 310 m (E-W), and a maximum height above the crater floor of 110 m. A substantial but uncertain amount of uplift of the entire crater floor was associated with the April extrusion, and some points on the crater floor spread away from the dome as much as 1.5 m, with most of the movement occurring during extrusion. One radial fissure exhibited about 55 cm of strike-slip movement during the episode. As of 5 May, only a few mm of additional deformation had taken place within the crater. No net deformation of the volcano as a whole has been associated with any of the extrusion episodes.

In the weeks following the April extrusion, characteristic low-level seismicity was sometimes correlated with witnessed bursts of steam emission. Simultaneous seismicity and ejection of steam containing a little ash occurred on 13 April at 0842; 14 April at 0950, 0953, and 1021; 17 April at 0958; and 24 April at 1018. Seismicity accompanied ejection of plumes of steam (without ash) on 25 April at 0921 and 26 April at 0821. A small amount of ash that fell about 50 km SE of Mt. St. Helens on 6 May between 1500 and 1530 may have been ejected during a period of seismicity at 1415.

Information Contacts: D. Swanson, C. Newhall, USGS, Vancouver, WA; C. Boyko, S. Malone, E. Endo, C. Weaver, Univ. of Washington; R. Tilling, USGS, Reston, VA.

5/81 (6:5) Small steam explosions, some ejecting a little ash, occurred intermittently through May. Explosions correlated with emergent low-frequency seismic signals occurred on 6 May at 1414; 11 May at 2140; 13 May at 0810 and 1059; 20 May at 0615; 22 May at 0247; 25 May at 2206; and 29 May at 1229. However, increased seismicity did not accompany an increase in steaming at 1040 on 13 May, or 3 ash puffs ejected between 1425 and 1431 the same day. Furthermore, seismic signals produced by steam explosions are often indistinguishable from those generated by rockfalls in the crater, so the seismic record alone cannot confirm the occurrence of either.

Rates of ground displacement within the crater and the volume of SO_2 emitted by the volcano both began to increase in late May. Until about 20 May, only very slow changes were noted in the position of the N crater rampart and in thrust faults surrounding the dome (fig. 12-16). Measurements 27 May showed an acceleration in the rate of displacement, and reoccupation of rampart stations 5 June showed outward movement of about 1 cm/day. The rate of rampart movement had increased

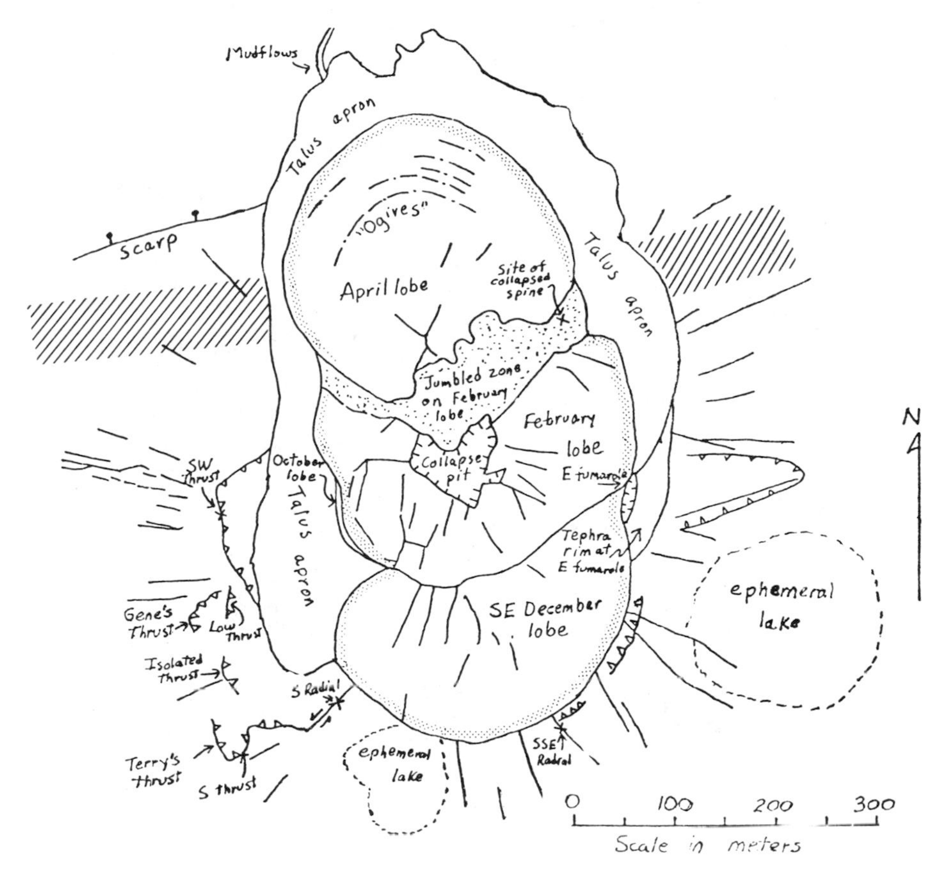

Fig. 12-16. Map of the May 1981 composite lava dome in the crater of Mt. St. Helens, prepared by D. Swanson. The margins of the dome are outlined by a stipple pattern. The approximate position of the crest of the N crater rampart is shown by a pattern of diagonal lines on each side of the April lobe. The base station for deformation measurements is located several kilometers to the N, off the map.

to about 2 cm/day between 11 and 15 June, and the S thrust fault (SE of the dome) moved 8.3 cm/day during the same period. Data telemetered 29 May-9 June by a newly installed bubble tiltmeter just NE of the dome showed substantial uplifts consistent with other deformation data.

The rate of SO_2 emission was measured by COSPEC from fixed-wing aircraft flying under the plume. Between 1 and 16 May, the 3-day moving average of SO_2 output decreased from 450 to 150 t/d. This trend reversed in late May, with emission rates rising from 190 t/d on 22 May to about 500 t/d by 11 June.

The USGS and the Univ. of Washington Geophysics Program issued a joint advisory 12 June stating that an eruption, probably of the dome-building type, was likely to begin within the next 1-2 weeks if ground deformation and gas emission trends continued.

Information Contacts: T. Casadevall, D. Dzurisin, C. Newhall, D. Swanson, USGS, Vancouver, WA; C. Boyko, S. Malone, E. Endo, C. Weaver, Univ. of Washington; R. Tilling, USGS, Reston, VA; UPI.

6/81 (6:6) The increased ground deformation and SO_2 emission described last month were followed by an episode of lava extrusion atop the pre-existing composite dome that probably began late 18 June. The new lobe was roughly comparable in volume to lobes emplaced during previous extrusion episodes in Oct 1980, Dec 1980-Jan 1981, Feb 1981, and April 1981. A final volume figure awaits analysis of aerial photographs.

Seismicity began to increase during the evening of 17 June, and by afternoon seismographs were recording several events per hour. The seismic events were impulsive and of higher frequency than those that had typically accompanied previous eruptive episodes, but were centered directly beneath the crater within 1 km of the surface. The increased seismicity prompted an advisory alert notice, issued by the USGS and Univ. of Washington at 1130 on 18 June, stating that a dome-building eruption was likely within the next day or two.

Between 1600 and 1700 on 18 June, the seismicity changed character to smaller, indistinct (non-impulsive) events, and the direction of tilt recorded by the single bubble tiltmeter (about 30 m from the NE margin of the dome) reversed. These changes were interpreted by the USGS as probably marking the beginning of lava extrusion, but cloudy weather prevented direct observation. As many as 12 of the indistinct seismic events, sometimes merging into bursts of noise, occurred each hour until about midnight, when the character of seismicity changed again to more typical low-frequency events with emergent arrivals. These events, some larger than those of the previous few hours, decreased gradually to number only a few per day by 22 June.

Poor weather prevented access to the crater until the afternoon of 19 June, when geologists observed new lava that had emerged from near the center of the pre-existing dome. The new lava covered an area roughly 300 m in both N-S and E-W dimensions, overriding portions of the lobes extruded in Feb and April and much of the talus at their margins. The June extrusion increased the height of the composite dome by around 50 m, but the top of the new lobe subsided 5-10 m between the afternoons of 19 and 20 June.

Between 29 May and the probable beginning of lava extrusion on 18 June, the bubble tiltmeter had recorded roughly 2000 μrad of tilt, in a direction consistent with other deformation data showing outward movement radial to the dome. Tilt rates increased

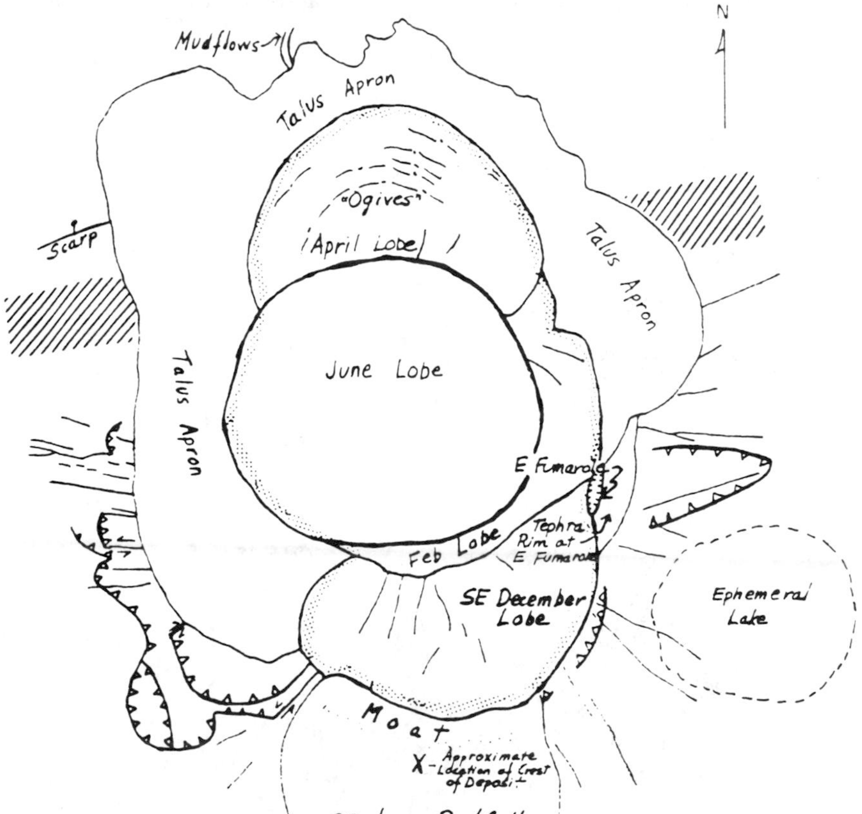

Fig. 12-17. Sketch map of the June 1981 composite dome in the crater, prepared by D. Swanson. The margins of the dome are outlined by a stipple pattern. The approximate position of the N crater rampart is shown by a pattern of diagonal lines on each side of the talus apron. The tiltmeter just NE of the dome was destroyed by a rockfall 23 June 1981.

from about 100 µrad/day during the first 10 days of measurements to 140 µrad/day on 16 June, and to roughly 140 µrad/hr during the final 3 hours before the tilt reversal that probably marked the beginning of lava extrusion. Between the time of the tilt reversal and the rockfall that ended telemetry early 23 June, roughly half of the relative uplift recorded 29 May-18 June had been lost. In contrast to the pre-extrusion period, the post-extrusion record showed numerous features including instantaneous offsets probably associated with rockfalls and short-term tilt fluctuations. Present plans call for the installation of a network of telemetered bubble tiltmeters in the crater.

Information Contacts: T. Casadevall, D. Dzurisin, C. Newhall, D. Swanson, USGS, Vancouver, WA; C. Boyko, S. Malone, E. Endo, C. Weaver, Univ. of Washington; R. Tilling, USGS, Reston, VA.

7/81 (6:7) The mid-June lobe (fig. 12-17) was roughly comparable in volume to lobes emplaced during previous episodes last Oct, Dec-Jan, Feb, and April that built the pre-existing composite dome. A continuously recording tiltmeter about 30 m from the NE margin of the dome provided data on pre-extrusion uplift and probably recorded the beginning of extrusion at about 1700 on 18 June (fig. 12-18). This tiltmeter was destroyed early 23 June, but 3 new tiltmeters were installed in early July, within 100 m of the NE, E, and SE sides of the dome. None showed any significant changes through July. Other July deformation measurements did not show the accelerating outward movement that has typically preceded extrusion episodes. The volume of SO_2 emissions, measured by COSPEC from fixed-wing aircraft flying under the plume, usually ranged from 100 to 300 t/d during July, averaging about 150 t/d. Through the end of July there was no suggestion of the increase in SO_2 emissions that preceded both the Dec 1980-Jan 1981 and the June 1981 lava extrusion episodes by several weeks. Poor weather precluded determination of SO_2 trends before other extrusion episodes.

Occasional steam and ash emissions were observed during July and early Aug. An ash-laden gas plume rose to nearly 3 km altitude at 1453 on 9 July, accompanied by seismicity. A small ash plume just cleared the rim of the crater at 1138 on 14 July, and other plumes, accompanied by seismicity, were seen by USGS field crews at about 0845, 0948, 1442, and 1805 on 15 July, the largest reaching about 3 km altitude. A plume emerging from the Feb lobe of the composite dome reached 3 km altitude at 1257 on 16 July. Light ashfall was reported at Cougar between 0800 and 0900 on 27 July; this ash may have been ejected during a period of seismicity recorded at 0750. Five minutes of low-level tremor accompanied weak gas emission at 1605 on 28 July. An ash-laden plume rose to more than 3 km altitude at about 1805 on 30 July, accompanied by a seismic event and followed by about 5 minutes of low-level tremor. Several episodes of very low-level tremor were recorded 1-2 Aug. A characteristic burst of seismicity accompanied a plume, recorded on USFS video equipment at 0735 on 2 Aug, that appeared to be ash-laden and rose to about 3.5 km altitude. Several moderate seismic bursts at about 1905 on 3 Aug accompanied a small ash plume that reached 3.5 km altitude according to Portland Airport radar; 7 min of moderate tremor followed this ash emission. A small ash-laden gas emission occurred at 1133 on 4 Aug.

Information Contacts: D. Dzurisin, C. Newhall, D. Swanson, USGS, Vancouver, WA; W. Rose, Michigan Technological Univ.; C. Boyko, S. Malone, E. Endo, C. Weaver, Univ. of Washington; R. Tilling, USGS, Reston, VA.

8/81 (6:8) Extrusion of a new lobe onto the NE portion of the composite lava dome started 6 Sept and had ended

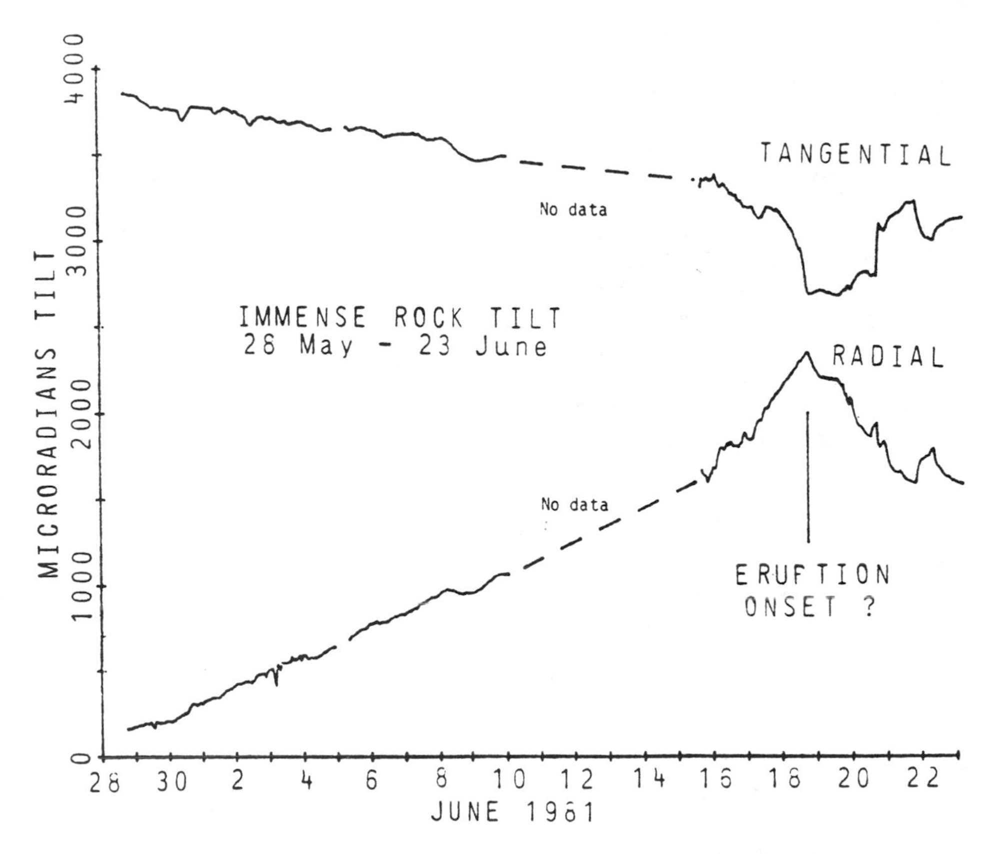

Fig. 12-18. Data from the tangential and radial components of the continuously recording bubble tiltmeter, about 30 m NE of the margin of the dome, between its installation 29 May and its destruction by a rockfall 23 June 1981. Courtesy of D. Dzurisin.

by noon on 11 Sept, after reaching a volume comparable to previous lobes. The eruption was preceded by increases in the crater displacement rate, SO_2 emission, and seismicity. This was the first extrusion episode at Mt. St. Helens in which weather conditions allowed observation of the crater immediately before and during the event.

On 5 July, the USGS installed 3 continuously recording bubble tiltmeters along a roughly N-S line within 150 m of the E side of the composite dome. The central tiltmeter, directly E of the middle of the dome, began to record gradual inflation on 8 July. The rate of inflation recorded by this tiltmeter increased systematically through July and most of Aug, reaching 235 μrad/day by 27 Aug and totaling about 2800 μrad for that month. Telemetry problems plagued the tiltmeter 175 m to the NE (at the same site as the instrument destroyed by a rockfall during the June extrusion). Field measurements of its output in July and data returned after resumption of telemetry 13 Aug showed a similar trend of uplift, but at only about half the rate recorded by the central tiltmeter. Before another rockfall interrupted telemetry 29 Aug, the N tiltmeter had detected about 800 μrad of inflation in 16 days. About 300 m SW of the central tiltmeter (directly E of the S end of the dome), the third tiltmeter, although functioning properly, recorded no significant net tilt through Aug.

Of the numerous thrust faults in the crater floor around the dome, the most vigorous showed rapid acceleration in late Aug. The highest measured rate of movement occurred about 100 m SW of the dome, along a fault first observed after the June extrusion. From 1.8 cm/day between 18 and 22 Aug, thrusting along this fault increased to 6.3 cm/day 27-29 Aug. Two other thrusts W of the dome showed similar accelerations but less total movement. Near the very active central tilt meter E of the dome, movement resumed along an older thrust 22 Aug and accelerated on the 29th.

However, most of the thrust faults that formed prior to the June extrusion remained inactive. The substantial differences in rates of thrust fault movement, combined with the variation in trends shown by the 3 bubble tiltmeters, indicated to USGS personnel that the crater floor was behaving as a group of independent blocks or plates on a scale of the order of 100 m rather than as a single coherent body.

Rates of displacement along the former N crater rampart varied considerably during Aug. At the most active site, the rate of outward movement increased to 2 cm/day between 12 and 18 Aug, declined to about 0.7 cm/day until late Aug, then resumed an irregular acceleration. A very gradual increase in the number of shallow volcanic earthquakes began about 29 July. Most of the events were centered near the former site of Goat Rocks Dome, slightly NW of the present composite dome. Several larger events occurred 13-16 Aug, then seismic activity remained relatively constant until early Sept. Small steam and ash plumes were occasionally ejected during Aug, sometimes accompanied by brief periods of harmonic tremor. The volume of SO_2 emissions, measured by COSPEC from fixed-wing aircraft flying under the plume, averaged 60 t/d 9-18 Aug, then increased sharply to about 360 t/d 19-24 Aug.

Increases in rates of tilting, outward movement of the former N crater rampart, movement along thrust faults, and SO_2 emission, prompted a USGS advisory on 26 Aug that stated in part "an eruption, probably of the dome-building type, will likely begin in 1-3 weeks". After this advisory was issued, activity in the crater continued to increase, although rates of SO_2 emission dropped to an average of 150 t/d 26 Aug-5 Sept. Between 3 and 5 Sept, outward movement of the former N crater rampart reached 10.4 cm/day at 1 site and 6.5-7 cm/day at 3 others. Thrusting along the most active fault about 100 m SW of the composite dome accelerated to 23 cm/day 1-3 Sept, 28 cm/day 3-4 Sept, and 48 cm/day 4-5 Sept. Movement along the older thrust fault near the central tiltmeter reached a rate of 7-8 cm/day by 4 Sept.

Beginning 2 Sept, USGS personnel working in the crater noted 1-2 rockfalls per hour and frequent audible and felt earthquakes. However, the earthquakes were probably very shallow, as no significant increase in seismicity was recorded by the Univ. of Washington seismic net through 4 Sept. Audible and felt earthquakes in the crater were nearly constant on 5 Sept, and rockfalls increased further, particularly from the overhanging NE portion of the June lobe. Recorded seismicity began to increase shortly after noon, and increased more rapidly during the predawn hours of 6 Sept, triggering a joint USGS-Univ. of Washington advisory at 0800 on 6 Sept that predicted a dome-building eruption within the next 12-48 hours.

During this period, sharply varying data were returned by the 3 continuously recording tiltmeters. After recording about 80 μrad/day of inflation 1-4 Sept, tilt at the N station reversed to relatively slow deflation on 5 Sept. Deflation continued on this instrument until its telemetry was ended by a rockfall during the afternoon of 6 Sept. Only 175 m to the SE, the central tiltmeter continued to record increasingly vigorous inflation, with rates reaching 700 μrad/hr on 6 Sept. This instrument recorded more than 10,000 μrad of inflation on 6 Sept before an incandescent boulder ended its telemetry during the afternoon. The S tiltmeter (about 300 m SW of the central instrument) had recorded no significant tilt previously but began to show deflation 5 Sept that continued through the 8th.

The seismic character changed to lower frequency events with emergent arrivals after dawn on 6 Sept, and epicenters moved to the area of the present dome. At about 1000 on the 6th, avalanche events began to dominate the seismic record, with only a few discrete low-frequency events appearing for the next several hours. USGS personnel working in the crater observed huge blocks falling from the NE portion of the June lobe, and were soon forced to retreat to a ridge N of the crater. Avalanche events peaked on the seismic record at about noon, but remained at high levels until about 1700. Clouds of dust from the frequent rockfalls made observation of the crater difficult, but by 1500-1530 it was evident to USGS personnel that the entire NE portion of the June lobe was breaking up. A bulge appeared to be developing on the E side of the lobe, but poor viewing conditions made this observation uncertain. By 1600-1700, an area of tens of cubic meters of fresh lava was clearly visible on the dome, and by 1830 many glowing rockfalls could be seen. . . . The number of seismic events began to decline after 1700. Significant numbers of low-frequency events resumed briefly about 2200, but seismicity dropped sharply at about 2330.

Aircraft crews monitoring the crater during the night of 6-7 Sept saw numerous glowing rockfalls, and by 0500 on 7 Sept a new lobe had been emplaced in the area formerly occupied by the NE portion of the June lobe. Most of the NE portion of the June lobe had fallen as talus, but from its high point to its SW margin, the June lobe remained intact. Slow aseismic growth and downslope spread of the new lobe continued through the afternoon of 10 Sept, but USGS field parties reported that growth had stagnated by noon 11 Sept. The new lobe was comparable in size to lobes extruded in previous episodes, but precise determination of its volume and daily growth rate await analysis of airphotos and reduction of field data.

Information Contacts: T. Casadevall, D. Dzurisin, C. Heliker, USGS, Vancouver, WA; C. Boyko, S. Malone, E. Endo, C. Weaver, Univ. of Washington; R. Tilling, USGS, Reston, VA.

9/81 (6:9) When USGS personnel arrived in the crater on the morning of 11 Sept, there was a characteristic area of smoother lava on the top of the new lobe. Similar features had marked the end of the Dec 1980-Jan 1981 and June 1981 extrusion episodes. No further growth was observed. The new lobe had a volume of about 5×10^6 m^3, comparable in size to previous lobes, and brought the total volume of the dome to about 30×10^6 m^3.

Poor weather plagued monitoring efforts after the extrusion episode. At 1559 on 10 Sept, just before extrusion ended, gas and fine ash rose to about 3 km altitude in a 15-minute eruptive episode accompanied by seismicity. Other gas emissions, all accompanied by seismicity, occurred at 0705 on 13 Sept, 1426 on 14 Sept, and 1028 on 16 Sept. No additional gas emissions were observed through the end of Sept.

Deformation within the crater showed a pattern similar to that of previous post-extrusion periods. The rate of thrust fault movement, which had accelerated to nearly 50 cm/day on the most active fault just prior to the Sept extrusion, decreased rapidly before stabilizing on 10 Sept. After the Sept extrusion ended, continued slow movement (about $^1/_2$ - 1 cm/day) was measured on some thrust faults around the dome, while other thrusts remained inactive. Similarly, outward movement of one station on the N crater rampart reached more than 10 cm/day before extrusion began; after the extrusion episode ended, rates of outward movement had dropped to 0.25-0.6 cm/day.

The volume of SO_2 emission peaked at 660 t/d during the afternoon of 6 Sept, just prior to the beginning of lava extrusion. During the extrusion episode, emission rates varied from 190 to 310 t/d, then dropped on 11 Sept to 70 t/d, the lowest measured rate for the month. SO_2 emission increased sharply in mid-Sept to 530 t/d on the 17th and dropped to 340 t/d on the 18th; then poor weather stopped data collection until the end of the month, when 2 days of measurements showed a rate of about 250 t/d.

Information Contacts: T. Casadevall, D. Dzurisin, D. Swanson, USGS, Vancouver, WA; C. Boyko, S. Malone, E. Endo, C. Weaver, Univ. of Washington; R. Tilling, USGS, Reston, VA.

10/81 (6:10) Lava extrusion that probably began 30 Oct added a new lobe to the composite dome in the crater, the 9th extrusive episode since the catastrophic eruption of 18 May 1980. After lava extrusion ended 10 Sept (6:8), rates of displacement in the crater remained low for several weeks, as they had after earlier extrusion episodes. Sulfur dioxide emission ranged from 70 to 190 t/d 9-24 Oct, but showed no particular trends.

The crater floor was not quiet during this period. Inflation of the dome has caused small thrust faults to form in the surrounding crater floor. In early Oct, the most active thrust, S of the dome, was moving at about 1.5 cm/day, and stations on the N crater rampart showed outward movement of about 0.5 cm/day. By 24 Oct, these rates had increased to 14.5 and 3.5-4 cm/day respectively, and levelling profiles perpendicular to the dome showed that crater floor tilt rates had reached 400-500 μrad/day, prompting the USGS to issue an advisory prediction of renewed lava extrusion within the next 2 weeks.

Eleven laser targets were installed on the dome and distances to the targets were measured from fixed points 150-1000 m away beginning 21-23 Oct. Rates of lateral expansion of the dome varied considerably, but were as much as 50 cm/day as of 25 Oct. Three continuously recording tiltmeters located 550, 850, and 1050 m N of the center of the dome began returning data 13 Oct. No significant tilt was detected by these instruments until about noon on 26 Oct, when all 3 began to record inflation at 25-50 μrad/day. The number of local earthquakes began to increase on 27 Oct, and gradually became more frequent through the 29th, but most of the events were extremely small. Seismicity increased substantially between midnight and 0700 on 30 Oct, then all but rockfall events ceased fairly abruptly shortly after 0730. During this period, there were a few events that reached roughly magnitude 1, but none were in the M 1.5-1.8 range typical of the larger events prior to past additions to the dome, and the total seismic energy release was much smaller than that associated with previous extrusion episodes. Inflation registered by the 3 continuously recording tiltmeters N of the dome flattened out on 29 Oct and at least 1 instrument began to show deflation by 0800 on the 30th.

Poor weather prevented any observations of the crater 30 Oct. When visibility improved early on the 31st, a substantial quantity of lava had already been extruded from the N part of the pre-existing dome's summit area. The new lobe advanced gradually down the N flank over portions of lobes extruded in April, June, and Sept. Geologists in the crater 31 Oct noted that several plastic airphoto targets N of the dome had melted and some metal fenceposts had been bent, although boards on the ground were not disturbed, indicating that a directed hot blast may have moved 200-300 m N from the dome. By the evening of 1 Nov, the new lava had advanced about halfway down the dome's N flank (fig. 12-19), but was moving more slowly than the previous day. The lobe front moved roughly 20-40 m downslope during the next 24 hours, $^1/_4$ to $^1/_2$ the previous rate of advance. As of late 4 Nov, lava had nearly reached the foot of the April lobe, where a substantial pile of talus had accumulated, and appeared to be advancing slowly, but lava extrusion had stopped by 5 Nov. The volume of the new lobe appeared to be roughly comparable to that of previous lobes (a few million cubic meters).

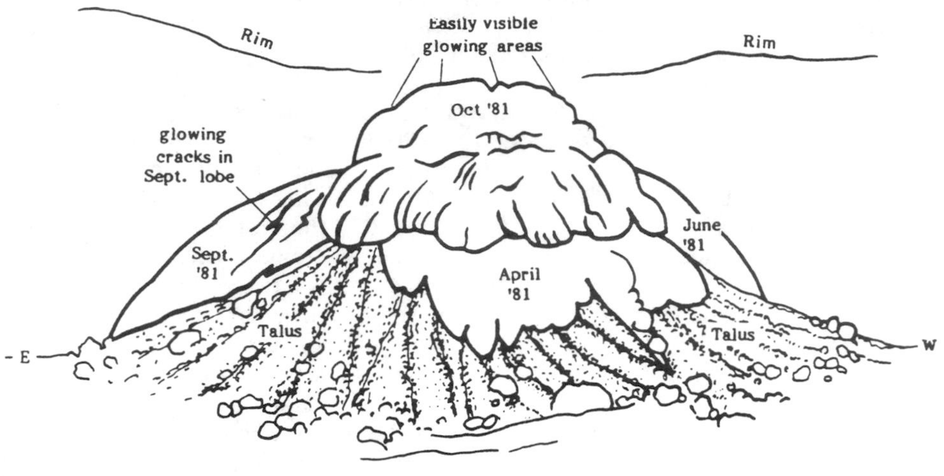

The extrusion was accompanied by major deformation of the dome, particularly at the S end, where dramatic changes took place. Large blocks were thrust upwards 10-20 m and steep outward tilting was obvious. On 31 Oct, the 4 remaining laser targets on the dome showed outward movements of 1-8.2 m since the previous measurements 6 days earlier. However, by 2 Nov, expansion of the pre-existing dome had slowed to only a few centimeters per day and some areas may have been showing subsidence. Large incandescent cracks radial to the center of the new extrusion extended from top to bottom of the Sept lobe, and the June lobe also showed large new cracks. As during previous extrusion episodes, thrust faults stopped moving (on 1 Nov) and significant deformation of the crater floor had not resumed as of 10 Nov. SO_2 emission during the extrusion peaked at 220 t/d (on 31 Oct) then dropped steadily, to 160 t/d on 3 Nov, 130 t/d on the 6th, and 80 t/d on the 10th.

Information Contacts: T. Casadevall, D. Dzurisin, D. Peterson, D. Swanson, USGS, Vancouver, WA; C. Boyko, S. Malone, E. Endo, C. Weaver, Univ. of Washington; R. Tilling, USGS, Reston, VA.

Fig. 12-19. Sketch by Bobbie Myers of the N side of the composite dome at 1400 on 1 Nov 1981, looking S from the mouth of the crater.

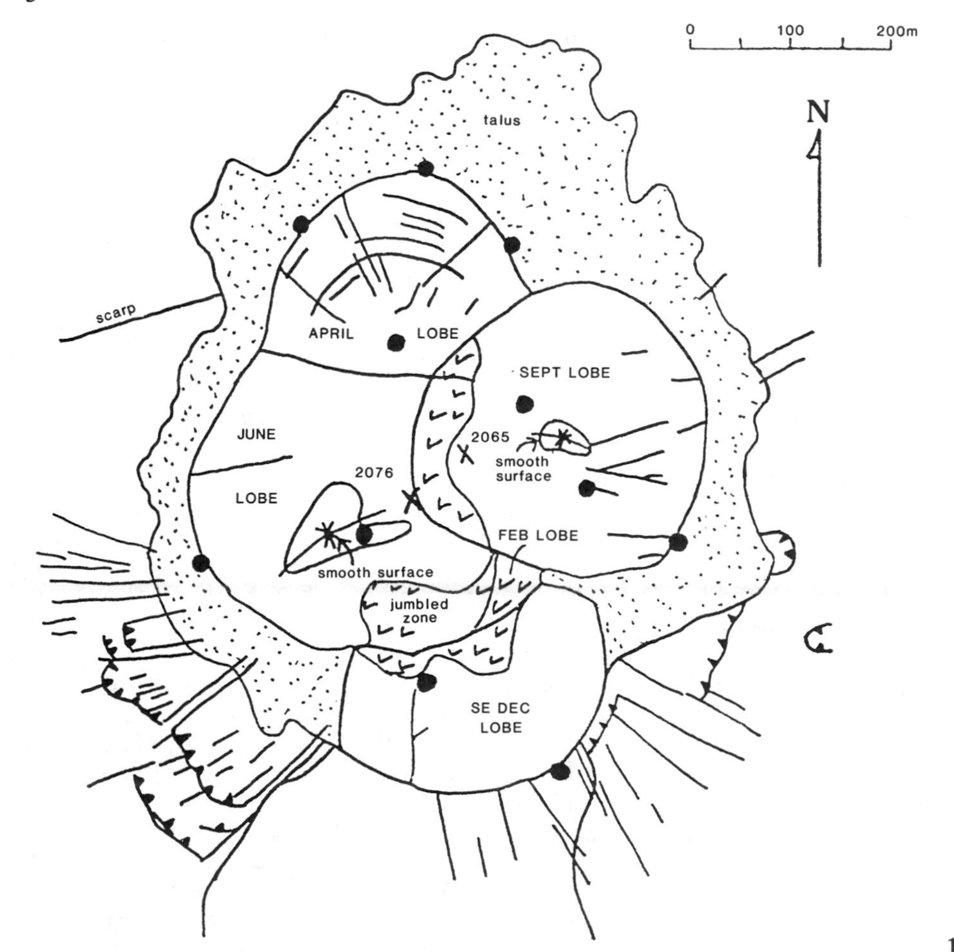

Fig. 12-20. Sketch map of the composite dome before extrusion of the Oct-Nov 1981 lobe, prepared by D. Swanson. The dark circles indicate the positions of the 11 laser targets installed on the dome in late Oct, before the latest lobe was extruded. Some of the small thrust faults in the crater floor around the dome are shown by hachured lines.

11/81 (6:11) Measurable downslope movement of the new lobe extruded onto the composite lava dome (fig. 12-20) had ended by 4 Nov. Distance measurements from fixed points on the crater floor to laser targets on the dome showed that very slow expansion of the entire dome continued through Nov. As of late Nov, the rate of expansion was stable at roughly $^1/_4$ to $^1/_3$ cm/day. No movement occurred during Nov along any of the small thrust faults formed in the crater floor by past expansion of the dome, nor was there any measureable outward movement of the N crater rampart. Gas sampling was infrequent because of poor weather, but no significant changes were evident in the rate of SO_2 emission.

During Nov, seismic instruments recorded 54 earthquakes in the Mt. St. Helens area (some were aftershocks from the Elk Lake earthquake of 13 Feb; 6:2), a typical level of seismicity for a month without an extrusion episode. The largest event was slightly stronger than M 1. Many of the earthquakes resembled steam plume or rock avalanche events, but weather conditions prevented visual confirmation. Medium-frequency events with sharp arrivals, similar to some of the Aug and early Sept earthquakes, were recorded at a rate of about 1/day in early Dec.

A series of seismic signals that indicated rock avalanching in the crater began with isolated 30-45-min periods of seismicity late 4 Dec. Activity peaked the next day, with 14 events of more than 45 seconds duration between 0930 and 1800, the longest a 220-second episode beginning at 1253. Seismic instruments continued to record occasional rock avalanche events during the night of 5-6 Dec. As of press time, weather conditions had prevented observations of any changes that may have occurred in the crater. None of the Dec rock avalanche events were as large as the seismic signal from the 2 Oct avalanche that deposited as much as 7 m of debris in the S part of the crater.

Information Contacts: D. Swanson, USGS [newly established] Cascades Volcano Observatory (CVO), Vancouver, WA; C. Boyko, S. Malone, E. Endo, C. Weaver, Univ. of Washington; R. Tilling, USGS, Reston, VA.

12/81 (6:12) After the entire dome showed very slow expansion through Nov, it subsided about 5 cm between early Dec and early Jan, and spread outward a roughly proportional amount (fig. 12-21). Deformation of the crater floor surrounding the dome has typically begun a few weeks after past lava extrusion episodes, but as of 11 Jan there had been no significant deformation of the crater or of the edifice as a whole. No movement occurred along any of the small crater-floor thrust faults produced by past expansion of the dome, and no outward movement of the N crater rampart was detected. No changes in crater floor tilt were recorded. The average quantity of SO_2 emitted daily has decreased steadily through 1981 and this trend continued in Dec and early Jan. Poor weather limited gas sampling to only 4 days in Dec, when 50-135 t/d of SO_2 were emitted, with an average of about 100 t/d. SO_2 emission rates were measured at 120 t/d on 7 Jan and 66 t/d on the 10th.

Seismicity remained at a low level through Dec. Most of the 64 discrete earthquakes in the Mt. St. Helens area were aftershocks of the 13 Feb Elk Lake event (6:2). Large rock avalanches from the crater walls occurred during periods of poor weather but were detected on seismographs. On 5 Dec, a 1-hour period of seismicity that began about 0200 included 5 events of more than 70 seconds duration, and 2 others of 40-70 seconds in length. Between 1200 and 1700 the same day, 7 more events lasting 70 or more seconds were recorded, the longest (150 seconds) at 1607. Geologists observed rockfall debris covering much of the NE floor of the crater after this seismic episode. Debris from another large rockfall, on the NW crater floor, was discovered 23 Dec after geologists had been prevented from entering the

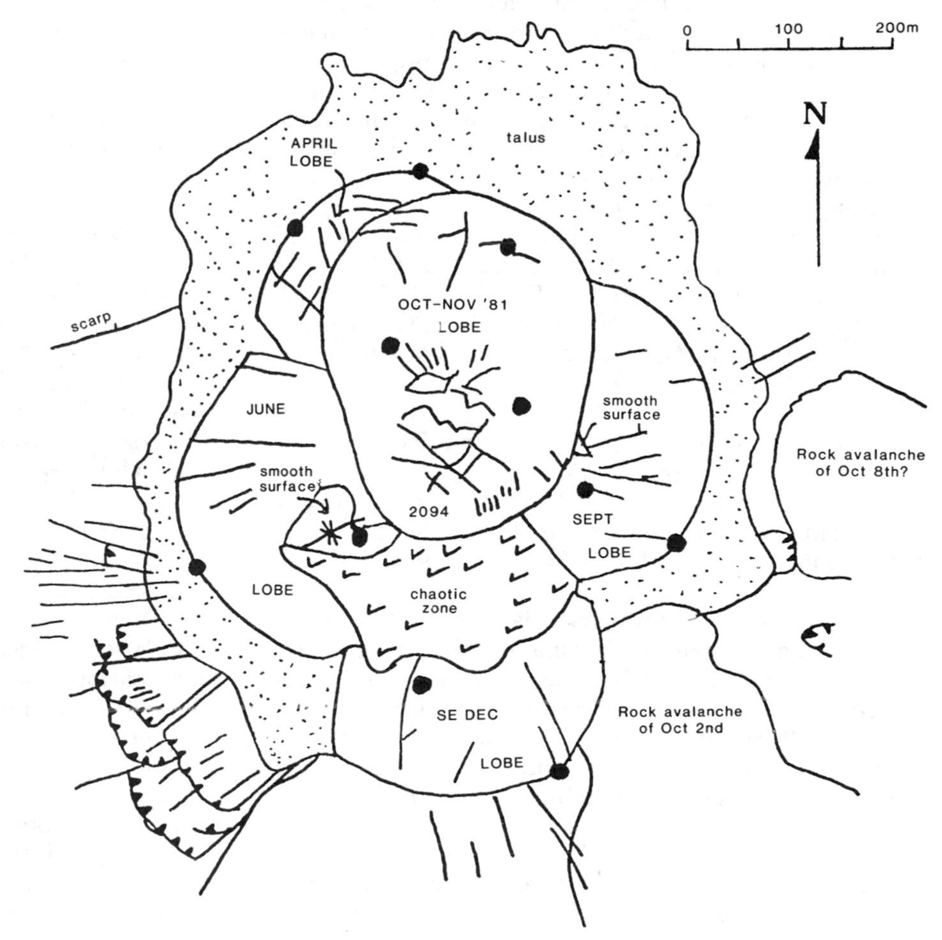

Fig. 12-21. Sketch map of the Jan 1982 composite dome in the crater, prepared by D. Swanson. The dark circles indicate the positions of the 11 laser targets installed on the dome in late Oct, before the latest lobe was extruded. Some of the small thrust faults in the crater floor around the dome are shown by hachured lines.

crater by a week of wet weather. Seismic records did not show any activity that could be unequivocally associated with this rockfall. Most rockfalls have occurred during or just after storms and have not apparently been correlated with volcanic activity.

Information Contacts: T. Casadevall, D. Swanson, USGS CVO, Vancouver, WA; C. Boyko, S. Malone, E. Endo, C. Weaver, Univ. of Washington; R. Tilling, USGS, Reston, VA.

1/82 (7:1) Although deformation of the crater floor has typically begun a few weeks after past extrusion episodes, USGS monitoring had shown few changes within the crater by early Feb. The small crater-floor thrust faults produced by past expansions of the dome had shown no significant movement as of 10 Feb, nor had there been any outward movement of the N crater rampart or inflation of the edifice as a whole. However, telemetry from a crater-floor tiltmeter 1 km N of the dome resumed 4 Feb after a 3-month hiatus and showed 17 μrad of inflation in 4 days. The dome continued to spread very slowly outward through Dec and Jan, at 1-3 mm/day. The rate of SO_2 emission remained low in Jan, usually ranging from 50 to 100 t/d, but increased steadily during the first 8 days of Feb from about 60 to 130 t/d.

During the first half of Jan, there were only 3 seismic events large enough to be recorded on more than 1 seismograph of the Mt. St. Helens net. However, during intermittent periods of increased seismicity in the last 2 weeks of Jan, several events per day were recorded. These active periods lasted as long as 3-4 days before seismicity declined to background levels. A total of 27 earthquakes in the Mt. St. Helens area, a few as large as magnitude 1, were recorded during the second half of the month. Similar activity continued into early Feb.

Information Contacts: D. Dzurisin, J. Ewert, D. Swanson, USGS CVO, Vancouver, WA; C. Boyko, S. Malone, E. Endo, C. Weaver, Univ. of Washington; R. Tilling, USGS, Reston, VA.

2/82 (7:2) No eruptive activity has occurred since measureable downslope movement of the newest lobe of the composite dome stopped 4 Nov. However, seismicity beneath the crater and deformation within the crater began in late Feb, and the USGS expected an eruption by late March.

Only 2 shallow earthquakes large enough to be detected by more than one instrument of the Mt. St. Helens net occurred 1-23 Feb. Seismic activity increased on 24 Feb. Since then, seismographs have recorded about 5 very small (negative magnitude) events per day centered below the crater at 5-12 km depth, and 10-12 somewhat larger (but usually less than M 1) shocks per day at 1-3 km depth. No earthquakes have been located in the zone between the groups of shallow and deeper foci, and no harmonic tremor has been recorded. The deeper earthquakes were the first since 1980, when they often followed (but never preceded) explosive activity.

Deformation within the crater began at about the same time as the increased seismicity. Outward movement of the N crater rampart started in late Feb and had increased to 2.4 cm/day by early March. Horizontal expansion of the lower portion of the composite lava dome increased to 1.6-1.7 cm/day by early March, apparently accompanied by uplift, but conditions prevented measurements of targets on the upper section of the dome. Movement also occurred along the small crater-floor thrust faults, produced by past expansions of the dome, but snow made quantitative monitoring difficult. A dry tilt station about 300 m N of the dome showed inflation during each of 3 measurement intervals (6-24 Feb, 24 Feb-5 March, and 5-8 March), but a tiltmeter 700 m farther N detected no inflation, suggesting to USGS geologists a shallow source near the first tilt station. In contrast, simultaneous inflation was measured at these 2 sites before the last extrusion episode Oct-Nov 1981. No expansion of the edifice as a whole had occurred as of early March. Measurements of SO_2 emission have been infrequent because of poor weather. The rate of SO_2 emission had increased in early Feb and reached 240 t/d on the 10th. Poor weather prevented further measurements until 21 Feb. Since then SO_2 emission has remained at background levels of 80-120 t/d.

On 12 March, the USGS and Univ. of Washington issued a joint extended outlook advisory stating that an eruption was likely within the next 3 weeks. Another dome-building episode was rated as the most probable eruption type, but because of the changed seismic pattern the advisory noted that explosions or lava flows were possible.

Information Contacts: D. Dzurisin, J. Ewert, D. Swanson, K. Cashman, USGS CVO, Vancouver, WA; C. Boyko, S. Malone, E. Endo, C. Weaver, Univ. of Washington; R. Tilling, USGS, Reston, VA.

3/82 (7:3) The first explosive eruption in 17 months ejected a tephra cloud that briefly rose to more than 13.5 km altitude on 19 March. A directed blast from near the base of the lava dome spawned a multilobate avalanche that flowed several kilometers down the volcano's N flank. A mudflow moved down the N fork of the Toutle River, but caused only minor damage. Clouds produced by explosions 20-21 March were much smaller and contained only a little tephra. Lava extrusion began 21 March, adding a new lobe to the SE side of the crater's composite dome. No injuries resulted. Smaller explosions 4-5 April were followed by the extrusion of a second lobe onto the N side of the dome.

Premonitory Activity 26 Feb-18 March. Seismic activity began to increase 3 weeks before the March eruption and included a substantial number of deeper events, in contrast to previous dome extrusion episodes, which were typically preceded by only a few days of shallow seismicity. Earthquakes occurred in 2 zones, at about 1-3 and 4-12 km depth (below average seismic station elevation of about 1 km above sea level). An average of 1 event per day stronger than magnitude 1.5 occurred in the shallow zone 26 Feb-12 March, with the rate of energy release remaining relatively constant. Most of the deeper events had negative magnitudes, and energy release from the deeper zone was about 2 orders of magnitude less than from the shallow zone. The end of deeper seismicity 12 March coincided with both an increase in the number of events (to an average of 3/day of m_b > 1.5 through 17 March) and a jump in the rate of energy release.

Deformation in the crater accelerated rapidly in mid-March. Between 17 and 18 March, uplift of an area near the SW base of the dome accompanied about 30 cm of movement along a nearby thrust, higher than any rate of crater-floor thrust movement previously measured at Mt. St. Helens. Outward displacement rate of the N crater rampart reached 32 cm/day, and a portion of the dome itself expanded 42 cm in the 24 hours ending shortly before the eruption. However, no deformation of the edifice as a whole was detected by measurements outside the crater. For the first 18 days of March, the rate of SO_2 emission averaged 110 t/d, remaining at about the same level as it has since the lava extrusion episode of Oct-Nov 1981.

After remaining approximately constant for several days, the rate of seismic energy release increased again about noon 18 March, and 14 events larger than magnitude 1.5 were recorded in the next 24 hours. A few brief (1-2 minutes or less) periods of low-level harmonic tremor were recorded during the afternoon of 19 March, as were 20 discrete events stronger than magnitude 1.5. SO_2 emission doubled to about 230 t/d. Tilt measured about 300 m N of the dome reversed about 1900 and seismic data showed that explosions began at 1928. After 2 minutes of initial seismicity there was a brief hiatus, followed by about 40 minutes of activity that declined gradually.

Explosive Eruption 19 March. A vertical tephra column, probably ejected from a vent near the center of the dome, reached its maximum altitude of more than 13.5 km (as measured by radar at Portland airport) at 1933 on 19 March. By 1950, radar data indicated that the altitude of the top of the column had dropped to 10.5 km. An infrared image returned at 2003 by a NOAA geostationary weather satellite showed a cloud-top temperature of -35°C, yielding an altitude of about 7 km. According to radar data, the eruption column contained 20-60 times less tephra than the cloud produced by the last significant explosion, in Oct 1980 [but see 7:4]. Ash blew SE at about 30 km/hr. Light ashfalls were reported as much as 80 km away, but caused only minor disruptions to auto travel. Bombs up to 3 m across fell 200-300 m from the dome. Frothy pumice (density about 0.8) fell 8 km away. Smaller explosions occurred at 0135 the next morning, when radar detected a cloud, containing a little ash, that rose to about 5.5 km altitude, and a small steam and ash column was ejected at 0415 on 21 March.

[Further investigation revealed a more complex sequence of events than was originally reported in the *Bulletin*. The following has been modified by R. Waitt and D. Swanson. A detailed description can be found in Waitt and others, 1983.] The initial avalanche apparently resulted from a directed blast that emerged from near the SW base of the dome. This blast destroyed the dome's SW margin, and struck the S wall of the crater, removing snow cover and rock. The resulting mixture of snow granules (0.5-2 mm in diameter), hot pumice, and lithic material [descended] the [E and] S crater walls, [flowed around] the E and W sides of the dome, joined N of it, then flowed out through the breach in the N side of the crater and continued for several kilometers down the N flank. Fed by water from the avalanche . . . and a [transient] pond [behind the dome], a complex mudflow sequence moved down the N fork of the Toutle River, which flows W . . . at the N foot of the volcano. Upstream deposits showed evidence of 2 distinct pulses, but gauges downstream registered only 1 well-defined peak. About 70 families were evacuated from the Toutle valley, but no major damage was reported. The mudflow buried trucks at an earthen flood-control dam and breached its S side. Three storms earlier this winter produced higher peak river stages at Castle Rock, roughly 70 km downstream. Floods produced by these storms had breached the N side of the dam and the combined damage has essentially destroyed the dam's effectiveness.

Lava Extrusion 20-24 March. Seismographs began to record rockfall events, probably associated with extrusion of a new lobe of lava, during the evening of 20 March. This activity slowly increased, and aerial observers first saw the new lobe during the night. It emerged from a vent at the top of the most recent lobe (extruded Oct-Nov 1981) and flowed down the SE side of the dome, barely reaching the crater floor. Growth was fairly rapid through 23 March, but there was little apparent increase in size between the 23rd and 24th, and the number of rockfall events was noticeably declining early 24 March. By the time growth slowed, the volume of new lava appeared to be greater than that for any previous lobe. SO_2 emission increased to 370 t/d on 21 March, about 3.5 times background levels, but had dropped to 90 t/d by 24 March.

However, before dawn on 24 March new glowing radial cracks were observed in older portions of the dome. The N crater rampart and the N side of the dome showed 12 cm of outward movement between the mornings of 23 and 24 March and 16-18 cm during field work 24 March. No unusual seismicity accompanied the movement, nor was any significant tilt measured N of the dome, but at similar stages of previous dome extrusion episodes, little or no deformation of any kind has been observed.

Poor weather prevented geologists from entering the crater again until early April. Seismicity remained at low levels through the end of March. SO_2 emission dropped to about background levels 24 March, but by the next measurement, on 28 March, had increased to about 200 t/d and reached a rate of 440 t/d during a small gas explosion. On 29 March, the rate was still high, at 180 t/d, but weather conditions prevented further measurements until a week later. Seismographs began to record a few very small, brief (20 seconds or less) harmonic events 1-2 April, and these became more numerous 3-4 April. Occasional low-frequency earthquakes began to appear on the seismic records 3 April. A few were recorded the next morning, then these events increased to about 2 per hour after 1400. A further increase in seismicity was noted in the early evening, and at about 2000, Univ. of Washington seismologists alerted USFS and Washington state officials that an eruption was imminent.

Renewed Explosions and Dome Growth 3-12 April.

[A large rock avalanche and] explosive activity began at 2052 on 3 April, and 3 seismic pulses occurred in 3 minutes. A plume containing a little ash rose to 8.5 km (altitude data from Portland airport radar) and drifted NE. Minor ashfall was reported in Packwood, 65 km away. Seismographs recorded pulsating activity for the next several hours, then a pair of stronger events at 0035 and 0039 that accompanied the ejection of an ash-poor cloud to almost 10 km altitude (as measured by Portland airport radar). A small mudflow emerged from the breach in the N side of the crater and flowed a short distance down the N flank. After 10-15 minutes, seismicity briefly dropped to background levels, but apparent harmonic tremor began about 0230 and continued for the next 14 hours. Gas and/or rockfall events began at roughly 0330 and became increasingly frequent during the next several hours. Before dawn, geologists observed a new lobe of lava on the N side of the composite dome. Growth of this lobe continued through 8 April, but had slowed considerably by the 9th. The April lava, perched on the N side of the dome, looked very similar to the Oct 1981 lobe but appeared to be smaller than any previously extruded. Gas emission events, including one that sent a plume to 7 km altitude at 1719 on 5 April, could be seen on seismic records, as well as large avalanche events as large chunks fell off the dome. Seismicity declined gradually as lava extrusion continued and had dropped to low levels by 12 April. By 10 April, deformation in the crater had decreased to levels typical of periods between extrusion episodes. As lava extrusion was beginning early 5 April, the rate of SO_2 emission increased to 900 t/d, dropping to 500 t/d during the afternoon, and to 390 t/d, a typical value during dome extrusion episodes, on 6 April. No gas data were available 7 April but SO_2 emission had returned to background levels 8-10 April.

Information Contacts: T. Casadevall, R. Janda, C. Newhall, D. Swanson, R. Waitt, USGS CVO, Vancouver, WA; C. Boyko, S. Malone, E. Endo, C. Weaver, Univ. of Washington; O. Karst, NOAA/NESS; D. Harris, Univ. of Alberta; R. Bailey, USGS, Reston, VA.

Further Reference: Waitt, R.B., Pierson, T.C., MacLeod, N.S., and Janda, R.J., 1983, Eruption-Triggered Avalanche, Flood, and Lahar at Mount St. Helens-Effects of Winter Snowpack; *Science*, v. 221, no. 4618, p. 1394-1396.

4/82 (7:4) The tephra volume of the 19 March eruption column was of the same order of magnitude as (but probably slightly less than) that of the Oct 1980 cloud (not 20-60 times less, as reported in 7:3). Extrusion of lava onto the N side of the composite dome stopped by 10 April and seismicity had dropped to low levels by the 12th, but another dome-building phase began on 14 May.

During the 19 March activity, the crater-floor thrust fault scarps near the dome were either buried or scoured away. Measurement of the rates of movement of these small faults had previously been an important deformation monitoring technique. By early May a number of small new thrusts had formed on the crater floor W and SW of the dome, but had not yet yielded useful deformation data. Data from a continuously recording tiltmeter NW of the dome varied considerably from dry tilt measurements made as little as 100 m away, indicating that (as in previous inter-eruption periods) the crater floor was deforming as a group of relatively small, independent blocks rather than as a coherent unit.

Rates of SO_2 emission have remained at about 2 times background levels since the end of lava extrusion 10 April. Ejection of small, ash-poor plumes from the dome's summit began 21 April, and geologists observed 1-2/day through early May. Many of the plumes could be correlated with seismic events, some of which were felt in the crater. Comparison of distance measurements made immediately before and after individual plume ejections showed that points on the dome had moved outward and upward by 3-5 cm after each small explosion.

Swelling of the composite dome was relatively slow until early May, then accelerated. By 11 May the SW portion of the dome was moving outward as much as 70 cm/day. Other areas of the dome were less active. Although the SW portion of the dome had also experienced the most rapid expansion prior to the Oct-Nov 1981 eruption, the lobe extruded during that episode emerged from the N part of the dome's summit area.

The number of local earthquakes began to build about 8 May and by 12 May had reached about 12 events per day (magnitude greater than about 1). All were shallow and had magnitudes of less than 2. Many were felt by geologists working near the dome.

Because of the increasing seismicity and deformation, the USGS and Univ. of Washington issued a joint extended outlook advisory late 11 May stating that an eruption was likely to begin within the next week, possibly within the next few days. The number of shallow, low-frequency events grew considerably late 13 May. Lava began to emerge from the summit of the composite dome between 0100 and 0200 on 14 May and began to flow down the dome's NE flank. Harmonic tremor increased substantially about 0200, and a steam plume rose about 2 km above the crater rim. Deformation and growth of the dome was continuing on 17 May, but seismicity had gradually decreased.

Information Contacts: T. Casadevall, C. Newhall, D. Peterson, D. Swanson, USGS CVO, Vancouver, WA; C. Boyko, S. Malone, E. Endo, C. Weaver, Univ. of Washington, D. Harris, Univ. of Alberta.

5/82 (7:5) Lava began to flow down the NE side of the dome 14 May, but the bulk of the new lava formed a lobe on the dome's NW flank 15-19 May. Since then, ejections of steam and ash, similar to those of July and Aug 1981, have occurred about once a day. Two of these, on 7 and 8 June, caused light ashfalls on Portland. Gas emission rates remained high through early June.

The crater-floor deformation before the extrusion was blocky and incoherent, as during previous pre-extrusion periods. While a continuously recording tiltmeter at a new site on the W crater floor recorded increasingly rapid subsidence (as did a dry tilt station at the same location), reoccupation of dry tilt stations less than 100 m away showed accelerating uplift. The development of a small thrust fault was observed between the dome and the continuously recording tiltmeter, leading Dan Dzurisin to suspect that thrusting was responsible for the different tilt directions at nearby sites.

Local seismicity had begun to increase on 8 May (7:4). In the 24 hours starting at 0700 on 13 May, 63 earthquakes were recorded (about twice the previous day's number) and some were felt by geologists working in the crater that day. Three radiating fractures, trending NE, N, and NW, were seen in the April lobe, on the N side of the dome. SO_2 emission remained at background levels of about 100 t/d. The rapid subsidence measured by the continuously recording tiltmeter stopped about midnight. Harmonic tremor started shortly thereafter, at 0055 on 14 May, and continued until about 0600. Bursts of seismic energy could be seen within the tremor. During an overflight at 0415, spectacular, nearly continuous cascades of incandescent material could be seen on the NE flank of the dome, but an hour later the rockfalls had ceased almost entirely. After dawn, a jumbled, blocky area could be seen on the dome's summit and upper NE flank, and there was a rockfall apron on the NE side of the dome. The jumbled area was larger by afternoon, but it was not certain whether it was new lava or scoriaceous older material being uplifted by endogenous dome growth. Episodic gas emission was observed on 14 May, and by afternoon the rate of SO_2 release had increased fourfold from the previous day, to about 400 t/d. The number of earthquakes decreased to 20 in the 24 hours starting at 0700 on 14 May.

On 15 May, rockfall activity continued on the dome's NE flank and rockfalls began on its NW side. The surface morphology of the N side of the April lobe was changed, but no movement was visible, and growth appeared to be occurring within the April lobe. However, by afternoon, flow texture had developed on a tiny lobe on the NE flank and a dominant lobe on the NNW flank. The number of recorded earthquakes declined to 11 between 0700 on 15 May and 0700 on the 16th. Downslope movement of the NNW flank lobe continued on 16 May, but the tiny NE flank lobe was stagnant. The number of earthquakes had dropped to only 1-2/day. Poor weather prevented observations 17-18 May, but seismographs detected numerous rockfalls. When geologists returned to the crater 19 May, rockfalls and some downslope movement of the NNW flank lobe were continuing, but little extrusion appeared to have occurred since the 16th. Only minor rockfalls were observed 20 May and these had stopped by evening.

The rate of SO_2 emission remained high 15-19 May, at 250-650 t/d. On 19 May, several fissures, surrounded by a small tephra blanket, were observed in the top of the March lobe. That day, a plume from these fissures quickly increased the SO_2 emission rate from 340 to 2600 t/d, but the rate dropped back to 500 t/d after about 50 minutes. An average of one large gas and ash ejection per day has occurred through early June. Many caused light dustings of ash near the volcano. The ash consisted of abraded and rounded lithic fragments and crystals, but included no fresh magma, although some of the larger fragments were hot. The highest observed plume rose to 5.5 km altitude late 6 June. Light ashfalls occurred in Portland early 7 and 8 June. SO_2 emission continued at an elevated level of 200-300 t/d through early June.

USGS analyses showed no significant chemical differences between the lobes extruded in 1981 and in March, April, and May 1982. All contain 62-63% SiO_2 and 40-42% phenocrysts (table 12-3).

Information Contacts: T. Casadevall, K. Cashman, D. Dzurisin, C. Newhall, USGS CVO, Vancouver, WA; C. Boyko, S. Malone, E. Endo, C. Weaver, Univ. of Washington.

6/82 (7:6) After downslope movement of lava extruded onto the N flank of the composite dome ended about 19 May, activity was limited to ejection of vapor plumes that sometimes contained tephra. Vigorous gas emission from the top of the dome produced plumes that lasted 5-45 minutes. Old material from the walls of the vent was carried upward by the gas, and blocks 20-30 cm across fell about 125 m from the dome. On 9 June at 1316, the largest of these rose to 6 km altitude, dropping ash a few kilometers to the N and larger tephra in the crater. The accompanying high-amplitude burst of seismicity saturated nearby seismographs but appeared to consist of a series of individual events. Before 9 June, most of the plume emissions were associated with lower amplitude seismic signals, but about half of those after 9 June were accompanied by bursts of stronger seismicity. Plumes were ejected about once a day until 22 June, but none has been observed since then. Before the May extrusion episode, similar gas and tephra ejections were typically preceded by 1-3 minutes of felt earthquakes, and electronic distance measurements showed that they were accompanied by several-centimeter expansions of the dome. However, neither precursory earthquakes nor deformation were associated with the post-extrusion plume emissions. The rate of SO_2 emission had remained high for several weeks after the May lava extrusion, but returned to the normal inter-eruption background level of about 100 t/d by early June.

Electronic distance measurements to targets on the

SiO_2	Al_2O_3	Fe_2O_3*	MgO	CaO	Na_2O	K_2O	TiO_2	P_2O_5	MnO	Total
62.5	17.8	5.28	2.27	5.40	4.42	1.28	0.72	0.16	0.08	99.91

*Total iron as Fe_2O_3

Table 12-3. Analysis from the USGS analytical laboratory, Denver, CO, averages 3 samples from the May 1982 lobe. Each sample was split into 3 groups before analysis and all were run with an internal standard. Data provided by Kathy Cashman.

dome and crater floor began to show [swelling] in mid to late June. The rate of [swelling] was a few millimeters per day in early July and was not accelerating significantly. Tilt stations on the crater floor also showed uplift, at a rate about twice as high 7 June-6 July as mid May-early June. The 7 June-6 July rate was similar to that observed several weeks before the May lava extrusion.

Information Contacts: T. Casadevall, W. Chadwick, D. Dzurisin, USGS CVO, Vancouver, WA; C. Boyko, S. Malone, E. Endo, C. Weaver, Univ. of Washington.

7/82 (7:7) Electronic distance measurements from sites on the crater floor to targets on the composite dome registered [swelling] beginning in mid-June. The rate of [swelling] remained a few millimeters/day until the end of July, when it accelerated to a few centimeters/day. On 27 July local seismicity increased to 3-4 events/day, then stabilized on 2-3 Aug to 1-2 events/day. By 13 Aug, 3-4 events/day were again occurring. The rate of SO_2 emission has remained at the inter-eruption background level of about 100 t/d.

The seismic pattern and the rise in displacement rate prompted the USGS and Univ. of Washington to issue an extended outlook advisory on 30 July, predicting an eruption within 3 weeks. Seismicity and displacement rates continued to increase. On 16 Aug an eruption was predicted within 4 days, then within 24 hours on 17 Aug.

The dome began to grow endogenously on 17 Aug. The W and SW sides were expanding at a rate of about 10 m/day. Numerous rockfalls were occurring, but without explosions or changes in rates of deformation, displacement, seismicity, or gas emission. Extrusion of a new lobe on top of the dome began during the day on 18 Aug, with lava flowing slowly onto the dome's W and S sides.

Information Contacts: W. Chadwick, C. Newhall, USGS CVO, Vancouver; S. Malone, Univ. of Washington.

8/82 (7:8) Lava extrusion that began early 18 Aug added a new lobe to the composite dome during the next 5 days. Seismicity and increasing deformation of the dome and the surrounding crater floor preceded the eruption, which included about a day of strong endogenous dome growth before lava appeared at the surface.

Distances and vertical angles were measured between 5 sites on the crater floor and targets (5-6/station) on the dome. Rapid acceleration of dome deformation started about 6 Aug, but was strongly asymmetrical, concentrated on the W side. Some targets on the W side of the dome were moving outward at 30 cm/day by 15 Aug, but distances between the crater floor and points on the E and N sides of the dome shortened by no more than 4 and 5 cm/day respectively, and no movement occurred there until a week before the eruption.

The number of earthquakes that could be located by the Univ. of Washington-USGS seismic net (usually events with magnitudes of 1 or greater) began to build in late July (7:7), then increased to an average of 6/day 11-13 Aug and about 12/day 14-16 Aug. All were centered within a few kilometers of the surface. Smaller events ($M \leq 0$) showed comparatively steady increases through this period.

Continuously recording tiltmeters 50-60 m W of the dome measured thousands of μrad of tilting. A tilt reversal had been noted a few hours before the start of some earlier extrusion episodes, but no such reversal preceded the Aug eruption. Rapid tilting was recorded until rockfalls from the new lobe ended transmission.

Rapid thrusting began in mid-Aug along cracks in the crater floor W of the dome, reaching rates of 2.5 m/day laterally and 1.5 m/day vertically along the most vigorous fault just before the eruption. The thrusting subdivided the W crater floor into numerous small blocks, and one dry tilt station was cut by a 3 m thrust scarp. Tilt stations will be re-established but on a part of the crater floor less prone to thrusting. Rapid endogenous growth of the W side of the dome began 17 Aug. Parts of the W side of the dome moved 12 m outward that day and 22 m of displacement (13 m [downward]) of one W side dome target was measured the next day. The character of seismicity began a gradual change on 17 Aug, when only 2 events occurred that were large enough to be located, and surface (rockfall or gas emission) events started to dominate the seismic records.

New lava was first seen on the surface of the dome at 1130 on 18 Aug, emerging from the NW side of the summit. After the new lava appeared, the June 1981 and March 1982 lobes began to crack apart, and large blocks rolled down the W side of the dome. When geologists left the crater at about 1800, the new lobe was still small and was confined to the summit area.

By the next morning, lava had flowed onto the NW flank and was moving down the talus slope at the base of the dome. Substantial growth of the new lobe continued through the 19th, and slower extrusion occurred for the next few days before ending between 22 and 23 Aug. Slow deformation of the E side of the dome (where targets remained in place during lava extrusion) continued through 20 Aug, then stopped, and had not resumed as of early Sept.

SO_2 emission remained at the background level of 80 ± 50 t/d through 17 Aug, but increased sharply, to about 500 t/d, when lava extrusion began early 18 Aug (fig. 12-22). After reaching a maximum of 530 ± 100 t/d that afternoon, the rate of SO_2 emission declined gradually to 150 t/d on the 22nd, then remained at roughly 130 t/d through the end of Aug.

Information Contacts: D. Dzurisin, C. Heliker, C. Newhall, R. Symonds, USGS CVO, Vancouver, WA; S. Malone, Univ. of Washington.

9/82 (7:9) As of 5 Oct, no significant deformation of the lava dome or the surrounding crater floor had been observed since the Aug extrusion episode, although some spreading and subsidence of the new lobe continued. SO_2 emission remained very uniform, ranging from 50 to 135 t/d and averaging 95 ± 25 t/d in Sept. Seismicity remained at background levels through early Oct.

Information Contacts: T. Casadevall, D. Swanson, USGS CVO, Vancouver, WA; C. Boyko, Univ. of Washington.

10/82 (7:10) Seismicity remained at background levels through early Nov, and no significant deformation has been measured. SO_2 emission ranged from 20 to 70 t/d in Oct, the lowest rate since the eruption of 18 May 1980.

Information Contacts: T. Casadevall, D. Swanson, USGS CVO, Vancouver, WA; S. Malone, Univ. of Washington.

11/82 (7:11) As of early Dec, no significant deformation had been measured. Snow conditions in the crater limited vertical angle and distance measurements between the dome and the crater floor, but comparison of 24 Nov and 4 Dec data showed no change. A continuously recording tiltmeter was installed 19 Oct on the E crater floor, within 25 m of the talus at the base of the dome. This instrument measured daily oscillations in tilt, but had recorded no deformation exceeding the diurnal variation as of 6 Dec. The rate of SO_2 emission declined to an average of about 35 ± 10 t/d in Nov, the lowest measured since 18 May 1980. Before the Aug extrusion episode, the average rate of SO_2 emission between eruptions had been roughly 100 t/d. Seismicity remained at background levels through Nov.

Information Contacts: T. Casadevall, D. Dzurisin, D. Swanson, USGS CVO, Vancouver, WA; R. Norris, Univ. of Washington.

12/82 (7:12) Gas emission, deformation, and seismicity remained quiet through the end of Dec. SO_2 emission ranged from 20 to 35 t/d, similar to the decreased amounts measured in Nov. No significant deformation of the composite lava dome or the edifice as a whole was detected. Some sagging of the dome continued, but at declining rates. Seismicity remained at background levels. Seismic records showed a few signals that may have been generated by small avalanches, but no gas emission events were detected.

Information Contacts: T. Casadevall, D. Swanson, USGS CVO, Vancouver, WA; S. Malone, Univ. of Washington.

Further Reference: Special Section: Mount St. Helens; *Science*, 1983, v. 221, p. 1369-1396 (9 papers).

1/83 (8:1) Increases in SO_2 emission, deformation, and seismicity preceded a series of small explosions and the extrusion of a new lobe onto the composite lava dome, the first since Aug 1982.

The rate of SO_2 emission, which had remained very low for several months, tripled between measurements 13 and 15 Jan and remained between 70 and 120 t/d through the end of the month. About 20 small shallow earthquakes were recorded 17-18 Jan, but seismicity declined and remained at background levels for the next 2 weeks. Heavy snow in the crater made deformation measurements impossible on the S and E sides of the dome, but very slow acceleration in the rate of outward movement of the dome's N side began in mid-Jan. A few small gas and ash emissions occurred in late Jan.

Gas monitoring on 30 Jan showed that SO_2 emission had increased to roughly twice the rate of the previous 2 weeks, and SO_2 flux ranged from 170 to 260 t/d through 7 Feb. On 31 Jan, a pronounced acceleration was measured in the outward movement of the N side of the dome. Points on the W side of the dome, usually the area of most rapid outward movement, showed little such activity, but sagged downward several tens of centimeters. A gradual, slight increase in the number of seismic events began 1 Feb, but seismicity remained relatively weak, reaching about the level of the 17-18 Jan activity.

At 2339 on 2 Feb and 0256 the next morning, explosions sent plumes containing small amounts of ash to about 6 km altitude. A pilot reported that the cloud top was at 8 km altitude at 0015 on 3 Feb. GOES East satellite images showed the plumes moving slowly NW. At 0430 a cloud about 150 km in diameter remained centered over the volcano, but it had begun to diffuse 30 minutes later and by 0530 had reached nearly to Puget Sound, about 100 km from the volcano. Ashfall was reported at Olympia, near the S end of Puget Sound. During a predawn flight 3 Feb, geologists observed that the explosions had created a small notch in the upper E

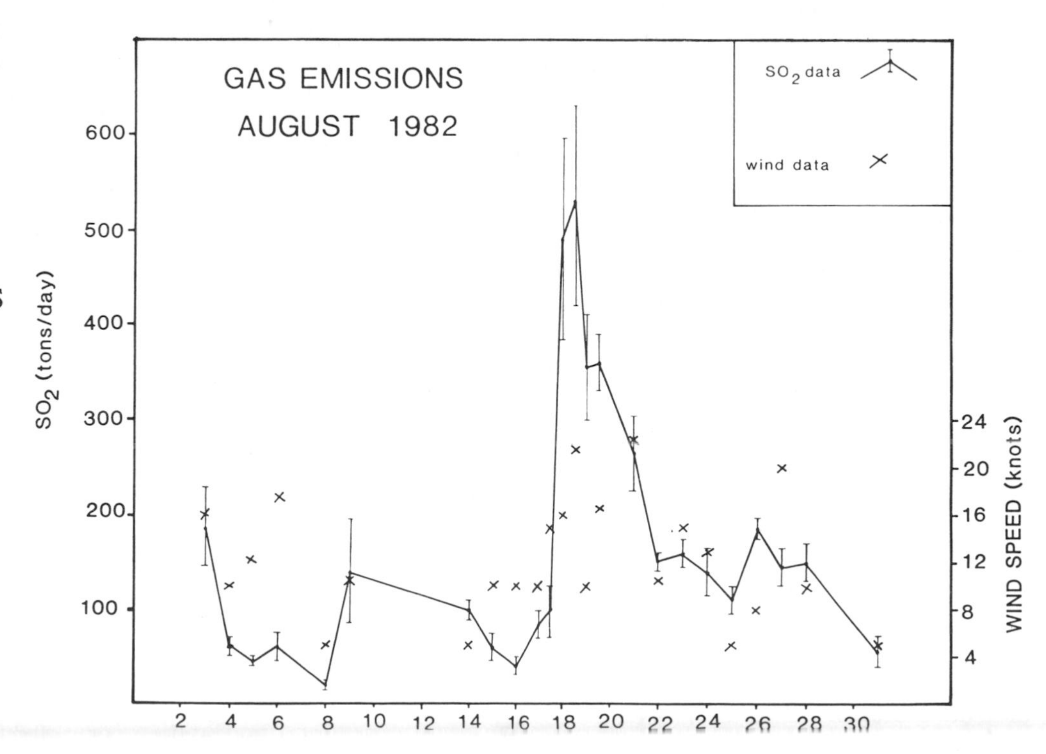

Fig. 12-22. Rates of SO_2 emission at Mt. St. Helens, Aug 1982. Courtesy of Robert Symonds.

flank of the dome. Within the crater, the deposits from these explosions showed a complex stratigraphy. Rare breadcrust bombs were found at the top of the deposits. A laterally directed component from one or both of the explosions melted snow on the E crater floor and wall, producing a mudflow that reached Spirit Lake. The ash column from a third explosion on 4 Feb at 1728 reached about 4.5 km altitude. This explosion enlarged the flank notch to 60-100 m deep and 80-100 m wide. Deformation data on 3 and 5 Feb showed continued acceleration of outward movement of the N side of the dome, reaching 5-6 cm/day by the 5th. Visual observations showed severe deformation of the E side of the dome, where a large wedge of rock just S of the notch had tipped up and out several meters. Locatable seismic events stopped 5 Feb, and only events typically associated with steam emissions and rockfalls were detected during the next several days.

Late 5 Feb, the USGS and Univ. of Washington issued an extended outlook advisory notice stating that an eruption was likely within the next week and could include some explosive activity. Poor weather prevented observations until about noon on 7 Feb, when geologists observed the extrusion of a new lobe of lava from the floor of the E flank notch. Lava advanced mainly toward the E, filling the notch, and by afternoon had reached the top of the talus pile at the base of the dome. From the air, geologists estimated that the new lobe extended roughly 100 m E-W and 50 m N-S. A small explosion occurred from the dome at 1640 on 7 Feb. Weather conditions prevented access to the crater for the next few days, but seismographs recorded rockfall events, suggesting that the new lobe continued to advance. Glimpses of the dome beneath low weather clouds 14 Feb indicated that the new lobe was still growing.

Information Contacts: T. Casadevall, C. Newhall, D. Swanson, S. Brantley, USGS CVO, Vancouver, WA; S. Malone, Univ. of Washington; D. Haller, NOAA/NESDIS.

2/83 (8:2) Extrusion of a new lobe onto the E flank of the composite lava dome stopped by 2 March. However, a renewed increase in seismicity was evident by 4 March and an eruption was expected by the end of the month. Poor weather hampered observations throughout Feb and early March, denying geologists access to the crater and views of the dome on most days.

Because of the weather, it was difficult to determine exactly when the Feb lava extrusion began. Between 30 Jan and 4 Feb, 21 gas and ash explosions were observed on seismic records and by FAA radar at Portland airport. Infrared photographs from an overflight during the night of 4-5 Feb did not appear to show new lava in the notch in the dome's upper E flank, but at 0930 on the 5th it contained a large smooth-sided creased rock (about 10 × 20 m in lateral dimension and 10 m high), probably the new lobe. An overflight during the night of 5-6 Feb showed that a substantial amount of new lava had been extruded. Univ. of Washington seismologists note that for most extrusion episodes at Mt. St. Helens, surface events (principally rockfalls) have begun to dominate the seismic record at about the time that lava extrusion started. Surface events began to increase noticeably late 4 Feb, although most were small and the number of surface events did not exceed the number of subsurface earthquakes until early 6 Feb. Gas emission events were also recorded [seismically], some of which were associated with [observations of] minor vapor and ash plumes. By the afternoon of 7 Feb, subsurface earthquake activity had decreased to 1-5 events per day and remained at that level through the end of the month.

After observing the new lobe on 7 Feb, geologists were next able to see it on the 11th, when it appeared to have grown in size by about 30%. A well-developed, smooth-sided crease was oriented along the long axis of the lobe's surface. Similar features observed during late stages of extrusions in Dec 1980, and Feb, June, and Sept 1981 were thought to represent the last material extruded from a vent and not disrupted by later flow. Although rockfall seismicity continued, indicating that the lobe was advancing, little growth was apparent between observations 11 and 15 Feb. Measurements on 23 Feb showed that the new lobe had advanced 23.5 m to the E since the 11th.

A type of seismicity not previously recorded at Mt. St. Helens was first detected 14 Feb. Hundreds of tiny events that were remarkably similar to each other (many were identical for as much as 20 cycles) occurred at an average interval of 40 seconds. In the 24 hours beginning at 0830, 559 of these events were recorded, but none was large enough to locate and their origin is uncertain.

Television footage 21 Feb showed a spine, not present on the 19th, growing from the center of the lobe. On 24 Feb, the spine was roughly 30 m tall, and by the 28th it had roughly doubled in height, extending about 20 m above the dome's summit. The relatively undisturbed growth of the spine indicated that little downslope movement of the lobe was occurring during this period. On 28 Feb, geologists noted that no rocks had fallen from the lobe front onto the previous night's snowfall. Observations 1 March indicated that extrusion had ended. The new lobe had filled all but 10-15 m of the 60-100 m-deep notch, oozed out its E end, and reached the E foot of the dome (fig. 12-23). Geologists estimated that it was roughly the same size as previous lobes. Total seismic energy released during the Feb extrusion episode was comparable to that associated with previous extrusions, but occurred over a longer time span.

Deformation data were limited, but indicated that little swelling of the dome was associated with the extrusion. Although the W side of the dome has usually been the area of most rapid outward movement, none was measured until 7 Feb, and only 0.7 cm of expansion occurred between then and 1 March. During the same period, the N side of the dome moved outward 7-8 cm, but deformation there began before the extrusion (8:1) and shortening of measured lines totaled about 20-25 cm. Visual observations indicated substantial deformation of the dome's E side, but no instrumental measurements were possible.

SO_2 emission peaked on 15 Feb, reaching 400 t/d, about twice the early Feb rate (8:1). By late Feb, SO_2 emission had dropped to slightly more than 100 t/d and the average for the month was about 170 t/d.

The number of surface seismic events remained steady through early March, but an increase in subsurface earthquakes was evident by 4 March. Six subsurface events were recorded on 1 March and 10 on the 3rd; the 12 events on 4 March were larger, so energy release was substantially higher. Energy release continued to accelerate significantly 5-6 March, and 25 subsurface events were recorded on the 6th. Because of the increased seismicity, the USGS and Univ. of Washington issued an advisory notice 6 March stating that renewed eruptive activity could be expected. Poor weather prevented deformation measurements that have previously been successfully used to predict the time of eruption onset.

Seismic energy release declined late 6 March, and only 12-15 events were recorded daily 7-9 March, but both values remained significantly above background levels. An updated advisory notice issued 8 March suggested that an eruption would begin within the next 3 weeks. On 9 March, observers in a helicopter saw that most of the spine had fallen, but did not report the presence of any additional new lava. Measurements on the N and W sides of the dome 10 March did not show large acceleration of displacement rates.

Information Contacts: T. Casadevall, C. Newhall, D. Swanson, B. Myers, S. Brantley, USGS CVO, Vancouver, WA; S. Malone, Univ. of Washington.

3/83 (8:3) CVO personnel report that frequent rockfalls from the toe of the Feb lobe and changes to its morphology, coupled with elevated SO_2 emission and small seismic events, may reflect continuing endogenous dome growth. Poor weather limited visibility and restricted access to the crater through March and early April.

A new incandescent radial fissure was observed on the NE side of the Feb lobe during a night overflight 23 March and remained visible through the end of the month. In the past, such fissures have typically formed during periods of rapid endogenous growth. By 31 March a mound of rubble estimated from brief aerial observations to be roughly 50-60 m in diameter and 20-30 m high had developed on the lobe, in the area where a spine had been extruded in late Feb. Frequent rockfalls from the E end of the lobe were continuing on 11 April, but conditions in the crater prevented geologists from mapping changes to the lobe front. Gas and ash explosions, some of moderate size, were observed or detected seismically roughly 1-2 times per day through early April. Most appeared to originate from fumaroles at the top of the dome. The main fumarole, a conical pit tens of meters in diameter at the surface and tens of meters deep, was located at the head of the E flank notch that was the source of the Feb lobe.

Rates of SO_2 emission remained elevated through early April, averaging 150 ± 95 t/d in March, only 20 t/d less than the Feb mean. Early April values ranged from 135 to 180 t/d. In contrast, rates in the months prior to Jan were only about 35 t/d. Deformation of the N and W sides of the dome continued but did not accelerate in March. Measurements of the deformation of the active E side of the dome have not been possible.

Extremely small events, felt by field crews in the crater but recorded only by a single seismometer 1 km N of the dome, continued through March. The number of larger earthquakes remained above background levels, averaging about half a dozen per day. This value oscillated considerably, without obvious trends or apparent correlations with variations in activity at the dome, but dropped substantially after the early March increase (8:3). Seismic energy release exceeded previous post-extrusion periods. In the past, 90-95% of cumulative energy release had occurred during extrusion episodes, but this pattern changed in Oct 1982, and seismic energy equivalent to 2 typical extrusions has been released since then. However, seismic energy accompanying extrusion of the Feb lobe was unusually low, comparable to that associated with smaller extrusions

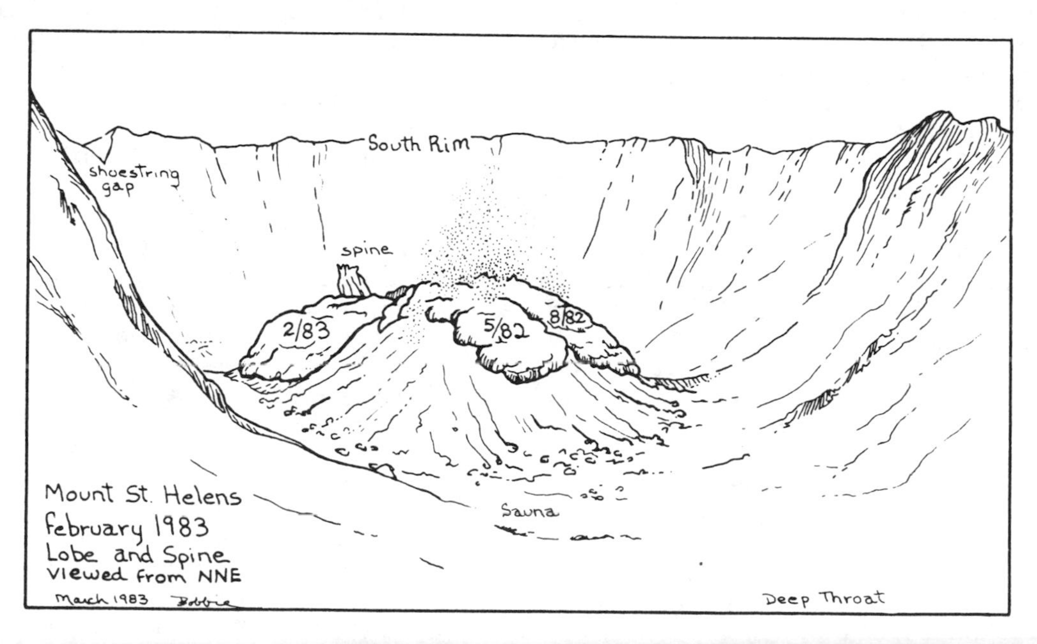

Fig. 12-23. Sketch by Bobbie Meyers showing the composite lava dome in the crater as viewed from the NNE in late Feb 1983. The Feb 1983 lobe is shown with its spine, as are lobes extruded in May and Aug 1982.

such as in Oct 1981. Energy release in early April exceeded Feb rates.

Information Contacts: T. Casadevall, C. Newhall, USGS CVO, Vancouver, WA; S. Malone, Univ. of Washington.

4/83 (8:4) Since early Feb, growth of the composite lava dome has been essentially continuous. Accelerating outward movement of the dome had preceded previous extrusion episodes, but stopped as lava reached the surface. However, substantial endogenous growth has continued throughout the current episode. Poor weather continued to hamper observations.

About 1 April, a broad stubby spine began to emerge at ~1 m/day from the center of the Feb lobe, reaching 30 m in N-S dimension, 20 m E-W, and about 25 m in height. Growth of this spine stopped about 15 April, and extrusion of another spine started about 70 m to the SE. The latter spine remained active until about 27 April, when at 60 m height it was the highest point on the dome and roughly the same size as the now-toppled Feb spine (8:2). Between visits to the crater 29 April and 4 May a new lobe began to grow high on the NE flank of the Feb lobe. This lava had a typical "spreading center" source and scoriaceous carapace. Extrusion continued as of 11 May, but the growth rate was slow and the lobe remained several times smaller than previous ones.

Dramatic deformation has continued on the E and particularly the NE sector of the dome since early-mid March. Because of frequent rockfalls, it was difficult to maintain targets on these areas of the dome, but rates of displacement reached 1.5 m/day, and averaged about 1 m/day over roughly 1-week periods. Between measurements 4 and 11 May, the NE margin of the dome moved 9 m outward and 2.5 m downward. Deformation on the N side of the dome was limited, but significant rate changes were observed. Through March the rate was constant at about 1.5 cm/day, but dropped to about 1 cm/day around 1 April as spine growth started. Deformation slowed further to 7-8 mm/day around the 15th as growth of one spine stopped and extrusion of another began, but returned to about 1 cm/day at the end of the month and remained at that rate as of 11 May. The W side of the dome, site of the most rapid deformation before many previous extrusion episodes, remained quite stable. No significant deformation of the edifice as a whole was detected.

Vapor and tephra emissions continued from the main vent near the source area of the Feb lobe but were relatively infrequent, occurring 1-3 times per day. Blocks up to 30 cm in diameter were ejected. Tephra could often be seen in the plumes, which sometimes rose to 1 km above the crater rim; the largest, 18 April at 1259, reached 6 km altitude. There was no apparent correlation between plume emissions and changes in extrusive activity or deformation.

SO_2 emission remained at roughly 150 t/d until about 27 April, when it dropped to 60-90 t/d. A similar rate was measured 30 April and 4 May, but SO_2 emission returned to 150 t/d 11 May.

Seismic activity remained elevated through April. All but 9 of the 243 events with locatable foci were of low-frequency with emergent onsets, a similar pattern to that seen in March. Between 1 and 12 April, daily earthquake totals commonly ranged from 4 to 8, increased to 8-12 events per day 13-24 April, then dropped slightly to an average of 8/day through the end of the month. Surface and avalanche events showed a similar pattern. Numerous gas emission events were recorded. Many began with a series of high-amplitude events that ended within a few minutes or gradually faded into brief periods of harmonic tremor. About 20 April, sequences of tiny discrete similar events, previously seen in Feb (8:2), reappeared on one seismometer, but these events could not be located and their significance remained uncertain. The start of extrusion of a new lobe of lava between visits to the crater 29 April and 4 May was not marked by an obvious change in seismicity. Geologists working in the crater 11 May heard loud but relatively small earthquakes, which had not been audible during previous extrusion episodes.

Information Contacts: D. Swanson, T. Casadevall, USGS CVO, Vancouver, WA; S. Malone, Univ. of Washington.

5/83 (8:5) New lava began to emerge high on the NE flank of the Feb lobe between 29 April and 4 May. Slow extrusion continued for about the next 3 weeks, but little new material appeared to have emerged onto the surface since then. Deformation measurements indicated that intrusive activity continued after extrusion stopped, as the new lobe continued to move outward at an average rate of 15-20 cm/day through early June. Little movement was noted on the N and W sides of the composite dome. The spine that grew on the Feb lobe from 15 April to about the end of the month and had formed the highest point on the composite dome toppled between visits to the crater 24 and 26 May.

The rate of SO_2 emission averaged 95 ± 35 t/d in May and remained similar in early June. This represented a decline from the April average of roughly 150 t/d, but remained substantially above the Sept 1982-mid Jan 1983 quiet rate of 25-50 t/d.

The number of earthquakes with locatable foci dropped from 243 in April to 155 in May. Nearly all were low-frequency events with emergent onsets, as has been observed since the early Feb explosions. However, strain release for surface events was nearly twice as high as for earthquakes, and nearly twice as many were recorded in May as in April. Most of the rockfalls observed by geologists in the crater occurred from the dome's active NE flank. Large . . . avalanche events were recorded 5, 15, and 18 May.

Information Contacts: T. Casadevall, E. Iwatsubo, USGS CVO, Vancouver, WA; C. Boyko, Univ. of Washington.

6/83 (8:6) Lava extrusion continued through early July. The rate of advance of the new lobe was about 4-5 m/day in June, but the net increase in length was substantially reduced by rockfalls from its front, which remained 75-100 m above the base of the dome in early July. Since extrusion began, the thickness of the lobe just downslope from the vent has increased from a few meters to at least 50 m.

Increasingly rapid deformation of the lava dome had preceded previous extrusion episodes but ended as lava reached the surface. However, deformation did not stop when extrusion began in Feb (8:1), and deformation measurements showed that intrusive activity has continued since then. Deformation remained strongest in the NE part of the dome, which moved outward an average of 1 m/day and downward about 30 cm/day in June. Both of these rates gradually decreased in June. In contrast, outward movement of the much less active SE sector of the dome increased from 4-5 cm/day through 28 June to 8 cm/day between then and the following measurement 5 July. Acceleration of the N side of the dome was also noted.

The SO_2 emission rate averaged about 65 t/d 3-16 June, increased by a factor of 3 to roughly 200 t/d 20-28 June, then dropped to about 75 t/d 4-6 July.

The number of earthquakes and the rate of seismic energy release in June were both similar to May values. Low-frequency shocks continued to dominate the seismic records as they have since extrusion began in Feb. Surface events, primarily rockfalls from the dome and walls of the crater, were more frequent in June. Much of the change can probably be attributed to warmer weather, which typically increases the number of rockfalls from the crater walls.

Information Contacts: D. Swanson, T. Casadevall, USGS CVO, Vancouver, WA; S. Malone, Univ. of Washington.

7/83 (8:7) Growth of the composite lava dome continued through early Aug (fig. 12-24). Net advance of the active lobe was reduced to roughly 10-20 m in July by frequent rockfalls from its front, but one area on the lobe thickened 11 m during the month. Deformation measurements showed continuing intrusive activity. The highest rates of expansion were on the NE side of the dome, where outward movement averaged 60-70 cm/day, although fluctuations by a factor of 2-3 were observed for periods lasting a maximum of 3-4 days. The station on the SE side of the dome typically moved outward 1-2 cm/day but this rate sometimes briefly increased to several centimeters per day. A short-term acceleration of endogenous growth in mid-July was accompanied by a slowing in extrusion of the new lobe.

Two vents on the summit of the dome were the sources of 3-6 small steam and ash explosions per day in July and early Aug. Plumes from most of the explosions barely cleared the crater rim, typically causing a fallout of fine ash within the crater, but little or none outside. A larger explosion in late July produced ballistic fragments 1 cm in diameter and a plume with considerable lightning. A steam and ash plume reached 4.5 km altitude 4 Aug and another rose to 3.5 km on 6 Aug.

The average rate of SO_2 emission in July was 120 ± 80 t/d. Measurements 4, 5, 6, and 10 July yielded mean values of 85 ± 15 t/d, then the emission rate increased 15, 17, and 21 July to 225 ± 70 t/d before dropping back to 80 ± 50 t/d 22-29 July and roughly 50 t/d 1-9 Aug. Mid-month increases in endogenous dome growth and seismicity were also noted (see first and last paragraphs of this report).

Measurements were made of the SO_2 content of 2 plumes produced by vapor and ash ejections. On 1 Aug,

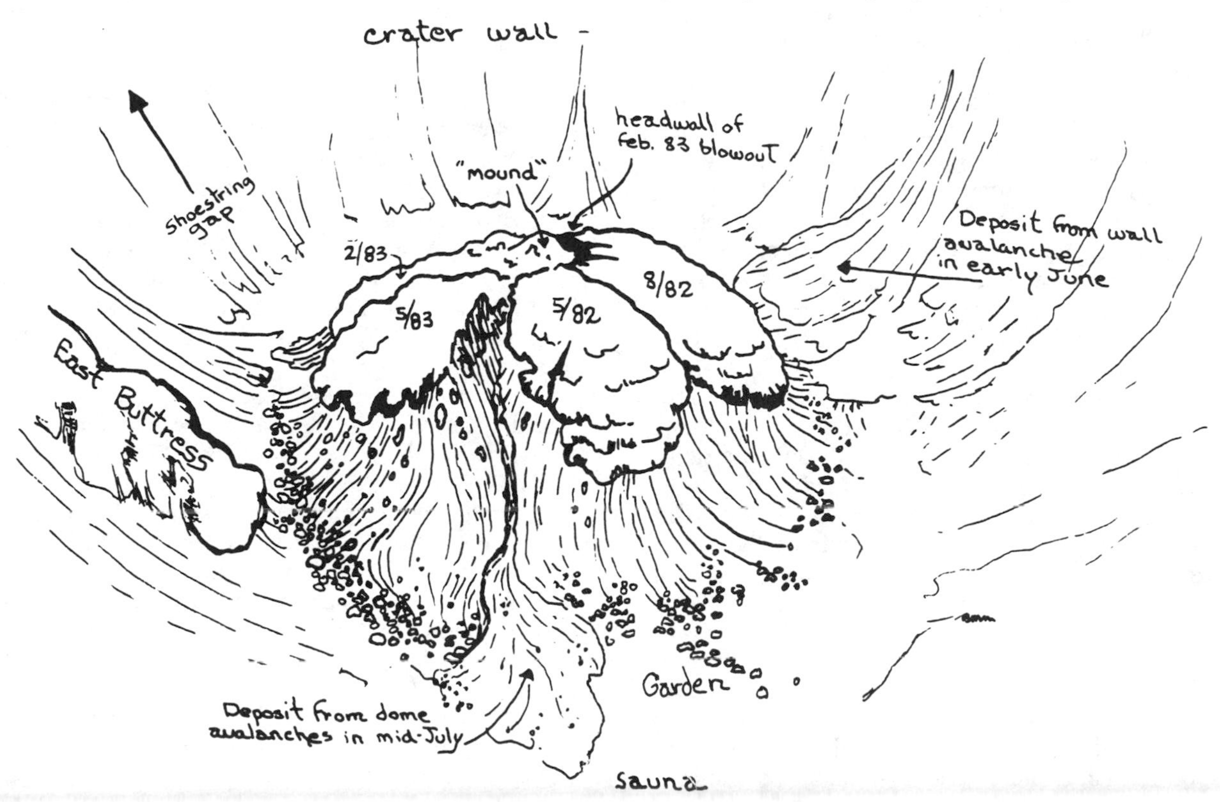

Fig. 12-24. Sketch by Bobbie Meyers of the composite lave dome, viewed from the N on 17 July 1983. Recent lobes are dated.

data collection began with the onset of the explosion and continued for 30 minutes after plume ejection ended. No change in the SO_2 emission rate (55 ± 5 t/d) was observed. Two measurements before a plume ejection on 9 Aug yielded rates of about 40 t/d. Values within the plume were initially 105 t/d before dropping gradually back to 40 t/d as it dissipated.

July seismic activity was similar to that of previous months. Vigorous surface activity was recorded, some of which was thought to be caused by avalanching from the crater walls on warm days. Several brief periods of increased seismicity were noted, and the largest, around the middle of the month, produced a noticeable step in the seismic energy release curve.

Information Contacts: D. Swanson, R. Symonds, E. Iwatsubo, B. Myers, USGS CVO, Vancouver, WA.

8/83 (8:8) New lava was still being added to the active lobe in early Sept and deformation of other parts of the dome was accelerating.The lobe front moved down the NE flank at about 1 m per day in Aug, roughly the same rate as in July. Rockfalls from the lobe's leading edge appeared to decline in July and Aug, but continued to remove some material, reducing the lava's net Aug advance to 20-25 m.

Rates of outward movement of survey targets on the S, SE, and N flanks of the dome began an irregular increase about 8 July and by early Sept had reached nearly 11 cm/day high on the S side. No acceleration of endogenous growth was observed in the area of most rapid deformation, below the active lobe on the NE flank, where rates averaged 60 cm/day. Movement of crater floor stations N and S of the dome was first detected around early Aug, gradually increasing into the millimeters per day range by early Sept. The pattern of increasing deformation was generally similar to periods that preceded extrusion of new lobes in 1981 and 1982. However, Donald Swanson noted that the irregular acceleration of endogenous growth contrasted with the quite steady increases measured before 1981-2 extrusion episodes, and that it was continuing after 2 months without the onset of new extrusion, exceeding the typical 1 month-6 week durations of the 1981-82 premonitory periods.

Numerous rockfalls, some quite large, occurred from a N flank notch that was propagating upslope toward the dome's extrusive vent. This activity built a large structurally unstable talus slope of hot blocks. Upon reaching the talus, some rockfalls became fluidized, probably by entrainment of heated air from between talus boulders. Early 12 Aug, Daniel Dzurisin observed a group of large boulders from the notch bounce onto the talus. A few seconds later, a second rockfall reached the talus and fluidized. An ash cloud quickly formed over the avalanche, and moved downslope at the same speed as the entrained boulders, stopping as they came to rest. The avalanche formed a lobate deposit with marginal levees less than 1 m high. Fine particles extended to roughly the distal end of the boulder deposit. Ash clouds formed by smaller avalanches were diffuse enough so that boulders could be seen rolling slowly downslope; these avalanches seemed to be only partially fluidized. The avalanches traveled no more than several hundred meters beyond the base of the talus, into the large breach on the N side of the crater. For several days after a large rockfall, avalanches occurred roughly every 2 hours, but declined to 1-2/day during quiet periods.

Occasional ejection of steam and ash plumes continued from several vents in the broad summit region of the dome. The number of plumes varied from day to day, but generally ranged from 3-6 daily and remained relatively unchanged through the summer. Plumes typically rose about 1 km above the dome and deposits were usually limited to the dome's summit area. No projectiles from these plumes reached the crater floor in Aug. Tom Casadevall reported that COSPEC measurements indicate that the volcano emits more SO_2 while plumes are being ejected than during quiet periods; on 18 Aug a plume briefly produced a 4-fold increase in SO_2 emission. However, plume events normally last only 15-20 min, and the excess SO_2 values decay exponentially, so they do not have a large effect on daily gas flux. The rate of SO_2 emission averaged 70 ±50 t/d in Aug, ranging from 40 to 90 t/d most of the month, but measurements 18-23 Aug yielded values of more than 150 t/d.

Aug seismic activity was generally similar to that of July. A substantial increase in surface events was recorded, but was thought to reflect increased avalanching from the crater walls as warm weather melted snow on the rim. For about 10 days in late Aug, the number of earthquakes and the rate of seismic energy release increased slightly, but declined to previous levels by early Sept.

Information Contacts: T. Casadevall, D. Dzurisin, D. Swanson, USGS CVO, Vancouver, WA; S. Malone, Univ. of Washington.

9/83 (8:9) Much of the NE flank lobe has been intruded just below the surface, carrying a thin rubbly carapace of older material downslope. New lava with visible flow structures has occasionally broken through to the surface. Expansion of the S and SE flanks continued to increase slowly until late Sept but accelerated more rapidly after the 28th and had reached 3-4 times early Sept values by 1 Oct. A USGS-Univ. of Washington advisory notice issued on that date predicted a new extrusion within 10 days.

Lava began to emerge from a new vent about 50 m S of the source of the NE flank lobe by 7 Oct. As with the NE flank lobe, little lava reached the surface, instead lifting a thin layer of older material to form a ridge roughly 300-400 m in E-W dimension, 100 m in N-S dimension, and 20 m thick, at the S edge of the NE flank lobe. The vent area was surmounted by a large pile of material, some of which appeared to be new lava, that by 12 Oct was 12-13 m higher than the previous summit of the dome. The advance of the NE portion of the May lobe slowed to less than 10% of its previous rate between 5 and 10 Oct. Deformation of the S portion of the dome stopped accelerating 5-6 Oct, also approximately coincident with the appearance of the new lava, but remained at rates as high as 105 cm/day. A small thrust fault began to form just beyond the talus pile at the dome's S side about 4-5 Oct and moved at an approximately constant rate of 2.5-3 cm/day 7-12 Oct.

Plumes of hot gas (a smell of H_2S was often noted by geologists in the crater) continued to be ejected 2-4 times per day. During a period of very clear weather and strong winds 20-22 Sept, plumes were observed on satellite imagery, rising roughly 2 km above the summit and extending 100-150 km. Most plumes contained little ash. No significant change in plume size, frequency, or density accompanied the appearance of the new lava. Rates of SO_2 emission averaged 100 ± 50 t/d in Sept, a slight increase over Aug values. Rates decreased slightly in mid-Sept to 70-90 t/d and usually remained in that range in early Oct. Since early Sept, seismic energy release has varied only slightly and there was no change in seismicity associated with the emergence of the new lava.

Information Contacts: D. Swanson, T. Casadevall, C. Newhall, S. Brantley, USGS CVO, Vancouver, WA; S. Malone, Univ. of Washington; M. Matson, S. Kusselson, E. Legg, NOAA/NESDIS.

10/83 (8:10) Growth of the composite lava dome continued through Oct. At the end of Sept, the pattern of lava extrusion began to change, with lava redirected southward along the margin of the active lobe. Examination of daily airphotos indicated that by 6 Oct a substantial area of uplift had developed along the S edge of the lobe and a spine was emerging from a point about 50 m WSW of the vent that had fed the active lobe since the beginning of May. The spine grew 1.5-2 m per day until 15 Oct, then its growth slowed to about $^1/_2$ m/day for the next 2 weeks. The spine reached 30 m in height, 17 m higher than any other point on the dome. From the spine, a sharp, lateral ridge extended about 100 m NE. By 31 Oct, this feature had crumbled. The spine remained nearly intact, but had stopped growing and some crumbling had occurred.

As the spine grew, lava emerged from a "spreading center" just to the S, pushing the spine slowly NW. Lava advanced S and SE along the periphery of the May lobe at roughly 1-2 m/day through Oct, thickening this portion of the lobe by a factor of 4-5. As this area of the lobe grew, advance of the NE end of the lobe slowed and had nearly stagnated by the end of Oct.

Although deformation of the S and SE flanks of the dome stopped accelerating about 6 Oct as the spine began to emerge, outward movement of this part of the dome continued at high but relatively stable rates of as much as 120 cm/day through Oct as the active lobe advanced and thickened. The crater floor adjacent to the S and SE flanks of the dome also continued to deform slightly through Oct, with a maximum uplift of 6-8 cm and maximum horizontal strain of about 20 cm during the month.

SO_2 emission averaged 75 ± 45 t/d in Oct. Several days of elevated SO_2 emission 1-8 Oct (reaching 210 t/d on the 1st) accompanied the onset of the changed lava extrusion pattern on the dome. From 8 Oct through the end of the month, rates dropped to 30-80 t/d. Several small gas and ash ejections from an explosion pit near the summit of the dome continued to occur daily, elevating SO_2 flux to 3-4 times background, usually for only a few minutes but occasionally for tens of minutes. Plumes were usually grayish-white and contained only a little tephra. Sand-size and occasionally cobble-size fragments fell near the vent, but only small quantities of very fine material were deposited on the crater rim.

Since Nov 1980, a drainage system has developed in the 1980 pyroclastic flow deposits. In the late spring of 1983, steam was noted in the drainage system for the first time. A zone of 6-8 small hot springs had developed near the N edge of the crater at the contact between hydrothermally altered ancestral dacite and the pyroclastic flow deposits. Flow rates were typically less than 1 liter per second. Temperatures measured at the springs in Sept and on 26 Oct ranged from 76° to 91°C, pH was 7.1-8.2, and specific conductance was 3300-5800 mhos. Travertine was being deposited at one of the springs.

Seismic energy release declined in mid-Sept, but a gradual increase began in early Oct, leveling off about 9-10 Oct at roughly twice the late Sept rate. A small decrease in the slope of the energy release curve occurred in late Oct but poor weather may have caused instrumental interference. By early Nov, the mid-Oct rates had been regained. Little change in surface events was observed in Oct. In the summer, many surface seismic events were the result of seasonal avalanching from the crater walls, but in Oct most were caused by steam ejections.

Information Contacts: D. Swanson, T. Casadevall, USGS CVO, Vancouver, WA; R. Norris, Univ. of Washington.

11/83 (8:11) Lava extrusion continued to extend the active lobe SSE through Nov. Lava advanced about 100 m across the dome's broad, gently sloping summit area during the month, approaching the break in slope at the top of its steep upper flank. The rate of outflow appeared to be roughly the same as in previous months. A new depression, about 50 m × 80 m in horizontal dimensions and 30-40 m deep, formed about 50 m SE of the "spreading center" that fed the lobe in Oct. The "spreading center" had stagnated and the new depression may have been the source of some of the Nov lava. Just N of the stagnant "spreading center," the spine extruded in Oct was crumbling, but was still the high point on the dome at the end of Nov.

Collection of deformation data was hampered by the loss or inaccessibility of targets on the most active parts of the dome. Outward movement of accessible targets on the dome's SE flank was about 55 cm/day between late Oct and early Nov, 35 cm/day 8-21 Nov, and 48-50 cm/day after 21 Nov. If deformation patterns remained similar to those measured before the most active targets were lost, the most rapidly moving areas on the dome may have been expanding at roughly 90-100 cm/day, down slightly from the 120 cm/day of early Oct. Deformation of the dome's NE side had been rapid as the active lobe advanced down the NE flank from early May through late Sept but was negligible in Nov.

Gas and ash plumes continued to be emitted 3-6 times daily. The plumes, dominantly gas but sometimes containing some tephra, rose a few hundred meters to

1 km above the dome. Poor weather limited airborne gas measurements to 5 in Nov. The rate of SO_2 emission averaged 70 ± 45 t/d, nearly identical to Oct values. However, the 21 Nov rate of 150 t/d may have been measured in the remnants of a gas and ash plume; without this figure, Nov SO_2 emission averaged 50 ± 20 t/d.

For most of Nov, seismicity continued at approximately the Oct rate. Both seismic energy release and the number of events increased at the end of the month to values higher than in Oct. Average daily earthquake counts ranged from 5-13/day through 18 Nov, 12-15/day 19-25 Nov, and 24-33/day 26 Nov-1 Dec.

Information Contacts: D. Swanson, C. Mullins, USGS CVO, Vancouver, WA; R. Norris, Univ. of Washington.

12/83 (8:12) Numerous rockfalls occurred through Dec as the lobe front broke up at the top of the S and SE flanks. Some rockfalls were hot enough to become fluidized, moving down a chute and spreading over a small area of the crater floor at the base of the dome. The fluidized rockfalls generated from the front of the NE flank lobe in Aug (8:8) were larger and had stronger seismic signatures. Maximum displacement rates measured on the S and SE sides of the dome dropped from about 50 cm/day in late Nov to about 20 cm/day by late Dec. Between 28 Dec and 4 Jan, values increased sharply, to as much as 60 cm/day.

New lava appeared on the NE flank in mid-Dec, forming jagged spires and ridges, but little downslope advance was observed. Instead, the lava appeared to have broken through the crust of the dome in a zone that extended about 200 m NE from the summit spine and was about 100 m wide (in a NW-SE direction). Night overflights showed an increased number of glowing cracks on the dome's NE flank. Deformation of the NE flank also began to accelerate in mid-Dec, increasing from 0.5-1 cm/day to a few centimeters per day by the end of the month.

The dome's summit spine crumbled rapidly, but continued to receive some new lava. As of early Jan, it remained the high point on the dome, but several meters of net height loss appeared to have occurred. Snow had accumulated in the depression that formed in Nov near the top of the dome. However, its NW rim tilted dramatically away from the remainder of the depression, and a new mound had grown in that area.

Gas and ash ejection from vents high on the dome continued to occur several times a day. On 16 Dec, a substantial number of breadcrust bombs with a maximum diameter of about 4 cm were found in the crater on snow that had probably fallen 2 days earlier. The bombs were more vesicular than any material ejected since the strong explosive activity of 19 March 1982 (7:3). Dec SO_2 emission averaged 105 ± 25 t/d, a 30-40% increase over Oct and Nov values, but similar to rates measured April-Sept.

Seismic events were slightly more numerous in Dec, but only minor growth was observed in the rate of energy release. Seismicity increased slightly in early Dec, dropped a little at mid-month, then showed a minor increase at month's end. There were no obvious changes in seismicity that could be correlated with changes in activity on the dome. The number of surface events decreased seasonally because winter conditions inhibit rock avalanching from the crater walls.

Information Contacts: D. Swanson, T. Casadevall, USGS CVO, Vancouver WA; R. Norris, Univ. of Washington.

Further Reference: Swanson, D.A., Dzurisin, D., Holcomb, R.T., Iwatsubo, E.Y., Chadwick, W.W., Jr., Casadevall, T.J., Ewert, J.W., and Heliker, C.C., 1987, Growth of the Lava Dome at Mount St. Helens, Washington (USA), 1981-1983 *in* Fink, J. (ed.) The Emplacement of Silicic Domes and Lava Flows; *Geological Society of America Special Paper* 212, p. 1-16.

1/84 (9:1) A new phase of the ongoing activity began with increasing deformation and seismicity in late Jan 1984, followed by the extrusion of a new lobe near the dome's summit.

Numerous rockfalls occurred in Dec after the active lobe had reached the top of the dome's steep S and SE flanks. Rockfalls from that area were infrequent in Jan, suggesting that the advance of the lobe had nearly stopped. Deformation of the dome's SE flank accelerated briefly in late Dec to more than 50 cm/day, but slowed by an order of magnitude in early Jan and remained at 5-6 cm/day through the end of the month. On the NE flank, rates of outward movement remained relatively high through 4 Jan. On 9 Jan, when weather next allowed access to the crater, NE flank deformation had slowed and a new mound was perched just E of the dome's summit (8:12). The surface of the mound was old material, but it was apparently cored by magma. It reached its maximum elevation 20 Jan, then subsided slowly. Earthquakes remained relatively numerous through 9 Jan, but the period 10-17 Jan was the quietest seismically since late Sept, with the number of events per day dropping from about 14 to 2-6 and energy release declining to near background level.

In mid-Jan, points on the floor of the crater's breach, about 1 km N of the dome, began to move outward, and medium-frequency events started to appear on seismic records. Energy release remained low but the number of medium-frequency events increased gradually through the end of Jan. Gas ejection episodes increased noticeably in duration and amplitude on seismic records. By 23 Jan, some were followed by several minutes of weak harmonic tremor. Vigorous plumes were observed and small blocks were deposited on the crater floor. SO_2 emission averaged 90 ± 40 t/d during the first 3 weeks of Jan (as compared to 105 ± 25 t/d in Dec), but dropped to 35 ± 20 t/d for the remainder of the month and was below detection limits 26-28 Jan. A fairly large gas and ash ejection on the 28th was not followed by the typical temporary several-fold increase in SO_2 flux.

New cracks were observed on top of the dome 29 Jan. A graben was evident on its SW side by 1 Feb, and radial cracks had appeared on the W side. By 3 Feb, the graben was a few tens of meters wide and a few meters deep, extending across the summit crater. Deformation began to accelerate rapidly. A point halfway up the N flank of the dome that had moved outward no more than a few centimeters per day through most of Jan showed rates of 11.6 cm/day on 30 Jan, 46 cm/day on 3 Feb and about 1 m/day by the 5th. Deformation changes on the SE flank were less dramatic, but rates also increased, from 6 cm/day through 3 Feb to 20 cm/day on 5 Feb (all rates are average daily changes since the previous measurement). This activity was accompanied by a rapid increase in the number of earthquakes and seismic energy release beginning 1 Feb. SO_2 emission increased to 55 t/d on 1 Feb and reached 140 t/d on the 6th.

When geologists arrived at the crater 6 Feb they observed a new mound filling the NW part of the dome's summit crater. The early Jan mound, about 100 m to the E, was subsiding. That evening, a small landslide from the E side of the dome moved 50-100 m to the main crater wall, causing minor snowmelt. The next day, the new mound had elongated and extended a short distance down the dome's N flank, while the Jan mound continued to subside. The surface of the new mound, like the Jan mound, was old material. No glow from the new mound was observed at night. N flank displacement was 2.4 m between 5 and 6 Feb, and rates of 3 m per day were measured on the 6th. SO_2 emission increased to 140 t/d on 6 Feb and 170 t/d the next day. Seismicity peaked during the evening of 7 Feb, dropping sharply in the next 24 hours to only slightly elevated levels. The number of rockfall events increased but they were smaller and less frequent than during previous extrusion episodes. Poor weather prevented access to the crater after 6 Feb. A brief glimpse of the dome on 10 Feb revealed a new lobe perched on its summit. . . . The new lobe appeared to be hot; snow had accumulated on the rest of the dome, but not on the new lobe.

Information Contacts: R. Holcomb, T. Casadevall, USGS CVO, Vancouver, WA; S. Malone, Univ. of Washington.

2/84 (9:2) Increasing seismicity, gas emission and deformation of the composite lava dome culminated in the extrusion of a small new lobe in early Feb. The extrusion was not observed because of poor weather, but was believed to have started between late 7 and early 8 Feb (when seismicity decreased sharply from peak levels reached during the evening of the 7th). The roughly circular new lobe, about 140 m in diameter and 25 m thick, spread N from the former apex of the Jan mound, just E of the dome's summit. The mound, first seen 9 Jan, had a surface of old material but was probably cored by magma. It had reached its maximum elevation 20 Jan, then subsided slowly. A new mound, almost as large as the Feb lobe, formed in the dome's summit crater after 6 Feb, probably just before the lobe was extruded. Observations of this feature have been limited because it has remained enveloped in vapor, but it appeared to be cored by a highly oxidized monolith or spine rising about 30-40 m from the crater floor and having a similar or slightly smaller width. The USGS estimated that the dome expanded by about 4×10^6 m^3 during the period of rapid deformation from late Jan through early Feb. The volume of the Feb lobe was estimated to be about an order of magnitude smaller.

After the end of the Feb extrusion episode, activity declined to the lowest level since continuous dome growth began in early 1983. From mid-Feb through early March, deformation on all sides of the dome has been less than 1 cm/day. Although one station moved 5 cm between 3 and 4 March, it was located close to the new lobe and may have been affected by local movement. Seismic energy release, which had increased in early Feb, dropped abruptly with the sharp decline in the number of events late 7-early 8 Feb. Two magnitude 2 events were recorded 10 and 12 Feb, but there were few earthquakes after the 12th and very little energy release. SO_2 emission dropped gradually through Feb, from 170 t/d on the 7th to slightly less than 100 t/d at the end of the month. Gas and ash ejections, which had occurred several times a day in past months, were not observed after 3 Feb. A wispy plume emitted continuously from the dome 3-7 Feb deposited a very thin layer of powdery ash nearby, but this had stopped by the end of the extrusion episode.

Information Contacts: R. Holcomb, C. Mullins, T. Casadevall USGS CVO, Vancouver, WA; S. Malone, Univ. of Washington.

3/84 (9:3) Strong seismicity and rapid deformation were followed in late March by the addition of a new lobe to the composite lava dome. Deformation, seismicity, and SO_2 emission declined after the extrusion of a small lobe in early Feb, and remained low through mid-March. Poor weather prevented access to the crater 22-27 March. Deformation on the 22nd was at low levels, but measurements on the 27th showed that targets on the N side of the dome had moved outward at an average rate of $^1/_2$ m/day since the 22nd. Instantaneous rates increased from 1.9 to 2.4 m/day during a 2-hour period on 27 March.

Seismicity began to increase late 22 March, and the number of events doubled each day 24-28 March. The swarm was characterized by type "M" (medium-frequency) events similar to those that preceded the Feb extrusion episode. The type "M" swarm peaked early 28 March, then declined rapidly, ending by midnight. As the type "M" events diminished, a swarm of small crater events began. By noon, 11 events were being recorded per minute, and by 1600 there were 15/minute. On more distant stations, the individual events could not be resolved, merging into a tremor-like signal with an amplitude that increased through the evening.

A continuously recording tiltmeter about 30 m from the base of the talus N of the dome began telemetering data 27 March. Within a day, outward tilt, initially 200 μrad/hr, increased to more than 400 μrad/hr, accelerating abruptly to more than 1000 μrad/hr during the afternoon of 28 March, then went off scale by 2000 that evening.

At 0320 on 29 March, an avalanche from the N side of the dome removed much of the Feb lobe and advanced 0.5-1 km onto the crater floor. Minor snowmelt occurred, but there was no significant mudflow. The March 1984 avalanche was similar in size to the 4 April 1982 avalanche (7:3). Fine particles from the avalanche rose to 4.5 km altitude, dusting the E and SE parts of the crater and flanks of the volcano. A large arcuate crack in the Feb lobe had been observed 27 March, and failure occurred along this crack.

By 29 March, the N side of the dome had decoupled from the rest of the structure and was moving very rapidly outward. The W and probably the SE sides of the dome were virtually stationary, but the N side had moved outward 42 m since 27 March, and instantaneous rates of 15 m per day were observed on the 29th. Since 22 March, the SE side of the dome had moved only a few centimeters outward, while the W side had expanded about 4 m. When the tiltmeter N of the dome was releveled early 29 March, the rate of outward tilt had dropped to 400 μrad/hr. By midnight, the tilt rate was decreasing rapidly.

Tremor gradually separated into individual events early 29 March. During the first couple of hours after the avalanche, large rockfalls were superimposed on the tremor. The tremor gradually evolved into an earthquake swarm that remained vigorous until midnight.

Poor weather prevented frequent measurements of SO_2 emission rates immediately before the extrusion episode. From 3 to 28 March, SO_2 emission remained relatively constant at about 80 t/d, increasing to 400 t/d on 29-30 March.

An overflight at about 2200 on 29 March confirmed that lava had reached the surface, emerging just W of the remnants of the Feb extrusion. The lobe eventually grew to nearly fill the crater at the top of the dome and reached the edge of the 29 March avalanche chute. Fragments spalled down the chute but lava did not flow beyond the edge of the dome's summit area. Weather conditions prevented direct observations of the extrusion, but deformation and seismic data suggested that lava production ended within a few days.

The earthquake swarm began to decline on 30 March, but did not reach background levels until 4 days later. This slow decline in seismicity contrasts with previous years when seismicity often dropped to background levels within hours of the onset of lava extrusion.

Total outward movement of the N flank was about 3.2 m between 30 March and 2 April, but deformation declined rapidly and had probably nearly stopped by 31 March. Between 22 March and 2 April, the N flank moved outward a total of 55 m. Tilting measured near the N foot of the dome had stopped by the morning of 31 March. Unlike the extrusion episodes of 1981-2, no tilt reversal was detected. Total tilt (assigning a rate of 400 μrad/hr while the instrument was off-scale late 28-early 29 March) was 28 milliradians, more tilt than had previously been recorded in association with an extrusion episode at Mt. St. Helens (20 milliradians of tilt preceded the Sept 1981 extrusion). Tiltmeters were redeployed 80 and 250 m N of the dome 6 April and had detected no tilt as of 16 April. Rates of outward movement of the dome were only ~5 mm/day in mid-April.

Information Contacts: S. Brantley, T. Casadevall, D. Dzurisin, C. Newhall, P. Otway, USGS CVO, Vancouver, WA; R. Norris, S. Malone, Univ. of Washington; D. Sowa, Northwest Orient Airlines.

4/84 (9:4) Deformation, seismicity, and gas emission declined to background levels after the end of the extrusion episode that began in late March. More than half of the April earthquakes were recorded on the 1st. Seismicity reached background level 3 April and remained low through early May. Deformation was very slow through early May. Rates were roughly half the background values measured before deformation began to accelerate prior to the March extrusion episode. Maximum rates were 6 mm/day on the SE side. However, at some points on the dome, no outward movement was measured. Three single-axis telemetering tiltmeters installed along a radial line at 50, 250, and 750 m N of the dome recorded little tilt in April. SO_2 emission averaged 85 ± 50 t/d in April.

. . . On 14 May at 0932 seismicity in the crater began to increase. A pronounced increase was detected at 0935 and a great increase at 0937. No plume was visible at 0934, but by 0939 a plume had risen to an altitude estimated at 5.5-6 km by ground observers. Airplane pilots reported that the eruption cloud had reached 7.5 km altitude at 0950 and 10.5 km at 1000-1005. Radar at Portland airport measured an elevation of 6 km at 0940. Light ashfall was reported at Mt. Rainier National Park, 65 km to the NNE.

In addition to a vertical plume, cold to hot (but not molten) material was ejected laterally from the upper SW portion of the dome. Some cleared the rim of the crater and fell on the outer W flank of the volcano. However, much of the fragmented dome material struck the W wall of the crater, melting heavy snow that had accumulated there during the winter. A flow consisting of about 90% snow and 10% sand-sized lithic fragments plus a few blocks to 1 m in diameter moved through the [crater and the] breach N of the crater. Near the crater, the flow was about 200 m wide. In the outer half of this zone, some melting occurred, but there was little erosion of the underlying snow. Deposits were typically 0.5-1 m thick but locally reached 2 m in thickness. The inner half of the zone was dominantly erosional. Approximately the outer 10 m of the erosional section consisted of fluted snow covered by a lag deposit of fine lithics a couple of centimeters thick. Most of the erosional section was eroded down to [1980 deposits]. The initial flow was apparently primarily snow, followed by [a slushy flow] and then (in the erosional zone) by [a watery flow].

The flow divided into 2 branches [on the pumice plain (on the volcano's N flank)]. The NE branch . . . reached Spirit Lake (about 5.5 km from the crater) at 0955. The main (NW) branch, a mudflow, entered the North Fork of the Toutle River. Addition of water to the flow caused degeneration to a hyperconcentrate (exhibiting Newtonian behavior, a transition that typically occurs when water concentration reaches about 30% of the flow by weight) by the time it reached Coldwater Lake (about 12 km from the crater). It reached the lake at 1020, carrying boulders to 2 m in diameter, but peak flow there did not occur until 1100. At N-1 debris dam (about 30 km from the crater), peak flow was 135 m^3/sec above background. Peak flow had attenuated to 27 m^3/sec at the mouth of the North Fork (about 90 km from the crater).

Information Contacts: D. Peterson, T. Casadevall, D.

Childers, N. MacLeod, C. Mullins, B. Myers, P. Otway, USGS CVO, Vancouver, WA; R. Norris, Univ. of Washington; M. Matson, NOAA/NESDIS; NWS, Portland, OR.

5/84 (9:5) Occasional ejections of gas and tephra from the upper W flank of the dome produced plumes and snow/water flows in late May and early June, but were much smaller than the 14 May episode.

Seismicity that was similar in duration and character but less vigorous than that associated with the 14 May episode began on 26 May at 0814 and lasted for about 20 minutes. A plume that contained little ash rose to about 6.5 km altitude. A snow/water flow moved down a channel 20-25 m wide that had been carved by the 14 May episode, and a small amount of material reached Spirit Lake (5.5 km from the crater). The outer portion of the flow was similar to a snow avalanche and the inner zone was more fluid. Although ejecta from the 26 May explosion did not strike the crater wall, the explosion was partly directed and apparently picked up snow as it moved across a deposit left on the crater floor by a large snow and rockfall from the crater wall that occurred on 22 May at 1516. An additional source of snow may have been the 1.5 m that had drifted into the area near the breach in the N side of the crater since 14 May. Warm rock from the interior of the dome was found in the 14 May flow deposit, but all of the rock in the 26 May flow deposit was cold and appeared to be from the exterior of the dome or the 22 May rockfall debris.

The source of the 14 and 26 May episodes was . . . a notch roughly 10-20 m deep and up to 50 m wide that extended roughly 200 m down the W flank from near the dome's summit. Most of the notch cut through the Aug 1982 lobe, but its head was in the W side of the March 1984 lobe. During an overflight in the evening of 1 June, a prominent long straight chain of glowing spots was visible extending from the head of the notch E across the summit of the dome and the March 1984 lobe. This feature was not observed during the previous night overflight on 17 May.

A 20-minute seismic signal weaker than that of 26 May began on the 27th at 1320. A small white plume rose roughly 600-900 m above the crater rim but was not detected by radar at Portland Airport. No mudflow was reported. An 18-minute seismic signal intermediate in strength between those of 26 and 27 May started at 0351 on 6 June. At 0415, radar at Portland airport detected a plume to about 6 km altitude. Light ashfall occurred on the NE side of the volcano and ash with ballistic fragments struck about 1/3 of the way up the crater wall. Minor snowmelt occurred, and relatively clear water flowed down the 14 and 26 May channel, causing a small increase in the normal flow into Spirit Lake. The source of the explosion was again the W side of the dome, where a crater had developed since the explosions of 26-27 May.

Outward movement of the dome remained minor but increased slightly, from 2 mm/day in early May to 4 mm/day in early June on the N side (most noticeably at the base of the dome), and also doubled in the same period on the dome's SE side, reaching 10 mm/day in early June. SO_2 emission measured 1-1 1/2 hours after the 14 May episode was 260 t/d. Two weeks later, rates of SO_2 emission were about 55-60 t/d.

Information Contacts: R. Holcomb, C. Mullins, P. Otway, R. Waitt, USGS CVO, Vancouver, WA; R. Norris, Univ. of Washington.

6/84 (9:6) A new lobe in the composite lava dome began to emerge in mid-June. Its location, on the W flank, was within the notch that was the source of explosions 14, 26, and 27 May, and 6 June as well as a similar event on 7 June. Growth of the W flank lobe had stopped by 1 July. Accelerating deformation on the dome's N side was measured in late June and early July, but rates of displacement began to decrease after several days and no lava reached the surface.

A small explosion occurred 7 June at 1720, when airline pilots observed an ash cloud that rose to about 9 km altitude. Increased water flow from the crater began at about 1740, and a mudflow about 3 m wide and 15 cm deep reached Spirit Lake at 1802.

Beginning 14-15 June, the number of earthquakes at the nearest crater seismometer (Yellow Rock) increased from about 20-30 to 50-60/day, and the number of surface (rockfall) events began to increase 16-17 June. The small new lobe was first seen during the afternoon of 17 June (by gas-monitoring aircraft pilot Allyn Merris), and was more clearly visible during an overflight at 2230 that night. The dome was photographed from the air 18 June (fig. 12-25).

The following information is from Peter Otway.

"When first tracked, on 18 June, the leading edge of the lobe was moving downslope at over 30 m/day although rates near the extrusion site were at least 50% higher. By 22 June, movement of the leading edge had slowed to 13 m/day, and by 25 June, to 6 m/day. When next observed, on 1 July, the flow had stopped.

"The lobe, which overfilled the notch formed by the the May-June explosions, was 60 m wide at its maximum and 150 m long. Its volume was estimated to be of the order of 0.2×10^6 m^3.

"The extrusion of 16-17 June was preceded by a slight acceleration of spreading rate on the SE side of the dome from less than 10 mm/day during late May to 13 mm/day between 10 and 19 June, declining to 5 mm/day by the end of June. Although monitoring of the W side has not been possible because of the gas blasts that commenced in May, targets 100 m N and NE of the June lobe moved radially away from the extrusion site at 60 mm/day prior to 22 June.

"During this time, targets on the N side of the dome moved outward at rates that steadily increased from 5 mm/day in early June to 60 mm/day by 25 June, but then accelerated rapidly to peak at 0.8 m/day between 30 June and 2 July. By 10 July, the rates had fallen to 20 mm/day. The direction of movement for all northern targets prior to 25 June was NNE, away from the mid-June extrusion, but swung to the N as the rates rapidly increased beginning 25 June, suggesting that an intrusive event occurred in late June at a site at least 100 m E of the mid-June extrusion. Targets on the NE side of the dome averaged a steady 13 mm/day NNE movement during June indicating that this area has decoupled from the mobile N sector."

Rates of SO_2 emission ranged from 15 to 35 t/d (near the detection limit) during the first 2 weeks of June. The next measurement, the morning of 18 June, was 105 t/d. Rates averaged about 100 t/d through 1 July, but had dropped to 40 t/d by 6 July.

Information Contacts: S. Brantley, T. Casadevall, E. Endo, C. Mullins, C. Newhall, P. Otway, USGS CVO, Vancouver, WA; R. Norris, Univ. of Washington.

7/84 (9:7) Seismicity, gas emission, and deformation of the composite lava dome declined after the extrusion of a new lobe in June and an intrusive event at the end of the month. The quoted material is from Peter Otway.

"Following late June's record deformation of up to 0.8 m/day at the upper stations on the N side of the dome, rates decreased quickly after 2 July and were down to background levels at a steady 10 mm/day on 16 July. No changes in the rates had been recorded as of early Aug. On the SE side, displacement rates maintained a constant 6 mm/day throughout July and on the S side remained steady at 1.2 mm/day.

"A number of new survey phototheodolite targets were installed across the top of the dome in July, resulting in a total of 32 targets being surveyed by theodolite several times a month, in addition to the 12 EDM targets tracked 2-3 times a week.

"During the installation of targets on the top of the dome on 30 July, several small steam emissions (accompanying a minor seismic signal) were witnessed at close range. They came from 2 separate locations: the top of the March lobe, and the old summit crater 80 m to the W. Some ash was present in the emission from the W side. The fume cloud rose less than 300 m from the dome and dispersed slowly. Such emissions appear to occur from time to time even when the general activity is low."

Rates of SO_2 emission increased before the lava extrusion in mid-June, then declined gradually, to 55 t/d on 13 July, and only 15 t/d on the 20th and 12 t/d on the 27th. Seismicity was elevated through early July as dome growth continued, then declined to background levels by the second week of the month and remained low through early Aug.

Information Contacts: P. Otway, C. Mullins, USGS CVO, Vancouver, WA; R. Norris, Univ. of Washington.

8/84 (9:8) Strong deformation, vigorous seismicity, and increased SO_2 emission were followed by extrusion of a new lobe on the NW side of the composite lava dome (fig. 12-26 is a profile showing the dome's 1980-1983 growth).

Deformation on the N side of the dome gradually accelerated from 1 cm/day on 15 Aug to about 14 cm/day by the end of the month. A slight increase in the number of recorded earthquakes started on 9 Aug and

Fig. 12-25. Photograph of the new lobe on the W side of the composite dome taken on 18 June 1984, early in the extrusion episode. By 1 July, the lobe had overfilled the 50 m-wide notch. Photo by T. Casadevall.

steepening of the seismic energy release curve was evident by the 12th as occasional events in the magnitude 1.5-2.3 range began to occur. Another slight upturn in earthquake counts started after 27 Aug. Rates of SO_2 emission remained low in Aug, ranging from 12 to 37 t/d.

Deformation accelerated rapidly in early Sept, rising to 40 cm/day by 2 Sept and 70 cm/day by the 5th. Instantaneous rates of 1.5 m/day were measured early 8 Sept and reached 2 m/day late that afternoon. An extensive crack system defined the S margin of the zone of large-scale deformation, which included the dome's NW sector from about due N to about N70°W. On 4 and 5 Sept, fissuring extended from the W side of the dome across the N end of the June lobe, passing just N of the dome's summit, but it did not continue across the E side of the dome. As deformation became more rapid, rockfalls began to fill the widening fissures.

Seismic instruments began to detect increased rockfall activity on 6 Sept at about 2200, and earthquakes started to increase shortly thereafter. The number of small earthquakes increased further during the day 8 Sept. Between 0300 and 0400 on 9 Sept, the number of low-frequency (type L) events increased, as did the number of very small events previously termed "peppercorns", which appeared to grade upward in magnitude into the type L's. A major increase in earthquake activity continued through the day, saturating many nearby seismic stations set at maximum attenuation. At 1830, the earthquake rate began to drop, and within an hour had declined from 1 event every few minutes to virtually none. Simultaneously, background tremor-like activity began to build, increasing strongly through the evening. Tremor reached maximum amplitude about 0300, saturating nearby stations, and was detected by seismometers on the S side of Mt. Rainier, about 60 km away. Tremor declined about dawn on the 10th and discrete earthquakes became visible again on seismic records. Because of the tremor saturation during the night, it was not possible to determine at what time the discrete earthquakes had resumed. The number of earthquakes gradually decreased from about 1 every 2 minutes to 5-10/hour by the 11th.

Deformation measurements suggested that maximum rates of internal dome growth approximately coincided with the strongest seismicity. Measurements of targets on the NW side of the dome early 10 Sept indicated that outward movement of almost 52 m had occurred since the previous afternoon. Instantaneous rates of 15 m/day were measured between 1000 and 1030, dropping to about 10 m/day in late afternoon and a few meters per day on the 11th.

Aerial observations before dawn 10 Sept also showed vigorous activity. Large incandescent block and ash avalanches originating from 3 areas, 2 on the dome's NW flank and 1 on the N flank, occurred about every 5 minutes. The avalanches moved quickly down the flank and several hundred meters onto the crater floor, building an extensive talus pile. Cascades of individual blocks were continuous, occurring at rates of hundreds per minute. Avalanches had been distinctly smaller and less frequent during an overflight 8 hours earlier, and none was seen on a flight before dawn on the 9th. No new areas of incandescence were seen on the E and S sectors of the dome, although pre-existing incandescent areas had brightened somewhat.

Rates of SO_2 emission also reached maximum values on 10 Sept, increasing from 14 t/d 30 Aug, to 54 t/d on 5 Sept, 440 t/d on the 8th, and 786 t/d on the 10th. Enhanced SO_2 emission was accompanied by increased fuming and small steam and ash ejections. The largest rose about 1 km above the crater rim early 9 Sept, dropping about 1 mm of ash in the crater and a trace SE of the volcano. All of the ash in this plume was from older dome rocks.

On 12 Sept, geologists observed that new lava had been extruded, extending about 300 m down the NW flank. By the next day, seismicity had declined to background levels and deformation had slowed.

Reference: Brantley, S. and Topinka, L. (eds.), 1984, Volcanic Studies at the U.S. Geological Survey's David A. Johnston Cascades Volcano Observatory, Van-

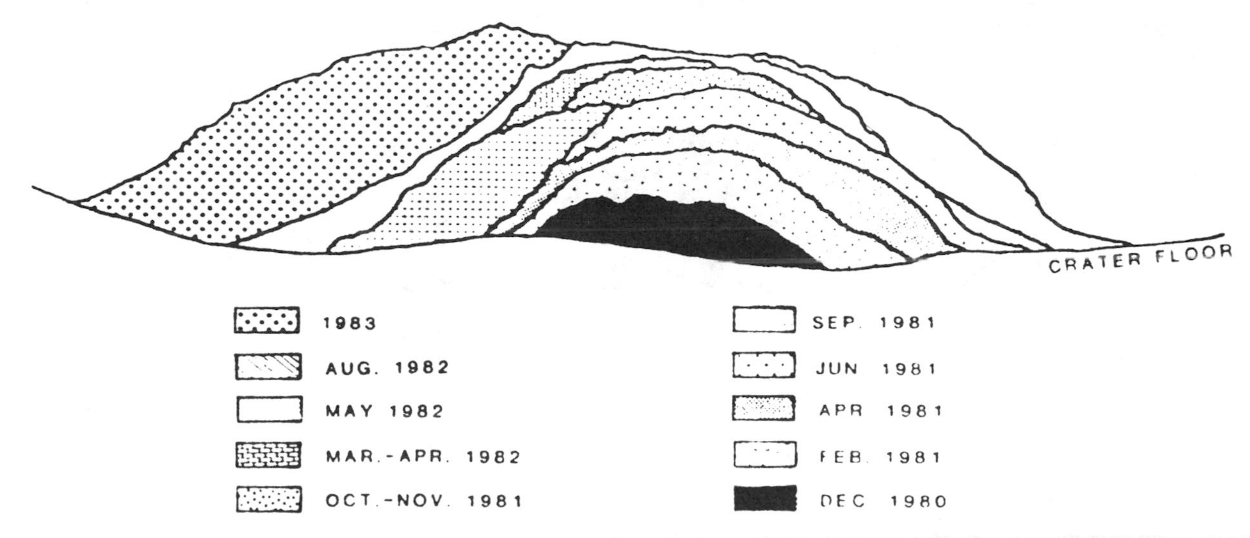

Fig. 12 26. E W profile of the composite lava dome, drawn from a succession of photographs taken from 1 km to the N. Successive lobes added to the dome Dec 1980-83 are shown. Growth since Oct 1983 is not shown. From Brantley and Topinka, 1984.

couver, Washington; *Earthquake Information Bulletin*, v. 16, no. 2, p. 43-122.

Information Contacts: M. Doukas, R. Holcomb, D. Swanson, J. Sutton, USGS CVO, Vancouver, WA; R. Norris, Univ. of Washington; UPI.

9/84 (9:9) The new lobe, which began to emerge during the night of 9-10 Sept and had stopped growing by the 14th, partially filled an elongate E-W-trending collapse zone, about 350 m long by 250 m wide, on the dome's upper W side. The collapse, in what had been the highest part of the dome, destroyed much of the March 1984 lobe and engulfed the notch produced by explosions in May (9:5). The SE wall of the collapse zone was defined by the zone of fissuring observed in early Sept. The volume of collapse was about 1.3×10^6 m^3.

The new lobe was about 250 m long by 220 m wide, with an average thickness of 35 m. The USGS calculated its volume at about 2×10^6 m^3. It was almost entirely confined within the collapse zone, although it protruded considerably above the rim of the collapse zone. Sublimate deposits were more extensive than on previous lobes, and a large area at the top of the new lobe was a dirty yellow color.

At least 3.4×10^6 m^3 of internal expansion of the dome took place during the period of accelerating deformation prior to extrusion of the new lobe. Between the afternoon of 9 Sept and the next morning, targets on the NW side of the dome moved outward as much as 52 m. Displacement rates dropped sharply on the 10th, and by 11 Sept outward movement was occurring at only centimeters per day. Deformation remained very slow through early Oct.

After increasing sharply to 786 t/d on 10 Sept shortly after the new lobe was first observed, rates of SO_2 emission declined to less than 300 t/d on the 12th, less than 200 t/d the next afternoon, and just under 100 t/d by the 15th. A period of higher SO_2 emission rates occurred in late Sept, with measurements on the 23rd, 26th, and 28th yielding values of 190, 135, and 148 t/d, but SO_2 values had dropped to less than 50 t/d on 2 Oct.

Seismicity gradually declined 11-12 Sept, from about 20 events per hour on the 11th to about 5 per hour late on the 12th, after very vigorous activity the previous 2 days (9:8). Most earthquakes had stopped by 13 Sept, but surface events, primarily caused by rockfalls, were frequent that day and remained frequent for the next week. The number of surface events declined by the last week in Sept and seismic activity remained low through early Oct.

Information Contacts: D. Swanson, C. Mullins, USGS CVO, Vancouver, WA; C. Jonientz-Trisler, Univ. of Washington.

10/84 (9:10) Since the extrusion of a new lobe onto the composite lava dome in early Sept (fig. 12-27), seismicity, deformation, and SO_2 emission have returned to near background levels. Poor weather has limited observation of the crater, slow settling of the lava dome has continued, and rates of displacement averaged 4-5 mm/day in Oct. No large landslides were evident from field work or seismic records. Little vapor emission from the dome has been visible, but seismic instruments recorded signals apparently caused by very small steam and ash events, with as many as 20 on a single day (27 Oct) [but see 9:11]. Field observations on 1 Nov found no new tephra in fresh snow layers in the crater. Rates of SO_2 emission averaged 75 ± 25 t/d in Oct. Very few earth-

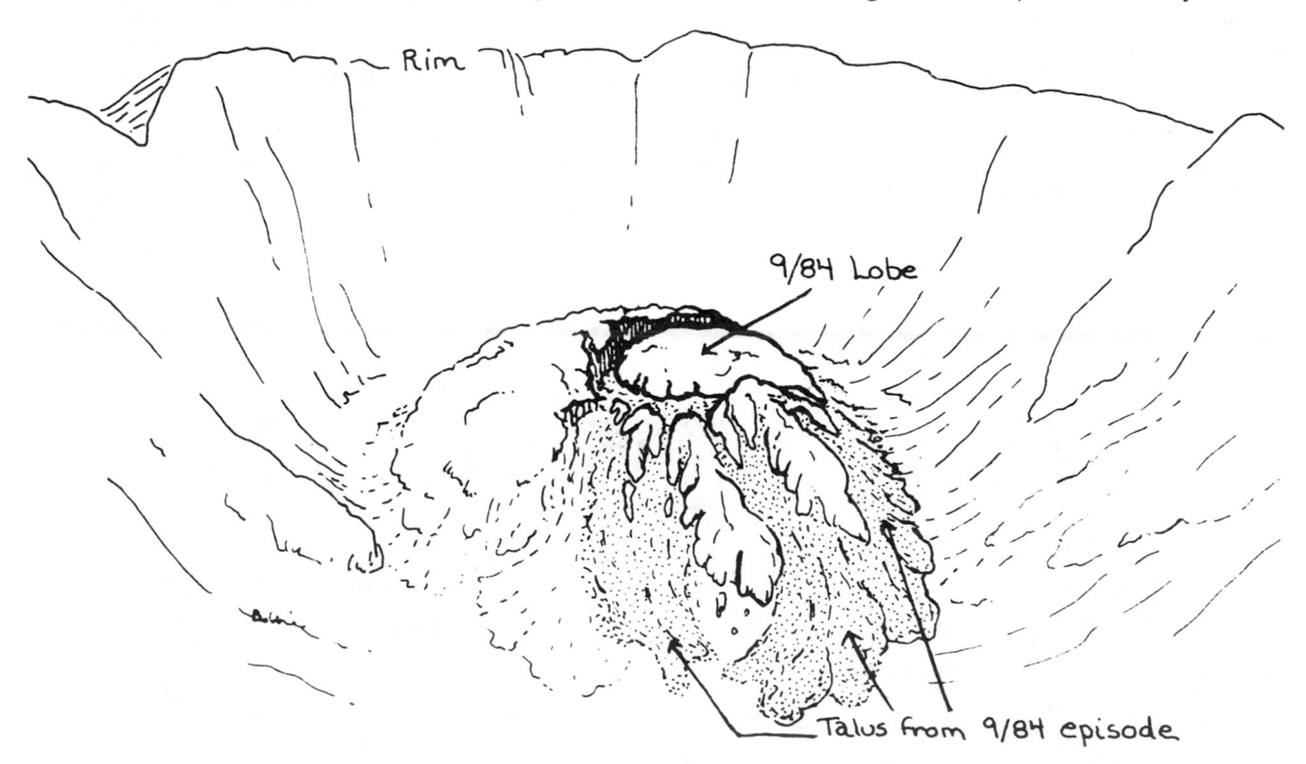

Fig. 12-27 Sketch by Bobbie Meyers, looking S, showing the dome after extrusion of the new lobe in early Sept 1984.

quakes were recorded in Oct.

Information Contacts: D. Swanson, C. Mullins, B. Myers, USGS CVO, Vancouver, WA; C. Jonientz-Trisler, Univ. of Washington.

11/84 (9:11) Seismicity, SO_2 emission, and deformation of the composite lava dome remained at background levels in Nov. The maximum displacement rate on the N side of the dome was 3 mm/day. SO_2 data were successfully collected from 4 flights in mid to late Nov. Rates of emission averaged 25 ± 10 t/d. Very few seismic events were recorded in Nov and early Dec.

It has proven difficult to discriminate between the seismic signatures produced by different types of surface events, including tephra emissions, snow avalanches, and rockfalls from the dome and crater walls. Field evidence indicates that the seismic signals reported in 9:10 were not produced by steam and ash emissions. Since 24 Sept, no emissions of tephra from the dome have been observed by field geologists, nor have any tephra layers been found in the snowpack that has been accumulating in the crater through Oct and Nov. USGS geologists noted that each of the 3 most recent large explosions (March 1982, Feb 1983, and May 1984) were preceded by a period (1½ - 5 months range) in which no tephra emission events were observed.

Information Contacts: D. Swanson, C. Mullins, USGS CVO, Vancouver, WA; C. Jonientz-Trisler, Univ. of Washington.

12/84 (9:12) Deformation of the composite lava dome, seismicity, and rates of SO_2 emission remained at background levels through early Jan. The maximum rate of outward movement on the N and S sides of the dome in Dec and early Jan was 5 mm/day. The rate of SO_2 emission, 10 t/d or less on 3 and 5 Dec, had increased slightly to 50 ± 10 t/d by the next measurement on 2 Jan and 25 ± 4 t/d 5 Jan. No tephra emissions have been observed by geologists working in the crater, nor have fresh tephra layers been found in the snowpack accumulating in the crater. Unusually good weather during the early winter has allowed the emplacement of 2 telemetering tiltmeters on the dome itself (tiltmeters had previously been installed on the crater floor near the dome), a telemetering strainmeter on a crack near the dome's summit, and a small trilateration network consisting of 2 benchmarks and 5 targets in the dome's summit area.

Information Contacts: D. Swanson; J. Sutton, USGS CVO, Vancouver, WA; C. Jonientz-Trisler, Univ. of Washington.

1/85 (10:1) Seismicity, SO_2 emission, and deformation of the composite lava dome remained at background levels through early Feb. The duration of the current quiet period was comparable to the longest previous interval between extrusion episodes, which separated the Aug 1982 extrusion from the onset of more than a year of continuous activity beginning in Feb 1983.

Very few earthquakes or surface events were recorded through early Feb. Rates of SO_2 emission measured on 5 days in Jan and early Feb ranged from 76 ± 4 t/d on 22 Jan to 12 ± 5 t/d on 4 Feb. Two telemetering H_2 monitors were installed on the dome in early Jan, one in a 70°C fumarole, the other in ambient air. Some variation in H_2 values was observed, but more background data will be necessary before the significance of such variations can be determined. No tephra emissions have occurred since 24 Sept. USGS geologists noted that each prolonged lull in tephra emissions (ranging from 6 to 20 weeks) between 1982 and 1984 was followed by a moderate explosion.

Information Contacts: Same as above.

2/85 (10:2) Seismicity and rates of displacement and SO_2 emission remained at background levels in Feb and early March. Maximum displacement rates on the dome were 2-3 mm/day. SO_2 emission averaged 50 ± 10 t/d in Feb and did not change significantly in early March. The highest temperature yet measured on the dome, 912°C, was recorded 18 Feb in a rubble-filled crack high on the N side of the dome. Glow at this site has been seen in night photos since early fall 1984, but no temperature measurements had previously been made there.

No tephra emission has occurred from the lava dome since 24 Sept, and the last extrusion episode was in early Sept (9:8-9). This is the longest quiet interval since the volcano became active in 1980.

Information Contacts: Same as above.

3/85 (10:3) SO_2 emission averaged 50 ± 10 t/d in March, identical to the Feb rate, and remained low in early April. Maximum displacement rates were 2-3 mm/day. No gas and ash emissions from the dome were observed, or detected by seismographs.

Information Contacts: Same as above.

4/85 (10:4) No gas and ash emissions from the dome were observed, and none was detected by seismographs through early May. Maximum displacement rates were 2-3 mm/day. SO_2 emission averaged 30 ± 5 t/d in April, lower than the average Feb and March rates of 50 ± 10 t/d.

The following is a report from Daniel Dzurisin and Roger Denlinger.

"Measurements of total magnetic field intensity on and near the lava dome are providing unique insights into the dome's internal structure and cooling history. Permanent changes in magnetic intensity occur on and near the dome during episodes of rapid growth, and secular increases occur on the dome as its exterior cools and becomes permanently magnetized.

"Magnetic intensity data with a precision of 0.25 gamma are collected simultaneously at 2 base stations on the volcano's flanks, and at fixed stations on the dome and surrounding crater floor. Identical proton precession total field magnetometers are used at the base stations and in the crater. The base stations are automated and transmit data to CVO once each minute; crater stations are measured sequentially by a field crew using portable instruments. The base station record is subtracted from the crater data to remove diurnal variations to an empirical accuracy of about 2 gammas.

"Two types of changes in magnetic field intensity have been detected since measurements began in March

1984. The first type occurs at stations near the dome during rapid endogenous or exogenous growth of the dome. Permanent decreases in magnetic intensity of a few to a few tens of gammas accompanied dome extrusions in March and Sept 1984. No comparable changes occurred during a relatively passive extrusion in June 1984 [but see 10:10], or during a period of rapid endogenous growth that followed in early July. Parts of the dome were displaced by several tens of meters during the March and Sept extrusions, but displacements during the June and July events were considerably smaller. We tentatively attribute this first type of magnetic change to large displacements of the cooled magnetic exterior of the dome, relative to nearby magnetic monitoring stations.

"A second type of change occurs only on the dome, where the magnetic field intensity at most stations has increased steadily since measurements began there in Dec 1984. Rates of increase vary from 0.1 to 2.2 gammas per day on different parts of the dome, but do not change significantly with time at any one station. We tentatively attribute these secular increases in field strength to cooling and magnetization of the outer parts of the dome.

"To better understand this second type of magnetic intensity change, we have made a preliminary magnetic intensity map of the dome, and have repeatedly measured several short magnetic profiles. The magnetic intensity map shows a broad positive anomaly of about 1000 gammas amplitude associated with the dome. A magnetic profile with more closely spaced stations across the Sept 1984 lobe reveals local anomalies with wavelengths of a few to a few hundred meters, with a maximum amplitude of about 700 gammas. We tentatively attribute the long-wavelength anomaly to a cooled magnetic exterior enclosing the lobe's hotter, non-magnetic interior. The strength of the long-wavelength anomaly increased by as much as 100 gammas from Feb to April 1985, presumably owing to continued cooling. More detailed magnetic profiles centered at each magnetic monitoring station on the dome tell a similar story. Measurements are made at 1-m intervals along N-S and E-W profiles about 20 m long, at sensor heights of 2.5, 3.7, and 5.0 m above the ground. Typically, short-wavelength (1-10 m) anomalies decay rapidly with increasing height, but they do not change significantly with time. Instead, the magnetic intensity along the entire profile increases uniformly with time, implying the growth of broader anomalies at depths of a few tens of meters.

"During the next year, we plan to improve our data base on the dome, and to begin quantitative modelling and interpretation of our results. The existing magnetic intensity map is not sufficiently detailed to distinguish the magnetic signatures of various lobes comprising the dome, so we will make a more detailed map this summer. We will also numerically model the development of magnetic anomalies associated with the Sept 1984 lobe and the dome as a whole, to estimate the downward migration rate of permanent magnetization, and the implied volumetric cooling rate of the dome. Combined with results from other concurrent geophysical studies, the goal of this research is to characterize the thermal structure of the lava dome and its temporal evolution. Our results may eventually bear on such diverse topics as the rheology of the dome and the volcanic hazards implications of its continued growth."

Information Contacts: D. Dzurisin, R. Denlinger, D. Swanson, J. Sutton, USGS CVO, Vancouver, WA; C. Jonientz-Trisler, Univ. of Washington.

5/85 (10:5) During mid-May, seismicity and rates of displacement began to increase from background levels. After 8 months of quiescence, a new lobe was extruded onto the composite lava dome, and a major intrusive event further enlarged the dome.

On 13 May, the number of medium- to high-frequency events recorded from the crater seismic station began to increase. Measurements on 16 May indicated that displacement rates on the dome had increased slightly while seismicity continued to build gradually. Several vigorous gas emissions, the first since 24 Sept 1984, occurred on 17 May from a vent in the NE part of the dome's summit area. No tephra was ejected, and the largest plume rose to just above the crater rim. Seismicity and displacement rates continued to increase, and on 20 May the USGS and Univ. of Washington issued a volcano advisory notice stating that recent changes at the volcano suggested "renewed eruptive activity will begin

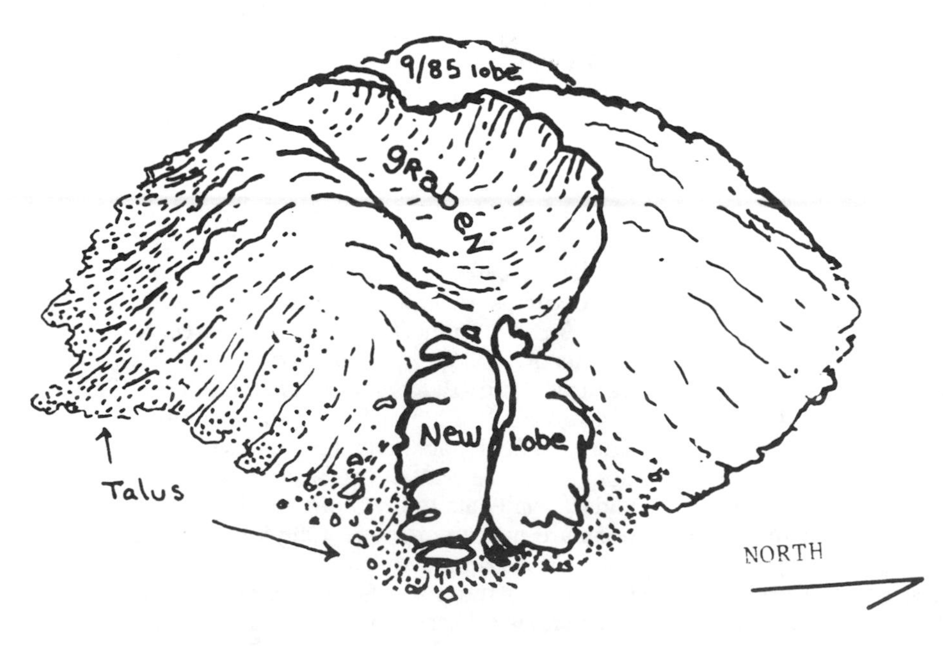

Fig. 12-28. Sketch by Bobbie Meyers, looking NW, showing the composite lava dome after formation of the graben and extrusion of the new lobe in May 1985.

within the next 2 weeks, possibly within the next few days."

By the 23rd, seismicity had reached high levels. On the 24th, very low-frequency events increased in size and number, although medium- to high-frequency events decreased. Observations made during a night overflight on the 25th at 0030 indicated increased glow from cracks on the dome. Later that morning, the low-frequency earthquakes that typically immediately precede and accompany extrusions had become more numerous, and deformation patterns suggested to USGS scientists that magma had nearly reached the surface of the dome. Observations during the next week were few because of poor weather conditions.

Between 21 and 23 May, a trough-like fracture system began to develop on top of the dome. By 27 May, this graben was 30-50 m wide and had cut across the entire S portion of the dome. Sometime between the 27th and 30th, a new lobe of gas-poor lava was extruded on the SE flank of the dome, near the E end of the newly formed graben. The 80 m-wide lobe extended out of the graben about 100 m down to the crater floor (fig. 12-28). By the 30th, the graben was 400 m long, 30 m deep, and averaged 90-100 m wide. The dome flank S of the graben had been displaced outward.

Seismicity continued at very high levels; on the 29th, magnitude 2.5-2.8 events were recorded every few minutes as deep as 1 km below the dome. During past extrusion episodes, seismicity typically decreased sharply when lava reached the surface, but on the 31st, seismicity remained vigorous. Late on 31 May or early on 1 June, a slow decrease in the number of seismic events began. By 5 June seismicity had decreased to moderate levels and had reached background levels by the 17th.

Measurements by USGS scientists on 16 June indicated that the S flank of the dome had moved outward about 70-100 m during the eruption, but that the current displacement rate had by then decreased to about a centimeter/day. A station on the crater floor S of the dome moved 46 m outward and 18 m upward between 17 May and 8 June, but had moved only a few centimeters more by the 16th. These measurements suggested to USGS scientists that about 6-8 $\times$ 10^6 m^3 of magma were intruded into the dome during this eruption, the largest single dome-building event since 12 June 1980.

After the vigorous gas emission events of 17 May, vapor emission continued from the same vent during the eruption. However, SO_2 emission rates averaged ~ 40 t/d during May, about the same as the average April rate of 30 $\pm$ 5 t/d. On 30 May, the rate increased from background levels to 90 $\pm$ 10 t/d, by 2 June to 160 $\pm$ 25 t/d, and by the 8th to 220 $\pm$ 10 t/d. SO_2 emission decreased to 165 $\pm$10 t/d on June 10th and to 60 $\pm$ 10 t/d on the 12th.

Equipment to measure H_2 emission was installed on top of the dome in Jan. After 3 months of very low measurements, ambient H_2 around the top of the dome began to increase dramatically on 24 May, continued to rise over the next few days, and reached a very high value before the station was destroyed early on the 29th. Because the instrument was not yet calibrated, the measurements cannot be quantified.

Information Contacts: D. Swanson, C. Newhall, S. Brantley, K. McGee, J. Sutton, B. Myers, USGS CVO, Vancouver, WA; C. Jonientz-Trisler, Univ. of Washington.

6/85 (10:6) After the extrusion of a new lobe and endogenous growth of the composite lava dome in late May-early June, seismicity, deformation, and SO_2 emission returned to background levels. Displacement rates began to decline on 10 June and had reached background levels by late June. As of early July the dome was stable, with maximum displacement rates of a few millimeters per day. After reaching a high of 220 $\pm$ 10 t/d on 8 June, SO_2 emission had decreased to a background level of 60 $\pm$ 5 t/d by the 12th, and remained between 25 $\pm$ 5 and 45 $\pm$ 5 t/d through early July.

Information Contacts: D. Swanson, J. Sutton USGS CVO, Vancouver, WA; C. Jonientz-Trisler, Univ. of Washington.

7/85 (10:7) Seismicity, SO_2 emission, and deformation remained at background levels. Maximum displacement rates on the dome were 1-2 mm/day, and SO_2 emissions continued to decline, averaging 28 $\pm$ 5 t/d in July, compared to 45 t/d in late June. No vigorous gas emissions from the dome have been observed since mid-May.

Information Contacts: D. Swanson, S. Brantley, USGS CVO, Vancouver, WA; C. Jonientz-Trisler, Univ. of Washington.

8/85 (10:8) Displacement rates in Aug and early Sept were 1 mm or less per day. SO_2 emissions averaged 30 $\pm$ 10 t/d. No gas or ash emissions were observed, and none was detected by seismographs.

Information Contacts: S. Brantley, E. Iwatsubo, USGS CVO, Vancouver, WA; C. Jonientz-Trisler, Univ. of Washington.

9/85 (10:9) Through early Oct, maximum displacement rates were 1-2 mm/day on the S side of the dome, the region where deformation has been concentrated for the past 6 months. The period includes several weeks preceding the May/June lobe extrusion on the S side of the dome (10:5). Three measurements during Sept showed an average rate of 30 $\pm$ 15 t/d of SO_2 emission. There were no tephra or energetic gas emissions during the month.

Information Contacts: D. Swanson; S. Brantley, USGS CVO, Vancouver, WA; S. Malone, Univ. of Washington.

10/85 (10:10) Seismicity, deformation, and SO_2 emissions were at background levels throughout Oct. Seven measurements of SO_2 yielded an average of 35 $\pm$ 10 t/d. Maximum displacement rates on the dome were 1-2 mm/day.

A continously recording stainless-steel wire strainmeter was installed on the dome on 24 Aug, spanning several cracks just north of the graben produced during the May-June 1985 dome building episode. Since 1 Sept the meter has shown a constant extension rate of 0.2 mm/day, thought to be caused by gravitational stresses.

The following is a report from Daniel Dzurisin and Roger Denlinger.

"In our last report (10:4), we discussed preliminary results of magnetic field intensity measurements since March 1984. Two types of temporal changes had been detected. On the crater floor near the dome, permanent decreases in magnetic intensity accompanied the March, June, and Sept 1984 extrusions (the report in 10:4 stated incorrectly that no changes were detected in June 1984). These changes were attributed to displacement of the N part of the dome during rapid endogenous growth. At some stations, smaller reversible changes accompanied the March and June 1984 extrusions. Their cause is not known, but stress-induced piezomagnetic effects could have played a role. On the dome itself, there had been a steady increase in magnetic intensity at most monitored sites since measurements began in Dec 1984. Those changes were attributed to cooling and magnetization of outer parts of the dome.

"There have been three significant developments since the April 1985 report. First, the largest dome-building eruption since activity began in 1980 occurred during May-June 1985. One magnetic site was destroyed and 2 others were significantly displaced when a graben split the S part of the dome during rapid endogenous growth. A small flow was eventually extruded from the graben floor. Large magnetic changes occurred only at those sites on the S part of the dome that were obviously displaced; elsewhere, earlier trends continued uninterrupted. On the crater floor, the field intensity at the station closest to the dome increased significantly during endogenous growth; other stations showed no changes. Large changes on the dome were almost surely caused by displacement of monitoring sites and nearby rock (one station rode the floor of the new graben downward almost 50 m). The cause of the intensity increase at one crater floor station is not known, but disruption of the dome could have been responsible.

"A second development is that rates of magnetization at stations on the N part of the dome decreased in early June 1985, and rates at stations on the S part of the dome increased. By Oct, there was a suggestion that rates at some N sites had started to increase again. Two explanations for the changes at N sites are under consideration: 1) a seasonal change in cooling and magnetization rates (less precipitation, less rapid cooling and magnetization in summer); and 2) a heat pulse re-

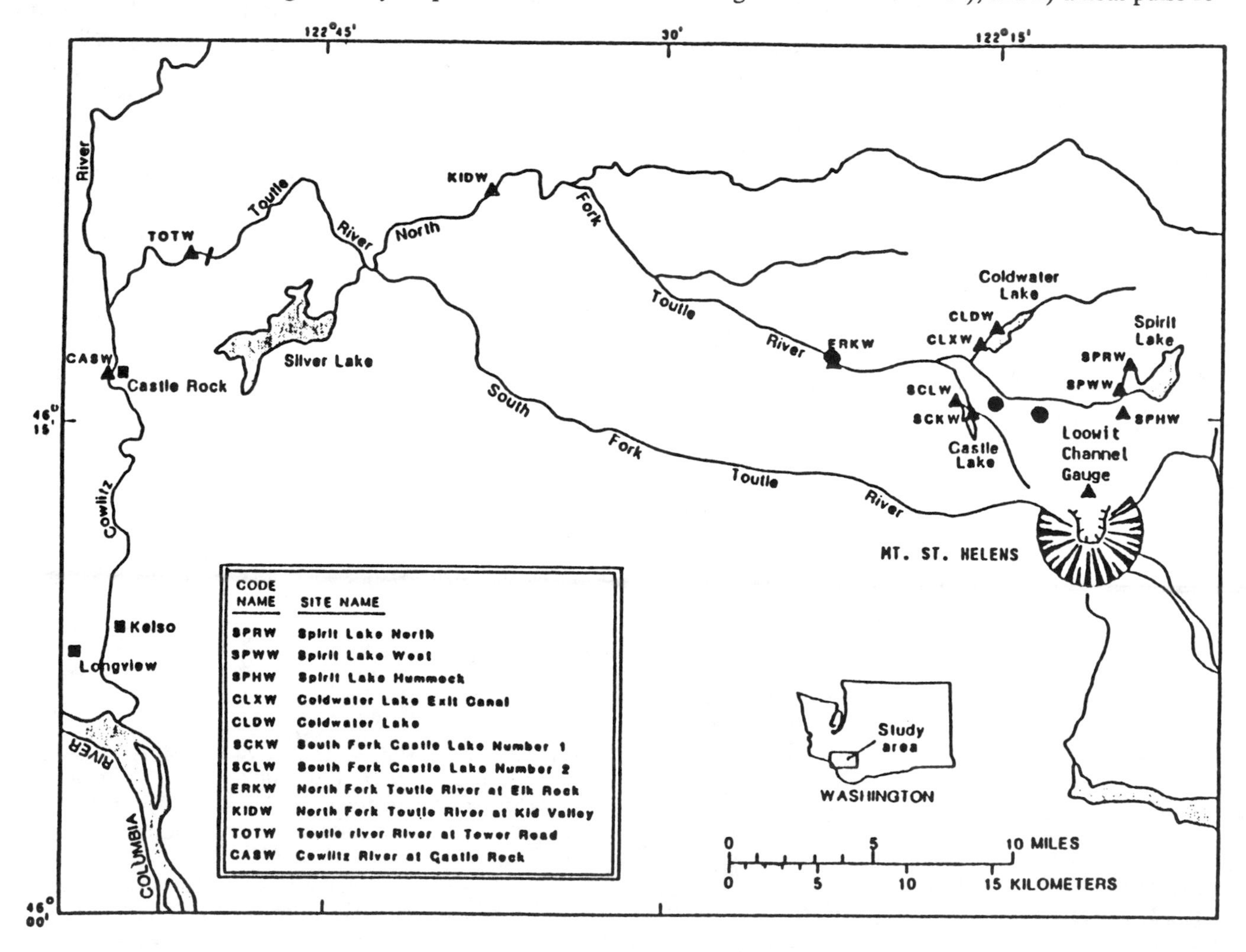

Fig. 12-29. Location of monitoring sites, Mt. St. Helens, Washington; circles indicate seismometers, triangles indicate gauges. Base map modified from Childers and Carpenter, 1985.

lated to the May-June episode of dome growth. Increased magnetization rates at S sites probably reflect rapid cooling of the walls of the new graben, or of newly-emplaced magma beneath the floor of the graben.

"Finally, a preliminary aeromagnetic survey was conducted above the dome in Oct, and additional surveys are planned this winter. Our goal is to model the internal thermal structure of the dome, to establish a better framework for interpreting temporal changes. Preliminary results suggest that the magnetic carapace of the dome is 1-10 m thick in most places, with large local variations."

Information Contacts: D. Dzurisin, R. Denlinger, D. Swanson, J. Sutton, USGS CVO, Vancouver, WA; C. Jonientz-Trisler, Univ. of Washington.

11/85 (10:11) Mt. St. Helens was quiet through Nov and early Dec, with seismic activity, SO_2 emission, and displacement rates remaining at background levels. Displacement rates on the dome were less than 1 mm/day; SO_2 levels continued to be low and quite variable.

The USGS's Water Resources Division has installed a new gauge along the crater's main drainage channel to improve warnings of mudflows and snow avalanches. The new gauge, just N of the crater (in Loowit Channel at The Steps), and an existing one 25 km downstream in the North Fork Toutle River (at Elk Rock) transmit to CVO via satellite every 5 minutes. These stations, in addition to seismometers and other gauges downstream, should provide an early warning of rapidly rising water levels (fig. 12-29).

Much of the network was installed prior to Nov 1982 to monitor levels of the lakes created when the 18 May 1980 avalanche blocked the outflow from Spirit Lake and downstream tributaries to the North Fork Toutle River. Major flooding could occur if one of the debris dams failed, and seismometers were installed in 1983 to supplement the gauges. A characteristic signature is produced by lahars flowing from the crater in a confined channel, as in May and June 1984. Although exact times and discharge volumes could not be determined from the seismic records for those events, future refinements may provide a relationship between seismic signals and discharge volumes (Brantley, et al., 1985).

References: Brantley, S., Power, J., and Topinka, L., 1985, Reports from the U.S. Geological Survey's Cascades Volcano Observatory at Vancouver, Washington; *Earthquake Information Bulletin*, v. 17, no. 1, p. 20-32.

Childers, D. and Carpenter, P.J., 1985, *International Symposium on Erosion, Debris Flow, and Disaster Prevention*; Sept 3-5, Tsukuba, Japan.

Information Contacts: G. Gallino, S. Brantley, E. Iwatsubo, USGS CVO, Vancouver, WA; C. Jonientz-Trisler, Univ. of Washington.

12/85 (10:12) Activity at Mt. St. Helens remained at background levels in Dec. For the 7th consecutive month, only minor seismicity, geodetic changes, and SO_2 emissions were detected. Displacement rates on the dome were approximately 2 mm/day, while the continuously recording strainmeter just north of the May-June 1985 graben showed steplike extension across cracks averaging 0.2 mm/day. Seismicity consisted mostly of surface events (primarily rockfalls). SO_2 emissions continued to be low and variable.

Information Contacts: D. Swanson, S. Brantley, USGS CVO, Vancouver, WA; C. Jonientz-Trisler, Univ. of Washington.

Further References: Keller, S.A.H. (ed.), 1986, *Mount St. Helens, Five Years Later: Proceedings of a Symposium at Eastern Washington University May 16-18, 1985*; Eastern Washington Univ. Press, 448 pp. (47 papers).

Manson, C.J., Messick, C.H., and Sinnott, G.M., 1987, Mount St. Helens-A Bibliography of Geoscience Literature, 1882-1986: *USGS Open-File Report* 87-292 (1600+ references).

Special Section: Mount St. Helens; *JGR*, 1987, v. 92, no. B10, p. 10149-10334 (12 papers).

MT. HOOD

Cascade Mountains, Oregon
45.37°N, 121.70°W, summit elev. 3424 m
Local time = GMT - 7 hours

Mid-19th century eruptions were also known from this prominent volcano, 80 km ESE of Portland.

General Reference: Crandell, D.R., 1980, Recent Eruptive History of Mount Hood, Oregon, and Potential Hazards from Future Eruptions; *USGS Bulletin* 1492, 81 pp.

7/80 (5:7) [Seismic data in the following report have been extensively modified using Rite and Iyer, 1981.] A series of earthquakes in the vicinity of Mt. Hood, about 100 km SSE of Mt. St. Helens, began with a magnitude [2.8] shock on 6 July at 1817. [Within 30 minutes, 7 events of magnitudes 1.6-2.8 occurred in the vicinity of the first shock. Depths were tightly clustered at 4-7 km.]

[Activity remained anomalously high for about 48 hours after the initial shock, with 40 events recorded on 3-9 stations in the first 24 hours (fig. 12-30). Epicenters were generally on the S flank (fig. 12-31).] A magnitude [2.44] earthquake at 0259 and a magnitude [2.25] event at 1315 were centered [near the summit] at [5.9 and 5.3] km depth. [Between 6 and 16 July, 143 small events were recorded on a single station (VHE) about 5 km SE of the summit. About 50 were correlated with nearby stump blasting, but the other 90 appeared to be swarm events, most with coda magnitudes less than 0.5.] . . . The swarm's total seismic energy release [was about 8×10^{15} ergs]. . . .

The USGS issued a hazard watch formally notifying government officials. On 11 July, the USGS and the Univ. of Washington installed 3 portable seismographs on Mt. Hood. No harmonic tremor has been recorded, and no new surface activity has been observed. Gas analyses made from aircraft 11 July showed no increase in atmospheric SO_2 or CO_2 above normal levels.

Because no additional seismicity and no new erup-

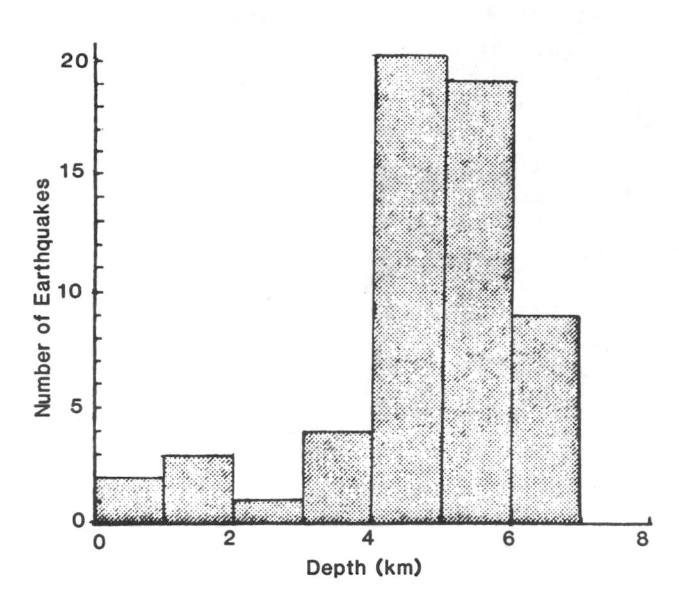

Fig. 12-30. Number of earthquakes vs. depth at Mt. Hood, 6-20 July 1980, after Rite and Iyer, 1981.

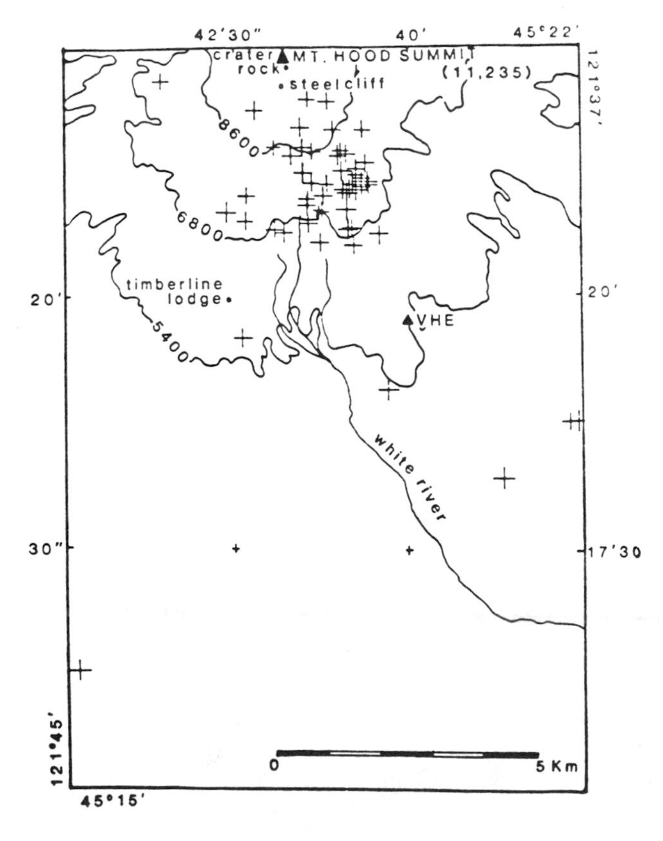

Fig. 12-31. Epicenters of most of the July 1980 events. Sizes of crosses are proportional to event magnitudes. After Rite and Iyer, 1981.

tive activity have occurred at Mt. Hood, the USGS ended the hazard watch on 5 Aug.

Information Contacts: R. Decker, USGS HVO, Hawaii; R. Tilling, C. Zablocki, USGS, Reston, VA; S. Malone, R. Crosson, E. Endo, Univ. of Washington.

Further Reference: Rite, A. and Iyer, H.M., 1981, July 1980 Mt. Hood Earthquake Swarm; *USGS Open File Report* 81-48, 22 pp.

MT. SHASTA

Cascade Mountains, California
41.40°N, 122.18°W, summit elev. 4317 m
Local time = GMT - 7 hrs

One of the largest of the Cascade strato-volcanoes, Mt. Shasta is 535 km directly S of Mt. St Helens and 425 km N of San Francisco. Its only known historic eruption was in 1786.

General Reference: Miller, C.D. 1980, Potential Hazards from Future Eruptions in the Vicinity of Mount Shasta, Northern California; *USGS Bulletin* 1503, 43 pp.

7/78 (3:7) An earthquake swarm in the vicinity of Mt. Shasta began at 0202 on 1 Aug. The first event, magnitude 4.2, was followed by six magnitude 3-4 events in the next 45 min. Between 35 and 40 shocks (M ≥ 2) were recorded on 1 Aug. The number of events declined slightly the next day and only about 10 had been recorded by midday on 3 Aug. About 20 of the total had magnitudes ≥ 3. Permanent seismographs were too distant for precise hypocenter determinations, but USGS personnel were bringing portable seismographs into the area.

An ash eruption, probably from Shasta, was sighted from a ship off the California coast in 1786.

Information Contact: R. Lester, USGS, Menlo Park, CA.

8/78 (3:8) The number of earthquakes declined to only a few events per day (M ≥ 2) by 11 Aug. However, on 12 Aug a magnitude 4.3-4.5 shock was followed by several similar events (approximately M 4), and others in the magnitude 3.5-4 range occurred on the 13th. Activity then declined again; on 23 Aug only 13 events of magnitude 2-2.5 were recorded, and by the end of Aug only about 6 events greater than magnitude 2 were being recorded daily.

Epicenters have been located along about 2 km of a pre-existing N-S-trending fault zone 28 km E of the summit. Uncertain crustal velocities for the area have made depth determinations difficult, but all events have been shallow, probably less than 5 km, and some may have been less that 1 km deep. No migration of events has been observed. New tensional fissures have been found in the epicentral area, but were not growing as of late Aug. Leveling, microearthquake studies, and gravity profiles are planned.

Information Contacts; R. Sherburne, California Div. of Mines & Geology; A. Walter, USGS, Menlo Park, CA.

9/78 (3:9) Seismic activity E of Mt. Shasta had declined by late Sept to about 6 locatable events per day, most stronger than magnitude 2. Hypocenters extended E from surface fissures 28 km E of Shasta, along a pre-existing N-S-trending fault zone. Focal depths were very shallow near the surface fissures, but increased to 4-6 km along an inclined seismic zone dipping 35°-45° E. The events have not been migrating, nor has there been any evidence of volcanic activity associated with the swarm.

Information Contact: R. Sherburne, California Div. of Mines & Geology.

Further Reference: Bennett, J.H., et al., 1979, Stephens Pass Earthquakes, Mount Shasta—August 1978; *California Geology*, Feb., 1979, p. 27-33.

LONG VALLEY CALDERA

Sierra Nevada, California
37.68°N, 118.86°W, summit elev. 2065 m
Local time = GMT - 8 hrs winter (- 7 hrs summer)

This caldera, at the E foot of the Sierra Nevada and 320 km E of San Francisco, has had no historic eruptions. It was formed 730,000 years ago during a major eruption that produced over 1,000 times the recent ash volume of Mt. St Helens and distributed it over much of the western United States.

General Reference: Bailey, R.A., Dalrymple, G.B., and Lanphere, M.A., 1976, Volcanism, Structure, and Geochronology of Long Valley Caldera, Mono County, California; *JGR*, v. 81, no.5, p. 725-744. (This issue of *JGR* includes 13 other papers on various aspects of Long Valley.)

5/82 (7:5) Since the four magnitude 5.5-6.1 earthquakes of 25-27 May, 1980, 8 seismic swarms have occurred in the S part of the caldera (table 12-4). Most lasted roughly 1-2 hours and consisted of several hundred events, including some B-type shocks. All of the swarms included periods of spasmodic tremor, produced by a succession of earthquakes too closely spaced to allow clear separation into discrete events on seismograph records. Some earthquakes were also recorded between swarms. In Jan 1982, a new group of fumaroles was discovered in the vicinity of Casa Diablo Hot Springs, about 2.5 km E of the earthquake epicenters, and existing fumaroles in the area had become more vigorous. The most recent swarm began 7 May at 0517 with a 40-minute burst of seismicity. Occasional 5-10-minute periods of spasmodic tremor were recorded during the next 24 hours. The swarm's strongest earthquake was a magnitude 4.3 event at 2047. Depths of the 7-8 May events ranged from 3.8 to 8.5 km below the surface, comparable to the 9-10 July 1981 swarm, but substantially shallower than others since May 1980. Epicenters were about 1/2 km N of those from previous swarms.

In Oct 1980, a survey along the highway that passes through the caldera showed as much as 25 cm of uplift of the caldera's resurgent dome, possibly within the previous 2 years. A 10 km-long geodimeter line over the dome, remeasured in May 1982, showed 50 cm of lateral spreading since 1978. Dry tilt stations, installed 8-9 May over the earthquake swarm epicentral area, showed no changes when reoccupied about 3 weeks later.

Miller et al. (1982) noted "A preliminary interpretation of this evidence is that magma at depth in the Long Valley Caldera moved upward at about the time of the May 1980 swarm of earthquakes. This caused bulging of the resurgent dome and opened fractures at depth in the S part of the caldera (R.A. Bailey and R. Cockerham, written communication, 1982), thereby allowing a tongue of magma to move toward the surface beneath the epicentral site. . . ." On 25 May 1982, the USGS issued a notice of potential volcanic hazard for the Long Valley area, adding to the earthquake hazard watch in effect since 27 May 1980.

Long Valley Caldera formed 0.7 million years ago during the eruption of the Bishop Tuff, a rhyolite ash flow deposit with a total volume exceeding 600 km^3. Since then, numerous post-caldera eruptions have occurred in the area, including resurgent doming 0.63 to roughly 0.51 million years ago. The most recent features in the area are the Mono craters, and the Inyo craters and domes (fig. 12-32), some of which have ^{14}C ages younger than 1000 years (Bailey et al., 1976).

Reference: Miller, C.D., Crandell, D.R., Mullineaux, D.R., Hoblitt, R.P., and Bailey, R.A., 1982, Preliminary Assessment of Potential Volcanic Hazards in the Long Valley-Mono Lake Area, East-Central California and Southwestern Nevada; *USGS Open File Report* 82-583.

Information Contacts: R. Bailey, USGS, Reston, VA; C.D. Miller, D. Mullineaux, USGS, Denver, CO; A. Ryall, Univ. of Nevada, Reno.

6/82 (7:6) Since the seismic swarm of 7–8 May, only 1 minor swarm of microearthquakes has been recorded in the epicentral area under the S side of the caldera. This swarm occurred 13 July at 1200 and consisted of 6 minor shocks of magnitude 1.5 or less during a span of 1 1/2 hours, at a depth of about 3.5-5.5 km. The swarm was not accompanied by spasmodic tremor. Seismicity otherwise has been at background levels.

Relevelling of U.S. route 395 across the resurgent dome in June 1982 revealed additional uplift since Sept 1980—a maximum of 8.2 cm at the crest of the dome, suggesting an average rate of uplift of about 0.5 cm/month during the 18-month time span.

Gas analyses and H_2 probe work that began in May have shown no systematic changes at Casa Diablo Hot Springs (2.5 km E of the epicentral area), where increased fumarolic activity was discovered in Jan. The USGS, University of Nevada, California Division of

Date			Average Depth (km)	Standard Error (km)	Largest Event Magnitude	No. of Events Located
1980	June	7	8.4	1.2	3.5	6
	July	2-3	8.0	1.0	4.2	14
	Aug	3	8.2	1.3	3.4	7
	Nov	25	7.8	1.3	3.4	5
1981	April	21	8.5	3.0	2.8	6
	July	9-10	5.9	1.3	3.2	30
	Aug	9	9.0	2.9	3.4	5

Table 12-4. Data from 7 seismic swarms at Long Valley Caldera, June 1980-Aug 1981, courtesy of Alan Ryall.

Mines and Geology, and USFS are continuing a coordinated geophysical and geochemical monitoring program at Long Valley.

Information Contacts: R. Bailey, USGS, Reston, VA; T. Casadevall, D. Dzurisin, USGS CVO, Vancouver, WA; A. Ryall, Univ. of Nevada, Reno; W. Duffield, R. Cockerham, USGS, Menlo Park, CA.

7/82 (7:7) Seismicity remained at background levels through mid-Aug. No changes occurred in fumarole activity at Casa Diablo Hot Springs.

Information Contact: B. Dalrymple, USGS, Menlo Park, CA.

8/82 (7:8) Seismic activity in the Long Valley area remained minor in Aug and early Sept. No spasmodic tremor was recorded, and since the 6 minor shocks of 13 July only occasional individual earthquakes have occurred in the epicentral area of previous swarms, on the S side of the caldera.

Francis Riley provided the following report.

"Preliminary review of spirit levels and electronic distance measurements obtained May–Aug 1982 in Long Valley Caldera suggests that uplift and horizontal extension in the W-central part of the caldera are continuing. Maximum values of uplift as defined by presently available data occur along the W side of the resurgent dome and amount to 33 cm since 1975. Maximum values of extensional strain since 1978 exceed 45 microstrain (change in length divided by original length x 10^6) and are associated with the resurgent dome and S caldera rim."

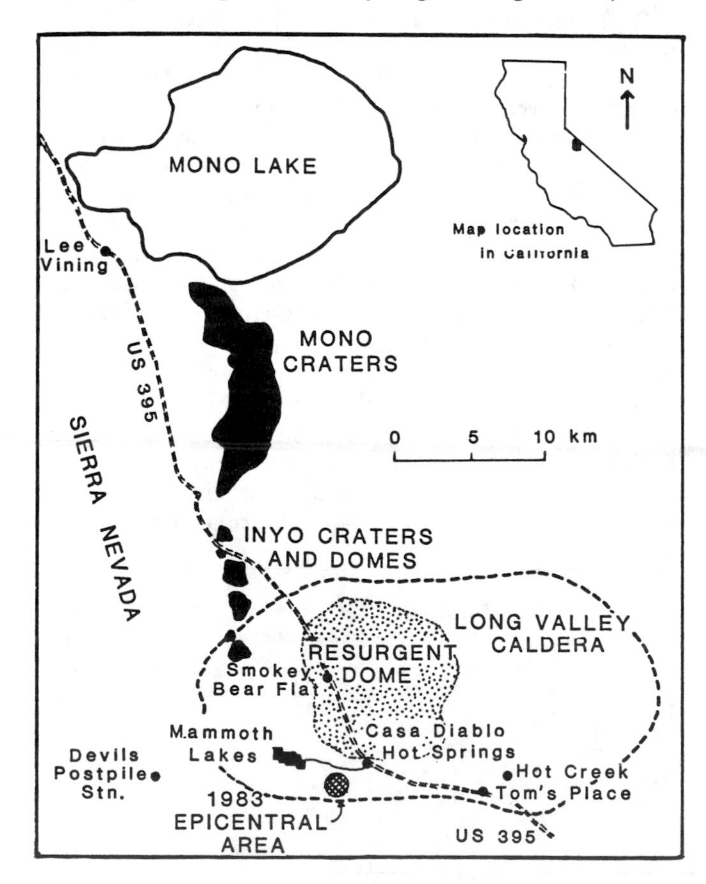

Fig. 12-32. Sketch map of Long Valley Caldera, after Bailey et al., 1976, and Miller, et al., 1982.

Information Contacts: A. Ryall, Univ. of Nevada, Reno; F. Riley, USGS, Denver, CO; R. Bailey, USGS, Reston, VA.

10/82 (7:10) Occasional earthquake swarms have been located in the S side of the caldera, but the region that had been most active seismically has remained quiet since the swarm of 7-8 May. The earthquake swarms since 7-8 May have not included spasmodic tremor. Hypocenters for the swarms have typically been restricted to pipe-like zones. The typical swarm pattern has begun with a burst of deep events, followed by a shallow burst, then a series of events beneath the two.

On 19 Sept, 40-50 events occurred at depths of 3-6.5 km along an intracaldera branch of the Hilton Creek Fault, on the edge of the resurgent dome about 4 km NE of Casa Diablo Hot Springs (fig. 12-32). The largest event, magnitude 3.2, occurred well after the beginning of the swarm. Roughly 9 km to the WSW, a swarm of more than 100 shocks started 18 Oct. Most of the events occurred the first day, but a few continued through 24 Oct. All ranged from 3.5 to 6.5 km depth (most 5-6 km) and the maximum magnitude was 2.5. On 3 Nov at 0945, a swarm of more than 100 events began near a thermal area about 2.5 km N of the 19 Sept epicenters. Depths ranged from roughly 3.5 to 10 km and the maximum magnitude was 2.9. Only very small events were recorded after 2113.

Information Contacts: R. Bailey, USGS, Reston, VA; R. Cockerham, USGS, Menlo Park, CA; A. Ryall, Univ. of Nevada, Reno.

11/82 (7:11) As of early Dec, no earthquake swarms had been reported in the caldera since the 3 Nov events. A network of 9 dry tilt stations in the S part of the caldera has been reoccupied about 6 times since May. The sensitivity limit of this network, estimated at about 10 μrad, is too large to allow monitoring of the continuing uplift of the resurgent dome at present (1980–1982) rates. However, these tilt stations would be sensitive to substantial increases in the overall rate of resurgence, or significant local deformation, neither of which has been detected. Borehole tiltmeters installed near 2 of the dry tilt sites 1 Nov are expected to be able to measure tilt changes several times as small and will telemeter data to the USGS.

Information Contacts: D. Dzurisin, USGS CVO, Vancouver, WA; R. Cockerham, USGS, Menlo Park, CA.

12/82 (7:12) The following is from the USGS.

"Earthquake swarms in the Long Valley area resumed in mid-Dec, after quiescence that lasted through most of Nov. On 14 Dec between 0056 and 0200, 200-300 small events were recorded, of which only about 10 could be located. These were centered at 2-3 km depth in the S part of the caldera, in the Casa Diablo epicentral area of many previous swarms. Spasmodic tremor was recorded for the first time since the 7-8 May swarm (7:5). Increased thermal activity was noted along Hot Creek and near the epicentral area a few days before this swarm. Geyser-like activity at one Hot Creek vent occasionally ejected hot water to about 10 m height and water from another vent surged intermittently to about 1.5 m height.

"On 21 Dec at 1428, two magnitude 3.3 earthquakes occurred at 6 km depth in the same epicentral area and were followed by a series of aftershocks. On 22 Dec between 2140 and 2200 about 100 events were recorded outside the caldera near Red Cones, 2 basaltic cinder cones about 9 km SW of the 14 and 21 Dec epicenters. Spasmodic tremor also accompanied this brief swarm.

"On 6 Jan at 1623, the most intense and prolonged swarm of earthquakes since May 1980 began in the S moat of the caldera. During the first 12 hours, more than 1000 events were recorded, most in the Casa Diablo epicentral area, but with a secondary concentration near the caldera wall at Convict Creek (about 10 km ESE) and with many distributed between. Strong spasmodic tremor was nearly continuous during the first 12 hours. Two particularly strong shocks, M 5.5 and 5.6 at 1738 and 1924, caused minor damage in Mammoth Lakes and disrupted electrical and telephone service for about an hour.

"During the first 36 hours, earthquakes of M ≥ 1 were occurring at a rate of 80-100/hr, those of M ≥ 3 at 1-5/hr. During the succeeding 36 hours, the number of earthquakes gradually declined to about 15/hr. Sporadic events of M 3-3.5 continued through 1200 on 10 Jan. As of 12 Jan, recorded events were continuing at a rate of 4-5/hr, still above the normal background of about 50/day. Hypocenters during the swarm ranged from 10 km to ≤ 3 km depth, with most between 4 and 7 km.

"Deformation (borehole tiltmeter, dry tilt, and geodimeter) measurements made during the swarm on 10-11 Jan suggest that uplift of the resurgent dome accompanied the swarm, but the exact amount awaits completion of remeasurement of selected parts of the leveling network. This, together with the concentration of seismicity in the S moat and the absence of significant seismicity in the Sierra block S of the caldera during this swarm, strongly suggests that the swarm was associated with magma movement at depth. Reoccupation of the geodimeter network in early Dec had shown no apparent change in deformation since the previous measurements in Aug (7:8)."

Information Contacts: R. Bailey, USGS, Reston, VA; R. Cockerham, USGS, Menlo Park, CA; F. Riley, USGS, Denver, CO.

1/83 (8:1) The following is from David Hill.

"The level of earthquake activity in the caldera continued to abate following the intense swarm that began 6 Jan. In mid-Feb, earthquakes of M ≥ 1 have occurred in the caldera at a rate of roughly 30/day, compared to rates of ~100/day toward the end of Jan and 1000/day on 7 Jan. Background activity for several months prior to the Jan swarm averaged 8-10 earthquakes/day of M ≥ 1 in the caldera. No events with M > 3 have been recorded in Long Valley since the brief flurry that included one M 3.5 and two M 4 earthquakes on 3-4 Feb.

"In the days immediately following the earthquake activity of 6 Jan, various deformation networks in the epicentral region were resurveyed. Preliminary analysis of laser-ranging measurements and precise leveling showed that strains of 3-4 ppm and uplift of up to 7 cm accompanied the earthquake activity. The deformation pattern generally resembled that observed since mid-1980, although recently determined changes appeared to be most pronounced in the epicentral region, 2-5 km E of the town of Mammoth Lakes.

"While no definitive statement can be made on the basis of available information, it appears that the deformation pattern can be explained by movement within the source region of the earthquakes. One preliminary model suggests up to 20 cm of right-lateral slip on the seismically defined fault zone, accompanied by 80 cm of opening within that zone. The right slip is consistent with seismically determined focal mechanisms for the earthquakes. Other evidence for extension at depth in the region comes from the re-analysis of the 2 largest May 1980 earthquakes, which can most simply be explained as the rapid opening of a tensile crack that is filled with fluid as it grows."

Information Contact: D. Hill, USGS, Menlo Park, CA.

2/83 (8:2) As of early March, an average of 10-30 events/day of M ≥ 1 continued to occur in the epicentral area of the major Jan earthquake swarm. Few larger events were recorded in Feb, but 5 shocks with M > 3 occurred 18-19 Feb and a M 4 earthquake was recorded 24 Feb in the Jan epicentral region. Heavy snows have severely limited deformation monitoring, but available data suggest that no major changes have occurred since Jan.

Information Contact: Same as above.

3/83 (8:3) Seismicity continued to decline in the epicentral area of the major Jan earthquake swarm. Brief bursts of small shocks were occasionally recorded, but by early April, daily earthquake counts were approaching the pre-January background levels of 2-5 events of magnitude ≥ 1 per day.

Information Contact: Same as above.

6/83 (8:6) Occasional periods of increased seismicity have continued in the Long Valley area. On 4 June a series of 100-150 small (maximum M 2.1) events occurred in the S part of the caldera, about 1 km N of the Jan epicentral zone. Focal depths of the June earthquakes ranged from about 3.8-4.9 km. Several kilometers S of the caldera rim, in the aftershock zone of the four M 5.5-6.1 earthquakes of 25–27 May 1980, a M 4 event occurred on 3 July at 0950, followed by a M 5 shock in the same area at 1140. About 150 aftershocks have been recorded at depths of 8-10 km, the largest a M 3.5.

Little measurable deformation has been detected during the past several months, but large areas of the caldera are inaccessible until summer because of heavy snow cover. Considerable uplift and horizontal extension had occurred between 1975 and mid-1982 in the W-central part of the caldera (7:8), but data from summer field work will be necessary before geophysicists can determine if this activity has continued.

Information Contacts: R. Cockerham, D. Hill, USGS, Menlo Park, CA; D. Dzurisin, USGS CVO, Vancouver, WA.

7/83 (8:7) As of early Aug, no significant earthquake swarms had been detected since 4 June. An average of 2-4 events/day larger than magnitude 1 were recorded in July. Completion of deformation measurements in the caldera is expected at the end of the summer.

The following is from Dartmouth College geologists:

"While conducting an extensive geochemical study of Rn and Hg° concentration at Long Valley, Dartmouth College geologists found evidence of activity in 2 previously unreported places. Very recent ground breakage (one large collapse pit 3 m long, 2.5 m wide, and 1.5 m deep; another pit 0.5 m in diameter and 0.5 m deep) was found along a fault sag ~0.5 km long bearing 340°. The fault is approximately 350 m W of Deer Mountain, the southernmost of the Inyo Domes in the NW part of the caldera (at UTM coordinates 321450 E and 4175800 N). A fumarole located in the Casa Diablo area (just S of the resurgent dome in the S part of the caldera) and associated with older alteration also appears to be previously unreported. New ground breakage may have allowed its formation (at UTM coordinates 333190 E and 4169370 N)."

Information Contacts: D. Hill, R. Cockerham, USGS, Menlo Park, CA; S. Williams, K. Hudnut, E. Lawrence, J. Lytle, Dartmouth College.

11/83 (8:11) A small earthquake swarm occurred in the S part of the caldera on 14 Nov. Between the start of the swarm at 1909 and its end about 40 minutes later, 40-50 events M > 0.8 were detected. The strongest shocks had magnitudes of 3.2 (at 1929) and 3.0 (at 1936). The 14 Nov events (centered about 6.5 km SE of the town of Mammoth Lakes) were within the epicentral zone of the major Jan swarm (7:12, 8:1), but about 3 km SE of the initial Jan swarm seismicity. Focal depths of the Nov earthquakes ranged from 2.5 to 4 km, making it the first swarm for which it was possible to confirm that no events were deeper than 4 km; depths of past swarms were typically 3-7 km. Some seismicity continued after the end of the main swarm at 1950; all events during the next 4 hours were at depths of 2-3.5 km.

Roughly 8 km to the W (under the E flank of Mammoth Mountain) a swarm of ~50 events had occurred 24 Sept between 0525 and 0630. The largest magnitude was 2.8. Seismic stations were too distant for precise depth determinations, but the events were roughly between 4 and 6.5 km depth. A slight increase in seismicity was noted in the same area at the end of Nov.

Information Contact: R. Cockerham, USGS, Menlo Park, CA.

8/84 (9:8) The following is from a USGS report.

"Following decay of the Jan 1983 earthquake swarm activity to background levels in the spring of 1983 (7:12, 8:1-3), seismic activity has persisted at a relatively low level. Within the caldera, this background level typically involves several earthquakes (M ≥ 1) per day and an occasional locally felt M 3 event. A M 4.2 earthquake on 28 April 1984 and a M 3.8 event in a swarm that began 16 July 1984 are the largest events to occur in the caldera since the Jan 1983 swarm. Deformation data show evidence for only modest inflation of the resurgent dome following the Jan 1983 swarm. Frequently repeated trilateration measurements, however, show continued extensional deformation within the caldera at rates of several microstrain units per year.

"With the reduced rates of seismic activity and ground deformation in the caldera during the last year, the possibility of an imminent eruption appears more remote than during the previous 2 years. In a letter dated 11 July 1984, the Director of the USGS advised California officials that a volcanic eruption does not pose an immediate threat to public safety in the Long Valley region and that the Long Valley region does not satisfy the imminent threat criteria for an official Hazard Warning for volcanic activity.

"The Long Valley region had been under a Notice of Potential Volcanic Hazard since May 1982. Prior to Sept 1983, this was the lowest of 3 levels used by the USGS to express the relative urgency of a potential geologic hazard when formally notifying the public and responsible officials. In Sept 1983, the 3-level terminology was changed to a single-level system termed a Hazard Warning. The USGS is continuing its intensive seismic, geophysical, and geodetic monitoring of Long Valley Caldera that was initiated in mid-1982."

Information Contact: D. Hill, USGS, Menlo Park, CA.

10/85 (10:10) The quoted material is from the USGS.

"Since the spring of 1983, seismicity has persisted at relatively low levels and regional extensional deformation and uplift centered over the resurgent dome have continued.

"Seismicity and geodetic measurements through Aug 1985 indicated that broad-scale deformation across the entire caldera was continuing. However, decreases have occurred in both seismicity within the caldera and the rate of deformation within the S moat, the region of the caldera between the resurgent dome and the S rim (fig. 12-32).

"No detectable earthquakes occurred within the caldera 1-23 July, and only 4 small ones (M < 2.2) were recorded during the last 8 days of June. On 24 July, however, the low-level seismicity characteristic of the previous several months (an average of 0.7 events with M ≥ 1 per day) resumed and continued through Aug. The largest July-Aug earthquake in the caldera was a magnitude 2.1 event on 12 Aug. Preliminary examination of more recent data suggests that this pattern of low-level seismicity continued through Sept and Oct.

"Recently completed annual measurements of the regional trilateration network and several of the major level lines indicated that regional extensional deformation and uplift centered over the resurgent dome continued at rates little changed from those measured from mid-1980 to mid-1984. Summer 1985 results for the level line along Highway 395, for example, showed that cumulative uplift for bench marks near Casa Diablo (at the S side of the resurgent dome) was 47 cm with respect to elevations measured in 1975 (fig. 12-33).

"Data from the frequently measured 2-color geodimeter network showed that local deformation within the S moat, including the epicentral region of the major Jan 1983 earthquake swarm, continued to slow and had virtually stopped along certain lines. July-Aug data from the telemetered tiltmeters are consistent with minimal local deformation within the caldera. Data from a continuously recording dilational strainmeter in a borehole ~3 km outside the caldera (at Devil's Postpile) showed that the strain rate apparently decreased somewhat abruptly on about 20 June, and remained flat until a small change about 28 Aug. Measurements of magnetic field variations at 2 locations in the SE and NW parts of the caldera (Hot Creek and Smokey Bear Flat) showed no changes through Aug.

"The geodetic evidence that broad-scale deformation across the entire caldera is continuing at a more or less uniform rate points to continuing inflation of the deeper section of the magma chamber (depths of 10-12 km). In contrast, the marked decrease in both the seismicity rate within the caldera and the rate of local deformation across the S moat indicate that stress differences in the upper crust associated with the Jan 1983 earthquake swarm (and possible related intrusive event) have substantially relaxed. However, as long as we see evidence for continuing inflation of the deeper magma chamber, we must regard the recurrence of another episode of unrest somewhere within the caldera as a distinct possibility."

In cooperation with the USGS, the California Division of Mines and Geology installed a system for monitoring seismicity within the caldera, with particular emphasis on low-frequency events. Stephen McNutt reported that no low-frequency events were recorded in the caldera between July 1984, when the system became operational, and Oct 1985.

Information Contacts: D. Hill, J. Savage, USGS, Menlo Park, CA; S. McNutt, California Div. of Mines & Geology.

Further References: Hill, D.P., Bailey, R.A., and Ryall, A.S. (eds.) 1984, Proceedings of Workshop XIX: Active Tectonic and Magmatic Processes Beneath Long Valley Caldera, Eastern California; *USGS Open File Report* 84-939, 942 pp. (32 papers).

Savage, J.C., Cockerham, R.S., and Estrem, J.E., 1987, Deformation Near the Long Valley Caldera, Eastern California, 1982-1986; *JGR*, v. 92, no. B3, p. 2721-2746.

Special Section — Long Valley Caldera, California; *JGR*, 1985, v. 90, no. B13, p. 11111-11252 (14 papers).

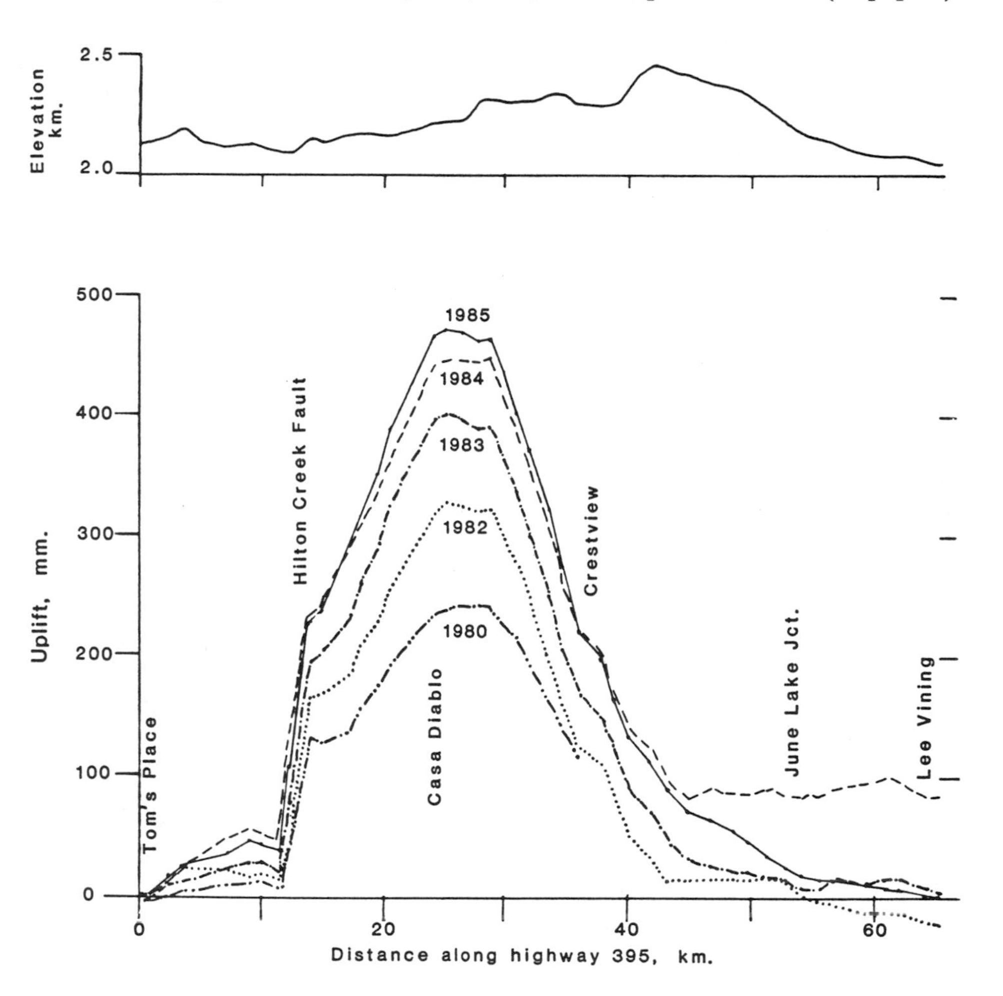

Fig. 12-33. Elevation differences in Long Valley Caldera for 1980-85, relative to the 1975 survey. Traverse is along Highway 395 between Tom's Place and Lee Vining. Courtesy of David Hill and James Savage.

CENTRAL PACIFIC OCEAN

LOIHI SEAMOUNT

Hawaii, USA
18.92°N, 155.25°W, summit depth 969 m

A future island is growing 28 km off the SE coast of Hawaii, the largest and most volcanically active island in the Hawaiian chain. Earthquake swarms recognized in 1971 first suggested contemporary volcanism that has more recently been confirmed by a variety of geophysical observations including exploration by manned submersible.

General References: Klein, F.W., 1982, Earthquakes at Loihi Submarine Volcano and the Hawaiian Hot Spot; *JGR*, v. 87, p. 7719-7726.

Loihi Seamount: Recent Results, 1987 (10 abstracts), *EOS*, v. 68, no. 44, p. 1553-1554.

See also 'General References' under Kilauea.

1/85 (10:1) The following is from the Hawaiian Volcano Observatory (HVO).

"Since Nov 1984, a renewed increase in earthquakes in the Loihi region has been recorded on the HVO seismic network. The recent activity occurred episodically in 3 minor bursts on 11 Nov and 10 and 21 Jan. From Nov 1984 to Jan 1985, 19 earthquakes of M ≥ 2.0 occurred in the Loihi region (fig. 13-1). Nine of the larger events ranged from M 3.0 to M 4.2.

"Although the current activity is relatively minor, swarm earthquakes characteristically associated with magmatic movement and volcanism in Hawaii have been known to gradually develop from such episodic increases. Significant earthquake swarms inferring magmatic activity at Loihi were detected in 1971-72 and 1975. The recurring episodes of earthquake swarms suggesting development of volcanism S of Hawaii are consistent with the pattern of progressive southeastward growth of the Hawaiian Islands in the context of plate tectonics and volcanism in the mid-Pacific region."

The nearest seismic station is about 30 km from Loihi on the S coast of the island of Hawaii. Seismic events at Loihi of M < 2 may be recorded by stations on Hawaii, but their locations cannot be determined as precisely as for larger events.

Information Contacts: R. Koyanagi, W. Tanigawa, USGS HVO.

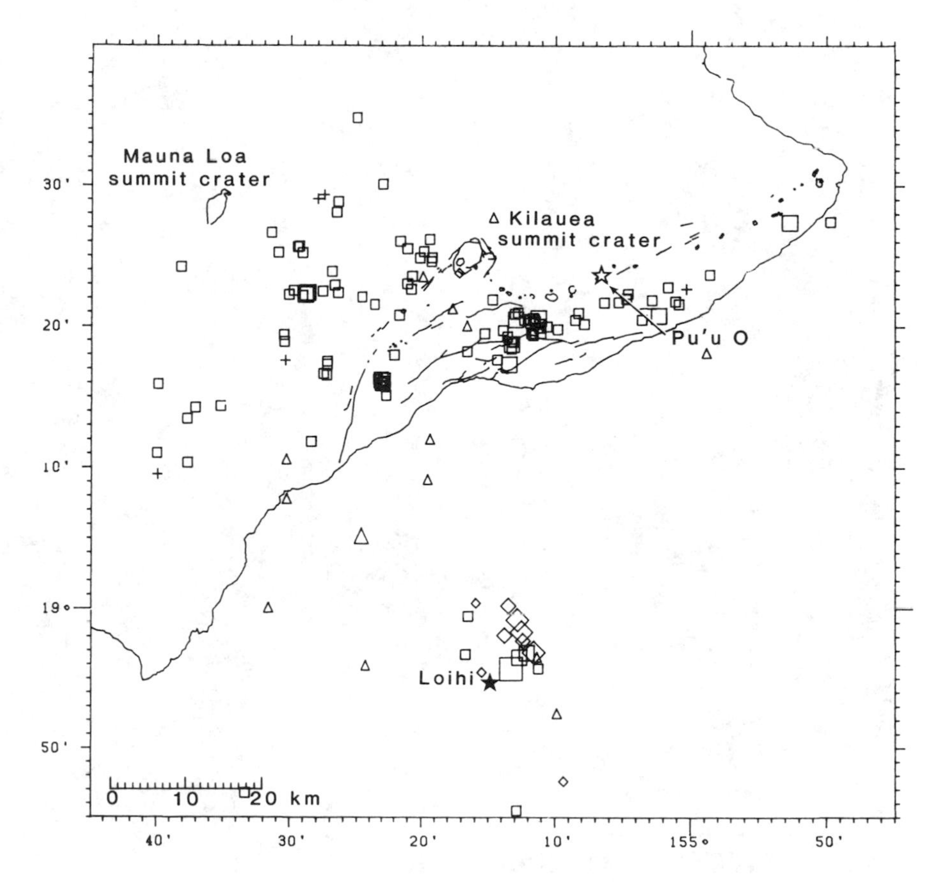

Fig. 13-1. Plot of earthquakes of M ≥ 2 in the Hawaii region, Nov 1984-Jan 1985. Squares indicate events with depths of 5-13 km, diamonds 13-20 km, and triangles > 20 km. Symbols of different sizes show the relative magnitudes of events; the largest event near Loihi was M 4.2.

KILAUEA

Hawaii, USA
19.425°N, 155.292°W, summit elev. 1222 m
Local time = GMT - 10 hrs

This basaltic shield, on the E flank of giant Mauna Loa (fig. 13-2), is perhaps the most extensively studied volcano in the world. Its lava lake eruptions of the 19th and early 20th centuries attracted widespread attention. Following a violent explosive eruption in 1790, 74 eruptions (including nearly half of the world's known lava lake eruptions) have been reported, some lasting many years.

The USGS's Hawaiian Volcano Observatory, located on the caldera rim since 1912, has pioneered many monitoring techniques, the results of which—over a particularly active decade—make up the second-largest (after Mt. St. Helens) set of reports in this book.

General References: Decker, R.W., Wright, T.L., and Stauffer, P.H. (eds.), 1987, Volcanism in Hawaii; *USGS Professional Paper* 1350, 1667 pp. (64 papers).

Dzurisin, D., Koyanagi, R.Y., and English, T.T.,

	Date	Time	Mag.	Lat.	Long.	Depth
a	29 Nov	0448	7.2 Ms	19.335°N	155.024°W	5 km
b	29 Nov	0448	7.2 Ms	19.35°N	155.02°W	8 km

Table 13-1. Seismic parameters of the 29 Nov 1975 Kalapana earthquake. *a,* reported in *SEAN Bulletin* 1:2, *b,* from *HVO Annual Summary* for 1975. The epicenter was 30 km ESE of Kilauea Caldera. Additional data on the earthquake may be found in Crosson and Endo (1981).

1984, Magma Supply and Storage at Kilauea Volcano, Hawaii, 1956-1983; *JVGR*, v. 21, no. 3/4, p. 177-207.

Heliker, C., Griggs, J.D., Takahashi, T.J., and Wright, T.L., 1986/7, Volcano Monitoring at the U.S. Geological Survey's Hawaiian Volcano Observatory; *Earthquakes and Volcanoes*, v. 18, no. 1, p. 3-71.

11/75 (1:2) At 0532 on 29 Nov, 44 minutes after a severe earthquake beneath Hawaii's SE coast (table 13-1), lava erupted from a N85E-trending fissure on the floor of Kilauea caldera. Lava fountains were 50 m high for the first 15 minutes but decreased to 5-10 m heights for the next 75 minutes before stopping. Eruptive activity resumed at about 0830 in Halemaumau pit crater and continued intermittently before ceasing around 2200, after 16 hrs of activity. Lava drained into the SW rift zone for days after the end of visible activity. Maximum horizontal displacement on the order of 3.5 m was involved in the seaward movement of Kilauea's S flank.

The earthquake was the strongest in Hawaii since at least 1868, and extensive damage was reported. On the SE shore, at Punaluu, a 6-m tsunami was observed. At Halape, 2 people died when a wave in excess of 7 m high [swept inland] and reached 16 m up a fault scarp perpendicular to the coast. The coastal area at Halape, the area of maximum subsidence, was permanently lowered 3 m. The tide gauge at Hilo recorded a 2.5 m wave.

Information Contact: R. Tilling, USGS HVO.

Further References: Crosson, R.S., Endo, E.T., 1981, Focal Mechanisms of Earthquakes Related to the 29 November 1975 Kalapana Hawaii Earthquake: The

Fig. 13-2. Space Shuttle photograph (no. 61C-41-050), taken 15 Jan 1986, showing most of the island of Hawaii. Mauna Loa is at center. On its SE flank is the much smaller Kilauea, partially obscured by clouds. A large plume extends from the active east rift crater, Pu'u O'o, following episode 40 of the eruption that began in Jan 1983. Courtesy of C. Wood.

Effect of Structural Models, *Bull. Seismol. Soc. Am.*, v. 71, p. 713-729.

Lipman, P.W., Lockwood, J.P., Okamura, R.T., et al., 1985, Ground Deformation Associated with the 1975 Magnitude 7.2 Earthquake and Resulting Changes in Activity of Kilauea Volcano, Hawaii; *USGS Professional Paper* 1276, 45 pp.

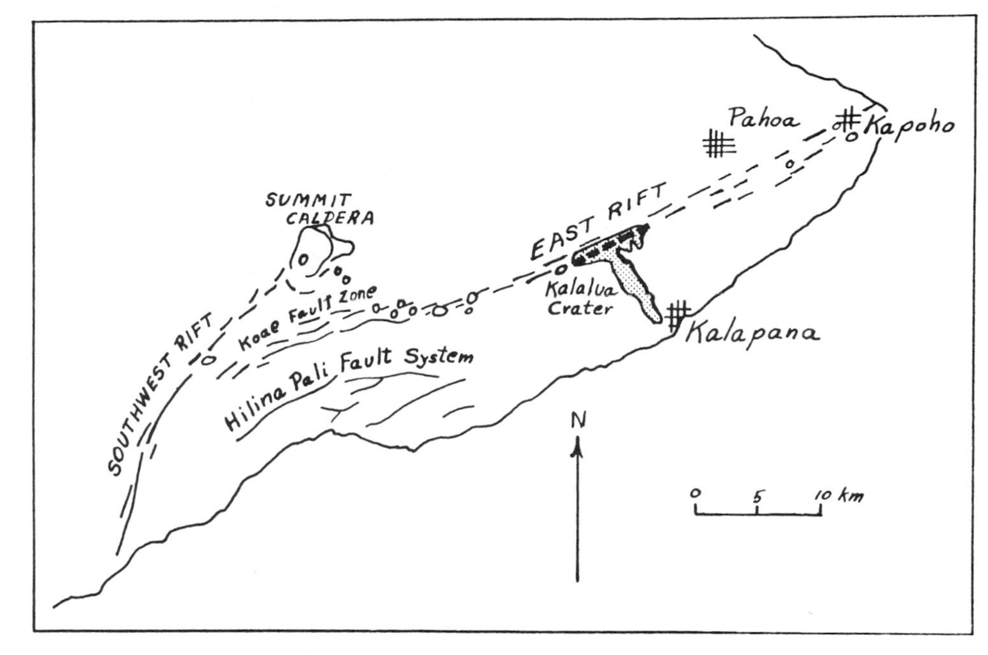

Fig. 13-3. Sketch map outlining the Sept 1977 lava flows (stipple pattern) and various features of Kilauea. Open circles are craters. Courtesy of Robert Tilling.

1/76 (1:4) The volcano has been inflating slowly since the 29 Nov eruption. Two earthquakes, one of M 4.0, shook Kilauea on 11 Jan, the largest events associated with the volcano in several months.

Information Contact: Same as above.

[Intrusive episodes on 21 June and 14 July 1976, not reported in the original *Bulletin,* are described in the comprehensive summary of Dzurisin et al., 1984.]

11/76 (1:14) A [gradual] 40 μrad deflation of the summit occurred [aseismically] during late Aug and Sept. In late Oct, a 15-cm dilation and a several-acre area of steam-killed trees were noted N of Kalapana on the E rift zone, ~ 25 km from the summit of Kilauea. It is assumed that the magma leaving the summit area migrated into the E rift zone, causing the effects observed near Kalapana.

Information Contacts: G. Eaton, USGS HVO; D. Shackelford, CA.

4/77 (2:4) A fascinating magma intrusion event at Kilauea on 8-9 Feb was unusually well documented. Continuously recording tiltmeters monitored a sharp summit deflation beginning 8 Feb at 1902, 5 hrs after the start of an earthquake swarm (M 3-4) on the upper E rift zone. A local magnetic anomaly (~10 gamma) also occurred in the upper E rift zone, and seismicity reached 200 events/hr with 3-7 km focal depths, but no eruption took place. Geodimeter surveys 1 day after the event showed extensions of up to 0.25 m across the upper E rift and electrical self-potential traverses add more documentation of magma migration. Similar events took place in June and July 1976, and HVO scientists suggest that magma is draining from beneath the summit area along subsurface paths created by the major earthquake of 29 Nov 1975. These drainage paths readily allow periodic intrusion into the E rift and are perhaps preventing major inflation of the summit reservoir.

Information Contact: USGS HVO.

9/77 (2:9) A substantial eruption on the E rift began from fissures near Kalalua Crater (fig. 13-3) on 13 Sept, after about 24 hrs of premonitory activity.

Field observations suggested interesting changes in lava composition during the course of the eruption. Lava erupted during the first 3 phases and about the first 1 1/2 days of the fourth phase apparently consisted of differentiated material containing plagioclase and pyroxene phenocrysts. This was succeeded on the 27th by more typical Hawaiian tholeiite basalt containing olivine phenocrysts.

Information Contacts: R. Tilling, USGS, Reston, VA; G. Eaton, USGS HVO; R. Fiske, SI.

10/77 (2:10) The following is from Gordon Eaton.

"At approximately 1930 on 13 Sept an eruption broke out on the central E rift, near Kalalua, a prehistoric cinder cone (see table 13-2 for detailed chronology). It followed a swarm of earthquakes that began on the previous day at 2130 on the upper E rift, near the young satellite shield, Mauna Ulu, and Makaopuhi crater. These earthquakes were accompanied by harmonic tremor and rapid summit deflation, indicating that magma was moving into the E rift in the subsurface. The deflation continued strongly for about a week and then tapered off gradually. A total of 90 μrad of tilt change was measured on the tiltmeters at Uwekahuna vault. Levelling later showed maximum summit subsidence of 44 cm.

"The initial active section of rift was approximately 5.5 km long, but fountaining at all times and locations was restricted to a few hundred meters of this length. The remainder of the rifted zone opened as a series of en echelon fractures and were sites of profuse steaming. Maximum fountain heights reached during the early phases of the eruption did not exceed 70-80 m. Flows at that time consisted chiefly of aa, with a maximum rate of advance of about 170 m/hr. By dawn on 15 Sept these

1977	Time	Activity
Sept 12	2130	Earthquake swarm began in the upper E rift
	2200	Summit deflation began
13	mid-morning	Earthquake hypocenters began to migrate E along the E rift
	1930	Fountaining began at newly opened fissures extending 3 km E from Kalalua Crater, accompanied by heavy harmonic tremor. Total summit deflation (as measured at Uwekahuna) had reached about 42 μrad.
14	0800	Fountaining was confined to the E $^1/_3$ of the new fissures, feeding a lava flow moving S. Summit deflation, 3.5 μrad/hr.
15	0200	Two areas of fountains, about 60 m high. Activity along remainder of the fissure was confined to low spattering. The deflation rate had declined to about 1 μrad/hr; total subsidence was about 75 μrad.
	late afternoon	First phase of the eruption ended, after the lava flow had advanced about 2.5 km. Earthquakes and harmonic tremor had declined. Total summit deflation was about 85 μrad.
	2400	Harmonic tremor ended.
16	0400	Renewed fountaining (phase 2), feeding a small flow [but see 2:10] parallel to the first flow. Fountains were discontinuous, rising to about 50 m from a vent area about 200 m long, slightly W of the earlier vents.
18	1530	The eruption had declined to weak, intermittent spattering, and the new flow had stopped less than 0.5 km from the vent. Harmonic tremor was still being recorded from the vent area, but not from the summit, where deflation had ended. Earthquakes had declined.
20	evening	Phase 2 activity ended [but see 2:10].
23	early afternoon	Minor fountaining (to 15 m) fed small flows, and ended by nightfall (Phase 3).
25	2350	Phase 4 began from a vent W of the earlier ones. During the next 24 hours, fountains rose 100 m, and discharge rates briefly reached an estimated 5-7 x 10^5 m^3/hr. Lava advanced SE at up to 300 m/hr.
29		Kalapana, a coastal village with population about 250, [but see 2:10] was evacuated.The flow front, several thousand meters from the village, was advancing toward it at about 150 m/hr down a steep slope. A transition from pahoehoe to aa flow types occurred at the edge of the steep slope.
30	0300	The fountains feeding the flow declined to 20-30 m and the flow had slowed to 60-90 m/hr after reaching a gentler slope. Summit tilt remained irregular, varying 2 μrad throughout phase 4.
Oct 1	1000	Flow advance had stopped 400 m from the nearest house in Kalapana. The pahoehoe to aa transition had retreated to the vent area. The flow front had thickened from 4.5 m to 12 m and had widened from 300 m to 900 m.
	1530	Harmonic tremor near the vent declined markedly.
	1625	Fountaining stopped, after building an irregular 100 m [but see 2:10] spatter cone.

Table 13-2. Chronology of the Sept-Oct 1977 eruption.

flows had slowed to 65 m/hr. They came to rest about 2.5 km from their source fountains, close to a papaya field and ranch.

"On 18 Sept new fountaining began uprift, immediately NW of Kalalua cone, several kilometers from the initial fountains. By late afternoon on 19 Sept this activity had decayed and flow movement was scarcely perceptible, but by midnight fountaining resumed. By 0900 on 20 Sept this phase of the eruption had ended.

"The next phase consisted of Strombolian activity at a small, new cone downrift in the afternoon of 23 Sept. The lava was highly viscous and was ejected sporadically in a series of taffy-like, irregular sheets and long clots. All lava to this point in the eruption was tholeiite rich in plagioclase microphenocrysts, presumably old and highly differentiated lava.

"The period 24-25 Sept was free of activity at the rift. Harmonic tremor decayed to very low levels. Just before midnight on 25 Sept, however, tremor resumed and strong glow was visible over the rift. Heat from the eruption domed a blanket of stratus clouds over the volcano into a huge cumulus cloud. Harmonic tremor amplitude rose at the 2 seismometers closest to the fountaining. Except for a 2-hour lull in the early afternoon of 26 Sept this fountaining continued until mid-afternoon on 1 Oct. Fountains played from heights of 20 m to as much as 300 m, lava production was copious at all times, and the new flows ran NE, ENE, and SE, but only the ENE flow eventually threatened populated areas. In the early hours of 28 Sept it turned away from the rift down which it had flowed for 1.5 km and started toward the village of Kalapana. Evacuation of Kalapana began at dawn 30 Sept and was completed by evening.

"On 1 Oct at 1530 tremor levels along the central E rift dropped dramatically. Fountaining had ceased by 1615. It did not resume again, although measurable tremor continued through 12 Oct.

"Evacuees returned 3 Oct, 38 hrs after cessation of fountaining. The source cone, named Pu'u Kia'i (Hill of the Guardian) is 250 m long, 140 m wide, and 35 m high."

Further Reference: Moore, R.B., Helz, R.T., et al., 1980, The 1977 Eruption of Kilauea Volcano, Hawaii; *JVGR*, v. 7, p. 189-210.

Information Contact: G. Eaton, USGS HVO.

11/79 (4:11) The following is from HVO.

"Kilauea erupted on its upper E rift zone for 22 hours on 16-17 Nov. Seismicity since the 1977 eruption was sustained at moderate to high levels at the summit and the E rift zone until the onset of the swarm of shallow earthquakes that preceded the eruption (fig. 13-4).

"The swarm began at 2100 on 15 Nov near Pauahi Crater, 7-8 km SE of the central caldera region, and within $^1/_2$ hour the summit tiltmeter indicated the onset of deflation (fig. 13-5). Simultaneously, 2 borehole tilt-

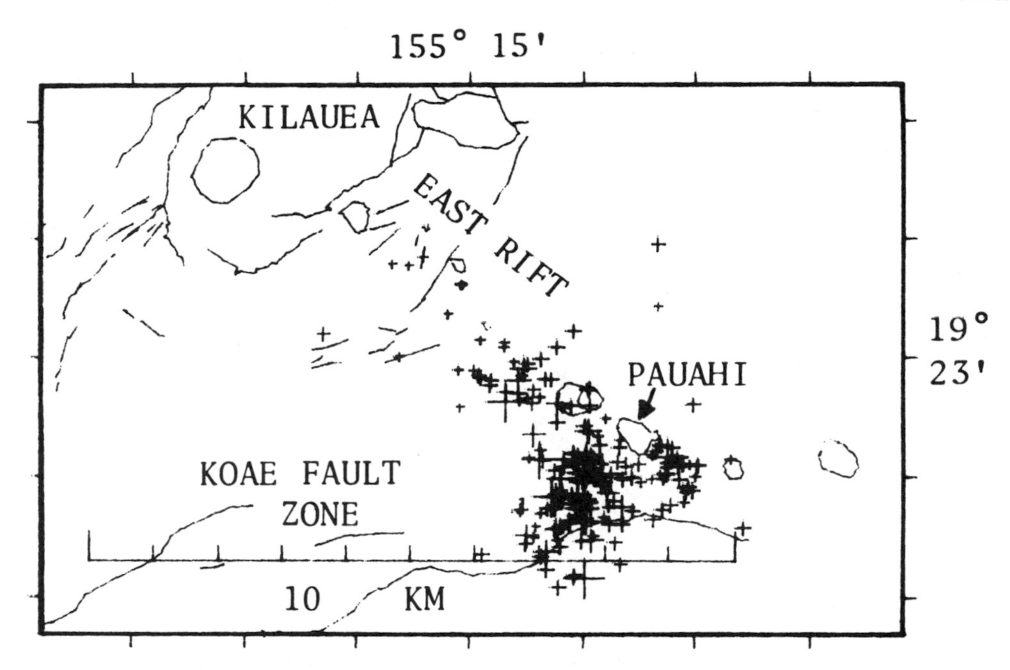

Fig. 13-4. Plot of shallow earthquake epicenters (1-5 km depth) associated with the 16-17 Nov 1979 eruption.

meters detected inflation in the upper part of the E rift zone (see table 13-3 for detailed chronology). Shallow volcanic tremor very local to Pauahi Crater (fig. 13-6) became strong at about 0700 the next morning as the number of earthquakes gradually decreased. At 0821, low fountaining (less than 10 m high) started in Pauahi Crater. Fifteen minutes later, observers arrived on the E side of the crater and found a curtain of fire 5-10 m high and 100 m long. These E vents ceased eruption less than 1 hr later. At 1130, 2 more vents opened in Pauahi Crater, W of the first Pauahi vent. Shortly thereafter, brief fountaining was observed N of the earlier E vents, followed by cessation of activity of the initial (eastern) Pauahi Crater vent. Over the next $1^1/_2$ hours, 6 more vents opened progressively to the W, one more in the crater and five W of the crater. Slightly before 1600, activity at the W vents began to wane and over the next hour fountaining ceased progressively eastward at the five W vents. Lava production in the 3 remaining vents in Pauahi Crater stayed relatively constant until 0100 on 17 Nov, gradually waned, then ceased at 0630, 22 hours after the eruption began.

"During the eruption, tremor amplitude fluctuated with the extrusion rate, and earthquakes continued to decline to a frequency only slightly higher than a normal background rate. Earthquakes immediately preceding and accompanying the eruption occurred within a roughly triangular zone bounded by the E rift, Koae fault zone, and a N-S line 1 km W of Pauahi (fig. 13-4).

"Unlike previous events, which presumably defined a downrift propagation of magma from the summit reservoir, epicenters during this eruption showed no downrift migration and tremor did not occur near stations uprift of the eruption. Furthermore, earthquakes migrated upward from 3 to 1 km in depth during the first 2 hrs of the swarm. It therefore seems probable that the eruption was fed by magma already stored beneath the rift zone and that magma drained from the summit reservoir to replace the magma mobilized in the E rift zone near the eruption site.

"Fig. 13-5, summarizing the pattern of summit infla-

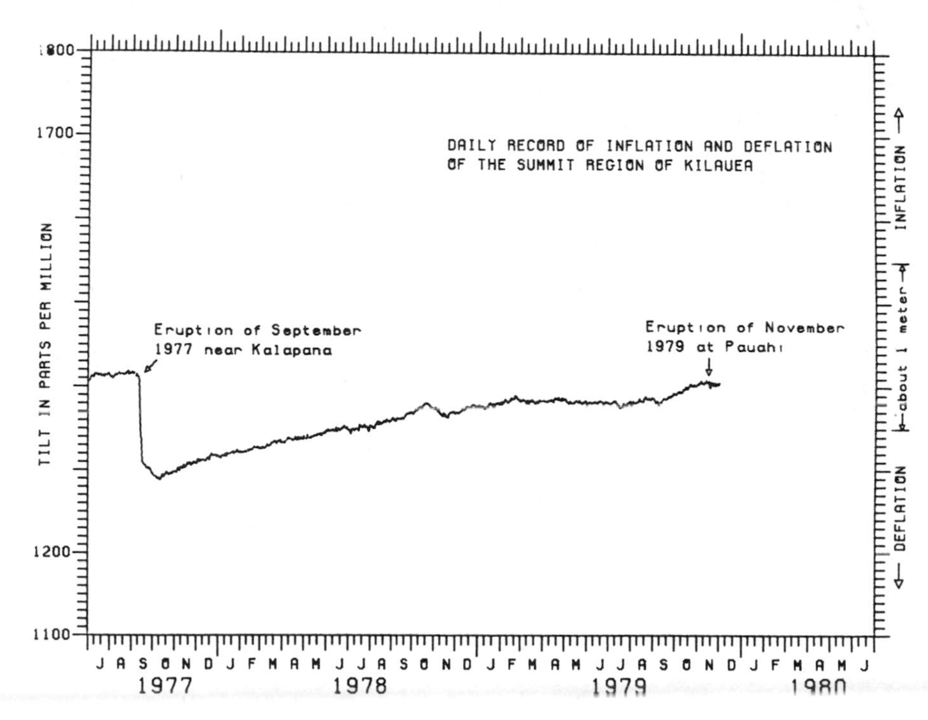

Fig. 13-5. Daily record of summit inflation of Kilauea Volcano 1977-79.

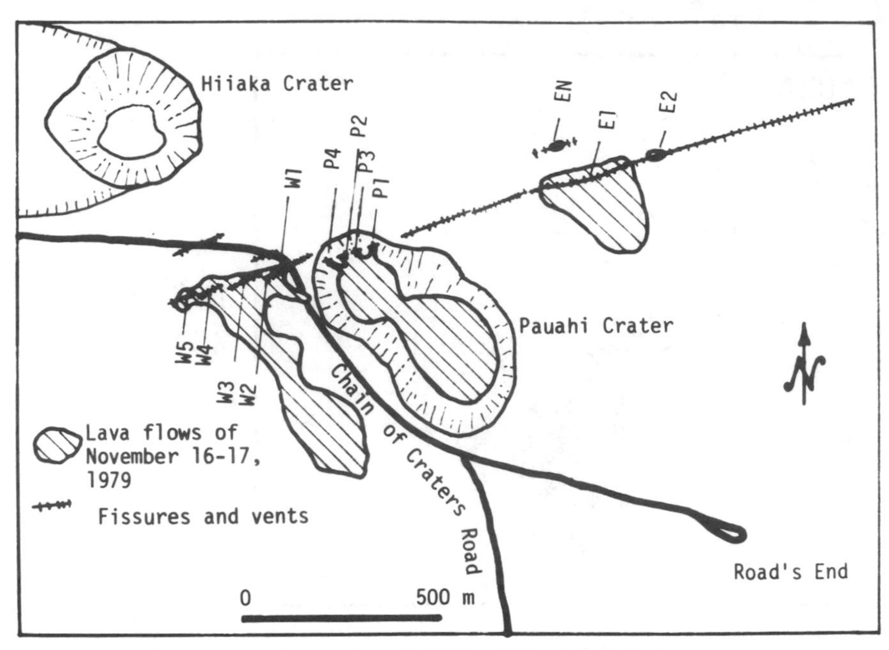

Fig. 13-6. Preliminary sketch map of the 16-17 Nov 1979 eruption site.

tion between the 1977 and 1979 eruptions, shows that throughout the month of Nov 1978, there was a slow deflation of the summit. Concurrently, an area approximately centered on the 1977 eruption site started to inflate. Except for the minor deflation centers along the upper E rift zone, the upper and lower parts of the E rift zone remained relatively undeformed during the period between eruptions.

"Summit SO_2 emission averaged 100-200 t/d over the 13 months preceding the eruption and peaked occasionally at 350 t/d. Ten days prior to the eruption a spike of 500 t/d was recorded, followed by a return to approximately normal daily emission throughout and following the eruption. Anomalous abundances of S were observed in condensates at 2 sites and of CO_2 at one site. The data base is too short to determine whether the observed variations of S and CO_2 are indicators of magmatic activity preceding the Nov eruption, or were coincident fluctuations unrelated to magma movement.

"The eruption was characterized by < 700,000 m^3 erupted volume, low fountains, generally low amounts of fume, viscous lava and spatter, and low temperatures (generally 1040°-1080°C with an infrared pyrometer). The overall pattern of the fissure system suggests that a left lateral shear couple was present during the eruption. The only obvious phenocryst observed by preliminary megascopic observation is olivine, $^1/_2$ - 2 mm in diameter. The earliest samples, especially those from the easternmost vents, appear to be poor in olivine (1-3%), whereas those from the later stages contain 3-8% olivine phenocrysts. Interpretations of this variation include: 1) eruption of a single fractionated magma body; 2) eruption of several discrete, compositionally variable, possibly fractionated magma bodies, and 3) an influx, during the later stages of the eruption, of more primitive (olivine-rich) magma that drained from the summit reservoirs.

"The seismology, petrology, and surface deformation suggest that the eruption was likely the result of a disturbance of a shallow local storage chamber where the magma had resided for several months or years. The immediate area had been the site of the Aug 1968, May 1973, and Nov 1973 eruptions and lies along the geologically and seismically defined shallow conduit system leading from the summit magma chamber.

"Summit deflation suggests that the volume of magma withdrawn from the summit reservoirs was 4-6 times that erupted. Presumably this excess magma is stored in conduits and reservoirs in the upper E rift zone. Fig. 13-5 indicates that the Nov eruption involved a trivial amount of the magma that has been supplied to the summit since the Sept 1977 eruption, and at present, the level of summit inflation is approximately that prior to the eruption."

Information Contacts: N. Banks, F. Klein, USGS HVO.

3/80 (5:3) Magma was intruded into the upper E rift zone on 2 occasions in March. On 2 March, a microearthquake swarm began near Pauahi Crater, 6 km downrift from the summit and site of a brief eruption in Nov 1979. Summit deflation, indicating draining of the magma chamber below, started within 7 minutes and a tiltmeter began to record deformation as magma was intruded downrift. About 2000 earthquakes, including several felt events, were recorded before the swarm ended 8 hrs later. Summit deflation, totaling only 8 mm, continued for another hour. On 10 March, a second intrusion took place a few kilometers farther down the E rift beneath Mauna Ulu, active 1969-74. Earthquakes began at 2157, summit deflation at 2206, and volcanic tremor at 2310. Eight cm of deflation, representing the draining of 8 × 10^6 m^3 of magma from the summit chamber, occurred before the end of summit deformation on 12 March at 1630. Volcanic tremor around Mauna Ulu waned slowly and had almost ceased by 13 March.

Inflation has resumed since the second intrusion. As of 2 April, the level of the summit was slightly higher than immediately prior to the Nov eruption.

Information Contact: USGS HVO.

8/80 (5:8) Magma was intruded into the upper E rift zone on 27 Aug, the first intrusive activity there since March. As in the two March intrusions, no eruption took place. An earthquake swarm began near Puhimau Crater (about 1.5 km SE of the caldera rim) at 1425 on 27 Aug. Within 6 minutes, the number of microearthquakes had increased to several per minute. Summit deflation started 30 minutes after the swarm, at 1455. The earthquakes migrated generally downrift at about 1 km/hr. Several were felt nearby, with the 3 highest magnitudes ranging from 3.6 to 4.0. Hundreds of M 1-3 events occurred at depths of 1-4 km. The seismographs closest to the swarm apparently registered some shallow volcanic tremor, but microearthquakes occurred so rapidly that tremor was obscured on the records. The number of microearthquakes per minute started to decrease at about 1830 and the swarm ended early the next morning. Summit deflation had stopped at about 1930 on 27 Aug, after 7.5 μrad of tilt had been recorded.

The USGS interpreted the activity as the formation of a dike estimated to be about 3 km long, 1-2 km high, and 1 m wide, located 1-3 km beneath the surface. About 4×10^6 m^3 of magma were calculated to have been injected into the dike between 1500 and 1930 on 27 Aug.

SO_2 emission was detected in the Puhimau thermal area (where the earthquake swarm began) on 28 Aug. The CO_2/SO_2 ratios of gases emitted by the summit fumaroles before and after the intrusion remained about the same as the previous week.

Information Contact: Same as 5:3.

10/80 (5:10) A minor intrusion of magma into the upper E rift zone was recorded on 22 Oct. At 1840, an earthquake swarm began near Pauahi Crater (6 km downrift from the summit), and summit deflation started 8 minutes later. Seismic activity declined about 2000, and deflation leveled off at about 2200, after 1.9 μrad of movement had been recorded. In the 24 hrs before the intrusion, 7 earthquakes with magnitudes of 3.1-4.2 had occurred at less than 5 km depth along the middle and lower E rift.

At press time, a second intrusion was recorded. An earthquake swarm began on 2 Nov at about 1415 and

1979	Time	Activity
Nov 15	2100	Seismic swarm began local to Pauahi station.
	2130	Deflation at the summit and inflation at the eruption site.
16	0005	Peak earthquake rate.
	0700	Strong tremor began local to Pauahi station.
	0805	Copious steam and fume emission began E of Pauahi Crater. The fissures occupied an old spatter rampart and never emitted lava, but the emitted gases were hot enough to ignite adjacent vegetation.
	0821	A sharp cracking sound accompanied the opening of a vent (P1, fig 13-6) on the NE wall of the NW lobe of Pauahi Crater; initial fountain heights were < 1 m.
	0836	Observers arriving at Road's End parking lot found fissures already erupting a low (5-10 m) curtain of fire (E1, fig 13-6) about 100 m long, 230 m E of the copiously fuming area noted at 0805. Fissuring migrated E, eventually producing a separate lava pad (E2). New fissures east of the E2 vent occupied the center of an old spatter rampart.
	0925	Eruptive activity on E1 and E2 fissures ceased.
	1130	A fissure opened about 70 m W of the still active P1 vent in Pauahi Crater, and almost immediately began to fountain to heights of 2-10 m on its W end. A few seconds later, 3 smaller vents began activity between the new fountain (P2) and P1. These vents collectively are labeled P3.
	1135	(time uncertain) – Brief eruption from a fissure (EN) N of the main E vents.
	1140	An eruptive fissure (P4) opened 20 m W of vent P2.
	1149	Activity at P1 vent abruptly decreased with concurrent increase of activity at P2, P3, P4.
	1150	P1 vent ceased activity.
	1155-1200	Flows produced by the now inactive P1 began to cascade down the mezzanine into the SE crater of Pauahi, followed by flows from P2-4.
	1203	Fissures migrating W of Pauahi crater cut the overlook parking lot and the Chain of Craters road.
	1214	Eruption began from E to W on 3 fissures (W1, W2, and W3) beginning just W of the Chain of Craters road; concurrently there was a temporary decrease in activity of vents P2-4 in Pauahi. Curtains of fire to 10 m high with spatter ejected to 30-40 m were soon established on 2 of the new vents (W2 and W3); weak spattering occurred at W1 vent.
	1239	Eruption began at W4 vent. Activity at W1, W2, and W3 vents decreased abruptly at the onset of this activity.
	1255	Eruption began from W5 vent, followed by roughly constant rates of effusion (about 50,000 m^3/hr) from all the W vents over the next $3\,^1/_2$ hours.
	1445	Tremor amplitude at seismic stations 6 km from the eruption site reached a peak. The earthquake rate dropped by an order of magnitude from its peak values at 0005.
	1542	Abrupt brief decrease in fountain height of all W vents followed by cessation of activity at W5 vent and slight decrease in activity of vents P2-4.
	1543-1631	Decrease and cessation of activity of W4 vent. Decrease in activity of W2 and W3 vents; nature of activity at W1 vent unknown but total emission from that vent was less than 15 m^3.
	1547-1555	Chain of Craters road cut by small flow lobes.
	1631-1648	Activity of W2 and W3 vents declined to sporadic spatter emission.
	1651	Cessation of activity at all W vents.
	1716	Increase in activity of P2, P3, and P4 vents; approximately constant combined effusion rate of P2-4 of 15,000-20,000 m^3/hr for the next 6 hours.
	1845	Activity of P4 vent decreased.
	2030	Activity of P4 vent essentially ceased.
17	0100-0409	Activity at P2 and P3 continually waned.
	0413	Continued decrease in activity of P2 and P3 vents; tilt at summit reversed; seismic tremor subsided.
	063	All but gas activity ceased in Pauahi crater vents.
	0809	Surface movement of red lava in channels in Pauahi ceased, followed over the next 2 hours by collapse of crust and levees and formation of slab pahoehoe.

Table 13-3. Chronology of the Nov 1979 eruption.

summit deflation started less than 30 minutes later. Between 1000 and 1500 earthquakes occurred at depths from 4 km to less than 1 km, migrating from just above Kokoolau Crater (about halfway between the caldera rim and Pauahi Crater) 2-3 km downrift to the Heake Crater area. About 4.5 μrad of deflation took place in the summit area before seismicity ended around 1700, and another 0.5 μrad of deflation were recorded in the succeeding 24 hrs. In contrast to the aftermath of the smaller 22 Oct intrusion, inflation did not resume immediately after deflation ended, tilt remaining essentially stable. During and just after the intrusion, no changes in amount or composition of gas emission were observed at various standard sites.

Information Contact: Same as 5:3.

5/81 (6:5) HVO has documented an intrusion into the SW rift in early 1981. This rift intrusion is the 13th since the major earthquake of Nov 1975, but the first since that date in the SW rift. On 20 Jan at ~0300 an intense earthquake swarm began just S of the caldera (fig. 13-7). A few earthquakes migrated northward into the caldera during the next 60 hrs, but most of the activity remained in the area where the swarm began, at depths of 2.5-3.5 km. Reoccupation of dry tilt stations 21 Jan revealed that significant inflation had taken place in the swarm area since the last tilt measurements 13 days earlier.

During the next few days, some epicenters advanced about 3 km down the SE side of the upper SW rift, while earthquakes continued in the initial swarm area. Migration of the seismicity was initially measured in hundreds of meters/day, more than an order of magnitude slower than the 500-1500 m/hr during previous intrusions into the E rift zone. Additional dry tilt measurements 28 Jan confirmed that the center of inflation had also moved SW, but to a point distinctly NW of the seismically active zone. Until early Feb, earthquakes remained within 3 km of the initial swarm area. Seismicity had been concentrated in the same region during the last intrusion and eruption episode in the SW rift, in Dec 1974.

On 6 Feb, seismic activity suddenly shifted to a zone about 17 km downrift from the caldera center and the rate of deflation increased sharply in the N caldera area, which had shown only minor tilt changes earlier in the episode. Intense seismicity was confined to less than 1 km of the rift for the next 2 days. Earthquakes propagated downrift 8-17 Feb, forming a narrow, tubular zone that rose (at an angle of about 40°) from depths of 7 to 2 km over about 3 km horizontal distance. These earthquakes eventually reached depths of only 1.5 km. Two clusters of less intense activity occurred uprift between this zone and the area of late Jan-early Feb activity.

Seismicity subsided considerably 17 Feb as inflation

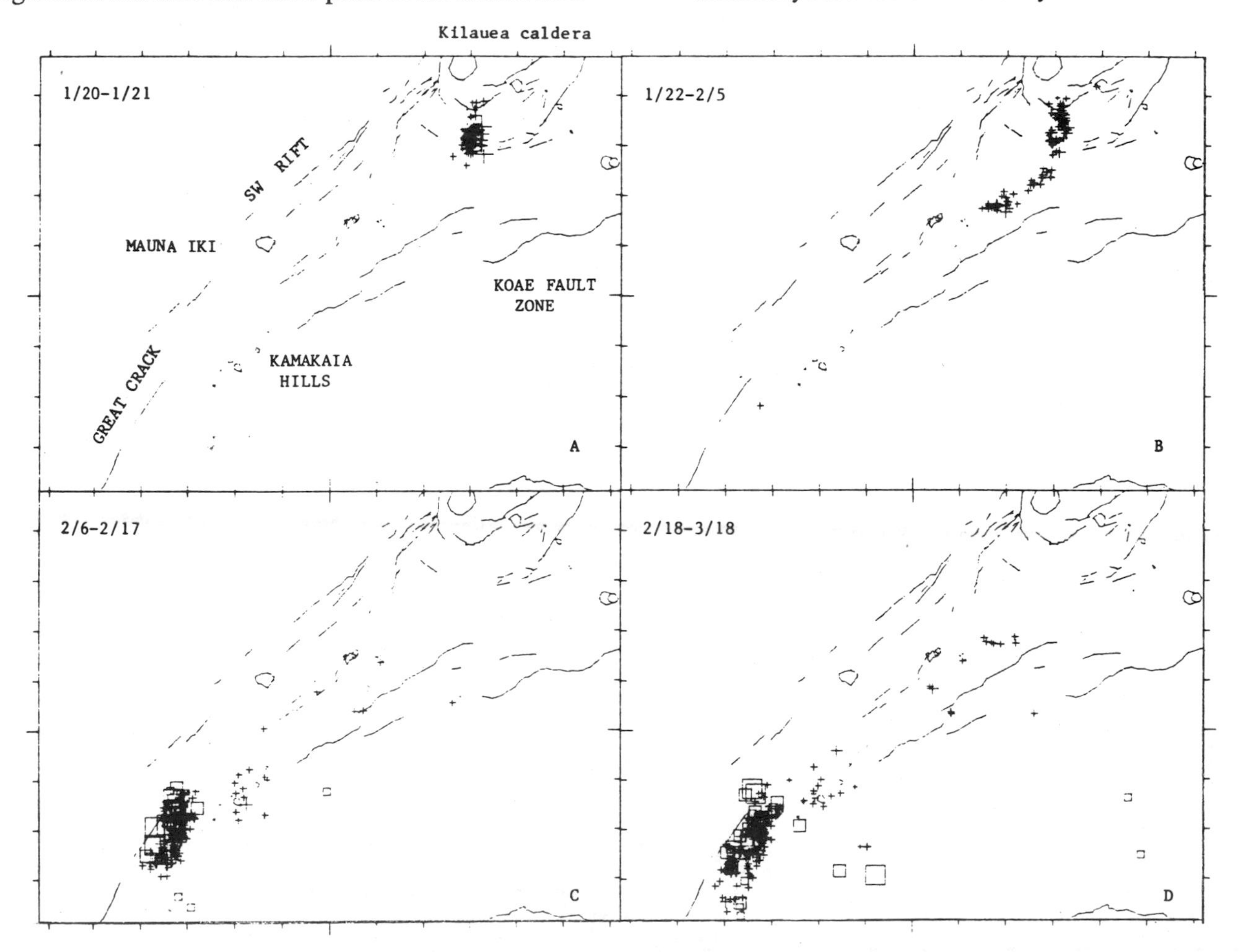

Fig. 13-7. Locations of shallow (< 7 km) SW rift earthquakes during 4 time periods, 20 Jan – 18 March 1981. Hypocenters < 5 km deep are shown by + signs; those > 5 km by □. Size of symbol is proportional to magnitude.

resumed in the N caldera area. By the end of the month much of the 15 μrad of deflation recorded on borehole tiltmeters since early Feb had been regained. The most active seismic zone continued to grow slowly (less than 100 m/day) downrift, reaching a length of 5 km by mid-March, but the number of events per day declined from a peak of 1800 in mid-Feb to 110-350 by late March. SW rift seismicity decreased gradually through April, reaching near-normal levels by the end of the month.

Information Contact: Same as 5:3.

6/81 (6:6) On 25 June, HVO recorded a small but rapid summit-area inflation episode accompanied by shallow microearthquakes and harmonic tremor. As rapid summit inflation started about 1500, a single seismic station in the SE part of the caldera began to record weak harmonic tremor. Six minutes later, tremor amplitude increased and scattered small individual earthquakes started to appear, still at only one seismic station. Tremor strengthened again and earthquakes became larger and more frequent at 1510, as instruments to a radius of 5 km began to register the individual events.

A magnitude 2.5 earthquake at 1517 was followed by relatively intense localized shallow activity, including tremor and more than 6 microearthquakes per minute between 1518 and 1530. Inflation ended about 1530, as harmonic tremor weakened and the number of microearthquakes started to decline. Between 1500 and 1530, 6 μrad of inflation had been registered by the E-W component of a continuously recording tiltmeter on the W rim of the caldera. No tremor was detected after 1600, but microearthquake activity remained high, gradually decreasing over the next 2 days. After inflation had ended, reoccupation of dry tilt stations near the caldera showed a maximum net tilt for the episode of 20 μrad (toward the NW) at a station near the SW rim.

The USGS's preliminary interpretation of this episode is that a small shallow intrusion [took place] beneath the S or SE part of the caldera. None of the hypocenters was shallower than 2 km, indicating that magma came no closer than that to the surface.

Information Contacts: R. Koyanagi, R. Decker, USGS HVO.

7/81 (6:7) The following is a report from HVO.

"On 10-11 Aug seismographs and tiltmeters at HVO recorded a moderate intrusion at Kilauea. The event was characterized by an earthquake swarm and harmonic tremor, accompanied by summit deflation and ground cracking. As of 0800 on 11 Aug, an estimated 30-50 × 10^6 m^3 of magma had intruded into the S summit and SW rift zones. The activity started with an increase of microearthquakes in the S summit area at 0330 on 10 Aug. Shortly before 0430 tiltmeters recorded the onset of the sharp summit deflation. By 0500 the seismic intensity increased and maintained a continuous state of activity. Microearthquakes and harmonic tremor less than 5 km in depth indicated that magma was migrating from the summit to the SW rift zone in the vicinity of the Kamakaia Hills nearly 20 km away. At mid-morning 11 Aug several thousand earthquakes of M_s ≤ 4.5 were detected, and monitoring instruments continued to record a diminishing pattern of seismicity and ground tilt."

Information Contact: R. Okamura, USGS HVO.

8/81 (6:8) An estimated 35 × 10^6 m^3 of magma were intruded into Kilauea's S summit region and SW rift 10-12 Aug, accompanied by seismicity, substantial summit deflation, and ground cracking in the SW summit area. Microearthquake activity in the S summit region began to increase about 0330 on 10 Aug, followed by the onset of rapid deflation about 1 hour later (fig. 13-8). Earthquakes began to migrate away from the summit into the SW rift (fig. 13-9) and by evening were concentrated about 17 km downrift, in approximately the same zone that had been seismically active during the later stages of the Jan-Feb intrusion (6:5). Seismic activity peaked during the evening of 10 Aug and gradually declined the next day. During the intrusion, seismic instruments detected thousands of small earthquakes, most shallower

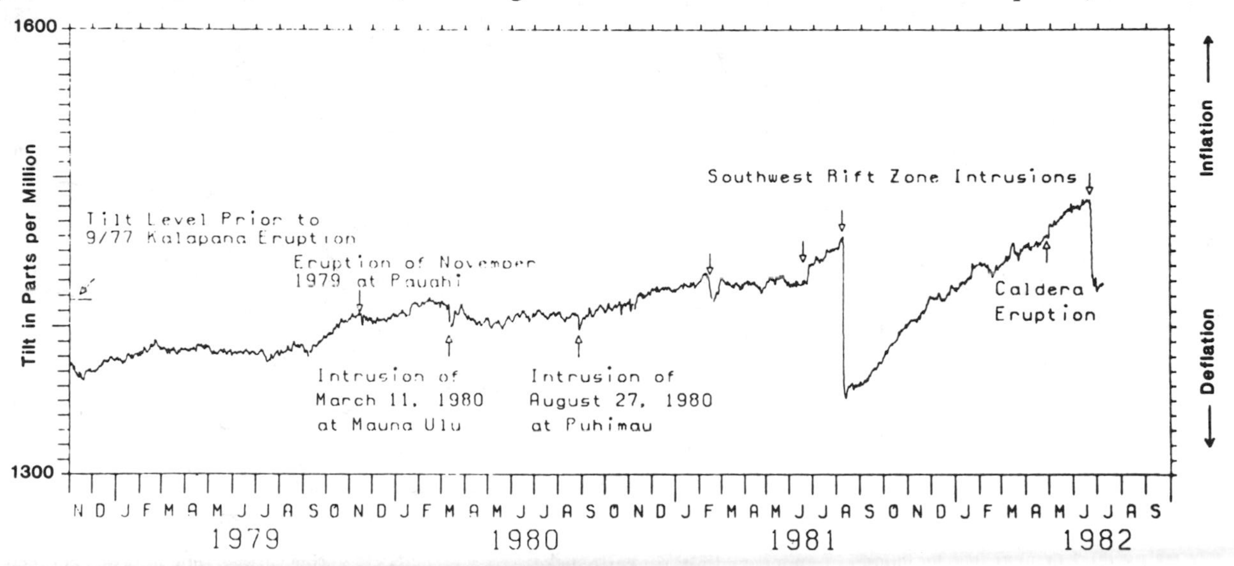

Fig. 13-8. Summit tilt measured by HVO Nov 1978 – June 1982. Tilt is recalculated to N 60°W from E-W and N-S components. Each unit mark on the x axis represents 10 μrad of tilt (up = inflation). [Originally in 7:6.]

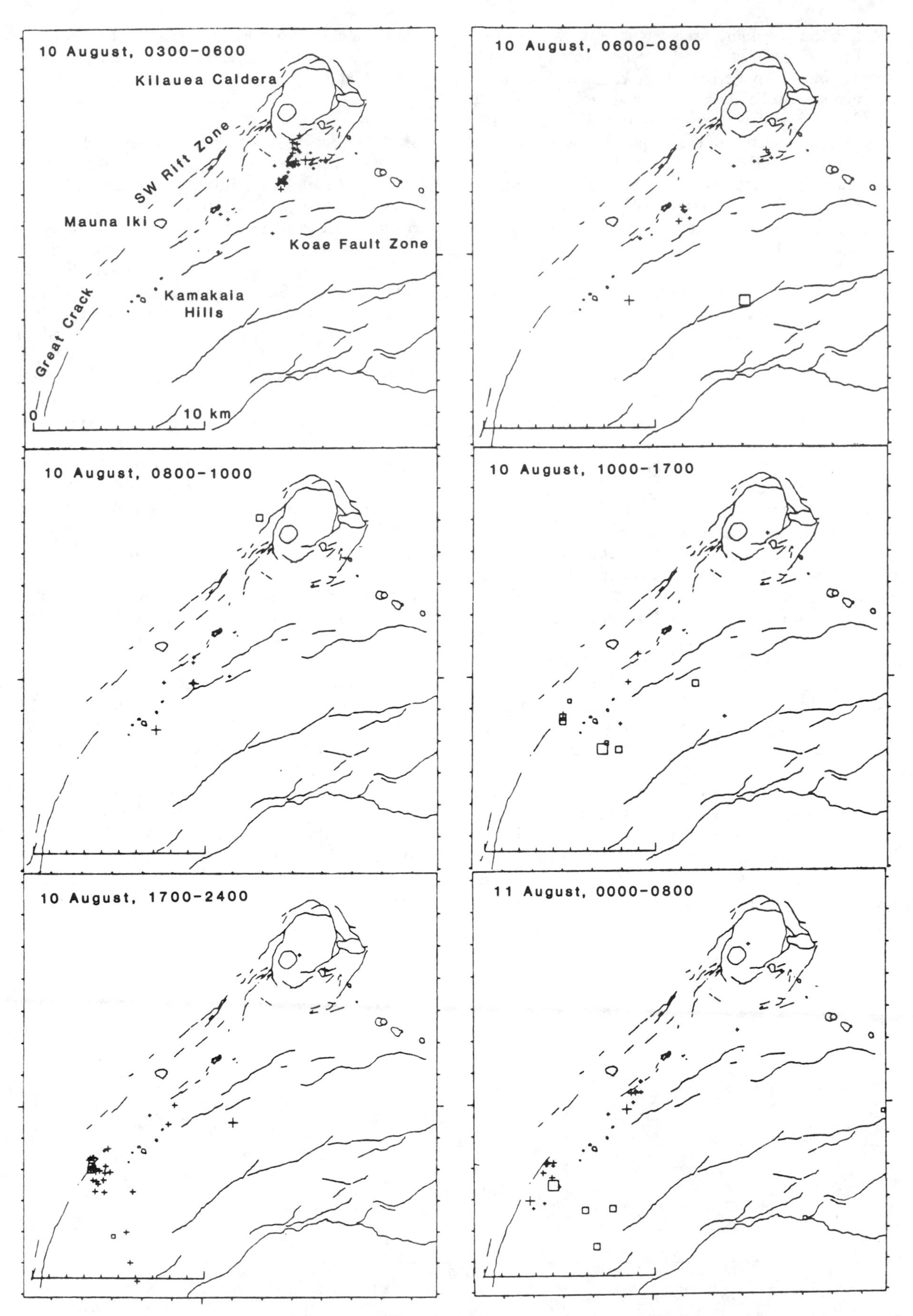

Fig. 13-9. Locations of shallow (< 7 km deep) summit and SW rift earthquakes recorded by HVO during various time intervals between 0330 on 10 Aug and 0800 on 11 Aug 1981. Hypocenters < 5 km deep are indicated by + signs; those deeper than 5 km by □. Size of symbol is proportional to magnitude.

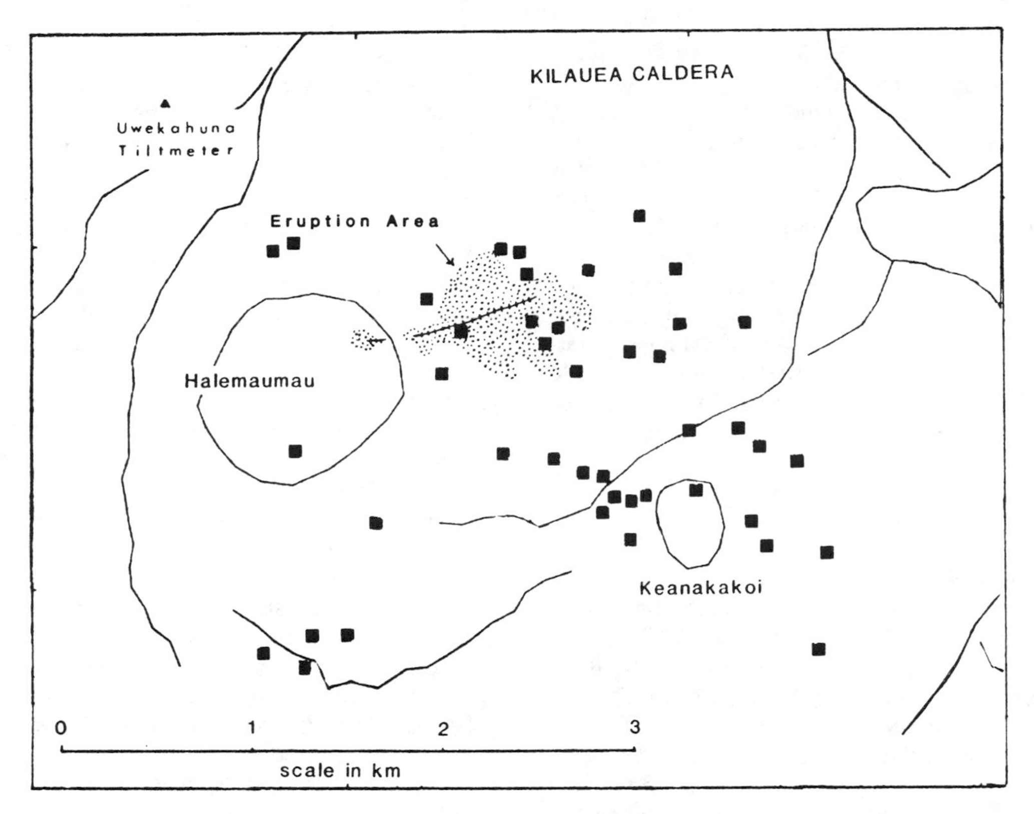

Fig. 13-10. Sketch map of Kilauea Caldera, showing locations of the 30 April – 1 May 1982 eruption fissures (hachured lines), the area covered by the new lava flows (stippled), and the earthquakes that preceded the eruption (solid squares).

than 5 km, plus several magnitude 3.5-4.5 events that were felt near the volcano. Summit deflation continued, but at a slowly decreasing rate, until the late evening of 12 Aug, when inflation resumed. Weak harmonic tremor accompanied the slow summit reinflation. SW rift seismicity remained slightly above average in late Aug, but was continuing to decline.

Information Contact: USGS HVO.

3/82 (7:3) Summit seismicity had increased to nearly normal daily counts by late Dec 1981. Since Jan, several very small intrusions (occasionally seismic but generally aseismic) have been detected by changes in tilt, gas emission, and fumarole temperatures in the E and SW rifts. By late March, tiltmeters showed that the summit area had recovered most of the roughly 100 μrad of deflation recorded during the intrusion of magma into the S summit region and SW rift 10-12 Aug (6:8). The inflation center was in the S caldera-upper SW rift area. A 45-minute swarm of 400-500 earthquakes that started about 1430 on 23 March indicated that magma was forcing open a new channel (or reopening an old one). The seismic swarm was not accompanied by any detectable ground deformation. Overall seismicity in the SW rift remained high in early April but seismicity in the E rift was still relatively unchanged.

Information Contact: N. Banks, USGS HVO.

4/82 (7:4) The following is from HVO.

"A summit eruption began at 1137 on 30 April and lasted about 19 hrs, preceded by a microearthquake swarm about 3 hrs long. Earthquakes at 1-3 km depth and magnitudes of less than 3 occurred in the S part of the caldera (fig. 13-10). Rapid inflation of 5.5 μrad was recorded at the Uwekahuna tiltmeter, on the NW rim of the caldera, during the earthquake swarm (fig. 13-11).

"Lava was erupted from an ENE-trending fissure approximately 1 km long. The fissure, on the S flank of the 1954 spatter ramparts, first opened near its E end about 2 minutes before spatter appeared. Steam emission closely followed opening of the crack as it extended in length both to the E and W. Preliminary data indicate that the line of vents propagated both E and W from the initial vent. Eastward propagation occurred at about 1-2 m/sec. The entire line of vents, including its westernmost part (in Halemaumau pit crater), was active within 25 minutes of the eruption's start, forming a nearly continuous and steady curtain of spatter on the order of 5-10 m high with bursts to 25-50 m. Lava flows to the N, E, and S, which eventually formed prominent lobes, were under way within minutes of the beginning of the eruption.

"A decrease in eruption rates near the ends of the line of vents was first recognized at approximately 1630, about 5 hrs from the beginning of the eruption. About an hour later (at 1740) the westernmost vents, except for those in Halemaumau, had shut off, and drainback was occurring in the area of the central vents, from which the relatively extensive flows to the N and S were fed. By 2100, only the central 150-200 m of the fissure system was active. At this group of vents, activity continued steadily until about midnight, after which the rate of fountaining decreased and the length of the active fissure system gradually diminished.

"A preliminary estimate suggests that the volume of new lava is about 0.5×10^6 m^3, primarily pahoehoe flows. A quick hand-lens inspection suggests that it is aphyric, or has at most rare olivine microphenocrysts. Relatively high temperatures, about 1135-1150°C, were measured by thermocouple and 2-wavelength infrared pyrometer. Minimal fume production, relatively high density of spatter, and low fountain heights suggest that the gas content of the lava was low. However, gas bursting increased in the last hours of the eruption as the rate of fountain activity waned at the remaining centrally located vents. Frothy pumiceous lapilli were ejected during this phase."

Information Contact: USGS HVO.

6/82 (7:6) Summit deflation and an earthquake swarm in the SW rift marked Kilauea's 16th intrusive event since the magnitude 7.2 earthquake of Nov 1975. A continuously recording tiltmeter high on the NW flank began to show deflation during the night of 21-22 June, and seismographs detected the gradual onset of earthquakes in the middle SW rift early 22 June. Low-level harmonic tremor began about 1800. Deflation reached its maximum rate of 1.05 μrad/hr (E-W component) between 2100 and 2200 on 23 June, then gradually decreased. By 0900 on the 24th, tremor had generally ceased. Deflation had nearly stopped by 2200 and the number of microearthquakes had dropped from a maximum of a few hundred per hour to about 1 every 2 minutes.

Early in the swarm, earthquakes were concentrated at 7-8 km depth about 10 km SW of the summit caldera (in the Koae fault system about 3 km S of Pu'u Koae). No significant downrift migration of the epicenters occurred. About 50 μrad of summit deflation (E-W and N-S tilt recalculated to N60°W) were recorded during the intrusion (fig. 13-7). Roughly 20×10^6 m^3 of magma were injected into the SW rift, about half the volume of the last SW rift intrusion in Aug 1981 (6:8-9). During Kilauea's 30 April-1 May 1982 eruption, about 0.5×10^6 m^3 of lava flowed onto the caldera floor.

Information Contact: Same as above.

9/82 (7:9) "An eruption in the S part of the caldera began at 1844:40 on 25 Sept, following nearly 2 hrs of a premonitory seismic swarm and an abrupt increase in summit tilt. The eruption continued for 15 hrs. Lava erupted from a kilometer-long set of left-stepping en echelon fissures in the the southernmost part of the caldera (fig. 13-12). The E and central parts of the vent system were oriented in the usual ENE direction, nearly parallel to the nearby caldera wall. In the W part, however, the fissure turned NW, as if to follow the arcuate circum-caldera fault system. At the NW end of this dogleg, the fissure intersected a circumcaldera fault, and a small isolated vent erupted weakly at the top of the fault scarp.

"The ENE-trending vents were fully active within about 3 minutes of the onset of the eruption and fountained vigorously and steadily through the night. General fountain height was estimated at 20-40 m, with Strombolian bursts that occasionally went as high as 50-70 m. Lava, fed primarily by these vents, rapidly filled a broad graben. At about 1930 the lava spilled southward through a gap in the caldera wall and fed an actively flowing channel that eventually extended more than 1.5 km to the S. Between 2100 and 2200, NE-flowing lava spilled

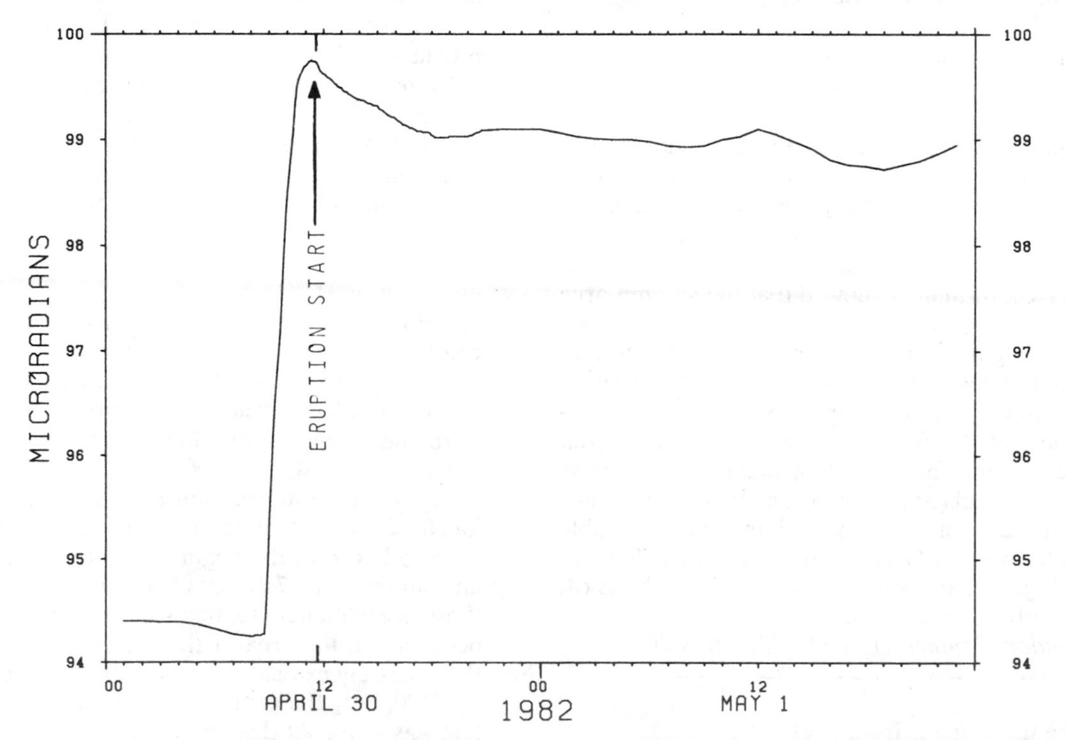

Fig. 13-11. Tilt measured 30 April – 1 May 1982 at Uwekahuna, on the NW rim of the caldera. Increasing tilt readings correspond to summit inflation (up to E).

into the interior of the caldera.

"The NW-trending vent segment was first recognized at about 1900. It extended NW to the caldera wall, which it reached at about 1945. The northwesternmost vent, on a segment of the caldera rim, opened at about 2100, and a small graben about 7 m wide and 20-40 cm deep extended E and W from the vent. The NW part of this line of vents had largely ceased erupting by 2300, but all other vents erupted with unchanged vigor until about 0500 on 26 Sept. At that time, the color changed from the normal yellow to an orange cast, and by 0600 most of the vents had shut down. Diminution of fountaining closely followed a marked decrease in tremor amplitude and a change in summit tilt from slow decrease to slow increase (see below). For the remainder of the eruption the only active vent was one in the central or S part of the NW-trending vent alignment. By 0830, eruption of new lava was about over, but loud explosive gas bursts continued until 0940.

"After the eruption was over, lava from the interior of the extensive pond surrounding the main vents drained back into the vents until 1800 on 27 Sept. The resulting collapse of the pond surface left a 'bathtub ring' on the order of 2-4 m high on the enclosing escarpments. An early estimate suggests that perhaps as much as 3-4 $\times$ 10^6 m^3 of pahoehoe lava was erupted. Of this, possibly as much as 1-2 $\times$ 10^6 m^3 drained back into the vent system.

"Temperatures measured during the eruption were ~1140-1145°C for flows (pahoehoe toes and lava channels), 1130-1160°C for the NW-trending line of vents (cooler to the NW), and 1170°C for the main, ENE-trending, vents. The latter vents fountained more vigorously, almost certainly erupted much more lava, were hotter, and emitted far more gas than the vents on the NW alignment.

"Gases from the erupting vents had atomic C/S ratios of 0.08-0.11 indicating considerable degassing of the magma prior to eruption. The gases were different from those of the 30 April 1982 eruption, which had a higher C/S ratio and were more oxidized. Hand lens inspection shows the new basalt to be almost aphyric. Olivine microphenocrysts are rare."

Kilauea's small eruption of 30 April-1 May, also within the caldera, was from fissures roughly 1.5 km N of the Sept vents (7:4).

Seismic Activity. "Earthquakes related to the eruption were centered in the S summit area (fig. 13-13), initially forming a NE-trending epicentral zone 3.5 km long. Hypocenters were concentrated in a zone ranging from about 0.5 to 4.0 km in depth (fig. 13-14). Located earthquakes ranged from 0.5 to 3.6 in magnitude. Several dozen were felt in the summit region. Nearly 100 earthquakes were located from the pre-eruption swarm, which started at 1650 when small earthquakes and weak harmonic tremor began to record on the summit seismographs. The seismic activity intensified rapidly so that within a few minutes stations 50 km away registered the signals and the number of recorded earthquakes reached 2-5/min. For the first $^3/_4$ hour, the onset of earthquake activity migrated from SW to NE at a rate of about 4 km/hr (fig. 13-15). Moderate-sized earthquakes continued repeatedly along the entire epicentral zone until the eruption began, when earthquake activity in the S caldera region virtually ceased and strong harmonic tremor started. During the eruption, detectable earthquakes were smaller and less numerous than during the pre-eruption swarm. By midnight, locatable earthquakes had decreased to about 10/hr and were mainly confined to the NE part of the seismic zone.

"Tremor intensity remained fairly steady through most of the eruption, but declined sharply between 0400 and 0500 on 26 Sept. By 0500, the tremor amplitude had declined almost to threshold level. As the eruptive activity waned during the morning of 26 Sept, earthquake activity renewed in the NE part of the seismic zone. Intermittent swarms accompanied minor surges of infla-

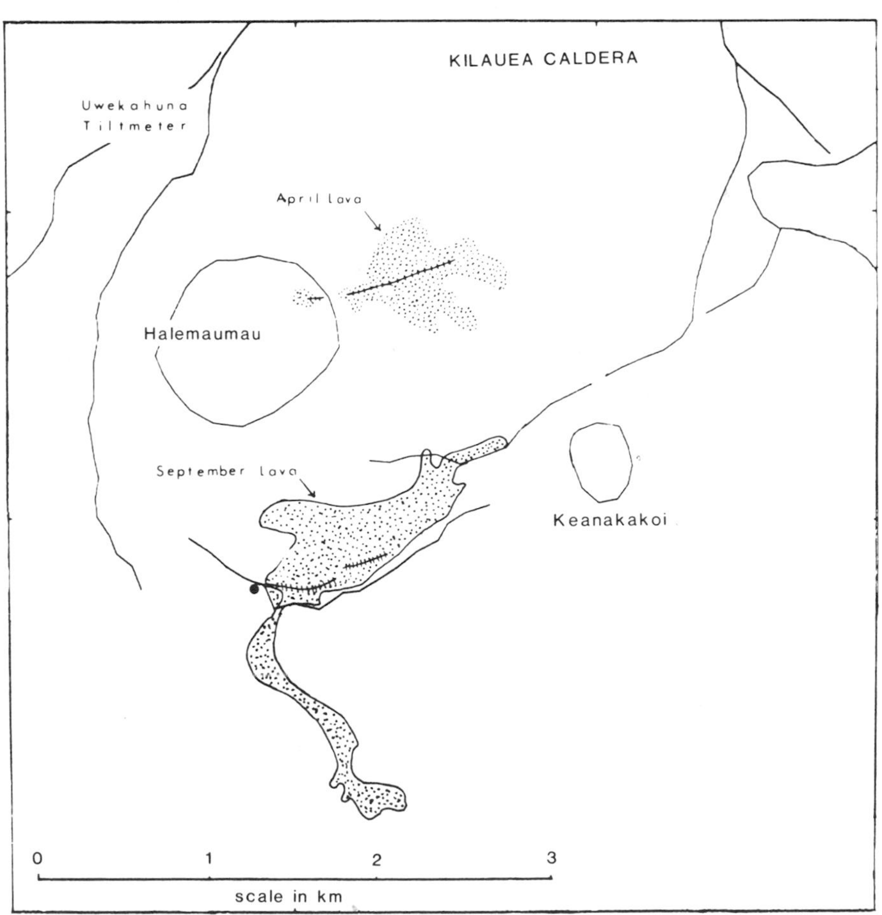

Fig. 13-12. Preliminary map of Kilauea caldera, showing the April and Sept 1982 lavas (stippled areas) and eruption fissures (cross-hatched lines). An isolated vent W of the Sept fissures is represented by a filled circle.

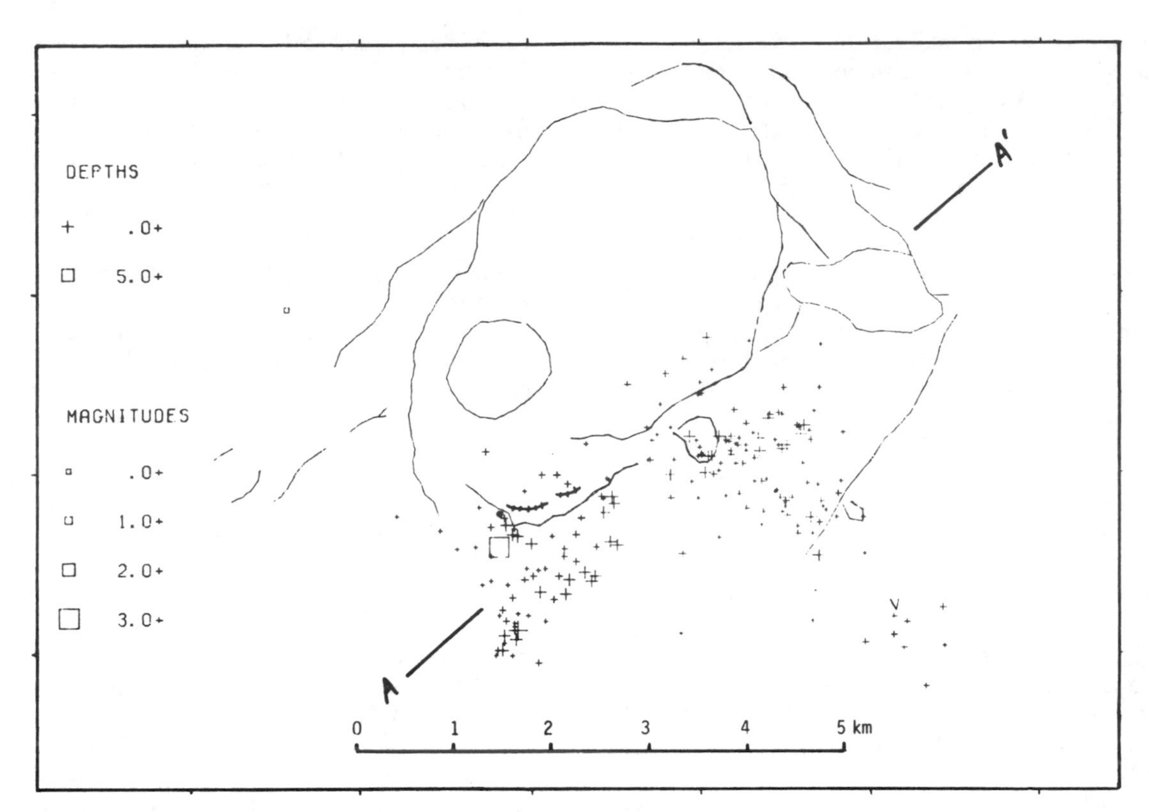

Fig. 13-13. Map of Kilauea caldera, with epicenters of earthquakes recorded 25-26 Sept 1982. Sept eruption fissures are shown as cross-hatched lines and an isolated vent just to the W is represented by a filled circle.

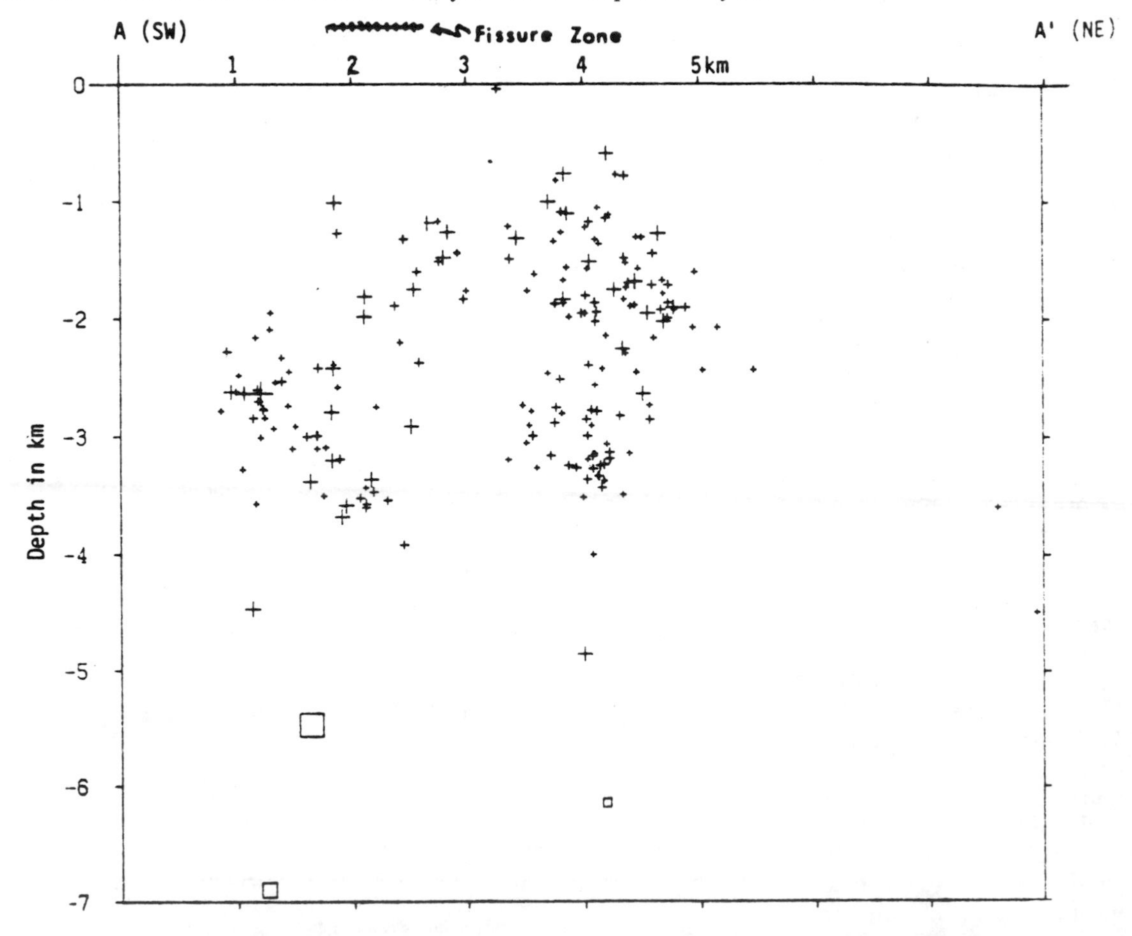

Fig. 13-14. Depths of 25-26 Sept 1982 earthquakes projected into line A-A′ from fig. 13-13. The position of the eruption fissures along A-A′ is represented by a cross-hatched line (as in fig. 13-13).

tion indicated by summit tiltmeters. Post-eruption earthquakes occurred vigorously for much of 26 Sept and activity decreased slowly over the next 2 days. As of the end of Sept, the number of microearthquakes beneath the summit and upper E rift remained higher than average.

Deformation. "As in the April eruption, a rapid increase in summit tilt, probably related at least in part to emplacement of the feeding dike, coincided closely with the pre-eruption earthquake swarm. Uplift had reactivated old cracks by the time (1715) the first observer arrived near the site of the eventual outbreak. The recorded change in tilt on the upper NW flank was about 30 μrad, down to the NNW (fig. 13-16). Termination of the rapid increase in tilt coincided approximately with the onset of strong harmonic tremor and the first appearance of lava. Most of the vigorous phase of the eruption was marked by gradual summit deflation. However, the tilt reversed at about 0450 on 26 Sept and the waning phase of the eruption was accompanied by gradual to moderate inflation that has continued intermittently since the eruption.

"A distance survey line across the zone where the new eruptive fissure formed was measured on 9 Sept, and repeated measurements were made during and after the eruption. Widening caused by the new dike was at least 705 mm and was accommodated in large part by contraction distributed across a broad zone on the flanks of the dike."

Information Contact: E. Wolfe, USGS HVO.

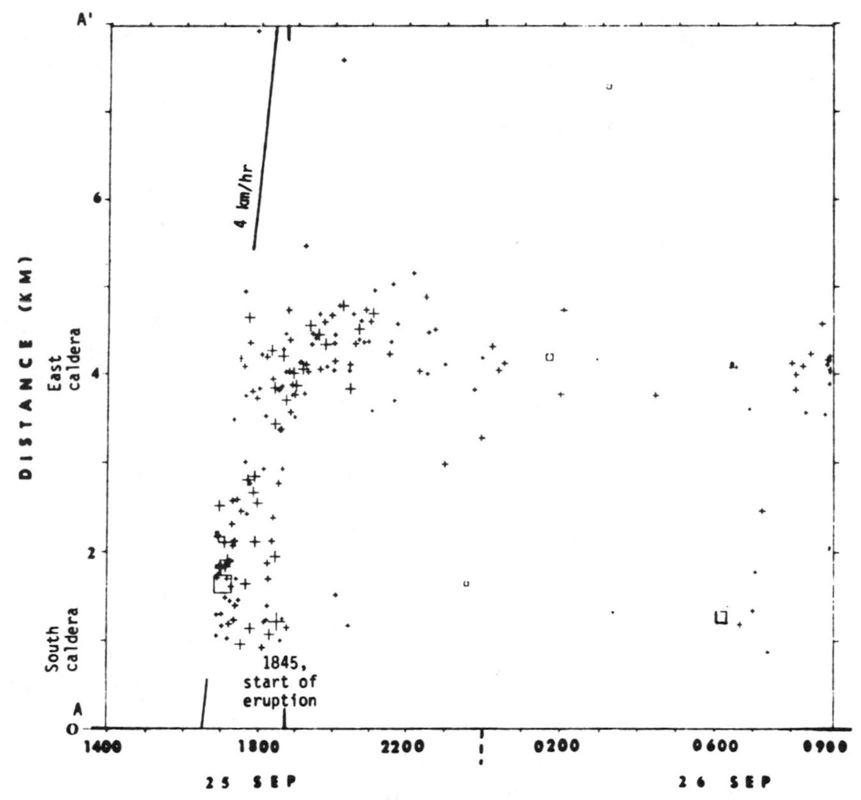

Fig. 13-15. Space-time diagram of earthquakes projected along line A-A´ from fig. 13-13, showing migration of epicenters from SW to NE at about 4 km/hr early in the pre-eruption swarm.

11/82 (7:11) A shallow earthquake swarm began in the summit region at about 1730 on 9 Dec. Epicenters soon migrated into the upper E rift, to the vicinity of Lua Manu and Kokoolau Craters (1.4-3.2 km from the caldera rim). Preliminary locations indicate depths from 0.5 to < 4 km beneath the surface. The largest earthquake had a magnitude of 3.5. No harmonic tremor was recorded.

Tiltmeter records from the NW caldera rim (Uwekahuna Vault) showed slow deflation starting about 1730 and accelerating about 1830. Just before 2000, the deflation rate decreased, and by 2030 slow inflation had resumed. Total deflation was about 5 μrad, suggesting that 2×10^6 m^3 of magma from the summit chamber was intruded into the E Rift.

As of early 10 Dec, earthquake counts were slightly elevated in the swarm area and slow summit inflation was continuing. This was Kilauea's 17th intrusive event since the M 7.2 earthquake of Nov 1975. Four eruptions have occurred since the 1975 earthquake.

Information Contact: Same as above.

12/82 and 1/83 (7:12 and 8:1) [Much of the information in 7:12 was repeated or updated in 8:1. We have therefore combined the reports.] "An eruption in the E rift zone

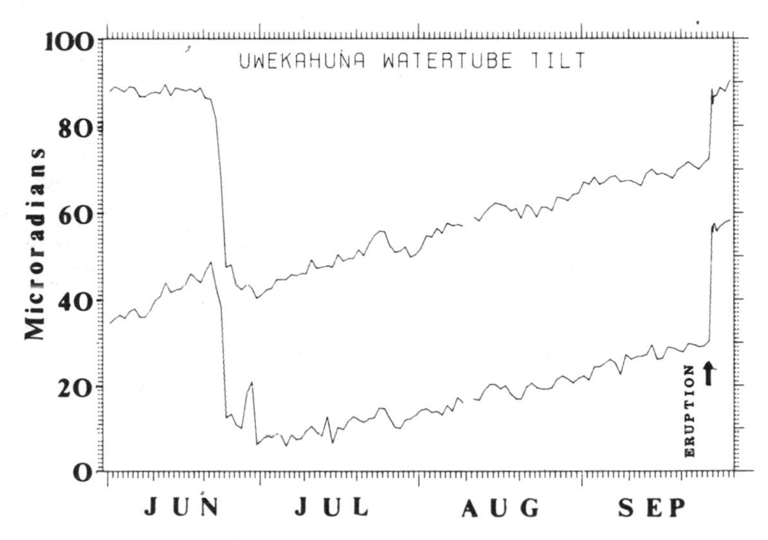

Fig. 13-16. E-W, *top*, and N-S, *bottom*, water tube tiltmeter records at Uwekahuna Vault, on the NW caldera rim. Large summit deflation in late June 1982 accompanied an intrusion into the SW Rift. The prominent inflation at right occurred mainly during the pre-eruption earthquake swarm 25 Sept 1982.

began at 0031 on 3 Jan. The outbreak began at Napau Crater (fig. 13-17), 14 km SE of the caldera rim, and extended progressively NE. By 0740, the line of eruptive fissures was 6 km long and its E end was about 0.7 km SE of Pu'u Kahauale'a (fig. 13-18). Fountaining and production of SE-moving lava flows of local extent continued until 1002. After a $4^1/_2$ hr pause, the eruption resumed at 1425 along a 100-m fissure at the NE (downrift) end of the vent system (about 600 m S of Pu'u Kahauale'a). This eruption lasted until 1535. During this first day's activity, fountains up to 80 m high produced an estimated 2-3 $\times$ 10^6 m^3 of lava.

"The volcano remained quiet for nearly 2 days. Eruptive activity resumed at 1123 on 5 Jan and continued with only brief interruptions until 2049 on 6 Jan. The eruptive activity jumped from one segment to another of a kilometer-long section of the vent system S of Pu'u Kahauale'a (including, at its E end, the easternmost vent of 3 Jan). A large amount of the lava poured into an open crack, parallel to the eruptive vents and about 0.1 km SE of their E ends. The crack was along a bounding fault at the NW edge of a prominent older graben. Minor aditional fountaining and flow production also occurred at, and within 1.5 km SW of, Pu'u Kamoamoa, along vents established on 3 Jan.

"On the morning of 7 Jan the main eruptive center shifted temporarily still farther NE to a 1 km-long line of vents approximately 1.4 km ESE of Pu'u Kahauale'a. After a brief introductory emission from 0957-0959, these vents erupted strongly from 1030-1557 on 7 Jan. During this period, they produced the highest fountains of the eruption; maximum sustained fountain heights of 80-100 m were estimated, with bursts sending fragmented spatter higher. This episode also fed a lava flow nearly 6 km long that extended E toward Kalalua then turned SE toward the coast. The flow, with an estimated volume of about 4 $\times$ 10^6 m^3, converted to aa as its front passed near Kalalua and stopped nearly 5 km from the coast. A second eruption from the same vents, from 1625 on 7 Jan to 0430 on 8 Jan, produced a smaller lava flow that overrode the near end of the first flow from the vents E of Pu'u Kahauale'a.

"Seven episodes of lava production ranging in duration from 8 minutes to 11 hrs occurred 8-15 Jan from a group of vents S and SW of Pu'u Kahauale'a that were active 5-6 Jan. The first and briefest of these (from 1446 to 1454 on 8 Jan) was from the E end of this group of vents. The remainder erupted from the W half of the kilometer-long line. Six of the 7 extrusive episodes occurred 8-11 Jan, the latest on 15 Jan. One additional extrusion occurred 8 days later, during the evening of 23 Jan, when approximately 7000-8000 m^3 of lava were erupted about 0.25 km E of Pu'u Kamoamoa following a magnitude 4.2 earthquake on the S flank.

"New lava covered an area of approximately 4.4 $\times$ 10^6 m^3. The erupted volume is estimated to be on the order of 10 $\times$ 10^6 m^3. Repeated measurements indicate an eruption temperature of about 1135°C. The basalt is slightly porphyritic with scattered small plagioclase and olivine phenocrysts.

"Strong emission of sulfur-rich gases has continued since the beginning of the eruption; widespread dispersal of the gases had caused vegetation damage over large areas of E Hawaii. Between extrusive events (and continuing at the time of this report) some vents S of Pu'u Kahauale'a remained at temperatures of about 1070°C and burning gases were visible at night. These vents periodically emit spatter composed, at least in part, of incandescent and partly melted fragments eroded from the vent walls.

Seismicity. "In the weeks prior to the eruption seismographs recorded increasing rates of microearthquakes in the E rift zone. At 0030 on 2 Jan the seismicity developed into a swarm of small shallow (depth < 5 km) earthquakes and weak harmonic tremor. The swarm started in the upper E rift near Mauna Ulu, increased in the early hours and migrated downrift about 9 km to Napau Crater. Tiny earthquakes were recorded at a rate of 3-5/min; seismic intensity peaked between 0040 and 0110, when several earthquakes, M 2.5 to 3.0, were felt in the Hawaii Volcanoes National Park area. From 0300 to 1300 2 Jan, the seismic zone spread farther downrift to beyond Pu'u Kamoamoa (fig. 13-18). From then until the eruptive outbreak, small earthquakes accompanied by harmonic tremor occurred at a nearly constant rate, mainly along a zone between Napau Crater and Pu'u Kamoamoa. When the eruption was sighted at Napau Crater by a ground crew at 0031 on 3 Jan, [instruments] started to record increasing harmonic tremor amplitudes and constant summit deflation.

"From 3 through 6 Jan the shallow earthquakes were broadly distributed from near Napau to beyond Pu'u Kahauale'a. Harmonic tremor, although continuous, waxed and waned in consonance with the production of lava. In the early morning hours of 7 Jan, however, the shallow earthquake activity precursory to the opening of the vents E of Pu'u Kahauale'a was concentrated in a zone that extended approximately from Pu'u Kahauale'a to about 2 km downrift from Kalalua. Since then earthquake activity in the eruptive zone has been relatively low. At the end of Jan, seismographs were still registering a moderate level of harmonic tremor originating from a source in the vicinity of the eruptive vents extending 2 or 3 km E of Pu'u Kamoamoa to near Pu'u Kahauale'a.

"Throughout the month thousands of shallow earthquakes were generated in the eruption zone. Of these, nearly 500, most from depths of less than 4 km and magnitude 0.5-4.0, were processed for size and location. The vast majority of the earthquakes probably record emplacement and establishment of the feeder dike system, a process that was completed on the morning of 7 Jan.

"Deeper earthquakes (5 to 13 km) that occurred throughout the month beneath the S flank were more or less representative of adjustment of the S flank that continues over the long term. However, a somewhat elevated frequency of occurrence of S flank earthquakes in the early part of the month was probably a direct response to increased stress from the magmatic activity in the rift zone.

Deformation. "Tiltmeters in the Uwekahuna Vault recorded a summit collapse of about 125 μrad (fig. 13-19), which represents an estimated volume loss of 50 $\times$ 10^6 m^3 from the summit region. Broken only by temporary inflation during the non-eruptive interval between 3 and 5 Jan, almost all of the summit deflation occurred at a high rate from early on 2 Jan to early on 8 Jan. The summit subsidence and the period of intense shallow seismic activity were coincident. Approximately $^3/_4$ of the volume of extruded lava was erupted during the same period. Minor, very slow deflation of the summit continued until about 18 Jan. Gradual recovery of about 4 μrad of tilt occurred during the remainder of the month.

"Geo-electric and recorded tilt changes in the E rift zone on 3 Jan strongly suggest that magma was intruding the rift as far down as Kalalua at that time. However, observations of ground cracking, tilt measurements, and electronic distance measurements show that major extension (> 2 m) perpendicular to the rift zone occurred N and NE of Kalalua late on 6 Jan and during 7 Jan. An electronically measured line across the eruptive fissures S of Pu'u Kahauale'a, initiated on 19 Jan, has shown a steady extension averaging 7 mm/day during the last part of Jan.

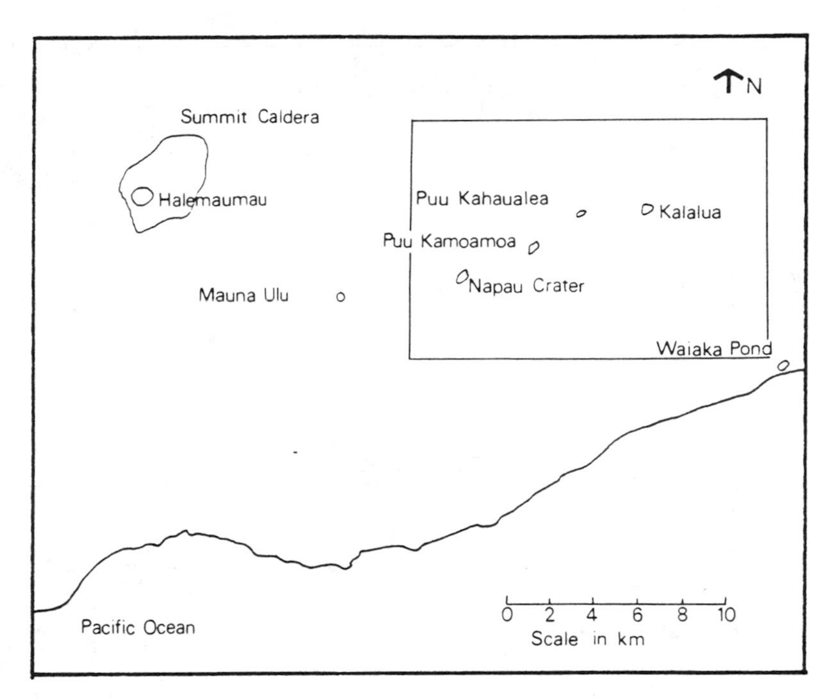

Fig. 13-17. Index map of the summit region and E rift. The area covered by fig. 13-18 is outlined.

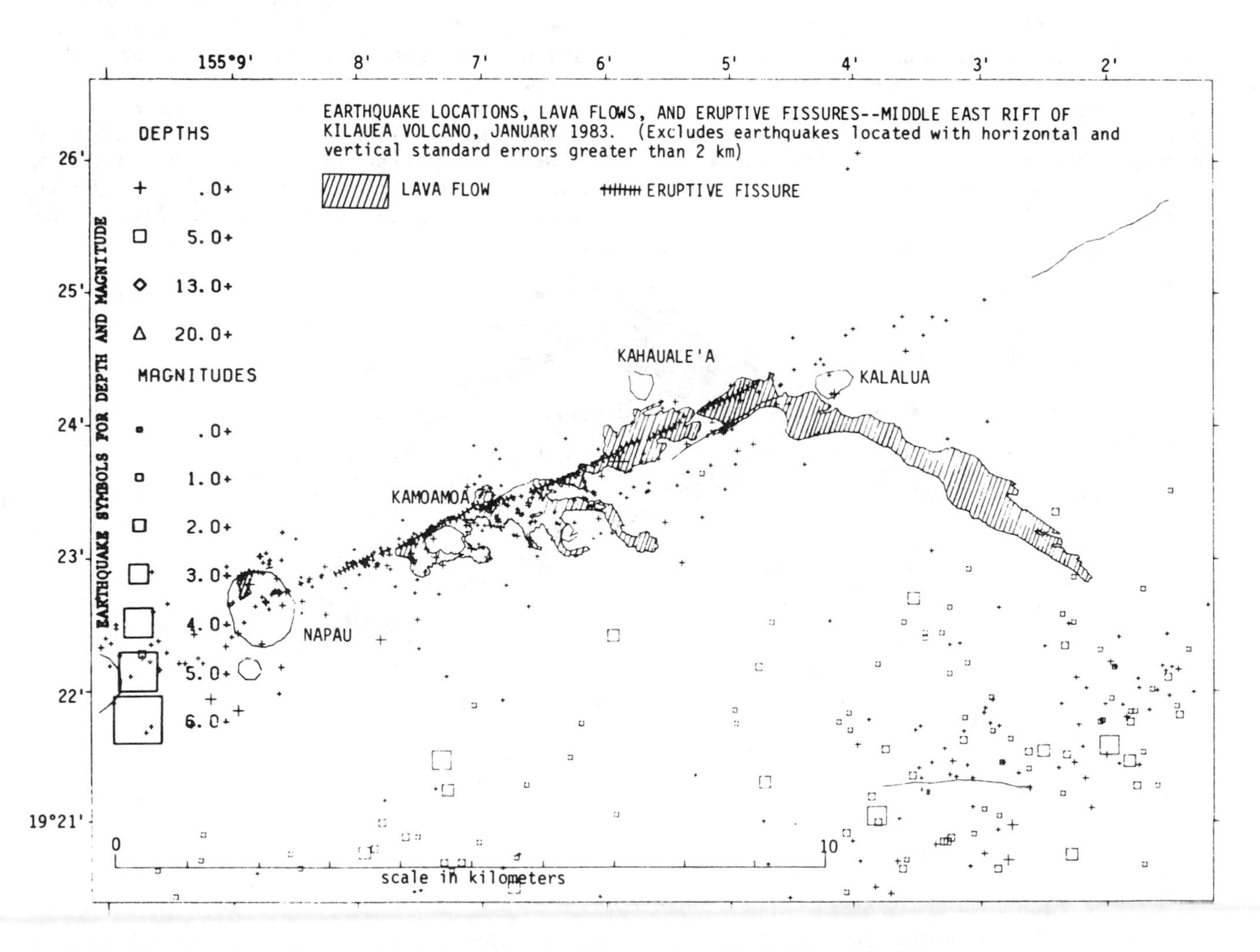

Fig. 13-18. Map of earthquake locations, lava flows, and eruptive fissures on the middle E rift zone, Jan 1983.

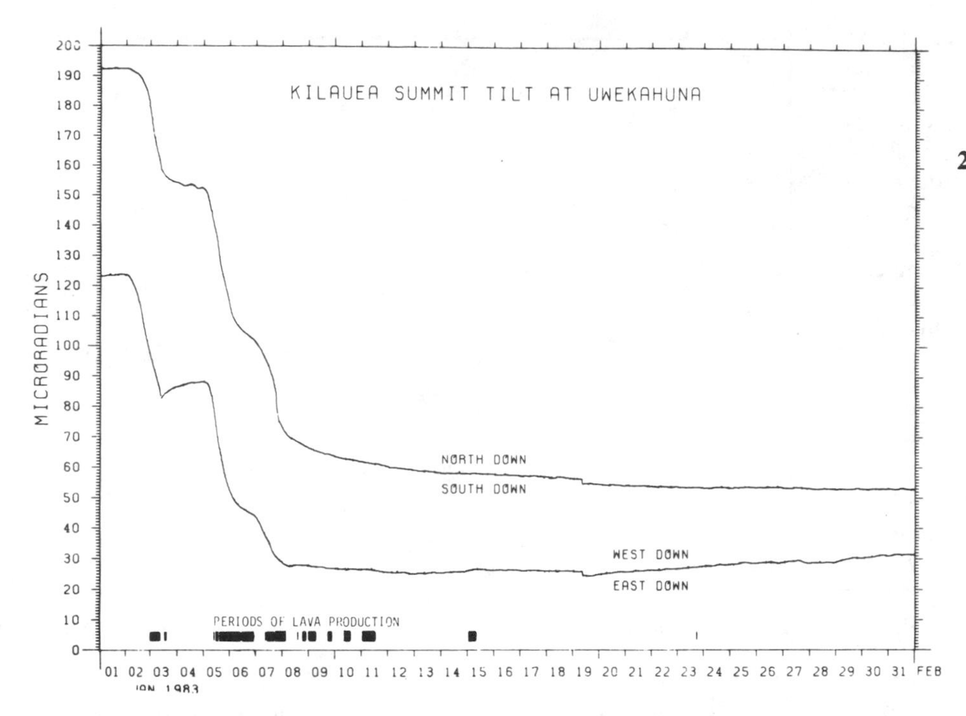

Fig. 13-19. N-S and E-W components of summit tilt measured at Uwekahuna Vault, NW caldera rim, Jan 1983. Periods of lava production are also shown.

Summary. "Although no major lava production has occurred since 15 Jan (a minor extrusion occurred on 23 Jan), the continuing steady harmonic tremor, voluminous gas emission, occurrence of incandescent and flaming vents, and extension across the recently active eruptive fissure S of Pu'u Kahauale'a indicate that the magmatic activity related to the Jan 1983 eruption has not yet ended."

Information Contacts: E. Wolfe, A. Okamura, R. Koyanagi, USGS HVO.

2/83 (8:2) "The E rift zone eruption of Kilauea that began on 3 Jan resumed on 10 Feb, and lava production continued until 4 March along eruptive fissures established during the initial outbreak on 3 Jan (fig. 13-20). The renewed eruption followed nearly a month in which vent activity was limited largely to incandescence and emission of burning gases along parts of a 2-km segment of the vent system that extended E from about 0.75 km NE of Pu'u Kamoamoa to the area S of Pu'u Kahauale'a. During the quiet period, a little incandescent spatter was ejected sporadically from the E vents; at least some of the ejecta consisted of wallrock remelted and eroded from the vents by vigorous emission of burning gas. There was no measurable production of new lava.

"Increased spatter production was first recognized on 10 Feb; a small (6 m-high) spatter cone had formed at the E vents (0.7 km S of Pu'u Kahaule'a). By 12 Feb, a second small spatter cone had formed, and a glowing crack extended tens of meters NE of the 2 cones. Sub-

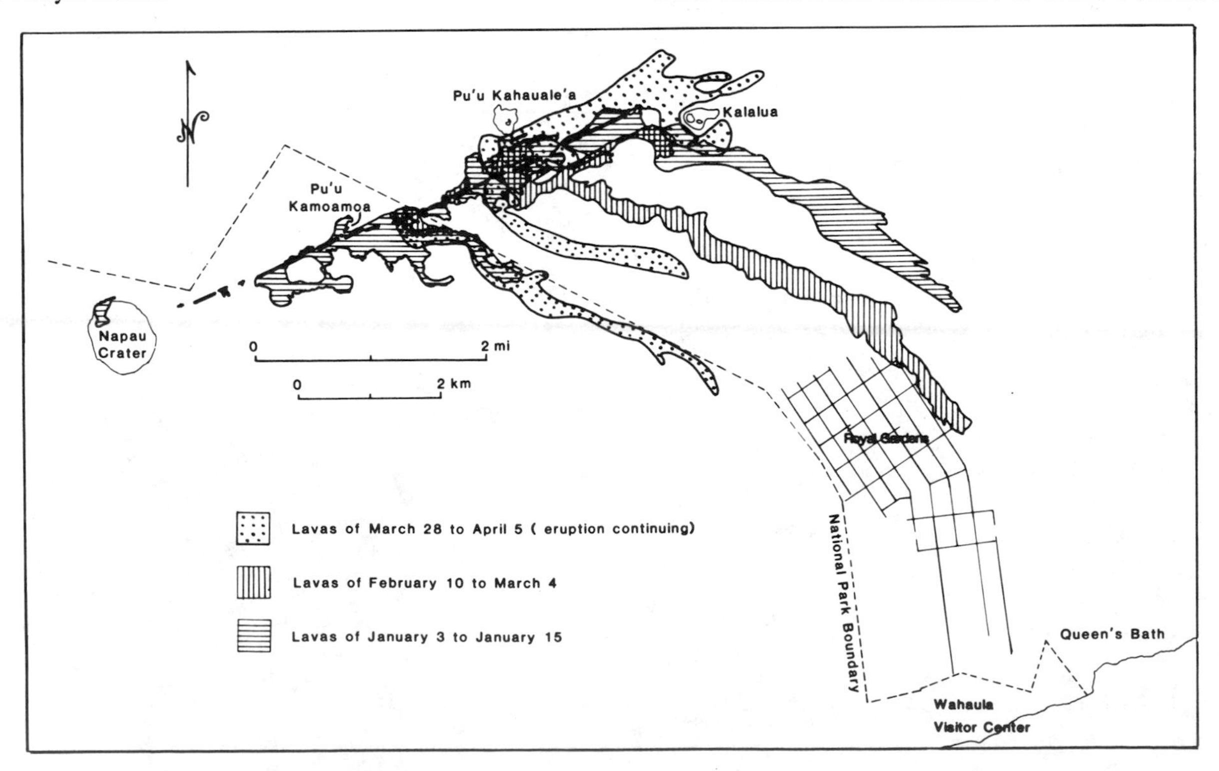

Fig. 13-20. Map of 1983 lava flows and eruptive fissures on the middle E rift zone for the periods 3-15 Jan (horizontal lines), 10 Feb – 4 March (vertical lines), and 28 March – 5 April (dotted patern). [For full 28 March – 8 April coverage see fig. 13-22.]

sequent intermittent production of low fountains and small lava flows through 24 Feb led to growth of a flat-topped shield estimated to be ~200 m long, 100 m wide, and 10 m thick. The shield was capped by a 170 m-long line of juxtaposed spatter cones ranging up to ~15 m high. Lava production during the 2-week shield-building period is estimated at 0.5×10^6 m^3. In addition, a short (probably 10-minute) episode of spatter production occurred at a vent just E of Pu'u Kamoamoa at about 2220 on 19 Feb. Gas emission during this period was low and was characterized by extremely low atomic C/S ratios (~0.05) suggesting that the near-surface magma had largely degassed during the non-eruptive interval.

"Beginning at 0145 on 25 Feb, fountaining and lava flow production increased in the W and central parts of the shield and a flow about a kilometer long extended NE. Gas composition also changed at this time, becoming more C-rich (C/S approximately 0.15) and generally reverting to a composition indistinguishable from that of the early Jan gases. Thirteen hours later, at about 1440 on 25 Feb, the main eruptive locus shifted about 100 m uprift and eruption from the shield vents soon terminated, at 1518. Fountains played continuously at this new locus until the end of the eruptive episode on 4 March. During this period, sporadic lava production also occurred from local vents as far uprift as 0.75 km NE of Pu'u Kamoamoa.

"During its week of sustained activity, the main fountain, about 0.75 km SSW of Pu'u Kahauale'a, was commonly 40-80 m high. Estimated to be about 30 m wide at its base, the fountain arose from a lava pond about 60 m in diameter. By the evening of 25 Feb, the S rim of the levee containing the lava pond had developed a spillway through which 2 major flows were supplied during the ensuing week. One of these moved NE 25-26 Feb within the same graben that contained the upper part of the 7 Jan flow. Following the path of that earlier flow, the new flow turned SE about 0.5 km W of Kalalua and stopped about 3 km from its source.

"By the morning of 27 Feb, the active lava river leading from the pond had been diverted SE, producing a flow that eventually extended more than 7 km from the vent to its terminus, about 3.8 km from the coast. This latest flow, parallel to and a kilometer SW of the 7 Jan flow, advanced slowly through the rain forest until 4 March, when lava production stopped. In the half nearest the vent, where the feeding channel was largely pahoehoe, the average velocity of the advancing flow front was about 90 m/hr. In the lower half, where the flow was dominantly aa, the front advanced episodically, but at an average rate of about 30 m/hr, even on the steepest (about 7°) slopes.

"In the early evening of 2 March, the advancing aa front, locally up to 10 m thick, entered the NE part of a sparsely populated subdivision on the S flank, just E of Hawaii Volcanoes National Park. Two dwellings were destroyed before lava production at the vent stopped at about 1500 on 4 March. Subsequent movement in the distal part of the flow was limited to minor adjustments that led to only a few meters of additional lava advance.

"Preliminary estimates, as yet without benefit of careful mapping or methodical thickness measurements, indicate that about 10×10^6 m^3 of lava were extruded 10 Feb – 4 March. Thermocouple measurements in pahoehoe toes gave lava temperatures of about 1112-1120°C, slightly cooler than in Jan. The basalt is sparsely porphyritic with scattered small phenocrysts of olivine and plagioclase.

Seismicity and Deformation. "Apparently because the feeder dike system had become fully established in early Jan and was maintained until eruptive activity resumed in Feb, the renewed lava emission was not accompanied by increased numbers of shallow earthquakes in either the summit or E rift zone. Harmonic tremor had declined by 30 Jan to about 10% of its high amplitude in early Jan; originating from a source beneath Feb's eruptive vents, the tremor slowly doubled in amplitude 30 Jan-25 Feb. From 25 Feb-4 March, average tremor amplitude was about 30% of the early Jan level. On 4 March, when lava production terminated, the tremor amplitude dropped abruptly to a low, but constant, level that was continuing as of 10 March.

"By 25 Feb, the E-W component tiltmeter in Uwekahuna Vault near the summit had recorded approximately 9 μrad of gradual summit re-inflation following the major subsidence of early Jan. The vigorous eruptive activity 25 Feb-4 March in the middle E rift zone coincided with an 11 μrad E-W deflation at Uwekahuna. Rapid summit re-inflation averaging about a μrad per day has occurred 4-10 March.

"No significant extension across the E rift has occurred in the vicinity of Kalalua since the major eruption of 7 Jan. However, a survey line across the eruptive fissure near Pu'u Kahauale'a showed extension 29 Jan-12 Feb of at least 13 mm/day. By 14 Feb, new lava had obstructed the line."

Information Contacts: Same as 7:12.

EPISODE 3

[With the following report, HVO scientists recognized that they were dealing with episodic eruptive behavior, which continued for the next three years. We have added numbered headers to help organize the reports. Originally referred to as "phases," the HVO staff later decided that the term "episode" was more appropriate (see 10:6). We have replaced the word "phase" throughout the text of earlier reports. Where the word "episode" was used in reports 8:3 to 10:6, we have substituted "period" to avoid confusion with the revised usage of "episode."]

3/83 (8:3) "The 1983 eruption entered its third major episode of lava production in the early morning of 28 March. A 3½-week quiescent period 4-28 March was interrupted only briefly by minor emission of spatter and pahoehoe lava on 21 March at vents S of Pu'u Kahauale'a.

"Initially, on 28 March, the major eruptive activity occurred at a vent 700 m NE of Pu'u Kamoamoa, just inside the Hawaii Volcanoes National Park boundary. Vigorous extrusive activity at this vent produced a flow that extended nearly 5 km SE along the National Park boundary (fig. 13-20). Eruption of this vent stopped at 2019 on 30 March.

"A vent S of Pu'u Kahauale'a that was the source of the major flows of late Feb and early March resumed erupting 28 March, sporadically at first. Its eruptive activity became steady at approximately 1800 on 29 March. From then through 5 April, when this report was prepared, it supplied a flow that slowly extended 4 km [revised to 3 km in 8:4] NE along the rift zone to the vicinity of Kalalua. Another flow from this vent moved about 3 km SE on 4 and 5 April. The vent continued as the single locus of lava production. Its vigorous fountain, commonly 100 m high and at times estimated as high as 300 m, was visible from a number of vantage points along the highway from the Hawaii Volcanoes National Park to Hilo. Frothy scoria fragments, bombs of spatter, and thin spatter-fed flows built a prominent cone 60 m high [note growth to 80 m in 8:4], and a thin airfall pyroclastic blanket extended more than 1 km from the cone. Pele's hair fell as far as 17 km from the vent.

"The volume of basalt extruded since the eruption began on 3 Jan was on the order of 30×10^6 m^3 [to at least 50×10^6 m^3 by end of episode 3; 8:4]. Flows of the present episode were dominantly aa. Lava temperatures ranging from 1112° to 1129°C were measured by thermocouple. Like the earlier 1983 lavas, those of late March-early April are slightly porphyritic, with scattered small plagioclase and olivine phenocrysts. The gas composition remained unchanged throughout the eruption, indicating that stored E rift magma remained the predominant source of the erupted lava."

Deformation and Seismicity. "The water tube tiltmeter at Uwekahuna Vault in the summit region showed the correspondence of summit subsidence with major extrusion episodes in the E rift zone (fig. 13-21). Moderate summit re-inflation followed the extrusive episodes of early Jan and early March. The tiltmeter data in combination with levelling results indicated a cumulative volume of at least 70×10^6 m^3 [to 80×10^6 m^3 by the end of episode 3; 8:4] for magma withdrawn from the shallow summit region since the beginning of the eruptive/intrusive activity in early Jan.

"Since cessation of the initial earthquake swarm in early Jan (8:1), seismicity in the eruptive zone was characterized by unceasing harmonic tremor that waxed and waned in amplitude in concert with the eruptive activity. As determined from a seismic station near Pu'u Kamoamoa and from portable seismometer traverses, the tremor originated from a source within a few kilometers of the surface in a zone between Pu'u Kamoamoa and the vents S of Pu'u Kahauale'a.

"Following the major outbreak of 25 Feb-4 March, tremor continued at a decreased level. On 21 March, the amplitude gradually increased from 0430 to 0630, remained moderately high for most of the day, and decreased to its previous low level on the following day.

"A gradual increase in amplitude occurred again beginning in the early morning of 27 March. By 0100 on 28 March the tremor amplitude increased by about 5 times at the Pu'u Kamoamoa station. Glow from active fountains was reported shortly thereafter. Tremor remained strong as of 5 April, at times reaching an amplitude greater than 10 times background for periods of a few minutes to several hours."

Addendum. Steven Brantley reported that lava fountaining from the vent S of Pu'u Kahauale'a stopped at 0257 [revised to 0247 in 8:4] on 9 April. By 0430, harmonic tremor had decreased to low levels, and the rate of summit deflation had decreased to < 0.05 μrad/hr. The SE flow from this vent entered the Royal Gardens subdivision on 8 April. Approximately 7 [6 confirmed in 8:4] structures were destroyed before the flow stopped late in the afternoon of 9 April.

Information Contacts: E. Wolfe, A. Okamura, R. Koyanagi, S. Brantley, USGS HVO; UPI.

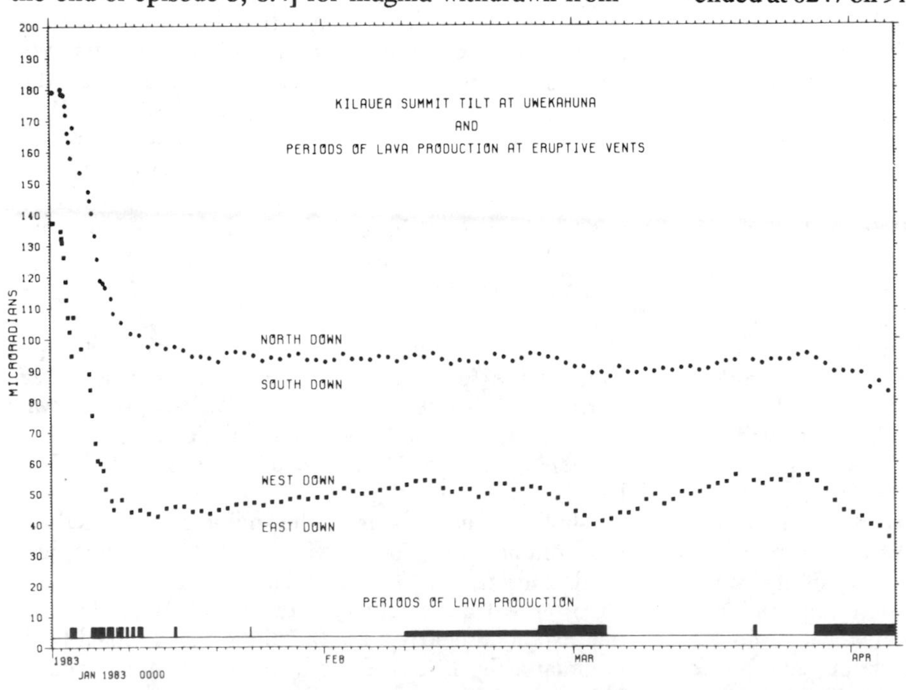

Fig. 13-21. N-S and E-W component of summit tilt measured at Uwekahuna Vault, NW caldera rim, Jan-April 1983. Periods of lava production are also shown.

4/83 (8:4) "The third major episode of the E rift eruption ended at 0247 on 9 April. The initial activity was at a vent that erupted 28-30 March, producing an aa flow that extended nearly 5 km SE along the Hawaii Volcanoes National Park boundary. A vent about 2 km to the ENE in the same location as the source of the lava flow that entered the Royal Gardens subdivision during episode 2 erupted steadily from 29 March-9 April. Marked by 2 fountains that were the sources of flows that exited N and S, the vent was the locus of all April lava production. These fountains, and in particular the steadily active and more prominent N fountain that was commonly 100 m high and at times estimated to be 300 m high, were the source of tephra for construction of an 80 m-high cinder cone.

"From 29 March to the end of the eruption, a flow composed of aa locally more that 12 m thick advanced slowly about 3 km NE to the vicinity of Kalalua. Another flow moved SE between earlier flows of 7 Jan and 27 Feb-4 March. Lava production at the S fountain changed from sporadic to steady on 4 April and at least 3 lava flows were fed in succession during the remainder of episode 3. These flows were ~ 3.0, 1.5, and 7.2 km long. The advance of each of the first two apparently stopped when the lava stream feeding it was diverted to form the next. The fronts of these aa flows advanced at average rates of approximately 100 m/hr and at times faster than 200 m/hr through gently sloping rain forests. The last and longest of the 3 flows began its advance on 6 April. It entered Royal Gardens subdivision on 8 April and destroyed 6 structures before the eruption stopped early on 9 April. Near the end of the eruption the flow front reached a velocity of 6 m/min as the narrow and elongate terminus, centered on a subdivision street, advanced down a steep slope. Decelerating advance of the flow front continued through at least 11 April.

"Jan-April lava covered about 15 km^2, approximately double the 7 km^2 that had been covered by the end of episode 2 on 4 March (fig 13-20). As a preliminary estimate, a minimum of 50 × 10^6 m^3 of lava had been extruded since eruptive activity began on 3 Jan. The most recent basalt is megascopically identical to the earlier lava. It is slightly porphyritic with scattered small phenocrysts of plagioclase and olivine. This, as well as continued relatively low lava temperatures (1112-1129°C) and unchanged composition of eruptive gases, implied that stored E rift magma had remained as the source for all Jan-April 1983 lava.

"The summit (Uwekahuna) water tube tiltmeter recorded nearly 30 µrad of decreasing tilt related to subsidence during the third episode of the eruption (fig. 13-21). Since the beginning of intrusive/eruptive activity on 2 Jan, the cumulative tilt decrease was nearly 200 µrad, suggesting a net volume loss in the summit area of at least 80 × 10^6 m^3. Harmonic tremor decreased significantly when lava fountaining stopped. As of 10 May, tremor continued at a very low level in the general area of the recently active vents.

"Although hot gases of magmatic origin were becoming progressively less concentrated and more oxidized, they continued to be emitted from the recently active part of the fissure system. Local but decreasing incandescence also continued. These observations combined with the observations of persistent low tremor and minor extension across the fissure system (10 mm of extension occurred by 29 April on a survey line established 8 days earlier) suggested that the feeder system was still active and renewal of eruptive activity was possible."

Information Contacts: Same as 8:3.

5/83 (8:5) "Although Kilauea did not erupt during May, the fissure system for the 1983 E rift zone eruption was still active. Incandescent fissures with temperatures around 800C were present at both vents that erupted from 28 March to 9 April. Steady slow dilation, averaging slightly less than 1 mm/day, was measured across the fissure system from Pu'u Kahaualc'a (near active vents about 17 km E of the summit caldera rim). Weak harmonic tremor continued throughout the month. These data suggest that slow intrusion was sustaining the feeder, and resumption of eruptive activity is possible."

Addendum. Eruptive activity accompanied by gradually increasing harmonic tremor resumed at about 0800 on 13 June, from the 28-30 March vent (near Pu'u Kamoamoa) about 15 km ESE of the summit caldera rim. Lava fountains 20-30 m high fed a flow that moved SE over the 28-30 March lava, along the Hawaii Volcanoes National Park boundary. The summit tiltmeter showed a deflation rate of 1 µrad/6 hrs. Moderately high and very steady harmonic tremor was recorded in the vicinity of the vent as the eruption continued 14 June.

Information Contacts: R. Decker, E. Wolfe, A. Okamura, R. Koyanagi, J. Nakata, USGS HVO.

EPISODE 4

6/83 (8:6) "The fourth and fifth major episodes of Kilauea's E rift zone eruption occurred during June and early July. The two eruptive events, each about 4 days long, produced 3 new major lava flows that extended SE down the S flank.

"The eruptive vent for both episodes was located just within the Hawaii Volcanoes National Park about 750 m NE of Pu'u Kamoamoa (fig. 13-22). The same vent had been active intermittently since early Jan; in late March it produced a 5 km-long lava flow that extended SE along the National Park boundary (8:3-4).

"Episode 4 lava fountains were first reported from a passing aircraft at 1025 on 13 June. When the first ground observers arrived at midday, a line of low fountains about 100 m long was feeding flows to both the NW and SE. The NE end of the vent quickly became the major locus of lava production, and an aa flow fed by a vigorous river of pahoehoe began extending SE, on top of and adjacent to the late March (episode 3) flow. A steep-sided spatter cone 30-40 m high was built at the source of the flow, which cascaded over a spillway $1/2$ to $2/3$ of the way up the S side of the cone. A low fountain, up to about 20 m high, played from the surface of the lava pond that filled the interior of the cone to the level of the spillway.

"Lava discharge was estimated at about 1 × 10^5 m^3/hr. The main flow extended about 7.5 km SE from the vent and covered approximately 1.5 × 10^6 m^2. Its front advanced at about 30-200 m/hr. Following the National Park boundary, the flow entered Royal Gardens subdivision only locally, and no homes were destroyed. Episode 4 ended abruptly at 1413 on 17 June, a little more than 4 days after it began.

"Like previous 1983 lavas, episode 4 basalt is slightly porphyritic, with scattered small phenocrysts of plagioclase and olivine. Lava temperatures, measured by thermocouple, ranged from 1115-1132°C.

EPISODE 5

"Episode 5 began on 29 June. At 1000, a pool of lava was seen slowly rising inside the main episode 4 vent. At about 1300 lava production became vigorous, and episode 5 lava cascaded over the earlier spillway and began flowing SE within the previously evacuated episode 4 channel. Lava production quickly reached and was maintained at a rate similar to that of episode 4, ~ 1 × 10^5 m^3/hr, and an aa flow began advancing SE over episode 3 and 4 basalts. The flow was fed by a vigorous pahoehoe channel that was generally bank-full and frequently overflowing. Advancing at average rates ranging from 80-165 m/hr, the flow front entered the NW part of Royal Gardens subdivision at 1919 on 1 July. It finally stopped 8 km from the vent at about 1030 on 3 July, more than 3 hrs after the vent had stopped erupting. Traversing the subdivision, it burned and crushed 7 dwellings and cut off 4 others from road access. The average velocity of the flow moving down the 4-8° slopes of the subdivision was 56 m/hr, but the actual velocity ranged from 0-30 m/min. Periods of stagnation up to a few hours long alternated with rapid surges that advanced the flow front by 100-300 m in 30 minutes.

"At about 1600 on 29 June a satellite vent on the W flank of the main vent began erupting. For the next 24 hrs it supplied local pahoehoe flows that extended about a kilometer N and NE of the vent. Then, in mid to late afternoon of 30 June, the satellite vent stopped feeding flows to the north and began to feed an aa flow that extended 5 km SE along the SW edge of the episode 3 and 4 flows. It, too, was fed by a pahoehoe channel; the front of this flow advanced at average rates of 70-110 m/hr.

"Fountain activity at the episode 5 vents constructed a pair of juxtaposed spatter cones about 40 m high. Lava pond surfaces within the 2 vents were 20-30 m above the bases of the cones. Fountains played from the ponds, and spatter was ejected to maximum heights of about 50 m above the pond surfaces. Fountaining was more vigorous than in episode 4, which suggested that the episode 5 magma may have been less depleted in gas. Lava production at the vents stopped at 0717 on 3 July, nearly 4 days after the eruption began.

"Thermocouple measurements gave lava temperatures of 1127-1129°C. Basalt collected near the end of episode 5 may be compositionally different from lavas erupted in previous episodes. Millimeter-size olivine phenocrysts are abundant, and plagioclase phenocrysts are rare. Unfortunately, no temperature measurements are specifically correlated with these samples."

Robert Symonds measured a rate of SO_2 emission from Kilauea of 8000 t/d from the ground on 30 June and the same flux from the air on 1 July.

Deformation and Seismicity. "Water-tube tilt measurements in the summit region (at Uwekahuna) showed small but distinct periods of summit deflation that correlated with episodes 4 and 5. Minimum volume loss at the summit was estimated to be about 14×10^6 m^3 for episodes 4 and 5 combined. Cumulative deflation since early Jan was ~ 235 μrad; a minimum volume loss

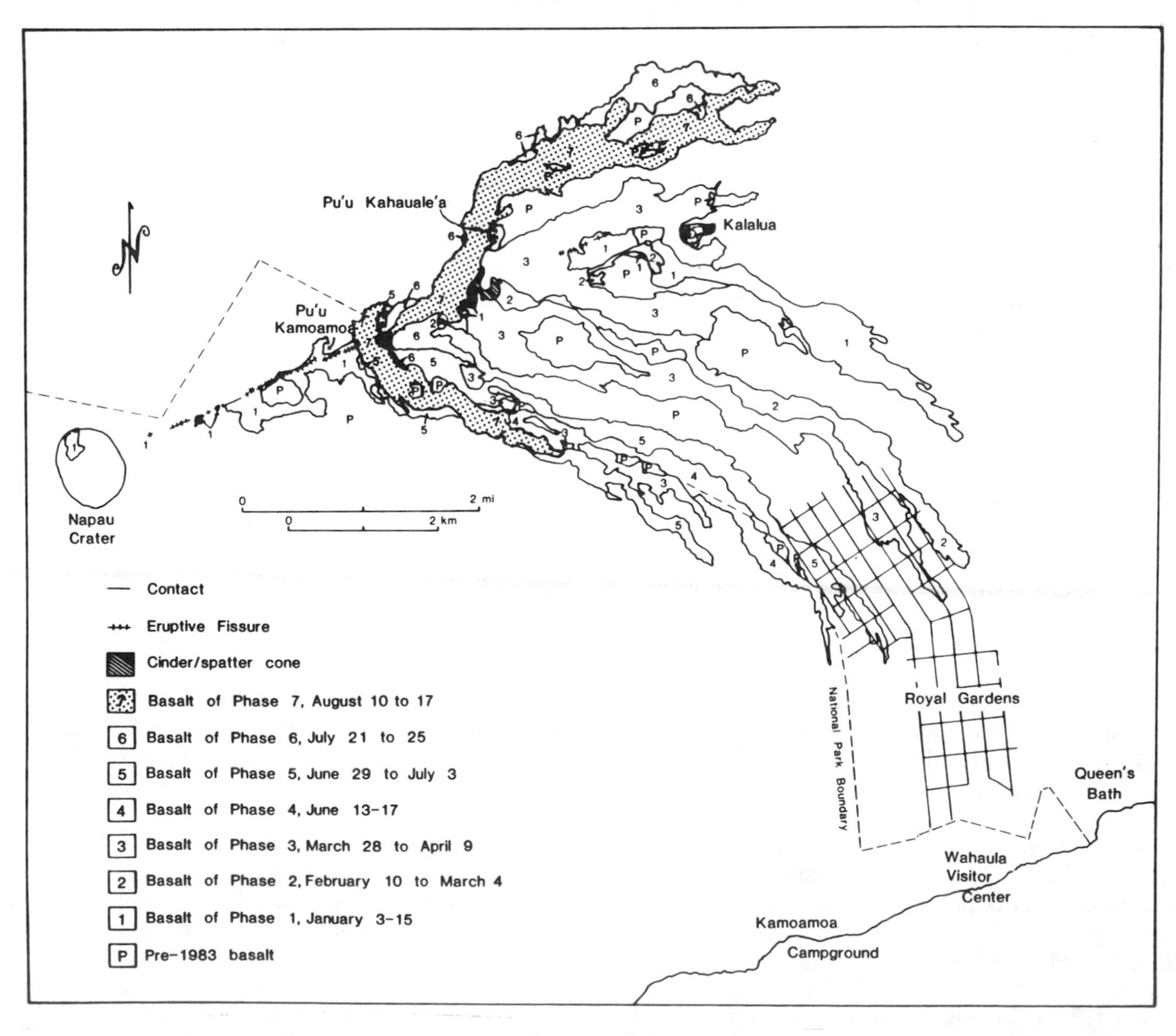

Fig. 13-22. Distribution of lavas and vent deposits for episodes 1-7. [Although the word "phase" in early reports has been changed to "episode" (see 8:3 and 10:6), we have not redrafted original figures to reflect this change.]

at the summit of about 95 $\times$ 10^6 m^3 is suggested.

"Very low-level harmonic tremor has characterized the periods between eruptive episodes. On 13 June, approximately coincident with the onset of episode 4, tremor increased from 0500 to about 1100. It remained constantly high until 17 June, when it declined rapidly from 1400-1600. Again, coincident with episode 5, tremor amplitude increased beginning at about 0900 on 29 June. It stayed high through the eruption, and, in concert with the end of lava production, the tremor dropped dramatically from 0713-0720 on 3 July."

Information Contacts: E. Wolfe, A. Okamura, R. Koyanagi, USGS HVO; R. Symonds, T. Casadevall, USGS CVO, Vancouver WA.

EPISODE 6

7/83 (8:7) "Kilauea's E rift zone eruption was in its sixth episode from 21 to 25 July. The eruptive vent was the same spatter cone, approximately 750 m NE of Pu'u Kamoamoa, that was the major source of episode 4 and 5 lavas (fig. 13-22). Between eruptive episodes the vent contained incandescent cracks and continued to emit magmatic gases.

"The first eruptive activity was seen from a passing aircraft at about 0600 on 21 July. From then until mid-afternoon the next day, extrusive activity consisted of cyclic filling and draining of the funnel-shaped interior of the spatter cone. Fountain activity at this stage ranged from intermittent bursts of spatter to the more steady play of a 3-4 m-high dome fountain.

"At approximately 1530 on 22 July, the pond filled to a depth of about 20 m, and lava spilled over low places on the rim of the spatter cone to begin feeding flows to the N, NE, and SE. Lava production rapidly increased, and a major aa flow advanced NE, fed by a vigorous pahoehoe river issuing from the pond at spillways on the N and NE flanks of the cone. Blocked from advancing SE by flows and vent deposits of earlier episodes, this flow moved ENE through the rain forest on the N side of Pu'u Kahauale'a. During the period of greatest lava discharge, estimated on 24 July at about 0.25 $\times$ 10^6 m^3/hr, the flow advanced at more than 200 m/hr. Ultimately it extended 6 km from the vent, covering an area of about 2 $\times$ 10^6 m^2 with a volume on the order of 10 $\times$ 10^6 m^3.

"During the period of active lava flow production, a fountain played continuously from the surface of the pond within the vent. The fountain was at its most vigorous on 23 July, when it often reached heights ranging from 50 to 150 m. It produced flows of spatter-fed pahoehoe and a local tephra blanket that extended SW from the vent. Subsequently the fountain was less vigorous and 30-60 m high, approximately 1.5 times its height at the end of episode 5.

"Lava temperatures measured by thermocouple ranged from 1128°C on 22 and 23 July to 1138°C (a new high temperature for the 1983 eruption) on 24 July. Hand-lens inspection indicated that episode 6 basalt contains scattered small phenocrysts of plagioclase and olivine. Thus, it generally resembles, except for the more olivine phyric basalt of episode 5, the earlier 1983 lavas.

"Harmonic tremor, which had persisted at a low level after the end of episode 5 eruptive activity on 3 July, began to fluctuate slightly and increase gradually during the period of cyclic filling and draining of the lava pond on 21 and 22 July. Shortly before 1600 on 22 July, tremor intensity increased rapidly in concert with increasing eruptive vigor. Tremor reached a high level at about 1800 and maintained it through the 3-day period of strong effusion. At about 1620 on 25 July, tremor decreased rapidly as the eruption came to an end. Low-level tremor, like that characteristic of other inter-eruptive periods, has continued in the vent area since 25 July.

"The Uwekahuna water-tube tiltmeter recorded about 17 μrad of summit deflation during episode 6. Following gradual inflation that had been underway since the end of episode 5, the summit of Kilauea began to deflate at about 1600 on 22 July, approximately coincident with the increases in tremor amplitude and lava emission. Summit deflation increased to a maximum rate of 3-4 μrad/hr on 23 and 24 July. The rate decreased thereafter until about 1800 on 25 July. Since then the summit has been gradually reinflating."

Addendum. Eruptive activity resumed at the episode 6 vent on 15 Aug. Harmonic tremor increased at approximately 0700; lava fountains 20-30 m high and a 1 km-long lava flow to the NE were reported by field crews that arrived at the vent at 0845. At press time on 17 Aug, Tina Neal reported that lava fountains had decreased to 5-10 m high, the lava flow extended about 6 km NE with lobes both N and S of the episode 6 flow, and summit deflation continued.

Information Contacts: E. Wolfe, A. Okamura, R. Koyanagi, T. Neal, USGS HVO.

EPISODE 7

8/83 (8:8) "The seventh major episode of extrusive activity in Kilauea's continuing middle E rift zone eruption occurred during the 3rd week of Aug. Lava flows extended NE and SE from the eruptive vent, but did not threaten any developed areas (fig. 13-22).

"After slightly more than 2 weeks of summit inflation, low-level extrusive activity was first observed on 10 Aug on the floor of the spatter cone (750 km NE of Pu'u Kamoamoa) that marks the vent for episodes 4-6 (8:6-7). Sporadic production of very small lava flows through 14 Aug, accompanied by weak spattering, produced enough new basalt to cover the 30 m-diameter floor of the crater with thin pahoehoe. During this period, the summit continued to inflate, and harmonic tremor, which had been low but steady in the eruptive zone since the end of episode 6, was marked by intermittent bursts of higher amplitude.

"Starting at about 0709 on 15 Aug, tremor intensity increased rapidly. Within an hour, the seismograph at Pu'u Kamoamoa registered a 10-fold increase in tremor amplitude. By the time the first observers reached the eruptive zone at 0850, fountains 50 m high were playing from the surface of a 20 m-deep pond within the crater. A rapidly moving lava flow had advanced 1 km NE on top of the voluminous episode 6 flow, and a smaller flow was moving slowly S. Steady lava production continued for about 57 hrs, ending abruptly at about 1600 on 17 Aug. An aa flow that had been fed by a vigorous pahoehoe river extended 6.4 km NE and buried a large portion of the episode 6 flow. A second aa flow that advanced less rapidly extended 2.9 km SE. Basalt samples contain scattered small olivine phenocrysts. Lava temperatures measured by thermocouple ranged from 1132° to 1141°C.

"Strong harmonic tremor persisted throughout the period of vigorous effusion. At about 1602 on 17 Aug, tremor decayed rapidly and within an hour assumed a fluctuating low level. Tremor amplitude slowly decreased even further over the next few days, then remained constant and low for the rest of the month.

"Summit deflation recorded by the Uwekahuna tiltmeter began at 1000 on 15 Aug, several hours after the onset of intense lava production. Nearly 20 μrad of summit collapse, which ended at 1900 on 17 Aug, suggests that a volume decrease of at least 8×10^6 m^3 occurred at the summit during [episode 7]. For the remainder of the month, the summit reinflated at an average rate of about 1 μrad/day. At least 110×10^6 m^3 of magma is estimated to have been lost from the summit reservoir since the E rift eruptive activity began in Jan 1983.

"On 2 Sept, field observations of the vent area indicated that small amounts of lava had been intermittently erupted within the main vent crater since the last field check on 29 Aug. There was some increase in the amplitude of low-level tremor during the night of 1-2 Sept, probably associated with the new lava emission.

EPISODE 8

"On the morning of 6 Sept the eighth major episode of the eruption began. At 0503, an increase in volcanic tremor was recorded on seismographs at HVO. By 0514 the tremor had increased to about 20 times its background level of the previous day. At 0530 a major fume cloud was visible from HVO and the roar of fountains was audible. At 0730 a field crew reported a large single fountain reaching approximately 50 m above the top of the spatter cone built in earlier episodes. A major flow was advancing NE on top of flows from the last 2 major eruptive episodes.

"The eruption was steady and vigorous throughout 6 Sept. Fountains reached heights of 140 m above the cone. Abruptly at 0525 on 7 Sept, fountaining ceased and tremor amplitude dropped to slightly above inter-eruptive background levels. Significant lava production during episode 8 lasted for approximately 24 hrs. Ten μrad of summit collapse were recorded."

Addendum. On 9 Sept at 0631 a magnitude 5.5 earthquake occurred at 9 km depth in the S flank, between the eruption vent and the sea coast. Low-level activity resumed at the episode 6-8 vent around midday on 15 Sept [episode 9]. Vigorous fountaining to ~100 m above the spatter cone began between 1700 and 1730 on 15 Sept, and produced a lava flow that extended 5 km NE on the N and W sides of the episode 6-8 flows. Fountaining ceased between 1900 and 1930 on 17 Sept.

Information Contacts: Same as 8:7.

9/83 (8:9) "Episodes 8 and 9 of Kilauea's long-lived middle E rift zone eruption occurred during Sept. All lava issued from the same vent, 750 m NE of Pu'u Kamoamoa, that has been the dominant locus of eruptive activity starting with episode 4 in mid-June. The major episode 8 and 9 flows (fig. 13-23) extended NE along the NW edge of the episode 6 and 7 flows.

Low-level Activity, 2-5 Sept. "In the pre-dawn hours of 2 Sept, following about 2 weeks of low but constant harmonic tremor in the eruptive zone and steady inflation of the shallow summit reservoir, approximately 4 days of low-level, intra-crater eruptive activity began at the vent. Harmonic tremor during this 4-day period, although still low, was characterized by intermittent episodes of slightly higher intensity. Prior to the renewal of activity, the crater floor was dominated by a steep-walled deep hole; an incandescent crack issued fume at the W base of the interior crater wall. By 0900 on 2 Sept, the deep central depression had been filled, and a 4 m-high mound of spatter surrounded a fuming and sporadically spattering conduit near the center of what had become a nearly flat, 30 m-diameter crater floor. At the W edge of the floor, several very small lava flows had been extruded from the area of the incandescent crack. Intermittent extrusion from this area continued through 4 Sept, producing about 300 m^3 of small solidified flows. On 5 Sept, a lava pond 5 m deep covered the crater floor, and a lava stream carrying an estimated 1000-2000 m^3/hr issued from the vent at the pond's W edge, disappearing downward near the pond's center. Although a 'bathtub ring' showed that the pond had earlier been as deep as 15 m, no significant change in the pond's volume was seen during 6 hrs of observation on 5 Sept. Minimal spatter production implied that the lava of this intra-crater activity was relatively degassed.

"Beginning at 0503 on 6 Sept, harmonic tremor rapidly increased. Within 15 minutes, tremor amplitude was an order of magnitude higher, and remained intense throughout [episode 8]. By about 0530 the roar from the fountaining vent could be heard and the eruption plume seen from the summit region.

"Although brief, the eruption was vigorous. Fountains rose 100-200 m from the surface of the pond within the crater. As in episodes 6 and 7, lava issued from a breach in the NE rim of the crater, producing an aa flow that extended NE for more than 4 km. A subordinate aa flow extended about 2 km SE.

"Harmonic tremor decreased rapidly from about 0520 on 7 Sept, and fountaining stopped at about 0525, approximately 24 hrs after it had begun. Steady low tremor continued until 14 Sept. Two open conduits were left extending steeply downward from the W edge and the center of the crater floor.

EPISODE 9

"After a week of quiescence, sporadic spattering began in the crater shortly after midnight on 14 Sept, and harmonic tremor gradually increased to mark the onset of episode 9. Lava began to cover the crater floor at about 1030 on 15 Sept. After an hour, when the pond was about 5 m deep, a few hundred m^3 of lava spilled through the deep breach in the NE crater rim. The remainder, about 3,000 m^3, drained back into the central conduit and disappeared at about 1300, 2 1/2 hrs after the pond began to form. Two hours later, a second pond, not deep enough to overflow the breach, formed and drained within 1/2 hr. Refilling of the pond began again about 5 minutes later, and within 15 minutes (by about 1540) lava had overtopped the spillway from the crater. A flow supplied at a rate of about 10,000 m^3/hr began advancing NE along the evacuated episode 8 channel. A small dome fountain 3-8 m high played on the pond surface above the central conduit, and there was almost no spatter. Beginning about 1700, harmonic tremor increased gradually. At about 1711, the first spatter was visible above the approximately 50 m-high W rim of the crater, and by 1730 the tremor had increased to the large amplitude typical of vigorous lava production. By 2130 the fountain was visible from the Wahaula Visitor Center near the coast (approximately 11 km SE of the active vent), which suggests a fountain height on the order of 300 m.

"Vigorous eruption continued for just over 2 days, ending at about 1915 on 17 Sept. The major flow of this episode, aa supplied during vigorous eruption at about 1-2 $\times 10^5$ m^3/hr, advanced more than 5 km NE through the rain forest along the NW edge of the episode 8 flow. Fountains were initially very high, but became erratic in height and direction, ranging up to about 200 m high. Rapidly changing in inclination and azimuth of trajectory, they heavily armored the growing spatter cone (about 60 m high) with spatter-fed flows.

"At the end of episode 9, 2 apparent conduits, much like those seen before it began, were visible in the crater floor. Near the center of the floor was a nearly cylindrical, vertical, fuming hole ~4 m in diameter. At the W edge of the floor was a fuming, glowing, elongated hole 1 m wide and 2-3 m long.

"Harmonic tremor in the eruption zone diminished rapidly at the end of the eruption. By 1930 on 17 Sept it had dropped to the low level characteristic of repose periods. Since then tremor has remained continuous and low.

"The volume of new basalt, mostly aa, produced during episodes 8 and 9 is $\sim 16 \times 10^6$ m^3, and the total volume of basalt produced since the beginning of episode 1 on 3 Jan is $\sim 106 \times 10^6$ m^3. The episode 8 and 9 basalts are slightly porphyritic, with scattered small olivine phenocrysts.

"Summit reservoir deflation and inflation associated with episodes 8 and 9 extended the strongly cyclic pattern that has been established in recent months. Deflation approximately coincident with the episodes was recorded at Uwekahuna vault by rapid inward tilt changes of 14.8 µrad for episode 8 and 16.2 µrad for episode 9. These correspond approximately to summit volume losses of 6 and 6.5 $\times 10^6$ m^3, respectively. The total deflationary volume loss at Kilauea's summit since the beginning of Jan is about 122 $\times 10^6$ m^3."

Addendum. The 10th major episode of eruptive activity began early 5 Oct following 3 days of minor and intermittent lava emission and drainback in the episode 4-9 vent. Tremor amplitude increased at 0327, and lava fountaining started by about 0430. Fountains rose to maximum heights of about 300 m, producing a substantial volume of airfall pumice, and fed 2 lava flows that extended 4 km NE and SE of the vent. Fountaining ceased at about 1700 on 7 Oct, and inflation of the summit region resumed. Extremely low-level harmonic tremor continued as of 11 Oct.

Information Contacts: Same as 8:7.

EPISODE 10

10/83 (8:10) "The 10th major episode of Kilauea's prolonged E rift zone eruption occurred in early Oct. As in episodes 4-9, the active vent was at the growing spatter cone, now tentatively named Pu'u O [later Pu'u O'o, see 10:4], approximately 750 m NE of Pu'u Kamoamoa (fig. 13-23).

"Low-level eruptive activity marking the onset of episode 10 began at about 0800 on 2 Oct when a small lava flow, about 300-500 m^3 in volume, was extruded onto the floor of the crater within Pu'u O. Some of the lava drained back into the open, glowing, 4 m-diameter conduit, which had been preserved near the center of the crater floor at the end of episode 9. No further activity was observed until 4 Oct when small bursts of spatter were intermittently emitted, most of which fell back into the conduit.

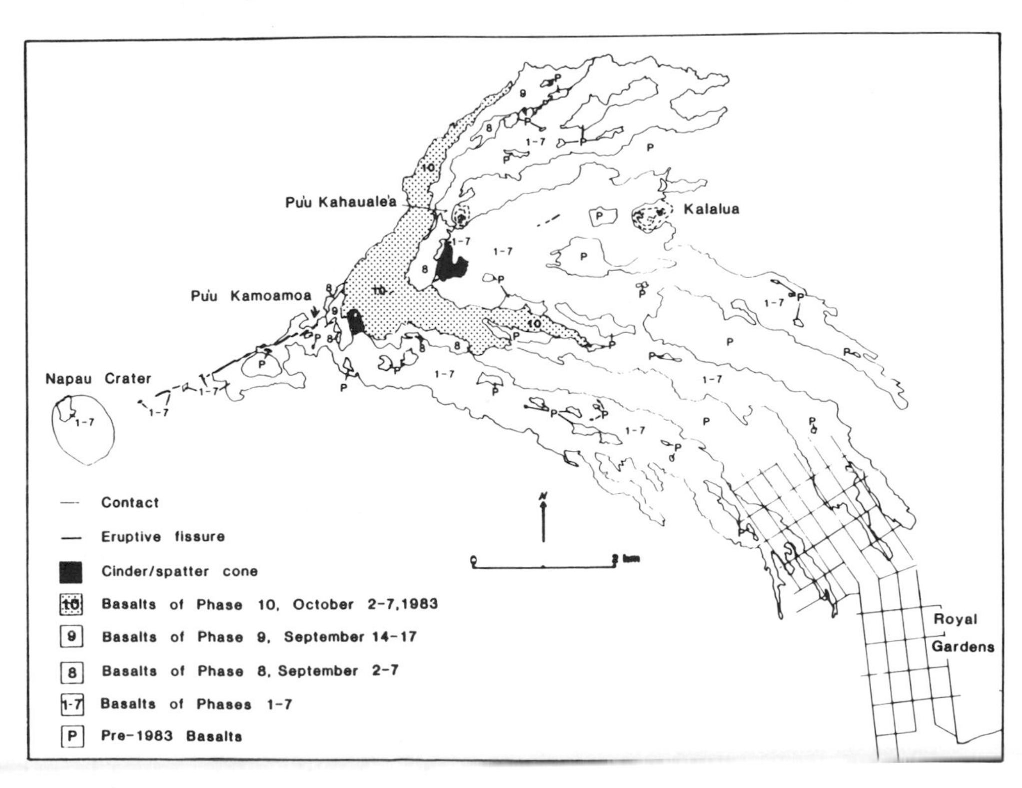

Fig. 13-23. Distribution of lavas and vent deposits of episodes 8, 9, and 10.

"Beginning at about 0100 on 5 Oct, harmonic tremor increased above the low repose-period background level, and lava began to spill through a deep breach in the NE wall of the spatter cone. Intensity of both harmonic tremor and eruptive activity increased over the next several hours. By 0330, tremor amplitude had increased by an order of magnitude, and by 0400 the glow was visible and the roar of the vent was audible 20 km away at HVO. Near dawn, the 300 m-high fountain was visible from the coast at the Wahaula Visitor Center (approximately 11 km S of the active vent). Late on 5 Oct, tremor amplitude increased to a maximum exceeding that of the previous eruptive episode, consistent with a trend of progressively increasing tremor that began with episode 4. High tremor occurred throughout vigorous episode 10 lava emission.

"After initially high activity, the fountains became erratic, changing rapidly in height from a few tens of meters to a maximum of about 250 m. At times, 3-4 distinct fountains were observed within the crater; sometimes high jetting fountains and a relatively low dome fountain played simultaneously from different parts of the lava pond surface.

"On 5-6 Oct, a thick slow-moving aa flow advanced 3.7 km ESE from Pu'u O, and rapidly moving pahoehoe formed superimposed sheets over an elongated area that extended about 2 km from the vent. By 7 Oct, the aa flow had stagnated, and the pahoehoe issuing to the NE had become confined to a channel in which a vigorously flowing lava river supplied a flow of pahoehoe and aa that eventually extended 4 km NE.

"Eruptive activity waned between about 1630 and 1700 on 7 Oct; harmonic tremor amplitude decreased rapidly at about 1650, and, after 1700, assumed the low level typical of repose-period activity. Such low-level tremor continued in the E rift zone for the remainder of Oct.

"As in the previous several episodes, this basalt is characterized by the occurrence of scattered small olivine phenocrysts. Lava temperatures measured by thermocouple ranged from 1134° to 1142°C; the latter is the highest lava temperature measured since the beginning of eruptive activity on 3 Jan.

"Episode 10 produced nearly 14×10^6 m^3 of lava for a total erupted volume since 3 Jan of approximately 120×10^6 m^3 distributed over about 23×10^6 m^2 of the rift zone. At the end of episode 10, Pu'u O was approximately 600 m in diameter and 80 m high. A chaotic jumble of disrupted blocks of agglutinated spatter covered the crater floor and the spillway through which the NE flow exited. The blocks, which were apparently transported as the last lava drained back into the eruptive conduit, choked the central conduit, which had stood open after episode 9. Oxidized fume issued from the margins of the crater floor and from the nearby parts of the interior walls of the crater but not from the rubble-covered floor.

"During episode 10, summit subsidence was recorded by nearly 19 μrad of ESE tilt change at the Uwekahuna vault on the NW rim of the caldera. This corresponds to a summit volume decrease of $\sim 7.5 \times 10^6$ m^3, bringing the total deflationary volume loss since Jan to about 130×10^6 m^3. Summit inflation from the end of episode 10 to the end of Oct was sufficient for the Uwekahuna tiltmeter to more than recover the deflationary tilt change of episode 10."

Addendum. Episode 11 began with an increase in harmonic tremor amplitude at 2314 on 5 Nov and the onset of eruptive activity around midnight. Summit deflation began at about 0200 on 6 Nov. An overflight at 0730 revealed a discontinuous line of fountains up to 40 m high extending ~ 200 m up and down rift from the spatter cone (Pu'u O) active in previous eruptive episodes. Vents high on the W flank of the spatter cone fed 2 lava flows that extended 4 and 1.5 km S before activity ceased there late on 6 Nov. The main vent, on the E summit and flank of the

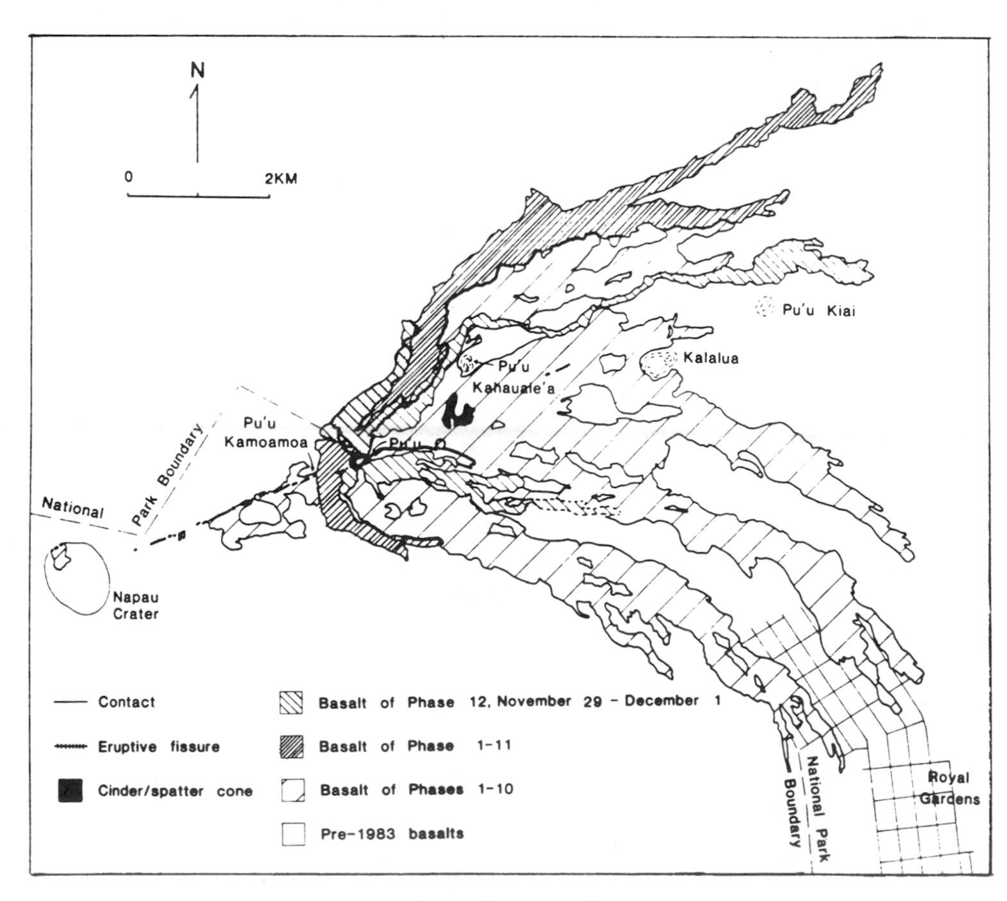

Fig. 13-24. Lava flows and vent deposits of episodes 11 and 12.

spatter cone, produced lava fountains 10-40 m high and fed the main lava flow that extended 8 km NE [it reached 9.5 km length; 8:11]. At 1840 on 7 Nov, the amplitude of harmonic tremor decreased sharply; lava fountaining from the main vent ceased later on 7 Nov, and inflation of the summit region resumed.

Information Contacts: E. Wolfe, A. Okamura, R. Koyanagi, USGS HVO.

EPISODE 11

11/83 (8:11) "The 11th and 12th eruptive episodes occurred on 6-7 Nov and 29 Nov-1 Dec. As in episodes 4-10, the growing spatter cone, Pu'u O, located just within Hawaii Volcanoes National Park, was the major eruptive locus.

"Unlike most of the previous eruptive episodes at Pu'u O, when vigorous lava production was preceded by several hours to several days of low-level activity within the crater, episode 11 began quickly. Harmonic tremor in the eruptive zone began to increase above normal repose-period background levels at 2314 on 5 Nov and reached a maximum amplitude at 0025 on 6 Nov. An electric trip wire, installed within the crater of Pu'u O to monitor the first appearance of lava, was broken between 2350 on 5 Nov, and 0000 on 6 Nov. By 0010, glow from the eruption was visible 20 km away at HVO. Between midnight and dawn on 6 Nov, low fountains were seen from Mountain View, 18 km to the N and from Kalapana, on the coast 14 km to the SE.

"When the first observers arrived at 0730, discontinuous low fountains played along a 700-m line approximately centered on Pu'u O. All of the fountains were relatively low. Those NE and SW of Pu'u O were about 10-20 m high. The most voluminous fountains rose about 40 m from the surface of a pond within the crater at Pu'u O; throughout the eruption they were barely visible above the 80 m-high crater rim.

"Pu'u O was the major locus of lava production. As in episodes 6-10, its lava pond overflowed through a breach in the NE crater wall, forming a vigorous lava river that eventually extended about 6 km before turning from pahoehoe to aa. By the end of episode 11, the river had fed a 9.5 km-long aa flow, the longest in the 1983 eruption, that advanced NE on top of earlier flows from Pu'u O, then extended an additional 4 km through dense rain forest (fig. 13-24). Vents 100-200 m SW of Pu'u O, in combination with a vent high on the W flank of the cone, produced pahoehoe flows that spread an estimated 1.2×10^6 m^3 of new basalt W and S of Pu'u O until the early morning of 7 Nov, when the W vents stopped erupting.

"During the 43 hrs of eruption, approximately 12.1×10^6 m^3 of lava was extruded, 85% in the long flow to the NE. In hand-lens view, the basalt is sparsely porphyritic with scattered olivine phenocrysts up to about 2 mm in diameter. Lava temperatures, measured by thermocouple in pahoehoe toes in flows from the W vents and in sustained overflows from the lava river, ranged from 1141 to 1144°C. Lower temperatures, 1133-1135°C, were measured at the river's edge.

"Harmonic tremor amplitude, sustained during the eruption at a level comparable to that of recent previous episodes, dropped rapidly between 1840 and 1850 on 7 Nov, at the end of the eruption.

"Summit subsidence began at about 0200 on 6 Nov, approximately 2 hrs after the onset of eruption, and continued until about half an hour after the eruption's end on 7 Nov. The deflationary tilt change of about 19 μrad at Uwekahuna corresponds with a summit volume loss of about 8×10^6 m^3.

16 Nov Earthquake; Repose Period Activity. "At 0613:01 on 16 Nov, a magnitude 6.7 earthquake occurred in the Kaoiki region (fig. 13-25) between Mauna Loa and Kilauea, an area of persistently high seismicity. Several tens of earthquakes are commonly recorded dai-

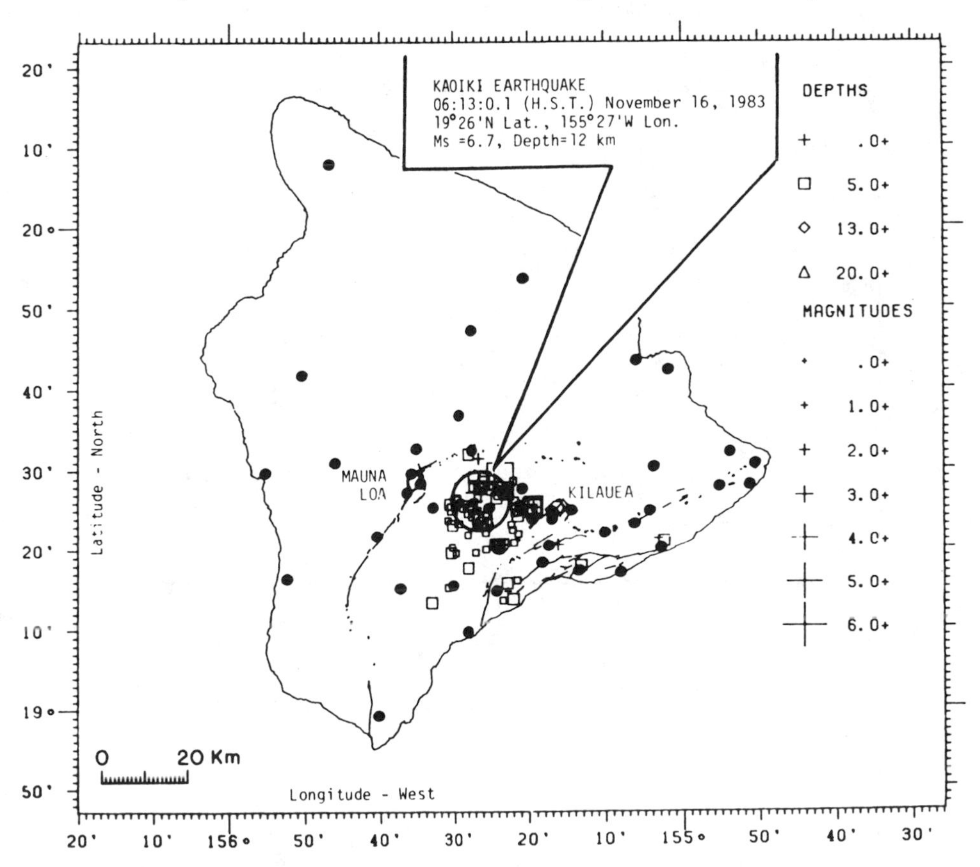

Fig. 13-25. Location of the M 6.7 Kaoiki earthquake and the initial 10 hrs of aftershocks. Local seismic data indicate strike-slip faulting with an E-W orientation of the pressure axis, and a N-S orientation of the tension axis. Darkened circles represent seismic stations that continuously telemeter signals to HVO.

ly by nearby seismic stations of HVO's 48-station island-wide network. Over the past 3 decades, about 35 earthquakes of M ≥ 4 have occurred in the Kaoiki region, two of the largest on 27 June 1962 (M 6.1), and on 30 Nov 1974 (M 5.5). The 16 Nov 1983 earthquake was highest in magnitude, ground breakage, and structural damage. Ground breakage related to shaking and gravitational adjustments on 16 Nov occurred along Kaoiki fault scarps and along Kilauea caldera faults. Landslides and rockfalls occurred on the steep caldera and crater walls, and on the (Hilina) escarpments on Kilauea's S flank. A 4.7 km-long zone of E-W-trending tension fissures has been identified as the possible main surface rupture. A N50°E line from the epicenter of the main shock intersects the fissure zone 5 km to the N. The fissure zone, ~50-100 m wide, is recognized as a series of en echelon offsets to the left. Thus far, 15-20 cm of southward extension has been measured on some E-W-trending extension cracks.

"The earthquake was 12 km deep, and a preliminary first motion analysis from local seismic data indicates a strike-slip mechanism. The faulting was a probable result of compressional stresses within the crust induced by magmatic activity generating lateral tensions from Mauna Loa and Kilauea. The NE-trending nodal plane with a right lateral strike-slip motion is preferred on the basis of aftershock distribution of past events.

"In 8 days following the main shock, more than 9000 earthquakes have been detected on stations a few kilometers from the aftershock zone. Over 800 of these, ranging in magnitude from about 1.0-4.3, were selected for processing by computer for location and magnitude. About 50 aftershocks were greater than M 3.0, and 3 were over M 4.0. Preliminary determinations of hypocenters during the initial 10 hrs of aftershock activity indicate an epicentral zone about 20 km in diameter centered between the summit areas of Mauna Loa and Kilauea. Focal depths are concentrated at about 10 km.

"Between episodes 11 and 12, low-level harmonic tremor typical of repose periods persisted in the eruption zone. Copious hot oxidized fume issued continuously from Pu'u O, maintaining incandescence in numerous openings. Neither tremor nor fume production changed recognizably in response to the 16 Nov earthquake. The earthquake, however, produced a major perturbation in the record of Kilauea summit tilt (fig. 13-26).

EPISODE 12

"Episode 12 began in earnest at 0447 on 30 Nov, and continued until 1545 on 1 Dec. Lava flows to the N, NE, and E originated from Pu'u O, where low fountains played on the S and N rims of the cone and within the breach on its NE flank. The main flow issued through the NE breach, which has been the principal spillway since late July. Advancing 8 km during the eruption, the front covered 1977 lavas S of the main episode 11 flow. Less voluminous flows issued from the S part of the vent and advanced about half the distance to the Royal Gardens subdivision.

"Lava apparently first reached the surface within the

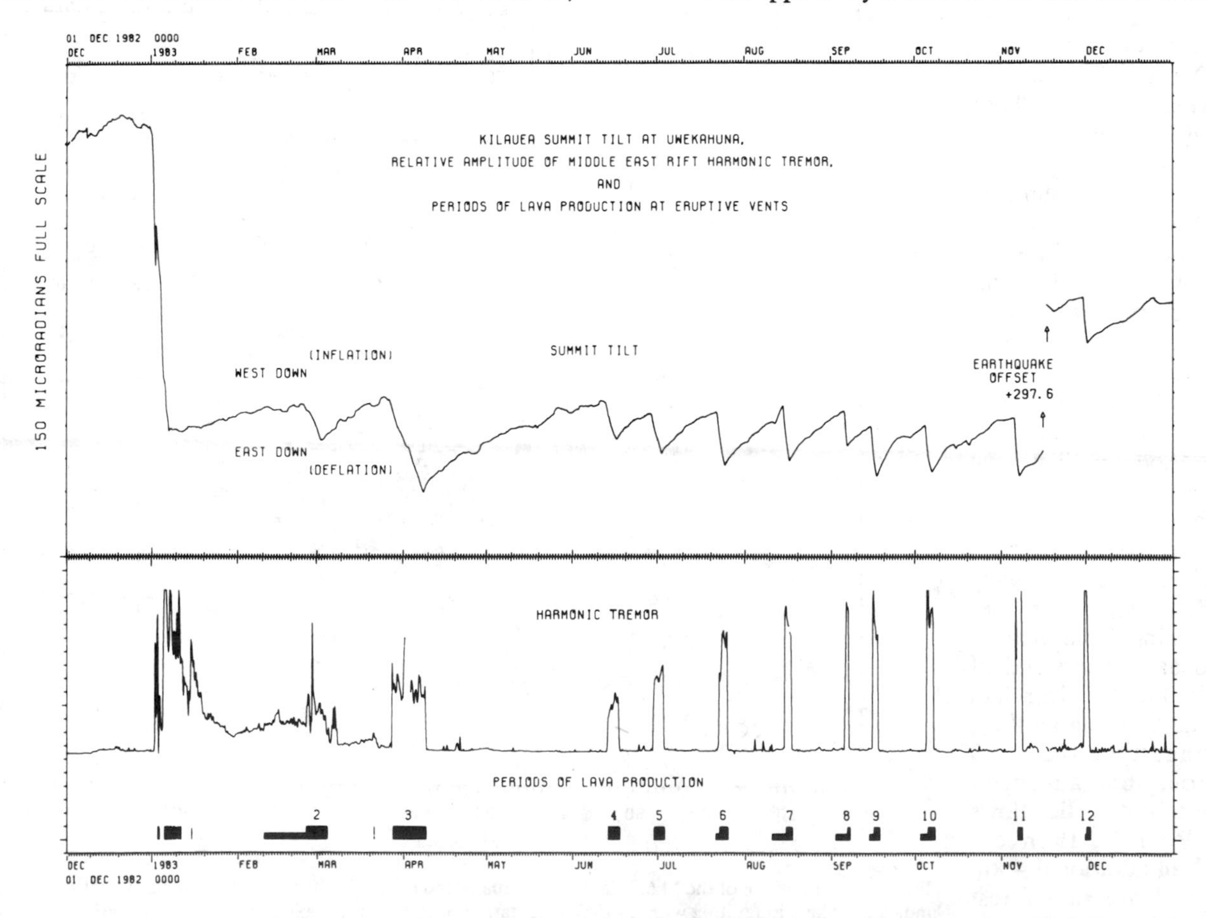

Fig. 13-26. Kilauea summit tilt, periods of lava production, and amplitude of harmonic tremor for episodes 1-12, 1983.

crater of Pu'u O at 1600 on 29 Nov, when an electric trip wire within the crater was cut. Faint glow in the eruptive zone was first reported at 2300. Intermittent low-level tremor near Pu'u O started to increase gradually at about that time (2300). The increase accelerated in the next few hours, and at 0447 on 30 Nov, a sharp increase was recorded, apparently marking the start of major lava emission. Visible fountains and audible roar from the vent were first reported at about 0450. Tremor peaked between 0530 and 0730 on 30 Nov with an amplitude more than an order of magnitude above background level. Tremor started to decrease rapidly at about 1545 on 1 Dec, marking the end of the eruptive episode. As of 5 Dec, the level of harmonic tremor in the E rift zone was very low but frequent microearthquakes were associated with near-surface activity at the vent."

Information Contacts: Same as 8:10.

12/83 (8:12) The 12th major episode ended on 1 Dec after 35 hrs of sustained lava production. Approximately 8.1 $\times$ 10^6 m^3 of lava was erupted, making episode 12 the second smallest of 1983, and bringing the total volume of lava erupted to about 140 $\times$ 10^6 m^3.

"Residents of Hilo (about 35 km NNE of the vent) first sighted visible fountains at about 0450 on 30 Nov, suggesting that fountains may have been higher during the early morning hours of 30 Nov than during the remainder of the eruption. When an HVO crew reached the vent at 0830, multiple low (10-60 m-high) fountains were playing within the crater of Pu'u O as well as on its N and S rims and in the breach in its NE rim. These changed little throughout the rest of episode 12. However, beginning shortly after midnight on 1 Dec, a vent just inside the N rim of Pu'u O produced a 50 m-high steam jet that distributed tephra about 1 km W and SW. The steam jet persisted until the end of episode 12 and was the first of its kind observed on Kilauea in 1983.

"A vigorous flow was fed by a lava river that issued steadily during the eruption from a major breach of the cone's NE flank, the dominant exit from the crater since the beginning of episode 6. Like the major flows of episodes 6-11, the main episode 12 flow advanced 2.5 km NE, then turned ENE along the S edge of episode 6-11 lavas, veered directly E, and eventually halted on top of 1977 lava 8.2 km from the vent. Smaller flows comprising about $^1/_3$ of the erupted volume extended N, S, and ESE of the cone (fig. 13-24).

"As the eruption stopped abruptly at 1545 on 1 Dec, harmonic tremor amplitude decreased (fig. 13-26). For the remainder of Dec, low harmonic tremor and shallow microearthquake activity persisted in the vicinity of Pu'u O. In early Jan, the interior of Pu'u O was a complex of craters separated by low septa. The cone was gradually cooling; incandescence that was conspicuous in numerous cracks and holes immediately after episode 12 had almost completely disappeared. A small amount of oxidized fume issued from the cone.

"As in previous recent episodes, hand samples of episode 12 are slightly porphyritic with scattered small olivine phenocrysts. Temperatures measured in actively advancing pahoehoe sheets near the vent ranged from 1135° to at least 1141°C.

"Minor summit subsidence as recorded by the Uwekahuna tiltmeter (fig. 13-26) began at 0200 on 30 Nov; vigorous subsidence began at about 0700, approximately 2 hrs after the onset of vigorous effusion at the vent. Summit deflation continued until about 1900 on 1 Dec. Total subsidence, measured E-W, was 13.9 μrad, suggesting a summit volume loss of about 6 $\times$ 10^6 m^3. By the end of Dec the E-W tiltmeter at Uwekahuna showed a net recovery of approximately 12 μrad, including a small deflationary episode late in the month.

"A 2 km-long zone of new steaming cracks parallel to the axis of the E rift was first observed on 17 Dec in the rain forest 7 km NE of Pu'u O. Ground observation suggested that at least 1 m of very recent extension perpendicular to the rift had occurred in the zone. The new cracks disappeared at their uprift end under episode 12 lava, 1 km NE of Pu'u Kiai, and the evidence suggests that the lava flow may have been broken locally by propagation of the cracks. However, the lava was also emplaced against fresh fault scarps that apparently had not moved subsequently. The cracks were within a narrow (approximately 25 m-wide) graben that has apparently been the locus of such extensional movements since prehistoric time. The zone of cracking was approximately on strike with the Jan 1983 eruptive fissure, but it was entirely downrift of the shallow earthquake swarm that recorded emplacement of the Jan 1983 eruptive dike."

Tom Casadevall and Barry Stokes made airborne measurements of SO_2 and CO_2 with a COSPEC and a Moran Infrared Spectrometer. On 2 Dec, about 16 hrs after the end of episode 12 lava production, SO_2 was being emitted at about 260 t/d from Pu'u O. On 9 Dec, the rate of SO_2 emission had dropped to 20 t/d. On both days, CO_2 production was below the 800-1000 t/d detection limit. Successive traverses through the plume in the south half of the summit crater yielded values of 3600 t/d of CO_2 and 300 $\pm$ 30 t/d SO_2 on 9 Dec.

Information Contacts: E. Wolfe, A. Okamura, R. Koyanagi, T. Neal, B. Stokes, USGS HVO; T. Casadevall, USGS CVO, Vancouver WA.

EPISODE 13

1/84 (9:1) The following reports are from HVO.

"Kilauea's episodic E rift zone eruption resumed in late Jan after 7 weeks of repose characterized by a desultory pattern of summit inflation, vanishingly weak harmonic tremor, and the virtual disappearance of magmatic gas at the vent. Episode 13 occurred 20-22 Jan, and episode 14 on 30-31 Jan. As in episodes 4-12, the eruptive vent for episodes 13 and 14 was at Pu'u O (fig. 13-27).

"Episode 12, which ended on 1 Dec 1983, left a complex of craters separated by low septa in the interior of Pu'u O. The largest and most central of the craters narrowed downward into a nearly vertical pipe about 20 m in diameter and at least 90 m long.

"On 20 Jan, harmonic tremor in the eruption zone increased gradually beginning at about 1030, and at 1117 HVO personnel first sighted moving lava about 50 m down in the pipe; by mid-afternoon the lava column could be seen slowly rising within the pipe. At 1724, the lava, which had by then filled the pipe and spread across the 30-40 m-wide floor of the crater, began to flow through the deep breach in the crater's NE rim. Supplied at a rate of approximately 10,000 m^3/hr (an order of magnitude less than normal minimum rates for vigorous eruption at Pu'u O), the lava began advancing NE in the channel evacuated at the end of episode 12. By this time, tremor amplitude had increased tenfold.

"As the lava rising within Pu'u O approached and overflowed the spillway, the lava surface, from which a low dome fountain repeatedly rose and fell, was agitated, resembling a rolling boil in a saucepan. Over the next few hours, the vigor of lava emission and fountain activity gradually increased, accompanied by increasingly intense harmonic tremor.

"Unlike some of the recent previous episodes, in which a complex of fountains formed within Pu'u O as well as on its rim and flanks, episode 13 developed a single fountain over the large central conduit. The production of lava, normally steady in previous episodes, pulsed strongly. At intervals on the order of $^1/_2$ - 1 minute the flux of lava in the channel and the height of the fountain waxed and waned. Hence, the fountain alternated rapidly from about the height of the crater rim (30-40 m above the pond surface within the crater) to as much as 40 or 50 m above the crater rim. The repeated increases in discharge of lava caused surges that advanced at about 10 m/sec down the first 100 m or so of the lava channel.

"Episode 13 effusion stopped temporarily on 22 Jan at about 0030. During this first 31-hr period the lava river debouching from Pu'u O fed a 7 km-long flow that advanced NE. It split into 2 lobes (at Pu'u Kahauale'a) that rejoined farther NE. Continuous eruptive activity resumed at about 0530; it was preceded by an hour of intermittent, low fountain activity. The fountain was visible above the rim of Pu'u O by 0550, and effusion accompanied by renewed intense harmonic tremor continued until the eruptive activity waned sporadically from 1115 to its termination at 1123. These 6 hours produced a second flow that advanced directly on top of the first; it followed the fork N of Pu'u Kahauale'a and stopped about 3 km NE of Pu'u O. The 2 effusive periods each terminated with [discontinuous] lava emission such that the fountain disappeared and reappeared repeatedly for a period of several minutes. Simultaneously, harmonic tremor decayed with marked alternations in amplitude.

Repose-Period Activity. "After episode 13, the interior of Pu'u O was a single bowl-like crater with a narrowing conduit, about 25 m wide at its mouth, extending downward from the crater floor. Throughout the entire repose period between episodes 13 and 14, the surface of a lava column was visible at depths of 0-25 m in the conduit. Sometimes fresh to barely crusted lava was exposed at the pond surface; at other times the pond surface was a solid frozen crust through which a small (0.5-3 m) orifice accommodated intermittent venting of magmatic gas, spatter, and small flows. At times, emission of spatter and gas through the small vent in the crust or alternating rise and fall of the fluid pond surface through a vertical interval of 10-15 m became strongly rhythmic with cycles about 4-6 minutes long. The activity closely resembled the gas-piston activity noted in Mauna Ulu lava ponds (1969–74). In a typical cycle at Pu'u O, the pond surface appeared deep in the conduit and slowly rose for about 4 minutes. It remained poised momentarily, became agitated, and the lava then drained rapidly out of sight in about $^1/_2$ minute. Draining was accompanied by emission of a plume of magmatic gas and a brief increase in tremor amplitude. After $1^1/_2$ minutes, lava would reappear deep in the conduit. When last visited in the early afternoon of 30 Jan, the lava pond had filled the steep-walled conduit and had begun to spread across the more gently sloping floor of the crater. About 30 m in diameter, the pond surface was about 5 m below the low point in the breach of the crater's NE rim. The surface was solid, and gas under pressure

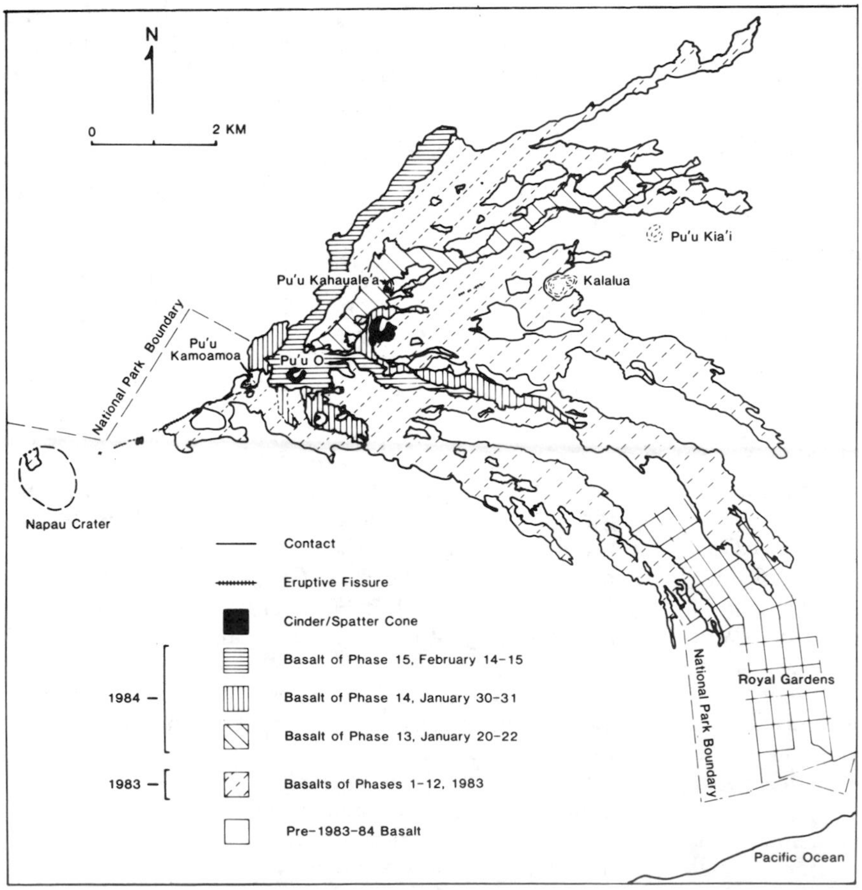

Fig. 13-27. Lava flows of episodes 13, 14, and 15.

along with minor spatter issued intermittently with a deafening roar from a 0.5 m-diameter vent in the crust. Gas samples from this vent showed that there had been no change in gas composition during more than a year of eruptive activity.

EPISODE 14

"At about 1030 on 30 Jan, the amplitude of harmonic tremor began a gradual and persistent increase that reached a high level by about 1830, when glow and visible fountains of episode 14 were first reported. Observations from a vantage point in the upper E rift zone as well as from the S coast of Kilauea indicate that in the evening hours the sustained height of the fountain at Pu'u O was 150-200 m above the rim of the cone, and bursts of spatter were rising as high as 300 m. By morning, when observers arrived at the vent, the single fountain, again centered over the large central conduit, was lower; through the remainder of episode 14 the fountain height fluctuated at intervals of 10-20 seconds from low (commonly 10-20 m above the rim of the cone) to high levels (up to about 80-100 m above the rim of the cone). The high bursts produced tephra plumes and short-lived spatter-fed flows on the flanks of Pu'u O.

"Flows to the N and E (fig. 13-27) were fed by distributary channels branching from the lava river that poured through the breached NE crater rim. The longest flow extended about 4 km from the vent. It turned SE and extended about halfway to Royal Gardens subdivision. A thick aa flow, fed entirely by spatter that cascaded over the S rim of the cone, advanced ~1.5 km S.

"Between 1315 and 1318 on 31 Jan lava emission declined, once again in spasmodic fits and starts. Harmonic tremor also decreased rapidly beginning at 1315, marking the eruption's end. In early Feb, the level of tremor, like that between episodes 13 and 14, was considerably higher than the background level during many of the previous repose periods.

"Following episode 14, the crater of Pu'u O was again a broad, steep-walled bowl from which a 20 m-diameter conduit extended nearly vertically downward. A 30-40 m-deep cleft in the NE crater rim marked the breach through which lava exited the crater.

Petrology. "Basalt of episodes 13 and 14 is sparsely porphyritic with scattered small olivine phenocrysts visible in hand-lens view. Lava temperatures measured by thermocouple were 1129-1131°C in the first few hours of episode 13. Subsequent episode 13 temperatures were 1140-1147°C, and include the highest temperature measured so far in this series of eruptions. During episode 14, temperatures, measured only at the edges of the widespread pahoehoe flow N of Pu'u O, were 1136-1137°C.

Deformation. "Rapid summit subsidence, as recorded by the Uwekahuna tiltmeter, began at 2100 on 20 Jan and again at 1930 on 30 Jan—in each case a short time after vigorous eruption was under way. Resumption of inflationary tilt followed the end of each eruption by several hours. E-W deflationary changes measured at Uwekahuna were 11 μrad for episode 13 and 10 μrad for episode 14. This suggests that a minimum of 8×10^6 m^3 of magma was withdrawn from the summit reservoir system during episodes 13 and 14."

Addendum: Episode 15 began at about 1945 on 14 Feb with a sharp increase in harmonic tremor. Summit deflation began about 2 hours later. During the evening, lava fountains rose as high as 300 m from the previously active vent at Pu'u O. Vigorous lava fountaining continued the next morning and lava flowed E and NE from Pu'u O. Eruptive activity ceased at about 1500 on 15 Feb.

Information Contacts: E. Wolfe, A. Okamura, R. Koyanagi, T. Duggan, R. Okamura, USGS HVO.

EPISODE 15

2/84 (9:2) The following report was received before the start of the 16th episode (see addendum).

"The 15th major episode of Kilauea's protracted series of middle E rift zone eruptions occurred on 14-15 Feb. As in recent previous eruptive episodes, the vent was at Pu'u O.

"Following episode 14, the more gently sloping floor of Pu'u O steepened downward near the crater's center to a nearly vertical 20 m-diameter conduit. Lava was not visible in the conduit on 1 Feb but was seen during aerial reconnaissance on 3 and 5 Feb. On 7 Feb, the partly crusted surface of a magma column in the conduit was 45 m below the crest of the spillway through the crater's breached NE rim. The column rose steadily at a rate of 4-5 m/day. On the morning of 13 Feb its upper surface was an actively roiled lava pond 18 m below the spillway. By the next morning, the lava column had risen enough for a 100-m flow to have spilled through the breach, after which it subsided to a level about 10-20 m below the spillway. A vigorous 19-hour eruption began during the evening of 14 Feb.

"Time-lapse film shows that steady low fountain activity within the crater began at about 1940 on 14 Feb. By 2000, the fountain height began to increase dramatically, and a vigorous lava flow issued NE through the breach. The fountain reached a maximum height of about 320 m at approximately 2100. It remained high for about 6 more hours and was easily visible from the coast S of Royal Gardens. The high fountain produced a thick tephra blanket on the SW side of the vent, and heavy tephra fall was reported by campers on the uprift side of Napau, about 5 km from the vent. The fountain height declined abruptly (by approximately 50%) at about 0330 on 15 Feb.

"When HVO observers arrived at about 0700 on 15 Feb, the fountain typically rose about 100 m above the rim of the cone. However, pulsations in its height ranged 20-200 m above the rim. The major lava flow, debouching through the long-lived breach in the crater's NE rim, extended 2 km to the NE. A stagnating aa flow, which had apparently been fed during the night by spatter falling over the S rim of Pu'u O during the more vigorous fountain activity, extended E about 3 km.

"By the end of episode 15 at 1501 on 15 Feb, the main flow had extended about 5 km NE along the NW edge of flows erupted July – Dec 1983. Approximately 8×10^6 m^3 of new basalt was extruded; the total volume emplaced at the surface since the beginning of Jan 1983 is about 165×10^6 m^3. As in recent previous episodes, the basalt is slightly porphyritic with scattered olivine phenocrysts of ~1 mm size. Lava temperatures measured by thermocouple (in thin pahoehoe overflows from the lava river that supplied the NE flow) were 1136°-1140°C.

"Steady, low harmonic tremor originating in the vent area was continuous after the end of episode 14. A slight increase in tremor amplitude related to the onset of episode 15 occurred from about 1939 to 1945 on 14 Feb. After a 1-min decrease, increase in tremor amplitude resumed at 1946. The amplitude increased rapidly to the maxima reached in previous eruptive episodes, and high tremor continued until the end of eruptive activity. After episode 15, low harmonic tremor continued steadily through the remainder of the month.

"Approximately 7 μrad of inflationary E-W tilt change was recorded in the NW part of Kilauea's summit region between the end of episode 14 and the onset of episode 15. Deflationary tilt change of 11 μrad, related to the extrusive activity, was recorded from 2030 on 14 Feb to 2100 on 15 Feb. Summit reinflation (of nearly 10 μrad) continued from 15 Feb through the end of the month.

"The morphology of Pu'u O's crater and vertical conduit was essentially unchanged by episode 15 eruptive activity. A partly crusted lava surface deep within the conduit was sighted from the air on 22 Feb. It was not visible again until 28 Feb, when it had risen to about 30 m below the surface of the spillway.

"Unlike many of the earlier repose periods, during which magmatic gases at the vent were highly oxidized or so dilute as to be practically undetectable, the repose periods preceding and following episode 15 have been characterized by the continuous emission of fresh, reduced, SO_2-rich magmatic gas. This may reflect the uninterrupted exposure of the active top of the magma column within the conduit (no plug of rubble or solidified lava above the column). Repose-period flux rates of such reduced gas are less than 1% of eruption-related flux rates."

Addendum: Episode 16 began at approximately 1500 on 3 March and ended at about 2230 on 4 March. Lava fountains rose as high as 200 m above the top of Pu'u O and produced a large amount of tephra. The main lava flow, fed by a vigorous pahoehoe river, extended 8 km ESE, reaching the NE corner of Royal Gardens subdivision atop episode 2 lava. Spatter-fed lava flows extended W and N of the vent.

Information Contacts: E. Wolfe, A. Okamura, R. Koyanagi, USGS HVO.

EPISODE 16

3/84 (9:3) "The 16th and 17th major episodes of the eruption occurred 3-4 and 30-31 March. Pu'u O was the eruptive locus for both episodes. Simultaneous eruptions on 30 March at Mauna Loa, Mt. St. Helens, Veniaminof, and Kilauea make this the first date known on which four U. S. volcanoes were erupting at the same time.

"Beginning on 28 Feb, lava was visible continuously within the upper 30 m of the vertical 20 m-diameter pipe that extended downward from the bowl-like crater within Pu'u O. Sometimes completely open and sometimes partly crusted, the surface of the magma column rose slowly over the next 4 days to the level of the spillway in the deep breach in the NE rim of Pu'u O. Minor amounts of spatter and small volumes of SO_2-rich gas issued from the lava surface.

"Time-lapse camera data indicate that the lava ponded within Pu'u O first overflowed the spillway on 3 March at about 1450. The vigor of fountain activity increased from that time, and the fountain became visible above the rim of Pu'u O at about 1520. By 1700, when HVO personnel arrived, the fountain rose 200-250 m above the lava pond (the crater rim was about 40 m above the pond). Glowing tephra was at times wafted to twice the height of the fountain by convective air currents rising over the cone. Erratic winds distributed tephra on all sides of the vent and at times during the night, intense tephra falls, including incandescent bombs up to 20 cm in diameter, rained on observers 750 m uprift of the vent. At about 0400 on 4 March, fountain activity diminished greatly and became sporadic. During the remainder of the eruption the fountain played to heights ranging from about 20-150 m above the pond.

"The intense fountain activity produced a thick spatter-fed aa flow that advanced about 1 km N; smaller spatter-fed flows extended E, SE, and W. The major flow, dominantly aa, was fed by a vigorous lava river debouching through the breach in the NE rim of Pu'u O at an average of about 0.25×10^6 m^3/hr. It advanced E and ESE, mostly on top of flows from earlier episodes. Approximately 6.4 km from the vent, it split into 2 lobes, the longer of which extended another 1.6 km SE across the NE corner of Royal Gardens subdivision. This lobe, however, was contained within the evacuated channel of the episode 2 flow that invaded the subdivision in March 1983, and caused no new damage. After a momentary pause at 2228, the eruption stopped abruptly at 2231 on 4 March. Occasional small bursts of spatter issued from the vent over the next 10 or 15 minutes.

"Basalt produced during episode 16, as in previous episodes, is sparsely porphyritic. In hand specimen, scattered, small (millimeter-size) olivine phenocrysts are visible. Lava temperatures, measured by thermocouple, were 1139-1142°C in pahoehoe within 1.5 km of the vent and 1135-1138°C at the advancing distal end of the main flow in aa and local breakouts of viscous pahoehoe.

"The volume of lava erupted during episode 16 was $\sim 12 \times 10^6$ m^3. The corresponding summit collapse caused almost 15 μrad of deflationary tilt change at Uwekahuna. As in previous episodes, the beginning and end of measurable summit deflation lagged slightly behind the onset and termination of lava production at the vent.

"Harmonic tremor associated with episode 16 began to increase gradually in amplitude on 3 March at 1435, and rapid increase began at 1504. Sustained, high-level tremor continued in the eruption area until 4 March at 2229 when the intensity began to pulsate and gradually decrease. Rapid decrease in tremor amplitude began at 2231, coincident with the end of lava production.

Repose-Period Activity. "The repose period between episodes 16 and 17 was marked by an unusually gradual reinflation of the summit, slightly over 9 μrad of inflationary tilt change in 3 1/2 weeks. The upper surface of the gradually rising magma column was first seen at the vent on 20 March, when it was approximately 50 m deep inside the pipe that extends downward from the crater floor, and 60 m below the spillway through the NE crater rim. Intermittent observations showed that the column rose slowly until the onset of vigorous lava production on 30 March.

EPISODE 17

"Harmonic tremor related to episode 17 began to increase 30 March at 0510, and glow from the eruptive area was visible from Kilauea's summit at 0515. A photographer camped near the vent reported seeing a low dome fountain in the crater and a short NE-moving lava flow at about 0520-0530. He estimated that the flow was about 0.5 km long by 0545, and that by about 0610 the fountain top was at the level of the rim of Pu'u O (about 40 m above the surface of the overflowing lava pond).

"HVO personnel arrived at about 1000. From then on, they observed the fountain playing in a rapidly pulsating fashion (up to 20 pulses per minute); the fountain height ranged about 40-160 m above the surface of the lava pond. A narrow lava flow, fed by a voluminous lava river discharging through the breach in the NE crater rim, was more than 1.5 km long at 1000. It extended ENE (to the spatter/cinder cone south of Pu'u Kahauale'a) then, following the episode 16 flow, turned ESE. During the day the flow continued advancing ESE on top of the episode 16 flow at 0.5 km/hr. During the night it followed the northern episode 16 lobe N of Royal Gardens. Near the terminus of that episode 16 lobe, the episode 17 flow turned SE and followed a gully parallel to and 1 km NE of the subdivision's E boundary. When seen the next morning, 3 1/2 hrs after the eruption's end, the still-advancing aa flow front was ~3 km from the ocean. A preliminary estimate suggests that this flow is > 10 km long, the longest of the 1983–84 eruptive series (fig 13-28). Episode 17 stopped on 31 March at 0324. The sustained high-amplitude tremor characteristic of Pu'u O eruptions began to diminish in intensity at 0316 and by 0324, when lava production stopped, the tremor amplitude had nearly reached the repose-period background level.

"The episode 17 basalt, like the episode 16 basalt, is slightly porphyritic with scattered millimeter-size olivine phenocrysts. However, thermocouple temperatures measured in overflows at the edge of the lava river range from 1131° to 1137°C, distinctly lower than the episode 16 temperatures.

"Measurable summit deflation began about 2 1/2 hours after the onset of lava production and ended nearly 4 hours after the termination of lava production. Deflationary tilt change at Uwekahuna was about 10 μrad. A preliminary guess is that the volume of episode 17 lava is significantly less than the episode 16 volume."

Information Contacts: Same as 9:2.

EPISODE 18

4/84 (9:4) "After 19 days of repose, the eruption resumed at Pu'u O at approximately 1800 on 18 April. Following nearly 60 hours of continuous lava emission, vigorous lava production stopped at 0533 on 21 April. Episode 18 was the longest eruptive episode of 1984 on Kilauea, and undoubtedly was one of the more voluminous of the 1983–84 series. About 200 × 10^6 m^3 of basalt has been produced since the eruption began on 3 Jan 1983.

Eruption Narrative. "After episode 17 ended on 31 March, the crater of Pu'u O was again a steep-walled, nearly flat-floored bowl about 50 m deep and 120 m across at the crater rim. As before, a nearly vertical 20 m-diameter pipe descended from near the center of the crater floor. Visibility down the pipe was obscured by fume in the days immediately following episode 17; exposed lava was first seen tens of meters down the pipe on 5 April. The lava surface, largely crusted much of the time, rose higher in the pipe over the following days; on 17 April, when last observed prior to episode 18, the crusted lava surface was < 10 m below the top of the pipe.

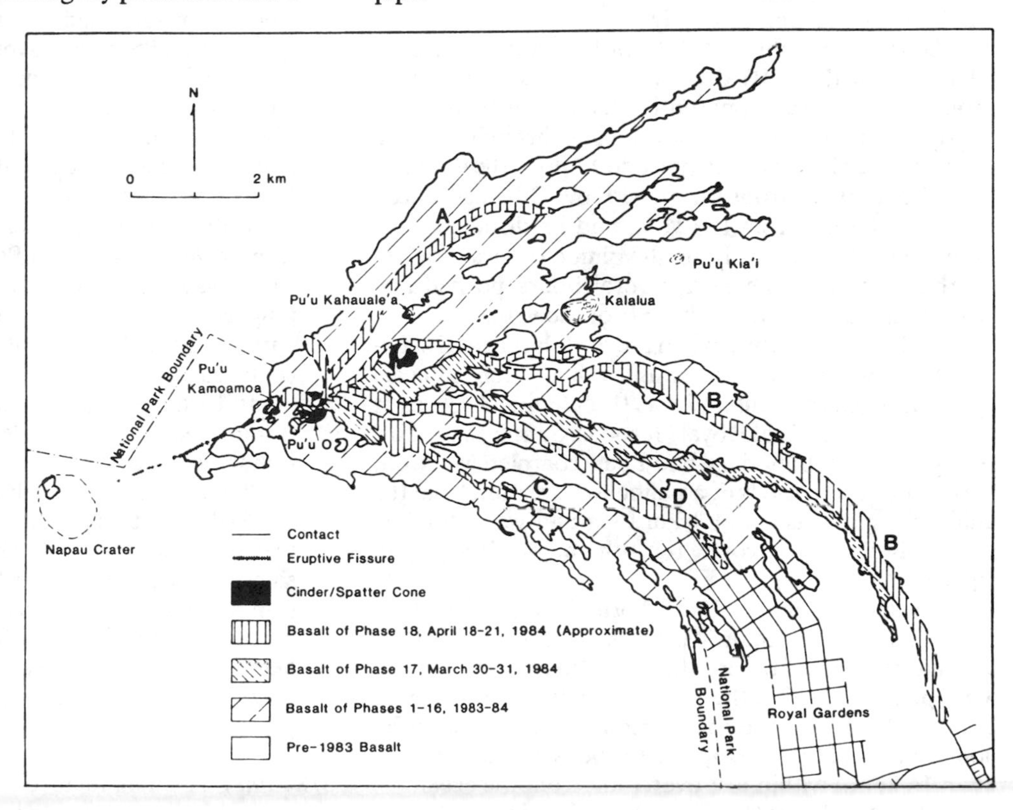

Fig. 13-28. Lava flows of episode 17 and approximate distribution of the 4 flows of episode 18.

"A time-lapse camera set at a 1-minute interval recorded the first low fountain activity within Pu'u O at 1751 on 18 April. At about 1800, lava was first recorded passing through the deep breach in the crater's NE rim. Intensity of activity increased, and by approximately 1900 the fountain crest was about 40 m above the surface of the lava pond within Pu'u O. Bad weather obscured visibility through much of the night, but frames exposed during brief clear periods suggest that the fountain was variable and at times was at least 200 m high. During the night, 10-20 cm of lapilli were deposited near Pu'u Kamoamoa (less than 1 km W of the vent), and Pele's hair fell at HVO, 20 km W of the vent. In the early morning the fountain height decreased, and for the rest of the eruption it was about 40-80 m above the pond. Major fountain activity and lava production stopped at 0533 on 21 April; emission of minor spatter continued for another hour.

"Episode 18 produced 4 major aa flows (fig. 13-28), all fed by distributaries of the lava river that debouched vigorously and steadily through the breach in the NE crater wall. Three flows were active when the first observers arrived at 0700 on 19 April. Flows A and B were receiving the bulk of the lava, and their active fronts were 3 and 4 km from the vent. A small flow (C), extending SE along the Hawaii Volcanoes National Park boundary, was about 1.5 km long.

"By 1800 on 19 April, flow A, 5 km long, had stopped, its supply cut off. Most of the lava was being channeled into flow B, which had become the main flow. It was about 8 km long and had advanced at an average rate of more than 300 m/hr to a point N of Royal Gardens subdivision. Flow C, along the park boundary, was about 3 km long and advancing slowly.

"During the night of 19-20 April, the rate of advance of the main flow (B) decreased, although the supply from the vent was apparently steady. By 0730 on 20 April, the flow front was advancing through the rain forest at the E edge of the episode 17 flow; the terminus was about 10 km from the vent. Flow C had advanced to about 4.5 km from the vent, and its distal end had stagnated. The relatively small volume of lava supplied to this flow over the next several hours ponded and spread laterally about 1.5 km SE of the vent.

"At 1730 on 20 April, when the day's last aerial reconnaissance was made, the main flow had extended nearly 14 km from the vent and was threatening 3 houses at the end of a road E of Royal Gardens. Supply to the flow along the park boundary had been completely cut off. However, just to the NE, a pahoehoe overflow from the main channel was about 1 km long and was extending SE. This was to become the 4th major flow (D) of episode 18.

"During the night of 20-21 April, the main flow slowly advanced more than 300 m and overran 2 of the 3 threatened houses before the eruption stopped at 0533. Creeping slowly after the eruption ended, the flow overran the 3rd house the following afternoon. This is the longest flow of the 18 eruptive episodes, extending more than 14 km to within just over 1 km from the ocean.

"The 4th flow (D), which had originated as a pahoehoe overflow in the late afternoon of 20 April, apparently captured a greater part of the supply during the night. Advancing at an average rate of between 400 and 500 m/hr, it extended more than 7 km and entered the N part of the Royal Gardens subdivision. Observed in Royal Gardens in the early morning hours, it was thin and fluid, moving more like pahoehoe than like aa.

"The episode 18 basalt is similar to that of other recent episodes in hand-lens view and in lava temperature. Scattered small olivine phenocrysts comprise less than 1% of the rock, which is otherwise aphyric. The maximum temperature measured by thermocouple in pahoehoe was 1144°C.

"At the eruption's end, debris from the crater walls plugged the opening of the pipe. However, by 23 April the pipe was once again open and vertical, and lava was seen deep (tens of meters) within it. Intermittent sightings of lava deep within the pipe have continued. Exact determinations of the depth to the lava are prevented by fume and by excessively hazardous access to the crater floor.

Seismicity and Deformation. "Harmonic tremor, centered near Pu'u O, remained at a low level from 1-18 April. During this time, periods (up to several days long) of constant low amplitude alternated with periods of variable amplitude that fluctuated on a time scale of a few minutes to several hours. Starting from about 0245 on 18 April, 1-2-minute bursts of stronger tremor occurred at intervals of 15-30 minutes. By 1200, these bursts were longer, and eventually the activity became constant. Tremor amplitude increased sharply from 1810. At 1848 the seismic alarm was triggered, and shortly thereafter the tremor had increased to levels comparable to earlier major episodes of the eruption. High tremor was sustained throughout the period of vigorous eruption. At 0535 on 21 April, tremor decreased abruptly, following about 15 minutes of spasmodic decay in amplitude. The seismicity resumed its typical low interphase level characterized by fluctuating weak tremor and tiny shocks, associated with near surface events at Pu'u O. From 0520 to 0545 on 22 April, a flurry of seismic events resembling rockfall signatures, and increased tremor were recorded at Pu'u Kamoamoa seismograph. The episode may record collapse of the rubble that plugged the pipe inside Pu'u O. This burst of activity occurred about 3 hours after two S flank earthquakes centered about 10 km SW of Pu'u O: at 0223, M 3.8, depth 10 km; and at 0226, M 3.9, depth 8 km. Harmonic tremor continued at low level near Pu'u O through the remainder of the month.

"Summit subsidence during episode 18, measured in the E-W direction by the Uwekahuna tiltmeter, was between 19 and 20 μrad. This is the largest subsidence since episode 3, 28 March-9 April 1983. Again, onset of summit deflation and resumption of inflation as recorded by the tiltmeter followed slightly behind the beginning and end of lava emission."

Information Contacts: Same as 9:2.

EPISODE 19

5/84 (9:5) "A small, relatively low-level eruption at Pu'u O on 16-17 May comprised an effusive episode in the eruption that began on 3 Jan 1983. Unlike the preceding eruptive episodes, which have been characterized by generally steady, voluminous lava emission and continuous vigorous fountaining, episode 19 was characterized by intermittent low-volume overflows from a lava pond in the crater of Pu'u O and a few brief periods of increased lava emission and high fountaining.

"After episode 18, and continuing through late April and the first half of May, the top of a magma column, partly crusted much of the time, remained tens of meters deep in the 20 m-diameter pipe that descends from the floor of the Pu'u O crater. At times, low-level spattering could be seen within the pipe from passing aircraft.

Eruption Narrative. An observer on the ground in the eruption area saw intermittent spatter at Pu'u O beginning shortly after 0200 on 16 May. At about 0500, the first of many small pahoehoe flows of episode 19 (fig. 13-29) spilled through the deep breach in the NE part of the crater rim. The flows typically lasted about 3-30 minutes and carried lava at about 10^3-10^5 m^3/hr. Intervals ranging from 4 minutes to several hours separated these small effusive events.

"During episode 19 an active lava lake was maintained within Pu'u O. At times the lake was confined to the upper part of the pipe; at other times it rose and spread across the crater floor. Flows issued quietly from the crater whenever the pond rose high enough to overflow the breach. A dome fountain played intermittently on the pond surface above the pipe, particularly during times of more rapid lava supply.

"In addition to the numerous, relatively quiet overflows of the pond, there were 4 periods, 1-3 hours long, of high fountaining with more vigorous lava production estimated at 1-2 × 10^5 m^3/hr. A fountain 30-100 m high played on the pond surface during these periods on 16 May at about 0940-1226, 1510-1608, 1840-2020, and 2355-0125 on 17 May. The first high fountain event was marked by rapid 2- to 3-fold changes, about a minute apart, in fountain height. These were followed within seconds by similar changes in lava discharge through the breach. In addition, this nearly 3-hour episode stopped and restarted 14 times. The pauses ranged in duration from a few seconds to 4 minutes. Vigorous lava emission and high fountaining would stop and start almost instantaneously at those times.

"During the evening of 17 May, the overflows from Pu'u O were smaller and more uniform in duration and spacing than previously. Several observations suggested that regular gas-piston activity had become established. During 3 hours of careful timing, the overflows averaged about 5 minutes in duration and occurred in cycles with an average length of 12 minutes. In each a [brief gas] burst was recognized immediately before the overflow stopped. Vigorous drainback into the pipe at the termination of overflow was seen within the crater in the late afternoon; presumably such drainback events, initiated by gas bursts, terminated each of the evening overflows. In addition, the distinctive bursts of tremor that we have come to associate with the drainback episode of gas-piston events were occurring with the same regularity as the overflows. At about 0030 on 18 May, the tremor bursts diminished in amplitude and became exceedingly regular, suggesting that the overflows probably stopped at that time.

"The repeated overflows from the crater built a relatively smooth apron of dense pahoehoe within a kilometer of the E base of Pu'u O. The higher lava emission associated with the periods of strong fountaining produced narrow, thin flows of pahoehoe and some aa that extended 2-3 km E and SE of the vent. Total estimated volume of new basalt is about 10^6 m^3.

"The lithology and lava temperature resembled those of previous episodes. The basalt is nearly aphyric with scattered small olivine phenocrysts. Temperatures measured by thermocouple in overflows were 1138°-1141°C.

"Measurements of SO_2 flux during episode 19 gave

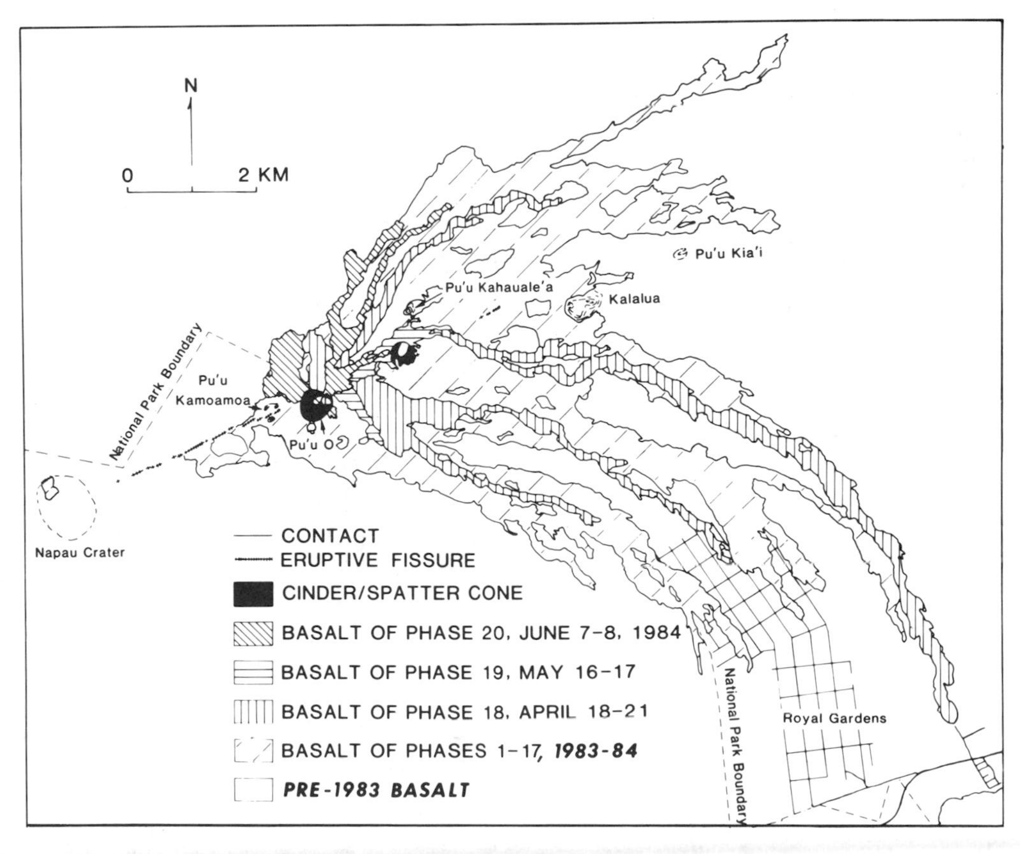

Fig. 13-29. Lava flows erupted since 3 Jan 1983 on the E rift zone. Flows from episodes 18-20 are shown separately.

surprising results. Even during the periods of high fountaining, it was comparable to that measured during repose periods when the magma column was visible within the pipe. That value was about 2 orders of magnitude less than the SO_2 flux measured during previous eruptive episodes.

Seismicity and Deformation. "After episode 18, harmonic tremor centered at Pu'u O remained low and continuous. Intermittent high tremor was associated with renewed fountaining during episode 19. On 16 May at 0640, background tremor began to increase rapidly, initiating a series of alternating [periods of] high- and low-amplitude tremor episodes at intervals of several minutes to several hours. Periods of relatively high and sustained tremor (0940-1255, 1423-1625, and 1857-2025 on 16 May; and 2351-0135 on 16-17 May) coincided closely with the intermittent high fountain events seen at Pu'u O. After the high fountain events, tremor of intermediate amplitude continued at irregular intervals, coincident with overflows at Pu'u O, until 1600 on 17 May.

"Tremor amplitude shown in fig. 13-30 is misleadingly low. Amplitudes during the high fountain events were comparable to those of previous episodes. However, the events were intermittent and of short duration, and they occurred between the programmed sampling intervals on which the tremor-amplitude plot was based.

"The Uwekahuna E-W tiltmeter, above the NW flank of Kilauea's summit reservoir, showed that slow deflation of the summit began at about 1100 on 15 May, 18 hrs before the first lava overflowed from the crater. Following a gradual 1.5-μrad deflationary tilt change, the rate of summit deflation increased, after high fountain activity had begun. By 0600 on 17 May, when deflation stopped, an additional 4 μrad of deflationary tilt change had been recorded. Gradual reinflation began at about 1100 on May 17, half a day before the overflows at the vent stopped.

Post-Episode 19 Activity. "From 17 May until the end of the month, the tremor pattern was dominated by cyclic low-level tremor. Half-minute bursts of higher amplitude alternated with 8-12-min intervals at lower amplitude. Repeated observation at the vent showed that the short bursts of higher tremor occurred during the gas burst and the accompanying vigorous drainback in gas-piston events.

"The pattern of periodic half-minute tremor bursts was temporarily interrupted by more continuous intermediate-level tremor from 1020 on 28 May to 2330 on 30 May. Observation at the vent on 29 and 30 May showed that the top of the magma column was open and smoothly boiling at the top of the pipe. Slowly rising and falling over a 2-m range, it occasionally spilled a minor flow onto the flat crater floor. There was no gas-piston activity.

"By 31 May, the top of the magma column was several meters down the pipe, and gas-pistoning had been re-established. Occasionally, the rise of the gas volume through the magma column caused a small pahoehoe flow to well out of the pipe and spread over

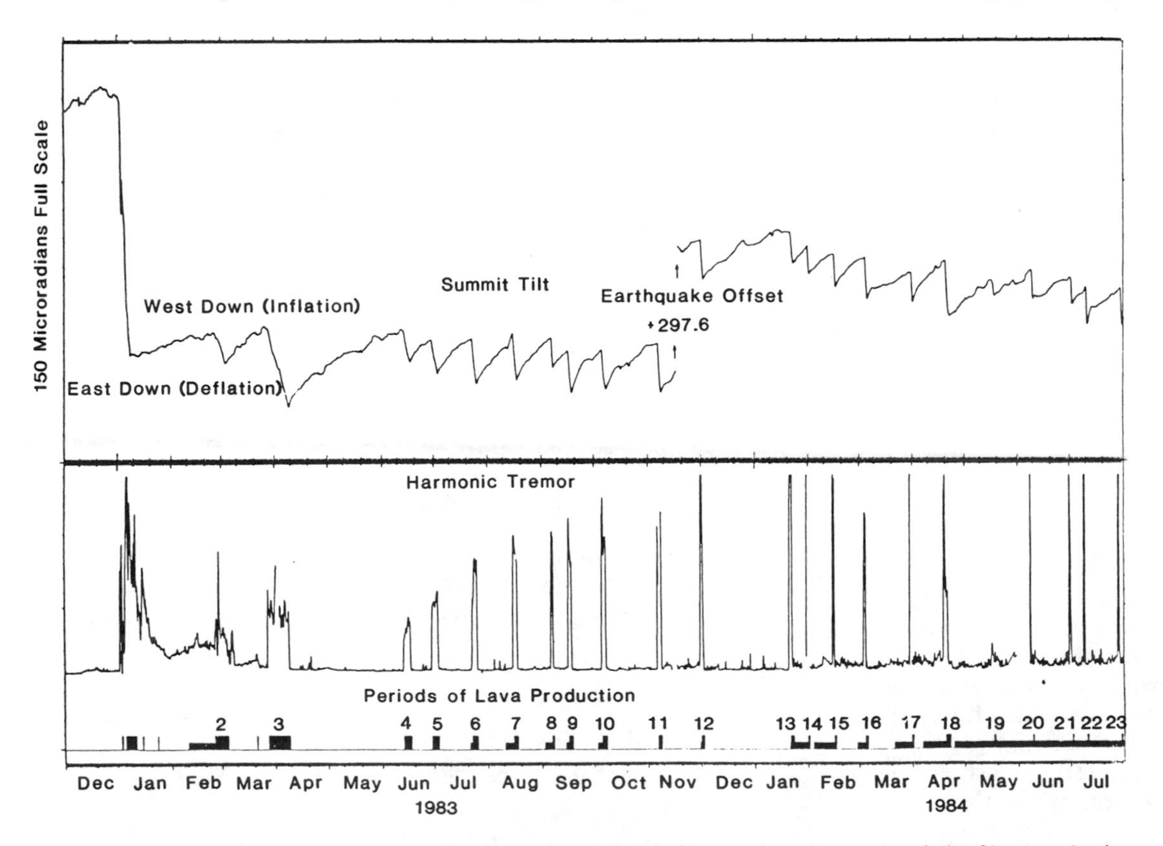

Fig. 13-30. Kilauea summit tilt (measured at Uwekahuna), amplitude of harmonic tremor, and periods of lava production for episodes 1-23. Tall bars represent periods of high fountaining and vigorous flow production; short bars represent periods of low-level activity including the visible presence of magma in the conduit. During episode 19, the brief periods of high tremor did not coincide with the programmed sampling intervals (9:5).

the nearby part of the nearly flat 100 m-diameter crater floor. When the gas accumulation broke through the magma column surface, a flare of burning hydrogen could be seen during the drainback.

"Just under 2 μrad of deflation were associated with the period of increased harmonic tremor on 28-30 May. As in episode 19, the onset and termination of deflation preceded by several hours the onset and termination of increased tremor."

Addendum: Episode 20 began at approximately 2115 on 7 June and ended at about 0625 on 8 June. Very vigorous lava fountains, visible for a short time from HVO, fed 2 lava flows that extended about 4 km NE of Pu'u O.

Information Contacts: Same as 9:2.

EPISODE 20

6/84 (9:6) "Two eruptive episodes occurred on the E rift zone in June [and one in early July]. Both produced relatively small volumes of lava (< 6 × 10^6 m^3 each) and unusually high fountains (in excess of 300 m) seen from HVO for the first time since the eruption began in Jan 1983. Episode 20 (7-8 June) and episode 21 (30 June) produced lava flows as long as 4.5 km, all on or NW of the E rift zone. Following episode 21, Pu'u O, the steep-sided spatter cone that has been the locus of eruptions for more than a year, stood about 140 m above the pre-1983 surface and was mantled by an extensive blanket of tephra. Between episodes, an active column of magma was visible in an open conduit several tens of meters below the floor of Pu'u O's summit crater. It periodically emitted gas and small particles of spatter in bursts of less than a minute, separated by intervals of 5-12 minutes. This gas pistoning was reflected in the tremor record from the seismic station near the eruption site.

"Low-level repose-period activity preceded Episode 20. The time-lapse film shows that intermittent spatter was visible above the spillway as early as 6 June at 2328 but was confined to the crater; it continued until 0448 on 7 June. For the next 11.5 hours only heavy fume was recorded.

"Episode 20 occurred largely at night, so the record consists primarily of time-lapse photography and the tiltmeter and seismic data described below. At 1911 on 7 June continuous low-level spatter activity commenced and sporadic fountaining began 65 minutes later. Lava first poured over the spillway at 2104, and from 2200 until 0624 on 8 June high fountaining was continuous. Maximum fountain heights were estimated at 300 m, because they were visible above the horizon from HVO. The last fountains recorded on film were at 0627; they were followed by intense fuming that decreased over the next several hours. During the period of vigorous eruption, 4 lava flows moved N and NE of Pu'u O, the longest reaching about 4 km from the vent (fig. 13-29).

EPISODE 21

"On 30 June, observers reached the eruption site at 1045, a few minutes after episode 21 lava topped the spillway and just before vigorous fountaining began. A maximum fountain height of approximately 390 m was measured by theodolite at 1115. This high level was maintained for about 4 hours after which fountaining decreased as flow apparently increased down the spillway. Early in the eruption, a fissure vent opened on the SW flank of Pu'u O, producing a line of low fountains and a pahoehoe flow that ponded on the S side of Pu'u Kamoamoa (about $^1/_2$ km to the W). A large lava flow that divided into several tongues extended 3-4.5 km NE of the vent, generally on top of episode 20 flows. A third flow of spatter-fed aa from the N rim of the crater overran a similar episode 20 flow for about 1 km N of Pu'u O. A thin blanket of tephra, including an early deposit of reticulite, was deposited SW of the vent. The eruption ended abruptly at 1827 with a simultaneous termination of fountaining and lava flow at the spillway. Heavy fuming and sporadic burning of hydrogen gas followed the eruption for several hours.

"Approximately 3.7 × 10^6 m^3 of new basalt was produced in episode 20 and a similar amount (mapping still in progress) was produced in episode 21.

EPISODE 22

"At the time of this writing, episode 22 has just ended, having produced 5 additional flows and another tephra blanket during activity that lasted from about 1930 on 8 July to 1017 on the 9th."

Deformation and Seismicity. "The E-W tilt measured at Uwekahuna increased irregularly after the brief episode 19 eruption of 16-17 May. On 7 June at about noon, deflation of the summit region began slowly (fig. 13-30), increasing to a rapid rate after 2100 when eruptive activity of episode 20 began. Subsidence continued through the eruption and for nearly 10 hours after fountaining activity had ceased. The post-eruptive deflation accounted for 30% of the total tilt change.

"Following episode 20, the return to summit inflation was unusually slow. For the first week, summit tilt increased by less than 2 μrad. During the next 10 days there was a 5-μrad increase in tilt followed by 5 days of little change. On 28 June, slow inflationary tilt resumed for about a day until the summit began to subside slowly. Rapid deflation commenced on 30 June at about 1100, shortly after the start of episode 21 lava production at Pu'u O. Deflation continued through the eruption and for nearly 9 hours afterwards. Post-eruption deflation accounted for nearly 40% of the total tilt change.

"Harmonic tremor at the Kamoamoa seismic station continued at low levels during the repose periods between brief eruptive stages. As in previous episodes, tremor increased to high amplitudes shortly before high fountaining began and returned to background levels when fountaining ended. High tremor of episode 20 began on 7 June at 2120 and ended on 8 June at 0627. Episode 21 tremor began on 30 June at 1028 and ended the same day at 1827. Small bursts of increased tremor during the repose periods coincided with brief intervals of gas-pistoning (9:5). This phenomenon of gas-pistoning (restricted to the vent conduit) has occurred regularly since episode 19. It became erratic several hours prior to the onset of episode 21, coincident with a slight increase in background tremor, and resumed its periodic pattern soon after episodes 20-21 ended.

Gas Measurements. "COSPEC measurements of SO_2 emissions during the repose period between episodes 20 and 21 averaged 1.4 metric tons/hr, rising to nearly 6 metric tons/hr during gas bursts. Measured SO_2 emission during episode 21 was ~1000 metric tons/hr."

Further Reference: Wolfe, E.W. (ed.), 1988, The Pu'u O'o Eruption of Kilauea Volcano, Hawaii, Episodes 1-20, January 3, 1983 Through June 8, 1984; *USGS Professional Paper* 1463 (8 papers).

Information Contacts: G. Ulrich, A. Okamura, R. Koyanagi, USGS HVO.

7/84 (9:7) "Two eruptive episodes occurred on the E rift zone in July. Episode 22 began about 1930 on 8 July, stopped abruptly at 1017 on the 9th, and produced approximately 7.7×10^6 m^3 of basalt (fig. 13-31).

EPISODE 23

"Episode 23 began at about noon on 28 July and lasted until about 0540 on the 29th (mapping in progress). Pu'u O (Enduring Hill in Hawaiian) was again the vent for lavas that flowed about 5 km to the NE, down the NW side of the rift zone (mostly outside the National Park). Both episodes were characterized by high fountain activity, short spatter-fed aa flows, and thin tephra blankets near the cone. The crater at Pu'u O has filled with solid basalt to near the level of the spillway that, following episode 23, was located on the N side of the cone and about 50 m below the highest point on the crater rim (on its SW side). More detailed surveying since June placed the summit elevation at 145 m above the pre-1983 surface and 80-100 m above surrounding accumulations of new basalt. The average diameter of the base of the cone was about 550 m. The basalt produced continued to be nearly aphyric with widely scattered phenocrysts of small (< 1 mm) olivine.

Deformation. "During the 8-day period after episode 21 (30 June), the E-W summit tilt (at Uwekahuna) increased at a rate of < 1 μrad/day (fig. 13-30). At 1200 on 8 July, the tilt started downward, indicating summit subsidence; eruptive activity at Pu'u O began 6 hrs later. The maximum rate of summit subsidence (1 μrad/hr) occurred 1.5 hours after the onset of episode 22. Summit subsidence continued until 1600 on 9 July, 6 hrs after lava production had ceased. The total E-W tilt change recorded was 13 μrad. Rapid inflationary tilt change during the 2 days following episode 22 totaled 5 μrad.

"Between 13 and 28 July, a slower rate of summit uplift was reflected in the net upward tilt change of 7 μrad. The amount of summit subsidence that occurred during episode 23 (28-29 July) corresponded to a downward tilt change of 13 μrad. At the end of July, Kilauea's summit was reinflating.

"The summit has undergone a net vertical subsidence of 50 mm since Jan 1984. Cross-rift deformation lines have been established 0.7 and 9 km uprift of Pu'u O. An additional telemetering tiltmeter has been installed about 7 km uprift of the 1983–84 eruption site.

Seismicity. "Harmonic tremor at the Kamoamoa seismic station was continuous and centered near Pu'u O. Tremor amplitudes were characteristically high during periods of vigorous lava output and low between the eruptive episodes. High tremor accompanying episode 22 started at 1859 on 8 July and ended at 1017 the next day.

"Tremor began to increase for episode 23 at 1158 on 28 July and declined at 0543 the next day. Associated with the accelerated summit deflation and toward the end of the eruptive episodes, a secondary source of intermediate-amplitude tremor in the shallow summit region was detected. Low interphase tremor near Pu'u O varied from cyclic bursts at about 10-min intervals to signals of uniform amplitude lasting up to several days."

Information Contacts: Same as above.

EPISODE 24

8/84 (9:8) "The 24th eruptive episode of the 1983–84 eruption occurred on 19-20 Aug. Lava was first seen flowing through the vent's NE spillway at 2127 on 19 Aug. Vigorous fountaining began at 2210, reached a maximum height of 315 m above the spillway at about 0100 on 20 Aug, and ended at 1721. One major flow moved through the N spillway (the dominant path from the vent) and advanced N and NE, overriding much of the lava from

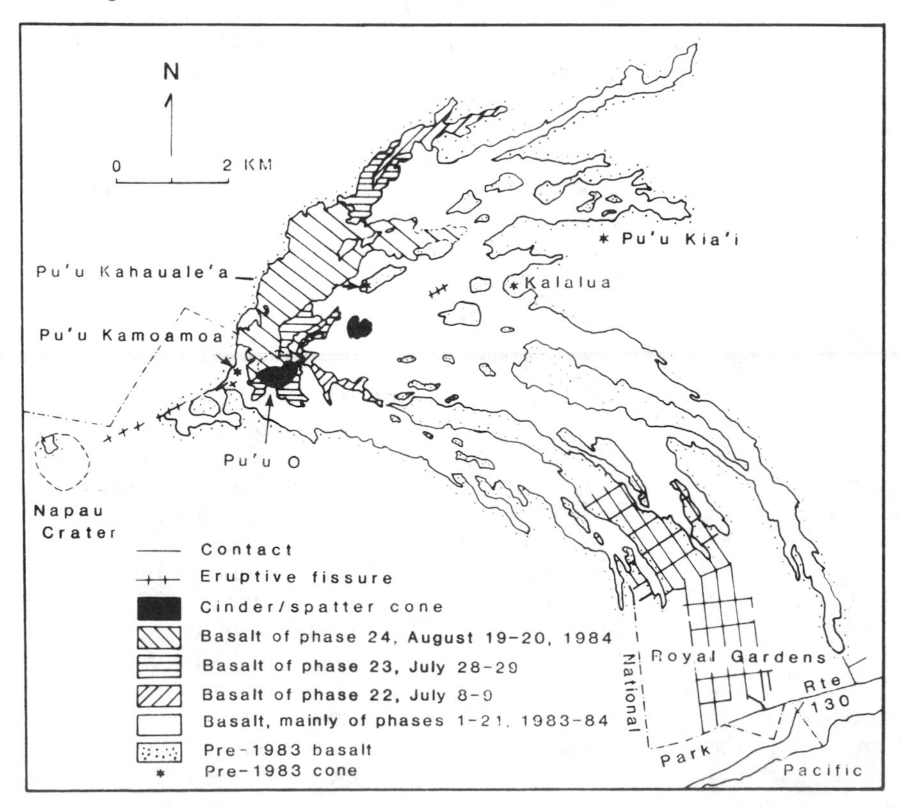

Fig. 13-31. Lava flows erupted since 3 Jan 1983. Flows from episodes 22-24 are shown separately.

episodes 22 and 23 (fig. 13-29). The volume of lava produced was approximately 11.6×10^6 m^3. A thin, uniform, tephra blanket was deposited over much of the area within a radius of about 2 km.

"The height of Pu'u O's summit increased by ~3.5 m and the cone had much the same form as in earlier episodes (see sketched profiles, fig. 13-32). The total volume of tholeiitic basalt (not corrected for void space) produced since Jan 1983 is slightly more than 0.25 km^3. The recent basalt, like that of earlier episodes, is nearly aphyric with widely scattered small (< 1 mm) phenocrysts of olivine.

Deformation. "During the 21-day repose prior to episode 24, the E-W summit tilt (measured at Uwekahuna bluff) increased by 10 μrad. Summit deflation, reflected by a downward tilt change of 14 μrad, began on 19 Aug at almost the same time as the onset of high fountaining and high-amplitude tremor at the eruption site. Deflation continued until 2000 on 20 Aug, ~2½ hrs after the end of lava production.

"Preliminary results from 2 horizontal deformation lines, installed across the E rift at points 0.7 and 9 km uprift from Pu'u O, show that changes since late June have been very small (< 20 mm), but systematic trends, similar on both lines, may be indicated. During the period of episodes 21-24, extension occurred on both sides of the rift axis between eruptive episodes and an almost equal contraction was measured following each episode. This cyclic extension and contraction suggests elastic behavior of the wall rock. Cumulative change in total line lengths has not occurred. The new telemetering tiltmeter installed near Makaopuhi Crater (about 6.5 km uprift from Pu'u O) has recorded no significant tilt changes since July.

Seismicity. "The Kamoamoa seismic station recorded continuous harmonic tremor through Aug. An increase in tremor amplitude near Pu'u O accompanied episode 24. A gradual buildup of high tremor started at 2125 on 19 Aug, 2 minutes before lava was first observed in the spillway; increased tremor continued throughout the episode and ended at 1721 the next day as fountaining stopped. Tremor of low to intermediate amplitude was also detected in the summit region. Low-level tremor with many sporadic bursts continued near Pu'u O for several days following the end of episode 24. The number of microearthquakes at the summit region fluctuated from low to average and was high in the E rift zone."

Information Contacts: Same as 9:6.

EPISODE 25

9/84 (9:9) "After a 30-day repose period, the 25th episode of the 1983–84 E rift zone eruption began at about 1604 on 19 Sept and ended at about 0532 the next day. Brief intervals of Strombolian activity and a small pahoehoe flow across the N spillway occurred over a 10-day period preceding episode 25. The largest flow exited through the N spillway and advanced on a broad front up to 3.3 km from Pu'u O (fig. 13-33). A small, short-lived, fissure eruption occurred ENE of Pu'u O early in the episode, producing a thin pahoehoe flow about 800 m long just N of the National Park boundary. The total volume of lava produced was approximately 11.1×10^6 m^3. An unusually thick tephra mantle, estimated to have a volume of about 1×10^6 m^3, fell early in this episode, accompanying the highest (460 m) fountains yet observed in the 1983–84 eruption.

"The summit of Pu'u O increased in height by 12 m during episode 25 and, within 11 days, had collapsed about 5 m, to stand 875 m above sea level and 155 m above the pre-1983 surface. The recent basalt, like that of earlier episodes, is nearly aphyric with widely scattered small (< 1 mm) phenocrysts of olivine.

Deformation. "During the 30-day repose period prior to episode 25, the E-W tilt recorded in the Uwekahuna vault increased by 17 μrad. Summit subsidence began around 1400 on 19 Sept. The net deflationary tilt change recorded during episode 25 was 15 μrad. Tiltmeters near the eruption site and at a site 6.5 km uprift recorded no significant tilt change during Sept. By the end of the month, slow reinflation of the summit was indicated by a recovery of 6 μrad.

Seismicity. "Prior to the onset of episode 25, tiny earthquakes, registered only at the Pu'u Kamoamoa station (about ½ km uprift from Pu'u O), started to increase after about 1500 on 19 Sept. At 1530, the level of harmonic tremor began to increase as well. Tremor increased rapidly at 1540, and was marked by a sharper increase at 1604. At 1613 the tremor alarm at HVO was triggered. At 1615, tremor amplitude peaked and remained at a high level until 0532 on 20 Sept. A rapid decrease in tremor was accompanied by intermittent high-amplitude bursts. For about 6 hours following the drop in tremor, there were many tiny high-frequency shocks recorded near Pu'u O.

"From about 2000 on 19 Sept to 0800 on 21 Sept, a

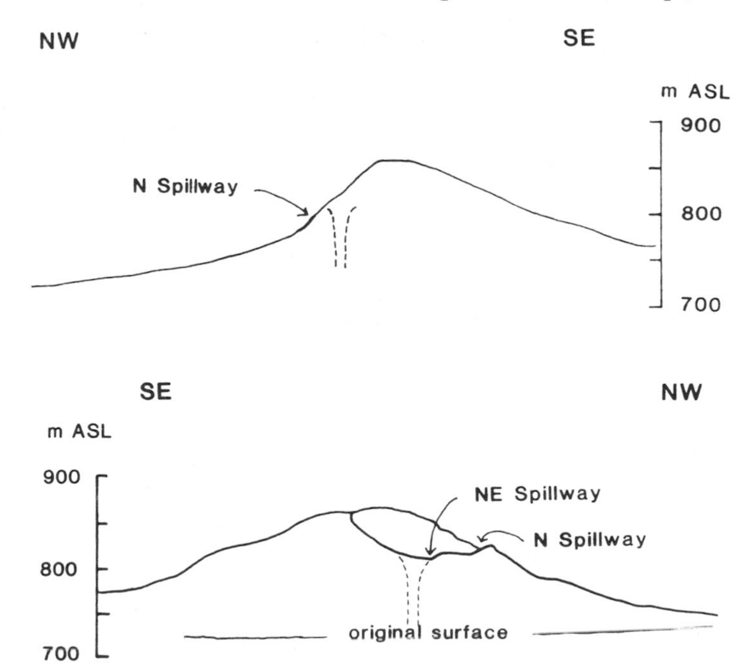

Fig. 13-32. Profiles of Pu'u O cinder/spatter cone. No vertical exaggeration. *Top,* view to the SW showing location of the N and NE spillways following episode 23. *Bottom,* view to the NE during episode 24 showing the asymmetry of cone shape and the slope of the N spillway, which has become the dominant path of lava from the vent.

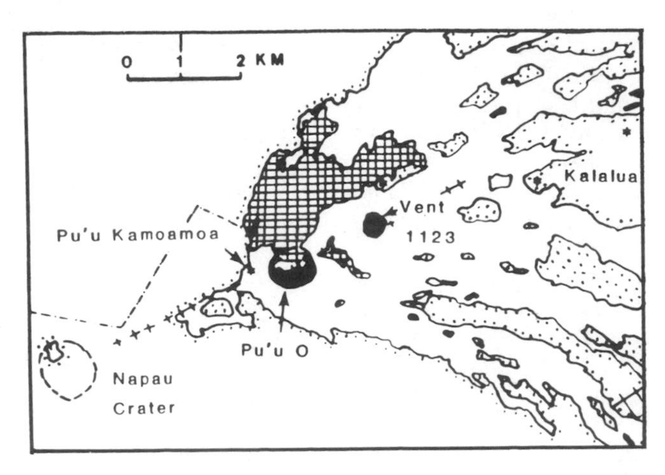

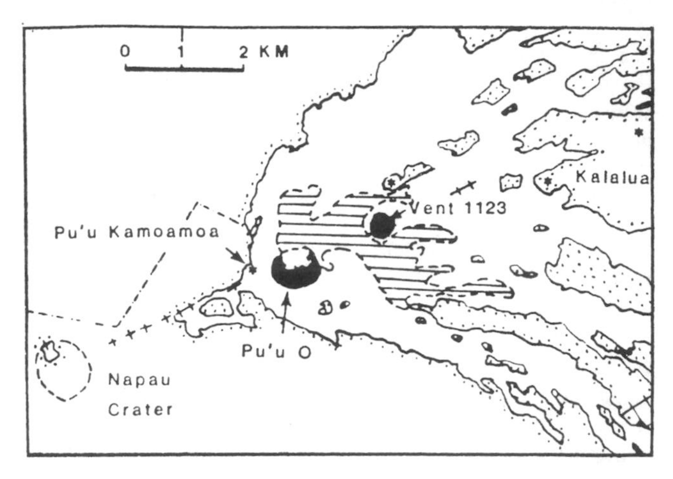

Fig. 13-33. Lava flows from episode 25, *left*, and episodes 25-27, *right*. [Scale same as fig. 13-31.]

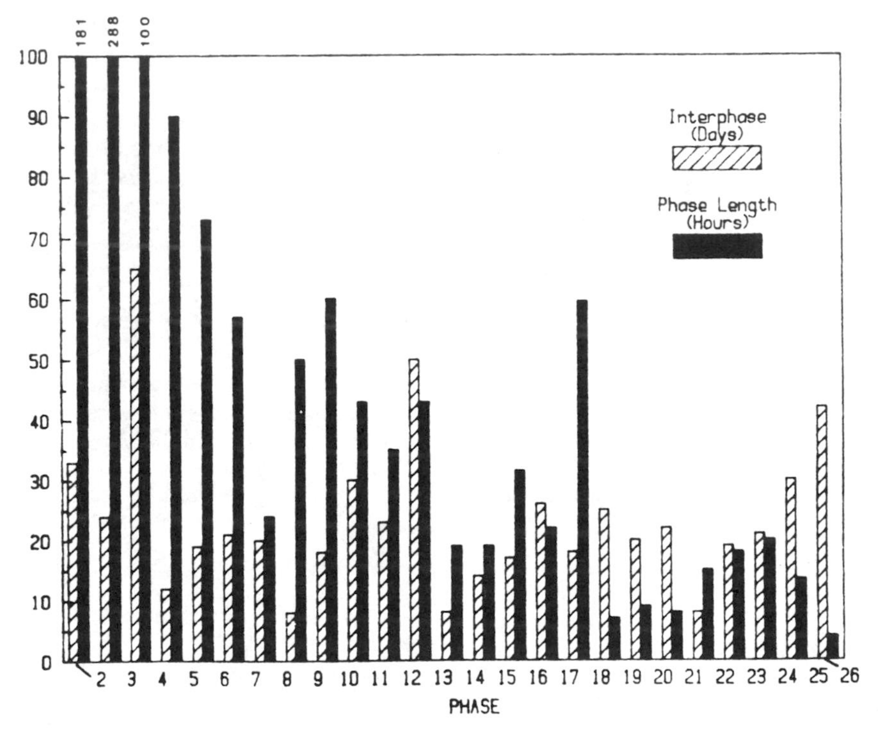

Fig. 13-34. Comparison of lengths of eruptive episodes (in hours) and preceding interphase periods (in days) for episodes 2-26 of the E rift zone eruption.

swarm of very small long-period earthquakes combined with intermediate-amplitude harmonic tremor, apparently related to the post-deflation adjustments of the summit storage system, was recorded in the North Pit station at the summit.

"At the end of Sept, the level of tremor at Pu'u O was constant and very low in amplitude. The number of microearthquakes was low in the summit region and moderate in the E rift zone."

Information Contacts: Same as 9:6.

10/84 (9:10) "By the end of Oct, 41 days had passed since the last vigorous production of lava and high fountaining on the E rift zone. However, low-level activity continued during this period. Magma was visible several tens of meters below the top of the conduit. Periodic to irregular gas-piston activity, consisting of slowly rising magma followed by low-level Strombolian activity and rapid drainback coincident with heavy degassing, occurred at intervals of minutes to hours. SO_2 emissions during the interphase period remained normal at about 100 t/d.

"On 25 Oct, lava ponded at the top of the conduit and occasionally spilled onto the floor of Pu'u O. On 27 Oct, a small pahoehoe flow, similar to that preceding episode 25, moved down the N spillway ~300 m from the conduit. During the next 5 days, a small pahoehoe shield, about 100 m in diameter and 10 m high, was built above the conduit and mainly within the rim of Pu'u O. A lava temperature of 1145°C, near the maximum measured during the 1983–84 eruption, was determined by thermocouple.

"On 2 Nov, a period of high fountaining, continuous lava production, and rapid summit deflation, designated as episode 26, occurred from 1140 to 1636. This was the shortest episode since the eruption began in Jan 1983. A comparison of the lengths of eruptive episodes and repose periods is shown in fig. 13-34.

Deformation. "Between the end of episode 25 (20 Sept) and 31 Oct, 17 μrad of E-W inflation were recorded at the summit (Uwekahuna). This increase brought the level of inflation to its highest point since episode 21 began in late June. Geodetic measurements near Pu'u O and about 9 km uprift indicated that very little net deformation was occurring across the rift zone. EDM measurements SW of Pu'u O show that the pattern of minor pre-eruption extension and post-eruption contraction reported in Aug (9:8) appears to be continuing. However, theodolite measurements NE of Pu'u O showed no movement above detection limits.

Seismicity. "Seismic activity was relatively low during Oct. In the summit area, the number of microearthquakes gradually increased corresponding to the pattern of inflation. In the middle E rift zone, the frequency of tiny shocks resulting from minor structural and thermal adjustments near Pu'u O also increased. Harmonic tremor varied in amplitude from constant, to erratic, to cyclic one-minute bursts characteristic of gas-piston activity within the Pu'u O vent."

Information Contacts: Same as 9:6.

EPISODES 26 AND 27

11/84 (9:11) "Episode 26 occurred 2 Nov between about 1140 and 1634. This episode of high fountaining from the Pu'u O vent lasted only 5 hrs, making it the shortest episode in 1983–84 eruption sequence. Lava exited through the N spillway of the cone and fanned out into broad aa flows that extended a maximum of 2.2 km NE and ESE from the vent. During the 17-day repose period following episode 26 there was little activity at the vent, unlike the period between episodes 25 and 26. Episode 27 began 20 Nov at 0005 and lasted 10 hrs. High fountains produced aa flows extending 3.5 km SE.

"Because poor weather hampered efforts to obtain aerial photographs, an accurate map of episode 26 flows was not completed before the flows were overrun by lava of episode 27 (fig. 13-33). The estimated volume of lava from the 2 episodes combined is approximately 16×10^6 m^3. The summit of Pu'u O, which increased in height by 10 m in Nov, was 886 m above sea level and 167 m above the pre-1983 surface after episode 27.

Deformation. "Summit subsidence (recorded by the Uwekahuna tiltmeter) associated with episode 26 began at 1100 on 2 Nov. The net E-W tilt change related to the subsidence was 7.5 μrad; more than half of the tilt change occurred after the end of episode 26. During the next 17 days, the summit tilt regained 7.7 μrad. Subsidence accompanying episode 27 started at about 0030 on 20 Nov and continued until 1630, resulting in a net tilt change of 12.3 μrad. By the end of the month, the summit tilt had recovered 8 μrad.

Seismicity. "Shallow seismic events correlated with the continuing volcanic activity on the middle E rift zone. Strong harmonic tremor was recorded during major outpourings of lava, and low-level tremor and numerous tiny shocks occurred near the active vent between eruptive episodes.

"Harmonic tremor associated with episode 26 started with bursts of increasing amplitude at 1004 on 2 Nov and developed into sustained high-level tremor by 1140. Tremor remained strong for the duration of the eruptive episode. At 1634, after the cessation of high fountaining, tremor decreased rapidly. Low tremor and tiny microearthquakes resumed in the middle E rift zone during the period of relative quiescence following episode 26.

"At 2101 on 19 Nov, harmonic tremor began to increase intermittently at Pu'u O. The intensity of tremor reached about an order of magnitude above background at 0005 on 20 Nov, triggering the HVO tremor alarm system. For the next 10 hours, the seismic signal was sustained at the high level characteristic of periods of high fountaining and continuous lava production. Tremor decreased rapidly between 1008 and 1012, following the end of episode 27.

"Seismic activity assumed a typical inter episode pattern for the remainder of the month. Low-level harmonic tremor continued in the middle E rift zone, varying from a pattern of constant amplitude to episodic short bursts indicative of gas-piston activity in the Pu'u O vent. The number of microearthquakes was generally below average in the summit region and above average in the E rift zone."

Addendum: Episode 28 began 3 Dec at about 1905. Vigorous fountaining fed lava flows to the N and SE during the 14-hour episode, which ended about 0941 the next morning.

Information Contacts: G. Ulrich, A. Okamura, R. Koyanagi, C. Heliker, USGS HVO.

EPISODE 28

12/84 (9:12) "Episode 28 began at 1905 on 3 Dec. High fountaining and continuous lava production lasted for 14.5 hours, ending at 0941 on 4 Dec. The lava spilled over the broad NE rim of Pu'u O and produced a broad short aa fan to the NE and 2 larger aa flows to the SE. [An area subsequently covered by flows of episode 29 (fig. 13-35).] The SE flows extended 4.7 km from the vent, ending about 1.2 km short of Royal Gardens subdivision. The high fountains were accompanied by heavy tephra fallout in a broad swath downwind of the vent. Episode 28 produced 12.3×10^6 m^3 of lava and approximately 0.2×10^6 m^3 of tephra (lava equivalent).

"Pu'u O was in repose for the remainder of the month. On 30 Dec, low fountains (3-10 m) produced thin pahoehoe overflows that reached the base of the cone. Similar activity occurred intermittently for the next 4 days until high fountains and continuous lava production signaled the start of episode 29.

Deformation. "Summit subsidence began at about 1900 on 3 Dec, at the start of episode 28. The Uwekahuna tiltmeter recorded continuous summit deflation until 1330 the next day, resulting in a net tilt change of 15.9 μrad. During the rest of the month the summit tilt recovered 17.7 μrad.

Seismicity. "Harmonic tremor associated with episode 28 at Pu'u O started to increase gradually at 1815 on 3 Dec. By 1905 tremor recorded at the Kamoamoa seismic station increased to a level characteristic of high lava fountaining. High-amplitude tremor continued for nearly 15 hrs. The sharp decrease in tremor following the end of the eruptive episode was recorded at 0941 on 4 Dec. For the rest of Dec, tremor was at low levels, varying from the fluctuating amplitude associated with gas-pistoning activity at Pu'u O, to continuous amplitude with no visible lava activity.

Addendum. The current Kilauea eruption sequence celebrated its 2nd anniversary on 3 Jan 1985 with the start of episode 29. Fountains up to 460 m high fed aa flows to the NE and SE during the 16-hour episode.

Information Contacts: C. Heliker, R. Koyanagi, A. Okamura, G. Ulrich, USGS HVO.

EPISODE 29

1/85 (10:1) "The E rift zone of Kilauea entered its 3rd year of repeated eruptive activity with its 29th major display of high fountaining and vigorous lava production beginning at 1315 on 3 Jan, exactly 2 years after the first eruption in 1983. This episode followed 29 days when the magma column was frequently visible within the upper 40 m of the 25 m-diameter conduit of Pu'u O. It was also preceded by 2 small events of fountaining 3-10 m high and pahoehoe spillovers from the Pu'u O conduit on 30 Dec and 2 Jan.

"Fountains briefly reached a maximum height of 460 m about 3 hours after episode 29 began, matching the previous record set early in episode 25. Very heavy tephra fall (estimated volume [0.15] $\times 10^6$ m^3) occurred on the SE side of Pu'u O and the cone height increased by 14 m, but about 8 m were lost by collapse during the next 2 weeks, after which Pu'u O's summit was about 195 m above the pre-1983 surface. A broad aa flow traveled 4 km S (fig. 13-35), burying much of the episode 28 flow and terminating about 1.7 km from Royal Gardens subdivision. Episode 29 ended abruptly at 0504 on 4 Jan. The total production of new lava was about 13 $\times$ 10^6 m^3. The recent basalt, like that of earlier episodes, is nearly aphyric with widely scattered small (< 1 mm) olivine phenocrysts.

Deformation. "The summit subsided rapidly during episode 29. The Uwekahuna (W-E) tiltmeter recorded subsidence from ~1100 on 3 Jan to 0800 on 4 Jan. The most rapid deflation recorded was between 1600 and 1800 on the 3rd and reached a value of nearly 1.3 μrad/hr (fig. 13-36). The net tilt change during episode 29 was 16 μrad. The summit had recovered approximately 13 μrad of inflationary tilt by the end of Jan.

Seismicity. "On 2 Jan between 1500 and 1950, a slight increase in tremor amplitude was recorded near Pu'u O. The seismic level decreased and remained relatively low until the following day at about 1100 (2 hrs and 15 minutes before the eruption began), when the amplitude of continuous tremor began to increase gradually. A rapid increase of tremor associated with the high fountaining of episode 29 started about 1340. Tremor amplitude remained strong throughout the eruptive episode. Tremor decreased sharply at 0504 on 4 Jan, at the end of major lava production.

"Seismicity then decreased to the normal background activity in the Kilauea region. Harmonic tremor in the middle E rift zone remained low, varying from relatively constant amplitude to periodic fluctuations in amplitude that were commonly correlated with gas-pistoning activity at the vent. Shallow microearthquakes were few in the summit region and varied from few to moderate in number in the E rift zone."

Addendum. At 0546 on 4 Feb, episode 30 began. Vigorous fountaining fed a major lava flow that extended 7 km SE from Pu'u O, before eruptive activity ended at about 0246 on the 5th.

Information Contacts: G. Ulrich, R. Koyanagi, R. Hanatani, USGS HVO.

EPISODE 30

2/85 (10:2) "Episode 30 began on 4 Feb at 0546. High fountains and continuous lava production from the Pu'u O vent continued for 21 hrs until 0246 on 5 Feb. Lava flows spilled over the NE rim of the cone and turned S to form one 8.3 km-long aa flow (fig. 13-35). The flow advanced rapidly through National Park forest and passed 0.7 km W of Royal Gardens subdivision. Nearly 2 km^2 of forest were destroyed by the lava flow, and grass fires at the distal end of the lava burned an additional 0.3 km^2 over the next 3 days. SW winds deposited tephra and Pele's hair in Hilo, 35 km from the vent.

"Episode 30 produced about 14 $\times 10^6$ m^3 of lava and a minimum of 0.09 $\times 10^6$ m^3 of tephra (dense rock equivalent). Dense basalt specimens from the episode 30 flow appear to contain a higher concentration of olivine microphenocrysts than those of previous episodes.

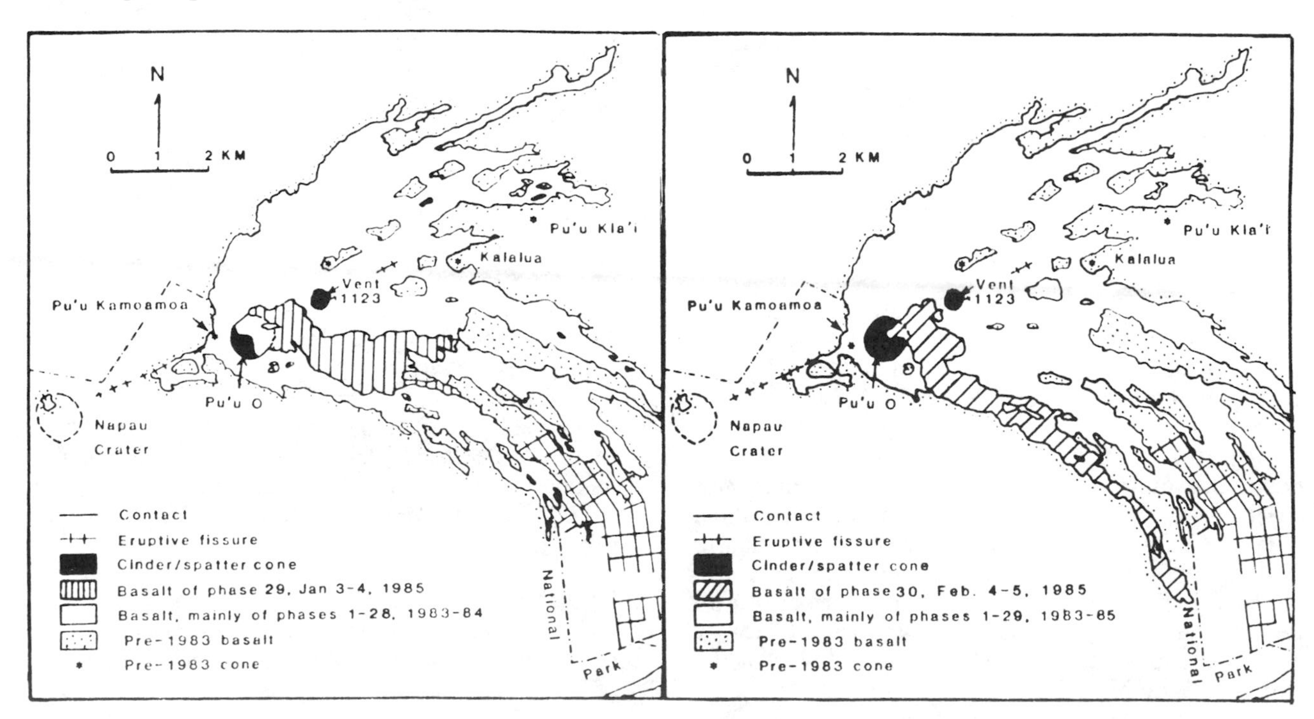

Fig. 13-35. Lava flows erupted since 3 Jan 1983 on the E rift zone. Flows from episode 29, *left*, and episode 30, *right*, are shown separately.

"Because of the unusual wind direction during episode 30, the elevation of Pu'u O's summit increased by only a meter to 913 m above sea level. The growth of the cone since episode 24, as seen from a photo station on vent 1123, is illustrated in fig. 13-37.

"Pu'u O was in repose for the remainder of Feb, although the magma column was intermittently visible at an estimated depth of 40-50 m in the 20 m-wide conduit.

Deformation. "Rapid summit subsidence associated with episode 30 began on 4 Feb at 0600, about $^{1}/_{4}$ hour after the onset of lava fountaining. The Uwekahuna (W-E) tiltmeter recorded continuous summit deflation until 0400 on 5 Feb, about $1^{1}/_{4}$ hours after vigorous lava production ended. The maximum deflation rate was 1.75 μrad/hr, the highest rate recorded since the initial episode of intrusion and eruption in Jan 1983. The net tilt change was -22 μrad. Reinflation began very slowly, but by the end of Feb, the summit tilt had recovered 10.4 μrad.

Seismicity. "Harmonic tremor recorded at the Kamoamoa seismic station near Pu'u O began to increase 4 Feb at 0552, a few minutes after lava emission began. After 0600, tremor increased rapidly to a high amplitude that was sustained until the early morning hours of 5 Feb, when the tremor level began to fluctuate. Tremor amplitude dropped to the characteristic inter-episode signal at 0247 as episode 30 ended. For the remainder of the month, harmonic tremor continued at variable low amplitudes in the middle E rift zone.

"Seismic activity was highlighted by a M 4.9 earthquake 9 km beneath the S flank on 21 Feb at 1948. The event was widely felt on the island and was followed by a series of aftershocks, but the earthquake had no immediate effect on the pattern of summit tilt or inter-episode activity at Pu'u O.

"Preliminary geodetic measurements near the epicenter of the 21 Feb earthquake showed that surface deformation was relatively local and did not extend beyond 4 km from the epicenter. Similarly, measurements made across the E rift zone at Pu'u O and near Mauna Ulu 4 days after the earthquake showed no significant deformation.

Addendum: On 13 March at 0720, episode 31 began. Lava fountaining fed flows that extended about 6 km SE from Pu'u O; lava flows entered Royal Gardens subdivision, but no structures were damaged. The 21-hour episode ended about 0455 on the 14th.

Information Contacts: C. Heliker, R. Hanatani, R. Koyanagi, USGS HVO.

EPISODE 31

3/85 (10:3) "The 1983–85 eruption produced the 31st episode of vigorous fountaining and continuous lava production on 13-14 March. Sustained high-level activity began on 13 March at 0720 and continued for almost 21.5 hours. Fountain heights reached a maximum of 340

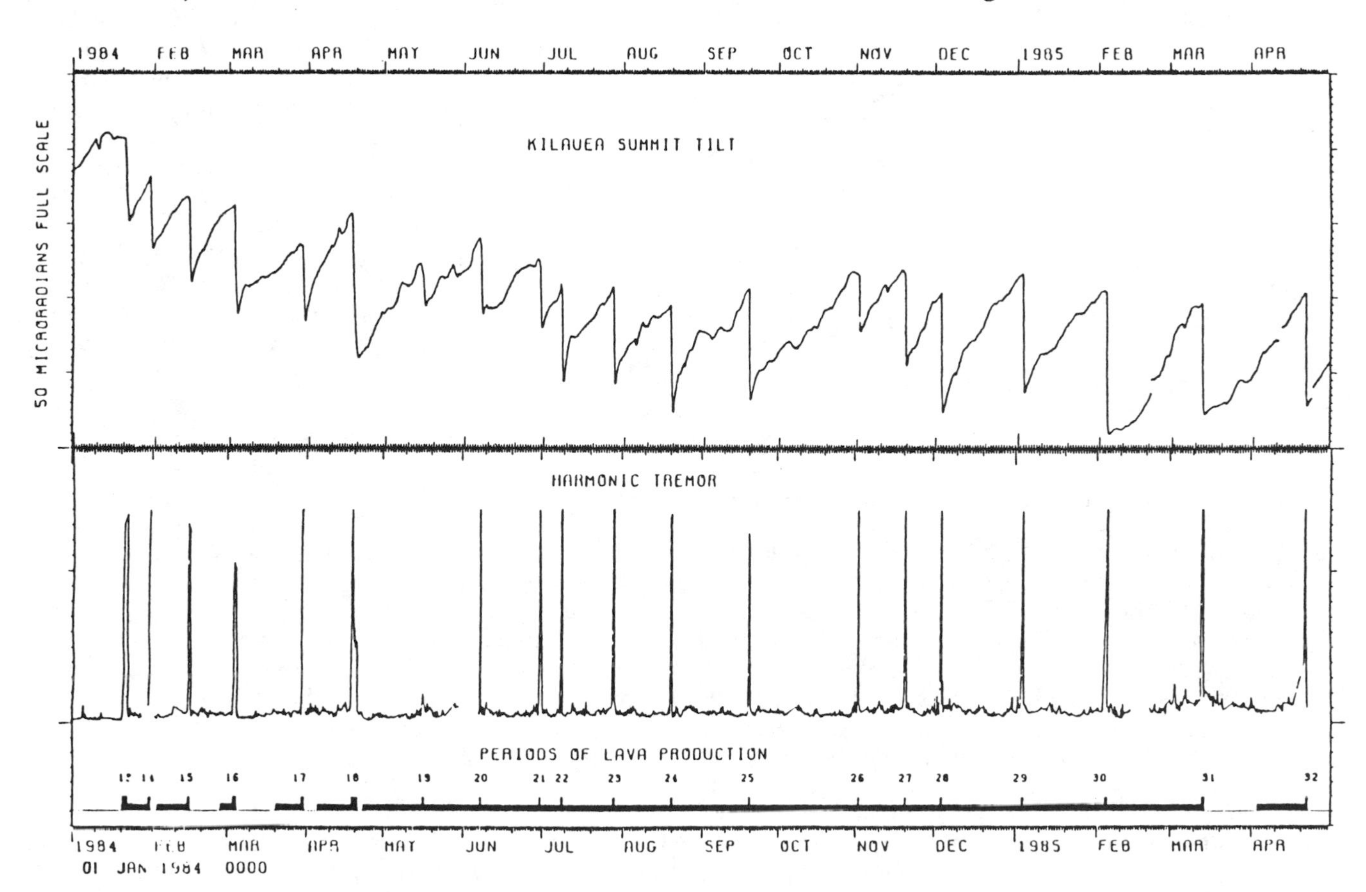

Fig. 13-36. Kilauea summit tilt, relative amplitude of harmonic tremor at the Pu'u Kamoamoa seismic station, and periods of lava production for Jan 1984–April 1985. In the lower plot, showing lava production, tall bars represent periods of high fountaining and vigorous flow production, short bars represent periods of low-level activity including magma frequently visible in the conduit. The gap in the tremor record following episode 30 is a result of the relocation of the Pu'u Kamoamoa seismic station.

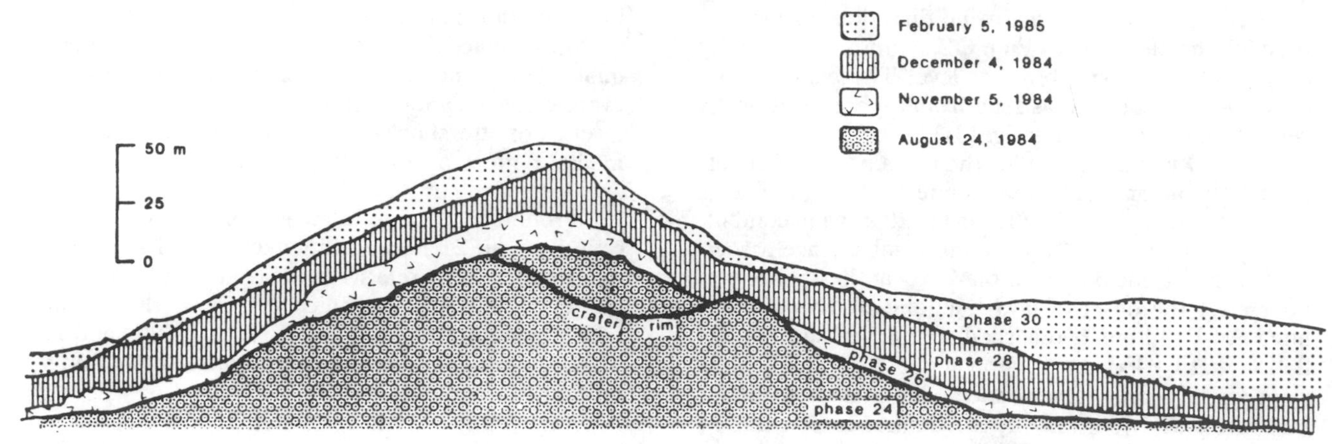

Fig. 13-37. Profiles of Pu'u O as seen from a photo station on vent 1123, 1500 m to the E, after episodes 24, 26, 28, and 30.

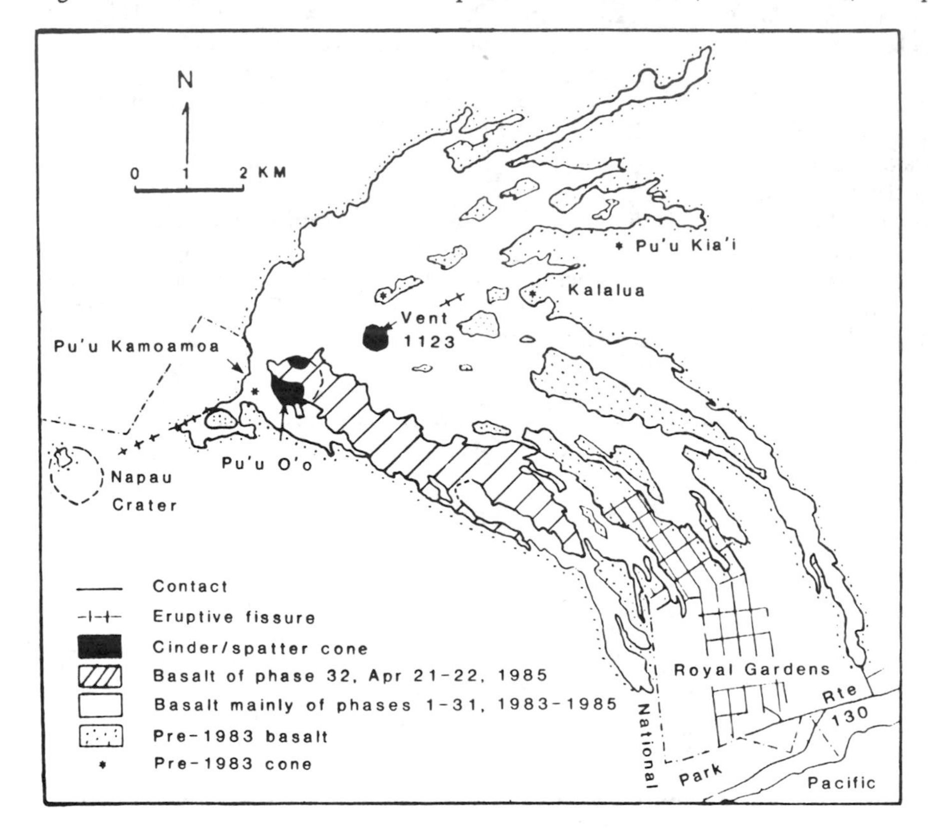

Fig. 13-38. Lava flows erupted since Jan 1983. The episode 32 flow is shown by diagonal lines.

m above the Pu'u O conduit 5 hours after the eruptive episode began. A single broad aa flow advanced 4 km SE before splitting into 2 lobes. The E lobe entered Royal Gardens subdivision on 14 March at 0100; no buildings were damaged, although the lava passed within 200 m of one house and eventually stagnated within 150 m of another. The W lobe extended 7.8 km from the vent, to about 90 m elevation (within the National Park).

"Heavy tephra fall occurred on the SW side of Pu'u O, adding 11 m to its height. At the end of episode 31, the summit elevation was 924 m above sea level and 205 m above the pre-1983 surface. The volume of lava produced during episode 31 was approximately 19×10^6 m^3, with an estimated tephra volume of 0.2×10^6 m^3 (dense rock equivalent). The basalt of episode 31, like that of recent episodes, is microporphyritic, with about 2.6 modal % olivine microphenocrysts (< 1 mm).

"Following episode 31, dense fume obscured the view into the conduit of Pu'u O, and the magma column was not seen for the remainder of March.

Deformation. "Rapid summit subsidence associated with episode 31 began at approximately 0700 on 13 March, about as high-level eruptive activity began. The Uwekahuna (W-E) tiltmeter recorded continuous summit subsidence until 0600 the next day. The net tilt loss was 20.4 μrad. The maximum rate of deflation, 2.3 μrad/hr recorded between 1200 and 1300 on the 13th, was the highest recorded since the episode 1 intrusion. As at the end of episode 30, reinflation of the summit began slowly, and by the end of March the summit tilt had recovered 4.7 μrad.

Seismicity. "The intensity of harmonic tremor recorded at the Kamoamoa seismic station near Pu'u O increased rapidly starting on 13 March at 0600. By the start of vigorous eruptive activity at 0720, tremor had reached a high amplitude that was sustained until the end of the episode at 0458 the next day. Following episode 31, tremor amplitude dropped to the low level characteristic of inter-episode periods. Fluctuating low-level tremor continued in the middle E rift zone for the rest of March, with variation in amplitudes occurring at intervals of a few minutes to several days."

Information Contacts: Same as 10:2.

EPISODE 32

4/85 (10:4) "The 32nd episode of the 1983–85 eruption occurred on 21-22 April, following 38 days of repose. Intermittent dome fountaining began at 1516 on the 21st and lasted for about 4 hrs. Continuous fountaining and lava production commenced at 1804. By 1910, tremor amplitude had increased and vigorous lava fountaining had begun. A broad aa flow advanced SE and split into 3 lobes about 3-4 km from the vent. The central lobe advanced the farthest, passing W of Royal Gardens subdivision, and stagnated 6 km from the vent (fig 13-38). Lava fountaining ceased at 0906 on the 22nd.

"Heavy tephra fallout added 9 m to the elevation of Pu'u O'o (a name recently suggested by local elders to replace the previously used Pu'u O) and blanketed the area SW of the vent. The summit collapsed 5 m during the following week.

"At the end of April, the cone was 928 m above sea level and 209 m above the pre-1983 surface. The total volume of lava produced through episode 31 was 354×10^6 m^3, slightly exceeding the total reported for the Mauna Ulu eruption of 1969–74.

"Kilauea remained quiet through the end of April. Heavy fume obscured the view into the vent, and the magma column was not observed.

Deformation. "The summit of Kilauea subsided rapidly during the 32nd episode of the eruption. The Uwekahuna (W-E) tiltmeter recorded continuous summit subsidence from 1700 on the 21st until 1100 the next day, about 2 hours after fountaining stopped. The maximum rate of subsidence occurred between 2200 and 2300 on 21 April and measured nearly 2.1 μrad per hour. The net tilt change was 17.0 μrad. Reinflation of the summit had reached 6 μrad by the end of April.

Seismicity. "Harmonic tremor in the middle E rift zone near Pu'u O'o increased gradually from about 1516 on 21 April, when intermittent dome fountaining began. Starting at 1843, tremor increased rapidly to an order of magnitude above background and remained high for the duration of the eruptive episode. At 0903 on the 22nd, about the time that lava fountaining ceased, the tremor level decreased rapidly, and by 1240 the intensity was down to nearly background levels.

"Following episode 32, harmonic tremor continued near Pu'u O'o in the characteristic pattern observed during previous inter-episode periods. The number of microearthquakes decreased to a low level in the summit region and E rift zone."

Information Contacts: C. Heliker, G. Ulrich, R. Hanatani, R. Koyanagi, USGS HVO.

5/85 (10:5) "For the first time since Oct 1984, Kilauea did not erupt during a calendar month. Mapping of episode 32 (21-22 April) eruptive products was completed in May and showed that 16.1×10^6 m^3 of lava and 0.25×10^6 m^3 of tephra (dense rock equivalent) were produced. Lava from the 1983–85 eruption covered 40 km^2 of the middle E rift zone.

"Following episode 32, magma was first visible in the conduit on 9 May, at ~50 m depth. By the end of May, the magma column had risen to within 30 m of the Pu'u O'o crater floor and remained mostly crusted over.

Deformation. "The summit of Kilauea continued to inflate during May. However, a period of significant summit subsidence occurred from 22 May at 1800 to 27 May at 0900. During this interval, the Uwekahuna tiltmeter recorded 2.8 μrad of summit deflation. The net inflation during May was 5.7 μrad (fig 13-39).

Seismicity. "Harmonic tremor continued at a fluctuating low level in the middle E rift zone near Pu'u O'o following episode 32. The number of shallow microearthquakes in the E rift zone and summit region varied from several tens to a few hundreds/day. Short-period summit earthquakes increased gradually in response to the slow rate of inflation. Frequent bursts of long-period events and harmonic tremor were recorded from intermediate to deep sources beneath Kilauea."

Addendum: Episode 33 of the middle E rift zone eruption occurred on 1213 June. Discontinuous low-level fountaining and spillover from the Pu'u O'o vent began at 0430 on the 12th. By 2306, fountaining had become vigorous and continuous. Lava flowed about 3.5 km to the SE over older lavas from Pu'u O'o before the episode ended at [0453] on the 13th.

Information Contacts: C. Heliker, G. Ulrich, R. Hanatani, R. Koyanagi, R. Okamura, USGS HVO.

EPISODE 33

[A footnote to the following report read, "USGS scientists at HVO have returned to the use of the word 'episode' rather than 'phase', a term that carries some connotation of evolutionary stages of development. The word 'episode' has been preferred in previous HVO discussions of Hawaiian volcanism such as D.A. Swanson et al. in *USGS Professional Paper* 1056."]

6/85 (10:6) "Two more episodes of the 2 1/2-year-old middle E rift eruption occurred 12-13 June and 6-7 July. Episode 33 was both preceded and followed by continuous low-level activity at the vent. Spattering was first observed on 1 June and continued intermittently until 12 June at 0430, when a low dome fountain produced a thin pahoehoe flow that extended 0.25 km from the vent. Fountaining ended after 20 minutes, but was followed by 20 more periods of low fountaining over the next 19 hours. At 2306 on the 12th, continuous high fountaining

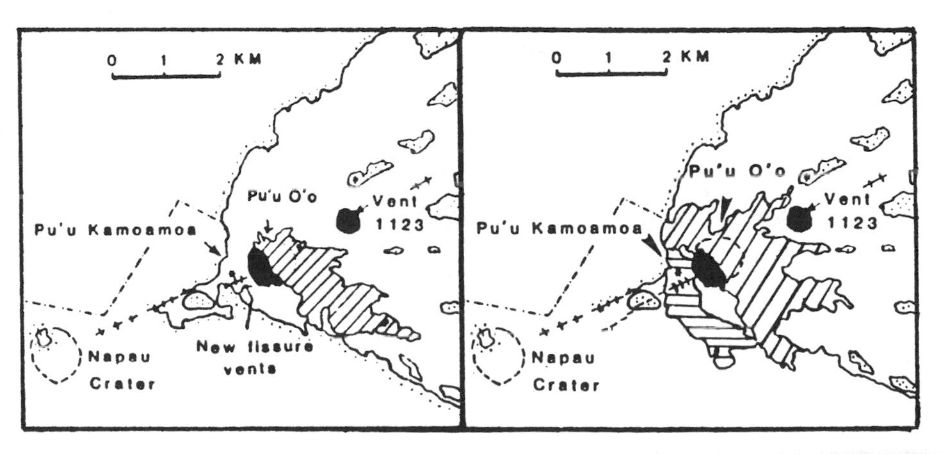

Fig. 13-39. *Left*, episode 34 flow is shown in a diagonal pattern. *Right*, the fissure-fed pahoehoe flows of episode 35 (horizontal lines) and the predominantly aa flows of episodes 35 and 36 (diagonal lines) are shown separately. [Scale same as fig. 13-38.]

began. The high fountains persisted for almost 6 hours and produced a single aa flow that extended 4.5 km SE from the vent, overriding an episode 32 flow.

"Following episode 33, magma was continuously visible in the conduit, and by 20 June spatter was again reaching the conduit rim. Episode 34 began on 6 July at [1903] and ended 14 hours later, producing an aa flow that terminated 3 km SE of the vent.

Deformation. "The Uwekahuna tiltmeter began recording subsidence on 12 June at 2200, ~1 hour before continuous high fountaining began. Summit subsidence continued for 17 hrs, resulting in a net loss of 7.0 μrad. Reinflation of the summit began slowly then increased rapidly after 22 June; by the end of the month, the summit tilt had regained 6.7 μrad. Deformation data for episode 34 had not been reduced at the time of this report.

Seismicity. "Harmonic tremor began with a minor increase in amplitude on 12 June at 0430, recorded by the Kamoamoa seismic station 1 km W of Pu'u O'o. Tremor amplitude increased rapidly at 2230 and remained high until 0452 on 13 June. The intensity of eruption tremor subsided rapidly thereafter, marking the end of vigorous lava fountaining. Moderate-amplitude tremor was recorded in the summit region until about 1600 on the 13th.

"The number of short-period caldera, or summit, earthquakes increased gradually in May then decreased from 1 June until the onset of the eruptive episode on 12 June. Intermediate-depth (5-13 km), long-period events were persistent throughout the month. Shallow long-period events at the summit accompanied the deflation event and eruptive episode on 12-13 June."

Accident. "On 12 June, George Ulrich, staff geologist at HVO, and Dario Tedesco, a volcanologist from the Osservatorio Vesuviano, were collecting lava samples and making temperature measurements at Pu'u O'o vent. A low dome fountain from Pu'u O'o was feeding a short flow to the SE. George was standing on the edge of the lava channel on solid but recently formed crust. At 1430, after completing a temperature measurement (1137°C), he walked onto the older lava to return the thermocouple and gather a sampling pick. When he returned to the lava channel to obtain a sample, the pahoehoe river had stopped moving, and he inadvertently went past the stable recent crust onto even more recent crust, which then gave way. He went in over his field boots, and was unable to pull himself out. In the attempt to reach for safety, his knees made further contact with the molten lava. Within an estimated 5 seconds, Dario pulled him to safety. His outer clothing (NOMEX suit) was then burning, but quickly extinguished itself. At the time of the accident, a helicopter piloted by Bill Lacey was landing to deliver an extra time-lapse camera. George got into the helicopter under his own power and Bill Lacey flew him directly to Hilo hospital. George was subsequently transferred to the Straub Clinic in Honolulu, where the following evaluation was made one week after the accident: first degree burns over parts of arms, hands, and face, healing normally; second degree burns from ankles to uppermost thighs on both legs; third degree burns on left kneecap, extending to upper thigh, and possible third degree burns on right kneecap. The long-term prognosis is for full recovery with no permanent disabilities, as none of the burns affected the musculature of the legs."

[George Ulrich returned to work part-time after a 2-month hospital recovery. After a few weeks, he resumed full-time responsibilities. Within 5 months, after achieving more than 90% recovery, he returned to full-time field work.]

Information Contacts: C. Heliker, R. Hanatani, R. Koyanagi, USGS HVO.

EPISODES 34 and 35

7/85 (10:7) "Kilauea's E rift eruption continued with 2 episodes of high fountaining at the Pu'u O'o vent in July. Episode 34 began 6 July at 1903 with a low dome fountain that gradually increased to 300-400 m in height over the next 4 hours. The activity ended after 14 hrs and produced an aa flow that extended 3 km SW (fig. 13-39). The estimated volume of the flow is 10.6×10^6 m^3 (including $\sim 0.4 \times 10^6$ m^3 of tephra reduced to lava equivalent).

"Episode 35 departed from the established pattern of Pu'u O'o eruptions. Fissure vents opened on the W flank of Pu'u O'o either shortly before or simultaneously with the start of fountaining at the Pu'u O'o vent, which began on 26 July at 0252 (observers were not in the area because of a hurricane warning). By 0700, the fissure vent was inactive after producing a pahoehoe flow that extended 1.5 km SE. High fountains continued at the Pu'u O'o vent until 0952 and produced short aa flows to the NW and a 2.2 km-long aa flow to the SW. Two hours later, a 50-m stretch of the fissures nearest Pu'u O'o reopened and a small volume of pahoehoe oozed to the surface. This activity ceased after an hour and a half but was followed late in the afternoon by extensive ground cracking that extended the fissure system 2.5 km up the rift (NW) from the base of Pu'u O'o.

"At 0414 the next morning, the eruption resumed along the 1 km of fissures nearest Pu'u O'o. By 28 July activity was localized in the eastern 400 m of the fissure system. Lava production in the first week of Aug was nearly continuous and characterized by low fountains that generated lava flows N and S, within 1 km of the 26 July fissure system. The fissure eruption was continuing 9 Aug with activity localized at 2 vents at the base of Pu'u O'o. When geologists returned to the vent on 12 Aug, no lava outpouring or spattering was observed. Late July-early Aug pahoehoe flows cover an area of about 400,000 m^2 with an estimated maximum depth near the vent of 15 m.

Deformation. "The summit rapidly subsided starting on 6 July at approximately 2000, as one of the events marking the beginning of the 34th episode of the eruption (fig. 13-40). The subsidence continued until approximately 1400 the next day. The Uwekahuna tiltmeter (at the summit of Kilauea) recorded a net loss of 12.0 μrad. The maximum rate of subsidence was 1.3 μrad/hr.

"Reinflation of the summit following episode 34 occurred at a fairly steady rate until 26 July at 0200, when rapid subsidence marked the beginning of episode 35. This second major subsidence of July continued for

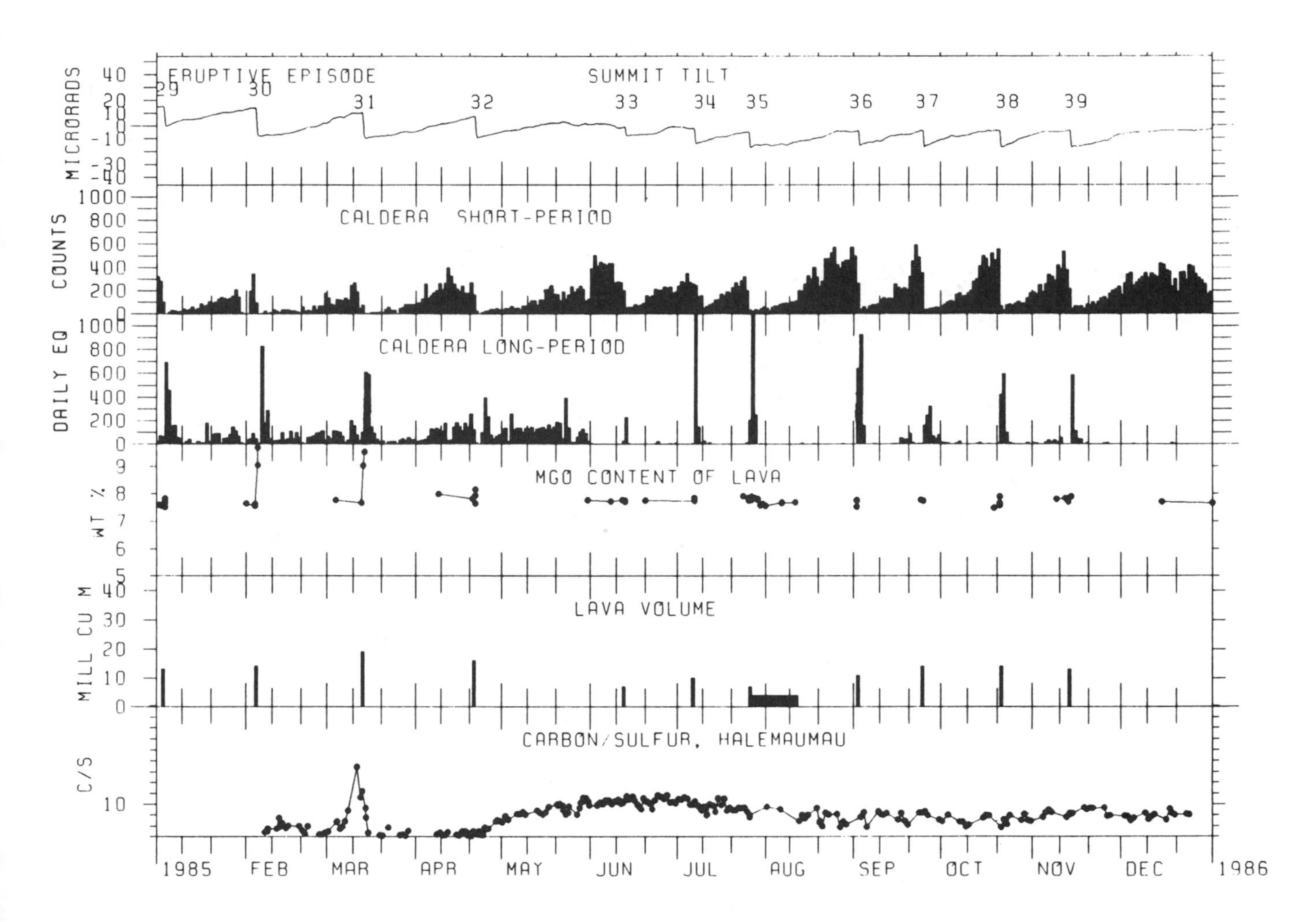

Fig. 13-40. Summary of 1985 Kilauea activity showing summit tilt, caldera short- and long-period earthquakes, C/S ratios of summit (Halemaumau) gases, MgO content of lava collected during and between eruptive episodes, and lava volumes.

about 12 hrs and resulted in the loss of 12.1 μrad of tilt. The maximum rate of summit subsidence associated with the 35th episode was 1.6 μrad/hr and occurred between 0400 and 0500. The summit briefly reinflated until 27 July when a fissure eruption began SW of Pu'u O'o. The Uwekahuna tiltmeter recorded little change from then until the end of July.

Seismicity. "Harmonic tremor associated with eruptive episode 34 started gradually on 6 July at 1903. The amplitude of tremor near Pu'u O'o increased several orders of magnitude above background after 1915. High tremor continued with minor fluctuations in amplitude until 0912 on 7 July.

"Episode 35 was preceded on 26 July at 0230 by an increase in the number of tiny earthquakes near Pu'u O'o. At about 0252, harmonic tremor increased in amplitude by nearly an order of magnitude and continued until 0952, when tremor decreased to moderate to low levels. At 0414 on 27 July, tremor increased to a moderate level, marking the start of renewed activity on the 26 July fissure system, and assumed a pattern of slightly fluctuating amplitudes until the end of the month. The number of micrcoearthquakes had decreased to a low level above the E rift zone and summit region at the month's end. In early Aug harmonic tremor continued with moderate amplitude near the eruption site."

Information Contacts: Same as 10:6.

8/85 (10:8) A decrease in harmonic tremor on 12 Aug at 0430 indicated the end of episode 35 lava production from fissures uprift of Pu'u O'o vent. The fissure activity formed a small lava shield at the W base of Pu'u O'o. The total volume of lava erupted during episode 35 was estimated to be 11.7 × 10^6 m^3. Kilauea's summit began to reinflate on 8 Aug, four days before the end of the fissure activity. By the onset of the next eruptive episode, on 2 Sept, the summit had recovered 12.9 μrad, compared with the 12.1 μrad of deflation that accompanied episode 35.

EPISODE 36

After only 21 days of repose, episode 36 began on 2 Sept at 1400. Fountaining at the Pu'u O'o vent increased slowly during the first 3 hours from 100 m to over 300 m high. The fountains fed a broad aa flow that extended 2.5 km SE on top of previous flows, and a small 1.8-km flow to the NE. Harmonic tremor was first recorded about 1 hour prior to the start of the episode. The seismic station at nearby Pu'u Kamoamoa was moved uprift during episode 35, so the tremor signal from episode 36

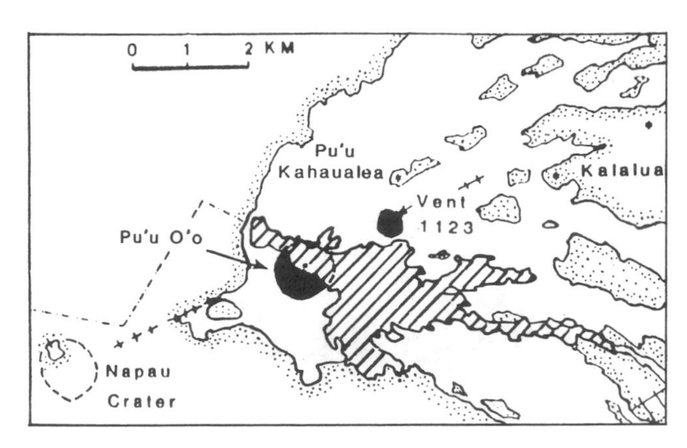

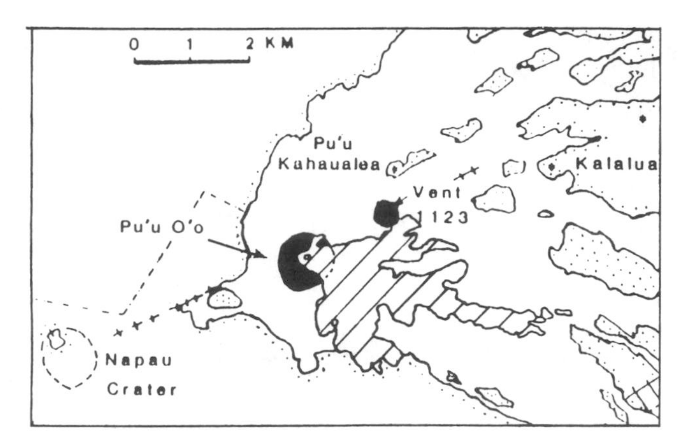

Fig. 13-41. Lava flows produced by episode 37, *left*, and 38, *right*.

was not as sensitive as in the past.

Summit deflation started on 2 Sept at 1300, 1 hour before fountaining began. Fountains stopped abruptly at 2335, after 9 1/2 hours of activity, but subsidence continued until 0300 on 3 Sept, totalling 11.7 μrad before inflation resumed (fig. 13-40). The episode 35 fissure system remained inactive during episode 36. The Sept activity extended the duration of the ongoing eruption, beginning in Jan 1983, to 33 months.

Information Contacts: G. Ulrich, C. Heliker, R. Hanatani, R. Koyanagi, USGS HVO.

9/85 (10:9) "Two episodes of high fountaining and lava production at the Pu'u O'o vent occurred in Sept. Episode 36 occurred on 2 Sept. The distribution of episode 36 lava flows and the partly covered flows of episode 35 (July–Aug) are shown in fig. 13-39. Episode 37 flows are [mapped in fig. 13-41]. The total volume of lava produced by episode 35 was 11.7×10^6 m^3. Episode 36 produced a lava volume of 11.5×10^6 m^3.

EPISODE 37

"After a 22-day repose period, when the magma column in the Pu'u O'o conduit was visible at depths of 50-80 m, episode 37 occurred. Preceded by several small pahoehoe spillovers from intermittent fountains, the fountaining became continuous on 24 Sept at 1808. The eruption increased slowly to a peak between midnight and 0200 when fountain heights of ~250 m were estimated. Episode 37 ended as the fountaining died at 0619, having lasted 12 hours. The basalt, as in recent episodes, is microporphyritic with abundant olivine crystals (< 0.5 mm in diameter). The composition has remained unchanged since episode 32 in April (fig. 13-41).

"Summit deflation during episode 37, as recorded by the Uwekahuna tiltmeter (W-E), commenced on 24 Sept at 1900 (about an hour after lava fountaining became continuous) and ended at 0900 the next day, resulting in a net loss of 13.2 μrad.

"Tremor amplitude at the newly located seismometer near Pu'u O'o (10:8) increased on 24 Sept at about 2050 (2 1/2 hours after the onset of continuous lava fountaining), activating the tremor alarm at HVO at 2214. Tremor level decreased rapidly at 0619, coinciding with the end of fountaining. The decreased tremor following episode 37 assumed a pulsating pattern characterized by one-minute bursts of moderate amplitude alternating with several minutes of lower amplitude. This gas-piston pattern continued for about a day then changed to a more continuous background of low-amplitude harmonic tremor for the remainder of the month.

Information Contacts: Same as above.

10/85 (10:10) "Episode 37 produced a lava volume of 14.7×10^6 m^3. During the 26-day repose period that followed episode 37, the summit tiltmeter recorded 13.0 μrad of inflation. Summit subsidence associated with episode 38 began on 20 Oct at 1900, nearly simultaneously with the onset of low-level, discontinuous activity at the Pu'u O'o vent. Intermittent dome

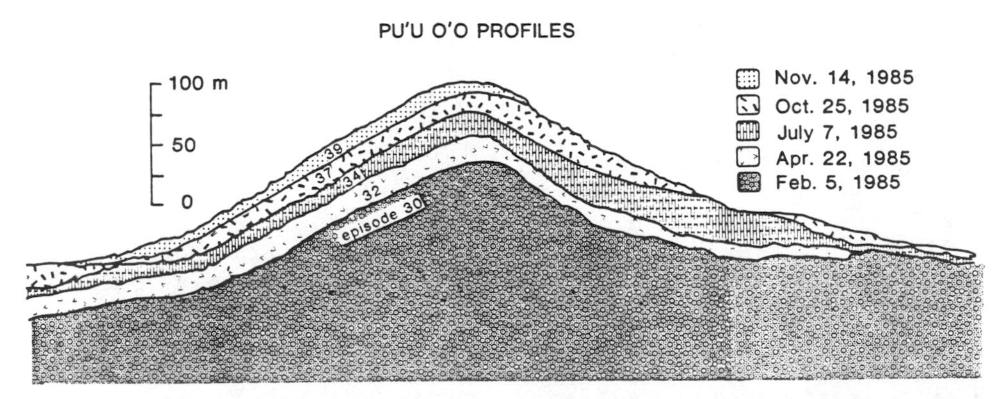

Fig. 13-42. Profiles showing the growth of Pu'u O'o since 5 Feb 1985, drawn from photographs taken from the 1123 cone, 1.5 km to the E.

fountaining produced thin pahoehoe spillovers for the next 8 hrs.

EPISODE 38

"Beginning at 0230 on 21 Oct, a gradual increase in tremor amplitude was recorded by the seismometer near Pu'u O'o. Continuous lava production began at 0301, signalling the official beginning of episode 38. Fountain heights reached approximately 300 m by 0607. Broad aa flows advanced SE, closely following the path of the episode 37 flows (fig. 13-41) and eventually stagnating 4.2 km from the vent.

"Harmonic tremor amplitude began to decrease rapidly at 1118, and the fountain died at 1120. By 1145 the tremor was at a non-eruptive level. Gas-piston activity followed the end of the eruptive episode for about 24 hrs, before continuous low-amplitude background tremor resumed. Summit subsidence continued until about 1500 on 21 Oct, resulting in a net deflation of 13.0 μrad. Following the eruption, the familiar pattern of summit reinflation resumed, and by the end of Oct, the summit tiltmeter had recorded a net gain of 9.5 μrad."

Addendum. On 13 Nov at 0530, low-level activity at Pu'u O'o signaled the onset of episode 39. At 1534 continuous lava production began with fountain heights reaching an estimated 260 m, producing lava flows that extended about 6 km E. Vigorous activity lasted nearly 10 hours, stopping at 0124 on 14 Nov.

Information Contacts: C. Heliker, G. Ulrich, R. Hanatani, J. Nakata, USGS HVO.

EPISODE 39

11/85 (10:11) "Kilauea's eruption continued in Nov with the 39th eruptive episode. On 12-13 Nov, the magma column was within a few meters of the conduit rim and spattering nearly continuously. At about 0530 on 13 Nov, 3 small vents opened on the S flank of Pu'u O'o. Two of these vents remained active through the day and produced a pahoehoe flow that extended approximately 600 m beyond the base of Pu'u O'o (fig. 13-43).

"The summit of Kilauea began to deflate on 13 Nov at 1430. An hour later, activity at the main Pu'u O'o vent increased, and a low fountain began to feed a thin pahoehoe flow. The fountains gradually increased in height, reaching 415 m by 1925. The vents on the S side of the cone probably died at about the same time. The high fountains continued until 0124 the next morning and produced an aa flow extending 6 km ESE of Pu'u O'o.

"Summit subsidence continued until 14 Nov at 0330, with a total loss of 13.9 μrad recorded by the summit tiltmeter. By the end of Nov, the summit had regained 9.6 μrad.

"Harmonic tremor associated with episode 39 increased gradually from 13 Nov at 1610, peaked in amplitude between 1725 and 0100, and decreased at 0126 on 14 Nov. Following episode 39, harmonic tremor persisted at background levels in the E rift zone near Pu'u O'o."

Information Contacts: C. Heliker, R. Hanatani, R. Koyanagi, USGS HVO.

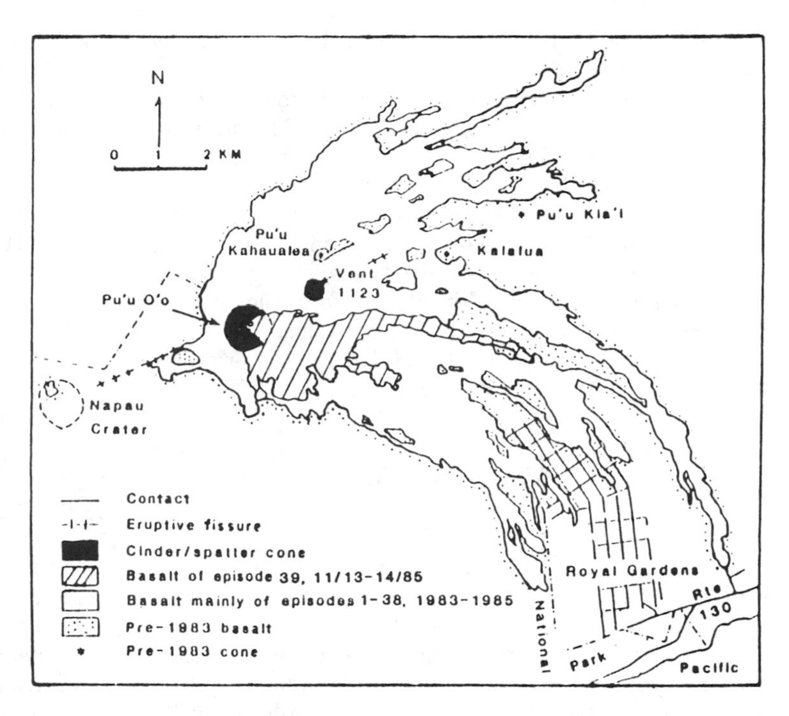

Fig. 13-43. Lava flows produced by the E rift zone eruption since Jan 1983. Episode 39 flows are marked by diagonal lines.

12/85 (10:12) "Dec ended the third year of Kilauea's ongoing E rift zone eruption. The month was uneventful, with Pu'u O'o maintaining an unusually long repose period after episode 39 on 13-14 Nov correlating with the low rate of inflation measured at the summit during most of the month.

"Harmonic tremor continued at a low level near Pu'u O'o. On 21 Dec a short burst of moderate-amplitude tremor that lasted about 30 minutes originated from a mantle source ~40 km deep, SW of Kilauea. The number of earthquakes of magnitude 2.5-3.5 on Kilauea's S flank increased during the last 2 weeks of Dec.

EPISODE 40

Addendum: Episode 40 arrived with the New Year, beginning at 1309 on 1 Jan and lasting 13.5 hours. A narrow aa flow passed through the NE corner of Royal Gardens subdivision, but remained on top of older lava flows. The flow eventually stagnated about 9 km from the vent.

1985 Summary. "Eleven eruptive episodes occurred during 1985, each characterized by high fountains from the Pu'u O'o vent and broad aa flows extending 3-7 km from the cone. Departures from the normal pattern were observed during the early stages of episodes 29, 35, and 39, when short-lived vents erupted at the base of the cone. During episode 35, fissure activity resumed after fountaining at the main Pu'u O'o vent had ceased and continued for 17 days, producing a pahoehoe shield 26 m high at the base of Pu'u O'o.

"Lava has covered 39 km^2 since Jan 1983, and the total volume of lava exceeds 0.46 km^3, surpassing all historic eruptions of Kilauea. The Pu'u O'o cone reached a height of 250 m in 1985 and is now the most prominent landmark on the E rift zone. The single conduit is 20 m in diameter, with its top 110 m below the summit of the asymmetric cone.

"Pu'u O'o lava compositions showed marked intra-episode variation during episodes 30 and 31, with more mafic lava erupted late in the episode. These variations suggest that during these episodes the eruption drew down a compositionally zoned magma chamber. Over the long term, the lava compositions appear to have stabilized at nearly constant values, suggesting that less mixing with differentiated melts stored in the rift zone is taking place compared to early episodes.

Information Contacts: C. Heliker, G. Ulrich, R. Koyanagi, R. Hanatani, USGS HVO.

MAUNA LOA

Hawaii, USA
19.48°N, 155.61°W, summit elev. 4170 m
Local time = GMT - 10 hrs

The summit caldera of this massive shield volcano (fig. 13-2) stands 9 km above the surrounding sea floor, making it the tallest historically active volcano in the world in height above base. It has had 40 known eruptions since the 18th century.

General Reference: Lockwood, J.P., Dvorak, J., English, T., Koyanagi, R., Okamura, A., Summers, M., and Tanigawa, W., 1987, Mauna Loa 1974–1984: A Decade of Intrusive and Extrusive Activity; *USGS Professional Paper* 1350, p. 537-570.

See also General References under Kilauea.

11/76 (1:14) A substantial increase in the number of earthquakes beneath Mauna Loa was recorded 20-24 Nov (table 13-4).

On the morning of 25 Nov, only 1 earthquake was recorded (beneath the NE rift). The USGS predicts a major eruption of Mauna Loa within the next 18 months.

Information Contact: UPI.

1/78 (3:1) The July 1975 eruption of Mauna Loa has been described by Lockwood and others (1976). Soon after this eruption, Mauna Loa began to inflate, and, chiefly on the basis of the historic patterns of activity, the USGS predicted that a second eruption, followed soon thereafter by a larger eruption from the NE rift, was likely to occur before July 1978. During the first half of 1977, however, local seismicity and inflation rates declined considerably, and in July 1977 the USGS issued a press release withdrawing the "before July 1978" date for the predicted eruption. Seismicity beneath the summit region has remained low, but measurements made in Dec 1977 indicate that the inflation rate for June–Dec 1977 had increased to that recorded during 1976. Furthermore, dense fume clouds have been emitted from the 1975 eruptive fissures since Oct 1977. The USGS has installed gas monitoring devices on one of these fumaroles and is continuing to monitor the volcano closely.

1976		Event Totals
Nov	20	350
	21	200-300
	22	200-300
	23	> 300
	24	< 100

Table 13-4. Earthquakes recorded at Mauna Loa 20-24 Nov 1976.

Reference: Lockwood, J., Koyanagi, R., Tilling, R., Holcomb, R., and Peterson, D., 1976, Mauna Loa Threatening; *Geotimes*, v. 21, no. 6, p. 12-15.

Information Contact: G. Eaton, USGS HVO.

5/83 (8:5) The following is from HVO.

"Mauna Loa last erupted in July 1975. That eruption was preceded by an increase in both shallow and intermediate-depth earthquakes, and by extension of survey lines across the caldera (fig. 13-44, left). Since 1980, and especially since early 1983, the number of shallow earthquakes beneath Mauna Loa has been increasing again. Intermediate-depth earthquakes have continued at a higher rate during the period from 1978 to present than during 1971–73, but have not shown the same pattern of increase as they did in 1974 (fig. 13-44, right). Figure 13-44 also shows a recent increase in the rate of extension of survey lines across the summit caldera.

"The recent rate of strain from apparent intrusion of magma beneath the summit region shows an increasing trend from both seismic and ground surface deformation data. The present strength of the summit region is not known, so no precise forecast of the next eruption can be made. However, the present seismic and deformation data indicate a significantly increased probability of eruption of Mauna Loa during 1983 or 1984."

Information Contacts: R. Decker, R. Koyanagi, J. Dvorak, USGS HVO.

3/84 (9:3) The following (except for the plume data) is from HVO. Times noted below are preliminary and subject to slight revision after later analysis.

"A long-expected flank eruption of Mauna Loa began on 25 March, and had ended by 14 April.

Background. "When summit seismic activity increased sharply in April 1974, Mauna Loa had not erupted since June 1950. Measurement of EDM lines across the summit caldera (Mokuaweoweo) in the summer of 1974 revealed significant extensions, monitoring capabilities were increased, and a forecast of renewed activity was issued (Koyanagi et al., 1975). The summit eruption of 5-6 July, 1975 lasted for less than 20 hrs, and only about 30 × 10^6 m^3 of lava were erupted. The eruption was identical to numerous other Mauna Loa summit eruptions that had been followed within 3 years by large flank eruptions. Given the historic record and continuing inflation, a forecast was made for renewed eruptive activity sometime before the summer of 1978 (Lockwood et al., 1976; see 3:1). The 1976 forecast was rescinded in 1977 but slow inflation continued and

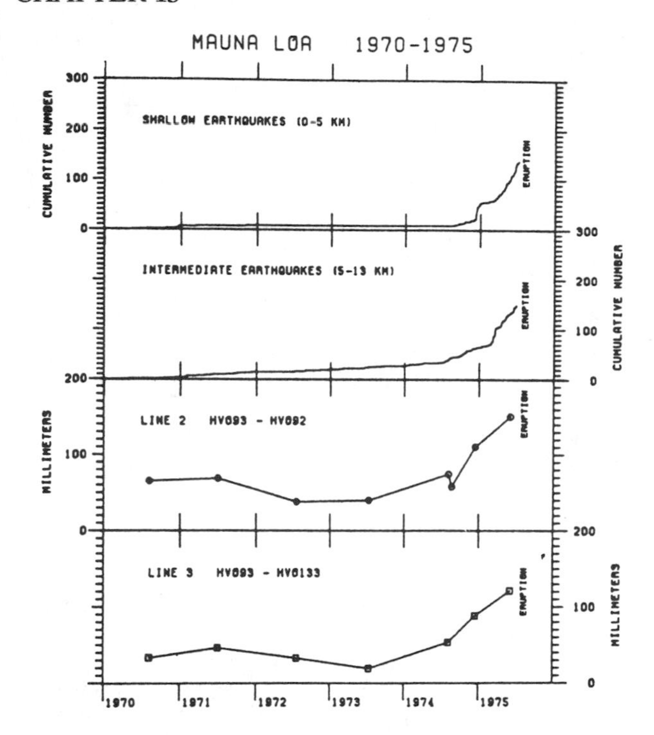

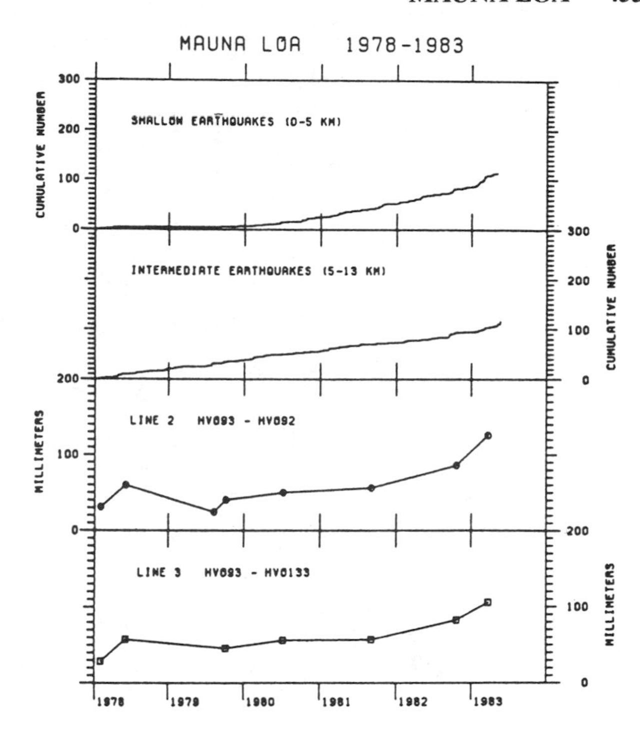

Fig. 13-44. Plot of cumulative number of local earthquakes at shallow (0-5 km) and intermediate (5-13 km) depths, and extension (in mm) on 2 survey lines across the summit caldera, for 1970–75, *left,* and 1978–83, *right.*

another forecast (based on an increase in the rate of geodetic change and seismic activity) was issued in 1983. This called attention 'to the increased probability of a Mauna Loa eruption within the next two years' (Decker et al., 1983), but see 8:5 in which the forecast was more specific: 'a significantly increased probability of eruption of Mauna Loa during 1983 or 1984'.

Premonitory Activity. "The 25 March outbreak gave almost no short-term instrumental warning. Seismic activity had been increasing gradually through March (fig. 13-45), but was relatively low immediately preceding the outbreak; only 29 microearthquakes were recorded beneath the summit caldera during the preceding 24 hrs (in contrast to 700 microearthquakes/day in Sept 1983).

"Several people saw probable fume clouds from the summit caldera and a camper at the summit noted small explosions from the 1975 eruptive fissures on 23 March. One hiker had reported seeing 'glowing cracks' near the 1940 cone on 18 March, but no anomalous activity was detected on a thermal probe in the 1975 fumaroles. Oxidation state and temperatures of fumarolic gases remained essentially unchanged prior to thc last satellite transmission about midnight on 24 March.

Eruption Narrative. "At 2255 on 24 March, a small earthquake swarm began directly beneath the summit. Weak harmonic tremor with an amplitude of about 1 mm was recorded at the summit station (WIL, fig. 13-46) at 2330. The number of small summit earthquakes increased at 2350. Tremor amplitude recorded at the summit increased to about 5 mm at 0015 on 25 March, remained high at 0051, and was recorded on all Mauna Loa and Kilauea summit area stations.

"At 0055 a magnitude 4.0 earthquake beneath the summit awoke geologists (from the University of Massachusetts) camped at Pu'u Ula'ula on the NE Rift Zone. At 0056, the telescope at the summit of Mauna Kea (42 km NNW of Mauna Loa) began high-amplitude oscillation, preventing astronomical observations for the next few hours. Between 0051 and 0210, 11 earthquakes

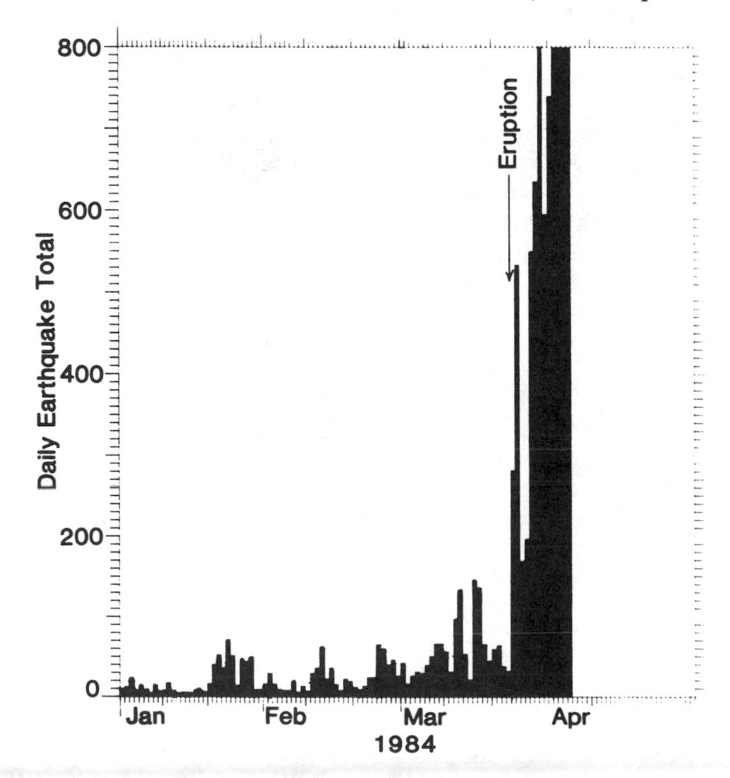

Fig. 13-45. Number of earthquakes per day at Mauna Loa, 1 Jan–5 April 1984. The start of the eruption is indicated by an arrow.

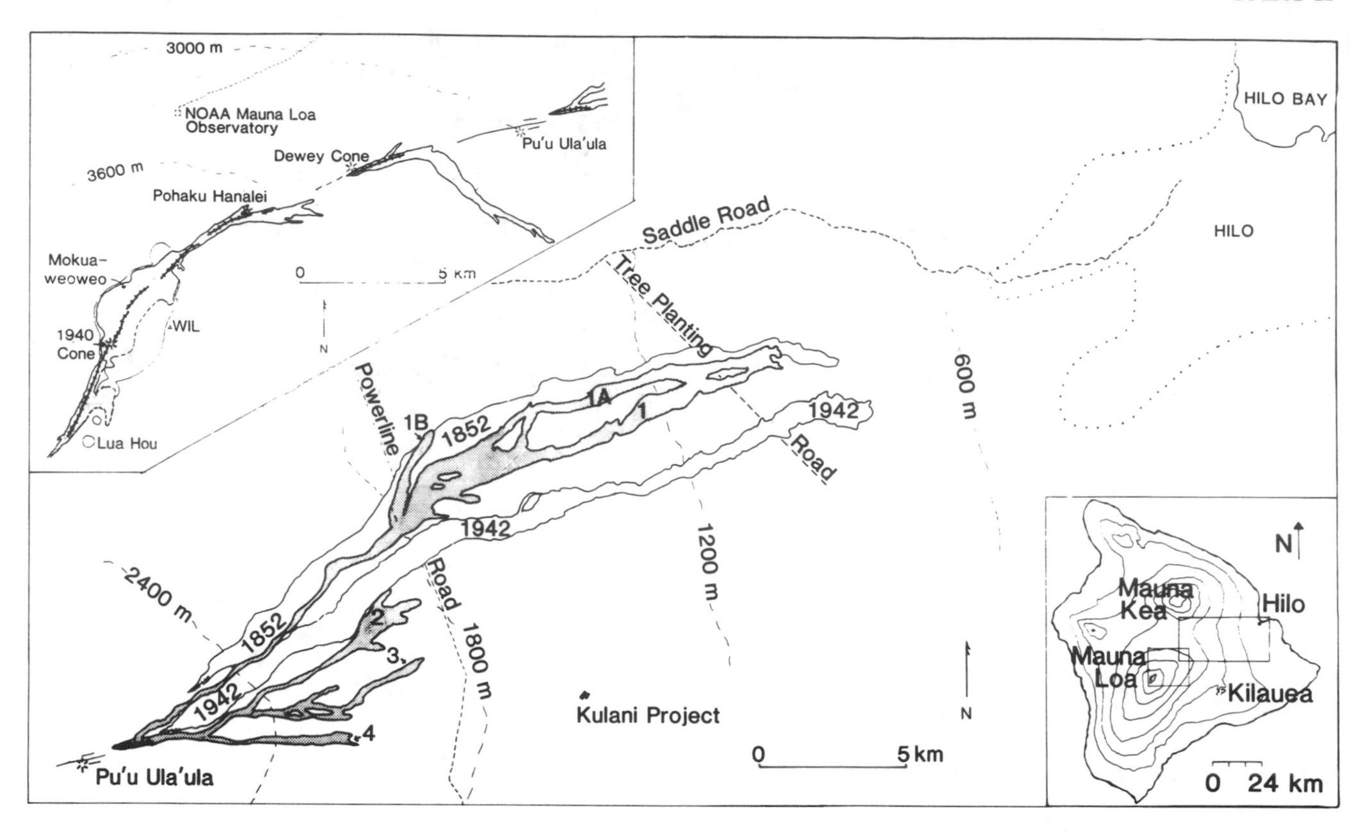

Fig. 13-46. Sketch maps of the NE rift zone and summit of Mauna Loa, showing positions of 1984 lava flows (stippled) as of 5 April 1984. Eruption fissures are indicated by hachured lines. The edge of the suburbs of Hilo is shown by a dotted line on the NE rift zone map. The areas covered are shown by the index map (inset).

with magnitudes between 2.0 and 4.1 were recorded beneath the summit. At 0100 borehole tiltmeters recorded the onset of rapid summit inflation.

"A military satellite detected a strong infrared signal from the summit at 0125. Glow was sighted in the SW portion (1940 cone area) of the summit caldera by an observer on the summit of Mauna Kea at 0129, by the geologists at Pu'u Ula'ula at 0130, and from Kilauea at 0140. At 0146, fountain reflection on fume clouds observed from HVO suggested that fountaining extended across much of SW Mokuaweoweo and was migrating down the SW Rift Zone.

"At 0232 the tops of fountains within Mokuaweoweo were seen from Pu'u Ula'ula, suggesting a height greater than 100 m. At approximately 0340, fountaining ceased on the SW Rift Zone. At 0357, 30 m-high fountains migrated out of Mokuaweoweo, down the upper NE Rift Zone. Lava flowed downrift and onto the SE flank.

"At approximately 0600, fountaining in the caldera gradually ended. At 0632, a new vent opened about 700 m E of Pohaku Hanalei and 8 minutes later another en echelon fissure began to erupt about 600 m downrift. Lava appearance was preceded by 3 minutes of copious white steam emission from the fissure. For the next 2 $^{1}/_{2}$ hours, activity waned.

"At 0905, profuse steaming appeared on a fracture at about 3510 m altitude, and at 0910 fountaining 15-40 m high began at 3410 m and migrated downrift. At 0930, fountains above Pohaku Hanalei died down as lava production increased to approximately 1-2 × 10^6 m^3/hr along a 2 km-long curtain of fire between about 3400 and 3470 m. The loci of most vigorous fountaining alternated along the 2-km fountain length. Much of the production from these vents was consumed by an open fissure parallel to and S of the principal fissure upslope, although an aa flow did move 5 km SE, S of an 1880 flow. During activity of these vents, episodic turbulent emissions of red and brown 'dust' from the eruptive fissures sent clouds to about 500 m height. At 1030, steaming was noted along a 1 km-long crack system extending from about 3260-3170 m, but there was no further downrift migration of eruptive vents for several hours. At about 1550, ground cracking extended below 3000 m, and at 1641 eruptive vents opened at about 2800 m and migrated both up- and downrift. At 1830 an eruptive vent extended about 1.7 km from about 2770-2930 m elevation. Fountains to 50 m height fed fast-moving flows to the E and NE. Activity waned at the 3400-m vents.

"By 0640 the next day, all lava production had ceased above 3000 m. Fountains (to 30 m height) were localized along a 500-m segment of the fissure that had opened the previous afternoon. The fastest moving flow cut the power line to the NOAA Mauna Loa Observatory shortly before dawn. At 0845, the E flows were spread out over a wide area above 1900 m elevation, but their advance slowed during the day. Four principal eruptive vents then developed along this fissure system. Two vents fed the NE flow (1), while the other two fed the S (2-4) flows. Flow 1 steadily advanced downslope 27-28 March (fig. 13-47), between the 1852 and 1942 lava flows. About 80% of the lava production fed flow 1. Flows 2-4 ceased significant advance by 28 March. The

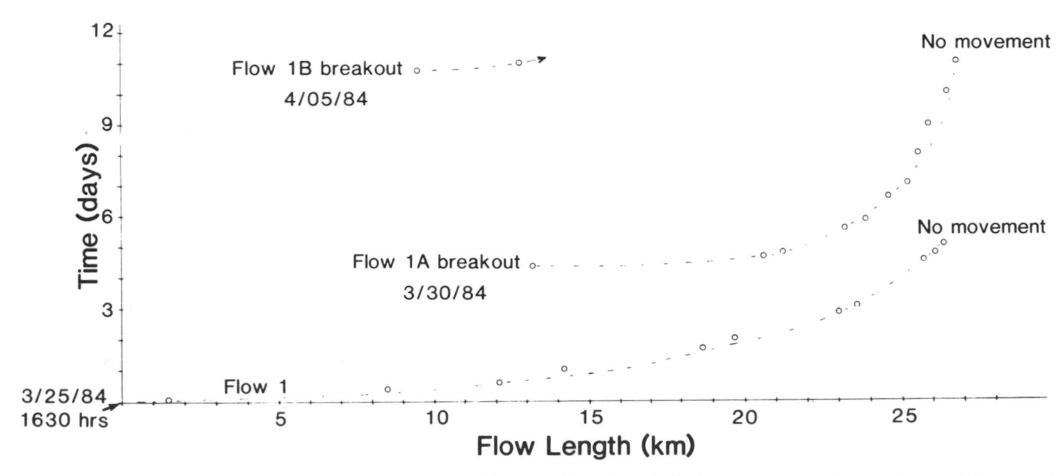

Fig. 13-47. Rates of movement of flows 1, 1A, and 1B in km/day. Small circles represent observations of flow positions. Courtesy of J.P. Lockwood.

terminus of flow 1 stopped significant advance by early 29 March, while production at the vents remained essentially constant. This suggested that a new branch flow had developed upslope. Bad weather prevented confirmation of the new branch until 30 March. This new flow (1A) moved rapidly downslope, N of flow 1.

"Phase 17 of Kilauea's E Rift Zone eruption began that morning but had no apparent effect on Mauna Loa activity. Likewise Kilauea tilt showed no deflection at the time of the Mauna Loa outbreak on 25 March.

"Flow 1A slowed on 31 March as the feeding channel became sluggish, and the flow thickened and widened upstream. At 1215 on 5 April, the flow was moving very slowly (18 m/hr) slightly below 900 m elevation. A major overflow at about 2000 m shut off most of its lava supply and created a fast-moving flow (1B), which advanced 3 km NE to about 1800 m elevation by 1700.

Deformation. "Much of the NE rift zone geodetic monitoring network was measured shortly before the 25 March outbreak, and EDM, tilt, and gravity stations were re-measured several times during the eruption. Although continuously recording tiltmeters at the summit showed sharp inflation (dike emplacement) immediately preceding the outbreak, major subsidence of the summit region accompanied eruptive activity along the NE rift. The center of subsidence, near the S edge of the summit caldera (fig. 13-48), was coincident with the center of uplift identified from repeated geodetic surveys between 1977 and 1983. The amount of summit deflation recorded by tilt and horizontal distance measurements exceeded the amount of gradual inflation of the volcano since the July 1975 summit eruption, suggesting substantial injection of magma into the summit area prior to this eruption, and possibly prior to the first EDM line across the summit caldera in 1964. Maximum vertical elevation change, inferred from repeated gravity measurements, is 500 mm.

"Large extensions occurred across the middle NE rift zone during dike emplacement on 25 March, but EDM monitor lines across this zone showed no significant change after the initial dilation. The rate of summit subsidence initially followed an exponential decay, similar to subsidence episodes in the summit region of Kilauea. Since 30 March, tilt and horizontal distance measurements have indicated a steady rate of deflation (figs. 13-49 and 13-50), although measurements on 6 April suggest decreasing deflation rates.

Dike Propagation. "All dikes were emplaced within the first 15 hours of the eruption. The eruptive fissure (surface expression of dikes) extended discontinuously along a 25-km zone from 3890 m on the SW rift zone to about 2770 m on the NE rift zone. Ground cracking along most of this zone demonstrates the continuity of the dike at shallow levels. Lateral propagation rates vary from > 2500 m/hr down the SW rift zone to about 1200 m/hr in lower parts of the NE rift zone (fig. 13-51).

Petrography, Lava Temperatures, Gas Measurements. "Hand specimens of the 1984 basalt are very fine-grained

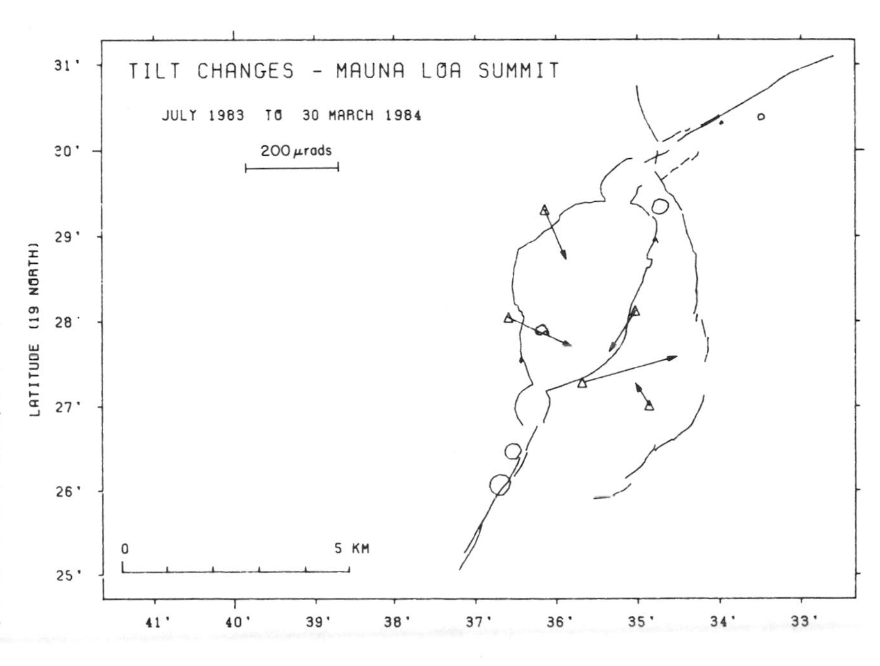

Fig. 13-48. Tilt changes near the summit of Mauna Loa, July 1983–30 March 1984.

Island	Distance	Visibility	1984
Johnston	1400 km WSW	6 km	2 Apr 2200-3 Apr 0200
Wake	3900 km W	1.6 km	2 Apr 1200-1700
Ponape	5000 km WSW	3.2 km	2 Apr 1000-1400

Table 13-5. Visibilities at airports on several islands affected by the plume from Mauna Loa (distances are from Mauna Loa). All times are Hawaiian Standard Time. Note that all except Johnston Is. are across the International Date Line from Hawaii. Data courtesy of NOAA.

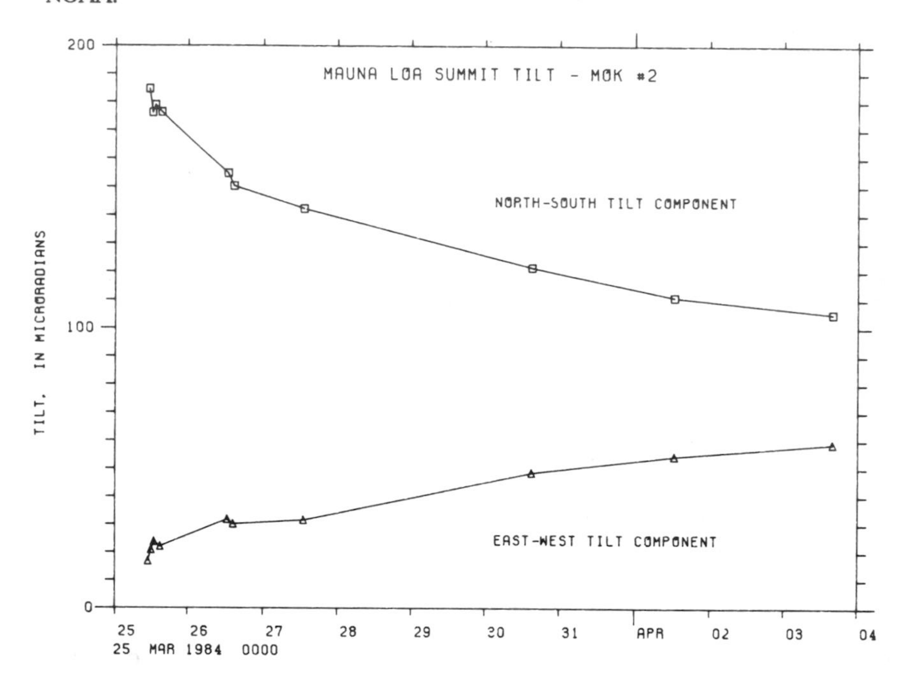

Fig. 13-49. Plot of summit tilt vs. time, 25 March – 3 April 1984.

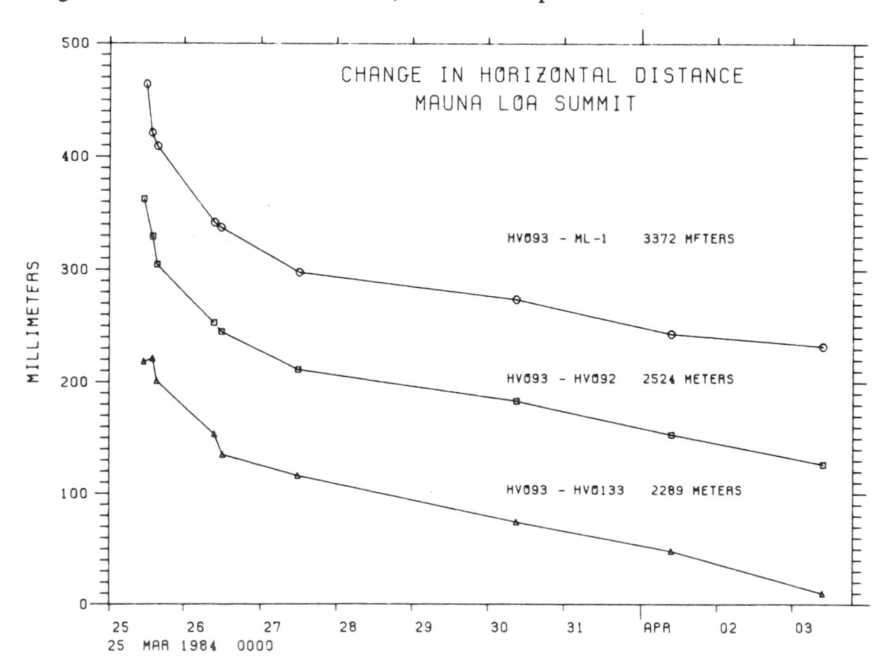

Fig. 13-50. Change in horizontal distance across the summit of Mauna Loa vs. time, 25 March – 3 April 1984.

with widely scattered (<1%) phenocrysts of olivine <3 mm in diameter and sparse microphenocrysts of plagioclase and clinopyroxene. Most olivines are anhedral, resorbed, commonly kinked, and surprisingly forsteritic (Fo 88-90). Plagioclase and clinopyroxene are barely resolvable in the groundmass. Maximum temperatures determined repeatedly by thermocouple and radiometer ranged from 1137° to 1141°C and had not changed as of 5 April.

"Eruptive gases have been extensively sampled and analyzed. Observed C/S ratios are much lower than expected in primitive Hawaiian tholeiite, suggesting extensive degassing in a shallow (<4 km deep) magma reservoir.

Geo-electric Studies. "One self-potential (SP) profile, first measured in July 1983, exists across the NE Rift Zone about 1 km W of the main erupting vents. The first complete reoccupation of the SP line 3 days after the eruption's start showed an amplitude increase slightly > 100 mV centered over a zone about 300 m wide across the 1.5 km-long crack zone N of Pu'u Ula'ula. VLF measurements show that the dike is located nearly in the center of the cracked zone, directly beneath the pre-existing SP maximum, at a very approximate depth of 150 m.

Areal Extent and Lava Volume. "As of 5 April, 25-30 km^2 of area was covered. The lava is mostly pahoehoe near the vents, but is mostly aa more than 2 km from the vents. The volume was estimated to be about 150 × 10^6 m^3 by 5 April."

Eruption Plume. The eruption produced a large gas plume that was carried thousands of kilometers to the W. The plume from the summit caldera activity was clearly visible from HVO. An airline pilot approaching Honolulu at dawn 25 March reported that the top of the plume was 10.7-11 km altitude and was drifting SW. Observers at Honolulu airport tower (300 km NW of Mauna Loa) reported that the top of a tall cumulus-like cloud became visible S of the airport just before dawn. There was no evidence that the plume reached the stratosphere; the tropopause on 25 March was at about 18 km altitude. The plume was carried W by trade winds. By 30 March, a haze layer was detected at Wake and Johnston Islands (table 13-4). Haze reached Kwajalein (4000 km WSW of Mauna Loa) the next day and reached Guam (6300 km WSW of Mauna Loa) by 2 April.

SO_2 emitted by Mauna Loa was detected by the TOMS instrument on the Nimbus-7 polar orbiting satellite, which passed over Hawaii daily at about 1200 (fig. 13-52). Although the TOMS instrument was designed to measure ozone, it is also sensitive to SO_2. An algorithm has been developed to isolate SO_2 values and calculate its approximate concentration within pixels (picture elements) ~50 km in diameter. Preliminary estimates of the total SO_2 in the Mauna Loa plume, using TOMS data, were roughly 130,000 metric tons on 26 March and 190,000 metric tons on 27 March.

References: Decker, R.W., Koyanagi, R.Y., Dvorak, J.J., Lockwood, J.P. Okamura, A.T. Yamashita, K.M., and Tanigawa, W.R., 1983, Seismicity and Surface Deformation of Mauna Loa Volcano, Hawaii, *EOS*, v. 64, no. 37, p. 545-547.

Koyanagi, R.Y., Endo, E.T., and Ebisu, J.S., 1975, Reawakening of Mauna Loa Volcano, Hawaii; A Preliminary Evaluation of Seismic Evidence, *Geophysical Research Letters,* v. 2, no. 9, p. 405-408.

Information Contacts: J. Lockwood and HVO staff, Hawaii; M. Rhodes, Univ. of Massachusetts; M. Garcia, Univ. of Hawaii; T. Casadevall, USGS CVO, Vancouver, WA; A. Krueger, NASA/GSFC; M. Matson, NOAA/NESDIS.

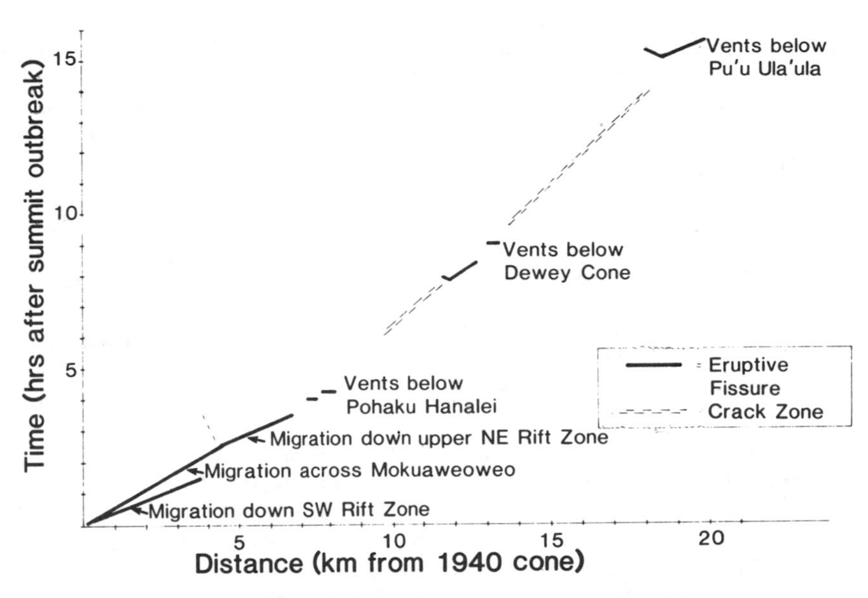

Fig. 13-51. Rate of propagation of eruption fissures, shown as distance from the 1940 cone (in the SW part of the summit caldera) vs. hours after the start of the eruption.

4/84 (9:4) "The NE Rift Zone eruption, which began on 25 March, ended early on the morning of 15 April. Lava output and fountain vigor steadily decreased during the last week of the eruption. As flow channels became blocked (by sluggish aa and channel collapse breccias) progressively farther upslope, flows terminated higher on Mauna Loa's NE flank. Many short overflows of viscous aa, up to 15 m thick, moved less than a few hundred meters from these points of channel blockage. By 10 April, no lava flowed below 2400 m. The total area

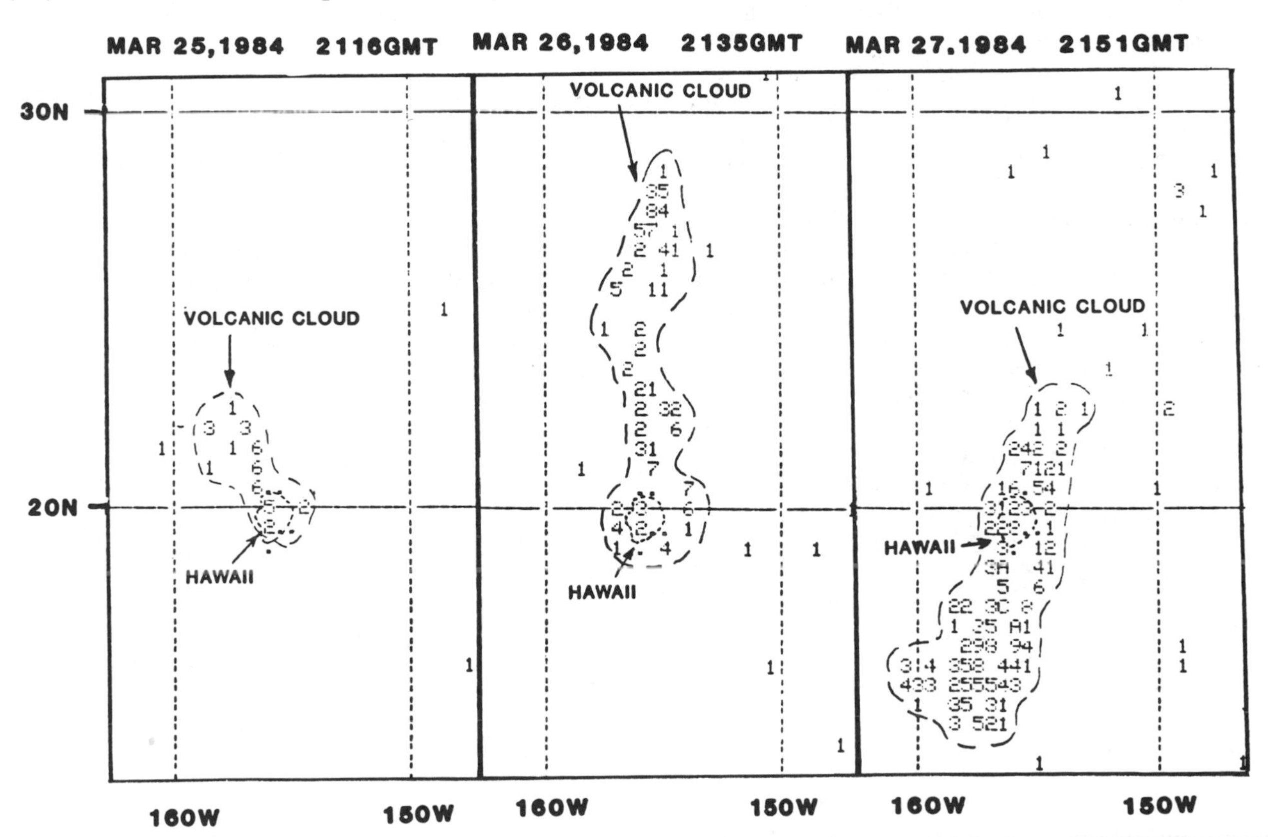

Fig. 13-52. Preliminary SO_2 data from the TOMS instrument on the Nimbus 7 satellite. All values less than 10 milliatmosphere-cm (100 ppm-meters) have been supressed. Each number or letter represents the average SO_2 value within an area 50 km across. 1 = 11-15 matm-cm = 101-150 ppm-m, 2 = 16-20 matm-cm = 151-200 ppm-m, etc.; 9 is followed by A, B, C, etc. Courtesy of Arlin Krueger.

covered by new lavas increased very little after 5 April, as multiple flows mostly piled on top of older flows. Total volume for this eruption was estimated at 180-250 $\times 10^6$ m^3."

Information Contacts: J. Lockwood, T. Wright, HVO, Hawaii.

MACDONALD SEAMOUNT

S-central Pacific Ocean
28.98°S, 140.25°W, depth ca 25 m
Local time = GMT - 10 hrs

Midway between Australia and South America, this volcano is 1590 km SE of Tahiti and 3000 km W of Easter Island. The Tubai (or Austral) chain of islands extend WNW parallel to the Hawaiian and other Pacific island chains. They are believed to be the progressively older products of a "hot spot," fixed in the mantle and feeding magma to the overlying Pacific plate as it moves slowly (10 cm/yr) WNW.

11/81 (6:11) Hydrophones recorded a submarine eruption from Macdonald Seamount 29 May 1967 (Johnson, 1970). After more than 10 years of apparent inactivity, acoustic waves (T-phase) from 6 eruptions were detected by French Polynesian seismic stations (at Tubuai, 23.3°S, 149.5°W; Tahiti, 17.6°S, 149.5°W; Moorea, 17.5°S, 149.8°W; Rangiroa 15.0°S, 147.7°W; Hao, 18.1°S, 141.0°W; and Rikitea, 23.1°S, 135.0°W) between Dec 1977 and Feb 1981. The acoustic record of each eruption began with intense explosive signals, followed by a few hours to a few days of amplitude-modulated noise that was frequently punctuated by brief periods of additional explosive activity, less commonly by strong (but apparently non-explosive) increases in noise amplitude.

The first eruption, which was the longest and most vigorous, was first detected at 1630 on 10 Dec 1977 and returned continuous noise for about 92 hours, during which nearly 50 periods of explosive activity could be recognized. Occasional bursts of noise continued for another 30 hours before activity ended at about 1830 on 15 Dec. A strong explosion recorded at 0246 on 30 Sept 1979 was followed by fewer than 5 hours of diminishing activity. About 20 minutes of frequent explosions first recorded at 1330 on 12 Feb 1980 were followed by roughly 12 hrs of eruption noise. Explosions that began to appear on the records at 0109 on 10 Nov 1980 remained strong and frequent for about 7 hrs and eruption noise continued for about 15 additional hrs. On 24 Dec 1980, seismic instruments detected intermittent explosions from 0610 until roughly 0900, continuous eruption noise for about 4 more hrs, then intermittent noise that lasted until early the next morning. Explosions at 0611 and 0620 on 15 Feb 1981 were followed by about 12 hrs of eruption noise. No additional activity had been recorded as of late Oct.

Macdonald was discovered after the 1967 eruption, and bathymetric work in Dec 1973 defined a submarine edifice reaching to within 49 m of the ocean surface (Johnson, 1980). More recent bathymetry by the French National Marine vessel *La Paimpolaise* indicates that further growth of Macdonald has occurred, bringing its summit to 23 m below sea level.

References: Johnson, R.H., 1970, Active Submarine Volcanism in the Austral Islands; *Science,* v. 167, p. 977-979.

Johnson, R.H., 1980, Seamounts in the Austral Islands Region; *National Geographic Society Research Reports,* v. 12, p. 389-405.

Information Contact: J. Talandier, Lab. de Géophysique, Tahiti.

Further Reference: Talandier, J. and Okal, E.A., 1982, Crises Sismiques au Volcan Macdonald (Ocean Pacifique Sud); *C.R. Acad. Sci. Paris,* ser. II, v. 295, p. 195-200.

2/82 (7:2) Renewed submarine activity at Macdonald Seamount was detected 1 March by Polynesian seismic stations. Acoustic waves (T-phase) were recorded at Moorea, Tubuai, Vaihoa at Rangiroa, and Rikitea in the Gambier Islands. Explosive signals were first recorded at 1337 and were followed by continuous noise of varying intensity. The initial phase, a few hours long, was succeeded by sporadic activity that lasted until about 0400 on 2 March. The latest activity is comparable in length, intensity and development to that of Feb 1980.

Bathymetric work that ended in Feb 1982 and included dredging rocks from the summit peak and adjacent plateau precisely located the submarine volcanic edifice. Its top was 27 m below sea level.

Information Contact: Same as above.

4/83 (8:4) On March, 4 RSP stations recorded acoustic waves (T-phase) from activity interpreted as a shallow submarine eruption at Macdonald Seamount. The signals were received at Moorea, Vaihoa on Rangiroa, Tubuai, and Rikitea [coordinates in 6:11].

Strong explosive signals began at 0914, and were followed by continuous noise of varying intensity. The activity began in the same manner as the 7 previous eruptions, and was probably comparable in strength and development to that of Feb 1981. However, high background noise levels caused by the passage of Cyclone Reva over Polynesia made interpretation difficult.

Information Contact: Same as above.

8/83 (8:8) In May, the RSP recorded seismicity from renewed eruptive activity at Macdonald. Its 8 previous eruptions had begun with explosive events, but the May activity did not, and probably was a continuation of the March eruption. Reconaissance by a Marine National Française vessel did not show a perceptible increase in the volcano's summit altitude since the bathymetric survey of Feb 1982.

[This same report carried observations of pumice both E of the Kermadecs and E of Tahiti (Tuamotu Archipelago). This report is reproduced following Monowai Seamount (Kermadecs) in Chapter 4 of Part II.]

Information Contact: Same as above.

Further Reference: Talandier, J. and Okal, E.A., 1984, New Surveys of Macdonald Seamount Following Volcanoseismic Activity, 1977–1983; *Geophysical Research Letters*, v. 1, no. 9, p. 813-816.

4/84 (9:4) Between May 1983 and Jan 1984, the RSP recorded acoustic waves (T-phase) from 3 shallow submarine eruptions at Macdonald Seamount. On 17 May 1983, eruptive activity began gradually with a few explosive sequences and lasted 4 1/2 days until 21 May. Activity resumed explosively on 27–28 Oct and continued for 15 hours with numerous explosive sequences. On 24 Dec, activity began gradually with no explosive sequences and continued for almost 10 days until 3 Jan 1984. This last event was the 12th and longest eruption recorded since the volcano was discovered in 1967.

Information Contact: Same as 6:11.

[Although all evidence of volcanism at Macdonald had been by indirect, geophysical means, its active status was dramatically confirmed in 1987, when the volcano erupted during surveys by the research ship *Melville* (*SEAN Bulletin* 12:9-10). There was no damage to ship or personnel.]

MEHETIA

Society Islands, French Polynesia
17.88°S, 148.07°W, summit elev. 435 m

Mehetia is a 1.5 km-diameter island 100 km E of Tahiti. No historic eruptions were known prior to the activity described here.

General Reference: Talandier, J. and Okal, E., 1987, Seismic Detection of Underwater Volcanism: The Example of French Polynesia; *Pure and Applied Geophysics*, v. 125, no. 6, p. 919-950.

10/81 (6:10) A swarm of earthquakes centered beneath Mehetia began suddenly on 6 March and was continuing in late Oct. After the first 2 days of the swarm, characterized by numerous weak events, seismographs began to record occasional larger shocks (fig. 13-53). The seismic energy released during the first week of the swarm greatly exceeded the previous total energy release detected by the Tahitian seismic net since the first stations were installed in 1962–63. Both the number of earthquakes (fig. 13-54) and energy release (fig. 13-55) varied considerably, with periods of increased activity separated by brief lulls. Epicenters were about 10 km SE of the crater. There were several groups of foci, which may indicate vertical migration, but it was impossible to compute depths of focus for most of the events. A temporary seismic station operated on Mehetia 25-30 March recorded local earthquakes at about 13 km depth.

By late Oct, more than 3000 local events of M_L 0.9 or stronger had been recorded. Of these, about 30 were stronger than M_L 3.0, including M_L 4.0 and 4.3 shocks. Because of the detection limits (usually about M_L 1.1, but weaker seismicity is sometimes recorded during periods of lesser microseismic noise) for events in the Mehetia area (about 140 km from the nearest Tahitian net station), harmonic tremor, if any, would probably not be recorded. Talandier's extrapolation of the well-defined frequency-magnitude relationship indicates that ~50,000 events stronger than M_L 0.1 have probably occurred during the swarm. Talandier also notes that the Mehetia swarm and seismicity associated with Hawaiian volcanoes showed similarities in number and magnitude of events and seismogram characteristics, as well as the detection of long-period waves that could be generated by events occurring beneath a magma chamber. No surface volcanic activity has been

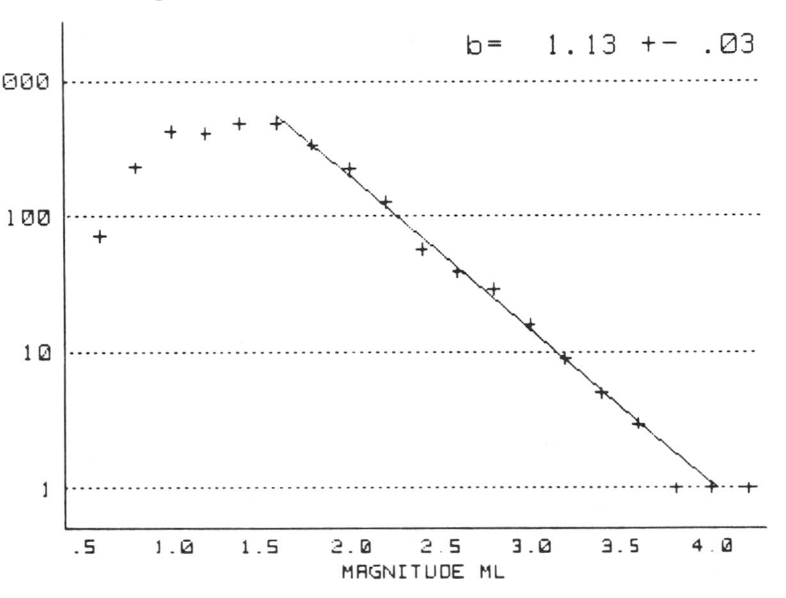

Fig. 13-53. Log plot of the number of events vs magnitude at intervals of M_L 0.2, Mehetia, 6 March – 15 Dec 1981. The b value (slope) of the regression line is 1.13 ± .03 for the entire swarm. For the first 2 days of the swarm, characterized by numerous weak events, b = [1.63]. Courtesy of J. Talandier.

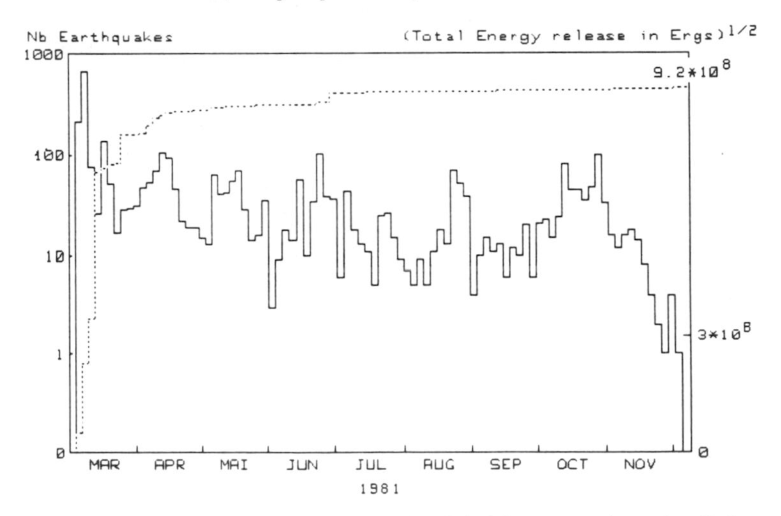

Fig. 13-54. Number of recorded earthquakes (M_L 0.9 or greater) per day (3-day means) at Mehetia, 6 March – early Dec 1981. The Tahitian seismic net records most local events of M_L 1.1 or greater, but fluctuating levels of microseismic noise prevent consistent recording of weaker events. Courtesy of J. Talandier.

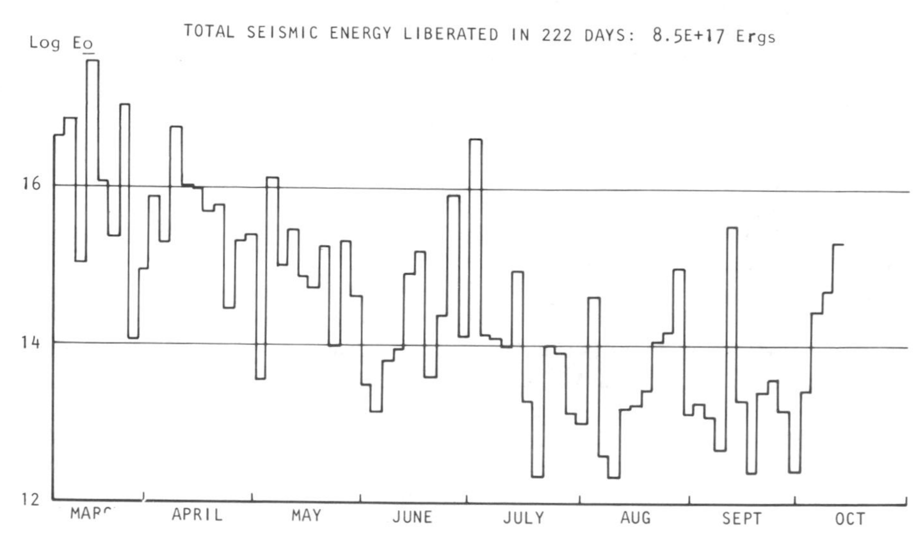

Fig. 13-55. Seismic energy released by the Mehetia swarm (3-day intervals), 6 March-mid Oct, 1981. Courtesy of J. Talandier.

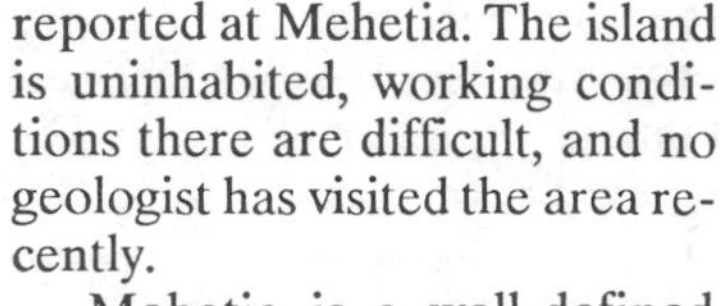
reported at Mehetia. The island is uninhabited, working conditions there are difficult, and no geologist has visited the area recently.

Mehetia is a well-defined cone about 1500 m in diameter and 435 m high, with a 200 m-diameter crater at the summit. No historic eruptions have been reported, but the limited erosion of the crater and flanks, lack of vegetation at the summit, and Tahitian legends of "big fires" all indicate that eruptions probably took place less than 2000 years ago.

Information Contact: J. Talandier, Lab. de Géophysique, Tahiti.

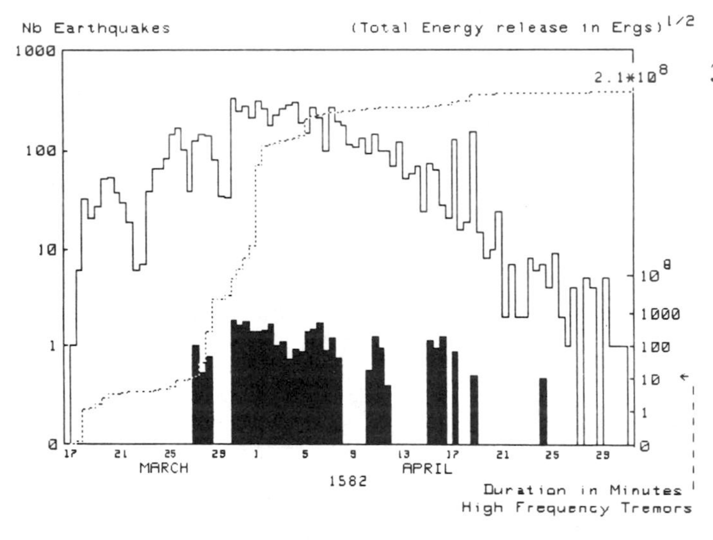

Fig. 13-56. Number of recorded earthquakes centered near Teahitia at 12-hour intervals (solid line), duration of high-frequency tremor in minutes (bars), and cumulative seismic energy release (dashed line, log scale), 17 March – 30 April 1982. Courtesy of J. Talandier.

3/82 (7:3) Seismic activity that began in March 1981 ceased in Dec. Only a few low-energy events/month have been recorded since. Bathymetric reconnaissance around the island found evidence of an elliptical opening at 1700 m below sea level on the SE flank, in the same location as the initial events of the earthquake swarm. RSP scientists interpreted the opening as a possible crater and the activity as a magmatic intrusion or eruption.

Information Contact: Same as above.

TEAHITIA

Society Islands, French Polynesia
17.57°S, 148.86°W, summit depth 2000 m

This deep seamount, 40 km NE of Tahiti, was not thought to be volcanically active before 1982.

General Reference: See Mehetia introduction.

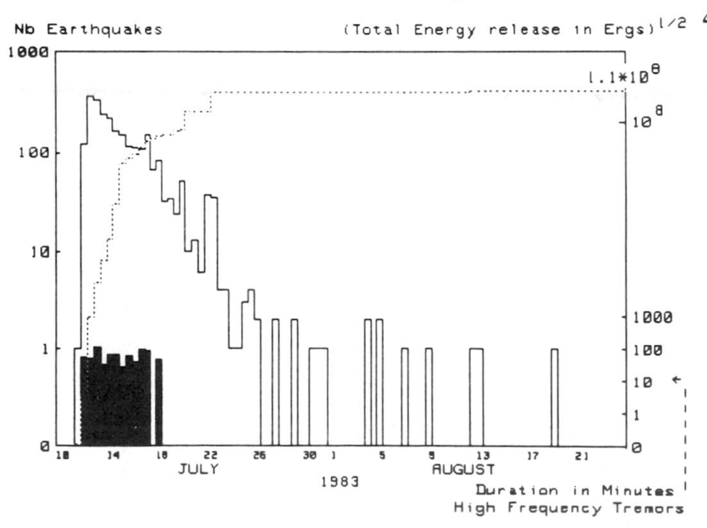

Fig. 13-57. Number of recorded earthquakes, as in fig. 13-56, 10 July – 24 Aug 1983. Courtesy of J. Talandier.

4/82 (7:4) On 14 March, the RSP began to record earthquakes in the vicinity of a seamount, with a summit at about 2 km below sea level, located about 3 km W of Rocard submarine volcano (17.640°S, 148.60°W) and 40 km NE of Tahiti's Taiarapu Peninsula. The increasing activity consisted only of low-magnitude earthquakes until 25 March, when the 5 Tahiti-Moorea seismic stations began to detect volcanic tremor of 5-10 Hz frequency (fig. 13-56). The Laboratoire de Géophysique attributed the tremor to magma movement and submarine eruptions. Periods of tremor, of variable duration but nearly continuous on some days, accompanied the increasingly numerous discrete earthquakes. Tremor began to decline after 9 April and none has been recorded since the 18th. As of 23 April, weak earthquakes and episodic 1.5-2 Hz seismic noise continued. More than 10,000 individual earthquakes (M_L > 0.9) were recorded during the swarm, the strongest of which had magnitudes of 3.5-4 (M_L) and were felt on Tahiti.

No volcanic activity had previously been known at

the seamount, for which the name Teahitia has been proposed. The activity was similar to that of Mehetia (90 km ESE), March – Dec 1981 (6:10, 7:3).

Information Contact: J. M. Talandier, Lab. de Géophysique, Tahiti.

8/83 (8:8) Between 11 and 20 July, the RSP recorded 3-4000 shallow earthquakes at Teahitia, accompanied by high-frequency volcanic tremor (fig. 13-57).

Information Contact: Same as above.

4/84 (9:4) From Aug 1983 to March 1984, the RSP recorded numerous sequences of low-frequency volcanic tremor and 2 seismic swarms associated with submarine eruptions at Teahitia. On 20 – 21 Dec, 300 very small earthquakes were recorded. From 3 March – 15 April 1984, approximately 9000 earthquakes were recorded, accompanied by low- and high-frequency spasmodic and harmonic tremor (fig. 13-58).

Information Contact: Same as above.

1/85 (10:1) Between 11 and 22 Jan, RSP stations on Tahiti recorded about 10,000 seismic events near Teahitia (fig. 13-59). From 15 to 19 Jan, about 50 seismic events stronger than M_L 3.5 were recorded. A number of those were felt by some of the inhabitants of Tahiti, 40 km SW, and three, of magnitudes 4.0, 4.2, and 4.4, were felt by the entire population. Very intense, high-frequency, long-duration volcanic tremors were recorded 19 Jan. Talandier noted that the intensity and duration of the tremors pointed to magmatic transfer. He also stated that this swarm certainly led to a submarine eruption, as did previous swarms. There were more events and greater seismic energy release in the Jan swarm than in previous swarms in April – May 1982, July and Dec 1983, and April – May 1984 (7:4, 8:8, 9:4). Only the 1982 swarm had stronger deep activity.

Information Contact: Same as above.

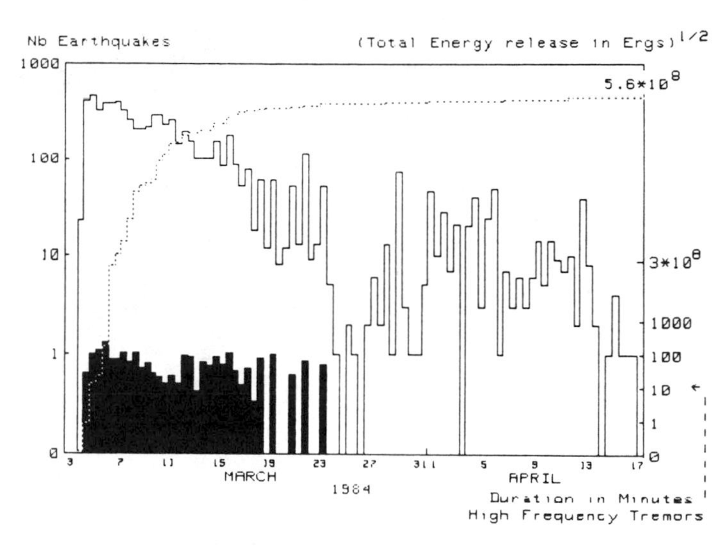

Fig. 13-58. Number of recorded earthquakes as in fig. 13-56, 3 March – 17 April 1984. Courtesy of J. Talandier.

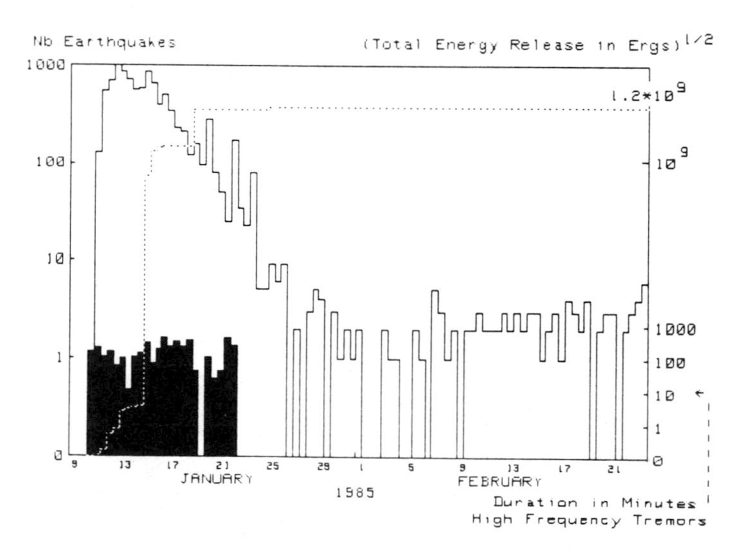

Fig. 13-59. Number of recorded earthquakes as in fig. 13-56, 5 Jan – 23 Feb 1985. Courtesy of J. Talandier.

MEXICO and CENTRAL AMERICA

COLIMA

Jalisco, México
19.42°N, 103.72°W, summit elev. 4100 m
Local time = GMT - 6 hrs

Colima is 145 km S of Guadalajara and 475 km W of México City. It has been México's most active volcano during historic time, with a record of 49 eruptions since 1560.

General References: Luhr, J.F. and Carmichael, I.S.E., 1980–82, The Colima Volcanic Complex: I, II, and III; *Contributions to Mineralogy and Petrology*, v. 71, p. 343-372; v. 76, p. 127-147; and v. 80, p. 262-275.

Medina Martínez, F., 1983, Analysis of the Eruptive History of the Volcán de Colima, México (1560–1980); *Instituto de Geofísica*, v. 22-2, p. 157-178.

Robin, C., Mossand, P., Camus, G., et al., 1987, Eruptive History of the Colima Volcanic Complex (México); *JVGR*, v. 31, no. 1/2, p. 99-113.

12/75 (1:3) On 1 Dec, 1975, the volcano began erupting. The summit crater was filled with viscous lava, and 2 flows descended the E flanks. There were hot avalanches, and eruptions were frequently visible at night from locations 20 km away. Colima, a strato-volcano, was last active in April, 1966.

Information Contact: S. de la Cruz-Reyna, Universidad Nacional Autonoma de México (UNAM).

1/76 (1:4) By the end of Dec the E lava flow had descended the flank of the cone, reaching the vegetation and a ravine at the base. The flow advanced about 2500 m in 60 days. On 28 Jan 1976 two new lava flows started to develop at the S and W sides of the dome [but see 1:6].

Information Contact: Same as above.

3/76 (1:6) The eruption decreased during Feb and (presumably) March. The lava flows that were thought to be developing on the W side of the dome in late Jan did not develop at all. The lava flow on the E flank, about 2 km long, was moving very slowly.

Other activity has almost stopped. de la Cruz last visited the volcano about 1 March 1976.

Information Contact: Same as above.

4/76 (1:7) Servando de la Cruz has not visited the volcano recently, but reports that he has received indicate that activity has almost completely ceased. He will visit the volcano again in mid-May.

Information Contact: Same as above.

5/76 (1:8) Servando de la Cruz did not visit the volcano in mid-May since the activity had already ceased. An observer located near the volcano will report any new activity directly to him.

Information Contact: Same as above.

9/81 (6:9) The following information is from visits to Colima Dec 1977–Jan 1978, Dec 1979, and Feb 1981 by James Luhr and others.

Since the extrusion of more than 10^8 m^3 of andesitic block lava between Nov 1975 and June 1976, activity has consisted of numerous brief ejections of ash and incandescent material, and several episodes of lava dome growth, all in the E part of the summit crater. A small steaming dome (about 100 m in diameter and extending about 15 m above the crater rim) was observed in the E part of the summit crater during the Dec 1977–Jan 1978 observations. When Luhr and others returned in Dec 1979, this dome had disappeared and the E part of the crater had a relatively flat floor, only about 2 m below the crater rim, containing numerous explosion vents 1-5 m in diameter. By Feb 1981, a new lava dome had been extruded into the E part of the summit crater. A steep-walled vent about 50 m across occupied the center of the dome, which was about 150 m in diameter and reached a height of about 50 m above the crater rim. The geologists interpreted the dome's smooth reddish SE flank (in the direction of the principal 1975–76 lava flows) as more likely to have been caused by slumping than by tephra accumulation. The remainder of the dome was composed of block lava. No information is presently available on post-Feb activity.

Information Contact: J. Luhr, Univ. of California, Berkeley.

1/82 (7:1) Colima began to erupt during the first week of Dec. Very viscous lava emerged from 2 vents in the dome that has covered the summit crater since the late 1950's. Aerial observations revealed that lava extruded from a vent in the E half of the dome flowed down its S flank. Collapse at the end of this lava flow produced small glowing avalanches, visible 30 km away during the night, that moved about 2 km downslope to the base of the cone. Less voluminous lava extrusion occurred from a vent in the W half of the dome. Activity was continuing as of 20 Jan.

No felt earthquakes have been reported, nor were volcanic earthquakes detected during 4 days of monitoring with 4 portable seismographs near Colima. However, small tremors of unspecified origin were detected.

The present eruption is similar to that of Nov 1975–June 1976 (1:3-4, 6-8). The most recent dome growth occurred between visits by geologists in Dec 1979 and Feb 1981.

Information Contacts: S. de la Cruz-Reyna, F. Medina, UNAM, México.

2/82 (7:2) The following is from William Rose, Jr.

"Colima was observed from the air for 3 hrs around

midday 20 Feb as part of the Research on Active Volcanic Emissions (RAVE) sampling mission. The volcano had an apparently active block lava flow descending from the summit down the S flank for several hundred meters. Below this level was active scree. The flow was gray, contrasting with the various browns of the earlier lava units. Heat waves emanated from the flow and rockfalls were rare, indicating a slow extrusion rate. The volcano had a significant gaseous plume that was sampled and measured by the RAVE Electra aircraft with a complete battery of atmospheric sampling gear. During the sampling, 1 short (30-second) burst of gaseous emission occurred, in which the emission rate approximately doubled."

Information Contacts: W. Rose, Jr., Michigan Tech. Univ.; T. Casadevall, USGS CVO, Vancouver, WA; W. Zoller, Univ. of Maryland; W. Fuller, NASA Langley Research Center.

Further Reference: Casadevall, T., Rose, W.I. Jr., Fuller, W., Hunt, W., Hart, M., Moyers, J., Woods, D., Chuan, R., and Friend, J., 1984, Sulfur Dioxide and Particles in Quiescent Volcanic Plumes from Poás, Arenal, and Colima Volcanoes, Costa Rica and México; *JGR*, v. 89, no. D6, p. 9633-9641.

3/82 (7:3) The following report from James Luhr supplements the report from Mexican scientists in 7:1.

"The andesitic block lava that began to flow from the summit crater dome in early Dec was the first to descend Colima's S flank for hundreds of years. Geologists from the Univ. of California at Berkeley observed the flow from the S side of the volcano starting 18 Jan, about the time of the report in 7:1. The new lava was moving down a polished avalanche chute with a slope of about 36°. On 20 Jan, the flow had a simple tongue shape and was some 600 m long. By 3 March, the lava had reached 1 km length. Block and ash flows were common from the uppermost margins of the lobe with surprisingly few from the flow front. In several instances, sizeable (2000 m^2 ?) areas on the flow surface suddenly shifted downslope 5-10 m, accompanied by only small amounts of ash and steam. This may be a major process of downslope movement of the flow. The active scree deposit below the lava contained blocks several meters in diameter, grading into a new sand and conglomerate wedge flooding the upper reaches of the Barranca Playa de Montegrande.

"Since the early part of Colima's lava eruption of 1975–76, through several episodes of dome growth, the andesitic magma has become progressively more basic. The latest lava continues this trend."

Information Contact: J. Luhr, Univ. of California, Berkeley.

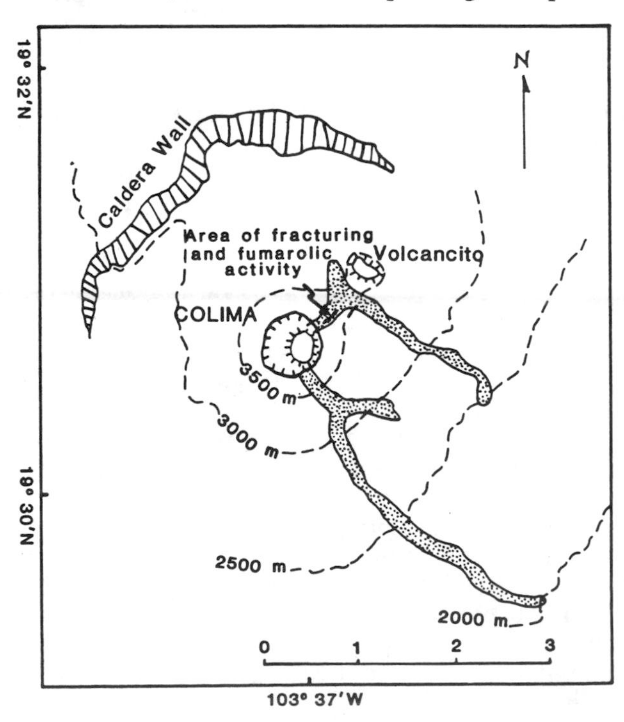

Fig. 14-1. Location of the area of active fracturing and intense fumarolic activity on Colima. The 1975–76 lava flow is shaded. Base map modified from Luhr, 1981.

2/83 (8:2) A French team reached the N rim of the summit cone in early Dec. Storm damage to trails prevented them from reaching the S side of the cone, so they were unable to see the S flank lava flow produced by the eruption that began in Dec 1981 (7:1-3). Only fumarolic activity was observed in the W part of the crater and on the N flank. Gas of essentially atmospheric composition was emitted at 500°C from the NE part of the cone and from a vent that had recently extruded a lava flow. Rockfalls occurred several times/day from the front of this flow and it may still have been advancing very slowly.

James Luhr and others visited Colima in mid-Jan and again in early Feb. The S flank lava flow appeared to have advanced very little since last observed by Luhr in March 1982. Residents of the area reported that incandescence had ended in June 1982. Plume emission continued in early 1983 at about the same intensity as a year earlier, but there were no episodic increases in intensity of plume emission as there had been in early 1982.

Information Contacts: J. Cheminée, IPG, Paris; J. Luhr, Univ. of California, Berkeley.

4/83 (8:4) "Colima was emitting a moderate-sized white plume when observed 25–28 April. Most of the plume came from the fumarole field on the NW side of the summit lava field. Temperatures were as high as 565°C in some instances. Preliminary estimates of SO_2 flux suggest a rate of up to 100 t/d."

Information Contacts: R. Stoiber, L. Benton, C. Connor, D. Douglass, D. Shumway, J. Swarts, Dartmouth College.

11/85 (10:11) "Dartmouth volcanologists visited Colima 26-27 Nov. Large linear fractures had opened on the NE flank of the volcano between the summit and Volcancito Cone, at an elevation of approximately 3700 m (fig. 14-1). These fractures strike N30°-45°E and are as much as 15 m deep, 5 m wide, and 50-100 m long. New arcuate fractures of comparable size are concave downslope and head across the linear fractures. Fumarolic activity in an area of 100 m^2 was intense with temperatures up to 800°C and several fumaroles hotter than 600°C. Fracturing and fumarolic activity persisted at higher elevations, but was concentrated below the summit. Hundreds of tremors were felt during a bivouac near the fumarole field on the night of 26 Nov. A small fissure was observed cutting across the summit of Volcancito roughly in line with those found at higher elevations. Temperatures in this fissure were 40°C. This is a marked change in activity since our group last visited Colima in April 1983 (8:4), when fractures and fumarolic activity were not observed in this area. The 1983 fumaroles were located at higher elevations and were much cooler; highest temperature was 565°C."

Reference: Luhr, J. 1981, Colima: History and Cyclicity of Eruptions; *Volcano News,* no. 7, p. 1-3.

Information Contacts: C. Connor, B. Gemmell, R. Stoiber, Dartmouth College.

PARICUTIN (Michoacán Volcanic Field)

Michoacán, México
19.48°N, 102.25°W, summit elev. 3170 m

Famous as the volcano born in a cornfield, this cinder cone grew to a height of 410 m from 1943 to 1952. It is 150 km E of Colima in the E-W Mexican volcanic belt.

4/83 (8:4) "Fumaroles at several localities were emitting small amounts of acid gases but there was no visible plume at the summit or elsewhere when visited 28 April – 1 May. Temperatures were mostly less than 150°C but some were much hotter, over 400°C."

Parícutin's 9-year eruption produced 1.3 km^3 of lava and 0.7 km^3 of tephra.

Information Contacts: R. Stoiber, L. Benton, C. Connor, D. Douglass, D. Shumway, J. Swartz, Dartmouth College.

11/85 (10:11) "Fumarolic activity persisted at Ahuan vent on the SW flank. When temperatures were measured at Ahuan vent on 29 Nov, the hottest fumarole was 473°C, 70° higher than in April 1983 (8:4), when Dartmouth scientists last measured temperatures at Parícutin. Several fumaroles over an area of about 50 m^2 were hotter than 300°C. No physical changes in the area were apparent since April 1983."

Information Contacts: C. Connor, B. Gemmell, R. Stoiber, Dartmouth College.

EL CHICHON

Chiapas, México
17.33°N, 93.20°W, summit elev. 1060 m
Local time = GMT - 6 hrs

Before 1982, only fumarolic activity had been recognized on this thickly-forested mountain. It is in the state of Chiapas, 670 km ESE of Mexico City, and part of the Central American belt of active volcanoes rather than the E-W Mexican belt.

General References: Duffield, W.A., Tilling, R.I., and Canul, R., 1984, Geology of El Chichón Volcano, Chiapas, México; *JVGR*, v. 20, p. 117-132.

Tilling, R.I., Rubin, M., Sigurdsson, H., Carey, S., Duffield, W.A., and Rose, W.I., 1984, Holocene Eruptive Activity of El Chichón Volcano, Chiapas, México; *Science*, v. 224, p. 747-749.

3/82 (7:3) After several weeks of local seismicity, explosions in late March and early April ejected a series of tephra columns, 2 of which penetrated well into the stratosphere. No previous historic eruptions are known from this volcano, SE of México's main volcanic belt. Officials reported that as many as 100 persons may have been killed by the eruption and associated seismic activity. Tephra falls were very heavy near the volcano, forcing tens of thousands of residents to flee their homes, and causing major damage to crops and livestock.

28 – 29 March. The eruption began 28 March at 2332 and NOAA geostationary weather satellite imagery showed that the eruption column was about 100 km in diameter 40 minutes later. Analysis of an infrared image returned at 0300 yielded a cloud top temperature of -75°C, corresponding to an altitude of 16.8 km, about 1 km above the tropopause. Surface and vault microbarographs and a KS36000 (SRO-type) seismograph operated by Teledyne Geotech near Dallas, Texas (1797 km from El Chichón) received 22 minutes of infrasonic signals generated by explosive activity. Nine distinct signals were recorded, including a strong gravity wave, indicating that the eruption column struck the tropopause. Instruments at McMurdo, Antarctica, 11865 km from El Chichón, recorded about 2 hours of infrasonic signals. Nine intensity peaks were detected, of which 5 were clearly from the eruption.

Vigorous feeding of the plume continued for several hours but had clearly ended by 0600. A dense tephra cloud drifted ENE from the volcano and a much more diffuse plume moved in roughly the opposite direction (fig. 14-2). By 0530 the next morning, satellite images showed the main plume extending from the Yucatán Peninsula, S of Cuba, to Haiti, and remnants of the more diffuse plume over the E Pacific Ocean at about 15°N, and 118-119°W. The U. S. National Weather Service analyzed wind directions and speeds at different altitudes near the volcano, and concluded that the ENE drift of the dense cloud indicated that it was in the upper troposphere, whereas the diffuse plume blown to the WSW was in the middle troposphere at roughly 6-7.5 km altitude. Initially, none of the tephra appeared to be drifting in a direction consistent with the lower stratospheric circulation, but significant aerosol development in the stratosphere is indicated by the lidar measurements described in the next-to-last paragraph of this report.

Heavy ashfall was reported from towns near the volcano. At Pichucalco, about 20 km NE of the summit, 15 cm of ash was reported, and 5 cm of ash fell at Villahermosa (population 100,000), 70 km NE of the volcano.

Residents of Nicapa, a village on the NE flank, took refuge in a church that was toppled by a M 3.5 earthquake, killing 10 people and injuring about 200. Initial estimates of the number of additional deaths varied, ranging as high as 100, and many more were probably killed on the SW flank during this or subsequent eruptions (see below). Most of the casualties on the N flank were reportedly caused by fires started by incandescent airfall tephra. Tens of thousands of people fled the area. The heavy ashfall forced the closure of roads and the airports at Villahermosa and Tuxtla Gutiérrez (~70 km S of the volcano). Cocoa, coffee, and banana crops were destroyed, and the cattlemen's association requested that animals from a wide area be transported for butchering because ashfall had made grazing impossible.

30 March – 3 April. A second but much smaller explosion was observed on the satellite imagery at about 0900 on 30 March. A thin plume drifted E about 120 km before dissipating. A somewhat larger explosion that was first visible at 1500 produced a cloud that rose into the mid-troposphere and moved about 350 km N. Activity was declining by 1900. Haze was widespread over central México, reducing visibility to about 8 km in México City (about 650 km WNW of the volcano) and to only about 3 km in Tampico (about 750 km NW of the volcano). A small explosion shortly before 1330 on 31 March produced a plume that reached the upper troposphere and blew to the E but dissipated quickly.

A small explosion during the early afternoon of 2 April ejected a mushroom-shaped cloud that rose to about 3.5 km altitude in 30 min. Satellite images showed renewed explosive activity early 3 April. An eruption column was emerging from the volcano by 0300 and blew to both the NE and SW. A series of gravity waves and acoustic signals from this activity were again recorded by Teledyne Geotech instruments near Dallas, Texas. The calculated start time for this activity was 0250 and signals continued for 14 minutes. As with the initial explosion 28 March, the powerful gravity waves generated by this event indicated that the eruption column struck the tropopause forcefully. Smaller explosions, calculated to have begun at 0312, generated acoustic waves and a single gravity wave that were received near Dallas for 10 minutes. During the next 5 hrs, ash drifted over N Guatemala and Belize. At Nicapa, on the NE flank, 7.5 cm of new ash was reported and a haze of SO_2 was visible during the day. Explosive activity resumed about 2000. Acoustic data recorded by Teledyne Geotech indicated that explosions probably occurred every 2-3 minutes, generating a few initial gravity waves and a complex series of acoustic waves that continued for 48 minutes. The total acoustic energy of this activity was significantly greater than that produced by the early morning explosions, and the eruption plume was denser and probably rose somewhat higher. It was initially elongate NE-SW and drifted over S México, N Guatemala and Belize. By noon the next day, a faint plume extended to about 25°N, 79°W, almost to Cuba, and lower altitude material, probably at only about 1.5 km, was drifting directly northward along the 95°W meridian.

4 April. A stronger explosion, possibly larger than the initial event 28 March, first appeared on the NOAA geostationary weather satellite image returned at 0530 on 4 April and was reported by ground observers to have started at 0522. An infrared image 3.5 hours later showed a temperature of -76°C at the top of the eruption cloud, corresponding to an altitude of 16.8 km, identical to the altitude measured from the 28 March plume. Wind speeds near the volcano apparently remained relatively low and most of the cloud remained over S México and N Guatemala more than 24 hrs later. In Pichucalco (about 20 km NE of the summit) incandescent tephra could be seen rising from the volcano and the ash cloud darkened the sky during the morning as though it were night. Felt earthquakes were also reported early 4 April. At Ixtacomitán, 18 km ENE of the summit, there was a

Fig. 14-2. NOAA geostationary weather satellite image returned 29 March 1982 at 1000, about 10.5 hours after El Chichón's initial explosion. A dense upper tropospheric eruption cloud drifts ENE to the edge of the Yucatan Peninsula, and a more diffuse cloud drifts WSW, probably in the mid-troposphere.

heavy fall of tephra no larger than 4 cm in diameter and the army was sent to evacuate 3000 residents. No casualties were reported. All villages within 15 km of the summit had previously been evacuated and tens of thousands of people had fled their homes. Government officials reported ashfall over an area of 24,000 km^2 and crop damage of $55,000,000.

A pumice flow deposit from the 4 April eruption extended about 5 km NE from the summit, terminating about 2 km from Nicapa. At its distal end, the deposit was about 100 m wide and 3 m thick and contained pumice blocks 1 m in diameter. Temperatures measured by a thermocouple at 40 cm depth on 8 April averaged 360°C, and were as high as 402°C. The pumice flow deposit appeared to have been emplaced as 2 separate events in rapid succession. Shortly afterward, an ash flow flattened trees in the valley surrounding the pumice flow deposit and left a relatively thin layer of ash that had a temperature of 94°C at 10 cm depth 3 days later.

Airfall tephra thickness in Nicapa, 7 km NE of the summit, totaled 25-40 cm [but see 7:4] after the 4 April eruption. Bombs as large as 50-60 cm in diameter had made numerous holes in the roofs of houses and many other roofs had collapsed. In hand specimen, the tephra appeared to be a crystal-rich andesite or dacite containing hornblende and considerable feldspar. In Ostuacán, 12.5 km NW of the summit, tephra was 15-20 cm thick after the 4 April eruption, including pumice as large as 15 cm in diameter. Many roofs had been destroyed. Extreme heat made it impossible to approach the village of Francisco León, 5 km SW of the summit. Midway between Ostuacán and Francisco León, a river was boiling and flattened trees could be seen upslope. Geologists thought it was likely that pyroclastic flows had moved through the area. Of the roughly 1000 residents of Francisco León, about half had reportedly left before the eruption because of the many felt earthquakes in Feb and March, but the remainder were missing in early April. A helicopter flight over the village during the first week in April revealed no signs of life. Because of the danger of mudflows when the rainy season begins around the end of April, authorities established a prohibited zone extending outward 10 km from the summit.

By 5 April, the low-altitude plume from the second 3 April explosion had reached the S Texas coast and Brownsville reported visibility of only 6.5 km in haze. A few flights into small S Texas airports were cancelled, but winds initially forced most of this material into the Gulf of México. Low-altitude (1.5-2 km) ejecta from the 4 April explosion also moved northward, and a slight change in wind direction blew the ash cloud further N and inland over Texas by late 7 April. A light ashfall occurred in Houston during the night of 7–8 April and samples were collected for analysis by NASA geologists.

5–11 April. A plume generated by a smaller explosion was observed on satellite imagery at 1130 on 5 April. Ground observers reported that the comparatively minor activity lasted about 3 hours and that no incandescent tephra was ejected. A similar but possibly slightly larger explosion could be seen on the satellite image returned at 0930 on 6 April. Geologists reported that earthquakes as strong as magnitude 1.5 were recorded about every 3 minutes on 6 April. Geologists working a few km NE of the summit reported that about 2 mm of wet ash fell at about 1000 on 8 April and 1130 on the 9th. Satellite images returned at 0728 on 9 April and 0238 on 10 April both showed small diffuse plumes, drifitng NNE and SSE respectively.

Data from laser radar (lidar) measurements at Mauna Loa Observatory, Hawaii (about 19.5°N, 155.6°W) during the nights of 9-10 and 10-11 April indicated that El Chichón had injected large quantities of volcanic material into the stratosphere. Several layers were detected, with strongest backscattering at an altitude of 25.7 km. Analysis of wind conditions at 25 km altitude in Hawaii and México indicated a likely drift of about 5-7 m/sec (roughly 430-600 km/day) towards the W, which would carry volcanic debris from El Chichón to Hawaii in 1 1/2 to 2 weeks. Inspection of a satellite image returned late 11 April showed a moderately dense cloud extending from México to just W of Hawaii, spreading from roughly 300 km wide near the Mexican coast to nearly 850 km near its distal end.

No previous eruptions of El Chichón are known in historic time. Before the 1982 eruption, the volcano was heavily forested, with a shallow crater, 1900 m × 900 m, elongate NNW-SSE. Solfataras and hot springs were present in the crater and on the flanks. Müllerried (1933) describes voluminous airfall deposits from previous eruptions that he believed to be post-Pleistocene.

Reference: Müllerried, F.K.G., 1933, El Chichón, Unico Volcán en Actividad en el Sureste de México; *Universidad de México*, v. 5, no. 27, p. 156-170.

Information Contacts: C. Lomnitz, S. de la Cruz-Reyna, F. Medina, UNAM, México; M. Krafft, Cernay; D. Haller, C. Kadin, M. Matson, NOAA/NESS; A. Krueger, NOAA/NWS; F. Mauk, Teledyne Geotech; C. Wilson, Univ. of Alaska; K. Coulson, T. DeFoor, MLO, Hawaii; C. Wood, NASA, Houston; Notimex Radio, México; *New York Times*; UPI.

4/82 (7:4) The 28 March explosion produced heavy tephra falls N of the volcano, but the initial press reports of the thickness of ash deposits, included last month, were exaggerated. A series of 3 explosions of increasing size occurred 3-4 April, the last of which, at 0522 on 4 April, was the largest of the eruption and produced a major stratospheric cloud. Conflicting reports persist about the fate of the approximately 1000 residents of a SW flank village that was apparently in the path of one or more pyroclastic flows ejected 28 March or 3-4 April.

Infrasonic Data. At College, Alaska (6634 km from the volcano), about an hour of acoustic signals were received from both the 28 March and 4 April explosions. Antipodal acoustic-gravity wave signals from the first 3 April and 4 April explosions were detected at Tennant Creek, Australia (19.52°S, 134.25°E).

Volume and Composition of Tephra. Tephra samples were collected from about 100 sites around El Chichón in mid-April. Near the volcano, 3 separate layers were evident, ejected by explosions 28 March, 3 April (at 2000) and 4 April. Farther away, only 28 March and 4 April tephra had been deposited. The axis of maximum deposition extended approximately N from the summit for the 28 March tephra and roughly E from the summit for the 4 April material. Both layers were normally graded but the 3 April layer, where present, consisted only of fine ash. James Luhr calculated that about [0.30] km^3 of tephra (converted to a density of 2.6 g/cm^3) had fallen within the 0.1 cm isopach (fig. 14-3). X-ray fluorescence analyses of pumice samples showed no significant variation in chemical composition, either within individual units or between units. The pumice, a porphyritic trachyandesite, has a whole-rock silica content of about 57.5%; silica content of the glass is ~61%.

Minor Activity April-May. No large explosions have taken place since 4 April, but occasional minor ash emission continued. The active crater, about 600 m in diameter in early May, was located within the pre-existing shallow summit crater that had dimensions of about 1900 m by 900 m. A 12-km prohibited zone around the summit remained in effect in early May. A 4-station seismic monitoring network operated by UNAM recorded 6-8 small earthquakes/day in early May, including some B-type events. The ejection of a minor ash column on 11 May was accompanied by a few small discrete earthquakes centered at about 2 km depth, and additional seismicity that may have been harmonic tremor.

Stratospheric Cloud. Careful inspection of visible satellite imagery from the NOAA 6 and 7 polar orbiters, the GOES East and West (U.S.), GMS (Japan), and Meteosat (Europe) geostationary weather satellites has permitted the tracking of the densest portion of the 4 April stratospheric cloud as it circled the globe from E to W. The cloud reached Hawaii by 9 April, Japan by 18 April, the Red Sea by 21 April, and had crossed the Atlantic Ocean by 26 April, dipping S to about 5°N at its W edge. Diffusion into higher latitudes appeared to be very limited. During its first circuit of the globe, the cloud could be seen in part (but usually not all) of the range 5°-30°N, sometimes occupying a band roughly 15°-20° wide. Tracking of the cloud after late April has been difficult, but careful work may allow the position of the cloud front to be established after that date. Ozone data from the Nimbus-7 polar orbiting satellite, available for the 2 weeks following the cloud's ejection, allowed its path to be clearly traced, and scientists at NASA's Goddard Space Flight Center hope to continue observations of its position as more satellite data arrives. A balloon flight from Laramie, Wyoming (41.33°N, 105.63°W) in mid-April detected a sharp peak at 17 km (just below the tropopause at 18 km).

Lidar stations in the U.S., Japan, and Europe recorded enhanced backscattering in the stratosphere at times that correlated well with satellite observations of the cloud's movement. Very strong signals were detected by stations at lower latitudes, while the cloud appeared to be present only intermittently and near the base of the stratosphere over mid-latitude stations. A possible northward diffusion of the cloud, probably on its second circuit of the globe, is shown by the sharply higher backscattering ratios detected at Fukuoka, Japan (33.65°N, 130.35°E) in May. At lower latitudes, the strongest layers were centered above 25 km. The highest layer detected was at 29.7 km altitude.

Persons in the SW U.S. observed phenomena that indicated the presence of stratospheric layers. A Bishop's Ring was first seen from Houston, Texas on 11 April and has been intermittently visible since. The 22° angular distance from the sun indicated a particle size of 0.7-0.9 μm. Unusual sunrises and sunsets have been reported from E Texas since 24 April. From Tucson,

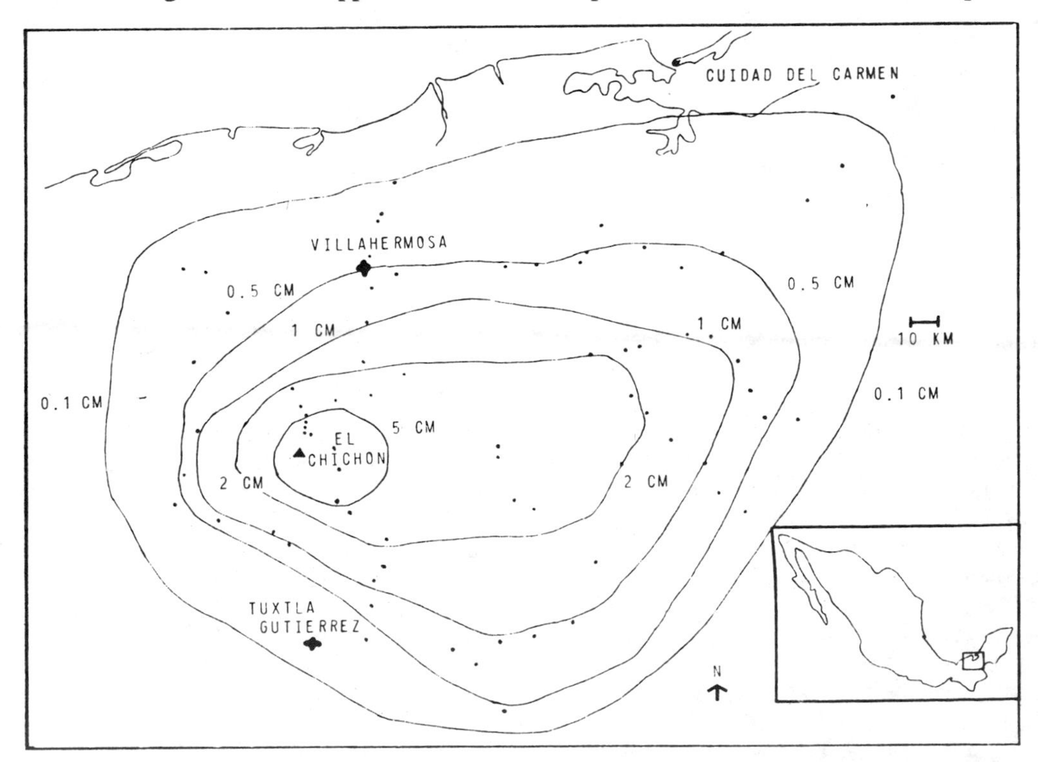

Fig. 14-3. Isopach map showing thickness of compacted ash at a density of 1.2 g/cm^3, for tephra ejected from El Chichón, 28 March – 4 April. Courtesy of James Luhr.

Arizona, Aden and Marjorie Meinel observed a primary scattering layer at 13.2 km and a weaker layer at 20 km around sunset on 30 April. By early May, the aerosol cloud had become extremely dense, with most of the material between 14 and 18 km and a trace to 20.5 km. The main body of the layer appeared to pass NW of Tucson during the evening of 7 May. Long windrows of aerosol were visible, similar to the phenomenon seen after the 1976 Augustine eruption. The conspicuous clouds had nearly disappeared by 9 May, but a strong aureole remained around the sun all day 10 May.

The high altitude of the cloud made direct sampling difficult and none has been possible in the densest portion above 25 km. Flights from the NASA Ames Research Center near San Francisco to about 23°N on 19 April and 5 May sampled the base of the cloud at about 19 km altitude. Optical depths of the cloud as measured with a sun photometer reached 0.3-0.4, increasing southward. Particles ranging in size from 0.1 to 3 μm were collected. Sulfuric acid droplets were common, but there were very few silicate particles. However, NaCl crystals were a significant component and salt has apparently never before been found in significant quantities in a volcanic cloud. Measurement of the degree of depolarization of layers detected at 15-16.5 km and 24-25.5 km on 22 April by lidar equipment at Nagoya University (about 35°N, 137°E) suggested that particles in the lower layer were strongly non-spherical, while those in the upper layer were mostly spherical droplets.

Information Contacts: F. Medina, UNAM, México; J. Luhr, Univ. of California, Berkeley; M. Matson, D. Haller, C. Kadin, NOAA/NESS; B. Mendonça, NOAA/ERL; K. Coulson, MLO, Hawaii; M. Hirono, Kyushu Univ.; W. Fuller, NASA Langley Research Center; Y. Iwasaka, Nagoya Univ.; R. Chuan, Brunswick Corp.; C. Wilson, Univ. of Alaska; D. Christie, Australian National Univ.; J. Rosen, Univ. of Wyoming; A. Meinel, M. Meinel, Univ. of Arizona; M. Helfert, NOAA, Houston TX; W. Evans, ARPX, Atmospheric Environment Service, Canada.

Further References: Krueger, A.J., 1983, Sighting of the El Chichón Sulfur Dioxide Clouds with the Nimbus-7 Total Ozone Mapping Spectrometer; *Science*, v. 220, p. 1377-1379.

Woods, D.C., Chuan, R.L., and Rose, W.I., 1985, Halite Particles Injected into the Stratosphere by the 1982 El Chichón Eruption; *Science,* v. 230, p. 170-172.

5/82 (7:5) The following summary was prepared by L. Silva, J.J. Cochemé, R. Canul, W. Duffield, and R. Tilling.

"Violent eruptions of the strato-volcano El Chichón destroyed its summit dome and formed a 1 km-wide crater. Field studies and eyewitness accounts indicate that the initial activity (28-29 March) was phreatomagmatic, and produced a Plinian column over 15 km high and tephra deposits extending more than 200 km downwind. More vigorous activity 3-4 April produced several pyroclastic flows, some more than 15 m thick, followed by 2 airfall deposits. Distal sections consist of 3 airfall layers whereas proximal sections include pyroclastic flows. The total volume of eruptive products is probably less than 0.5 km^3, much of which is juvenile pumice, which is highly porphyritic with plagioclase, amphibole, and clinopyroxene as major phenocrysts. Petrographic and chemical data suggest an alkali-rich 'andesitic' composition. The high alkali content of the pumice, occurrence of anhydrite in tephra, and presence of halite in the stratospheric cloud reflect contamination by evaporites. Villages within a 7-km radius were entirely destroyed or heavily damaged. Pyroclastic flows dammed a river and created a 5 km-long lake of hot water; the failure of the natural dam on 26 May caused a destructive flood. Study of pre-1982 deposits indicates that El Chichón has developed by several cycles of pyroclastic eruptions, with or without a subsequent growth of domes, with the last pre-1982 pyroclastic eruption about 130 years ago. The current activity may continue and could include dome emplacement."

Premonitory Seismicity. During field work at El Chichón between Dec 1980 and Feb 1981, more than 1 year before the eruption, Rene Canul heard loud noises and felt small earthquakes near the central dome, and could also feel some events while on the flanks of the volcano. People living near the volcano reported felt earthquakes several months before the eruption.

March-April Explosions. The March and April explosions destroyed most of the former central lava dome and formed a new crater, about 1 km in diameter and slightly elongate NW-SE. In early June, there were several explosion pits on the floor of the new crater, all of which were filled with boiling water or mud and were emitting vapor. Although heavy rains had compacted the ash, 3-4 m remained at the rim of the new crater.

The army reported that 187 deaths were caused directly by the eruption. Among the deaths were one geologist and 32 soldiers sent to the village of Francisco León, about 6 km SW of the summit, after the 28-29 March explosion. The pyroclastic flow that traveled through Francisco León left only a thin deposit, but of the structures in the village, only one wall of the church, parallel to the direction of the pyroclastic flow's movement, remained standing.

All of the volcano's major drainages contained pyroclastic flow deposits, which were more than 15 m thick in some of the deeper valleys. These deposits were still hot in late May and were occasionally the source of small secondary explosions. In the 2 months since the March-April explosions, as much as 20 m of erosion has taken place in some areas and fan deposits have formed at the base of the volcano.

May Activity and Flood. Lakes formed behind natural dams of new pyroclastic flow deposits at several sites around the volcano. The largest lake, in the valley of the Río Magdalena at the SW foot of the volcano, grew about 1 m deeper each day until 26 April, then more slowly, eventually reaching 5 km in length and several million m^3 in volume. Late 26 May, the pyroclastic dam holding back this water failed. Seismographs recorded the draining of the lake over a period of about 1 hour, sending a flood of hot water downstream. At Ostuacán, more than 10 km from the dam, the water temperature was measured at 82°C. Most residents of low-lying areas had been evacuated, but at a hydroelectric project 35 km downstream 1 worker was killed and 3 were badly burned by 52°C water. The flood also destroyed a bridge several kilometers from the pyroclastic dam. Geologists inspecting the former lake bed in early June saw a series of strand lines several meters high, indicating that the lake had been draining slowly before the dam failed.

No large explosions have occurred at El Chichón since 4 April. Minor ash emission continued through early May but none has been reported since the 11th. A 4-station Instituto de Geofísica seismic net N of the volcano recorded 4-7 very small events/day in late May.

By late May, significant revegetation had begun in some areas devastated by the eruption. Near Nicapa (about 7 km NE of the summit), coconut trees totally denuded by the 4 April explosion showed new leafy growth. Some residents had returned to Nicapa and cattle were grazing in the area. Closer to the summit, fields that had been completely buried contained tufts of grass about 1/3 m high.

Stratospheric Cloud. The major stratospheric cloud ejected by El Chichón has remained concentrated over lower northern latitudes, but lidar data and observations of brilliant sunsets appeared to indicate the beginning of significant northward dispersal in early June. However, with NASA's SAGE satellite no longer functional, determination of the extent of the cloud at any given time is very difficult. Through mid-May, wind data from Hilo, Hawaii showed a strong (up to 240 km/hr), steady W to E flow between 10 and 20-22 km altitude, and a steady, 50-60 km/hr E to W flow above 25-26 km. Between these levels, winds were light and variable. No significant N-S component had developed above 10 km since the El Chichón eruption. Lidar at Mauna Loa, Hawaii and Fukuoka, Japan continued to detect dense layers of stratospheric material through early June, at altitudes of as much as 32 km over Hawaii. University of Wyoming balloon flights from Laredo, Texas 17-19 May passed through 2 primary layers, at 15-20 km and 24-27 km altitude. In the upper layer there were more than 500 particles of less than 0.01 $\mu m/cm^3$ and about 20 particles larger than 1 $\mu m/cm^3$. April 1982 NOAA 7 satellite data between 120°E and 122.5°W showed an apparent increase in albedo (visible band) and an apparent decrease in outgoing longwave radiation (thermal infrared band) between 15°N and 35°N, peaking at 23°-26°N, when compared with the zonal average from the previous 4 years. Further analysis of samples collected 5 and 7 May during NASA Ames Research Center flights south from San Francisco shows particles ranging from < 0.1 µm to several µm at the base of the cloud (~19 km altitude). Silicates and halite crystals of several µm in size were found. Halite concentration was only a few percent of the amount of H_2SO_4 sampled, but H_2SO_4 was not as dominant as in many previous volcanic clouds. Geologists suggested that the halite sampled by NASA and anhydrite found in tephra near the volcano are probably the result of contamination by evaporites, which were found in bedrock penetrated by 2 Petroleos Mexicanos drillholes near El Chichón.

From Tucson, Arizona on 14 May, Aden and Marjorie Meinel observed a roughly 40 × 400 km band of smoky clouds pass overhead during the afternoon, but at sunset these clouds appeared to be at an altitude of only 8 km. A dense veil covered the sky 15-16 May. A brilliant fiery red glow appeared 35 min after sunset on 16 May. The top of this glow was at 24 km, the highest altitude observed from Tucson thus far. A feature similar to Bishop's Ring was observed 17 May and windrows of aerosol moved over Tucson later that afternoon. A dense veil was present 18 May but was nearly gone on the 19th and skies were almost normal early 20 May. From Kitt Peak National Observatory near Tucson, extinction of 2 times normal at 3900 Å was measured in mid-May. Scattering was about equal at all wavelengths except near-infrared, where the cloud was more transparent. Unusual twilight colors were observed through mid-May from Flagstaff, Arizona, and extinction coefficients measured there were about 3 times the normal value [Livingston and Lockwood, 1983]. A strong haze has been present over Houston, Texas since early May. Spectacular sunsets were observed there in late May and early June and the haze blotted out stars during the night of 2-3 June. Similar conditions plus a Bishop's Ring were seen from Austin, Texas during the night of 5-6 June. From Norwich, England, H.H. Lamb observed a rose-colored pillar of light at sunset 27 April and 9 May. At sunset 10-12 May, this phenomenon was accompanied by brilliant orange to fiery red diffused light that extended to 3-4 sun diameters. Strongly diffused light extended 2-3 diameters from a fiery red sun on 23 May. From the dimensions of the extended twilight illumination, Lamb estimated that the layer was at very roughly 20 km altitude. Lamb saw no abnormal effects on other evenings, although poor weather frequently made observations impossible.

By early June, lidar observations and reports of unusual sunsets indicated that the cloud was beginning to move northward. Lidar operated by NASA at Hampton, Virginia (37.1°N, 76.3°W), began to detect layers at higher altitudes in early June. After a brilliant sunset on 14 June, a dense layer at 20.2-23.1 km was accompanied by material at 26.5 km. However, as of early June, lidar stations in Italy and West Germany had not detected layers at these altitudes. Enhanced sunsets with definite striations began 4 June in Boulder, Colorado and continued for the next several days. Residents of Jacksonville, Florida also began to see brilliant sunsets 4 June.

[Details of the continuing dispersal of the stratospheric cloud are reported in Chapter 19.]

Information Contacts: L. Silva M., J-J. Cochemé, S. de la Cruz-Reyna, F. Medina, M. Mena, J. Havskov, S. Singh, UNAM, México; R. Canul D., Comisión Federal de Electricidad, Morelia; R. Tilling, USGS, Reston, VA; W. Duffield, USGS, Menlo Park, CA; W. Fuller, NASA Langley Research Center; T. DeFoor, MLO, Hawaii; M. Hirono, Kyushu Univ.; B. Mendonça, NOAA/ERL; J. Rosen, Univ. of Wyoming; R. Chuan, Brunswick Corp.; M. Matson, NOAA/NESS; A. & M. Meinel, W. Livingston, Univ. of Arizona; B. Skiff, Lowell Observatory; M. Helfert, NOAA, Houston, TX; H. Lamb, Univ. of East Anglia; J. Nania, Deaconess Hospital, Spokane, WA; *Numero Uno*, Tuxtla Gutiérrez.

Further References: Guerrero, J.C. (ed.), 1983, *El Volcán Chichonal*; Instituto de Geología, Universidad Nacional Autónoma de México, 100 pp. (9 papers).

Havskov, J., de la Cruz-Reyna, S., Singh, S.K.,

Medina, F., and Gutiérrez, C., 1983, Seismic Activity Related to the March-April, 1982 Eruptions of El Chichón Volcano, Chiapas, México; *Geophysical Research Letters*, v. 10, no. 4, p. 293-296.

Livingston, W. and Lockwood, G.W., 1983, Volcanic Ash over Arizona in the Spring of 1982: Astronomical Observations; *Science*, v. 220, p. 300-302.

Luhr, J.F. and Varekamp, J.C. (eds.), 1984, El Chichón Volcano, Chiapas, México; *JVGR*, v. 23, no. 1/2, p. 1-191 (8 papers).

Sigurdsson, H., Carey, S.N., and Fisher, R.V., 1987, The 1982 Eruptions of El Chichón Volcano, México (3): Physical Properties of Pyroclastic Flows; *BV*, v. 49, p. 467-488.

6/82 (7:6) No large explosions have occurred at El Chichón since 4 April and weak ash emission was last observed 11 May. Minor microseismic activity was continuing in early July. The large stratospheric cloud ejected by the 4 April explosion remained dense over lower northern latitudes, but lidar measurements indicated that gradual northward dispersal was continuing.

Pyroclastic Flows and Casualties. Major erosion of pyroclastic flow deposits around El Chichón has taken place since the eruption. Some small accumulations of water remain associated with these deposits, but there have been no recent observations of large lakes such as the one that produced a fatal flood 26 May. The largest eruption killed many people in and near the village of Francisco León (about 5 km SW of the summit; fig. 14-4), but initial reports that all of its residents died were incorrect, according to an American missionary who had lived in the village for many years. Many villagers who had fled the heavy tephra falls from the initial explosions 28-29 March, however, returned a few days later. About 140 residents of the village itself and a similar number from the countryside nearby were killed by the pyroclastic flow that destroyed the village 4 April.

Cloud Sampling. A late April-early May NASA flight collected stratospheric material at altitudes above 18 km over the W U.S. (including Alaska). The flight crew reported unambiguous evidence of the cloud as far N as the US-Canada border and estimated that it reached more than 21 km altitude. A preliminary examination of the material collected showed that it was a well-sorted assemblage of 5-10 μm plagioclase crystals and silicic glass, with a small amount of a mafic mineral (probably an amphibole) and traces of a Ca and S-rich mineral (probably a Ca sulfate). . . .

Information Contacts: S. de la Cruz-Reyna, L. Silva M., UNAM, México; R. Tilling, USGS, Reston, VA; W. Wonderly, Albuquerque, NM; J. Gooding, NASA, Houston.

8/82 (7:8) Seismicity continued to decline in Aug and no eruptive activity has been reported. There was a large increase in the volume of radon emitted after the initial explosion 28 March, but it has been declining since then.

Information Contact: S. de la Cruz-Reyna, UNAM, México.

9/82 (7:9) A small ash ejection, lasting only a few minutes, occurred on 11 Sept, the first eruptive activity reported since minor ash emission stopped in early May. Instruments near the volcano recorded a slight increase in seismicity beginning 26 Aug, to a few very small events per day, and seismic activity remained at this level through late Sept. There have been no reports or evidence of lava dome extrusion (a large summit dome was destroyed by the March-April explosions). By Sept, the 3 lakes that covered much of the crater floor after the March-April explosions had coalesced into 2 larger lakes. Heavy rains have washed out most of the roads near the volcano and have made field work extremely difficult. In late Oct, after the end of the rainy season, geologists will climb the volcano and sample gases.

Residents of the municipio of Francisco León, destroyed by a pyroclastic flow 4 April, estimate that about 400 people in the area were killed by the eruption, and the Catholic diocese of Tuxtla Gutiérrez has a list of more than 1000 persons believed to have died. The eruption was also reported to have killed most of the birds near the volcano. As a result, insects multiplied and devoured crops planted at the beginning of the rainy season, leaving many farmers without food. Coffee, which normally provides a cash crop, survived the eruption but appears unlikely to produce any beans in 1982. The Mexican government has resettled people from heavily damaged villages onto land in other parts of the state of Chiapas. By the end of July, few of the 60,000 refugees from the eruption remained in temporary camps.

Information Contacts: S. de la Cruz-Reyna, UNAM, México; W. Wonderly, Albuquerque, NM; R. Engel, Instituto Lingüístico del Verano, México.

10/82 (7:10) As of early Nov, no new eruptive activity has been reported from El Chichón since the small ash ejection of 11 Sept described last month. Mexican and

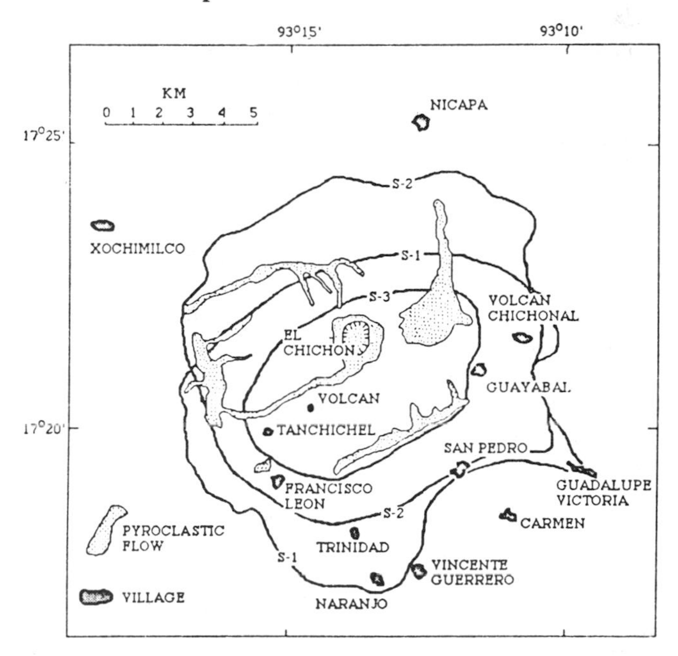

Fig. 14-4. Areal distribution of pyroclastic flows and the 3 major surge deposits from the March-April 1982 eruption. Surges S-1 and S-2 were associated with the first 4 April explosion and the smaller S-3 surge with the second 4 April explosion. From Sigurdsson, Carey, and Fisher, 1987.

1982	Time	Temperature	Altitude	Tropopause
March 29	0230	-75.2°	16.31	16.5
April 3	0400	-71.2°	15.1	17
	2200	-78.2°	*	16.9
4	0744	-83.0°	*	16.9

1982	Time	Alititude (km)	Direction of Drift
March 29	1500	0-5.6	SW
		10.4-15.2	NE
		24	SW
April 3	0900	0-4.9	SW
		13.7-16.3	NE
4	0100	13.7-18.5	NE
		20.7-24	SW
4	1730	9.2-20.7	NE
		24-31	SW

Table 14-1. Determinations from satellite and radiosonde data of initial cloud top altitudes, *top*, and later downwind drift directions in different altitude ranges, *bottom*, from the 4 largest explosions of El Chichón, March-April 1982. Altitudes could not be determined for the 2 largest clouds (these positions in the table are marked by a *) because they yielded unrealistically low temperatures, indicating that they did not radiate as perfect blackbodies, perhaps because of high tephra content. [Carey and Sigurdsson (1986) calculate average column altitudes of 20, 24, and 22 km respectively for the 29 March and the two 4 April explosions.]

American geologists flew over the volcano 4 Nov (during the NASA mission to observe the stratospheric cloud ejected 4 April; see Chapter 19). Little new vegetation could be seen on the volcano and lahar deposits were evident in stream valleys. The 2 crater lakes observed in Sept had combined into a single bright yellow-green lake that covered more than 80% of the crater floor. A sulfur slick covered the lake and sulfur deposits were visible along its shore. Vapor was emitted from 6-8 vents distributed around the shore and churning of the lake water indicated that gases were also emerging from sources beneath its surface. The gas emission fed a relatively diffuse, low-altitude plume that was visible within about 15 km of the volcano. Filter samples collected from the NASA aircraft showed that the plume was relatively poor in ash. The ratio of NO (measured by a chemoluminescent technique) to total sulfur gases (analyzed by flame photometry) was an order of magnitude higher in the El Chichón plume than at Arenal, Poás, Colima, and Mt. St. Helens during similar sampling in Feb. H_2S made up more than 95% of the sulfur gases in the plume.

Michael Matson provided additional data on the initial heights and subsequent directions of movement of different layers of the eruption clouds from the largest El Chichón explosions (table 14-1). Blackbody temperatures of the tops of dense eruption columns are determined from GOES East geostationary weather satellite digital data, and are compared to temperature/altitude profiles from radiosondes launched from Veracruz, México (19.15°N, 96.12°W) at 0600 and 1800 daily. Note that the cloud top temperatures are usually for eruption columns that have just been ejected from the volcano and thus have not yet reached their maximum altitudes. Accurate temperatures cannot be assigned to diffuse plumes because data from the plume are mixed with data from the underlying terrain. However, their direction and speed of drift can be measured on a series of satellite images and then compared with known wind directions and speeds at various heights (in this case from the Veracruz radiosondes) to estimate altitudes.

Information Contacts: J. Friend, Drexel Univ.; S. de la Cruz-Reyna, UNAM, México; M. Matson, NOAA/NESS.

Further Reference: Carey, S.N. and Sigurdsson, H., 1986, The 1982 Eruptions of El Chichón Volcano México (2): Observations and Numerical Modelling of Tephra-Fall Distribution; *BV*, v. 48, no. 2/3, p. 127-141.

1/83 (8:1) Scientists from UNAM's Instituto de Geofísica, Michigan Technological Univ., the Univ. of Maryland, and the USGS visited El Chichón 26-29 Jan 1983, the first time that observations were made within the active crater since the devastating March–April 1982 eruption. The expedition was made with helicopter support provided by the governments of the states of Tabasco and Chiapas. The following is from the scientific team's report.

"The crater was geologically mapped and rocks and gases were sampled extensively. There was no evidence of any eruptive activity since at least 3 Nov, when excellent NASA photography was obtained. Extensive gas emission was still occurring from vents under and adjacent to the crater lake. The emissions were H_2S-rich, apparently partly because the lake (temperature 52°C, pH 0.56) and the ground water were selectively extracting SO_2. Most of the fumaroles were drowned by ground water, had temperatures between 90° and 115°C, and were audibly emitting steam. One fumarole on the SW side of the crater had a temperature of 446°C. The crater as a whole typically had H_2S concentrations of 2-6 ppm, which required special precautions for the scientists. Seismic activity and landsliding were at a very low level.

"Geologic observations document two pre-1982 dome extrusion periods and 2 periods of pyroclastic flows in the crater wall stratigraphy, as well as domal units extruded on the somma or ring fracture around the crater.

"The level of the crater lake was receding of 2-3 cm/day during the period of observations. The high water mark of the lake coincided with the NASA flight in early Nov (Chapter 19; 7:9-11) when it was 60 cm higher than in Jan. The recession correlated with a deepening of color in the lake and may in part have been due to the end of the rainy season."

Information Contacts: S. de la Cruz-Reyna, R. Mota P., M. Mena J., UNAM, México; W. Rose, Jr., T. Bornhorst, S. Halsor, P. Plumley, W. Capaul, Michigan Tech. Univ.; W. Zoller, Univ. of Maryland; T. Casadevall, USGS CVO, Vancouver, WA.

2/83 (8:2) Geologists who left the volcano 10 Feb reported that activity at that time was limited to vapor emission from vents in and around the crater lake. No eruptive activity had been reported as of early March.

Information Contact: H. Sigurdsson, Univ. of Rhode Island.

4/83 (8:4) UNAM geologists visited El Chichón 20-24 April. Fumarolic activity continued but there was no evidence of any recent ash ejection. Variations observed in the size of the fumarole plumes were thought to be caused by changes in temperature and humidity. Portable seismometers on the crater rim 21-23 April detected only 2 small events in 26 hrs. The shore of the crater lake had receded 38 m (horizontally) since Nov 1982 (when the lake was roughly 500 m long, 300 m wide, and a maximum of 50 m deep), and about 30 m of the shrinkage had taken place since the previous measurements in Jan (8:1). Less than a week after geologists left the volcano major additional shrinkage of the lake was reported, and as of early May the lake occupied less than half the area it had covered 24 April. Pilots overflying the volcano were asked to report further changes in the lake, but no reports had been received as of 11 May. Data from a telemetering seismometer 16 km SSE of the volcano showed no change in seismicity through 4 May.

Information Contact: S. de la Cruz-Reyna, UNAM, México.

10/83 (8:10) A team of scientists descended into the crater during the week of 16 Oct. They found no new tephra or other signs of recent eruptive activity. The temperature of the crater lake had decreased to 42°C from 52-58°C during the previous descent in late Jan (8:1). The lake had receded substantially between the late Jan visit and observations from the crater rim 21 April (8:4) but lake level was not appreciably lower in Oct. Average fumarole temperatures had dropped from 116°C in Jan to 99° in Oct. In the talus rampart just above the crater floor, a small low-pressure fumarole that had registered a temperature of 446°C in Jan was no longer active in Oct. No other notable changes in fumarole position were noted. No significant seismic activity has been recorded at El Chichón in recent months although some tectonic seismicity has occurred SE of the volcano.

Information Contact: Same as above.

Further Reference: Casadevall, T., de la Cruz-Reyna, S., Rose, W.I. Jr., Bagley, S., Finnegan, D.L., and Zoller, W.H., 1984, Crater Lake and Post-Eruption Hydrothermal Activity, El Chichón Volcano, México; *JVGR*, v. 23, p. 163-191.

2/84 (9:2) A team of scientists from UNAM and elsewhere visited El Chichón's crater lake on 15 Feb. There was no evidence of recent eruptive activity. Temperature (43°C) and pH (1.9) of the lake had changed little since the previous visit on 21 Oct 1983 (42°C and 1.8; 8:10). High concentrations of H_2S forced the geologists to leave after 40 minutes. Photographs taken in Dec 1983 showed little change in lake level since Oct; no information on lake level was available from the Feb visit.

Information Contacts: S. de la Cruz-Reyna, UNAM, México; T. Casadevall, USGS CVO, Vancouver, WA.

3/84 (9:3) Weather satellite images 3-4 April showed a series of plume-like features originating from the vicinity of El Chichón. Data from the TOMS instrument on the Nimbus-7 polar orbiting satellite 9 1/2 hours after the last plume observation showed no enhancement in SO_2 over the area. However, cloud elevations were estimated at 13.5 km and this prompted an on-site investigation.

Servando de la Cruz visited the volcano and found no evidence that an eruption had occurred. There was no change in the appearance of the crater, crater lake, or outer flanks, and no sign of any recent tephra deposits. Since the 15 Feb visit, temperatures of the crater lake (36-37°C) and fumaroles (96-98°C) had declined slightly, and the crater lake level was somewhat lower after several weeks of very little rainfall. De la Cruz noted that smoke produced by the centuries-old practice of burning the remains of the corn plants after the harvest might have looked like eruption plumes on the satellite imagery, although corn is not the most common crop in the immediate vicinity of the volcano.

Information Contacts: S. de la Cruz-Reyna, UNAM, México; A. Krueger, NASA/GSFC; M. Matson, NOAA/NESDIS; T. Casadevall, USGS CVO, Vancouver, WA; R. Tilling, USGS, Reston, VA.

[The 1982 eruption of El Chichon had a stronger effect on the atmosphere than any other event of the decade. These effects have been reported in many *SEAN Bulletins* throughout the decade, but are included here in chapter 19 (devoted specifically to Atmospheric effects).]

SANTIAGUITO DOME (SANTA MARIA)

Guatemala
14.76°N, 91.55°, summit elev. ca. 2500 m
Local time = GMT - 6 hrs

Santa María's first historic eruption, in 1902, was the second largest explosive event of this century. It came just 6 months after the twin eruptions of Mont Pelée and Soufrière of St. Vincent on the opposite side of the Caribbean. Santiaguito is a complex dacitic lava dome that has been growing since 1922 within the large SW flank crater formed in 1902. The volcano is 335 km SE of El Chichón and 125 km WNW of Guatemala City.

7/76 (1:10) Richard Stoiber of Dartmouth College visited the volcano in July, and reported that the Caliente crater was the site of small explosions every 15-60 minutes. Incandescent blocks were thrown up during these small explosive eruptions, which are more or less continuous. Robert Decker noted that the volcano has been active like this for years, but now the explosions seem to be bigger and more regular. [Rose, in 2:5, places the start of this increased activity in April 1975.]

Information Contact: R. Decker, Dartmouth College.

12/76 (1:15) Dartmouth College geologists visited Santa María in late Nov and early Dec. Ash eruptions, containing some incandescent material, occurred at intervals of 30 minutes or less from Caliente crater. The eruption clouds reached a maximum height of 1.5 km. Steam was emitted between ash eruptions. Considerable landsliding occurred from El Brujo Dome. The activity was comparable to that of July 1976 (1:10), but there was less incandescence than in Jan 1976.

Information Contact: Dept. of Earth Sciences, Dartmouth College.

2/77 (2:2) Beginning about 25 Jan, ash emission from Santiaguito increased significantly. On 8 Feb, an ash cloud reduced visibility in Quetzaltenango (12 km NNE) to about 10 m and coated vegetation and roofs. On 9 Feb, ash was still falling "incessantly" on Quetzaltenango. The ashfall zone extends at least 70 km to the Pacific coast, where an "extensive zone" is reportedly affected. The ashfall has been annoying, but no damage or casualties have been reported.

Information Contact: Guatemalan press.

5/77 (2:5) The Caliente vent at Santiaguito continues to be in a state of unusually strong pyroclastic activity, a condition that began in April 1975 and has been confirmed by every reported observation since that date. The intensity and number of the explosions has varied, but observations are too infrequent to be sure of trends. The volcano's lack of visibility from inhabited locations has limited recorded observations to about 30 different days since early 1975. On all of these dates, pyroclastic activity from the Caliente vent was noted. The frequency of explosions was typically 0.2-4.0/hr and the heights of ash clouds ranged from 300 to more than 6000 m.

Especially large ash eruptions, with clouds to heights of more than 6 km, were observed on 7 and 9 May, and in early June (exact date unrecorded) 1976, and on 9 and 21 Feb and 14 and 19 March 1977. Most of these larger events resulted in ash fallout at nearby towns and cities. No nuée ardente activity has been reported in the recent activity period. Lava and/or dome extrusion at the El Brujo vent has continued, but has slowed since 1975.

Information Contact: W. Rose, Jr., Michigan Tech. Univ.

9/78 (3:9) The Guatemalan press reports that blocks and ash erupted from Santiaguito on 23 July dammed the headwaters of 3 S flank rivers; the Nimá I and II and the Tambor, forming a large lake. The breakup of these temporary dams on 24 July produced mudflows that damaged farms and destroyed bridges, isolating some villages. Damage was estimated at about $1 million, but no casualties were reported. Ash emission was continuing on 28 July. Another mudflow, on 2 Sept, killed 1 person and caused further damage.

Santiaguito has shown continuous extrusion of dacite lava since 1922, with periodic ash eruptions and nuées ardentes.

Information Contact: *Diario el Gráfico*, Guatemala.

11/78 (3:11) A joint expedition of geology students and professors from Dartmouth College and Michigan Tech. Univ. visited Santiaguito in late Nov. The following report, by William I. Rose, Jr. and Richard Stoiber, is based on their observations on 22 and 23 Nov.

"Volcanic activity at Santiaguito was concentrated at Caliente Vent (fig. 14-5), with no activity occurring at El Brujo. Steam and ash explosions occurred at regular intervals of $1^1/_2$ to 2 hrs. They lasted 1-2 min and usually produced an ash cloud ~1 km above the vent. Blocks and bombs, seen to be incandescent at night, were thrown 200 m from the vent. This is very similar to Caliente Vent activity reported several times since 1975.

"Lava extruded in 1976 from Caliente Vent and material washed by torrential rains from the walls of Santa María have helped to produce a flat floor in the 1902 crater, and easy access to the E flanks of Santiaguito for the first time in many years. Observations of Caliente Vent from the E show that it is now surrounded by a 50 m-high cone of ash, blocks, and bombs."

The Dartmouth-Michigan Tech. team could see burned vegetation extending at least 2 km down the valley of the Río Nimá II (on the S flank), indicating that ashflows have descended from the dome since the rainy season ended in Oct.

Reference: Rose, W. Jr., Pearson, T., and Bonis, S., 1976/7, Nuée Ardente Eruption from the Foot of a Dacite Lava Flow, Santiaguito Volcano, Guatemala; *BV*, v. 40, no. 1, p. 23-38.

Information Contact: W. Rose, Jr., Michigan Tech. Univ.

8/79 (4:8) Press reports state that rumbling and seismic activity began at Santiaguito before dawn on 23 Aug, followed by a fallout of fine ash on Quetzaltenango (12 km NNE) and vicinity. At 1300, ash mixed with rain severely obscured visibility in the area and covered many sectors of Quetzaltenango with mud.

Information Contact: P. Newton, Antigua.

11/79 (4:11) The following report was received from Dartmouth College geologists who observed the volcano 11-24 Nov 1979.

"Activity was confined to the crater of Caliente dome, the oldest dome in the complex. Periodic pyroclastic eruptions were the predominant type of activity, occurring on average every 30 min (standard deviation = 24 min for n=67). These eruptions lasted an average of 130 sec (standard deviation = 150 sec for n=72). The eruption cloud in most instances rose about 1500 m above the Caliente summit with some rising 1900 m (500 m above the summit of Santa María). At night, incandescent material was visible within the eruption columns. The pyroclastic activity has produced a horseshoe-shaped cone in the summit of Caliente vent which is open to the SSW. Ash generally blew NW and fell as far away as 3 km, where the leaves of plants were covered.

"COSPEC measurements of SO_2 emission during the pyroclastic events show that 10-20 t/d of SO_2 were being emitted from Santiaguito. The range of emission values is due to variations in the recorded eruption rates from day to day.

"Observations from the Finca Florida overlook S of the dome showed that there was a viscous flow moving out of the summit cone towards the SSW. The flow had proceeded perhaps $^{1}/_{4}$ of the way down the side of the dome. Periodically there were rockfalls off the front of the flow that rolled down the flanks of the dome into a barranca (dry valley). At night these rockfalls were often spectacularly incandescent.

"Nuées ardentes that glowed at night were observed. These appeared to erupt from the toe of the Caliente vent lava flow, perhaps generated when rock fell off the front of the flow exposing hot material beneath. The nuées traveled the same general path as the hot rock avalanches, SW into the barranca. Sporadic observations suggest that large nuées possibly occurred twice a day.

"Geologists ascending the dome made measurements on some of the fumaroles on Caliente. Most of the fumarolic vents seemed to be cooling off. Sapper fumarole was measured at around 82°C, significantly cooler than the temperatures of 170°-300°C reported by Stoiber and Rose (1970) for the period of 1965–69. On the other hand, the von Türkheim fumarole seemed to have increased slightly in temperature to 120°C. There also appeared to be deposition of sulfur minerals at von Türkheim where previously only anhydrite was being deposited."

Reference: Stoiber, R.E. and Rose, W.I., Jr., 1970, The Geochemistry of Central American Volcanic Gas Condensates; *GSA Bulletin,* v. 81, no. 10, p. 2891-2912.

Information Contacts: R. Stoiber, L. Malinconico, R. Naslund, S. Williams, Dartmouth College.

2/80 (5:2) The following is a report from W.I. Rose, Jr., based on air and ground observations 22 Jan – 10 Feb.

"Activity was similar to that of the past 5 years, characterized by steady weak gas emission from Caliente vent, punctuated by ash explosions at intervals of $^{1}/_{2}$-6 hours. Ash from some of the explosions reached heights of more than 2.5 km above the crater. Particularly large blasts were observed on 22 Jan at 0945, 26 Jan at 1500, and 6 Feb at 1110. On 26 Jan, a dusting of ash fell on Quetzaltenango.

"Caliente vent was surrounded by a cone of debris that by 28 Jan had reached the height of the highest spines (2500 m above sea level) on the dome. The cone was breached on the S side, and a 400 m-long blocky lava flow descended from the mouth of the vent down the talus slope to the S, where it broke up into hot, dusty avalanches. These avalanches occurred almost continuously and larger ones resembled small nuées ardentes. No activity (other than fumarolic) was observed anywhere else on the dome. The combination of lava flow activity and ash explosions at Caliente vent is similar to activity described by Von Türkheim in the 1930's. At that time, nuées ardentes became quite prominent."

Information Contact: W. Rose, Jr., Michigan Tech. Univ.

12/80 (5:12) Ash and gas eruptions from Caliente vent occurred irregularly over the 3-day period of observation, with intervals of $^{1}/_{2}$-4 hours between eruptions. Most eruptions lasted 2-3 minutes and sent ash and gas col-

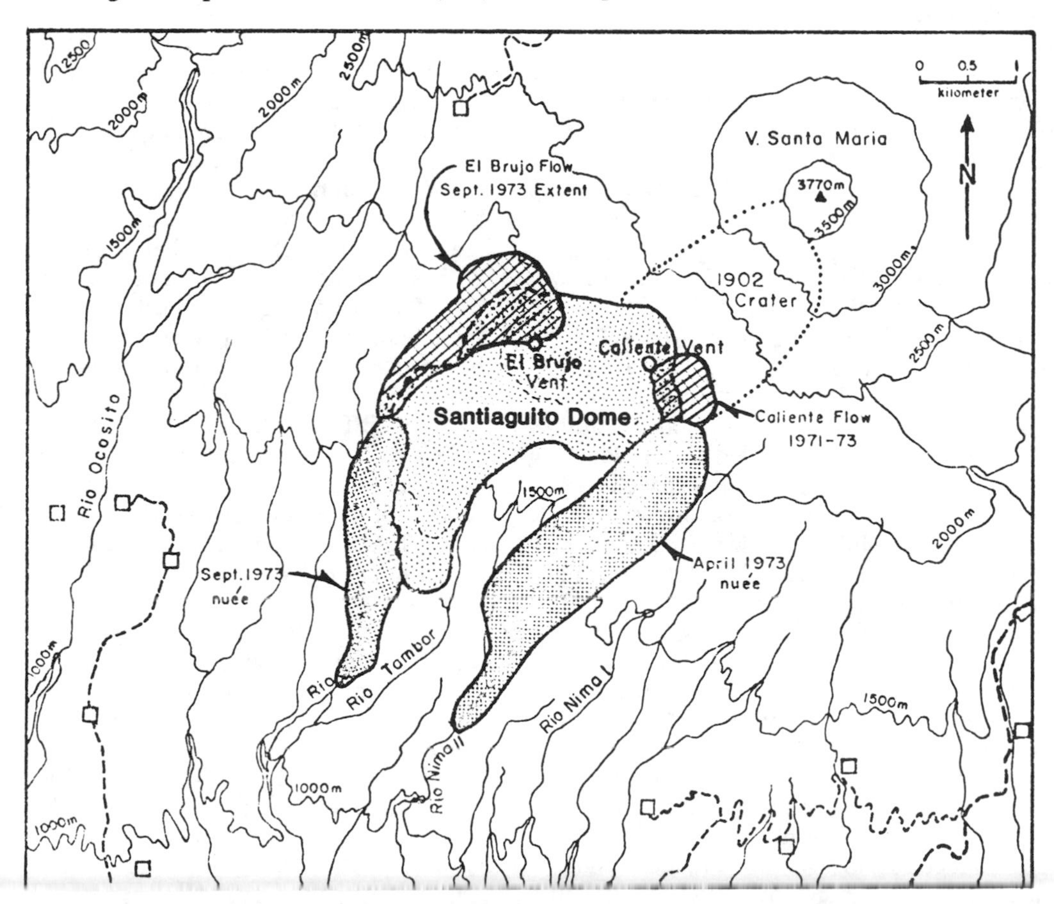

Fig. 14-5. Map of Santiaguito and vicinity showing areas devastated by nuées ardentes in April and Sept 1973. Squares represent inhabited areas. From Rose, Pearson, and Bonis (1976–77).

umns to heights of several hundred meters to 1 km above the vent. Five mm of ash accumulated at the foot of the dome over one 12-hour period. Eruptions occasionally threw 10-cm blocks several hundred meters and ejected tephra to well above the summit of Santa María. Although not directly observed, the plug dome and blocky lava flow that were seen being extruded from Caliente vent in Feb were apparently still very active. Large avalanches of glassy material could be heard from Caliente vista many times per hour. Debris from these avalanches was visible in the barranca below Santiaguito.

Information Contacts: R. Stoiber, S. Williams, R. Naslund, M. Conrad, L. Malinconico, Dartmouth College; S. Bonis, Instituto Geográfico Nacional (IGN); A. Aburto, D. Fajardo, Instituto de Investigaciones Sísmicas, Nicaragua.

2/81 (6:2) Three geologists from Michigan Tech. Univ. spent 12 Feb on Santiaguito Dome. At 1410 an explosion at Caliente Vent sent up a 400 m-high vertical column of fine ash. It was the only explosion in 8 hours of observation, but 2 increases in the vent's vapor plume indicated additional gas emissions during that time. The vent was more active late last year when other geologists visited it.

Large dust clouds in the early morning suggested that avalanching was continuing down the SE slope of the dome (5:2). Fine ash coating the leaves and the ground was notable in the area NW of the volcano.

Information Contacts: W. Rose, Jr., T. Bornhorst, C. Chesner, Michigan Tech. Univ.

3/81 (6:3) On several occasions between 17 Feb and 2 March, R. W. Hodder and a group of students observed explosive activity at Santiaguito Dome. They saw morning eruptions from Caliente Vent on 17, 23, 24, 26, 27, and 28 Feb and 2 March, and late afternoon eruptions on 17 Feb and 2 March.

The group climbed Santiaguito on 23 and 24 Feb. During one eruption, accretionary lapilli fell, followed by raindrops coated with fine ash. About 1000 on 26 Feb, a large 30-minute eruption of gas with very little ash occurred from Caliente Vent. The eruption column rose to about 1800 m, reaching a diameter of about 500 m (much larger than any other observed by Hodder's group) and forming a well-developed anvil-shaped top. At its maximum, the upper $^1/_4$ of the column was ash-poor, nearly white vapor while the lower $^3/_4$ darkened downward to a light brown (lighter colored than the 12 Feb eruption column described last month).

The group saw eruptions at 1000 and 1115 on 28 Feb during 5 hrs of observations. The first consisted of a single 10-minute pulse that sent a vapor column to about 500 m above the vent. The second comprised 4 pulses in 30 minutes. Each pulse began with a white-topped column that developed a light tan base and an anvil-shaped top as it rose as much as 1500 m above the vent. Between each pulse there was intense fuming.

Dartmouth College scientists climbed to the summit of Santa María on the morning of 24 March. They provided the following report.

"The plug dome previously observed in the crater of Caliente Vent was clearly visible and appeared to be covered with huge blocks of light gray lava. Four eruptions occurred within 3 hours with repose periods of 20 minutes, 1 hour, and 1 hour 40 minutes. Each was ash-rich and clearly audible from the summit (a distance of 2.8 km). All rose in the gas-thrust phase to approximately the elevation of the summit (a vertical distance of 1272 m) and beyond convectively.

"Avalanches in the crater and down the SW flank occurred every 5-15 minutes suggesting nearly continuous activity of the dome. The several hundred meter-long lava flow, visible on the SW flank in Feb 1980 (5:2) was not visible from the summit but avalanche clouds rising from that area suggested that it was still active there.

"One large fumarole in the NW part of the plug dome was continuously and very vigorously degassing, remaining essentially unchanged even during eruptions. All four eruptions began in the NE and E region of the crater and lasted 2-4 minutes."

Information Contacts: R. Hodder, Univ. of Western Ontario; T. Bornhorst, C. Chesner, W. Rose, Jr., Michigan Tech. Univ.; S. Williams, M. Mort, Dartmouth College.

2/82 (7:2) "Santiaguito was observed on 10 and 11 Feb during excellent weather conditions. A group of scientists climbed the volcano on 10 Feb in an attempt to sample gases at Caliente Vent (at the E end of the dome). All activity was from Caliente Vent. The tuff ring surrounding the vent was breached to the S and a block lava flow was actively descending a 25° slope, terminating after 300 m in an active scree flow. Avalanching occurred several times per hour.

"Ash eruptions occurred at 1- to 2-hr intervals from Caliente Vent. These reached altitudes of 3.5 km, 1 km above the vent. Fine light brown ash fallouts resulted and the top of the dome had a thick ash mantle, which made walking much easier than in previous years. Some of these eruptions lasted 15 minutes, most only 2-5 minutes. High-frequency noise similar to jet engines pulsated and changed frequency during the eruptions. All of the activity was similar to previous observations. The ash eruptions were identical to those seen consistently since 1975. The block lava flow represented a low rate of lava extrusion, which has been occurring for at least the past 2 years."

Information Contacts: W. Rose, Jr., Michigan Tech. Univ.; T. Casadevall, USGS CVO, Vancouver, WA; W. Zoller, Univ. of Maryland.

9/82 (7:9) The newspaper *El Gráfico* reported 26 Aug that activity from Santiaguito had forced the evacuation of hundreds of residents of towns on and near its flanks. The activity caused the overflow of the Nimá River, leaving hundreds of families in the Nimá valley homeless. No additional information was available.

Information Contacts: *Diario El Gráfico*, Guatemala City; P. Newton, Antigua.

1/83 (8:1) On 29 and 30 Jan, Maurice Krafft flew over Santiaguito. Explosions about once every 2 hours from

Caliente Vent ejected gray, relatively ash-poor plumes. On 29 Jan at 1114, an eruption column rose to about 4.5 km altitude, 2 km above the vent, and another column reached about 3 km altitude the next day at 0946. No rockfalls were noted at the fronts of viscous block lava flows that had been active in previous years, and pilots reported that the flow fronts had also been quiet a few weeks earlier. High above the dome, frequent rockfalls occurred from the NE part of Santa Maria's crater, breached during the major eruption of 1902.

Information Contact: M. Krafft, Cernay, France.

11/83 (8:11) "A group from Michigan Tech. and INSIVUMEH visited Santiaguito 15-16 Nov to observe its activity and recent mudflow damage. Activity was very similar to that of Feb 1982 (7:2) with small vertical ash eruptions occurring at 30-minute intervals. Typical eruption cloud heights were 200-300 m above Caliente Vent, and tephra has built an ash cone around it to an elevation equal to the highest spines of the dome (2500 m). Occasionally in the last few months much larger explosions, to elevations of at least 1 km above the vent, have been observed by residents of the coastal slope. Extrusion of lava continues from the area near Caliente Vent, producing incandescent rockfalls into the Río Nimá II and other nearby rivers (fig. 14-6).

"The combined effects of continued lava extrusion and avalanching, the incessant ash emissions, and an unusually wet rainy season have increased the hazard of mudflows in the Río Nimá II, Río Tambor, and Río Concepción south of Santiaguito. Around and just below the dome at 1700-2000 m, deep canyons have been incised into the volcanic debris. On the N side of the dome, the 1902 crater has been breached and a deep canyon separates the dome from the Casita Base Camp. The formerly stabilized talus slopes of the dome have become active again because of the erosion from below. Beginning in late June and continuing into Aug, the Río Nimá II was especially active, aggrading several tens of meters at Finca La Florida (900 m altitude) and downstream at the town of El Palmar (680 m altitude). The situation at El Palmar was complicated by the confluence of Ríos Nimá I and II. The debris in the Nimá II eventually dammed the Nimá I, creating a natural reservoir which was becoming larger quickly. More and more mudflow and fluvial debris choked the previous mouth of the Nimá I. A channel was dredged through the debris but before it could be sufficiently deepened the Nimá I broke out of its reservoir and suddenly overflowed into the southern part of El Palmar. Evacuation of several hundred people prevented loss of life there, but several dozen houses were completely destroyed by mudflows created by the reservoir waters and the volcanic debris. As the rains lessened, the rivers seemed to be downcutting into the new laharic material and the danger of mudflows seemed less."

Information Contacts: W. Rose, Jr., S. Halsor, T. Bornhorst, Michigan Tech. Univ.; E. Quevec Robles, C. Martinez, The Instituto Nacional de Sismología Vulcanología, Meteorología, e Hidrología (INSIVUMEH), Guatemala City.

Further Reference: Rose, W.I., 1987, Volcanic Activity at Santiaguito Volcano, 1976–1984; *Geological Society of America Special Paper* 2121, p. 17-27.

2/85 (10:2) INSIVUMEH geologists observed explosive activity and seismicity at Santiaguito during visits in late Jan and early Feb. Seven gas and ash ejections occurred on 24 Jan between 1230 and 1730, and 9 were seen between 0945 and 1030 the next day. Eruption clouds rose 1-2 km above Caliente vent. Similar activity has been observed during the past few months. Vigorous fumaroles were present near the vent. At El Brujo vent, on the NW side of the dome, only fumarolic activity was observed. On 8-9 Feb, gas and ash ejections again occurred from Caliente vent but were weaker and less frequent than in late Jan.

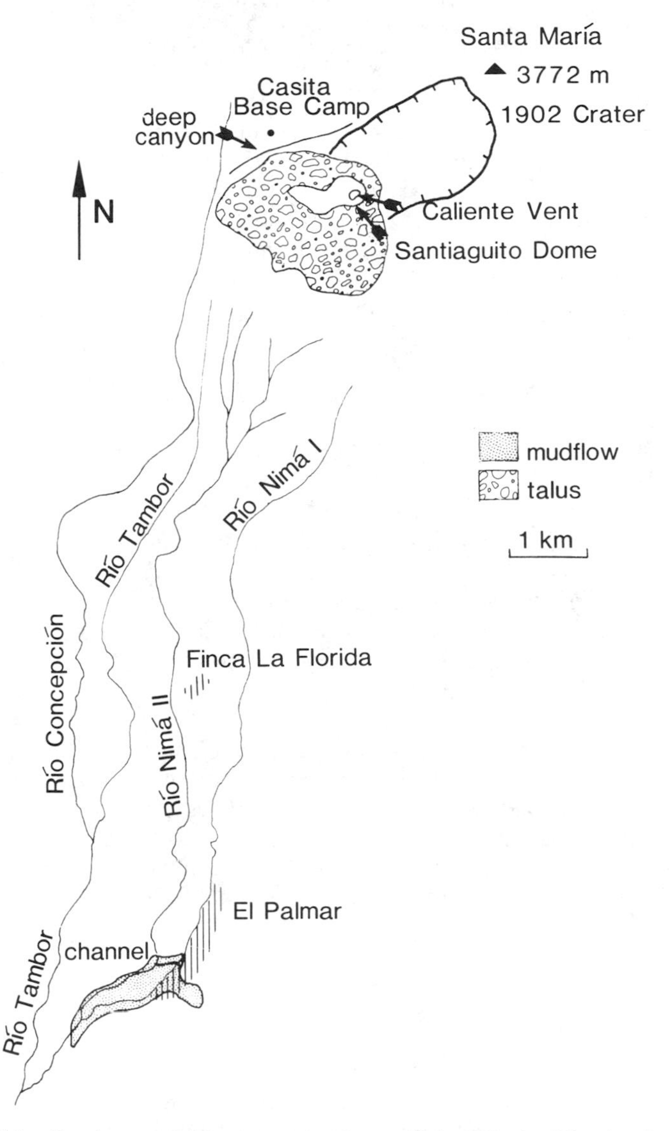

Fig. 14-6. Sketch map of the vent area and part of the S flank of Santiaguito, showing drainage affected by mudflows. The area occupied by houses at El Palmar and Finca La Florida is indicated by vertical ruling. Other inhabited areas are not shaded.

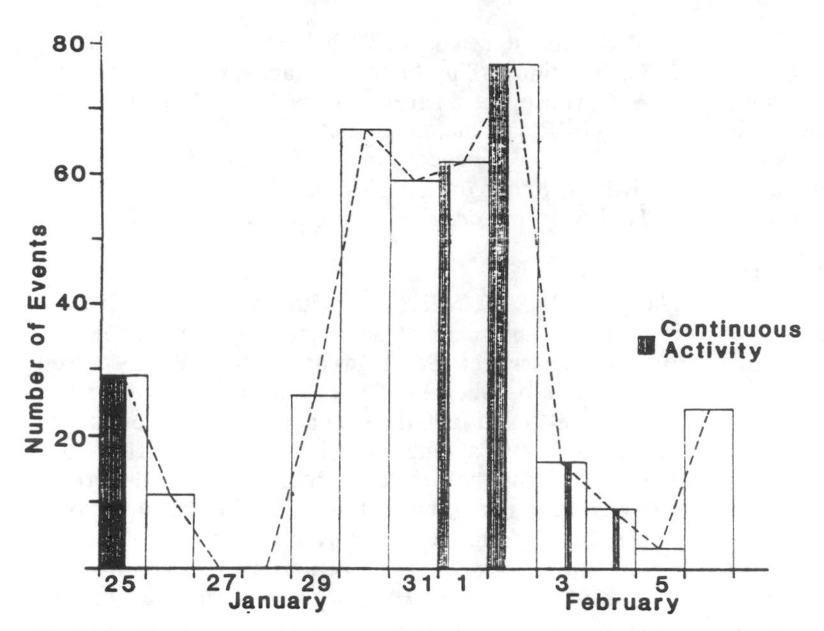

Fig. 14-7. Number of recorded seismic events/day at Santiaguito, 25 Jan – 6 Feb, 1985. Periods of continuous seismicity are shaded (25 Feb, 0115-1230; 1 March, 0145-0540; 2 March, 0120-0930; 3 March, 1335-1615; and 4 March, 1300-1600). Courtesy of Edgar Quevec.

A portable seismograph was installed N of the volcano (near the "Hotel Magermann") on 25 Jan. Very shallow B-type earthquakes accompanied the eruptions. The number of events recorded each day ranged from none 27-28 Jan to 77 on 2 Feb. Continuous seismicity was recorded for more than 11 hours on 25 Jan and for periods of several hours each 1-4 Feb (fig. 14-7).

Information Contact: E. Quevec Robles, INSIVUMEH, Guatemala City.

ACATENANGO

Guatemala
14.50°N, 90.87°W, summit elev. 3976 m

One of the highest volcanoes in Central America, Acatenango is 125 km SE of Santa María and 55 km WSW of Guatemala City. Historic eruptions are known from 1924 – 27 and 1972.

3/81 (6:3) Geologists visited the summit 16, 17, and 18 Feb. There was no visible fumarolic activity around the summit, or in the explosion craters from the volcano's last eruption in 1972. The geologists smelled a strong sulfur odor in the immediate vicinity of the summit craters.

Information Contacts: T. Bornhorst, C. Chesner, Michigan Tech. Univ.

FUEGO

Guatemala
14.48°N, 90.88°W, summit elev. 3763 m
Local time = GMT - 6 hrs

With its summit only 3 km S of Acatenango's, Fuego has been Guatemala's most active volcano since 1524. It has erupted 61 times, and 3 of these were responsible for fatalities. Antigua, the former capital of Guatemala, lies nestled among strato-volcanoes 17 km NE of Fuego (fig. 14-8).

General Reference: Chesner, C.A. and Rose, W.I., 1984, Geochemistry and Evolution of the Fuego Volcanic Complex, Guatemala; *JVGR*, v. 21, p. 25-44.

10/75 (1:1) Unusual activity on Fuego, aside from minor vapor emissions, started the night of Sept 19. Since that time there have been occasional emissions of dark gray to black ash clouds. The ash was either dissipated as dust in the atmosphere or fell on the upper slopes of the cone. A trace fell on some populated areas but was insufficient for collection. A large ash cloud was observed on 2 Oct at 0830, and on 11 – 12 Oct ash, brownish-gray in color, fell on Antigua.

Information Contacts: D. Willever, Antigua; S. Bonis, IGN.

Further Reference: Yuan, A.T.E., McNutt, S.R., and Harlow, D.H., 1984, Seismicity and Eruptive Activity at Fuego Volcano, Guatemala: February 1975 – January 1977; *JVGR*, v. 21, p. 277-298.

3/77 (2:3) On 3 March, a steam plume was emitted for 5-10 minutes from a vent about 200 m below the summit crater on the S flank. The plume first appeared to be dust-colored then turned white. About 2 weeks earlier, road workers on the N flank had reported a small saline brook originating below the summit crater.

Information Contact: P. Newton, Antigua.

4/77 (2:4) A small ash eruption from Fuego occurred at about 1100 on 19 April.

Information Contact: Same as above.

9/77 (2:9) Steam clouds containing a little ash rose about 1000 m above the summit, beginning before dawn on 11 Sept. The eruption was preceded by felt earthquakes on 6 Sept at 2206 and 9 Sept at 0420 and was accompanied by harmonic tremor. On 12 Sept, emission of voluminous brown to black clouds was nearly continuous, frequently broken by 10-20-second intervals of quiescence. At 2130, incandescent bombs and ash were ejected. Ash emission was continuous on 13 Sept and harmonic tremor amplitude increased. Some ash fell on the W flank, at Yepocapa. The eruption had ended by 14 Sept [but see 2:10].

Information Contacts: S. Bonis, IGN; R. Stoiber, Dartmouth College; D. Harlow, USGS, Menlo Park, CA; P. Newton, Antigua; D. Shackelford, CA.

10/77 (2:10) Steam and ash emission has continued intermittently through late Oct. A substantial increase in activity 19-20 Oct was reported. Columns of dark gray ash were ejected at ~1-hr intervals on 24 Sept, accompanied by rumbling. Activity declined the next day but similar ash ejections resumed on 26 Sept, lasting until evening. At 2000, loud rumbling was succeeded by a small eruption of incandescent ash and bombs, accompanied by hot avalanches that moved a short distance down canyons on the upper flanks. Persons near the volcano felt an earthquake 2 hours after the eruption and several more the next day. Incandescent ash was ejected on 28 Sept until 2200 and thick ash, accompanied by loud explosions, was emitted for several hours the next day. A moderate ash eruption, including small hot avalanches, was visible through thick cloud cover on 2 Oct. Two days later, several explosions at ~1630 produced black ash clouds. About 3-4 brown to black ash puffs per hour were reported on 7 Oct, accompanied by a red glow at night. Incandescence was also reported during the night of 9 Oct.

Explosions increased markedly on 19 Oct, producing minor local ashfalls from clouds that rose 1500 m above the vent, and continuous small hot avalanches, which were again restricted to canyons on Fuego's upper slopes. Maximum activity was on 20 Oct, when a trace of ash fell on villages SW of the vent. Intermittent small steam, ash, and ash flow eruptions continued on a reduced scale as of 26 Oct.

Information Contacts: S. Bonis, IGN; R. Stoiber, Dartmouth College.

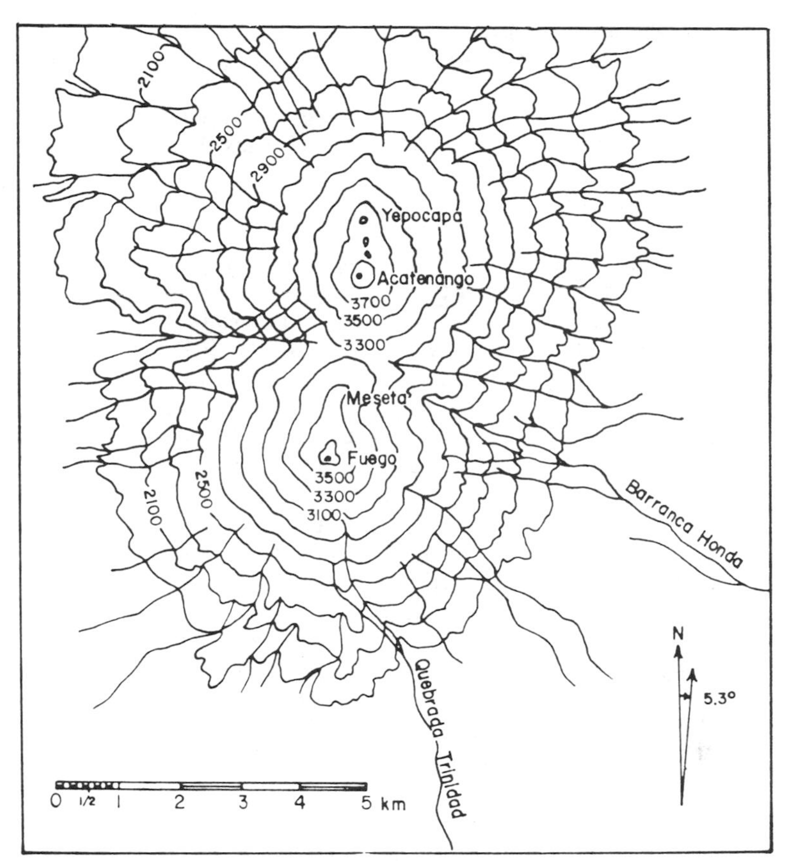

Fig. 14-8. Topographic map of the Fuego and Acatenango volcanic complexes, adapted by Chesner and Rose, 1984, from Alotenango and Chimaltenango 1:50,000 quadrangle maps, Instituto Geográfico Nacional, Guatemala.

11/77 (2:11) Increased activity was continuing in mid-Nov. Only occasional observation of the volcano has been possible because of heavy cloud cover. 29 Oct: Weak ejection of incandescent material at 2230; no hot avalanches observed. 3 Nov: Emission of large volumes of steam under considerable pressure. 5 Nov: Moderate steam emission from the summit crater and a thin, 60 m- high steam column issued from a vent high on the NE flank. 7-9 Nov: Weak steaming from both vents 7 Nov, increased slightly on the 8th. Ash ejection 9 Nov was mainly from the summit crater. 12 Nov: Steam and ash emission from the summit crater and steaming from the NE vent. 13 Nov: Intermittent 250-300-m steam and ash clouds from the summit crater during the day. Between 1915 and 1930, bright red incandescent material was ejected. 14 Nov: Steaming during the morning, succeeded by intermittent ejection of dark gray and black ash in the afternoon. 15 Nov: Steam emission from the summit crater, and clouds of steam from the upper portion of a canyon on the E flank, possibly from a fresh hot avalanche deposit. 16 Nov: Weak felt earthquake at 1455, accompanied by noise. 17 Nov: Intermittent bursts of dark gray and black ash from the main crater visible during the morning.

Information Contact: P. Newton, Antigua.

12/77 (2:12) Cloudy weather continued to hamper observations through Dec, but black ash emission was frequently visible during the day and glow or small incandescent eruptions at night.

Brief, intermittent ejection of black ash occurred 20 Nov and was occasionally visible through clouds 24-28 Nov. Weak glow during the night of 20 Nov brightened the next night and was weak again on 22 Nov. Felt earthquakes on 23 Nov preceded incandescent tephra ejection during the night.

On 29 Nov, activity had declined to steaming within the crater without night glow, which persisted until the afternoon of 2 Dec, when felt earthquakes preceded black ash ejection. Intermittent, 5-minute incandescent eruptions were seen that night and more ash was ejected the next morning. Glow was seen on 6 Dec and weak black ash emission was occasionally visible through the 9th.

Visibility was poor during the next 10 days. Black ash was seen on 12 and 14 Dec, but the volcano was only steaming on the 17th. A glow was observed on 18 Dec and intermittent black ash puffs were seen during the next 3 days. Ash clouds rose more than 650 m above the crater during the mornings of 22 and 23 Dec, but activity had weakened by the afternoon of the 23rd. On 24 Dec, 200-m ash clouds were emitted in the afternoon and there was a glow in the crater at night. The next day, intermittent ash puffs reached 600 m above the crater and incandescent eruptions, accompanied by a small glowing avalanche, occurred at night. Ash puffs continued through 28 Dec, and small incandescent eruptions were seen that night.

Information Contact: Same as above.

1/78 (3:1) Intermittent activity continued through late Jan, culminating in the largest eruption since 1974. The 29

Dec – 28 Jan information is from Paul Newton, and the 29 Jan – 1 Feb information was reported by Samuel Bonis to Richard Stoiber.

29 Dec – 15 Jan: Black (primarily) to gray ash clouds were emitted at low velocity from at least 3 vents within the summit crater. Ash fell mainly on the upper N flank, but some fine material fell several kilometers downwind. Incandescent ejecta was visible above the summit on 30 Dec, and 1, 5, 7, 9, and 15 Jan. Incandescent activity was weak except on the 15th, when bombs were ejected and ash fell on the E flank. Weak steaming from a small vent on the upper E flank was observed on 9 Jan. 16-17 Jan: Mostly steaming, with a little black ash emission. 18-22 Jan: Low-velocity black ash emission to a maximum height of about 370 m above the crater. 23-28 Jan: Activity decreased markedly to weak steam emission.

Weak earthquakes, accompanied by rumbling, were felt at 2015 on 29 Dec, 2130 and 2200 on 5 Jan, between 1100 and 1115 on 9 Jan (2 events), 1100 and 1215 on 15 Jan (lasting about 4 sec each), and 0600 on 20 Jan (followed by black ash emission). Larger events were felt on 19 Jan at 2015 (15 sec) and 2320 (about 4 sec).

"Fuego had its strongest eruption since 1974 from 29-31 Jan. The eruption column reached about 4000 m above the crater, and 1 mm of ash blanketed the ground 5 km SW of the vent. Small nuées ardentes flowed down the canyon of the Río Taniluya, on the SW flank of the volcano. Fuego was inactive at sunrise on 1 Feb."

Information Contacts: P. Newton, Antigua; S. Bonis, IGN; R. Stoiber, Dartmouth College.

2/78 (3:2) More prominent Vulcanian activity 28-31 Jan was followed in Feb by a lava flow, uncommon in Fuego's history. Since Fuego's first recorded event, in 1524, less than one in four eruptions have been accompanied by a lava flow.

The following information, provided by Paul Newton, supplements the report last month.

Weak glow in Fuego's summit crater was observed at 1800 on 28 Jan, beginning 3 days of strong explosive activity. By 2000, bright orange-red incandescent material could be seen in the crater and glowing avalanches flowed down the flanks, accompanied by large incandescent blocks. The next day, black to gray-black ash was emitted steadily to several hundred meters above the crater. Strong, low-pitched rumbling was heard and a few weak felt shocks occurred. At night, glowing avalanches were again visible and incandescent material rose a short distance above the crater rim. On 30 Jan, frequent strong rumbling accompanied emission of black to gray-black ash to an estimated height of 2300 m above the crater. After sunset, incandescent ejecta including large light orange to bright red blocks rose as much as 250 m above the crater and large glowing avalanches flowed down the flanks. The eruption began to weaken after dawn on the 31st. Ash ejection had become intermittent by 1630 and had stopped by 2200. Crater glow had dimmed to a dull red. The next morning, activity was limited to emission of a small brown-white to white plume.

The following is a report by William I. Rose, Jr.

"Throughout Feb, Fuego has been repeatedly visited by a plume-sampling aircraft from NCAR. Lava flows were first observed on 9 Feb in two 30 m-diameter craters 0.1 km SW of Fuego's summit. The flows had advanced 0.5 km down the SW flank of the cone by 11 Feb and the lava craters showed continuous incandescence. Hot debris slides from the toes of the 2 flows had cascaded more than 3 km down the valley of the Río Taniluya by 22 Feb. Small (100-500 m-high) ash eruptions (3-15/hr) occurred from the summit crater as the lava flow activity continued. On 24 Feb the 3 vents were observed in simultaneous activity. The summit crater was erupting a continuous stream of gas and fine ash, while the 2 lava vents exhibited lava fountaining interspersed with dark ash explosions. The rate of lava flow extrusion appeared to be increasing."

Information Contacts: W. Rose, Jr., Michigan Tech. Univ.; P. Newton, Antigua.

Further Reference: Rose, W.I., Jr., Chuan, R.L., Cadle, R.D., and Woods, D.C., 1980, Small Particles in Volcanic Eruption Clouds; *American Journal of Science*, v. 280, p. 671-696.

3/78 (3:3) Emission of black to gray ash continued through March. Cloudy weather prevented observations on 9 days during March, and no report on the lava flows was available. Intermittent incandescence was seen during the evening of 26 Feb, but no incandescence has been observed since then. Activity was limited to weak steaming on 28 Feb and 22, 25, 27, and 29 March.

Strong activity occurred on 25 Feb, and 2, 4, 23, and 30 March. Ash columns rose about 1000 m above the summit at 1700 on 25 Feb, and from 0600 until clouds obscured the volcano at 1100 on 2 March. On 4 March, black ash was emitted to 1250 m above the vent for over 3 hours, beginning at 0600, but the column had become lighter colored by 1100 when clouds prevented further observation. Between 0730 and 1030 on 23 March, fine ash and vapor reached 1800 m above the vent, while heavy ashfall occurred on the upper N flank. At 0600 on 30 March, emission of black ash from the main crater was accompanied by a vapor plume from the N vent. Shortly before 0800, a 2300-m black column was emitted, appearing to originate from both vents, and a hot avalanche flowed down a canyon on the E flank. The eruption could be seen above the clouds until 1100, when visibility became completely obscured.

Earthquakes were felt in Antigua, 17 km NE of Fuego, on 4 March at 0550 (moderately strong) and 1137 (weaker); 13 March at 0130 (weak); 22 March at 0340 (weak); and 30 March at 1330 (strong, lasting 4 sec).

Information Contact: P. Newton, Antigua.

4/78 (3:4) Emission of black to gray ash continued through late April. Clouds and haze completely obscured the volcano on 9 days during April and limited visibility to only a few hours on most other days.

Activity usually consisted of mild to moderate emission of gray ash. Stronger activity occurred on 2 April, when black ash rose about 1000 m during the morning, but activity had declined by late afternoon. Dark gray ash emission was visible between 0730 and 0830 on the 20th (to 1000 m above the summit) and at about 1030 on the 23rd (to 550 m). Small amounts of fine ash fell on Antigua on 17, 18, 19, 21, and 25 April. Incandescence was seen before dawn on 15 and 20 April.

At 1120 on 15 April an earthquake centered about 15 km NE of Fuego was felt at intensity 3 in Antigua, 3 km from the epicenter, for about 10 sec. A small 1 $^1/_2$-sec event was also felt in Antigua at 1822 on 26 April.

Information Contact: Same as 3:3.

5/78 (3:5) Intermittent emission of gray to black ash clouds continued through late May. Clouds and dense haze totally obscured the volcano for 11 days of the 25-day reporting period (28 April–22 May), and visibility was limited to only a few hours on many other days.

Small incandescent eruptions occurred after sunset on 3 May, the only incandescent activity seen during the reporting period. Fine ash fell in Antigua on 7 May, then after 2 days of weak steam emission (8-9 May) and a cloudy day (10 May), ashfall in Yepocapa, 8.5 km NW of the summit, was reported on the 11th and 12th. Ash rose to more than 1400 m early on 13 May and to about 1100 m the next day, when simultaneous activity from 3 vents was observed and several small directed blasts from a fourth vent sent hot avalanches down a canyon on the SE flank. Ash fell on Antigua each day between 15 and 20 May, but the volcano was visible only for brief periods; maximum observed ash cloud height was about 1500 m, at 0900 on the 18th. Only 1 earthquake was felt in Antigua during the report period, a small shock at 1145 on 5 May.

Information Contact: Same as 3:3.

6/78 (3:6) A column of incandescent ejecta was seen above the summit after sunset on 22 May, but observed activity declined between 23 and 28 May to emission of steam and ash plumes (usually gray to dark gray) reaching a maximum height of 900 m. Ash fell on Antigua on 29 May, but clouds limited observation until evening. During the night of 29-30 May a red glow was seen over the summit, accompanied by occasional incandescent eruptions. Glowing avalanches flowed down canyons on the NE and SE flanks. After sunrise on the 30th, black ash could be seen discoloring a cloud bank 1000 m above the summit before visibility was totally obscured at mid-morning. Activity the next day was weak and no incandescence was observed.

The following information is from Samuel Bonis, David Harlow, and Keith Priestly (see 3:7 for 1-14 June observations).

Fuego's eruption intensified on 15 June. Two viscous flows were extruded from the summit area and traveled slowly down canyons on the E and SE flanks, accompanied by vigorous ash ejection. This activity continued the next day, accompanied by almost constant rumbling. The ash cloud rose an estimated 3 km on the 16th and was seen by airline pilots from as far away as the Mexican border, about 170 km NW. At Yepocapa, 1-2 mm of ash fell in about 1 $^1/_2$ hrs. By 17 June, lava extrusions had ended, and explosive activity was confined to weak ash emission, which was continuing on 19 June.

Information Contacts: P. Newton, Antigua; S. Bonis, IGN; D. Harlow, USGS, Menlo Park, CA; K. Priestly, Univ. of Nevada, Reno.

7/78 (3:7) Between 1 and 6 June, activity was confined to weak steam and ash emission from the summit crater and vapor emission from a small subsidiary vent near the summit. During the late afternoon of 7 June, steam and ash clouds rose 1100 m, and after sunset loud rumbling was heard for several hours. Activity was very weak on 8 June and clouds obscured the volcano on the 9th.

Several incandescent avalanches were observed flowing down a canyon on Fuego's E flank before dawn 10 June, and that evening more incandescent material was seen in the same canyon, as much as 725 m below the summit. Similar activity was observed 11-12 June, but during the morning of the 13th, only weak steaming was visible.

A lava flow moved slowly down the E flank canyon after sunset on the 13th, to about 550 m below the summit. The volcano was obscured by clouds on the 14th and much of the 15th, but lava was visible during the evening of the 15th. Activity intensified on 16 June. At 0700, 3 lava flows were moving down the E and NE flanks, and at 1100 a hot avalanche traveled several kilometers down the NE flank as ash rose 1900 m from the summit crater. Felt earthquakes accompanied by loud rumbling occurred at 1400, followed by a 2300-m ash cloud and hot avalanches at 1600. Throughout the evening, several streams of lava could be seen descending the E and NE flanks. Lava extrusion had ended by the next morning, but a few hot avalanches were observed.

Activity between 18 and 25 June consisted primarily of weak steam and ash emission, but some incandescence was seen above the summit on 19, 21, 24, and 25 June. A hot avalanche was observed early on 26 June accompanied by 1200-m ash clouds. Incandescent material, possibly another lava flow, reached a level more than 1000 m below the summit in 4 hrs of evening activity. Incandescent ash rose about 800 m above the summit and loud rumbling was heard. Similar activity continued through the end of June, with frequent hot avalanches and accompanying vertical emissions rising as much as 2200 m.

Powerful explosions were heard at 2230 on 30 June and at 0215, 0700, and 2300 on 1 July. The 3rd explosion produced a 2200-m ash cloud; the 4th caused a strong air shock at the base of the volcano, and yellow to red ejecta including large blocks rose a short distance above the crater.

Poor weather conditions prevented observation of Fuego during much of July. The volcano was inactive during brief periods of visibility on 4 and 5 July, but incandescence was observed above the crater on the 6th and a glowing avalanche moved down the NE flank. Some incandescence was visible above the crater on 8 July. Activity was limited to weak steaming and a few ash puffs between 12 and 19 July. A hot avalanche flowed about 800 m down an E flank canyon on 21 July and ash rose about 1000 m on the 21st and 22nd. On 23 July, the last day of observations reported here, only weak ash emission was observed.

Information Contacts: P. Newton, P. Alquijay, D. Willever, Antigua.

8/78 (3:8) Intermittent ash emission continued through late Aug, but no new lava flows were reported. The maximum observed ash cloud height was about 1650 m, on 18 Aug. Summit glow and ejections of incandescent material were often visible at night.

Numerous hot avalanches moved down the NE, E, and SE flanks during the late afternoon and evening of 24 July. Small incandescent ejections were seen the next night and dark gray ash rose to 1100 m the morning of the 26th. Activity had declined by that evening, and was characterized through the end of July by the emission of small, dense, brown to black ash puffs. Occasional jets of incandescent material from the main crater and fumarolic activity from a vent near the main crater were seen 29 July.

Activity declined further in early Aug. Very weak black ash emission during the morning of 1 Aug was succeeded by steaming that continued intermittently through the next day. Ash emission resumed 3 Aug and jets of incandescent ejecta were seen at night. Ash clouds rose about 800 m on 4 Aug; each ejection lasted 3-4 minutes followed by about 30 minutes of quiescence. Clouds obscured the volcano 5-9 Aug.

A series of loud rumbling sounds was heard in Antigua during the late afternoon and evening of 9 Aug. Another rumble was heard early 10 Aug, accompanying the ejection of an 1100-m black ash cloud. Intermittent ejection of black ash continued for the next several days. Clouds rose to 1200 m on the 12th and 1600 m on the 13th, and a small earthquake was felt in Antigua at 2214 on 12 Aug. Incandescent ejecta were seen after sunset on 10, 11, 13, 14, and 15 Aug. A hot avalanche traveled about 600 m down the E flank on 17 Aug.

Immediately after a M 5.4 earthquake at 0936 on 18 Aug, 51 km W of Fuego, ash emission became more voluminous. That evening ash clouds reached about 1650 m above the summit. Early on 19 Aug, ash rose 1100 m above the summit and a hot avalanche moved about 550 m down the SE flank late in the afternoon. Incandescent material was occasionally seen during the evenings of 19-21 Aug, but clouds prevented daylight observations. Activity weakened 22-23 Aug (the last days of observations reported here), with weak steaming and ash emission from several summit area vents.

Information Contact: P. Newton, Antigua.

9/78 (3:9) Intermittent emission of ash and some hot avalanches persisted through Sept. Ash clouds reached a maximum observed height of 5500 m, on 3 Sept.

The weak activity of 22-23 Aug continued on the 24th. Activity strengthened on 25 Aug, and between the 25th and 28th varied from emission of thin grayish clouds about 1000 m high, to ejection of larger, more ash-saturated clouds about 1500 m high. On 29 Aug, intermittent bursts of dense black ejecta were thrown slightly above the crater rim during the morning, and incandescent ejecta rising to about 550 m were seen after sunset. Weak ash emission was briefly visible through clouds the next day, then clouds and rain obscured the volcano until 2 Sept. That evening, incandescent material, including large blocks, was thrown to about 550 m. An avalanche of yellow-orange to red material, originating from the toe of what appeared to be a short lava flow just below the crater rim, traveled down a canyon on the E flank. After midnight, a nuée ardente traveled down the same E flank canyon to the foot of the volcano. Between 0630 and 0930 on 3 Sept, a cauliflower-shaped eruption cloud rose to 5500 m above the crater and nuées ardentes flowed down the E flank canyon. Ashfall was confined to Fuego's flanks. Rumbling and felt seismicity accompanied the activity during the night of 2-3 Sept and the morning of the 3rd. Clouds obscured the volcano for much of 3 Sept, but a 2200-m eruption cloud and a large red glow were visible above the crater in the early evening.

Activity had declined to weak steam and ash emission early on 4 Sept, although incandescence was seen through clouds that night. From 4 to 14 Sept, clouds of varying ash contents reached a maximum observed height of 1100 m (on the 8th) but most rose only a few hundred meters and there were periods of quiescence. A small earthquake was felt for about 3 sec in Antigua at 1305 on 7 Sept, and a much larger, magnitude 5.5 (m_b) event, 67 km WNW of Fuego, was felt there at 1724 on 10 Sept, continuing for nearly 1 minute. A rounded dome or cone could be seen protruding over the top of the crater rim.

Glow was seen over the crater after sunset on 14 Sept and intermittent incandescent activity occurred the next night. On 16 Sept, voluminous dark gray ash clouds were emitted to 1100 m and a thunderstorm generated large lightning bolts that struck Fuego's crater just before sunset. Similar ash ejection occurred on 17 Sept. Incandescence was seen above the crater that night, as hot avalanches traveled down a canyon on the E flank. Black and gray ash rose as much as 2000 m on the 18th and intermittent incandescence was visible at night. Rain and clouds prevented further Sept observations.

Information Contact: Same as above.

10/78 (3:10) Fuego's eruption declined substantially in Oct. The summit dome noted last month continued to grow. Fuego was steaming weakly on 23 Sept and a brief break in the clouds on the 24th revealed no activity. Voluminous steaming had begun by the morning of the 25th. That night, a glowing avalanche traveled ~550 m down a canyon on the E flank. After dawn on 26 Sept, steam rose slowly to about 1100 m, then clouds prevented further observation. Weak steaming was seen after dawn the next day. Beginning before dawn on 28 Sept and continuing through the evening, nuées ardentes moved down the E flank canyon. Red glow and some small incandescent eruptions were visible at night. Heavy cloudiness prevented observations on 29 Sept, but brief glimpses of the volcano on 30 Sept and during the day on 1 Oct showed no explosive activity. Weak ejections of incandescent material after sunset on 1 Oct were separated by long periods of quiescence. After 1 Oct, no ash or incandescence was seen. Although the summit dome was clearly growing, other activity was confined to weak or moderate steaming from the summit crater, and occasionally from 2 smaller vents near the summit. The maximum steam plume height during the period was about 1000 m. Small earthquakes were felt in Antigua at 1209 on 4 Oct and at about 0245 on 22 Oct.

Information Contact: Same as 3:8.

11/78 (3:11) After a month of dome growth and weak steaming, lava flow extrusion began in late Oct and was continuing in late Nov.

The following is from Paul Newton.

Fuego's activity was limited to weak steaming from 24 Oct until the evening of 28 Oct, when a loud explosion was heard at 1930. About 30 minutes later, avalanches of glowing ash and some large blocks began to travel down Barranca Honda (fig. 14-8), a canyon on Fuego's E flank. Hot avalanches continued through at least 1130 the next day, accompanied by weak steaming. The SE side of the summit dome appeared "distorted" from Antigua. Clouds then obscured the volcano until 1830 when a glow was visible over the summit but hot avalanches had ceased.

Only minor steaming was observed between 30 Oct and 4 Nov. The vapor column reached 700 m in height on 5 Nov. Incandescent avalanches resumed by the evening of 6 Nov, stopped by the next morning, then began again during the evening of the 7th and continued through 12 Nov, the last day of Newton's observations. Weak steaming and occasional emission of dark gray ejecta to a maximum observed height of 500 m accompanied the avalanches. Earthquakes were felt in Antigua at 1224 and 1312 on 30 Oct, and at 0030 on 31 Oct.

The following report, by Dennis Martin, is based on observations from the E side of Fuego 19-20 Nov, and from Alotenango (at Fuego's NE foot) beginning 21 Nov, plus conversations with local residents.

"An incandescent lava flow about 300 m long extruded from a vent just E of and below the summit was flowing down Barranca Honda. Where the slope within the barranca steepens to more than 60°, the flow formed a spectacular lava cascade where incandescent lava blocks dropped hundreds of meters. Extending about 3 km further down the barranca to elevations of about 2 km was a block avalanche fed by accumulations of incandescent blocks from above. This avalanche continually oversteepened and produced its own incandescent cascades. At night, incandescence from the block avalanche disappeared below about 3 km elevation. Because of accumulations high in Barranca Honda, a block avalanche had began to advance down an adjacent canyon (N of Barranca Honda) in mid-Nov. Both avalanches had reached a point about 6 km from National Route 14, where the nearest people live. Weak, sporadic lava fountaining at the summit mound and gas emission from an apparent 500 m-long N-S fissure accompanied the lava flow.

"A moderate Vulcanian eruption, similar in intensity to the 29-31 Jan events (3:1-2) began at 0030 on 21 Nov. The eruption cloud rose more than 2 km and was blown WSW and W. Incandescent blocks were thrown to a height of a few hundred meters. An apparent ash flow descended the Barranca Honda at 0430. The loudest of nearly continuous audible explosions heard at Antigua occurred at 0600, but by 0900 the eruption was clearly decreasing in intensity and continued to do so for the rest of the day. By the evening of 22 Nov, the Vulcanian activity had declined to 1-2-minute explosions at about 2-hr intervals. The lava flow activity of the previous weeks continued during the Vulcanian eruption."

Information Contacts: D. Martin, W. Rose, Jr., Michigan Tech. Univ.; P. Newton, Antigua.

12/78 (3:12) Information for the period 23 Nov through 5 Dec is from Don Willever and Paulino Alquijay, compiled by Paul Newton. Observations from 6 through 18 Dec are by Paul Newton. Clouds frequently obscured the volcano, permitting only brief views of the summit area on many days.

After the moderate Vulcanian eruption of 21 Nov, activity had declined to occasional explosions and intermittent ejection of small amounts of incandescent material by the night of 22 Nov. No activity was visible when clouds briefly dissipated on 23 Nov. Between 24 Nov and the evening of 2 Dec, steam and ash clouds, usually white or gray but occasionally black, were ejected intermittently from the main crater, and steam was emitted from the N-S-trending fissure extending northwards from the summit. The steam and ash clouds usually rose only a few hundred meters, but an 1100-m column was seen early on 28 Nov. A weak glow could be seen over the crater at night, but no lava fountaining or flows were reported.

Lava fountaining from the summit crater was observed after sunset on 2 Dec. The next night, incandescent material was visible in Barranca Honda, the principal lava flow channel on Fuego's E flank. Intermittent steam and ash emission, similar to the 24 Nov-2 Dec activity, was reported on 4-5 Dec. Clouds prevented observations on the 6th, but lava was seen in Barranca Honda on the morning of 7 Dec. Only weak steaming occurred on 8 Dec. Cloudiness obscured the volcano until after sunset on the 10th, when incandescent material rose a short distance above the summit. Low but voluminous black clouds were ejected the next day, then weather clouds prevented observations on 12 Dec. Activity was weak during the morning of the 13th, but frequent bursts of incandescent ejecta were thrown more than 500 m above the crater after sunset. Ejection of incandescent material was visible each night through 18 Dec, the last day of observations. Dense gray ash clouds rose about 700 m above the summit. Block avalanches, or rising dust produced by them, were seen in Barranca Honda on 15 and 16 Dec.

The following is from a report by Dennis Martin.

"By 16 Dec, the summit mound seemed to have grown markedly, perhaps more than 100 m in height since 21 Nov. Between 11 and 16 Dec, ash-laden clouds were ejected at intervals of a few seconds to a few minutes, to a maximum height of approximately 1 km. Most clouds rose only a few hundred meters and were blown away by predominantly westerly winds. The duration of explosive intervals varied from 1 sec to 2 min. At night, fountains of molten lava thrusting up to a maximum of 750 m could be seen within the ash-laden gray clouds. These incandescent blocks cascaded in a spectacular display from 100 m-1000 m down the flank. However, the incandescent lava flow and resulting block avalanche down Barranca Honda were no longer visible.

A walk up the Barranca Honda on 13 Dec confirmed that the block avalanche had stopped at approximately 2600 m elevation. The new lava is a vesicular, olivine-bearing basalt like previous Fuego magmas."

Information Contacts: P. Newton, P. Alquijay, D. Willever, Antigua; D. Martin, Michigan Tech. Univ.

1/79 (4:1) Sporadic ejection of ash clouds, sometimes accompanied by incandescent material, continued through early Jan. Lava flows and block avalanches moved down the E flank 6-8 Jan, then a pattern of intermittent ash emission and occasional fountaining resumed, continuing through 21 Jan. Fuego's summit was only briefly visible through clouds on many days of the 19 Dec–21 Jan observation period, but was totally obscured only on 3-4 and 12 Jan.

A thin ash column rose more than 1300 m during the late afternoon of 19 Dec and a small ash flow moved down Barranca Honda the next morning. Low incandescent ejections were visible during the night of 21-22 Dec, but ash rose only about 1/2 km the next day. On the 23rd, voluminous ash clouds were ejected to about 1 km above the summit and there was a slight ashfall in Antigua. That night incandescent ejecta was again visible, and ash and small blocks flowed about 500 m down Barranca Honda. Ash clouds as high as 1600 m were emitted on 24 Dec, but activity declined to minor steaming between the 25th and 31st. Ejection of incandescent material after sunset on 31 Dec and 1 Jan was succeeded by weak steaming through 5 Jan.

Before dawn on 6 Jan, lava began to flow down Fuego's E flank, dividing into separate streams that moved down Barranca Honda and several other canyons with a rolling, tumbling motion (see the description of lava cascades at Fuego in 3:11). Large blocks fell away from the main body of the flow, and a persistent brighter red area could be seen about 200 m below its source. On the 7th, a dense dust cloud rose from Barranca Honda and remained above it through evening, as lava continued to flow down the E flank. At 0400 on 8 Jan, a loud explosion was heard in Antigua and bursts of incandescent ejecta were thrown about 1/2 km above the crater. After dawn, ash clouds could be seen rising 2 km. Some of the ash was ejected from the base of the summit mound at an angle of about 30° from the vertical. Nuées ardentes flowed down the E flank. By late morning, the height of the ash columns was decreasing and ash emission had ended by late afternoon.

Several streams of lava continued to flow down the E flank through the 8th and block avalanching was prominent that night. However, only weak steaming was observed from 9 Jan until the night of 14 Jan, when small bursts of incandescent material were ejected from the summit crater and red glow in Barranca Honda was barely visible through dense haze. Weak ash emission on the 15th was succeeded by low but more voluminous ash clouds on the 16th, accompanied by a brown dust cloud that was present over the entire visible length of Barranca Honda. That evening, low but dense black clouds dropped ash onto the NW flank and brilliant bursts of lava and blocks were ejected nearly vertically from the foot of the summit mound to a height of about 0.3 km. A block avalanche descended Barranca Honda, then clouds prevented further observations. Similar but slightly stronger activity was seen on 17 Jan. Only weak steaming was visible during the day on the 18th, but at night bursts of lava fed block avalanches that traveled almost 1 km down Barranca Honda. Intermittent dark gray to black ash columns rose more than 1100 m on 19 Jan. A light fall of fine ash on Antigua obscured the volcano from view on 20 Jan. The next morning, low lava fountaining was visible before dawn. Ash clouds rose more than 2 km by the evening of the 21st, accompanied by moderate ash emission from the base of the summit mound, at an angle of 40° from the vertical.

One large and several small earthquakes were felt in Antigua during the observation period. The largest was a magnitude 5.5 event centered 67 km SE of Fuego on 12 Jan.

Information Contact: P. Newton, Antigua.

2/79 (4:2) After considerable lava extrusion and ash emission in early to mid-Jan, Fuego's activity was confined to weak steaming between 24 Jan and 7 Feb. Clouds obscured the volcano 22-23 Jan and from 8 Feb until sunset the next day.

Renewed visibility late on 9 Feb revealed red to bright orange-yellow block lava flowing down Barranca Honda, a canyon on the E flank. Large blocks bouncing from wall to wall of the canyon could be seen from Antigua. The next morning, dense black ash clouds were ejected, obscuring the summit area, and white vapor clouds reached a height of 2.5-3 km above the summit. By nightfall, black ash emitted from 2 vents formed a thin column about 1 km high and white vapor moved upwards about an additional 1.3 km. Bursts of incandescent material were thrown more than 200 m above these 2 vents. A large amount of block lava flowed down Barranca Honda, but blocks did not bounce out of the flow as they had the previous night. Lava also flowed into a second canyon N of Barranca Honda. Later in the evening, less lava was moving down Barranca Honda, but large numbers of incandescent blocks were thrown onto the N shoulder, then disappeared from view down the W flank (not visible from Antigua).

Clouds prevented night observations 11-18 Feb, and no lava was seen in Barranca Honda in the daytime during this period. However, ash emission continued, reaching a maximum observed height of 1.5 km and frequently rising about 1 km. Eruption clouds decreased in height and their ash contents declined 16-17 Feb, and no ash rose above low summit-area weather clouds on the 18th.

About midday on 19 Feb, block lava could again be seen flowing down Barranca Honda, and low clouds of black ash were ejected from the vents. After dark, several streams of lava were visible, flowing down Barranca Honda, a canyon to the N, and an open area on the NE flank. Glowing blocks tumbled away from the flows. On 20 Feb, the last day of observations reported here, lava extrusion persisted and dense black ash clouds rose about 0.5 km.

No earthquakes were felt in Antigua during the observation period, but persons nearer the volcano felt some shocks and heard numerous rumblings.

Information Contacts: P. Newton, Antigua; UPI.

3/79 (4:3) The extrusion of block lava observed 19-20 Feb was no longer visible by sunrise on the 21st. Dark gray to black clouds were ejected sporadically that morning, rising as much as 500 m, but ash contents declined in the afternoon, and by 1930 no activity was visible. Only intermittent emission of vapor, sometimes containing a little ash, occurred 22-24 Feb, with columns reaching a maximum estimated height of slightly more than 1000 m. Activity declined further, to low but voluminous steaming from vents on the E and SW sides of the summit cinder cone, on 25 Feb. Mild steaming followed for 2 days. Slumping on the SE side of the summit cone was first observed on the 27th. After clouds obscured the volcano on 28 Feb, sporadic ejection of gray steam and ash columns could be seen on 1 March. The columns rose as much as 1400 m on the 1st and more than 2000 m on the 3rd. The summit showed no activity early on 4 March, after which a period of cloudiness prevented observations until the 9th.

Activity 9-13 March ranged from periods of very mild steaming to emission of ash clouds that rose to a maximum observed height of 1500 m on the 10th. Significant quantities of ash could be seen falling on Fuego's N and E flanks on 13 March. The slumped area on the summit cone had become a clearly-defined, wedge-shaped fissure by mid-March, extending from near the summit to somewhat less than halfway down the SE side of the cone.

Extrusion of block lava resumed on 17 March, after 3 days of weak steaming. The flow first became clearly visible after sunset, but probably began at or before daybreak, appearing to originate from the base of the summit cone fissure. Lava moved about km down Barranca Honda. Many large blocks tumbled down the canyon. At 1915 on 18 March, an explosion ejected incandescent tephra, including large blocks, to a height of about 500 m. Clouds obscured the volcano on 19, 20, and most of 21 March; continuing lava extrusion was briefly visible on the 21st. Explosive activity on 22 March, the last day of observations reported here, produced ash clouds that rose about 2500 m, accompanied by spectacular lightning. Fine ash fell on Antigua.

Information Contact: P. Newton, Antigua.

4/79 (4:4) Fuego's pattern of frequent ash emission and occasional extrusion of block lava continued through late April. Cloudiness obscured the volcano frequently during this period, as Guatemala's rainy season approached.

The strong explosive activity observed on 22 March had declined somewhat by the following morning, but ash emission remained voluminous for the next several days. Dense black ash rose as much as 0.7 km and vapor reached a height of 1.3 km. Ash fell on Antigua on 23 and 25 March.

Activity intensified on 29 March. During the early morning, a black eruption column was ejected to more than 1 km over the summit. By late afternoon (when cloud conditions again permitted observations) the height of the eruption column had decreased but heavy ashfall was occurring on the flanks. Sporadic ejection of incandescent tephra was visible after sunset, and a glowing avalanche traveled down the E flank. Block lava extrusion began the next day, probably by morning, when dust could be seen rising from Barranca Honda. By evening, slow-moving lava had flowed more than $^1/_2$ km down the canyon, and some large blocks had tumbled from its upper end. Small amounts of incandescent tephra could be seen emerging from a vent W of the summit cone.

Clouds obscured the volcano for most of the period 31 March – 5 April. A dark gray eruption column, about $^1/_2$ km high, was briefly visible on 31 March, and mild to moderate steam emission could sometimes be seen in the mornings during the cloudy period. On 6 April, a glowing avalanche descended Barranca Honda, and steam and ash were sporadically ejected from the summit. Stronger ash emission occurred the next day, causing heavy ashfall near the summit. Vapor clouds rose higher than the ash, to 1.1-1.7 km. The fissure on the summit cone had been substantially filled by ash.

Small bursts of incandescent tephra, separated by 5-18-minute periods of quiescence, were seen during the evening of 7 April. Activity was limited to mild steaming during the day on the 8th, but low, sporadic ejection of incandescent tephra could be seen after sunset. A 4-5-sec, very low-pitched rumble was heard in Antigua just before midnight. Ash fell on Antigua during each of the next 3 days, while the volcano was hidden by clouds. Between 12 and 14 April, gray ash and vapor were intermittently ejected, rising as much as 750 m. Ash content of the eruption clouds increased on 15 April. A new fissure that had formed in the top of the summit cone was the source of low emission of black ash.

At 0820 on 16 April, deep rumbling was followed by an explosion that produced a 2.1 km-high ash cloud. Poor weather prevented further observations on the 16th, but by the next morning, maximum ash cloud heights had declined to about $^1/_2$ km. Some incandescent tephra was visible, falling a few hundred meters from the summit cone.

Although clouds obscured the volcano from 18-20 April, ashfalls in Antigua on each of these 3 days indicated continuing activity. When visible through clouds 21-23 April, ash emission was moderate, rising about $^1/_2$-1 km. Better visibility on 24 April revealed ash ejection from the fissure at the top of the summit cone; ash rose about $^1/_2$ km and white vapor as much as 2 km. Ash again fell in Antigua on 23-24 April, the last days of observations reported here.

Information Contact: Same as above.

5/79 (4:5) Activity observed at Fuego during the 26 April – 24 May reporting period was limited to intermittent and usually weak ash ejection. All of the ash appeared to originate from the summit cone fissure, but Newton suggests that significant quantities may also be ejected from vents on the W side of the summit area, not visible from his vantage point. Clouds prevented any observations on 25, 27, and 28 April and 2, 5, and 19-23 May, as well as limiting visibility to brief glimpses on many other days.

From 26 April through 9 May, Fuego's eruption column was gray to black, usually rising 0.5 km or less. Ash ejection was often intermittent, as on 3 May when low but voluminous ash clouds were emitted for 4- to 7-minute periods, separated by 5-8 minutes of quiescence.

Windless conditions allowed the ash column to rise an estimated 1.4 km on 6 May and 1.6 km on 9 May. Fine ash fell on Antigua, 17 km NE of Fuego, on 1, 7, and 8 May. A hot avalanche appeared to descend about 1/2 km down an E flank canyon on 30 April.

A small earthquake, lasting less than 2 sec, was felt in Antigua at 1700 on 9 May. Rumbles were heard there the next morning, at 0300, 0400, 0600, and 0630, the last of which was quite loud. After dawn, thick and voluminous ash clouds were ejected to about 1/2 km above the crater. By midmorning the plume was white, rising a maximum of 1.4 km in still air. That night, the only incandescent material observed during the reporting period was ejected to a short distance above the summit.

Activity similar to that of 26 April–9 May resumed on 11 May. Intermittent bursts of black ejecta rose about 1/2 km, accompanied by steady steaming. Larger ash clouds were ejected on the 15th (to 1.6 km) and 16th (more than 2 km), then activity declined on 17 May to mild steaming and a single burst of gray ash to about 1/2 km above the crater. During the 2 hrs of visibility the next day, no activity was observed. What appeared to be dense white gas was twice seen flowing down an E flank canyon during this period, on 14 and 17 May.

Clouds obscured the volcano 19-23 May. A mild, 4-second earthquake was felt in Antigua at 0300 on the 20th and a single sharp shock at 2315 on the 21st. Residents of Alotenango, at the base of Fuego, had reported numerous rumblings and earthquakes during a 15-day period prior to 17 May.

Although fine ash fell on Antigua on 22 and 23 May, only a very small plume, rising less than 0.3 km, was seen on 24 May, the last day of observations reported here.

Information Contact: Same as 4:3.

6/79 (4:6) Activity at Fuego remained intermittent and relatively weak during most of the 25 May–27 June reporting period, although vigorous ejection of incandescent tephra occurred on 20 June. Rainy season clouds often made observation of the volcano impossible, particularly in early June.

During periods of visibility between 25 and 30 May, Fuego's activity was usually confined to mild steaming from the summit. However, voluminous black ejecta rose about 3/4 km early on 25 May and about 1/2 km in the late afternoon of 30 May, the latter followed by a light ashfall on Antigua. On 2 June at 1615, a brown tephra cloud was ejected at high velocity to 2.5-3 km above the summit. The volcano was briefly visible late the next afternoon, but no activity was taking place.

Except for a 30-minute period on 7 June (when it was inactive) Fuego was obscured by clouds 4-11 June. Weak vapor emission was glimpsed through clouds on 12, 13, and 15 June. Ash ejection was seen twice between the 16th and 19th, to 0.3 km height early 16 June and to about 0.5 km 24 hours later, but the volcano was otherwise inactive.

On 20 June, weak emission of light brown ash to 400 m height was followed at 0930 by ejection of incandescent tephra to more than 0.8 km above the summit. Large yellow, orange, and red blocks fell on the flanks before clouds prevented further observations at 1030. Intermittent voluminous ash emission could be seen for the next 4 days, some originating from a vent W of the summit. Maximum cloud heights were about 0.7 km. When the volcano was next (briefly) visible, on 27 June, activity had declined to occasional steaming.

Information Contact: Same as 4:3.

7/79 (4:7) Activity at Fuego increased during July. Voluminous ash clouds frequently rose more than 1 km above the summit, and incandescent tephra was seen on several occasions, but no new lava flows were reported.

During the afternoon of 29 June, an explosion projected a black eruption cloud to about 2.5 km height, but activity quickly declined to intermittent gray steam and ash emission. When the volcano was next visible, on 3 July, occasional low-velocity bursts of brown to black tephra rose about 0.7 km. Tephra rose only about 0.5 km the next day, but incandescent material was ejected a short distance above the summit that night. After Fuego was obscured by 2 days of clouds, ash emission was voluminous and steady early on 7 July, rising to about 1.5 km. Shortly after sunset, incandescent tephra rose about 0.3 km, and large blocks fell on the flank. This activity intensified 45 minutes later, some tephra reaching an estimated 1.8 km above the summit, then subsided after about 10 minutes. The next day a brief explosion produced a 2 km-high voluminous gray-black cloud, but otherwise only weak steaming was visible. During the early evening 3 loud explosions were heard from Antigua but the volcano could not be seen through clouds. Several explosions, similar to the one observed on 8 July, took place on the 9th, with clouds reaching maximum heights of 1.5 km. Ash fell on the flanks of Fuego and adjacent Acatenango. The S end of the summit fissure could be seen to have widened.

Activity ceased the evening of 9 July, and only weak steam emission was seen 10-13 July. A single explosion produced a 3/4 km-high gray cloud on the 14th. Clouds obscured the volcano the next day. On 16 July, weak steaming during the day was succeeded by 15 minutes of thick intermittent bursts of black ejecta, rising about 0.5 km. Two small incandescent ejections were seen after nightfall.

Clouds prevented observations 17-18 July. Brown eruption columns were seen above the clouds on three occasions 19 July, rising an estimated 1.5-2 km. After another cloudy day on the 20th, a 0.6-km black ash column was seen in a brief clear period on the 21st. On 22 July, sporadic bursts of brown tephra produced dense clouds that rose 1-2 km. Eruption cloud densities declined 23-24 July, and reached heights of 1 km or less.

Information Contact: Same as 4:3.

8/79 (4:8) The intensity and frequency of ash emission at Fuego began to decline in late July. By mid-Aug only weak steam emission was visible, a condition that continued through 21 Aug.

On 25 July, 4 small explosions in a 45-minute period produced dark gray ash clouds that rose < 0.6 km, after 1 hr of inactivity earlier that morning. A similar pattern of activity was observed on 26 and 27 July, but ash clouds were more voluminous, and reached 0.8 km on the 27th. Only steaming took place on 28 July until late afternoon, when 3-4-minute bursts of gray ejecta, rising to more than 1 km above the summit, occurred at intervals of 10-30 minutes. The next day, long periods of quiescence separated voluminous, but ash-poor emissions lasting 2-4 minutes each. The volcano was inactive on 30 July and only weakly active the following morning, but at 1330 on 31 July, the largest explosion of the reporting period took place, producing a cloud that rapidly rose to more than 2 km. Weak steam and ash emission followed this explosion, and similar activity was visible on 2 and 3 Aug (after a cloudy day on 1 Aug).

From 4-7 Aug, steam and ash emission was very weak, and by 8 Aug was interspersed with periods of complete inactivity. No activity was occurring during 1-hr breaks in cloud cover early on 9 and 10 Aug. Clouds obscured the volcano 11-12 Aug, but clear visibility 13-15 Aug revealed no activity. Fumarolic emission from a subsidiary vent near the summit resumed on 16 Aug and was continuing on 21 Aug, the last day of observations reported here.

Information Contact: Same as 4:3.

9/79 (4:9) From late Aug through late Sept the summit was usually obscured by clouds. When the volcano was visible (26-29 Aug, 4-7 and 11-13 Sept), only steaming took place, from some or all of several vents high on the E and ENE flanks. Two fissures, of approximately equal depth, emitted steam S of and slightly below the top of the summit cone. Although winds for the past 2 months have frequently blown from Fuego toward Antigua, no ashfalls have occurred there during that time, in contrast to earlier, more active periods.

Unusually heavy rains have caused landslides and widespread flooding throughout Guatemala, resulting in many deaths, major crop losses, and closures of main road and rail transportation routes. Secondary mudflows formed from tephra on Fuego's flanks have been particularly destructive, as far downstream as the coastal plain.

Information Contact: Same as 4:3.

10/79 (4:10) Activity was again limited to steaming from the summit crater and several other vents high on the flanks between late Sept and late Oct. Steaming was occasionally quite voluminous early in the reporting period, but had weakened considerably and was interspersed with periods of inactivity by late Oct. A clear view of the summit on 18 Oct revealed that a new crater had formed a short distance from 2 fissures seen the previous month.

Information Contact: Same as 4:3.

2/80 (5:2) Paul Newton and Paulino Alquijay have continued to monitor Fuego from Antigua since last Oct. Intermittent vapor emission from the summit and several other vents high on the flanks continued through Jan. Vapor occasionally rose more than 1 km above the summit, but activity was often weak and there were periods of as much as 4 days when no vapor emission was visible from Antigua.

The following report from William I. Rose, Jr., is based on air and ground observations between 22 Jan and 10 Feb.

"Fuego was in a state of continous gas emission, mainly from vents around the edges of the summit crater. The plume varied considerably in intensity, but on its most impressive days it extended many tens of kilometers. The intense fumarolic activity at and around Fuego's summit has produced a broad white and yellow zone of encrustations above about 3500 m elevation. The crater itself was 35 m in diameter, symmetrical, and about 20 m deep."

Information Contacts: P. Newton, P. Alquijay, Antigua; W. Rose, Jr., Michigan Tech. Univ.

3/81 (6:3) On 16, 17, and 18 Feb geologists visited the summits of Fuego and Acatenango Volcanoes. Comparisons of photographs of Fuego taken on this expedition to ones taken by W. I. Rose, Jr., in Feb 1980 showed no striking physical changes in the summit region. The main areas of gas emission, on the N and the SE sides of the main crater, were the same as in 1980. The SE area is a spatter vent from Fuego's last eruption in 1977–79. During the group's visit, gas was being emitted at a moderate, steady rate, as in early 1980. On 21 Feb, however, the group observed that there was a clear pulsation in the rate of emission, with a period of about 2 minutes. A light wind on the 21st allowed the gas plume to rise nearly vertically about 400 m above the crater. Around the crater rim there were only a few fumaroles in contrast to many in early 1980. New fumaroles had appeared around and atop an older irregular domal protrusion on the W flank of the summit.

Information Contacts: T. Bornhorst, C. Chesner, Michigan Tech. Univ.

PACAYA

Guatemala
14.38°N, 90.60°W, summit elev. 2552 m
Local time = GMT - 6 hrs

Pacaya is a complex of cones and lava domes 30 km SE of Fuego and SSW of Guatemala City. A total of 22 eruptions are known since 1565.

2/76 (1:5) This is a report by Richard Birnie, Sam Bonis, and Bradley Dean.

"Pacaya, inactive since the Oct 1975 lava flow, resumed activity for at least 1 day prior to the 4 Feb earthquake. Black steam clouds from MacKenney Crater were seen in Guatemala City. A group camped at Pacaya on the morning of the 4 Feb earthquake reported that a black steam cloud was discharged from MacKenney Crater immediately after the event. Eruptions continued until sunrise. Primarily ash and steam with very few small blocks were discharged. At sunrise the eruptions ceased and Pacaya remained quiet until sunset when similar activity resumed. A group from Dartmouth College and IGN visited Pacaya on 16 Feb. There were no eruptions, but earth tremors of several seconds duration were felt during the 18 hrs that the party was on the summit. Gas condensate samples were collected from a fumarole with a temperature of 138°C on the rim of MacKenney Crater. The summit crater of Pacaya was cut by at least 7 steaming fractures that described a radial pattern. These fractures had been noted on visits to the summit in Nov 1975 and Jan 1976. A thermal infrared map was made of the W flank of Pacaya on the early morning of 18 Feb from a station near El Patrocinio. A large zone approximately 100 m square, at an elevation of 2250 m, contained apparent surface temperatures 10° above ambient with a maximum apparent surface temperature of 40° above ambient. This zone bears watching as the possible locus of a future lava flow. An earth tremor of 10 sec duration was felt while making the thermal map at 0620. At 0903 a dark cloud was erupted from MacKenney Crater. Six more similar eruptions took place in the next hour before the summit was obscured by clouds. Despite the absence of a favorable wind direction, ash from the eruptions fell near El Patrocinio, 2200 m from the vent. The eruption clouds rose approximately 500 m above the summit and rocks were heard cascading down the slopes."

Information Contacts: S. Bonis, IGN; R. Birnie, B. Dean, Dartmouth College.

5/76 (1:8) Residents living near the volcano reported that in mid-May Pacaya's activity consisted of seismic movements and eruptions of moderate amounts of ash.

Information Contact: S. Bonis, IGN.

8/77 (2:8) After a month of increasing steam emission, a 1-day series of ash eruptions occurred on 27 Aug from MacKenney Crater, the active vent of the 1965–75 eruption. Although the eruptions were conspicuous, the amount of ash ejected was not large, and was carried W by the wind. Small eruptions were also reported 19 Aug (black ash) and 20 Aug, accompanied by a strong sulfur odor near the volcano.

Information Contacts: S. Bonis, IGN; R. Stoiber, Dartmouth College; P. Newton, Antigua.

10/77 (2:10) Steam and ash emission from Pacaya increased in Oct, and was continuing at the end of the month.

Information Contacts: S. Bonis, IGN; R. Stoiber, Dartmouth College.

11/78 (3:11) The strong continuous steam emission that began in late 1977 from MacKenney Crater was still underway when Pacaya was visited on 16 Nov 1978, unabated in intensity since Feb.

Information Contact: W. Rose, Jr., Michigan Tech. Univ.

2/80 (5:2) Pulsating gas emission from MacKenney Crater was continuous during observations 22 Jan–10 Feb 1980. The activity was similar in intensity to that observed Jan-March and Nov, 1978.

Information Contact: Same as above.

12/80 (5:12) A very small cinder cone had grown inside MacKenney Crater in the last 2 months. A large gas plume rose continuously from the summit.

Information Contacts: R. Stoiber, S. Williams, H. R. Naslund, L. Malinconico, M. Conrad, Dartmouth College; S. Bonis, IGN.

2/81 (6:2) Pacaya displayed weak Strombolian activity during a visit by Michigan Tech. Univ. geologists on 14 Feb, the first Strombolian activity observed at Pacaya since 1975. Gas emissions have characterized the activity since late 1977 (2:8, 10; 3:11; 5:2).

Lava was fountaining to 200 m at 10-sec to 1-min intervals from 2 coalesced spatter vents in the center of MacKenney Crater, high on the WNW flank. Four subsidiary vents, 2 N of the spatter vents and 2 W of them (in the direction of the volcano summit), also ejected lava. New pahoehoe lava flows, some of which were moving, had filled the N half of the crater floor to the rim. The fountaining was interspersed with intense, pulsating gas emission from the spatter vents.

By 20 Feb, when Robert Hodder climbed Pacaya, one lava flow had traveled a quarter of the way (about 200 m) down the N flank of MacKenney Crater cone, over one of the Sept 1969 flows. Within the crater, cracks and pressure ridges in the lava crust indicated continued lava movement. Strombolian activity was occurring at about 30-min intervals. Patches of sublimate were visible on the SE crater wall.

During a second climb on 28 Feb Hodder observed that aa lava had flowed about 750 m from the crater rim to the base of MacKenney Crater cone, into the trough between it and the rim of the older Pacaya edifice. The level of lava in the crater had risen. The 2 vents observed on 14 Feb had totally coalesced and had built cones about 15 m high. The lava crust seemed solid, but incandescence showed through surface cracks at night. Strombolian activity occurred about every 20 min. Large cow-dung bombs, hurled as high as 100 m, fell onto the cones and the lava crust. Bomb ejection was sometimes preceded by a puffy steam cloud at least 300 m high. Sublimate solidly coated the SE crater wall. Hodder noted that this eruption seemed similar to that of 1969.

Information Contacts: W. Rose, Jr., T. Bornhorst, C. Chesner, Michigan Tech. Univ.; R. Hodder, Univ. of Western Ontario.

3/81 (6:3) Newspapers reported that vigorous magmatic activity had begun at Pacaya on 9 Feb. According to newspapers, activity peaked 18 Feb at 1730 as lava overflowed the N rim of the crater and began to move down Pacaya's N flank.

Michigan Tech. Univ. geologists climbed Pacaya again on 5 March. Since their last visit 14 Feb, the level of the lava lake in MacKenney Crater had risen considerably, the 2 coalesced spatter vents had grown, and an additional small spatter vent had formed in the S part of the crater on the lava lake surface. The new S vent continuously extruded 2 pahoehoe lava flows but was not the source of any Strombolian explosions. A few small pahoehoe flows were also moving across the E side of the crusted lava lake surface. Nearly continuous weak Strombolian activity occurred from the 2 older vents. The smaller N vent had many small Strombolian explosions at intervals of 10-20 seconds. From the larger vent, activity was cyclical, consisting of a 1-5-second explosion that ejected spatter to 200-300 m above the vent, followed almost instantly by a large increase in gas emission that peaked in about 1 minute, decreased slowly, then dropped sharply about 30 seconds before the next explosion. Gases above the vent had an intense blue color. The alignment of the 3 vents in MacKenney Crater indicated that the activity may have been from a fissure trending approximately N-S.

The geologists estimated that the lava flow descending the N flank had a volume of about 2×10^4 m^3 on 5 March. They estimated the total volume of 1981 lava at about 1×10^6 m^3, for an average eruption rate of about 4×10^4 m^3/day. The lava was petrologically similar to lavas from eruptions since 1970, consisting of basalt with abundant plagioclase phenocrysts and sparse olivine phenocrysts.

Information Contacts: T. Bornhorst, C. Chesner, Michigan Tech. Univ.

2/82 (7:2) Geologists climbed to Pacaya's summit on 12 Feb. During their observations, lava from a 10 m-high hornito at about 2100 m altitude on the SW flank flowed into a lava tube that extended about 1/2 km downslope. Lava emerged from the end of the tube and continued several kilometers farther down the SSW flank, forming an aa flow about 10 m wide, incandescent in daylight, that was bordered by levees. Two smaller subflows branched from the main flow 1/2-1 km below the end of the lava tube. Lava flowing through the tube appeared to be of pahoehoe type when viewed through a skylight just upslope from its distal end. A considerable quantity of blue fume escaped through the skylight. Near the active hornito, 2 smaller ones, each a few meters high, were no longer emitting lava. Loud booming from the area of the hornitos was heard every few seconds. According to a security guard working on Pacaya, activity from the hornitos had begun about 3 weeks earlier.

Since geologists visited Pacaya in March 1981, several additional lava flows had spilled over the rim and moved down the flank. Residents of the area said that lava effusion had stopped a few months prior to the Feb 1982 visit but Strombolian explosions had continued until activity began at the SW flank hornitos in early 1982. In Feb 1982, MacKenney Crater contained a single spatter cone, elongate NE-SW, 50-60 m in diameter, and 30-35 m high, breached on the NE side by a fissure that was vigorously emitting vapor. Lava had apparently flowed through this breach in the past. No incandescent rock was visible within the spatter cone, which had a crater no more than 50 m deep. Considerable sublimate deposition had occurred on the spatter cone's NW side. Intense booming from within the spatter cone occurred every 10-15 seconds, occasionally causing felt shaking on parts of the cone for a few seconds, but was not accompanied by any obvious increase in vapor emission.

Information Contact: T. Casadevall, USGS CVO, Vancouver, WA.

3/82 (7:3) Rodolfo Alvarado reported that on 4 March lava continued to flow from a hornito on the upper SW flank.

Information Contacts: R. Alvarado, Inst. Nacional de Electrificación; T. Casadevall, USGS CVO, Vancouver, WA.

1/83 (8:1) Maurice Krafft climbed Pacaya 28-29 Jan and observed lava emerging sluggishly from the SW flank vent that had been much more vigorously active in early 1982. The lava formed a single flow, about 3-4 m thick and less than 50 m long, that advanced only about 1 m/day. When Alfredo MacKenney had climbed the volcano 3 weeks earlier, the same vent had been feeding 3 aa flows, each about 2-3 m wide and 20 m long. None of these tiny flows were still active in late Jan. Moderate degassing continued from summit vents.

Information Contacts: M. Krafft, Cernay, France; A. MacKenney, Guatemala City.

9/83 (8:9) In March 1983, small Strombolian explosions began in MacKenney Crater, 50 m in diameter and 30 m deep. Similar activity continued through June. Activity increased on 17 July when explosions were heard and glow seen from nearby towns. By 31 July, the crater was completely filled with blocks and ash. A pyramidal cone 30 m in diameter and 20 m high, with a small upper crater, was present within the summit crater on 21 Aug. On 4 Sept, activity increased again as lava flowed from a hornito on the upper S flank of the newly formed cone. On 11 Sept, the cone was 25 m high and 50 m in diameter with 3 active vents trending N and 2 lava flows emerging from its W flank. The 2 upper vents produced continuous explosions on 15 Sept, and lava flowed about 200 m downslope from a hornito on the N flank of the summit crater.

Information Contact: A. MacKenney, Guatemala City.

11/83 (8:11) Geologists from Michigan Tech. Univ. and the Instituto Geográfico Militar (IGM) visited the volcano 13 and 16 Nov. Cloudy weather on the 13th allowed only brief glimpses of the crater; activity appeared to be weak but some spattering may have been occurring. On 16 Nov, the geologists noted that the active cone, centered on the S edge of MacKenney Crater, had grown to more than 100 m in height and filled 1/3 - 1/2 of the crater. Several Strombolian bursts per minute occurred from a vent at the top of the cone, ejecting spatter to 20 m height. Bombs reached 30 cm in maximum dimension. Aa lava emerged from below the base of the active cone, flowed down the crater's NW flank, and was beginning to pond in the saddle between MacKenney Crater and an older cone (Cerro Chino).

Information Contact: S. Halsor, Michigan Tech. Univ.

8/84 (9:8) During Dec 1983 and the next 2 months, activity increased and the new cone grew in height and diameter. Many lava flows emerged from below its N base. In March, two lava flows were extruded from the cone's S base while activity declined in its upper crater. On 15

May, people living near the volcano heard strong explosive activity. When the volcano was visited on 20 May, half of the active cone had been destroyed. Small pyroclastic explosions were occurring and a great quantity of vapor was being emitted. On 10 June, several lava flows were emerging in the saddle between the active crater and Cerro Chino (fig. 14-9). In July, lava continued to fill this area and flowed W. On 5 Aug, Strombolian activity had increased in the upper crater. In Aug and early Sept lava emerged slightly SE of the June-July vents. On 2 Sept, minor pyroclastic activity was occurring from MacKenney Crater.

Information Contact: A. MacKenney, Guatemala City.

3/85 (10:3) Lava flows from the saddle vent that began erupting on 5 Aug 1984 remained active until 6 Feb 1985 and built a cone approximately 50 m high.

On 1 Jan, minor pyroclastic activity, which could only be seen from the crater rim, resumed at MacKenney Crater. By 10 Feb, a steep-sided lava cone about 6 m high had developed within the crater. Pyroclastic material was ejected from the top of the small cone, a lava flow emerged from its E flank, and a lava spine had grown on its W flank. On 17 March, the cone was approximately 20 m high and another explosion crater had formed on its E flank. By mid-March, the activity had begun to be visible from Guatemala City (about 40 km E) on clear days.

Information Contact: Same as above.

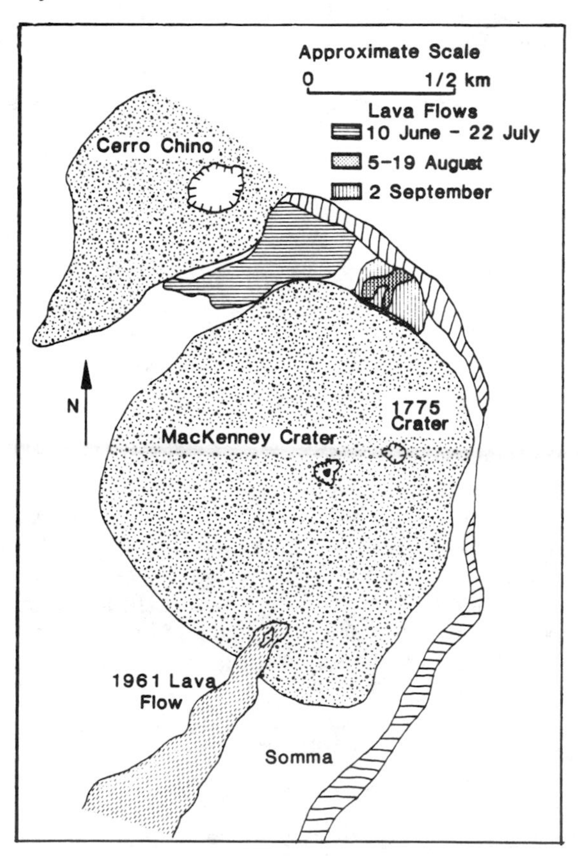

Fig. 14-9. Sketch map of the summit area of Pacaya, showing 1984 lava flows. Courtesy of Alfredo MacKenney.

4/85 (10:4) David Harlow climbed Pacaya during the night of 10 Feb and observed a slow-moving lava flow emerging from a vent between MacKenney Crater and the Aug 1984 lava NNE of the crater. When Enrique Molina and Randy White visited the volcano at the end of Feb, two vents were active in MacKenney Crater: the W vent had built a hornito and was spattering every second or two; only steam was emerging from the E vent, but it was glowing red at 6 m depth. By 17 March, a new cone over these vents had grown to 20 m height.

Enrique Molina and Michael Carr climbed Pacaya 24 April and observed continuing activity from the new cone. Small Strombolian explosions ejected tephra to 20 m height about every 7 minutes from one or both of a pair of vents 6-7 m apart. Activity was more vigorous from the W vent; only minor spattering occurred from the E vent. A steep lava tower, 3-4 m high, had grown over the E vent. Most of the ejecta were black scoria, as large as 1/3 m in diameter, that contained a few small plagioclase, olivine, and pyroxene phenocrysts. Considerable quantities of Pelé's Hair were found near the vents. The lava flow that was emerging from the E flank of the cone in Feb was no longer active. Farmers reported an active lava flow on the W flank of Pacaya, but the geologists did not visit the area.

Information Contacts: E. Molina, INSIVUMEH, Guatemala City; M. Carr, Rutgers Univ.; D. Harlow, R. White, USGS, Menlo Park, CA.

Further Reference: Eggers, A.A., 1983, Temporal Gravity and Elevation Changes at Pacaya Volcano, Guatemala; *JVGR*, v. 19, p. 223-238.

SANTA ANA

El Salvador
13.85°N, 89.63°W, summit elev. 2382 m

The first historic eruption of this strato-volcano, in 1520, is the second oldest in Central America. Twelve eruptions have followed; the most recent in 1920. Santa Ana is 120 km ESE of Pacaya and 55 km WNW of the capital city San Salvador.

12/80 (5:12) Observations were made during a flight over the country. A moderate plume rose from a bank of fumaroles on the SE wall of the inner crater, very similar to its appearance in Nov 1978.

Information Contacts: R. Stoiber, S. Williams, R. Naslund, L. Malinconico, M. Conrad, Dartmouth College.

TECAPA

El Salvador
13.50°N, 88.50°W, summit elev. 1592 m
Local time = GMT - 6 hrs

Only fumarolic activity has been reported from this strato-volcano 40 km ESE of Santa Ana.

5/85 (10:5) The following is from David Harlow.

"An earthquake swarm occurred by surface faulting on 21 April in the Chinameca Complex, near Tecapa, a volcano with no known historic eruption. An m_b 4.7 earthquake, the largest so far, occurred on 23 April at 0916. Four days later, on 27 April, during the peak activity (in numbers of recorded and felt events per day), an 8 km-long graben formed on the NW flank of the volcano. The N40°W-striking graben is 0.8-1 km wide, with a vertical offset of 30 cm. Through 6 June, 170 earthquakes were felt in the area. A very preliminary analysis of the earthquake location data implies that activity began on a tectonic fault some 15 km NW of the volcanic summit, and moved nearer to the volcano at the time the graben formed. Since 3 June, 5-20 earthquakes per day were recorded by a high-gain seismograph station 50 km from the volcano, down from 300 events per day on 26, 27, and 28 April and 21 and 23 May."

Information Contacts: J. González, Centro de Investigaciones Geotécnicas, San Salvador; D. Harlow, USGS, Menlo Park, CA.

SAN MIGUEL

El Salvador
13.44°N, 88.27°W, summit elev. 2132 m

San Miguel, 25 km ESE of Tecapa, has erupted 32 times since 1586. The only Salvadoran volcano to have erupted more frequently in historic time is Izalco (perennially active until a hotel was built to view its pyrotechnics in 1957).

12/76 (1:15) An explosive eruption of San Miguel was reported to have begun during the morning of 2 Dec. "Smoke," loud noises, and a sulfur smell were reported from the nearby town of San Miguel. By 9 Dec, the eruption cloud was visible from a considerable distance and ashfalls had caused some crop damage. A few persons were evacuated. Although "fire" was reported by the press, only a white cloud was seen by Richard Stoiber during a 9 Dec overflight. The last activity at San Miguel was the eruption of more than 75,000 m^3 of ash beginning 30 March, 1970.

Information Contacts: AFP; Sercano Radio Network; R. Stoiber, Dartmouth College.

Further Reference: Martínez, M.A., 1977, *The Eruption of 2 December, 1976 of San Miguel Volcano, Republic of El Salvador, Central America*; Centro de Investigaciones Geotécnicas, San Salvador, 4 pp.

12/80 (5:12) During a flight over El Salvador by Dartmouth geologists, a small, continuous vapor plume rose from the summit crater.

Information Contacts: R. Stoiber, S. Williams, R. Naslund, L. Malinconico, M. Conrad, Dartmouth College.

COSIGUINA

Nicaragua
12.97°N, 87.58°W, summit elev. 859 m

Cosigüina is a basaltic shield volcano with a caldera lake, 190 km ESE of San Miguel. It has erupted only twice in historic time; a violent eruption in 1835 and a small explosion in 1852. The volcano forms the NW tip of Nicaragua.

12/80 (5:12) No fumarolic activity was visible from the rim [during a visit between mid-Nov. and early Dec.]

Information Contacts: R. Stoiber, S. Williams, H.R. Naslund, L. Malinconico, M. Conrad, Dartmouth College; S. Bonis, IGN, Guatemala; A. Aburto, D. Fajardo, Instituto de Investigaciones Sísmicas (IIS).

SAN CRISTOBAL

Nicaragua
12.70°N, 87.02°W, summit elev. 1745 m
Local time = GMT - 6 hrs

A strato-volcano 70 km ESE of Cosigüina, San Cristóbal has erupted 9 times since the Spanish conquest. Summit fumarolic activity has been strong since 1971.

3/76 (1:6) A team from Dartmouth College visited the volcano from 12 to 30 March and supplied the following information.

The level of activity was very low and static. There was no regular activity, and no big changes have occurred. Every few days there was a little shudder of superficial tremor in the volcano, and perhaps a short, spasmodic earthquake, but nothing very important. There was nothing out of character with the gradual buildup of activity that has been observed over the past couple of months.

On 9-10 March an eruption of ash drifted into the town of Chinandega, and there was a small ash eruption, mainly confined to the crater area, on 16 March at 1130. The team collected some sand from the latter eruption. Because the gases were too thick and heavy the crater could not be entered. When viewed from the air, sulfur coated everything in the crater to the point where it was actually overlapping the rim on the leeward side. The team did not get the impression that any inordinate activity was imminent at San Cristóbal.

Information Contact: Dept. of Earth Sciences, Dartmouth College.

4/76 (1:7) It is believed that the activity during April remained static. The gas output was about five times more than it was before the 2 small ash eruptions in March, and this involves a sulfur output of thousands of t/d.

Information Contact: R. Hazlett, Dartmouth College.

8/76 (1:11) On 29 Aug San Cristóbal erupted briefly with a minor ejection of ash — the eruption being another signal in the rise of the volcano's thermal gradient as evidenced by earlier eruptions in May 1971 and 9-10 March 1976, both of which were minor. The volcano has been quiet since the most recent eruption.

According to Richard Stoiber, both gas emission and its SO_2 content increased from early 1971 to March 1976, but had declined by mid-July 1976.

The ash ejection of 29 Aug has increased anticipation of major activity. Richard Stoiber has been monitoring the volcano for the past 6 months, collecting condensed water samples from active fumaroles. Analysis of those samples has not yet been completed.

Information Contact: A. Aburto Q., IIS.

12/76 (1:15) The fume cloud from San Cristóbal was still very large in Dec, but the SO_2 content of the gas had declined since the summer.

Information Contact: R. Stoiber, Dartmouth College.

10/77 (2:10) A small ash eruption from San Cristóbal began on 16 Oct at 0800 and lasted about 45 minutes. According to local estimates, the eruption column rose about 1500 m above the summit. Ashfall at Chinandega, near the volcano, was very slight.

Information Contact: Same as above.

2/80 (5:2) A large white plume was observed from the air. Remote sensing of SO_2 revealed very low levels as compared to records from monitoring over the last 8 years.

Information Contacts: R. Stoiber, S. Williams, M. Bruzga, Dartmouth College.

7/80 (5:7) A moderate vapor plume was emitted. Remote sensing showed that the SO_2 content of the plume was low in comparison to recent years.

Information Contacts: R. Stoiber, S. Williams, Dartmouth College; M. Carr, J. Walker, Rutgers Univ.; A. Creusot, Instituto Nicaraguense de Energía.

12/80 (5:12) A moderate-sized vapor plume rose continuously from the summit. Remote sensing of SO_2 revealed increased flux since June 1980, but SO_2 emission remained far below the levels of the mid-1970's.

Information Contacts: R. Stoiber, S. Williams, H.R. Naslund, L. Malinconico, M. Conrad, Dartmouth College; A. Aburto Q., D. Fajardo B., IIS.

Further Reference: Zapata, R., 1981, La Actividad del Volcán San Cristóbal (Nicaragua), Iniciada el 24 de Agosto de 1980; *Boletín de Vulcanología* (Universidad Nacional, Heredia, Costa Rica), no. 11, p. 16-18.

1/81 (6:1) The gas plume released essentially continuously since 1971 has become intermittent. Periods of energetic gas release of less than 1 hr duration were separated by periods (measured in hours) of only low fumarolic release. Shallow seismic activity continued at levels above background.

Information Contacts: R. Stoiber, S. Williams, Dartmouth Colege; D. Jerez, Instituto Nicaraguense de Recursos Naturales (IRENA), Managua; D. Fajardo B., IIS.

3/81 (6:3) The following is a report from Stanley Williams and Richard Stoiber.

"A trend of decreasing SO_2 emissions had been evident since the small ash eruptions of March 1976. However, San Cristóbal has suddenly reversed this trend, after being in a heightened state of seismic activity since Aug 1980. In late Feb, SO_2 output increased by approximately an order of magnitude to the several thousand t/d level of the mid-1970's. Flights over the crater in mid-Feb and mid-March showed evidence of considerable recent slumping in the crater formed by the 1976 eruptions, especially on the N and NW walls. Fumarolic activity was evident all over the crater but was most concentrated in the S and SE margins of the floor and in the lower parts of the walls. No new fumaroles or fissures were observed outside the 1976 crater. Night observation revealed extensive incandescence over much of the crater, even more than that observed in Dec at Momotombo (5:12). High gas concentrations and unstable footing prevented measurement of any fumarole temperatures. Seismic activity continued at high levels, with almost continuous harmonic tremor and at least one earthquake with magnitude greater than 2 (which occurred one week before the elevated SO_2 emission was detected)."

Information Contacts: R. Stoiber, S. Williams, Dartmouth College; D. Jerez, IRENA; D. Fajardo B., IIS.

7/81 (6:7) Between 14 June and 11 July, personnel from PIRPSEV, CNRS, and the volcano observation section of IPG sampled gases from 5 Central American volcanoes. Medium- to high-temperature gases from San Cristóbal were analyzed (table 14-2). Maximum gas temperatures measured were 525°C.

Information Contacts: H. Delorme, Univ. de Paris; J. L. Cheminée, IPG, Paris.

12/81 (6:12) The following 3 reports are from Dartmouth College geologists.

"The volcano continued to emit a small white irregular gas plume. SO_2 emission rates were measured at very low levels, similar to those reported for Nov 1980 and Jan 1981. Some caving had occurred in the summit crater since March 1981, and incandescence was again observed."

Information Contacts: R. Stoiber, S. Williams, R. Naslund, J.B. Gemmell, D. Sussman, Dartmouth College; D. Fajardo B., IIS.

1/82 (7:1) "COSPEC measurements indicated that very little SO_2 was being emitted in the vapor plume. Levels were lower than any measured since San Cristóbal began emitting gases in 1971. Seismicity beneath the volcano has diminished to normal levels, but incandescence could still be observed in the crater."

Information Contacts: R. Stoiber, S. Williams, Dartmouth College; D. Fajardo B., Instituto Nicaraguense de Estudios Territoriales (INETER).

3/82 (7:3) "During a crater visit 9 March we found that the small white vapor plume was almost entirely made up of water vapor, with little acid gas content. We were unable to reach the fumaroles, but Bruce Gemmell of Dartmouth College measured fumarole temperatures as high as 590°C in Dec 1981.

Information Contact: S. Williams, R. Stoiber, Dartmouth College, I. Menyailov, V. Shapar, IVP, Kamchatka; D. Fajardo B., INETER.

8/82 (7:8) Most of the following is from Douglas Fajardo B., with additional notes from Roderic Parnell.

A gas column, usually drifting SW, was continuously present over the volcano. The strongest activity was concentrated in the E part of the summit crater, where several incandescent areas were observed. A temperature of 565°C was measured in July. Landslides within the crater enlarged its E portion. Since the strong seismic activity of March 1981 (6:3), only occasional small earthquakes have been recorded, normal at San Cristóbal.

Information Contacts: D. Fajardo B., INETER; R. Parnell, Jr., Dartmouth College.

4/83 (8:4) Temperatures continued to increase slowly, to 620°C on 2 Feb. A strong gas column was blown W. New N-S-trending fissures were observed. During Feb, 13-14 seismic events were recorded per day, but could only be detected by very sensitive instruments.

Information Contact: D. Fajardo B., INETER.

8/85 (10:8) Space shuttle astronauts photographed plumes from two volcanoes on 2 Sept. A whitish, rather wispy plume about 70 km long was sighted over San Cristóbal.

Information Contact: C. Wood, NASA, Houston.

SO_2	65.40%	CO	0.02%
CO_2	33.93%	CH_4	65 ppm
H_2	0.31%	He	12 ppm
H_2S	0.33%	COS	27 ppm
N_2	4.35% in air		

Table 14-2. Analytical mean of 3 mid-1981 gas samples (dry, HCl excluded) from San Cristóbal.

Information Contacts: R. Gleason, R. Stoiber, Dartmouth College.

TELICA

Nicaragua
12.60°N, 86.87°W, summit elev. 1010 m
Local time = GMT - 6 hrs

Telica's eruptive history is similar to San Cristóbal's, 20 km to the WNW. Fourteen small to moderately explosive eruptions are known since 1527, 5 of them since 1965.

11/76 (1:14) Telica emitted large dark clouds during the morning of 3 and 4 Nov.

12/76 (1:15) During the morning of 18 or 19 Nov, there was a little dark material in the small fume cloud that is usually present over Telica. Alain Creusot suggested that landsliding from the crater walls may have been responsible.

Information Contact: R. Stoiber, Dartmouth College.

11/77 (2:11) The quoted material in Telica reports is by Dartmouth College geologists.

"During the early morning of 25 Nov, our group observed a series of ultravulcanian ash eruptions from Telica. An explosion was observed at 0605 from León, 19 km from Telica, producing a light brown ash cloud that rose 0.91 km above the summit. A dispersed cloud of ash from an earlier explosion was noted SW of the volcano. Four more explosions were observed, at 0617, 0622, 0626, and 0644. For the rest of the day, only intermittent observations were possible, but explosions were noted at 0740, 0835, and 1530. Ash fell in a zone extending about 5 km SSW of the volcano. Four kilometers from the source, the ash formed a discontinuous coating < 1 mm thick. Samples were collected for analysis at Dartmouth. A similar explosion, sending ash 1.6 km above the cone, was seen at 0605 on 26 Nov. At 0920 on 27 Nov, a dust cloud, presumably produced by another explosion, was seen from the air. The group reports that other Nicaraguan volcanoes show their normal level of fumarolic activity."

Information Contacts: R. Stoiber, R. Birnie, S. Self, Dartmouth College.

12/77 (2:12) Increased activity began 11 Nov, when sizeable explosions began to occur at a rate of about 1/hr, in contrast to about 1/month during the preceding year. Ash from some of the larger explosions reached the Pacific Ocean, about 35 km from the volcano. By early Jan, explosion frequency had declined to 1-2/day and ashfall was no longer reaching the ocean.

Information Contacts: A. Aburto Q., IIS; D. Harlow, USGS, Menlo Park, CA.

2/80 (5:2) A minor white plume was observed in early Feb.

Information Contacts: R. Stoiber, S. Williams, M. Bruzga, Dartmouth College.

7/80 (5:7) When geologists visited Telica in June, a small steam plume continued to be emitted.

Information Contacts: R. Stoiber, S. Williams, Dartmouth College; M. Carr, J. Walker, Rutgers Univ.; A. Creusot, Instituto Nicaraguense de Energía.

12/80 (5:12) In late 1980, a moderate-sized but continuous vapor plume rose from the summit crater. SO_2 flux was remotely measured and found to be about 150 t/d.

Information Contacts: R. Stoiber, S. Williams, H.R. Naslund, L. Malinconico, M. Conrad, Dartmouth College; A. Aburto Q., D. Fajardo B., IIS.

1/81 (6:1) A small-volume plume of vapor was intermittently released in early 1981. Shallow seismicity was regularly observed in the vicinity.

Information Contacts: R. Stoiber, S. Williams, Dartmouth College; D. Jerez, IRENA; D. Fajardo B., IIS.

3/81 (6:3) "Two flights were made over the summit crater of Telica, in mid-Feb and mid-March. Two large holes (each with a diameter of approximately 20-30 m) occurred high on the NW wall of the crater. They are reported (by Alain Creusot, Instituto Nicaraguense de Energía) to coalesce at depth. One or both of them emitted a continuous vapor plume. Occasional minor ash eruptions were reported by local people."

Information Contacts: Same as above.

12/81 (6:12) "A small continuous plume of white vapor and occasional ash was observed in late Nov. The 2 summit vents on the W wall of the crater had merged into one."

Information Contacts: R. Stoiber, S. Williams, H.R. Naslund, B. Gemmell, D. Sussman, Dartmouth College; D. Fajardo B., IIS.

1/82 (7:1) "A series of small gas and ash eruptions took place in late Dec and early Jan. Ash fell only very close to the crater. Seismicity was above normal throughout the period and remained so in late Jan."

Information Contacts: R. Stoiber, S. Williams, Dartmouth College; D. Fajardo B., IIS.

2/82 (7:2) "On 12 Feb between 1100 and 1200 Telica had its largest eruption since 1976. It began after $1^1/_2$ days of relative seismic quiet and lasted about 45 minutes. Blocks and scoria 2-3 m in diameter were thrown over an area extending 200 m from the crater rim. Minor collapse occurred on the crater rim. Fires were ignited in grass up to 1 km from the crater, especially to the E. At 5-6 km downwind, moist sand-sized ash accumulated to 2-3 cm depth. Ash fell to the SW as far away as Corinto (45 km from Telica), on the Pacific coast."

While flying over the Pacific coast of Nicaragua in a commercial aircraft on 12 Feb at 1200, William I. Rose, Jr. saw a vertical ash column penetrate the top of the cloud deck (at about 1.8-2.4 km altitude) over the Telica area. After about 50 seconds, the eruption column reached its maximum altitude of 3.7-4.3 km. Rose observed a second, smaller eruption column at 1207 before moving out of view of the area at 1215.

"The eruption followed several months of minor gas and ash ejection and notably heightened levels of seismic activity. Eruptive activity had apparently increased in duration and frequency during the week preceding 12 Feb." Dull red incandescence had been visible in the deepest portion of the vent for weeks and new bright incandescence was noted on 8 Feb in a fumarole at a much higher elevation, after a day of eruption. A second eruption from 1545 to 1610 on 12 Feb also ejected blocks and scoria but was much smaller.

La Prensa reported that between 12 and 14 Feb at least 50 people (of the roughly 8,000 living nearby) had been evacuated from the S side of the volcano, and the Red Cross established an aid station. Some farm birds were killed and small cotton fields on the flanks of Telica were coated with ash. Wild birds and animals also fled the area. The Red Cross reported strong activity 14 Feb and ejection of ash and blocks at 0930 on 15 Feb.

"On 19 Feb at 1030 a second large eruption sent ash as far to the WSW as Chinandega (35 km away)." William Rose was again flying along the coast of Nicaragua, about 160 km offshore, and observed 2 distinct explosions, the first starting at about 1030, the second between 1100 and 1115. Tephra clouds rose to more than 3.5 km altitude and black ash blew to the NW. "In the following week, 2-3 small eruptions each day sent ash and gas over the immediate area.

"The day geologists visited the crater, 24 Feb, there were no large eruptions and there had not been since 20 Feb. An impressive bomb and block field was laid down on the N side of the crater, extending approximately 300 m to the N. Craters 1-1.5 m across were created by low-angle impact of bombs 50 cm in diameter at the 300 m range. At about 100 m from the crater rim, the surface was 100% cratered and at 80 m was 100% covered with debris. Maximum thickness was 50 cm at the crater rim. The deposit was strongly inversely graded. Fresh bread-crusted scoria bombs up to 80 cm in diameter were widely scattered throughout the deposit. The active vent had increased in size by about 2 times. It was elongated N-S and was approximately 75 m by 50 m. No incandescence was visible during the day. Slumping of the wall above the vent may have removed about 10 m from the rim of the crater.

"Gas emission between eruptions on 24 Feb was very small, as it was on 28 Jan. The gas was bluish in color and smelled like SO_2 and HCl, but not H_2S. Samples were taken on both days. Volcanic tremor occurred in irregular pulses only. The volcano has been relatively quiet seismically since the eruption of 12 Feb. In Jan and early Feb, there had been several magnitude 2-3 earthquakes each day."

Information Contacts: R. Stoiber, S. Williams, W. Crenshaw, D. Sussman, Dartmouth College; W. Rose, Jr., Michigan Tech. Univ.; D. Fajardo B., INETER; A. Creusot, Instituto Nicaraguense de Electricidad; *La Prensa*.

3/82 (7:3) "The eruption sequence that began in mid-Dec 1981 appears to have drawn to a close. The last confirmed eruption occurred at approximately noon on 2 March, sending ash to Corinto and beyond. Since then the volcano has also been seismically quiet. A crater visit on 19 March revealed continued collapse of the crater walls. The vent was clogged with boulders and a ring of strongly jetting fumaroles was established around its margins."

Information Contacts: S. Williams, R. Stoiber, Dartmouth College; I. Menyailov, V. N. Shapar, IVP, Kamchatka; D. Fajardo B., INETER.

Further Reference: Williams, S.N., 1985, La Erupción del Volcán Telica, Nicaragua, 1982; *Boletín de Vulcanología* (Universidad Nacional, Heredia, Costa Rica), no. 15, p. 10-19.

8/82 (7:8) The diameter of the E part of the crater increased by about 10 m. As of early Aug no seismicity was being recorded but fumarolic activity was stronger than it had been before the Dec–March eruption.

Information Contacts: D. Fajardo B., INETER; R. Parnell, Jr., Dartmouth College.

4/83 (8:4) After a tephra eruption ended 1 March 1982 (7:1-3), seismicity returned to normal. Through Nov 1982 about 14 events were recorded/day. On 10 Feb 1983 a portable seismograph installed on the flank recorded 66 earthquakes. This increase was paralleled by stronger fumarolic activity on the S, SE, and SW part of the volcano.

Information Contact: D. Fajardo B., INETER.

11/85 (10:11) Although normal activity continued, Telica continued to show stronger seismic activity than other Nicaraguan volcanoes. A seismograph registered 24 low-frequency microearthquakes in a 5-hour period on 9 July.

Information Contacts: Same as above.

CERRO NEGRO

Nicaragua
12.50°N, 86.70°W, summit elev. 675 m

This is the youngest of 4 basaltic cinder cones clustered along a N-S line and 20 km ESE of Telica. It has erupted at least 17 times since its birth in 1850.

2/80 (5:2) A small white plume formed by several small fumaroles was visible from the air in early Feb.

Information Contacts: R. Stoiber, S. Williams, M. Bruzga, Dartmouth College.

7/80 (5:7) When geologists visited Cerro Negro in June, fumaroles on the N side of the crater reached 320°C. A new, low-temperature, fumarolic area had formed 1 km N of Cerro Negro on the E flank of Cerro La Mula. A linear zone of recently killed vegetation extended 200 m along a NNW trend. Cerro Negro last erupted in Feb 1971, ejecting about 4.5×10^6 m^3 of tephra.

Information Contacts: R. Stoiber, S. Williams, Dartmouth College; M. Carr, J. Walker, Rutgers Univ.; A. Creusot, Instituto Nicaraguense de Energía.

12/80 (5:12) In late 1980, summit crater fumaroles remained at temperatures as high as 300°C. A small vapor plume was intermittently visible. Seismic activity had dropped from the high level of June.

Information Contacts: R. Stoiber, S. Williams, H.R. Naslund, L. Malinconico, M. Conrad, Dartmouth College; A. Aburto Q., D. Fajardo B., IIS.

3/81 (6:3) The following 3 reports are from Dartmouth College geologists.

"A flight over the crater in mid-March revealed 1 area of minor fumarolic activity in the SW center region of the crater. No significant seismicity has occurred recently."

Information Contacts: S. Williams, R. Stoiber, Dartmouth College; D. Jerez, IRENA; D. Fajardo B., IIS.

12/81 (6:12) "A small region of fumarolic activity was observed high on the W wall of the crater in late Nov."

Information Contacts: R. Stoiber, S. Williams, H.R. Naslund, J.B. Gemmell, D. Sussman, Dartmouth College; D. Fajardo B., IIS.

3/82 (7:3) "A very small gas plume was being emitted from a group of fumaroles on the NW inner crater wall. Maximum fumarole temperatures of 505°C were measured on 3 March."

Information Contacts: S. Williams, R. Stoiber, Dartmouth College; I. Menyailov, V. Shapar, IVP, Kamchatka; D. Fajardo B., INETER.

8/82 (7:8) Minor fumarolic activity continued. The maximum temperature was 320°C and has not varied since 1980.

Information Contacts: D. Fajardo B., INETER; R. Parnell, Jr., Dartmouth College.

4/83 (8:4) A small column of gases was observed in March, but there was no seismic activity. A fumarole temperature of 337°C was measured, 17° higher than in mid-1982.

Information Contact: D. Fajardo B., INETER.

LAS PILAS

Nicaragua
12.48°N, 86.68°W, summit elev. 938 m

Only 5 km SE of Cerro Negro, but substantially larger, this composite volcano's only known eruptions were in 1952 and 1954.

7/80 (5:7) Minor fumarolic activity continued in June, and an area of recently killed vegetation was observed on the NNE flank. The Las Pilas complex last erupted 29-31 Oct 1954, producing explosions from the central crater El Hoyo.

Information Contacts: R. Stoiber, S. Williams, Dartmouth College; M. Carr, J. Walker, Rutgers Univ.; A. Creusot, Instituto Nicaraguense de Energía.

12/80 (5:12) In late 1980 a small continuous vapor plume was still being emitted from the top of the kilometer-long crack in the summit.

Information Contacts: R. Stoiber, S. Williams, H.R. Naslund, L. Malinconico, M. Conrad, Dartmouth College; A. Aburto, D. Fajardo B., IIS.

MOMOTOMBO

Nicaragua
12.42°N, 86.55°W, summit elev. 1191 m

A strato-volcano on the N shore of Lake Managua and 20 km ESE of Cerro Negro, Momotombo has erupted 9 times since the 16th century. Activity since 1905 has been largely fumarolic, and geothermal power production began in 1982.

7/80 (5:7) Temperatures of 465°-780°C were measured in summit crater fumaroles in June. However, there was little gas emission. Momotombo's last eruption, in 1905, ejected tephra and lava flows from the S crater.

Information Contacts: R. Stoiber, S. Williams, Dartmouth College; M. Carr, J. Walker, Rutgers Univ.; A. Creusot, Instituto Nicaraguense de Energía.

12/80 (5:12) The summit crater fumaroles remained very hot in late 1980 with temperatures measured up to 735°C and reported to > 900°C. A small vapor plume continued and remote sensing revealed very low rates of SO_2 emission. Portions of the crater were seen to glow red and orange when observed at night, with the highest temperatures on the steep S wall of the crater. No seismic activity has occurred recently at Momotombo.

Information Contacts: R. Stoiber, S. Williams, H.R. Naslund, L. Malinconico, M. Conrad, Dartmouth College; A. Aburto Q., D. Fajardo B., IIS.

1/81 (6:1) "A small, continuous vapor plume was visible in early 1981. No shallow seismicity was observed around Momotombo."

Information Contacts: R. Stoiber, S. Williams, Dartmouth College; D. Jerez, IRENA, Managua; D. Fajardo B., IIS.

SO_2	24.02%	CO	0.07%
CO_2	64.86%	CH_4	21 ppm
H_2	4.89%	He	18 ppm
H_2S	6.15%	COS	18 ppm
N_2	4.93% in air		

Table 14-3. Analytical means of 4 mid-1981 gas samples from Momotombo (dry, HCl excluded).

3/81 (6:3) "A small continuous plume continued to be released in early 1981. No new measurements were made. No significant seismicity has occurred recently."

Information Contacts: R. Stoiber, S. Williams, Dartmouth College; D. Jerez, IRENA; D. Fajardo B., IIS.

7/81 (6:7) Between 14 June and 11 July, personnel from PIRPSEV, CNRS, and the volcano observation section of IPG sampled medium- to high-temperature gases from Momotombo (table 14-3). Maximum gas temperatures measured were 640°C.

Information Contacts: H. DeLorme, Univ. de Paris; J. Cheminée, IPG, Paris.

12/81 (6:12) The following 3 reports are from Dartmouth College geologists.

"A very small continuous vapor plume was observed in late Nov. Summit fumarole temperatures seemed lower than in March 1981."

Information Contacts: R. Stoiber, S. Williams, H.R. Naslund, J.B. Gemmell, D. Sussman, Dartmouth College; D. Fajardo B., IIS.

1/82 (7:1) "A very small plume was continuously emitted in late Dec and early Jan."

Information Contacts: R. Stoiber, S. Williams, Dartmouth College; D. Fajardo B., IIS.

3/82 (7:3) "Temperatures of the crater fumaroles, measured on 13 March, were as high as 800°C. Heating has occurred since Dec 1981, but it was not apparent whether this was a result of dry season effects or was a true increase in heat. A small gas plume was continuously emitted."

Information Contacts: S. Williams, R. Stoiber, Dartmouth College; I. Menyailov, V. Shapar, IVP, Kamchatka; D. Fajardo B., INETER.

Further Reference: Menyailov, I.A., Nikitina, L.P., Shapar, V.N., Grinenko, V.A., Buachidze, G.I., Stoiber, R., and Williams, S., 1986, The Chemistry, Metal Content, and Isotope Composition of Fumarolic Gases from Momotombo Volcano, Nicaragua, in 1982; *Volcanology and Seismology*, no. 2, p. 60-70.

8/82 (7:8) A small vapor plume was still being emitted in Aug and incandescence could be seen in the summit crater. The maximum fumarole temperature was 825°C, 25° higher than in March. Tremor and small-magnitude discrete earthquakes continued.

Information Contacts: D. Fajardo B., INETER; R. Parnell Jr., Dartmouth College.

11/82 (7:11) Continued heating of the entire fumarole field has been observed in late 1982. No significant volume change has occurred in the small plume of volcanic gas released continuously. Seismic activity remained at low levels.

Information Contacts: S. Williams, R. Stoiber, Dartmouth College; G. Hodgson V., D. Fajardo B., INETER.

4/83 (8:4) Fumarolic activity increased over that of 1981, especially in the S part of the crater. Temperatures continued to increase, reaching 855°C in March, 30° higher than in mid-1982. Seismographs recorded magmatic movement but no discrete earthquakes.

Information Contact: D. Fajardo B., INETER.

5/83 (8:5) The following report is from Róger Argeñal A.

"In mid-March 1983 a joint Soviet-Nicaraguan team began a 3-month program of continuous observations of fumarolic activity in Momotombo's summit crater. The observations included temperature measurements and gas sampling from the highest temperature fumarole (N 9; fig. 14-10). Immediately after sampling, the gas samples were analyzed in a field chemical laboratory near the volcano. CH_2, O_2, and N_2 contents were determined and the S/Cl, F/Cl, and H_2S/SO_2 ratios were calculated, as well as the analyses shown in fig. 14-11.

"During the initial period of observations (27 March–5 April) a gradual temperature rise from 851° to 858°C was noted, as well as a negligible increase in H_2O and a more appreciable increase in reduced gases (H_2S, H_2, and CO). The seismic station registered volcanic tremor that appeared 17 March and stopped 4 April. When the tremor disappeared the temperature of fumarole N 9 immediately fell sharply, dropping 15°C in 4 days and continuing downward to 815°C by 22 April (fig. 14-11). Simultaneous with the falling temperature, an increase in H_2O content from 94.1 to 96.7 mole % and decreasing proportions of CO_2, H_2, CO, and other gases were registered. After 15 April the H_2O content gradually decreased, reaching previous values at the end of April. The concentrations of other gases also reached previous values by that time.

"Although the variations in the temperature and gas compositions of N 9 fumarole were very appreciable, the S/Cl, F/Cl, and H_2S/SO_2 ratios fluctuated inside very narrow ranges of values. This fact gave us the chance to assume that no eruptive activity occurred at Momotombo during the period of observations."

Information Contacts: I. Menyailov, V. Shapar, L. Gartzeva, V. Pilipenko, IVP, Kamchatka; R. Argeñal, D. Fajardo B., R. Espinoza, INETER, I. Vallecillo, Proyecto Geotérmico.

5/85 (10:5) Summit fumarole temperatures have remained high, with small fluctuations prohably related to seasonal rainfall variations. The maximum temperature

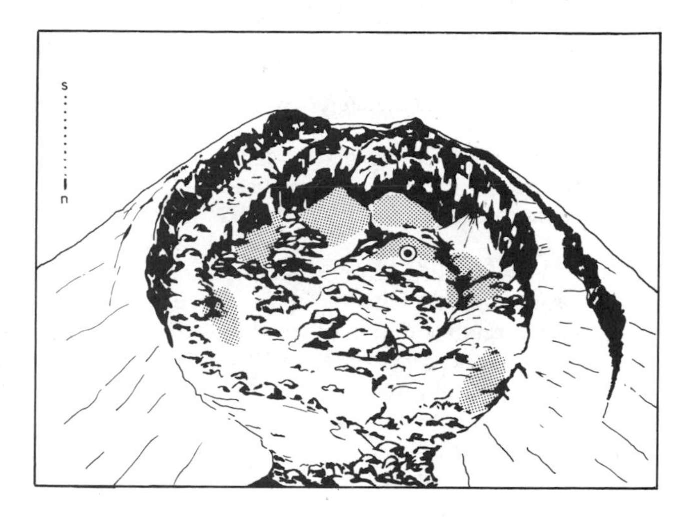

Fig. 14-10. Sketch of the summit area of Momotombo. Areas of fumarolic activity are shown by a stipple pattern. Temperatures were measured and gases sampled at the hottest fumarole (N 9), marked

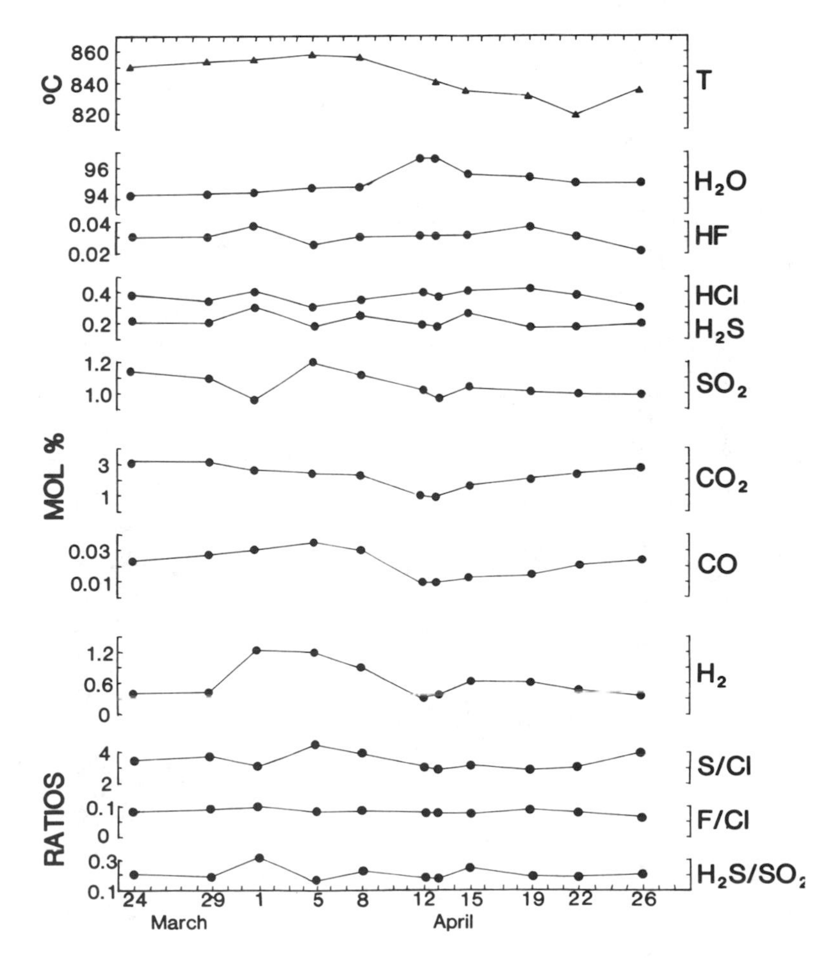

Fig. 14-11. Temperature in °C, *top*, gas concentrations in mole %, *center*, and gas concentration ratios, *bottom*, at Momotombo's fumarole N 9 (fig. 14-10), 24 March–26 April 1983.

Month	Temp	Month	Temp
Jan	870°	Jun	875°
Feb	870°	Jul	875°
Mar	875°	Aug	875°
Apr	879°	Sep	874°
May	882°	Oct	874°

Table 14-4. Fumarole temperatures at Momotombo, 1985.

in 1984 was 895°C in Aug; the same fumarole had a temperature of 875°C when measured in 1985.

Information Contact: D. Fajardo B., INETER.

11/85 (10:11) On 30 May, Momotombo was monitored with 3 sensitive portable seismographs. Three microearthquakes were recorded, one of which was detected by all 3 instruments, and was located at 12.455°N, 86.547°W at 1.27 km depth. Fumarole temperatures remained high (table 14-4), but were slightly lower than the 895°C measured in Aug 1984.

Information Contacts: Same as above.

Further Reference: Menyailov, I.A., Nikitina, L.P., Shapar, V.N., and Pilipenko, V.P., 1986, Temperature Increase and Chemical Change of Fumarolic Gases at Momotombo Volcano, Nicaragua, in 1982–1985: Are These Indicators of a Possible Eruption?; *JGR*, v. 91, no. B12, p. 12199-12214.

MASAYA

SE of Managua, Nicaragua
11.95°N, 86.15°W, summit elev. 635 m
Local time = GMT - 6 hrs

Masaya is a 6 × 11 km caldera 20 km SE of Managua and 65 km SSE of Momotombo. It became Nicaragua's first National Park in 1979. Masaya has been active almost continually since first documented in 1524, and has exhibited an active lava lake from 1965 to 1979. Its prehistoric record indicates explosive violence unusual in basaltic volcanoes.

4/78 (3:4) The past 4 months have produced a gradual increase in the intensity of activity at Masaya. Fissures

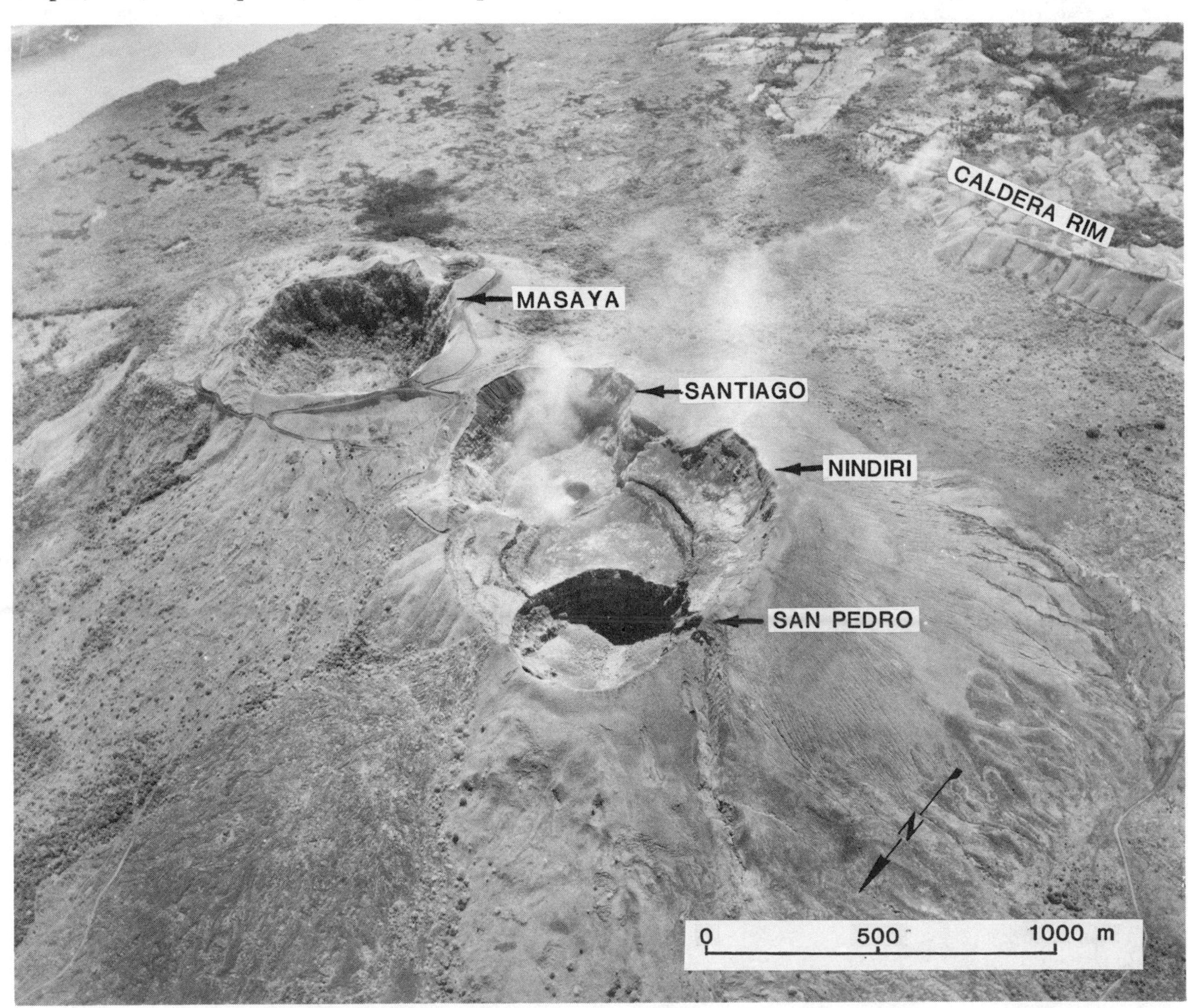

Fig. 14-12. Oblique airphoto of the Masaya Complex, taken by IGN, 6 Nov 1975. Courtesy of Jaime Incer.

have appeared in the floor of Santiago Crater (fig. 14-12) a collapse feature that formed, along with neighboring San Pedro crater, in 1858. The vent opening about 100 m below the rim of Santiago's pit crater has widened to about 3 times its size of a few months ago. The persistent lava lake inside the pit crater is usually not visible from Santiago's rim, but splashes of lava can occasionally be seen and minor amounts of lava clots are sometimes thrown from the vent. When the volcano was visited in late March, rare bursts of scoria reached the rim of the pit crater. Gas emission was strong, but has not seriously damaged nearby coffee trees.

Information Contacts: D. Jerez, Parque Nacional Volcán Masaya; D. Shackelford, CA.

2/80 (5:2) In early Feb, the plume appeared larger than any observed between 1968 and 1977. The diameter of the active lava lake in the pit crater was several times larger than in 1977 and the level of lava has dropped since then.

Information Contacts: R. Stoiber, S. Williams, M. Bruzga, Dartmouth College.

7/80 (5:7) Emission of a very large plume continued in June. Remote sensing of SO_2 gas revealed high output rates. The gas plume allowed only brief glimpses of the small pit crater in which an active lava lake had been observed on many occasions since 1970. The lake was not seen during the brief clear moments, nor did a glow appear in photographs of the pit. The lake's characteristic roaring noise, if present, was masked by the sounds created by gas emission. There were no night observations at Masaya.

Information Contacts: R. Stoiber, S. Williams, Dartmouth College; M. Carr, J. Walker, Rutgers Univ.; A. Creusot, INETER.

12/80 (5:12) Emission of a very large gas plume has continued without interruption since fall, 1979. Remote sensing of SO_2 revealed continued high flux, with a 1500-2000 t/d average for the entire year. The hole through the surface of the lava lake was larger than in previous years and a great deal of sublimation was occurring around its edge. No lava or red glow was visible during daylight. Acid gas and rain continued to cause considerable damage downwind.

Information Contacts: R. Stoiber, S. Williams, H.R. Naslund, L. Malinconico, M. Conrad, Dartmouth College; A. Aburto Q., D. Fajardo B., IIS.

1/81 (6:1) The gas emission event that began in fall 1979 continued with a steady release of very large amounts of SO_2 in early 1981. Strong winds carried the gas plume onto populated areas at high elevations. A day of notable rockfall activity in the crater was followed for 1 day by a significantly larger rate of gas release.

Information Contacts: R. Stoiber, S. Williams, Dartmouth College; D. Jerez, IRENA, Managua; D. Fajardo B., IIS.

3/81 (6:3) "Scientists from Dartmouth College, IRENA, and the Instituto de Investigaciones Sísmicas report the following based on their continuing cooperative observation of Nicaraguan volcanoes.

"The fourth gas emission crisis of this century continued unabated. Extensive remote measurement of SO_2 output (by COSPEC) has revealed a greater variability in emission rates than had previously been recognized (several hundred to several thousand t/d). The pit crater from which the gas is emitted continued to increase slowly in diameter and was strongly elongate in the NW-SE direction. Night observation of the activity was possible and confirmed the complete absence of any incandescence in the pit where lava was visible as recently as Nov 1978."

Information Contacts: S. Williams, R. Stoiber, Dartmouth College; D. Jerez, IRENA; D. Fajardo B., IIS.

12/81 (6:12) Quoted material following 6 reports is from Dartmouth College geologists.

"Emission of a very large white vapor plume continued in late Nov. SO_2 emission rates, measured using COSPEC, were at the same high levels reported since Feb 1980. Acid rain and gas fumigation continued to cause problems downwind. Incandescence was seen in the bottom of the inner crater through the crust on the surface of Santiago Crater lava lake on 29 Nov. Park rangers reported that this incandescence has been visible since Sept 1981, but it was not noted by several observers who specifically looked for it while working around the crater 25-29 Nov. The roaring sound of gas emission (or possibly lava splashing) may have been louder than in March 1981."

Information Contacts: R. Stoiber, S. Williams, H.R. Naslund, J.B. Gemmell, D. Sussman, Dartmouth College; D. Fajardo B., IIS.

1/82 (7:1) "A small eruption occurred from the hole in Santiago Crater lava lake in the early evening of 16 Dec. No one witnessed the event, but people living S of the caldera reported hearing an explosion that was followed by ashfall several kilometers to the S. Highly vesiculated scoria fragments up to 20 cm in diameter fell as much as 200 m S of Santiago pit crater. As of late Jan, no subsequent explosive activity had been observed. A very large plume was still being continuously emitted. Incandescence was not readily visible during the day but was evident at night."

Information Contacts: R. Stoiber, S. Williams, Dartmouth College; D. Fajardo B., IIS.

3/82 (7:3) "Bright yellow incandescence was plainly visible at night in Santiago Crater in early March. No change had occurred except for a small collapse of the inner crater walls. The huge gas plume still poured out continuously."

Information Contacts: S. Williams, R. Stoiber, Dartmouth College; I. Menyailov, V. Shapar, IVP, Kamchatka; D. Fajardo B., INETER.

8/82 (7:8) Emission of acid gas from Santiago Crater was strong and continuous in early Aug. Incandescence from the small pit in Santiago Crater's lava lake was visible at night. The gas and associated acid rain affected vegetation downwind. An explosive gas emission event occurred 6 June at 1622. As of early Aug, seismographs recorded constant tremor.

Information Contacts: D. Fajardo B., INETER; R. Parnell, Jr., Dartmouth College.

10/82 (7:10) A small, brief, explosive eruption from the bottom of the lava lake in Santiago Crater occurred at dawn on 7 Oct. Tephra, including blocks with volumes to 55 cm^3, fell 300 m SE and covered an area of 150,000 m^2. The eruption killed a few trees and animals near the summit. Heat from the ejecta melted asphalt on a road, which was also slightly damaged by impact from larger tephra. Rumbling and explosion sounds were heard through the day. After the initial explosion, no additional tephra was ejected, but gas emission increased considerably, forming wide vapor columns that reached high altitudes.

The eruption was preceded by a change in the pattern of seismicity and accompanied by a M 2.3 event lasting 3.7 seconds. After the eruption, small earthquakes occurred about every 6 minutes until 1100 on 8 Oct.

The 7 Oct eruption was larger than Masaya's previous explosion on 26 Dec 1981 (7:1). Strong vapor emission has made observations of the bottom of the crater difficult, obscuring any changes that may have occurred to the lava lake.

Information Contact: G. Hodgson V., INETER.

11/82 (7:11) "The approximately 3-year gas emission crisis from Santiago Crater continued in late 1982. Total SO_2 flux was apparently reduced from the very large levels reported before. Incandescence within the inner crater was dull red-orange, as compared to the brilliant orange observed in Feb, 1982. The 7 Oct, 1982 explosion threw out abundant, juvenile, highly vesiculated scoria, which was often flattened and oxidized against the ground surface. Numerous fragments of sublimate minerals torn from the lip of the inner crater were ejected with the juvenile scoria. No new explosions have thrown debris out of Santiago Crater but several gas bursts have been reported by Park guards."

Information Contacts: S. Williams, R. Stoiber, Dartmouth College; G. Hodgson V., D. Fajardo B., INETER.

4/83 (8:4) A strong gas column was observed. Through 9 March, incandescence was noted.

Information Contact: D. Fajardo B., INETER.

5/85 (10:5) In Dec 1983, 4993 microearthquakes were recorded at Masaya. Early that month, a very large gas column was continuing to emerge from Santiago Crater. A series of small ash eruptions occurred in April 1984. There was a small gas explosion on 23 Jan 1985, and another ash eruption occurred in April 1985.

Information Contact: Same as above.

8/85 (10:8) An apparent volcanic plume roughly 400 km long was sighted in the Masaya area, extending W over the Pacific Ocean. Its exact source could not be clearly ascertained as Masaya was obscured by clouds. It was broader and appeared to be more dense then the San Cristóbal plume but was still relatively diffuse.

Information Contact: C. Wood, NASA, Houston.

11/85 (10:11) Heights of the gas column above the rim of Santiago Crater (485 m above sea level) were measured on three occasions in 1985: 22 Jan (48 m), 17 June (78 m), and 21 Oct (78 m).

On 3 Dec, during re-entry from mission 61-B, Space Shuttle pilot Brian O'Connor took three 35-mm photographs (nos. 61B-12-020, 021, and 022) of Masaya. These showed a large white plume extending at least 25 km due W toward the Pacific Ocean.

Information Contacts: D. Fajardo B., INETER; C. Wood, M. Helfert, NASA, Houston.

Further Reference: Stoiber, R.E., Williams, S.N., and Huebert, B.J., 1986, Sulfur and Halogen Gases at Masaya Caldera Complex, Nicaragua: Total Flux and Variations with Time; *JGR*, v. 91, no. B12, p. 12215-12232.

MOMBACHO

Nicaragua
11.83°N, 85.98°W, summit elev. 1345 m

On the NW shore of Lake Nicaragua, 25 km SE of Masaya, Mombacho's last activity was in 1570.

12/80 (5:12) In late 1980 a small, intermittent plume was visible, rising from the SE section of the summit.

Information Contacts: R. Stoiber, S. Williams, H.R. Naslund, L. Malinconico, M. Conrad, Dartmouth College; M. Carr, J. Walker, Rutgers Univ.; A. Creusot, Instituto Nicaraguense de Energía.

3/82 (7:3) The following is a report from Dartmouth College geologists.

"Paolo Pisani, a consultant to the Instituto Nicaraguense de Electricidad, reported finding 4 previously unknown low-temperature hot springs on the S side of Mombacho. These are not believed to be new, however."

Information Contacts: S. Williams, R. Stoiber, Dartmouth College; I. Menyailov, V. Shapar, IVP, Kamchatka; D. Fajardo B., INETER.

CONCEPCION

Isla de Ometepe, Nicaragua
11.53°N, 85.65°W, summit elev. 1610 m
Local time = GMT - 6 hrs

This strato-volcano forms the NW half of an island near the center of Lake Nicaragua and 75 km SE of

Masaya. Seventeen eruptions have been recorded since 1883.

4/77 (2:4) A 1-2-minute eruption from Concepción, which lit the sky "like daylight," began at 2326 on 4 April. Earthquakes were felt at about $1\,^1/_2$ hours and at 2 minutes prior to the eruption (2156 and 2324 on 4 April) and about 9 hours afterwards (0822 on 5 April). During the next several weeks, frequent small ash eruptions, separated by periods of gas emission, caused light ashfalls on Isla de Ometepe.

Sixteen separate explosions, some sending incandescent ash more than 1500 m above the summit, occurred between the early afternoon of 29 April and the morning of 1 May. A burst of seismicity accompanied each explosion. Ash fell intermittently at Rivas, 25 km SW of Concepción. A few minor ash clouds were reported on 3 May.

Local seismicity had begun to increase in Oct 1976 with many events occurring in Dec 1976 and March 1977. Between 1 and 27 April, 145 local earthquakes and many brief (a few seconds to a few minutes) periods of low-frequency tremors were recorded.

Information Contacts: D. Harlow, USGS, Menlo Park, CA; A. Aburto Q., IIS.

4/78 (3:4) The newspaper *La Prensa* reports that strong ash eruptions from Concepción began in late March. Ashfall made life "intolerable" for persons living near the volcano, and was reported as far away as Belén, on the Pacific Coast 28 km SW of Concepción.

Information Contacts: *La Prensa*; D. Jerez, Parque Nacional Volcán Masaya.

3/82 (7:3) The following 2 reports are from Dartmouth College geologists.

"A series of small steam and ash eruptions occurred from mid-Jan to mid-Feb. During flights over Concepción on 18 Feb and 4 March we saw a moderate-sized continuous white vapor plume being emitted from the crater."

Information Contacts: S. Williams, R. Stoiber, Dartmouth College; I. Menyailov, V. Shapar, IVP, Kamchatka; D. Fajardo B., INETER.

11/82 (7:11) "We observed Concepción for a short time in early Dec. It appeared to be emitting a small gas plume and no explosive activity has been reported recently."

Information Contacts: S. Williams, R. Stoiber, Dartmouth College; G. Hodgson V., D. Fajardo B., INETER.

4/83 (8:4) At the beginning of March, earthquakes were recorded and unusually strong gas columns were ob-

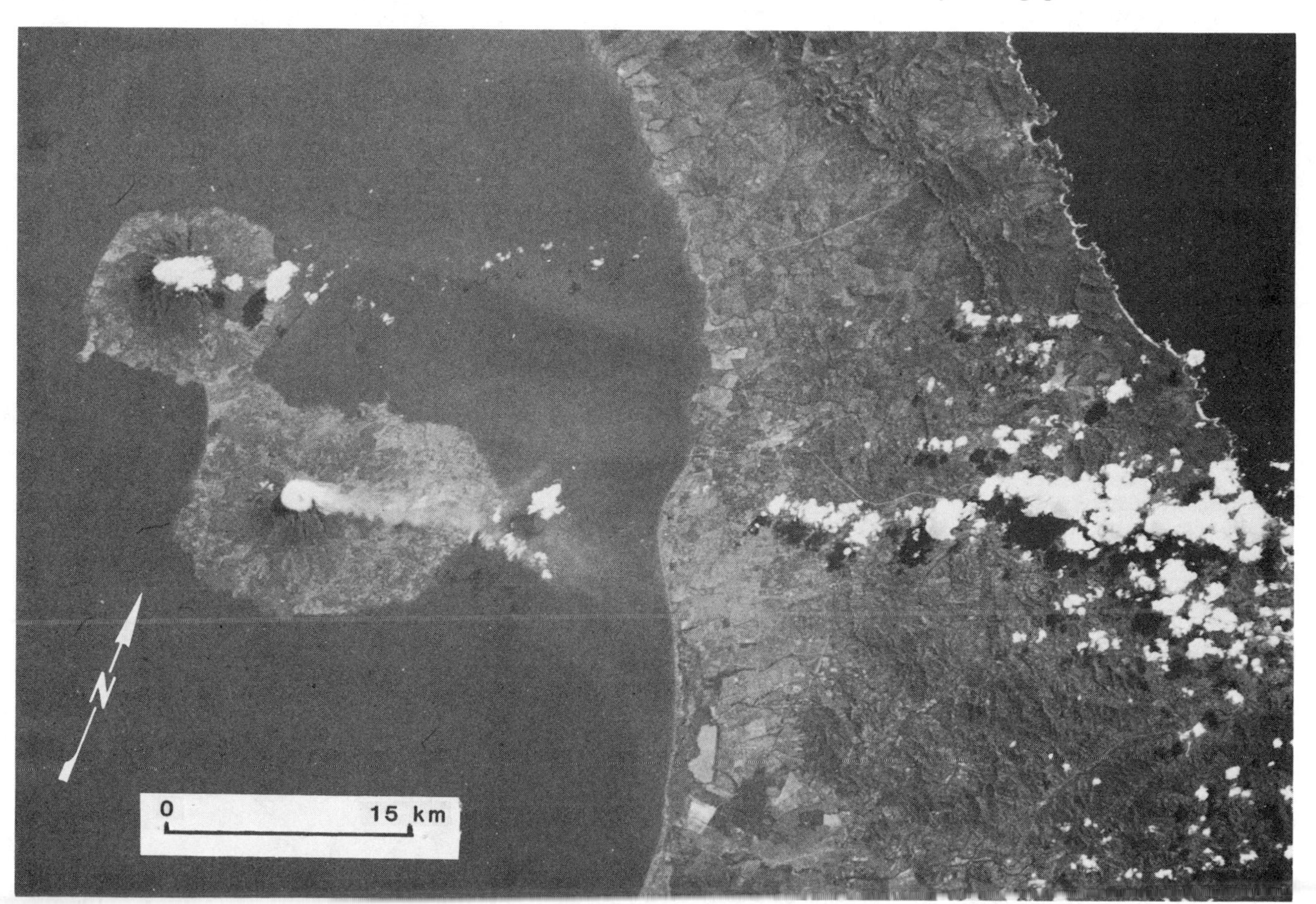

Fig. 14-13. Space Shuttle photograph (no. 61B-39-066) of Concepción, taken Dec 1985, showing plume over volcano. Courtesy of C. Wood.

served. A strong ash eruption began on 15 March and continued until the 24th, but caused no damage. Earthquakes accompanied the activity, but there were very few compared to the usual seismicity at the volcano. The ash eruptions were smaller than those of 1977 (2:4).

Information Contact: D. Fajardo B., INETER.

5/85 (10:5) A series of small ash eruptions occurred in Dec 1984.

Information Contact: Same as above.

11/85 (10:11) A violent tephra eruption occurred on 2 Jan 1985, accompanied by strong rumbling. Ashfall damaged tobacco and sesame crops, and fell on the towns of Esquipulas, Los Angeles, and to a lesser extent, Moyogalpa (at the W foot of the volcano). As the eruption began, Esquipulas was celebrating the festival of its patron saint, but heavy ashfall forced the festival's suspension. Blocks fell on the volcano's flanks. The eruption was comparable to that of 1977 (2:4).

During Space Shuttle mission 61B (27 Nov–3 Dec 1985), astronauts photographed Concepción 3 times. The first 2 photographs (nos. 61B-121-101 and 61B-36-086), taken on 28 Nov at 1300 and 30 Nov at 1230, showed no plume or other signs of volcanic activity. A photograph (no. 61B-39-066; fig. 14-13) taken later in the mission, however, exhibited a buff-brown plume over the volcano, which could be traced clearly for 6.6 km to the SW with possible wispy material for an additional 10 km. Space Shuttle photographs are available through USGS, EROS Data Center, Sioux Falls SD.

Information Contacts: D. Fajardo B., INETER; C. Wood, M. Helfert, NASA, Houston.

RINCON DE LA VIEJA

NW Costa Rica
10.83°N, 85.33°W, summit elev. 1895 m

This remote volcano, 85 km SE of Concepción and 30 km S of Lake Nicaragua, has erupted at least 12 times since 1860.

General References (Costa Rican Volcanoes): Barquero, J., et al., 1978–1986, Estado de los Volcanes de Costa Rica (15 annual or semi-annual reports); *Boletín de Vulcanología*, nos. 2-13 and 15-17.

Garcia, M.O. and Malavassi, E. (eds.), 1983, *Memoir, USA-Costa Rica Joint Seminar in Volcanology, San José, January 1982*; Universidad Nacional, Heredia, 155 pp. (18 papers).

11/82 (7:11) The following is a report from Dartmouth College geologists.

"In Nov the odor of H_2S was present from the summit crater, but there was no visible gas plume and no eruptive activity. Low-temperature steam vents and mudpots persisted at Estación las Pailas, at the foot of the volcano."

Information Contacts: R. Stoiber, S. Williams, H.R. Naslund, C. Connor, J. Prosser, J.B. Gemmell, Dartmouth College; E. Malavassi, J. Barquero H., Univ. Nacional, Heredia.

3/83 (8:3) The following paragraph is primarily from a report by Jorge Barquero H. and Juan de Dios Segura.

During the night of 6 Feb, residents of towns (Dos Ríos de Upala, Colonia Blanca, and Colonia Libertad) 8 km N and NE of the volcano heard strong rumblings and observed the rise of a large eruption column from the crater. Personnel from the Proyecto de Investigaciones Vulcanológicas climbed the volcano 19 Feb. The odor of sulfur was stronger than it had been during their previous ascent in Nov 1982. Phreatomagmatic eruptions had ejected bombs, lapilli, and ash, as well as blocks 10 cm-1 m in diameter that formed impact craters. Tephra fell SE, S, and SW of the vent to a distance of about 1.5 km. Destruction, primarily to vegetation, was greatest to the SE and S. The tephra had a high water content because the vent contained a lake. Strong rains and rapid erosion since the eruption made it difficult to calculate the original depth of the airfall deposits, although in some places SE of the vent they were 4 cm thick. The eroded ash washed into a ravine, producing a small mudflow in a NE flank river (Río Pénjamo), causing the deaths of thousands of fish 7-8 Feb, possibly because of the acidity of the water. The pH of the cold lake was 3.5 on 19 Feb and 4.1 on 5 March.

Jorge Barquero H., J. Bruce Gemmell, and Jerry Prosser climbed the volcano on Nov 1982, and Gemmell provided the following report.

"Rincón de la Vieja is a large composite volcano with a series of collapse craters aligned ENE-WSW. Its main cone is covered with thick vegetation but 3 craters to the W are not vegetated. The most recently active crater (250 m in diameter) is 1 km NW of the main cone. No activity or gas emissions were seen in this crater and a cold yellowish-green lake covered the crater floor. No steam was rising from the lake but 2 areas of brown discoloration near its center may have indicated subaqueous vents.

"The area around the summit craters was covered with accessory blocks of andesitic lava and tuff breccias, in addition to juvenile andesitic breadcrust bombs, lapilli, and ash from the most recent recorded eruptions in 1966–70. Numerous mudpots, hot springs, and steam vents occurred in 2 main areas (Aguas Termales and Sitio Hornillas), on the S flank at about 900 m elevation."

Information Contacts: J. Barquero H., J. de Dios Segura, Univ. Nacional, Heredia; J.B. Gemmell, J. Prosser, Dartmouth College; R. Sáenz R., Ministro de Energía y Minas; *La República*, San José.

Further Reference: Barquero, J., and de Diós Segura, J., 1983, La Actividad del Volcán Rincón de la Vieja; *Boletín de Vulcanología*, no. 13, p. 5-10.

ARENAL

NW Costa Rica
10.48°N, 84.72°W, summit elev. 1552 m

Arenal's first historic eruption in 1968 devastated its W flank and killed 78 people. It has been active ever since, emitting viscous lava flows and Strombolian explosions. This conical volcano is 40 km SE of Rincón de la Vieja.

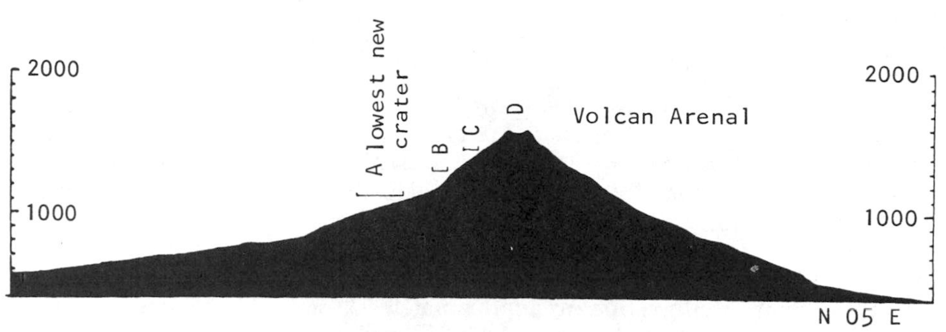

Fig. 14-14. N85°E section through Arenal, showing locations of craters (no vertical exaggeration). After Melson and Sáenz (1973).

10/76 (1:13) Arenal Volcano was very active during observations 12-22 Oct. Thick block flows were being emitted from the summit crater (D in fig. 14-14) which has been greatly enlarged and now opens to the W. The flows were advancing NW. Avalanches from the flow fronts near the crater were audible about once/minute. Since the initial explosions of July–Aug 1968, a very large lava "delta" of basaltic andesite ($\sim$0.1 km^6 in volume) has been built, covering most of the W slope of the volcano. An additional hot avalanche was emitted in June 1975.

The activity has sequentially migrated from the lower new explosion crater (A in fig. 14-14) to the middle slope craters (B and C), and is presently from the summit crater (D). Arenal's activity presents an excellent opportunity to observe the eruption and dynamics of thick block flows.

Reference: Melson, W.G. and Saenz, R., 1973, Volume, Energy, and Cyclicity of Eruptions of Arenal Volcano, Costa Rica; *BV*, v. 37, p. 416-437.

Information Contact: W. Melson, SI.

1/77 (2:1) Arenal volcano, observed 11 Jan, 1977, remained as active as it was 12-22 Oct 1976 (figs. 14-15 and 14-16). The only change is that an active block flow is now moving from the summit crater (D) towards the SW, overriding 1974 flows. On 29 Jan, it had descended to about 1000 m elevation, burying seismic and dry tilt stations recently installed on the rim of the lower crater (A; fig. 14-17).

Information Contact: Same as above.

Fig. 14-15. N side of Arenal, showing the still-growing N lava field. Crater D (summit crater) has greatly enlarged since the beginning of current eruptions (1968). Photo taken by W. G. Melson, Oct 1976.

11/78 (3:11) The following is from *Boletín de Volcanología* No. 2 of the Volcanology Section, School of Geographical Sciences, National University of Costa Rica.

Lava extrusion and gas emission at Arenal continued through mid-Oct. A block lava flow descended NW from Crater C, near the summit at about 1400 m altitude. By 15 July, this flow had bifurcated at about 1100 m altitude, one arm moving N, the other W. Three months later, on 14 Oct, the front of the W arm had traveled 500-600 m from the point of bifurcation, to approximately 830 m altitude, and was about 15 m thick. The N arm had only advanced about 150 m because of damming by older flows. A large gas column from Crater C was also observed on 14 Oct.

William Melson notes that Crater C has merged with Crater D, the old summit crater, over about the past 5 years, forming a single much-enlarged summit crater breached on the NW. This crater is rimmed elsewhere by nearly vertical cliffs, up to an estimated 100 m high on the E side.

Information Contacts: J. Barquero H., Univ. Nacional, Heredia; W. Melson, SI.

2/79 (4:2) In Oct 1976, 4 dry tilt stations were installed along a radial line on the W side of the volcano. Since then, these stations have been relevelled at about 1-month intervals. Throughout the 2 years of measurement, Arenal has been emitting block lava flows from its summit crater, at largely unknown but widely ranging rates of emission. During this period each station has shown, on the average, continuing deflation. Fig. 14-18 shows deflation along the about 40 m-long radial line at each station. As expected, stations closest to the volcano show the largest rate of deflation. Periods of rapid deflation in some cases correlate with times of high rates of lava emission.

W. Melson notes that these data are consistent with a very shallow magma chamber 2 km below the surface, following the Mogi model of surface deformation over a spherical magma chamber and assuming maximum tilt at Station A. The extensive tilt measurements at Kilauea show a quite different pattern than at Arenal. Specifically, deflation is extremely rapid, occurring over 12-hour to 6-month intervals, often accompanied by flank eruptions. At Arenal, more or less continuous deflation is accompanied by nearly continuous, long-term eruption of lava from the summit crater. At Hawaii, rates of lava emission are much higher than at Arenal, where basaltic andesite lavas (55% SiO_2) contain about 50% phenocrysts and are clearly more viscous than the basalts of Hawaii. It seems reasonable to conclude that these different deflation behaviors are related directly to the different viscosities.

Fig. 14-16. SW side of Arenal, showing now inactive Crater A and approximate outline of active SW moving flow as of 29 Jan 1977. Photo taken by W. G. Melson, 11 Jan 1977.

Access to Station A was blocked in Feb, 1977 by a new lava flow. Stations A, C, and D (fig. 14-17) are tilt vectors oriented very close to, or on, the radial line. Station B is along a sharp ridge crest and only one component of tilt, along the ridge crest, (roughly on the radial line) can be measured.

The installation and relevelling were done by ICE topographers with advice from W. G. Melson and R. S. Fiske, in accordance with similar successful stations on Hawaii. Relevelling is done using a Wilde N3 level and precision stadia. The benchmarks at each station are in unconsolidated airfall, lahar, and avalanche deposits. Stability was obtained by driving 3 m-long, 2.5 cm-diameter steel rods into the ground. In most cases, it proved possible to drive these completely into the ground. The upper soil zones around each stake were excavated to about 1 m and filled with concrete. The top of each stake was then rounded, and between measurements was coated with grease and capped to prevent rusting.

Information Contacts: W. Melson, SI; J. Umaña, E. Evans, ICE.

3/79 (4:3) The following 6 reports are from *Boletín de Vulcanología* Nos. 3-8, respectively.

The block lava that had been flowing NW from the summit crater since mid-1978 stopped in Nov. However, a new block flow from the summit crater headed NE, reaching 1100 m altitude (~500 m below the summit) by 2 Dec. This flow continued to advance in Jan and Feb.

Information Contact: J. Barquero H., Univ. Nacional, Heredia.

7/79 (4:7) A team from the Institute of Volcanology climbed Arenal via the W and N flanks on 18 and 19 May. A new block lava flow from the summit crater was moving slowly down the W flank, where its front had reached 1300 m altitude (about 330 m below the summit). The NE flank flow described in 4:3 had stopped at 1000 m elevation, about 100 m below its altitude of early Dec 1978. Glow had been visible over the NE flank in March and April.

Strong fumarolic activity obscured the summit crater. Within about 70 m of the summit on the N flank, numerous fumaroles emitted vapors ranging in temperature from 55° to 95°C.

Information Contact: J. Barquero H., Univ. Nacional, Heredia.

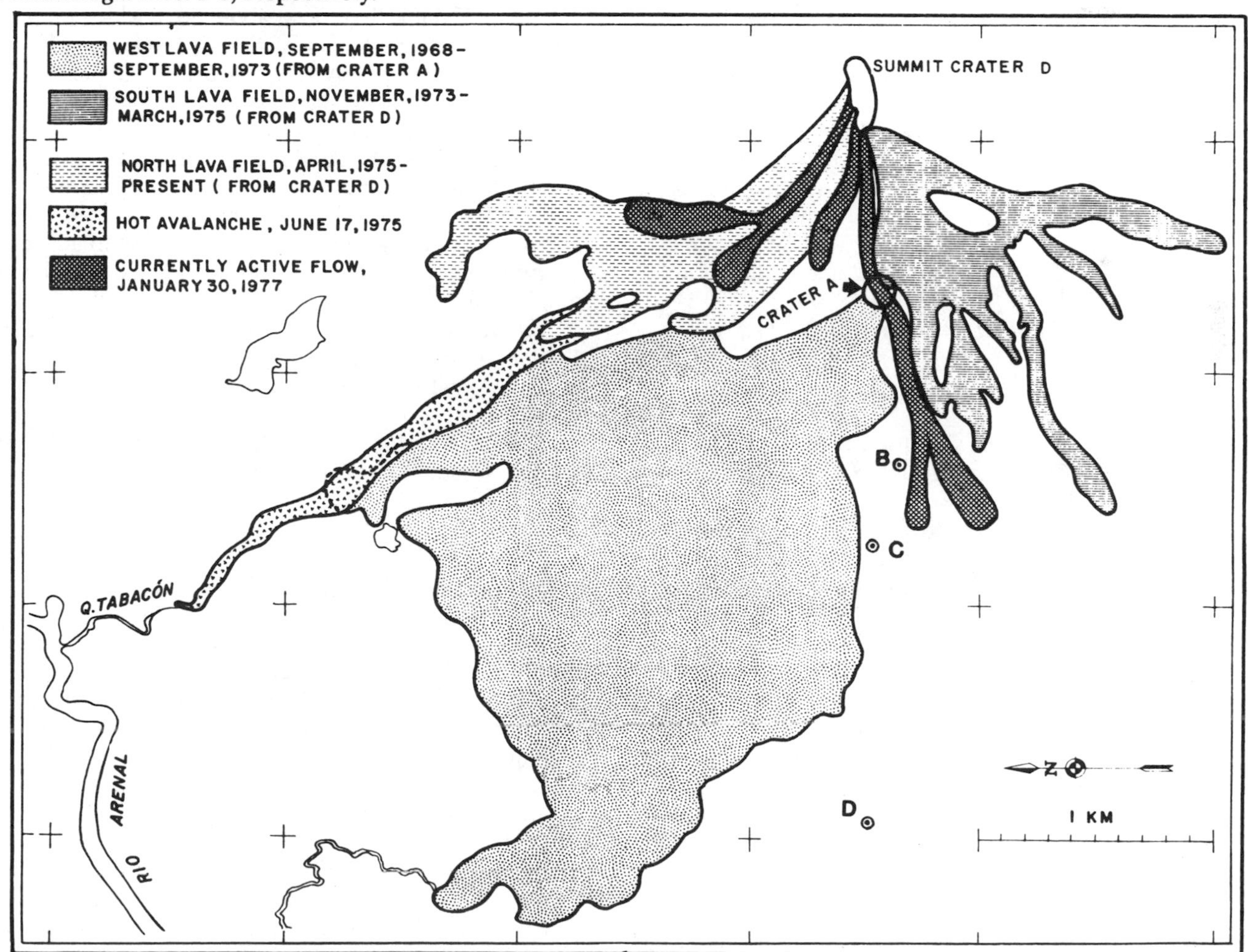

Fig. 14-17. Preliminary sketch map of the N, S, and W lava fields, showing lava flows and the hot avalanche deposit of 17 June 1975. Dates are approximate. Tilt stations B, C, and D are shown by circles. Grid markings refer to the 1000 Meter Transverse Mercator Grid, Zone 16, Clarke 1866 Spheroid (Fortuna, Costa Rica 1:50,000 Quadrangle). Adapted from a map prepared by Hugo T. Tims and provided by Jorge Umaña, Instituto Costarricense de Electricidad. Tilt station A, not shown in the map, is about 100 m S of the center of Crater A. Courtesy of W.G. Melson.

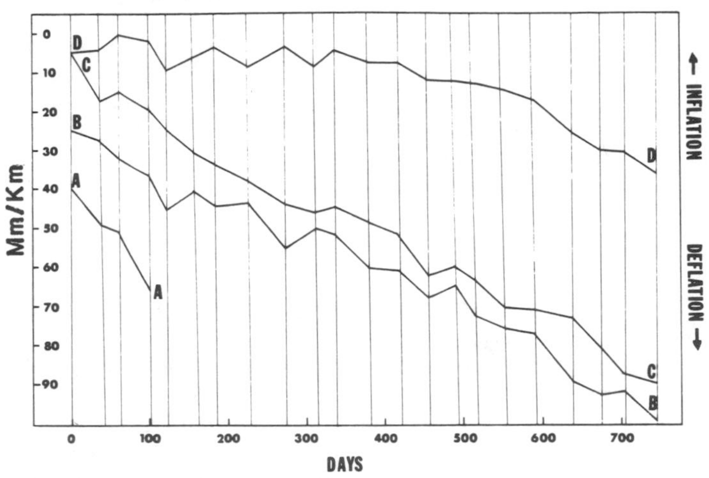

Fig. 14-18. Inflation/deflation radial to Arenal's summit, in µrad, at tilt stations A, B, C, and D (fig. 14-17), from 19 Oct 1976 (day 0) to 3 Nov 1978.

10/79 (4:10) A team from the Institute of Volcanology climbed the W flank on 20 Sept. The active block lava flow observed during the previous ascent on 18-19 May had stopped when its front reached an altitude of about 1130 m on the W flank. A new block flow from the same crater (Crater C, at the W end of the elliptical summit crater area) had moved to about 900 m altitude on the SW flank. From there, a small lobe flowed W, reaching 700 m altitude by 20 Sept, with an approximate thickness of 20 m at its front.

During the team's ascent, a series of white vapor ejections occurred from Crater C, producing clouds that rose about 200 m above the summit. Each ejection was accompanied by loud noise. Collapse appeared to have taken place in the summit crater's W wall, which separates it from the active Crater C. Collapse debris fell above Crater C.

Information Contact: J. Barquero H., Univ. Nacional, Heredia.

1/80 (5:1) During a late Dec visit to Arenal by Guillermo Avila and Jorge Barquero H., the SW flank lava flow described in 4:10 continued to advance at 20-25 m/day, more slowly than when the flow was higher, on steeper slopes. The flow, originating from Crater C, was channeled and composed of blocks about m across. Vegetation surrounding the flow front was ignited by the heat of the lava. Since the last ascent of Arenal on 20 Sept, a new lobe had separated from the main flow and was moving downslope to the S.

Whitish vapor ejections from Crater C remained frequent, accompanied by loud noises and a very strong sulfurous odor. The strong fumarolic emanations forced Avila and Barquero to wear gas masks. A red glow was visible in the summit area at night. Major fissures in the upper wall of the summit crater threaten collapse of part of this wall into the vent area.

Information Contacts: G. Avila, ICE; J. Barquero H., Univ. Nacional, Heredia.

4/80 (5:4) The block lava flow described in 4:10 and 5:1 continued to advance down the SW flank through the end of Feb. An inspection of the N flank on 23 Jan revealed no new lava flows on that side of the volcano.

The rate of extrusion from Crater C had diminished in late Jan to the extent that it was difficult to see blocks moving in the central channel of the SW flank flow. However, the extrusion rate increased in Feb, and lava overflowed its channel at about 1200 m above sea level. By 28 Feb, the new lobe had reached an altitude of about 1000 m, with a front 50 m wide and 20 m high, partially covering an older flow. Emission of white vapor from Crater C also continued, accompanied by loud noise.

Information Contact: J. Barquero H., Univ. Nacional, Heredia.

8/80 (5:8) The block lava flow extruded from the W end of the elliptical summit crater since mid-1979 continued to descend the SW flank through April, causing small forest fires. During a period of increased extrusion in Feb, when lava had overflowed its channel, it divided into 7 subflows, partially covering 3 earlier flow deposits.

In early May, lava from the same vent began moving down the W flank. This flow, about 60 m wide and 20 m thick, reached the N rim of Crater A (about 1100 m above sea level) by early July. An arm of the flow at 900 m altitude was still advancing on 10 July, although feeding from the vent had stopped.

Another flow, the 33rd since nearly continuous extrusion of block lava began in 1968, started to descend the NW flank in early July. On the 11th, its front was at about 1200 m elevation and was continuing to advance.

Information Contact: J. Barquero H., Univ. Nacional, Heredia.

Further References: Guendel, F. and Malavassi, E., 1980, La Actividad del Volcán Arenal Entre los Días 15 al 20 de Agosto de 1980; *Boletín de Vulcanología*, no. 9, p. 3-4.

Wadge, G., 1983, The Magma Budget of Volcán Arenal, Costa Rica from 1968–1980; *JVGR*, v. 19, p. 281-302.

2/81 (6:2) The following information is from Jorge Barquero Hernández.

The vent located at 1450 m altitude at the W end of the elliptical summit crater area continued to emit block lava and vapor. The lava flow that began to descend the NW flank in early July had reached 1100 m by Nov and continued to advance. Two other flows that had been active in July on the SW and W flanks had stopped advancing by Nov. A newer flow, the 34th since 1968, descended the W flank to 1300 m altitude, where it bifurcated into lobes moving W and NW over the channels of older flows. The front of the W lobe was at 800 m altitude on 11 Nov, and the other (NW) lobe had reached 1200 m altitude by 12 Nov. A mean velocity of 1.5 km/hr was measured on blocks in the central flow channel on the upper W flank.

The vapor emissions observed 15-20 Aug were a little more voluminous than normal, included small quantities of ash, and were accompanied by rumblings. The constant noise from the violently escaping gases was occasionally loud enough to be heard in nearby villages. Vegetation on the upper part of the volcano's E flank had been burned by the effects of the vapor eruptions. The loss of vegetation had noticeably augmented fluvial erosion.

In a separate communication, Jorge Umaña reports that as of early Feb Arenal continued to emit lava and vapor from the summit area. The gases had a high chlorine content.

Information Contacts: J. Barquero H., Univ. Nacional, Heredia; J. Umaña, ICE.

3/81 (6:3) Lava flow 34 continued to descend the W flank. By mid-March, the flow had divided into 5 lobes. Geologists noted an increase in the chlorine content of gas emitted from the summit area.

Information Contact: J. Barquero H., Univ. Nacional, Heredia.

5/81 (6:5) The following is from *Boletin de Vulcanologia* no. 10, March 1981.

Lava extrusion continued from the vent at 1450 m altitude at the W end of the summit crater. Portions of multi-lobed flow 34 were still advancing in early 1981. One lobe partially covered the W flank explosion crater that had formed in 1968 and extruded lava until 1973. At about 1000 m altitude, this lobe separated into 3 sublobes, all of which flowed generally westward, reaching an altitude of about 900 m by late Feb. Two other lobes had halted, one in Dec 1980 at 700 m above sea level, the second in Jan 1981 at 750 m altitude.

Information Contact: Same as above.

8/81 (6:8) In June 1979 and again in April 1980, the active vent was obstructed by a lava dome. Lava extruded from the dome flowed down the flanks, and gases (table 14-5) with temperatures between 930° and 954°C escaped at low pressure from radial fissures in the dome. Beginning in March 1981, violent degassing of the upper part of the conduit ejected large vapor columns, accompanied by strong detonations heard in La Fortuna, 6 km to the E. The vapor columns occasionally included fine tephra and bombs, which fell as far as 100 m from the crater rim. Incandescent tephra was sometimes observed at night.

In May 1981, the dome deflated, reopening the active vent's central conduit. At about the same time a multi-lobed flow, fed by the dome for the past several months, stopped advancing down the W flank. Later in May, block lava overflowed the S wall of the vent and began to move downslope at 50 m/hr in a channel 20 m wide. The flow split into 2 parallel lobes at 1200 m altitude. The front of one lobe halted at 500 m above sea level on the S flank, but the second lobe continued down the SW flank. Vapor emission continued to kill vegetation on the N and E flanks, resulting in increased erosion that carved large gullies.

An ascent to the crater area 25 June revealed a new flow (the 36th since 1968) descending the W flank. Continuous vigorous emission of gas and fresh tephra (basaltic andesite with plagioclase phenocrysts) was punctuated 1-3 times/day by stronger explosions that ejected bombs and blocks to roughly 200 m above the crater rim. The stronger explosions, which typically occurred after periods of decreased degassing but without immediate warning, kept geologists from getting close to the crater. A network of 5 radon-monitoring stations were installed on the volcano, in cooperation with Michel Monnin, Univ. of Clermont-Ferrand.

Information Contacts: J. Barquero H., E. Malavassi R., Univ. Nacional, Heredia; J. Cheminée, IPG, Paris; H. Delorme, Univ. de Paris; G. Avila, F. Guendel, ICE.

12/81 (6:12) Block lava emission continued from the single crater near the summit at a rate of approximately 1 m/hr. By 3 Dec this flow had reached a cliff SW of the crater, and incandescent material cascaded in spectacular avalanches from the flow front. A continuous white gas column was only briefly observed due to poor weather conditions at the summit.

Information Contacts: R. Stoiber, S. Williams, H.R. Naslund, J.B. Gemmell, D. Sussman, Dartmouth College; E. Malavassi R., J. Barquero H., Univ. Nacional, Heredia.

5/82 (7:5) Lava continued to emerge from the active vent (Crater C). In Jan, a new lava flow, the 38th since 1968, started to move down the SW flank. By April, the flow front had reached an altitude of about 700 m. Continuous gas emission produced a vapor column that was always present over the active vent.

Information Contact: J. Barquero H., Univ. Nacional, Heredia.

11/82 (7:11) The following is a report from Dartmouth College geologists.

"A blocky lava flow continued to descend the upper slopes. On 17 Nov, a short, very slow-moving flow descended barely a hundred meters from the summit on the W side. This flow was much like others of recent months, slow-moving and not far-reaching, and the summit elevation had been built up as a result. By 27 Nov, a new flow had moved 250 m down the NE slope toward the head of the Río Tabacón, where a small nuée ardente eruption killed one and injured another in 1975. This flow was moving at a velocity more like that observed in Nov, 1981 (approximately 1 m/hr; 6:12).

"Numerous loud gas explosions could be heard dur-

H_2	8%	CO	0.27%
CO_2	50%	H_2S	900 ppm
SO_2	37%	CH_4	120 ppm

Table 14-5. Analysis of the anhydrous component (H_2O = 94%) of one of 60 samples of gases collected from fissures in Arenal's dome between June 1979 and Oct 1980. Courtesy of J.L. Cheminée.

ing several days of observation. SO_2 output, as determined by COSPEC, was extremely small considering the active state of the volcano. Total SO_2 output was about a quarter of the 200 ± 30 t/d reported by Casadevall, et al. (1984; see Further Reference under Colima, 7:2) for Arenal in Feb 1982. Large areas of the volcano's lower slopes were covered by a thin haze of blue gas suspected of being HCl-rich."

Information Contacts: R. Stoiber, S. Williams, H. Naslund, C. Connor, J. Prosser, J. Gemmell, Dartmouth College; E. Malavassi R., J. Barquero H., Univ. Nacional, Heredia.

12/82 (7:12) The lava flow that had advanced down the SW flank since Jan (7:5) stopped in Aug. A new lava flow, the 39th since 1968, began descending the SW flank in Aug. In early Sept, the front of its main lobe stopped at an altitude of about 700 m, but another lobe continued to move SW until Nov. In mid-Nov, another new flow began advancing NW. On 18 Nov, its front had reached an altitude of about 1300 m. Vapor emissions and gas explosions continued in the active crater.

Information Contacts: J. Barquero H., E. Malavassi R., Univ. Nacional, Heredia.

3/83 (8:3) By late March, the lava flow that had been advancing down the NW flank along the Río Tabacón had stopped. However, a new flow was emerging from the active crater and had advanced a short distance over its predecessor. Lava extrusion from the summit area has produced more than 40 discrete flows.

Information Contacts: E. Malavassi R., Univ. Nacional, Heredia; J. Prosser, Dartmouth College.

4/83 (8:4) The lava flow that began to move down the NW flank in mid-Nov continued to advance very slowly. By 5 April, its front was at 700 m altitude, but it was being overridden by a new faster-moving flow (the 41st since 1968) which had reached 1200 m above sea level. Gas emission remained constant from the active crater.

Information Contacts: J. Barquero H., J. de Dios Segura, Univ. Nacional, Heredia.

10/83 (8:10) The lava flow that was advancing rapidly down the NW flank in April stopped in July with its front at an altitude of 625 m above sea level. A new flow, the 42nd since 1968, began to emerge in July and by Sept had reached 1380 m altitude. Gas emission was continuous and strong rumblings were heard.

Information Contacts: J. Barquero H., E. Fernández S., Univ. Nacional, Heredia.

3/84 (9:3) Lava extrusion continued from the vent at 1450 m altitude. The lava flow that had been active in Sept 1983 stopped advancing in Oct. During the same month, a new flow (the 43rd since 1968) began to emerge, moving NW before halting at 980 m above sea level in Nov. Another flow (no. 44) started to advance NW in Dec, remaining active until Feb, and still another flow moved N between Jan and March. Extrusion of flow no. 46 started in March and it continued to travel westward late in the month. Rumblings, or sounds similar to those produced by jet aircraft, were often heard in the crater.

Information Contacts: J. Barquero H., E. Fernández S., Univ. Nacional, Heredia.

7/84 (9:7) The following is a report from Rodolfo Van der Laat.

"An increase in the activity of Arenal, in the form of eruptions of ash and large pyroclastics, was observed beginning the first week in June.

"A small increase in the crystallinity and quantity of silica in the lava, signs of a greater although low viscosity, may explain the increase in the intensity and frequency of eruptions of ash and bombs originating in the active crater. This coincides with a clear diminuition in the movement of lava flow number 49 with an active front on 10-11 July at about 1100 m above sea level (400 m below the level of the active crater, fig. 14-18).

"The frequency of the eruptions was approximately every 30 minutes. All emitted gases and about every fourth one produced ash and bombs accompanied by rumbling. The size of the ash eruptions was medium to small, reaching a mean maximum height of 800 m, and ashfall extended some 10 km W to SW of the active crater. The large pyroclastics did not reach farther than 1 km (on the flanks of the volcano, fig. 14-18).

"Seismic data obtained in 3 visits to the volcano revealed a nearly total absence of volcanic earthquakes (other than seismicity accompanying the eruptions) but showed drastic changes in the eruptive character. Abundant on the seismograms were volcanic tremors and sonic waves that accompanied the eruptions of ash and bombs.

"Because bombs did not reach inhabited or farming-grazing areas and thanks to the great quantity of rain that continuously washes the ash, the risk at this time is minimal."

Strombolian activity was occurring during an 11-14 July visit to the volcano by Michael Carr. The following is from his report.

"Explosions usually occurred erratically, but occasionally in a regular pattern with small explosions spaced at 10 ± 2-min intervals. Weak explosions and 10-50-second periods of continuous pulsing gas emission occurred sporadically. Most of the explosions produced large blocks, which were red at night. Ashfall was intermittent and very light, even immediately downwind of the vent. A portable seismometer about 2.5 km from the crater, operated by Luis Diego Morales and Walter Montero of the Universidad Nacional, recorded B-type earthquakes during each explosion. The first seismic wave arrivals preceded the appearance of the eruption column by 2 seconds, indicating that the explosions originated at least 50 m below the crater rim. Until recently, lava flows were being nearly continuously produced from the crater, but the level of magma is now lower."

Information Contacts: R. Van der Laat, Univ. Nacional, Heredia; M. Carr, Rutgers Univ.

11/84 (9:11) The Strombolian activity that began to accompany the extrusion of block lava flows in June was continuing in mid-Nov. The eruptions were accompanied by strong rumblings. Blocks and bombs fell as much as 300 m from the crater. Ash was carried by the wind toward the W, to a distance of 5 km. No losses have occurred to agriculture or livestock. Nevertheless, because of the action of acid rain and ash, some vegetable species have chlorosis symptoms and fungus proliferation, both of which have affected plant development.

Lava flow number 49 ended its advance in Oct. A new flow (no. 50) began to advance toward the SW in Oct, stopping in Nov. Flow no. 51, descending toward the W, was still active in mid-Nov. Portable seismic stations have been operated periodically. They have not registered type A or B volcanic earthquakes, but only events produced by the explosions, and volcanic tremors of different character.

Information Contact: Observatorio Vulcanológico y Sismológico, Univ. Nacional, Heredia.

12/84 (9:12) The quoted material is from Dartmouth College geologists.

"Dartmouth geologists visited the volcano 3-4 Dec. Lava flows had reached Arenal's lower flanks earlier in the eruption, but lava now being extruded was restricted to the summit. Eruption noises were heard at approximately half-hour intervals, some as far as 5 km from the summit. They were accompanied by sounds of ejecta falling on the upper slopes. Light ashfall was observed 2 km from the summit. A local farmer reported that eruptions (perhaps mud pots?) had occurred in a small lake near La Fortuna, 6 km E of Arenal, but the report was not substantiated."

Information Contacts: B. Barreiro, R. Naslund, R. Stoiber, P. Turner, Dartmouth College.

2/85 (10:2) In late Feb, a new lava flow was descending Arenal's N flank, filling gullies that had been formed by erosion. Summit area Strombolian activity (9:7) was continuing, although explosions were not frequent. Explosions typically began with strong rumblings, followed a few seconds later by the ejection of bombs, blocks, and ash from the upper crater to a height of 2000 m. The ash was carried by the prevailing winds toward the W, to a distance of 4 km from the crater. Blocks and bombs fell as far as 500 m from the crater. Some blocks reached 40 cm in diameter. The N and E flanks of the volcano are covered by trees, and falling tephra caused small fires that burned vegetation. Gas emission continued at the same rate as before.

Information Contacts: J. Barquero H., E. Fernández S., Univ. Nacional, Heredia.

7/85 (10:7) During July, two lava flows descended from the active vent. On the NE flank a flow that originated in April continued to advance. A new flow (the 55th since 1968) began erupting in July. Strombolian eruptions continued but were becoming less frequent. Accompanied by very strong rumblings, they produced small quantities of ash that were carried 3 km W and SW from the crater. Gas emission was continuous.

Information Contacts: Same as above.

11/85 (10:11) Lava production continued through early Nov from the active vent (Crater C), feeding lava flows that advanced NW, W, and S. A dome has formed on the W rim of Crater C. Gas emission was continuous. The ejections of pyroclastic materials that began in June 1984 were also continuing in early Nov. Geologists noted that when the advance of the lava flows was slow, the explosions were strong and very frequent, whereas when much lava was flowing from the crater there were almost no pyroclastic ejections. There were strong explosions during the first week in Oct. A 50 × 60 cm bomb fell 1 km from the crater, forming an impact crater that measured 80 × 50 cm and 20 cm deep. A block with dimensions of 90 × 30 × 20 cm formed a 90 × 100 cm crater that was 30 cm deep; this block came to rest 20 m from its initial impact crater. Small quantities of ash were emitted. Prevailing winds carried the ash about 3 km W and SW.

Information Contacts: Same as above.

Further Reference: Alvarado, I.G.E. and Barquero, R., 1987, *Las Señales Sísmicas del Volcán Arenal (Costa Rica) y su Relación con las Fases Eruptivas (1968–1986)*; Instituto Costarricense de Electricidad, Departamento de Geología, San José, 33 pp.

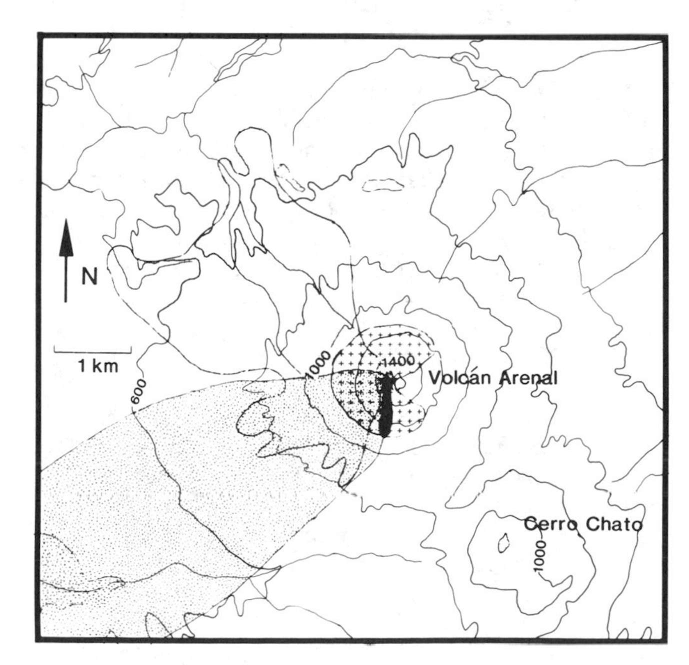

Fig. 14-19. Map of Arenal and vicinity. Contour interval is 100 m. Lakes are indicated by a horizontal line pattern. The area covered by lava since the eruption started in 1968 is shown by vertical lines, the most recent lava flow by heavy shading, and the zones of June–July 1984 ash and bomb deposition by stipple pattern and crosses. Courtesy of R. Van der Laat.

POAS

Central Costa Rica
10.18°N, 84.22°W, summit elev. 2704 m
Local time = GMT - 6 hrs

Poás is 65 km ESE of Arenal and 30 km NW of San José. One of its twin crater lakes is clear, but the northern one is distinctive in both behavior and composition. Most of the volcano's 34 eruptions since 1828 have been from this crater lake, and the 1910 event created a geyser over 4 km high.

General References: Casertano, L., Borgia, A., Cigolini, C., Morales, L.D., Montero, W., Gómez, M., and Fernández, J.F., 1985, Investigaciones Geofísicas y Caracteristicas Geoquímicas de las Aguas Hidrotermales: Volcán Poás, Costa Rica; *Geofísica Internacional*, v. 24, p. 315-332.

Prosser, J., 1985, Geology and Medium-Term Temporal Magmatic Variation Found at the Summit Region of Poás Volcano, Costa Rica; *Boletín de Vulcanología*, no. 15, p. 21-39.

7/76 (1:10) The news media reported that Poás Volcano was "intensely active" on the morning of 23 July, but Rodrigo Sáenz reported that there is no evidence to support this. He is not aware of any recent activity at the volcano, but noted that some minor activity could have taken place.

Information Contact: R. Sáenz R., Dirección de Geología, Minas, y Petroleo.

12/77 (2:12) Local Civil Defense officials reported that Poás erupted early in the week of 18 Dec, for the first time in 1977. A column of water, ash, and mud was ejected, rose more than 1 km above the vent, then fell on the flanks of the volcano, covering a large area with mud.

Information Contact: Sercano Broadcast Network.

11/78 (3:11) The following is from *Boletín de Vulcanología* nos. 1 and 2.

Two periods of increased activity have been observed at Poás since Feb 1978, one in April and May, the other beginning in late Sept. Ejection of sulfur-rich gray clouds of ash and mud to a height that fluctuated between 5 and 100 m began on 10 April and continued through May. Nine to 50 eruptions occurred per day at intervals of 5-30 min. Ejecta fell within the crater or nearby.

Activity was confined to steady gas emission from June until late Sept. On 22 Sept explosions resumed, somewhat larger than those of April and May. Gray ash and mud columns were thrown 40-300 m above 2 vents, one in the center of the crater lake, the other at its S end. Explosions were somewhat less frequent than in April and May; there were up to 30/day at 10- to 45-minute intervals. Much of the ejecta fell within the crater but Volcanology Section personnel visiting Poás on 4 Oct observed ash deposits averaging 15 cm thick on the E side of the crater. The deposits also contained fine- and medium-grained lapilli and blocks of various sizes. The crater lake, occasionally turquoise green in the past, was entirely gray, and its level had dropped about 3 m, as compared to a 0.5-m drop during the April–May activity. Block ejection and landsliding had produced a vertical wall at the S end of the lake by 4 Oct.

The following average crater lake temperatures were recorded: 22 March, 50°C; 8 June, 56°; 27 June, 48°; 8 Sept, 58°; 4 Oct, 70°.

Information Contact: J. Barquero H., Univ. Nacional, Heredia.

3/79 (4:3) The following information is from *Boletín de Vulcanología* no. 3.

The period of mud and tephra ejection that began 22 Sept 1978 declined in Nov and ended in Dec. Eruption columns, containing ash and small lapilli (up to 0.4 cm), did not exceed 25 m in height during Nov, and only sporadic very small eruptions occurred in Dec.

The temperature of the crater lake water fell from 70°C on 4 Oct to 50° on 7 Dec, and decreased further, to 40°, in Feb.

Fumarolic activity from the central "dome" [the eroded cone from the 1953–55 eruption at the S end of the lake] continued, with vapor temperatures ranging from 60 to 90°C during both the Dec and Feb visits.

Information Contact: Same as above.

12/79 (4:12) Small phreatic explosions from the crater lake of Poás resumed in Sept after about 9 months of quiet. Water and tephra formed frequent mushroom- or pine-tree-shaped clouds that rose 15-300 m. Tephra fall was confined to the crater lake. Vapor emission was continuous, occurring in about equal quantities from the lake and the N wall of the [1953–55 cone].

On 29 Nov, personnel from the Volcanology Project, National University, descended into the crater. Explosions, accompanied by strong noise, occurred at 20-minute intervals. The mean lake temperature had risen to 65°C, from 30-40°C in Aug. Some morphological changes had taken place, for which the Volcanology Project personnel suggested 2 causes: the force of the explosions had dislodged material from the crater walls, especially at the S coast of the lake, and tephra fall into the lake produced wave surges, which caused landslides when they struck the crater walls. The level of the lake had risen about 2 m because of heavy rainfall.

Information Contact: Same as above.

1/80 (5:1) *Boletín de Vulcanolgía* no. 6 provided the following.

A resurgence of phreatic activity began on 8 Sept. Poás National Park personnel reported that a pine-tree-shaped eruption column containing tephra and a large quantity of water rose about 400 m, before falling back into the crater lake. Only fumarolic activity was observed 9-14 Sept. From 15 Sept until the end of the month, up to 10 small phreatic explosions occurred per day, ejecting material to only 15-50 m above the lake surface.

Geysering to heights of 150 m continued through Oct. Ejecta fell back into the lake or onto the beach at the E end of the crater. On 9 Oct, volcanologists measured an average lake water temperature of 60°C, and temperatures of 80-90°C on the [1953–55 cone]. The lake level had risen. During the volcanologists' visit, ejections of water and tephra took place from the central part of the lake. Eruption columns rose a maximum of 300 m in Nov, and Dec activity was similar.

The following is from Guillermo Avila.

Steam clouds produced by explosions from the crater lake reached heights of 100 m or less in Jan. The mottled brown color of the lake indicated the presence of active fumaroles on the lake bottom. Water samples from the lake are being chemically analyzed.

Information Contacts: J. Barquero H., Univ. Nacional, Heredia; G. Avila, ICE.

Further Reference: Francis, P.W., Thorpe, R.S., Brown, G.C., and Glasscock, J., 1980, Pyroclastic Sulfur Eruption at Poás Volcano, Costa Rica; *Nature*, v. 283, p. 754-756.

4/80 (5:4) The following information is from *Boletín de Vulcanolgía* no. 7.

By Jan, the small phreatic explosions that began in Sept 1979 had ended. Activity in Jan and Feb was limited to emission of gases with a strong sulfur odor from the N wall of the [1953–55 cone]. Fumarole temperatures ranged from 56° to 95°C. Mean lake water temperature dropped from 65°C in Nov 1979 to 55° in Jan and 50° in Feb. The lake was gray with green patches in Jan, but had turned to a turquoise color in Feb. The water level dropped about 30 cm between the Jan and Feb surveys.

Information Contact: J. Barquero H., Univ. Nacional, Heredia.

2/81 (6:2) Fumarolic activity continued at Poás during Aug and early Sept. Sulfurous vapors emitted under pressure from the N wall of the dome in the crater lake rose noisily in an almost continuous column about 200 m high. The lake color was turquoise green. Temperatures registered 40°C in the N part of the lake, 45° in the S part near the [1953–55 cone], and 70-90° in accessible fumaroles on the dome.

On 11 Sept at 0950 an explosion from the S part of the lake (near the [1953–55 cone]) produced a 250 m-high column of lake water laden with ash, sand, and small blocks rich in mineralized sulfur. The ejecta fell back into the lake and onto the E shore where they covered an area of 50 m^2. A landslide that originated from the NE part of the [1953–55 cone], the area of greatest fumarole activity, deposited debris in the lake and changed the morphology of the E sector of the crater.

The initial activity was followed by similar explosions throughout Sept and Oct [but see 6:3 and 6:5]. Institute of Volcanology scientists had predicted resumption of phreatic activity from the thermal behavior of the lake, which had been similar to the pattern observed before previous such eruptions. Temperatures declined slightly in Oct, to 40°C in the NE part of the lake from 45° in Sept, and to 45° in the SE sector (near the Sept explosion site) from 50° in Sept. Temperatures of the accessible fumaroles on the dome continued to oscillate between 70° and 90° in Sept and Oct.

Information Contact: Same as above.

3/81 (6:3) Activity at Poás had increased, with explosions observed 11 Sept and 26 Dec 1980. As of mid-March, ICE and the Univ. Nacional were keeping the volcano under continuous observation. The temperature of the [1953–55 cone] in the crater lake was 650-750°C and some red areas were seen along fissures in the [cone]. Lake water temperatures were 50°C, similar to temperatures in the fall of 1980. The pH of the lake had decreased to 0.1. Fumaroles emitted large quantities of water vapor and SO_2. Many landslides had occurred in the walls of the main crater.

Information Contacts: G. Avila, ICE; J. Barquero H., Univ. Nacional, Heredia.

5/81 (6:5) About dawn on 26 Dec 1980, an explosion from the S portion of the crater lake ejected ash, small blocks, and hot water that rose above the level of the crater rim (about 320 m above the lake surface). This was the first such activity since a similar explosion on 11 Sept from the same area, near the wall of the [cone] in which fumarolic activity has been concentrated. Both the volume of vapor emitted and the amount of SO_2 and other toxic gases in the vapor have increased since mid-1980, forcing the use of gas masks near the fumaroles. In mid-Jan 1981, Parque Nacional Poás employee Geiner Chacón reported seeing incandescence during the night in the N wall of the [cone]. In Feb, volcanologists descended to the [cone] and observed incandescent areas, estimating temperatures of 700-800°C from their color. Significant quantities of sulfur had been deposited, some of which had been melted and turned a yellow-orange color by the heat. Many landslides had occurred in the crater walls, forming talus slopes at their base. The incandescence was continuing as of late May.

Information Contact: J. Barquero H., Univ. Nacional, Heredia.

7/81 (6:7) Between 14 June and 11 July, personnel from PIRPSEV, CNRS, and the volcano observation section of IPG sampled and analyzed medium-to-high temperature gases from Poás, (table 14-6). Maximum gas temperatures measured were 940°C.

Information Contacts: H. Delorme, Univ. de Paris; J. L. Cheminée, IPG, Paris.

8/81 (6:8) In Jan 1981, incandescent fissures were observed for the first time in the eroded cone that formed during the 1953–55 eruption. Although the hottest fissures were inaccessible, geologists measured temperatures of 875°C on the cone in Feb. Between March and June, maximum temperatures generally oscillated between

SO_2	55.79%	CO	0.24%
CO_2	26.06%	CH_4	84.3 ppm
H_2	17.90%	He	52 ppm
H_2S	0.52%	COS	25.8 ppm
N_2	1.98% in air		

Table 14-6. Analytical means of 28 gas samples from Poás (dry, HCl excluded).

950° and 1000°, although a temperature of 1020° was recorded on 28 April. New fissures formed, widened, and grew hotter; on the E part of the cone, a fissure that on 4 March was 2 cm wide and had a temperature of 350°, had by 19 April heated to 910°, and on 28 June was 20 cm wide and had reached 940°. Fumaroles reappeared in an area on the S portion of the cone where they had died out in 1979. Lake water temperatures increased from 43° on 19 April to 51° on 17 June. Lake level lowered 1.3 m between Jan and June; such lowerings are typical early in the year, but are usually reversed by rains in May.

Vapor emission was continuous and the more vigorous activity was occasionally visible from the area surrounding the volcano. The larger vapor emissions were typically accompanied by weak rumbling that could be heard only from within the crater. On 4 May a vapor column originating in the fumarolic area of the cone reached an estimated height of 2 km.

A seismograph was installed at Poás on 19 March. Through May, the instrument recorded harmonic tremor at frequencies of 3-4 Hz for a few minutes to a few hours daily; discrete events caused by internal rupturing; signals produced by degassing; explosion events accompanying vapor eruptions; very shallow (about 1 km deep) B-type earthquakes; and a very few A-type earthquakes centered at depths of 1-10 km. An inverse relationship was evident between the daily duration of harmonic tremor and the daily number of discrete earthquakes, a phenomenon that had also been observed at Arenal in 1975 (T. Matsumoto, personal communication).

Information Contacts: J. Barquero H., E. Malavassi R., Univ. Nacional, Heredia; J. L. Cheminée, IPG, Paris; H. Delorme, Univ. de Paris; G. Avila, F. Guendel, ICE; T. Matsumoto, Univ. of Texas, Austin.

Date		Fumarole Temperature	Crater Lake Temperature	Lake Water pH
1982	Jan	860°	47°	
	Feb	883°	48°	
	Mar	887°	47°	
	Apr	873°	50°	
	Dec	731°	40°	
1983	Jan	753°	56°	
	Feb	732°	60°	
	Mar	780°	60°	
	Apr	818°	60°	
	May	832°	57°	0.06
	Jun	834°	57°	0.1
	Jul	822°	57°	0.2
	Aug	806°	58°	0.2
	Sep	810°	58°	0.3
	Oct	801°	60°	0.6
	Nov	750°	58°	–
	Dec	750°	56°	0.07
1984	Jan	690°	51°	0.7
	Feb	650°	51°	–
	Mar	570°	54°	0.7
	Apr	570°	49°	
	May	586°	50°	
	Jun	603°	50°	
	Jul	602°	48°	
	Aug	530°	48°	
	Sep	500°	50°	
	Oct	490°	48°	
	Nov	510°	48°	
	Dec	515°	48°	
1985	Jan	550°	44°	
	Feb	560°	44°	
	Mar	584°	44°	
	Apr	568°	44°	
	May	490°	44°	
	Jun	420°	48°	
	Jul	316°	46°	
	Aug	295°	45°	
	Sep	294°	45°	
	Oct	310°	45°	

Table 14-7. Temperatures (in °C) measured in a fissure on the E side of the eroded 1953-55 cone and in the crater lake, Jan 1982 – Oct 1985.

12/81 (6:12) A large white vapor plume was continuously emitted from the eroded cone at the S end of the crater lake. SO_2 emission rates measured by COSPEC were at average to high levels. Bright orange-red incandescence was visible in cracks within 20 cm of the surface. Temperatures of up to 873°C were measured in these cracks. The crater lake had a temperature of 40°C and abundant sulfur bubbles were floating on its surface.

Information Contacts: R. Stoiber, S. Williams, H.R. Naslund, J.B. Gemmell, D. Sussman, Dartmouth Coll., NH; E. Malavassi R., J. Barquero H., Univ. Nacional, Heredia.

5/82 (7:5) High temperatures and incandescence continued to be observed through April at the eroded cone at the S end of the crater lake. Water temperatures in the crater lake remained relatively high (table 14-7).

Information Contact: J. Barquero H., Univ. Nacional, Heredia.

11/82 (7:11) The following is a report from Dartmouth College geologists.

"The volcano has cooled over the past year, with maximum fumarole temperatures on the cone of 575°C. Only very dull, brown incandescence could be occasionally seen, whereas one year ago bright red was seen easily. Water temperature was 41°C and the pH ~0.

"COSPEC measurements of SO_2 flux suggest levels very far below the approximately 800 t/d measured in Feb 1982 by Casadevall and others (1984; see further reference under Colima, 7:2). A portable seismograph indicated very little seismic activity in 2 days of recording in Nov."

Information Contacts: R. Stoiber, S. Williams, H.R. Naslund, C. Connor, J. Prosser, J.B. Gemmell, Dartmouth College; E. Malavassi R., J. Barquero H., Univ. Nacional, Heredia.

	Poás			Irazú		Turrialba	
	Jan 1982	Mar 1982	Dec 1982	1981	1982	1981	1982
SO_2 %	66.4	59.8	66.1	0.106	0.007	0.016	0.018
CO_2 %	20.11	20.9	21.8	98.02	99.8	99.95	99.93
H_2S %	0.02	0	0.246	1.860	0.119	0.003	0.008
CO %	0.265	0.160	0.180	0.003	0.002	0.001	0
CH_4 %	-	-	-	0	0	0	0
H_2 %	13.1	13.5	11.5	0.015	0	0.022	0.047
He ppm	128	40	38	8.4	6.0	4.7	1.7

Table 14-8. Average values of gases sampled at Poás, Irazú, and Turrialba, 1981–82. Late 1982 sampling was by a team from PIRPSEV (CNRS) and a volcanological team from the Universidad Nacional de Costa Rica.

12/82 (7:12) Fumarolic activity continued in the eroded cone at the S end of the crater lake, with temperatures ranging from 652 to 873°C. Temperatures were also measured in the hot crater lake. Landslides continued from the N wall of the eroded cone.

Information Contacts: J. Barquero H., E. Malavassi R., Univ. Nacional, Heredia.

1/83 (8:1) Between 5 Dec and 20 Dec, 1982, a team from PIRPSEV (CNRS) and a volcanological team from the Universidad Nacional de Costa Rica sampled gases from Poás (table 14-8). The gas temperatures have been variable but generally decreasing: 940°C in June 1981 (6:7), about 870° between Jan and April 1982 (7:5), about 790° between April and Nov 1982, and 731° on 17 Dec 1982. Since June 1981, 20 measurements have been collected from the control fissure with the aid of a silica rod of new design: the decrease in temperature at the end of the rod being only 10%. Since Dec 1981, the ratio S/C appeared to have stabilized at approximately 3, a higher value than in other available data.

Information Contacts: J. Cheminée, IPG, Paris, M. Javoy, H. Delorme, Univ. de Paris.

3/83 (8:3) A fissure extending from the E side across the summit of the eroded cone at the S end of the crater lake emitted gases that were hotter in late March than several months earlier. On 22 March, Jerry Prosser measured a temperature of 800°C on the side of the cone and 890°C at the summit. Cheminée and others had reported variable but generally falling temperatures between June 1981 (940°C) and Dec 1982 (731°C) (8:1). Prosser also noted that the crater lake was consistently warmer than 60°C, its highest temperature since just prior to the 1978 eruption.

Information Contact: J. Prosser, Dartmouth College.

4/83 (8:4) Fumarolic activity continued from the eroded cone formed during the 1953–55 eruption. Fumarole and crater lake temperatures generally increased through the 5-month period. The water level in the lake has dropped 5 m during the current dry season. The lake's pH was 0.2.

Information Contacts: J. Barquero H., J. de Dios Segura, Univ. Nacional, Heredia.

10/83 (8:10) Fumarolic activity continued with strong continuous gas emission from the eroded cone. The crater lake remained hot and the water level descended about 10 m.

Information Contacts: J. Barquero H., E. Fernández S., Univ. Nacional, Heredia.

Further Reference: Barquero, J., 1983, Termometría de la Fumarola del Volcán Poás; *Boletín de Vulcanología*, no. 13, p. 11-12.

3/84 (9:3) Fumarolic activity continued with vigorous gas emission, but temperatures declined at a fumarole on the eroded cone.

Information Contact: Same as above.

12/84 (9:12) "Dartmouth geologists visited Poás on 6 and 10 Dec. Extensive fumarolic activity continued. Fumaroles had lower temperatures than measured on previous trips, not exceeding 525°C. Lake water sampled 100 m NW of the fumarole bank had a temperature of 36°C and a pH of 0.22 at the surface, and a temperature of 38°C and a pH of 0.08 at 4 m depth."

Information Contacts: B. Barreiro, H.R. Naslund, R. Stoiber, P. Turner, Dartmouth College.

7/85 (10:7) Activity was characterized by the emission of gas from fumaroles on the eroded 1953–55 cone. Fumarole temperatures continued to drop, while crater lake temperature remained constant.

Information Contacts: J. Barquero H., E. Fernández S., Univ. Nacional, Heredia.

11/85 (10:11) Fumarolic activity continued, with constant gas emission. Fumarole temperatures have dropped substantially since early 1985.

Information Contact: Same as above.

IRAZU

Costa Rica
9.98°N, 83.85°W, summit elev. 3432 m

The highest volcano in Costa Rica, Irazú also has the earliest historic eruption, in 1722. It is 25 km E of San José, and ashfalls from its largest eruption, in 1963–65, were common in the capital city.

General Reference: Barquero, J., 1976, *El Volcán Irazú y su Actividad*; Escuela de Ciencias Geográficas, San José, 63 pp.

12/81 (6:12) "The summit was visited on 19 Nov and 5 Dec. No activity was observed [but see 7:5 and 7:11]."

Information Contacts: R. Stoiber, S. Williams, H.R. Naslund, J.B. Gemmell, D. Sussman, Dartmouth College; D. Fajardo, IIS, Nicaragua; E. Malavassi R., J. Barquero H., Univ. Nacional, Heredia.

5/82 (7:5) Fumarolic activity continued on the NW flank, where temperatures varied between 70° and 90°C in March. No activity was observed in the main crater.

Information Contact: J. Barquero H., Univ. Nacional, Heredia.

11/82 (7:11) The following is a report from Dartmouth College geologists.

"Some indications that Irazú may be becoming more active were noted in several trips to the summit. The smell of H_2S, only occasionally noticeable Nov–Dec 1981 and again in Jan 1982, was always detectable in the summit region. On 22 Nov a small flow of molten sulfur was observed to have advanced several meters from a fumarole at the base of the N wall of the active crater. Park guards believed that it was new that day.

"Faulting and slumping of scarps were reported to have occurred over a large area of the summit during Nov. This was largely confined to a belt oriented NE-SW and approximately 3-4 km from the summit crater. Park guards reported a single day in Sept in which 20-30 earthquakes were felt."

Information Contacts: R. Stoiber, S. Williams, H.R. Naslund, C. Connor, J. Prosser, J.B. Gemmell, Dartmouth College; E. Malavassi R., J. Barquero H., Univ. Nacional, Heredia.

12/82 (7:12) Fumarolic activity was limited to the NW flank of the volcano. Temperatures oscillated between 78°C and 85°C. No activity was observed in either the principal crater or Diego de la Haya, just to the SE.

Activity on a local fault caused an earthquake swarm 4 June. The swarm generated landslides in the wall that divides the principal crater from Diego de la Haya. Another stronger swarm occurred 23–24 Sept affecting the area between Irazú and Turrialba.

Information Contacts: R. Stoiber, S. Williams, R. Naslund, C. Connor, J. Prosser, J. Gemmell, Dartmouth Coll., NH; E. Fernández S., J. Barquero H., Univ. Nacional, Heredia.

1/83 (8:1) The low 1982 temperatures do not vary from June 1981. [For gas analyses, see table 14-8.]

Information Contacts: J. Cheminée, IPG, Paris; M. Javoy, H. Delorme, Univ. de Paris.

4/83 (8:4) Fumarolic activity continued on the NW flank, where temperatures oscillated between 74° and 81°C. The main crater and Diego de la Haya crater remained inactive.

InformationContacts: J. Barquero H., J. de Dios Segura, Univ. Nacional, Heredia.

10/83 (8:10) Fumarolic activity continued on the NW flank with a mean temperature of 77°C. No activity was observed in the craters.

Information Contacts: J. Barquero H., E. Fernández S., Univ. Nacional, Heredia.

7/85 (10:7) Fumarolic activity continued on the N flank, with temperatures averaging about 60°C. A lake that started to form in the main crater in Sept 1984 was still present in July 1985. Bubbles from the gas emission could be seen rising through the lake water.

Information Contact: Same as above.

TURRIALBA

Costa Rica
10.03°N, 83.77°W, summit elev. 3335 m

Turrialba is 35 km ENE of San José. It is now thickly vegetated but was quite active from the mid 18th to mid 19th centuries.

5/82 (7:5) Fumarole temperatures averaged 86°C.

Information Contact: J. Barquero H., Univ. Nacional, Heredia.

12/82 (7:12) Activity remained the same as in previous years. The mean temperature was 86°C in the central crater and 89°C in the W crater. The E crater showed no activity.

Information Contacts: J. Barquero H., E. Malavassi R., Univ. Nacional, Heredia.

1/83 (8:1) [See table 14-8 for gas analyses.]

10/83 (8:10) Fumarolic activity continued in the central and W craters, where a mean temperature of 89°C was measured.

Information Contacts: J. Barquero H., E. Fernández S., Univ. Nacional, Heredia.

7/85 (10:7) Fumarolic activity continued, with temperatures averaging 90°C in the W crater and 85°C in the central crater.

Information Contacts: Same as above.

SOUTH AMERICA

LA LORENZA

NW Colombia
9°N, 76°W
Local time = GMT - 5 hrs

Mud volcanoes do not form from deep-seated igneous processes and are normally excluded from volcanological compilations, but these distinctions are often lost on concerned neighbors and the press, as illustrated by the following reports from SEAN's first year. La Lorenza is on the Caribbean coast, 440 km N of Nevado del Ruiz, the nearest historically active volcano.

10/76 (1:13) A possible volcanic eruption began during the afternoon of Oct 21. Forty families were evacuated. No human casualties were reported, but 100 head of cattle have been killed and another 100 trapped by the activity.

A lava flow was reported, but it should be noted that La Lorenza is on the Caribbean coast, in an area where mud volcanoes are common.

Information Contact: AFP.

11/76 (1:14) The possible eruption reported in 1:13 from La Lorenza was mud volcano activity, as suspected. At 0800 on 21 Oct, ambient-temperature gray mud was ejected. At 0900, the petroleum gases (largely methane) emitted by the mud volcano ignited, producing a flame several hundred meters high, which lasted for several days. Contrary to press reports, no lava was emitted. People and cattle fled the area, but there were no deaths. Some livestock were injured.

Information Contact: E. Ramirez, Univ. Javeriana, Bogotá.

RUIZ

W Colombia
4.88°N, 75.37°W, summit elev. 5389 m
Local time = GMT - 5 hrs

In the Andes 150 km WNW of Bogotá, Ruiz is the northernmost and highest Colombian volcano with historic activity. Its first major historic eruption, in 1595, caused damaging mudflows and those of 1845 resulted in fatalities presaging the tragic events of 1985.

5/85 (10:5) The following is a report from Minard L. Hall, John Tomblin, and Omar Gómez.

"Since late Nov 1984, local earthquakes have been felt near the summit. On 22 Dec, stronger earthquakes were detected, followed by a half hour of apparent harmonic tremor. During a visit to the crater in early Jan 1985, increased fumarolic activity, evidence of phreatic explosions, and the wide deposition of sulfur salts over the adjacent snowcap were noted. At times, a thin layer of ash had been ejected, which was analyzed by J. Tomblin and found to consist of alteration products and sulfur.

"Seismic activity continued, with 17 felt earthquakes in March and 18 in April. There are no operating seismographs in the region. Abnormal fumarolic activity also continued. The one hot spring with frequent temperature monitoring, NW of the crater, had not shown any variation in temperature.

"Ruiz is a glacier-clad volcano (bordering Tolima and Caldas Departments). . . . Colombian officials have begun the necessary studies."

Information Contacts: M. Hall, Escuela Politécnica, Ecuador; J. Tomblin, UNDRO; O. Gómez, Civil Defense Coordinator, Manizales.

7/85 (10:7) Increased thermal and seismic activity have continued. Scientists from the Central Hidroeléctrica de Caldas visited the crater on 8 July and found evidence of increased activity since their previous visit on 22 Feb. They reported intense noise from the fumaroles as well as increased fuming which frequently made breathing difficult. Sulfur deposits were more extensive than those noted on their previous visit. On 22 Feb, there had been only a thin film of sulfur covering the surface near the fumaroles, but by July sand-sized material on the inner slopes of the crater was impregnated with sulfur deposits, creating a crust 10 cm thick. A new crack was observed near the crater rim; it was 1.5 m in length, 8 cm wide, and emitted hot gas and vapors. Ground temperatures were measured at various locations as the team descended into the crater. Isotopic studies are being conducted on waters collected at thermal vents on the flanks of the volcano.

Much of the bottom of the crater was covered by a green lake that emitted hot steam from its surface. On 22 Feb, the lake had a pH of 0.2. The water in the lake had risen at least 1 m since Feb, covering a mud pool and its surrounding "mud volcano" seen on 22 Feb, and also nearby fumaroles. Projecting into the SE side of the lake on 8 July was a peninsula that was thought to be the remnant of these features. An unusual thaw of the glacial ice that covers much of the summit area seemed to have contributed to the rise in the water level. Further evidence for thawing came from the presence of large blocks of ice scattered near the base of the crater. A major increase in fumarolic activity producing an enormous yellow vapor cloud was reported on 23 July.

Information Contacts: B. Salazar A., M. Calvache V., N. Garciá P., Central Hidroeléctrica de Caldas.

8/85 (10:8) Ash emission began 11 Sept at about 1300 and lasted up to 7 hours, accompanied by a persistent roaring noise and electrical discharges. A few millimeters of ash fell on the cities of Manizales and Chinchiná (about

30 km NW and WNW of the volcano) and ashfall was reported to have locally reached 2 cm. A mudslide blocked a road on the E flank of the volcano. Before the eruption, tremors were recorded at about 1.5-hour intervals by seismographs about 3 km N, S, E, and W of the volcano's summit. Tremor was continuous during the ashfall, punctuated by discrete earthquakes.

On 13 Sept, morphological changes in the glaciers on the W side of Ruiz were reported by ground and air observers. New fissures were also reported on the icecap N of and below the active crater near a small growing pond on the glacier surface.

Nevado del Ruiz is a broad shield-shaped volcano covering more than 200 km^2, composed of an extensive sequence of hypersthene-augite andesite, hornblende andesite, basaltic andesite, and dacite. Its 1595 eruption produced ash and lapilli falls and lahars that spilled down the Río Gualí and Río Lagunillas valleys N and E of the volcano. Eruptions were reported in 1828 and 1829, but no descriptions exist. Apparently these events produced no tephra. Ruiz is also reported to have been "smoking" in 1831 and 1833 (Herd, 1982).

Reference: Herd, D.G., 1982, Glacial and Volcanic Geology of the Ruiz-Tolima Volcanic Complex, Cordillera Central, Colombia; *Publicaciones Geológicas Especiales del INGEOMINAS*, no. 8, 48 pp.

Information Contacts: B. Salazar A., Central Hidroeléctrica de Caldas; Germán Mejía, Univ. Nacional, Manizales.

9/85 (10:9) After a series of magnitude-4 earthquakes on 22 Dec marked the onset of stronger seismicity (10:5-7), 25-30 events were felt each month until an episode of strong phreatic activity on 11 Sept. Seismologists have located about 100 earthquakes, centered 1-2 km NW of the active crater at average depths of 0-2 km below sea level. Increased thermal and phreatic activity began in early 1985 and continued after the 11 Sept episode into early Oct. No juvenile material has been recognized in the 1985 tephra.

The vigorous 11 Sept phreatic activity began at 1330 from the summit crater (Arenas) and had ended by the next morning. Ashfall was 1 cm or less near the volcano and a trace of ash reached Manizales and Chinchiná. Lithic blocks were deposited on snowfields as much as 2 km from the crater. Thunderous detonations and summit-area lightning accompanied the activity. A small- to moderate-sized lahar began at 1830, advancing 27 km down the valley of the NE flank's Río Azufrado, from 4700 m to 3000 m altitude. As it travelled at an estimated 10-30 km/hour, the lahar left the river channel at various locations, particularly along curves, and rose as high as 10-20 m up canyon walls. Valley residents were placed on alert but have not been evacuated. A hazard map has been prepared by the international team studying the volcano and distributed to Red Cross and Civil Defense officials [fig. 15-4].

For 5 days preceding the 11 Sept activity, seismographs registered a very regular pattern consisting of 15 minutes of strong high-frequency tremor every hour. Although similar seismicity was recorded during 3 other periods of a day or less in Aug and Sept, none of these episodes was as intense or long-lasting.

Phreatic activity was continuous for the rest of Sept, emitting variable amounts of ash, typically darkening the snow to a few kilometers from the crater. Heavier emissions occurred on 23, 24, and 29 Sept, producing dense dark plumes that deposited trace ashfalls more than 10 km from the crater and lithic blocks on nearby snowfields. Activity had declined by early Oct. A steam plume 1-3 km high was visible daily but contained no obvious ash. Seismographs recorded 5-10 microseismic events/day and an irregular harmonic tremor that may have been related to the steam plume. Geologists visited the crater at the beginning of Oct, noting a slight decline in fumarolic activity and little ash emission. Glaciers seemed unchanged since Sept.

Information Contacts: L. Jaramillo, INGEOMINAS, Bogotá; A. Rivera, Univ. de Caldas, Manizales; G. Duque, Univ. Nacional, Manizales; A. César, Central Hidroeléctrica de Caldas; A. Solano, Univ. Nacional de Colombia, Bogotá; M. Hall, Escuela Politécnica, Ecuador; D. Herd, USGS, Reston, VA.

10/85 (10:10) An explosive eruption on 13 Nov melted ice and snow in the summit area, generating lahars that flowed tens of kilometers down flank river valleys, killing more than [22,000] people. [Other estimates range to 25,000.] This is history's fourth largest single-eruption death toll, behind only Tambora in 1815 (92,000), Krakatau in 1883 (36,000) and Mt. Pelée in May 1902 (28,000). The following briefly summarizes the very preliminary and inevitably conflicting information that had been received by press time. [An 11-page Appendix that originally appeared at the end of this issue presented a more detailed but very preliminary chronology.]

After the moderate explosive activity (and 27-km lahar) of 11 Sept, there were several smaller ash ejection episodes through the end of Sept. Activity declined by early Oct to emission of a 1-3 km steam plume that contained no obvious ash. Low-level steaming with very little ash continued into late Oct, accompanied by occasional earthquakes, apparently tectonic, at about 12 km depth. Shallow (0-5 km) earthquakes, as many as 5/day, were centered about $^1/_2$ km N of the active crater. Three dry tilt stations were established in late Oct, N, S, and W of the summit, and 11 days of data (before 5 Nov) suggested that some deflation was occurring.

Seismicity began to increase by 7 Nov, characterized by a series of high-frequency seismic swarms, although fewer than were associated with the 11 Sept ash emission. Continuous volcanic tremor began 10 Nov, but was weaker than the 11 Sept tremor.

No additional changes in seismicity were apparent before the onset of eruptive activity at about 1545 on 13 Nov. Ashfall reportedly began about 1600 in heavy rain at Mariquita (59 km NE of the summit) and lapilli with ash started falling about 1730 in Armero (46 km to the ENE; fig. 15-1). At 2109, Bernardo Salazar, tending seismic equipment 9 km from the summit, heard the start of strong explosions, much louder than those on 11 Sept, that shook the building and lit up the rain clouds "like a lamp." Heavy rains prevented Salazar from seeing the eruption column, but pumice to 15 cm in diameter began to fall by 2137, and 1-cm lapilli fell 18 km [W of] the summit. Ashfall at Armero increased about 2200. At 2220 a Caribbean Air Lines cargo plane flew through the eruption cloud at 8 km altitude. Pilot Manuel Cervera reported that "smoke" and the smell of sulfur filled the cabin, and his windows were etched. Two attempts to land at Bogotá failed, but he landed safely at Cali by putting his head out a side window.

Lahars traveled down 11 flank valleys, the most destructive inundating the city of Armero, where an estimated [21,000] of 25,000 residents died. The first mud, which reached Armero about [2335], was cold [reports that the mud became increasingly hot were incorrect]. Resident E. Nieto described "A frightening noise and then a blast of wind hit us and we saw fire falling from the sky." On the volcano's W flank, low-lying neighborhoods of Chinchiná were also buried by mud, and officials estimated that [1000] died there. Geologists estimated that the lahars advanced at 30-35 km/hour; Armero, at about 300 m altitude, is 5 vertical km below the summit of Ruiz.

No pyroclastic flows were observed during the eruption, but geologists noted deposits with cross-stratification typical of those from surges. The Refugio (about 2 km NW of the summit) was knocked down and its walls scattered radially downslope.

In hand specimen, the 13 Nov tephra appeared to be a dacitic andesite, about 30-40% phenocrysts, with plagioclase dominant, more hornblende than pyroxene, and a trace of biotite. No estimates of tephra thickness, volume, or extent were available at press time, but ashfall was reported as far away as Táchira, Venezuela, 500 km to the NE. Tephra fell on all but the S and SE flanks.

NASA's Nimbus-7 polar orbiting satellite passed over Ruiz on 14 Nov at 1151. Using data from its TOMS instrument, Arlin Krueger calculated that a cloud containing more than 0.5×10^6 metric tons of SO_2 covered a 650,000 km^2 zone extending NE and slightly SW of the volcano (fig. 15-2). Bernardo Londoño reported that the cloud extended to 10.5 km above the crater, but it remains uncertain whether the cloud reached the stratosphere. Heavy weather clouds prevented observation of the eruption on NOAA's weather satellite images. Lack of nearby wind data prevented precise altitude determination by correlation with cloud movement as indicated by Nimbus-7, but the available data suggested that it was upper tropospheric.

As of 20 Nov, no additional strong explosions had occurred. Geologists who flew over the volcano reported that the crater had enlarged to about 300 m in diameter and 200 m deep, and that new fumaroles had developed 500-900 m from the crater, but they observed no lava flow. Newly installed telemetering seismometers recorded harmonic tremor of varying amplitude beginning 18 Nov at 0200 that preceded ash emission at 0600. An episode of stronger tremor that lasted from about 0600

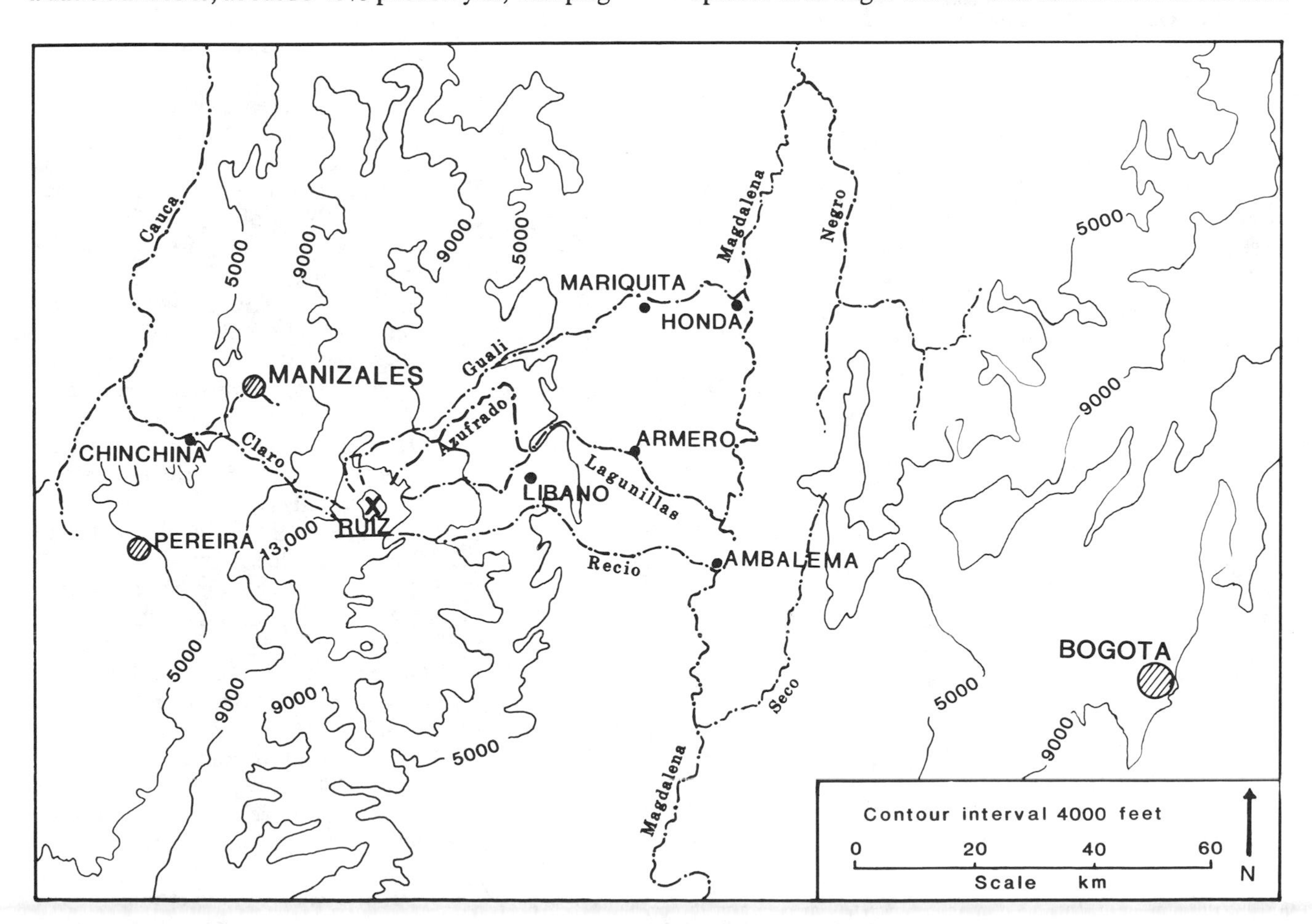

Fig. 15-1. Map showing Ruiz volcano and its major drainages. Contour interval is 4000 feet (about 1200 m).

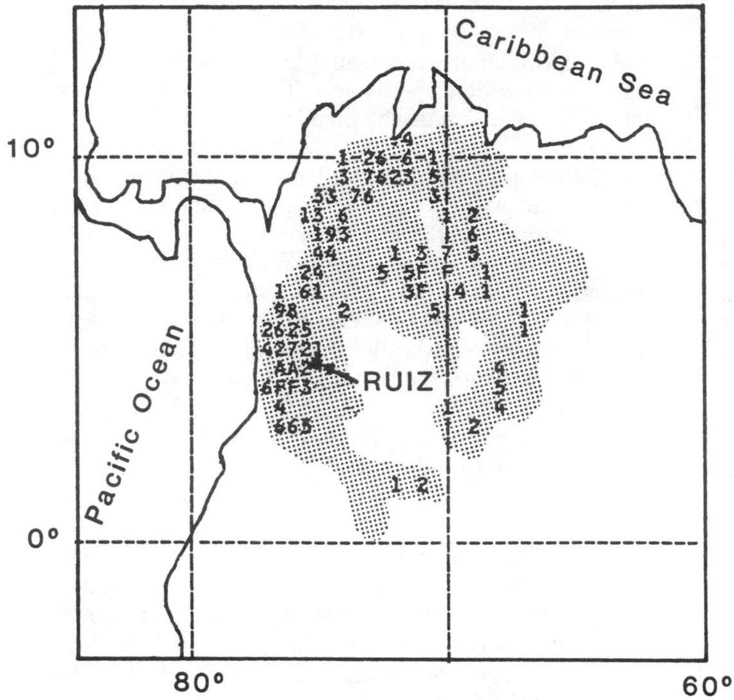

Fig. 15-2. Preliminary SO_2 data from the TOMS instrument of the Nimbus-7 satellite at 1151 on 14 Nov. Each number or letter represents the average SO_2 value within an area 50 km across. Values above 9 are represented by A, B, C, etc. The area of perturbed values is shaded. For a more detailed derivation of these values, see fig. 13-52. Courtesy of Arlin Krueger.

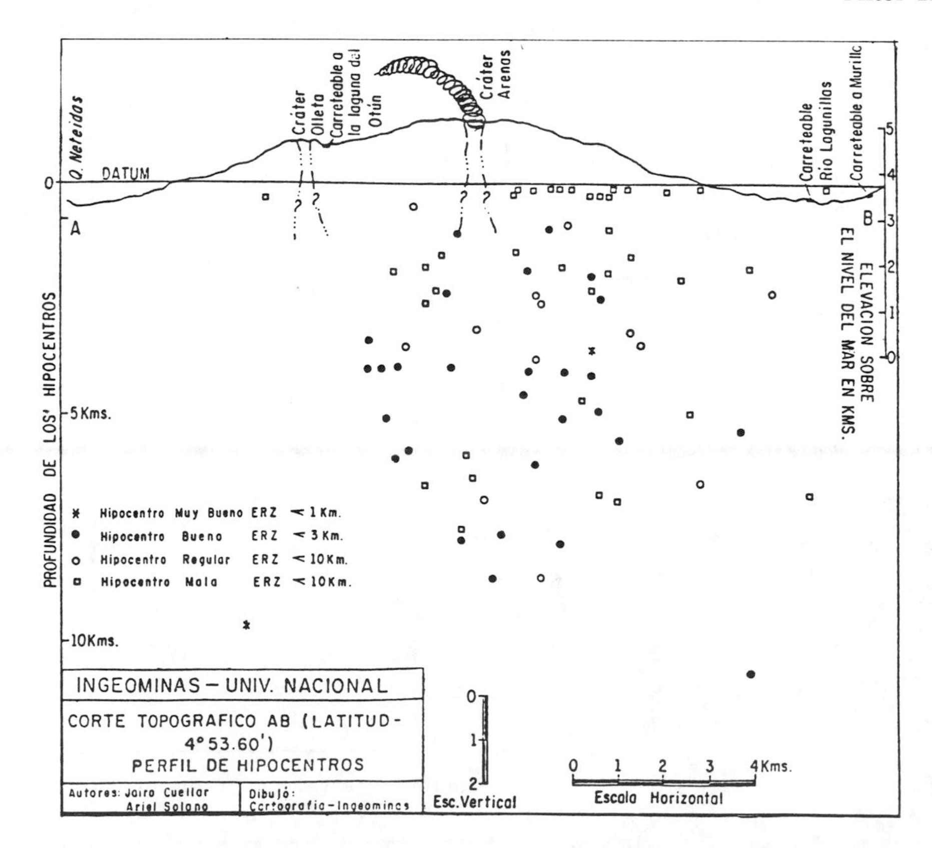

Fig. 15-3. Pre-13 Nov earthquake hypocenters projected onto a vertical E-W plane passing through the summit of Ruiz. Depths are below a datum at 3.8 km altitude. No vertical exaggeration. Reprinted from *Estudio de los Riesgos Potenciales del Volcán Nevado del Ruiz; Informe de las Actividades Desarolladas (Período Octubre 8 – Noviembre 10, 1985)*; INGEOMINAS, Bogotá,

to 0945 the next day was followed by a small explosion at 1030.

Information Contacts: B. Salazar A., Central Hidroeléctrica de Caldas; B. Londoño, Comité de Estudios Vulcanológicos, Manizales; D. Herd, R. Tilling, USGS, Reston, VA; N. Banks, USGS CVO, Vancouver, WA; M. Matson, W. Gould, NOAA/NESDIS; A. Krueger, NASA/GSFC; J. Tomblin, UNDRO; J. Barquero, R. van der Laat, Univ. Nacional, Heredia, Costa Rica; AP.

11/85 (10:11) Since the 13 Nov eruption, activity at Ruiz has been limited to emission of a vapor plume and a few seismic swarms, one accompanied by measureable inflation. Work by numerous geologists has yielded new information on the 13 Nov eruption, its products, and pre-eruption activity.

Pre-13 Nov Activity. The most vigorous seismic energy release occurred in the days preceding the 11 Sept ash emission. The rate of energy release increased prior to the 13 Nov eruption, but more gradually than before the 11 Sept activity. Hypocenters were concentrated N and NE of the summit with best-located events concentrated at depths 0-1 km below sea level (fig. 15-3).

The quoted material below is from a report from Rodolfo Van der Laat, Eduardo Parra, and Heyley Vergara.

"After 11 Sept, when there was a significant ash emission (10:8,9), activity at Ruiz had decreased notably through 10 Nov. The activity caused concern in Manizales (30 km NW of Arenas Crater), but the presentation by INGEOMINAS of a preliminary volcanic risk map (fig. 15-4) calmed the population.

"Seismic activity reached a maximum of 60 events/day 19-21 Oct, declining by 3 Nov to 3-5 daily locatable events. An increase in temperature of the thermal vent 'La Hedionda' (on the NE flank) may have been related to the increase in seismic activity.

"The height of the plume during this period decreased from about 3 km at the end of Sept and the beginning of Oct to about 800 m, with an occasional nucleus of ash 200-300 m high. There were 2 main fumarolic vents: one yellowish (sulfur), the smaller one gray/coffee-colored (ash derived from mud).

"A tilt network was established, detecting a general deflation 26 Oct – 3 Nov, with small pulses of inflation of the order of 5-10 µrad/day. At the beginning of Nov, the first measurement was made of a geodesic net to monitor horizontal deformation by the Instituto Geográfico Agustín Codazzi."

13 Nov Eruption and Products. Details of the 13 Nov eruption sequence remain uncertain and field investigations were still in progress at press time. An initial explosion at 1530 deposited a very thin, fine-grained layer of ash around the summit and NNE of the volcano. The main explosion started at 2108 or 2109 and continued for 20-30 min. Five km from the crater, tephra from the main explosion was 7 cm thick and included 30-cm pumice fragments, but the deposit thinned rapidly and was only 1-2 mm thick at Armero with similar amounts at Mariquita and Honda (75 km NE). Preliminary estimates by Haraldur Sigurdsson and Steven Carey place the volume of tephra at roughly 39×10^6 m^3. Cloudy weather and lack of nearby wind data on 13 Nov impeded determination of the height of the Ruiz eruption column. Based on the position of tephra diameter isopleths, Sigurdsson and Carey inferred that the top of the eruption cloud reached [31] km altitude, but emphasized that most of the tephra probably remained in the upper troposphere [Naranjo et al., 1986].

Mudflows that moved E down the valleys of the Lagunillas and Azufrado rivers and inundated Armero were overlain by airfall tephra within 5-10 km of the volcano. However, the mudflow that moved W down the Río Claro valley to Chinchiná contained fresh pumice, and the fluid mudflow that traveled down the Gualí river washed tephra off vegetation, suggesting that both were generated after tephra ejection. The Armero mudflows emerged from both the Lagunillas and Azufrado valleys, which join upstream from the city. The first wave of mud, probably from the Lagunillas, was apparently colder, lighter colored, more water-rich, and formed a more extensive deposit than the second wave, probably from the Azufrado, which was hotter, coarser, and darker colored. Donald Lowe estimated that outflow from the mouth of the Río Lagunillas reached about 47,500 m^3/sec. Preliminary calculations by Sigurdsson and Carey yield a volume of about 30-60 $\times 10^6$ m^3 for the deposits of the Armero, and Gualí and Chinchiná valley mudflows, plus about 30-90 $\times 10^6$ m^3 of water, roughly 6-18% of the pre-eruption volume of the summit ice cap. The Lagunillas mudflow probably included water from a lake that had been trapped behind a debris dam in that valley's headwaters for at least several months. Other estimates suggested that about 5% of the summit ice was removed during the 13 Nov eruption.

Preliminary chemical analyses of a few samples of the 13 Nov pumice suggest that it is a hypersthene andesite, very similar to an earlier pumice that was probably from Ruiz's last large eruption, in 1595. Little

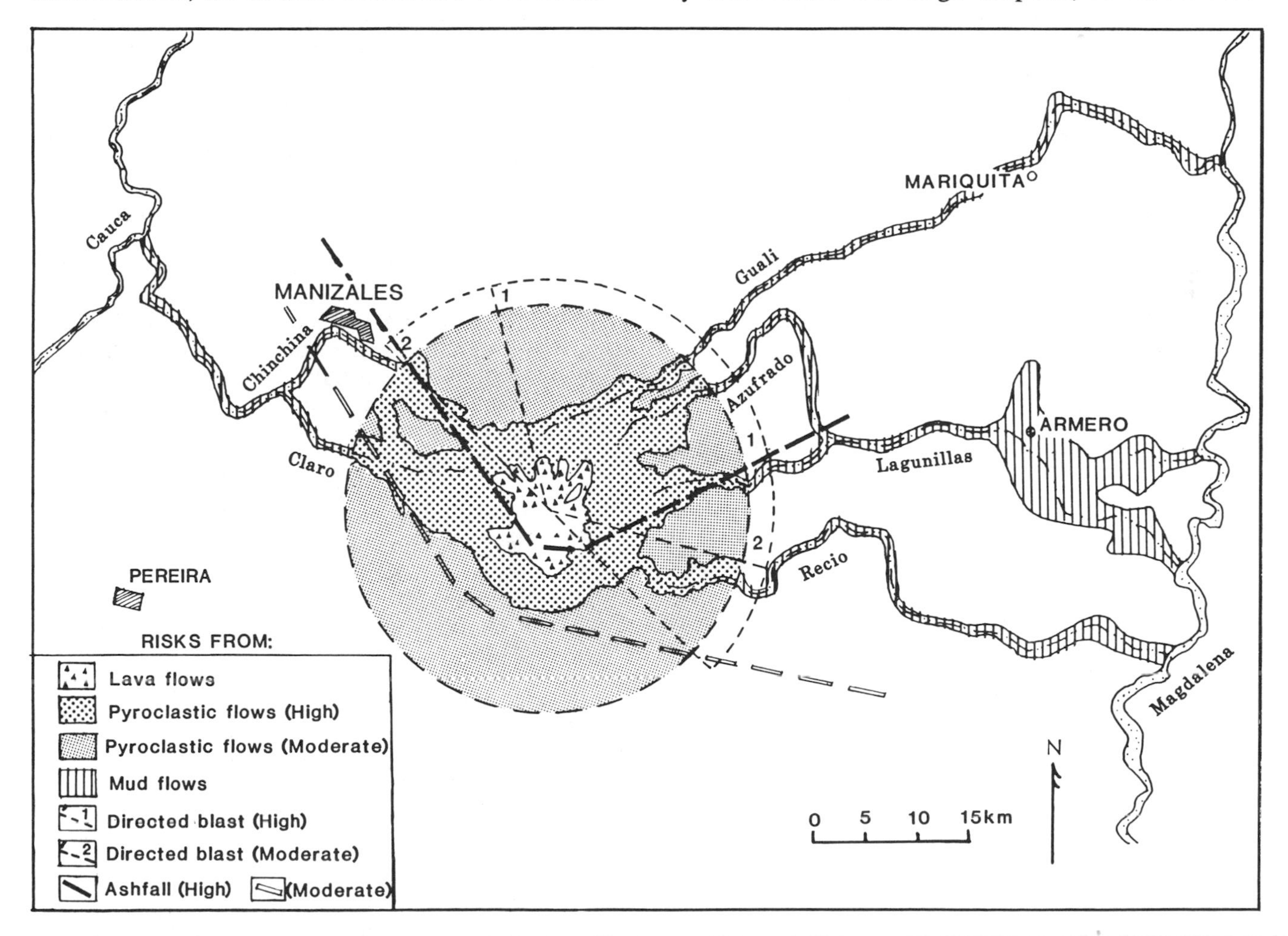

Fig. 15-4. Volcanic risk map presented by geologists from INGEOMINAS and the Univ. de Caldas to officials and the press before the 13 Nov eruption. Some redrafting has been done to facilitate reproduction in the *Bulletin,* but boundaries of hazard zones are unchanged.

No.	SiO_2	Al_2O_3	FeO*	MgO	CaO	K_2O	Na_2O	TiO_2	P_2O_5	MnO	Total
1	59.31	16.83	5.87	5.40	6.30	1.87	3.80	0.82	0.30		100.50
2	58.69	16.81	5.72	5.13	6.04	1.85	3.79	0.81	0.28		99.12
3	59.5	15.7	5.94	4.94	6.11	3.67	2.07	0.67	0.19	0.09	98.9
4	61.5	15.2	5.44	3.98	5.43	3.66	2.45	0.65	0.19	0.09	98.6
5	65.36	16.01	3.98	1.51	3.81	3.47	4.12	0.77	0.30		99.33
6	63.97	16.33	4.14	1.54	4.24	3.15	4.22	0.77	0.29		98.65

1 and 5: 13 Nov 1985 pumice collected by Stanley Williams. USNM116158.
2 and 6: Probable 1595 pumice collected by Stanley Williams. USNM116159.

Table 15-1. Preliminary analyses of bulk compositions of Ruiz pumice (1-4) and glass septa (5-6). Nos. 1, 2, 5, and 6 are from electron microprobe analyses by William Melson and Deborah Reid Jerez; 3 & 4 are X-ray fluorescence analyses by Joseph Taggart.

systematic variation was found in different-colored samples that had suggested mixed magma in hand specimen (table 15-1).

Post-13 Nov Activity. No significant eruptive activity occurred in the succeeding weeks. The vapor column varied in height from 200-300 m to 1-1.4 km. Rates of SO_2 emission measured by COSPEC were 200 t/d on 18 Nov, 50 t/d on the 19th, and several thousand t/d on 22 Nov. Possible new fissures have been observed near the summit along with possible development of a depression SW of the summit. However, the fissures may have been pre-existing features exposed by clearer weather and seasonal snowmelt. Slight advances of some of the summit glaciers have been noted, but no large-scale ice movements were apparent and there was no evidence of significant melting from below.

Six telemetering seismometers have been installed, ringing the summit at altitudes of 4000-4500 m, supplementing the 4-station seismic net that was in place before 13 Nov. Telemetering tiltmeters were emplaced at 4200 m altitude on the NW flank, 4600 m altitude on the NNW flank, and on the NE flank, and 8-10 EDM lines have been established, in addition to the dry tilt network installed on the N flank in Oct.

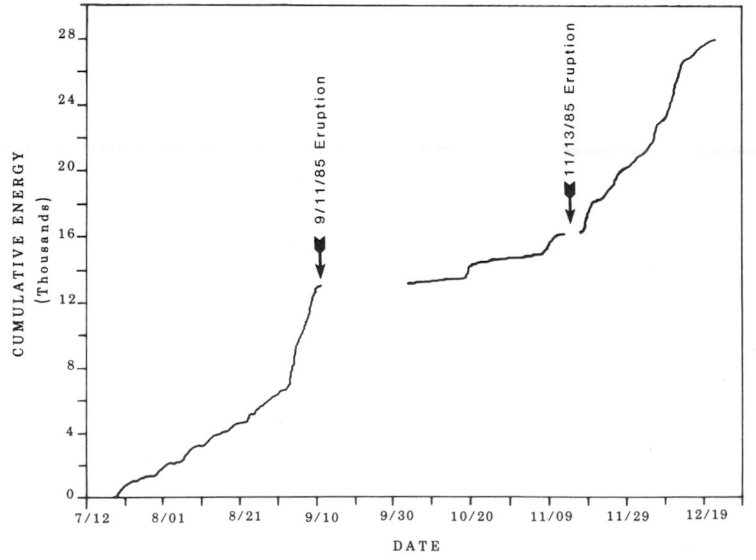

Fig. 15-5. Seismic energy release at Ruiz volcano, 20 July – 19 Dec 1985. Timing of eruptions on 11 Sept and 13 Nov are shown. Courtesy of the Comité de Estudios Vulcanológicos.

Seismic energy release was at relatively low levels shortly after the 13 Nov eruption, but the slope of the energy release curve steepened in the succeeding weeks. Earthquake swarms that were small but of increasing energy occurred 19-20 and 27 Nov, and 6-7 Dec. Maximum magnitudes were 2.5-3 in the Nov swarms; the 6-7 Dec activity included two magnitude 3-3.5 shocks. Locations were available for only a few events, which were centered along a generally N-S trend, usually somewhat N of the crater. The swarms were not accompanied by measurable tilt episodes or obvious changes to the plume. The rate of seismic energy release doubled during the first day of a stronger swarm 12-13 Dec and Civil Defense personnel were put on alert. The same day, the NW flank tiltmeter recorded a 5 µrad tilt event, the first change recorded in the weeks since it was installed, and a NW flank EDM line shortened 14 cm between measurements 11 and 13 Dec. However, seismicity declined 13 Dec, and the seismic energy release curve was nearly flat 14-17 Dec.

Information Contacts: P. Medina, Comité de Estudios Vulcanológicos, Manizales; A. López R., INGEOMINAS, Bogotá; R. Van der Laat, Univ. Nacional, Heredia, Costa Rica; E. Parra, INGEOMINAS, Medellín; H. Vergara, INGEOMINAS, Tolima; H. Sigurdsson, S. Carey, Univ. of Rhode Island; S. Williams, D. Lowe, Louisiana State Univ.; A. Londoño C., Univ. Nacional, Manizales; Néstor Garcia P., Industria Licorera de Caldas, Manizales; R. Stoiber, B. Gemmell, Dartmouth College; D. Harlow, USGS, Menlo Park, CA; C. Hearn, D. Klick, D. Herd, R. Tilling, USGS, Reston, VA; J. Taggart, Jr., USGS, Denver, CO; W. Melson, D. Jerez, SI; P. Clemente-Colón, NOAA/NESDIS.

12/85 (10:12) Explosive activity on 4 Jan ejected a small amount of ash and was accompanied by vigorous seismicity. The 4 Jan activity did not generate mudflows or cause any apparent changes in river flow, but residents of low-lying areas were temporarily evacuated as a precautionary measure.

A series of earthquake swarms followed the 13 Nov eruption, including a strong swarm 12-13 Dec that was accompanied by deformation (fig. 15-5). Seismicity then declined briefly, followed by a period of stronger seismicity 22-24 Dec, then diminished again at the end of Dec to ~10 events/day (M > 0) and brief bursts of tremor. Epicenters were generally S of the active crater, extending E and W under the flanks. Before the 4 Jan eruption, focal depths decreased from 4-8 km to 2-5 km (below a datum at 3.8 km above sea level). EDM lines on the SW, N, NE, and E flanks began to show changes in the rate and/or direction of deformation between 19 and 24 Dec. Equipment problems prevented remeasurement of EDM lines immediately before the 4 Jan eruption, so the amount of pre-eruption inflation is uncertain. The net change in the lengths of several radial lines (of 5 km average length) measured 3 days after the eruption was about 10 cm, but this figure probably included substantial post-eruption deflation. By 28-30 Dec, small but distinct changes in rate or direction of tilt had begun to appear on all 4 electronic tilt stations (at 4100 m altitude on the NE flank, 4800 m on the W flank, about 3900 m on the NW flank, and 4600 m on the SE flank).

Movement of cracks in summit glaciers continued through Dec and early Jan at roughly constant rates. Extensional changes of 5-10 cm/day were measured near the head of the Azufrado valley, and both extensional and compressional motion of a few millimeters to 5 cm/day elsewhere. Little baseline data exist on typical rates of glacier advance on Ruiz.

Strong seismicity began 3 Jan at about 2320, and was saturating seismographs within less than an hour. The seismicity was initially characterized by superimposed high- and low-frequency tremor, but tremor amplitude declined somewhat around 0115 and low-frequency (2-2.5 Hz) tremor began to dominate the seismic records at 0128. B-type earthquakes and explosion events accompanied the tremor. Darkness initially prevented direct observations of the summit, but ash began falling about 0300. The eruption cloud was small, generally 300-600 m high, occasionally rising to 1 km above the summit. Ashfalls were minor, concentrated around the summit and in a narrow zone to the WNW. Several hundred meters from the vent, new ash was only about 7 mm thick; 3 km downwind the deposit was only 2 mm deep; and only traces of ash were found more than 10 km away. Vigorous seismicity continued until about noon, then declined slowly until the eruption ended in mid-afternoon.

Evacuations of about 15,000 people from low-lying areas of the valleys of the Azufrado, Lagunillas, Recio, Gualí, Sabandija, and Chinchiná rivers began 4 Jan at about 0600. Most residents returned to their homes shortly after the eruption, but about 2,000 people remained evacuated 10 days later.

Smaller earthquake swarms occurred 5-7 Jan, then seismicity declined to about 1-2 A- or B-type events per hour, generally with magnitudes of 0 or less. No additional explosions or major increases in seismicity had occurred as of mid-Jan.

Information Contacts: P. Medina, Comité de Estudios Vulcanológicos, Manizales; N. Banks, USGS CVO, Vancouver, WA; AP.

Further References: Herd, D.G. and Comité de Estudios Vulcanológicos, 1986, The 1985 Ruiz Volcano Disaster; *EOS*, v. 67, p. 457-460.

Katsui, Y., Takahashi, H., Egashira, S., Kawachi, S., and Watanabe, H., 1986, The 1985 Eruption of Nevado del Ruiz Volcano and Associated Mudflow Disaster; *Rep. Natur. Disast. Sci. Res.*, B-60-7, p. 1-102.

Naranjo, J.L., Sigurdsson, H., Carey, S., and Fritz, W.J., 1986, The November 13, 1985 Eruption of Nevado del Ruiz Volcano, Colombia: Tephra Fall and Lahars; *Science*, v. 233, p. 961-963.

Thouret, J.C., 1986, L'éruption du 13 Novembre 1985 au Nevado El Ruiz: L'originalité du Dynamisme Eruptif Phréato-magmatique et Plinien sur une Calotte Glaciaire aux Latitudes Equatoriales; *Revue de Géographie Alpine,* v. 74, no. 4, p. 373-391.

Valdiri Wagner, J. (ed.), 1987, Memorias del Simposio Internacional Sobre Neotectónica y Riesgos Volcánicos (Bogotá, Colombia, 1-3 Diciembre, 1986); *Revista del Centro Interamericano de Fotointerpretación*, v. 11, nos. 1-3, p. 1-399 (23 papers).

Williams, S.N., Stoiber, R.E., García, P.N., et al., 1986, Eruption of the Nevado del Ruiz Volcano, Colombia, on November 13, 1985: Gas Flux and Fluid Geochemistry; *Science*, v. 233, p. 964-967.

PURACE

W Colombia
2.37°N, 76.38°W, summit elev. 4600 m
Local time = GMT - 5 hrs

This complex strato-volcano, 300 km SSW of Ruiz and 115 km S of Cali, has erupted 23 times since 1827.

3/77 (2:3) An eruption on 19 March deposited fine gray ash as far as 7 km from the crater according to press reports. No casualties or damage were reported.

Information Contact: J.E. Ramirez, Univ. Javeriana, Bogotá.

4/77 (2:4) The eruption of a black and gray ash cloud began at 0545 on 19 March from 2 new vents. Fine gray ash was deposited as far as 7 km away. The volcano was visited a few days later by Guillermo Cajino, who noted a small tremor and rumbling noises while 5 km from Puracé at 2300 on 24 March. The next day, he observed the emission of a gas column from the 2 vents, which scattered ash SE over the flanks. By 28 March the fume clouds rose only 200 m. Puracé last erupted in 1957.

Information Contact: U.S. Dept. of State.

GALERAS

SW Colombia
1.22°N, 77.30°W, summit elev. 4482 m

Galeras, with 16 eruptions since 1535, is 165 km SW of Puracé near the Ecuadorian border. Minor explosive activity was reported there throughout 1977 and had been continuing since 1974 (*Bulletin of Volcanic Eruptions* no. 17, 1979).

REVENTADOR

N-central Ecuador
00.07°S, 77.67°W, summit elev. 3485 m
Local time = GMT - 5 hrs

This equatorial volcano, 90 km ENE of Quito, has erupted at least 24 times since 1541. It is E of the Andean crest and perennially obscured by clouds.

General References (Ecuador): Hall, M.L., 1977, *El Volcanismo en el Ecuador*; Instituto Panamericano de Geografía e Historia, Quito, 120 pp.

1/76 (1:4) An explosive eruption began during the early morning of 4 Jan. At dawn an ash column 1 km high was observed. Fine ash was carried W and SW over the Andes, dusting Quito through 10 Jan. Bombs blown 100 m vertically from the crater were large enough to be seen by the naked eye from a distance of 3 km. Strange seismic signatures, detected 90 km away (at Quito) and attributed to Reventador, began at 0115 on 4 Jan and continued until 0900 on 9 Jan. A portable seismograph 10 km from the cone measured continuous harmonic tremor.

Two lava flows descended from the breached crater and divided into 3 lobes at the base of the cone. During the first 40 hours flows traveled approximately 1.5 km E at 37 m/hr. By 9 Jan the 3 lobes had traveled 2.5 km and were advancing approximately 5 m/hr over lahar deposits and jungle. As of 27 Jan the lava flows had stopped, but infrequent explosive activity, including nuées ardentes, was continuing. The continual ash column had terminated by Jan 25. The flows were a black basaltic andesite with olivine, augite, hypersthene, and oxyhornblende.

Reventador had similar eruptions in July 1972 and Nov 1973. When last visited before the current eruption, on 10 Dec 1975, it was producing a large steam column.

Information Contact: M. Hall, Escuela Politécnica, Quito.

Further Reference: Hall, M.L., 1980, El Reventador, Ecuador; Un Volcán Activo de los Andes Septentrionales; *Revista Politécnica*, v. 5, no. 2, p. 123-136.

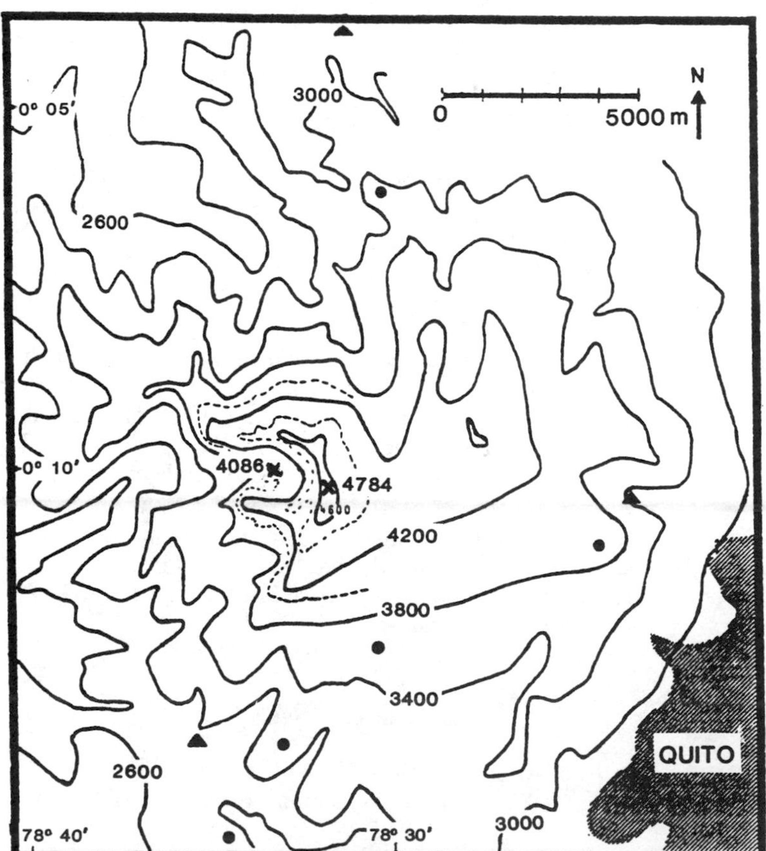

Fig. 15-6. Map of Guagua Pichincha. Contour interval is 400 m except in the summit area, where 200 m contours are shown by dashed lines. The positions of dry tilt stations are shown by solid circles, and seismographs are indicated by solid triangles. The area shaded with diagonal lines at lower right is the city of Quito.

GUAGUA PICHINCHA

N-central Ecuador
0.17°S, 78.60°W, summit elev. 4794 m
Local time = GMT - 5 hrs

The capital city of Quito, with a population of 900,000, occupies the E foot of this large, caldera-topped strato-volcano. Although 20 eruptions have been reported since 1533, 8 from the 16th century are uncertain and none has caused fatalities.

9/81 (6:9) A small phreatic explosion that probably occurred in mid-Aug deposited fine tephra as much as 1 km SE of 3 new vents (3-8 m in diameter) in the summit crater. The new vents formed just E of a lava dome, about [625] m in diameter, emplaced in the center of the summit crater, probably in 1660. Pichincha's horseshoe-shaped summit crater, about 2 km in diameter and 600 m deep, occupies the W end of a 9 km-long massif and is breachéd to the W, in the opposite direction from Ecuador's capital Quito, at the E foot of the volcano (fig. 15-6). Aerial observers reported increased fumarolic activity in the summit crater about 20 Aug. Plume heights of as much as several hundred meters were reported in mid-Aug and a group that climbed the volcano 11-13 Sept observed a 200-300 m-high plume, but vapor emission had declined to only 2-3 times its normal level by early Oct. Temperatures of summit crater fumaroles in early Oct were 88-90°C, comparable to those recorded in 1976.

Seismographs at Quito and at Cotopaxi volcano (60 km to the SSE) recorded a series of earthquakes, some of which were large enough to be felt (fig. 15-7). However, the volcano is in a tectonically active zone and none of these events were large enough to be detected by the WWSSN. Earthquakes on 12 Aug at 0804 (probably centered near Quito) and 21 Aug at 0718 (probably centered about 40 km S of the volcano) had intensities of MM III-IV in Quito. Smaller events recorded on 25 Aug at 0651 and 26 Aug at 1311, both apparently centered about 40 km S of the volcano, were not felt, but residents of Quito noticed an event on 28 Aug at 1822 that probably had a nearby epicenter. Seismographs installed on the N, E, and S flanks 25-27 Sept had recorded no local seismicity (magnitude threshold about 1.5) as of 7 Oct. Dry tilt stations were emplaced beginning 28 Sept at sites 11.25 km NNE, 9 km E, and 7.25 km SSW of the central dome.

A UNDRO volcanological team of John Tomblin, Karl Grönvold, and J. C. Sabroux assessed volcanic hazards in Ecuador 1-12 Oct. Chemical analyses of gas samples collected by Sabroux at Guagua Pichincha on 5 Oct will be compared to his analyses of gases collected from the same fumaroles in 1976.

The last major eruption from Guagua Pichincha occurred in 1660, when 40 cm of ash fell on Quito and nuées ardentes flowed down the W flank. Several minor phreatic eruptions were reported in the 19th century, the most recent in 1881.

Information Contacts: M. Hall, Escuela Politécnica, Quito; J. C. Sabroux, CNRS, Gif-sur-Yvette, France; USGS/NEIS, Denver, CO.

11/81 (6:11) In late Aug or early Sept, one or more small phreatic explosions ejected about 5000 m^3 of ash from 5 new vents, 2-12 m in diameter, on the NE flank of the central lava dome. Ash thicknesses decreased from > 1 m adjacent to the vents to about 1 cm at 500 m to the E, and traces extended about halfway up the E inner wall of the crater. There was no evidence of any fresh magma in the ash. By early Oct, vapor emission from the explosion vents and a group of new fumaroles at the base of the S inner wall of the crater had declined to only 2-3 times its normal level.

Seismographs installed on the N, E, and S flanks of the volcano 25-27 Sept detected no local earthquakes until early Oct, but recorded several events in most 5-day periods between 1 Oct and 15 Nov. The number of recorded events peaked at about 2/day in early Nov and had declined slightly to 4-6/week by early Dec. Three dry tilt stations were established on the flanks of the volcano in late Sept (6:9) and 2 more were added in mid-Nov, but no significant change in tilt had been measured as of early Dec.

Geologists climbed to the crater rim 3 Dec and noted that a small amount of ash had been deposited near the explosion vents since the area was last observed from the ground in mid-Oct. The volume of new ash appeared to be less than the estimated 5000 m^3 ejected in late Aug or early Sept. Steam emission had become more voluminous and 2 new vents opened since Oct.

Information Contacts: M. Hall, Escuela Politécnica, Quito; J. Tomblin, UNDRO.

6/82 (7:6) The following is from Minard Hall.

"Since the initiation of activity in Aug 1981 (6:9) there had been a progressive increase, although irregular, in fumarolic activity within the summit crater. Prior to this, the volcano had apparently been quiet since 1881.

"The major fumarolic activity remained associated with the 625 m-diameter dacitic dome, which covers much of the crater floor. The new vent area, on the NE side of the dome, continued as one of the principal steam vents. Originally consisting of several small vents in line, the NE vents had grown to form one principal crater, approximately 50 m in diameter, plus another crater. Occasional phreatic explosions ejected ash and rock fragments. There was no evidence of new magmatic material in the ejected debris.

"Two older vents on the dome also remained active, producing individually about the same volume of steam as the NE vent. During a 16 May visit considerably more steam was escaping via many small fumaroles along the S talus slope of the dome. Apparently no material was being ejected by the older vents. Two other major fumaroles, at the foot of the S crater wall, also had notable steam emission.

"A 3-station seismograph net, in continuous operation since 25 Sept, registered a fairly constant level of 4-10 local events per week. Three dry tilt stations were established in late Sept, and 2 others in Nov (fig. 15-6). These stations were re-levelled during Nov, Jan and Feb, but unfortunately have not been visited during the past 4 months. No change in tilt (threshold value about 5 μrad) was observed in the sporadic data."

Preliminary analysis of data from the 3-station seismic net shows hypocenters concentrated in 2 areas, both at roughly 4-5 km depth. Most of the events (maximum magnitude about 1) were centered 5-7 km SE of the summit, but some occurred below the caldera.

A geologist climbed to the summit 10 June and observed phreatic explosion deposits not present during the 16 May visit. The explosion appeared to have been

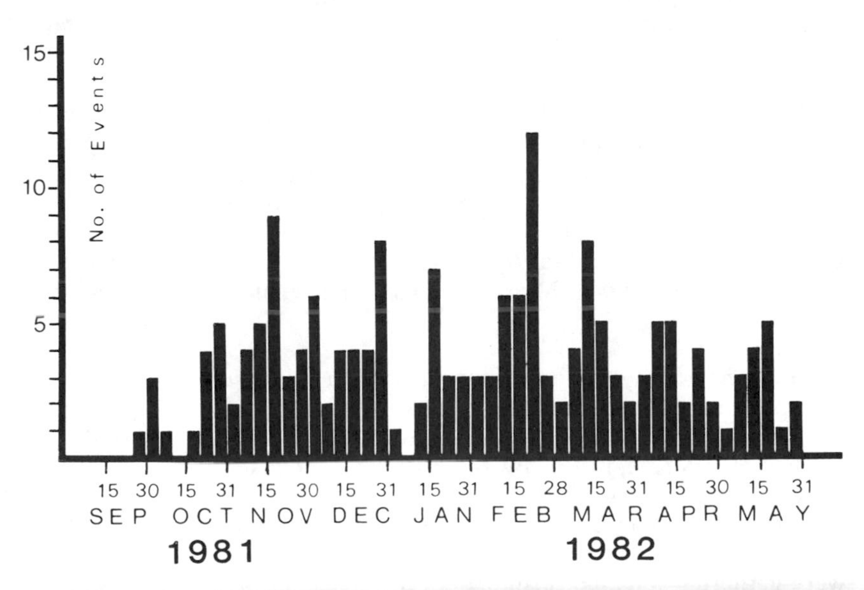

Fig. 15-7. Number of volcanic earthquakes recorded 25 Sept 1981 – 31 May 1982 by a 3-station seismic net at Guagua Pichincha. Courtesy of Minard Hall.

directed toward the N, where ash coated recently stripped tree trunks to a maximum thickness of 5 cm. Ash deposition was followed by ejection of blocks up to 1 m across that produced impact craters as much as 4 m in diameter and 1.5 m deep. A zone of intense cratering extended 150 m from the vent and blocks 20 cm across were found as far as 350 m away. Hall's field work in the caldera 26-27 June revealed no evidence of additional explosions. No fracturing or other indications of uplift were observed on the top of the lava dome. Fumarole vapors had very little odor and were less voluminous than in previous visits. The 26-27 June visit followed 3 days without precipitation on the normally damp summit, and Hall is investigating the correlation between local rainfall and steam emission.

Information Contact: M. Hall, Escuela Politécnica, Quito.

Further Reference: Salazar Medina, E., 1985, Riesgo Volcánico de los Volcanes Guagua Pichincha, Cotopaxi, y Tungurahua; *Riesgo Volcánico en el Ecuador*, INEMIN, Ministerio de Recursos Naturales, Quito, p. 1-24.

COTOPAXI

N-central Ecuador
00.65°S, 78.43°W, summit elev. 5897 m
Local time = GMT - 5 hrs

This classic strato-volcano, 50 km S of Quito, has erupted 59 times since 1532. Along with Java's Kelut, Cotopaxi leads the world's volcanoes in the number of historic eruptions producing mudflows (27), and 4 of these have resulted in fatalities.

General References: Hall, M., 1987, Peligros Potenciales de las Erupciones Futuras del Volcán Cotopaxi; *Revista Politécnica*, v. 12, p. 41-80.

Miller, C.D., Mullineaux, D.R., and Hall, M.L., 1978, Reconaissance Map of Potential Volcanic Hazards from Cotopaxi Volcano, Ecuador; *USGS Misc. Investigations Series, Map* I-1072.

10/75 (1:1) On 23 July 1975 small grayish puffs of smoke were observed emanating from the crater. In mid-Sept a 300-m vapor plume rose above the crater, and a small earthquake shook the volcano on 24 Sept. In mid-Oct vapor plumes were reported to be increasing in volume and frequency. Cotopaxi last erupted in 1944.

Information Contact: M. Hall, Escuela Politécnica, Quito.

11/75 (1:2) The activity continued during Nov. The amount of vapor was about the same as in Oct, but daily cloud activity was observed. There was increased fumarolic activity on the W side, just below the crater. Earthquakes were felt in the town of Mulaló around 0000, 11 Nov, and at 0431, 12 Nov, and again on 14 Nov. Three portable seismographs were placed around the volcano, and were gathering valuable seismic data by the end of the month.

Information Contact: Same as above.

12/75 (1:3) Activity declined during Dec. The crater steam clouds of previous months were observed only once, on 12 Dec, since 28 Nov. Seismic activity was not strong and focal depths were at least 20 km below the volcano. Relevelling of dry tilt stations installed 7 Nov showed no inflation of the volcano. However, a USGS team including Donal Mullineaux and C. Dan Miller reached Ecuador in late Dec to assess volcanic hazards and assist the local authorities in contingency planning.

Information Contact: Same as above.

SANGAY

Central Ecuador
2.03°S, 78.33°W, summit elev. 5230 m
Local time = GMT - 5 hrs

Commonly cloud-covered, this high cone E of the Andean crest has been in virtually continuous Strombolian eruption since at least 1934. Its first documented eruption was in 1628, and it lies 150 km S of Cotopaxi.

7/76 (1:10) A recent expedition, 28 July–9 Aug, to the strato-volcano Sangay reports the following activity.

Mild explosive activity occurred at intervals of 6-12 hrs with the expulsion of white, vapor-rich, sulfurous plumes that rose approximately 100 m. Very acidic rains were falling W of the cone. The NW side of the cone was covered by still-hot lava flows (basaltic andesite) from the last few years. A new lava flow was leaving the S crater and has descended W several hundred meters. A light coating of ash covered the snow on the SW side of the cone. No other activity was observed on the other sides of the volcano. It appeared that this activity has continued steadily from last year. A small parasitic cone of approximately 50 m height was recently discovered on the lower E flank. It is not presently active.

On 10 Aug an independent British team, which apparently included no geologist, reached the 3700 m basecamp level on the volcano. Two days later 6 members ascended to within 300 m of the summit. At 1230 an explosion produced a black mushroom cloud that reached an estimated 300 m above the volcano and dropped ejecta (to 35 cm) on the group. Later, search parties found 3 injured, 1 dead, and another had not been found 5 days after the accident. Helicopter rescue attempts were abandoned on 18 Aug after 2 days of heavy snowfall.

Information Contacts: M. Hall, Escuela Politécnica, Quito; J. Aucott, British Embassy, Quito.

7/83 (8:7) . . . During overflights on 4 and 6 Aug, Maurice Krafft observed frequent ash emission from 1 of 4 WSW-ENE-trending vents in the summit area. The westernmost vent was filled by a blocky lava dome 15-20 m in diameter, partially covered by ash. ENE of the dome, explosions at least every 10 min from a 15 m-diameter crater produced thick black cauliflower-shaped ash columns 100-300 m high. Winds blew ash from these explosions to the SW, toward the dome. Each explosion also triggered small ash avalanches from deposits on the upper W and SW flanks. The largest of the 4 vents, ENE of the active crater, was 80-100 m across and contained 2 fumaroles that were emitting vapor. The fourth vent, 20-30 m in diameter and slightly N of the trend of the other 3 vents, was not active during the overflights.

Minard Hall reported that activity was generally similar when he visited the volcano in 1976. Although lava was oozing from the westernmost vent at that time, it had not yet built a dome.

Information Contacts: M. Krafft, Cernay, France; M. Hall, Escuela Politécnica, Quito.

FERNANDINA

Galápagos Islands, Ecuador
0.37°S, 91.55°W, summit elev. 1495 m
Local time = GMT - 6 hrs

Fernandina is 1450 km W of Quito on the W edge of the Galápagos archipelago. Like its neighboring volcanoes, Fernandina is a basaltic shield with a central caldera. It is the most active Galápagos volcano, with 20 eruptions known since 1813, but reporting has been poor in this uninhabited part of the archipelago, and even a 1981 eruption (see 9:3) was not witnessed at the time. In 1968 the caldera floor dropped 350 m following a major explosive eruption.

***General References** (Galapagos):* McBirney, A. R., and Williams, H., 1968; Geology and Petrology of the Galápagos Islands, *Geological Society of America/ Memoir* 118, 197 pp.

Simkin, T., 1984, Geology of Galápagos Islands; *in* Perry, R. (ed.), *Galápagos*; Pergamon Press, Oxford, p. 15-41.

3/77 (2:3) A 4-day eruption began 23 March from fissures at the SE end of Fernandina caldera. As in the similar eruptions of 1972 and 1973, lava flowed down the inner caldera wall into the large (2 km-diameter) caldera lake. The following report is compiled largely from information provided by Dagmar Werner and the Charles Darwin Research Station (CDRS).

A red glow over Fernandina's summit was first noticed at 2140 on 23 March by Werner, who was camped at the coast, 16 km WSW of the eruption site. Later inspection of seismograms at CDRS, 140 km ESE of Fernandina, showed 3 small events ($M \leq 3$) between 1831 and 1852 that same evening, but no tremors were felt by Werner. A light ashfall dusted her camp that night and heavier ashfall was experienced twice while climbing to the caldera rim the next morning. Reaching the rim at 1300, she observed low fountaining from fissures along the W half of the prominent bench 300 m below the caldera's SE rim. Lava cascaded over 500 m to the lake (formed by the 1968 caldera collapse) and steaming was localized around a growing lava delta, forming there. This activity continued through the night and little change was observed before Dr. Werner departed the rim at 0700 on 25 March. Glow was again observed that night and a cloud was visible as she sailed away on the morning of 26 March.

A separate group including David Doubilet (National Geographic) and Jerry Wellington (University of California, Santa Barbara) was working near Fernandina and on the evening of 26 March observed a bright red glow that increased in intensity until midnight. By dawn, however, the intensity had decreased greatly and the eruption was essentially over [when] they reached the caldera rim at 2000 on 27 March. On 31 March a CDRS team was on its way to study the eruption's products and effects on the lake.

Information Contacts: D. Werner, C. MacFarland, CDRS, Galápagos; T. Simkin, SI.

8/78 (3:8) A new eruption began on 8 Aug and had apparently ended on 19 Aug when the last observers left the island. An earthquake (m_b 4.5) at 0955 was located by USGS/NEIS 43 km NE of Fernandina caldera (easily within hypocenter location error for an event of this size at Fernandina) and appears to have triggered the eruption. One hour later an eruptive cloud was first noticed over the volcano and at 1230-1240 cloud heights of 4500 and 6000 m were independently estimated from distant parts of the archipelago. At this time (1231) a NOAA infrared satellite image recorded an irregular cloud roughly 55 km in diameter with its SE boundary over Fernandina. One hour later the cloud was still 55 km in maximum dimension, but had become wedge shaped with its apex over Fernandina and measured 27 km across at its WNW end. After another hour, at 1432, the plume had narrowed to 13 km some 46 km from its apex, and the next hourly image (1533) showed only the indistinct hint of a plume. That night, glow was observed over Fernandina several times and lightning was seen for $^1/_2$ hr. On 9 Aug a large cloud remained over Fernandina, growing during the afternoon, but it has not been recognized on satellite imagery.

A group including Tui Deroy Moore and Howard Snell reached the caldera rim on 10 Aug as the eruption entered its third full day. They observed smoking vents along a fissure on the NW bench of the caldera, roughly 300 m below the rim [see also 3:9]. This bench, isolated by a prehistoric collapse of the elliptical 4 × 6 km caldera, is symmetrically opposite the SE bench that has been the site of eruptions in 1972, 1973, and 1977. The SE end of the caldera also experienced maximum subsidence (350 m) in the major caldera collapse of 1968, while the NW end was unaffected both then and in the 10 years since. Scoria was recognized down to 200 m elevation on the NW flank and Pele's hair reached the NW coast of the island, 12 km from the caldera rim.

Remaining on the rim for nearly 3 days, the group observed intermittent fountaining along the 1-km fissure feeding lava to the caldera lake 500 m below and 2 km to the SE. One strong tremor was felt and rockfalls on the caldera walls were nearly continuous. Activity had declined when they left on 13 Aug, but another group, including Dee Boersma and Bob Tindle, reached the volcano on the night of 16 Aug and observed glow over the caldera. They climbed to the rim on the 18th and observed fountaining to 100 m from vents at the back of the bench, some 400 m NW of the vents active on 10-13 Aug [but see 3:9]. Fresh spatter was on caldera walls 200 m above the vents, and lava was flowing to the steaming lake. Activity declined in the early morning of 19 Aug and only vapor issued from the vents as the group left the rim. Neither vapor nor glow were seen over the volcano on 20 and 21 Aug, and no more recent reports are available.

Information Contacts: H. Hoeck, R. Tindle, H. Snell, CDRS, Galápagos; T. Moore, Academy Bay, Galápagos; A. Kreuger, NOAA; D. Boersma, Univ. of Washington; USGS/NEIS, Denver, CO.

9/78 (3:9) The eruption reported last month took place along a fault on the caldera's NW bench. This pre-1947 fault dropped the SE edge of the bench by about 80 m. The 3 active vents viewed 18-19 Aug were located on this fracture, not at the back of the bench as reported last month. Although the witnessed cessation of lava venting during the early morning of 19 Aug appeared to be the end of the eruption, on 24 and 26 Aug explosive "popping" sounds were heard by Robert Tindle 15 km from the caldera at the island's NE coast. These explosions lasted about 2 hours and were accompanied by "diffuse, smoky haze" drifting down the upper slopes of the volcano. From 27 Aug until he left the island 5 Sept, Tindle heard no other activity and saw no other clouds over the NW caldera rim. Visual observations of the caldera on 6 and 19 Sept likewise showed no signs of activity.

Information Contacts: R. Tindle, H. Hoeck, P. Ramón, CDRS, Galápagos.

Further Reference: Moore, T. De Roy, 1980, *Galápagos: Islands Lost in Time,* Viking, New York, 161 pp.

3/84 (9:3) At 0500 on 30 March, Oswaldo Chapi and Fausto Cepeda (of the Galápagos National Park) heard noise from Fernandina Caldera, 22 km SW of their position at Tagus Cove. Glow was visible over the NW end of the caldera and a cloud was seen issuing from the same location after sunrise. The eruption was described as being smaller than the Volcán Wolf eruption of 1982 (Wolf, 7:8).

On 1 and 2 April, the TOMS instrument in the Nimbus-7 polar orbiting satellite detected SO_2 produced by the eruption (fig. 15-8). No data were available 30-31 March, and SO_2 had dropped below the detection threshold by 3 April. Strongest values on 1 April were directly over the volcano and a preliminary estimate of total SO_2 was 60,000 metric tons. No eruption cloud was evident on NOAA weather satellite imagery.

On the afternoon of 4 April, the cruise ship *Santa Cruz* reported a long vapor plume coming from the caldera, but apparently decreasing in size. They looked for glow over the volcano that night but reported none.

On 11 April Fernandina was climbed from the NW by David Day [and others], who reported an apparently inactive lava flow reaching from the W side of the caldera (near the site of the major eruption of 1968) to the lake. At 0650 the next morning, [Day's group] heard a noise "like a large landslide" from their camp near the W caldera rim. Within 30 sec, they reached the rim in

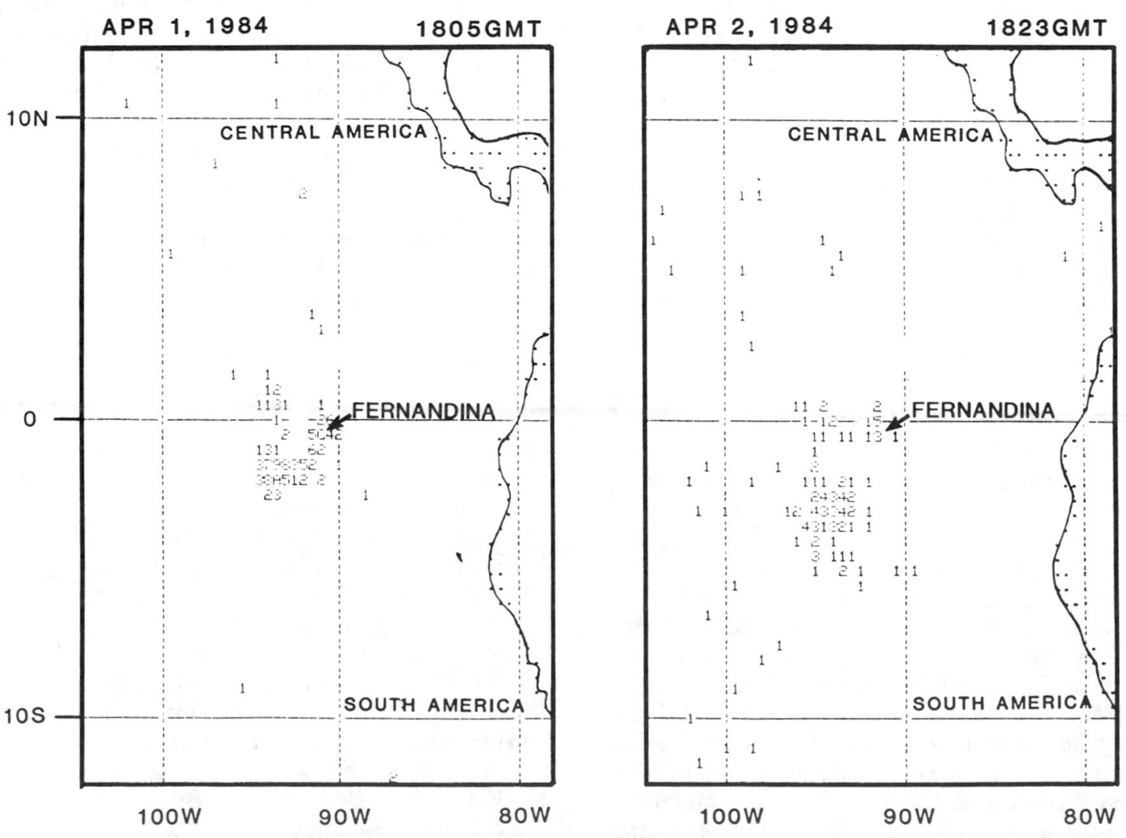

Fig. 15-8. Preliminary map of SO_2 data from the TOMS instrument on the Nimbus-7 satellite. Each number or letter represents the average SO_2 value within an area 50 km across. Values above 9 are represented by A, B, C, etc. For a more detailed derivation of these values, see fig. 13-52. Courtesy of Arlin Krueger.

time to see what Day described as a nuée ardente that had already moved from the vent area halfway to the lake. They left the rim, and observers from Punta Espinoza (17 km to the NE) described an eruptive cloud rising at 0655 to an estimated height of about 7 km. At 0704, [Day's group] was overtaken by an ash rain described as "raindrops with ash" and total darkness persisted until 0720. A thickness of 3 mm of tephra accumulated during that period at their rim camp. By 0725 it was clear enough to see into the caldera. Tephra covered the new lava on the caldera floor with the exception of an area a few hundred meters across in which molten lava could be seen. [The group] left the rim at 1030 and no further volcanism had been witnessed at the time of their radio report, at 1500 on 13 April, from Punta Espinoza. [A substantial part of the caldera wall collapsed into the 1984 vent area on 11 April, and was responsible for most, if not all, of the phenomena witnessed by Day and his group.]

This is the 6th known eruption of Fernandina since the major explosive eruption and massive caldera collapse of 1968. The last eruption was not recognized in the Galápagos, but its products are visible in an aerial photograph taken 26 March 1982. From a 900 m-long circumferential fissure on the S rim of the caldera, flows moved both inward (N) down the caldera wall and over a high topographic bench, and outward (S) where the flow ponded behind another row of circumferential vents. The eruption had not yet taken place when Tom Simkin and others passed this area on 4 December 1980.

Information Contacts: G. Reck, CDRS, Galápagos; L. Maldonado, Quito, Ecuador; D. Day, Isla Santa Cruz, Galápagos; A. Krueger, NASA/GSFC; M. Matson, NOAA/NESDIS.

WOLF

Galápagos Islands, Ecuador
0.02°N, 91.35°W, summit elev. 1710 m
Local time = GMT - 6 hrs

Wolf is the archipelago's highest volcano, 50 km NNE of Fernandina. It straddles the equator at the N end of Isabela, the largest island in Galápagos. Wolf's 1797 eruption was the first documented in the islands and has been followed by at least 9 more. It should not be confused with the small island of Wolf, or Wenman, 160 km to the NNW.

8/82 (7:8) Eruption clouds began to emerge from the volcano during the afternoon of 28 Aug. Plume emission was first detected on visible band satellite images between 1300 and 1400, and feeding continued until nightfall. The plume, which drifted W, could not be seen on infrared imagery, indicating that it remained at low altitudes. Observers on a tour ship first saw clouds issuing from the summit about 1430-1500 and reported strong summit glow that night. Activity on the SE flank was first observed at 0830 the next morning, but heavy weather clouds had obscured this area the previous afternoon, and flank vents may have been active then as well.

Tui DeRoy Moore and others arrived at the SE flank late 31 Aug. Lava fountained from a radial fissure that extended 1 km or more downslope from 875 m altitude. Fresh lava covered the area near the fissure. Lava flowed SE then turned toward the E, reaching about 280 m altitude. This flow had stopped advancing by 1 Sept and fountaining had ended by that evening, although some SE flank glow remained visible. As the flank activity declined, summit activity increased. Summit glow had been visible since 28 Aug, but strengthened during the night of 1-2 Sept and a large convecting cloud was present over the caldera. Moore reached the caldera rim on 3 Sept and found several vents active in the caldera, the strongest on its floor at the base of the steep SW wall. Lava fountaining from this vent was continuous and the fountains occasionally rose as high as the caldera rim, approximately 700 m above the floor. Intermittent, relatively weak fountaining (less than 50 m estimated height) occurred from 4 small vents along a 100-200 m-long fissure on the S caldera floor. Thick-looking pahoehoe lava covered slightly more than half of the caldera floor, or ~6 km^2, and was mainly on the N and NW side. Gases emerging from the base of the convecting eruption cloud formed a haze that drifted W, away from the observers. By early 4 Sept, when Moore left the volcano, a cone had begun to form around the main vent. Activity appeared to be dominated by scoria ejection, with little lava being added to the caldera floor flows. No earthquakes were felt by the observers, and Moore reported that there seemed to be little effect on the flora and fauna. Glow was still visible in the eruption cloud late 5 Sept and airplane passengers saw a strong plume 6 Sept.

Wolf has been one of the more active Galápagos volcanoes. Flank eruptions from the same SE vent area took place in 1948 and 1963, but summit caldera activity had not been documented since 1800. A probable SE flank eruption was heard but not seen in 1973. The present eruption is a two-hemisphere event, with the caldera lying mostly N of the equator and the SE vent, less than 10 km distant, in the Southern Hemisphere.

Information Contact: T. Moore, Isla Santa Cruz, Galápagos.

Further Reference: Schatz, H. and Schatz, I., 1983, Der Ausbruch des Vulkanes Wolf (Inseln Isabela, Galápagos-Inseln Ecuador) im Jahre 1982 – Ein Augenzeugenbericht; *Ber. Nat. Med. Verein Innsbruck*; v. 70, p. 17-28.

SIERRA NEGRA

Galápagos Islands, Ecuador
0.83°S, 91.17°W, summit elev. 1490 m
Local time = GMT - 6 hrs

At the opposite end of Isabela, 100 km S of Wolf, Sierra Negra's 7 × 10 km caldera is the largest in the islands. Eleven historic eruptions are attributed to Sierra Negra, but several poorly described events in the last century may have been from neighboring volcanoes.

General Reference: Delaney, J.R., Colony, W.E., Gerlach, T.M., and Nordlie, B.E., 1973, Geology of the

Volcán Chico Area on Sierra Negra Volcano, Galápagos Islands; *GSA Bulletin*, v. 84, p. 2455-2470.

11/79 (4:11) An eruption began on 13 Nov at Sierra Negra, the only historically active volcano in the Galápagos Islands that is presently inhabited. The location was Volcán Chico, a circumferential fissure zone 1 km N of, and 100-200 m below, the caldera rim of Sierra Negra. Volcán Chico was also the site of the last 2 eruptions from this volcano, in 1963 and 1953, and has been described by Delaney and others (1973). At the time of the last report lava was still flowing down the volcano's uninhabited N slope 3 weeks after it began. The residents of the S slope, evacuated on the first day of the eruption, had returned to find damaged crops and livestock.

At 0730 on 13 Nov a local earthquake was registered on the CDRS seismograph (90 km E of Sierra Negra on Isla Santa Cruz), part of the WWSSN maintained by the USGS [see also 4:12]. A second, larger, earthquake followed in 20 min, and at 0845 the first explosion was heard by people living on the S flank. Within 20 min of the first explosion, tephra (including scoria and Pele's hair) fell on the villages of Santo Tomás and Villamil, 13 and 26 km SE of the eruption site. Residents of Santo Tomás began evacuation to the coastal village of Villamil.

The eruptive cloud was large enough to be seen on NOAA's SMS-1 weather satellite, which transmits images of the whole hemisphere every 30 minutes from its equatorial geostationary orbit. The cloud first appeared on the 0930 image and grew rapidly to an area estimated to be 220 × 130 km within 2 hours. The cloud soon separated into 2 lobes and measurements on the infrared imagery by Arthur F. Krueger indicate a maximum elevation of at least 14 km for the main lobe at 1500. This lobe moved SE at 25 km/hr while the other lobe, estimated to be about 8 km high, moved SW at 30 km/hr. By 2200, the main lobes were quite diffuse, but a low, thin plume (near the 4-8 km resolution of the imagery) extended about 50 km S from Sierra Negra. No prominent eruptive clouds were visible on SMS-1 imagery during the following days.

A CDRS team (including P. Ramón, J. Budris, and T. & A. Moore) reached Volcán Chico 24 hours after the eruption began. Upon arrival at Villamil, they noted a thin coating of tephra with some light scoria fragments to 5 cm in diameter. Tephra at the caldera rim was 2-3 cm thick, with some 15-20-cm bombs, but nothing larger than 1 cm was falling at the time. Up to 20 vents were active along the 1-2-km [8 km, see 5:1] circumferential fissure and individual fountains reached 100-m heights. As many as 13 lava flows coalesced downslope to the N and probably reached the sea on the first day. Billowing clouds were visible at the coast of Elizabeth Bay on the morning of 15 Nov, and 2 biological groups were reported to be investigating the lava/sea interface.

At 1500 on 14 Nov, a higher resolution NOAA satellite, VHRR Tiros 6, returned an infrared image showing the lava flow on the N flank and a high vapor plume heading S. The CDRS seismograph recorded 8 local events on 14 Nov and 6 on the following day. Observers at the site felt many tremors, but noted that eruptive intensity seemed to decrease after a peak around 2200-2300 on 14 Nov. Local haze increased during the next 2 days and vent activity declined steadily. Only 4 vents were active and fountaining had almost stopped by sunset on 16 Nov. At 0400 the next morning, however, observers were awakened by the largest explosion of their 5 days at the site. Fountaining increased immediately and continued to build to a peak around 5 hours later. That same night the wind changed, carrying haze ENE over Isla Santa Cruz and reducing visibility there to 2 km on 19 Nov.

Activity at the vents remained low on 18 and 19 Nov, when the group left. The flows continued, however, and on the morning of 19 Nov, NASA's Landsat C satellite returned a high-resolution near-infrared image clearly showing the main flow as a 100 m band bearing 030° for about 12 km from the vent area (fig. 15-9). On that same day, tephra fragments to 5 mm in diameter were collected on the ship *Delfin*, over 100 km E of the volcano.

As the CDRS group left the volcano, they noted an average tephra thickness of 3-5 cm on the caldera rim and serious damage to the biota on the upper SE flank. Acid rain and haze killed much vegetation and the group noted dead birds and rats. Residents of the S flank, who returned to their farms after the eruption's first day, reported damage to crops and livestock.

H. Hoeck was on the volcano 1-2 Dec and reported that the vegetation was beginning to recover. One large lava flow continued down the N slope, and several vents were steaming. Six small vapor vents were observed on the S floor of the caldera nearly 10 km S of the eruptive fissure. This caldera, the widest (10 × 7 km) and shallowest in the Galápagos, was not otherwise affected by the eruption, and Arnaldo Tupiza (CDRS representative on Isabela and very familiar with the volcano) recognizes these 6 vents as new.

None of the eruption earthquakes have yet been located on the WWSSN [but see 4:12], but a magnitude 5.0 event was located in the Galápagos at 1006 on 28 Oct. Its preliminary epicenter determination is 60 km NW of the eruption site.

Information Contacts: P. Ramón, J. Budris, H. Hoeck, P. López, J. Villa, CDRS, Galápagos; M. Hall, Escuela Politécnica, Quito; A. Krueger, NOAA/NESS; C. Wood, NASA/GSFC; B. Presgrave, USGS/NEIS, Denver, CO.

12/79 (4:12) The fissure eruption was continuing on 5 Dec. LANDSAT imagery on 7 Dec, however, showed no thermal anomaly like the lava flow on the 19 Nov LANDSAT image (fig. 15-9), although a vapor cloud similar to that of 19 Nov was present. When Arnaldo Tupiza visited the eruption site on 29 Dec, the lava flow had ceased, but the vents continued to steam and the solfataric area had grown. The new fumaroles on the S caldera floor remained active.

Two earthquakes connected with this eruption have been located by USGS/NEIS. The first, at 0751 on 13 Nov [4:11], was M 4.4 and its hypocenter (0.89°S, 91.23°W) was located 18 km SW of the eruption site. This was followed at 0817 by another earthquake that wrote a very similar, and only slightly smaller, record on the CDRS seismograph 90 km to the E. This event has not been located by NEIS, but was followed in less than 30 minutes by the first explosion recognized on the island. Seismicity increased on the eruption's second afternoon, and at 1858 on 15 Nov, an M 4.8 event was recorded. Its hypocenter (0.96°S, 91.25°W) was located by NEIS 24 km SSW of the eruption site. These hypocenters are in the area of the Feb 1979 flank eruption of Cerro Azul [4:1-2], but study of the 1968 caldera collapse earthquakes from Fernandina (Filson et al., 1973), where true epicenters were known, showed that hypocenter mislocations for events of this magnitude range averaged about 50 km.

Reference: Filson, J., Simkin, T., and Leu, L.-K., 1973, Seismicity of a Caldera Collapse: Galápagos Islands 1968; *JGR*, v. 78, no. 35, p. 8591-8622.

Information Contacts: A. Tupiza, H. Hoeck, CDRS, Galápagos; USGS/NEIS, Denver, CO.

1/80 (5:1) On 14 Jan the new lava delta, formed at Bahia Elizabeth by the 13 Nov eruption, was visited for the first time in 2 months. Hendrik Hoeck, inspecting the flow at night with binoculars, saw red spots along the flow length and some active feeding at vents 12 km inland. Motion was not visible along the 1-1 1/2-km flow front in daylight, but water temperatures of 50°-60°C at distances of 30 m from the shore provide further evidence that the flow remains active. Pelicans and Franklin's gulls were concentrated in the area, presumably feeding on killed marine organisms, and 2 dead green turtles were observed. A pack of feral dogs appears to have been cut off by the new flows from more vegetated parts of Sierra Negra, and it is feared that they may cross the barren older lava flows to reach volcán Alcedo to the N. Alcedo has not been threatened with feral dogs and it is the home of the largest population of giant tortoises in the Galápagos.

Arnaldo Tupiza walked about 6 km inland on 14 Jan, noting flow thickness of 3-5 m, two 10 m-diameter "pools" of molten lava, and another major flow diverging toward the NE. The most active vents were farther W, along the circumferential fissure, than any previously reported, and it is not yet clear whether activity has resumed after migrating westward or whether the

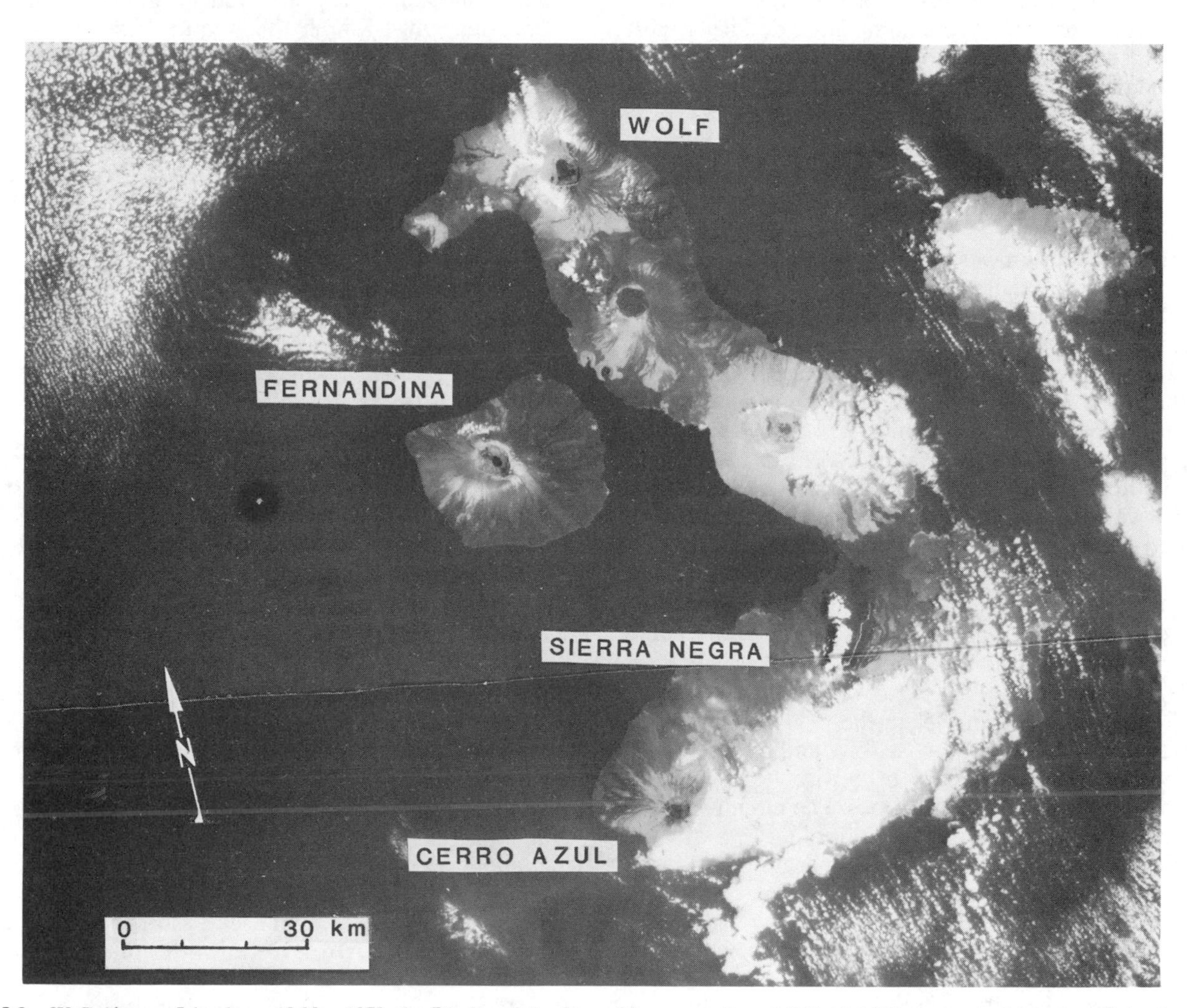

Fig. 15-9. W Galápagos Islands on 19 Nov 1979, the 7th day of the Sierra Negra eruption. A 130 × 165 km area is shown in a [joined pair of Landsat C images]. The large cloud bank near the base of the figure obscures the summit and S flank of Sierra Negra, but a prominent white eruptive cloud trends NNE from the eruption site near the volcano's N rim and a lava flow appears just E of the cloud as a thin white band trending NE for 12 km. The near-infrared image (Band 7) shows the vegetated portions of these volcanic islands as light gray, in contrast to the dark young lavas. Summit calderas of most volcanoes are visible. Images provided by NASA/GSFC. They may be ordered from EROS Data Center, Sioux Falls, SD 57198 (scene ID's: PAO file #E-931-225BN and E-932-225BN).

westerly vents were not recognized during previous observations from the vigorously steaming E end of the fissure system. The mid-Jan observations indicate an active fissure length of 8 km.

Information Contacts: H. Hoeck, A. Tupiza, CDRS, Galápagos.

CERRO AZUL

Galápagos Islands, Ecuador
0.90°S, 91.42°W, summit elev. 1690 m
Local time = GMT - 6 hrs

Forming the SW end of "J"-shaped Isla Isabela, and 30 km WSW of Sierra Negra, Cerro Azul has erupted 9 times since 1932.

1/79 (4:1) At 2000 on 2 Feb red glow was sighted from a fishing boat E of Isla Isabela. The source has since been confirmed as 2 small craters on the E flank of Cerro Azul. Red glow was seen from CDRS, [110] km to the E, at 0100 on 5 Feb. By this date, 2 groups were investigating the eruption on foot and a more detailed report will be included next month.

The last certain eruption of this volcano was 29 June 1959, but there were unconfirmed reports of activity in 1968 synchronous with the major explosive eruption and caldera collapse at Fernandina, 65 km to the N.

Information Contact: H. Hoeck, CDRS, Galápagos.

2/79 (4:2) The eruption was at an elevation of about 300 m on the E flank of Cerro Azul, a young shield volcano. The earliest report of activity is from fishing boats that were anchored in Iguana Cove, 5 km W of the volcano's caldera, on the night of 29 Jan. They experienced substantial ashfall overnight, but the first reported glow seen from the village of Villamil, 23 km E of the eruption, was not until 2 nights later. The eruption site was seen on the morning of 2 Feb by Howard Snell and Hans Kruuk, who landed on the S coast that afternoon and reached a good viewing position 3-4 km from the vent on 3 Feb. Lava fountaining reached 300 m above the 20-50-m new crater rim, and fed a lava flow moving SE toward the sea. The main vent was 6 km E of the caldera center, near the low point between Cerro Azul and Sierra Negra volcanoes. This area was the site of recent eruptions (apparently in 1952 or the 1940's) that fed lava flows toward the sea both to the NW and SE, forming a narrow lava barrier on the otherwise vegetated low landscape between the 2 volcanoes. The main 1979 vent was several hundred meters into this older lava, and the new flow moved SE along the SW margin of the older flow. The flow front advanced an estimated 1-2 km during the night of 4-5 Feb. Another vent, roughly 400-500 m E of the main one, was recognized on 6 or 7 Feb and appeared to be a later development.

Cloud cover over the eruption area is normally heavy and was particularly bad during the eruption. On the first clear day (8 Feb), Snell described a "thunderhead" cloud to an estimated 5-7 km above the vent with substantial amounts of tephra being carried downwind. No other cloud data have been reported. . . .Snell's last view of the eruption was 10 Feb, when fountaining remained vigorous and the flow had traveled an estimated 8 km toward the sea. Geologist Patricio Ramón reached the eruption site that day and reported that the eruption was from a radial fissure 600-700 m long, with lava flowing from its full length. Four principal vents were active, but the main one described by Snell was the most vigorous. Eruptive intensity declined during the next 5 days. Although quite active on 14 Feb, there was a strong decrease on the 15th when Ramón left to return to CDRS. During Ramón's time at the eruption, the flow advanced no closer to the sea, remaining 5-6 km from the shore, but became much broader. Snell estimated that the maximum flow width was approximately 1 km whereas it was 3-4 km wide when Ramón left nearly a week later.

There were no reports of the eruption for the following 2 weeks (the site is not directly visible from the SE side of Sierra Negra, the only inhabited part of Isla Isabela, and is distant from all normal tour routes in the archipelago), but on the night of 28 Feb Captain Bernardo of the *San Juan* observed red glow over the eruption site. On 4 March the weather was clear and an eruptive cloud was seen by Snell and H. Hoeck from the opposite side of Cerro Azul, roughly 15 km away on the NW coast.

Information Contacts: H. Snell, P. Ramón, H. Hoeck, CDRS, Galápagos.

Further Reference: Moore, T. De Roy, 1980, The Awakening Volcano; *Pacific Discovery*, v. 33, no. 4, p. 25-31.

EL MISTI

S Peru
16.30°S, 71.41°W, summit elev. 5825 m

In southern Peru, this beautifully symmetrical stratovolcano is on the W side of the Andes, W of Lake Titicaca and 765 km SE of Lima. The city of Arequipa is on its flank. Several eruptions have been reported since the mid-15th century, the most recent in 1870, but few are well documented.

5/84 (9:5) The following is a report from Alberto Parodi Isolabella.

"An increase in the normal emission of vapor from the volcano has been noted since the beginning of April. At times the column rose to about 1 km above the crater and was ejected from the dome (150 m in diameter) that is about 250 m below the outer rim of the crater. A temperature of 125°C has been registered for some years in the dome's fumarolic fissures. [At the dome, blocks of andesite are covered with sulfur, gypsum, anhydrite, and ralstonite.]

"It is possible that this increase in vapor (no such increase had been noted since April 1971) is caused by the evaporation of water from the rains which have been intense this year and abundant for 3 months, Jan-March.

As the rains ended and the clouds disappeared, the impressive and sometimes intermittent vapor column from the volcano was visible from Arequipa (1,000,000 inhabitants including the suburbs). It cannot be excluded that a notable increase in natural degassing has also been occurring, in combination with the evaporation of atmospheric water.

"Persons who intended to go to the bottom of the crater 29 April to collect samples were prevented from doing so because gases irritated noses, throats, and eyes. They said that they also observed flaring, perhaps caused by the combustion of hydrogen or other gases.

"The Observatorio de Chacarato of the Instituto Geofísico, Universidad Nacional de San Agustín de Arequipa (10 km from the city and about 20 km from the volcano's crater) equipped with seismographs and magnetometers, had not registered any changes as of 8 May that could be attributed to perturbations of volcanic origin." . . .

Information Contact: A. Parodi Isolabella, Arequipa.

12/85 (10:12) Inside the SE rim of El Misti's [690 × 935]-m summit crater is a younger cinder cone, about [545] m wide at the top and having an inner crater [198] m deep, with a flat floor [158] m across. On 7 and 8 Aug geologists observed vigorous fumaroles, which had not been active a few months earlier, on the N side of the cinder cone floor. High-pressure degassing, as "noisy as a reaction motor," emitted white-gray vapor from 6 vents. There were red sulfur deposits inside the vents, yellow sulfur outside them. Fumaroles were still visible on the N rim of the crater.

The last strong eruption of El Misti occurred between 1438 and 1471 (the reign of the Inca Pachacutec); several weeks of vigorous tephra emission forced residents of the region to flee. Several smaller explosive eruptions have been reported since then, but some were probably only periods of increased fumarolic activity [such as reports from 1878, 1901, 1906, 1929, 1949, and 1971].

Information Contact: M. Decobecq Dominique, Univ. Paris Sud.

UBINAS

S Peru
16.36°S, 70.90°W, summit elev. 5672 m

Ubinas is 55 km E of El Misti and has erupted 17 times since the 16th century. In 1969, ashfall caused crop damage.

12/85 (10:12) When geologists visited Ubinas on 12 Aug, fumarolic activity was weak and emissions were dilute. Some noise was coming from a pit about 300 m in diameter in the N side of the 1-km summit crater. . . .

Information Contact: M. Decobecq Dominique, Univ. Paris Sud.

GUALLATIRI

N Chile
18.42°S, 69.10°W, summit elev. 6060 m
Local time = GMT - 4 hrs

Guallatiri is the highest volcano to have erupted during the report decade. It is 340 km SE of El Misti in northernmost Chile and has erupted at least 4 times since the first half of the 19th century.

The activity described below was initially attributed to Acotango in the Quimsachata Volcano Group. It is now thought to have originated from Guallatiri, further along the same line of sight as Acotango (fig. 15-10). Investigation of Acotango volcano in 1987, by Oscar González-Ferrán, revealed no evidence of a recent eruption (12:12).

11/85 (10:11) The quoted material is a report from Robert Koeppen.

"On 1 Dec at 0750, Robert Koeppen (USGS), Walter West (U. S. Embassy, La Paz, Bolivia), and Jaime Jauregui (Geological Survey of Bolivia) observed steam plumes erupting from a source near the crest of Nevados de Quimsachata. The observation was made from a point about 25 km to the NNE, near the W base of Volcán Nevado Sajama, Bolivia. Visibility was initially excellent, but within about 45 minutes cloud cover completely obscured the mountain's summit. Intervening peaks prevented exact determination of the vent, but the eruption appeared to originate from Cerro Acotango, either from the main summit or possibly immediately to the NW. If the eruption plume came from Volcán Guallatiri, the next possible source area along the line of sight, then it would represent a significantly larger eruption (fig. 15-10).

"The eruptions produced white, billowing clouds that rose vertically above the Quimsachata crest, perhaps about 500 m, then drifted W. The eruptions occurred in episodic bursts, and, based on first sightings of the plumes, at intervals ranging from 45-75 sec. One large plume drifted more NW and appeared to trail a curtain, possibly of ash fallout. Several bursts seemed particularly vigorous and appeared to consist of several plumes coalescing at higher levels."

Chilean forest service personnel based near Lago Chungara, roughly 20 km to the NW, reported that they had seen no activity, but visibilities in the area are frequently poor.

No eruptions are known in historic time from Cerro Acotango, but Yoshio Katsui and Oscar González-Ferrán (1968) mapped it as Holocene. Four historic eruptions bave been reported from Guallatiri, most recently in Dec 1960. Chilean geologists note that fumarolic activity from Guallatiri is apparently continuous but less vigorous than the plume emission observed 1 Dec.

Reference: Katsui, Y., and González-Ferrán, O., 1968, Geología del Area Neovolcánica de los Nevados

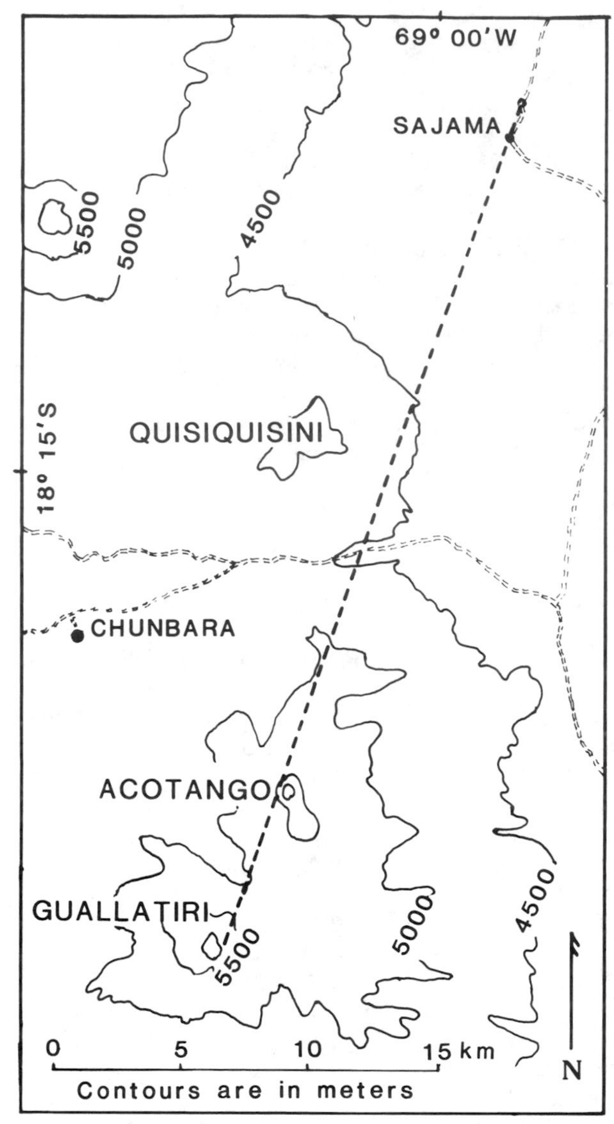

Fig. 15-10. Sketch map of the Nevados de Quimsachata area. Volcano names are in large type, town names in smaller type. Roads are indicated by pairs of dashed lines. Contour interval is 500 m. Robert Koeppen's line of sight from his observation point near Sajama, Bolivia, is shown by a dashed line. The border between Chile and Bolivia is not shown, but follows the crest of the Andes.

de Payachata; *Publicaciones del Instituto de Geología, Universidad de Chile,* no. 29, 161 pp.

Information Contacts: R. Koeppen, USGS, Reston, VA; L. López E., Univ. de Chile, Santiago.

TUPUNGATITO

Central Chile
33.40°S, 69.80°W, summit elev. 5640 m
Local time = GMT - 3 hrs

Tupungatito is a strato-volcano 80 km ENE of Santiago, and has erupted 18 times since the early 1800s. It lies 1670 km S of Guallatiri. No other N Chile volcanoes were reported active during the decade. A volcano gap of nearly 1000 km separates Tupungatito from Llullaillaco, the next historically active volcano to the N (and the world's highest, at 6723 m).

2/80 (5:1) A series of subterranean noises at 0623 on 10 Jan was followed by an explosion that ejected a 1500 m-high cloud of gas and ash from Tupungatito's SW crater. The next day, Oscar González-Ferrán and Sergio Barrientos flew over the volcano, observing that ash covered the snow NE of the vent and that the eruption was continuing, but with decreased intensity.

The Seismologic Service of the Geophysics Department, University of Chile, recorded considerable seismic activity near Tupungatito. The principal earthquake, at 1851 on 14 Jan, was a shallow event with an epicenter calculated at 33.2°S, 69°W, 78 km NE of the volcano. It was felt at intensity 3 in Santiago, about 150 km from the calculated epicenter. In the next 2 hrs, 17 similar events were recorded, of which 3 were located with the same epicenter. Between 2100 on 14 Jan and 0100 on 16 Jan, 13 more local events were recorded, one of which was fairly large. As of 18 Jan, both seismic and eruptive activity had diminished. Tupungatito's last eruption, in 1964, consisted of explosions from the central crater.

Information Contact: O. González-Ferrán, Univ. de Chile, Santiago.

DESCABEZADO GRANDE

Central Chile
35.58°S, 70.75°W, summit elev. 3830 m

The only historic eruption of this volcano, 260 km S of Tupungatito, was in 1932.

3/82 (7:3) Fumarolic activity was observed on the morning of 19 March. A white plume was rising from the summit crater during the 3 hours the observer was on Nevados de Chillán Volcano, 160 km to the S. The only recorded eruption at Descabezado Grande, in 1932, was from a crater at its NE foot. Weak fumarolic activity has been reported on the W slope at about 3500 m, but none had previously been observed in the main crater.

Information Contact: H. Moreno R., Univ. de Chile, Santiago.

NEVADOS DE CHILLAN

Central Chile
36.85°S, 71.40°W, summit elev. 3089 m

This strato-volcano has erupted 17 times since the 17th century.

7/79 (4:7) When visited by Oscar González-Ferrán on 21 Feb 1979, the eruption of Nevados de Chillán that began in July 1973 was continuing. An explosion lasting $1^1/_2$ hrs produced a cloud, containing bombs, [blocks], and ash, that rose almost 2 km before reaching a windy layer that prevented further upward movement. The new cone had grown to about the same height as the adjacent 1906 cone [Volcán Nuevo], where fumarolic activity persisted (fig. 15-11).

[Hugo Moreno reports that by 1983 the phreatomagmatic eruption had almost ended. From 1983 to 1987, only a few explosions have been reported (about 1 every 2 months). These generated small pyroclastic flows, by column collapse, over the snow cover. By late 1987, the dome extruded earlier in the eruption had been covered by tephra that built a new cone (named Tata) about 30 m higher than neighboring Volcán Nuevo.]

Information Contact: O. González-Ferrán, Univ. de Chile, Santiago.

Further Reference: Deruelle, B., 1977, New Activity of the Nevados de Chillán; *C.R. Acad. Sci. Paris*, serie D, v. 284, p. 1651-1654.

CALLAQUI

Central Chile
37.92°S, 71.42°W, summit elev. 3164 m

Callaqui is 115 km S of Nevados de Chillan. A small phreatic ash emission from its summit was observed in October 1980 (H. Moreno, 1985, personal communication). Although this strato-volcano has youthful cones and lava flows, its only other historic activity is "smoke" reported in the middle of the 18th century.

LLAIMA

S Chile
38.70°S, 71.70°W, summit elev. 3124 m
Local time = GMT - 4 hrs

Llaima has erupted 35 times since 1640, second only to Villarrica among Chilean volcanoes. It is 90 km S of Callaqui.

10/79 (4:10) After a series of subterranean noises and local earthquakes, an eruption from Llaima's central crater began at about 0200 on 15 Oct, with intermittent explosions ejecting gas and ash. A group of schoolteachers and students 7 km to the W saw explosions beginning at about 0800 that projected an ash column to about 1 km above the central crater. Police reports state that the eruption was briefly visible from San Patricio, 35 km to the W, at 0857. After 0900, steam and ash emission rapidly declined to fumarolic activity. Ash could be seen on the snow-covered slopes of the volcano. Tourists and

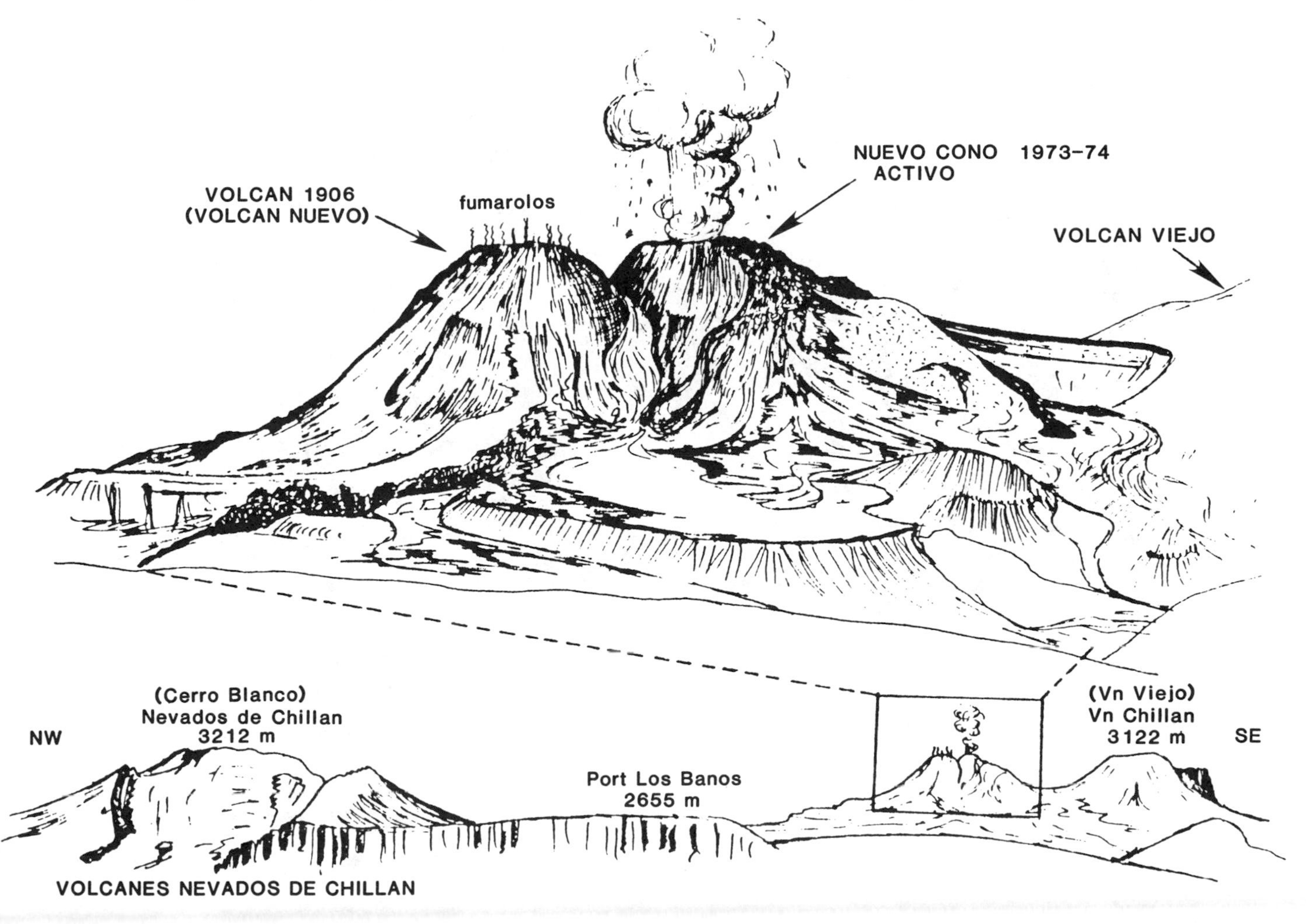

Fig. 15-11. Sketch by Oscar González-Ferrán of the Nevados de Chillán complex. Upper sketch shows detail of the area active on 21 Feb 1979.

residents of the area immediately around the volcano were evacuated.

[Hugo Moreno notes that reports of lava extrusion following the explosive activity were incorrect. Three lava debris flows/pyroclastic flows were generated by column collapse during the main explosion.]

A brief period of lava extrusion from the central crater followed the explosive activity. The lava melted some of the ice near the summit, generating small lahars that moved N and E. Chilean Air Force personnel saw 2 lobes of lava on Llaima's W slopes during an overflight at 1200.

Hugo Moreno flew over the volcano the next day, observing 3 [black] flows, and . . . deposits of ash and lapilli to the NE (fig. 15-12). Fine ash covered the slopes of Sierra Nevada volcano (20 km NE) and nearly reached Lonquimay village (45 km NE). Dense white vapor emerged from the central crater.

Llaima last erupted during the austral summer of 1971–72. Since the lava and ash eruptions of 1957 and 1964, active fumaroles have been present in the summit crater.

Information Contacts: H. Moreno R., O. González-Ferrán, Univ. de Chile, Santiago.

12/79 (4:12) H. Moreno and C. Marangunic visited the volcano on 28 Oct. They found the flow fronts to be composed of fresh lava [debris] and pyroclastic fragments mixed with older material, forming a kind of "lava debris flow" [and pyroclastic flow] that moved rapidly over the ice. The continuous fumarolic activity that followed the eruption was quite intense on the 28th.

Pedro Riffo A. reports that 2 new explosions from the summit crater took place at 1226 on 24 Nov. A steam and ash column rose about 2 km. Minor emission of fine dark gray to brownish ash followed the main explosions.

Information Contacts: H. Moreno R., Univ. de Chile, Santiago; P. Riffo A., Univ. de Chile, Temuco.

7/84 (9:7) The following is a report from Hugo Moreno.

"An eruption from Llaima's central crater was observed between 0800 and 1800 on 20 April, with emission of dense columns of dark ash that fell around the crater. The activity could be seen from Temuco, 70 km W of the volcano. Tourists and residents of the area immediately around the volcano were evacuated by police and forest guards, and vehicle traffic was restricted.

Information Contact: H. Moreno R., Univ. de Chile, Santiago.

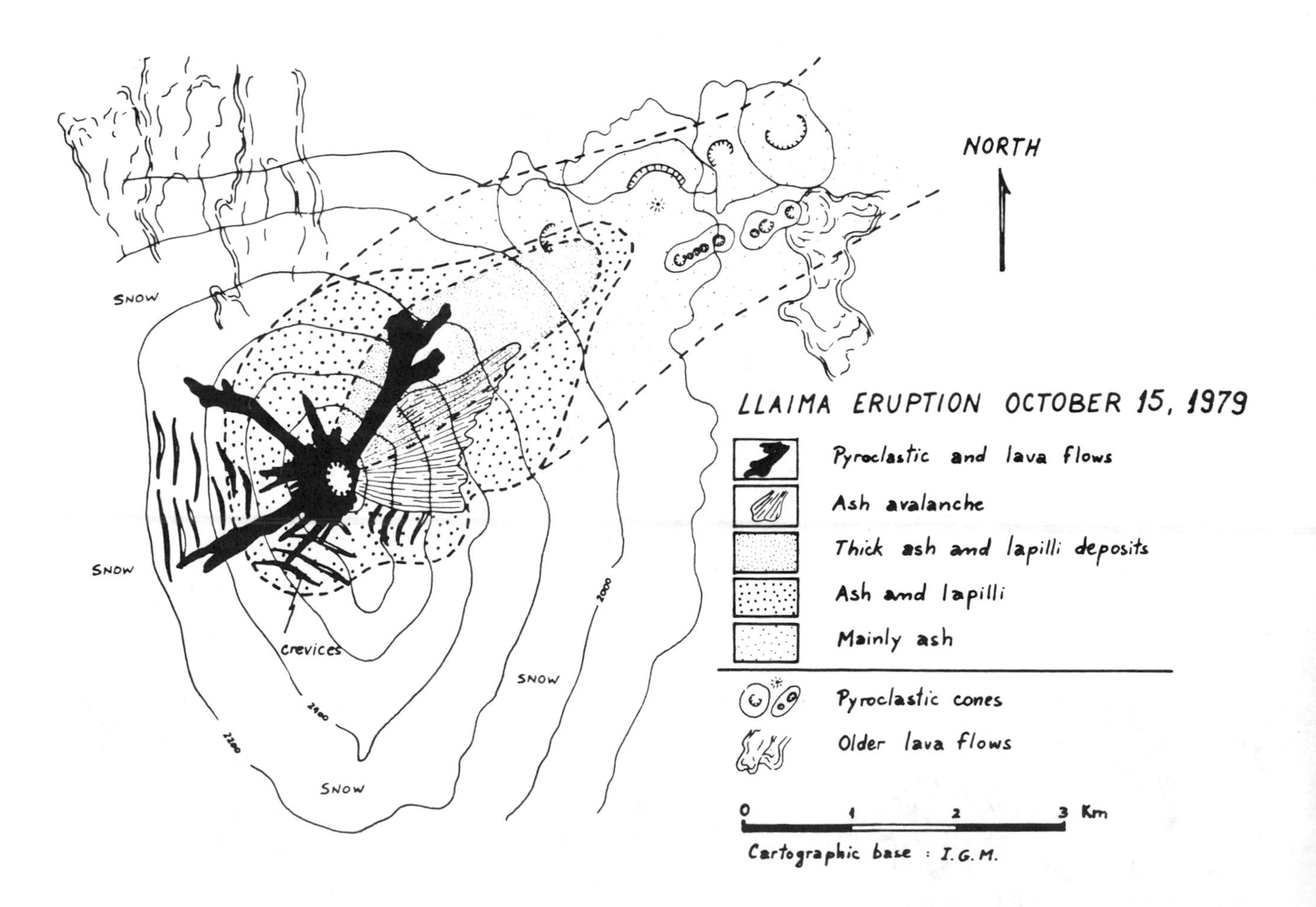

Fig. 15-12. Sketch map of Llaima by Hugo Moreno, showing the deposits from the 15 Oct 1979 eruption.

VILLARRICA

S Chile
39.42°S, 71.95°W, summit elev. 2840 m
Local time = GMT - 3 hrs winter (-4 hrs summer)

This strato-volcano has erupted 51 times since 1558, and mudflows from its steep, ice-covered slopes caused fatalities in 4 eruptions this century. Villarrica is 85 km S of Llaima and 120 km ENE of the port city of Valdivia.

10/79 (4:10) A notable increase in fumarolic activity in the central crater was observed beginning 25 Sept, after about 8 months of quiet. Villarrica last erupted in 1971, extruding large amounts of lava.

Information Contact: O. González-Ferrán, Univ. de Chile, Santiago.

6/80 (5:6) On 20 June, AFP reported that Villarrica ejected "dense smoke and abundant ashes." During the activity, continuous underground noise was audible nearby.

Information Contact: AFP.

9/80 (5:9) After a long period of fumarolic activity, a series of explosions that began on 19 Sept at 2200 ejected ash from the main crater. The next morning, a long, dark-colored pyroclastic flow could be seen on the NW flank. On 24 Sept at 0800 fine ash was ejected briefly, covering Villarrica's snowy slopes.

Information Contact: H. Moreno R., Univ. de Chile, Santiago.

10/83 (8:10) The following is a report from Hugo Moreno.

"Forest guards in the Villarrica National Park reported that the volcano entered into a remarkable eruptive stage on 14 Oct, after a long period of moderate activity. Continuous explosions with tephra emissions and some black pyroclastic flows over the ice-covered slopes have been observed. By night, a red glow over the summit indicates that a lava fountain is filling the crater. Since the big lava and pyroclastic eruptions of Oct – Dec 1971, active fumaroles have been present in the main crater."

Information Contact: Same as above.

Further Reference: Muñoz, M., 1984, Probabilidad de Erupción en el Volcán Villarrica en los Próximos Años; *Tralka*, v. 2, no. 3, p. 323-325.

9/84 (9:9) The following is from Hugo Moreno.

"A brief eruption from Villarrica's central crater was reported during the afternoon of 11 Aug. After 3 strong explosions and underground rumbling, a dark ash column was seen rising more than 200 m above the summit. Winds blew the ash over the snow-covered SE flank. Seismometers operated by the Universidad de la Frontera at Temuco (100 km NW of the volcano) recorded an increase in seismicity during the event."

Information Contact: Same as above.

11/84 (9:11) The first paragraph is from a report from Oscar González-Ferrán. The quoted material is from a report from Hugo Moreno, Leopoldo López Escobar, Pedro Riffo A., and Gustavo Fuentealba.

Villarrica began to erupt on 30 Oct. Activity was generally similar to that of the 1971 – 72 eruption. A very fluid basaltic lava column ascended the central crater without the emission of pyroclastics. Gases escaped freely, generating explosions in the crater that ejected lava spatter to 20-100 m in height, forming a spatter cone. Lava flowed NE from the base of this cone over the snow and ice that cover the upper flanks, excavating a channel and generating a large column of vapor. The Emergency Office took preventive measures to protect the population against possible avalanches. As of mid-Nov, the level of lava in the central crater continued to rise.

"On 30 Oct at 1745, authorities 16 km N of the summit (in Pucón) reported that explosions were occurring in the central crater and a small lava flow was pouring out from the NNE side of the crater through a small V-shaped opening left by the 1971 fissure eruption. The lava moved across the ice, quenched, and generated an avalanche mixed with ice and snow that reached 5 km from the summit (phase 1, fig. 15-13).

"Lava was emitted continuously from the central crater, advancing toward the NNE, where it melted the ice cover and formed a channel that was estimated to be 30-40 m deep, 50 m wide, and 1 km long during aerial observations 2-3 Nov. The central crater was occupied by a small flat spatter cone showing weak Strombolian activity. Lava from a small lake at the NNE foot of the spatter cone poured into the ice channel. Over the flat bottom of the channel, formed by solidified black lava, two narrow red lava flows were observed. Voluminous quantities of water vapor emerged from the area where the lava flow front was in contact with the channel's steep ice wall. Numerous fissures were present in the ice surrounding the channel, and on 3 Nov the ice cover on the SW flank also showed several deep fissures (phase 2, fig. 15-13).

"Weather conditions obscured the volcano 4-5 Nov, but seismometers operated by the Universidad de la Frontera at Temuco recorded intense shallow seismicity (0-1 km depth), [tremors] and B-type [earthquakes] (table 15-2).

"There were no signs of eruptive activity 6-9 Nov, al-

Time	B-Type Events	Tremors
2100 - 0100	4	
0100 - 0200	5	
0200 - 0300	9	
0300 - 0400	8	
0400 - 0500	3	4
0500 - 0600	8	2
0600 - 0635	9	

Table 15-2. Number of B-type events and [tremors] recorded by seismometers at Temuco, from 4 Nov, 2100, to 0635 the next day.

though tremors and underground rumbling were reported at Pucón. Strombolian activity at the small spatter cone in the central crater resumed 10-12 Nov, and lava flowed NNE into the ice channel. The speed of the flow was estimated at 10 m/sec.

"As of midday on 13 Nov, almost 2×10^6 m^3 of lava had flowed into the ice channel, most of which was concentrated at the lava front under the ice cover. At 1350

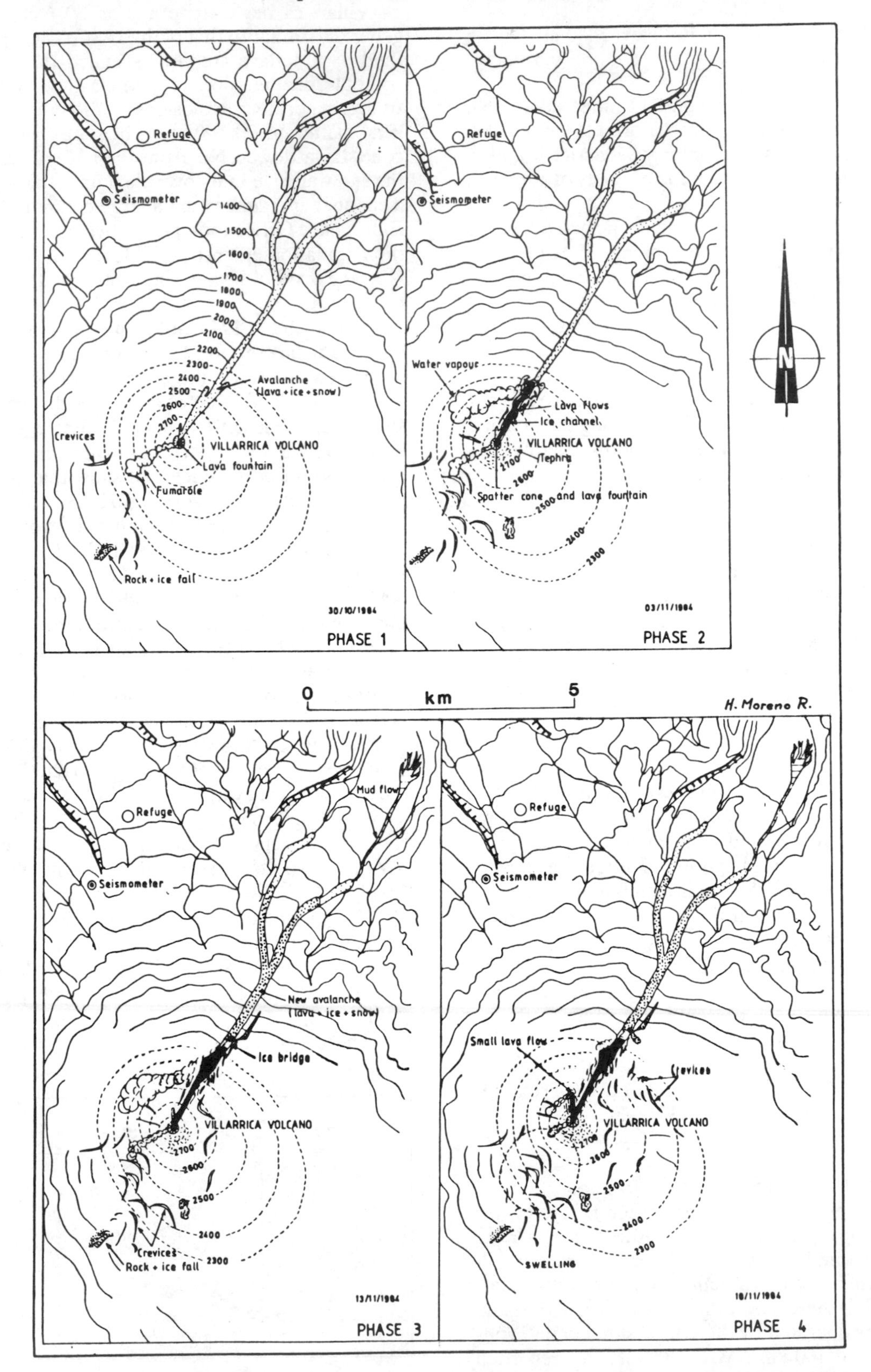

Fig. 15-13. Sketch map of the summit area and N flank of Villarrica on 30 Oct and 3, 13, and 18 Nov 1984. Courtesy of Hugo Moreno.

on the 13th, the lava front emerged onto the surface, generating a 3 km-long avalanche of lava blocks, ice, and snow, and leaving behind a 150 m-long ice bridge. Since only a very small mudflow moved downstream, it seems that most of the water generated by melting of the ice evaporated. New fissures were observed on the ice-covered SW, E, and NE flanks (phase 3, fig. 15-13).

"Eruptive activity decreased 14-17 Nov, with only weak fumarolic emissions seen at the spatter cone and water vapor emission at the ice bridge. [Earthquakes] were reported 13 km E and 10 km SW of the volcano (at Palguin and Chaillupen).

"On 18 Nov, a clear deformation of the SW slope was observed between 2200 and 2800 m above sea level. Weak lava production from the central crater opened a new small ice channel, about 200 m long and 50 m wide, toward the N (phase 4, fig. 15-13).

"Villarrica's Oct-Nov eruptive behavior is quite similar to the 1971 eruptive cycle that ended in a big lava effusion (29 Dec at 2345). Villarrica last erupted 11 Aug (9:9) and during the first week in Sept, with small explosions and tephra emissions."

Information Contacts: O. González-Ferrán, H. Moreno R., L. López E., Univ. de Chile, Santiago; P. Riffo A., G. Fuentealba C., Univ. de la Frontera, Temuco.

1/85 (10:1) Between 16 Nov and 1 Dec, activity remained constant. The lava column maintained its pressure and level in the central crater, and there were small explosions and gas emissions. Strombolian activity increased 1-6 Dec. Tephra was ejected to about 100 m height every 10 minutes. Lava from the lake in the central crater continued to pour out slowly through the initial NE flank channel.

On 6 Dec between 1200 and 1500 there was a violent increase in the rate of lava production. Lava flowed out through a new channel NE toward the Río Correntoso (which turns NW and flows about 20 km into Lake Villarrica), reaching the base of the volcano. The activity generated a small lahar that flattened a small wooden bridge and affected houses beside the river. The volume of water returned to its normal level after 24 hrs. An overflight of the crater revealed that the level of the lava lake was higher than before and the pyroclastic cone had grown higher than the central crater rim. Intense Strombolian activity continued 7-10 Dec. Pyroclastic material was ejected to 50-100 m height. Very liquid lava continued to pour out of the crater to the NE. Activity decreased gradually 11-19 Dec. Small explosions occurred 20 Dec but the pyroclastic cone and lava lake collapsed and effusive activity ended.

Explosive activity resumed 12 Jan between 1015 and 2300. A column of pyroclastics reached about 400 m height. A sequence of explosions ejected incandescent material to 100 m. On 18 Jan, the pyroclastic cone and lava lake in the central crater had completely collapsed.

Information Contact: O. González-Ferrán, Univ de Chile, Santiago.

11/85 (10:11) The following is a report from Gustavo Fuentealba Cifuentes.

"When the last eruptive cycle of Villarrica Volcano (30 Oct – 26 Feb, 9:11, 10:1) began to decay in Jan 1985, seismic activity also decreased. Between Jan and June 1985, the seismograph located on the N flank of the volcano recorded a monthly average of 15 volcanic earthquakes (Minakami's B-type). In Feb, only 5 seismic events were recorded with very little harmonic tremor. However, since June 1985 volcano-seismic activity has increased significantly. At the same time, notable harmonic tremor was observed. Figure 15-14 shows monthly seismic activity between Jan and Nov 1985. This situation was continuing as of 25 Nov, with a small gap in mid-late Nov. On 19 Nov at 0700, harmonic tremor stopped abruptly, and only apparently very shallow seismic activity was recorded. On 21 Nov at 1000, harmonic tremor activity resumed.

"According to personal observations and reports from Pucón, a town at the N foot of the volcano, an increase in fumarolic activity and lava fountaining with weak explosions and very small ash emissions have been registered since April. A red glow has been seen at night since late Sept."

Information Contacts: G. Fuentealba C., P. Riffo A., P. Acevedo, Univ. de La Frontera, Temuco; H. Moreno R., Univ. de Chile, Santiago.

Further References: Acevedo, P. and Fuentealba, G., 1987, Antecedentes de la Actividad Microsísmica del Volcán Villarrica Relacionada con la Erupción de Octubre de 1984; *Boletín de Vulcanología* (Universidad Nacional, Heredia, Costa Rica), no. 18, p. 13-17.

Moreno, H., Fuentealba, G., and Riffo, P. (in press), The 1984–1985 Eruption of Villarrica, Southern Andes of Chile (39°21'S): Basaltic Lava Flows Furrowed the Ice Cap.

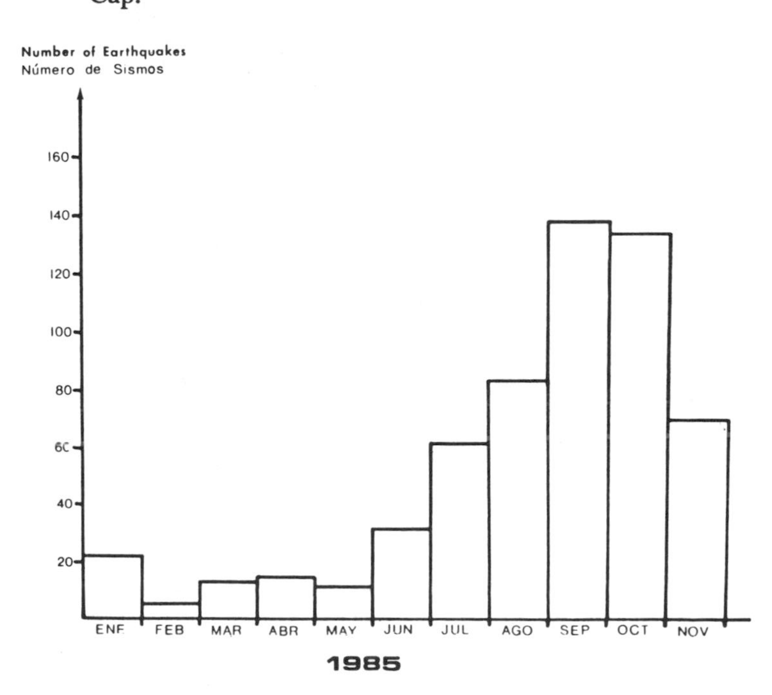

Fig. 15-14. Number of earthquakes/month at Villarrica, Jan – Nov 1985. Courtesy of Gustavo Fuentealba C.

CARRAN VOLCANIC GROUP

S Chile
40.35°S, 72.05°W, summit elev. 1114 m
Local time = GMT - 4 hrs

The Carrán and Los Venados volcano groups are contiguous scoria cones and maars ESE of Lake Ranco and 100 km S of Villarica. Río Nilahue runs through the area (fig. 15-15), and the name Nilahue was used for a maar formed in 1955. Nilahue was also used for the group in *CAVW*, *Volcanoes of the World*, and elsewhere. However, "Carrán Volcano Group" is the name now preferred by Chilean volcanologists, and it was used in the original SEAN reports below.

4/79 (4:4) An eruption of pyroclastics from Mirador, one of the craters in the Carrán volcanic group (fig. 15-15), began at 0200 on [14] April, preceded by local seismicity. The column of gas and ash reached altitudes of 3-4 km, and was accompanied by gas explosions in the mouth of the crater. After about 40 hrs, lava began to flow SSW, traveling about 500 m by the 17th. Voluminous ash emission persisted, with explosions occurring every 5 minutes, and local seismicity was continuous. The activity continued to increase through 18 April. Ash covered agricultural land near the volcano. No one has been killed, but authorities have evacuated 125 persons from the area.

The most recent previous activity in the area was an explosive eruption in 1955, in which 2 persons were killed. At least 2 of the volcanic features in the Carrán group are maars.

Information Contact: O. González-Ferrán, Univ. de Chile, Santiago.

5/79 (4:5) Local authorities report that the eruption of Mirador crater resumed on 12 May [see also 4:7] and was continuing 24 hrs later. Incandescent material rose 30 m above the vent and a series of low-intensity tremors occurred. Subterranean rumbling, followed by several hours of earthquakes, was reported from Río Buenos, about 40 km away. Carrán's vigorous mid-April eruption had stopped after a few days.

Information Contact: Latin Radio Network, Buenos Aires, Argentina.

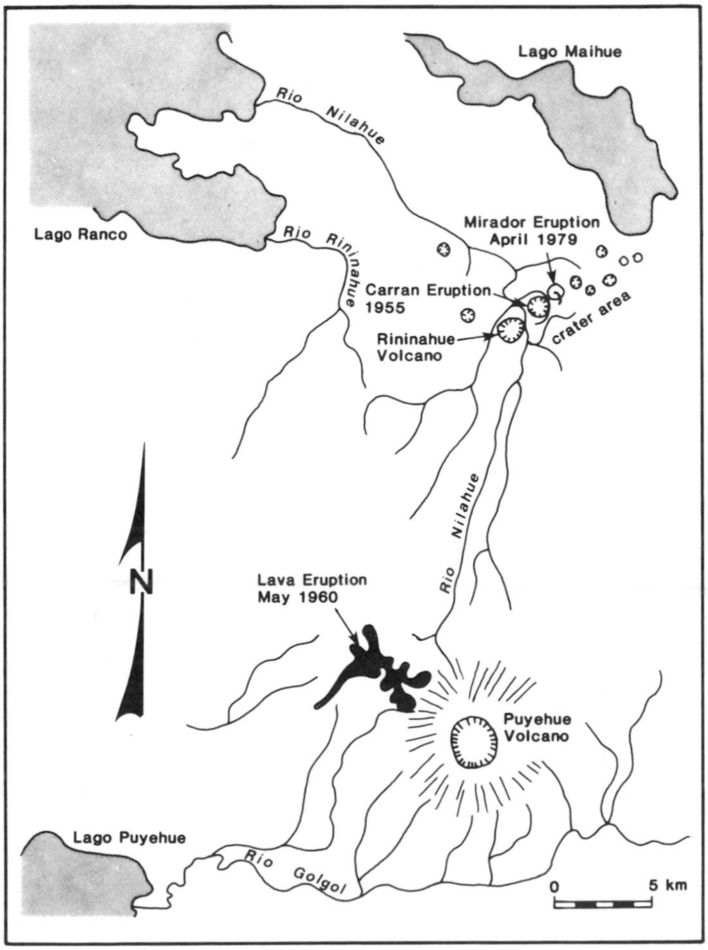

Fig. 15-15. Sketch map of Mirador and vicinity, from López and Moreno, 1981, showing eruption sites and flows from the 1979 eruption. Six-digit numbers mark sample localities.

7/79 (4:7) The eruption of Mirador began violently on [14] April, but activity declined quickly to more moderate levels. Explosions and accompanying seismicity then gradually decreased in frequency and intensity until the eruption ended on 25 May. Ash deposits were 2-3 cm thick at the base of Mirador and ash fell as much as 20 km away. A small amount of lava (andesitic basalt in hand specimen) flowed NE and SE.

Information Contact: O. González-Ferrán, Univ. de Chile, Santiago.

Further References: López, L. and Moreno, H., 1981, Erupción de 1979 del Volcán Mirador, Andes del Sur, 40°21'S: Características Geoquímicas de las Lavas y Xenolitos Graníticos; *Revista Geológica de Chile*, no. 13-14, p. 17-33.

Moreno, H., 1980, La Erupción del Volcán Mirador en Abril-Mayo de 1979, Lago Ranco-Rininahue, Andes del Sur; *Comunicaciones, Universidad de Chile*, no. 28, p. 1-23.

12/79 (4:12) Fumarolic activity has persisted from Mirador crater since its ash and lava eruption of April-May. A dense, yellowish-white steam cloud emerged from Mirador's summit during a visit by L. López, A. Lahsen, and H. Moreno on 1 Nov. The surfaces of 2 lava flows extruded on 12 May were covered by a yellowish salt deposit composed mainly of iron chlorides.

Information Contact: H. Moreno R., Univ. de Chile, Santiago.

WEST INDIES

SOUFRIERE DE GUADELOUPE

Guadeloupe
16.05°N, 61.67°W, summit elev. 1467 m
Local time = GMT - 4 hrs

Guadeloupe is the largest island in the Lesser Antilles. This strato-volcano, making up the island's SW part, has erupted 9 times since 1660.

4/76 (1:7) The recent seismic crisis began in Nov 1975 when 25 tremors were recorded. Thirty tremors were registered in Dec, 36 in Jan 1976, 93 in Feb, and 680 in March (of which $^1/_3$ were felt by inhabitants very near the volcano). The crisis was declining at the end of April. The foci of the tremors were under the NW part of Soufrière, and depths of focus varied from 1 to 5 km.

No new surface manifestations accompanied or followed these tremors (up to 9 April). There were no modifications of temperatures nor of the composition of gases emitted by the fumaroles. An overflight of the volcano did not indicate any sign of abnormal activity on the surface. The volcano had previous historic eruptions in 1956, 1903, 1899, 1896, 1879, 1837–38, 1809–12, 1797–99, 1696, 1680, and 1645.

Local earthquake swarms are rather frequent at volcanoes of this type. They have taken place approximately once per year on Guadeloupe since 1962, but none have reached the frequency of March 1976. The same phenomenon has previously occurred on neighboring islands (Montserrat 1966, St. Kitts in 1950 and 1961, and St. Vincent 1947) without any eruption.

J. Dorel believes that, for the present time, there is no important risk of an eruption of this volcano, but that it is impossible to make a long-term (several months or years) prognostication on the evolution of such volcanoes.

Information Contact: J. Dorel, IPG, Paris.

6/76 (1:9) On 8 July at approximately 0900, a phreatic eruption began at Soufrière (fig. 16-1). A new crack, 300 m long [but see 1:10] and up to 10 m wide, opened down the SE side of the summit dome. Steam and gas with ash and lava fragments were blowing through the crack on that day, but on 9 July there were only steam and gas with no solids. The eruption appeared to be purely phreatic, with no molten material reported.

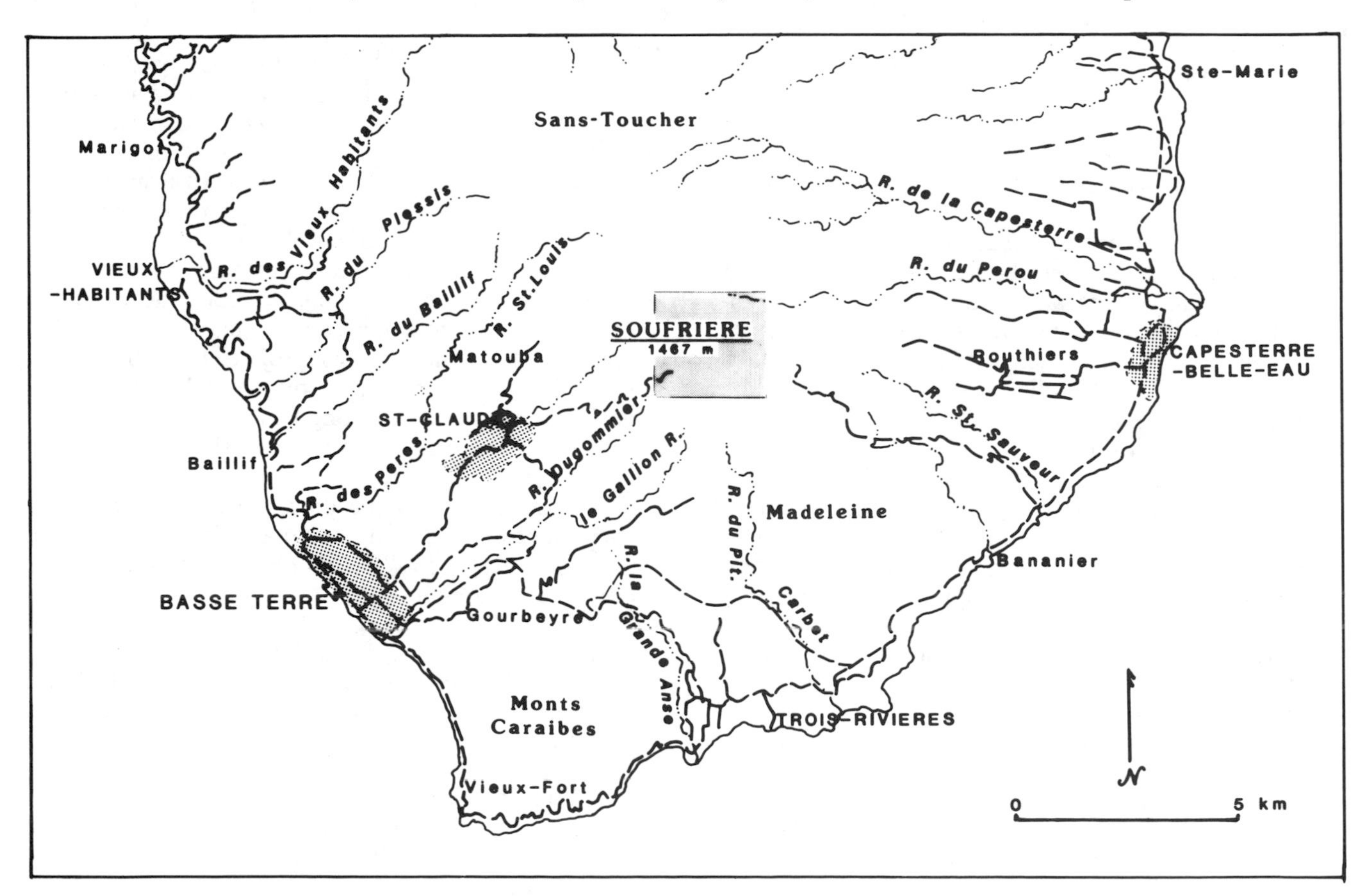

Fig. 16-1. Map of the S part of Basse-Terre (one of the islands that comprise Guadeloupe). Soufrière's summit area, within the shaded box, is shown in fig. 16-2. After Guadeloupe Tourist Map, 1:100,000, Institut Géographique National, 1982.

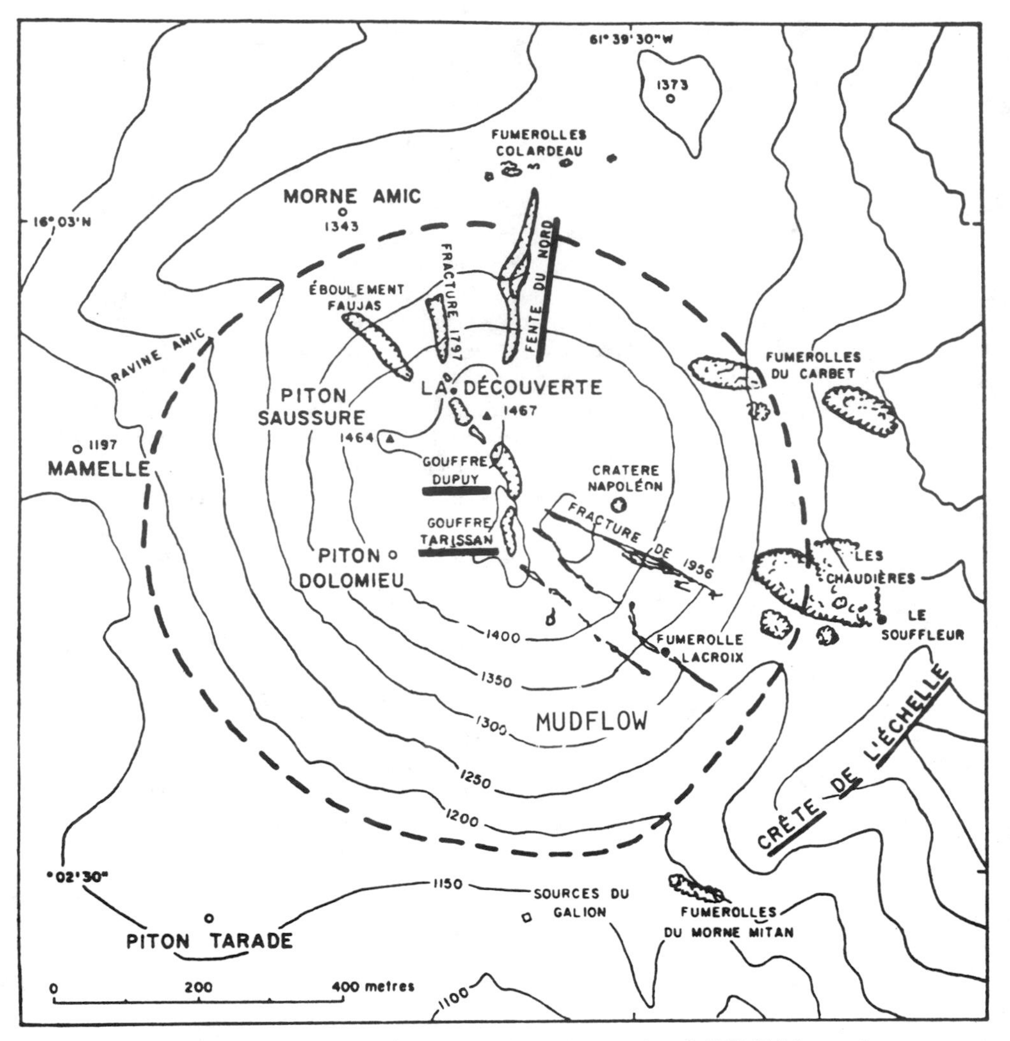

Fig. 16-2. Map of the summit of Soufrière volcano, Guadeloupe, after *CAVW*. Heights are in meters.

of the Observatory of St. Claude, Guadeloupe. The earthquakes numbered about 607 in March, 747 in April, 611 in May, and 668 in June. Most foci were distributed between 2 and 6 km under the volcano, along a NW-SE axis. Many earthquakes were felt by people living near Soufrière. The maximum magnitude was 3.8.

On 8 July at 0856, the volcano threw out a column of fine sand and dust, plunging the surrounding regions into darkness for 40 minutes. This fine volcanic material settled over the W side of the mountain and extended to the towns of Basse-Terre and Vieux-Habitants, on the W coast of Guadeloupe 8 km away. The thickness of this grey tephra was very irregular because of the influence of winds: about 8 cm near the summit to 1 or 2 cm near Basse-Terre. This eruption was followed by steam and gases rising from a new fracture 400 m long on the SE flank of the cone. A little avalanche (800 m long) of mud and rocks from this fracture accompanied the eruption.

Since 8 July the pressure of the gases has decreased and the temperature has remained around 96°C, but the volcanic seismicity remains important (300 earthquakes registered 9-20 July). A group of scientists from IPG, Commissariat à l'Energie Atomique, and the Université de Paris Sud, is investigating this eruption. They concluded that the eruption is phreatic (like the eruptions in 1956, [1837–38], and 1797). The chemical composition of the volcanic sand is the same as that of 1956 and corresponds to the composition of the old material of the cone (SiO_2 = 57.8%). No new magma was emitted during this eruption.

Addendum: Summary of Eruptive and Seismic Activity 25 July–1 Sept. Earthquake hypocenters are given in kilometers below sea level unless otherwise noted. [Corrections and additions to this addendum supplied by J. Gautheyrou and LANL, originally printed in 1:11, have been incorporated here.]

25 July: Significant ash eruption.

9 Aug: Ash eruption, causing a small river to stop flowing for 24 hours. When flow resumed, the water was very muddy.

12 Aug: Powerful new ash eruption containing juvenile material [but see 1:12]. Evacuation of 72,000

On the 9th the gas and steam were venting at a "vast rate," and the column rose about 300 m above the crack. Steam and gas venting was continuing as of this report, but the activity appeared to be slowing down. It was not possible to ascertain the height of the steam column because of clouds. It is believed that local earthquakes occurred just before the eruptive activity began; longer term premonitory seismicity was described in 1:8.

A phreatic eruption occurred in 1956 from a fracture across the top of the dome (fig. 16-2), but the current eruption is bigger in that the crack is longer and the volume of products is greater. There were comparable phreatic eruptions in [1837–38] and 1797, each lasting a few days. The present activity appears to be following the pattern of these previous eruptions. Radiocarbon evidence indicates that there was a large magmatic eruption about 1500 A.D.

In general characteristics Soufrière is comparable to the Martinique and St. Vincent volcanoes, which are capable of extremely violent activity.

Information Contacts: J. Tomblin, Univ. of the West Indies; Lab. de Physique du Globe.

7/76 (**1:10**) Since March 1976 a great number of volcanic earthquakes have been registered by the seismic array persons began (fig. 16-3); evacuation was completed 15 Aug.

16 Aug: Very fine ash erupted from the fissure that opened 8 July. At 1730 a magnitude 4.6 earthquake with a hypocenter at about 2.5 km occurred, followed by 20 minutes of very frequent tremors that saturated a detector 8 km from the volcano for about 10 minutes.

17 Aug: Approximately 900 earthquakes were recorded in the 24 hrs after the beginning of the 16 Aug seismic crisis. Depths of focus ranged from 1 to 3 km. An earthquake registering about 4.0 occurred at 2315. An expedition to the volcano noted weaker gas pressure and a new 5-cm ashfall at the foot of the dome. Mudflows were forming from the new ash because of heavy rainfall. At 1700 the principal vent was puffing and a plume of fine particles was being emitted from the N fracture. The 1500 residents of Vieux-Fort were evacuated. Microprobe analysis of earlier 1976 ejecta indicated andesitic magma at approximately 900°C.

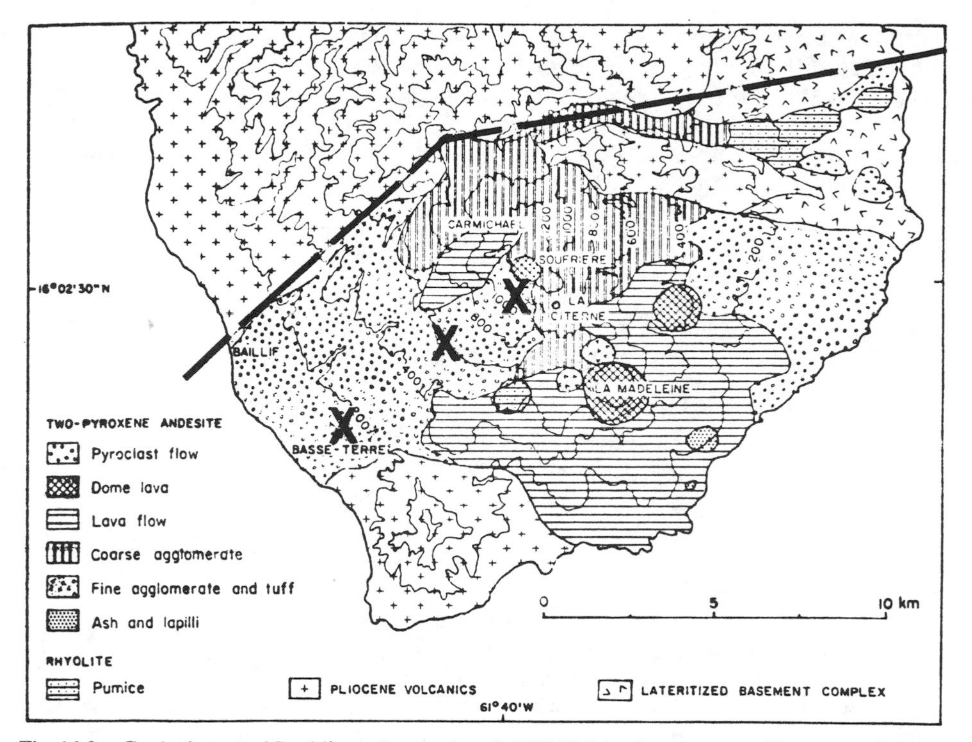

Fig. 16-3. Geologic map of Soufrière volcano, after *CAVW*. Heights are in meters. The area below the dashed line is the section of Basse-Terre that was evacuated due to the possibility of a major eruption. The crosses indicate the locations of the USGS dry tilt stations at Savane à Mulets (closest to summit), Parnasse, and Bonne Terre (farthest from summit).

18 Aug: The volcano was quiet enough to allow the temporary return of farmers. Earthquake foci were about 2.5 km deep.

19 Aug: About 1 mm of fine dust was found on the dome. 95% of the erupted material was less than 0.2 mm in diameter; the coarsest was 2 mm mushroom-colored glass. Dust made breathing difficult in a sector from Basse Terre to Bouillante. Earthquake hypocenters were at about 2.5 km depth.

20 Aug: There was a steady, non-explosive emission of steam rising to less than 300 m.

21 Aug: A major increase in the SO_2 content of gases sampled was followed in about 8 hrs by 11 minutes of harmonic tremor, beginning at 2326. All instruments were saturated for the first 3 minutes. An ash eruption, probably coinciding with the harmonic tremor, left a few centimeters of ash at the base of the dome, 3 mm $^1/_3$ of the way up the mountain, and about 1 mm on the S coast. An expedition the following day discovered that the SE fracture had been widened and that blocks up to 1 m across had been blown 400 m to the old rim (L'Echelle). Hypocenters of recorded earthquakes were at or above sea level.

22 Aug: More than 100 of the 156 recorded earthquakes occurred by 0800. Activity then declined. Emission of steam and other gases occurred from vents south of the summit dome (Gouffre Tarissan, Pont Chinois and Gouffre Dupuy). Rain caused a mudflow in the Galion River.

23 Aug: An increase in radon emission during the afternoon predicted an increase in activity; minor dust emissions were reported.

24 Aug: A period of major seismic activity began at 0256 with a magnitude 4.1 event centered 2 km under the W side of Soufrière. More than 1000 earthquakes were recorded by 1530 (fig. 16-4), 2 of which were felt in Pointe-a-Pitre, 20 km away. Intervals between events and earthquake amplitudes were both decreasing by afternoon. Radon percentage in the gas increased with increased seismicity. SO_2 had returned to normal. Minor dust emission continued. 50 cm of dust, largely composed of fresh glass, was present at the S side of the summit dome (Gouffre Tarissan).

25 Aug: The volume of ash erupted increased significantly and the eruption clouds were much darker after 1530. Ash emission was from the S and E vents (Gouffre Tarissan and Dent de l'Est). Study of the tephra erupted since the 24 Aug seismic crisis indicates both an increase in the percentage of fresh glass and an increase (to 2 mm) in the size of the glass shards. The percentage of chemical weathering products in the ejecta also increased. Between 2000 on 25 Aug and 0600 on 26 Aug, 120 earthquakes were recorded, some having hypocenters as shallow as 300 m above sea level. Gas analyses showed lower percentages of SO_2, SO_3, and H_2S than for previous days. Radon again increased during seismic crises.

26 Aug: A 100-200 m column of ash was erupted continuously from Gouffre Tarissan (the main vent at the top of the dome) and from the upper end of the SE fracture. The remainder of the SE fracture emitted steam that was almost entirely of meteoric origin. The 26 Aug ash was dry, and therefore paler than the wet ash of 25

Aug. The ash was blown W; 2 cm were deposited on the W slopes of the dome and 1-2 mm on the W coast towns. Approximately 50% of the ash was juvenile glass [but see 1:12]. Ash emission ended in the morning. The magmatic component of the emitted gases was low.

27 Aug: Considerable steam was emitted, along with some dust. Fresh glass was erupted from Pont Chinois and the N fracture. SO_2 and H_2S were present in small amounts in the volcanic gases.

28 Aug: An expedition to the dome found 1 m of very fine dust around Gouffre Tarissan and a few centimeters of dust along the flanks of the dome. One block a few centimeters across was found near the top of the dome.

29 Aug: A 10 μrad inflation of the volcano had occurred since 28 Aug. Eruptive activity was limited to minor ash emission. Earthquake hypocenters 27-29 Aug were all shallower than 3 km; all but 4 were shallower than 2 km, and one was above sea level. No ash emission occurred 28-29 Aug. The gases analyzed were almost entirely of meteoric origin.

30 Aug: A small eruption occurred at 1015 from the main vent, accompanied by 22 minutes of harmonic tremor, the first 13 minutes of which saturated the instruments. A party of 7 volcanologists was showered by blocks up to 1 m across; 4 were injured but none seriously. 209 earthquakes were recorded. The ash, analyzed by the LANL, is 95% alteration products (fine-grained clayey material and pyrite) and 5% rock and crystal fragments (andesite, plagioclase, and pyroxene). Volatile content was high, and there was no fresh glass.

31 Aug: Large volumes of steam with some admixed ash erupted from all of the previously active vents. The steam cloud from the main vent reached 2000 m altitude in the evening due to light winds. 6-8 cm of juvenile ash (80% glass), which flocculated into 5 mm flakes as it fell, was deposited near the main vent, as were blocks up to 20 cm across that were torn from the vent and blown up to 400 m away. A new earthquake swarm began at 1035 and lasted for 6-8 hrs, accounting for most of the 249 seismic events recorded 31 Aug. A total of 5989 earthquakes were recorded in Aug, 43 of which were felt.

1 Sept: Strong steam emission continued, producing cauliflower clouds that reached a height of 300 m before being blown W. A grayish-brown haze of very fine ash was falling on the top 1/4 of the volcano.

Information Contacts: J. Dorel, IPG, Paris; J. Gautheyrou, ORSTOM; J. Tomblin, Univ. of the West Indies; M. Feuillard, Lab. de Physique du Globe; LANL.

8/76 (1:11) *Summary of Eruptive and Seismic Activity 2-15 Sept.* 2 Sept: Heavy steam emission from all vents continued after the 30 Aug eruption. Ash emission was slight. Several loud, booming reports were heard in rapid succession at 1600.

3-7 Sept: Surface activity diminished. No significant ash eruption occurred. Tilt measurements showed a very slight movement downwards to the NW through 5 Sept; a 2-3 μrad deflation was noted between 6 and 7 Sept. Mean earthquake focal depth increased to about 3 km.

7-9 Sept: Eruptive activity was limited to mild steaming. Gas analysis yielded low values for acid gases (SO_2, F_2, and fluorides). The magnetic anomaly increased slightly. Mean focal depths were 2.5 km (based on 4 events) on 8 Sept, increasing to 3.4 km (based on 3 events) on 9 Sept.

10-13 Sept: Mild steaming continued; only small amounts of gases other than H_2O were present. At 0418 on 13 Sept a M 3 earthquake occurred, centered under St. Claude (SW of the summit).

14 Sept: A summit eruption, lasting 5-9 minutes, began at 1922, 15.4 days after the last significant eruption. Juergen Kienle (University of Alaska) had predicted an eruption between 14 and 18 Sept, a period centered around the lunar last quarter on 16 Sept. Immediately prior to the eruption, tilt at the summit went off scale. Loud noise was heard, but cloud cover at the

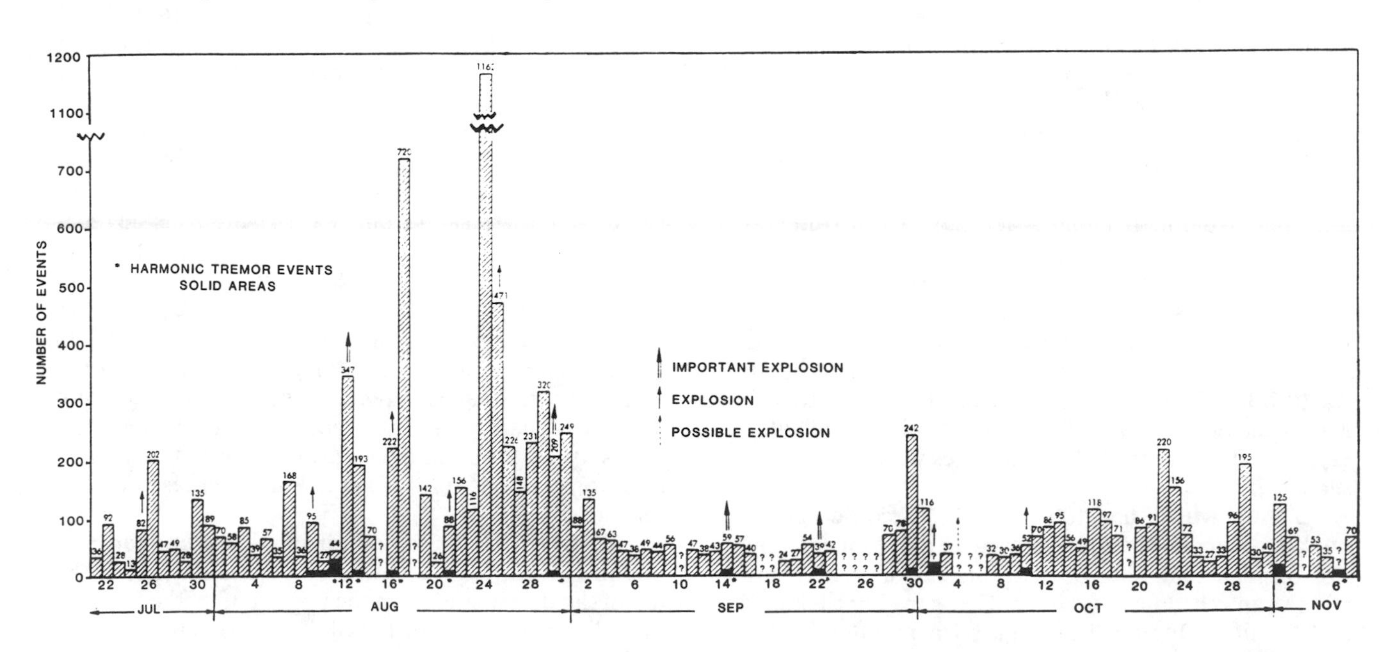

Fig. 16-4. Number of recorded seismic events per day, 21 July – 7 Nov 1976.

summit prevented visual observation. Harmonic tremor accompanied the eruption, saturating seismographs for the first 4 minutes. Inspection of the summit area revealed airfall ash as far away as St. Claude (4.5 km SW of the summit) and Vieux Habitants (11 km W of the summit). Tom McGetchin reported that a small directed blast, originating from a fissure just S of the main vent, defoliated vegetation up to 500 m away.

15 Sept: Eruptive activity was confined to steaming and minor ash emission.

Local seismic energy release from mid-July to mid-Sept was calculated at the University of the West Indies (table 16-1).

Dry tilt stations were installed 28 Aug by the USGS (fig. 16-3) and are being relevelled daily by George Jezouin, government engineer (table 16-2).

Raw data from the LANL borehole tiltmeter array appear to be generally compatible with the dry tilt figures. 5-10 μrad tilt events occurring over about 10-minute time spans have been recorded simultaneously at more than one tilt station.

Seismic velocity profiles (18 shot points on land, 2-3 in the ocean) were being run 16 and 17 Sept by a French team. Velocities previously used to calculate focal depths were estimates which may have been too high. The French scientific team is responsible for gas sampling, magnetic readings, and seismic recording, as well as overseeing all scientific work on the volcano. The LANL is operating a camera station at Fort St. Charles and plans to establish another at Pointe-a-Pitre. Juergen Kienle is running a daily gravity survey. Michael Sheridan (Arizona State Univ.) is studying pyroclastic flow deposits from past eruptions.

Information Contacts: J. Tomblin, W. Aspinall, Univ. of the West Indies; T. McGetchin, LANL.

1976	Total Local Earthquake Energy (Ergs)
8-17 July	3.0×10^{17} (Initial steam explosion 8 July)
18-27 July	6.3×10^{17}
28 July-6 Aug	1.6×10^{18}
7-16 Aug	2.27×10^{19} (Evacuation began 12 Aug)
27 Aug-5 Sept	1.72×10^{19}
6-13 Sept	4.63×10^{16} (8-day period)

Table 16-1. Total local earthquake energy for each of six 10-day periods and one 8-day period, 8 July-13 Sept 1976. Calculated using the formula $\log E = 9.4 + 2.14\,M + 0.054\,M^2$ by the Seismic Research Unit, University of the West Indies.

9/76 (1:12) 16 Sept: Heavy steaming occurred from the S and SE sides of the summit dome. Differential magnetometer readings increased at the summit and decreased at Matouba with respect to a base station at Basse Terre.

17-19 Sept: Large inflation was shown by LANL tilt stations to continue through the 18th. Deflation was observed by the same stations on the 19th. Sulfur was noted in the gases on the 18th. Eruptive activity was minor.

20 Sept: A seismic event occurred at 0100. LANL reports that the tilt, gravity, differential magnetometry, and seismicity each showed distinctive patterns prior to this event, similar to the patterns noted before the 14 Sept activity. These data have yet to be interpreted in terms of a magma chamber model. LANL noted approximately 10 μrad inflation at the summit, and about 5 μrad deflation at lower tilt stations during the the 24 hrs after the event.

21 Sept: Eruptive activity was minor.

22 Sept: An ash eruption, producing a 3500-m cloud, occurred at 0645, accompanied by 18 minutes of harmonic tremor, 6 minutes of which saturated seismographs. About 1 cm of ash fell downwind at Baillif-Vieux-Habitants. Ash was sampled from a 1-mm deposit near the S edge of the ashfall area.

23 Sept: Strong steaming, possibly with some ash emission, was occasionally visible through heavy clouds.

24-27 Sept: The activity stabilized to moderate steaming and minor ash emission. LANL recorded a large tilt episode, beginning at 2300 on the 27th and lasting several hours.

28 Sept: A 300-m plume was visible for much of the day. At 1010 an earthquake occurred that was felt in Pointe-a-Pitre. Seventy minutes later, LANL recorded the beginning of deflation. The tilt and seismic data are thought by LANL to indicate cavity collapse.

29 Sept: LANL reported the beginning of inflation at 1445, then deflation at 1745.

30 Sept: A 20-minute period of tremors began at 0002. Two large and 4 small events were recorded. Cavity collapse was again suggested by LANL as an explanation for the seismic episode.

In other tilt experiments, R.S. Fiske and W.T. Kinoshita, supported by the USGS, SI, and OFDA, established 4 dry tilt arrays on the SW flank of Soufrière in late Aug. These 4 arrays are at about 1, 3, 4, and 5 km from the summit. More than 25 reoccupations of these arrays (29 Aug-30 Sept) by French technicians have indicated daily station "noise" sometimes exceeding 5 μrad, but the suggestion of true volcano inflation has been recorded only at the station 5 km from the sum-

	Bonne Terre	Parnasse	Savane à Mulets
Azimuth to Summit	N47°E	N66°E	N10°E
Distance to Summit	5 km	3 km	1 km
Aug 29	10 μrad inflation; no individual station data		
31	-4.1	-7.0	+32.0
Sept 2	0.0	+5.0	-9.0
3	+3.0	-4.0	-6.0
4	-3.0	-6.0	-4.5
6	+5.0	-4.0	+2.5
7	0.0	-7.0	–
8	-3.5	-5.0	0.0
9	+2.3	+3.8	–
10	Deflation; no figures		
12	Inflation; no figures		
13	Inflation; no figures		
14	+5.0	0.0	0.0

Table 16-2. Data from 3 USGS dry tilt stations on the SW flank of Soufrière, Aug-Sept 1976. All values are expressed in μrad vectored towards or away from the summit. Positive values represent inflation.

Components %	30 Sept	15 Oct
Altered and unaltered plagioclase, ortho- and clinopyroxene	18	16
Altered and unaltered lithic fragments (mostly porphyritic, glassy dacite)	18	20
Altered colorless glass	4	1
Matrix; <2 μm altered glass and mineral phases, clay, and celadonite (?)	55	60

Table 16-3. Modal analyses by LANL of unsieved 30 Sept and 15 Oct 1976 ejecta from Soufrière.

mit. The bulk of the dry tilt data gathered to date therefore do not indicate that the volcano is being deformed as a result of the inflation of a magma reservoir. This tentative conclusion is supported by measurements of P.A. Blum, who installed 2 pendulum-type tiltmeters at the now-abandoned volcano observatory about 3 km from the summit. These tiltmeters, capable of detecting changes in ground tilt of a few parts in 10^8, have indicated tilt changes of only fractions of a μrad from 25 Sept, when measurements were begun, to 29 Sept. Neither of these observations, however, has been made over times that would record the short periods of tilt reported by LANL during the hour preceding small eruptions.

[Magnetic variations measured on the volcano in Sept were positive at all but one station, and intense, reaching 6 nT at the end of the month (Zlotnicki, 1986).]

Earlier reports of fresh volcanic glass in the ejected ash from La Soufrière are now regarded to have been in error. There is no confirmed evidence that juvenile magma has reached the surface.

Information Contacts: W. Aspinall, J. Tomblin, Univ. of the West Indies; LANL; J. Gautheyrou, ORSTOM; R. Fiske, SI.

10/76 (1:13) 2 Oct: At about 1630, a steam and ash eruption began, producing a cloud that reached a maximum of 1000-1500 m height. The cloud deposited a fine veneer of ash on vegetation as far as 7 km downwind. Maximum ejecta diameter on the summit dome was 1-2 cm. Mudflows accompanied the eruption. Harmonic tremor events of 10 and 13 minutes respectively saturated seismographs at 1615 and 1654.

3 Oct: Normal activity resumed (200-300-m sulfurous steam plumes).

1976		Wt.% S
Aug	12	1.7
	22	2.3 (1.1, 3.4; two locations)
	26	2.7
	27	3.6 (3.0, 4.1; two locations)
	30	4.3

Table 16-4. Sulfur content of several ash samples from Aug 1976 eruptions of Soufrière. Analyzed by D. Curtis, LANL, using the thermal neutron-prompt gamma activation method. [Originally in 1:12.]

4 Oct: Steam clouds, fluctuating in volume and ash content, were emitted from the dome fracture system. Activity was greatest near the S flank. Ash slurries, 4-5 m wide, flowed downslope from the dome fractures. Two ash samples and one mudflow sample were taken by LANL the next day.

5 Oct: Summit fumarolic activity was visible in the early morning and late afternoon. Little or no ash was present in the steam plumes.

6-7 Oct: Clouds and heavy rain precluded observations of the summit area.

8 Oct: Moderate steam plumes, sometimes containing small amounts of ash, were emitted during the morning. A significant increase in activity began at 1237, about 2.5 hours after unusual heavy rains on the summit area. The steam plumes reached 50 m height.

9 Oct: Continued moderate steam emission, with varying but minor amounts of ash in the plumes.

10 Oct: At 1111, a 13-minute period of harmonic tremor began, coincident with a pronounced increase in ash emission from the summit, which lasted for 20-30 minutes. Four mm of ash were deposited at the summit parking lot and at Marigot (5.5 km W of the summit), considerably more than was deposited from the larger ash clouds of 2 and 4 Oct.

11 Oct: A pronounced increase in seismic activity began at about 0500 and continued through about 1100. No simultaneous increase in eruptive activity was noted.

12 Oct: Eruptive activity was limited to weak steaming. Plumes rose only 10-12 m above the summit.

13-14 Oct: Steam emission returned to the "moderate" level of 8-9 Oct.

15 Oct: LANL scientists returned to Los Alamos leaving borehole tiltmeters in concrete building foundations for continued use by the French team.

21 Oct: 65 earthquakes were recorded.

22 Oct: Earthquake activity increased sharply. Between 1330 and 1610, 182 earthquakes larger than magnitude 2.2 were recorded. Eruptive activity increased during the afternoon and evening.

23 Oct: A harmonic tremor event occurred at about 0530, saturating instruments for 1 minute. Steam and "smoke" emission had declined since the previous evening.

Continuing dry tilt measurements by French technicians have indicated no coherent pattern of deformation 1-16 Oct. It is probable that the small tilts observed can be attributed to the near-surface effects of variation in rainfall, sunlight, and evaporation at each of the 4 stations.

Analysis of unsieved material by LANL from the 30 Sept and 15 Oct eruptions detected mainly hydrothermally altered lithic and mineral fragments. No fresh glass was found (table 16-3).

Sulfur contents of tephra samples increased over the period mid-Aug to early Sept, according to LANL analyses (table 16-4).

Information Contacts: LANL; J. Gautheyrou, ORSTOM; AFP.

11/76 (1:14) 30 Oct: A 9-minute phreatic explosion occurred at about 1000. At 2238 a period of volcanic tremor began. 1 Nov: Phreatic explosion at about 1100. 6 Nov: Phreatic explosion at about 2200. The periods between phreatic explosions were characterized by mild steaming from fissures on the summit dome.

On 15-18 Nov, CNRS convened a committee of scientists in Paris to review the Soufrière situation and to assess the hazard posed by the volcano. This committee was chaired by Frank Press of the U.S., and included the following members: Shigeo Aramaki, Japan; Franco Barberi, Italy; Jean Coulomb, France; Richard S. Fiske, U.S.; Paolo Gasparini, Italy; Claude Guillemin, France; and Gudmundur Sigvaldason, Iceland.

The committee concluded that as of mid-Nov there was low probability of a dangerous eruption at La Soufrière, but that the situation could change. In order to detect these changes, the committee recommended an augmented program of study that included increased geological and petrological observation of the volcano and its ejecta, expansion of the deformation studies (precise levelling and tiltmetry), monitoring of volcanic gases, and expansion and improvement of seismic monitoring procedures. The committee report also outlined a possible model explaining the volcano's behavior. The model calls for the tectonic fracturing of hot rock or partly solidified magma beneath the volcano. This fracturing permitted the rapid influx of groundwater, of both meteoric and marine origin, which was then heated and driven off in copious outpourings of steam from the summit of the volcano. At times, steam was expelled with such violence that particles of pre-existing, altered rock were torn from the interior of the volcano and ejected to form airborne clouds of ash.

Several hours after the release of the committee report on 18 Nov, the French Government announced that the evacuation on Guadeloupe would end, and that the last of the 72,000 evacuees could return to their homes on 1 Dec.

Information Contacts: M. Feuillard, Lab. de Physique du Globe; J. Dorel, IPG, Paris; R. Fiske, SI.

12/76 (1:15) Activity has declined to mild steaming with an occasional weak phreatic explosion.

Information Contact: J. Gautheyrou, ORSTOM.

1/77 (2:1) Significant steam and ash explosions after 11 Oct occurred 30 Oct, 1, 6, and 7 Nov, and 5, 13, 14, 15, 17, 19, and 30 Jan [but see 2:2]. Two harmonic tremor events were recorded on 1 Nov and one on 6 Nov. No information on harmonic tremor is available after 7 Nov. [Both the number of earthquakes and seismic energy release declined after mid-Nov (table 16-5).]

Information Contact: M. Feuillard, Lab. de Physique du Globe.

1976		Number of Earthquakes	Total Seismic Energy (x 10^{13} ergs)
Oct	1-10	412	446
	11-20	895	532
	21-31	1008	929
Nov	1-10	489	523
	11-20	235	402
	21-30	316	168
Dec	1-10	204	225
	11-20	102	165
	21-31	93	41

Table 16-5. Number of earthquakes and total seismic energy release, Oct-Dec 1976.

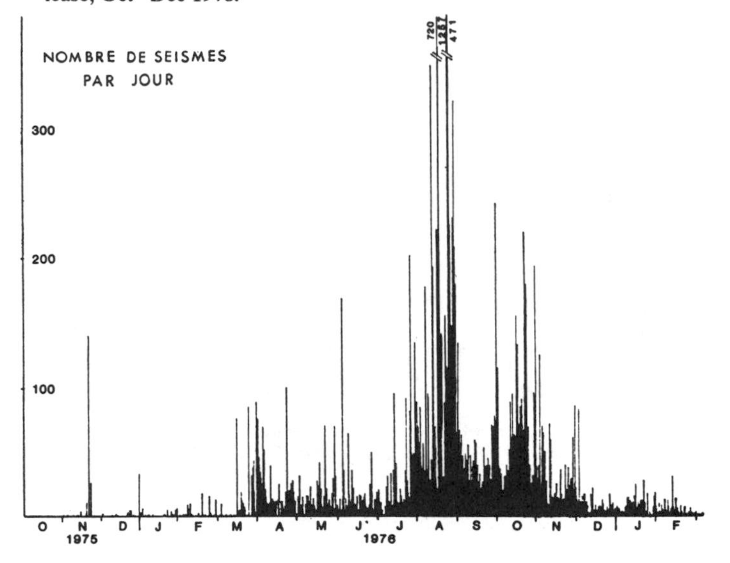

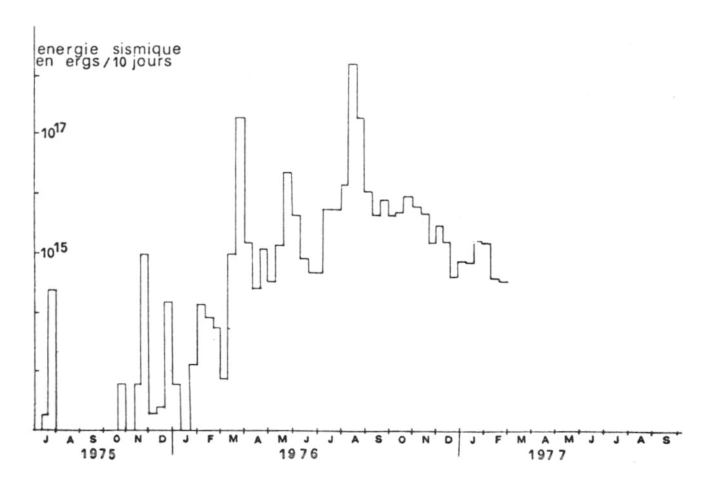

Fig. 16-6. Seismic energy release in ergs for 10-day periods, July 1975-Feb 1977.

2/77 (2:2) The following report is by Richard Fiske.

"Phreatic activity has continued at La Soufrière through 23 Feb. The largest explosion of the 1976-77 activity occurred at 1911 on 29 Jan 1977 and was heard by people living as far as 10 km from the summit. Lithic blocks as large as 0.5 m in diameter were thrown 1 km from the summit fissures. Five cm of ash was deposited at Savane à Mulets and 5 mm of ash fell on Matouba, 3 km W of the summit. Smaller phreatic explosions occurred on 13 and 15 Feb, and measurable amounts of ash were erupted on 5, 7, and 10-15 Feb. On 23 Feb only minor amounts of steam were being emitted from the volcano.

"In contrast to the eventful phreatic activity of recent weeks, the local seismicity has continued to decline. Only 6-10 earthquakes/day were detected 1-23 Feb and

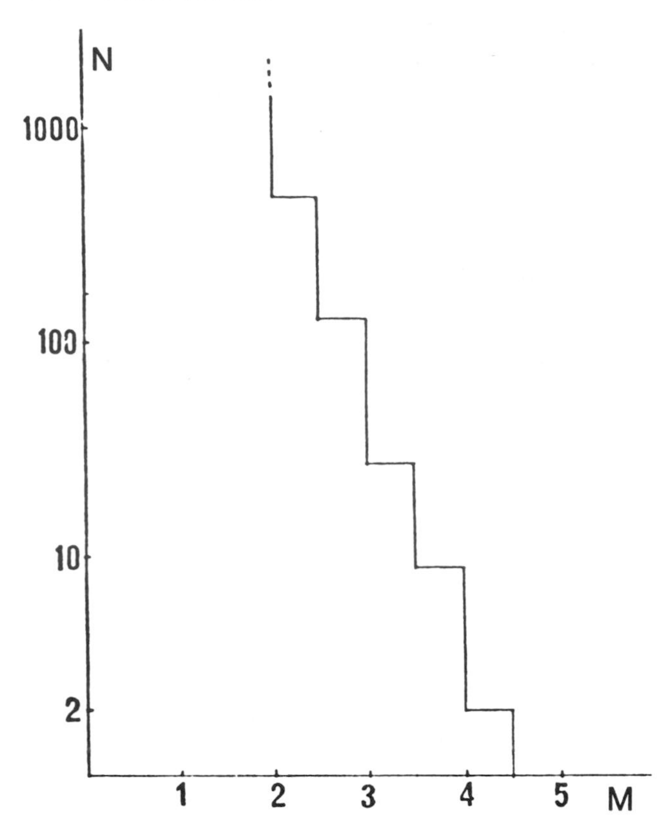

Fig. 16-7. Number of recorded earthquakes at Soufrière in magnitude increments of 0.5, late 1975 – Feb 1977.

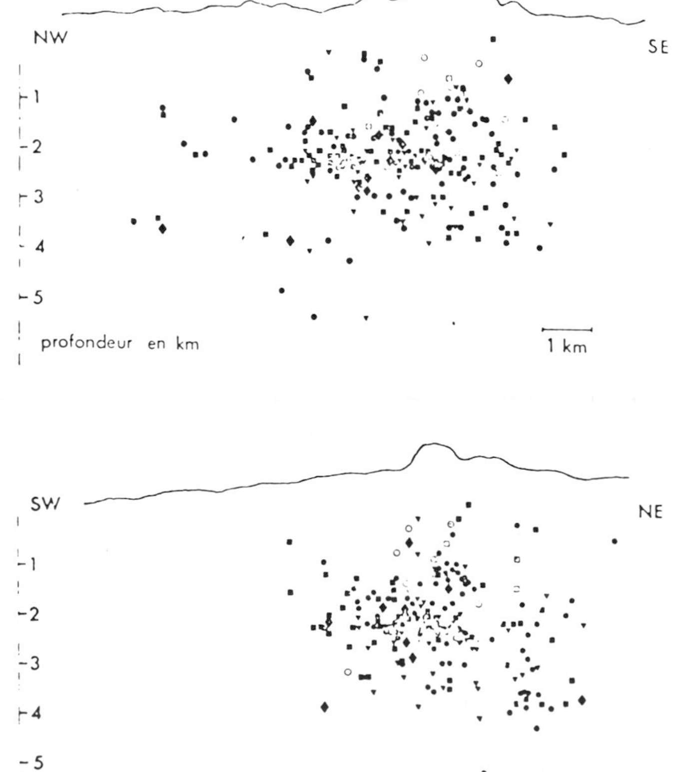

Fig. 16-8. NW-SE and NE-SW cross-sections showing earthquake hypocenters.

a felt event occurs only once in every 10 days.

"Petrographic study continues to confirm that ejecta from Soufrière consist only of older, altered material. No fresh glass has been observed. As before, ground deformation measurements with various types of tiltmeters (pendulum, borehole, dry tilt) have indicated no significant changes in the overall shape of the volcano.

Information Contact: R. Fiske, SI.

3/77 (2:3) The following report is by Jacques Dorel.

"Daily recorded earthquakes at Soufrière had decreased from a maximum of 1257 on 24 Aug 1976 to one per day by the first week in March 1977 (fig. 16-5). Fig. 16-6 shows the seismic energy released during 10-day periods. Total energy release from 1 Nov 1975 through Feb 1977 was about 1.9×10^{18} ergs. Fig. 16-7 shows the number of earthquakes in various magnitude ranges. The largest magnitude was 4.2 (4.9 by NEIS determination) on 16 Aug 1976.

Hypocenter determination has been made for about 350 events. Fig. 16-8 shows NW-SE and NE-SW vertical sections (under the volcano) onto which the foci are projected. Different symbols are used for different time series, but the variation with time of the depths of quakes during the crisis is not meaningful. Mean depth is about 2.5 km.

"Volcanic tremor, the duration of which varied from a few minutes to a few tens of minutes, has been recorded by the seismographs. The tremor events correspond to such important surface manifestations as steam, solid ejecta, and fracture opening. With its 16,000 earthquakes and a total energy release $> 10^{18}$ ergs, the Soufrière crisis is the most important in the Carribean region in the past 40 years."

Information Contact: J. Dorel, Univ. Pierre et Marie Curie, Paris.

4/77 (2:4) Surface and seismic activity have declined considerably at Soufrière since the peak of the volcano-seismic crisis in Aug 1976 (table 16-6). The most recent phreatic explosion occurred on 1 March, after which solfataric activity continued into April from the uppermost craters but diminished on the flank and in the Col de l'Echelle, just SE of the summit dome. Ash was emitted for portions of 20 days 1 Jan-15 April.

Information Contact: M. Feuillard, Lab. de Physique du Globe.

5/77 (2:5) Surface activity at La Soufrière remained entirely fumarolic, and was strongest from the fissures toward the top of the summit dome. [Seismicity continued an irregular decline (table 16-6).]

Information Contact: M. Feuillard, Lab. de Physique du Globe.

6/77 (2:6) Table 16-7 summarizes seismicity at Soufrière from mid-1975 through May 1976.

Information Contact: M. Feuillard, Lab. de Physique du Globe.

Further References: Feuillard, M., Allegre, C.J., Brandeis, G., Gaulon, R., et al., 1983, The 1975 – 1977 Crisis of La Soufrière de Guadeloupe (F.W.I.): A Still-born Magmatic Eruption; *JVGR*, v. 16, p. 317-334.

Smith, A.L. (ed.), 1980, Special Issue on Circum-Caribbean Volcanism; *BV*, v. 43, no. 2 (5 papers on Soufrière de Guadeloupe), p. 383-411 and 419-452.

10/81 (6:10) IPG geophysicists have installed 3 telemetering magnetometers and established a network of 20 additional non-telemetering magnetometer stations on the flanks of Soufrière. Although significant changes in the magnetic field were recorded over periods of a few [months] during the phreatic and seismic activity of 1976–77, no variations of more than 1.5 nT were detected between Aug 1980 and March 1981.

IPG geophysicists also drilled 2 boreholes, to 77 m depth in the summit dome, and to 97 m depth immediately beside the dome. The temperature at the bottom of dome borehole decreased steadily, from 97.5°C in April 1978 to 87°C in Dec 1980. . . . No significant temperature variation was recorded in the borehole near the dome, which remained at about 19°C through 1980. The temperature of a fumarole (Lacroix) about 80 m below the summit dropped from about 160°C in July 1978 to just over 100° by Jan 1979 and has remained between 96° and 98°C in 1981.

Information Contacts: J. Pozzi, J. Zlotnicki, G. Simon, J. Le Mouel, J. Cheminée, IPG, Paris.

Further Reference: Zlotnicki, J., 1986, Magnetic Measurements on La Soufrière Volcano, Guadeloupe (Lesser Antilles), 1976–1984: A Re-examination of the Volcanomagnetic Effects Observed During the Volcanic Crisis of 1976–1977; *JVGR*, v. 30, p. 83-116.

1/82 (7:1) Since Jan 1979, samples have been collected 2-3 times/month from 6 springs with water temperatures of 25-69°C, 0.5-3.5 km from the summit dome. Water temperatures and chemistry remained nearly constant at 5 of the 6 springs, as shown by the following maximum variations recorded at any one spring in the nearly 3 years of sampling: temperature (°C) 4%, pH 5%, Cl 20%, HCO_3 25%, F 25%. These variations were independent of rainfall and apparently random.

However, Carbet l'Echelle spring, the hottest and nearest to the summit (0.5 km) showed appreciable changes during the survey period. Its water temperature dropped from 69°C in Jan 1979 to 57° in Nov 1981, while the concentration of HCO_3 ions approximately doubled and large variations in Cl concentrations occurred (fig. 16-9).

Information Contacts: S. Bigot, Lab. de Géochemie, Paris; G. Hammouya, Lab. de Physique du Globe; J. Le Mouel, J. Cheminée, IPG, Paris.

3/83 (8:3) A study of the geochemical evolution of Carbet l'Echelle spring, continued during 1981 and 1982 (fig. 16-10). The water temperature decreased from 62.5°C in Jan 1981 to

1977		Number of Earthquakes	Total Seismic Energy ($\times 10^{13}$ ergs)
Jan	1-10	69	76
	11-20	126	71
	21-31	116	154
Feb	1-10	76	152
	11-20	86	38
	21-28	17	32
Mar	1-10	23	4
	11-20	65	54
	21-31	66	52
April	1-10	12	2
	11-20	28	6.9
	21-30	6	1.1
May	1-10	5	0.9
	11-20	9	1.7

Table 16-6. Number of earthquakes and total seismic energy release, 1 Jan–20 May, 1977. [11 April–20 May data from 2:5.]

Days of Month:		1-10	11-20	21-end of month
1975	July	0	0	23
	Aug	0	0	0
	Sept	0	0	0
	Oct	0	0	1
	Nov	0	1	91
	Dec	0	0	14
1976	Jan	0	0	1
	Feb	1	7	5
	March	1	100	18000
	April	175	26	125
	May	35	150	4600
	June	380	88	64
	July	80	510	768
	Aug	1544	33831	20010
	Sept	1027	469	949
	Oct	446	532	930
	Nov	523	402	168
	Dec	225	165	41
1977	Jan	76	71	154
	Feb	152	38	33
	March	4	54	52
	April	2	7	1
	May	1	2	6

Table 16-7. Total seismic energy release ($\times 10^{13}$ ergs) per 1/3-month period, July 1975-May 1977.

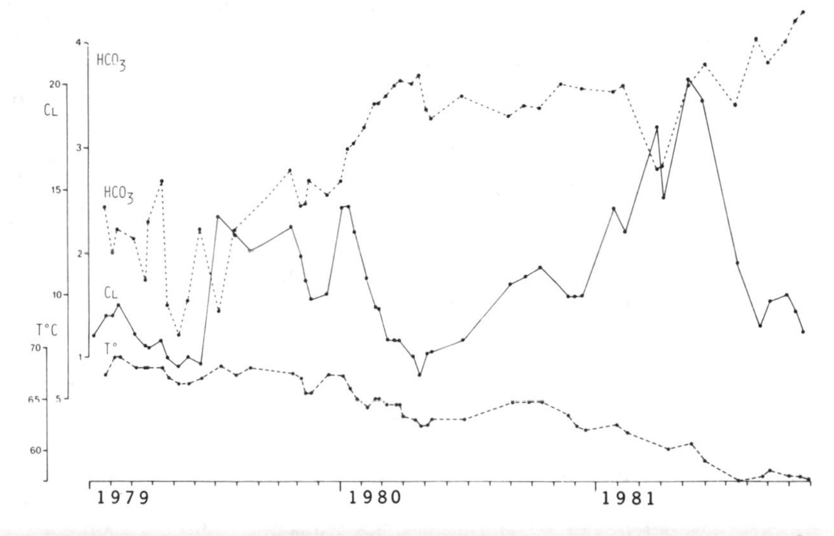

Fig. 16-9. Changes in temperature (°C, dashed line), Cl concentration (10^{-3} moles/liter, solid line) and HCO_3 concentration (10^{-3} moles/liter, dotted line) for Carbet l'Echelle spring, 0.5 km from the summit, Jan 1979–Oct 1981.

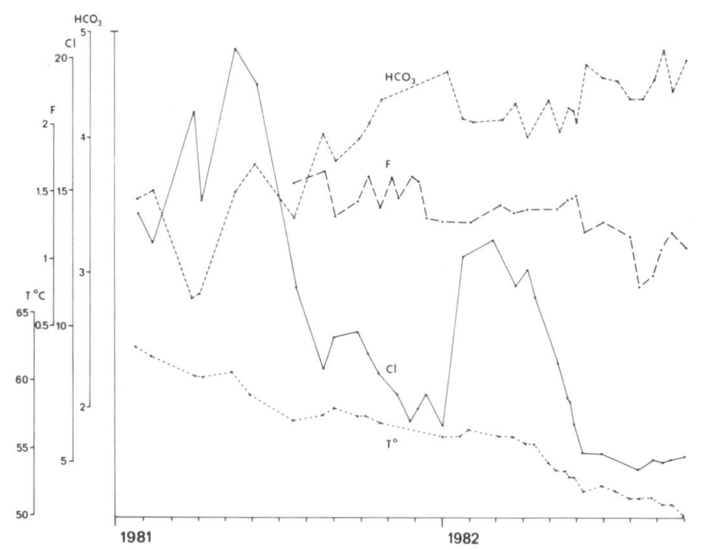

Fig. 16-10. Changes in temperature (°C), HCO_3 and Cl concentration (10^3 moles/liter), and F concentration (10^5 moles/liter) for Carbet l'Echelle spring, 0.5 km from the summit, Jan 1981 – Sept 1982.

50° in Sept 1982, while the concentration of HCO_3 ions continued to increase (+ 30%). Moreover, although large variations in Cl concentration have been observed since 1979, the background level of Cl began to fall in Sept 1981 and was reduced by half in one year. Likewise, the concentration of F ions, which has oscillated around a constant value since the beginning of the survey, decreased slightly in late 1982. Water temperatures and chemistry remained nearly constant at the other springs monitored by this study.

Information Contacts: S. Bigot, Lab. de Géochemie, Paris; G. Hammouya, Lab. de Physique du Globe; J. Le Mouel, J. Cheminée, IPG, Paris.

Further References: Bigot, S. and Hammouya, G., 1987, Surveillance Hydrogéochimique de la Soufrière de Guadeloupe, 1979 – 1985: Diminuition d'activité ou Confinement?; *C.R. Acad. Sci. Paris*, t. 304, Série II, no. 13, p. 757-760.

Observations Volcanologiques: Rapport d'Activite des Observatoires Volcanologiques des Antilles (Guadeloupe-Martinique) - Nov 1976 – April 1977, April 1977 – Dec 1978, 1979, 1980, 1981, 1982, and 1983 – 1984; Institut de Physique du Globe, Paris.

MICOTRIN

Dominica
15.30°N, 61.30°W, summit elev. 1221 m
Local time = GMT - 4 hrs

This volcano in SW Dominica is between Guadeloupe and the island of Martinique, site of the 1902 Mont Pelée eruption that killed 28,000 in hot pyroclastic flows. Micotrin itself is a prominent lava dome 9 km NE of Dominica's capital city, and is believed to be the source for the Roseau ash, the largest recent tephra deposit in the West Indies. The volcano's only historic eruption (at the Valley of Desolation solfatara field) was in 1880, but there have been several episodes of seismic unrest. The original *Bulletin* reports simply list the 1976 activity under "Dominica."

Reference: Carey, S.C. and Sigurdsson, H., 1980, The Roseau Ash: Deep-sea Tephra Deposits from a Major Eruption on Dominica, Lesser Antilles Arc; *JVGR* v. 7, p 67-86.

4/76 (1:7) On 10 Feb a local earthquake swarm began on the island of Dominica. Only 12 events were recorded during the next month, but they increased sharply following a regional earthquake (M 5.9) 170 km to the N on 10 March. The largest event of this swarm was 14 March at 2115, but was not recorded on the WWSSN despite a locally estimated intensity of MM VI. On 30 March a 5-station seismograph net was established by a team from the Seismic Research Unit of the University of the West Indies, Trinidad. They have since located over 40 earthquakes with focal depths of generally less than 3 km. Activity was generally concentrated in a belt extending beneath Roseau, capital of the island, and up to 2 km offshore to the SW.

. . . There has been neither increased steam venting at the nearby Watten Waven Soufrière (inspected on 11 April by John Tomblin, Director of the Seismic Research Unit) nor additional indications of volcanic activity reported elsewhere on the island. Furthermore, earthquake frequency was decreasing toward the end of April and focal depths appeared to be increasing. Similar earthquake swarms in Dominica have been recorded in 1974 and 1971, but they were felt less in the capital city and were therefore less conspicuous. Apart from a minor phreatic explosion in 1880, there has been no historic volcanism on Dominica. Tomblin interprets the recent earthquake swarm as subsurface migration of magma, but feels that the risk of an eruption in the near future is decreasing. The Seismic Research Unit team is continuing to keep a close watch on geophysical activity in Dominica.

Information Contacts: J. Tomblin, Univ. of the West Indies; W. Person, USGS/NEIS, Denver, CO.

5/76 (1:8) The trend of decreasing earthquake frequency reported last month did not continue into early May nor did the apparent trend toward increasing focal depths. Seismographs recorded 1-3 events/day in early May with depths of 1.2-2.2 km. Scientists from the University of West Indies' Seismic Research Unit have continued monitoring the activity with a telemetered network.

Information Contact: Seismic Research Unit, Univ. of the West Indies.

8/76 (1:11) The increased seismic activity that began 10 Feb gradually declined and had returned to normal by late May.

Information Contact: W. Aspinall, Univ. of the West Indies.

SOUFRIERE OF ST. VINCENT

St. Vincent
13.33°N, 61.18°W, summit elev. 1178 m
Local time = GMT - 4 hrs

Soufrière means "sulfur," and this Soufrière is 305 km S of the one on Guadeloupe. It has a history of violent eruptions, including those in 1718, 1812, and 1902, when 1600 people were killed only hours before the disastrous Pelée eruption 165 km to the N. This history, and the controversial activity at Guadeloupe 3 years earlier, were much on the minds of those dealing with the 1979 eruption.

4/79 (4:4) A series of powerful explosions from Soufrière produced large ash clouds and several pyroclastic avalanches, forcing the evacuation of more than 17,000 persons from the N end of St. Vincent. This eruption is particularly noteworthy because of the wide variety of observations made by various scientific teams (from land, low-flying aircraft, a high-altitude research plane, and from satellites). When these data are analyzed and integrated, the geophysical community can look forward to an unusually well-documented account of an episode of explosive island-arc volcanism.

The first pre-eruption seismic event, telemetered to the Seismic Research Unit of the University of the West Indies by seismometers 3 and 9 km from Soufrière's summit, was a strong local earthquake at 1106 on 12 April, within 1 hour of the fortnightly earth tide maximum (calculated by F.J. Mauk). Seismic activity gradually increased through the day, and by 1900 about 15 clearly identifiable earthquakes, apparently B-type, were occurring per hour. Continuous harmonic tremor began to build an hour later, and within 2 hrs was saturating the seismometers.

A team of volcanologists and seismologists from the Seismic Research Unit arrived on St. Vincent 13 April, and were later joined by researchers from several other institutions. The first explosive activity was observed at dawn (about 0500) on 13 April. Subsequent explosive events were reported at 1115, 1700, and 2050 on the 13th, 0300 and 1200 on 14 April, 1705 on 17 April, 0635 on 22 April, and 2355 on 25 April. Analysis of infrared imagery from NOAA's SMS-1 weather satellite indicates that most of these high eruption clouds were fed briefly (< $^1/_2$ hour) by the volcano. The largest of the clouds, from the 17 April explosion, reached an estimated height of 18 km and ultimately grew to a diameter of 140 km. The noon explosion on 14 April produced a 100 km-diameter cloud, and 2 explosions on 13 April also produced sizeable clouds, 60 km-diameter at 1700 and 40 km at 2050. The first 2 explosion clouds on 13 April were smaller, and the explosion reported at 0300 on 14 April produced no infrared signature. Most of the explosions occurred close to diurnal earth tide maxima, and the 17 April explosion also fell on the fortnightly earth tide minimum (calculations from F.J. Mauk).

A NASA P-3 Electra aircraft, equipped with lidar, a 10-stage quartz-crystal microbalance, and an NO_2 remote-sensing instrument, flew through the periphery of the 17 April cloud minutes after it was ejected. Ash was collected and photographs were taken. The next evening, the NASA aircraft's lidar detected significant quantities of ejecta in the stratosphere, in patches 0.5-3 km thick with a base altitude of about 18.7 km. Scientists at NCAR, Kyushu University, and other institutions, will search at higher latitudes for stratospheric aerosols from this event.

Hot pyroclastic avalanches have accompanied at least 3 of the explosions. The largest avalanche flowed down the Larikai River valley (fig. 16-11) at noon on 14 April and continued beyond the mouth of the river (3 km W of the crater) several kilometers out to sea. Its deposit at the coast was 1.5 m thick, about 300 m wide, and contained scoria blocks up to 60 cm in diameter. The surface temperature of this deposit was well in excess of 100°C when it was inspected 28 hrs after emplacement. A hot avalanche from one of the earlier explosions traveled 2.5 km SE from the summit, down the Rabacca Valley, and the 17 April explosion produced numerous hot avalanches that moved down several valleys on the flanks of the cone.

Ashfalls from most of the explosions were limited to a few centimeters in the N portion of St. Vincent, and small amounts on Barbados, about 150 km E. However, during the 22 April eruption (fig. 16-12) abnormal winds dropped 4 mm of ash on Kingstown (the capital of St. Vincent), on the S end of the island.

The character of the seismic activity varied considerably during the eruption. Harmonic tremor con-

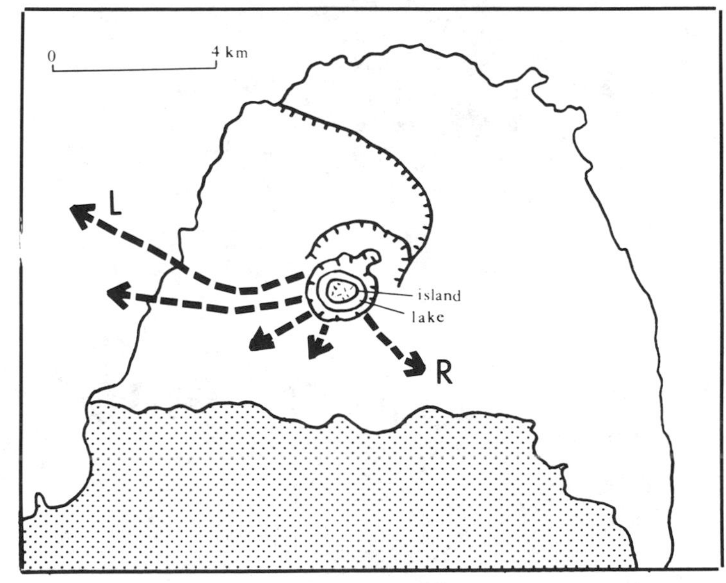

Fig. 16-11. Sketch map of Soufrière and the N portion of St. Vincent, based on the map in *CAVW.* Arrows show routes taken by pyroclastic avalanches, which traveled down valleys on Soufrière's flanks during at least 3 of the explosions. Lengths are approximately proportional to distances traveled by the hot avalanches. Arrows labeled L and R indicate avalanches in the Larikai and Rabacca valleys. The lake and island (a lava dome extruded during the 1971 eruption) shown in Soufrière's summit crater were both destroyed during the present eruption. Rocks older than the Soufrière edifice are shown by a stippled pattern.

Fig. 16-12. Photograph taken shortly after the onset of the 22 April explosion. Height of the eruption cloud is estimated to be about 6 km. Photo by R. Fiske.

tinued to saturate the instruments from 2200 on 12 April through 1600 on 14 April. Beginning at about 0900 on the 15th, a decline in tremor amplitude made it possible to identify more than 100 individual events per hour at the station 9 km from the summit. The number of individual events declined irregularly through the morning of 16 April. At about 2000 on the 16th, the seismicity changed entirely to rock fracture events, which continued until the explosion of 17 April. After this explosion, the seismicity consisted primarily of 30-50 small explosion events per hour, which could be correlated with the ejection of steam puffs that sometimes contained a little ash. Most of the later explosions were preceded by brief periods (up to 3 hrs) of seismic quiet. Seismicity ended almost completely at about 0300 on 29 April. With the exception of a 20-minute period of tremor during the night of 29-30 April, the volcano has remained seismically quiet through early 3 May.

Soufrière of St. Vincent has a history of violent eruptions, the most devastating of which occurred in 1902, killing more than 1600 persons. The amount of tephra produced by the current eruption is about 2 orders of magnitude less than in 1902. Soufrière's most recent activity occurred in 1971, when a lava dome was extruded aseismically into the summit crater lake. Both dome and lake have been destroyed by the present eruption.

Information Contacts: J. Tomblin, W. Aspinall, K. Rowley, J. Shepherd, Univ. of the West Indies; A. Kreuger, R. LaPorte, NOAA/NESS; M.P. McCormick, NASA Langley Research Center; F. Mauk, Univ. of Michigan; R. Fiske, SI.

5/79 (4:5) A new phase of activity from Soufrière began in late April. At 0300 on 29 April, the seismicity that had accompanied the eruption since 12 April ended almost completely, and there have been virtually no local earthquakes since then. Katia Krafft climbed to the crater rim on 3 May and observed a small new lava dome growing in the base of the crater, the same area that had been occupied by the center of the now-destroyed 1971 dome. A sample of the new dome collected by John Tomblin in mid-May was described in hand specimen as a basaltic andesite, similar to the 1971 dome.

By the end of May, the new dome had grown to about 500 m in diameter and 60 m in height. The dome's growth rate was difficult to estimate, but was probably less than 0.5×10^6 m^3/day in late May. Virtually all of the recent expansion of the relatively fluid dome has been lateral rather than vertical.

A few small explosions have taken place from the dome, but none were large enough to project material above the rim (about 300 m above the crater floor). Fumarolic activity from the top of the dome was moderately strong during the early stages of its growth, but had stopped by 20 May. In late May, fuming was still fairly vigorous around the dome, on the floor of the crater. A team from Guadeloupe's volcanological observatory measured a temperature of 239°C from the fumaroles in late May, and collected samples from the dome.

The April explosions deposited a total of about 30 cm of ash on the zone within 1 km of the crater rim. Large numbers of massive dense remnants of the 1971 dome and scoria blocks up to 60 cm in diameter were also found in this zone.

H.H. Lamb's preliminary estimate of Dust Veil Index (Lamb, 1970) for this eruption is from 3 to 9, probably nearer to 3. The Dust Veil Index for the 1902 eruption of Soufrière was calculated at 300; the Krakatau 1883 eruption = 1000.

Much of the crater floor is presently mantled with rubble. The new dome had spread over the remnants of the old crater lake by about 20 May, but a new lake had begun to form at the beginning of June because of the onset of the rainy season.

Most of the more than 17,000 evacuees were allowed to return to their homes on 14 May, including residents of the 2 largest towns near the volcano, Georgetown and Chateaubelair. Restrictions remain in effect for the 4,000 persons who live in the zone devastated by the 1902 eruption, N of the Rabacca and Wallibou Rivers.

Reference: Lamb, H.H., 1970, Volcanic Dust in the Atmosphere; *Philosophical Transactions of the Royal Society of London*; series A, v. 266, no. 1178, p. 425-533.

Information Contacts: W. Aspinall, K. Rowley, J. Shepherd, J. Tomblin, Univ. of the West Indies; Katia Krafft, Cernay, France; H. Lamb, Univ. of East Anglia.

6/79 (4:6) A period of summit crater lava extrusion, accompanied by little or no seismicity, began in late April and was continuing at the end of June.

Richard Fiske and Haraldur Sigurdsson descended into Soufrière's crater on 18 June. Lava extrusion continued from the same vent that produced the 1971 dome and the 1979 explosions. Blocky lava had flowed to the N wall of the crater, reaching maximum dimensions of 725 m across and 110 m high according to their tape and compass survey. The small number of flow front rock avalanches observed during their 4-hr stay may indicate a rate of extrusion substantially lower than in late April. Heavy steaming in the crater limited visibility.

Since the recent beginning of the rainy season, large quantities of tephra have been eroded from Soufrière's flanks. Major mudflows that traveled down the larger valleys (mostly on the W side of the volcano) carved deep, narrow canyons, dramatically exposing pre-1979 valley fill deposits. The mudflows disrupted road crossings in the Rabacca Valley (E flank) but otherwise did little property damage. Revegetation of the areas devastated by the 1979 eruption had begun by mid-June.

The zone of destruction from the 1902 eruption (N of the Wallibou and Rabacca Rivers) remains partially evacuated. Some of the several thousand evacuees work in the area during the day, but most leave at night.

Information Contacts: R. Fiske, SI; H. Sigurdsson, Univ. of Rhode Island.

7/79 (4:7) Lava extrusion at Soufrière continued through July. Between surveys on 2 July and 4 Aug, the lava had expanded 30-50 m horizontally (except on the N side where it had reached the crater wall, 4:6) and about 6 m vertically, to a mean diameter of 820 m and a mean height of 85 m. Assuming 45° sides, its volume was 36.5

1979	Daily Volume Increase (x 10^6 m^3)
25 May–2 July	.30
2 July–4 Aug	.24
4-21 Aug	.17
21 Aug–23 Sept	.06
23 Sept–2 Oct	.10
2-25 Oct	.01

Table 16-8. Rate of lava extrusion in the summit crater of Soufrière, 25 May–25 Oct 1979, calculated by John Tomblin.

x 10^6 m^3 on 4 Aug, having increased an average of 0.36 x 10^6 m^3/day in July (similar to the June rate).

During the Aug survey, some pulsing steam emission took place, mostly at the S edge of the lava, but no explosions were observed. The summit seismic station recorded 50-200 small earthquakes per day during July, but none of these were detected by instruments on the flanks.

Information Contact: J. Tomblin, Univ. of the West Indies.

8/79 (4:8) Lava extrusion continued through Aug. However the extrusion rate has decreased during the past 3 months, as shown in table 16-8. The mean diameter of the lava extrusion increased from 820 m on 4 Aug to 832 m on 21 Aug, but the highest point has remained at 130 m above the crater floor since 11 Aug, after 13 m of vertical growth from 10 July to 11 Aug.

Between 50 and 200 very small local earthquakes continued to be recorded daily by the summit seismic station, but very few were detected by seismometers on the flanks.

When the summit was not obscured by clouds, gentle but continuous steam emission was visible. Steam frequently rose slightly above the crater rim. Water was sometimes present in the crater, especially after heavy rains, but the volume of water remained small and water depths did not exceed about 1 m.

Information Contact: Same as above.

9/79 (4:9) Extrusion of lava into Soufrière's summit crater continued through Sept. However, the rate of extrusion continued to decrease; the late Aug-late Sept rate was an order of magnitude less than peak values observed in mid-May.

Between 21 Aug and 23 Sept, the mean diameter of the lava body grew from 832 m to 840 m, covering about 60% of the crater floor, and its maximum height increased 1 m to 131 m. Its volume on 23 Sept was calculated at 37.6 x 10^6 m^3.

Small seismic events continued to be recorded by the summit seismograph. There has been no significant change in seismicity since a several-day increase in the number of events in late June and early July.

Information Contact: Same as above.

10/79 (4:10) By late Oct, extrusion of lava into Soufrière's central crater had virtually stopped. The Oct extrusion rate was an order of magnitude less than that of late Sept and 2 orders of magnitude less than the May rate.

The mean diameter of the lava extrusion increased by only 1.5 m between 2 and 25 Oct, to 870 m, and the maximum height remained ~130 m. Vigorous steaming from the lava was continuing in late Oct, but the number of small local earthquakes recorded by the summit seismograph had declined markedly since early Oct.

Information Contact: Same as above.

11/79 (4:11) No lava has been extruded into Soufrière's central crater since the survey of 25 Oct (4:10). However, monitoring of the volcano by the Seismic Research Unit, University of the West Indies, continues. Data from 4 seismometers are telemetered continuously to Trinidad, and scientists visit the volcano every weekend.

Information Contact: Same as above.

Further References: Fiske, R. and Sigurdsson, H. (eds)., 1982, Soufrière Volcano, St. Vincent: Observations of its 1979 Eruption from the Ground, Aircraft, and Satellites; *Science*, v. 216, no. 4550, p. 1105-1126 (11 papers).

Shepherd, J.B., Aspinall, W.P., Rowley, K.C., et al., 1979, The Eruption of Soufrière Volcano, St. Vincent April–June 1979; *Nature*, v. 282, p. 24-28.

Shepherd, J.B. and Sigurdsson, H., 1982, Mechanism of the 1979 Explosive Eruption of Soufrière Volcano, St. Vincent; *JVGR*, v. 13, p. 119-130.

Sparks, R.S.J. and Wilson, L., 1982, Explosive Volcanic Eruptions - V. Observations of Plume Dynamics During the Soufrière Eruption, St. Vincent; *Geophysical Journal of the Royal Astronomical Society*, v. 69, p. 551-570.

KICK-'EM-JENNY

N of Grenada
12.30°N, 61.63°W, summit depth -160 m

This submarine volcano, 10 km N of Grenada and 125 km S of Soufrière St. Vincent, has erupted at least 8 times since 1939. Most eruptions were detected by seismic monitoring. The last event occurred in 1977, when strong acoustic (T-phase) signals indicated probable submarine eruptions between 0449 and 0530 on 14 Jan (H. Sigurdsson, 1979, personal communication).

ICELAND and JAN MAYEN

KATLA

beneath Myrdalsjökull, S Iceland
63.63°N, 19.03°W, summit elev. 1363 m

Only 20 km from Iceland's southernmost coast, and 155 km ESE of Reykjavik, Katla is covered by an icesheet 30 km in diameter. Seventeen eruptions are known since the 10th century.

11/77 (2:11) Fig. 17-1 shows annual seismic strain release from 1970 – Sept 1977 in the Myrdalsjökull area. After 1977, the annual strain release was similar to that of 1970 – 74. Katla's subglacial eruptions have caused major floods (jökulhlaups) including a catastrophic event in 1755 – 56. The last substantial jökulhlaup in this area occurred in 1918, but a small flood in 1955 may have been caused by minor volcanic activity.

Information Contact: R. Stefánsson, Icelandic Meteorological Office.

HEKLA

S Iceland
63.98°N, 19.70°W, summit elev. 1491 m
Local time = GMT

Hekla's large (VEI 5) eruption in 1104 was the first in Iceland's recorded history. Eighteen eruptions have followed, and (like Katla's 50 km to the SW) many have been violent. Property damage has resulted from 12 eruptions and fatalities from 3.

8/80 (5:8) The following is a report from Karl Grönvold and Sigurdur Thorarinsson.

"Hekla started erupting at 1328 on 17 Aug. Small earthquakes were recorded on local seismographs for 20-25 minutes prior to the first explosions but these precursors were not noticed until later.

"This was a mixed eruption starting in the summit area and quickly extending to the full fissure length of 7 km, longer than observed in the 1947 and 1970 eruptions. The initial activity at 1320 was a steam column, then a dark tephra column started between 1327 and 1330. The main tephra fall lasted about 2 hours and extended NNE, and the eruption column reached about 15 km altitude. [Atmospheric effects that may have been caused by Hekla are described in Chapter 19, 5:9.] The maximum tephra thickness 10 km N of the summit was 20 cm, and at the N coast, about 230 km distant, 1 mm or less. The fluorine content in many grazing districts is above danger level, causing problems for livestock, especially sheep. Lava began flowing shortly after the beginning of the eruption. The first flows appeared near the summit, then lava eruption extended to the lower parts of the fissure. Most of the lava issued within 12 hours and nearly all within 24 hours, forming 4 main separate lava flows, covering an area of about 22 km^2. The volume of the lava is estimated at about 0.1 km^3 and the tephra somewhat less than in the 1970 eruption, which produced 0.07 km^3.

"Glowing scoria was last observed in the early morning of 20 Aug. Steam emission was continuing as of late Aug. For the first few days following the eruption this steam column was often darkened at the base, but no glow was observed.

"Preliminary chemical analyses of tephra and lava show composition similar to that of the 1970 eruption products. The last Hekla eruption took place 5 May – 5 July 1970, making this period of quiescence of only 10 years the shortest recorded for Hekla since 1104. The shortest previous period between eruptions was 1206-1222 and the second shortest was between the 1947 and 1970 eruptions. If the present eruption episode is over now, the behavior is highly unusual. All previous known eruptions have lasted from 2 months to 2 years. Initially there have been a few days of major activity, followed by a few almost quiet days, then renewed explosive activity and lava eruption concentrated on small parts of the fissure."

Information Contacts: K. Grönvold, Nordic Volcanological Institute (NVI); S. Thorarinsson, Univ. of Iceland.

3/81 and 4/81 (6:3 and 6:4) [The report in 6:4 repeats and expands upon the preliminary information received at press time for 6:3. These reports are combined here.]

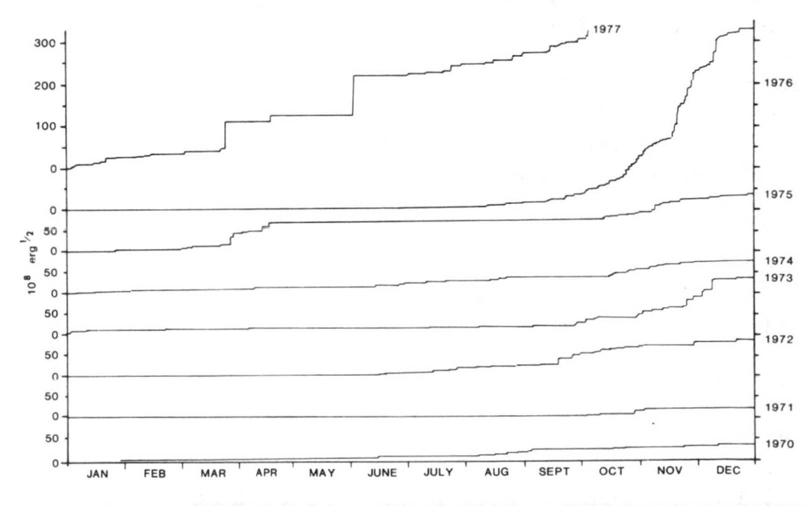

Fig. 17-1. Annual seismic strain release in the Myrdalsjökull area, Feb 1970 through Sept 1977. From *Skjálftabref* (published by the Icelandic Meterological Office), no. 26, Oct 1977. Courtesy of Ragnar Stefánsson.

An eruption of Hekla was first observed about 0300 on 9 April. The initial activity consisted of ejection of ash columns. Pilots said that ash reached 4.2 km altitude and radar registered the top of an eruption column at 6.6 km above sea level. Ash was blown N, toward the interior of Iceland, falling at least 30 km from the volcano. Lava extrusion from a newly formed summit crater began later on 9 April, reached a peak during the afternoon, then declined gradually until the eruption ended on 16 April. Three flows were extruded, the 2 largest moving down the N side of the main Hekla ridge, the smallest down the S side. The lava covered an area of 5-6 km^2 and extended a maximum of 4.5 km from the crater, reaching the base of Hekla Ridge on both the N and S sides. After the initial ash ejection, activity was dominantly effusive, but a small amount of ash fell E and NE of the volcano later in the eruption. The area has remained inaccessible, but ash amounts were estimated to be only a few millimeters thick. No earthquakes were recorded prior to the eruption, and distinct harmonic tremor did not appear on seismic records until after the eruption began.

Volcanologists interpreted the April activity as a continuation of the much larger eruption of Aug 1980.

Information Contacts: K. Gronvold,NVI; S. Thorarinsson, Univ. of Iceland.

Further Reference: Grönvold, K., Larsen, G., Einarsson, P., Thorarinsson, S., et al., 1983, The Hekla Eruption of 1980 – 1981; *BV*, v. 46, p. 349-364.

5/81 (6:5) The COSPEC detected SO_2 N and SW of Hekla's April 1981 lava flows.

Information Contacts: K. Grönvold; NVI; R. Stoiber, S. Williams, Dartmouth College.

GRIMSVOTN

S Iceland
64.42°N, 17.33°W, summit elev. 1719 m

Under Vatnajökull, Iceland's largest icesheet, Grímsvötn is 125 km ENE of Hekla. More subglacial eruptions are known from this caldera than from any other volcano in the world, but only one (1629) has caused fatalities. The first historic eruption (of about 48) was in 1332.

General Reference: Björnsson, H., Björnsson, S., and Sigurgeirsson, Th., 1982, Penetration of Water into Hot Rock Boundaries of Magma at Grímsvötn; *Nature*, v. 295, p. 580-581.

2/82 (7:2) The following is a report from Sigurdur Thorarinsson, Helgi Björnsson, and Sigurjon Rist.

"A glacier burst (jökulhlaup) from Grímsvötn caldera in Vatnajökull glacier started 28 Jan. It increased slowly, with a discharge on 4 Feb of 420 m^3/sec. It culminated on 11 Feb at 2000 m^3/sec and ended 21 Feb. Total discharge is estimated at 1.3 km^3 and the lowering of the ice level in the caldera at ~50 m. The volume of the last jökulhlaup from Grímsvötn, in Sept 1976, was estimated at 2.4 km^3 and that of 1972 at 3.2 km^3."

[Most of the] documented jökulhlaups from Vatnajökull between 1332 and 1934 were probably caused by eruptions of Grímsvötn, under the glacier. Since then, Icelandic volcanologists believe that steady heat production from Grímsvötn has caused a gradual buildup of meltwater beneath the glacier. Occasional sudden failures have produced large outflows from beneath Vatnajökull.

Information Contacts: S. Thorarinsson, H. Björnsson, Univ. of Iceland; K. Grönvold, NVI; S. Rist, National Energy Authority.

5/83 (8:5) The following is a report from Karl Grönvold, Páll Einarsson, and Helgi Björnsson.

"A subglacial volcanic eruption started in Grímsvötn below the W part of the ice cap Vatnajökull on 28 or 29 May. The central part of the Grímsvötn area is a caldera of about 35 km^2, one of the most active volcanic and geothermal areas in Iceland.

"Melting of ice due to intense geothermal activity and continuous inflow of ice causes a gradual buildup of meltwater below the 200 m-thick floating ice shelf in the caldera. This buildup culminates in a jökulhlaup when the level of the lake reaches a certain threshold value. The water drains in a catastrophic flood beneath the ice, 50-60 km into the rivers on the Skeidarársandur outwash plain S of the ice cap. The jökulhlaups last about 3 weeks and occur about once to twice every decade, the last one in Jan – Feb 1982.

"The last definite volcanic eruption in the caldera was in 1934 but small eruptions may have occurred since without being noticed due to the remoteness of the area. In most cases known previous eruptions have been accompanied by jökulhlaups. The new eruption was preceded by an intense earthquake swarm that began at about 0400 on 28 May. The largest earthquakes were in the magnitude range 3-3.5. Earthquake activity declined at about 1000 and soon after that bursts of volcanic tremor began to appear on the seismograms. The tremor amplitude increased at about 1500 and intense bursts of tremor were recorded for the rest of that day and the next day. During the following days the tremor gradually decreased in amplitude.

"The first definite observation of the eruption was on 29 May at 1030 from an aircraft, diverted to fly over Grímsvötn by request from seismologists. At that time the eruption had broken through the ice and produced a 5 km-long very thin ash fan downwind on the ice cap S of Grímsvötn. Steam clouds were observed in the direction of Grímsvötn on 28 May at about 2115. Weather satellite images on 29 May show a long narrow cirrus cloud that almost certainly originated at Grímsvötn during the morning of 29 May, at 0300 at the latest.

"When the eruption was observed on 29 May an opening had formed in the ice shelf inside the caldera near the SW wall. This lake was oval-shaped, about 300 m in diameter, and during 29 May covered by raft ice from the overhanging caldera wall. Explosions were observed in the lake at varying time intervals. The highest explosions reached about 50 m but the accompanying steam columns reached 1-2 km. During the next few days, weather conditions prevented direct observations except the height of the steam column. On 30 May, a maximum height of 6000-7000 m was observed, on the 31st 7000-8000 m, and on 1 June about 5000 m. The steam column was intermittent, never continuous. After that, no activity has been observed, but on 5 June a small island was observed in the steaming lake.

"No change has been observed in the rivers that drain the glacier. At this time, it is not known whether the eruption has affected the lake level within the Grímsvötn caldera."

Information Contacts: K. Grönvold, NVI, P. Einarsson, H. Björnsson, Univ. of Iceland.

Further References: Einarsson, P. and Brandsdóttir, B., 1984, Seismic Activity Preceding and During the 1983 Volcanic Eruption in Grímsvötn, Iceland; *Jökull*, v. 34, p. 13-23.

Grönvold, K. and Jóhannesson, H., 1984, Eruption in Grímsvötn 1983; Course of Events and Chemical Studies of the Tephra; *Jökull*, v. 34, p. 1-11.

KRAFLA

N Iceland
65.71°N, 16.75°W, summit elev. 818 m
Local time = GMT

This large caldera is 150 km N of Grímsvötn and only 50 km from Iceland's N coast. A series of eruptions from 1724 to 1729 comprised its only recorded history before the events described below. SEAN's early reports listed the activity under Leirhnjúkur, the principal vent system of the 1725–29 activity, on the caldera's W side.

General Reference: Jacoby, W., Björnsson, A., and Möller, D. (eds.) 1980, Iceland: Evolution, Active Tectonics, and Structure; *Journal of Geophysics*, v. 47, nos. 1-3, 277 pp. (33 papers; 6 on Krafla).

12/75 (1:3) A minor eruption consisting mainly of steam began at about 1120 on 20 Dec. A small lava flow also occurred. The eruption was preceded by minor local earthquakes that began at about 1000. A 2 km-long fissure opened, and lava was erupted from 3 vents. The main lava flow lasted for 1 hour, and the eruption was essentially over by 1700 that same day.

Information Contact: H. Sigtryggsson, Icelandic Meteorological Office.

9/76 (1:12) More than 2 m of subsidence occurred in a 2-3 km^2 roughly circular area, immediately E of Leirhnjúkur (about 4 km W of Krafla; fig. 17-2), between 20 Dec 1975, the first day of a minor eruption of Leirhnjúkur, and the end of Jan, 1976. Subsidence was followed by uplift of about 6.5 mm/day, beginning in Feb and continuing to the present (fig. 17-3).

The area of subsidence and subsequent uplift is thought to be underlain by a magma chamber about 3 km in diameter with its root at about 3 km depth. Assuming that the cause of uplift was flow of magma from below into the chamber, velocity of flow was of the order of 4-4.5 m^3/sec.

The number of earthquakes per day in the area (fig. 17-3) has increased considerably since early July, although remaining an order of magnitude less than the 1500/day experienced in Jan, then decreased rapidly in the last days of Sept.

Information Contact: H. Sigtryggsson, Icelandic Meteorological Office.

Reference: Björnsson, A., Saemundsson, K., Einarsson, P., Tryggvason, E., and Grönvold, K., 1977, Current Rifting Episode in North Iceland; *Nature*, v. 266, p. 318-323. [Originally cited in 2:3.]

1/77 (2:1) On 20 Jan a remarkable intrusion of magma was documented in N Iceland, one of a series of similar events during the last year [Sept and Oct/Nov] that are being carefully studied by virtually all geoscientists in Iceland. For a detailed account see Björnsson et al, 1977. The following summary was provided by Gudmundur Sigvaldason.

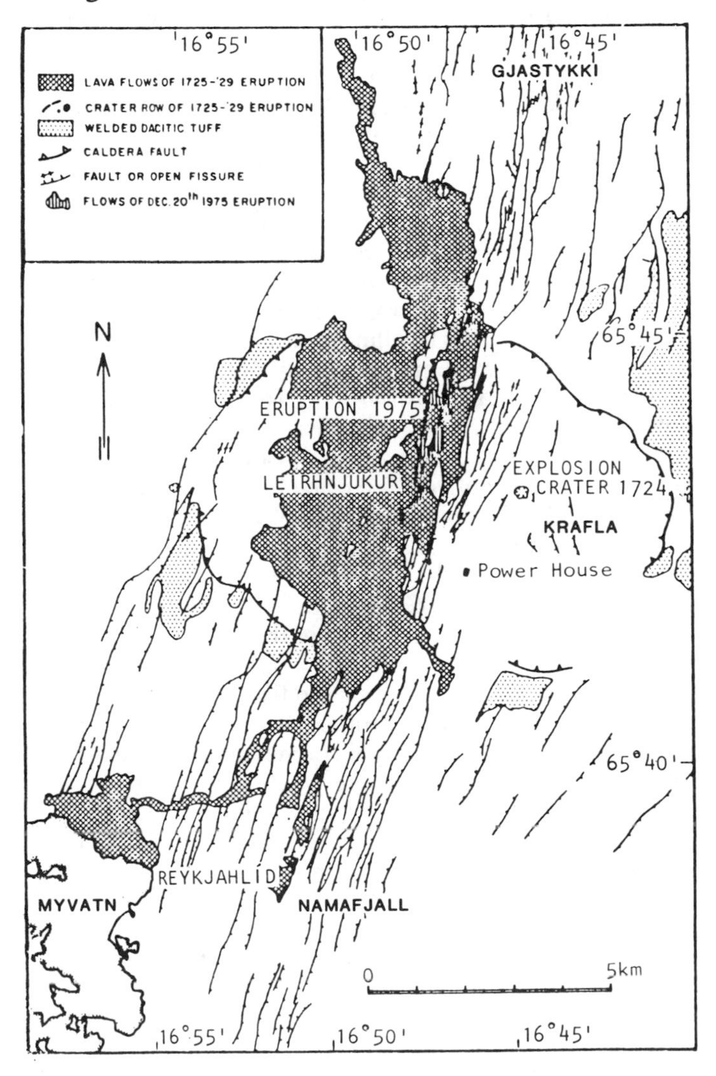

Fig. 17-2. Outline geologic map of Krafla caldera and the associated fault swarm. Note that Leirhnjúkur is a feature within the caldera and that Lake Mývatn intercepts a portion of the fault swarm. From Björnsson, et al., 1977.

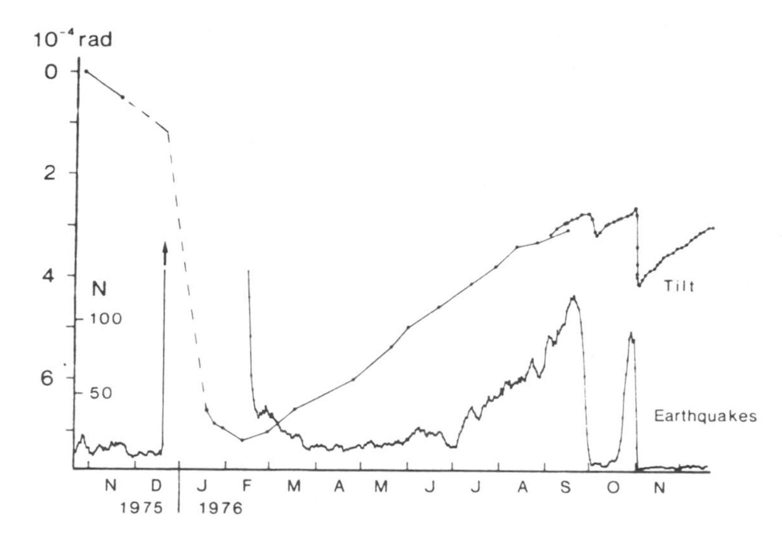

Fig. 17-3. Deformation and seismicity at Krafla caldera, Nov 1975 – Nov 1976. *a*, N-S component of tilt at the Krafla power house; increasing numbers mean tilt down towards the N. *b*, 5-day running average of the number of earthquakes per day recorded at the seismic station near Reykjahlíd. Tilt was measured by precision levelling until Aug 1976, after which a conventional "wet" tiltmeter (79 m base) was used. From Björnsson et al., 1977.

On 20 Dec 1975, intense earthquake activity began in Krafla caldera, 10 km N of Lake Mývatn in N Iceland. Within half an hour a volcanic eruption had occurred, producing a minor lava flow (1:1). During the following 2 hours, earthquake activity propagated along 40 km of a fissure system N of the caldera. Horizontal ground movement of the order of 1 m and vertical displacement of the same magnitude were recorded within the fissure zone. Within Krafla caldera, a subsidence of 2.5 m was measured by precision levelling relative to Lake Mývatn. A high level of earthquake activity (maximum magnitude 6.3) continued within and N of Krafla caldera until mid Feb. In March, significant inflation was recorded within Krafla caldera, and drastic chemical changes were noted in both drillhole and fumarole emissions from the caldera's geothermal field. While inflation continued at a steady rate, earthquake activity remained low and constant until July, when the number of seismic events started to increase. This activity culminated at the end of Sept, with rapid deflation of the Krafla area, accompanied by harmonic tremor and migration of hypocenters 20 km northward along the fissure system.

Seismic activity discontinued with this deflation event, but inflation started again immediately. When the ground level reached the same position as before the deflation event, seismic activity rapidly increased again. This cycle culminated, at the end of Oct, with another (and much greater) deflation event, accompanied by the same seismic behavior as before.

As with the 2 earlier deflations, inflation started immediately, and at the same rate as before. This rate, when combined with the measured amount of deflation, enabled geologists to estimate exactly the timing of the third event, which started shortly after midnight on 20 Jan 1977. Tilt measurements indicated a deflation of the order of 100 μrad during 12 hours, accompanied by continuous harmonic tremor and frequent earthquakes that migrated in the same direction as on previous occasions. This deflation, however, lasted for < 24 hours (in contrast to 5 days for the Sept and 2 days for the Oct deflation) and only 12 km of the fissure system was active this time. Surface fissures to 40 cm in width were measured in the snow, and fuming (water only) was observed along the active zone. After deflation, surface geothermal activity was limited to the end of the 12 km fissure system farthest from Krafla caldera.

The last major eruption in the area, the Mývatn fires, occurred in 1724 – 29. The activity culminated during the last 2 years of the 5-year period with voluminous outpouring of basaltic lava covering 35 km^2. Contemporary descriptions of these events indicate some similarities with the current behavior at Krafla. Using the same "inflation" approach they used successfully in Nov, Icelandic geologists are predicting the next deflation event on 20 March 1977.

Information Contact: G. Sigvaldason, NVI.

3/77 (2:3) After the deflation event of 20 Jan 1977, the Krafla caldera started inflating again. The rate of uplift near the center was ~ 10 mm/day. Earthquake activity was at a minimum after the 20 Jan deflation. During the first week of March the earthquake activity increased again, and during March there have been 100-150 earthquakes/day compared to 100/day prior to the Jan deflation event, all confined within the caldera. The largest earthquakes are of M 3.6. This increase in seismic activity was predicted 2 weeks in advance based on the inflation rate of 20 Jan. One of the fissures monitored during the last few months showed significant widening during the last 2 inflation periods and contraction during deflation. This fissure is now widening at a slowly increasing rate (Björnsson et al. 1977).

Information Contact: K. Grönvold, NVI.

4/77 (2:4) Inflation at Krafla continued irregularly until late April, while 100-130 earthquakes were recorded per day.

Harmonic tremor began at 1317 on 27 April, followed at about 1400 by series of earthquakes centered in a fissure swarm S of Krafla. About one M 3.5-4.5 event occurred per minute during the swarm, which culminated at 1830. An eruption from a discontinuous fissure extending 3 km N from Leirhnjúkur (about 4 km W of Krafla) had begun before 1600, when a minor ashfall was recorded in the Mývatn area (about 10 km SW of Krafla). A 200 m × 40 m lava flow was extruded from the N end of the fissure and steam and mud along the rest of its length. Tilt measurements indicate a 1 m subsidence of the caldera bottom in 17 hours, then renewed inflation after subsidence ended. New 2 m-wide fissures opened and fumarolic activity began in the Mývatn area, where more than 1 m of vertical displacement occurred, causing damage at a factory. Earthquakes continued on 1 May, but were declining.

Information Contacts: G. Sigvaldason, NVI; P. Einarsson, Univ. of Iceland; H. Sigtryggsson, Icelandic Meteorological Office.

5/77 (2:5) Caldera floor inflation of about 1 cm/day resumed on 29 April, 2 days after the 27 April deflation event. Seismicity has declined to 400-1000 events per day, in contrast to rapid declines after previous deflation episodes to < 10 events/day (fig. 17-4). Hydrothermal activity remained higher than before 27 April.

Information Contacts: G. Sigvaldason, NVI; P. Einarsson, Univ. of Iceland.

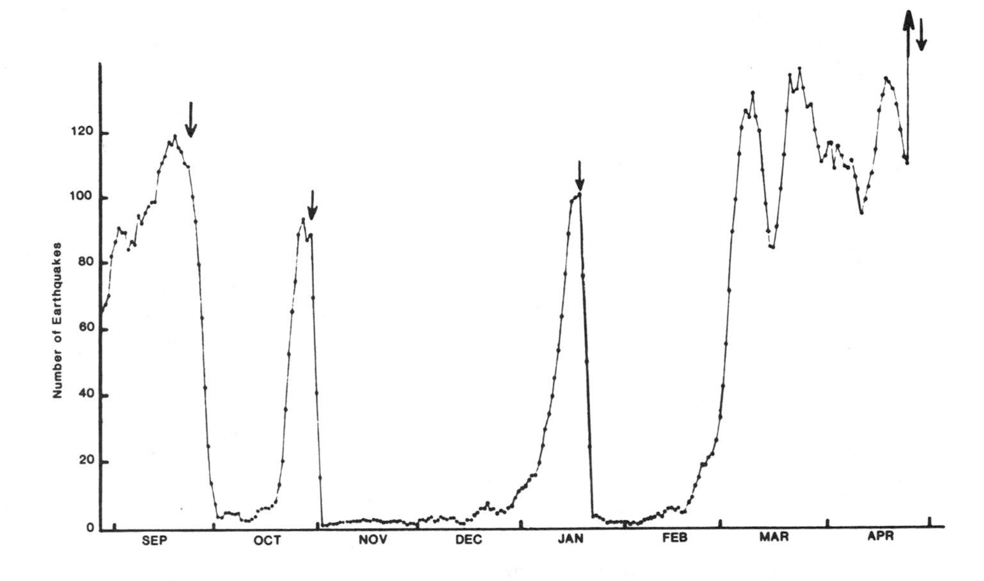

Fig. 17-4. Five-day running average of the number of earthquakes per day at Krafla caldera, Sept 1976–April 1977. Deflation events are shown by ↓. Courtesy of Páll Einarsson.

6/77 (2:6) Inflation has continued at a uniform rate. Seismicity has declined from 400-1000 earthquakes/day in late May to 100-150/day in late June. All earthquakes have been located S of the caldera, where maximum deformation occurred during the 27 April deflation event.

Information Contact: G. Sigvaldason, NVI.

7/77 (2:7) The following reports are from Karl Grönvold.

"Inflation in the central part of the Krafla caldera continues. The rate is 6-7 mm/day, similar to that of previous inflation periods. In the fault swarm S of the caldera (Námafjall), where maximum deformation occurred during the 27 April deflation, seismicity has decreased still further. Significant increase in gas discharge, accompanied by an increase in CO_2, has taken place. So far, there are no signs that the activity in the Krafla area is decreasing."

Information Contact: K. Grönvold, NVI.

8/77 (2:8) "Inflation continues in the Krafla caldera at a similar rate as before, with seismic activity still at a minimum. During the evening of 2 Aug, steam explosions started at the northern end of Leirhnjúkur, near the site of the 20 Dec, 1975 eruption (1:3)."

Information Contact: Same as above.

9/77 (2:9) Krafla caldera deflated on 8 Sept, accompanied by a short basaltic eruption and large-scale movement on faults S of the caldera. Since the 27 April deflation (2:4) the center of the caldera had inflated about 6-7 mm/day, and by 8 Sept had returned to approximately the pre-27 April elevation. An increase in the number of earthquakes within the caldera, which had preceded earlier deflation events, did not occur prior to the 8 Sept deflation.

Chronology of Events. 1547: First appearance on seismographs of continuous harmonic tremor, coinciding with the beginning of deflation.

1800: Effusive eruption began from a 900-m fissure N of the caldera rim, just N of the 27 April eruption site. By just after 2000, lava flows had reached their maximum extent, covering about 0.8 km^2. The eruption had ceased by [2230].

[2240]: Beginning of movement on faults in the Námafjall area, S of the caldera and about 14 km from the eruption site. The earthquakes were centered 3-4 km closer to the caldera than during the 27 April event, but many of the same faults were active. A few cubic meters of extremely glassy vesicular scoria were erupted from an 1134-m drillhole in the Námafjall area during the earthquake swarm.

After 2 days of deflation, inflation of the caldera resumed and was continuing in late Sept at the pre-deflation rate. The fault zone in the Námafjall area has widened > 1 m since Aug. Old thermal fields show increased activity and new fields have developed. Parts of the diatomite factory near Námafjall were severely damaged and some of the steam drillholes were seriously affected, but there were no accidents or injuries.

Information Contact: G. Sigvaldason, NVI.

Further References: Brandsdottir, B. and Einarsson, P., 1979, Seismic Activity Associated with the September 1977 Deflation of the Krafla Central Volcano in Northeastern Iceland; *JVGR*, v. 6, p. 197-212.

Larsen, G., Grönvold, K., and Thorarinsson, S., 1979, Volcanic Eruption Through a Geothermal Borehole at Námafjall, Iceland; *Nature*, v. 278, p. 707-710.

10/77 (2:10) After the eruption and deflation of 8 Sept, inflation of Krafla caldera resumed, and has continued at the pre-8 Sept rate. Dry tilt measurements indicated a complicated pattern of deformation. Repeated levelling showed that the caldera bottom had reached the pre-8 Sept elevation by mid-Oct.

During the 8 Sept event, magma was injected less than 1200 m beneath the thermal areas, as shown by one drill hole that ejected tephra for half an hour. Since then, thermal activity at Námafjall has increased conspicuously, making access to the area difficult. This fall farmers in the area literally harvested boiled potatoes!

Remeasuring of geodimeter lines in late Oct indicated continued rifting of the fissure system, amounting to 20 cm since 15 Sept. Seismic activity is presently at a minimum, as was the case before the 8 Sept event. Renewed volcanic activity is expected before the end of Nov.

Information Contact: Same as 2:9.

11/77 (2:11) The following is a report from Gudmundur Sigvaldason.

"Two continuously recording tiltmeters, of a new type designed and built by the NVI, have been installed in the Krafla-Mývatn area. One, close to the power house in Krafla, has been in operation since Aug, and replicates the water tiltmeter which has been operated at this location for over a year. A second tiltmeter was installed in late Oct close to a hotel at Reykjahlid and the recorders are placed at a nearby temporary observatory. This tiltmeter shows inflation towards the SE, confirming inflation between Reykjahlid and Hverfjall (just off fig. 17-2, SE of Reykjahlid and SW of Námafjall) discovered by dry tilt levelling in Oct.

"At 0600 on 2 Nov, all seismometers started to show harmonic tremor, the continuous tiltmeter in Krafla showed deflation towards the NW, and the Reykjahlid tiltmeter a slight deflation towards the SE. At about noon, both harmonic tremor and deflation in Krafla had stopped. Total deflation within the caldera was about 3 cm and had been fully reversed by 5 Nov. The 8 Sept tilt change at the Krafla power house should be fully reversed by early Dec. Only a few small earthquakes were recorded, none of which was felt. Increased thermal activity in Gjástykki, N of the 8 Sept eruption site (2:9), indicated magma intrusion N of the caldera.

"At Lake Mývatn, geothermal activity has increased and steam causes considerable problems in the area. The main road to eastern Iceland passes through the steam field, where visibility is usually only a couple of meters. The Mývatn area is enveloped in fog in calm weather, and, with the advent of winter, telephone and power lines are continuously threatened by accumulating ice. The diatomite factory near Mývatn has now been encircled by a dam 3-4 m high in order to divert lava flows. Equipment for water pumping has been installed on the dam for additional security. Civil defense is in a state of top alertness, access roads have been partially rebuilt to insure that they can be kept passable under any weather conditions, and the telecommunication system has been greatly improved."

Information Contact: Same as 2:9.

12/77 (2:12) Inflation continued at Krafla during Dec, at a rate similar to that recorded during the previous month. However, the focus of inflation has changed, and, although the amount of inflation significantly exceeds that which triggered earlier deflation events, no deflation had occurred by early Jan and seismicity remained at a minimum.

Information Contact: Same as 2:9.

1/78 (3:1) Renewed deflation at Krafla began on 6 Jan and continued through 22 Jan. Uplift then resumed and was continuing in early Feb.

On 6 Jan at 1700 the continuously recording tiltmeter at Krafla started to show slight deflation towards the N. The next morning at 0700, the rate of deflation increased and half an hour later harmonic tremor began. The rate of deflation (9.4 μrad/hr) was considerably slower than in previous events. In the following hours, it became evident from earthquake locations that magma injection into the fissure system N of Krafla was occurring. Epicenters migrated to 30 km N of Krafla on 7 Jan and had migrated 10 km farther N by the 9th, affecting the Dec 1975 rift zone (Björnsson, et al., 1977). The number of earthquakes gradually decreased 8-10 Jan, but magnitudes increased, reaching a maximum of 4.8.

Deflation ended 22 Jan after subsidence equivalent to about 110 cm had been recorded. Uplift then resumed, but has been interrupted several times, coinciding with periods of volcanic tremor. Earthquakes, many with $M > 4$, were continuing in early Feb from the rift zone, 25-50 km N of Krafla. The amount of rifting has not been measured, but about 2 m were added to telephone lines crossing the rift zone. Roads crossing the rift zone became impassable on several occasions, and one farmhouse was seriously damaged due to the opening of a ground fissure beneath it. No marked change in geothermal activity has been observed, and no visible eruption has occurred.

Information Contact: Same as 2:9.

2/78 (3:2) The following reports are from Karl Grönvold.

"Since 22 Jan, the center of Krafla caldera has been inflating at a similar rate as before. It is expected that the level at which previous deflation events were triggered could be reached near the end of June. The center of the Jan rifting is ~40 km N of the center of Krafla caldera. Large-scale rifting took place as in previous events, but weather conditions have prevented detailed measurements. The volume of magma estimated to have left the central reservoirs below Krafla caldera during this event is 80×10^6 m^3. This remained below the surface, since no eruption took place. The site of the main rifting is just S of the area where the main rifting took place following the initial outbreak in Dec 1975, but is about 10 km N of other events since 1975. This is a marked change in behavior, as all other rifting events along the fault swarm have been centered closer to the caldera. According to newspaper reports, the geothermal power station recently built within Krafla caldera has now started electricity production at 7-8 megawatts."

Information Contact: K. Grönvold, NVI.

4/78 (3:4) Inflation continued through March and April at about the same rate as before.

Information Contact: Same as above.

5/78 (3:5) "Inflation continued in the caldera at a similar rate as before, with an estimated magma inflow of ~5 m^3/sec. In a recently published prediction by the geologists working in the area the following points were made:

"A. Timing of the next deflation event: If inflation continues at the present rate, the ground level reached before the last 3 deflation events will be reached by 20 June. If inflation slows when this level is approached, as has often happened, the last week of July is the likeliest deflation date. Experience shows that once this level is reached, new events may come immediately or after a maximum delay of 2 months.

"B. Location of the next rifting: Large parts of the fault swarm have already been rifted. There are, however, 2 segments that still have not been rifted or where lesser rifting has been noted. One is a 10-15 km-long segment centered 20-25 km N of the caldera center. The other is the S part of the fault swarm, from 15 km S of the caldera center onwards.

"C. The probability of a lava eruption: The rifting is most likely approaching its final stages. The flow of magma into the holding chambers beneath the caldera continues at a constant rate. The probability of a basaltic lava eruption is therefore increasing."

Information Contact: Same as 3:2.

6/78 (3:6) By early July, inflation at Krafla had become irregular, as it neared the level at which previous deflation events were triggered. A new deflation event is expected in the near future, but it is not possible to predict its exact timing.

Information Contact: Same as 3:2.

7/78 (3:7) The following is a report from Karl Grönvold and Páll Einarsson.

"Krafla deflated again 10-13 July. After the deflation event of 6-21 Jan, the Krafla area inflated again at the previous rate of about 6-7 mm/day. About 10 June, the inflation slowed down and became somewhat irregular, as had happened shortly before most previous deflation events. Earthquake activity remained low. A new deflation event was anticipated in late June or the following weeks.

"On 10 July at about 1400 slow deflation was recorded at the tiltmeters. The deflation rate increased markedly shortly before 1700 and continuous tremor appeared on seismographs a few minutes later. Maximum deflation rate (11 μrad/hr) and tremor amplitude were reached about 2000. Earthquake activity increased and epicenters migrated northward along the Krafla fault swarm. Maximum earthquake activity occurred early 11 July in the uninhabited area 10-30 km N of Krafla caldera. The largest earthquakes reached magnitude 4.0 and only a few earthquakes were felt.

"The deflation rate decreased gradually during the next 2 days, and by 2000 on 13 July inflation had resumed. Total subsidence at the center of deflation (Krafla caldera) was about 60 cm. This subsidence is caused by movement of magma from the magma reservoir below the Krafla caldera to a fault swarm to the N. The amount of rifting in the fault swarm remains to be measured but movements on faults of the order of 0.5-1.5 m were observed. No new steam fields were formed but a very substantial increase was observed in the emission of a steam field formed in 1976. The increase occurred in the time interval 0300-0600 on 11 July.

"Inflation now continues at a similar rate as before. Based on previous experience, the next deflation event can be expected in late Oct or the following weeks."

Information Contacts: K. Grönvold, NVI; P. Einarsson, Univ. of Iceland.

8/78 (3:8) Inflation has continued through Aug at the same rate as before (6-7 mm/day).

Information Contact: K. Grönvold, NVI.

9/78 (3:9) The following reports are from Karl Grönvold.

"Since the end of the deflation event of 10-13 July, inflation within Krafla caldera has continued at a similar rate as before. The rate of inflation as recorded on the tiltmeters was very uniform until about 20 Sept. Since then the inflation has been slower and more irregular. Similar irregularities have generally been observed a few weeks before previous deflation events. It is expected that the land elevation will reach the previous level about the second week of Oct. After that, a deflation event can be expected within a few weeks.

"Where the magma will go this time is more uncertain. Very significant rifting has now taken place on most parts of the fault swarm N of Krafla and on the S part next to the caldera. Recent predictions have therefore emphasized the possibility of a rifting event in the southernmost part of the fault swarm. However, the possibility of another rifting event to the N, or an eruption, can in no way be excluded."

Information Contact: Same as above.

10/78 (3:10) "Inflation in the Krafla caldera continues at a somewhat reduced rate. The land height at the center of the uplift is now higher than at any time previously and deflation is expected within the next few weeks. A geologist is now permanently positioned in the area to advise the Civil Defense authorities. The seismometers and the continuously recording tiltmeters are now watched hourly day and night."

Information Contact: Same as above.

11/78 (3:11) The following report is from Karl Grönvold and Páll Einarsson.

"The deflation expected in the Krafla caldera started on 10 Nov. Deflation was first noted on tiltmeters at the Krafla power house at about 1000 and about 20 minutes later continuous tremor began on seismometers. The maximum deflation rate ($\sim$12 μrad/hr) was reached at about 1500.

"Earthquake epicenters moved northward along the fault swarm and the earthquake activity increased markedly at about 2400. Most of the activity occurred 15-25 km N of the caldera center. The deflation rate decreased gradually, and by midday 15 Nov inflation had started. At the same time, the earthquake activity began decreasing.

"The maximum subsidence at the center of the caldera was about 70 cm and the volume of magma that moved into the fault swarm from the reservoir below Krafla is estimated at about 40×10^6 m^3. Due to difficult weather conditions, it was not possible to look for surface faulting in the active zone and measurements of reference lines await better conditions. Although this deflation event involves twice the volume of the most damaging rifting event of 8 Sept 1977, it was not felt or noted by the population because of its location in an uninhabited area. It was therefore only observed instrumentally.

"Inflation continues presently at a similar rate as before and the next deflation event is expected in 4 to 6 months."

Information Contacts: K. Grönvold, NVI; P. Einarsson, Univ. of Iceland.

12/78 (3:12) After the deflation event of 10-15 Nov, inflation at Krafla resumed and was continuing at a similar rate as before.

Information Contact: K. Grönvold, NVI.

1/79 (4:1) The following reports are from Karl Grönvold.

"The ground inflation at the Krafla caldera continues as before. The ground level that preceded the Nov 1978 event will most probably be reached in early March. After that, a deflation event can be expected, accompanied by rifting as in earlier events, and possibly by a volcanic eruption."

Information Contact: Same as above.

3/79 (4:3) "The behavior of ground movement in the Krafla area is similar to that shortly before earlier deflation events. In early March, ground level reached the same height as before previous events, then inflation slowed markedly. Earthquake activity remains low, but the number recorded has increased significantly. A new deflation event with associated magma movement is now expected at any time."

Information Contact: Same as above.

5/79 (4:5) The following is a report from Karl Grönvold and Páll Einarsson.

"Deflation of Krafla took place 13-18 May. The main features are similar to the previous deflation event. Before this deflation, the volcano had inflated for 2 months beyond previous maximum levels and a deflation event had been anticipated since March. Earthquake activity above the magma reservoirs increased significantly during this time. This earthquake activity stopped with the first sign of deflation during the early hours of 13 May. The deflation rate increased only gradually, reaching a maximum (about 5 μrad/hr at the Krafla powerhouse) during the afternoon of 14 May.

"Small earthquakes occurred N of the volcano and the epicenters moved northward along the fault swarm. The seismic activity increased markedly on 14 May, shortly before midnight. The earthquakes were associated with extensive rifting in the fault swarm 10-20 km N of Krafla. Geodimeter lines in this part of the fault swarm extended up to 1.5 m during the event. Some of those lines have now extended 3.5 m in less than 1 year, in 3 rifting events.

"Total subsidence of the center of the deflation is estimated at 70-80 cm, corresponding to about 40×10^6 m^3 of removed magma. Inflation started again at about 1600 on 18 May and continues. Land elevation is expected to reach the previous maximum in 4-5 months."

Information Contacts: K. Grönvold, NVI; P. Einarsson, Univ. of Iceland.

8/79 (4:8) The following reports are from Karl Grönvold.

"Krafla continues to inflate. As in many previous inflation periods, the rate of inflation as recorded in the Krafla power house was greatest during the weeks following the May deflation event. The rate then slowed, and during Aug was ~1 μrad/day, $1/3$ of the initial rate. The ground level just prior to the May deflation event will at this rate be reached during the second half of Oct."

Information Contact: K. Grönvold, NVI.

9/79 (4:9) The rate of inflation during Sept was similar to that of Aug. The ground level at which previous deflation events have been triggered should be reached in mid-Oct.

Information Contact: Same as above.

11/79 (4:11) "About the middle of Oct, the inflation of Krafla reached the level at which deflation previously occurred. Since then, the number of earthquakes has increased gradually, but no deflation had begun as of 5 Dec. The rate of inflation during the last 3 months has been about $1/3$ of the rate during the initial inflation period, or less than 1 μrad/day."

Information Contact: Same as above.

12/79 (4:12) "The relatively slow but steady inflation of Krafla was interrupted by a minor deflation event 3-10 Dec. This deflation was very slow and no volcanic tremor was observed. Inflation resumed on 10 Dec and on 15 Dec the previous ground level had been reached.

"The number of earthquakes in the vicinity of the magma chambers had increased gradually in Nov, but during the mini-deflation earthquake activity decreased to background levels. When the previous ground level was exceeded in the second half of Dec, the number of recorded earthquakes started to increase again. The largest, of estimated magnitude 4, took place on 3 Jan and was widely felt in the area."

Information Contact: Same as above.

1/80 (5:1) Slow, steady inflation has continued through Jan. Earthquake activity remained at a level similar to that of late Dec.

Information Contact: Same as above.

2/80 (5:2) The following is a report from Karl Grönvold and Páll Einarsson.

"The gradual inflation after the minor Dec event was reversed about 1 Feb, when very slow deflation began and the earthquake activity associated with inflation stopped. As measured at the Krafla power house, the initial deflation rate of about 0.5 μrad/day increased on 7 Feb to about 1.3 μrad/day. In the evening of 10 Feb, a dense microearthquake swarm began near the S rim of the Krafla caldera and started to propagate southward. At about the same time, the deflation rate increased at Krafla, reaching a maximum of 7 μrad/day on 12 Feb. The rate of deflation decreased gradually thereafter; on 17 Feb it stopped and on 21 Feb inflation started again. The total deflation of 40 μrad is somewhat larger than the Dec deflation but still only about 1/6 of the last major deflation-rifting event in May 1979 (4:5).

"The active zone extended 7 km S from the rim of the caldera. The same part of the fault swarm was activated during 2 deflation/rifting events in 1977. This time, the earthquakes were significantly deeper in the crust than in the earlier events in that area. No earthquakes reached M 3 but many were felt and heard in villages near Mývatn (about 10 km SW of Krafla). Movements of faults over the seismically active area were very minor and the geodimeter lines near the S end of the active zone showed no significant movement."

Information Contacts: K. Grönvold, NVI; P. Einarsson, Univ. of Iceland.

3/80 (5:3) The following reports are from Karl Grönvold.

"After a small deflation event in early Feb Krafla showed a slow but continuous inflation until 16 March, when a new deflation and rifting took place, accompanied by an eruption. At about 1515 the continuously recording tiltmeters showed very rapid deflation and at the same time the seismometers showed intense volcanic tremor.

"The eruption started at about 1620 and the beginning was observed by 2 NVI staff members doing geodimeter work a few kilometers N of the eruption site. Most of the eruption, which produced very fluid basaltic lava, took place during the first 2 hours and the final flow faded about 2230 the same night.

"Eight separate lava flows were erupted, covering a total area of about 1.3 km^2 with average thickness of about 2 m. The total length of the active fissure is about 4.5 km and the longest continuous eruptive section about 800 m. The northernmost lava extends about 2 km N of the eruption site of Sept 1977. This fissure, active in April and Sept 1977, erupted again, but the southernmost new lava came from the fissure active in Dec 1975.

"Preliminary chemical analyses show a pattern similar to that observed earlier, with more evolved basalt erupted towards the S but less evolved to the N.

"The zone of new rifting, about 15 km long and 1-2 km wide, extends right through the Krafla caldera. Preliminary estimates indicate about 1 m widening of this zone. Most of the rifting apparently took place during the first hours of the deflation but the maximum earthquake activity took place towards the evening in the S parts of the rifted zone towards Námafjall.

"The maximum subsidence over the magma chambers is ~40 to 50 cm and the rate of deflation was greater than that observed in any previous deflation event.

"Inflation resumed early on 17 March at a rate exceeding any previous inflation rate. The rate is still higher than usual and about half the land subsidence was already recovered on 1 April."

Information Contact: K. Grönvold, NVI.

4/80 (5:4) The inflation rate had slowed by early May, but remained unusually high.

Information Contact: Same as above.

5/80 (5:5) "Krafla continued to inflate in early June, but at a slower rate than initially after the eruption and deflation of 16 March. Ground level over the magma chambers has regained its previous height. From experience of earlier events, another rifting event or eruption can now be expected, possibly within the next few weeks."

Information Contact: Same as above.

6/80 (5:6) "The inflation of Krafla continued during June, but at much reduced rates. On 15 June phreatic explosions started at Leirhnjúkur near the S end of the eruptive fissures, last active on 16 March. Similar phreatic activity has taken place previously at a similar phase in the inflation cycle."

Information Contact: Same as above.

7/80 (5:7) "By the end of May, the land level over the magma reservoirs at Krafla had regained the height reached prior to the 16 March deflation and eruption. A new event was therefore expected as tiltmeters showed continuous height increase indicating further magma inflow into the reservoirs.

"On 10 July at about 0800, the continuously recording tiltmeters started to show rapid deflation and about 1 hour later volcanic tremor was noted on the seismometers. Earthquake activity was soon seen to migrate northwards along the Krafla fault swarm. At about 1245 an eruption started about 8 km N of the magma reservoirs. At least 4 groups of fissures were active, extending over about 4 km. Activity was greatest during the first day but then decreased and concentrated in the northernmost crater group. The eruption then continued with decreasing vigor until 18 July when activity faded away in the early morning.

"The eruption took place far away from habitation so damage was minimal. The beginning of the eruption was observed from the air and on the ground by an unsuspecting tourist.

"The eruption was accompanied by large-scale fault movements close to the eruption site. Possibly the most spectacular sight was when large volumes of lava disappeared down into one of the faults about 3 km from the crater. This lasted for a few hours and the width of the falls was about 200 m.

"The lava covers 5-6 km^2 and is much thicker than in previous eruptions. Preliminary volume estimates indicate that the volume of erupted lava exceeds that which left the central magma reservoirs. This is the first time during the Krafla rifting episode when no magma appears to have been added to the rift zone.

"Inflation of the magma reservoirs started again on 16 July, before the eruption was over. Inflation now continues at a high rate so that the Krafla rifting and magmatic episode continues."

Information Contact: Same as 5:3.

8/80 (5:8) "Inflation of the magma reservoirs at Krafla, as monitored by ground movement, continues. At present,inflation is occurring at about the rate measured at similar ground levels during previous inflation periods, suggesting that the rate of magma supply remains similar. About 70% of the July deflation has now been recovered and the previous maximum ground level is likely to be reached after mid-Oct."

Information Contact: Same as 5:3.

9/80 (5:9) Continued inflation of the magma reservoirs beneath Krafla had raised the ground level by mid-Sept to nearly the height at which previous deflation events and eruptions were triggered.

Information Contact: Same as 5:3.

10/80 (5:10) "Inflation of the magma chambers, as monitored by crustal movement, continued after the eruption of 10-18 July. The previous ground levels were reached about the middle of Oct when 2 very minor deflations took place. Early in the morning of 18 Oct, continuously recording tiltmeters close to the crater row (about 600 m) showed rapid inflation while tiltmeters farther away showed no unusual behavior.

"At about 1842 the 3 tiltmeters close to the magma reservoirs suddenly started to show very rapid deflation and continuous tremor started on the seismometers at 1847.

"At 2203 a fissure eruption started about 1 km N of Leirhnjúkur, near the center of Krafla caldera. The eruptive fissure quickly extended N about 6 km and S towards Leirhnjúkur. The eruptive fissures were discontinuous and extended over a 7 km-long segment of the fissure system. In the early morning of 19 Oct activity decreased in the S part of the fissure system and finally concentrated on a 200-300 m-long segment near the northernmost original fissure.

"Activity decreased gradually after that until between 1500 and 1600 on 23 Oct when lava production ceased somewhat abruptly. Some activity remained in the craters until about midnight. During the eruption, tiltmeters near the magma reservoirs showed little movement but almost simultaneously with the ceasing of the lava production rapid inflation started again.

"The new lava covers an area of 11-12 km^2. The absolute amount of subsidence is not known yet but the amount of tilt recorded by the tiltmeters is only about half of that accompanying the July eruption. This indicates that again the bulk of the magma drained from the magma reservoirs reached the surface. The relatively small deflation and present high inflation rate suggest that previous ground levels could be reached during Nov or early Dec, with at least similar probability of a renewed eruption.

"The beginning of the eruption was witnessed by a few observers. The eruption site was closer to the populated area and the Krafla power plant than in previous eruptions and the population had been warned before the eruption started. But with the activity concentrating in the N the hazard quickly ceased."

Information Contact: Same as 5:3.

11/80 (5:11) "After the Oct eruption the magma reservoirs at Krafla inflated rapidly until the last week of Nov. Ground-level monitoring indicates that at that time land over the magma reservoirs was higher than before the Oct eruption. During the week or so prior to 3 Dec, the rate of inflation has been slower and more irregular.

"From the pattern of behaviour so far, an eruption can be expected to take place soon. Evacuation plans and civil defence measures have been strengthened in case of an eruption in the S part of the fissure system, closer to the village near Lake Mývatn."

Information Contact: Same as 5:3.

12/80 (5:12) Ground level remained about the same until 25 Dec when 4 days of very slow deflation began. Total deflation was about $^1/_3$ the amount recorded during the Oct eruption (5:10). At the end of the deflation episode, a swarm of small indistinct earthquakes was recorded, with epicenters about 10 km N of the caldera center. Inflation resumed on 29 Dec, and by 8 Jan the caldera had nearly returned to its pre-deflation ground level.

Information Contact: G. Sigvaldason, NVI.

1/81 (6:1) The following report is from Karl Grönvold and Páll Einarsson.

". . . The ground level at which previous deflation events and eruptions were triggered was again reached about 10 Jan, but inflation continued.

"On 30 Jan at about 0700, slow deflation of the magma reservoirs started, as recorded by tiltmeters at the Krafla power plant. The rate of deflation rapidly increased and at about 0730 tremor appeared on seismometers. Deflation rate and tremor amplitude reached a maximum at about 0900 and declined very gradually thereafter. The earthquake epicenters indicated movement of magma along the fault swarm toward the N. Soon after 1400, a fissure eruption started in the fault swarm 8-9 km N of the center of the magma reservoirs. The fissure soon extended to 2 km length and the lava front quickly moved toward the N. The eruption site was close to those of July and Oct, 1980 and the eruptive behavior was broadly similar. In the morning of 31 Jan, the fissure had shortened to about 300-400 m and the lava production rate had decreased somewhat.

"The eruption and very slow deflation were continuing on 2 Feb. The eruption site is in an uninhabited area and poses no danger to the local population. Observations are hampered due to remoteness and difficult weather conditions."

Information Contacts: K. Grönvold, NVI; P. Einarsson, Univ. of Iceland.

2/81 (6:2) The following reports are from Karl Grönvold.

"The initial vigorous phase of the eruption lasted until the early morning of 31 Jan. Then activity began to decrease, with shortening of the crater row that initially extended 2 km, then decreasing activity in the craters and declining lava production.

"The final activity in the craters died out just after 1400 on 4 Feb. During the eruption, slow deflation over the Krafla magma reservoirs, 8-9 km to the S, was observed, but inflation started again at about the same time as the eruption ceased. The lava covered 6.3 km^2 and appeared to be similar in volume to the 2 previous eruptions in July and Oct 1980 (5:7,10).

"Considerable movement of faults extending about 1 km N of the main lava (about 8 km N of the craters) was observed. Large volumes of steam emitted from these faults suggest that lava again forced its way down into the faults and then northward. Renewed earthquake activity in this region on 1 Feb was possibly associated with this fault movement.

"By carly March the inflation of the magma reservoirs had regained over half of the deflation that accompanied the eruption. Experience indicates that previous ground levels will be reached about the end of March to early April."

Information Contacts: K. Grönvold, NVI.

3/81 (6:3) "As of 2 April, land elevation over the Krafla magma reservoirs, as indicated by tiltmeters, was about the same as at the time of the eruption that began 30 Jan. Inflation continued and from past experience an eruption can be expected to take place in the near future, possibly within the next few weeks."

Information Contact: Same as above.

4/81 (6:4) Very slow inflation was continuing at Krafla as of 7 May. Tilt measurements indicated that the ground level was substantially higher than it had been just before the 30 Jan eruption.

Information Contact: Same as above.

5/81 (6:5) As of 10 June, slow inflation continued at Krafla.

Dartmouth College geologists visited Iceland 11-19 May and measured SO_2 emissions with the cooperation of NVI personnel. At Krafla, SO_2 was remotely monitored by COSPEC in the Leirhnjúkur area WNW of Krafla, where the Oct 1980 fissure eruption occurred (5:10). Significant concentrations of SO_2 were directly detected in the area of the Jan 1981 eruption fissures. The COSPEC detected SO_2 north and SW of Hekla's April 1981 lava flows.

Information Contacts: K. Grönvold, NVI; R. Stoiber, S. Williams, Dartmouth College.

8/81 (6:8) Inflation has continued through early Sept. Previous periods of inflation had been characterized by a single center of uplift beneath the caldera, but data gathered by tiltmeters since 4 Feb has been more complex and may indicate multiple centers of uplift. Because of the changed pattern of inflation, the rate of magma inflow from depth can no longer be calculated, nor can the timing of future deflation events or eruptions be predicted.

Information Contact: K. Grönvold, NVI.

11/81 (6:11) The following report is from Karl Grönvold and Páll Einarsson.

"The 30 Jan-4 Feb eruption (6:1-2) was followed by a rapid recovery of land elevation. Tiltmeters indicated that previous land levels had been reached by the end of March. After that the rate of tilt change was very slow. There are, however, indications that the rate of tilt change as recorded by the continuously recording tiltmeters did not reflect the land elevation changes as closely as in earlier inflation periods.

"On 18 Nov at 0036 seismometers started to show continuous tremor, and the continuously recording tiltmeters simultaneously showed the onset of very rapid deflation. Warnings of an imminent eruption were issued, and at 0152 an eruption broke out. Weather conditions were unfavourable, with strong wind and poor visibility. The eruption started about 1.5 km N of the hill Leirhnjúkur (near the center of Krafla caldera), the same place as the Oct 1980 eruption (5:10).

"In about 25 minutes the eruption extended S to Leirhnjúkur, about $^1/_2$ km farther S than in previous eruptions. At the same time the eruptive fissure extended N. When the first planes flew over the area about 0400, the eruptive fissures had extended to 8 km, and were displaying a more or less continuous curtain of fire. The fissure consisted of several segments, slightly offset with respect to each other, but the width of the fissure zone was always less than 100 m. The craters were situated very close to or coincided with those of the Oct 1980 eruption. The rate of deflation reached a maximum between 0115 and 0135, and the rate of lava production had apparently already reached maximum by 0400. Most of the lava flowed W (fig. 17-5). Lava production was highest from the N half of the fissure.

"The initial tremor amplitude was higher in the caldera region than in most previous eruptions. The tremor was mixed with earthquakes, but low-frequency events began about 1 hour after the onset of tremor. The characteristics of the seismic activity changed when the eruption broke out. The irregular, high-frequency, intrusion tremor gave way to more regular, low-frequency, eruption tremor. Low-frequency earthquakes continued for about 1 hour after the outbreak of the eruption, but after that the eruption tremor was the only seismic disturbance seen on the seismographs.

"By about 0600 the force of the eruption was already decreasing. In the afternoon land deflation stopped and inflation resumed. Segments of the fissure stopped activity and the active segments gradually shortened. Activity in individual craters or crater groups decreased and became more pulsating. Substantial lava production lasted longest in craters about 5 km N of Leirhnjúkur. Activity on these craters increased rather suddenly on 20 Nov, doubling from an estimated 50 m^3/sec. During the same afternoon tiltmeters showed deflation lasting a few hours. But inflation resumed and about noon on 21 Nov lava production decreased rather suddenly to about 30 m^3/sec. By the morning of 22 Nov activity in these craters ceased. Activity continued in the initial S crater until 23 Nov, but lava production was minimal. About 0900 the activity stopped but resumed for about 15-20 minutes around 1700. After that no activity was recorded.

"After inflation resumed, the tiltmeters showed inflation rates similar to those following previous eruptions. The area covered by the new lava is estimated at about 15 km^2 and the maximum land subsidence at 45-50 cm."

Information Contacts: K. Grönvold, NVI; P. Einarsson, Univ. of Iceland.

12/81 (6:12) The following is a report from Karl Grönvold.

"According to tiltmeters, the inflation of the Krafla magma reservoirs had by early Jan reached similar levels as before the Nov eruption (6:11). The inflation rate was slowing somewhat as is usually observed after the initial rapid recovery. This recovery rate was similar to the fastest rate observed after previous eruptions and deflations. There were therefore still no signs that the Krafla episode is nearing its end or final phases."

Information Contact: K. Grönvold, NVI.

Further Reference: Tryggvason, E., 1984, Widening of the Krafla Fissure Swarm During the 1975 – 1981 Volcano-Tectonic Episode; *BV*, v. 47, p. 47-71.

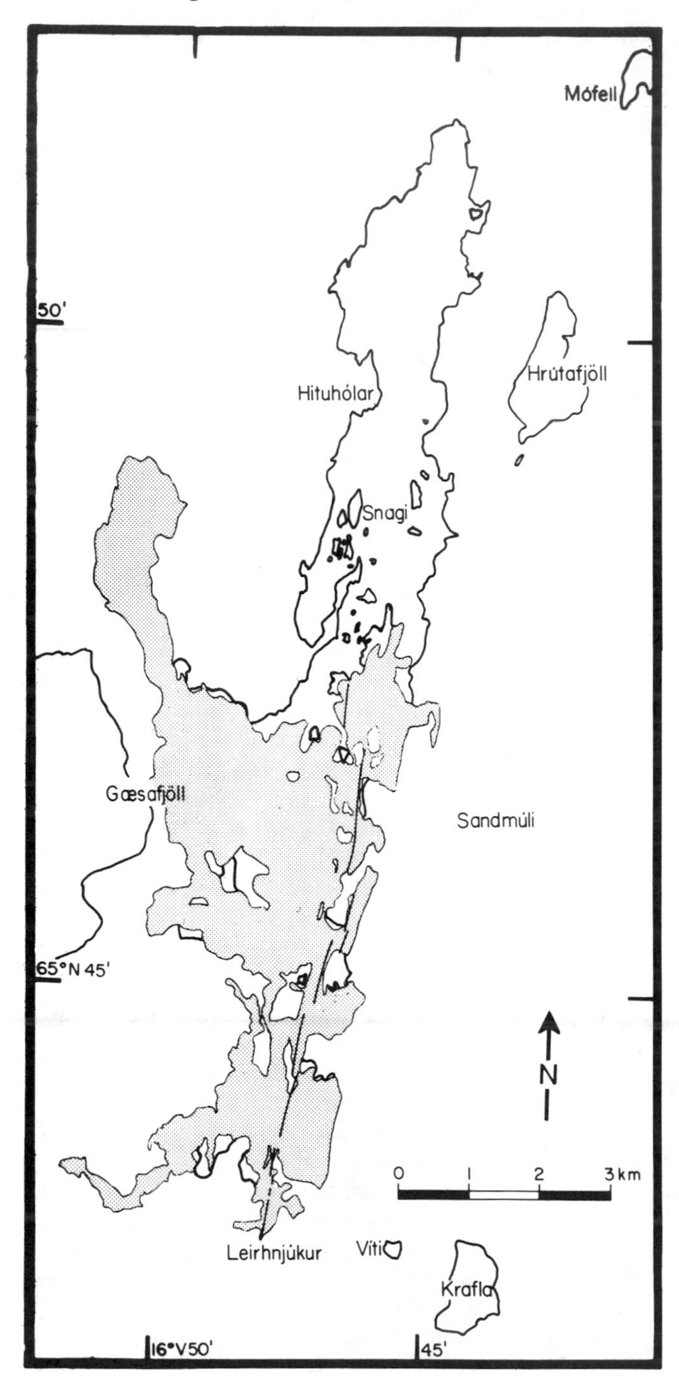

Fig. 17-5. 16-23 Nov 1981 lava (stippled area) and eruption fissures (N-S-trending lines within stippled area). The area covered by lava from the Dec 1975 – Jan 1981 eruptions is outlined. [Originally in 7:4.]

4/82 (7:4) The following is a report from Karl Grönvold and Kristjian Saemundsson.

"Rifting in the Krafla area with associated magmatic and eruptive activity has continued since Dec 1975. During the early phase of the episode, rifting and underground magma movement were the main features. Rifting occurred in individual relatively short events, the most recent in March 1980. Some of the rifting events were accompanied by eruptions but the volumes of the lavas were very small compared with the volumes of magma moving underground to fill the rifting segments. After March 1980 the character of the activity changed. Basaltic fissure eruptions are now the major feature but rifting is much less significant. A zone that now extends over 80 km has been rifted about 5-7 m.

"Since July 1980 there have been 4 basaltic fissure eruptions, each lasting 5-7 days. The lavas cover 30 km^2 and their estimated volume is about 0.2 km^3. The eruptive fissures extend 11 km N from Leirhnjúkur, which is situated right above the magma reservoirs and the center of inflation. In each eruption, only part of the eruptive fissures are active. Lava production has so far been dominantly from the N half of the fissure system. The most productive part of the fissure system is therefore situated near the center of the 80 km-long rifted zone, but the magma reservoirs that deflate during the eruptions are situated at the S end of the eruptive fissures. The chemical composition of the glasses indicates a more complex pattern than simple magma reservoirs draining magma to the surface.

"The pattern of inflation of the magma reservoirs between eruptions, and deflation during eruptions, has remained very regular (fig. 17-6). Eruptions take place only after the inflation of the magma reservoirs has reached well over the level of the previous inflation. After the most recent eruption in Nov 1981 the magma reservoirs inflated again and the previous ground levels were reached in early Jan. At present, inflation continues very slowly and an eruption at any time during the next few months would come as no surprise."

Information Contacts: K. Grönvold, NVI; K. Saemundsson, National Energy Authority.

5/82 (7:5) Ground level over Krafla's magma reservoirs remained high as of 4 June. No eruption has occurred since lava emerged from fissures N of the caldera 18-23 Nov (6:11).

Information Contact: K. Grönvold, NVI.

6/82 (7:6) Slow inflation continued through the spring. As of early July, however, no change in tilt had occurred over the magma reservoirs for several weeks and no eruption had taken place.

Information Contact: Same as 7:5.

8/84 (9:8) The following is a report from Karl Grönvold and Páll Einarsson.

"After a quiet interval of 2 years and 9 months, an eruption broke out at Krafla on 4 Sept 1984. The eruption in Nov 1981 (6:11) and associated deflation of magma reservoirs below Leirhnjúkur were followed by inflation, reaching previous levels in early 1982. Since then slow and intermittent inflation has continued, accompanied by earthquakes in the reservoir roof.

"Rapid deflation over the magma reservoirs followed by volcanic tremor began 4 Sept at about 2025, but the eruption broke out at 2349. The beginning of the eruption was observed from the air by alerted scientists and a television reporter. The first fissure segment opened about 6 km N of Leirhnjúkur, followed within a minute by another about 3 km to the S. The fissures quickly joined and in 1 hour reached their full length of 8.5 km, extending from Leirhnjúkur to the N. During the first hours lava was erupted along the whole fissure, advancing on broad fronts.

"Already in the early hours of the morning, the ac-

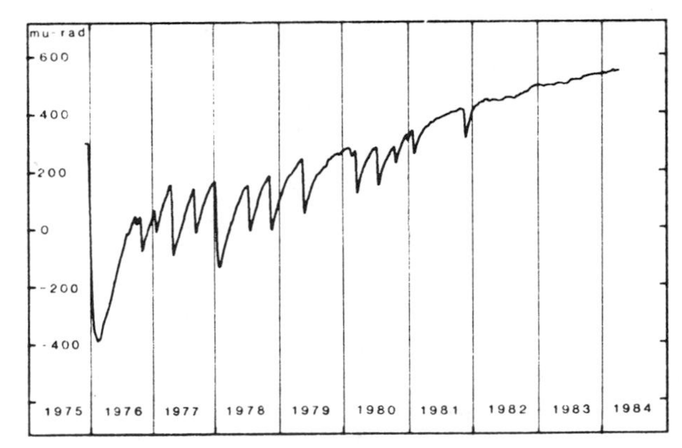

Fig. 17-6. Inflation/deflation sequence in the Krafla region, 1975–82, as measured by the N-S component of the water tube tiltmeter in the Krafla power station, about 1.5 km S of the center of inflation/deflation. One µrad of tilt corresponds to ~3.4 mm of vertical ground displacement at the center of inflation/deflation. From Tryggvason, 1984. [Not in original *Bulletin*.]

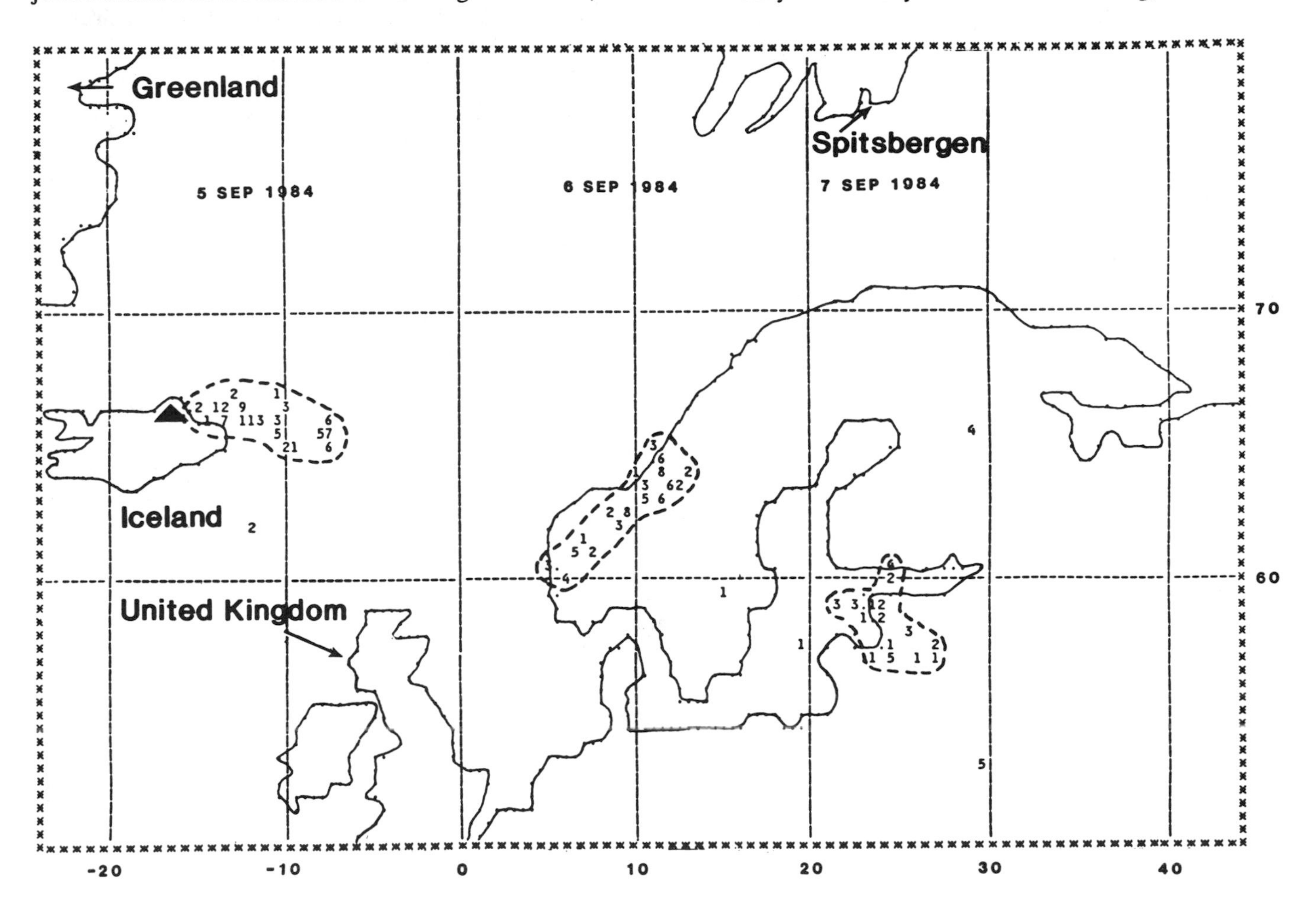

Fig. 17-7. Preliminary SO_2 data from the TOMS instrument on the Nimbus-7 polar orbiting satellite. The areas of enhanced SO_2 values at local noon on 5, 6, and 7 Sept 1984 are outlined. Krafla is marked by a triangle. NW Europe, Iceland, and E Greenland are outlined. Each number represents the average SO_2 value within an area 50 km across. Values above 9 are represented by A, B, etc. For a more detailed discussion of these values, see fig. 13-52. Courtesy of Arlin Krueger.

tivity had decreased and lava production on various sections of the fissures had faded out. By midday on 5 Sept, inflation had resumed. By 6 Sept, eruption on the S part of the fissure had ceased except on 1 crater, which then changed into phreatic activity.

"By 8 Sept, inflation rates over the magma chambers diminished and slow deflation started [but see 9:11]. On the northernmost part of the fissure, activity continued as of the morning of 12 Sept with significant lava production.

"As in previous eruptions, lava production was highest on the northern part of the fissure and has so far not constituted any threat to inhabited areas."

Information Contacts: K. Grönvold, NVI; P. Einarsson, Univ. of Iceland.

10/84 (9:10) The eruption continued in the northernmost part of the fissure system for nearly 2 weeks, before stopping on 18 Sept between 1600 and 1700. Inflation began again only a few hours after the end of the eruption.

Plumes from the eruption were detected by the NOAA 6 polar orbiting satellite and by the TOMS instrument on the Nimbus-7 polar orbiter. A NOAA 6 visible-band image 5 Sept at 0841, about 9 hours after the eruption started, shows a plume extending about 180 km E from Krafla. From a point source, the plume broadened, reaching about 35 km width roughly 90 km from the volcano and remaining about that wide farther downwind.

The TOMS instrument detected large areas of SO_2 enhancement at approximately local noon on 5, 6, and 7 Sept (fig. 17-7). The plume extended ESE from the volcano on the 5th. Twenty-four hours later, it was detected as a coherent body over Norway (roughly 1300 km ESE of Krafla), and another 24 hours later over Estonia (roughly 2200 km ESE of Krafla). NASA's preliminary estimate of the amount of SO_2 in the plume is 35,000 metric tons. An SO_2 ground monitor at Norrköping, Sweden (58°N, 16°E, about 1900 km ESE of Krafla) detected a 40 milliatmosphere-cm anomaly on 7 Sept. The TOMS instrument detected smaller areas of SO_2 enhancement on 10 Sept—again extending ESE from the vicinity of Krafla—and 18 Sept—NW of and detached from the volcano (fig. 17-8). On other days between 4 and 19 Sept there were no SO_2 enhancements detected by TOMS near Iceland.

Information Contacts: A. Krueger, NASA/GSFC; M. Matson, NOAA/ NESDIS; K. Grönvold, NVI.

11/84 (9:11) The following is a report from Karl Grönvold and Páll Einarsson.

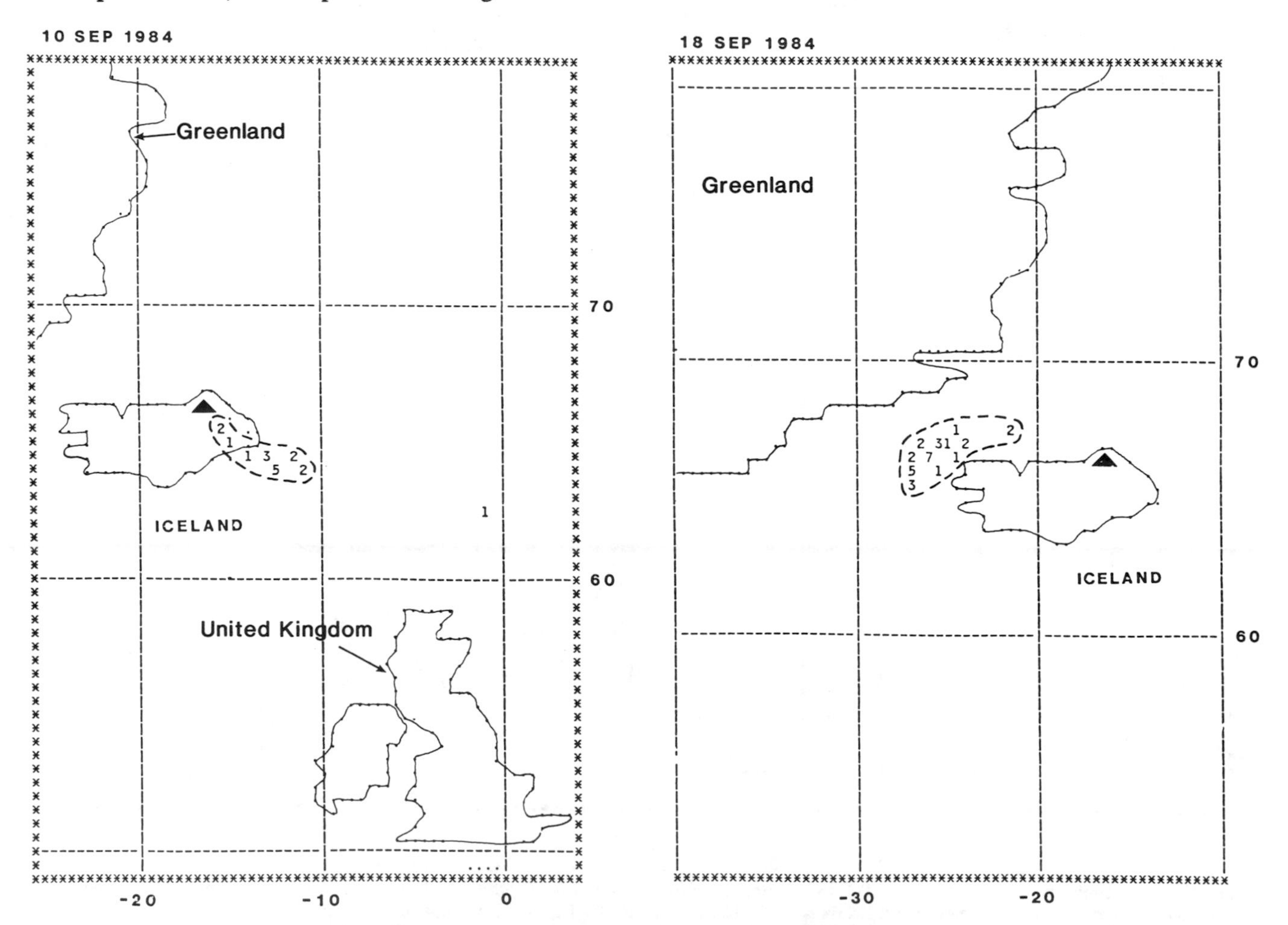

Fig. 17-8. Areas of enhanced SO_2 values detected in the vicinity of Iceland 10 and 18 Sept 1984 by the TOMS instrument on the Nimbus-7 satellite. Values of SO_2 enhancements are indicated as in fig. 17-7 and Krafla is again marked by a triangle.

"The early phase of the Sept eruption (9:8) was very similar to previous eruptions. Maximum eruptive activity was reached within about 1 hour, and while the whole 8.5 km-long fissure was active thin fluid lava advanced along the entire zone (fig. 17-9). In the early hours of the eruption's first morning, 5 Sept, activity decreased rapidly, dying out on many of the fissure segments, especially in the southern half. By 6 Sept only one crater remained active, just S of the N end of the original fissure.

"By midday 5 Sept, while activity was dying out along most of the fissure, deflation over the magma reservoirs below Leirhnjúkur stopped and inflation began (fig. 17-10). This situation changed on 9 Sept when deflation resumed. Around the same time, activity at the one remaining crater increased. Deflation and relatively vigorous eruption continued until the afternoon of 18 Sept when activity at the crater died out and inflation resumed. The inflation pattern since then is similar to that following previous eruptions.

"The new lava covers about 24 km^2 and the total area now covered by Krafla lava is about 36 km^2. As in some of the previous eruptions lava was observed flowing into older fissures, causing secondary rifting. Sometimes this lava reemerged farther along the fissures. Flames from burning gas were widely observed near the flow margin."

Information Contacts: K. Grönvold, G. Sigvaldason, NVI; P. Einarsson, Univ. of Iceland.

Further Reference: Tryggvason, E., 1986, Multiple Magma Reservoirs in a Rift Zone Volcano: Ground Deformation and Magma Transport During the September 1984 Eruption of Krafla, Iceland; *JVGR*, v. 28, p. 1-44.

12/85 (10:12) The following is from the NVI.

"The inflation of Krafla after the Sept 1984 eruption (9:8, 10-11) apparently came to a halt during Jan 1985. Recording tiltmeters at Krafla and Víti showed no significant inflation from Jan through late May although the conventional winter noise would have prevented detection of a miniscule inflation. After the winter noise subsided in April 1985, a very slight tilt, indicating deflation, prevailed in the Krafla power station, where the tilt rate was approximately 6 µrad/month.

"On 21 May, however, significant inflation started at Krafla with a tilt rate of about 10 µrad/ month. This corresponds to an inflation rate of about 1 mm per day at Leirhnjúkur. This inflation continued until 1 July. A noticeable subsidence was recorded 1-3 July on the Krafla and Víti tiltmeters. The tilt at Krafla in the conventional subsidence direction was about 14-17 µrad, corresponding to land subsidence of about 5 cm at Leirhnjúkur. The Víti tiltmeter showed about 12 µrad of tilt toward the WSW, the usual subsidence tilt direction. This corresponds to a removal of 2×10^6 m^3 of magma from the Krafla magma reservoir.

"After 3 July no noticeable inflation or deflation was observed at Krafla or Víti, although minimal inflation may have occurred during Nov. An apparent deflation

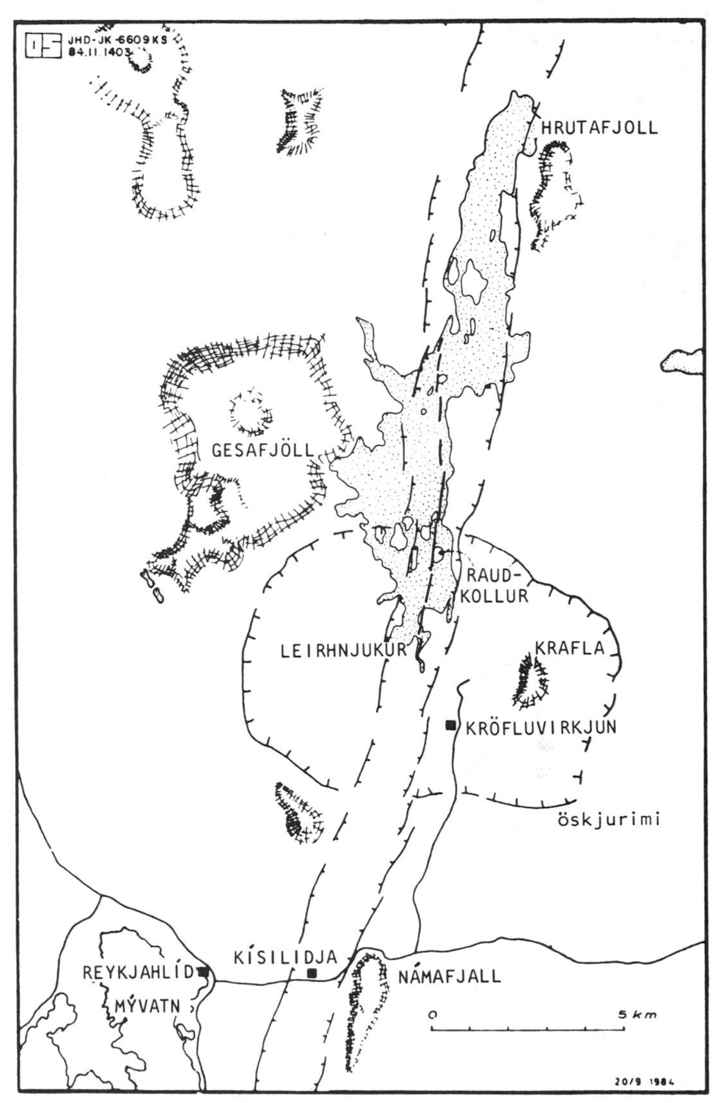

Fig. 17-9. Sketch map of the Krafla area by Kristjan Saemundsson (NVI) made with the aid of airphotos by Oddur Sigurdsson at the same institute. Lava flows from the Sept 1984 eruption are shown in a stipple pattern. Courtesy of Karl Grönvold.

at Krafla in July and early Aug was probably caused by thermal stress, but similar tilt was observed in July and Aug 1983 and 1984. This signifies an annual cycle in tilt at this station. Dry tilt measurements during the last days of May and again in late Oct indicate subsidence centered near Leirhnjúkur between those dates. The maximum subsidence was about 4.5 cm, similar to that indicated by the Krafla tiltmeter on 1-3 July.

"Although obvious ground deformation occurred 21 May–3 July, the net ground movement throughout the year was near zero and only further measurements in 1986 will show if any measurable deformation is in progress. Measurements of ground deformation at Krafla in 1985 do not allow any conclusion regarding the present progress or expected continuation of the activity. The inflation, if any, is slower than during any previous one-year period after 1975. The small subsidence event of 1-3 July is different from all earlier events, as no inflation was observed after the subsidence. Thus the behavior of Krafla is greatly different from what it has been

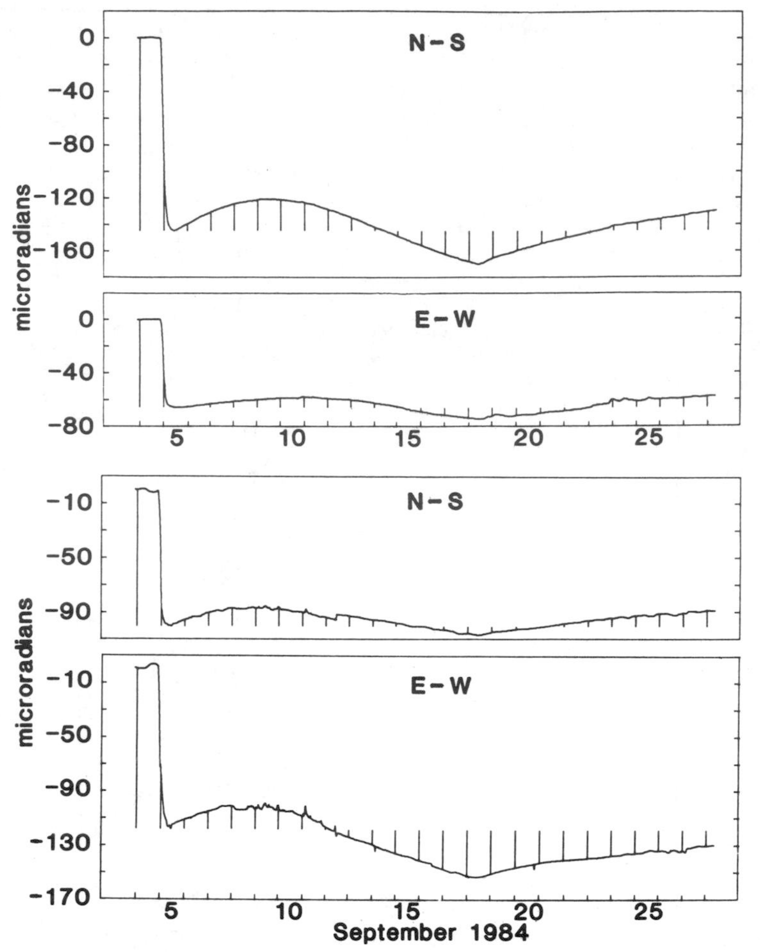

Fig. 17-10. Tilt at the Krafla powerhouse (Kröfluvirkjun), *top,* and Víti (about 3 km NE of the powerhouse), *bottom,* 4-27 Sept 1984. Vertical lines are time signals, at midnight. Courtesy of Gudmundur Sigvaldason.

during the previous 9 years and the experience from those years is of no use in predicting the continuation of activity at Krafla."

Information Contact: E. Tryggvason, Univ. of Iceland.

BEERENBERG

Jan Mayen Island
71.08°N, 8.17°W, summit elev. 2277 m
Local time = GMT + 1 hr.

This strato-volcano dominates Jan Mayen, a small Norwegian island 600 km NNE of Iceland's nearest point. It is the northernmost subaerial volcano in the world with historic volcanism (submarine eruptions have been reported from a site in the Arctic Ocean, and Holocene activity has been recognized in Spitsbergen). Beerenberg has erupted 8 times since 1732.

General Reference: Imsland, P., 1984, *Petrology, Mineralogy, and Evolution of the Jan Mayen Magma System*; Vísindafélag Islendinga, Reykjavík, 382 pp.

12/84 (9:12) A NE flank eruption of Beerenberg accompanied by recorded and felt seismicity occurred 6-9 Jan. First reports indicated that seismic swarm activity began 5 Jan at 2329 with typical low-frequency volcanic earthquakes. A later report put the start of the swarm at 0044 on the 6th, with 1-4 events, mostly explosions, recorded per minute and said that "normal" earthquakes did not precede the activity. Earthquakes were initially no stronger than magnitude 3, but magnitude 5.4 and 5.0 events were felt at 1122 and 1332 at the Norwegian meteorological station, about 30 km from the summit. Later that afternoon, at 1725, red clouds and smoke were observed above and N of Beerenberg from the meteorological station. Local topography made visual inspection of the volcano impossible. Seismic activity reached a level of about 100 events/hr on the 6th and decreased to about 5 events/hr on the 7th.

At 0645 on 7 Jan an eruption with lava flows was observed on the N part of the volcano from a commercial aircraft. At 0915, observers in another airplane saw lava entering the sea. At 1009, a thermal infrared image from the NOAA 6 polar-orbiting satellite showed a wispy plume extending about 60 km NE from the N end of Jan Mayen and a bright spot caused by heat from the lava flow. Weather clouds had obscured the volcano the 2 previous days.

At 1520, scientists from the NVI flew over the eruption and saw an E-W fissure extending into the sea from about 200 m altitude at the NE end of the island, very close to the site of the last lava-producing eruption (1970). The fissure was somewhat less than 1 km long, but during their flight only about 30 m of its length was active, emitting very low lava [spatter] and gas flames. Lava covered an area estimated at 1 km × 300 m and ash about 1.5 km^2. A dark ash column rose to a height of about 1 km; the plume was initially blown E, and then curved gently S, extending for several tens of kilometers.

At 1600 on the 8th, personnel from the Norwegian Coast Guard cutter *Senja* inspected the eruption area by helicopter and reported a 400 m-diameter crater. A 1 km-long lava flow moved NNE from the crater into the sea, which was "boiling" up to 150 m from the shore. The next day, the crater was again observed from a helicopter and located at 71.147°N, 7.998°W, about 10 km NE of the summit. The lava flow had almost stopped and "boiling" seawater was still present to 150 m from the shore. By 10 Jan the seismic activity had almost returned to normal levels. The preliminary location of most of the crater seismic activity was 71.13°N and 7.98°W, about 9 km NE of the summit, at depths of about 12 km.

Beerenberg's Sept 1970 eruption began with lava emission from a 6 km-long NE-trending fissure that extended almost to sea level on the NE flank. Activity soon concentrated at 5 major vents; lava flows reached the

Fig. 17-11. 7 April 1985 photograph showing steam emission from Beerenberg's new central crater vent, looking N from the opposite side of the central crater rim (fig. 17-12).

sea and formed a delta that was about 4 km long and 1 km wide before it was partially eroded (Sylvester, 1975). Since then, phreatic eruptions occurred in March 1971 from a SW flank crater, in Oct 1971 from the central crater, and in late 1972-early 1973 from a NE flank crater. [Imsland (1986, personal communication) attributes the 1971–73 activity to secondary windblown tephra and evaporation of the island's plentiful precipitation on contact with the cooling 1970 lava flows.]

Reference: Sylvester, A.G., 1975, History and Surveillance of Volcanic Activity on Jan Mayen Island; *BV*, v. 39, p. 313-335.

Information Contacts: K. Sandvik, Loran Station, Jan Mayen; J. Havskov, Univ. of Bergen; K. Grönvold, G. Sigvaldason, P. Imsland, NVI; W. Gould, NOAA/NESDIS.

Further References: Birkenmajer, K., 1972, Geotectonic Aspects of the Beerenberg Volcano Eruption 1970, Jan Mayen Island; *Acta Geologica Polonica*, v. 12, p. 1-16.

Imsland, P., 1986, The Volcanic Eruption on Jan Mayen, January 1985: Interaction Between a Volcanic Island and a Fracture Zone; *JVGR*, v. 28, p. 45-54.

Siggerud, T., 1972, The Volcanic Eruption on Jan Mayen 1970; *Norsk Polarinstitutt, Arbok 1970*, p. 5-18.

3/85 (10:3) Weather conditions make observation of Beerenberg rare and difficult, but during clear weather on 3-4 April a steam column could be seen rising 400-500 m above the central crater rim. An expedition from the Jan Mayen Loran station climbed the volcano 7 April. Steam was rising from a new 300 m-wide subglacial vent (fig. 17-11) within the central crater at 71.09°N, 8.17°W, in Weyprectbreen (Weyprect Glacier) near Gjuvtinden crag at 2113 m altitude (fig. 17-12). No activity was recorded by the Jan Mayen seismic array in early April. During an overflight 17 April, similar steam emission was observed. The pilot of the aircraft reported a strong sulfur smell. Two small craters in the Jan eruption area were emitting "smoke".

Loran station personnel estimated that the new central crater vent had developed within recent weeks. Very slight and apparently constant steam emission from Gjuvtinden has been observed since Sept 1971.

Information Contacts: K. Sandvik, P. Dalheim, R. Kirkemo, F. Moen, M. Gundersen, T. Eliseussen, Loran Station, Jan Mayen.

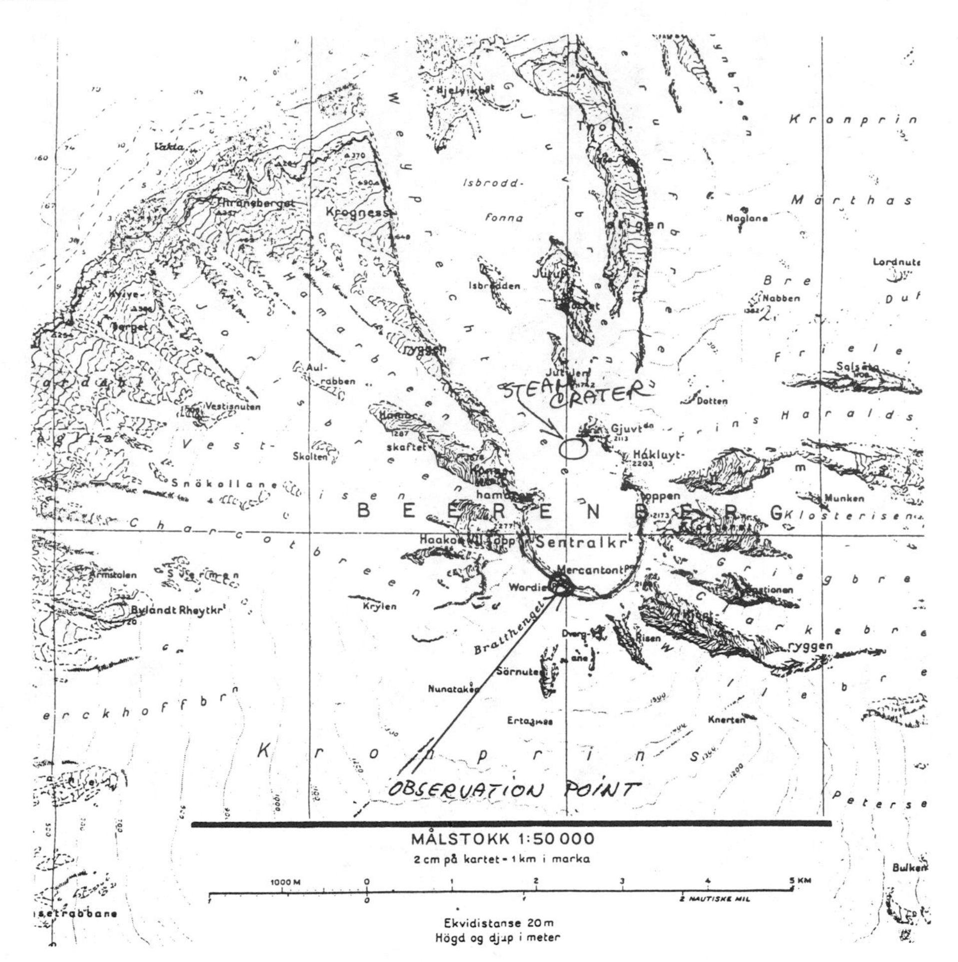

Fig. 17-12. Portion of topographic map showing Beerenberg's central crater and parts of its flanks. The position of the new vent is shown. Fig. 17-11 was taken from the "observation point."

ANTARCTICA

General Reference: LeMasurier, W.E. and Thomson, J.W. (eds.), 1989, *Antarctic Volcanoes: Late Cenozoic Volcanism On and Near the Antarctic Plate;* American Geophysical Union Monograph, Washington, DC (in press).

MOUNT MELBOURNE

Antarctica
74.35°S, 164.70°E, summit elev. 2733 m

Only fumarolic activity has been reported from this large volcano in Victoria Land, but the steam builds dramatic ice towers in the frigid climate. It is 365 km N of Ross Island and 3000 km directly S of New Zealand.

3/83 (8:3) The following is a report from Philip Kyle.

"Mt. Melbourne, one of two known active volcanoes in the Ross Sea embayment, is a composite volcano composed predominantly of lavas ranging from trachyandesite to trachyte. Observations in Dec 1972 indicated 3 main areas of steaming ground in which temperatures as high as 59°C were recorded at depths of 0.25 m. Numerous fumarolic ice towers were scattered throughout the summit area (Lyon and Giggenbach, 1974). Two members of the U.S. Antarctic Research Program climbed the mountain in Jan 1983. No change was noted in the number, size, and distribution of ice towers and steaming ground from the 1972 reports. Measured ground temperatures were also similar to those in 1972. There was no evidence of any change in the activity over the last 10 years."

Reference: Lyon, G.L. and Giggenbach, W.F., 1974, Geothermal Activity in Victoria Land, Antarctica; *New Zealand Journal of Geology and Geophysics*, v. 17, p. 511-521.

Information Contact: P. Kyle, New Mexico Inst. of Mining and Tech.

Further Reference: Keys, J.R., McIntosh, W.C., and Kyle, P.R., 1983, Volcanic Activity of Mount Melbourne, North Victoria Land; *Antarctic Journal of the United States*, 1983 review, v. 18, no. 5, p. 10-11.

MOUNT EREBUS

Ross Island, Antarctica
77.58°S, 167.17°E, summit elev. 3794 m

The southernmost volcano in the world with historic activity, Erebus is still 1381 km from the S Pole. This ice-covered strato-volcano has an active lava lake in its summit crater, and was erupting when first sighted in 1841. Erebus is the highest peak on Ross Island, site of the largest permanent base (McMurdo) on the continent.

3/76 (1:6) The following is from the Nov/Dec issue of the *Antarctic Journal of the United States.*

"The lava lake in Mount Erebus' inner crater (fig. 18-1) has increased in size since the 1974–75 austral summer. This observation was made by 4 New Zealand scientists who camped near the volcano's summit 30 Nov–7 Dec 1975.

"Last season the lava covered only part of the N half of the inner crater, moving along a curved path originating at the E end and disappearing in a tunnel at the W end. A 14-person French-New Zealand-U.S. team made the observations during 1974–75, and attempted unsuccessfully to enter the inner crater.

"According to P.R. Kyle, and W.F. Giggenbach, lava now fills the entire N half of the inner crater with the previously circular movement having been replaced by a series of overlapping areas where lava is welling up. An average of about 2 explosions per day were heard in the camp area; they appeared to be less violent, and were characterized by a more prolonged, whooshing noise compared to the short, sharp bangs heard during last season's visit.

"Only small pools of lava were observed 3 years ago. Since then a steady expansion in the area of exposed lava has taken place."

Information Contacts: P. Kyle, Ohio State Univ.; W. Giggenbach, DSIR, New Zealand.

1/77 (2:1) On 20 Dec 1976, a team visited the crater of Mt. Erebus. They reported the continued existence of the anorthoclase phonolite lava lake first discovered in Dec 1972. During the 4 years of its existence, the lava lake has slowly increased in size and is now approximately 100 m in diameter. A temperature of 980°C ± 20°C was obtained using an optical pyrometer. Vigorous convection occurred over part of the lava lake.

Information Contact: P. Kyle, Ohio State Univ.

12/77 (2:12) Philip Kyle and 2 associates visited the summit crater on 28 Nov. The persistent anorthoclase phonolite lava lake did not appear to have changed substantially since the previous observation in Dec 1976. The semicircular lake was about 130 m long and covered by an olive green crust, which was continuously being disturbed by minor upwellings. No major convective flow was noted.

Two small Strombolian eruptions originating from the lava lake and a small adjacent vent occurred during 5 hours of observation. One bomb was projected onto the main crater floor, 100 m above the inner crater floor. Large fresh bombs, up to 2 m long, were numerous around the NE main crater rim and were randomly scattered across the entire main crater floor, indicating that moderate Strombolian eruptions had occurred recently, probably during the previous 2-3 weeks. Further observations are planned for Jan 1978.

Information Contact: Same as above.

5/78 (3:5) Geologists from France, New Zealand, and the U.S. conducted studies on Mt. Erebus between 2 and 17

Jan. The anorthoclase phonolite lava lake, which has persisted since its discovery in Dec 1972, occupied the entire N half of the inner crater. There were 2 zones of upwelling in the 130 m-long, oval-shaped lake: one at the extreme SW end (often obscured by fumes) and a second about 30 m from the E end. Doming of the lava lake surface was occasionally observed, including one large blister that grew to ~80 m height before bursting.

Between 2 and 6 moderate Strombolian eruptions occurred daily (54 during the 16-day observation period) from a vent about 30 m S of the lava lake. The eruptions lasted 1-15 seconds and were frequently followed by emission of fume clouds containing Pelé's hair up to 15 mm long and 3 mm in diameter. Bombs up to 0.3 m in diameter were thrown over the main crater rim (about 270 m higher than the vent) and bombs as large as 10 m in diameter were found near the vent. The vent had built a small spatter cone with an orifice 2-3 m across containing a small lava pool (briefly absent after eruptions).

Oscillations of the lava lake level were observed 3 times on 16 Jan. The oscillations were periodic, with an amplitude of ~2 m and a period of 14-18 minutes. There was no apparent correlation with explosive activity.

Geothermal activity had changed little from previous observations. However, large cracks were developing above geothermal features on the N wall of the main crater and may lead to collapse in this area. The cracks, 30-100 mm wide and 1.5 m deep in many places, were traced ~200 m around the N main crater rim. Material falling from this area would land in the lava lake.

Information Contact: B. Scott, NZGS, Rotorua.

7/78 (3:7) The following is excerpted from Kyle and McIntosh, 1978.

"The most significant change in the activity during 1977–78 was an increase in explosive Strombolian eruptions. In the 1976–77 field season (22-31 Dec 1976) no explosive eruptions were heard, although in previous years they were frequent (Kyle, et al., 1982). Compared to the 1972–75 period, eruptions during 1977–78 were more frequent and possibly larger in size. It is believed that most of the eruptions originated from the Active Vent. The number and size of the bombs were greater than observed in any previous season since observations began in 1971–72. Pelé's hair up to 100 mm in length was particularly common and was found lying on the snowy N flanks over 3 km from the crater.

"The persistent lava lake showed no apparent increase in size over that observed last year (fig. 18-2). Subsidence along a large depression parallel to the ridge that divides the Inner Crater in half may be due to a slight lowering in the level of the magma column. On the E end of the lava lake, 2 raised benches of consolidated lava are also suggestive of a slight lowering in the lava lake.

"Activity within the lava lake consists of 3 main features: (1) lava upwelling in nearly circular areas; (2) small bubble-like degassing eruptions; and (3) downwelling of the consolidated crust along planar troughs or 'subduction zones'. While there was moderate and steady downwelling along the N wall, downwelling along central 'subduction' cracks was more rapid or at least more apparent. Migration of subduction zones themsel-

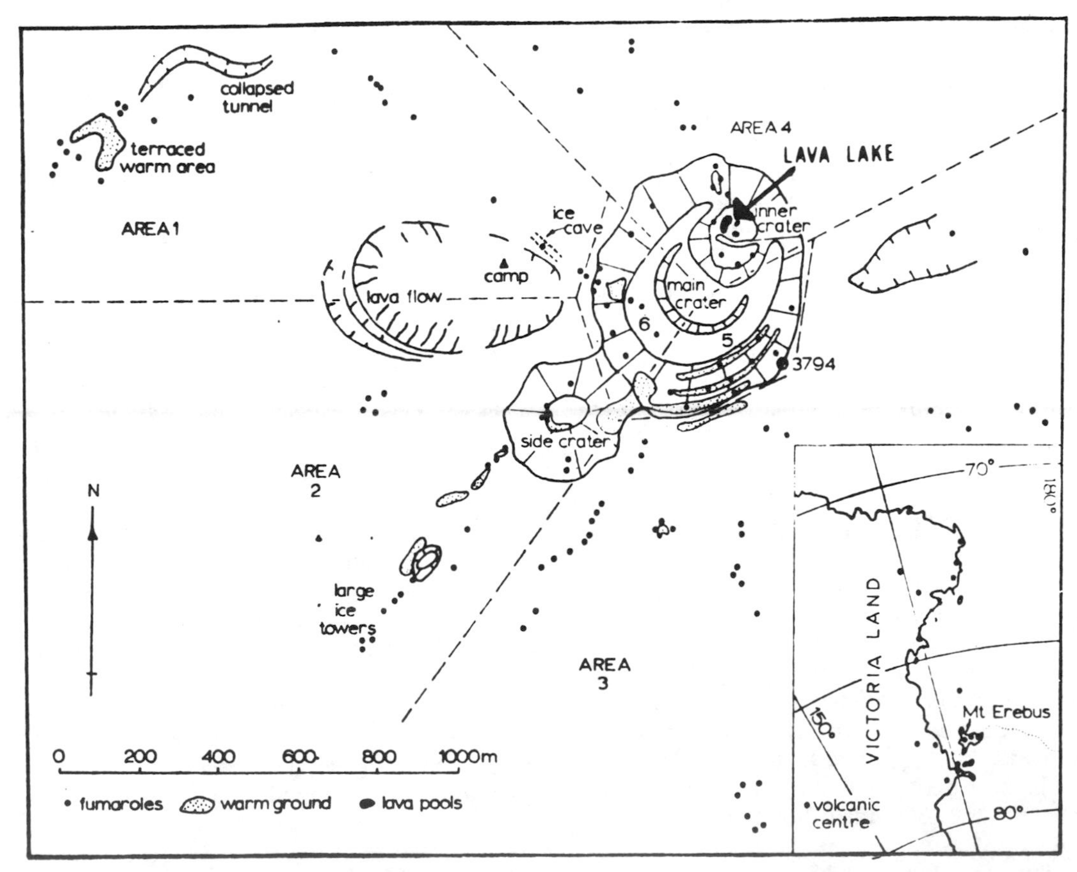

Fig. 18-1. Summit region of Mt. Erebus, and its location. Ice towers are marked as fumaroles. After Lyon and Giggenbach, 1974. [Originally in 2:1.]

ves was the most easily observed surface movement. Upwelling was difficult to detect as it did not usually involve bubble formation; instead, a slight increase of incandescence and cracking of the lava crust was the main surface manifestation.

"Major element analyses of 4 bombs collected in 1977–78 are indistinguishable (within analytical error) from analyses of ejecta collected in previous years. The magma column is therefore not undergoing rapid changes due to crystal fractionation or influxes of new magma. Apparently the lava lake is the surface expression of a stable convecting column of magma."

References: Kyle, P.R. and McIntosh, W., 1978, Obervations of Volcanic Activity at Mt. Erebus, *Antarctic Journal of the United States*, v. 13, no. 4, p. 32-34.

Kyle, P.R., Dibble, R.R., Giggenbach, W., and Keys, J., 1982, Volcanic Activity Associated with the Anorthoclase Phonolite Lava Lake, Mount Erebus, Antarctica; *in* Craddock, C. (ed.), *Antarctic Geoscience*; Univ. of Wisconsin Press, Madison, p. 735-745.

Information Contact: P. Kyle, Ohio State Univ.

10/78 (3:10) The following is from Philip Kyle.

"The persistent anorthoclase phonolite lava lake at Mt. Erebus was observed on 26 Oct during a 5-hour visit to the summit crater by Philip Kyle and others. First discovered in Dec 1972 by Giggenbach and others, the lava lake gradually increased in size over a 3-year period, but has shown little change for the last 2 years. Two small congealed lava flows were visible on the S side of the inner crater. One flow originated from a small vent in the SW quadrant of the inner crater, the other flow apparently came from the lava lake.

"The general appearance of the lava lake was one of increased heat flow. The consolidated crust on the lake appeared to be less continuous and was broken by numerous E-W cracks. Larger areas of incandescent lava were visible compared to observations made in Jan 1978. The lava lake appeared to have increased in temperature, but no quantitative measurements are available. Small bubble-like degassing eruptions were common over most of the lake, with a greater concentration toward the W. The level of the lava lake appeared similar to that in Jan, although the surface area may have been reduced due to the collapse of the N crater wall. Minor collapse is common around the whole inner crater wall.

"Only a few small bombs were found on the crater rim and none were observed on the main crater floor, which suggests that there has been a considerable decrease in Strombolian eruptions compared to Jan, when up to 6 eruptions per day occurred. Further observations at Erebus are planned for Dec [see 4:5]."

Information Contact: Same as above.

5/79 (4:5) The following is excerpted from Kyle, 1979.

"An average of 1.6 eruptions/day occurred during Dec 1978, compared to 3.6/day in Jan 1978 and an average of ~2.5 daily between Dec 1972 and Jan 1976.

"During an attempt on 23 Dec 1978 by a joint U.S.-New Zealand team to get a member into the Inner Crater, an eruption occurred from a small vent (termed the active vent) adjacent to the lava lake. Lava was observed to rise rapidly in the active vent, and as it reached the rim the explosion occurred. At the time of the eruption, New Zealand volcanologist W. F. Giggenbach had descended to about 25 m above the Inner Crater floor. One small bomb hit him above his knee, burning his woolen pants but causing no injury. Other bombs rose about 200 m above the Main Crater rim, many landing on the Main Crater floor. After the eruption there was a drop in the general level of the lava lake, which

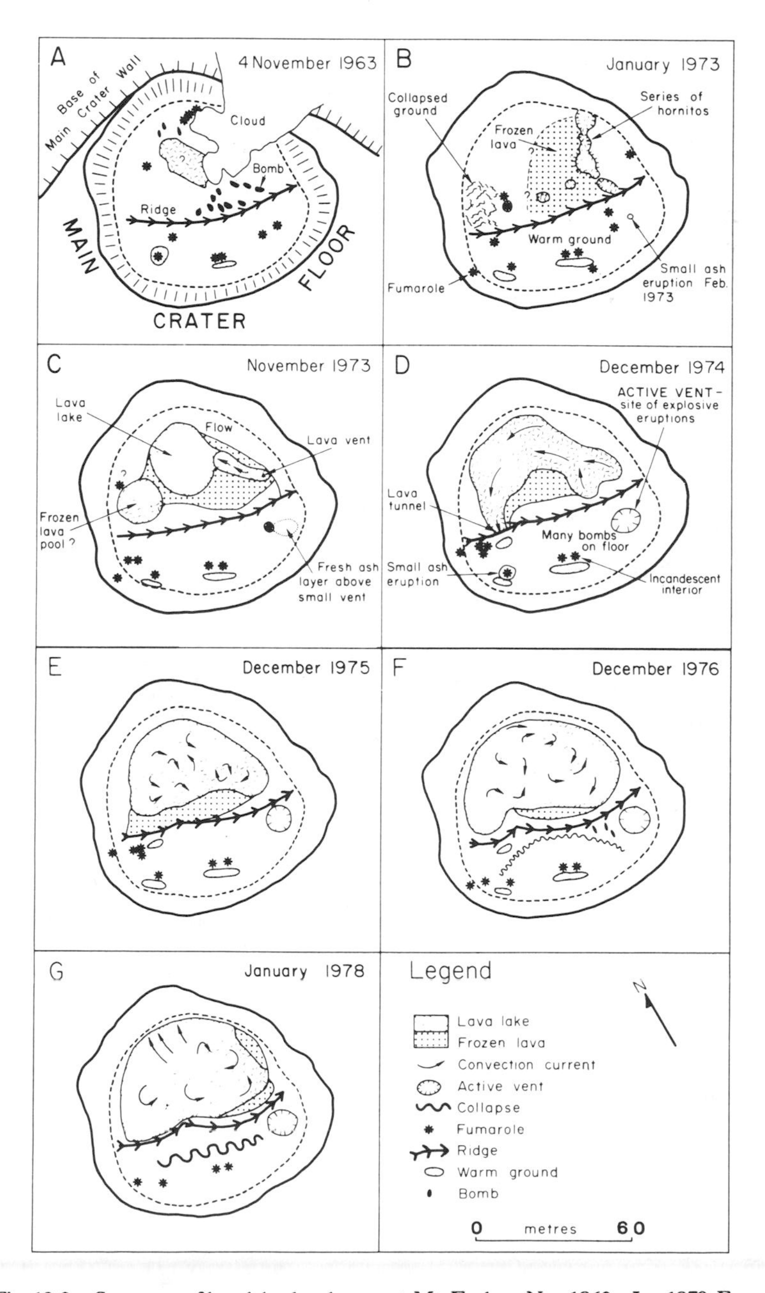

Fig. 18-2. Summary of lava lake development, Mt. Erebus, Nov 1963 – Jan 1978. From Kyle and McIntosh, 1978.

returned to its original level in about 15 minutes. It is believed that a subterranean connection exists between the active vent and the lava lake.

"The overall pattern of activity at Mt. Erebus is unchanged. The small Strombolian eruptions are likely to continue as long as the active vent is open and connected to the main magma column. Since 1976 there has been no increase in the size of the lava lake; collapse of the N crater wall has probably reduced the overall size. It is apparent from observations made inside the Inner Crater that the lava is perched above the floor in the Inner Crater. A ridge and small levee retain the lava to the N half of the Inner Crater."

Reference: Kyle, P.R., Volcanic Activity of Mt. Erebus, 1978/79; *Antarctic Journal of the United States,* v. 14, no. 5, p. 35-36.

Information Contact: Same as 3:7.

3/80 (5:3) The following reports are from Philip Kyle.

"The summit was visited by a Japan-New Zealand-U.S. group in Dec. The pattern of volcanic activity remains similar to that observed a year ago. Almost no change in the shape of the persistent anorthoclase phonolite (Kenyte) lava lake was noted. . . . The lake is oval, about 100 m long and 60 m wide, with total surface area approximately 4500 m^2 [but see 9:1]. The lava lake level appears to be slightly lower, by about 3-5 m, compared to 1 year ago. A small distinct bench, 1-5 m wide, occurs at several places around the perimeter of the lake and probably marks a former lake level, possibly that of last year [see 9:1]. Five conspicuous fumarole vents were observed on the bench. Each vent was zoned by concentric bands of differently colored sublimates. Incandescent lava was observed beneath each vent.

"Convection of the lava lake was variable, with upwelling occurring about $^1/_3$ of the distance from both ends of the lake. During periods of moderate convection, degassing was quiet and spread more evenly over the lava lake. Degassing bubbles were fewer, larger, and reached about 5-10 m in diameter. However, during stages of more active convection, degassing of the lava lake was noisy with more rapid upwelling. Explosive eruptions from the active vent continued at a frequency of about 2/day. One eruption was seen to eject bombs about 300 m above the vent. Minor ash eruptions occurred on 2 occasions.

"A tripartite seismic array consisting of 1 horizontal and 2 vertical seismometers was operated for 1 week at the summit. The system detected explosive earthquakes as well as other shallow events not accompanied by explosive activity. Some of the larger explosions could be seen on the WWSSN station at Scott Base, 30 km away.

"Major element analyses of whole rock and interstitial glasses indicate no variation in chemical composition of the anorthoclase phonolite between Dec 1972 and Dec 1978. The lava lake probably represents the top of an extremely stable convection system, presumably connected at depth to a major magma chamber. No attempt was made to descend into the Main or Inner Craters this year."

Information Contact: Same as 3:7.

2/81 (6:2) "The summit crater was visited by Japanese, New Zealand and U.S. scientists during late Dec and early Jan. A 1-day visit was also made in Nov. The anorthoclase phonolite lava lake was still present, although its level may have been slightly lower than that observed over the last 2 years [but see 9:1]. The 120 m-long oval-shaped lava lake still showed a simple convection pattern with lava apparently welling up from 2 centers about $^1/_3$ of the way from each end.

"Small Strombolian eruptions continued at a frequency of 2-6/day. The noise associated with the eruptions consisted of a long drawn-out roar, contrasted with the strong explosive eruptions heard in previous years. Although no eruptions were witnessed they are believed to occur from the small Active Vent, adjacent to the lava lake. Very few bombs were found on the main crater rim during Dec, but in Jan there were a few sharper explosive eruptions and these ejected material onto the rim.

"The 3 nations commenced a new project and installed 3 permanent seismometers on the mountain. The seismometers have radio-telemetry links with Scott Base (the New Zealand research station on Ross Island). Two seismometer stations are on the W flank of the volcano at altitudes between 1500 and 1900 m about 5 km from the crater rim. It is anticipated that these stations will run until April, when darkness sets in. The stations should be reactivated in Oct, when new batteries will be installed and the solar panels will function. The third station is at the summit, and has its batteries buried in warm ground. It is hoped that it will operate all year round. The summit station is also transmitting the output from an acoustic sensor (a microphone which monitors the sounds of volcanic eruptions) and a large wire loop around the crater, which monitors induced currents. A fourth permanent seismic station will be installed in Dec 1981.

"Preliminary observations from the seismic network, which can detect events with magnitudes less than 1, indicate a surprisingly high level of microearthquake activity, with up to 10 events per day. Some of these are apparently tectonic earthquakes occurring some distance from Mt. Erebus. Antarctica has been considered aseismic, but this is apparently not the case, at least not for microearthquakes with magnitudes < 3."

Information Contact: Same as 3:7.

3/82 (7:3) "The summit crater was visited by New Zealand and U.S. scientists during late Nov and Dec 1981, and on one day in late Jan 1982. The anorthoclase phonolite lava lake was still present and the pattern of activity was similar to that observed over the last 5 years.

"The lake was undergoing simple convection. Small Strombolian explosions continued at a frequency of 4-6/day. The eruptions were believed to originate from the Active Vent, adjacent to the lava lake. Many fresh bombs were found on the crater rim, suggesting that the eruptions were the strongest observed in the last 3 years. This may reflect an increase in distance between the lip of the Active Vent and the underlying magma level.

"The lava lake grew from small hornitos in 1972 to a semi-circular lake about 100 m long by 1976. Since then there has been little change in surface area, but a slight lowering in the lake level has occurred. No measurements of the magma column withdrawal were available but it was small, perhaps 5-10 m over the last 3 years [but see 9:1]. The withdrawal was possibly equivalent to the amount of material ejected by the small Strombolian eruptions. A deformation survey pattern set up in Dec 1980 was remeasured in Dec 1981; . . . data indicate [little change in the width] of the crater rim, [despite the] lowering of the column. Withdrawal was [however] suggested by the development of a semi-radial fracture, on the main crater floor, that parallels the inner crater rim."

Information Contacts: P. Kyle, New Mexico Inst. of Mining & Tech.; P. Otway, NZGS, Wairakei.

Further References: Dibble, R.R., Kienle, J., Kyle, P.R., and Shibuya, K., Geophysical Studies of Erebus Volcano, Antarctica, from 1974 December to 1982 January; *in* Lynch, R.P. (ed.), 1984, *Tenth Antarctic Issue, New Zealand Journal of Geology and Geophysics*, v. 27, no. 4, p. 425-455.

Wiesnet, D.R. and D'Aguanno, J., 1982, Thermal Imagery of Mount Erebus from the NOAA-6 Satellite; *Antarctic Journal of the United States*, v. 17, no. 5, p. 32-34.

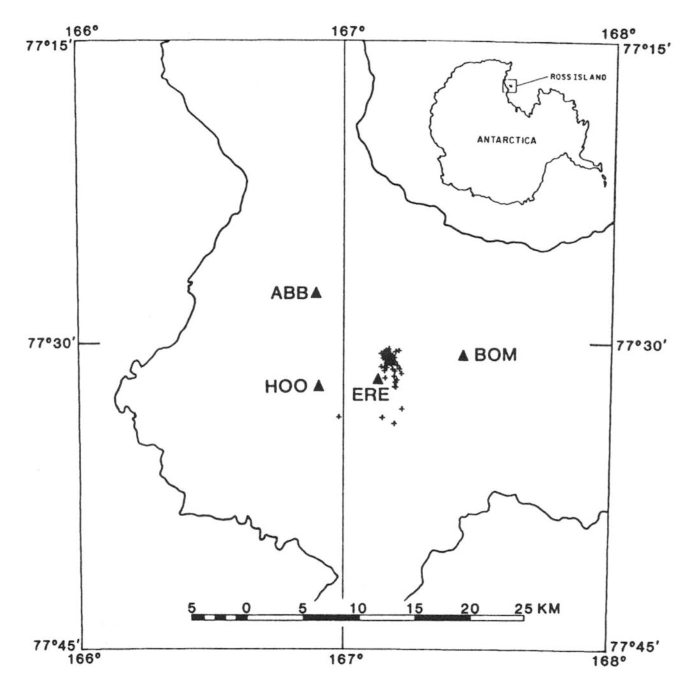

Fig. 18-3. Locations of 75 earthquakes recorded 23 Dec 1981 – 18 Jan 1982 in map view. Large triangles indicate positions of seismic stations. Inset shows the position of Ross Island. [New determination of the velocity structure of Mt. Erebus will cause significant revision in earthquake locations.]

3/83 (8:3) The following report from Juergen Kienle summarizes work by U.S., New Zealand, and Japanese scientists.

"Surveillance of the activity associated with the anorthoclase phonolite lava lake continued during the 1982–83 austral field season. The summit crater was visited by New Zealand, U.S., and Japanese scientists on several occasions between Nov 1982 and Feb 1983. The ~100 m-long semicircular lava lake was still present. Its level had dropped by about 3 m, a loss of roughly 9000 m^3 of lava since the previous visit a year earlier. Since 1976, the lake area has stayed fairly constant. However, its level has been dropping at an average rate of about 2-3 m over the past 4 years [but see 9:1].

"An original tripartite array of short-period seismic stations was installed in Dec 1980 (6:2). During the 1981–2 field season, this array was expanded by 2 stations. All stations have single-component vertical seismometers. The summit station also transmits acoustic data to monitor explosive gas discharge from the lava lake. Another data channel is used to monitor electromagnetic signals induced in a wire loop laid around the summit crater by the eruption of conducting magma in the static field of the earth.

"Over the past 2 years many of the microearthquakes we recorded were located immediately beneath the summit lava lake. For example, fig. 18-3 shows the epicenters and hypocenters of events located between Dec 1981 and Jan 1982. All but one of these events were explosion earthquakes that positively correlated with acoustic and sometimes electromagnetic signals. Explosive gas discharges from the lava lake typically occurred 2-6 times per day. Observers living in the summit hut commonly reported hearing several explosions per day.

"Over the past 2 years we have also recorded microearthquake swarms that show a negative correlation with acoustic and electromagnetic signals. Typical daily counts of about 20 events during quiet periods rise by factors of about 5-10 in swarm periods. Again, most of these events were located beneath the summit region at depths shallower than 3 km (fig. 18-4.)

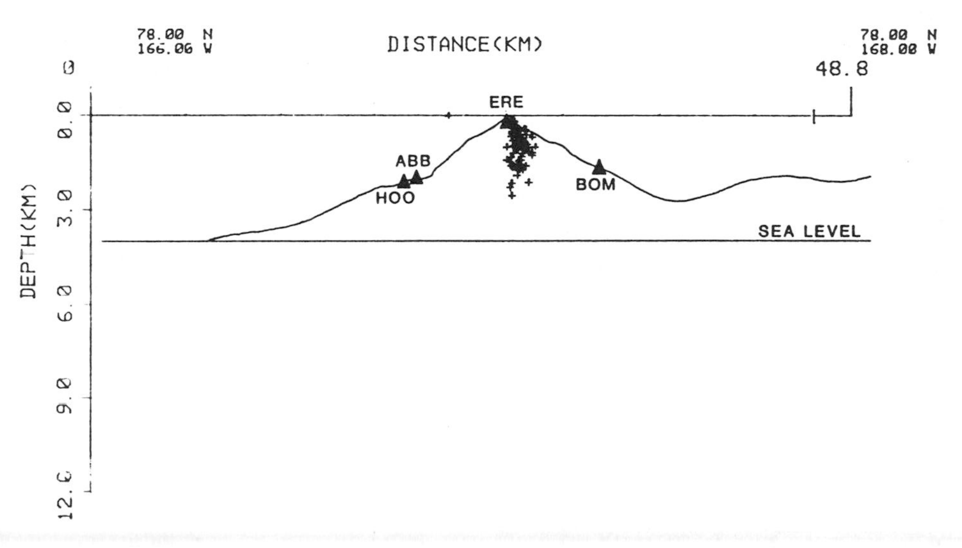

Fig. 18-4. Cross-section of earthquakes in fig. 18-3. All shallow (< 3 km) events were associated with an acoustic signal.

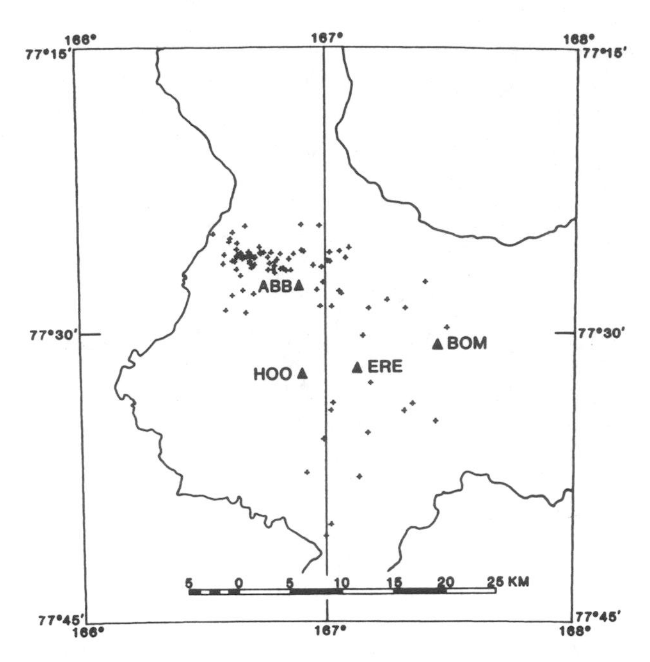

Fig. 18-5. Epicenters of 92 earthquakes recorded 4-13 Oct 1982. [New determination of the velocity structure of Mt. Erebus will cause significant revision in earthquake locations.]

"On 8 Oct 1982 an unusual earthquake swarm was recorded from a new source region on Ross Island. On that day almost 700 events occurred near Abbott Peak, a station 10 km NNE of the summit of Mt. Erebus. At this time we do not have reliable magnitudes for the events, but the fact that some of them were recorded at Scott Base and Mt. Terror suggest that the largest events had a local magnitude of 2-3. Fig. 18-5 shows epicenters and a hypothetical cross section of the Oct events. It is interesting to note that the epicentral area is located halfway between Mt. Bird and Mt. Erebus and roughly correlates with an area that apparently was hydrothermally active in 1908. Professor T. W. Edgeworth David, R. Priestley, and J. Murray, all members of the 1908 Shackelton expedition, reported steam clouds in April 1908 and a major steam eruption ('geysir') on 17 June 1908, rising from a source region at the 600-m level on the SSW slope of Mt. Bird. A tall jet of steam erupted from the same place on 8 Sept 1908. Philip Kyle has investigated rock outcrops in this area in recent years but could not find any sign of hydrothermal activity. The 8 Oct earthquake swarm may be related to renewed magma movement (dike injection?) at depth between Mt. Erebus and Mt. Bird. Preliminary hypocentral determinations suggest a clustering of events at 12-15 km depth. Fig. 18-5 is a compilation of data by the Japanese participants in the International Mt. Erebus Seismic Study (IMESS) and shows the number of earthquakes recorded per hour and day at Abbott Peak, Sept-Nov 1982. On 8-9 Oct, the Abbott seismic station also recorded volcanic tremor from presently unknown depths."

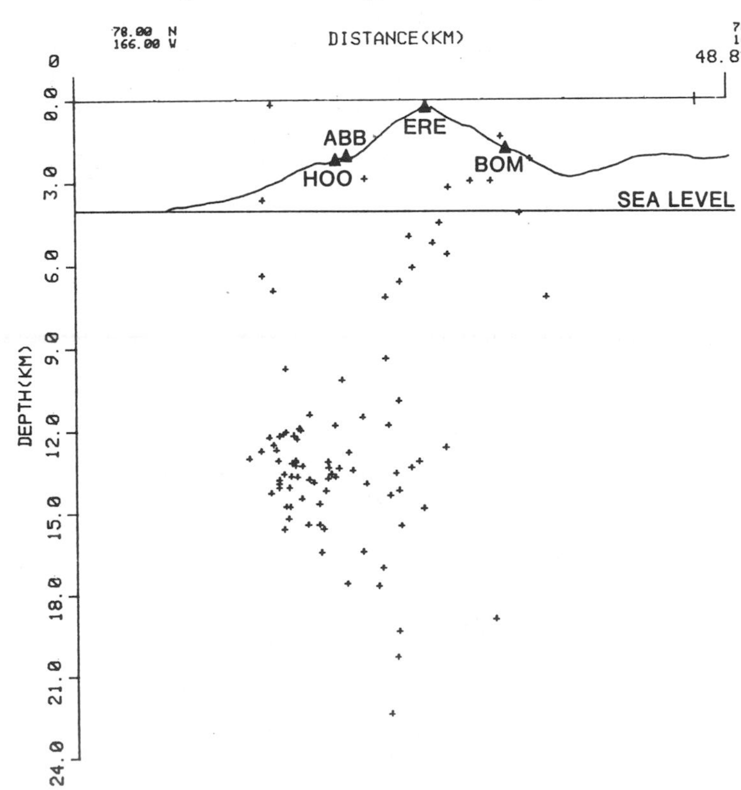

Fig. 18-6. Cross-section of earthquakes in fig. 18-5.

Information Contacts: J. Kienle, D. Marshall, Univ. of Alaska; P. Kyle, New Mexico Inst. of Mining & Tech.; K. Kaminuma, Nat. Inst. of Polar Research, Tokyo; R. Dibble, Victoria Univ., New Zealand.

1/84 (9:1) The following is a report from Philip Kyle and William I. Rose.

"Volcanic activity at Mt. Erebus was observed in Nov and Dec 1983 by scientists from New Zealand and the U.S. The persistent convecting anorthoclase phonolite lava lake still existed and displayed activity similar to that observed over the last few years. Since 1976 its surface area has remained relatively constant. A recent airphoto showing the lava lake allowed a more precise determination of its size. The photo showed that the lake was ovoid and about 60 m long × 45 m wide, considerably smaller than other ground-based observations have suggested.

"It has been previously reported that the level of the lava lake was dropping. A recent analysis of ground-based photographs makes it difficult to confirm a substantial lowering in the lake surface relative to other features seen on the Inner Crater floor. A small amount of lowering was documented in the 1979-80 austral summer field season, when a distinct bench with fumaroles rimmed the lake (5:3). However, the volcanic gas plume from the lake makes it extremely difficult to determine the lava lake height using simple visual estimates from the crater rim. Deformation studies currently in progress by NZGS personnel should give a better indication of the changes in the lava lake.

"The seismic network on Mt. Erebus was expanded by University of Alaska and Japanese personnel. Six radio-telemetry stations consisting of a single vertical-component seismometer are now situated on the flanks of the mountain. An additional station near the summit is presently inoperative. Three other stations are placed at sites 30-40 km from the summit of Mt. Erebus. Preliminary analysis of the seismic records show the level of seismicity to be normal.

"Airborne observations made using a C130 aircraft included sampling the plume for particle size distribution, and remote sensing using a COSPEC to determine SO_2 flux, which was 230 ± 90 t/d [on 19 Dec]."

Information Contacts: P. Kyle, New Mexico Inst. of Mining & Tech.; W. Rose, Michigan Tech. Univ.

Further Reference: Chuan, R.L., Palais, J., and Rose, W.I., 1986, Fluxes, Sizes, Morphology, and Compositions of Particles in the Mt. Erebus Volcanic Plume, December 1983; *Journal of Atmospheric Chemistry*, v. 4, p. 467-477.

9/84 (9:9) The following is a report from Philip Kyle and Juergen Kienle.

"Brief reports from technical staff operating seismic instruments at Scott Base and infrasonic equipment at McMurdo Sound indicate a significant change in eruptive activity at Mt. Erebus.

"Activity associated with the lava lake has consisted of quiet degassing with emission of about 230 t/d of SO_2 and 21 t/d of aerosol particles. Two to six small Strombolian eruptions occurred per day, often ejecting bombs of anorthoclase phonolite onto the crater rim, about 220 m above the lava lake.

"The reports indicate that starting on 13 Sept a number of large explosions were recorded by the IMESS network situated on the volcano, by infrasonic detectors in Windless Bight (about 29 km away), by the WWSSN seismograph at Scott Base (37 km distant), and by a tidal gravimeter at South Pole station (about 1400 km from Mt. Erebus). Previous Strombolian activity has generally been too weak to record except on the IMESS seismic stations.

"From 13 to 19 Sept, the volcano was very active with 8-19 large explosions (recorded on WWSSN, IMESS, and infrasound instruments) per day, decreasing to 2-8 per day 20-25 Sept, then increasing again to 12-27 explosions per day 26-29 Sept. Numerous mushroom-shaped clouds were reported, and were estimated to rise as much as 2 km above the summit. Observers at McMurdo, 37 km SW of the volcano, reported hearing explosions on 16 Sept at 0459, and 26 Sept at 1133 and 1135. Slight earth tremors were also felt there. On 17 Sept at 1010, a bright summit glow was observed from McMurdo Sound. Six minutes later, incandescent bombs were ejected to about 600 m above the summit; observers at Butter Point, 70 km from the volcano, reported seeing incandescent tephra from this explosion, which produced one of the larger infrasonic and seismic signals of the eruption sequence.

"Ash covered the NW side of the volcano down to 3400 m elevation. Fumaroles around the summit crater showed a substantial increase in activity. A 300-500 m-high very narrow plume was observed lower on the E flank (1800 m?). Observers suggested that it might have been a geyser."

Information Contacts: P. Kyle, New Mexico Inst. of Mining & Tech.; J. Kienle, C. Wilson, Univ. of Alaska.

10/84 (9:10) Vigorous explosive activity continued through the end of Oct. During the first half of Oct, there were about 15 large explosions per day, most of which were recorded by infrasonic detectors at Windless Bight and the WWSSN seismograph at Scott Base, as well as by the seismograph network on the volcano. Most explosions ejected a small number of very vesicular bombs that were typically 2-4 m in diameter but sometimes reached 6-8 m. Some of the bombs were ejected to heights of 500 m or more, possibly to as much as 1 km. Most fell on the upper 150 m of the outer crater rim. Continuous daylight in the area has prevented distant observers from seeing ejections of incandescent tephra after the early part of the increased activity.

The summit crater lava lake was about 100 m below the rim and actively convecting in late 1983. When Philip Kyle flew over the crater on 20 Oct, the lava lake surface had frozen and domed upward to within roughly 30 m of the rim, piling up against the N wall of the crater and sloping S. [Subsequent work has shown that the lake surface had not been domed upward. Instead, the crater had been partially filled by ejecta from the increased activity.] The bench area and fumaroles that formerly occupied the S part of the crater were covered. A small vent containing incandescent material was present near the center of the uplift and there were scattered fumaroles in the frozen lake surface. Kyle saw no explosions during his overflight but 500-1000 bombs had accumulated on the outer crater rim in the 3-4 days since a heavy snowfall had covered earlier ejecta. Prior to the increased activity in Sept, few of the bombs ejected by the small Strombolian explosions that have accompanied lava lake activity since 1972 have reached the crater rim. Microprobe analyses of glass from bombs sampled by Kyle on 20 Oct were identical in composition to glasses in the anorthoclase phonolite bombs ejected since 1972.

Information Contact: P. Kyle, New Mexico Inst. of Mining & Tech.

3/85 (10:3) During Nov and Dec, about 4-12 explosions occurred per day. Many threw bombs over the crater rim, especially on the N side. The character of the activity changed in early Dec to high-velocity gas streaming episodes that lasted about 30 seconds each. A few bombs were entrained with the gas. Analysis of the bombs showed that they were of identical composition to the anorthoclase phonolite of previous years; no changes were noted in glass composition or SO_2 content of the magma. No blocks were ejected. The explosions occurred from a small new vent in the center of the inner crater, on the S edge of the former lava lake. The vent seemed to increase in size, reaching about 30 m in diameter by late Dec. Sounds from the vent in late Dec were similar to those previously produced by degassing from the lava lake. Lava was not visible, but it was not possible to see the bottom of the vent.

In previous years, an active vent had been located just S of the SE edge of the lava lake. This vent was not present in Nov but reappeared in early Dec in a similar position. Gas streaming episodes were not observed in late Dec. However, the 8-16 explosion events/day recorded in Feb by the seismic network on the volcano

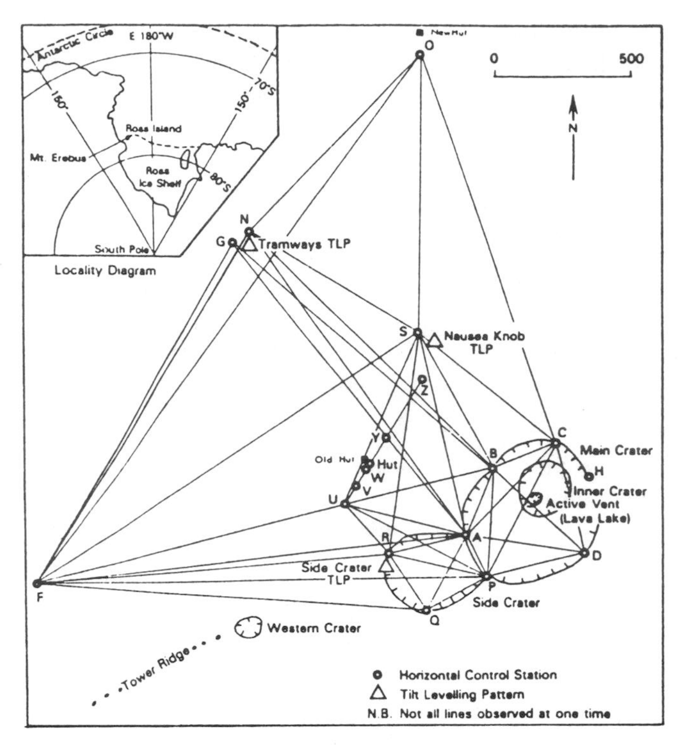

Fig. 18-7. Deformation network and craters at Mt. Erebus. From Blick, 1987.

were accompanied by about a dozen periods of seismicity that lasted > 1 minute and appeared to be records of gas streaming episodes. Magnitudes were increasing in Feb after a decrease during the austral summer.

SO_2 measurements using an airborne COSPEC were unsuccessful because the plume hugged the ground, dropping over the crater rim. Geologists estimated that a minimum of about 30 t/d of SO_2 were being emitted, much less than the 230 t/d measured the previous austral summer. The odor of SO_2 in the plume had decreased substantially from the previous year. . . . Reoccupation of a network of stations around the main

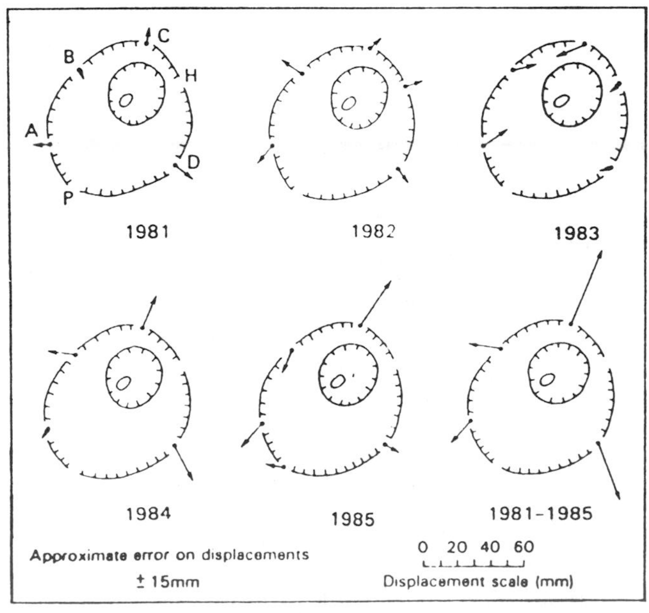

Fig. 18-8. Displacements of stations around the main crater of Mt. Erebus, 1981–85. From Blick, 1987.

crater rim (fig. 18-7) suggested that there had been [minor but significant inflation (23 mm)] in the past year [bringing the total over the last 4 years to 45 mm (fig. 18-8). No tilt has been detected from arrays on the outer slopes of the crater rim during that time.]

Information Contacts: P. Kyle, New Mexico Inst. of Mining & Tech.; J. Kienle, Univ. of Alaska; W. Rose, Michigan Tech. Univ.; P. Otway, DSIR, New Zealand.

Further References: Blick, G.H., 1987, Mt. Erebus, Antarctica: Volcanological Observations and Deformation Survey; *New Zealand Volcanological Record*, v. 15, p. 52-55.

Blick, G.H., Otway, P.M., and Scott, B.J., 1988?, Deformation Monitoring and Volcanic Activity of Mt. Erebus, Antarctica, 1980–1985; *BV*, in press.

Summaries of activity at Mt. Erebus in the *Antarctic Journal of the United States* (year described: citation); 1975: v. 11, no. 4, p. 270-271; 1977–78: v. 13, no. 4, p. 32-34; 1978–79: v. 14, no. 5, p. 35-36; 1980–81: v. 16, no. 5, p. 34; 1981–82: v. 17, no. 5, p. 29-31; 1982–83: v. 18, no. 5, p. 41-44; 1983–84: v. 19, no. 5, p. 25-27; 1984–85: v. 20, no. 5, p. 25-28.

SEAL NUNATAKS GROUP

Larsen Ice Shelf, Antarctica
65.03°S, 60.05°W, summit elev. 368 m

The Antarctic Peninsula extends toward S America and the Larsen Ice Shelf lines its E side. Frequent eruptions are known from Deception Island, 235 km N of the Seal Nunataks, but only one eruption (in 1893) has been reported from this volcano group, 1080 km S of Cape Horn.

9/82 (7:9) In Jan 1982, geologists from the Universidad de Chile found evidence of recent volcanism at 2 sites in the Seal Nunataks Group that had not previously been identified as recent volcanic centers. At Dallman (65.02°S, 60.32°W), very fresh-looking basaltic lava had emerged from the central crater and flowed to the NW foot of the volcano. Active fumaroles were observed at the N side of the summit. Abundant basaltic lapilli and ash covered wide areas of the Larsen Ice Shelf in the vicinity of Murdoch Volcano (65.03°S, 60.03°W). Portions of these deposits were covered by fresh snow. Fumaroles were active in a small parasitic cinder cone SE of Murdoch.

In Dec 1893, C.A. Larsen saw an eruption of black ash from Lindenberg (64.92°S, 59.70°W) and solfataric activity at Christensen (65.10°S, 59.57°W), but no activity was observed at either of these volcanoes in 1982.

Reference: Larsen, C.A. 1894, The Voyage of the "Jason" to the Antarctic Regions; *Geographical Journal*, v. 4, no. 4, p. 333-344.

Information Contact: O. González-Ferrán, Univ. de Chile.

Further Reference: González-Ferrán, O., 1983, The Seal Nunataks: an Active Volcanic Group on the Larsen Ice Shelf, West Antarctica; *in* Oliver and Jay, (eds.), *Antarctic Earth Science*; Cambridge University Press, p. 334-338.

ATMOSPHERIC EFFECTS

The enormous aerosol cloud from the March-April 1982 eruption of Mexico's El Chichón persisted for years in the stratosphere, and led to the Atmospheric Effects section becoming a regular feature of the *Bulletin.* Descriptions of the initial dispersal of major eruption clouds remain with individual eruption reports, but observations of long-term stratospheric aerosol loading will be found in this section.

Only observations of aerosols from an unknown source or from multiple sources were listed in this section until mid-1982. Earlier data on the atmospheric effects of volcanism were included within reports of individual eruptions.

General References: Keen, R.A., 1983, Volcanic Aerosols and Lunar Eclipses; *Science*, v. 222, p. 1011-1013.

Kent, G.S. and McCormick, M.P., 1984, SAGE and SAM II Measurements of Global Stratospheric Aerosol Optical Depth and Mass Loading; *JGR (Oceans & Atmospheres)*, v. 89, no. D4, p. 5303-5314.

Meinel, A.B. and Meinel, M.P., 1983, *Sunsets, Twilights and Evening Skies*; Cambridge University Press, Cambridge, England.

Pollack, J.B. and McCormick, M.P. (eds.), 1981, Special Issue on Aircraft and Spacecraft Measurements of Stratospheric Aerosols and Their Implications; *Geophysical Research Letters*, v. 8, p. 1-28 (9 papers).

9/80 (5:9) In the Tucson, Arizona area (32.25°N, 110.95°W), Marjorie and Aden Meinel report strongly enhanced sunset glows beginning 24 Aug. That evening, silvery, undulate striae, similar to phenomena associated with the 1974 Fuego eruption, were seen at sunset against a glowing background. As the sun set, striae and background passed simultaneously through the same sequence of colors, indicating that they were at the same altitude, calculated by the Meinels to be about 19 km. The striae were not visible the following evening, but the strong background glow recurred and a secondary glow was visible until about 70 minutes after sunset, as after the eruptions of Krakatau in 1883 and Agung in 1963. The glow became fainter on succeeding evenings, but a strong enhancement occurred 16 Sept and the glow was still nearly as bright several days later. The Meinels believe that the Aug phenomena were probably caused by material injected into the upper atmosphere by the 17 Aug eruption of Hekla [5:8], while the Sept enhancement could have been due either to a return of the Hekla material or normal seasonal trends in glow intensity.

In England, H.H. Lamb observed a week or more of reddened sunsets, culminating in a colored ring around the sun seen from Ketteringham, Norfolk before sunset on 12 June. The same evening lidar observations confirmed the presence of a dust veil over Garmisch-Partenkirchen, West Germany (47.5°N, 11.0°W). Lamb saw no apparent abnormal coloring the following week, nor was any observed during the few breaks in bad weather from late June through mid-July. On 14 July, unusual light diffusion in brownish layers above cumulonimbus clouds was followed by a shower that deposited russet-colored mud. Meteorological data suggest transit from the Arctic. By early Sept, clear skies showed a brownish coloration at sunset and more than usually diffused light around the sun, interpreted by Lamb to indicate a rather thin dust veil. [See Gareloi, 5:11 for aerial observations of an extensive but poorly documented plume erupted in Aug 1980.]

Information Contacts: M. and A. Meinel, Institute for Physics & Astronomy, Taiwan; H. Lamb, Univ. of East Anglia.

5/81 (6:5) A thick zone of probable volcanic material was observed in the upper troposphere and lower stratosphere during the night of 8-9 June. NASA's lidar at Hampton, Virginia (37.1°N, 76.3°W), operating at the ruby wavelength of 0.6943 μm, recorded several layers with scattering ratios greater than the normal background reading of 1.1 (scattering ratio = 1 + aerosol scatter/molecular scatter). Thin layers were centered at altitudes of 17.4 km (about 1 km thick, scattering ratio 1.5) and 16.5 km (about $^1/_2$ km thick, scattering ratio 1.3). A much broader layer extended downward from 16 km through the tropopause at 13.3 km to 12 km altitude. Within this broad layer, clearly-defined scattering peaks were located at 14.9 km and 14.1 km, both with scattering ratios of 2. Residual material from Mt. St. Helens remains in the stratosphere, raising the background scattering ratio to 1.2.

Weather balloons launched from Laramie, Wyoming (41.33°N, 105.63°W) began to detect volcanic material 16 May. Since then, a thick zone showing some variation in structure has remained between 12 and 18 km altitude. During the night of 8-9 June, an intense new volcanic layer between 11 and 14 km altitude joined the post-26 May material over Laramie and nearby Boulder, Colorado. The previous material, concentrated between 16 and 18 km, was truncated by a reversal in wind direction at 18 km altitude.

The 2 most likely source volcanoes are Alaid and Pagan, which both injected eruption columns into the stratosphere, at the end of April and on 15 May respectively. No other major explosive eruptions have been reported since then.

Information Contacts: M.P. McCormick, NASA Langley Research Center; D. Hofmann, J. Rosen, Univ. of Wyoming.

7/81 (6:7) High-altitude aerial sampling 9, 10, and 13 July revealed an extensive zone of sulfate aerosols and silicate fragments just below the tropopause at high northern latitudes.

On 9 July, instruments aboard an LANL B-57 research aircraft, flying at an altitude of about 13.5 km from Seattle, Washington (47.5°N, 122.5°W) about 2300 km to Anchorage, Alaska (62°N, 149°W), sampled a very constant sulfate concentration of 0.7-0.8 mg/m^3, well above the normal mid-latitude background of about 0.1 mg/m^3. The next day, flying directly N from Anchorage at a constant 12 km altitude from 62°N to 75°N (a distance of more than 1500 km) along the 145° meridian, sampling instruments measured sulfate concentrations of about 1.5 mg/m^3. A few silicate particles, larger than 1 mm and probably coated with acid, were also recovered. At 75°N, the aircraft climbed through the tropopause, just above the 12 km altitude of the northbound flight path, to about 16.5 km, then flew S at that altitude back to 62°N. The zone of high sulfate and silicate concentration terminated sharply at the tropopause and no unusual concentrations were recorded in the stratosphere during the return flight.

A U-2 aircraft operating from the NASA Ames Research Center near San Francisco (37.33°N, 121.92°W) flew at gradually increasing altitude to about 50°N, 155°W on 13 July. As the aircraft climbed toward the tropopause, marked by the polar jet stream, concentrations of sulfate increased gradually from about 2 mg/m^3 to about 4.5 mg/m^3. Some silicate particles were also collected. Sulfate concentrations of 2 mg/m^3 or higher were measured for about 2 hours of flight time, representing a lateral distance of about 1800 km. As in the 9-10 July flights, the zone of high sulfate and silicate concentration was truncated abruptly at the tropopause. While descending on the return leg of the flight, a sulfate concentration of 5 mg/m^3 was recorded at the top of the polar jet stream.

While the sulfates and silicates sampled from the aircraft were almost certainly of volcanic origin, it is not yet possible to pinpoint their source, nor their time of eruption [but see 6:10]. Airmass movement in the days prior to 9 July will be analyzed to help locate a probable source area and the presence or absence of significant variation in silicate chemistry should help determine whether all are from a single eruption or whether multiple sources are likely.

Information Contacts: R. Chuan, Brunswick Corp.; W. Rose, Jr., Michigan Tech. Univ.

10/81 (6:10) Filter samples from LANL high-altitude aircraft and data from NASA's SAGE satellite provided information about the height and dispersal of the eruption cloud ejected by Ulawun in Oct 1980, and the LANL aircraft also collected tephra probably produced by the May 1981 eruption of Pagan.

Ulawun's brief but powerful eruption took place 6-7 Oct 1980, producing a cloud estimated by ground observers to have reached 7-10 km in height [see Ulawun, 5:10]. On 24 Oct, the LANL aircraft sampled the lower stratosphere at about 19 km altitude, between the equator and 5°N at about 80°W (just S of Panama]. Data from these samples indicated atmospheric concentrations of as much as 6 parts per billion (ppb) of sulfate by mass, of which only 25-50% could be attributed to the 18 May eruption of Mt. St. Helens. The SAGE satellite detected tephra from Mt. St. Helens N of 40°N in Oct 1980, but also detected a large cloud of new material from 26°N to 10°S (data collection was truncated at 10°S) between 125°W and the International Date Line. SAGE next collected data from the equatorial region in mid-Nov, when a zone of significant particle enhancement (roughly 5 times background), extending upward from the tropopause (about 16.5 km altitude at the equator) to about 22 km, circled the globe in an irregular band 10°-20° wide between 20°S and 10°N.

Pagan's eruption began 15 May. Japanese weather radar recorded the top of the eruption column at 18-20 km altitude and weather satellite images showed that the high-altitude cloud traveled SSE [see Pagan, 6:4-5]. Ten weeks later, filter samples collected by the LANL aircraft just S of Panama (from the equator to 5°N) on 24 July showed lower stratospheric sulfate concentrations of about 5 ppb by mass at altitudes of 18.2, 19, and 19.2 km. Between 5°N and 35°N at 16.8 km altitude (above the tropopause), sulfate concentrations were as high as 4 ppb by mass; only about 1 ppb could be attributed to material remaining from the Mt. St. Helens eruption 14 months earlier. High concentrations of sulfate aerosols and silicate particles collected at mid to high northern latitudes in early July are probably from the late April-early May eruption of Alaid (6:4-5). From these data, the average lower stratospheric sulfate concentration over the entire Northern Hemisphere in July 1981 was calculated to be about 2.5 ppb by mass, primarily contributed by the eruptions of Pagan and Alaid but including a little material from the Mt. St. Helens eruption. The same calculations made from July 1980 data yielded a slightly lower concentration, about 2.3 ppb by mass, with Mt. St. Helens as the dominant source. In the last decade, only 2 relatively brief periods can be identified as showing "background" sulfate concentrations, without a substantial volcanic component: mid-1973 through mid-1974 (about 0.34 ppb sulfate by mass), and late 1978 through late 1979 (about 0.47 ppb sulfate by mass).

Information Contacts: W. Sedlacek, LANL; M.P. McCormick, NASA Langley Research Center.

1/82 (7:1) A sudden increase in stratospheric aerosols was recorded on 23 Jan at 1200 GMT by the Nd-YAG lidar, wavelength 1.06 μm, operated by Kyushu Univ., Fukuoka, Japan (33.65°N, 130.35°E). The scattering ratio at 17 km altitude (about 4) was about 20 times the normal average value. The same equipment detected a strong aerosol layer at 11-17 km on 30 Jan and a very strong layer at 10-17 km on 2 Feb. Peak concentrations were about the same as those recorded 1 month after the 18 May 1980 eruption of Mt. St. Helens [Hirono et al. (1981)]. NOAA's lidar unit on Mauna Loa, Hawaii (19.5°N, 155.6°W) detected a several kilometer-thick layer centered at 17 km on 28 Jan, and its next reading, on 4 Feb, showed 2-3 different layers between 17 and 20 km altitude. No unusual atmospheric debris had been detected during the previous measurement by this instrument on 19 Jan. Lidar data gathered during clear weather the last week in Jan from Wallops Island, Virginia (37.9°N, 75.5°W) revealed no notable stratospheric material. However, ground-based ruby lidar at Garmisch-Partenkirchen, West Germany measured strong aerosol layers at 13-16 km on 2 Feb and 15-17 km the next night.

Motokazu Hirono interpreted the fine structures of the Kyushu Univ. lidar profiles to indicate a volcanic source. However, SEAN has no recent report of a large explosive eruption and the source of the stratospheric aerosols is not yet known [but see 7:3].

Reference: Hirono, M., Fujiwara, M., Shibata, T., and Kugimiya, N., 1981, Lidar Observations of Volcanic Clouds in the Stratosphere over Fukuoka caused by Eruptions of Mt. St. Helens in May 1980; *Geophysical Research Letters*, v. 8, no. 9, p. 1019-1022.

Information Contacts: M. Hirono, Kyushu Univ.; B. Mendonça, NOAA/Air Resources Lab (ARL); M.P. McCormick, NASA Langley Research Center; R. Reiter, Garmisch-Partenkirchen, W Germany.

2/82 (7:2) A widely-distributed and voluminous cloud of aerosols remained in the upper troposphere in early March. Aircraft observations indicated that aerosols had been disseminated over broad areas of middle and lower northern latitudes by mid-Feb. Although the cloud was clearly of volcanic origin, no eruption has been unequivocally identified as its source.

Solar Irradiance Measurements. On 11 Jan a pyrheliometer (which measures direct solar irradiance over a broad spectrum at the earth's surface) at Mauna Loa Observatory, Hawaii detected a substantial decrease in solar radiation, nearly as large as that measured after the major eruption of Agung in 1963. A small decrease measured by this instrument during its preceding reading 4 Jan was within the noise level. Lidar data collected at Mauna Loa indicated that material in the upper troposphere, between 10 and 13 km altitude, was responsible for the initial decrease. Solar irradiance has remained low at Mauna Loa, as it did for about 3 years after the Agung eruption. A pyrheliometer at Aspendale Observatory, Australia (about 38°S, 145°E) had shown no major increase in atmospheric turbidity as of early March. No other atmospheric data are presently available from the Southern Hemisphere.

Lidar and Balloon Measurements. Stratospheric aerosols were first detected by lidar on 23 Jan at Kyushu Univ., Fukuoka, Japan as a thin layer centered at about 17 km altitude. Scattering ratios were at background levels during their next observation 26 Jan but each measurement since then, beginning 30 Jan, has recorded backscatter from the cloud, with peak concentrations on 2 Feb. Their most recent data, on 3 March, indicated the presence of an aerosol layer between 9 and 17 km altitude.

No stratospheric layer was evident on the Mauna Loa lidar until 28 Jan (9 days after the preceding measurement). It continued to detect stratospheric material through late Feb and on the 26th the aerosol layer was about 3 km thick, centered at 18 km altitude. Lidar data collected during poor conditions 5 March showed an apparently weak layer centered at 17.3 km. No aerosol layer was detected during a balloon flight from Laramie, Wyoming 5 Feb, but instruments aboard balloons launched 17 and 27 Feb measured aerosol concentrations between 13.5 and 18 km altitude that were similar to those observed following the 18 May, 1980 eruption of Mt. St. Helens. William Fuller reports that lidar data obtained 26 Jan from Hampton, Virginia indicated that the stratosphere at this latitude was reasonably clear, with peak backscattering ratios at about background levels. Measurements made with the NASA Langley Research Center airborne lidar 10 Feb from the ground at Wallops Island, Virginia showed a considerable increase in stratospheric material. The base of the layer was at 12 km and it extended to 18 km with a peak backscattering ratio 4-5 times greater than normal. Stratospheric lidar measurements will continue on a regular basis from Hampton with a ground-based 48-inch lidar and possibly with the airborne lidar. During the night of 8-9 March, the ground-based lidar at Hampton detected the aerosol layer between 12.9 and 16.5 km.

Airborne Lidar Measurements. By mid-Feb, stratospheric aerosols were distributed over a broad area of the Northern Hemisphere, as described in the following report from William Fuller.

"A flight was conducted 13 Feb from Wallops Island to San José, Costa Rica (9.9°N, 84.1°W) with the lidar on board the NASA Wallops Electra aircraft. Lidar measurements indicated that the very intense stratospheric layer was present along the entire flight path. As the aircraft proceeded S, the layer became very intense until the maximum scattering ratio occurred at 21°N, 87.1°W, just E of the Yucatán Peninsula. The stratospheric layer remained strong to 10°N, 82-83°W where the southernmost measurement was made. On the return flight 21 Feb, lidar measurements were made from New Orleans (30.0°N, 90.05°W) to Wallops Island and the stratospheric layer continued to be present."

Aerial Sampling. On 6 March at about 1900 GMT, impact samples of the aerosol cloud were collected at about 20°N, 96.2°W (SW Gulf of Mexico) on a grid carried by a NASA U-2 aircraft. The samples were similar to those collected from the Mt. St. Helens aerosol cloud 1-2 weeks after the 18 May, 1980 eruption. More than 70 particles/cm^3 greater than 0.6 μm in diameter, all liquid droplets, were recovered at 16.75 km altitude. Preliminary analysis by the Atmospheric Experiments Branch of the NASA Ames Research Center showed that substantial quantities of H_2SO_4 were present in the cloud, indicating that its source was volcanic. Virtually no mineral grains were recovered. Hand-held geiger counters in the aircraft recorded radioactivity at only $^1/_2$ of normal background. Additional analyses of the 6 March samples are planned, as are a series of additional flights.

Unusual Sunsets. Aden and Marjorie Meinel saw sunset glows from Tucson, Arizona beginning 23 Jan, when late twilight coloration indicated possible enhancement in the region between 18 and 20 km altitude. Glow was stronger the next night and on 15 Jan glowset was 43 minutes after sunset. The glow was weaker 26 and 27 Jan, and after cloudy weather 28-29 Jan, no enhancement was visible 30 Jan. The Meinels observed a faint glow 19 Feb, similar to that of 24 Jan, but after 2 nights of cloudiness saw none 22 Feb. Glow was next visible from Tucson on 2 March and was present for the next 2 nights.

Information Contacts: W. Fuller, M.P. McCormick, M. Fujiwara, NASA Langley Research Center; B. Ragent, G. Ferry, V. Overbeck, K. Snetsinger, D. Hayes,

NASA Ames Research Center; K. Coulson, Mauna Loa Observatory (MLO); B. Mendonça, NOAA/ARL; A. and M. Meinel, Univ. of Arizona; J. Rosen, Univ. of Wyoming; M. Hirono, Kyushu Univ.; J. Gras, CSIRO, Australia.

3/82 (7:3) The widely distributed volcanic aerosol cloud remained in the lower stratosphere through early April. Since 29 Jan, each lidar measurement at MLO has detected the cloud. As of 9 April, it was centered at about 18 km altitude (with a peak backscattering ratio of 1.6) and was about 2 km thick. A balloon flight the first week in April from Laramie, Wyoming showed a broad layer centered at 18 km altitude. From Hampton, Virginia, lidar data [17 March] showed a 3 km-thick layer centered at about 17 km altitude (backscattering ratio about 1.6). The cloud has also been intermittently present over Toronto, Canada (43.6°N, 70.5°W) since early March.

A NASA sampling aircraft flew S from San Francisco 18 March, and collected about 20 times the normal concentration of H_2SO_4 from a layer at the base of the stratosphere. Silicate particles about 0.25 m in diameter were present both as discrete fragments and within the acid droplets. Chemical analysis of these particles showed that they contained no Na, and their Si/Al ratio was consistent with a basaltic composition. Additional sampling flights are planned in mid-April by NASA and LANL.

No eruption can be unequivocally identified as the source for the cloud. Careful inspection of satellite images has yielded no large eruption clouds that had gone unreported from the ground, but cloudy weather often obscured volcanically active areas of the world. The best candidate appears to be Pagan (18.13°N, 145.80°E), where moderate explosive activity was reported in early Jan. However, no ground observations are available between 6 Jan and 8 Feb, and the source eruption for the cloud probably occurred in mid-Jan. Careful inspection of images from the Japanese Geostationary Meteorological Satellite by Yosihiro Sawada showed a possible volcanic cloud from Pagan 14 Jan at 1900 local time (0900 GMT), but interference from weather clouds made this impossible to confirm. Sawada observed a similar feature on an image returned at 2200 local time 19 Jan 1981, the same day that visiting islanders reported explosive activity (Pagan, 6:11).

[Unpublished data from NASA's Total Ozone Mapping Spectrometer (TOMS), which is sensitive to the SO_2 that is emitted by most eruptions, strongly suggest that this cloud was ejected by Nyamuragira (Zaire) during the initial explosive phase of its Dec 1981 – Jan 1982 eruption.]

Information Contacts: R. Chuan, Brunswick Corp.; Y. Sawada, Meteorological Research Inst., Japan; N. Banks, USGS HVO, Hawaii; K. Coulson, T. DeFoor, MLO; W. Fuller, NASA Langley Research Center; D. Hofmann, Univ. of Wyoming; B. Ragent, NASA Ames Research Center; W. Evans, ARPX-AES, Downsview, Canada.

[The initial dispersal of the major stratospheric cloud from the March-April, 1982 eruption of El Chichón is described in the El Chichón section of 7:3-5. Its persistent atmospheric effects are reported below. The following report, and those from 7:8 through 8:5, were originally part of the El Chichón reports, the "Atmospheric Effects section no becoming a regular feature of the *Bulletin* until 8:6. Lidar data originally presented each month in lengthy tables have been summarized in figs. 19-1 and 19-6 through 19-13.]

General References: Galindo, I., Hofmann, D.J., and McCormick, M.P. (eds.), 1984, Atmospheric Effects of the Volcanic Eruption of El Chichón; *Geofísica Internacional*, v. 23, nos. 2-3, p. 113-448 (22 papers).

Pollack, J.B., Toon, O.B., Danielsen, E.F., Hofmann, D.J., and Rosen, J.M. (eds.), 1983, Climatic Effects of the Eruption of El Chichón; *Geophysical Research Letters*, v. 10, no. 11, p. 989-1060 (18 papers).

6/82 (7:6) *Lidar measurements.* Data collected in June at Mauna Loa, Hawaii showed backscattering that typically increased from near the base of the stratosphere to a peak at 26-27 km altitude, with significantly enhanced values to 33-34 km. Data were less variable from night to night and layering within the cloud was less distinct than in May. The cloud above Hawaii has apparently affected incoming solar radiation, measured for about 50 years by the Hawaiian Sugar Planters Association. Although mean daily solar radiation would normally have been about 110% of the long-term average (because precipitation in May at the primary station was only 27% of normal), the measured value for the month was only 92% of average.

Lidar at Fukuoka, Japan showed decreased backscattering from the 21-29 km layer in late June, but backscattering increased again in early July. The less dense layer at 18.5 km remained stable through this period. In early July, backscattering detected by lidar at Hampton, Virginia increased sharply for the (highest) layer centered at about 25 km altitude, approaching values measured at lower latitudes for the first time.

To assess latitudinal variation in the stratospheric cloud, a lidar-equipped NASA aircraft flew from Wallops Island, Virginia to Puerto Rico during the night of 8-9 July, to about 12°N (near the coast of Venezuela) 10 July, and from Puerto Rico to the vicinity of Albany, New York (about 42°N) 11 July. From 25-30°N to the southern limit of the flight, preliminary data show greatly enhanced backscattering from a dense layer between 21 and 33 km altitude. Some material was present below 21 km, but it was much less dense. Strong local variation in the cloud was observed. Although the cloud diminished in density N of 25-30°N, significantly enhanced stratospheric backscattering was detected to the N limit of the flight.

Brilliant Sunrises and Sunsets. Weather satellite images first showed the front of the 4 April stratospheric cloud (visible over water during the day) over the Red Sea on 21 April. Edward Brooks, who has made frequent sunrise and sunset observations from Jeddah, Saudi Arabia (21.5°N, 39.16°E) saw WNW-ESE-trending bands of haze in the WNW sky after sunset 20 April and similar bands before dawn the next morning. The twilight of 24 April was a brilliant pink from bands and patches of WSW-ENE-trending aerosol. During the next several weeks, volcanic cloud effects could be seen in the sky around sunrise and sunset most days, often as bands of material oriented within 45° of E-W. Brooks observed a layer 10° above the horizon at twilight 18 May and calculated its altitude at roughly 20-25 km. Multiple layers began to be visible in June. At sunrise on 5 June, criss-cross bands of aerosol trended SW-NE and SSE-NNW. Brilliant sunrises and sunsets were common in mid-June. Beginning on 19 June, many sunrises illuminated 2 distinct layers, at about 30-minute intervals. This effect weakened later in the month and by 30 June the higher layer (illuminated earlier in the morning) had disappeared. Brooks and others also noted that on 6 July roughly the upper half of the eclipsed moon was considerably darker than the lower half, which Brooks interprets as indicating the presence of volcanic aerosols in the atmosphere of the earth's northern (but not southern) hemisphere.

Information Contacts: W. Fuller, M.P. McCormick, NASA Langley Research Center; T. DeFoor, MLO; M. Hirono, Kyushu Univ.; E. Brooks, Saudi Arabia; K. How, Hawaiian Sugar Planters Assoc.; M. Matson, NOAA/NESS.

7/82 (7:7) The major stratospheric cloud remained dense over lower northern latitudes. It has been estimated to cover the earth from S of the equator to as far N as Japan between 21 and 33 km altitude and to average 9.6 km thick. Lidar measurements and reports from England indicated that gradual northward dispersal was continuing.

At Mauna Loa, Hawaii, lidar measurements showed that cloud material was densest at 25-27 km altitude. Lidar at Fukuoka, Japan showed increasing concentrations during July at 22-25 km altitude. Measurements of well-resolved fine structures on 26 July showed a more dense layer at 22-24 km and a less dense one at 28-29 km. The backscattering ratio of 42 at 25 km altitude detected 1 July by lidar at Hampton, Virginia was the highest ever observed in the stratosphere from there. Measurements from the 8-13 July NASA flight revealed several separate layers of material, with greatest concentrations between 24 and 26 km altitude.

As observed from Norwich, England (52.5°N, 1°E) the cloud was a thin veil that was not always present. On most evenings when the sun was visible, a round area of diffused pale bluish-white light appeared around the sun, extending 20-30° out from it and remaining obvious after sunset. On 28 and 29 May a brownish band appeared around the perimeter of the bright area, separating it from blue sky; this was interpreted as Bishop's Ring. A similar ring and a prominent sun pillar were noted around the midnight sun on 13 June at latitude 67.5°N between Bodø and the Lofoten Islands, Norway.

Information Contacts:: W. Fuller, NASA Langley Research Center; M. Hirono, Kyushu Univ.; H. Lamb, Univ. of East Anglia.

8/82 (7:8) Lidar measurements from several locations in the northern hemisphere suggest that the bulk of the El Chichón stratospheric cloud remains confined to lower northern latitudes. Measurements at MLO, Hawaii showed that the cloud remained densest at 25-27 km altitude, but seemed more uniform than last month. Lidar at Fukuoka, Japan showed the aerosols in multi-layer structures above 21 km, in the easterly winds, but well-mixed below that altitude, in the westerly winds. Over Hampton, Virginia, multiple layers were detected between 16 and 31 km, but lower backscattering ratios indicated that the aerosols were much less dense than over Hawaii or Japan. Farther N, at L'Aquila, Italy (42.37°N, 13.4°E) large aerosol enhancements were detected 28 July, and 5, 11-21, and 25-30 Aug. Layers 3-4 km thick with backscattering ratios as large as 10-12 were observed ~25 km altitude, and on 26 Aug there was a layer at 32 km altitude with a backscattering ratio of 2.

Scientists from the Department of Physics and Astronomy, Univ. of Wyoming studied the cloud during 2 unmanned balloon flights from S Texas (27.3° and 27.7°N) and 7 flights from SE Wyoming (41°N). Optical particle counter measurements of the concentration of particles with radii > 0.15 μm revealed 2 stratospheric layers. The lower layer was 2-3 km thick with peak concentrations, at least 40 times background levels, at about 18 km. The upper was about 3 km thick with peak concentrations, at least 200 times background levels, at about 25 km. Ten sublayers were identified in the upper layer over Texas in May, but only some were found over Wyoming before July.

Since early Sept brilliant colors persisting at least 30 minutes after sunset have been observed from Norwich, England. At that time on 7 Sept the brightest area was faintly rimmed with a brown band, interpreted as Bishop's Ring. On 8 Sept, long smooth streaks at an altitude of 20-25 km were illuminated until 30-35 minutes after sunset.

Information Contacts: T. DeFoor, MLO; M. Hirono, Kyushu Univ.; M.P. McCormick, W. Fuller, NASA Langley Research Center; G. Visconti, Univ. L'Aquila; D. Hofmann, Univ. of Wyoming; H. Lamb, Univ. of East Anglia.

9/82 (7:9) Satellite, lidar, and balloon data continue to indicate that little latitudinal movement of the stratospheric cloud has occurred in recent months. Mean monthly sea surface temperatures determined from the NOAA 7 satellite have been as much as 3°C lower in some regions than the actual ocean temperatures measured at the same time and place by ships, apparently because of interference from the El Chichón cloud. Previous work had shown that the satellite temperatures are normally quite accurate, varying from ship measurements by a maximum of about 0.5°C. Significant discrepancies between actual temperatures and satellite temperatures have remained between 10°S and 30°N, with maximum variations from 15-20°N. No substantial northward movement has been detected through Sept. Variations peaked in June and July, declining somewhat in Aug and Sept.

Sept lidar data showed gradual vertical expansion of the cloud but little evidence of large-scale northward movement. Layers were detected below the tropopause in Hawaii, but this material may be from Galunggung or some other source, rather than from El Chichón. Balloon soundings from Wyoming (41°N) continued to show a very broad layer around 18 km altitude and some layers as high as 25 km during Sept, but there were large variations between soundings. The upper layers were not as broad as they had been over Texas (about 27.5°N) during previous measurements (7:7). When samples collected from these layers by the Wyoming sondes were heated to 150°C, 98% of the material volatilized, a result consistent with an H_2SO_4-H_2S composition. The composition of the particulate matter, the 2% that did not volatilize, has not been determined.

Brilliant sunsets continued to be seen by H.H. Lamb from Norwich, England through mid-Sept. [Optical effects were observed 7-8 Sept (7:8)], but low-level haze prevented any useful observations 9-10 Sept. No anomalous features were present at sunset or during twilight on the 11th, but a brightly-colored layer could be seen around sunset 12-14 Sept. The timing of the end of illumination of this layer indicated that it was at about 21 km altitude. When dark, it showed a structure of long streaks.

In late Oct, government and university scientists will begin a satellite and field experiment designed to determine the key radiative, dynamical, and chemical properties of the El Chichón stratospheric eruption cloud. A NASA Electra aircraft, equipped with a number of remote sensors (including 2-wavelength lidar, 13-channel sun photometer, 4-channel direct diffuse photometer, and Brewer spectrometer) will collect data over the US, the Caribbean, and Central and South America between 19 Oct and 5 Nov. The flight plan will be coordinated with a number of concurrrent in-situ and satellite (SME, NOAA/TIROS N, Nimbus-7, and GOES) measurements. Coordinated rendezvous are planned with high-flying aircraft and with balloon experiments in Texas, New Mexico, and Wyoming. Comprehensive data sets from this research will be made readily available to the scientific community. For more information, contact Pat McCormick or George Maddrea at NASA Langley Research Center.

Information Contacts: W. Fuller, M.P. McCormick, G. Maddrea, NASA Langley Research Center; T. DeFoor, MLO; M. Hirono, Kyushu Univ.; A. Strong, M. Matson, NOAA/NESS; D. Hofmann, Univ. of Wyoming; B. Mendonça, NOAA/ARL; H. Lamb, Univ. of East Anglia.

10/82 (7:10) In late Oct and early Nov, a NASA Electra aircraft with a package of remote sensing instruments on board gathered data on El Chichón's stratospheric cloud from about 46°N to 46°S. Only very preliminary results were available at press time. Lidar profiles were collected over the entire flight path, optical depths of the atmosphere were determined by sun photometry at 13 wavelengths, and the total SO_2 and O_3 columns over the aircraft were measured. No dramatic changes in the position or morphology of the cloud appeared to have occurred since a similar flight (N Hemisphere only) 8-13 July (7:7). Stratospheric material was detected over the entire flight path, and M.P. McCormick believes that ejecta from El Chichón has reached both poles. As in July, there was a distinct boundary between the much more dense cloud at lower northern latitudes and more diffuse material farther from the volcano. The densest portion of the cloud extended only a few degrees farther N in Nov than in July [see also 7:11]. A similar boundary was found at lower southern latitudes. The strongest concentration of aerosols was typically at 23-24 km altitude, with the base of the cloud at about 21 km, but considerable variation in its morphology was observed.

From lower northern latitudes, ground-based lidar continued to detect stratospheric debris from El Chichón, but peak backscattering ratios were smaller and occurred at slightly lower altitudes in Oct than in Sept. Farther north, however, new layers have been observed since 3 Nov over Garmisch-Partenkirchen, West Germany.

Edward M. Brooks resumed daily sunrise and sunset observations from Jeddah, Saudi Arabia on 26 Aug. Between April and June, brilliant colors had been visible 40 minutes from sunrise and sunset, indicating material at 20-25 km altitude, but by Oct, only faint remnants of color could be seen that long before sunrise and after sunset. However, Brooks continued to see unusual colors, although somewhat nearer to sunrise and sunset, and at dawn on 21 Oct observed a brown NNW-SSE-trending band that looked like volcanic ash, at 10° above the ESE horizon. In early Nov, the timing of the appearance and disappearance of color indicated that the layers being illuminated may have only been in the upper troposphere. H. H. Lamb observed brilliant sunsets from Norwich, England in early and mid-Sept (7:9), but reported that sunset colors were less spectacular, although still usually abnormal, by late Oct.

Information Contacts: M.P. McCormick, NASA Langley Research Center; T. DeFoor, MLO; M. Hirono, Kyushu Univ.; S. Hayashida, A. Ono, Nagoya Univ.; R. Reiter, Garmisch-Partenkirchen, W Germany; E. Brooks, Saudi Arabia; H. Lamb, Univ. of East Anglia.

11/82 (7:11) Data from ground-based lidar distant from El Chichón indicated that the dense portion of the stratospheric cloud ejected 4 April was spreading slowly northward. However, at low latitudes, where dense aerosols have been observed since shortly after the eruption, both the altitudes and concentrations of the strongest layers have decreased noticeably in Nov.

During the late Oct-early Nov NASA flight between 46°N and 46°S the dense portion of the cloud terminated at 6-10°S and 30-37°N. In some areas, the edge was quite abrupt, almost cliff-like, but in other regions, it was more gradual. The N boundary of the dense aerosols was at 35-37°N in late Oct, but at the time of the return flight in early Nov an arctic air mass had pushed it back to ~30°N, its approximate July position. Peak backscattering ratios were about half of those measured in July, but the cloud had become more homogeneous. By mid to late Nov, however, ground-based lidar showed that the dense cloud had advanced significantly farther N.

The following is from a report from Reinhold Reiter.

"Short-interval lidar observations by the Fraunhofer-Institut für Atmospharische Umweltforschung, Garmisch-Partenkirchen, West Germany have been made since Oct 1976. Our first lidar sighting of an aerosol cloud attributed to the El Chichón activity was on 3 May at 15-16 km altitude (circles in fig. 19-1). This layer, with a scattering ratio of 3 (ratio of total to molecular backscattering) at 0.69 μm (ruby wavelength) was clearly distinguishable from the layer produced by the as yet unknown eruption in late Dec 1981 or early Jan 1982 (the so-called 'mystery cloud'). From 11 May on, a broad layer between 10 and 20 km had developed with a backscattering maximum at 18 km. This layer remained very steady throughout the summer months. The maximum scattering ratio of 8 was observed 16 May; later on the layer showed values of 2-3. This layer, transported by the stratospheric westerlies, was joined by aerosol layers after the end of May that were carried by the stratospheric summer easterlies above 20 km. In the height range 20-25 km, the aerosol concentrations fluctuated drastically and not before mid-Aug had a rather homogeneous aerosol layer developed. The highest scattering ratio of 14 was observed with a height resolution of 600 m at 24.6 km on 1 Aug. After mid-Oct the double structure merged into a broad layer between 13 and 25 km, which can be explained by the change of the stratospheric wind pattern to the winter regime with westerlies throughout this height range.

"In Nov the stratospheric aerosol again changed its structure because of the arrival of the uppermost El Chichón clouds. Above 25 km, 2 new layers could be observed with scattering ratios of about 3, one centered at about 27 km, the other centered at about 30 km, tailing off to 35 km."

In Hawaii, lidar measurements showed a gradual decrease in total integrated backscatter until 3 Dec, when a hole appeared in in the aerosol layers at 26 km altitude, dropping integrated backscatter by nearly a factor of 2. Preliminary data from 10 Dec measurements indicates a return to a pattern similar to that of late Nov. The lower stratospheric layer first observed there in Oct was very strong in mid-Nov, but had nearly disappeared by the end of the month. In Hampton, Virginia increasingly strong backscattering ratios (twice those previously recorded at this latitude) at high altitudes were measured in Nov, showing that northward spread of the dense portion of the cloud had resumed, but lower stratospheric layers like those detected from Hawaii were not observed.

Unusual sunrises and sunsets were seen in Japan, Saudi Arabia, and Germany. Toshio Fujita reported fiery red glows and unusual twilights from various locations in Japan since 17 May. These were seen as far N as Sendai (38.27°N, 140.90°E) in late Oct. During the autumn, evening glows have been more brilliant than usual at Sapporo (43.05°N, 141.33°E). Many of these observations corresponded with detection of increased aerosols at 25-27 km altitude by the Meteorological Research Institute lidar (36°N, 140°E). Backscattering ratios ranged from 30 to 150. Small pale brownish halos often surrounded the moon, with diffused light extending 3-5 diameters. In Saudi Arabia, Edward Brooks continued to see long and unusually colored dawns and twilights through early Dec, but noted considerable day-to-day variation. He could often see distinct bands of volcanic material (27 and 29 Oct, 5, 6, 11, 14-15, 17, 24, and 27 Nov, and 3-4 Dec). After sunset on 5 Nov, a SSW-NNE band of ash was illuminated after cirrus clouds (in the upper troposphere) had lost their illumination, indicating that the ash was at stratospheric altitudes. Separate tropospheric and stratospheric layers were illuminated before sunrise 3 Dec.

Information Contacts: M.P. McCormick, W. Fuller, NASA Langley Research Center; R. Reiter, Garmisch-Partenkirchen, W Germany; T. DeFoor, MLO; M. Hirono, Kyushu Univ.; S. Hayashida, Nagoya Univ.; T. Fujita, Meteorological Research Inst., Japan; E. Brooks, Saudi Arabia; A. Strong, NOAA/NESDIS.

12/82 (7:12) Atmospheric data indicated continued disper-

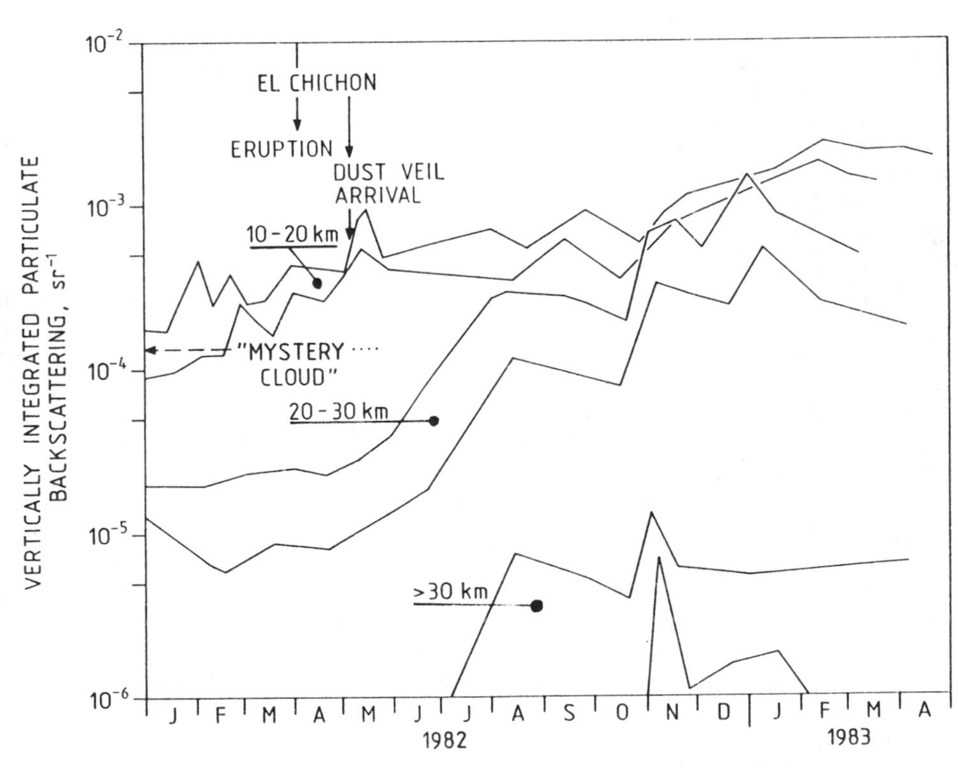

Fig. 19-1. Lidar observations of volcanic aerosols over West Germany, Jan 1982-April 1983, showing the arrival dates of layers at different altitudes from 2 eruptions. Data include the passage of material from the "Mystery Cloud" [now believed to have been ejected by Nyamuragira (Zaire) in late Dec 1981; see 7:3] and the aerosols from the March-April 1982 eruption of El Chichón. A 3-dimensional plot of these (and later) data appears in Fig. 19-7. Vertically integrated particulate backscatter intensity (sr^{-1}) [see discussions in 6:5 and 9:10] is subdivided into 10-20, 20-30, and > 30 km altitude ranges. Measurements were made by the 694.3 nm pulsed ruby laser (height resolution 600 m) at Garmisch-Partenkirchen. [Illustrations throughout this chapter have been rearranged and captions revised for clarity.]

sal of the dense part of the stratospheric aerosol cloud ejected by El Chichón's 4 April eruption. Balloon data from Wyoming (41°N) began to show a few isolated layers of the dense portion of the cloud in July and Aug. More layers gradually appeared and by 30 Nov about 80% of the zone between the tropopause (roughly 10 km altitude) and 30 km altitude contained aerosols from the cloud. Particle concentration dropped sharply above 30 km altitude and was 2 orders of magnitude lower at 32 km. By the next balloon launch on 9 Dec, no major gaps were evident in the aerosols within this 20 km-thick region. Data collected 30 Dec were similar. Particles in the upper half of the cloud averaged about 0.3 μm in diameter, as compared to a mean particle size of about 0.1 μm in the lower half, more typical of volcanic clouds. From the balloon data, the total mass of the El Chichón cloud was estimated to be 8-10 megatons, about 40 times that of the cloud ejected 18 May 1980 from Mt. St. Helens. Most of the mass of the El Chichón cloud was concentrated in its upper layers. No decay of the cloud was evident from the balloon data, indicating that the rate of particle settling did not yet exceed the rate of gas to particle conversion.

Gas and particle samples were collected for LANL between the tropopause and 20 km altitude from a WB57-F aircraft that flew from the equator to 75°N in April-May, July-Aug, and Oct. The maximum lower stratospheric sulfate concentration detected was 167 ppb by mass from a sample taken 20 April. Because the 4 April cloud had probably not reached the sampling area by then, this material is thought to have been collected from the smaller 29 March cloud. The average lower stratospheric concentrations over the entire Northern Hemisphere were calculated to be 11.85 ppb by mass in April-May, 9.27 ppb in July-Aug, and 7.54 ppb in Oct, in contrast to the July 1981 value of 2.5 ppb (primarily from Alaid and Pagan) and the July 1980 value of 2.3 ppb (mostly from Mt. St. Helens, 6:10). However, unlike the 1980 and 1981 eruptions, the bulk of the material from El Chichón's explosions reached altitudes higher than 20 km, so the 1982 concentration figures represent only $^1/_3$-$^1/_5$ of the total cloud mass, estimated at about 5 megatons from the aircraft data.

Comparison of Oct and Dec data collected from a NASA Ames Research Center Convair 990 aircraft showed that considerable mass transport of the El Chichón cloud took place in the Northern Hemisphere during the autumn. Visible wavelength optical depth measurements indicated that large quantities of material had reached 54°N, the northern limit of the flight, by mid-Dec. Significant variation in particle size distribution at different locations was detected. Measurements were also made at many infrared wavelengths. Data reduction was not yet complete, but spectral resolution of about 1.5% will allow isolation of the effects from volcanic components in the stratosphere from the effects of other material such as ozone. Samples were collected during U-2 flights to 21 km altitude. The concentrations of SO_2 and condensation nuclei (CN) had declined since the summer and new particle formation in the lower stratosphere appeared to have stopped.

Lidar data from Hawaii indicated that most of the aerosol material was between 16 and 30 km although some enhancement in backscattering was detected to 38 km. Total integrated backscatter varied, but was generally less than in Nov. Maximum backscatter measured from Fukuoka, Japan dropped in mid-Dec, but returned to Nov levels 5 days later. The amount of stratospheric material over Hampton, Virginia increased in Dec, but peak values for individual layers have not been as large as those measured over lower latitudes before significant lateral dispersal of the cloud began.

During the lunar eclipse of 30 Dec, the moon was much darker than normal, with only the extreme S limb showing substantial light. Edward Brooks notes that this relatively uniform darkening of most of the moon during the total phase of the eclipse suggests that the El Chichón cloud was present over all but extreme southern latitudes. Darkening of the moon during the 6 July total eclipse was asymmetrical, concentrated on its N half.

Brooks continued to see some brilliant dawns and twilights from Saudi Arabia, but noted considerable variation in their intensity and length. SSW-NNE bands of ash were visible low in the ESE sky near sunrise on 14 Dec. Beginning in mid-Dec, long dawns and twilights indicated the presence of high-altitude material over the area. From Boulder, Colorado (40°N, 105.2°W), Richard Keen observed brick-red color to $1^1/_2$ hours after sunset 3-4 Nov and 11-12 Jan. This corresponds to a solar depression angle of about 18°, indicating that the volcanic material reached an altitude of about 40 km [but see 8:1]. On other nights, the glow has persisted only until a solar depression angle of 6-7° was reached, yielding a maximum cloud height of about 23 km.

Information Contacts: D. Hofmann, Univ. of Wyoming; W. Sedlacek, LANL; J. Pollack, NASA Ames Research Center; T. DeFoor, MLO; M. Hirono, Kyushu Univ.; W. Fuller, NASA Langley Research Center; E. Brooks, Saudi Arabia; R. Keen, Univ. of Colorado; M. Matson, A. Strong, NOAA/NESDIS.

1/83 (8:1) *Aerosol Cloud-Instrumental Observations.* Lidar data continued to show a gradual decrease in both the altitude and intensity of backscatter from the aerosol cloud's densest layer. From Fukuoka, Japan, no notable peak was found above 21 km altitude after 24 Jan. A single broad layer was detected, with maximum backscatter at 18-19 km. The strongest layer over Mauna Loa, Hawaii remained at about 22 km altitude through early Feb while the cloud's integrated backscatter declined slowly. A layer between 16.8 and 17.4 km measured 26 Jan may not have been El Chichón material; it accounted for about 2% of the 26 Jan integrated backscatter. A 12-16 km layer was detected the same night from Wallops Island, Virginia.

Between 27 Jan and 5 Feb, a NASA P-3 Electra aircraft collected aerosol data from 27°N-76°N. The cloud was quite homogeneous from 27-38°N, with peak ruby lidar scattering ratios of roughly 8 at about 20 km altitude. To 55°N, both the upper and lower altitude portions of the aerosol cloud continued to be present, in similar concentrations. As the aircraft approached Greenland at about 55°N and entered the polar vortex, a system circulating air southward from the polar region, the upper aerosol layers disappeared fairly abruptly. However, lower stratospheric material remained, with scattering ratios of 3-8 at 15-16 km altitude, values similar to those at the same altitudes S of 55°N. As the aircraft flew W at 76°N, a similar pattern persisted until it exited the polar vortex at about 100°W, when the upper layer reappeared; scattering ratios ranged from 2-4 at 18-23 km altitude and some material was detected to 30 km altitude. M.P. McCormick noted that these data support information from the SAM II satellite indicating that the lower stratospheric aerosols from El Chichón moved fairly rapidly to the poles but material at higher altitudes has yet to fully penetrate the polar regions.

David Hofmann reported that late Jan-early Feb balloon data from Laramie, Wyoming revealed an extensive cloud of aerosols at higher altitudes than previously observed. The base of the layer, at about 29 km altitude, was marked by a boundary zone that was only about 50 m thick. Particle concentrations on 28 Jan exceeded $600/cm^3$ at 29 km, compared to normal background values of $1\text{-}2/cm^3$ at that altitude. Enhanced concentrations were measured to about 35 km altitude. A second balloon flight, on 1 Feb, again penetrated the cloud. The aerosol particles were about 0.02 μm in diameter, too small to be detected by lidar. They had no non-volatile cores and were probably H_2SO_4 droplets formed in the north polar region from SO_2 ejected by El Chichón. Given a wind speed of about 80 km/hr (from the E) at these altitudes, the cloud was at least 8000 km in lateral extent. By a third flight on 4 Feb, the high-altitude cloud was greatly attenuated.

Gas and particle samples were collected for LANL between the tropopause and 20 km altitude from a WB57-F aircraft that flew from the equator to 75°N in April-May, July-Aug, and Oct. Eugene Mroz reports that calculations based on information from these samples, combined with data from balloon launches to 30 km altitude at 33°N and limited sampling to 10°S, yield a mass of [7.6] $\times 10^{12}$ g of sulfate injected into the stratosphere by El Chichón's explosions. Using the same methods, the mass of sulfate in the "mystery cloud" ejected in early Jan by a volcano that remains unidentified [but see 7:3] was calculated to be 0.85×10^{12} g.

Unusual Sunrises and Sunsets. Brilliant sunrises and sunsets continued to be reported from England in early Jan and Saudi Arabia in late Jan. However, no unusual colors were seen from Wyoming in Jan, or from Colorado after mid-Jan, and sunset colors in New Jersey weakened considerably in late Jan and early Feb.

In mid-Jan, H. H. Lamb reported that during clear days for the previous 3-4 weeks the sun over Norwich, England had increasingly appeared to be surrounded by a white sheen of diffused light extending to about a 20° radius, although there was little apparent diminuition of solar brightness. On 16 Dec at 1605 GMT, with the sun about 6° below the horizon, a roughly round, vivid purple patch was seen at 20-25° elevation, indicating to Lamb that the layer was at 20-24 km altitude. Twenty minutes later, the W sky was a brilliant orange, changing to fiery red nearer the horizon during the next 20 minutes. A brilliant afterglow continued until 1700, which Lamb interpreted to presumably indicate the presence of aerosol material to 34 km altitude. There were no unusual sunset colors for the next several nights, and cloudy weather made observations after 24 Dec difficult. Glow was stronger than usual through breaks in the clouds 26 Dec and on the 28th the increased spread of diffuse white light around the sun throughout the day began to be obvious. On 9 Jan at 1630-1635 GMT, with the sun about 5° below the horizon 22-27 minutes after sunset, a vivid magenta-purple area developed at 15-27° elevation above a greenish-pale yellow sky, suggesting an aerosol layer at about 18-20 km altitude. During the next 15-20 minutes, the sky changed to a more normal appearance, but at 1700-1705 a fainter purple patch appeared at 10-15° elevation, suggesting a possible second aerosol layer at about 35 km altitude.

From Jeddah, Saudi Arabia, Edward Brooks saw several sunrises in early Jan that were preceded by two distinct periods of unusual colors. SW-NE-trending bands of volcanic aerosols were seen at dawn 1 Jan and the next day 2 periods of dawn color were separated by the appearance of dull reddish SSW-NNE-trending volcanic aerosol layers. Similar layers were seen 5 and 6 Jan in association with 2-stage dawns, and after a period of cloudy weather, on 13 Jan. Several long-lasting and bright-colored dawns and twilights were observed during the next several days. On 21 Jan, the second part of a 2-stage dawn included faint N-S bands of volcanic aerosols. When weather conditions permitted, bright dawn and twilight colors were visible until 29 Jan, then were succeeded by several days of little or no color. A brilliant twilight 2 Feb was followed by the observation of NNE-SSW-trending bands after sunset 3-4 Feb that may have been volcanic aerosols. An early dawn 5 Feb indicated the presence of high-altitude aerosols, but the later stage of dawn color was absent, indicating that no lower altitude material was present.

From Boulder, Colorado, Richard Keen reported that a salmon-colored primary twilight glow visible to solar depression angles of 6-7° preceded the brick-red secondary glow that persisted to 1 1/2 hours after sunset 11-13 Jan [7:12; note that 13 Jan has been added]. Keen noted that Volz (1969) described similar double twilights after the 1963 Agung eruption and showed that the later glow can be produced by secondary illumination of the same single layer. Keen therefore suggests that the double twilights that he observed in Nov and Jan were caused by a particularly thick layer at about 23 km altitude and that there probably was no 40 km layer. To produce a double twilight, the 23 km layer would have to extend at least 1500 km W of Boulder. Since 13 Jan, he has seen no unusual twilights. From Laramie, Wyoming, David Hofmann observed no unusual twilight colors since about early Jan. Fred Schaaf observed numerous double twilights from Millville, New Jersey (39.4°N, 74.9°W) through mid-Jan, but secondary glow was not present on 19 Jan and other twilight colors were much weaker. Twilight color remained subdued or absent for the next several days. A rather strong double twilight was visible 28 Jan, but colors were weaker on succeeding days. Daytime aerosol effects also seemed weaker in Jan than in Dec.

Reference: Volz, F. E., 1969, Twilights and Stratospheric Dust Before and After the Agung Eruption; *Applied Optics*, v. 8, p. 2507-2517.

Information Contacts: M. Hirono, Kyushu Univ.; T. DeFoor, MLO; M.P. McCormick, W. Fuller, NASA Langley Research Center; D. Hofmann, Univ. of Wyoming; E. Mroz, LANL; H. Lamb, Univ. of East Anglia; E. Brooks, Saudi Arabia; R. Keen, Univ. of Colorado; F. Schaaf, Millville, NJ.

2/83 (8:2) *Lidar Data*. Lidar measurements from Nagoya, Japan (35.13°N, 136.88°E) 2, 6, 13, 19, and 27 Dec showed similar altitudes and peak backscattering ratios, but small secondary peaks were detected at 35-36 km only on the 13th and 19th. Both the altitude and the strength of peak backscatter measured at Garmisch-Partenkirchen, Germany were significantly higher on 1 Jan than for the very consistent readings of 10, 18, and 29 Jan. Late Feb-early March lidar data from Mauna Loa, Hawaii, Fukuoka, Japan, and Hampton, Virginia were similar to data at the same locations a month earlier. Integrated aerosol backscatter was considerably higher in Virginia than in Hawaii, suggesting that the bulk of the cloud had moved from the low latitudes where it was concentrated for several months after the March-April 1982 eruption.

Unusual Sunrises and Sunsets. From Tsukuba, Japan (36°N, 140°E) Toshio Fujita observed unusual twilight glows through Jan. Evening glows in Dec appeared redder than those in Nov, but by mid-Jan the red twilight colors were rapidly becoming lighter. A twilight photograph taken 24 Jan showed much paler colors than one taken 7 Dec at a time of similar solar depression angle. Despite the difference in color, the peak backscattering ratio measured 26 Jan was very close to the 8 Dec value. The maximum backscattering ratio increased from 17 (at 23 km altitude) on 8 Dec, to 28 on 28 Dec, but both heights of the strongest aerosol layers and their peak backscattering ratios were gradually descending by mid-Jan, and maximum backscattering was 16 on the 26th. Fujita attributed the differences in color at times of similar lidar readings to varying turbidity in the lower atmosphere. In mid-Feb, lidar at Tsukuba again measured relatively weak backscattering and the strongest aerosol layer had descended farther to about 20 km altitude.

Edward Brooks noted considerable variation in dawn and dusk colors from Jeddah, Saudi Arabia in Feb. Colorful early dawns 6-9 Feb (8:1 for his observations 1-5 Feb) indicated the presence of higher stratospheric aerosols, but the absence of unusual late dawn colors suggested that lower stratospheric aerosols were absent. This pattern reversed early 10 Feb, when no early dawn was evident but a colorful late dawn resulted from illumination of volcanic layers near the tropopause, visible as faint N-S bands. Similar bands were visible that evening, when aerosols could be observed both at the tropopause and higher in the stratosphere. Few unusual colors were visible early 11 Feb, but on the 12th both the upper and lower aerosol layers were illuminated. For the next several days, bands of material, generally trending SSW-NNE, were often observed at dawn and twilight with both early and late colors. Only higher aerosols were illuminated at dawn 19 Feb; some lower-altitude material appeared to be present early 20 Feb, but no unusual colors were evident that evening.

From Norwich, England, H. H. Lamb reported that on all cloudless days the sun continued to be surrounded by a white sheen of diffused light that seemed to be increasing steadily in extent, from about 20° in angular radius in mid-Jan (8:1) to 25-30° as of 10 Feb. The sun itself often appeared nearly white at elevations of 5-15° and on partly cloudy days the sky was a paler gray than usual. Richard Keen observed no unusual sunsets from Boulder, Colorado between 13 Jan and 17 Feb.

Fred Schaaf observed increased optical effects in Feb from Millville, New Jersey after a notable weakening in Jan. By 15 Feb, late dawn colors (lower altitude aerosols) had returned to moderate levels. During the afternoon of 18 Feb, the sun was surrounded by a red-brown ring with a radius of about 30° that remained visible until shortly after sunset, when weather clouds obscured further observations. The next evening, twilight glow was stronger and Schaaf calculated that later glows seen for the first time since 31 Jan were produced by aerosols as high as about 16 km. Twilights were less impressive for the next few days, but similar effects were seen 23 Feb. On the 26th, the length of twilight glows indicated aerosols to 16-19 km. On 2 March, moderate to strong early twilight colors from material at about 16 km were followed by a very weak secondary glow that may have indicated the presence of aerosols to 32-40 km.

Balloon Data. David Hofmann reported that balloon launches from Laramie, Wyoming continued to penetrate remnants of the extensive cloud of tiny aerosols at 29-35 km altitude first detected 28 Jan, and encountered a newly formed cloud of similar particles 2 March. The average radius of the 28 Jan particles was about 0.015 μm, with a few as large as 0.05-0.06 μm. Given the particle size distribution and a 30-50% drop in electrical conductivity measured within the cloud, Hofmann calculated a particle concentration of about $1200/cm^3$. During the next week, this concentration dropped to about $100/cm^3$, a rate of decay corresponding closely to the expected rate of particle coagulation. About 90% of the particles disappeared when heated to 150°C, indicating that the cloud was composed of H_2SO_4 and H_2O droplets. Remnants of this cloud were still present 2 March but had diffused and coagulated considerably, extending from 25-35 km altitude with a maximum concentration of about 50 particles per cm^3. Although coagulation had increased the size of individual particles, they remained too small to be detected by lidar at these concentrations. Superimposed on the remnants of the 28 Jan cloud, a new cloud was detected 2 March. Sharply constrained between 31 and 34 km altitude, the new cloud reached concentrations of ~300 particles per cm^3, and appeared to be about 4-5 days old.

Hofmann noted that between 25 and 35 km altitude, liquid H_2SO_4 is vulnerable to vaporization if its temperature is raised slightly. If cooled again, it would then recondense into tiny droplets. High-altitude wind data indicated that the 28 Jan cloud originated in the Alaska-Siberia area, in a zone of 30-40°C stratospheric warming. From this warm area, the cloud reached Wyoming in about 30 hours, carried by 200 km/hr winds. Cooling of about 40°C probably occurred during transport, sufficient to recondense the H_2SO_4. Similar clouds have been detected for the past several years, usually in the spring, but the 28 Jan cloud had particle concentrations 15 times as high as clouds seen in 1982, which in turn were 5 times as concentrated as 1981 clouds (Rosen and Hofmann, 1983).

The much larger particles from the original El Chichón cloud remained evident over Wyoming. During the most recent fully analyzed sounding, on 11 Feb, a very broad layer extended from the tropopause (11 km) to about 27 km with a peak concentration of 10 particles (larger than 0.15 μm) per cm^3 at 19 km altitude. A sounding in an equatorial airmass, on 10 March, showed a similar profile, with a peak concentration of 8 particles per cm^3 at 19-20 km altitude and the top of the cloud at roughly 27 km. The profile included a 1 km-thick zone of very clean air (about $^1/_2$ particle per cm^3) centered at about 15 km altitude, probably tropospheric in origin.

Reference: Rosen, J. M. and Hofmann, D. J., 1983, Unusual Behavior in the Condensation Nuclei Concentration at 30 km; *JGR*, v. 88, p. 3725-3731.

Information Contacts: D. Hofmann, Univ. of Wyoming; W. Fuller, NASA Langley Research Center; T. DeFoor, MLO; M. Hirono, Kyushu Univ.; R. Reiter, Garmisch-Partenkirchen, W Germany; E. Brooks, Saudi Arabia; H. Lamb, Univ. of East Anglia; T. Fujita, Meteorological Research Inst., Japan; S. Hayashida, Nagoya Univ.; F. Schaaf, Millville NJ; R. Keen, Univ. of Colorado.

3/83 (8:3) Lidar data from Fukuoka, Japan showed a significant decrease in peak values during the limited intervals when weather permitted observations. Broad, almost monolayer profiles were obtained. On 22 March, lidar at Hampton, Virginia showed a broader peak than it had on the 3rd, but about the same total amount of aerosol. From Mauna Loa, Hawaii, lidar detected only minor variations in total aerosol through March. In late April and early May, a lidar-equipped NASA aircraft will collect data on the El Chichón aerosols from high northern to high southern latitudes, and will coordinate with balloon launches from Palestine, Texas.

David Hofmann reported that a balloon launch from Laramie, Wyoming early 8 April detected remnants of the extensive cloud of tiny aerosols observed 28 Jan (8:1-2). About 20 particles per cm^3 remained between 25 and 33 km altitude. A new layer of similar particles, probably about 1 week old, was observed at 20 km, an unusually low altitude. Particle concentrations were about $125/cm^3$, but the layer was only 200 m thick. The arctic airmass over Wyoming on 8 April lowered the tropopause to 9-10 km altitude, so the densest layers of the main El Chichón cloud were lower than usual. Counts of particles larger than 0.15 μm reached $13/cm^3$ at 12 km and were still $7/cm^3$ at 20 km. The layer terminated rather abruptly at 23 km.

Edward Brooks reported brilliant dawns and twilights and visible bands of volcanic aerosols over Jeddah, Saudi Arabia during several periods between late Feb and late March. In addition to colors observed shortly before sunrise and soon after sunset, caused by illumination of aerosols in the lower stratosphere, the presence of higher layers often resulted in unusual colors long before sunrise and after sunset. Early and late colors were both visible near dawn 21 Feb, but remained feeble. Brightly colored sunsets were observed 21-22 Feb, and another 2-stage dawn the 23rd. That evening, brown volcanic aerosols formed a layer at about 6° above the horizon. Clouds obscured the sky for the next several days, but many N-S bands of aerosols were visible at 1-3° elevation in the E sky early 28 Feb. During the first week in March, both early and late dawn colors were usually faint and were sometimes entirely absent. N-S bands of volcanic aerosols were present early 4 March at about 5° elevation. Clouds made observations difficult 10-14 March, but the return of clear weather revealed more bands of aerosols 15-19 and 22 March accompanying long, brilliant dawns and twilights.

Fred Schaaf saw several examples of Bishop's Ring in March from Millville, New Jersey, but frequent cloudiness limited his observations. Before sunset on 13 March, the sun was surrounded by a band about 6° wide that formed a ring with a radius of about 24-30°. On 15 March, the sun's brightness was considerably diminished by a haze that could not be accounted for by weather conditions or local pollution. A milky area bordered by a Bishop's Ring that was again about 24-30° in radius was visible an hour before sunset 20 March. A similar Bishop's Ring was observed before sunset 30 March. Sunset glows through the month were only weak to moderate and there was only one weak example of late glow indicating illumination of higher aerosols. Richard Keen reported that he had observed no unusual twilights from Boulder, Colorado since mid-Jan.

Information Contacts: M. Hirono, Kyushu Univ.; M. Osborn, NASA Langley Research Center; T. DeFoor, MLO; D. Hofmann, Univ. of Wyoming; E. Brooks, Saudi Arabia; F. Schaaf, Millville, NJ; R. Keen, Univ. of Colorado.

4/83 (8:4) Atmospheric scientists continued to monitor the stratospheric cloud ejected by El Chichón's March-April 1982 explosions. Poor weather plagued attempts to gather lidar data in Hawaii and Japan, but a few measurements were obtained. In Hawaii, declines were recorded in April for both peak and integrated backscatter, indicating decreases in the aerosol concentration within the densest layers and the cloud as a whole. In Japan, however, limited data showed a recovery of peak backscattering ratios to March levels, after a decline in early April.

Pyrheliometer data from Japan and Colorado show substantial reductions in direct solar radiation after the March-April 1982 injection of stratospheric aerosols. Japanese stations between 26.3°N and 43.3°N showed significant increases in turbidity and decreases in transmissivity since last autumn. These effects seemed to occur earlier at the southernmost station. Direct solar radiation first began to show a slight decline at Boulder, Colorado in July 1982 (about 2%) and was about 6% below the 1978 value in Sept (table 19-1). The major decrease in direct solar radiation occurred in late Oct

	1982	1978	Ratio	Hours
Mar	1005	1000	1.005	13/10
Apr	1000	1049	0.953	16/ 6
May	929	1007	0.922	5/ 8
Jun	943	950	0.993	13/ 9
Jul	938	957	0.980	17/15
Aug	931	959	0.97	16/16
Sep	902	963	0.936	9/32
Oct	915	979	0.935	13/26
Nov	830	958	0.866	19/15
Dec	789	992	0.795	14/22
	1983	**1979**	**Ratio**	**Hours**
Jan	808	999	0.810	11/10
Feb	849	1021	0.832	11/ 6
Mar	916	1006	0.911	6/18
Apr	900	985	0.914	6/10

Table 19-1. Comparison of direct solar radiation (in watt-hours/m^2) at Boulder, Colorado (40°N, 105.2°W) before and after the arrival of El Chichón aerosols. The dense cloud of H_2SO_4 droplets scatters sunlight, measureably reducing the amount that reaches the earth's surface. Ratios (1982-83 divided by 1978-79 values) show little change before the appearance of the dense aerosol cloud over Colorado in late Oct 1982. A slight increase in direct solar radiation Jan-April 1983 suggests a slow diminuition in aerosol density. The right-hand column shows the number of hours of measurements each month, 1982-83 and 1978-79 (left and right of slash respectively). Courtesy of Edwin Flowers.

(around the 27th), with Nov and Dec 1982 having average values 13% and 20% less than the 1978 means. However, Jan-April 1983 data show increasing direct solar radiation, indicating a slow diminuition in cloud density over Boulder. Total radiation data (direct plus diffuse on a horizontal surface) show a much smaller effect. Changes to these data first became apparent in Nov with a decrease of slightly more than the 2% measurement/analytical noise, and the Dec 1982 values were about 3% below the Dec 1978 mean. Edwin Flowers noted that reductions in direct solar radiation caused by previous volcanic aerosol clouds were usually of short duration, interspersed with periods of normal transmission. However, once the effects of the El Chichón aerosols began to be observed, solar radiation values remained depressed, without periods of normal transmission, indicating the cloud's strong lateral uniformity.

A cooperative NASA-NOAA effort, using real-time TOMS data from the Nimbus-7 polar orbiting satellite, identified episodes of tropospheric-stratospheric "folding" that brought stratospheric air to within 3 km of sea level in the arctic. During one of these events, on 23 March, a NOAA P3 Orion aircraft sampled stratospheric aerosols along the coast of Greenland on a flight between Thule (77.5°N, 69.3°W) and Söndre Strömfjord (67°N, 50.6°W). In 6 hours of sampling, mainly at about 4.5 km altitude, aerosols were collected with 9 different filter systems. Concentrations of 0.1 and 1.0 μm particles exceeded 2000/cm^3 with occasional peak values to 5000/cm^3. These particles were predominantly droplets, probably of H_2SO_4. Larger (1-5 μm) fragmented particles were present in concentrations about 1/1000 of the smaller ones and were predominantly composed of Si, lesser amounts of Fe, and traces of Al. In scanning electron micrographs, these particles appeared similar to El Chichón ash collected on the ground shortly after the March-April 1982 explosions. Although the sampling system was designed to operate at 4.5 km altitude or lower, some aerosols were also collected at about 7.5 km, also in stratospheric air. Along with the droplets and fragmented particles, some NaCl was recovered, although in concentrations only 1/1000-1/10,000 of the droplets. Salt had also been collected at 4.5 km, but Russell Schnell noted that at the lower altitude the source of the NaCl could have been sea water, extremely unlikely at 7.5 km. NaCl crystals were collected by NASA aircraft from the El Chichón cloud in April and May 1982 (El Chichón, 7:4). In tropospheric air, aerosol concentrations were 100/cm^3 or less and included very few particles that looked volcanic.

H. H. Lamb reported increased evidence of stratospheric aerosols over Norwich, England. Since early March normal blue skies have been absent; clear sky colors range from milky blue at considerable distance from the sun to strongly increasing whiteness within 30°-50° of arc from the sun. Sky seen in gaps between clouds within 15°-30° of the sun has invariably been virtually white. During many evenings, particularly in March but continuing as of mid-April, the dominant sky color after sunset has been sepia brown to bronze. From Jeddah, Saudi Arabia, Edward Brooks occasionally observed bands of volcanic aerosols from late March through late April, although brilliant dawn and twilight colors were less frequent than in previous months. In addition to colors observed shortly before sunrise and shortly after sunset, caused by illumination of aerosols in the lower stratosphere, material higher in the stratosphere sometimes resulted in colors long before sunrise and after sunset. On 23 March, SSW-NNE bands of aerosols were visible to 7° altitude in the WSW sky around sunset, and faint SSE-NNW aerosol bands below 5° in the ENE were illuminated during a 2-stage dawn the next day. Many sunrises and sunsets in late March were pale to nearly colorless. A brilliant sunset 4 April was followed by a 2-stage dawn on the 5th. Although dawn colors were nearly absent 7 April, faint brown, narrow, closely-spaced N-S bands of volcanic aerosols were seen at 15° altitude in the E; a colorful dusk that evening was also accompanied by brown N-S-trending aerosols. More aerosol bands were visible the next morning. A dense haze, probably stratospheric, was visible early 9 April as were gray N-S bands of haze that evening, accompanying chalky-appearing dawn and dusk colors that extended to 30-35° altitude. Pale dawns and twilights were observed when cloud conditions permitted 10-15 April, but brilliant colors reappeared at sunset 16 April and bands of aerosols were present at 3° altitude late on the 18th and early on the 19th. Neither bright colors nor aerosol bands were observed from sunset 19 April through the 21st.

Information Contacts: R. Schnell, NOAA/GMCC; E. Flowers, NOAA/ERL; T. Yamauchi, JMA; T. Fujita, Meteorological Research Inst., Japan; M. Hirono, Kyushu Univ.; T. DeFoor, MLO; E. Brooks, Saudi Arabia; H. Lamb, Univ. of East Anglia.

5/83 (8:5) The following paragraph is from Alan Strong.

"Stratospheric aerosols from El Chichón continue to produce offsets in satellite measurements of sea-surface temperatures. NOAA has used these offsets to monitor the month-to-month evolution of the aerosol cloud over the northern and southern hemispheres. Revised analyses agree well with data from the Solar Mesospheric Explorer satellite and show spreading S of the equator by June 1982, followed by a more extensive southward push during Aug, when the northern limit of the aerosols was rarely beyond 30°N. During autumn, material began diffusing northward over the Pacific Ocean and by Dec was extensive over North America. The NOAA 7 satellite measurements continued to show large sea surface temperature offsets (i.e. high aerosol concentrations) into Jan, but by Feb this higher-latitude material appeared to migrate away from North America over the North Pacific. In April, although some material was indicated between 10°S and 20°N, with some increases being seen in the southern hemisphere, the highest concentrations appeared to be north of 50°N over the Pacific."

David Hofmann reported that data from a balloon launched at Palestine, Texas (31.6°N, 96.5°W) on 16 May showed a peak concentration of about 5.2 particles (larger than 0.15 μm) per cm^3 at about 20 km altitude. Remnants of the extensive cloud of tiny aerosols first detected above Wyoming on 28 Jan (8:1-2) were encountered between 25 km and the upper limit of reliable data collection at 35 km altitude, peaking at 12 particles per cm^3 at 30 km. Background at this altitude is about 1-2 particles per cm^3. Similar concentrations of these tiny particles were measured from Laramie, Wyoming on 25 May. The larger (> 0.15 μm) aerosols were again centered at about 20 km altitude. Concentrations of $7.3/cm^3$ were detected, more than in Texas 9 days earlier, but slightly less than Jan-Feb peak values of 8-$10/cm^3$ over Wyoming.

Lidar observations from Fukuoka, Japan showed a slight decrease in both the altitude and density of the strongest layer of El Chichón aerosols in early May. From Mauna Loa, Hawaii a new layer that straddled the troposphere/stratosphere boundary (tropopause heights during May lidar measurements were 15.6-15.9 km) was first noted 11 May between 15.2 and 16.5 km altitude and was quite strong during the next observation 25 May, ranging from 13-16.4 km. This layer was weaker during subsequent lidar measurements, but some material was still present at these altitudes 8 June. The new material slightly boosted the integrated backscatter over Mauna Loa, but little change was noted in aerosol layers higher in the stratosphere.

JMA has collected pyrheliometer data from a number of stations since 1959. A marked increase in atmospheric turbidity has been evident since autumn 1982 in the combined data (fig. 19-2), while data from individual locations shows effects beginning earlier at southern stations (fig. 19-3).

José Caburian reported gaudy dawns and twilights in the Philippines in early 1983. Dawns and twilights at low latitudes are usually relatively brief, but colors have been observed for an hour or more before sunrise and after sunset. From Jeddah, Saudi Arabia, Edward Brooks observed pale dawn and twilight colors 22-25 April but on the 26th, colors were present only in the sky

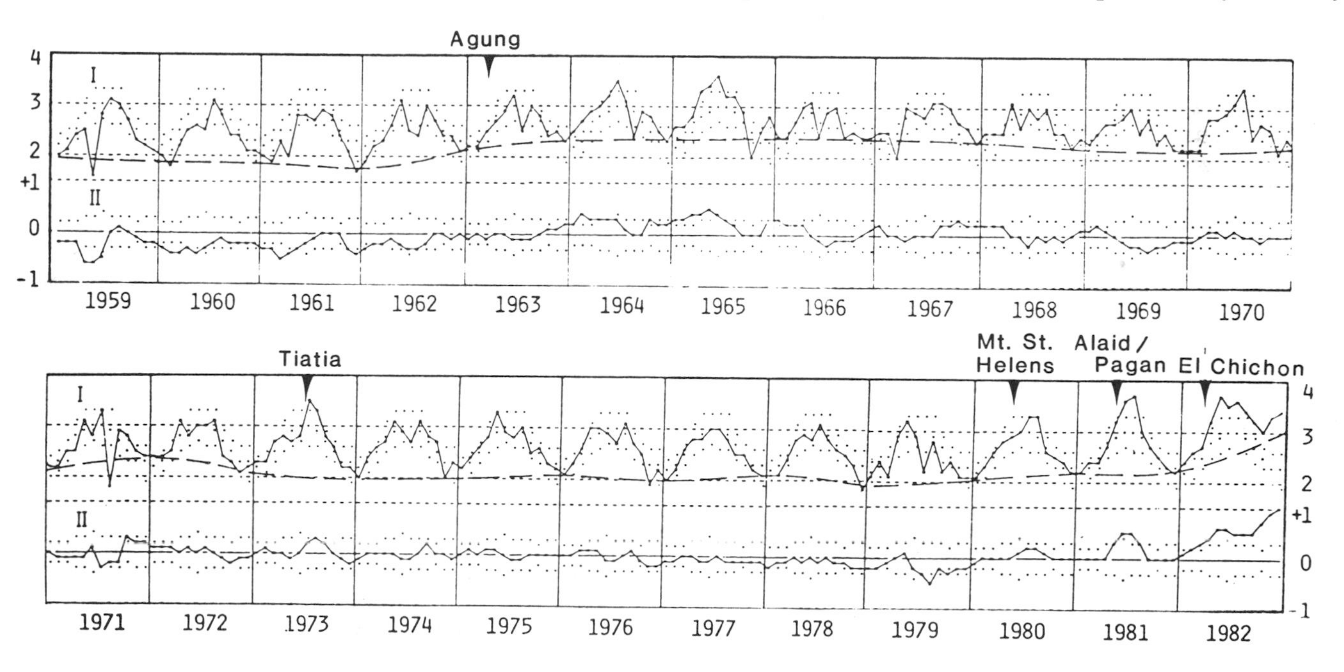

Fig. 19-2. Variations in atmospheric turbidity over Japan, 1959-82, comparing the effects of several large eruptions in two plots. The increase after the El Chichón eruption is substantially larger than those following major explosive events at Tiatia, Agung, Mt. St. Helens, Alaid, and Pagan. *Above,* (I): Monthly minimum turbidity coefficients (connected by the solid line) from pyrheliometer observations at JMA stations were used to reduce local tropospheric effects. The dashed line connects winter values only; T. Yamauchi cautions that abundant water vapor over Japan during the summer might disturb turbidity observations. Three other points are plotted for each month: the middle point is the mean of the 1959-82 minimum values for that month; the upper and lower points define the range of 1 standard deviation from this "normal" value. *Below,* (II): The 3-month running means of deviations from the "normal" value are connected by the solid line. The dots again show one standard deviation for each month. The abnormally high turbidity after several of the large eruptions, especially El Chichón, is emphasized by this plot.

N of the sunrise and sunset points. Late dawn and early dusk colors were visible 27-28 April, indicating aerosols at lower altitudes, but material at higher altitudes produced colors long before sunrise 29 April-1 May and well after sunset 29 April. Bands of aerosols were observed early 1 May and late 2 May. Only late dawn colors were seen 4-5 May but [NNW-SSE] bands of aerosols were present early on the 5th. Overcast weather limited observations 7-10 May but little color was apparent through breaks in the clouds. Two-stage dawns, indicative of aerosols in both the lower stratosphere and at higher altitudes, were seen 11-12 May; early dawn on the 12th was fiery red, and long dark [NNE-SSW] bands of aerosols were visible nearer to sunrise. Only late dawn colors were noted 13-15 May, but there were broad N-S bands of aerosols on the 14th. Sunrises and sunsets were nearly colorless 16-19 May, but brilliant 2-stage dawns and twilights with N-S bands of aerosols occurred 20-21 May. Long crepuscular rays extended to 30° above the horizon early 21 May. Fred Schaaf noted that twilight effects at Millville, New Jersey continued to diminish in April and early May. An aureole around the sun remained prominent on some days, although it was far weaker than in previous months. A Bishop's Ring of 20-30° radius was visible for 1 or more hours before sunset in late April. From a good viewing site 22 April, Schaaf observed 10 minutes of purple light that faded at 5-10° altitude at 1910 local time, indicating material only to roughly 9.5-13 km altitude.

Information Contacts: A. Strong, NOAA/NESDIS; T. Yamauchi, JMA; T. Fujita, Meteorological Research Inst., Japan; D. Hofmann, Univ. of Wyoming; E. Brooks, Saudi Arabia; F. Schaaf, Millville, NJ; J. Caburian, Manila; T. DeFoor, MLO; M. Hirono, Kyushu Univ.

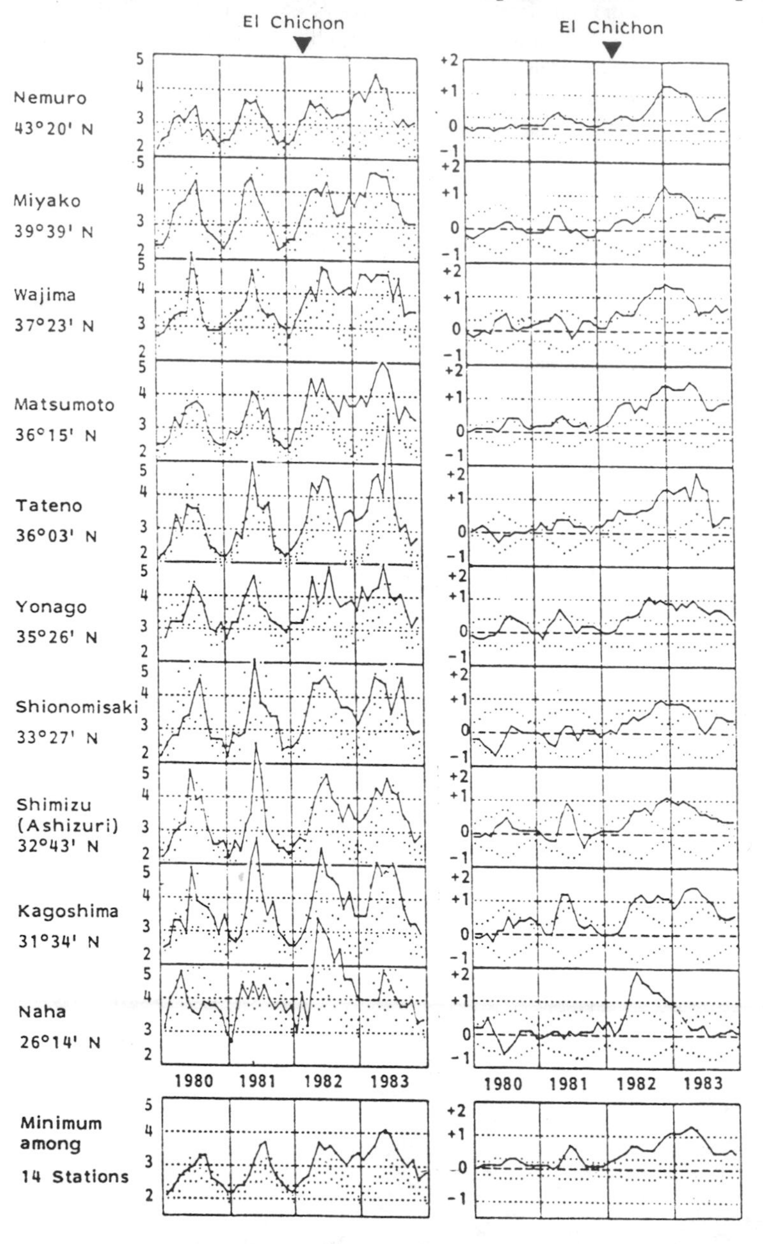

Fig. 19-3. Atmospheric turbidity measured at 10 stations in Japan, 1980-83, comparing the cloud from El Chichón with those produced by eruptions from Mt. St. Helens, Alaid, and Pagan. Note that El Chichón aerosols were first detected at the southernmost stations, nearest in latitude to the volcano. *Left,* monthly mimimum (solid lines) of Feussner-Dubois' turbidity factor (τ_0) from pyrheliometer observations (at 0900, 1200, and 1500 local time) by JMA stations. As in fig. 19-2, the middle dot is the mean minimum monthly value, 1951-80; the top and bottom dots define one standard deviation from the mean. *Right,* the solid line shows the deviation of the 3-month running mean of the monthly minimum turbidity factor (τ_0) from the 1951-80 mean monthly minimum. Dots again mark a range of 1 standard deviation. Courtesy of T. Yamauchi and H. Shimura.

6/83 (8:6) Reinhold Reiter provided the following summary of the results of lidar measurements from Garmisch-Partenkirchen, West Germany, Jan 1982-April 1983. A more detailed analysis [appears in Reiter, et al., 1983].

"The stratospheric dust veil from the early 1982 'Mystery Cloud' (7:1-3) caused from Jan on (before the El Chichón eruption) a clearly increased background level of particle backscattering at 10-20 km altitude, but not at higher levels. We observed the arrival of the dust veil from El Chichón on 3 May 1982 at 10-20 km altitude (peak at 15 km). In this height interval the aerosol backscatter intensity rose until March 1983, with fluctuations of a factor of 5 (weakened by preloading from the 'Mystery Cloud'). We noted the strongest increase, by about a factor of 40, in Jan 1983 in the height interval 20-30 km. Since Aug 1982, a clearly increased aerosol backscatter from the El Chichón dust veil can also be seen at levels higher than 30 km. In contrast to the other dust veils that we traced in the stratosphere (Mt. St. Helens and Alaid), the aerosol backscatter intensity increased, as fig. 19-3 shows, up to the recent past, i.e. over many months. Only in the 20-30 km range can a slight decrease be recognized from Jan 1983 on; at this level sedimentation of particles seems now to prevail over their influx. The time history of the stratospheric aerosol loading can be explained by El Chichón's position near the equator. There was only a slow meridional transport and a poleward homogenization of the stratospheric material in the northern hemisphere. Surely, part of the material also entered the stratosphere in the southern hemisphere.

"Important for an assessment of possible global climate affects is the change in the optical thickness. Before the eruption of Mt. St. Helens, its value was about 0.002, before the eruption of El Chichón, 0.022, and presently it is about 0.18. This means a considerable increase which, according to known models, suggests most probably a cooling of the earth's surface by some tenths of a degree within the next 1-2 years. In the stratosphere, however, warming is to be expected."

The following is from M.P. McCormick.

"A dedicated El Chichón survey mission was flown on the NASA CV990 aircraft 8-20 May, 1983, from 71°N to 56°S (California-Alaska-S of New Zealand, and return to California). Previous missions were conducted on the NASA Electra aircraft in July and Oct-Nov 1982, and Jan-Feb 1983 (7:7, 9-11, and 8:1) and on the CV990 in Dec 1982 (7:12). These flights were coordinated with measurements by satellites or balloons, or with in situ measurements aboard other aircraft.

"During the May 1983 flight, lidar profiles were obtained over the full latitude range, showing a definite peak in integrated stratospheric backscattering (related to optical depth or mass of stratospheric aerosols) at about 50-55°N. This agreed with lidar data collected during the Jan-Feb 1983 arctic mission (8:1). The May lidar profiles showed a well-defined southern boundary of the dense stratospheric aerosols between 40° and 50°N, with a minimum near 20°N, a small maximum near the equator, and a definite falloff to even lower values near 10°S. These data are preliminary at this time and are being processed for future publication. In addition to the airborne lidar measurements, other experiments conducted during the May 1983 flight included measurements of optical depth at visible and infrared wavelengths, radiation flux, and total SO_2 and O_3 column content."

Ground-based lidar in Hawaii and Virginia showed a continuing gradual (although somewhat irregular) decrease in total integrated backscatter. Peak backscattering remained similar.

Edward Brooks reported that bands of aerosols and brilliant dawns and twilights were visible on some days through mid-June from Jeddah, Saudi Arabia. Aerosol bands were seen early 22-24 and 26 May and impressive sunrises and sunsets were frequent during this period, some showing 2 distinct stages caused by illumination of unusually high layers. Optical effects declined 16-20 June, but weather conditions made observations difficult. From Millville, New Jersey, Fred Schaaf reported that only occasional observations were possible because of cloudy weather. Aerosols at moderate altitudes were illuminated 10 May and early sunset color was slightly enhanced 11 and 13 May. José Caburian noted that fiery red sunrises and sunsets, seen in the Philippines for the previous several months, were not evident at the end of May and early June.

Recently erupted volcanic material from an unknown source was collected at 18-19 km altitude over the western US during a series of flights by a NASA U-2 aircraft 22-29 April. Samples from a 22 April mission flown at 37°N from near San Francisco (about 37.7°N, 122.5°W) to Topeka, Kansas (39.02°N, 95.68°W) included particles ranging from less than 0.1 μm to 20-30 μm in diameter: fragments of magnesian olivine, Cl-rich agglomerates, and H_2SO_4 droplets, plus a few glass shards, fragments of SiO_2, and particles of copper and zinc oxide. The copper oxide-zinc oxide particles were similar to those found in previous aircraft samples of volcanic debris, with a characteristic 2:1 Cu to Zn ratio. During a flight from Topeka southward to Palestine, Texas 28 April, concentrations of volcanic material were less than on 22 April but remained above background levels. However, flying northward from Topeka to the U.S.-Canada border (49°N) the next day, particle concentrations were substantially higher, and the debris included many large fragments.

The source of the volcanic material remains uncertain. The samples were not similar to ejecta collected from the El Chichón aerosol cloud, and Raymond Chuan added that the large size of some of the particles suggests that the eruption probably occurred no more than about 2 months before the late April flights. Some large particles had been found in July 1982 samples from the El Chichón stratospheric cloud (Labitzke et al., 1983), but none were recovered during flights in Nov 1982. Chuan also noted that a mid to high northern latitude source for the April 1983 material is suggested by the higher concentration of volcanic debris on more northern flight paths. No observations of early 1983 eruption clouds large enough to penetrate the stratosphere have been reported to SEAN. However, the largest clouds from the 8 April eruption of Asama (Japan, 36.40°N, 138.53°E) were produced before dawn and were not observed (8:3-4). Ashfall from this eruption extended more than 250 km ENE, reaching the coast of Honshu. Large plumes from Etna (Italy, 37.73°N, 15.00°E) have been observed from the ground and on satellite imagery during the eruption that began 28 March (8:3-5).

Lidar data from Mauna Loa, Hawaii showed an aerosol layer straddling the tropopause on 11 May that had not been present during the previous measurement on 4 May. This layer was narrow and well-defined, extending from 15.2-16.4 km altitude on 11 May and from 15.6-16.4 km on 28 May. Lidar data on 25 May seemed to show some mixing between the new layer and the base of the El Chichón material. Lesser amounts of aerosol were detected by lidar at 13-15.6 km on 8 June and 11.4-15.2 km on 29 June, but no upper tropospheric material was present 22 June. Clouds that were probably at too high an altitude to be cirrus persisted over Hawaii for several weeks and a corona and brilliant white aureole sometimes surrounded the sun. A relatively weak layer peaking at 13.5 km altitude was detected by lidar at Hampton, Virginia on 23 June but none was present at that level 22, 26, 27, or 30 June. Lidar data were not collected at Hampton between 22 March and 22 June. The relationship between these lidar observations and the fresh volcanic material collected during the NASA flight on 22 April is uncertain.

Reference: Labitzke, K., Naujokat, B., and McCormick, M.P., 1983, Temperature Effects on the Stratosphere of the April 4, 1982 Eruption of El Chichón,

Mexíco; *Geophysical Research Letters*, v. 10, no. 1, p. 24-26.

Information Contacts: R. Reiter, Garmisch-Partenkirchen, W Germany; M.P. McCormick, D. Woods, W. Fuller, NASA Langley Research Center; T. DeFoor, MLO; E. Brooks, Saudi Arabia; F. Schaaf, Millville, NJ; J. Caburian, Manila, Philippines; R. Chuan, Brunswick Corp.

Further Reference: Reiter, R., et al., 1983, The El Chichón Cloud over Central Europe Observed by Lidar at Garmisch-Partenkirchen During 1982; *Geophysical Research Letters*, v. 10, no. 11, p. 1001-1004.

7/83 (8:7) Lidar data showed a continuing gradual decline in backscattering ratios from El Chichón aerosols. However, lidar in Hampton, Virginia and Fukuoka, Japan detected layers near the local tropopause that may have been from another eruption. A weak layer peaking at 13.5 km altitude was present over Hampton 23 June and aerosols were also observed at 9-14 km on 7 July and 11-13.5 km on 26 July. In July, this material was associated with a "double tropopause", a condition in which 2 temperature inversions were found instead of the single one that typically marks the boundary between the troposphere (where temperature decreases with altitude) and the lower stratosphere (where temperature increases with altitude). No aerosols were detected below the El Chichón material 12 and 27 July. From Fukuoka, new thin layers were observed 1-5 Aug between 12 and 16 km altitudes, at or below the local tropopause. The layers had fine structures of 100 m and were much more stable than the often-observed cirrus clouds in that altitude range, indicating the possibility that they were volcanic ejecta. Their peak scattering ratios were about 3-30 using YAG lidar. Aerosol layers were observed below the El Chichón material at Mauna Loa, Hawaii on some nights in May and June (8:5-6), but none were detected in July and early Aug. The source of this material and its relationship to the recently-erupted volcanic material collected in late April by a NASA aircraft over the central United States remain uncertain. Although the eruption of Una Una, Indonesia (0.17°S, 120.61°E) that began 18 July probably injected tephra into the stratosphere, meteorologists anticipated little northward migration of this material until autumn.

From Norwich, England, H. H. Lamb observed little change in optical phenomena. Skies in the direction of the sun continued to be whitish with much diffuse radiation; broken clouds seen against this background appeared an unusually pale gray. Clear twilight skies still produced abnormal colors, often bronze or sepia near the horizon, and whitish shades sometimes with a magenta or purplish patch above, in the direction of the sun. Before sunset, a fan-shaped area of brilliant bluish-white glow above the sun was common. On 30 June between 2245 and 2315 GMT (and probably for some time before and after) noctilucent clouds, structured like dense cirrus, were seen glowing strongly with a soft bluish-white light against the background of the brightest part of the twilight sky, between the horizon and about 9° elevation. The implied altitude of the clouds was about 80 km. Noctilucent clouds are rare at England's latitude and are seen only within about 2-3 weeks of the summer solstice. Lamb last observed noctilucent clouds a year or 2 after the 1963 Agung eruption, which injected large quantities of aerosols into the stratosphere.

Richard Keen reported that enhanced twilights returned to Boulder, Colorado on 14 June, after an absence of 5 months. Unusual twilights were observed 14 and 17 June, and 2-5, 13-14, 17, 22, and 24 July. The twilights were salmon-colored, with brightest and most pronounced coloring at solar depression angle (SDA) of 4°, disappearing on the horizon at SDA 5-6°. In addition, the 3, 13, and 17 July twilights included fainter purplish color that continued to an SDA of about 11°. None of the twilights were as bright as those seen in Jan, and they ended at somewhat smaller SDA's, suggesting that they were produced by aerosols at somewhat lower altitudes. Cloudy weather Feb-May made observations difficult but no unusual twilights were observed on clear evenings. Raymond Chuan began to see similar salmon-colored twilights about 10 July from Costa Mesa, California (33.65°N, 117.93°W) and these continued through the week of 18 July, but no enhanced colors were present in early Aug.

From Millville, New Jersey, Fred Schaaf observed twilight glows in June, July, and early Aug that were the strongest since Jan. Skies were whitened by volcanic aerosols 9 June and a 20° Bishop's Ring was seen the next evening. Purple light after sunset indicated aerosols to more than 8 km altitude 13-14 June. Faint but very late color 22 June was followed by a 2-stage twilight the next evening suggesting the presence of both lower altitude material and aerosols extending to 24 km. Colors caused by high-altitude material were not evident 24-25 June but returned on the 26th, from aerosols at 24-27 km. Sunsets showed no evidence of high-altitude material for the rest of June, but Bishop's Ring was seen on the 30th. Poor weather prevented additional observations until 10 July, when a 2-stage twilight again indicated aerosols at low and high altitudes; if aerosols were being directly illuminated, continued color at 2200 suggests material extended to 40-48 km. Twilight was spectacular the next evening. Shortly after sunset, an intense narrow red band was observed, indicating strong low-altitude aerosols, but late color was also the strongest since Jan; if caused by direct illumination, aerosols reached more than 50 km altitude. Twilights were similar 12-13 July. Poor weather limited observations during the next 2 weeks. Moderately colored 2-stage twilights were observed 22 and 26 July, with a mid-level layer at 13-19 km altitude. This layer was still present 27 July, but higher material seemed absent. Dawn observations 1, 2, and 5 Aug continued to show mid-level aerosols but no high-altitude material was observed.

Late June observations by Edward Brooks of dawns and twilights from Jeddah, Saudi Arabia were hampered by haze, sand, and occasional clouds. However, very early dawns, although often nearly colorless, suggested the presence of high-altitude aerosols. Brilliant dusk colors were observed 3 July and and faint bands of volcanic aerosols were observed with a very early dawn the next day.

Information Contacts: T. DeFoor, MLO; W. Fuller, NASA Langley Research Center; M. Hirono, Kyushu Univ.; R. Reiter, Garmisch-Partenkirchen, W Germany; H. Lamb, Univ. of East Anglia; F. Schaaf, Millville, NJ; R. Keen, Univ. of Colorado; E. Brooks, Saudi Arabia; R. Chuan, Brunswick Corp.

8/83 (8:8) Balloon data from Laramie, Wyoming in June and July showed little change in the El Chichón stratospheric aerosol cloud. Maximum concentrations of about 5-6 particles (> 0.15 μm)/cm^3 were measured at about 20 km altitude. A few small layers were also detected near the tropopause, perhaps from recent eruptions.

The profile of stratospheric aerosols from the March-April 1982 eruption of El Chichón remained almost constant during Aug over Fukuoka, Japan. Integrated backscatter of the aerosol cloud over Hampton, Virginia declined slightly in Aug from July values. New layers near the local tropopause, occasionally observed in recent months, were not reported in Aug.

From Millville, New Jersey, Fred Schaaf continued to observe moderate to strong twilight glows in Aug and early Sept. Primary glow shortly after sunset often blended with somewhat later purple light but faded quickly, indicating aerosol layers that reached a maximum altitude of 8-13 km. Two-stage twilights 14, 15, and 24 Aug were characterized by long intervals between illumination of the lower and higher layers. If late colors were caused by direct illumination of high-altitude layers, aerosol material was present to roughly 32-40 km altitude. Late colors were usually present, but were not observed 5-6 Sept. Schaaf noted that their occasional absence during evenings when early colors were visible suggests that they are produced by primary illumination of high-altitude layers rather than secondary glow from material at lower latitudes.

Information Contacts: W. Fuller, NASA Langley Research Center; M. Hirono, Kyushu Univ.; D. Hofmann, Univ. of Wyoming; F. Schaaf, Millville, NJ.

9/83 (8:9) Lidar data from Fukuoka, Japan and Hampton, Virginia showed little change in the remnants of the stratospheric cloud ejected by the March-April 1982 eruption of El Chichón.

Edward Brooks resumed dawn and twilight observations from Jeddah, Saudi Arabia in late Aug. Bright colors were seen during some dawns and dusks in late Aug but colors became more intense in early Sept. NNW-SSE bands of aerosols were seen low in the W sky during brilliant dusks 12 and 13 Sept. During the evening of the 14th, faint N-S aerosol bands were visible, but their rapid fading suggested that they were at relatively low altitudes. Aerosol bands, trending N-S and NNW-SSE, were seen in the E sky at dawn 18-19 Sept. Colors remained brilliant through 24 Sept. Brooks noted that the very late twilight illumination at high sky angles observed at times since April 1982 would require particles to be present at altitudes near 100 km if caused by primary scatter. He suggests instead that secondary scattering from aerosols near 25 km altitude was responsible for late, high-angle illumination.

From Millville, New Jersey, Fred Schaaf observed moderate to strong early and late twilight colors during the first half of Sept. After the arrival of a northern air mass 15 Sept, both early and late colors became weak. Moderately strong early and late colors returned 6 Oct. Schaaf noted that if late colors were caused by direct illumination of high-altitude aerosols (timing of illumination would indicate altitudes in excess of 70 km), they should have been unaffected by the northern airmass; the near disappearance of late colors in mid-Sept suggests that they are the result of secondary scattering from lower altitude material.

Information Contacts: E. Brooks, Saudi Arabia; F. Schaaf, Millville, NJ; W. Fuller, NASA Langley Research Center; M. Fujiwara, M. Hirono, Kyushu Univ.

10/83 (8:10) Lidar data from Hampton, Virginia showed a sharp increase in backscattering from aerosols in the lower stratosphere on 27 Oct. Accompanying the enhanced aerosols was a lowered, multiple tropopause, with several temperature inversions (the strongest at 9.4 km) instead of the single one that usually marks the boundary between the troposphere and the stratosphere. Poor weather had prevented observations since 3 Oct, when backscattering integrated from the tropopause to 30 km altitude was only half the 27 Oct value. Four nights later, lower stratospheric backscattering had declined but was still somewhat enhanced, and the altitude of the tropopause was 2 km higher. Preliminary analysis of 8 Nov data indicated an additional decrease in the amount of lower stratospheric aerosols and a further rise in the tropopause altitude. The source of the enhanced aerosols could not be determined but NASA scientists suggested either a recent eruption (perhaps the 3 Oct Miyakejima activity) or El Chichón material transported southward by an arctic air mass from high latitudes, where it has recently been concentrated (8:6). No lower stratospheric layers were reported in Oct from Fukuoka, Japan.

From Jeddah, Saudi Arabia, Edward Brooks reported variable dawns and dusks; some were brief because aerosols were at lower elevation than before. Dusk on 28 Sept ended quickly and was the same color as the sun, indicating that the scatterers were at low altitude and relatively large. Two distinct periods of enhanced colors were observed at dusk 2 Oct and dawn on the 8th, suggesting illumination of layers at 2 altitudes. Dusks 10-11 Oct and dawn on the 11th ended quickly, indicating that the scattering layer was at low altitude. Dusks 29 Sept and 9 Oct produced no significant color enhancement.

Information Contacts: W. Fuller, M.P. McCormick, NASA Langley Research Center; M. Fujiwara, M. Hirono, Kyushu Univ.; E. Brooks, Saudi Arabia.

11/83 (8:11) *Balloon Data.* Univ. of Wyoming atmospheric scientists launched a balloon from McMurdo, Antarctica (77.85°S, 166.62°E) on 27 Oct. Aerosols with a size and altitude distribution typical of remnants of debris injected by El Chichón were detected between the tropopause (at 12 km) and 19 km altitude. The layer's broad peak was centered on 15 km, where concentrations of particles larger than 0.15 μm were about

$6/cm^3$, similar to values observed over Laramie, Wyoming at the same latitude 6 days earlier (see below).

From 22 km to the top of the sounding at 33 km, the balloon passed through a zone of much smaller particles, reminiscent of the cloud of tiny condensation nuclei observed over Wyoming in early 1983 (8:1-2) and (in lesser quantities) during previous springtimes. Maximum concentrations of particles larger than 0.01 μm reached $100/cm^3$ at 25 km altitude, an order of magnitude less than the maximum in the early 1983 cloud over Wyoming. A secondary peak was measured at 30 km altitude ($20/cm^3$), and concentrations were increasing again (to $10/cm^3$) shortly before the balloon burst at about 33 km.

Stratospheric temperatures were unusually warm during the 27 Oct balloon mission, rising from -75°C at the tropopause, to about -50 to -60°C at 25 km, (the altitude of peak CN concentrations), to -25°C at 33 km. Comparison of the peak CN concentration during the 27 Oct balloon launch with a particle coagulation curve suggests that roughly 16 days had elapsed since the evaporation event. Temperature profile data in the South Polar region is sparse, but weather soundings at the Pole show a warming from about -50°C at roughly 30 km altitude (10 millibars) on 6 Oct, to -30°C at the same altitude on 11 Oct. This temperature change is consistent with those observed in conjunction with CN events in the arctic.

Balloons were launched from Laramie, Wyoming on 21 Oct and 22 Nov. A double tropopause was detected during the 21 Oct mission, with temperature reversals at 11 and 15 km altitude. Each reversal was associated with an aerosol layer. The denser layer was at 15 km, where particle concentrations were $6/cm^3$ larger than 0.15 μm and $4/cm^3$ larger than 0.25 μm. On 22 Nov, the aerosol layer extended from the tropopause, at 7.5 km altitude in arctic air, to 22 km, reaching peak particle concentrations of $9\text{-}10/cm^3$ (> 0.15 μm) at 13 km. A secondary peak of $0.3/cm^3$ at 30 km (a high concentration for that altitude), suggested that equatorial air was present above the arctic air. Ratios of particles larger than 0.15 μm to those larger than 0.25 μm were similar to a month earlier, remaining at about 1.5, a value characteristic of the El Chichón aerosols but in distinct contrast with ratios of 4-5 for most volcanic aerosol clouds. Hofmann noted that multiple aerosol layers and complex boundaries between troposphere and stratosphere are often seen over mid-latitudes in the autumn, when air from polar regions is present at low altitudes while higher altitude air has been transported from equatorial regions.

Lidar Data. From Mauna Loa, Hawaii aerosol profile shape and backscattering values became increasingly variable in the autumn. The 13 Sept and 19 Oct profiles were jagged, with many small peaks. The 10 Aug profile was smooth; the 6 Oct pattern was similar but with a single new peak superimposed. On 9 Nov, a layer only 30-60 m thick was present between 15 and 16 km, just below the tropopause. On 30 Nov, backscattering increased very steeply from the base of the aerosol layer, dropping off above 21 km. At Hampton, Virginia, lidar data in late Oct and Nov showed variable aerosol profiles. When the jet stream moved south of Hampton, it transported polar air containing lower stratospheric aerosols into the region. Bases of aerosol layers in both the polar and tropical air masses were near the tropopause, several kilometers lower in polar air than in tropical air (roughly 10 km vs 15 km). "Polar" aerosols were emplaced beneath the higher-altitude "tropical" layers, approximately doubling integrated backscattering values over Hampton in polar airmasses. A distinct layer at 18.5-19.8 km observed 29 and 30 Nov did not appear to be part of this pattern and its source is uncertain. This layer was not present 8 Dec. Late Nov lidar data from Fukuoka, Japan were very similar to late Oct results.

Unusual Sunrises and Sunsets. From Norwich, England, H. H. Lamb provided the following report.

"Abnormal coloured sunset and twilight glows presumed associated with the El Chichón eruption's stratospheric cloud have continued to be observed on almost any clear evening. In July, Aug, and Sept, the most regularly noticeable anomalies were cold yellow to sepia or stone-coloured twilights, later turning fiery red at the horizon. Frequently a nearly circular beautiful purple to rose patch of softer colour, some 10° of arc or a little more in diameter, was present above the horizontal yellow-brown layers of colour. The timing of the decline in elevation and ultimate disappearance of the purple/rose patch consistently indicated a height of about 20 km for the layer responsible throughout the period from late July to 14 Oct. On 5, 6, and 25 Sept and again on 7 Oct the purple/rose patch merged or extended into crepuscular rays in the same colour, the timing of which was used to derive the probable height of the layer.

"Since mid-Oct the displays have become more brilliant again and greater heights have been indicated for the active layer. On 21 Oct, the after-sunset glow was a brilliant fiery red along the horizon and at 1750 GMT a roughly circular purple patch of light developed strongly to over 20° elevation. This indicates a layer above 50 km height unless some secondary scattering effect was responsible. But on 24, 29, and 30 Oct a purple patch which later turned to rose pink became strongly illuminated about 1700 GMT up to 30° elevation, declining in elevation to 20° by 1707 and to 10° as it faded quickly about 1710. Over these dates a height of about 27 km is consistently indicated by these timings.

"It is not known whether this renewal of the optical effects and evidence of greater heights should be attributed to fresh dispersal of El Chichón material from other latitudes and/or other heights by the seasonal changes of the winds or whether a new stratospheric veil is involved."

Richard Keen continued to observe colorful twilights from Boulder, Colorado. An enhanced, salmon-colored twilight with maximum brightness and coloration at an SDA of 4° occurred 28 July. Similar twilights were observed 3 and 11-13 Sept (no observations were made in Aug). These included a fainter purplish glow to an SDA of 9°, similar to twilights of 3, 13, and 17 July (8:7). Strong double twilights, comparable to those seen from Boulder in Nov 1982 and Jan 1983 (8:1), occurred 15-16 and 21 Sept, 3-6, 22, and 24-26 Oct, and 3, 6, 12, 14, and 23 Nov. They typically consisted of an enhanced salmon-colored 4° SDA primary twilight followed by a deep red secondary twilight to an SDA of about 17°. Keen had earlier noted that an aerosol layer at about 20 km altitude would have to extend about 1500 km W of a given observation point for secondary twilight illumination to occur. He suggests that a second requirement appears to be the absence of cirrus clouds in the same range. Satellite imagery showed that middle and high-altitude weather clouds were absent for 1500 km W from Boulder during all such observations, and these conditions were successfully used to predict double twilights.

Edward Brooks continued to observe some bright dawns and dusks through mid-Nov from Jeddah, Saudi Arabia. Colors were often strong in late Oct and sometimes reached 40° elevation, but illumination was usually most intense within a few degrees of the horizon. In early Nov, dawn and dusk colors were typically present only near the horizon; on the 8th, the absence of early dawn colors and the brief late dawn illumination low in the sky suggested that the scattering layer was at lower altitude than before. In mid-Nov, the onset of variable autumn weather coincided with frequent changes in dawns and dusks, which ranged from prolonged and bright-colored to brief and subdued.

From Millville, New Jersey, Fred Schaaf observed strong twilight colors in Oct. The timing and extent of illumination 10 Oct suggested an aerosol altitude of little more than 8 km. Strong 2-stage twilights were seen 27-28 Oct. Twilight effects weakened suddenly 30 Oct as arctic air moved into the area, as had occurred in mid-Sept (8:9), and remained weak through early Nov. Double twilights returned 8 Nov, when first-stage aerosol illumination ended at a time indicating that the top of the layer was at about 16-19 km. On the 12th, the arrival of arctic air again coincided with the weakening of twilight colors. In late Nov, colors were often intense but the aerosols appeared to be lower, with the top of the illuminated layer estimated at 11-13 km on the 29th.

Information Contacts: D. Hofmann, Univ. of Wyoming; R. Reiter, Garmisch-Partenkirchen, W Germany; T. DeFoor, MLO; W. Fuller, M.P. McCormick, NASA Langley Research Center; M. Fujiwara, M. Hirono, Kyushu Univ.; H. Lamb, Univ. of East Anglia; R. Keen, Univ. of Colorado; E. Brooks, Saudi Arabia; F. Schaaf, Millville, NJ.

12/83 (8:12) *Balloon Data.* Particle counters on a balloon flight from Laramie, Wyoming 14 Dec detected a layer of tiny condensation nuclei (CN), centered at about 30 km altitude. Particle concentrations reached about $200/cm^3$, compared to background values of about $10/cm^3$ at that altitude. Temperatures at 30 km altitude had warmed to -30 to -35°C on 10-11 Dec. David Hofmann noted that particle concentrations observed on the 14th were consistent with values that would be expected in a CN cloud produced a few days earlier.

Data on larger particles were collected during a balloon flight 21 Dec. The base of the aerosol layer was just above the tropopause (at about 8 km altitude, indicating an arctic air mass). Maximum particle concentrations of about $10/cm^3$ (larger than 0.15 μm) were detected at about 12.5 km, declining gradually to about half that value at 18 km and 10% of the peak at 23 km. No distinct sublayers were observed.

Lidar Data. In Dec, lidar at Mauna Loa, Hawaii detected slightly less stratospheric aerosol, as measured by total integrated backscattering, than in Oct and Nov. On 7 Dec, major backscattering peaks, marking the strongest aerosol layers, were separated by only small decreases in values; the 23.2 km layer was the strongest among several sharp peaks. The 2 major layers were well-defined on 14 Dec. On the 28th, each had a broad, smooth shape and backscattering values dropped sharply between them. At Hampton, Virginia, peak backscattering remained somewhat enhanced over lowest summer values. The aerosol profile was considerably more irregular on 3 Jan than during the previous reading 7 Dec. The 12 Jan data at Fukuoka, Japan showed a broad monolayer that included most aerosols below 21 km.

Unusual Sunrises and Sunsets. From Jeddah, Saudi Arabia, Edward Brooks reported that dawn and twilight colors diminished in late Nov. Dawns showed no evidence of the presence of a volcanic aerosol layer over nearby Saudi Arabia 28 Nov-6 Dec. Colored but rather subdued dawns indicated that aerosols returned during the second week in Dec. On 18 Dec, only the late dawn had significant color, indicating that the aerosol layer was at relatively low altitude. Separate early and late dawn illumination 24 Dec showed the presence of 2 scattering layers, at high and low altitudes respectively.

Information Contacts: D. Hofmann, Univ. of Wyoming; T. DeFoor, MLO; W. Fuller, M. Osborn, NASA Langley Research Center; M. Fujiwara, M. Hirono, Kyushu Univ.; E. Brooks, Saudi Arabia.

1/84 (9:1) *Airborne Lidar.* The following is a report from M.P. McCormick.

"An airborne lidar mission was flown 19-28 Jan on the NASA Electra aircraft from 37°N to the North Pole, via Goose Bay, Labrador, and Söndre Strömfjord and Thule airbases, Greenland. The primary objective of the mission was to provide correlative stratospheric aerosol measurements for the SAM II (Stratospheric Aerosol Measurement) satellite, successfully accomplished on 3 separate satellite underflights.

"The El Chichón cloud was very consistent from 37°-76°N, generally as a single broad layer with very little structure. From Thule a flight was conducted 24 Jan to the North Pole along the 60°W meridian to determine the northerly extent of the El Chichón material and to search for polar stratospheric clouds (PSC's, see below). As the aircraft proceeded north from Thule, considerable structure and varied intensity were observed in the El Chichón cloud, with an increase in peak scattering ratio.

"First detected by the SAM II satellite, PSC's are thought to be ice clouds that form during the Arctic and Antarctic winter by freezing of diluted sulfuric acid-water aerosol droplets at temperatures less than about -80° to -85°C, followed by rapid growth by sublimation (McCormick et al., 1982 and Steele et al., 1983). Nacreous or mother-of-pearl clouds are thought to be subsets of PSC's. From 85°N to the Pole, PSC's were detected for the first time by a remote sensor other than SAM II.

They occurred at 19-21 km altitude, above the main El Chichón layer.

"A second flight was conducted 25 Jan from Thule to 86°N and PSC's were again detected within the temperature region of -85°C from about 81°-86°N. A third mission was flown 27 Jan from Thule to 87°N over the same flight path (60° meridian) and somewhat to the east. The El Chichón layer decreased in intensity and was less structured. There were no indications of PSC's, correlating with stratospheric temperature data, which showed that the low-temperature region had moved over the north pole toward Siberia. Returning to Virginia on 28 Jan, the same consistency was observed in the stratospheric layer as on the earlier northbound flight, with a slight decrease in peak scattering ratio.

"In addition to the uplooking airborne lidar, a downlooking lidar was used to study the tropospheric aerosols. In-situ measurements of aerosol mass and number density, CO_2 and O_3 were made over the full flight range at various altitudes during the mission."

Lidar Data. At Mauna Loa, Hawaii, a distinct double aerosol layer was observed 3 Jan, similar to the pattern observed in Dec. No strong upper peak was present a week later, but numerous small layers were detected above the main lower peak. On 19 Jan, enhanced higher altitude layers were no longer detected. Two distinct layers were observed over Fukuoka, Japan on 10 Feb, in contrast to the broad monolayer present a month earlier.

Lidar at Garmisch-Partenkirchen, West Germany continued to detect aerosols from the March-April 1982 eruption of El Chichón. Altitude and values of peak backscattering were slightly lower than in the summer (8:11) but secondary peaks at higher altitudes were sometimes detected. Integrated backscattering between 1 km above the tropopause and the top of the aerosol layer, about 20% below expected values in the second half of Sept and the first half of Oct, rose to 70% above expected values 14 Nov as the tropopause altitude dropped to 9.9 km in arctic air and backscattering was enhanced between 9 and 13 km.

Unusual Sunrises and Sunsets. Edward Brooks reported that dawns and dusks indicated a variable and often weak aerosol layer over Jeddah, Saudi Arabia in late Dec and early Jan. Morning and evening colors on 26 Dec suggested that few scattering particles were present at either low or high altitudes. Stronger sunrises and sunsets were observed 27-30 Dec, but only late dawn colors, illuminating low-altitude aerosols, were visible 31 Dec. Dawns were essentially colorless 3 and 7-9 Jan; only late dawn colors (aerosols at low altitude) were visible on the 4th, while only early dawn colors (high-altitude material) were seen 2 days later.

From Millville, New Jersey, Fred Schaaf reported that twilight colors were usually weak to moderate in Dec and early Jan. On 16 Dec, early twilight colors merged into a strong crimson glow at 12° altitude, suggesting that the top of the scattering layer was at 13-18 km altitude. Colors were strong again the following evening and included secondary illumination to a solar depression angle of about 12°. As in previous months, the arrival of arctic air 24 Dec weakened twilight colors. On 8 Jan, weak to moderate colors were replaced suddenly by a milky area at 12° altitude, indicating the top of the aerosol layer at 13-16 km. The same evening, a secondary glow remained visible to an SDA of 10°. On 12 Jan, the time that illumination ended suggested that highest aerosols were at 11-13 km. Only early twilight colors were visible 19 Jan. A week later, timing of the end of purplish illumination indicated that aerosols reached 16-19 km altitude and a deep-hued secondary glow persisted for much longer. However, colors were very weak the following night. On 1 Feb, moderate colors disappeared at a time indicating aerosols reached 11-16 km.

References: McCormick, M.P., Steele, H.M., Hamill, P., Chu, W.P., and Swissler, T.J., 1982, Polar Stratospheric Cloud Sightings by SAM II; *Journal of the Atmospheric Sciences*, v. 39, no. 6, p. 1387-1397.

Steele, H.M., Hamill, P., McCormick, M.P., and Swissler, T. J., 1983, The Formation of Polar Stratospheric Clouds; *Journal of the Atmospheric Sciences*, v. 40, no. 8, p. 2055-2067.

Information Contacts: M.P. McCormick, NASA Langley Research Center; R. Reiter, Garmisch-Partenkirchen, W Germany; T. DeFoor, MLO; M. Fujiwara, M. Hirono, Kyushu Univ.; E. Brooks, Saudi Arabia; F. Schaaf, Millville, NJ.

2/84 (9:2) *Lidar Data.* Lidar from Mauna Loa, Hawaii showed a substantial decrease in total aerosol backscatter between measurements on 1 and 15 Feb. Total backscatter as well as altitudes and values of peak backscatter were nearly identical on 15, 22, and 29 Feb but shapes of aerosol profiles varied somewhat. A broad, single layer without fine structures was detected over Fukuoka, Japan on 11 March. Two distinct layers had been seen from Fukuoka a month earlier. Total backscatter over Hampton, Virginia remained at roughly double the Mauna Loa values. Three distinct layers were observed 8 Feb, but the aerosol profile showed a single broad layer on 21 Feb and 1 March.

Unusual Sunrises and Sunsets. From Boulder, Colorado, Richard Keen reported that unusual twilights were not seen during Dec. However, Keen notes that clear skies for 1500 km to the west of the observation site are necessary to generate double twilights, and Dec weather was usually poor. Double twilights, with salmon color at 4° SDA followed by persistent deep red secondary color, were observed on 2, 4, 6, and 7 Jan. Double twilights were also observed on 24, 28, and 30 Jan, and 5-6 Feb, but were not as bright as those in late 1983 (8:11) or early Jan 1984. The secondary twilights of late Jan-Feb were also not as persistent; at the end of red secondary glow, SDA was 16° on 4 Jan, 14° on 28 Jan, 13° on 30 Jan and 12° on 6 Feb. Keen notes that these observations imply a thinning and/or lowering of the aerosol layer. No double twilights have occurred at Boulder from 6 Feb through early March, but skies west of Boulder were again frequently cloudy.

Optical effects at Jeddah, Saudi Arabia were often weak in late Jan and Feb. From sunset 25 Jan to sunrise 28 Jan, dusk and dawn observations by Edward Brooks indicated that aerosol scatterers were nearly absent.

Only early dawn colors (higher altitude aerosols) were seen 1, 2, and 4 Feb. Of the twilights between 1 and 9 Feb, only on the 6th did colors indicate a strong scattering layer. Brighter dusks and dawns were more frequent from mid-Feb, but scattering effects remained inconsistent. No dawn colors were visible 21, 25, and 27 Feb and colors were subdued on many other mornings and evenings late in the month.

Fred Schaaf reported that twilight effects at Millville, New Jersey in Feb and early March were the weakest since aerosols from the March-April 1982 El Chichón eruption first arrived over the area. Frequent cloudy weather limited observations. Little or no high-altitude aerosol material appeared to be present and colors produced by lower altitude material had diminished.

Information Contacts: T. DeFoor, MLO; W. Fuller, NASA Langley Research Center; M. Fujiwara, M. Hirono, Kyushu Univ.; R. Keen, Univ. of Colorado; E. Brooks, Saudi Arabia; F. Schaaf, Millville, NJ.

3/84 (9:3) *Unusual Sunrises and Sunsets.* Paul Handler observed brilliant twilights 11-18 March from Guana Island, British Virgin Islands (18.50°N, 64.62°W). Skies 30-45° above the horizon were tinted lavender pink and colors remained for 36-37 minutes after sunset, suggesting the presence of aerosols to about 18 km altitude (Meinel and Meinel, 1983) [see chapter introduction]. Yellowish-green illumination was observed one evening, and green clouds were seen around sunset during the week before Handler's visit. The source of the aerosols was unknown.

From Jeddah, Saudi Arabia, Edward Brooks observed few colorful dawns and twilights in early March. Little stratospheric aerosol material appeared to be present, and the bright yellow dawns of 9 and 12 March were the only colorful ones observed during the first half of the month. Effects of stratospheric aerosols were occasionally observed in late March, but colors were not usually strong. Pale colors were seen 17-21 March. The absence of late dusk illumination 22 March indicated that there were no significant aerosols in the stratosphere, and scatterers appeared scarce the next morning. The brief dusk color sequence 30 March indicated a thin layer of aerosols. Dawns were bright and began early 31 March and 1 April, suggesting the presence of high aerosol layers. Dusks and dawns were strong during the first 3 days of April.

Lidar Data. Lidar from Fukuoka, Japan showed two backscattering peaks 28 and 29 March; on the 29th the lower peak was below the local tropopause. Layer altitudes and peak backscattering were very similar 1, 7, and 14 March at Hampton, Virginia, but integrated values were substantially lower on the 7th. More structure was evident on 2 April and integrated backscattering had dropped again to below 7 March levels. Data collection from Mauna Loa, Hawaii was curtailed by the onset of the eruption 25 March. Integrated backscattering has remained very similar since 16 Feb. Two layers were evident 7 and 13 March, but a broad, multiple-peaked layer was present on the 21st.

Information Contacts: P. Handler, Univ. of Illinois; E. Brooks, Saudi Arabia; M. Fujiwara, M. Hirono, Kyushu Univ.; T. DeFoor, MLO; W. Fuller, NASA Langley Research Center.

4/84 (9:4) Lidar data from Fukuoka, Japan and Hampton, Virginia showed generally weaker stratospheric aerosol backscattering in April. On 22 April, the lidar profile at Fukuoka showed a broad single layer that decreased gradually upward from 17 km. At Hampton, integrated backscattering on 25 and 26 April was only half that on the 11th, and lower than values measured in previous months. Lava from the March-April eruption of Mauna Loa cut electric power lines to Mauna Loa Observatory and no lidar data has been collected there since 21 March.

Edward Brooks reported that dawn and dusk colors at Jeddah, Saudi Arabia were variable but generally unimpressive in April, continuing the trend of recent months. Dawn colors were bright 1-3 April and remained bright on the 4th and 5th, but no late dusk colors were observed, indicating the absence of high-altitude aerosols. Dawn was colorless 7 April and faint 8-11 April. A bright dusk 11 April and dawn the next morning were followed by cloudy weather through the 14th. Dawn and dusk were colorless on the 15th, but moderately strong from the 16th through dawn on the 21st.

Information Contacts: W. Fuller, NASA Langley Research Center; M. Fujiwara, M. Hirono, Kyushu Univ.; E. Brooks, Saudi Arabia.

5/84 (9:5) Lidar continued to detect remnants of the aerosol cloud produced by the March-April 1982 explosions from El Chichón, México. Peak backscattering ratios decreased at Fukuoka, Japan, where 2 layers were evident. Altitudes and peak backscattering ratios of the main layers at Hampton, Virginia were similar in April and May. However, measurements made in an Arctic airmass 16 May yielded a total backscatter nearly twice as large as it had been a week earlier in tropical air. Aerosols were relatively dense down to the tropopause, at 11 km altitude on the 9th and 9.6 km on the 16th. On 24 May, many small layers were present, the tropopause was much higher (13.5 km), and total backscatter had decreased to late April levels. Little structure was evident in the aerosol material 4 June. Values from Garmisch-Partenkirchen, Germany showed minor variation between Jan and April, but were generally slightly lower than in late 1983 (9:1). Lidar measurements resumed 30 May at Mauna Loa Observatory. Peak and total backscattering had decreased slightly since late March.

From Jeddah, Saudi Arabia, dawns and dusks observed by Edward Brooks varied in intensity. Moderately strong colors observed 16-21 April persisted through dawn on the 23rd, but were only intermittently present through the end of the month. Weak dawns and twilights from 30 April through the dawn of 6 May suggested that little or no aerosol material was present. Moderate colors returned late 6 May and dusk on the 9th was long and strongly illuminated, indicating the presence of high-altitude aerosols. Dawns and twilights were colorful through 14 May, and faint N-S bands of high-altitude aerosols were visible at dawn on the 11th. Morning and evening colors were again absent 15-18 May, then clouds prevented observations through the 22nd. Moderate colors were present 23-24 May but were weak the following 2 days.

Information Contacts: R. Reiter, Garmisch-Partenkirchen, W Germany; W. Fuller, NASA Langley Research Center; M. Fujiwara, M. Hirono, Kyushu Univ.; T. DeFoor, MLO; E. Brooks, Saudi Arabia.

6/84 (9:6) Toyotaro Yamauchi and Hidehiro Shimura report that atmospheric turbidity over Japan (fig. 19-4) reached a maximum between late autumn 1982 and spring 1983 (8:5). Atmospheric turbidity has decreased gradually since then, but was still considerably higher than normal during the winter of 1983-84. At noon on clear days in Dec 1982, direct solar radiation (I) and global solar radiation (G) showed decreases of 0.17 kW/m^2 (20%) and 0.02 kW/m^2 (3%), and diffuse solar radiation (D) a 0.07 kW/m^2 (74%) increase at Tsukuba, Japan (36.17°N, 141.09°E) as compared to a normal year. In Dec 1983, I was 0.05 kW/m^2 (6%) lower and D was 0.02 kW/m^2 (26%) higher than a normal year at Tsukuba, while G was approximately normal (fig. 19-5).

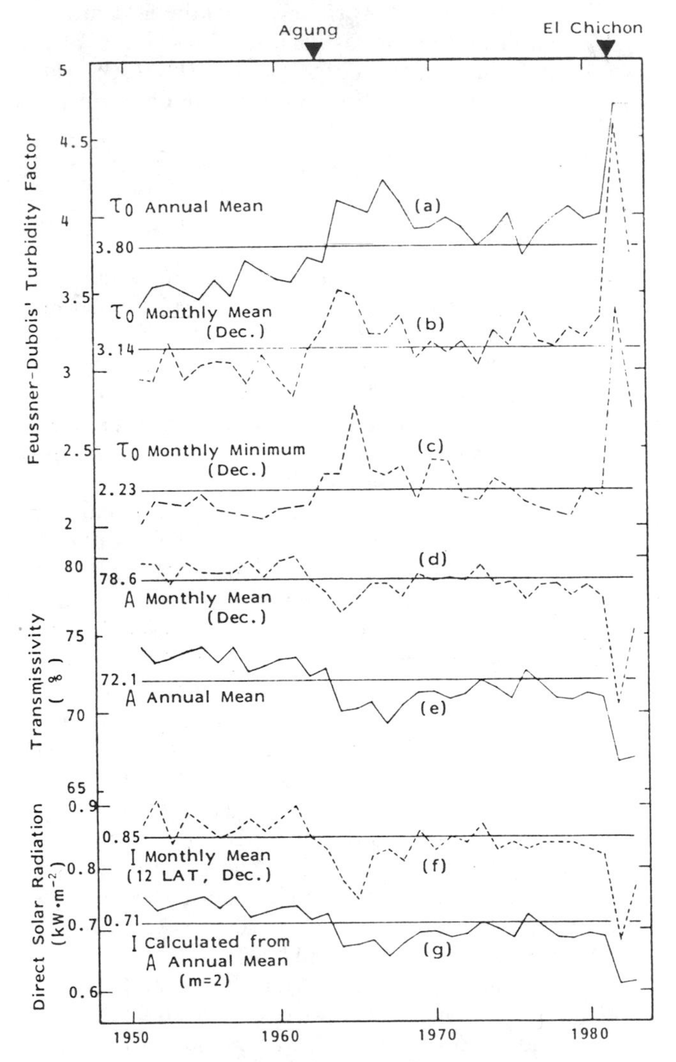

Fig. 19-4. Three measures of aerosol content of the atmosphere; variation in atmospheric turbidity (τ_0) *top*, transmissivity (A) *center*, and direct solar radiation (I) *bottom* for all stations in Japan, 1950-83. The largest changes followed the eruptions of Agung in 1963 and El Chichón in 1982. Courtesy of T. Yamauchi and H. Shimura.

Lidar at Mauna Loa, Hawaii continued to detect stratospheric aerosols from the March-April 1982 eruption of El Chichón. Altitude and peak backscatter of the main layer remained similar through June. A small peak in the upper troposphere or lower stratosphere was also observed throughout the month. Particularly pronounced on 14 June, it was distinctly weaker although still evident on the 29th. There was no obvious single source for this layer, although moderate explosions had occurred at low latitudes at several volcanoes in previous weeks (9:4-5), including Pagan (Mariana Islands), Soputan (Indonesia), and Manam (Papua New Guinea). A similar layer was present over Hampton, Virginia in late May, but a distinct lower layer was not reported from there in June. Altitude and backscattering of the main layer were relatively uniform through the month, but a lower tropopause on the 25th (10.5 km) was accompanied by an increase in total backscatter.

Information Contacts: T. Yamauchi, H. Shimura, JMA; T. DeFoor, MLO; W. Fuller, NASA Langley Research Center.

7/84 (9:7) July lidar data indicated the continuing presence of stratospheric aerosols from El Chichón. A small layer

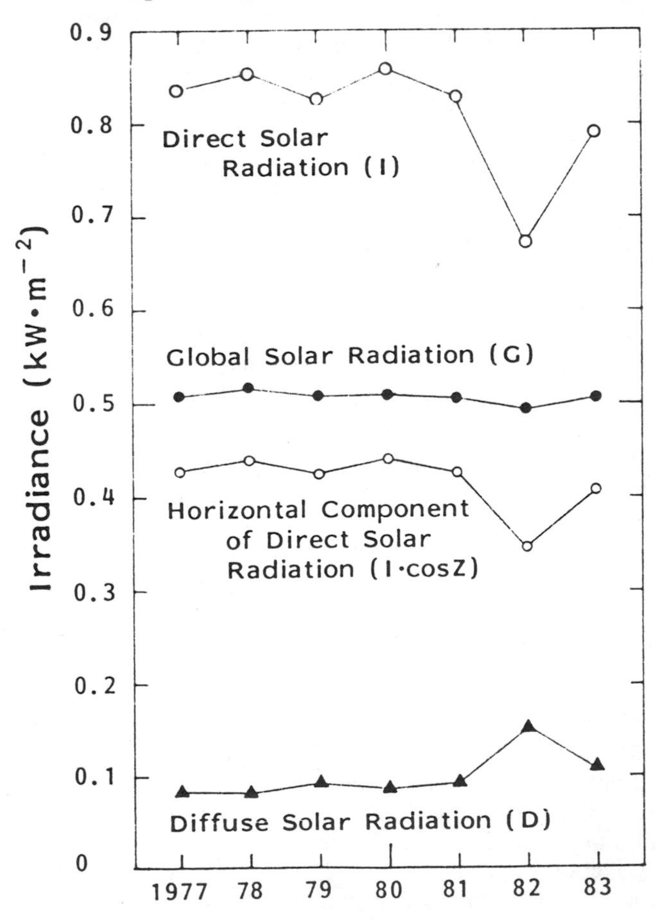

Fig. 19-5. Monthly means of direct, diffuse, and global solar radiation observed at noon on clear days at Tsukuba, Japan. [Note that the decline in direct solar radiation after the 1982 El Chichón eruption is mirrored by an increase in diffuse radiation as the aerosol cloud scatters sunlight.] Courtesy of Minoru Obata, Aerological Observatory.

in the upper troposphere or lower stratosphere over Mauna Loa, Hawaii in June was still present in July, although somewhat weaker. Layer altitudes, and values of peak and integrated backscatter at Hampton, Virginia in early July and early Aug were similar to late June figures. Collection of lidar data resumed at Fukuoka, Japan in mid-July. A single layer was reported in contrast to the pair of layers observed in May.

Brightness and duration of July twilights at Boulder, Colorado were the strongest since Jan, but Richard Keen noted that it was difficult to determine whether the cause was less cloudiness west of Boulder or the presence of thicker and/or higher aerosols. Bright salmon-pink to pinkish-lavender twilights, peaking in color intensity at an SDA of 4-5°, were observed 20 March, 14-15 April, 7 May, 1-5 and 16-18 July, and 2 Aug. Rather dim reddish secondary twilight colors persisted to an SDA of 9° on 14-15 April. Brighter red secondary twilights remained visible to an SDA of 9° on 4 July, 11° on the 5th and 16th, 12° on the 17th, and 13° on the 18th. Keen related the general lack of twilight effects March-June to frequent high-level coloudiness over the western United States in the spring. Dates in July without twilight effects corresponded to dates of thunderstorm activity over the Great Basin. Late May-early June dawns observed by Edward Brooks from Jeddah, Saudi Arabia were later than those of the previous year, indicating a significant reduction in stratospheric aerosols. Brooks noted that observations during the second week in June showed no firm evidence of volcanic aerosols in the stratosphere. Fred Schaaf noted that twilight glows were sometimes visible from Millville, New Jersey in June and July, after having been nearly absent March-May. Timing and position of the glow suggested maximum aerosol altitudes of little more than 8 km.

Information Contacts: R. Keen, Univ. of Colorado; E. Brooks, Saudi Arabia; F. Schaaf, Millville, NJ; M. Fujiwara, M. Hirono, Kyushu Univ.; T. DeFoor, MLO; W. Fuller, NASA Langley Research Center.

8/84 (9:8) H. H. Lamb's report describes observations from Holt, England (52.9°N, 1.1°E), about 30 km N of Norwich, his previous observing site. "Twilight optical effects presumably attributable to the remnants of the El Chichón aerosol remained visible on clear evenings in Nov, 1983, but then became less noticeable and in early 1984 seemed generally weak. After a long period in which an untrained observer would surely have noticed nothing unusual about the twilight glows, on 19 Aug pink crepuscular rays reached 20-22° elevation at 1945-50 GMT, about ½ hour after sunset, indicating an illuminating layer at about 25-27 km altitude."

Little variation was seen in aerosol profiles measured by lidar at Mauna Loa, Hawaii in Aug and at Hampton, Virginia in Aug and early Sept. Integrated backscattering at Mauna Loa was consistently higher in Aug than in July; values dropped somewhat at Hampton. The lidar profile at Mauna Loa suggested that aerosols extended downward several kilometers into the troposphere from the stratospheric layer. Fig. 19-6 shows changes in stratospheric aerosols over Hampton during the past 10 years, with timing of large explosive eruptions.

Information Contacts: H. Lamb, Univ. of East Anglia; W. Fuller, NASA Langley Research Center; T. DeFoor, MLO.

9/84 (9:9) Lidar and balloon data showed that aerosols from the March-April 1982 eruption of El Chichón remain in the stratosphere. Despite the 15-km eruption clouds reported during the Mayon eruption, no new aerosol layers were detected. At Mauna Loa, Hawaii, integrated backscattering values were lower in Sept than in Aug. Cirrus clouds were abundant throughout the month, masking any upper tropospheric aerosols that might have been present. A sunrise during clear weather in early Oct did not show any apparent aerosol layers. At Hampton, Virginia, a very broad backscattering peak and somewhat increased integrated backscatter were measured 6 Sept, but a profile with a narrower peak on 19 Sept (the last measurement of the month) yielded integrated values lower than in Aug. Sept values from Fukuoka, Japan were similar to those from the last measurement in mid-July. April-June lidar data from Garmisch-Partenkirchen, Germany show a gradual decrease in backscattering from stratospheric aerosols. A 3-dimensional representation of the structure of the aerosol profile since early 1982 is shown in fig. 19-7. Balloon data from Laramie, Wyoming have continued to

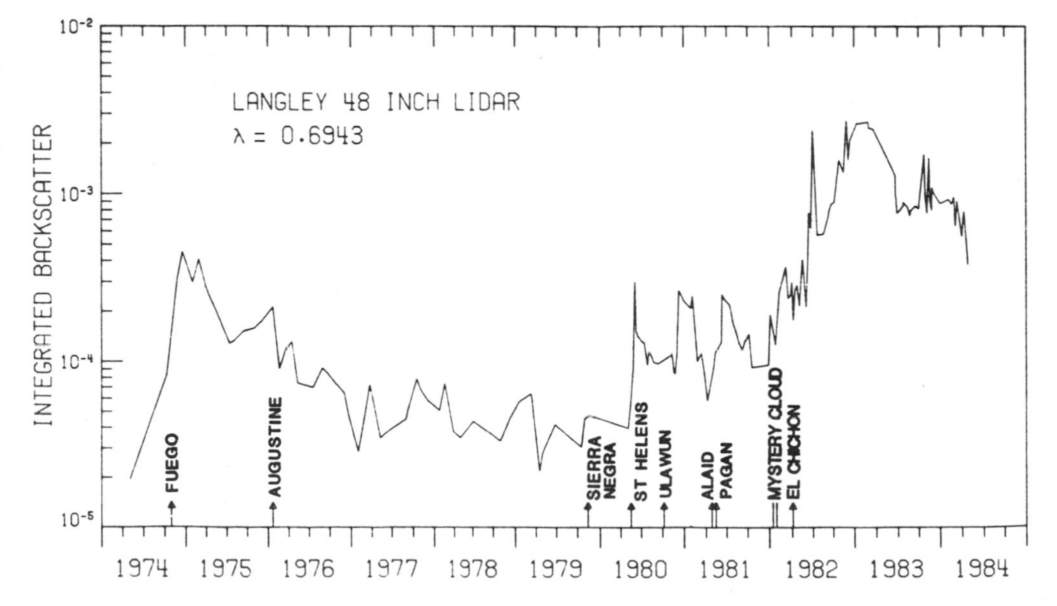

Fig. 19-6. Atmospheric effects of some large explosive eruptions, 1974-84. Integrated backscattering measured by the 48-inch lidar at Hampton, Virginia. Courtesy of the NASA Langley Research Center.

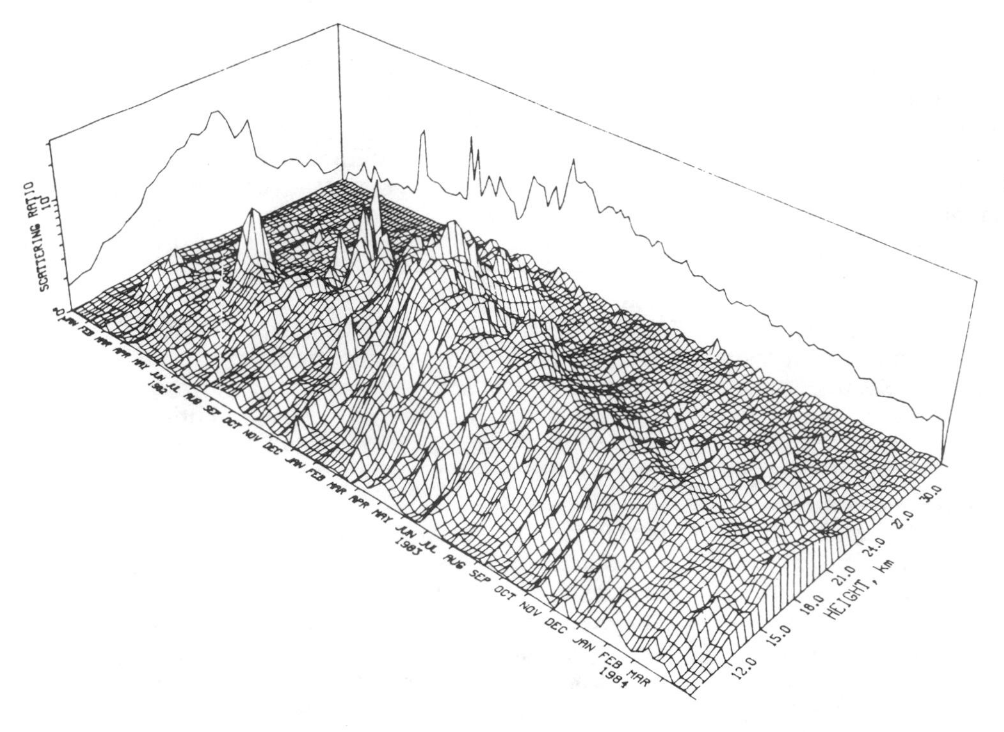

Fig. 19-7. Development and gradual decay of the El Chichón aerosol cloud over Garmisch-Partenkirchen, West Germany, Jan 1982-March 1984. The aerosol profile structure is represented in 3 dimensions. Courtesy of Reinhold Reiter.

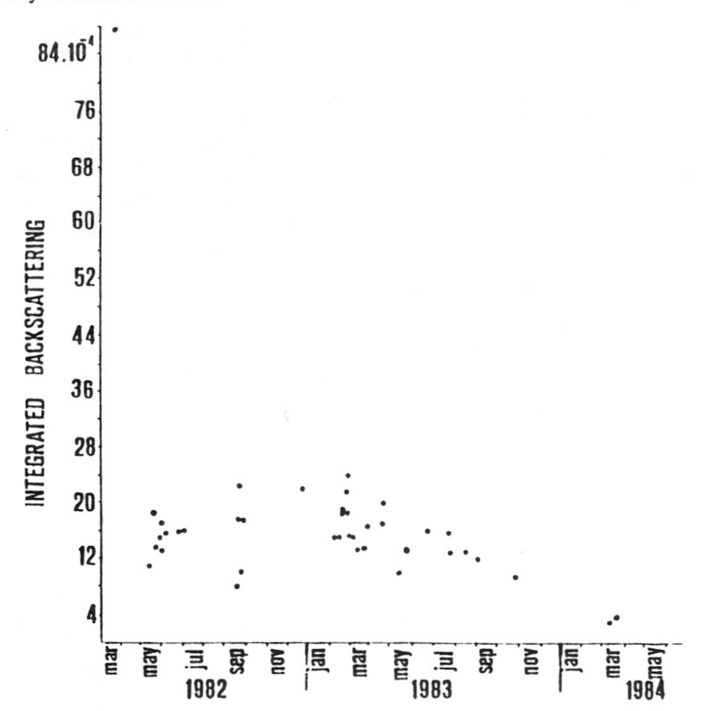

Fig. 19-8. Arrival and decay of the El Chichón aerosol cloud over Firenze, Italy, April 1982-March 1984. Integrated backscattering measured by Nd-YAG lidar (with second harmonic generator, producing a wavelength of 0.53 microns). Receiving optics were redesigned and new acquisition and processing software written in fall, 1982. Prior data were from a system designed for detecting lower tropospheric aerosols and are thus less reliable. Courtesy of Leopoldo Stefanutti.

show slow decay of the El Chichón aerosols. No material from Mayon was evident during a flight 6 Oct.

Information Contacts: R. Reiter, Garmisch-Partenkirchen, W Germany; T. DeFoor, MLO; M. Osborn, NASA Langley Research Center; M. Fujiwara, M. Hirono, Kyushu Univ.; J. Rosen, Univ. of Wyoming.

10/84 (9:10) Lidar at Garmisch-Partenkirchen, Germany continued to detect remnants of the stratospheric aerosols from the El Chichón eruption. Peak backscattering ratios were somewhat lower in the summer and early autumn than they had been in the spring. At Firenze, Italy (43.8°N, 15.25°E), lidar data were collected April 1982–March 1984 by the Istituto di Ricerca Sulle Onde Electromagnetiche (figs. 19-8 and 19-9). Integrated backscattering increased from just after the El Chichón eruption through early 1983, then declined gradually. At Hampton, Virginia, integrated backscattering was about the same in late Oct as in mid-Sept. A relatively weak secondary layer appeared to be present above the main layer of El Chichón material. Integrated backscattering varied considerably in Oct at Mauna Loa, Hawaii. An intense layer observed 30 Oct between a double tropopause at 15 and 15.3 km was probably cirrus cloud; below 13 km several layers appeared on the lidar data and cirrus were visible to the naked eye. At Fukuoka, Japan, increases in peak and integrated backscattering were noted for several days beginning 8 Oct between 9 and 22 km. Values rose to about 1.8 times seasonal means, then returned to previous levels.

M. Patrick McCormick and Thomas Swissler provided the following information about the relationship between backscattering ratios measured by lidar at wavelengths generated by ruby (0.6943 μm), and Nd-YAG (1.064 and 0.532 μm) laser transmitters. For any two wavelengths, the relationship can be expressed as:

$$(R_1 - 1) = (\lambda_1/\lambda_2)^{4-x}\,(R_2 - 1) \qquad (1)$$

where R_1 and R_2 are backscattering ratios produced by lidar operating at wavelengths λ_1 and λ_2. The value of x varies with the aerosol size distribution at a given time. Using techniques described in McCormick et al., (1984), McCormick and Swissler calculated values of x for a typical background aerosol size distribution (no significant volcanic contribution) from Russell et al. (1981) and for the aerosols measured by Hofmann using a 6-channel dustsonde on 24 Aug and 21 Dec, 1983 (about 17 and 21 months after the March-April 1982 eruption of El Chichón). After calculating x, equation (1) can be simplified to:

$$(R_1-1) = k(R_2-1) \quad (2)$$

By substituting k values into equation (2) from the appropriate model in table 19-2, lidar data of different frequencies can be made approximately comparable.

From Millville, New Jersey, Fred Schaaf continued to observe unusual twilight colors. From mid-July through mid-Sept, weak to moderate primary glows were usually present, and purple and crimson colors were often observed for somewhat longer after sunset. Timing of the disappearance of later colors suggested that aerosols were present to at least 8-13 km. Strong crepuscular rays were observed during the evenings of 20 Aug and 11 Sept. On a few evenings, little or no glow was evident. In late Sept and early Oct, weak secondary illumination was visible, and the timing of primary colors suggested that aerosols were present to 13-19 km altitude on 26 Sept. From about 38°N, 75.5°W (Maryland-Virginia border), Schaaf saw moderate colors and many crepuscular rays on 28 Oct. In arctic air over New Jersey 7-8 Nov, colors were relatively weak and faded quickly.

References: McCormick, M.P., Swissler, T.J., Fuller, W.H., Hunt, W.H., and Osborn, M.T., 1984, Airborne and Ground-based Lidar Measurements of the El Chichón Stratospheric Aerosol from 90°N to 56°S; *Geofísica Internacional*, v. 23, no. 2, p. 187-221.

Russell, P.B., Swissler, T.J., McCormick, M.P., Chu, W.P., Livingston, J.M., and Pepin, T.J., 1981, Satellite and Correlative Measurements of the Stratospheric Aerosol. I: An Optical Model for Data Conversions; *Journal of the Atmospheric Sciences*, v. 38, no. 6, p. 1279-1294.

Information Contacts: L. Stefanutti, Isto. di Ricerca Sulle Onde Electromagnetiche; R. Reiter, Garmisch-Partenkirchen, W Germany; M.P. McCormick, T. Swissler, W. Fuller, M. Osborn, NASA Langley Research Center; T. DeFoor, MLO; M. Fujiwara, M. Hirono, Kyushu Univ.; F. Schaaf, Millville, NJ.

11/84 (9:11) Since the El Chichón eruption cloud was first detected over Mauna Loa, Hawaii in April 1982, the aerosols measured by lidar there have extended downward from the stratosphere into the upper troposphere, without a sharply-defined base. Aerosol concentrations in the upper troposphere decreased gradually with decreasing altitude. Nov lidar data showed a return to typical pre-El Chichón profiles in the upper troposphere, with few aerosols and cleanest air at the top of the troposphere, near the tropopause.

The Mauna Loa lidar cannot reliably measure aerosol concentrations above about 30 km altitude, but the presence of a distinct break in slope in the recorded profile at roughly 39 km in past months has suggested enhanced aerosol concentrations to that altitude. However, lidar measurements on 15 and 27 Nov showed no structure between 30 and 40 km altitude, suggesting that no aerosols were present. The 27 Nov data also showed a substantial decrease in the integrated aerosol backscattering, suggesting a decline in the stratospheric aerosol load. The decrease in integrated backscattering appeared real and the instrument signal looked typical, but the presence of a heavy cirrus layer at the altitude where the instrument is usually normalized may have distorted the data. No similar decreases in aerosol backscattering were observed at Fukuoka, Japan or Hampton, Virginia.

Aerosol model	x	k(r,y)	k(g,y)	k (g,r)
Hirono (7:5)	1.85	0.40	0.23	0.56
Russell et al. 1981	1.60	0.36	0.19	0.53
Hofmann 24 Aug & 21 Dec 1983	0.90	0.27	0.12	0.44

Table 19-2. For each of three aerosol models, values of k relating pairs of lidar frequencies are shown. Values of x used to derive k for each model are also shown. Subscripts of k show the two frequencies being compared, where r = ruby (0.6943 μm), y = Nd YAG (1.064 μm), and g = Nd YAG 2nd harmonic (0.532 μm). In addition to the Russell and Hofmann models, Hirono's value of 0.4 for k(r,y) is extrapolated for k(g,y) and k(g,r).

Lidar data at Firenze, Italy showed no firm evidence of aerosols from the eruption of Mayon (Philippines) in Sept, but integrated backscattering increased from the end of Oct through the end of Nov. Aerosol loading seemed quite continuous from about 14 km to 22-23 km altitude. Measurements at the end of Nov showed more evidence of inhomogeneity of aerosol distribution with

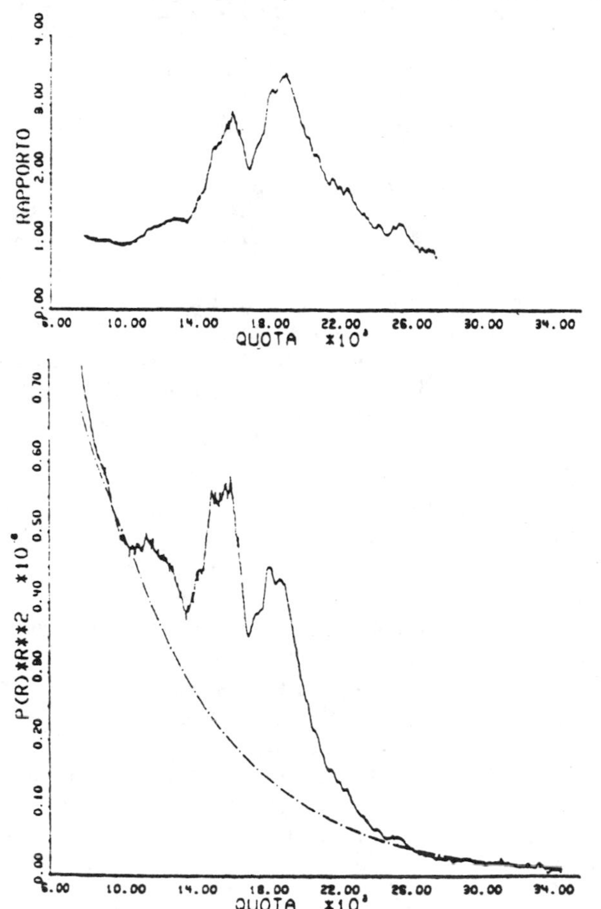

Fig. 19-9. Profile of volcanic aerosol layers over Firenze, Italy, 18 Feb 1983, about the time of maximum enhancement (fig. 19-8). Backscattering ratios, *top*, and coefficients, *bottom*, vs altitude, from Nd-YAG lidar data. The dashed line on the lower graph represents the US standard atmosphere. Courtesy of Leopoldo Stefanutti.

height. William Fuller reports that a ground truth measurement experiment took place over Laramie, Wyoming 29-30 Nov during the overflight of the newly-launched SAGE II satellite. Sun photometer and airborne lidar measurements were conducted on board the Ames Research Center CV 990 aircraft. Aerosol, water vapor, ozone, and NO_2 measurements were made using balloon-borne samplers. Excellent data sets were obtained on each day.

Information Contacts: T. Defoor, MLO; W. Fuller, NASA Langley Research Center; M. Fujiwara, M. Hirono, Kyushu Univ.; L. Stefanutti, Isto. di Ricera Sulle Onde Elettromagnetiche.

12/84 (9:12) H. H. Lamb reported a resumption of strong sunset colors in late Nov in the vicinity of Holt, England. The effects were stronger than they had been in more than 12 months. The timings and elevations of the optical phenomena seemed to suggest the illumination of an aerosol layer at about 22-25 km altitude. Lamb noted that for both the March-April 1982 eruption of El Chichón, and the Aug 1883 eruption of Krakatau, the first strong optical effects over England were at about the same time of year. On 25 Nov, Lamb noted a strong purple patch that extended to more than 20° elevation and appeared about 30 minutes after sunset. At sunset on 28 Nov, the slightly greenish orange sun was surrounded by strong orange diffused light, and a purplish patch more prominent than on the 25th was present 50-60 minutes after sunset. However, no anomalous colors or other optical effects were seen at sunset in clear weather on 30 Nov.

During clear weather in early Dec, optical phenomena similar to those of late 1982 and early 1983 yet stronger than those of late 1983 were consistently observed by Lamb and his colleague Michael Kelly. The 6 Dec effects were very similar to the most impressive twilight seen in Nov 1982. As the sun set on 6 Dec, it was surrounded by orange light to about 3 solar diameters. From 10 to 20 minutes after sunset, a horizontal pinkish purple band appeared, evidently from the illumination of a layer above the surface haze. This band gradually climbed to 10° elevation, becoming broader and more diffuse. At 1600, 20 minutes after sunset, the entire western sky was a brilliant yellow to 35° elevation, edged by a brown layer along the horizon. By 1610, a purple patch had developed from 10-30° elevation above a shield-shaped area of bright white sky. At 1616, the maximum elevation of the purple glow was 20° and the sky from the horizon to 3° was a fiery deep brownish red. The next morning, the rising sun was pale yellow and surrounded by orange diffused light to 4 solar diameters. Twilight observations that evening were similar to those the previous day, but the purple patch at 1615 was asymmetrical, roughly triangular, with a vertical northern edge. It then narrowed to a broad column of light, which reached only 10-12° elevation by 1622 and faded fast. Sunrise on 8 Dec was obscured by clouds. That evening, purple light development was less pronounced than on the 6th and 7th, and gradually changed to a dirty gray. A fiery red band was present along the horizon at 1630. After sunset on 10 Dec, the purple patch was especially beautifully colored and in the form of crepuscular rays, reaching 25° elevation at 1615 and fading soon after 1620, when the maximum elevation was about 18°.

Richard Keen reported that brightness and duration of twilights at Boulder, Colorado showed a noticeable and steady decrease from late Aug through early Dec, indicative of a continued thinning and/or lowering of the aerosol layer. Enhanced salmon-pink to lavender twilights, peaking in color intensity at an SDA of 4°, were observed on either the mornings or evenings of 26, 27, 29, and 30 Aug; 2, 3, 5, and 15 Sept, 7 Oct, and 4-6 Dec. Extended lavender to purplish twilights were visible to an SDA of 11° in either the morning or evening sky on all these dates except 26 and 29 Aug and 5 Dec. Keen related the occurrence of extended twilights on individual dates to the absence of cirrus clouds for 1000-2000 km in the direction of the sun.

The shapes of lidar profiles and total aerosol backscattering at Mauna Loa, Hawaii varied considerably in Dec. Data on 11 and 19 Dec were similar to those of Oct and early Nov (9:10-11), with higher integrated backscattering than in late Nov and early Dec. Breaks in slope in recorded profiles on those dates suggested that aerosols were present to 34 and 39 km altitudes. Aerosol concentrations and maximum layer altitudes decreased again at the end of Dec.

Information Contacts: H. Lamb, Univ. of East Anglia; R. Keen, Univ. of Colorado; T. DeFoor, MLO.

1/85 (10:1) Significant concentrations of aerosols from El Chichón remained in the stratosphere at the beginning of 1985. Major stratospheric warming in late Dec and Jan may have evaporated and recondensed the El Chichón aerosols over a large portion of mid and high northern latitudes.

A small stratospheric warming event started about 7 Dec over the Aleutians. Air circulation at the 30 km level carried air from the zone of warming toward the western United States, cooling the air in transport. Balloon-borne particle counters detected increased numbers of tiny condensation nuclei (CN) at about 30 km altitude over Laramie, Wyoming on 14 and 18 Dec. The bases of Dec particle count profiles were relatively smooth, suggesting that the CN droplets were about 1-2 weeks old.

A much larger and more intense stratospheric warming began in late Dec and increased stratospheric temperatures persisted through Jan. Labitzke et al. (1985) note that "The evolution of this warming was very unusual. Only the development of the winter of 1956/57 appears to be similar, although few stratospheric data are available from that event". They report that stratospheric warming was first observed 26 Dec over Sable Island (44°N, 60°W) where the temperature at about 30 km altitude (10 hPa) was -24°C. Within a few days, the entire Arctic had warmed, resulting in a complete reversal of the stratospheric and mesospheric circulation over high latitudes and a breakdown of the polar vortex. By 29 Dec, temperatures at 30 km over the Labrador Sea area were 55° higher than they had been 5 days earlier and effects extended over much of eastern and central

North America. Satellite data showed the rapid disappearance of the polar vortex over Europe and radiosonde measurements over Berlin (52.32°N, 13.25°E) on 2 Jan showed that intense warming had occurred in the 3 days since the previous measurement. The lower and middle stratosphere remained very disturbed in mid-Jan and a new warming pulse was developing over Labrador. Balloon soundings from Laramie on 9, 24, and 31 Jan measured CN concentrations as high as $100/cm^3$ at 30 km altitude, compared to background values of $2\text{-}3/cm^3$. In contrast to previous years, meteorological data suggested that Laramie was within the zone of warming, so the Jan soundings may have sampled ongoing or very fresh CN events.

Jan lidar data from Mauna Loa, Hawaii showed no major changes from the previous month. The profile on 15 Jan showed a pronounced peak, whereas the layer was much broader on the 22nd, but integrated backscattering values on the two nights were very similar. Early Feb data from Hampton, Virginia were similar to those from previous measurements in Nov.

Reference; Labitzke, K., Lenschow, R., Naujokat, B., and Petzoldt, K., 1985, First Note on the Major Stratospheric Warming at the End of December 1984; *Beilage zur Berliner Wetterkarte* SO 1/85, Met. Inst. Free University of Berlin. A shortened version has been submitted to the Map Newsletter.

Information Contacts: D. Hofmann, Univ. of Wyoming; T. DeFoor, MLO; W. Fuller, NASA Langley Research Center.

2/85 (10:2) The 20 Feb lidar profile at Hampton, Virginia showed quite uniform aerosol density to 23.5 km altitude. By 6 March, the lidar profile had returned to a more normal shape with a distinct peak. At Mauna Loa, Hawaii, lidar data indicated that the aerosol layer on 19 Feb terminated at a much lower altitude and had smaller peaks than 5 days earlier; integrated backscattering was nearly halved. No significant increase in backscattering was observed from Garmisch-Partenkirchen, Germany Oct 1984 – Jan 1985. A second higher-altitude layer was detected in Dec and Jan, but maximum backscattering ratios did not increase.

Information Contacts: R. Reiter, Garmisch-Partenkirchen, W Germany; T. DeFoor, MLO; W. Fuller, NASA Langley Research Center.

3/85 (10:3) Integrated aerosol backscattering remained about the same over Mauna Loa, Hawaii and decreased slightly over Hampton, Virginia in March. Lidar data from Garmisch-Partenkirchen, Germany showed little change in altitudes of aerosol layers or peak backscattering ratios during winter 1984–85. No new aerosol layers were reported.

Information Contacts: T. DeFoor, MLO; M. Osborn, NASA Langley Research Center; R. Reiter, Garmisch-Partenkirchen, W Germany.

4/85 (10:4) An airborne lidar mission supporting the SAGE II satellite detected substantially smaller amounts of stratospheric aerosol at low latitudes than at mid-latitudes. Maximum backscattering ratios off the east coasts of Central America and Brazil were similar to those measured over southern California. However, the aerosol layers off Central America and Brazil were narrower, truncated at the base by higher tropopauses characteristic of the tropics, and integrated backscattering at low latitudes was only about [$^1/_2$] that over California. Sun photometer data were also collected from the aircraft, and balloons and ozone-sensing rockets were launched from Natal, Brazil.

Data from Mauna Loa, Hawaii showed a continuing gradual decline in stratospheric aerosols, with integrated backscattering at the end of April dropping to little more than half the early March values. Peak backscattering over Fukuoka, Japan declined sharply in early April, but had returned to near late March values by mid-April.

Information Contacts: W. Fuller, NASA Langley Research Center; M. Fujiwara, Kyushu Univ.; T. DeFoor, MLO.

5/85 (10:5) Stratospheric aerosol concentrations continued to decline through mid-May at Mauna Loa, Hawaii, but a new layer was detected just above the tropopause late in the month. The new layer was only a small anomaly on the 23 May lidar profile, but was somewhat stronger by the time of the next measurement on 30 May. It was not sharply defined, suggesting that it was at least a month old, although it had not been previously observed at Mauna Loa. Small layers of this type can be produced by recirculation of existing aerosols as well as by a new injection of material. No such layer was detected by the SAGE II support mission in late March and early April. Lidar at Garmisch-Partenkirchen, Germany continued to detect remnants of the El Chichón stratospheric aerosol cloud. Layer altitudes and peak backscattering ratios changed little from Feb through April.

Information Contacts: R. Reiter, Garmisch-Partenkirchen, W Germany; T. DeFoor, MLO; W. Fuller, NASA Langley Research Center.

6/85 (10:6) Persistent remnants of the El Chichón stratospheric aerosol cloud were measured over Virginia and Hawaii. Tropospheric aerosol layers were also detected by lidar over Virginia in late June and early July, and by an airline pilot at about 11 km altitude over Tennessee in late June. No volcanic source was recognized for the tropospheric aerosols, but many forest fires were burning in the western US during this period. In early July, smoke from one major fire rose to 5.5 km altitude. Enhanced sunrises and sunsets have been reported from Boulder, Colorado, roughly 1,500 km away.

Information Contacts: W. Fuller, NASA Langley Research Center; T. DeFoor, MLO; P. Handler, Univ. of Illinois; M. Matson, NOAA/NESDIS.

7/85 (10:7) Stratospheric aerosols produced by the 1982 eruption of El Chichón continued to be measured over Virginia and Hawaii in July. A tropospheric layer, perhaps smoke from major forest fires in the western United States was detected over Virginia on 9 July, but

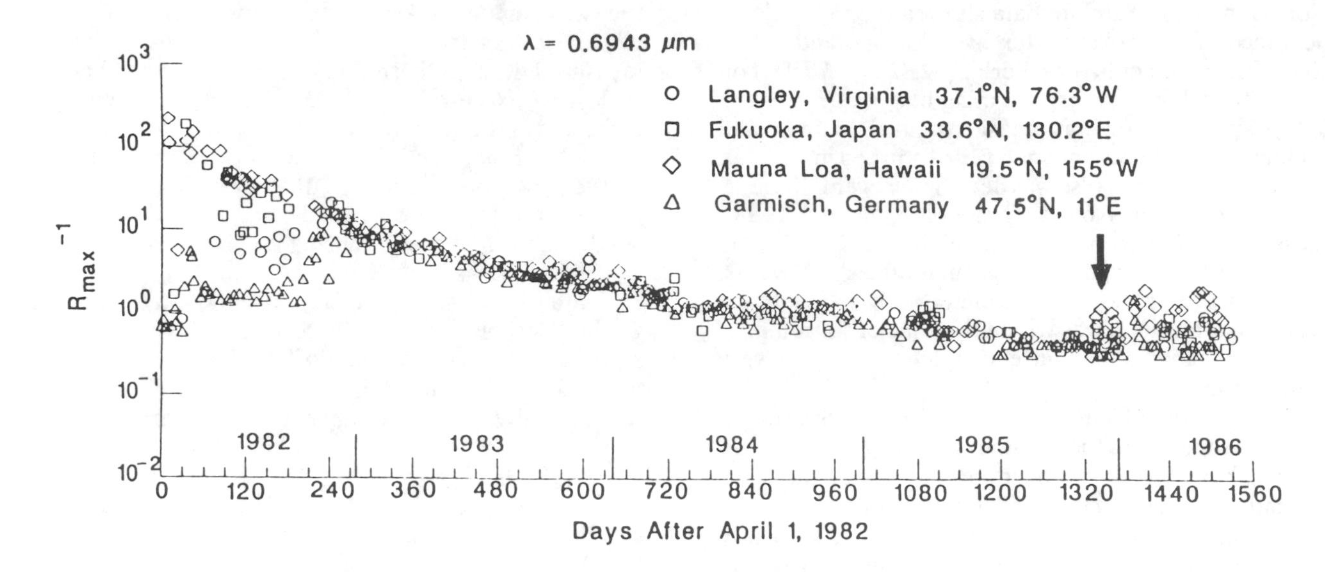

Fig. 19-10. Aerosol layers from the eruptions of El Chichón and Ruiz as monitored from 4 lidar sites, April 1982-mid 1986. El Chichón aerosols arrived first at lower latitude sites, but by early 1983, measurements were very similar at low and mid-latitudes. Values were much smaller for the Ruiz aerosols (to the right of the arrow). Peak backscatter data (x axis; note logarithmic scale) were normalized to ruby values (see 9:10). Courtesy of M.P. McCormick.

no such layer was present 9 days later. Peak and integrated backscattering remained similar to values of the past few months, but were distinctly lower than the beginning of this year.

Information Contacts: T. DeFoor, MLO; W. Fuller, NASA Langley Research Center.

8/85 (10:8) Stratospheric aerosols from the El Chichón eruption weakened in Aug over Japan, Hawaii, and Virginia.

The following is a report from William Fuller.

"An airborne lidar mission supporting the SAGE II/SAM II correlative measurement experiment was flown on the NASA Wallops P-3 aircraft on 7, 8, and 9 Aug 1985 from Fairbanks, Alaska. SAGE II missions were flown on 7 and 9 Aug just south and north of Fairbanks (64.8°N, 147.9°W) and SAM II measurements were conducted to 74°N. Other correlative measurements supporting the experiments were as follows: in-situ balloon-borne aerosol, water vapor, ozone measurements, and higher-altitude sampling from a NASA U-2 aircraft. Sun photometer data were also obtained on the P-3 flights. One set of data was taken at the Wallops flight facility on 2 Aug prior to departing for Alaska. Peak scattering ratios from 63°N to 74°N were approximately the same as at 38°N latitude, on the order of 1.4. The peaks occurred at lower altitudes because of the lower tropopause heights at high latitudes (10 km at 64°N, 16 km at 38°N)."

Information Contacts: W. Fuller, NASA Langley Research Center; M. Hirono, Kyushu Univ.; T. DeFoor, MLO.

9/85 (10:9) Lidar measurements indicated that the stratospheric aerosol cloud from the 1982 eruption of El Chichón had weakened very slightly over Germany, Japan, and Hawaii.

Information Contacts: R. Reiter, Garmisch-Partenkirchen, W Germany; T. DeFoor, MLO; M. Fujiwara, Kyushu Univ.

10/85 (10:10) Lidar data from Japan, Hawaii, and Virginia showed the continuing presence of aerosols from the 1982 eruption of El Chichón (fig. 19-10). Although both peak and integrated backscattering values remained very uniform over Hawaii from measurement to measurement, the lidar profiles showed substantial variation. The 1 Oct profile was relatively smooth; 2 small peaks were apparent on the 16 Oct data; a zone of sharply decreased aerosol concentration was detected between 20.5 and 23 km altitude on 23 Oct; and there was a step increase (in contrast to the usual gradual increase) in backscattering at the base of the stratosphere on the 30 Oct profile.

From Holt, England, H. H. Lamb has observed weakening of optical phenomena associated with stratospheric aerosols since his report of strong effects in Nov and Dec 1984 (9:12). On 16 Feb, the clear sky appeared dirty at twilight and a distinct purple patch developed 20 minutes after sunset. Strong optical effects and colors were observed after sunset on 4-6 and 13 March, and 21 April, with measurements indicating an aerosol height of 20-25 km, as in Dec. Lamb's next detailed sunset observations were on 1 Sept and 13 Oct, with a characteristic fiery red layer along the horizon being particularly notable on the latter date. Measurements of the elevation of the top of the illuminated patch in Sept and Oct suggested that the aerosol layer was at 15-18 km altitude.

Information Contacts: H. Lamb, Univ. of East Anglia; T. DeFoor, MLO; M. Fujiwara, Kyushu Univ.; M. Osborn, NASA Langley Research Center.

11/85 (10:11) Data from Hawaii and Wyoming suggest that

aerosol material, perhaps from the 13 Nov eruption of Ruiz volcano, Colombia, has recently been injected into the stratosphere. Through 22 Nov, lidar measurements at Mauna Loa, Hawaii continued to show only remnants of the 1982 El Chichón aerosol cloud. On 26 and 27 Nov, a distinct new layer centered at 25-25.5 km appeared on Mauna Loa lidar profiles (fig. 19-11), accompanied by a substantial increase in total backscatter. This layer was absent over Mauna Loa on 3 Dec, but new layers centered at 15, 16.8, and 18.6 km altitude (tropopause altitude was 16.5 km) were detected and total backscatter remained elevated. Preliminary data 7 and 10 Dec showed some apparent new material, but less distinctly than on the 3rd. No new material was evident 10 Dec over Hampton, Virginia. While flying from Honolulu to Los Angeles on 27 Nov at approximately 22°N, 140°W, David Hofmann saw a cloud of particles above the aircraft at an estimated elevation of 12 km. The gray haze and large ring around the sun were very similar to what Hofmann observed shortly after the eruption of Fuego volcano, Guatemala in Oct 1974, suggesting that this cloud was also of volcanic origin.

On 5 Dec, balloon-borne particle counters detected a 10-fold increase in condensation nuclei (H_2SO_4 droplets with diameters $<< 0.1$ μm) between 15 and 17 km altitude over Laramie, Wyoming. Particle counts were 400/cm^3 compared to background values of 40 particles per cm^3 at that altitude [see 10:12]. This CN event is distinguishable from El Chichón aerosols by both its altitude (El Chichón aerosols are found at 17-19 km) and particle size (El Chichón particles are now > 0.1 μm). A subsequent balloon flight on 11 Dec showed only

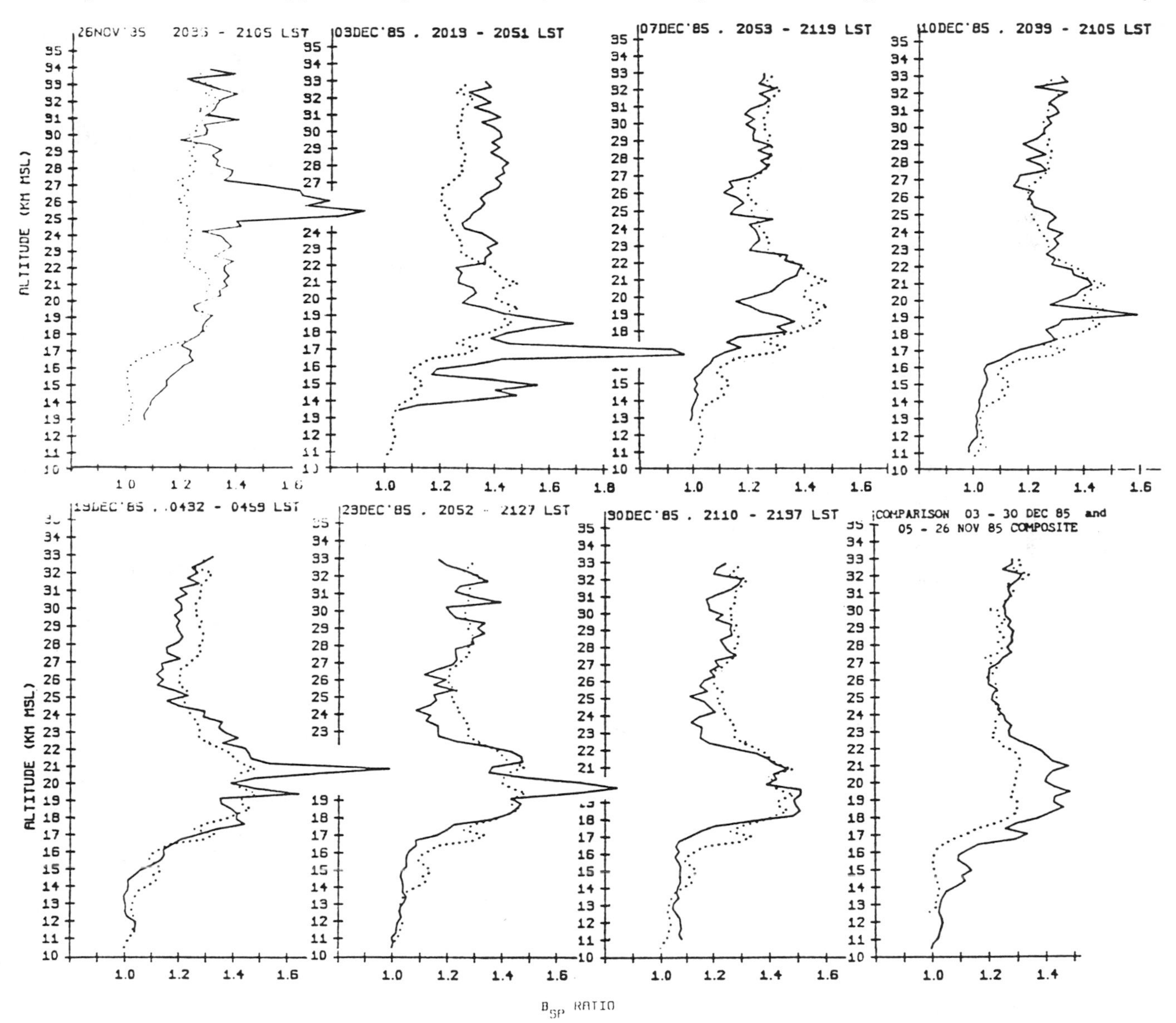

Fig. 19-11. Arrival of Ruiz aerosols at different altitudes, shown by preliminary lidar profiles from Mauna Loa, Hawaii, 26 Nov–30 Dec 1985. Backscattering ratios are on the x axis, altitudes above sea level on the y axis. In the 6 dated graphs, the solid line records the data collected that night, the dotted line represents the average Dec profile. The final graph compares the average Dec profile (solid line) containing new Ruiz aerosols with the average Nov profile (dotted line), primarily aerosols remaining from the 1982 eruption of El Chichón. Courtesy of Thomas DeFoor.

background CN concentrations at these elevations.

Information Contacts: D. Hofmann, J. Rosen, Univ. of Wyoming; T. DeFoor, MLO; R. Reiter, Garmisch-Partenkirchen, W Germany; W. Fuller, NASA Langley Research Center.

12/85 (10:12) Lidar instruments in Hawaii and Japan detected new stratospheric aerosol layers that may have been produced by the 13 Nov eruption of Ruiz volcano, Colombia. Lidar profiles at Mauna Loa, Hawaii showed a distinct new layer centered at 25-25.5 km altitude on 26 and 27 Nov. That layer was not detected during the next measurement on 3 Dec, but distinct new upper tropospheric and lower stratospheric material was evident that night, and apparent new layers centered at 18-22 km were present during the rest of Dec (figs. 19-11 and 19-13). At Fukuoka, Japan, a relatively strong scattering layer appeared at 16.9 km (0.2 km above the tropopause observed at the Fukuoka Meteorological Observatory, 7 km from the lidar site) on 28 Nov, but there were two probable cirrus cloud layers at 6-15 km and it was not possible to confirm that the layer was volcanic. The next night, a very thin scattering layer was present at 18.4 km (3 km above the tropopause); the 29 Nov lidar profile was of a type not observed except after major volcanic eruptions. Layers were observed at about the same altitude during most Dec lidar measurements at Fukuoka. Lidar at the National Institute for Environmental Studies at Tsukuba, Japan detected an aerosol layer about 1 km thick at 18 km altitude on 11 and 12 Dec (fig. 19-12). No such layer had been observed through Nov. The layer became more obscure 13 and 16 Dec. Another small layer was detected at about 22 km through Dec, but it was not certain whether it had been present in previous months. Weak layers centered at 25.8 and 24.5 km were detected from Hampton and Wallops Island, Virginia in Dec.

The following is from Ram Krishna.

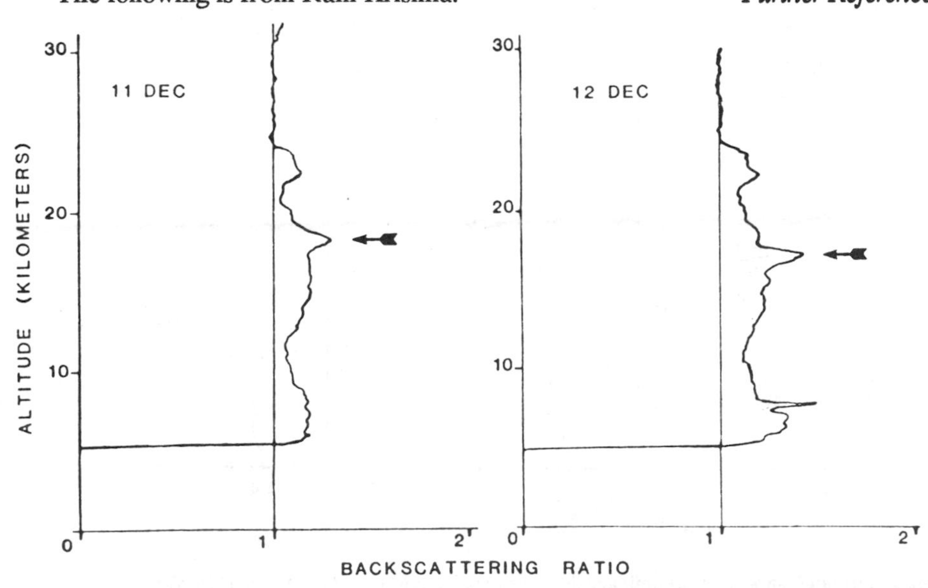

Fig. 19-12. New aerosol layer at 18 km altitude (arrow) detected on preliminary lidar profiles from Tsukuba, Japan, 11 and 12 Dec 1985, 1 month after the Ruiz eruption. Backscattering ratios are on the x axis, with the solid vertical line representing a ratio of 1, and altitudes are on the y axis. Courtesy of Sachiko Hayashida.

"An unusual, heavy haze was observed at Nadi, Fiji (17.78°S, 177.48°E) from 20 through 22 Nov. The haze significantly reduced the very good visibility normally encountered in this area, and, according to reports by pilots, it extended through the boundary layer to the inversion and was evident for many tens of kilometers across Viti Levu to Vanua Levu. Winds were light throughout the period and meteorological analyses could not provide reliable back trajectories. The appearance, density, and spatial extent of the haze suggest that it was an aerosol formed from volcanic sulfur-containing gas emission not too far upstream. Volcanic activity in Vanuatu has been implicated in previous haze episodes and is a likely explanation for the present episode, but this could not be confirmed. A similarly heavy haze was observed on 21 Dec, but it did not persist beyond that date."

Four balloon-borne aerosol observations were made over Laramie, Wyoming during Dec, showing enhanced concentrations of condensation nuclei (with radii between 0.01 and 0.1 μm) above background, probably from the 13 Nov eruption of Ruiz (table 19-3 and fig. 19-14). The strongest enhancement was a 10-fold increase at 15-18 km on 5 Dec, when smaller increases were also measured at 23 and 28 km altitudes (data in table 19-3 replace the preliminary 5 Dec values in 10:11). Only one weak layer was detected on 11 Dec, but flights on 18 and 31 Dec showed several zones of enhanced aerosol concentrations. Increased concentrations of optically active aerosols (larger than 0.15 μm) were not present in any of the samplings. Preliminary data from the 10 Jan flight showed no substantial CN enhancement.

Information Contacts: T. DeFoor, MLO; M. Fujiwara, Kyushu Univ.; S. Hayashida, National Inst. for Environmental Studies, Japan; W. Fuller, M. Osborn, NASA Langley Research Center; R. Krishna, Fiji Meteorological Service; D. Hofmann, Univ. of Wyoming.

Further References:
DeLuisi, J., DeFoor, T., et al., 1984, 1985, 1986, Lidar Observations of Stratospheric Aerosol over Mauna Loa Observatory: 1974-1981, 1982-1983, 1984-1985; *NOAA Data Reports* ERL ARL 4, 5, 9; 107, 78, 66 pp.

Fujita, T., 1985, The Abnormal Temperature Rises in the Lower Stratosphere after the 1982 Eruptions of the Volcano El Chichón, México; *Papers in Meteorology and Geophysics*, v. 36, no. 2, p. 47-60.

Lockwood, G.W. and Thompson, D.T., 1986, Atmospheric Extinction: The Ordinary and Volcanically Induced Variations, 1972-1985; *Astronomical Journal*, v. 92, no. 4, p. 976-985.

Reiter, R. and Jäger, H., 1986, Results of 8-Year Continuous Measurements of Aerosol Profiles in the Stratosphere with Discussion of the Importance of

Stratospheric Aerosols to an Estimate of Effects on the Global Climate; *Meteorology and Atmospheric Physics*, v. 35, p. 19-48.

Sedlacek, W.A., Mroz, E.J., Lazrus, A.L., and Gandrud, B.W., 1983, A Decade of Stratospheric Sulfate Measurements Compared with Observations of Volcanic Eruptions; *JGR*, v. 88, no. C6, p. 3741-3776.

1985	Altitude (km)	Concentration (/cm^3)
Dec 5	15-18	600/50
	23	30/ 7
	28	10/ 4
11	20	18/ 7
18	12-14	70/20
	16	35/15
	18	45/ 8
31	13-17	80/30
	19	40/10
	20	15/ 7
	22-25	30/ 7

Table 19-3. Zones of enhanced condensation nuclei concentration detected by balloon-borne instruments over Laramie, Wyoming, 5-31 Dec. Concentrations are expressed as counts per cm^3; normal background concentrations for each altitude are given after the slash. Data courtesy of David Hofmann.

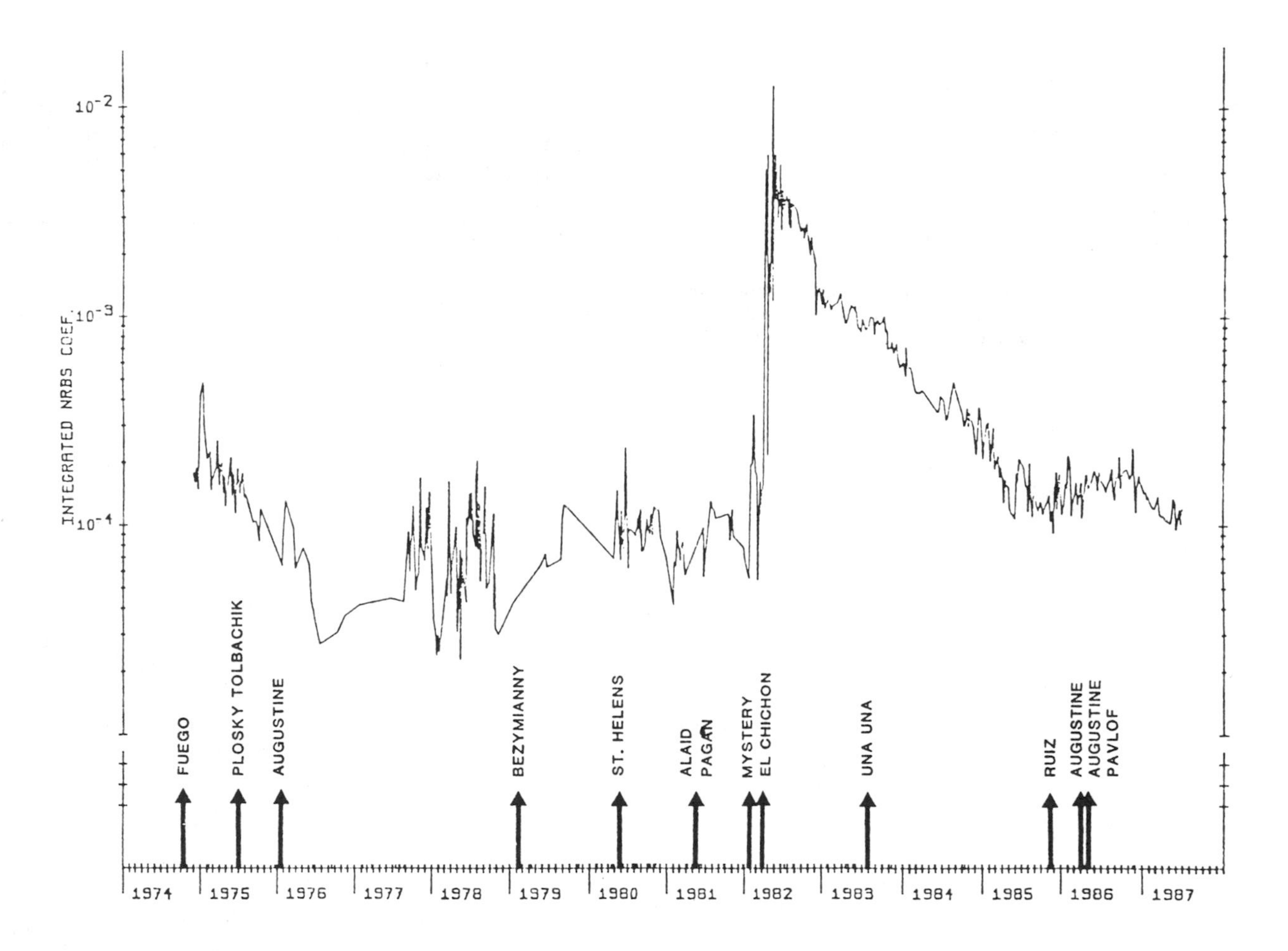

Fig. 19-13. Timing of large explosive eruptions, 1974-86, compared to lidar measurements of stratospheric aerosols over Mauna Loa, Hawaii. Non-Rayleigh backscattering coefficients are integrated over the lower stratosphere. Courtesy of Thomas DeFoor and Elmer Robinson.

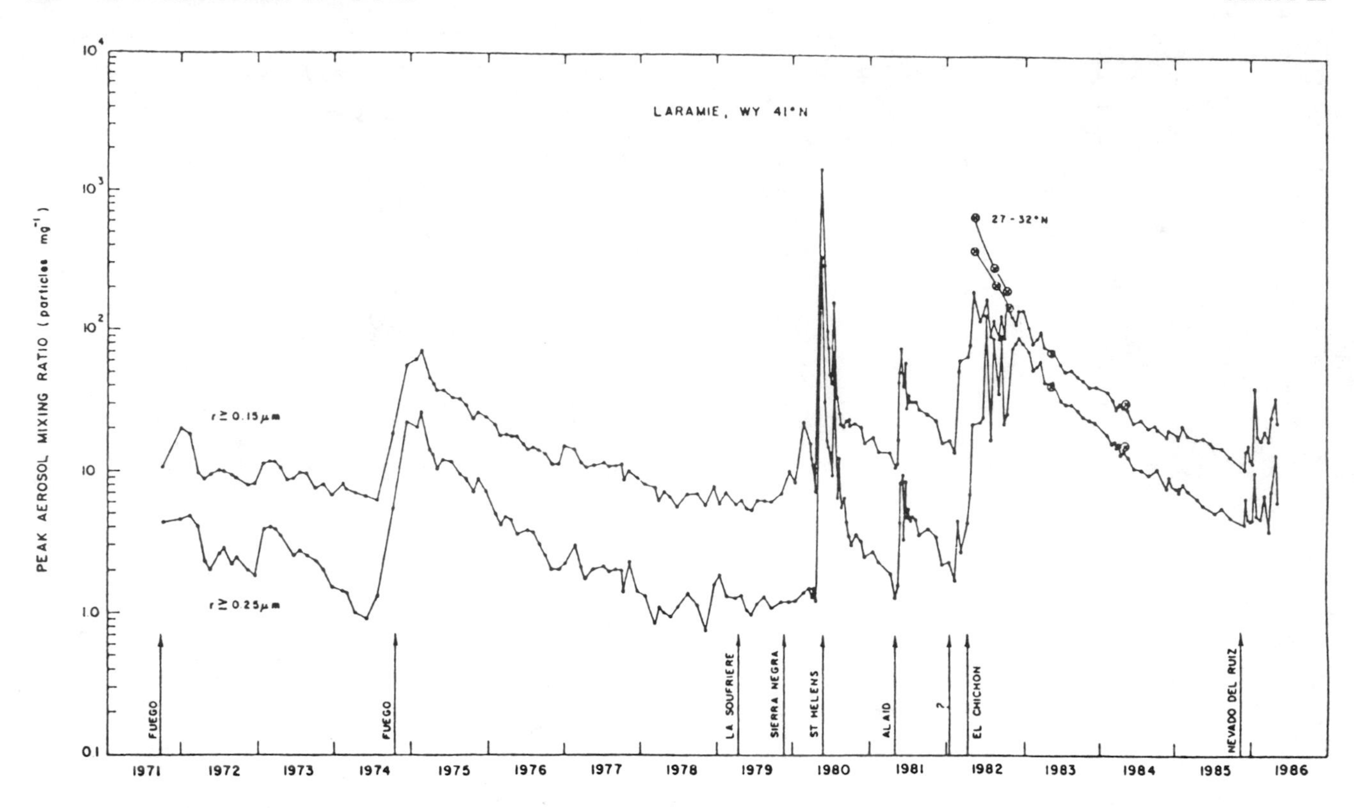

Fig. 19-14. Timing of large explosive eruptions, 1971-85, compared to particle concentrations (per milligram of ambient air) measured during 200 balloon soundings from Laramie, Wyoming and 5 from S Texas (circled crosses). Data were collected for 2 particle sizes (radii 0.25 and 0.15 microns) at the stratospheric maximum (18-22 km altitude). Courtesy of David Hofmann.

IO (A Moon of Jupiter)

3/79 (4:3) Cameras on the Voyager 1 spacecraft have recorded clear images of volcanic eruptions on Jupiter's moon Io, the first extraterrestrial body on which eruptions have been observed. Linda Morabito, a Jet Propulsion Laboratory (JPL) engineer, discovered the activity on fig. 20-1, taken 8 March at a distance of 4.5×10^6 km. Inspection of other images, many taken at much closer range, revealed 8-10 active vents, several of which erupted repeatedly during the 4-day period when Io was in view of Voyager's cameras. All of the eruptions identified on Voyager imagery were explosive, with ejection velocities of 1000-2000 km/hr, producing irregular to mushroom-shaped clouds 100-400 km high. Although Io has no atmosphere to frictionally slow the rising ejecta, most seemed to fall back to the surface within 5-10 minutes after eruption.

No active lava flows were identified on the Voyager 1 images. However, features that appeared to be volcanoes at the center of fresh radial lava flows were observed, as in fig. 20-2. Images of these features from Voyager 2, scheduled to pass Io in early July, will be carefully compared with those from Voyager 1, with the hope of discovering new flows.

Morphology of the active vents varied substantially, from small dark-centered areas resembling calderas to larger doughnut-shaped bright spots. The volcanic features tended to be concentrated in Io's equatorial region. Only a few seemed to form a linear pattern.

Infrared instruments measured temperatures in excess of 0°C around some of the caldera-like features, as compared to Io's average surface temperature of about 120°K. An eruption cloud temperature of almost 100°C was also recorded.

Information Contacts: L. Soderblom, USGS, Flagstaff, AZ; F. Bristow, JPL.

Further Reference: Stone, E.C. and Lane, L.A. (eds.), 1979, Voyager 1 Encounter with the Jovian System; *Science*, v. 204, p. 945-1008 (14 papers).

7/79 (4:7) The Voyager 2 spacecraft flew past Jupiter and its moons in early July, about 4 months after the Voyager 1 encounter. Voyager 2's cameras viewed Io intermittently for 5 days, including continuous imaging for about 7 hours around the closest encounter on 9 July. Resolutions ranged from about 100 km to about 20 km. Of the 8 plumes seen on Voyager 1 imagery, 6 were still active,

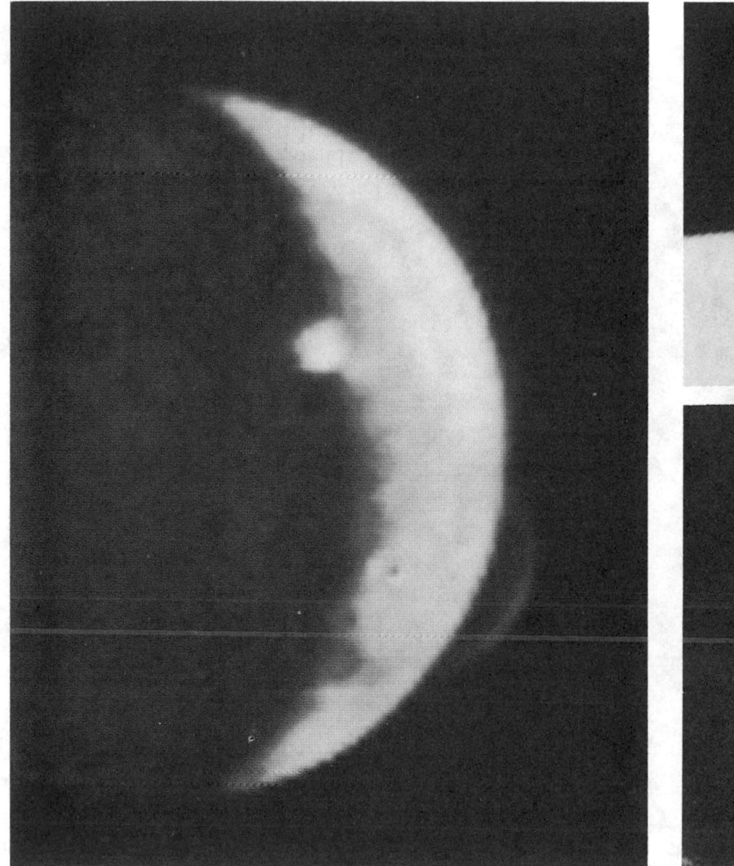

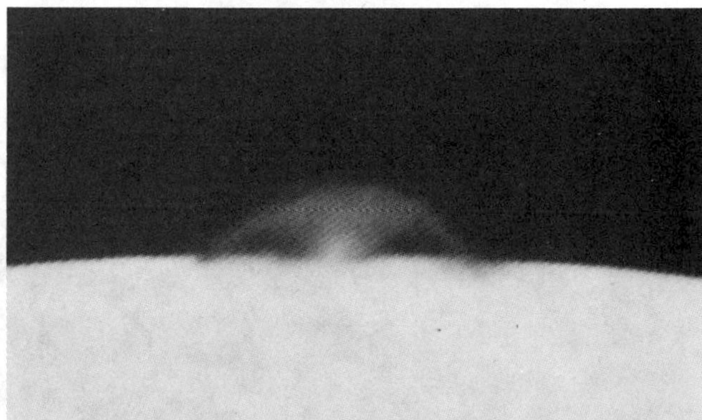

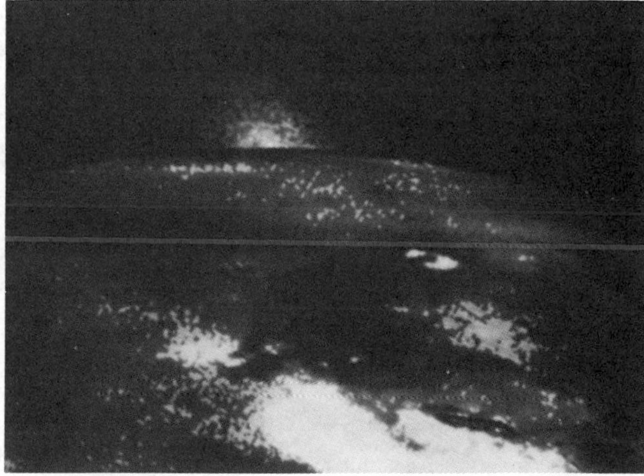

Fig. 20-1. *Left,* view of Io from the Voyager 1 spacecraft 8 March 1979 from a distance of 4.5×10^6 km. Two eruptions are visible: the cloud rising from the limb of Io, at lower right, rises > 250 km from the surface; a second cloud is lit by the sun at the upper center of the terminator (the boundary between night and day). *Upper right,* mushroom-shaped eruption cloud rising > 100 km from Io's surface on 4 March. The small dark spot at lower left is used to align Voyager's 1 camera. *Lower right,* 150 km-high eruption cloud silhouetted against black space, from 490,000 km on 4 March 1979. The cloud is greenish-white, originating from a complex circular structure consisting of a bright ring about 300 km in diameter and a central region of irregular dark and light patterns. All images courtesy of Frank Bristow.

Fig. 20-2. Dark spot, probably a volcano, surrounded by a radial pattern of flows. This image, taken 5 March from a distance of 128,500 km, is about 500 km wide. In the color version, the dark spot is black, but the flows are bright red-orange with a few darker longitudinal streaks. Courtesy of Frank Bristow.

one (the largest seen by Voyager 1, 280 km high) was definitely quiescent, and the state of activity of another could not be determined. Most of the plumes seemed to be about the same size as on Voyager 1 imagery, but the largest currently active plume had grown to roughly 150 km from its approximately 100 km height of 4 months earlier. No new eruption sites were discovered, but preliminary analysis indicates that some fairly large-scale changes in surface morphology (visible at resolutions of 40-50 km) have taken place in the past 4 months, both around active vents and in other areas.

Information Contact: R. Strom, JPL.

Further References: Stone, E.C. and Lane, L.A. (eds.), 1979, Voyager 2 Imaging Science Results; *Science*, v. 206, p. 925-996 (14 papers).

Special supplement, 1979, Voyager 1 Encounter with Jupiter and Io; *Nature*, v. 280, p. 725-806 (28 papers, 15 on Io).

11/79 (4:11) On 11 June 1979, William Sinton observed what may have been an eruption from Io, using infrared photometry from the 2.2-m telescope on Mauna Kea, Hawaii. Within the region on Io where Sinton believes the eruption may have occurred, a caldera was discovered on Voyager 2 imagery (early July) that was not present on imagery from Voyager 1 (early March). An earlier observation of what may also have been an eruption is described in Witteborn et al. (1979). Sinton plans further observations in Dec.

References: Sinton, W.M., 1980, Io's 5 Micron Variability; *Astrophysical Journal Letters*, v. 235, p. L49-L51.

Witteborn, F. C., Bergman, J. D., and Pollack, J. B., 1979, Io: An Intense Brightening near 5 Micrometers; *Science*, v. 203,. p. 653-646.

Information Contact: W. Sinton, Inst. for Astronomy, Honolulu.

Further References: Carr, M.C., 1986, Silicate Volcanism on Io; *JGR*, v. 91, no. B3, p. 3521-3532.

Johnson, T.V. et al., 1984, Volcanic Hot Spots on Io: Stability and Longitudinal Distribution; *Science,* v. 226, p. 134-137.

Kieffer, S.W., 1982, Dynamics and Thermodynamics of Volcanic Eruptions: Implications for the Plumes on Io; *in* Morrison, D. (ed.), *Satellites of Jupiter*; Univ. of Arizona Press, Tucson, p. 647-723.

McEwen, A.S., Matson, D. L., Johnson, T.V., and Soderblom, L.A., 1985, Volcanic Hot Spots on Io: Correlation with Low-Albedo Calderas; *JGR*, v. 90, no. B14, p. 12345-12380.

Strom, R.G., Schneider, N. M., Terrile, R.J., et al., 1981, Volcanic Eruptions on Io; *JGR (Space Physics)*, v. 86, p. 8593-8620.

Wilson, L. and Head, J.W., 1983, A Comparison of Volcanic Eruption Processes on Earth, Moon, Mars, Io, and Venus; *Nature*, v. 302, p. 663-669.

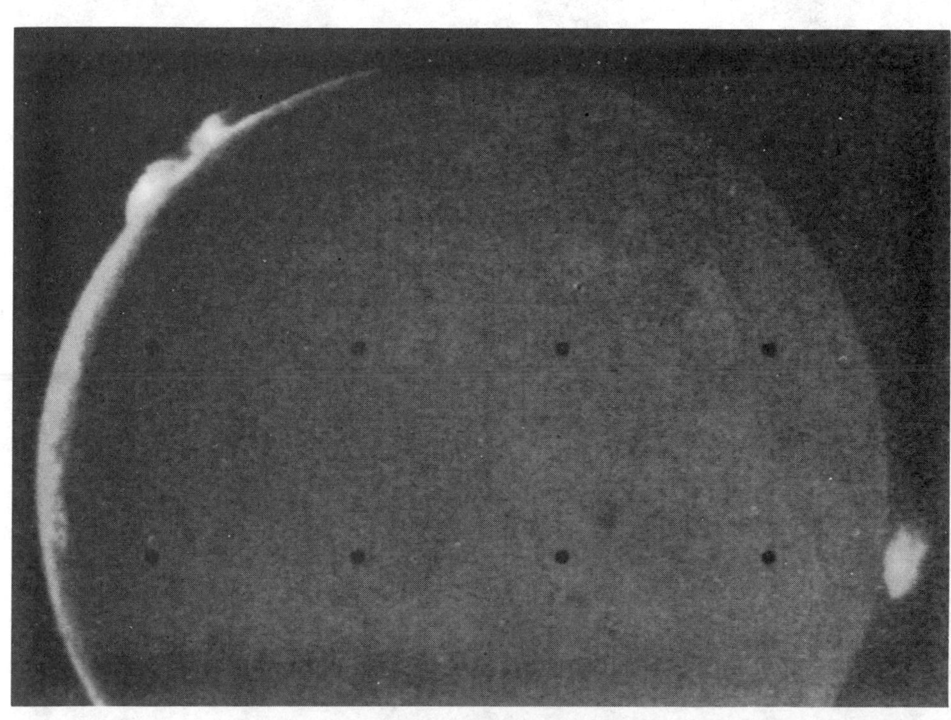

Fig. 20-3. Voyager 2 image of Io taken 10 July 1979 from 1.2 million km. The sunlit crescent of Io is at the left; the rest of Io is dimly illuminated by light reflected from Jupiter. Three volcanic plumes discovered by Voyager 1 four months earlier are visible: the two on the bright limb are about 100 km high, while the plume on the dark limb is about 185 km high and 325 km wide, about 1 $^1/_2$ times its dimensions during the Voyager 1 encounter. Courtesy of Frank Bristow. [Originally in 4:8.]

PART III

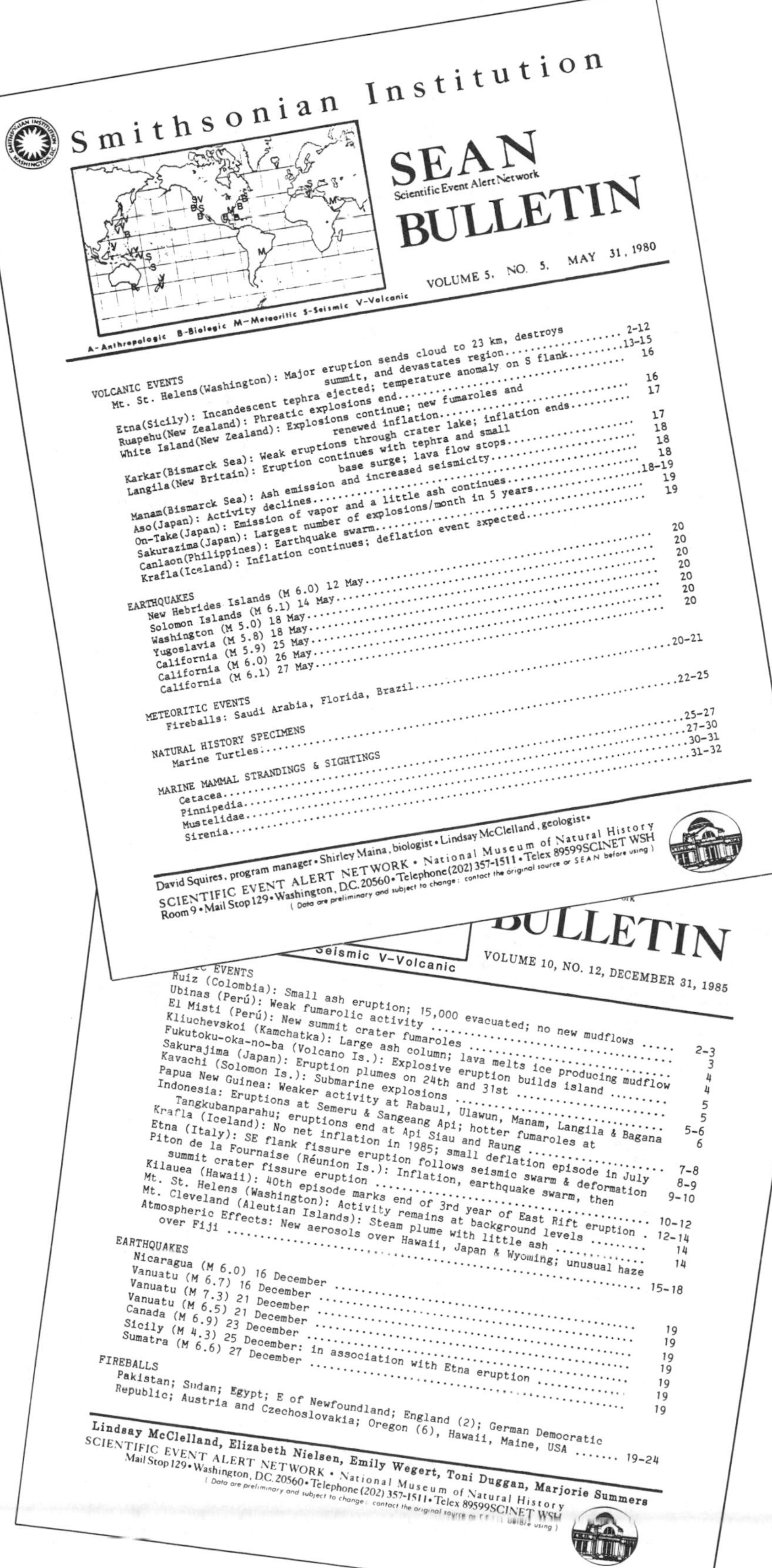

Smithsonian Institution

SEAN
Scientific Event Alert Network
BULLETIN

VOLUME 5, NO. 5, MAY 31, 1980

A-Anthropologic B-Biologic M-Meteoritic S-Seismic V-Volcanic

VOLCANIC EVENTS
Mt. St. Helens (Washington): Major eruption sends cloud to 23 km, destroys summit, and devastates region 2-12
Etna (Sicily): Incandescent tephra ejected; temperature anomaly on S flank 13-15
Ruapehu (New Zealand): Phreatic explosions end 16
White Island (New Zealand): Explosions continue; new fumaroles and renewed inflation 16
Karkar (Bismarck Sea): Weak eruptions through crater lake; inflation ends 17
Langila (New Britain): Eruption continues with tephra and small base surge; lava flow stops 17
Manam (Bismarck Sea): Ash emission and increased seismicity 18
Aso (Japan): Activity declines 18
On-Take (Japan): Emission of vapor and a little ash continues 18
Sakurazima (Japan): Largest number of explosions/month in 5 years 18-19
Canlaon (Philippines): Earthquake swarm 19
Krafla (Iceland): Inflation continues; deflation event expected 19

EARTHQUAKES
New Hebrides Islands (M 6.0) 12 May 20
Solomon Islands (M 6.1) 14 May 20
Washington (M 5.0) 18 May 20
Yugoslavia (M 5.8) 18 May 20
California (M 5.9) 25 May 20
California (M 6.0) 26 May 20
California (M 6.1) 27 May 20

METEORITIC EVENTS
Fireballs: Saudi Arabia, Florida, Brazil 20-21

NATURAL HISTORY SPECIMENS
Marine Turtles 22-25

MARINE MAMMAL STRANDINGS & SIGHTINGS
Cetacea 25-27
Pinnipedia 27-30
Mustelidae 30-31
Sirenia 31-32

David Squires, program manager • Shirley Maina, biologist • Lindsay McClelland, geologist •
SCIENTIFIC EVENT ALERT NETWORK • National Museum of Natural History
Room 9 • Mail Stop 129 • Washington, D.C. 20560 • Telephone (202) 357-1511 • Telex 89599SCINET WSH
(Data are preliminary and subject to change; contact the original source or SEAN before using)

BULLETIN

Seismic V-Volcanic

VOLUME 10, NO. 12, DECEMBER 31, 1985

EVENTS
Ruiz (Colombia): Small ash eruption; 15,000 evacuated; no new mudflows 2-3
Ubinas (Perú): Weak fumarolic activity 3
El Misti (Perú): New summit crater fumaroles 4
Kliuchevskoi (Kamchatka): Large ash column; lava melts ice producing mudflow 4
Fukutoku-oka-no-ba (Volcano Is.): Explosive eruption builds island 5
Sakurajima (Japan): Eruption plumes on 24th and 31st 5
Kavachi (Solomon Is.): Submarine explosions 5-6
Papua New Guinea: Weaker activity at Rabaul, Ulawun, Manam, Langila & Bagana 6
Indonesia: Eruptions at Semeru & Sangeang Api; hotter fumaroles at Tangkubanparahu; eruptions end at Api Siau and Raung 7-8
Krafla (Iceland): No net inflation in 1985; small deflation episode in July 8-9
Etna (Italy): SE flank fissure eruption follows seismic swarm & deformation 9-10
Piton de la Fournaise (Réunion Is.): Inflation, earthquake swarm, then summit crater fissure eruption 10-12
Kilauea (Hawaii): 40th episode marks end of 3rd year of East Rift eruption 12-14
Mt. St. Helens (Washington): Activity remains at background levels 14
Mt. Cleveland (Aleutian Islands): Steam plume with little ash 14
Atmospheric Effects: New aerosols over Hawaii, Japan & Wyoming; unusual haze over Fiji 15-18

EARTHQUAKES
Nicaragua (M 6.0) 16 December 19
Vanuatu (M 6.7) 16 December 19
Vanuatu (M 7.3) 21 December 19
Vanuatu (M 6.5) 21 December 19
Canada (M 6.9) 23 December 19
Sicily (M 4.3) 25 December: in association with Etna eruption 19
Sumatra (M 6.6) 27 December 19

FIREBALLS
Pakistan; Sudan; Egypt; E of Newfoundland; England (2); German Democratic Republic; Austria and Czechoslovakia; Oregon (6), Hawaii, Maine, USA 19-24

Lindsay McClelland, Elizabeth Nielsen, Emily Wegert, Toni Duggan, Marjorie Summers
SCIENTIFIC EVENT ALERT NETWORK • National Museum of Natural History
Mail Stop 129 • Washington, D.C. 20560 • Telephone (202) 357-1511 • Telex 89599SCINET WSH
(Data are preliminary and subject to change; contact the original source or SEAN before using)

ABBREVIATIONS/ACRONYMS

AGU	American Geophysical Union
AID	Agency for International Development (US Dept. of State)
AP	Associated Press
AFP	Agence France-Presse
ARL	Air Resources Laboratory (NOAA)
AZAP	Agence Zairoise de Presse (Zaire)
BBC	British Broadcasting Corporation
BP	Before Present
BRGM	Bureau de Recherches Géologiques et Minières
BV	*Bulletin of Volcanology* (Springer International) continuation of *Bulletin Volcanologique* (IAVCEI)
BVE	*Bulletin of Volcanic Eruptions* (Volcanological Society of Japan)
CAVW	*Catalogue of the Active Volcanoes of the World* (IAVCEI)
CN	Condensation Nuclei
CNET	Centre National d'Etudes des Télécommunications (France)
CNRS	Centre National de la Recherche Scientifique (France)
CDRS	Charles Darwin Research Station (Galápagos)
COMVOL	Philippine Commission on Volcanology; predecessor of PHIVOLCS
COSPEC	Correlation Spectrometer
CSLP	Center for Short-Lived Phenomena (Smithsonian)
CVO	Cascades Volcano Observatory (USGS)
DNS	Domestic News Service (various countries)
DPA	Deutche-Presse Agentur
DRM	Délégation aux Risques Majeurs (France)
DRS	Domestic Radio Service (various countries)
DSIR	Department of Scientific and Industrial Research (New Zealand)
EDM	Electronic Distance Measurement
ERI	Earthquake Research Institute, Univ. of Tokyo
ERL	Environmental Research Laboratory (NOAA)
FAA	Federal Aviation Administration (US)
FSD	Full-Scale Deflection (seismograph)
GMS	Geostationary Meteorological Satellite
GMT	Greenwich Mean Time
GOES	Geostationary Orbiting Earth Satellite
GSA	Geological Society of America
GSFC	Goddard Space Flight Center (NASA)
GSI	Geological Survey of Indonesia
HVO	Hawaiian Volcano Observatory (USGS)
IATA	International Air Transport Association
IAVCEI	International Association of Volcanology and Chemistry of the Earth's Interior
ICAO	International Civil Aviation Organization
ICE	Instituto Costarricense de Electricidad
IGM	Instituto Geográfico Militar (Guatemala)
IGN	Instituto Geográfico Nacional, predecessor of IGM; Institut Géografique National (France)
IIS	Instituto de Investigaciones Sísmicas (Nicaragua)
IIV	Istituto Internazionale di Vulcanologia (Italy)
IMESS	International Mt. Erebus Seismic Study
INE	Instituto Nicaraguense de Energía
INETER	Instituto Nicaraguense de Estudios Territoriales
INGEOMINAS	Instituto Nacional de Investigaciones Geológico-Mineras (Colombia)
INSIVUMEH	Instituto Nacional de Sismología, Vulcanología, Meteorología, e Hidrología (Guatemala)
IPG	Institut de Physique du Globe (France)
IRENA	Instituto Nicaraguense de Recursos Naturales
IRS	Institut de Recherches Scientifiques Afrique Centrale (Zaire)
IVP	Institute of Volcanology, Petropavlovsk (USSR)
JGR	*Journal of Geophysical Research* (AGU)
JMA	Japan Meteorological Agency
JMSA	Japan Maritime Safety Agency
JMSDF	Japan Maritime Self-Defense Force
JPL	Jet Propulsion Laboratory (US)
JSC	Johnson Space Center (NASA)
JSDF	Japan Self-Defense Force
JVGR	*Journal of Volcanology and Geothermal Research* (Elsevier)
LANL	Los Alamos National Laboratory (formerly Los Alamos Scientific Laboratory)
LDGO	Lamont-Doherty Geological Observatory
LSNR	Lands Survey and Natural Resources (Tonga)
M	Magnitude (earthquake)
m_b	Body-wave magnitude
M_L	Long-wave magnitude
MLO	Mauna Loa Observatory (NOAA, Hawaii)
MM	Modified Mercalli (earthquake intensity scale)
MRF	Modified Rossi-Forel (earthquake intensity scale)
MRI	Meteorological Research Institute (Tokyo)
M_s	Surface-wave magnitude
NASA	National Aeronautics and Space Administration (US)
NCAR	National Center for Atmospheric Research (US)
NESDIS	National Environmental Satellite Data and Information Service (US)
NEIC	National Earthquake Information Center (US)
NEIS	National Earthquake Information Service (US); name changed to NEIC
NESS	National Earth Satellite Service (US); predecessor of NESDIS
NOAA	National Oceanographic and Atmospheric Administration (US)
NVI	Nordic Volcanological Institute (Iceland)
NWS	National Weather Service (US)
NZGS	New Zealand Geological Survey
OFDA	Office of Foreign Disaster Assistance (US)
ORSTOM	Office de la Recherche Scientifique et Technique Outre-Mer (France)
OV	Osservatorio Vesuviano (Italy)
OVPDLF	Observatoire Volcanologique du Piton de la Fournaise
PHIVOLCS	Philippine Institute of Volcanology and Seismology
PIRPSEV	Programme Interdisciplinaire de Recherches sur la Prevision et la Surveillance des Eruptions Volcaniques (France)
PSC	Polar Stratospheric Cloud
RAVE	Research on Active Volcanic Emissions
RNZAF	Royal New Zealand Air Force
RSP	Réseau Sismique Polynésien (Tahiti)
RVO	Rabaul Volcano Observatory (Papua New Guinea)
SAGE	Stratospheric Aerosol Gas Experiment (satellite)
SAM	Stratospheric Aerosol Measurement (satellite)
SAO	Smithsonian Astrophysical Observatory
SDA	Solar Depression Angle
SI	Smithsonian Institution
SME	Solar Mesospheric Explorer (satellite)
SOLAIR	Solomon Island Airways
SPIA	South Pacific Island Airways (American Samoa)
TOMS	Total Ozone Mapping Spectrometer
UNAM	Universidad Nacional Autónoma de México
UNDRO	United Nations Disaster Relief Organization
UPI	United Press International
US	United States
USAF	US Air Force
USCG	US Coast Guard
USCGC	US Coast Guard Cutter
USFS	US Forest Service
USGS	US Geological Survey
USN	US Navy
USNRL	US Naval Research Laboratory
UW	University of Washington, Seattle
VEI	Volcanic Explosivity Index
VRF	Volcano Reference File (Smithsonian)
VSI	Volcanological Survey of Indonesia (Direktorat Vulkanologi)
WWSSN	Worldwide Seismic Station Network (USGS)

WEIGHT/MEASURE UNITS

°C	Centigrade degrees ($^5/_9$ · (Fahrenheit-32))
cm	centimeter (0.3937 inches)
g	gram (0.0353 avoirdupois ounces)
hPa	hecto Pascal (10^2 Pa)
hr	hour
kg	kilogram (2.205 pounds)
km	kilometer (0.6214 statute miles; 0.5396 nautical miles)
m	meter (3.2808 feet)
matm-cm	milliatmosphere-centimeters
min	minute
mm	millimeter (0.0394 inches)
µrad	microradian (1 mm elev. difference over 1 km)
µm	micrometer (micron; 10^{-6} m; 0.00004 inches)
nT	nano Tesla
pH	measure of acidity (the negative logarithm of hydrogen-ion concentration)
ppm	parts per million
rms	root mean square
sec	second
t/d	metric tons per day (1.1023 short tons/day; 2205 pounds/day)
hectare	(10,000 m^2, 2.4710 acres; 0.0039 $miles^2$)

DATES AND DIRECTIONS

Jan	January
Feb	February
Aug	August
Sept	September
Oct	October
Nov	November
Dec	December

Other months spelled out.

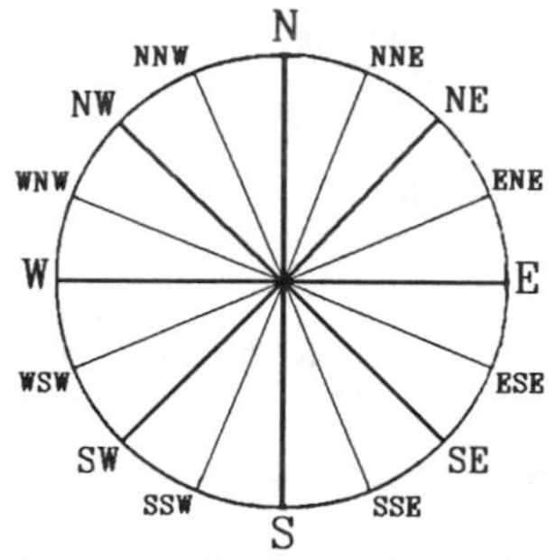

Compass Direction Abbreviations

US STATE ABBREVIATIONS

AK	Alaska
AZ	Arizona
CA	California
CO	Colorado
DC	District of Columbia
FL	Florida
HI	Hawaii
IA	Iowa
ID	Idaho
IL	Illinois
LA	Louisiana
MA	Massachusetts
MD	Maryland
MI	Michigan
MN	Minnesota
MO	Missouri
MS	Mississippi
NC	North Carolina
NH	New Hampshire
NJ	New Jersey
NM	New Mexico
NV	Nevada
NY	New York
OH	Ohio
OK	Oklahoma
OR	Oregon
PA	Pennsylvania
RI	Rhode Island
TX	Texas
VA	Virginia
WA	Washington
WI	Wisconsin
WV	West Virginia
WY	Wyoming

INFORMATION CONTACTS

Abiko, T.
Dept. of Applied Chemistry
Muroran Institute of Technology
Mizumoto-chu, Muroran 050, Japan

Aburto, Arturo
Instituto de Investigaciones Sísmicas
Apartado Postal 1761
Managua, Nicaragua

Acevedo, Patricio
Depto. de Ciencias Físicas, Area Geofísica
Universidad de la Frontera
Casilla 54-D, Temuco, Chile

Adams, A. B.
Botany Dept.
University of Washington
Seattle, WA 98195 USA

Adkerson, Roy, Lt. Cmdr.
COMNAVMAR, Code N3
FPO San Francisco 96630 USA
(current address unknown)

Agence France-Presse
11 Place de La Bourse
75002 Paris, France

Almond, Richard
Bureau of Mineral Resources
P. O. Box 378
Canberra, A.C.T., Australia

Alquijay, Paulino
2^A^ Calle Poniente No. 5
Antigua, Guatemala

Alvarado, Rodolfo
Instituto Nacional de Electrificación
7^A^ Avenida 1-17, Zona 4
Guatemala City, Guatemala

Andal, Gregorio (retired)
Commission on Volcanology
5th Floor, Hizon Bldg
29 Quezon Avenue
Quezon City, Philippines

Anderson, Diane
Public Relations, Singapore Airlines
8350 Wilshire Boulevard
Beverly Hills, CA 90211 USA

Antara News Agency
Jalan Medan Merdeka Selatan 17
P. O. Box 257
Jakarta, Indonesia

Archambault, C.
Centre National d'Etudes des Télécommunications (CNET)
22301 Lannion Cedex, France

Argeñal A., Róger
Dirección de Geología y Geofísica
Instituto de Estudios Territoriales
Apartado 1761
Managua, Nicaragua

Arnette, Steve
Synoptic Analysis Branch
NOAA/NESDIS
Room 401, World Weather Bldg.
Washington, DC 20233 USA

Aspinall, William
Yew Tree Cottage
5 Woodside Close
Beaconsfield, Bucks. HP9 1JQ
England

Associated Press
50 Rockefeller Plaza
New York, NY 10020 USA

Aucott, J. W.
c/o British Embassy
Quito, Ecuador

Avila, Guillermo
Depto. de Geología
Instituto Costarricense de Electricidad
Apartado 10032
San José, Costa Rica

AZAP News agency (Zaire)

Baade, J. H.
Gulf Oil
Douala, Cameroon

Bachelery, Patrick
Laboratoire de Géologie
Université de la Réunion
B.P. 5, 94790 Ste. Clothilde
Réunion Island

Bailey, Roy
USGS, 345 Middlefield Road
Menlo Park, CA 94025 USA

Baldwin, Thomas
Synoptic Analysis Branch
NOAA/NESDIS
Room 401, World Weather Bldg.
Washington, DC 20233 USA

Banks, Norman G.
USGS Cascades Volcano Observatory
5400 MacArthur Boulevard
Vancouver, WA 98661 USA

Barberi, Franco
Dipartimento de Scienze della Terra
Università degli Studi di Pisa
Via S. Maria 53
56100 Pisa, Italy

Barnes, John
New Zealand Defense Staff
1601 Connecticut Ave. NW
Washington, DC 20036 USA

Barquero H., Jorge
Observatorio Vulcanológico y Sismológico de Costa Rica
Universidad Nacional
Heredia, Costa Rica

Barreiro, Barbara A.
Dept. of Earth Sciences
Dartmouth College
Hanover, NH 03755 USA

Bartaire, N.
Université de Paris VI
4 Avenue de Neptune
94100 St. Maur-des-Fossés, France

Basile, R.
Gruppo Ricerca Speleologica
Acireale, Italy

Beauchamp, George, Jr. (retired)
AID/OFDA
Department of State
Washington, DC 20523 USA

Becker, Lt.
U.S. Air Force, Shemya, AK USA
(current address unknown)

Beech, Edwin, Lt.
U.S. Navy, Adak Island, AK USA
(current address unknown)

Beetham, R. D.
New Zealand Geological Survey
Turangi, New Zealand

Benson, Marie
USAID
Arusha, Tanzania

Benton, L.
Basin Research Institute
Louisiana State University
Baton Rouge, LA 70803 USA

Berruti, A.
Percy Fitzpatrick Institute
University of Cape Town
Rondebosch 7700, South Africa

Betah, Dr.
Ministère des Mines et Energie
B. P. 70, Yaoundé, Cameroon

Bigot, S.
Laboratoire de Géochimie Comparée et Systématique
4 Place Jussieu
75230 Paris Cedex 05, France

Billington, Selena
NOAA/CIRES
University of Colorado
Boulder, CO 80302 USA

Birnie, Richard
Dept. of Earth Sciences
Dartmouth College
Hanover, NH 03755 USA

Björnsson, Helgi
Science Institute
University of Iceland
Reykjavík, Iceland

Blong, Russell J.
School of Earth Sciences
Macquarie University
North Ryde, NSW 2113, Australia

Bloom, Justin L.
Scientific and Technical Affairs
(U. S. Embassy, Tokyo)
APO San Francisco, 96503 USA

Blum, P. A.
Institut de Physique du Globe
4 Place Jussieu
75252 Paris Cedex 05, France

Boeing Commercial Airplane Co.
P. O. Box 3707
Seattle, WA 98124 USA

Boersma, Dee
Institute for Environmental Studies
University of Washington
Seattle, WA 98195 USA

Bogoyavlenskaya, G. Ye.
Institute of Volcanology
Piip Avenue 9
Petropavlovsk, Kamchatka 683006 USSR

Bonis, Samuel
Instituto Geográfico Militar
Avenida las Américas, 5-76, Zona 13
Guatemala City, Guatemala

Bonneville, Alain
Centre Géologique et Géophysique
Université des Sciences et Techniques du Languedoc
Place Eugène Bataillon
34060 Montpellier Cedex, France

Borneman, Richard (retired)
NOAA/NESDIS
Washington, DC 20233 USA

Bornhorst, Theodore J.
Dept. of Geology & Geol. Engineering
Michigan Technological University
Houghton, MI 49931 USA

Boudon, G.
Observatoire Volcanologique de la Montagne Pelée
Fond, 11 Denis
97250 St Pierre, Martinique

Bougainville Copper Ltd.
Panguna, Solomon Islands.

Bougeres, J.
Laboratoire de Géographie
Université de la Réunion
B.P. 5, 94790 Ste. Clothilde
Réunion Island

Boyko, Christina
Lamont-Doherty Geological Observatory
Palisades, NY 10964 USA

Brantley, Steven
USGS Cascades Volcano Observatory
5400 MacArthur Boulevard
Vancouver, WA 98661 USA

Briole, Pierre
PIRPSEV
77, Avenue Denfert-Rochereau
75014 Paris, France

Bristow, Frank E.
Office of Public Information
Jet Propulsion Laboratory
4800 Oak Grove Drive
Pasadena, CA 91103 USA

BBC Radio and Head Office
Broadcasting House, Portland Place
London W1, England

British Consulate
Douala, Cameroon

Brooks, Edward M.
321 Kendrick St.
Newton, MA 01258 USA

Brussels Domestic Service
Belgium

Bruzga, M.
1245 Bear Mountain Court
Boulder, CO 80303 USA

Budris, J.
Charles Darwin Research Station
Galápagos Islands, Ecuador
(current address unknown)

Bull, J.
Terres Australes et Antarctiques Françaises
34 Rue des Renaudes
75017 Paris, France

Caburian, José
c/o Marsman & Co.
P. O. Box 297
Manila, Philippines

Calvache V., Marta
Observatorio Vulcanológico de Colombia
INGEOMINAS, Apartado 1296
Manizales, Colombia

The Cameroon Tribune
P. O. Box 23
Yaoundé, Cameroon

Camus, Guy
Dept. de Géologie et Mineralogie
5 Rue Kessler
63000 Clermont-Ferrand, France

Canul D., René
Comisión Federal de Electricidad
Gerencia de Geotermia
Morelia, México

Capaul, W.
Idaho Geological Survey
Room 332, Morris Hall
University of Idaho
Moscow, ID 83843 USA

Carapezza, Marcello (deceased)
Università di Palermo, Italy

Carey, Steven
School of Oceanography
University of Rhode Island
Kingston, RI 02881 USA

Carney, John
Geological Survey, British Residency
Vila, Vanuatu

Carr, Michael
Dept. of Geology
Rutgers University
New Brunswick, NJ 08903 USA

Casadevall, Thomas
USGS Cascades Volcano Observatory
5400 MacArthur Boulevard
Vancouver, WA 98661 USA

Cashman, Kathy
Dept. of Geological & Geophysical Sciences
Princeton University
Princeton, NJ 08544 USA

Centre National de la Recherche Scientifique (CNRS)
91190 Gif-Sur-Yvette, France

Centre National d'Etudes des Télécommunications (CNET)
22301 Lannion, France

César, Ariel
Central Hidroeléctrica de Caldas, S.A.
Apartado Aereo 83
Manizales, Colombia

Chadwick, William
USGS, 345 Middlefield Rd.
Menlo Park, CA 94025 USA

Cheminée, Jean-Louis
Institut de Physique du Globe
Direction des Observatoires Volcanologiques,
4, Place Jussieu
75252 Paris Cedex 05, France

Cherkis, Norman
Code 8110
U. S. Naval Research Laboratory
Washington, DC 20375 USA

Chesner, Craig
Dept. of Geology and Geological Engineering
Michigan Technological University
Houghton, MI 49931 USA

Childers, Dallas
USGS Cascades Volcano Observatory
5400 MacArthur Boulevard
Vancouver, WA 98661 USA

Chong, Francisco (retired)
Disaster Control Officer, Saipan
Office of the Governor
Saipan, Northern Marianas 96950

Christiansen, Robert
USGS, 345 Middlefield, Road
Menlo Park, CA 94025 USA

Christie, D.R.
Research School of Earth Sciences
Australian National University
Canberra, ACT, Australia

Christoffel, D.
Victoria University
Private Bag, Wellington, New Zealand

Chuan, Raymond
Brunswick Corporation,
Costa Mesa, CA 92626 USA

Clark, R. H. (deceased)
Victoria University, New Zealand

Clark, Richard
Box R
Blaine, WA 98230 USA

Clarke, Warren, General Consul
American Embassy, Djibouti
Republic of Djibouti

Clemente-Colón, Pablo
Ocean Sciences Branch
NOAA/NESDIS, Room 0310
Suitland Professional Center
Washington, DC 20233 USA

Clocchiatti, R.
Centre National de la Recherche Scientifique
15, Quai Anatole France
75700 Paris, France

Cochemé, Jean-Jacques
Instituto de Geología
Universidad Nacional Autónoma de México
Ciudad Universitaria
04510 México, D.F., México

Cockerham, Robert
USGS 345 Middlefield Road
Menlo Park, CA 94025 USA

Cody, Ashley D.
New Zealand Geological Survey
P.O. Box 499
Rotorua, New Zealand

Cole, James W.
Geology Dept.
University of Canterbury
Christchurch 1, New Zealand

Compton, M.,
USCGC Morgenthau,
APO, San Francisco, CA
(current address unknown)

Condarelli, D.
Istituto Internazionale di Vulcanologia
Viale Regina Margherita, 6
95123 Catania, Italy

Connor, Charles B.
Dept. of Geology
Florida International University
Miami, FL 33199 USA

Conrad, Mark
628 Arkansas
San Francisco, CA 94107 USA

Consoli, O.
Istituto Internazionale di Vulcanologia
Viale Regina Margherita, 6
95123 Catania, Italy

Continental Air Micronesia
P. O. Box 138
Saipan, Mariana Islands 96950 USA

Cooke, Robin J. S. (deceased)
Rabaul Volcano Observatory

Cooper, Eric
Solomon Islands Airways, Ltd.
P. O. Box 23
Honiara, Solomon Islands

Corriere della Sera
Via Solferino 28
20121 Milano, Italy

Cosentino, M.
Istituto di Scienze della Terra
Catania, Italy

Coulson, Frank I.
T C Mapping Project, Western Solomons
Ministry of Natural Resources
Honiara, Solomon Islands

Coulson, Kinsell (retired)
Mauna Loa Observatory
P. O. Box 275
Hilo, HI 96720 USA

Crandell, Dwight R. (retired)
USGS Engineering Geology Branch,
Box 25046, Denver Federal Center
Denver, CO 80225 USA

Crenshaw, William
Northwestern University Medical Center
303 E. Chicago Avenue
Chicago, IL 60657 USA

Creusot, Alain
División Hidrotérmico
Instituto Nicaraguense de Energía
Managua, Nicaragua

Cristofolini, R.
Istituto di Scienze della Terra
Catania, Italy

Crosson, Robert S.
Geophysics Program
University of Washington
Seattle, WA 98195 USA

Dahl, Arthur Lyon
Les Allues
73250 Saint-Pierre D'Albigny
France

Dailey, William
NASA Johnson Space Center
Houston, TX 77058 USA

Dalheim, Per
Loran Station-Jan Mayen
N-8013 Jan Mayen, Norway

Dalrymple, Brent
USGS, 345 Middlefield Road
Menlo Park, CA 94025 USA

Dalton, Russell
Alexandria Health Dept.
517 N St. Asaph Street
Alexandria, VA 22314 USA

Davies, J. N.
Geophysical Institute
University of Alaska
Fairbanks, AK 99775-0800 USA

Davis, William J.
New Zealand Geological Survey
Rotorua, New Zealand
(current address unknown)

Day, David
Isla Santa Cruz, Galápagos Islands
Ecuador

Dean, Bradley
4411 Iberville Street
New Orleans, LA 70119 USA

Dean, Roger H.
Cold Bay, Alaska USA
(current address unknown)

Decker, Robert
USGS, 345 Middlefield Road
Menlo Park, CA 94025 USA

Decobecq M., Dominique
Laboratoire de Petrologie et
Volcanologie
Université Paris Sud, Bot 504
91405 Orsay Cedex, France

de Dios Segura, Juan
Observatorio Vulcanológico y
Sismológico de Costa Rica
Universidad Nacional
Heredia, Costa Rica

Defense Scientific Establishment
Auckland, New Zealand

DeFoor, Thomas
Mauna Loa Observatory
P. O. Box 275
Hilo, HI 96720 USA

de la Cruz-Reyna, Servando
Instituto de Geofísica
Universidad Nacional Autónoma de
México
Ciudad Universitaria
04510 México, D.F. México

de Larouzière, F. D.
Institut Geologique Albert de Lapparent
21 Rue d'Assas
75270 Paris Cedex 06, France

Delarue, Jean-François
Observatoire Volcanologique du Piton
de la Fournaise
14 R.N. 3, 27ème km
97418 La Plaine des Cafres
Réunion Island

Delattre, Jean Noel
Institut de Physique du Globe
Université de Paris VI
4 Place Jussieu
75252 Paris Cedex 05, France

Delmotte, Catherine
Laboratoire de Pétrologie et
Volcanologie
Université Paris Sud, Bot 504
91405 Orsay Cedex, France

Delorme, Hughes
Observatoire Volcanologique du Piton
de la Fournaise
14 R.N. 3, 27 ème km
97418 La Plaine des Cafres
Réunion Island

De Nève, George A.
Volcanological Survey of Indonesia
Jalan Diponegoro 57
Bandung, Indonesia

Denlinger, Roger
USGS Cascades Volcano Observatory
5400 MacArthur Boulevard
Vancouver, WA 98661 USA

De Roy, Tui
Isla Santa Cruz, Galápagos Islands
Ecuador

Déruelle, Bernard
Département de Science de la Terre
Université de Yaoundé
B.P. 812, Yaoundé, Cameroon

de Saint Ours, Patrice
Rabaul Volcano Observatory
P. O. Box 386
Rabaul, Papua New Guinea

Deutche Presse-Agentur
866 United Nations Plaza, Rm 4056
New York, NY 10017 USA

Dibble, Raymond R.
Dept. of Geology
Victoria University
Private Bag, Wellington, New Zealand

DiFrancesco, M.
Istituto di Scienze della Terra
Università di Catania, Italy

Dion, Gerald
P. O. Box 417
Kenwood, CA 95452 USA

Dipartimento di Scienze della Terra
via S. Maria 53
Pisa, Italy

Donn, William (deceased)
Lamont-Doherty Geological
Observatory

Dorel, Jacques
Institut de Physique du Globe de Paris
4 Place Jussieu
75252 Paris Cedex 05, France

Doubik, Yuri M.
Institute of Volcanology
Piip Avenue 9
Petropavlovsk, Kamchatka 683006 USSR

Doukas, Michael
USGS Cascades Volcano Observatory
5400 MacArthur Boulevard
Vancouver, WA 98661 USA

Douglass, D.
2309 Naomi, Apt. F
Burbank, CA 91504 USA

Duffield, Wendell
USGS, 2255 N Gemini Drive
Flagstaff, AZ 86001 USA

Duggan, Toni A.
13323 Schwenger Place
Herndon, VA 22070 USA

du Plessis, M.
Geological Survey
Private Bag X112
Pretoria 0001, South Africa

Duque, Gonzalo
Universidad Nacional
Manizales, Colombia

Dvorak, John
USGS Hawaiian Volcano Observatory
P. O. Box 51
Hawaii National Park, HI 96718 USA

Dzurisin, Daniel
USGS Cascades Volcano Observatory
5400 MacArthur Boulevard
Vancouver, WA 98661 USA

Eaton, Gordon
Office of the President
Iowa State University
Ames, IA 50011 USA

Eggers, Albert
Dept. of Geology
University of Puget Sound
Tacoma, WA 98416 USA

Einarsson, Páll
Science Institute
University of Iceland
Dunhagi 3
107 Reykjavík, Iceland

Eissen, J. P.
ORSTOM, Centre de Noumea
B.P. A5
Noumea Cedex, New Caledonia

Eldredge, L. G.
U.N. Food & Agriculture Organization
Via delle Terme di Caracalla
01000 Roma, Italy

El Gráfico
14 Avenida 4-33, Zona 1
Guatemala City, Guatemala

Eliseussen, Tor
Loran Station-Jan Mayen
N-8013 Jan Mayen, Norway

Ellis, P. R.
Magadi Soda Company PLC
Magadi, Kenya

Emanuel, R.
USGS, 4200 University Drive
Anchorage, AK 99508 USA

Endo, Elliot
USGS Cascades Volcano Observatory
5400 MacArthur Boulevard
Vancouver, WA 98661 USA

Engel, Ralph
Instituto Linguístico del Verano
Apartado Postal 22067
14000 México D.F., México

Espinoza, R.
INETER, Dir. de Geología y Geofísica
Apartado 1761
Managua, Nicaragua

Evans, D.
U.S. Geological Survey
Adak Island, AK USA

Evans, E.
Instituto Costarricense de Electricidad
Apartado 10032
San José, Costa Rica

Evans, W. F. J.
ARPX-AES
4905 Dufferin St.
Downsview, Ontario M3H 5T4, Canada

Everingham, I. B.
Mineral Resources Department
Private Mail Bag, G.P.O.
Suva, Fiji

Ewert, J.
USGS Cascades Volcano Observatory
5400 MacArthur Boulevard
Vancouver, WA 98661 USA

Fajardo, Douglas
Sección Geológica
Instituto de Estudios Territoriales
Apartado Postal 1761
Managua, Nicaragua

Falsaperla, Susanna
Istituto Internazionale di Vulcanologia
Viale Regina Margherita, 6
95123 Catania, Italy

Federal Aviation Administration
800 Independence Avenue, SW
Washington, DC 20591 USA

Fedotov, Sergei
Institute of Volcanology
Piip Avenue 9
Petropavlovsk, Kamchatka 683006 USSR

Fegley, R. W.
NOAA/Air Resources Laboratory
325 Broadway
Boulder, CO 80303 USA

Fernández Soto, Erick
Observatorio Vulcanológico y Sismológico de Costa Rica
Universidad Nacional
Heredia, Costa Rica

Ferry, Guy
Atmospheric Experiments Branch
MS 245-5, NASA Ames Research Center
Moffett Field, CA 94035 USA

Feuillard, Michel
Observatoire Volcanologique de la Soufrière
97120 Le Pasnasse St. Claude, Guadeloupe

Fifita, Mr.
Chief Meteorological Officer
Nuku'alofa, Tonga

The Fiji Times
20 Gordon Street, GPO Box 1167
Suva, Fiji

Fiocco, Giorgio
Istituto di Fisica dell 'Atmosfera
P. Le Luigi Sturzo 31
00144 Roma, Italy

Fiske, Richard
NHB MS 119
Smithsonian Institution
Washington, DC 20560 USA

Fitton, Godfrey
Department of Geology
University of Edinburgh
Edinburgh EH9 3JW, Scotland

Flowers, Edwin
Solar Radiation Facility, NOAA/ARL
Boulder, CO 80303 USA

Forbes, Robert (retired)
Dept. of Geology and Geophysics
University of Alaska
Fairbanks, AK 99775 USA

Frazetta, G.
Istituto Internazionale di Vulcanologia
Viale Regina Margherita, 6
95123 Catania, Italy

Friend, James
Chemistry Dept.
Drexel University
Philadelphia, PA 19104 USA

Fuentealba Cifuentes, Gustavo
Depto. de Ciencias Físicas, Area Geofísica
Universidad de la Frontera
Casilla 54-D, Temuco, Chile

Fujita, Toshio
Minami-Yukigawa 4-7-15
Ohta-Ku, Tokyo 145, Japan

Fujiwara, Motowo
Dept. of Physics
Fukuoka University
Jonan-ku, Fukuoka 814-01, Japan

Fuller, William
NASA, Langley Research Center
Hampton, VA 23665 USA

Gallagher, Captain
Polynesian Airlines, Ltd.
P. O. Box 599
Apia, Western Samoa

Gallino, Gary
USGS Cascades Volcano Observatory
5400 MacArthur Boulevard
Vancouver, WA 98661 USA

Garcia, Michael
Dept. of Geology and Geophysics
University of Hawaii
Honolulu, HI 96822 USA

Garcia P., Nestor
Derivados del Azufre, S.A.
Apartado Aereo 590
Manizales, Colombia

Gartzeva, L.
Institute of Volcanology
Piip Avenue 9
Petropavlovsk, Kamchatka 683006 USSR

Gautheyrou, J.
ORSTOM, Centre des Antilles
B.P. 504
97165 Pointe-a-Pitre, Guadeloupe

Gemmell, J. Bruce
Geology Department
University of Tasmania
GPO Box 252C
Hobart, Tasmania 7001, Australia

Giggenbach, Werner
DSIR, P.O. Box 8005
Wellington, New Zealand

Giret, A.
Expéditions Polaires Françaises
47 Av. du Marechal Fayolle
Paris, France

Gleason, Richard
Exploration Department
Callahan Mining Co.
2452 US 41 West
Marquette, MI 49855

Gómez, Omar
Civil Defense Coordinator,
Carrera 13A No. 12A-17
Manizales, Colombia

Gonzalez, José
Centro de Investigaciones Geotécnicas
Depto. de Sismología
Apartado Postal 109
San Salvador, El Salvador

González-Ferrán, Oscar
Depto. de Geología y Geofísica
Universidad de Chile,
Casilla 27116,
27 Santiago, Chile

Gooding, James
Code SN2, NASA
Johnson Space Center
Houston, TX 77058 USA

Gould, Will
Satellite Data Services Division
Room 100, World Weather Bldg.
Washington, DC 20233 USA

Gras, John
CSIRO, Div. of Atmospheric Research
Private Bag 1
Mordialloc, Victoria 3195 Australia

Greene, H. Gary
USGS Branch of Pacific Marine Geology
345 Middlefield Road
Menlo Park, CA 94025 USA

Greenhall, F.
Park Board
Ohakune, New Zealand

Gresta, S.
Istituto di Scienze della Terra
Catania, Italy

Grönvold, Karl
Nordic Volcanological Institute
University of Iceland
101 Reykjavík, Iceland

Guendel, Federico
Sección de Sismología
Instituto Costarricense de Electricidad
Apartado 10032
San José, Costa Rica

Guest, John E.
University of London Observatory
Mill Hill Park
London NW7 2QS, England

Gundersen, Magne
Loran Station-Jan Mayen
N-8013 Jan Mayen, Norway

Gunther, G.
Lamont-Doherty Geological Observatory
Palisades, NY 10964 USA

Hadikusumo, Djayadi (deceased)
Geological Survey of Indonesia

Hadisantono, Rudy
Volcanological Survey of Indonesia
Jalan Diponegoro 57
Bandung, Indonesia

Hair, John
Marine Environmental Branch
P. O. Box 3-5000 (MEP)
Juneau, AK 99802 USA

Halbwachs, M.
Université de Chambéry
Chambéry, France

Hall, Minard L.
Inst. Geofísica, Escuela Politécnica
Casilla 2759
Quito, Ecuador

Haller, Dennis
Forecast Branch
National Meteorological Center
Room 402, World Weather Bldg.
Washington, DC 20233 USA

Halsor, Sid P.
Dept. of Earth & Environmental Sciences
Wilkes College
Wilkes-Barre, PA 18766 USA

Hammouya, G.
Laboratoire de Physique du Globe
97120 Saint Claude, Guadeloupe

Hanatani, Ronald
USGS Hawaiian Volcano Observatory
P. O. Box 51
Hawaii National Park, HI 96718 USA

Handler, Paul
Physics Dept.
University of Illinois
Urbana, IL 61801 USA

Hardy, E. F.
Victoria University
Private Bag, Wellington, New Zealand

Harlow, David
USGS, 345 Middlefield Road
Menlo Park, CA 94025 USA

Harvey, Samuel
Civil Defence
Whakatane, New Zealand

Hauksson, Egill
Lamont-Doherty Geological Observatory
Palisades, NY 10964 USA

Hauptmann, J.
Lamont-Doherty Geological Observatory
Palisades, NY 10964 USA

Havskov, Jens
Seismological Observatory
Universetetet I Bergen
Allegt. 41, 5000 Bergen, Norway

Hawkins, Jamie
NOAA/NESDIS
Room 2064, Federal Bldg. no. 4
Washington, DC 20233 USA

Hay, Richard
Dept. of Geology
University of Illinois
Urbana, IL 61801 USA

Hayashida, Sachiko
Water Research Institute
Nagoya University
Chikusa-ku, Nagoya 464 Japan

Hayes, Dennis
Atmospheric Experiments Branch
MS 245-5, NASA Ames Research Center
Moffett Field, CA 94035 USA

Hazlett, Richard
Dept. of Geology
Pomona College
Claremont, CA 91711 USA

Hearn, B. Carter, Jr.
USGS National Center, MS 951
Reston, VA 22092 USA

Heckman, Gary
NOAA/Space Environment Services Center
325 Broadway
Boulder, CO 80303 USA

Heiken, Grant H.
Los Alamos National Laboratory
Mail Stop 575
Los Alamos, NM 87545 USA

Helfert, Michael
NOAA, 1050 Bay Area Boulevard
Houston, TX 77058 USA

Heliker, Christina
USGS Hawaiian Volcano Observatory
P. O. Box 51
Hawaii National Park, HI 96718 USA

Hemond, C.
Observatoire Volcanologique du Piton de la Fournaise
14 R.N. 3, 27 ème km
97418 La Plaine des Cafres
Réunion Island

Herd, Darrell G.
USGS National Center, MS 922
Reston VA 22092 USA

Hide, C. G.
Office of the Scientific Counsellor
South African Embassy
Suite 300, 2555 M St. NW
Washington, DC 20037 USA

Hill, David
USGS, 345 Middlefield Road, MS 77
Menlo Park, CA 94025 USA

Hirn, Alfred
Institut de Physique du Globe
4 Place Jussieu
75252 Paris Cedex 05, France

Hirono, Motokazu (retired)
Dept. of Physics
Kyushu University
Fukuoka 812, Japan

Hoadley, Donald
Reeve Aleutian Airways
Dutch Harbor, AK 99692 USA

Hodder, Robert W
Dept. of Geology
University of Western Ontario
London, Ontario, N6A 5B7 Canada

Hodgson V., Glen
INETER, Depto. de Geología
Apartado 1761
Managua, Nicaragua

Hoeck, Hendrik
Fac. Biologie
Universität Konstanz
Postfach 7733
Konstanz, West Germany

Hofmann, David J.
Dept. of Physics and Astronomy
University of Wyoming
Laramie, WY 82071 USA

Holcomb, Robin
(USGS) School of Oceanography
University of Washington
Seattle, WA 98195 USA

Honma, Kenneth
USGS Hawaiian Volcano Observatory
P. O. Box 51
Hawaii National Park, HI 96718 USA

Hooper, Earl
Synoptic Analysis Branch, NOAA/NESDIS
Room 401, World Weather Building
Washington, DC 20233 USA

Hopson, Clifford
Dept. of Geological Sciences
University of California at Santa Barbara
Goleta, CA 93106 USA

Horan, Hume, Ambassador
American Embassy
Rue Nachtigal
B.P. 817, Yaoundé, Cameroon

Houghton, Bruce F.
New Zealand Geological Survey
P. O. Box 499
Rotorua, New Zealand

How, Karl
Hawaiian Suger Planters Association
P. O. Box 1057
Aiea, HI 96701 USA

Hudnut, Ken
Lamont-Doherty Geological Observatory
Palisades, NY 10964 USA

Hughes, D.
Portsmouth Polytechnic
Museum Road
Portsmouth PO1 2QQ, England

Humphrey, T. J.
Gulf Oil
Douala, Cameroon

Il Progresso Italo-Americano
15 Bland Street
Emerson, NJ 07630 USA

Imsland, Páll
Nordic Volcanological Institute
University of Iceland
101 Reykjavík, Iceland

Indonesia Times
Jalan Letjen S. Parman Kav. 72, Slipi
Jakarta, Indonesia

INGEOMINAS
Apartado Aereo 1296
Manizales, Colombia

Institut de Physique du Globe de Paris
Direction des Observatoires Volcanologiques
4 Place Jussieu
75252 Paris Cedex 05, France

Institute de Recherches Scientifiques Afrique Centrale
Lwiro, D/S Bukavu, Zaire

Istituto di Geochimica Applicata
Università di Palermo,
Palermo, Italy

Istituto Internazionale di Vulcanologia
Viale Regina Margherita, 6
95123 Catania, Italy

Ivanov, Boris V.
Institute of Volcanology
Piip Avenue 9
Petropavlovsk, Kamchatka 683006 USSR

Iwasaka, Yasunobu
Water Research Institute
Nagoya University
Chikusa-ku, Nagoya 264, Japan

Iwatsubo, Eugene
USGS Cascades Volcano Observatory
5400 MacArthur Boulevard
Vancouver, WA 98661 USA

Izumo, Akiko
Yokohama Science Center
5-2-1 Yokodai, Isogo-ku
Yokohama 235, Japan

Jäger, Horst
Fraunhofer-Institut für Atmosphärische Umweltforschung
Kreuzeckbahnstrasse 19
D-8100 Garmisch-Partenkirchen
West Germany

Jakarta Domestic Radio Service
Jakarta, Indonesia

Janda, Richard
USGS Cascades Volcano Observatory
5400 MacArthur Boulevard
Vancouver, WA 98661 USA

Japan Meteorological Agency
Volcanological Center
1-3-4 Ote-machi, Chiyoda-ku
Tokyo 100, Japan

Japan Times
4-5-4 Shibaura, Minato-ku
Tokyo 108, Japan

Jaramillo, Luis
INGEOMINAS
Diagonal 53, no. 34-53
Bogotá, Colombia

Javoy, M.
Lab. de Géochimie Isotopique
Université de Paris VII
2 Place Jussieu
75221 Paris Cedex 05, France

Jerez, Debbie Reid
Roy F. Weston, Inc.
955 L'Enfant Plaza, 8th Floor
Washington, DC 20024 USA

Jezek, Peter
167 The Strand
Winthrop, MA 02152 USA

Johnson, H. M.
Appl. Lab., NOAA/NESDIS
Washington, DC 20233 USA

Johnston, David A. (deceased)
USGS Cascades Volcano Observatory

Jones, Peter
c/o Olduvai Gorge
P. O. Box 7
Ngorongoro, Tanzania

Jones, Robert
Air Tonga,
Nuku'alofa, Tonga

Jonientz-Trisler, Chris
Geophysics Program
University of Washington
Seattle, WA 98195 USA

Kadin, Charles
Synoptic Analysis Branch
NOAA/NESDIS
Room 401, World Weather Bldg.
Washington, DC 20233 USA

Kamid, Mohammad
Volcanological Survey of Indonesia
Jalan Diponegoro 57
Bandung, Indonesia

Kaminuma, Katsutada
Nat. Inst. of Polar Research
Kaga 1-9-10
Itabashi-ku, Tokyo, Japan

Kampala News Service
Kampala, Uganda

Karst, Otto
Synoptic Analysis Branch
NOAA/NESDIS
Room 401, World Weather Bldg.
Washington, DC 20233 USA

Kasser, M.
Observatoire Volcanologique du Piton de la Fournaise
14 R.N. 3, 27 ème km
97418 La Plaine des Cafres
Réunion Island

Katsui, Yoshio
Dept. of Geology and Mineralogy
Hokkaido University
Sapporo, Japan

Keen, Richard
34296 Gap Road
Golden, CO 80403 USA

Kelly, Madelon
The Grier School
Tyrone, PA 16686 USA

Kergomard, Dr.
Elf-Serepca
B.P. 2214
Doula, Cameroon

Khudyakov, V.
Sovetskaya Rossiya
Moscow, USSR

Kieffer, Guy
Dépt. de Géologie et Minéralogie
5, Rue Kessler
63038 Clermont-Ferrand Cedex, France

Kienle, Jürgen
Geophysical Institute
University of Alaska
Fairbanks AK 99701 USA

Kigali News Service
Rwanda Office of Information
P. O. Box 83, Kigali, Rwanda

Kilburn, Christopher
Dipartimento di Geofisica e Vulcanologia
Università degli Studi di Napoli
Largo S. Marcellino 10
80138 Napoli, Italy

Kirkemo, Rolf
Loran Station-Jan Mayen
N-8013 Jan Mayen, Norway

Kirkwood, Bruce
Solomon Islands Airways, Ltd.
P. O. Box 23
Honiara, Solomon Islands

Kirsanov, I. T.
Institute of Volcanology
Piip Avenue 9
Petropavlovsk, Kamchatka 683006 USSR

Klein, Fred
345 Middlefield Road
Menlo Park, CA 94025 USA

Klick, Donald
DOSECC
Suite 700, 1755 Massachusetts Ave. NW
Washington, DC 20036 USA

Kobayashi, Takeshi
Toyama University
Toyama, Japan

Koeppen, Robert
USGS National Center, MS 954
Reston, VA 22092 USA

Komiya, Manabu
Volcanological Center
Japan Meteorological Agency
1-3-4 Ote-machi, Chiyoda-ku
Tokyo 100, Japan

Kompas Newspaper
Jalan Pal Merah Selatan, no. 26/28
P. O. Box 615 DAK
Jakarta, Indonesia

Koreis, Rocke
2401 Larson Road
Yakima, WA 98908 USA

Kossarias, Tom
Federal Aviation Administration
Washington, DC 20591 USA

Koyanagi, Robert Y.
USGS Hawaiian Volcano Observatory
P. O. Box 51
Hawaii National Park, HI 96718 USA

Kozhemyaka, N. N.
Institute of Volcanology
Piip Avenue 9
Petropavlovsk, Kamchatka 683006 USSR

Krafft, Maurice and Katia
Equipe Vulcain
B.P. 5
68700 Cernay, France

Krishna, Ram
Fiji Meteorological Service
Nandi Airport, Fiji

Krueger, Arlin
Code 614
NASA Goddard Space Flight Center
Greenbelt, MD 20771 USA

Krueger, Arthur
Rt. 1, Box 711-5
Accokeek, MD 20607 USA

Kusselson, Sheldon
Synoptic Analysis Branch
NOAA/NESDIS
Room 401, World Weather Bldg.
Washington, DC 20233 USA

Kyle, Philip
Dept. of Geoscience
New Mexico Inst. of Mining and Tech.,
Socorro NM 87801 USA

Kyodo Radio and News Service
2-2-5 Toranomon, Minato-ku
Tokyo 107, Japan

Lahr, J.
USGS, 345 Middlefield Road
Menlo Park, CA 94025 USA

Lalanne, F. X.
Observatoire Volcanologique du Piton de la Fournaise
14 R.N. 3, 27 'eme km
97418 La Plaine des Cafres,
Réunion Island

Lamb, Hubert H. (retired)
Climatic Research Unit
School of Environmental Sciences
University of East Anglia
Norwich NR4 7TJ, England

LaPorte, H. Ross
NOAA/NESDIS
World Weather Building
Washington, DC 20233 USA

La Stampa
Via Marenco 32,
10126 Torino, Italy

Latin Radio Network
Buenos Aires, Argentina

Latter, John H.
DSIR Geophysics Division
P. O. Box 1320
Wellington, New Zealand

Lawrence (Dittrick), Elizabeth A.
P. O. Box 501
Burton, OH 44201 USA

Legg, E.
NOAA/NESDIS, Synoptic Analysis Branch
Room 401, World Weather Building
Washington, DC 20233 USA

LeGuern, François
CNRS
91190 Gif-Sur-Yvette, France

Le Mouel, J.
Direction des Observatoires Volcanologiques
Institut de Physique du Globe
4, Place Jussieu
75252 Paris Cedex 05, France

Lenat, Jean-François
Centre de Recherches Volcanologiques
5 Rue Kessler
63038 Clermont-Ferrand Cedex, France

Les Nouvelles Caledoniennes
34 Rue de la Republique
Noumea, New Caledonia

Lester, Richard
345 Middlefield Road
Menlo Park, CA 94025 USA

Lewis, Glyn
Sengo Safaris Ltd.
P. O. Box 180
Arusha, Tanzania

Livingston, William
Kitt Peak National Observatory
Box 26732
Tucson, AZ 85726 USA

Lloyd, E. F.
New Zealand Geological Survey
P. O. Box 499
Rotorua, New Zealand

Lockwood, John P.
USGS Hawaiian Volcano Observatory
P. O. Box 51
Hawaii National Park, HI 96718 USA

Lombardo, G.
Istituto di Scienze della Terra
Catania, Italy

Lomnitz, Cinna
Instituto de Geofísica
Universidad Nacional Autónoma de México
Ciudad Universitaria
04510 México, D.F. México

Londoño C., Adela
Universidad Nacional de Colombia
Seccional Manizales
Manizales, Colombia

Londoño, Bernardo
Observatorio Vulcanológico de Colombia
INGEOMINAS
Apartado Aereo 1296
Manizales, Colombia

Lonne, Dr.
Ardic Cameroun Aérienne Cie.
Douala, Cameroon

Lopes, Rosaly
Planetary Image Centre
University of London Observatory
33-35 Daws Lane, Mill Hill,
London NW7 4SD, England

López E., Leopoldo
Depto. de Geología y Geofísica
Universidad de Chile
Casilla 27116, 27 Santiago, Chile

López, P.
Charles Darwin Research Station
Galápagos Islands, Ecuador
(current address unknown)

López R., Alfonso
INGEOMINAS, Diagonal 53, No. 34-53
Bogotá, Colombia

Lorfsher, Dr.
Geophysics Division, Elf-Serepca
B.P. 2214, Doula, Cameroon

Los Alamos National Laboratory
P. O. Box 1663
Los Alamos, NM 87545 USA

Lowe, Donald R.
Dept. of Geology
Louisiana State University
Baton Rouge, LA 70803 USA

Lowenstein, Peter
Rabaul Volcano Observatory
P. O. Box 386
Rabaul, Papua New Guinea

Luhr, James
Dept. of Earth and Planetary Sciences
Campus Box 1169
Washington University
St. Louis, MO 63130 USA

Lum, J.
Ministry of Energy and Mineral Resources
Private Mail Bag
Suva, Fiji

Luongo, Giuseppe
Osservatorio Vesuviano
Largo San Marcellino 10
80138 Napoli, Italy

Lytle, Joseph P.
Samson Resources Co.
2 W 2nd Street
Tulsa, OK 74103 USA

MacFarland, Craig
836 Mabelle
Moscow, ID 83843 USA

MacFarlane, A.
Dept. of Geology, Mines, and Rural Water Supplies
GPO, Port Vila, Vanuatu
(current address unknown)

Machta, Lester
NOAA/Air Resources Laboratory
Silver Spring, MD 20910 USA

MacKenney, Alfredo
2[A] Avenida 10-60, Zona 1
Guatemala City, Guatemala

MacLeod, Norman (retired)
USGS Cascades Volcano Observatory
5400 MacArthur Boulevard
Vancouver, WA 98661 USA

Maddrea, George
NASA, Langley Research Center
Hampton, VA 23665 USA

Mahieu, L. J.
Division of Oceanographic Research
Royal Netherlands Meteorological Institute
Postbus 201, 3730 AE De Bilt, Netherlands

Maillet, Patrick
ORSTOM, Centre de Noumea
B.P. A5
Noumea Cedex, New Caledonia

Malavassi R., Eduardo
Observatorio Vulcanológico y Sismológico de Costa Rica
Universidad Nacional
Heredia, Costa Rica

Maldonado, Lucho
Metropolitan Touring
P. O. Box 2542, Avenida Amazonas 239
Quito, Ecuador

Malin, Michael
Department of Geology
Arizona State University
Tempe, AZ 85281 USA

Malinconico, Lawrence L.
Dept. of Geology
Southern Illinois University
Carbondale, IL 62901 USA

Malone, Steven
Geophysics Program
University of Washington
Seattle, WA 98195 USA

Mandraguna, O.
Volcanological Survey of Indonesia
Jalan Diponegoro 57
Bandung, Indonesia

Marshall, Dianne
Geophysical Institute
University of Alaska
Fairbanks, AK 99701 USA

Martin, Dennis
Minerals Division
Minnesota Dept. of Natural Resources
1525 E 3rd Ave.
Hibbing, MN 55746 USA

Martinez, Carlos
INSIVUMEH
7[A] Avenida 14-57, Zona 13
Guatemala City, Guatemala

Matahelumual, J.
Volcanological Survey of Indonesia
Jalan Diponegoro 57
Bandung, Indonesia

Matson, Michael
Interactive Processing Branch
NOAA/NESDIS
Room 510, World Weather Bldg.
Washington, DC 20233 USA

Matsumoto, T.
Institute for Geophysics
University of Texas
Austin, TX 78759 USA

Matteson, Lois
American Consulate
21 Avenue du General de Gaulle
B. P. 4006, Douala, Cameroon

Mauk, Fred
5907 Swiss Ave.
Dallas, TX 75214 USA

McCormick, M. Patrick
NASA, Langley Research Center
Hampton, VA 23665 USA

McCue, Kevin
Bureau of Mineral Resources Room 111
P. O. Box 378
Canberra A.C.T. 2601 Australia

McCutchan, A.
Dept. of Geology, Mines, and Rural Water Supplies
Port Vila, Vanuatu

McGee, Kenneth
USGS Cascades Volcano Observatory
5400 MacArthur Boulevard
Vancouver, WA 98661 USA

McGetchin, T. (deceased)
Los Alamos National Laboratory

McKee, C. O.
Rabaul Volcano Observatory
P. O. Box 386
Rabaul, Papua New Guinea

McNutt, Stephen
Division of Mines and Geology
630 Bercutt St.
Sacramento, CA 95814 USA

Medina, Francisco
Instituto de Geofísica
Universidad Nacional Autónoma de México
Ciudad Universitaria
04510 México D.F., México

Medina, Pablo
Oficina Nacional-Atención de Emergencias
Presidencia de la República
Casa Narino
Bogotá, D.E. Colombia

Meinel, Aden and Marjorie
MS 186-134, Jet Propulsion Laboratory
4800 Oak Grove Drive
Pasadena, CA 91109 USA

Mejía, Germán
Departamento de Agronomía
Universidad Nacional
Manizales, Colombia

Melson, William G.
NHB MS 119
Smithsonian Institution
Washington, DC 20560 USA

Melbourne Overseas Service
Australian Broadcasting Commission
Melbourne, Australia

Mena J., Manuel
Instituto de Geofísica
Universidad Nacional Autónoma de México
Ciudad Universitaria
04510 México, D.F., México

Mendonça, Bernard
NOAA/ERL, Air Res. Lab.
Code RF 3292
325 Broadway
Boulder, CO 80303 USA

Menyailov, I. A.
Institute of Volcanology
Piip Avenue 9
Petropavlovsk, Kamchatka 683006 USSR

Meyer, James
COMNAVMAR
FPO San Francisco, CA 96630 USA
(current address unknown)

Miller, C. Dan
USGS Cascades Volcano Observatory
5400 MacArthur Boulevard
Vancouver, WA 98661 USA

Miller, David C., Ambassador
American Embassy
Dar Es Salaam, Tanzania

Miller, Thomas
USGS, 4200 University Drive
Anchorage, AK 99508 USA

Mitchell, Murray
NOAA, Code R-32
8060 13th Street
Silver Spring, MD 20910 USA

Mitrohartono, F.X. Suparban
Volcanological Survey of Indonesia
Jalan Diponegoro 57
Bandung, Indonesia

Modjo, Subroto
Volcanological Survey of Indonesia
Jalan Diponegoro 57
Bandung, Indonesia

Moen, Finn
Loran Station-Jan Mayen
N-8013 Jan Mayen, Norway

Molina, Enrique
INSIVUMEH
7^A Avenida 14-57, Zona 13
Guatemala City, Guatemala

Monoarta, Muslim
Volcanological Survey of Indonesia
Jalan Diponegoro 57
Bandung, Indonesia

Montaggioni, L.
Laboratoire de Géologie
Université de la Réunion
B.P. 5, 94790 Ste. Clothilde
Réunion Island

Monzier, M.
ORSTOM, Centre de Noumea
B.P. A5
Noumea Cedex, New Caledonia

Moore, James G.
USGS, 345 Middlefield Road
Menlo Park, CA 94025 USA

Moreno R., Hugo
Depto. de Geología y Geofísica
Universidad de Chile
Casilla 13518, Correo 21
Santiago, Chile

Morgan, Guy
Peninsula Airways
Box 32
Cold Bay AK 99571 USA

Mori, James
USGS Office of Earthquakes, Volcanoes, and Engineering
525 S Wilson Ave.
Pasadena, CA 91106 USA

Morris, R. D.
P. O. Box 22468
G.M.F. Guam 96921 USA

Mort, Mona A.
Division of Biological Sciences
University of Michigan
Ann Arbor, MI 48109 USA

Moscow Television Service
Moscow, USSR

Mota P., R.
Instituto de Geofísica
Universidad Nacional Autónoma de México
Ciudad Universitaria
04510 México, D.F., México

Mousnier-Lompré, P.
Servizio Sismico Regionale
Messina, Italy

Mroz, Eugene J.
Los Alamos National Laboratory
P. O. Box 1663
Los Alamos, NM 87545 USA

Mullineaux, Donal (retired)
USGS, Denver Federal Center
Box 25046
Denver, CO 80225 USA

Mullins, Clint
USGS Cascades Volcano Observatory
5400 MacArthur Boulevard
Vancouver, WA 98661 USA

Muñoz, Robert
NASA, Ames Research Center
Moffett Field, CA 94035 USA

Murray, John B.
"Field End", Marshall's Heath
Wheathampshead, Herts., England

Mutschlecner, Paul
Los Alamos National Laboratory
ESS-5 Mail Stop F665
Los Alamos, NM 87545 USA

Myers, Bobbie
USGS Cascades Volcano Observatory
5400 MacArthur Boulevard
Vancouver, WA 98661 USA

Nairn, Ian A.
New Zealand Geological Survey
P. O. Box 499
Rotorua, New Zealand.

Nakamura, Kazuaki (deceased)
Earthquake Research Institute, Tokyo

Nakata, Jennifer
USGS Hawaiian Volcano Observatory
P. O. Box 51
Hawaii National Park, HI 96718 USA

Nania, James
Deaconess Hospital
West 800 5th Avenue
Spokane, WA 99210 USA

Nappi, G.
Istituto Internazionale di Vulcanologia
Viale Regina Margherita, 6
95123 Catania, Italy

Naslund, H. Richard
Dept. of Geological Sciences
State University of New York
Binghamton, NY 13901 USA

National Earthquake Information Center
USGS, Denver Federal Center, Box 25046
Denver, CO 80225 USA

Neal, Christina
Dept. of Geological Sciences
University of California at Santa Barbara
Goleta, CA 93106 USA

Nelson, Jerry
Travis AFB CA
(current address unknown)

Nelson, S.
USGS, 4200 University Drive
Anchorage, AK 99508 USA

Neri, Giancarlo
Istituto Internazionale di Vulcanologia
Viale Regina Margherita, 6
95123 Catania, Italy

Nercessian, A.
Institut de Physique du Globe
Direction des Observatoires Volcanologiques
4, Place Jussieu
75252 Paris Cedex 05, France

Newhall, Chris
USGS National Center, MS 905
Reston, VA 22092 USA

Newport Geophysical Observatory
USGS, Newport, WA 99156 USA

Newton, Paul S. (deceased)
Antigua, Guatemala

New York Times
229 W. 43rd Street
New York, NY 10036 USA

New Zealand Geological Survey
P. O. Box 499
Rotorua, New Zealand

Ngungi, Dr.
Ministère des Mines et Energie
B. P. 70, Yaoundé, Cameroon

Nolf, Bruce
Central Oregon College
Bend, OR 97701 USA

Norris, Robert
Geophysics Program
University of Washington
Seattle, WA 98195 USA

Notimex Radio
Morena 110, Col. Vel Valle
03100 México D.F., México

Numero Uno
Avenida Norte 1^A
Tuxtla Gutiérrez, México

Nunnari, G.
Istituto Internazionale di Vulcanológia
Viale Regina Margherita, 6
95123 Catania, Italy

Nyamweru, Celia
Dept. of Geography
Kenyatta University
P. O. Box 43844
Nairobi, Kenya

Observatoire Volcanologique du Piton de la Fournaise
14 R.N. 3, 27 ème km
97418 La Plaine des Cafres
Réunion Island

Observatorio Vulcanológico de Colombia
INGEOMINAS
Apartado Aereo 1296
Manizales, Colombia

Observatorio Vulcanológico y Sismológico de Costa Rica
Universidad Nacional
Heredia, Costa Rica

O'Donnell, W.
University of London Observatory
Mill Hill Park
London NW7 2QS, England

Okamura, Arnold
USGS Hawaiian Volcano Observatory
P. O. Box 51
Hawaii National Park, HI 96718 USA

Okamura, Reginald T.
USGS Hawaiian Volcano Observatory
P. O. Box 51
Hawaii National Park, HI 96718 USA

Ono, Akira
Water Research Institute
Nagoya University
Chikusa-ku, Nagoya 464 Japan

Osborn, Mary
NASA, Langley Research Center
Hampton, VA 23665 USA

Osservatorio Vesuviano
Largo S. Marcellino 10,
80138 Napoli, Italy

Otway, Peter M.
New Zealand Geological Survey
Wairakei, Taupo, New Zealand

Overbeck, Vernon
Atmospheric Experiments Branch
MS 245-5, NASA Ames Research Center
Moffett Field, CA 94035 USA

Palmer, H. W.
New Zealand Post Office
Rotorua, New Zealand

Pambrun, C.
Centre National d'Etudes des Télécommunications
22301 Lannion, France

Paquette, John
Synoptic Analysis Branch
NOAA/NESDIS
Room 401, World Weather Building
Washington, DC 20233 USA

Pardyanto, Liek
Volcanological Survey of Indonesia
Jalan Diponegoro 57
Bandung, Indonesia

Parmenter, Frances
Satellite Applications Laboratory
NOAA/NESDIS
Room 601, World Weather Bldg.
Washington, DC 20233 USA

Parnell, Jr., Roderic A.
Dept. of Geology
Northern Arizona University
Flagstaff, AZ 88011 USA

Parodi Isolabella, Alberto
P. O. Box 208
Arequipa, Perú

Parra, Eduardo
Observatorio Vulcanológico de Colombia
INGEOMINAS
Apartado Aereo 1296
Manizales, Colombia

Patané, G.
Istituto di Scienze della Terra
Catania, Italy

Patterson, J. A.
Medical School, Box 150, HB 7000
Dartmouth College
Hanover, NH 03755 USA

Pece, R.
Osservatorio Vesuviano
Largo San Marcellino 10
80138 Napoli, Italy

Peña, Olimpio
Philippine Institute of Volcanology
5th Floor, Hizon Bldg.
29 Quezon Ave.
Quezon City, Philippines

Person, Waverly
USGS NEIC
MS 967, Denver Federal Center
Box 25046
Denver, CO 80225 USA

Peterson, Donald
USGS, 345 Middlefield Road
Menlo Park, CA 94025 USA

Philippine Institute of Volcanology
5th Floor, Hizon Bldg.
29 Quezon Avenue
Quezon City, Philippines

Pilipenko, V.
Institute of Volcanology
Piip Avenue 9
Petropavlovsk, Kamchatka 683006 USSR

Plumley, P.
239 Fairway Court
Santa Rosa, CA 95405 USA

PNA Radio
Manila, Philippines

Polian, Dr.
Centre National de la Recherche Scientifique
91190 Gif-Sur-Yvette, France

Pollack, James P.
NASA Ames Research Center
Moffett Field, CA 94035 USA

Portland International Airport
Portland, OR 97218 USA

Pottier, Yves
Laboratoire de Volcanologie
Bat. 504-Faculté des Sciences
Université Paris-Sud,
91405 Orsay, France

Pozzi, J. P.
Laboratoire de Magnétisme
Institut de Physique du Globe
4 Place Jussieu
75252 Paris Cedex 05, France

Presgrave, Bruce
USGS NEIS
MS 967, Denver Federal Center
Box 25046
Denver, CO 80225 USA

Priestly, Keith
Seismology Laboratory
University of Nevada
Reno, NV 89507 USA

Proffett, J.
Anaconda Co.
Anchorage, AK 99501 USA

Programa de Investigaciones Vulcanológicas y Sísmicas
Universidad Nacional
Heredia, Costa Rica

Programme Interdisiplinaire de Recherche sur la Prévision et la Surveillance des Eruptions Volcaniques
77, Avenue Denfert-Rochereau
75014 Paris, France

Prosser, Jerome T.
Texas Gulf
239 S Elliot Road
Chapel Hill, NC 27514 USA

Pullen, A.
Department of Geology
Imperial College
London NW7 2BP, England

Punongbayan, Raymundo
Philippine Institute of Volcanology
5th Floor, Hizon Bldg
29 Quezon Avenue
Quezon City, Philippines

Quevec Robles, Edgar René
INSIVUMEH
7ᴬ Avenida 14-57, Zona 13
Guatemala City, Guatemala

Rabaul Volcano Observatory
P. O. Box 386
Rabaul, Papua New Guinea

Ragent, Boris
Atmospheric Experiments Branch
MS 245-5, NASA Ames Research Center
Moffett Field, CA 94035 USA

Ramírez, J. Emilio (deceased)
Universidad Javeriana, Columbia

Ramón, Patricio
c/o Inst. Geofísica, Escuela Politécnica
Casilla 2759
Quito, Ecuador

Rampino, Michael
NASA
2880 Broadway
New York, NY 10025 USA

Reck, Gunther
Charles Darwin Research Station
Isla Santa Cruz, Galápagos Islands
Ecuador

Reeder, John W.
Alaska Div. of Geological and Geophysical Surveys
P. O. Box 772116
Eagle River, AK 99577 USA

Reiter, Reinhold
Consulting Bureau Reiter
Fritz-Müller Strasse 54
8100 Garmisch-Partenkirchen
West Germany

La República
P. O. Box 2130
San José, Costa Rica

Réseau Sismique Polynésien
Laboratoire de Géophysique
B. P. 640
Papéete, Tahiti

Reuters News Service
1700 Broadway
New York, NY 10019 USA

Rhodes, J. Michael
Dept. of Geology
University of Massachusetts
Amherst, MA 01003 USA

Richmond, Ronald N.
Mineral Resources Department
Suva, Fiji
(current address unknown)

Riehle, James
USGS, 4200 University Drive
Anchorage, AK 99508 USA

Riffo A., Pedro
Depto. de Ciencias Físicas, Area Geofísica
Universidad de la Frontera
Casilla 54-D, Temuco, Chile

Riley, Francis
USGS Water Resources Division
Denver Federal Center, Box 25046
Denver, CO 80225 USA

Rist, Sigurjon
National Energy Authority
Reykjavík, Iceland

Rivera, Antonio
Depto. de Geología
Universidad de Caldas
Manizales, Colombia

Roberts, G.
Cold Bay Weather Station,
Cold Bay, AK 99571 USA

Rodda, Peter
Mineral Resources Department
Private Mail Bag
Suva, Fiji

Romano, Romolo
Istituto Internazionale di Vulcanologia
Viale Regina Margherita, 6
95123 Catania, Italy

Rose, William I.
Dept. of Geology & Geological Engineering
Michigan Technological University
Houghton, MI 49931 USA

Rosen, James
Dept. of Physics and Astromony
University of Wyoming
Laramie, WY 82071 USA

Rousset, Dr.
Geophysics Division, Elf-Serepca
B. P. 2214
Doula, Cameroon

Rowley, Keith
Seismic Research Unit
University of the West Indies
St. Augustine, Trinidad

Russell, Philip B.
Atmospheric Science Center
SRI International
333 Ravenswood Avenue
Menlo Park, CA 94025 USA

Russell, Shaun
Institute of Environmental Sciences
University of Orange Free State
Bloemfontein 9300, South Africa

Russell-Robinson, Susan
USGS National Center, MS 951
Reston, VA 22092 USA

Ryall, Alan
Dept. of Geological Sciences
University of Nevada
Reno, NV 89507 USA

Sabroux, J. C.
Centre des Faibles Radioactivités
91190 Gif-Sur-Yvette, France

Saemundsson, Kristján
Division of Natural Heat
National Energy Authority
Grensásvegi 9
IS-108 Reykjavík, Iceland

Saenz R., Rodrigo
Dir. de Geología, Minas, y Petroleo
Ministerio de Energía y Minas
MEIC Apt. 10216
San José, Costa Rica

Salazar A., Bernardo
Central Hidroeléctrica de Caldas, S.A.
Apartado Aereo 83
Manizales, Colombia

Sanders, John E.
Dept. of Geology, Barnard College
Columbia University
606 W 12th St.
New York, NY 10027 USA

Sanderson, T.
Department of Geology
Imperial College
London NW7 2BP, England

Sandvik, Kaare
Loran Station-Jan Mayen
N-8013 Jan Mayen, Norway

Saos, J. L.
Mineral Resources Dept.
B. P. 637
Port Vila, Vanuatu

Savage, James C.
USGS, 345 Middlefield Road, MS 77
Menlo Park, CA 94025 USA

Sawada, Yoshiro
Seis. and Volc. Management Division
Japan Meteorological Agency
1-3-4 Ote-machi, Chiyoda-ku
Tokyo 100, Japan

Scalia, S.
Gruppo Ricerca Speleologica
Acireale, Italy

Scandone, Roberto
Dipartimento di Geofisica e Vulcanologia
Università degli Studi di Napoli
Largo S. Marcellino 10
80138 Napoli, Italy

Scarpinati, G.
Gruppo Ricerca Speleologica
Acireale, Italy

Schaaf, Fred
R.D. 2, Box 248
Millville, NJ 08332 USA

Schmincke, Hans-Ulrich
Institut für Mineralogie
Ruhr-Universität Bochum
Postfach 102148
D-4630 Bochum, West Germany

Schnell, Russell
GMCC/Analysis and Interpretation
NOAA, 325 Broadway
Boulder, CO 80303 USA

Scott, Brad J.
New Zealand Geological Survey
P. O. Box 499, Rotorua, New Zealand

Scott, S.
University of London Observatory
Mill Hill Park
London, NW7 2QS England

Sedlacek, William A.
Los Alamos National Laboratory
P. O. Box 1663, Mail Stop 514
Los Alamos, NM 85545 USA

Self, Stephen
Dept. of Geology
University of Texas
Arlington, TX 76019 USA

Sercano Radio Network (defunct)
San Salvador, El Salvador

Shackelford, Daniel
3124 E. Yorba Linda Boulevard, Apt. H-33
Fullerton, CA 92631 USA

Shapar, V. N.
Institute of Volcanology
Piip Avenue 9
Petropavlovsk, Kamchatka 683006 USSR

Shaw, Robert R. R.
International Air Transport Association
2000 Peel Street
Montreal H3A 2R4 Canada

Shepherd, John
Seismic Research Unit
University of the West Indies
St. Augustine, Trinidad

Shepherd, P. J. R.
1 SQD ALM LDR
Royal New Zealand Air Force
Whenuapai, Auckland, New Zealand

Sheppard, D. S.
DSIR, Chemistry Division
Gracefield, New Zealand

Sherburn, Steven
DSIR, Geophysics Division
Private Bag
Wairakei, Taupo, New Zealand

Sherburne, Roger
California Div. of Mines and Geology
2815 O Street
Sacramento, CA 95816 USA

Sheridan, Michael
Department of Geology
Arizona State University
Tempe, AZ 85281 USA

Shimozuru, Daisuke
Physics Dept.
Tokyo University of Agriculture
Sakura-oka 1-1-1, Setagayu-ku
Tokyo 156, Japan

Shimura, Hidehiro
Observation Division
Japan Meteorological Agency
1-3-4 Ote-machi, Chiyoda-ku
Tokyo 100, Japan

Shumway, D.
9946 Victor
Hesperia, CA 92345 USA

Siebert, Lee
NHB MS 119
Smithsonian Institution
Washington, DC 20560 USA

Sigtryggsson, Hlynur
Icelandic Meteorological Office
Bustadavegur 9
105 Reykjavík, Iceland

Sigurdsson, Haraldur
Graduate School of Oceanography
University of Rhode Island
Kingston, RI 02881 USA

Sigvaldason, Gudmundur E.
Nordic Volcanological Institute
University of Iceland
101 Reykjavík, Iceland

Silva M., Luís
Instituto de Geología
Universidad Nacional Autónoma de México
Ciudad Universitaria
04510 México, D.F., México

Simkin, Tom
NHB MS 119
Smithsonian Institution
Washington, DC 20560 USA

Simon, G.
Laboratoire de Magnétisme
Institut de Physique du Globe
4 Place Jussieu
75252 Paris Cedex 05, France

Simpson, Barbara McG.
GCNZ Consultants, Ltd.
P. O. Box 8441
Symonds Str.
Auckland 3, New Zealand

Sinar Harapan
Jalan Dewi Sartika 136D. Cawang
P. O. Box 260
Jakarta, Indonesia

Singh, Shri Krishna
Instituto de Ingenería
Universidad Nacional Autónoma de México
Ciudad Universitaria
04510 México, D.F., México

Sinton, William
Institute for Astronomy
2680 Woodlawn Drive
Honolulu, HI 96822 USA

Siswowidjoyo, Suparto
Volcanological Survey of Indonesia
Jalan Diponegoro 57
Bandung, Indonesia

Skiff, Brian
San Francisco Mtn. Cosmographic Group
421 W. Aspen
Flagstaff, AZ 86001 USA

Smigielski, Frank (retired)
Synoptic Analysis Branch
NOAA/NESDIS
Washington, DC 20233 USA

Smith, Art
NOAA/NESDIS
Washington, DC 20233 USA
(current address unknown)

Smith, Warwick
DSIR, Seismological Laboratory
Wellington, New Zealand

Smith, William S.
Space Science & Applications Subcommittee
House Science, Space & Technology Committee
2324 Rayburn House Office Bldg.
Washington, DC 20515 USA

Snape, Robert
Solomon Islands Airways, Ltd.
P. O. Box 23
Honiara, Solomon Islands

Snell, Howard
Dept. of Biology
University of New Mexico
Albuquerque, NM 87131 USA

Snetsinger, Kenneth
Atmospheric Experiments Branch
MS 245-5, NASA Ames Research Center
Moffett Field, CA 94035 USA

Soderblom, Lawrence
USGS, 2255 N Gemini Drive
Flagstaff, AZ 86001 USA

Solano, Ariel E.
Universidad Nacional de Colombia
Apartado Aereo 14490
Ciudad Universitaria
Bogotá, Colombia

Solomon Islands Airways (Solair) Ltd.
P. O. Box 23
Honiara, Solomon Islands

Sorrell, G.
DSIR, Geophysics Division
P. O. Box 8005,
Wellington, New Zealand

South Pacific Island Airways
P. O. Box 400
Pago Pago, American Samoa 96799

Sovetskaya Rossiya Radio
Moscow, USSR

Sowa, Daniel
Chief Meteorologist
Northwest Orient Airlines
Minneapolis-St.Paul Airport
St. Paul, MN 55111 USA

Stefánsson, Ragnar
Icelandic Meteorological Office
P. O. Box 5330
Reykjavík, Iceland

Stefanutti, Leopoldo
Istituto di Ricerca Sulle Onde Elettromagnetiche
Via Panciatichi 64,
50127 Firenze, Italy

Steinberg, Genrich S.
Sakhalin Complex Scientific Research Institute
Novoalexandrovsk, Sakhalin 694050
USSR

Steuer, R.
Lamont-Doherty Geological Observatory
Palisades, NY 10964 USA

Stieltjes, Laurent
BRGM, Service Géologique Régional
B. P. 1206, 97484 Saint Denis
Réunion Island

Stirrat, Andrew
Outward Bound
Box 250
Long Lake, MN 55356 USA

Stoiber, Richard
Dept. of Earth Sciences
Dartmouth College
Hanover, NH 03755 USA

Stokes, Barry
USGS Hawaiian Volcano Observatory
P. O. Box 51
Hawaii National Park, HI 96718 USA

Stoschek, J.
Centre National d'Etudes des Télécommunications
22301 Lannion, France

Strom, Robert
Jet Propulsion Laboratory
4800 Oak Grove Drive
Pasadena, CA 91103 USA

Strong, Alan
Ocean Sciences Branch
NOAA/NESDIS
Room 0310, Suitland Professional Center
Washington, DC 20233 USA

Sudradjat, Adjat
Directorate General of Geology and Mineral Resources
Jalan Jenderal Gatot Subroto Kav. 49
Jakarta, Indonesia

Suratman
Volcanological Survey of Indonesia
Jalan Diponegoro 57
Bandung, Indonesia

Sussman, David
Geothermal Division
UNOCAL Corp.
P. O. Box 6854
Santa Rosa, CA 95406 USA

Sutherland, F.
Polynesian Airlines, Ltd.
P. O. Box 599
Apia, Western Samoa

Sutton, Jeffrey
National Center
Reston, VA 22092 USA

Sventek, Paul
11685 Alief Road, #53
Houston, TX 77082 USA

Swan, Peter
Hotel 77
Arusha, Tanzania

Swanson, Donald
USGS Cascades Volcano Observatory
5400 MacArthur Boulevard
Vancouver, WA 98661 USA

Swarts, J.
Geographic Data Technology, Inc.
13 Dartmouth College Hwy
Lyme, NH 03768 USA

Swissler, Thomas
NASA Langley Research Center
Hampton, VA 23665 USA

Symonds, Robert
USGS Cascades Volcano Observatory
5400 MacArthur Boulevard
Vancouver, WA 98661 USA

Taber, John
Lamont-Doherty Geological Observatory
Palisades, NY 10964 USA

Taggart, Jr., Joseph
USGS, Denver Federal Center
Box 25046
Denver, CO 80225 USA

Talai, Ben
Rabaul Volcano Observatory
P. O. Box 386
Rabaul, Papua New Guinea

Talandier, Jacques
Laboratoire de Géophysique
Commiss. à l'Energie Atomique
B. P. 640, Papéete, Tahiti

Tanguy, J. C.
Laboratoire de Géomagnétisme
Université de Paris VI
4 Avenue de Neptune
94100 St. Maur-des-Fossés, France

Tanigawa, Will
USGS Hawaiian Volcano Observatory
P. O. Box 51
Hawaii National Park, HI 96718 USA

Tappin, David
Ministry of Lands, Survey, and Natural Resources
P. O. Box 5
Nuku'alofa, Tonga

Tarits, Pascal
Observatoire Volcanologique du Piton de la Fournaise
14 R.N. 3, 27 ème km
97418 La Plaine des Cafres
Réunion Island

TASS (Soviet News Agency)
50 Rockefeller Plaza
New York, NY 10020 USA

Taylor, Paul
School of Earth Sciences
Macquarie University
North Ryde, New South Wales 2113
Australia

Tazieff, Haroun
Centre National de la Recherche Scientifique
15, Quai Anatole France
75700 Paris, France

Tchoua, Felix
Dept. de Science de la Terre
Université de Yaoundé
B. P. 812, Yaoundé, Cameroon

Telegadas, Gus
Rm. 617, NOAA/ARL
Silver Spring, MD 20910 USA

Thompson, R. B. M. (retired)
Ministry of Natural Resources
G.P.O. Box G24, Honiara
Guadalcanal, Solomon Islands

Thorarinsson, Sigurdur (deceased)
University of Iceland

Tiba, Tokiko
Dept. of Geology
National Science Museum
3-23-1 Hyakunin-cho, Shinjuku-ku
Tokyo 160, Japan

Till, A.
USGS, 4200 University Drive
Anchorage, AK 99508 USA

Tilling, Robert
USGS, 345 Middlefield Road
Menlo Park, CA 94025 USA

Tindle, Robert
26 Lammermuir Gardens
Bear's Den
Glasgow, Scotland

Tokarev, P. I.
Institute of Volcanology
Piip Avenue 9
Petropavlovsk, Kamchatka 683006 USSR

Tomblin, John
UNDRO, Palais de Nations
CH-1211 Genève 10, Switzerland

Tongilava, Sione L.
Lands Survey and Natural Resources
Nuku'alofa, Tonga

Tryggvason, Eysteinn
Nordic Volcanological Institute
University of Iceland
101 Reykjavík, Iceland

Tuni, Deni
Ministry of Lands, Energy, and Natural Resources
P. O. Box G 24
Honiara, Solomon Islands

Tupiza, Arnaldo
Charles Darwin Research Station
Isla Santa Cruz, Galápagos Islands
Ecuador

Turner, P. A.
730 S. Jackson
Denver, CO 80209 USA

Twining, Charles
American Consulate
21 Avenue du General de Gaulle
B. P. 4006
Douala, Cameroon

Ueki, Sadato
Earthquake Prediction Observation Center
Facuty of Science, Tohoku University
Sendai 980, Japan

Ulrich, George
USGS Mapping Division
National Center
Reston, VA 22092 USA

Umaña, Jorge
Depto. de Geología
Instituto Costarricense de Electricidad
Apartado 10032
San José, Costa Rica

Underwood, Richard
Code JL, NASA
Johnson Space Center
Houston, TX 77058 USA

UNDRO News
UNDRO, Palais de Nations
CH-1211 Genève 10, Switzerland

United Press International
1400 I Street NW
Washington, DC 20005 USA

United States Geological Survey
National Center
12201 Sunrise Valley Drive
Reston, VA 22092 USA

United States Department of State
Office of Foreign Disaster Assistance
Room 1262A State Dept. Building
Washington, DC 20523 USA

Université de Paris VI
4 Avenue de Neptune
94100 St. Maur-des-Fossés, France

Université de la Réunion
Laboratoire de Géologie
B.P. 5, 94790 Ste. Clothilde
Réunion Island

Vakin, Dr.
Institute of Volcanology
Piip Avenue 9
Petropavlovsk Kamchatka 683006 USSR

Vallecillo, I.
Proyecto Geotérmico
Managua, Nicaragua

van der Laat, Rodolfo
Observatorio Vulcanológico y
Sismológico de Costa Rica
Universidad Nacional
Heredia, Costa Rica

van der Werff, P.
New Zealand Geological Survey
P. O. Box 499
Rotorua, New Zealand

van Zant, Charles
511 Lawnmeadow
Richardson, TX 75080 USA

Varne, R.
Dept. of Geology, Box 252-C
University of Tasmania
Hobart, Tasmania 7001 Australia

Velardita, R.
Istituto Internazionale di Vulcanologia
Viale Regina Margherita, 6
95123 Catania, Italy

Vergara, Heyley
INGEOMINAS
Apartado Aereo 916
Ibague, Tolima, Colombia

Vie le Sage, R.
Observatoire Volcanologique du Piton
de la Fournaise
14 R.N. 3, 27 ème km
97418 La Plaine des Cafres
Réunion Island

Viglianisi, A.
Istituto di Scienze della Terra
Catania, Italy

Villa, J.
Charles Darwin Research Station
Galápagos Islands, Ecuador
(current address unknown)

Villari, L.
Instituto di Scienze de la Terra
Università di Messina
Via dei Verdi, 75-98100 Messina, Italy

Villari, P.
Istituto di Scienze della Terra
Catania, Italy

Vincent, Pierre M.
Département de Géologie
Université de Clermont
5 Rue Kessler
63038 Clermont-Ferrand Cedex, France

Vinogradov, A. P.
Institute of Volcanology
Piip Avenue 9
Petropavlovsk, Kamchatka 683006 USSR

Visconti, Guido
Istituto di Fisica
Università L'Aquila
L'Aquila, Italy

Volcanological Survey of Indonesia
Jalan Diponegoro 57
Bandung, Indonesia

Waitt, Richard
USGS Cascades Volcano Observatory
5400 MacArthur Boulevard
Vancouver, WA 98661 USA

Walker, James, Cmdr.
COMNAVMAR
FPO San Francisco, 96630 USA
(current address unknown)

Walker, James
Dept. of Geology
Northern Illinois University
DeKalb, IL 60115 USA

Wallace, David A.
Bureau of Mineral Resources
P. O. Box 378
Canberra A.C.T., Australia

Walter, Allan
USGS, 345 Middlefield Road
Menlo Park, CA 94025 USA

Ward, William T. T.
HQ 13th Air Force (PACAF)
Attn: 13 AF/HO
APO San Francisco, 96274 USA
(current address unknown)

Watanabe, Mitsugi
Volcanological Center
Japan Meteorological Agency
1-3-4 Otemachi, Chiyoda-ku
Tokyo 100, Japan

WCBS Radio
New York, NY USA

Weaver, Craig S.
Geophysics Program, AK-50
University of Washington
Seattle, WA 98195 USA

Werner, Dagmar
Smithsonian Tropical Research Institute
APO Miami, FL 34002-0011 USA

The Whakatane Beacon
Whakatane, New Zealand

White, Randy
USGS, 345 Middlefield Road
Menlo Park, CA 94025 USA

Whitney, Linda
Off. of Representative to the U.S.
Commonwealth of the Northern
Mariana Islands
2121 R St. NW
Washington, DC 20036 USA

Willever, Don
7^A Calle Oriente, no. 18
Antigua, Guatemala

Williams, Stanley N.
Dept. of Geology
Louisiana State University
Baton Rouge, LA 70803 USA

Wilson, Charles
Geophysical Institute
University of Alaska
Fairbanks, AK 99701 USA

Wolfe, Edward
USGS Hawaiian Volcano Observatory
P. O. Box 51
Hawaii National Park, HI 96718 USA

Wolfe, John A.
Pan Asean Technical Services Inc.
MCC P. O. Box 1868
Makati, Metro Manila, Philippines

Wonderly, William
4209 San Pedro NE, Apt. 322
Albuquerque, NM 87109 USA

Wood, Charles A.
Code SN-6, NASA Johnson Space
Center
Houston, TX 77058 USA

Wood, C. P.
New Zealand Geological Survey
P.O. Box 499, Rotorua, New Zealand

Woodhall, D.
Mineral Resources Department
Suva, Fiji

Woods, David
NASA, Langley Research Center
Hampton, VA 23665 USA

Wright, Thomas L.
USGS Hawaiian Volcano Observatory
P. O. Box 51
Hawaii National Park, HI 96718 USA

Yamauchi, Toyotara
Observation Division
Japan Meteorological Agency
1-3-4 Ote-machi, Chiyoda-ku
Tokyo 100, Japan

Yaoundé Domestic Radio Service
Yaoundé, Cameroon

Yokoyama, I.
Dept. of Geophysics, Faculty of Science
University of Hokkaido
Sapporo, Japan

Younker, W.
NOAA/NESS, SFSS,
Anchorage, AK 99501 USA

Yount, M. Elizabeth
USGS, 4200 University Drive
Anchorage, AK 99508 USA

Zablocki, Charles
USGS, Denver Federal Center
Box 25046
Denver, CO 80225 USA

Zana, Ndontoni
Département de Géophysique
Institut de Recherches Scientifiques
Lwiro, D/S Bukavu, Zaire

Zlotnicki, J.
Laboratoire de Magnétisme
Institut de Physique du Globe
4 Place Jussieu
75252 Paris Cedex 05, France

Zoller, William
Chemistry Dept., BG-10
University of Washington
Seattle, WA 98195 USA

INDEX

The following index combines references to subjects, people, organizations, places, and even a few acronymns.

The reader should use the Chronology (pages 16-29), to find eruptions of a particular type (e.g., fissure, submarine) or those that exhibit another of the eruptive characteristics described on pages 12-14 (e.g., pyroclastic flow, dome). Page numbers are shown for each eruption in the Chronology, and only general discussions of eruptive characteristics are given page references in the index below. However, as described on pages 34-35 ("Eruption Measurements"), actual measurements, such as lava flow length and eruptive cloud height, were captured during the indexing process and are indexed using ranges of values. For example, the Chronology shows which eruptions have produced lava flows, while the index provides references to lava flows of a particular length or velocity range. When several values are listed in a report, as for the length of a growing lava flow, only the largest is indexed, so page references are indicated as "maximum" ranges under the appropriate subheading.

Several headings (e.g., "seismicity") are followed by a daunting number of page references. However, the reader interested in a particular geographic area can use the page number range for that region to select a smaller number of appropriate references.

Place names, including volcanoes, are indexed by country or geographic region. We have tended to index using geographic rather than political boundaries, but include cross references under appropriate national entries. Names of volcanoes with activity in 1975-85 are shown in bold, capitalized type (e.g., **ST. HELENS, MOUNT**). In most cases we have used the volcano name as given in *Volcanoes of the World* (page 35), but have cross referenced other names. Page references to the primary reports of each volcano are also shown in bold type, while references to this volcano in Part I or other volcano reports are shown in normal type. A volcano's subfeatures are indexed if its reports span many pages in this volume, but if the report is short we have not indexed separate geographic elements or subfeatures.

Information contacts – individuals and organizations – are indexed in bold type (e.g., **Rose, William I.**, **Japan Meteorological Agency**, **United Press International**). Individuals or organizations referred to in the text, but not listed as information contacts, appear in normal type. Most organizations are indexed geographically, and appear before the place references. For example, the Volcanological Survey of Indonesia is indexed as "Indonesia, Volcanological Survey of" before other "Indonesia" references.

Publications (primarily newspapers) that have served as sources of information for SEAN reports are shown in bold italics (e.g., ***New York Times***). Other publications appear in italics (e.g., *Catalogue of the Active Volcanoes of the World*) as they do in the text. Ship names are given in italics with quotation marks.

Literature references are indexed by the first author's name followed by the publication year (in parentheses). When there is only one coauthor, a cross-reference is given under the second author's name. References with three or more authors are only indexed under the first author. References indexed under "Anonymous" are followed by the volcano name in which the reference appears and the year of publication [e.g., Anonymous:Etna (1979)]. Compilations of papers (with no named editor) are indexed under the journal or publication title.

As an aid to the reader, we have included many cross references(using *see* and *see also*). Within these cross references, colons separate different index levels, and commas separate multiple references. For example, under the main heading "hazard," the cross reference "*see also* destruction, lava flow:diversion" refers the reader to the "destruction" and "lava flow" main headings. The abbreviation "*q.v.*" (quod vide, or "which see") follows entries under a main heading when that entry appears elsewhere as its own main heading.

A

B

C

D

E

H

I

J

K

L

M

N

O

P

T

U

V

W

Y

AUTHORS/EDITORS

The principal authors of this book are the SEAN network correspondents; the 696 information contacts listed on pages 628-40 and named at the end of each report. Without their timely reports, often produced under difficult and stressful conditions, there would be no SEAN.

The editors involved in producing this book, and in shaping most of the reports that make the book, are listed below:

Lindsay McClelland worked part time on the Smithsonian Institution's Volcano Reference File while a student at George Washington University and full time after receiving a BS in geology in 1973. His interest in volcanology then led him to Dartmouth College, where he received an MA in geology in 1976. He returned to the Smithsonian as a geologist with SEAN, where he has gathered and reported information on contemporary volcanism and other geophysical activity since 1976.

Tom Simkin, Curator of Petrology and Volcanology, received a BS in civil engineering from Swarthmore College in 1955 and a PhD in geology from Princeton University in 1965. He was at the University of Chicago and SUNY-Binghamton before joining the Smithsonian in 1967. He heads the Museum's Global Volcanism Program, and other research interests include the Galápagos Islands and the Isle of Skye, Scotland. One of the founders of SEAN, he has been final editor of its volcano and earthquake reports since the *Bulletin's* first issue.

Marjorie Summers received a BA in geology from the University of Hawaii in 1976 and, after additional studies in art and drafting, joined the U.S. Geological Survey's Hawaiian Volcano Observatory. She was involved in field and laboratory studies of the Mauna Loa and Kilauea eruptions, and production of several publications. She came to the Smithsonian in 1985, focusing on the preparation of this book, while reporting on current volcanism, seismicity, and meteoritic events for the *SEAN Bulletin*.

Elizabeth Nielsen came to the Smithsonian after receiving a BS/BA in geology and mathematics at Southern Methodist University (1980) and an MA in geology from the University of California at Santa Barbara (1982). She collaborated on (and indexed) two other major books, and reported on volcanic activity for SEAN. She also participated in research on volcano deformation and developed the Smithsonian Volcano Archives from 1985 until moving to the Cascades Volcano Observatory in 1988.

Thomas Stein came to the Smithsonian after earning a BA degree in earth and planetary sciences from Washington University (1987), where his undergraduate studies included 2 years at the NASA Regional Planetary Image Facility. At the Smithsonian, his work has been dominated by computer applications in volcanology, particularly the Volcano Reference File, documenting 10,000 years of global activity. He has taken a leading role in developing the program's use of desktop publishing and museum exhibits.

Additional contributors to this book are named under "Acknowledgments" (p. 4-5), and the contributions of former SEAN staff members are discussed under "SEAN History" (p. 5-7).

Because these reports are the product of many correspondents, we request that citations be to the network rather than the editors. Please reference this book as:

> Smithsonian Institution/SEAN, 1989, *Global Volcanism 1975-1985;* Prentice Hall, Englewood Cliffs, NJ, and American Geophysical Union, Washington, DC, 657 pp.

SEAN depends on prompt communication from volcano watchers around the world. Please help by sending news of current volcanism to SEAN via telephone (202/357-1511), telex (USA 89599 answerback SCINET WSH), telefax (202/357-2476), or airletter (NHB MRC 129, Smithsonian Institution, Washington, DC 20560, USA).

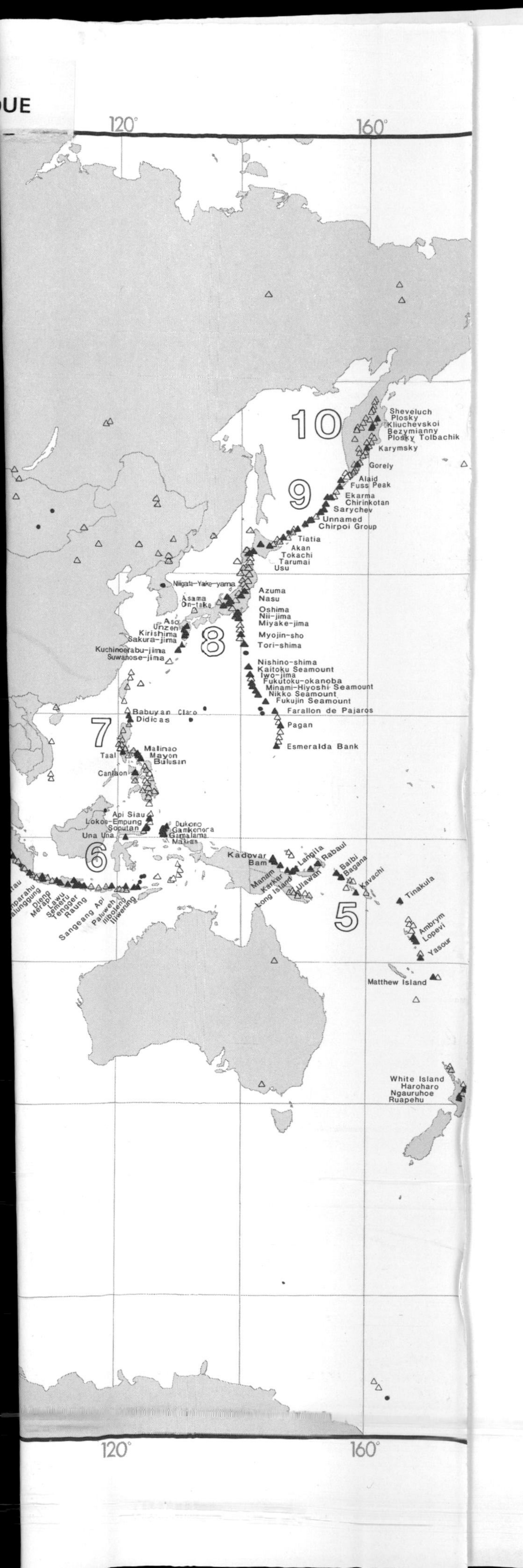

OUE
120°
160°
10
Sheveluch
Plosky
Kliuchevskoi
Bezymianny
Plosky Tolbachik
Karymsky
Gorely
9
Alaid
Fuss Peak
Ekarma
Chirinkotan
Sarychev
Unnamed
Chirpoi Group
Tiatia
Akan
Tokachi
Tarumai
Usu
Niigata-Yake-yama
Asama
On-take
Azuma
Nasu
Oshima
Nii-jima
Miyake-jima
Aso
Unzen
Kirishima
Sakura-jima
8
Myojin-sho
Tori-shima
Kuchinoerabu-jima
Suwanose-jima
Nishino-shima
Kaitoku Seamount
Iwo-jima
Fukutoku-okanoba
Minami-Hiyoshi Seamount
Nikko Seamount
Fukujin Seamount
Farallon de Pajaros
Babuyan Claro
Didicas
7
Pagan
Esmeralda Bank
Taal
Malinao
Mayon
Bulusan
Canlaon
Api Siau
Lokon-Empung
Soputan
Una Una
Dukono
Gamkonora
Gamalama
Makian
6
Kadovar
Bam
Manam
Karkar
Long Island
Langila
Ulawun
Rabaul
Balbi
Bagana
Kavachi
5
Tinakula
Ambrym
Lopevi
Yasour
Matthew Island
Sangeang Api
Raung
Tengger
Semeru
Lawu
Merapi
Dieng
Galunggung
Tangkuban Parahu
Paluweh
Iliboleng
Iliwerung
White Island
Haroharo
Ngauruhoe
Ruapehu
120°
160°

DATE D
75°
40°
60°
40°
20°
0°
20°
40°
60°
70°
40°
80°
1
2
3
Campi Flegrei
Stromboli
Vulcano
Etna
Erta Ale
Ardoukoba
Lake Monoun
Mt. Cameroon
Nyamuragira
Nyiragongo
Ol Doinyo Lengai
Karthala
Piton de la Fournaise
Marion Island
Heard Island

40°
80°
120°
160°
75°
60°
40°
20°
0°
20°
40°
60°
70°
1
Campi Flegrei
Stromboli
Vulcano
Etna
2
Erta Ale
Ardoukoba
Lake Monoun
Mt. Cameroon
Nyamuragira
Nyiragongo
Ol Doinyo Lengai
3
Karthala
Piton de la Fournaise
Marion Island
Heard Island
10
Sheveluch
Plosky
Kliuchevskoi
Bezymianny
Plosky Tolbachik
Karymsky
Gorely
Alaid
Fuss Peak
9
Ekarma
Chirinkotan
Sarychev
Unnamed
Chirpoi Group
Tiatia
Akan
Tokachi
Tarumai
Usu
Niigata-Yake-yama
Azuma
Nasu
Asama
On-take
Oshima
Nii-jima
Miyake-jima
Aso
Unzen
Kirishima
Sakura-jima
Myojin-sho
Tori-shima
8
Kuchinoerabu-jima
Suwanose-jima
Nishino-shima
Kaitoku Seamount
Iwo-jima
Fukutoku-okanoba
Minami-Hiyoshi Seamount
Nikko Seamount
Fukujin Seamount
Farallon de Pajaros
Pagan
Esmeralda Bank
7
Babuyan Claro
Didicas
Malinao
Mayon
Bulusan
Taal
Canlaon
Api Siau
Lokon-Empung
Soputan
Una Una
Dukono
Gamkonora
Gamalama
Makian
6
Marapi
Kaba
Krakatau
Tangkubanparahu
Galunggung
Dieng
Merapi
Kelut
Semeru
Tengger
Raung
Sangeang Api
Paluweh
Iliboleng
Iliwerung
Kadovar
Bam
Manam
Karkar
Long Island
Langila
Ulawun
Rabaul
Balbi
Bagana
Kavachi
5
Tinakula
Ambrym
Lopevi
Yasour
Matthew Island
White Island
Haroharo
Ngauruhoe
Ruapehu